# CHARACTERIZATION
# OF MATERIALS

# CHARACTERIZATION OF MATERIALS

## Volume 2

**Elton N. Kaufmann,** *(Editor-in-Chief)*
Argonne National Laboratory
Argonne, IL

*Characterization of Materials* is available Online in full color
at www.mrw.interscience.wiley.com/com.

**WILEY-INTERSCIENCE**
A John Wiley and Sons Publication

*Library of Congress Cataloging in Publication Data is available.*

Characterization of Materials, 2 volume set
Elton N. Kaufmann, editor-in-chief

ISBN: 0-471-26882-8 (acid-free paper)

Printed in the United States of America

10  9  8  7  6  5  4  3  2  1

# CHARACTERIZATION
# OF MATERIALS

# OPTICAL IMAGING AND SPECTROSCOPY

## INTRODUCTION

All materials absorb and emit electromagnetic radiation. The characteristics of a material frequently manifest themselves in the way it interacts with radiation. The fundamental basis for the interaction varies with the wavelength of the radiation, and the information gleaned from the material under study will likewise vary. Materials interact with radiation in some fashion across the entire electromagnetic spectrum, which can be thought of as spanning the frequency realm from near DC oscillations found in circuits ($<1$ Hz) through the extremely high frequency oscillations typical of gamma radiation ($>10^{19}$ Hz). Shown in the Table 1 is a representation of the electromagnetic spectrum, giving the correspondence of wavelength, frequency, and common terminology applied to the sub-bands.

Because human vision depends on the interaction of light with the rods and cones of the retina in the range of photon wavelengths from about 0.7 to 0.4 micrometers, this "visible" spectrum is particularly important to us. Materials are directly observed using instrumentation that magnifies and records visible light; this is the subject of the optical microscopy units in this chapter. As an adjunct to the value of direct observation of the structure of materials, often, a desirable material property arises from its interaction with visible light. The visible and ultraviolet regions are the basis for the units in this chapter, and, as seen in Table 1, they occupy the midpoint on the energy scale. Yet, it is valuable to note the interactions of electromagnetic radiation from the full-spectrum perspective, both for understanding the role that optical methods play and for realizing their relationship to those presented in other chapters throughout *Characterization of Materials*. Thus, we briefly consider here how all radiation—from DC to x ray—interacts with materials.

In the most general sense, radiation can be considered to interact with materials either by absorption, reflection, or scattering; this is the basis for Kirchoff's law, which is discussed in COMMON CONCEPTS IN MATERIALS CHARACTERIZATION.

As indicated in Table 1, the nature of the interaction between photons and matter governs the observed effects. At very low frequencies, we think of the photons more as a time-dependent electric field than as a packet of energy. Interactions with matter in this regime are exploited in the electrical and electronic characterization of transients and charge carriers in materials. This is the topic of many of the units in ELECTRICAL AND ELECTRONIC MEASUREMENTS. At higher frequencies, the nuclear (RF) and electronic (microwave) Zeeman effects become observable by virtue of transitions between states in the presence of the internal quadrupole field of the nuclei (nuclear quadrupole resonance) and/or in the presence of an externally applied field (nuclear magnetic resonance and electron spin resonance, see RESONANCE METHODS).

The microwave region is also an important regime for materials characterization, as the measurement of dielectric properties such as permittivity and dielectric loss tangent, surface resistance, and skin depth become feasible using radiation in this region. Some of these measurements are discussed in detail in ELECTRICAL AND ELECTRONIC MEASUREMENTS. As we continue up in frequency, we enter the domain of the lowest-frequency molecular vibrations in materials, lattice vibrations, or phonons, which couple to the oscillating electric field of the probe radiation. Phonons can also be probed using neutron scattering, a complementary method described in detail in NEUTRON TECHNIQUES, and by inelastic scattering of radiation (Brillouin spectroscopy).

The next jump in frequency enters the infrared radiation region, in which the electromagnetic field is coupled with the intermolecular motions of chemical bonds in the material structure. Also known as vibrational spectroscopy, the study of the spectra of materials in this regime provides information on chemical composition, as several characteristic frequency bands can be directly related to the atom and bond types involved in the vibrational motion. An important detail deserves mention here, as a shift occurs at this point in convention and nomenclature. Researchers in the lower frequency fields (DC through far infrared) typically report their measurements in frequency units. In infrared work the most common unit employed is the wavenumber, or reciprocal centimeter, although there is an increasing use of wavelength (reported in microns) in some disciplines. Analogous to Brillouin spectroscopy is Raman spectroscopy, which measures vibrational transitions by observing inelastically scattered radiation. Raman spectroscopy is complementary to infrared spectroscopy by virtue of the selection rules governing transitions between vibrational states—whereas a change in overall dipole moment in the molecule gives rise to infrared absorption, a change in overall polarizability of the molecule gives rise to the vibrational structure observed in Raman spectroscopy.

At still higher frequencies, we access transitions between electronic states within molecules and atoms, and again the conventions change. Here, convention favors units of wavelength. The underlying phenomena studied range from overtones of the molecular vibrations (near-infrared) to vibrational-electronic (vibronic) transitions, to charge transfer transitions and pure electronic spectra (visible through ultraviolet spectroscopy). Electronic spectroscopy, like vibrational spectroscopy, also reflects the chemical makeup of the material. The intensity of electronic transitions, a function of the type of molecular orbitals

**Table 1. The Electromagnetic Spectrum: A Materials Perspective**[a]

| Spectral Region | Frequency Range (Hz) | Wavelength Range (m) | Unit Conventions | Interactions with Matter | Representative Materials Properties Accessible |
|---|---|---|---|---|---|
| Extremely low frequency | $<3 \times 10^{2}$ | $>1 \times 10^{6}$ | Hz | Electron transport | Conductivity, free carrier concentration |
| Radiofrequency | | | | | |
| Very low frequency (VLF) | $3 \times 10^{2}$ to $3 \times 10^{4}$ | $1 \times 10^{6}$ to $1 \times 10^{4}$ | Hz | Nuclear magnetic resonance; nuclear quadrupole resonance | Microscopic structure; chemical composition |
| Low frequency (LF) | $3 \times 10^{4}$ to $3 \times 10^{5}$ | $1 \times 10^{4}$ to $1 \times 10^{3}$ | kHz | | |
| Medium frequency (MF) | $3 \times 10^{5}$ to $3 \times 10^{6}$ | $1 \times 10^{3}$ to $1 \times 10^{2}$ | kHz | | |
| High frequency (HF) | $3 \times 10^{6}$ to $3 \times 10^{7}$ | $1 \times 10^{2}$ to $1 \times 10^{1}$ | kHz | | |
| Very high frequency (VHF) | $3 \times 10^{7}$ to $3 \times 10^{8}$ | $1 \times 10^{1}$ to $1 \times 10^{0}$ | MHz | | |
| Ultra high frequency | $3 \times 10^{8}$ to $3 \times 10^{9}$ | $1 \times 10^{0}$ to $1 \times 10^{-1}$ | MHz | | |
| Microwave | | | | | |
| Super high frequency (SHF) | $3 \times 10^{9}$ to $3 \times 10^{10}$ | $1 \times 10^{-1}$ to $1 \times 10^{-2}$ | GHz | Electron paramagnetic resonance; molecular rotations | Microscopic structure; surface conductivity |
| Extremely high frequency (EHF) | $3 \times 10^{10}$ to $3 \times 10^{11}$ | $1 \times 10^{-2}$ to $1 \times 10^{-3}$ | GHz | | |
| Submillimeter | $3 \times 10^{11}$ to $3 \times 10^{12}$ | $1 \times 10^{-3}$ to $1 \times 10^{-4}$ | THz | | |
| Infrared | | | | | |
| Far infrared (FIR) | $\sim 3 \times 10^{12}$ to $1.2 \times 10^{13}$ | $\sim 1 \times 10^{-4}$ to $2.5 \times 10^{-5}$ | THz, cm$^{-1}$, kiloKayser (kK, $= 1000$ cm$^{-1}$), micron (μm) | Librations, molecular vibrations, vibrational overtones, vibronic transitions | Microscopic and macroscopic structure, phases, chemical composition |
| Intermediate infrared | $1.2 \times 10^{13}$ to $3 \times 10^{14}$ | $2.5 \times 10^{-5}$ to $1 \times 10^{-6}$ | | | |
| Near infrared (NIR) | $3 \times 10^{14}$ to $3.8 \times 10^{14}$ | $1 \times 10^{-6}$ to $7.9 \times 10^{-7}$ | | | |
| Visible | $3.8 \times 10^{14}$ to $7.5 \times 10^{14}$ | $7.9 \times 10^{-7}$ to $4 \times 10^{-7}$ | kK, nm, μm | Valence shell, π bonding electronic transitions | Chemical composition and concentration |
| Ultraviolet | | | | | |
| Near ultraviolet | $7.5 \times 10^{14}$ to $1.5 \times 10^{15}$ | $4 \times 10^{-7}$ to $2 \times 10^{-7}$ | nm | Inner shell, σ bonding electronic transitions | Chemical composition, bond types, and bond strength |
| Vacuum ultraviolet | $1.5 \times 10^{14}$ to $3 \times 10^{16}$ | $2 \times 10^{-7}$ to $1 \times 10^{-8}$ | eV | | |
| X ray | $3 \times 10^{16}$ to $3 \times 10^{19}$ | $1 \times 10^{-8}$ to $1 \times 10^{-11}$ | keV | Core shell electronic transitions, nuclear reactions | Elemental analysis, nature of chemical bonding |
| γray | $> \sim 3 \times 10^{19}$ | $< \sim 1 \times 8 10^{-11}$ | MeV | Nuclear transitions | Elemental analysis, profiling |

[a]Note that values preceded by represent "soft definitions," i.e., frequencies that bridge disciplines and may be defined differently depending upon the field of study. In fact, many frequency definitions, especially at the boundaries between spectral regions, are "soft."

in the material, govern the appearance of the observed spectra. Absorption of radiation in this regime is often accompanied by electronic relaxation processes, and, depending on the nature of the electronic states involved, can cause fluorescence and/or phosphorescence, providing another window on the chemical makeup and structure of the material.

Visible and ultraviolet spectroscopies are frequently coupled with other measurement techniques to provide additional insight into the makeup and nature of materials. Changes in materials induced by electrochemical reactions, for instance, can change the observed spectra, or even lead directly to chemiluminescence (ELECTROCHEMICAL TECHNIQUES). Of particular interest to semiconductor studies is the luminescence (photoluminescence, or PL) that follows the promotion of electrons into the valence band of the material by incident photons (ELECTRICAL AND ELECTRONIC MEASUREMENTS). Electrons may even be ejected completely from the material, providing a measure of the work function (the photoelectric effect).

At the high end of the ultraviolet range (vacuum UV) and beyond, yet another shift in convention occurs. This regime traditionally employs the unit of electron volts (eV). Phenomenology of interactions ranges from inner-shell electronic transitions, as observed in ultraviolet and x-ray photoelectron spectroscopy, to nuclear rearrangement, as observed in gamma ray emission in nuclear reactions (ION-BEAM TECHNIQUES). Emitted electrons themselves carry spectroscopic information as kinetic energy; this is the basis for ESCA (electron spectroscopy for chemical analysis), energy dispersive spectroscopy, and Auger electron spectroscopy (ELECTRON TECHNIQUES).

The Mössbauer effect relies on the recoil-free emission of nuclear transition gamma radiation from the nuclei in a solid (the source) and its resonance absorption by identical nuclei in another solid (the absorber). Because recoilless resonance absorption yields an extremely sharp resonance, Mössbauer spectroscopy is a very sensitive probe of the slightest energy level shifts and therefore of the local environment of the emitting and absorbing atoms.

The units collected in this chapter present methods that are useful in the optical characterization of materials. The preceding discussion illustrates how spectroscopy permeates the entire *Characterization of Materials* volume—called upon where appropriate to specific materials studies. Spectroscopy, as a general category of techniques, will certainly remain central to the characterization of materials as materials science continues its advance.

ALAN C. SAMUELS

# OPTICAL MICROSCOPY

## INTRODUCTION

Among the numerous investigative techniques used to study materials, optical microscopy, with its several diverse variations, is important to the researcher and/or materials engineer for obtaining information concerning the structural state of a material. In the field of metallurgy, light optical metallography is the most widely used investigative tool, and to a lesser degree its methods may be extended to the investigation of nonmetallic materials. The state of microstructure of a metal or alloy, or other engineering material, whether ceramic, polymer or composite, is related directly to its physical, chemical, and mechanical properties as they are influenced by processing and/or service environment. To obtain qualitative, and in many cases quantitative information about the microstructural state of a material, it is often necessary to perform a sequence of specimen-preparation operations, which will reveal the microstructure for observation. Selecting a specimen, cutting it from the bulk material, grinding and polishing an artifact-free surface, then etching the surface to reveal the microstructure, is the usual sequence of specimen-preparation operations (see SAMPLE PREPARATION FOR METALLOGRAPHY). This scheme is appropriate for metals and alloys as well as some ceramic, polymer, and composite materials. Specimens thus prepared are then viewed with a reflected-light microscope (see REFLECTED-LIGHT OPTICAL MICROSCOPY). The microstructure of transparent or translucent ceramic and polymer specimens may be observed using a transmitted-light microscope when appropriately thinned sections are prepared. In all cases, observation of the microstructure is limited by the relatively shallow depth of focus of optical microscopes, the depth of focus being inversely related to the magnification.

Optical microscopy as related to the study of engineering materials can be divided into the categories of reflected-light microscopy for opaque specimens and transmitted-light microscopy for transparent specimens. On the whole, reflected light microscopy is of more utility to materials scientists and engineers because the vast bulk of engineering materials are opaque. Within each of the categories, there are a number of illumination variations (bright field, dark field, polarized light, sensitive tint, phase contrast, and differential interference contrast), each providing a different or enhanced observation of microstructural features. It is fortunate that a majority of illumination variations of reflected-light microscopy parallel those of transmitted-light microscopy.

In general, one does not set out to employ optical microscopy to measure a material property or set of properties. Information gained using optical microscopy is more often than not of a qualitative nature, which, to a skilled observer, translates to a set of correspondences among a material's physical, mechanical, chemical properties as influenced by chemical composition, processing and service history. One can use the measurement methods of quantitative microscopy to follow the progress of microstructural evolution during processing or during the course of exposure to a service environment. Such techniques may be used to follow grain growth, the coarsening of precipitates during a process, or the evolution of microstructure that would lead to catastrophic failure under service conditions. Such a case might be the evolution of the microstructure of steam pipes carrying superheated steam at high pressure that will ultimately fail by creep rupture.

In comparison with other instruments used to produce images of microstructure—transmission electron microscope (TRANSMISSION ELECTRON MICROSCOPY and SCANNING TRANSMISSION ELECTRON MICROSCOPY), scanning electron microscope (SCANNING ELECTRON MICROSCOPY), field ion microscope, scanning tunneling microscope (SCANNING TUNNELING MICROSCOPY), atomic force microscope, and electron beam probe—the light-optical microscope is the instrument of choice for a majority of studies. The reasons are as follows: (1) instrument cost is relatively low, (2) less technician training is required, (3) specimen preparation is relatively straightforward, and (4) the amount of information that can be obtained by an experienced microscopist is large. Because the maximum resolution is limited by the wavelength of light, useful magnification is limited to ~1600 times. At these upper limits of resolution and magnification, microstructural features as small as ~325 nm may be observed. Most of the other instruments are better suited to much higher magnifications, but the area of specimen surface is severely limited, so sampling of the microstructure state is better using optical microscopy.

## PRACTICAL ASPECTS OF THE METHOD

### The Basic Optical Microscope

Early microscopes, though rudimentary compared to today's instruments, provided suitable magnifications, image resolution, and contrast to give early researchers and engineers suitably detailed visual information necessary for the advancement of science and materials application. Modern instruments are indeed sophisticated and incorporate a number of devices designed to enhance image quality, to provide special imaging techniques, and to record images photographically or digitally. Regardless of the sophistication of an instrument, the fundamental components that are of particular importance are a light source, an objective lens, and an ocular or eyepiece.

Significant development of the microscope was not forthcoming until the wave nature of light became known and Huygens proposed a principle that accounted for the propagation of light based on an advancing front of wavelets (Jenkins and White, 1957). Rayleigh was the first to develop a criterion for resolution, which was later refined by Abbe, who is often credited as the "father of optical microscopes" (Gifkins, 1970). Abbe's theory of the microscope was founded upon the concept that every subject consisted of a collection of diffracting points like pinholes, each pinhole behaving like a Huygens light source (Jenkins and White, 1957). It is the function of the objective lens to collect diffracted rays from an illuminated specimen to form an image by constructive and destructive interference from a collection of Huygens sources. Therefore, the function of a lens through which the specimen is viewed is to gather more diffraction information than could be gathered by the unaided eye. This means that an objective lens must be of such geometric configuration as to gather as many orders of diffracted light as possible, to give an image of suitable resolution. The more orders of diffracted rays that can be collected by the objective lens, the greater

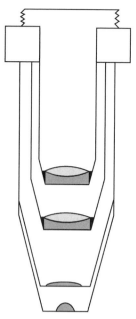

**Figure 1.** Schematic diagram of a typical objective lens. Note that the plano-convex lens is the "front" lens element of the objective and faces the specimen.

is the resolution of the lens, and generally the higher the magnification attainable.

By its most basic definition, a lens is a piece of glass or other transparent material bounded by two surfaces, each of a different and/or opposite curvature. From a geometric point of view, light rays passing through a lens either converge or diverge. Any device that can be used to concentrate or disperse light by means of refraction can be called a lens. Objective lenses usually contain a first element (nearest the specimen) that is plano-convex, having a plane front surface and a spherical rear surface (Fig. 1). Additionally, the objective lens may contain a number of elements arranged in several groups, secured in a lens barrel to keep the elements and groups in proper spatial relationship to one another. Likewise the eyepiece or ocular (Fig. 2), usually consists of two lenses, or two or more groups of lenses, each group containing two or more elements.

In a microscope the objective lens, acting like a projector lens, forms a real image of the subject near the top of the microscope tube. Because the image is formed in air, it is often called the aerial image (Restivo, 1992). The eyepiece acts to magnify this real image as a virtual image which is projected by the cornea on the retina of the eye, or it may be projected on the plane of a photographic film or on the surface of some other image recording device. The final magnification observed is the product of the magnification of the objective lens times the magnification of the ocular, except when the microscope contains a zoom magnification device or a magnification multiplier.

To a microscopist, the properties of the objective and the ocular lenses are a first consideration in setting out to use a microscope; therefore, two important properties of objective lenses, magnification and numerical aperture, are usually engraved on the barrel of objective lenses. Like-

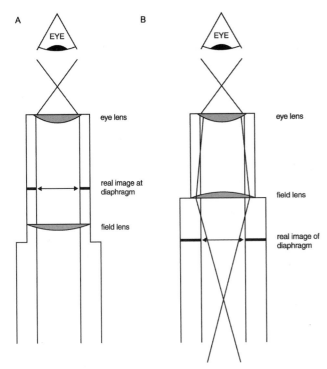

**Figure 2.** Schematic diagrams of (**A**) a Huygenian ocular and (**B**) a Ramsden ocular. Note the relative orientation of the field lens and the field diaphragms.

wise, magnification is engraved on the ocular. For an objective and ocular to work together properly, their focal lengths must be such that they are separated by a specific distance, so that a virtual image is produced at the eye or a real image is produced at the viewing plane. This distance is the tube length, defined as the distance between the rear focal plane of the objective and the front focal plane of the ocular. It is universally accepted that for any given microscope the objectives and oculars are designed to work together. Objectives and oculars designed for an instrument must be designed for the same tube length. As a practical rule, objective lenses and oculars should not be mixed or transferred to other microscopes. A number of recent instruments are infinity-corrected, meaning that the eyepieces must also be infinity corrected. Infinity-corrected systems allow for variations in the length of the optical path, so that other devices can be inserted for special purposes.

### Characteristics of the Microscope

The characteristics of the microscope that are of the greatest importance to the user are the magnification, resolution, and the absence of lens aberrations.

While *magnification* appears to be a simple concept, without concurrent consideration of resolution it becomes relatively unimportant. Magnification alone cannot assist in revealing detail unless the system also possesses resolving power. The microscopist defines resolution as the distance between two closely spaced points in the subject that can be made observable; thus, resolution, expressed as a linear measure, is better the smaller the distance between the points. Attainable resolution is established by the

properties of the objective lens, the wavelength of the illumination, and the index of refraction of the medium between the objective lens and the specimen. When the magnification is great enough for the resolution to be perceived by the eye, no amount of additional magnification will cause the image to have more resolution; no more detail will be visible. Regardless of the resolution of photographic films or charge-coupled devices, the resolution of the microscope, primarily that of the objective lens, is controlling in terms of image detail that can be detected.

Calibration of magnification is altogether straightforward. A ruled scale, usually in tenths and hundredths of a millimeter, is placed on the stage of the microscope and brought into sharp focus. It is customary to photograph such a scale at the full series of magnifications of which the instrument is capable. By measuring the photographic images of the magnified scale, one can easily determine the precise magnification.

The degree of resolution is dependent upon the amount of diffraction information that can be gathered by the objective lens. This factor is most conveniently expressed in terms of the numerical aperture (N.A.) of the objective lens. The N.A. of an objective lens is given by N.A. $= n \sin \mu$, where $n$ is the index of refraction of the medium between the specimen and the lens (for air, $n = 1$), and $\mu$ is the half-angle of the cone of light capable of being accepted by the objective (Gifkins, 1970).

Resolution ($R$) is dependent upon the wavelength of light ($\lambda$) illuminating the specimen, and the N.A. of the objective lens: $R = \lambda/(2 \times \text{N.A.})$ (Gifkins, 1970). The larger the numerical aperture and/or the shorter the wavelength of light, the finer the detail that can be resolved. This expression for resolution is used expressly for reflected-light microscopy, where the objective lens carries light to the specimen, then acts again to collect light reflected from the specimen. Resolution of other reflected-light microscopes, particularly those that operate at low magnification and utilize an off-axis external light source, is not a major concern, because the low-power objectives have small N.A. values.

In transmitted-light microscopy a separate condenser lens supplies light through the specimen; thus, the expression becomes $R = \lambda/(\text{N.A.}_{\text{condenser}} + \text{N.A.}_{\text{objective}})$. By way of example, if one illuminates a specimen with green light ($\lambda = 530$ nm) and uses an objective with a 0.65 N.A., then the $R = 407.7$ nm ($4.077 \times 10^{-4}$ mm). $R$ is the distance between two points on the specimen that can be resolved by the microscope. Even though these points may be resolved by the microscope, they will not appear as separate points unless the magnification of the microscope is sufficient to place them within the resolving power of the human eye. Most people can detect an angular separation of from 1 to 2 min of arc. This amounts to a distance between points of ~0.11 mm at a viewing distance of 250 mm. Therefore, by dividing 0.11 mm by $4.077 \times 10^{-4}$ mm, one obtains a system magnification of ~270×. At least this amount of magnification is required in order to make the two points perceptible to the eye. Any magnification greater than this only makes the image more easily perceived by the eye, but the resolution is not further enhanced.

In practice, the largest numerical aperture is ∼1.6. Using this value for numerical aperture and again using green light, one finds that the maximum resolution occurs at a magnification of the order of 650×. A magnification of 1000 or 1600 times may be justified by the increased ease of perception. Some microscope manufacturers suggest, by "rule of thumb," that the practical magnification should be 1000 times the numerical aperture. Other sources recommend lower magnifications depending on the type and magnification of the objective.

Because the resolving power of a microscope is improved as the wavelength of the illumination decreases, it is evident that blue light will give slightly better resolution than green light. Wavelengths of <400 nm have been used successfully to enhance resolution. Because conventional optical glasses absorb wavelengths below ∼300 nm, the use of special optical elements of quartz, fluorite, calcite, or combinations thereof are required. Thus, all wavelengths down to ∼200 nm can be transmitted through the microscope. Mercury vapor lamps produce a large amount of short-wavelength illumination and have proven suitable for ultraviolet (UV) microscopy. Because of the shortness of the wavelength, images can be damaging to the eye if viewed directly; therefore the UV part of the spectrum must be filtered out. Photomicroscopy presents a significant problem in focusing the image. If the image is focused at the film plane with the ultraviolet wavelengths filtered out, the image produced by the shorter wavelengths will not be in focus at the film plane. In 1926, Lucas developed a method for focusing the image on a fluorescent screen. Although resolution can theoretically be enhanced by as much as 60%, a more practical enhancement of 33% can be obtained using a wavelength of 365 nm rather than 540 nm (Kehl, 1949). Ultraviolet microscopy has not become a major technique in materials science because of the great amount of skill required to produce exceptionally flat and artifact-free surfaces. Also, the development of scanning electron micro-scopy (see SCANNING ELECTRON MICROSCOPY) has made it possible to get exceptional resolution, with great depth of field, without great concern for sample preparation.

A general knowledge of lens aberrations is important to the present-day microscopist. While some modern instruments are of very high quality, the kinds of objectives and oculars that are available are numerous, and it must be understood what aberrations are characteristic of each. In addition, various oculars (or projector lenses) are available for correction of aberrations introduced by objective lenses.

Lens aberrations are of two general classes: spherical and chromatic (Gifkins, 1970; Vander Voort, 1984; Kehl, 1949). Spherical aberrations are present even with monochromatic illumination. Its subclasses are spherical aberration itself, coma, curvature of field, astigmatism, and distortion. Chromatic aberration is primarily associated with objectives.

Spherical aberration occurs when a single lens is ground with truly spherical surfaces. Parallel rays come to focus at slightly different points according to their path through the lens—through the center portion of the lens, or through its outer portions. Thus, the image of a point source at the center of the field becomes split into a series of overlapping point source images resulting in an unsharp circle of confusion instead of a single sharp point source image. For spherical positive (convex or convergent) lenses, the spherical aberration is exactly the opposite of that produced by a spherical negative (concave or divergent) lens ground to the same radius of curvature. Correction for spherical aberration can take two forms. The first involves the use of a doublet (two lenses cemented together—one of positive curvature and one of negative curvature). The development of lens formulas for a doublet is rather more complex than it would seem, for should both the positive and negative lenses of the doublet be ground to the exact same radius, then the magnification of the doublet would be 1×. A second method of correcting for spherical aberration is to use a single lens with each face ground to a different radius; such a lens is said to be aspherical.

Coma affects those portions of the image away from the center. Differences in magnification resulting from ray paths that meet the lens at widely differing angles cause the image of a point to appear more like a comma. Lens formulas for the correction of coma provide for the sine of the angle made by each incident ray with the front surface of a lens to have a constant ratio to the sine of each corresponding refracted ray.

Curvature of field arises because the sharpest image of an extended object lies on a curved surface rather than a flat plane. While slight curvature of field is not a serious problem for visual work, because the eye can partially adjust for the visual image and fine focus adjustments can overcome that for which the eye cannot compensate— it becomes a serious problem for photographic work. At high magnifications, the center of the image may be sharp, with sharpness falling off radially away from the center. Attempts to focus the outer portions of the image only cause the central portion of the image to go out of focus. Correction for curvature of field is done by introducing correction in the ocular with the addition of specially designed lens groups. For this reason, it must be understood that oculars are designed to work with a specific objective or set of objectives. One should not attempt to use oculars in another instrument or with objectives for which they are not designed.

Astigmatism is a defect that causes points to be imaged as lines (or discs). Lenses designed to correct for astigmatism are called anastigmats. Correction for astigmatism is done by introducing an additional departure from spherical lens surfaces, together with the matching of refractive indices of the lens components.

Distortion is mainly an aberration found in oculars, where straight lines appear as curved lines in the outer portion of the image. The curvature may be either positive or negative and can often be observed when viewing the image of a stage micrometer. Again it is produced by unequal magnifications from various zones of the lenses in an ocular. Most modern microscopes have oculars that are well corrected for distortion.

Chromatic aberration arises because all wavelengths of light are not brought to focus at the same distance from the optical center of a lens (longitudinal chromatic aberration). If longitudinal chromatic aberration is serious and broad wavelength spectrum illumination is used, one can observe color fringes around the image of a subject. If a black-and-white photograph is taken of the image, it will appear slightly out of focus. Lateral chromatic aberration results from a lens producing differing magnifications for differing wavelengths of light. A simple solution is to use a narrow-band-pass filter to make the illumination nearly monochromatic; however, this precludes color photography, particularly at higher magnifications. Fortunately, lens formulas exist for the correction of chromatic aberration. Such formulas employ optical glasses of different indices of refraction.

The cost of making lenses corrected to various degrees for chromatic aberrations increases dramatically as the amount of correction increases. From the amount of correction one finds that there are several classes or names for objective lenses: nonachromatic, achromatic, semi-apochromatic, and apochromatic. For work in the materials science and engineering field, non-achromatic lenses are not used.

Achromatic lenses are corrected for spherical aberration such that all rays within a limited wavelength range of the spectrum (yellow-green: 500 to 630 nm) are brought into focus essentially at the same position on the optical path. It is common practice to make chromatic correction for two specific wavelengths within this range. Correction for spherical aberration is also made for one wavelength within the range. Achromatic lenses are useful at low to medium magnifications up to 500× magnification, provided that for photography one uses black-and-white film and also a green filter to provide illumination in the yellow-green portion of the spectrum.

Improved correction for chromatic aberration can be obtained by including fluorite (LiF) lens elements in the objective (Kreidl and Rood, 1965). LiF is used because of its higher index of refraction than the normal borosilicate crown glass elements. The effect of proper optical design using fluorite elements in conjunction with borosilicate elements is to shift the focal plane of the longer wavelengths toward that of the image formed by the shorter wavelengths. Objective lenses thus corrected for chromatic aberrations are called fluorites or semiapochromatics. Such objectives provide good correction over the spectral range of 450 to 650 nm.

Apochromatic lenses are corrected for chromatic aberrations at three specific wavelengths and simultaneously for spherical aberration at two specific wavelengths. Wavelengths in the range of 420 to 720 nm are brought to the same focus. To provide optimal correction for spherical aberration, apochromatic lenses are somewhat undercorrected for color. Thus, they should be used with compensating oculars. Apochromatic lenses are of special importance to high magnifications (>800×) and for color photography.

No discussion of objective lenses is complete without comments on working distance and on depth of focus. Normal working distances (the distance between the front element of an objective lens and the specimen) are rather small even for relatively low-magnification objectives. The working distance may range from ∼1 cm for a 5× objective to a fraction of a millimeter for a 100× objective.

Objective lenses designed to have larger working distances for special applications are available, but they are resolution-limited due to their smaller numerical apertures. Consequently, long-working-distance objectives are not employed for high-magnification work.

Depth of focus is a property of an objective lens related to its numerical aperture. The depth of focus is the distance normal to the specimen surface that is within acceptable focus when the microscope is precisely focused on the specimen surface. Depth of focus is of concern when the specimen surface is not perfectly flat and perpendicular to the optic axis. In practice, the preparation of specimens to obtain flatness, or the depth of etching, become important factors in the utility of a given objective lens. Based on a criterion devised by Rayleigh, the depth of focus is given by $d \approx \pm n\lambda/(\text{N.A.})^2$, where $d$ is the depth of focus, $n$ is the index of refraction for the medium occupying the space between the objective lens and the specimen, $\lambda$ is the wavelength of illumination, and N.A. is the numerical aperture (Gifkins, 1970).

Thus, for a 0.65 N.A. objective operating in air with illumination of 530-nm wavelength, the depth of focus is ∼2.35 times the wavelength of light. This means that the surface of the specimen must be flat to within ∼4.7 times the wavelength of light or, in this case, 2500 nm.

Oculars (eyepieces) used in modern microscope are of four main types: Huygenian, Ramsden (orthoscopic), compensating, and wide-angle (Restivo, 1992).

Huygenian oculars (Fig. 2A) consist of two simple lenses fitted with a fixed-field diaphragm located between the field lens and the eye lens. A real image is focused at the plane of the field diaphragm. Such oculars are basic low-magnification eyepieces with little correction for objective lens aberrations.

Ramsden oculars (Fig. 2B) contain the same components; however the field lens is in an inverted position and the field diaphragm is located below the field lens. As in the Huygenian ocular, a real image is located at the plane of the field diaphragm. Ramsden oculars are normally used for intermediate magnifications.

Compensating oculars have the greatest amount of correction for chromatic variations in magnification, as well as curvature of field. These eyepieces are recommended for use at high magnification and for color photomicrography with apochromatic objectives.

Wide-angle oculars in modern instruments are also designed as high-eyepoint lenses, giving a much larger relief distance between the eye lens and the eye. They contain many lens elements and their design may vary considerably from manufacturer to manufacturer. Being highly corrected for chromatic aberration and for distortion, they are often the only oculars supplied with a microscope, particularly when the instrument is fitted with flat-field objectives.

### Practical Adjustment of the Microscope

Simplified schematic diagrams of transmitted- and reflected-light microscopes are given in Figure 3. Experienced microscopists have learned that the most successful use of any microscope begins by following the light path

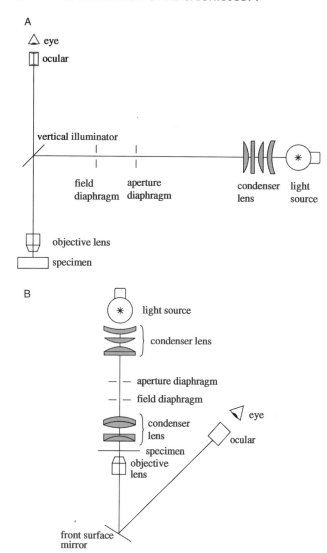

**Figure 3.** Schematic diagrams of (**A**) inverted reflected-light and (**B**) transmitted-light microscopes. The important components are labeled. Note that the transmitted-light microscope (panel B) does not have a vertical illuminator.

diaphragm. In some instruments, the light path is sufficiently open so that one can visually inspect the focus of the lamp on a closed-aperture diaphragm. In closed systems, specific instructions for obtaining proper focus are usually provided with the instrument. Centering of the light source is accomplished by adjustments that laterally translate the lamp in two directions. A focusing adjustment moves the lamp closer to or further away from the condenser to focus the lamp at the plane of the aperture diaphragm.

The aperture diaphragm must be properly adjusted to obtain the optimum resolution commensurate with image contrast, for either a transmission- or reflected-light microscope. The full numerical aperture of the objective lens must be utilized to obtain optimum resolution. From a practical standpoint, the aperture diaphragm is interposed in the optical path to control the diameter of the pencil of light so that it just fills the full circular area of the rear lens group of the objective lens. In a transmitted-light microscope the aperture diaphragm is located in the condenser lens barrel, while in a reflected-light microscope the aperture diaphragm is located in the optical path just beyond the condenser (collector) lens of the light source. In both instruments, the image of the diaphragm is focused in a plane coincident with the rear element of the objective lens. The aperture diaphragm is always partially closed to reduce internal light scattering within the microscope, a circumstance that degrades contrast in the image. In practice, the method for obtaining the optimum setting of the aperture diaphragm is to open the diaphragm fully, bring the specimen into focus, and then close the diaphragm slowly until the image begins to dim perceptibly. At this point the aperture is opened slightly to bring the image back to full brightness.

The image of the field diaphragm is focused on a plane coincident with the plane of the specimen. Its function is to reduce extraneous light scatter between the specimen surface and the front element of the objective in order to enhance contrast. To adjust the field diaphragm, one starts with it in the full-open position and gradually closes it while viewing the focused specimen either through the eyepieces or on a photographic viewing screen. When the image of the diaphragm is observed to just intrude into the field of view or into the photographic field, it should be opened slightly.

Vertical illuminators are required for reflected-light microscopes, because the specimens are opaque and the objective lens must act both to supply light to, and to collect reflected light from, the specimen. The light path must be passed to the specimen through the objective lens, then allow the reflected image-forming light to pass back through the objective to the ocular. From Figure 3A, note that the light path is diverted 90° from the horizontal to the specimen, which rests inverted on the stage (inverted microscope). This arrangement is of particular utility in using specimens that must have prepared surfaces, because it does not require that the specimen have a back surface exactly parallel to the prepared surface. Several schemes have been used for vertical illuminators, including various prism configurations and thin, optically flat glass plates. Of these, the thin, optically flat glass plate

through the instrument and making necessary adjustments, starting with the source of illumination. By consistently following the sequence, the chance of failure in the form of a poor photomicrograph or poor visual image is greatly reduced.

Illumination of the specimen is accomplished by two or more components including the lamp and a condenser or collector lens. Sometimes, the aperture diaphragm, field diaphragm, heat-absorbing filter, narrow band-pass filter, and neutral-density filter are also included as components of the illumination system. This author chooses not to include the additional components as parts of the illumination system, although some microscope makers mount them in a single modular unit of the instrument.

Proper centering and focusing of the light from the illuminator is essential to obtaining satisfactory results with any optical microscope. The image of the lamp filament or arc must be focused at the plane of the aperture

(or disc) is almost universally used in recent instruments. Generally, vertical illuminators have no adjustment controls except for a provision for dark field illumination, a topic that is discussed in REFLECTED-LIGHT OPTICAL MICROSCOPY.

The objective lens requires only two adjustments: centering the optic axis of the lens with the optic axis of the instrument and focusing the lens by moving it closer to or further away from the specimen. Focusing the objective is as simple as moving the focusing adjustments while viewing the specimen through the eyepieces and observing when the specimen appears to be sharp. A number of instruments have their objective lens mounted in a turret to facilitate changing from one objective to another. Many instruments are said to have "parafocaled" objectives. If the objectives are parafocaled, one may readily switch objectives without having to refocus the instrument. In practice, however, it is wise to use the fine-focus adjustment each time an objective is changed in order to assure that the optimum focus is obtained.

Centering of the objective lens on an instrument with an objective turret is accomplished by centering on the objective through the centering of the axis of rotation of the turret. This involves moving the axis of rotation of the turret by shifting it in one or two orthogonal directions of a plane normal to the axis of rotation. Such an operation is not recommended except when performed by competent service personnel. Fortunately, once a turret is centered and locked into the instrument, frequent centering is not required. Other instruments, in which objectives are placed in the instrument individually, require that each lens be centered individually by adjustment set screws on the mounting plate of each lens. Often the instrument contains one "standard" objective that is centered at the factory in the final fitting stages of instrument assembly. The mounting plate of this objective has no centering adjustments. Other objectives may be centered individually using the "standard" lens as a reference, provided the instrument is fitted with a rotating stage that is also centered with the optic axis of the instrument. A simple method involves starting with the "standard" objective, aligning a reflective cross-hair (or reflective point target) on the stage with the center of the field as viewed through the ocular, and making certain that the stage is itself centered. The stage is then rotated to ascertain that the cross-hair stays in the center of the field as the stage is rotated. Without moving the stage or the cross-hair, a centerable objective is substituted for the "standard" objective. If the objective is centered, the cross-hair will again fall in the center of the field and will remain there when the stage is rotated. To center the objective, it will be necessary to use the adjustment set screws on the mounting plate to move the optic axis of the objective in $x$ and $y$ directions in a plane normal to the optic axis. A number of instruments include special accessories for centering the objectives.

Ocular adjustment amounts only to focusing the ocular. Instruments with a single eyepiece generally do not have an adjustment for focus, because the eyepiece is set at its proper focal distance from the objective. Since most microscopes now have binocular eyepieces, some care must be taken to assure that the eyepieces are correctly focused to match the eyes of the user. One eyepiece has a focusing ring, while the other is fixed. Depending on the instrument, focusing methods may vary; however, they are all variations on the method outlined in the following. A stage micrometer, or other similar device, is placed on the stage and brought into focus with the fixed eyepiece using the objective focusing adjustment. Once this is done, the eyepiece focusing ring should be used to focus the image for the other eye. This method accounts for differences between the two eyes, and is necessary to reduce eye strain from long sessions with the microscope.

It is important to emphasize that if the microscopist normally wears eyeglasses, they should be worn while using a microscope. In recent years wide-field objective lenses and matching wide-field, high-eyepoint oculars have been introduced. Matching optics of this type provide a wider field of view of the specimen and are also particularly suited to users who wear glasses. Multifocal eyeglasses (bifocal or trifocal) present special problems that can be overcome by the adaptability of the microscopist.

Autofocusing has yet to be applied to microscopes except in the case of instruments designed to make precise measurements. Such devices are only of use when one of several turret-mounted objective lenses is brought into sharp focus. The autofocusing feature will then cause each of the other lenses in the turret to be brought into sharp focus when they are rotated into the optic axis of the microscope. Critical focusing depends on the skill of the microscopist in being able to judge when sharp focus is attained. Focusing for edge sharpness could probably be developed for instruments used in conjunction with image analysis computers, wherein the focusing would be accomplished by feedback of the contrast gradient at a sharp edge. However, obtaining the best overall focus for the entire field of view might be difficult, either because of a limited range of the gray scale or because of the lack of flatness of specimens.

Devices for calibrating resolution or the depth of focus are generally not used. Observable degradation of the image is a signal to inspect the optical components for contaminants and for physical defects such as scratched or etched surfaces. Regular annual service by a competent service technician is recommended, to maintain the microscope in optimum working condition.

Damage from contamination and damage to objectives is, for the most part, up to the microscopist in maintaining a clean environment for the microscope. The use of appropriate lubricants for the mechanical parts of the instrument is of paramount importance. Physical damage of the front element of the objective lens comes primarily from two sources: (1) contact of the front element with the specimen and (2) etching of the front element by chemical reagents used in specimen preparation. Careful operation by the microscopist is the most effective way to keep the objective lens from contacting the specimen; however, some microscopes, usually upright instruments, employ a stage lock to prevent the objective from contacting the specimen. To prevent contamination of the objective lenses with specimen-preparation reagents, it is essential that specimens be thoroughly cleaned and dried prior to placing them on the microscope. This is particularly important when HF or other corrosive reagents have been used in specimen preparation.

In the past, special stages for providing special specimen environments have been available as regular production accessories for some instruments. However, with the advent of the scanning electron microscope (SEM), most studies that require hot or cold environments are now undertaken in the SEM. Studies in liquid environments are especially difficult, particularly at high magnifications, because the index of refraction of the liquid layer interposed between the specimen and the objective lenses is different from that of air. Therefore, some corrections can be designed into objectives for certain biological microscopes. Objectives designed for use with glass slides and coverslips are also corrected for the presence of the thin glass layer of the coverslip, which is of the order of 0.1 mm thickness. A number of specialized environmental stages have been designed and constructed by microscopists.

Photomicrography, on film and with charge-coupled devices, as well as specimen preparation, are of paramount importance to successful optical microscopy for the materials scientist or engineer. These topics are to be covered in subsequent units of this chapter.

## LITERATURE CITED

Gifkins, R. C. 1970. Optical Microscopy of Metals. Elsevier Science Publishing, New York.

Jenkins, F. A. and White H. E. 1957. Fundamentals of Optics, 3rd ed. McGraw-Hill, New York.

Kehl, G. L. 1949. The Principles of Metallography Practice. McGraw Hill, New York.

Kreidl, N. and Rood, J. 1965. Optical Materials. Vol. I. Applied Optics and Optical Engineering. Academic Press, New York and London.

Restivo, F. 1992. A Simplified Approach to the Use of Reflected Light Microscopes. Form No. 200-853, LECO Corporation, St. Joseph, MI.

Vander Voort, G. 1984. Metallography Principles and Practice. McGraw-Hill, New York.

## KEY REFERENCES

Gifkins, 1970. See above.

*An excellent reference for microscope principles and details of techniques, using copious illustrations.*

Jenkins and White, 1957. See above.

*A classic text on optics that has been a mainstay reference for students of physics and engineering.*

Kehl, 1949. See above.

*A standard text on metallography used extensively for many years. It contains much useful information concerning microscopy of opaque specimens.*

Kreidl and Rood, 1965. See above.

*Contains a compilation of the optical properties of a large variety of materials.*

Restivo, 1992. See above.

*While this is a brief treatment of the use of reflected light microscopes, primarily from the standpoint of service technicians, it contains excellent schematic drawings with simplified text.*

Vander Voort, 1984. See above.

*This is the current standard text for students of metallography and a valuable reference for practicing metallographers. Chapter 4, dealing with light microscopy, is a superior overview containing 100 references.*

## INTERNET RESOURCES

http://www.carlzeiss.com
*Carl Zeiss Web site.*

http://www.olympus.com
*Olympus Web site.*

http://www.nikon.com
*Nikon Web site.*

http://www.LECO.com
*LECO Corporation Web site listing special Olympus reflected-light microscopes.*

http://microscopy.fsu.edu
*This site provides a historical review of the development of the microscope.*

RICHARD G. CONNELL, JR.
University of Florida
Gainesville, Florida

# REFLECTED-LIGHT OPTICAL MICROSCOPY

## INTRODUCTION

Reflected-light microscopes were developed for imaging the surfaces of opaque specimens. For illumination of the specimen, early microscopes used sunlight played on the specimen surface by mirrors and/or focused by simple condenser lenses. Optical microscopes began to evolve toward modern instruments with the understanding that engineering properties of materials were dependent upon their microstructures. Current higher-magnification instruments ($50\times$ to $2000\times$), for obtaining images of polished and etched specimens, employ vertical illumination devices to supply light to the specimen along the optical axis of the microscope. Vertical illumination overcomes the shadowing of surface features, which are enhanced by off-axis illumination, and improves the attainable resolution. On the other hand, lower-power ($5\times$ to $150\times$) stereo microscopes, most commonly used for imaging bulk specimens such as microelectronic circuits or fracture surfaces, utilize off-axis lighting supplied by focused lamps or fiber-optic light systems.

Microscopy, regardless of the instrumental technique, has as its sole objective the production of visual images that can be further analyzed by counting measurements to quantify any of a number of microstructural properties including grain size, the volume fractions of phases, count (or identification) of microconstituents or microstructural features such as twins, number of inclusions, surface area of grains and second phases, and more esoteric microstructural measurements such as the total mean curvature of the grain boundary in a given specimen. Quantification of microstructural features may be accomplished by manual counting while viewing the image through the microscope, taking measurements from photographic or digital images, or using a computer for digital image analysis. Obtaining quantitative microstructural information constitutes the specialized field of quantitative microscopy or stereology, based upon geometric and topological considerations and statistics. Stereology is a branch of quantitative microscopy that provides for the inference of three-dimensional structure information from two-dimensional measurements. The reader is referred to DeHoff and Rhines (1968), Gifkins (1970), and Vander Voort (1984) as primary references on quantitative microscopy.

While reflected-light microscopy alone is not intended for the measurement of material properties, coupled with the techniques of quantitative microscopy it does provide the connecting links among properties, chemical composition, and processing parameters. However, an experienced microscopist can obtain much useful qualitative information related to processing and properties. For instance, one may assess the degree of annealing of a metal or alloy by the amount of the microstructure that appears to consist of recrystallized grains. In a production setting, a microscopist may quickly determine if a material meets specified properties merely by observing the microstructure and judging it on the basis of experience.

It is the intent of this unit to describe the more common techniques of reflected-light microscopy, but not to delve into microstructural property measurement. Emphasis will be placed on the use of the higher-power instruments (metallographs), because of the variety of useful illumination modes and image-forming techniques that are available on these instruments.

## PRACTICAL ASPECTS OF THE METHOD

### Reflected-light Microscopes

The reflected-light microscope, shown by the light-path diagram in Figure 1, contains a vertical illuminator, in this case employing a half-mirror to direct light to the

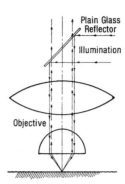

**Figure 2.** Light-path diagram for a plain-glass reflector for a reflected-light microscope (courtesy LECO Corporation).

specimen and to allow light from the objective lens to pass to the ocular (eyepiece). In arrangement of the specimen stage, with respect to the axis of the horizontal incoming light path, there are two types of microscopes—upright and inverted. Generally, the inverted instrument is preferred because it ensures that a specimen surface is always normal to the optic axis of the objective lens. Further, it allows for easier observation of large unmounted specimens and specimens of irregular shape that have a prepared surface (Vander Voort, 1984).

Vertical illuminators have taken several forms during their evolution. For the most part, two types of vertical illuminators are in use today; the plain-glass reflector and the half-mirror reflector. The plain-glass reflector consists of a very thin, optically flat glass plate inclined at 45° with respect to the path of the incoming illumination (Fig. 2). Part of the illumination is reflected at 90° through the objective to illuminate the specimen. A portion of the light that returns from the specimen through the objective lens passes through the glass plate on to the ocular. Note that the objective first acts as an illuminating (condenser) lens and second as the image-forming lens. When the plain-glass reflector is properly centered, plain axial illumination is the result and no shadows of specimen surface relief are formed. While the loss of light intensity is significant, this type of reflector can be used with any objective. The half-mirror reflector is a totally reflecting surface, usually silvered glass, with a thin clear coated glass area in its center. The clear central area is elliptical in shape, so that when the reflecting plate is placed at 45° in the optical path a round beam of light can be reflected (Restivo, 1992). An advantage of this type of reflector is that it is particularly well suited for dark-field illumination, and it also works well with bright-field illumination and polarized light (see the discussion below under "Illumination Modes and Image-Enhancement Techniques" for a description of these illumination modes). Light rays reflected back through the objective lens from the specimen surface pass through the central clear region of the reflector to form an aerial image that is further magnified by the ocular. Because the reflecting surface reflects all of the incident light, there is no loss of light intensity as with the plane glass reflector (Restivo, 1992). Figure 3 shows the light path for the half-mirror reflector as it is used in dark-field illumination mode. Other vertical illuminator/

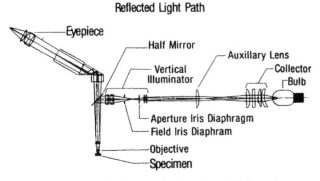

**Figure 1.** Light path diagram for a reflected-light microscope (courtesy LECO Corporation).

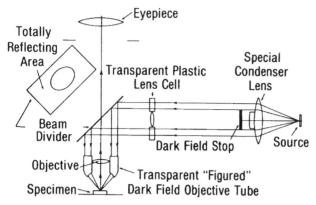

**Figure 3** Light path diagram for a half-mirror reflector for a reflected-light microscope. (courtesy LECO Corporation).

reflector schemes have been used with some success; however, they have fallen out of favor because of limitations or expense in producing suitable reflectors.

### Illumination Modes and Image-Enhancement Techniques

Provisions for a number of variations in illumination modes and image enhancement techniques may be included as accessories on many of the current production microscopes. Among the more common of these are: bright-field illumination, oblique-light illumination, dark-field illumination, polarized-light, sensitive-tint, phase-contrast, and differential-interference-contrast. Each of these will be discussed in the following paragraphs.

Bright-field illumination is the most widely used mode of illumination for metal specimens that have been prepared by suitable polishing and etching. The incident illumination to the specimen surface is on axis with the optic axis through the objective lens, and is normal to the specimen surface. As expected, those surfaces of the specimen that are normal or nearly so to the optic axis reflect light back through the objective lens, giving rise to bright portions of the image. Those features of the microstructure that have surfaces that are not normal to the optic axis appear dark and thus give the image contrast (Fig. 4). Image contrast also occurs in bright-field illumination because of variations in reflectivity within a specimen. For example, in properly prepared specimens of irons and steels that contain MnS (manganese sulfide)

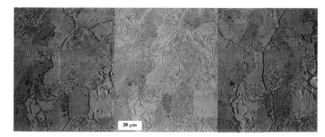

**Figure 5.** Bright-field photomicrograph (center) of pearlite and proeutectoid cementite in an annealed high-carbon steel. The left and right panels show oblique-light photomicrographs of the same field, with the right panel illuminated with light from the left and the left panel illuminated with light from the right.

as a microconstituent, the MnS constituent appears somewhat darker than the surrounding $\alpha$-ferrite iron, and possibly lighter than $Fe_3C$ (cementite). The image contrast in this case is due to the optical properties of the constituents that render their polished surfaces more or less reflective. In fact, if the specimen preparation is very well done, the MnS appears to have a smooth dark-slate color. Certain specimen-preparation techniques and spectral image enhancements provide a variety of colors to establish image contrast. The vast majority of metallographic work utilizes bright-field illumination. A typical bright-field photomicrograph appears in Figure 5 (center panel).

Oblique-light illumination is a variation of bright-field illumination wherein the condenser system is shifted off axis. The result is that relief in the specimen surface is shadowed. While the image appears to possess enhanced contrast, and detailed features become more easily observed, the optical resolution is decreased; fine details become broadened. In addition, by shifting the incoming light beam off axis, the effective numerical aperture of the objective lens is decreased, thereby reducing resolution (see OPTICAL MICROSCOPY). To some extent, compensation for this deficiency is made by opening the aperture diaphragm, but at some sacrifice to contrast. On some microscopes oblique illumination is obtained by shifting only the aperture diaphragm off axis, an adjustment that requires that the aperture diaphragm be opened, or else the specimen will be unevenly illuminated. Oblique illumination is illustrated in the photomicrographs of Figures 5, left and right. These photomicrographs are of the same specimen and of the same area as that of the bright-field photomicrograph of Figure 5 (center). Both the left and right panels of Figure 5 are taken under conditions of oblique light, but the off-axis shifts are in opposite directions. Note how certain microstructural features appear raised in one photomicrograph, while the same features appear depressed in the other.

Dark-field illumination is a useful technique to enhance image contrast by making the normally bright areas of an image dark while making dark features light. Dark-field illumination is effective because of the perception characteristics of human vision. It has been demonstrated that a small white disc on a black background is more easily perceived than a small black disc on a white background. The size limit of the white disc on the black background

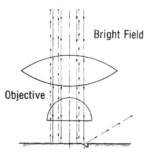

**Figure 4.** Light path diagram for bright-field illumination (courtesy LECO Corporation).

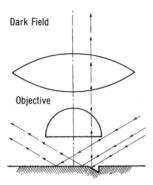

**Figure 6** Light path diagram for dark-field illumination (courtesy LECO Corporation).

depends entirely on its reflected luminous flux, or the total light reflected per unit area (Gifkins, 1970).

Dark-field illumination occurs when the light incident on the specimen arrives at a low angle. If the surface is featureless, reflected light does not reenter the objective lens and the image is dark. However, any feature on the surface that is inclined a relatively small angle can reflect light back through the objective lens to form an image. To effect dark-field illumination, a small circular stop is inserted in the light path ahead of the vertical illuminator to produce an annulus of light larger in inside diameter than the back element of the objective lens. The objective lens is surrounded by a "shaped" or "figured" transparent tube that directs the incident light to the specimen at a relatively low angle to the specimen surface. Reflected light from tilted features of the specimen reflects light back into the objective lens (Fig. 6). Figures 7 A and B are bright-field and dark-field photomicrographs, respectively, of a specimen of low-carbon iron. Note that the grain boundaries and inclusions are enhanced in dark-field.

Polarized-light microscopy has significant use in imaging microstructural details in some materials, notably, alloys of uranium, zirconium, beryllium and, from the author's experience, high-purity aluminum (Connell, 1973). Specimens of these materials are difficult to etch to reveal microstructural details. Before the invention of the electron-beam probe, polarized light was used as an aid in identifying inclusions in alloys. Also, polarized light has been of paramount importance in mineralogical work for the identification of various minerals.

Polarized-light microscopy depends upon a characteristic of light that cannot be detected by the human eye. When considering the electromagnetic wave character of light, it is easy to understand that, the greater the amplitude of the wave, the brighter the light. Likewise, variations in wavelength produce variations in color; the shorter the wavelength, the more blue the light, while the longer the wavelength, the more red the light. Two other characteristics of light—phase and polarization—cannot be directly detected by the eye. From an elementary point of view, one may consider that light waves travel in a straight line, and the light waves oscillate (vibrate) along the line in random directions around the line. Under certain conditions of reflection defined by Brewster's law, and when light passes through certain transparent crystalline

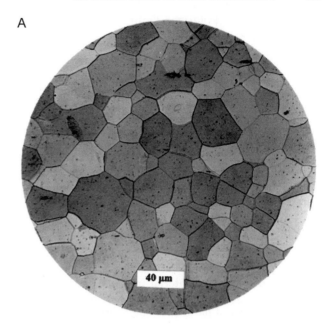

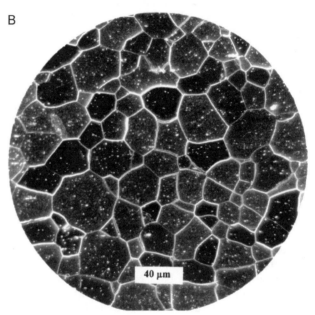

**Figure 7.** Photomicrographs of ferrite in low-carbon iron: (**A**) bright field; (**B**) dark field.

materials, some of the vibration directions around the light axis become attenuated, and there may be rotation of other directions, so that certain directions of vibration are predominant (Jenkins and White, 1957; Gifkins, 1970). From a practical standpoint, when such interaction of light waves with matter occurs, vibrations of the light are confined to one plane, and the light is said to be plane-polarized. It is to be further stated that, light being electromagnetic waves, it is the electrical or magnetic interaction of the light waves with matter that result in polarization (Gifkins, 1970).

In microscopy, the usual way to produce polarized light from a normal source is to pass the light through a transparent plate of some specific optically active material,

## Polarized Light

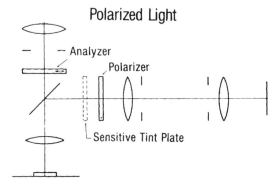

**Figure 8.** Light path diagram for a reflected-light microscope fitted with polarizer and analyzer. Note also the position of the sensitive-tint plate (courtesy LECO Corporation).

called a polarizer. Most polarizers in use in microscopes are made of synthetic material in sheet form, called by its proprietary name, Polaroid. If two sheets of polarizing material are placed parallel to one another separated by some distance (4 or 5 cm), and light is passed through them, one can observe that light passes through the first although it is polarized. The light may also pass through the second, but by rotation of the second sheet, one will observe that the light no longer passes through it. This is the case of extinction, and one can say that the polarizers are crossed. The second sheet, called the analyzer, will not allow the polarized light through, because its plane of polarization is 90° to the first sheet, the polarizer.

Both a polarizer and an analyzer are employed in a polarizing microscope. The polarizer is placed in the light path ahead of the vertical illuminator and the analyzer is placed in the light path between the vertical illuminator and the ocular (Fig. 8). The polarizer usually is fixed so that it will not rotate. When polarized light is incident on the reflective surface of an isotropic specimen at a right angle, the polarization of the reflected light will not be altered (there is no polarization dependence of reflectivity in isotropic specimens). If one rotates an analyzer located in the light path at the back end of the objective lens, extinction will occur when the polarizer and analyzer are crossed. The image turns dark and indicates that the originally polarized light has not been altered by the reflecting surface. If the specimen is replaced with an anisotropic material such as zirconium, whose reflectivity depends on orientation of polarization with respect to crystal axes, then it can be observed that some grains appear dark while others appear in various shades of gray all the way to white, indicating that each grain has a different crystallographic orientation. When the specimen stage is rotated, the orientation of individual grains will change with respect to the polarized light and go through several extinctions as the stage is rotated through 360°. In this way, the microscopist can observe the grain structure in a material for which etching to reveal the grain structure is difficult. An example of a photomicrograph of martensite taken under polarized light is given in Figure 9. Martensite, being body-centered tetragonal, is anisotropic and therefore responds to the polarization of the light according to its crystal orientation.

**Figure 9** Polarized-light photomicrograph of martensite in a hardened steel specimen.

Isotropic metals can be examined by polarized light if they can be etched to develop regular patterns of closely spaced etch pits or facets on each grain. Anodizing the polished specimen surface of isotropic metals can also make the material respond to polarized light. In the case of aluminum, a relatively thick transparent film is formed by anodizing. The observed polarizing effect results from double reflection from the surface irregularities in the film (Vander Voort, 1984). The author has used this technique to observe the subgrain structure in creep-deformed high-purity aluminum. Angular misorientations as small as 2.5° between adjacent subgrains have been observed (Connell, 1973).

Sensitive tint is an image-enhancement technique used with polarized light. Because the human eye is more sensitive to minute differences in color than to minute differences in gray scale, the sensitive-tint technique is a valuable tool in viewing and photographing images that lack contrast. Therefore, preferred orientation may be qualitatively assessed, counts to determine grain size are more easily made, and the presence of twins is more easily detected.

A sensitive-tint plate is placed in the light path of the microscope between the polarizer and the analyzer (Fig. 8). The material of the sensitive plate is a birefrigent. Birefringence occurs in certain anisotropic crystals when these are cut along specific crystallographic planes to produce double refraction (Jenkins and White, 1957). One sees a double image when viewing through such a plate. The sensitive-tint plate (also called magenta-tint or whole-wave plate) is made of quartz sliced parallel to its optic axis. The thickness of the plate is such that it is equivalent to a whole number of wavelengths of light, where the wavelength is in the middle of the visual spectrum (green, 5400 Å). Rays of plane-polarized light pass through the sensitive-tint plate and become divided

## Phase Contrast

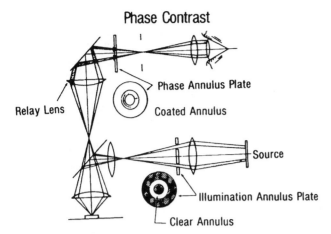

**Figure 10.** Light path diagram for phase-contrast microscopy. Note the positions of the illumination- and phase annulus plates in a reflected-light microscope (courtesy LECO Corporation).

into ordinary rays and extraordinary rays (Gifkins, 1970), the latter being refracted at some angle to the former. When the ordinary and extraordinary rays emerge from the plate, they are out of phase by one wavelength. The two rays recombine to form a single plane-polarized ray that is extinguished by the analyzer, while all other wavelengths pass through the analyzer. The result is that white light minus the green light excluded by the analyzer produces light that is magenta in color. Typically, the hue imparted to sensitive-tint images have a magenta hue, with differences in grain orientation displayed as different colors. Rotation of the specimen causes each grain to change color. Images are striking and usually aesthetically pleasing. Color control can be obtained by rotation of the sensitive-tint plate through a range of ~15° to 20°.

Phase-contrast illumination is an image enhancement that is dependent upon another property of light that cannot be observed by the eye. Subtle variations in polished and/or etched metallographic specimens sometimes are so minute that phase differences from the difference in optical path produce no image contrast. Differences in height as little as 5 nm can be observed using phase contrast (Vander Voort, 1984; Restino, 1992).

The physical arrangement of the microscope is seen in Figure 10, where an illumination annulus and a phase annulus are inserted in the optical path as shown. The illumination annulus is opaque, with a clear central annulus, while the phase annulus is a clear glass onto which two separate materials have been applied by vacuum coating. One material is a thin, semitransparent metallic film (antimony) designed to reduce transmission of the annulus, while the other (magnesium fluoride) is for the purpose of introducing a 1/4 wavelength difference between the light that passes through the annulus versus the light that passes through the other part of the phase plate. An alternate construction for the phase plate is to grind a shallow annular groove in the glass (Gifkins, 1970).

In using phase-contrast illumination, proper and precise adjustment of the instrument is critical for obtaining satisfactory images. Some instruments have sets of matching illumination and phase annuli for use with different

objective lenses. It is critical that the phase annulus and illumination annulus match. It is generally not sufficient merely to focus on the specimen and insert the annuli to obtain enhanced contrast, because the specimen surface is not generally exactly normal to the optic axis. To make the appropriate adjustment, the specimen must be sharply focused and the image of the illuminating annulus must be centered with the phase annulus. This requires replacing the ocular with an auxiliary lens (Bertrand lens) that is focused on the phase annulus. Once the ocular is reinstalled, phase contrast is realized. The technique is at best fussy, because specimen translation often throws the annuli out of coincidence and the alignment must be done all over again.

Phase-contrast illumination gained some popularity in the early 1950s for metallurgical work, but today it has lost favor because of the critical and frequent instrumental adjustments required. The technique remains viable for image enhancement in transmission microscopy, particularly for biological specimens.

Differential interference contrast (DIC; also known as Nomarski optics), because of its versatility, has become an image-enhancement technique highly favored by microscopists. A number of other interference techniques have been developed over the years, some requiring special instruments or elaborate accessories attached to rather standard instruments, for example two-beam interference microscopy. Of all these, the device that produces contrast by differential interference (Padawar, 1968) appears to be the most readily adaptable to standard instruments, because it requires no internal optically flat reference surfaces. Images produced by DIC display some of the attributes of other illumination modes including dark-field, polarized light, and sensitive-tint. Nonplanar surfaces in specimens are contrast-enhanced, grains in anisotropic materials display extinction behavior, and the color can be of benefit in enhancing perceived contrast.

Differential interference contrast can be classified as an application of polarizing interference, wherein the underlying principle is image duplication in some birefringent material. The image consists of two parts, one formed by the ordinary ray and a second formed by the extraordinary ray.

A Wollaston prism (Jenkins and White, 1957; Gifkins, 1970) is the birefringent unit used in DIC. This device consists of two quartz plates, $w_1$ and $w_2$, cut so that their optic axes are at right angles to each other and so that both make a right angle with the optic axis of the microscope (Fig. 11). The Wollaston prism is located at the focal plane of the objective and can be moved laterally across the optic axis. In operation, the incident light is plane polarized after passing through the polarizer, and the critical setting of the illuminating system is such that the beam is focused in $w_1$ at point $x$. The reflected light from the specimen is focused in $w_2$ at point $z$. Points $x$ and $z$ are symmetrical in the Wollaston prism; thus the path-length difference, $d$, of the ordinary and extraordinary rays in their passage through the prism to the specimen is the same as that produced on their return passage through the prism, but it is of opposite sign. So the path-length difference after passing through the Wollaston prism the second time, $D$,

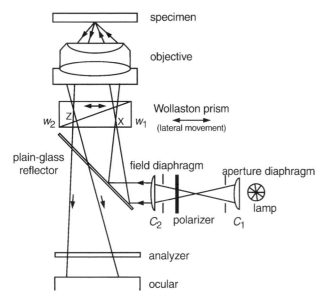

**Figure 11.** Light path diagram for differential interference contrast (DIC) microscopy. Note that the two sections ($w_1$ and $w_2$) of the Wollaston prism have optic axes that are normal to each other, and both optic axes are normal to the light path of the microscope. The axis in $w_1$ is normal to the plane of the figure, while that in $w_2$ is in the plane as shown. Points $x$ and $z$ are symmetrically located about the interface between $w_1$ and $w_2$.

becomes zero. With a crossed polarizing analyzer in place, and the equalization of path-length differences between the ordinary and extraordinary rays ($D = 0$), a dark field results. The analyzer suppresses both the ordinary rays and the extraordinary rays. Should a specimen contain tilted surfaces such as those that exist at grain boundaries, twins, or slip bands, the reflected rays will pass through the Wollaston prism at nonsymmetrical positions, and the path-length differences between the ordinary and extraordinary rays will not be equalized ($D \neq 0$); therefore, the ordinary and extraordinary rays will not be suppressed by the analyzer. Any path-length difference between the ordinary and extraordinary rays is proportional to the amount of tilt, so that the intensity of the light passed to the ocular is also proportional the amount of tilt.

An important feature of DIC is the ability to introduce color to the image. By laterally shifting the Wollaston prism across the optical path, the path-length difference, $D$, can be set to values other than zero, and the background takes on a hue dependent upon the path-length difference. For example, if the position of the prism is set so that $D = 0.565$ μm (a wavelength in the yellow region of the spectrum), the background takes on a purple color—i.e., white light minus yellow. Portions of the specimen surface that are tilted will impose other path-length differences, which will result in other colors.

Figure 12A and B are photomicrographs that illustrate a comparison between bright-field illumination and DIC, respectively.

### Sample Preparation

Successful use of reflected light microscopes depends to a large measure on specimen preparation. Without a

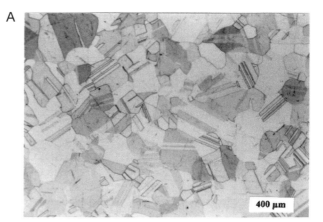

**Figure 12.** Photomicrographs of α-brass (**A**) bright-field, (**B**) DIC. Grain and twin boundaries are easily seen in the DIC image making it possible to determine grain size easily.

thorough discussion of specimen-preparation techniques, it is sufficient to make the reader aware that opaque specimens must have a properly prepared surface from which light may be reflected. Such a surface is usually prepared by mechanical grinding and polishing using appropriate abrasives until the surface is scratch-free and as specular as a mirror (see SAMPLE PREPARATION FOR METALLOGRAPHY). In most cases it is necessary to chemically (or electrochemically) etch the surface to reveal the details of the microstructure.

Metallographic preparation techniques have been developed for a wide variety of materials, but with some the metallographer must use his intuition and experience. An excellent source related to specimen preparation is Vander Voort's (1984) book.

### PROBLEMS

When inadequate results are obtained with the use of reflected-light microscopy, they are most often associated either with specimen preparation or the inappropriate setting of the controls of the instrument. Improper or poor specimen preparation may introduce artifacts such as surface pitting or inclusion of abrasive particles in the surface. Such artifacts can be confused with second-phase precipitates. When some portion of the image appears to be in focus, while others appear to be poorly focused, the problem is usually associated with lack of flatness of the specimen surface. In addition, one should pay attention to the action of polishing abrasives on the specimen surface. Some combinations of abrasive materials result in undesired chemical attack of the specimen surface, and the true structure of the specimen may not be revealed. Poor etching (over- or underetching) is a major concern to the metallographer. Overetching often causes what appears to be lack of resolution; fine features can be difficult or impossible to resolve. Underetching may leave some fine details not revealed at all. A better understanding of some aspects of metallographic specimen preparation is given in SAMPLE PREPARATION FOR METALLOGRAPHY.

A practical guide to the setting of the controls is given in OPTICAL MICROSCOPY. Some of the common errors in making settings that degrade resolution of the image are: (1) too small an aperture diaphragm setting, (2) incorrect selection of objective lens, and (3) too high a magnification. Closing the aperture diaphragm beyond the point where the rear element of the objective lens is just covered with light will not allow the objective lens to collect all the diffraction information needed to realize the full resolution capabilities of the lens. The numerical aperture (N.A.) of the objective lens must be consistent with the intended magnification of the final image. As a practical upper limit, the final magnification should be no greater than 1000 times the numerical aperture of the objective lens. For a 0.65 N.A. objective lens, the final magnification should be $650\times$ or less.

Dirty or damaged optical components seriously degrade image quality. Dirt and dust on lens, mirror, or prism surfaces, as well as degradation of lens coatings, are of primary concern. It is wise not to attempt optical surface cleaning other than that which can be done with a soft brush or light blast of air. It is well worth the expense to have a skilled service person thoroughly clean the microscope on an annual basis.

The forgoing discussions are not intended to be a comprehensive treatment of reflected-light microscopy. A number of interesting special techniques have been excluded because they require specialized instrumentation. Almost all of the illumination modes and image-enhancement techniques discussed can be done with current research-grade metallographs. One feature that most modern instruments do not include is a precision rotating stage that allows for rotation of the specimen without disturbing the focus of the instrument. Such a stage was incorporated on instruments manufactured through the mid-1970s. Stage rotation, wherein the specimen remains normal to the optic axis of the instrument, greatly enhances the ability to use polarized light, sensitive tint, and differential interference contrast.

## LITERATURE CITED

Connell, R. G. 1973. Microstructural Evolution of High-Purity Aluminum During High-Temperature Creep. Ph.D. thesis, University of Florida, Gainesville, Fla.

DeHoff, R. T. and Rhines, F. N. 1968. Quantitative Microscopy. McGraw-Hill, New York.

Gifkins, R. C. 1970. Optical Microscopy of Metals. Elsevier Science Publishing, New York.

Jenkins, F. A. and White, H. E. 1957. Fundamentals of Optics. McGraw-Hill, New York.

Padawar, J. 1968. The Nomarski interference-contrast microscope, an experimental basis for image interpretation. *J. R. Microsc. Soc.* 99:305-3–49.

Restivo, F. A. 1992. Simplified Approach to the Use of Reflected-Light Microscopes. Form No. 200–853. LECO Corporation, St. Joseph, Mich.

Vander Voort, G. F. 1984. Metallography, Principles and Practice. McGraw-Hill, New York.

## KEY REFERENCES

DeHoff and Rhines, 1968. See above.

*A textbook of the well-established fundamentals underlying the field of quantitative microscopy, which contains the details of applied measurement techniques.*

Gifkins, 1970. See above.

*Chapter 6 of this book provides an excellent treatment of interference techniques.*

Kehl, G.L. 1949. The Principles of Metallographic Laboratory Practice. McGraw-Hill, New York.

*Excellent reference on microscopes and photomicrography for use by technicians, although some of the instrumentation is dated.*

Vander Voort, 1984. See above.

*Chapter 4 of this book is a particularly good overview reference for reflected-light microscopy.*

## INTERNET RESOURCES

http://microscopy.fsu.edu/primer/resources/tutorials.html
*List of educational resources for microscopy.*

http://www.people.virginia.edu/jaw/mse310L/w4/mse4-1.html
*Optical micrography reference.*

RICHARD G. CONNELL, JR.
University of Florida
Gainesville, Florida

# PHOTOLUMINESCENCE SPECTROSCOPY

## INTRODUCTION

Photoluminescence (PL) spectroscopy is a powerful technique for investigating the electronic structure, both intrinsic and extrinsic, of semiconducting and semi-insulating materials. When collected at liquid helium temperatures, a PL spectrum gives an excellent picture of overall crystal quality and purity. It can also be helpful in determining impurity concentrations, identifying defect complexes, and measuring the band gap of semiconductors. When performed at room temperature with a tightly focused laser beam, PL mapping can be used to measure micrometer-scale variations in crystal quality or, in the case of alloys and superlattices, chemical composition.

The information obtainable from PL overlaps with that obtained by absorption spectroscopy, the latter technique being somewhat more difficult to perform for several reasons. First, absorption spectroscopy requires a broadband excitation source, such as a tungsten-halogen lamp, while PL can be performed with any above-band-gap laser source. Second, for thin-film samples absorption spectroscopy requires that the substrate be etched away to permit transmission of light through the sample. Third, for bulk samples the transmitted light may be quite weak, creating signal-to-noise ratio difficulties. On the other hand, absorption spectroscopy has the advantage of probing the entire sample volume, while PL is limited to the penetration depth of the above-band-gap excitation source, typically on the order of a micrometer.

The areas of application of PL also overlap somewhat with Raman spectroscopy (RAMAN SPECTROSCOPY OF SOLIDS), but the latter is much harder to perform because of the inherent weakness of nonlinear scattering processes and because of its sensitivity to crystallographic orientation. On the other hand, this sensitivity to orientation gives Raman spectroscopy a capability to detect variations in crystallinity that PL lacks.

There are also more elaborate techniques, such as photoluminescence excitation (PLE), time-resolved PL, and resonant Raman scattering, but these techniques are more useful for basic scientific investigations than for routine crystal characterization.

## PRINCIPLES OF THE METHOD

When a semiconductor is illuminated by a light source having photon energy greater than the band-gap energy ($E_g$) of the material, electrons are promoted to the conduction band, leaving holes behind in the valence band. When an electron-hole pair recombines, it emits a photon that has a wavelength characteristic of the material and the particular radiative process. Photoluminescence spectroscopy is a technique for extracting information about the electronic structure of the material from the spectrum of light emitted.

In the basic process of PL at low temperature (liquid nitrogen temperature and below), an incoming photon is absorbed to create a free electron-hole pair. This transition must be essentially vertical to conserve crystal momentum, since the photon momentum is quite small. At cryogenic temperatures the electrons (holes) rapidly thermalize to the bottom (top) of the conduction (valence) band by emitting phonons. For this reason there is typically no observable luminescence above the band-gap energy. Once they have relaxed to the band edges, an electron and hole can bind together to form an exciton, which then recombines radiatively as described below. Alternatively, the electron or hole (or both) can drop into a defect state by nonradiative recombination and then undergo radiative recombination.

## Excitons and Exciton-Polaritons

It is impossible to analyze luminescence spectra without considering the role of excitons. Formally an exciton must be thought of as an additional bound state of the crystal that appears when the Coulomb interaction between electrons and holes is incorporated into the crystal Hamiltonian, but there is also a simple, mechanistic picture that leads to the same result. The exciton can be thought of as a hydrogenlike atom that results when the electron and hole bind together. The binding energy of an exciton is then given by

$$E_n = -\frac{E_0}{n^2} \qquad E_0 > 0 \qquad n = 1, 2, 3 \qquad (1)$$

where $E_0$ is the free exciton binding energy. The total energy of the exciton is $E_g + E_n < E_g$, so when the exciton recombines, the energy of the emitted photon is below $E_g$.

This simple, hydrogenlike picture of the exciton is actually an oversimplification. In reality, the free exciton should be thought of as a polarization wave that can propagate through the crystal. This polarization wave can interact with an electromagnetic wave to produce *exciton-polariton* states. The dispersion curve for these states has two branches, as illustrated in many texts in solid-state and semiconductor physics (e.g., Yu and Cardona, 1996, p. 273). Given this continuum of states, the polariton emission spectrum can be quite complex, the peak positions being determined by the relative lifetimes of states at various positions on the dispersion curve. In particular, polaritons near the transverse free exciton energy have particularly long lifetimes, creating a "bottleneck" that leads to accumulation of exciton population. Hence there is typically a peak at the free exciton energy, which can be thought of as a "pure" free exciton peak and is often labeled as such in the literature and in this unit.

## Bound Excitons

At low temperatures excitons can bind to donor or acceptor sites via van der Waals forces to form states of lower energy than the polariton states. Since bound exciton states are localized, they have no dispersion. The dissociation energy for a bound exciton is dependent on the species of donor or acceptor; this fact is sometimes helpful in identifying impurities. Figure 1 shows donor- and acceptor-bound exciton luminescence for CdTe.

## Alloy Broadening

In an alloy, random microscopic composition variations in the crystal cause a broadening in the energy spectrum of electronic states. Figure 2 shows the exciton spectrum for $Cd_{0.96}Zn_{0.04}Te$. Much of the fine structure that was apparent for pure CdTe has been obscured by alloy broadening.

## Phonon Replicas

There may also be phonon replicas of any of the aforementioned peaks. That is, the system may emit part of its energy as one or more phonons—typically longitudinal-

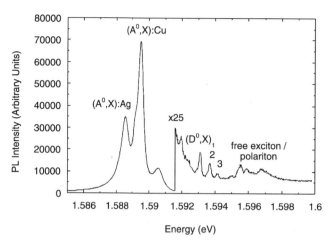

**Figure 1.** Free and bound exciton luminescence of CdTe at 4.2 K showing bound exciton luminescence from two acceptor species.

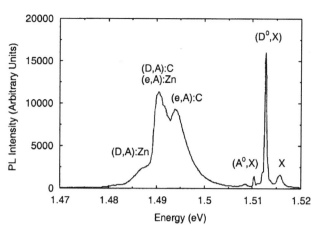

**Figure 3.** PL spectrum of semi-insulating GaAs at 4.2 K showing separate donor-acceptor and band-to-acceptor peaks for C and Zn.

optical (LO) phonons—and the remainder as a photon. The emitted photon is thus of longer wavelength than that for the no-phonon process. In some cases the emitted phonon corresponds to a local vibrational mode of the defect to which the exciton is bound; this provides another means for identifying defects. In Figure 2, the first phonon replica of the free exciton luminescence is visible. The phonon replicas are less well-resolved than the no-phonon lines, because of the slight dispersion in the LO phonon energy. The LO phonon energy can be read from the spacing of the peaks as ~22 meV.

### Donor-Acceptor and Free-to-Bound Transitions

Some electrons and holes drop by nonradiative processes from the conduction and valence bands into donor and acceptor states, respectively. When these charge carriers recombine, they emit photons of lower energy than do excitonic or direct band-to-band transitions. Possible transitions include those in which a conduction band electron

recombines with a hole in an acceptor level (e,A), an electron in a donor level combines with a hole in the valence band (D,h), and an electron in a donor level recombines with a hole in an acceptor level (D,A). The (D,A) luminescence is below the band gap by the sum of the donor and acceptor binding energies minus the Coulomb energy of attraction between the ionized donor and acceptor in the final state:

$$hv = E_g - (E_A + E_D) + \frac{e^2}{\varepsilon r_n} \quad \text{(Gaussian units)} \qquad (2)$$

where $r_n$ is the distance between the $n$th nearest neighbor donor and acceptor sites and can take on only those values consistent with the crystal lattice. The discrete values of $r$ imply that there are a large number of separate (D,A) peaks, but for relatively large values of $n$, they merge into a single broad band. With high-resolution measurements on high-quality crystals at low temperature, it is sometimes possible to detect separate peaks for low values of $n$ (Yu and Cardona, 1996, p. 346).

There can also be multiple donor or acceptor levels in a given material, each exhibiting its own set of peaks in the spectrum. A well-known example is C and Zn in GaAs, which give rise to two sets of overlapping (e,A) and (D,A) peaks (Stillman et al., 1991). A representative 4.2 -K GaAs spectrum is shown in Figure 3.

### Transitions Involving Defect Levels

Many kinds of defects produce energy states closer to the middle of the band gap than the shallow donor and acceptor states mentioned above. Recombination processes involving these defect levels result in emission of longer-wavelength photons. Because these defect states are localized, they have broad energy spectra. Also, they tend to be strongly coupled to the lattice and hence have a large number of phonon replicas that often smear together into a single broad band. An example CdZnTe spectrum is shown in Figure 4. The band at ~1.4 eV is believed to be related to the A-center (Cd vacancy plus donor), and the band at 1.1 eV is believed to be related to Te vacancies.

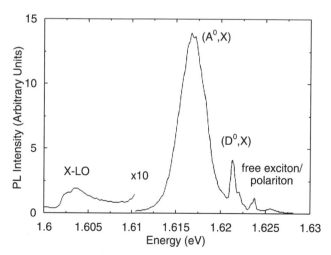

**Figure 2.** Exciton luminescence of $Cd_{0.96}Zn_{0.04}Te$ at 4.2 K. Many of the peaks that are present in the CdTe spectrum are not resolvable because of alloy broadening.

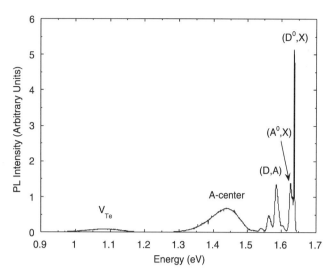

**Figure 4.** PL spectrum of $Cd_{0.9}Zn_{0.1}Te$ at 4.2 K showing excitonic, donor-acceptor, and deep-level luminescence.

## Band-to-Band Recombination

At room temperature there is a nonzero occupancy of electron states near the bottom of the conduction band and hole states near the top of the valence band. Hence, electron-hole pairs can recombine to emit photons over a range of energies, producing a broad band rather than a sharp peak. Donor, acceptor, and defect states are generally fully ionized at room temperature and do not contribute significantly to the observed spectrum.

Band-to-band recombination produces a luminescence spectrum that is the product of the joint density of states and the Fermi-Dirac distribution (see Bhattacharya, 1994):

$$I(E) \propto \frac{\sqrt{E - E_\mathrm{g}}}{1 + \exp[(E - E_\mathrm{g})/kT]} \qquad (3)$$

Observed room temperature luminescence typically does not follow Equation 3 except at the upper end of the energy range. Equation 3 has a sharp lower cutoff at the band-gap energy, while the measured spectrum has a long tail below the peak. This is an indication that the spectrum is only partly due to band-to-band recombination and also includes excitonic and phonon-assisted transitions (Lee et al., 1994b).

## PRACTICAL ASPECTS OF THE METHOD

### Experimental Setup

The apparatus for low-temperature PL can be set up easily on an optical table. The sample is enclosed in an optical cryostat and illuminated with an above-band-gap light source, in most cases a laser. An optical band-pass filter is typically used to remove unwanted plasma lines from the laser emission, and a low-pass filter is often placed at the entrance of the monochromator to remove stray light at shorter wavelengths than the range of interest. Both of these sources of extraneous light can show up in the

spectrum in second order. The light emitted by the sample is collected with a collimating lens and then focused onto the entrance slits of the monochromator with a second lens.

A photon detector, typically either a photomultiplier tube (PMT) or a positive-intrinsic-negative (PIN) diode detector, is placed at the exit of the monochromator. The output signal of the detector is routed either to a photon counter or to a lock-in amplifier. In the latter case, a light chopper is used to synchronize the signal with a reference. In the photon-counting mode, it is desirable to cool the detector with liquid nitrogen to reduce noise; $N_2$ gas must then be allowed to flow over the face of the detector to keep it from frosting.

The instrumentation for room temperature PL mapping is similar, with the addition of an x-y translation stage. Generally, the optics are kept fixed and the sample translated. The excitation source is focused onto the sample with a microscope objective, which also serves to collimate the luminescence. A beam splitter or a mirror containing a small hole is used to allow transmission of the laser beam while reflecting the luminescence into the monochromator.

An alternative scheme for PL uses an optical multichannel analyzer. In this case the monochromator disperses the light onto an array of charge-coupled devices (CCDs) so that the entire spectrum is captured at once. This configuration has the advantage of greater speed, which is especially valuable in room temperature mapping, but it generally suffers from reduced wavelength resolution.

Following is a list of instrumentation and representative manufacturers for low-temperature PL along with some of the requirements and considerations for each apparatus:

1. *Detectors* (Hamamatsu, Phillips, Burr-Brown, Electron Tubes, Oriel, EG&G, Newport)—The principal trade-off is between bandwidth and sensitivity. The need for cooling is a further consideration. Photomultiplier tubes are unmatched in sensitivity but typically have limited long-wavelength response. Most PMTs can be operated at room temperature in the current mode but may require cooling for optimal performance in the photon-counting mode. Narrow-band-gap semiconductor detectors, such as germanium PIN diodes, have better long-wavelength response but poorer sensitivity at shorter wavelengths. They generally must be cooled with liquid nitrogen.

2. *Spectrometers / monochromators* (Jobin Yvon-Spex, Oriel, Jarrell Ash, Acton Research, McPherson, Ocean Optics, CVI Laser)—For high resolution, a large, double-grating spectrometer is required. For lower resolution applications, compact, inexpensive prism spectrometers with integrated CCD detectors are available.

3. *Lasers* (Coherent, Uniphase, Kimmon, Spectra-Physics, Newport, Melles Griot, Anderson Lasers)—The principal requirement is that the photon energy be higher than the band gap of the material

at the temperature at which the measurements are made. For direct band-gap materials, output power of a few milliwatts is typically sufficient. For indirect band-gap materials, higher power may be desirable. Single-line operation is preferred, since it permits the use of a laser line filter to remove extraneous lines that can appear in the measured spectrum in second order.

4. *Lock-in amplifiers/photon counters* (EG&G, Stanford Research, Oriel)—Both analog and digital lock-in amplifiers are available, the digital instruments generally having superior signal-to-noise ratios.

5. *Complete PL spectroscopy/mapping/imaging systems* (Phillips Analytical, Perkin Elmer, Oriel, ChemIcon)—Some systems are configurable with various laser modules, gratings, and detectors to be optimized for the application.

### Use of Low-Temperature PL for Crystal Characterization

The most basic use of low-temperature PL spectra is as an indicator of overall crystal quality. The principal indicator of quality is the ratio of excitonic luminescence to donor-acceptor and deep-level luminescence. Figure 5 shows PL

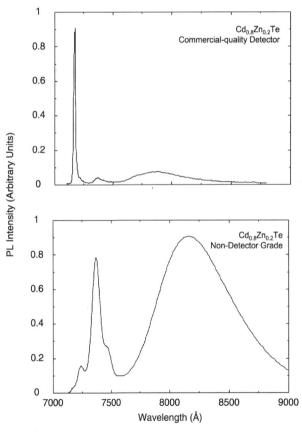

**Figure 5.** The 4.2-K PL spectra for $Cd_{0.8}Zn_{0.2}Te$ samples of differing quality. In the high-quality (commercial-quality detector grade) sample (top), the spectrum is dominated by excitonic luminescence. In the poor-quality (non-detector grade) sample the donor-acceptor and deep-level luminescence dominate.

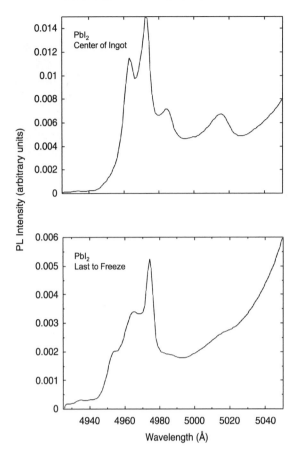

**Figure 6.** 4.2 K PL spectra for $PbI_2$ samples from center (top) and last-to-freeze portions of ingot. The spectrum for the sample from the middle of the ingot is sharper, indicating better crystal quality.

spectra for two CdZnTe samples of different quality. In the high-quality sample, the spectrum is dominated by a donor-bound exciton peak $(D^0,X)$, while in the poor-quality sample the excitonic luminescence is outweighed by donor-acceptor and defect bands.

A second measure of crystal quality is the sharpness of the spectrum—that is, the extent to which adjacent lines can be resolved. Figure 6 shows near-band-edge PL spectra for two $PbI_2$ samples, one from the center of the ingot and the other from the last-to-freeze portion. The spectrum of the sample from the center is sharper, indicating a more uniform crystal potential on a microscopic scale.

Aside from indicating overall crystallinity, the low-temperature PL spectrum can sometimes be helpful in identifying specific impurities. The example of the C and Zn (D,A) bands in GaAs was given in Figure 4, but the bound exciton luminescence can also be used. For example, in the CdTe spectrum shown in Figure 1, the $(A^0,X)$ peak at 1.5895 eV is due to Cu, while that at 1.5885 eV is due to Ag (Molva et al., 1984).

The ratio of bound to free exciton luminescence can also be used to measure impurity concentrations, but this application requires careful control over excitation intensity and sample temperature (Lightowlers, 1990). It also requires that calibration curves be generated based on chemical analysis.

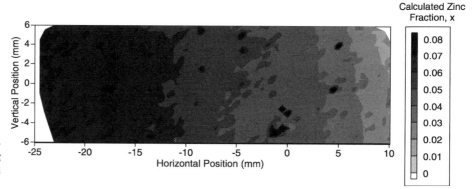

**Figure 7.** Room temperature PL composition map of $Cd_{1-x}Zn_xTe$ showing segregation of zinc along the growth axis.

## Uses of Room Temperature PL Mapping

One application of room temperature PL is composition mapping of alloys. Figure 7 shows a PL composition map for CdZnTe grown by a high-pressure Bridgman method. Segregation of zinc along the growth axis is clearly visible.

In addition to the systematic variation in composition along the growth axis, local variations are sometimes seen. Figure 8 shows an example in which minute striations are present, perhaps due to temperature fluctuations during growth. Note that the dark bands are only ~20 μm wide, demonstrating that high spatial resolution is important.

Photoluminescence mapping is also useful for studying structural defects. PL maps of CnZnTe often contain isolated dark spots that can be attributed to inclusions, possibly of tellurium or zinc. High-spatial-resolution PL maps of these inclusions show a region of near-zero PL intensity surrounded by a region in which the band gap is shifted to higher energy—hence the dark spots in the larger-scale maps. It is unclear whether the shift is caused by gettering of zinc around the inclusion or by strain.

## METHOD AUTOMATION

Complete automated systems for PL spectroscopy and mapping are now commercially available. If a system is

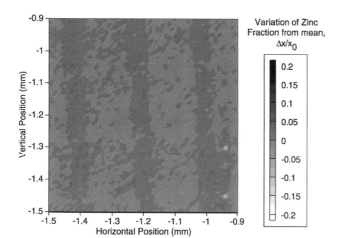

**Figure 8.** Room temperature PL composition map of $Cd_{1-x}Zn_xTe$ showing growth striations.

constructed from scratch, it is a relatively simple matter to integrate GPIB-controllable equipment using general-purpose laboratory software packages.

## DATA ANALYSIS AND INITIAL INTERPRETATION

### Identification of Bands in a Low-Temperature PL Spectrum

Qualitative interpretation of a low-temperature PL spectrum is quite simple if there are reliable sources in the literature to identify the origins of the various bands. If there are bands of uncertain origin, the intensity and temperature dependence of the spectrum can give clues to their nature.

As the sample temperature is raised, the luminescence is quenched as the centers involved in the transition are ionized by thermal excitation. The activation energy or energies of the level can then be determined by a curve fit to the following equation, which follows from Boltzmann statistics (Bimberg et al., 1971):

$$I(T) = I(0)\left[1 + C_1\exp\left(-\frac{E_1}{kT}\right) + \cdots + C_n\exp\left(-\frac{E_n}{kT}\right)\right]^{-1}$$

(4)

where $I(T)$ is the integrated intensity of the band at absolute temperature $T$, $k$ is Boltzmann's constant, $E_i$ is the activation energy of the $i$th process involved in the dissociation, and $C_i$ is formally the degeneracy of the level but in practice is left as a fitting parameter. Most processes can be modeled as one- or two-step dissociations. An example of a one-step dissociation is that of a free exciton into a free electron and hole, whereas the prototypical two-step dissociation is bound exciton quenching, in which the exciton first dissociates from the donor or acceptor to form a free exciton, which then dissociates in a second step. This difference in temperature dependence is one means of distinguishing between free and bound excitons in an unfamiliar spectrum.

Temperature dependence can also be used to distinguish between (D,A) and (e,A) transitions, since the former, which involve a shallow donor, will be quenched quickly as the donors are ionized.

The dependence of the spectrum on the excitation intensity can be used to distinguish between excitonic and donor-acceptor bands. In both cases the luminescence intensity varies with the excitation intensity as a power law, but for excitonic lines the exponent is >1.0, whereas for donor-acceptor and free-to-bound transitions it is <1.0 (Schmidt et al., 1992).

One additional technique that can help to identify peaks is infrared quenching. As the sample is subjected to below-band-gap infrared radiation in addition to laser excitation, donors and acceptors are ionized, causing quenching of $(D^0,X)$ and $(A^0,X)$ transitions and an increase in $(D^+,X)$ luminescence. It has been shown that $(A^-,X)$ complexes cannot exist in most semiconductors (Hopfield, 1964), so that this provides another way of distinguishing between donor- and acceptor-bound excitons. Infrared quenching has also been used quantitatively to measure impurity concentrations (Lee et al., 1994a).

### Curve Fitting in Room Temperature PL Mapping

Extracting material parameters from a room temperature PL spectrum in a systematic way requires that the calculated line shapes for various types of transitions (Bebb and Williams, 1972) be combined. This approach has been used with some success in cadmium telluride (Lee et al., 1994b), and in cadmium zinc telluride (Brunnett et al., 1999). In the alloy, variation in the band gap and exciton and phonon energies introduces additional uncertainty in the extracted composition. The composition determined from this procedure cannot be regarded as highly accurate, but variations can be detected with precision.

### SAMPLE PREPARATION

Photoluminescence generally requires no special sample preparation, other than to provide a reasonably smooth surface so that a large fraction of the light is not scattered away. Thin-film samples generally can be analyzed as grown, whereas bulk samples that have been cut typically require etching to remove the damaged layer produced by the cutting.

### SPECIMEN MODIFICATION

Low-temperature PL requires relatively low optical power density, so that the risk of damaging the material is minimal. Room temperature mapping, however, does use rather high power densities that cause significant sample heating. Most group II–VI and III–V compounds produce adequate luminescence at low enough intensity to avoid damaging the sample, but with less robust materials, such as $PbI_2$ and $HgI_2$, room temperature mapping may not be practical.

### POTENTIAL PROBLEMS

A difficulty with extracting accurate composition information from room temperature PL spectra is that the peak position depends on the excitation wavelength and inten-

sity. The spectrum shifts to lower energy as the intensity is raised, and the peak position is at lower energy for the shorter-wavelength light. The explanation for this effect probably lies in local heating of the sample by the rather high intensities used. The heating is greater for higher-energy (blue) photons because a greater fraction of each photon's energy is converted to phonons.

### LITERATURE CITED

Bebb, H. B. and Williams, E. W. 1972. Photoluminescence I: Theory. *In* Semiconductors and Semimetals, Vol. 8 (R. K. Willardson and A. C. Beer, eds.) pp. 183–217. Academic Press, New York.

Bhattacharya, P. 1994. Semiconductor Optoelectronic Devices. Prentice Hall, Englewood Cliffs, N.J.

Bimberg, D., Sondergeld, M., and Grobe, E. 1971. Thermal dissociation of excitons bound to neutral acceptors in high-purity GaAs. *Phys. Rev. B* 4:3451–3455.

Brunnett, B. A., Schlesinger, T. E., Toney, J. E., and James, A. B. 1999. Room-temperature photoluminescence mapping of CdZnTe detectors. *Proc. SPIE 3768.* In press.

Hopfield, J. J. 1964. The quantum chemistry of bound exciton complexes. *In* Proceedings of 7th International Conference on the Physics of Semiconductors (M. Hulin, ed.) pp. 725–735. Dunod, Paris.

Lee, J., Giles, N. C., Rajavel, D., and Summers, C. J. 1994b. Room-temperature band-edge photoluminescence from cadmium telluride. *Phys. Rev. B* 49:1668–1676.

Lee, J., Myers, T. H., Giles, N. C., Dean, B. E., and Johnson, C. J. 1994a. Optical quenching of bound excitons in CdTe and $Cd_{1-x}Zn_xTe$ alloys: A technique to measure copper concentration. *J. Appl. Phys.* 76:537–541.

Lightowlers, E. C. 1990. Photoluminescence characterization. *In* Growth and Characterization of Semiconductors (R. A. Stradling and P. C. Klipstein, eds.) pp. 135–163. Adam Hilger, Bristol, U.K.

Molva, E., Pautrat, J. L., Saminadayar, K., Milchberg, G., and Magnea, N. 1984. Acceptor states in CdTe and comparison with ZnTe. General trends. *Phys. Rev. B* 30:3344–3354.

Schmidt, T., Lischka, K., and Zulehner, W. 1992. Excitation-power dependence of the near-band-edge photoluminescence of semiconductors. *Phys. Rev. B* 45:8989–8994.

Stillman, G. E., Bose, S. S., and Curtis, A. P. 1991. Photoluminescence characterization of compound semiconductor optoelectronic materials. *In* Advanced Processing and Characterization Technologies, Fabrication and Characterization of Semiconductor Optoelectronic Devices and Integrated Circuits (P. H. Holloway, ed.) pp. 34–37. American Institute of Physics, Woodbury, N.Y.

Yu, P. Y. and Cadona, M. 1996. Fundamentals of Semiconductors. Springer-Verlag, Berlin.

### KEY REFERENCES

Bebb and Williams, 1972. See above.

*A thorough, mathematical treatment of the theory of PL.*

Demtroder, W. 1996. Laser Spectroscopy, 2nd ed. Springer-Verlag, Berlin.

*Contains a detailed discussion of instrumentation used in optical spectroscopy.*

Perkowitz, E. 1993. Optical Characterization of Semiconductors. Academic Press, London.

*Contains a good summary of the general principles of PL and some helpful case studies dealing with stress and impurity emission.*

Peyghambarian, N., Loch, S. W., and Mysyrowicz, A. 1993. Introduction to Semiconductor Optics. Prentice-Hall, Englewood Cliffs, N.J.

*A thorough, pedagogical introduction to optics of semiconductors with emphasis on properties of excitons.*

Swaminathan, V. and Macrander, A. T. 1991. Materials Aspects of GaAs and InP Based Structures. Prentice-Hall, Englewood Cliffs, N.J.

*Gives a fine, conceptual introduction to PL.*

JAMES E. TONEY
Spire Corporation
Bedford, Massachusetts

# ULTRAVIOLET AND VISIBLE ABSORPTION SPECTROSCOPY

## INTRODUCTION

Ultraviolet and visible (UV-Vis) absorption spectroscopy is the measurement of the attenuation of a beam of light after it passes through a sample or after reflection from a sample surface. This unit will use the term UV-Vis spectroscopy to include a variety of absorption, transmittance, and reflectance measurements in the ultraviolet (UV), visible, and near-infrared (NIR) spectral regions. These measurements can be at a single wavelength or over an extended spectral range. As presented here, UV-Vis spectroscopy will be found under other names, including UV-Vis spectrometry, UV-Vis spectrophotometry, and UV-Vis reflectance spectroscopy. This unit provides an overview of the technique and does not attempt to provide a comprehensive review of the many applications of UV-Vis spectroscopy in materials research. In this regard, many of the references were chosen to illustrate the diversity of applications rather than to comprehensively survey the uses of UV-Vis spectroscopy. A unifying theme is that most of the measurements discussed in this unit can be performed with simple benchtop spectrometers and commercially available sampling accessories.

The UV-Vis spectral range is approximately 190 to 900 nm. This definition originates from the working range of a typical commercial UV-Vis spectrometer. The short-wavelength limit for simple UV-Vis spectrometers is the absorption of UV wavelengths <180 nm by atmospheric gases. Purging a spectrometer with nitrogen gas extends this limit to 175 nm. Working beyond 175 nm requires a vacuum spectrometer and a suitable UV light source and is not discussed in this unit. The long-wavelength limit is usually determined by the wavelength response of the detector in the spectrometer. Higher end commercial UV-Vis spectrometers extend the measurable spectral range into the NIR region as far as 3300 nm. This unit includes the use of UV-Vis-NIR instruments because of the importance of characterizing the NIR properties of materials for lasers, amplifiers, and low-loss optical fibers for fiber-optic communications and other applications. Table 1 summarizes the wavelength and energy ranges of the UV-Vis and related spectral regions.

Ultraviolet-visible spectroscopy is one of the more ubiquitous analytical and characterization techniques in science. There is a linear relationship between absorbance and absorber concentration, which makes UV-Vis spectroscopy especially attractive for making quantitative measurements. Ultraviolet and visible photons are energetic enough to promote electrons to higher energy states in molecules and materials. Figures 1 and 2 illustrate typical absorption spectra for the absorption processes for molecules and semiconductor materials, respectively. Therefore, UV-Vis spectroscopy is useful to the exploration of the electronic properties of materials and materials precursors in basic research and in the development of applied materials. Materials that can be characterized by UV-Vis spectroscopy include semiconductors for electronics, lasers, and detectors; transparent or partially transparent optical components; solid-state laser hosts; optical fibers, waveguides, and amplifiers for communication; and materials for solar energy conversion. The UV-Vis range also spans the range of human visual acuity of approximately 400 to 750 nm, making UV-Vis spectroscopy useful in characterizing the absorption, transmission, and reflectivity of a variety of technologically important materials, such as pigments, coatings, windows, and filters.

The use of UV-Vis spectroscopy in materials research can be divided into two main categories: (1) quantitative measurements of an analyte in the gas, liquid, or solid phase and (2) characterization of the optical and electronic properties of a material. The first category is most useful as a diagnostic tool for the preparation of materials, either to quantitate constituents of materials or their precursors or as a process method to monitor the concentrations of reactants or products during a reaction (Baucom et al., 1995; Degueldre et al., 1996). In quantitative applications it is often only necessary to measure the absorbance or reflectivity at a single wavelength. The second more

**Table 1. Approximate Wavelengths, Energies, and Type of Excitation for Selected Spectral Regions**

| Spectral Region | Wavelength Range (nm) | Energy Range (cm$^{-1}$) | Energy Range (eV) | Types of Excitation |
|---|---|---|---|---|
| Vacuum-UV | 10–180 | $1 \times 10^6$–55,600 | 124–6.89 | Electronic |
| UV | 200–400 | 50,000–25,000 | 6.89–3.10 | Electronic |
| Visible | 400–750 | 25,000–13,300 | 3.10–1.65 | Electronic |
| Near IR | 750–2500 | 13,300–4000 | 1.65–0.496 | Electronic, vibrational overtones |
| IR | 2500–25,000 | 4000–400 | 0.496–0.0496 | Vibrations, phonons |

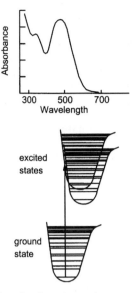

**Figure 1.** Typical molecular absorption spectrum and a schematic of the absorption process in a molecule.

qualitative application usually requires recording at least a portion of the UV-Vis spectrum for characterization of the optical or electronic properties of materials (Kovacs et al., 1997). Much of the discussion of UV-Vis spectroscopy will follow these two main applications of quantitative measurements and materials characterization.

## Materials Properties Measurable by UV-Vis Spectroscopy

Electronic excitations in molecules usually occur in the UV and visible regions. Charge transfer bands can occur from the UV to the NIR. The band gap of a semiconductor depends on the specific material and physical dimensions and can range from the UV to NIR. A reduction in semiconductor particle size or dimensions (<10 nm) will shift the

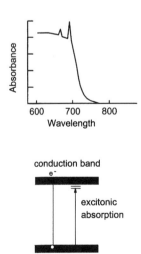

**Figure 2.** Typical absorption spectrum for a semiconductor and a schematic of the absorption process in a direct-bandgap semiconductor.

band edge to shorter wavelength (higher energy) due to quantum confinement effects (Bassani et al., 1997). The NIR light excites vibrational overtones and combination bands in molecules. Although these transitions are very weak, they become important in certain cases such as optical fibers in which a very long path length is of interest.

In organic molecules and polymers, the UV-Vis spectrum can help to identify chromophores and the extent of electronic delocalization (Yoshino et al., 1996). Similarly, absorption measurements can be correlated with bulk physical properties (Graebner, 1995). For inorganic complexes, the UV-Vis spectrum can provide information about oxidation states, electronic structure, and metal-ligand interactions. For solid materials, the UV-Vis spectrum can measure the band gap and identify any localized excitations or impurities (Banerjee et al., 1996).

Optical materials that rely on interference can be utilized throughout the UV-Vis-NIR spectral range. Collective excitations that determine the absorption or reflectivity of a material likewise occur throughout this range. For example, the peak absorption of the plasmon resonance absorption of thin films of aluminum, gold, and semiconductors occurs in the UV, visible, and NIR regions, respectively, For small metal particles, the plasmon resonance absorption also varies due to particle size (Ali et al., 1997).

## Competitive and Related Techniques for Qualitative and Quantitative Analysis

The spectrum of a molecule depends on its energy-level structure, and the UV-Vis absorption spectrum can sometimes be useful for identifying molecular compounds. However, UV-Vis spectra have broad features that are of limited use for sample identification. The broad lines do facilitate making accurate quantitative measurements. Lanthanide ions and some transition metals in molecular complexes have sharp UV-Vis bands and are easily identified by their absorption spectrum. Qualitative identification and structural analysis of molecules usually require multiple analytical methods, including C, H, and O elemental analysis; infrared (IR) absorption; Raman spectroscopy (see RAMAN SPECTROSCOPY OF SOLIDS); nuclear magnetic resonance (NMR) spectroscopy (see NUCLEAR MAGNETIC RESONANCE IMAGING); mass spectrometry; size exclusion chromatography; and x-ray diffraction (XRD, see X-RAY TECHNIQUES). The elemental composition of solids can be determined by x-ray fluorescence, microprobe techniques (see X-RAY MICROPROBE FOR FLUORESCENCE AND DIFFRACTION ANALYSIS), and surface spectroscopies such as x-ray photoelectron spectroscopy (XPS), Auger electron spectroscopy (AES, see AUGER ELECTRON SPECTROSCOPY), or secondary-ion mass spectrometry (SIMS). For crystalline or polycrystalline solids, x-ray (X-RAY TECHNIQUES), electron (ELECTRON TECHNIQUES), and neutron diffraction (NEUTRON TECHNIQUES) are the most reliable means of structure identification.

Quantitative measurements with benchtop UV-Vis spectrometers are most commonly performed on molecules and inorganic complexes in solution. Photoluminescence for fluorescent analytes is usually more sensitive than absorption measurements, because the fluorescence signal is measured on a nominally zero-noise background. Absorbance measurements require that the change in light power due to absorption be greater than the noise of the light source. Quantitating gas phase species or analytes on surfaces or in thin films using UV-Vis spectrometers is possible for high concentrations or strongly absorbing species (Gregory, 1995). In practice, polymer or Langmuir-Blodgett films are more often measured using IR absorption or some type of laser spectroscopy (Al-Rawashdeh and Foss, 1997). In addition to UV-Vis reflectance spectroscopy, inorganic surfaces often require analysis with multiple spectroscopies such as Raman spectroscopy, extended x-ray absorption fine structure (EXAFS, see XAFS SPECTROSCOPY), XPS, AES, and SIMS.

Laser spectroscopic absorption techniques can provide a variety of advantages compared to conventional UV-Vis spectroscopy. Lasers have much higher intensities and can provide much shorter time resolution than modulated lamp sources for measuring transient phenomena. The recent commercialization of optical parametric oscillators (OPOs) alleviates the limited scan range of dye lasers. Cavity-ringdown laser absorption spectroscopy (Paul and Saykally, 1997; Scherer et al., 1997) and intracavity laser absorption (Stoeckel and Atkinson, 1985; Kachanov et al., 1997) are much more sensitive than measurements using UV-Vis spectrometers. Similarly, photoacoustic and photothermal spectroscopy can provide very high sensitivity absorption measurements of opaque samples and analytes on surfaces (Bialkowski, 1995; Childers et al., 1986; Palmer, 1993; Welsch et al., 1997). Nonlinear laser spectroscopies, such as second-harmonic generation (SHG), can be surface selective and provide enhanced selectivity and sensitivity for analytes on surfaces compared to absorption or fluorescent measurements (Di Bartolo, 1994).

### Competitive and Related Techniques for Material Characterization

Because of the diverse applications to which UV-Vis spectroscopy can be applied, there are a multitude of related techniques for characterizing the electronic and optical properties of materials. Characterizing complex materials or processes often requires the results of multiple analytical techniques (see, e.g., Lassaletta et al., 1995; Takenaka et al., 1997; Weckhuysen et al., 1996). Electronic properties are often measured by a variety of methods, and UV-Vis spectral results are correlated with electrical measurements, surface spectroscopies, and electrochemical measurements (Ozer et al., 1996; Yoshino et al., 1996). Characterization of the electronic and optical properties of semiconductors often requires time-resolved laser spectroscopic methods, photoluminescence spectroscopy (see PHOTOLUMINESCENCE SPECTROSCOPY), and sophisticated modulation techniques such as piezospectroscopy (Ramdas and Rodriguez, 1992) and modulated photoreflectance (Glembocki and Shanabrook, 1992).

Measuring the transmittance or reflectance of most optical materials in the UV-Vis region rarely requires more sophisticated instrumentation than a commercial spectrometer. Measuring an extremely small absorbance can require some type of laser-based method. Measuring

polarization properties such as optical rotation and circular dichroism of materials also requires more sophisticated techniques and is usually performed on dedicated polarimetry, circular dichroism, or ellipsometry instruments. Many technologically important materials, such as light-emitting diodes (LEDs), optical amplifiers, laser materials, and lamp and display phosphors, require characterization of the emission properties as well as the absorption properties. For these types of materials, measurements of the photoluminescence or electroluminescence spectrum and the fluorescence quantum efficiency are as important as characterizing the absorption spectrum and absorbance.

### PRINCIPLES OF THE METHOD

Ultraviolet-visible spectroscopy measures the attenuation of light when the light passes through a sample or is reflected from a sample surface. The attenuation can result from absorption, scattering, reflection, or interference. Accurate quantitation requires that the measurement record the attenuation due only to absorption by the analyte, so the spectroscopic procedure must compensate for the loss of light from other mechanisms. The cause of the attenuation is often not important for many optical materials, and the total resulting transmittance or reflectance is sufficient to determine the suitability of a material for a certain application.

Experimental measurements are made in terms of transmittance $T$:

$$T = P/P_0 \qquad (1)$$

where $P$ is the radiant power (radiant energy on unit area in unit time) after it passes through the sample and $P_0$ is the initial radiant power. This relationship will also be found in terms of light intensities:

$$T = I/I_0 \qquad (2)$$

Percent transmittance is simply $T \times 100\%$ ($\%T$). The parameters $P$ and $P_0$ are not always well-defined and can depend on the UV-Vis application. Figures 3 and 4 illustrate the differences that are encountered in defining $P$ and $P_0$ in quantitative and qualitative characterization measurements.

### Quantitative Analysis

Within certain limits, the absorbance of light by a sample is directly proportional to the distance the light travels through the sample and to the concentration of the absorbing species. This linear relationship is known as the Beer-Lambert law (also called the Beer-Lambert-Bouguer law or simply Beer's law) and allows accurate concentration measurements of absorbing species in a sample. The general Beer-Lambert law is usually written as

$$A = a \times b \times c \qquad (3)$$

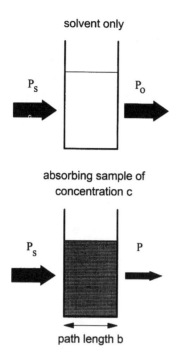

solvent only

$P_s$   $P_0$

absorbing sample of
concentration c

$P_s$   P

path length b

**Figure 3.** Schematic of a transmittance measurement for the quantitative determination of an analyte in solution.

where $A$ is the measured absorbance, $a$ is a wavelength-dependent absorptivity coefficient, $b$ is the path length, and $c$ is the analyte concentration. When working in concentration units of molarity, the Beer-Lambert law is written as

$$A = \varepsilon \times b \times c \qquad (4)$$

where $\varepsilon$ is the wavelength-dependent molar absorptivity coefficient with units of reciprocal molarity per centimeter and $b$ has units of centimeters. If multiple species that absorb light at a given wavelength are present in a sample, the total absorbance at that wavelength is the sum due to all absorbers:

$$A = (\varepsilon_1 \times b \times c_1) + (\varepsilon_2 \times b \times c_2) + \cdots \qquad (5)$$

where the subscripts refer to the molar absorptivity and concentration of the different absorbing species that are present.

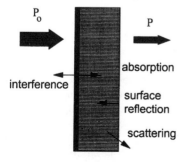

$P_0$   P

absorption

interference

surface
reflection

scattering

**Figure 4.** Schematic of a transmittance measurement of an optical component such as a dielectric-coated filter.

The linear relationship of the Beer-Lambert law depends on several conditions. The incident radiation must be monochromatic and collimated through a sample. Samples, including calibration standards, must be homogeneous without irreproducible losses due to reflection or scattering. For analytes in a matrix, such as a solution or a glass, the absorber concentration must be low enough so that the absorbing species do not interact. Interactions between absorbing species can lead to deviations in the absorptivity coefficient (at a given wavelength) as a function of concentration. More details on the most common limitations are described below (see Problems).

The relationship between absorbance $A$ and the experimentally measured transmittance $T$ is

$$A = -\log T = -\log(P/P_0) \qquad (6)$$

where $T$ and $A$ are both unitless. Absorption data and spectra will often be presented using $A$, $T$, $\%T$, or $1 - T$. An absorption spectrum that uses $1 - T$ versus wavelength will appear similar to a plot of absorbance $A$ versus wavelength, but the quantity $1 - T$ does not directly correspond to absorber concentration as does $A$.

Figure 3 shows the quantitative measurement of $P$ and $P_0$ for an analyte in solution. In this example, $P_s$ is the source light power that is incident on a sample, $P$ is the measured light power after passing through the analyte, solvent, and sample holder, and $P_0$ is the measured light power after passing through only the solvent and sample holder. The measured transmittance in this case is attributed to only the analyte.

If the absorptivity coefficient of an absorbing species is not known, an unknown concentration can be determined using a working curve of absorbance versus concentration derived from a set of standards of known concentration. Calibration with standards is almost always necessary for absorbance measurements made in a reflectance geometry or with a fiber-optic light delivery system.

### Material Characterization

Figure 4 illustrates the measurement of $P$ and $P_0$ to determine the transmittance of a dielectric coating on a substrate. This example is a typical measurement to characterize the optical properties of a material such as a window, filter, or other optical component. In this case $P$ is the light power after the sample and $P_0$ is the light power before the sample. Figure 5 shows two examples of transmittance spectra for optical components. The transmittance of the colored glass filter (top spectrum) does not reach 1.0 at wavelengths >600 nm due to surface reflection losses. The oscillations in the bottom spectrum arise from the interference origin of the transmittance properties. In these examples, the measured transmittance is due to all sources of light attenuation, which for practical applications is often unimportant compared to knowing the resulting transmittance at any given wavelength. These example spectra show the transmittance spectrum for an incidence angle of $0°$ through an optical component. Depending on the application, the transmittance or reflectance spectrum as a

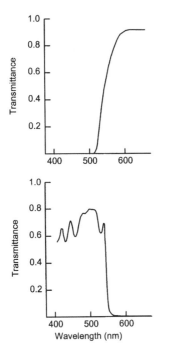

**Figure 5.** Transmittance spectra of a colored glass filter (top) and a dielectric-coated interference filter (bottom).

function of the angle the light beam makes with the optical component might be of more interest.

## PRACTICAL ASPECTS OF THE METHOD

### General Considerations

Most commercial UV-Vis absorption spectrometers use one of three overall optical designs: a scanning spectrometer with a single light beam and sample holder, a scanning spectrometer with dual light beams and dual sample holders for simultaneous measurement of $P$ and $P_0$, or a nonscanning spectrometer with an array detector for simultaneous measurement of multiple wavelengths. Most commercial instruments use a dispersive design, but Fourier transform UV-Vis spectrometers find applications in high-speed, low-resolution and other specialized applications (Thorne, 1991). In single- and dual-beam spectrometers, the light from a lamp is dispersed before reaching the sample cell. In an array detector instrument, all wavelengths pass through the sample and the dispersing element is between the sample and the array detector. The specific designs and optical components vary widely depending on the required level of spectrometer performance. The following paragraph describes common features of the UV-Vis instruments followed by descriptions of the three different designs and sampling options. Specific details about UV-Vis instruments can be found in the manufacturers' literature.

### Common Spectrometer Components

The light source is usually a deuterium discharge lamp for UV measurements and a tungsten-halogen lamp for visible and NIR measurements. The instruments automatically swap lamps when scanning between the UV and visible regions. The wavelengths of these continuous light sources are typically dispersed by a holographic grating in a single or double monochromator or spectrograph. The spectral bandpass is then determined by the monochromator slit width or by the array element width in array detector spectrometers. Spectrometer designs and optical components are optimized to reject stray light, which is one of the limiting factors in quantitative absorbance measurements. The detector in single-detector instruments is a photodiode, phototube, or photomultiplier tube (PMT). The UV-Vis-NIR spectrometers utilize a combination of a PMT and a Peltier-cooled PbS IR detector. The light beam is redirected automatically to the appropriate detector when scanning between the visible and NIR regions, and the diffraction grating and instrument parameters, such as slit width, can also change automatically.

### Single-Beam Spectrometer

The simplest UV-Vis spectrometer design has one light beam and uses one sample cell. These instruments are simple and relatively inexpensive and are useful for routine quantitative measurements. Making a quantitative measurement requires calibrating the instrument at the selected wavelength for 0 and 100% transmittance (or a more narrow transmittance range if accurate standards are available). As in Figure 3, the 100% calibration is done with the sample cell containing only the solvent for quantitative measurements in solution. These measurements are stored by the instrument, and the spectrometer readout can also display in absorbance units (A.U.).

Absorbance and reflectivity measurements can be made with discrete light sources without dispersing optics in specialized applications. One example of a very small and robust optical spectrometer is the Mars Oxidant Experiment, which was developed for the Mars96 mission. This spectrometer consisted of LED light sources, a series of thin-film sensors, a photodiode array detector, detection electronics, and optical fibers to get light from an LED to a sensor and from that sensor to a pixel of the array detector. A variety of sensor films provide a measure of the oxidizing power of the Martian surface and atmosphere. The reflectivity of the sensors changed depending on the chemical environment, as recorded by the attenuation of the LED power reaching the detector (Grunthaner et al., 1995).

### Dual-Beam Spectrometer

Spectra can be recorded with a single-beam UV-Vis instrument by manually recording absorbance measurements at different wavelength settings. Both a reference and the sample must be measured at each wavelength, which makes recording a wide spectral range or a small step size a tedious procedure. The dual-beam design greatly simplifies this process by simultaneously measuring $P$ and $P_0$ of the sample and reference cells, respectively. Most spectrometers use a mirrored rotating chopper wheel to alternately direct the light beam through the sample and reference cells. The detection electronics or software program can then manipulate the $P$ and $P_0$ values as the wavelength scans to produce the spectrum of absorbance or transmittance as a function of wavelength.

## Array Detector Spectrometers

In a large number of applications absorbance spectra must be recorded very quickly, such as in process monitoring or kinetics experiments. Dispersing the light after it passes through a sample allows the use of an array detector to simultaneously record the transmitted light power at multiple wavelengths. These spectrometers use photodiode arrays (PDAs) or charge-coupled devices (CCDs) as the detector. The spectral range of these array detectors is typically 200 to 1000 nm. Besides allowing rapid spectral recording, these instruments are relatively small and robust and have been developed into PC-card portable spectrometers that use optical fibers to deliver light to and from a sample. These instruments use only a single light beam, so a reference spectrum is recorded and stored in memory to produce transmittance or absorbance spectra after recording the raw sample spectrum.

## Sample Handling

Most benchtop UV-Vis spectrometers contain a 10- to 200-mm-wide sample compartment through which the collimated light beam(s) pass. Standard mounts in this sample compartment hold liquid sample cuvettes or optical components and can include thermostatted temperature regulation. Custom mounts can hold transparent solid samples of arbitrary shapes. The standard sample holder is a 10-mm-path-length quartz cuvette, which can be purchased in matched sets for use in dual-beam spectrometers. Micro-cuvettes and capillary cuvettes hold less than 10 μL of sample for limited amounts of liquid samples. These micro-cuvettes have shorter path lengths than the standard sample cells, and the smaller volume can introduce more error into a quantitative measurement. Sample cells and mounts with path lengths up to 100 mm are also available. Measuring longer path lengths requires conventional or fiber-optic accessories to deliver and re-collect the light beam outside of the sample compartment.

For samples that are opaque or strongly absorbing, the light absorption must be measured in a reflectance geometry. These measurements are done with reflectance accessories of various designs that fit internally or externally to the sample compartment. Variable-angle specular reflectance measurements can measure bulk optical properties of thin films, such as the refractive index or film thickness (Lamprecht et al., 1997; McPhedran et al., 1984; Larena et al., 1995). For quantitative or spectral characterization, diffuse reflectance spectra are recorded with an integrating sphere. A variation of a reflectance sampling method is attenuated total reflectance (ATR). In ATR the light beam travels through a waveguide or crystal by total internal reflection. The evanescent wave of the light extends beyond the surface of the waveguide and is attenuated by strongly absorbing samples on the outer surface of the waveguide. An ATR accessory can be used remotely by transferring the light to and from the ATR waveguide with optical fibers.

The use of optical fibers, or optical fiber bundles, allows the delivery of a light beam to and from a sample that is remote from the spectrometer sample compartment. This delivery mechanism is extremely useful for process control or to make remote measurements in clean rooms or reaction chambers. These accessories are usually used without a reference sample and can be used with single-beam, dual-beam, or array detector instruments. The transmission properties of optical fibers can limit the wavelength range of a single spectral scan, and complete coverage of the UV-Vis region can require measurements with multiple fibers. The delivery and return fibers or bundles can be arranged in a straight-through transmission geometry, a folded sample compartment geometry that uses a mirror in a compact probe head, or a reflectance mode to measure nontransparent samples.

## METHOD AUTOMATION

Most modern dual-beam UV-Vis absorption spectrometers include an integrated microprocessor or are controlled by a software program running on a separate microcomputer. Spectrometer parameters, such as slit width (spectral bandpass), scan range, scan step size, scan speed, and integration time, are set by the program. These parameters can be stored to disk as a method and recalled later to make the same types of measurements. Because UV-Vis spectroscopy is often used for repetitive measurements, instrument manufacturers offer a variety of automated sample-handling accessories. These automated accessories are controlled by the software programs, which also automatically store the data to disk after measurement. Most of these accessories are designed for liquid-containing sample cells, but some of the accessories can be adapted for measuring solid samples. Automated cell changers and multicell transporters can cycle eight to sixteen samples through the spectrometer. These units are useful for repetitive measurements on a small number of samples or to follow spectral changes in a set of multiple samples. As an example, the automated cell transporter can cycle through a set of systematically varied samples to determine the reaction kinetics as a function of reactant concentrations or reaction conditions.

Sipper systems use a peristaltic pump to deliver liquid sample to the sample cell in the spectrometer. A sipper system can automate repetitive measurements and also reduces the error in absorbance measurements due to irreproducibility in positioning the sample cell. Autosamplers are available to automatically measure large numbers of samples. An autosampler can use a sipper system to sequentially transfer a series of samples to the sample cell with intermediate rinsing or to sequentially place a fiber-optic probe into a set of sample containers. Autosamplers are used for repetitive measurements on the order of 100 to 200 different liquid samples.

## DATA ANALYSIS AND INITIAL INTERPRETATION

### Quantitative Analysis

An unknown concentration of an analyte can be determined by measuring the amount of light that a sample absorbs and applying the Beer-Lambert law as described above (see Principles of the Method). Modern UV-Vis absorption spectrometers will display the spectral data in a variety of formats, with the most common data presentation being in transmittance, percent transmittance, absorbance, reflectance, or percent reflectance (%R). For quantitative measurements, instrument control programs will store measurements of standards and automatically calculate the concentrations of unknowns from the absorbance measurement, including multicomponent mixtures. Spectra will usually be plotted versus wavelength (in angstroms, nanometers, or micrometers) or energy (in electron volts, or wavenumbers, in reciprocal centimeters). The data can be processed and displayed after a variety of processing methods, such as first to fourth derivative, the Kubelka-Munk function for turbidity measurements (Vargas and Niklasson, 1997), Savitzky-Golay smoothing (Savitzky and Golay, 1964; Barak, p. 1995), peak fitting, and others.

Standards are available from the National Institute of Standards and Technology (NIST) Standards Reference Materials Program to calibrate the transmittance, absorbance, wavelength, and reflectance of UV-Vis spectrometers. Many of the instrument manufacturers offer calibration standards and validation procedures for their UV-Vis spectrometers, as do third-party vendors (see the *Anal. Chem. Labguide* listed in Key References for more information). Table 4 in Dean (1995) gives the percent transmittance and absorbance values from 220 to 500 nm for a laboratory-prepared $K_2CrO_4$ standard solution, which is useful for checking the calibration of UV-Vis spectrometers. This handbook also lists other useful data such as the UV cutoff of common solvents and UV-Vis absorption methods for the analysis of metals and nonmetals (Dean, 1995).

## Material Characterization

Several manufacturers offer software packages for specific applications. Many of these programs are for biological applications, but they can be adapted to materials work. For example, an enzyme kinetics program that works in conjunction with an automated multicell holder can be adapted for monitoring the kinetics of the solution-phase synthesis of materials. Software modules are also available to automatically calculate the refractive index or thickness of a thin film from the interference pattern in transmittance or reflectance measurements. Another materials application is the characterization of color, including analysis of the CIE (1986) tristimulus values in diffuse reflectance measurements (e.g., CHROMA software package from Unicam UV-Visible Spectrometry). These colorimetry standards are also applicable in transmission experiments (Prasad et al., 1996).

Two examples are presented here to illustrate the use of UV-Vis spectrometry for materials characterization. In the first example, Figure 6 shows the UV-Vis absorption spectra of two samples of CdS quantum dots (Tsuzuki and McCormick, 1997).

These materials were prepared by a mechanochemical reaction:

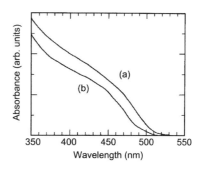

**Figure 6.** Absorption spectra of CdS quantum dots produced using 12.7-mm balls (curve *a*, average dot diameter 8.3 nm) and 4.8-mm balls (curve *b*, 4.3 nm average dot diameter). (Adapted with permission from Tsuzuki and McCormick, 1997. Copyright © 1997 by Springer-Verlag.)

$$CdCl_2 + Na_2S \rightarrow CdS + 2NaCl \qquad (7)$$

After reaction in a ball mill for 1 h, the NaCl is removed by washing and the CdS quantum dots are dispersed in $(NaPO_3)_6$ solution to obtain an absorption spectrum. Curve *a* in Figure 6 is the spectrum of 8.3-nm-diameter quantum dots produced using 12.7-mm balls in the ball mill, and curve *b* is the spectrum of 4.3-nm quantum dots produced using 4.8-mm balls. The smaller ball size produces lower collision energies, which results in smaller crystallite size. The absorption threshold is taken as the inflection point in the spectra. The 8.3-nm CdS has an absorption threshold of 510 nm, which is close to the threshold of 515 nm for bulk CdS. The blue shift due to quantum confinement is very evident for the 4.3-nm quantum dots, which exhibits an absorption threshold of 470 nm. Determining the absorption edge provides a convenient means of monitoring particle size for semiconductor and metal particles (Ali et al., 1997).

In the second example, Figure 7 shows reflectance spectra of GaN films under different uniaxial stresses in the *c* plane (Yamaguchi et al., 1997). The spectra show a narrow

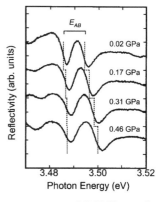

**Figure 7.** Reflectance spectra of GaN films under different uniaxial stresses in the *c* plane. The dashed lines indicate exciton energies as determined by theoretical fitting. The symbol $E_{AB}$ refers to the energy separation between the A and B excitons. (Adapted with permission from Yamaguchi et al., 1997. Copyright © 1997 by the American Institute of Physics.)

spectral region (3.47 eV = 357.3 nm, 3.52 eV = 352.2 nm) that encompasses two of three dispersive-shaped exciton absorption bands that occur near the valence band maximum. The closely spaced excitons lead to a large density of states at the valence band maximum, which results in a high threshold current for blue GaN laser diodes. Theory predicts that breaking the symmetry in the $c$ plane should increase the energy splitting of the A and B bands and decrease the density of states. The dashed lines in Figure 7 show that the energy splitting does increase with increasing stress. The dependence of $E_{AB}$ on the uniaxial stress, with other polarization measurements, allowed the authors to determine deformation potentials to be $D_4 = -3.4$ eV and $D_5 = -3.3$ eV. These results confirm the theoretical predictions that uniaxial strain should reduce the density of states and ultimately improve the performance of GaN laser diodes by reducing the lasing threshold current.

## SAMPLE PREPARATION

Sample preparation for UV-Vis spectroscopy is usually straightforward. Absorbing analytes that are in solution or can be dissolved are measured in an appropriate solvent or pH buffer. The pH and competing equilibria must be controlled so that the absorbing species are present in the same form in all samples and standards. Metals and nonmetals are quantitated by adding reagents that produce absorbing complexes in solution (Dean, 1995). To avoid saturation of the absorbance, strongly absorbing samples must be diluted, measured in a reflectance or ATR geometry, or measured in a short-path-length cell (Degiorgi et al., 1995). Because UV-Vis absorption bands are typically broad, quantitative analysis requires that a sample contain only one or a few absorbing species. Very complex mixtures usually require separation before making quantitative measurements. Similarly, particulates that scatter light should be removed by centrifugation or filtration to avoid interfering with the absorbance measurement.

Solid samples require clean and polished surfaces to avoid contaminant and scattering losses. If the solid has nonparallel surfaces, it will act as a prism and bend the light beam. For regular shaped samples, a compensating optic can redirect the light beam to its original path. Fiber-optic light delivery and collection provides a simple means of directing the light through irregular shaped samples. Preparing samples for reflectance measurements can be much more difficult than transparent samples. The reflectivity of a surface or film can depend on a variety of material parameters, including the presence of absorbing species, surface roughness, particle size, and packing density (Childers et al., 1986).

Specular reflectance is usually treated similarly to transmittance measurements, i.e., as the fraction or percentage of reflected light power compared to the incident light power. Diffuse reflectance is usually described as the relative reflectance $r_\infty$, where $r_\infty = r_{\text{sample}}/r_{\text{reference}}$. The absorptive and scattering components can be extracted

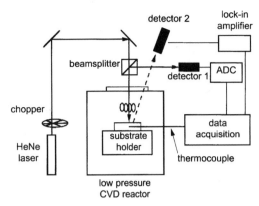

**Figure 8.** Experimental schematic of a CVD chamber that uses reflectivity (detector 1) and scattering (detector 2) to monitor diamond film growth on a Ni substrate. (Adapted with permission from Yang et al., 1997. Copyright © 1997 by the American Institute of Physics.)

from the relative reflectance using the Kubelka-Munk equation:

$$\frac{k}{s} = \frac{(1 - r_\infty)^2}{2r_\infty} \tag{8}$$

where $k$ is the absorption coefficient and $s$ is the scattering coefficient (Childers et al., 1986).

The reflectivity dependence on morphology and composition can be used to monitor the growth or processing of a thin film or surface. Figure 8 shows an experimental layout to use reflectivity and scattering to monitor diamond film growth on Ni in a low-pressure chemical vapor deposition (CVD) chamber (Yang et al., 1997). The 632.8-nm beam of a HeNe laser is incident on the substrate/film surface, and detector 1 monitors reflectivity and detector 2 monitors scattered light. The optical monitoring used bandpass interference filters in front of the detectors and a chopped light source with phase-sensitive detection to discriminate against the large background light emission from the hot filament. Figure 9 shows typical changes in reflected and scattered light as the substrate temperature increases. Correlation of these curves with scanning electron micrographs indicated that the time to change the reaction conditions from pretreatment of the diamond seeds to diamond growth should occur immediately after the drop in scattered light intensity.

## SPECIMEN MODIFICATION

Specimen modification rarely occurs in commercial UV-Vis spectrometers due to the fairly low intensity of UV and visible light that passes through or strikes the sample. Some samples can be very susceptible to photodegradation; obvious examples are photocurable polymers and photoresist materials. To minimize photochemical degradation, most instruments disperse the light before it passes through the sample, so that wavelengths that are not being measured are also not irradiating the sample. Photodegradation is more problematic in diode array and

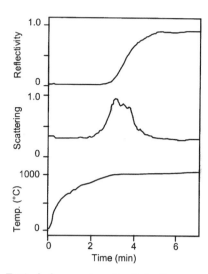

**Figure 9.** Typical changes in reflectivity (top curve, arbitrary units) and scattering (middle curve, arbitrary units) of a HeNe laser beam from a growing diamond film as the substrate temperature increases (bottom curve). (Adapted with permission from Yang et al., 1997. Copyright © 1997 by the American Institute of Physics.)

Fourier transform spectrometers since all wavelengths of light pass through the sample before dispersion and detection. Photosensitive samples can be protected to some degree by placing a filter between the light source and the sample to block shorter wavelengths that are not of interest in the measurement.

## PROBLEMS

The problems that can arise in UV-Vis spectroscopy can be divided into sample problems and instrumental limitations. These problems mostly affect the linear dependence of absorbance versus concentration in quantitative measurements, but in severe cases they can alter the qualitative nature of a UV-Vis spectrum. Sample problems can include the presence of absorbing or scattering impurities and degradation of very photochemically active molecules or materials. Very intense fluorescence from the analyte or other sample components is usually not a problem in scanning spectrometers but can cause problems in array detector spectrometers.

Problems due to chemical effects such as analyte-solvent interaction, pH effects, and competing equilibria can be minimized by making measurements of all samples under identical conditions. Quantitative measurements are usually calibrated with external or internal standards, rather than relying only on published absorptivity coefficients. For cases where the chemical behavior varies in different samples, the measurement can be calibrated using the standard addition method (Ingle and Crouch, 1988).

The two most common instrumental factors that affect quantitative measurements are the amount of stray light reaching the detector and the non-monochromatic nature of the light beam. The amount of stray light depends on the quality of the spectrometer components and the shielding of the sample compartment or fiber-optic probe. It is usually fixed for a given instrument, although it can vary significantly between the UV-Vis and NIR spectral regions. Maintaining measurement accuracy of approximately 1% requires stray light of <0.01% (Ingle and Crouch, 1988). The effects of stray light can also be minimized by measuring analytes at low concentration so that $P$ and $P_0$ remain large compared to the stray light power. Rule-of-thumb ranges for quantitative absorbance measurements are 0.1 to 2.5 A.U. for a high-quality spectrometer and 0.1 to 1.0 A.U. for a moderate-quality spectrometer (Ingle and Crouch, 1988; also see manufacturer's literature).

The unavoidable use of a fixed bandpass of light, which is non-monochromatic, results in a detector response due to multiple wavelengths around the central wavelength at which the spectrometer is set. The total amount of light reaching the detector will depend on the absorptivities of all wavelengths within the spectral bandpass and will vary nonlinearly with analyte concentration. This problem is eliminated by making measurements at a level part in the absorption spectrum with a spectrometer bandpass that is small compared to the width of the absorption line. The rule-of-thumb for accurate measurements is to use a spectrometer bandpass-to-linewidth ratio of approximately 0.1 (Ingle and Crouch, 1988). This criterion is easily met for most molecular absorption bands, which tend to be broad. Recording spectra with a bandpass that is larger than the absorption linewidths can drastically alter the qualitative appearance of a spectrum.

## ACKNOWLEDGMENTS

The author thanks Professor Paul Deck for critically reading the manuscript and gratefully acknowledges support from a National Science Foundation Career Award (CHE-9502460) and a Research Corporation Cottrell Scholars Award.

## LITERATURE CITED

Ali, A. H., Luther, R. J., and Foss, C. A., Jr. 1997. Optical properties of nanoscopic gold particles adsorbed at electrode surfaces: The effect of applied potential on plasmon resonance absorption. *Nanostruct. Mater.* 9:559–562.

Al-Rawashdeh, N. and Foss, C. A., Jr. 1997. UV/visible and infrared spectra of polyethylene/nanoscopic gold rod composite films: Effects of gold particle size, shape and orientation. *Nanostruct. Mater.* 9:383–386.

Banerjee, S., Chatterjee, S., and Chaudhuri, B. K. 1996. Optical reflectivity of superconducting $Bi_{3.9}Pb_{0.1}Sr_3Ca_3Cu_4O_x$ obtained by glass-ceramic route in 0.6–6.2 eV range. *Solid State Commun.* 98:665–669.

Barak, P. 1995. Smoothing and differentiation by an adaptive-degree polynomial filter. *Anal. Chem.* 67:2758–2762.

Bassani, F., Arnaud d'Avitaya, F., Mihalcescu, I., and Vial, J. C. 1997. Optical absorption evidence of quantum confinement in $Si/CaF_2$ multilayers grown by molecular beam epitaxy. *Appl. Surf. Sci.* 117: 670–676.

Baucom, K. C., Killeen, K. P., and Moffat, H. K. 1995. Monitoring of MOCVD reactants by UV absorption. *J. Electron. Mater.* 24: 1703–1706.

Bialkowski, S. E. 1995. Photothermal Spectroscopy Methods for Chemical Analysis. John Wiley & Sons, New York.

Childers, J. W., Rohl, R., and Palmer, R. A. 1986. Direct comparison of the capabilities of photoacoustic and diffuse reflectance spectroscopies in the ultraviolet, visible, and near-infrared regions. *Anal. Chem.* 58:2629–2636.

Commission Internationale de l'Eclairage (CIE, International Commission on Illumination). 1986. Colorimetry, 2nd ed. Publication CIE 15.2-1986. CIE, Vienna, Austria.

Dean, J. A. 1995. Analytical Chemistry Handbook. McGraw-Hill, New York.

Degiorgi, E., Postorino, P., and Nardone, M. 1995. A cell of variable thickness for optical studies of highly absorbing liquids. *Meas. Sci. Technol.* 6:929–931.

Degueldre, C., O'Prey, S., and Francioni, W. 1996. An in-line diffuse reflection spectroscopy study of the oxidation of stainless steel under boiling water reactor conditions. *Corros. Sci.* 38:1763–1782.

Di Bartolo, B. and Bowlby, B. (eds). 1994. Nonlinear spectroscopy of solids: Advances and applications. *In:* NATO ASI Series B: Physics, Vol. 339. Plenum, New York.

Glembocki, O. J. and Shanabrook, B. V. 1992. Photoreflectance spectroscopy of microstructures. *In* The Spectroscopy of Semiconductors (D. G. Seiler and C. L. Littler, eds.). pp. 221–292. Academic Press, San Diego.

Graebner, J. E. 1995. Simple correlation between optical absorption and thermal conductivity of CVD diamond. *Diam. Relat. Mater.* 4: 1196–1199.

Gregory, N. W. 1995. UV-vis vapor absorption spectrum of antimony(III) chloride, antimony(V) chloride, and antimony(III) bromide. The vapor pressure of antimony(III) bromide. *J. Chem. Eng. Data* 40:963–967.

Grunthaner, F. J., Ricco, A. J., Butler, M. A., Lane, A. L., McKay, C. P., Zent, A. P., Quinn, R. C., Murray, B., Klein, H. P., Levin, G. V., Terhune, R. W., Homer, M. L., Ksendzov, A., and Niedermann, P. 1995. Investigating the surface chemistry of Mars. *Anal. Chem.* 67:605A–610A. (Author's note: Unfortunately the Mars96 orbiter did not arrive at Mars.)

Ingle, J. D. and Crouch, S. R. 1988. Spectrochemical Analysis. Prentice-Hall, New York.

Kachanov, A. A., Stoeckel, F., and Charvat, A. 1997. Intracavity laser absorption measurements at ultrahigh spectral resolution. *Appl. Opt.*36:4062–4068.

Kovacs, L., Ruschhaupt, G., and Polgar, K. 1997. Composition dependence of the ultraviolet absorption edge in lithium niobate. *Appl. Phys. Lett.* 70:2801–2803.

Lamprecht, K., Papousek, W., and Leising, G. 1997. Problem of ambiguity in the determination of optical constants of thin absorbing films from spectroscopic reflectance and transmittance measurements. *Appl. Opt.* 36:6364–71.

Larena, A., Pinto, G., and Millan, F. 1995. Using the Lambert-Beer law for thickness evaluation of photoconductor coatings for recording holograms. *Appl. Surf. Sci.* 84:407–411.

Lassaletta, G., Fernandez, A., and Espinos, J. P. 1995. Spectroscopic characterization of quantum-sized $TiO_2$ supported on silica: Influence of size and $TiO_2$-$SiO_2$ interface composition. *J. Phys. Chem.* 99:1484–1490.

McPhedran, R. C., Botten, L. C., and McKenzie, D. R. 1984. Unambiguous determination of optical constants of absorbing films by reflectance and transmittance measurements. *Appl. Opt.* 23:1197–1205.

Ozer, N., Rubin, M. D., and Lampert, C. M. 1996. Optical and electrochemical characteristics of niobium oxide films prepared by sol-gel process and magnetron sputtering. A comparison. *Sol. Energ. Mat. Sol. Cells* 40:285–296.

Palmer, R. A. 1993. Photoacoustic and photothermal spectroscopies. In Physical Methods of Chemistry, 2nd ed., Vol. 8 (Determination of Electronic and Optical Properties) (B. W. Rossiter and R. C. Baetzold, eds.) pp. 61–108. John Wiley & Sons, New York.

Paul, J. B. and Saykally, R. J. 1997. Cavity ringdown laser absorption spectroscopy. *Anal. Chem.* 69:287A–292A.

Prasad, K. M. M. K., Raheem, S., Vijayalekshmi, P., and Kamala Sastri, C. 1996. Basic aspects and applications of tristimulus colorimetry. *Talanta* 43:1187–1206.

Ramdas, A. K. and Rodriguez, S. 1992. Piezospectroscopy of semiconductors. *In* The Spectroscopy of Semiconductors (D. G. Seiler and C. L. Littler, eds.) Academic Press, San Diego.

Savitzky A. and Golay, M. J. E. 1964. Smoothing and differentiation of data by simplified least squares procedures. *Anal. Chem.* 36:1627–1639.

Scherer, J. J., Paul, J. B., O'Keefe, A., and Saykally, R. J. 1997. Cavity ringdown laser absorption spectroscopy: History, development, and application to pulsed molecular beams. *Chem. Rev.* 97:25–51.

Stoeckel, F. and Atkinson, G. H. 1985. Time evolution of a broadband quasi-cw dye laser: Limitations of sensitivity in intracavity laser spectroscopy. *Appl. Opt.* 24:3591–3597.

Takenaka, S., Tanaka, T., and Yamazaki, T. 1997. Structure of active species in alkali-ion-modified silica- supported vanadium oxide. *J. Phys. Chem. B* 101:9035–9040.

Thorne, A. P. 1991. Fourier transform spectrometry in the ultraviolet. *Anal. Chem.* 63:57A–65A.

Tsuzuki, T. and McCormick, P. G. 1997. Synthesis of CdS quantum dots by mechanochemical reaction. *Appl. Phys. A* 65:607–609.

Vargas, W. E. and Niklasson, G. A. 1997. Applicability conditions of the Kubelka-Munk theory. *Appl. Opt.* 36:5580–5586.

Weckhuysen, B. M., Wachs, I. E., and Schoonheydt R. A. 1996. Surface chemistry and spectroscopy of chromium in inorganic oxides. *Chem. Rev.* 96:3327–3349.

Welsch, E., Ettrich, K., and Blaschke, H. 1997. Investigation of the absorption induced damage in ultraviolet dielectric thin films. *Opt. Eng.* 36:504–514.

Yamaguchi, A. A., Mochizuki, Y., Sasaoka, C., Kimura, A., Nido, M., and Usui, A. 1997. Reflectance spectroscopy on GaN films under uniaxial stress. *Appl. Phys. Lett.* 71:374–376.

Yang, P. C., Schlesser, R., Wolden, C. A., Liu, W., Davis, R. F., Sitar, Z., and Prater, J. T. 1997. Control of diamond heteroepitaxy on nickel by optical reflectance. *Appl. Phys. Lett.* 70:2960–2962.

Yoshino, K., Tada, K., Yoshimoto, K., Yoshida, M., Kawai, T., Zakhidov, A., Hamaguchi, M., and Araki, H. 1996. Electrical and optical properties of molecularly doped conducting polymers. *Syn. Metals* 78:301–312.

## KEY REFERENCES

*Analytical Chemistry.* American Chemical Society.

*This journal publishes several special issues that serve as useful references for spectroscopists. The June 15 issue each year is a set of reviews, and the August 15 issue is a buyer's guide*

*(Labguide) of instruments and supplies. The most relevant reviews to this unit are Coatings and Surface Characterization reviews in the Application Reviews issue (published odd years) and Ultraviolet and Light Absorption Spectrometry review in the Fundamental Reviews issue (published even years). An on-line version of the Labguide is currently available at http://pubs.acs.org/labguide/.*

Demtröder, W. 1996. Laser Spectroscopy: Basic Concepts and Instrumentation, 2nd ed. Springer-Verlag, Berlin.

*A graduate-level text on laser spectroscopy, including theory, laser-spectroscopic techniques, and instrumentation. Provides an overview of advanced absorption spectroscopies and competitive techniques such as photoacoustic spectroscopy.*

Hollas, J. M. 1996. Modern Spectroscopy, 3rd ed. John Wiley & Sons, New York.

*A graduate-level text that covers the principles and instrumentation of spectroscopy. Specific topics include quantum mechanics and interaction of light and matter, molecular symmetry, rotational spectroscopy, electronic spectroscopy, vibrational spectroscopy, photoelectron and surface spectroscopies, and laser spectroscopy.*

Ingle and Crouch, 1988. See above.

*A graduate-level text that covers the principles and instrumentation of spectroscopy. Specific techniques include atomic absorption and emission spectroscopy, molecular absorption and fluorescence spectroscopy, infrared absorption spectroscopy, and Raman scattering.*

Settle, F. A. 1997. Handbook of Instrumental Techniques for Analytical Chemistry. Prentice-Hall, New York.

*A broad comprehensive handbook of analytical techniques. It covers separation techniques, optical spectroscopy, mass spectrometry, electrochemistry, surface analysis, and polymer analysis.*

## APPENDIX:
## GLOSSARY OF TERMS AND SYMBOLS USED

| | |
|---|---|
| $a$ | Absorptivity |
| $A$ | Absorbance, $= abc = \varepsilon bc$ |
| A.U. | Absorbance unit |
| $b$ | Path length |
| $c$ | Analyte concentration; also velocity of light in vacuum ($2.99795 \times 10^8$ m/s), $= \lambda \nu$ |
| $D_4, D_5$ | Deformation potentials |
| $E$ | Energy of a photon, $= h\nu$ |
| $E_{AB}$ | Energy separation of A and B bands |
| $h$ | Planck's constant ($6.626 \times 10^{-34}$ Js) |
| $I$ | Light intensity |
| $I_0$ | Initial light intensity |
| $k$ | Absorption coefficient |
| $P$ | Radiant power (radiant energy on unit area in unit time) |
| $P_0$ | Initial radiant power; power transmitted through sample holder and solvent only |
| $P_s$ | Power of source light |
| $r_\infty$ | Relative reflectance, $= r_{sample}/r_{reference}$ |
| $R$ | Reflectance (includes diffuse and specular reflectance) |
| $\%R$ | Percent reflectance |
| $s$ | Scattering coefficient |
| $T$ | Transmittance |
| $\%T$ | Percent transmittance |

| | |
|---|---|
| $\varepsilon$ | Molar absorptivity |
| $\lambda$ | Wavelength |
| $\nu$ | Frequency |

Brian M. Tissue
Virginia Polytechnic Institute
and State University
Blacksburg, Virginia

# RAMAN SPECTROSCOPY OF SOLIDS

## INTRODUCTION

Raman spectroscopy is based on the inelastic scattering of light by matter and is capable of probing the structure of gases, liquids, and solids, both amorphous and crystalline. In addition to its applicability to all states of matter, Raman spectroscopy has a number of other advantages. It can be used to analyze tiny quantities of material (e.g., particles that are ~1 μm on edge), as well as samples exposed to a variety of conditions such as high temperature and high pressure and samples embedded in other phases, so long as the surrounding media are optically transparent.

Succinctly stated, Raman scattering results from incident radiation inducing transitions in the atoms/molecules that make up the scattering medium. The transition can be rotational, vibrational, electronic, or a combination (but first-order Raman scattering involves only a single incident photon). In most studies of solids by Raman spectroscopy, the transitions observed are vibrational and these will be the focus of this unit.

In a Raman experiment, the sample is irradiated with monochromatic radiation. If the sample is transparent, most of the light is transmitted, a small fraction is elastically (Rayleigh) scattered, and a very small fraction is inelastically (Raman) scattered. The inelastically scattered light is collected and dispersed, and the results are presented as a Raman spectrum, which plots the intensity of the inelastically scattered light as a function of the shift in wavenumber of the radiation. (The wavenumber of a wave is the reciprocal of its wavelength and is proportional to its momentum in units of reciprocal centimeters.) Each peak in the spectrum corresponds to one or more vibrational modes of the solid. The total number of peaks in the Raman spectrum is related to the number of symmetry-allowed, Raman active modes. Some of the modes may be degenerate and some may have Raman intensities that are too low to be measured, in spite of their symmetry-allowed nature. Consequently, the number of peaks in the Raman spectrum will be less than or equal to the number of Raman active modes. The practical usefulness of Raman spectroscopy resides largely in the fact that the Raman spectrum serves as a fingerprint of the scattering material. In fact, Raman activity is a function of the point group symmetry of a molecule and the space group symmetry of a crystalline solid; it can provide a range of information, including the strength of interatomic and intermolecular bonds, the mechanical strain present in a solid, the composition of multicomponent matter, the degree of crystallinity of a solid, and the effects of pressure and temperature on phase transformations.

## Competitive and Related Techniques

As mentioned above, Raman spectroscopy is extremely versatile and can be used to investigate the structures of solids, liquids, and gases. This unit is concerned with the analyses of solids. In general, two major types of experimental techniques are used to obtain information about the structures of solids: spectroscopy and diffraction. Spectroscopy is based on the interaction of electromagnetic radiation with matter. Diffraction occurs whenever a wave is incident on an array of regularly spaced scatterers in which the wavelength of the wave is similar to the spacing of the scatterers.

X rays, electrons, and neutrons with wavelengths similar to the interatomic spacing of a crystal will be diffracted by the crystal. The diffraction pattern will provide detailed information about, e.g., the crystal structure, crystal size, interplanar spacings, long- and short-range order, and residual lattice strain. While diffraction can provide some information about gases, liquids, and amorphous solids, the greatest amount of information is obtained for crystals.

Spectroscopic methods can probe the electronic, vibrational, rotational, and nuclear states of matter. In one family of spectroscopic techniques, the material absorbs electromagnetic radiation of energy equivalent to the difference in energy between particular states of the material. For example, transitions between adjacent electronic states generally require energies equivalent to that of ultraviolet and visible radiation (UV-vis electronic spectroscopy; see ULTRAVIOLET AND VISIBLE ABSORPTION SPECTROSCOPY). Vibrational transitions typically make use of infrared radiation (infrared absorption spectroscopy, IRAS). In the gas-phase molecules may absorb microwave radiation and undergo a change in rotational state. Finally, in nuclear magnetic resonance (NMR) the nuclei of atoms with angular momentum (spin) and magnetic moment absorb energy in the radio frequency region while subjected to magnetic fields that are alternating synchronous with the natural frequencies of the nucleus (see NUCLEAR MAGNETIC RESONANCE IMAGING).

Raman spectroscopy, unlike the above-mentioned spectroscopies, involves the scattering of electromagnetic radiation rather than absorption. In classical terms, the energy of the Raman scattered radiation is shifted from that of the incident radiation because of modulation by the vibrations of the scattering medium. Thus, Raman spectroscopy (RS), like IRAS, probes the vibrational spectra of materials. Because IRAS involves absorption of radiation and RS involves inelastic scattering of radiation, the two techniques are complementary. Since Raman spectroscopy makes use of optical radiation, it is a highly flexible technique and can probe the structures of materials in relatively inaccessible locations, such as in high-pressure chambers, high-temperature furnaces, and even aqueous solutions. All that is necessary is the ability to get laser light onto the sample and to collect the scattered light.

For materials that are both Raman active and IR active, the Raman scattering cross-section is much smaller than the IR absorption cross-section so the incident radiation needs to be more intense and the detectors need to be more sensitive for Raman spectroscopy than for IR spectroscopy. Because of the fundamental differences between absorption and scattering of electromagnetic radiation, some vibrational modes of a material may be Raman active but not IR active. In particular, for a molecule with a center of symmetry, vibrational modes that are Raman active are IR inactive and vice versa. Consequently, when deciding between IR and Raman spectroscopies, the point group or space group symmetry of the material needs to be taken into account.

A major advantage of Raman spectroscopy compared to x-ray diffraction is the ability of Raman spectroscopy to provide detailed structural information of amorphous materials. The Raman spectrum of an amorphous solid, for example, is directly related to its complete density of (vibrational) states. However, all metals are Raman inactive so Raman spectroscopy cannot be used for the structural analyses of this entire class of materials.

## PRINCIPLES OF THE METHOD

### Theoretical Background

This discussion of Raman spectroscopy will begin with a description of the fundamental phenomenon that underlies the scattering of light by matter, namely the generation of electromagnetic radiation by the acceleration of charged particles. As a light-scattering process, Raman scattering is rigorously described by treating both the incident and scattered radiation and the scattering medium quantum mechanically. In this unit a full quantum mechanical description of Raman scattering is not given. Instead, Raman scattering is treated less rigorously but in a manner that allows all of the practically important phenomena associated with Raman spectroscopy of solids to be identified. Specifically, a combination of classical physics and semiclassical physics (treating the scattering medium quantum mechanically while approximating light as a sinusoidally varying electric field) is used to describe Raman scattering.

**Generation of Electromagnetic Radiation—Classical Physics.** Classical physics dictates that accelerating charged particles emit electromagnetic radiation whose electric field $\varepsilon$ (a vector quantity) is directly proportional to the particle's acceleration $\mathbf{a}$ (a vector quantity; see Fig. 1; Rossi, 1957):

$$\varepsilon_{\mathbf{r}}(r, \theta, \phi, t) = \varepsilon_{\phi}(r, \theta, \phi, t) = 0 \tag{1a}$$

and

$$\varepsilon_{\theta}(r, \theta, \phi, t) = \left(\frac{q \sin \theta}{4\pi c^2 r}\right) a \tag{1b}$$

where $c$ is the velocity of light; $a$ is the magnitude of the acceleration of the charged particle of charge $q$; $r$, $\theta$, and $\phi$ are the spherical coordinates with the particle located at

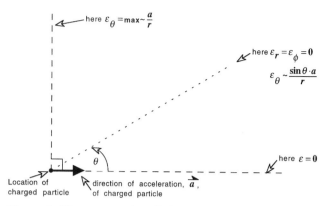

**Figure 1.** Electromagnetic radiation associated with an accelerating charged particle.

the origin and; $t$ is time. The term $\varepsilon_i(r, \theta, \phi, t)$ is the $i$th component of the electric field at the point $(r, \theta, \phi)$ and time $t$. The other terms are defined in Figure 1.

The optical behavior of matter, e.g., the emission of sharp spectral lines by elemental gases excited by an electric discharge, requires that matter, when treated by classical physics, be composed of charged particles that behave as harmonic oscillators when displaced from their equilibrium positions. When matter is illuminated by monochromatic optical radiation, the electrons are driven from their equilibrium positions by the sinusoidally varying electric field of light, creating a time-varying electric polarization **p** (a vector quantity) that is directly proportional to the acceleration of the electrons. Approximating the polarized molecule by a dipole, **p** is given by

$$\mathbf{p} = -e\mathbf{x} = -e\mathbf{x}_0 \cos \omega_{inc} t \tag{2a}$$

where **x** is the displacement of the center of negative charge of the molecule from the center of positive charge and $\mathbf{x}_0$ is the maximum value of this displacement.

$$\mathbf{p} = -\left(\frac{e}{\omega_{inc}^2}\right)\mathbf{a} \tag{2b}$$

where $-e$ is the electric charge of the electron and $\omega_{inc}$ is the frequency of oscillation of the electron, which matches the frequency of the incident light. The electrons thereby emit radiation whose electric field is proportional to the polarization of the molecule:

$$\varepsilon_\theta\left(\frac{\omega_{inc}^2 \sin \theta}{4\pi\varepsilon_0 c^2 r}\right)|\mathbf{p}| \tag{3}$$

where **x** is the displacement of the center of negative charge of the molecule from the center of positive charge and $\mathbf{x}_0$ is the maximum value of this displacement.

$$S_r = \left(\frac{\omega_{inc}^4 \sin^2 \theta}{16\pi^2\varepsilon_0^2 c^4 r^2}\right)|\mathbf{p}|^2 \tag{4}$$

where $\varepsilon_0$ is the dielectric strength of the medium in which the radiation is propagating and $S_r$ is the radial component of the Poynting vector, which is the only nonzero com-

ponent of the Poynting vector. The Poynting vector is the energy flux of the light wave. It is equal to the amount of energy flowing through a unit area in unit time.

When analyzing the effects of irradiating a molecule or a large quantity of matter with electromagnetic radiation, it is not necessary to treat each electron individually. Rather, it is possible to model the molecule(s) by atoms that are composed of heavy, positively charged ion cores and a mobile negative charge, which together form an electric dipole. If the centers of positive and negative charges are not coincident, the molecule has a permanent dipole moment. If the negative charge is displaced from its equilibrium position, e.g., by colliding with another particle, it will harmonically oscillate about its equilibrium position and emit electromagnetic radiation until all of the energy it gained in the collision is radiated away. Similarly, if the molecule is irradiated with an electromagnetic wave, the negative charge will be driven away from its equilibrium position by the applied electric field and will harmonically oscillate about its equilibrium position. The oscillating electric dipole will emit electromagnetic radiation with an electric field given by Equation 4 with $|\mathbf{p}| = p_0 \cos \omega_{inc} t$, where $\omega_{inc}$ is the frequency of the incident radiation and $p_0$ is the magnitude of the polarization when the centers of positive and negative charges experience their maximum separation.

The polarization **p** induced by the applied electric field of the incident electromagnetic radiation varies with time at the same frequency as the applied electric field. There is a second contribution to the time dependency of the material's polarization, namely that caused by the time-dependent nuclear displacements (thermally excited molecular vibrations), which, as illustrated in Figure 2, act to modulate the radiation generated by the electrons oscillating under the influence of the incident radiation (Long, 1977). As a result, there are three frequency components to the scattered light, which is the radiation emitted by the time-varying, molecular electric dipoles. One component (the Rayleigh line) has the same frequency, $\omega_{inc}$ as the incident radiation and the other two components have a slightly lower and a slightly higher frequency, $\omega_{inc} + \omega_v$ and $\omega_{inc} - \omega_v$, respectively, where $\omega_v$ is the vibrational frequency of the atoms/molecules making up the scattering medium. These last two components represent the Stokes and anti-Stokes components, respectively, of the Raman scattered radiation.

Classically, the expression for the intensity of the Raman scattered radiation is

$$I = \langle|S|\rangle\frac{dA}{d\Omega} \tag{5}$$

where $A$ is the surface area of the sphere centered at the source of the Raman scattered radiation, $\Omega$ is the solid angle through which the radiation passes, and $\langle|S|\rangle$ is the time-averaged value of the Poynting vector of the Raman scattered radiation. Its form is identical to that given above for $S_r$ except that the time-varying part of **p** (see Equation 4) is now given by

$$\mathbf{p} \sim \left(\frac{\partial\alpha}{\partial\mathbf{Q}_k}\right)_0 \mathbf{Q}_k\varepsilon \tag{6}$$

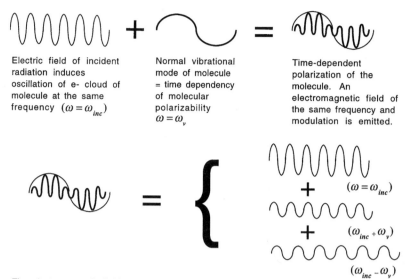

Electric field of incident radiation induces oscillation of e- cloud of molecule at the same frequency $(\omega = \omega_{inc})$

Normal vibrational mode of molecule = time dependency of molecular polarizability $\omega = \omega_v$

Time-dependent polarization of the molecule. An electromagnetic field of the same frequency and modulation is emitted.

$(\omega = \omega_{inc})$

$(\omega_{inc} + \omega_v)$

$(\omega_{inc} - \omega_v)$

The electromagnetic field may be subdivided into 3 components of frequency $\omega_{inc}$, $\omega_{inc} + \omega_v$, $\omega_{inc} - \omega_v$.

**Figure 2.** Polarization of vibrating molecule irradiated by light.

where $\underline{\alpha}$ (a tensor quantity) is the polarizability of the material, $\mathbf{Q}_k$ (a vector quantity) is the normal coordinate of the $k$th vibrational mode, and $(\partial\underline{\alpha}/\partial\mathbf{Q}_k)_0$ (a tensor quantity) is the differential polarizability, which indicates the substance's change in polarizability at its equilibrium position during the $k$th normal mode of vibration. Consequently, the material property that is directly probed by Raman spectroscopy is the differential polarizability $(\partial\underline{\alpha}/\partial\mathbf{Q}_k)_0$. It is a symmetric tensor quantity, and at least one component must be nonzero for Raman scattering to occur. Knowing the point group symmetry of a molecule or the space group symmetry of a crystalline solid, it is possible to use group theoretical techniques to determine if the symmetry permits nonzero components of the differential polarizability tensor. The use of molecular and crystal symmetry greatly simplifies the analyses of Raman scattering by a material, as will be demonstrated below (see Overview of Group Theoretical Analysis of Vibrational Raman Spectroscopy).

**Light Scattering—Semiclassical Physics.** From the perspective of semiclassical physics, Raman scattering results from the incident radiation inducing a transition (considered here to be vibrational) in the scattering entity, as schematically illustrated in Figure 3. The material is placed by the sinusoidally varying electric field into a higher energy state, termed a virtual state, which may be thought of as the distortion of the molecule by the electric field. As shown in Figure 3, when the molecule returns to a lower energy state, it emits radiation of frequency $\omega_{inc}(=>$ Rayleigh, or elastic scattering), $\omega_{inc} - \omega_v(=>$ Stokes Raman scattering), or $\omega_{inc} + \omega_v$ (=> anti-Stokes Raman scattering). The probability of these transitions occurring is, according to quantum mechanical perturbation theory, given by $a_{mn}^2$, where

$$a_{mn} \sim \int \psi_m^* \mathbf{p} \psi_n dV \qquad (7)$$

Here, $\psi_i$ is the wave function of the $i$th state and $\psi_i^*$ is its complex conjugate.

$$\mathbf{p} = \underline{\alpha}\varepsilon \qquad (8)$$

When expressed in terms of the individual components, $p_i$, Equation 8 becomes

$$p_i = \sum \alpha_{ij}\varepsilon_j \qquad (9)$$

where $\alpha_{ij}$ is the $ij$ matrix component of the polarizability, $\varepsilon_j$ is the $j$ component of the electric field, and the differential $dV$ includes differential displacements of all quantum vibrational states. For Rayleigh scattering, $m = n$ is the vibrational ground state; for Stokes Raman scattering, $n$ is the vibrational ground state and $m$ is the first vibrational excited state; for anti-Stokes Raman scattering, $n$

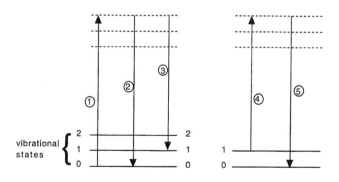

vibrational states

1+2 = Raleigh Scattering;
1+3 = Stokes Raman Scattering;
4+5 = anti-Stokes Raman Scattering

**Figure 3.** Rayleigh and Raman scattering. $1 + 2$, Rayleigh scattering; $1 + 3$, Stokes Raman scattering; $4 + 5$, anti-Stokes Raman scattering.

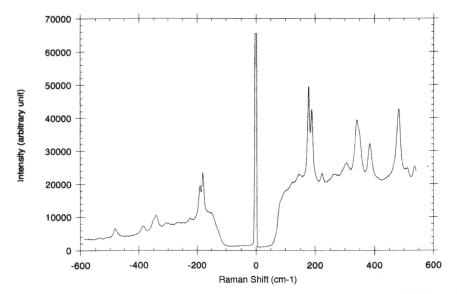

**Figure 4.** Raman spectrum of impure, polyphase (monoclinic+tetragonal) $ZrO_2$. (C. S. Kumai, unpublished results, 1998.)

is the first vibrational excited state and $m$ is the vibrational ground state.

Since the population of energy levels in a collection of atoms/molecules is governed by the Maxwell-Boltzmann distribution function, the population of the vibrational ground state will always be greater than the population of the first vibrational excited state. As a result, the intensity of the Stokes-shifted Raman scattering will always be greater than the intensity of the anti-Stokes-shifted Raman scattered radiation. This effect, which is predicted quantum mechanically but not classically, is illustrated in Figure 4, which presents the Raman spectrum of polycrystalline $ZrO_2$ (mixed monoclinic and tetragonal). The Raman spectrum is depicted in terms of the shift (in wavenumbers) in the wavelength of the scattered radiation with respect to that of the incident light. Thus the Rayleigh peak is seen in the center, and the anti-Stokes and Stokes lines are to the left and right, respectively. Note that for each peak in the Stokes Raman section of the spectrum there is a corresponding peak of lower intensity in the anti-Stokes Raman section.

### Overview of Group Theoretical Analysis of Vibrational Raman Spectroscopy

The use of group theory in determining whether or not a given integral can have a value of zero may be illustrated using the ground vibrational state $\psi_0(x)$ and first vibrational excited state $\psi_1(x)$ of a harmonic oscillator. The equations describing these two vibrational states are, respectively,

$$\psi_0(x) = \exp\left(-\frac{2\pi m\omega x^2}{\hbar}\right) \tag{10}$$

$$\psi_1(x) = \left(\frac{4\pi^{1/2}m\omega}{\hbar}\right)^{1/2} \exp\left(-\frac{2\pi m\omega x^2}{\hbar}\right) \tag{11}$$

where $\hbar = h/2\pi$ and $h$ is Planck's constant. These two vibrational states are plotted in Figure 5.

By inspection,

$$\int_{-\infty}^{\infty} \psi_1(x)dx = 0 \tag{12}$$

That is, performing the integration by summing up the infinite number of terms given by

$$\sum \psi_1(x_n)\Delta x_n \tag{13}$$

(where $\Delta x_n = x_n - x_{n-1}$), it is seen that for every positive contribution (to the sum) given by $\psi_1(x_i)\Delta x_i$ there is a term equal in magnitude but opposite in sign [$\psi_1(x_{-i}) \Delta x_{-i} = -\psi_1(x_i)\Delta x_i$].
Thus, the entire sum in Equation 13 is zero.

Similarly, by inspection,

$$\int_{-\infty}^{\infty} \psi_0(x)dx \neq 0 \tag{14}$$

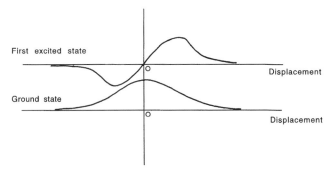

**Figure 5.** Ground-state vibrational wave function and first excited-state vibrational wave function of harmonic oscillator.

That is, every term $\psi_0(x_i)\Delta x_i$ in the sum $\sum \psi_0(x_n)\Delta x_n$ has the identical sign and, hence,

$$\int_{-\infty}^{\infty} \psi_0(x)dx \approx \sum \psi_0(x_n)\Delta x_n \neq 0 \qquad (15)$$

The above analysis of the integrals in Equations 12 and 13 can be repeated using group theory. To begin, there needs to be a mathematical description of what is meant by the symmetry of a function. The difference between the symmetry of $\psi_1(x)$ and $\psi_0(x)$ can be expressed as follows. Imagine that there is a mirror plane parallel to the YZ plane and that passes through the origin, (0, 0, 0); then the operation of this mirror plane on the function $\psi_0(x)$ is described by the phrase "a mirror plane acting on $\psi_0(x)$ yields $\psi_0(x)$."

If the effect of the mirror plane acting on $\psi_0(x)$ is represented by a matrix $[R]$, then

$$\text{Effect of mirror plane acting on } \psi_0(x) = [R][\psi_0(x)] \quad (16)$$

In this case, $[R] = [I]$ is the identity matrix.

Similarly, for $\psi_1(x)$, the effect of a mirror plane acting on $\psi_1(x)$ equals $-\psi_1(x)$. In matrix notation, this statement is given as

$$\text{Effect of mirror plane acting on}$$
$$\psi_1(x) = [R]\psi_1(x) = -[I]\psi_1(x) \qquad (17)$$

Note that the effect of the mirror operation on $x$ is to produce $-x$ and that $\psi(-x) = -\psi(x)$. The character of the one-dimensional matrix in Equation 16 is $+1$ and that in Equation 17 is $-1$. Consider now a molecule that exhibits a number of symmetry elements (e.g., mirror planes, axes of rotation, center of inversion). If the operation of each of those on a function leaves the function unchanged, then the function is said to be "totally symmetric" with respect to the symmetry elements of the molecule (or, in other words, the function is totally symmetric with respect to the molecular point group, which is the collection of all the molecule symmetry elements). For the totally symmetric case, the character of each one-dimensional (1D) matrix representing the effect of the operation of each symmetry element on the function is 1 for all symmetry operations of the molecule. For such a function, the value of the integral over all space of $\psi(x)dx$ will be nonzero.

If the operation of any symmetry element of the molecule (in the molecular point group) generates a matrix that has a character other than 1, the function is not totally symmetric and the integral of $\psi(x)dx$ over all space will be zero.

Now consider the application of group theory to vibrational spectroscopy. A vibrational Raman transition generally consists of the incident light causing a transition from the vibrational ground state to a vibrational excited state. The transition may occur if

$$\int_{-x}^{x} \psi_1(x)\alpha_{ij}\psi_0(x)dx \neq 0 \qquad (18)$$

**Table 1. Character Table of Point Group $C_{2v}$**

| $C_{2v}$ | $E$ | $C_2(z)$ | $\sigma_v(xz)$ | $\sigma'_v\,(yz)$ | Basis Functions |
|---|---|---|---|---|---|
| $A_1$ | $+1$ | $+1$ | $+1$ | $+1$ | $z;\ x^2;\ y^2;\ z^2$ |
| $A_2$ | $+1$ | $+1$ | $-1$ | $-1$ | $xy$ |
| $B_1$ | $+1$ | $-1$ | $+1$ | $-1$ | $x;\ xz$ |
| $B_2$ | $+1$ | $-1$ | $-1$ | $+1$ | $y;\ yz$ |

That is, the integral may have a nonzero value if the integrand product is totally symmetric over the range of integration. The symmetry of a function that is the product of two functions is totally symmetric if the symmetries of the two functions are the same. In addition, for the sake of completeness, the last statement will be expanded using group theoretical terminology, which will be defined below (see Example: Raman Active Vibrational Modes of $\alpha$-Al$_2$O$_3$). The symmetry species of a function that is the product of two functions will contain the totally symmetric irreducible representation if the symmetry species of one function contains a component of the symmetry species of the other.

The ground vibrational state is totally symmetric. Hence, the integrand is totally symmetric and the vibrational mode is Raman active if the symmetry of the excited vibrational state $\psi_1(x)$ is the same as, or contains, the symmetry of the polarizability operator $\alpha_{ij}$.

As shown below (see Appendix), the symmetry of the operator $\alpha_{xy}$ is the same as that of the product $xy$. Thus, the integrand $\psi_1\alpha_{xy}\psi_0$ is totally symmetric if $\psi_1$ has the same symmetry as $xy$ (recall that $\psi_0$, the vibrational ground state, is totally symmetric).

A molecular point group is a mathematical group (see Appendix) whose members consist of all the symmetry elements of the molecule. Character tables summarize much of the symmetry information about a molecular point group. Complete sets of character tables may be found in dedicated texts on group theory (e.g., Bishop, 1973). The character table for the point group $C_{2v}$ is presented in Table 1. In the far right-hand column are listed functions of interest in quantum mechanics and vibrational spectroscopy in particular. The top row lists the different symmetry elements contained in the point group and illustrated in Figure 6. Here, $C_2$ is a twofold rotational axis that is parallel to the $z$ axis. The parameters $\sigma_{xz}$ and $\sigma_{yz}$ are mirror planes parallel to the $xz$ and $yz$ planes, respectively. In each row beneath the top row are listed a set of numbers. Each number is the character of the matrix that represents the effect of the symmetry operation acting on any of the functions listed in the right-hand end of the same row. In the point group $C_{2v}$, e.g., the functions $z$, $x^2$, $y^2$, and $z^2$ have the same symmetry, which in the terminology of group theory is identified as $A_1$. The functions $y$ and $yz$ have the same symmetry, $B_2$.

As shown in Figure 6, the water molecule, H$_2$O, exhibits the point group symmetry $C_{2v}$. There are three atoms in H$_2$O and hence there are $3N - 6 = 3$ normal vibrational modes, which are depicted in Figure 7. These exhibit symmetries $A_1$, $A_1$, and $B_2$. That is, the vibrational wave functions of the first excited state of each of these modes

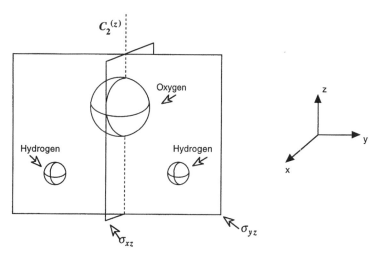

**Figure 6.** Symmetry operations of $H_2O$ (point group $C_{2v}$).

possess the symmetries $A_1$, $A_1$, and $B_2$, respectively. Since $x^2$, $y^2$, and $z^2$ exhibit $A_1$ symmetry, so too would $\alpha_{xx}$, $\alpha_{yy}$ and $\alpha_{zz}$. Hence the product of $\alpha_{xx}$, $\alpha_{yy}$ and $\alpha_{zz}$ with $\psi_1$ for the vibrational modes with $A_1$ symmetry would be totally symmetric, meaning that the integral over the whole space $\psi_{A1}\alpha_{xx(or\,yy,zz)}\psi_0 dV$ may be nonzero, meaning that this vibrational mode is Raman active (the integrand $\psi_{A1}\alpha_{xx(or\,yy,zz)}\psi_0 dV$ is the same as the integrand in Equation 7).

Now consider the Raman vibrational spectra of solids. One mole of a solid will have $3 \times 6.023 \times 10^{23} - 6$ normal modes of vibration. Fortunately, it is not necessary to consider all of these. As demonstrated below (see Appendix), the only Raman active modes will be those near the center of the Brillouin zone (BZ; $k = 0$). Solids with only one atom per unit cell have zero modes at the center of the BZ and hence are Raman inactive (see Appendix). Solids with

more than one atom per unit cell, e.g., silicon and $Al_2O_3$, are Raman active.

The symmetry of a crystal is described by its space group. However, as demonstrated below (see Appendix), for the purposes of vibrational spectroscopy, the symmetry of the crystal is embodied in its crystallographic point group. This is an extremely important and useful conclusion. As a consequence, the above recipe for deciding the Raman activity of the vibrational modes of $H_2O$ can be applied to the vibrational modes of any solid whose unit cell is multiatomic.

## PRACTICAL ASPECTS OF THE METHOD

The weakness of Raman scattering is the characteristic that most strongly influences the equipment and experimental techniques employed in Raman spectroscopy.

A conventional Raman facility consists of five major components: (1) a source of radiation, (2) optics for illuminating the sample and collecting the Raman scattered radiation, (3) a spectrometer for dispersing the Raman scattered radiation, (4) a device for measuring the intensity of the Raman scattered light, and (5) a set of components that control the polarization of the incident radiation and monitor the polarization of the Raman scattered radiation (Chase, 1991; Ferraro and Nakamoto, 1994).

### Sources of Radiation

Prior to the introduction of lasers, the primary source of monochromatic radiation for Raman spectroscopy was the strongest excitation line in the visible region (blue, 435.83 nm) of the mercury arc. The power density of the monochromatic radiation incident on the sample from a mercury arc is very low. To compensate, relatively large samples and complicated collection optics are necessary to, respectively, generate and collect as many Raman scattered photons as possible. Collecting Raman photons that are scattered in widely different directions precludes investigating the direction and polarization characteristics of the

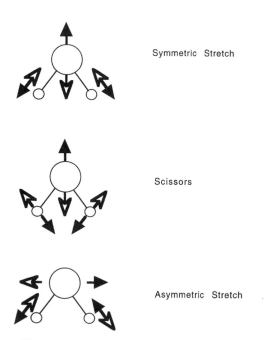

**Figure 7.** Normal vibrational modes of $H_2O$.

Raman scattered radiation. This is valuable information that is needed to identify the component(s) of the crystal's polarizability tensor that is (are) responsible for the Raman scattering (Long, 1977).

In general, laser radiation is unpolarized. However, in ionized gas lasers, such as argon and krypton ion lasers, a flat window with parallel faces and oriented at the Brewster angle ($\theta_{\text{Brewster}}$) relative to the incident beam is positioned at the end of the tube that contains the gas. This window is called a Brewster window. The Brewster angle is also called the polarizing angle and is given by the arctangent of the ratio of the indices of refraction of the two media that form the interface (e.g., for air/glass, $\theta_{\text{Brewster}} = \tan^{-1} 1.5 \approx 57°$). If an unpolarized beam of light traveling in air is incident on a planar glass surface at the Brewster angle, the reflected beam will be linearly polarized with its electric field vector parallel to the plane of incidence. (The plane of incidence contains the direction of propagation of the light and the normal to the interface between the two media.) A beam of light that is linearly polarized with its electric field vector in the plane of incidence and is incident on a Brewster window at $\theta_{\text{Brewster}}$ will be entirely transmitted. Thus, the light that exits from a laser tube that is capped with a Brewster window is linearly polarized (Fowles, 1989; Hecht, 1993).

Lasers have greatly increased the use of Raman spectroscopy as a tool for both research and chemical identification. In particular, continuous-wave gas lasers, such as argon, krypton, and helium-neon, provide adequately powered, monochromatic, linearly polarized, collimated, small-diameter beams that are suitable for obtaining Raman spectra from all states of matter and for a wide range of sample sizes using relatively simple systems of focusing and collection optics.

## Optics

As the light exits the laser, it generally first passes through a filter set to transmit a narrow band of radiation centered at the laser line (e.g., an interference filter). This reduces the intensity of extraneous radiation, such as the plasma lines that also exit from the laser tube. A set of mirrors and/or lenses will make the laser light incident on the sample at the desired angle and spot size.

The focusing lens reduces the beam's diameter to

$$d = \frac{4\lambda f}{\pi D} \qquad (19)$$

where $\lambda$ is the wavelength of the laser radiation, $f$ is the focal length of the lens, and $D$ is the diameter of the unfocused laser beam. Note that for a given focusing lens and incident light, the size of the focused spot is inversely proportional to the size of the unfocused beam. Thus, a smaller spot size can be generated by first expanding the laser beam to the size of the focusing lens. The distance $l$ over which the beam is focused is proportional to the square of the focused beam's diameter (Hecht, 1993):

$$l = \frac{16\lambda f^2}{\pi D^2} = \frac{\pi d^2}{\lambda} \qquad (20)$$

Thus, to reduce the diameter of the focal spot from an argon laser ($\lambda = 514.5$ nm) to 1 μm, the distance between the focusing lens and the sample must be accurate to within $\pm \frac{1}{2}\pi(1 \text{ μm})^2/(0.5145 \text{ μm}) \approx 3$ μm.

The specularly reflected light is of no interest in Raman spectroscopy. The inelastically scattered light is collected using a fast lens (i.e., low $f$ number, defined as the focal length per diameter of the lens). Often, the lens from a 35-mm camera is a very cost-effective collection lens. The collection lens collimates the Raman scattered light and transmits it toward a second lens that focuses the light into the entrance slit of the spectrometer. The $f$ number of the second lens should match that of the spectrometer. Otherwise, the spectrometer's grating will either not be completely filled with light, which will decrease the resolving power (which is directly proportional to the size of the grating), or be smaller than the beam of radiation, which will result in stray light that may reflect off surfaces inside the spectrometer and reduce the signal-to-noise ratio.

There are many sources of stray light in a Raman spectrum. One of the major sources is elastically scattered light (it has the same wavelength as the incident radiation), caused by Rayleigh scattering by the molecules/atoms responsible for Raman scattering and Mie scattering from larger particles such as dust. Rayleigh scattering always accompanies Raman scattering and is several orders of magnitude more intense than Raman scattering. The Rayleigh scattered light comes off the sample at all angles and is captured by the collection optics along with the Raman scattered light and is directed into the spectrometer. This would not be a problem if the Rayleigh scattered light exited the spectrometer only at the $0$-cm$^{-1}$ position. However, because of imperfections in the gratings and mirrors within the spectrometer, a portion of the Rayleigh scattered light exits the spectrometer in the same region as Raman scattered radiation that is shifted in the range of 0 to $\pm 250$ cm$^{-1}$ from the incident radiation. Even though only a fraction of the Rayleigh scattered light is involved in this misdirection, the consequence is significant because the intensity of the Rayleigh component is so much greater than that of the Raman component. Raman peaks located within $\approx 250$ cm$^{-1}$ of the laser line can only be detected if the intensity of the Rayleigh scattered light is strongly reduced.

The intensity of stray light that reaches the detector can be diminished by passing the collimated beam of elastically and inelastically scattered light collected from the sample into a notch filter. After exiting the notch filter, the light enters a lens that focuses the light into the entrance of the spectrometer.

A notch filter derives its name from the narrow band of wavelengths that it filters out. In a Raman experiment, the notch filter is centered at the exciting laser line and is placed in front of the spectrometer. (Actually, as mentioned above and explained below, the notch filter should be located in front of the lens that focuses the scattered light into the spectrometer.) The notch filter then removes the Rayleigh scattered radiation (as well as Brillouin scattered light and Mie scattering from dust particles). Depending on the width of the notch, it may also remove a significant amount of the Raman spectrum.

Dielectric notch filters consist of multiple, thin layers of two materials with different indices of refraction. The materials are arranged in alternating layers so that the variation of index of refraction with distance is a square wave. Typically, dielectric filters have a very wide notch with diffuse edges and have nonuniform transmission in the region outside the notch. As a consequence, a dielectric notch filter removes too much of the Raman spectrum (the low-wavenumber portion) and distorts a significant fraction of the portion it transmits. In contrast, holographic notch filters have relatively narrow widths, with sharp edges and fairly uniform transmission in the region outside the notch (Carrabba et al., 1990; Pelletier and Reeder, 1991; Yang et al., 1991; Schoen et al., 1993).

A holographic notch filter (HNF) is basically an interference filter. In one manufacturing process (Owen, 1992), a photosensitive material consisting of a dichromate gelatin film is placed on top of a mirror substrate. Laser light that is incident on the mirror surface interferes with its reflected beam and forms a standing-wave pattern within the photosensitive layer. The angle between the incident and reflected beams determines the fringe spacing of the hologram. If the incident beam is normal to the mirror, the fringe spacing will be equal to the wavelength of the laser light.

Chemical processing generates within the hologram-exposed material an approximately sinusoidal variation of index of refraction with distance through the thickness of the layer. The wavelength of the modulation of the index of refraction determines the central wavelength of the filter. The amplitude of the modulation and the total thickness of the filter determine the bandwidth and optical density of the filter.

The precise shape and location of the "notch" of the filter is angle tunable so the filter should be located in between the lens that collects and collimates the scattered light and the lens that focuses the Raman scattered radiation into the spectrometer such that only collimated light passes through the filter.

A holographically generated notch filter can have a relatively narrow, sharp-edged band of wavelengths that will not be transmitted. The band is centered at the wavelength of the exciting laser line and may have a total width of $\approx 300$ cm$^{-1}$. The narrow width of the notch and its high optical density permit the measurement of both the Stokes and anti-Stokes components of a spectrum, as is illustrated in the Raman spectrum of zirconia presented in Figure 4.

In addition to a notch filter, it is also possible to holographically generate a bandpass filter, which can be used to remove the light from the plasma discharge in the laser tube (Owen, 1992). In contrast to dielectric filters, holographic filters can have a bandpass that is five times more narrow and can transmit up to 90% of the laser line. Dielectric bandpass filters typically transmit only 50% of the laser line.

### Optical Alignment

Alignment of the entire optical system is most easily accomplished by use of a second laser, which need only have an intensity of $\approx 1$ mW. The spectrometer is set to transmit the radiation of the second laser, which is located at the exit slit of the spectrometer. For convenience, instead of having to remove the light detector, the light from the second laser can enter through a second, nearby port and made to hit a mirror that is inserted inside the spectrometer, just in front of the exit slits. The light reflected from the mirror then passes through the entire spectrometer and exits at the entrance slits, a process termed "back illumination." The back-illuminated beam travels the same path through the spectrometer (from just in front of the exit slits to the entrance slits) that the Raman scattered radiation from the sample will travel in getting to the light detector. Consequently, the system is aligned by sending the back-illuminated beam through the collection optics and into coincidence with the exciting laser beam on the sample's surface.

### Spectrometer

The dispersing element is usually a grating, and the spectrometer may typically have one to three gratings. Multiple gratings are employed to reduce the intensity of the Rayleigh line. That is, light that is dispersed by the first grating is collimated and made incident on a second grating. However, the overall intensity of the dispersed light that exits from the spectrometer decreases as the number of gratings increases. The doubly dispersed light has a higher signal-to-noise ratio, but its overall intensity is significantly lower than singly dispersed light. Since the intensity of Raman scattered radiation is generally very weak (often by one or more orders of magnitude), there is an advantage to decreasing the intensity of the Rayleigh light that enters the spectrometer by using a single grating in combination with a notch filter rather than by using multiple gratings (see Optics).

Modern spectrometers generally make use of interference or holographic gratings. These are formed by exposing photosensitive material to the interference pattern produced by reflecting a laser beam at normal incidence off a mirror surface or by intersecting two coherent beams of laser light. Immersing the photosensitized material into a suitable solvent, in which the regions exposed to the maximum light intensity experience either enhanced or retarded rates of dissolution, produces the periodic profile of the grating. The surface of the grating is then typically coated with a thin, highly reflective metal coating (Hutley, 1982).

The minimum spacing of grooves that can be generated by interference patterns is directly proportional to the wavelength of the laser light. The minimum groove spacing that can be formed using the 458-nm line of an argon ion laser is 0.28 μm (3500 grooves/mm). The line spacing of a grating dictates one of the most important characteristics of the grating, its resolving power.

The resolving power of a grating is a measure of the smallest change of wavelength that the grating can resolve:

$$\frac{\lambda}{\Delta\lambda} = \frac{mW}{d} \tag{21}$$

where $\lambda$ is the wavelength at which the grating is operating, $m$ is the diffraction order, $W$ is the width of the grating, and $d$ is the spacing of the grating grooves (or steps). This relation indicates the importance of just filling the grating with light and the influence of the spacing of the grating grooves, which is generally expressed in terms of the number of grooves per millimeters, on its resolving power.

A second parameter of interest for characterizing the performance of a grating is its absolute efficiency, which is a measure of the fraction of the light incident on the grating that is diffracted into the required order. It is a function of the shape of the groove, the angle of incidence, the wavelength of the incident light, the polarization of the light, the reflectance of the material that forms the grating, and the particular instrument that houses the gratings (Hutley, 1982). The efficiency of a grating can be measured or calculated with the aid of a computer. The measurement of the absolute efficiency is performed by taking the ratio of the flux in the diffracted beam to the flux in the incident beam. Generally, it is not necessary for a Raman spectroscopist to measure and/or calculate the absolute efficiencies of gratings. Such information may be available from the manufacturer of the gratings.

### Measurement of the Dispersed Radiation

A device for measuring the intensity of the dispersed radiation will be located just outside the exit port of the spectrometer. Since the intensity of the Raman scattered radiation is so weak, a very sensitive device is required to measure it. Three such devices are a photomultiplier tube (PMT), a photodiode array, and a charge-transfer device.

The key elements in a PMT are a photosensitive cathode, a series of dynodes, and an anode (Long, 1977; Ferraro and Nakamoto, 1994). The PMT is located at the exit of the spectrometer, which is set to transmit radiation of a particular wavelength. For Stokes-shifted Raman spectroscopy, the wavelength is longer than that of the exciting laser. As photons of this energy exit the spectrometer, they are focused into the photocathode of the PMT. For each photon that is absorbed by the photocathode, an electron is ejected and is accelerated toward the first dynode, whose potential is $\approx 100$ V positive with respect to the photocathode. For each electron that hits the first dynode, several are ejected and are accelerated toward the second dynode. A single photon entering the PMT may cause a pulse of $10^6$ electrons at the anode. Long (1977) describes the different procedures for relating the current pulse at the anode to the intensity of the radiation incident on the PMT.

The efficiency of the PMT is increased through the use of a photoemission element consisting of a semiconductor whose surface is coated with a thin layer ($\approx 2$ nm) of material with a low work function (e.g., a mixture of cesium and oxygen; Fraser, 1990). For a heavily doped $p$-type semiconductor coated with such a surface layer, the electron affinity in the bulk semiconductor is equal to the difference between its band-gap energy and the surface layer work function. If the work function is smaller than the band gap, the semiconductor has a negative electron affinity. This means that the lowest energy of an electron in the conduction band in the bulk of the semiconductor is higher than the energy of an electron in vacuum and results in an increase in the number of electrons emitted per absorbed photon (i.e., increased quantum efficiency). The gain of the PMT can be increased by treating the dynodesurface in a similar fashion. A dynode with negative electron affinity emits a greater number of electrons per incident electron. The overall effect is a higher number of electrons at the anode per photon absorbed at the cathode.

It is important to note that the PMT needs to be cooled to reduce the number of thermally generated electrons at the photocathode and dynodes, which add to the PMT "dark count." Heat is generally extracted from the PMT by a thermoelectric cooler. The heat extracted by the cooler is generally conducted away by flowing water.

A PMT is well suited to measuring the weak signal that exits a spectrometer that has been set to pass Raman scattered radiation of a single energy. The entire Raman spectrum is measured by systematically changing the energy of the light that passes through the spectrometer (i.e., rotating the grating) and measuring its intensity as it exits from the spectrometer. For a double monochromator fitted with a PMT, it may take minutes to tens of minutes to generate a complete Raman spectrum of one sample.

In a PMT, current is generated as photons hit the cathode. In contrast, each active segment in a photodiode array (PDA) and a charge-coupled device (CCD) stores charge (rather than generating current) that is created by photons absorbed at that location. The Raman spectrum is generated from the spatial distribution of charge that is produced in the device. In both a PDA and CCD, photons are absorbed and electron-hole pairs are created. The junction in which the electron-hole pairs are created are different in the two devices. In a PDA, the junction is a reverse-biased $p$-$n$ junction. In a CCD, the junction is the depletion zone in the semiconductor at the semiconductor-oxide interface of a metal-oxide-semiconductor.

When the reverse-biased $p$-$n$ junction is irradiated by photons with an energy greater than the band gap, electron-hole pairs are generated from the absorbed photons. Minority carriers from the pairs formed close (i.e., within diffusion distance) to the charge-depleted layer of the junction are split apart from their complementary, oppositely charged particles and driven in the opposite direction by the electric field in the junction. The increase in reverse saturation current density is proportional to the light intensity hitting the photodiode. To minimize the "dark counts", the photodiode must be cooled to reduce the number of thermally generated electron-hole pairs.

If the junction is at equilibrium and is irradiated with photons of energy greater than the band gap, electron-hole pairs generated in the junction are separated by the built-in electric field. The separated electrons and holes lower the magnitude of the built-in field. If an array of diodes is distributed across the exit of the spectrometer, the distribution of charge that is created in the array provides a measure of the intensity of the radiation that is dispersed by the spectrometer across the focal plane for the exiting radiation.

The time to measure a Raman spectrum can be greatly lowered by measuring the entire spectrum at once, rather than one energy value at a time, as is the case with a PMT. Multichannel detection is accomplished by, e.g., placing a PDA at the exit of a spectrometer. There are no slits at the exit, which is filled by the PDA. Each tiny diode in the array measures the intensity of the light that is dispersed to that location. Collectively, the entire array measures the whole spectrum (or a significant fraction of the whole spectrum) all at once. Generally, an individual photodiode in a PDA is not as good a detector as is a PMT. However, for some experiments the multichannel advantage compensates for the lower quality detector. Unfortunately, a PDA do not always provide the sensitivity at low light intensities that are needed in Raman spectroscopy.

A CCD is a charge-transfer device that combines the multichannel detection advantage of a PDA with a sensitivity at low light intensities that rivals that of a PMT (Bilhorn et al., 1987a,b; Epperson, 1988). Charge-coupled devices are made of metal-oxide-semiconductor elements and, in terms of light detection, function analogously to photographic film. That is, photons that hit and are absorbed at a particular location of the CCD are stored at that site in the form of photogenerated electrons (for $p$-type semiconductor substrate) or photogenerated holes (for $n$-type semiconductor substrate).

One example of a CCD is a heavily doped, $p$-type silicon substrate whose surface is coated with a thin layer of $SiO_2$ on top of which are a series of discrete, closely spaced metal strips, referred to as gates. The potential of each adjacent metal strip is set at a different value so that the width of the depletion layer in the semiconductor varies periodically across its surface. When a positive potential is applied to a metal strip, the mobile holes in the $p$-type silicon are repelled from the surface region, creating a depleted layer. If a high enough potential is applied to a metal strip, significant bending of the bands in the region of the semiconductor close to the oxide layer causes the bottom of the conduction band to approach the Fermi level. Electrons occupy states in the conduction band near the surface forming an inversion layer, i.e., an $n$-type surface region in the bulk $p$-type semiconductor. Photons that are absorbed in the surface region generate electron-hole pairs in which the electrons and holes are driven apart by the field in the depletion layer. The holes are expelled from the depletion layer and the electrons are stored in the potential well at the surface. As the intensity of light increases, the number of stored electrons increases. The light exiting the spectrometer is dispersed and so hits the CCD at locations indicative of its wavelength. The dispersed light creates a distribution of charge in the depletion layer from which the sample Raman spectrum is generated. The distribution of charge corresponding to the Raman spectrum is sent to a detector by shifting the periodic variation in potential of the metal strips in the direction of the detector. In this manner the charge stored in the depletion layer of the semiconductor adjacent to a particular metal strip is shifted from one strip to the next. Charge arrives at the detector in successive packets, corresponding to the sequential distribution of the metal gates across the face of the CCD.

An illustration of the increased efficiency of measuring Raman spectra that has emerged in the past 10 years is provided by a comparison of the time required to measure a portion (200 to 1500 $cm^{-1}$) of the surface-enhanced Raman spectrum (see the following paragraph) of thin passive films grown on samples of iron immersed in aqueous solutions. Ten to 15 min (Gui and Devine, 1991) was required to generate the spectrum by using a double spectrometer (Jobin-Yvon U1000 with 1800 grooves/mm holographic gratings) and a PMT (RCA c31034 GaAs), while the same spectrum can now be acquired in $\approx 5$ s (Oblonsky and Devine, 1995) using a single monochromator (270M Spex), in which the stray light is reduced by a notch filter (Kaiser Optics Super Holographic Notch Filter centered at 647.1 nm), and the intensity of the dispersed Raman scattered radiation is measured with a CCD (Spectrum One, $298 \times 1152$ pixels). The enhanced speed in generating the spectrum makes it possible to study the time evolution of samples.

In the previous paragraph, mention was made of surface-enhanced Raman scattering (SERS), which is the greatly magnified intensity of the Raman scattered radiation from species adsorbed on the surfaces of specific metals with either roughened surfaces or colloidal dimensions (Chang and Furtak, 1982). SERS is not a topic of this unit, but a great deal of interest during the past 15 years has been devoted to the use of SERS in studies of surface adsorption and surface films.

### Polarization

By controlling the polarization of the incident radiation and measuring the intensity of the Raman scattered radiation as a function of its polarization, it is possible to identify the component(s) of the polarizabilty tensor of a crystal that are responsible for the different peaks in the Raman spectrum (see Example: Raman Active Vibrational Modes of $\alpha$-$Al_2O_3$, below). This information will then identify the vibrational mode(s) responsible for each peak in the spectrum. The procedure is illustrated below for $\alpha$-$Al_2O_3$ (see Example: Raman Active Vibrational Modes of $\alpha$-$Al_2O_3$). The configuration of the polarizing and polarization measuring components are discussed in Scherer (1991).

### Fourier Transform Raman Spectroscopy

The weak intensity of Raman scattering has already been mentioned as one of the major shortcomings of Raman spectroscopy. Sample heating and fluorescence are two other phenomena that can increase the difficulty of obtaining Raman spectra. Sample heating is caused by absorption of energy during illumination with laser beams of high power density. Fluorescence is also caused by absorption of laser energy by the sample (including impurities in the sample). When the excited electrons drop back down to lower energy levels, they emit light whose energy equals the difference between the energies of the initial (excited) and final states. The fluorescence spectrum is (practically speaking) continuous and covers a wide range of values. The intensity of the fluorescence can be much greater than that of the Raman scattered radiation.

Both sample heating and fluorescence can be minimized by switching to exciting radiation whose energy is too low to be absorbed. For many materials, optical radiation in the red or near-IR range will not be absorbed. If such radiation is to be used in Raman spectroscopy, two consequences must be recognized and addressed. First, since Raman scattering is a light-scattering process, its intensity varies as $\lambda^{-4}$. Hence, by switching to longer wavelength exciting radiation, the intensity of the Raman spectrum will be significantly lowered. Second, the sensitivity of most PMTs to red and especially near-infrared radiation is very low. Consequently, the use of long-wavelength radiation in Raman spectroscopy has been coupled to the use of Fourier transform Raman spectroscopy (FTRS) (Parker, 1994).

In FTRS, an interferometer is used in place of a monochromator. The Raman scattered radiation is fed directly into a Michelson interferometer and from there it enters the detector. The Raman spectrum is obtained by taking the cosine-Fourier transform of the intensity of the radiation that reaches the detector, which varies in magnitude as the path difference between the moving mirror and the fixed mirror within the interferometer changes. Since the Raman scattered radiation is not dispersed, a much higher Raman intensity reaches the detector in FTRS than is the case for conventional Raman spectroscopy. The much higher intensity of the Raman scattered radiation in FTRS permits the use of longer wavelength incident radiation (e.g., $\lambda = 1064$ nm of Nd-YAG), which may eliminate fluorescence.

## DATA ANALYSIS AND INITIAL INTERPRETATION

Raman scattering may result from incident radiation inducing transitions in electronic, vibrational, and rotational states of the scattering medium. Peaks in the Raman spectrum of a solid obtained using visible exciting radiation are generally associated with vibrational modes of the solid. The vibrations may be subdivided into internal modes that arise from molecules or ions that make up the solid and external modes that result from collective modes ($k = 0$) of the crystal. In either case, the presence of a peak in the Raman spectrum due to the vibration requires that the Raman transition be symmetry allowed, i.e., that the integrand in Equation 7 have a nonzero value. In addition, for a measurable Raman intensity, the integral in Equation 7 must have a large enough magnitude. It is generally difficult to calculate the intensity of a Raman peak, but it is rather straightforward to use group theoretical techniques to determine whether or not the vibrational mode is symmetry allowed.

### Example: Raman Active Vibrational Modes of α-Al₂O₃

The extremely useful result that the crystallographic point group provides all the information that is needed to know the symmetries of the Raman active vibrational modes of the crystal is derived below (see Appendix). The left-hand column in the character table for each point group lists a set of symbols such as $A_1$, $B_1$, and $B_2$ for the point group $C_{2v}$. The term $A_1$ is the list of characters of the matrices that represent the effect of each symmetry operation of the point group on any function listed in the far right-hand column of the character table. A remarkable result of group theory is that the list of characters of the matrices that represent the effect of each symmetry operation of the point group on "any" function can be completely described by a linear combination of the few lists presented in the character table. For example, for the point group $C_{2v}$, the symmetry of the function $x^2$ is described by $A_1$. This means that when the effect of each of the symmetry operations operating on $x^2$ is represented by a matrix, the characters of the four matrices are 1, 1, 1, 1. In fact, within the context of the point group $C_{2v}$, the symmetry of any function is given by $aA_1 + bB_1 + cB_2$, where $a$, $b$, and $c$ are integers. In the language of group theory, $A_1$, $B_1$, and $B_2$ are called the irreducible representations of the point group $C_{2v}$. The linear combination of irreducible representations that describe the symmetry of a function is called the symmetry species of the function.

The irreducible representations of the normal vibrational modes of a crystal are most easily identified by the correlation method (Fately et al., 1971). The analysis for the lattice vibrations of α-Al₂O₃ will be summarized here. The details of the analysis are contained in Fately et al. (1971).

The space group of α-Al₂O₃ is $D_{3d}^6$. There are two molecules of α-Al₂O₃ per Bravais cell. Each atom in the Bravais cell has its own symmetry, termed the site symmetry, which is a subgroup of the full symmetry of the Bravais unit cell. The site symmetry of the aluminum atoms is $C_3$ and that of the oxygen atoms is $C_2$. From the character table for the $C_3$ point group, displacement in the $z$ direction is a basis for the $A$ irreducible representation. Displacements parallel to the $x$ and $y$ axes are bases for the $E$ irreducible representation. The correlation tables for the species of a group and its subgroups (Wilson et al., 1955) correlate the $A$ species of $C_3$ to $A_{1g}$, $A_{2g}$, $A_{1u}$, and $A_{2u}$ of the crystal point group $D_{3d}$.

Displacement of the oxygen atom in the $z$ direction is a basis of the $A$ irreducible representation of the $C_2$ point group, which is the site symmetry of the oxygen atoms in α-Al₂O₃. Displacements of the oxygen atom that are parallel to the $x$ and $y$ axes are bases for the $B$ irreducible representation of $C_2$. The correlation table associates the $A$ species of $C_2$ to $A_{1g}$, $E_g$, $A_{1u}$, and $E_u$ of $D_{3d}$. The $B$ species of $C_2$ correlates to $A_{2g}$, $E_g$, $A_{2u}$, and $E_u$ of $D_{3d}$.

After accounting for the number of degrees of freedom contributed by each irreducible representation of $C_2$ and $C_3$ to an irreducible representation of $D_{3d}$ and removing the irreducible representations associated with the rigid translation of the entire crystal, the irreducible species for the optical modes of the corundum crystal are determined to be

$$2A_{1g} + 2A_{1u} + 3A_{2g} + 2A_{2u} + 5E_g + 4E_u \qquad (22)$$

Consulting the character table of the point group $D_{3d}$ indicates that the Raman active vibrational modes would span

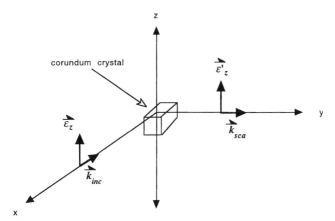

**Figure 8.** Directions and polarizations of incident and scattered radiation for the $x(zz)y$ configuration.

the symmetry species $A_{1g}$ and $E_g$. The basis functions for these symmetry species are listed in the character table:

$$x^2 + y^2 \quad \text{and} \quad z^2 \quad \text{for } A_{1g} \qquad (23)$$

and

$$(x^2 - y^2, xy) \quad \text{and} \quad (xz, yz) \quad \text{for } E_g \qquad (24)$$

Thus there are a total of seven Raman active modes so there may be as many as seven peaks in the Raman spectrum. The question now is which peaks correspond to which vibrational modes.

Figure 8 depicts a Cartesian coordinate system, and a crystal of corundum is imagined to be located at the origin with the crystal axes parallel to $x$, $y$, and $z$. If light is incident on the crystal in the $-x$ direction and polarized in the $z$ direction, and scattered light is collected in the $y$ direction after having passed through a $y$ polarizer, then the experimental conditions of incidence and collection are described by the Porto notation (Porto and Krishnan, 1967) as $x(zz)y$. The first term inside the parentheses indicates the polarization of the incident light and the second term denotes the polarization of the scattered light that is observed. In general, if $z$ polarized light is incident on a crystal in the $x$ direction, then the induced polarization of the molecule will be given by $[p_x, p_y, p_z] = [\alpha_{xz}E_z, \alpha_{yz}E_z, \alpha_{zz}E_z]$. Light scattered in the $y$ direction would be the result of the induced polarization components $p_x$ and $p_z$. If the light scattered in the $y$ direction is passed through a $z$ polarizer, then the source of the observed radiation would be the induced polarization along the $z$ direction. Consequently, only the $\alpha_{zz}$ component of the polarizability contributes to scattered light in the case of $x(zz)y$. Thus, peaks present in the Raman spectrum must originate from Raman scattering by vibrational modes that span the $A_{1g}$ irreducible species.

Figure 9 presents the Raman spectra reported by Porto and Krishnan (1967) for single-crystal corundum under a variety of incidence and collection conditions. Figure 9A is the Raman spectrum in the $x(zz)y$ condition that was

discussed above so the two peaks present in the spectrum correspond to vibrational modes with $A_{1g}$ symmetry. The peaks present in the spectrum in Figure 9B can originate from vibrational modes with either $A_{1g}$ or $E_g$ symmetries. Consequently, the peaks present in Figure 9B but missing in Figure 9A have $E_g$ symmetry. The spectra in Figures 9C-E exhibit only peaks with $E_g$ symmetry. Collectively, the spectra reveal all seven Raman active modes and distinguish between the $A_{1g}$ modes and the $E_g$ modes.

The magnitude of the Raman shift in wavenumbers of each peak with respect to the exciting laser line can be related to the energy of the vibration. Combined with the knowledge of the masses of the atoms involved in the vibrational mode and, e.g., the harmonic oscillator assumption, it is possible to calculate the force constant associated with the bond between the atoms participating in the vibration. If one atom in the molecule were replaced by another without changing the symmetry of the molecule, the peak would shift corresponding to the different strength of the bond and the mass of new atom compared to the original atom. This illustrates why the Raman spectrum can provide information about bond strength and alloying effects.

### Other Examples of the Use of Raman Spectroscopy in Studies of Solids

Raman spectroscopy can provide information on the amorphous/crystalline nature of a solid. Many factors can contribute to the widths of peaks in the Raman spectra of a solids. The greatest peak broadening factor can be microcrystallinity. For a macroscopic-sized single crystal of silicon at room temperature, the full width at half-maximum (FWHM) of the peak at 522 cm$^{-1}$ is $\approx$3 cm$^{-1}$ (Pollak, 1991). The breadth of the peak increases dramatically as the size of the crystal decreases below 10 nm. The cause of peak broadening in microcrystals is the finite size of the central peak in the Fourier transform of a phonon in a finite-size crystal. In an infinitely large crystal, the Fourier transform of a phonon consists of a single sharp line at the phonon frequency. The peak width $\Delta k$ (i.e., FWHM) in a crystal of dimension $L$ is given approximately by

$$\Delta k = \frac{2\pi(v/c)}{L} \qquad (25)$$

where $v$ is phonon velocity and $c$ the velocity of light. Assuming $v = 1 \times 10^5$ cm/s, $\Delta k = 20$ cm$^{-1}$ for $L = 10$ nm and $\Delta k = 200$ cm$^{-1}$ for $L = 1$ nm. In the extreme case, the silicon is amorphous, in which case the $k = 0$ selection rule breaks down and all phonons are Raman active. In that case, the Raman spectrum resembles the phonon density of states and is characterized by two broad peaks centered at 140 and 480 cm$^{-1}$ (Pollak, 1991). Consequently, the Raman spectrum can distinguish between the crystalline and amorphous structures of a solid and can indicate the grain size of microcrystalline solids.

Raman spectroscopy can be used to nondestructively measure the elastic stress in crystalline samples. Elastic straining of the lattice will alter the spring constant of a chemical bond and, hence, will shift the frequency of the

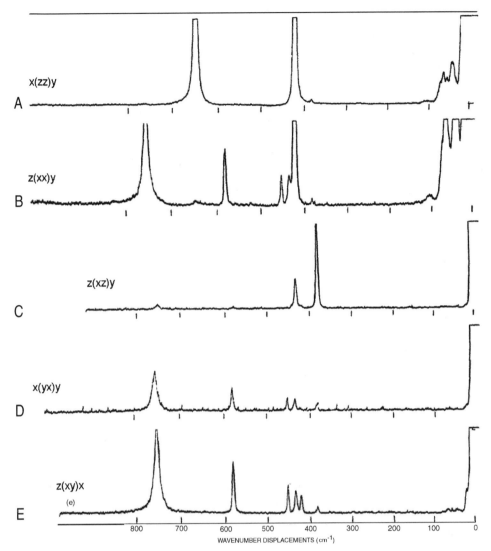

**Figure 9.** Raman spectra of corundum as a function of crystal and laser light polarization orientations (Porto and Krishnan, 1967).

vibrational mode associated with the strained bond. The direction and magnitude of the shift in peak location will depend on the sign (tensile or compressive) and magnitude, respectively, of the strain. Raman spectroscopy has been used to measure the residual stresses in thin films deposited on substrates that result in lattice mismatches and in thermally cycled composites where the strains result from differences in thermal expansion coefficients of the two materials (Pollak, 1991).

Another example of the use of Raman spectroscopy to nondestructively investigate the mechanical behavior of a solid is its use in studies of $Al_2O_3$-$ZrO_2$ composites (Clarke and Adar, 1982). These two-phase structures exhibit improved mechanical toughness due to transformation of the tetragonal $ZrO_2$ to the monoclinic form. The transformation occurs in the highly stressed region ahead of a growing crack and adds to the energy that must be expended in the propagation of the crack. Using Raman

microprobe spectroscopy, it was possible to measure, with a spatial resolution of $\approx 1$ μm, the widths of the regions on either side of a propagating crack in which the $ZrO_2$ transformed. The size of the transformed region is an important parameter in theories that predict the increased toughness expected from the transformation of the $ZrO_2$.

Phase transformations resulting from temperature or pressure changes in an initially homogeneous material have also been studied by Raman spectroscopy (Ferraro and Nakamoto, 1994). Raman spectra can be obtained from samples at temperatures and pressures that are markedly different from ambient conditions. All that is needed is the ability to irradiate the sample with a laser beam and to collect the scattered light. This can be accomplished by having an optically transparent window (such as silica, sapphire, or diamond) in the chamber that houses the sample and maintains the nonambient conditions. Alternatively, a fiber optic can be used to transmit the incident light to the sample and the scattered light to the spectrometer.

The influence of composition on the Raman spectrum is in evidence in semiconductor alloys such as $Ga_{1-x}AL_xAs$ $(0 < x < 1)$, which exhibit both GaAs-like and AlAs-like longitudinal optical (LO) and transverse optical (TO) phonon modes. The LO modes have a greater dependence on composition than do the TO modes. The AlAs-like LO mode shifts by 50 cm$^{-1}$ and the GaAs-like mode shifts by $\approx$35 cm$^{-1}$ as $x$ varies from 0 to 1. Thus the Raman spectrum of $Ga_{1-x}Al_xAs$ $(0 < x < 1)$ can be used to identify the composition of the alloy (Pollak, 1991).

Similarly, the Raman spectra of mixtures of HfO$_2$ and ZrO$_2$ are strong functions of the relative amounts of the two components. Depending on the particular mode, the Raman peaks shift by $\approx$7 to $\approx$60 cm$^{-1}$ as the amount of HfO$_2$ increases from 0 to 100% spectroscopy (Ferraro and Nakamoto, 1994).

Raman spectroscopy has long been used to characterize the structure of polymers: identifying functional groups, end groups, crystallinity, and chain orientation. One word of caution: As might be expected of any Raman study of organic matter, fluorescence problems can arise in Raman investigations of polymers (Bower and Maddams, 1989; Bulkin, 1991; Rabolt, 1991).

### Raman Spectroscopy of Carbon

Because carbon can exist in a variety of solid forms, ranging from diamond to graphite to amorphous carbon, Raman spectroscopy is the single most effective characterization tool for analyzing carbon. Given the wide spectrum of structures exhibited by carbon, it provides a compelling example of the capability of Raman spectroscopy to analyze the structures of a large variety of solids. The following discussion illustrates the use of Raman spectroscopy to distinguish between the various structural forms of carbon and describes the general procedure for quantitatively analyzing the structure of a solid by Raman spectroscopy.

Figure 10, which was originally presented in an article by Robertson (1991), who compiled the data of a number of researchers, presents the Raman spectra of diamond, large crystal graphite, microcrystalline graphite, glassy carbon, and several forms of amorphous carbon. Diamond has two atoms per unit cell and therefore has a single internal vibrational mode ($3N - 5 = 1$). This mode is Raman active and is located at 1332 cm$^{-1}$. Graphite, with four atoms per unit cell, has six internal vibrational modes ($3N - 6 = 6$), two of which are Raman active. The rigid-layer mode of graphite spans the symmetry species $E_{2g}$ and occurs at 50 cm$^{-1}$, which is too close to the incident laser line to be accessible with most Raman collection optics and spectrometers. The second Raman active mode of graphite is centered at 1580 cm$^{-1}$, and it too spans the symmetry species $E_{2g}$. The relative displacements of the carbon atoms during this in-plane mode are presented next to the Raman spectrum in Figure 10. Thus, Raman spectroscopy can easily distinguish between diamond and large-crystal graphite.

The third spectrum from the top of Figure 10 was obtained from microcrystalline graphite. The next lower spectrum is similar and belongs to glassy carbon. There

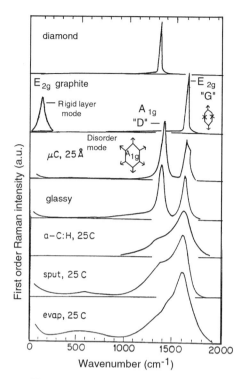

**Figure 10.** First-order Raman spectra of diamond, highly oriented pyrolytic graphite (hopg), microcrystalline graphite, glassy C, plasma-deposited a-C:H, sputtered a-C, and evaporated a-C. (From Robertson, 1991.)

are two peaks in these spectra, one at 1580 cm$^{-1}$ and the other at 1350 cm$^{-1}$. Given these peak positions, originally there was some thought that the structures were a mixture of graphitelike ($sp^2$ bonding) and diamondlike ($sp^3$ bonding) components. However, experiments that revealed the effects of heat treatments on the relative intensities of the two peaks clearly showed that the material was polycrystalline graphite and the peak at 1350 cm$^{-1}$ is a consequence of the small size of the graphite crystals. The new peak at 1350 cm$^{-1}$ is labeled the disorder, or "D," peak, as it is a consequence of the breakdown of the crystal momentum conservation rule. In the Raman spectrum of an infinite crystal, the peak at 1350 cm$^{-1}$ is absent. This peak results from the vibrational mode sketched next to the spectrum for microcrystalline graphite in Figure 10, and it spans the symmetry species $A_{1g}$. Its Raman activity is symmetry forbidden in a crystal of infinite size. Finite crystal size disrupts translational symmetry, and so the D mode is Raman active in microcrystalline graphite. The ratio of the integrated intensities of the peaks associated with the D and G modes is proportional to the size of the crystal:

$$\frac{I(D \text{ mode})}{I(G \text{ mode})} = \frac{k}{d} \tag{26}$$

where $d$ is the crystal size. This relationship may be used, for example, to calculate the relative sizes of two microcrystals and the relative mean grain diameters of two

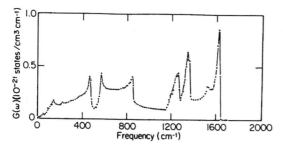

**Figure 11.** Phonon density of states for graphite.

polycrystalline aggregates, so long as the grain size is greater than 12 Å. The behavior of crystals smaller than 12 Å begins to resemble that of amorphous solids.

An amorphous solid of $N$ atoms may be thought of as a giant molecule of $N$ atoms. All $3N - 6$ of its internal vibrational modes are Raman active, and the Raman spectrum is proportional to the phonon density of states. This is demonstrated by a comparison of the bottom three spectra in Figure 10 with the calculated phonon density of states of graphite in Figure 11.

At this point several important distinctions between x-ray diffraction and Raman spectroscopy can be appreciated. Like Raman spectroscopy, x-ray diffraction is also able to distinguish between crystals, microcrystals, and amorphous forms of the same material. Raman spectroscopy not only is able to distinguish between the various forms but also readily provides fundamental information about the amorphous state as well as the crystalline state. This is due to the fact that x-ray diffraction provides information about the long-range periodicity of the arrangement of atoms in the solid, while Raman spectroscopy provides information concerning the bonds between atoms and the symmetry of the atomic arrangement. On the other hand, while x-ray diffraction is in principle applicable to every crystal, Raman spectroscopy is only applicable to those solids with unit cells containing two or more atoms.

### Quantitative Analysis

The quantitative analysis of a spectrum requires fitting each peak to an expression of Raman scattered intensity vs. frequency of the scattered light. Typically, a computer is used to fit the data to some mathematical expression of the relationship between the intensity of the Raman scattered radiation as a function of its frequency with respect to that of the incident laser line. One example of a peak-fitting relationship is the expression (DiDomenico et al., 1968)

$$\frac{dI}{d\omega} = \frac{(\text{const})\Gamma\omega_0}{\{\omega_0^2 - (\Delta\omega)^2\}^2 + 4\Gamma^2\omega_0^2(\Delta\omega)^2} \qquad (27)$$

where $dI/d\omega$ is the average Raman scattered intensity per unit frequency and is proportional to the strength of the signal at the Raman shift of $\Delta\omega$; $\Gamma$ is the damping constant, $\omega_0$ is the frequency of the undamped vibrational mode, and $\Delta\omega$ is the frequency shift from the laser line. The calcu-

lated fit yields the values of $\Gamma$ and $\omega_0$ for each peak in the spectrum.

An example of the quantitative analysis of a Raman spectrum is provided in the work of Dillon et al. (1984), who used Raman spectroscopy to investigate the growth of graphite microcrystals in carbon films as a function of annealing temperature. The data for the $D$ and $G$ peaks of graphite were fitted to Equation 27, which was then integrated to give the integrated intensity of each peak. The ratio of the integrated intensity of the $D$ mode to the integrated intensity of the $G$ mode was then calculated and plotted as a function of annealing temperature. The ratio increased with annealing temperature over the range from 400 to 600°C, and the widths of the two peaks decreased over the same range. The results suggest that either the number or the size of the crystal grains was increasing over this temperature range. The integrated intensity ratio of the $D$-to-$G$ mode reached a maximum and then decreased with higher temperatures, indicating growth in the grain size during annealing at higher temperatures. Dillon et al. also used the peak positions defined by Equation 27 to monitor the shift in peak locations with annealing temperature. The results indicated that the annealed films were characterized by threefold rather than fourfold coordination.

In summary, a single-phase, unknown material can be identified by the locations of peaks and their relative intensities in the Raman spectrum. The location of each peak and its breadth can be defined by fitting the peaks to a quantitative expression of peak intensity vs. Raman shift. The same expression will provide the integrated intensity of each peak. Ratios of integrated intensities of peaks belonging to different phases in a multiphase sample will be proportional to the relative amounts of the two phases.

### PROBLEMS

Many problems can arise in the generation of a Raman spectrum. Peaks not part of the spectrum of the sample can sometimes appear. Among the factors that can produce spurious peaks are plasma lines from the laser and cosmic rays hitting the light detector. The occurrence of peaks caused by cosmic rays increases with the time required for acquisition of the spectrum. The peaks from cosmic rays can be identified by their sharpness and their nonreproducibility. Peaks attributed to plasma lines are also sharp, but not nearly as sharp as those caused by cosmic rays. A peak associated with a plasma line will vanish when the spectrum is generated using a different laser.

An all too common problem encountered in Raman spectroscopy is the occurrence of fluorescence, which completely swamps the much weaker Raman signal. The fluorescence can be caused by the component of interest or by impurities in the sample. Fluorescence can be minimized by decreasing the concentration of the guilty impurity or by long-time exposure of the impurity to the exciting laser line, which "burns out" the fluorescence. Absorption of the exciting laser line by the impurity results in its thermal decomposition.

Fluorescence is generally red shifted from the exciting laser line. Consequently, the anti-Stokes Raman spectrum may be less affected by fluorescence than the Stokes-Raman spectrum. The intensity of the anti-Stokes Raman spectrum is much weaker than that of the Stokes Raman so this is not always a viable approach to avoiding the fluorescence signal.

Increasing the wavelength of the exciting laser line may also reduce fluorescence, as was mentioned above (see Fourier Transform Raman Spectroscopy). One potential problem with the latter approach is tied to the wavelength dependence of the intensity of the scattered light, which varies inversely as the fourth power of the wavelength. Shifting to exciting radiation with a longer wavelength may reduce fluorescence but it will also decrease the overall intensity of the scattered radiation.

Fluctuations in the intensity of the incident laser power during the generation of a Raman spectrum are particularly problematic for spectra measured using a single-channel detector. Errors in the relative intensities of different peaks can result from fluctuations in the laser power. Similarly, fluctuations in laser power will cause variations in the intensities of successive measurements of the same spectrum using a multiple-channel detector. These problems will be largely averted if the laser can be operated in a constant-light-intensity mode rather than in a controlled (electrical) power input mode.

If the sample absorbs the incident radiation, problems may occur even if the absorption does not result in fluorescence. For example, if an argon laser ($\lambda = 514.5$ nm) were used to generate the Raman spectrum of an adsorbate on the surface of a copper or gold substrate, significant reduction in the Raman intensity would result because of the strong absorption by copper and gold of green light. In this case, better results would be obtained by switching to a krypton laser ($\lambda = 647.1$ nm) or helium-neon laser ($\lambda = 632.8$ nm).

Absorption of the incident radiation can also lead to increases in the temperature of the sample. This may have a major deleterious effect on the experiment if the higher temperature causes a change in the concentration of an adsorbate or in the rate of a chemical or electrochemical reaction. If optically induced thermal effects are suspected, the spectra should be generated using several different incident wavelengths.

Temperature changes can also cause significant changes in the Raman spectrum. There are a number of sources of the temperature dependency of Raman scattering. First, the intensity of the Stokes component relative to the anti-Stokes component decreases as temperature increases. For studies that make use of only the Stokes component, an increase in temperature will cause a decrease in intensity. A second cause of the temperature dependency of a Raman spectrum is the broadening of peaks as the temperature increases. This cause of peak broadening is dependent on the decay mechanism of the phonons. Broadening of the Raman peak can also result from hot bands. Here the higher temperature results in a higher population of the excited state. As a result, the incident photon can induce the transition from the first excited state to the second excited state. Because of anharmonic effects,

the energy required for this transition is different from the energy required of the transition from the ground state to the first excited state. As a result, the peak is broadened. Hence, Raman peaks are much sharper at lower temperatures. Consequently, it may be necessary to obtain spectra at temperatures well below room temperature in order to minimize peak overlap and to identify distinct peaks.

Temperature changes can also affect the Raman spectrum due to the temperature dependencies of the phonons themselves. For example, the phonon itself will have a strong temperature dependency as the temperature is raised close to that of a structural phase transition if at least one component of the oscillation coincides with the displacement during the phase transition. Temperature changes can also affect the phonon frequency. As the temperature increases, the anharmonic character of the vibration causes the well in the plot of potential energy vs. atomic displacement to be more narrow than is the case for a purely harmonic oscillation. The atomic displacements are therefore smaller, which results in a higher frequency since the amplitude of displacement is inversely proportional to frequency. In addition, thermal expansion increases the average distance between the atoms as temperature increases, leading to a decrease in the strength of the interatomic interactions. This results in a decrease in the phonon frequency with increasing temperature. In summary, superior Raman spectra are generally obtained at lower temperatures, and it may be necessary in some cases to obtain spectra at temperatures that are considerably lower than room temperature.

Changes in the temperature of the room in which the Raman spectra are measured can cause changes in the positions of optical components (lenses, filters, mirrors, gratings) both inside and outside the spectrometer. Such effects are important when attempting to accurately measure the locations of peaks in a spectrum, or, e.g., stress-induced shifts in peak locations.

One of the more expensive mistakes that can be made in systems using single-channel detectors is the inadvertent exposure of the PMT to the Rayleigh line. The high intensity of the Rayleigh scattered radiation can "burn out" the PMT, resulting in a significant increase in PMT dark counts. Either closing the shutter in front of the PMT or shutting off the high voltage to the PMT will prevent damage to the PMT from exposure to high light intensity. Typically, accidents of this type occur when the frequency of the incident radiation is changed and the operator neglects to note this change at the appropriate point in the software that runs the experiment and controls the operation of the PMT shutter.

## ACKNOWLEDGMENTS

It is a pleasure to thank J. Larry Nelson and Wylie Childs of the Electric Power Research Institute for their long-term support and interest in the use of Raman spectroscopy in corrosion investigations. In addition, Gary Chesnut and David Blumer of ARCO Production and Technology have encouraged the use of surface-enhanced Raman spectroscopy in studies of corrosion inhibition.

Both organizations have partially supported the writing of this unit.

The efforts and skills of former and current graduate students, in particular Jing Gui, Lucy J. Oblonsky, Christopher Kumai, Valeska Schroeder, and Peter Chou, have greatly contributed to my continually improved understanding and appreciation of Raman spectroscopy.

## LITERATURE CITED

Altmann, S. L. 1991. Band Theory of Solids: An Introduction from the Point of View of Symmetry (see p. 190). Oxford University Press, New York.

Atkins, P. W. 1984. Molecular Quantum Mechanics. Oxford University Press, New York.

Bilhorn, R. B., Epperson, P. M., Sweedler, J. V., and Denton, M. B. 1987b. Spectrochemical measurements with multichannel integrating detectors. *Appl. Spectrosc.* 41:1125–1135.

Bilhorn, R. B., Sweedler, J. V., Epperson, P. M., and Denton, M. B. 1987a. Charge transfer device detectors for analytical optical spectroscopy—operation and characteristics. *Appl. Spectrosc.* 41:1114–1125.

Bishop, D. 1973. Group Theory and Chemistry. Clarendon Press, Oxford.

Bower, D. I. and Maddams, W. F. 1989. The Vibrational Spectroscopy of Polymers. Cambridge University Press, Cambridge.

Bulkin, B. J. 1991. Polymer applications. *In* Analytical Raman Spectroscopy (J. G. Grasselli and B. J. Bulkin, eds.) pp. 45–57. John Wiley & Sons, New York.

Carrabba, M. M., Spencer, K. M., Rich, C., and Rauh, D. 1990. The utilization of a holographic Bragg diffraction filter for Rayleigh line rejection in Raman spectroscopy. *Appl. Spectrosc.* 44:1558–1561.

Chang, R. K. and Furtak, T. E. (eds.). 1982. Surface Enhanced Raman Scattering. Plenum Press, New York.

Chase, D. B. 1991. Modern Raman instrumentation and techniques. *In* Analytical Raman Spectroscopy (J. G. Grasselli and B. J. Bulkin, eds.) pp. 45–57. John Wiley & Sons, New York.

Clarke, D. R. and Adar, F. 1982. Measurement of the crystallographically transformed zone produced by fracture in ceramics containing tetragonal zirconia. *J. Am. Ceramic Soc.* 65:284–288.

DiDomenico, M., Wemple, S. H., Perto, S. P. S., and Bauman, R. P. 1968. Raman spectrum of single-domain $BaTiO_3$. *Phys. Rev.*

Dillon, R. O., Woollam, J. A., and Katkanant, V. 1984. Use of Raman scattering to investigate disorder and crystallite formation in As-deposited and annealed carbon films. *Phys. Rev. B.*

Epperson, P. M., Sweedler, J. V., Bilhorn, R. B., Sims, G. R., and Denton, M. B. 1988. Applications of charge transfer devices in spectroscopy. *Anal. Chem.* 60:327–335.

Fateley, W. G., McDevitt, N. T., and Bailey, F. F. 1971. Infrared and Raman selection rules for lattice vibrations: The correlation method. *Appl. Spectrosc.* 25:155–173.

Ferraro, J. R. and Nakamoto, K. 1994. Introduction to Raman Spectroscopy. Academic Press, San Diego, Ca.

Fowles, G. R. 1989. Introduction to Modern Optics. Dover Publications, Mineola, N.Y.

Fraser, D. A. 1990. The Physics of Semiconductor Devices. Oxford University Press, New York.

Gui, J. and Devine, T. M. 1991. *In-situ* vibrational spectra from the passive film in iron in buffered borate solution. *Corr. Sci.* 32:1105–1124.

Hamermesh, M. 1962. Group Theory and Its Application to Physical Problems. Dover Publications, New York.

Hecht, J. 1993. Understanding Lasers: An Entry-Level Guide. IEEE Press, Piscataway, N.J.

Hutley, M. C. 1982. Diffraction Gratings. Academic Press, New York.

Kettle, S. F. A. 1987. Symmetry and Structure. John Wiley & Sons, New York.

Koster, G. F. 1957. Space groups and their representations. *In* Solid State Physics, Advances in Research and Applications, Vol. 5 (F. Seitz and D. Turnbull, eds.) 173–256. Academic Press, New York.

Lax, M. 1974. Symmetry Principles in Solid State and Molecular Physics. John Wiley & Sons, New York.

Leech, J. W. and Newman, D. J. 1969. How To Use Groups. Methuen, London.

Long, D. A. 1977. Raman Spectroscopy. McGraw-Hill, New York.

Mariot, L. 1962. Group Theory and Solid State Physics. Prentice-Hall, Englewood Cliffs, N.J.

Meijer, P. H. E. and Bauer, E. 1962. Group Theory—The Application to Quantum Mechanics. North-Holland Publishing, Amsterdam, The Netherlands.

Oblonsky, L. J. and Devine, T. M. 1995. Surface enhanced Raman spectroscopic study of the passive films formed in borate buffer on iron, nickel, chromium and stainless steel. *Corr. Sci.* 37:17–41.

Owen, H. 1992. Holographic optical components for laser spectroscopy applications. *SPIE* 1732:324–332.

Parker, S. F. 1994. A review of the theory of Fourier-transform Raman spectroscopy. *Spectrochim. Acta* 50A:1841–1856.

Pelletier, M. J. and Reeder, R. C. 1991. Characterization of holographic band-reject filters designed for Raman spectroscopy. *Appl. Spectrosc.* 45:765–770.

Pollak, F. H. 1991. Characterization of semiconductors by Raman spectroscopy. *In* Analytical Raman Spectroscopy (J. G. Grasselli and B. J. Bulkin, eds.) pp. 137–221. John Wiley & Sons, New York.

Porto, S. P. S. and Krishnan, R. S. 1967. Raman effect of corundum. *J. Chem. Phys.* 47:1009–10012.

Rabolt, J. F. 1991. Anisotropic scattering properties of uniaxially oriented polymers: Raman studies. *In* Analytical Raman Spectroscopy (J. G. Grasselli and B. J. Bulkin, eds.) pp. 45–57. John Wiley & Sons, New York.

Robertson, J. 1991. Hard amorphous (diamond-like) carbons. *Prog. Solid State Chem.* 21:199–333.

Rossi, B. 1957. Optics. Addison-Wesley, Reading, Mass.

Scherer, J. R. 1991. Experimental considerations for accurate polarization measurements. *In* Analytical Raman Spectroscopy (J. G. Grasselli and B. J. Bulkin, eds.) pp. 45–57. John Wiley & Sons, New York.

Schoen, C. L., Sharma, S. K., Henlsley, C. E., and Owen, H. 1993. Performance of a holographic supernotch filter. *Appl. Spectrosc.* 47:305–308.

Weyl, H. 1931. The Theory of Groups and Quantum Mechanics. Dover Publications, New York.

Wilson, E. B., Decius, J. C., and Cross, P. C. 1955. Molecular Vibrations: The Theory of Infrared and Raman Vibrational Spectra. McGraw-Hill, New York.

Yang, B., Morris, M. D., and Owen, H. 1991. Holographic notch filter for low-wavenumber stokes and anti-stokes Raman spectroscopy. *Appl. Spectrosc.* 45:1533–1536.

## KEY REFERENCES

Altmann, 1991. See above.

*In Chapter 11, Bloch sums are used for the eigenvectors of a crystal's vibrational Hamiltonian.*

Atkins, 1984. See above.

*Provides a highly readable introduction to quantum mechanics and group theory.*

Fateley et al., 1971. See above.

*Recipe for determining the symmetry species of vibrational modes of crystals.*

Ferraro and Nakamoto, 1994. See above.

*Provides a more current description of equipment and more extensive and up-to-date discussion of applications of Raman spectroscopy than are provided in the text by Long (see below).*

Long, 1977. See above.

*Provides a thorough introduction to Raman spectroscopy.*

Pollak, 1991. See above.

*A comprehensive review of the use of Raman spectroscopy to investigate the structure and composition of semiconductors.*

## APPENDIX:
## GROUP THEORY AND VIBRATIONAL SPECTROSCOPY

This appendix was written, in part, to provide a road map that someone interested in vibrational spectroscopy might want to follow in migrating through the huge number of topics that are covered by the vast number of textbooks on group theory (see Literature Cited). Although necessarily brief in a number of areas, it does cover the following topics in some depth, selected on the basis of their importance and/or the rareness with which they are fully discussed elsewhere: (1) How character tables are used to identify the Raman activity of vibrational modes is demonstrated. (2) That the *ij* component, $\alpha_{ij}$, of the polarizability tensor has the same symmetry as the product function $x_i x_j$ is demonstrated. Character tables list quadratic functions such as $x^2$, $yz$ and identify the symmetry species (defined below) spanned by each. The same symmetry species will be spanned by the Raman active vibrational modes. (3) The proper rotations found in crystallographic point groups are identified. (4) A general approach, developed by Lax (1974) and using multiplier groups, is presented that elegantly establishes the link between all space groups, both symmorphic and nonsymmorphic, and the crystallographic point groups.

### Vibrational Selection Rules

Quantum mechanically a vibrational Raman transition may occur if the integral in Equation 7 has a nonzero value. This can quickly be determined by the use of group theoretical techniques, which are summarized in the form of a Raman selection rule. The Raman selection rule is a statement of the symmetry that a vibrational mode must possess in order that it may be Raman active.

Before discussing the symmetry-based selection rules for Raman scattering, there is also a restriction based on the principles of energy and momentum conservation

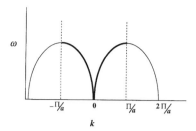

**Figure 12.** Dispersion of longitudinal wave in a linear monoatomic lattice (nearest-neighbor interactions only).

that is especially helpful in considering the Raman activity of solids. The requirements of energy and momentum conservation are expressed as follows:

$$\hbar\omega_i = \hbar\omega_s \pm \hbar\omega_v \qquad (28)$$

$$\hbar k_i = \hbar k_s \pm \hbar k_v \qquad (29)$$

where $\hbar k_i$, $\hbar k_s$, and $\hbar k_v$ are the momentum vectors of the incident and scattered radiation and the crystal phonon, respectively. For optical radiation $|k_i|$ is on the order of $10^4$ to $10^5$ cm$^{-1}$. For example, for $\lambda = 647.1$ nm, which corresponds to the high-intensity red line of a krypton ion laser,

$$k = \frac{2\pi}{\lambda} = 9.71 \times 10^3 \text{ cm}^{-1} \qquad (30)$$

For a crystal with $a_0 = 0.25$ nm, $k_{v,\text{max}} \approx 2.51 \times 10^8$ cm$^{-1}$. Hence, for Raman scattering, $k_v \ll k_{v,\text{max}}$. In other words, $k_v$ must be near the center of the Brillouin Zone (BZ).

The phonon dispersion curves for a simple, one-dimensional, monoatomic lattice and a 1D diatomic lattice are presented in Figures 12 and 13. Since there are no available states at the center of the BZ in a monatomic lattice [for a three-dimensional (3D) lattice as well as a one-dimensional lattice], such structures are not Raman active. Consequently, all metals are Raman inactive. Diatomic lattices, whether homonuclear or heteronuclear, do possess phonon modes at the center of the BZ (see Fig. 13) and are Raman active. Thus, crystals such as diamond, silicon, gallium nitride, and aluminum oxide are all Raman active.

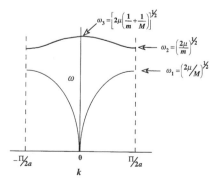

**Figure 13.** Dispersion of longitudinal wave in a linear diatomic lattice.

A Raman spectrum may consist of several peaks in the plot of the intensity of the scattered radiation versus its shift in wavenumber with respect to the incident radiation, as illustrated in Figure 4. If the immediate task is to identify the material responsible for the Raman spectrum, then it is only necessary to compare the measured spectrum with the reference spectra of candidate materials. Since the Raman spectrum acts as a fingerprint of the scattering material, once a match is found, the unknown substance can be identified.

Often the identity of the sample is known, and Raman spectroscopy is being used to learn more about the material, e.g., the strength of its atomic bonds or the presence of elastic strains in a crystalline matrix. Such information is present in the Raman spectrum but it must now be analyzed carefully in order to characterize the structure of the material.

Once a vibrational Raman spectrum is obtained, the first task is to identify the vibrational modes that contribute to each peak in the spectrum. Group theoretical techniques enormously simplify the analyses of vibrational spectra. At this point, the rules of group theory that are used in vibrational spectroscopy will be cited and examples of their use will then be given. In doing so, it is worth recalling the warning issued by David Bishop in the preface to his textbook on Group Theory (Bishop, 1973, p. vii): "The mathematics involved in actually applying, as opposed to deriving group theoretical formulae is quite trivial. It involves little more than adding and multiplying. It is in fact possible to make the applications, by filling in the necessary formulae in a routine way, without even understanding where the formulae have come from. I do not, however, advocate this practice."

Although the approach taken in the body of this unit ignores Bishop's advice, the intent is to demonstrate the ease with which vibrational spectra can be analyzed by using the tools of group theory. One researcher used the automobile analogy to illustrate the usefulness of this approach to group theory. Driving an automobile can be a very useful practice. Doing so can be accomplished without any knowledge of the internal combustion engine. Hopefully, the elegance of the methodology, which should be apparent from reading this appendix, will encourage the reader to acquire a firm understanding of the principles of group theory by consulting any number of texts that provide an excellent introduction to the subject (see Literature Cited, e.g., Hammermesh, 1962; Koster, 1957; Leech and Newman, 1969; Mariot, 1962; Meijer and Bauer, 1962; Weyl, 1931).

### Group Theoretical Tools for Analyzing Vibrational Spectra

**Point Groups and Matrix Representations of Symmetry Operations.** The collection of symmetry operations of a molecule make up its point group. A point group is a collection of symmetry operations that are linked to one another by the following rules: (1) The identity operation is a member of the set. (2) The operations multiply associatively. (3) If $R$ and $S$ are elements, then so is $RS$. (4) The inverse of each element is also a member of the set. In satisfying these requirements, the point groups meet the criteria

for a mathematical group so that molecular symmetry can be investigated with the aid of group theory.

At this juncture, point groups, which describe the symmetry of molecules, and the use of group theoretical techniques to identify the selection rules for Raman active, molecular vibrations will be described. Although the symmetry of crystals is described by space groups, it will be demonstrated below that the Raman active vibrational modes of a crystal can be identified with the aid of the point group associated with the crystal space group. That is, the Raman spectrum of a crystal is equated to the Raman spectrum of a single unit cell, which is treated like a molecule. Specifically, the symmetry of the unit cell is described by one of 32 point groups. All that is presented about point groups and molecular symmetry will be directly applicable to the use of symmetry and group theory for analyses of vibrational spectra of solids. After completing the discussion of point groups and molecular symmetry, the link between point groups and vibrational spectroscopy of solids will be established.

There are a total of fourteen different types of point groups, seven with one principal axis of rotation; groups within these seven types form a series in which one group can be formed from one other group by the addition of a symmetry operation. The seven other types of point groups do not constitute a series and involve multiple axes of higher order. The symmetries of the vast number of different molecules can be completely represented by a small number of point groups.

The water molecule belongs to the point group $C_{2v}$, which consists of the symmetry elements $\sigma_{v(xz)}$, $\sigma_v(yz)$, and $C_2$, as illustrated in Figure 6, plus the identity operation $E$. Each symmetry operation can be represented by a matrix that describes how the symmetry element acts on the molecule. First, a basis function that represents the molecule is identified. Examples of basis functions for the water molecule are: (1) the $1s$ orbitals of the two hydrogen atoms and the $2s$ orbital of the oxygen atom; (2) the displacement coordinates of the molecule during normal-mode vibrations; and (3) the $1s$ orbitals of the hydrogen atoms and the $2p_{x,y,z}$ orbitals of oxygen. In fact, there are any number of possible basis functions, although some are more efficient and effective than others in representing the molecule. The number of functions in the basis defines the dimensions of the matrix representations of the symmetry operations. In the case of a basis consisting of three components, e.g., the two $1s$ orbitals of hydrogen atoms and the $2s$ orbital of oxygen, each matrix would be $3 \times 3$. The action of the symmetry operation on the molecule would then be provided by an equation of the form

$$[f_1 f_2 f_3] = \begin{pmatrix} d_{11} & d_{12} & d_{13} \\ d_{21} & d_{22} & d_{23} \\ d_{31} & d_{32} & d_{33} \end{pmatrix} \begin{bmatrix} 2s \\ 2s_a \\ 1s_b \end{bmatrix} \quad (31)$$

where $[f_1, f_2, f_3]$ represents the basis function following the action of the symmetry operation and $1s_a$ represents the $1s$ orbital on the "a" hydrogen atom. If the symmetry operation under consideration is the twofold axis of rotation,

$$D(R) = \begin{bmatrix} a & b & o \\ c & d & o \\ o & o & e \end{bmatrix} = \begin{bmatrix} a & b \\ c & d \end{bmatrix} + [e]$$

**Figure 14.** Matrix representation of the symmetry operation $R$ is given as $D(R)$. In the form depicted, $D(R)$ can be visualized as the "direct sum" of a $2 \times 2$ and a $1 \times 1$ matrix.

then $[f_1, f_2, f_3]$ is $[2s, 1s_b, 1s_a]$ and the matrix representation of the twofold axis of rotation for this basis function is

$$D(C_2) = \begin{pmatrix} 1 & 0 & 0 \\ 0 & 0 & 1 \\ 1 & 0 & 0 \end{pmatrix} \tag{32}$$

In instances in which one component of the basis function is unchanged by all of the symmetry operations, each matrix representation has the form depicted in Figure 14. The matrix can be thought of as consisting of the sum of two matrices, a $1 \times 1$ matrix and a $2 \times 2$ matrix. This type of sum is not an arithmetic operation and is referred to as a "direct sum." The process of converting all of the matrices that represent each of the symmetry operations for a given basis to the direct sum of matrices of smaller dimensions is referred to as "reducing the representation." In some cases each $2 \times 2$ matrix representation of the symmetry operations is also diagonalized, and the representation can be reduced to the direct sum of two $1 \times 1$ matrices. When the matrix representations have been converted to the direct sum of matrices with the smallest possible dimensions, the process of reducing the representation is complete.

Obviously, there are an infinite number of matrix representations that could be developed for each point group, each arising from a different basis. Fortunately, and perhaps somewhat surprisingly, most of the information about the symmetry of a molecule is contained in the characters of the matrix representations. Knowledge of the full matrix is not required in order to use the symmetry of the molecule to help solve complex problems in quantum mechanics. Often, only the characters of the matrices representing the group symmetry operations are needed.

The many different matrix representations of the symmetry operations can be reduced via similarity transformations to block diagonalized form. In block diagonalized form, each matrix is seen to consist of the direct sum of a number of irreducible matrices. An irreducible matrix is one for which there is no similarity transformation that is capable of putting it into block diagonalized form. A reduced representation consists of a set of matrices, each of which represents one symmetry operation of the point group and each of which is in the identical block diagonalized form.

Now it turns out that the character of a matrix representation is the most important piece of information concerning the symmetry of the molecule, as mentioned above. The list composed of the characters of the matrix representations of the symmetry operations of the group is referred to as the symmetry species of the representation. For each point group there is a small number of

unique symmetry species that are composed of the characters of irreducible matrices representing each of the symmetry operations. These special symmetry species are called irreducible representations of the point group. For each point group the number of irreducible representations is equal to the number of symmetry classes.

Remarkably, of the unlimited number of representations of a point group, corresponding to the unlimited number of bases that can be employed, the symmetry species of each representation can be expressed by a linear combination of symmetry species of irreducible representations. The list of the symmetry species of each irreducible representation therefore concisely indicates what is unique about the symmetry of the particular point group. This information is provided in character tables of the point groups.

**Character Tables.** The character table for the point group $C_{2v}$ is presented in Table 1. The top row lists the symmetry elements in the point group. The second through the fourth rows list the point group irreducible representations. The first column is the symbol used to denote the irreducible representation, e.g., $A_1$. The numbers in the character tables are the characters of the irreducible matrix representations of the symmetry operations for the various irreducible symmetry species. The right-hand column lists basis functions for the irreducible symmetry species. For example, if $x$ is used as a basis function for the point group $C_{2v}$, the matrix representations of the various symmetry operations will all be $1 \times 1$ matrices with character of 1. The resultant symmetry species, therefore, is 1, 1, 1, 1 and is labeled $A_1$. On the other hand, if $xy$ is the basis function, the matrix representations of all of the symmetry operations will consist of $1 \times 1$ matrices. The resultant symmetry species is 1, 1, $-1$, $-1$ and is labeled $A_2$. If a function is said to have the same symmetry as $x$, then that function will also serve as a basis for the irreducible representation $B_1$ in the point group $C_{2v}$. There are many possible basis functions for each irreducible representation. Those listed in the character tables have special significance for the analyses of rotational, vibrational, and electronic spectra. This will be made clear below. When a function serves as a basis of a representation, it is said to span that representation.

**Vibrational Selection Rules.** Where group theory is especially helpful is in deciding whether or not a particular integral, taken over a symmetric range, has a nonzero value. It turns out that if the integrand can serve as a basis for the $A_1$, totally symmetric irreducible representation, then the integral may have a nonzero value. If the integrand does not serve as a basis for the $A_1$ irreducible representation, then the integral will necessarily be zero.

The symmetry species that is spanned by the products of two functions is obtained by forming the "direct product" of the symmetry species spanned by each function. Tables are available that list the direct products of all possible combinations of irreducible representations for each point group (see, e.g., Wilson et al., 1955; Bishop, 1973; Atkins, 1984; Kettle, 1987).

Thus, if we want to know whether or not a Raman transition from the ground-state vibrational mode $n$ to the first vibrationally excited state $m$ can occur, we need to examine the six integrals of the form

$$\int \psi_m^* \alpha_{ij} \psi_n dV \qquad (33)$$

one for each of the six independent values of $\alpha_{ij}$ (or fewer depending on the symmetry of the molecule or crystal). If any one is nonzero, the Raman transition may occur. If all are zero, the Raman transition will be symmetry forbidden.

The Raman selection rule states that a Raman transition between two vibrational states $n$ and $m$ is allowed if the product $\psi_m^* \psi_n$ of the two wave functions describing the $m$ and $n$ vibrational states has the same symmetry species as at least one of the six components of $\alpha_{ij}$. This rule reflects the fact that for the integral $\int f_1 f_2 f_3 \, dV$ to have a nonzero value when taken over the symmetric range of variables $V$, it is necessary that the integrand, which is the product of the three functions $f_1$, $f_2$, and $f_3$, must span a symmetry species that is or contains the totally symmetric irreducible representation of the point group.

The symmetry species spanned by the product of three functions is obtained from the direct product of the symmetry species of each function (see, e.g., Wilson et al., 1955; Bishop, 1973; Atkins, 1984; Kettle, 1987). For the symmetry species of a direct product to contain the totally symmetric irreducible representation, it is necessary that the two functions span the same symmetry species or have a common irreducible representation in their symmetry species. The function $x^2$ acts as a basis for the $A_1$ irreducible representation of the point group $C_{2v}$. A basis for the irreducible representation $B_1$ is given as $x$. The function $x^2 x$ would not span the irreducible representation $A_1$, and so its integral over all space would be zero. The function $xz$ is also a basis for the irreducible representation $B_1$ in the point group $C_{2v}$. Hence, the function $xxz (=x^2 z)$ spans the irreducible representation $A_1$ and its integral over all space could be other than zero. Thus, for the integrand to span the totally symmetric irreducible representation of the molecular point group, it is necessary that the representation spanned by the product $\psi_m^* \psi_n$ be the same as the representation spanned by $\alpha_{ij}$. This last statement is extremely useful for analyzing the probability of a Stokes Raman transition from the vibrational ground state $n$ to the first vibrational excited state $m$ or for an anti-Stokes Raman transition from the first vibrational excited state $m$ to the vibrational ground state $n$. The vibrational ground state spans the totally symmetric species. Consequently, the direct product $\psi_m^* \psi_n$ spans the symmetry species of the first vibrational excited state $m$. Hence, the Raman selection rule means that the symmetry species of $\alpha_{ij}$ must be identical to the symmetry species of the first vibrational excited state.

The symmetry of $\alpha_{ij}$ can be obtained by showing that it transforms under the action of a symmetry operation of the point group in the same way as a function of known symmetry. Here, $\alpha_{ij}$ relates $p_i$ to $\alpha_{ij}$. If a symmetry operation $R$ is applied to the molecule/crystal, then the applied electric field is transformed to $R = \mathbf{D}(R)$, where $\mathbf{D}(R)$ is the matrix representation of $R$. The polarizability, which expresses the proportionality between the applied electric field and the induced polarization of the molecule/crystal, must transform in a manner related to but in general different from that of the applied electric field because the direction of $\mathbf{p}$ will generally be different from the direction of $\varepsilon_j$. The polarization of the crystal induced by the applied electric field is transformed to $Rp = D(R)p$ as a result of the action of the symmetry operation $R$. Consequently, the polarizability transforms as $RR\alpha = D(R)D(R)\alpha$. The transformation of $\alpha$ can be derived as follows (taken from Bishop, 1973).

If the symmetry operation $R$ transforms the coordinate system from $x$ to $x'$, then (summations are over all repeated indices)

$$p_i(x') = \sum \alpha_{ij}(x')\varepsilon_j(x')$$
$$= \sum \alpha_{ij}(x') \sum D_{jk}(R)\varepsilon_k(x) \qquad (34)$$

$$p_l(x) = \sum D_{ml}(R)p_m(x') \qquad (35)$$

$$p_l(x) = \sum D_{ml}(R) \sum \alpha_{im}(x') \sum D_{jk}(R) _k(x) \qquad (36)$$

Substituting into the last equation the expression

$$p_l(x) = \sum \alpha_{lk}(x) _k(x) \qquad (37)$$

gives

$$\alpha_{lk}(x) = \sum \sum D_{ml}(R)D_{jk}(R)\alpha_{mj}(x') \qquad (38)$$

This describes the action of the symmetry operation $R$ on the polarizability $\alpha$.

Now,

$$x_l = \sum D_{ml}(R)x_m \qquad (39)$$

and

$$x_k = \sum D_{jk}(R)x_j \qquad (40)$$

so

$$x_l x_k = \sum \sum D_{ml}(R)D_{jk}(R)x_m x_j \qquad (41)$$

Comparison of Equations 19 and 22 indicates that $\alpha_{lk}$ transforms as the function $x_l x_k$. Any symmetry species spanned by the function $x_m x_j$ will also be spanned by $\alpha_{mj}$. Thus, any vibrational mode that spans the same symmetry species as $x_l x_k$ will be Raman active. This information is included in the character tables of the point groups. In the column that lists the basis functions for the various symmetry species of the point group are listed quadratic functions such as $x^2, yz, x^2 + y^2, x^2 + y^2 - 2z^2, \ldots$.

By way of illustration, the character table indicates that if a molecule with point group symmetry $C_{2v}$ has a vibrational mode that spans the symmetry species $A_1$, it will be Raman active since this symmetry species is also spanned by the quadratic function $x^2$. Any vibrational mode that spans a symmetry species that has at least one quadratic function (e.g., $x^2$, $yz$) as a basis will be Raman active.

### Raman Active Vibrational Modes of Solids

It is now possible to illustrate how Raman spectroscopy can identify the symmetry species of the vibrational modes of a single crystal. To begin, it is important to recognize that the symmetry of a molecule may be lowered when it is part of a crystal. This can be appreciated by considering a simpler case of symmetry reduction: the change in symmetry of a sulfate anion as a consequence of its adsorption on a solid surface. As illustrated in Figure 15, the sulfate anion consists of oxygen nuclei located on the four corners of a tetrahedron and the sulfur nucleus positioned at the geometric center of the tetrahedron. The anion exhibits $T_d$ point group symmetry. If the sulfate is adsorbed on the surface of a solid through a bond formed between the solid and one of the oxygen anions of the sulfate, then the symmetry of the sulfate is lowered to $C_{3v}$. If two oxygen-surface bonds are formed, in either a bridging or nonbridging configuration, the symmetry of the sulfate is further lowered to $C_{2v}$. When present in a crystal, a molecule or anion will exhibit either the same or lower symmetry than the free molecule or anion.

The distinctive feature of the symmetry of a crystal is the presence of translational symmetry elements. In fact, it is possible to generate the entire lattice of a crystal by the repeated operation of the three basic primitive translation vectors that define the unit cell. When all possible arrays of lattice points in three dimensions are considered, it turns out that there are only fourteen distinctive types of lattices, called Bravais lattices. The Bravais lattice may not completely describe the symmetry of a crystal. The unit cell may have an internal structure that is not completely specified by the symmetry elements of the bare lattice. The symmetry of the crystal is completely specified by its space group, which is a mathematical group that contains translational operations and that may contain as many as three additional types of symmetry elements: rotations (proper and improper), which are the point group elements, screw axes, and glide planes. The latter two elements may be conceptually subdivided into a rotation (proper = screw axis; improper = glide plane) plus a rational fraction of a primitive lattice translation.

Lattices can be brought into themselves by the operations of certain point groups. Each Bravais lattice type is compatible with only a particular set of point groups, which are 32 in number and are referred to as crystallographic point groups. The appropriate combinations of the 14 Bravais lattices and the 32 crystallographic point groups result in the 230 three-dimensional space groups.

The restrictions that a lattice imposes on the rotational symmetry elements of the space group can be readily illustrated. Each lattice point is located at the tip of a primitive lattice vector given by

$$\mathbf{T} = n_1\mathbf{t}_1 + n_2\mathbf{t}_2 + n_3\mathbf{t}_3 \qquad (42)$$

where the $\mathbf{t}_i$ are the basic primitive translation vectors and the $n_i$ are positive and negative integers including zero. The rotation of the lattice translates the points to new locations given by

$$\mathbf{T}' = \mathbf{R}T = R_{ij}n_j\mathbf{t}_j \qquad (43)$$

where $R_{ij}n_j$ must be integers for all values of $n_j$, which are integers. Hence, the $R_{ij}$ must be integers as well. For a proper or improper rotation of $\theta$ about the $z$ axis,

$$\mathbf{R} = \begin{bmatrix} \cos\theta & \sin\theta & 0 \\ -\sin\theta & \cos\theta & 0 \\ 0 & 0 & \pm1 \end{bmatrix} \qquad (44)$$

The requirement of integral values for each element $R_{ij}$ means, in particular, that

$$\text{Tr } \mathbf{R} = 2\cos\theta \pm 1 = \text{integer} \qquad (45)$$

The plus sign corresponds to a proper rotation, while the minus sign corresponds to an improper rotation (e.g., reflection or inversion). Thus, the requirement of integral values of $R_{ij}$ limits the possible rotation angles $\theta$ to

$$\cos\theta = (n - \pm1)/2 \Rightarrow \theta = 0, \pi/3, \pi/2, 2\pi/3, 2\pi \qquad (46)$$

Crystallographic point groups may therefore contain the following proper rotations: $C_1$, $C_2$, $C_3$, $C_4$, and $C_6$, where $C_n$ indicates a rotation angle of $180°/n$ about an axis of the crystal.

Within each space group is an invariant subgroup consisting of the primitive translational operations. An invariant group is one in which a conjugate transformation

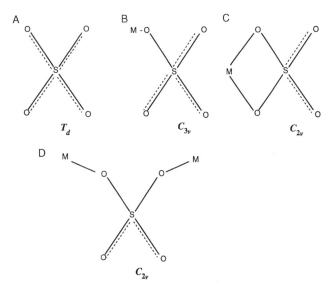

**Figure 15.** Influence of bonding on the symmetry of sulfate: (**A**) free sulfate; (**B**) unidentate sulfate; (**C**) bidentate sulfate; (**D**) bidentate sulfate in the bridging configuration.

produces another member of the group; i.e., if for all elements $X$ and $t$ of the group $G$, $X^{-1}tX = t'$, where $t$ and $t'$ are elements of the same subgroup of $G$, then this subgroup is an invariant subgroup of $G$.

A translational operation is represented by the symbol $\underline{\varepsilon}|\mathbf{t}$, which is a particular case of the more general symbol for a rotation of $\alpha$ followed by a translation of $\mathbf{a}$, $\underline{\alpha}|\mathbf{a}$. The notation $\underline{\varepsilon}$ represents a rotation of $0°$. The invariance of the translational subgroup is demonstrated by

$$(\underline{\alpha}|\mathbf{a})\mathbf{x} = \underline{\alpha}\mathbf{x} + \mathbf{a}$$
$$(\underline{\beta}|\mathbf{b})(\underline{\alpha}|\mathbf{a})\mathbf{x} = \underline{\beta}\underline{\alpha}\mathbf{x} + \underline{\beta}\mathbf{a} + \mathbf{b}$$
$$= (\underline{\beta}\underline{\alpha}|\underline{\beta}\mathbf{a} + \mathbf{b})\mathbf{x}$$

Since $\quad (\underline{\alpha}|\mathbf{a})^{-1}(\underline{\alpha}|\mathbf{a}) = (\underline{\varepsilon}|0)$

$(\underline{\alpha}|\mathbf{a})^{-1}$ must be given by $\quad (\underline{\alpha}^{-1}| - \underline{\alpha}^{-1}\mathbf{a})$

$$(\underline{\alpha}|\mathbf{a})^{-1}(\underline{\varepsilon}|\mathbf{t})(\underline{\alpha}|\mathbf{a})\mathbf{x} = (\underline{\alpha}|\mathbf{a})^{-1}(\underline{\varepsilon}|\mathbf{t})(\underline{\alpha}\mathbf{x} + \mathbf{a})$$
$$(\underline{\alpha}|\mathbf{a})^{-1}(\underline{\alpha}\mathbf{x} + \mathbf{t} + \mathbf{a}) = \underline{\alpha}^{-1}\underline{\alpha}\mathbf{x} + \underline{\alpha}^{-1}\mathbf{t} + \underline{\alpha}^{-1}\mathbf{a} - \underline{\alpha}^{-1}\mathbf{a}$$
$$= \mathbf{x} + \underline{\alpha}^{-1}\mathbf{t}$$
$$= (\underline{\varepsilon}|\underline{\alpha}^{-1}\mathbf{t})\mathbf{x}$$
$$= (\underline{\varepsilon}|\mathbf{t}')\mathbf{x} \tag{47}$$

Thus, a conjugate transformation of a translation operator $\underline{\varepsilon}|\mathbf{t}$ using a general rotation plus translation operator $\underline{\alpha}|\mathbf{a}$ that is also a member of the group $G$ produces another simple translation operator $\underline{\varepsilon}|\mathbf{t}'$, demonstrating that the translational subgroup is invariant to the conjugacy operation.

The invariant property of the translational subgroup is exploited quite heavily in developing expressions for the irreducible representations of space groups. The difficulty in dealing with the symmetry of a crystal and in analyzing the vibrational modes of a crystal stems from the fact that the translational subgroup is infinite in size. Stated slightly differently, a crystal consists of a large number of atoms so that the number of normal vibrational modes is huge, i.e., $\approx 10^{23}$. Fortunately, group theory provides a method for reducing this problem to a manageable size.

Most textbooks dealing with applications of group theory to crystals develop cofactor expressions for space groups. The cofactor group is isomorphic with the crystallographic point group providing the space group is symmorphic. A symmorphic space group consists of rotational and translational symmetry elements. A nonsymmorphic space group contains at least one symmetry element $(\underline{\alpha}|\mathbf{a})$ in which the rotation $\alpha$, all by itself, and/or the translation $\mathbf{a}$, all by itself, are not symmetry elements of the space group. Of the 230 three-dimensional space groups, only 73 are symmorphic. Consequently, the use of factor groups is of no benefit for analyzing the vibrational modes of crystals in the 157 nonsymmorphic space groups.

Since the use of cofactor groups is so common, albeit of limited value, the approach will be summarized here and then a more general approach that addresses both symmorphic and nonsymmorphic space groups will be developed.

The translational factor group $G|T$ of the symmorphic space group $G$ consists of

$$G|T = T + \sum s_i T \qquad (i = 2, 3, \ldots, n) \tag{48}$$

where the $s_i$ are the nontranslational symmetry elements except the identity element $(s_1 = E)$ of the symmorphic space group $G$ whose crystallographic point group is of order $n$; $T$ is the translational symmetry elements of $G$. The translational factor group is distinct from the group $G$ because $T$ is treated as a single entity, rather than as the infinite number of lattice translational operations that it is in the group $G$. The factor group has a different multiplication rule than the space group and $T$ acts as the identity element. As long as $G$ is symmorphic, the factor group $G|T$ is obviously isomorphic with the crystallographic point group of $G$. Because of the isomorphic relationship of the two groups, the irreducible representations of the crystallographic point group will also serve as the irreducible representations of the factor group $G|T$. The irreducible representations of the factor group provide all the symmetry information about the crystal that is needed to analyze its vibrational motion. On the other hand, if $G$ is nonsymmorphic, then it contains at least one element $s_j$ of the form $[\underline{\alpha}|\mathbf{v}(\alpha)]$, where $\alpha$ and/or $\mathbf{v}(\alpha)$ are not group elements. As a consequence, $G|T$ is no longer symmorphic with any crystallographic point group.

The value of this approach, which makes use of cofactor groups, is that the total number of group elements have been reduced from $g \times 10^{23}$, where $g$ is the number of point group operations, to $g$. Furthermore, the factor group $G|T$ is isomorphic with the crystallographic point group. Consequently, the irreducible representations of the point group serve as irreducible representations of the factor group. How does this help in the analyses of vibrational modes? Recall that the only Raman active modes are characterized by $k = 0$ (i.e., approximately infinite wavelength) and nonzero frequency. The modes of $k = 0$ consist of corresponding atoms in each unit cell moving in phase. The number of normal modes of this type of motion is given by $3N$, where $N$ is the number of atoms in the unit cell. The symmetry of the unit cell is just that of the factor group. Thus, the irreducible representations of the crystallographic point group provide the irreducible representations of the Raman active vibrational modes of the crystal.

As mentioned above, a more general approach to the group theoretical analysis of crystal vibrational modes will be developed in place of the use of factor groups. This approach is enunciated by Lax (1974).

The group multiplication of symmetry elements of space groups is denoted by

$$[\underline{\alpha}|\mathbf{v}(\alpha)][\underline{\beta}|\mathbf{v}(\beta)]\mathbf{r} = [\underline{\alpha}|\mathbf{v}(\alpha)][\underline{\beta}\mathbf{r} + \mathbf{v}(\beta)]$$
$$= \underline{\alpha}\underline{\beta}\mathbf{r} + \underline{\alpha}\mathbf{v}(\beta) + \mathbf{v}(\alpha)$$
$$= \underline{\alpha}\underline{\beta}\mathbf{r} + \underline{\alpha}\mathbf{v}(\beta) + \mathbf{v}(\alpha) + \mathbf{v}(\alpha\beta) - \mathbf{v}(\alpha\beta)$$
$$\underline{\alpha}\underline{\beta}\mathbf{r} + \mathbf{v}(\alpha\beta) + \underline{\alpha}\mathbf{v}(\beta) + \mathbf{v}(\alpha) - \mathbf{v}(\alpha\beta)$$
$$= \alpha\beta\mathbf{r} + \mathbf{v}(\alpha\beta) + \mathbf{t}$$
$$\text{where} \quad \mathbf{t} = \underline{\alpha}\mathbf{v}(\beta) + \mathbf{v}(\alpha) - \mathbf{v}(\alpha\beta)$$
$$= (\underline{\varepsilon}|\mathbf{t})(\underline{\alpha}\underline{\beta}|\mathbf{v}(\alpha\beta)) \tag{49}$$

Next, consider the operation of $(\varepsilon|\mathbf{t})[\underline{\alpha\beta}|\mathbf{v}(\alpha\beta)]$ on a Bloch function $\exp(i\mathbf{k}\cdot\mathbf{r})u_k(\mathbf{r})$. Functions of this form serve as eigenvectors of a crystal's vibrational Hamiltonian that is expressed in normal-mode coordinates with $u_k(\mathbf{r})$ representing the displacement in the $k$th normal vibrational mode of the harmonic oscillator at $\mathbf{r}$ (Altmann, 1991). Therefore, the operation of two successive symmetry elements $[\underline{\alpha}|\mathbf{v}(\alpha)]$ and $[\underline{\beta}|\mathbf{v}(\beta)]$ on the Bloch function is given by

$$(\varepsilon|\mathbf{t})[\underline{\alpha\beta}|\mathbf{v}(\alpha\beta)]\exp(i\mathbf{k}\cdot\mathbf{r})u_k(\mathbf{r}) \qquad (50)$$

Applying $[\underline{\alpha\beta}|\mathbf{v}(\alpha\beta)]$ to both functions that constitute the Bloch function converts the above expression to

$$(\varepsilon|\mathbf{t})\{\exp[i\mathbf{k}\cdot(\underline{\alpha\beta}|\mathbf{v}(\alpha\beta))^{-1}\cdot\mathbf{r}](\underline{\alpha\beta}|\mathbf{v}(\alpha\beta))u_k(\mathbf{r})\} \qquad (51)$$

Now focus attention on the argument of the exponential in the previous expression. It may be rewritten as

$$
\begin{aligned}
&[i\mathbf{k}\cdot((\underline{\alpha\beta})^{-1}|-(\underline{\alpha\beta})^{-1}\mathbf{v}(\alpha\beta))\cdot\mathbf{r}]\\
&=[i\mathbf{k}\cdot(\underline{\alpha\beta})^{-1}\cdot[\mathbf{r}-\mathbf{v}(\alpha\beta)]]\\
&=[i(\underline{\alpha\beta})\mathbf{k}\cdot(\mathbf{r}-\mathbf{v}(\alpha\beta))]
\end{aligned}
\qquad (52)
$$

(1) Substituting this expression for the argument of the exponential back into Equation 51, (2) operating $\underline{\alpha\beta}|\mathbf{v}(\alpha\beta)$ on $\mathbf{r}$ in $u_k(\mathbf{r})$, and (3) including $\varepsilon|\mathbf{t}$ in the arguments of both functions of the Bloch function convert Equation 51 to

$$\exp\{i(\underline{\alpha\beta})\mathbf{k}\cdot(\varepsilon|\mathbf{t})^{-1}\cdot[\mathbf{r}-\mathbf{v}(\alpha\beta)]\}u_k\{(\varepsilon|\mathbf{t})^{-1}[\underline{\alpha\beta}|\mathbf{v}(\alpha\beta)]^{-1}\mathbf{r}\} \qquad (53)$$

Now,

$$u_k\{(\varepsilon|\mathbf{t})^{-1}[\underline{\alpha\beta}|\mathbf{v}(\alpha\beta)]^{-1}\mathbf{r}\}=u_k\{[\underline{\alpha\beta}|\mathbf{v}(\alpha\beta)]^{-1}\mathbf{r}\} \qquad (54)$$

because

$$u_k(\mathbf{r}-\mathbf{t})=u_k(\mathbf{r}) \qquad (55)$$

where $\mathbf{t}$ is the lattice vector as $u_k(\mathbf{r})$ has the periodicity of the lattice. Equation 53 now becomes

$$\exp\{i(\underline{\alpha\beta})\mathbf{k}\cdot[\mathbf{r}-\mathbf{v}(\alpha\beta)-\mathbf{t}]\}u_k\{[\underline{\alpha\beta}|\mathbf{v}(\alpha\beta)]^{-1}\mathbf{r}\}$$

$$\exp[-i(\underline{\alpha\beta})\mathbf{k}\cdot\mathbf{t}]\exp\{i(\underline{\alpha\beta})\mathbf{k}\cdot[\mathbf{r}-\mathbf{v}(\alpha\beta)]\}u_k\{[\underline{\alpha\beta}|\mathbf{v}(\alpha\beta)]^{-1}\mathbf{r}\}$$

$$\exp[-i(\underline{\alpha\beta})\mathbf{k}\cdot\mathbf{t}](\underline{\alpha\beta}\mathbf{v}(\alpha\beta))\psi(\mathbf{r},\mathbf{k}) \qquad (56)$$

where $\psi(\mathbf{r},\mathbf{k})$ is the Bloch function. At this point it is no longer possible to continue in a completely general fashion. To include in Equation 56 all possible values of $\mathbf{k}$, it would be necessary to develop two separate expressions of Equation 56: one for symmorphic space groups and the other for nonsymmorphic space groups. Alternatively, a single expression for Equation 56 can be developed for both symmorphic and nonsymmorphic space groups if we consider only one value of $\mathbf{k}$: the long-wavelength limit, $\mathbf{k}=0$. For $\mathbf{k}=0$, the above group multiplication becomes

$$(\underline{\alpha}|\mathbf{v}(\underline{\alpha}))(\underline{\beta}|\mathbf{v}(\underline{\beta}))=(\underline{\alpha\beta}|\mathbf{v}(\underline{\alpha\beta})) \qquad (57)$$

which is the same group multiplication rule obeyed by symmetry elements of point groups (i.e., groups consisting of elements of the form $[\underline{\alpha}|0]$). Thus, the representations of space group elements $[\underline{\alpha}|\mathbf{v}(\alpha)]$ are identical to the representations of the point group elements $[\underline{\alpha}|0]$.

Thomas M. Devine
University of California
Berkeley, California

# ULTRAVIOLET PHOTOELECTRON SPECTROSCOPY

## INTRODUCTION

### Photoemission and Inverse Photoemission

Ultraviolet photoelectron spectroscopy (UPS) probes electronic states in solids and at surfaces. It relies on the process of photoemission, in which an incident photon provides enough energy to bound valence electrons to release them into vacuum. Their energy $E$, momentum $\hbar\mathbf{k}$, and spin $\sigma$ provide the full information about the quantum numbers of the original valence electron using conservation laws. Figure 1 depicts the process in an energy diagram. Essentially, the photon provides energy but negligible momentum (due to its long wavelength $\lambda=2\pi/|\mathbf{k}|$), thus shifting all valence states up by a fixed energy ("vertical" or "direct" transitions). In addition, secondary

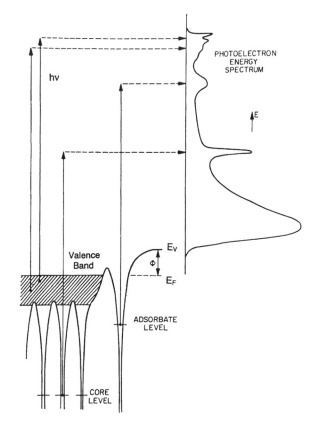

**Figure 1.** Photoemission process (Smith and Himpsel, 1983).

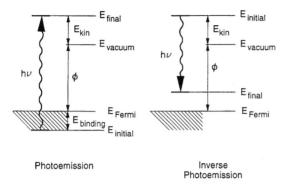

**Figure 2.** Photoemission and inverse photoemission as probes of occupied and unoccupied valence states: $\phi$ = work function (Himpsel and Lindau, 1995).

processes, such as energy loss of a photoelectron by creating plasmons or electron-hole pairs, produce a background of secondary electrons that increases toward lower kinetic energy. It is cut off at the vacuum level $E_V$, where the kinetic energy goes to zero. The other important energy level is the Fermi level $E_F$, which becomes the upper cutoff of the photoelectron spectrum when translated up by the photon energy. The difference $\phi = E_V - E_F$ is the work function. It can be obtained by subtracting the energy width of the photoelectron spectrum from the photon energy. For reviews on photoemission, see Cardona and Ley (1978), Smith and Himpsel (1983), and Himpsel and Lindau (1995). Information specific to angle-resolved photoemission is given by Plummer and Eberhardt (1982), Himpsel (1983), and Kevan (1992).

Photoemission is complemented by a sister technique that maps out unoccupied valence states, called inverse photoemission or bremsstrahlung isochromat spectroscopy (BIS). This technique is reviewed by Dose (1985), Himpsel (1986), and Smith (1988). As shown in Figure 2, inverse photoemission represents the reverse of the photoemission process, with an incoming electron and an outgoing photon. The electron drops into an unoccupied state and the energy is released by the photon emission.

Both photoemission and inverse photoemission operate at photon energies in the ultraviolet (UV), starting with the work function threshold at ~4 eV and reaching up to 50- to 100-eV photon energy, where the cross-section of valence states has fallen off by an order of magnitude and the momentum information begins to get blurred. At kinetic energies of 1 to 100 eV, the electron mean free path is only a few atomic layers, making it possible to detect surface states as well as bulk states.

### Characterization of Valence Electrons

In atoms and molecules, the important parameters characterizing a valence electron are its binding energy (or ionization potential) plus its angular momentum (or symmetry label in molecules). This information is augmented by the vibrational and rotational fine structures, which shed light onto the interatomic potential curves (Turner et al., 1970). In a solid, it becomes necessary to consider not only energy but also momentum. Electrons in a crystalline solid are

completely characterized by a set of quantum numbers. These are energy $E$, momentum $\hbar\mathbf{k}$, point group symmetry (i.e., angular symmetry), and spin. This information can be summarized by plotting $E(\mathbf{k})$ band dispersions with the appropriate labels for point group symmetry and spin. Disordered solids such as random alloys can be characterized by average values of these quantities, with disorder introducing a broadening of the band dispersions. Localized electronic states (e.g., the $4f$ levels of rare earths) exhibit flat $E(\mathbf{k})$ band dispersions. Angle-resolved photoemission, combined with a tunable and polarized light source, such as synchrotron radiation, is able to provide the full complement of quantum numbers. In this respect, photoemission and inverse photoemission are unique among other methods of determining the electronic structure of solids.

Before getting into the details of the technique, its capabilities are illustrated in Figure 3, which shows how much information can be extracted by various techniques about the band structure of Ge. Optical spectroscopy integrates over the momentum and energy of the photoelectron and leaves only the photon energy $h\nu$ as variable. The resulting spectral features represent regions of momentum and energy space near critical points at which the density of transitions is high (Fig. 3A). By adjusting an empirical band structure to the data (Chelikowski and Cohen, 1976; Smith et al., 1982), it is possible to extract rather accurate information about the strongest critical points. Angle-integrated photoemission goes one step further by detecting the electron energy in addition to the photon energy (Fig. 3B). Now it becomes possible to sort out whether spectral features are due to the lower state or the upper state of the optical transition. To extract all the information on band dispersion, it is necessary to resolve the components of the electron momentum parallel and perpendicular to the surface, $\mathbf{k}^{\parallel}$ and $k^{\perp}$, by angle-resolved photoemission with variable photon energy (Fig. 3C).

Photoemission and inverse photoemission data can, in principle, be used to derive a variety of electronic properties of solids, such as the optical constants [from the $E(\mathbf{k})$ relations; see Chelikowski and Cohen (1976) and Smith et al. (1982)], conductivity (from the optical constants), the electron lifetime [from the time decay constant $\tau$ or from the line width $\delta E$; see Olson et al. (1989) and Haight (1995)], the electron mean free path [from the attenuation length $\lambda$ or from the line width $\delta k$; see Petrovykh et al. (1998)], the group velocity of the electrons [via the slope of the $E(\mathbf{k})$ relation; see Petrovykh et al. (1998)], the magnetic moment (via the band filling), and the superconducting gap (see Olson et al., 1989; Shen et al., 1993; Ding et al., 1996).

### Energy Band Dispersions

How are energy band dispersions determined in practice? A first look at the task reveals that photoemission (and inverse photoemission) provides just the right number of independent measurable variables to establish a unique correspondence to the quantum numbers of an electron in a solid. The energy $E$ is obtained from the kinetic energy of the electron. The two momentum components parallel to

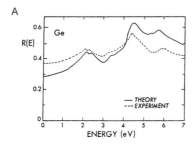

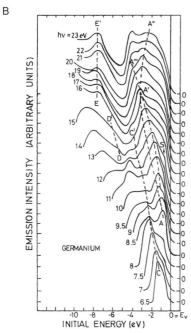

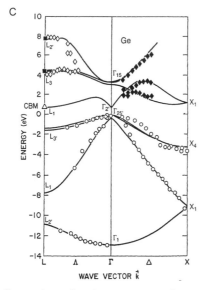

**Figure 3.** Comparison of various spectroscopies applied to the band structure of germanium. (**A**) The optical reflectivity $R(E)$ determines critical points near the band gap, (**B**) angle-integrated photoemission adds critical points farther away, and (**C**) angle-resolved photoemission and inverse photoemission provide the full $E(\mathbf{k})$ band dispersion (Phillip and Ehrenreich, 1963; Grobman et al., 1975; Chelikowski and Cohen, 1976; Wachs et al., 1985; Hybertsen and Louie, 1986; Ortega and Himpsel, 1993).

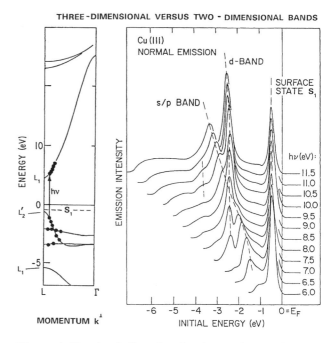

**Figure 4.** Mapping bulk and surface bands of copper by angle-resolved photoemission. Variable photon energy from monochromatized synchrotron radiation makes it possible to tune the **k** component perpendicular to the surface $k^{\perp}$ and to distinguish surface states from bulk states by their lack of $k^{\perp}$ dispersion (Knapp et al., 1979; Himpsel, 1983).

the surface, $\mathbf{k}^{\parallel}$, are derived from the polar and azimuthal angles $\vartheta$ and $\phi$ of the electron. The third momentum component, $k^{\perp}$, is varied by tuning the photon energy $h\nu$. This can be seen in Figure 4, in which angle-resolved photoemission spectra at different photon energies $h\nu$ are plotted together with the relevant portion of the band structure. The parallel momentum component $\mathbf{k}^{\parallel}$ is kept zero by detecting photoelectrons in normal emission; the perpendicular component $k^{\perp}$ changes as the photon energy of the vertical interband transitions is increased. A complete band structure determination requires a tunable photon source, such as synchrotron radiation (or a tunable photon detector in inverse photoemission). Surface states, such as the state near $E_{\mathrm{F}}$, labeled $S_1$ in Figure 4, do not exhibit a well-defined $k^{\perp}$ quantum number. Their binding energy does not change when the $k^{\perp}$ of the upper state is varied by changing $h\nu$. This lack of $k^{\perp}$ dispersion is one of the characteristics of a surface state. Other clues are that a surface state is located in a gap of bulk states with the same $\mathbf{k}^{\parallel}$ and the surface state is sensitive to contamination. To complete the set of quantum numbers, one needs the point group symmetry and the spin in ferromagnets. The former is obtained via dipole selection rules from the polarization of the photon, the latter from the spin polarization of the photoelectron.

For two-dimensional (2D) states in thin films and at surfaces, the determination of energy bands is almost trivial since only $E$ and $\mathbf{k}^{\parallel}$ have to be determined. These quantities obey the conservation laws

$$E_1 = E_{\mathrm{u}} - h\nu \tag{1}$$

and

$$\mathbf{k}_l^{\parallel} - \mathbf{k}_u^{\parallel} - \mathbf{g}^{\parallel} \qquad (2)$$

where $\mathbf{g}^{\parallel}$ is a vector of the reciprocal surface lattice, u denotes the upper state, and l the lower state. These conservation laws can be derived from the invariance of the crystal with respect to translation in time and in space (by a surface lattice vector). For the photon, only its energy $h\nu$ appears in the balance because the momentum of a UV photon is negligible compared to the momentum of the electrons. The subtraction of a reciprocal lattice vector simply corresponds to plotting energy bands in a reduced surface Brillouin zone, i.e., within the unit cell in $\mathbf{k}^{\parallel}$ space. For a three-dimensional (3D) bulk energy band, the situation becomes more complicated since the momentum component perpendicular to the surface is not conserved during the passage of the photoelectron across the surface energy barrier. However, $k^{\perp}$ can be varied by changing the photon energy $h\nu$, and extremal points, such as the $\Gamma$ and $L$ points in Figure 4, can thus be determined. A discussion of the practical aspects and capabilities of various band mapping schemes is given below (see Practical Aspects of the Method) and in several reviews (Plummer and Eberhardt, 1982; Himpsel, 1983; Kevan, 1992). Experimental energy bands are compiled in Landolt-Börnstein (1989, 1994).

In ferromagnets the bands are split into two subsets, one with majority spin, the other with minority spin (see Fig. 5). The magnetic exchange splitting $\delta E_{ex}$ between majority and minority spin bands is the key to magnetism. It causes the majority spin band to become filled more than the minority band, thus creating the spin imbalance that produces the magnetic moment. An overview of the band structure of ferromagnets and magnetic thin-film structures is given in Himpsel et al. (1998).

## Competitive and Related Techniques

The relation of UPS to optical spectroscopy has been mentioned in the context of Figure 3 already. Several other spectroscopies involve core levels but provide information about valence electrons as well. Figure 6 shows the simplest processes, which involve transitions between two levels only, one of them a core level. More complex phenomena involve four levels, such as Auger electron spectroscopy (see AUGER ELECTRON SPECTROSCOPY) and appearance potential spectroscopy.

Core-level photoelectron spectroscopy determines the energy of a core level relative to the Fermi level $E_F$ (Fig. 6A) and is known as x-ray photoelectron spectroscopy (XPS) or electron spectroscopy for chemical analysis (ESCA). A core electron is ionized, and its energy is obtained by subtracting the photon energy

## Core Level Spectroscopies

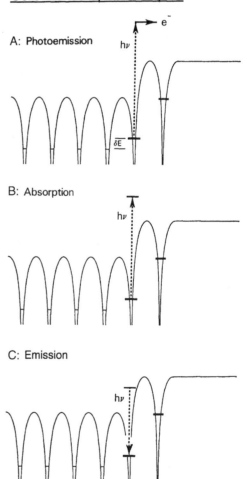

**Figure 6.** Spectroscopies based on core levels. (**A**) X-ray photoelectron spectroscopy measures the binding energy of core levels, (**B**) core-level absorption spectroscopy detects transitions from a core level into unoccupied valence states, and (**C**) core-level emission spectroscopy detects transitions from occupied valence states into a core hole. In contrast to UPS, these spectroscopies are element-specific but cannot provide the full momentum information.

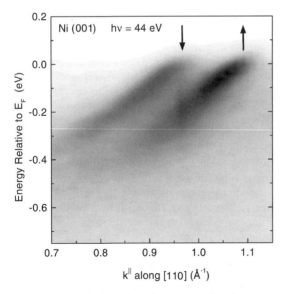

**Figure 5.** Using multidetection of energy and angle to map out the $E(\mathbf{k})$ band dispersion of Ni near the Fermi level $E_F$ from a Ni(001) crystal. The majority and minority spin bands are split by the ferromagnetic exchange splitting $\delta E_{ex}$. High photoemission intensity is shown in dark (Petrovykh et al., 1998).

from the kinetic energy of the photoelectron. The energy shifts of the substrate and adsorbate levels are a measure of the charge transfer and chemical bonding at the surface.

Core-level absorption spectroscopy determines the pattern of unoccupied orbitals by exciting them optically from a core level (Fig. 6B). It is also known as near-edge x-ray absorption fine structure (NEXAFS; see XAFS SPECTROSCOPY) or x-ray absorption near-edge structure (XANES). Instead of measuring transmission or reflectivity, the absorption coefficient is determined in surface experiments by collecting secondary products, such as photoelectrons, Auger electrons, and core-level fluorescence. The short escape depth of electrons compared to photons provides surface sensitivity. These spectra resemble the density of unoccupied states, projected onto specific atoms and angular momentum states. In a magnetic version of this technique, magnetic circular dichroism (MCD), the difference in the absorption between parallel and antiparallel alignment of the electron and photon spin is measured. Compared to UPS, the momentum information is lost in core-level absorption spectroscopy, but element sensitivity is gained. Also, the finite width of the core level is convolved with the absorption spectrum and limits the resolution.

Core-level emission spectroscopy can be viewed as the reverse of absorption spectroscopy (Fig. 6C). Here, the valence orbital structure is obtained from the spectral distribution of the characteristic x rays emitted during the recombination of a core hole with a valence electron. The core hole is created by either optical or electron excitation. As with photoemission and inverse photoemission, core-level emission spectroscopy complements core-level absorption spectroscopy by mapping out occupied valence states projected onto specific atoms.

## PRINCIPLES OF THE METHOD

### The Photoemission Process

The most general theory of photoemission is given by an expression of the golden rule type, i.e., a differential cross-section containing a matrix element between the initial and final states $\psi_i$ and $\psi_f$ and a phase space sum (see Himpsel, 1983; Dose, 1985; Smith, 1988):

$$\frac{d\sigma}{d\Omega} \sim \sqrt{E_{kin}} \sum_i \langle \Psi_f | \mathbf{A} \cdot \mathbf{p} + \mathbf{p} \cdot \mathbf{A} | \Psi_i \rangle|^2 \times \delta(E_f - E_i - h\nu)$$

(3)

The factor $\sqrt{E_{kin}}$ containing the kinetic energy $E_{kin}$ of the photoelectron represents the density of final states, the scalar product of the vector potential $\mathbf{A}$ of the photon and the momentum operator $\mathbf{p} = -i\hbar\partial/\partial\mathbf{r}$ is the dipole operator for optical excitation, and the $\delta$ function represents energy conservation. This is often called the one-step model of photoemission, as opposed to the three-step model, which approximates the one-step model by a

sequence of three simpler steps. For practical purposes (see Practical Aspects of the Method; Himpsel, 1983), we have to consider mainly the various selection rules inherent in this expression, such as the conservation of energy (Equation 1), parallel momentum (Equation 2), and spin, together with the point group selection rules. Since there is no clear-cut selection rule for the perpendicular momentum, it is often determined approximately by using a nearly free electron final state.

While selection rules provide clear yes-no decisions, there are also more subtle effects of the matrix element that can be used to bring out specific electronic states. The atomic symmetry character determines the energy dependence of the cross-section (Yeh and Lindau, 1985), allowing a selection of specific orbitals by varying the photon energy. For example, the $s,p$ states in transition and noble metals dominate the spectra near the photoelectric threshold, while the $d$ states turn on at 10 eV above threshold. It takes photon energies of 30 eV above threshold to make the $f$ states in rare earths visible. Resonance effects at a threshold for a core-level excitation can also enhance particular orbitals. Conversely, the cross-section for states with a radial node exhibits so-called Cooper minima, at which the transitions become almost invisible.

The wave functions of the electronic states in solids and at surfaces are usually approximated by the ground-state wave functions obtained from a variety of schemes, e.g., empirical tight binding and plane-wave schemes (Chelikowski and Cohen, 1976; Smith et al., 1982) and first-principles local density approximation (LDA; see Moruzzi et al., 1978; Papaconstantopoulos, 1986). Strictly speaking, one should use the excited-state wave functions and energies that represent the hole created in the photoemission process or the extra electron added to the solid in the case of inverse photoemission. Such quasiparticle calculations have now become feasible and provide the most accurate band dispersions to date, e.g., the so-called $GW$ calculations, which calculate the full Green's function $G$ of the electron/hole and the fully screened Coulomb interaction $W$ but still neglect vertex and density gradient corrections (Hybertsen and Louie, 1986; see also BONDING IN METALS). Particularly in the case of semiconductors the traditional ground-state methods are unable to determine the fundamental band gap from first principles, with Hartree-Fock overestimating it and LDA underestimating it, typically by a factor of 2. The band width comes out within 10% to 20 % in LDA calculations.

Various types of wave functions can be involved in the photoemission process, as shown in Figure 7. The states that are propagating in the solid lead to vertical transitions that conserve all three momentum components and are being used for mapping out bulk bands. Evanescent states conserve the parallel momentum only and are more sensitive to the surface. At elevated temperatures one has to consider phonon-assisted transitions, which scramble the momentum information about the initial state completely.

The materials property obtained from angle-integrated UPS is closely related to the density of states and is often interpreted as such. Strictly speaking, one measures

WAVE FUNCTIONS AT THE SURFACE

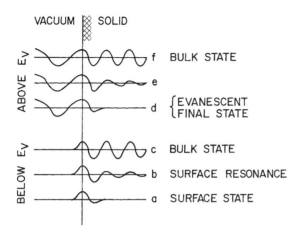

**Figure 7.** Various types of wave functions encountered in UPS. Propagating bulk states give rise to transitions that conserve all three **k** components; evanescent states ignore the $k^\perp$ component (Smith and Himpsel 1983).

the energy distribution of the joint density of initial and final states, which includes the optical matrix element (Grobman et al., 1975). Angle-resolved UPS spectra provide the $E(\mathbf{k})$ band dispersions that characterize electrons in a solid completely. Since the probing depth is several atomic layers in UPS, it is possible to determine these properties for the bulk as well as for the surface.

### Energy Bands in Solids

The mapping of energy bands in solids and at surfaces is based on the conservation laws for energy and parallel momentum (Equations 1 and 2). Figures 3, 4, and 5 give examples. Presently, energy bands have been mapped for practically all elemental solids and for many compounds. Such results are compiled in Landolt-Börnstein (1989, 1994). For the ferromagnets, see Himpsel et al. (1998). Since energy band dispersions comprise the full information about electrons in solids, all other electronic properties can in principle be obtained from them. Such a program has been demonstrated for optical properties (Chelikowsky and Cohen, 1976; Smith et al., 1982). For ferromagnetism, the most significant band parameters are the exchange splitting $\delta E_{ex}$ between majority and minority spin bands and the distance between the top of the majority spin $d$ bands and the Fermi level (Stoner gap). The exchange splitting is loosely related to the magnetic moment (~1 eV splitting per Bohr magneton), the Stoner gap to the minimum energy for spin flip excitations (typically from a few tenths of an electron volt down to zero).

### Surface States

The electronic structure of a surface is characterized by 2D energy bands that give the relation between $E$ and $\mathbf{k}^\|$. The momentum perpendicular to the surface, $k^\perp$, ceases to

be a good quantum number. The states truly specific to the surface are distinguished by their lack of interaction with bulk states, which usually requires that they are located at points in the $E(\mathbf{k}^\|)$ diagram where bulk states of the same symmetry are absent. This is revealed in band diagrams where the $E(\mathbf{k}^\|)$ band dispersions of surface states are superimposed on the regions of bulk bands projected along $k^\perp$.

On metals, one finds fairly localized, $d$-like surface states and delocalized $s,p$-like surface states. The $d$ states carry most of the spin polarization in ferromagnets. Their cross-section starts dominating relative to $s,p$ states as the photon energy is increased. An example of an $s,p$-like surface state is given in Figure 4. On the Cu(111) surface a $p_z$-like surface state appears close to the Fermi level ($S_1$). It is located just above the bottom of the $L'_2$-$L_1$ gap of the bulk $s,p$ band. A very basic type of $s,p$ surface state is the image state in a metal. The negative charge of an electron outside a metal surface induces a positive image charge that binds the electron to the surface. In semiconductors, surface states may be viewed as broken bond orbitals.

### Electronic Phase Transitions

The electronic states at the Fermi level are a crucial factor in many observed properties of a solid. States within a few $kT$ of the Fermi level determine the transport properties, such as electrical conductance, where $k$ is the Boltzmann constant and $T$ the temperature in kelvin. They also drive electronic phase transitions, such as superconductivity, magnetism, and charge density waves. Typically, these phase transitions open a gap at the Fermi level of a few multiples of $kT_C$, where $T_C$ is the transition temperature. Occupied states near the Fermi level move down in energy by half the gap, which lowers the total energy. In recent years, the resolution of UPS experiments has reached a level where sub-$kT$ measurements have become nearly routine. The best-known examples are measurements of the gap in high-temperature superconductors (Olson et al., 1989; Shen et al., 1993; Ding et al., 1996). Compared to other techniques that probe the superconducting gap, such as infrared absorption and tunneling, angle-resolved photoemission provides its **k** dependence.

### Atoms and Molecules

Atoms and molecules in the gas phase exhibit discrete energy levels that correspond to their orbitals. There is an additional fine structure due to transitions between different vibrational and rotational states in the lower and upper states of the photoemission process. This fine structure disappears when molecules are adsorbed at surfaces. Nevertheless, the envelope function of the molecular orbitals frequently survives and facilitates the fingerprinting of adsorbed species. A typical molecular photoelectron spectrum is shown in Figure 8 (Turner et al., 1970), a typical spectrum of an adsorbed molecular fragment in Figure 9 (Sutherland et al., 1997).

Ultraviolet photoelectron spectroscopy has been used to study surface reactions in heterogeneous catalysis and

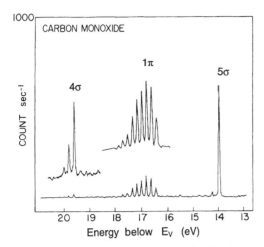

**Figure 8.** UPS spectrum of the free CO molecule, showing several molecular orbitals and their vibrational fine structure. Note that the reference level is the vacuum level $E_V$ for atoms and molecules, whereas it is the Fermi level $E_F$ for solids (Turner et al., 1970).

in semiconductor surface processing. For example, the methyl and ethyl groups adsorbed on a silicon surface in Figure 9 exhibit the same C 2s orbital splittings as in the gas phase (compare Pireaux et al., 1986), allowing a clear identification of the fragments that remain on the surface after reaction with dimethylsilane and diethylsilane. Gen-

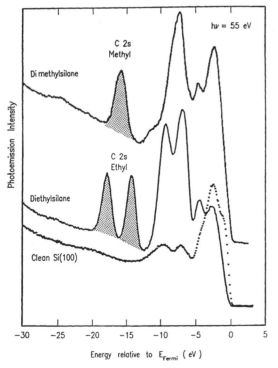

**Figure 9.** Fingerprinting of methyl and ethyl groups deposited on a Si(100) surface via reaction with dimethylsilane and diethylsilane. The lowest orbitals are due to the C 2s electrons. The number of carbon atoms in the chain determines the number of C 2s orbitals (Sutherland et al., 1997).

erally, one distinguishes the weak adsorption on inert surfaces at low temperatures (physisorption regime) and strong adsorption on reactive surfaces at room temperature (chemisorption). During physisorption, molecular orbitals are shifted rigidly toward higher energies by 1 to 2 eV due to dielectric screening of the valence holes created in the photoemission process. Essentially, surrounding electrons in the substrate lower their energy by moving toward the valence hole on the adsorbate molecule, generating extra energy that is being transferred to the emitted photoelectron. In the chemisorption regime, certain molecular orbitals react with the substrate and experience an additional chemical shift. A well-studied example is the adsorption of CO on transition metals, where the chemically active 5σ orbital is shifted relative to the more inert 4σ and 1π orbitals. The orientation of the adsorbed molecules can be determined from the angle and polarization dependence of the UPS spectra using dipole selection rules. If the CO molecule is adsorbed with its axis perpendicular to the surface and photoelectrons are detected along that direction, the σ orbitals are seen with the electric field vector **E** of the light along the axis and the π orbitals with **E** perpendicular to it. Similar selection rules apply to mirror planes and make it possible to distinguish even and odd wave functions.

## PRACTICAL ASPECTS OF THE METHOD

### Light Sources

To be able to excite photoelectrons across a valence band that is typically 10 to 20 eV wide, the photon energy has to exceed the valence band width plus the work function (typically 3 to 5 eV). This is one reason conventional lasers have been of limited applicability in photoelectron spectroscopy, except for two-photon pump-probe experiments with short pulses (Haight, 1995). The most common source of UV radiation is based on a capillary glow discharge using the He(I) line. It provides monochromatic radiation at 21.2 eV with a line width as narrow as 1 meV (Baltzer et al., 1993; a powerful electron cyclotron resonance source is marketed by Gammadata). This radiation originates from the 2p-to-1s transition in neutral He atoms. Emission can also be produced from He ions, whose primary He(II) emission line is at an energy of 40.8 eV. Similarly, emission lines of Ne(I) at 16.8 eV and Ne(II) at 26.9 eV are being used in photoelectron spectroscopy.

During the last two decades, synchrotron radiation has emerged both as a powerful and convenient excitation source in photoelectron spectroscopy (Winick and Doniach, 1980; Koch, 1983; Winick et al., 1989, some synchrotron light sources are listed in Appendix A). Synchrotron radiation is emitted by relativistic electrons kept in a circular orbit by bending magnets. In recent years, undulators have been developed that contain 10 to 100 bends in a row. For a well-focused electron beam and a highly perfect magnetic field, photons emitted from all bends are superimposed coherently. In this case, the amplitudes add up, as opposed to the intensities, and a corresponding increase in spectral brilliance by 2 to 4 orders of magnitude is achievable. The continuous synchrotron radiation spectrum shifts its weight toward higher energies with increasing energy of the stored electrons and increasing magnetic fields. A variety of electron storage rings exists worldwide that are dedicated to synchrotron radiation. Three spectral regions can be distinguished by their scientific scope: For the photon energies of 5 to 50 eV required by UPS, storage ring energies of 0.5 to 1 GeV combined with undulators are optimum. For core-level spectroscopies at 100- to 1000-eV photon energy, a storage ring energy of 1.5 to 2 GeV is the optimum. For studies of the atomic structure of solids, even shorter wavelengths are required, which correspond to photon energies of 5 to 10 keV and storage ring energies of 6 to 8 GeV combined with an undulator. Bending magnets emit higher photon energies than undulators due to their higher magnetic field. Synchrotron radiation needs to be monochromatized for photoelectron spectroscopy. The typical range of UPS is covered best by normal-incidence monochromators, which provide photon energies up to ~30 eV with optimum intensity and up to ~50 eV with reduced intensity, while suppressing higher energy photons from higher order diffracted light and harmonics of the undulators. Resolutions of better than 1 meV are routinely achievable now. Synchrotron radiation has a number of desirable properties. The most widely utilized properties are a tunable photon energy and a high degree of polarization. Tunable photon energy is necessary for reaching the complete momentum space by varying $k^\perp$, for distinguishing surface from bulk states, and for adjusting relative cross-sections of different orbitals. Linear polarization (in the plane of the electron orbit) is critical for determining the point group symmetry of electrons in solids. Circular polarization (above and below the orbit plane) allows spin-specific transitions in magnetic systems.

Some synchrotron facilities are listed in Appendix A, and figures of merit to consider regarding light sources and detectors for UPS are listed in Appendix B.

## Electron Spectrometers

The most common type of photoelectron spectrometers have been the cylindrical mirror analyzer (CMA) for high-throughput, angle-integrated UPS and the hemispherical analyzer for high-resolution, angle-resolved UPS. Popular double-pass CMAs were manufactured by PHI. With hemispherical spectrometers, resolutions of <3 meV are achievable with photoelectrons. A recent high-resolution model is the Scienta analyzer developed in Uppsala (Martensson et al., 1994; marketed by Gammadata). As the energy resolution of spectrometers improves, the sample temperature needs to be reduced to take full advantage of that resolution for solid-state samples. For example, the thermal broadening of the Fermi edge is ~0.1 eV at room temperature. Temperatures below 10 K are required to reach the limits of today's spectrometers. With increasing energy and angular resolution, the throughput of electron spectrometers drops dramatically. This can be offset by parallel detection. Energy multidetection in hemispherical analyzers has become a widely available option. It is possible to detect an angle together with the energy in hemi-spherical and toroidal analyzers. Such a 2D scheme consists of a channel plate electron multiplier with optical readout. An example is given in Figure 5, where the band structure of Ni is mapped using a hemispherical Scienta analyzer with multidetection of the energy and the polar angle (Petrovykh et al., 1998). Other designs detect two angles simultaneously and display the angular distribution of monochromatized photoelectrons directly on a channel plate using an ellipsoidal electron mirror combined with a spherical retarding grid (Eastman et al., 1980). Photoemission intensity distributions over an 85° emission cone can be acquired in a few seconds at an undulator synchrotron light source.

To detect the spin of photoelectrons, one typically measures the right/left asymmetry in the electron-scattering cross-section at heavy nuclei (Mott scattering) after accelerating photoelectrons to 20 to 100 keV (Burnett et al., 1994). Less bulky detectors use scattering at the electron cloud surrounding a heavy nucleus, typically at 100- to 200-eV electron energy. This can be achieved by random scattering or by diffraction at a single crystal. Such detectors lack the absolute calibration of a Mott detector. Generally, spin detection costs 3 to 4 orders of magnitude in detection efficiency.

## Alignment Procedures

The electron spectrometer needs to be aligned with respect to the light source (such as a synchrotron beam) and the sample with respect to the electron spectrometer. These seemingly straightforward steps can take a substantial amount of time and may lead to artifacts by inexperienced users, such as a strongly energy- and angle-dependent transmission of the spectrometer. Residual electric and magnetic fields in the sample region make the alignment more difficult and prevent low-energy electrons of a few electron volts from being detected. Electrostatic deflection plates in the spectrometer lens speed up the alignment.

First, the light beam needs to intersect the spectrometer axis at the focal point of the spectrometer. A course alignment can be achieved by using a laser beam along the axis of the spectrometer and zero-order light at a synchrotron. Hemispherical spectrometers often have small holes in the hemispheres for that purpose. The two beams can be brought to intersect at a target sample. If they do not cross each other, there will not be any sample position for a reasonable spectrum. To move the intersection point to the focal point, it is useful to have mechanical alignment markers, such as wire cross-hairs for sighting the proper distance from the front lens of the spectrometer.

Second, the sample needs to be placed at the intersection of the light beam and the analyzer axis. Again, the laser and the light beam can be utilized for this purpose. If they are not available (e.g., when bringing in a new sample into a prealigned spectrometer), a systematic optimization of the photoemission intensity can be used: First, set the sample at the focal distance from the analyzer, as determined by the cross-hairs. Then translate the sample in the two directions orthogonal to the analyzer axis and determine the cutoff points for the photoemission intensity at opposite edges of the sample and place the sample half way in between.

These alignment procedures provide a first approximation, but residual fields and an uncertain position of the ultraviolet photon beam on the sample require a symmetry check for acquiring accurate angle-resolved UPS data. Mirror symmetry with respect to crystallographic planes provides the exact normal emission geometry and the azimuthal orientation of the sample.

### Useful Relations

Some relations occur frequently when processing UPS data. In the following we list the most common ones:

The energy $h\nu$ of a photon is given (in electron volts) by its wavelength $\lambda$ (in angstroms) via

$$\lambda = \frac{12399}{h\nu} \tag{4}$$

The wave vector $k = |\mathbf{k}|$ of an electron is given (in reciprocal angstroms) by its kinetic energy $E_{\text{kin}}$ (in electron volts) via

$$k = 0.51\sqrt{E_{\text{kin}}} \tag{5}$$

The inverse of this relation is

$$E_{\text{kin}} = 3.81 \times k^2 \tag{6}$$

For determining $k^\perp$, the upper energy bands are often approximated by the "empty lattice" bands, which are given by

$$E = V + 3.81 \times (k \pm |g|)^2 \tag{7}$$

where $k$ is in reciprocal angstroms and $E$ in electron volts, $V$ is the inner potential (typically $-10$ eV relative to the vacuum level or $-5$ eV relative to the Fermi level), and $\mathbf{g}$ (also in reciprocal angstroms) is a reciprocal lattice vector determined by the spacing $d_{\text{plane}}$ of equivalent atomic planes:

$$|\mathbf{g}| = \frac{2\pi}{d_{\text{plane}}} \quad \mathbf{g} \perp \text{planes} \tag{8}$$

Likewise, a reciprocal surface lattice vector $\mathbf{g}^{\parallel}$ is determined by the spacing $d_{\text{row}}$ of equeivalent atomic rows:

$$|\mathbf{g}| = \frac{2\pi}{d_{\text{row}}} \quad \mathbf{g} \perp \text{rows} \tag{9}$$

The empty lattice approximation used in Figure 10 and Equation 7 improves at higher photon energies ($>30$ eV), where the kinetic energy of the photoelectrons becomes large compared to the Fourier components of the lattice

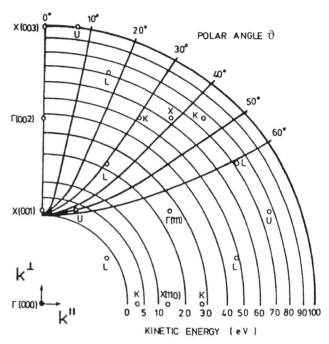

**Figure 10.** Location of direct transitions in $\mathbf{k}$ space, shown for a Ni(001) surface along the [110] emission azimuth. The circular lines represent the location of possible upper states at a given kinetic energy using empty lattice bands. The lines fanning out toward the top represent different emission angles from the [001] sample normal (Himpsel, 1983).

potential. Various "absolute" methods are available to refine the $\mathbf{k}$ values. These methods determine the $k^\perp$ component by triangulation from different crystal faces or by mirror symmetry properties of high-symmetry lines (see Plummer and Eberhardt, 1982; Himpsel, 1983; Kevan, 1992).

The thermal broadening of the Fermi function that cuts off photoelectron spectra at high energy is proportional to $kT$. Taking the full width at half-maximum of the derivative of the Fermi function, one obtains a width $3.5kT$, which is 0.09 eV at room temperature and 1 meV at liquid He temperature. The half-height points of the derivative correspond to the 15% and 85% levels of the Fermi function.

### Sensitivity Limits

A typical measurement volume in UPS is an area 1 mm$^2$ times a depth of 1 nm. In photoelectron microscopes, it is possible to reduce the sampling area to dimensions of $<0.1$ μm. Several types of such microscopes have been tested in recent years. Imaging devices accelerate the photoelectrons to $\sim20$ keV and use regular electron microscope optics to form an image. In scanning microscopes, the sample is scanned across a finely focused ultraviolet light and soft x-ray source. In both cases, the photoelectron energy analysis has been rather rudimentary so far. Most contrast mechanisms utilized to date involved core levels or work function differences, and not fine structure in the valence band.

The sensitivity to dilute species is rather limited in UPS due to the substantial background of secondary electrons (see Fig. 1). A surface atom density of $10^{14}$ cm$^{-2}$ (about a tenth of a monolayer) is relatively easy to detect, a concentration of $10^{13}$ cm$^{-2}$ is the limit. A much higher signal-to-background ratio can be achieved by core-level emission spectroscopy (see Fig. 6C).

The time resolution can be as short as tens of femtoseconds in pump-probe two-photon photoemission experiments using fast lasers (Haight, 1995).

## METHOD AUTOMATION

While the sample preparation is highly individual in UPS, the actual data acquisition can be automated easily. A few voltages are sufficient to control a photoelectron spectrometer, such as the kinetic energy and the pass energy. The photon energy is controlled by a monochromator at a synchrotron light source, which in most cases is computer controlled. Labview has become a widely used control system. The angular parameters can be set by stepping motors that drive goniometers for the sample and the spectrometer, with a maximum of five axes, three for the sample and two for the spectrometer. Typically, the acquisition of a series of spectra is preprogrammed. They start out with a wide kinetic energy range at high photon energy, which gives an overview of all valence states, plus some shallow core levels for elemental characterization. Subsequent spectra zero in on specific states, such as adsorbate orbitals or Fermi-level crossings. For each of them, one wants to step through a variety of emission angles and photon energies to vary the parallel and perpendicular momenta, respectively.

## DATA ANALYSIS AND INITIAL INTERPRETATION

The first look at UPS spectra from a new sample is typically reserved for making sure that the surface is not contaminated. For example, oxygen contamination generates a characteristic O $2p$ peak at $\sim 6$ eV below the Fermi level in metals. Then one might want to compare the spectrum to a set of published reference spectra from related materials or molecules. This "fingerprinting" often gives a quick overview of the energies of specific orbitals, such as $d$ states in transition metals, $f$ states in rare earths, and adsorbate orbitals. When in doubt, the photon energy dependence of the cross-section can be utilized to enhance certain orbitals by resonant photoemission at core-level thresholds (e.g., $f$ levels) or to quench them at Cooper minima (e.g., orbitals with radial nodes).

The next step is often a determination of the Fermi-level position from the high-energy cutoff of a metallic sample, such as Au or Pt. A fit of the Fermi-Dirac function convoluted by a Gaussian resolution function over the uppermost few tenths of an electron volt gives an accurate Fermi-level position.

After characterizing the sample by angle-integrated UPS, the full $E(\mathbf{k})$ band dispersion is obtained from the variation with angle and photon energy. Usually, the starting point is normal emission at various photon energies, which provides the $E(k^\perp)$ relation along a high-symmetry line. This geometry features the strictest selection rules and, therefore, is the easiest to interpret. An example is given in Figure 4, where a series of normal emission spectra at various photon energies (on the right side) is converted into a set of energy bands (on the left side). The $k$ values of the data points on the left are obtained from the peak positions on the right (dashed lines) after adding the photon energies and converting the resulting upper state energies $E_u$ into $k^\perp$ using the calculated upper band on the left (cf. Equation 1). Alternatively, one could use the somewhat cruder empty lattice bands from Equation 7.

By switching the polarization of the electric field vector of the photons between parallel and perpendicular to the surface (the latter asymptotically), different orbital orientations can be selected, such as $p_{x,y}$ and $p_z$, respectively. In normal emission, for example, the $p_{x,y}$ bands are excited by the component of the electric field vector in the $x,y$ directions (parallel to the sample surface), and the $p_z$ states are excited by the $z$ component normal to the surface.

Having covered normal emission, it is useful to vary the polar angle within a mirror plane perpendicular to the surface, where the wave functions still exhibit even/odd symmetry properties. Here, orbitals that are even with respect to the mirror plane are excited by the in-plane electric field component, and odd orbitals are excited by the component perpendicular to the mirror plane.

Off-normal photoemission peaks can be assigned to transitions at specific $\mathbf{k}$ points by using Equations 5 to 7, as in Figure 10. Taking the data in Figure 5 as example, let us follow a typical processing sequence. The data in Figure 5 are taken from a Ni(001) crystal in the (110) emission plane using a photon energy $h\nu = 44$ eV. In the raw data, the Fermi-level crossings occur at a voltage $V_{spectr} \approx 40$ V on the electron spectrometer and at a polar angle $\vartheta \approx 18°$ from the sample normal. The spectrometer voltage yields a kinetic energy $E_{kin} \approx 40$ eV when discounting work function differences between spectrometer and sample (see below). From $E_{kin}$ and $\vartheta$, one obtains a parallel wave vector $k^\parallel = k \sin\vartheta \approx 1$ Å$^{-1}$ via Equation 5. Going with $E_{kin}$ and $k^\parallel$ into Figure 10, one finds the location of the transition in $\mathbf{k}$ space as the intersection of the circle for $E_{kin} \approx 40$ eV and the vertical line for $k^\parallel = 1$ Å$^{-1}$. The $k$ scale in Figure 10 is in units of $2\pi/a = 1.8$ Å$^{-1}$ ($a$ = cubic lattice constant of Ni). Consequently, the transition happens about half way on a horizontal line between $\Gamma(002)$ and $K$, where a line for $\vartheta = 18°$ would intersect the circle for $E_{kin} = 40$ eV.

It should be noted that the energy scale provided by the voltage reading of an electron spectrometer is referred to the vacuum level of the spectrometer and not to that of the sample. A contact potential develops spontaneously between the sample and the spectrometer that compensates for the difference in work functions and accelerates or retards the photoelectrons slightly toward the analyzer. Such a contact potential needs to be compensated for accurate angular measurements of slow photoelectrons (a few electron volts). A typical spectrometer work function is 4 eV, so samples with smaller work functions require negative bias and samples with larger work function positive bias. The optimum bias is determined such that the vacuum-level cutoff coincides with a zero voltage reading on the spectrometer. A negative sample bias of a few volts is often used for moving the spectrum up in energy when determining the sample work function from the difference between the photon energy and the width of the photoelectron spectrum (Fig. 2).

With two-dimensional detectors, such as the $E$, $\vartheta$ detection of the hemispherical analyzer (Martensson et al., 1994; Petrovykh et al., 1998; Fig. 5) and the $\vartheta$, $\varphi$ detection of the ellipsoidal mirror analyzer (Eastman et al., 1980), one needs to divide by the two-dimensional transmission function of the spectrometer to obtain quantitative data. An approximate transmission function can be achieved by increasing the photon energy at identical spectrometer settings, such that secondary electrons are detected whose angular distribution is weak. Careful alignment of sample and light source with respect to the spectrometer and thorough shielding from magnetic and electric fields minimize transmission corrections. After obtaining the 2D intensity distributions $I(E, \vartheta)$ and $I(\vartheta, \varphi)$ and converting $\vartheta$, $\varphi$ into $\mathbf{k}^{\parallel}$, one is able to generate $I(\mathbf{k}^{\parallel})$ distributions. These are particularly useful at the Fermi level, where the features are sharpest and the the electronic states relevant for electronic transport, magnetism, and superconductivity are located.

## SAMPLE PREPARATION

Well-characterized surfaces are essential for reliable UPS data. In general, data from single crystals (either in bulk form or as epitaxial films) are more reliable than from polycrystalline materials. Single-crystal surfaces can be cleaned of adsorbed molecules and segregated bulk impurities more thoroughly and can be characterized by diffraction methods, such as low-energy electron diffraction (LEED) and reflection high-energy electron diffraction (RHEED). Surface impurities are monitored by Auger electron spectroscopy or by various core-level spectroscopies, including photoemission itself (see Fig. 6). A wide variety of methods have been developed for preparing single-crystal surfaces, with treatments specific to individual materials and even specific crystal faces. Many of them are compiled by Musket et al. (1982; see also SAMPLE PREPARATION FOR METALLOGRAPHY).

Metals are typically sputter annealed, whereby impurities are removed from the surface by $Ar^+$ ion bombardment (typically 1 keV at a few microamperes of current and an Ar pressure of $5 \times 10^{-5}$ Torr). Bulk impurities continue to segregate to the surface and are sputtered away one monolayer at a time, a process that can take months to complete. Therefore, many chemical methods have been developed to deplete the bulk more quickly of impurities. For example, the typical impurities C, N, O, and S can be removed from a catalytically active surface, such as Fe, by heating in 1 atm of hydrogen, which forms volatile $CH_4$, $NH_3$, $H_2O$, and $H_2S$. Careful electropolishing can also go a long way in removing damage from mechanical polishing and producing a highly perfect surface after just a few sputter

anneals, particularly for the soft noble metals. Highly reactive metals such as rare earths, actinides, and early transition metals are prepared in purest form as epitaxial films on a W(110) surface, where they can also be removed by simple heating to make room for a new film.

Semiconductors can be cleaved or heated above the desorption point of the native oxide. Only the crystal orientations with the lowest surface energy cleave well, such as Ge(111), Si(111), and GaAs(110). On silicon, the native oxide film of $\sim$10 Å thickness desorbs at $\sim$1050°C. Dipping into ultrapure 10% HF solution removes the oxide and leaves a mostly H-terminated surface that can be cleaned at 850°C and below. If carbon is present at the surface, it forms SiC particles upon heating that require a flash to 1250°C for diffusing carbon away from the surface into the bulk.

Insulator surfaces are most perfect if they can be obtained by cleavage, which produces the electrically neutral surfaces, e.g., (100) for the NaCl structure and (111) for the $CaF_2$ structure. Sputter annealing tends to deplete ionic insulators of the negative ion species.

Layered compounds such as graphite, $TaS_2$, or certain high-temperature superconductors are easiest to prepare. A tab is glued to the front surface with conductive, vacuum-compatible epoxy and knocked off in vacuum. A Scotch tape will do the same job but needs to be kept out of the ultrahigh-vacuum chamber.

If only polycrystalline samples are available, one sometimes needs to resort to cruder cleaning methods, such as filing in vacuum with a diamond file. It is not clear, however, how much of the surface exposed by filing consists of grain boundaries with segregated contamination.

Sample mounting needs to be matched to the preparation method. Samples can be clamped down by spring clips or spotwelded to wires. The material for the springs or wires should not diffuse along the sample surface. Molybdenum is a widely useful spring material that remains elastic even at high temperatures. Tantalum spotwelds very well and diffuses little. Heating can be achieved ohmically or by electron bombardment. The former is better controllable; the latter makes it easier to reach high temperatures. It is particularly difficult to mount samples that require high-temperature cleaning (such as W) but have to be cooled and insulated electrically as well. In such a case, the sample is attached to a cryostat via a sapphire plate that insulates electrically but conducts the heat at low temperature. Most of the temperature drop during heating occurs at W heating wires that attach the sample to a copper plate on top of the sapphire. For high-resolution measurements, it is helpful to minimize potential fluctuations around the sample by painting the area surrounding the sample with graphite using an Aquadag spray. This method is also used to coat the surfaces of electron spectrometers seen by the electrons. Insulating samples need special mounting to avoid charging. Often, the sample potential stabilizes itself after a few seconds of irradiation and leads to a shifted spectrum with reduced kinetic energies. Mounting an electron flood gun near the sample can provide zero-kinetic-energy electrons that compensate the charging. Alternatively, mild heating of the sample may create thermal carriers to provide sufficient conductivity.

Molecules for adsorption experiments are admitted through leak valves that allow gas dosing with submonolayer control. A dosage of 1 s at $10^{-6}$ Torr (= 1 Langmuir = 1 L) corresponds to every surface atom being exposed to about one gas molecule. The probability of an arriving molecule to stick to the surface (the sticking coefficient) ranges from unity down to $10^{-3}$ in typical cases.

## PROBLEMS

Potential problems can arise mainly due to insufficient sample preparation and mounting. Ultraviolet photoelectron spectroscopy is a surface-sensitive technique that probes only a few atomic layers and thus detects surface contamination and structural defects with uncanny clarity. A thorough literature search is advisable on preparation methods for the particular material used. Insulating samples need charging compensation. Certain semiconductor surfaces produce a significant photovoltage at low temperature, which leads to an energy shift. The determination of the emission angle from purely geometrical sample alignment often is not accurate due to deflection of the photoelectrons by residual fields. A symmetry check is always advisable.

On the instrumentation side, there are a variety of lesser pitfalls. If the sample and spectrometer are not aligned properly to the light beam, the spectra can be highly distorted by energy- and angle-dependent transmission. A typical example is the loss of intensity for slow electrons near the vacuum level. It is important to have a clearly thoughtout alignment procedure, such as the one described above. Residual magnetic and electric fields have the same effect, e.g., due to insufficient magnetic shielding and charging insulators in the sample region, respectively. Another artifact is the flattening of intense photoemission peaks due to a saturation of the electron detector. The channeltron or channelplate multipliers used for electron detection have a rather clearly defined saturation point where the voltage drop induced by the signal current reduces the applied high voltage and thus reduces the gain. Low-resistance "hot" channel plates have higher saturation count rates, which can be as large as $10^7$ counts/s for large plates. The high-resolution capability of hemispherical analyzers is easily degraded by incorrect electrical connections, such as ground loops or capacitive pickup. It may be necessary to use a separation transformer for the electronics.

Finally, the photoemission process itself is not always straightforward to interpret. Many different types of wave functions involved are in UPS (see Fig. 7). Some of them conserve all $\mathbf{k}$ components (direct transitions between bulk states), others conserve $\mathbf{k}^{\parallel}$ only (evanescent states), and phonon-assisted transitions do not conserve $\mathbf{k}$ at all. That leaves quite a few options for the assignment of spectral features that have led to long-standing controversies in the past, such as direct versus indirect transitions. Many years of experience with well-defined single-crystal surfaces and support from calculations of photoemission spectra have made it possible to sort out the various processes, as long as a full set of data at various photon energies, angles, and surface orientations is available. Such complexity is inherent in a process that is capable of determining all the details of an electron wave function.

## ACKNOWLEDGMENTS

This work was supported in part by the National Science Foundation under Award No. DMR-9531009 (the Synchrotron Radiation Center at the University of Wisconsin at Madison) and by the Department of Energy under Contract No. DE-AC03-76SF00098 (the Advanced Light Source at the Lawrence Berkeley National Laboratory).

## LITERATURE CITED

Baltzer, P., Karlsson, L., Lundqvist, M., and Wannberg, B. 1993. *Rev. Sci. Instrum.* 62:2174.

Burnett, G. C., Monroe, T. J., and Dunning, F. B. 1994. *Rev. Sci. Instrum.* 65:1893.

Cardona, M. and Ley, L. 1978. Photoemission in Solids I, General Principles, Springer Topics in Applied Physics, Vol. 26. Springer-Verlag, Berlin.

Chelikowski, J. R. and Cohen, M. L. 1976. *Phys. Rev. B* 14:556.

Ding, H., Yokoya, T., Campuzano, J. C., Takahashi, T., Randeria, M., Norman, M. R., Mochiku, T., Hadowaki, K., and Giapintzakis, J. 1996. *Nature* 382:51.

Dose, V. 1985. *Surf. Sci. Rep.* 5:337.

Eastman, D. E., Donelon, J. J., Hien, N. C., and Himpsel, F. J. 1980. *Nucl. Instrum. Methods* 172:327.

Grobman, W. D., Eastman, D. E., and Freeouf, J. L. 1975. *Phys. Rev. B* 12:4405.

Haight, R. 1995. *Surf. Sci. Rep.* 21:275.

Himpsel, F. J. 1983. *Adv. Phys.* 32:1–51.

Himpsel, F. J. 1986. *Comments Cond. Mater. Phys.* 12:199.

Himpsel, F. J. and Lindau, I. 1995. Photoemission and photoelectron spectra. *In* Encyclopedia of Applied Physics, Vol. 13 (G. L. Trigg and E. H. Immergut, eds.) pp. 477–495. VCH Publishers, New York.

Himpsel, F. J., Ortega, J. E., Mankey, G. J., and Willis, R.F. 1998. *Adv. Phys.* 47:511–597.

Hybertsen, M. S. and Louie, S. G. 1986.. *Phys. Rev. B* 34:5390.

Kevan, S. D. (ed.) 1992. Angle-Resolved Photoemission. Elsevier, Amsterdam, The Netherlands.

Knapp, J. A., Himpsel, F. J., and Eastman, D. E. 1979. *Phys. Rev. B* 19:4952.

Koch, E. E. (ed.). 1983. Handbook on Synchrotron Radiation. North-Holland Publishing, Amsterdam, The Netherlands.

Landolt-Börnstein, 1989, 1994. Electronic structure of solids: Photoemission spectra and related data. *In* Numerical Data and Functional Relationships in Science and Technology, New Series, Group III, Vol. 23 a,b (A. Goldmann and E.-E. Koch, eds.). Springer-Verlag, Berlin.

Martensson, N., Baltzer, P., Brühwiler, P. A., Forsell, J.-O., Nilsson, A., Stenborg, A., and Wannberg, B. 1994. *J. Electron. Spectrosc.* 70:117.

Moruzzi, V. L., Janak, J. F., and Williams, A. R. 1978. Calculated Electronic Properties of Metals. Pergamon, New York.

Musket, R. G., McLean, W., Colmenares, C. A., Makowiecki, S. M., and Siekhaus, W. J. 1982. *Appl. Surf. Sci.* 10:143.

Olson, C. G., Liu, R., Yang, A.-B., Lynch, D. W., Arko, A. J., List, R. S., Veal, B. W., Chang, Y. C., Jiang, P. Z., and Paulikas, A. P. 1989. *Science* 245:731.

Ortega, J. E. and Himpsel, F. J. 1993. *Phys. Rev. B* 47:2130.

Papaconstantopoulos, D. A. 1986. Handbook of the Band Structure of Elemental Solids. Plenum, New York.

Petrovykh, D. Y., Altmann, K. N., Höchst, H., Laubscher, M., Maat, S., Mankey, G. J., and Himpsel, F. J. 1998. *Appl. Phys. Lett.* 73:3459.

Phillip, H. R. and Ehrenreich, H. 1963. *Phys. Rev.* 129:1550.

Pireaux, J. J., Riga, J., Thiry, P. A., Caudano, R., and Verbist, J. J. 1986. *Phys. Scr. T* 13:78.

Plummer, E. W. and Eberhardt, W. 1982. *Adv. Chem. Phys.* 49:533.

Shen, Z.-X., Dessau, D. S., Wells., B. O., King, D. M., Spicer, W. E., Arko, A. J., Marshall, D., Lombardo, L. W., Kapitulnik, A., Dickinson, P., Doniach, S., DiCarlo, J., Loeser, A. G., and Park, C. H. 1993. *Phys. Rev. Lett.* 70:1553.

Smith, N. V. 1988. *Rep. Prog. Phys.* 51:1227.

Smith, N. V. and Himpsel, F. J. 1983. Photoelectron spectroscopy. *In* Handbook on Synchrotron Radiation, Vol. 1 (E.E. Koch, ed.) pp. 905–954. North-Holland Publishing, Amsterdam, The Netherlands.

Smith, N. V., Lässer, R., and Chiang, S. 1982. *Phys. Rev. B* 25:793.

Sutherland, D. G. J., Himpsel, F. J., Terminello, L. J., Baines, K. M., Carlisle, J. A., Jimenenz, I., Shuh, D. K., and Tong, W. M. 1997. *J. Appl. Phys.* 82:3567.

Turner, D. W., Baker, C., Baker, A. D., and Brundle, C. R. 1970. Molecular Photoelectron Spectroscopy. Wiley, New York.

Wachs, A. L., Miller, T., Hsieh, T. C., Shapiro, A. P., and Chiang, T. C. 1985. *Phys. Rev. B* 32:2326.

Winick, H. and Doniach, S. (eds.). 1980. Synchrotron Radiation Research. Plenum, New York.

Winick, H., Xian, D., Ye, M.-H., and Huang, T. (eds.). 1989. Applications of Synchrotron Radiation. Gordon and Breach, New York.

Yeh, J. J. and Lindau, I. 1985. *Atomic Data Nucl. Data Tables* 32:1.

## KEY REFERENCES

Cardona and Ley, 1978. See above.

*An overview of UPS, provides chapters on selected areas of research.*

Gaarenstroom, S. W. and Hecht, M.H. (eds.) *Surface Science Spectra.* American Vacuum Society, Research Triangle Park, N.C.

*A journal that publishes UPS spectra of specific materials, including documentation about the measurement.*

Himpsel, 1983. See above.

*Presents principles of angle-resolved photoemission with emphasis on band structure determination.*

Kevan 1992. See above.

*Comprehensive book containing articles on various topics in angle-resolved photoemission.*

Landolt-Börnstein 1989, 1994. See above.

*Compilations of UPS data about the electronic structure of solids, combining photoemission spectra with optical data and band structure calculations.*

Plummer and Eberhardt, 1982. See above.

*Principles of angle-resolved photoemission with emphasis on adsorbate states.*

Smith and Himpsel, 1983. See above.

*Brief introduction to UPS and its applications*

## APPENDIX A: SYNCHROTRON LIGHT SOURCES

A complete list of the dozens of synchrotron radiation sources worldwide would go beyond the scope of this unit. The following list gives typical machines that are optimized for either valence states (10- to 100-eV photon energy, "VUV") or shallow core levels (100- to 1000-eV photon energy, "Soft X-ray"):

United States: ALS, Berkeley, CA (Soft X-ray); NSLS UV ring, Brookhaven, NY (VUV); SRC, Madison, WI (VUV).

Europe: BESSY, Berlin, Germany (VUV, Soft X-ray); ELETTRA, Trieste, Italy (Soft X-ray); MAX-Lab, Lund, Sweden (VUV, Soft X-ray).

Asia: Photon Factory, Tsukuba, Japan (Soft X-ray); SRRC, Hsinchu, Taiwan (Soft X-ray).

### Some Companies that Carry UPS Equipment

In general, companies that specialize in equipment for ultra-high-vacuum and surface science carry UPS equipment as well, such as He lamps and spectrometers. Companies that provide equipment at the high end of the line are (1) Gammadata (formerly Scienta), PO Box 15120, S-750 15 Uppsala, Sweden, http://www.gammadata.se (also has the unique ECR He lamp) and (2) Omicron (bought out VSW, a previous supplier of hemispherical analyzers), Idsteiner Str. 78, D-65232 Taunusstein, Germany, http:// www.omicron.de. Companies that have equipment in the middle range are (3) PHI (in the United States, used to sell CMAs, also selling hemispherical analyzers) and (4) Staib (in Germany, sells a CMA).

The problem with naming particular companies in this field is the volatile nature of this business. Companies are bought out and spun off frequently. Up-to-date information can usually be found at trade shows associated with conferences such as the AVS (American Vacuum Society) fall conference and the APS (American Physical Society) March meeting.

## APPENDIX B: FIGURE OF MERIT FOR LIGHT SOURCES AND DETECTORS

Liouville's theorem of the conservation of the volume in phase space, $(\delta x\, \delta y\, \delta z)(\delta k_x\, \delta k_y\, \delta k_z)$, dictates the ultimate limitations of light sources and electron spectrometers. Applying this theorem to electron and photon beams, one realizes that reducing the cross-section of a beam causes its angular divergence to increase (at fixed energy). Likewise, when electrons are retarded to a low-pass energy in a spectrometer to obtain better energy resolution, the effective source size goes up (with a hemispherical retarding grid) or the divergence increases (in a retarding lens).

Applied to light sources, one can define spectral brilliance (or brightness) as the photon flux divided by the emission solid angle, source area, and the energy band width $\delta E/E$. Typical bending magnet values are $10^{13}$ photons/(s $\times$ mrad$^2$ $\times$ mm$^2$ $\times$ 0.1%). Undulators provide $10^2$ to $10^4$ times the spectral brilliance of bending magnets. Applied to electron storage rings for synchrotron radiation, one uses the emittance (cross-section times angular spread) of the stored electron beam as a measure of beam quality, with the lowest emittance electron beam providing the highest gain in spectral brilliance at an undulator. Brilliance is important for microresolved UPS. Flux is often the relevant quantity for ordinary UPS, particularly when spin resolved, in which case only a good photon energy resolution is required. To achieve this, light needs to be squeezed only in one dimension through the entrance slit of a monochromator.

For electron spectrometers, the relevant figure of merit is étendue, i.e., the source area times the solid angle accepted by the spectrometer at a given resolving power $E/\delta E$. Normal-incidence electron optics provides a large étendue (see Eastman et al., 1980). Likewise, large physical dimensions increase the étendue (Baltzer et al., 1993).

F. J. HIMPSEL
University of Wisconsin
Madison, Wisconsin

# ELLIPSOMETRY

## INTRODUCTION

Ellipsometry, also known as reflection polarimetry or polarimetric spectroscopy, is a classical and precise method for determining the optical constants, thickness, and nature of reflecting surfaces or films formed on them. Ellipsometry derives its name from the measurement and tracking of elliptically polarized light that results from optical reflection. Typical experiments undertaken with this technique consist of measuring the changes in the state of polarization of light upon reflection from the investigated surfaces. Since this technique is very sensitive to the state of the surface, many experiments that involve such surface changes can be detected and followed. Hence ellipsometry finds applicability in a wide variety of fields, such as physical and chemical adsorption; corrosion; electrochemical or chemical formation of passive layers on metals, oxides, and polymer films; dissolution; semiconductor growth; and immunological reactions (Hayfield, 1963; Archer, 1964; Arsov, 1985; Hristova et al., 1997).

The main strength of this technique lies in its capability not only to allow *in situ* measurements and provide information about the growth kinetics of thin films but also simultaneously to allow the determination of many or all of the optical parameters necessary to quantify the system (Martens et al., 1963; Zaininger and Revesz, 1964; Schmidt, 1970). Ellipsometry can be used to detect and follow film thickness in the monoatomic range, which is at least an order of magnitude smaller than what can be studied by other optical methods (e.g., interferometry and reflection spectroscopy). The basic theory and principles of ellipsometry have been described by a number of authors. The original equations of ellipsometry given by Drude (1889) and various other authors have been reviewed by Winterbottom (1946), who has described straightforward procedures for determining the thickness and optical constants of films on reflecting surfaces. Heavens (1955) and Vasicek (1960) have also discussed the exact and approximate solution procedures for measuring optical constants.

When linearly polarized light having an arbitrary electric field direction is reflected from a surface, the resultant light is elliptically polarized due to the difference in phase shifts for the components of the electric field that are parallel and perpendicular to the plane of incidence. According to the principle of reversibility, light of this resultant ellipticity, when incident on a reflecting surface, should produce linearly polarized light. This is the principle under which the ellipsometer operates. The angle of incidence of light, its wavelength, and the ellipticity of the incident light are related by means of certain theoretical relations to the optical parameters of the substrate and any film that might exist on it. By knowing the optical properties of the substrate and by carrying out ellipsometric measurements, one can find the thickness of a film and its refractive index with relative ease based on certain well-known relations that will be described in the next section.

Ellipsometry can be broadly divided into two categories. One class of ellipsometry, called null ellipsometry, concerns itself with performing a zero signal intensity measurement of the light beam that is reflected from the sample. In the other class of ellipsometry, measurements of the ellipticity and intensity of light reflected from the sample are performed and correlated to the sample properties. This technique is referred to as photometric ellipsometry. The techniques and methodology described in this unit will concentrate mainly on the principles of null ellipsometry. However, references to the photometric techniques will be provided in those cases where they would provide an advantage over nulling methods.

Two parameters that are measured in ellipsometry are the change in relative amplitude and relative phase of two orthogonal components of light due to reflection. The relative, rather than absolute, measurement is one reason for the high resolution of ellipsometry. The other arises from the measured quantities usually being azimuth angles (rotation around an optical axis) that can easily be measured with high resolution (0.01°). The availability of high-speed computers and comprehensive software algorithms have also generated interest in ellipsometry measurements (Cahan and Spainer, 1969; Ord, 1969).

## PRINCIPLES OF THE METHOD

### Reflecting Surfaces

Let us consider a thin-film reflecting surface as shown in Figure 1. Consider that a monochromatic light beam is incident on the substrate at an angle of 45°. This incident beam can be resolved into two orthogonal components, one

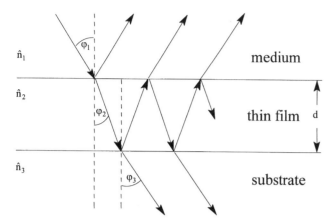

**Figure 1.** Reflection and refraction of light at a planar thin-film covered metal surface ($d$ = film thickness).

in the plane of incidence, denoted by $p$, and the other perpendicular to it, denoted by $s$. The beam obtained after the reflection from the substrate (and film) will be shifted in phase and amplitude in both the $p$ and $s$ components. If we denote the phase shift in the parallel component by $\delta^s p$ and the perpendicular component by $\delta^S$, the phase shift between the $p$ and $s$ components can then be written as

$$\Delta = \delta^p - \delta^s \tag{1}$$

The amplitudes of these two waves will also be different from their initial amplitudes, and their ratio can be written as

$$\tan\psi = \frac{E_r^p}{E_r^s} \tag{2}$$

It is these two parameters, the relative phase difference ($\Delta$) and the ratio of the relative amplitudes ($\tan\psi$) between the parallel and perpendicular components of an electromagnetic wave that is reflected from a film-covered surface, that one obtains from ellipsometric experiments. Both the phase and amplitude shifts occur at each boundary between regions of different optical properties. The final amplitude of the wave that is observed by an analyzer is a sum from multiple reflecting layers. Hence, the reflected wave will depend on the properties of both the thin film and the substrate.

The sum of the difference in amplitudes between the parallel and perpendicular components of a multiply reflected beam is obtained using the well-known Fresnel reflection coefficients (Azzam and Bashara, 1977).

For Figure 1, the Fresnel coefficients can be calculated as follows. From the known value of the incidence angle $\varphi_1$ at the boundary between air (or a suitable medium in which the sample is immersed during the experiment) and the film, the cosine of refraction angle $\varphi_2$ can be calculated by Snell's law (Menzel, 1960):

$$\cos\varphi_2 = \left[1 - \left(\frac{\hat{n}_1}{\hat{n}_2}\sin\varphi_1\right)^2\right]^{0.5}, \tag{3}$$

where $\hat{n}_1$ and $\hat{n}_2$ are the complex refractive indices of media 1 and 2, respectively (Brown, 1965). One can then determine a complex reflection coefficient at each interface, 1-2 and 2-3, as the ratio of the magnitude of the electric vectors in a single orientation (either parallel or perpendicular) before and after reflection. It can be seen that there will be two reflection coefficients at each interface, one for the parallel and another for the perpendicular component of the reflected and incident light. These two components for the reflection between interfaces 1 and 2 are

$$r_{1,2,p} = \frac{E_r^p}{E_1^p} = \frac{\hat{n}_2\cos\varphi_1 - \hat{n}_1\cos\varphi_2}{\hat{n}_2\cos\varphi_1 + \hat{n}_1\cos\varphi_2} \tag{4}$$

$$r_{1,2,s} = \frac{E_r^s}{E_1^s} = \frac{\hat{n}_1\cos\varphi_1 - \hat{n}_1\cos\varphi_2}{\hat{n}_1\cos\varphi_1 + \hat{n}_1\cos\varphi_2} \tag{5}$$

In a similar manner, the refraction angle at the boundary between the film and the substrate can be calculated as

$$\cos\varphi_3 = \left[1 - \left(\frac{\hat{n}_1}{\hat{n}_3}\sin\varphi_1\right)^2\right]^{0.5} \tag{6}$$

and

$$r_{2,3,p} = \frac{\hat{n}_3\cos\varphi_2 - \hat{n}_2\cos\varphi_3}{\hat{n}_3\cos\varphi_2 + \hat{n}_2\cos\varphi_3} \tag{7}$$

$$r_{2,3,s} = \frac{\hat{n}_2\cos\varphi_2 - \hat{n}_3\cos\varphi_3}{\hat{n}_2\cos\varphi_2 + \hat{n}_3\cos\varphi_3} \tag{8}$$

The four reflection coefficients given above are called Fresnel's reflection coefficients, after Augustin Fresnel. These reflection coefficients can be combined to give two total reflection coefficients, one for the parallel and another for the perpendicular component:

$$r_p = \frac{r_{1,2,p} + r_{2,3,p}\exp(-2i\delta)}{1 + r_{1,2p}r_{2,3,p}\exp(-2i\delta)} \tag{9}$$

and

$$r_s = \frac{r_{1,2,s} + r_{2,3,s}\exp(-2i\delta)}{1 + r_{1,2s}r_{2,3,s}\exp(-2i\delta)} \tag{10}$$

where

$$\delta = 2\pi\hat{n}_2\frac{d_2}{\lambda}\cos\varphi_2 \tag{11}$$

These total reflection coefficients $r_p$ and $r_s$ also include the contributions of partial reflections from the immersion medium/film and film/substrate. The ratio of these two reflection coefficients is denoted as $\rho$ and is also related to the relative phase shift parameter $\Delta$ and the relative amplitude attenuation parameter $\psi$ by the following fundamental equation in ellipsometry:

$$\rho = \frac{r_p}{r_s} = \tan\psi\exp(i\Delta) \tag{12}$$

Hence, combining the above two equations, one can relate the parameters $\psi$ and $\delta$ into the optical parameters of the surface and the substrate as

$$\tan\psi\exp(i\Delta) = \frac{r_{1,2,\mathrm{p}} + r_{2,3,\mathrm{p}}\exp(-2i\delta)}{1 + r_{1,2,\mathrm{p}}r_{2,3,\mathrm{p}}\exp(2i\delta)}\frac{1 + r_{1,2,\mathrm{s}}r_{2,3,\mathrm{s}}\exp(-2i\delta)}{r_{1,2,\mathrm{s}} + r_{2,3,\mathrm{s}}\exp(2i\delta)}$$

(13)

The values of $\Delta$ and $\psi$ are determined from the ellipsometric experiments and can hence be used to determine the ratio of the reflection coefficients. Subsequently, they can be used through Equation 13 to determine the optical properties of the surface or film of interest.

The following paragraphs provide some equations that can be used to determine various physical parameters that are of interest in a sample, such as reflectivity, refractive index, dielectric constants, and optical conductivity. The derivations for these properties are not presented since it is beyond the scope of this unit.

### Reflectivity

Reflectivity is defined as the quotient of reflected energetic flux to incident energetic flux, and is given as

$$R_{\mathrm{s}} = \frac{(p - n_0\cos\varphi)^2 + q^2}{(p + n_0\cos\varphi)^2 + q^2}$$

(14)

and

$$R_{\mathrm{p}} = R_{\mathrm{s}}\tan^2\psi$$

(15)

where

$$p = n_0\tan\varphi\sin\varphi\frac{\cos 2\psi}{1 + \sin 2\varphi\cos\Delta}$$

(16)

and

$$p = n_0\tan\varphi\sin\varphi\frac{\sin 2\psi\sin\Delta}{1 + \sin 2\psi\cos\Delta}$$

(17)

Here, $R_p$ and $R_s$ denote the reflectivity along the parallel and perpendicular components, respectively, $\varphi$ is the angle of incidence of the light beam, and $n_0$ is the refractive index of the surrounding media in which the reflected surface is located (for air $n_0 = 1$).

### Optical Constants

The complex refractive index $\hat{n}$ is represented as

$$\hat{n} = n - ik$$

(18)

Here $n$ and $k$ represent the index of refraction (real part) and the index of extinction (imaginary part), respectively. The following equations can be used for the calculation of $n$ and $k$ through ellipsometric parameters $\Delta$, $\psi$ and the angle

of incidence $\varphi$ for bare surfaces without the presence of any film or impurities at the surface:

$$n^2 - k^2 = n_0^2\sin^2\varphi\left[1 + \frac{\tan^2\varphi(\cos^2 2\psi - \sin^2\psi\sin^2\Delta)}{(1 + \sin 2\psi\cos\Delta)^2}\right]$$

(19)

and

$$2nk = \frac{n_0^2\sin^2\varphi\tan^2\varphi\sin 4\psi\sin\Delta}{(1 + \sin 2\psi\cos\Delta)^2}$$

(20)

### Dielectric Constants

The complex dielectric function $\hat{\varepsilon}$ is given as

$$\hat{\varepsilon} = \varepsilon_1 + i\varepsilon_2$$

(21)

where, again, $\varepsilon_1$ and $\varepsilon_2$ are real and imaginary parts, respectively. These values can be calculated by the equations

$$\varepsilon_1 = n^2 - k^2$$

(22)

$$\varepsilon_2 = 2nk$$

(23)

or

$$\varepsilon_2 = 2pq$$

(24)

where

$$pq = nk$$

(25)

### Optical Conductivity

The complex optical conductivity is

$$\hat{\sigma} = \sigma_1 + i\sigma_2$$

(26)

where $\sigma_1$ and $\sigma_2$ are real and imaginary parts, respectively, and are calculated by the values of dielectric constants:

$$\sigma_1 = \frac{\varepsilon_2\omega}{4\pi}$$

(27)

$$-\sigma_2 = \frac{(\varepsilon_1 - 1)\omega}{4\pi}$$

(28)

The angular velocity $\omega = 2\pi\upsilon$ and $\upsilon = c/\lambda$ is the frequency (in inverse seconds).

### PRACTICAL ASPECTS OF THE METHOD

A schematic of a nulling ellipsometer is shown in Figure 2. It consists of two arms, the first containing the components

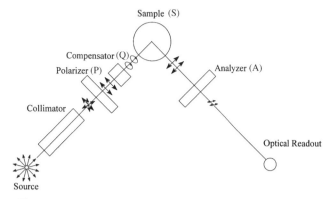

**Figure 2.** Schematic representation of a nulling ellipsometer.

necessary for sending polarized light into the sample and the second for analyzing the light reflected from the sample. There are two fundamental sets of component arrangements that are prevalent in ellipsometers. The first and the most widely used configuration consists of a polarizer (P), a compensator (Q), the sample (S), and an analyzer (A) arranged in that order, also referred to as a PQSA or PCSA arrangement. The other set consists of an arrangement in which the compensator follows the sample, thereby giving the name PSQA or PSCA. There are significant differences in the principle and the equations for computing the surface properties between these two arrangements. The most widely used PQSA alignment will be treated here.

In the PQSA alignment, light from a source passes through a polarizer. The polarizer typically converts the unpolarized incident light into linearly polarized light. This light then passes through the compensator that converts it into elliptically polarized light. The light then falls on the sample at a known angle of incidence $\varphi$ with respect to the sample normal. The light reflected from the sample then passes through an analyzer, which is set at an angle so as to extinguish the reflected light. The criterion of light extinction is verified by either the naked eye or a photosensor coupled with a photomultiplier. For a specific incident wavelength and angle of incidence, there are only two readings that are obtained from the ellipsometer, the angle of the polarizer and that of the analyzer. These two readings can then be used to calculate $\psi$ and $\Delta$ and subsequently calculate the optical properties of either a bare surface or a coating on top of it. The following paragraphs provide a more detailed description of the components in a generic ellipsometer.

### Light Source

In theory, any wave that can be polarized, i.e., any electromagnetic radiation, can be used as a source for ellipsometric detection. However, most experimental systems use a light source in the visible range because of the availability of lenses and detectors for efficiently polarizing and detecting light. The source can be either monochromatic by itself, such as a low-power helium-neon laser or a gas discharge lamp that can be filtered using an external monochromator.

### Polarizer and Analyzer

The function of the polarizer is to produce a linearly polarized light as the output to an incident light of arbitrary polarization. A birefringent polarizer can be used to obtain linear polarization to analyze. A typical birefringent polarizer converts the incident light into two beams, both of which are linearly polarized and one of which is rejected by making it undergo total internal reflection. A Glan-Thompson prism is an example of a birefringent polarizer that is commonly used.

### Compensator

The compensator is typically a quarter-wave plate that converts the incident linearly polarized light into elliptically polarized light of known orientation. The quarter-wave plate introduces a relative phase shift of 1/4 between the fast- and slow-axis components of the light. A Babinet-Soleil compensator, the principle of which is explained by Azzam and Bashara (1977), is common in modern-day instruments.

### Alignment

A vertical alignment of the polarizer and analyzer arm is typically the first step in ensuring that the system is properly aligned before the experiment is conducted. Manual ellipsometers often have some adjustable height mechanism such as adjustable feet or simple screws. It should also be verified that the geometric axes of the two arms are perpendicular to the central axis of the ellipsometer system.

Once the three crucial axes are aligned, the next step is to align and calibrate the ellipsometer angles. Conventionally, the polarizer and analyzer angles are measured as being positive counterclockwise from the plane of incidence when looking into the light beam. The alignment of the polarizer and analyzer angles in the ellipsometer is a very critical step in determining the optical properties of the system. There are several methods for arriving at a null point in order to calibrate the polarizer and analyzer angles (McCrackin et al., 1963; Archer, 1968; Azzam and Bashara, 1977). The method for step-by-step alignment for a manual ellipsometer given by McCrackin et al. (1963) has been used extensively by many manual ellipsometer users. The alignment must be made such that the polarizer and analyzer scales read zero when the plane of transmission of the films are parallel to the plane of incidence. For the case where there is ellipticity in either the polarizer or the analyzer, the calibrations account for both true calibration errors and imperfection parameters in the polarizer or the analyzer. The reader is referred to specific literature for the effect and the means to counteract these imperfections (Ghezzo, 1969; Graves, 1971; Steel, 1971).

After the alignment, the actual readings that are to be taken are the values of the polarizer and the analyzer angles $P$ and $A$, respectively. Initially, the quarter-wave plate is set either at $Q = +45°$ or $Q = −45°$. There is a multiplicity of readings for $P$ and $A$ for both these settings for the compensator (McCrackin et al., 1963). It has been shown that for two positions of the compensator lens, at

±45° there exist 32 readings for the polarizer and analyzer combination, for which there is a minimum obtained in the photomultiplier intensity. This whole data range falls into four zones, two with the compensator at +45° and two at −45°. Experiments are carried out typically in two out of the four zones and averaged in order to reduce azimuth errors. Although a complete zonal average as mentioned by McCrackin et al. (1963) need not be done for each experiment, it is advisable to be aware of the zone in which the experimental angles are being recorded.

## METHOD AUTOMATION

It takes several minutes to perform the manual nulling of an ellipsometer reading by changing the values of the polarizer and analyzer angles P and A. However, in most cases, as when *in situ* studies are being carried out or when there are multiple states that are to be studied in a very short time, it is necessary to have some form of automation in order to give accurate and fast results. Hence there has been interest in developing automatic ellipsometers. These ellipsometers can be divided into two broad categories, those that still use the principle of nulling the light exiting from the sample and others that measure the intensity of light coming out of the analyzer and correlate this intensity and ellipticity into the surface properties.

### Automatic Null Ellipsometers

The first ing automatic ellipsometers used motors to change the polarizer and analyzer angles P and A to drive them to null. Initially, the systems that used motor-driven nulling also required visual reading of the P and A values. This process was time consuming and very inefficient. This led to instruments that avoided mechanical nulling. One of the most widely used automatic nulling ellipsometers uses Faraday coils between the polarizer-compensator and sample-analyzer. One such automatic null ellipsometer using Faraday coil rotators was built by Mathieu et al. (1974). One Faraday coil rotates the polarization of light before it enters the compensator. The Faraday coil that is placed before the analyzer compensates for the polarized light by rotating to a certain degree, until a null point is reached. The DC current levels in the two Faraday coils provide an accurate measure of the degree of rotation of these coils. These values can be directly used for the calculation of the physical properties or can be fed to a computer program that computes the necessary optical parameters. The rate of data acquisition is very high since there are no mechanically moving parts. This technique has seen widespread use in systems in which *in situ* measurements are done. In such a case, the polarizer or the analyzer is kept at a fixed azimuth and the other is driven to null by use of the Faraday coil, and the current is recorded as a function of time.

### Automatic Intensity-Measuring Ellipsometer

Another major class of ellipsometers measures the intensity and the polarization state of light after reflection from the sample. These instruments fall into two categories:

(1) rotating-analyzer ellipsometer and (2) polarization-modulated ellipsometer. The rotating-analyzer ellipsometer (Hauge and Dill, 1973), as the name indicates, uses a continuously rotating analyzer to carry out the measurement. Light passes through a fixed monochromator, a fixed polarizer, an adjustable sample, and the rotating analyzer before being detected by the detector as a beam of cyclically varying intensity. The output light intensity is digitized as a function of discrete angular positions of the analyzer. From the digital output light intensity, the output polarization state of the light is obtained. This is subsequently used along with other experimental parameters, such as the angle of incidence and polarizer angle, and fed into a numerical computational algorithm that yields the film thickness and refractive index.

The polarization-modulated ellipsometer uses a piezo-birefringent element to produce a periodic, known relative phase shift in the amplitudes of the perpendicular incident light components. This light input is also converted into digital data typically using an analog-to-digital converter (ADC), and the output polarization state is determined as a function of each discrete input state. This is subsequently fed into a numerical algorithm to calculate the required optical parameters. Many different variations of these polarization-modulated ellipsometers have been developed over the years (Jasperson and Schnatterly, 1969; Drevillon et al., 1982) and have found widespread use because of their accuracy and high-speed measurement capabilities.

There have also been many other combinations used to improve the speed and accuracy of the ellipsometric technique. Some of these ellipsometers are the rotating-polarizer ellipsometer (Faber and Smith, 1968), the rotating-detector ellipsometer (Nick and Azzam, 1989), and the comparison ellipsometer.

## DATA ANALYSIS AND INITIAL INTERPRETATION

### Calculation of $\Delta$ and $\psi$ from Ellipsometer Readings P and A

The experimental data typically consist of two angles, those of the polarizer, P, and the analyzer, A. When using a manual ellipsometer, one should be careful to cover the zone in which the readings are taken. Since there are errors in azimuth values that might arise out of measurements in different zones, it is preferred to either take the whole sequence of data from a single zone or average each data point in two or more zones. For an example of multiple zones, consider the perpendicular Fresnel reflection coefficient for one interface in Equation 10. The value for this expression is the same for $\delta$, $\delta + 2\pi$, $\delta + 4\pi, \ldots, \delta + 2n\pi$. Hence, the curve for $\Delta$ vs. $\psi$ repeats when $\delta = 2\pi$. It is therefore helpful to have an advance estimate of the thickness in ellipsometric experiments. The reader is referred to the original paper by McCrackin et al. (1963) for a detailed procedure for the zonal analysis of a manual ellipsometer.

Once confirmed that the whole data set is in the same zone or that it is properly averaged, the calculation of the phase and amplitude change parameters $\Delta$ and $\psi$ for the system is straightforward. The relative phase change between the parallel and perpendicular polarized light, $\Delta$, and the relative amplitude change between these two components, $\psi$, are related to $P$ and $A$ by

$$\Delta = 2P + \frac{1}{2}\pi \qquad (29)$$

$$\psi = A \qquad (30)$$

when the compensator is kept at $+45°$. Recall that the above relations assume the polarizer, compensator, and analyzer are ideal instruments arranged in a PCSA configuration. If the compensator follows the sample, appropriate adjustments must be made.

Smith (1969) has derived a set of exact and approximate equations for determining the values for the shift parameters $\Delta$ and $\psi$ that is independent of the compensator transmission properties. For such a set of calculations, one has to obtain readings in at least two of the four zones and then use the following formulas to obtain $\Delta$ and $\psi$. For a PSCA arrangement,

$$\tan^2 \psi = -\tan p_1 \tan p_2 \qquad (31)$$

$$\cos \Delta = \frac{1 - \cot p_1 \tan p_2}{2 - (\cot p_1 \tan p_2)^{1/2}} \frac{\cos (A_1 + A_2)}{\sin (A_2 - A_1)} \qquad (32)$$

The subscripts 1 and 2 indicate the readings obtained in zones 1 and 2 for $P$ and $A$, respectively. Similar exact equations for the PCSA arrangement are

$$\tan^2 \psi = -\tan A_1 \tan A_2 \qquad (33)$$

$$\tan \Delta = Y \sin \Delta_e \tan (P_1 + P_2) \qquad (34)$$

where

$$Y = \frac{2[\cot(p_1 - \pi/4)\cot(\pi/4 - p_2)]^{1/2}}{1 + \cot(P_1 - \pi/4)\cot(\pi/4 - P_2)} \qquad (35)$$

$\sin \Delta_e =$

$$\left[\frac{1 + [1 - \tan(P_1 - \pi/4)\cot(\pi/4 - P_2)]^2}{4\tan(P_1 - \pi/4)\cot(\pi/4 - P_2)} \frac{\sin^2(A_2 + A_1)}{\sin^2(A_2 - A_1)}\right]^{1/2} \qquad (36)$$

Modern automatic ellipsometers use this or other similar formulas to account for the nonidealities while determining the values of these two ellipsometer parameters.

### Calculation of Optical Properties from Values of $\Delta$ and $\psi$

The optical properties of interest in the system can be determined using values of $\Delta$ and $\psi$.

The parameters are functions of many physical and material properties. Some of these dependencies can be written in the form

$$\Delta_i = f_i(n_1, k_1, n_2, k_2, n_3, k_3, d, \varphi, \lambda) \qquad (37)$$

$$\psi_i = f_i(n_1, k_1, n_2, k_2, n_3, k_3, d, \varphi, \lambda) \qquad (38)$$

Most often, some form of previous measurements can determine some of these unknown parameters. For

example, $\varphi$ can be adjusted beforehand, and $\lambda$ can be fixed by knowing or determining the wavelength of the incident light. If the surrounding medium is air, then $n_1 = 1$ and $k_1 = 0$, or if the medium is some totally transparent liquid, $k_1 = 0$ and $n_1$ can be determined by refractometry.

The equations used to compute these optical properties are very complicated. Hence, some sort of computer algorithm is usually necessary. One of the first computer programs for ellipsometric calculations was written by McCrackin (1969). In this program, the calculations can be made for various options pertaining to the physical properties of the measured specimen. The program calculates the properties of, e.g., multiple films, inhomogenous films, and films composed of a mixture of materials. For the case of a substrate with a very thin surface oxide film, Shewchun and Rowe (1971) provided a method for calculation of the apparent (substrate-plus-film) and true (substrate-only) values of $\Delta$ and $\psi$ by varying the incident angle. These authors have shown a flow chart for a portion of the computer program used to determine the substrate and film optical constants and thickness. Joseph and Gagnaire (Gagnaire, 1983) utilized a variance analysis using the least-squares method for application to anodic oxide growth on metallic surfaces. They have developed a method of analysis that permits the simultaneous calculation of complex refractive indices of the film and the substrate and the film thickness.

By fixing the values of the refractive indices and all other properties in a system, one can theoretically compute a $\Delta$-$\psi$ curve. Figure 3 gives such a theoretical $\Delta$-$\psi$ curve for $\varphi = 70°$, $\lambda = 546.1$ nm, $n_1 = 1.337$, and $\hat{n}_3 = 2.94 - 3.57i$, $k_1 = k_2 = 0$, for various $n_2$ and $\delta$ values. Each curve in this figure shows the locus of points for increasing film

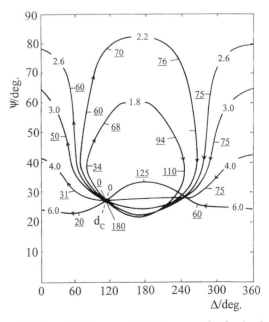

**Figure 3.** Theoretically computed $\delta$–$\psi$ curves for fixed values of $n_1 = 1.337$, $k_1 = 0$, $\hat{n}_3 = 2.94 - 3.57i$, and $k_2 = 0$ and various values of $n_2$ from 1.8 to 6. Underlined numbers correspond to values of $\delta$.

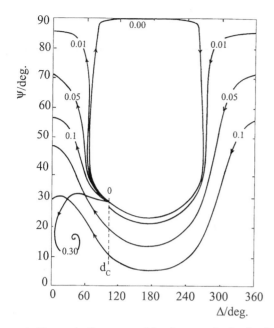

**Figure 4.** Theoretically computed $\delta - \psi$ curves for fixed values of $n_1 = 1.337$, $k_1 = 0$, $n_2 = 2.38$, and $\hat{n}_3 = 2.94 - 3.57i$ for various values of $k_2$ from 0 to 0.3.

thickness of fixed $n_2$ or $k_2$. The arrows show the direction of increasing thickness and the nonunderlined numbers are the values of the indices of refraction of the film.

For a nonabsorbing film, $k_2 = 0$, and $\Delta$ and $\psi$ are cyclic functions of thickness for each $n_2$ value. The curves repeat periodically with every 180° in $\delta$. Figure 4 shows a similar figure obtained for an absorbing film, where $k_2 > 0$. It can be seen that the curves do not quite have a periodic form. The curve front moves toward a lower $\psi$ value with increasing cycle number. In the case where $k_2$ is high (0.3), the curves show a spiral shape. To determine the value of the refractive index $n_2$ for a nonabsorbing film $k_2 = 0$, a large number of theoretical curves have to be drawn and compared with the experimental values. The data-fitting procedure is much more complicated for absorbing films that have a complex refractive index. In such a case, both $n$ and $k$ values have to be changed, and families of curves have to be drawn for the relationship between $\Delta$ and $\psi$. From these curves, the one that best fits the experimentally observed curves can be used to determine $n_2$ and $k_2$. One can use any standard fitting tool for determining the minimum error between the experimental and predicted values or use some of the ellipsometry-specific methods in the literature (Erfemova and Arsov, 1992).

## SAMPLE PREPARATION

Ellipsometric measurements can be carried out only with specimens that have a high degree of reflectivity. The degrees of smoothness and flatness are both factors that must be considered in the selection of a substrate. An indication of the extent of regularity of the whole surface is obtained by determination of $P$ and $A$ values at several points on the specimen. Various techniques have been used for the preparation and cleaning of the substrate surface, such as evaporation in high vacuum, mechanical polishing, chemical polishing, electrochemical polishing, and cathodic cleaning of surfaces by dissolution of natural oxide films. Other relatively less common methods include sputtering with argon atoms and thermal annealing.

### Evaporation in Vacuum

Metal specimens with high reflectivity and purity at the surface can be prepared as films by vapor deposition in vacuum. The substrates generally used are glass microscope slides, which are first cleaned by washing and vapor phase degreasing before being introduced into the vacuum chamber, where they are cleaned by ion bombardment with argon. Adjusting parameters such as the rate and time of evaporation regulates the thickness of the evaporated film. Many other variables have an impact on the properties of evaporated films. They include the nature and pressure of residual gas in the chamber, intensity of the atomic flux condensing on the surface, the nature of the target surface, the temperature of the evaporation source and energy of the impinging atoms, and contamination on the surface by evaporated supporting material from the source.

### Mechanical Polishing

Mechanical polishing is one of the most widely used methods in surface preparation. Generally, mechanical polishing consists of several steps. The specimens are initially abraded with silicon carbide paper of various grades, from 400 to 1000 grit. After that, successive additional fine mechanical polishing on silk disks with diamond powders and sprays (up to 1/10 μm grain size) is practiced to yield a mirror finish. This results in a bright and smooth surface suitable for optical measurements. But, it is well known that mechanical treatments change the structure of surface layers to varying depths depending on the material and the treatment. Mechanically polished surfaces do not have the same structural and optical properties as the bulk material. As a result of mechanical polishing, surface crystallinity is impaired. In addition, there exists a thin deformed layer extending from the surface to the interior due to the friction and heating caused by mechanical polishing. Such changes are revealed by the deviations in the optical properties of the surface from those of the bulk. Hence mechanical polishing is utilized as a sole preparation method, primarily if the specimen will be used as a substrate for ellipsometric study of some other surface reactions, e.g., adsorption, thermal, or electrochemical formation of oxide films and polymer electrodeposition.

### Chemical Polishing

This method of surface preparation is generally used after mechanical polishing in order to remove the damaged layer produced by the mechanical action. A specimen is immersed in a bath for a controlled time and subsequently ultrasonically washed in an alcohol solution. The final cleaning is extremely important to remove bath components that may otherwise be trapped on the surface.

Various baths of different chemical compositions and immersion conditions (temperature of bath and time of treatment depending on the nature of the metal specimen) are described in the literature (Tegart, 1960). During chemical polishing, oxygen evolution is possible along with the chemical dissolution of the metal substrate. In such cases, the polished surfaces may lose their reflectivity, becoming less suitable for optical measurements. A major problem associated with chemical polishing is the possible formation of surface viscous layers that result in impurities on the surface. Care must be taken to identify and remove impurities like S, Cl, N, O, and C that might be formed by the use of chemicals.

### Electrochemical Polishing

Electrochemical polishing is a better way to prepare metal surfaces than mechanical and chemical polishing. The metal is placed in an electrochemical cell as an anode, while a Pt or carbon electrode with a large surface area is used as the cathode. Passage of current causes the dissolution of parts of the metal surface, the peaks dissolving at a greater rate than the valleys, which leads to planarization and a subsequently reflecting surface.

### Electrochemical Dissolution of Natural Oxide Films

Dissolution of the naturally formed oxide films can be carried out by cathodic polarization. It is very important to determine an optimum value of the cathodic potential for each metal in order to minimize the thickness of the oxide layer. For potentials more anodic than the optimum value, the dissolution is very slow and the natural oxide film is not dissolved completely. Potentials higher than this optimum increase the risk for hydrogen contamination at the surface, its penetration in the substrate, and its inclusion in the metal lattice. The formation of thin hydrated layers is also possible.

### PROBLEMS

There are various sources, some from the equipment and others from the measuring system or the measurement process itself, that can introduce errors and uncertainties during an ellipsometric measurement. Most of these errors are correctable with certain precautions and corrections during or after the experiment.

A frequent source of error is the presence of a residual oxide layer on the surface of the film under study. The presence of even a very thin film (50 Å) can cause the refractive index to be markedly different than the actual substrate value (Archer, 1962). Hence, in the case of an absorbing substrate and a nonabsorbing thin film, four optical parameters need to be determined: the refractive index and the thickness of the film, $n_2$ and $d_2$, and the complex refractive index coefficients of the substrate, $n_3$ and $k_3$. Since only two parameters are determined, $\Delta$ and $\psi$, it is not possible to determine all four parameters using a single measurement. To estimate the optical properties of the substrate, multiple measurements are made (So and Vedam, 1972) using (1) various film thickness values, (2) various angles of incidence, or (3) various incident media.

One set of substrate optical properties consistent with all the multiple readings can then be found.

Errors in measurement also occur due to surface roughness in the substrate (or the film). Predictions of the optical parameters for surfaces with very minor surface roughness (50 Å) have shown large errors in the determined refractive indices (Festenmaker and McCrackin, 1969). Hence, it is imperative that one has a good knowledge of the degree of impurities or abnormalities that a sample might have. Sample preparation is crucial to the accuracy of the results, and care must be taken to determine a method most suited for the particular application.

The polarizer, compensator, and analyzer are all subject to possibly incorrect settings and hence azimuth errors. The analysis can be freed from azimuth errors by performing the analysis and averaging the results with components set at angles that are 90° apart, described as zone averaging by McCrackin et al. (1963).

The magnitudes of $\Delta$ and $\psi$ vary as a function of the angle of incidence. There are several types of systematic errors encountered during the angle-of-incidence measurements that produce errors in $\Delta$ and $\psi$. Many studies suggest that the derived constants are not true constants, but are dependent on the incidence angle (Vasicek, 1960; Martens et al., 1963). The beam area divided by the cosine of the angle of incidence gives the area of the surface under examination. Hence, there is a significant increase in the area under examination for a small change in the incident angle. A change in the angle of incidence simultaneously changes the area of the surface under examination. As a matter of convenience, it is suggested that the ellipsometric measurements be made with an angle of incidence near the principal angle, i.e., the angle at which the relative phase shift between the parallel and perpendicular components of reflected light, $\Delta$, is 90°. The value of this principal angle depends mainly on the refractive index of the substrate, and is between 60° and 80° for typical metal substrates.

Other sources of error include those arising from the deflection of the incident light beam due to an imperfection in the polarizer, multiply reflected beams, and incompletely polarized incident light. The importance of reliable and accurate optical components should therefore be stressed, since most of these errors are connected to the quality of the equipment one procures.

It should also be noted that the steps after the experimental determination of $\Delta$ and $\psi$ are also susceptible to error. One typically has to input the values of $\Delta$ and $\psi$ into a numerical algorithm to compute the optical properties. One should have an idea of the precision of the computation accuracy of the input data. There are also cases in which the approximations in the relations used to compute the physical parameters are more extensive than necessary and introduce more uncertainty than a conservative approach would cause.

### LITERATURE CITED

Archer, R. J. 1962. Determination of the properties of films on silicon by the method of ellipsometry. *J. Opt. Soc. Am.* 52:970.

Archer, R. J. 1964. Measurement of the physical adsorption of vapors and the chemisorption of oxygen and silicon by the method of ellipsometry. *In* Ellipsometry in the Measurements of Surfaces and Thin Films (E. Passaglia, R.R. Stromberg, and J. Kruger, eds.) pp. 255–272. National Bureau of Standards, Washington, DC, Miscellaneous Publication 256.

Archer, R. J., 1968. Manual on Ellipsometry. Gaertner Scientific, Chicago.

Arsov, Lj. 1985. Dissolution electrochemique des films anodiques du titane dans l'acide silfurique. *Electrochim. Acta* 30:1645–1657.

Azzam, R. M. A. and Bashara, N. M. 1977. Ellipsometry and Polarized Light. North-Holland Publishing, New York.

Brown, E. B. 1965. Modern Optics. Reinhold Publishing, New York.

Cahan, B. D. and Spainer, R. F. 1969. A high speed precision automatic ellipsometer. *Surf Sci.* 16:166.

Drevillon, B., Perrin, J., Marbot, R., Violet, A., and Dalby, J. L. 1982. Fast polarization modulated ellipsometer using a microprocessor system for digital Fourier analysis. *Rev. Sci. Instrum.* 53(7):969–977.

Drude, P. 1889. Ueber oberflachenschichten. *Ann. Phys.* 272:532–560.

Erfemova, A. T. and Arsov, Lj. 1992. Ellipsometric in situ study of titanium surfaces during anodization. *J. Phys. France* II:1353–1361.

Faber, T. E. and Smith, N. V. 1968. Optical measurements on liquid metals using a new ellipsometer. *J. Opt. Soc. Am.* 58(1):102–108.

Festenmaker, C. A. and McCrackin, F. L. 1969. Errors arising from surface roughness in ellipsometric measurement of the refractive index of a surface. *Surf. Sci.* 16:85–96.

Gagnaire, J. 1983. Ellipsometric study of anodic oxide growth: Application to the titanium oxide systems. *Thin Solid Films* 103:257–265.

Ghezzo, M. 1960. Method for calibrating the analyser and the polarizer in an ellipsometer. *Br. J. Appl. Phys. (J. Phys. D)* 2:1483–1485.

Ghezzo, M. 1969. Method for calibrating the analyzer and polarizer in an ellipsometer. *J. Phys. Ser. D* 2:1483–1490.

Graves, R. H. W. 1971. Ellipsometry using imperfect polarizers. *Appl. Opt.* 10:2679.

Hauge, P. S. and Dill, F. H. 1973. Design and operation of ETA, and automated ellipsometer. *IBM J. Res. Dev.* 17:472–489.

Hayfield, P. C. S. 1963. American Institute of Physics, Handbook. McGraw-Hill, New York.

Heavens, O. S., 1955. Optical Properties of Thin Solid Films. Dover, New York.

Hristova, E., Arsov, Lj., Popov, B., and White, R. 1997. Ellipsometric and Raman spectroscopic study of thermally formed films on titanium. *J. Electrochem. Soc.* 144:2318–2323.

Jasperson, S. N. and Schnatterly, S. E. 1969. An improved method for high reflectivity ellipsometry based on a new polarization modulation technique. *Rev. Sci. Instrum.* 40(6):761–767.

Martens, F. P., Theroux, P., and Plumb, R. 1963. Some observations on the use of elliptically polarized light to study metal surfaces. *J. Opt. Soc. Am.* 53(7):788–796.

Mathieu, H. J., McClure, D. E., and Muller, R. H. 1974. Fast self-compensating ellipsometer. *Rev. Sci. Instrum.* 45:798–802.

McCrackin, F. L. 1969. A FORTRAN program for analysis of ellipsometer measurements. Technical note 479. National Bureau of Standards, pp. 1–76.

McCrackin, F. L., Passaglia, E., Stromberg, R. R., and Steinberg, H. L. 1963. Measurement of the thickness and refractive index of very thin films and the optical properties of surfaces by ellipsometry. *J. Res. Natl. Bur. Stand.* A67:363–377.

Menzel, H. D. (ed.). 1960. Fundamental Formulas of Physics, Vol. 1. Dover Publications, New York.

Nick, D. C. and Azzam, R. M. A. 1989. Performance of an automated rotating-detector ellipsometer. *Rev. Sci. Instrum.* 60(12):3625–3632.

Ord, J. L. 1969. An elliposometer for following film growth. *Surf. Sci.* 16:147.

Schmidt, E. 1970. Precision of ellipsometric measurements. *J Opt. Soc.* 60(4):490–494.

Shewchun, J. and Rowe, E. C. 1971. Ellipsometric technique for obtaining substrate optical constants. *J. Appl. Phys.* 41(10):4128–4138.

Smith, P. H. 1969. A theoretical and experimental analysis of the ellipsometer. *Surf. Sci.* 16:34–66.

So, S. S. and Vedam, K. 1972. Generalized ellipsometric method for the absorbing substrate covered with a transparent-film system. Optical constants for silicon at 3655 Å. *J. Opt. Soc. Am.* 62(1):16–23.

Steel, M. R. 1971. Method for azimuthal angle alignment in ellipsometry. *Appl. Opt.* 10:2370–2371.

Tegart, W. J. 1960. Polissage electrolytique et chemique des metaux au laboratoire et dans l'industry. Dunod, Paris.

Vasicek, A. 1960. Optics of Thin Films. North-Holland Publishing, New York.

Winterbottom, A. W. 1946. Optical methods of studying films on reflecting bases depending on polarization and interference phenomena. *Trans. Faraday Soc.* 42:487–495.

Zaininger, K. H. and Revesz, A. G. 1964. Ellipsometry—a valuable tool in surface research. *RCA Rev.* 25:85–115.

## KEY REFERENCES

Azzam, R. M. A. (ed.). 1991. Selected Papers on Ellipsometry; SPIE Milestone Series, Vol. MS 27. SPIE Optical Engineering Press, Bellingham, Wash.

*A collection of many of the path-breaking publications up to 1990 in the field of ellipsometry and ellipsometric measurements. Gives a very good historical perspective of the developments in ellipsometry.*

Azzam and Bashara, 1977. See above.

*Gives an excellent theoretical basis and provides an in-depth analysis of the principles and practical applications of ellipsometry.*

McCrackin et al., 1963. See above.

*Provides a good explanation of the practical aspects of measuring the thickness and refractice indices of thin films. Includes development of a notation for identifying different null pairs for the polarizer/analyzer rotations. Also provides a method to calibrate the azimuth scales of the ellipsometer divided circles.*

Passaglia, E., Stromberg, R.R., and Kruger, J. (eds.). 1964. Ellipsometry in the Measurement of Surfaces and Thin Films. Symposium Proceedings, Washington, DC, 1963. National Bureau of Standards Miscellaneous Publication 256.

*Presents practical aspects of ellipsometric measurements in various field.*

## APPENDIX:
## GLOSSARY OF TERMS AND SYMBOLS

| | |
|---|---|
| $A$ | Analyzer settings with respect to the plane of incidence, deg |
| $C$ | Compensator settings with respect to the plane of incidence, deg |
| $E_i$ | Amplitude of the incident electric wave |
| $E_r$ | Amplititude of the reflected electric wave |
| $d$ | Film thickness, cm |
| $E_x$ | Amplitude of the electric wave in the $x$ axis |
| $i$ | $\sqrt{-1}$ |
| $k$ | Absorption index (imaginary part of the refractive index) |
| $n$ | Real part of the refractive index |
| $\hat{n}$ | Complex refractive index |
| $P$ | Polarizer settings with respect to the plane of incidence, deg |
| $Q$ | Compensator settings with respect to the plane of incidence, deg |
| $R$ | Reflectivity |
| $r_p$ | Fresnel reflection coefficient for light polarized parallel to the plane of incidence |
| $r_s$ | Fresnel reflection coefficient for light polarized perpendicular to the plane of incidence |
| $\Delta$ | Relative phase change, deg |
| $\delta$ | Change of phase of the beam crossing the film |
| $\hat{\varepsilon}$ | Complex dielectric function |
| $\lambda$ | Wavelength, cm |
| $\nu$ | Frequency, $s^{-1}$ |
| $\rho$ | Ratio of complex reflection coefficients |
| $\hat{\sigma}$ | Complex optical conductivity |
| $\varphi$ | Angle of incidence, deg |
| $\tan\psi$ | Relative amplitude attenuation, deg |
| $\omega$ | Angular velocity, rad |

LJ. ARSOV
University Kiril and Metodij
Skopje, Macedonia

M. RAMASUBRAMANIAN
B. N. POPOV
University of South Carolina
Columbia, South Carolina

# IMPULSIVE STIMULATED THERMAL SCATTERING

## INTRODUCTION

Impulsive stimulated thermal scattering (ISTS) is a purely optical, non-contacting method for characterizing the acoustic behavior of surfaces, thin membranes, coatings, and multilayer assemblies (Rogers et al., 2000a), as well as bulk materials (Nelson and Fayer, 1980). The method has emerged as a useful tool for materials research in part because: (1) it enables accurate, fast, and nondestructive measurement of important acoustic (direct) and elastic (derived) properties that can be difficult or impossible to evaluate in thin films using other techniques; (2) it can be applied to a wide range of materials that occur in microelectronics, biotechnology, optics, and other areas of technology; and (3) it does not require specialized test structures or impedance-matching fluids which are commonly needed for conventional mechanical and acoustic tests. Further, recent advances in experimental design have simplified the ISTS measurement dramatically, resulting in straightforward, low-cost setups, and even in the development of a commercial ISTS photoacoustic instrument that requires no user adjustments of lasers or optics. With this tool, automated single-point measurements, as well as scanning-mode acquisition of images of acoustic and other physical properties, are routine.

ISTS, which is based on transient grating (TG) methods (Nelson and Fayer, 1980; Eichler et al., 1986), is a spectroscopic technique that measures the acoustic properties of thin films over a range of acoustic wavelengths. It uses mild heating produced by crossed picosecond laser pulses to launch coherent, wavelength-tunable acoustic modes and thermal disturbances. The time-dependent surface ripple produced by these motions diffracts a continuous-wave probing laser beam that is overlapped with the excited region of the sample. Measuring the temporal variation of the intensity of the diffracted light yields the frequencies and damping rates of acoustic waves that propagate in the plane of the film. It also determines the arrival times of acoustic echoes generated by subsurface reflections of longitudinal acoustic wavepackets that are launched at the surface of the film. The wavelength dependence of the acoustic phase velocities (i.e., product of the frequency and the wavelength) of in-plane modes, which is known as the acoustic dispersion, is determined either from a single measurement that involves the excitation of acoustic waves with a well defined set of wavelengths, or from the combined results of a series of measurements that each determine the acoustic response at a single wavelength. Interpreting this dispersion with suitable models of the acoustic waveguide physics yields viscoelastic (e.g., Young's modulus, Poisson's ratio, acoustic damping rates, stress, etc.) and/or other physical properties (e.g., density, thickness, presence or absence of adhesion, etc.) of the films. The acoustic echoes, which are recorded in the same measurements, provide additional information that can simplify this modeling. The ISTS data also generally contain information from nonacoustic (e.g., thermal, electronic, etc.) responses. This unit, however, focuses only on ISTS measurement of acoustic motions in thin films. It begins with an overview of other related measurement techniques. It then describes the ISTS acoustic data and demonstrates how it can be used to determine: (1) the stress and flexural rigidity in thin membranes (Rogers and Bogart, 2000; Rogers et al., 2000b; Rogers and Nelson, 1995); (2) the elastic constants of membranes and supported films (Rogers and Nelson, 1995; Duggal et al., 1992; Shen et al., 1996; Rogers and Nelson, 1994; Rogers et al., 1994a); and (3) the thicknesses of single or multiple films in multilayer stacks (Banet et al., 1998; Gostein et al., 2000).

### Competitive and Related Techniques

Acoustic properties of thin films are most commonly evaluated with conventional ultrasonic tests, which involve a source of ultrasound (e.g., a transducer), a propagation path, and a detector. The ability to excite and detect acoustic waves with wavelengths that are short enough for thin film evaluation (i.e., wavelengths comparable to or smaller than the film thickness) requires high-frequency transducers/detectors fabricated directly on the sample, or coupled efficiently to it with impedance-matching liquids or gels. Both of these approaches restrict the range of structures that can be examined; they limit the usefulness of conventional acoustic tests for thin film measurement. Photoacoustic methods overcome the challenge of acoustic coupling by using laser light to excite and probe acoustic disturbances without contacting the sample. In addition to ISTS, there are two other general classes of photoacoustic techniques for measuring acoustics in thin films. In the first, a single excitation pulse arrives at the surface of the sample and launches, through mild heating, a longitudinal (i.e., compressional) acoustic wavepacket that propagates into the depth of the structure (Thomsen et al., 1986; Eesley et al., 1987; Wright and Kawashima, 1991). Parts of this acoustic disturbance reflect at buried interfaces, such as the one between the film and its support or between films in a complex multilayer stack. A variably delayed probe pulse measures the time dependence of the optical reflectivity or the slope of the front surface of the sample in order to determine the time of arrival of the various acoustic echoes. Data from this type of measurement are similar in information content to the acoustic echo component of the ISTS signal. In both cases, the data reveal the out-of-plane longitudinal acoustic velocities when the thicknesses of the films are known. The measured acoustic reflectivity can also be used to determine properties (e.g., density) that are related to the change in acoustic impedance that occurs at the interface. This technique has the disadvantage that it requires the excitation pulses to be strongly absorbed by the sample. It also typically relies on expensive and complex femtosec laser sources and relatively slow detection schemes that use probe pulses. Although it can be used to measure thicknesses accurately, it does not yield information on transversely polarized (i.e., shear) acoustic waves or on modes that propagate in the plane of the film. As a result, the only elastic property that can be evaluated easily is the out-of-plane compressional modulus.

Another method uses a cylindrically-focused excitation pulse as a broadband line source for surface propagating waves (Neubrand and Hess, 1992; Hess, 1996). Examining the changes in position, intensity, or phase of at least one other laser beam that strikes the sample at a location spatially separated from the excitation region provides a means for probing these waves. The data enable reliable measurement of surface acoustic wave velocities over a continuous range of acoustic wavelengths (i.e., the dispersion) when the separation between the excitation and probing beams (or between the probing beams themselves) is known precisely. The measured dispersion can be used, with suitable models for the acoustic physics, to extract other properties (e.g., density, thickness, elastic properties) of the samples. This technique has the disadvantage that out-of-plane acoustic properties are not probed directly. Also, data that include multiple acoustic veloci-

ties at a single wavelength (e.g., multiple modes in an acoustic waveguide) can be difficult to interpret. We note that this method and the one described in the previous paragraph share the "source–propagation path–receiver" approach of conventional acoustic testing techniques. They are similar also in their generation of an essentially single-cycle acoustic pulse or "wavepacket" that includes a wide range of wavevectors and corresponding frequencies. The combined information from these two techniques is present in the ISTS data, in a separable and more easily analyzed form.

Although it is not strictly a photoacoustic method, surface Brillouin scattering (SBS; Nizzoli and Sandercock, 1990) is perhaps more closely related to ISTS than the techniques described above. An SBS experiment measures the spectral properties of light scattered at a well-defined angle from the sample. The spectrum reveals the frequencies of incoherent, thermally populated acoustic modes with wavelengths that satisfy the associated phase-matching condition for scattering into the chosen angle. The information obtained from SBS and ISTS measurements is similar. An important difference is that the ISTS technique uses coherent, laser-excited phonons rather than incoherent, thermally populated ones. ISTS signals are therefore much stronger than those in SBS and they can be detected rapidly in the time domain. This form of detection enables acoustic damping rates, for example, to be evaluated accurately without the deconvolution procedures that are necessary to interpret spectra collected with the sensitive Fabry-Perot filters that are commonly used in SBS. Also, with ISTS it is possible simultaneously to excite and monitor acoustic modes with more than one wavelength, to determine their phases, and to measure them in real-time as they propagate across the sample. These and other capabilities, which are useful for accurately evaluating films or other structures with dimensions comparable to the acoustic wavelength, are absent from traditional forms of SBS.

Finally, in some cases, certain properties that can be derived from ISTS measurements (e.g., elastic constants, density, thickness, stress, etc.) can be determined with other, nonacoustic methods. Although complete descriptions of all of the possible techniques is beyond the scope of this unit, we list a few of the more established methods.

1. Elastic constants can be determined with uniaxial pull-testers, nanoindenters (Pharr and Oliver, 1992), and specialized micromechanical test structures (Allen et al., 1987).

2. Stress, and sometimes elastic constants, are typically measured with tests that use deflections or vibrations of drumhead membranes (Maden et al., 1994; Vlassak and Nix, 1992) or cantilevered beams (Mizubayashi et al., 1992), or these are inferred from strain evaluated using X-ray diffraction (Clemens and Bain, 1992; also see X-RAY AND NEUTRON DIFFUSE SCATTERING MEASUREMENTS).

3. Thicknesses of transparent films are often determined with reflectometers or ellipsometers (See ELLIPSOMETRY). For opaque films, thickness is evaluated

using stylus profilometry or grazing incidence x-ray reflection; in the case of conducting films, it is determined indirectly from measurements of sheet resistance. Vinci and Vlassak (1996) present a review of techniques for measuring the mechanical properties of thin films and membranes.

## PRINCIPLES OF THE METHOD

As mentioned above (see Introduction), ISTS uses short (relative to the time-scale of the material response of interest) pulses of light from an excitation laser to stimulate acoustic motions in a sample. The responses are measured with a separate probing laser. Figure 1 schematically illustrates the mechanisms for excitation and detection in the simplest case. Here, a single pair of excitation pulses crosses at an angle $\theta$ at the surface of a thin film on a substrate. The optical interference of these pulses produces a sinusoidal variation in intensity with a period $\Lambda$, given by

$$\Lambda = \frac{\lambda_e}{2\sin(\theta/2)} \tag{1}$$

### (1):  Excitation

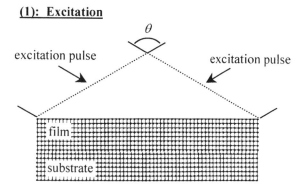

### (2):  Detection

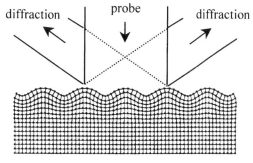

**Figure 1.** Schematic illustration of the ISTS measurement. Crossed laser pulses form an optical interference pattern and induce coherent, monochromatic acoustic and thermal motions with wavelengths that match this pattern. The angle between the pulses determines the wavelength of the response. A continuous wave probing laser diffracts from the ripple on the surface of the sample. Measuring the intensity of the diffracted light with a fast detector and transient recorder reveals the time dependence of the motions.

The wavelength of the excitation light ($\lambda_e$) is chosen so that it is partly absorbed by the sample; this absorption induces heating in the geometry of the interference pattern. The resulting spatially periodic thermal expansion launches coherent, monochromatic counter-propagating surface acoustic modes with wavelength $\Lambda$. It also simultaneously generates acoustic waves that propagate into the bulk of the sample and produce the acoustic echoes mentioned previously (Shen et al., 1996; Crimmins et al., 1998). In thin films, the former and latter responses typically occur on nanosecond and picosecond time scales, respectively. On microsecond time scales, the thermally induced strain slowly relaxes via thermal diffusion. The temporal nature of the surface motions that result from these responses can be determined in real time with each shot of the excitation laser by measuring the intensity of diffraction of a continuous wave probe laser with a fast detector and transient recorder. For motions that exceed the bandwidth of conventional detection electronics (typically ~1 to 2 GHz), it is possible to use either variably delayed picosecond probing pulses to map out the response in a point-by-point fashion (Shen et al., 1996; Crimmins et al., 1998) or a continuous wave probe with a streak camera for ultrafast real-time detection. It is also possible to use, instead of the streak camera, a scanning spectral filter to resolve the motions in the frequency domain (Maznev et al., 1996). This unit focuses primarily on the in-plane acoustic modes, partly because they can be excited and detected rapidly in real-time using commercially available, low-cost laser sources and electronics. Also, these modes provide a route to measuring a wide range of thin film elastic constants and other physical properties.

The following functional form describes the response, $R(t)$, when the thermal and the in-plane acoustic motions both contribute

$$R(t) \propto A_{\text{therm}}e^{-\Gamma_t} + \sum_i B_i e^{-\gamma_i t}\cos\omega_i t \tag{2}$$

where $A_{\text{therm}}$ and $B_i$ are the amplitudes of the thermal and acoustic responses, respectively, and the $\omega_i$ are the frequencies of the acoustic modes. The summation extends over the excited acoustic modes, $i$, of the system. The thermal and acoustic decay rates are, respectively, $\Gamma_t$ and $\gamma_i$. In the simplest diffraction-based detection scheme, the measured signal is proportional to the diffraction efficiency, which is determined by the square of the $R(t)$. These responses, as well as the acoustic echoes, are all recorded in a single measurement of the area of the sample that is illuminated by the excitation and probing lasers. The ISTS measurement provides spatial resolution that is comparable to this area, which is typically a circular or elliptical region with a characteristic dimension (e.g., diameter or major axis) of 50 to 500 μm. In certain situations, the effective resolution can be significantly better than this length scale (Gostein et al., 2000).

Figure 2 shows data from a polymer film on a silicon substrate. The onset of diffraction coincides with the arrival of the excitation pulses at $t = 0$. The slow decay of signal is associated with thermal diffusion (Fig. 2B); the oscillations in Fig. 2A are due to acoustic modes that propagate

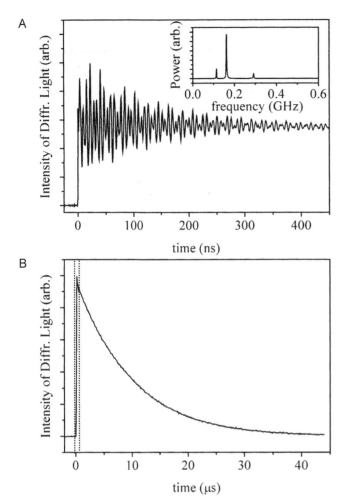

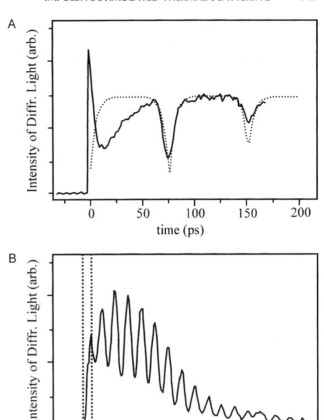

**Figure 2.** Typical ISTS data from a thin polymer film (thickness ~4 μm) on a silicon substrate. Part (**A**) shows the in-plane acoustic response, which occurs on a nanosec time scale and has a single well defined wavelength (8.32 μm) determined by the color and crossing angle of the excitation pulses. The oscillations in the signal reveal the frequencies of the different acoustic waveguide modes that are excited in this measurement. The inset shows the power spectrum. The frequencies are determined by the acoustic wavelength and the mechanical properties of the film, the substrate, and the nature of the interface between them. The acoustic waves eventually damp out and leave a nonoscillatory component of signal that decays on a microsec time scale (**B**). This slow response is associated with the thermal grating. Its decay rate is determined by the wavelength of the response and the thermal diffusivity of the structure. The dashed lines in (B) indicate the temporal range displayed in (A).

**Figure 3.** Typical ISTS data from an ultrathin metal film (thickness ~200 nm) on a silicon substrate. Part (**A**) shows the arrival of two acoustic echoes produced by longitudinal acoustic wavepackets that are launched at the surface of the film and reflect at the interface between the film and the substrate and at the interface between the film and the surrounding air. Propagation of acoustic modes in the plane of the film causes oscillations in the signal on a nanosec time scale (**B**). Thermal diffusion produces the overall decay in signal that occurs on a nanosec time scale. The dashed lines in (B) indicate the temporal range displayed in (A).

in the plane of the film. The frequencies and damping rates of the acoustic modes, along with the acoustic wavelength determined from Equation 1, define the real and imaginary parts of the phase velocities. On a typically faster (picosecond) time scale it is also possible to resolve responses due to longitudinal waves that reflect from the film/substrate interface. Figure 3 shows both types of acoustic responses evaluated in a single measurement on a thin metal film on a silicon substrate.

Although the measured acoustic frequencies and echoes themselves can be important (e.g., for filters that use sur-

face acoustic waves or thin film acoustic resonances, respectively), the intrinsic elastic properties of the films are often of interest. The acoustic echoes yield, in a simple way, the out-of-plane compressional modulus, $c_o$, when the density, $\rho$, and thickness, $h$, are known. The measured roundtrip time in this case defines, with the thickness, the out-of-plane longitudinal acoustic velocity, $v_o$; the modulus is $c_0 = \rho v_0^2$.

Extracting moduli from the in-plane acoustic responses is more difficult because thin films form planar acoustic waveguides that couple in- and out-of-plane compressional and shearing motions (Farnell and Adler, 1972; Viktorov, 1976). An advantage of this characteristic is that, in principle, the dispersion of the waveguide modes can be used to determine a set of anisotropic elastic constants as well as film thicknesses and densities. Determining these properties requires an accurate measurement of the dispersion of the waveguide and a detailed understanding of how

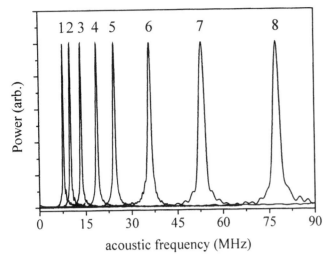

**Figure 4.** Power spectra from data collected in ISTS measurements on a thin membrane at several different acoustic wavelengths between ~30 μm and ~6 μm and labeled 1 to 8. The variation of the frequency with wavelength defines the acoustic dispersion. Analyzing this dispersion with physical models of the waveguide acoustics of the membrane yields the elastic constants and other properties.

acoustic waves propagate in layered systems. There are two approaches for determining the dispersion. One involves a series of measurements with different angles between the excitation pulses to determine the acoustic response as a function of wavelength. Figure 4 shows the results of measurements that determine the dispersion of a thin unsuppported membrane using this method (Rogers and Bogart, 2000). The other approach uses a single measurement performed with specialized beam-shaping optics that generate more than two excitation pulses (Rogers, 1998). Figure 5 shows data that were collected in an

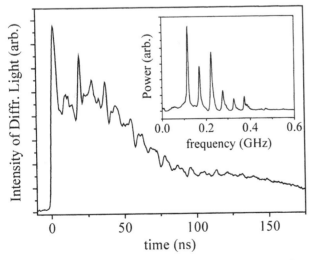

**Figure 5.** ISTS data from a thin film of platinum on a silicon wafer. This measurement used a system of optics to generate and cross six excitation pulses at the surface of the sample. The complex acoustic disturbance launched by these pulses is characterized by six different wavelengths. The power spectrum reveals the frequency components of the signal. A single measurement of this type defines the acoustic dispersion.

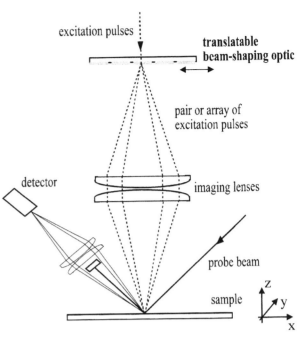

**Figure 6.** Optical system for ISTS measurements. Pulses from an excitation laser pass through a beam-shaping optic that splits the incident pulse into an array of diverging pulses. This optic contains a set of patterns, each of which produces a different number and distribution of pulses. A pair of imaging lenses crosses some of the pulses at the surface of a sample. The optical interference pattern produced in this way defines the wavelengths of acoustic motions stimulated in the sample. Diffraction of a probe laser is used to monitor the time dependence of these motions. A pair of lenses collects the diffracted signal light and directs it to a fast photodetector. The wavelengths of the acoustic waves can be changed simply by translating the beam-shaping optic.

ISTS measurement with six crossed excitation pulses. The response in this case contains contributions from acoustic modes with six different wavelengths, which are defined by the geometry of the six-pulse interference pattern. In both cases, the size of the excitation region is chosen, through selection of appropriate lenses, to be at least several times larger than the longest acoustic wavelength. This geometry ensures precise definition of the spatial periods of the interference patterns and, therefore, of the acoustic wavelengths.

The simplicity and general utility of the modern form of ISTS derives from engineering advances in optical systems that enable rapid measurement of acoustic dispersion with the two approaches described above (Rogers et al., 1997; Maznev et al., 1998; Rogers and Nelson, 1996; Rogers, 1998). Figure 6 schematically illustrates one version of the optical setup. Specially designed, phase-only beam shaping optics produce, through diffraction, a pair or an array of excitation pulses from a single incident pulse. A selected set of the diffracted pulses pass through a pair of imaging lenses and cross at the sample surface. Their interference produces simple or complex intensity patterns with geometries defined by the beam shaping optic and the imaging lenses. For the simple two-beam interference described by Equation 1, this optic typically consists of a binary phase grating optimized for diffraction at the excitation wavelength. Roughly 80% of the light that passes through this type of grating diffracts into the +1 and −1 orders. The imaging lenses cross these two orders at the sample to produce a sinusoidal intensity pattern with periodicity $\Lambda$ when the magnification of the lenses is unity and the period of the grating is $2\Lambda$. More complex beam-shaping optics produce more than two excitation pulses, and, therefore, interference patterns that are characterized by more than one period. In either case, a useful beam-shaping optic contains many (20 to 50) different spatially separated diffracting patterns. The excitation geometry can then be adjusted simply by translating this optic so that the incident excitation pulses pass through different patterns. With this approach, the response of the sample at many different wavelengths can be determined rapidly, without moving any of the other elements in the optical system.

Detection is accomplished by imaging one or more diffracted probe laser beams onto the active area of a fast detector. When the intensity of these beams is measured directly, the signal is proportional to the product of the square of the amplitude of the out-of-plane acoustic displacements (i.e., the square of Equation 2) and the intensity of the probing light. Heterodyne detection approaches, which measure the intensity of the coherent optical interference of the signal beams with collinear reference beams generated from the same probe laser, provide enhanced sensitivity. They also simplify data interpretation since the heterodyne signal, $S(t)$, is linear in the material response. In particular, in the limit that the intensity of the reference beam, $I_r$, is large compared to the diffracted signal

$$S(t) \propto |\sqrt{I_p}R(t) + \sqrt{I_r}e^{i\varphi}| \approx I_r + 2\sqrt{I_p I_r}R(t)\cos\varphi \quad (3)$$

where $\varphi$ is the phase difference between the reference and diffracted beams and $I_p$ is the intensity of the probing beam. A general scheme for heterodyne detection that uses the beam-shaping optic to produce the excitation pulses, as well as to generate the reference beam, is a relatively new and remarkably simple approach that makes this sensitive detection method suitable for routine use (Maznev et al., 1998; Rogers and Nelson, 1996). Figure 7 schematically illustrates the optics for measurements on transparent samples; a similar setup can be used in reflection mode. With heterodyne and nonheterodyne detection, peaks in the power spectrum of the signal define frequencies of the acoustic responses. In the case of nonheterodyne signals, sums and differences and twice these frequencies (i.e., cross terms that result from squaring Equation 2) also appear. Figure 8 compares responses measured with and without heterodyne detection.

The measured dispersion can be combined with models of the waveguide acoustics to determine intrinsic material properties. The general equation of motion for a nonpiezoelectric material is given by

$$\frac{\partial^2 \mathbf{u}_j}{\partial t^2} = \frac{\mathbf{c}_{ijkl}}{\rho}\frac{\partial^2 \mathbf{u}_k}{\partial x_i \partial x_l} + \frac{\partial}{\partial x_i}\left(\sigma_{ik}^{(r)}\frac{\partial \mathbf{u}_j}{\partial x_k}\right) \quad (4)$$

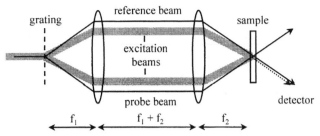

**Figure 7.** Optical system for ISTS measurements with heterodyne detection. Both the probe and the excitation laser beams pass through the same beam-shaping optic. The excitation pulses are split and recombined at the surface of the sample to launch the acoustic waves. The probe light is also split by this optic into more than one beam. For this illustration, the beam shaping optic produces a single pair of probing and excitation pulses. One of the probe beams is used to monitor the material response. The other emerges collinear with the signal light and acts as a reference beam for heterodyne amplification of signal generated by diffraction of the beam used for probing. The beam-shaping optic and imaging lenses ensure overlap of the signal and reference light. An optical setup similar to the one illustrated here can be used for measurements in reflection mode.

where $\mathbf{c}$ is the elastic stiffness tensor, $\mathbf{u}$ is the displacement vector, $\rho$ is the density of the material and $\sigma$ is the residual stress tensor. Solutions to this equation, with suitable boundary conditions at all material interfaces, define the dispersion of the velocities of waveguide modes in arbitrary systems (Farnell and Adler, 1972; Viktorov, 1976). Inversion algorithms based on these solutions can be used to determine the elastic constants, densities, and/or film thicknesses from the measured dispersion. For thickness determination, the elastic constants and densities are typically assumed to be known; they are treated as fixed parameters in the inversion. Similarly, for elastic constant evaluation, the thicknesses and densities are treated as fixed parameters. It is important to note, however, that only the elastic constants that determine the velocities of modes with displacement components in the vertical (i.e., sagittal) plane can be determined, because ISTS probes only these modes. Elastic constants that govern the propagation of in-plane shear acoustic waves polarized in the plane of the film, for example, cannot be evaluated. Figure 9 illustrates calculated distributions of displacements in the lowest six sagittal modes for a polymer film that is strongly bonded to a silicon substrate. As this figure illustrates, the modes involve coupled in- and out-of-plane shearing and compressional motions in the film and the substrate. The elastic constants and densities of the film and substrate materials and the thickness of the film determine the spatial characters and velocities of the modes. Their relative contributions to the ISTS signal are determined by their excitation efficiency and by their diffraction efficiency; the latter is dictated by the amount of surface ripple that is associated with their motion.

## PRACTICAL ASPECTS OF THE METHOD

The setup illustrated in Figure 6 represents the core of the experimental apparatus. The entire system can either

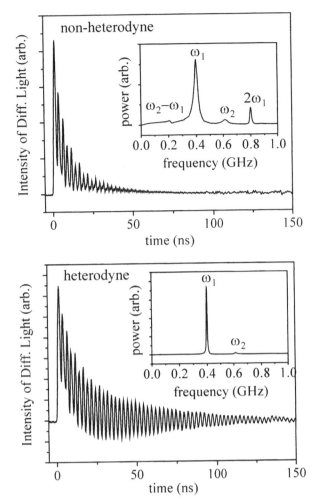

**Figure 8.** ISTS signal and power spectrum of the signal from an electroplated copper film (thickness ∼1.8 μm) on a layer of silicon dioxide (thickness ∼0.1 μm) on a silicon substrate at an acoustic wavelength ∼7 μm. Part (**A**) shows the signal obtained without heterodyne amplification. The fact that the signal is quadratic with respect to the material displacement results in what appears as a fast decay of the acoustic oscillations. This effect is caused by the decay of the background thermal response rather than a decay in the acoustic component of the signal itself. Part (**B**) shows the heterodyne signal; the spectrum contains two peaks corresponding to Rayleigh and Sezawa waves, i.e., the two lowest-order modes of the layered structure. The signal without heterodyning shows these frequencies and combinations of them, due to the quadratic dependence of the signal on the response. The peak width in the spectrum of the signal in (**A**) is considerably larger than that in (**B**). This effect is due to the contribution of the thermal decay to the peak width in the nonheterodyne case. Note also that the heterodyne signal appears to drop below the baseline signal measured before the excitation pulses arrive. This artifact results from the insensitivity of the detector to DC levels of light.

be obtained commercially or most of it can be assembled from conventional, off-the-shelf components (lenses, lasers, etc.). The beam-shaping optics can be obtained as special order parts from commercial digital optics vendors. Alternatively, they can be custom built using relatively simple techniques for microfabrication (Rogers, 1998).

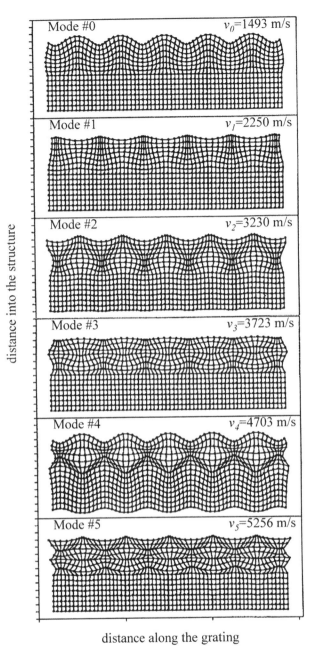

distance along the grating

**Figure 9.** Computed displacement distributions for the lowest six waveguide modes in a polymer film supported by a silicon substrate. These modes are known as Rayleigh or Sezawa modes. They are excited and probed efficiently in ISTS measurements because they involve sagittal displacements that couple strongly both to the thermal expansion induced by the excitation pulses and to surface ripple, which is commonly responsible for diffracting the probe laser. The spatial nature of these modes and their velocities are defined by the mechanical properties of the film and the substrate, the nature of the interface between them, and the ratio of the acoustic wavelength to the film thickness.

The excitation pulses are typically derived from a flash lamp or diode-pumped Nd:YAG laser, although any source of laser pulses with good coherence and mode properties can be used. The pulse duration must be short compared to the temporal period of the excited acoustic wave. In many instances, pulses shorter than ∼300 psec are suitable. Pulse energies in the range of one or a few μJ provide adequate excitation for most samples. Nonlinear crystals can be used to double, triple, quadruple, or even quintuple the output of the Nd:YAG (wavelength ~1064 nm) to produce 532, 355, 266, or 213 nm light. The color is selected so that a sufficient fraction of the incident light is absorbed by the sample. Ultraviolet (UV) light is a generally useful wavelength in this regard, although the expense and experimental complexity of using multiple nonlinear crystals represent a disadvantage of UV light when a Nd:YAG laser is used. Nonlinear frequency conversion can be avoided by depositing thin absorbing films onto samples that are otherwise too transparent or reflective to examine directly at the fundamental wavelength of the laser. The probe laser can be pulsed or continuous wave. For most of the data presented in this unit, we used a continuous-wave infrared (850 nm) diode laser with a power of ~200 mW. Its output is electronically gated so that it emits only during the material response, which typically lasts no longer than 100 μsec after the excitation pulses strike the sample.

Alignment of the optics (which is performed at the factory for commercial instruments) ensures that the probing beam overlaps the crossed excitation beams and that signal light reaches the detector. The size of the beams at the crossing point is typically in the range of one or several hundred microns. The alignment of the probe laser can be aided by the use of a pinhole aperture to locate the crossing point of the excitation beams. Routine measurement is accomplished by placing the surface of the sample at the intersection of the excitation and probing laser beams, moving the beam-shaping optic to the desired pattern(s), recording data, and interpreting the data. The sample placement is most easily achieved by moving the sample along the $z$ direction (Fig. 6) to maximize the measured signal, which can be visualized in real time on a digitizing oscilloscope. The surface normal of the sample is adjusted to be parallel to the bisector of the excitation beams.

Recorded data typically consist of an average of responses measured for a specified number of excitation-probing events. The strength of the signal, the required precision, and the necessary measurement time determine the number of averages: between one hundred and one thousand is typical. This averaging requires <1 sec with common excitation lasers (repetition rates ~1 kHz). Translating this optic so that the excitation laser beam passes through different diffracting patterns changes the acoustic wavelength(s). Single or multiple measurements of this type define acoustic response frequencies as a function of wavelength, which, in most cases, varies from 2 to 200 μm. The optics and the wavelengths of the lasers determine the range that is practical.

The measurement procedures described above apply to a wide variety of samples. Figures 10 to 13 show, respectively, data from a film of lead zirconium titanate (PZT) on silicon, a thin layer of nanoporous silica glass on silicon, a layer of paint on a plastic automobile bumper part, and a film of cellulose. In the last example, the data were collected with a single shot of the laser. In each case, the oscillations in the signal are produced by surface acoustic waveguide modes that involve coupled displacements in

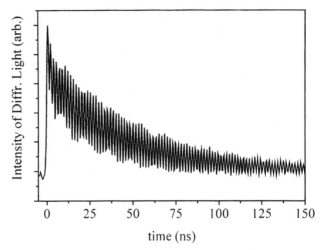

**Figure 10.** ISTS data from a thin film of lead zirconium titanate on a layer of platinum on a silicon wafer. Crossed laser pulses coherently stimulate the lowest-order Rayleigh waveguide acoustic mode, with a wavelength defined by the optical interference pattern. The motions induce ripple on the surface of the structure; this ripple diffracts a probing laser beam that is overlapped with the excited region of the sample. The acoustic waveguide motions involve coupled displacements in the film and underlying substrate. In this case, the overall decay of the signal caused by thermal diffusion occurs on a time scale comparable to the acoustic decay time.

the outer film, and, in general, in all material layers beneath it. Their frequencies are therefore functions of the material properties of each component of the structure and the nature of the boundary conditions between them. These and other waveguide properties can be calculated

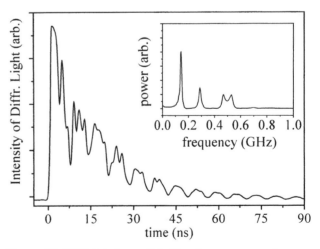

**Figure 11.** ISTS data from a thin film of a nanoporous glass on a silicon substrate. A thin overlayer of aluminum was added in this case to induce optical absorption at the wavelength of the excitation laser (1064 nm). The measurement stimulates numerous acoustic waveguide modes, each with a wavelength equal to the period of the optical interference pattern formed by the crossed laser pulses. The power spectrum in the inset shows the various frequency components of the signal. In this case, the overall decay of the signal caused by thermal diffusion occurs on a time scale comparable to the decay time of the lowest order acoustic waveguide mode.

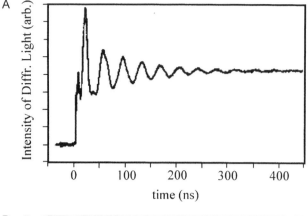

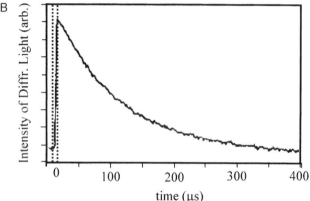

**Figure 12.** ISTS data from a layer of paint on a plastic automobile bumper part. Part (**A**) shows the acoustic component of the signal. Part (**B**) shows the thermal decay, which occurs on a microsec time scale. The vertical dashed lines in the bottom frame indicate the temporal window of the signal shown in the upper frame. The acoustic and thermal properties of paint can be relevant to developing materials that resist delamination and maintain their mechanical flexibility even after exposure to harsh environmental conditions. The dashed lines in (B) indicate the temporal range illustrated in (A).

from first principles using standard procedures (Farnell and Adler, 1972; Viktorov, 1976).

The results in Figures 10 to 13 are representative of the range of acoustic responses that are typically encountered. The PZT film on silicon (Fig. 10) only supports a single waveguide mode at this acoustic wavelength, because its thickness is small compared to the wavelength. The response consists, therefore, of a single frequency whose value is determined primarily by the density and thickness of the film and by the mechanical properties of the silicon. In contrast, the nanoporous glass on silicon (Fig. 11) supports many acoustic waveguide modes because the acoustic wavelength is comparable to the film thickness. The data, as a result, contain many frequency components. Several of these modes involve motions confined largely to the film; their characteristics are determined primarily by the film mechanical properties and its film thickness. The layer of paint on the plastic bumper part (Fig. 12) is thick compared to the acoustic wavelength and it therefore supports many waveguide modes, like the nanoporous glass. The excitation in this case, however, is localized to

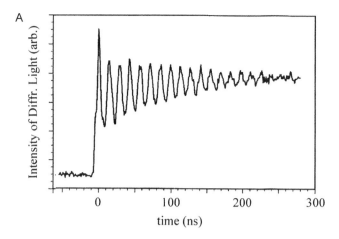

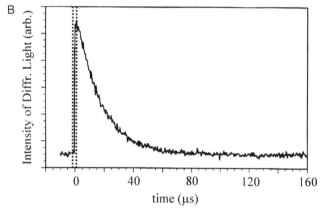

**Figure 13.** ISTS data collected in a single shot of the laser from a film of cellulose. Parts (**A**) and (**B**) show the acoustic and thermal components of the signal that was recorded in this measurement. The dashed lines in (B) indicate the temporal range illustrated in (A).

the very top surface of the sample. Only the lowest waveguide mode, which is known as the Rayleigh mode, is stimulated and detected efficiently. The frequency of this mode is determined entirely by the mechanical properties of the film. It is independent of the film thickness and of the mechanical properties of the substrate. Finally, the cellulose sample (Fig. 13) is a thick free-standing film. The excited mode, in this case, is similar, in its physical characteristics and in its sensitivity to film properties, to the Rayleigh mode observed in the paint sample.

The precision (i.e., repeatability) of the frequencies measured from data like those shown in Figs. 10 to 13 can be better than 0.01% to 0.001%. The acoustic wavelengths are determined by the experimental apparatus: the geometry of the beam-shaping optic, its position relative to the imaging lenses, and the magnification of these lenses. This information, or a direct measurement of the excitation beam crossing angles, can be used to evaluate the wavelengths. Measurements with high accuracy are typically achieved by measuring response frequencies from standard samples or by using a microscope to view the interference pattern directly. With these latter two techniques the wavelengths can be determined to an accuracy of better than 0.1%.

## METHOD AUTOMATION

Many aspects of the measurement can be automated. The beam-shaping optic in a commercial machine, for example, is mounted on a computer-controlled motorized translation stage. This motor, along with the transient recorder (i.e., digitizing oscilloscope), a variable neutral density filter for adjusting the intensity of the excitation light, and a precision stage that supports the sample, are all controlled by a single computer interface. Robotic handlers are also available to load and unload samples into and from the machine. An automated focusing operation positions the sample along the $z$ direction to maximize the signal. The computer controls data collection at user-defined acoustic wavelengths and also computes the power spectrum from the time domain data. It fits peaks in this spectrum to define the acoustic frequencies, and, in the case of film thickness, uses these frequencies (and the known acoustic wavelength, elastic constants, and densities) to evaluate the unknown layer(s). This fully automated procedure can be coordinated with motion of the sample stage to yield high-resolution (better than 10-μm in certain cases) images of the acoustic frequency, other aspects of the ISTS signal (e.g., amplitude, acoustic damping rate, thermal decay time, etc.), and/or computed material characteristics (e.g., film thickness). In a noncommercial ISTS apparatus for general laboratory use, the stages for the sample, the beam-shaping optic, the neutral density filter, etc., are manually controlled. Data downloaded from an oscilloscope in such a setup are analyzed using a separate computer equipped with software for the fitting and modeling procedures.

## DATA ANALYSIS AND INITIAL INTERPRETATION

As discussed in the previous sections, the ISTS measurements provide information on the acoustic properties of the sample. The acoustic frequencies can be obtained directly by fitting the time domain ISTS data to expressions given in Equation 2 or Equation 3. They can also be determined by Fourier transformation. In the latter case, the positions of the peaks in the power spectrum are typically defined by fitting them to Lorentzian or Gaussian line shapes. The experimental apparatus defines the crossing angles of the excitation pulses and, therefore, the acoustic wavelength(s).

Inverse modeling with the acoustic frequencies and wavelengths determines elastic and other properties. In this procedure, the elastic constants, densities and thicknesses of the "unknown" layer(s) are first set to some approximate initial values. The dispersion is then calculated with Equation 4 and the known properties of the other layers. The sum of squared differences, $\chi^2$, between the computed and measured modal phase velocities provides a metric for how well the modeling reproduces the observed dispersion. In a generally nonlinear iterative search routine, the properties of the unknown layers are adjusted and $\chi^2$ is calculated for each case. The properties that minimize $\chi^2$ represent best fit estimates for the intrinsic characteristics of the "unknown" layers. For the simple case of determining the thickness of a single layer in a film stack whose other properties are known, this procedure reduces to a straightforward non-iterative calculation that yields the thickness from a single measured phase velocity. Clearly, for these types of fitting procedures to be successful, the number of parameters to be determined cannot exceed the number of measured velocities. The uncertainties in the fitted parameters will typically decrease as the number of measured velocities, and the range of wavelengths increase. These uncertainties can be quantitatively estimated using established statistical analysis such as the $F$ test (Beale, 1960). Below we examine this inverse modeling in some detail, with illustrations for rigidity and stress determination in thin membranes, elastic moduli determination in supported and unsupported films, and thickness evaluation of films in multilayer stacks.

We begin with the simple case of a thin unsupported membrane. When the thickness ($h$) of the membrane is much smaller than the acoustic wavelength, then it is possible to model the dispersion of the drumhead mode (i.e., the lowest-order waveguide mode, which is typically easy to observe in an ISTS measurement) with small-deflection plate theory. We also assume that the ISTS excitation region is large enough to ignore the dimension perpendicular to the interference fringes (i.e., the $y$ axis is Fig. 6). In these limits, Equation 4 reduces to (Rogers and Bogart, 2000)

$$\frac{Eh^2}{12(1-v^2)}\frac{\partial^4 \mathbf{u}}{\partial x^4} + \rho \frac{\partial^2 \mathbf{u}}{\partial t^2} = \sigma \frac{\partial^2 \mathbf{u}}{\partial x^2} \qquad (5)$$

where $E$ is Young's modulus, $v$ is the Poisson ratio, $\rho$ is the density of the film, and $\sigma$ is the residual stress. The coordinate $x$ lies along the interference fringes. The dispersion of the phase velocity ($v_\varphi$) determined with Equation 5 can be written

$$v_\varphi = \sqrt{\frac{E}{12\rho(1-v^2)}(kh)^2 + \frac{\sigma}{\rho}} = \sqrt{\frac{D}{\rho h}k^2 + \frac{\sigma}{\rho}} \qquad (6)$$

where $D$ is the flexural rigidity. This simple equation can be used with ISTS measurements of the dispersion of the lowest-order Lamb mode to determine, for example, the flexural rigidity and stress when the thickness and density of the membrane are known. Figure 14 shows ISTS data from an unsupported bilayer membrane of tungsten (W; ~25 nm)/silicon nitride (SiN; ~150 nm) and best fit curves that use Equation 6. The results confirm the validity of the plate mode approximations for this sample. The stress and rigidity measured for this sample are 241 ± 1 MPa and 253 ± 1 GPa respectively. The uncertainties in this case are dominated by uncertainties in the density of the membrane. We note that the ISTS results agree with independent evaluation of these quantities by the resonant frequency (RF) method and the bulge test, respectively (Rogers and Bogart, 2000; Rogers et al., 2000b). Unlike ISTS, however, both of these methods require specialized test structures. Also, they do not offer spatial resolution

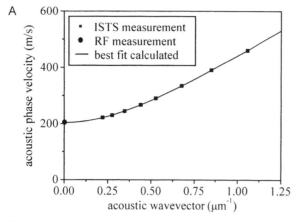

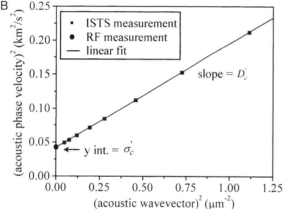

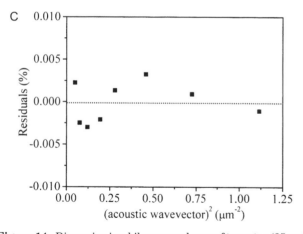

**Figure 14.** Dispersion in a bilayer membrane of tungsten (25 nm) and silicon nitride (150 nm) measured by ISTS. Part (**A**) shows the measured (symbols) and best fit calculated (line) variation of the acoustic phase velocity with wavevector. Part (**B**) shows the expected linear behavior of the square of the velocity with the square of the wavevector. The slope of the line that passes through these data points defines the ratio of the composite flexural rigidity to the product of the composite density and the membrane thickness. Its intercept determines the ratio of the composite residual stress (tensile) to the composite density. Part (**C**) illustrates the percent deviation of the data shown in part (**B**) from the best fit line. The small size of these residuals provides some indication of (1) the accuracy of the acoustic frequency measured by ISTS, (2) the accuracy of the measured wavelengths, and (3) the validity of the plate theory approximations. The large circular symbols in parts (**A**) and (**B**) are data from independent resonant frequency measurements on this sample.

and they cannot easily evaluate in-plane anisotropies in the stress or rigidity.

As the acoustic wavelength approaches the thickness of the membrane, the plate theory ceases to be valid, and a complete waveguide description based on Equation 4 is required. In this regime, multiple dispersive modes known as Lamb modes are possible (Farnell and Adler, 1972; Viktorov, 1976). Their velocities, $v_i$, at an acoustic wavelength, $\Lambda$, are defined by positions of zeroes in the determinant of a matrix defined by the boundary conditions appropriate for the system. If this determinant is denoted by the function $G$, then for a mechanically isotropic, stress-free membrane we can write

$$G(v_i, \Lambda, h, v_{tr}^{(f)}, v_{lg}^{(f)}) = 0 \qquad (7)$$

where $v_{tr}^{(f)}$ and $v_{lg}^{(f)}$ are the intrinsic transverse and longitudinal velocities of the film (i.e., the velocities that would characterize acoustic propagation in a bulk sample of the film material). Simple equations relate these intrinsic velocities to the Young's modulus and Poisson ratio (Fetter and Walecka, 1980). Figure 15 shows the measured dispersion of a set of Lamb modes in an unstressed polymer membrane and calculations based on Equation 7 with best fit estimates for the intrinsic velocities: $v_{tr}^{(f)} = 1126 \pm 4$ m/sec, and $v_{lg}^{(f)} = 2570 \pm 40$ m/sec (Rogers et al., 1994b). A general feature of the dispersion is that the mode velocities scale with the product of the acoustic wavevector, $k = 2\pi/\Lambda$, and the thickness, $h$, as expected based on the plate mode result of Equation 6. Plate-like behavior of the lowest-order mode results when the acoustic wavelength is long compared to the film thickness (i.e., $kh$ is small). In this limit, the second-lowest-order mode acquires a velocity slightly smaller than the intrinsic longitudinal velocity of the film, $v_{lg}^{(f)}$. When $kh$ is large, the velocities of the two lowest modes approach the Rayleigh wave velocity of the film (i.e., the Rayleigh mode is a surface localized wave that propagates on a semi-infinite substrate). All

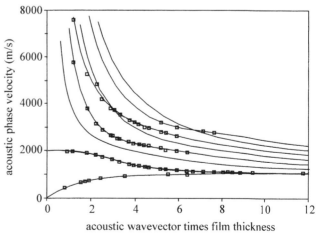

**Figure 15.** Dispersion in a thin unsupported polymer film measured by ISTS (symbols). The lines represent best fit calculations that allowed the intrinsic longitudinal and transverse acoustic velocities of the film to vary. With the density, simple expressions can be used to relate these velocities to the elastic moduli.

other mode velocities approach the intrinsic transverse velocity of the film, $v_{tr}^{(f)}$. These and other characteristic features of the dispersion illustrate how the elastic properties (i.e., intrinsic velocities) of a thin film can be determined from the measured variation of waveguide mode velocities with wavelength. Iterative fitting routines achieve this determination in an automated fashion that also allows for statistical estimation of uncertainties. Similar procedures can be used to evaluate membranes with anisotropic elastic constants (Rogers et al., 1994a).

When the film is supported by a substrate or when it is part of a multilayer stack, the associated boundary conditions are functions of the elastic properties, densities, and thicknesses of all constituent layers and the supporting substrate. In the general case, the boundary condition determinant can be written

$$G(v_i, \Lambda, h^{(f1)}, v_{tr}^{(f1)}, v_{lg}^{(f1)}, \rho^{(f1)}, h^{(f2)}, v_{tr}^{(f2)}, v_{lg}^{(f2)}, \rho^{(f2)}, \dots,$$
$$v_{tr}^{(s)}, v_{lg}^{(s)}, \rho^{(s)}) = 0 \qquad (8)$$

where $v_{tr}^{(fi)}$ and $v_{lg}^{(fi)}$ are the intrinsic transverse and longitudinal velocities of the films, $h^{(i)}$ are their thicknesses, and $\rho^{(fi)}$ are their densities. The corresponding quantities for the substrate (assumed to be semi-infinite) are $v_{tr}^{(s)}$, $v_{lg}^{(s)}$ and $\rho^{(s)}$. Figure 16 shows the best fit computed and measured dispersion for a thin film of nanoporous silica on a silicon wafer (Rogers and Case, 1999). The intrinsic velocities determined from this fitting are $v_{lg} = 610 \pm 50$ m/sec and $v_{tr} = 400 \pm 30$ m/sec. The uncertainties in this case are dominated by uncertainties in the thickness and density of a thin overlayer of aluminum that was used to

induce slight absorption at 1064 nm, the wavelength of the excitation pulses in these measurements. Note that the overall behavior of the dispersion in this case is much different from that for an unsupported film (Fig. 15). The velocity of the lowest mode approaches the Rayleigh velocities of the film and substrate at large and small values of $kh$, respectively. All other modes approach $v_{tr}^{(f)}$ at large $kh$. As $kh$ decreases, each of these modes is, at a certain value of $kh$, "cut off" at $v_{tr}^{(s)}$. Above this velocity they are no longer strictly guided modes. In this regime, some of their energy "leaks" into bulk transverse waves in the semi-infinite substrate.

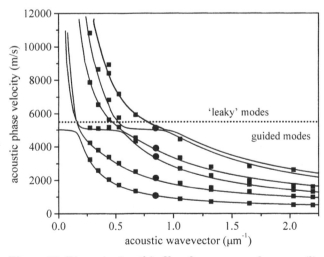

**Figure 16.** Dispersion in a thin film of nanoporous glass on a silicon substrate measured by ISTS (symbols). In this case, a thin film of aluminum (thickness ~75 nm) was deposited on top of the nanoporous glass in order to induce absorption at the wavelength (1064 nm) of the excitation laser used for these measurements. The lines represent best fit calculations that allowed the intrinsic longitudinal and transverse acoustic velocities of the nanoporous glass to vary. The fitting used literature values for the densities and elastic properties of the aluminum and the silicon. The thickness of the aluminum film was fixed to its nominal value. The circular symbols represent data illustrated in Figure 8b.6.11.

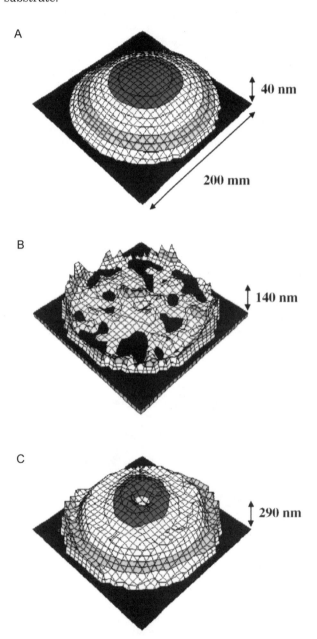

**Figure 17.** ISTS measured maps of thickness in thin films of copper on silicon wafers with 200 mm diameters. Parts (**A**), (**B**), and (**C**) show, respectively, copper deposited by sputter deposition (average thickness ~180 nm), electroplating (average thickness ~1.6 μm), and an electroplated copper film after chemical-mechanical polishing (average thickness ~500 nm).

The dependence of the waveguide mode velocities on film thickness provides the basis for thickness determination with ISTS (Banet et al., 1998). Often, in practice a single measurement of a mode velocity coupled with the properties of the other components of the structure determines the thickness. The accuracy of the measurement is, of course, highest when the velocity of the waveguide mode depends strongly on the thickness and when the properties of the other components of the structure are known accurately. In a silicon supported system, the velocity of the lowest-order mode (typically excited and probed efficiently in an ISTS measurement) in most cases exhibits sufficient variation with thickness when $kh$ is not large. For films with thicknesses less than $\sim$100 μm, and with properties that are reasonably dissimilar to those of the substrate, it is possible to tune $k$ into a range that allows good sensitivity. Figures 17 and 18 show several examples of thickness evaluation in thin metal films used in microelectronics. Note that the accuracy of the measured thickness depends on the assumed elastic properties, densities, and thicknesses of the substrate and other films in the stack. For many systems, literature values for the physical properties yield acceptable accuracy. In other cases, procedures that calibrate these properties through ISTS analysis of samples with known thicknesses are necessary. For complex structures, such as copper damascene features in microelectronics (i.e., lines of copper embedded in layers of oxide), effective medium models are often preferred to the complex calculations that are required to simulate these systems using first-principles physics based on Equation 4. Figure 19 shows the results of line scans across copper damascene structures that use these types of models for computing thicknesses. With calibrated tools for measuring these and other systems for microelectronics, thickness

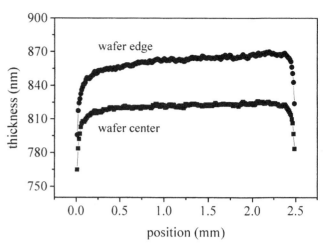

**Figure 19.** Thickness profiles of copper damascene arrays consisting of copper lines (width 0.5 μm, length $\sim$1 mm and average thickness $\sim$800 nm) separated by 0.5 μm and embedded in a layer of silicon dioxide. Acoustic wave propagation direction is along the trenches (i.e., fringes of the excitation interference pattern lie perpendicular to the long dimension of the copper lines). In this case, effective elastic properties of the copper damascene structure needed for the thickness calculation can be determined by averaging elastic constants of copper and silicon dioxide using a parallel spring model. Thickness values determined using these effective properties were in agreement with SEM measurements. The data illustrate a major problem in the copper interconnect technology, i.e., nonuniformities in the chemical-mechanical polishing process used to form these structures.

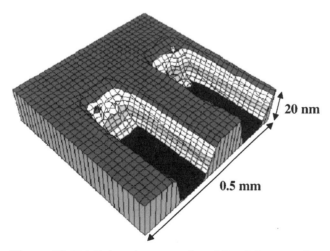

**Figure 18.** Detailed contour map of an 0.8 × 0.8–mm region showing variations in the thickness of a layer of tantalum in a stack structure of 200 nm copper/Ta (patterned)/300 nm silicon dioxide/silicon. The measurements clearly resolve the regions of the structure where there is no tantalum. The thickness of tantalum in the other areas is $\sim$25 nm. In order to measure the thickness of this buried layer, thicknesses and properties of the other layers, and the mechanical properties and density of the tantalum, were assumed to be known.

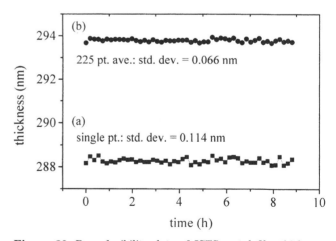

**Figure 20.** Reproducibility data of ISTS metal film thickness measurement obtained on a commercially available system (Philips Analytical's PQ Emerald). The sample (a silicon wafer with 200 mm diameter and coated with 300 nm tantalum on 300 nm silicon dioxide) was loaded and unloaded from the system for each measurement. The signal was averaged over 300 laser shots (measurement time $\sim$1 sec per point). (**A**) Single-point data. (**B**) Average of a 225-point wafer map. In both cases, the reproducibility is on the order or less than 0.1 nm, roughly a single atomic diameter.

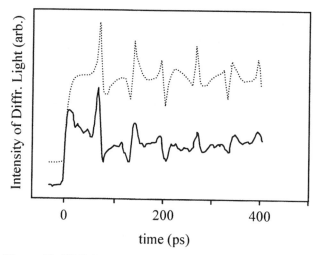

**Figure 21.** ISTS data on a picosec time scale from a multilayer stack of aluminum (thickness ∼100 nm) on titanium-tungsten (thickness ∼75 nm) on a silicon wafer. The dashed line shows a simulated response. The features in the data correspond to reflections of longitudinal acoustic wavepackets generated at the surface of the sample by the crossed excitation pulses. The timing of these echoes, with the out-of-plane longitudinal acoustic velocities of the materials, yields the thicknesses of both of the films.

effects of parasitic scatter. Spatial filtering can also help to reduce the amount of scatter that reaches the detector. From a sample-preparation point of view, scatter can be reduced in many cases by adding an index-matching fluid, or by mechanical polishing.

As mentioned above (see Practical Aspects of the Method), in order to ensure adequate ISTS signals, excitation light must produce some mild amount of heating in the sample. This heating can be achieved by using an excitation wavelength that is absorbed directly by the sample, or by adding a thin absorbing overlayer. Generally, the materials used for these overlayer films and their thicknesses are chosen to minimize their effects on the acoustic response of the sample under test. Alternatively, the acoustic influence of these films can be accounted for directly through calculation. Thin aluminum coatings are useful because they provide sufficient absorption to allow high quality ISTS measurements with excitation pulse energies in the range of a few μJ and wavelengths in the infrared (e.g., 1064 nm). Aluminum also has the advantage that it is a low-density material (small acoustic loading for thin films) that can be easily deposited in thin film form by thermal evaporation. Coatings with ∼50 to 75 nm thickness provide sufficient absorption, and they also only minimally affect the acoustic response of most samples.

## PROBLEMS

Problems that can arise with ISTS evaluation fall into two classes, those associated with acquiring data and those related to interpreting it. For interpretation, multilayer samples can provide the biggest challenge, because uncertainties in the properties of any layer in the stack can significantly affect the accuracy for evaluating the "unknown" layer(s). Accurate measurements of the dispersion of multiple modes over a wide range of $k$ reduces the significance of this problem because it decreases the number of assumptions necessary for the properties of the other layers in the stack.

For data collection, the sample must absorb some reasonable fraction of the excitation light. Convenient laser sources for excitation (e.g., pulsed Nd:YAG lasers) provide direct output in the near-IR range (e.g., 1064 nm). Because of the relatively high peak power, nonlinear optical crystals can be used to generate harmonics of this fundamental frequency. Gold, for example, reflects too efficiently at 1064 nm to allow for ISTS measurement with the fundamental output of a Nd:YAG laser. It absorbs, however, enough frequency-doubled light at 532 nm to allow for easy measurement. Generally, the fourth harmonic of these types of lasers (266 nm) provides a useful wavelength for measurements on a wide range of samples.

With a suitable wavelength for excitation, high-quality signals can be obtained for most samples that do not scatter strongly at the wavelength of the probing laser. Scattered light that reaches the detector can interfere with data collection either by simply adding noise or, when it coherently interferes with diffracted signal light, by distorting the shape of the measured waveform. The former problem can, in most cases, be eliminated by carefully

precision and accuracy are both in the range of a few angstroms. Figure 20 illustrates the repeatability for a typical film.

When the thicknesses of more than one film in a multilayer stack are required, additional data, such as (1) velocities of more than one waveguide mode, (2) the velocity of a single mode at more than one $k$, (3) information other than the acoustic frequency (e.g., acoustic damping, thermal or electronic response, etc.), or (4) the out-of-plane acoustic echo component of the ISTS signal can be employed. Figure 21 illustrates the last approach. The precision and accuracy (again, with proper calibration) are both in the angstrom range with this method. Commercial ISTS tools use a combination of the thermal and acoustic components of the data to determine thicknesses of two films in a stack (Gostein et al., 2000). Here, the precision and accuracy are in the range of 0.05 to 0.15 nm. The particular technique that is most convenient for a given measurement application depends on the sample.

## SAMPLE PREPARATION AND SPECIMEN MODIFICATION

The primary constraints on the sample are that (1) it be sufficiently smooth that it does not scatter enough light to interfere with measurement of the diffracted signal, and (2) it absorb a sufficient fraction of the excitation light to yield a measurable response. From a practical standpoint, the intensity of scattered light that is spatially close to or overlapping the signal should not be large compared to the intensity of the signal itself. Typical diffraction efficiencies from acoustic motions in an ISTS experiment are in the range of $10^{-6}$. Heterodyne detection schemes significantly amplify the signal levels, thereby reducing the

blocking the scattered light so that it does not reach the detector. The latter problem cannot be handled in the same way because these effects are produced by scattered light that spatially overlaps the signal light. This interference essentially causes uncontrolled heterodyne amplification. It is often difficult to simulate accurately this interference and its subtle effects on the accuracy of procedures for extracting the response frequency (e.g., Fourier transformation followed by peak fitting), because the phase and intensity of the scattered light are not generally known. These effects can be minimized, however, by employing heterodyne detection and adjusting the intensity of the reference beam so that its interference with the signal dominates the measured response. In any case, the precision of the measurement can be easily evaluated by repeated measurement of a single spot on a sample. The standard deviation of a set of frequencies measured in this fashion should be less than $\sim$0.1% for most samples.

## LITERATURE CITED

Allen, M. G., Mehregany, M., Howe, R. T., and Senturia, S. D. 1987. Microfabricated structures for the in situ measurement of residual stress, Young's modulus, and ultimate strain of thin films. *Appl. Phys. Lett.* 51:241–243.

Banet, M. J., Fuchs, M., Rogers, J. A., Reinold, J. H., Knecht, J. M., Rothschild, M., Logan R., Maznev, A. A., and Nelson, K. A. 1998. High-precision film thickness determination using a laser-based ultrasonic technique. *Appl. Phys. Lett.* 73:169–171.

Beale, E. M. L. 1960. Confidence regions in non-linear estimation. *J. Roy. Stat. Soc. B.* 22:41–76.

Clemens, B. M., and Bain, J. A. 1992. Stress determination in textured thin films using X-ray diffraction. *M. R. S. Bulletin* July:46–51.

Crimmins, T. F., Maznev, A. A., and Nelson, K. A. 1998. Transient grating measurements of picosecond acoustic pulses in metal films. *Appl. Phys. Lett.* 74:1344–1346.

Duggal, A. R., Rogers, J. A., and Nelson, K. A. 1992. Real-time optical characterization of surface acoustic modes of polyimide thin-film coatings. *J. Appl. Phys.* 72:2823–2839.

Eesley, G. L., Clemens, B. M., and Paddock, C. A. 1987. Generation and detection of picosecond acoustic pulses in thin metal films. *Appl. Phys. Lett.* 50:717–719.

Eichler, H. J., Gunter, P., and Pohl, D. W. 1986. Laser-Induced Dynamic Gratings. Springer-Verlag, New York.

Farnell, G. W. and Adler, E. L. 1972. Elastic wave propagation in thin layers. *In* Physical Acoustics, Principles and Methods, Vol. IX (W. P. Mason and R. N. Thurston, eds.) pp. 35–127. Academic Press, New York.

Fetter, A. L. and Walecka, J. D. 1980. Theoretical Mechanics of Particles and Continua. McGraw-Hill, New York.

Gostein, M., Banet, M. J., Joffe, M., Maznev, A., Sacco, R., Rogers, J. A. and Nelson, K. A. 2000. Opaque film metrology using transient-grating optoacoustics (ISTS). *In* Handbook of Silicon Semiconductor Metrology (A. Diebold, ed.) Marcel-Dekker, New York.

Hess, P. 1996. Laser diagnostics of mechanical and elastic properties of silicon and carbon films. *Appl. Surf. Sci.* 106:429–437.

Maden, M. A., Jagota, A., Mazur, S., and Farris, R. J. 1994. Vibrational technique for stress measurement in thin films. 1. Ideal membrane behavior. *J. Am. Ceram. Soc.* 77:625–635.

Maznev, A. A., Nelson, K. A., and Yagi, T. 1996. Surface phonon spectroscopy with frequency-domain impulsive stimulated light scattering. *Sol. St. Comm.* 100:807–811.

Maznev, A. A., Rogers, J. A., and Nelson, K. A. 1998. Optical heterodyne detection of laser-induced gratings. *Opt. Lett.* 23:1319–1321.

Mizubayashi, H., Yoshihara, Y., and Okuda, S. 1992. The elasticity measurements of aluminum-nm films. *Phys. Status Solidi A* 129:475–481.

Nelson, K. A. and Fayer, M. D. 1980. Laser induced phonons: A probe of intermolecular interactions in molecular solids. *J. Chem. Phys.* 72:5202–5218.

Neubrand, A. and Hess, P. 1992. Laser generation and detection of surface acoustic-waves: Elastic properties of surface layers. *J. Appl. Phys.* 71:227–238.

Nizzoli, F. and Sandercock, J. R. 1990. Surface Brillouin scattering from phonons. *In* Dynamical Properties of Solids Vol. 6 (G. K. Horton and A. A. Maradudin, eds.) pp. 281–335. Amsterdam, North-Holland.

Pharr, G. M. and Oliver, W. C. 1992. Measurement of thin film mechanical properties using nanoindentation. *M. R. S. Bulletin* July, 28–33.

Rogers, J. A. 1998. Complex acoustic waveforms excited with multiple picosecond transient gratings formed using specially designed phase-only beam-shaping optics. *J. Acoust. Soc. Amer.* 104:2807–2813.

Rogers, J. A. and Nelson, K. A. 1994. Study of Lamb acoustic waveguide modes in unsupported polyimide films using real-time impulsive stimulated thermal scattering. *J. Appl. Phys.* 75: 1534–1556.

Rogers, J. A. and Nelson, K. A. 1995. Photoacoustic determination of the residual stress and transverse isotropic elastic moduli in thin films of the polyimide PMDA/ODA. *IEEE Trans. UFFC* 42:555–566.

Rogers, J. A. and Nelson, K. A. 1996. A new photoacoustic/photothermal device for real-time materials evaluation: An automated means for performing transient grating experiments. *Physica B* 219–220:562–564.

Rogers, J. A. and Case, C. 1999. Acoustic waveguide properties of a thin film of nanoporous silica on silicon. *Appl. Phys. Lett.* 75: 865–867.

Rogers, J. A. and Bogart, G. R. 2000. Optical evaluation of the flexural rigidity and residual stress in thin membranes: picosecond transient grating measurements of the dispersion of the lowest order Lamb acoustic waveguide mode. *J. Mater. Res.* 16:217–225.

Rogers, J. A., Dhar, L., and Nelson, K. A. 1994a. Noncontact determination of transverse isotropic elastic moduli in polyimide thin films using a laser based ultrasonic method. *Appl. Phys. Lett.* 65:312–314.

Rogers, J. A., Yang, Y., and Nelson, K. A. 1994b. Elastic modulus and in-plane thermal diffusivity measurements in thin polyimide films using symmetry selective real-time impulsive stimulated thermal scattering. *Appl. Phys. A* 58:523–534.

Rogers, J. A., Fuchs M., Banet, M. J., Hanselman, J. B., Logan, R., and Nelson, K. A. 1997. Optical system for rapid materials characterization with the transient grating technique: Application to nondestructive evaluation of thin films used in microelectronics. *Appl. Phys. Lett.* 71:225–227.

Rogers, J. A., Maznev, A. A., Banet, M. J., and Nelson, K. A. 2000a. Optical generation and characterization of acoustic waves in thin films: fundamentals and applications. *Ann. Rev. Mat. Sci.* 30:117–157.

Rogers, J. A., Bogart, G. R., and Miller, R. E. 2000b. Noncontact quantitative spatial mapping of stress and flexural rigidity in thin membranes using a picosecond transient grating photoacoustic technique. *J. Acoust. Soc. Amer.* 109:547–553.

Shen, Q., Harata, A., and Sawada, T. 1996. Theory of transient reflecting grating in fluid/metallic thin film/substrate systems for thin film characterization and electrochemical investigation. *Jpn. J. Appl. Phys.* 35:2339–2349.

Thomsen, C., Grahn, H. T., Maris, J. H., and Tauc, J. 1986. Surface generation and detection of phonons by picosecond light pulses. *Phys. Rev. B* 34:4129–4138.

Viktorov, I. A. 1976. Rayleigh and Lamb Waves. Plenum, New York.

Vinci, R. P. and Vlassak, J. J. 1996. Mechanical behavior of thin films. *Ann. Rev. Mater. Sci.* 26:431–462.

Vlassak, J. J. and Nix, W. D. 1992. A new bulge test technique for the determination Youngs modulus and Poisson ratio of thin films. *J. Mater. Res.* 7:3242–3249.

Wright, O. B., and Kawashima, K. 1991. Coherent phonon detection from ultrafast surface vibrations. *Phys. Rev. Lett.* 69:1668–1671.

## KEY REFERENCES

Eichler et al., 1986. See above.

*Provides a review of transient grating methods and how they can be used to measure nonacoustic properties.*

Rogers et al., 2000a. See above.

*Provides a review of ISTS methods with applications to acoustic evaluation of thin films, membranes and other type of microstructures.*

## APPENDIX: VENDOR INFORMATION

Currently there is a single vendor that sells complete ISTS instruments for metrology of thin metal films found in microelectronics:

Philips Analytical
Worldwide headquarters:
Lelyweg 1, 7602 EA Almelo,
The Netherlands
Tel.: +31 546 534 444
Fax: +31 546 534 598

In the U.S.:
Philips Analytical
12 Michigan Dr.
Natick, Mass. 01760

Tel.: (508) 647–1100
Fax: (508) 647–1111
*http://www.analytical.philips.com*

The names of the two free-standing tools are Impulse 300 and PQ Emerald.

There are also integrated metrology solutions for Cu interconnect process tools (e.g.: Paragon electroplating system from Semitool).

Semitool
655 West Reserve Dr.
Kalispell, Montana 59901
Tel.: (406) 752-2107
Fax: (406) 752-5522

Philips Analytical's Impulse 300 and PQ Emerald are fully automated all-optical metal thin film metrology systems for metal interconnect process control in semiconductor integrated circuit manufacturing. Impulse 300 is a versatile system for wafers up to 300 mm in size. PQ Emerald is a small-footprint ($1.2 \times 1.53-$m) system for wafer size up to 200 mm. Both systems have measurement spot size $25 \times 90$ microns and pattern-recognition capabilities enabling measurements on product wafers. Measurements can be done on test pads as well as on high-density arrays of submicron structures, e.g., damascene line arrays.

Typical reproducibility: 1 to 2 Å (plasma vapor deposited Cu, Ta), 10 to 50 Å (thick electroplated Cu). Measurement time 1 sec /site, throughput up to 70 wafers per hr.

Key applications are copper interconnect process control, including seed deposition, electroplating and chemical-mechanical polishing.

Philips Analytical also offers integrated metrology solutions for Cu PVD, electroplating and polishing process tools.

JOHN A. ROGERS
Bell Laboratories, Lucent
    Technologies
Murray Hill, New Jersey

ALEX MAZNEV
Philips Analytical
Natick, Massachusetts

KEITH A. NELSON
Massachusetts Institute of
    Technology
Cambridge, Massachusetts

# RESONANCE METHODS

## INTRODUCTION

This chapter shows how nuclear and electron resonance spectroscopies can help solve problems in materials science. The concept of a probe, located centrally within a material, is common to all resonance spectroscopy techniques. For example, the nucleus serves as the probe in nuclear magnetic resonance (NMR), nuclear quadrupole resonance (NQR), and Mössbauer spectrometry, while a paramagnetic atom (i.e., one containing unpaired electrons) often serves as a probe in electron spin resonance (ESR). Sometimes the interest is in measuring the total numbers of probes within a material, or the concentration profiles of probe nuclei as in the case for basic NMR imaging. More typically, details of the measured energy spectra are of interest. The spectra provide the energies of photons that are absorbed by the probe, and these are affected by the electronic and/or nuclear environment in the neighborhood of the probe. It is often a challenge to relate this local electronic information to larger features of the structure of materials.

The units in this chapter describe how local electronic and magnetic structure can be studied with resonance techniques. The energy levels of the probe originate with fundamental electrostatic and magnetic interactions. These are the interaction of the nuclear charge with the local electron density, the electric quadrupole moment of the nucleus with the local electric field gradient, and the spin of the nucleus or spin of the electrons with the local magnetic field. For magnetic interactions in NMR, for example, the experimental spectra depend on how the energy of the probe differs for different orientations of the nuclear spin in the local magnetic field. In general these interactions involve some constant factors of the probe itself, such as its spin and gyromagnetic ratio. The energy level of the probe is the product of these constant factors and a local field. The resonance condition is such that a photon is absorbed, which causes the probe to undergo a transition to a state that differs in energy from its initial state by the energy of the absorbed photon. Since the parameters intrinsic to the probe are known for both states, the experimental spectrum provides the spectrum of the local field quantity characteristic of the material. Continuous wave methods excite probes with specific local fields. Pulse techniques excite all available probes, and differentiate them later by the phases of their precession in their local fields.

The energy differences of the probe levels are small. Radio transmitters provide the excitation photons in the case of NMR and NQR, and microwave generators provide the photons in ESR. The energy from these excitations is eventually converted into heat in the material. For Mössbauer spectrometry and other nuclear methods such as perturbed angular correlation spectroscopy (PACS), the photon, a $\gamma$ ray, is provided by a radioisotope source. The photon energy is absorbed by exciting internal nuclear transitions. The subsequent decay occurs through the emission of ionizing radiation.

The ultimate energy resolution of these resonance spectroscopies is the energy precision, $\Delta E$, that is set by the uncertainty principle

$$\Delta E \cong \frac{\hbar}{\Delta t} \qquad (1)$$

where $\Delta t$ is the lifetime of the excited state. [The $\Delta t$ can also be imposed experimentally by exposing the sample to an additional radiofrequency pulse(s).] It is fortunate for the nuclear techniques of NMR, NQR, and Mössbauer spectrometry that $\Delta t$ is relatively long, so the energy resolution, $\Delta E$, is high. This makes nuclear spectra useful for studies of materials, because the weak hyperfine interactions between the nucleus and the surrounding electrons provide information about the electrons in the solid. The sampling of the local field is modified when the probe atom undergoes diffusive motions during the time $\Delta t$. Local probe methods provide unique information on the jumping frequencies of the probe atoms, and sometimes on the jump directions.

In spite of the high energy resolution of resonance spectroscopies, they are not typically known for their spatial resolution. Microscopies, as those for electrons or light, have not been developed for resonance spectroscopies because their photons cannot be focused. Especially for NMR and NQR, however, control over magnetic fields and the imposed pulse sequences can be used to make images of the concentrations of resonant probes within the material. These methods are under rapid development, driven largely by market forces outside materials science.

The units in this chapter enumerate some requirements for samples, but an obvious requirement is that the material must contain the probe itself. A set of allowed nuclei is listed in the units of NMR, NQR, and Mössbauer spectrometry, and the unit on ESR discusses unpaired electrons in materials. It is rare for the sample to have too high a concentration of probes; usually the difficulty is that the sample is too dilute in the nuclear or electron probe. Sensitivity is often a weakness of resonance spectroscopies, and measurements may require modest sample sizes and long data acquisition times to achieve statistically reliable spectra. An increased data acquisition rate is possible if the resonance frequencies are known in advance, and data need be acquired at only one point of the spectrum. Such specialized methods can be useful for routine characterizations of materials, but are usually not the main mode of operation for laboratory spectrometers.

Sometimes the probe is peripheral to the item of interest, which may be an adjacent atom, for example. Useful data can sometimes be obtained from measurements where the probe is a nearby spectator. As a rule of thumb, however, resonant probe experiments tend to be most informative when the probe is itself at the center of the action, or at least is located at the atom of interest in the material.

Nuclear and electron resonance spectroscopies are specialized techniques. They are not methods that will be used by many groups in materials science. Especially when capital investment is high, as in the case of NMR imaging, it makes sense to contact an expert in the field to assess feasibility and reliability of measurements. Even when the capital investment in equipment is low, as for Mössbauer spectrometry, novices can be misled in several aspects of data interpretation. Collaborations are fairly commonplace for practitioners of resonance spectroscopies. The viewpoint of a material from a resonant probe is a small one, so resonance spectroscopists are accustomed to obtaining complementary experimental information when solving problems in materials science. Nevertheless, the viewpoint of a material from a resonant probe is unique, and can provide information unavailable by other means.

BRENT FULTZ

# NUCLEAR MAGNETIC RESONANCE IMAGING

## INTRODUCTION

Nuclear magnetic resonance imaging is a tomographic imaging technique that produces maps of the nuclear magnetic resonance signal in a sample. The signal from a volume element (voxel) in the sample is represented as an intensity of a picture element (pixel) in an image of the object. Nuclear magnetic resonance imaging has developed as an excellent noninvasive medical imaging technique, providing images of anatomy, pathology, circulation, and functionality that are unmatched by other medical imaging techniques. Nuclear magnetic resonance imaging is also an excellent nondestructive analytical tool for studying materials. Applications of nuclear magnetic resonance imaging in the field of materials science have also been developed, but to a lesser extent. The most probable reason why the development of materials applications of nuclear magnetic resonance imaging has lagged behind that of clinical applications is the cost-to-information ratio. The 1998 price of a clinical magnetic resonance imager is between one and two million U.S. dollars.

Before beginning a discussion of nuclear magnetic resonance imaging, it is useful to review the field of magnetic resonance (MR). Magnetic resonance is the general term that encompasses resonance processes associated with magnetic moments of matter. Please refer to Figure 1 throughout this introductory discussion. The two major branches of magnetic resonance are electron paramagnetic resonance (EPR), or electron spin resonance (ESR), as it is sometimes called (see ELECTRON PARAMAGNETIC RESONANCE SPECTROSCOPY), and nuclear magnetic resonance (NMR). A much smaller third component of magnetic resonance is muon spin resonance (μSR). EPR and ESR measure properties of the magnetic moment of the electrons or electron orbitals, while NMR is concerned with the magnetic

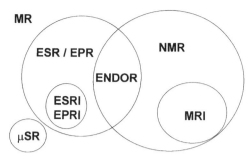

**Figure 1.** A Venn diagram of the field of magnetic resonance (MR). Abbreviations: ENDOR, electron nuclear double resonance; EPR, electron paramagnetic resonance; EPRI, electron paramagnetic resonance imaging; ESR, electron spin resonance; ESRI, electron spin resonance imaging; μSR, muon spin resonance; MRI, magnetic resonance imaging; NMR, nuclear magnetic resonance.

moment of nuclei with unpaired protons and neutrons. Electron nuclear double resonance (ENDOR) examines the interactions between the nuclear and electron moments. All of the above techniques are used to measure properties of atoms and molecules in materials.

Within the fields of EPR and NMR are imaging disciplines. The term magnetic resonance imaging (MRI) should be used to describe imaging techniques based on both NMR and EPR or ESR. For historical reasons, MRI has been used for nuclear magnetic resonance imaging, the mapping of the spatial distribution of the NMR signals in matter. EPR imaging (EPRI) and ESR imaging (ESRI) have been used to describe imaging of the spatial distribution of electron and electron orbital magnetic moments in matter. This unit will concentrate on MRI, as previously defined, of materials.

MRI is used to image both large- and small-diameter objects. When imaging small objects, <1 cm in diameter, the imaging process is typically called NMR microscopy. When imaging larger objects, the procedure is generally called MRI. In reality, the imaging procedures are the same. The distinction arises from the type of instrument used, rather than the dimensions of the material. NMR microscopy is performed on an NMR spectrometer equipped with imaging hardware. Samples that do not fit into the microscope are imaged on a small-bore imager or clinical whole-body imager. No distinction will be made in this unit between MRI and NMR microscopy of materials.

MRI has been applied in many disciplines to study materials. MRI, as a nondestructive analytical technique, has very few competing techniques. X-ray and ultrasound imaging techniques come close to providing similar information in a nondestructive manner, but do not have the sensitivity to many of the fundamental properties of matter that MRI has. MRI may be used to measure distance, flow, diffusion coefficients, concentration, microscopic viscosity, partial pressure, pH (Metz et al., 1998), temperature (Aime et al., 1996; Zuo et al., 1996), magnetic susceptibility, and the dielectric constant. A few good MRI reviews are (Komoroski, 1993; Callaghan, 1991; Blumich et al., 1994; Hall et al., 1990; Gibbs and Hall, 1996; and Attard et al., 1991).

In food science, MRI has been used to study the consistency (Duce et al., 1990), baking (Duce and Ablett, 1995), cooking, curing (Guiheneuf et al., 1996; Guiheneuf, Gibbs, and Hall, 1997), storage, spoilage (Guiheneuf et al., 1997), hydration (Duce and Hall, 1995), and gelling (Holme and Hall, 1991) of food. MRI can be used to examine the ripening of fruits and vegetables, as well as damage by disease, insects, and bruising. MRI can determine the moisture and oil content, distribution, and transport in harvested grains and seeds (Duce et al., 1994). Lumber grading, or the mapping of knots, defects, and disease in lumber (Pearce et al., 1997), can be achieved by MRI. MRI has been used in botany to study disease and the flow and diffusion of water in plants.

In soil science, MRI can be used to map the water content and transport in soil. Soil (Amin et al., 1994, 1996), rock cores (Horsfield et al., 1990; Fordham et al., 1993a; Fordham et al., 1995), and oil well cores have been imaged to determine hydrocarbon and water content, as well as porosity, diffusion (Fordham et al., 1994; Horsfield, 1994), capillary migration (Carpenter et al., 1993), and flow characteristics.

In civil engineering, MRI has been used to study the curing and porosity of cement as well as water transport in concrete (Link et al., 1994) and other building materials (Pel et al., 1996). MRI can determine the effectiveness of paints and sealers in keeping out moisture.

In chemistry, MRI has been used to image chemical waves associated with certain chemical reactions (Armstrong et al., 1992), hydrogen adsorption into palladium metal (McFarland et al., 1993), processes in chromatographic columns (Ilg et al., 1992), as well as in reactions and crystals (Komoroski, 1993). MRI has been used to characterize ceramics (Wang et al., 1995)—flaws (Karunanithy and Mooibrook, 1989), cracks, voids (Wallner and Ritchey, 1993), and binder distribution (Wang et al., 1993) in ceramics have been imaged. MRI has been used to noninvasively study the homogeneity of solid rocket propellants (Maas et al., 1997; Sinton et al., 1991).

Polymers have been the subject of many MRI investigations (Blumich and Blumler, 1993). MRI has been used to study the uptake of water into the polymer coating used on silicon chips (Hafner and Kuhn, 1994) and polyurethane dispersion coating (Nieminen and Koenig, 1990). MRI has been used to study curing (Jackson, 1992), vulcanization (Mori and Koenig, 1995), solvent ingress (Webb and Hall, 1991), inhomogeneities (Webb et al., 1989), diffusion (Webb and Hall, 1990), multicomponent diffusion (Grinsted and Koenig, 1992), cyclic sorption-desorption (Grinsted et al., 1992), aging (Knorgen et al., 1997), filler inhomogeneities (Sarkar and Komoroski, 1992), adhesion (Nieminen and Koenig, 1989), and water uptake (Fyfe et al., 1992) in polymers.

MRI is also used to image flow (Amin et al., 1997; Seymour and Callaghan, 1997), diffusion (Callaghan and Xia, 1991), and drainage (Fordham et al., 1993b) in materials, and hence gain a better understanding of the internal microscopic structure of the materials.

There have been two ways in which MRI has been used to study materials. One approach images the material, and the other images the absence of material. The first approach is to directly image the NMR signal from a signal-bearing substance in the sample. The second is to introduce a signal-bearing substance into voids within the material and image the signal from the signal-bearing substance. The latter procedure is good for porous samples that do not have a measurable NMR signal of their own. The latter approach is more common because it can be performed on clinical magnetic resonance imagers with routinely available clinical imaging sequences. The former approach typically requires either faster imaging sequences, or different hardware and imaging sequences capable of looking at a solid-state NMR signal.

## PRINCIPLES OF THE METHOD

Nuclear magnetic resonance imaging is based on the principles of nuclear magnetic resonance. The choice of imaging parameters necessary to produce a high quality MR image requires knowledge of many of the microscopic properties of the system of spins being imaged. It is therefore necessary to understand the principles of NMR before those of MRI can be introduced.

### Theory of NMR

NMR is based on a property of the nucleus of an atom called spin. The material presented in this unit pertains to spin 1/2 nuclei. Spin can be thought of as a simple magnetic moment. When a spin 1/2 nucleus is placed in an external magnetic field, the magnetic moment can take one of two possible orientations, one low-energy orientation aligned with the field and one high-energy orientation opposing the field. A photon with an amount of energy equal to the energy difference between the two orientations or states will cause a transition between the states. The greater the magnetic field, the greater the energy difference, and hence the greater the frequency of the absorbed photon. The relationship between the applied magnetic field $B$ and the frequency of the absorbed photon $\nu$ is linear.

$$\nu = \gamma B \qquad (1)$$

The proportionality constant $\gamma$ is called the gyromagnetic ratio. The gyromagnetic ratio is a function of the magnitude of the nuclear magnetic moment. Therefore, each isotope with a net nuclear spin possesses a unique $\gamma$. The $\gamma$ value of some of the more commonly imaged nuclei are listed in Table 1. In NMR, $B$ fields are typically 1 to 10 tesla (T); therefore $\nu$ is in the MHz range.

**Table 1. Gyromagnetic Ratio of Some Commonly Imaged Nuclei**

| Nucleus | Gyromagnetic Ratio (MHz/T) |
| --- | --- |
| $^1$H | 42.58 |
| $^{13}$C | 10.71 |
| $^{17}$O | 5.77 |
| $^{19}$F | 40.05 |
| $^{23}$Na | 11.26 |
| $^{31}$P | 17.24 |

NMR has the ability to distinguish between nuclei in different chemical environments. For example, the resonance frequency of hydrogen in water is different than that for the hydrogen nuclei in benzene. This difference is called the chemical shift of the nucleus, and is usually reported in ppm relative to a reference. In hydrogen NMR this reference is tetramethylsilane (TMS). The chemical shift for a freely tumbling nucleus $i$ in a nonviscous solution is

$$\delta_i = \frac{(\nu_{TMS} - \nu_i)}{\nu_{TMS}} \times 10^6 \qquad (2)$$

In solids and some viscous liquids, $\delta$ may be anisotropic due to the fixed distribution of orientations of the nuclei in the molecule.

Molecules with interacting spins will display spin-spin coupling, or a splitting of the energy levels due to interactions between the nuclei. The spin-spin interaction is orientation-dependent. Once again, in nonviscous liquids, this orientation-dependence is averaged out and seen as one average interaction. In liquids, multiple resonance lines may be observed in an NMR spectrum, depending on the magnitude of the spin-spin coupling. In solids and viscous liquids, the distribution of orientations results in broad spectral absorption peaks. Imaging solids will require different and less routine techniques than are used in imaging nonviscous liquids.

The remainder of this description of NMR will adopt a more macroscopic perspective of the spin system in which a static magnetic field, $B_0$, is applied along the $z$ axis. Groups of nuclei experiencing the exact same $B_0$ are called spin packets. When placed in a $B_0$ field, spin packets precess about the direction of $B_0$ just as a spinning top precesses about the direction of the gravitational field on earth. The precessional frequency, also called the Larmor frequency, $\omega$, is equal to $2\pi\nu$. Adapting the conventional magnetic resonance coordinate system, the direction of the precession is clockwise about $B_0$, and the symbol $\nu_0$ is reserved for spin packets experiencing exactly $B_0$ (Fig. 2). It is often helpful in NMR and MRI to adopt a rotating frame of reference to describe the motion of magnetization vectors. This frame of reference rotates about the $z$ axis at $\nu_0$. The coordinates in the rotating frame of reference are referred to as $z, x'$, and $y'$. An excellent animated depiction

of the motion of a magnetization vector in a magnetic field can be found in the Web-based hypertext book, *The Basics of MRI*, by J. P. Hornak (*http://www.cis.rit.edu/htbooks/mri/*).

An NMR sample contains millions of spin packets, each with a slightly different Larmor frequency. The magnetization vectors from all these spin packets form a cone of magnetization around the $z$ axis. At equilibrium, the net magnetization vector **M** from all the spins in a sample lies in the center of the cone along the $z$ axis. Therefore, the longitudinal magnetization $M_z$ equals $M$ and the transverse magnetization $M_{xy}$ equals zero at equilibrium. Net magnetization perturbed from its equilibrium position will want to return to its equilibrium position. This process is called spin relaxation.

The return of the $z$ component of magnetization to its equilibrium value is called spin-lattice relaxation. The time constant that describes the exponential rate at which $M_z$ returns to its equilibrium value $M_{z0}$ is called the spin-lattice relaxation time, $T_1$. Spin-lattice relaxation is caused by time-varying magnetic fields at the Larmor frequency. These variations in the magnetic field at the Larmor frequency cause transitions between the spin states and hence change $M_z$. Time-varying fields are caused by the random rotational and translational motions of the molecules in the sample possessing nuclei with magnetic moments. The frequency distribution of random motions in a liquid varies with temperature and viscosity. In general, relaxation times tend to get longer as $B_0$ increases. $T_1$ lengthens because there are fewer relaxation-causing frequency components present in the random motions of the molecules as $\nu$ increases.

At equilibrium, the transverse magnetization, $M_{xy}$, equals zero. A net magnetization vector rotated off of the $z$ axis creates transverse magnetization. This transverse magnetization decays exponentially with a time constant called the spin-spin relaxation time $T_2$. Spin-spin relaxation is caused by fluctuating magnetic fields that perturb the energy levels of the spin states and dephase the transverse magnetization. $T_2$ is inversely proportional to the number of molecular motions less than and equal to the Larmor frequency. Specialists in NMR break down $T_2$ further into a pure $T_2$ due to molecular interactions and one due to inhomogeneities in the $B_0$ field. The overall $T_2^*$ is referred to as "$T_2$ star."

$$\frac{1}{T_2^*} = \frac{1}{T_{2\,molec}} + \frac{1}{T_{2\,inhomogeneous}} \qquad (3)$$

In pulsed NMR and MRI, radiofrequency (RF) energy is put into a spin system by sending RF into a resonant LC circuit, the inductor of which is placed around the sample. The inductor must be oriented with respect to the $B_0$ magnetic field so that the oscillating RF field created by the RF flowing through the inductor is perpendicular to $B_0$. The RF magnetic field is called the $B_1$ magnetic field. When the RF inductor, or coil as it more often called, is placed around the $x$ axis, the $B_1$ field will oscillate back and forth along the $x$ axis.

In pulsed NMR spectroscopy, it is the $B_1$ field that is pulsed. Turning on a $B_1$ field for a period of time, $\tau$, will

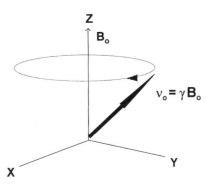

**Figure 2.** Clockwise precession of the net magnetization vector in an *xyz* coordinate system.

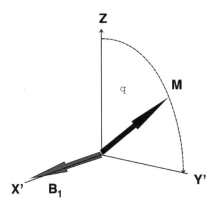

**Figure 3.** Rotation of net magnetization vector $M$ by $B_1$ field in a rotating frame of reference.

cause the net magnetization vector to precess in ever widening circles around the $z$ axis. Eventually the vector will reach the $xy$ plane. If $B_1$ is left on longer, the net magnetization vector will reach the negative $z$ axis. In the rotating frame of reference this vector appears to rotate away from the $z$ axis, as depicted in Figure 3. The rotation angle $\theta$, which is measured clockwise about the direction of $B_1$ in radians, is proportional to $\gamma$, $B_1$, and $\tau$

$$\theta = 2\pi\gamma B_1\tau \tag{4}$$

Any transverse magnetization, $M_{xy}$, will precess about the direction of $B_0$. An NMR signal is generated from transverse magnetization rotating about the $z$ axis. This magnetization will induce a current in a coil of wire placed around the $x$ or $y$ axis. As long as there is transverse magnetization that is changing with respect to time, there will be an induced current in the coil. For a group of nuclei with one identical chemical shift, the signal will be an exponentially decaying sine wave. The sine wave decays with time constant $T_2^*$. It is predominantly the inhomogeneities in $B_0$ that cause the spin packets to dephase.

Net magnetization, which has been rotated away from its equilibrium position along the $z$ axis by exactly $180°$, will not create transverse magnetization and hence not give a signal. The time-domain signal from a net magnetization vector in the $xy$ plane is called a free induction decay (FID). This time-domain signal must be converted to a frequency-domain spectrum to be interpreted for chemical information. The conversion is performed using a Fourier transform. The hardware in most NMR spectrometers and magnetic resonance imagers detects both $M_x$ and $M_y$ simultaneously. This detection scheme is called quadrature detection. These two signals are equivalent to the real and imaginary signals; therefore the input to the Fourier transform will be complex. Sampling theory tells us that one need only digitize the FID at a frequency of $f$ complex points per second in order to obtain a spectrum of frequency width $f$, in Hz.

In pulsed Fourier transform NMR spectroscopy, short bursts of RF energy are applied to a spin system to induce a particular signal from the spins within a sample. A pulse sequence is a description of the types of RF pulses used and the response of the magnetization to the pulses. The simplest and most widely used pulse sequence for routine NMR spectroscopy is the 90-FID pulse sequence. As the name implies, the pulse sequence is a $90°$ pulse followed by the acquisition of the time-domain signal called an FID. The net magnetization vector, which at equilibrium is along the positive $z$ axis, is rotated by $90°$ down into the $xy$ plane. The rotation is accomplished by choosing an RF pulse width and amplitude so that the rotation equation for a $90°$ pulse is satisfied. At this time the net magnetization begins to precess about the direction of the applied magnetic field $B_0$. Assuming that the spin-spin relaxation time is much shorter than the spin-lattice relaxation time ($T_2 \ll T_1$), the net magnetization vector begins to dephase as the vectors from the individual spin packets in the sample precess at their own Larmor frequencies. Eventually, $M_{xy}$ will equal zero and the net magnetization will return to its equilibrium value along $z$. The signal that is detected by the spectrometer is the decay of the transverse magnetization as a function of time. The FID or time-domain signal is Fourier transformed to yield the frequency domain NMR spectrum.

### Theory of MRI

The basis of MRI is Equation 1, which states that the resonance frequency of a nucleus is proportional to the magnetic field it is experiencing. The application of a spatially varying magnetic field across a sample will cause the nuclei within the sample to resonate at a frequency related to their position.

For example, assume that a one-dimensional linear magnetic field gradient $G_z$ is set up in the $B_0$ field along the $z$ axis. The resonant frequency, $\nu$, will be equal to

$$\nu = \gamma(B_0 + zG_z) \tag{5}$$

The origin of the $xyz$ coordinate system is taken to be the point in the magnet where the field is exactly equal to $B_0$ and spins resonate at $\nu_0$. This point is referred to as the isocenter of the magnet. Equation 6 explains how a simple one-dimensional imaging experiment can be performed. The sample to be imaged is placed in a magnetic field $B_0$. A $90°$ pulse of RF energy is applied to rotate magnetization into the $xy$ plane. A one-dimensional linear magnetic field gradient $G_z$ is turned on after the RF pulse and the FID is immediately recorded. The Fourier transform of the FID yields a frequency spectrum that can be converted to a spatial, $z$, spectrum as a function

$$z = \frac{(\nu - \nu_0)}{\gamma G_z} \tag{6}$$

This simple concept of a one-dimensional image can be expanded to a two-dimensional image by employing the concept of back-projection imaging similar to that used in computed tomography (CT) imaging. (Hornak, 1995). If a series of one-dimensional images, or projections of the signal in a sample, are recorded for linear one-dimensional magnetic field gradients applied along several different trajectories in a plane, the spectra can be transformed into a two-dimensional image using an inverse

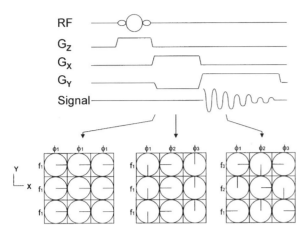

**Figure 4.** Timing diagram for a 90-FID imaging sequence and the behavior of nine spin vectors in the imaged plane at three times in the imaging sequence.

radon transform (Sanz, 1988) or back projection algorithm (Lauterbur, 1973). This procedure is often used in materials imaging where short $T_2$ values prevent the use of the more popular Fourier-based method (Kumar and Ernst, 1975; Smith, 1985). A Fourier-based imaging technique collects data from the $k$ space of the imaged object. This $k$ space is the two-dimensional Fourier transform of the image. Figure 4 depicts a timing diagram for a Fourier imaging sequence using simple 90-FID sequence. The timing diagram describes the application of RF and three magnetic-field gradients called the slice selection ($G_s$), phase encoding ($G_p$), and frequency encoding ($G_f$) gradients. The first step in the Fourier imaging procedure is slice selection of those spins in the object for which the image is to be generated.

Slice selection is accomplished by the application of a magnetic field gradient at the same time as the RF pulse is applied. An RF pulse, with frequency width $\Delta\nu$ centered at $\nu$, will excite, when applied in conjunction with a field gradient $G_z$, spins centered at $z = (\nu - \nu_0)/\gamma G_z$ with a spread of spins at $z$ values $\Delta z = \Delta\nu/\gamma G_z$. Spins experiencing a magnetic field strength not satisfying the resonance condition will not be rotated by the RF pulses, and hence slice selection is accomplished. The image slice thickness (Thk) is given by $\Delta z$. For a clean slice, i.e., where all spins along the slice thickness are rotated by the prescribed rotations, the frequency content of the pulse must be equal to a rectangular-shaped function. Therefore, the RF pulse must be shaped as a sinc function ($\sin x/x$) in the time domain.

The next step in the Fourier imaging procedure is to encode some property of the spins as to the location in the selected plane. One could easily encode spins as to their $x$ position by applying a gradient $G_x$ after the RF pulse and during the acquisition of the FID. The difficulty is in encoding the spins with information as to their $y$ location. This is accomplished by encoding the phase of the precessing spin packets with $y$ position (Fig. 4). Phase encoding is accomplished by turning on a gradient in the $y$ direction immediately after the slice-selection gradient is turned off and before the frequency-encoding gradient

is turned on. The spins in the excited plane now precess at a frequency dependent on their $y$ position. After a period of time, the gradient is turned off and the spins have acquired a phase equal to

$$\phi = 2\pi\gamma\tau y G_y \qquad (7)$$

Figure 4 describes, for nine magnetization vectors, the effect of the application of a phase-encoding gradient, $G_y$, and a frequency-encoding gradient, $G_x$. The phase-encoding gradient assigns each $y$ position a unique phase. The frequency-encoding gradient assigns each $x$ position a unique frequency. If one had the capability of independently assessing the phase and frequency of a spin packet, one could assign its position in the $xy$ plane. Unfortunately, this cannot be accomplished with a single pulse and signal. The amplitude of the phase-encoding gradient must be varied so a $2\pi$ radian phase variation between the isocenter and the first resolvable point in the $y$ direction can be achieved, as well as a $256\pi$ radian variation from center to edge of the imaged space for 256-pixel resolution in the phase-encoding direction. The result is to traverse, line by line, the $k$ space of the image. The negative lobe on the frequency-encoding gradient (Fig. 4.), which was not previously described, shifts the center of $k$ space to the center of the signal acquisition window.

The $B_0$ field is created by a large-diameter, solenoidal-shaped, superconducting magnet. The gradient fields are created by room temperature gradient coils located within the bore of the magnet. These coils are driven by high-current audiofrequency amplifiers. The $B_1$ field is introduced into the sample by means of a large LC circuit that surrounds the object to be imaged. The same or a separate LC circuit is used to detect the signals from the precessing spins in the body.

The field of view (FOV) is dependent on the quadrature sampling rate, $R_s$ during the $G_f$ and on the magnitude of $G_f$.

$$\text{FOV} = R_s/G_f \qquad (8)$$

The two-dimensional $k$ space data set is Fourier transformed and the magnitude image generated from the real and imaginary outputs of the Fourier transform.

The number of pixels across the field of view, $N$, is typically 256 in both directions of the image. Assuming a 20-cm FOV and 3-mm slice thickness, the in-plane resolution is ~0.8 mm. The volume of a voxel is therefore ~2 mm$^2$. When imaging features occupying fractions of a voxel, each voxel comprises more than one substance. As a consequence, the NMR signal from a voxel is a sum of the NMR signals from the substances found in the voxel. Variations in the signal from a voxel, due to the relative amounts of the components found in a voxel, is referred to as a partial volume effect.

A magnetic resonance image can be thought of as a convolution of a two-dimensional NMR spectrum of the spin-bearing substance with the spin-concentration map from the imaged object. This is better visualized in one dimension, where $x$ is the imaged dimension. If $f_i(\nu G_x/\gamma)$ is the NMR spectrum of the spin type converted to distance units, and $g_i(x)$ is the distribution of the spins, the one-dimensional image is the convolution of these two functions. Since there may be more than one component in a voxel, the one-dimensional image, $h(x)$, becomes

$$h(x) = \sum_i g_i(x) \otimes f_i(\nu G_x/\gamma) \qquad (9)$$

Defining $\Gamma_i$ as the full line width at half height of component $i$ in Hz, $\Gamma_i G_x/\gamma$ for the largest $\Gamma_i$ must be less than FOV/N for optimum resolution in $h(x)$.

The most commonly imaged spin-bearing nucleus is hydrogen. The two most commonly imaged molecules containing hydrogen are hydrocarbons and water. These hydrogens yield one signal in the image, as chemical shift and spin-spin splitting information is generally not utilized. Occasionally the different chemical shifts for water and hydrocarbon hydrogens can lead to an artifact in an image called a chemical-shift artifact. Other hydrogens associated with many samples have very short $T_2$ values and do not contribute directly to the signal.

In the example of Figure 4, the slice-selection gradient was applied along the $z$ axis and the phase and frequency-encoding gradients along the $y$ and $x$ axes, respectively. In practice, the gradients can be applied along any three orthogonal directions, with the only restrictions being that the slice selection gradient be perpendicular to the imaged plane.

The most routinely used imaging sequence is the spin-echo. Its popularity is attributable to its ability to produce images that display variations in $T_1$, $T_2$, and spin concentration of samples. This sequence consists of 90° and 180° RF pulses repeated every TR seconds (Fig. 5). These pulses are applied in conjunction with the slice selection gradients.

The phase-encoding gradient is applied between the 90° and 180° pulses. The frequency-encoding gradient is turned on during the acquisition of the signal. The signal is referred to as an echo because it comes about from the refocusing of the transverse magnetization at time TE after the application of the 90° pulse. The signal, when TE ≪ TR, from a voxel will be equal to a sum over all the different types of spins, $i$, in the voxel.

$$S(\text{TE}, \text{TR}) = k \sum_i \rho_i (1 - e^{-\text{TR}/T_{1i}}) e^{-\text{TE}/T_{2i}} \qquad (10)$$

Spin density, $\rho$, is the number of spins per voxel, and $k$ is a proportionality constant.

## PRACTICAL ASPECTS OF THE METHOD

The signal in a magnetic resonance image is a function of $T_1$, $T_2$, $T_2^*$, $\rho$, the diffusion coefficient ($D$), the velocity ($V$), and chemical shift ($\delta$) of the spins. The various imaging sequences are designed to produce a signal that is proportional to one of these parameters. A weighted image is one in which the signal intensity is dependent on one of these properties. For example, a $T_1$-weighted image is one in which the image intensity displays differences in $T_1$ of the sample.

All images are generated with the aid of an imaging pulse sequence similar to those presented in Figures 4 and 5. There are hundreds of imaging sequences, but only a fraction of these are routinely used. A few are mentioned here, and others can be found in the literature. Because the implementation of these is imager-dependent, readers are encouraged to refer to their instrument manuals for specific details. When searching the scientific literature for information on pulse sequences, the reader is reminded that much MRI development work has been published by scientists in the clinical literature.

There are two goals in MRI of materials. One is to visualize structure, texture, or morphology in samples. The second is to visualize spatial or temporal variations in some property of the sample. In the former, the MRI researcher needs to maximize the contrast-to-noise ratio between features in the sample. The imaging parameters are chosen to maximize the difference in signal intensity of two adjacent regions in the image of the sample. For example, Figures 6 and 7 indicate how contrast can be achieved based on the choice of TE and differences in $T_2$, and the choice of TR and variations in $T_1$. If the visualization of spatial or temporal variations is the goal, the absolute intensity of a pixel is important and the choice of imaging parameters is generally chosen to maximize the signal-to-noise ratio.

Before any of the following imaging sequences may be implemented, $T_1$ and $T_2$ of the sample at the field strength of the imager must be known. These values are needed to determine the optimum values of TR, TE, FOV, Thk, and so on.

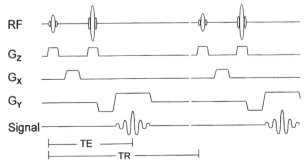

**Figure 5.** Timing diagram for a spin-echo imaging sequence.

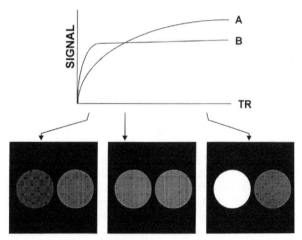

**Figure 6.** Signal of two substances, A and B, as a function of TR, and their corresponding simulated images at the indicated TR values.

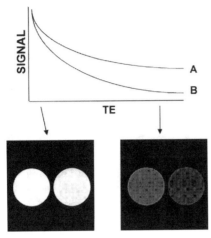

**Figure 7.** Signal of two substances, A and B, as a function of TE, and their corresponding simulated images at the indicated TE values.

### Spin-Echo Sequence

The spin-echo sequence is one of the most common imaging sequences. This sequence was described earlier (see discussion of Theory of MRI), and its timing diagram can be found in Figure 5. This sequence is used to produce images whose signal intensity is a function of the $T_1$, $T_2$, and $\rho$ of the sample as described by Equation 10.

A spin-echo sequence will produce a $T_1$ weighted image when $\text{TR} < T_1$ and $\text{TE} \ll T_2$. A $T_2$-weighted image can be produced by a spin-echo sequence when $\text{TR} \gg T_1$ and $\text{TE} > T_2$. A spin-echo sequence will produce a $\rho$-weighted image when $\text{TR} \gg T_1$ and $\text{TE} \ll T_2$. Due to time constraints, these conditions are rarely precisely met, so weighted images will have some dependence on the remaining two properties.

### Gradient Recalled Echo Sequence

A gradient recalled echo sequence is essentially the 90-FID sequence described earlier in the theory section. The signal from the gradient-recalled echo sequence is given by the following equation, where $\theta$ is the rotation angle of the RF pulse

$$S(\text{TE}, \text{TR}) = k \sum_i \rho_i \frac{(1 - e^{\text{TR}/T_{1i}})\sin\theta\, e^{-\text{TE}/T_{2i}^*}}{(1 - \cos\theta\, e^{-\text{TR}/T_{1i}})} \quad (11)$$

The repetition time, TR, has the same definition as in the spin-echo sequence. A maximum in the signal is produced because the frequency-encoding gradient is first turned on negative so as to position the start of the signal at the edge of $k$ space. Therefore, the echo time, TE, is the time between the RF pulse and the maximum signal. The signal has a $T_2^*$ dependence instead of a $T_2$ dependence. The rotation angle $\theta$ is often set to a value less than 90° so that equilibrium is reached more quickly in the period TR. $T_1$-, $T_2^*$-, and $\rho$-weighted images can be made with a gradient-recalled sequence.

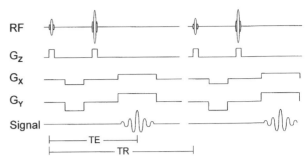

**Figure 8.** Timing diagram for a back-projection imaging sequence.

### Back-Projection Imaging Sequence

The back-projection imaging sequence is useful when $T_2$ values are shorter than 10 ms. The principle is based on the acquisition of several one-dimensional images of the sample (Lauterbur, 1973). The direction of the one-dimensional image is varied by 360° around the imaged object. Figure 8 depicts the timing diagram for a spin-echo back-projection sequence. A slice selection gradient is applied in conjunction with the RF pulses so as to select the desired image plane. In this example an $xy$ plane is imaged.

A readout gradient, $G_R$, is applied during the acquisition of the signal such that the angle ($\alpha$) subtended by its direction relative to the $x$-axis varies between 1° and 360° as given by the following equations

$$G_x = G_R\cos(\alpha) \quad (12)$$
$$G_y = G_R\sin(\alpha) \quad (13)$$

Each of these one-dimensional images represents the projection of the NMR signal across the image perpendicular to $G_R$. $G_R$ is also applied between the 90° and 180° pulses so as to position the echo properly in $k$ space. This set of projection images is back-projected through space using an inverse radon transform to produce an image (Sanz et al., 1988). The advantage of this sequence is that TE can be made less than the corresponding TE in a conventional spin-echo sequence. Therefore samples with shorter $T_2$ values can be imaged. Since this back-projection sequence utilizes a spin-echo pulse sequence, $S(\text{TR},\text{TE})$ is given by Equation 10. Since

$$\Gamma_i = \frac{1}{\pi T_{2i}} \quad (14)$$

and being mindful of Equation 9, gradient strengths must be greater for samples with shorter $T_2$ values.

### Echo-Planar Imaging Sequence

Echo-planar imaging is a rapid magnetic resonance imaging technique, capable of producing images at video rates. The technique records an entire image in a single TR period. Echo-planar imaging therefore measures all lines of $k$ space in a single TR period. A timing diagram for an echo-planar imaging sequence is depicted in Figure 9.

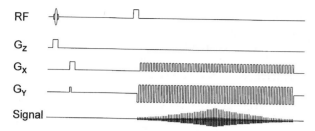

**Figure 9.** Timing diagram for an echo-planar imaging sequence.

A 90° slice-selective RF pulse is applied in conjunction with a slice-selection gradient. Initial phase-encoding gradient and frequency-encoding gradient pulses position the spins at the corner of $k$ space. A 180° pulse follows, which is not slice selective. Next, the phase- and frequency-encoding gradients are cycled so as to raster through $k$ space. This is equivalent to putting 256 phase- and frequency-encoding gradients in the usual period when the echo is recorded.

The rate at which $k$ space is traversed is so rapid that it is possible to obtain 30 images per second. This implies that echo-planar imaging is suitable for studying dynamic processes in materials such as reaction kinetics. Echo-planar images require large and faster-than-normal gradients. As a consequence, echo-planar images are prone to severe magnetic susceptibility artifacts.

### Diffusion Imaging Sequence

Diffusion imaging is used to produce images whose intensity is related to the diffusion coefficient, $D$, of the spins being imaged. The pulsed-gradient spin-echo imaging sequence is used to produce diffusion-weighted images. The timing diagram for this sequence is presented in Figure 10. The signal from a pulsed-gradient spin-echo imaging sequence, $S$, as a function of TR, TE, diffusion encoding gradient ($G_j$) in the $j$ direction, width of the diffusion gradient ($\delta$), and separation of the diffusion gradients ($\Delta$) is given by Equation 15.

$$S(\text{TR}, \text{TE}, G_j, \delta, \Delta) = k \sum_i \rho_i (1 - e^{-\text{TR}/T_{1i}})$$
$$\times\, e^{-\text{TE}/T_{2i}} e^{-(G_j \gamma \delta)^2 D_{j,i}(\Delta - \delta/3)} \quad (15)$$

The summation is over the $i$ components in the voxel. The function of the diffusion gradient pulses is to dephase magnetization from spins that have diffused to a new location in the period $\Delta$. These pulses have no effect on stationary spins, as can be seen by Equation 15 reducing to Equation 10 when the diffusion coefficient is zero.

An image map of the diffusion in one dimension ($j$) may be calculated from diffusion images recorded at different $G_j$ values. Data of $S(G_j)$ as a function of $G_j^2$ is fit on a pixel-by-pixel basis to obtain $D_j$ for each pixel in the image space. Diffusion coefficient images can be measured for $10^{-9} < D_j < 10^{-4}$ cm$^2$/s.

### Flow Imaging Sequence

A flow imaging sequence is similar to that used for diffusion imaging, except that the motion-encoding gradients are tuned to pick up very fast motion. Figure 11 depicts the timing diagram for a phase-contrast imaging sequence implemented with a spin-echo sequence for imaging flow in the $z$ direction. The basis of the phase-contrast sequence is a bipolar magnetic field gradient ($G_{\text{BP}}$) pulse. A bipolar gradient pulse is one in which the gradient is turned on and off in one direction for a period of time, and then turned on and off in the opposite direction for an equivalent amount of time. A positive bipolar gradient pulse has the positive lobe first and a negative bipolar gradient pulse has the negative lobe first. The area under the first lobe of the gradient pulse must equal that of the second. A bipolar gradient pulse has no net effect on stationary spins. The bipolar gradient pulse will affect spins that have a velocity component in the direction of the gradient.

If a bipolar gradient pulse is placed in an imaging sequence, it will not affect the image since all we have done is impart a phase shift to the moving spins. Since an image is a magnitude representation of the transverse magnetization, there is no effect. However, if two imaging sequences are performed in which the first has a positive bipolar gradient pulse and the second a negative bipolar gradient pulse, and the raw data from the two are subtracted, the signals from the stationary spins will cancel and the flowing spins will add. The vectors from the stationary spins cancel and those from the moving spins have a net magnitude. The net effect is an image of the flowing spins.

The direction of the bipolar gradient yields signal only from those spins with a flow component in that direction. The timing diagram in Figure 11 will measure flow in the $z$

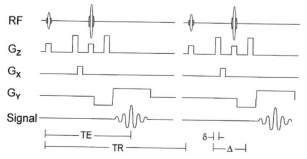

**Figure 10.** Timing diagram for a diffusion imaging sequence.

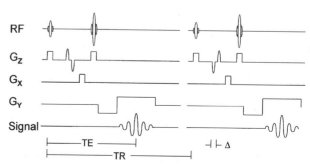

**Figure 11.** Timing diagram for a flow imaging sequence.

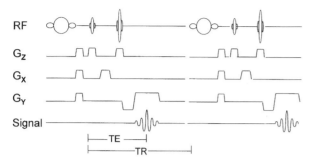

**Figure 12.** Timing diagram for a chemical shift imaging sequence that saturates the spins at a specific chemical shift.

direction because the bipolar gradient pulses are applied by $G_z$. Images of the flow rate between 5 and 400 cm/s may be recorded.

### Chemical Shift Imaging Sequence

A chemical shift imaging (CSI) sequence is one that can map the spatial distribution of a narrow band of chemical shift components found in a sample. There are several possible CSI sequences. The one presented here consists of the modified spin-echo sequence in Figure 12. This sequence has the usual slice-selective $90°$ and $180°$ pulses, and gradients. It has an additional one or more saturation pulses before the normal spin-echo sequence designed to saturate the spins at specific chemical shift values. The frequency of the $90°$ and $180°$ pulses are tuned to the resonance frequency of the desired chemical shift component. The frequency of the saturation pulse is tuned to the resonance frequency of the chemical shift component that is to be eliminated. The frequency, bandwidth, and amplitude of the saturation pulse must be set to effectively saturate the undesired chemical shift components. This technique works best when the $T_1$ of the saturated component is long, thus preventing the generation of transverse magnetization before the application of the spin-echo sequence. Magnetic field gradients are occasionally applied between the saturation pulse and the spin echo sequence to dephase any residual transverse magnetization after the saturation pulse. Images of chemical shift differences as low as 1 ppm are attainable.

### Three-Dimensional Imaging

With the development of fast desktop computers with high-resolution video or holographic monitors, three-dimensional images are readily viewed and manipulated. MRI can readily produce three-dimensional images by acquiring contiguous tomographic slices through an object. To minimize the time necessary to acquire this amount of data, a fast imaging sequence, such as a gradient-recalled echo or echo-planar is typically used. Alternatively, a multislice sequence, which utilizes the dead time between repetitions to acquire data from other slices, can be utilized.

### Solid-State Imaging

The NMR spectra of solids typically consist of very broad absorption lines due to anisotropic chemical shifts and spin-spin (or dipole-dipole) interactions. Recalling the limitations imposed by Equation 14 and Equation 9, this would require extremely large and fast gradients, as well as fast digitizers. Solid-state magic-angle spinning NMR techniques can be used to decrease the line width (Fukushima, 1981). This task is accomplished by spinning the sample about an axis oriented at $54.7°$ to the applied $B_0$ field. This angle is referred to as the magic angle because at this angle the dipole interaction contribution to the line width equals zero. Unfortunately, two-dimensional imaging becomes difficult due to the need to spin the sample at the magic angle. A one-dimensional image may be recorded of a sample spinning at the magic angle by applying a frequency-encoding gradient along the direction of the magic-angle spinning axis.

The dipolar interaction may also be minimized using line-narrowing pulse sequences. The specifics of these sequences are beyond the scope of this unit. The reader is directed to the relevant literature (Chingas et al., 1986; McDonald et al., 1987; Rommel et al., 1990) and instrument manuals for specifics.

### Instrument Setup and Tuning

Certain aspects of the imager should be tuned for each sample. These aspects include the sample probe or imaging coil, the power required to produce a $90°$ pulse, the receiver gain, and the homogeneity of the $B_0$ magnetic field. The reader is directed to a more detailed work (Fukushima and Roeder, 1981) for more information than is presented here.

The imaging coil sends an RF magnetic field into the sample and detects an RF response from the sample. The homogeneity of the transmitted RF field as well as that of the sensitivity will directly influence image quality. A coil with maximal field homogeneity, or a technique for correcting the inhomogeneity in the measured property (Li, 1994), is needed.

The imaging coil is an LC circuit which is designed to resonate at the operating frequency of the imager. This resonance frequency changes when the sample is placed inside the imaging coil. A tuning capacitor on many imaging coils allows the user to change the resonance frequency of the coil. A second capacitor allows the user to adjust the impedance of the coil to that of the imager, typically 50 Ω.

The amount of power necessary to produce a $90°$ pulse, $P_{90}$, will vary from sample to sample and from sample coil to sample coil. $P_{90}$ is proportional to the coil volume, $V_c$, and the bandwidth, BW, of the imaging coil. Therefore the more closely matched the coil is to the size of the sample, the less power is required to produce a $90°$ pulse. The BW of the imaging coil will depend on the electrical quality of the coil and the dielectric constant and conductivity of the sample. A salt water solution sample will cause the imaging coil to have a larger BW than a hydrocarbon sample, and hence require more transmitter power to produce a $90°$ pulse. The amount of power necessary to produce a $90°$ pulse is determined by observing the signal from the sample as the transmitter power is increased from zero. The first maximum in the signal corresponds to the amount of power necessary to produce a 90° pulse. Optimizing $V_c$ and BW is also important because the signal is proportional to the fraction of the imaging coil occupied by the imaged object, and inversely proportional to the BW of the coil.

The receiver gain of the imager must be adjusted such that the amplitude of the greatest signal is less than that of the dynamic range of the digitizer. Each imager has a slightly different procedure for achieving this step. In general, the amplitude of the time-domain signal from the center of $k$ space must be less than the dynamic range. To set this, the image acquisition parameters, such as TR, TE, FOV, and Thk, are first set and a time-domain signal is recorded from the imaging sequence with the phase-encoding gradient turned off. Amplifier gains are adjusted to keep this signal from saturating the amplifier.

The quality of the images produced by an MR imaging device is directly related to the homogeneity of the $B_0$ magnetic field across the sample. The $B_0$ field from the imaging magnet may be extremely homogeneous in the absence of a sample. When a sample is placed in the magnetic field, it distorts the fields around and within the sample due to its magnetic susceptibility and geometry. The homogeneity of the $B_0$ field in the sample is optimized using a set of shim coils found on the imager. The coils are set up to superimpose small, temporally static, spatial magnetic field gradients on the $B_0$ field. There are often ten or twenty different functional gradient forms (i.e., $z$, $z^2$, $x$, $x^2$, $y$, $y^2$, $xy$, etc.) which can be superimposed on $B_0$. These fields can be adjusted to cancel out spatial variations in the $B_0$ field across the sample. The FID from a 90-FID pulse is monitored as the gradients are varied. Maximizing the height and length of the FID optimizes the $B_0$ field.

## Safety

Imaging magnets are typically 1 to 5 tesla (T). These magnets, especially those with a large-diameter bore, tend to have large magnetic fields that extend out around the magnet. These magnetic fields can cause pacemakers to malfunction, erase magnetic storage media, and attract ferromagnetic objects into the bore of the magnet. Persons with pacemakers must not be allowed to stray into a magnetic field of greater than $5 \times 10^{-4}$ T. A magnetic field of approximately $5 \times 10^{-3}$ T will erase the magnetically encoded material on credit cards and floppy disks. Ferromagnetic materials should be kept clear of the magnet, for these objects can be attracted to the magnet and damage experimental apparatus or injure persons in their path. They will also cause damage to the internal support structure of the superconducting magnet and could cause the magnet to quench.

## DATA ANALYSIS AND INITIAL INTERPRETATION

The raw data from a magnetic resonance imager is the $k$ space data of the image. These data are converted to the image with the use of a two-dimensional Fourier transform. The raw $k$ space data are occasionally multiplied by an exponential with a time constant less than $T_2^*$ to reduce noise in the image. This is equivalent to convolving the image with a Lorentzian function, but less time-consuming. This procedure for reducing the noise in an image will also reduce resolution if the time constant for the exponential multiplier is less than $T_2^*$.

Most magnetic resonance imaging data are 16-bit unsigned integer numbers and many images have signals that span this range. Displaying these data in the form of a pixel in an image for interpretation by the human eye necessitates the windowing and leveling of the data. This process takes the useable portion of the 16-bit data and converts it to 8-bit data for interpretation by the human eye-brain system. All imagers have software to perform this conversion. The level is translated to a brightness, and the window to the contrast, on a monitor.

Often the goal is to determine the concentration of a signal-bearing substance in the sample as a function of distance. When TE is much less than the minimum $T_2$ value and TR is much greater than the maximum $T_1$ in the sample, the image intensity from a spin-echo sequence is approximately equal to the concentration. Often it is not possible or practical to record an image under these conditions. In these cases, the spatial variation in $T_1$ and $T_2$ must be measured before the spin density can be determined.

There are several MR imagers on the market ranging in price from one to two million U.S. dollars. As with any instrument, there is a range of available sophistication. An expensive unit may be purchased that has much of the hardware and software necessary to image any material. Less expensive instruments are available that require the user to write some specific pulse sequence computer code or build specific application imaging coils. Details on imaging coil construction can be found in the literature (Hornak et al., 1986, 1987, 1988; Marshall et al., 1998; Szeglowski and Hornak, 1993).

When imaging a pure substance, $T_1$, $T_2$, and $\rho$ images may be calculated from spin-echo data. $T_1$ is best calculated from $S(\text{TR})$ at constant TE using a fast least-squares technique (Gong, 1992). $T_2$ is calculated from $S(\text{TE})$ at constant TR using a nonlinear least-squares approach (Li and Hornak, 1993). The spin density may then be calculated from $S(\text{TR,TE})$, $T_1$, $T_2$, and Equation 10. Samples with multiple components require multiexponential techniques for calculating $T_{1i}$, $T_{2i}$, and $\rho_i$ (Windig et al., 1998).

The concentration of some substances may be imaged indirectly through their effect on $T_1$ and $T_2$. For example, paramagnetic substances will shorten $T_1$ and $T_2$ of water. The relaxation rates ($1/T_1$ and $1/T_2$) of water are inversely proportional to the concentration of the paramagnetic species. Therefore, if images of $T_1$ or $T_2$ can be generated, the concentration of the paramagnetic substance can be determined. The concentration profiles of paramagnetic substances that have been studied by this technique include oxygen and transition metal ions, such as Ni, Cu, and Mn (Antalek, 1991). Other properties of materials, such as pH and temperature, may be measured if they cause a change in resonant frequency of the imaged nucleus. A change in the frequency will cause a change in the phase in the image. Phase information is typically discarded when a magnitude image is taken, so this information must be extracted before the magnitude calculation.

## SAMPLE PREPARATION

In nuclear MRI, sample preparation is minimal, as the goal is to image samples in their natural state. Samples must be of a size that will fit inside of the RF sample coil of the imager. Sample containers must be made of materials that are transparent to RF. The dielectric constant of the sample container must be between that of air and that of the sample, so as to minimize reflection of RF. Polyvinyl chloride or polyethylene work well for most aqueous and hydrocarbon-based signals. The magnetic susceptibility of the sample container, and the sample, must be approximately equal to that of air to minimize susceptibility distortions. Glass containers produce large susceptibility distortions.

MRI of materials under extreme conditions of pressure and temperature are possible, but sample containers and hardware for containing the extreme pressures and temperatures are not commercially available and must be made by the researcher.

## PROBLEMS

The major reason for not seeing a magnetic resonance image or for seeing one with a poor signal-to-noise ratio (SNR) is insufficient signal intensity. Occasionally, other hardware problems may be the cause. Both of these causes are discussed in this section.

Recall from Equation 8 that the signal intensity in an imaging sequence is proportional to the spin density. If the number of spins in a voxel is insufficient to produce a measurable signal, a poor SNR will result. The spin density may be increased by increasing the concentration of the signal-bearing nuclei or by increasing the voxel size. The voxel size may be increased by increasing the Thk, decreasing the number of pixels across the image, or by increasing the FOV. Increasing the voxel size will, however, result in a loss of image resolution.

Equation 10 also tells us that if $T_2$ is small compared to TE, or if $T_1$ is long compared to TR, we will see a small signal. These conditions are often the case in many solids. A short $T_2$ will cause the transverse magnetization to decay before a signal can be detected. A long $T_1$ will not allow the net magnetization to recover to its equilibrium value along the positive $z$ axis before the next set of RF pulses.

Occasionally, a sample may contain ferromagnetic particles that destroy the homogeneity of the $B_0$ field in the sample and either distort the image or destroy the signal entirely. Certain concrete samples contain iron filings, which prevent one from seeing an image of free water in concrete. Some samples contain high concentrations of paramagnetic substances that shorten the $T_2$ of water to value less than TE, thus destroying the signal.

Two additional properties that may perturb the NMR signal are conductivity and dielectric constant (Roe et al., 1996). A high conductivity will cause a skin effect artifact in the image. A high conductivity causes the RF magnetic fields to diminish with depth into the sample. Therefore, the outside of the object appears brighter than the inside of the object. A sample with a high dielectric constant causes the opposite artifact. Standing waves are set up within the imaged object when the dimension of the object approaches a half wavelength of the RF in the object. If concentration maps are desired, it may be possible to compensate for the effects of the spatially inhomogeneous $B_1$ field (Li, 1994).

The magnetic resonance imager is a multicomponent, complex imaging system consisting of RF, gradient, magnet, computer, etc. subsystems. All of these subsystems must be operating properly for optimal signal. As with any complex instrument, it is recommended that test standards be imaged before each run to test the operation of the subsystem components individually and as a whole, integrated system. The subcomponent that the user has the most control over is the imaging coil. The imaging coil sends out the oscillating $B_1$ field into the sample and detects the oscillating field from the magnetization in the sample. The imaging coil is an electrical LC circuit that must be tuned to the resonant frequency of the spins in the sample at the magnetic field strength of the imager, and impedance matched to that of the transmission line from the imager. The conductivity and dielectric constant of the sample affect the resonant frequency of the imaging coil and the matching. When high-$Q$ imaging coils are used, the resonant frequency of the coil must be tuned for and the impedance matched for each sample. Failure to tune and match the coil for each sample may result in large amounts of reflected power from the imaging coil, which will diminish signal and possibly damage the RF transmission and detection circuitry. Tuning and matching the coil is typically accomplished with tuning and matching variable capacitors located on the imaging coil.

Differences in the magnetic susceptibility of the sample may distort the image. Adjusting the shim settings may minimize the susceptibility artifact.

## LITERATURE CITED

Aime, S., Botta, M., Fasano, M., Terreno, E., Kinchesh, P., Calabi, L., and Paleari, L. 1996. A new chelate as contrast agent in chemical shift imaging and temperature sensitive probe for MR spectroscopy. *Magn. Reson. Med.* 35:648–651.

Amin, M. H. G., Hall, L. D., and Chorley, R. J. 1994. Magnetic resonance imaging of soil-water phenomena. *Magn. Reson. Imag.* 12:319.

Amin, M. H. G., Richards, K. S., and Hall, L. D. 1996. Studies of soil-water transport by MRI. *Magn. Reson. Imag.* 14:879.

Amin, M. H. G., Gibbs, S. J., and Hall, L. D. 1997. Study of flow and hydrodynamic dispersion in a porous medium using pulsed-field-gradient magnetic resonance. *Proc. Math. Phys. Eng. Sci.* 453:489.

Antalek, B. J. 1991. MRI studies of porous materials. M. S. Thesis, Rochester Institute of Technology.

Armstrong, R. L., Tzalmona, A., Menzinger, M., Cross, A., and Lemaire, C. 1992. *In* Magnetic Resonance Microscopy (B. Blumich and W. Kuhn, eds.) pp. 309–323. VCH Publishers, Weinheim, Germany.

Attard, J., Hall, L., Herrod, N., and Duce, S. 1991. Materials mapped with NMR. *Physics World.* 4:41.

Blumich, B. and Blumler, P. 1993. NMR imaging of polymer materials. *Makromol. Chem.* 194:2133.

Blumich, B., Blumler, P., and Weigand, F. 1994. NMR imaging and materials research. *Die Makromolekulare Chemie Macromolecular Symp.* 87:187.

Callaghan, P. T., 1991. Principles of Nuclear Magnetic Resonance Microscopy, Clarendon, Oxford.

Callaghan, P. T. and Xia, Y. 1991. Velocity and diffusion imaging in dynamic NMR microscopy. *J. Magn. Reson.* 91:326.

Carpenter, T. A., Davies, E. S., and Hall, C. 1993. Capillary water migration in rock: Process and material properties examined by NMR imaging. *Mater. Struct.* 26:286.

Chingas, G. C., Miller, J. B., and Garroway, A. N. 1986. NMR images of solids. *J. Magn. Reson.* 66:530–535.

Duce, S. L. and Hall, L. D. 1995. Visualisation of the hydration of food by nuclear magnetic resonance imaging. *J. Food Engineering.* 26:251.

Duce, S. L., Carpenter, T. A., and Hall, L. D. 1990. Nuclear magnetic resonance imaging of chocolate confectionery and the spatial detection of polymorphic states of cocoa butter in chocolate. *Lebensm. Wiss. Technol.* 23:545.

Duce, S. L., Ablett, S., and Hall, L. D. 1994. Quantitative determination of water and lipid in sunflower oil and water and meat/fat emulsions by nuclear magnetic resonance imaging. *J. Food Sci.* 59:808.

Duce, S. L., Ablett, S., and Hall, L. D. 1995. Nuclear magnetic resonance imaging and spectroscopic studies of wheat flake biscuits during baking. *Cereal Chem.* 72:05.

Fordham, E. J., Horsfield, M. A., and Hall, L. D. 1993a. Depth filtration of clay in rock cores observed by one-dimensional $^1$H NMR imaging. *J. Colloid Interface Sci.* 156:253.

Fordham, E. J., Hall, L. D., and Ramakrishnan, T. S. 1993b. Saturation gradients in drainage of porous media: NMR imaging measurements. *Am. Inst.. Chem. Eng. J.* 39:1431.

Fordham, E. J., Gibbs, S. J., and Hall, L. D. 1994. Partially restricted diffusion in a permeable sandstone: Observations by stimulated echo PFG NMR. *Magn. Reson. Imag.* 12:279.

Fordham, E. J., Sezginer, A., and Hall, L. D. 1995. Imaging multiexponential relaxation in the (y,log 3 T1) plane, with application to clay filtration in rock cores. *J. Magn. Reson. A* 113:139.

Fukushima, E. and Roeder, S. B. W. 1981. Experimental Pulse NMR: A Nuts and Bolts Approach. Addison-Wesley, Reading, Mass.

Fyfe, C. A., Randall, L. H., and Burlinson, N. E. 1992. Observation of heterogeneous trace (0.4% w/w) water uptake in bisphenol a polycarbonate by NMR imaging. *Chem. Mater.* 4:267.

Grinsted, R. A. and Koenig, J. L. 1992. Study of multicomponent diffusion into polycarbonate rods using NMR imaging. *Macromolecules* 25:1229.

Grinsted, R. A., Clark, L., and Koenig, J. L. 1992. Study of cyclic sorption-desorption into poly(methyl methacrylate) rods using NMR imaging. *Macromolecules* 25:1235.

Gibbs, S. J. and Hall, L. D. 1996. What roles are there for magnetic resonance imaging in process tomography? *Meas. Sci. Technol.* 7:827.

Gong, J. and Hornak, J. P. 1992. A fast $T_1$ algorithm. *J. Magn. Reson. Imag.* 10:623–626.

Guiheneuf, T. M., Tessier, J.-J., and Hall, L. D. 1996. Magnetic resonance imaging of meat products: Automated quantitation of the NMR relaxation parameters of cured pork, by both "bulk" NMR and MRI methods. *J. Sci. Food Agric.* 71:163.

Guiheneuf, T. M., Couzens, P. J., and Hall, L. D. 1997. Visualisation of liquid triacylglycerol migration in chocolate by magnetic resonance imaging. *J. Sci. Food Agric.* 73:265.

Guiheneuf, T. M., Gibbs, S. J., and Hall, L. D. 1997. Measurement of the inter-diffusion of sodium ions during pork brining by one-dimensional $^{23}$Na magnetic resonance imaging (MRI). *J. Food Eng.* 31:457.

Hafner, S. and Kuhn, W. 1994. NMR-imaging of water content in the polymer matrix of silicon chips. *Magn. Reson. Imag.* 12:1075–1078.

Hall, L. D., Hawkes, R. C., and Herrod, N. J. 1990. A survey of some applications of NMR chemical microscopy. *Philos. Trans. Phys. Sci.* 333:477.

Holme, K. R. and Hall, L. D. 1991. Chitosan derivatives bearing c10-alkyl glycoside branches: a temperature-induced gelling polysaccharide. *Macromolecules* 24:3828.

Hornak, J. P. 1995. Medical imaging technology. *In* Kirk-Othmer Encyclopedia of Chemical Technology, 4th ed, John Wiley & Sons, New York.

Hornak, J. P., Ceckler, T. L., and Bryant, R. G. 1986. Phosphorus-31 NMR spectroscopy using a loop-gap resonator. *J. Magn. Reson.* 68:319–322.

Hornak, J. P., Szumowski, J., and Bryant, R. G. 1987. Elementary single turn solenoids used as the transmitter and receiver in magnetic resonance imaging. *J. Magn. Res. Imag.* 5:233–237.

Hornak, J. P., Marshall, E., Szumowski, J., and Bryant, R. G. 1988. MRI of extremities using perforated single turn solenoids. *Magn. Reson. Med.* 7:442–448.

Horsfield, M. A., Hall, C., and Hall, L. D. 1990. Two-species chemical-shift imaging using prior knowledge and estimation theory: Application to rock cores. *J. Magn. Reson.* 87:319.

Ilg, M., Maier-Rosenkrantz, J., Muller, W., Albert, K., and Bayer, E. 1992. Imaging of the chromatographic process. *J. Magn. Reson.* 96:335–344.

Jackson, P. 1992. Curing of carbon-fiber reinforced epoxy resin; non-invasive viscosity measurement by NMR imaging. *J. Mater. Sci.* 27:1302.

Karunanithy, S. and Mooibroek, S. 1989. Detection of physical flaws in alumina reinforced with SiC fibers by NMR imaging in the green state. *J. Mater. Science.* 24:3686.

Komoroski, R. A. 1993. NMR imaging of materials. *Anal. Chem.* 65:1068A.

Knorgen, M., Heuert, U., and Kuhn, W. 1997. Spatially resolved and integral NMR investigation of the aging process of carbon black filled natural rubber. *Polymer Bull.* 38:101.

Kumar, A., Welti, D., and Ernst, R. E. 1975. NMR Fourier zeugmatography. *J. Magn. Reson.* 18:69–83 (also *Naturwiss.* 62:34).

Lauterbur, P. G. 1973. Image formation by induced local interactions: examples employing nuclear magnetic resonance. *Nature* 242:190–191.

Li, X. 1994. Tissue parameter determination with MRI in the presence of imperfect radiofrequency pulses. M. S. Thesis, Rochester Institute of Technology.

Li, X. and Hornak, J. P. 1993. Accurate determination of $T_2$ images in MRI. *Imag. Sci. Technol.* 38:154–157.

Link, J., Kaufmann, J., and Schenker, K. 1994. Water transport in concrete. *Magn. Reson. Imag.* 12:203–205.

Maas, W. E., Merwin, L. H., and Cory, D. G. 1997. Nuclear magnetic resonance imaging of solid rocket propellants at 14.1 T. *J. Magn. Reson.* 129:105.

Marshall, E. A., Listinsky, J. J., Ceckler, T. L., Szumowski, J., Bryant, R. G., and Hornak, J. P. 1998. Magnetic resonance imaging using a ribbonator: Hand and wrist. *Magn. Reson. Med.* 9:369–378.

McDonald, P. J., Attard, J. J., and Taylor, D. G. 1987. A new approach to NMR imaging of solids. *J. Magn. Reson.* 72:224–229.

McFarland, E. W. and Lee, D. 1993. NMR microscopy of absorbed protons in palladium. *J. Magn. Reson.* A102:231–234.

Metz, K. R., Zuo, C. S., and Sherry, A. D. 1998. Rapid simultaneous temperature and pH measurements by NMR through thulium complex. 39th Experimental NMR Conference, Asilomar, Calif., 1998.

Mori, M. and Koenig, J. L. 1995. Solid-State C-13 NMR studies of vulcanized elastomers XIII: TBBS accelerated, sulfur-vulcanization of carbon black filled natural rubber. *Rubber Chem. Technol.* 68:551.

Nieminen, A. O. K. and Koenig, J. L. 1989. NMR imaging of the interfaces of epoxy adhesive joints. *J. Adhesion* 30:47.

Nieminen, A. O. K. and Koenig, J. L. 1990. Evaluation of water resistance of polyurethane dispersion coating by nuclear magnetic resonance imaging. *J. Adhesion* 32:105.

Pearce, R. B., Fisher, B. J., and Hall, L. D. 1997. Water distribution in fungal lesions in the wood of sycamore, *Acer pseudoplatanus*, determined gravimetrically and using nuclear magnetic resonance imaging. *New Phytol.* 135:675.

Pel, L., Kopinga, K., and Brocken, H. 1996. Determination of moisture profiles in porous building materials by NMR. *Magn. Reson. Imag.* 14:931.

Roe, J. E., Prenttice, W. E., and Hornak, J. P. 1996. A multipurpose MRI phantom based on a reverse micelle solution. *Magn. Reson. Med.* 35:136–141.

Rommel, E., Hafner, S., and Kimmich, R. 1990. NMR imaging of solids by jenner-broekaert phase encoding. *J. Magn. Reson.* 86:264–272.

Sanz, J. L. C., Hinkle, E. B., and Jain, A. K. 1988. Radon and Projection Transform-Based Computer Vision, Springer-Verlag, Berlin.

Sarkar, S. N. and Komoroski, R. A. 1992. NMR imaging of morphology, defects, and composition of tire composites and model elastomer blends. *Macromolecules*, 25:1420–1426.

Seymour, J. D. and Callaghan, P. T. 1997. Generalized approach to NMR analysis of flow and dispersion in porous media. *Am. Inst.. Chem. Eng. J.* 43:2096.

Sinton, S. W., Iwamiya, J. H., Ewing, B., and Drobny, G. P. 1991. NMR of solid rocket fuel. *Spectroscopy* 6:42–48.

Smith, S. L. 1985. Nuclear magnetic resonance imaging. *Anal. Chem.* 57:A595–A607.

Szeglowski, S. D. and Hornak, J. P. 1993. Asymmetric single-turn solenoid for MRI of the wrist. *Magn. Reson. Med.* 30:750–753.

Wang, P.-S., Malghan, S. G., and Raman, R. 1995. NMR characterization of injection-moulded alumina green compact. Part II. T2-weighted proton imaging. *J. Mater. Sci.* 30:1069.

Wang, P.-S.., Minor, D. B., and Malghan, S. G. 1993. Binder distribution in Si3N4 ceramic green bodies studied by stray-field NMR imaging. *J. Mater. Sci.* 28:4940.

Wallner, A. S. and Ritchey, W. M. 1993. Void distribution and susceptibility differences in ceramic materials using MRI. *J. Mater. Res.* 8:655.

Webb, A. G. and Hall, L. D. 1990. Evaluation of the use of nuclear magnetic resonance imaging in the study of Fickian diffusion in rubbery polymers. 1. Unicomponent solvent ingress. *Polymer Commun.* 31:422.

Webb, A. G. and Hall, L. D. 1990. Evaluation of the use of nuclear magnetic resonance imaging in the study of Fickian diffusion in rubbery polymers. 2. Bicomponent solvent ingress. *Polymer Commun.* 31:425.

Webb, A. G. and Hall, L. D. 1991. An experimental overview of the use of nuclear magnetic resonance imaging to follow solvent ingress into polymers. *Polymer Bull.* 32:2926.

Webb, A. G., Jezzard, P., and Hall, L. D. 1989. Detection of inhomogeneities in rubber samples using NMR imaging. *Polymer Commun.* 30:363.

Windig, W., Hornak, J. P., and Antalek, B. J. 1998. Multivariate image analysis of magnetic resonance images with the direct exponential curve resolution algorithm (DECRA). Part 1: algorithm and model study. *J. Magn. Reson.* 132:298–306.

Zuo, C. S. Bowers, J. L., Metz, K. R., Noseka, T., Sherry, A. D., and Clouse, M. E. 1996. TmDOTP$^{5-}$: a substance for NMR temperature measurements. *In Vivo Magn. Reson. Med.* 36:955–959.

## KEY REFERENCES

Stark, D. D. and Bradley, W. G. 1988. Magnetic Resonance Imaging. Mosby, St. Louis, Mo.

*A comprehensive source of information on MRI as applied to clinical magnetic resonance imaging.*

Callaghan, P. T. 1991. Principles of Nuclear Magnetic Resonance Microscopy, Clarendon, Oxford.

## INTERNET RESOURCES

http://www.cis.rit.edu/htbooks/mri/

*J. P. Hornak, 1996. The Basics of MRI. An excellent hypertext resource describing the theory of magnetic resonance imaging with animated diagrams.*

http://www.cis.rit.edu/htbooks/nmr/

*J. P. Hornak, 1998. The Basics of NMR. An excellent hypertext resource describing the theory of nuclear magnetic resonance spectroscopy with animated diagrams.*

## APPENDIX:
## GLOSSARY OF TERMS AND SYMBOLS

| | |
|---|---|
| $\alpha$ | angle between readout gradient and $x$ axis |
| $B_0$ | static magnetic field |
| $B_1$ | magnitude of the RF magnetic field |
| BW | bandwidth |
| $C$ | capacitance |
| CSI | chemical shift imaging |
| $D_i$ | diffusion coefficient of component $i$ |
| $\delta$ | width of the diffusion encoding gradient in a pulse sequence |
| $\Delta$ | separation of the diffusion encoding gradients in a pulse sequence |
| $\delta_i$ | chemical shift of component $i$ |
| ESR | electron spin resonance |
| EPR | electron paramagnetic resonance |
| FOV | field of view |
| $f$ | frequency width of an NMR spectrum |
| $\phi$ | phase angle |
| $\gamma$ | gyromagnetic ratio |
| $G_{BP}$ | bipolar magnetic field gradient pulse |
| $G_i$ | magnetic field gradient in the i direction |
| $G_R$ | readout gradient |
| $G_s$ | slice selection gradient |
| $G_p$ | phase encoding gradient |
| $G_f$ | frequency encoding gradient |

$\Gamma_i$    full width at half height of a spectral absorption line for component $i$

$k$    proportionality constant used in NMR signal equations to include amplifier gains

$L$    inductance

MRI    magnetic resonance imaging

$\mathbf{M}$    net magnetization vector

$\mathbf{M}_{xy}$    transverse component of magnetization

$\mathbf{M}_z$    longitudinal component of magnetization

$\mu$SR    muon spin resonance

NMR    nuclear magnetic resonance

$\nu_i$    frequency of $i$

$P_{90}$    RF power necessary to produce a $90^\circ$ pulse

$\rho_i$    spin density of $i$

RF    radio frequency

$S$    signal

$T_{1i}$    spin-lattice relaxation time or component $i$

$T_{2i}$    spin-spin relaxation time of component $i$

$T_{2i}$    *$T_2$ star of component $i$

Thk    slice thickness

TR    repetition time

TE    echo time

$\tau$    width of an RF or phase encoding gradient pulse

TMS    the NMR standard tetramethylsilane

$\theta$    rotation angle of magnetization by the RF pulse

$V$    velocity

$V_c$    imaging coil volume

$\otimes$    convolution symbol

JOSEPH P. HORNAK
Rochester Institute of Technology
Rochester, New York

# NUCLEAR QUADRUPOLE RESONANCE

## INTRODUCTION

Nuclear quadrupole resonance (NQR) was once written off as a "dead" field, but has recently had a modest rebirth, and has been applied with success to several areas of materials science, most notably the high-$T_c$ superconductors. The name itself is a misnomer—NQR is really (NMR) at zero field. For that reason, it has many of the same disadvantages as the more familiar NMR spectroscopy (NUCLEAR MAGNETIC RESONANCE IMAGING)—e.g., poor sensitivity and complications arising from molecular dynamics, but because it does not require a superconducting magnet, NQR is generally cheaper and can be applied to a far greater number of nuclei. It is inherently noninvasive and nondestructive, and can be applied both to pure substances and to materials in situ. It is currently, for example, being used as an antiterrorist technique to screen persons and baggage for the presence of explosives, using receiver coils that are of the order of 1 m in radius. The principal disadvantage of NQR, other than sensitivity, is that, as a radiofrequency (RF) technique, it is incompatible with conducting or ferromagnetic materials; however, for the somewhat narrow range of materials to which it can be applied, NQR gives information which is otherwise unavailable.

The NQR properties of a material are primarily defined by a transition frequency or set of transition frequencies, which can be used to determine two parameters—the quadrupole coupling constant or q.c.c., and the asymmetry parameter, or $\eta$. These quantities are often quite characteristic of particular materials—as when they are used to detect drugs or explosives—but are also sensitive to temperature and pressure, and can therefore be used as a noninvasive probe of these properties. Moreover, by applying external spatially varying magnetic or radiofrequency fields, the NQR signal can be made position-dependent, allowing the distribution of the NQR-active species in the material to be imaged. Finally, the relaxation times of NQR nuclei are highly sensitive to dynamics, and so can be used as a local probe of molecular mobility.

In general, NQR imaging gives the same information about the distribution of NMR active species as NMR imaging, but applies to nuclei with spin $>1/2$ rather than spin $1/2$ nuclei.

The internal structure of materials can also be imaged by x-ray tomography and ultrasonic methods, which depend on the existence of inhomogeneities in the distribution of x-ray absorbers, or acoustic properties, respectively. The stress and pressure dependences of the NQR frequency are generally much stronger than those of NMR imaging, and therefore NQR is a superior method for mapping the temperature or stress distribution across a sample. Such stress distributions can be determined for optically transparent material by using polarized light; for opaque material, there may be no alternative methods.

This unit aims to be a general and somewhat cursory review of the theory and practice of modern one-dimensional (1D) and two-dimensional (2D) NQR and NQR imaging. It will certainly not be comprehensive. For a more general grounding in the theory of the method, one may best go back to the classic works of Abragam (1961) and Das and Hahn (1958); more modern reviews are referenced elsewhere in the text.

## PRINCIPLES OF THE METHOD

### Nuclear Moments

Elementary electrostatics teaches us that the nucleus, like any electromagnetic distribution, can be physically treated from the outside as a series of electric and magnetic moments. The moments of the nucleus follow a peculiar alternating pattern; nuclei in general have an electric monopole (the nuclear charge), a magnetic dipole moment, electric quadrupole moment, magnetic octopole moment, and electric hexadecapole moment. The last two, while they have been observed (Liao and Harbison, 1994) have little practical significance, and will not be examined further. The magnetic dipole moment of the nucleus enables NMR, which is treated elsewhere (NUCLEAR MAGNETIC RESONANCE IMAGING). It is the electric quadrupole moment of the nucleus that enables NQR.

Whether or not a nucleus has an electric quadrupole moment depends on its total spin angular momentum, $\mathbf{I}$, a property that is quantized in half-integer multiples of the quantum of action, $h$ (Planck's constant), and often referred to in shorthand form as the nuclear spin. Nuclei with spin zero, such as $^{12}$C, have neither a magnetic dipole nor an electric quadrupole moment. Nuclei with spin 1/2 have magnetic dipole moments but not electric quadrupole moments, while nuclei with spin 1 or higher have both magnetic dipole moments and electric quadrupole moments. For reasons connected with the details of nuclear structure and particularly the paucity of odd-odd nuclei (those with odd numbers of both protons and neutrons) there are few nuclei with integer spin: the naturally occurring spin 1 nuclei are $^{2}$H, $^{6}$Li, and $^{14}$N, while $^{10}$B, $^{40}$K, and $^{50}$V are the single examples of spin 3, 4, and 6 respectively. The quadrupole moments of $^{2}$H and $^{6}$Li are extremely small, making direct nuclear quadrupole resonance impossible except through the use of SQUIDs (superconducting quantum interference devices, see Practical Aspects of the Method and (TECHNIQUES TO MEASURE MAGNETIC DOMAIN STRUCTURES). The higher integer spin nuclei are of no experimental importance, and therefore the only integer spin nucleus that has been studied much by NQR is $^{14}$N.

In contrast, half-integer spin nuclei are liberally spread across the periodic table, and they have therefore been the subject of most NQR work. They range from the very common spin 3/2, for which there are scores of different stable isotopes, through the comparatively rare spin 9/2, which is present in a few, such as $^{87}$Sr and $^{209}$Bi.

Nuclear electric quadrupole moments are typically about eight orders of magnitude smaller than molecular electric quadrupole moments. In Figure 1, they are listed in SI units of $C \cdot m^2$, though it is common to divide out the charge of the electron and give them in units of area. They vary from the tiny moment of $-1.6 \times 10^{-50}$ $C \cdot m^2$ for $^{6}$Li to the comparatively huge value of $6.1 \times 10^{-47}$ $C \cdot m^2$ in $^{179}$Hf. They can be directly measured by atomic beam methods, but are frequently known only with very rough precision; for example, recent measurements of the moment of $^{137}$Ba have varied from $Q = 0.228$ barns (1 barn $= 10^{-28}$ $m^2$)to $Q = 0.35$ barns (to convert $Q$ in barns to $eQ$ in $C \cdot m^2$ multiply by $1.602 \times 10^{-47}$). An extensive tabulation of measurements of nuclear moments is available in Raghavan (1989). As will be discussed below, such moments cannot easily be independently determined by NQR.

## Nuclear Couplings

**The Electric Field Gradient.** Group theory dictates that nuclear moments couple only with the derivative of the local electric or magnetic field that possesses the same symmetry. This means that the nuclear magnetic dipole couples only with the magnetic field, and the electric quadrupole couples with the electric field gradient. In NMR, the magnetic field is usually externally applied and in most cases dwarfs the local field from the paramagnetism of the sample. However, in NQR, the local electric field gradient dwarfs any experimentally feasible external field gradient. As a ball park estimate of the local field gradient ($\nabla \mathbf{E}$) in an ionic crystal, we can compute the magnitude of the field gradient at a distance of $r = 0.2$ nm from a unipositive ion ($q_r$), using

$$\nabla \mathbf{E} = -\frac{(3\hat{\mathbf{r}}\hat{\mathbf{r}} - 1)q_r}{4\pi\varepsilon_0 r^3} \tag{1}$$

where $\varepsilon_0$ is the permitivity of the vacuum and $\hat{\mathbf{r}}$ is a unit vector. This yields a gradient of $3.6 \times 10^{20}$ $V \cdot m^{-2}$, or between 8 and 9 orders of magnitude greater than the external field gradients that can currently be produced in the laboratory. NQR is therefore carried out using only the local field gradient of the sample.

The electric field gradient $\nabla \mathbf{E}$ is a traceless symmetric second-rank tensor with five independent elements that can be expressed in terms of two magnitudes and three

| IA | IIA | IIIB | IVB | VB | VIB | VIIB | — | VIIIB | — | IB | IIB | IIIA | IVA | VA | VIA | VIIA | 0 |
|---|---|---|---|---|---|---|---|---|---|---|---|---|---|---|---|---|---|
| $^{2}$D<br>1<br>4.58(2) | | | | | | | | | | | | | | | | | He |
| $^{6}$Li<br>3/2<br>-1.3(1)<br>$^{7}$Li<br>3/2<br>-64.2(2) | Be | | | | | | | | | | | $^{10}$B<br>3<br>135.5(1)<br>$^{11}$B<br>3<br>65.0(4) | C | $^{14}$N<br>1<br>32.36(16) | $^{17}$O<br>5/2<br>-41.3(2) | F | Ne |
| $^{23}$Na<br>3/2<br>161(3) | $^{25}$Mg<br>5/2<br>322(5) | | | | | | | | | | | $^{27}$Al<br>5/2<br>224(10) | Si | P | $^{33}$S<br>3/2<br>-102(16) | $^{35}$Cl<br>3/2<br>-132.16(3)<br>$^{37}$Cl<br>3/2<br>-104.03(3) | Ar |
| $^{39}$K<br>3/2<br>96.2(6) | $^{43}$Ca<br>7/2<br>-78(8) | $^{45}$Sc<br>7/2<br>-346(2) | $^{47}$Ti<br>5/2<br>466(16)<br>$^{49}$Ti<br>7/2<br>384(16) | $^{51}$V<br>7/2<br>-83(16) | $^{53}$Cr<br>3/2<br>-240(80) | $^{55}$Mn<br>5/2<br>528(16) | Fe | $^{59}$Co<br>5/2<br>672(48) | $^{61}$Ni<br>3/2<br>260(24) | $^{63}$Cu<br>3/2<br>-352(24)<br>$^{65}$Cu<br>3/2<br>-312(6) | $^{67}$Zn<br>5/2<br>240(24) | $^{69}$Ga<br>3/2<br>269(2)<br>$^{71}$Ga<br>3/2<br>169(2) | $^{73}$Ge<br>9/2<br>-277(42) | $^{75}$As<br>3/2<br>503(10) | Se | $^{79}$Br<br>3/2<br>530(6)<br>$^{81}$Br<br>3/2<br>442(6) | Kr |
| $^{85}$Rb<br>5/2<br>439(3)<br>$^{87}$Rb<br>3/2<br>211(2) | $^{87}$Sr<br>9/2<br>536(32) | Y | $^{91}$Zr<br>5/2<br>-330(16) | $^{93}$Nb<br>3/2<br>-512(32) | $^{95}$Mo<br>5/2<br>-320(128)<br>$^{97}$Mo<br>5/2<br>408(21 | Tc | $^{99}$Ru<br>3/2<br>126.57<br>$^{101}$Ru<br>5/2<br>732.19 | Rh | $^{105}$Pd<br>5/2<br>1057(18) | Ag | Cd | $^{115}$In<br>9/2<br>1298(2) | Sn | $^{121}$Sb<br>5/2<br>-577(64)<br>$^{123}$Sb<br>7/2<br>-785(80) | Te | $^{127}$I<br>7/2<br>-1264(2) | $^{131}$Xe<br>3/2<br>-192(19) |
| $^{133}$Cs<br>7/2<br>-5.94(22) | $^{135}$Ba<br>3/2<br>256(5)<br>$^{137}$Ba<br>3/2<br>392(6) | $^{139}$La<br>7/2<br>320(16) | $^{177}$Hf<br>7/2<br>5378(46)<br>$^{179}$Hf<br>9/2<br>6077(53) | $^{181}$Ta<br>7/2<br>5079(32) | W | $^{185}$Re<br>5/2<br>3492(32)<br>$^{187}$Re<br>5/2<br>3316(32) | $^{189}$Os<br>3/2<br>1371(45) | $^{191}$Ir<br>3/2<br>1307(14)<br>$^{193}$Ir<br>3/2<br>1203(14) | Pt | $^{197}$Au<br>3/2<br>876(26) | $^{201}$Hg<br>3/2<br>616(64) | Tl | Pb | $^{209}$Bi<br>9/2<br>-801(16) | Po | At | Rn |

**Figure 1.** Spin and quadrupolar moments (in $C \cdot m^2$ \times $10^{50}$) of quadrupolar nuclei.

Euler angles defining its directionality. The magnitude of the field gradient tensor along its $zz$ direction in its principal axis system $(\nabla E)_{zz}$ is often given the shorthand notation $eq$, with $e$, the electric charge, factored out. The asymmetry parameter $\eta$ is defined by

$$\eta = \frac{(\nabla E)_{yy} - (\nabla E)_{xx}}{(\nabla E)_{zz}} \qquad (2)$$

Electric field gradients can be computed for nuclear sites in ionic lattices and in covalently bound molecules. For lattice calculations, point-charge models are rarely adequate, and in fact often give results that are entirely wrong. In most cases, we find that fixed and induced dipole and quadrupole moments need to be introduced for each atom or residue in order to get accurate values. For example, in calculations of the field gradient at the $^{14}N$ and cationic sites in $Ba(NO_3)_2$ and $Sr(NO_3)_2$ (Kye, 1998), we include the charges, the quadrupole moment of the nitrate ion (treated as a single entity and calculated by *ab initio* methods), the anisotropic electric polarizability of the nitrate ion (obtained from the refractive indices of a series of nitrates via the Clausius-Mosotti equation), and the quadrupole polarizability of the cations (calculated by other authors *ab initio*). We omit the electric polarizability of the cations only because they sit at a center of crystal symmetry and the net dipole moment must inevitably be zero; otherwise, this would be an important contribution. Briefly, after calculating the field and field gradient for each site using point charges, we compute the induced dipole at the nitrate as a vector using the nitrate polarizability tensor, and recalculate the field iteratively until convergence is obtained. We repeat this procedure again after introducing the quadrupole moment of the nitrate and quadrupole polarizabilities of the cations. Convergence of the lattice sums may be facilitated by an intelligent choice of the unit cells. The contributions of the various terms to the total field gradient at the nitrate in $Ba(NO_3)_2$ and $Pb(NO_3)_2$ are shown in Figure 2.

In covalent molecules, field gradients at the nucleus may be adequately obtained using computational modeling methods.

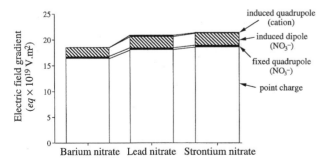

**Figure 2.** Contributions of the point charges, induced dipole moments, fixed quadrupolar and induced quadrupole moments to the total electric field gradient in ionic nitrates (from Kye and Harbison, 1998).

**The Electric Quadrupole Hamiltonian.** The coupling between the electric quadrupole moment and the ambient electric field gradient is expressed mathematically by the electric quadrupole coupling Hamiltonian.

$$H = \frac{e^2 qQ}{4hI(2I-1)}\left[ 3I_z^2 - I(I+1) + \frac{\eta}{2}(I_+^2 + I_-^2)\right] \qquad (3)$$

The quantity $e^2qQ/h$ is the so-called electric quadrupole coupling constant, and is a measure of the strength of the coupling, dimensioned, as is usual in radiofrequency spectroscopy, in frequency units. $I_z$, $I_+$, and $I_-$ are nuclear spin operators defined in the eigenbasis of the field gradient tensor.

**The Sternheimer Effect and Electron Deformation Densities.** One major complication of lattice calculations of electric field gradients is the Sternheimer antishielding phenomenon. Briefly, an atom or ion in the presence of a field gradient deforms into an induced quadrupole. The induced quadrupole does not merely perturb other atoms in the lattice; it also changes the field gradient within the electron distribution, usually increasing it, and often by a large amount. The field experienced by the nucleus is related to the external field by

$$\nabla \mathbf{E}_{\text{nuclear}} = (1 - \gamma_\infty)\nabla \mathbf{E}_{\text{external}} \qquad (4)$$

where $\gamma_\infty$ is the Sternheimer antishielding factor. Antishielding factors can be calculated by *ab initio* methods, and are widely available in the literature. They vary from small negative values for ions isoelectronic with helium, to positive values of several hundred for heavy ions. A representative selection of such factors for common ions is given in Table 1.

### Energy Levels at Zero Field

**Spin 1.** At zero field, the electric quadrupole coupling is generally the only important component of the spin Hamiltonian. If the asymmetry parameter is zero, the Hamiltonian is already diagonal in the Cartesian (field gradient) representation, and yields nuclear energy levels of

$$E_{\pm 1} = \frac{e^2qQ}{4h}; \qquad E_0 = -\frac{e^2qQ}{2h} \qquad (5)$$

There are two degenerate transitions, between $m = -1 \rightarrow m = 0$ and $m = 0 \rightarrow m = +1$, appearing at a frequency

$$\omega_Q = E_{\pm 1} - E_0 = \frac{3e^2qQ}{4h} \qquad (6)$$

If $\eta \neq 0$, the degeneracy is lifted and the Hamiltonian is no longer diagonal in any Cartesian representation, meaning that the nuclear spin states are no longer pure eigenstates of the angular momentum along any direction, but rather are linear combinations of such eigenstates. The three energies are given by the equation

$$E_{\pm} = \frac{e^2qQ}{4h}(1 \pm \eta); \qquad E_0 = -\frac{e^2qQ}{2h} \qquad (7)$$

**Table 1. Sternheimer Quadrupole Antishielding Factors for Selected Ions**

| Ion | Free Ion | Crystal Ion |
|---|---|---|
| $Li^+$ | $0.261^a$, $0.257^{b,c}$, $0.256^d$, $0.248^e$ | $0.282^a$, $0.271^f$ |
| $B^{3+}$ | $0.146^a$, $0.145^{b,c,d}$, $0.142^e$ | $0.208^a$, $0.189^b$ |
| $N^{5+}$ | $0.101^a$ | $0.110^a$ |
| $Na^+$ | $-5.029^a$, $-5.072^b$, $4.5\pm0.1^{c,d,e}$ | $-7.686^a$, $-4.747^f$ |
| $Al^{3+}$ | $-2.434^a$, $-2.462^b$, $-2.59^d$, $-2.236^c$ | $-5.715^a$, $-3.217^f$ |
| $Cl^-$ | $-82.047^a$, $-83.5^b$, $-53.91^c$, $-49.28^g$, $-63.21^h$ | $-38.915^a$, $-27.04^f$ |
| $K^+$ | $-18.768^a$, $-19.16^b$, $-17.32^i$, $-12.17^c$, $-12.84^g$, $-18.27^h$ | $-28.701^a$, $-22.83^f$ |
| $Ca^{2+}$ | $-14.521^a$, $-13.95^b$, $-12.12^b$, $-13.32^h$ | $-25.714^a$, $-20.58^f$ |
| $Br^-$ | $-195.014^a$, $-210.0^b$, $-99.0^g$, $-123.0^i$ | $-97.424^a$ |
| $Rb^+$ | $-51.196^a$, $-54.97^b$, $-49.29^g$, $-47.9^j$ | $-77.063^a$ |
| $Nb^{5+}$ | $-22.204^a$ | $-28.991^a$ |
| $I^-$ | $-331.663^a$, $-396.60^b$, $-178.75^g$, $-138.4^k$ | $-317.655^a$ |

[a]Sen and Narasimhan (1974).
[b]Feiock and Johnson (1969).
[c]Langhoff and Hurst (1965).
[d]Das and Bersohn (1956).
[e]Lahiri and Mukherji (1966).
[f]Burns and Wikner (1961).
[g]Wikner and Das (1958).
[h]Lahiri and Mukherji (1967).
[i]Sternheimer (1963).
[j]Sternheimer and Peierls (1971).
[k]Sternheimer (1966).

And there are now three observable transitions, two of which correspond to the degenerate transitions for $\eta = 0$, and one which generally falls at very low frequency.

$$\omega_\pm = \frac{3e^2qQ}{4h}\left(1 \pm \frac{\eta}{3}\right); \qquad \omega_0 = \frac{e^2qQ\eta}{2h} \tag{8}$$

Observation of a single NQR transition at zero field for $^{14}N$ is an indication that the asymmetry parameter is zero (or that one transition has been missed). Observation of a pair of transitions for any species permits the q.c.c. and $\eta$ to be extracted, and thence all of the magnitude information for the tensor to be obtained. The low-frequency transition is seldom observed, and in any case contains redundant information.

**Spin 3/2.** The energy levels of spin 3/2 nuclei are given by

$$E_{\pm1/2} = -\frac{e^2qQ}{4h}\left(1 + \frac{\eta^2}{3}\right)^{1/2} : E_{\pm3/2} = \frac{e^2qQ}{4h}\left(1 + \frac{\eta^2}{3}\right)^{1/2} \tag{9}$$

As can be seen, the $1/2$ and $-1/2$ states are degenerate, as are the $3/2$ and $-3/2$ states. This degeneracy is not lifted by a nonzero asymmetry parameter, which mixes the degenerate states without splitting them. The single quadrupolar frequency is

$$\omega_Q = \frac{e^2qQ}{2h}\left(1 + \frac{\eta^2}{3}\right)^{1/2} \tag{10}$$

Since this frequency is a function of both $e^{2qQ/h}$ and $\eta$, those quantities cannot be separately measured by simple NQR, though they can be obtained by a variety of two-dimensional methods (see Practical Aspects of the Method).

**Spin 5/2.** At $\eta = 0$, the energy levels of the spin 5/2 nucleus are given by

$$E_{\pm1/2} = \frac{e^2qQ}{5h}; \quad E_{\pm3/2} = -\frac{e^2qQ}{20h}; \quad e_{\pm5/2} = \frac{e^2qQ}{4h} \tag{11}$$

Transitions between the $m = \pm1/2$ and $\pm3/2$ states, and between the $\pm3/2$ and $\pm5/2$ states are allowed. These appear at

$$\omega_1 = \frac{3e^2qQ}{20h}; \qquad \omega_2 = \frac{3e^2qQ}{10h} \tag{12}$$

and obviously lead to a 2:1 ratio of frequencies.

If the asymmetry parameter is nonzero, there are still two transition frequencies, but these deviate from a 2:1 ratio. Solving for them requires finding the roots of a cubic secular equation. The algebraic solutions have been published (Creel et al., 1980; Yu, 1991); more useful are the q.c.c. and $\eta$ in terms of the measured frequencies

$$\frac{e^2qQ}{h} = \frac{20}{3}\sqrt{\frac{(\nu_2^2 + \nu_2\nu_1 + \nu_1^2)}{7}}\cos(\phi/3)$$

$$\eta = \sqrt{3}\tan(\phi/3) \tag{13}$$

$$\cos\phi = \left(\frac{7}{(\nu_2^2 + \nu_2\nu_1 + \nu_1^2)}\right)^{3/2}\frac{(\nu_2 + 2\nu_1)(\nu_2 - \nu_1)(2\nu_2 + \nu_1)}{20}$$

with $\nu_{1,2} = \omega_{1,2}/2\pi$. Thus, measurement of both frequencies for a spin 5/2 nucleus such as $^{127}$I gives a complete set of magnitude information about the quadrupolar tensor.

In addition to these two transitions, for $\eta \neq 0$ the $\pm 1/2 \rightarrow \pm 5/2$ transition becomes weakly allowed, and can occasionally be observed. However, if the two single-quantum frequencies are already known, this nearly forbidden transition gives only redundant information.

**Higher-Spin Nuclei.** Spin 7/2 nuclei have three, and spin 9/2 four observable transitions at zero field. At $\eta = 0$, these frequencies are in simple algebraic ratio, and can be easily obtained from the Hamiltonian given in Abragam (1961). If $\eta \neq 0$, the situation is more complicated, although analytical expressions for the spin 7/2 frequencies are still available by solution of a quartic secular equation (Creel, 1983). Any pair of frequencies (if assigned) is sufficient to give $e^2qQ/h$ and $\eta$.

## First-Order Zeeman Perturbation and Zeeman Line Shapes

Most NQR spectra are collected not at zero field, but in the earth's field. In practice, in most cases, the effect of this small field is unobservable, save for certain systems with extremely narrow NQR lines. In these cases, most frequently highly ordered ionic crystals, Zeeman effects may be observed.

An external field introduces a preferential external reference from of the system, and for a single crystal (the simplest case) it therefore makes the spectrum dependent on the relative orientation of magnetic field and $\nabla\mathbf{E}$. For spin 3/2 systems (which are most frequently studied) the effect is to split the single transition into a pair of doublets whose splittings depend on the size of the magnetic field and its orientation with respect to the field-gradient tensor. Analytical expressions for the positions of these lines have been determined (Creel, 1983); in addition, certain preferred orientations of the tensor relative to field give no splitting, and the loci of these cones of zero splitting can be used to get the field-gradient tensor orientation, which is not directly available from NQR at zero field (Markworth et al., 1987).

In powders, Zeeman-perturbed spectra of spin 3/2 nuclei have a characteristic lineshape with four maxima; these line shapes can be used to determine the asymmetry parameter (Morino and Toyama, 1961).

Figure 3 shows the Zeeman-perturbed spectra of potassium chlorate ($KClO_3$), an excellent test sample for NQR, which has a $^{35}$Cl resonance at 28.08 MHz. The spectra were obtained in a probe with an external field coil oriented parallel to the Earth's field, and shown for a series of coil currents. As can be seen, the minimum linewidth is obtained with a current of 0.2 A, where the external field presumably nulls out the Earth's field precisely. At higher or lower values of the current, characteristic Zeeman-perturbed line shapes are obtained.

## Spin Relaxation in NQR

As in NMR, NQR relaxation is generally parameterized in terms of two relaxation times: $T_1$, the longitudinal relaxation time, which is the time constant for relaxation

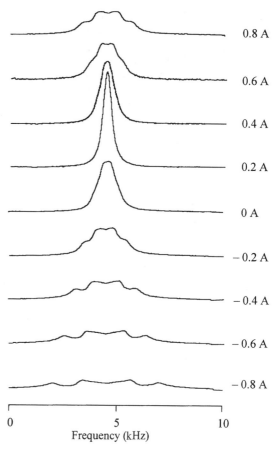

**Figure 3.** Effect of small external magnetic fields (expressed in terms of external coil current) on the lineshape of potassium chlorate resonance recorded at 28.85 MHz. A negative current corresponds to an applied field in the direction of the terrestrial magnetic field.

of the populations of the diagonal elements of the electric quadrupole coupling tensor in its eigenbasis, and $T_2$, the transverse relaxation time, which governs the relaxation of off-diagonal elements. Except in the case of spin 3/2, more than one $T_1$ and $T_2$ are generally observed.

The longitudinal relaxation time is probably in most cases dominated by relaxation via very high-frequency motions of the crystal, typically librations of bonds or groups. Phenomenologically, it is generally observed to have an Arrhenius-like dependence on temperature, even in the absence of a well defined thermally activated motional mechanism.

The transverse relaxation time, $T_2$, must be distinguished from $T_2^*$, the reciprocal of the NQR line width. The latter is generally dominated by inhomogeneous broadening, which is a significant factor in all but the most perfectly crystalline systems. The processes giving rise to $T_2$ in NQR have, to our knowledge, not been systematically investigated, but it has been our observation that, in many cases, $T_2$ is similar in value to $T_1$, and is therefore likely a result of the same high-frequency motions.

$T_1$ and $T_2$ may be measured by the same pulse sequences used in high-field NMR; $T_1$ by an inversion recovery pulse sequence $(\pi - t - \pi/2 - \text{acquire})$; $T_2$ using a single spin echo or train of spin echoes. $T_1$ values at room temperature range from hundreds of microseconds to tens of milliseconds for half-integer spin nuclei with appreciable quadrupole coupling constants, such as $^{35}Cl$, $^{63}Cu$, $^{79}Br$, or $^{127}I$; for nuclei with smaller quadrupole moments, such as $^{14}N$, they can range from a few milliseconds to (in rare cases) hundreds of seconds. $T_1$ values generally increase by one or two orders of magnitude on cooling to 77 K.

One phenomenon in which spin-relaxation times in NQR can be used to get detailed dynamical information is the "bleach-out" effect seen for trichloromethyl groups in the solid state. At low temperatures, three chlorine resonances are usually seen for such groups; however, as the temperature is raised, the resonances broaden and weaken, eventually disappearing, as a result of reorientation of the group about its three-fold axis. Depending on the environment of the grouping, "bleach-out temperatures" can be as low as $-100°C$ or as high at $70°C$, as is observed in the compound $p$-chloroaniline trichloroacetate (Markworth et al., 1987). Such exchange effects are quantum-mechanically very different from exchange effects in NMR, since reorientation of the group dramatically reorients the axis of quantization of the nuclear spin Hamiltonian, and is therefore in the limit of the "strong collision" regime. Under these circumstances, the relaxation time $T_1$ is to a very good approximation equal to the correlation time for the reorientation, making extraction of that correlation time trivial.

Similar phenomena are expected for other large-angle, low-frequency motions, but have otherwise been reported only for $^{14}N$.

## PRACTICAL ASPECTS OF THE METHOD

### Spectrometers for Direct Detection of NQR Transitions

**Frequency-Swept Continuous Wave Detection.** For its first 30 years, the bulk of NQR spectroscopy was done using superregenerative spectrometers, either home-built or commercial (Decca Radar). The superregenerative oscillator was a solution to the constraint that an NMR spectrometer must be capable of sweeping over a wide range of frequencies, while in the 1950s and 1960s most radiofrequency devices (e.g., amplifiers, phase modulators and detectors, and receivers) were narrow-band and tunable. Briefly, the superregenerative circuit is one where the sample coil forms the inductive element in the tank circuit of the primary oscillator, which is frequency-swept using a motor-driven variable capacitor. Absorption of radiofrequency by the NQR spins causes a damping of the oscillator, which can be detected by a bridge circuit. The result is a frequency-swept spectrum in which NQR transitions are detected as the second derivative of their lineshape. The superregenerative circuit has the virtues of cheapness and robustness; however it has all the disadvantages of continuous wave detection—it does not lend itself easily to signal averaging, or to the more sophisticated 2D approaches discussed below. It is probably useful only for pure samples of molecular weights < 1000 Da.

**Frequency-swept Fourier-transform Spectrometers.** Most modern NQR spectroscopy likely is done on commercial

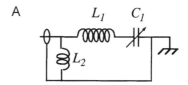

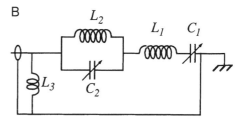

**Figure 4.** Probes for Fourier transform NQR. (**A**) Inductively matched series tank single-resonance probe; (**B**) inductively matched series/parallel tank double-resonance probe.

solid-state NMR instruments, as the equipment requirements (save for the magnet) are very similar to those of NMR. Critical elements are a fast digitizer (<2 μs/point) and high-power amplification (preferably capable of 2 kW over a range of 5 to 200 MHz, although this depends on the sample).

Standard static single-resonance solid-state NMR probes can be used for NQR; however, since there is no necessity that the probe be inserted in a magnet bore, some efficiencies can be achieved by building a dedicated probe. The circuit in Figure 4, panel A, has been successfully used for 10 years in the authors' laboratory. $L_1$, the sample coil, is typically a 20-turn solenoid wound from 26 AWG (American Wire Gauge) magnet wire, with dimensions of 4 cm long by 1.2 cm diameter; it holds a sample volume of $\sim$4 ml. It is tuned by a series capacitance $C_1$; we generally employ a 1.5 to 30 pF Jennings vacuum capacitor rated at 15 kV. Matching to 50 $\Omega$ is accomplished by a 3 turn, 1 cm diameter inductor $L_2$. In general, inductive matches, while somewhat more cumbersome to adjust, are much more broad-banded than capacitive matches.

The probe gives a 1 to 2 μs $\pi/2$ pulse for most nuclei over a frequency range of $\sim$20 to 45 MHz at a pulse power level of 500 W; additional series capacitance can be used to lower the frequency range, while going higher requires a smaller sample coil. The probe can simply be built inside an aluminum box, though somewhat better ringdown times are achieved with stainless steel. Acoustic ringing is however much less of a problem than in solid-state NMR, since the magnetoacoustic contribution is not present. If low temperatures are desired, the sample coil can be immersed in a coolant such as liquid nitrogen.

For double-resonance experiments, such as are required to correlate connected NQR transitions, we use the probe design in Figure 4, panel B. These double-resonance techniques involve irradiation of a spin >3/2 at two frequencies corresponding to distinct but connected NQR resonances. Double-resonance methods have not yet found significant application in materials research, and so are not described at length here; the interested reader is referred to Liao and

Harbison (1994). The circuit consists of a series and a parallel tank in series. In principle, either coil could contain the sample; in practice we use the series inductor $L_1$. A typical probe configuration uses coil inductances of $L_1 = 2.3$ µH and $L_2 = 1.2$ µH, with a small two- or three-turn matching inductor $L_3$. The circuit is essentially a pair of tank circuits, whose coupling varies with the separation of their resonance frequencies. Again, high-voltage 1.5 to 30 pF variable capacitors are used for $C_1$ and $C_2$. The two-fold difference between $L_1$ and $L_2$ ensures that the coupling between the two tanks is only moderate, so that resonance $\nu_1$ primarily resides in tank circuit $L_1 C_1$ and $\nu_2$ in $L_2 C_2$. With this configuration, $\nu_1$ tunes between 17 MHz and 44 MHz and $\nu_2$ between 47 and 101 MHz, although the two ranges are not entirely independent.

$L_3$ acts as a matching inductor for both resonances; differential adjustment of the matching inductance can be achieved by adjusting the coupling between the two inductors, either by partially screening the two coils by an intervening aluminum plate, or even by adjusting their relative orientation with a pair of pliers.

One notable feature that the probe lacks is any isolation between the two resonances; in fact, it has but a single port. This is because, unlike high-field NMR, it is seldom necessary to pulse one resonance while observing a second. We can therefore combine the output of both radio frequency channels before the final amplifier (which is broad-band and linear) and send the single amplifier output to a standard quarter-wave duplexer, which in practice is broad-band enough to handle the typical 2 : 1 frequency ratio used in NQR double-resonance experiments. Using single output state amplification and omitting isolation makes it far easier to design a probe to tune both channels over a wide frequency range.

**SQUID Detectors.** To overcome the sensitivity problems of NQR at low frequency, a DC superconducting quantum interference device (SQUID)—based amplifier can be used (Clarke, 1994; TonThat and Clarke, 1996). Using this device allows one to detect magnetic flux directly rather than differentially. This makes the intensity of the signal a linear function of the frequency instead of a quadratic function, greatly improving sensitivity at low frequencies compared to conventional amplification. SQUID NMR spectrometers can operate in both the continuous wave (Connor et al., 1990) or pulsed (TonThat and Clarke, 1996) modes. The latest spectrometers can attain a bandwidth up to 5 MHz (TonThat and Clarke, 1996). One of the drawbacks of the technique is cost, a result of the more complex design of the probe and spectrometer. SQUID spectrometers are not yet commercially available.

Experiments are performed at 4.2 K in liquid helium. Examples of nuclei observed using this technique are $^{27}Al$ in $Al_2O_3[Cr^{3+}]$ at 714 kHz (TonThat and Clarke, 1996), $^{129}Xe$ in laser-polarized solid xenon (Ton That et al., 1997), and $^{14}N$ in materials such as amino acids (Werner-Zwanzinger et al., 1994), cocaine hydrochloride (Yesinowski et al., 1995), and ammonium perchlorate (Clarke, 1994). Note that the detection of nitrogen transitions can be performed by cross-relaxation to protons coupled to nitrogen by the dipole-dipole interaction (Werner-Zwanzinger et al., 1994; Yesinowski et al., 1995).

### Indirect Detection of NQR Transitions by Field Cycling Methods

Where relaxation properties of a material are favorable (long intrinsic proton relaxation time, short quadrupole relaxation time), indirect detection methods give a significant improvement in signal-to-noise ratio, since detection is via the proton spins, which have a large gyromagnetic ratio and relatively narrow linewidth. Such methods are therefore often favored over direct detection methods for spins with low NQR frequencies and low magnetic moments, e.g., $^{14}N$ or $^{35}Cl$.

The term "field cycling" comes from the fact that the sample experiences different magnetic fields during the course of the experiment. The field intensity is therefore cycled between transients. Field cycling techniques were first developed during the 1950s (Pound, 1951). During a simple cycle, the sample is first magnetized at a high field, $B_0$, then rapidly brought adiabatically to a low field or to zero field. During the evolution period, the magnetization is allowed to oscillate under local interactions. The sample is finally brought back to high field, where the signal is detected. The field strength during the detection period can be identical to that for the preparation period, which is convenient but it can also be lower, as in "soak field" techniques (Koening and Schillinger, 1969). Also, other cycles involving more than two different field strengths have been developed, such as the "zero-field technique" (Bielecki et al., 1983). Field switching can be implemented either by electronic switches or by mechanically moving the sample. The latter is more simple and can be easily implemented on any spectrometer, but is limited for experiments necessitating very rapid field switching. Electronic switches are limited by Faraday's induction law, which dictates the maximum value of $(dB_0/dt)$ for a given static $B_0$ field.

Field cycling can be applied in the solution state and in liquid crystals as well as in the solid state. As a competitive technique to NQR, quadrupolar nuclei can be observed directly; some examples are, but are not limited to, deuterium (Thayer and Pines, 1987) and nitrogen-14 (Selinger et al., 1994). Indirect detection ("level crossing") is also feasible by observing the signal of the abundant nucleus (typically protons) dipolar-coupled to the quadrupolar nucleus (Millar et al., 1985). An example of level crossing is given in a latter section to record the nutation spectrum of *tris*-sarcosine calcium chloride at 600 kHz (Blinc and Selinger, 1992). The large number of applications of field cycling NQR is outside the scope of this unit, and more detailed information can be found in reviews by Kimmich (1980) and Noack (1986); see Literature Cited for additional information.

### One-Dimensional Fourier Transform NQR: Implementation

The hardest task in Fourier transform NQR is to find the resonance. While the chemistry of the species being studied sometimes gives one insight about where the resonances might lie—e.g., an organochlorine can be expected

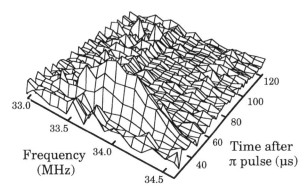

**Figure 5.** Frequency-swept spin-echo NQR. Each slice in the time dimension is a spin echo, recorded at a particular frequency; a slice in the frequency dimension parallel to the frequency axis gives the NQR spectrum.

to fall between 30 and 40 MHz—in general, this range will be far wider than the bandwidth of an NQR probe or the effective bandwidth of a high-power pulse. Some frequency sweeping will therefore have to be done. In a modern solid-state NMR spectrometer, the output power will usually be fairly constant over a range of 10 to 20 MHz, and the other components will generally be broad-band. Sweeping frequency, therefore, involves incrementing the spectrometer frequency by some fairly large fraction of the probe bandwidth, retuning the probe to that frequency, and acquiring a free-induction decay or preferably a spin echo at that frequency. Spectrometer frequencies are generally under computer control; the task of retuning the probe can be accomplished manually, or preferably automatically, by controlling the probe's tuning capacitor from the spectrometer via a stepper motor, and using this motor to minimize reflected power. The result of such a sweep is a series of spin echoes collected as a function of spectrometer bandwidth; these echoes, either phased or in magnitude mode, can be stacked in an array such as shown in Figure 5, which is a set of spin echoes collected for the $^{35}$Cl signal of the polymer poly-4-chlorostyrene. A slice along the spine of the spin echo is the NQR spectrum collected at the resolution of the frequency sweep. If the line is narrower than this step frequency, the individual spin-echo slice in which the line appears may be Fourier transformed; if it is broad, the slice through the spin echoes suffices.

### Two-dimensional Zero-Field NQR

**Zeeman-Perturbed Nuclear Resonance Spectroscopy ZNQRS.** For 3/2 spins, the $\pm 3/2 \rightarrow \pm 1/2$ transition depends on both the quadrupolar coupling constant, $e^2qQ/h$, and the asymmetry parameter $\eta$. The resonance frequency of the single line is given by

$$\nu_Q = \frac{1}{2}\left[\frac{e^2qQ}{h}\right]\left(1 + \frac{\eta^2}{3}\right)^{1/2} \tag{14}$$

That single-resonance frequency is insufficient to determine these two parameters separately. To overcome the problem, Ramachandran and Oldfield (1984) applied a

small static Zeeman field, $H_0$, parallel to $H_1$, the radiofrequency field, as a perturbation to remove the degeneracy of the quadrupolar levels. The resulting splitting creates singularities, first described by Morino and Toyama (1961), at

$$\nu_Q \pm (1 + \eta)\nu_0 \tag{15}$$

and

$$\nu_Q \pm (1 - \eta)\nu_0 \tag{16}$$

where $\nu_0$ is the Larmor frequency.

Note that applying the static field $H_0$ perpendicular to $H_1$ does not change the position of the singularities but affects the intensity distribution. By measuring the splitting, the asymmetry parameter can be determined.

Ramachandran and Oldfield's two-dimensional version of this experiment was implemented using both one pulse and spin echo (see Fig. 6, panels A and B). The field is turned on during the preparation period, during which the spins evolve under both the Zeeman and quadrupole interactions. The field is turned off during acquisition. The 2D spectrum is obtained by incrementing $t_1$ in regular intervals followed by 2D Fourier transform. The projection

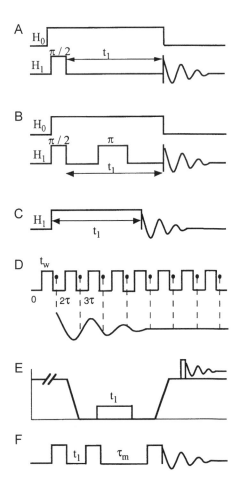

**Figure 6.** 2D NQR pulse sequences: (**A**) Zeeman-perturbed NQR one-pulse; (**B**) Zeeman-perturbed NQR spin echo; (**C**) zero-field nutation; (**D**) RF pulse train method; (**E**) level-crossing double resonance NQR nutation; (**F**) 2D exchange NQR.

on the $\omega_2$ axis shows the zero-field NQR spectrum. The Zeeman NQR powder pattern is observed by projection along the $\omega_1$ axis, which allows the determination of the asymmetry parameter $\eta$. The limitation of the method is that it requires a rapidly switchable homogeneous Zeeman field.

**Zero-Field Nutation Nuclear Resonance Spectroscopy.** Another method to determine the asymmetry parameter has been described (Harbison et al., 1989; Harbison and Slokenbergs, 1990) in which no Zeeman field is applied. The absence of a static field makes the frequency spectrum orientation independent. To determine $\eta$, it is necessary to obtain an orientation-dependent spectrum, but to do so, it is unnecessary to introduce an extra perturbation such as the Zeeman field. The sample radio frequency coil itself introduces an external preferential axis and thus an orientation dependence.

During an RF pulse in a zero-field NQR experiment, a 3/2 spin undergoes nutation about the unique axis of the EFG (Bloom et al., 1955). The strength of the effective $H_1$ field depends on the relative orientation of the coil and quadrupolar axes and goes to zero when the two axes are parallel. The nutation frequency is given by

$$\omega_N = (\sqrt{3}\omega_R \sin\theta)/2 \qquad (17)$$

where $\omega_R = \gamma H_1$, $\theta$ is the angle between the coil axis and the unique axis of the electric field gradient tensor, and $\gamma$ is the gyromagnetic ratio of the nucleus.

The voltage induced in the coil by the precessing magnetization after the pulse is proportional to $\sin\theta$. Figure 6, panel C, shows the 2D nutation NQR pulse sequence. For a single crystal, the NQR free precession signal is

$$F(t_1, t_2, \theta) \propto \sin\theta \sin(\sqrt{3}\omega_R t_1 \sin\theta/2)\sin(\omega_Q t_2) \qquad (18)$$

where $\omega_Q$ is the quadrupolar frequency, $t_1$ is the time the RF pulse is applied and $t_2$ is the acquisition time.

For an isotropic powder, the nutation spectrum is obtained by powder integration over $\theta$, followed by complex Fourier transformation in the second dimension and an imaginary Fourier transform in the first

$$G(\omega_1, \omega_2) \propto \int_0^\pi \int_{-\infty}^\infty \int_{-\infty}^\infty \sin^2\theta \, \sin(\sqrt{3}\omega_R t_1 \sin\theta/2)$$
$$\times \sin(\omega_Q t_2)e^{i\omega_1 t_1}e^{i\omega_2 t_2}\,dt_2\,dt_1\,d\theta \qquad (19)$$

The angular factor $\sin\theta$ must be replaced by a factor $R(\theta, \phi)$ where $\theta$ and $\phi$ are the polar angles relating the coil axis and the quadrupolar tensor (Pratt et al., 1975). For an axially asymmetric tensor

$$R(\theta, \phi) = \sqrt{4\eta^2\cos^2\theta + [9 + \eta^2 + 6\eta\cos(2\phi)]\sin^2\theta} \qquad (20)$$

The 2D frequency-domain spectrum becomes

$$G(\omega_1, \omega_2) \propto \int_0^\pi \int_{-\infty}^\infty \int_{-\infty}^\infty \int_{-\infty}^\infty \sin\theta R(\theta, \phi)\sin(\omega_R t_1 R(\theta, \phi)/2\sqrt{3}\rho)$$
$$\times \sin(\omega_Q t_2)e^{i\omega_1 t_1}e^{i\omega_2 t_2}dt_2\,dt_1\,d\theta\,d\phi \qquad (21)$$

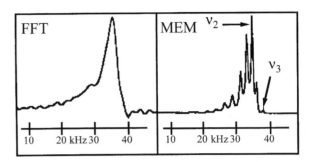

**Figure 7.** Fourier transform of the on-resonance nutation spectrum and corresponding MEM spectrum (the slices through the 2D spectrum parallel to the $F_1$ axis at $F_2 = \nu_{NQR}$) obtained at 77 K for $\nu_{NQR} = 36.772$ in $C_3N_3Cl_3$. The linear prediction filter used to obtain the MEM spectrum from $N = 162$ time-domain data points was $m = 0.95$ N (from Mackowiak and Katowski, 1996).

The on-resonance nutation spectrum in the $\omega_1$ dimension (see Fig. 7) shows three singularities, $\nu_1$, $\nu_2$, and $\nu_3$, whose frequencies are given by

$$2\pi\nu_1 = \frac{\eta\omega_R}{\sqrt{3}\sqrt{1 + \frac{1}{3}\eta^2}}, \quad 2\pi\nu_2 = \frac{(3 - \eta)\omega_R}{2\sqrt{3}\sqrt{1 + \frac{1}{3}\eta^2}},$$

$$\text{and} \quad 2\pi\nu_3 = \frac{(3 + \eta)\omega_R}{2\sqrt{3}\sqrt{1 + \frac{1}{3}\eta^2}} \qquad (22)$$

The asymmetry parameter can be determined from the position of $\nu_2$ and $\nu_3$ alone (Harbison et al., 1989)

$$\eta = \frac{3(\nu_3 - \nu_2)}{\nu_3 + \nu_2} \qquad (23)$$

Off-resonance effects can be calculated using a similar procedure (Mackowiak and Katowski, 1996). The nutation frequency is given by

$$\nu_N^{off} = \xi/\pi = \sqrt{(\nu_N^{on})^2 + (\Delta\nu)^2} \qquad (24)$$

where $\nu_N^{on}$ is the on-resonance nutation frequency and $\Delta\nu$ is the resonance offset.

The asymmetry parameter is determined from the relation

$$\eta = \frac{3\left[\sqrt{(\nu_3^{off})^2 - (\Delta\nu)^2} - \sqrt{(\nu_2^{off})^2 - (\Delta\nu)^2}\right]}{\sqrt{(\nu_3^{off})^2 - (\Delta\nu)^2} + \sqrt{(\nu_2^{off})^2 - (\Delta\nu)^2}} \qquad (25)$$

Another way to do an off-resonance nutation experiment is to start acquisition at some constant delay, $t_o$, after the RF pulse procedure (Sinyavski, 1991; Mackowiak and Katowski, 1996). The nutation spectrum consists of three lines. The line at $\Delta\nu$ is independent of the EFG parameters and the asymmetry parameter can be calculated from any of the two other lines using

$$\eta = \frac{3\left[\sqrt{(\nu_3^{off})^2 \pm 2(\Delta\nu)^2} - \sqrt{(\nu_2^{off})^2 \pm 2(\Delta\nu)^2}\right]}{\sqrt{(\nu_3^{off})^2 \pm 2(\Delta\nu)^2} + \sqrt{(\nu_2^{off})^2 \pm 2(\Delta\nu)^2}} \quad (26)$$

The spectral resolution for an off-resonance experiment decreases with the frequency offset up to ~80% for a 100 kHz offset.

For experimental nutation spectra, the data are occasionally truncated and noisy. Therefore the position of the singularities is often poorly resolved, which makes an accurate determination of $\eta$ difficult. As an alternative to the 2D Fourier transform, the maximum entropy method (MEM) can be used for processing the time-domain data (Sinjavski, 1991; Mackowiak and Katowski, 1993, 1996). The basis of the maximum entropy method is to maximize the entropy of the spectrum while maintaining correlation between the inverse Fourier transform of the spectrum and the free induction decay (FID). The process works by successive iterations in calculating the $\chi^2$ correlation between the inverse Fourier transform of the trial spectrum and the FID. The spectrum is modified so its entropy increases without drastically increasing $\chi^2$, until convergence is reached. One of the inherent advantages to this method is that the MEM spectrum is virtually noiseless and free of artifacts. Mackowiak and Katowski used the Burg algorithm to determine the lineshape that maximizes the entropy for the nutation experiment (Stephenson, 1988; Mackowiak and Katowski, 1993, 1996). In this unit we do not intend to discuss the mathematics involved in MEM, but we can point readers to a review by D.S. Stephenson about linear prediction and maximum entropy methods used in NMR spectroscopy (Stephenson, 1988; also see references therein). MEM has been very successfully applied in high-field NMR for both 1D and 2D data reconstruction.

Figure 7 shows on-resonance simulated and experimental 2D Fourier transform (FT) and MEM spectra. The noise and resolution improvements for the MEM method are clearly visible. The singularities $\nu_2$ and $\nu_3$ can be easily measured from the MEM spectrum. The asymmetry parameter is determined as previously using Equation 23.

The quadrupolar frequency is very temperature sensitive. Rapid recycling can cause significant sample heating and introduce unwanted frequency shifts during the nutation experiment. Thus, recycle delays much longer than necessary for spin relaxation are used (Harbison et al., 1989; Harbison and Slokenbergs, 1990; Mackowiak and Katowski, 1996), together with active gas cooling. Changes in the RF excitation pulse length between slices can cause undesirable effects. Also, for weak NQR lines, acquiring the 2D NQR nutation experiment can take a long time. To reduce the experiment time, Sinyavski et al. (1996) reported a sequence of identical short RF pulses separated by time intervals $\tau$.

The NQR signal induced in the RF coil just after turnoff of the $n$th pulse is

$$G(nt_w) = \langle I_x\rangle_n \sin\theta\cos\phi + \langle I_y\rangle_n \sin\theta\sin\phi + \langle I_z\rangle_n \cos\theta$$

$$(27)$$

where $\langle I_{x,y,z}\rangle_n$ are the expectation values of the magnetization along the coordinate axis and are calculated using a wave function approach (Pratt et al., 1975; Sinyavski et al., 1996).

In the general case, the expression in Equation 27 is a sum of signals with various phases. For the case when

$$-\omega_0\tau + \Delta\omega t_w = 2\pi k \quad\text{with}\quad k = 0, \pm 1, \pm 2\dots \quad (28)$$

that is, when the phase accumulated due to resonance offset is an integer multiple of $2\pi$, the free induction decay obtained is identical to that found in the two-dimensional nutation experiment. Here, $t_w$ is the time delay between pulses in the nutation pulse train.

In addition, if the NQR signal is measured at some constant delay, $t_d$, after each pulse in the sequence, then the signal after synchronous detection is

$$G(nt_w) = \alpha R^2(\theta,\phi)$$
$$\times \left[\begin{array}{l}\sin(2n\xi t_w)\cos(\Delta\omega t_d + \Psi) \\ +\Delta\omega[1 - \cos(2n\xi t_w)]\sin(\Delta\omega t_d + \Psi)/2\xi\end{array}\right]\Big/2\xi \quad (29)$$

$R(\theta,\phi)$ is defined in Equation 20, $\Delta\omega$ is the frequency offset, $\Psi$ is a constant phase shift of the reference signal for the synchronous detector, and

$$2\xi = \sqrt{(\Delta\omega)^2 + 4m^2} \quad\text{with}\quad m = (\gamma H_1/4)R(\theta,\phi)/\sqrt{3 + \eta^2}$$

$$(30)$$

Equation 29 describes the "discrete interferogram" of a single nutation line at the frequency $2\xi$ (Sinyavski et al., 1996). The pulse sequence for the train-pulse method is shown in Figure 6, panel D.

For the on-resonance condition ($\Delta\omega = 0$), $t_w$ and $\tau$ can easily be chosen so that condition in Equation 28 is satisfied. For $\Delta\omega \neq 0$, Equation 28 can also be satisfied. If the sum of the intervals $t_w + \tau$ in the pulse train is an integer multiple of the data sampling period, the nutation interferogram $D(i)$ can be reconstructed from the raw data by the simple algorithm

$$D(i) = D[c + (i - 1)(p \backslash n)] \quad (31)$$

where $i = 1, 2, \dots, n, n$ is the number of pulses, $p$ the number of data points, the backslash denotes an integer division (e.g., $a \backslash b$ is the largest integer less than $a/b$), and $c$ is a constant that can be determined by trial (Sinyavski et al., 1996).

If the NQR line is broad, dead-time problems occur and a spin echo signal can be used. The equivalent of the expression in Equation 29 for the echo signal is

$$G(nt_w) = \alpha R^2(\theta,\phi)(1 - [\Delta\omega/2\xi]^2)[1 - \cos(2\xi t'_w)]\times$$
$$[-\sin(2n\xi t_w)\cos\Psi + \Delta\omega[1 - \cos(2n\xi t_w)]\sin\Psi/(4\xi)]/4\xi \quad (32)$$

where $t'_w$ is the delay between nutation pulses.

The maximum time-saving factor can be roughly estimated as equal to the number of pulses contained in a pulse train (Sinyavski et al., 1996).

**Level-Crossing Double Resonance NQR Nutation Spectroscopy.** When the NQR signal of the 3/2 spin nuclei is weak (low natural abundance) or if the quadrupolar frequency is too low to be observable, a double-resonance technique known as level-crossing double resonance NQR nutation (Blinc and Selinger, 1992) can be used to retrieve the nutation spectrum. In this experiment, the quadrupolar nuclei are observed via their effect on signals from the other nuclei (typically protons). Dipolar coupling must be present between the two nuclei and spin-lattice relaxation times have to be relatively long. Figure 6, panel E, shows the field cycling and pulse sequence used for the experiment. The sample is first polarized in a strong magnetic field, $H_0$. The system is then adiabatically brought to zero magnetic field (either by decreasing $H_0$ or by moving the sample out of the magnet). At a specific field strength, level crossing occurs between the Zeeman levels of the 1/2 spin (protons) and the Zeeman-perturbed quadrupolar levels (Blinc et al., 1972; Edmonds and Speight, 1972; Blinc and Selinger, 1992). The level crossing polarizes the quadrupolar nuclei (decreasing the population difference $N_{\pm 1/2} - N_{\pm 3/2}$) and decreases the magnetization of the protons. An RF pulse of length $t_1$ is applied at zero field with frequency of $\Omega = \omega_Q + \delta$, where $\omega_Q$ is the pure quadrupolar frequency. The proton system is remagnetized and the proton FID acquired after a 90° pulse. The 2D spectrum is acquired by varying $t_1$ and $\Omega$. The proton signal $S_H(t_1)$ is proportional to the population difference (Blinc and Selinger, 1992).

$$S_H(t_1) \propto (N_{\pm 1/2} - N_{\pm 3/2}) = A \cos[\omega(\theta, \varphi, \delta)t_1] \qquad (33)$$

where $A$ is a constant and $\omega(\theta, \varphi, \delta) = \gamma_Q H_1 R(\theta, \varphi)/(2\sqrt{3} + \eta)$ is the nutation frequency of the quadrupolar nuclei. For a powder, Equation 33 is integrated over the solid angle $\Omega$

$$S_H(t_1) = \frac{S_H(0)}{4\pi} \int \cos[\omega(\theta, \varphi, \delta)t_1]d\Omega \qquad (34)$$

The Fourier transform of $S_H(t_1)$ gives the nutation spectrum with singularities given in Equation 22. The $\omega_2$ dimension shows the zero-field NQR spectra. This technique has been successfully applied to determine $e^2qQ/h$ and $\eta$ for NQR frequencies as low as 600 kHz (Blinc and Selinger, 1992) for *tris*-sarcosine calcium chloride.

**2D Exchange NQR Spectroscopy.** 2D exchange NMR spectroscopy was first suggested by Jeener et al. (1979) The technique has been widely used in high field. Rommel et al. (1992a) reported the first NQR application of the 2D exchange experiment.

The 2D exchange experiment (Rommel et al., 1992a; Nickel and Kimmich, 1995) consists of three identical RF pulses separated by interval $t_1$ and $\tau_m$ (Fig. 6, panel F). The first pulse produces spin coherence evolving during $t_1$; the second pulse partially transfers the magnetization components orthogonal to the rotating frame RF field into components along the local direction of the EFG. The mixing time, $\tau_m$, is set to be much longer than $t_1$, so most of the exchange occurs during this interval. The

last pulse generates the FID. The NQR signal oscillates along the axis of the RF coil. By incrementing $t_1$, and after double Fourier transform, the 2D spectrum $S(\nu_1, \nu_2, \tau_m)$ shows diagonal and cross-peaks, the latter corresponding to nuclei undergoing exchange.

The theoretical treatment is based on the fictitious spin 1/2 formalism (Abragam, 1961; Goldman, 1990; Nickel and Kimmich, 1995). The matrix treatment of the high field 2D exchange NMR (Ernst et al., 1987) has to be modified for the specifics of the NQR experiment. During the exchange process, in contrast to its high-field equivalent (where only the resonance frequency changes), the direction of quantization (given by the direction of the EFG) can also change. This reduces the cross-peak intensities by projection losses from the initial to the final direction of quantization of the exchange process (Nickel and Kimmich, 1995).

It is important for the 2D exchange that all resonances participating in the exchange be excited simultaneously. Phase-alternating pulses producing tunable sidebands can overcome this problem (Rommel et al., 1992a). If the exchanging resonances fall outside the feasable single resonance bandwidth, a double-resonance configuration could be used instead.

### Spatially Resolved NQR and NQR Imaging

A logical step following 2D NQR experiments is to perform spatially resolved NQR and NQR imaging. This would be of particular value for the characterization of solid materials. Imaging is a very recent technique in NQR spectroscopy that only a handful of research groups are developing. The main restriction to performing NQR imaging is that even if the quadrupole resonances are narrow, it is practically impossible to shift the NQR frequency, since it only depends on the interaction between the quadrupole moment and the electric field gradient at the nuclei.

**Magnetic Field Gradient Method.** The first NQR experiment that allowed retrieval of spatial information was performed recently by Matsui et al. (1990). If the NQR resonance is relatively sharp, the half-width of a Zeeman-perturbed spectrum is proportional to the strength of the small static Zeeman field applied (Das and Hahn, 1958). Thus, the height of the pattern at the resonance frequency is reduced proportionally to the Zeeman field strength (Matsui et al., 1990). Figure 3 shows the changes in the spectrum lineshape of potassium chlorate due to different Zeeman field strengths. The sample coil is placed between two Helmholtz coils oriented perpendicular to the terrestrial magnetic field. Changing the current applied through the coil changes the field strength. Note that because the Zeeman field is parallel to the terrestrial field, a shimming effect appears for a current of 0.2 A, reducing the half-width by approximately a factor of two (unpublished results). This effect was not observed in the work of Matsui et al. (1990).

If a magnetic field gradient is applied, the reduction effect can be used for imaging since it depends on spatial location (Matsui et al., 1990). Given $N$ discrete quadrupolar spin densities $\rho(X_n)$, the observed signal for the one-dimensional experiment will be the sum of all the powder patterns affected by the field gradient. The spectral height of the spectrum at the resonance frequency $\omega_0$ becomes

$$H(X_{n_0}) = \sum_{n=1}^{N} W(X_n - X_{n_0})\rho(X_n) \qquad (35)$$

The function $W(X_n - X_{n_0})$ represents the spectral height reduction at $X_n$. The reduction function can be measured by applying uniform Zeeman fields of known strength. The $N$ discrete spin densities can be determined by solving the system of $N$ linear equations given by Equation 21 at the resonance frequency (Matsui et al., 1990).

As the field gradient increases, the plot becomes similar to the projection of the sample. This is because the signal contribution of the neighboring locations decreases with increasing gradients. Determining the spin densities is a deconvolution, and small experimental errors (e.g., amplitude and location) are enhanced by the conversion.

**Rotating Frame NQR Imaging.** Rotating frame NQR imaging ($\rho$ NQRI) is similar to the rotating frame zeugmatography proposed by Hoult (1979). This technique has the advantage of using pure NQR without any magnetic field or magnetic-field gradient, in contrast to the previous method. The rotating frame zeugmatography is a flip-angle encoding technique used in NMR imaging. Nonuniform radiofrequency fields are applied so that the flip angle of an RF pulse depends on the position with respect to the RF field gradient. The RF coils are designed to produce constant field gradients. For NQR, only the amplitude-encoding form of the method is applicable. In NQR, the transverse magnetization oscillates rather than precesses, making the phase encoding (Hoult, 1979) variant of the method inapplicable.

The first application of $\rho$ NQRI was developed by Rommel et al. (1991b) in order to determine the one-dimensional profile of solid samples. In this experiment, an anti-Helmholtz coil (transmitter coil) produces the RF whose distribution is given by

$$B_1(z) = \frac{\mu_0 I}{2}\left[\frac{R^2}{\left(R^2 + (z+z_0)^2\right)^{3/2}} - \frac{R^2}{\left(R^2 + (z-z_0)^2\right)^{3/2}}\right] \qquad (36)$$

where $R$ is the coil radius, $z_0 = R\sqrt{3}/2$ is half the distance between the coils and $2I$ is the current amplitude through the coils. The receiver coil is placed coaxial to the transmitter coil (Rommel et al., 1991b). With this arrangement, transmitter and receiver coils are electrically decoupled. Later coil designs (Nickel et al., 1991) prefer the use of a surface coil acting as both a transmitter and a receiver coil. The sample is placed in only one-half of the transmitter coil.

The pulse sequence for the experiment is very similar to the 2D-nutation experiment (Harbison et al., 1989) described earlier in this unit. Free induction decay signals are excited with increasing RF pulse length.

Since $\rho$ NQRI is restricted to solid samples, a simple way to access the second dimension is to rotate the sample by small angle increments, the rotation axis being perpendicular to the RF field gradient (Nickel et al., 1991; Kimmich et al., 1992; Rommel et al., 1992b). Using a surface coil, where the sample is placed outside the coil, creates the RF field gradient and allows two-dimensional spatial encoding. The 2D image is reconstructed using the projection/reconstruction procedure proposed by Lauterbur (1973). A stepping motor allows the sample to rotate, producing the 2D image. The surface coil is placed perpendicular to the rotation axis. The spatial information is amplitude encoded in the FID signals by the gradient of the RF pulse $B_1$. The RF gradient ($G_1$) is aligned along the coil $z$ axis and considered as constant (Nickel et al., 1991).

$$G_1(z) = \frac{\partial B_1(z)}{\partial z} \qquad (37)$$

The excitation pulse is characterized by the effective pulse length: $t_p = t_w \alpha$, where $\alpha$ is the transmitter attenuation factor and $t_w$ is the proper pulse length. $t_p$ can be varied by varying either $t_w$ or $\alpha$.

For 3/2 spins, the "pseudo" FID given in Equation 21 for the nutation experiment (Harbison et al., 1989) can be rewritten as

$$S(t_p) = \int_0^\infty \frac{1}{\pi}\tilde{\rho}(z)dz \int_0^\pi d\theta \int_0^{2\pi} \frac{\sqrt{3}}{2}R(\theta,\phi)\sin\theta$$
$$\times \sin(\sqrt{3}t_p\omega_1(z)R(\theta,\phi)/2\sqrt{1+\eta^2/3}) \qquad (38)$$

or

$$S(t_p) = c\int_0^\infty \frac{1}{\pi}\tilde{\rho}(z)f(z,t_p)dz \qquad (39)$$

The proper distribution is $\tilde{\rho}(z) = g(\omega,z)\rho(z)$, where $g(\omega,z)$ is the local line shape acting as a weighting factor. $\omega_1(z) = \gamma B_1(z)$ and $R(\theta,\phi)$ are defined by Equation 6. For constant gradient, $\omega_1(z). = G_1$.

Introducing $t_p = t_d e^u$, $z = z_0 e^{-v}$ and $\xi = u - v$ leads to

$$\tilde{S}(u) \equiv S(t_d e^u) = c\int_0^\infty f_a(v)f_b(u-v)dv \qquad (40)$$

The deconvolution can be performed using the following expression

$$f_a(v) = F_u^{-1}\left\{\frac{F_u\{S(t_d e^u)\}}{F_\xi\{f_b(\xi)\}}\right\} \qquad (41)$$

$f_a(v)$ can be derived numerically, since both $f_b(\xi)$ and $S(t_d e^u)$ are known. The profile is given by $\tilde{\rho}(z) = f_a(v)/z$ and $v = \ln(z_0/z)$.

The 2D image is reconstructed from the profiles of all the orientations using the following steps (Nickel et al., 1991; Kimmich et al., 1992; Rommel et al., 1992b). The pseudo FIDs are baseline- and phase-corrected so that the imaginary part can be zeroed with no loss of information, and the profiles are determined. The profiles are centered with respect to the rotation axis. If $G_1(z)$ is not constant, the profiles must be stretched or compressed by a coordinate transformation so that the abscissae are linearly related to the space coordinate. The profile ordinates must be corrected by a factor $G_1(z)/B_1(z)$ because the RF pickup sensitivity of the coil depends on $z$ in the same way as $B_1$ and because the signal intensity is weighted by a factor $1/G_1(z)$. Finally, the image is reconstructed by the back-projection method (Lauterbur, 1973).

Robert et al. (1994) used the maximum entropy method (MEM) as an alternative to the Fourier transform to process the pseudo FIDs. The MEM procedure shows better resolution and avoids the noise and artifacts of the Fourier method. The drawback is the much longer time required for the MEM method to process the data ($\sim$30 s per FID compare to the millisecond scale of the fast Fourier transform or FFT). A Hankel transform can also be applied instead of FFT or MEM (Robert et al., 1994).

To reduce the acquisition time, a procedure similar to the 2D nutation NQR train pulse method (Sinyavski et al., 1996), discussed previously, can be used (Robert et al., 1996). Theoretical calculations show that the pseudo-FIDs recorded using this single experiment (SEXI) method (Robert et al., 1996) are identical to those obtained using the multiple-experiment procedure. Data processing and image reconstruction are performed as described previously using MEM processing and the back-projection algorithm. The SEXI method reduces acquisition time by a factor of $\sim$50. So far, theoretical treatments and experiments using the SEXI method are only applied for 3/2 spins, while the multiple-experiment method also can be applied for 5/2 spins (Robert et al., 1996).

The combination of $\rho$ NQRI and Matsui's Zeeman perturbed NQR technique, described earlier, allows adding the third dimension to the imaging process by adding slice selection (Robert and Pusiol, 1996a,b). The width of the Zeeman-perturbed NQR spectrum is proportional to the local Zeeman field (Matsui et al., 1990). In a zero-crossing magnetic field gradient, the nuclei in the zero-field region show sharp resonance while nuclei experiencing a field show weak and broad resonances. Therefore, the sample slice localized within the $B = 0$ plane will show an unchanged spectrum, while the other slices away from the zero-field plane show broad or even vanishing spectra. If the field gradient is large enough, signals outside the zero-field slice virtually disappear. Two Helmholtz coils of similar diameter, separated by a distance $L$ and carrying current in opposite directions, produce the magnetic field gradient. Obviously, other coil designs can be used to produce the magnetic field gradient. Varying the electric current ratio between the Helmholtz coils shifts the zero-field plane. Each slice can be obtained as described previously using the SEXI method.

Robert and Pusiol (1996a,b) applied the $\rho$ NQRI with slice-selection method on a test object filled with $p$-dichlorobenzene. Results are shown in Figure 8 for two different orientations of the test object. In both cases, the geometry of the test objects is well resolved and slice-selection appears to be a nice improvement of the $\rho$ NQRI method.

Recently, a new imaging technique developed by Robert and Pusiol (1997) was reported. This method, instead of

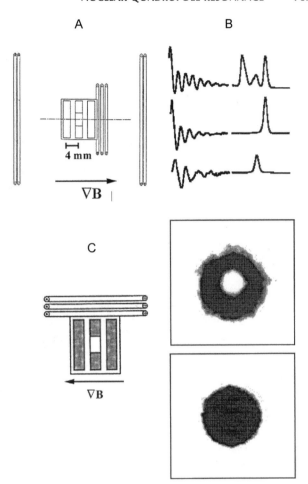

**Figure 8.** (**A**) Schematic arrangement of the surface RF and the selective magnetic gradient coils together with the cross-section of the object used for the test experiment. The selective static field gradient ($\nabla B$) is applied in a direction normal to the desired image plane. (**B**) Top: Pseudo-FID and profile along the cylindrical symmetry axis of the object imaged without external magnetic field. Middle: Magnetic gradient selecting the right hand cylinder. Bottom: After shifting the $B = 0$ plane towards the central cylinder. (**C**) Coil and test object arrangement for the 2D imaging experiment and cross image for the selection of the central cylinder (top) and of the external paradichlorobenzene disks (bottom) From Robert and Pusiol (1996a).

retrieving the second spatial dimension by rotating the sample, used a second surface RF coil perpendicular to the first one. In the "bidimensional rotating frame imaging technique 2D $\rho$ NQRI," orthogonal RF gradients along the $x$ and $y$ axis are applied. A constant time interval $T$ is introduced between the two pulses $t_x$ and $t_y$. This is necessary to remove the transverse coherence created by the first pulse (Robert and Pusiol, 1997). The 2D experiment is carried out by incrementing $t_x$ and $t_y$. Alternatively, the second pulse can be replaced by the SEXI train pulse described previously (Robert et al., 1996), and the complete 2D signal is obtained by incrementing only $t_x$. The resulting FIDs depend on both the $x$ and $y$ coordinates

$$F(t_x, t_y) \approx \int_0^\infty dx \int_0^\infty H_d\rho(x,y)S_1(t_x)S_2(t_y)\,dy \qquad (42)$$

where

$$S_1(t_x) = \int_0^{2\pi} d\phi_1 \int_0^\pi \sin\theta_1 \cos[\omega_1\lambda(\theta_1,\phi_1)t_x]\, d\theta_1 \qquad (43)$$

and

$$S_2(t_y) = \int_0^{2\pi} d\phi_2 \int_0^\pi \sin\theta_2\lambda(\theta_2,\phi_2)\sin[\omega_2\lambda(\theta_2,\phi_2)t_y]\, d\theta_2$$
$$(44)$$

$\lambda(\theta,\phi)$ is proportional to $R(\theta,\phi)$ from Equation 20.

The corresponding 2D image is produced by determining the two-dimensional spin density function $\rho(t_x,t_y)$ The 2D image reconstruction is performed by a "true" 2D version of maximum entropy method and after RF field correction (Robert and Pusiol, 1997).

**Temperature, Stress, and Pressure Imaging.** As mentioned earlier, temperature strongly affects the quadrupolar frequency. Although this is an inconvenience for the spatially resolved imaging techniques, where the temperature must remain constant during acquisition to avoid undesired artifacts or loss in resolution, the temperature shift, as well as the effect of applying pressure or stress to the sample, can in fact be useful to detect temperature and pressure gradients, giving spectroscopic resolution in the second dimension. The spatial resolution in the first dimension can be performed using the initial $\rho$ NQRI technique (Rommel et al., 1991a; Nickel et al., 1994) with no rotation of the sample and with a surface coil to produce the RF field gradient.

To study the effect of the temperature (Rommel et al., 1991b), a temperature gradient can be applied by two water baths at different temperature on each side of the sample. Data are collected in the usual way and are processed by 2D fast Fourier transform (FFT; Brigham, 1974). Figure 9, panel A, represents the $\rho$ NQRI image of the test object and clearly shows the temperature gradients on the three sample layers through their corresponding frequency shifts. Using this technique, temperature gradients can be resolved to ~1°C/mm. Using MEM and deconvolution methods can highly improve the resolution, and experimental times can be greatly reduced by using the pulse-train method (Robert et al., 1996) instead of the multiple-experiment method.

To study pressure and stress, a similar procedure can be applied (Nickel et al., 1994). The probe contains two sample compartments, the first one serving as a reference while pressure is exerted on the second. Images are recorded for different applied pressures. If the sample is "partially embedded" in soft rubber or other soft matrices, broadening due to localized distribution of stress does not appear and only frequency shifts are observed in the image. If the pressure coefficient (in kHz/MPa) is known, pressures can be calculated from the corresponding frequency shifts. Figure 9, panel B, shows the shifting effect of applied pressure on the test sample. In the case of harder matrices or pure crystals, the frequency shift is accompanied by a line broadening caused by the local

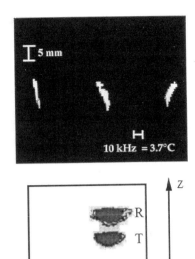

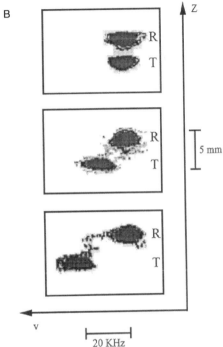

**Figure 9.** (**A**) Two-dimensional representation of temperature imaging experiments. The vertical axis represents the spacial distribution of the (cylindrical) sample, whereas the horizontal axis is the line shift corresponding to the local temperature (from Rommel et al., 1991a). (**B**) Spatial distribution of the $^{127}$I NQR spectra of lithium iodate embedded in rubber in a two-compartment sample arrangement. Top: no applied pressure. Middle and bottom: applied pressure of 7 MPa and 14 MPa, respectively (from Nickel et al.,1994).

stress distribution within the sample. This broadening can even hide the shifting effect of the stress or pressure applied and the image only shows a decrease in intensity (Nickel et al., 1994).

**Field Cycling Methods.** Field cycling techniques are of potential interest for the imaging of materials containing quadrupolar nuclei dipolar-coupled to an abundant 1/2 nucleus (i.e., protons), and for detecting signals in the case where the quadrupolar frequency is too low or the signal too weak to be directly observed by pure NQR spectroscopy. Recently, Lee et al. (1993) applied field cycling to the one-dimensional $^{11}$B imaging of a phantom of boric acid. The experiment consisted of a period at high field, during which the proton magnetization is built up, followed by a rapid passage to zero field, during which the magnetization is transferred to the quadrupolar nuclei. RF irradiation at the quadrupolar frequency is applied and the remaining magnetization is transferred back to the proton by rapidly bringing the sample back to high field. Finally, the proton FID is obtained after solid echo. Spatial resolution is obtained by translating along the sample the small irradiation RF coil that produced the field gradient (Lee et al., 1993). The percentage of "recovered magnetization" is obtained from two reference experiments—maximum and minimum values of the "recovered magnetization" corresponding to "normal" and long residence times at zero field with no irradiation. Despite the low resolution, the image accurately represents the object and the experiment demonstrated the applicability of the technique. Using more elaborate coil designs, the same research group (Lee and Butler, 1995) produced $^{14}$N 2D images spatially resolved in one dimension and frequency-resolved in the other. The image was produced by swapping both the RF coil position and the irradiation frequency at zero field.

It may be worth noting that field cycling imaging can possibly be achieved by performing the "level crossing double resonance NQR nutation spectroscopy" experiment (Blinc and Selinger, 1992) as described earlier, with a surface RF coil arrangement to access the spatial information as in the ρ NQRI techniques. If applicable, this would lead to improved image resolution.

So far, in contrast to its high field equivalent, and especially compared to medical imaging, the spatial resolution of the NQR imaging techniques is still fairly poor (∼1 mm). Nevertheless, the recent advances are very convincing for the future importance of NQR imaging in materials science. Resolution improvements can be achieved by new coil arrangement designs and deconvolution algorithms. The limiting factor in resolution is ultimately the intrinsic linewidth of the signal. The future of NQR imaging may reside in the development of line-narrowing techniques.

## DATA ANALYSIS AND INITIAL INTERPRETATION

Most modern NMR spectrometers contain a full package of one and two-dimensional data processing routines, and the interested reader is referred to the spectrometer documentation. Very briefly, processing time domain NMR data (free induction decays or FIDs) is done in five steps.

1. Baseline correction is applied to remove the DC offset in the FID.

2. Usually, the first few points of the free-induction decays are contaminated by pulse ringdown and give rise to artifacts. They must be removed by shifting the all data sets by several points.

3. Exponential multiplication is usually performed to increase the signal-to-noise ratio. This is performed by multiplying the real and imaginary part by the function $(-Ct)$, where $t$ is the time in seconds and $C$ is a constant in Hz. The larger the $C$, the faster the decay and the larger will be the broadening in the spectrum after the Fourier transform. This is

equivalent to convolving the frequency domain spectrum with a Lorentzian function. Other apodization functions, such as sinebell and Gaussian multiplication, can also be used.

4. Spectral resolution can also be improved by doubling the size of the data set. This is done by filling the second half of the time-domain data with zeros.

5. Fourier transform allows transformation from the time to the frequency domain where the spectrum is analyzed. The theory of the fast Fourier transform (FFT) has been described in detail elsewhere and will not be discussed here. We refer the reader instead to Brigham (1974).

Two-dimensional data processing in general repeats this procedure in a second dimension, although there are some mathematical niceties involved. Any modern handbook of NMR spectroscopy, such as Ernst et al. (1987), deals with the details.

## PROBLEMS

### Sensitivity

As with most radiofrequency spectroscopy methods, the overriding problem is sensitivity. Boltzmann populations and magnetic moments of quadrupolar nuclei are small, and their direct detection either by nuclear induction or by SQUID methods often stretches spectrometers to the limit. Offsetting these sensitivity difficulties is the fact that longitudinal relaxation times are often short, and so very rapid signal averaging is often possible; in fact, typical data acquisition rates are limited not by relaxation but by thermal heating of the sample by the RF pulses.

Sensitivity depends on a host of factors—linewidth, relaxation times, the quadrupole constant and gyromagnetic ratio of the nucleus, the amount of sample, and, of course, the quality of the spectrometer. As an order-of-magnitude estimate, a "typical" NMR nucleus ($^{63}$Cu) with a quadrupole coupling constant in the range 20 to 25 MHz, can probably be detected in amounts of 1 to 10 μM.

### Spurious Signals

As in any form of spectroscopy, it is essential to distinguish between signals arising from the sample and artifactual resonances from the sample container, probe, or elsewhere in the experimental apparatus. Artifacts extraneous to the sample can, generally, easily be detected by adequate controls, and with experience can be recognized and discounted; our probes, for example, have an impressive-looking NQR signal at 32.0 MHz, apparently due to a ceramic in the capacitors! A harder problem to deal with is identifying and assigning multiple NQR resonances in the same material; in contrast to NMR, at zero field the identity of a resonance cannot simply be determined from its frequency. The particular nucleus giving rise to an NQR signal can be identified with certainty if there exists another isotope of that nucleus. For example, natural-abundance copper has two isotopes—$^{63}$Cu and $^{65}$Cu—with a well-known ratio of quadrupole moments. Similar pairs are $^{35}Cl$ and $^{37}Cl$, $^{79}Br$ and $^{81}Br$, $^{121}Sb$ and $^{123}Sb$, etc. Where a second isotope does not exist, or where additional confirmation is required, the gyromagnetic ratio of the nucleus can be measured with a nutation experiment (see Practical Aspects of the Method).

Two signals from the same nucleus and arising from different chemical species are the most difficult problem to deal with. Such signals generally cannot be assigned by independent means, but must be inferred from the composition and chemistry of the sample. This is an area where much more work needs to be done.

## Dynamics

The broadening and disappearance of NQR lines caused by dynamical processes on a time scale comparable to the NQR frequency is a useful means of quantifying dynamics, if such processes are recognized; however, they may also cause expected NQR signals to be very weak or absent. We suspect the tendency of NQR signals to be very idiosyncratic in intensity to be largely due to this effect. If NQR lines are missing, or unaccountably weak, and dynamics is suspected to be the culprit, cooling the sample will often alleviate the problem.

## Heterogeneous Broadening

Finally, another limitation on the detectability of NQR samples is heterogeneous broadening. In pure samples of perfect crystals, NQR lines can often be very narrow, but such systems are seldom of interest to the materials scientist; more often the material contains considerable physical or chemical heterogeneity. The spectrum of poly-4-chlorostyrene in Figure 5 is an example of chemically heterogeneous broadening; the coupling constants of the chlorine are distributed over ~1 MHz due to local variations in the chemical environment of the aromatic chlorines. Similar broadening is observed, for example, in the copper NQR signals of high $T_c$ superconductors. Physical heterogeneous broadening is induced by variations in temperature or strain over the sample. Simply acquiring NQR data too fast will generally induce inhomogenous RF heating of the sample, and since NQR resonances are exquisitely temperature dependent, the result is a highly broadened and shifted line. Strain broadening can be demonstrated simply by grinding a crystalline NQR sample in a mortar and pestle; the strain induced by grinding will often broaden the lines by tens or hundreds of kilohertz. Obviously, the former effect can be avoided by limiting the rate of data acquisition, and the latter by careful sample handling. When working with crystalline solids, we avoid even pressing the material into the sample tube, but pack it instead by agitation.

## ACKNOWLEDGMENTS

We thank Young-Sik Kye for permission to include unpublished results. The data in Figure 5 were collected by Solomon Arhunmwunde, who died under tragic circumstances in 1996. This project was funded by NSF under grant number MCB 9604521.

## LITERATURE CITED

Abragam, A. 1961. The Principles of Nuclear Magnetism. Oxford University Press, New York.

Bielecki, A., Zax, D., Zilm, K., and Pines, A. 1983. Zero field nuclear magnetic resonance. *Phys. Rev. Lett.* 50:1807–1810.

Blinc, R. and Selinger, J. 1992. 2D methods in NQR spectroscopy. *Z. Naturforsch.* 47a:333–341.

Blinc, R., Mali, M., Osredkar, R., Prelesnik, A., Selinger, J., Zupancic, I., and Ehrenberg, L. 1972. Nitrogen-14 NQR (nuclear quadrupole resonance) spectroscopy of some amino acid and nucleic bases via double resonance in the laboratory frame. *J. Chem. Phys.* 57:5087.

Bloom, M., Hahn, E. L., and Herzog, B. 1955. Free magnetic induction in nuclear quadrupole resonance. *Phys. Rev.* 97:1699–1709.

Brigham, E. O. 1974. The Fast Fourier Transform. Prentice-Hall, Englewood Cliffs, N.J.

Burns, G. and Wikner, E. G. 1961. Antishielding and contracted wave functions. *Phys. Rev.* 121:155–158.

Clarke, J. 1994. Low frequency nuclear quadrupole resonance with SQUID amplifiers. *Z. Naturforsch.* 49a:5–13.

Connor, C., Chang, J., and Pines, A. 1990. Magnetic resonance spectrometer with a D. C. SQUID detector. *Rev. Sci. Instrum.* 61:1059–1063.

Creel, R. B. 1983. Analytic solution of fourth degree secular equations: I=3/2 Zeeman-quadrupole interactions and I=7/2 pure quadrupole interaction. *J. Magn. Reson.* 52:515–517.

Creel, R. B., Brooker, H. R., and Barnes, R. G. 1980. Exact analytic expressions for NQR parameters in terms of the transition frequencies. *J. Magn. Reson.* 41:146–149.

Das, T. P. and Bersohn, R. 1956. Variational approach to the quadrupole polarizability of ions. *Phys. Rev.* 102:733–738.

Das, T. P. and Hahn, E. L. 1958. Nuclear Quadrupole Resonance Spectroscopy. Academic Press, New York.

Edmonds, D. T. and Speight, P. A. 1972. Nuclear quadrupole resonance of $^{14}N$ in pyrimidines purines and their nucleosides. *J. Magn. Reson.* 6:265–273.

Ernst, R. R., Bodenhausen, G., and Wokaun, A. 1987. Principles of Nuclear Magnetic Resonance in One and Two Dimensions. Clarendon, Oxford.

Feiock, F. D. and Johnson, W. R. 1969. Atomic susceptibilities and shielding factors. *Phys. Rev.* 187:39–50.

Goldman, M. 1990. Spin 1/2 description of spins 3/2. *Adv. Magn. Reson.* 14:59.

Harbison, G. S. and Slokenbergs, A. 1990. Two-dimensional nutation echo nuclear quadrupole resonance spectroscopy. *Z. Naturforsch.* 45a:575–580.

Harbison, G. S., Slokenbergs, A., and Barbara, T. S. 1989. Two-dimensional zero field nutation nuclear quadrupole resonance spectroscopy. *J. Chem. Phys.* 90:5292–5298.

Hoult, D. I. 1979. Rotating frame zeugmatography. *J. Magn. Reson.* 33:183–197.

Jeener, J., Meier, B. H., Bachmann, P., and Ernst, R. R. 1979. Investigation of exchange processes by two-dimensional NMR spectroscopy. *J. Chem. Phys.* 71:4546.

Kimmich, R. 1980. Field cycling in NMR relaxation spectroscopy: Applications in biological, chemical and polymer physics. *Bull. Magn. Res.* 1:195.

Kimmich, R., Rommel, E., Nickel, P., and Pusiol, D. 1992. NQR imaging. *Z. Naturforsch.* 47a:361–366.

Koening, S. H. and Schillinger, W. S. 1969. Nuclear magnetic relaxation dispersion in protein solutions. *J. Biol. Chem.* 244: 3283–3289.

Kye, Y.-S. 1998 The nuclear quadrupole coupling constant of the nitrate ion. Ph.D. Thesis. University of Nebraska at Lincoln.

Lahiri, J. and Mukherji, A. 1966. Self-consistent perturbation. II. Calculation of quadrupole polarizability and shielding factor. *Phys. Rev.* 141:428–430.

Lahiri, J. and Mukherji, A. 1967. Electrostatic polarizability and shielding factors for ions of argon configuration. *Phys. Rev.* 155:24–25.

Langhoff, P. W. and Hurst, R. P. 1965. Multipole polarizabilities and shielding factors from Hartree-Fock wave functions. *Phys. Rev.* 139:A1415–A1425.

Lauterbur, P. C. 1973. Image formation by induced local interactions: Examples employing nuclear magnetic resonance. *Nature* 242:190–191.

Lee, Y. and Butler, L. G. 1995. Field cycling $^{14}$N imaging with spatial and frequency resolution. *J. Magn. Reson.* A112:92–95.

Lee, Y., Michaels, D. C. and Butler, L. G. 1993. $^{11}$B imaging with field cycling NMR as a line narrowing technique. *Chem. Phys. Lett.* 206:464–466.

Liao, M. Y. and Harbison, G. S. 1994 The nuclear hexadecapole interaction of iodine-127 in cadmium iodide measured using zero-field two dimensional nuclear magnetic resonance. *J. Chem. Phys.* 100:1895–1901.

Mackowiak, M. and Katowski, P. 1993. Application of maximum entropy methods in NQR data processing. *Appl. Magn. Reson.* 5:433–443.

Mackowiak, M. and Katowski, P. 1996. Enhanced information recovery in 2D on- and off-resonance nutation NQR using the maximum entropy method. *Z. Naturforsch.* 51a:337–347.

Markworth, A., Weiden, N., and Weiss, A. 1987. Microcomputer controlled 4-Pi-Zeeman split NQR spectroscopy of Cl-35 in single-crystal para-chloroanilinium trichloroacetate—crystal-structure and activation-energy for the bleaching-out process. *Ber. Bunsenges. Phys. Chem.* 91:1158–1166.

Matsui, S., Kose, K., and Inouye, T. 1990. An NQR imaging experiment on a disordered solid. *J. Magn. Reson.* 88:186–191.

Millar, J. M., Thayer, A. M., Bielecki, A., Zax, D. B., and Pines, A. 1985. Zero field NMR and NQR with selective pulses and indirect detection. *J. Chem. Phys.* 83:934–938.

Morino, Y. and Toyama, M. 1961. Zeeman effect of the nuclear quadrupole resonance spectrum in crystalline powder. *J. Chem. Phys.* 35:1289–1296.

Nickel, P. and Kimmich, R. 1995. 2D exchange NQR spectroscopy. *J. Mol. Struct.* 345:253–264.

Nickel, P., Rommel, E., Kimmich, R., and Pusiol, D. 1991. Two-dimensional projection/reconstruction rotating-frame NQR imaging (ρ NQRI). *Chem. Phys Lett.* 183:183–186.

Nickel, P., Robert, H., Kimmich R., and Pusiol D. 1994. NQR method for stress and pressure imaging. *J. Magn. Reson.* A111:191–194.

Noack, F. 1986. NMR field cycling spectroscopy: Principles and applications. *Prog. NMR Spectrosc.* 18:171–276.

Pound, R. V. 1951. Nuclear spin relaxation times in a single crystal of LiF. *Phys. Rev.* 81:156–156.

Pratt, J. C., Raganuthan, P., and McDowell, C. A. 1975. Transient response of a quadrupolar system in zero applied field. *J. Magn. Reson.* 20:313–327.

Raghavan, P. 1989. Table of Nuclear Moments. *Atomic Data and Nuclear Data Tables* 42:189–291.

Ramachandran, R. and Oldfield, E. 1984. Two dimensional Zeeman nuclear quadrupole resonance spectroscopy. *J. Chem. Phys.* 80:674–677.

Robert, H. and Pusiol, D. 1996a. Fast ρ -NQR imaging with slice selection. *Z. Naturforsch.* 51a:353–356.

Robert, H. and Pusiol, D. 1996b. Slice selection in NQR spatially resolved spectroscopy. *J. Magn. Reson.* A118:279–281.

Robert, H. and Pusiol, D. 1997. Two dimensional rotating-frame NQR imaging. *J. Magn. Reson.* 127:109–114.

Robert, H., Pussiol, D., Rommel, E., and Kimmich R. 1994. On the reconstruction of NQR nutation spectra in solid with powder geometry. *Z. Naturforsch.* 49a:35–41.

Robert, H., Minuzzi, A., and Pusiol, D. 1996. A fast method for the spatial encoding in rotating-frame NQR imaging. *J. Magn. Reson.* A118:189–194.

Rommel, E., Kimmich, R., and Pusiol, D. 1991a. Spectroscopic rotating-frame NQR imaging (ρ NQRI) using surface coils. *Meas. Sci. Technol.* 2:866–871.

Rommel, E., Nickel, P., Kimmich, R., and Pusiol, D. 1991b. Rotating-frame NQR imaging. *J. Magn. Reson.* 91:630–636.

Rommel, E., Nickel, P., Rohmer, F., Kimmich, R., Gonzales, C., and Pusiol, D. 1992a. Two dimensional exchange spectroscopy using pure NQR. *Z. Naturforsch.* 47a:382–388.

Rommel, E., Kimmich, R., Robert, H., and Pusiol, D. 1992b. A reconstruction algorithm for rotating-frame NQR imaging (ρ NQRI) of solids with powder geometry. *Meas. Sci. Technol.* 3:446–450.

Selinger, J., Zagar, V., and Blinc, R. 1994. $^{1}$H-$^{14}$N nuclear quadrupole double resonance with multiple frequency sweeps. *Z. Naturforsch.* 49a:31–34.

Sen, K. D. and Narasimhan, P. T. 1974. Polarizabilities and anti-shielding factors in crystals. *In* Advances in Nuclear Quadrupole Resonance, Vol. 1 (J. A. S. Smith, ed.). Heyden and Sons, London.

Sinyavski, N., Ostafin, M., and Mackowiak, M. 1996. Rapid measurement of nutation NQR spectra in powder using an rf pulse train. *Z. Naturforsch.* 51a:363–367.

Stephenson, D. S. 1988. Linear prediction and maximum entropy methods in NMR spectroscopy. *Prog. Nucl. Magn. Reson Spectrosc.* 20:516–626.

Sternheimer, R. M. 1963. Quadrupole antishielding factors of ions. *Phys Rev.* 130:1423–1424.

Sternheimer, R. M. 1966. Shielding and antishielding effects for various ions and atomic systems. *Phys. Rev.* 146:140.

Sternheimer, R. M. and Peierls, R. F. 1971. Quadrupole antishielding factor and the nuclear quadrupole moments of several alkali isotopes. *Phys. Rev.* A3:837–848.

Thayer, A. M and Pines, A. 1987. Zero field NMR. *Acc. Chem. Res.* 20:47–53.

TonThat, D. M. and Clarke, J. 1996. Direct current superconducting quantum interference device spectrometer for pulsed magnetic resonance and nuclear quadrupole resonance at frequencies up to 5 MHz. *Rev. Sci. Instrum.* 67:2890–2893.

TonThat, D. M., Ziegeweid, M., Song, Y. Q., Munson, E. G., Applelt, S., Pines, A., and Clarke, J. 1997. SQUID detected NMR of laser polarized Xenon at 4.2 K and at frequencies down to 200 Hz. *Chem. Phys. Lett.* 272:245–249.

Werner-Zwanzinger, U., Zeigeweid, M., Black, B., and Pines, A. 1994. Nitrogen-14 SQUID NQR of L-Ala-L-His and of serine. *Z. Naturforsch.* 49a:1188–1192.

Wikner, E. G. and Das, T. P. 1958. Antishielding of nuclear quadrupole moment of heavy ions. *Phys. Rev.* 109:360–368.

Yesinowski, J. P., Buess, M. L., Garroway, A. N., Zeigeweid, M., and Pines A. 1995. Detection of $^{14}N$ and $^{35}Cl$ in cocaine base and hydrochloride using NQR, NMR and SQUID techniques. *Anal. Chem.* 67:2256–2263.

Yu, H.-Y. 1991. Studies of NQR spectroscopy for spin-5/2 systems. M. S. Thesis, SUNY at Stony Brook.

## KEY REFERENCES

Abragam, 1961. See above.

*Still the best comprehensive text on NMR and NQR theory.*

Das, T. P. and Hahn, E. L. 1958. Nuclear Quadrupole Resonance Spectroscopy. Academic Press, New York.

*A more specialized review of the theory of NQR: old but still indispensible.*

Harbison et al., 1989. See above.

*The first true zero-field multidimensional NMR experiment.*

Robert and Pusiol, 1997. See above.

*A good source for references on NQR imaging.*

## APPENDIX:
## GLOSSARY OF TERMS AND SYMBOLS

| | |
|---|---|
| $eq$ | the most distinct principal value of the electric field gradient at the nucleus |
| $eQ$ | electric quadrupole moment of the nucleus |
| $h$ | Planck's constant |
| $H_1$ | applied radiofrequency field in tesla |
| $I$ | nuclear spin |
| $T_1, T_2$ | longitudinal and transverse relaxation times |
| $\gamma$ | gyromagnetic ratio of the nucleus |
| $\gamma_\infty$ | Sternheimer antishielding factor of the atom or ion |
| $\eta$ | asymmetry parameter of the electric field gradient at the nucleus |
| $\nu_Q$ | NQR resonance frequency in Hz |
| $\omega_Q$ | NQR resonance frequency in rad s$^{-1}$ |
| $\omega_R$ | intrinsic precession frequency in rad s$^{-1}$; $\omega_R = \gamma H_1$ |

Bruno Herreros
Gerard S. Harbison
University of Nebraska
Lincoln, Nebraska

# ELECTRON PARAMAGNETIC RESONANCE SPECTROSCOPY

## INTRODUCTION

Electron paramagnetic resonance (EPR) spectroscopy, also called electron spin resonance (ESR) or electron magnetic resonance (EMR), measures the absorption of electromagnetic energy by a paramagnetic center with one or more unpaired electrons (Atherton, 1993; Weil et al., 1994). In the presence of a magnetic field, the degeneracy of the electron spin energy levels is removed and transitions between

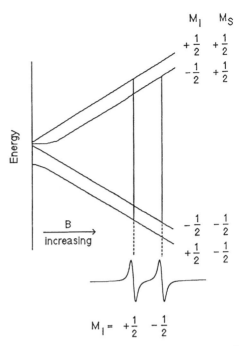

**Figure 1.** Energy-level splitting diagram for an unpaired electron in the presence of a magnetic field, interacting with one nucleus with $I = 1/2$. The energy separation between the two levels for the unpaired electron is linearly proportional to the magnetic field strength, $B$. Coupling to the nuclear spin splits each electron spin energy level into two. Transitions between the two electron energy levels are stimulated by microwave radiation when $h\nu = g\beta B$, where $\beta$ is the electron Bohr magneton. Only transitions with $\Delta m_S = \pm 1, \Delta m_I = 0$ are allowed, so interaction with nuclear spins causes the signal to split into $2nI + 1$ lines, where $n$ is the number of equivalent nuclei with spin $I$. The microwave magnetic field, $B_1$ is perpendicular to $B$. If the line shape is determined by relaxation, it is Lorentzian. The customary display in EPR spectroscopy is the first derivative of the absorption line shape. Sometimes the dispersion signal is detected instead of the absorption because the dispersion signal saturates less readily than the absorption. The dispersion signal is related to the absorption signal by the Kramers-Kronig transform.

the energy levels can be caused to occur by supplying energy. When the energy of the microwave photons equals the separation between the energy levels of the unpaired electrons, there is absorption of energy by the sample and the system is said to be at "resonance" (Fig. 1). The fundamental equation that describes the experiment for a paramagnetic center with one unpaired electron is $h\nu = g\beta B$, where $h$ is Planck's constant, $\nu$ is the frequency of the microwaves, $g$ is a characteristic of the sample, $\beta$ is the Bohr magneton, and $B$ is the strength of the magnetic field in which the sample is placed. Typically the experiment is performed with magnetic fields such that the energies are in the microwave region. If the paramagnetic center is tumbling rapidly in solution, $g$ is a scalar quantity. When the paramagnetic center is immobilized, as in solid samples, most samples exhibit $g$ anisotropy and $g$ is then represented by a matrix. Hyperfine splitting of the signal occurs due to interaction with nuclear spins and can therefore be used to identify the number and types of nuclear spins in proximity to the paramagnetic center.

The information content of EPR arises from the ability to detect a signal and from characteristics of the signal, including integrated intensity, hyperfine splitting by nuclear spins, g value, line shape, and electron spin relaxation time. The characteristics of the signal may depend on environmental factors, including temperature, pressure, solvent, and other chemical species. Various types of EPR experiments can optimize information concerning these observables. The following list includes some commonly asked materials science questions that one might seek to answer based on these observables and some corresponding experimental design considerations. Detailed background information on the physical significance and theoretical basis for the observable parameters is provided in the standard texts cited (Poole, 1967, 1983; Pake and Estle, 1973; Eaton and Eaton, 1990a, 1997a; Weil et al., 1994). Some general information and practical considerations are given in this unit.

Whenever unpaired electrons are involved, EPR is potentially the best physical technique for studying the system. Historically, the majority of applications of EPR have been to the study of organic free radicals and transition metal complexes (Abragam and Bleaney, 1970; Swartz et al., 1972; Dalton, 1985; Pilbrow, 1990). Today, these applications continue, but in the context of biological systems, where the organic radicals are naturally occurring radicals, spin labels, and spin-trapped radicals, and the transition metals are in metalloproteins (Berliner, 1976, 1979; Eaton et al., 1998). Applications to the study of materials are extensive, but deserve more attention than they have had in the past. Recent studies include the use of EPR to monitor the age of archeological artifacts (Ikeya, 1993), characterize semiconductors and superconductors, measure the spatial distribution of radicals in processed polymers, monitor photochemical degradation of paints, and characterize the molecular structure of glasses (Rudowicz et al., 1998).

One might seek to answer the following types of questions.

*1. Are there paramagnetic species in the sample?*

The signal could be due to organic radicals, paramagnetic metal ions, defects in materials (Lancaster, 1967) such as dangling bonds or radiation damage, or paramagnetic species intentionally added as probes. The primary concern in this case would be the presence or absence of a signal. The search for a signal might include obtaining data over a range of temperatures, because some signals can only be detected at low temperatures and others are best detected near room temperature.

*2. What is the concentration of paramagnetic species?* Does the concentration of paramagnetic species change as a function of procedure used to prepare the material or of sample handling?

Spectra would need to be recorded in a quantitative fashion (Eaton and Eaton, 1980, 1990a, 1992, 1997a).

*3. What is the nature of the paramagnetic species? If it is due to a paramagnetic metal, what metal? If it is due to an organic radical, what is the nature of the radical? Is more than one paramagnetic species present in the sample?*

One would seek to obtain spectra that are as well resolved as possible. To examine weak interactions with nuclear spins that are too small compared with line widths to be resolved in the typical EPR spectrum, but which might help to identify the species that gives rise to the EPR signal, one might use techniques such as ENDOR (electron nuclear double resonance; Atherton, 1993) or ESEEM (electron spin echo envelope modulation; Dikanov and Tsvetkov, 1992).

*4. Does the sample undergo phase transitions that change the environment of the paramagnetic center?*

Experiments could be run as a function of temperature, pressure, or other parameters that cause the phase transition.

*5. Are the paramagnetic centers isolated from each other or present in pairs, clusters, or higher aggregates?*

EPR spectra are very sensitive to interactions between paramagnetic centers (Bencini and Gatteschi, 1990). Strong interactions are reflected in line shapes. Weaker interactions are reflected in relaxation times.

*6. What is the mobility of the paramagnetic centers?*

EPR spectra are sensitive to motion on the time scale of anisotropies in resonance energies that arise from anisotropies in g values and/or electron-nuclear hyperfine interactions. This time scale typically is microseconds to nanoseconds, depending on the experiment. Thus, EPR spectra can be used as probes of local mobility and microviscosity (Berliner and Reuben, 1989).

*7. Are the paramagnetic centers uniformly distributed through the sample or spatially localized?*

This issue would be best addressed with an EPR imaging experiment (Eaton and Eaton, 1988a, 1990b, 1991, 1993a, 1995a, 1996b, 1999a; Eaton et al., 1991; Sueki et al., 1990).

Bulk magnetic susceptibility (GENERATION AND MEASUREMENT OF MAGNETIC FIELDS & MAGNETIC MOMENT AND MAGNETIZATION) also can be used to study systems with unpaired electrons, and can be used to determine the nature of the interactions between spins in concentrated spin systems. Rather large samples are required for bulk magnetic susceptibility measurements, whereas EPR typically is limited to rather small samples. Bulk susceptibility and EPR are complementary techniques. EPR has the great advantage that it uniquely measures unpaired electrons in low concentrations, and in fact is most informative for systems that are magnetically dilute. It is particularly powerful for identification of paramagnetic species present in the sample and characterization of the environment of the paramagnetic species.

In this unit, we seek to provide enough information so that a potential user can determine whether EPR is likely to be informative for a particular type of sample, and which type of EPR experiment would most likely be useful.

## PRINCIPLES OF EPR

The fundamental principles of EPR are similar to those of nuclear magnetic resonance (NMR; Carrington and

McLachlan, 1967) and magnetic resonance imaging (MRI), which are described elsewhere in this volume (NUCLEAR MAGNETIC RESONANCE IMAGING). However, several major differences between the properties of unpaired electron spins and of nuclear spins result in substantial differences between NMR and EPR spectroscopy. First, the magnetogyric ratio of the electron is 658 times that of the proton, so for the same magnetic field the frequency for electron spin resonance is 658 times the frequency for proton resonance. Many EPR spectrometers operate in the 9 to 9.5 GHz frequency range (called "X-band"), which corresponds to resonance for an organic radical ($g \sim 2$) at a magnetic field of 3200 to 3400 gauss [G; 1 G=$10^{-4}$ tesla (T)]. Second, electron spins couple to the electron orbital angular momentum, resulting in shorter relaxation times than are observed in NMR. Electron spin relaxation times are strongly dependent on the type of paramagnetic centers. For example, typical room temperature electron spin relaxation times range from $\sim 10^{-6}$ s for organic radicals to as short as $10^{-12}$ s for low-spin Fe(III). Third, short relaxation times can result in very broad lines. Even for organic radicals with relatively long relaxation times, EPR lines frequently are relatively broad due to unresolved coupling to neighboring nuclear spins. Line widths for detectable EPR signals range from fractions of a gauss to tens or hundreds of gauss. Since 1 G corresponds to about 2.8 MHz, these EPR line widths correspond to $10^6$ to $10^8$ Hz, which is much greater than NMR line widths. Fourth, coupling to the electron orbital angular momentum also results in larger spectral dispersion for EPR than for NMR. A room temperature spectrum of an organic radical might extend over 10 to 100 G. However, the spectrum for a transition metal ion might extend over one hundred to several thousand gauss. Finally, the larger magnetic moment of the unpaired electron also results in larger spin–spin interactions, so high-resolution spectra require lower concentrations for EPR than for NMR. These differences make the EPR measurement much more technically challenging than the NMR measurement. For example, pulsed Fourier transform NMR provides major advances in sensitivity; however pulsed Fourier transform spectroscopy has more restricted applicability in EPR because a pulse of finite width cannot excite the full spectrum for many types of paramagnetic samples (Kevan and Schwartz, 1979; Kevan and Bowman, 1990).

In a typical EPR experiment the sample tube is placed in a structure that is called a resonator. The most common type of resonator is a rectangular cavity. This name is used because microwaves from a source are used to set up a standing wave pattern. In order to set up the standing wave pattern, the resonator must have dimensions appropriate for a relatively narrow range of microwave frequencies. Although the use of a resonator enhances the signal-to-noise (S/N) ratio for the experiment, it requires that the microwave frequency be held approximately constant and that the magnetic field be swept to achieve resonance. The spectrometer detects the microwave energy reflected from the resonator. When spins are at resonance, energy is absorbed from the standing wave pattern, the reflected energy decreases, and a signal is detected. In a typical continuous wave (CW) EPR experiment, the

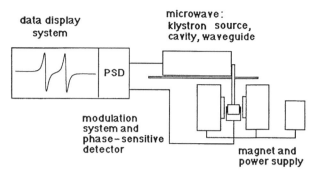

**Figure 2.** Block diagram of an EPR spectrometer. The fundamental modules of a CW EPR spectrometer include: the microwave system comprising source, resonator, and detector; the magnet system comprising power supply and field controller; the magnetic field modulation and phase-sensitive detection system; and the data display and manipulation system. Each of these subsystems can be largely independent of the others. In modern spectrometers there is a trend toward increasing integration of these units via interfaces to a computer. The computer is then the controller of the spectrometer and also provides the data display and manipulation. In a pulse system a pulse timing unit is added and magnetic field modulation is not used.

magnetic field is swept and the change in reflected energy (the EPR signal) is recorded. To further improve S/N, the magnetic field is usually modulated (100 kHz modulation is commonly used), and the EPR signal is detected using a phase-sensitive detector at this modulation frequency. This detection scheme produces a first derivative of the EPR absorption signal. A sketch of how magnetic-field modulation results in a derivative display is found in Eaton and Eaton, (1990a, 1997a). Because taking a derivative is a form of resolution enhancement and EPR signals frequently are broad, it is advantageous for many types of samples to work directly with the first-derivative display. Thus, the traditional EPR spectrum is a plot of the first derivative of the EPR absorption as a function of magnetic field. A block diagram for a CW EPR spectrometer is given in Figure 2.

**Types of EPR Experiments**

One can list over 100 separate types of EPR spectra. One such list is in an issue of the *EPR Newsletter* (Eaton and Eaton, 1997b). For this unit we focus primarily on experiments that can be done with commercial spectrometers. For the near future, those are the experiments that are most likely to be routinely available to materials scientists. In the following paragraphs we outline some of the more common types of experiments and indicate what characteristics of the sample might indicate the desirability of a particular type of experiment. Experimental details are reserved for a later section.

**1. CW Experiments at X-band with a Rectangular Resonator**

This category represents the vast majority of experiments currently performed, whether for routine analysis or for research, in materials or in biomedical areas. In this experiment the microwaves are on continuously and the magnetic field is scanned to achieve resonance. This is the experiment that was described in the introductory

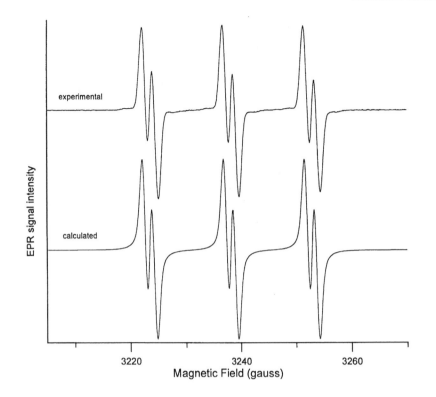

**Figure 3.** Room temperature CW spectrum of a nitroxyl radical obtained with a microwave frequency of 9.09 GHz. The major three-line splitting is due to coupling to the nitroxyl nitrogen ($I = 1$) and the small doublet splitting is due to interaction with a unique proton ($I = 1/2$). The calculated spectrum was obtained with hyperfine coupling to the nitrogen of 14.7 G, coupling to the proton of 1.6 G, and a Gaussian line width of 1.1 G due to additional unresolved proton hyperfine coupling. The $g$ value (2.0056) is shifted from 2.0023 because of spin-orbit coupling involving the nitrogen heteroatom.

paragraphs and which will be the focus of much of this unit. Examples of spectra obtained at X-band in fluid solution and for an immobilized sample are shown in Figures 3 and 4. The parameters obtained by simulation of these spectra are indicated in the figure captions.

## 2. CW Experiments at Frequencies Other than X-band

Until relatively recently, almost all EPR was performed at approximately 9 GHz (X-band). However, experiments at

lower frequencies can be advantageous in dealing with larger samples and samples composed of materials that absorb microwaves strongly at X-band, as well as for resolving nitrogen hyperfine splitting in certain Cu(II) complexes (Hyde and Froncisz, 1982).

Experiments at higher microwave frequencies are advantageous in resolving signals from species with small $g$-value differences (Krinichnyi, 1995; Eaton and Eaton, 1993c, 1999b), in obtaining signals from

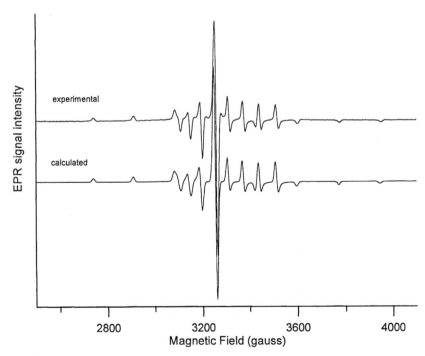

**Figure 4.** Room temperature CW spectrum of a vanadyl porphyrin in a 2:1 toluene:chloroform glass at 100 K, obtained with a microwave frequency of 9.200 GHz. The anisotropy of the $g$ and hyperfine values results in different resonant conditions for the molecules as a function of orientation with respect to the external field. For this random distribution of orientations, the EPR absorption signal extends from about 2750 to 3950 G. The first derivative display emphasizes regions of the spectrum where there are more rapid changes in slope, so "peaks" in the first derivative curve occur at extrema in the powder distribution. These extrema define the values of $g$ and $A$ along the principal axes of the magnetic tensors. The hyperfine splitting into eight approximately equally spaced lines is due to the nuclear spin ($I = 3.5$) of the vanadium nucleus. The hyperfine splitting is greater along the normal to the porphyrin plane ($z$ axis) than in the perpendicular plane. The calculated spectrum was obtained with $g_x = 1.984$, $g_y = 1.981$, $g_z = 1.965$, $A_x = 55 \times 10^{-4}$ cm$^{-1}$, $A_y = 53 \times 10^{-4}$ cm$^{-1}$, and $A_z = 158 \times 10^{-4}$ cm$^{-1}$.

paramagnetic centers with more than one unpaired electron, and in analyzing magnetic interactions in magnetically concentrated samples (Date, 1983). In principle, experiments at higher microwave frequency should have higher sensitivity than experiments at X-band, due to the larger difference in the Boltzmann populations of the electron spin states. Sensitivity varies with (frequency)$^{11/4}$ if the microwave magnetic field at the sample is kept constant and the size of the resonator is scaled inversely with frequency (Rinard et al., 1999). This is a realistic projection for very small samples, as might occur in some materials research applications. To achieve the expected increase in sensitivity (Rinard et al., 1999) will require substantial engineering improvements in the sources, resonators and detectors used at higher frequency.

Computer simulations of complicated spectra as a function of microwave frequency provide a much more stringent test of the parameters than can be obtained at a single frequency.

### 3. Electron-Nuclear Double Resonance (ENDOR)

Due to the large line widths of many EPR signals, it is difficult to resolve couplings to nuclear spins. However, these couplings often are key to identifying the paramagnetic center that gives rise to the EPR signal. By simultaneously applying radio frequency (RF) and microwave frequency energies, one can achieve double resonance of both the electron and nuclear spins. The line widths of the signals in this experiment are much narrower than typical EPR line widths and the total number of lines in the spectrum is smaller, so it is easier to identify the nuclear spins that are interacting with the electron spin in an ENDOR spectrum (Box, 1977; Schweiger, 1982; Atherton, 1993; Piekara-Sady and Kispert, 1994; Goslar et al., 1994) than in an experiment without the RF frequency—i.e., a "normal" CW spectrum.

### 4. Pulsed/Fourier Transform EPR

Measurements of relaxation times via pulsed or time-domain techniques typically are much more accurate than continuous wave measurements. Pulse sequences also can be tailored to obtain detailed information about the spins. However, the much shorter relaxation times for electron spins than for nuclear spins limit pulse sequences to shorter times than can be used in pulsed NMR (Kevan and Schwartz, 1979). A technique referred to as electron spin echo envelope modulation (ESEEM) is particularly useful in characterizing interacting nuclear spins, including applications in materials (Kevan and Bowman, 1990). Interaction between inequivalent unpaired electrons can cause changes in relaxation times that depend on distance between the spins (Eaton and Eaton, 1996a).

### 5. Experiments with Resonators Other than the Rectangular Resonator

New lumped-circuit resonators, such as the loop-gap resonator (LGR) and others designed on this principle (Hyde and Froncisz, 1986) can be adapted to a specific experiment to optimize aspects of the measurement (Rinard et al., 1993, 1994, 1996a, 1996b). These structures are particularly important for pulsed experiments and for experiments at frequencies lower than X-band.

## PRACTICAL ASPECTS OF THE CW METHOD

The selection of the microwave power and modulation amplitude for recording CW EPR signals is very important in obtaining reliable results. We therefore give a brief discussion of issues related to the selection of these parameters. Greater detail concerning the selection of parameters can be found in Eaton and Eaton (1990a, 1997a).

### Microwave Power

If microwave power is absorbed by the sample at a rate that is faster than the sample can dissipate energy to the lattice (i.e., faster than the electron spin relaxation rate), the EPR signal will not be proportional to the number of spins in the sample. This condition is called saturation. The power that can be used without saturating the sample depends on the relaxation rate. To avoid saturation of spectra, much lower powers must be used for samples that have long relaxation times, such as organic radicals, than for samples with much shorter relaxation times, such as transition metals. Relaxation times can range over more than 18 orders of magnitude, from hours to $10^{-14}$ s, with a wide range of temperature dependence, from $T^9$ to temperature-independent, and a wide variety of magnetic field dependence and/or dependence on position in the EPR spectrum (Du et al., 1995). One can test for saturation by recording signal amplitude at a series of microwave powers, $P$, and plotting signal amplitude as a function of $\sqrt{P}$. This plot is called a power-saturation curve. The point where the signal dependence on $\sqrt{P}$ begins to deviate from linearity is called the onset of saturation. Data must be recorded at a power level below the onset of saturation in order to use integrated data to determine the spin concentration in the sample. Higher powers also cause line broadening, so powers should be kept below the onset of saturation if data analysis is based on line shape information. Above the onset of saturation, the signal intensity typically increases with increasing power and goes through a maximum. If the primary goal is to get maximum S/N at the expense of line shape or quantitation, then one can operate at the power level that gives maximum signal amplitude, but under such conditions quantitation of spins can be erroneous.

### Modulation Amplitude

Provided that the modulation amplitude is less than 1/10 of the peak-to-peak line width of the first-derivative EPR signal to be recorded, increasing modulation amplitude improves S/N without distorting line shape. However, as the modulation amplitude is increased further, it causes broadening and distortion (Poole, 1967). If the key information to be obtained from the spectrum is based on the line shape, care must be taken in selecting the modulation amplitude. Unlike the problems that occur when too high a microwave power is used to record the spectrum, increasing the modulation amplitude does not change the linear relationship between integrated signal intensity and modulation amplitude. Thus, if the primary information to be obtained from a spectrum is the integrated signal intensity (to determine spin concentration), it can be useful to increase modulation amplitude, and thereby improve S/N, at the expense of some line shape distortion. The maximum spectral amplitude, and therefore the best S/N, occurs at a peak-to-peak modulation amplitude ∼2 times the peak-to-peak width of the derivative EPR spectrum.

## Sensitivity

The most important issue concerning practical applications of EPR is sensitivity (Eaton and Eaton 1980, 1992; Rinard et al., 1999). In many analytical methods, one can describe sensitivity simply in terms of minimum detectable concentrations. The problem is more complicated for EPR because of the wide range of line widths and large differences in power saturation among species of interest. Vendor literature cites sensitivity as about $0.8 \times 10^{10}/\Delta B$ spins/G, where $\Delta B$ is the line width of the EPR signal. This is a statement of the minimum number of detectable spins with a S/N of 1. It is based on the assumption that spectra can be recorded at a power of 200 mW, with a modulation amplitude of 8 G, and that there is no hyperfine splitting that divides intensity of the signal into multiple resolved lines. To apply this statement to estimate sensitivity for another sample requires taking account of the actual conditions appropriate for a particular sample. Those conditions depend upon the information that one requires from the spectra. One can consider two types of conditions as described below.

**Case 1.** Selection of microwave power, modulation amplitude, and time constant of the signal detection system that provide undistorted line shapes. This mode of operation is crucial if one wishes to obtain information concerning mobility of the paramagnetic center or partially resolved hyperfine interactions. Typically this would also require a S/N significantly greater than 1. This case might be called "minimum number of detectable spins with desired spectral information." It would require the use of a microwave power that did not cause saturation and a modulation amplitude that did not cause line broadening.

**Case 2.** Selection of microwave power, modulation amplitude, and time constant of the signal detection system that provide the maximum signal amplitude, at the expense of line shape information. This mode of operation might be selected, for example, in some spin trapping experiments, where one seeks to determine whether an EPR signal is present and one can obtain adequate information for identification of the species from observed hyperfine couplings even in the presence of line shape distortions that come from power broadening or overmodulation. This case might be called "minimum number of detectable spins."

To correct the minimum number of detectable spins for microwave power, multiply $0.8 \times 10^{10}/\Delta B$ by

$$\sqrt{\frac{P_{sample}}{200}} \qquad (1)$$

where $P_{sample}$ is the power incident on the critically coupled resonator, in mW. To correct for changes in modulation amplitude (MA), multiply by $MA_{sample}$ (in gauss)/8 (G). For samples with narrow lines and long relaxation times, the minimum number of detectable spins may be higher than $0.8 \times 10^{10}/\Delta B$ by several orders of magnitude.

## Calibration

Quality assurance requires a major effort in EPR (Eaton and Eaton 1980, 1992). Because the properties of the sample affect the characteristics of the resonator, and because the microwave magnetic field varies over the sample, selecting a standard for quantitation of the number of spins requires understanding of the spin system and of the spectrometer [Poole, 1967, 1983; Wilmshurst, 1968 (see especially THERMAL ANALYSIS); Alger 1968; Czoch and Francik, 1989]. In general, the standard should be as similar as possible to the sample to be measured, in the same solvent or solid host, and in a tube that is as close as possible to the same size. The EPR tube has to be treated as a volumetric flask, and calibrated accordingly. Determining the number of spins in the sample requires subtraction of any background signal, and then integrating twice (derivative → absorption → area). If there is background slope, the double integration may not be very accurate. All of this requires that the EPR spectrum be in digital form. If the line shape is the same for the sample and the reference, then peak heights can be used instead of integrated areas. This is sometimes difficult to ensure, especially in the wings of the spectrum, which may constitute a lot of the area under the spectrum. However, for fairly widespread applications, such as spin trapping and spin labeling, reasonable estimates of spin concentration can be obtained by peak-height comparison with gravimetrically prepared standard solutions. Very noisy spectra with well defined peak positions might best be quantitated using simulated spectra that best fit the experimental spectra.

Another important calibration is of magnetic field scan. The best standard is an NMR gaussmeter, but note that the magnetic field varies across the pole face, and the gaussmeter probe has to be as close to the sample as possible—ideally inside the resonator. A few carefully-measured samples that can be prepared reproducibly have been reported in the literature (Eaton and Eaton 1980, 1992) and are good secondary standards for magnetic field calibrations.

The magnetic field control device on electromagnet-based EPR spectrometers is a Hall probe. These are inherently not very accurate devices, but EPR field control units contain circuitry to correct for their inaccuracies. Recent Bruker field controllers, for example, use a ROM containing the characteristics of the specific Hall probe with which it is matched. Such systems provide sufficient accuracy for most EPR measurements.

## METHOD AUTOMATION

Early EPR spectrometers did not include computers. Many, but not all of those systems, have been retrofitted with computers for data acquisition. Historically, the operator spent full time at the EPR spectrometer console while the spectrometer was being operated. In the next generation, a computer was attached to the EPR spectrometer for data acquisition, but the spectrometer was still substantially manually operated. Even when computers were built into the spectrometer, there was a control console at which the operator sat. The newer spectrometers are operated from the computer, and have a console of electronic modules that control various functions of the magnet, microwave system, and data acquisition. The latest pulsed EPR spectrometers are able to execute long (including overnight) multipulse experiments without operator intervention. The most recent cryogenic temperature-control systems also can be programmed and can control the cryogen valve in addition to the heater. There is one research lab that has highly automated ENDOR spectrometers (Corradi et al, 1991). Except for these examples, EPR has not been automated to a significant degree.

For many samples, the sample positioning is sufficiently critical, as is the resonator tuning and coupling, that automation of sample changing for unattended operation is less feasible than for many other forms of spectroscopy, including NMR. Automation would be feasible for multiple samples that shared the same geometry, dielectric constant, and microwave loss factor, if they could be run near room temperature. The only currently available spectrometer with an automatic sample changer is one designed by Bruker for dosimetry. This system is customized to handle a particular style of alanine dosimeter and includes software to quantitate the radical signal and to prepare reports. References to computer use for acquisition and interpretation are given by Vancamp and Heiss (1981) and Kirste (1994).

## DATA ANALYSIS AND INITIAL INTERPRETATION

The wide range of queries one may make of the spin system, listed elsewhere in this unit, makes the data analysis and interpretation diverse. An example of a spectrum of an organic radical in fluid solution is shown in Figure 3. Computer simulation of the spectrum gives the $g$ value for the radical and the nuclear hyperfine splittings which can be used to identify the nuclear spins with the largest couplings to the unpaired electrons. An example of a spectrum of a vanadyl complex immobilized in 1:1 toluene:chloroform at 100 K is shown in Figure 4. The characteristic 8-line splitting patterns permit unambiguous assignment as a vanadium species ($I = 3.5$). Computer simulation of the spectrum gives the anisotropic components of the $g$ and $A$ values, which permits a more detailed characterization of the electronic structure of the paramagnetic center than could be obtained from the isotropic averages of these parameters observed in fluid solution.

As discussed above, under instrument calibration, a first stage in data analysis and interpretation is to determine how many spins are present, as well as the magnitude of line widths and hyperfine splittings. The importance of quantitation should not be underestimated. There are many reports in the literature of detailed analysis of a signal that accounts for only a very small fraction of the spins in the sample, the others having been overlooked. The problem is particularly acute in samples that contain both sharp and broad signals. The first-derivative display tends to emphasize the sharp signals, and it is easy to overlook broad signals unless spectra are recorded under a wide range of experimental conditions.

The software available as part of the spectrometer systems is increasingly powerful and versatile. Additional software is available, some commercially and some from individual labs (easily locatable via the EPR Society software exchange; *http://www.ierc.scs.uiuc.edu*), for simulation of more specialized spin systems. The understanding of many spin systems is now at the stage where a full simulation of the experimental line shape is a necessary step in interpreting an EPR spectrum. For $S = 1/2$ organic radicals in fluid solution, one should be able to fit spectra within experimental error if the radical has been correctly identified. However, there will always remain important problems for which the key is to understand the spin system, and for these systems simulation may push the state of the art. For example, for many high-spin Fe(III) systems, there is little information about zero-field splitting (ZFS) terms, so one does not even know which transitions should be included in the simulation.

In pulsed EPR measurements, the first step after quantitation is to determine the relaxation times, and, if relevant, the ESEEM frequencies in the echo decay by Fourier transformation of the time-domain signal. Software for these analyses and for interpretation of the results is still largely resident within individual research groups, especially for newly evolved experiments.

## SAMPLE PREPARATION

Samples for study by EPR can be solid, dissolved solid, liquid, or, less frequently, gas. In the gas phase the electron spin couples to the rotational angular momentum, so EPR spectroscopy has been used primarily for very small molecules. Thus, the focus of this unit is on samples in solid or liquid phases. For most EPR resonators at X-band or lower frequency, the sample is placed in a 4-mm outside diameter (o.d.) quartz tube. Usually, synthetic fused silica is used to avoid paramagnetic metal impurities. The weaker the sample signal, the higher the quality of quartz required. Pyrex or Kimax (trade names of Corning and Kimble) have very strong EPR signals.

Often, oxygen must be removed from the sample because the sample reacts with oxygen, or because the paramagnetic $O_2$ broadens EPR signals by Heisenberg exchange during collisions. The oxygen broadening is a problem for high-resolution spectra of organic radicals in solution and for the study of carbonaceous materials. This problem is so severe that one can turn this around and use oxygen broadening as a measure of oxygen concentration. EPR oximetry is a powerful analytical tool (Hyde and Subczynski, 1989).

For some problems one would want to study the concentrated solid, which may not exist in any other form. Maybe one wants to examine the spin-spin interaction as a function of temperature. In this case, one would put the solid directly into the EPR tube, either as a powder or as a single crystal, which might be oriented. There are a number of caveats for these types of samples. First, one must avoid overloading the resonator. More sample is not always better. With a "standard" $TE_{102}$ (transverse electric 102 mode cavity) rectangular resonator (which until recently was the most common EPR resonator), one cannot use more than $\sim 10^{17}$ spins (Goldberg and Crowe, 1977). If the number of spins is too high, one violates the criterion that the energy absorption on resonance is a small perturbation to the energy reflected from the resonator. This is not an uncommon problem, especially with "unknown" samples, which might contain, for example, large quantities of iron oxide.

A similar problem occurs with samples that are lossy—i.e., that absorb microwaves other than by magnetic resonance (Dalal et al., 1981). Water is a prime example. For aqueous samples, and especially biological samples, there are special flat cells to keep the lossy material in a nodal plane of the microwave field in the resonator. There are special resonators to optimize the S/N for lossy samples. Usually, these operate with a TM (transverse magnetic) mode rather than a TE mode. Lower frequency can also be advantageous for lossy samples. There are many examples of lossy materials commonly encountered in the study of materials. Carbonaceous materials often have highly conducting regions. Some semiconductors are sufficiently conducting that only small amounts of sample can be put in the resonator without reducing the $Q$ (the resonator quality factor) so much that the spectrometer cannot operate.

These problems are obvious to the experienced operator, because general-purpose commercial EPR spectrometers provide a readout of the microwave power as a function of frequency during the "tune" mode of setting up the measurement. The power absorbed by the resonator appears as a "dip" in the power versus frequency display. The broader the dip, the lower the resonator $Q$. One useful definition of $Q$ is the resonant frequency divided by the half-power bandwidth of the resonator, $Q = v/\Delta v$. A lossy or conducting sample lowers $Q$, which is evident to the operator as a broadening of the dip. If the sample has much effect on $Q$, care must be taken in quantitation of the EPR signal, since the signal amplitude is proportional to $Q$. Ideally, one would measure the $Q$ for every sample if spin quantitation is a goal. No commercial spectrometer gives an accurate readout of $Q$, so the best approach is to make the $Q$ effect of the samples under comparison as precisely the same as possible.

Magnetically concentrated solids can exhibit the full range of magnetic interactions, including antiferromagnetism, ferromagnetism, and ferrimagnetism, in addition to paramagnetism, many of which are amenable to study by EPR (Bencini and Gatteschi, 1990; Stevens, 1997). Because of the strength of magnetic interactions in magnetically concentrated solids, ferrimagnetic resonance and ferromagnetic resonance have become separate subfields of magnetic resonance. High-field EPR is especially valuable to studies of magnetically concentrated solids, because it is possible in some cases to achieve external magnetic fields on the order of or greater than the internal magnetic fields.

If the goal is the study of isolated spins, they need to be magnetically dilute. In practice, this means that the spin concentration in a doped solid or liquid or solid solution should be less than $\sim 1$ mM ($\sim 6 \times 10^{17}$ spins per $cm^3$). For narrow-line spectra, such as defect centers in solids or organic radicals, either in liquid or solid solution, concentrations less than 1 mM are required to achieve minimum line width. For accurate measurement of relaxation times, concentrations have to be lower than those required for accurate measurement of line widths in CW spectra. For some materials, it is convenient to prepare a solution and cool it to obtain an immobilized sample. If this is done, it is important to select a solvent or solvent mixture that forms a glass, not crystals, when it is cooled. Crystallization excludes solute from the lattice and thereby generates regions of locally high concentrations, which can result in poor resolution of the spectra.

The discussion of lossy samples hints at part of the problem of preparing fluid solution samples. The solvent has to be nonlossy, or the sample has to be very small and carefully located. Some solvents are more lossy at given temperatures than others. Most solvents are nonlossy when frozen. Broadening by oxygen is usually not detectable in frozen solution line widths, but there still can be an impact on relaxation times. Furthermore, the paramagnetic $O_2$ also can yield an EPR signal in frozen solution (it is detectable in the gas phase and in the solid, but not in fluid solution), and can be mistaken for signals in the sample under study.

Care must be taken when examining powdered crystalline solids to get a true powder (random orientation) line shape. If the particles are not small enough they may preferentially orient when placed in the sample tube. At high magnetic fields, magnetically anisotropic solids can align in the magnetic field. Either of these effects can yield spectra that are representative of special crystal orientations, rather than a powder (spherical) average. The combination of these effects can dominate attempts to obtain spectra at high magnetic fields, such as at 95 GHz and above, where very small samples are needed and magnetic fields are strong enough to macroscopically align microcrystals.

Although experiments are more time consuming, very detailed information concerning the electronic structure of the paramagnetic center can be obtained by examining single crystals. In these samples, data are obtained as a function of the orientation of the crystal, which then provides a full mapping of the orientation dependence of the g-matrix and of the electron-nuclear hyperfine interaction (Byrn and Strouse, 1983; Weil et al., 1994).

## SPECIMEN MODIFICATION

The issue of specimen modification is inherent in the discussion of sample preparation. EPR is appropriately described as a nondestructive technique. If the sample fits the EPR resonator, it can be recovered unchanged.

However, some of the detailed sample preparations discussed above do result in modification of the sample. Grinding to achieve a true powder average spectrum destroys the original crystal. Sometimes grinding itself introduces defects that yield EPR signals, and this could be a study in itself. Sometimes grinding exposes surfaces to reactants that convert the sample to something different. Similarly, dissolving a sample may irreversibly change it, due to reactivity or association or dissociation equilibria. Cooling or heating a single crystal sometimes causes phase transitions that shatter the crystal. Unfortunately freezing a solvent may shatter the quartz EPR tube, due to differences in coefficient of expansion, causing loss of the sample.

These potential problems notwithstanding, EPR is a nondestructive and noninvasive methodology for study of materials. Consider the alternatives to EPR imaging to determine spatial distribution of paramagnetic centers. Even using EPR to monitor the radical concentration, one would grind or etch away the surface of the sample, monitoring changes in the signal after each step of surface removal. The sample is destroyed in the process. With EPR imaging, the sample is still intact and can be used for a further study, e.g., additional irradiation or heat treatment (Sueki et al., 1995).

## PROBLEMS

### How to Decide Whether and How EPR Should be Used

What do you want to learn about the sample? Is EPR the right tool for the task? This section attempts to guide thinking about these questions.

Is there an EPR signal? The first thing we learn via EPR is whether the sample in a particular environment exhibits resonance at the selected frequency, magnetic field, temperature, microwave power, and time after a pulse. This is a nontrivial observation, and by itself may provide almost the full answer to a question about the material. One should be careful, however, about jumping to conclusions based on only qualitative EPR results. With care it is possible to find an EPR signal in almost any sample, but that signal may not even be relevant to the information desired about the sample. There are spins everywhere. Dirt and dust in the environment, even indoors, will yield an EPR signal, often with strong Mn(II) signals and large broad signals due to iron oxides. Quartz sample tubes and Dewars contain impurities that yield EPR signal close to $g = 2$ and slightly higher. Most often, these are "background" interferences that have to be subtracted for quantitative work.

What if the sample does not exhibit a nonbackground EPR signal? Does this mean that there are no unpaired spins in the sample? Definitely not! The experienced spectroscopist is more likely to find a signal than the novice, because of the extensive number of tradeoffs needed to optimize an EPR spectrum (sample positioning, microwave frequency, resonator coupling, magnetic field sweep, microwave power, modulation frequency and amplitude, and time constant). The sample might have relaxation times such that the EPR signal is not observable at the

temperature selected (Eaton and Eaton 1990a). In general, relaxation times increase with decreasing temperature, and there will be some optimum temperature at which the relaxation time is sufficiently long that the lines are not immeasurably broad, but not so long that the line is saturated at all available microwave power levels.

Furthermore, the failure to observe an EPR signal could be a fundamental property of the spin system. If the electron spin, $S$, for the paramagnetic center of interest is half-integer, usually at least the transition between the $m_s = \pm 1/2$ levels will be observable at some temperature. However, for integer spin systems ($S = 1, 2, 3$) the transition may not be observable at all in the usual spectrometer configuration, even if the right frequency, field, temperature, and so forth are chosen (Abragam and Bleaney, 1970; Pilbrow, 1990). This is because such spin transitions are forbidden when the microwave magnetic field, $B_1$, is perpendicular to the external magnetic field, $B_0$, i.e., in the usual arrangement. For integer spin systems one needs $B_1$ parallel to $B_0$, and a special cavity is required. Fortunately, these are commercially available at X-band. Thus, if the question is "does this sample contain paramagnetic Ni(II)," a special parallel mode resonator is needed. An overview of EPR-detectable paramagnetic centers and corresponding parameters can be found in Eaton and Eaton (1990a).

Another reason for not observing an EPR signal, or at least the signal from the species of interest, is that spin-spin interaction resulted in a singlet-triplet separation such that only one of these states is significantly populated under the conditions of measurement. A classic example is dimeric Cu(II) complexes, which have singlet ground states, so the signal disappears as the sample is cooled to He temperatures. Similarly, in fullerene chemistry it was found that the EPR signal from $C_{60}^{2-}$ is due to a thermally excited triplet state and could only be seen in a restricted temperature range (Trulove et al., 1995).

Even if the goal is to measure relaxation times with pulsed EPR, one would usually start with CW survey spectra to ensure the integrity of the sample. Usually, pulsed EPR is less sensitive on a molar spin basis than CW EPR when both are optimized, because of the better noise filtering that can be done in CW EPR. However, if the relaxation time is long, it may not be possible to obtain a valid slow-passage CW EPR spectrum, and alternative techniques will be superior. Sometimes rapid-passage CW EPR, or dispersion spectra instead of absorption spectra, could be the EPR methodology of choice, especially for broad signals. However, for EPR signals that are narrow relative to the bandwidth of the resonator, pulsed FT EPR may provide both better S/N and more fidelity in line shape than CW EPR. This is the case for the signal of the E' center in the $\gamma$-irradiated fused quartz sample that has been proposed as a standard sample for time-domain EPR (Eaton and Eaton, 1993b; available from Wilmad Glass).

How many species are there? In many forms of spectroscopy one can monitor isosbestic points to determine limits on the number of species present. Note that, since the common CW EPR spectrum is a derivative display, one should in general not interpret spectral positions that are constant as a function of some variable as isosbestic points. Because of the derivative display they are isoclinic points—points of the same slope, not points of the same absorption. Integrated spectra need to be used to judge relative behavior of multiple species in the way that has become common in UV-visible electronic spectroscopy.

Relatively few paramagnetic species yield single-line EPR spectra, due to electron-nuclear coupling and electron-electron coupling. If the nuclear coupling is well resolved in a CW spectrum, it is a straightforward exercise to figure out the number of nuclear spins of each nuclear moment. However, unresolved coupling is common, and more powerful (and expensive) techniques have to be used. Electron-nuclear double resonance (ENDOR), which uses RF at the nuclear resonant frequency simultaneous with microwaves resonant at particular EPR transition, is a powerful tool for identifying nuclear couplings not resolved in CW EPR spectra (Kevan and Kispert, 1976; Box, 1977; Dorio and Freed, 1979; Schweiger, 1982; Piekara-Sady and Kispert, 1994; Goslar et al., 1994). A good introduction is given in (Atherton, 1993). Nuclear couplings too small to be observed in ENDOR commonly can be observed with spin echo EPR, where they result in electron spin echo envelope modulation (ESEEM, also abbreviated ESEM; Dikanov and Tsvetkov, 1992). This technique requires a pulsed EPR spectrometer. ESEEM is especially powerful for determining the nuclei in the environment of an electron spin, and is extensively used in metalloprotein studies.

Where are the species that yield the EPR signals? The ENDOR and ESEEM techniques provide answers about the immediate environments of the paramagnetic centers, but there may be a larger-scale question: are the paramagnetic species on the surface of the sample (Sueki et al., 1995), in the center, uniformly distributed, or localized in high-concentration regions throughout the sample? These questions also are answerable via various EPR techniques. EPR imaging can tell where in the sample the spins are located (Eaton et al., 1991). EPR imaging is analogous to NMR imaging, but is most often performed with slowly stepped magnetic field gradients and CW spectra, whereas most NMR imaging is done with pulsed-field gradients and time-domain (usually spin echo) spectra (see NUCLEAR MAGNETIC RESONANCE IMAGING). The differences in technique are due to the differences in spectral width and relaxation times. Low-frequency (L-band) spectrometers are available from JEOL and Bruker.

Uniform versus nonuniform distributions of paramagnetic centers can be ascertained by several observations. Nonuniformity is a more common problem than is recognized. Dopants (impurities) in solids usually distort the geometry of the lattice and sometimes the charge balance, so they are often not distributed in a truly random fashion. Frozen solutions (at whatever temperature) often exhibit partial precipitation and/or aggregation of the paramagnetic centers. The use of pure solvents rarely gives a good glass. For example, even flash freezing does not yield glassy water under normal laboratory conditions. A cosolvent such as glycerol is needed to form a glass. If there is even partial devitrification, it is likely that the paramagnetic species (an impurity in the solvent) will be localized

at grain boundaries. Local high concentrations sometimes can be recognized by broadening of the spectra (if one knows what the "normal" spectrum should look like), or even exchange narrowing of the spectra when concentrations are very high. Time-domain (pulse) spectra are more sensitive than CW spectra to these effects, and provide a convenient measure through a phenomenon called instantaneous diffusion (Eaton and Eaton, 1991), which is revealed when the two-pulse echo decay rate depends on the pulse power. The observation of instantaneous diffusion is an indication that the spin concentration is high. If the bulk concentration is known to be low enough that instantaneous diffusion is not likely to be observed (e.g., less than ~1 mM for broad spectra and less than ~0.1 mM for narrow spectra), then the observation of instantaneous diffusion indicates that there are locally high concentrations of spins—i.e., that the spin distribution is microscopically nonuniform.

Multiple species can show up as multiple time constants in pulsed EPR, but there may be reasons other than multiple species for the multiple time constants. Many checks are needed, and the literature of time-domain EPR should be consulted (Standley and Vaughan, 1969; Muus and Atkins, 1972; Kevan and Schwartz, 1979; Dikanov and Tsvetkov, 1992; Kevan and Bowman, 1990).

One of the most common questions is "how long will it take to run an EPR spectrum?" The answer depends strongly on what one wants to learn from the sample, and can range from a few minutes to many weeks. Even the simple question as to whether there are any unpaired electrons present may take quite a bit of effort to answer, unless one already knows a lot about the sample. The spins may have relaxation times so long that they are difficult to observe without saturation (e.g., defect centers in solids) or so short that they cannot be observed except at very low temperature, where the relaxation times become longer [e.g., high-spin Co(II) in many environments]. At the other extreme, column fractions of a nitroxyl-spin-labeled protein can be monitored for radicals about as fast as the samples can be put in the spectrometer. This is an example of an application that could be automated. Similarly, alanine-based radiation dosimeters, if of a standard geometry, can be loaded automatically into a Bruker spectrometer designed for this application, and the spectra can be run in a few minutes. If, on the other hand, one wants to know the concentration of high-spin Co(II) in a sample, the need for quantitative sample preparation, accurate cryogenic temperature control, careful background subtraction, and skillful setting of instrument parameters leads to a rather time-consuming measurement.

Current research on EPR is usually reported in the *Journal of Magnetic Resonance, Applied Magnetic Resonance, Chemical Physics Letters, Journal of Chemical Physics,* and journals focusing on application areas, such as *Macromolecules, Journal of Non-Crystalline Solids, Inorganic Chemistry,* and numerous biochemical journals. Examination of the current literature will suggest applications of EPR to materials science beyond those briefly mentioned in this unit. The *Specialist Periodical Report* on ESR of the Royal Society of Chemistry is the best source of annual updates on progress in EPR.

Reports of two workshops on the future of EPR (in 1987 and 1992) and a volume celebrating the first 50 years of EPR (Eaton et al., 1998) provide a vision of the future directions of the field (Eaton and Eaton, 1988b, 1995b).

Finally, it should be pointed out that merely obtaining an EPR spectrum properly, and characterizing its dependence on the factors discussed above, is just the first step. Its interpretation in terms of the physical properties of the material of interest is the real intellectual challenge and payoff.

## LITERATURE CITED

Abragam, A. and Bleaney, B. 1970. Electron Paramagnetic Resonance of Transition Ions. Oxford University Press, Oxford.

Alger, R. S. 1968. Electron Paramagnetic Resonance: Techniques and Applications. Wiley-Interscience, New York.

Atherton, N. M. 1993. Principles of Electron Spin Resonance. Prentice Hall, London.

Baden-Fuller, A. J. 1990. Microwaves: An Introduction to Microwave Theory and Techniques, 3rd ed. Pergamon Press, Oxford.

Bencini, A. and Gatteschi, D. 1990. EPR of Exchange Coupled Systems. Springer-Verlag, Berlin.

Berliner, L. J., ed., 1976. Spin Labeling: Theory and Applications. Academic Press, New York.

Berliner, L. J., ed., 1979. Spin Labeling II. Academic Press, New York.

Berliner, L. J. and Reuben, J., eds., 1989. Spin Labeling: Theory and Applications. Plenum, New York.

Box, H. C. 1977. Radiation Effects: ESR and ENDOR Analysis. Academic Press, New York.

Byrn, M. P. and Strouse, C. E. 1983. *g*-Tensor determination from single-crystal ESR data. *J. Magn. Reson.* 53:32–39.

Carrington, A. and McLachlan, A. D. 1967. Introduction to Magnetic Resonance. Harper and Row, New York.

Corradi, G., Söthe, H., Spaeth, J.-M., and Polgar, K. 1991. Electron spin resonance and electron-nuclear double resonance investigation of a new $Cr^{3+}$ defect on an Nb site in $LiNbO_3$: Mg:Cr. *J. Phys. Condens. Matter* 3:1901–1908.

Czoch, R. and Francik, A. 1989. Instrumental Effects of Homodyne Electron Paramagnetic Resonance Spectrometers. Wiley-Halsted, New York.

Dalal, D. P., Eaton, S. S., and Eaton, G. R. 1981. The effects of lossy solvents on quantitative EPR studies. *J. Magn. Reson.* 44:415–428.

Dalton, L. R., ed., 1985. EPR and Advanced EPR Studies of Biological Systems. CRC Press, Boca Raton, Fla.

Date, M., ed., 1983. High Field Magnetism. North-Holland Publishing Co., Amsterdam.

Dikanov, S. A. and Tsvetkov, Yu. D. 1992. Electron Spin Echo Envelope Modulation (ESEEM) Spectroscopy. CRC Press, Boca Raton, Fla.

Dorio, M. M. and J. H. Freed, eds., 1979. Multiple Electron Resonance Spectroscopy. Plenum Press, New York.

Du, J.-L., Eaton, G. R., and Eaton, S. S. 1995. Temperature, orientation, and solvent dependence of electron spin lattice relaxation rates for nitroxyl radicals in glassy solvents and doped solids. *J. Magn. Reson.* A115:213–221.

Eaton, S. S. and Eaton, G. R. 1980. Signal area measurements in EPR. *Bull. Magn. Reson.* 1:130–138.

Eaton, G. R. and Eaton, S. S. 1988a. EPR imaging: progress and prospects. *Bull. Magn. Reson.* 10:22–31.

Eaton, G. R. and Eaton S. S. 1988b. Workshop on the future of EPR (ESR) instrumentation: Denver, Colorado, August 7, 1987. *Bull. Magn. Reson.* 10:3–21.

Eaton, G. R. and Eaton, S. S. 1990a. Electron paramagnetic resonance. *In* Analytical Instrumentation Handbook (G. W. Ewing, ed.) pp. 467–530. Marcel Dekker, New York.

Eaton, S. S. and Eaton, G. R. 1990b. Electron spin resonance imaging. *In* Modern Pulsed and Continuous-Wave Electron Spin Resonance (L. Kevan and M. Bowman, eds.) pp. 405–435 Wiley Interscience, New York.

Eaton, G. R., Eaton, S., and Ohno, K., eds. 1991. EPR Imaging and in Vivo EPR, CRC Press, Boca Raton, Fla.

Eaton, S. S. and Eaton, G. R. 1991. EPR imaging, *In* Electron Spin Resonance Specialist Periodical Reports (M. C. R. Symons, ed.) 12b:176–190. Royal Society of London, London.

Eaton, S. S. and Eaton, G. R. 1992. Quality assurance in EPR. *Bull. Magn. Reson.* 13:83–89.

Eaton, G. R. and Eaton, S. S. 1993a. Electron paramagnetic resonance imaging. *In* Microscopic and Spectroscopic Imaging of the Chemical State (M. Morris, ed.) pp. 395–419. Marcel Dekker, New York.

Eaton, S. S. and Eaton, G. R. 1993b. Irradiated fused quartz standard sample for time domain EPR. *J. Magn. Reson.* A102:354–356.

Eaton, S. S. and Eaton, G. R. 1993c. Applications of high magnetic fields in EPR spectroscopy. *Magn. Reson. Rev.* 16:157–181.

Eaton, G. R. and Eaton, S. S. 1995a. Introduction to EPR imaging using magnetic field gradients. *Concepts in Magnetic Resonance* 7:49–67.

Eaton, S. S. and Eaton, G. R. 1995b. The future of electron paramagnetic resonance spectroscopy. *Bull. Magn. Reson.* 16:149–192.

Eaton, S. S. and Eaton, G. R. 1996a. Electron spin relaxation in discrete molecular species. *Current Topics in Biophysics* 20:9–14.

Eaton, S. S. and Eaton, G. R. 1996b. EPR imaging. *In* Electron Spin Resonance Specialist Periodical Reports (M. C. R. Symons, ed.) 15:169–185. Royal Society of London, London.

Eaton, S. S. and Eaton, G. R. 1997a. Electron paramagnetic resonance. *In* Analytical Instrumentation Handbook (G. W. Ewing, ed.), 2nd ed. pp. 767–862. Marcel Dekker, New York.

Eaton, S. S. and Eaton, G. R., 1997b, EPR methodologies: Ways of looking at electron spins. *EPR Newsletter* 9:15–18.

Eaton, G. R., Eaton, S. S., and Salikhov, K., eds. 1998. Foundations of Modern EPR. World Scientific Publishing, Singapore.

Eaton, G. R. and Eaton, S. S. 1999a. ESR imaging. *In* Handbook of Electron Spin Resonance (C. P. Poole, Jr., and H. A. Farach, eds.), vol 2. Springer-Verlag, New York.

Eaton, S. S. and Eaton, G. R. 1999b. Magnetic fields and high frequencies in ESR spectroscopy. *In* Handbook of Electron Spin Resonance (C. P. Poole, Jr., and H. A. Farach, eds.), vol 2. Springer-Verlag, New York.

Goldberg, I. B. and Crowe, H. R. 1977. Effect of cavity loading on analytical electron spin resonance spectrometry. *Anal. Chem.* 49:1353.

Goslar, J., Piekara-Sady, L. and Kispert, L. D. 1994. ENDOR data tabulation. *In* Handbook of Electron Spin Resonance (C. P. Poole, Jr., and H. A. Farach, eds.). AIP Press, NY.

Hyde, J. S. and Froncisz, W. 1982. The role of microwave frequency in EPR spectroscopy of copper complexes. *Ann. Rev. Biophys. Bioeng.* 11:391–417.

Hyde, J. S. and Froncisz, W. 1986. Loop gap resonators. *In* Electron Spin Resonance Specialist Periodical Reports (M. C. R. Symons, ed.) 10:175–185. Royal Society, London.

Hyde, J. S. and Subczynski, W. K. 1989. Spin-label oximetry. *In* Spin Labeling: Theory and Applications, (L. J. Berliner, and J. Reuben, eds.) Plenum, New York.

Ikeya, M. 1993. New Applications of Electron Spin Resonance. Dating, Dosimetry, and Microscopy. World Scientific, Singapore.

Kevan, L. and Kispert, L. D. 1976. Electron Spin Double Resonance Spectroscopy. John Wiley & Sons, New York.

Kevan, L. and Bowman, M. K., eds. 1990. Modern Pulsed and Continuous-Wave Electron Spin Resonance. Wiley-Interscience, New York.

Kevan, L. and Schwartz, R. N., eds. 1979. Time Domain Electron Spin Resonance. John Wiley & Sons, New York.

Kirste, B., 1994. Computer techniques. *In* Handbook of Electron Spin Resonance (C. P. Poole, Jr., and H. A. Farach, eds.). AIP Press, NY.

Krinichnyi, V. I. 1995. 2-mm Wave Band EPR Spectroscopy of Condensed Systems. CRC Press, Boca Raton, Fla.

Lancaster, G., 1967. Electron Spin Resonance in Semiconductors. Plenum, New York.

Muus, L. T. and Atkins, P. W., eds. 1972. Electron Spin Relaxation in Liquids, Plenum Press, New York.

Pake, G. E. and Estle, T. L. 1973. The Physical Principles of Electron Paramagnetic Resonance, 2nd ed. W. A. Benjamin, Reading, Mass.

Piekara-Sady, L. and Kispert, L. D. 1994. ENDOR spectroscopy. *In* Handbook of Electron Spin Resonance (C. P. Poole, Jr., and H. A. Farach, eds.). AIP Press, NY.

Pilbrow, J. R., 1990. Transition Ion Electron Paramagnetic Resonance. Oxford University Press, London.

Poole, C. P., Jr., 1967. Electron Spin Resonance: A Comprehensive Treatise on Experimental Techniques, pp. 398–413. John Wiley & Sons, New York.

Poole, C. P., Jr., 1983. Electron Spin Resonance: A Comprehensive Treatise on Experimental Techniques, 2nd ed. Wiley-Interscience, New York.

Quine, R. W., Eaton, G. R., and Eaton, S. S. 1987. Pulsed EPR spectrometer. *Rev. Sci. Instrum.* 58:1709–1724.

Quine, R. W., Eaton, S. S., and Eaton, G. R. 1992. A saturation recovery electron paramagnetic resonance spectrometer. *Rev. Sci. Instrum.* 63:4251–4262.

Quine, R. W., Rinard, G. A., Ghim, B. T., Eaton, S. S. and Eaton, G. R. 1996. A 1-2 GHz pulsed and continuous wave electron paramagnetic resonance spectrometer. *Rev. Sci. Instrum.* 67:2514–2527.

Rinard, G. A., Quine, R. W., Eaton, S. S. and Eaton, G. R. 1993. Microwave coupling structures for spectroscopy. *J. Magn. Reson.* A105:134–144.

Rinard, G. A., Quine, R. W., Eaton, S. S., Eaton, G. R., and Froncisz, W. 1994. Relative benefits of overcoupled resonators vs. inherently low-Q resonators for pulsed magnetic resonance. *J. Magn. Reson.* A108:71–81.

Rinard, G. A., Quine, R. W., Ghim, B. T., Eaton, S. S., and Eaton, G. R. 1996a. Easily tunable crossed-loop (bimodal) EPR resonator. *J. Magn. Reson.* A122:50–57.

Rinard, G. A., Quine, R. W., Ghim, B. T., Eaton, S. S., and Eaton, G. R. 1996b. Dispersion and superheterodyne EPR using a bimodal resonator. *J. Magn. Reson.* A122:58–63.

Rinard, G. A., Eaton, S. S., Eaton, G. R., Poole, C. P., Jr., and Farach, H. A., 1999. Sensitivity of ESR Spectrometers: Signal, Noise, and Signal-to-Noise. *In* Handbook of Electron Spin Resonance (C. P. Poole, Jr., and H. A. Farach, eds.), vol. 2. Springer-Verlag, New York.

Rudowicz, C. Z., Yu, K. N., and Hiraoka, H., eds. 1998. Modern Applications of EPR/ESR: From Biophysics to Materials Science. Springer-Verlag, Singapore.

Schweiger, A. 1982. Electron Nuclear Double Resonance of Transition Metal Complexes with Organic Ligands, Structure and Bonding vol. 51. Springer-Verlag, New York.

Standley, K. J. and Vaughan, R. A. 1969. Electron Spin Relaxation Phenomena in Solids. Plenum Press, New York.

Stevens, K. W. H. 1997. Magnetic Ions in Crystals. Princeton University Press, Princeton, N.J.

Sueki, M., Eaton, G. R., and Eaton, S. S. 1990. Electron spin echo and CW perspectives in 3D EPR imaging. *Appl. Magn. Reson.* 1:20–28.

Sueki, M., Austin, W. R., Zhang, L., Kerwin, D. B., Leisure, R. G., Eaton, G. R., and Eaton, S. S. 1995. Determination of depth profiles of E' defects in irradiated vitreous silica by electron paramagnetic resonance imaging. *J. Appl. Phys.* 77:790–794.

Swartz, H. M., Bolton, J. R., and Borg, D. C., eds. 1972. Biological Applications of Electron Spin Resonance, John Wiley & Sons, New York.

Trulove, P. C., Carlin, R. T., Eaton, G. R., and Eaton, S. S. 1995. Determination of the Singlet-Triplet Energy Separation for $C_{60}^{2-}$ in DMSO by Electron Paramagnetic Resonance. *J. Am. Chem. Soc.* 117:6265–6272.

Vancamp, H. L. and Heiss, A. H. 1981. Computer applications in electron paramagnetic resonance. *Magn. Reson. Rev.* 7:1–40.

Weil, J. A., Bolton, J. R., and Wertz, J. E. 1994. Electron Paramagnetic Resonance: Elementary Theory and Practical Applications, John Wiley & Sons, New York.

Wilmshurst, T. H. 1968. Electron Spin Resonance Spectrometers, Plenum, N. Y., see especially ch. 4.

## KEY REFERENCES

Poole, 1967. See above.

Poole, 1983. See above.

*The two editions of this book provide a truly comprehensive coverage of EPR. They are especially strong in instrumentation and technique, though most early applications to materials science are cited also. This is "the bible" for EPR spectroscopists.*

Eaton and Eaton, 1990a. See above.

Eaton and Eaton, 1997a. See above.

*These introductory chapters in two editions of the Analytical Instrumentation Handbook provide extensive background and references. They are a good first source for a person with no background in the subject.*

Weil et al., 1994. See above.

Atherton, 1993. See above.

*These two very good comprehensive textbooks have been updated recently. They assume a fairly solid understanding of physical chemistry. Detailed reviews of these texts are given in:*

Eaton, G.R. 1995. *J. Magn. Reson.* A113:135-136.

*Detailed review of Principles of Electron Spin Resonance (Atherton, 1993).*

Eaton, S.S. 1995. *J. Magn. Reson.* A113:137.

*Detailed book review of Electron Paramagnetic Resonance: Theory and Practical Applications (Weil et al., 1994).*

## INTERNET RESOURCES

http://ierc.scs.uiuc.edu

*The International EPR (ESR) Society strives to broadly disseminate information concerning EPR spectroscopy. This Web site contains a newsletter published for the Society by the staff of the Illinois EPR Research Center, University of Illinois. Supported by National Institutes of Health.*

http://www.biophysics.mcw.edu

*EPR center at the Medical College of Wisconsin. Supported by National Institutes of Health.*

http://spin.aecom.yu.edu

*EPR center at Albert Einstein College of Medicine. Supported by National Institutes of Health.*

epr-list@xenon.che.ilstu.edu

*E-mail service for rapid communication, run by Professor P.D. Morse, Illinois State University.*

http://www.du.edu/~seaton

*Web page for the authors' lab. The above Centers and the authors' lab attempt to link to each other and collectively provide many references to recent papers, notices of meetings, and background on magnetic resonance. There now exist many electronic data bases that one can search to find recent papers on EPR. However, only a small portion of the relevant papers will be found by using just "EPR" as the search term. Our experience is that at least the following set of keywords have to be used to retrieve the bulk of the papers: ESR, EPR, electron spin, electron paramag\*, ESEEM, ENDOR, and echo modulation. Other relevant papers will be found using keywords for the application field, since words relating to EPR may not occur in the title or abstract of the paper even if EPR is a primary tool of the study.*

## APPENDIX:
## INSTRUMENTATION

EPR measurements require a significant investment in instrumentation. There are three fundamental parts to an EPR spectrometer: the microwave system, the magnet system, and the data acquisition and analysis system (Figure 2). Brief discussions of spectrometer design are in introductory texts (Weil et al., 1994, Eaton and Eaton, 1990, 1997). Extensive discussions of microwaves relevant to EPR are available (Poole, 1967, 1983; Baden-Fuller, 1990). Some recent papers give details on pulsed EPR design (Quine et al., 1987, 1992, 1996). The magnets used in EPR range from ~3-in. to 12-in. diameter pole faces, the larger magnets providing better magnetic field homogeneity for larger gaps. Larger magnet gaps permit a wider range of accessory instrumentation for various measurements, such as controlling the sample temperature. A 3-in. magnet weighs roughly 300 pounds, and a 12-in. magnet weighs roughly 4000 pounds. Power supplies for the magnet, and cooling water systems, which optimally are temperature-controlled-recirculating deionized water systems, also require an investment in floor space. The microwave system usually sits on a table supported on the magnet for stability.

EPR spectrometers are expensive to purchase and to operate. They require a well trained operator, and except for a few tabletop models, they take up a lot of floor space and use substantial electrical power and cooling water for the electromagnet. Newer EPR spectrometer systems are operated via a computer, so the new operator who is familiar with use of computers can learn operation much faster than in the older equipment where remembering a critical sequence of manual valves and knobs was part of doing spectroscopy. Even so, the use of a high-$Q$ resonator results in much stronger interaction between the sample and the spectrometer than in most other analytical techniques, and a lot of judgment is needed to get useful, especially quantitatively meaningful results.

### Commercial Instruments

Most materials science needs will be satisfied by commercial spectrometers. Some references are provided below to recent spectrometers built by various research labs. These provide some details about design that may be helpful to the reader who wants to go beyond the level of this article. The following brief outline of commercial instruments is intended to guide the reader to the range of spectrometers available. The largest manufacturers—Bruker Instruments EPR Division, and JEOL—market general-purpose spectrometers intended to fulfill most analytical needs. The focus is on X-band (~9 to 10 GHz) CW spectrometers, with a wide variety of resonators to provide for many types of samples. Accessories facilitate control of the sample temperature from <4 to ~700 K. Magnets commonly range from 6- to 12-in. pole face diameters. Smaller tabletop spectrometers are available from Bruker, JEOL, and Resonance Instruments. Some of these have permanent magnets and sweep coils for applications that focus on spectra near $g = 2$, but some have electromagnets permitting wide field sweeps. Bruker makes one small system optimized for quantitation of organic radicals and defect centers, such as for dosimetry.

Bruker and JEOL market pulsed, time-domain spectrometers as well as CW spectrometers. These suppliers also market spectrometers for frequencies lower than X-band, which are useful for study of lossy samples. Bruker and Resonance Technologies market high-frequency, high-field EPR spectrometers, which require superconducting magnets, not resistive electromagnets.

There are numerous vendors of accessories and supplies essential for various aspects of EPR measurements, including Oxford Instruments (cryostats), Cryo Industries (cryostats), Wilmad (sample tubes, quartz Dewars, standard samples), Oxis (spin traps), Medical Advances (loop gap resonators), Research Specialties (accessories and service), Summit Technology (small spectrometers, accessories), Scientific Software Services (data acquisition), Cambridge Isotope Labs (labeled compounds), and CPI (klystrons). Addresses and up-to-date information can be accessed via the EPR Newsletter.

SANDRA S. EATON
GARETH R. EATON
University of Denver
Denver, Colorado

# CYCLOTRON RESONANCE

## INTRODUCTION

Cyclotron resonance (CR) is a method for measuring the effective masses of charge carriers in solids. It is by far the most direct and accurate method for providing such information. In the simplest description, the principle of the method can be stated as follows. A particle of effective mass $m^*$ and charge $e$ in a DC magnetic field $B$ executes a helical motion around $B$ with the cyclotron frequency $\omega_c = eB/m^*$. If, at the same time, an AC electric field of frequency $\omega = \omega_c$ is applied to the system, perpendicular to $B$, the particle will resonantly absorb energy from the AC field. Since $B$ and/or $\omega$ can be continuously swept through the resonance and known to a very high degree of accuracy, $m^*$ can be directly determined with high accuracy by $m^* = eB/\omega$.

In crystalline solids, the dynamics of charge carriers (or Bloch electrons) can be most conveniently described by the use of the effective mass tensor, defined for simple nondegenerate bands as

$$(\bar{m})_{\mu v} = \left( \frac{1}{\hbar^2} \frac{\partial^2 E(\vec{k})}{\partial k_\mu \partial k_v} \right)^{-1}, \qquad (\mu, v = 1, 2, 3) \qquad (1)$$

where $E = E(\vec{k})$ is the energy dispersion relation near the band edge ($\vec{p} = \hbar \vec{k}$ is the crystal momentum). Hence, the primary purpose of CR in solids is to determine the components of the effective mass tensor, or the curvature of the energy surface, at the extrema of the conduction and valence bands (or at the Fermi surface in the case of a metal). In the 1950s, the first CR studies were carried out in germanium and silicon crystals (Dresselhaus et al., 1953, 1955; Lax et al., 1953), which, in conjunction with the effective-mass theory (e.g., Luttinger and Kohn, 1955; Luttinger, 1956), successfully determined the band-edge parameters for these materials with unprecedented accuracy. Since then CR has been investigated in a large number of elementary and compound materials and their alloys and heterostructures.

Quantum mechanically, the energy of a free electron in a magnetic field is quantized as $E_N = (N + 1/2)\hbar\omega_c$, where $N = 0, 1, 2, \ldots \infty$ and $\hbar\omega_c$ is called the cyclotron energy. These equally-spaced energy levels, or an infinite ladder, are the well-known Landau levels (see, e.g., Kittel, 1987). At high magnetic fields and low temperatures ($T$), where $k_B T < \hbar\omega_c$ ($k_B$ is the Boltzmann constant), this magnetic quantization becomes important, and CR may then be viewed as a quantum transition between adjacent Landau levels ($\Delta N = \pm 1$).

In real solids, Landau levels are generally not equally-spaced since the energy dispersion relation, $E$ versus $\vec{k}$, is not generally given by a simple parabolic dispersion relation $E(\vec{k}) = \hbar^2 |\vec{k}|^2 / 2m^*$. The degree of nonparabolicity and anisotropy depends on the material, but, in general, the effective mass is energy-dependent as well as direction-dependent. Landau levels for free carriers in solids cannot be obtained analytically, but several useful approximation models exist (e.g., Luttinger, 1956; Bowers and Yafet,

1959; Pidgeon and Brown, 1966). These calculations could be complex, especially when one is concerned with degenerate bands such as the valence bands of group IV, III-V and II-VI semiconductors. However, a detailed comparison between theory and experiment on quantum CR can provide a critical test of the band theory of solids (e.g., Suzuki and Hensel, 1974; Hensel and Suzuki, 1974).

As a secondary purpose, one can also use CR to study carrier scattering phenomena in solids by examining the scattering lifetime $\tau$ (the time between collisions, also known as the collision time or the transport/momentum relaxation time), which can be found from the linewidth of CR peaks. In the classical regime, where most electrons reside in states with high Landau indices, $\tau$ is directly related to the static (or DC) conductivity in the absence of the magnetic field. The temperature dependence of $\tau$ then shows markedly different characteristics for phonon scattering, impurity scattering, and scattering from various imperfections. However, in the quantum regime, where most electrons are in the first few Landau levels, the effect of the magnetic field on scattering mechanisms is no longer negligible and $\tau$ loses its direct relationship with the DC mobility (for theory, see, e.g., Kawabata, 1967).

As the frequency of scattering events (or the density of scatterers) increases, CR linewidth increases, and, eventually, CR becomes unobservable when scattering occurs too frequently. More quantitatively, in order to observe CR, $\tau$ must be long enough to allow the electron to travel at least $1/2\pi$ of a revolution between two scattering events, i.e.,

$$\tau > \frac{T_c}{2\pi} = \frac{1}{\omega_c}$$

or

$$\omega_c \tau = \frac{eB}{m^*} \tau = \frac{e\tau}{m^*} B = \mu B > 1 \qquad (2)$$

where $T_c$ is the period of cyclotron motion and $\mu = e\tau/m^*$ is the DC mobility of the electron. Let us examine this CR observability condition for a realistic set of parameters. If $m^* = 0.1 m_0$ (where $m_0 = 9.1 \times 10^{-31}$ kg) and $B = 1$ Tesla, then $\omega_c = eB/m^* \approx 2 \times 10^{12}$ sec$^{-1}$. Thus one needs a microwave field with a frequency of $f_c = \omega_c/2\pi \approx 3 \times 10^{11}$ Hz = 300 GHz (or a wavelength of $\lambda_c = c/f_c \approx 1$ mm). Then, in order to satisfy Equation 2, one needs a minimum mobility of $\mu = 1$ m$^2$/V-sec = $1 \times 10^4$ cm$^2$/V-sec. This value of mobility can be achieved only in a limited number of high-purity semiconductors at low temperatures, thereby posing a severe limit on the observations of microwave CR.

From the resonance condition $\omega = \omega_c$, it is obvious that if a higher magnetic field is available (see GENERATION AND MEASUREMENT OF MAGNETIC FIELDS) one can use a higher frequency (or a shorter wavelength), which should make Equation 2 easier to satisfy. Hence modern CR methods almost invariably use far-infrared (FIR) [or Terahertz (THz)] radiation instead of microwaves. Strong magnetic fields are available either in pulsed form (up to $\sim 10^3$ T) or in steady form by superconducting magnets (up to $\sim 20$ T), water-cooled magnets (up to $\sim 30$ T), or hybrid magnets (up to $\sim 45$ T). In these cases, even at room temperature, Equation 2 may be fulfilled. Here we are only concerned with the methods of FIR-CR. The reader particularly interested in microwave CR is referred to Lax and Mavroides (1960).

Although this unit is mainly concerned with the simplest case of free carrier CR in bulk semiconductors, one can also study a wide variety of FIR magneto-optical phenomena with essentially the same techniques as CR. These phenomena ("derivatives" of CR) include: (a) spin-flip resonances, i.e., electron spin resonance and combined resonance, (b) resonances of bound carriers, i.e., internal transitions of shallow impurities and excitons, (c) polaronic coupling, i.e., resonant interactions of carriers with phonons and plasmons, and (d) 1-D and 2-D magneto-plasmon excitations. It should be also mentioned that in 2-D systems in the magnetic quantum limit, there are still unresolved issues concerning the effects of disorder and electron-electron interactions on CR (for a review, see, e.g., Petrou and McCombe, 1991; Nicholas, 1994).

It is important to note that all the early CR studies were carried out on semiconductors, not on metals. This is because of the high carrier concentrations present in metals, which preclude direct transmission spectroscopy except in the case of very thin films where thickness is less than the depth of penetration (skin depth) of the electromagnetic fields. In bulk metals, special geometries are thus required to detect CR, the most important of which is the Azbel-Kaner geometry (Azbel and Kaner, 1958). In this geometry, both the DC magnetic field $\vec{B}$ and the AC electric field $\vec{E}$ are applied parallel to the sample surface, either $\vec{B}//\vec{E}$ or $\vec{B} \perp \vec{E}$. The electrons then execute a spiral motion along $\vec{B}$, moving in and out of the skin depth, where $\vec{E}$ is present. Thus, whenever the electron enters the skin depth, it is accelerated by $\vec{E}$ and if the phase of $\vec{E}$ is the same every time the electron enters the skin depth, then the electron can resonantly absorb energy from the AC field. The condition for resonance here is $n\omega_c = \omega (n = 1, 2, 3, \ldots)$. For more details on CR in metals, see, e.g., Mavroides (1972).

Many techniques can provide information on effective masses, but none can rival CR for directness and accuracy. Effective masses can be estimated from the temperature dependence of the amplitude of the galvanomagnetic effects, i.e., the Shubnikov–de Haas and de Haas–van Alphen effects. Interband magneto-optical absorption can determine the reduced mass $\mu = (1/m_e + 1/m_h)^{-1}$ of photo-created electrons and holes. Measurements of the infrared Faraday rotation effect due to free carriers can provide information on the anisotropy of elliptical equi-energy surfaces. The temperature dependence of electronic specific heat provides a good measure of the density of levels at the Fermi level, which in turn is proportional to the effective mass. Nonresonant free carrier absorption (see CARRIER LIFETIME: FREE CARRIER ABSORPTION, PHOTOCONDUCTIVITY, AND PHOTOLUMINESCENCE) can be used to estimate effective masses, but, of course, this simply represents a tail or shoulder of a CR absorption curve.

It is worth pointing out here that several different definitions of effective masses exist in the literature and care must be taken when one discusses masses. The band-edge mass, defined as in Equation 1 at band extrema (e.g., $\vec{k} = 0$ in most semiconductors), is the most important band parameter to characterize a material. The specific heat mass is directly related to the density of states at the Fermi level, and is thus also called the density-of-states mass. The cyclotron mass is defined as $m_c^* = (\hbar^2/2\pi)\partial A/\partial E$, where $A$ is the momentum-space area enclosed by the cyclotron orbit; this definition follows naturally from calculation of $\omega_c$ in momentum space. The spectroscopic mass can be defined for any resonance peak, and is identical to the cyclotron mass when the resonance is due to free-carrier CR (also see Data Analysis and Initial Interpretation).

The basic theory and experimental methods of cyclotron resonance is presented in this unit. Basic theoretical background will first be presented (see Principles of the Method). A detailed description will be given of the actual experimentation procedures (see Practical Aspects of the Method). Finally, typical data analysis procedures are presented (see Data Analysis and Initial Interpretation).

## PRINCIPLES OF THE METHOD

As described in the Introduction, the basic physics of CR is the interaction of electromagnetic (EM) radiation with charge carriers in a magnetic field. Here, more quantitative descriptions of this physical phenomenon will be presented, based on (1) a semiclassical model and (2) a quantum-mechanical model. In analyzing CR data, judicious combination, modification, and refinement of these basic models are necessary, depending upon the experimental conditions and the material under study, in order to obtain the maximum amount of information from a given set of data.

The most commonly used method for describing the motion of charge carriers in solids perturbed by external fields is the effective-mass approximation (EMA), developed by many workers in the early history of the quantum theory of solids. The beauty of this method lies in the ability to replace the effect of the lattice periodic potential on electron motion by a mass tensor, the elements of which are determined by the unperturbed band structure. In other words, instead of considering electrons in a lattice we may consider the motion of effective-mass particles, which obey simple equations of motion in the presence of external fields. Rigorous treatments and full justification of the EMA can be found in the early original papers (e.g., Wannier, 1937; Slater, 1949; Luttinger, 1951; Luttinger and Kohn, 1955).

### Semiclassical Drude Description of CR

In many cases it is satisfactory to use the semiclassical Drude model (e.g., Aschcroft and Mermin, 1976) to describe the conductivity tensor of free carriers in a magnetic field (see MAGNETOTRANSPORT IN METALS AND ALLOYS). In this model each electron is assumed to independently obey the equation of motion

$$\overleftrightarrow{m}^* \cdot \frac{d\vec{v}}{dt} + \overleftrightarrow{m}^* \cdot \frac{\vec{v}}{\tau} = e(\vec{E} + \vec{v} \times \vec{B}) \tag{3}$$

where $\overleftrightarrow{m}^*$ is the effective-mass tensor (see, e.g., Equation 1 for nondegenerate bands), $\vec{v}$ is the drift velocity of the electrons, $\tau$ is the scattering lifetime (which is assumed to be a constant), $\vec{E}$ is the AC electric field, and $\vec{B}$ is the DC magnetic field. The complex conductivity tensor $\overleftrightarrow{\sigma}$ is then defined by $\vec{J} = ne\vec{v} = \overleftrightarrow{\sigma} \cdot \vec{E}$, where $\vec{J}$ is the current density and $n$ is the carrier density. Assuming that the AC field and the drift velocity have the harmonically varying form, i.e., $\vec{E}(t) = \vec{E}_0 \exp(-i\omega t)$, $\vec{v}(t) = \vec{v}_0 \exp(-i\omega t)$, one can easily solve Equation 3. In particular, for cubic materials and for $\vec{B} \parallel \hat{z}$, $\overleftrightarrow{\sigma}$ is given by

$$\overleftrightarrow{\sigma} = \begin{pmatrix} \sigma_{xx} & \sigma_{xy} & 0 \\ \sigma_{yx} & \sigma_{yy} & 0 \\ 0 & 0 & \sigma_{zz} \end{pmatrix} \tag{4a}$$

$$\sigma_{xx} = \sigma_{yy} = \sigma_0 \frac{i\omega\tau + 1}{(i\omega\tau + 1)^2 + \omega_c^2\tau^2} \tag{4b}$$

$$\sigma_{xy} = -\sigma_{yx} = \sigma_0 \frac{\omega_c\tau}{(i\omega\tau + 1)^2 + \omega_c^2\tau^2} \tag{4c}$$

$$\sigma_{zz} = \sigma_0 \frac{1}{i\omega\tau + 1} \tag{4d}$$

$$\sigma_0 = ne\mu = \frac{ne^2\tau}{m^*} \tag{4e}$$

where $\mu$ is the carrier mobility and $\sigma_0$ is the DC conductivity.

Once we know the conductivity tensor, we can evaluate the power $P$ absorbed by the carriers from the AC field as

$$P = \langle \vec{J}(t) \cdot \vec{E}(t) \rangle = \frac{1}{2} Re(\vec{j} \cdot \vec{E}^*) \tag{5}$$

where $\langle \ldots \rangle$ represents the time average and $\vec{E}^*$ is the complex conjugate of $\vec{E}$. For an EM wave linearly polarized in the $x$-direction, i.e., $\vec{E} = (E_x, 0, 0)$, Equation 5 simplifies to

$$P = \frac{1}{2} Re(J_x E_x^*) = \frac{1}{2} |E_0|^2 Re(\sigma_{xx}) \tag{6}$$

Substituting part b of Equation 4 into Equation 6, we obtain

$$P(\omega) = \frac{1}{2} |E_0|^2 \sigma_0 Re\left\{ \frac{i\omega\tau + 1}{(i\omega\tau + 1)^2 + \omega_c^2\tau^2} \right\}$$
$$= \frac{1}{4} |E_0|^2 \sigma_0 \left[ \frac{1}{(\omega - \omega_c)^2\tau^2 + 1} + \frac{1}{(\omega + \omega_c)^2\tau^2 + 1} \right] \tag{7}$$

This is plotted in Fig. 1 for different values of the parameter $\omega_c\tau$. It is evident from this figure that the absorption peak occurs when $\omega = \omega_c$ and $\omega_c\tau > 1$.

Note that Equation 7 contains two resonances—one at $\omega = \omega_c$ and the other at $\omega = -\omega_c$. These two resonances

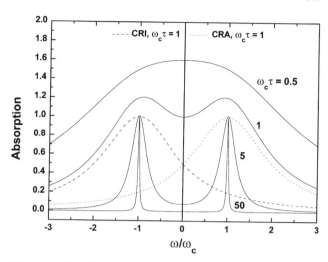

**Figure 1.** The CR absorption power versus $\omega$ for different values of $\omega_c\tau$. The traces are obtained from Equation 7. CR occurs at $\omega = \omega_c$ when $\omega_c\tau > 1$. The absorption is expressed in units of $|E_0|^2\sigma_0/4$.

correspond to the two opposite senses of circular polarization. It can be shown (see, e.g., Palik and Furdyna, 1970) that in the Faraday geometry $\vec{q} \parallel \vec{B} \parallel \hat{z}$, where $\vec{q}$ is the wave-vector of the EM wave) a 3-D magneto-plasma can support only two propagating EM modes represented by

$$\vec{E}_\pm = E_0 \frac{1}{\sqrt{2}} (\vec{e}_x \pm i\vec{e}_y) \exp[i(q_\pm z - \omega t)] \tag{8}$$

where $\vec{e}_x$ and $\vec{e}_y$ are the unit vectors in the $x$ and $y$ directions, respectively. The dispersion relations, $q_\pm$ versus $\omega$, for the two modes are obtained from

$$q_\pm = \frac{\omega N_\pm}{c} \tag{9a}$$

$$\begin{cases} N_\pm^2 = \kappa_\pm = \kappa_{xx} \pm ik_{xy} & (9b) \\ \kappa_{xx} = \kappa_l + \frac{i}{\omega\varepsilon_0}\sigma_{xx} = \kappa_l + \frac{i\sigma_0}{\omega\varepsilon_0}\frac{i\omega\tau + 1}{(i\omega\tau + 1)^2 + \omega_c^2\tau^2} & (9c) \\ \kappa_{xy} = \frac{i}{\omega\varepsilon_0}\sigma_{xy} = \frac{i\sigma_0}{\omega\varepsilon_0}\frac{\omega_c\tau}{(i\omega\tau + 1)^2 + \omega_c^2\tau^2} & (9d) \end{cases}$$

where $N_\pm$ is the complex refractive index, $\kappa_l$ is the relative dielectric constant of the lattice (assumed to be constant in the FIR), and $\kappa_{xx}$ and $\kappa_{xy}$ are components of the generalized dielectric tensor $\overleftrightarrow{\kappa} = \kappa_l \overleftrightarrow{I} + (i/\omega\varepsilon_0)\overleftrightarrow{\sigma}$, where $\overleftrightarrow{I}$ is the unit tensor and $\overleftrightarrow{\sigma}$ is the conductivity tensor (Equation 4). The positive sign corresponds to a circularly-polarized FIR field, rotating in the same sense as a negatively-charged particle, and is traditionally referred to as cyclotron-resonance-active (CRA) for electrons. Similarly, the negative sign represents the opposite sense of circular polarization, cyclotron-resonance-inactive (CRI) for electrons. The CRA mode for electrons is the CRI mode for holes, and vice versa. For linearly-polarized FIR radiation, which is an equal-weight mixture of the two modes, both terms in Equation 7 contribute to the absorption curve, as represented in Figure 1.

## Quantum Mechanical Description of CR

According to the EMA, if the unperturbed energy-momentum relation for the band $n$, $E_n(\vec{p})$, is known, then the allowed energies $E$ of a crystalline system perturbed by a uniform DC magnetic field $\vec{B} = \vec{\nabla} \times \vec{A}$ (where $\vec{A}$ is the vector potential) are given approximately by solving the effective Schrödinger equation

$$\hat{H} F_n(\vec{r}) = \hat{E}_n(-i\hbar\vec{\nabla} + e\vec{A}) F_n(\vec{r}) = E F_n(\vec{r}) \quad (10)$$

Here the operator $\hat{E}_n(-i\hbar\vec{\nabla} + e\vec{A})$ means that we first replace the momentum $\vec{p}$ in the function $E_n(\vec{p})$ with the kinematic (or mechanical) momentum $\vec{\pi} = \vec{p} + e\vec{A}$ (see, e.g., Sakurai, 1985) and then transform it into an operator by $\vec{p} \rightarrow -i\hbar\vec{\nabla}$. The function $F_n(\vec{r})$ is a slowly varying "envelope" wavefunction; the total wavefunction $\Psi(\vec{r})$ is given by a linear combination of the Bloch functions at $\vec{p} = 0$ (or the cell-periodic function) $\psi_{n0}(\vec{r})$

$$\Psi(\vec{r}) = \sum_n F_n(\vec{r}) \Psi_{n0}(\vec{r}) \quad (11)$$

For simplicity let us consider a 2-D electron in a conduction-band with a parabolic and isotropic dispersion $E_n(\vec{p}) = |\vec{p}|^2/2m^* = (P_x^2 + P_y^2)/2m^*$. The Hamiltonian in Equation 10 is then simply expressed as

$$\hat{H}_0 = \frac{|\vec{\pi}|^2}{2m^*} = \frac{1}{2m^*}(\pi_x^2 + \pi_y^2) \quad (12)$$

These kinematic momentum operators obey the following commutation relations

$$[x, \pi_x] = [y, \pi_y] = i\hbar \quad (13a)$$

$$[x, \pi_y] = [y, \pi_x] = [x, y] = 0 \quad (13b)$$

$$[\pi_x, \pi_y] = -i\hbar^2/\ell^2 \quad (13c)$$

where $\ell = (\hbar/eB)^{1/2}$ is the magnetic length, which measures the spatial extent of electronic (envelope) wavefunctions in magnetic fields—the quantum mechanical counterpart of the cyclotron radius $r_c$. Note the non-commutability between $\pi_x$ and $\pi_y$ (Equation 13, line c), in contrast to $p_x$ and $p_y$ in the zero magnetic-field case.

We now introduce the Landau level raising and lowering operators

$$\hat{a}_\pm = \frac{\ell}{\sqrt{2}\hbar}(\pi_x \pm i\pi_y) \quad (14)$$

for which we can show that

$$[\hat{a}_-, \hat{a}_+] = 1 \quad (15a)$$

$$[\hat{H}_0, \hat{a}_\pm] = \pm\hbar\omega_c\hat{a}_\pm \quad (15b)$$

Combining Equations 13 to 15, we can see that $\hat{a}_\pm$ connects the state $|N\rangle$ and the states $|N \pm 1\rangle$ and that $\hat{H}_0 = \hbar\omega_c(\hat{a}_+\hat{a}_- + 1/2)$, and hence the eigenenergies are

$$E = \left(N + \frac{1}{2}\right)\hbar\omega_c, \quad (N = 0, 1, \ldots) \quad (16)$$

These discrete energy levels are well-known as Landau levels.

If we now apply a FIR field with CRA (CRI) polarization (Equation 8) to the system, then, in the electric dipole approximation, the Hamiltonian becomes

$$\hat{H} = \frac{1}{2m^*}|\vec{p} + e\hat{A} + e\hat{A}'_\pm|^2 \approx \frac{1}{2m^*}|\vec{\pi}|^2 + \frac{e}{m^*}\vec{\pi} \cdot \vec{A}'_\pm = \hat{H}_0 + \hat{H}' \quad (17)$$

where, from Equation 8, the vector potential for the FIR field $\vec{A}'_\pm$ is given by

$$\vec{A}'_\pm = -\frac{E_0}{\sqrt{2}i\omega}(\vec{e}_x \pm i\vec{e}_y)\exp(-i\omega t) \quad (18)$$

Thus the perturbation Hamiltonian $\hat{H}'$ is given by

$$\begin{aligned}
\hat{H}' &= \frac{e}{m^*}(\pi_x A'_{\pm,x} + \pi_y A_{\pm,y}) \\
&= -\frac{eE_0}{\sqrt{2}i\omega m^*}(\pi_x \pm i\pi_y)\exp(-i\omega t) \\
&= -\frac{e\hbar E_0}{i\omega m^*\ell}\hat{a}_\pm\exp(-i\omega t) \quad (19)
\end{aligned}$$

We can immediately see that this perturbation, containing $\hat{a}_\pm$, connects the state $|N\rangle$ with the states $|N \pm 1\rangle$, so that a sharp absorption occurs at $\omega = \omega_c$.

## PRACTICAL ASPECTS OF THE METHOD

CR spectroscopy, or the more general FIR magneto-spectroscopy, is performed in two distinct ways—Fourier-transform magneto-spectroscopy (FTMS) and laser magneto-spectroscopy (LMS). The former is wavelength-dependent spectroscopy and the latter is magnetic-field-dependent spectroscopy. Generally speaking, these two methods are complimentary.

Narrow spectral widths and high output powers are two of the main advantages of lasers. The former makes LMS suitable for investigating spectral features that cannot be resolved by using FTMS, and the latter makes it suitable for studying absorption features in the presence of strong background absorption or reflection. It is also much easier to conduct polarization-dependent measurements by LMS than by FTMS. Moreover, LMS can be easily combined with strong pulsed magnets. Finally, with intense and short-pulse lasers such as the free-electron laser (FEL; see e.g., Brau, 1990), LMS can be extended to nonlinear and time-resolved FIR magneto-spectroscopy.

On the other hand, FTMS has some significant advantages with respect to LMS. First, currently available FIR laser sources (except for FELs) can produce only discrete wavelengths, whereas FTMS uses light sources that produce continuous spectra. Second, it is needless to say that LMS can only detect spectral features that are magnetic-field-dependent, so that it is unable to generate zero-field spectra. Third, LMS often overlooks or distorts features that have a very weak field dependence, in which case only FTMS can give unambiguous results, the

$1s \rightarrow 2p_-$ transition of shallow neutral donors being a good example (e.g., McCombe and Wagner, 1975). Finally, for studying 2-D electron systems, spectroscopy at fixed filling factors, namely at fixed magnetic fields, is sometimes crucial. In this section, after briefly reviewing FIR sources, these two modes of operation—FTMS and LMS—will be described in detail. In addition, short descriptions of two other unconventional methods—cross-modulation (or photoconductivity), in which the sample itself is used as a detector, and optically detected resonance (ODR) spectroscopy, which is a recently developed, highly sensitive method—will be provided. For the reader interested in a detailed description of FIR detectors and other FIR techniques, see, e.g., Kimmitt (1970), Stewart (1970), and Chantry (1971).

### Far-Infrared Sources

The two "classic" sources of FIR radiation commonly used in FTMS are the globar and the Hg arc lamp. The globar consists of a rod of silicon carbide usually ~2 cm long and 0.5 cm in diameter. It is heated by electrical conduction; normally 5 A are passed through it, which raises its temperature to ~1500 K. The globar is bright at wavelengths between 2 and 40 μm, but beyond 40 μm its emissivity falls slowly, although it is still sufficient for spectroscopy up to ~100 μm. The mercury lamp has higher emissivity than the globar at wavelengths longer than 100 μm. It is normally termed a "high-pressure" arc, although the actual pressure is only 1 to 2 atm. (Low-pressure gaseous discharges are not useful here because they emit discrete line spectra.) It is contained in a fused quartz inner envelope. At the shorter wavelengths of the FIR, quartz is opaque, but it becomes very hot and emits thermal radiation. At the longer wavelengths, radiation from the mercury plasma is transmitted by the quartz and replaces the thermal radiation. Originally used by Rubens and von Baeyer (1911), the mercury arc lamp is still the most widely employed source in the FIR.

Three types of laser sources currently available to FIR spectroscopists are molecular-gas lasers, the FEL, and the $p$-type germanium laser. The most frequently used among these are the hundreds of laser lines available from a large number of molecular gases. The low-pressure gas consisting of HCN, $H_2O$, $D_2O$, $CH_3OH$, $CH_3CN$, etc., flows through a glass or metal tube, where population inversion is achieved either through a high-voltage discharge or by optical excitation with a $CO_2$ laser. Output powers range from a few μW to several hundred mW, depending on the line, gas pressure, pumping power, and whether continuous or pulsed excitation is used. The FEL, first operated in 1977, is an unusual laser source which converts the kinetic energy of free electrons to EM radiation. It is tunable in a wide range of frequencies, from millimeter to ultraviolet. An FEL consists of an electron gun, an accelerator, an optical cavity, and a periodic array of magnets called an undulator or wiggler. The wavelength of the output optical beam is determined by (1) the kinetic energy of the incident electrons, (2) the spatial period of the wiggler, and (3) the strength of the wiggler magnets, all of which are continuously tunable. With FEL's enormously high peak powers (up to ~1 GW) and short pulse widths (down to ~200 fsec), a new class of nonequilibrium phenomena is currently being explored in the FIR. The $p$-Ge laser is a new type of tunable solid-state FIR laser (for a review, see Gornik, 1991; Pidgeon, 1994). Its operation relies on the fact that streaming holes in the valence band in crossed strong electric and magnetic fields can result in an inverted hot-carrier distribution. Two different lasing processes, employing light hole–light hole and light hole–heavy hole transitions, respectively, have been identified (the former is the first realization of a CR laser). The lasing wavelength is continuously tunable, by adjusting the electric and magnetic fields. Lasing in a wide spectral range (75 to 500 μm) with powers up to almost 1 W has been reported.

Nonlinear optical effects provide tunable sources in the FIR. Various schemes exist, but the most thoroughly studied method has been Difference Frequency Mixing using the 9.6 and 10.6 μm lines from two $CO_2$ lasers (for a review, see, Aggarwal and Lax, 1977). InSb is usually used as the mixing crystal. Phase matching is achieved either through the use of temperature dependence of the anomalous dispersion or by using the free-carrier contribution to the refractive index in a magnetic field. As the $CO_2$ laser produces a large number of closely spaced lines at 9.6 and 10.6 μm, thousands of lines covering the FIR region from ~70 μm to mm can be produced. However, the efficiency of this method is very low: considerable input laser powers are necessary to obtain output powers only in the μW range. Another important method for generating FIR radiation is Optical Parametric Oscillation (OPO; see, e.g., Byer and Herbst, 1977). In this method, a birefringent crystal in an optical cavity is used to split the pump beam at frequency $\omega_3$ into simultaneous oscillations at two other frequencies $\omega_1$ (idler) and $\omega_2$ (signal), where $\omega_1 + \omega_2 = \omega_3$. This is achieved either spontaneously (through vacuum fluctuation) or by feeding a beam at frequency $\omega_1$. The advantages of OPO are high efficiency, wide tuning range, and an all solid-state system. The longest wavelength currently available from OPO is ~25 μm. More recently, the remarkable advances in high-speed opto-electronic and NIR/visible femtosecond laser technology have enabled generation and detection of ultrashort pulses of broadband FIR radiation (more frequently referred to as THz radiation or "T Rays" in this context). The technique has proven to be extremely useful for FIR spectroscopic measurements in the time domain. Many experiments have shown that ultrafast photoexcitation of semiconductors and semiconductor heterostructures can be used to generate coherent charge oscillations, which emit transient THz EM radiation. This is currently an active topic of research, and the interested reader is referred to, e.g., Nuss and Orenstein (1998) and references therein.

### Fourier Transform FIR Magneto-Spectroscopy

A Fourier-transform spectrometer is essentially a Michelson type two-beam interferometer, the basic components of which are collimating optics, a fixed mirror, a beam splitter, and a movable mirror. The basic operation principle can be stated as follows. IR radiation emitted from the light source is divided by the beam splitter into two beams with approximately the same intensity. One of the beams reaches the fixed mirror, and the other reaches the movable mirror. The two beams bounce back from the two mirrors and recombine at the beam splitter. When the movable mirror is at the zero-path-difference (ZPD) position, the output of the light intensity becomes maximum, since the two beams constructively interfere at all wavelengths. When the path difference, $x$, measured from the ZPD position is varied, an interference pattern as a function of $x$, called an interferogram, is obtained that is the FT of the spectrum of the light passing through the interferometer. Hence, by taking the inverse Fourier transform of the interferogram using a computer, one obtains the spectrum.

Two different types of FT spectrometers exist: (1) "slow-scan" (or step-scan) and (2) "fast-scan" spectrometers. In a slow-scan FT spectrometer, a stepping motor drives the movable mirror. A computer controls the step size in multiples of the fundamental step size, the dwell time at each mirror position, and the total number of steps. The product of the step size and the number of steps determines the total path difference, and hence the spectral resolution. A mechanical chopper (see Fig. 2) usually chops the FIR beam. The AC signal at this frequency from the detector is fed into a lock-in amplifier and the reference signal from the chopper into the reference input of the lock-in amplifier. The data acquisition occurs at each movable-mirror position, and thus an interferogram is constructed as the magnitude of the output versus the position of the movable mirror. Computer Fourier analysis with a Fast Fourier Transform algorithm converts the interferogram into an intensity versus frequency distribution—the spectrum. Rapid-scan FT spectrometers operate quite differently, although the basic principles are the same. The movable mirror of a rapid-scan FT machine is driven at a constant velocity. Instead of using a mechanical chopper, the constant velocity of the mirror produces a sinusoidal intensity variation with a unique frequency for each spectral element $\omega$. The modulation frequency is given by

$\vartheta = 2V\omega$, where $V$ is the velocity of the mirror. High precision of parallel alignment between the two mirrors and the constant velocity of the moving mirror is provided *in situ* by a dynamic feedback controlling system. The signal sampling takes place at equally spaced mirror displacements, and is determined by the fringes of a He-Ne laser reference.

A slow-scan FTMS system for transmission CR studies is schematically shown in Fig. 2. FIR radiation generated by a Hg-arc lamp inside the spectrometer is coupled out by a Cassegrain output mirror and guided through a 3/4-inch ("oversized") brass light-pipe to a 45° mirror. The beam reflected by the mirror is then directed down and passes through a white polyethylene window into a sample probe, which consists of a stainless-steel light pipe, a sample-holder/heater/temperature-sensor complex, metallic light cones, and a detector. The probe is sealed by the white polyethylene window and a stainless steel vacuum jacket, and inserted into a superconducting-magnet cryostat. The beam is focused by a condensing cone onto the sample located at the end of the light-cone at the center of the field. A black polyethylene filter is placed in front of the sample in order to filter out the high frequency part ($\geq 500$ cm$^{-1}$) of the radiation from the light source. The FIR light transmitted through the sample is further guided by light-pipe/light-cone optics into a detector, which is placed at the bottom of the light-pipe system, far away from the center of the magnet. If a cancellation coil is available, the detector is placed at the center of the cancellation coil where $B = 0$. The sample and detector are cooled by helium exchange gas contained in the vacuum jacket of the probe.

Figures 3A and 3B show a typical interferogram and spectrum, respectively. The spectrum obtained contains not only the spectral response (transmittance in this case) of the sample but also the spectral response of combined effects of any absorption, filtering, and reflection when the light travels from the source to the detector, in addition to the output intensity spectrum of the radiation source. Therefore, in most experimental situations, spectra such as those obtained Fig. 3B are ratioed to an appropriate background spectrum taken under a different condition such as a different magnetic field, temperature, optical pumping intensity, or some other parameter that would only change the transmittance of the sample. In CR studies, spectra are usually ratioed to a zero-magnetic-field spectrum. In this way all the unwanted field-insensitive spectral structures are canceled out.

## Laser FIR Magneto-Spectroscopy

LMS is generally easier to carry out than FTMS, although the experimental setup is almost identical to that of FTMS (the only difference is that the FT spectrometer is replaced by a laser in Fig. 2). This is partly because of the high power available from lasers compared with conventional radiation sources employed in FTMS, and also because mathematical treatments of the recorded data are not required. The data acquisition simply consists of monitoring an output signal from the detector that is proportional to the amount of FIR light transmitted by the sample while

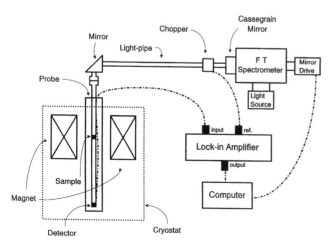

**Figure 2.** Schematic diagram of an experimental setup for CR studies with a (step-scan) Fourier transform spectrometer.

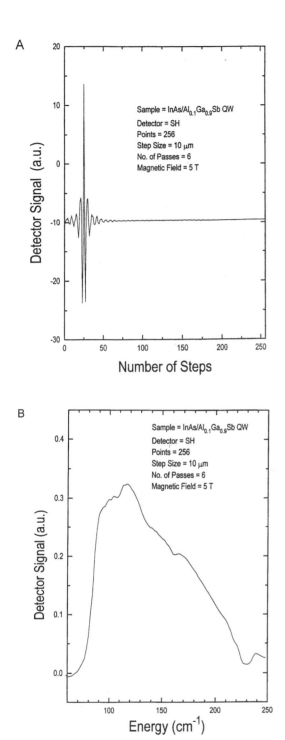

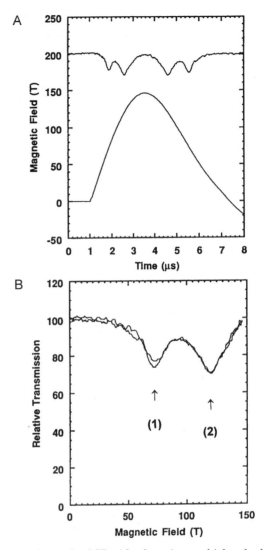

**Figure 3.** (**A**) An interferogram obtained with the FTMS setup shown in Fig. 2. (**B**) Spectrum obtained after Fourier transforming the interferogram in (A). This spectrum contains the output intensity spectrum of the Hg arc lamp, the spectral response of all the components between the source and the detector (filters, lightpipes, light cones, etc.), the responsivity spectrum of the detector (Ge:Ge extrinsic photoconductor), as well as the transmission spectrum of the sample.

the magnetic field is swept. The signal changes with magnetic field, decreasing resonantly and showing minima at resonant magnetic fields (see Fig. 4). Only magnetic-field-dependent features can thus be observed. If the laser out-

**Figure 4.** Example of CR with a laser in very high pulsed magnetic fields. (**A**) FIR transmission and magnetic field as functions of time. (**B**) Replot of the transmission signal as a function of magnetic field where the two traces arise from the rising and falling portions of the field strength shown in (A). Data obtained for $n$-type 3C-SiC.

put is stable while the field is being swept, no ratioing is necessary.

A very important feature of LMS is that it can easily incorporate pulsed magnets, thus allowing observations of CR at very high magnetic fields (see Miura, 1984). Pulsed magnetic fields up to 40 to 60 Tesla with millisecond pulse durations can be routinely produced at various laboratories. Stronger magnetic fields in the megagauss (1 megagauss = 100 Tesla) range can be also generated by special destructive methods (see, e.g., Herlach, 1984) at some institutes. These strong fields have been used to explore new physics in the ultra-quantum limit (see, e.g., Herlach, 1984; Miura, 1984) as well as to observe CR in wide-gap, heavy-mass, and low-mobility materials unmeasurable by other techniques (see, e.g., Kono et al., 1993). The cyclotron energy of electrons in megagauss fields can exceed other characteristic energies in solids such as binding energies of impurities and excitons, optical phonon energies, plasma energies, and even the fundamental band-gap in some materials, causing strong modifications in CR spectra.

An example of megagauss CR is shown in Fig. 4. Here, the recorded traces of the magnetic field pulse and the transmitted radiation from an $n$-type silicon carbide sample are plotted in Fig. 4A. The transmitted signal is then plotted as a function of magnetic field in Fig. 4B. The 36-$\mu$m line from a water-cooled $H_2O$ vapor pulse laser and a Ge:Cu impurity photoconductive detector are used. The magnetic field is generated by a destructive single-turn coil method (see, e.g., Herlach, 1984), in which one shot of large current ($\sim$2.5 MA) from a fast capacitor bank (100 kJ, 40 kV) is passed through a thin single-turn copper coil. Although the coil is destroyed during the shot due to the EM force and the Joule heat, the sample can survive a number of such shots, so that repeated measurements can be made on the same sample. The resonance absorption is observed twice in one pulse, in the rising and falling slopes of the magnetic field, as seen in Fig. 4. It should be noted that the coincidence of the two traces (Fig. 4B) confirms a sufficiently fast response of the detector system.

### Cross-Modulation (or Photoconductivity)

It is well known that the energy transferred from EM radiation to an electron system due to CR absorption induces significant heating in the system (free-carrier or Drude heating). This type of heating in turn induces a change in the conductivity tensor (e.g., an increase in $1/\tau$ or $m^*$), which causes (third-order) optical nonlinearity and modulation at other frequencies, allowing an observation of CR at a second frequency. Most conveniently, the DC conductivity $\sigma(\omega = 0)$ shows a pronounced change at a resonant magnetic field. This effect is known as Cross-Modulation or CR-induced photoconductivity, and has been described as a very sensitive technique to detect CR (Zeiger et al., 1959; Lax and Mavroides, 1960). The beauty of this method is that the sample acts as its own detector, so that there is no detector-related noise. The disadvantage is that the detection mechanism(s) is not well understood, so that quantitative lineshape analysis is difficult, unlike direct absorption. Either a decrease or increase in conductivity is observed, depending on a number of experimental conditions. Although many suggestions concerning the underlying mechanism(s) have been proposed, a complete understanding has been elusive.

Contrary to the situation of CR, which is a free carrier resonance, the mechanism of photoconductivity due to bound carrier resonances is much more clearly understood (see CARRIER LIFETIME: FREE CARRIER ABSORPTION, PHOTOCONDUCTIVITY, AND PHOTOLUMINESCENCE). For example, a resonant absorption occurs due to the $1s$ to $2p$ hydrogenic impurity transition in zero magnetic field in GaAs. Although the $2p$ state is below the continuum, an electron in the $2p$ state can be thermally excited into the conduction band, increasing conductivity (photo-thermal ionization). Other $p$-like excited states also can be studied in this manner, and these states evolve with increasing magnetic fields. As a result, one can map out the hydrogenic spectrum of the GaAs impurities simply by studying the photoconductivity of the sample. Since this is a null technique (i.e., there is photoresponse only at resonances), it is much more sensitive than transmission studies.

### Optically-Detected Resonance Spectroscopy

Recently, a great deal of development work has centered on a new type of detection scheme, Optically-Detected Resonance (ODR) spectroscopy in the FIR. This novel technique possesses several significant advantages over conventional CR methods, stimulating considerable interest among workers in the community. With this technique, FIR resonances are detected through the change in the intensity of photoluminescence (PL; see CARRIER LIFETIME: FREE CARRIER ABSORPTION, PHOTOCONDUCTIVITY, AND PHOTOLUMINESCENCE) while the magnetic field is swept, rather than measuring FIR absorption directly. This technique, originally developed for the microwave region, was extended to the FIR by Wright et al. (1990) in studies of epitaxial GaAs, and subsequently by others. Remarkable sensitivity in comparison with conventional FIR methods has been demonstrated in studies of CR (e.g., Ahmed et al., 1992), impurity transitions (Michels et al., 1994; Kono et al., 1995), and internal transitions in excitons (Cerne et al., 1996; Salib et al., 1996). Since carriers are optically created, ODR enables detection of CR in "clean" systems with no intentional doping, with increased scattering time, and in materials for which doping is difficult. Furthermore, with ODR it is possible to select a specific PL feature in the spectrum, among various band-edge features, as the detection "channel." It is then possible to use the specificity of the FIR spectrum to obtain information about recombination mechanisms and the interactions that give rise to the various lines.

Figure 5A is a schematic of the experimental apparatus used for an ODR study (Kono et al., 1995). The sample is mounted in the Faraday geometry in a FIR lightpipe at the center of a 9-T superconducting magnet cooled to 4.2 K. PL is excited with the 632.8 nm line of a He-Ne laser via a 600-$\mu$m-diameter optical fiber. The signals are collected with a second 600-$\mu$m fiber, and analyzed with 0.25-m single-grating spectrometer/Si diode detector combination. A $CO_2$-pumped FIR laser is used to generate FIR radiation. The FIR laser power supply is chopped with the lock-in amplifier referenced to this chopped signal. A computer is used to simultaneously record the magnetic field values and the detected changes in the PL, and to step the monochromator to follow the center of the desired PL peak as it shifts with magnetic field. Two scans of the change in PL intensity as a function of magnetic field (ODR signal) for a FIR laser line of 118.8 $\mu$m are presented in Fig. 6. The upper trace shows a positive change (i.e., increase) in the intensity of the free exciton PL from a well-center-doped GaAs quantum well, whereas the lower trace shows a negative change (i.e., decrease) in the intensity of the bound exciton luminescence, demonstrating spectral specificity. For both scans, four different FIR resonances are clearly seen with an excellent signal-to-noise ratio (the sharp feature near 6T is the electron CR, and the other features are donor-related features).

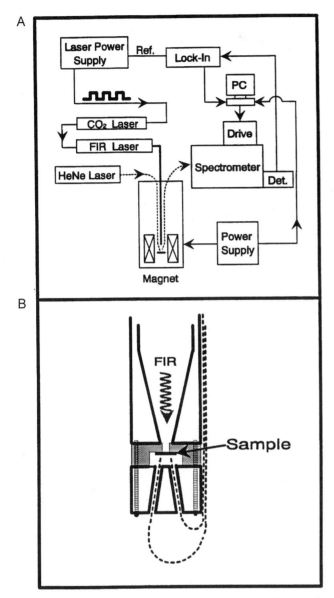

**Figure 5.** (**A**) Schematic diagram of an experimental setup for ODR spectroscopy. (**B**) Diagram of the FIR light cone, sample, and optical fiber arrangement used.

## DATA ANALYSIS AND INITIAL INTERPRETATION

After minimizing noise to get as clean a spectrum as possible, and making sure that the spectrum is free from any artifacts and multiple-reflection interference effects, one can analyze resonance "positions" (i.e., magnetic fields in LMS and frequencies in FTMS). For each resonance feature, an effective mass $m^*$ in units of $m_0$ (free electron mass in vacuum) can be obtained in different unit systems as follows

$$\frac{m^*}{m_0} = \frac{eB}{m_0\omega} = \frac{0.1158 \cdot B[\text{T}]}{\hbar\omega[\text{meV}]} = \frac{0.9337 \cdot B[\text{T}]}{\tilde{\nu}[\text{cm}^{-1}]}$$

$$= \frac{27.99 \cdot B[\text{T}]}{f[\text{GHz}]} = 93.37 \cdot B[\text{T}] \cdot \lambda[\text{m}] \qquad (20)$$

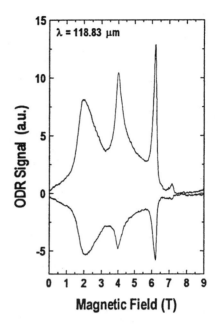

**Figure 6.** Two ODR spectra for a well-center-doped GaAs quantum well at a FIR wavelength of 118.8 μm. The upper trace shows a positive change in the intensity of the free exciton PL, whereas the lower trace shows a negative change in the intensity of the bound exciton luminescence, demonstrating the spectral specificity of ODR signals.

Note that one can obtain an effective mass (a spectroscopic mass) for any resonance. For example, for the resonances (1) and (2) for SiC in Fig. 4B one can obtain $m_1 = 0.247m_0$ and $m_2 = 0.406m_0$, and for the resonances (a) – (d) for doped GaAs quantum wells in Fig. 5C one can obtain $m_a = 0.023m_0$, $m_b = 0.044m_0$, $m_c = 0.069m_0$, and $m_d = 0.079m_0$, irrespective of their origins. However, the spectroscopic mass is identical to the cyclotron mass only when the feature is due to free carrier CR; bound-carrier resonances such as the donor-related features in Fig. 6, have different spectroscopic masses at different frequencies/ fields. Hence one needs to know which spectroscopic feature(s) arises from free-carrier CR. This can be found by two methods: temperature dependence and magnetic field (or frequency) dependence.

Examining the temperature dependence of a spectral feature is the easiest way to check the origin of the feature. As a rule of thumb, features associated with bound carriers increase in intensity with decreasing temperature at the expense of free carrier resonances. This is because in bulk semiconductors with doping level below the Mott condition, free carriers freeze out onto impurities, leaving no electrical conductivity at the lowest temperature. Thus, free-carrier CR grows with increasing temperature, but it broadens with the resulting increase of carrier scattering.

A more stringent test of whether a particular feature originates from CR is the response of its frequency versus magnetic field. In the case of LMS, one can take data only at discrete frequencies, but in the case of FTMS one can generate a continuous relationship between frequency and magnetic field. This is one of the advantages of FTMS over LMS as discussed earlier. Therefore, LMS is usually performed to study, at a fixed wavelength with a higher resolution than FTMS and with circular polarizers if necessary, those spectral features whose magnetic-field-dependence is already known by FTMS. The resonance frequency versus magnetic field thus obtained for CR should show a straight line passing through the origin, if the non-parabolicity of the material is negligible and there are no localization effects (e.g., Nicholas, 1994). The slope of this straight line then provides the spectroscopic mass, which is constant and identical to the cyclotron mass and is also equal to the band-edge mass in this idealized situation. In the case of a nonparabolic band, the cyclotron mass gradually increases with magnetic field. This means that the slope versus $B$ for CR becomes smaller (i.e., the line bends down) with increasing $B$.

Impurity-related lines, on the other hand, extrapolate to *finite* frequencies at zero magnetic field, corresponding to the zero-field binding energies of impurities. Different transitions have different slopes versus $B$, but all transitions originating from the same impurity atoms converge to an approximately common intersect at zero field. The most dominant donor-related line, the $1s$ to $2p_+$ transition, is sometimes called impurity-shifted CR (ICR). This is because its slope versus $B$ becomes nearly equal to that of free electron CR in the high-field limit, i.e., $\hbar\omega_c \gg R_y^*$, where $R_y^*$ is the binding energy of the impurity at zero field (McCombe and Wagner, 1975).

In many cases, multiple CR lines are observed in one spectrum (see, e.g., Dresselhaus et al., 1955; Otsuka, 1991; Petrou and McCombe, 1991; Nicolas, 1994). This generally indicates the existence of multiple types of carriers with different masses. Possible origins include: multi-valley splitting in the conduction band, light holes and heavy holes in the valence band, splitting due to resonant electron-phonon coupling, nonparabolicity-induced spin splitting, Landau-level splitting (see THEORY OF MAGNETIC PHASE TRANSITIONS), and population of two sub-bands in a quantum well. Explanation of each of these phenomena would be beyond the scope of this unit.

As discussed in the Introduction, CR linewidth is a sensitive probe for studying carrier scattering phenomena. In general, the linewidth of a CR line is related to the scattering lifetime, and the integrated absorption intensity of a CR line is related to the density of carriers participating in CR. Thus, if carrier density is constant, the product of absorption linewidth and depth is constant, even though the width and depth are not individually constant. More quantitatively, if the observed lineshape is well fitted by a Lorentzian, in the small absorption and reflection approximation it may be compared with

$$\frac{T(\omega, B)}{T(\omega, 0)} = 1 - A(\omega, B) = 1 - \frac{1}{2}\frac{dne^2\tau}{c\varepsilon_0\kappa_l^{1/2}m^*}\frac{1}{1 + (\omega - \omega_c)^2\tau^2}$$

(21)

where $T$ is transmission, $A$ is absorption, $d$ is the sample thickness, $c$ is the speed of light, and the other symbols have been defined earlier. The half-width at half maximum (HWHM) is thus equal to $1/\tau$. The peak absorption depth gives an estimate of the carrier density. For a

more detailed and complete description on carrier transport studies using CR, see Otsuka (1991).

## SAMPLE PREPARATION

Cyclotron resonance does not require any complicated sample preparation unless it is combined with other experimental techniques. The minimum sample size required depends on the design of the FIR optics used, i.e., how tightly the FIR beam can be focused onto the sample. Highly absorptive samples and samples with high carrier concentrations need to be polished down so that they are thin enough for transmission studies. In any case, wedging the sample substrates $2°$ to $3°$ is necessary to avoid multiple-reflection interference effects. Choosing the right sample is crucial for the success of a CR study. Samples with the highest possible carrier mobility are always preferable, if available. The DC mobility and density (see CONDUCTIVITY MEASUREMENT) can provide a rough estimate for the CR lineshape expected.

## PROBLEMS

Generally speaking, the FIR (or THz) frequency regime, where CR is usually observed, is a difficult spectral range in which to carry out sophisticated spectroscopy. This range lies in the so-called "technology-gap" existing between electronics ($\leq 100$ GHz) and optics ($\geq 10$ THz) frequencies. The well-developed NIR/visible technology does not extend to this range; sources are dimmer and detectors are less sensitive. In addition, because of the lack of efficient non-linear crystals, there exist no amplitude or phase modulators in the FIR, except for simple mechanical choppers. Therefore, in many experiments, one deals with small signals having large background noise. In steady-state experiments with a step-scan FT spectrometer, lock-in techniques are always preferable. Modulating a property of the sample that only changes the size of the signal of interest—e.g., modulating the carrier density with tunable gate electrodes—has proven to be a very efficient way to detect small signals. The cross-modulation (or photoconductivity) technique is also frequently used to detect small signals since it is a very sensitive method, as discussed earlier.

Aside from this signal-to-noise problem inherent in the FIR, there are some additional problems that CR spectroscopists might encounter. A problem particularly important in bulk semiconductors is the carrier freeze-out effect mentioned earlier. In most semiconductors, low-temperature FIR magneto-spectra are dominated by impurity transitions. At high temperatures, free carriers are liberated from the impurities, but at the same time CR often becomes too broad to be observed because of the increased scattering rate. So one has to be careful in choosing the right temperature range to study CR. In very pure semiconductors, the only way to get any CR signal is by optical pumping. In Si and Ge, whose carrier lifetimes are very long (~msec in high-quality samples), one can create a large number of carriers sufficient for steady-state

FIR absorption spectroscopy. In direct-gap semiconductors such as GaAs, carrier lifetimes are very short ($\leq 1$ nsec), so that it is nearly impossible to create enough carriers for steady-state FIR experiments, although short-pulse NIR-FIR two-color spectroscopy with an FEL is able to capture transient FIR absorption by photo-created nonequilibrium carriers. In low-dimensional semiconductor systems, so-called modulation doping is possible, where carriers can be spatially separated from their parent impurities so that they do not freeze out even at the lowest temperature.

The use of a strong magnet introduces a new class of problems. As we have seen above, in all CR studies, either in the form of FTMS or LMS, the transmission through the sample at finite magnetic field is compared with the transmission at zero magnetic field. The success of this ratioing relies on the assumption that it is only the sample that changes transmissivity with magnetic field. If anything else in the system changes some property with magnetic field, this method fails. Therefore, great care must be taken in order to make sure that no optical components have magnetic-field-dependent characteristics, that the FIR source and detector are not affected by the magnetic field, and that no component moves with magnetic field.

## ACKNOWLEDGMENTS

The author would like to thank Prof. B. D. McCombe for useful discussions, comments, and suggestions. He is also grateful to Prof. N. Miura for critically reading the article, Prof. R. A. Stradling and Prof. C. R. Pidgeon for useful comments, and G. Vacca and D. C. Larrabee for proofreading the manuscript. This work was supported in part by NSF DMR-0049024 and ONR N00014-94-1-1024.

## LITERATURE CITED

Aggarwal, R. L. and Lax, B. 1977. Optical mixing of $CO_2$ lasers in the far-infrared. *In* Nonlinear Infrared Generation—Vol. 16 of Topics in Applied Physics (Y.-R. Shen, ed.) pp. 19–80. Springer-Verlag, Berlin.

Ahmed, N., Agool, I. R., Wright, M. G., Mitchell, K., Koohian, A., Adams, S. J. A., Pidgeon, C. R., Cavenett, B. C., Stanley, C. R., and Kean, A. H. 1992. Far-infrared optically detected cyclotron resonance in GaAs layers and low-dimensional structures. *Semicond. Sci. Technol.* 7:357–363.

Aschcroft, N. W. and Mermin, N. D. 1976. Solid State Physics. Holt, Rinehart and Winston, Philadelphia.

Azbel, M. Ya. and Kaner, E. A. 1958. Cyclotron resonance in metals. *J. Phys. Chem. Solids* 6:113–135.

Bowers, R. and Yafet, Y. 1959. Magnetic susceptibility of InSb. *Phys. Rev.* 115:1165–1172.

Brau, C. A. 1990. Free-Electron Lasers. Academic Press, San Diego.

Byer, R. L. and Herbst, R. L. 1977. Parametric oscillation and mixing. *In* Nonlinear Infrared Generation—Vol. 16 of Topics in Applied Physics (Y.-R. Shen, ed.) pp. 81–137. Springer-Verlag, Berlin.

Cerne, J., Kono, J., Sherwin, M. S., Sundaram, M., Gossard, A. C., and Bauer, G. E. W. 1996. Terahertz dynamics of excitons in GaAs/AlGaAs quantum wells. *Phys. Rev. Lett.* 77:1131–1134.

Chantry, G. W. 1971. Submillimeter Spectroscopy. Academic Press, New York.

Dresselhaus, G., Kip, A. F., and Kittel, C. 1953. Observation of cyclotron resonance in germanium crystals. *Phys. Rev.* 92:827.

Dresselhaus, G., Kip, A. F., and Kittel, C. 1955. Cyclotron resonance of electrons and holes in silicon and germanium crystals. *Phys. Rev.* 98:368–384.

Gornik, E. 1991. Landau emission. *In* Landau Level Spectroscopy—Vol. 27.2 of Modern Problems in Condensed Matter Sciences (G. Landwehr and E. I. Rashba, eds.) pp. 912–996. Elsevier Science, Amsterdam.

Herlach, F. ed. 1984. Strong and Ultrastrong Magnetic Fields and Their Applications—Vol. 57 of Topics in Applied Physics. Springer-Verlag, Berlin.

Hensel, J. C. and Suzuki, K. 1974. Quantum resonances in the valence bands of germanium. II. Cyclotron resonances in uniaxially stressed crystals. *Phys. Rev. B* 9:4219–4257.

Kawabata, A. 1967. Theory of cyclotron resonance linewidth. *J. Phys. Soc. Japan* 23:999–1006.

Kimmitt, M. F. 1970. Far-infrared techniques. Pion Limited, London.

Kittel, C. 1987. Quantum Theory of Solids. John Wiley & Sons, New York.

Kono, J., Miura, N., Takeyama, S., Yokoi, H., Fujimori, N., Nishibayashi, Y., Nakajima, T., Tsuji, K., and Yamanaka, M. 1993. Observation of cyclotron resonance in low-mobility semiconductors using pulsed ultra-high magnetic fields. *Physica B* 184:178–183.

Kono, J., Lee, S. T., Salib, M. S., Herold, G. S., Petrou, A., and McCombe, B. D. 1995. Optically detected far-infrared resonances in doped GaAs quantum wells. *Phys. Rev. B* 52:R8654–R8657.

Lax, B. and Mavroides, J. G. 1960. Cyclotron resonance. *In* Solid State Physics Vol. 11 (F. Seitz and D. Turnbull, eds.) pp. 261–400. Academic Press, New York.

Lax, B, Zeiger, H. J., Dexter, R. N., and Rosenblum, E. S. 1953. Directional properties of the cyclotron resonance in germanium. *Phys. Rev.* 93:1418–1420.

Luttinger, J. M. 1951. The effect of a magnetic field on electrons in a periodic potential. *Phys. Rev.* 84:814–817.

Luttinger, J. M. 1956. Quantum theory of cyclotron resonance in semiconductors: General theory. *Phys. Rev.* 102:1030–1041.

Luttinger, J. M. and Kohn, W. 1955. Motion of electrons and holes in perturbed periodic fields. *Phys. Rev.* 97:869–883.

Mavroides, J. G. 1972. Magneto-optical properties. *In* Optical Properties of Solids (F. Abeles, ed.) pp. 351–528. North-Holland, Amsterdam.

McCombe, B. D. and Wagner, R. J. 1975. Intraband magneto-optical studies of semiconductors in the far-infrared. *In* Advances in Electronics and Electron Physics (L. Marton, ed.) Vol. 37, pp. 1–78 and Vol. 38, pp. 1–53. Academic Press, New York.

Michels, J. G., Warburton, R. J., Nicholas, R. J., and Stanley, C. R. 1994. An optically detected cyclotron resonance study of bulk GaAs. *Semicond. Sci. Technol.* 9:198–206.

Miura, N. 1984. Infrared magnetooptical spectroscopy in semiconductors and magnetic materials in high pulsed magnetic fields. *In* Infrared and Millimeter Waves Vol. 12 (K. J. Button, ed.) pp. 73–143. Academic Press, New York.

Nicholas, R. J. 1994. Intraband optical properties of low-dimensional semiconductor systems. *In* Handbook on Semiconductors Vol. 2 "Optical Properties" (M. Balkanski, ed.) pp. 385–461. Elsevier, Amsterdam.

Nuss, M. C. and Orenstein, J. 1998. Terahertz time-domain spectroscopy. *In* Millimeter and Submillimeter Wave Spectroscopy of Solids (G. Grüner, ed.) pp. 7–44. Springer-Verlag, Berlin.

Otsuka, E. 1991. Cyclotron resonance. *In* Landau Level Spectroscopy—Vol. 27.1 of Modern Problems in Condensed Matter Sciences (G. Landwehr and E. I. Rashba, eds.) pp. 1–78. Elsevier Science, Amsterdam.

Palik, E. D. and Furdyna, J. K. 1970. Infrared and microwave magnetoplasma effects in semiconductors. *Rep. Prog. Phys.* 33: 1193–1322.

Petrou, A. and McCombe, B. D. 1991. Magnetospectroscopy of confined semiconductor systems. *In* Landau Level Spectroscopy—Vol. 27.2 of Modern Problems in Condensed Matter Sciences (G. Landwehr and E. I. Rashba, eds.) pp. 679–775. Elsevier Science, Amsterdam.

Pidgeon, C. R. 1994. Free carrier Landau level absorption and emission in semiconductors. *In* Handbook on Semiconductors Vol. 2 "Optical Properties" (M. Balkanski, ed.) pp. 637–678. Elsevier, Amsterdam.

Pidgeon, C. R. and Brown, R. N. 1966. Interband magneto-absorption and Faraday rotation in InSb. *Phys. Rev.* 146:575–583.

Rubens, H. and von Baeyer, O. 1911. On extremely long waves, emitted by the quartz mercury lamp. *Phil. Mag.* 21:689–695.

Sakurai, J. J. 1985. Modern Quantum Mechanics. Addison-Wesley Publishing Co., Redwood City, California.

Salib, M. S., Nickel, H. A., Herold, G. S., Petrou, A., McCombe, B. D., Chen, R., Bajaj, K. K., and Schaff, W. 1996. Observation of internal transitions of confined excitons in GaAs/AlGaAs quantum wells. *Phys. Rev. Lett.* 77:1135–1138.

Slater, J. C. 1949. Electrons in perturbed periodic lattices. *Phys. Rev.* 76:1592–1601.

Stewart, J. E. 1970. Infrared Spectroscopy. Marcel Dekker, New York.

Suzuki, K. and Hensel, J. C. 1974. Quantum resonances in the valence bands of germanium. I. Theoretical considerations. *Phys. Rev. B* 9:4184–4218.

Wannier, G. H. 1937. The structure of electronic excitation levels in insulating crystals. *Phys. Rev.* 52:191–197.

Wright, M. G., Ahmed, N., Koohian, A., Mitchell, K., Johnson, G. R., Cavenett, B. C., Pidgeon, C. R., Stanley, C. R., and Kean, A. H. 1990. Far-infrared optically detected cyclotron resonance observation of quantum effects in GaAs. *Semicond. Sci. Technol.* 5:438–441.

Zeiger, H. J., Rauch, C. J., and Behrndt, M. E. 1959. Cross modulation of D. C. resistance by microwave cyclotron resonance. *Phys. Chem. Solids* 8:496–498.

## KEY REFERENCES

Dresselhaus et al.,1955. See above.

*This seminal article is still a useful reference not only for CR spectroscopists but also for students beginning to study semiconductor physics. Both experimental and theoretical aspects of CR in solids as well as the band structure of these two fundamental semiconductors (Si and Ge) are described in great detail.*

Landwehr, G. and Rashba, E. I. eds. 1991. Landau Level Spectroscopy—Vol. 27.1 and 27.2 of Modern Problems in Condensed Matter Sciences. Elsevier Science, Amsterdam.

*These two volumes contain a number of excellent review articles on magneto-optical and magneto-transport phenomena in bulk semiconductors and low-dimensional semiconductor quantum structures.*

Lax and Mavroides, 1960. See above.

*This review article provides a thorough overview on early, primarily microwave, CR studies in the 1950s. A historical discussion is given also to CR of electrons and ions in ionized gases, which had been extensively investigated before CR in solids was first observed.*

McCombe and Wagner, 1975. See above.

*Describes a wide variety of far-infrared mangeto-optical phenomena observed in bulk semiconductors in the 1960s and 1970 s. Detailed describtions are given to basic theoretical formulations and experimental techniques as well as extensive coverage of experimental results.*

J. KONO
Rice University
Houston, Texas

# MÖSSBAUER SPECTROMETRY

## INTRODUCTION

Mössbauer spectrometry is based on the quantum-mechanical "Mössbauer effect," which provides a nonintuitive link between nuclear and solid-state physics. Mössbauer spectrometry measures the spectrum of energies at which specific nuclei absorb $\gamma$ rays. Curiously, for one nucleus to emit a $\gamma$ ray and a second nucleus to absorb it with efficiency, the atoms containing the two nuclei must be bonded chemically in solids. A young R. L. Mössbauer observed this efficient $\gamma$-ray emission and absorption process in $^{191}$Ir, and explained why the nuclei must be embedded in solids. Mössbauer spectrometry is now performed primarily with the nuclei $^{57}$Fe, $^{119}$Sn, $^{151}$Eu, $^{121}$Sb, and $^{161}$Dy. Mössbauer spectrometry can be performed with other nuclei, but only if the experimenter can accept short radioisotope half-lives, cryogenic temperatures, and the preparation of radiation sources in hot cells.

Most applications of Mössbauer spectrometry in materials science utilize "hyperfine interactions," in which the electrons around a nucleus perturb the energies of nuclear states. Hyperfine interactions cause very small perturbations of $10^{-9}$ to $10^{-7}$ eV in the energies of Mössbauer $\gamma$ rays. For comparison, the $\gamma$ rays themselves have energies of $10^4$ to $10^5$ eV. Surprisingly, these small hyperfine perturbations of $\gamma$-ray energies can be measured easily, and with high accuracy, using a low-cost Mössbauer spectrometer.

Interpretations of Mössbauer spectra have few parallels with other methods of materials characterization. A Mössbauer spectrum looks at a material from the "inside out," where "inside" means the Mössbauer nucleus. Hyperfine interactions are sensitive to the electronic structure at the Mössbauer atom, or at its nearest neighbors. The important hyperfine interactions originate with the electron density at the nucleus, the gradient of the electric field at the nucleus, or the unpaired electron spins at the nucleus. These three hyperfine interactions are called the "isomer shift," "electric quadrupole splitting," and "hyperfine magnetic field," respectively.

The viewpoint from the nucleus is sometimes too small to address problems in the microstructure of materials.

Over the past four decades, however, there has been considerable effort to learn how the three hyperfine interactions respond to the environment around the nucleus. In general, it is found that Mössbauer spectrometry is best for identifying the electronic or magnetic structure at the Mössbauer atom itself, such as its valence, spin state, or magnetic moment. The Mössbauer effect is sensitive to the arrangements of surrounding atoms, however, because the local crystal structure will affect the electronic or magnetic structure at the nucleus. Different chemical and structural environments around the nucleus can often be assigned to specific hyperfine interactions. In such cases, measuring the fractions of nuclei with different hyperfine interactions is equivalent to measuring the fractions of the various chemical and structural environments in a material. Phase fractions and solute distributions, for example, can be determined in this way.

Other applications of the Mössbauer effect utilize its sensitivity to vibrations in solids, its timescale for scattering, or its coherence. To date these phenomena have seen little use outside the international community of a few hundred Mössbauer spectroscopists. Nevertheless, some new applications for them have recently become possible with the advent of synchrotron sources for Mössbauer spectrometry.

There have been a number of books written about the Mössbauer effect and its spectroscopies (see Key References). Most include reviews of materials research. These reviews typically demonstrate applications of the measurable quantities in Mössbauer spectrometry, and provide copious references. This unit is not a review of the field, but an instructional reference that gives the working materials scientist a basis for evaluating whether or not Mössbauer spectrometry may be useful for a research problem.

Recent research publications on Mössbauer spectrometry of materials have involved, in descending order in terms of numbers of papers: oxides, metals and alloys, organometallics, glasses, and minerals. For some problems, materials characterization by Mössbauer spectrometry is now "routine." A few representative applications to materials studies are presented. These applications were chosen in part according to the taste of the author, who makes no claim to have reviewed the literature of approximately 40,000 publications utilizing the Mössbauer effect (see Internet Resources for Mössbauer Effect Data Center Web site).

## PRINCIPLES OF THE METHOD

### Nuclear Excitations

Many properties of atomic nuclei and nuclear matter are well established, but these properties are generally not of importance to materials scientists. Since Mössbauer spectrometry measures transitions between states of nuclei, however, some knowledge of nuclear properties is necessary to understand the measurements.

A nucleus can undergo transitions between quantum states, much like the electrons of an atom, and doing so involves large changes in energy. For example, the first excited state of $^{57}$Fe is 14.41 keV above its ground state. The Mössbauer effect is sometimes called "nuclear resonant $\gamma$-ray scattering" because it involves the emission of a $\gamma$ ray from an excited nucleus, followed by the absorption of this $\gamma$ ray by a second nucleus, which becomes excited. The scattering is called "resonant" because the phase and energy relationships for the $\gamma$-ray emission and absorption processes are much the same as for two coupled harmonic oscillators.

The state of a nucleus is described in part by the quantum numbers $E$, $I$, and $I_z$, where $E$ is energy and $I$ is the nuclear spin with orientation $I_z$ along a $z$ axis. In addition to these three internal nuclear coordinates, to understand the Mössbauer effect we also need spatial coordinates, X, for the nuclear center of mass as the nucleus moves through space or vibrates in a crystal lattice. These center-of-mass coordinates are decoupled from the internal excitations of the nucleus.

The internal coordinates of the nucleus are mutually coupled. For example, the first excited state of the nucleus $^{57}$Fe has spin $I = 3/2$. For $I = 3/2$, there are four possible values of $I_z$, namely, $-3/2$, $-1/2$, $+1/2$, and $+3/2$. The ground state of $^{57}$Fe has $I = 1/2$ and two allowed values of $I_z$. In the absence of hyperfine interactions to lift the energy degeneracies of spin levels, all allowed transitions between these spin levels will occur at the same energy, giving a total cross-section for nuclear absorption, $\sigma_0$, of $2.57 \times 10^{-18}$ cm$^2$. Although $\sigma_0$ is smaller by a factor of 100 than a typical projected area of an atomic electron cloud, $\sigma_0$ is much larger than the characteristic size of the nucleus. It is also hundreds of times larger than the cross-section for scattering a 14.41-keV photon by the atomic electrons at $^{57}$Fe.

The characteristic lifetime of the excited state of the $^{57}$Fe nucleus, $\tau$, is 141 ns, which is relatively long. An ensemble of independent $^{57}$Fe nuclei that are excited simultaneously, by a flash of synchrotron light, for example, will decay at various times, $t$, with the probability per unit time of $\frac{1}{\tau}\exp(-1/\tau)$. The time uncertainty of the nuclear excited state, $\tau$, is related to the energy uncertainty of the excited state, $\Delta E$, through the uncertainty relationship, $\hbar \sim \Delta E \tau$. For $\tau = 141$ ns, the uncertainty relationship provides $\Delta E = 4.7 \times 10^{-9}$ eV. This is remarkably small—the energy of the nuclear excited state is extremely precise in energy. A nuclear resonant $\gamma$-ray emission or absorption has an oscillator quality factor, $Q$, of $3 \times 10^{12}$. The purity of phase of the $\gamma$ ray is equally impressive. In terms of information technology, it is possible in principle to transmit high-quality audio recordings of all the Beethoven symphonies on a single Mössbauer $\gamma$-ray photon. The technology for modulating and demodulating this information remains problematic, however.

In the absence of significant hyperfine interactions, the energy dependence of the cross-section for Mössbauer scattering is of Lorentzian form, with a width determined by the small lifetime broadening of the excited state energy

$$\sigma_j(E) = \frac{\sigma_0 p_j}{1 + \left(\frac{E - E_j}{\Gamma/2}\right)^2} \quad (1)$$

where for $^{57}$Fe, $\Gamma = \Delta E = 4.7 \times 10^{-9}$ eV, and $E_j$ is the mean energy of the nuclear level transition ($\sim$14.41 keV). Here $p_j$ is the fraction of nuclear absorptions that will occur with energy $E_j$. In the usual case where the energy levels of the different Mössbauer nuclei are inequivalent and the nuclei scatter independently, the total cross section is

$$\sigma(E) = \sum_j \sigma_j(E) \qquad (2)$$

A Mössbauer spectrometry measurement is usually designed to measure the energy dependence of the total cross-section, $\sigma(E)$, which is often a sum of Lorentzian functions of natural line width $\Gamma$.

A highly monochromatic $\gamma$ ray from a first nucleus is required to excite a second Mössbauer nucleus. The subsequent decay of the nuclear excitation need not occur by the reemission of a $\gamma$ ray, however, and for $^{57}$Fe only 10.9% of the decays occur in this way. Most of the decays occur by "internal conversion" processes, where the energy of the nuclear excited state is transferred to the atomic electrons. These electrons typically leave the atom, or rearrange their atomic states to emit an x ray. These conversion electrons or conversion x rays can themselves be used for measuring a Mössbauer spectrum. The conversion electrons offer the capability for surface analysis of a material. The surface sensitivity of conversion electron Mössbauer spectrometry can be as small as a monolayer (Faldum et al., 1994; Stahl and Kankeleit, 1997; Kruijer et al., 1997). More typically, electrons of a wide range of energies are detected, providing a depth sensitivity for conversion electron Mössbauer spectrometry of $\sim$100 nm (Gancedo et al., 1991; Williamson, 1993).

It is sometimes possible to measure coherent Mössbauer scattering. Here the total intensity, $I(E)$, from a sample is not the sum of independent intensity contributions from individual nuclei. One considers instead the total wave, $\Psi(\mathbf{r},E)$, at a detector located at $\mathbf{r}$. The total wave, $\Psi(\mathbf{r},E)$, is the sum of the scattered waves from individual nuclei, $j$

$$\Psi(\mathbf{r},E) = \sum_j \Psi_j(\mathbf{r},E) \qquad (3)$$

Equation 3 is fundamentally different from Equation 2, since wave amplitudes rather than intensities are added. Since we add the individual $\Psi_j$, it is necessary to account precisely for the phases of the waves scattered by the different nuclei.

### The Mössbauer Effect

Up to this point, we have assumed it possible for a second nucleus to become excited by absorbing the energy of a $\gamma$ ray emitted by a first nucleus. There were a few such experiments performed before Mössbauer's discovery, but they suffered from a well recognized difficulty. As mentioned above, the energy precision of a nuclear excited state can be on the order of $10^{-8}$ eV. This is an extremely small energy target to hit with an incident $\gamma$ ray. At room temperature, for example, vibrations of the nuclear center

of mass have energies of $2.5 \times 10^{-2}$ eV/atom. If there were any change in the vibrational energy of the nucleus caused by $\gamma$-ray emission, the $\gamma$ ray would be far too imprecise in energy to be absorbed by the sharp resonance of a second nucleus. Such a change seems likely, since the emission of a $\gamma$ ray of momentum $p_\gamma = E_\gamma/c$ requires the recoil of the emitting system with an opposite momentum (where $E_\gamma$ is the $\gamma$-ray energy and c is the speed of light). A mass, $m$, will recoil after such a momentum transfer, and the kinetic energy in the recoil, $E_{\text{recoil}}$, will detract from the $\gamma$-ray energy

$$E_{\text{recoil}} = \frac{p_\gamma^2}{2m} = \left( \frac{E_\gamma^2}{2mc^2} \right) \qquad (4)$$

For the recoil of a single nucleus, we use the mass of a $^{57}$Fe nucleus for $m$ in Equation 4, and find that $E_{\text{recoil}} = 1.86 \times 10^{-3}$ eV. This is again many orders of magnitude larger than the energy precision required for the $\gamma$ ray to be absorbed by a second nucleus.

Rudolf Mössbauer's doctoral thesis project was to measure nuclear resonant scattering in $^{191}$Ir. His approach was to use thermal Doppler broadening of the emission line to compensate for the recoil energy. A few resonant nuclear absorptions could be expected this way. To his surprise, the number of resonant absorptions was large, and was even larger when his radiation source and absorber were cooled to liquid nitrogen temperature (where the thermal Doppler broadening is smaller). Adapting a theory developed by W. E. Lamb for neutron resonant scattering (Lamb, 1939), Mössbauer interpreted his observed effect and obtained the equivalent of Equation 9, below. Mössbauer further realized that by using small mechanical motions, he could provide Doppler shifts to the $\gamma$-ray energies and tune through the nuclear resonance. He did so, and observed a spectrum without thermal Doppler broadening. In 1961, R. L. Mössbauer won the Nobel prize in physics. He was 32.

Mössbauer discovered (Mössbauer, 1958) that under appropriate conditions, the mass, $m$, in Equation 4 could be equal to the mass of the crystal. In such a case, the recoil energy is trivially small, the energy of the outgoing $\gamma$ ray is precise to better than $10^{-9}$ eV, and the $\gamma$ ray can be absorbed by exciting a second nucleus. The question is now how the mass, $m$, could be so large. The idea is that the nuclear mass is attached rigidly to the mass of the crystal. This sounds rather unrealistic, of course, and a better model is that the $^{57}$Fe nucleus is attached to the crystal mass by a spring. This is the problem of a simple harmonic oscillator, or equivalently the Einstein model of a solid with Einstein frequency $\omega_E$. The solution to this quantum mechanical problem shows that some of the nuclear recoils occur as if the nucleus is indeed attached rigidly to the crystal, but other $\gamma$-ray emissions occur by changing the state of the Einstein oscillator.

Nearly all of the energy of the emitted $\gamma$ ray comes from changes in the internal coordinates of the nucleus, independently of the motion of the nuclear center of mass. The concern about the change in the nuclear center of mass coordinates arises from the conservation of the momentum during the emission of a $\gamma$ ray with momentum $p_\gamma = E_\gamma/c$. Eventually, the momentum of the $\gamma$-ray emission will be taken up by the recoil of the crystal as a whole. However, it is possible that the energy levels of a simple harmonic oscillator (comprising the Mössbauer nucleus bound to the other atoms of the crystal lattice) could be changed by the $\gamma$-ray emission. An excitation of this oscillator would depreciate the $\gamma$-ray energy by $n\hbar$ if $n$ phonons are excited during the $\gamma$-ray emission. Since $\hbar\omega_E$ is on the order of $10^{-2}$ eV, any change in oscillator energy would spoil the possibility for a subsequent resonant absorption. In essence, changes in the oscillator excitation (or phonons in a periodic solid) replace the classical recoil energy of Equation 4 for spoiling the energy precision of the emitted $\gamma$ ray. We need to calculate the probability that phonon excitation will not occur during $\gamma$-ray emission.

Before $\gamma$-ray emission, the wavefunction of the nuclear center of mass is $\psi_i(X)$, which can also be represented in momentum space through the Fourier transformation

$$\phi_i(p) = \frac{1}{\sqrt{2\pi\hbar}} \int_{-\infty}^{\infty} \exp\left(\frac{-ipX}{\hbar}\right) \psi_i(X') \, dX' \qquad (5)$$

or

$$\psi_i(X) = \frac{1}{\sqrt{2\pi\hbar}} \int_{-\infty}^{\infty} \exp\left(\frac{ipX}{\hbar}\right) \phi_i(p) \, dp \qquad (6)$$

The momentum space representation can handily accommodate the impulse of the $\gamma$-ray emission, to provide the final state of the nuclear center of mass, $\psi_f(X)$. Recall that the impulse is the time integral of the force, $F = dp/dt$, which equals the change in momentum. The analog to impulse in momentum space is a translation in real-space, such as $X \rightarrow X - X_0$. This corresponds to obtaining a final state by a shift in origin of an initial eigenstate. With the emission of a $\gamma$ ray having momentum $p_\gamma$, we obtain the final state wave function from the initial eigenstate through a shift of origin in momentum space, $\phi_i(p) \rightarrow \phi_i(p - p_\gamma)$. We interpret the final state in real-space, $\psi_f(X)$, with Equation 6

$$\psi_f(X) = \frac{1}{\sqrt{2\pi\hbar}} \int_{-\infty}^{\infty} \exp\left(\frac{i(p + p_\gamma)X}{\hbar}\right) \phi_i(p) dp \qquad (7)$$

Now, substituting Equation 5 into Equation 7

$$\psi_f(X) = \frac{1}{2\pi\hbar} \int_{-\infty}^{\infty} \exp\left(\frac{i(p + p_\gamma)X}{\hbar}\right) \\ \times \int_{-\infty}^{\infty} \exp\left(\frac{-ipX'}{\hbar}\right) \psi_i(X') dX' dp \qquad (8)$$

Isolating the integration over momentum, $p$

$$\psi_f(X) = \frac{1}{2\pi\hbar} \int_{-\infty}^{\infty} \exp\left(\frac{ip_\gamma X}{\hbar}\right) \psi_i(X') \left[ \int_{-\infty}^{\infty} \exp\left(\frac{ip(X - X')}{\hbar}\right) dp \right] dX' \qquad (9)$$

because this integration over p gives a Dirac delta function (times $2\pi\hbar$)

$$\psi_f(X) = \int_{-\infty}^{\infty} \exp\left(\frac{ip_\gamma X}{\hbar}\right) \psi_i(X') \delta(X - X') \, dX' \qquad (10)$$

$$\psi_f(X) = \exp\left(\frac{ip_\gamma X}{\hbar}\right) \psi_i(X) \qquad (11)$$

The exponential in Equation 11 is a translation of the eigenstate, $\psi_i(X)$, in position for a fixed momentum transfer, $-p_\gamma$. It is similar to the translation in time, t, of an eigenstate with fixed energy, E, which is $\exp(-iEt/\hbar)$ or a translation in momentum for a fixed spatial translation, $X_0$, which is $\exp(-ipX_0/\hbar)$. (If the initial state is not an eigenstate, $p_\gamma$ in Equation 11 must be replaced by an operator.)

For the nuclear center-of-mass wavefunction after $\gamma$-ray emission, we seek the amplitude of the initial state wavefunction that remains in the final state wavefunction. In Dirac notation

$$\langle i|f \rangle = \int_{-\infty}^{\infty} \psi_i^*(X) \psi_f(X) \, dX \qquad (12)$$

Substituting Equation 11 into Equation 12, and using Dirac notation

$$\langle i|f \rangle = \langle i|\exp\left(\frac{ip_\gamma X}{\hbar}\right)|i \rangle \qquad (13)$$

Using the convention for the $\gamma$-ray wavevector, $k_\gamma \equiv 2\pi\nu/c = E_\gamma/\hbar c$

$$\langle i|f \rangle = \langle i|\exp(ik_\gamma X)|i \rangle \qquad (14)$$

The inner product $\langle i|f \rangle$ is the projection of the initial state of the nuclear center of mass on the final state after emission of the $\gamma$ ray. It provides the probability that there is no change in the state of the nuclear center of mass caused by $\gamma$-ray emission. The probability of this recoilless emission, $f$, is the square of the matrix element of Equation 14, normalized by all possible changes of the center-of-mass eigenfunctions

$$f = \frac{|\langle i|\exp(ik_\gamma X)|i \rangle|^2}{\sum_j |\langle j|\exp(ik_\gamma X)|i \rangle|^2} \qquad (15)$$

$$f = \frac{|\langle i|\exp(ik_\gamma X)|i \rangle|^2}{\sum_j \langle i|\exp(-ik_\gamma X)j \rangle \langle j|\exp(+ik_\gamma X)|i \rangle} \qquad (16)$$

Using the closure relation $\Sigma_j |j \rangle \langle j| = 1$ and the normalization $\langle i|i \rangle = 1$, Equation 16 becomes

$$f = |\langle i|\exp(ik_\gamma X)|i \rangle|^2 \qquad (17)$$

The quantity $f$ is the "recoil-free-fraction." It is the probability that, after the $\gamma$ ray removes momentum $p_\gamma$ from the nuclear center of mass, there will be no change in the lattice state function involving the nuclear center of mass. In other words, $f$ is the probability that a $\gamma$ ray will be emitted with no energy loss to phonons. A similar factor is required for the absorption of a $\gamma$ ray by a nucleus in a second crystal (e.g., the sample). The evaluation of $f$ is straightforward for the ground state of the Einstein solid. The ground state wavefunction is

$$\psi_{CM}^0(X) = \left(\frac{m\omega_E}{\pi\hbar}\right)^{1/4} \exp\left(-\frac{m\omega_E X^2}{2\hbar}\right) \quad (18)$$

Inserting Equation 18 into Equation 17, and evaluating the integral (which is the Fourier transform of a Gaussian function)

$$f = \exp\left(-\frac{\hbar^2 k_\gamma^2}{2m\hbar\omega_E}\right) = \exp\left(-\frac{E_R}{\hbar\omega_E}\right) = \exp\left(-k_\gamma^2 \langle X^2\rangle\right) \quad (19)$$

where $E_R$ is the recoil energy of a free $^{57}$Fe nucleus, and $\langle X^2\rangle$ is the mean-squared displacement of the nucleus when bound in an oscillator. It is somewhat more complicated to use a Debye model for calculating $f$ with a distribution of phonon energies (Mössbauer, 1958). When the lattice dynamics are known, computer calculations can be used to obtain $f$ from the full phonon spectrum of the solid, including the phonon polarizations. These more detailed calculations essentially confirm the result of Equation 19. The only nontrivial point is that low-energy phonons do not alter the result significantly. The recoil of a single nucleus does not couple effectively to long wavelength phonons, and there are few of them, so their excitation is not a problem for recoilless emission.

The condition for obtaining a significant number of "recoilless" $\gamma$-ray emissions is that the characteristic recoil energy of a free nucleus, $E_R$, is smaller than, or on the order of, the energy of the short wavelength phonons in the solid. These phonon energies are typically estimated from the Debye or Einstein temperatures of the solid to be a few tens of meV. Since $E_R = 1.86 \times 10^{-3}$ eV for $^{57}$Fe, this condition is satisfied nicely. It is not uncommon for most of the $\gamma$-ray emissions or absorptions from $^{57}$Fe to be recoil-free. It is helpful that the energy of the $\gamma$ ray, 14.41 keV, is relatively low. Higher-energy $\gamma$ rays cause $E_R$ to be large, as seen by the quadratic relation in Equation 4. Energies of most $\gamma$ rays are far greater than 14 keV, so Mössbauer spectrometry is not practical for most nuclear transitions.

### Overview of Hyperfine Interactions

Given the existence of the Mössbauer effect, the question remains as to what it can do. The answer is given in two parts: what are the phenomena that can be measured, and then what do these measurables tell us about materials? The four standard measurable quantities are the recoil-free fraction and the three hyperfine interactions: the isomer shift, the electric quadrupole splitting, and the hyperfine magnetic field. To date, the three hyperfine interactions have proved the most useful measurable quantities for the characterization of materials by Mössbauer spectrometry. This overview provides a few rules of thumb as to the types of information that can be obtained from hyperfine interactions. The section below (see More Exotic Measurable Quantities) describes quantities that are measurable, but which have seen fewer applications so far. For specific applications of hyperfine interactions for studies of materials, see Practical Aspects of the Method.

The isomer shift is the easiest hyperfine interaction to understand. It is a direct measure of electron density, albeit at the nucleus and away from the electron density responsible for chemical bonding between the Mössbauer atom and its neighbors. The isomer shift changes considerably with the valence of the Mössbauer atom in the cases of $^{57}$Fe and $^{119}$Sn. It is possible to use the isomer shift to estimate the fraction of Mössbauer isotope in different valence states, which may originate from different crystallographic site occupancies or from the presence of multiple phases in a sample. Valence analysis is often straightforward, and is probably the most common type of service work that Mössbauer spectroscopists provide for other materials scientists. The isomer shift has proven most useful for studies of ionic or covalently bonded materials such as oxides and minerals. Unfortunately, although the isomer shift is in principle sensitive to local atomic coordinations, it has usually not proven useful for structural characterization of materials, except when changes in valence are involved. The isomer shifts caused by most local structural distortions are generally too small to be useful.

Electric field gradients (EFG) are often correlated to isomer shifts. The existence of an EFG requires an asymmetric (i.e., noncubic) electronic environment around the nucleus, however, and this usually correlates with the local atomic structure. Again, like the isomer shift, the EFG has proven most useful for studies of oxides and minerals. Although interpretations of the EFG are not so straightforward as the isomer shift, the EFG is more capable of providing information about the local atomic coordination of the Mössbauer isotope. For $^{57}$Fe, the shifts in peak positions caused by the EFG tend to be comparable to, or larger than, those caused by the isomer shift.

While isomer shifts are universal, hyperfine magnetic fields are confined to ferro-, ferri-, or antiferromagnetic materials. However, while isomer shifts tend to be small, hyperfine magnetic fields usually provide large and distinct shifts of Mössbauer peaks. Because their effects are so large and varied, hyperfine magnetic fields often permit detailed materials characterizations by Mössbauer spectrometry. For body-centered cubic (bcc) Fe alloys, it is known how most solutes in the periodic table alter the magnetic moments and hyperfine magnetic fields at neighboring Fe atoms, so it is often possible to measure the distribution of hyperfine magnetic fields and determine solute distributions about $^{57}$Fe atoms. In magnetically ordered Fe oxides, the distinct hyperfine magnetic fields allow for ready identification of phase, sometimes more readily than by x-ray diffractometry.

Even in cases where fundamental interpretations of Mössbauer spectra are impossible, the identification of the local chemistry around the Mössbauer isotope is often possible by "fingerprint" comparisons with known standards. Mössbauer spectrometers tend to have similar instrument characteristics, so quantitative comparisons with published spectra are often possible. A literature search for related Mössbauer publications is usually enough to locate standard spectra for comparison. The Mössbauer Effect Data Center (see Internet Resources) is another resource that can provide this information.

### Recoil-Free Fraction

An obvious quantity to measure with the Mössbauer effect is its intensity, given by Equation 19 as the recoil-free fraction, $f$. The recoil-free fraction is reminiscent of the Debye-Waller factor for x-ray diffraction. It is large when the lattice is stiff and $\omega_E$ is large. Like the Debye-Waller factor, $f$ is a weighted average over all phonons in the solid. Unlike the Debye-Waller factor, however, $f$ must be determined from measurements with only one value of wavevector $k$, which is of course $k_\gamma$.

It is difficult to obtain $f$ from a single absolute measurement, since details about the sample thickness and absorption characteristics must be known accurately. Comparative studies may be possible with in situ experiments where a material undergoes a phase transition from one state to another while the macroscopic shape of the specimen is unchanged.

The usual way to determine $f$ for a single-phase material is by measuring Mössbauer spectral areas as a function of temperature, $T$. Equation 19 shows that the intensity of the Mössbauer effect will decrease with $\langle X^2 \rangle$, the mean-squared displacement of the nuclear motion. The $\langle X^2 \rangle$ increases with $T$, so measurements of spectral intensity versus $T$ can provide the means for determining $f$, and hence the Debye or Einstein temperature of the solid.

Another effect that occurs with temperature provides a measure of $\langle v^2 \rangle$, where v is the velocity of the nuclear center of mass. This effect is sometimes called the "second order Doppler shift," but it originates with special relativity. When a nucleus emits a $\gamma$ ray and loses energy, its mass is reduced slightly. The phonon occupation numbers do not change, but the phonon energy is increased slightly owing to the diminished mass. This reduces the energy available to the $\gamma$-ray photon. This effect is usually of greater concern for absorption by the specimen, for which the energy shift is

$$E_{\text{therm}} = -\frac{1}{2}\frac{\langle V^2 \rangle}{c^2} E_0 \qquad (20)$$

The thermal shift scales with the thermal kinetic energy in the sample, which is essentially a measure of temperature. For $^{57}$Fe, $E_{\text{therm}} = -7.3 \times 10^{-4}$ mm/s K.

### Isomer Shift

The peaks in a Mössbauer spectrum undergo observable shifts in energy when the Mössbauer atom is in different materials. These shifts originate from a hyperfine interaction involving the nucleus and the inner electrons of the atom. These "isomer shifts" are in proportion to the electron density at the nucleus. Two possibly unfamiliar concepts underlie the origin of the isomer shift. First, some atomic electron wavefunctions are actually present inside the nucleus. Second, the nuclear radius is different in the nuclear ground and excited states.

In solving the Schrödinger equation for radial wavefunctions of electrons around a point nucleus, it is found that for $r \rightarrow 0$ (i.e., toward the nucleus) the electron wavefunctions go as $r^l$, where $l$ is the angular momentum quantum number of the electron. For $s$ electrons (1s, 2s, 3s, 4s, etc.) with $l = 0$, the electron wavefunction is quite large at $r = 0$. It might be guessed that the wavefunctions of $s$ electrons could make some sort of sharp wiggle so they go to zero inside the nucleus, but this would cost too much kinetic energy. The $s$ electrons (and some relativistic $p$ electrons) are actually present inside the nucleus. Furthermore, the electron density is essentially constant over the size of the nucleus.

The overlap of the $s$-electron wavefunction with the finite nucleus provides a Coulomb perturbation which lowers the nuclear energy levels. If the excited state and ground-state energy levels were lowered equally, however, the energy of the nuclear transition would be unaffected, and the emitted (or absorbed) $\gamma$ ray would have the same energy. It is well known that the radius of an atom changes when an electron enters an excited state. The same type of effect occurs for nuclei—the nuclear radius is different for the nuclear ground and excited states. For $^{57}$Fe, the effective radius of the nuclear excited state, $R_{\text{ex}}$, is smaller than the radius of the ground state, $R_g$, but for $^{119}$Sn it is the other way around. For the overlap of a finite nucleus with a constant charge density, the total electrostatic attraction is stronger when the nucleus is smaller. This leads to a difference in energy between the nuclear excited state and ground state in the presence of a constant electron density $|\psi(0)|^2$. This shift in transition energy will usually be different for nuclei in the radiation source and nuclei in the sample, giving the following shift in position of the absorption peak in the measured spectrum

$$\Delta E_{\text{IS}} = \left[ CZe^2(R_{\text{ex}}^2 - R_g^2) \right] \left[ |\psi_{\text{sample}}(0)|^2 - |\psi_{\text{source}}(0)|^2 \right] \qquad (21)$$

The factor $C$ depends on the shape of the nuclear charge distribution, which need not be uniform or spherical. The sign of Equation 21 for $^{57}$Fe is such that with an increasing $s$-electron density at the nucleus, the Mössbauer peaks will be shifted to more negative velocity. For $^{119}$Sn, the difference in nuclear radii has the opposite sign. With increasing $s$-electron density at a $^{119}$Sn nucleus, the Mössbauer peaks shift to more positive velocity.

There remains another issue in interpreting isomer shifts, however. In the case of Fe, the 3$d$ electrons are expected to partly screen the nuclear charge from the 4$s$ electrons. An increase in the number of 3$d$ electrons at an $^{57}$Fe atom will therefore increase this screening, reducing the $s$-electron density at the $^{57}$Fe nucleus and causing a more positive isomer shift. The $s$-electron density at the nucleus is therefore not simply proportional to the number of valence $s$ electrons at the ion. The effect of this $3d$ electron screening is large for ionic compounds (Gütlich, 1975). In these compounds there is a series of trend lines for how the isomer shift depends on the $4s$ electron density, where the different trends correspond to the different number of $3d$ electrons at the $^{57}$Fe atom (Walker et al., 1961). With more $3d$ electrons, the isomer shift is more positive, but also the isomer shift becomes less sensitive to the number of $4s$ electrons at the atom. Determining the valence state of Fe atoms from isomer shifts is generally a realistic type of experiment, however (see Practical Aspects of the Method).

For metals it has been more recently learned that the isomer shifts do not depend on the $3d$ electron density (Akai et al., 1986). In Fe alloys, the isomer shift corresponds nicely to the $4s$ charge transfer, in spite of changes in the $3d$ electrons at the Fe atoms. For the first factor in Equation 21, a proposed choice for $^{57}$Fe is $\left[ CZe^2(R_{ex}^2 - R_g^2) \right] = -0.24\, a_0^3$ mm/s (Akai et al., 1986), where $a_0$ is the Bohr radius of 0.529 Å.

## Electric Quadrupole Splitting

The isomer shift, described in the previous section, is an electric monopole interaction. There is no static dipole moment of the nucleus. The nucleus does have an electric quadrupole moment that originates with its asymmetrical shape. The asymmetry of the nucleus depends on its spin, which differs for the ground and excited states of the nucleus. In a uniform electric field, the shape of the nuclear charge distribution has no effect on the Coulomb energy. An EFG, however, will have different interaction energies for different alignments of the electric quadrupole moment of the nucleus. An EFG generally involves a variation with position of the $x$, $y$, and $z$ components of the electric field vector. In specifying an EFG, it is necessary to know, for example, how the $x$ component of the electric field, $V_x \equiv \partial V/\partial x$ varies along the $y$ direction, $V_y \equiv \partial^2 V/\partial y \partial x$ [here $V(x,y,z)$ is the electric potential]. The EFG involves all such partial derivatives, and is a tensor quantity. In the absence of competing hyperfine interactions, it is possible to choose freely a set of principal axes so that the off-diagonal elements of the EFG tensor are zero. By convention, we label the principal axes such that $|V_{zz}| > |V_{yy}| > |V_{xx}|$. Furthermore, because the Laplacian of the potential vanishes, $V_{xx} + V_{yy} + V_{zz} = 0$, there are only two parameters required to specify the EFG. These are chosen to be $V_{zz}$ and an asymmetry parameter, $\eta \equiv (V_{xx} - V_{yy})/V_{zz}$.

The isotopes $^{57}$Fe and $^{119}$Sn have an excited-state spin of $I = 3/2$ and a ground-state spin of $1/2$. The shape of the excited nucleus is that of a prolate spheroid. This prolate spheroid will be oriented with its long axis pointing along the $z$ axis of the EFG when $I_z = \pm 3/2$. There is no effect from the sign of $I_z$, since inverting a prolate spheroid does not change its charge distribution. The $I_z = \pm 3/2$ states have a low energy compared to the $I_z = 1/2$ orientation of the excited state. In the presence of an EFG, the excited-state energy is split into two levels. Since $I_z = \pm 1/2$ for the ground state, however, the ground state

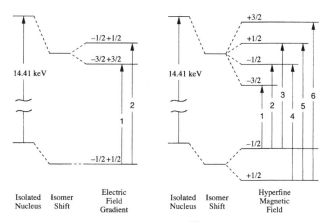

**Figure 1.** Energy level diagrams for $^{57}$Fe in an electric field gradient (EFG; left) or hyperfine magnetic field (HMF; right). For an HMF at the sample, the numbers 1 to 6 indicate progressively more energetic transitions, which give experimental peaks at progressively more positive velocities. Sign convention is that an applied magnetic field along the direction of lattice magnetization will reduce the HMF and the magnetic splitting. The case where the nucleus is exposed simultaneously to an EFG and HMF of approximately the same energies is much more complicated than can be presented on a simple energy level diagram.

energy is not split by the EFG. With an electric quadrupole moment for the excited state defined as $Q$, for $^{57}$Fe and $^{119}$Sn the quadrupole splitting of energy levels is

$$\Delta E_q = \frac{\pm 1}{4} eQV_{zz}\left(1 + \frac{\eta^2}{3}\right)^{1/2} \qquad (22)$$

where often there is the additional definition $eq \equiv V_{zz}$. The energy level diagram is shown in Figure 1. By definition, $\eta < 1$, so the asymmetry factor can vary only from 1 to 1.155. For $^{57}$Fe and $^{119}$Sn, for which Equation 22 is valid, the asymmetry can usually be neglected, and the electric quadrupole interaction can be assumed to be a measure of $V_{zz}$. Unfortunately, it is not possible to determine the sign of $V_{zz}$ easily (although this has been done by applying high magnetic fields to the sample).

The EFG is zero when the electronic environment of the Mössbauer isotope has cubic symmetry. When the electronic symmetry is reduced, a single line in the Mössbauer spectrum appears as two lines separated in energy as described by Equation 22 (as shown in Fig. 1). When the $^{57}$Fe atom has a $3d$ electron structure with orbital angular momentum, $V_{zz}$ is large. High- and low-spin Fe complexes can be identified by differences in their electric quadrupole splitting. The electric quadrupole splitting is also sensitive to the local atomic arrangements, such as ligand charge and coordination, but this sensitivity is not possible to interpret by simple calculations. The ligand field gives an enhanced effect on the EFG at the nucleus because the electronic structure at the Mössbauer atom is itself distorted by the ligand. This effect is termed "Sternheimer antishielding," and enhances the EFG from the ligands by a factor of about 7 for $^{57}$Fe (Watson and Freeman, 1967).

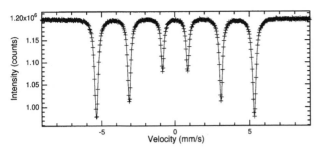

**Figure 2.** Mössbauer spectrum from bcc Fe. Data were acquired at 300 K in transmission geometry with a constant acceleration spectrometer (Ranger MS900). The points are the experimental data. The solid line is a fit to the data for six independent Lorentzian functions with unconstrained centers, widths, and depths. Also in the fit was a parabolic background function, which accounts for the fact that the radiation source was somewhat closer to the specimen at zero velocity than at the large positive or negative velocities. A $^{57}$Co source in Rh was used, but the zero of the velocity scale is the centroid of the Fe spectrum itself. Separation between peaks 1 and 6 is 10.62 mm/s.

### Hyperfine Magnetic Field Splitting

The nuclear states have spin, and therefore associated magnetic dipole moments. The spins can be oriented with different projections along a magnetic field. The energies of nuclear transitions are therefore modified when the nucleus is in a magnetic field. The energy perturbations caused by this HMF are sometimes termed the "nuclear Zeeman effect," in analogy with the more familiar splitting of energy levels of atomic electrons when there is a magnetic field at the atom.

A hyperfine magnetic field lifts all degeneracies of the spin states of the nucleus, resulting in separate transitions identifiable in a Mössbauer spectrum (see, e.g., Fig. 2). The $I_z$ range from $-I$ to $+I$ in increments of 1, being $\{-3/2, -1/2, +1/2, +3/2\}$ for the excited state of $^{57}$Fe and $\{-1/2, +1/2\}$ for the ground state. The allowed transitions between ground and excited states are set by selection rules. For the M1 magnetic dipole radiation for $^{57}$Fe, six transitions are allowed: $\{(-1/2 \rightarrow -3/2)\ (-1/2 \rightarrow -1/2)\ (-1/2 \rightarrow +1/2)\ (+1/2 \rightarrow -1/2)\ (+1/2 \rightarrow +1/2)\ (+1/2 \rightarrow +3/2)\}$. The allowed transitions are shown in Figure 1. Notice the inversion in energy levels of the nuclear ground state.

In ferromagnetic iron metal, the magnetic field at the $^{57}$Fe nucleus, the HMF, is 33.0 T at 300 K. The enormity of this HMF suggests immediately that it does not originate from the traditional mechanisms of solid-state magnetism. Furthermore, when an external magnetic field is applied to a sample of Fe metal, there is a decrease in magnetic splitting of the measured Mössbauer peaks. This latter observation shows that the HMF at the $^{57}$Fe nucleus has a sign opposite to that of the lattice magnetization of Fe metal, so the HMF is given as $-33.0$ T.

It is easiest to understand the classical contributions to the HMF, denoted $H_{mag}$, $H_{dip}$ and $H_{orb}$. The contribution $H_{mag}$ is the magnetic field from the lattice magnetization, $M$, which is $4\pi M/3$. To this contribution we add any magnetic fields applied by the experimenter, and we subtract the demagnetization caused by the return flux. Typically,

$H_{mag} < +0.7$ T. The contribution $H_{dip}$ is the classical dipole magnetic field caused by magnetic moments at atoms near the Mössbauer nucleus. In Fe metal, $H_{dip}$ vanishes owing to cubic symmetry, but contributions of $+0.1$ T are possible when neighboring Fe atoms are replaced with nonmagnetic solutes. Finally, $H_{orb}$ originates with any residual orbital magnetic moment from the Mössbauer atom that is not quenched when the atom is a crystal lattice. This contribution is about $+2$ T (Akai, 1986), and it may not change significantly when Fe metal is alloyed with solute atoms, for example. These classical mechanisms make only minor contributions to the HMF.

The big contribution to the HMF at a Mössbauer nucleus originates with the "Fermi contact interaction." Using the Dirac equation, Fermi and Segre discovered a new term in the Hamiltonian for the interaction of a nucleus and an atomic electron

$$\hbar_{FC} = -\frac{8\pi}{3} g_e g_N \mu_e \mu_N \mathbf{I} \cdot \mathbf{S}\, \delta(\mathbf{r}) \qquad (23)$$

Here $\mathbf{I}$ and $\mathbf{S}$ are spin operators that act on the nuclear and electron wavefunctions, respectively, $\mu_e$ and $\mu_N$ are the electron and nuclear magnetons, and $(\mathbf{r})$ ensures that the electron wavefunction is sampled at the nucleus. Much like the electron gyromagnetic ratio, $g_e$, the nuclear gyromagnetic ratio, $g_N$, is a proportionality between the nuclear spin and the nuclear magnetic moment. Unlike the case for an electron, the nuclear ground and excited states do not have the same value of $g_N$; that of the ground state of $^{57}$Fe is larger by a factor of $-1.7145$. The nuclear magnetic moment is $g_N\, \mu_N I$, so we can express the Fermi contact energy by considering this nuclear magnetic moment in an effective magnetic field, $H_{eff}$, defined as

$$H_{eff} = \frac{8\pi}{3} g_e \mu_e S |\psi(0)|^2 \qquad (24)$$

where the electron spin is $\pm 1/2$, and $|\psi(0)|^2$ is the electron density at the nucleus. If two electrons of opposite spin have the same density at the nucleus, their contributions will cancel and $H_{eff}$ will be zero. A large HMF requires an unpaired electron density at the nucleus, expressed as $|S| > 0$.

The Fermi contact interaction explains why the HMF is negative in $^{57}$Fe. As described above (see Isomer Shift), only $s$ electrons of Fe have a substantial presence at the nucleus. The largest contribution to the $^{57}$Fe HMF is from $2s$ electrons, however, which are spin-paired core electrons. The reason that spin-paired core electrons can make a large contribution to the HMF is that the $2s\uparrow$ and $2s\downarrow$ wavefunctions have slightly different shapes when the Fe atom is magnetic. The magnetic moment of Fe atoms originates primarily with unpaired $3d$ electrons, so the imbalance in numbers of $3d\uparrow$ and $3d\downarrow$ electrons must affect the shapes of the paired $2s\uparrow$ and $2s\downarrow$ electrons.

These shapes of the $2s\uparrow$ and $2s\downarrow$ electron wavefunctions are altered by exchange interactions with the $3d\uparrow$ and $3d\downarrow$ electrons. The exchange interaction originates with the Pauli exclusion principle, which requires that a multielectron wavefunction be antisymmetric under the exchange of electron coordinates. The process of antisymmetrization of a multielectron wavefunction produces an energy contribution from the Coulomb interaction between electrons called the "exchange energy," since it is the expectation value of the Coulomb energy for all pairs of electrons of like spin exchanged between their wavefunctions.

The net effect of the exchange interaction is to decrease the repulsive energy between electrons of like spin. In particular, the exchange interaction reduces the Coulomb repulsion between the $2s\uparrow$ and $3d\uparrow$ electrons, allowing the more centralized $2s\uparrow$ electrons to expand outward away from the nucleus. The same effect occurs for the $2s\downarrow$ and $3d\downarrow$ electrons, but to a lesser extent because there are fewer $3d\downarrow$ electrons than $3d\uparrow$ electrons in ferromagnetic Fe. The result is a higher density of $2s\downarrow$ than $2s\uparrow$ electrons at the $^{57}$Fe nucleus. The same effect occurs for the $1s$ shell, and the net result is that the HMF at the $^{57}$Fe nucleus is opposite in sign to the lattice magnetization (which is dominated by the $3d\uparrow$ electrons). The $3s$ electrons contribute to the HMF, but are at about the same mean radius as the $3d$ electrons, so their spin unbalance at the $^{57}$Fe nucleus is smaller. The $4s$ electrons, on the other hand, lie outside the $3d$ shell, and exchange interactions bring a higher density of $4s\uparrow$ electrons into the $^{57}$Fe nucleus, although not enough to overcome the effects of the $1s\downarrow$ and $2s\downarrow$ electrons. These $4s$ spin polarizations are sensitive to the magnetic moments at nearest neighbor atoms, however, and provide a mechanism for the $^{57}$Fe atom to sense the presence of neighboring solute atoms. This is described below (see Solutes in bcc Fe Alloys).

## More Exotic Measurable Quantities

**Relaxation Phenomena.** Hyperfine interactions have natural time windows for sampling electric or magnetic fields. This time window is the characteristic time, $\tau_{hf}$, associated with the energy of a hyperfine splitting, $\tau_{hf} = \hbar/E_{hf}$. When a hyperfine electric or magnetic field undergoes fluctuations on the order of $\tau_{hf}$ or faster, observable distortions appear in the measured Mössbauer spectrum. The lifetime of the nuclear excited state does not play a direct role in setting the timescale for observing such relaxation phenomena. However, the lifetime of the nuclear excited state does provide a reasonable estimate of the longest characteristic time for fluctuations that can be measured by Mössbauer spectrometry.

Sensitivity to changes in valence of the Mössbauer atom between Fe(II) and Fe(III) has been used in studies of the Verwey transition in $Fe_3O_4$, which occurs at $\sim$120 K. Above the Verwey transition temperature the Mössbauer spectrum comprises two sextets, but when $Fe_3O_4$ is cooled below the Verwey transition temperature the spectrum becomes complex (Degrave et al., 1993).

Atomic diffusion is another phenomenon that can be studied by Mössbauer spectrometry (Ruebenbauer et al., 1994). As an atom jumps to a new site on a crystal lattice, the coherence of its $\gamma$-ray emission is disturbed. The shortening of the time for coherent $\gamma$-ray emission causes a broadening of the linewidths in the Mössbauer spectrum. In single crystals this broadening can be shown to occur by different amounts along different crystallographic directions, and has been used to identify the atom jump directions and mechanisms of diffusion in Fe alloys (Feldwisch et al., 1994; Vogl et al., 1994; Sepiol et al., 1996).

Perhaps the most familiar example of a relaxation effect in Mössbauer spectrometry is the superparamagnetic behavior of small particles. This phenomenon is described below (see Crystal Defects and Small Particles).

**Phonons.** The phonon partial density of states (DOS) has recently become measurable by Mössbauer spectrometry. Technically, nuclear resonant scattering that occurs with the creation or annihilation of a phonon is inelastic scattering, and therefore not the Mössbauer effect. However, techniques for measuring the phonon partial DOS have been developed as a capability of synchrotron radiation sources for Mössbauer scattering. The experiments are performed by detuning the incident photon energies above and below the nuclear resonance by 100 meV or so. This range of energy is far beyond the energy width of the Mössbauer resonance or any of its hyperfine interactions. However, it is in the range of typical phonon energies. The inelastic spectra so obtained are called "partial" phonon densities of states because they involve the motions of only the Mössbauer nucleus. The recent experiments (Seto et al., 1995; Sturhahn et al., 1995; Fultz et al., 1997) are performed with incoherent scattering (a Mössbauer $\gamma$ ray into the sample, a conversion x ray out), and are interpreted in the same way as incoherent inelastic neutron scattering spectra (Squires, 1978). Compared to this latter, more established technique, the inelastic nuclear resonant scattering experiments have the capability of working with much smaller samples, owing to the large cross-section for nuclear resonant scattering. The vibrational spectra of monolayers of $^{57}$Fe atoms at interfaces of thin films have been measured in preliminary experiments.

**Coherence and Diffraction.** Mössbauer scattering can be coherent, meaning that the phase of the incident wave is in a precise relationship to the phase of the scattered wave. For coherent scattering, wave amplitudes are added (Equation 3) instead of independent photon intensities (Equation 2). For the isotope $^{57}$Fe, coherency occurs only in experiments where a 14.41 keV $\gamma$ ray is absorbed and a 14.41 keV $\gamma$ ray is reemitted through the reverse nuclear transition. The waves scattered by different coherent processes interfere with each other, either constructively or destructively. The interference between Mössbauer scattering and x-ray Rayleigh scattering undergoes a change from constructive in-phase interference above the Mössbauer resonance to destructive out-of-phase interference below. This gives rise to an asymmetry in the peaks measured in an energy spectrum, first observed by measuring a Mössbauer energy spectrum in scattering geometry (Black and Moon, 1960).

Diffraction is a specialized type of interference phenomenon. Of particular interest to the physics of Mössbauer diffraction is a suppression of internal conversion processes when diffraction is strong. With multiple transfers of energy between forward and diffracted beams, there is a nonintuitive enhancement in the rate of decay of the nuclear excited state (Hannon and Trammell, 1969; van Bürck et al., 1978; Shvyd'ko and Smirnov, 1989), and a broadening of the characteristic linewidth. A fortunate consequence for highly perfect crystals is that with strong Bragg diffraction, a much larger fraction of the reemissions from $^{57}$Fe nuclei occur by coherent 14.41 keV emission. The intensities of Mössbauer diffraction peaks therefore become stronger and easier to observe. For solving unknown structures in materials or condensed matter, however, it is difficult to interpret the intensities of diffraction peaks when there are multiple scatterings. Quantification of diffraction intensities with kinematical theory is an advantage of performing Mössbauer diffraction experiments on polycrystalline samples. Such samples also avoid the broadening of features in the Mössbauer energy spectrum that accompanies the speedup of the nuclear decay. Unfortunately, without the dynamical enhancement of coherent decay channels, kinematical diffraction experiments on small crystals suffer a serious penalty in diffraction intensity. Powder diffraction patterns have not been obtained until recently (Stephens et al., 1994), owing to the low intensities of the diffraction peaks. Mössbauer diffraction from polycrystalline alloys does offer a new capability, however, of combining the spectroscopic capabilities of hyperfine interactions to extract a diffraction pattern from a particular chemical environment of the Mössbauer isotope (Stephens and Fultz, 1997).

## PRACTICAL ASPECTS OF THE METHOD

### Radioisotope Sources

The vast majority of Mössbauer spectra have been measured with instrumentation as shown in Figure 3. The spectrum is obtained by counting the number of $\gamma$-ray photons that pass through a thin specimen as a function of the $\gamma$-ray energy. At energies where the Mössbauer effect is strong, a dip is observed in the $\gamma$-ray transmission. The $\gamma$-ray energy is tuned with a drive that imparts a Doppler shift, $\Delta E$, to the $\gamma$ ray in the reference frame of the sample:

$$\Delta E = \frac{v}{c} E_\gamma \qquad (25)$$

where $v$ is the velocity of the drive. A velocity of 10 mm/s will provide an energy shift, $\Delta E$, of $4.8 \times 10^{-7}$ eV to a 14.41 keV $\gamma$ ray of $^{57}$Fe. Recall that the energy width of the Mössbauer resonance is $4.7 \times 10^{-9}$ eV, which corresponds to 0.097 mm/s. An energy range of 10 mm/s is usually more than sufficient to tune through the full Mössbauer energy spectrum of $^{57}$Fe or $^{119}$Sn. It is conventional to present the energy axis of a Mössbauer spectrum in units of mm/s.

The equipment required for Mössbauer spectrometry is simple, and adequate instrumentation is often found in instructional laboratories for undergraduate physics students. In a typical coursework laboratory exercise, students learn the operation of the detector electronics and the spectrometer drive system in a few hours, and complete a measurement or two in about a week. (The under-

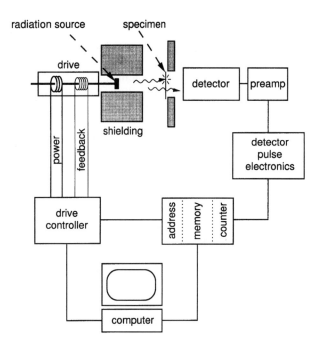

**Figure 3.** Transmission Mössbauer spectrometer. The radiation source sends $\gamma$ rays to the right through a thin specimen into a detector. The electromagnetic drive is operated with feedback control by comparing a measured velocity signal with a desired reference waveform. The drive is cycled repetitively, usually so the velocity of the source varies linearly with time (constant acceleration mode). Counts from the detector are accumulated repetitively in short time intervals associated with memory addresses of a multichannel scaler. Each time interval corresponds to a particular velocity of the radiation source. Typical numbers are 1024 data points of 50-μs time duration and a period of 20 Hz.

standing of the measured spectrum typically takes much longer.) Most components for the Mössbauer spectrometer in Figure 3 are standard items for x-ray detection and data acquisition. The items specialized for Mössbauer spectrometry are the electromagnetic drive and the radiation source. Abandoned electromagnetic drives and controllers are often found in university and industrial laboratories, and hardware manufactured since about 1970 by Austin Science Associates, Ranger Scientific, Wissel/Oxford Instruments, and Elscint, Ltd. are all capable of providing excellent results. Half-lives for radiation sources are: $^{57}$Co, 271 days, $^{119m}$Sn, 245 days, $^{151}$Sm, 93 years, and $^{125}$Te, 2.7 years. A new laboratory setup for $^{57}$Fe or $^{119}$Sn work may require the purchase of a radiation source. Suppliers include Amersham International, Dupont/NEN, and Isotope Products. It is also possible to obtain high-quality radiation sources from the former Soviet Union. Specifications for the purchase of a new Mössbauer source, besides activity level (typically 20 to 50 mCi for $^{57}$Co), should include linewidth and sometimes levels of impurity radioisotopes.

The measured energy spectrum from the sample is convoluted with the energy spectrum of the radiation source. For a spectrum with sharp Lorentzian lines of natural linewidth, $\Gamma$ (see Equation 1), the convolution of the source and sample Lorentzian functions provides a measured

Lorentzian function of full width at half-maximum of 0.198 mm/s. An excellent $^{57}$Fe spectrum from pure Fe metal over an energy range of 10 mm/s may have linewidths of 0.23 mm/s, although instrumental linewidths of somewhat less than 0.3 mm/s are not uncommon owing to technical problems with the purity of the radiation source and vibrations of the specimen or source.

Radiation sources for $^{57}$Fe Mössbauer spectrometry use the $^{57}$Co radioisotope. The unstable $^{57}$Co nucleus absorbs an inner-shell electron, transmuting to $^{57}$Fe and emitting a 122-keV γ ray. The $^{57}$Fe nucleus thus formed is in its first excited state, and decays about 141 ns later by the emission of a 14.41-keV γ ray. This second γ ray is the useful photon for Mössbauer spectrometry. While the 122-keV γ ray can be used as a clock to mark the formation of the $^{57}$Fe excited state, it is generally considered a nuisance in Mössbauer spectrometry, along with emissions from other contamination radioisotopes in the radiation source. A Mössbauer radiation source is prepared by diffusing the $^{57}$Co isotope into a matrix material such as Rh, so that atoms of $^{57}$Co reside as dilute substitutional solutes on the fcc Rh crystal lattice. Being dilute, the $^{57}$Co atoms have a neighborhood of pure Rh, and therefore all $^{57}$Co atoms have the same local environment and the same nuclear energy levels. They will therefore emit γ rays of the same energy. Although radiation precautions are required for handling the source, the samples (absorbers) are not radioactive either before or after measurement in the spectrometer.

Enrichment of the Mössbauer isotope is sometimes needed when, for example, the 2.2% natural abundance of $^{57}$Fe is insufficient to provide a strong spectrum. Although $^{57}$Fe is not radioactive, material enriched to 95% $^{57}$Fe costs approximately $15 to $30 per mg, so specimen preparation usually involves only small quantities of isotope. Biochemical experiments often require growing organisms in the presence of $^{57}$Fe. This is common practice for studies on heme proteins, for example. For inorganic materials, it is sometimes possible to study dilute concentrations of Fe by isotopic enrichment. It is also common practice to use $^{57}$Fe as an impurity, even when Fe is not part of the structure. Sometimes it is clear that the $^{57}$Fe atom will substitute on the site of another transition metal, for example, and the local chemistry of this site can be studied with $^{57}$Fe dopants. The same approach can be used with the $^{57}$Co radioisotope, but this is not a common practice because it involves the preparation of radioactive materials. With $^{57}$Co doping, the sample material itself serves as the radiation source, and the sample is moved with respect to a single-line absorber to acquire the Mössbauer spectrum. These "source experiments" can be performed with concentrations of $^{57}$Co in the ppm range, providing a potent local probe in the material. Another advantage of source experiments is that the samples are usually so dilute in the Mössbauer isotope that there is no thickness distortion of the measured spectrum. The single-line absorber, typically sodium ferrocyanide containing 0.2 mg/cm$^2$ of $^{57}$Fe, may itself have thickness distortion, but it is the same for all Doppler velocities. The net effect of absorber thickness is a broadening of spectral features without a distortion of intensities.

## Synchrotron Sources

Since 1985 (Gerdau et al., 1985), it has become increasingly practical to perform Mössbauer spectrometry measurements with a synchrotron source of radiation, rather than a radioisotope source. This work has become more routine with the advent of Mössbauer beamlines at the European Synchrotron Radiation Facility at Grenoble, France, the Advanced Photon Source at Argonne National Laboratory, Argonne, Illinois, and the SPring-8 facility in Harima, Japan. Work at these facilities first requires success in an experiment approval process. Successful beamtime proposals will not involve experiments that can be done with radioisotope sources. Special capabilities that are offered by synchrotron radiation sources are the time structure of the incident radiation, its brightness and collimation, and the prospect of measuring energy spectra off-resonance to study phonons and other excitations in solids.

Synchrotron radiation for Mössbauer spectrometry is provided by an undulator magnet device inserted in the synchrotron storage ring. The undulator has tens of magnetic poles, positioned precisely so that the electron accelerations in each pole are arranged to add in phase. This provides a high concentration of radiation within a narrow range of angle, somewhat like Bragg diffraction from a crystal. This highly parallel radiation can be used to advantage in measurements through narrow windows, such as in diamond anvil cells. The highly parallel synchrotron radiation should permit a number of new diffraction experiments, using the Mössbauer effect for the coherent scattering mechanism.

Measurements of energy spectra are impractical with a synchrotron source, but equivalent spectroscopic information is available in the time domain. The method may be perceived as "Fourier transform Mössbauer spectrometry." A synchrotron photon, with time coherence less than 1 ns, can excite all resonant nuclei in the sample. Over the period of time for nuclear decay, 100 ns or so, the nuclei emit photon waves with energies characteristic of their hyperfine fields. Assume that there are two such hyperfine fields in the solid, providing photons of energy $E_\gamma^0 + \varepsilon_1$ and $E_\gamma^0 + \varepsilon_2$. In the forward scattering direction, the two photon waves can add in phase. The time dependence of the photon at the detector is obtained by the coherent sum as in Equation 3

$$T(t) = \exp[i(E_\gamma^0 + \varepsilon_1)t/\hbar] + \exp[i(E_\gamma^0 + \varepsilon_2)t/\hbar] \quad (26)$$

The photon intensity at the detector, $I(t)$, has the time dependence

$$I(t) = T^*(t)T(t) = 2\{1 + \cos[\varepsilon_2 - \varepsilon_1]t/\hbar]\} \quad (27)$$

When the energy difference between levels, $\varepsilon_2 - \varepsilon_1$, is greater than the natural linewidth, $\Gamma$, the forward scattered intensity measured at the detector will undergo a number of oscillations during the time of the nuclear decay. These "quantum beats" can be Fourier transformed to provide energy differences between hyperfine levels of the nucleus (Smirnov, 1996). It should be mentioned that forward scattering from thick samples also shows a phenomenon of "dynamical beats," which involve energy interchanges between scattering processes.

## Valence and Spin Determination

The isomer shift, with supplementary information provided by the quadrupole splitting, can be used to determine the valence and spin of $^{57}$Fe and $^{119}$Sn atoms. The isomer shift is proportional to the electron density at the nucleus, but this is influenced by the different σ- and π-donor acceptance strengths of surrounding ligands, their electronegativities, covalency effects, and other phenomena. It is usually best to have some independent knowledge about the electronic state of Fe or Sn in the material before attempting a valency determination. Even for unknown materials, however, valence and spin can often be determined reliably for the Mössbauer isotope.

The $^{57}$Fe isomer shifts shown in Figure 4 are useful for determining the valence and spin state of Fe ions. If the $^{57}$Fe isomer shift of an unknown compound is +1.2 mm/s with respect to bcc Fe, for example, it is identified as high-spin Fe(II). Low-spin Fe(II) and Fe(III) compounds show very similar isomer shifts, so it is not possible to distinguish between them on the basis of isomer shift alone. Fortunately, there are distinct differences in the electric quadrupole splittings of these electronic states. For low-spin Fe(II), the quadrupole splittings are rather small, being in the range of 0 to 0.8 mm/s. For low spin Fe(III) the electric quadrupole splittings are larger, being in the range 0.7 to 1.7 mm/s. The other oxidation states shown in Figure 4 are not so common, and tend to be of greater interest to chemists than materials scientists.

The previous analysis of valence and spin state assumed that the material is not magnetically ordered. In cases where a hyperfine magnetic field is present, identification of the chemical state of Fe is sometimes even easier. Table 1 presents a few examples of hyperfine magnetic fields and isomer shifts for common magnetic oxides and oxyhydroxides (Simmons and Leidheiser, 1976). This table is given as a convenient guide, but the hyperfine parameters may depend on crystalline quality and stoichiometry (Bowen et al., 1993).

**Table 1. Hyperfine Parameters of Common Oxides and Oxyhydroxides**[a]

| Compound (Fe Site) | HMF (T) | Q.S. | I.S. (vs. Fe) | Temp. (K) |
|---|---|---|---|---|
| α-FeOOH | 50.0 | −0.25 | | 77 |
| α-FeOOH | 38.2 | −0.25 | +0.61 | 300 |
| β-FeOOH | 48.5 | 0.64 | +0.38 | 80 |
| β-FeOOH | 0 | 0.62 | +0.39 | 300 |
| γ-FeOOH | 0 | 0.60 | +0.38 | 295 |
| δ-FeOOH (large cryst.) | 42.0 | | +0.35 | 295 |
| FeO | | 0.8 | +0.93 | 295 |
| $Fe_3O_4$ (Fe(III), A) | 49.3 | | +0.26 | 298 |
| $Fe_3O_4$ (Fe(II, III), B) | 46.0 | | +0.67 | 298 |
| α-$Fe_2O_3$ | 51.8 | +0.42 | +0.39 | 296 |
| γ-$Fe_2O_3$(A) | 50.2 | | +0.18 | 300 |
| γ-$Fe_2O_3$(B) | 50.3 | | +0.40 | 300 |

[a]Abbreviations: HMF, hyperfine magnetic field; I.S., isomer shift; Q.S., quadrupole splitting; T, tesla.

The isomer shifts for $^{119}$Sn compounds have a wider range than for $^{57}$Fe compounds. Isomer shifts for compounds with Sn(IV) ions have a range from −0.5 to +1.5 mm/s versus $SnO_2$. For Sn(II) compounds, the range of isomer shifts is +2.2 to +4.2 versus $SnO_2$. Within these ranges it is possible to identify other chemical trends. In particular, for Sn compounds there is a strong correlation of isomer shift with the electronegativity of the ligands. This correlation between isomer shift and ligand electronegativity is especially reliable for Sn(IV) ions. Within a family of Sn(IV) compounds of similar coordination, the isomer shift depends on the electronegativity of the surrounding ligands as −1.27 χ mm/s, where χ is the Pauling electronegativity. The correlation with Sn(II) is less reliable, in part because of the different coordinations found for this ion.

Finally, it should be mentioned that there have been a number of efforts to correlate the local coordination of $^{57}$Fe with the electric quadrupole splitting. These correlations are often reliable within a specific class of compounds, typically showing a semiquantitative relationship between quadrupole splitting and the degree of distortion of the local atomic structure.

## Phase Analysis

When more than one crystallographic phase is present in a material containing $^{57}$Fe or $^{119}$Sn, it is often possible to determine the phase fractions at least semiquantitatively. Usually some supplemental information is required before quantitative information can be derived. For example, most multiphase materials contain several chemical elements. Since Mössbauer spectrometry detects only the Mössbauer isotope, to determine the volume fraction of each phase, it is necessary to know its concentration of Mössbauer isotope. Quantitative phase analysis tends to be most reliable when the material is rich in the Mössbauer atom. Phase fractions in iron alloys, steels, and iron oxides can often be measured routinely by Mössbauer

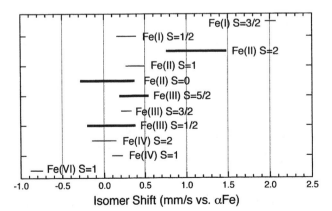

**Figure 4.** Ranges of isomer shifts in Fe compounds with various valences and spin states, with reference to bcc Fe at 300 K. Thicker lines are more common configurations (Greenwood and Gibb, 1971; Gütlich, 1975).

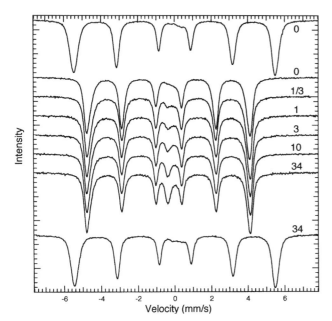

**Figure 5.** Mössbauer spectra of an alloy of Fe−8.9 atomic % Knee. The initial state of the material was ferromagnetic bcc phase, shown by the six-line spectrum at the top of the figure. This top spectrum was acquired at 23°C. The sample was heated in situ in the Mössbauer spectrometer to 600°C, for the numbers of hours marked on the curves, to form increasing amounts of fcc phase, evident as the single paramagnetic peak near −0.4 mm/s. This fcc phase is stable at 500°C, but not at 23°C, so the middle spectra were acquired at 500°C in interludes between heatings at 600°C for various times. At the end of the high-temperature runs, the sample temperature was again reduced to 23°C, and the final spectrum shown at the bottom of the figure showed that the fcc phase had transformed back into bcc phase. A trace of oxide is evident in all spectra as additional intensity around +0.4 mm/s at 23°C.

spectrometry (Schwartz, 1976; Simmons and Leidheiser, 1976; Cortie and Pollak, 1995; Campbell et al., 1995).

Figure 5 is an example of phase analysis of an Fe-Knee alloy, for which the interest was in determining the kinetics of fcc phase formation at 600°C (Fultz, 1982). The fcc phase, once formed, is stable at 500°C but not at room temperature. To determine the amount of fcc phase formed at 600°C it was necessary to measure Mössbauer spectra at 500°C without an intervening cooling to room temperature for spectrum acquisition. Mössbauer spectrometry is well suited for detecting small amounts of fcc phase in a bcc matrix, since the fcc phase is paramagnetic, and all its intensity appears as a single peak near the center of the spectrum. Amounts of fcc phase ("austenite") of 0.5% can be detected in iron alloys and steels, and quantitative analysis of the fcc phase fraction is straightforward.

The spectra in Figure 5 clearly show the six-line pattern of the bcc phase and the growth of the single peak at −0.4 mm/s from the fcc phase. These spectra show three other features that are common to many Mössbauer spectra. First, the spectrum at the top of Figure 5 shows a broadening of the outer lines of the sextet with respect to the inner lines (also see Fig. 2). This broadening originates with a distribution of hyperfine magnetic fields in alloys.

The different numbers of Knee neighbors about the various $^{57}$Fe atoms in the bcc phase cause different perturbations of the $^{57}$Fe HMF. Second, the Curie temperature of bcc Fe−8.9 atomic % Knee is ∼700°C. At the Curie temperature the average lattice magnetization is zero, and the HMF is also zero. At 500°C the alloy is approaching the Curie temperature, and shows a strong reduction in its HMF as evidenced by the smaller splitting of the six line pattern with respect to the pattern at 23°C. Finally, at 500°C the entire spectrum is shifted to the left towards more negative isomer shift. This is the relativistic thermal shift of Equation 20.

To obtain the phase fractions, the fcc and bcc components of the spectrum were isolated and integrated numerically. Isolating the fcc peak was possible by digital subtraction of the initial spectrum from spectra measured after different annealing times. The fraction of the fcc spectral component then needed two correction factors to convert it into a molar phase fraction. One factor accounted for the different chemical compositions of the fcc and bcc phases (the fcc phase was enriched in Knee to about 25%). A second factor accounted for the differences in recoil-free fraction of the two phases. Fortunately, the Debye temperatures of the two phases were known, and they differed little, so the differences in recoil-free fraction were not significant. The amount of fcc phase in the alloy at 500°C was found to change from 0.5% initially to 7.5% after 34 h of heating at 600°C.

### Solutes in bcc Fe Alloys

The HMF in pure bcc Fe is −33.0 T for every Fe atom, since every Fe atom has an identical chemical environment of 8 Fe first-nearest-neighbors (1nn), 6 Fe 2nn, 12 Fe 3nn, etc. In ferromagnetic alloys, however, the $^{57}$Fe HMF is perturbed significantly by the presence of neighboring solute atoms. In many cases, this perturbation is about +2.5 T (a reduction in the magnitude of the HMF) for each 1nn solute atom. A compilation of some HMF perturbations for 1nn solutes and 2nn solutes is presented in Figure 6. These data were obtained by analysis of Mössbauer spectra from dilute bcc Fe-X alloys (Vincze and Campbell, 1973; Vincze and Aldred, 1974; Fultz, 1993).

In general, the HMF perturbations at $^{57}$Fe nuclei from nearest-neighbor solute atoms originate from several sources, but for nonmagnetic solutes such as Si, the effects are fairly simple to understand. When the Si atom substitutes for an Fe atom in the bcc lattice, a magnetic moment of 2.2 $\mu_B$ is removed (the Fe) and replaced with a magnetic hole (the Si). The 4s conduction electrons redistribute their spin density around the Si atom, and this redistribution is significant at 1nn and 2nn distances. The Fermi contact interaction and $H_{eff}$ (Equation 24) are sensitive to the 4s electron spin density, which has finite probability at the $^{57}$Fe nucleus. Another important feature of 3p, 4p, and 5p solutes is that their presence does not significantly affect the magnetic moments at neighboring Fe atoms. Bulk magnetization measurements on Fe-Si and Fe-Al alloys, for example, show that the magnetic moment of the material decreases approximately in proportion to the fraction of Al or Si in the alloy. The core electron

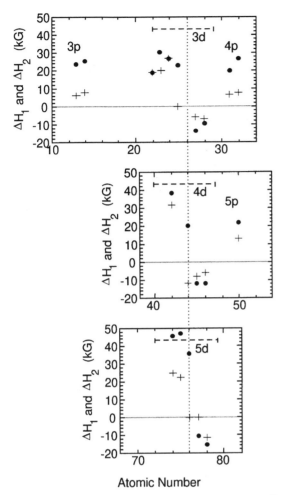

**Figure 6.** The hyperfine magnetic field perturbation, $\Delta H_1^X$, at a Fe atom caused by one 1nn solute of type X, and the 2nn perturbation, $\Delta H_2^X$, versus the atomic number of the solute. The vertical line denotes the column of Fe in the periodic table.

polarization, which involves exchange interactions between the unpaired $3d$ electrons at the $^{57}$Fe atom and its inner-shell $s$ electrons, is therefore not much affected by the presence of Si neighbors. The dominant effect comes from the magnetic hole at the solute atom, which causes the redistribution of $4s$ spin density. Figure 6 shows that the nonmagnetic $3p$, $4p$, and $5p$ elements all cause about the same HMF perturbation at neighboring $^{57}$Fe atoms, as do the nonmagnetic early $3d$ transition metals.

For solutes that perturb significantly the magnetic moments at neighboring Fe atoms, especially the late transition metals, the core polarization at the $^{57}$Fe atom is altered. There is an additional complicating effect from the matrix Fe atoms near the $^{57}$Fe atom, whose magnetic moments are altered enough to affect the $4s$ conduction electron polarization at the $^{57}$Fe (Fultz, 1993).

The HMF distribution can sometimes provide detailed information on the arrangements of solutes in nondilute bcc Fe alloys. For most solutes (that do not perturb significantly the magnetic moments at Fe atoms), the HMF at a $^{57}$Fe atom depends monotonically on the number of solute atoms in its 1nn and 2nn shells. Hyperfine magnetic field perturbations can therefore be used to measure the chemical composition or the chemical short-range order in an alloy containing up to 10 atomic % solute or even more. In many cases, it is possible to distinguish among Fe atoms having different numbers of solute atoms as first neighbors, and then determine the fractions of these different first neighbor environments. This is considerably more information on chemical short-range order (SRO) than just the average number of solute neighbors, as provided by a 1nn Warren-Cowley SRO parameter, for example.

An example of chemical short range order in an Fe–26 atomic % Al alloy is presented in Figure 7. The material was cooled at a rate of $10^6$ K/s from the melt by piston-anvil quenching, producing a polycrystalline ferromagnetic alloy with a nearly random distribution of Al atoms on the bcc lattice. With low-temperature annealing, the material evolved toward its equilibrium state of D0$_3$ chemical order. The Mössbauer spectra in Figure 7A change significantly as the alloy evolves chemical order. The overlap of several sets of six line patterns does confuse the physical picture, however, and further analysis requires the extraction of an HMF distribution from the experimental data. Software packages available for such work are described below (see Data Analysis). Figure 7B shows HMF distributions extracted from the three spectra of Figure 7A. At the top of Figure 7B are markers indicating the numbers of Al atoms in the 1nn shell of the $^{57}$Fe nucleus associated with the HMF. With low-temperature annealing, there is a clear increase in the numbers of $^{57}$Fe atoms with 0 and 4 Al neighbors, as expected when D0$_3$ order is evolving in the material. The perfectly ordered D0$_3$ structure has two chemical sites for Fe atoms, one with 0 Al neighbors and the other with 4 Al neighbors, in a 1:2 ratio. The HMF distributions were fit to a set of Gaussian functions to provide data on the chemical short range order in the alloys. These

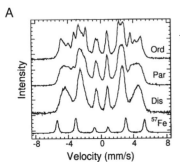

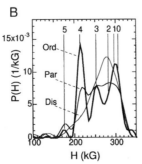

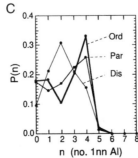

**Figure 7.** (**A**) Conversion electron Mössbauer spectra from a specimen of bcc $^{57}$Fe and three specimens of disordered, partially ordered, and D0$_3$-ordered $^{57}$Fe$_3$Al. (**B**) HMF distribution of the $^{57}$Fe$_3$Al specimens. Peaks in the HMF distribution are labeled with numbers indicating the different numbers of 1nn Al neighbors about the $^{57}$Fe atom. (**C**) Probabilities for the $^{57}$Fe atom having various numbers, $n$, of Al atoms as 1nn.

data on chemical short-range order are presented in Figure 7C.

### Crystal Defects and Small Particles

Since Mössbauer spectrometry probes local environments around a nucleus, it has often been proposed that Mössbauer spectra should be sensitive to the local atomic structures at grain boundaries and defects such as dislocations and vacancies. This is in fact true, but the measured spectra are an average over all Mössbauer atoms in a sample. Unless the material is chosen carefully so that the Mössbauer atom is segregated to the defect of interest, the spectral contribution from the defects is usually overwhelmed by the contribution from Mössbauer atoms in regions of perfect crystal. The recent interest in nanocrystalline materials, however, has provided a number of new opportunities for Mössbauer spectrometry (Herr et al., 1987; Fultz et al., 1995). The number of atoms at and near grain boundaries in nanocrystals is typically 35% for bcc Fe alloys with crystallite sizes of 7 nm or so. Such a large fraction of grain boundary atoms makes it possible to identify distinct contributions from Mössbauer atoms at grain boundaries, and to identify their local electronic environment.

Mössbauer spectrometry can provide detailed information on some features of small-particle magnetism (Mørup, 1990). When a magnetically ordered material is in the form of a very small particle, it is easier for thermal energy to realign its direction of magnetization. The particle retains its magnetic order, but the change in axis of magnetization will disturb the shape of the Mössbauer spectrum if the magnetic realignment occurs on the time scale $t_s$, which is $\hbar$ divided by the hyperfine magnetic field energy (see Relaxation Phenomena for discussion of the time window for measuring hyperfine interactions). An activation energy is associated with this "superparamagnetic" behavior, which is the magnetocrystalline anisotropy energy times the volume of the crystallite, $\kappa V$. The probability of activating a spin rotation in a small particle is the Boltzmann factor for overcoming the anisotropy energy, so the condition for observing a strong relaxation effect in the Mössbauer spectrum is

$$t_s = A \exp(-\kappa V/k_B T_b) \tag{28}$$

The temperature, $T_b$, satisfying Equation 28 is known as the "blocking temperature." The prefactor of Equation 28, the attempt frequency, is not so well understood, so studies of superparamagnetic behavior often study the blocking temperature versus the volume of the particles. In practice, most clusters of small particles have a distribution of blocking temperatures, and there are often interactions between the magnetic moments at adjacent particles. These effects can produce Mössbauer spectra with a wide variety of shapes, including very broad Lorentzian lines.

At temperatures below $T_b$, the magnetic moments of small particles undergo small fluctuations in their alignment. These small-amplitude fluctuations can be considered as vibrations of the particle magnetization about an average orientation, which serve to reduce the HMF by a modest amount. At increasing temperatures around $T_b$, however, large fluctuations occur in the magnetic alignment. The result is first a severe uncertainty of the HMF distribution, leading to a very broad background in the spectrum, followed by the growth of a paramagnetic peak near zero velocity. All of these effects can be observed in the spectra shown in Figure 8. Here, the biomaterial samples

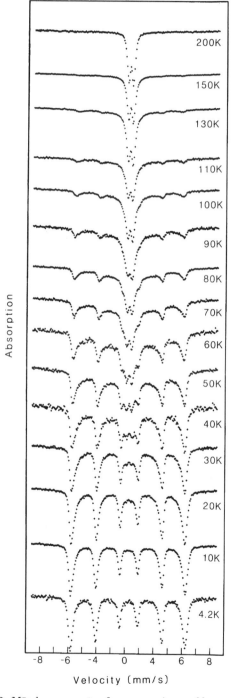

**Figure 8.** Mössbauer spectra from a specimen of haemosiderin, showing the effects of superparamagnetism with increasing temperature (Bell et al., 1984; Dickson, 1987).

comprised a core of haemosiderin, an iron storage compound, encapsulated within a protein shell. A clear six-line pattern is observed at 4.2 K, but the splitting of these six lines is found to decrease with temperature owing to small amplitude fluctuations in magnetic alignment. At temperatures around 40 to 70 K, a broad background appears under the measured spectrum, and a paramagnetic doublet begins to grow in intensity with increasing temperature. These effects are caused by large thermal reorientations of the magnetization. Finally, by 200 K, the thermally induced magnetic fluctuations are all of large amplitude and of characteristic times too short to permit a HMF to be detected by Mössbauer spectrometry.

## DATA ANALYSIS AND INITIAL INTERPRETATION

Mössbauer spectra are often presented for publication with little or no processing. An obvious correction that can be applied to most transmission spectra is a correction for thickness distortion (see Sample Preparation). This correction is rarely performed, however, in large part because the thickness of the specimen is usually not known or is not uniform. The specimen is typically prepared to be thin, or at least this is assumed, and the spectrum is assumed to be representative of the Mössbauer absorption cross-section.

A typical goal of data analysis is to find individual hyperfine parameters, or more typically a distribution of hyperfine parameters, that characterize a measured spectrum. For example, the HMF distribution of Figure 2 should resemble a delta function centered at 330 kG. On the other hand, the HMF distribution of Figure 7B shows a number of peaks that are characteristic of different local chemical environments. Distributions of electric quadrupole splittings and isomer shifts are also useful for understanding heterogeneities in the local atomic arrangements in materials.

Several software packages are available to extract distributions of hyperfine parameters from Mössbauer spectra (Hesse and Rutbartsch, 1974; Le Car and DuBoise, 1979; Brand and Le Caër, 1988; Lagarec and Rancourt, 1997). These programs are often distributed by their authors who may be located with the Mössbauer Information eXchange (see Internet Resources). The different programs extract hyperfine distributions from experimental spectra with different numerical approaches, but all will show how successfully the hyperfine distribution can be used to regenerate the experimental spectrum.

In the presence of statistical noise, the reliability of these derived hyperfine distributions must be considered carefully. In particular, over small ranges of hyperfine parameters, the hyperfine distributions are not unique. For example, it may be unrealistic to distinguish one Lorentzian-shaped peak centered at a particular velocity from the sum of several peaks distributed within a quarter of a linewidth around this same velocity. This nonuniqueness can lead to numerical problems in extracting hyperfine distributions from experimental data. Some software packages use smoothing parameters to penalize the algorithm when it picks a candidate HMF distribution with

sharp curvature. When differences in hyperfine distributions are small, there is always an issue of their uniqueness. Sometimes the data analysis cannot distinguish between different types of hyperfine distributions. For example, a spectrum that has been broadened by an EFG distribution, or even an HMF distribution, can be fit perfectly with an IS distribution. The physical origin of hyperfine distributions may not be obvious, especially when the spectrum shows little structure. Application of an external magnetic field may be helpful in identifying a weak HMF, however.

In general, distributions of all three hyperfine parameters (IS, EFG, HMF) will be present simultaneously in a measured spectrum. These parameters may be correlated; for example nuclei having the largest HMF may have the largest (or smallest) IS. Sorting out these correlations is often a research topic in itself, although the software for calculating hyperfine distributions typically allows for simple linear correlations between the distributions.

Both the EFG and the HMF use an axis of quantization for the nuclear spin. However, the direction of magnetization (for the HMF axis) generally does not coincide with the directions of the chemical bonds responsible for the EFG. The general case with comparable hyperfine interaction energies of HMFs and EFGs is quite complicated, and is well beyond the scope of this unit. Some software packages using model spin Hamiltonians are available to calculate spectra acquired under these conditions, however. In the common case when the HMF causes much larger spectral splittings than the EFG, with polycrystalline samples the usual effect of the EFG is a simple broadening of the peaks in the magnetic sextet, with no shifts in their positions.

With even modest experimental care, Mössbauer spectra can be highly reproducible from run to run. For example, the Mössbauer spectrum in Figure 2 was repeated many times over a time period of several years. Almost all of these bcc Fe spectra had data points that overlaid on top of each other, point by point, within the accuracy of the counting statistics. Because of this reproducibility, it is tempting and often appropriate to try to identify spectral features with energy width smaller than the characteristic linewidth. An underexploited technique for data analysis is "partial deconvolution" or "thinning." Since the lineshape of each nuclear transition is close to a Lorentzian function, and can be quite reproducible, it is appropriate to deconvolute a Lorentzian function from the experimental spectrum. This is formally the same as obtaining an IS distribution, but no assumptions about the origin of the hyperfine distributions are implied by this process. The net effect is to sharpen the peaks from the experimental Mössbauer spectrum, and this improvement in effective resolution can be advantageous when overlapping peaks are present in the spectra. The method does require excellent counting statistics to be reliable, however.

Finally, in spite of all the ready availability of computing resources, it is always important to look at differences in the experimental spectra themselves. Sometimes a digital subtraction of one normalized spectrum from another is an excellent way to identify changes in a material. In any event, if no differences are detected by direct inspection of the data, changes in the hyperfine distributions obtained by computer software should not be believed. For this reason it is still necessary to show actual experimental spectra in research papers that use Mössbauer spectrometry.

## SAMPLE PREPARATION

A central concern for transmission Mössbauer spectrometry is the choice and control of specimen thickness. The natural thickness of a specimen is $t$

$$t \equiv (f_a n_a \sigma_a)^{-1} \qquad (29)$$

where $f_a$ is the recoil-free fraction of the Mössbauer isotope in the specimen, $n_a$ is the number of Mössbauer nuclei per $cm^3$, and $\sigma_a$ is the cross-section in units of $cm^2$. The $f_a$ can be estimated from Equation 19, for which it is useful to know that the $\gamma$-ray energies are 14.41 keV for $^{57}$Fe, 23.875 keV for $^{119}$Sn, and 21.64 keV for $^{151}$Eu. To obtain $n_a$ it is important to know that the natural isotopic abundance is 2.2% for $^{57}$Fe, 8.6% for $^{119}$Sn, and 48% for $^{151}$Eu. The cross-sections for these isotopes are, in units of $10^{-19}$ $cm^2$, 25.7 for $^{57}$Fe, 14.0 for $^{119}$Sn, and 1.14 for $^{151}$Eu. Finally, natural linewidths, $\Gamma$, of Equation 1, are 0.097 mm/s for $^{57}$Fe, 0.323 mm/s for $^{119}$Sn, and 0.652 mm/s for $^{151}$Eu.

The observed intensity in a Mössbauer transmission spectrum appears as a dip in count rate as the Mössbauer effect removes $\gamma$ rays from the transmitted beam. Since this dip in transmission increases with sample thickness, thicker samples provide better signal-to-noise ratios and shorter data acquisition times. For quantitative work, however, it is poor practice to work with samples that are the natural thickness, $t$, or thicker owing to an effect called "thickness distortion." In a typical constant-acceleration spectrometer, the incident beam will have uniform intensity at all velocities of the source, and the top layer of sample will absorb $\gamma$ rays in proportion to its cross-section (Equation 2). On the other hand, layers deeper within the sample will be exposed to a $\gamma$-ray intensity diminished at velocities where the top layers have absorbed strongly. The effect of this "thickness distortion" is to reduce the overall sample absorption at velocities where the Mössbauer effect is strong. Broadening of the Mössbauer peaks therefore occurs as the samples become thicker. This broadening can be modeled approximately as increasing the effective linewidth of Equation 1 from the natural $\Gamma$ to $\Gamma(1 + 0.135t/t)$. However, it is important to note that in the tails of the Mössbauer peaks, where the absorption is weak, there is less thickness distortion. The peak shape in the presence of thickness distortion is therefore not a true Lorentzian function. Numerical corrections for the effects of thickness distortion are sometimes possible, but are rarely performed owing to difficulties in knowing the precise sample thickness and thickness uniformity. For quantitative work the standard practice is to use samples of thickness/2 or so.

We calculate an effective thickness, t, for the case of natural Fe metal, which is widely used as a calibration standard for Mössbauer spectrometers. If there are no

impurities in the Fe metal, its Mössbauer spectrum has sharp lines as shown in Figure 2. The recoil-free-fraction of bcc Fe is 0.80 at 300 K, and other quantities follow Equation 29

$$= \left( 0.80 \times 0.022 \times \frac{7.86\,g}{cm^3} \times \frac{1\,mol}{55.85\,g} \times \right.$$
$$\left. \frac{6.023 \times 10^{23}\,atoms}{mole} \times 25 \times 10^{-19}\,cm^2 \frac{1}{4} \right)^{-1} \qquad (30)$$
$$= 11 \times 10^{-4}\,cm = 11\,\mu m \qquad (31)$$

The factor of 1/4 in Equation 30 accounts for the fact that the absorption cross-section is split into six different peaks owing to the hyperfine magnetic field in bcc Fe. The strongest of these comprises 1/4 of the total absorption. Figure 2 was acquired with a sample of natural bcc Fe 25 μm in thickness. The linewidths of the inner two peaks are 0.235 mm/s whereas the outer two are 0.291 mm/s. Although the outer two peaks are broadened by thickness distortion, effects of impurity atoms in the Fe are equally important. The widths of the inner two lines are probably a better measure of the spectrometer resolution.

## LITERATURE CITED

Akai, H., Blügel, S., Zeller, R., and Dederichs, P. H., 1986. Isomer shifts and their relation to charge transfer in dilute Fe alloys. *Phys. Rev. Lett.* 56:2407–2410.

Bell, S. H., Weir, M. P., Dickson, D. P. E., Gibson, J. F., Sharp, G. A., and Peters, T. J. 1984. Mössbauer spectroscopic studies of human haemosiderin and ferritin. *Biochim. Biophys. Acta.* 787:227–236.

Black, P. J., and Moon, P. B. 1960. Resonant scattering of the 14-keV iron-57 γ-ray, and its interference with Rayleigh scattering. *Nature (London)* 188:481–484.

Bowen, L. H., De Grave, E., and Vandenberghe, R. E. 1993. Mössbauer effect studies of magnetic soils and sediments. *In* Mössbauer Spectroscopy Applied to Magnetism and Materials Science Vol. I. (G. J. Long and F. Grandjean, eds.). pp. 115–159. Plenum, New York.

Brand, R. A. and Le Car, G. 1988. Improving the validity of Mössbauer hyperfine parameter distributions: The maximum entropy formalism and its applications. *Nucl. Instr. Methods Phys. Res. B* 34:272–284.

Campbell, S. J., Kaczmarek, W. A., and Wang, G. M. 1995. Mechanochemical transformation of haematite to magnetite. *Nanostruct. Mater.* 6:35–738.

Cortie, M. B. and Pollak, H. 1995. Embrittlement and aging at 475°C in an experimental ferritic stainless-steel containing 38 wt.% chromium. *Mater. Sci. Eng. A* 199:153–163.

Degrave, E., Persoons, R. M., and Vandenberghe, R. E. 1993. Mössbauer study of the high temperature phase of co-substituted magnetite. *Phys. Rev. B* 47:5881–5893.

Dickson, D. P. E. 1987. Mössbauer Spectroscopic Studies of Magnetically Ordered Biological Materials. *Hyperfine Interact.* 33: 263–276.

Faldum, T., Meisel, W., and Gütlich, P. 1994. Oxidic and metallic Fe/Knee multilayers prepared from Langmuir-Blodgett films. *Hyperfine Interact.* 92:1263–1269.

Feldwisch, R., Sepiol, B., and Vogl, G. 1994. Elementary diffusion jump of iron atoms in intermetallic phases studied by Mössbauer spectroscopy 2. From order to disorder. *Acta Metall. Mater.* 42:3175–3181.

Fultz, B. 1982. A Mössbauer Spectrometry Study of Fe-Knee-X Alloys. Ph.D. Thesis. University of California at Berkeley.

Fultz, B. 1993. Chemical systematics of iron-57 hyperfine magnetic field distributions in iron alloys. *In* Mössbauer Spectroscopy Applied to Magnetism and Materials Science Vol. I. (G. J. Long and F. Grandjean, eds.). pp. 1–31. Plenum Press, New York.

Fultz, B., Ahn, C. C., Alp, E. E., Sturhahn, W., and Toellner, T. S. 1997. Phonons in nanocrystalline $^{57}$Fe. *Phys. Rev. Lett.* 79: 937–940.

Fultz, B., Kuwano, H., and Ouyang, H. 1995. Average widths of grain boundaries in nanophase alloys synthesized by mechanical attrition. *J. Appl. Phys.* 77:3458–3466.

Gancedo, J. R., Gracia, M. and Marco, J. F. 1991. CEMS Methodology. *Hyperfine Interact.* 66:83–94.

Gerdau, E., Rüffer, R., Winkler, H., Tolksdorf, W., Klages, C. P., and Hannon, J. P. 1985. Nuclear Bragg diffraction of synchrotron radiation in yttrium iron garnet, *Phys. Rev. Lett.* 54: 835–838.

Greenwood, N. N. and Gibb, T. C. 1971. Mössbauer Spectroscopy. Chapman & Hall, London.

Gütlich, P. 1975. Mössbauer spectroscopy in chemistry. *In* Mössbauer Spectroscopy. (U. Gonser, ed.). Chapter 2. Springer-Verlag, New York.

Hannon, J. P. and Trammell, G. T. 1969. Mössbauer diffraction. II. Dynamical Theory of Mössbauer Optics. *Phys. Rev.* 186:306–325.

Herr, U., Jing, J., Birringer, R., Gonser, U., and Gleiter, H. 1987. Investigation of nanocrystalline iron materials by Mössbauer spectroscopy. *Appl. Phys. Lett.* 50:472–474.

Hesse, J. and Rutbartsch, A. 1974. Model independent evaluation of overlapped Mössbauer spectra. *J. Phys. E: Sci. Instrum.* 7: 526–532.

Kruijer, S., Keune, W., Dobler, M., and Reuther, H. 1997. Depth analysis of phase formation in Si after high-dose Fe ion-implantation by depth-selective conversion electron Mössbauer spectroscopy. *Appl. Phys. Lett.* 70:2696–2698.

Lagarec, K. and Rancourt, D. G. 1997. Extended Voigt-based analytic lineshape method for determining n-dimensional correlated hyperfine parameter distributions in Mössbauer spectroscopy. *Nucl. Inst. Methods Phys. Res. B* 128:266–280.

Lamb, W. E. Jr. 1939. Capture of neutrons by atoms in a crystal. *Phys. Rev.* 55:190–197.

Le Caër, G. and Duboise, J. M. 1979. Evaluation of hyperfine parameter distributions from overlapped Mössbauer spectra of amorphous alloys. *J. Phys. E: Sci. Instrum.* 12:1083–1090.

Mørup, S. 1990. Mössbauer effect in small particles. *Hyperfine Interact.* 60:959–974.

Mössbauer, R. L. 1958. Kernresonanzfluoreszenz von Gammastrahlung in Ir$^{191}$. *Z. Phys.* 151:124.

Ruebenbauer, K., Mullen, J. G., Nienhaus, G. U., and Shupp, G. 1994. Simple model of the diffusive scattering law in glass-forming liquids. *Phys. Rev. B* 49:15607–15614.

Schwartz, L. H. 1976. Ferrous alloy phase transformations. *In* Applications of Mössbauer Spectroscopy, Vol. 1. pp. 37–81. (R. L. Cohen, ed.). Academic Press, New York.

Seto, M., Yoda, Y., Kikuta, S., Zhang, X. W., and Ando, M. 1995. Observation of nuclear resonant scattering accompanied by phonon excitation using synchrotron radiation. *Phys. Rev. Lett.* 74:3828–3831.

Sepiol, B., Meyer, A., Vogl, G., Ruffer, R., Chumakov, A. I., and Baron, A. Q. R. 1996. Time domain study of Fe-57 diffusion using nuclear forward scattering of synchrotron radiation. *Phys. Rev. Lett.* 76:3220–3223.

Shvyd'ko, Yu. V., and Smirnov, G. V. 1989. Experimental study of time and frequency properties of collective nuclear excitations in a single crystal. *J. Phys. Condens. Matter* 1:10563–10584.

Simmons, G. W. and Leidheiser, Jr., H. 1976. Corrosion and interfacial reactions. *In* Applications of Mössbauer Spectroscopy, Vol. 1. pp. 92–93. (R. L. Cohen, ed.). Academic Press, New York.

Squires, G. L. 1978. Introduction to the Theory of Thermal Neutron Scattering. p. 54. Dover Publications, New York.

Stahl, B. and Kankeleit, E. 1997. A high-luminosity UHV orange type magnetic spectrometer developed for depth-selective Mössbauer spectroscopy. *Nucl. Instr. Meth. Phys. Res. B* 122: 149–161.

Stephens, T. A. and Fultz, B. 1997. Chemical environment selectivity in Mössbauer diffraction from $^{57}$Fe$_3$Al. *Phys. Rev. Lett.* 78:366–369.

Stephens, T. A., Keune, W., and Fultz, B. 1994. Mössbauer effect diffraction from polycrystalline $^{57}$Fe. *Hyperfine Interact.* 92: 1095–1100.

Sturhahn, W., Toellner, T. S., Alp, E. E., Zhang, X., Ando, M., Yoda, Y., Kikuta, S., Seto, M., Kimball, C. W., and Dabrowski, B. 1995. Phonon density-of-states measured by inelastic nuclear resonant scattering. *Phys. Rev. Lett.* 74:3832–3835.

Smirnov, G. V. 1996. Nuclear resonant scattering of synchrotron radiation. *Hyperfine Interact.* 97/98:551–588.

van Bürck, U., Smirnov, G. V., Mössbauer, R. L., Parak, F., and Semioschkina, N. A. 1978. Suppression of nuclear inelastic channels in nuclear resonance and electronic scattering of $\gamma$-quanta for different hyperfine transtions in perfect $^{57}$Fe single crystals. *J. Phys. C Solid State Phys.* 11:2305–2321.

Vincze, I. and Aldred, A. T. 1974. Mössbauer measurements in iron-base alloys with nontransition elements. *Phys. Rev. B* 9:3845–3853.

Vincze, I. and Campbell, I. A. 1973. Mössbauer measurements in iron based alloys with transition metals. *J. Phys. F Metal Phys.* 3:647–663.

Vogl, G. and Sepiol, B. 1994. Elementary diffusion jump of iron atoms in intermetallic phases studied by Mössbauer spectroscopy 1. Fe-Al close to equiatomic stoichiometry. *Acta Metall. Mater.* 42:3175–3181.

Walker, L. R., Wertheim, G. K., and Jaccarino, V. 1961. Interpretation of the Fe$^{57}$ isomer shift. *Phys. Rev. Lett.* 6:98–101.

Watson, R. E. and Freeman, A. J. 1967. Hartree-Fock theory of electric and magnetic hyperfine interactions in atoms and magnetic compounds. *In* Hyperfine Interactions. (A. J. Freeman and R. B. Frankel, eds.) Chapter 2. Academic Press, New York.

Williamson, D. L. 1993. Microstructure and tribology of carbon, nitrogen, and oxygen implanted ferrous materials. *Nucl. Instr. Methods Phys. Res. B* 76:262–267.

## KEY REFERENCES

Bancroft, G. M. 1973. Mössbauer Spectroscopy: An Introduction for Inorganic Chemists and Geochemists. John Wiley & Sons, New York.

Belozerski, G. N. 1993. Mössbauer Studies of Surface Layers. Elsevier/North Holland, Amsterdam.

Cohen, R. L. (ed.). 1980. Applications of Mössbauer Spectroscopy, Vols. 1, 2. Academic Press, New York.

*These 1980 volumes by Cohen contain review articles on the applications of Mössbauer spectrometry to a wide range of materials and phenomena, with some exposition of the principles involved.*

Cranshaw, T. E., Dale, B. W., Longworth, G. O., and Johnson, C. E. 1985. Mössbauer Spectroscopy and its Applications. Cambridge University Press, Cambridge.

Dickson, D. P. E. and Berry, F. J. (eds.). 1986. Mössbauer Spectroscopy. Cambridge University Press, Cambridge.

Frauenfelder, H. 1962. The Mössbauer Effect: A Review with a Collection of Reprints. W. A. Benjamin, New York.

*This book by Frauenfelder was written in the early days of Mössbauer spectrometry, but contains a fine exposition of principles. More importantly, it contains reprints of the papers that first reported the phenomena that are the basis for much of Mössbauer spectrometry. It includes an English translation of one of Mössbauer's first papers.*

Gibb, T. C. 1976. Principles of Mössbauer Spectroscopy. Chapman and Hall, London.

Gonser, U. (ed.). 1975. Mössbauer Spectroscopy. Springer-Verlag, New York.

Gonser, U. (ed.). 1986. Microscopic Methods in Metals, Topics in Current Physics, 40, Springer-Verlag, Berlin.

Gruverman, I. J. (ed.). 1976. Mössbauer Effect Methodology, Vols. 1-10. Plenum Press, New York.

Gütlich, P., Link, R., and Trautwein, A. (eds.). 1978. Mössbauer Spectroscopy and Transition Metal Chemistry. Springer-Verlag, Berlin.

Long, G. J. and Grandjean, F. (eds.). 1984. Mössbauer Spectroscopy Applied to Inorganic Chemistry, Vols. 1-3. Plenum Press, New York.

Long, G. J. and Grandjean, F. (eds.). 1996. Mössbauer Spectroscopy Applied to Magnetism and Materials Science, Vols. 1 and 2. Plenum Press, New York.

*These 1996 volumes by Long and Grandjean contain review articles on different classes of materials, and on different techniques used in Mössbauer spectrometry.*

Long, G. J. and Stevens, J. G. (eds.). 1986. Industrial Applications of the Mössbauer Effect. Plenum, New York.

May, L. (ed.). 1971. An Introduction to Mössbauer Spectroscopy. Plenum Press, New York.

Mitra, S. (ed.). 1992. Applied Mössbauer Spectroscopy: Theory and Practice for Geochemists and Archaeologists. Pergamon Press, Elmsford, New York.

Thosar, B. V. and Iyengar, P. K. (eds.). 1983. Advances in Mössbauer Spectroscopy, Studies in Physical and Theoretical Chemistry 25. Elsevier/North Holland, Amsterdam.

Wertheim, G. 1964. Mössbauer Effect: Principles and Applications. Academic Press, New York.

## INTERNET RESOURCES

http://www.kfki.hu/~mixhp/

*The Mössbauer Information eXchange, MIX, is a project of the KFKI Research Institute for Particle and Nuclear Physics, Budapest, Hungary. It is primarily for scientists, students, and manufacturers involved in Mössbauer spectroscopy and other nuclear solid-state methods.*

http://www.unca.edu/medc

medc@unca.edu

*The Mössbauer Effect Data Center (University of North Carolina; J.G.. Stevens, Director) maintains a library of most publications involving the Mössbauer effect, including hard-to-access publications. Computerized databases and database search services are available to find papers on specific materials.*

BRENT FULTZ
California Institute of Technology
Pasadena, California

# X-RAY TECHNIQUES

## INTRODUCTION

X-ray scattering and spectroscopy methods can provide a wealth of information concerning the physical and electronic structure of crystalline and noncrystalline materials in a variety of external conditions and environments. X-ray powder diffraction, for example, is generally the first, and perhaps the most widely used, probe of crystal structure. Over the last nine decades, especially with the introduction of x-ray synchrotron sources during the last 25 years, x-ray techniques have expanded well beyond their initial role in structure determination.

This chapter explores the wide variety of applications of x-ray scattering and spectroscopy techniques to the study of materials. Materials, in this sense, include not only bulk condensed matter systems, but liquids and surfaces as well. The term "scattering," is generally used to include x-ray measurements on noncrystalline systems, such as glasses and liquids, or poorly crystallized materials, such as polymers, as well as x-ray diffraction from crystalline solids. Information concerning long-range, short-range, and chemical ordering as well as the existence, distribution and characterization of various defects is accessible through these kinds of measurements. Spectroscopy techniques generally make use of the energy dependence of the scattering or absorption cross-section to study chemical short-range order, identify the presence and location of chemical species, and probe the electronic structure through inelastic excitations.

X-ray scattering and spectroscopy provide information complementary to several other techniques found in this volume. Perhaps most closely related are electron and neutron scattering methods described in Chapters 11 and 13, respectively. The utility of all three methods in investigations of atomic scale structure arises from the close match of the wavelength of these probes to typical interatomic distances (a few Ångstroms). Of the three methods, the electron scattering interaction with matter is the strongest, so this technique is most appropriate to the study of surfaces or very thin samples. In addition, samples must be studied in an ultrahigh-vacuum environment. The relatively weak absorption of neutrons by most isotopes allows investigations of bulk samples. The principal neutron scattering interaction involves the nuclei of constituent elements and the magnetic moment of the outer electrons. Indeed, the cross-section for scattering from magnetic electrons is of the same order as scattering from the nuclei, so that this technique is of great utility in studying magnetic structures of magnetic materials. Neutron energies are typically on the order of a few meV to tens of meV, the same energy scale as many important elementary excitations in solids. Therefore inelastic neutron scattering has become a critical probe of elementary excitations including phonon and magnon dispersion in solids. The principal x-ray scattering interaction in materials involves the atomic electrons, but is significantly weaker than the electron-electron scattering cross-section. X-ray absorption by solids is relatively strong at typical energies found in laboratory apparatus (8 keV), penetrating a few tens of microns into a sample. At x-ray synchrotron sources, x-ray beams of energies above 100 keV are available. At these high energies, the absorption of x-rays approaches small values comparable to neutron scattering. The energy dependence of the x-ray absorption cross-section is punctuated by absorption edges or resonances that are associated with interatomic processes of constituent elements. These resonances can be used to great advantage to investigate element-specific properties of the material and are the basis of many of the spectroscopic techniques described in this chapter.

Facilities for x-ray scattering and spectroscopy range from turn-key powder diffraction instruments for laboratory use to very sophisticated beam lines at x-ray synchrotron sources around the world. The growth of these synchrotron facilities in recent years, with corresponding increases in energy and scattering angle resolution, energy tunability, polarization tunability, and raw intensity has spurred tremendous advances in this field. For example, the technique of x-ray absorption spectroscopy (XAS) requires a source of continuous radiation that can be tuned over approximately 1 keV through the absorption edge of interest. Although possible using the bremsstrahlung radiation from conventional x-ray tubes, full exploitation of this technique requires the high intensity, collimated beam of synchrotron radiation. As another example, the x-ray magnetic scattering cross-section is quite small (but finite) compared to the charge scattering cross-section. The high flux of x-rays available at synchrotron sources compared to neutron fluxes from reactors, however, allows x-ray magnetic scattering to effectively complement neutron magnetic scattering in several instances. Again, largely due to high x-ray fluxes and strong collimation, energy resolution in the meV range are achievable at current third generation x-ray sources, paving the way for new x-ray inelastic scattering studies of elementary excitations. Third generation sources include the Advanced Photon Source (APS) at the Argonne National Laboratory and the Advanced Light Source (ALS) at the Lawrence Berkeley National Laboratory in the United States, the European Synchrotron Radiation Facility (ESRF) in Europe, and Spring-8 in Japan.

ALAN I. GOLDMAN

## X-RAY POWDER DIFFRACTION

### INTRODUCTION

X-ray powder diffraction is used to determine the atomic structure of crystalline materials without the need for large (~100-μm) single crystals. "Powder" can be a misnomer; the technique is applicable to polycrystalline phases such as cast solids or films grown on a substrate. X-ray powder diffraction can be useful in a wide variety of situations. Below we list a number of questions that can be effectively addressed by this technique. This is an attempt to illustrate the versatility of x-ray powder diffraction and not, by any means, a complete list. Six experiments (corresponding to numbers 1 through 6 below), described later as concrete examples (see Practical Aspects of the Model), constitute an assortment of problems of varying difficulty and involvement that we have come across over the course of several years.

1. The positions and integrated intensities of a set of peaks in an x-ray powder diffraction pattern can be compared to a database of known materials in order to identify the contents of the sample and to determine the presence or absence of any particular phase.

2. A mixture of two or more crystalline phases can be easily and accurately analyzed in terms of its phase fractions, whether or not the crystal structures of all phases are known. This is called quantitative phase analysis, and it is particularly valuable if some or all of the phases are chemically identical and hence cannot be distinguished while in solution.

3. The crystal structure of a new or unknown material can be determined when a similar material with a known structure exists. Depending on the degree of similarity between the new and the old structure, this can be fairly straightforward.

4. The crystal structure of a new or unknown material can be solved *ab initio* even if no information about the material other than its stoichiometry is known. This case is significantly more difficult than the previous one, and it requires both high-resolution data and a significant investment of time and effort on the part of the investigator.

5. Phase transitions and solid-state reactions can be investigated in near real time by recording x-ray powder diffraction patterns as a function of time, pressure, and/or temperature.

6. Subtle structural details such as lattice vacancies of an otherwise known structure can be extracted. This also usually requires high-resolution data and a very high sample quality.

The variation of peak positions with sample orientation can be used to deduce information about the internal strain of a sample. This technique is not covered in this unit, and the interested reader is directed to references such as Noyan and Cohen (1987). Another related technique, not covered here, is texture analysis, the determination of the distribution of orientations in a polycrystalline sample.

X-ray powder diffraction is an old technique, in use for most of this century. The capabilities of the technique have recently grown for two main reasons: (1) development of x-ray sources and optics (e.g., synchrotron radiation, Göbel mirrors) and (2) the increasing power of computers and software for analysis of powder data. This unit discusses the fundamental principles of the technique, including the expression for the intensity of each diffraction peak, gives a brief overview of experimental techniques, describes several illustrative examples, gives a description of the procedures to interpret and analyze diffraction data, and discusses weaknesses and sources of possible errors.

## Competitive and Related Techniques

### Alternative Methods

1. Single-crystal x-ray diffraction requires single crystals of appropriate size (10 to 100 μm). Solution of single-crystal structures is usually more automated with appropriate software than powder structures. Single-crystal techniques can generally solve more complicated structures than powder techniques, whereas powder diffraction can determine the constituents of a mixture of crystalline solid phases.

2. Neutron powder diffraction generally requires larger samples than x rays. It is more sensitive to light atoms (especially hydrogen) than x rays are. Deuterated samples are often required if the specimen has a significant amount of hydrogen (see NEUTRON POWDER DIFFRACTION). Neutrons are sensitive to the magnetic structure (see MAGNETIC NEUTRON SCATTERING). Measurements must be performed at a special facility (reactor or spallation source).

3. Single-crystal neutron diffraction requires single-crystal samples of millimeter size, is more sensitive to light atoms (especially hydrogen) than are x rays, and often requires deuterated samples if the specimen has a significant amount of hydrogen. As in the powder case, it is sensitive to the magnetic structure and must be performed at a special facility (reactor or spallation source).

4. Electron diffraction can give lattice information for samples that have a strain distribution too great to be indexed by the x-ray powder technique. It provides spatial resolution for inhomogeneous samples and can view individual grains but requires relatively sophisticated equipment (electron microscope; see LOW-ENERGY ELECTRON DIFFRACTION).

## PRINCIPLES OF THE METHOD

If an x ray strikes an atom, it will be weakly scattered in all directions. If it encounters a periodic array of atoms, the waves scattered by each atom will reinforce in certain directions and cancel in others. Geometrically, one may imagine that a crystal is made up of families of lattice planes and that the scattering from a given family of planes will only be strong if the x rays reflected by each plane arrive at the detector in phase. This leads to a relationship between the x-ray wavelength $\lambda$, the spacing $d$ between lattice planes, and the angle of incidence $\theta$ known as Bragg's law, $\lambda = 2d\sin\theta$. Note that the angle of deviation of the x ray is $2\theta$ from its initial direction. This is fairly restrictive for a single crystal: for a given $\lambda$, even if the detector is set at the correct 2θ for a given $d$ spacing within the crystal, there will be no diffracted intensity unless the crystal is properly aligned to both the incident beam and the detector. The essence of the powder diffraction technique is to illuminate a large number of crystallites, so that a substantial number of them are in the correct orientation to diffract x rays into the detector.

The geometry of diffraction in a single grain is described through the reciprocal lattice. The fundamental concepts are in any of the Key References or in introductory solid-state physics texts such as Kittel (1996) or Ashcroft and Mermin (1976); also see SYMMETRY IN CRYSTALLOGRAPHY. If the crystal lattice is defined by three vectors **a**, **b**, and **c**, there are three reciprocal lattice vectors defined as

$$\mathbf{a}^* = \frac{2\pi \mathbf{b} \times \mathbf{c}}{\mathbf{a} \cdot \mathbf{b} \times \mathbf{c}} \qquad (1)$$

and cyclic permutations thereof for **b**\* and **c**\* (in the chemistry literature the factor $2\pi$ is usually eliminated in this definition). These vectors define the reciprocal lattice, with the significance that any three integers $(hkl)$ define a family of lattice planes with spacing $d = 2\pi/|h\mathbf{a}^* + k\mathbf{b}^* + l\mathbf{c}^*|$, so that the diffraction vector $\mathbf{K} = h\mathbf{a}^* + k\mathbf{b}^* + l\mathbf{c}^*$ satisfies $|\mathbf{K}| = 2\pi/d = 4\pi\sin\theta/\lambda$ (caution: most chemists and some physicists define $|\mathbf{K}| = 1/d = 2\sin\theta/\lambda$).

The intensity of the diffracted beam is governed by the unit cell structure factor, defined as

$$F_{hkl} = \sum_j e^{i\mathbf{K}\cdot\mathbf{R}_j} f_j e^{-2W} \qquad (2)$$

where $R_j$ is the position of the $j$th atom in the unit cell, the summation is taken over all atoms in the unit cell, and $f_j$ is the atomic scattering factor, tabulated in, e.g., the International Tables for Crystallography (Brown et al., 1992), and is equal to the number of atomic electrons at $2\theta = 0$, decreasing as a smooth function of $\sin\theta/\lambda$ (there are "anomalous" corrections to this amplitude if the x-ray energy is close to a transition in a target atom). The Debye-Waller factor $2W$ is given by $2W = K^2 u_{\text{rms}}^2/3$, where $u_{\text{rms}}$ is the (three-dimensional) root-mean-square deviation of the atom from its lattice position due to thermal and zero-point fluctuations. Experimental results are often quoted as the thermal parameter $B$, defined as $8\pi^2 u_{\text{rms}}^2/3$, so that the Debye-Waller factor is given by $2W = 2B\sin^2\theta/\lambda^2$. Note that thermal fluctuations of the atoms about their average position weaken the diffraction lines but do not broaden them. As long as the diffraction is sufficiently weak (kinematic limit, assumed valid for most powders), the diffracted intensity is proportional to the square of the structure factor. In powder diffraction, it is always useful to bear in mind that the positions of the observed peaks indicate the geometry of the lattice, both its dimensions and any internal symmetries, whereas the intensities are governed by the arrangement of atoms within the unit cell.

In a powder experiment, various factors act to spread the intensity over a finite range of the diffraction angle,

and so it is useful to consider the integrated intensity (power) of a given peak over the diffraction angle,

$$P_{hkl} = \left[\frac{P_0 \lambda^3 R_e^2 l}{16\pi R} \frac{V_s}{V^2}\right] M_{hkl} F_{hkl}^2 \left(\frac{1+\cos^2 2\theta}{\sin 2\theta \sin\theta}\right) \qquad (3)$$

where $P_0$ is the power density of the incident beam, $R_e = 2.82$ fm is the classical electron radius, $l$ and $R$ are the width of the receiving slit and the distance between it and the sample, and $V_s$ and $V$ are the effective illuminated volume of the sample and the volume of one unit cell. The term $M_{hkl}$ is the multiplicity of the $hkl$ peak, e.g., 8 for a cubic $hhh$ and 6 for a cubic $h00$, and $(1+\cos^2 2\theta)/(\sin\theta \sin 2\theta)$ is called the Lorentz polarization factor. The numerator takes the given form only for unpolarized incident radiation and in the absence of any other polarization-sensitive optical elements; it must be adapted, e.g., for a synchrotron-radiation source or a diffracted-beam monochromator. There are considerable experimental difficulties with measuring the absolute intensity either of the incident or the diffracted beam, and so the terms in the square brackets are usually lumped into a normalization factor, and one considers the relative intensity of different diffraction peaks, or more generally, the spectrum of intensity vs. scattering angle.

An inherent limitation on the amount of information that can be derived from a powder diffraction spectrum arises from possible overlap of Bragg peaks with different $(hkl)$, so that their intensities cannot be determined independently. This overlap may be exact, as in the coincidence of cubic (511) and (333) reflections, or it may allow a partial degree of separation, as in the case of two peaks whose positions differ by a fraction of their widths. Because powder diffraction peaks generally become broader and more closely spaced at higher angles, peak overlap is a factor in almost every powder diffraction experiment. Perhaps the most important advance in powder diffraction during the last 30 years is the development of whole-pattern (Rietveld) fitting techniques for dealing with partially overlapped peaks, as discussed below.

Diffraction peaks acquire a nonzero width from three main factors: instrumental resolution (not discussed here), finite grain size, and random strains. If the grains have a linear dimension $L$, then the full-width at half-maximum (FWHM) in 2θ, expressed in radians, of the diffraction line is estimated by the well-known Scherrer equation, $\text{FWHM}_{2\theta} = 0.89\lambda/L\cos\theta$. This is a reflection of the fact that the length $L$ of a crystal made up of atoms with period $d$ can only be determined to within $d$. One should not take the numerical value too seriously, as it depends on the precise shape of the crystallites and dispersion of the crystallite size distribution. If the crystallites have a needle or plate morphology, the size can be different for different families of lattice planes. On the other hand, if the crystallites are subject to a random distribution of lattice fractional strains having a FWHM of $e_{\text{FWHM}}$, the FWHM of the diffraction line will be $2\tan\theta e_{\text{FWHM}}$. It is sometimes asserted that size broadening produces a Lorentzian and strain a Gaussian lineshape, but there is no fundamental reason for this to be true, and counterexamples are frequently observed. If the sample peak width exceeds the instrumental resolution, or can be corrected for that effect, it can be informative to make a plot (called a Williamson-Hall plot) of FWHMcosθ vs. sinθ. If the data points fall on a smooth curve, the intercept will give the particle size and the limiting slope the strain distribution. Indeed, the curve will be a straight line if both effects give a Lorentzian lineshape, because the shape of any peak would be the convolution of the size and strain contributions. If the points in a Williamson-Hall plot are scattered, it may give useful information (or at least a warning) of anisotropic size or strain broadening. More elaborate techniques for the deconvolution of size and strain effects from experimental data are described in the literature (Klug and Alexander, 1974; Balzar and Ledbetter, 1993).

There are two major problems in using powder diffraction measurements to determine the atomic structure of a material. First, as noted above, peaks overlap, so that the measured intensity cannot be uniquely assigned to the correct Miller indices (hkl). Second, even if the intensities were perfectly separated so that the magnitudes of the structure factors were known, one could not Fourier transform the measured structure factors to learn the atomic positions because their phases are not known.

## PRACTICAL ASPECTS OF THE METHOD

The requirements to obtain a useful powder diffraction data set are conceptually straightforward: allow a beam of x rays to impinge on the sample and record the diffracted intensity as a function of angle. Practical realizations are governed by the desire to optimize various aspects of the measurement, such as the intensity, resolution, and discrimination against undesired effects (e.g., background from sample fluorescence).

Most laboratory powder x-ray diffractometers use a sealed x-ray tube with a target of copper, molybdenum, or some other metal. About half of the x rays from such a tube are in the characteristic $K_\alpha$ line ($\lambda = 1.54$Å for Cu, $\lambda = 0.70$Å for Mo), and the remainder are in other lines and in a continuous bremsstrahlung spectrum. Rotating-anode x-ray sources can be approximately ten times brighter than fixed targets, with an attendant cost in complexity and reliability. In either case, one can either use the x rays emitted by the anode directly (so that diffraction of the continuous component of the spectrum contributes a smooth background under the diffraction peaks) or select the line radiation by a crystal monochromator (using diffraction to pass only the correct wavelength) or by an energy-sensitive detector. The $K_\alpha$ line is actually a doublet (1.54051 and 1.54433 Å for Cu), which can create the added complication of split peaks unless one uses a monochromator of sufficient resolving power to pass only one component. Synchrotron radiation sources are finding increasing application for powder diffraction, due to their high intensity, intrinsically good collimation (~0.01° in the vertical direction) of x-ray beams, and tunability over a continuous spectrum and the proliferation of user facilities throughout the world.

There are a large number of detectors suitable for powder x-ray diffraction. Perhaps the simplest is photographic film, which allows the collection of an entire diffractogram at one time and, with proper procedures, can be used to obtain quantitative intensities with a dynamic range up to 100:1 (Klug and Alexander, 1974). An updated form of photographic film is the x-ray imaging plate, developed for medical radiography, which is read out electronically (Miyahara et al., 1986; Ito and Amemiya, 1991). The simplest electronic detector, the Geiger counter, is no longer widely used because of its rather long dead time, which limits the maximum count rate. The gas-filled proportional counter offers higher count rates and some degree of x-ray energy resolution. The most widely used x-ray detector today is the scintillation counter, in which x rays are converted into visible light, typically in a thallium-doped NaI crystal, and then into electronic pulses by a photomultiplier tube. Various semiconductor detectors [Si:Li, positive-intrinsic-negative (PIN)] offer energy resolutions of 100 to 300 eV, sufficient to distinguish fluorescence from different elements and from the diffracted x rays, although their count rate capability is generally lower than that of scintillation counters. There are various forms of electronic position-sensitive detectors. Gas-filled proportional detectors can have a spatial resolution of a small fraction of a millimeter and are available as straight-line detectors, limited to several degrees of 2θ by parallax, or as curved detectors covering an angular range as large as 120°. They can operate at a count rate up to $10^5$ Hz over the entire detector, but one must bear in mind that the count rate in one individual peak would be significantly less. Also, not all position-sensitive detectors are able to discriminate against x-ray fluorescence from the sample, although there is one elegant design using Kr gas and x rays just exceeding the Kr $K$ edge that addresses this problem (Smith, 1991). Charge-coupled devices (CCDs) are two-dimensional detectors that integrate the total energy deposited into each pixel and therefore may have a larger dynamic range and/or a faster time response (Clarke and Rowe, 1991).

Some of the most important configurations for x-ray powder diffraction instruments are illustrated in Figure 1. The simple Debye-Scherrer camera in (A) records a wide range of angles on curved photographic film but suffers from limited resolution. Modern incarnations include instruments using curved position-sensitive detectors and imaging plates and are in use at several synchrotron sources. It generally requires a thin rod-shaped sample either poured as a powder into a capillary or mixed with an appropriate binder and rolled into the desired shape. The Bragg-Brentano diffractometer illustrated in (B) utilizes parafocusing from a flat sample to increase the resolution available from a diverging x-ray beam; in this exaggerated sketch, the distribution of Bragg angles is 3°, despite the fact that the sample subtends an angle of 15° from source or detector. The addition of a diffracted-beam monochromator illustrated in (C) produces a marked improvement in performance by eliminating x-ray fluorescence from the sample. For high-pressure cells, with limited access for the x-ray beam, the energy-dispersive diffraction approach illustrated in (D) can be an attractive

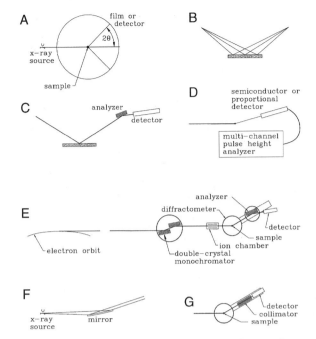

**Figure 1.** Schematic illustration of experimental setups for powder diffraction measurements.

these mechanisms will have an energy that is identical to or generally indistinguishable from the primary beam, and so it is not possible to eliminate these effects by using energy-sensitive detectors. Another important source of background is x-ray fluorescence, the emission of x rays by atoms in the target that have been ionized by the primary x-ray beam. X-ray fluorescence is always of a longer wavelength (lower energy) than the primary beam and so can be eliminated by the use of a diffracting crystal between sample and detector or an energy-sensitive detector (with the precaution that the fluorescence radiation does not saturate it) or controlled by appropriate choice of the incident wavelength.

It is not usually possible to predict the shape of this background, and so, once appropriate measures are taken to reduce it as much as practical, it is empirically separated from the relatively sharp powder diffraction peaks of interest. In data analysis such as the Rietveld method, one may take it as a piecewise linear or spline function between specified points or parameterize and adjust it to produce the best fit. There is some controversy about how to treat the statistical errors in a refinement with an adjustable background.

## DATA ANALYSIS AND INTERPRETATION

A very important technique for analysis of powder diffraction data is the whole-pattern fitting method proposed by Rietveld (1969). It is based on the following properties of x-ray (and neutron) powder diffraction data: a powder diffraction pattern usually comprises a large number of peaks, many of which overlap, often very seriously, making the separate direct measurement of their integrated intensities difficult or impossible. However, it is possible to describe the shape of all Bragg peaks in the pattern by a small (compared to the number of peaks) number of profile parameters. This allows the least-squares refinement of an atomic model combined with an appropriate peak shape function, i.e., a simulated powder pattern, directly against the measured powder pattern. This may be contrasted to the single-crystal case, where the atomic structure is refined against a list of extracted integrated intensities. The Rietveld method is an extremely powerful tool for the structural analysis of virtually all types of crystalline materials not available as single crystals.

The parameters refined in the Rietveld method fall into two classes: those that describe the shape and position of the Bragg peaks in the pattern (profile parameters) and those that describe the underlying atomic model (atomic or structural parameters). The former include the lattice parameters and those describing the shape and width of the Bragg peaks. In x-ray powder diffraction, a widely used peak shape function is the pseudo-Voigt function (Thompson et al., 1987), a fast-computing approximation to a convolution of a Gaussian and a Lorentzian (Voigt function). It uses only five parameters (usually called $U, V, W, X,$ and $Y$) to describe the shape of all peaks in the powder pattern. In particular, the peak widths are a smooth function of the scattering angle $2\theta$. Additional profile parameters are often used to describe the peak

solution. A polychromatic beam is scattered through a fixed angle, and the energy spectrum of the diffracted x rays is converted to $d$ spacings for interpretation.

A typical synchrotron powder beamline such as X7A or X3B1 at the National Synchrotron Light Source (Brookhaven National Laboratory) is illustrated in (E). One particular wavelength is selected by the dual-crystal monochromator, and the x rays are diffracted again from an analyzer crystal mounted on the 2θ arm. Monochromator and analyzer are typically semiconductor crystals with rocking-curve widths of a few thousandths of a degree. Besides its intrinsically high resolution, this configuration offers the strong advantage that it truly measures the angle through which a parallel beam of radiation is scattered and so is relatively insensitive to parallax or sample transparency errors. The advantage of a parallel incident beam is available on laboratory sources by the use of a curved multilayer (Göbel) mirror (F), currently commercialized by Bruker Analytical X-ray Systems. One can also measure the diffracted angle, free from parallax, by use of a parallel-blade collimator as shown in (G). This generally gives higher intensity than an analyzer crystal at the cost of coarser angular resolution (0.1° to 0.03°), but this is often not a disadvantage because diffraction peaks from real samples are almost always broader than the analyzer rocking curve. However, the analyzer crystal also discriminates against fluorescence.

There are a number of sources of background in a powder diffraction experiment. X rays can be scattered from the sample by a number of mechanisms other than Bragg diffraction: Compton scattering, thermal diffuse scattering, scattering by defects in the crystal lattice, multiple scattering in the sample, and scattering by noncrystalline components of the sample, by the sample holder, or even by air in the x-ray beam path. The x rays scattered by all of

asymmetry at low angles due to the intersection of the curved Debye-Scherrer cone of radiation with a straight receiving slit and corrections for preferred orientation. The structural parameters include the positions, types, and occupancies of the atoms in the structural model and isotropic or anisotropic thermal parameters (Debye-Waller factors). The power of the Rietveld method lies in simultaneous refinement of both profile and atomic parameters, thereby maximizing the amount of information obtained from the powder data.

The analysis of x-ray powder diffraction data can be divided into a number of separate steps. While some of these steps rely on the correct completion of the previous one(s), they generally constitute independent tasks to be completed by the experimenter. Depending on the issues to be addressed by any particular experiment, one, several, or all of these tasks will be encountered, usually in the order in which they are described here.

## Crystal Lattice and Space Group Determination from X-Ray Powder Data

Obtaining the lattice parameters (unit cell) of an unknown material is always the first step on the path to a structure solution. While predicting the scattering angles of a set of diffraction peaks given the unit cell is trivial, the inverse task is not, because the powder method reduces the three-dimensional atomic distribution in real space to a one-dimensional diffraction pattern in momentum space. In all but the simplest cases, a computer is required and a variety of programs are available for the task. They generally fall into two categories: those that attempt to find a suitable unit cell by trial and error (also often called the "semiexhaustive" method) and those that use an analytical approach based on the geometry of the reciprocal lattice. Descriptions of these approaches can be found below [see Key References, particularly Bish and Post (1989, pp. 188–196) and Klug and Alexander (1974, Section 6)]. Some widely used auto-indexing programs are TREOR (Werner et al., 1985), ITO (Visser, 1969), and DICVOL (Boultif and Louër, 1991). All of these programs take either a series of peak positions or corresponding $d$ spacings as input, and their accuracy is crucial for this step to succeed. The programs may tolerate reasonably small random errors in the peak position; however, even a small systematic error will prevent them from finding the correct (or any) solution. Hence, the data set used to extract peak positions must be checked carefully for systematic errors such as the diffractometer zero point, for example, by looking at pairs of Bragg peaks whose values of $\sin\theta$ are related by integer multiples.

Once the unit cell is known, the space group can be determined from systematic absences of Bragg peaks, i.e., Bragg peaks allowed by the unit cell but not observed in the actual spectrum. For example, the observed spectrum of a face-centered-cubic (fcc) material contains only Bragg peaks ($hkl$) for which $h$, $k$, and $l$ are either all odd or all even. Tables listing possible systematic absences for each crystal symmetry class and correlating them with space groups can be found in the International Tables for Crystallography (Vos and Buerger, 1996).

## Extracting Integrated Intensities of Bragg Peaks

The extraction of accurate integrated intensities of Bragg peaks is a prerequisite for several other steps in x-ray powder diffraction data analysis. However, the fraction of peaks in the observed spectrum that can be individually fitted is usually very small. The problem of peak overlap is partially overcome in the Rietveld method by describing the shapes of all Bragg peaks observed in the x-ray powder pattern by a small number of profile parameters. It is then possible to perform a refinement of the powder pattern without any atomic model using (and refining) only the lattice and profile parameters to describe the position and shape of all Bragg peaks and letting the intensity of each peak vary freely; two variations are referred to as the Pawley (1981) and LeBail (1988) methods. While this does not solve the problem of exact peak overlap [e.g., of the cubic (511) and (333) peaks], it is very powerful in separately determining the intensities of clustered, partially overlapping peaks. The integrated intensities of individual peaks that can be determined in this manner usually constitute a significant fraction of the total number of allowed Bragg peaks, and they can then be used as input to search for candidate atom positions. Many Rietveld programs (see The Final Rietveld Refinement, below) feature an intensity extraction mode that can be used for this task. The resulting fit can also serve as an upper bound for the achievable goodness of fit in the final Rietveld refinement. EXTRA (Altomare et al., 1995) is another intensity extraction routine that interfaces directly with SIRPOW.92 (see Search for Candidate Atom Positions, below), a program that utilizes direct methods. The program EXPO consists of both the EXTRA and SIRPOW.92 modules.

## Search for Candidate Atom Positions

When a suitable starting model for a crystal structure is not known and cannot easily be guessed, it must be determined from the x-ray powder data before any Rietveld refinements can be performed. The Fourier transform of the distribution of electrons in a crystal is the x-ray scattering amplitude. However, only the scattering intensity can be measured, and hence all phase information is lost. Nevertheless, the Fourier transform of the intensities can be useful in finding some of the atoms in the unit cell. A plot of the Fourier transform of the measured scattering intensities is called a Patterson map, and its peaks correspond to translation vectors between pairs of atoms in the unit cell. Obviously, the strongest peak will be at the origin, and depending on the crystal structure, it may be anywhere from simple to impossible to deduce atom positions from a Patterson map. Cases where there is one (or few) relatively heavy atom(s) in the unit cell are most favorable for this approach. Similarly, if the positions of most of the atoms in the unit cell (or at least of the heavy atoms) are already known, it may be reasonable to guess that the phases are dominated by the known part of the atomic structure. Then the differences between the measured intensities and those calculated from the known atoms together with the phase information calculated from the known atoms can be used to obtain a difference Fourier map that, if successful, will indicate the positions of the remaining atoms. The ability to calculate and plot Patterson and Fourier maps is included in common Rietveld packages (see below).

Another approach to finding atom candidate positions is the use of direct methods originally developed (and widely used) for the solution of crystal structures from single-crystal x-ray data. A mathematical description is beyond the scope of this unit; the reader is referred to the literature (e.g., Giacovazzo, 1992, pp. 335–365). In direct methods, an attempt is made to derive the phase of the structure factor directly from the observed amplitudes through mathematical relationships. This is feasible because the electron density function is positive everywhere and consists of discrete atoms. These two properties, "positivity" and "atomicity," are then used to establish likely relationships between the phases of certain groups of Bragg peaks. If the intensities of enough such groups of Bragg peaks have been measured accurately, this method yields the positions of some or all the atoms in the unit cell. The program SIRPOW.92 (Altomare et al., 1994) is an adaptation of direct methods to x-ray powder diffraction data and has been used to solve a number of organic and organometallic crystal structures from such data.

We emphasize that these procedures are not straightforward, and even for experienced researchers, success is far from guaranteed. However, the chance of success increases appreciably with the quality of the data, in particular with the resolution of the x-ray powder pattern.

### The Final Rietveld Refinement

Once a suitable starting model is found, the Rietveld method allows the simultaneous refinement of structural parameters such as atomic positions, site occupancies, isotropic or anisotropic Debye-Waller factors, along with lattice and profile parameters, against the observed x-ray powder diffraction pattern. Since the refinement is usually performed using a least-squares algorithm [chi-square ($\chi^2$) minimization], the result will be a local minimum close to the set of starting parameters. It is the responsibility of the experimenter to confirm that this is the global minimum, i.e., that the model is in fact correct. There are criteria based on the goodness-of-fit of the final refinement that, if they are not fulfilled, almost certainly indicate a wrong solution. However, such criteria are not sufficient in establishing the correctness of a model. For example, if an inspection of the final fit shows that a relatively small amount of total residual error is noticeably concentrated in a few peaks rather than being distributed evenly over the entire spectrum, that may be an indication that the model is incorrect. More details on this subject are given in the International Tables for Crystallography (Prince and Spiegelmann, 1992) and in Young (1993) and Bish and Post (1989).

A number of frequently updated Rietveld refinement program packages are available that can be run on various platforms, most of which are distant descendants of the original program by Rietveld (1969). Packages commonly in use today and freely distributed for noncommercial use by their authors include—among several—GSAS (Larson and Von Dreele, 1994) and FULLPROF (Rodriguez-Carvajal, 1990). Notably, both include the latest descriptions of the x-ray powder diffraction peak shape functions (Thompson et al., 1987), analytical descriptions of the peak shape asymmetry at low angles (Finger et al., 1994), and an elementary description for anisotropic strain broadening for different crystal symmetries (Stephens, 1999). Both programs contain comprehensive documentation, which is no substitute for the key references discussing the Rietveld method (Young, 1993; Bish and Post, 1989).

### Estimated Standard Deviations in Rietveld Refinements

The assignment of estimated standard deviations (ESDs) of the atomic parameters obtained from Rietveld refinements is a delicate matter. In general, the ESDs calculated by the Rietveld program are measures of the precision (statistical variation between equivalent experiments) rather than the accuracy (discrepancy from the correct value) of any given parameter. The latter cannot, in principle, be determined experimentally, since the "truly" correct model describing experimental data remains unknown. However, it is the accuracy that the experimenter is interested in, and it is desired to make the best possible estimate of it, based both on statistical considerations and the experience and judgment of the experimenter.

The Rietveld program considers each data point of the measured powder pattern to be an independent measurement of the Bragg peak(s) contributing to it. This implicit assumption holds only if the difference between the refined profile and the experimental data is from counting statistics alone (i.e., if $\chi^2 = 1$ and the points of the difference curve form a random, uncorrelated distribution with center 0 and variance $N$). However, this is not true for real experiments in which the dominant contributions to the difference curve are an imperfect description of the peak shapes and/or an imperfect atomic model. Consequently, accepting the precision of any refined parameter as a measure of its accuracy cannot be justified and would usually be unreasonably optimistic.

A number of corrections to the Rietveld ESDs have been proposed, and even though all are empirical in nature and there is no statistical justification for these procedures, they can be very valuable in order to obtain estimates. If one or more extra parameters are introduced into a model, the fit to the data will invariably improve. The $F$ statistic can be used to determine the likelihood that this improvement represents additional information (i.e., is statistically significant) and is not purely by chance (Prince and Spiegelmann, 1992). This can also be used to determine how far any given parameter must be shifted away from its ideal value in order to make a statistically significant difference in the goodness of fit. This can be used as an estimate of the accuracy of that parameter that is independent of its precision as calculated by the Rietveld program. There exists no rigid rule on whether precision or accuracy should be quoted when presenting results obtained from x-ray powder diffraction; it certainly depends on the motivation and outcome of the experiment. However, it is very important to distinguish between the two. We note that the International Union of Crystallography has sponsored a Rietveld refinement round robin, during which identical samples of $ZrO_2$ were measured and Rietveld refined by 51 participants (Hill and Cranswick, 1994), giving an empirical indication of the achievable accuracy from the technique.

### Quantitative Phase Analysis

Quantitative phase analysis refers to the important technique of determining the amount of various crystalline phases in a mixture. It can be performed in two ways, based on the intensities of selected peaks or on a multiphase Rietveld refinement.

In the first case, suppose that the sample is a flat plate that is optically thick to x rays. In a mixture of two phases with mass absorption coefficients $\mu_1^*$ and $\mu_2^*$ ($cm^2/g$), the intensity of a given peak from phase 1 will be reduced from its value for a pure sample of phase 1 by the ratio

$$\frac{I_1(x_1)}{I_1(\text{pure})} = \frac{x_1\mu_1^*}{x_1\mu_1^* + (1-x_1)\mu_2^*} \qquad (4)$$

where $x_1$ is the weight fraction of phase 1. Note that the intensity is affected both by dilution and by the change of the sample's absorption constant. When the absorption coefficients are not known, they can be determined from experimental measurements, such as by "spiking" the mixture with an additional amount of one of its constituents. Details are given in Klug and Alexander (1974) and Bish and Post (1989).

If the atomic structures of the constituents are known, a multiphase Rietveld refinement (see above) directly yields scale factors $s_j$ for each phase. The weight fraction $w_j$ of the $j$th phase can then be calculated by

$$w_j = \frac{s_j Z_j m_j V_j}{\sum_i s_i Z_i m_i V_i} \qquad (5)$$

where $Z_j$ is the number of formula units per unit cell, $m_j$ is the mass of a formula unit, and $V_j$ is the unit cell volume.

In either case, the expressions above are given under the assumption that the powders are sufficiently fine, i.e., the product of the linear absorption coefficient $\mu$ ($cm^{-1}$, equal to $\mu^*\rho$) and the linear particle (not sample) size $D$ is small ($\mu D < 0.01$). If not, one must make allowance for microabsorption, as described by Brindley (1945).

It is worth noting that the sensitivity of an x-ray powder diffraction experiment to weak peaks originating from trace phases is governed by the signal-to-background ratio. Hence, in order to obtain maximum sensitivity, it is desirable to reduce the relative background as much as possible.

## SAMPLE PREPARATION

The preparation of samples to avoid unwanted artifacts is an important consideration in powder diffraction experiments. One issue is that preferred orientation (texture) should be avoided or controlled. The grains of a sample may tend to align, especially if they have a needle or platelike morphology, so that reflections in certain directions are enhanced relative to others. Various measures such as mixing the powder with a binder or an inert material chosen to randomize the grains or pouring the sample sideways into the flat plate sample holder are in common use. [More details are given in, e.g., Klug and Alexander (1974), Bish and Post (1989), and Jenkins and Snyder (1996).] It is also possible to correct experimental data if one can model the distribution of crystallite orientations; this option is available in most common Rietveld programs (see The Final Rietveld Refinement, above).

A related issue is that there must be a sufficient number of individual grains participating in the diffraction measurement to ensure a valid statistical sample. It may be necessary to grind and sieve the sample, especially in the case of strongly absorbing materials. However, grinding can introduce strain broadening into the pattern, and some experimentation is usually necessary to find the optimum means of preparing a sample. A useful test of whether a specimen in a diffractometer is sufficiently powdered is to scan the sample angle $\theta$ over several degrees while leaving $2\theta$ fixed at the value of a strong Bragg peak, in steps of perhaps $0.01°$; fluctuations of more than a few percent indicate trouble. It is good practice to rock (flat plate) or twirl (capillary) the sample during data collection to increase the number of observed grains.

## PROBLEMS

X rays are absorbed by any material through which they pass, and so the effective volume is generally not equal to the actual volume of the sample. Absorption constants are typically tabulated as the mass absorption coefficient $\mu/\rho$ ($cm^2/g$), which makes it easy to work out the attenuation length in a sample as a function of its composition. A plot of the absorption lengths (in micrometers) of some "typical" samples vs. x-ray energy/wavelength is given in Figure 2. To determine the relative intensities of different

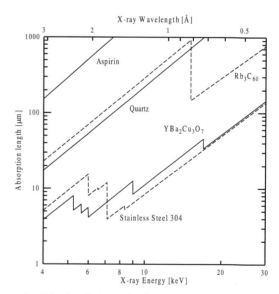

**Figure 2.** Calculated absorption lengths for several materials as a function of x-ray wavelength or energy.

peaks, it is important to either arrange for the effective volume to be constant or be able to make a quantitative correction. If the sample is a flat plate in symmetrical geometry, as illustrated in Figure 1B, the effective volume is independent of diffraction angle as long as the beam does not spill over the edge of the sample. If the sample is a cylinder and the absorption constant is known, the angle-dependent attenuation constant is tabulated, e.g., in the International Tables for Crystallography (Maslen, 1992). It is also important to ensure that the slits and other optical components of the diffractometer are properly aligned so that the variation of illuminated sample volume with $2\theta$ is well controlled and understood by the experimenter.

Inasmuch as a powder diffraction experiment consists of "simply" measuring the diffracted intensity as a function of angle, one can classify sources of systematic error as arising from measurement of the intensity or the angle or from the underlying assumptions. Errors of intensity can arise from detector saturation, drift or fluctuations of the strength of the x-ray source, and the statistical fluctuations intrinsic to counting the random arrival of photons.

Error of the angle can arise from mechanical faults or instability of the instrument or from displacement of the sample from the axis of the diffractometer (parallax). A subtle form of parallax can occur for flat-plate samples that are moderately transparent to x rays, because the origin of the diffracted radiation is located below the surface of the sample by a distance that depends on the diffraction angle. Another effect that can give rise to an apparent shift of diffraction peaks is the asymmetry caused by the intersection of the curved Debye-Scherrer cone of radiation with a straight receiving slit. This geometric effect has been discussed by several authors (Finger et al., 1994, and references therein), and it can produce a significant shift in the apparent position of low-angle peaks if fitted with a model lineshape that does not properly account for it. A potential source of angle errors in very high precision work is the refraction of x rays; to an x ray, most of the electrons of a solid appear free, and if their number density is $n$, the index of refraction is given by

$$1 - \left(\frac{e^2}{mc^2}\right)\frac{n\lambda^2}{2\pi} \qquad (6)$$

so that the observed Bragg angle is slightly shifted from its value inside the sample.

The interpretation of powder intensities is based on a number of assumptions that may or may not correspond to experimental reality in any given case.

The integrated intensity is proportional to the square of the structure factor only if the diffracted radiation is sufficiently weak that it does not remove a significant fraction of the incident beam. If the individual grains of the powder sample are too large, strong reflections will be effectively weakened by this "extinction" effect. The basic phenomenon is described in any standard crystallography text, and a treatment specific to powder diffraction is given by Sabine (1993). This can be an issue for grains larger than $\sim 1$ μm, which is well below the size that can be easily guaranteed by passing the sample through a sieve. Consequently, it is

not easy to assure *a priori* that extinction is not a problem. One might be suspicious of an extinction effect if a refinement shows that the strongest peaks are weaker than predicted by a model that is otherwise satisfactory.

To satisfy the basic premise of powder diffraction, there must be enough grains in the effective volume to produce a statistical sample. This may be a particularly serious issue in highly absorbing materials, for which only a small number of grains at the surface of the sample participate. One can test for this possibility by measuring the intensity of a strong reflection at constant $2\theta$ as a function of sample rotation; if the intensity fluctuates by more than a few percent, it is likely that there is an insufficient number of grains in the sample. One can partly overcome this problem by moving the sample during the measurement to increase the number of grains sampled. A capillary sample can be spun around its axis, while a flat plate can be rocked by a few degrees about the dividing position or rotated about its normal (or both) to achieve this aim.

If the grains of a powder sample are not randomly oriented, it will distort the powder diffraction pattern by making the peaks in certain directions stronger or weaker than they would be in the ideal case. This arises most frequently if the crystallites have a needle- or platelike morphology. [Sample preparation procedures to control this effect are described in, e.g., Klug and Alexander (1974), Bish and Post (1989), and Jenkins and Snyder (1996).] It is also possible to correct experimental data if one can model the distribution of crystallite orientations.

There are a number of standard samples that are useful for verifying and monitoring instrument performance and as internal standards. The National Institute of Standards and Technology sells several materials, such as SRM 640 (Silicon Powder), SRM 660 ($LaB_6$, which has exceedingly sharp lines), SRM 674a (a set of five metal oxides of various absorption lengths, useful for quantitative analysis standards), and SRM 1976 (corundum $Al_2O_3$ plate, with somewhat sharper peaks than the same compound in SRM 674a and certified relative intensities). Silver behenate powder has been proposed as a useful standard with a very large lattice spacing of $c = 58.4$ Å (Blanton et al., 1995).

## EXAMPLES

### Comparison Against a Database of Known Materials

An ongoing investigation of novel carbon materials yielded a series of samples with new and potentially interesting properties: (1) the strongest signal in a mass spectrometer was shown to correspond to the equivalent of an integer number of carbon atoms, (2) electron diffraction indicated that the samples were at least partially crystalline, and (3) while the material was predicted to consist entirely of carbon, the procedures for synthesis and purification had involved an organic Li compound. The top half of Figure 3 shows an x-ray powder diffraction pattern of one such sample recorded at $\lambda = 0.7$ Å at beamline X3B1 of the National Synchrotron Light Source. The peak positions can be indexed (Werner et al., 1985) to a monoclinic unit cell: $a = 8.36$ Å, $b = 4.97$ Å, $c = 6.19$ Å, and $\beta = 114.7°$.

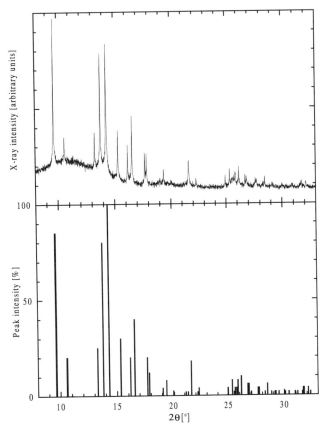

**Figure 3.** X-ray powder diffraction pattern of "unknown" sample (top) and relative peak intensities of $Li_2O_3$ as listed in the PDF database (bottom), both at $\lambda = 0.7$ Å.

The Powder Diffraction File (PDF) database provided by the International Center for Diffraction Data (ICDD, 1998) allows for searching based on several search criteria, including types of atoms, positions of Bragg peaks, or unit cell parameters. Note that it is not necessary to know the unit cell of a compound to search the database against its diffraction pattern, nor is it necessary to know the crystal structures (or even the unit cells) of all the compounds included in the database.

Searches for materials with an appropriate position of their strong diffraction peaks containing C only, or containing some or all of C, N, O, or H, did not result in any candidate materials. However, including Li in the list of possible atoms matched the measured pattern with the one in the database for synthetic $Li_2CO_3$ (zabuyelite). The peak intensities listed in the database, converted to $2\theta$ for the appropriate wavelength, are shown in the bottom half of Figure 3, leading to the conclusion that the only crystalline fraction of the candidate sample is $Li_2CO_3$, not a new carbon material.

This type of experiment is fairly simple, and both data collection and analysis can be performed quickly. Average sample quality is sufficient, and usually so are relatively low-resolution data from an x-ray tube. In fact, this test is routinely performed before any attempt to solve the structure of an unknown and presumably new material.

## Quantitative Phase Analysis

The qualitative and quantitative detection of small traces of polymorphs plays a major role in pharmacology, since many relevant molecules form two or more different crystal structures, and phase purity rather than chemical purity is required for scientific, regulatory, and patent-related legal reasons. Mixtures of the $\alpha$ and $\gamma$ phases of the anti-inflammatory drug indomethacin ($C_{19}H_{16}ClNO_4$) provide a good example for the detection limit of x-ray powder diffraction experiments in this case (Dinnebier et al., 1996).

Mixtures of 10, 1, 0.1, 0.01, and 0 wt% $\alpha$ phase (with the balance $\gamma$ phase) were investigated. Using its five strongest Bragg peaks, 0.1% $\alpha$ phase can easily be detected, and the fact that the "pure" $\gamma$ phase from which the mixtures were prepared contains traces of $\alpha$ became evident, making the quantification of the detection limits below 0.1% difficult. Figure 4 shows the strong (120) peak of the $\alpha$ phase and the weak (010) peak (0.15% intensity of its strongest peak) of the $\gamma$ phase for three different concentrations of the $\alpha$ phase.

While these experiments and the accompanying data analysis are not difficult to perform per se, it is probably necessary to have access to a synchrotron radiation source with its high resolution to obtain detection limits comparable to or even close to the values mentioned. The data of this example were collected on beamline X3B1 of the National Synchrotron Light Source.

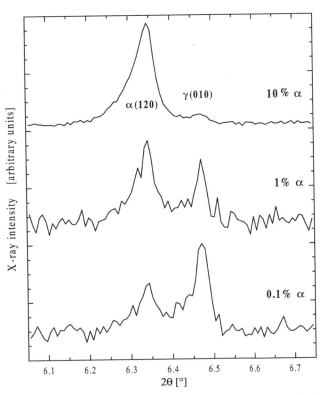

**Figure 4.** Strong (120) peak of the $\alpha$ phase and weak (010) peak of the $\gamma$ phase of the drug indomethacin for different concentrations of $\alpha$ phase.

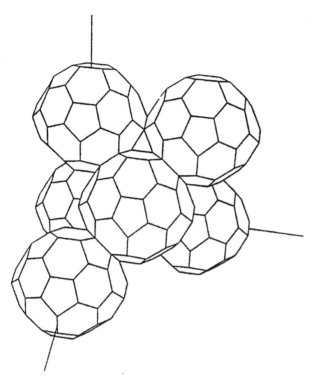

**Figure 5** Atomic clusters R(Al-Li-Cu) exhibiting nearly icosahedral symmetry.

## Structure Determination in a Known Analog

The icosahedral quasicrystal phase $i$(Al-Li-Cu) is close to a crystalline phase, $R$(Al-Li-Cu), which has a very similar local structure, making the crystal structure of the latter an interesting possible basis for the structure of the former. The crystal symmetry [body-centered-cubic (bcc)], lattice parameter ($a = 13.89$ Å), space group ($Im\bar{3}$), and composition ($Al_{5.6}Li_{2.9}Cu$) suggest that the crystal structure of $R$(Al-Li-Cu) is similar to that of $Mg_{32}(Zn, Al)_{49}$ (Bergman et al., 1952, 1957). However, using the atomic parameters of the latter compound does not give a satisfactory fit to the data of the former. On the other hand, a Rietveld analysis (see Data Analysis and Interpretation, above) using those parameters as starting values converges and gives a stable refinement of occupancies and atom positions (Guryan et al., 1988).

Such an experiment can be done with a standard laboratory x-ray source. This example was solved using Mo $K_\alpha$ radiation from a rotating-anode x-ray generator. Figure 5 shows the nearly icosahedral symmetry of the atomic clusters of which this material is comprised.

## *Ab Initio* Structure Determination

The Kolbe-Schmitt reaction, which proceeds from sodium phenolate, $C_6H_5ONa$, is the starting point for the synthesis of many pigments, fertilizers, and pharmaceuticals such as aspirin. However, replacing Na with heavier alkalis thwarts the reaction, a fact known since 1874 yet not understood to date. This provides motivation for the solution of the crystal structure of potassium phenolate, $C_6H_5OK$ (Dinnebier et al., 1997).

A high-resolution x-ray powder diffraction pattern of a sample of $C_6H_5OK$ was measured at room temperature, $\lambda = 1.15$ Å, and $2\theta = 5°-65°$ with step size 0.005° for a total counting time of 20 hr on beamline X3B1 of the National Synchrotron Light Source. Low-angle diffraction peaks had a FWHM of 0.013°. Following are the steps that were necessary to obtain the structure solution of this compound.

The powder pattern was indexed (Visser, 1969) to an orthorhombic unit cell, $a = 14.10$ Å, $b = 17.91$ Å, and $c = 7.16$ Å. Observed systematic absences of Bragg peaks, e.g., of all ($h0l$) peaks with $h$ odd, were used to determine the space group, $Pna2_1$ (Vos and Buerger, 1996), and the number of formula units per unit cell ($Z = 12$) was determined from geometrical considerations.

Next, using the LeBail technique (Le Bail et al., 1988), i.e., fitting the powder pattern to lattice and profile parameters without any structural model, ~300 Bragg peak intensities were extracted. Using these as input for the direct-methods program SIRPOW.92 (Altomare et al., 1994), it was possible to deduce the positions of potassium and some candidate oxygen atoms but none of the carbon atoms. In the next stage, the positions of the centers of the phenyl rings were searched by replacing them with diffuse pseudoatoms having the same number of electrons but a high temperature factor. After Rietveld refining this model (see Data Analysis and Interpretation, above) one pseudoatom at a time was back substituted for the corresponding phenyl ring, and the orientation of each phenyl ring was found via a grid search of all of its possible orientations. Upon completion of this procedure, the coordinates of all nonhydrogen atoms had been obtained.

The final Rietveld refinement (see Data Analysis and Interpretation) was performed using the program GSAS (Larson and Von Dreele, 1994), the hydrogen atoms were included, and the model remained stable when all structural parameters were refined simultaneously. Figure 6 shows a plot of the final refinement of $C_6H_5OK$, and Figure 7 shows the crystal structure obtained.

This example illustrates the power of high-resolution x-ray powder diffraction for the solution of rather complicated crystal structures. There are 24 nonhydrogen atoms in the asymmetric unit, which until recently would have been impossible to solve without the availability of a single crystal. At the same time, it is worth emphasizing that such an *ab initio* structure solution from x-ray powder diffraction requires both a sample of extremely high quality and an instrument with very high resolution. By comparison, it was not possible to obtain the crystal structure of $C_6H_5OLi$ from a data set with approximately 3 times fewer independently observed peaks due to poorer crystallinity of the sample. Even with a sample and data of sufficient quality, solving a structure *ab initio* from a powder sample (unlike the single-crystal case) requires experience and a significant investment of time and effort on the part of the researcher.

## Time-Resolved X-Ray Powder Diffraction

Zeolites are widely used in a number of applications, for example, as catalysts and for gas separations. Their

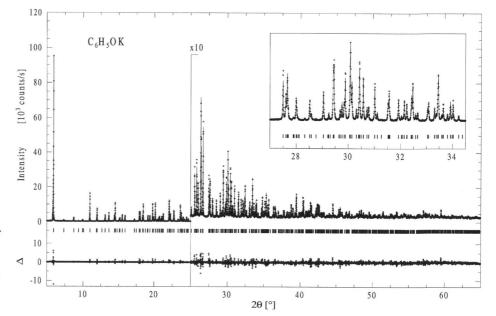

**Figure 6.** Rietveld refinement of potassium phenolate, $C_6H_5OK$. Diamonds denote x-ray powder diffraction data; solid line denotes atomic model. Difference curve is given below.

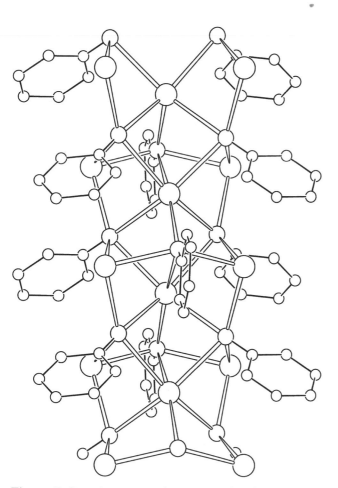

**Figure 7.** Crystal structure of potassium phenolate, $C_6H_5OK$, solved *ab initio* from x-ray powder diffraction data.

structural properties, particularly the location of the extra-framework cations, are important in understanding their role in these processes.

Samples of CsY zeolite dehydrated at 300° and at 500°C show differences in their diffraction patterns (Fig. 8). Rietveld analysis (see Data Analysis and Interpretation, above) shows that the changes occur in the extra-framework cations. The basic framework contains large so-called supercages and smaller sodalite cages, and in dehydration at 300°C, almost all of the Cs cations occupy the supercages while the Na cations occupy the sodalite cages. However, after dehydration at 500°C, a fraction of both Cs and Na ions will populate each other's cage position, resulting in mixed occupancy for both sites (Poshni et al., 1997; Norby et al., 1998).

To observe this transition while it occurs, the subtle changes in the CsY zeolite structure were followed by *in situ* dehydration under vacuum. The sample, loosely packed in a 0.7-mm quartz capillary, was ramped to a final temperature of 500°C over a period of 8 hr. Figure 9 shows the evolution of the diffraction pattern as a function of time during the dehydration process (Poshni et al., 1997). The data were collected at beamline X7B of the National Synchrotron Light Source with a translating image plate data collection stage.

To investigate phase transitions and solid-state reactions in near real time by recording x-ray powder diffraction patterns as a function of time, pressure, and/or temperature, a number of conditions must be met. For data collection, a position-sensitive detector or a translating image plate stage is required. Also, since there can be no collimator slits or analyzer crystal between the sample and the detector/image plate, such an experiment is optimally carried out at a synchrotron radiation source. These have an extremely low vertical divergence, allowing for a reasonably good angular resolution in this detector

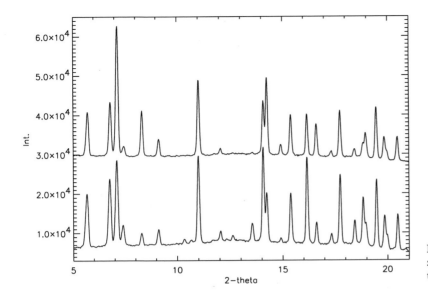

**Figure 8.** X-ray powder diffraction patterns of CsY zeolite dehydrated at 500°C (top) and 300°C (bottom).

geometry. The sample environment must be designed with care in order to provide, e.g., the desired control of pressure, temperature, and chemical environment compatible with appropriate x-ray access and low background. Examples include a cryostat with a Be can, a sample heater, a diamond-anvil pressure apparatus, and a setup that allows a chemical reaction to take place inside the capillary.

### Determination of Subtle Structural Details

The superconductor $Rb_3C_{60}$ has $T_c \approx 30$ K. The fullerenes form an fcc lattice, and of the three $Rb^+$ cations, one occupies the large octahedral ($O$) site at $(\frac{1}{2}, 0, 0)$ and the remaining two occupy the small tetrahedral ($T$) sites at $(\frac{1}{4}, \frac{1}{4}, \frac{1}{4})$. The size mismatch between the smaller $Rb^+$ ions and the larger octahedral site led to the suggestion that the $Rb^+$ cations in the large octahedral site could be displaced from the site center, supported by some nuclear

magnetic resonance (NMR) and extended x-ray absorption fine structure (EXAFS) experiments (see XAFS SPECTROSCOPY). If true, such a displacement would have a significant impact on the electronic and superconductin properties. By comparison, the Rb–C distances of the $Rb^+$ cation in the tetrahedral site are consistent with ionic radii.

In principle, a single x-ray pattern cannot distinguish between static displacement of an atom from its site center (as proposed for this system) and dynamic thermal fluctuation about a site (as described by a Debye-Waller factor). Hence, to address this issue, an x-ray powder diffraction study of $Rb_3C_{60}$ at various temperatures was carried out at beamline X3B1 of the National Synchrotron Light Source (Bendele et al., 1998).

At each temperature the data were Rietveld refined (see Data Analysis and Interpretation, above) against two competing (and otherwise identical) structural models: (1) The octahedral $Rb^+$ cation is fixed at $(\frac{1}{2}, 0, 0)$ and its isotropic

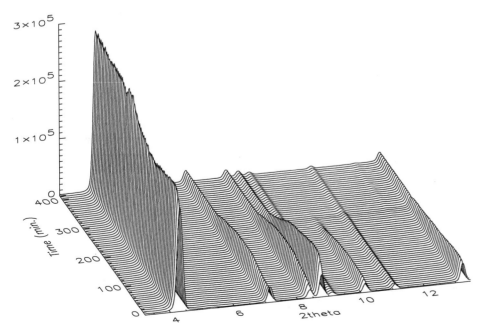

**Figure 9.** Evolution of the diffraction pattern during the *in situ* dehydration of CsY zeolite.

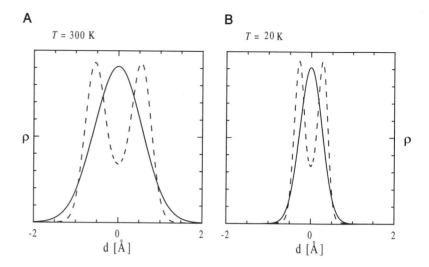

**Figure 10.** Electron distributions obtained from x-ray powder diffraction data in the octahedral site of $Rb_3C_{60}$ for $T = 300$ K (**A**) and $T = 20$ K (**B**).

Debye-Waller factor $B_o$ is refined and (2) the octahedral $Rb^+$ cation is shifted an increasing distance $\varepsilon$ away from $(\frac{1}{2}, 0, 0)$ until the refined $B_o$ becomes comparable to the temperature factor of the tetrahedral ion, $B_t$. In each case, both models result in an identical goodness of fit, and the direction of the assumed displacement $\varepsilon$ has no effect whatsoever. Figure 10 shows the electron distributions in the octahedral site for both models (in each case convoluted with the appropriate Debye-Waller factor). While the width of this distribution (static or dynamic) is 1.32 Å at room temperature, it has decreased to 0.56 Å upon cooling the sample to $T = 20$ K.

If the octahedral $Rb^+$ cations were truly displaced away from their site center (either at all temperatures or below some transition temperature $T_{tr}$), the amount of that displacement would act as an order parameter, i.e., increase or saturate with lower temperature and not decrease monotonically to zero as the data show it does. Hence, any off-center displacement of the octahedral $Rb^+$ cations can be excluded from this x-ray powder diffraction experiment. Consequently, other possible causes for the features seen in NMR and EXAFS experiments that had given rise to this suggestion must be researched. For such an experiment to be valid and successful, it must be possible to put very tight limits on the accuracy of the measured structural parameters. This requires that two conditions are met. First, the sample must have the highest possible quality, in terms of both purity and crystallinity, and the data must have both high resolution and very good counting statistic. Second, the accuracy for each measured parameter must be judged carefully, which is not trivial in x-ray powder diffraction. Note, in particular, the discussion under Estimated Standard Deviations in Rietveld Refinements, above.

## LITERATURE CITED

Altomare, A., Burla, M. C., Cascarano, G., Giacovazzo, C., Guagliardi, A., Moliterni, A. G. G., and Polidori, G. 1995. Extra: A program for extracting structure-factor amplitudes from powder diffraction data. *J. Appl. Crystallogr.* 28:842–846. See Internet Resources.

Altomare, A., Cascarano, G., Giacovazzo, C., Guagliardi, A., Burla, M. C., Polidori, G., and Camalli, M. 1994. SIRPOW.92—a program for automatic solution of crystal structures by direct methods optimized for powder data. *J. Appl. Crystallogr.* 27:435–436. See Internet Resources.

Ashcroft, N. W. and Mermin, N. D. 1976. Solid State Physics. Holt, Rinehart and Winston, New York.

Balzar, D. and Ledbetter, H. 1993. Voigt-function modeling in Fourier analysis of size- and strain-broadened X-ray diffraction peaks. *J. Appl. Crystallogr.* 26:97–103.

Bendele, G. M., Stephens, P. W., and Fischer, J. E. 1998. Octahedral cations in $Rb_3C_{60}$: Reconciliation of conflicting evidence from different probes. *Europhys. Lett.* 41:553–558.

Bergman, G., Waugh, J. L. T., and Pauling, L. 1952. Crystal structure of the intermetallic compound $Mg_{32}(Al,Zn)_{49}$ and related phases. *Nature* 169:1057.

Bergman, G., Waugh, J. L. T., and Pauling, L. 1957. The crystal structure of the metallic phase $Mg_{32}(Al, Zn)_{49}$. *Acta Crystallogr.* 10:254–257.

Bish, D. L. and Post, J. E. (eds.). 1989. Modern Powder Diffraction. Mineralogical Society of America, Washington, D.C.

Blanton, T. N., Huang, T. C., Toraya, H., Hubbard, C. R., Robie, S. B., Louër, D., Göbel, H. E., Will, G., Gilles, R., and Raftery T. 1995. JCPDS—International Centre for Diffraction Data round robin study of silver behenate. A possible low-angle X-ray diffraction calibration standard. *Powder Diffraction* 10:91–95.

Boultif, A. and Louër, D. 1991. Indexing of powder diffraction patterns for low symmetry lattices by the successive dichotomy method. *J. Appl. Crystallogr.* 24:987–993.

Brindley, G. W. 1945. The effect of grain or particle size on x-ray reflections from mixed powders and alloys, considered in relation to the quantitative determination of crystalline substances by x-ray methods. *Philos. Mag.* 36:347–369.

Brown, P. J., Fox, A. G., Maslen, E. N., O'Keefe, M. A., and Willis, B. T. M. 1992. Intensity of diffracted intensities. *In* International Tables for Crystallography, Vol. C (A. J. C. Wilson, ed.) pp. 476–516. Kluwer Academic Publishers, Dordrecht.

Clarke, R. and Rowe, W. P. 1991. Real-time X-ray studies using CCDs. *Synchrotron Radiation News* 4(3):24–28.

Dinnebier, R. E., Pink, M., Sieler, J., and Stephens, P. W. 1997. Novel alkali metal coordination in phenoxides: Powder diffraction results on $C_6H_5OM$ [M = K, Rb, Cs]. *Inorg. Chem.* 36:3398–3401.

Dinnebier, R. E., Stephens, P. W., Byrn, S., Andronis, V., and Zografi, G. 1996. Detection limit for polymorphs in drugs using powder diffraction (J. B. Hastings, ed.). p. B47. *In* BNL-NSLS 1995 Activity Report. National Synchrotron Light Source, Upton, N.Y.

Finger, L. W., Cox, D. E., and Jephcoat, A. P. 1994. A correction for powder diffraction peak asymmetry due to axial divergence. *J. Appl. Crystallogr.* 27:892–900.

Giacovazzo, C. (ed.) 1992. Fundamentals of Crystallography. Oxford University Press, Oxford.

Guryan, C. A., Stephens, P. W., Goldman, A. I., and Gayle, F. W. 1988. Structure of icosahedral clusters in cubic $Al_{5.6}Li_{2.9}Cu$. *Phys. Rev. B* 37:8495–8498.

Hill, R. J. and Cranswick, L. M. D. 1994. International Union of Crystallography Commission on Powder Diffraction Rietveld Refinement Round Robin. II. Analysis of monoclinic $ZrO_2$. *J. Appl. Crystallogr.* 27:802–844.

International Center for Diffraction Data (ICDD). 1998. PDF-2 Powder Diffraction File Database. ICDD, Newtown Square, Pa. See Internet Resources.

Ito, M. and Amemiya, Y. 1991. X-ray energy dependence and uniformity of an imaging plate detector. *Nucl. Instrum. Methods Phys. Res.* A310:369–372.

Jenkins, R. and Snyder, R. L. 1996. Introduction to X-ray Powder Diffractometry. John Wiley & Sons, New York.

Kittel, C. 1996. Introduction to Solid State Physics. John Wiley & Sons, New York.

Klug, H. P. and Alexander, L. E. 1974. X-ray Diffraction Procedures for Polycrystalline and Amorphous Materials. John Wiley & Sons, New York.

Larson, A. C. and Von Dreele, R. B. 1994. GSAS: General Structure Analysis System. Los Alamos National Laboratory, publication LAUR 86–748. See Internet Resources.

Le Bail, A., Duryo, H., and Fourquet, J. L. 1988. Ab-initio structure determination of $LiSbWO_6$ by X-ray powder diffraction. *Mater. Res. Bull.* 23:447–452.

Maslen, E. N. 1992. X-ray absorption. *In* International Tables for Crystallography, Vol. C (A. J. C. Wilson, ed.). pp. 520–529. Kluwer Academic Publishers, Dordrecht.

Miyahara, J., Takahashi, K., Amemiya, Y., Kamiya, N., and Satow, Y. 1986. A new type of X-ray area detector utilizing laser stimulated luminescence. *Nucl. Instrum. Methods Phys. Res.* A246:572–578.

Norby, P., Poshni, F. I., Gualtieri, A. F., Hanson, J. C., and Grey, C. P. 1998. Cation migration in zeolites: An in-situ powder diffraction and MAS NMR study of the structure of zeolite Cs(Na)-Y during dehydration. *J. Phys. Chem. B* 102:839–856.

Noyan, I. C. and Cohen, J. B. 1987. Residual Stress. Springer-Verlag, New York.

Pawley, G. S. 1981. Unit-cell refinement from powder diffraction scans. *J. Appl. Crystallogr.* 14:357–361.

Poshni, F. I., Ciraolo, M. F., Grey, C. P., Gualtieri, A. F., Norby, P., and Hanson, J. C. 1997. An in-situ X-ray powder diffraction study of the dehydration of zeolite CsY (J. B. Hastings, ed.). p. B84. *In* BNL-NSLS 1996 Activity Report. National Synchrotron Light Source, Upton, N.Y.

Prince, E. and Spiegelmann, C. H. 1992. Statistical significance tests. *In* International Tables for Crystallography, Vol. C (A. J. C. Wilson, ed.). pp. 618–621. Kluwer Academic Publishers, Dordrecht.

Rietveld, H. M. 1969. A profile refinement method for nuclear and magnetic structures. *J. Appl. Crystallogr.* 2:65–71.

Rodriguez-Carvajal, R. 1990. FULLPROF: A program for Rietveld refinement and pattern matching analysis. *In* Abstracts of the Satellite Meeting on Powder Diffraction of the XV Congress of the IUCr, p. 127. IUCr, Toulouse, France. See Internet Resources.

Sabine, T. M. 1993. The flow of radiation in a polycrystalline material. *In* The Rietveld Method (R. A. Young, ed.). pp. 55–61. Oxford University Press, Oxford.

Smith, G. W. 1991. X-ray imaging with gas proportional detectors. *Synchrotron Radiation News* 4(3):24–30.

Stephens, P. W. 1999. Phenomenological model of anisotropic peak broadening in powder diffraction. *J. Appl. Crystallogr.* 32:281–289.

Thompson, P., Cox, D. E., and Hastings. J. B. 1987. Rietveld refinement of Debye-Scherrer synchrotron X-ray data from $Al_2O_3$. *J. Appl. Crystallogr.* 20:79–83.

Visser, J. W. 1969. A fully automatic program for finding the unit cell from powder data. *J. Appl. Crystallogr.* 2:89–95.

Vos, A. and Buerger, M. J. 1996. Space-group determination and diffraction symbols. *In* International Tables for Crystallography, Vol. A (T. Hahn, ed.) pp. 39–48. Kluwer Academic Publishers, Dordrecht.

Werner, P.-E., Eriksson, L., and Westdahl, M. 1985. TREOR, a semi-exhaustive trial-and-error powder indexing program for all symmetries. *J. Appl. Crystallogr.* 18:367–370.

Young, R. A. (ed.) 1993. The Rietveld Method. Oxford University Press, Oxford.

## KEY REFERENCES

Azaroff, L. V. 1968. Elements of X-ray Crystallography. McGraw-Hill, New York.

*Gives comprehensive discussions of x-ray crystallography albeit not particularly specialized to powder diffraction.*

Bish and Post, 1989. See above.

*An extremely useful reference for details on the Rietveld method.*

Giacovazzo, 1992. See above.

*Another comprehensive text on x-ray crystallography, not particularly specialized to powder diffraction.*

International Tables for Crystallography (3 vols.). Kluwer Academic Publishers, Dordrecht.

*An invaluable resource for all aspects of crystallography. Volume A treats crystallographic symmetry in direct space, and it includes tables of all crystallographic plane, space, and point groups. Volume B contains accounts of numerous aspects of reciprocal space, such as the structure-factor formalism, and Volume C contains mathematical, physical, and chemical information needed for experimental studies in structural crystallography.*

Jenkins and Snyder, 1996. See above.

*Offers a modern approach to techniques, especially of phase identification and quantification; however, there is very little discussion of synchrotron radiation or of structure solution.*

Klug and Alexander, 1974. See above.

*The classic text about many aspects of the technique, particularly the analysis of mixtures, although several of the experimental techniques described are rather dated.*

Langford, J. I. and Louër, D. 1996. Powder diffraction. *Rep. Prog. Phys.* 59:131–234.

*Offers another approach to techniques that includes a thorough discussion of synchrotron radiation and of structure solution.*

Young, 1993. See above.
*Another valuable reference for details on the Rietveld refinements.*

## INTERNET RESOURCES

http://www.ba.cnr.it/IRMEC/SirWare.html

http://ww.iccd.com

http://www.ccp14.ac.uk

*An invaluable resource for crystallographic computing that contains virtually all freely available software for power (and also single-crystal) diffraction for academia, including every program mentioned in this unit.*

ftp://ftp.lanl.gov/public/gsas

ftp://charybde.saclay.cea.fr/pub/divers/fullp/

PETER W. STEPHENS
State University of New York
Stony Brook, New York

GOETZ M. BENDELE
Los Alamos National Laboratory
Los Alamos, New Mexico, and
State University of New York
Stony Brook, New York

# SINGLE-CRYSTAL X-RAY STRUCTURE DETERMINATION

## INTRODUCTION

X-ray techniques provide one of the most powerful, if not the most powerful, methods for deducing the crystal and molecular structure of a crystalline solid. In general, it is a technique that provides a large number of observations for every parameter to be determined and results obtained thereby can usually be considered very reliable. Significant advances have taken place, especially over the last twenty years, to make the collection of data and the subsequent steps to carry out the rest of the structure determination amenable to the relative novice in the field. This has led to the adoption of single-crystal diffraction methods as a standard analytical tool. It is no accident that this has paralleled the development of the digital computer—modern crystallography is strongly dependent on the use of the digital computer. Typically, a small single crystal (a few tenths of a millimeter in dimension) of the material to be investigated is placed on a diffractometer and data are collected under computer-controlled operation. A moderate-sized unit cell (10 to 20 Å on edge) can yield 5000 or so diffraction maxima, which might require a few day's data collection time on a scintillation counter diffractometer or only hours of data collection time on the newer charge-coupled device (CCD) area detectors.

From these data the fractional coordinates describing the positions of the atoms within the cell, some characteristics of their thermal motion, the cell dimensions, the crystal system, the space group symmetry, the number of formula units per cell, and a calculated density can be obtained. This "solving of the structure" may only take a few additional hours or a day or two if no undue complications occur.

We will concentrate here on the use of X-ray diffraction techniques as would be necessary for the normal crystal structure investigation. A discussion of the basic principles of single-crystal X-ray diffraction will be presented first, followed by a discussion of the practical applications of these techniques. In the course of this discussion, a few of the more widely used methods of overcoming the lack of measured phases, i.e., the "phase problem" in crystallography, will be presented. These methods are usually sufficient to handle most routine crystal structure determinations. Space will not permit discussion of other more specialized techniques that are available and can be used when standard methods fail. The reader may wish to refer to other books devoted to this subject (e.g., Ladd and Palmer, 1985; Stout and Jensen, 1989; Lipscomb and Jacobson, 1990) for further information on these other techniques and for a more in-depth discussion of diffraction methods.

## Competitive and Related Techniques

Single-crystal X-ray diffraction is certainly not the only technique available to the investigator for the determination of the structure of materials. Many other techniques can provide information that is often complementary to that obtained from single-crystal X-ray experiments. Some of these other techniques are briefly described below.

Crystalline powders also diffract X rays. In fact, it is convenient to view powder diffraction as single-crystal diffraction integrated over angular coordinates, thus retaining only $\sin \theta$ dependence (assuming the powder contains crystallites in random orientation). Diffraction maxima with essentially the same $\sin \theta$ values combine in the powder diffraction pattern. This results in fewer discrete observations and makes the unit cell determination and structure solution more difficult; this is especially true in lower symmetry cases. The fewer number of data points above background translates into larger standard deviations for the determined atomic parameters. Since the angular information is lost, obtaining an initial model is often quite difficult if nothing is known about the structure. Therefore, the majority of quantitative powder diffraction investigations are done in situations where an initial model can be obtained from a related structure; the calculated powder pattern is then fit to the observed one by adjusting the atomic parameters, the unit cell parameters, and the parameters that describe the peak shape of a single diffraction maximum. (This is usually termed Rietveld refinement; see X-RAY POWDER DIFFRACTION and NEUTRON POWDER DIFFRACTION) If an appropriate single crystal of the material could be obtained, single-crystal diffraction would be the preferred approach. Powder diffraction sometimes provides an alternate approach when only very small crystals can be obtained.

Neutrons also undergo diffraction when they pass through crystalline materials. In fact, the theoretical descriptions of X-ray and neutron diffraction are very similar, primarily differing in the atomic scattering factor used (see NEUTRON POWDER DIFFRACTION and SINGLE-CRYSTAL

NEUTRON DIFFRACTION). The principal interaction of X rays with matter is electronic, whereas with neutrons two interactions predominate—one is nuclear and the other is magnetic (Majkrzak et al., 1990). The interaction of the neutron with the nucleus varies with the isotope and typically does not show the larger variation seen with X rays, where the scattering is proportional to the atomic number. Therefore neutrons can be useful in distinguishing between atoms of neighboring atomic number or light atoms, especially hydrogen in the presence of heavy atoms. The neutron scattering factors for hydrogen and deuterium are also markedly different, which opens up a number of interesting possible structural studies, especially of organic and biological compounds. As noted above, neutrons are also sensitive to magnetic structure (see MAGNETIC NEUTRON SCATTERING). In fact, neutron scattering data constitute the most significant body of experimental evidence regarding long-range magnetic ordering in solids. In practice, with the neutron sources that are presently available, it is necessary to use a considerably larger sample for a neutron diffraction investigation than is necessary for an X-ray investigation. Single-crystal X-ray studies are usually carried out initially, followed by the single-crystal neutron investigation to obtain complementary information where sample preparation is not a problem. (It might be noted that for powder studies neutron diffraction can have some advantages over X-ray diffraction in terms of profile pattern fitting due to the Gaussian character of the neutron diffraction peaks and the enhanced intensities that can be obtained at higher $\sin \theta$ values; see NEUTRON POWDER DIFFRACTION).

Electron diffraction can also provide useful structural information on materials. The mathematical description appropriate for the scattering of an electron by the potential of an atom has many similarities to the description used in discussing X-ray diffraction. Electrons, however, are scattered much more efficiently than either X rays or neutrons, so much so that electrons can penetrate only a few atomic layers in a solid and are likely to undergo multiple scattering events in the process. Hence, electron diffraction is most often applied to the study of surfaces (low-energy electron diffraction; LEED), in electron-microscopic studies of microcrystals, or in gas-phase diffraction studies.

X-ray absorption spectroscopy (EXAFS) (Heald and Tranquada, 1990) provides an additional technique that can be used to obtain structural information (see XAFS SPECTROSCOPY). At typical X-ray energies, absorption is a much more likely event than the scattering process that we associate with single-crystal diffraction. When an X-ray photon is absorbed by an electron in an atom, the electron is emitted with an energy of the X-ray photon minus the electron-binding energy. Thus some minimum photon energy is necessary, and the X-ray absorption spectrum shows sharp increases in absorption as the various electron-binding energies are crossed. These sharp steps are usually termed X-ray edges and have characteristic values for each element. Accurate measurements have revealed that there is structure near the edge. This fine structure is explained by modulations in the final state of the photoelectron that are caused by backscattering from the sur-

rounding atoms. The advent of synchrotron X-ray sources has accelerated the use of EXAFS since they provide the higher intensities that are necessary to observe the fine structure. The local nature of EXAFS makes it particularly sensitive to small perturbations within the unit cell and can thus complement the information obtained in diffraction experiments. With EXAFS it is possible to focus in on a particular atom type, and changes can be monitored without determining the entire structure. It can be routinely performed on a wide variety of heavier elements. When long-range order does not exist, diffraction methods such as X-ray diffraction have less utility. EXAFS yields an averaged radial distribution function pertinent to the particular atom and can be applied to amorphous as well as crystalline materials. EXAFS can be considered as being complementary to single-crystal diffraction.

Solid-state nuclear magnetic resonance (Hendrichs and Hewitt, 1980) and Mössbauer spectroscopy (Berry, 1990; see MOSSBAUER SPECTROMETRY) can sometimes provide additional useful structural information, especially for non-single-crystal materials.

## PRINCIPLES OF X-RAY CRYSTALLOGRAPHY

When a material is placed in an X-ray beam, the periodically varying electric field of the X-ray accelerates the electrons into periodic motion; each in turn emits an electromagnetic wave with a frequency essentially identical to that of the incident wave and with a definite phase relation to the source that we will take as identical to that of the incident wave. To the approximation usually employed, all electrons are assumed to scatter with the same amplitude and phase and to do so independently of all other electrons. (The energy of the X ray, approximately 20 keV for Mo Kα, is usually large in comparison with the binding energy of the electron, except for the inner electrons of the heavier atoms. It should be noted, however, that the anomalous behavior of first shell, or first- and second-shell electrons in heavy elements, can provide a useful technique for phase determination in some cases and has proven to be especially valuable in protein crystallography.)

Interference can occur from electrons occupying different spatial positions, providing the spatial difference is comparable to the wavelength being employed. Interference effects can occur within an atom, as is shown in Figure 1, where the variation of scattering with angle for

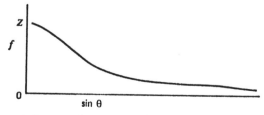

**Figure 1.** Decrease of amplitude of scattering from an atom of atomic number $Z$ as a function of scattering angle $\theta$. The function $f$ is the atomic scattering factor. All atoms show similar shaped functions, the rate of decrease being somewhat less for atoms of larger $Z$; their inner electrons are closer to the nucleus.

a typical atom is plotted. These interference effects cause $f$, the atom's amplitude of scattering, to decrease as the scattering angle increases. Interference effects can also occur between atoms, and it is the latter that permits one to deduce the positions of the atoms in the unit cell, i.e., the crystal structure.

It is mathematically convenient to use a complex exponential description for the X-ray wave,

$$E = E_0 \exp[2\pi i(\nu t - x/\lambda + \delta)] \tag{1}$$

where $x$ is the distance, $\lambda$ is the wavelength, $\nu$ is the frequency, $t$ is the time, and $\delta$ is a phase factor for the wave at $x = 0$ and $t = 0$. Anticipating that we will be referring to intensities (i.e., $EE^*$, $E^*$ being the complex conjugate of $E$) and only be concerned with effects due to different spatial positions, we will use the simpler form

$$E = E_0 \exp(-2\pi i x/\lambda) \tag{2}$$

For two electrons displaced in one dimension the intensity of scattering would be given by

$$\begin{aligned} I = EE^* &= |E_1 + E_2|^2 \\ &= |E_{0_1} \exp(-2\pi i x_1/\lambda) + E_{0_2} \exp(-2\pi i x_2/\lambda)|^2 \end{aligned} \tag{3}$$

Furthermore, since $E_{0_1}$ would be expected to equal $E_{0_2}$,

$$\begin{aligned} I &= |E_0[\exp(-2\pi i x_1/\lambda) + \exp(-2\pi i x_2/\lambda)]|^2 \\ &= E_0^2|[\{1 + \exp[-2\pi i(x_2 - x_1)/\lambda]\}\exp(-2\pi i x_1/\lambda)]|^2 \\ &= E_0^2|\{1 + \exp[-2\pi i(x_2 - x_1)/\lambda]\}|^2 \end{aligned} \tag{4}$$

Generalizing to three dimensions, consider a wave (Fig. 2) incident upon an electron at the origin and a second electron at $\mathbf{r}$. Then the difference in distance would be given by $\mathbf{r} \cdot (\mathbf{s}_0 - \mathbf{s})$, where $\mathbf{s}_0$ and $\mathbf{s}$ represent directions of the incident and scattered rays and $\mathbf{r}$ is the distance between electrons

$$I = E_0^2|1 + \exp[2\pi i \mathbf{r} \bullet (\mathbf{s} - \mathbf{s}_0)/\lambda]|^2 \tag{5}$$

or for $n$ electrons

$$I = E_0^2 \left| \sum_{j=0}^{n-1} \exp[2\pi i \mathbf{r}_j \bullet (\mathbf{s} - \mathbf{s}_0)/\lambda] \right|^2 \tag{6}$$

Since we will only require relative intensities, we define

$$F = \sum_{j=0}^{n-1} \exp[2\pi i \mathbf{r}_j \bullet (\mathbf{s} - \mathbf{s}_0)/\lambda] \tag{7}$$

where $F$ is termed the structure factor and

$$I \propto |F|^2 \tag{8}$$

The structure factor expression can be further simplified by replacing the summation by an integral,

$$F = \int_V \rho(\mathbf{r})\exp[2\pi i \mathbf{r} \bullet (\mathbf{s} - \mathbf{s}_0)/\lambda]dV \tag{9}$$

where $\rho(\mathbf{r})$ is the electron density. If the assumption is made that the electron density is approximated by the sum of atom electron densities, then

$$F = \sum_{j=0}^{N} f_j \exp[2\pi i \mathbf{r}_j \bullet (\mathbf{s} - \mathbf{s}_0)/\lambda)] \tag{10}$$

where $f_j$ is the atomic scattering factor, i.e., the scattering that would be produced by an isolated atom of that atomic number, convoluted with effects due to thermal motion.

To simplify this equation further, we now specify that the sample is a single crystal, and $\mathbf{r}_j = (x_j + p)\mathbf{a}_1 + (y_j + m)\mathbf{a}_2 + (z_j + n)\mathbf{a}_3$. The vectors $\mathbf{a}_1$, $\mathbf{a}_2$, and $\mathbf{a}_3$ describe the repeating unit (the unit cell), $x_j, y_j, z_j$ are the fractional coordinates of atom $j$ in the cell, and $p$, $m$, and $n$ are integers. Since $\mathbf{r}_j \bullet (\mathbf{s} - \mathbf{s}_0)/\lambda$ must be dimensionless, the $(\mathbf{s} - \mathbf{s}_0)/\lambda$ quantity must be of dimension reciprocal length and can be represented by

$$(\mathbf{s} - \mathbf{s}_0)/\lambda = h\mathbf{b}_1 + k\mathbf{b}_2 + l\mathbf{b}_3 \tag{11}$$

where $\mathbf{b}_1$, $\mathbf{b}_2$, and $\mathbf{b}_3$ are termed the reciprocal cell vectors. They are defined such that $\mathbf{a}_i \cdot \mathbf{b}_i = 1$ and $\mathbf{a}_i \cdot \mathbf{b}_j = 0$. (It should be noted that another common convention is to use $\mathbf{a}$, $\mathbf{b}$, $\mathbf{c}$ to represent unit dimensions in direct space and $\mathbf{a}^*$, $\mathbf{b}^*$, $\mathbf{c}^*$ to represent the reciprocal space quantities.)

By substituting into the structure factor expression and assuming at least a few hundred cells in each direction, it can be shown (Lipscomb and Jacobson, 1990) that the restriction of $p$, $m$, and $n$ to integer values yields nonzero diffraction maxima only when $h$, $k$, and $l$ are integer. Hence the structure factor for single-crystal diffraction is usually written as

$$F_{hkl} = s \sum_{j=1}^{N} f_j \exp[2\pi i(hx_j + ky_j + lz_j)] \tag{12}$$

where the sum is now over only the atoms in the repeating unit and the other terms (those involving $p$, $m$, and $n$) contribute to a constant multiplying factor, $s$.

**Figure 2.** (**A**) Unit vectors $\mathbf{s}_0$ and $\mathbf{s}$ represent directions of the incident and scattered waves, respectively. (**B**) Scattering from a point at the origin and a second point at $\mathbf{r}$. The point of observation $P$ is very far away compared with the distance $\mathbf{r}$.

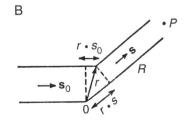

A

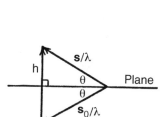

B

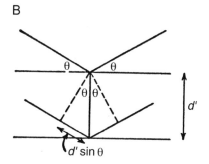

**Figure 3.** **(A)** Graphical representation of $\mathbf{s}/\lambda - \mathbf{s}_0/\lambda = \mathbf{h}$. **(B)** Path difference between successive planes a distance $d'$ apart is $2d'' \sin \theta$, which yields a maximum when equal to $n\lambda$.

We now define a reciprocal lattice vector $\mathbf{h}$ as $\mathbf{h} = h\mathbf{b}_1 + k\mathbf{b}_2 + l\mathbf{b}_3$.

Also from Equation 11,

$$\mathbf{h}\lambda = \mathbf{s} - \mathbf{s}_0 \tag{13}$$

and this is termed the Laue equation.

The diffraction pattern from a single crystal therefore can be viewed as yielding a reciprocal lattice pattern ($h$, $k$, $l$ all integer) that is weighted by $|F_{hkl}|^2$.

It can be readily shown that $\mathbf{h}$ is a vector normal to a plane passing through the unit cell with intercepts $\mathbf{a}_1/h$, $\mathbf{a}_2/k$, and $\mathbf{a}_3/l$. The integers $h$, $k$, and $l$ are termed crystallographic indices (Miller indices if they are relatively prime numbers; see SYMMETRY IN CRYSTALLOGRAPHY). If $2\theta$ is defined as the angle between $\mathbf{s}$ and $\mathbf{s}_0$, then the scalar equivalent of the Laue equation is

$$\lambda = 2d\sin\theta \tag{14}$$

i.e., Bragg's law, where $|\mathbf{h}| = 1/d$, the distance between planes. (If Miller indices are used, a factor of $n$, the order of the diffraction, is included on the left side of Equation 14.) A convenient graphical representation of the Laue equation is given in Figure 3 and alternately in what is termed the Ewald construction in Figure 4. These are

especially useful when discussing diffractometers and other devices for the collection of X-ray data. Diffraction maxima will only occur when a reciprocal lattice point intersects the sphere of diffraction.

The scattering from an isolated atom represents a combination of the scattering of an atom at rest convoluted with the effects due to thermal motion. These can be separated, and for an atom undergoing isotropic motion, a thermal factor expression

$$T_j = \exp(-B_j \sin^2\theta/\lambda^2) \tag{15}$$

can be used, where $B_j = 8\pi^2\mu^2$, $\mu^2$ being the mean-square amplitude of vibration.

For anisotropic motion, we use

$$T_j = \exp[-2\pi^2(U_{11}h^2\mathbf{b}_1^2 + U_{22}k^2\mathbf{b}_2^2 + U_{33}l^2\mathbf{b}_3^2 + 2U_{12}hk\mathbf{b}_1 \bullet \mathbf{b}_2 \\ + 2U_{13}hl\mathbf{b}_1 \bullet \mathbf{b}_3 + 2U_{23}kl\mathbf{b}_2 \bullet \mathbf{b}_3)] \tag{16a}$$

or the analogous expression

$$T_j = \exp[-(\beta_{11}h^2 + \beta_{22}k^2 + \beta_{33}l^2 + 2\beta_{12}hk \\ + 2\beta_{13}hl + 2\beta_{23}kl)] \tag{16b}$$

The $U$'s are thermal parameters expressed in terms of mean-square amplitudes in angstroms, while the $\beta$'s are the associated quantities without units.

The structure factor is then written as

$$F_{hkl} = s \sum_{j=1}^{N} f_j \exp[2\pi i(hx_j + ky_j + lz_j)]T_j \tag{17}$$

As we have seen, one form of the structure factor expression (Equation 9) involves $\rho(\mathbf{r})$, the electron density function, in a Fourier series. Therefore, it would be expected that there exists an inverse Fourier series in which the electron density is expressed in terms of structure factor quantities. Indeed, the electron density function for the cell can be written as

$$\rho(\mathbf{r}) = \frac{1}{V} \sum_{h=-\infty}^{\infty} \sum_{k=-\infty}^{\infty} \sum_{l=-\infty}^{\infty} F_{hkl} \exp[-2\pi i(hx + ky + lz)] \tag{18}$$

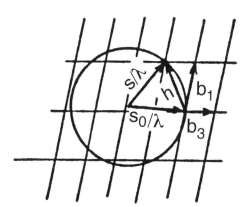

**Figure 4.** Ewald construction. Assume the reciprocal lattice net represents the $h0l$ zone of a monoclinic crystal. Then the $\mathbf{a}_2$ axis [010] is perpendicular to the plane of the paper. As the crystal is rotated about this axis, various points of the reciprocal lattice cross the circle of reflection; as they do so, the equation $\mathbf{s}/\lambda - \mathbf{s}_0/\lambda = \mathbf{h}$ is satisfied, and a diffracted beam occurs. Diffraction by the [10 1] plane is illustrated.

In this expression $V$ is the volume of the cell and the triple summation is in theory from $-\infty$ to $\infty$ and in practice over all the structure factors.

This series can be written in a number of equivalent ways. The structure factor can be expressed in terms of the real and imaginary components $A$ and $B$ as

$$F_{hkl} = A_{hkl} + iB_{hkl} \tag{19}$$

$$= s \sum_j f_j \cos 2\pi(hx_j + ky_j + lz_j)T_j$$

$$+ is \sum_j f_j \sin 2\pi(hx_j + ky_j + lz_j)T_j \tag{20}$$

and

$$\rho(xyz) = \frac{1}{V} \sum_{h=-\infty}^{\infty} \sum_{k=-\infty}^{\infty} \sum_{l=-\infty}^{\infty} [A_{hkl} \cos 2\pi(hx + ky + lz)$$

$$+ B_{hkl} \sin 2\pi(hx + ky + lz)] \tag{21}$$

Another form for the structure factor is

$$F_{hkl} = |F_{hkl}| \exp(2\pi i \alpha_{hkl}) \tag{22}$$

Then

$$\alpha_{hkl} = \tan^{-1}(B_{hkl}/A_{hkl}) \tag{23}$$

and

$$|F_{hkl}| = (A_{hkl}^2 + B_{hkl}^2)^{1/2} \tag{24}$$

The quantity $\alpha_{hkl}$ is termed the crystallographic phase angle. It could also be noted that, using $\bar{h}$ to designate $-h$

$$A_{\overline{hkl}} = A_{hkl} \tag{25}$$

and

$$B_{\overline{hkl}} = -B_{hkl} \tag{26}$$

if the atomic scattering factor $f_j$ is real. Also,

$$|F_{hkl}|^2 = |F_{\overline{hkl}}|^2 \tag{27}$$

Equation 27 is termed Friedel's law and as noted holds as long as $f_j$ is real.

Equation 18 implies that the electron density for the cell can be obtained from the structure factors; a knowledge of the electron density would enable us to deduce the location and types of atoms making up the cell. The magnitude of the structure factor can be obtained from the intensities (Equation 8); however, the phase is not measured, and this gives rise to what is termed the phase problem in crystallography. Considerable effort has been devoted over the last half century to developing reliable methods to deduce these phases. One of the most widely used of the current methods is that developed in large part by J. Karle and H. Hauptman (Hauptman and Karle,

1953; Karle and Hauptman, 1956; Karle and Karle, 1966), which uses a statistical and probability approach to extract phases directly from the magnitudes of the structure factors. We will give a brief discussion of the background of such "direct methods" as well as a discussion of the heavy-atom method, another commonly employed approach. It should be noted, however, that a number of other techniques could be used if these methods fail to yield a good trial model; these will not be discussed here due to space limitations.

### Symmetry in Crystals

Symmetry plays an important role in the determination of crystal structures; only those atomic positional and thermal parameters that are symmetry independent need to be determined. In fact, if the proper symmetry is not recognized and attempts are made to refine parameters that are symmetry related, correlation effects occur and unreliable refinement and even divergence can result.

What restrictions are there on the symmetry elements and their combinations that can be present in a crystal? (We will confine our discussion to crystals as generated by a single tiling pattern and exclude cases such as the quasi-crystal.) Consider a lattice vector **a** operated on by a rotation axis perpendicular to the plane of the paper, as shown in Figure 5. Rotation by $\alpha$ or $-\alpha$ will move the head of the vector to two other lattice points, and these can be connected by a vector **b**. Since **b** must be parallel to **a**, one can relate their lengths by $b = ma$, where $m$ is an integer, or by $b = 2a \cos \alpha$. Thus $m = 2 \cos \alpha$ and

$$\alpha = \cos^{-1}(m/2) \tag{28}$$

Obviously the only allowed values for $m$ are $0, \pm 1, \pm 2$, which yield $0, 60°, 90°, 120°, 180°, 240°, 270°$, and $300°$ as possible values for the rotation angle $\alpha$. This in turn implies rotation axes of order 1, 2, 3, 4, or 6. No other orders are compatible with the repeating character of the lattice. A lattice can also have inversion symmetry, as would be implied by Friedel's law (Equation 27). Coupling the inversion operation with a rotation axis produces the rotatory inversion axes: $\bar{1}, \bar{2}, \bar{3}, \bar{4}$, and $\bar{6}$. (These operations can be viewed as a rotation followed by an inversion. The $\bar{1}$ is just the inversion operation and $\bar{2}$ the mirror operation.)

Various combinations of these symmetry elements can now be made. In doing so, 32 different point groups are produced (see SYMMETRY IN CRYSTALLOGRAPHY). Since in a lattice description the repeating unit is represented by a point and the lattice has inversion symmetry, those point

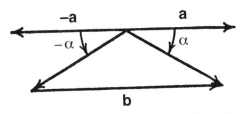

**Figure 5.** Rotation axis $A(\alpha)$ at a lattice point. Note that the vector **b** is parallel to **a**.

**Table 1. Fourteen Bravais Lattices**

| Symmetry at a Lattice Point | Name | Lattice Points Per Unit Cell | Elementary Cell Relations |
|---|---|---|---|
| $C_i - 1$ | Triclinic | 1 | |
| $C_{2h} - 2/2m$ | Primitive monoclinic | 1 | $\alpha_1 = \alpha_3 = 90°$ |
| | End-centered monoclinic $C$ | 2 | |
| $D_{2h}$ | Primitive orthorhombic | 1 | $\alpha_1 = \alpha_2 = \alpha_3 = 90°$ |
| | End-centered $C, A, B$ | 2 | |
| | Face-centered $F$ | 4 | |
| | Body-centered $I$ | 2 | |
| $D_{4h} - 4/m\,mm$ | Primitive tetragonal (equivalent to end-centered) | 1 | $\alpha_1 = \alpha_2 = \alpha_3 = 90°$ $a_1 = a_2$ |
| | Body-centered (equivalent to face-centered) $I$ | 2 | |
| $D_{3d} - 3m$ | Trigonal | 1 | $\alpha_1 = \alpha_2 = \alpha_3$ $a_1 = a_2 = a_3$ (rhombohedral axes, for hexagonal description, see below) |
| $D_{6h} - 6/m\,mm$ | Heagonal | 1 | $\alpha_1 = \alpha_2 = 90°$ $\alpha_3 = 120°$ $a_1 = a_2$ |
| $O_h - m3m$ | Primitive cubic | 1 | $\alpha_1 = \alpha_2 = \alpha_3 = 90°$ $a_1 = a_2 = a_3$ |
| | Face-centered cubic $F$ | 4 | |
| | Body-centered cubic $I$ | 2 | |

groups that differ only by an inversion center can be grouped together. This produces the seven crystal systems given in Table 1. For a number of crystal systems, in addition to the primitive cell (one lattice point per cell), there are also centered cell possibilities. (Such cells are not allowed if redrawing the centered cell to obtain a primitive one can be done without loss of symmetry consistent with the crystal system. Also, if redrawing or renaming axes converts one type to another, only one of these is included.) The notation used to describe centered cells is as follows: $A$—centered on the $\mathbf{a}_2$, $\mathbf{a}_3$ face; $B$—centered on the $\mathbf{a}_1$, $\mathbf{a}_3$ face; $C$—centered on the $\mathbf{a}_1$, $\mathbf{a}_2$ face; $I$—body centered; and $F$—face centered.

As noted in the table, trigonal cells can be reported using a primitive rhombohedral cell or as a hexagonal cell. If the latter, then a triply primitive R-centered cell is used. R centering has lattice points at (1/3, 2/3, 2/3) and (2/3, 1/3, 1/3) as well as at (0, 0, 0).

The next task is to convert these 32 point groups into the related space groups. This can be accomplished by generating not only space groups having these point group symmetries but also groups in which the point group operation is replaced by an "isomorphous" operation that has a translational element associated with it. However, this must be done in such a way as to preserve the multiplicative properties of the group.

The symmetry operation isomorphous to the rotation axis is the screw axis. They are designated by $N_m$, where the rotation is through an angle $2\pi/N$ and the translation is $m/N$, in terms of fractions of the cell, and is parallel to the rotation axis. Thus $2_1$ is a 180° rotation coupled with a 1/2 cell translation, $4_1$ is a 90° rotation coupled with 1/4 cell translation, and $6_2$ represents a 60° rotation coupled

with a 1/3 cell translation. Note that $2_1 \cdot 2_1$ yields a cell translation along the rotation axis direction and hence produces a result isomorphous to the identity operation; performing a $6_2$ six times yields a translation of two cell lengths.

Mirror planes can be coupled with a translation and converted to glide planes. The following notation is used: $a$ glide, translates 1/2 in $\mathbf{a}_1$; $b$ glide, translates 1/2 in $\mathbf{a}_2$; $c$ glide, translates 1/2 in $\mathbf{a}_3$; $n$ glide, translates 1/2 along the face diagonal; and $d$ glide, translates 1/4 along the face or body diagonal.

Let us digress for a moment to continue to discuss notation in general. When referring to space groups, the Hermann-Mauguin notation is the one most commonly used (see SYMMETRY IN CRYSTALLOGRAPHY). The symbol given first indicates the cell type, followed by the symmetry in one or more directions as dictated by the crystal system.

For the triclinic system, the only possible space groups are $P1$ and $P\bar{1}$.

For the monoclinic system, the cell type, $P$ or $C$, is followed by the symmetry in the $b$ direction. (The $b$ direction is typically taken as the unique direction for the monoclinic system, although one can occasionally encounter $c$-unique monoclinic descriptions in the older literature.) If a rotation axis and the normal to a mirror plane are in the same direction, then a slash (/) is used to designate that the two operations are in the same direction. For the point group $C_{2h}$, the isomorphous space groups in the monoclinic system include $P2/m$, $P2_1/m$, $P2/c$, $P2_1/c$, $C2/m$ and $C2/c$.

In the orthorhombic system, the lattice type is followed by the symmetry along the $\mathbf{a}$, $\mathbf{b}$, and $\mathbf{c}$ directions. The space groups $P222$, $Pca2$, $P2_12_12_1$, $Pmmm$, $Pbca$, $Ama2$,

*Fdd*2, and *Cmcm* are all examples of orthorhombic space groups.

The notation for the higher symmetry crystal systems follows in the same manner. The symmetry along the symmetry axis of order greater than 2 is given first followed by symmetry in the other directions as dictated by the point group. For tetragonal and hexagonal systems, the symmetry along the **c** direction is given, followed by that for **a** and then for the **ab** diagonal, when isomorphous with a *D*-type point group. For cubic space groups the order is symmetry along the cell edge, followed by that along the body diagonal, followed by the symmetry along the face diagonal, where appropriate.

A space group operation can be represented by

$$\begin{pmatrix} x' \\ y' \\ z' \end{pmatrix} = \mathbf{R} \begin{pmatrix} x \\ y \\ z \end{pmatrix} + \mathbf{t} \qquad (29)$$

where **R** is a $3 \times 3$ matrix corresponding to the point group operation and **t** is a column vector containing the three components of the associated translation. As noted above, group properties require that operations in the space group combine in the same fashion as those in the isomorphous point group; this in turn usually dictates the positions in the cell of the various symmetry elements making up the space group. Consider, for example, the space group $P2_1/c$ in the monoclinic system. This commonly occurring space group is derived from the $C_{2h}$ point group. Since in the point group a twofold rotation followed by the reflection is equivalent to the inversion, in the space group the $2_1$ operation followed by the *c* glide must be equivalent to the inversion. The usual convention is to place an inversion at the origin of the cell. For this to be so, the screw axis has to be placed at $z = 1/4$ and the *c* glide at $y = 1/4$. Therefore, in $P2_1/c$, the following relations hold:

$$x, y, z \xrightarrow{i} -x, -y, -z \qquad (30a)$$

$$x, y, z \xrightarrow{2_1} -x, 1/2 + y, 1/2 - z \qquad (30b)$$

$$x, y, z \xrightarrow{c} x, 1/2 - y, 1/2 + z \qquad (30c)$$

and these four equivalent positions are termed the general positions in the space group. In many space groups it is also possible for atoms to reside in "special positions," locations in the cell where a point group operation leaves the position invariant. Obviously these can only occur for symmetry elements that contain no translational components. In $P2_1/c$, the only special position possible is associated with the inversion. The number of equivalent positions in this case is two, namely, 0, 0, 0 and 0, 1/2, 1/2 or any of the translationally related pairs.

By replacing any point group operation by any allowable isomorphous space operation, the 32 point groups yield 230 space groups. A complete listing of all the space groups with associated cell diagrams showing the placement of the symmetry elements and a listing of the general and special position coordinates can be found in Volume A, International Tables for Crystallography, 1983.

Examination of the symmetry of the diffraction pattern (the Laue symmetry) permits the determination of the crystal system. To further limit the possible space groups within the crystal system, one needs to determine those classes of reflections that are systematically absent (extinct).

Consider the *c*-glide plane as present in $P2_1/c$. For every atom at $(x, y, z)$, there is another at $(x, 1/2-y, 1/2+z)$. The structure factor can then be written as

$$F_{hkl} = \sum_{j=1}^{N/2} f_j \left\{ \exp[2\pi i(hx_j + ky_j + lz_j)] \right.$$
$$\left. + \exp\left[2\pi i\left(hx_j + k\left(\frac{1}{2} - y_j\right) + l\left(\frac{1}{2} - z_j\right)\right)\right] \right\} \qquad (31)$$

For those reflections with $k = 0$,

$$F_{h0l} = \sum_{j=1}^{N/2} f_j \, \exp[2\pi i(hx_j + lz_j)][1 + (-1)^l] \qquad (32)$$

Thus, for $h0l$ reflections, all intensities with $l$ odd would be absent. Different symmetry elements give different patterns of missing reflections as long as the symmetry element has a translational component. Point group elements show no systematic missing reflections.

Table 2 lists the extinction conditions caused by various symmetry elements. In many cases an examination of the extinctions will uniquely determine the space group. In others, two possibilities may remain that differ only by the presence or absence of a center of symmetry. Statistical tests (Wilson, 1944) have been devised to detect the presence of a center of symmetry; one of the most reliable is that developed by Howells et al. (1950). It should be cautioned, however, that while such tests are usually reliable, one should not interpret their results as being 100% certain.

Further discussion of space group symmetry can be found in SYMMETRY IN CRYSTALLOGRAPHY as well as in the literature (International Tables for Crystallography, 1983; Ladd and Palmer, 1985; Stout and Jensen, 1989; Lipscomb and Jacobson, 1990).

### Crystal Structure Refinement

After data have been collected and the space group determined, a first approximation to the crystal structure must then be found. (Methods to do so are described in the next section.) Such a model must then be refined to obtain a final model that best agrees with the experimental data. Although a Fourier series approach could be used (using electron density map or difference electron density map calculations), this approach has the disadvantage that the structural model could be affected by systematic errors due to series termination effects, related to truncation of data at the observation limit. Consequently, a least-squares refinement method is employed instead. Changes in the positional and thermal parameters are calculated that minimize the difference between $|\mathbf{F}_{hkl}^{\text{obs}}|$ and $|\mathbf{F}_{hkl}^{\text{calc}}|$, or alternatively $\mathbf{I}_{hkl}^{\text{obs}}$ and $\mathbf{I}_{hkl}^{\text{calc}}$.

**Table 2. Extinctions Caused by Symmetry Elements[a]**

| Class of Reflection | Condition for Nonextinction ($n$, an integer) | Interpretation of Extinction | Symbol for Symmetry Element |
|---|---|---|---|
| $hkl$ | $h+k+l=2n$ | Body-centered lattice | $I$ |
| | $h+k=2n$ | $C$-centered lattice | $C$ |
| | $h+l=2n$ | $B$-centered lattice | $B$ |
| | $k+l=2n$ | $A$-centered lattice | $A$ |
| | $h+k,h+l$ | Face-centered lattice | $F$ |
| | $h+l=2n$ | | |
| | $-h+k+l=3n$ | Rhombohedral lattice indexed on hexagonal lattice | $R$ |
| | $h+k+l=3n$ | Hexagonal lattice indexed on rhombohedral lattice | $H$ |
| $0kl$ | $k=2n$ | Glide plane $\perp a$, translation $b/2$ | $b$ |
| | $l=2n$ | Glide plane $\perp a$, translation $c/2$ | $c$ |
| | $k+l=2n$ | Glide plane $\perp a$, translation $(b+c)/2$ | $n$ |
| | $k+l=4n$ | Glide plane $\perp a$, translation $(b+c)/4$ | $d$ |
| $h0l$ | $h=2n$ | Glide plane $\perp b$, translation $a/2$ | $a$ |
| | $l=2n$ | Glide plane $\perp b$, translation $c/2$ | $c$ |
| | $h+l=2n$ | Glide plane $\perp b$, translation $(a+c)/2$ | $n$ |
| | $h+k=4n$ | Glide plane $\perp a$, translation $(a+b)/4$ | $d$ |
| $hhl$ | $l=2n$ | Glide plane $\perp (a-b)$, translation $c/2$ | $c$ |
| | $2h+l=2n$ | Glide plane $\perp (a-b)$, translation $(a+b+c)/2$ | $n$ |
| | $2h+l=2n$ | Glide plane $\perp (a-b)$, translation $(a+b+c)/4$ | $d$ |
| $h00^{b}$ | $h=2n$ | Screw axis $\parallel a$, translation $a/2$ | $2_1, 4_2$ |
| | $h=4n$ | Screw axis $\parallel a$, translation $a/4$ | $4_1, 4_3$ |
| $00l$ | $l=2n$ | Screw axis $\parallel c$, translation $c/2$ | $2_1, 4_2, 6_3$ |
| | $l=4n$ | Screw axis $\parallel c$, translation $c/4$ | $4_1, 4_3$ |
| | $i=3n$ | Screw axis $\parallel c$, translation $c/3$ | $3_1, 3_2, 6_2, 6_4$ |
| | $l=6n$ | Screw axis $\parallel c$, translation $c/6$ | $6_1, 6_5$ |
| $hh0$ | $h=2n$ | Screw axis $\parallel (a+b)$, translation $(a+b)/2$ | $2_1$ |

[a]Adapted from Buerger, 1942.
[b]Similarly for $k$ and $b$ translations.

It is common practice to refer to one portion of the data set as observed data and the other as unobserved. The terminology is a carry-over from the early days of crystallography when the data were measured from photographic film by visual matching to a standard scale. Intensities could be measured only to some lower limit, hence the "observed" and "unobserved" categories. Although measurements are now made with digital counters (scintillation detectors or area detectors), the distinction between these categories is still made; often $I > 3\sigma(I)$ or a similar condition involving $|F|$ is used for designating observed reflections. Such a distinction is still valuable when refining on $|F|$ (see below) and when discussing $R$ factors.

Crystallographers typically refer to a residual index ($R$ factor) as a gauge for assessing the validity of a structure. It is defined as

$$R = \frac{\sum_{\mathbf{h}} \| F_{\mathbf{h}}^{\text{obs}} | - | F_{\mathbf{h}}^{\text{obs}} \|}{\sum_{\mathbf{h}} |F_{\mathbf{h}}^{\text{obs}}|} \quad (33)$$

Atoms placed in incorrect positions usually produce $R > 0.40$. If positional parameters are correct, $R$'s in the 0.20 to 0.30 range are common. Structures with refined isotropic thermal parameters typically give $R \sim 0.15$, and least-squares refinement of the anisotropic thermal parameters will reduce this to $\sim 0.05$ or less for a well-behaved structure.

Positional and thermal parameters occur in trigonometric and exponential functions, respectively. To refine such nonlinear functions, a Taylor series expansion is used. If $f$ is a nonlinear function of parameters $p_1, p_2, \ldots, p_n$, then $f$ can be written as a Taylor series

$$f = f^0 + \sum_{i=1}^{n} \left( \frac{\partial f}{\partial p_i} \right)_{p_j} \Delta p_i + \text{higher order terms} \quad (34)$$

where $f^0$ represents the value of the function evaluated with the initial parameters and the $\Delta p_i$'s are the shifts in these parameters. In practice, the higher order terms are neglected; if $\Delta_j$ for observation $j$ is defined as

$$\Delta_j = f_j - f_j^0 - \sum_{i=1}^{n} \left( \frac{\partial f}{\partial p_i} \right)_{p_j} \Delta p_i \quad (35)$$

then $\sum_j w_j \Delta_j^2$ can be minimized to obtain the set of best shifts of the parameters in a least squares sense. These in turn can be used to obtain a new set of $f^0$s, and the process repeated until all the shifts are smaller than some prescribed minimum. The weights $w_j$ are chosen such as to be reciprocally related to the square of the standard deviation associated with that observation.

In the X-ray case, as noted above, $f$ can be $I_{hkl}$ or $|F_{hkl}|$. If the latter, then

$$|F_{hkl}^c| = F_{hkl}^c \exp(-2\pi i \alpha_{hkl})$$
$$= A_{hkl}^c \cos \alpha_{hkl} + B_{hkl}^c \sin \alpha_{hkl} \quad (36)$$

In this case, only observed reflections should be used to calculate parameter shifts since the least-squares method assumes a Gaussian distribution of errors in the observations. (For net intensities that are negative, i.e., where the total intensity is somewhat less than the background measurement, $|F^0_{hkl}|$ is not defined. After refinement, structure factors can still be calculated for these unobserved reflections.)

The set of least-squares equations can be obtained from these $\Delta_j$ equations as follows for $m$ observations. Express

$$\begin{pmatrix} \frac{\partial f_1}{\partial p_1} & \frac{\partial f_1}{\partial p_2} & \frac{\partial f_1}{\partial p_3} & \cdots & \frac{\partial f_1}{\partial p_n} \\ \vdots & \vdots & \vdots & \cdots & \vdots \\ \frac{\partial f_m}{\partial p_1} & \frac{\partial f_m}{\partial p_2} & \frac{\partial f_m}{\partial p_3} & \cdots & \frac{\partial f_m}{\partial p_n} \end{pmatrix} \begin{pmatrix} \Delta p_1 \\ \Delta p_2 \\ \vdots \\ \Delta p_n \end{pmatrix} = \begin{pmatrix} f_1 - f_1^0 \\ f_2 - f_2^0 \\ \vdots \\ f_m - f_m^0 \end{pmatrix} \quad (37)$$

It is convenient to represent these matrices by $\mathbf{MP} = \mathbf{Y}$. Then multiplying both sides by the transpose of $\mathbf{M}$, $\mathbf{M^T}$, gives $\mathbf{M^TMP} = \mathbf{M^TY}$. Let $\mathbf{A} = \mathbf{M^TM}$ and $\mathbf{B} = \mathbf{M^TY}$. Since $\mathbf{A}$ is now a square matrix, $\mathbf{P} = \mathbf{A^{-1}B}$. Least-squares methods not only give a final set of parameters but also provide

$$s^2(p_j) = a^{jj} \left( \frac{\sum_i w_i \Delta j^2}{m - n} \right) \quad (38)$$

where $s$ is the standard deviation associated with the parameter $p_j$, $a^{jj}$ is the $j$th diagonal matrix element of the inverse matrix, $m$ is the number of observations, and $n$ is the number of parameters. The square root of the quantity in parentheses is sometimes referred to as the standard deviation of an observation of unit weight, or the goodness-of-fit parameter. If the weights are correct, i.e., if the errors in the data are strictly random and correctly estimated, and if the crystal structure is properly being modeled, then the value of this quantity should approximately equal unity.

## PRACTICAL ASPECTS OF THE METHOD

In practice, the application of single-crystal X-ray diffraction methods for the determination of crystal and molecular structure can be subdivided into a number of tasks. The investigator must: (i) select and mount a suitable crystal; (ii) collect the diffracted intensities produced by the crystal; (iii) correct the data for various experimental effects; (iv) obtain an initial approximate model of the structure; (v) refine this model to obtain the best fit between the observed intensities and their calculated counterparts; and (vi) calculate derived results (e.g., bond distances and angles and stereographic drawings including thermal ellipsoid plots) from the atomic parameters associated with the final model.

For further discussion of the selection and mounting of a crystal see Sample Preparation, below; for further details on methods of data collection, see Data Collection Techniques.

The fiber or capillary holding the crystal is placed in a pin on a goniometer head that has $x, y, z$ translations. (Preliminary investigation using film techniques—precession or Weissenberg—could be carried out at this stage, if desired, but we will not discuss such investigations here.) The goniometer head is then placed on the diffractometer and adjusted using the instrument's microscope so that the center of the crystal remains fixed during axial rotations.

The next task is to determine the cell dimensions, the orientation of the cell relative to the laboratory coordinate system, and the likely Bravais lattice and crystal system. This process is usually termed indexing.

To accomplish this, an initial survey of the low to mid angle ($\theta$) regions of reciprocal space is carried out. This can be done via a rotation photograph using a Polaroid cassette—the user then inputs the $x, y$ positions of the diffraction maxima and the diffractometer searches in the rotation angle—or by initiating a search procedure that varies a variety of the angles on a four-circle diffractometer. Any peaks found are centered. Peak profiles give an indication of crystal quality—one would expect Gaussian-like peaks from a crystal of good quality. Sometimes successful structure determinations can be carried out on crystals that give split peaks or have a significant side peak, but it is best to work with crystals giving well-shaped peaks if possible. With an area detector, reflections can be collected from a few frames at different rotation angles. An indexing program is then employed that uses this information to determine the most probable cell and crystal system. (Although these indexing programs are usually quite reliable, the user must remember that the information is being obtained from a reasonably small number of reflections and some caution is advisable. The user should verify that all or almost all of the reflections were satisfactorily assigned indices by the program.)

Computer programs (Jacobson, 1976, 1997) typically produce a reduced cell as a result of indexing. The reduced cell is one that minimizes the lengths of $\mathbf{a}_1$, $\mathbf{a}_2$, and $\mathbf{a}_3$ that describe the repeating unit (and hence give angles as close to 90° as possible) (Niggli, 1928; Lawton and Jacobson, 1965; Santoro and Mighell, 1970). It also orders the axes according to a convention such as $|\mathbf{a}_1| \leq |\mathbf{a}_2| \leq |\mathbf{a}_3|$. This is followed by a Bravais lattice determination in which characteristic length and angle relationships are used to predict the presence of centered cells and the axes are rearranged to be consistent with the likely crystal system. The next step would then be to measure intensities of some sets of reflections that would be predicted to be equal by the Laue group symmetry (to obtain additional experimental verification as to the actual Laue group of the crystal or a possible subgroup).

Once the likely Laue symmetry has been determined, the data collection can begin. This author usually prefers to collect more than just a unique data set; the subsequent data averaging gives a good estimate of data quality. Data are typically collected to a $2\theta$ maximum from 50 to 60° with Mo $K_\alpha$ radiation, the maximum chosen depending on the crystal quality and the amount of thermal motion associated with the atoms. (Greater thermal motion means weaker average intensities as $2\theta$ increases.)

Once data collection has finished, the data will need to be corrected for geometric effects such as the Lorentz effect (the velocity with which the reciprocal lattice point moves through the sphere of reflection) and the polarization effect; the X-ray beam becomes partially polarized in the course of the diffraction process. Usually data will also have to be corrected for absorption since as the beam traverses the crystal, absorption occurs, as given by the relation $e^{-\mu t}$, where $\mu$ is the absorption coefficient and $t$ is the path length. "Analytical" absorption correction can be made if the orientations of crystal faces and their distances from the center of the crystal are carefully measured. Often empirical absorption corrections are employed instead. For data collected on a four-circle diffractometer, a series of psi scans can be used for this purpose. These are scans at different $\phi$ angle positions for reflections close to $\chi = 90°$. (For a reflection at $\chi = 90°$, diffraction is $\phi$ independent and any variation in observed diffraction can be attributable to effects due to absorption.) For data collected on an area detector, computer programs determine the parameters associated with an analytic representation of the absorption surface using the observed differences found between the intensities of symmetry-related reflections. Corrections can also be made for a decrease in scattering caused by crystal decomposition effects, if necessary. This composite set of corrections is often referred to as data reduction.

Following data reduction, the data can be averaged and then the probable space group or groups found by an examination of systematic absences in the data. Statistical tests, such as the Howells, Philips, and Rogers tests for the possible presence of a center of symmetry, can be carried out if more than one space group is indicated.

Next the investigator typically seeks an initial trial model using a direct method program, or if a heavy-atom approach is appropriate, the analysis of peaks between symmetry-related atoms found in the Patterson calculation might be used as discussed below. An initial electron density map would then be calculated. The bond distances and bond angles that are associated with these atom position possibilities and the residual index often indicate whether a valid solution has been found. If so, the inclusion of additional atoms found on electron density or difference electron density maps and subsequent refinement usually leads to a final structural model. Final bond distances, angles, and their standard deviations can be produced along with a plot of the atom positions (e.g., an ORTEP drawing); see Data Analysis and Initial Interpretations, Derived Results.

If a valid initial model is not obtained, the number of phase set trials might be changed if using a direct method program. Different Patterson peaks could be tried. Other space group or crystal system possibilities could be investigated. Other structure solution methods beyond the scope of this chapter might have to be employed to deduce an appropriate trial model. Fortunately, if the data are of good quality, most structures can be solved using the standard computer software available with commercial systems.

## Data Collection Techniques

Film has long been used to record X-ray diffraction intensities. However, film has a number of disadvantages; it is relatively insensitive, especially for the shorter X-ray wavelengths, and does not have a large dynamic range. Gas-filled proportional counters can be used to detect X rays. These detectors can attain high efficiencies for longer wavelength radiations; with the use of a center wire of high resistance, they can be used to obtain positional information as well. The curved one-dimensional versions are used for powder diffraction studies, and the two-dimensional counterparts form the basis for some of the area detectors used in biochemical applications. They operate at reasonably high voltages (800 to 1700 V), and the passage of an X-ray photon causes ionization and an avalanche of electrons to occur that migrate to the anode. The number of electrons is proportional to the number of ion pairs, which is approximately proportional to the energy of the incident X-ray photon.

The scintillation counter is composed of a NaI crystal that has been activated by the addition of some Tl. An absorbed X-ray produces a photoelectron and one or more Auger electrons; the latter energize the Tl sites, which in turn emit visible photons. The crystal is placed against a photomultiplier tube that converts the light into a pulse of electrons. The number of color centers energized are approximately dependent on the initial X-ray energy.

Scintillation detectors are typically used on four-circle diffractometers. The four-circle diffractometer is one of the most popular of the counter diffractometers. As the name suggests, this diffractometer has four shafts that can be moved usually in an independent fashion. Such a diffractometer is shown schematically in Figure 6. The $2\theta$ axis is used to move the detector parallel to the basal plane, while the remaining three axes position the crystal (Hamilton, 1974). The $\omega$ axis rotates around an axis that is perpendicular to the basal plane, the $\chi$ axis rotates around an axis that is perpendicular to the face of the circle, and the $\phi$ axis rotates around the goniometer head mount. To collect data with such an instrument, these angles must be positioned such that the Laue equation, or its scalar equivalent, Bragg's law, is satisfied (see DYNAMICAL DIFFRACTION).

Electronic area detectors have become more routinely available in the last few years and represent a major new advance in detector technology. Two such new detectors are the CCD detector and the imaging plate detector. In a typical CCD system, the X-ray photons impinge on a phosphor such as gadolinium oxysulfide. The light

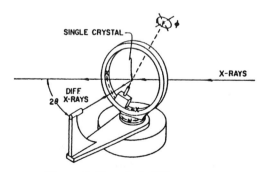

**Figure 6.** Four-circle diffractometer.

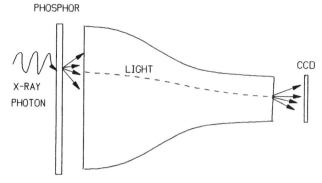

**Figure 7.** A CCD area detector. An X-ray photon arriving from the left is converted to light photons at the phosphor. Fiber optics conduct these to the CCD chip on the right.

produced is conducted to a CCD chip via fiber optics, as shown in Figure 7, yielding about eight electrons per X-ray photon. The typical system is maintained at $-50°C$ and has a dynamic range of 1 to $> 10^5$ photons per pixel. Readout times are typically 2 to 9 s depending on the noise suppression desired.

In a typical imaging plate system, the X-ray photons impinge on a phosphor (e.g., europium-doped barium halide) and produce color centers in the imaging plate. Following the exposure, these color centers are read by a laser and the light emitted is converted to an electronic signal using a photomultiplier. The laser action also erases the image so that the plate can be reused. In practice, usually one plate is being exposed while another is being read. The readout time is somewhat longer than with a CCD, typically 4 min or so per frame. However, the imaging plate has excellent dynamic range, 1 to $>10^6$ photons per pixel, and a considerably larger active area. It operates at room temperature.

As noted above, imaging plate systems usually provide a larger detector surface and slightly greater dynamic range than does the CCD detector, but frames from the latter system can be read much more rapidly than with imaging plate systems. Crystallographers doing small-molecule studies often tend to prefer CCD detectors, while imaging plate systems tend to be found in those laboratories doing structural studies of large biological molecules. Both types of area detectors permit the crystallographer to collect data in a manner almost independent of cell size. With the older scintillation counter system, a doubling of the cell size translates into a doubling of the time needed to collect the data set (since the number of reflections would be doubled). On the other hand, with an area detector, as long as the detector does not have to be moved further away from the source to increase the separation between reflections, the time for a data collection is essentially independent of cell size. Furthermore, if the crystal is rotated around one axis (e.g., the vertical axis) via a series of small oscillations and the associated frame data are stored to memory, most of the accessible data within a theta sphere would be collected. The remaining "blind region" could be collected by offsetting the crystal and performing a second limited rotation. Many commercial instruments use a four-circle diffractometer to provide maximum rotational capability. Due to the high sensitivity of the detector, a full data collection can be readily carried out with one of these systems in a fraction of a day for moderate-size unit cells.

## METHOD AUTOMATION

Single-crystal X-ray methods involve the collection and processing of thousands to tens of thousands of X-ray intensities. Therefore, modern instrumentation comes with appropriate automation and related computer software to carry out this process in an efficient manner. Often, user input is confined to the selection of one of a few options at various stages in the data-collection process. The same is true of subsequent data processing, structure solution, and refinement.

## DATA ANALYSIS AND INITIAL INTERPRETATION

### Obtaining the Initial Model

There are a number of approaches that can be utilized to obtain an initial model of the structure appropriate for input into the refinement process. Many are specialized and tend to be used to "solve" particular kinds of structures, e.g., the use of isomorphous replacement methods for protein structure determination. Here we will primarily confine ourselves to two approaches: direct methods and the heavy-atom method.

Direct methods employ probability relations relying on the magnitudes of the structure factors and a few known signs or phases to determine enough other phases to reveal the structure from an electron density map calculation. The possibility that phase information could be extracted from the structure factor magnitudes has intrigued investigators throughout the history of crystallography. Karle and Hauptman (1956), building on the earlier work of Harker and Kasper (1948) and Sayre (1952), developed equations that have made these methods practical. A much more complete discussion of direct methods can be found in the Appendix.

One of the most important of the equations developed for the direct methods approach is the sigma-two equation:

$$E_{\mathbf{h}} \approx \langle E_{\mathbf{h}} E_{\mathbf{h-k}} \rangle_{\mathbf{k}} \qquad (39)$$

where $E_{\mathbf{h}}$ is derived from $F_{\mathbf{h}}$ and $\langle \rangle_{\mathbf{k}}$ represents the average over the $\mathbf{k}$ contributors. In practice, only those $E$'s whose sign or phase has been determined are used in the average on the right. Since $|E_{\mathbf{h}}|$ is known, the equation is used to predict the most probable sign or place of the reflection.

The probability of $|E_{\mathbf{h}}|$ being positive is given by

$$P_+(\mathbf{h}) = \frac{1}{2} + \frac{1}{2}\tanh \sigma_3 \sigma_2^{-3/2} |E_{\mathbf{h}}| \sum_{\mathbf{k}} E_{\mathbf{k}} E_{\mathbf{h-k}} \qquad (40)$$

For the non-centrosymmetric case

$$\tan\alpha_{\mathbf{h}} = \frac{\sum_{\mathbf{k}} |E_{\mathbf{h}} \| E_{\mathbf{h-h}}| \sin(\alpha_{\mathbf{k}} + \alpha_{\mathbf{h-k}})}{\sum_{\mathbf{k}} |E_{\mathbf{k}} \| E_{\mathbf{h-k}}| \cos(\alpha_{\mathbf{k}} + \alpha_{\mathbf{h-k}})} \quad (41)$$

Direct method computer programs use these equations and other related ones as described in the Appendix to attempt to obtain an approximate set of signs (or phases) directly from the magnitudes of the structure factors. These are then used to calculate an electron density map, which could reveal the approximate positions of most of the atoms in the structure. Further details on two of the more commonly used direct methods programs can be found in the Appendix.

Direct methods can be expected to be the most reliable when applied to structures in which most of the atoms are approximately equal in atomic number. In structures where one or a few atoms are present that have appreciably larger atomic numbers than the rest, the heavy-atom method is typically the approach of choice.

The basic assumption of the heavy-atom method is that these atoms alone provide good enough initial phasing of an electron density map that other atoms can be found and included, and the process repeated, until all atoms are located. See the Appendix for a more complete theoretical treatment of the heavy-atom method. The initial positions for the heavy atoms can come from direct methods or from an analysis of the Patterson function (see Appendix).

### Derived Results

As a result of a crystal structure determination, one usually wishes to obtain, e.g., bond distances, bond angles, and least-squares planes, along with associated standard deviations. These can be determined from the least-squares results in a straightforward fashion.

Consider the bond distance between two atoms A and B. In vector notation, using $\mathbf{a}_1$, $\mathbf{a}_2$, and $\mathbf{a}_3$ to represent the unit cell vectors.

$$\begin{aligned}
d_{AB} = |\mathbf{r}_B - \mathbf{r}_A| &= [(\mathbf{r}_B - \mathbf{r}_A) \bullet (\mathbf{r}_B - \mathbf{r}_A)]^{1/2} \\
&= [(x_B - x_A)^2 a_1^2 + (y_B - y_A)^2 a_2^2 \\
&\quad + (z_B - z_A)^2 a_3^2 + 2(x_B - x_A)(y_B - y_A)\mathbf{a}_1 \bullet \mathbf{a}_2 \quad (42) \\
&\quad + 2(y_B - y_A)(z_B - z_A)\mathbf{a}_2 \bullet \mathbf{a}_3 \\
&\quad + 2(x_B - x_A)(z_B - z_A)\mathbf{a}_1 \bullet \mathbf{a}^3]^{1/2}
\end{aligned}$$

For atoms $A$, $B$, and $C$, the bond angle A–B–C can be obtained from

$$\text{Angle} = \cos^{-1}\left\{ \frac{(\mathbf{r}_A - \mathbf{r}_B) \bullet (\mathbf{r}_C - \mathbf{r}_B)}{d_{AB} \bullet d_{BC}} \right\} \quad (43)$$

Furthermore, the standard deviation in the bond distances can be obtained from

$$\sigma^2(d) = \sum_j \left( \frac{\partial d}{\partial p_j} \right) \sigma^2(p_j) \quad (44)$$

if the parameters are uncorrelated. The parameters in this case are the fractional coordinates ($x_A$, $y_A$, $z_A$, $x_B$, $y_B$, $z_B$)

and their standard deviations ($\sigma(x_A)$, $\sigma(y_A)$, $\sigma(z_A)$, $\sigma(x_B)$, $\sigma(y_B)$, $\sigma(z_B)$) and the cell dimensions and their standard deviations. [Usually the major contributors to $\sigma(d)$ are from the errors in the fractional coordinates.] A similar expression can be written for the standard deviation in the bond angle. One can also extend these arguments to e.g., torsion angles and least-squares planes.

As noted above, anisotropic thermal parameters are usually obtained as part of a crystal structure refinement. One should keep in mind that $R$ factors will tend to be lower on introduction of anisotropic temperature factors, but this could merely be an artifact due to the introduction of more parameters (five for each atom). One of the most popular ways of viewing the structure, including the thermal ellipsoids, is via plots using ORTEP [a Fortran thermal ellipsoid plot program for crystal structure illustration (Johnson, 1965)]. ORTEP plots provide a visual way of helping the investigator decide if these parameters are physically meaningful. Atoms would be expected to show less thermal motion along bonds and more motion perpendicular to them. Atoms, especially those in similar environments, would be expected to exhibit similar thermal ellipsoids. Very long or very short principal moments in the thermal ellipsoids could be due to disordering effects, poor corrections for effects such as absorption, or an incorrect choice of space groups, to mention just some of the more common causes.

Once data collection has finished, the data will need to be corrected for geometric effects such as the Lorentz effect (the velocity with which the reciprocal lattice point moves through the sphere of reflection) and the polarization effect; the X-ray beam becomes partially polarized in the course of the diffraction process. Usually data will also have to be corrected for absorption since as the beam traverses the crystal, absorption occurs, as given by the relation $e^{-\mu t}$, where $\mu$ is the absorption coefficient and $t$ is the path length. "Analytical" absorption correction can be made if the orientations of crystal faces and their distances from the center of the crystal are carefully measured. Often empirical absorption corrections are employed instead. For data collected on a four-circle diffractometer, a series of psi scans can be used for this purpose. These are scans at different $\phi$ angle positions for reflections close to $\chi = 90°$. (For a reflection at $\chi = 90°$, diffraction is $\phi$ independent and any variation in observed diffraction can be attributable to effects due to absorption.) For data collected on an area detector, computer programs determine the parameters associated with an analytic representation of the absorption surface using the observed differences found between symmetry-related reflections. Corrections can also be made for a decrease in scattering caused by crystal decomposition effects, if necessary. This composite set of corrections is often referred to as data reduction.

Following data reduction, the data can be averaged, if desired, and then the probable space group or groups found by an examination of systematic absences in the data. Statistical tests such as the Howells, Philips, and Rogers tests can be carried out to check for the possible presence of a center of symmetry; if more than one space group is indicated.

Next the investigator typically seeks an initial trial model using a direct method program, or if a heavy-atom approach is appropriate, the analysis of peaks between symmetry-related atoms found in the Patterson approach might be used. An initial electron density map would then be calculated. The bond distances and bond angles that are associated with these atom position possibilities and sometimes the residual index often indicate whether a valid solution has been found. If so, the inclusion of additional atoms found on electron density or difference electron density maps and subsequent refinement usually leads to a final structural model. Final bond distances, angles, and their standard deviations can be produced along with a plot of the atom positions (e.g., an ORTEP drawing).

If a valid initial model is not obtained, the number of phase set trials might be changed if using a direct method program. Different Patterson peaks could be tried. Other space group or crystal system possibilities could be investigated. Other structure solution methods beyond the scope of this chapter might have to be employed to deduce an appropriate trial model. Fortunately, if the data are of good quality, most structures can be solved using the standard computer software available with commercial systems.

### Example

The compound $(C_5H_5)(CO)_2MnDBT$, where DBT is dibenzothiophene, was synthesized in R. J. Angelici's group as part of their study of prototype hydrodesulfurization catalysts (Reynolds et al., 1999). It readily formed large, brown, parallelepiped-shaped crystals. A crystal of approximate dimensions $0.6 \times 0.6 \times 0.5$ mm was selected and mounted on a glass fiber. (This crystal was larger than what would normally be used, but it did not cleave readily and the absorption coefficient for this material is quite small, $\mu = 9.34\,cm^{-1}$. A larger than normal collimator was used.) It was then placed on a Bruker P4 diffractometer, a four-circle diffractometer equipped with a scintillation counter. Molybdenum $K_\alpha$ radiation was used as a source from a sealed tube target and a graphite monochromator ($\lambda = 0.71069$ Å).

A set of 42 reflections was found by a random-search procedure in the range $10 \leq 2\theta \leq 25°$. These were carefully centered, indexed, and found to fit a monoclinic cell with dimensions $a = 13.076(2)$, $b = 10.309(1)$, $c = 23.484(3)$ Å, and $\beta = 92.93(1)°$. Based on atomic volume estimates and a formula weight of 360.31 g/mol, a calculated density of 1.514 g/cm³ was obtained with eight molecules per cell. Systematic absences of $h0l$: $h + l \neq 2n$ and $0k0$: $k \neq 2n$ indicated space group $P2_1/n$, with two molecules per asymmetric unit, which was later confirmed by successful refinement.

Data were collected at room temperature (23°C) using an $\omega$-scan technique for reflections with $2\theta \leq 50°$. A hemisphere of data was collected ($+h, \pm k, \pm l$), yielding 12,361 reflections. Symmetry-related reflections were averaged (5556 reflections) and 4141 had $F > 3\sigma(F)$ and were considered observed. The data were corrected for Lorentz and polarization effects and were corrected for absorption using an empirical method based on psi scans. The transmission factors ranged from 0.44 to 0.57.

**Table 3. Fractional Coordinates (Nonhydrogen Atoms) for One Molecule of $Mn(SC_{12}H_9)(CO)_2(C_5H_5)$**

| Atom | $X$ | $Y$ | $Z$ | $B_{eq}{}^a$ |
|------|------|------|------|------|
| Mn1 | .0356(1) | .0387(1) | .14887(6) | 3.18(4) |
| S1 | .0970(2) | −.1032(2) | .2155(1) | 3.38(6) |
| C1 | .4723(8) | .2495(9) | .2839(4) | 3.2(3) |
| C2 | .4098(8) | .1431(9) | .2962(4) | 3.5(3) |
| C3 | .4538(10) | .0169(10) | .2953(4) | 4.7(3) |
| C4 | .5588(11) | .0055(12) | .2818(5) | 5.0(4) |
| C5 | .6168(10) | .1102(11) | .2692(5) | 4.7(3) |
| C6 | .5742(9) | .2367(11) | .2701(4) | 4.1(3) |
| C7 | .7902(8) | .1841(10) | .8059(4) | 3.4(3) |
| C8 | .3040(8) | .1813(9) | .3090(4) | 3.4(3) |
| C9 | .2202(9) | .1044(11) | .3220(5) | 4.5(3) |
| C10 | .1274(10) | .1599(13) | .3303(5) | 5.0(3) |
| C11 | .6169(10) | .2017(14) | .8285(5) | 5.5(4) |
| C12 | .6989(8) | .1259(11) | .8158(5) | 4.4(3) |
| C13 | .0334(10) | .1004(11) | .0620(4) | 4.6(3) |
| C14 | .9012(10) | .0103(13) | .9320(5) | 5.1(4l) |
| C15 | .9462(10) | .1140(11) | .9141(4) | 4.5(3) |
| C16 | .0649(10) | .0685(11) | .9097(4) | 4.3(3) |
| C17 | .9348(10) | .0674(12) | .0753(4) | 4.6(3) |
| C18 | .9439(10) | .1038(11) | .1963(5) | 4.4(3) |
| C19 | .1292(10) | .1583(11) | .1670(4) | 4.3(3) |
| O1 | .8838(7) | .1452(9) | .2255(4) | 6.6(3) |
| O2 | .1901(7) | .2394(8) | .1754(4) | 6.0(3) |

$^aB_{eq} = (8\pi^2/3)\sum_{i=1}^{3}\sum_{j=1}^{3} u_{ij}a_i^* a_j^* a_i \bullet a_j$.

The SHELX direct method procedure was used and yielded positions for the heavier atoms and possible positions for many of the carbon atoms. An electron density map was calculated using the phases from the heavier atoms; this map yielded further information on the lighter atom positions, which were then added. Hydrogen atoms were placed in calculated positions and all nonhydrogen atoms were refined anisotropically. The final cycle of full-matrix least-squares refinement converged with an unweighted residual $R = \sum \| F_0| - |F_c \| / \sum |F_0|$ of 3.3% and a weighted residual $R_w = [\sum w\ (|F_0| - |F_c|)^2 / \sum w F_0^2]^{1/2}$, where $w = 1/\sigma^2(F)$, of 3.9%. The maximum peak in the final electron density difference map was $0.26\ e/\text{Å}^3$. Atomic scattering factors were taken from Cromer and Weber (1974), including corrections for anomalous dispersion. Fractional atomic coordinates for one of the molecules in the asymmetric unit are given in Table 3, and their anisotropic temperature factors in Table 4.

The two independent molecules in the asymmetric unit have essentially identical geometries within standard deviations. A few selected bond distances and angles are given in Table 5 and Table 6, respectively. A thermal ellipsoid plot of one of the molecules is shown in Figure 8.

### SAMPLE PREPARATION

The typical investigation starts obviously with the selection of a crystal. This is done by examining the material under a low- or moderate-power microscope. Samples that display sharp, well-defined faces are more likely to be single crystals. Crystals with average dimensions of a

**Table 4. Anisotropic Thermal Parameters (Non-Hydrogen Atoms)\* for One Molecule of $Mn(SC_{12}H_9)(CO)_2(C_5H_5)$**

| Atom | $U_{11}$ | $U_{22}$ | $U_{33}$ | $U_{23}$ | $U_{13}$ | $U_{12}$ |
|------|----------|----------|----------|----------|----------|----------|
| Mn1 | 5.11(10) | 3.41(8) | 3.56(7) | −.07(6) | .38(6) | .33(7) |
| S1 | 5.5(2) | 3.5(1) | 3.8(1) | −.1(1) | −.1(1) | .4(1) |
| C1 | 5.6(7) | 4.1(5) | 2.6(4) | .0(4) | −.3(4) | −.2(5) |
| C2 | 6.1(7) | 3.5(5) | 3.6(5) | −.6(4) | −.6(5) | −.1(5) |
| C3 | 8.8(9) | 3.8(6) | 5.1(6) | −.6(5) | −1.3(6) | .6(6) |
| C4 | 7.8(9) | 5.8(7) | 5.4(6) | −1.5(6) | .0(6) | 1.5(7) |
| C5 | 6.6(8) | 5.7(7) | 5.7(6) | −1.8(6) | .0(6) | .7(6) |
| C6 | 5.2(7) | 6.1(7) | 4.2(5) | −.7(5) | .4(5) | −.3(6) |
| C7 | 4.6(6) | 4.1(5) | 4.1(5) | .1(4) | −.6(5) | .6(5) |
| C8 | 5.6(6) | 3.6(5) | 3.7(5) | −.1(4) | −.5(5) | −.6(5) |
| C9 | 6.5(8) | 5.4(7) | 5.2(6) | −.1(5) | −.5(6) | −1.8(6) |
| C10 | 5.6(8) | 6.5(8) | 6.9(7) | −.1(6) | .4(6) | −.7(6) |
| C11 | 5.9(8) | 7.9(9) | 6.8(7) | −.6(7) | −.4(6) | .4(7) |
| C12 | 4.6(7) | 5.9(7) | 6.2(7) | .4(6) | −.1(5) | −.6(6) |
| C13 | 7.9(9) | 6.1(7) | 3.6(5) | .7(5) | .4(5) | −.4(7) |
| C14 | 7.3(8) | 8.0(9) | 4.3(6) | −.8(6) | .9(6) | .0(7) |
| C15 | 7.5(8) | 6.2(7) | 3.4(50) | −1.1(5) | −.3(5) | .5(7) |
| C16 | 7.3(8) | 5.3(6) | 3.6(5) | −.5(5) | .0(5) | −.2(6) |
| C17 | 7.3(8) | 6.1(7) | 3.9(5) | .3(5) | −.4(5) | .2(6) |
| C18 | 6.4(8) | 4.4(6) | 5.7(6) | −.2(5) | −.9(6) | .2(6) |
| C19 | 6.7(8) | 4.5(6) | 5.2(6) | .4(5) | .8(6) | 1.1(6) |
| O1 | 8.0(7) | 9.1(7) | 8.4(6) | −2.1(6) | 3.1(5) | 3.0(6) |
| O2 | 8.1(7) | 5.1(5) | 9.6(7) | −.6(5) | .2(5) | −1.7(5) |

\**U* values are scaled by 100.

few tenths of a millimeter are often appropriate—somewhat larger for an organic material. (An approximately spherical crystal with diameter $2/\mu$, where $\mu$ is the absorption coefficient, would be optimum.) The longest dimension should not exceed the diameter of the X-ray beam, i.e., it is important that the crystal be completely bathed by the beam if observed intensities are to be compared to their calculated counterparts.

The crystal can be mounted on the end of a thin glass fiber or glass capillary if stable to the atmosphere using any of a variety of glues, epoxy, or even fingernail polish. If unstable, it can be sealed inside a thin-walled glass capillary. If the investigation is to be carried out at low temperature and the sample is unstable in air, it can be immersed in an oil drop and then quickly frozen; the oil should contain light elements and be selected such as to form a glassy solid when cooled.

**Table 5. Selected Bond Distances (Å) for One Molecule of $Mn(SC_{12}H_9)(CO)_2(C_5H_5)$**

| | | | |
|------|----------|--------|----------|
| Mn1–S1 | 2.255(1) | C4–C5 | 1.381(5) |
| Mn1–C13 | 2.116(3) | C5–C6 | 1.387(4) |
| Mn1–C14 | 2.139(3) | C7–C8 | 1.397(4) |
| Mn1–C15 | 2.151(3) | C7–C12 | 1.376(4) |
| Mn1–C16 | 2.148(3) | C8–C9 | 1.391(4) |
| Mn1–C17 | 2.123(3) | C9–C10 | 1.376(5) |
| Mn1–C18 | 1.764(3) | C10–C11 | 1.405(5) |
| Mn1–C19 | 1.766(3) | C11–C12 | 1.377(5) |
| S1–C1 | 1.770(3) | C13–C14 | 1.424(5) |
| S1–C7 | 1.769(3) | C13–C17 | 1.381(5) |
| C1–C2 | 1.398(4) | C14–C15 | 1.412(5) |
| C1–C6 | 1.380(4) | C15–C16 | 1.393(4) |
| C2–C3 | 1.405(4) | C16–C17 | 1.422(4) |
| C2–C8 | 1.459(4) | C18–O1 | 1.166(4) |
| C3–C4 | 1.365(5) | C19–O2 | 1.164(4) |

## SPECIMEN MODIFICATION

In most instances, little if any specimen modification occurs, especially when area detectors are employed for data collection, thereby reducing exposure times. In some cases, the X-ray beam can cause some decomposition of the crystal. Decomposition can often be reduced by collecting data at low temperatures.

## PROBLEMS

Single-crystal X-ray diffraction has an advantage over many other methods of structure determination in that there are many more observations (the X-ray intensities) than parameters (atomic coordinates and thermal parameters). The residual index (R-value) therefore usually serves as a good guide to the general reliability of the solution. Thermal ellipsoids from ORTEP should also be examined, as well as the agreement between distances involving similar atoms, taking into account their standard deviations. Ideally, such distance differences should be less than 3 sigma, but may be somewhat greater if packing or

**Table 6. Selected Bond Angles (deg) for One Molecule of $Mn(SC_{12}H_9)(CO)_2(C_5H_5)$**

| | | | |
|-------------|----------|--------------|----------|
| S1– Mn1 –C18 | 92.5(1) | C1–C2 –C8 | 112.3(2) |
| S1– Mn1 –C19 | 93.3(1) | S1–C7 –C8 | 112.2(2) |
| C18– Mn1 –C19 | 92.8(1) | C2 –C8 –C7 | 112.3(2) |
| Mn1– S1 –C1 | 113.5(1) | Mn1 –C18 –O1 | 177.6(3) |
| Mn1– S1 –C7 | 112.7(1) | Mn1 –C19 –O2 | 175.0(3) |
| S1–C1 –C2 | 112.1(2) | | |

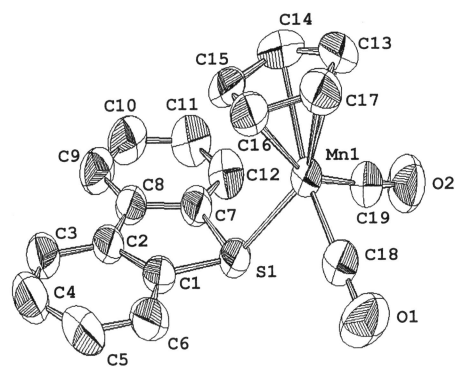

**Figure 8.** ORTEP of $Mn(SC_{12}Ha)(CO)_2(C_5H_5)$.

systematic error effects are present. If thermal ellipsoids are abnormal, this can be an indication of a problem with some aspect of the structure determination. This might be due to a poor correction for absorption of X rays as they pass through the crystal, a disordering of atoms in the structure, an extinction coefficient, or an incorrect choice of space group.

Approximations are inherent in any absorption correction, and therefore it is best to limit such effects by choosing a smaller crystal or using a different wavelength where the linear absorption coefficient would be less. The optimum thickness for a crystal is discussed above (see Sample Preparation).

If an atom is disordered over a couple of sites in the unit cell, an appropriate average of the electron density is observed and an elongation of the thermal ellipsoid may be observed if these sites are only slightly displaced. Sometimes it is possible to include multiple occupancies in the refinement to reasonably model such behavior with a constraint on the sum of the occupancies.

If a crystal is very perfect, extinction effects may occur, i.e., a scattered X-ray photon is scattered a second time. If this occurs in the same "mosaic block," it is termed primary extinction; if it occurs in another "mosaic block," it is termed secondary extinction. The effect is most noticeable in the largest intensities and usually manifests itself in a situation in which ten or so of the largest intensities are found to have calculated values that exceed the observed ones. Mathematical techniques exist to approximately correct for such extinction effects, and most crystallographic programs include such options.

Problems can be encountered if some dimension of the crystal exceeds the beam diameter. The crystal must be completely bathed in a uniform beam of X rays for all orientations of the crystal, if accurate intensities are to be predicted for any model.

The crystal should not decompose significantly (less than about 10%) in the X-ray beam. Some decomposition effects can be accounted for as long as they are limited.

## ACKNOWLEDGEMENTS

This work was supported in part by the U.S. Department of Energy, Office of Basic Energy Sciences, under contract W-7405-Eng-82.

## LITERATURE CITED

Berry, F. J. 1990. Mössbauer Spectroscopy. *In* Physical Methods of Chemistry. Determination of Structural Features of Crystalline and Amorphous Solids, Vol. V, 2nd ed. (B. W. Rossiter and J. F. Hamilton, eds.). pp. 273–343. John Wiley & Sons, New York.

Buerger, M. J. 1942. X-Ray Crystallography. John Wiley & Sons, New York.

Buerger, M. J. 1959. Vector Space. John Wiley & Sons, New York.

Cochran, W. and Woolfson, M. M. 1955. The theory of sign relations between structure factors. *Acta Crystallogr.* 8:1–12.

Cromer, D. T. and Weber, J. T. 1974. International Tables for X-Ray Crystallography, Vol. IV. Kynoch Press, Birmingham, England, Tables 2.2 A, 2.3.1.

Germain, G., Main, P., and Woolfson, M. M. 1970. On the application of phase relationships to complex structures. II. Getting a good start. *Acta Crystallogr.* B26:274–285.

Hamilton, W. C. 1974. Angle settings for four-circle diffractometers. *In* International Tables for X-ray Crystallography, Vol. IV (J. A. Ibers and W. C. Hamilton, eds.). pp. 273–284. Kynoch Press, Birmingham, England.

Harker, D. 1936. The application of the three-dimensional Patterson method and the crystal structures of provstite, $Ag_3A_5S_3$ and pyrargyrite, $Ag_3SbS_3$. *J. Chem. Phys.* 4:381–390.

Harker, D. and Kasper, J. S. 1948. Phase of Fourier coefficients directly from crystal diffraction data. *Acta Crystallogr.* 1:70–75.

Hauptman, H. and Karle, J. 1953. Solution of the Phase Problem I. Centrosymmetric Crystal. American Crystallographic Association, Monograph No. 3. Polycrystal Book Service, Pittsburgh, Pa.

Heald, S. M. and Tranquada, J. M. 1990. X-Ray Absorption Spectroscopy: EXAFS and XANES. *In* Physical Methods of Chemistry. Determination of Structural Features of Crystalline and Amorphous Solids, Vol. V, 2nd ed. (B. W. Rossiter and J. F. Hamilton, eds.). pp. 189–272. John Wiley & Sons, New York.

Hendrichs, P. M. and Hewitt, J. M. 1980. Solid-State Nuclear Magnetic Resonance. *In* Physical Methods of Chemistry. Determination of Structural Features of Crystalline and Amorphous Solids, Vol. V, 2nd ed. (B. W. Rossiter and J. F. Hamilton, eds.). pp. 345–432. John Wiley & Sons, New York.

Howells, E. R., Phillips, D. C., and Rogers, D. 1950. The probability distribution of x-ray intensities. II. Experimental investigation and the x-ray detection of centres of symmetry. *Acta Crystallogr.* 3:210–214.

International Tables for Crystallography. 1983. Vol. A: Space-Group Symmetry. D. Reidel Publishing Company, Dordrecht, Holland.

Jacobson, R. A. 1976. A single-crystal automatic indexing procedure. *J. Appl. Crystallogr.* 9:115–118.

Jacobson, R. A. 1997. A cosine transform approach to indexing. *Z. Kristallogr.* 212:99–102.

Johnson, C. K. 1965. ORTEP: A Fortran Thermal-Ellipsoid Plot Program for the Crystal Structure Illustrations. Report ORNL-3794. Oak Ridge National Laboratory, Oak Ridge, TN.

Karle, J. and Hauptman, H. 1956. A theory of phase determination for the four types of non-centrosymmetric space groups $1P222, 2P22, 3P_12,3P_22$. *Acta Crystallogr.* 9:635–651.

Karle, J. and Karle, I. L. 1966. The symbolic addition procedure for phase determination for centrosymmetric and noncentrosymmetric crystals. *Acta Crystallogr.* 21:849–859.

Ladd, M. F. C. and Palmer, R. A. 1980. Theory and Practice of Direct Methods in Crystallography. Plenum, New York.

Ladd, M. F. C. and Palmer, R. A. 1985. Structure Determination by X-Ray Crystallography. Plenum Press, New York.

Lawton, S. L. and Jacobson, R. A. 1965. The Reduced Cell and Its Crystallographic Applications. USAEC Report IS-1141. Ames Laboratory, Ames, IA.

Lipscomb, W. N. and Jacobson, R. A. 1990. X-Ray Crystal Structure Analysis. *In* Physical Methods of Chemistry. Determination of Structural Features of Crystalline and Amorphous Solids, Vol. V, 2nd ed. (B. W. Rossiter and J. F. Hamilton, eds.). pp. 3–121. John Wiley & Sons, New York.

Majkrzak, C. F., Lehmann, M. S., and Cox, D. E. 1990. The Application of Neutron Diffraction Techniques to Structural Studies, Physical Methods of Chemistry. Determination of Structural Features of Crystalline and Amorphous Solids, Vol. V, 2nd ed. (B. W. Rossiter and J.F. Hamilton, eds.). pp. 123–187.

Niggli, P. 1928. Handbuch der Experimentalphysike 7 Part 1. Akademische Verlagsgesellschaft, Leipzig.

Patterson, A. L. 1935. A direct method for the determination of components of interatomic distances in crystals. Z. Kristallogr. A90:517–542; Tabulated data for the seventeen phase groups. *Z. Kristallogr.* A90:543–554.

Reynolds, M. A., Logsdon, B. C., Thomas, L. M., Jacobson, R. A., and Angelici, R. J. 1999. Transition metal complexes of Cr, Mo, W and Mn containing $h^1(S)$-2,5-dimethylthiophene, benzothiophene, and dibenzothiophene ligands. *Organometallics* 18: 4075–4081.

Richardson, Jr., J. W. and Jacobson, R. A. 1987. Computer-aided analysis of multi-solution Patterson superpositions. *In* Patterson and Pattersons (J. P. Glusker, B. K. Patterson, and M. Rossi, eds.). p. 311. Oxford University Press, Oxford.

Santoro, A. and Mighell, A. D. 1970. Determination of reduced cells. *Acta Crystallogr.* A26:124–127.

Sayre, D. 1952. The squaring method: A new method for phase determination. *Acta Crystallogr.* 5:60–65.

Sheldrick, G. M. 1990. Phase annealing in SHELX-90: Direct methods for larger structures. *Acta Crystallogr.* A46:467–473.

Stout, G. H. and Jensen, L. H. 1989. X-Ray Crystal Structure Determination. John Wiley & Sons, New York.

Wilson, A. J. C. 1944. The probability distribution of x-ray intensities. *Acta Crystallogr.* 2:318–321.

Woolfson, M. M. and Germain, G. 1970. On the application of phase relationships to complex structures. *Acta Crystallogr.* B24:91–96.

## KEY REFERENCES

Lipscomb and Jacobson, 1990. See above.

*Provides a much more detailed description of the kinematic theory of X-ray scattering and its application to single-crystal structure determination.*

Ladd and Palmer, 1985. See above.

*Another good text that introduces the methods of crystallography, yet does so without being too mathematical; also provides a number of problems following each chapter to facilitate subject comprehension.*

Stout and Jensen, 1989. See above.

*A good general text on X-ray diffraction and one that emphasizes many of the practical aspects of the technique.*

## APPENDIX

### Direct Methods

There were a number of early attempts to use statistical approaches to develop relations that could yield the phases (or signs for the centrosymmetric case) of some of the reflections. Harker and Kasper (1948), for example, developed a set of inequalities based on the use of Cauchy's inequality. A different approach was suggested by Sayre (1952). Although later superceded by the work of Karle and Hauptman (1956), this work is still of some interest since it affords some physical insight into these methods.

Sayre (1952) noted that $\rho^2$ would be expected to look much like $\rho$, especially if the structure contains atoms that are all comparable in size. (Ideally, $\rho$ should be nonnegative; it should contain peaks where atoms are and should equal zero elsewhere, and hence its square should have the same characteristics with modified peak shape.) One can express $\rho^2$ as a Fourier series,

$$\rho^2(xyz) = \frac{1}{V}\sum_h\sum_k\sum_l G_{hkl}\exp[-2\pi i(hx+ky+lz)] \quad (45)$$

where the $G$'s are the Fourier transform coefficients.

Since $\rho^2$ would be made up of "squared" atoms, $G_{hkl}$ can be expressed as

$$G_{hkl} = \sum_{j=1}^n g_j\exp[2\pi i(hx_j+ky_j+lz_j)] \quad (46)$$

or

$$G_{hkl}\approx\frac{g}{f}F_{hkl} \quad (47)$$

if all atoms are assumed to be approximately equal. Therefore

$$\rho^2(xyz) = \frac{1}{V}\sum_h\sum_k\sum_l\frac{g}{f}F_{hkl}\exp[2\pi i(hx+ky+lz)] \quad (48)$$

but also

$$\rho^2(xyz) = \frac{1}{V^2}\left(\sum_h\sum_k\sum_l F_{hkl}\exp[2\pi i(hx+ky+lz)]\right)$$
$$\times\left(\sum_H\sum_K\sum_L\frac{g}{f}F_{HKL}\exp[2\pi i(Hx+Ky+Lz)]\right)$$
$$= \frac{1}{V^2}\sum_h\sum_k\sum_l\sum_H\sum_K\sum_L F_{hkl}\exp\{2\pi i[(h+H)x$$
$$+(k+K)y+(l+L)z]\} \quad (49)$$

Comparing the two expressions for $\rho^2$ and letting $\mathbf{h} = \mathbf{k} + \mathbf{H}$ yield

$$\frac{g}{f}F_{\mathbf{h}} = \frac{1}{V}\sum_{\mathbf{k}}F_{\mathbf{h}-\mathbf{k}}F_{\mathbf{k}} \quad (50)$$

Sayre's relation therefore implies that the structure factors are interrelated. The above relation would suggest that in a centrosymmetric structure, if $F_{\mathbf{k}}$ and $F_{\mathbf{h}-\mathbf{k}}$ are both large in magnitude and of known sign, the sign of $F_{\mathbf{h}}$ would likely be given by the product $F_{\mathbf{k}}F_{\mathbf{h}-\mathbf{k}}$ since this term will tend to predominate in the sum. Sayre's relation was primarily of academic interest since in practice few $F$'s with known signs have large magnitudes, and the possibilities for obtaining new phases or signs are limited.

Karle and Hauptman (1956), using a different statistical approach, developed a similar equation along with a number of other useful statistical-based relations. They did so using $E_{hkl}$'s that are related to the structure factor by defining

$$|E_{hkl}|^2 = |F_{hkl}|^2/\varepsilon\sum f_j^2 \quad (51)$$

where $f_j$ is the atomic scattering factor corrected for thermal effects and $\varepsilon$ is a factor to account for degeneracy and is usually unity except for special projections.

For the equal atom case, a simple derivation can be given (Karle and Karle, 1966). Then

$$|E_{\mathbf{h}}|^2 = f^2\left|\sum_j\exp(2\pi i\mathbf{h}\bullet\mathbf{r}_j)\right|^2/Nf^2 \quad (52)$$

or

$$E_{\mathbf{h}} = \frac{1}{N^{1/2}}\sum_{j=1}^N\exp(2\pi i\mathbf{h}\bullet\mathbf{r}_j) \quad (53)$$

Now consider

$$E_{\mathbf{k}}E_{\mathbf{h}-\mathbf{k}} = \frac{1}{N}\left[\sum_j\exp(2\pi i\mathbf{k}\bullet\mathbf{r}_j)\right]$$
$$\times\left[\sum_j\exp(2\pi i(\mathbf{h}-\mathbf{k})\bullet\mathbf{r}_j)\right]$$
$$= \frac{1}{N}\left\{\sum_j\exp(2\pi i\mathbf{h}\bullet\mathbf{r}_j)\right.$$
$$\left.+\sum_i\sum_j\exp[2\pi i\mathbf{k}\bullet(\mathbf{r}_j-\mathbf{r}_i)][\exp(2\pi i\mathbf{h}\bullet\mathbf{r}_i)]\right\}$$
$$(54)$$

If an average is now taken over a reasonably large number of such pairs, keeping $h$ constant, then

$$\langle E_{\mathbf{k}}E_{\mathbf{h}-\mathbf{k}}\rangle\approx\frac{1}{N}\sum\exp(2\pi i\mathbf{h}\bullet\mathbf{r}_i)$$
$$\approx\frac{1}{N^{1/2}}E_{\mathbf{h}} \quad (55)$$

(The second term in Equation 54 will tend to average to zero.) Since the magnitudes are measured and our concern is with the phase, we write

$$E_{\mathbf{h}}\propto\langle E_{\mathbf{k}}E_{\mathbf{h}-\mathbf{k}}\rangle_{\mathbf{k}} \quad (56)$$

Karle and Hauptman showed, using a more sophisticated mathematical approach, that for the unequal atom ease, a similar relation will hold. Equation 56 is often referred to as the sigma-two ($\sum_2$) relation.

For the non-centrosymmetric case, this is more conveniently written (Karle and Hauptman, 1956) as

$$\tan\alpha_{\mathbf{h}} = \frac{\sum_{\mathbf{k}}|E_{\mathbf{h}}\|E_{\mathbf{h}-\mathbf{h}}|\sin(\alpha_{\mathbf{k}}+\alpha_{\mathbf{h}-\mathbf{k}})}{\sum_{\mathbf{k}}|E_{\mathbf{k}}\|E_{\mathbf{h}-\mathbf{k}}|\cos(\alpha_{\mathbf{k}}+\alpha_{\mathbf{h}-\mathbf{k}})} \quad (57)$$

Karle and Hauptman argued that the relation should hold for an average using a limited set of data if the $|E|$'s are large, and Cochran and Woolfson (1955) obtained an approximate expression for $|E_{\mathbf{h}}|$ being positive from the $\sum_2$ relation,

$$P_+(\mathbf{h}) = \frac{1}{2}+\frac{1}{2}\tanh\sigma_3\sigma_2^{-3/2}|E_{\mathbf{h}}|\sum_{\mathbf{k}}E_{\mathbf{k}}E_{\mathbf{h}-\mathbf{k}} \quad (58)$$

where, $Z_j$ being the atomic number of atom $j$

$$\sigma_n = \sum_j Z_j^n \qquad (59)$$

The strategy then is to start with a few reflections whose phases are known and extend these using the $\sum_2$ relation accepting only those indications that have high probability (typically >95%) of being correct. The process is then repeated until no significant further changes in phases occur.

However, to employ the above strategy, phases of a few of the largest $|E|$'s must be known. How does one obtain an appropriate starting set? If working in a crystal system of orthorhombic symmetry or lower, three reflections can be used to specify the origin. Consider a centrosymmetric space group. The unit cell has centers of symmetry not only at $(0, 0, 0)$ but also halfway along the cell in any direction. Therefore the cell can be displaced by 1/2 in any direction and an equally valid structural solution obtained (same set of intensities) using the tabulated equivalent positions of the space group. If in the structure factor expression all $x_j$ were replaced by $1/2+x_j$, then

$$E_{hkl}^{\text{new}} = (-1)^h E_{hkl}^{\text{old}} \qquad (60)$$

Therefore the signs of the structure factors for $h$ odd would change while the signs for the even would not. Thus specifying a sign for an $h$ odd reflection amounts to specifying an origin in the $x$ direction. A similar argument can be made for the $y$ and $z$ directions as long as the reflection choices are independent. Such choices will be independent if they obey

$$\mathbf{h}_1 + \mathbf{h}_2 + \mathbf{h}_3 \neq (e, e, e) \qquad (61a)$$
$$\mathbf{h}_1 + \mathbf{h}_2 \neq (e, e, e), \cdots \qquad (61b)$$
$$\mathbf{h}_1 \neq (e, e, e), \cdots \qquad (61c)$$

where $e$ indicates an even value of the index.

Another useful approach to the derivation of direct method equations is through the use of what are termed structure semivariants and structure invariants.

Structure semivariants are those reflections or linear combination of reflections that are independent of the choice of permissible origins, such as has just been described above.

Structure invariants are certain linear combinations of the phases whose values are determined by the structure alone and are independent of the choice of origin. We will consider here two subgroups, the three-phase (triplet) structure invariant and the four-phase (quartet) structure invariant.

A three-phase structure invariant is a set of three reciprocal vectors $\mathbf{h}_1$, $\mathbf{h}_2$, and $\mathbf{h}_3$ that satisfy the relation

$$\mathbf{h}_1 + \mathbf{h}_2 + \mathbf{h}_3 = 0 \qquad (62)$$

Let

$$A = \frac{\sigma_3}{\sigma_2^{3/2}} (|E_{\mathbf{h}_1} \| E_{\mathbf{h}_2} \| E_{\mathbf{h}_3}|) \qquad (63)$$

and, in terms of phases $\varphi$

$$\varphi_3 = \varphi_{\mathbf{h}_1} + \varphi_{\mathbf{h}_2} + \varphi_{\mathbf{h}_3} \qquad (64)$$

then $\varphi_3$ is found to be distributed about zero with a variance that is dependent on $A$—the larger its value, the narrower the distribution. It is obviously then another version of the $\sum_2$ relation.

A four-phase structure invariant is similarly four reciprocal vectors $\mathbf{h}_1$, $\mathbf{h}_2$, $\mathbf{h}_3$, and $\mathbf{h}_4$ that satisfy the relation

$$\mathbf{h}_1 + \mathbf{h}_2 + \mathbf{h}_3 + \mathbf{h}_4 = 0 \qquad (65)$$

Let

$$B = \frac{\sigma_3}{\sigma_2^{3/2}} (|E_{\mathbf{h}_1} \| E_{\mathbf{h}_2} \| E_{\mathbf{h}_3} \| E_{\mathbf{h}_4}|) \qquad (66)$$

and

$$\varphi_4 = \varphi_{\mathbf{h}_1} + \varphi_{\mathbf{h}_2} + \varphi_{\mathbf{h}_3} + \varphi_{\mathbf{h}_4} \qquad (67)$$

If furthermore three additional magnitudes are known, $|E_{\mathbf{h}_1+\mathbf{h}_2}|$, $|E_{\mathbf{h}_2+\mathbf{h}_3}|$, and $|E_{\mathbf{h}_3+\mathbf{h}_1}|$, then, in favorable cases, a more reliable estimate of $\varphi_4$ may be obtained, and furthermore, the estimate may lie anywhere in the interval 0 to $\pi$. If all seven of the above $|E|$'s are large, then it is likely that $\varphi_4 = 0$. However, it can also be shown that for the case where $|E_{\mathbf{h}_1}|, |E_{\mathbf{h}_2}|, |E_{\mathbf{h}_3}|$, and $|E_{\mathbf{h}_4}|$ are large but $|E_{\mathbf{h}_1+\mathbf{h}_2}|, |E_{\mathbf{h}_2+\mathbf{h}_3}|$ and $|E_{\mathbf{h}_3+\mathbf{h}_1}|$ are small, the most probable value of $\varphi_4$ is $\pi$. The latter is sometimes referred to as the negative-quartet relation. (For more details see Ladd and Palmer, 1980.)

Normally additional reflections will be needed to obtain a sufficient starting set. Various computer programs adopt their own specialized procedures to accomplish this end. Some of the more common approaches include random choices for signs or phases for a few hundred reflections followed by refinement and extension; alternately, phases for a smaller set of reflections, chosen to maximize their interaction with other reflections, are systematically varied followed by refinement and extension (for an example of the latter, see Woolfson and Germain, 1970; Germain et al., 1970). Various figures of merit have also been devised by the programmers to test the validity of the phase sets so obtained.

Direct method approaches readily lend themselves to the development of automatic techniques. Over the years, they have been extended and refined to make them generally applicable, and this has led to their widespread use for deducing an initial model.

Computer programs such as the SHELX direct method programs (Sheldrick, 1990) use both the $\sum_2$ (three-phase structure invariant) and the negative four-phase quartet to determine phases. Starting phases are produced by random-number generation techniques and then refined using

$$\text{New } \varphi_h = \text{phase of } [\alpha - \eta] \qquad (68)$$

where $\alpha$ is defined by

$$\alpha = 2|E_{\mathbf{h}}|E_{\mathbf{k}}E_{\mathbf{h-k}}/N^{1/2} \qquad (69)$$

and $\eta$ is defined by

$$\eta = g|E_{\mathbf{h}}|E_{\mathbf{k}}E_{\mathbf{l}}E_{\mathbf{h-k-l}}/N \qquad (70)$$

$N$ is the number of atoms (for an equal atom case), and $g$ is a constant to compensate for the smaller absolute value of $\eta$ compared to $\alpha$. In the calculation of $\eta$, only those quartets are used for which all three cross terms have been measured and are weak. An $E$ map is then calculated for the refined phase set giving the best figure of merit.

### Heavy-Atom Method

Direct methods can be expected to be the most reliable when applied to structures in which most of the atoms are approximately equal in atomic number. In structures where one or a few atoms are present that have appreciably larger atomic numbers than the rest, the heavy-atom method is typically the approach of choice.

Assume a structure contains one atom in the asymmetric unit that is "heavier" than the rest. Also assume for the moment that the position of this atom ($r_{\mathrm{H}}$) is known. The structure factor can be written as

$$F_{hkl} = f_H \exp(2\pi i \mathbf{h} \bullet \mathbf{r}_H) + \sum_{j=2}^{N} f_j \exp(2i\mathbf{h} \bullet \mathbf{r}_j) \qquad (71)$$

Alternately the structure factor can be written as

$$F_{hkl} = F_{hld}^{\text{heavy}} + F_{hkl}^{\text{other}} \qquad (72)$$

where "heavy" denotes the contributors to the structure factor from the heavy atom(s) and "other" represents the contribution to the structure factor from the remaining atoms in the structure. The latter contains many small contributions, which can be expected to partially cancel one another in most reflections. Therefore, if the "other" contribution does not contain too large a number of these smaller contributors, one would expect the sign (or the phase) of the observed structure factor to be approximately that calculated from $F_{hkl}^{\text{heavy}}$, although the agreement in terms of magnitudes would likely be poor. Thus the approach is to calculate $F_{hkl}^{\text{heavy}}$ and transfer its sign or phase to the $|F_{hkl}^{\text{obs}}|$, unless $|F_{hkl}^{\text{heavy}}|$ is quite small. Use these phased $|F_{hkl}^{\text{obs}}|$ to calculate an electron density map and examine this map to attempt to find additional positions. (As a general rule of thumb, atoms often appear with ~1/3 of their true height if not included in the structure factor calculation.) Those atoms that appear, especially if they are in chemically reasonable positions, are then added to the structure factor calculation, giving improved phases, and the process is repeated until all the atoms have been found. For this approach to be successful, it is usually necessary for $\sum Z_{\mathrm{H}}^2 \geq \sum Z_{\mathrm{L}}^2$.

How does one go about finding the position of the heavy atom or atoms, assuming the statistics indicate that the heavy-atom approach could be successful? One possibility is to use direct methods as discussed earlier. Although direct methods may not provide the full structure, the largest peaks on the calculated map may well correspond to the positions of the heavier atoms.

An alternate approach is to deduce the position of the heavy atom(s) through the use of a Patterson function (Patterson, 1935). The Patterson function is an autocorrelation function of the electron density function:

$$P(\mathbf{u}) = \int \rho(\mathbf{r}) \bullet \rho(\mathbf{r} + \mathbf{u})d\tau \qquad (73)$$

If the Fourier series expressions for $\rho$ are substituted into this equation, it can be readily shown that

$$P(\mathbf{u}) = \frac{1}{V}\sum_h \sum_k \sum_l |F_{hkl}|^2 \cos 2\pi \mathbf{h} \bullet \mathbf{u} \qquad (74)$$

The Patterson function therefore can be directly calculated from the intensities. From a physical point of view, a peak would be expected in the Patterson function anytime two atoms are separated by a vector displacement $\mathbf{u}$. Since a vector could be drawn from A to B or B to A, a centrosymmetric function would be expected, consistent with a cosine function. Moreover, the heights of the peaks in the Patterson function should be proportional to the products of the atomic electron densities, i.e., $Z_iZ_j$. If a structure contained nickel and oxygen atoms, the heights of the Ni-Ni, Ni-O, and O-O vectors would be expected to be in the ratio 784:224:64 for peaks representing single interactions.

Most materials crystallize in unit cells that have symmetry higher than triclinic. This symmetry can provide an additional very useful tool in the analysis of the Patterson function. All Patterson functions have the same symmetry as the Laue group associated with the related crystal system. Often peaks corresponding to vectors between symmetry-related atoms occur in special planes or lines. This is probably easiest seen with an example. Consider the monoclinic space group $P2_1/c$. The Patterson function would have $2/m$ symmetry—the symmetry of the monoclinic Laue group. Moreover, because of the general equivalent positions in this space group, namely, $(x,y,z)(x,1/2-y,1/2+z)(\bar{x},1/2+y,1/2-z)(\bar{x},\bar{y},\bar{z})$, peaks between symmetry-related atoms would be found at $(0,1/2-2y,1/2)(0,1/2+2y,1/2)$ $(-2x,1/2,1/2-2z)$ $(2x, 1/2,1/2+2z)$ $(-2x,2y,-2z)$ $(2x,2y,2z)$ $(-2x,2y,-2z)$ and $(2x,-2y,2z)$ as deduced by obtaining all the differences between these equivalent positions. Such vectors are often termed Harker vectors (Harker, 1936; Buerger, 1959). In the case of the first four peaks, two differences give the same values and yield a double peak on the Patterson function. This set of Patterson peaks can be used to determine the position of a heavy atom. First we should again note that the peaks corresponding to vectors between heavy atoms should be larger than the other types. For this space group, one can select those peaks with $u = 0$ and $w = 1/2$ and, from their $v$ coordinate, determine $y$ possibilities. In a similar fashion, by selecting those large peaks with $v = 1/2$ (remember peaks in both of these categories must be double), an $x$ and $z$ pair can be determined. These results can be combined together to predict the $2x$, $2y$, $2z$ type peak position, which can then be tested to see if the combination is a valid one. The same process can be carried out if more than one heavy atom is present per asymmetric unit—the analysis just becomes somewhat more complicated. [There is another category of methods termed Patterson superposition methods, which are designed to systematically break down the multiple images of the structure present in the Patterson function to obtain essentially a single image of the structure. Because of space limitations, they will not be discussed here. An interested reader should consult Richardson and Jacobson (1987) and Lipscomb and Jacobson (1990) and references therein for further details.] It may also be appropriate here to remind the reader that the intensities are invariant if the structure is acted upon by any symmetry element in the Laue group or is displaced by half-cell displacements. Two solutions differing only in this fashion are equivalent.

ROBERT A. JACOBSON
Iowa State University
Ames, Iowa

# XAFS SPECTROSCOPY

## INTRODUCTION

X rays, like other forms of electromagnetic radiation, are both absorbed and scattered when they encounter matter. X-ray scattering and diffraction are widely utilized structural methods employed in thousands of laboratories around the world. X-ray diffraction techniques are undeniably among the most important analysis tools in nearly every physical and biological science. As this unit will show, x-ray absorption measurements are achieving a similar range of application and utility. Absorption methods are often complementary to diffraction methods in terms of the area of application and the information obtained.

The basic utility of x-ray absorption arises from the fact that each element has characteristic absorption energies, usually referred to as absorption edges. These occur when the x rays exceed the energy necessary to ionize a particular atomic level. Since this is a new channel for absorption, the absorption coefficient shows a sharp rise. Some examples are shown in Figure 1A for elements in a high-temperature superconductor. Note that the spectra for each element can be separately obtained and have distinctly different structure. Often the spectra are divided into two regions. The region right at the edge is often called the XANES (x-ray absorption near-edge structure), and the region starting ~20 or 30 eV past its edge is referred to as EXAFS (extended x-ray absorption fine structure). The isolated EXAFS structure is shown in Figure 1B. Recent theories have begun to treat these in a more unified manner, and the trend is to refer to the entire spectrum as the XAFS (x-ray absorption fine structure)

and the technique in general as XAS (x-ray absorption spectroscopy). Historically the principal quantum number of the initial state atomic level is labeled by the letters $K$, $L$, $M$, $N$, ... for $n = 1, 2, 3, 4, ...$, and the angular momentum state of the level is denoted as subscripts $1, 2, 3, 4, ...$ for the $s$, $p_{1/2}$, $p_{3/2}$, $d_{3/2}$, ... levels. The most common edges used for XAFS are the $K$ edge or $1s$ initial state (the subscript 1 is omitted since it is the only possibility) and the $L_3$ edge or $2p_{3/2}$ initial state.

The utility of XAS is demonstrated in Figure 1. The different elements have different fine structure because they have different local atomic environments. In this unit it will be shown how these spectra can be analyzed to obtain detailed information about each atom's environment. This includes the types of neighboring atoms, their distances, the disorder in these distances, and the type of bonding. The near-edge region is more sensitive to chemical effects and can often be used to determine the formal valence of the absorbing atom as well as its site symmetry.

In the simplest picture, the spectra are a result of quantum interference of the photoelectron generated by the absorption process as it is scattered from the neighboring atoms. This interference pattern is, of course, related to the local arrangement of atoms causing the scattering. As the incoming x-ray energy is changed, the energy of the photoelectron also varies along with its corresponding

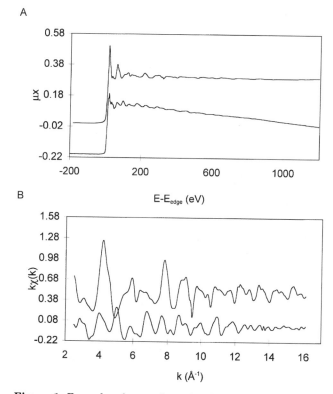

**Figure 1.** Examples of x-ray absorption data from the high-temperature superconductor material $YBa_2Cu_3O_7$. (**A**) X-ray absorption. (**B**) Normalized extended fine structure vs. wave vector extracted from the spectra in (A). The top spectra in both plots are for the Y $K$ edge at 17,038 eV, and the bottom spectra are for the Cu $K$ edge at 8,980 eV. Both edges were obtained at liquid nitrogen temperature.

wavelength. Therefore, the interference from the different surrounding atoms goes in and out of phase, giving the oscillatory behavior seen in Figure 1B. Each type of atom also has a characteristic backscattering amplitude and phase shift variation with energy. This allows different atom types to be distinguished by the energy dependence of the phase and amplitude of the different oscillatory components of the spectrum. This simple picture will be expanded below (see Principles of the Method) and will form the basis for the detailed analysis procedures used to extract the structural parameters. It is important to emphasize that the oscillations originate from a local process that does not depend on long-range order. XAFS will be observed any time an atom has at least one well-defined neighbor and, in addition to well-ordered crystalline or polycrystalline materials, has been observed in molecular gases, liquids, and amorphous materials.

The widespread use of x-ray absorption methods is intimately connected to the development of synchrotron radiation sources. The measurements require a degree of tunability and intensity that is difficult to obtain with conventional laboratory sources. Because of this need and of advances in the theoretical understanding that came at about the same time as the major synchrotron sources were first developed, the modern application of absorption methods for materials analysis is only ~25 years old. However, it has a long history as a physical phenomenon to be understood, with extensive work beginning in the 1930s. For a review of the early work see Azaroff and Pease (1974) and Stumm von Bordwehr (1989).

This unit will concentrate on the modern application of XAS at synchrotron sources. Conducting experiments at remote facilities is a process with its own style that may not be familiar to new practitioners. Some issues related to working at synchrotrons will be discussed (see Practical Aspects of the Method).

### Comparison to Other Techniques

There are a wide variety of structural techniques; they can be broadly classified as direct and indirect methods. Direct methods give signals that directly reflect the structure of the material. Diffraction measurements can, in principle, be directly inverted to give the atomic positions. Of course, the difficulty in measuring both the phase and amplitude of the diffraction signals usually precludes such an inversion. However, the diffraction pattern still directly reflects the underlying symmetry of the crystal lattice, and unit cell symmetry and size can be simply determined. More detailed modeling or determination of phase information is required to place the atoms accurately within the unit cell. Because there is such a direct relation between structure and the diffraction pattern, diffraction techniques can often provide unsurpassed precision in the atomic distances.

Indirect methods are sensitive to structure but typically require modeling or comparison to standards to extract the structural parameters. This does not mean they are not extremely useful structural methods, only that the signal does not have a direct and obvious relation to the structure. Examples of such methods include Mössbauer spec-

troscopy (MOSSBAUER SPECTROMETRY), nuclear magnetic resonance (NMR; NUCLEAR MAGNETIC RESONANCE IMAGING), electron paramagnetic resonance (EPR; ELECTRON PARAMAGNETIC RESONANCE SPECTROSCOPY), and Raman spectroscopy (RAMAN SPECTROSCOPY OF SOLIDS). The XANES part of the x-ray absorption signal can be fit into this category. As will be shown, strong multiple-scattering and chemical effects make direct extraction of the structural parameters difficult, and analysis usually proceeds by comparison of the spectra with calculated or measured models. On the other hand, as is the case with many of the other indirect methods, the complicating factors can often be used to advantage to learn more about the chemical bonding.

The indirect methods usually share the characteristic that they are atomic probes. That is, they are foremost measures of the state of an individual atom. The structure affects the signal since it affects the atomic environment, but long range order is usually not required. Also, atomic concentration is not important in determining the form of the signal, only its detectability. Therefore, exquisite sensitivity to low concentrations is often possible. This is also true for x-ray absorption. Methods that allow detection of structural information to concentrations for which diffraction techniques would be useless will be described under Detection Methods, below.

In some respects, the EXAFS part of the absorption signal shares the characteristics of both direct and indirect structural methods. The absorbing atom acts as both the source and detector of the propagating photoelectrons. The measured signal contains both phase and amplitude information and, as in direct methods, can be inverted to obtain structural information. However, the photoelectrons interact strongly with the surrounding atoms, which modifies both their amplitude and phase. Therefore, a direct inversion (Fourier transform) of the spectrum gives only a qualitative view of the local structure. This qualitative view can be informative, but modeling is usually required to extract detailed structural parameters.

## PRINCIPLES OF THE METHOD

### Single-Scattering Picture

When an x ray is absorbed, most of the time an electron is knocked out of the atom, which results in a photoelectron with energy $E = E_x - E_b$, where $E_x$ is the x-ray energy and $E_b$ is the binding energy of the electron. The x-ray edge occurs when $E_x = E_b$. The photoelectron propagates as a spherical wave with a wave vector given by

$$k = \frac{\sqrt{2m_e E}}{\hbar} \qquad (1)$$

As shown in Figure 2, this wave can be scattered back to the central atom and interfere with the original absorption. In a classical picture this can seem like a strange concept, but this process must be treated quantum mechanically, and the absorption cross-section is governed by Fermi's golden rule:

$$\sigma_K = 4\pi^2 \alpha \hbar \omega \sum_j |\langle \Psi_f | \mathbf{e} \cdot \mathbf{r} | \Psi_K \rangle|^2 \times \delta(E_f + E_K - \hbar\omega) \qquad (2)$$

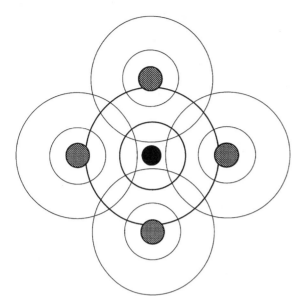

**Figure 2.** Two-dimensional schematic of the EXAFS scattering process. The photoelectron originates from the central (black) atom and is scattered by the four neighbors. The rings represent the maxima of the photoelectron direct and scattered waves.

where **e** is the electric field vector, **r** is the radial distance vector, $\alpha$ is the fine structure constant, $\omega$ is the angular frequency of the photon, and $\Psi_K$ and $\Psi_f$ are the initial state (e.g., the $1s$ state in the $K$ shell) and the final state, respectively. The sum is over all possible final states of the photoelectron.

It is necessary to consider the overlap of the initial core state with a final state consisting of the excited atom and the outgoing and backscattered photoelectron. Since the photoelectron has a well-defined wavelength, the backscattered wave will have a well-defined phase relative to the origin (deep core electrons characteristic of x-ray edges are highly localized at the center of the absorbing atom). This phase will vary with energy, resulting in a characteristic oscillation of the absorption probability as the two waves go in and out of phase. Using this simple idea and considering only single scattering events gives the following simple expression (Sayers et al., 1971; Ashley and Doniach, 1975; Lee and Pendry, 1975):

$$\chi(k) = \frac{\mu(k) - \mu_0(k)}{\mu_0(k)}$$
$$= \sum_j \frac{N_j}{kR_j^2} A_j(k) p_j(\mathbf{e}) \sin[2kR_j + \Psi_j(k)] \quad (3)$$

where $\chi(k)$ is the normalized oscillatory part of the absorption determined by subtracting the smooth part of the absorption, $\mu_0$, from the total absorption, $\mu$, and normalizing by $\mu_0$. The sum is over the neighboring shells of atoms each of which consists of $N_j$ atoms at a similar distance $R_j$. The $R_j^2$ factor accounts for the fall-off of the spherical photoelectron wave with distance.

The sine term accounts for the interference. The total path length of the photoelectron as it travels to a shell and back is $2kR_j$. The factor $p_j(\mathbf{e})$ accounts for the

polarization of the incoming x-rays. The remaining factors, $A_j(k)$ and $\Psi_j(k)$, are overall amplitude and phase factors. It is these two factors that must be calibrated by theory or experiment. The amplitude factor $A(k)$ can be further broken down to give

$$A_j(k) = S_0^2 F_j(k) Q_j(k) e^{-2R_j/\lambda} \quad (4)$$

The amplitude reduction factor $S_0^2$ results from multielectron excitations in the central atom (the simple picture assumes only a single photoelectron is excited). This factor depends on the central atom type and typically has nearly $k$ independent values from 0.7 to 0.9. The magnitude of the complex backscattering amplitude factor $f_j(k, \pi)$, $F_j(k)$ depends on the type of scattering atom. Often its $k$-dependence can be used to determine the type of backscattering atom. The factor $Q_j(k)$ accounts for any thermal or structural disorder in the $j$th shell of atoms. It will be discussed in more detail later. The final exponential factor is a mean free path factor that accounts for inelastic scattering and core hole lifetime effects.

The phase term can be broken down to give

$$\Psi_j(k) = 2\phi_c + \theta_j(k) + \varphi_j(k) \quad (5)$$

where $\theta_j(k)$ is the phase of the backscattering factor $f_j(k, \pi)$, $\phi_c$ is the $p$-wave phase shift caused by the potential of the central atom, and $\varphi_j(k)$ is a phase factor related to the disorder of the $j$th shell.

In the past, the phase and amplitude terms were generally calibrated using model compounds that had known structures and chemical environments similar to those of the substance being investigated. The problem with this approach is that suitable models were often difficult to come by, since a major justification for applying XAFS is the lack of other structural information. In recent years the theory has advanced to the state where it is often more accurate to make a theoretical determination of the phase and amplitude factors. Using theory also allows for proper consideration of multiple scattering complications to the simple single scattering picture presented so far. This will be discussed below (see Data Analysis and Initial Interpretation). Figures 3, 4, and 5 show some calculated

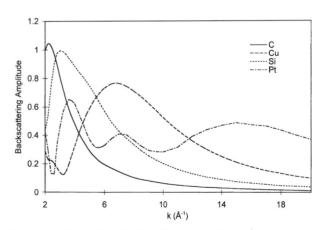

**Figure 3.** Theoretical backscattering amplitude versus photoelectron wave vector $k$ for some representative backscattering atoms. Calculated using FEFF 5.0 (Zabinsky et al., 1995).

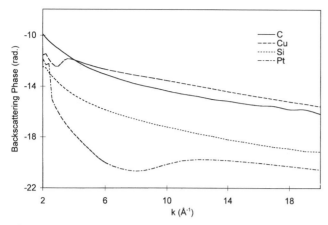

**Figure 4.** Theoretical backscattering phase shift versus photoelectron wave vector $k$ for some representative backscattering atoms. Calculated using FEFF 5.0 (Zabinsky et al., 1995).

amplitudes and phases, illustrating two main points. First, there are significant differences in the backscattering from different elements. Therefore, it is possible to distinguish between different types of atoms in a shell. For neighboring atoms such as Fe and Co, the differences are generally too small to distinguish them reliably, but if the atomic numbers of atoms in a shell differ by more than ~10%, then their contributions can be separated. The figures nicely illustrate this. Platinum is quite different from the other elements and is easily distinguished. Silicon and carbon have similar backscattering amplitudes that could be difficult to distinguish if the additional amplitude variations from disorder terms are not known. However, the backscattering phase is about a factor of π different, which means their oscillations will be out of phase and easily distinguished. Similarly, copper and carbon have almost the same phase but a distinctly different amplitude.

The second point is that the phase shifts are fairly linear with $k$. This means that the sinusoidal oscillations in Equation 3 will maintain their periodic character. The additional phase term will primarily result in a constant frequency shift. Each shell of atoms will still have a characteristic frequency, and a Fourier transform of the spec-

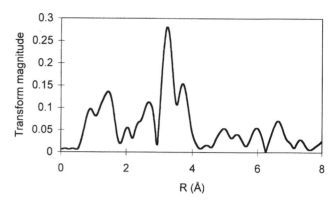

**Figure 6.** Magnitude of the $k$-weighted Fourier transform of the Cu spectrum in Figure 1B. The $k$ range from 2.5 to 16 Å$^{-1}$ was used.

trum can be used to separate them. This is shown in Figure 6, which is a Fourier transform of the Cu data in Figure 1B. The magnitude of the Fourier transform is related to the partial radial distribution function of the absorbing atoms. The peaks are shifted to lower $R$ by the phase shift terms, and the amplitude of the peaks depends on the atomic type, number of atoms in a shell, and disorder of the shell. This is a complex structure and there are no well-defined peaks that are well separated from the others. Methods that can be used to extract the structural information are described below (see Data Analysis and Initial Intepretation). This is high-quality data and all the structure shown is reproducible and real. However, above ~4 Å the structure is a composite of such a large number of single- and multiple-scattering contributions that it is not possible to analyze. This is typical of most structures, which become too complex to analyze at distances above 4 to 5 Å. It is also important to note that because of the additional phase and amplitude terms, the transform does not give the true radial distribution. It should only be used as an intermediate step in the analysis and as a method for making a simple qualitative examination of the data. It is very useful for assessing the noise in the data and for pointing out artifacts that can result in nonphysical peaks.

## EXTENSIONS TO THE SIMPLE PICTURE

### Disorder

The x-ray absorption process is energetic and takes place in a typical time scale of $10^{-15}$ sec. This is much faster than the typical vibrational periods of atoms and means that the x-ray absorption takes a snapshot of the atomic configuration. Even for an ideal shell of atoms all at a single distance, thermal vibrations will result in a distribution about the average $R_j$. In addition, for complex or disordered structures there can be a structural contribution to the distribution in $R_j$. This results in the disorder factors $Q_j(k)$ and $\varphi_j(k)$ given above. To a good approximation these factors are given by the Fourier transform of the real-space distribution $P_j(r)$:

$$Q_j(k)\exp[i\varphi_j(k)] = \int dr\, P_j(r)\exp[i2k(r - R_j)] \quad (6)$$

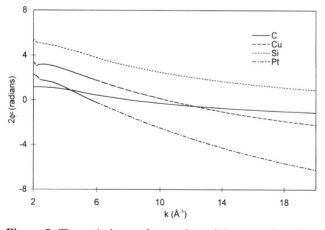

**Figure 5.** Theoretical central atom phase shift versus photoelectron wave vector $k$ for some representative absorbing atoms. Calculated using FEFF 5.0 (Zabinsky et al., 1995).

The simplest case is for a well-ordered shell at low temperature where the harmonic approximation for the vibrations applies. Then the distribution is Gaussian, and the transform is also Gaussian. This results in $\varphi_j(k) = 0$ and a Debye-Waller-like term for $Q : Q_j(k) = \exp(-2k^2\sigma_j^2)$, where $\sigma_j^2$ is the mean-square width of the distribution. This simple approximation is often valid and provides a good starting point for analysis. However, it is a fairly extreme approximation and can break down. If $\sigma^2 >\sim 0.01^2$, more exact methods should be used that include the contribution to the phase (Boyce et al., 1981; Crozier et al., 1988; Stern et al., 1992). One commonly used method included in many analysis packages is the cumulant expansion (Bunker, 1983).

The changes in the Debye-Waller term with temperature can often be determined relatively accurately. It can be quite sensitive to temperature, and low-temperature measurements will often produce larger signals and higher data quality.

Although the EXAFS disorder term is often called a Debye-Waller factor, this is not strictly correct. The EXAFS Debye-Waller factor is different from that determined by x-ray diffraction or Mössbauer spectroscopy (Beni and Plantzman, 1976). Those techniques measure the mean-square deviation from the ideal lattice position. The EXAFS term is sensitive to the mean-square deviation in the bond length. Long-wavelength vibrational modes in which neighboring atoms move together make relatively little contribution to the EXAFS Debye-Waller term but can dominate the conventional Debye-Waller factor. Thus, the temperature dependence of the EXAFS Debye-Waller term is a sensitive meaure of the bond strength since it directly measures the relative vibrational motion of the bonded atoms.

## L Edges

The preceding discussion is limited to the $K$ edge since that has a simple $1s$ initial state. For an $L$ or higher edge the situation can be more complicated. For the $L_2$ or $L_3$ edge, the initial state has $2p$ symmetry and the final state can have either $s$ or $d$ symmetry. It is also possible to have mixed final states where an outgoing $d$ wave is scattered back as an $s$ state or vice versa. The result is three separate contributions to the function $\chi(k)$ that can be denoted $\chi_{00}$, $\chi_{22}$, and $\chi_{20}$, where the subscripts refer to the angular momentum state of the outgoing and backscattered photoelectron wave. Each of these can have different phase and amplitude functions. The total EXAFS can then be expressed as follows (Lee, 1976; Heald and Stern, 1977):

$$\chi = \frac{M_{21}^2\chi_{22} + M_{01}^2\chi_{00} + 2M_{01}M_{21}\chi_{20}}{M_{21}^2 + M_{01}^2/2} \qquad (7)$$

The matrix element terms (e.g., $M_{01}$) refer to radial dipole matrix elements between the core wave function with $l = 1$ and the final states with $l = 0$ and 2. For the $K$ edge there is only one possibility for $M$ that cancels out. The $L_1$ edge, which has a $2s$ initial state, has a similar cancellation and can be treated the same as the $K$ edge.

This would seriously complicate the analysis of $L_{2,3}$ edges, but for two fortuitous simplifications. The most important is that $M_{01}$ is only $\sim 0.2$ of $M_{21}$. Therefore, the $M_{01}^2$ terms can be ignored. The cross-term can still be $\sim 40\%$ of the main term. Fortunately, for an unoriented polycrystalline sample or a material with at least cubic symmetry, the angular average of this term is zero. The cross-term must be accounted for when the sample has an orientation. This most commonly occurs for surface XAFS experiments, which often employ $L$ edges and which have an intrinsic asymmetry.

## Polarization

At synchrotron sources the radiation is highly polarized. The most common form of polarization is linear polarization, which will be considered here (it is also possible to have circular polarization). For linear polarization, the photoelectron is preferentially emitted along the electric field direction. At synchrotrons the electric field is normally oriented horizontally in a direction perpendicular to the beam direction. For $K$ and $L_1$ shells the polarization factor in Equation 3 is $p_j(\mathbf{e}) = 3\langle\cos^2\theta\rangle$, where

$$\langle\cos^2\theta\rangle = \frac{1}{N_j}\sum_i\cos^2\theta_i \qquad (8)$$

The sum is over the $i$ individual atoms in shell $j$. If the material has cubic or higher symmety or is randomly oriented, then $\langle\cos^2\theta\rangle = \frac{1}{3}$, and $p_j(\mathbf{e}) = 1$. A common situation is a sample with uniaxial symmetry such as a layered material or a surface experiment. If the symmetry in the plane is 3-fold, then the signal does not depend on the orientation of $\mathbf{e}$ within the plane. The signal only depends on the orientation of the electric field vector with respect to the plane, $\Theta$, and $\chi(\Theta) = 2\chi_0\cos^2\Theta/3 + \chi_{90}\sin^2\Theta/3$, where $\chi_0$ and $\chi_{90}$ are the signals with the field parallel and perpendicular to the plane, respectively.

For $L_{2,3}$ edges the situation is more complicated (Heald and Stern, 1977). There are separate polarization factors for the three contributions:

$$p_{22}^j = \frac{1}{2}(1 + 3\langle\cos^2\theta\rangle_j)$$
$$p_{00}^j = \frac{1}{2}$$
$$p_{02}^j = \frac{1}{2}(1 - 3\langle\cos^2\theta\rangle_j) \qquad (9)$$

As mentioned previously, the 00 case can be ignored. For the unoriented case $\langle\cos^2\theta\rangle_j = 1/3$, giving $p_{22} = 1$ and $p_{02} = 0$. This is the reason the cross-term can often be ignored.

## Multiple Scattering

The most important extension to the simple single-scattering picture discussed so far is the inclusion of multiple-scattering paths. For example, considering the first shell of atoms, the photoelectron can scatter from one first-shell atom to a second first-shell atom before being scattered back to the origin. Since in an ordered structure there

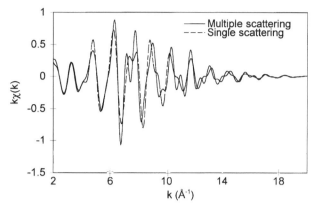

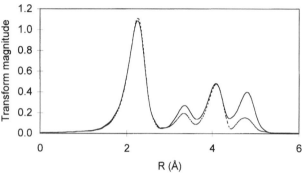

**Figure 7.** Comparison of the theoretical multiple- and single-scattering spectra for Cu metal at 80 K. The calculation only included the first four shells of atoms, which show up as four distinct peaks in the transform.

are typically 4 to 12 first-shell atoms, there are a large number of equivalent paths of this type, and the contribution can potentially be comparable to the single-scattering signal from the first shell. Two facts rescue the single-scattering picture. First, it is obvious that the multiple-scattering path is longer than the single-scattering path. Therefore, the multiple-scattering contribution would not contaminate the first-shell signal. It can, however, complicate the analysis of the higher shells. The second important point is that the scattering of the photoelectron is maximum when the scattering angle is either 0° (forward scattering) or 180° (backscattering). For the example given, there are two scattering events at intermediate angles and the contribution to the signal is reduced.

The single-scattering picture is still valid for the first shell and provides a reasonably accurate representation of the next one or two shells. For best results, however, modern analysis practice dictates that multiple scattering should be taken into account whenever shells past the first are analyzed, and this will be discussed further in the analysis section.

The enhancement of the photoelectron wave in the forward direction is often referred to as the focusing affect. It results in strong enhancement of shells of atoms that are directly in line with inner shells. An example is a face-centered-cubic (fcc) material such as copper. The fourth-shell atoms are directly in line with the first. This gives a strong enhancement of the fourth-shell signal and significantly

changes the phase of the signal relative to a single-scattering calculation. Figure 7 compares the $\chi(k)$ and the Fourier transforms of calculated spectra for copper in the single-scattering and multiple-scattering cases. Sometimes this focusing effect can be used to advantage in the study of lattice distortions. Because the focusing effect is strongly peaked in the forward direction, small deviations from perfect colinearity are amplified in the signal, allowing such distortions to be resolved even when the change in atomic position of the atoms is too small to resolve directly (Rechav et al., 1994). The focusing effect has even allowed the detection of hydrogen atoms, which normally have negligible backscattering (Lengeler, 1986).

## XANES

The primary difference in the near-edge region is the dominance of multiple-scattering contributions (Bianconi, 1988). At low energies the scattering is more isotropic, and many more multiple-scattering paths must be accounted for, including paths that have negligible amplitude in the extended region. This complicates both the quantitative analysis and calculation of the near-edge region. However, the same multiple scattering makes the near-edge region much more sensitive to the symmetry of the neighboring atoms. Often near-edge features can be used as indicators of such things as tetrahedral versus octahedral bonding. Since the bond symmetry can be intimately related to formal valence, the near edge can be a sensitive indicator of valence in certain systems. A classic example is the near edge of Cr, where, as shown in Figure 8, the chromate ion has a very strong pre-edge feature that makes it easy to distinguish from other types. This near-edge feature is much larger than any EXAFS signal and can be used to estimate the chromate-to-total-Cr ratio in many cases for which the EXAFS would be too noisy for reliable analysis.

While great strides have been made in the calculation of near-edge spectra, the level of accuracy often is much less than for the extended fine structure. In addition to the multiple scattering approach discussed, attempts have been made to relate the near-edge structure to the projected density of final states obtained from a band-structure-type calculation. In this approach, for the $K$ edge,

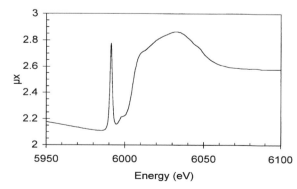

**Figure 8.** Near-edge region for the Cr $K$ edge in a 0.5 M solution of $KCrO_4$.

the $p$ density of states would be calculated. Often there is quite good qualitative agreement with near-edge features that allows them to be correlated to certain electronic states. This approach has some merit since one approach to band structure calculation is to consider all possible scattering paths for an electron in the solid, a quite similar calculation to that for the multiple scattering of the photo-electron. However, the band structure calculations do not account for the ionized core hole created when the photo-electron is excited. If local screening is not strong, the core hole can significantly affect the local potential and the resulting local density of states. While the band structure approach can often be informative, it cannot be taken as a definitive identification of near-edge features without supporting evidence.

It is important to point out that the near edge has the same polarization dependence as the rest of the XAFS. Thus, the polarization dependence of the XANES features can be used to associate them with different bonds. For example, in the $K$ edge of a planar system a feature associated with the in-plane bonding will have a $\cos^2\Theta$ dependence as the polarization vector is rotated out of the plane by $\Theta$. For surface studies this can be a powerful method of determining the orientation of molecules on a surface (Stohr, 1988).

## PRACTICAL ASPECTS OF THE METHOD

### Detection Methods

The simplest method for determining the absorption coefficient $\mu$ is to measure the attenuation of the beam as it passes through the sample:

$$I_t = I_0 e^- \quad \text{or} \quad \mu x = \ln(I_0/I_t) \qquad (10)$$

Here, $I_0$ is measured with a partially transmitting detector, typically a gas ionization chamber with the gas chosen to absorb a fraction of the beam. It can be shown statistically that the optimum fraction of absorption for the $I_0$ detector is $f = 1/(1 + e^{\mu x/2})$, or $\sim 10\%$ to $30\%$ for typical samples (Rose and Shapiro, 1948). The choice of the optimum sample thickness depends on several factors and will be discussed in more detail later. There are two methods for measuring the transmission: the standard scanning method and the dispersive method.

The scanning method uses a monochromatic beam that is scanned in energy as the transmission is monitored. The energy scanning can be either in a step-by-step mode, where the monochromator is halted at each point for a fixed time, or the so-called quick XAFS method, where the monochromator is continuously scanned and data are collected "on the fly" (Frahm, 1988). The step-scanning mode has some overhead in starting and stopping the monochromator but does not require as high a stability of the monochromator during scanning.

The dispersive method uses a bent crystal monochromator to focus a range of energies onto the sample (Matsushita and Phizackerley, 1981; Flank et al., 1982). After the sample, the focused beam is allowed to diverge. The constituent energies have different angles of divergence

and eventually separate enough to allow their individual detection on a linear detector. The method has no moving parts, and spectra can be collected in a millisecond time frame. The lack of moving apparatus also means that a very stable energy calibration can be achieved, and the method is very good for looking at the time response of small near-edge changes. With existing dispersive setups, the total energy range is limited, often to a smaller range than would be desired for a full EXAFS measurement. Also, it should be noted that, in principle, the dispersive and quick XAFS methods have the same statistical noise for a given acquisition time, assuming they both use the same beam divergence from the source. For $N$ data points, the dispersive method uses $1/N$ of the flux in each data bin for the full acquisition time, while the quick XAFS method uses the full flux for $1/N$ of the time.

The transmission method is the preferred choice for concentrated samples that can be prepared in a uniform layer of the appropriate thickness. As the element of interest becomes more dilute, the fluorescence technique begins to be preferred. In transmission the signal-to-noise ratio applies to the total signal, and for the XAFS of a dilute component the signal-to-noise ratio will be degraded by a factor related to the diluteness. With the large number of x rays ($10^{10}$ to $10^{13}$ photons/s) available at synchrotron sources, if the signal-to-noise ratio was dominated by statistical errors, then the transmission method would be possible for very dilute samples. However, systematic errors, beam fluctuation, nonuniform samples, and amplifier noise generally limit the total signal-to-noise ratio to $\sim 10^4$. Therefore, if the contribution of an element to the total absorption is only a few percent, the XAFS signal, which is only a few percent of that element's absorption, will be difficult to extract from the transmission signal.

In the fluorescence method, the fluorescence photons emitted by the element of interest are detected (Jacklevic et al., 1977). The probability of fluorescence is essentially constant with energy, and the fluorescence intensity is, therefore, directly proportional to the absorption of a specific atomic species. The advantage is that each element has a characteristic fluorescence energy, and in principle the fluorescence from the element under study can be measured separately from the total absorption. For an ideal detector the signal-to-noise ratio would then be independent of the concentration, and would depend only on the number of photons collected. Of course, the number of fluorescence photons that can be collected does depend on concentration, and there would be practical limits. In actuality the real limits are determined by the efficiency of the detectors. It is difficult to achieve a high degree of background discrimination along with a high efficiency of collection. Figure 9 shows a schematic of the energy spectrum from a fluorescence sample. The main fluorescence line is $\sim 10\%$ lower than the edge energy. The main background peak is from the elastic scattered incident photons. There are also Compton-scattered photons, which are energy-shifted slightly below the elastic peak. The width and shape of the Compton scattering peak is determined by the acceptance angle of the detector. In multicomponent samples there can also be fluorescence lines from other sample components.

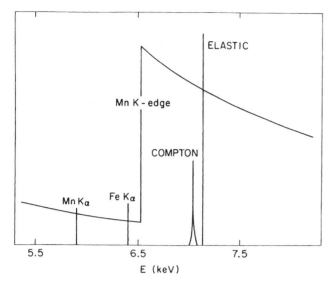

ELASTIC

Mn K - edge

COMPTON

Mn Kα    Fe Kα

5.5            6.5            7.5

E (keV)

**Figure 9.** Schematic of the fluorescence spectrum from an Fe-containing sample with the incident energy set at the Fe $K$ edge. The details of the Compton peak depend on the solid angle and location of the detector. This does not include complications such as fluorescence lines from other components in the sample. Also shown is the Mn $K$-edge absorption as an example of how filters can be used to reduce the Compton and elastic background.

To obtain a reasonable fluorescence spectrum, it is typically necessary to collect at least $10^6$ photons in each energy channel. This would give a fractional statistical error of $10^{-3}$ for the total signal and 1% to 10% in the EXAFS. If the fluorescence signal $N_f$ is accompanied by a background $N_b$, which cannot be separated by the detector, then the signal-to-noise ratio is equivalent to a signal $N_f/(1+A)$, where $A = N_b/N_f$. Therefore, a detector that is inefficient at separating the background can substantially increase the counts required.

There are three basic approaches to fluorescence detection: solid-state detectors, filters and slits (Stern and Heald, 1979), and crystal spectrometers. High-resolution Ge or Si detectors can provide energy resolution sufficient to separate the fluorescence peak. The problem is that their total counting rate (signal plus all sources of background) is limited to somewhere in the range of $2 \times 10^4$ to $5 \times 10^5$ counts per second depending on the type of detector and electronics used. Since the fluorescence from a dilute sample can be a small fraction of the total, collecting enough photons for good statistics can be time consuming. One solution has been the development of multichannel detectors with up to 100 individual detector elements (Pullia et al., 1997).

The other two approaches try to get around the count rate limitations of solid-state detectors by using other means of energy resolution. Then a detector such as an ion chamber can be used that has essentially no count rate limitations (saturation occurs at signal levels much greater than from a typical sample). The principle behind filters is shown in Figure 9. It is often possible to find a material whose absorption edge is between the fluorescence line and the elastic background peak. This will selectively attenuate the background. There is one problem

with this approach: the filter will reemit the absorbed photons as its own fluorescence. Many of these can be prevented from reaching the detector by using appropriate slits, but no filter slit system can be perfect. The selectivity of a crystal spectrometer is nearly ideal since Bragg reflection is used to select only the fluorescence line. However, the extreme angular selectivity of Bragg reflection means that it is difficult to collect a large solid angle, and many fluorescent photons are wasted. Thus, the filter with a slit detector trades off energy selectivity to collect large solid angles, while the crystal spectrometer achieves excellent resolution at the expense of solid angles.

Which approach is best will depend on the experimental conditions. For samples with moderate diluteness (cases where the background ratio $A < \sim 20$), the filter slit is an excellent choice, providing simple setup and alignment, ease of changing the fluorescence energy, and good collection efficiency. For more dilute samples or samples with multiple fluorescence lines that cannot be filtered out, the standard method is to use a multielement detector. These can often collect a substantial solid angle and with the right control software can be fairly easy to set up. As sources become more intense, the count rate limitations of solid-state detectors become more serious. This has spurred renewed interest in developing either faster detectors or those with more elements (Gauthier et al., 1996; Pullia et al., 1997) and more efficient crystal detectors. The potential user would be advised to discuss with facility managers the performance of the detectors they have available. Since x-ray cross-sections are well known, it is usually possible to estimate the relative merits of the different detection methods for a particular experiment.

The emission of a fluorescence photon is only one possible outcome of an absorption event. Another, which is dominant for light elements, is the emission of Auger electrons (see AUGER ELECTRON SPECTROSCOPY). The electron yield detection method takes advantage of this. There are several schemes for detecting the electron emission (Stohr, 1988), but the total electron yield method is by far the most widely used. This involves measuring the total electron current emitted by the sample, usually by using a positively biased collecting electrode. It is also possible to measure the sample drain current, which in effect uses the sample chamber as the collecting electrode. The electron yield method can be performed in a vacuum, which makes it an attractive choice for surface XAFS experiments carried out in ultra-high-vacuum (UHV) chambers. The yield signal is composed of the excited photoelectron, directly emitted Auger electrons, and numerous secondary electrons excited by the primary Auger or photoelectrons. Since the electron mean free path in solids is very short, all of these electron signals originate from the surface region (typically within 10 to 100 nm of the surface). This effectively discriminates against absorption occurring deep within a sample and allows for monolayer sensitivity in UHV surface experiments.

For samples that do not need UHV conditions, it is also possible to use a gas such as He or $H_2$ to enhance the electron signal (Kordesch and Hoffman, 1984). In this case, the emitted electrons ionize the gas atoms or molecules, which are then detected as in an ionization chamber. It typically requires $\sim$30 eV to ionize a gas atom, and the electrons can have energies up to several keV. Thus many gas ions can be produced for each absorption event, giving an effective amplification of the signal current. The total yield method is often applied to concentrated samples that are difficult to prepare in a thin uniform manner. It is especially useful for avoiding thickness effects in samples with strong white lines, as will be discussed below (see Data Analysis and Initial Interpretation).

### Synchrotron Facilities

Most XAFS experiments are carried out at synchrotron beamlines over which the user typically has little direct control. To carry out an effective experiment, it is important to understand some of the common beamline characteristics. These must be measured or obtained from the beamline operator.

### Energy Resolution

Bragg reflection by silicon crystals is the nearly universal choice for monochromatizing the x rays at energies >2 keV. These crystals are nearly always perfect and, if thermal or mounting strains are avoided, have a very well defined energy resolution. For Si(111), the resolution is $\Delta E/E = 1.3 \times 10^{-4}$. Higher-order reflections such as Si(220) and Si(311) can have even better resolution ($5.6 \times 10^{-5}$ and $2.7 \times 10^{-5}$, respectively). This is the resolution for an ideally collimated beam. All beamlines have some beam divergence, $\Delta\theta$. Taking the derivative of the Bragg condition, this will add an additional contribution $\Delta E = \Delta\theta \cot(\theta_b)$ where $\theta_b$ is the monochromator crystal Bragg angle. Typically, these two contributions can be added in quadrature to obtain the total energy spread in the beam. The beam divergence is determined by the source and slit sizes and can also be affected by any focusing optics that precede the monochromator.

At lower energies, other types of monochromators will likely be employed. These are commonly based on gratings but can also use more exotic crystals such as quartz, InSb, beryl, or $YB_{66}$. In these cases, it is generally necessary to obtain the resolution information from the beamline operator.

### Harmonic Content

As the later discussion on thickness effects will make clear, it is extremely important to understand the effects of harmonics on the experiment planned. These are most serious for transmission experiments but can be a problem for the other methods as well. Both crystals and gratings can diffract energies that are multiples of the fundamental desired energy. These generally need to be kept to a low level for accurate measurements. Often the beamline mirrors can be used to reduce harmonics if their high-energy cutoff is between the fundamental and harmonic energies. When this is not possible (or for unfocused beamlines), detuning of crystal monochromators is another common method. Nearly all scanning monochromators use two or more crystal reflections to allow energy selection while avoiding wide swings in direction of the output beam.

Typically the harmonic reflection curves are much narrower than for the fundamental. This means that by setting the diffraction angles of the two crystals to slightly different values the intensity of the harmonic reflections can be dramatically reduced while only losing 20% to 50% of the fundamental intensity. Nearly all XAFS double-crystal monochromators allow for this possibility.

Detuning only works well for ideal crystals and for higher energies. It is generally good practice to determine the harmonic content of the beam if it is not already known. The simplest method is to insert a foil whose edge is near the harmonic energy. For Si(111) the second harmonic is forbidden, so this would involve choosing a material whose edge is $\sim$3 times the desired energy. When the monochromator is scanned through the desired energy range, the harmonic content can be accurately estimated from the edge step of this test sample. Other methods include using an energy-analyzing detector such as a solid-state detector or Bragg spectrometer.

### Energy Calibration

All mechanical systems can have some error, and it is important to verify the energy calibration of the beamline. For a single point this can be done using a standard foil containing the element planned for study. It is also important that the energy scale throughout the scan is correct. Most monochromators rely on gear-driven rotation stages, which can have some nonlinearity. This can result in nonlinearity in the energy scale. To deal with this, many modern monochromators employ accurate angle encoders. If an energy scale error is suspected, a good first check is to measure the energy difference between the edges of two adjacent materials and compare with published values. Unfortunately it is sometimes difficult to determine exactly which feature on the edge corresponds to the published value. Another approach is the use of Bragg reflection calibrators, available at some facilities.

### Monochromator Glitches

An annoying feature of perfect crystal monochromators is the presence of sharp features in the output intensities, often referred to as "glitches." These are due to the excitation of multiple-diffraction conditions at certain energies and angles. Generally, these are sharp dips in the output intensity due to a secondary reflection removing energy from the primary. For ideal detectors and samples these should cancel out, but for XAFS measurements cancellation the $10^{-4}$ level is needed to make them undetectable. This is difficult to achieve. It is especially difficult for unfocused beamlines since the intensity reduction may only affect a part of the output beam, and then any sample nonuniformity will result in noncancellation. A sharp glitch that affects only one or two points in the spectrum can generally be handled in the data analysis. Problematic glitches extend over a broader energy range. These can be minimized only by operating the detectors at optimum linearity and making samples as uniform as possible. Usually a particular set of crystals will have only a few problematic glitches, which should be known to the beamline operator. The presence of glitches makes it essential that the $I_0$ spectra be recorded separately. Then it can be assured that weak features in the spectra are not spurious by checking for correlation with glitches.

## METHOD AUTOMATION

Most XAFS facilities are highly automated. At synchrotron facilities the experiments are typically located inside radiation enclosures to which access is forbidden when the beam is on. Therefore, it is important to automate sample alignment. Usually, motorized sample stages are provided by the facility, but for large or unusual sample cells these may not be satisfactory. It is important to work with the facility operators to make sure that sample alignment needs are provided for. Otherwise tedious cycles of entering the enclosure, changing the sample alignment, interlocking the radiation enclosure, and measuring the change in the signal may be necessary.

Since beamlines are complex and expensive, it is usually necessary that the installed control system be used. This may or may not be flexible enough to accommodate special needs. While standard experiments will be supported, it is difficult to anticipate all the possible experimental permutations. Again, it is incumbent upon the users to communicate early with the facility operators regarding special needs. This point has already been made several times but, to avoid disappointment, cannot be emphasized enough.

## DATA ANALYSIS AND INITIAL INTERPRETATION

XAFS data analysis is a complex topic that cannot be completely covered here. There are several useful reviews in the literature (Teo, 1986; Sayers and Bunker, 1988; Heald and Tranquada, 1990). There are also a number of software packages available (see Internet Resources listing below), which include their own documentation. The discussion here will concentrate on general principles and on the types of preliminary analysis that can be conducted at the beamline facility to assess the quality of the data.

There are four main steps in the analysis: normalization, background removal, Fourier transformation and filtering, and data fitting. The proper normalization is given in Equation 3. In practice, it is simpler to normalize the measured absorption to the edge step. For sample thickness $x$,

$$\chi_{\exp}(k) = \frac{\mu(k)x - \mu_0(k)x}{\Delta\mu(0)x} \qquad (11)$$

where $\Delta\mu(0)x$ is the measured edge step. This can be determined by fitting smooth functions (often linear) to regions above and below the edge. This step normalization is convenient and accurate when it is consistently applied to experimental data being compared. To compare with theory, the energy dependence of $\mu_0(k)$ in the denominator must be included. This can be determined from tabulated absorption coefficients (McMaster et al., 1969; see Internet Resources). For data taken in fluorescence or electron

yield, it is also necessary to compensate for the energy dependence of the detectors. For a gas-filled ionization chamber used to monitor $I_0$, the absorption coefficient of the gas decreases with energy. This means the $I_0$ signal would decrease with energy even if the flux were constant. Again, tabulated coefficients can be used. The fluorescence detector energy dependence is not important since it is detecting the constant-energy fluorescence signal. This correction is not needed for transmission since the log of the ratio is analyzed. For this case, the detector energy dependence becomes an additive factor, which is removed in the normalization and background removal process.

Background subtraction is probably the most important step. The usual method is to fit a smooth function such as a cubic spline through the data. The fitting parameters must be chosen to only fit the background while leaving the oscillations untouched. This procedure has been automated to a degree in some software packages (Cook and Sayers, 1981; Newville et al., 1993), but care is needed. Low-$Z$ backscatterers are often problematic since their oscillation amplitude is very large at low $k$ and damps out rapidly. It is then difficult to define a smooth background function that works over the entire $k$ region in all cases. Often careful attention is needed in choosing the low-$Z$ termination point. The background subtraction stage is also useful in pointing out artifacts in the data. The background curve should be reasonably smooth. Any unusual oscillation or structure should be investigated. One natural cause of background structure is multielectron excitations. These can be ignored in many cases but may be important for heavier atoms.

Once the background-subtracted $\chi(k)$ is obtained, its Fourier transform can be taken. As shown in Figure 6, this makes visible the different frequency components in the spectra. It also allows a judgment of the success of the background subtraction. Poor background removal will result in strong structure at low $r$. However, not all structure at low $r$ is spurious. For example, the rapid decay of low-$Z$ backscatterers can result in a low-$r$ tail on the transform peak. To properly judge the source of the low-$r$ structure, it is useful to transform theoretical calculations of similar cases for comparison.

Before Fourier transforming, the data is usually multiplied by a factor $k^w$, $W = 1$, 2, or 3. This is done to sharpen the transform peaks. Generally the transform peaks will be sharpest if the oscillation amplitude is uniform over the data window being transformed. Often, $W = 1$ is used to compensate for the $1/k$ factor seen in Equation 3. For low-$Z$ atoms and systems with large disorder, higher values of $W$ can be used to compensate for the additional $k$-dependent fall-off. High values of $W$ also emphasize the higher-$k$ part of the spectrum where the heavier backscatterers have the largest contribution. Thus, comparing transforms with different $k$ weighting is a simple qualitative method of determining which peaks are dominated by low- or high-$Z$ atoms. In addition to $k$ weighting, other weighting can be applied prior to transforming. The data should only be transformed over the range for which there is significant signal. This can be accomplished by truncating $\chi(k)$ with a rectangular window prior to transforming. A sharp rectangular window can induce truncation ripples

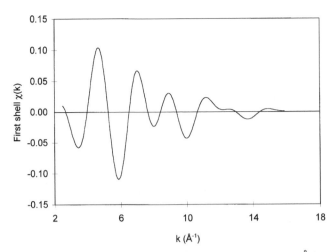

**Figure 10.** Inverse Fourier transform of the region 0.8 to 2 Å for the Cu data in Figure 6. This region contains contributions from three different Cu-O distances in the first coordination sphere.

in the transform. To deal with this, most analysis packages offer the option of various window functions that give a more gradual truncation. There are many types of windows, all of which work more or less equivalently.

The final stage of analysis is to fit the data either in $k$ or $r$ space using theoretical or experimentally derived functions. If $k$-space fitting is done, the data are usually filtered to extract only the shells being fit. For example, if the first and second shells are being fit in $k$ space, an $r$-space window is applied to the $r$-space data, which includes only the first- and second-shell contributions, and then the data are transformed back to $k$ space. This filtered spectra only includes the first and second shell contributions, and it can be fit without including all the other shells. An example of filtered data is shown in Figure 10. An important concept in fitting filtered data is the number of independent data points (Stern, 1993). This is determined by information theory to be $N_1 = 2\Delta k \Delta r/\pi + 2$, where $\Delta k$ and $\Delta r$ are the $k$- and $r$-space ranges of the data used for analysis. If fitting is done in $r$ space, $\Delta r$ would be the range over which the data are fit, and $\Delta k$ is the $k$ range used for the transform. The number of independent parameters $N_I$ is the number that are needed to completely describe the filtered data. Therefore, any fitting procedure should use fewer than $N_I$ parameters.

The relative merits of different types of fitting are beyond the scope of this unit, and fitting is generally not carried out at the beamline while data is being taken. It is important, however, to carry the analysis through the Fourier transform stage to assess the data quality. As mentioned, systematic errors often show up as strange backgrounds or unphysical transform peaks, and random statistical errors will show up as an overall noise level in the transform.

The most widely used theoretical program is FEFF (Zabinsky et al., 1995). It is designed to be relatively portable and user friendly and includes multiple scattering. It is highly desirable for new users to obtain FEFF or an equivalent package (Westre et al., 1995) and to calculate some example spectra for the expected structures.

The most common causes of incorrect data can all be classified under the general heading of thickness effects or thickness-dependent distortions. A good experimental check is to measure two different thicknesses of the same sample: if thickness effects are present, the normalized XAFS amplitude for the thicker sample will be reduced. Thickness effects are caused by leakage radiation, such as pinholes in the sample, radiation that leaks around the sample, or harmonic content in the beam. All of these are essentially unaffected by the sample absorption. As the x-ray energy passes through the edge, the transmission of the primary radiation is greatly reduced while the leakage is unchanged. Thus, the percentage of leakage radiation in the transmitted radiation increases in passing through the edge, and the edge step is apparently reduced. The greater the absorption, the greater is this effect. This means the peaks in the absorption are reduced in amplitude even more than the edge if leakage is present. In the normalized spectrum such peaks will be reduced. The distortion is larger for thick samples, which have the greatest primary beam attenuation. It can be shown statistically that the optimum thickness for transmission samples in the absence of leakage is about $\mu x = 2.5$. In practice, a $\mu x \sim 1.5$ is generally a good compromise between obtaining a good signal and avoiding thickness effects.

An analogous effect occurs in fluorescence detection. It is generally referred to as self-absorption. As the absorption increases at the edge, the penetration of the x rays into the sample decreases, and fewer atoms are available for fluorescing. Again, peaks in the absorption are reduced. Obviously this is a problem only if the absorption change is significant. For very dilute samples the absorption change is too small for a noticeable effect. Self-absorption is the reason that fluorescence detection should not be applied to concentrated samples. In this context "concentrated" refers to samples for which the absorption step is $>\sim 10\%$ of the total absorption. When sample conditions preclude transmission measurements on a concentrated sample, electron yield detection is a better choice than fluorescence detection. For electron yield detection, the signal originates within the near-surface region where the x-ray penetration is essentially unaffected by the absorption changes in the sample. Thus, in electron yield detection, self-absorption can almost always be ignored.

## SAMPLE PREPARATION

XAFS samples should be as uniform as possible. This can be a significant challenge. Typical absorption lengths for concentrated samples are from 3 to 30 μm. Foils or liquid solutions are a simple approach but cannot be universally applied. It can be difficult to grind and disperse solid particles into uniform layers of appropriate thickness. For example, to achieve a 10-μm layer of a powdered sample, the particle size should be of order 2 to 3 μm. This is difficult to achieve for many materials. Some common sample preparation methods include rubbing the powder into low-absorption adhesive tape or combining the powder with a low-absorption material such as graphite or BN prior to pressing into pellets. The tape method has been especially successful, probably because the rubbing process removes some of the larger particles leaving behind a smaller average particle size adhered to the tape. Nevertheless, if less than about four layers of tape are required to produce a reasonable sample thickness, then the samples are likely to be unacceptably nonuniform. XAFS signals can be significantly enhanced by cooling samples to liquid nitrogen temperatures. This can also be a constraint on sample preparation since tape-prepared samples can crack, exposing leakage paths, upon cooling. This can be avoided by using Kapton-based tape, although the adhesive on such tapes can have significant absorption.

Materials with layered or nonsymmetric structures can result in anisotropic particles. These will tend to be oriented when using the tape or pellet method and result in an aligned sample. Since all synchrotron radiation sources produce a highly polarized beam, the resulting data will show an orientation dependence. This can be accounted for in the analysis if the degree of orientation is known, but this can be difficult to determine. For layered materials, one solution is to orient the sample at the "magic" angle (54.7°) relative to the polarization vector. This will result in a signal equivalent to an unoriented sample for partially oriented samples.

Sample thickness is less of a concern for fluorescence or electron yield samples, but it is still important that the samples are uniform. For electron yield, the sample surface should be clean and free of thick oxide layers or other impurities. Sample charging can also be a factor. This can be a major problem for insulating samples in a vacuum. It is less important when a He-filled detector is used. Also, the x-ray beam is less charging than typical electron beam techniques, and fairly low conductivity samples can still be successfully measured.

## SPECIMEN MODIFICATION

In general, x rays are less damaging than particle probes such as the electrons used in electron microscopy. However, samples can be modified by radiation damage. This can be physical damage resulting in actual structural changes (usually disordering) or chemical changes. This is especially true for biological materials such as metalloproteins. Metals and semiconductors are generally radiation resistant. Many insulating materials will eventually suffer some damage, but often on timescales that are long compared to the measurement. Since it is difficult to predict which materials will be damaged, for potentially sensitive materials, it is wise to use other characterization techniques to verify sample integrity after exposure or to look for an exposure dependence in the x-ray measurements.

## PROBLEMS

The two most common causes of incorrect data are sample nonuniformity and thickness effects, which have already been discussed. Other problems can occur that are exacerbated by the need to run measurements at remote facilities under the time constraints of a fixed schedule. The critical role played by beam harmonics has already been discussed. Eliminating harmonics requires the correct setting of beamline crystals or mirrors. It is easy to do this incorrectly at a complex beamline. Another area of concern is the correct operation of detectors. The high fluxes at synchrotrons mean that saturation or deadtime effects are often important. For ion-chamber detectors it is important to run the instrument at the proper voltage and to use the appropriate gases for linear response. Similarly, saturation must be avoided in the subsequent amplifier chain. Single-photon detectors, if used, are often run at count rates for which deadtime corrections are necessary. For fluorescence experiments this means that the incoming or total count rate should be monitored as well as the fluorescence line(s) of interest.

Guidance on many of these issues can be obtained from the beamline operators for typical running conditions. However, few experiments are truly typical in all respects, and separate tests may need to be made. In contrast to laboratory experiments, the time constraints of synchrotron experiments make it tempting to skip some simple tests needed to verify that undistorted data is being collected. This should be avoided if at all possible.

## LITERATURE CITED

Ashley, C. A. and Doniach, S. 1975. Theory of extended x-ray absorption edge fine structure (EXAFS) in crystalline solids. *Phys. Rev. B* 11:1279–1288.

Azaroff, L. V. and Pease, D. M. 1974. X-ray absorption spectra. *In* X-ray Spectroscopy (L. V. Azaroff, ed.). McGraw-Hill, New York.

Beni, G. and Platzman, P. M. 1976. Temperature and polarization dependence of extended x-ray absorption fine-structure spectra. *Phys. Rev. B* 14:1514–1518.

Bianconi, A. 1988. XANES spectroscopy. *In* X-ray Absorption (D. C. Koningsberger and R. Prins, eds.). pp. 573–662. John Wiley & Sons, New York.

Boyce, J. B., Hayes, T. M., and Mikkelsen, J. C. 1981. Extended-x-ray-absorption-fine-structure of mobile-ion density in superionic AgI, CuI, CuBr, and CuCl. *Phys. Rev. B* 23:2876–2896.

Bunker, G. 1983. Application of the ratio method of EXAFS analysis to disordered systems. *Nucl. Instrum. Methods* 207:437–444.

Cook, J. W. and Sayers, D. E. 1981. Criteria for automatic x-ray absorption fine structure background removal. *J. Appl. Physiol.* 52:5024–5031.

Crozier, E. D., Rehr, J. J., and Ingalls, R. 1988. Amorphous and liquid systems. *In* X-ray Absorption (D. C. Koningsberger and R. Prins, eds.). pp. 373–442. John Wiley & Sons, New York.

Flank, A. M., Fontaine, A., Jucha, A., Lemonnier, M., and Williams, C. 1982. Extended x-ray absorption fine structure in dispersive mode. *J. Physique* 43:L-315-L-319.

Frahm, R. 1988. Quick scanning EXAFS: First experiments. *Nucl. Instrum. Methods* A270:578–581.

Gauthier, Ch., Goulon, J., Moguiline, E., Rogalev, A., Lechner, P., Struder, L., Fiorini, C., Longoni, A., Sampietro, M., Besch, H.,

Pfitzner, R., Schenk, H., Tafelmeier, U., Walenta, A., Misiakos, K., Kavadias, S., and Loukas, D. 1996. A high resolution, 6 channels drift detector array with integrated JFET's designed for XAFS spectroscopy: First x-ray fluorescence excitation recorded at the ESRF. *Nucl. Instrum. Methods* A382:524–532.

Heald, S. M. and Stern, E. A. 1977. Anisotropic x-ray absorption in layered compounds. *Phys. Rev. B* 16:5549–5557.

Heald, S. M. and Tranquada, J. M. 1990. X-ray absorption spectroscopy: EXAFS and XANES. *In* Physical Methods of Chemistry, Vol. V (Determination of Structural Features of Crystalline and Amorphous Solids) (B.W. Rossiter and J.F. Hamilton, eds.). pp. 189–272. Wiley-Interscience, New York.

Jacklevic, J., Kirby, J. A., Klein, M. P., Robinson, A. S., Brown, G., and Eisenberger, P. 1977. Fluorescence detection of EXAFS: Sensitivity enhancement for dilute species and thin films. *Sol. St. Commun.* 23:679–682.

Kordesch, M. E. and Hoffman, R. W. 1984. Electron yield extended x-ray absorption fine structure with the use of a gas-flow detector. *Phys. Rev. B* 29:491–492.

Lee, P. A. 1976. Possibility of adsorbate position determination using final-state interference effects. *Phys. Rev. B* 13:5261–5270.

Lee, P. A. and Pendry, J. B. 1975. Theory of the extended x-ray absorption fine structure. *Phys. Rev. B* 11:2795–2811.

Lengeler, B. 1986. Interaction of hydrogen with impurities in dilute palladium alloys. *J. Physique* C8:1015–1018.

Matsushita, T. and Phizackerley, R. P. 1981. A fast x-ray spectrometer for use with synchrotron radiation. *Jpn. J. Appl. Phys.* 20:2223–2228.

McMaster, W. H., Kerr Del Grande, N., Mallett, J. H., and Hubbell, J. H. 1969. Compilation of x-ray cross sections, LLNL report, UCRL-50174 Section II Rev. 1. National Technical Information Services L-3, U.S. Department of Commerce.

Newville, M., Livins, P., Yacoby, Y., Stern, E. A., and Rehr, J. J. 1993. Near-edge x-ray-absorption fine structure of Pb: A comparison of theory and experiment. *Phys. Rev. B* 47:14126–14131.

Pullia, A., Kraner, H. W., and Furenlid, L. 1997. New results with silicon pad detectors and low-noise electronics for absorption spectrometry. *Nucl. Instrum. Methods* 395:452–456.

Rechav, B., Yacoby Y., Stern, E. A., Rehr, J. J., and Newville, M. 1994. Local structural distortions below and above the antiferrodistortive phase transition. *Phys. Rev. Lett.* 69:3397–3400.

Rose, M. E. and Shapiro, M. M. 1948. Statistical error in absorption experiments. *Physiol. Rev.* 74:1853–1864.

Sayers, D. E. and Bunker, B. A. 1988. Data analysis. *In* X-ray Absorption (D. C. Koningsberger and R. Prins, eds.). pp. 211–253. John Wiley & Sons, New York.

Sayers, D. E., Stern, E. A., and Lytle, F. W. 1971. New technique for investigating non-crystalline structures: Fourier analysis of the extended x-ray absorption fine structure. *Phys. Rev. Lett.* 27:1204–1207.

Stern, E. A. 1993. Number of relevant independent points in x-ray-absorption fine-structure spectra. *Phys. Rev. B* 48:9825–9827.

Stern, E. A. and Heald, S. M. 1979. X-ray filter assembly for fluorescence measurements of x-ray absorption fine structure. *Rev. Sci. Instrum.* 50:1579–1582.

Stern, E. A., Ma, Y., Hanske-Pettipierre, O., and Bouldin, C. E. 1992. Radial distribution function in x-ray-absorption fine structure. *Phys. Rev. B* 46:687–694.

Stohr, J. 1988. SEXAFS: Everything you always wanted to know. *In* X-ray Absorption (D. C. Koningsberger and R. Prins, eds.). pp. 443–571. John Wiley & Sons, New York.

Stumm von Bordwehr, R. 1989. A history of x-ray absorption fine structure. *Ann. Phys. Fr.* 14:377–466.

Teo, B. K. 1986. EXAFS: Basic Principles and Data Analysis. Springer-Verlag, Berlin.

Westre, T. E., Dicicco, A., Filipponi, A., Natoli, C. R., Hedman, B., Soloman, E. I., and Hodgson, K. O. 1995. GNXAS, a multiple-scattering approach to EXAFS analysis—methodology and applications to iron complexes. *J. Am. Chem. Soc.* 117:1566–1583.

Zabinsky, S. I., Rehr, J. J., Ankudinov, A., Albers, R. C., and Eller, M. J. 1995. Multiple scattering calculations of x-ray absorption spectra. *Phys. Rev. B* 52:2995–3009.

## KEY REFERENCES

Goulon, J., Goulon-Ginet, C., and Brookes, N. B. 1997. Proceedings of the Ninth International Conference on X-ray Absorption Fine Structure, *J. Physique* 7 Colloque 2.

*A good snapshot of the current status of the applications and theory of XAFS. See also the earlier proceedings of the same conference.*

Koningsberger, D. C. and Prins, P. 1988. X-ray Absorption: Principles, Applications, Techniques of EXAFS, SEXAFS and XANES. John Wiley & Sons, New York.

*A comprehensive survey of all aspects of x-ray absorption spectroscopy. Slightly dated on some aspects of the calculation and analysis of multiple-scattering contributions, but a very useful reference for serious XAFS practitioners.*

Stohr, J. 1992. NEXAFS Spectroscopy. Springer-Verlag, New York.

*More details on the use and acquisition of near-edge spectra, especially as they apply to surface experiments.*

## INTERNET RESOURCES

http://ixs.csrri.iit.edu/index.html

*International XAFS Society homepage, containing much useful information including upcoming meeting information, an XAFS database, and links to many other XAFS-related resources.*

http://www.aps.anl.gov/offsite.html

*A list of synchrotron facility homepages maintained by the Advanced Photon Source at Argonne National Laboratory, one of several similar Web sites.*

http://www-cxro.lbl.gov/optical_constants/

*Tabulation of absorption coefficients and other x-ray optical constants maintained by the Center for X-Ray Optics at the Lawrence Berkeley National Laboratory.*

http://www.esrf.fr/computing/expg/subgroups/theory/xafs/xafs_software.html

*A compilation of available XAFS analysis software maintained by the European Synchrotron Radiation Facility (ESRF).*

STEVE HEALD
Pacific Northwest National Laboratory
Richland, Washington

# X-RAY AND NEUTRON DIFFUSE SCATTERING MEASUREMENTS

## INTRODUCTION

Diffuse scattering from crystalline solid solutions is used to measure local compositional order among the atoms, dynamic displacements (phonons), and mean species-dependent static displacements. In locally ordered alloys, fluctuations of composition and interatomic distances break the long-range symmetry of the crystal within local regions and contribute to the total energy of the alloy (Zunger, 1994). Local ordering can be a precursor to a lower temperature equilibrium structure that may be unattainable because of slow atomic diffusion. Discussions of the usefulness of local chemical and displacive correlations within alloy theory are given in Chapter 2 (see PREDICTION OF PHASE DIAGRAMS and COMPUTATION OF DIFFUSE INTENSITIES IN ALLOYS). In addition to local atomic correlations, neutron diffuse scattering methods can be used to study the local short-range correlations of the magnetic moments. Interstitial defects, as opposed to the substitutional disorder defects described above, also disrupt the long-range periodicity of a crystalline material and give rise to diffusely scattered x rays, neutrons, and electrons (electron scattering is not covered in this unit; Schweika, 1998).

Use of tunable synchrotron radiation to change the x-ray scattering contrast between elements has greatly improved the measurement of bond distances between the three types of atom pairs found in crystalline binary alloys (Ice et al., 1992). The estimated standard deviation of the first-order (first moment) mean static displacements from this technique approaches $\pm 0.001$ Å (0.0001 nm), which is an order of magnitude more precise than results obtained with extended x-ray absorption fine structure (EXAFS; XAFS SPECTROSCOPY) measurements. In addition, both the radial and tangential displacements can be reliably determined to five or more near-neighbor shells (Jiang et al., 1996). In a binary A-B alloy, the number of A or B near-neighbor atoms to, for example, an A atom can be determined to even less than 1 atom in 100. The second moment of the static displacements, which gives rise to Huang scattering, is also measurable (Schweika, 1998). Measurements of diffuse scattering can also reveal the tensorial displacements associated with substitutional and interstitial defects. This information provides models of the average arrangements of the atoms on a local scale.

An example of chemical local ordering is given in Figure 1, where the probability $P_{lmn}^{AB}$ of finding a B atom out to the sixth $lmn$ shell around an A atom goes from a preference for A atoms (clustering) to a preference for B atoms (short-range order) for a body-centered cubic (bcc) $A_{50}B_{50}$ alloy. The real-space representation of the atom positions is derived from a Monte Carlo simulation of the $P_{lmn}^{AB}$ values (Gehlen and Cohen, 1965) from the measurement of the intensity distribution in reciprocal space (Robertson et al., 1998). In the upper panel, the probability of finding a B atom as the first neighbor to an A atom is 40% (10% clustering; $P_{111}^{AB} = 0.4$). Atoms are located on a (110) plane so that first-neighbor pairs are shown. The middle panel depicts the random alloy where $P_{lmn}^{AB} = -0.5$ for the

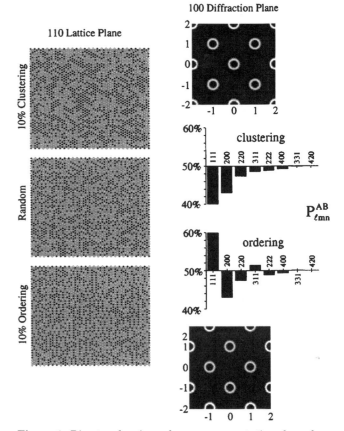

**Figure 1.** Direct and reciprocal space representations for a clustering, a random, and an ordering $A_{50}B_{50}$ bcc alloy. Courtesy of Robertson et al. (1998).

first six shells $(lmn)$. The lower panel shows the case where $P_{lmn}^{AB} = 0.6$ (a preference for unlike atom pairs). The intensity distribution in the (100) plane of reciprocal space (with the fundamental Bragg maxima removed) is shown in the right column of Figure 1. Note that a preference for like nearest neighbors causes the scattering to be centered near the fundamental Bragg maxima, such as at the origin, 0,0. A preference for unlike first-neighbor pairs causes the diffuse scattering to peak at the superlattice reflections for an ordered structure. Models, such as those shown in Figure 1, are used to understand materials properties and their response to heat treatment, mechanical deformation, and magnetic fields. These local configurations are useful to test advances in theoretical models of crystalline alloys as discussed in COMPUTATION OF DIFFUSE INTENSITIES IN ALLOYS.

The diffraction pattern from a crystalline material with perfect periodicity, such as nearly perfect single-crystal Si, consists of sharp Bragg maxima associated with long-range periodicity. With Bragg's law, we can determine the size of the unit cell. Because of thermal motion, atom positions are smeared and Bragg maxima are reduced. In alloys with different atomic sizes, static displacements will also contribute to this reduction. This intensity, which is lost from the Bragg reflections, is diffusely distributed.

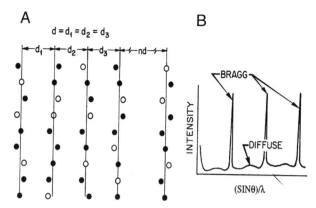

**Figure 2.** Displacements about the average lattice preserve the regular spacing between atomic planes such that $d = d_1 = d_2 = d_3 = \ldots$. The average lattice is obtained from the positions of the sharp Bragg reflections (**B**). Information about short-range correlations among the atoms is contained in the diffusely distributed intensity between the Bragg peaks. Courtesy of Ice et al. (1998).

Shown schematically in Figure 2A is a solid solution of two kinds of atoms displaced from the sites of the average lattice in such a way that the average plane of atoms is regularly spaced with a constant "$d$" spacing existing over hundreds of planes. As shown schematically in Figure 2B, there is weak diffuse scattering but no broadening of the fundamental Bragg reflections, as would be the case for more extended defects such as stacking faults, high dislocation densities, displacive transformations, and incoherent precipitates, among others (Warren, 1969). In cases where the fundamental Bragg reflections are broadened, our uncertainty in the size of the average lattice increases and the precision of the measured pair separation is reduced.

This unit will concentrate on the use of diffuse x-ray and neutron scattering from single crystals to measure local chemical correlations and chemically specific static displacements. Particular emphasis will be placed on the use of resonant (anomalous) x-ray techniques to extract information on atomic size from binary solid solutions with short-range order. Here the alloys have a well-defined average lattice but have local fluctuations in composition and displacements from the average lattice. In stoichiometric crystals with long-range periodicity, sharp superlattice Bragg reflections appear. If the compositional order is correlated only over short distances, the superlattice reflections are so broadened that measurements throughout a symmetry related volume in reciprocal space are required to determine its distribution. In addition, the displacement of the atom pairs (e.g., the A-A, A-B, and B-B pairs in a binary alloy) from the sites of the average lattice because of different atom sizes also contributes to the distribution of this diffuse scattering. By separating this diffuse intensity into its component parts—that associated with the chemical preference for A-A, A-B, and B-B pairs for the various near-neighbor shells and that associated with the static and dynamic displacements of the atoms from the sites of the average lattice—we are able to

recover pair correlation probabilities for the three kinds of pairs in a binary alloy. The interpretation of diffuse scattering associated with dynamic displacements of atoms from their average crystal sites will be discussed only briefly in this unit.

## Competitive and Related Techniques

Other techniques that measure local chemical order and bond distances exist. In EXAFS, outgoing photoejected electrons are scattered by the surrounding near neighbors (most from first and to a lesser extent from second nearest neighbors). This creates modulations of the x-ray absorption cross-section, typically extending for $\sim$1000 eV above the edge, and gives information about both local chemical order and bond distances (see XAFS SPECTROSCOPY for details). Usually, the phase and amplitudes for the interference of the photo-ejected electrons must be extracted from model systems, which necessitates measurements on intermetallic (ordered) compounds of known bond distances and neighboring atoms. The choice of an incident x-ray energy specific to an elemental absorption edge makes EXAFS information specific to that elemental constituent. For an alloy of A and B atoms, the EXAFS for an absorption edge of an A atom would be sensitive to the A and B atoms neighboring the A atoms. Separation of the signal into A-A and B-B pairs is typically done by using dilute alloys containing 2 at.% or less of the constituent of interest (e.g., the A atoms). The EXAFS signal is then interpreted as arising from the predominately B neighborhood of an A atom, and analyzed in terms of the number of B first and second neighbors and their bond distances from the A atoms. Claims for the accuracy of the EXAFS method vary between 0.01 and 0.02 Å for bond distance and $\sim$10% for the first shell coordination number (number of atoms in the first shell; Scheuer and Lengeler, 1991). For crystalline solid-solution alloys, the crystal structure precisely determines the number of neighbors in each shell but not the kinds for nondilute alloys. For most alloys, the precision achieved with EXAFS is inadequate to determine the deviations of the interatomic spacings from the average lattice. Whenever EXAFS measurements are applicable and of sufficient precision to determine the information of interest, the ease and simplicity of this experiment compared with three-dimensional diffuse scattering measurements makes it an attractive tool. An EXAFS study of concentrated Au-Ni alloys revealed the kind of information available (Renaud et al., 1988).

Mössbauer spectroscopy is another method for obtaining near-neighbor information (MOSSBAUER SPECTROMETRY). Measurements of hyperfine field splitting caused by changes in the electric field gradient or magnetic field because of the different charge or magnetic states of the nuclear environments give information about the near-neighbor environments. Different chemical and magnetic environments of the nucleus produce different hyperfine structure, which is interpreted as a measure of the different chemical environments typical for first and second neighbors. The quantitative interpretation of Mössbauer spectra in terms of local order and bond distances is often ambiguous. The use of Mössbauer spectroscopy is limited to those few alloys where at least one of the constituents is a Mössbauer-active isotope. This spectroscopy is complimentary but does not compete with diffuse scattering measurements as a direct method for obtaining detailed information about near-neighbor chemical environments and bond distances (Drijver et al., 1977; Pierron-Bohnes et al., 1983).

Imaging techniques with electrons, such as $Z$ contrast microscopy and other high-resolution electron microscopy techniques (see Chapter 11), do not have the resolution for measuring the small displacements associated with crystalline solid solutions. Imaging for high-resolution microscopy requires a thin sample about a dozen or more unit cells thick with identical atom occupations, and precludes obtaining information about short-range order and bond distances. Electron diffuse scattering measurements are difficult to record in absolute units and to separate from contributions to the diffuse background caused by straggling energy loss processes. Electron techniques provide extremely useful information on more extended defects as discussed in Chapter 11.

Field ion microscopy uses He or Ne gas atoms to image the small radius tip of the sample. Atom probes provide direct imaging of atom positions. Atoms are pulled from a small radius tip of the sample by an applied voltage and mass analyzed through a small opening. The position of the atom can be localized to ~5 Å. Information on the species of an atom and its neighbors can be recovered. Reports of successful analysis of concentration waves and clusters in phase separating alloys have occurred where strongly enriched clusters of like atoms are as small as 5 Å in diameter (Miller et al., 1996). However, information on small displacements cannot be obtained with atom probes. Scanning tunneling (see SCANNING TUNNELING MICROSCOPY) and atomic force microscopy can distinguish between the kinds of atoms on a surface and reveal their relative positions.

## PRINCIPLES OF THE METHOD

In this section, we formulate the diffraction theory of diffuse scattering in a way that minimizes assumptions and maximizes the information obtained from a diffraction pattern without recourse to models. This approach can be extended with various theories and models for interpretation of the recovered information. Measurements of diffusely scattered radiation can reveal the kinds and number of defects. Since different defects give different signatures in diffuse scattering, separation of these signatures can simplify recovery of the phenomenological parameters describing the defect.

Availability of intense and tunable synchrotron x-ray sources, which allow the selection of x-ray energies near absorption edges, permits the use of resonant scattering techniques to separate the contribution to the diffuse scattering from different kinds of pairs (e.g., the A-A, A-B, and B-B pairs of a binary alloy). Near an x-ray $K$ absorption edge, the x-ray scattering factor of an atom can change by ~8 electron units (eu) and allows for scattering contrast control between atoms in an alloy. Adjustable contrast,

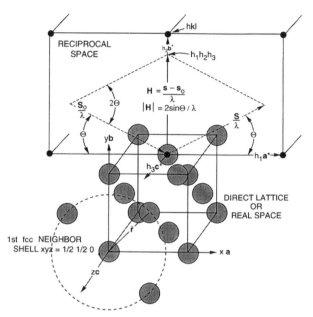

**Figure 3.** The atom positions for a face-centered cubic (fcc) structure are used to illustrate the notation for the real-space lattice. The unit cell has dimensions $\mathbf{a} = \mathbf{b} = \mathbf{c}$. The corresponding reciprocal space lattice is $\mathbf{a}^*, \mathbf{b}^*, \mathbf{c}^*$. A position in reciprocal space at which the scattered intensity, $I(\mathbf{H})$, is measured for an incoming x ray in the direction of $\mathbf{S}_0$ of wavelength $\lambda$ and detected in the outgoing direction of $\mathbf{S}$ would be $\mathbf{H} = (\mathbf{S} - \mathbf{S}_0)/\lambda = h_1\mathbf{a}^* + h_2\mathbf{b}^* + h_3\mathbf{c}^*$. At Bragg reflections, $h_1h_2h_3$ are integers and are usually designated $hkl$, the Miller indices of the reflection. This notation follows that used in the International Tables for Crystallography. Courtesy of Sparks and Robertson (1994).

either through resonant (anomalous) x-ray scattering or through isotopic substitution for neutrons, allows for precision measurement of chemically specific local-interatomic distances within the alloy. Figure 3 gives the real-space notation used in describing the atom positions and its reciprocal space notations used in describing the intensity distribution.

### X Rays Versus Neutrons

The choice of x rays or neutrons for a given experiment depends on instrumentation and source availability, the constituent elements of the sample, the information sought, the temperature of the experiment, the size of the sample, and isotopic availability and cost, among other considerations. Neutron scattering is particularly useful for measurements of low-$Z$ materials, for high-temperature measurements, and for measurements of magnetic ordering. X-ray scattering is preferred for measurements of small samples, for measurement of static displacements, and for high $\mathbf{H}$ resolution.

### Chemical Order

Recovery of the local chemical preference for atom neighbors has been predominately an x-ray diffuse scattering measurement, although x-ray and neutron measurements are complimentary. More than 50 systems have been studied with x rays and around 10 systems with neutrons.

The choice between x-ray and neutron methods often depends upon which one gives the best contrast between the constituent elements and which one allows the greatest control over contrast-isotopic substitution for contrast change with neutrons (Cenedese et al., 1984) or resonant (synonymous with dispersion and anomalous) x-ray techniques with x-ray energies near to absorption edge energies (Ice et al., 1994). In general, x-ray scattering is favored by virtue of its better intensity and collimation, which allow for smaller samples and better momentum-transfer resolution. Neutron diffuse scattering has the advantage of discriminating against thermal diffuse scattering (TDS); there is a significant change in energy (wavelength) when neutrons are scattered by phonons (see PHONON STUDIES). For example, phonons in the few tens of millielectron volt energy range make an insignificant change in the energy of kiloelectron volt x rays but make a significant change in the $\sim$35-meV energy of thermal neutrons, except near Bragg reflections. Thus, TDS of neutrons is easily rejected with crystal diffraction or time-of-flight techniques even at temperatures approaching and exceeding the Debye temperature. With x rays, however, TDS can be a major contribution and can obscure the Laue scattering. Magnetic short-range order can also be studied with neutrons in much the same way as the chemical short-range order. However, when the alloy is magnetic, extra effort is needed to separate the magnetic scattering from the nuclear scattering that gives the information on chemical pair correlations.

Neutrons can be used in combination with x rays to obtain additional scattering contrast. The x-ray scattering factors increase with the atomic number of the elements (since it increases with the number of electrons), but neutron scattering cross-sections are not correlated with the atomic number as they are scattered from the nucleus. When an absorption edge of one of the elements is either too high for available x-ray energies or too low in energy for the reciprocal space of interest, or if enough different isotopes of the atomic species making up the alloy are not available or are too expensive, then a combination of both x-ray and neutron diffuse scattering measurements may be a way to obtain the needed contrast. A more in-depth discussion of the properties of neutrons is given in Chapter 13. Information on chemical short-range order obtained with both x rays and neutrons and a general discussion of their merits is given by Kostorz (1996).

Local chemical order among the atoms including vacancies has been measured for $\sim$70 metallic binary alloys and a few oxides. Only two ternary metallic systems have been measured in which the three independent pair probabilities between the three kinds of atoms have been recovered: an alloy of $Cr_{21}Fe_{56}Ni_{23}$ with three different isotopic contents studied with neutron diffuse scattering measurements (Cenedese et al., 1984) and an alloy $Cu_{47}Ni_{29}Zn_{34}$ studied with three x-ray energies (Hashimoto et al., 1985).

## Bond Distances

In binary alloys, recovery of the bond distances between A-A, A-B, and B-B pairs requires measurement of the diffuse intensity at two different scattering contrasts to separate the A-A and B-B bond distances (as shown, later the A-B bond distance can be expressed as a linear combination of the other two distances to conserve volume). Such measurements require two matched samples of differing isotopic content to develop neutron contrast, whereas for x rays, the x-ray energy need only be changed a fraction of a kilovolt near an absorption edge to produce a significant contrast change (Ice et al., 1994). Of course, one of the elemental constituents of the sample needs to have an absorption edge energy above $\sim$5 keV to obtain sufficient momentum transfer for a full separation of the thermal contribution. However, more limited momentum-transfer data can be valuable for certain applications. Thus binary solid solutions where both elements have atomic numbers lower than that of Cr ($Z=24$) may require both an x-ray and a neutron measurement of the diffuse intensity to obtain data of sufficient contrast and momentum transfer. To date, there have been about 10 publications on the recovery of bond distances. Most have employed diffuse x-ray scattering measurements, but some employed neutrons and isotopically substituted samples (Müller et al., 1989).

### Diffuse X-ray (Neutron) Scattering Theory for Crystalline Solid Solutions

In the kinematic approximation, the elastically scattered x-ray (neutron) intensity in electron units per atom from an ensemble of atoms is given by

$$I(\mathbf{H})_{\text{Total}} = \left| \sum_p f_p e^{2\pi i \mathbf{H} \bullet \mathbf{r}_p} \right|^2 = \sum_p \sum_q f_p f_q^* e^{2\pi i \mathbf{H} \bullet (\mathbf{r}_p - \mathbf{r}_q)} \quad (1)$$

Here $f_p$ and $f_q^*$ denote the complex and complex-conjugate x-ray atomic scattering factors (or neutron scattering lengths), $p$ and $q$ designate the lattice sites from 0 to $N-1$, $\mathbf{r}_p$ and $\mathbf{r}_q$ are the atomic position vectors for those sites, and $\mathbf{H}$ is the momentum transfer or reciprocal lattice vector $|\mathbf{H}| = (2 \sin \theta)/\lambda$ (Fig. 3). For crystalline solid solutions with a well-defined average lattice (sharp Bragg reflections), the atom positions can be represented by $\mathbf{r} = \mathbf{R}+\delta$, where $\mathbf{R}$ is determined from the lattice constants and $\delta$ is both the thermal and static displacement of the atom from that average lattice. Equation 1 can be separated into terms of the average lattice $\mathbf{R}$ and the local fluctuations $\delta$,

$$I(\mathbf{H})_{\text{Total}} = \sum_p \sum_q f_p f_q^* e^{2\pi i \mathbf{H} \bullet (\delta_p - \delta_q)} e^{2\pi i \mathbf{H} \bullet (\mathbf{R}_p - \mathbf{R}_q)} \quad (2)$$

We limit our discussion of the diffraction theory to crystalline binary alloys of A and B atoms with atomic concentration $C_A$ and $C_B$, respectively, and with complex x-ray atomic scattering factors of $f_A$ and $f_B$ (neutron scattering lengths $b_A$ and $b_B$). Since an x-ray or neutron beam of even a millimeter diameter intercepts $>10^{20}$ atoms, the double sum in Equation 2 involves $>10^{21}$ first-neighbor atom pairs (one at $p$, the other at $q$); the sum over all the atoms is a statistical description, which includes all possible atom pairs that can be formed (i.e., A-A, A-B, B-A, B-B;

Warren, 1969). A preference for like or unlike neighboring pairs is introduced by the conditional probability term $P_{pq}^{AB}$. This term is defined as the probability for finding a B atom at site $q$ after having found an A atom at site $p$ (Cowley, 1950). The probability for A-B pairs is $C_A\,P_{pq}^{AB}$, which must equal $C_B\,P_{pq}^{BA}$, the number of B-A pairs. Also, $P_{pq}^{BB} = 1 - P_{pq}^{BA}$, $P_{pq}^{AA} = 1 - P_{pq}^{AB}$; $C_A + C_B = 1$. With the Warren-Cowley definition of the short-range order (SRO) parameter (Cowley, 1950), $\alpha_{pq} \equiv 1 - P_{pq}^{AB}/C_B$. Spatial and time averages taken over the chemically distinct A-A, A-B, or B-B pairs with relative atom positions $p-q$, produce the total elastically and quasielastic (thermal) scattered intensity in electron units for a crystalline solid solution of two components as

$$
\begin{aligned}
I(\mathbf{H})_{\text{Total}} = \sum_p \sum_q &\Big[ (C_A^2 + C_A C_B \alpha_{pq})|f_A|^2 \langle e^{2\pi i \mathbf{H}\bullet(\delta_p - \delta_q)} \rangle^{AA} \\
&+ C_A C_B (1 - \alpha_{pq}) f_A f_B^* \Big( \langle e^{2\pi i \mathbf{H}\bullet(\delta_p - \delta_q)} \rangle^{BA} \\
&+ \langle e^{2\pi i \mathbf{H}\bullet(\delta_p - \delta_q)} \rangle^{AB} \Big) + (C_B^2 + C_A C_B \alpha_{pq})|f_B|^2 \\
&\times \langle e^{2\pi i \mathbf{H}\bullet(\delta_p - \delta_q)} \rangle^{BB} \Big] \times e^{2\pi i \mathbf{H}\bullet(\mathbf{R}_p - \mathbf{R}_q)}
\end{aligned} \tag{3}
$$

where $|f_A|$ and $|f_B|$ denote the absolute value or moduli of the complex amplitudes.

From the theoretical development given in Appendix A, a complete description of the diffusely distributed intensity through the second moment of the displacements is given as

$$
\frac{I(\mathbf{H})_{\text{Diffuse}}}{N} = \frac{I(\mathbf{H})_{\text{SRO}}}{N} + \frac{I(\mathbf{H})_{j=1}}{N} + \frac{I(\mathbf{H})_{j=2}}{N} \tag{4}
$$

where

$$
\frac{I(\mathbf{H})_{\text{SRO}}}{N} = \sum_{lmn} C_A C_B |f_A - f_B|^2 \alpha_{lmn} \cos \pi (h_1 l + h_2 m + h_3 n) \tag{5}
$$

$$
\begin{aligned}
\frac{I(\mathbf{H})_{j=1}}{N} = &-\text{Re}\big[f_A(f_A^* - f_B^*)\big]\big[h_1 Q_x^{AA} + h_2 Q_y^{AA} + h_3 Q_z^{AA}\big] \\
&+\text{Re}\big[f_B(f_A^* - f_B^*)\big]\big[h_1 Q_x^{BB} + h_2 Q_y^{BB} + h_3 Q_z^{BB}\big]
\end{aligned} \tag{6}
$$

and

$$
\begin{aligned}
\frac{I(\mathbf{H})_{j=2}}{N} = &\ |f_A|^2 (h_1^2 R_X^{AA} + h_2^2 R_Y^{AA} + h_3^2 R_Z^{AA}) \\
&+ f_A f_B^* (h_1^2 R_X^{AB} + h_2^2 R_Y^{AB} + h_3^2 R_Z^{AB}) \\
&+ |f_B|^2 (h_1^2 R_X^{BB} + h_2^2 R_Y^{BB} + h_3^2 R_Z^{BB}) \\
&+ |f_A|^2 (h_1 h_2 S_{XY}^{AA} + h_1 h_3 S_{XZ}^{AA} + h_2 h_3 S_{YZ}^{AA}) \\
&+ f_A f_B^* (h_1 h_2 S_{XY}^{AB} + h_1 h_3 S_{XZ}^{AB} + h_2 h_3 S_{YZ}^{AB}) \\
&+ |f_B|^2 (h_1 h_2 S_{XY}^{BB} + h_1 h_3 S_{XZ}^{BB} + h_2 h_3 S_{YZ}^{BB})
\end{aligned} \tag{7}
$$

Here the individual terms are defined in Appendix A. As illustrated in Equations 4, 5, 6, and 7, local chemical order (Warren-Cowley $\alpha$'s) can be recovered from a crystalline binary alloy with a single contrast measurement of the diffuse scattering distribution, provided the displacement

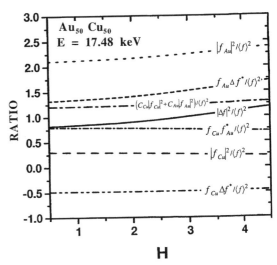

**Figure 4.** Variation in the ratio of the x-ray atomic scattering factor terms as a function $\mathbf{H}$. The divisor $\langle f \rangle^2 = |C_{Cu} f_{Cu} + C_{Au} f_{Au}|^2$ was chosen to reduce the $\mathbf{H}$ dependence of all the terms for an incident energy of Mo $K_\alpha = 1.748$ keV. The relatively larger x-ray atomic scattering factor of Au, $f_{Au} = 79$ versus Cu, $f_{Cu} = 29$ at $\mathbf{H} = 0$, would require a divisor more heavily weighted with $f_{Au}$, such as $|f_{Au}|^2$ to reduce the $\mathbf{H}$ dependence of those terms.

contributions are negligible. This was the early practice until a series of papers used symmetry relationships among the various terms to remove the $I(\mathbf{H})_{j=1}$ term in two dimensions (Borie and Sparks, 1964), in three dimensions (Sparks and Borie, 1965), and to second moment in all three dimensions: $I(\mathbf{H})_{\text{SRO}}$, $I(\mathbf{H})_{j=1}$, $I(\mathbf{H})_{j=2}$ (Borie and Sparks, 1971), henceforth referred to as BS. The major assumption of the BS method is that the x-ray atomic scattering factor terms $|f_A - f_B|^2$, $\text{Re}[f_A(f_A^* - f_B^*)]$, $\text{Re}[f_B(f_A^* - f_B^*)]$, $|f_A|^2$, $|f_B|^2$, and $f_A f_B^*$ of Equations 4, 5, 6, and 7 have a similar $\mathbf{H}$ dependence so that a single divisor renders them independent of $\mathbf{H}$. With this assumption, the diffuse intensity can be written as a sum of periodic functions given by Equation 34. For neutron nuclear scattering, this assumption is excellent; neutron nuclear scattering cross-sections are independent of $\mathbf{H}$, and in addition, the TDS terms C and D can be filtered out. Even with x rays, the BS assumption is generally a good approximation. For example, as shown in Figure 4 for Mo $K_\alpha$ x rays and even with widely separated elements such as Au-Cu, a judicious choice of the divisor allows the BS method to be applied as a first approximation over a large range in momentum transfer. In this case, division by $f_{Au}(f_{Au} - f_{Cu}) = f_{Au}\Delta f$ would be a better choice since the Au atom is the major scatterer. Iterative techniques to further extend the BS method have not been fully explored. This variation in the scattering factor terms with $\mathbf{H}$ has been proposed as a means to recover the individual pair displacements (Georgopoulos and Cohen, 1977).

Equations 5, 6, and 7 are derived from the terms first given by BS, but with notation similar to that used by Georgopoulos and Cohen (1977). There are 25 Fourier series in Equations 5, 6, and 7. For a cubic system with centrosymmetric sites, if we know $Q_X^{AA}$, then we know $Q_Y^{AA}$ and

$Q_Z^{AA}$. Similarly, if we know $Q_X^{BB}$, $R_X^{AA}$, $R_X^{BB}$, $R_X^{AB}$, $S_{XY}^{AA}$, $S_{XY}^{BB}$, and $S_{XY}^{AB}$, then we know all the $Q$, $R$, and $S$ parameters. Thus with the addition of the α series, there are nine separate Fourier series for cubic scattering to second order.

As described in Appendix A (Derivation of the Diffuse Intensity), the nine distinct correlation terms from the α, Q, R, and S series can be grouped into four unique **H**-dependent functions, A, B, C, D within the BS approximation. By following the operations given by BS, we are able to recover these unique **H** dependent functions and from these the nine distinct correlation terms. For a binary cubic alloy, one x-ray map is sufficient to recover A, B, C, and D and from A($h_1h_2h_3$), the Warren-Cowley α's. Measurements at two x-ray energies with sufficient contrast are required to separate the A-A and B-B pair contributions to the B($h_1h_2h_3$) terms, and three x-ray energies for the A-A, A-B, and B-B displacements given in Equation 29 and contained in the terms C and D of Equation 34.

In an effort to overcome the assumption of **H** independence for the x-ray atomic scattering factor terms and to use that information to separate the different pair contributions, Georgopoulos and Cohen (1977), henceforth GC, included the variation of the x-ray scattering factors in a large least-squares program. Based on a suggestion by Tibballs (1975), GC used the **H** dependence of the three different x-ray scattering factor terms to separate the first moment of the displacements for the A-A, A-B, and B-B pairs. Results from GC's error analysis (which included statistical, roundoff, x-ray scattering factors, sample roughness, and extraneous backgrounds) showed that errors in the x-ray atomic scattering factors had the largest effect, particularly on the Q terms. They concluded, based on an error analysis of the BS method by Gragg et al. (1973), that the errors in the GC method were no worse than for the BS method and provided correct directions for the first moment displacements.

Improvements in the GC method, with the use of Mo $K_\alpha$ x rays to obtain more data and the use of a Householder transformation to avoid matrix inversion and stabilization with ridge-regression techniques, still resulted in unacceptably large errors on the values of the $R$ and $S$ parameters (Wu et al., 1983). To date, there have been no reported values for the terms $R$ and $S$ that are deemed reliable. However, the Warren-Cowley α's are found to have typical errors of ~10% or less for binary alloys with a preference for unlike first-neighbor pairs with either the BS or GC analysis. For clustering systems, the BS method was reported to give large errors of 20% to 50% of the recovered α's (Gragg et al., 1973). Smaller errors were reported on the α's for clustering systems with the GC method (Wu et al., 1983). With increasing experience and better data from intense synchrotron sources, errors will be reduced for both the BS and GC methods.

Another methodology to recover the pair correlation parameters uses selectable x-ray energies (Ice et al., 1992). Of most practical interest are the α's and the first moment of the static displacements as given in Equations 27 and 28. When alloys contain elements that are near one another in the periodic table, the scattering factor term $f_A - f_B$ can be made to be nearly zero by proper choice of x-ray energy nearby to an x-ray absorption. In this way,

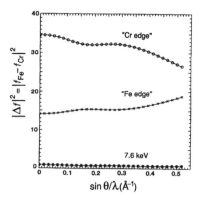

**Figure 5.** For elements nearby in the periodic table, x-ray energies can be chosen to obtain near null Laue scattering to separate intensity arising from quadratic and higher moments in atomic displacements. Courtesy of Reinhard et al. (1992).

the intensities expressed in Equations 27 and 28 are made nearly zero and only that intensity associated with Equation 29 remains. Then, the term $I(\mathbf{H})_{j=2}$ can be measured and scaled to diffuse scattering measurements made at other x-ray energies (which emphasize the contrast between the A and B atoms) and subtracted off. This leaves only the $I(\mathbf{H})_{SRO}$ term, Equation 5, and the first moment of the static displacements $I(\mathbf{H})_{j=1}$, Equation 6. Shown in Figure 5 are the values of $|f_{Fe} - f_{Cr}|^2$ for x-ray energies selected for maximum contrast at 20 eV below the Fe $K$ and Cr $K$ edges. The near null Laue energy, or energy of minimum contrast, was 7.6 keV.

The major assumption in this null Laue or 3 λ method is that the $I(\mathbf{H})_{j=2}$ and higher moment terms scale with x-ray energy as $|C_A f_A + C_B f_B|^2$, which implies that the A and B atoms have the same second and higher moment displacements or that the different elements have the same x-ray atomic scattering factors. This assumption is most valid for alloys of elements with similar atomic numbers, which have similar masses (similar thermal motion), atom sizes (small static displacement), and numbers of electrons (similar x-ray scattering factors). This 3 λ method has been used to analyze four different alloys, $Fe_{22.5}Ni_{77.5}$ (Ice et al., 1992), $Cr_{47}Fe_{53}$ (Reinhard et al., 1992), $Cr_{20}Ni_{80}$ (Schönfeld et al., 1994), and $Fe_{46.5}Ni_{53.5}$ and recalculated $Fe_{22.5}Ni_{77.5}$ (Jiang et al., 1996). An improvement in the null Laue method by Jiang et al. (1996) removed an iteration procedure to account for the residuals left by the fact that $f_A - f_B$ was not strictly zero over the measured volume.

The same techniques used for x-ray diffuse scattering analysis can also be applied to neutron scattering measurements. Neutrons have the advantage (and complication) of being sensitive to magnetic order as described in Appendix B. This sensitivity to magnetic order allows neutron measurements to detect and quantify local magnetic ordering but complicates analysis of chemical ordering.

Error analysis of the null Laue method has been given by Jiang et al. (1995) and by Ice et al. (1998). The statistical uncertainties of the recovered parameters can be estimated by propagating the standard deviation $\pm\sqrt{n}$ of the total number of counts $n$ for each data point through the

**Table 1. Contributions to the Uncertainties in the Short-Range-Order Parameter, of $Fe_{46.5}Ni_{53.5}$[a] ($\pm 1\sigma$)**

| $lmn$ | $\alpha_{lmn}$ ($\sigma_{Total}$) | $\sigma(\sqrt{n})$ | $\sigma(f')$ $\pm 0.2$ eu | $\sigma(P_0)$ $\pm 1\%$ | $\sigma$(RRS) $\pm 1$ eu | $\sigma_{Compton}$ | $\sigma(C_A)$ $\pm 0.3$ at.% |
|---|---|---|---|---|---|---|---|
| 000 | 1.0000(100) | 0.0024 | 0 | 0 | 0 | 0 | 0 |
| 110 | −0.0766 (54) | 0.0018 | 0.0010 | 0.0048 | 0 | 0.0006 | 0.0011 |
| 200 | 0.0646 (28) | 0.0017 | 0.0003 | 0.0016 | 0.0008 | 0.0013 | 0.0003 |
| 211 | −0.0022 (15) | 0.0014 | 0 | 0.0004 | 0.0001 | 0.0002 | 0.0001 |
| 220 | 0.0037 (14) | 0.0013 | 0.0002 | 0.0003 | 0.0003 | 0.0003 | 0.0001 |
| 310 | −0.0100 (11) | 0.0011 | 0.0001 | 0.0002 | 0.0001 | 0.0001 | 0.0001 |
| 222 | 0.0037 (12) | 0.0011 | 0 | 0.0002 | 0.0002 | 0.0003 | 0 |
| 321 | −0.0032 (19) | 0.0009 | 0 | 0.0001 | 0.0001 | 0.0001 | 0.0001 |
| 400 | 0.0071 (12) | 0.0011 | 0.0002 | 0.0001 | 0.0003 | 0.0004 | 0 |
| 330 | −0.0021 (9) | 0.0008 | 0.0001 | 0 | 0.0003 | 0.0001 | 0 |
| 411 | 0.0007 (7) | 0.0007 | 0 | 0 | 0 | 0.0002 | 0 |
| 420 | 0.0012 (8) | 0.0007 | 0.0002 | 0 | 0.0004 | 0.0001 | 0 |
| 332 | −0.0007 (7) | 0.0007 | 0 | 0 | 0 | 0.0001 | 0 |

[a]For statistical and possible systematic errors associated with counting statistics $n$, the real part of the resonant x-ray scattering factor $f'$ the scaling parameter $P_0$ to absolute intensities, inelastic resonant Raman scattering (RRS) and Compton contributions, and concentration $C_A$. Total error is shown in parentheses and 0 indicates uncertainties < 0.00005 Å.

nonlinear least-squares processing of the data. Systematic errors can be determined by changing the values of input variables such as the x-ray atomic scattering factors, backgrounds, and composition; then the data is reprocessed and the recovered parameters are compared.

Because the measured pair correlation coefficients are very sensitive to the relative and to a lesser degree the absolute intensity calibration of data sets collected with varying scattering contrast, the addition of constraints greatly increases reliability and reduces uncertainties. For example, the uncertainty in recovered parameters due to scaling of the measured scattering intensities is determined as input parameters are varied. Each time, the intensities are rescaled so that the $I_{SRO}$ values are everywhere positive and match values at the origin of reciprocal space measured by small-angle scattering. The inte-

grated Laue scattering over a repeat volume in reciprocal space is also constrained to have an average value of $C_A C_B |f_A - f_B|^2$ (i.e., $\alpha_{000} = 1$). These two constraints eliminate most of the systematic errors associated with converting the raw intensities into absolute units (Sparks et al., 1994). The intensities measured at three different energies are adjusted to within ~1% on a relative scale and the intensity at the origin is matched to measured values. For these reasons, the systematic errors for $\alpha_{000}$ are estimated at ~1%.

For the null Laue method, errors on the recovered $\alpha$'s and $\Delta X$'s arising from statistical and various possible systematic errors in the measurement and analysis of diffuse scattering data are given in Tables 1 and 2 for the $Fe_{46.5}Ni_{53.5}$ alloy (Jiang et al., 1995; Ice et al., 1998). Details of the conversion to absolute intensity units are given

**Table 2.  Standard Deviation of $\pm 1\sigma$ of $x$, $y$, and $z$ Components of the Pair Fe-Fe Displacements $\delta$ Fe−Fe[a]**

| $lmn$ | $\Delta X(\sigma_{Total})$ (Å) | $\sigma\sqrt{n}$ | $\sigma(f')$ $\pm 0.2$ eu | $\sigma(P_0)$ $\pm 1\%$ | $\sigma$(RRS) $\pm 1$ eu | $\sigma_{Compton}$ | $\sigma(C_A)$ $\pm 0.3$ at.% |
|---|---|---|---|---|---|---|---|
| 110 | 0.0211 (25) | 0.0002 | 0.0023 | 0.0007 | 0.0002 | 0.0004 | 0.0004 |
| 200 | −0.0228 (14) | 0.0004 | 0.0010 | 0.0007 | 0.0002 | 0.0004 | 0.0002 |
| 211 | 0.0005 (2) | 0.0002 | 0 | 0.0001 | 0.0001 | 0 | 0 |
| 121 | 0.0014 (4) | 0.0001 | 0.0003 | 0.0001 | 0.0002 | 0 | 0 |
| 220 | 0.0030 (7) | 0.0002 | 0.0006 | 0.0001 | 0.0003 | 0.0001 | 0 |
| 310 | 0.0022 (3) | 0.0002 | 0.0001 | 0.0001 | 0.0002 | 0.0001 | 0 |
| 130 | 0.0009 (2) | 0.0002 | 0.0001 | 0 | 0.0001 | 0 | 0 |
| 222 | 0.0003 (3) | 0.0002 | 0.0002 | 0 | 0.0001 | 0 | 0 |
| 321 | 0.0011 (2) | 0.0001 | 0.0001 | 0 | 0.0002 | 0 | 0 |
| 231 | 0.0001 (1) | 0.0001 | 0 | 0 | 0.0001 | 0 | 0 |
| 123 | 0.0008 (4) | 0.0001 | 0.0001 | 0 | 0 | 0 | 0 |
| 400 | −0.0019 (6) | 0.0004 | 0.0002 | 0.0001 | 0.0003 | 0.0001 | 0 |
| 330 | 0.0011 (4) | 0.0002 | 0.0001 | 0 | 0.0003 | 0 | 0 |
| 411 | −0.0008 (3) | 0.0002 | 0.0002 | 0 | 0.0002 | 0 | 0 |
| 141 | −0.0001 (2) | 0.0001 | 0.0001 | 0 | 0.0001 | 0 | 0 |

[a]For the various atom pairs of $Fe_{46.5}Ni_{53.5}$ for statistical and possible systematic errors described in the text. Total error is shown in parentheses and 0 indicates uncertainties <0.00005 Å.

elsewhere (Sparks and Borie, 1965; Ice et al., 1994; Warren, 1969; Reinhard et al., 1992). A previous assessment of the systematic errors, without the constraint of forcing $\alpha_{000} = 1$ and keeping the intensity at the origin and fundamentals a positive match to known values, resulted in estimated errors approximately two to five times larger than those reported here (Jiang et al., 1995). Parameters necessary to the analysis of the data (other than well-known physical constants) with our best estimate of their standard deviations and their contributing standard deviations to the $\alpha$ and $DX$ parameters are listed in Tables 1 and 2. From a comparison of theoretical and measured values, we estimate a 0.2-eu error on the real part of the x-ray atomic scattering factors, a 1% error in the $P_0$ calibration for converting the raw intensities to absolute units (eus), a 1-eu error in separating the inelastic resonant Raman scattering (RRS; Sparks, 1974), a 0- to 1-eu **H** dependent Compton scattering error (Ice et al., 1994), and an error of $\pm 0.3$ at.% in composition (Ice et al., 1998). Systematic errors are larger than the statistical errors for the first three shells.

The asymmetric contribution of the first moment of the static displacements, $I_{j=1}$, Equation 13, to the diffuse intensity $I_{SRO} + I_{j=1}$ for an $Fe_{63.2}Ni_{36.8}$ alloy is displayed in Figure 6. Without static displacements the $I_{SRO}$ maxima would occur at the (100) and (300) superlattice positions. The static atomic displacements for the alloy are similar to those given in Table 2. Such large distortions of the short-range order diffuse scattering caused by displacements of <0.02 Å (0.002 nm) emphasizes the sensitivity of this technique. With a change in the x-ray energy from 7.092 to 8.313 keV, $f_{Ni}$ becomes smaller than $f_{Fe}$. Figure 6 displays a reversal in the shift of the position of the diffuse scattering maxima. Two of these x-ray energies

for the $3\lambda$ method are chosen to emphasize this contrast and a third nearest the null Laue energy for removal of the TDS. The total estimated standard deviation on the values of the $\alpha$'s and in particular the $\Delta X$'s give unprecedented precision for the displacements with errors $\pm 0.003$ Å and less.

## PRACTICAL ASPECTS OF THE METHOD

Local chemical order (Warren-Cowley $\alpha$'s) from a crystalline binary alloy can be recovered with a single contrast measurement of the diffuse scattering distribution. Recovery of the two terms of the first moment of the static displacements $\langle \Delta X_{lmn}^{AA} \rangle$ and $\langle \Delta X_{lmn}^{BB} \rangle$ requires two measurements of sufficiently different contrast in $f_A \Delta f^*$ and $f_B \Delta f^*$ to separate those two contributions (Ice et al., 1992). Scattering contrast can be controlled in at least three ways: (1) by selecting x-ray energies near to and far from an absorption edge energy (resonance or anomalous x-ray scattering, Ice et al., 1994); (2) by measuring diffuse scattering over a wide $Q$ range where there is a significant change in the atomic form factors (Georgopoulos and Cohen, 1977); or (3) with neutrons by isotopic substitution (Cenedese et al., 1984).

The measurement of weak diffuse scattering normally associated with local order and displacements requires careful attention to possible sources of diffusely distributed radiation. Air scatter and other extraneous scattering from sources other than the sample, inelastic contributions such as Compton and resonant Raman, surface roughness attenuation, and geometrical factors associated with sample tilt must be removed. These details are important for placing the measured diffuse scattering in absolute units: a necessary requirement for the recovery of the $\alpha$'s and displacements.

### Measurement of Diffuse X-ray Scattering

Methods for collecting diffuse x-ray scattering data from crystalline solid solutions have been discussed by Sparks and Borie (1965), Warren (1969), and Schwartz and Cohen (1987). As demonstrated in Equations 18 and 20, the x-ray scattering intensity from a solid solution alloy contains components arising from long-range periodicity of the crystalline lattice, correlations between the different atom types, and displacements of the atoms off the sites of the average lattice.

Because of the average periodicity of the crystalline solid solution lattice, the x-ray scattering repeats periodically in reciprocal space. The equations for recovering the various components are best conditioned when the data are collected in a volume of reciprocal space that contains at least one repeat volume for the diffuse scattering (Borie and Sparks, 1971). The volume required depends on the significance of the displacement scattering. If static displacements can be ignored and thermal scattering is removed, the smallest repeat volume for SRO is sufficient. This minimum volume increases as higher-order displacement terms become important, but in no case exceeds one-fourth of the volume bounded by $\Delta h_1 = 1$, $\Delta h_2 = 1$, and $\Delta h_3 = 1$.

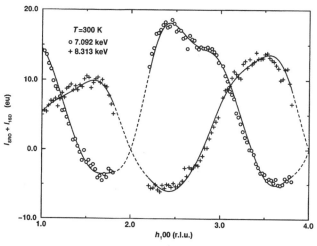

**Figure 6.** Diffusely scattered x-ray intensity from an $Fe_{63.2}Ni_{36.8}$ Invar alloy associated with the chemical order $I_{SRO}$ and the first moment of the static displacements $I_{J=1}$ versus $h$ in reciprocal lattice units (r.l.u.) along the $[h_1 00]$ direction. A major intensity change is affected by the choice of two different x-ray energies. The solid lines calculated from the $\alpha$ and $\delta$ parameters recovered from the $3\lambda$ data sets closely fit the observed data given by o and +. The dashed lines are calculated intensity through the fundamental reflections. Courtesy of Ice et al. (1998).

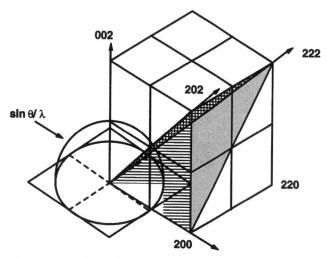

**Figure 7.** The diffuse intensity is mapped in a volume of reciprocal space bounded by three mirror planes that contain all the information available for a cubic alloy.

A conservative approach is to measure the scattered intensity in a volume of reciprocal space that extends from the origin to as far out in $|\mathbf{H}|$ space as possible, but contains the minimum repeat volume for cubic symmetry. As illustrated in Figure 7, the repeat volume for cubic crystals is 1/48 of the reciprocal space volume limited by the maximum momentum transfer, $\sin\theta/\lambda = 1/\lambda$. This repeat unit contains all the accessible information about the structure of a crystal with average cubic symmetry. As it takes several days to prepare for the experiment and its setup, actual collection of the data is not the time-limiting step with x rays, and as much data as possible should be collected in an effort to accumulate 1000 or more counts at each of several thousand positions.

The various points $h_1$, $h_2$, $h_3$ in reciprocal space are measured by orienting the sample and detector arm with a four-circle diffractometer. Diffuse scattering data are typically collected at intervals of $\Delta h = 0.1$ in a volume of reciprocal space bounded by $4 \geq h_1 \geq h_2 \geq h_3 \geq 1$. Regions of detailed interest are measured at intervals of $\Delta h = 0.05$. There are on the order of 7000 data points collected for a diffuse volume at each x-ray energy. Angular dependence of the absorption corrections is eliminated by measuring diffuse scattering in a bisecting geometry where the redundancy of the three sample-orienting circles is used to maintain the same incident and exit angle for the radiation with respect to the surface.

Shown in Figure 8 is the typical optical train for the measurement of diffuse x-ray scattering. Every effort should be made to ensure that the detector receives the radiation from the sample with the same collection efficiency regardless of the sample orientation. The sample needs to be replaced with a well-calibrated standard to convert the flux incident on the sample to absolute units. As all scattering measurements are made for fixed $I_0$ monitor counts, any changes to the optical train such as slit positions (excepting scatter slit) and sizes, distances, detectors, and changes in energy require replacing the sample with the scattering standard for recalibration of the incident flux against the $I_0$ monitor counts.

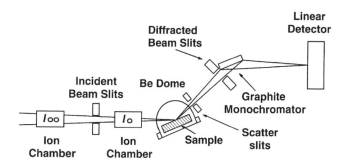

**Figure 8.** Optical setup for resonant diffuse x-ray scattering measurement.

New challenges arise from the application of resonant (anomalous) x-ray scattering to the study of local order in crystalline solid solutions: (1) the need to work near absorption edges that can create large fluorescent and resonant Raman backgrounds, and (2) the need to know the resonant (anomalous) scattering factors and absorption cross-sections to ~1%, especially at x-ray energies near absorption edges. Background problems due to inelastic scattering are exacerbated. Experimental measurement to recover the elastic scattering from these inelastic contributions (Compton, fluorescence, and resonant Raman) requires a combination of spectroscopy and diffraction.

### Removal of Inelastic Scattering Backgrounds

Photoelectric absorption is the dominant x-ray cross-section for elements with $Z > 13$ and x-ray energies $E < 20$ keV. The resultant fluorescence is typically orders of magnitude larger than the diffuse elastic scattering. Fluorescence can be removed by the use of an energy-sensitive detector when the incident x-ray energy is far enough above the photoabsorption threshold that the detector's energy resolution is adequate for separation.

Maximum change in scattering amplitude is obtained when measurements are made with the x-ray energy very near and then far from an absorption edge, as shown in Figure 9. The size of the $f'$ component roughly doubles as the energy gap between the incident x-ray energy $E_I$ and the absorption edge $E_K$ is halved. Near an edge, the size of the inelastic background grows rapidly due to RRS (Sparks, 1974; Eisenberger et al., 1976a,b; Åberg and Crasemann, 1994). RRS is interpreted as the filling of a virtual hole by a bound electron. As with fluorescence, the x-ray energy spectra is distinctive with peaks corresponding to the filling of the virtual hole, say in the $K$ shell, by various higher lying shells: $K$ filled from $L$, or $K$ filled from $M$ shells, often referred to as $K$-$L$ and $K$-$M$ RRS. Unlike fluorescence, the energy of the RRS peaks shift with incident x-ray energy, and the energy of the nearest RRS $K$-$L$ line is only a few tens of electronvolts from the incident x-ray energy (Åberg and Crasemann, 1994). This large inelastic background must either be removed experimentally or be calculated and subtracted.

The resolution of a solid-state detector at ~150 eV is inadequate to resolve all the RRS $K$-$M$ component from the elastic peak when excited near the threshold and can

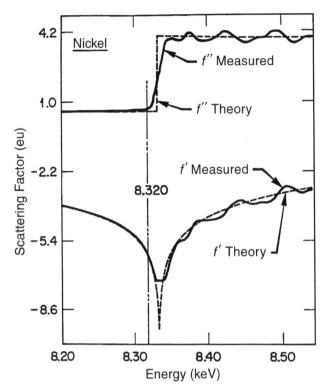

**Figure 9.** Variation of $f'$ and $f''$ near the $K$ absorption edge of nickel. Dashed lines are the theoretical calculation from Cromer and Liberman (1970; 1981). Courtesy of Ice et al. (1994).

only resolve the Compton inelastic scattering at high scattering angles. A typical energy spectrum excited by 8.0-keV x rays on a $Ni_{77.5}Fe_{22.5}$ single crystal measured with a Si(Li) detector is shown in Figure 10. Near and below the Fe edge of 7.112 keV, the RRS $K$-$M$ component cannot be resolved from the elastic scattering peak. The RRS $K$-$M$ component can be removed by measuring the RRS $K$-$L$ component and assuming the $K$-$L$/$K$-$M$ ratio remains constant. The Compton scattering component at low scattering angles is removed by using theoretical tables.

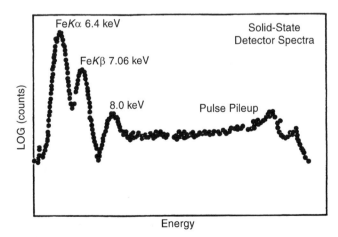

**Figure 10.** Energy spectrum measured with a solid-state detector from Ni-Fe alloy excited by 8.0-keV x rays. Courtesy of Ice et al. (1994).

Disadvantages to the use of solid-state detectors include the statistical and theoretical uncertainty of the inelastic contributions. Another disadvantage is the large deadtime imposed by the fluorescence $K$ signal in a solid-state detector, which restricts the useful flux on the sample.

Resolution on the order of 10 to 30 eV is necessary to cleanly separate the resonant Raman and Compton scattering from elastic scattering. A crystal spectrometer improves the energy resolution beyond that available with a solid-state detector. Perfect crystal spectrometers are highly inefficient compared to mosaic crystal spectrometers due to their smaller integrated reflectivity. This inefficiency is usually unacceptable for the typical weak diffuse x-ray scattering. A mosaic crystal x-ray spectrometer (Ice and Sparks, 1990) has been found to be a more practical device. The advantages of a mosaic crystal spectrometer is that it is possible to obtain energy resolutions similar to that of a perfect crystal spectrometer, but with an overall angular acceptance and efficiency similar to those of a solid-state detector. A schematic is shown in Figure 11A.

Figure 11B,C illustrates the ability of the graphite spectrometer to resolve Ni $K_{\alpha_1}$ from $K_{\alpha_2}$ and to resolve RRS from elastic scattering near the Ni $K$ edge. Good efficiency is possible if the x-ray energy, $E_I$, lies within a bandpass, $\Delta E$, set by the crystal Bragg angle $\theta_B$ and the rocking curve width $\Delta\omega_R$; $\Delta E = \Delta\omega_R E \cot \theta_B$. The diffracted beam is parafocused onto a linear detector and the beam position is correlated to x-ray energy. The energy resolution and energy scale of the spectrometer are determined by varying the energy of the incident beam and observing the peak position of the elastic peak. At 8 keV, the bandpass, $\Delta E$, of a graphite crystal with a 0.4° full-width at half-maximum (FWHM) mosaic spread is ~250 eV. The energy resolution is limited by the effective source size viewed by the energy analyzer and by imperfections in the crystal. Energy resolutions of 10 to 30 eV are typical with a 0.3- to 1.5-mm high beam in the scattering plane at the sample.

Elastic scattering is resolved from the Compton scattering at higher scattering angles and $K$-$L$ RRS at all angles. When an x-ray energy is selected near but below the absorption edge of the higher $Z$ element, the lower $Z$ element will fluoresce. The mosaic crystal spectrometer can discriminate against this fluorescence and avoid deadtime from this emission. This graphite monochromator gives an overall decrease of 3 to 4 in counting efficiency compared with a solid-state detector, but provides a much cleaner signal with greatly reduced deadtime.

A consideration when using a crystal spectrometer is the sensitivity of the energy resolution to the effective source size. As the source size increases, the energy resolution decreases and increasingly small incident beams are required for good energy resolution (Ice and Sparks, 1990). In addition, the effective source size as viewed by the crystal spectrometer depends on the spread of the incident beam on the sample and the angle of the detector axis to the sample surface. These geometrical factors are governed by the scattering angle, $\theta$, and the chi tilt, $\chi$, as shown in Figure 12.

Diffuse scattering is normally collected in the bisecting mode as intended here. The size and shape of the beam

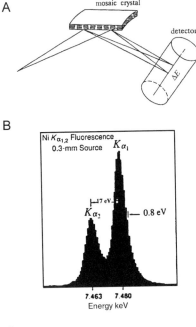

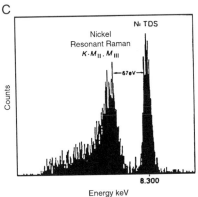

**Figure 11.** (**A**) Scattered radiation is energy analyzed with a mosaic graphite crystal dispersing radiation along a position sensitive detector to resolve (**B**) fluorescence and (**C**) resonant Raman and Compton scattering. Courtesy of Ice and Sparks (1990).

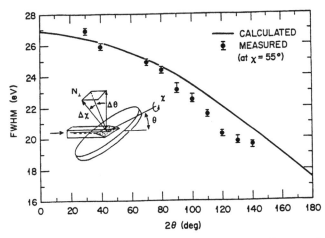

**Figure 12.** Incident beam spread on the sample depends on $\chi$ and $\theta$, which orient the surface normal with respect to the incident beam. As the beam spread on the sample is proportional to the effective source size viewed by the spectrometer, the energy resolution changes as $\chi$ and $\theta$ change. The measured energy resolution (points) is plotted for $\chi = 55°$ and compared with the theoretical prediction (line).

intercept with the sample surface is determined by the beam height $h$, width $w$, and the sample angles $\theta$ and $\omega$. As shown in the insert of Figure 12, the intercept is a parallelogram that extends along the beam direction by $h/\sin \theta$ due to the beam height and $w \tan(\omega - \pi/2)/\sin \theta$ due to the beam width. In the reference frame of the detector, the length of the parallelogram is projected into a root-mean-square source height of

$$\sigma_s = \frac{\sqrt{[2w \tan(\chi - \pi/2)\cos \theta]^2 + h^2}}{12} \qquad (8)$$

The measured energy resolution for a $1.5 \times 1.5$-mm$^2$ beam at $\chi = 55°$ is plotted in Figure 12 as a function of scattering angle $\theta$. Deviations from $\chi = 90°$, where the sample normal lies in the scattering plane, are held to a minimum by choosing a crystal surface normal centered in the volume of reciprocal space to be measured. This reduces

surface roughness corrections and maintains good energy resolution. Actual spectra collected during a diffuse scattering measurement are shown in Figure 13. Crystal analyzers will perform even better with third-generation synchrotron sources that have smaller source size and higher flux. Measurement of the fluorescent intensity or RRS throughout the volume used for the data with the identical optical train provides a correction for this beam spread, sample roughness, and alignment errors.

A model is used to describe the energy distributions of the RRS shown in Figures 13A,B as observed with the graphite spectrometer. The model contains a truncated Lorentzian centered at the energy for the RRS peak (Sparks, 1974; Ice et al., 1994). The high-energy cutoff is determined from energy conservation. The spectra are corrected for the graphite monochromator efficiency, which has a Gaussian distribution centered on the elastic scattering peak. The spectra are also corrected for the finite spectrometer resolution by convolving the Gaussian shape of the elastically scattered peak. The simple model for the resonant Raman peak shape allows for a good fit to the experimental resonant Raman peak observed with the graphite monochromator, as shown in Figures 13A,B.

Compton scattering can be removed from the elastic scatter by subtracting tabulated theoretical Compton intensities. It is possible to experimentally separate the elastic scattering peak from the Compton peak except at the lowest scattering angles. A correction for overlap of the two peaks at small angles is achieved by modeling the energy dependence of the Compton profile. The doubly differential Compton scattering cross-section is calculated using the impulse approximation (IA; Carlsson et al., 1982; Biggs et al., 1975). The Compton scattering is calculated for each subshell, and energy conservation is used to restrict the high-energy tail from each shell. The total spectrum is determined by adding the contribution from

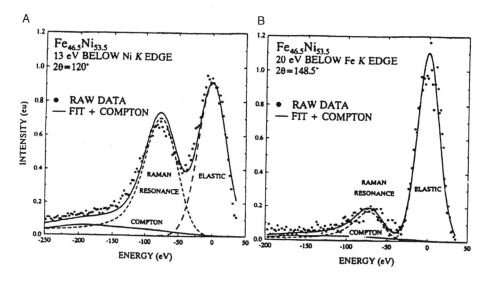

**Figure 13.** Energy spectrum of scattered radiation when the incident energy is (**A**) 13 eV below the Ni K edge and (**B**) 20 eV below the Fe K edge of a Fe-Ni crystal. Courtesy of Ice et al. (1994).

each shell and from each atom type. This cross-section is subtracted from the measured data normalized to absolute electron units per atom, which leaves only a resonant Raman peak and an elastic scattering peak. The slight overlap of the Compton peak with the elastic peak is typically small compared with statistical uncertainties.

Comparison of the calculated to the observed Compton spectrum can be achieved with x-ray energies sufficiently removed from an edge that the resonant Raman contribution is negligible. At 8.000 keV, the resonant Raman contribution from an Fe-Ni sample is centered far below the elastic peak and outside the spectrometer window. The IA-calculated Compton profiles are observed to be in qualitative agreement with the observed spectra, but the intensity is overestimated at low angles and underestimated at high angles. Particularly noticeable is a low-energy tail at high scattering angles. A more exact theory without the impulse approximation might improve matters.

### Determination of the Resonant Scattering Terms $f'$ and $f''$

The widely used Cromer-Liberman tabulation (Cromer and Liberman, 1981; Sasaki, 1989) of $f'$ and $f''$ explicitly ignores the presence of pre-edge unfilled bound states (bound-to-bound transitions), lifetime broadening of the inner-shell hole, and x-ray absorption fine structure (XAFS; XAFS SPECTROSCOPY). These assumptions are justified ~100 eV below and ~1 keV above an edge, but not near an absorption edge (Lengeler, 1994; Chantier, 1994). Of particular concern is the underestimation of the absorption coefficient and $f'$ just below an absorption edge due to the Lorentzian hole width of an inner shell. This is illustrated in Figure 14. An inner-shell hole with a 2-eV broadened lifetime and a K edge jump ratio of 8 will increase the absorption cross-section by ~11% at 20 eV below the nominal edge and by 2% at 100 eV below the edge. Theoretical tabulations that ignore the lifetime broadened hole width must be corrected, as they underestimate the absorption coefficient (and $f'$) just below an edge and overestimate $f''$ above the edge. Theoretical tabulations also do

not include unfilled pre-edge states or absorption fine structure that is highly sample dependent. Thus, it is necessary to determine the photoelectric absorption cross-section, $\mu/\rho$, experimentally for each sample, to calculate $f''$ with the optical theorem (James, 1948), and then to calculate $f'$ from $f''$ with the Kramers-Kronig relationship.

The practical method of measuring the sample-specific absorption cross-section is to measure the relative absorption cross-section across the edge of each of the elements of the sample over a range of ~100 to 1000 eV. These data are normalized to theoretical data far from the edge (Kawamura and Fukamachi, 1978; Dreier et al., 1983; Hoyt et al., 1984). Measurements are made with a thin foil of the sample in transmission geometry. The measured value of $f'$ is obtained by adding the difference integration to tabulated values of $f'$, as shown in Figure 9.

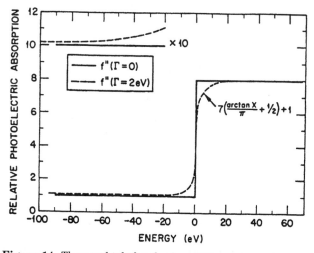

**Figure 14.** The usual tabulated values of the resonant (anomalous) scattering terms are not corrected for hole width (lifetime), which causes a Lorentzian broadening of the absorption edge and affects the values of $f'$ and $f''$ near the edge. Courtesy of Ice et al. (1994).

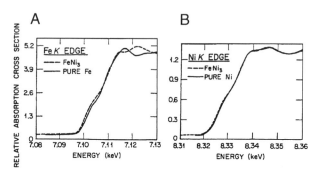

**Figure 15.** The absorption edge energy shifts are very small for metallic alloys with differing nearest neighbors. Courtesy of Ice et al. (1994).

We find $f'$ is 5 to 10 times less sensitive than $f''$ to lifetime broadening of the inner-shell hole. For example, the effect of a 2-eV lifetime on $f'$ at 20 eV below the edge is only 2% and at $-100$ eV the effect is only 0.3%. The value of $f'$ is sensitive to the position of the average inflection point of the absorption edge. A shift of 5 eV results in a 4% to 5% change in $f'$ at 20 eV below the edge. Errors in the absolute energy calibration are removed when the energy of the incident radiation is fixed to the same absorption edge energy as the calculation of $f'$ and $f''$.

As shown in Figure 15A,B, the absorption edge energies for Fe and Ni in a fully ordered $FeNi_3$ foil do not shift compared to the absorption edge energies of pure Fe or Ni foils. However, the local environment of the Fe atoms in $FeNi_3$ are sufficiently different that calculated or measured values of $f'$ and $f''$ for pure Fe would be in error close to the Fe absorption edge. For samples where there is a large charge transfer (change in oxidation state), this difference becomes even larger.

### Absolute Calibration of the Measured Intensities

Conversions to absolute units depend on previously calibrated standards to place the intensity in electron units (Suortti et al., 1985). Calibrated powder standards account for the monitor efficiency, the beam path transmission, and the efficiency and solid angle of the detector. The largest uncertainty is in the values of the linear x-ray absorption cross-sections, $\mu$, near an absorption edge for both the powder and the sample. This problem is reduced by using a powder similar in elemental composition to the sample or by careful calibration of the sample absorption. Comparison between standardizations with various powder samples are consistent to within 3% to 5% (Suortti et al., 1985).

The relative scaling factors between different energy sets can be refined with great sensitivity by restricting the short-range-order intensity as discussed previously and as described below. For alloys that cluster, the Laue scattering can be obscured by proximity of the fundamental Bragg peaks. Higher-resolution measurements such as small-angle x-ray scattering (SAXS) techniques may then be required to recover the Laue scattering. SAXS can be used to measure the total SRO scattering at the origin, and the relative scaling factors of data sets can be adjusted so that $\alpha_{000}=1$, and the fitted SRO diffuse scattering,

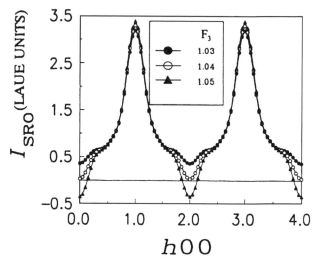

**Figure 16.** Fitted $I_{SRO}$ along the $h00$ line for three relative scale factors on the near zero contrast data of a $Ni_{77.5}Fe_{22.5}$ sample. With a scale factor of 1.04, $I_{SRO}$ is near zero at the origin and fundamental Bragg peaks as measured by SAXS. Courtesy of Sparks et al. (1994).

approach the SAXS value obtained near the origin. This scaling method is illustrated in Figure 16 for a $Ni_{77.5}Fe_{22.5}$ alloy. As shown in Figure 16, a small change in the relative scaling (here the zero contrast or near null Laue scale factor) makes a big change in the SRO scattering near the origin. Fe-Ni alloys are known to show negligible scattering near the origin (Simon and Lyon, 1991). Scale factors are adjusted to set the SRO scattering to near zero at the Bragg peaks. Nonlinear fitting routines that refine the relative intensities and include these and other restrictions may be possible with more reliable data sets from third-generation synchrotron sources.

### DATA ANALYSIS AND INITIAL INTERPRETATION

Most x-ray detectors are count-rate limited, and measured x-ray scattering intensities must be corrected for detector-system deadtime (Ice et al., 1994). With proportional counters and solid-state detectors, the measured deadtime is typically 3 times the amplifier shaping constant. For most measurements, this results in deadtimes of $\sim$1 to 10 $\mu$s/count.

Detector survival can also be challenged by intense x-ray beams and requires that the sample orientation and detector position be controlled such that a Bragg reflection from the cyrstal does not enter the detector. Bragg reflections can contain in excess of $10^9$ x rays per second at synchrotron sources, which can paralyze or damage detectors. Position-sensitive wire proportional counters are especially vulnerable, and count rates below $10^4$ counts/s are generally advisable to prevent damage to the wire or coated filament. Just falshing throug a Bragg reflection when changing orientation can damage the wire anode of a linear-senitive proportional counter, which materially degrades its spatial resolution. Extreme caution is necessary when measurements are taken near Bragg reflections with flux-sensitive detectors.

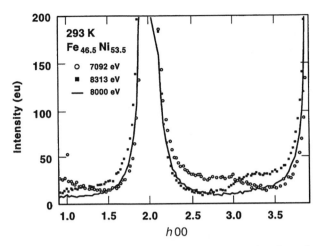

**Figure 17.** Total elastically scattered x-ray intensity along the $\langle h00 \rangle$ measured at 293 $K$ for the three x-ray energies listed. Note the shift in contrast for intensities measured with energies 20 eV below the Fe $K$ edge at 7092-eV and the Ni $K$ edge at 8313 eV, which changes the sign of $\mathrm{Re}(f_{\mathrm{Ni}} - f_{\mathrm{Fe}})$. The outlying data point at the {100} position is from harmonic energy contamination of the incident radiation and such points are removed before processing. Courtesy of Jiang et al. (1996).

Mirrors and crystal monochromators can pass a significant number of harmonics (multienergy x rays), which are then diffracted by the sample at positions of $h/n$, $k/n$, $l/n$, where $n$ is an integer and $hkl$ are the Miller indices of the Bragg reflections. Any sharp spikes observed in the measurement of diffuse intensities at these positions are suspect and must be removed before processing the data to recover the local correlations. At the position (100) in Figure 17, we note an outlying data point that can be attributed to the Bragg diffraction from the (200) reflection of an x-ray energy twice that of the nominal energy. Such spurious data can also be caused by surface films left from chemical treatment or oxidation.

An example of the raw data measured for three different x-ray energies from an $Fe_{46.5}Ni_{53.5}$ alloy single crystal is shown in Figure 17 (Jiang et al., 1996). The solid line in Figure 17 is the near null Laue measurement, which can be used to remove the quadratic and higher-order displacement terms from the other data sets. The assumption made is that for elements with similar masses and small static displacements the second and higher moment terms are the same for both atoms species so that for $I_{j \geq 2}$, the

intensity scales as $\langle f \rangle^2$ This method avoids the need to calculate thermal scattering from a set of force constants with the Born-von Karman central forces model that assumes harmonic vibrations and Hooke's law forces between the atoms independent of their environment (Warren, 1969). Theoretically calculated TDS increasingly deviates from measured TDS on approach to the Brillouin zone boundaries. This is not unexpected as local structure is important near the zone boundaries. A comparison of x-ray null Laue results with neutron diffuse scattering measurements, which are not complicated by TDS, gives very similar $\alpha$'s for $Fe_3Ni$ (Jiang et al., 1996).

An example of the measured diffuse scattering data in the $h_1h_20$ plane (labeled here as $H_1$, $H_2$, 0) for three x-ray energies is shown in Figure 18 for an alloy of $Fe_{27.5}Ni_{77.5}$ (Ice et al., 1992). The x-ray energy at 8.000 keV (Fig. 18B) is the near null Laue energy ($f_{\mathrm{Ni}} - f_{\mathrm{Fe}} \sim 0$), where $I_{\mathrm{SRO}}$ is nearly zero compared with the intense SRO maxima such as (100), (110), and (210) at 7.092 keV of Figure 18A. With the data in electron units and with inelastic scattering removed, the data are now processed to recover the $\alpha$ and $\delta$ values. As discussed previously (see Principles of the Method), we have the choice of different processing methods.

1. The null Laue method (also referred to as the $3\lambda$ method). This method is used when the elements of an alloy are sufficiently near each other in the periodic table that their dynamic displacements are similar and $I_{j \geq 2}$ scales as $(C_A f_A + C_B f_B)^2$ for the different energies; $I_{j \geq 2}$ can be experimentally measured at a low x-ray scattering contrast and then substracted from the diffuse scattering with high elemental contrast (Ice et al., 1992; Reinhard et al., 1992; Schoenfeld et al., 1994; Jiang et al., 1996). The null Laue method is implemented using a nonlinear least-squares approach.

2. Collection of sufficient data such that the BS separation technique can first be used to recover $I_{\mathrm{SRO}}$, $I_{j=1}$, $I_{j=2}$, and higher terms separately for each of the three x-ray energies. Then a least-squares program is used to recover the $\alpha$'s from $I_{\mathrm{SRO}}$. The $\alpha$'s are used to recover the displacements from $I_{j=1}$ and the second moments from $I_{j=2}$ as given in Equations 4, 5, 6, and 7. As the assumption is made that the x-ray atomic scattering factors are independent of **H**, an interactive technique is required to remove that

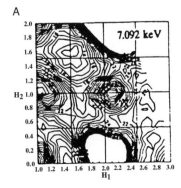
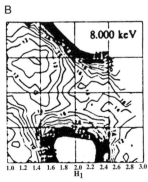
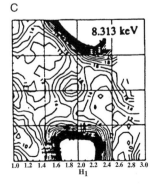

**Figure 18.** Diffuse x-ray scattering intensities from $Fe_{22.5}Ni_{72.5}$ in the $h_3 = 0$ plane collected with x-ray energies of (**A**) 7.092, (**B**) 8.000, and (**C**) 8.313 keV. Courtesy of Ice et al. (1992).

assumption. This is possibly the most robust of the methods as it does not have the assumption of method 1. Therefore it can be extended to include the $I_{j=3}$ and $I_{j=4}$ terms to account for higher moments of the displacements including second-order TDS, which becomes important for measurements made at temperatures approaching the Debye temperature. Furthermore, data for each energy can be separately into $I_{SRO}$ and displacement scattering and checked for the correct normalization factors before different energy sets are subjected to a simultaneous least-squares program to recover the A-A, A-B, and B-B pair correlations.

3. The GC method of analysis, which uses data measured at only one energy for 25 symmetry-related points for each of the 25 terms expressed in Equations 4, 5, 6, and 7. These symmetry-related points are chosen such that the values of $I_{SRO}$, and the $Q$, $R$, and $S$ terms are of the same or of opposite magnitude. Only their scattering factors and $\mathbf{H}$ dependence differ. In this way, the terms of Equations 4, 5, 6, and 7 are obtained from a system of linear equations stabilized by a ridge regression technique. These terms are then inverted to recover their Fourier coefficients. Results with this technique have been mixed. An analysis of $AuCu_3$ data (Butler and Cohen, 1989) concluded that the Au-Au bond distance was shorter than that for Cu-Cu. This result is contrary to the experimental findings that ordering of $AuCu_3$ reduces the lattice constant as more first-neighbor Au-Cu pairs are formed and that the addition of the $\sim$14% larger Au atoms to Cu increases the lattice constant because of the larger Au-Au bond distance. Theoretical considerations have also concluded that the Au-Au bond distance is the largest of the three kinds (Chakraborty, 1995; Horiuchi et al., 1995). Apparently, the $\mathbf{H}$ variation of $f_{Au}\Delta f^*$ and $f_{Cu}\Delta f^*$ shown in Figure 4 is not sufficiently different to provide for a meaningful separation of the Au-Au and Cu-Cu bond distances. In a direct comparison with the $3\lambda$ technique on an alloy of $Ni_{80}Cr_{20}$, the GC result gave a Ni-Ni bond distance with a different sign, which was contrary to other information (Schöenfeld et al., 1994). In addition, published GC values for $\langle(\Delta X)^2\rangle$ coefficients are not reliable (Wu et al., 1983). Though first-order TDS is included in the separation, higher-order TDS is calculated from force constants and subtracted.

### Interpretation of Recovered Static Displacements

The displacements are defined as deviations from the average lattice and are given by

$$\mathbf{r}_p - \mathbf{r}_q = (\mathbf{R}_p - \mathbf{R}_q) + (\boldsymbol{\delta}_p - \boldsymbol{\delta}_q) \tag{9}$$

As we can move the frame of reference so that its origin always resides on one of the atoms of the pair, such that $\mathbf{r}_p \equiv \mathbf{r}_0$, $\mathbf{R}_p \equiv \mathbf{R}_0$, and $\boldsymbol{\delta}_p \equiv \boldsymbol{\delta}_0$, then

$$\mathbf{r}_p - \mathbf{r}_q = \mathbf{r}_0 - \mathbf{r}_q = \mathbf{r}_0 - \mathbf{r}_{lmn} = -\mathbf{r}_{lmn} \tag{10}$$

and with the atom pair identified by $ij$

$$\mathbf{r}_{lmn}^{ij} = \mathbf{R}_{lmn} + \boldsymbol{\delta}_{lmn}^{ij} \tag{11}$$

where $\mathbf{R}_{lmn}$ is independent of the kinds of atom pairs since it is defined by the average lattice (i.e., Bragg reflection positions). The average value of the measured $\mathbf{r}_{lmn}$ for all the $N$ pairs contributing to the measured intensity is

$$\langle \mathbf{r}_{lmn}^{ij} \rangle = \frac{1}{N^{ij}} \sum_{ij} \langle \mathbf{R}_{lmn} + \boldsymbol{\delta}_{lmn}^{ij} \rangle = \mathbf{R}_{lmn} + \langle \boldsymbol{\delta}_{lmn}^{ij} \rangle \tag{12}$$

Here $\langle \boldsymbol{\delta}_{lmn}^{ij} \rangle$ is the variable recovered from the diffuse scattering. As shown in Equation 25, we recover the Cartesian coordinates of the average displacement vector,

$$\langle \boldsymbol{\delta}_{lmn}^{ij} \rangle \equiv \langle \Delta X_{lmn}^{ij} \rangle \mathbf{a} + \langle \Delta Y_{lmn}^{ij} \rangle \mathbf{b} + \langle \Delta Z_{lmn}^{ij} \rangle \mathbf{c} \tag{13}$$

For cubic systems, when the atom has fewer than 24 neighboring atoms in a coordination shell (permutations and combinations of $\pm l$, $\pm m$, $\pm n$), $\langle \boldsymbol{\delta}_{lmn}^{ij} \rangle$ must be parallel to the lattice vector $\mathbf{R}_{lmn}$. This maintains the statistically observed long-range cubic symmetry even though on a local scale this symmetry is broken. For $lmn$ multiplicities $\geq 24$, the displacements on average need not be parallel to the average interatomic vector $\mathbf{R}_{lmn}$ to preserve cubic symmetry (Sparks and Borie, 1965).

Measurements of diffuse scattering from single crystals provides the components of the atomic displacements $\langle \Delta X \rangle$, $\langle \Delta Y \rangle$, and $\langle \Delta Z \rangle$ whereas the spherical average usually obtained from EXAFS and x-ray measurements on amorphous materials and crystalline powders gives only the magnitude of the radial displacements. Thus, diffuse x-ray scattering from single crystals provides new information about the vector displacements associated with near-neighbor chemistry.

Measured displacements such as those presented in Table 2 provide unique insight into how atoms move off their lattice sites when local symmetry is broken. Local symmetry is broken when a multicomponent crystalline material is above the ordering temperature (with less-than-perfect long-range order) and/or off stoichiometry. With perfect long-range order the atoms are constrained to lie precisely on the sites of the average lattice by balanced forces. In alloys, where the local symmetry is broken, we gain new insights into the chemically distinct bonding, including the interatomic bond distances and whether the displacements have both radial and tangential components. With reference to Figure 23, the displacement for the [110] nearest-neighbor atoms is on average radial with a magnitude given by $|\langle \boldsymbol{\delta}_{110} \rangle| = \sqrt{2}|\Delta X_{110}|$.

We note that the Fe-Fe first-neighbor pair distances given in Table 2 are 0.021(3) Å $\times \sqrt{2} = 0.030(4)$ Å further apart then the average lattice and that second neighbor pairs are closer by $(-)$ 0.023(1) Å. Average bond distances along the interatomic vector between nearest-neighbor pairs for this fcc lattice are obtained by adding the $\sqrt{2}|\Delta X_{110}|$ to the average interatomic vector $\mathbf{R}_{110}$, as defined in Figure 19. The parameter $|\mathbf{R}_{110}|$ is just the

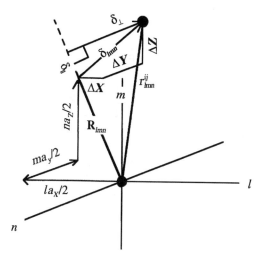

**Figure 19.** Construction of the vectors recovered from diffuse scattering measurements on single crystals. The parameter $\mathbf{R}_{lmn}$ is obtained from the lattice parameter $|\mathbf{a}|$, and the average components of the displacement $\delta_{lmn}^{ij}$ are recovered from measurements of the diffuse scattering. Courtesy of Ice et al. (1998).

cubic lattice constant $|\mathbf{a}|$ times $1/\sqrt{2}$. From the construction shown in Figure 19, it follows that the vector distance between a pair of atoms, $\mathbf{r}_{lmn}^{ij}$, has radial and tangential displacement components with magnitudes given by

$$\left|\boldsymbol{\delta}_{lmn}^{ij}\right|_{\parallel} = \frac{\boldsymbol{\delta}_{lmn}^{ij} \cdot \mathbf{R}_{lmn}}{|\mathbf{R}_{lmn}|} \qquad (14)$$

and

$$\left|\boldsymbol{\delta}_{lmn}^{ij}\right|_{\perp} = \sqrt{|\boldsymbol{\delta}_{lmn}^{ij}|^2 - |\boldsymbol{\delta}_{lmn}^{ij}|_{\parallel}^2} \qquad (15)$$

The radial ($\parallel$) and tangential ($\perp$) components of the displacements recovered from diffuse scattering measurements on single crystals are shown in Figure 19.

As the $Fe_{46.5}Ni_{53.5}$ alloy is cubic (face centered), the $\Delta Y$ and $\Delta Z$ displacements are derived from the $\Delta X$'s given in Table 2 by permutation of the indices. (Henceforth, we will omit the $<>$ on the displacements for simplicity.) For example, $\Delta X_{321}$ has the identical value as $\Delta Y_{231}$ and $\Delta Z_{123}$, and $\Delta X_{321}=\Delta X_{312}=\Delta Y_{231}=\Delta Y_{132}=\Delta Z_{123}=\Delta Z_{213}$. In addition, $\Delta X_{321}=-\Delta X_{\bar{3}21}$ and similarly for the other combinations as illustrated in Figure 20. The nearest-atom pairs that could have, on average, nonradial components are those in the third neighboring shell, $lmn=211$ (Sparks and Borie, 1965). If the displacements between atom pairs is on average along their interatomic vector, then $\Delta X_{211} = 2\,\Delta X_{121}$ (Fig. 20). For the Fe-Fe pair displacements given in Table 2, $\Delta X_{211}=0.0005(2)\text{Å}$ and $2\,\Delta X_{121}=0.0028(8)\text{Å}$; thus the (211) Fe-Fe pair displacements have a significant tangential component. From Equations 14 and 15, the magnitude of the displacement between (211) Fe-Fe pairs along the radial direction $|\boldsymbol{\delta}_{211}^{Fe\text{-}Fe}|_{\parallel}$ is 0.0016 (7)Å and $|\boldsymbol{\delta}_{211}^{Fe\text{-}Fe}|_{\perp}$ tangential is 0.0013(7)Å. Thus the (211) Fe-Fe neighbors have a similar radial and tangential component to their displacements. For the (310) Fe-Fe pair displacements,

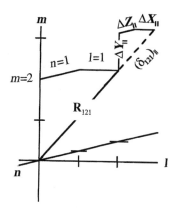

**Figure 20.** Radial displacements (parallel to the interatomic vector $\mathbf{R}_{lmn}$) between the atom pairs require that the relative magnitudes of the displacement components be in the same proportion as the average lattice vector; $\Delta X : \Delta Y : \Delta Z = l:m:n$. As shown for $lmn = 211$, a radial displacement requires $|\Delta X| = 2|\Delta Y|$ and $|\Delta X| = 2|\Delta Z|$. For $lmn = 121$, $|\Delta X| = |\Delta Y|/2$ and $|\Delta X| = |\Delta Z|$. For a cubic lattice, we can interchange $l$, $m$, and $n$ and similarly $\Delta X$, $\Delta Y$, and $\Delta Z$. Thus there is only one value $\Delta X$ for $lmn$ multiplicities $<24$ (i.e., 110, 200, and 222), two values for $\Delta X$ when $lmn$ has multiplicities equal to 24 ($l \neq m$ and $l = m, n$), and three values for $\Delta X$ with multiplicities equal to 48. Courtesy of Ice et al. (1998).

$\Delta X_{310} \cong 3\ \Delta X_{130}$ within the total estimated error, and on average (310) displacements are predominantly radial. These measured displacements provide new information not obtained in other ways about the local atomic arrangements in crystalline solid solutions.

Results for the few crystalline binary alloys that have had their individual pair displacements measured with this $3\lambda$ technique are summarized in Figure 21. Here the $\Delta X$ static displacements are plotted as a function of the radial distance $\frac{1}{2}\sqrt{l^2 + m^2 + n^2}$. When there is more than one value for $\Delta X$, the plots show the various values. Most striking is the observation that for the three ordering alloys the near-neighbor Fe-Ni and Cr-Ni bond distances are the smallest of the three possible pairs (Fig. 21). However, for the clustering $Cr_{47}Fe_{53}$ alloy the Cr-Cr nearest-neighbor bond distances are closest and the Cr-Fe are furthest apart. More details, including the short-range-order parameters $\alpha$ and numerical values of the displacements for each shell, are given in the original papers. These pair displacement observations provide a more rigid test of theoretical predictions than variations of the average lattice parameter with concentrations (Froyen and Herring, 1981; Mousseau and Thorpe, 1992).

## METHOD AUTOMATION

Because of the large number of data that is collected ($\sim$7000 data points per diffuse scattering volume, each consisting of a multichannel energy spectrum with 200 to 500 channels), data collection is under computer control. The program most widely used for converting the reciprocal space coordinates to Eulerian angles and then stepping the three- (or four-) axis diffractometer to this list of

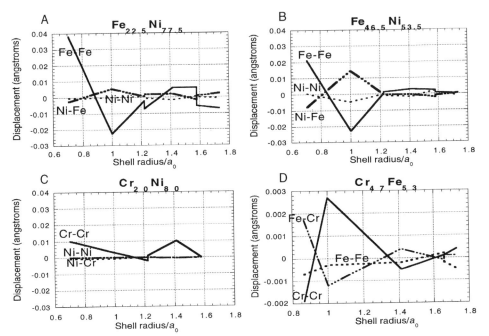

**Figure 21.** Displacement from the average lattice sites for chemically specific pairs. Shell radius divided by the lattice parameter $a_0$ becomes 1 for second neighbors (seperated by the cube edge). Courtesy of Ice et al. (1998).

coordinates is SPEC (Certified Scientific Software). Because the diffuse scattering is a slowly varying function without sharp peaks, it is fastest to take the data while the diffractometer is moving and avoid the time-consuming starting and stopping of the diffractometer. The reciprocal space coordinate is estimated from the midrange of the angular scans during data collection. When using the BS separation procedure, it is necessary to have the diffuse scattering data on a uniform cubic grid in reciprocal space. Least-squares procedures do not have this requirement. Since the experiment is enclosed in a radiation hutch, computerized control of all slits, sample positioning, and electronic components in the hutch is desirable.

## SAMPLE PREPARATION

The most detailed measurements of diffuse x-ray scattering are made on homogeneous single crystals with polished and etched surfaces. These samples must be carefully prepared and handled to minimize surface roughness effects, surface contamination, and inhomogeneities. To prepare a sample, a single crystal of the intended composition and ≥99.9% purity metals should be homogenized for ≥3 days at ~50° to 100°C below the crystal melting point in an atmosphere that protects its composition. This step is intended to create a single crystal with uniform chemical composition. For a crystal with cubic symmetry, the crystal should be oriented and then cut with a surface near the <421> normal. This orientation minimizes the goniometer χ range, improves the energy resolution of measurements with a crystal spectrometer, and improves access to a cubic symmetry volume of reciprocal space. The surface of the crystal should be ~20 mm in diameter to ensure that the

incident beam is completely intercepted by the sample at all orientations. Electrical discharge machining or chemical sawing minimizes surfaces strains and distortions and is preferred to standard machining. After chemical polishing to remove distorted metal, the crystals should be heat treated to the desired state of thermodynamic equilibrium before quenching. Metallographic polishing is followed by chemical dissolution or electropolishing to produce a mirror smooth surface with nondistorted metal (see SAMPLE PREPARATION FOR METALLOGRAPHY). Verification of negligible distortion is obtained by optical microscopy of subgrain boundary images. Because surface roughness reduces x-ray scattering intensity at low angles, smooth surfaces reduce the corrections to the absolute intensity measurements. The effect of sample roughness on intensity can be checked with fluorescence or RRS measurements to determine if there is any angular variation with respect to **H** in the bisecting geometry (Sparks et al., 1992). The acceptable mosaic spread of the crystals measured from x-ray rocking curves depends on the **H** resolution necessary to resolve the features of interest, but can be as large as ~2 degrees. Verification of the composition to ± 0.2 at.% or less is needed to reduce the uncertainty in the data analysis. A check of the weight of the raw materials against the weight of the single-crystal ingot will determine if more extensive chemical analysis is necessary.

## PROBLEMS

The fundamental problem of diffuse x-ray scattering is that the measurements must be made in absolute units and must be consistent over large ranges in reciprocal space and with different experimental conditions (x ray, neutron, energy, etc.). Even small systematic uncertainties increase the uncertainties in the recovered local correlations. Special procedures can sometimes be used to calibrate the relative normalization between different experimental conditions, but the reality remains that great care is required to collect meaningful data. For example, near an absorption edge, uncertainties in the scattered intensity can occur because the values of $f'$ and $f''$ change rapidly with small fluctuations in incident photon energy, as seen in Figure 9. The intensities of RRS also increases as the incident photon energy approaches an absorption edge. As a compromise, the incident energies are usually chosen $\sim20$ eV below an absorption edge. This requires that the x-ray optics, which select the x-ray energy from the white synchrotron spectrum, be very stable so as to control the selected energy fluctuations to less than about $\pm1$ eV.

Similarly, environments must maintain a clean surface free of condensates and oxidation that can add unwanted scattered intensity to measurements. Extremely thin coverages of a few tens of monolayers can contribute noticeable intensity. Equilibrium temperatures can sometimes be difficult to define on quenched samples. For example, quenching to a desired equilibrium local-ordered state may not be possible due to rapid kinetics; quenched-in vacancies enhance diffusion, which can alter chemical order at the actual temperature of interest for local order. Kinetic studies or measurements at the higher temperature are required to ensure that equilibrium order is achieved. Even if diffusion, which changes the $\alpha$'s on cooling, is not a problem, changes in lattice constant on cooling may affect the static displacements. It is better to have the sample in a known state of equilibrium so that the recovered parameters can be directly compared to theory. A discussion of the effect of quenching parameters on the diffuse scattering from an Al-Cu sample has been given by Epperson et al. (1978), and a study of the kinetics of short-range ordering in $Ni_{0.765}Fe_{0.235}$ has been given by Bley et al. (1988).

The complexities of actually performing and interpreting a three-dimensional diffuse scattering experiment require a major commitment of time and resources. To ensure success, we suggest that beginners collaborate with one of the referenced authors who has experience in this science and who has access to the specialized instrumentation and software required for successful diffuse scattering experiments.

## ACKNOWLEDGMENTS

We wish to express our appreciation to the early pioneers of diffuse scattering and to our many contemporaries who have contributed to this subject. Research was sponsored by the Division of Materials Sciences, U.S. Department of Energy, under contract DE-AC05-96OR22464 with Lockheed Martin Energy Research. Research was performed at the Oak Ridge National Laboratory Beamline X-14 at the National Synchrotron Light Source, Brookhaven National Laboratory, sponsored by the Division of Materials Sciences and Division of Chemical Sciences, U.S. Department of Energy.

## LITERATURE CITED

Åberg, T. and Crasemann, B. 1994. Radiative and radiationless resonant Raman scattering. *In* Resonant Anomalous X-Ray Scattering: Theory and Applications (G. Materlik, C. J. Sparks, and F. Fischer, eds.). pp. 431–448. Elsevier Science Publishing, New York.

Biggs, F., Mendelshohn, L. B., and Mann, J. B. 1975. Hartree-Fock Compton profiles for the element. *Atomic Data Nucl. Tables* 16:201–309.

Bley, F., Amilius, Z., and Lefebvre, S. 1988. Wave vector dependent kinetics of short-range ordering in $^{62}Ni_{0.765}Fe_{0.235}$, studied by neutron diffuse scattering. *Acta Metall. Mater.* 36:1643–1652.

Borie B. and Sparks C. J. 1964. The short-range structure of copper-16 At. % aluminum. *Acta. Crystallogr.* 17: 827–835.

Borie, B. and Sparks, C. J. 1971. The interpretation of intensity distributions from disordered binary alloys. *Acta Crystallogr.* A27:198–201.

Butler, B. D. and Cohen, J. B. 1989. The structure of $Cu_3Au$ above the critical temperature. *J. Appl. Phys.* 65:2214–2219.

Carlsson, G. A., Carlsson, C. A., Berggren, K., and Ribberfors, R. 1982. Calculation of cattering cross sections for increased accuracy in diagnostic radiology. 1. Energy broadening of Compton-scattered photons. *Med. Phys.* 9:868–879.

Cenedese, P., Bley, F., and Lefebvre, S. 1984. Diffuse scattering in disordered ternary alloys: Neutron measurements of local order in stainless steel $Fe_{0.56}Cr_{0.21}Ni_{0.23}$. *Acta Crystallogr.* A40: 228–240.

Chantier, C. T. 1994. Towards improved form factor tables. *In* Resonant Anomalous X-Ray Scattering: Theory and Applications (G. Materlik, C. J. Sparks, and K. Fischer, eds.). pp. 61–78. Elsevier Science Pubishers, New York.

Chakraborty, B. 1995. Static displacements and chemical correlations. *Eur. Phys. Lett.* 30:531–536.

Cowley, J. M. 1950. X-ray measurement of order in single crystals of $Cu_3Au$. *J. Appl. Phys.* 21:24–30.

Cromer, D. T. and Liberman, D. 1970. Relativistic calculations of anomalous scattering factors for x rays. *J. Chem. Phys.* 53:1891–1898.

Cromer, D. T. and Liberman, D. 1981. Anomalous dispersion calculations near to and on the long-wavelength side of an absorption edge. *Acta. Crystallogr.* A37:267–268.

Dreier, P., Rabe, P., Malzfeldt, W., and Niemann, W. 1983. Anomalous scattering factors from x-ray absorption data by Kramers-Kronig analysis. *In* Proceedings of the International Conference on EXAFS and Near Edge Structure, Frascati, Italy, Sept. 1982 (A. Bianconi, L. Incoccia, and S. Stipcich, eds.). pp. 378–380. Springer-Verlag, New York and Heidelberg.

Drijver, J. W., van der Woude, F., and Radelaar, S. 1977. Mössbauer study of atomic order in $Ni_3Fe$. I. Determination of the long-range-order parameter. *Phys. Rev. B* 16:985–992.

Eisenberger, P., Platzman, P. M., and Winick, H. 1976a. X-ray resonant Raman scattering: Observation of characteristic radiation narrower than the lifetime width. *Phys. Rev. Lett.* 36:623–625.

Eisenberg, P., Platzman, P. M., and Winick, H. 1976b. Resonant x-ray Raman scattering studies using synchrotron radiation. *Phys. Rev. B.* 13:2377–2380.

Epperson, J. E., Fürnrohr, P., and Ortiz, C. 1978. The short-range-order structure of α-phase Cu-Al alloys. *Acta Crystallogr.* A34:667–681.

Froyen, S. and Herring, C. 1981. Distribution of interatomic spacings in random alloys. *J. Appl. Phys.* 52:7165–7173.

Gehlen, P. C. and Cohen, J. B. 1965. Computer simulation of the structure associated with local order in alloys. *Phys. Rev. A* 139:884–A855.

Georgopoulos, P. and Cohen, J. B. 1977. The determination of short range order and local atomic displacements in disordered binary solid solutions. *J. Phys. (Paris) Colloq.* 38:C7-191–196.

Georgopoulos, P. and Cohen, J. B. 1981. The defect arrangement in (non-stoichiometric) β'-NiAl. *Acta Metall. Mater.* 29:1535–1551.

Gragg. J. E., Hayakawa, M., and Cohen, J. B. 1973. Errors in quantitative analysis of diffuse scattering from alloys. *J. Appl. Crystallogr.* 6:59–66.

Hashimoto, S., Iwasaki, H., Ohshima, K., Harada, J., Sakata, M., and Terauchi, H. 1985. Study of local atomic order in a ternary $Cu_{0.47}N_{0.29}Zn_{0.24}$ alloy using anomalous scattering of synchrotron radiation. *J. Phys. Soc. Jpn.* 54:3796–3807.

Horiuchi, T., Takizawa, S., Tomoo, S., and Mohri, T. 1995. Computer simulation of local lattice distortion in Cu−Au solid solution. *Metall. Mater. Trans. A* 26A:11–19.

Hoyt, J. J., Fontaine, D. D., and Warburton, W. K. 1984. Determination of the anomalous scattering factors for Cu, Ni and Ti using the dispersion relation. *J. Appl. Crystallogr.* 17:344–351.

Ice, G. E. and Sparks, C. J. 1990. Mosaic crystal x-ray spectrometer to resolve inelastic background from anomalous scattering experiments. *Nucl. Instrum. Methods* A291:110–116.

Ice, G. E., Sparks, C. J., Habenschuss, A., and Shaffer, L. B. 1992. Anomalous x-ray scattering measurement of near-neighbor individual pair displacements and chemical order in $Fe_{22.5}Ni_{77.5}$. *Phys. Rev. Lett.* 68:863–866.

Ice, G. E., Sparks, C. J., Jiang, X., and Robertson, L. 1998. Diffuse scattering measurements of static atomic displacements in crystalline binary solid solutions. *J. Phase Equilib.* 19:529–537.

Ice, G. E., Sparks, C. J., and Shaffer, L. B. 1994. Chemical and displacement atomic pair correlations in crystalline solid solutions recovered by resonant (anomalous) x-ray scattering. *In* Resonant Anomalous X-Ray Scattering: Theory and Applications (G. Materlik, C. J. Sparks, and K. Fischer, eds.), pp. 265–294. Elsevier Science Publishing, New York.

James, R.W. 1948. The Optical Principles of the Diffraction of X-Rays. Cornell University Press, Ithaca, N.Y.

Jiang, X., Ice, G. E., Sparks, C. J., Robertson, L., and Zachack, P. 1996. Local atomic order and individual pair displacements of $Fe_{46.5}Ni_{53.5}$ and $Fe_{22.5}Ni_{77.5}$ from diffuse x-ray scattering studies. *Phys. Rev. B* 54:3211–3226.

Jiang, X., Ice, G. E., Sparks, C. J., and Zschack, P. 1995. Recovery of SRO parameters and pairwise atomic displacements in a $Fe_{46.5}Ni_{53.5}$ alloy. *In* Applications of Synchrotron Radiation Techniques to Materials Science. *Mater. Res. Soc. Symp. Proc.* 375:267–273.

Kawamura, T. and Fukamachi, T. 1978. Application of the dispersion relation to determine the anomalous scattering factors. Proceedings of the International Conference on X-ray and VUV Spectroscopies, Sendai, Japan. *J. Appl. Phys.*, Suppl. 17-2:224–226.

Kostorz, G. 1996. X-ray and neutron scattering. *In* Physical Metallurgy, 4th and revised and enhanced edition. (R. W. Cahn and P. Hoosen, eds.). pp. 1115–1199. Elsevier Science Publishing, New York.

Lengeler, B. 1994. Experimental determination of the dispersion correction $f'(E)$ of the atomic scattering factor. *In* Resonant Anomalous X-Ray Scattering: Theory and Applications (G. Materlik, C. J. Sparks, and K. Fischer, eds.). pp. 35–60. Elsevier Science Publishing, New York.

Miller, M. K., Cerezo, A., Hetherington, M. G., and Smith, G. D. W. 1996. Atom Probe Field Ion Microscopy. Oxford University Press, Oxford.

Mousseau, N. and Thorpe, M. F. 1992. Length distributions in metallic alloys. *Phys. Rev. B* 45:2015–2022.

Müller, P. P., Schönfeld, B., Kostorz, G., and Bührer, W. 1989. Guinier-Preston I zones in Al-1.75 at.% Cu single crystals. *Acta Metall. Mater.* 37:2125–2132.

Pierron-Bohnes, V., Cadeville, M. C., and Gautier, F. 1983. Magnetism and local order in dilute Fe-C alloys. *J. Phys. F* 13:1689–1713.

Reinhard, L., Robertson, J. L., Moss, S. C., Ice, G. E., Zschack, P., and Sparks, C. J. 1992. Anomalous-x-ray scattering study of local order in BCC $Fe_{0.53}Cr_{0.47}$. *Phys. Rev B* 45:2662–2676.

Renaud, G., Motta, N., Lançon, F., and Belakhovsky, M. 1988. Topological short-range disorder in $Au_{1-x}Ni_x$ solid solutions: An extended x-ray absorption fine structure spectroscopy and computer-simulation study. *Phys. Rev. B* 38:5944–5964.

Robertson, J. L., Sparks, C. J., Ice, G. E., Jiang, X., Moss, S. C., and Reinhard, L. 1998. Local atomic arrangements in binary solid solutions studied by x-ray and neutron diffuse scattering from single crystals. *In* Local Structure from Diffraction: Fundamental Materials Science Series (M. F. Thorpe and S. Billinge, eds.).. Plenum, New York. In press.

Sasaki, S. 1989. Numerical tables of anomalous scattering factors calculated by the Cromer and Liberman's method. *KEK Report* 88–14.

Scheuer, U. and Lengeler, B. 1991. Lattice distortion of solute atoms in metals studied by x-ray absorption fine structure. *Phys. Rev. B* 44:9883–9894.

Schönfeld, B., Ice, G. E., Sparks, C. J., Haubold, H.-G., Schweika, W., and Shaffer, L.B. 1994. X-ray study of diffuse scattering in Ni-20 at% Cr. *Phys. Status Solidi B* 183:79–95.

Schwartz, L. H. and Cohen, J. B. 1987. Diffraction from Materials. Springer-Verlag, New York and Heidelberg.

Schweika, W. 1998. Disordered Alloys: Diffuse Scattering and Monte Carlo Simulations, Vol. 141. Springer-Verlag, New York and Heidelberg.

Simon, J. P. and Lyon, O. 1991. The nature of the scattering tail in Cu-Ni-Fe and invar alloys investiaged by anomalous small angle x-ray scattering. *J. Appl. Crystallogr.* 24:1027–1034.

Sparks, C. J. 1974. Inelastic resonance emission of x rays: Anomalous scattering associated with anomalous dispersion. *Phys. Rev. Lett.* 33:262–265.

Sparks, C. J. and Borie, B. 1965. Local atomic arrangements studied by x-ray diffraction. *In* AIME Conference Proceedings 36 (J.B. Cohen and J.E. Hilliard, eds.). pp. 5–50. Gordon and Breach, New York.

Sparks, C. J., Ice, G. E., Shaffer, L. B., and Robertson, J. L. 1994. Experimental measurements of local displacement and chemical pair correlations in crystalline solid solutions. *In* Metallic Alloys: Experimental and Theoretical Perspectives (J. S. Faulkner and R. G. Jordan, eds.). pp. 73–82. Kluwer Academic Publishers, Dordrecht, The Netherlands, NATO Vol. 256.

Sparks, C. J. and Robertson, J. L. 1994. Guide to some crystallographic symbols and definitions with discussion of short-range correlations. *In* Resonant Anomalous X-ray Scattering: Theory and Applications (G. Materlik, C. J. Sparks, and K. Fischer, eds.). pp. 653–664. Elsevier Science North-Holland, Amsterdam, The Netherlands.

Suortti, P., Hastings, J. B., and Cox, D. E. 1985. Powder diffraction with synchrotron radiation. I. Absolute measurements. *Acta Crystallogr.* A41:413–416.

Tibballs, J. E. 1975. The separation of displacement and substitutional disorder scattering: A correction for structure factor ratio variation. *J. Appl. Crystallogr.* 8:111–114.

Warren, B. E. 1969 (reprinted) 1990. *In* X-Ray Diffraction. Dover Publications, New York.

Warren, B. E., Averbach, G. L., and Roberts, B. W. 1951. Atomic size effect in the x-ray scattering by alloys. *J. Appl. Phys.* 22:1493–1496.

Welberry, T. R. and Butler, B. D. 1995. Diffuse x-ray scattering from disordered crystals. *Chem Rev.* 95:2369–2403.

Wu, T. B., Matsubara, E., and Cohen, J. B. 1983. New procedures for quantitative studies of diffuse x-ray scattering. *J. Appl. Crystallogr.* 16:407–414.

Zunger, A. 1994. First-principles statistical mechanics of semiconductor alloys and intermetallic compounds. *In* Statics and Dynamics of Alloy Phase Transitions, Vol. B319 of NATO ASI Series B (P.E.A. Turchi and A. Gonis, eds.). p. 361. Plenum, New York.

## KEY REFERENCES

Cowley, J.M. 1975. Diffraction Physics. North-Holland, Amsterdam, The Netherlands.

*This reference is a good basic introduction into diffraction physics and x-ray techniques.*

Materlik, G., Sparks, C. J., and Fisher, K. (eds.).. 1994. Resonant Anomalous X-ray Scattering. North-Holland, Amsterdam, The Netherlands.

*This reference is a collection of recent work on the theory and application of resonant x-ray scattering techniques. It provides the most complete introduction to the application of anomalous (resonant) scattering techniques and how the field has been revolutionized by the availability of intense synchrotron radiation sources.*

Schwartz and Cohen, 1987. See above.

*This is another introductory text on x-ray diffraction with an emphasis on the application to materials. The section on diffuse x-ray scattering is especially strong and the notation is the same as used in this unit.*

Schweika, 1998. See above.

*This monograph provides an excellent reference to the interplay between experiment and theory in the field of local atomic order in alloys.*

Warren, 1969 (reprinted 1990). See above.

*This reference provides an authoritative treatment of all phases of diffuse x-ray scattering, including thermal diffuse scattering, TDS, short-range order, and atomic size displacement scattering. Although it does not include a modern outlook on the importance of resonant scattering, it provides a clear foundation for virtually all modern treatments of diffuse scattering from materials.*

## APPENDIX A:
## DERIVATION OF THE DIFFUSE INTENSITY

Our interest is in the diffusely distributed intensity. To separate Equation 3 into an intensity that may be sharply peaked and one that is diffusely distributed, we follow the method of Warren (1969). This method expands $e^{2\pi i \mathbf{H} \bullet (\delta_p - \delta_q)}$ in a Taylor series about $\delta_p - \delta_q$ and examines the displacement terms as the separation of atom pairs becomes large, $p-q \to \infty$. As the x-ray or neutron beam intercepts many atoms and the atoms undergo many thermal vibrations during the period of the intensity measurement, both a spatial and a time average are taken. These are indicated by $< >$. The spatial and time average of the $j$th-order Taylor series expansion of the exponential displacement term is

$$\langle e^{e\pi i \mathbf{H}} \bullet (\delta_p - \delta_g) \rangle \equiv \langle e^{iX_{pq}} \rangle = 1 + i \langle X_{pq} \rangle - \frac{\langle X_{pq}^2 \rangle}{2} - \frac{i \langle X_{pq}^3 \rangle}{3!} + \cdots + \frac{i^j \langle X_{pq}^j \rangle}{j!} \quad (16)$$

The time average for harmonic thermal displacements causes odd-order terms to vanish. With the definition $\langle X_{pq} \rangle = \langle X_p - X_q \rangle$, so that $\langle (X_p - X_q)^2 \rangle = \langle X_p^2 \rangle + \langle X_q^2 \rangle - 2 \langle X_p X_q \rangle$, and for sharply peaked Bragg reflections, where $p-q \to \infty$, the displacements become uncorrelated so that $\langle X_p X_q \rangle = 0$. Therefore, $\langle X_{pq}^2 \rangle_{p-q \to \infty} = \langle X_p^2 \rangle + \langle X_q^2 \rangle$, and with the harmonic approximation we can estimate the long-range dynamical displacement term by

$$\langle e^{iX_{pq}} \rangle_{p-q \to \infty} \cong e^{-\frac{1}{2}(\langle x_p^2 \rangle + \langle x_q^2 \rangle)} = e^{-(M_p + M_q)} \quad (17)$$

Here $M_p$ is the usual designation for the Debye-Waller temperature factor (Warren, 1990, p. 35). The subject is treated by Chen in KINEMATIC DIFFRACTION OF X RAYS. We also include the mean-square static displacements in $M$. Experience has shown the validity of the harmonic approximation in Equation 17 to account for the reduction in the intensity of the Bragg reflections as a function of $\mathbf{H}$. With the understanding that when there is an A atom at site $p$ or $q$, $M_p$ or $M_q$ is written as $M_A$, and similarly $M_B$ when there is a B at $p$ or $q$, the fundamental Bragg intensity for an alloy is given by the substitution of Equation 17 into Equation 3 as

$$I(\mathbf{H})_{\mathbf{Fund}} = |C_A f_A e^{-M_A} + C_B f_B f_B e^{-M_B}|^2 \sum_p \sum_q e^{2\pi i \mathbf{H} \bullet (\mathbf{R}_p - \mathbf{R}_q)} \quad (18)$$

This expression accounts for the reduced intensity of the fundamental Bragg reflections due to thermal motion and static displacements of the atoms. Fundamental reflections scale as the average scattering factor and are insensitive to how the chemical composition is distributed on the lattice sites. When the alloy has long-range order among the kinds of atoms, the $\alpha_{pq}$'s do not converge rapidly with larger $p$, $q$ and account for the superstructure Bragg reflections that depend on how the atoms are distributed among the sites. We are now concerned with the distribution of this thermal and static scattering. To recover the diffuse intensity, we subtract $I(\mathbf{H})_{\mathrm{Fund}}$ from $I(\mathbf{H})_{\mathrm{Total}}$ in Equation 3. To avoid making the harmonic approximation of Equation 17, we subtract $I(\mathbf{H})_{\mathrm{Fund}}$ term by term and take the limit as $p-q \to \infty$. For example, to second order we have

$$\langle e^{iX_{pq}} \rangle_{p-q \to \infty} = 1 - \frac{1}{2}\left(\langle X_p^2 \rangle + \langle X_2^q \rangle\right) \quad (19)$$

By substitution of Equation 19 for each of the $\langle e^{iX_{pq}} \rangle$ terms in Equation 3 and assignment of the proper atom identity for different pairs, and recalling that as $p-q \to \infty$, $\to \infty$, $\alpha_{pq}=0$, we subtract this expression for $I(\mathbf{H})_{\text{Fund}}$ from $I(\mathbf{H})_{\text{Total}}$ and write the diffuse scattering to second order in the displacements as

$$\begin{aligned}
I(\mathbf{H})_{\text{Diffuse}} = \sum_p \sum_q \Big\{ &\left(C_A^2 + C_A C_B \alpha_{pq}\right)|f_A|^2 \\
&\times \left(1 + i\langle X_p^A - X_q^A \rangle - \frac{1}{2}\langle (X_p^A - X_q^A)^2 \rangle\right) \\
&- C_A^2|f_A|^2\left(1 - \langle X^2 \rangle^A\right) + C_A C_B(1 - \alpha_{pq})f_A f_B^* \\
&\times \left(1 + i\langle X_p^B - X_p^A \rangle - \frac{1}{2}\langle (X_p^B - X_q^A)^2 \rangle\right) \\
&- 2C_A C_B f_A f_B^*\left(1 - \frac{1}{2}\langle X^2 \rangle^A - \frac{1}{2}\langle X^2 \rangle^B\right) \\
&+ C_A C_B(1 - \alpha_{pq})f_A f_B^* \\
&\times \left(1 + i\langle X_p^A - X_q^B \rangle - \frac{1}{2}\langle (X_p^A - X_q^B)^2 \rangle\right) \\
&+ \left(C_B^2 + C_A C_B \alpha_{pq}\right)|f_B|^2 \\
&\times \left(1 + i\langle X_p^B - X_q^B \rangle - \frac{1}{2}\langle (X_p^B - X_q^B) \rangle\right) \\
&- C_B^2|f_B|^2\left(1 - \langle X^2 \rangle^B\right) \Big\} e^{2\pi i \mathbf{H} \bullet (\mathbf{R}_p - \mathbf{R}_q)}
\end{aligned}$$
$$(20)$$

Our use of the double sum requires that the pairs of atoms be counted in both directions, that is, a $p, q$ pair will become a $q, p$ pair such that $i(\boldsymbol{\delta}_p - \boldsymbol{\delta}_q) = -i(\boldsymbol{\delta}_q - \boldsymbol{\delta}_q)$ as shown in Figure 22. As seen in Equation 2, the $I(\mathbf{H})_{\text{Total}}$ double sum is made up of $p, q$ elements that are the product of four complex numbers: the two complex scattering factors and the two complex phase factors. For every $p, q$ pair there is a corresponding $q, p$ pair where the two scattering factors and the two phase factors are the $p, q$ pair complex conjugates; hence the $p, q$ and $q, p$ elements add up to a real number. This means that the terms in the series expansion of the fluctuation displacements must add in

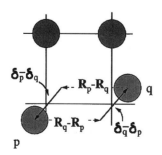

**Figure 22.** Schematic illustration showing that since all the pairs of atoms are counted as to kind and displacement in both directions, odd-power terms in the displacements are replaced with their negatives $i(\boldsymbol{\delta}_p - \boldsymbol{\delta}_q) = -i(\boldsymbol{\delta}_q - \boldsymbol{\delta}_q)$ and $\mathbf{R}_p - \mathbf{R}_q = -(\mathbf{R}_q - \mathbf{R}_p)$ so that the imaginary terms cancel.

pairs to give real intensity and from Equation 20, the $j =$ odd order displacement terms have a $\sin 2\pi(\mathbf{H} \bullet \mathbf{R})$ $\langle (X_p^A - X_q^A)^j \rangle$ dependence and the $j =$ even-order terms have a $\cos 2\pi(\mathbf{H} \bullet \mathbf{R})$ $\langle (X_p^A - X_q^A)^j \rangle$ dependence; the imaginary components cancel.

From the definition of an average lattice, the weighted average of the displacements for all the kinds of pairs formed for any coordination shell is zero (Warren et al., 1951),

$$\begin{aligned}
2(\alpha_{pq} - 1)\langle \boldsymbol{\delta}_p^A - \boldsymbol{\delta}_q^B \rangle = &\left(\frac{C_A}{C_B} + \alpha_{pq}\right)\langle \boldsymbol{\delta}_p^A - \boldsymbol{\delta}_q^A \rangle \\
&+ \left(\frac{C_B}{C_A} + \alpha_{pq}\right)\langle \boldsymbol{\delta}_p^B - \boldsymbol{\delta}_q^B \rangle \quad (21)
\end{aligned}$$

If a crystal structure has more than one kind of site symmetry (sublattices with different site symmetries), than Equation 21 may be true for only that sublattice with all the same site symmetries. Different-sized atom species will most likely have a preference for the symmetry of a particular sublattice. This preference could produce long-range correlations and superstructure reflections from which site occupation preferences can be recovered. Disorder among the atoms on any one sublattice and between sublattices will produce scattering from which pair correlations can be recovered. Discussion of this issue is beyond the scope of this unit. A study of a partially ordered nonstoichiometric bcc NiAl crystal, where one of the two sublattices with the same site symmetry is partly occupied by vacancies, has been discussed in Georgopoulos and Cohen (1981). A general review of the application of diffuse scattering measurements to more complicated structures with different site symmetries shows a wider use of models to reduce the number of variables necessary to describe the local pair correlations (Welberry and Butler, 1995).

Equation 20 can be expressed in a more tractable form through three steps: (1) Replace the double sum over atomic sites $p$ and $q$ by $N$ single sums around an origin site where the relative sites are identified by lattice difference $lmn$ such that $\mathbf{R}_p - \mathbf{R}_q = \mathbf{R}_0 - \mathbf{R}_{lmn}$. This approximation neglects surface effects. (2) Use trigonometric functions to simplify the phase factors. (3) Substitute Equation 21 into Equation 20. With these steps, the diffuse intensity can be expressed as,

$$\begin{aligned}
\frac{I(\mathbf{H})_{\text{Diffuse}}}{N} = &\sum_{lmn} C_A C_B |f_A - f_B|^2 \alpha_{lmn} \cos 2\pi \mathbf{H} \bullet (\mathbf{R}_0 - \mathbf{R}_{lmn}) \\
&- \sum_{lmn}[(C_A^2 + C_A C_B \alpha_{lmn}) \text{Re}(f_a(f_A - f_B)^*) \\
&\times \langle X_0^A - X_{lmn}^A \rangle - (C_B^2 + C_A C_B \alpha_{lmn}) \text{Re}(f_B(f_A - f_B)^*) \\
&\times \langle X_0^B - X_{lmn}^B \rangle] \sin 2\pi \mathbf{H} \bullet (\mathbf{R}_0 - \mathbf{R}_{lmn}) \\
&+ \sum_{lmn}\Big[ C_A^2|f_A|^2\left(\langle X^2 \rangle^A - \left(1 + \frac{C_B}{C_A}\alpha_{lmn}\right)\right. \\
&\times \frac{1}{2}\langle (X_0^A - X_{lmn}^A)^2 \rangle\Big) + C_A C_B f_A f_B^* (\langle X^2 \rangle^A + \langle X^2 \rangle^B \\
&- (1 - \alpha_{lmn})\langle (X_0^B - x_{lmn}^A)^2 \rangle) + C_B^2|f_B|^2\Big(\langle X^2 \rangle^B \\
&- \left(1 + \frac{C_A}{C_B}\alpha_{lmn}\right)\frac{1}{2}\langle (X_0^B - X_{lmn}^B)^2 \rangle\Big) \Big] \\
&\times \cos 2\pi \mathbf{H} \bullet (\mathbf{R}_0 - \mathbf{R}_{lmn}) \quad (22)
\end{aligned}$$

Equation 22 is a completely general description of the diffuse scattering from any crystal structure through the second moment of the static and thermal displacements. We now apply this to binary solid solutions, which have received the most attention among crystalline solid solutions.

It is helpful to choose real-space basis vectors **a**, **b**, **c** that reflect the long-range periodicity of the crystal structure. This long-range periodicity in turn is reflected in the intensity distribution in the reciprocal space lattice with basis vectors $\mathbf{a}^* = 1/\mathbf{a}$, $\mathbf{b}^* = 1/\mathbf{b}$, and $\mathbf{c}^* = 1/\mathbf{c}$, as shown in Figure 3. For an alloy that is on average statistically cubic, such as that shown in Figure 3, we define

$$\mathbf{R}_0 - \mathbf{R}_{lmn} \equiv \frac{l}{2}\mathbf{a} + \frac{m}{2}\mathbf{b} + \frac{n}{2}\mathbf{c}, \ \mathbf{H} \equiv h_1\mathbf{a}^* + h_2\mathbf{b}^* \, h_3\mathbf{c}^* \quad (23)$$

so that

$$2\pi\mathbf{H} \bullet (\mathbf{R}_0 - \mathbf{R}_{lmn}) = \pi(h_1 l + h_2 m + h_3 n) \quad (24)$$

In addition,

$$\boldsymbol{\delta}_{lmn} \equiv \Delta X_{lmn}\mathbf{a} + \Delta Y_{lmn}\mathbf{b} + \Delta Z_{lmn}\mathbf{c} \quad (25)$$

so that

$$X_0 - X_{lmn} \equiv 2\pi\mathbf{H} \bullet (\boldsymbol{\delta}_0 - \boldsymbol{\delta}_{lmn}) \equiv 2\pi[h_1(\Delta X_0 - \Delta X_{lmn}) + h_2$$
$$+ (\Delta Y_0 - \Delta Y_{lmn}) + h_3(\Delta Z_0 - \Delta Z_{lmn})] \quad (26)$$

This definition of $\mathbf{R}_0 - \mathbf{R}_{lmn}$ causes the continuous variables $h_1 h_2 h_3$ in reciprocal space to have the integer values of the Miller indices at reciprocal lattice points. We further specify that the site symmetry is cubic such as for the bcc Fe structure and fcc Cu structure. With these definitions, the various diffuse x-ray scattering terms in Equation 22 can be written, starting with the local chemical order term as

$$\frac{I(\mathbf{H})_{\text{SRO}}}{N} = \sum_{lmn} C_A C_B |f_A - f_B|^2 \alpha_{lmn} \cos \pi(h_1 l + h_2 m + h_3 n)$$
$$(27)$$

which was first given by Cowley (1950). For bcc Fe and fcc Cu structures, $\alpha_{lmn} = \alpha_{\overline{lmn}} = \alpha_{\overline{l}mn} = \alpha_{lm\overline{n}} = \alpha_{l\overline{m}n} = \alpha_{\overline{l}\overline{m}n} = \alpha_{\overline{l}m\overline{n}} = \alpha_{lm\overline{n}}$ and the cosine term takes the form $\cos \pi h_1 l \cos \pi h_2 m \cos \pi h_3 n$.

For the first-order displacement term,

$$\frac{I(\mathbf{H})_{j=1}}{N} = -\text{Re}\left(f_A (f_A^* - f_B^*)\right)\left[h_1 Q_x^{AA} + h_2 Q_Y^{AA} + h_3 Q_Z^{AA}\right]$$
$$+ \text{Re}\left(f_B(f_A^* - f_B^*)\right)\left[h_1 Q_X^{BB} + h_2 Q_Y^{BB} + h_3^{BB}\right]$$
$$(28)$$

where

$$Q_X^{AA} = 2\pi \sum_{lmn} \left(C_A^2 + C_A C_B \alpha_{lmn}\right) \langle \Delta X_{lmn}^A \rangle_0^A \sin \pi h_1 l$$
$$\times \cos \pi h_2 m \cos \pi h_3 n$$

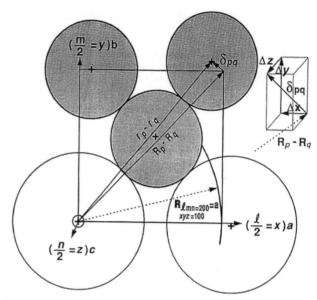

**Figure 23.** The rectangular square of solid lines is the average lattice about which the atom centers (+) are displaced by the amount $\delta_{pq}$. Shown in the smaller box on the right are the rectangular components of the displacement, $\Delta x$, $\Delta y$, and $\Delta z$. Courtesy of Jiang et al. (1996).

and

$$Q_X^{BB} = 2\pi \sum_{lmn} \left(C_B^2 + C_A C_B \alpha_{lmn}\right) \langle \Delta X_{lmn}^B \rangle_0^B \sin \pi h_1 l$$
$$\times \cos \pi h_2 m \cos \pi h_3 n$$

and similarly for the other terms as given by Borie and Sparks (1964) and Georgopoulos and Cohen (1977). Where

$$\langle \Delta X_{lmn}^A \rangle_0^A = \langle \Delta X_0^A - \Delta X_{lmm}^A \rangle$$

Equation 28 is a result given by Borie and Sparks (1964) that avoids an earlier assumption of radial displacements first given by Warren et al. (1951). Shown in Figure 23 is a schematic of the displacements described by Equation 28.

Diffuse scattering from the second-order displacement term can be expanded as,

$$\frac{I(\mathbf{H})_{j=2}}{N} = |f_A|^2 \left(h_1^2 R_X^{AA} + h_2^2 R_Y^{AA} + h_3^2 R_Z^{AA}\right)$$
$$+ f_A f_B \left(h_1^2 R_X^{AB} + h_2^2 R_Y^{AB} + h_3^2 R_Z^{AB}\right)$$
$$+ |f_B|^2 \left(h_1^2 R_X^{BB} + h_2^2 R_Y^{BB} + h_3^2 R_Z^{BB}\right)$$
$$+ |f_A|^2 \left(h_1 h_2 S_{XY}^{AA} + h_1 h^3 S_{XZ}^{AA} + h_2 h_3 S_{YZ}^{AA}\right)$$
$$+ f_A f_B^* \left(h_1 h_2 S_{XY}^{AB} + h_1 h_3 S_{XZ}^{AB} + h_2 h_3 S_{YZ}^{AB}\right)$$
$$+ |f_B|^2 \left(h_1 h_2 S_{XY}^{BB} + h_1 h_3 S_{XZ}^{BB} + h_2 h_3 S_{YZ}^{BB}\right) \quad (29)$$

where

$$R_X^{AA} = 4\pi^2 C_A^2 \sum_{lmn} \left\{ \langle \Delta X^2 \rangle^A - \left(1 + \frac{C_B}{C_A}\alpha_{lmn}\right)\right.$$
$$\times \left[\langle(\Delta X^2)_{lmn}^A\rangle_0^A - \langle \Delta X_0^A \Delta X_{lmn}^A\rangle\right]\right\}$$
$$\times \cos \pi(h_1 l + h_2 m + h_3 n)$$

and similarly for $R_Y$ and $R_Z$ with $Y$ and $Z$ replacing $X$, respectively, or $R_X^{AA}(h_1h_2h_3) = R_Y^{AA}(h_2h_3h_1)$. The $R^{BB}$ terms are given by

$$R_X^{BB} = 4\pi^2 C_B^2 \sum_{lmn} \left\{ \langle \Delta X^2 \rangle^B - \left(1 + \frac{C_A}{C_B}\alpha_{lmn}\right) \right.$$
$$\times \left. \left[ \langle(\Delta X^2)_{lmn}^B \rangle_0^B - \langle \Delta X_0^B \Delta X_{lmn}^B \rangle \right] \right\}$$
$$\times \cos\pi(h_1l + h_2m + h_3n) \tag{30}$$

For the $R^{AB}$ terms, we have

$$R_X^{AB} = 4\pi^2 C_A C_B \sum_{lmn} \left[ \langle \Delta X^2 \rangle^A + \langle \Delta X^2 \rangle^B - (1 - \alpha_{lmn}) \right.$$
$$\times \left. (\langle(\Delta X^2)_{lmn}^A\rangle_0^B + \langle(\Delta X^2)_0^B\rangle_{lmn}^A - 2\langle \Delta X_0^B \Delta X_{lmn}^A \rangle) \right]$$
$$\times \cos\pi(h_1l + h_2m + h_3n) \tag{31}$$

and similar terms for $R_Y^{AB}$ with $Y$ replacing $X$ and for $R_Z^{AB}$ with $Z$ replacing $X$. For the cross terms

$$S_{XY}^{AA} = 4\pi^2 C_A^2 \sum_{lmn} \left(1 + \frac{C_B}{C_A}\alpha_{lmn}\right) \langle \Delta X_0^A - \Delta X_{lmn}^A \rangle$$
$$\times \langle \Delta Y_0^A - \Delta Y_{lmn}^A \rangle \cos\pi(h_1l + h_2m + h_3n) \tag{32}$$

and similarly for $S_{XZ}^{AA}$ and $S_{YZ}^{AA}$ with $XY$ replaced with $XZ$ and $YZ$, respectively. The terms $S_{XY}^{BB}$, $S_{XZ}^{BB}$, and $S_{YZ}^{BB}$ are derived from the $S^{AA}$ terms by replacing AA with BB and by replacing $C_A^2$, with $C_B^2$ and $C_A/C_B$ with $C_B/C_A$. For the term $S_{XY}^{AB}$, we write

$$S_{XY}^{AB} = 8\pi^2 C_A C_B \sum_{lmn} (1 - \alpha_{lmn}) \langle (\Delta X_0^B - \Delta X_{lmn}^A)$$
$$\times (\Delta Y_0^B - \Delta Y_{lmn}^A)\rangle \cos\pi(h_1l + h_2m + h_3n) \tag{33}$$

and similarly for $S_{XZ}^{AB}$ and $S_{YZ}^{AB}$, where $XZ$ and $YZ$ replace $XY$, respectively. The periodicity of the terms in Equations 27, 28, and 29, and the assumption that the scattering factor terms can be made independent of $\mathbf{H}$, permit us to write their sum as

$$\frac{I(\mathbf{H})_{\text{Diffuse}}}{N|f(\mathbf{H})^2|} = A(h_1h_2h_3) + h_1 B(h_1h_2h_3) + h_2 B(h_2h_3h_1)$$
$$+ h_3 B(h_3h_1h_2) + h_1^2 C(h_1h_2h_3) + h_2^2 C(h_2h_3h_1)$$
$$+ h_3^2 C(h_3h_1h_2) + h_1h_2 D(h_1h_2h_3)$$
$$+ h_1h_3 D(h_2h_3h_1) + h_2h_3 D(h_3h_1h_2) \tag{34}$$

where $A(h_1, h_2, h_3)$ is given by Equation 27 $\div |f(\mathbf{H})|^2$, and $B(h_1h_2h_3)$ contains the two terms $-f_A\Delta f^* Q_X^{AA} + f_B\Delta f^* Q_X^{BB}$ given by Equation 28 $\div |f(\mathbf{H})|^2$, and likewise for the other terms.

## APPENDIX B:
## NEUTRON MAGNETIC DIFFUSE SCATTERING

The elastic diffuse scattering of neutrons from binary alloys with magnetic short-range order is composed of three parts: the nuclear scattering, the magnetic scattering, and the nuclear magnetic interference scattering. The nuclear scattering length for neutrons is analogous to the x-ray atomic scattering factor for x rays. Thus the information obtained is the same for chemical short-range order and displacements as discussed for x rays. Because neutrons have a magnetic moment, there is also magnetic scattering associated with the unpaired electron spins (see MAGNETIC NEUTRON SCATTERING). If the direction of magnetization is perpendicular to the scattering plane, the magnetic scattering cross-section (in barns) for an A-B alloy of atomic concentration $C_A$ of A atoms is given by

$$I_M(\mathbf{H}) = C_A C_B T(\mathbf{H})(0.270)^2 \tag{35}$$

Here, $T(\mathbf{H})$ is the moment-moment correlation function expressed in terms of $\mu_{lmn} f_{lmn}(\mathbf{H})$, where $\mu_{lmn}$ is the magnetic moment of the atom on site $lmn$ and $f_{lmn}(\mathbf{H})$ its magnetic form factor (the magnetic scattering comes from the unpaired electrons rather than the nucleus so that it has a scattering angle dependent form factor much like that for x rays):

$$\mu_{lmn} = \mu_{lmn} f_{lmn}(\mathbf{H}) \tag{36}$$
$$C_A C_B T(\mathbf{H}) = \sum_{lmn} e^{2\pi i \mathbf{H}\cdot\mathbf{R}} \langle \mu_{lmn}(\mathbf{H})[\mu_{000}(\mathbf{H}) - \langle\mu(\mathbf{H})\rangle]\rangle \tag{37}$$

The nuclear magnetic interference term $I_{NM}(\mathbf{H})$ is proportional to a site occupation-magnetic moment correlation $M(\mathbf{H})$. For a magnetization perpendicular to the scattering plane, we have

$$I_{NM}(\mathbf{H}) = C_A C_B \Delta b(0.540) M(\mathbf{H}) \tag{38}$$

and

$$C_A C_B M(\mathbf{H}) = \sum_{lmn} e^{2\pi i \mathbf{H}\cdot\mathbf{R}} \langle(\alpha_{lmn} - 1)\mu_{000}(\mathbf{H})\rangle \tag{39}$$

Here the quantity in $<>$'s represents the average increase in the moment of the atom at the origin (000) due to the atomic species of the atom located at site $lmn$.

While the terms $I_M(\mathbf{H})$ and $I_{NM}(\mathbf{H})$ allow one to study the magnetic short-range order in the alloy, they also complicate the data analysis by making it difficult to separate these two terms from the chemical SRO. One experimental method for resolving the magnetic components is to use polarization analysis where the moment of the incident neutron beam is polarized to be either parallel ($\varepsilon = 1$) or antiparallel ($\varepsilon = -1$) to the magnetization. The total scattering for each case can now be written as

$$I_\varepsilon(\mathbf{H}) = I_N(\mathbf{H}) + \varepsilon I_{NM}(\mathbf{H}) + I_M(\mathbf{H}) \tag{40}$$

The intensity difference between the two polarization states gives

$$\Delta I_{\text{Total}}(\mathbf{H}) = 2I_{NM}(\mathbf{H}) \tag{41}$$

and the sum gives

$$\sum_\varepsilon I_\varepsilon(\mathbf{H}) + 2I_N(\mathbf{H}) + 2I_M(\mathbf{H}) \tag{42}$$

If $I_N(\mathbf{H})$ is known from a separate measurement with x rays, all three components of the scattering can be separated from one another.

One of the greatest difficulties in studying magnetic short-range order comes when the moments of the atoms cannot be aligned in the same direction with, for example, an external magnetic field. In the above analysis, it was assumed that the moments are perpendicular to the scattering vector, $\mathbf{H}$. The magnetic scattering cross-section is reduced by the sine of the angle between the magnetic moment and the scattering vector. Thus if the magnetization is not perpendicular to the scattering vector, the moments on the atoms must be reduced by the appropriate amount. When the spins are not aligned, the sine of the angle between the moment and the scattering vector for each individual atom must be considered. In this case, it becomes necessary to construct computer models of the spin structure to extract $M(\mathbf{H})$ and $T(\mathbf{H})$. More in-depth discussion is given in MAGNETIC NEUTRON SCATTERING.

GENE E. ICE
JAMES L. ROBERTSON
CULLIE J. SPARKS
Oak Ridge National Laboratory
Oak Ridge, Tennessee

# RESONANT SCATTERING TECHNIQUES

## INTRODUCTION

This unit will describe the principles and methods of resonant (anomalous) x-ray diffraction as it is used to obtain information about the roles of symmetry and bonding on the electronic structure of selected atoms in a crystal. These effects manifest themselves as crystal orientation-dependent changes in the diffracted signal when the x-ray energy is tuned through the absorption edges for those atoms. Applications of the method have demonstrated it is useful in: (1) providing site-specific electronic structure information in a solid containing the same atom in several different environments; (2) determining the positions of specific atoms in a crystal structure; (3) distinguishing electronic from magnetic contributions to diffraction; (4) isolating and characterizing multipole contributions to the x-ray scattering that are sensitive to the local environment of atoms in the solid; and (5) studying the ordering of $3d$ electron orbitals in the transition metal oxides.

These effects share a common origin with x ray resonant magnetic scattering. Both are examples of anomalous dispersion, both are strongest near the absorption edge of the atoms involved, and both display interesting dependence on x-ray polarization.

Resonant electronic scattering is a microscopic analog of the optical activity familiar at longer wavelengths. Both manifest as a polarization-dependent response of x-ray or visible light propagation through anisotropic media. The former depends on the local environment (molecular point symmetry) of individual atoms in the crystalline unit cell, while the latter is determined by the point symmetry of the crystal as a whole (Belyakov and Dmitrienko, 1989, Section 2). This analogy led to a description of "polarized anomalous scattering" with optical tensors assigned to individual atoms (Templeton and Templeton, 1982). These tensors represent the anisotropic response when scattering involves electronic excitations between atomic core and valence states that are influenced by the point symmetry at the atomic site. In the simplest cases (dipole-dipole), these tensors may be visualized as ellipsoidal distortions of the otherwise isotropic x-ray form factor. Symmetry operations of the crystal space group that affect ellipse orientation can modify the structure factors, leading to new reflections and angle dependencies.

## Materials Properties that Can Be Measured

Resonant x-ray diffraction is used to measure the electronic structure of selected atoms in a crystal and for the direct determination of the phases of structure factors. Results to date have been obtained in crystalline materials possessing atoms with x-ray absorption edges in an energy range compatible with Bragg diffraction. As shown by examples below (see Principles of the Method), resonant diffraction is a uniquely selective spectroscopy, combining element specificity (by selecting the absorption edge) with site selectivity (through the choice of reflection). The method is sensitive to the angular distribution of empty states near the Fermi energy in a solid. Beyond distinguishing common species at sites where oxidation and/or coordination may differ, it is sensitive to the orientation of molecules, to deviations from high symmetry, to the influence of bonding, and to the orientation (spatial distribution) of $d$ electron orbital moments when these valence electrons are ordered in the lattice (Elfimov et al., 1999; Zimmermann et al., 1999).

The second category of measurement takes advantage of differences in the scattering from crystallographically equivalent atoms. Because of its sensitivity to molecular orientation, resonant scattering can lead to diffraction at reflections "forbidden" by screw-axis and glide-plane crystal symmetries (SYMMETRY IN CRYSTALLOGRAPHY). Such reflections can display a varying intensity as the crystal is rotated in azimuth about the reflection vector. This results from a phase variation in the scattering amplitude that adds to the position-dependent phase associated with each atom. This effect can lead to the determination of the position-dependent phase in a manner analogous to the multiple-wavelength anomalous diffraction (MAD) method of phase determination.

Secondary applications involve the formulation of corrections used to analyze anomalous scattering data. This includes accounting for orientation dependence in MAD measurements (Fanchon and Hendrickson, 1990) and for the effects of birefringence in measurements of radial distribution functions of amorphous materials (Templeton and Templeton, 1989).

## Comparison with Other Methods

The method is complementary to standard x-ray diffraction because it is site selective and can provide direct information on the phase of structure factors (Templeton and

Templeton, 1992). A synchrotron radiation source is required to provide the intense, tunable high-energy x-rays required. Highly polarized beams are useful but not required.

Perhaps the technique most closely related to resonant diffraction is angle-dependent x-ray absorption. However, as pointed out by Brouder (1990), except in special cases, the absorption cross-section reflects the full crystal (point) symmetry instead of the local symmetry at individual sites in the crystal. In diffraction, the position-dependent phase associated with each atom permits a higher level of selectivity. This added sensitivity is dramatically illustrated by (but not limited to) reflections that violate the usual extinction rules.

Resonant diffraction thus offers a distinct advantage over spectroscopy methods that provide an average picture of the material. It offers site specificity, even for atoms of a common species, through selection of the diffraction and/or polarization conditions. When compared to electron probe spectroscopies, the method has sensitivity extending many microns into the material, and does not require special sample preparation or vacuum chambers.

Drawbacks to the method include the fact that signal sizes are small when compared to standard x-ray diffraction, and that the experimental setup requires a good deal of care, including the use of a tunable monochromator and in some cases a polarization analyzer.

### General Scope

This unit describes resonant x-ray scattering by presenting its principles, illustrating them with several examples, providing background on experimental methods, and discussing the analysis of data. The Principles of the Method section presents the classical picture of x-ray scattering, including anomalous dispersion, followed by a quantum-mechanical description for the scattering amplitudes, and ends with a discussion on techniques of symmetry analysis useful in designing experiments. Two examples from the literature are used to illustrate the technique. This is followed by a discussion of experimental aspects of the method, including the principles of x-ray polarization analysis. A list of criteria is given for selecting systems amenable to the technique, along with an outline for experimental design. A general background on data collection and analysis includes references to sources of computational tools and expertise in the field.

## PRINCIPLES OF THE METHOD

Resonant x-ray diffraction is used to refine information on the location of atoms and the influence of symmetry and bonding for specific atoms in a known crystal structure. Measurements are performed in a narrow range of energy around the absorption edges for the atoms of interest. The energy and crystal structure parameters determine the range of reflections (regions in reciprocal space) accessible to measurement. The point symmetry at the atom location selects the nature (i.e., the multipole character) of the resonance-scattering amplitude for that site. This amplitude is

described by a symmetric tensor, and the structure factor is constructed by summing these tensors, along with the usual position-dependent phase factor, for each site in the unit cell. The tensor structure factor is sensitive to the nonisotropic nature of the scattering at each site and this leads to several important phenomena: (1) the usual extinction rules that limit reflections (based on an isotropic response) can be violated; (2) the x-ray polarization may be modified by the scattering process; (3) the intensity of reflections may vary in a characteristic way as the crystal is rotated in azimuth (maintaining the Bragg condition) about the reflection vector; and (4) these behaviors may have a strong dependence on x-ray energy near the absorption edge.

### Theory

The classical theory of x-ray scattering, carefully developed in James' book on diffraction (James, 1965), makes clear connections between the classical and quantum mechanical descriptions of anomalous scattering. The full description of resonant (anomalous) scattering requires a quantum-mechanical derivation beginning with the interaction between electrons and the quantized electromagnetic field, and yielding the Thomson and resonant electronic scattering cross-sections. An excellent modern exposition is given by Blume (1985). Several good discussions on the influence of symmetry in anomalous scattering, including an optical model of the anistropy of x-ray susceptibility, are given by Templeton (1991), Belyakov and Dmitrienko (1989), and Kirfel (1994). The quantum mechanical description of symmetry-related effects in anisotropic scattering is included in the work by Blume (1994) on magnetic effects in anomalous dispersion. This has been extended in a sophisticated, general group-theoretical approach by Carra and Thole (1994).

We describe x-ray diffraction by first considering the atomic scattering factor in classical terms and the corrections associated with anomalous scattering. This is followed by the quantum description, and finally the structure factor for resonant diffraction is calculated.

### Classical Picture

The scattering of x rays from a single free electron is given by the Thomson scattering cross-section; the geometry is illustrated in Figure 1.

The oscillating electric field ($\mathbf{E}$) of the incident x-ray, with wave vector $\mathbf{k}$ and frequency $\omega$, causes a sinusoidal

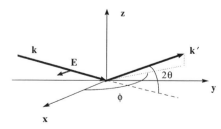

**Figure 1.** Geometry of x-ray scattering from a single electron. The notation is explained in the text.

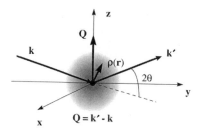

**Figure 2.** The geometry used in the calculation of the atomic scattering form factor for x rays. The diffraction vector is also defined and illustrated.

acceleration of the electron (with charge $-e$) along the field direction. This produces a time-dependent dipole moment of magnitude $e^2/(m\omega^2)$, which radiates an electromagnetic field polarized so that no radiation is emitted along the direction $\mathbf{E}$. This polarization effect (succinctly given in terms if the incident and scattered polarization directions as $\varepsilon \bullet \varepsilon'$) yields a $\sin \phi$ dependence where $\phi$ is the angle between the incident polarization and the scattered wave vector $\mathbf{k}'$. The Thomson scattered radiation is 180° out of phase with the incident electric field, and has an intensity

$$I_{\text{Thomson}} = I_0[(e^2/mc^2)\sin \phi/R]^2 \qquad (1)$$

where the incident intensity $I_0 = |\mathbf{E}|^2$, $e$, and $m$ are the electron charge and mass, $c$ the speed of light, and $R$ is the distance between the electron and field point where scattered radiation is measured.

When electrons are bound to an atom, the x-ray scattering is changed in several important ways. First, consider an electron cloud around the nucleus as shown in Figure 2. James treats the cloud as a continuous, spherically symmetric distribution with charge density $\rho(\mathbf{r})$ that depends only on distance from the center. The amplitude for x rays scattered from an extended charge distribution into the $\mathbf{k}'$ direction is calculated by integrating over the distribution, weighting each point with a phase factor that accounts for the path length differences between points in the cloud. This is given by

$$f_0 = \int \rho(\mathbf{r})e^{i\mathbf{Q}\cdot\mathbf{r}}dV \qquad (2)$$

with $\mathbf{Q}$ the reflection vector of magnitude $|\mathbf{Q}| = 4\pi \sin \theta/\lambda$, and $2\theta$ and $\lambda$ the scattering angle and wavelength, respectively.

The atomic scattering form factor, $f_0$, defined by this integral, is equal to the total charge in the cloud when the phase factor is unity (in the forward scattering direction), and approaches zero for large $2\theta$. Values of $f_0$ versus $\sin \theta/\lambda$ for most elements are provided in tables and by algebraic expressions (MacGillavry and Rieck, 1968; Su and Coppens, 1997). The form factor is given in electron units, where one electron is the scattering from a single electron.

An important assumption in this chapter is that the x-ray energy is unchanged by the scattering (i.e., elastic or coherent). Warren (1990) describes the relation between

the total intensity scattered from an atom and the coherent scattering. He defines the "modified" (i.e., incoherent or Compton) intensity as the difference between the atomic number and the sum of squares of form factors, one for each electron in the atom.

A second effect of binding to the nucleus must be considered in describing x-ray scattering from atomic electrons. This force influences the response to the incident electric field. James gives the equation of motion for a bound electron as

$$d^2\mathbf{x}/d^2t + \kappa d\,\mathbf{x}/dt + \omega_s^2\mathbf{x} = (e/m)\mathbf{E}e^{i\omega t} \qquad (3)$$

where $\mathbf{x}$, $\kappa$, $\omega_s$ are, respectively, the electron position, a damping factor, and natural oscillation frequency for the bound electron. Using the oscillating dipole model, he gives the scattered electric field at unit distance in the scattering plane (i.e., normal to the oscillator) as

$$A = e^2/(mc^2)[\omega^2 E/(\omega_s^2 - \omega^2 + i\kappa\omega)] \qquad (4)$$

The x-ray scattering factor, $f$, is obtained by dividing $A$ by $-e^2/(mc^2)E$, the scattering from a single electron. This factor is usually expressed in terms of real and imaginary components as

$$f = \omega^2/(\omega^2 - \omega_s^2 - i\kappa\omega) \equiv f' + if'' \qquad (5)$$

This expression, and its extension to the many-electron atom discussed by James, are the basis for a classical description of resonant x-ray scattering near absorption edges. The dispersion of x rays is the behavior of $f$ as a function of $\omega$ (where the energy is $\hbar\omega$), and "anomalous dispersion" refers to the narrow region near resonance ($\omega = \omega_s$) where the slope $df/d\omega$ is positive. Templeton's review on anomalous scattering describes how $f'$ and $f''$ are related to refraction and absorption, and gives insight into the analytic connection between them through the Kramers-Kronig relation.

Returning to the description of scattering from all the electrons in an atom, it is important to recognize that the x-ray energies required in most diffraction measurements correspond to $\omega \gg \omega_s$ for all but the inner-shell electrons of most elements. For all but these few electrons, $f$ is unity and their contribution to the scattering is well described by the atomic form factor. The scattering from a single atom is then given by

$$f = f_0 + f' + if'' \qquad (6)$$

To this approximation, the dependence of scattering amplitude on scattering angle comes from two factors. The first is the polarization dependence of Thomson scattering, and second is the result of interference because of the finite size of the atomic charge density which effects $f_0$. This is the classical description assuming the x-ray wavelength is larger than the spatial extent of the resonating electrons. We will find a very different situation with the quantum-mechanical description.

## Quantum Picture

The quantum mechanical description of x-ray scattering given by Blume may be supplemented with the discussion by Sakurai (1967). Both authors start with the Hamiltonian describing the interaction between atomic electrons and a transversely polarized radiation field

$$H = \sum_j \left[ (\mathbf{p}_j - (e/c)\mathbf{A}(\mathbf{r}_j))^2/(2m) + \sum_i V(\mathbf{r}_{ij}) \right] \quad (7)$$

where $\mathbf{p}_j$ is the momentum operator and $\mathbf{r}_j$ is the position of the $j$th electron, $\mathbf{A}(\mathbf{r}_j)$ is the electromagnetic vector potential at $\mathbf{r}_j$, and $V(\mathbf{r}_{ij})$ is the Coulomb potential for the electron at $\mathbf{r}_j$ due to all other charges at $\mathbf{r}_i$. We have suppressed terms related to magnetic scattering and to the energy of the radiation field. Expanding the first term gives

$$H' = \sum_j [e^2/(2mc^2)\mathbf{A}(\mathbf{r}_j)^2 - e/(mc)\mathbf{A}(\mathbf{r}_j) \cdot \mathbf{p}_j] \quad (8)$$

The interaction responsible for the scattering is illustrated in Figure 3, and described as a transition from an initial state $|\mathbf{I}\rangle = |\,a; \mathbf{k}, \varepsilon\rangle$ with the atom in state $a$ (with energy $E_a$) and a photon of wave vector $\mathbf{k}$, polarization $\varepsilon$, and energy $\hbar\omega_k$ to a final state $|F\rangle$ with the atom in state $b$ (of energy $E_b$) and a photon of wave vector $\mathbf{k}'$, polarization $\varepsilon'$, and energy $\hbar\omega_{k'}$.

The transition probability per unit time is calculated using time-dependent perturbation theory (Fermi's golden rule) to second order in $\mathbf{A}$ (Sakurai, 1967, Sections 2-4,2-5)

$$W = (2\pi/\hbar) \left| e^2/(2mc^2) \sum_j \langle F/\mathbf{A}(\mathbf{r}_j) \cdot \mathbf{A}(\mathbf{r}_j)|I\rangle + (e/mc)^2 \right.$$
$$\left. \sum_n \left\{ \frac{\langle F| \sum_j \mathbf{A}(\mathbf{r}_j) \cdot \mathbf{p}_j|n\rangle \langle n| \sum_l \mathbf{A}(\mathbf{r}_l) \cdot \mathbf{p}_{lrml}|I\rangle}{E_I - E_n} \right\} \right|^2 \delta(E_I - E_F) \quad (9)$$

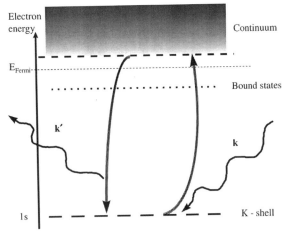

**Figure 3.** Schematic diagram illustrating the electronic transitions responsible for resonant scattering. An x-ray photon of wave vector $\mathbf{k}$ excites a core electron to an empty state near the Fermi level. The deexcitation releases a photon of wave vector $\mathbf{k}'$.

where the delta function insures that energy is conserved between the initial state of energy $E_I = E_a + \hbar\omega_k$, and the final one with $E_F = E_b + \hbar\omega_{k'}$, $|n\rangle$ and $E_n$ are the wave functions and energies of intermediate (or virtual) states of the system during the process.

To get the scattered intensity (or cross-section), $W$ is multiplied by the density of allowed final states and divided by the incident intensity. Blume uses a standard expansion for the vector potential, and modifies the energy resonant denominator for processes that limit the scattering amplitude. Sakurai introduces the energy width parameter, $\Gamma$, to account for "resonant fluorescence" (Sakurai, 1967). In the x-ray case, the processes that limit scattering are both radiative (such as ordinary fluorescence), and nonradiative (like Auger emission). They are discussed by many authors (Almbladh and Hedin, 1983). When the solid is returned to the initial state $|a\rangle$, the differential scattering cross-section, near resonance, is

$$d^2\sigma/d\Omega\,dE = e^2/(2mc^2) \left| \langle a| \sum_j e^{i\mathbf{Q}\cdot\mathbf{r}_j}|a\rangle \varepsilon' \cdot \varepsilon - \hbar/m \right.$$
$$\left. \sum_n \left[ \frac{\langle a| \sum_j (\varepsilon' \cdot \mathbf{p}_j e^{i\mathbf{k}'\cdot\mathbf{r}j})|n\rangle \langle n| \sum_l (\varepsilon \cdot \mathbf{p}_j e^{i\mathbf{k}\cdot\mathbf{r}_l})|a\rangle}{E_a - En + \hbar\omega_k - i(\Gamma/2)} \right] \right|^2 \quad (10)$$

where $\mathbf{Q} = \mathbf{k} - \mathbf{k}'$. It is useful to express the momentum operator in terms of a position operator using a commutation relation (Baym, 1973, Equations 13-96). The final expression, including an expansion of the complex exponential is given by Blume (1994) as

$$d^2\sigma/(d\Omega\,dE)$$
$$= e^2/(2mc^2) \left| \langle a| \sum_j e^{i\mathbf{Q}\cdot\mathbf{r}_j}|a\rangle \varepsilon' \cdot \varepsilon \frac{m}{\hbar^2} \frac{\sum_n (E_a - E_n)^3}{\hbar\omega} \right.$$
$$\left. \times \frac{\langle a| \sum_j (\varepsilon' \cdot \mathbf{r}_j)(1 - i\mathbf{k}' \cdot \mathbf{r}_j/2)|n\rangle \langle n| \sum_l (\varepsilon \cdot r_l)(1 + i\mathbf{k} \cdot \mathbf{r}_l/2)|a\rangle}{E_a - E_n + \hbar\omega_k - i(\Gamma/2)} \right|^2 \quad (11)$$

The amplitude, inside the squared brackets, should be compared to the classical scattering amplitude given in Equation 6. The first term is equivalent to the integral which gave the atomic form factor; the polarization dependence is that of Thomson scattering. The second term, the resonant amplitude, is the same form as Equation 5

$$\omega^2/\omega_k^2 - \omega_{ca}^2 - i(\Gamma/\hbar\omega)_k \quad (12)$$

Now, however, the ground-state expectation value,
$\langle a|$ transition operator $|a\rangle$, can have a complicated polarization and angular dependence.

The transition operator is a tensor that increases in complexity as more terms in the expansion of the exponentials (Equation 10) are kept. Blume (1994) and Templeton (1998) identify the lowest-order term (keeping only the 1 yields a second-rank tensor) as a dipole excitation and dipole deexcitation. Next in complexity is the dipole-quadrupole terms (third-rank tensors); this effect has been observed by Templeton and Templeton (1994). The next order (fourth-rank tensor) in scattering can be dipole-octapole or quadrupole-quadrupole; the last term has been measured by Finkelstein (Finkelstein et al., 1992). An important point is that different multipole contributions make the scattering experiment sensitive to different features of the resonating atoms involved. The atom site symmetry, the electronic structure of the outer orbitals, and the orientational relationship between symmetry equivalent atoms combine to influence the scattering.

## A METHOD FOR CALCULATION

Two central facts pointed out by Carra and Thole (1994) guide our evaluation of the resonant scattering. The first is a selection rule; the scattering may exist only if the transition operator is totally symmetric (i.e., if it is invariant under all operations which define the symmetry of the atomic site). This point is explained in Landau and Lifshitz (1977) and many books on group theory (Cotton, 1990). The second point is that the Wigner-Eckart Theorem permits separation of the scattering amplitude into products of angle- and energy-dependent terms. The method outlined below involves an evaluation of this angular dependence. We cannot predict the strength of the scattering (the energy-dependent coefficients); this requires detailed knowledge of electronic states that contribute to the scattering operator.

The first task is to determine the multipole contributions to resonant scattering that may be observable at sites of interest in the crystal; the selection rule provides this information. The rule may be cast as an orthogonality relation between two representations of the point symmetry. Formula 4.3–11 in Cotton (1990) gives the number of times the totally symmetric representation $(A_1)$ exists in a representation of this point symmetry. It is convenient to choose spherical harmonics as basis functions for the representation because harmonics with total angular momentum $L$ represent tensors of rank $L$.

First calculate the trace $\chi(\mathbf{R}i)$ of each symmetry operation, $\mathbf{R}i$, of the point group. Because the trace of any symmetry operation of the totally symmetric representation equals 1, Cotton's formula reduces to

$$N(A_1) = (1/h) \sum_i [\chi(\mathbf{R}i)] \qquad (13)$$

where $N(A_1)$ is the number of times the totally symmetric representation is found in the candidate representation, $h$ is the number of symmetry operations in the point group, and the sum is over all group operations. If the result is nonzero, this number equals the number of independent tensors (of rank $L$) needed to describe the scattering.

The second task is to construct the angle-dependent tensors that describe the resonant scattering. There are a number of approaches (Belyakov and Dmitrienko, 1989; Kirfel, 1994; Blume, 1994; Carra and Thole, 1994), but in essence this problem is familiar to physicists, being analogous to finding the normal modes of a mechanical system, and to chemists as finding symmetry-adapted linear combinations of atomic orbitals that describe states of a molecule. To calculate where, in reciprocal space, this scattering may appear, and to predict the polarization and angular dependence of each reflection, we use spherical harmonics to construct the angle-dependent tensors. This choice is essential for the separation discussed above, and simplifies the method when higher-order (than dipole-dipole) scattering is considered.

Our method uses the projection operator familiar in group theory (see Chapter 5 in Cotton, 1990; also see Section 94 in Landau and Lifshitz, 1977). We seek a tensor of rank $L$, which is totally symmetric for the point group of interest. The projection operator applied to find a totally symmetric tensor is straightforward; we average the result of applying each symmetry operation of the group to a linear combination of the $2L + 1$ spherical harmonics with quantum number $L$, i.e.,

$$f(L) = (1/h) \sum_n \left[ \mathbf{R}(n) \left\{ \sum_m a_m Y(L,m) \right\} \right] \qquad (14)$$

where $a_m$ is a constant which may be set equal to 1 for all $m$, and $Y(L,m)$ is the spherical harmonic with quantum numbers $L$ and $m$. If our "first task" indicated more than one totally symmetric contribution to the scattering amplitude, the other tensors must be orthogonal to $f(L)$ and again totally symmetric.

The third task is to calculate the contribution this atom makes to the structure factor. In calculating $f(L)$, we assumed an oriented coordinate system to evaluate the symmetry operations. If this orientation is consistent with the oriented site symmetry for this atom and space group in the International Tables (Hahn, 1989), then the listed symmetry operations can be used to transform the angle-dependent tensor from site to site in the crystal. The structure factor of the atoms of site $i$, for reflection $\mathbf{Q}$ is written

$$F(i; \mathbf{Q}) = \sum_j \left[ \left\{ f_{0j} + \sum_p f_{pj}(L) \right\} e^{i\mathbf{Q} \cdot \mathbf{r}_j} \right] \qquad (15)$$

where $f_{0j}$ is the nonresonant form factor for this atom, and $f_{pj}(L)$ is the $p$th resonant scattering tensor for the atom at $\mathbf{r}_j$; both terms are multiplied by the usual position-dependent phase factor. The tensor transformations are familiar in Cartesian coordinates as a multiplication of matrices (Kaspar and Lonsdale, 1985, Section 2.4). Here, having used spherical harmonics, it is easiest to use standard formulas (Sakurai, 1985) for rotations and simple rules for reflection and inversion given in APPENDIX A of this unit.

When more then one atom contributes to the scattering, these structure factors are added together to give a total $F(\mathbf{Q})$.

The final task is to calculate the scattering as a function of polarization and crystal orientation. For a given $L$, our resonant $F_L(\mathbf{Q})$ is a compact expression

$$F_L(\mathbf{Q}) = \sum_m [b_m Y(L,m)] \qquad (16)$$

with $b_m$ the sum of contributions to the scattering with angle dependence $Y(L,m)$. To evaluate the scattering, we express each $Y(L,m)$ as a sum of products of the vectors in the problem. The products are obtained by standard methods using Clebsch-Gordan coefficients, as described in Sakurai (1985, Section 3.10). For $L = 2$ (dipole-dipole) scattering, the two vectors are the incident and scattered polarization, for $L = 3$ a third vector (the incident or outgoing wave vector) is included, and for $L = 4$ both wave vectors are used to characterize the scattering. The crystal orientation (relative to these defining vectors) is included by applying the rotation formula (see APPENDIX C in this article) to $F_L(\mathbf{Q})$.

### Illustrations of Resonant Scattering

The following two examples illustrate the type of information available from these measurements; the first describes an experiment sensitive to $L = 2$ scattering, while the second presents results from $L = 4$ work.

***L = 2* measurements.** Wilkinson et al. (1995) illustrate how resonant scattering can be used to separately measure parameters connected to the electronic state of gold at distinct sites. The diffraction measurements were performed at the gold $L_{III}$ edge in Wells' salt (Cs$_2$[AuCl$_2$] [AuCl$_4$]). Both sites, I and III in Figure 4, have $D_{4h}$ symmetry, but the first has a linear coordination of neighbors (along the tetragonal $c$ axis), while the coordination is square planar (parallel to the crystalline $a$-$b$ plane) at the second site.

Application of Equation 13 shows that one $L = 2$ tensor is required (for each of the two sites) to describe the anisotropic scattering. The construction formula, Equation 14, gives the angular dependence, which is proportional to $Y(2,0)$. This agrees with the anisotropic (the traceless symmetric) part of the matrix given (in Cartesian form) by

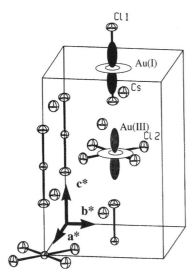

**Figure 4.** The crystal structure of Wells' salt contains two sites for gold; the labeling indicates the formal valence of the two atoms. At each of these two sites, the lobe-shaped symbol indicates the angular dependence for resonant scattering of rank $L = 2$.

Wilkinson et al. (1995) in their Equation 1. Subtracting one-third of the trace of this matrix from each component gives $f_{xx} = f_{yy} = -2 \cdot f_{zz}$, which is equivalent to $Y(2,0) \propto 2z^2 - (x^2 + y^2)$. Further discussion on the angular dependence for dipole scattering is found in APPENDIX B of this article. The angular function at each gold site is represented by the lobe-shaped symbols in Figure 4. This angular dependence is then used to calculate the structure factor, which is a sum of three terms: one for each of the two gold atomic sites and a third for the remaining atoms in the unit cell. The 16 symmetry operations of this space group (#139 in Hahn, 1989) leave the angular function unchanged, so the structure factor for the two gold sites may be written

$$\mathbf{F_{gold}} = [\mathbf{Q}(h, k, l)] \propto \cos[(\pi/2)(h + k + l)]$$
$$\times \{2f_0\mathbf{I}\cos[\pi l/2] + \mathbf{F_I}e^{-i(\pi l/2)} + \mathbf{F_{III}}e^{+i(\pi l/2)}\} \quad (17)$$

where $\mathbf{F_{gold}}$, $\mathbf{F_I}$, $\mathbf{F_{III}}$, and $\mathbf{I}$ are second-rank tensors and the reciprocal lattice vector $\mathbf{Q}(h, k, l) = h\mathbf{a}^* + k\mathbf{b}^* + l\mathbf{c}^*$. $\mathbf{I}$, the diagonal unit tensor, represents the isotropic nature of the gold atom form factor. Wilkinson et al. (1995) point out that the resonant scattering is proportional to the sum (difference) of $\mathbf{F_I}$, $\mathbf{F_{III}}$ for reflections with l = even (odd), so that separation of the two quantities requires data for both types of reflections. Figure 4 shows that $\mathbf{F_I}$ and $\mathbf{F_{III}}$ share a common orientation in the unit cell. The tensor structure factor is thus easily described in a coordinate system that coincides with the reciprocal lattice. Wilkinson et al. (1995) use $\sigma$ for the direction along $\mathbf{c}^*$ and $\pi$ for directions in the basal plane. To avoid confusion with notation for polarization, we will use $s$ and $p$ instead of $\sigma$ and $\pi$.

With the crystal oriented for diffraction, the resonant scattering amplitude depends on the orientation of the tensors with respect to the polarization vectors. $\mathbf{F_{gold}}(\mathbf{Q})$ should be re-expressed in the coordinate system where these vectors are defined (see APPENDIX C in this article). The amplitude is then written

$$A = \varepsilon' \cdot \mathbf{F}_{lab}(\mathbf{Q}) \cdot \varepsilon \quad (18)$$

where $\varepsilon'$ and $\varepsilon$ are the scattered and incident polarization vectors and matrix multiplication is indicated. The amplitude expressed in terms of the four (2-incoming, 2-outgoing) polarization states is shown in Table 1, where $F_{ij}$ are cartesian tensor components in the new (laboratory) frame, and $2\theta$ is the scattering angle.

This scheme for representing the scattering amplitude makes it easy to predict basic behaviors. Consider a vertical scattering plane with $\mathbf{Q}$ pointing up, and the incident

**Table 1. Amplitude Expressed in Terms of the Four (2-Incoming, 2-Outgoing) Polarization States[a]**

|  | $\varepsilon = \sigma$ | $\varepsilon = \pi$ |
|---|---|---|
| $\varepsilon' = \sigma'$ | $F_{11}$ | $F_{13}$ |
| $\varepsilon' = \pi''$ | $F_{31}\cos(2\theta)-$ | $F_{21}\sin(2\theta)-$ |
|  | $F_{33}\cos(2\theta)$ | $F_{23}\sin(2\theta)$ |

[a]$\sigma(\pi)$ points normal (parallel) to the scattering plane.

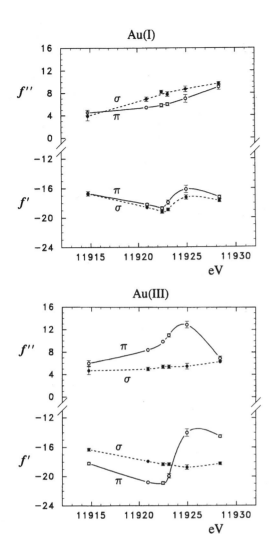

**Figure 5.** The energy dependence of the real and imaginary parts of the anomalous scattering from the two gold sites in Wells' salt is given in electron units. The symbols $\sigma$ and $\pi$ (referred to in the text as $s$ and $p$) correspond to the tensor components along the $\mathbf{c}^*$ axis and in the basal plane in Figure 4.

Their results, plotted versus energy in Figure 5, show a clear distinction between the two sites. The gold (III) site exhibits a strongly anisotropic near-edge behavior (the difference $|f_{\mathrm{p}} - f_{\mathrm{s}}|$ is >15% of the total resonant scattering) which the authors suggest may result from an electronic transition between the $2p_{3/2}$ and empty $3d_{x^2-y^2}$ orbitals. The gold(I) site on the other hand would (formally) have a filled $d$ shell, and in fact shows a much smaller anisotropic difference (<4%).

***L = 4* measurements**. The second example illustrates $L = 4$ scattering where polarization analysis is critical to measurement success.

The measurements were performed on a single crystal of hematite ($Fe_2O_3$), at x-ray energies corresponding to the iron K-edge (Finkelstein et al., 1992). The results show: (1) resonant scattering is a sensitive measure of crystal field splitting of the iron $3d$ level; (2) an intensity variation with azimuth identifies the symmetry at the iron site and separates chemical from magnetic contributions to the scattering; and (3) that polarization analysis is a powerful tool for selection and quantitative evaluation of these signals.

The structure of hematite is trigonal (space group #167; Hahn, 1989), with a single iron site of symmetry $C_3$, positioned at $c/6$ separation along the crystalline $c$ axis. The structure, illustrated in Figure 6, indicates that the oxygen atoms of each molecule are mirror reflected from site to site by 3-fold vertical glide planes.

A single reflection, the $(00.3)_{\mathrm{Hexagonal}}$ (probing lattice planes of $c/3$ spacing normal to the $c$ axis), was measured

polarization ($\sigma$) normal to the plane. For any $\mathbf{Q}(h,k,l)$, one can rotate the crystal around that reflection vector until the donut shaped $p$ lobe of the tensor is along the incident polarization; the measurement is then most sensitive to this feature. Only for $\mathbf{Q}(h,k,0)$ reflections can scattering from the $s$ lobe be isolated. This analysis also suggests the benefits of selectively measuring the polarization of the scattered beam. By looking only at the $\sigma'$-scattered polarization, one can, in the first case, completely separate out $p$-lobe scattering; in the second case both contributions can be determined (by the azimuth scan; see Practical Aspects of the Method) at a single reflection. By selecting the $\pi'$-scattered polarization, one eliminates the Thomson ($f_0$) contribution at allowed reflections.

Wilkinson et al. (1995) do not use polarization analysis, but instead measure a series of reflections, at angular orientations that favor either tensor component, and extract four values for the anomalous scattering at each gold site (Templeton and Templeton, 1982, 1985, 1992).

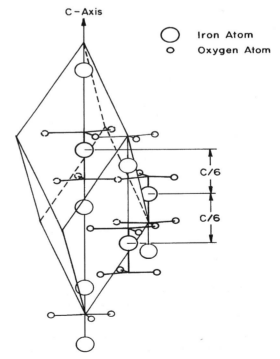

**Figure 6.** The crystal structure of hematite is displayed with a rhombohedral unit cell. Each iron atom is surrounded by an approximate octahedral arrangement of oxygen atoms.

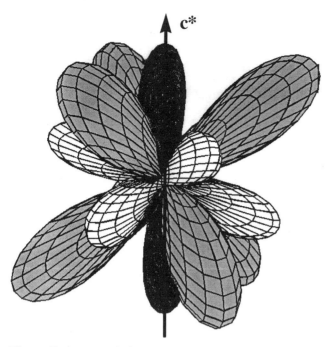

**Figure 7.** A geometrical representation of the angular dependence for $L = 4$ resonant scattering in $O_h$ symmetry. The object has 3-fold rotational symmetry about the **c**$^*$ axis. The black lobes are rotationally symmetric; the scattering amplitude associated with gray and white lobes is opposite in sign.

at a series of energies and azimuth angles around this axis. This reflection is "forbidden" by general conditions of the space group, but magnetically "allowed" because the iron atoms are antiferromagnetically ordered.

The iron has a formal valence of $+3$, implying a half-filled $3d$ shell. The site symmetry is very close to cubic ($O_h$), and to this approximation the $3d$ levels are crystal field split into $e_g$ bonding states (with electron density directed toward oxygen neighbors), and $t_{2g}$ antibonding states (with the density directed between these neighbors). $1s \rightarrow 3d$ resonant scattering is sensitive to the filling and directionality of these states.

In $O_h$ symmetry Equation 10b.3.13, excludes $L = 2$ and 3, but permits $L = 4$ scattering with one tensor contributing. Equation 14 gives an angle-dependent tensor proportional to

$$\sqrt{(7/10)}Y(4,0) + Y(4,3) - Y(4,-3) \qquad (19)$$

The angular properties of this tensor are illustrated in Figure 7.

The shape of this object is a geometrical measure of the sensitivity which this type of scattering has to electronic transitions between the $1s$ and $3d$ states. By considering this object positioned at each iron site, the explanation for 3 out of 4 experimental observations is immediately apparent: (1) the "forbidden" reflection is allowed because the orientation of oxygen atoms changes the sign of the scattering amplitude from site to site; (2) this amplitude changes sign every $60°$ as the crystal is rotated about the $c$ axis; (3) this behavior depends on electronic transi-

tions (with $\Delta L = 2$; quadrupole excitation and deexcitation) and it therefore occurs only at a well-defined resonance energy.

A fourth experimental fact, that the resonant behavior was visible as a rotation of the x-ray polarization (from primarily $\sigma$ to $\pi'$), can be explained when the tensor is expanded using the methods in Sakurai (1985). The quantitative details, worked out by Hamrick (1994) and by Carra and Thole (1994), show that the energy dependences of this resonance are exceedingly sensitive to the magnitude of the crystal field splitting. The resonant scattering amplitude was equivalent to less than about 0.08 electrons! However, by analyzing the scattered beam for the rotated polarization, the signal intensity was found to be 45 times that of the background.

## PRACTICAL ASPECTS OF THE METHOD

### Instrumentation

These measurements are made with synchrotron radiation because an intense, energy-tunable incident x-ray beam is required. It is useful to work in a vertical scattering geometry to take advantage of the small divergence and high degree of horizontal, linear polarization available from synchrotron sources. A standard silicon (111) double crystal monochromator is commonly used for energy tuning. Focusing optics, when used, can deliver substantially more flux to small sample crystals, but can also lead to several problems. In particular, strong horizontal focusing (i.e., sagittal focusing) increases the beam divergence normal to the scattering plane and therefore increases the likelihood of multiple-beam (Renninger) reflection contamination in the scattered signal. In addition, when polarization analysis is employed to improve sensitivity, the analyzer may not fully accept the increased divergence. The use of a vertical reflection mirror for harmonic rejection and focusing (if available) is generally an advantage. Many experiments have been performed with bend magnet radiation, and most of these with an incident energy resolution not better than $1 - 5 \times 10^{-4}$. The ideal monochromator would be tunable, and provide a beam with an energy width and stability matched to the core-hole energy associated with the absorption edge under study.

A four-circle diffractometer is essential for access to a wide range of reflections from one sample. With appropriate software, it permits flexible scanning of reciprocal space, including the important azimuthal scan discussed below. For more challenging experiments, the diffractometer may need to accommodate a polarization analyzer. The technique of polarization analysis is also discussed in this section. It is often useful to simultaneously monitor fluorescence from the sample as diffraction data are collected. In so doing, one may: (1) reduce errors caused by energy drifts (through periodic energy scans that monitor the position of the absorption edge) and (2) conveniently obtain a measurement of $f''$ useful for signal normalization; also (3) as discussed earlier, measurement of the angular behavior of $f''$ gives an essentially independent measure of the average site anisotropy for the absorbing species (Brouder, 1990).

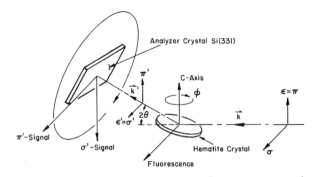

**Figure 8.** The experimental arrangement for resonant scattering measurements may include an analyzer for selecting polarization in the signal beam.

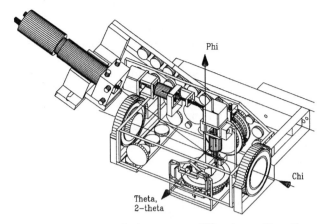

**Figure 9.** An x-ray polarimeter uses diffraction, with a Bragg angle (theta) near 45° to measure the direction of linear polarization in a beam. The device is rotated (in Chi) about the beam for this purpose.

Figure 8 illustrates the geometry of the scattering experiment, including the placement of polarization analyzer and fluorescence detector. The samples are usually single crystals mounted on a eucentric goniometer for convenient orientation of crystal faces. The azimuth scan is performed to measure the anisotropic nature of the atomic environment. The sample is oriented so that the Bragg angle is maintained as it is rotated about the axis of the reflection vector. Because tensor scattering effects are small, one often chooses weak or space-group forbidden reflections for these studies. In this case an appropriate choice of azimuth is important for eliminating the influence of Renninger reflections, which are localized and predictable (Cole et al., 1962) in both energy and angle.

## Polarization Analysis

The principles and methods of x-ray polarization analysis are discussed by several authors (Blume and Gibbs, 1988; Belyakov and Dmitrienko, 1990; Shen and Finkelstein, 1993; Finkelstein et al., 1994). Although tensor scattering effects can appear even with an unpolarized incident beam, it is important to know the beam polarization for accurate analysis of the data. Further, it has been shown (Finkelstein et al., 1992) that polarization analysis of the diffraction signal can significantly improve sensitivity and provide a detailed characterization of the scattering effects. In particular, resonant scattering is one of only a few effects that rotate the incident polarization; thus a search for resonant scattering is facilitated by a polarized incident beam and an analyzer set to accept the orthogonal polarization state.

The linear polarization analyzer, illustrated in Figure 9, is basically a 3-circle diffractometer mounted on the 2-theta (detector) arm of the 4-circle instrument. The heart of the analyzer is a crystal that intercepts signal from the sample and diffracts it into a detector set near 90° to the signal beam. The analyzer can be rotated about the beam by changing chi. When the analyzer and sample diffraction planes are parallel, the measurement is sensitive to σ polarization. When rotated 90° about chi, the analyzer passes the rotated or π polarized radiation. The linear polarization state (relative to these planes) is characterized by the Stokes-Poincaré parameter

$$P_1 = (I_\sigma - I_\pi)/(I_\sigma + I_\pi) \qquad (20)$$

where $P_1$ is $+1$ for a beam fully σ-polarized, and $P_1$ is equal to $-1$ if the beam is fully π polarized. Three parameters are required to completely characterize the beam polarization, but for the purposes of these experiments, knowledge of the incident and scattered $P_1$ is usually sufficient.

Besides possessing high reflectivity, the polarization-analyzing crystal should have an angular acceptance matched to the signal beam and a Bragg angle close to 45° for the energy being measured. If the first condition is not met, the analyzer must be scanned through its reflection (rocking) curve to ensure accurate measurement. When set to pass π-polarized radiation, the analyzer is considerably less sensitive to noise sources. This may help in the search for resonant signals, but is of only limited value if the incident beam horizontal divergence is large. In the case where an energy scan, at constant sample **Q** (also known as a DAFS scan; Stragier, et al., 1992) is required, the monochromator, sample and detector arm (2-theta) angles, as well as the analyzer crystal angle must move in synchronization.

## Experiment Preparation

Resonant scattering experiments are generally nontrivial to perform and require significant advanced planning. No well established techniques exist for estimating signal size, but the method for calculation outlined earlier can provide a map for candidate reflections. Several recent attempts have demonstrated how difficult it is to explore higher multipole scattering processes without employing the methods of polarization analysis. Some general rules, gleaned from the literature, may help define appropriate characteristics for systems under consideration for study.

Sample size and crystal perfection have not generally limited these measurements. Crystal structures with compound symmetry elements (glide planes, and screw axes) are good candidates for study. The resonant scattering from atoms related by these elements may differ because of differences in orientation from atom to atom.

Structure-factor relationships are an important experimental consideration. Forbidden reflections, which are excited at resonance (by the previous mechanism), are a sensitive measure of the local environment. Weak allowed reflections reveal tensor scattering effects through azimuthal and polarization anisotropy. $L = 3$ scattering may occur only from sites with no center of symmetry. Templeton (1994) has shown that this leads to an interesting selectivity of the scattering process.

The width, strength, and orientational dependence of absorption edge features influence the resonant scattering amplitude. Because absorption ($f''$) and refraction ($f'$) are related through the Kramers-Kronig dispersion relation, x-ray dichroism (polarization-dependent absorption) is a good predictor of birefringence (polarization-dependent scattering). In general, the higher the edge energy, the broader the intrinsic (core-hole) width and the weaker the resonant response. $M$ and $L$ edges have yielded the largest signals, followed by first row transition metal $K$ edges. In this regard, another predictor of resonant behavior in diffraction comes from examining the optical activity of the molecules that compose the crystal. The highest-order multipole scattering observed corresponds to $L = 4$. This effect was measured to be less than 0.08 electrons in amplitude, but was visible through polarization analysis. $L = 3$ scattering has ranged in size from 0.5 to 3 electrons, while $L = 2$ scattering amplitudes (including the isotropic component) as large as 30 electrons have been observed.

Multiple beam reflections must be accounted for in experiment design and analysis. It is important to note that this effect can also be responsible for a rotation of the scattered beam polarization.

## Experiment Design

A reasonable approach to the design of these experiments is to begin by constructing maps of reciprocal space. Strong, weak, and forbidden nonresonant reflections should be noted. This is followed by adding the location of resonant reflections and their associated multipole order. The map is then sectioned into regions consistent with the resonant energy and range of scattering angles available. Principle directions (faces of the crystal) should be indicated to help define orientations for azimuthal scanning.

For polarization analysis, the appropriate analyzer crystal will have a strong reflection with approximately a 45° Bragg angle at the resonant energy. The angular acceptance of this crystal should be consistent with beam divergence and it should not fluoresce at the energies of interest. The analyzer should be used with a high count rate detector (ion chamber) to measure the incident beam polarization and can also determine energy resolution and divergence, as well as the absolute energy (*http://www.chess.cornell.edu/technical/hardware/special_setups/energy_analyzer.html*). The incident beam is also used to align the detector for accurate measurement of both polarization states. A low-count-rate detector (scintillation counter) is then installed for the diffraction measurements.

## METHOD AUTOMATION

Single-crystal data are collected with a 4-circle diffractometer and an energy scanning monochrometer. Automated procedures for controlling the diffractometer allow the experimenter to: (1) collect a large numbers of reflections, one energy at a time; (2) perform a scan in azimuth around a reciprocal lattice vector set to maintain the Bragg diffraction condition; and (3) coordinate the diffractometer and monochromator in such a way that an energy scan is performed while one reflection is excited (i.e., at constant **Q**). These capabilities are available through software available at most synchrotron radiation beamlines set up for 4-circle diffraction, with an energy-tunable monochromator.

The first data collection scheme, applied to resonant scattering in work at a synchrotron, was by Phillips (Phillips, et al., 1978) using a CAD-4 diffractometer. Methods of azimuth scanning were developed for surface x-ray diffraction studies (Mochrie, 1988) and are built into 4-circle control programs such as FOURC (Certified Scientific Software). An implementation of this method, used with the FOURC program, is documented on the CHESS World Wide Web site available at *http://www.chess.cornell.edu/CHESS_spec/azimesh.html*. The third data collection mode (sometimes referred to as a DAFS scan) is again built into some commercial control programs. In FOURC the monochromator energy scan ("Escan") permits the optional argument "fixQ" which moves the 4-circle to maintain a constant reflection condition. For information on this scan see *http://www.chess.cornell.edu/Certif/spec/help/mono.html*.

## DATA ANALYSIS AND INITIAL INTERPRETATION

The tools required for a preliminary analysis of resonant scattering measurements have generally been developed by individual investigators. A good, general-purpose scientific graphics and analysis package, which includes a capability for nonlinear least-squares fitting, will suffice for all but the most specialized data manipulations.

Basic corrections to the data include the scaling of signals to account for: (1) changes of incident beam intensity with time and energy; (2) sharp variations in sample absorption because measurements are always taken near absorption edges; (3) standard geometrical corrections to the scattering from different reflections; and (4) efficiency (throughput) of the polarization analyzer. It is also crucial to set up criteria for eliminating data that are too weak (often due to misalignments in monochromator, diffractometer, or analyzer) or too strong (a typical effect when working close to multiple-beam reflections) to be considered reliable.

Accurate measurements of absorption, obtained by monitoring fluorescence from the sample, are extremely useful for several purposes. First, because they may be combined with numerical values for absorption coefficients far from these edges (calculated in programs like those by Cromer and Liberman (Cromer, 1983) available on the WWW at *http://www.cindy.lbl.gov/optical_constants/*)

to calculate the energy dependence of $f'$. These methods are discussed in the Templeton (1991) review of anomalous scattering. Second, the angular dependence of $f''$ (Brouder, 1990) is an important source of supplementary information.

Problems with multiple beam reflections (when measuring weak or "forbidden" reflections) can be reduced by mapping their locations before data are collected.

The work on hematite, discussed earlier, shows how important it is to know the polarization content and energy resolution of the incident beam. The former is required when an analyzer is used to determine polarization-dependent cross-sections. The latter is needed for determining information from resonance widths.

Finally, the angular dependence of azimuthal scans at resonant reflections may be compared to models derived using the calculational method (discussed earlier) by non-linear least-squares analysis.

## ACKNOWLEDGMENTS

The author is indebted to B.T. Thole, R. Lifshitz, M. Hamrick, M. Blume, and S. Shastri for helping him appreciate the theory of resonant scattering. He also thanks A.P. Wilkinson, L.K. Templeton, D.H. Templeton, A. Kirfel, Q. Shen, and E. Fontes, for contributions to the manuscript, and is grateful to E. Pollack, L. Pollack, and A.I. Goldman for their guidance in improving the manuscript. This work is supported by the National Science Foundation through CHESS, under Grant No. DMR 93-11772.

## LITERATURE CITED

Almbladh, C. and Hedin, L. 1983. Beyond the one-electron model: Many-body effects in atoms, molecules, and solids. *In* Handbook on Synchrotron Radiation, Vol. 1B, (E. E. Koch, ed.) North-Holland Publishing, New York.

Baym, G. 1973. Lectures on Quantum Mechanics. Benjamin/Cummings, Reading, Mass.

Belyakov, V. A. and Dmitrienko, V. E. 1989. Polarization phenomena in x-ray optics. *Sov. Phys. Usp.* 32:697–718.

Blume, M. 1985. Magnetic scattering of x-rays. *J. Appl. Phys.* 57:3615–3618.

Blume, M. 1994. Magnetic effects in anomalous scattering. *In* Resonant Anomalous X-Ray Scattering (G. Materlik, C. J. Sparks, and K. Fischer, eds.). pp. 495–512. North-Holland, Amsterdam, The Netherlands.

Blume, M. and Gibbs, D. 1988. Polarization dependence of magnetic x-ray scattering. *Phys. Rev. B* 37:1779–1789.

Brouder, C. 1990. Angular dependence of x-ray absorption spectra. *J. Phys. Condens. Matter* 2:701–738.

Busing, W. R. and Levy, H. A. 1967. Angle calculations for 3- and 4-circle x-ray and neutron diffractometers. *Acta Crystallogr.* 22:457–464.

Carra, P. and Thole, B. T. 1994. Anisotropic x-ray anomalous diffraction and forbidden reflections. *Rev. Mod. Phys.* 66:1509–1515.

Cole, H., Chambers, F. W., and Dunn, H. M. 1962. Simultaneous diffraction: Indexing Umweganregung peaks in simple cases. *Acta Crystallogr.* 15:138–144.

Cotton, F. A. 1990. Chemical Applications of Group Theory 3rd Edition. John Wiley & Sons, New York.

Cromer, D. T. 1983. Calculation of anomalous scattering factors at arbitrary wavelengths, *J. Appl. Crystallogr.* 16:437.

Elfimov, I. S. Anisimov, V. I., and Sawatzky, G. A. 1999. Orbital ordering, Jahn-Teller distortion, and anomalous x-ray scattering in manganates. *Phys. Rev. Lett.* 82:4264–4267.

Fanchon, E. and Hendrickson, W. A. 1990. Effects of the anisotropy of anomalous scattering on the MAD phasing method. *Acta Crystallogr.* A46:809–820.

Finkelstein, K. D., Shen, Q., and Shastri, S. 1992. Resonant x-ray diffraction near the iron k-edge in hematite ($Fe_2O_3$). *Phys. Rev. Lett.* 69:1612–1615.

Finkelstein, K., Staffa, C., and Shen, Q. 1994. A multi-purpose polarimeter for x-ray studies. *Nucl. Instr. Meth.* A 347:124–127.

Hahn, T. (ed.) 1989. International Tables for Crystallography, Vol. A: Space-Group Symmetry, 2nd ed. Kluwer Academic Publishers, Dordrecht, The Netherlands.

Hamrick, M. D. 1994. Magnetic and Chemical Effects in X-ray Resonant Exchange Scattering in Rare Earths and Transition Metal Compounds. Ph.D. thesis, Rice University, Houston, Tex.

James, R. W. 1965. The Optical Principles of the Diffraction of X-rays. Cornell University Press, Ithaca, N.Y.

Kaspar, J. S. and Lonsdale, K. (eds.). 1985. International Tables for X-Ray Crystallography, Vol. II. Reidel Publishing Company, Dordrecht, The Netherlands.

Kirfel, A. 1994. Anisotropy of anomalous scattering in single crystals in Resonant Anomalous X-ray Scattering: Theory and Applications, (G. Materlik, C. J. Sparks, and K. Fischer, eds.).. Elsevier Science Publishing, New York.

Landau, L. D. and Lifshitz, E. M. 1977. Quantum Mechanics (Non-relativistic Theory). Pergamon Press Ltd., Elmsford, New York.

MacGillavry, C. H. and Rieck, G. D. (eds.). 1968. International Tables for X-ray Crystallography, Vol. III: Physical and Chemical Tables. Kynoch Press, Birmingham, U.K.

Mochrie, S. G. J. 1988. Four-circle angle calculations for surface diffraction. *J. Appl. Crystallogr.* 21:1–3.

Nowick, A. S. 1995. Crystal Properties Via Group Theory. Cambridge University Press, Cambridge.

Phillips, J. C., Templeton, D. H., Templeton, L. K., and Hodgson, K. O. 1978. LIII-edge anomalous x-ray scattering by cesium measured with synchrotron radiation. *Science* 201:257–259.

Sakurai, J. J. 1967. Advanced Quantum Mechanics, Chapter 2. Addison-Wesley, Reading, Mass.

Sakurai, J. J. 1985. Modern Quantum Mechanics. Benjamin/Cummings, Menlo Park, Calif.

Shen, Q. and Finkelstein, K. 1993. A complete characterization of x-ray polarization state by combination of single and multiple bragg reflections. *Rev. Sci. Instrum.* 64:3451–3455.

Stragier, H., Cross, J. O., Rehr, J. J., and Sorensen, L. B. 1992. Diffraction anomalous fine structure: A new x-ray structural technique. *Phys. Rev. Lett.* 69:3064–3067.

Su, Z. and Coppens, P. 1997. Relativistic x-ray elastic scattering factors. *Acta. Crystallogr.* A53:749–762.

Templeton, D. H. 1991. Anomalous Scattering. *In* Handbook on Synchrotron Radiation, Vol. 3, (G. Brown and D. E. Moncton, eds.).. Elsevier Science Publishing, New York.

Templeton, D. H. 1998. Resonant scattering tensors in spherical and cubic symmetries. *Acta Crystallogr.* A54:158–162.

Templeton, D. H. and Templeton, L. K. 1982. X-ray dichroism and polarized anomalous scattering of the uranyl ion. *Acta Crystallogr.* A38:62–67.

Templeton, D. H. and Templeton, L. K. 1985. Tensor x-ray optical properties of the bromate ion. *Acta Crystallogr.* A41:133–142.

Templeton, D. H. and Templeton, L. K. 1989. Effects of x-ray birefringence on radial distribution functions for amorphous materials. *Phys. Rev. B* 40:6506–6508.

Templeton, D. H. and Templeton, L. K. 1992. Polarized dispersion, glide-rule forbidden reflections and phase determination in barium bromate monohydrate. *Acta Crystallogr.* A48:746–751.

Templeton, D. H. and Templeton, L. K. 1994. Tetrahedral anisotropy of x-ray anomalous scattering. *Phys. Rev.* B49:14850–14853.

Warren, B. E. 1990. X-ray Diffraction, Dover Publications, New York.

Wilkinson, A. P., Templeton, L. K., and Templeton, D. H. 1995. Probing local electronic anisotropy using anomalous scattering diffraction: $Cs_2[AuCl_2][AuCl_4]$. *J. Solid State Chem.* 118:383–388.

Zimmermann, M. v., Hall, J. P., Gibbs, Doon, Blume, M., Casa, D., Keimer, B., Murakami, Y., Tomioka, Y., and Tokura, Y. 1999. Interplay between charge, orbital, and magnetic order in $Pr_{1-x}Ca_xMnO_3$. *Phys. Rev. Lett.* 83:4872–4875.

## KEY REFERENCES

Belyakov and Dmitrienko, 1989. See above.

*A good early review of resonant scattering and more general polarization-related x-ray phenomena. Section 2 provides physical insight and a general discussion on anisotropy of x-ray susceptibility for atoms in diffracting crystals.*

Blume, 1994. See above.

*This presentation is comprehensive enough to include all the basic concepts and references. All tensor components up to fourth rank are given in a Cartesian (as opposed to spherical) representation.*

Sakurai, 1967. See above.

*An excellent, primarily nonrelativistic description of the quantum theory used to describe scattering of the type discussed here. Of particular relevence are the discussions on the Kramers-Heisenberg formula and radiation damping.*

## INTERNET RESOURCES

CHESS World Wide Web site

*http://www.chess.cornell.edu/technical/hardware/special_setups/energy_analyzer.html*

LBNL's X-Ray Interactions with Matter

*http://cindy.lbl.gov/optical_constants*

## APPENDIX A:
## TRANSFORMING THE TENSOR STRUCTURE FACTORS

The tensor structure factor is constructed by adding the partial structure factors for each unique site in the unit cell. Each of these contributions has a form given by Equation 15, above. To calculate each contribution, the tensor derived using Equation 14 is transformed from site to site according to the space group symmetry operations.

Assume a tensor is written as a sum of spherical harmonics ($|L, m\rangle$) and defined in a Cartesian coordinate system ($x, y, z$). A general rotation is performed using the rotation operator $\mathbf{D}(\alpha, \beta, \gamma)$ and Euler angles ($\alpha, \beta, \gamma$). See Sakurai (1985), Sections 3.3, 5, and 9. This is represented as

$$\mathbf{D}(\alpha, \beta, \gamma)|L, m\rangle = \sum_{m'} D^{(L)}_{m',m}(\alpha, \beta, \gamma)|L, m'\rangle \quad (21)$$

where

$$D^{(L)}_{m',m}(\alpha, \beta, \gamma)$$
$$= e^{-i(m'\alpha + m\gamma)} \sum_k (-1)^{k-m-m'}$$
$$\times \frac{\sqrt{[(L+m)!(L-m)!(L+m')!(L-m')!]}}{(L+m-k)!k!(L-k-m')!(k-m+m')!}$$
$$\times (\cos \beta/2)^{2L-2k+m+m'}(\sin \beta)^{2k-m-m'} \quad (22)$$

For inversions the rule is

$$|L, m\rangle \rightarrow (-1)^L |L, m\rangle \quad (23)$$

For reflections in a plane normal to the $z$ axis: $z \rightarrow -z$, so

$$|L, m\rangle \rightarrow (-1)^{(L+M)}|L, m\rangle \quad (24)$$

For reflections in a plane parallel to the $z$ axis and at angle to the $x$ axis, the azimuthal angle of the harmonic $\phi \rightarrow 2\delta - \phi$, so

$$|L, m\rangle \rightarrow (-1)^m e^{i2m\delta}|L, -m\rangle \quad (25)$$

## APPENDIX B:
## ANGULAR DEPENDENT TENSORS FOR L=2

It should be pointed out that the angular function $Y(2,0)$ describes $L = 2$ anisotropic scattering for all seven tetragonal and all twelve trigonal/hexagonal point groups. In fact, one can find the form of $L = 2$ angular dependence for all 32 point groups in character tables that list the "binary products of coordinates" for the totally symmetric representation. Such tables indicate there are four other possible $L = 2$ angular functions, one for each of the remaining families (triclinic, monoclinic, orthorhombic, and cubic). One may also see that the $L = 2$ angular dependence for all cubic groups is isotropic (i.e., $\propto x^2 + y^2 + z^2$), which means that resonant scattering from these sites cannot display anisotropy. More information on these functions can be found in books that discuss the properties of crystals via group theory (Nowick, 1995).

## APPENDIX C:
## TRANSFORMATION OF COORDINATES

Describing the scattering experiment involves transforming the resonant scattering structure factor, usually

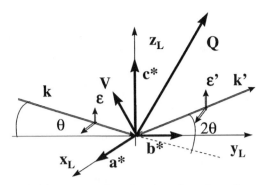

**Figure 10.** Two vectors describe the orientation of a crystal. If the reciprocal lattice of the crystal parallels the "laboratory" coordinate system, then the transformation formula (below) will reorient $\mathbf{Q}$ to be along $\mathbf{z}_L$ for diffraction and set $\mathbf{V}$ at azimuth $\psi$.

defined in a coordinate system fixed to the crystal, into a system fixed with respect to the scattering geometry (i.e., the "laboratory frame"). The methods involved are found in references used to calculate diffractometer angle settings in 4-circle crystallography (Busing and Levy, 1967). Here, these ideas merit special consideration because the tensor nature of the structure factors lead to an important additional dependence of the scattering on crystal orientation.

Figure 10 shows a rectilinear reciprocal lattice (a crystal coordinate system) coincident with a laboratory coordinate system. The orientation of vectors representing the scattering geometry are tied to the laboratory system.

Two vectors in the crystal system are needed to describe its orientation; the reflection vector $\mathbf{Q}(h,k,l)$, and a noncollinear reference vector $\mathbf{V}(h_V, k_V, l_V)$. The condition for Bragg diffraction ($\sin\theta_B = \lambda Q/2$) and the constraint that $\mathbf{Q}$ points along $\mathbf{z}_L$, allows any orientation of $\mathbf{V}$ on a cone about $\mathbf{Q}$. The azimuth angle ($\psi$) fixes the crystal orientation by specifying the projection of $\mathbf{V}$ normal to $\mathbf{z}_L$; $\psi = 0$ when this projection is along $-\mathbf{y}_L$ and increases with clockwise rotation about $\mathbf{Q}$.

The tensor structure factor (usually derived in crystal coordinates) is expressed in the laboratory frame through a coordinate transformation based on Euler angles (Sakari 1985, Section 3.3). In matrix form the transformation is written

$$
\begin{bmatrix}
\cos\alpha\cos\beta\cos\gamma & \sin\alpha\cos\beta\cos\gamma & -\sin\beta\cos\gamma \\
-\sin\alpha\sin\gamma & +\cos\alpha\sin\gamma & \\
-\cos\alpha\cos\beta\sin\gamma & -\sin\alpha\,\beta\sin\gamma & \sin\gamma\sin\beta \\
-\sin\alpha\cos\gamma & +\cos\alpha\cos\gamma & \\
\sin\beta\cos\alpha & \sin\beta\sin\alpha & \cos\beta
\end{bmatrix}
\quad (26)
$$

where $\alpha$ and $\beta$ (the usual azimuthal and polar angles of spherical coordinates) describe the orientation of $\mathbf{Q}$ relative to crystal axes. These angles may be expressed in components of the reflection vectors as $\cos\beta = 1c^*/Q$, and

$\tan\alpha = (k\,b^*)/(h\,a^*)$. The angle $\gamma$ is related to the experimental azimuth, $\psi$, through the relation

$$
\sin(\gamma - \psi) = (l_v c^* - (\mathbf{V}\cdot\mathbf{Q}/Q)\cos\beta)/[(\mathbf{V}\cdot\mathbf{Q}/Q - V)\sin\beta]
$$
$$(27)$$

KENNETH D. FINKELSTEIN
Cornell University
Ithaca, New York

# MAGNETIC X-RAY SCATTERING

## INTRODUCTION

For an x ray incident on a solid, the dominant scattering process is Thomson scattering, which probes the charge density of that solid. It is this interaction that provides the basis for the vast majority of x-ray scattering experiments, supplying a wealth of information on crystal structures and the nature of order in solid-state systems (see, e.g., XAFS SPECTROSCOPY, X-RAY AND NEUTRON DIFFUSE SCATTERING and RESONANT SCATTERING TECHNIQUES). However, in addition to such scattering, there are other terms in the cross-section that are sensitive to the magnetization density. Scattering associated with these terms gives rise to new features in the diffraction pattern from magnetic systems. This is magnetic x-ray scattering. For antiferromagnets, for which the magnetic periodicity is different from that of the crystalline lattice, such scattering gives rise to magnetic superlattice peaks, well separated from the charge Bragg peaks. For ferromagnets, the magnetic scattering is coincident with the charge scattering from the underlying lattice, and difference experiments must be performed to resolve the magnetic contribution.

The magnetic terms in the cross-section are much smaller than the Thomson scattering term, and pure magnetic scattering is typically $10^4$ to $10^6$ times weaker than the charge Bragg peaks. Nevertheless, this scattering is readily observable with synchrotron x-ray sources, and successful experiments have been performed on both antiferromagnets and ferromagnets. The technique has undergone a rapid growth in the last decade as a result of the availability of synchrotrons and the consequent access to intense, tunable x-ray beams of well-defined energy and polarization.

The basic quantity measured in a magnetic x-ray scattering experiment is the instantaneous magnetic moment-moment correlation function. In essence, the technique takes a snapshot of the Fourier transform of the magnetic order in a system. This means that the principal properties measured by magnetic scattering are quantities connected with the magnetic structure—including the ordering wave-vector, the magnetic moment direction, the correlation length, and the order parameter characterizing the phase (for example, this latter might be the staggered magnetization in the case of an antiferromagnet or the uniform magnetization for a ferromagnet). These are obtained from the position, intensity, and lineshape of the magnetic peaks. The scattering is very weak and therefore not affected by extinction effects, so that the temperature dependence of the order parameter may be measured very precisely, leading, for example, to an accurate determination of the order parameter critical exponent, β. However, used alone, x-ray magnetic scattering can typically determine the order parameter only to within an arbitrary scale factor, because of the difficulty in obtaining an absolute scale for the scattered intensities.

There are two regimes of importance in magnetic x-ray scattering, characterized by the incident photon energy relative to the positions of the absorption edges in the system under study. These are the nonresonant and resonant limits to the cross-section.

Nonresonant magnetic scattering is observed when the incident photon energy is far from any absorption edges. It is energy independent, and has the advantages of being (in principle) present for all magnetic ions and of having a simple cross-section. The non-resonant cross-section contains terms due to both the orbital, $\vec{Z}$, and spin, $\vec{S}$, components of the magnetic moment. These each contribute, with different polarization dependences, to the scattering, and it is therefore possible to separate the two contributions empirically. This ability is one of the technique's unique strengths. However, nonresonant magnetic scattering is extremely weak, and for the signal to be observable samples of high crystallographic quality are required in order to match the high collimation of the synchrotron source. The size of the signal varies as the square of the moment; therefore large-moment systems are good candidates for study.

Conversely, resonant magnetic scattering is observed when the incident photon energy is tuned to an absorption edge of an ion with which there is an associated magnetic polarization. It arises from second-order scattering processes. Under certain circumstances, enhancements in the magnetic scattering may be observed at such edges. These enhancements can be very large; increases of a factor of 50 have been observed at the L edges of the rare earths, and of several orders of magnitude at the M edges of the actinides. Such enhancements make the technique particularly useful for the study of weakly scattering systems—e.g., thin films, very small samples, critical fluctuations, and magnetic scattering from surfaces. Resonant scattering is sensitive to the total moment at a given site and it is not possible to extract $\vec{L}$ and $\vec{S}$ in the same way as it is for non-resonant scattering.

The resonant energy depends on the atomic species, and therefore resonant magnetic scattering is element specific—by tuning the incident photon energy to the absorption edge of a particular element, one can study the magnetic order of that sublattice with very little contribution from other magnetic sublattices. This is a particular strength of resonant scattering, in contrast to magnetic neutron scattering (Magnetic neutron scattering), which is sensitive to the total moment of ++ all sublattices contributing at a particular wave-vector. In addition, resonant scattering is also orbital specific—i.e., the magnetic polarization of different electronic orbitals may be probed independently. For example, in the elemental rare earths, both the (itinerant) 5d orbitals and the (localized) 4f

orbitals carry an ordered moment. At an $L_2$ or $L_3$ edge, these states are reached through dipole or quadrupole transitions, respectively. Typically the dipole and quadrupole resonances differ in energy by 5 to 10 eV (Bartolomé et al., 1997), and have different polarization dependencies. They may therefore be separated empirically in a relatively straightforward manner. Recent experiments, for example, have exploited these features to elucidate the RKKY interaction in multicomponent systems (Pengra et al., 1994; Everitt et al., 1995; Langridge et al., 1999; Vigliante et al., 1998).

Both nonresonant and resonant magnetic scattering techniques offer extremely high reciprocal space resolution as a result of the intrinsic collimation of the synchrotron radiation. Typically, the resolution widths are $\sim 10^{-4}$ $\text{Å}^{-1}$. Thus, magnetic x-ray scattering is useful for detailed quantitative measurements of such properties as incommensurabilities, correlation lengths, and scattering lineshapes. In addition, the poor energy resolution of x-ray scattering experiments, relative to the relevant magnetic excitations, ensures that the quasi-elastic approximation is rigorously fulfilled and that the instantaneous correlation function is measured.

In many ways, x-ray magnetic scattering is complementary to the more established technique of neutron magnetic scattering (Magnetic neutron scattering). This latter probe measures many of the same properties of a system, and each technique has its own strengths and weaknesses. It is frequently of value to perform experiments using both techniques on the same sample.

Neutron scattering offers bulk penetration and a large magnetic cross-section, making it more generally applicable than x-ray magnetic scattering. In addition, inelastic magnetic neutron scattering is readily observable (in samples of sufficient volume) allowing the study of spin dynamics. Absolute intensities may be obtained, so it is also possible to measure ordered moments. As a result, unless there are special circumstances, it is generally preferable to solve the magnetic structure using neutron scattering. Here, powder neutron diffraction (Neutron powder diffraction) is particularly powerful. In contrast, powder magnetic x-ray diffraction is, typically, prohibitively small. However, neutron scattering is not inherently element specific, nor can it resolve $\vec{L}$ and $\vec{S}$ contributions simply, or study small-volume samples. Further, it requires a reactor or spallation source to perform the experiments. The reciprocal space resolution is typically lower by a factor of 10 than that achievable with x-rays, and there are some elements that have high neutron absorption cross-sections (e.g., Gd and Sm), which are difficult to study with neutrons but that have relatively large x-ray magnetic cross-sections. In addition, in some cases, the incident neutron flux will activate the sample, which can cause problems for subsequent experiments on the same sample. Finally, for high-quality samples, the magnetic elastic intensity can be affected by extinction, making quantitative determinations of the order parameter and lineshape difficult.

In summary, magnetic x-ray scattering is particularly suitable when high-resolution, or elemental or orbital-specific information is required, or when small samples are involved, or if one needs to separate the $\vec{L}$ and $\vec{S}$ contributions to the ordered moment.

There are other techniques that can measure quantities related to those obtained by x-ray magnetic scattering. These include x-ray magnetic circular dichroism (XMCD), discussed in X-RAY MAGNETIC CIRCULAR DICHROISM, which is closely related to resonant magnetic x-ray scattering. XMCD is particularly useful for obtaining moment directions and $\vec{L}$ and $\vec{S}$ contributions in ferromagnets. Like resonant magnetic scattering it is both elemental and orbital specific. In addition, bulk techniques (Chapter 6) such as SQUID (superconducting quantum interference device; THERMOMAGNETIC ANALYSIS) magnetometry, and heat capacity measurements can provide insight into the magnetic behavior of a particular system, including the total (uniform) magnetization and the existence of phase transitions. These techniques also have the advantage of being available in a home laboratory, and are less demanding in their crystallographic quality requirements. However, such techniques cannot provide the microscopic information on the magnetic order that is available from scattering (x-ray and neutron) techniques.

In any discussion of magnetic x-ray scattering, one is inevitably led to considering the polarization states of the photon. The two important cases are linear and circular polarization. The degree of linear polarization is characterized by

$$P_\zeta = \frac{I_{\text{hor}} - I_{\text{vert}}}{I_{\text{hor}} + I_{\text{vert}}} \qquad (1)$$

where $I_{\text{hor}}$ is the horizontally polarized and $I_{\text{vert}}$ the vertically polarized intensity. Similarly, the degree of circular polarization is defined by

$$P_\eta = \frac{I_{\text{left}} - I_{\text{right}}}{I_{\text{left}} + I_{\text{right}}} \qquad (2)$$

where $I_{\text{left}}$ is the intensity of the left circularly polarized light and $I_{\text{right}}$ the intensity of the right circularly polarized light. In the case of linear polarization, the polarization state is often described relative to the scattering plane. Polarization perpendicular to the scattering plane is referred to as $\sigma$ polarization; polarization parallel to the scattering plane is called $\pi$ polarization.

It is also useful to review the nomenclature for absorption edges, which are labeled by the core hole created in the absorption process. The principal quantum number ($n = 1,2,3,4\ldots$) of the core-hole is labeled by K,L,M,N,$\ldots$ and the angular momentum $s$, $p_{1/2}$, $p_{3/2}$, $d_{3/2}$, $d_{5/2}, \ldots$ by $1,2,3,4,5\ldots$ Thus the $L_3$ edge corresponds to the excitation of the $2p_{3/2}$ electron.

This unit provides an overall background to the theoretical and experimental principles of nonresonant and resonant magnetic x-ray scattering. The more general applications of the resonant terms in the cross-section will not be discussed here; some of them are covered in RESONANT SCATTERING TECHNIQUES and X-RAY MAGNETIC CIRCULAR DICHROISM. The experimental techniques are very similar to standard synchrotron x-ray diffraction experiments; therefore discussion of the beamlines and spectrometers will be limited, with only those elements particular to magnetic scattering, such as polarization analyzers, discussed in any detail. At high incident photon energy, it is possible to perform inelastic experiments utilizing the nonresonant magnetic scattering terms in the cross-section. This allows for the measurement of magnetic Compton profiles (projections of the momentum density of the magnetic electrons, along a given direction). Magnetic Compton scattering may only be applied to ferromagnets, because of the incoherent nature of the inelastic scattering, and the technique differs in many aspects of its practical implementation from the diffraction experiments discussed here. It will therefore not be covered in this unit. Reviews may be found in Schülke (1991) and Lovesey and Collins (1996).

The history of the field will not be covered in this unit. Brief reviews have been given in Brunel and de Bergevin (1991), Gibbs (1992), and McMorrow et al. (1999).

## PRINCIPLES OF THE METHOD

### Theoretical Concepts

As mentioned in the introduction, magnetic x-ray scattering arises from terms in the scattering cross-section that are dependent on the magnetic moment. For nonresonant scattering, these terms can be most simply be understood using a classical picture of scattering from free electrons (de Bergevin and Brunel, 1981a,b). The different classical scattering mechanisms are shown in Figure 1. Charge (Thomson) scattering is illustrated in the uppermost (a) diagram of Figure 1, in which the electric field of the

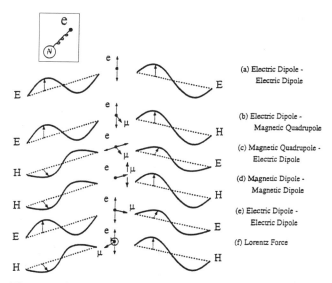

**Figure 1.** Classical picture of the dominant mechanisms for an x-ray scattering from an electron ($e$). In (**a**) the x-ray is scattered from the electronic charge, while in (**b-f**) the magnetic moment of the electron is involved. Mechanisms (**b-d**) result in spin-dependent scattering, while (**e-f**) are for a moving magnetic dipole with finite orbital angular momentum. Mechanism (**a**) represents resonant scattering from an electron bound to an atomic nucleus ($N$) as shown in the inset. Figure adapted from McMorrow et al. (1999).

incident photon accelerates the electronic charge, leading to electric dipole radiation in the form of the scattered photon. The terms that give rise to nonresonant magnetic scattering are illustrated in Figure 1, diagrams b to f. These arise from the (weaker) interaction between the electric and magnetic fields of the photon and the magnetic moment of the electron. The scattering due to nonresonant interactions is reduced relative to charge scattering, by a factor proportional to $(\hbar\omega/mc^2)^2$, which is $\approx 10^{-4}$ for an incident photon energy of $\hbar\omega = 10$ keV (here $h$ is Planck's constant, $\hbar = h/2\pi$, $\omega$ is the frequency, $m$ is the electron mass, and $c$ is the speed of light).

Resonant scattering arises from the fact the electrons are not free, but bound to the nucleus. The resulting quantized bound states give rise to well defined transition energies, and resonant scattering results when these transitions are selectively excited. Maintaining the classical picture, this may be visualized with a ball and spring model and its associated resonant frequencies (see insert over diagram "a" of Fig. 1).

The scattering mechanisms of Figure 1 may be calculated quantum mechanically within second-order perturbation theory with the appropriate choice of interaction Hamiltonian (Blume, 1985, 1994).

The cross-section for elastic scattering of x-rays is

$$\left(\frac{d\sigma}{d\Omega}\right) = r_0^2 \left| \sum_n e^{i\vec{Q}\cdot\vec{r}_n} f_n(\vec{k},\vec{k}',\hbar\omega) \right|^2 \qquad (3)$$

where $\vec{Q}$ is the momentum transfer in the scattering process, $r_0 = 2.8 \times 10^{-15}$ m is the classical electron radius, $\vec{k}$ and $\vec{k}'$ are the incident and scattered wave-vectors of the photon, $d\sigma$ is the differential cross-section of the scattering center for scattering into the solid angle element $d\Omega$. The scattered amplitude from site $\vec{r}_n$ is $f_n$. It may be expressed as the sum of several contributions

$$f_n(\vec{k},\vec{k}',\hbar\omega) = f_n^{\text{charge}}(\vec{Q}) + f_n^{\text{non-res}}(\vec{Q},\vec{k},\vec{k}') + f_n^{\text{res}}(\vec{k},\vec{k}',\hbar\omega) \qquad (4)$$

where, for incident and scattered photon polarization states described by the unit vectors $\hat{\varepsilon}$ and $\hat{\varepsilon}'$, respectively

$$f_n^{\text{charge}}(\vec{Q}) = -\rho_n(\vec{Q})\hat{\varepsilon}\cdot\hat{\varepsilon}' \qquad (5)$$

is the Thomson scattering, which is proportional to the Fourier transform of the charge density, $\rho(\vec{Q})$. The dot product of the polarization vectors requires that the polarization state not be rotated by the scattering. The nonresonant magnetic scattering amplitude, $f_n^{\text{non-res}}$ and the resonant magnetic scattering amplitude, which is contained in the anomalous (energy dependent) terms, $f_n^{\text{res}}$, are considered in the following section.

**Nonresonant Scattering.** The nonresonant scattering amplitude may be written (Blume, 1985)

$$f_n^{\text{non-res}}(\vec{Q}) = -i\left(\frac{\hbar\omega}{mc^2}\right)\left[\frac{1}{2}\vec{L}_n(\vec{Q})\cdot\vec{A} + \vec{S}_n(\vec{Q})\cdot\vec{B}\right] \qquad (6)$$

where $\frac{1}{2}\vec{L}_n(\vec{Q})$ and $\vec{S}_n(\vec{Q})$ are proportional to the Fourier transforms of the orbital and spin magnetization densities, respectively. The vectors $\vec{A}$ and $\vec{B}$ contain the polarization dependences of the two contributions to the scattering. Importantly, these two vectors are not equal and have distinct $\vec{Q}$ dependences. Hence, by measuring the polarization of a scattered beam for a number of magnetic reflections, one may obtain the ordered $\vec{L}$ and $\vec{S}$ moments for the system. For synchrotron diffraction experiments with linear incident polarization, $\vec{A}$ and $\vec{B}$ are most usefully expressed in terms of linear polarization basis states (Blume and Gibbs, 1988). In this basis, the charge scattering may be written as

$$\langle M_{\text{charge}}\rangle = \rho(\vec{Q})\begin{pmatrix} 1 & 0 \\ 0 & \cos 2\theta \end{pmatrix} \qquad (7)$$

where $2\theta$ is the angle between $\vec{k}$ and $\vec{k}'$. Note, this matrix is diagonal, reflecting the fact that the polarization state is not rotated by the charge scattering process. In this same basis the non-resonant magnetic scattering operator is

$$\langle M_{\text{non-res}}\rangle = \begin{pmatrix} \langle M_{\sigma\sigma}\rangle & \langle M_{\pi\sigma}\rangle \\ \langle M_{\sigma\pi}\rangle & \langle M_{\pi\pi}\rangle \end{pmatrix} = -i\left(\frac{\hbar\omega}{mc^2}\right)\times$$
$$\begin{pmatrix} \sin 2\theta S_2 & -2\sin^2\theta[\cos\theta(L_1+S_1) - \sin\theta S_3] \\ 2\sin^2\theta[\cos\theta(L_1+S_1) + \sin\theta S_3] & \sin 2\theta[2\sin^2\theta L_2 + S_2] \end{pmatrix} \qquad (8)$$

in the coordinate system of Figure 2. Here and throughout this unit, the $\langle\ \rangle$ symbol represents the ground state expectation value of an operator. Note that non-resonant magnetic scattering is only sensitive to the components of the orbital angular momentum perpendicular to the

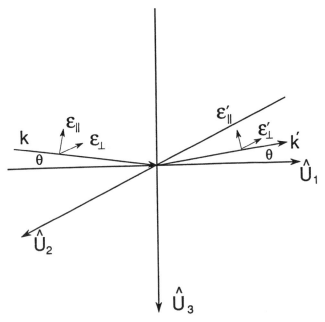

**Figure 2.** The coordinate system used in this unit, after Blume and Gibbs (1988). Note that the momentum transfer $\vec{Q}$ is along $-\hat{U}_3$.

momentum transfer (i.e., the components $L_1$ and $L_3$). For pure magnetic scattering, then

$$\frac{d\sigma}{d\Omega} = r_o^2 \left(\frac{\hbar\omega}{mc^2}\right)^2 \text{tr}\langle M_{\text{non-res}}\rangle\rho\langle M_{\text{non-res}}^\dagger\rangle \qquad (9)$$

where tr is the trace operator, $\langle M_{\text{non-res}}^\dagger\rangle$ is the Hermitian adjoint of $\langle M_{\text{non-res}}\rangle$ and $\rho$ is the density matrix for the incident polarization. For example, for linear incident polarization

$$\rho = \frac{1}{2}\begin{pmatrix} 1+P_\zeta & 0 \\ 0 & 1-P_\zeta \end{pmatrix} \qquad (10)$$

The general case is given in Blume and Gibbs (1988). With this form, the $\sigma \to \sigma$ scattering is given by $|\langle M_{\sigma\sigma}\rangle|^2$, the $\sigma \to \pi$ by $|\langle M_{\sigma\pi}\rangle|^2$ and so on.

Equation 9 allows for the straightforward determination of the polarization dependence of the scattering for a given scattering geometry and magnetic structure. As an illustrative example, consider the case of $MnF_2$, an Ising antiferromagnet for which the spins are oriented along the $c$ axis, and the orbital moment is quenched, $L_1 = L_2 = L_3 = 0$. The ordering wave vector is perpendicular to the $c$ axis. If the scattering geometry is arranged such that the $c$ axis is perpendicular to the scattering plane, then $S_2 = S$ and $S_1 = S_3 = 0$. The scattering matrix, Equation 8, is then diagonal, and for a $\sigma$ incident beam, all the scattering is into the $\sigma$ channel. If the sample is then rotated about the scattering vector such that the spins are then in the scattering plane, i.e., $S_1 = S$ and $S_2 = S_3 = 0$, then the diagonal elements are zero and all the scattering rotates the photon polarization state by 90°, i.e., the scattering is entirely $\sigma \to \pi$ and $\pi \to \sigma$. Further, the scattered intensity is reduced by a factor of $\sin^2\theta$, relative to the previous geometry. In typical experiments, this is a small number, and, as a result, non-resonant magnetic scattering is largely sensitive to the component of the spin perpendicular to the scattering plane.

These predictions have been largely experimentally verified in $MnF_2$ (Goldman et al., 1987; Hill et al., 1993; Brückel et al., 1996).

An estimate of the size of pure non-resonant magnetic scattering, relative to that of charge scattering (ignoring polarization factors) may be obtained from Blume (1985)

$$\frac{\sigma_{\text{mag}}}{\sigma_{\text{charge}}} = \left(\frac{\hbar\omega}{mc^2}\right)^2 \left(\frac{N_m}{N}\right)^2 \left(\frac{f_m}{f}\right)^2 \langle\mu^2\rangle \qquad (11)$$

where $N_m/N$ is the ratio of the number of magnetic electrons to the total number of electrons, $f_m/f$ is the ratio of the magnetic to charge form factors, and $\mu^2$ is the average magnetic moment. For Fe and 10 keV photons

$$\frac{\sigma_{\text{mag}}}{\sigma_{\text{charge}}} \sim 4 \times 10^{-6}\langle\mu^2\rangle \qquad (12)$$

In fact, this ratio may be further reduced for a particular Bragg magnetic reflection because the magnetic form factor falls off faster than the charge form factor. While,

to a large extent, the small amplitude of the non-resonant scattering may be countered by the use of the intense incident x-ray flux available at synchrotrons, it is generally true that samples of relatively small mosaic spread (e.g., <0.1°) are required in order to make full use of this incident intensity. Nevertheless, even quite small moments have been observed. For example, non-resonant magnetic scattering from the spin density wave in chromium, which has an ordered moment of 0.4 $\mu_B$/atom, gave rise to count rates on the order of 10 counts per sec at a bending magnet beamline at the National Synchrotron Light Source (Hill et al., 1995a) and >1000 counts per second at an undulator beamline at the European Synchrotron Radiation facility (Mannix, 2001).

**Resonant Scattering.** The resonant terms arise from second-order scattering processes which may, in a loose sense, be thought of as the absorption of an incident photon, the creation of a short lived intermediate (excited) state, and the decay of that state, via the emission of an elastically scattered photon, back into the ground state. For Bragg scattering, this process is coherent and the phase relationship between the incident and scattered photons is preserved. A schematic energy level diagram illustrating this process for a rare-earth ion is shown in Figure 3.

The resonant scattering amplitude is (Blume, 1994)

$$f_n^{\text{res}}(\vec{k},\vec{k}',\hbar\omega) = -\frac{1}{m}\sum_c\left(\frac{E_a - E_c}{\hbar\omega}\right)\frac{\langle a|O_\lambda(\vec{k})|c\rangle\langle c|O_{\lambda'}^\dagger(\vec{k}')|a\rangle}{E_a - E_c - \hbar\omega}$$
$$+\frac{1}{m}\sum_c\left(\frac{E_a - E_c}{\hbar\omega}\right)\frac{\langle a|O_{\lambda'}^\dagger(\vec{k}')|c\rangle\langle c|O_\lambda(\vec{k})|a\rangle}{E_a - E_c + \hbar\omega - i\frac{\Gamma}{2}}$$
$$(13)$$

where $E_a$, $|a\rangle$, and $E_c$, $|c\rangle$, are the energies and wavefunctions of the initial and intermediate states, respectively, $\Gamma$ is the inverse lifetime of the intermediate state, and the operator $O_\lambda(\vec{k})$ is given by

$$O_\lambda(\vec{k}) = \sum_n e^{i\vec{k}\cdot\vec{r}_n}(\vec{P}_n \cdot \hat{\varepsilon}_\lambda - i\hbar\hat{\varepsilon}_\lambda \cdot (\vec{k} \times \vec{S}_n)) \qquad (14)$$

The two terms of Equation 13 give rise to anomalous dispersion, i.e., they contain all the energy dependence of the scattering. They have come to play an increasingly important role in x-ray scattering experiments with the availability of energy tunable x-ray sources at synchrotron beamlines. Some of these applications are discussed in the units on resonant scattering techniques (RESONANT SCATTERING TECHNIQUES). In the limit of zero momentum transfer, these terms give rise to XMCD effects (X-RAY MAGNETIC CIRCULAR DICHROISM).

In a resonant experiment, the incident photon energy, $\hbar\omega$ is tuned to the absorption edge, such that $\hbar\omega = E_c - E_a$, and the second of the two terms in Equation 13 is enhanced. In this limit, it is common to ignore the first, non-resonant term of Equation 13 (Blume, 1994). That resonant effects may contain some sensitivity to the magnetization was first suggested by Blume (1985) and observed by Namikawa et al. (1985). It was explored in

## $L_{III}$ Edge

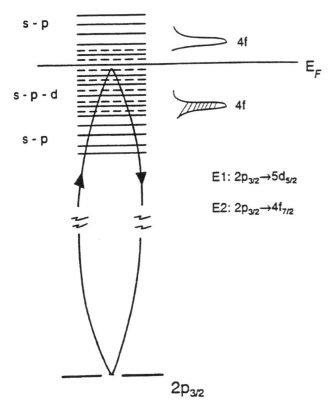

**Figure 3.** A schematic single-electron picture of the resonant scattering process, illustrated for a rare earth, at the $L_3$ edge. A $2p_{3/2}$ electron is excited into an unoccupied state above the Fermi level by the incident photon. The $5d$ states are reached through electric dipole transitions and the $4f$ through electric quadrupole transitions. In the elemental rare earths, the $5d$ states form delocalized bands, polarized through an exchange interaction with the localized, magnetic $4f$ electrons. Resonant elastic scattering results when the virtually excited electron decays by filling the core hole and emitting a photon. Figure taken from Gibbs (1992).

detail by Hannon et al. (1988) following the experimental work of Gibbs et al. (1988).

The magnetic sensitivity arises from spin-orbit correlations, which must be present in at least one of the two levels involved in the resonant process, and from exchange effects. "Exchange effects" refers to any phenomena originating from the anti-symmetry of the wave function under the exchange of electrons. This includes the exclusion principle, which only allows transitions to unoccupied spin orbitals in partially filled spin polarized valence states, as well as the exchange interaction between electrons in different orbitals. Resonant magnetic scattering is often referred to as x-ray resonant exchange scattering (XRES), as a result of the requirement that exchange effects be present.

Almost all of the experimental phenomena observed in resonant magnetic scattering can be explained in terms of a one-electron picture of electric multipole transitions

(Hannon et al., 1988; Luo et al., 1993). Magnetic multipole transitions are smaller by a factor of $\hbar\omega/mc^2$ and have not been considered in the literature to date. In this picture, the size of the resonant enhancement of the magnetic scattering depends on the radial matrix elements ($\langle c|r^L|a\rangle$, where $L = 1, 2..$ for dipole, quadrupole... transitions), the magnetic polarization of the intermediate state, the value of the intermediate-state lifetime, and the energy width of the incident photon. Thus the largest enhancements are expected when large matrix elements couple states with strong polarizations.

Such a situation occurs at the $M_4$ and $M_5$ edges in the actinides, for which dipole allowed transitions couple to the magnetic $5f$ electrons. Enhancements of several orders of magnitude are typical at such edges, allowing extremely small moments to be studied. For example, in URu$_2$Si$_2$, scattering from an ordered moment of 0.02$\mu_B$ was observed (Isaacs et al., 1990).

Conversely, at the $L_2$ and $L_3$ edges of the rare-earths, the dipole transitions are $2p \leftrightarrow 5d$. The $5d$ electrons are polarized through an exchange interaction with the localized, magnetic $4f$ electrons, and the induced moment is relatively small, $\approx 0.2\mu_B$. In addition, the radial matrix elements are 5 to 10 times smaller than at the $M_{4,5}$ edges (Hamrick, 1994) and smaller enhancements are therefore expected. Enhancement factors of $\sim 50$ were observed at the Ho $L_3$ edge (Gibbs et al., 1988, 1991). Quadrupolar transitions at the rare-earth $L_{2,3}$ edges, however, couple to the magnetic $4f$ levels, $2p \leftrightarrow 4f$. To some extent the greater polarization of the $4f$ orbitals compensates for the much smaller quadrupole matrix elements, and this scattering is also observable. In Ho, it is $\approx 5$ times weaker than the dipole scattering at the first harmonic, after correction for absorption (Gibbs et al., 1988). Table 1 summarizes the absorption edges for various elements and approximate enhancement factors, together with a typical reference.

One property of resonant scattering is that it does not contain a magnetic form factor, in contrast with non-resonant scattering and magnetic neutron diffraction. In essence, this is because of the two-step nature of the scattering event. The lack of a magnetic form factor is in fact beneficial when observing magnetic reflections at large $\vec{Q}$, something that is often necessary when studying off-specular peaks in a reflection geometry.

In order to extract quantities such as ordered moments and moment directions, the cross-section needs to be considered in more detail (Hannon et al., 1988; Carra et al., 1993; Luo et al., 1993). Absolute moments of one species have been estimated in multicomponent systems by comparison with signals from a known moment in the other species, measured in the same experiment, e.g., in alloy systems such as Dy/Lu (Everitt et al., 1995). However, the large number of estimates and approximations that are required, especially when the intermediate states are delocalized and band structure calculations must be invoked, mean that this method is at best semiquantitative. Work along these lines in Ho/Pr alloys (Vigliante et al., 1998) highlights another problem with extracting such quantitative information, i.e., a breakdown of the one-electron, atomic picture. This is manifest most

**Table 1. Resonant Energies and Approximate Enhancements for Various Atomic Absorption Edges**

| Series | Edge | Energy Range | Transition | Enhancement | Reference |
|---|---|---|---|---|---|
| 3d transition metals | $L_{2,3}$ | 400–1000 eV | E1 $2p \rightarrow 3d$ | $\geq 10^2$ | Tonnerre et al., 1995 |
| | K | 4.5–9.5 keV | E1 $1s \rightarrow 4p$ | ~2 | Namikawa et al., 1985 |
| | | | E2 $1s \rightarrow 3d$ | ~2 | Hill et al., 1996; Neubeck et al., 2001 |
| 5d transition metals | $L_{2,3}$ | 5.8–14 keV | E1 $2p \rightarrow 5d$ | ~$10^2$ | de Bergevin et al., 1992 |
| Rare earths | $L_{2,3}$ | 5.8–10.3 keV | E1 $2p \rightarrow 5d$ | 50 | Gibbs et al., 1988 |
| | | | E2 $2p \rightarrow 4f$ | | |
| Actinides | $M_{4,5}$ | 3.5–4.7 keV | E1 $3d \rightarrow 5f$ | $10^7$ | Isaacs et al., 1989 |
| | $M_3$ | 4.3 keV | E1 $3p \rightarrow 6d$ | < 2 | McWhan et al., 1990 |
| | | | E2 $3p \rightarrow 5f$ | | |
| | $L_{2,3}$ | 7–20 keV | E1 $2p \rightarrow 6d$ | ~$10^2$ | —[a] |
| | | 7–20 keV | E2 $2p \rightarrow 5f$ | | |

[a] D. Wermeille, C. Vettier, N. Bernhoeft, N. Stunault, P. Lejay, and J. Flouquet, unpub. observ.

obviously in the branching ratio (the ratio of scattered intensities obtained at a pair of resonances resulting from excitations of two spin-orbit split core-holes, e.g., the $L_2$ and $L_3$ or the $M_4$ and $M_5$ edges). In the simplest theories (Hannon et al., 1988), these are predicted to be near unity. In fact, empirically, they are known to vary across the rare-earth series, with the heavy rare earths supporting $L_2/L_3 < 1$ and the light rare earths having $L_2/L_3 > 1$ (Gibbs et al., 1991; Hill et al., 1995b; D. Watson et al., 1996; Vigliante et al., 1998). Further, the ratio has even been observed to be temperature dependent. These results point to a need for more detailed theoretical work (e.g., van Veenendaal et al., 1997). However, until a full understanding of the scattered amplitude (and electronic structure used in the calculations) is obtained, the resonant scattered intensity must be interpreted with some caution. A very recent example of an attempt along these lines is the work of Neubeck et al. (2001).

The polarization dependence of the resonant scattering is a function of moment direction and scattering geometry. For dipole (E1) scattering, it is

$$f_{nE1}^{\text{X RES}} = F^{(0)} \begin{pmatrix} 1 & 0 \\ 0 & \cos 2\theta \end{pmatrix} - iF^{(1)} \begin{pmatrix} 0 & z_1\cos\theta + z_3\sin\theta \\ z_3\sin\theta - z_1\cos\theta & -z_2\sin 2\theta \end{pmatrix}$$
$$+ F^{(2)} \begin{pmatrix} z_2^2 & -z_2(z_1\sin\theta - z_3\cos\theta) \\ +z_2(z_1\sin\theta + z_3\cos\theta) & -\cos^2\theta(z_1^2\tan^2\theta + z_3^2) \end{pmatrix}$$

$$(15)$$

in the same coordinate system and basis as previously (Hannon et al., 1988; Pengra et al., 1994; Hill and McMorrow, 1996). The $z_i$ are the components of a unit vector aligned along the quantization axis at site $n$ and the $F^{(L)}$ are the resonant matrix elements. The first term contains no dependence on the magnetic moment and therefore only contributes to the charge Bragg peaks. The second term is linear in $z$ and for an antiferromagnet will produce new

scattering at the magnetic wave-vector. The final term is quadratic in $z$. This will produce scattering at twice the ordering wave vector. For an incommensurate antiferromagnet, this will give rise to a second-order resonant harmonic.

By way of an example, consider the case of a basal plane spiral, such as occurs in Ho, Tb, and Dy. In this structure the moments are confined to the a-b plane of the hcp crystal structure, in ferromagnetic sheets. The direction of the magnetization rotates from basal plane to basal plane, creating a spiral structure propagating along the c axis. The modulation vector then lies along (00L). The magnetic moment takes the form

$$\hat{z}_n = \cos(\tau \cdot \mathbf{r}_n)\hat{U}_1 + \sin(\tau \cdot \mathbf{r}_n)\hat{U}_2 \qquad (16)$$

where $\tau$ is the period of the spiral, $\mathbf{r}_n$ the coordinate of the $n$th atom, and $\hat{U}_1$ and $\hat{U}_2$ are unit vectors defined in Figure 2.

Now, the full x-ray scattering cross-section may be written as

$$\frac{d\sigma}{d'} = \sum_{\lambda\lambda'} P_\lambda |\langle\lambda'|\langle M_{\text{charge}}\rangle|\lambda\rangle - i\frac{\hbar\omega}{mc^2}\langle\lambda'|\langle M_{\text{non-res}}\rangle|\lambda\rangle$$
$$+ \langle\lambda'|\langle M_{E1}^{\text{X RES}}\rangle|\lambda\rangle + \langle\lambda'|\langle M_{E2}^{\text{X RES}}\rangle|\lambda\rangle + \cdots I^2 \qquad (17)$$

where $\lambda$, $\lambda'$ are incident and scattered polarization states and $P_\lambda$ the probability for incident polarization $\lambda$. As before, $\langle\ \rangle$ represents the ground-state expectation value of the respective operators i.e.,

$$\langle M_{E1}^{\text{X RES}}\rangle = \langle a|\sum_n e^{i\mathbf{Q}\cdot\mathbf{r}_n} f_{nE1}^{\text{X RES}}|a\rangle \qquad (18)$$

and E1 and E2 are the dipole and quadrupole terms, respectively. For the purposes of this example, we assume that the dipole contribution is dominant and ignore any interference effects. This will be valid at resonance and away from charge Bragg peaks. Then from Equation 15

$$M_{E1}^{\text{X RES}}$$

$$= \sum_n e^{i\mathbf{Q}\cdot\mathbf{r}_n} \begin{pmatrix} z_2^2 F^{(2)} & -iz_1\cos\theta F^{(1)} - z_2 z_1\sin\theta F^{(2)} \\ iz_1\cos\theta F^{(1)} + z_2 z_1\sin\theta F^{(2)} & iz_2\sin 2\theta F^{(1)} - z_1^2\sin^2\theta F^{(2)} \end{pmatrix}$$

$$(19)$$

For simplicity, we take $P_\lambda = \delta_{\lambda\sigma}$, i.e., an incident beam of linear polarization perpendicular to the scattering plane. Then, only $\sigma \to \sigma$ and $\sigma \to \pi$ terms contribute and

$$\frac{d\sigma}{d\Omega} = |\langle a| \sum_n e^{i\mathbf{Q}\cdot\mathbf{r}_n} z_2^2 F^{(2)} |a\rangle|^2$$

$$+ |\langle a| \sum_n e^{i\mathbf{Q}\cdot\mathbf{r}_n} (iz_1\cos\theta F^{(1)} + z_2 z_1\sin\theta F^{(2)}) |a\rangle|^2 \quad (20)$$

Substituting $z_1 = \cos(\tau \cdot \mathbf{r}_n), z_2 = \sin(\tau \cdot \mathbf{r}_n)$ and writing the sine and cosine terms as the sums of complex exponentials, we obtain the cross-section for a flat spiral with incident radiation perfectly polarized perpendicular to the scattering plane

$$\left.\frac{d\sigma}{d\Omega}\right)_{E1}^{\text{X RES}} = \frac{1}{4} F^{2(2)}\delta(\mathbf{Q} - \mathbf{G}) + \frac{1}{4}\cos^2\theta F^{2(1)}\delta(\mathbf{Q} - \mathbf{G} \pm \tau)$$

$$+ \frac{1}{16}(1 + \sin^2\theta)F^{2(2)}\delta(\mathbf{Q} - \mathbf{G} \pm 2\tau) \quad (21)$$

where $\mathbf{G}$ is a reciprocal lattice vector. The intensity of the observed scattering will only be proportional to the above expression if both polarization components of the scattered beam are collected with equal weight. This is the case if no analyzer crystal is employed. If one is used, then the polarization component that is in the scattering plane of the analyzer must be weighted by a factor $\cos^2 2\theta_A$, where $\theta_A$ is the Bragg angle of the analyzer crystal. From Equation 21, we see that in addition to producing scattering at the Bragg peak, the resonant dipole contribution produces two magnetic satellites at $\pm\tau$ and $\pm 2\tau$ on either side of the Bragg peak. The scattering at $\pm 2\tau$ are the resonant harmonics. Note that these are distinct from any diffraction harmonics that may arise from a nonsinusoidal modulation of the magnetization density.

For quadrupole transitions, the situation is still more complex (Hannon et al., 1988; Hill and McMorrow, 1996) and terms up to $O(z^4)$ are obtained, giving rise to up to four orders of resonant harmonics. The presence of third and fourth resonant harmonics is a clear indication of quadrupole transitions.

Finally, the original formulation of the resonant cross-section (Hannon et al., 1988) has been re-expressed in terms of effective scattering operators following theoretical work performed for the XMCD cross-section (Thole et al., 1992; Carra et al., 1993; Luo et al., 1993; Carra and Thole, 1994). These scattering operators are constituted from spin-orbital moment operators of the valence shell involved in the scattering process and are valid in the short collision time approximation. They offer the possibility of relating resonant elastic and inelastic scattering measurements directly to physically interesting properties of the system.

**Observation of Ferromagnetic Scattering.** Given the presence of terms in the cross-section sensitive to the magnetization density, the observation of pure magnetic scattering is, in principle, straightforward for antiferromagnets because the superlattice peaks are well separated from the much stronger charge Bragg peaks of the underlying chemical lattice. For ferromagnets, however, the situation is complicated by the fact that the magnetic and charge scattering are superposed.

At first glance, the situation might seem hopeless, since one is apparently required to resolve a signal of one part in $10^6$. However, the presence of the charge scattering at the same position does allow for the possibility of observing the interference between the charge and magnetic scattering amplitudes. For example, if the charge and magnetic scattering were in phase, then

$$\frac{d\sigma}{d\Omega} = |f_c + f_m|^2 = f_c^2 + f_m^2 + 2f_c f_m \quad (22)$$

where $f_m(f_c)$ is the magnetic (charge) scattering amplitude. The interference term is linear in $f_m$ and not quadratic as for pure magnetic scattering, and this scattering is therefore much stronger, on the order of $10^{-3}$ of the charge scattering. In addition, it changes sign when the sign of the magnetization is changed, allowing the magnetic scattering to be distinguished from the charge scattering in a difference experiment. Unfortunately, the charge and magnetic scattering are in fact precisely 90° out of phase for plane-polarized incident light (in centrosymmetric structures), and thus there is no interference between them.

The interference can be recovered, however, if there is a phase shift between the magnetic and charge amplitudes—e.g., by utilizing resonant effects, or, in non-centrosymmetric structures, by use of circularly polarized light. For non-resonant magnetic scattering, the last option is the only generally applicable alternative. Circularly polarized light also allows for the possibility of reversing the sign of the magnetic scattering by "flipping" the helicity of the incident beam, as well as by reversing the magnetization.

In practice, difference experiments are performed in which the scattered intensity is measured in each of the two configurations and the asymmetry ratio

$$R = \frac{I_+ - I_-}{I_+ + I_-} \quad (23)$$

is calculated, where $I_+$ and $I_-$ are the intensities measured in the two configurations. This approach eliminates many sources of systematic errors.

For the observation of non-resonant ferromagnetic scattering, the preferred incident beam has some degree of linear polarization as well as a finite circular polarization. The asymmetry ratio is then maximized by scattering in the plane of the linear polarization of the beam at a scattering angle of $2\theta = 90°$, thus minimizing the charge scattering related to the linear polarization of the incident beam. In this limit, for non-resonant magnetic scattering (Lovesey and Collins, 1996)

$$R = -2\left(\frac{\hbar\omega}{mc^2}\right)\frac{P_\eta}{1+P_\eta}\left(\frac{F_S}{F_c}\sin\alpha + \frac{F_L}{F_c}(\sin\alpha + \cos\alpha)\right) \quad (24)$$

where $F_S$, $F_L$, and $F_c$ are the structure factors for the spin, orbital and charge densities respectively, $\alpha$ is the angle between the spin and orbital moments (assumed collinear) and the incident beam, $P_\zeta$ is the degree of linear polarization (also see Equation 1), and $P_\eta$ is the degree of circular polarization (also see Equation 2). Since the scattering is within the plane of linear polarization, $P_\zeta < 0$, and therefore $R$ is maximized by maximizing $P_\zeta$ while maintaining a non-zero circular polarization. By measuring the asymmetry ratio for two different angles, $\alpha$, and taking the ratio of these quantities, then the only unknown is the ratio of the spin to orbital structure factors. This data may then be used together with polarized beam neutron experiments, which determine $(F_S + F_L)$, to obtain $F_S$ and $F_L$ separately, without the need to characterize the polarization state.

**Observation of Ferromagnetic Scattering: Resonance Effects.** The anomalous dispersion effects associated with absorption edges open up new possibilities for inducing interference between charge and magnetic scattering amplitudes, and thus for studying ferrimagnetism and ferromagnetism. Such techniques can be placed into two classes: the interference of non-resonant magnetic scattering with resonant charge scattering and the interference of resonant magnetic scattering with non-resonant charge scattering. Lovesey and Collins (1996) refer to these as magnetic-resonant charge interference scattering and resonant magnetic-charge interference scattering, respectively. In each case, the phase shift is induced through the anomalous terms, and thus circularly polarized light is not required. In the former case, for which the non-resonant magnetic scattering terms are utilized, it is still possible, in principle, to separate the $\vec{L}$ and $\vec{S}$ contributions to the scattering.

In order for the results of such experiments to be cleanly interpreted, it is important that the experiment be performed in the appropriate limit of the magnetic scattering terms (i.e., resonant or non-resonant). For example, in order to ensure that non-resonant magnetic scattering is dominant, but still be able to take advantage of the resonant phase shift in the charge scattering amplitude, one could use an absorption edge of a nonmagnetic ion, or move far enough above an edge associated with a weak enhancement (such as a K edge in a transition metal), so that the resonant magnetic scattering is negligible. This latter approach was utilized in the first magnetic scattering experiments on ferromagnets (de Bergevin and Brunel, 1981a,b; Brunel, 1983).

Alternatively, the different polarization dependence of resonant and non-resonant magnetic scattering may be used to select a particular channel, utilizing the fact that for interference to occur it must be allowed by both magnetic and charge scattering. Recall that charge scattering (Equation 7) permits only $\sigma \to \sigma$ and $\pi \to \pi$ processes and that in a difference experiment, the only terms which contribute to the observed signal are odd in $z_n$. Now, for dipole resonant scattering, $\sigma \to \sigma$ processes are forbidden in the first-order term (Equation 15) and therefore resonant magnetic-charge interference is only observed in the $\pi \to \pi$ channel (ignoring second-order terms). Conversely, non-resonant magnetic scattering may contribute to both $\sigma \to \sigma$ and $\pi \to \pi$ channels (Equation 15) and so, for example, the $\sigma \to \sigma$ channel may be used to select the non-resonant magnetic-charge interference.

## PRACTICAL ASPECTS OF THE METHOD

### Experimental Hardware

Although the first x-ray magnetic scattering experiments were performed with laboratory sources, the inherent advantages of synchrotron sources—the high flux and collimation, energy tunability, and the possibility of manipulating the incident polarization state—make such facilities nearly ideal for the study of magnetic phenomena with x rays. For all practical purposes, therefore, a synchrotron source is a requirement for a successful experiment. There are three sources of synchrotron radiation in current storage rings. These are the dipole bending magnet, which applies a circular acceleration to the electron beam, and the so-called insertion devices: wigglers and undulators. These latter are linear periodic arrays of alternating bending magnets. A wiggler of $N$ poles will provide $N$ times the flux of a single bending magnet of the same field strength, with the same spectral distribution. An undulator is similar to a wiggler except that the amplitude of the oscillations is much smaller. The amplitude is constrained such that, when viewed far from the source, the radiation from each oscillation is coherent with the others. In such a limit, the flux scales as $N^2$, and emission is peaked around harmonics of a fundamental (Brown and Lavender, 1991).

The experimental hardware may be broken down into three broad classes of components:

1. *The beamline.* This delivers the x-rays to the sample and is comprised of various optical elements, including monochromator and (typically) focusing mirror, together with evacuated beam transport tubes and possibly a further optical element to alter the polarization state, such as a phase plate.
2. *The spectrometer.* This instrument manipulates the sample and the detector, providing the ability to determine the angle of incidence and exit of the x rays, to a precision of ~0.001°, allowing the experimenter to explore reciprocal space, often as a function of incident energy. The spectrometer may also include a polarization analyzer for detecting the polarization state of the scattered beam.
3. *The sample environment.* Frequently in magnetic experiments this is a cryostat of some kind.

In the following, the practical aspects of the beamline and the spectrometer are discussed.

**The Beamline.** Synchrotron beamlines are large, expensive pieces of equipment. They must accept the incident white light spectrum and its associated heat load from the x-ray source and deliver a monochromatic, stable beam to the sample. Optical elements upstream of and including the monochromator must be cooled to prevent problems associated with heating, such as degradation of the optical elements and the loss of flux delivered to the sample due to mismatches of Bragg reflecting crystals that are experiencing different heat loads. At second-generation sources this is typically accomplished with water cooling. The third-generation sources such as the European Sychrotron Radiation Facility (ESRF) in France, the Advanced Photon Source (APS) in the U.S.A., and the Spring-8 synchrotron in Japan) experience greater heat loads and power densities, and cooling with liquid nitrogen is a common solution.

Typical beamlines suitable for magnetic x-ray scattering are composed of a number of components. The mirror (or pair of mirrors) acts as a low-pass filter, eliminating the high-energy component of the white beam; the mirrors are frequently used as focusing elements. They are operated in a grazing incidence geometry such that the angle of incidence is below the critical angle of reflection for the x-ray energies of interest, and these energies are passed with a reflectivity approaching unity. However, for the higher energies, which have a lower critical angle, the angle of incidence will exceed the corresponding critical angle and photons of these wavelengths will be increasingly attenuated.

The high-energy part of the spectrum is particularly troublesome for a magnetic scattering experiment because of the property whereby the monochromator passes higher-order components of the beam. A typical monochromator, e.g., Si(111) or Ge(111), set to pass $\lambda$ will also pass $\lambda/2$, $\lambda/3 \ldots$ (in varying amounts), since the Bragg conditions for these energies will be met simultaneously by the (222), (333)... reflections. These components of the incident beam will produce charge scattering at apparent fractional Bragg positions, when indexed based on a wavelength of $\lambda$. Such harmonics may superpose magnetic scattering for commensurate magnetic systems. Typically, this problem is overcome through a combination of one or more of the following techniques. Firstly, the mirror and inherent energy dependence of the incident spectrum reduce the incident intensity at the higher energies. Secondly, the (222) reflection is all but forbidden in the diamond structure, so that after two reflections from a Si(111) or Ge(111) double-bounce monochromator, and (if used) a (111) reflection in an analyzer crystal, it is greatly reduced. Thirdly, the energy resolution of the detector can further discriminate between the $\lambda$ and $\lambda/n$ components (though if the harmonic content of the signal is too great, the detector will saturate, and this approach will not work). Fourth, the monochromator may be detuned. This final method is based on the fact that the rocking widths (Darwin curves) of the higher-order reflections are sharper than that of the fundamental. Thus, if one operates the monochromator with the second crystal slightly rotated away from the $\theta_B$, the reflectivity for the higher-order components may be greatly reduced with only a slight loss of intensity of the fundamental.

For resonant scattering experiments, the monochromator needs to be a scanning, fixed exit monochromator. The energy resolution of the monochromator is crucial in determining the size of any resonant enhancement, and, to maximize the enhancement, it should be approximately matched with the width of the resonance, i.e., to the core-hole lifetime of the associated transition. For $3d$ transition metal K edges, this is 1 to 2 eV, and for the rare earth L edges and the actinide M edges it is $\approx 5$ eV.

Other elements in the beamline include various defining slits. Those upstream of the monochromator and mirrors determine the fraction of the full beam accepted by the optical elements. This is useful in quantitative polarization analysis experiments, where the vertical and horizontal divergences must be closely controlled. A final set of slits before the sample are used to trim the beam of diffuse scattering arising from imperfections in the optical elements and to define the footprint of the beam on the sample. This is of importance when measuring integrated intensities of a number of Bragg peaks. The last component of the beamline, immediately preceding the sample, is an incident beam monitor that is used to normalize scattered count rates.

Bending magnets, wigglers, and undulators are all planar devices—i.e., the electron orbit is only perturbed in the horizontal plane, and the radiation produced is therefore horizontally polarized in the same plane. The degree of polarization represents a figure of merit for performing polarization analysis experiments. It is largely dependent on the type and stability of the source, although subsequent optics can modify it. For example, for a bending magnet source at the National Synchrotron Light Source (NSLS; Brookhaven National Laboratory, U.S.A.), $P_\zeta = 0.9$. For a wiggler $P_\zeta = 0.99$, and for an undulator source at the ESRF, $P_\zeta = 0.998$. This greater degree of polarization, together with smaller horizontal and vertical divergences at the third-generation sources, make undulator beamlines significantly better for quantitative polarization analysis. The degree of linear polarization can be further improved by choosing a monochromator Bragg reflection close to 45°—or by inserting a second (polarizing) monochromator just upstream of the sample, also with a $\theta_B \approx 45°$. This will further reduce the $\pi$ component of the incident polarization. This approach has been employed successfully in the study of ferromagnets.

While a linearly polarized beam is often sufficient for magnetic scattering, as discussed above (see Principles of the Method, Theoretical Concepts) the observation of ferromagnetic order via interference between non-resonant magnetic and charge scattering requires circular polarization. There are three main methods for obtaining circularly polarized light in the hard x-ray regime at a synchrotron (see also X-RAY MAGNETIC CIRCULAR DICHROISM).

1. Firstly, although bending magnet radiation is horizontally polarized on-axis, it is circularly polarized out of the plane of the synchrotron. This results from the fact that the electron motion appears elliptical when viewed from above or below the plane, and thus the off-axis radiation is elliptically polarized, with opposite polarization states being

obtained above and below the plane. Techniques based on this fact have been relatively popular, because of the ready access to bending magnet sources. The degree of circular polarization increases rapidly from zero to near unity as the linear, horizontal polarization drops to zero. The disadvantage of this method is that the flux also drops very rapidly, and even at the small viewing angles involved (a few tenths of a milliradian) intensity losses at factors of 5 to 10 must be suffered to gain appreciable circular polarization. In addition, the technique is more sensitive to beam motion, since such motions cause a change in both intensity and polarization. Finally, in order to switch the handedness of the polarization, the whole experiment must be realigned from above the plane to below the plane. This is a cumbersome procedure at best, and one liable to incur systematic errors.

2. A second source of circularly polarized light in the hard x-ray regime is to use purpose-built insertion devices. Planar wigglers and undulators do not produce circularly polarized light when viewed from out of the plane of the orbit, because they are comprised of equal numbers of left and right oscillations and the net helicity is zero. However, it is possible to design periodic magnet structures such that circularly polarized light is produced on-axis. These include crossed undulator pairs (i.e., a vertical undulator followed by a horizontal undulator close enough together for the radiation to remain coherent) and elliptical wigglers for which the electron motion is helical. These devices produce intense beams with high degrees of polarization, combined with the ability, in some cases, to rapidly switch the polarization state. However, they are expensive, and it can be difficult to preserve the circular polarization through the optical elements. To date there have been only a few magnetic scattering beamlines built around such devices in the hard x-ray regime.

3. A third device converts linear polarization into circular polarization, with some cost in intensity. This is a quarter wave plate. In the optical regime, such devices are constituted from birefringent materials—i.e., optically anisotropic materials whose thickness is chosen such that a linearly polarized wave sent in at 45° to the optical axes has its two components phase-shifted by $\pi/2$ relative to one another, thus producing circularly polarized radiation. In the hard x-ray regime, a similar result may be achieved by utilizing the fact that dynamical diffraction is birefringent. In this case, birefringent refers to the fact that the refractive index is different for polarizations parallel or perpendicular to the diffracting planes. Thus, by arranging the Bragg planes to be at 457° with respect to the plane of linear polarization, an x-ray phase plate may be constructed. The most effective devices employ a transmission Bragg geometry, for which the phase shift between the $\pi$ and $\sigma$ components is inversely proportional to the angle of offset from the Bragg peak. The resulting

transmitted beam is circularly polarized, and the circularly polarized flux is maximized by minimizing absorption losses. X-ray phase plates have been constructed from silicon, germanium, and diamond (Hirano et al., 1993, 1995). For sufficiently thin crystals, transmission factors of 10% to 20% are achievable. Since these devices may be employed on-axis of insertion device beamlines, a relatively high circularly polarized flux is obtained. These devices are particularly well suited for use with undulator radiation, where the collimation of the beam does not compromise the polarization by providing a spread in angles incident on the phase plate. For example, a diamond (111) phase plate 0.77 mm thick, installed at the undulator beamline, ID10, at the ESRF, achieved a degree of circular polarization $P_\eta = 0.96$, with a transmission of 17% (Sutter et al., 1997). The crystal was offset 0.016° from the Bragg peak. Typically, such phase plates are the last optical component before the sample.

Finally, we note that for experiments performed at the M edges of the actinides, the problem of absorption becomes acute, and this must be reflected in the beamline design. The number and total thickness of all x-ray windows should be minimized. At 3.7 keV, the absorption length of Be is $\Lambda = 0.5$ mm, and for Kapton $\Lambda = 0.05$ mm. For comparison, the same numbers at 8 keV are $\Lambda(\text{Be}) = 5$ mm and $\Lambda(\text{Kapton}) = 0.7$ mm. This puts constraints on the design of the experiment. For example, in their work on uranium arsenides, McWhan et al. (1990) utilized a beamline configuration with only one Be window and a further 500 μm of Be and 39 μm of Kapton between the source and detector. In addition, a helium-filled bag was placed over the spectrometer, so that the beam path was either through vacuum or helium.

**The Spectrometer.** Often, no special equipment is required for the magnetic scattering spectrometer. For the study of antiferromagnets, in particular, a standard x-ray diffraction spectrometer, such as those manufactured by Huber GmbH (*http://www.xhuber.com*) is sufficient. Larger, more robust spectrometers have also been made for magnetic scattering beamlines by, for example, Franke GmbH (X22C, NSLS; *http://www.franke-gmbh.de*) and Micro-Control (ID20, ESRF). These have the advantage that they can support large cryostats, for example. The latter design also allows for diffraction in both $\sigma$-incident and $\pi$-incident geometries with the same spectrometer.

Two types of detector are commonly used. The NaI photomultiplier tube is frequently the workhorse of such experiments, with a relatively low dark current (<0.1 counts per sec) and a high efficiency in the regimes of interest. However, it has poor energy resolution ($\Delta E/E \sim 0.3$). This is a particular problem for resonant experiments if an analyzer is not used, because of the large background arising from the fluorescent decays associated with the creation of the core hole. In addition, harmonic rejection of such detectors is not always sufficient. In such circumstances, a solid-state detector is used. This has an energy

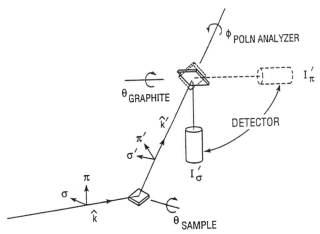

**Figure 4.** Linear polarization analyzer. When the analyzer, with a Bragg angle of 45°, is set to diffract in the scattering plane (solid lines), the σ′ component of the scattered beam is detected and the π′ component suppressed. Rotating the analyzer crystal by 90° about the scattered photon wave-vector, $\vec{k}'$, ($\phi_{poln\ analyzer} = 90°$) selects the π′ component of the scattered beam (dotted lines). Figure taken from Gibbs et al. (1988).

**Table 2. Selected Crystals and the Energy for Which $\theta_{anal} = 45°$**

| Crystal | Energy (keV) | Crystal | Energy (keV) |
|---------|--------------|---------|--------------|
| Al(111) | 3.75 | PG(006) | 7.85 |
| LiF(200) | 4.35 | Ge(333) | 8.05 |
| Cu(200) | 4.85 | Si(333) | 8.39 |
| LiF(220) | 6.16 | Cu(222) | 8.41 |
| Cu(220) | 6.86 | LiF(400) | 8.71 |
| Be(11.0) | 7.64 | Cu(400) | 9.70 |

resolution, $\Delta E/E$, of approximately 0.02, which allows for the removal of the fluorescent background with electronic discrimination. Often the fluorescence is in fact recorded simultaneously with the elastic signal by means of a multichannel analyzer. This then provides an on-line calibration of the energy. In addition, this signal is directly proportional to the intensity incident on the sample and may be used to normalize the scattered count rate. This method is of particular value when measuring the integrated intensity for a number of Bragg peaks with different beam footprints on the sample.

For polarization analysis experiments, a polarimeter must be mounted on the 2θ arm of the spectrometer. A number of ways of measuring the polarization have been developed, including methods to determine the complete polarization state with multiple beam diffraction methods (Shen et al., 1995). However, for analyzing linear polarization, the simplest method is to choose an analyzer crystal with a Bragg angle of $\theta_A = 45°$. This geometry is illustrated in Figure 4, taken from Gibbs et al. (1988). With the analyzer crystal set to diffract in the scattering plane, the linear component perpendicular to the diffraction plane (σ component) is reflected, while the parallel (π) component is suppressed. If the analyzer crystal is rotated by 90° about the scattered beam, the π component is reflected and the σ component is suppressed. Several devices have been built based on such an approach (see, e.g., Gibbs et al., 1989, and Brückel et al., 1996).

In order to perform quantitative polarization analysis, one must measure the integrated intensities of both the σ and π scattered components. How this is achieved depends on the relative divergences in the vertical and horizontal directions (which depend on the sample and beamline optics) and the mosaic of the analyzer crystal. For example, in the early work on holmium (Gibbs et al., 1991), the perpendicular and parallel divergences of the scattered beam were 0.05° and 0.03°, respectively,

and a pyrolytic graphite crystal was used as an analyzer ($\theta_A = 45°$ at $E = 7.85$ keV). This has a mosaic of 0.3° and therefore effectively integrates over the scattered beam. In this limit, it is sufficient to rock the sample in both configurations to obtain integrated intensities.

The choice of analyzer crystal is dictated by the need to obtain a reflection close to 45° at the energy of interest, and to match the experimental conditions. A number of crystals are listed in Table 2.

Inevitably, a compromise must be made, typically in the Bragg angle of the crystal. The leakage rate, i.e., the fraction of intensity from the other polarization channel that is reflected, is equal to $\cos^2(2\theta_A)$. For a Bragg angle of $\theta_A = 43°$, the leakage is $\approx 0.5\%$, so such a crystal is relatively tolerant of errors (the actual leakage also depends on the analyzer mosaic and the beam divergence).

**Specific Experimental Examples**

X-ray magnetic scattering offers a wealth of information on the magnetic order of a system, and the combination of resonant and non-resonant techniques make it widely applicable. It is therefore not possible to adequately represent all the practical aspects of the various implementations of the principles discussed above (see Principles of the Method). In the following, four broad classes of experiments are identified. In each case, a few examples are given to illustrate some of the strengths of the particular technique, together with some of the different experimental realizations.

**Non-resonant Scattering: Antiferromagnets.** The implementation for non-resonant magnetic scattering from antiferromagnets is relatively straightforward. The experiments are done at fixed incident energy, well away from any absorption edges; the magnetic signal is well separated from charge scattering (even for long period incommensurate structures) and the scattering may be interpreted directly in terms of magnetic densities. The principal difficulty lies in the weakness of the signal and the need to separate it from the background (diffuse) charge scattering. This is simplest to achieve in samples of high crystallographic quality, well matched to the incident beam divergence. For systems with long-range magnetic order, this will maximize the peak count rate. The situation may be further improved by employing an analyzer crystal on the 2θ arm, in front of the detector. This sharpens the longitudinal resolution and discriminates against any broad diffuse background. In addition, the effective energy resolution is then $\sim$10 eV, and this will eliminate any fluorescence background. For samples of reasonable quality (mosaic $<$0.1°), the use of a Ge(111) analyzer, for example, will dramatically improve the signal-to-noise ratio of a magnetic peak with long-range order, at a cost of a factor of $\sim$5 in the counting rate. If there is a significant amount of charge scattering present (for example, when studying thin films, reflectivity from the various interfaces may be the dominant contribution to the background), then it may be preferable to employ a polarization analyzer and record only the photons of rotated polarization. Depending on the degree of incident polarization, as well as the magnetic structure, this can provide gains in data quality. For non-resonant scattering, one is free to pick the energy at which the experiment is performed, such that an appropriate analyzer crystal has a Bragg angle of exactly 45°.

As mentioned in the introduction, the weakness of the scattering can also be an advantage, because it ensures that extinction effects, which can prevent reliable determination of intensities, are negligible. This has been exploited in a number of experiments in which precise measurements of the evolution of the order parameter are required.

One such example is the series of experiments performed on dilute antiferromagnets in an applied field. These are believed to be realizations of the random field Ising model and have attracted a great deal of interest, including numerous neutron scattering investigations. These latter, however, are hindered by extinction effects and by an inability to study the long length scale behavior. In addition, the samples must be doped with non-magnetic ions in order to induce the required phenomena and concentration gradients in the samples can then be a serious problem, masking the intrinsic behavior close to the phase transition. This illustrates another strength of x-ray magnetic scattering for these experiments, that is the small penetration depths ($\sim$5 μm in this case), which significantly reduce the effects of any concentration gradients. The experiments were performed on $MnF_2$ and $FeF_2$, doped with Zn, utilizing a superconducting magnet to apply fields up to 10 T (Hill et al., 1993, 1997). The cryostat constrained the scattering plane to be horizontal, that is a $\pi$ incident geometry was utilized.

One problem in this work that is worth mentioning (Hill et al., 1993) arose because the crystals order as commensurate antiferromagnets, but have two magnetic ions per unit cell, so that the periodicity is not doubled by the magnetic order. Magnetic scattering can then only be observed at reflections for which the charge structure factor is identically zero. Under these conditions, multiple charge scattering can be a problem (see Problems), and efforts must be taken to reduce such scattering.

Non-resonant experiments to separate $\vec{L}$ and $\vec{S}$ contributions to the ordered moment in an antiferromagnet have also been performed. These require measuring the integrated magnetic intensities, in both scattered polarization channels, for a number of Bragg peaks. Integrated intensities are obtained by rocking the sample and/or the polarization analyzer in an appropriate manner (see, e.g., Langridge et al., 1997). These must be corrected for any

differences in the rocking widths of the analyzer in the two configurations. Any charge scattering background may be subtracted off by performing the same measurements above the Nel temperature. The integrated intensity must also be corrected for absorption. The magnetic structure must already be known and Equation 6 is then used to relate the intensities to the quantities $\vec{L}$ and $\vec{S}$. Typically, intensities are not put on an absolute scale because of the difficulty in making extinction corrections for the strong charge Bragg peaks. Thus, often it is the ratio $\vec{L}/\vec{S}$ which is determined, or more precisely, $(\mu_L f_L(\vec{Q}))/(\mu_S f_S(\vec{Q}))$, where $f_{L,S}(\vec{Q})$ are the orbital and spin form factors, and $\mu_{L,S}$ the atomic orbital and spin moments. Such experiments require a high degree of incident polarization in order to simplify the analysis and thus benefit greatly from the third-generation sources. Examples of measurements of this type that have been carried out on antiferromagnets include work on holmium (Gibbs et al., 1988, 1991; C. Sutter, unpublished observation), uranium arsenide (McWhan et al., 1990; Langridge et al., 1997), and NiO (Fernandez et al., 1998).

The non-resonant cross-section, (Equation 6) is energy independent, and this may be exploited to work at very high incident photon energies ($\hbar\omega \geq 80$ keV). While experiments at these energies bring new difficulties, there are obvious advantages, primarily the greater penetration depth ($\sim$mm rather than $\sim$μm), which ensures that bulk-like properties are measured. In addition, the larger penetration depth can provide enhancements in the scattered count rate, relative to that at medium x-ray energies. This is because for weak reflections that are limited by absorption, such as magnetic peaks, the integrated scattered intensity is proportional to the scattering volume. Thus, quite large increases can be realized in going from medium to high energies due to this volume enhancement (in contrast to the case for strong charge reflections, for which the intensity is limited by extinction effects). However, care must be taken to match the instrumental resolution (which is typically extremely good at high x-ray energies) with the quality of the sample to take advantage of any increases.

The principles of high-energy x-ray magnetic scattering experiments are identical to those described previously; however, there are some differences in implementation. Firstly, the need for specialized x-ray windows is eliminated and simple Al windows on cryostats may be used (e.g., at 80 keV, 10 mm of Al results in a 40% loss of intensity). This can bring improvements in parameters such as temperature stability. The experiments are typically performed in transmission, and the need to worry about sample surface preparation is eliminated. At high energies, the Bragg angles are correspondingly reduced; for low index reflections, the non-resonant cross-section greatly simplifies. Neglecting terms containing $\sin^2\theta$, then

$$\langle M_m \rangle = i\left(\frac{\hbar\omega}{mc^2}\right)\begin{pmatrix} \sin2\theta S_2 & 0 \\ 0 & \sin2\theta S_2 \end{pmatrix} \quad (25)$$

and the experiments are only sensitive to the spin component perpendicular to the scattering plane. For linear incident polarization, the scattering does not alter the polarization state. In this limit, one cannot distinguish the magnetic scattering from charge scattering through the use of polarization analysis. On the other hand, one obtains the spin density, without the need for polarization analysis to separate the orbital contribution. High-energy experiments can then be combined, in principle, with neutron diffraction experiments to derive $\vec{L}$ and $\vec{S}$.

A disadvantage of working at such high energies is the limited number of beamlines capable of providing a high flux at the relevant wavelengths. In addition, higher-order contamination can be a significant problem because of the lack of x-ray mirrors, and multiple scattering, if present, is significantly worse at high energies. Nevertheless, successful experiments have been performed (Lippert et al., 1994; Strempfer et al., 1996; 1997; Vigliante et al., 1997; Strempfer et al., 2001).

**Non-resonant Scattering: Ferromagnets.** Non-resonant magnetic scattering can be used to determine $\vec{L}$ and $\vec{S}$ in ferromagnets. One particular implementation for such an experiment, first proposed by Laundry et al. (1991), utilizes a white, elliptical polarized beam incident on a single crystal and an energy dispersive detector positioned at a scattering angle of 90°. The advantages of this method are that it is relatively insensitive to sample motion upon reversal of the magnetization, that the scattering is performed at the optimum angle of 90°, and that several reflections may be collected at once. The disadvantages are that strong fluorescence lines may superpose some Bragg peaks and that high count rates may saturate the detector, giving rise to dead-time errors. If measurements are made for two different magnetization angles, $\alpha$ (Equation 24), then the $\vec{L}$ and $\vec{S}$ contributions may be separated. In an experiment on HoFe$_2$, Collins et al. (1993) chose $\alpha = 0$ and 90°, as shown in Figure 5. The ratio of the asymmetry ratios then determines the spin-orbital ratio

$$\frac{F_S}{F_L} = \frac{R(\alpha = 90°)}{R(\alpha = 0)} - 1 \tag{26}$$

The measurements of Collins et al. (1993) were in approximate agreement with band-structure calculations. The disadvantage of this method is the need for two easy magnetization axes 90° apart and the requirement that the material be magnetically soft. It should also be mentioned that in cases for which absorption edges are accessible, the technique of XMCD (X-RAY MAGNETIC CIRCULAR DICHROISM) can provide an alternative to such approaches and can determine $\langle L_z \rangle$ and $\langle S_z \rangle$ independently, in an element-specific manner, in ferromagnets.

**Resonant Scattering: Antiferromagnets.** The principal difference between a resonant and a non-resonant magnetic scattering experiment is the need for a scanning monochromator. The incident energy resolution should match the width of the resonance to attain the greatest enhancement, or exceed it, if information on lifetimes is required. In order to compare the experimentally measured resonances with the theoretical predictions of Equation 15, integrated magnetic intensities must be recorded at

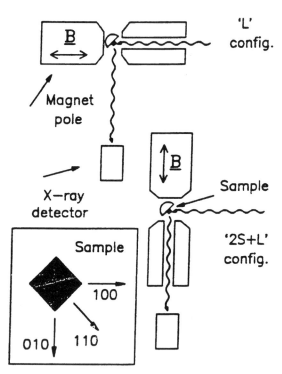

**Figure 5.** Schematic of the diffraction geometry used in Collins et al. (1993), for scattering from ferrimagnetic HoFe$_2$. In the top configuration, with the sample magnetized along the incident beam direction, the scattering is only sensitive to the orbital moment density. In the second configuration (bottom), the sample is magnetized perpendicular to the incident beam and contributions from both $\vec{L}$ and $\vec{S}$ are observed. The sample is a single crystal and the incident beam is "white" (unmonochromatized).

each energy and corrected for absorption. While calculations for the energy dependence of the absorption exist (Cromer and Lieberman, 1981), these are not reliable in the vicinity of the edge, where they are most important. Ideally, to obtain the appropriate corrections, the absorption would be measured directly through a thin film of the sample material. By reference to calculations made far from the edge, such data may be brought to an absolute scale and $\mu(E)$ obtained. However, this is not always possible, and various schemes have been attempted to circumvent this problem. One is to obtain $\mu$ from measurements of the fluorescence intensity, $I_f$ using the relation

$$I_f(E) = C \frac{\mu_R(E)}{\mu_S(E_f) + \mu_S(E)\dfrac{\sin\psi}{\sin\phi}} \tag{27}$$

where $\mu_S(E)$ is the absorption of the sample at photon energy $E$, $E_f$ is the fluorescence energy, $\mu_R$ is the absorption of the resonance ion and $\psi$ and $\phi$ are the angles of incidence and fluorescent beams, respectively. The fluorescence is measured as a function of incident energy by placing an open detector at 90° to the scattering plane (to minimize charge scattering) and with the discriminator set to select the fluorescent energy. The constant, $C$, may

UAs (0,0,5/2)

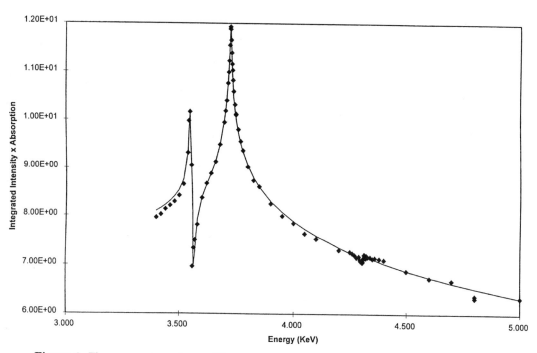

**Figure 6.** The energy dependence of the magnetic scattering at the (0,0,2.5) antiferromagnetic Bragg peak in uranium arsenide as the incident energy is scanned through the U $M_5$, $M_4$, and $M_3$ edges. Each data point represents an integrated intensity and is corrected for absorption. The solid line is a fit to three dipole oscillators (McWhan, 1998).

be obtained by normalizing the data to the calculated values far from the edges (see, e.g., McWhan et al., 1990; McWhan, 1998). This method has its limitations, and other approaches have also been tried (e.g., Tang et al., 1992; Sanyal et al., 1994). For a Bragg peak along the surface normal, and a flat sample, the absorption correction corresponds to simply multiplying the integrated intensity by $\mu$. The angular factors of the cross-section, Equation 15, will also contribute some energy dependence, together with the Lorentz factor $1/\sin 2\theta$.

Figure 6 shows the absorption-corrected energy dependence of the (0,0,2.5) magnetic Bragg peak in uranium arsenide on scanning the incident energy through the U $M_5$, $M_4$, and $M_3$ edges (McWhan et al., 1990). The solid line is a fit to three coherent dipole resonances, together with the appropriate angular factors. These data were taken with an incident energy resolution of 1 eV, and the authors were thus able to extract branching ratios and lifetimes for the various resonances, though these are sensitive to the absorption correction and are therefore only semiquantitative. The overall energy dependence illustrated in Figure 6 is also sensitive to the exchange splitting present in the excited state, and these data led McWhan and co-workers to conclude that it was smaller than 0.1 eV.

The position of the resonance is a function of atomic species and valence state, and thus this technique also provides a means for investigating magnetic order in mixed valent compounds. For example, McWhan et al. (1993) studied TmSe. In this material both $Tm^{2+}$ and $Tm^{3+}$ config-

urations exist, each of which may carry a magnetic moment. The resonance energies of the two ions differ by 7 eV, as determined by x-ray absorption spectroscopy (XAS) which is comparable to the energy width of the resonances. Thus, by tuning the incident photon energy to one or other of the dipole transitions of the two ionic configurations, the dominant contribution to the cross-section may be varied from one configuration to the other, and the magnetic order of the two valences may be studied separately. The (003) magnetic reflection showed two peaks as a function of energy, demonstrating that both valence states exhibit long-range magnetic order.

Both non-resonant and resonant magnetic x-ray scattering may be used to determine moment directions in ordered structures, and this has been demonstrated for a number of antiferromagnets. To do this, integrated intensities of a number of magnetic Bragg peaks are measured, and then the $Q$ dependence of the intensities is fit to expressions such as Equation 9 and Equation 15. Detlefs et al. (1996, 1997) demonstrated this approach in studies of $RNi_2B_2C$, with R = Gd, Nd, and Sm. For R=Gd and Sm, the structure could not be solved using neutron scattering techniques because of the high neutron absorption of the Gd and Sm, together with the B. The different polarization dependences of the non-resonant, resonant dipole, and resonant quadrupole scattering were exploited to solve the structures from a relatively small data set. Before a direct comparison with the cross-section can be made, a variety of other angular factors must be included. For a Bragg peak at an angle $\alpha$ away from the surface

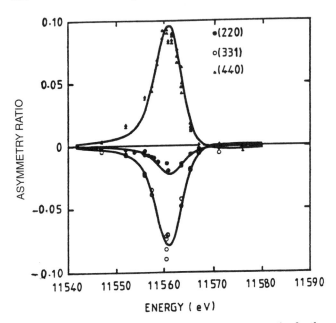

**Figure 7.** The energy dependence of the asymmetry ratios for the (220), (331) and (440) Bragg peaks of the ferromagnet, CoPt$_3$. Data show a large resonant enhancement at the Pt L$_3$ edge and are well fit by a single $2p \rightarrow 5d$ dipole transition. Taken from de Bergevin et al. (1992).

normal and scattering from a flat fully illuminated sample, then

$$I \propto \left| \sum_n e^{i\vec{Q} \cdot \vec{r}_n} f_n^{res} \right|^2 \frac{\sin(\theta + \alpha)\sin(\theta - \alpha)}{2\mu \sin\theta \, \cos\alpha \, \sin 2\theta} \quad (28)$$

where the Lorentz factor has also been included. A more detailed study has been carried out on pure Nd (D. Watson et al., 1998).

These measurements are made simpler by the lack of a magnetic form factor for resonant scattering, and the fact that the $F^{(L)}$ of Equation 15 are independent of the scattering angle. Apart from an overall scale factor, then, the only unknowns are the spin components, $z_1$, $z_2$, and $z_3$. A coordinate transformation, which depends on $\theta$, is then required to rotate into the coordinate system of the sample, obtaining the moment directions relative to the crystallographic axes.

Other experiments on antiferromagnets have studied thin films, ultra-small samples, multilayers, and compounds with more than one magnetic species. These take advantage of the small penetration depth, high-resolution, and element specificity of the technique. However, in implementation, they do not involve concepts different from those already discussed.

**Resonant Scattering: Ferromagnets.** The first experimental observation of magnetic scattering with x-rays from ferro- and ferrimagnets is credited to de Bergevin and Brunel (1981a,b), who utilized Cu K$_\alpha$ radiation to study powdered samples of zinc-substituted magnetite, Fe$_{2-x}$ Zn$_x$O$_4$, and Fe. Interference was induced through the proximity to the iron K edge. Studies of interfacial magnetism were first performed in ferromagnetic multilayers, namely Gd-

Y (Vettier et al., 1986), utilizing non-resonant magnetic-charge interference scattering. The interference was induced by tuning through the Gd L$_2$ and L$_3$ edges.

More recently, de Bergevin et al. (1992) carried out a study of CoPt$_3$ utilizing resonant magnetic scattering at the Pt L$_3$ edge. The asymmetry ratio was determined for three Bragg peaks, and as a function of energy (Fig. 7). The data are well fit by a single dipole ($2p \leftrightarrow 5d$) resonance, with a peak scattering amplitude of $0.8r_0$.

**Surface Magnetic Scattering.** The large resonant enhancements in the cross-section allow the measurement of the very weak signals associated with the surface magnetization density profile. To a first approximation, the experimental techniques for such work are identical to those developed for surface x-ray scattering (SURFACE X-RAY DIFFRACTION), with the additional complication of a strong energy dependence to the scattering.

Briefly, for a semi-infinite crystal, the diffraction pattern consists of rods of scattering, known as truncation rods, which pass through the Bragg points and which are parallel to the surface normal (Fig. 8A). Characterizing the intensity along such rods provides information on the decay of the charge density near the surface. Similarly, magnetic truncation rods are present when a magnetic lattice is terminated, and the decay of the magnetic intensity is related to the profile of that magnetic interface. For a step function interface, the diffraction pattern is the convolution of a reciprocal lattice of delta function Bragg peaks with the Fourier transform of the step function, $1/q$. This gives a $1/(\Delta q_z)^3$ broadening of each Bragg point, where $\Delta q_z$ is the deviation from the Bragg peak, along the surface normal. If a broader interface is present, the Fourier transform will be sharper than $1/q$ and the intensity fall-off around each Bragg point will be faster (Robinson, 1991). Thus, measuring the intensity fall-off measures the Fourier transform of the magnetic interface profile, which may in principle be different from the chemical interface.

For an antiferromagnet, there is both a magnetic contribution to the charge rods, as well as pure magnetic rods, as shown in Figure 8A, taken from G. Watson et al. (1996b; 2000). In addition to scans around Bragg points, the surface magnetic behavior may also be probed explicitly by employing the grazing incidence scattering geometry shown in Figure 8B. $\alpha_j$ and $\alpha_s$ are the angles of the incident and scattered beams with respect to the surface, respectively. In the figure, most of the momentum transfer is in the plane of the crystal $\vec{q}_\parallel$, with a small component along the surface normal, $\vec{q}_z$. For example, such a situation occurs for scans along the (0,1,L) magnetic rod, with L small. In such a geometry, it is possible to take advantage of the well known enhancement in the scattering, when $\alpha_i = \alpha_s = \alpha_c$, the critical angle. Such techniques have been demonstrated in the antiferromagnet UO$_2$, working at the U $M_4$ edge (G. Watson et al., 1996b; 2000). The energy dependence of scans taken along a magnetic rod (0,1,L) are shown in Figure 9 (G. Watson et al., 1996b; 2000). These data illustrate that the critical angle, for which the intensity is a maximum, is also a function of the incident energy, since $\alpha_c^2 \sim (Z + f'')\rho\lambda$, where $Z$ is the total charge, $\rho$ is the average atomic density and, $f''$ the

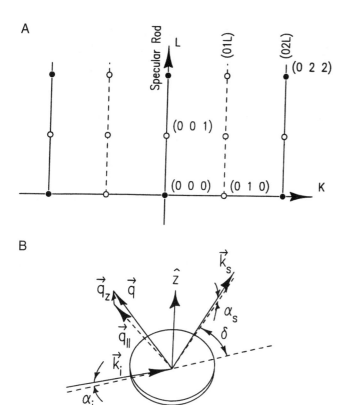

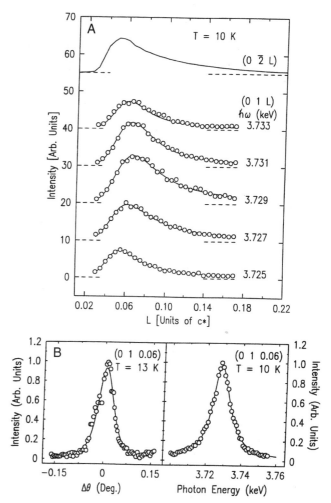

**Figure 8.** (**A**) Schematic of reciprocal space for a semiinfinite antiferromagnet, illustrated for UO$_2$. In addition to the charge (closed circles) and antiferromagnetic (open circles) bulk Bragg peaks, there are rods of scattering connecting the Bragg points and running parallel to the surface (vertical lines). The dotted lines are pure magnetic rods of scattering. The specular rod (00L) corresponds to scattering in which there is no in-plane momentum transfer, and the angle of incidence equals the angle of reflection. (**B**) Scattering geometry for surface magnetic scattering. $\alpha_i$ and $\alpha_S$ are the angles of incidence and exit, normal to the surface, and $\delta$ is the in-plane scattering angle. In this diagram, the momentum transfer $\vec{q}$ is largely in-plane, $\vec{q}_{\parallel}$, with a small component normal to the surface, $\vec{q}_z$. Figure taken from G. Watson et al. (1996b).

**Figure 9.** (**A**) Energy dependence of scans along the (0,1,L) magnetic rod in UO$_2$. The data exhibit an enhancement at an energy dependent critical angle, $\alpha_c$. For comparison, the same scan is shown for along the (0,2,L) charge rod, solid line. (**B**) Transverse scan across the (0,1,L) magnetic rod. (**C**) Intensity of the magnetic rod as a function of incident photon energy. Figure taken from G. Watson et al. (1996b).

real part of $f_n^{\mathrm{res}}$ of Equation 4. Measurements of $\alpha_c$ thus provide an independent measure of $f'$. Scans across the magnetic rod provide information on the disorder of the magnetic layer normals. In UO$_2$, such scans were found to be identical to those of the charge rods (Fig. 9). The scattered intensity may be described by

$$\frac{I(k_i, k_s)}{I_0} = \frac{A}{A_0} \sin\alpha_i \ \sin\alpha_s |T(\alpha_i)|^2 |T(\alpha_s)|^2$$
$$\times \left| \sum_{n=0}^{\infty} r_0 f_n^{\mathrm{res}} \rho_n^{\mathrm{M}}(q_{\parallel}) e^{iq_z z_n} \right|^2 \quad (29)$$

where $T(\alpha)$ is the Fresnel transmission coefficient, $A/A_0$ the ratio of the illuminated area to the cross-sectional area of the beam, $I_0$ the incident flux, $r_0 f_n^{\mathrm{res}}$ the magnetic scattering amplitude, and $\rho_n^{\mathrm{M}}$ the magnetic density in the $n$th layer with wave-vector $q_{\parallel}$. The fits to this form in Figure 9A assume a perfectly sharp magnetic interface, and describe the data well. This technique can be used to

determine the temperature and depth dependence of the magnetic order parameter profile. By varying L, the effective penetration depth varies. For example, in UO$_2$, for $L = 0.075, 0.15,$ and 1, the corresponding penetration depth was 50, 120, and 850Å, respectively.

Similar studies of surface magnetism have been carried out in ferromagnets, for which the asymmetry ratio is measured along the truncation rod. This was first done in the soft x-ray regime at the Fe L$_{2,3}$ edges (Kao et al., 1990) and Co L$_{2,3}$ edges (Kao et al., 1994) in which the specular magnetic reflectivity was measured (i.e., (0,0,L) scans). This describes the magnetic order normal to the surface. Recent studies of Co multilayers have focused on the off-specular diffuse scattering, which describes the roughness of the interface (Mackay et al., 1996). In this work, the magnetic interfaces were found to be less rough than the structural interfaces. Finally, working at the Pt L$_3$ edge, Ferrer et al. (1996) have measured complete magnetic truncation rods in ferromagnetic Co$_3$Pt(111) and Co/Pt(111) ultra-thin films (Ferrer et al., 1997). The advantage of resonant magnetic scattering in such studies is the technique's element specificity, allowing multicomponent systems to be resolved and buried interfaces to be studied, in contrast to scanning probe techniques. In addition, it is essential to characterize the structural interface(s) in order to fully understand the magnetic interface. X-ray scattering techniques are able to characterize both the structural and magnetic interfaces with the same depth sensitivity and resolution, in the same experiment.

## DATA ANALYSIS AND INITIAL INTERPRETATION

Quantities such as ordering wave vector, ordered moment, and moment direction may be obtained from the position and integrated intensity of magnetic reflections. For more detailed information on the magnetic correlations, e.g., to extract forms for the moment-moment correlation function, as well as parameters such as correlation lengths and critical exponents, then it is necessary to fit the lineshapes of the magnetic peaks. The observed intensity, $I\vec{Q}$, is given by the convolution of the cross-section, $d\sigma(\vec{Q})/d$, and the instrumental resolution $R(\vec{q})$

$$I(\vec{Q}) = \int R(\vec{q}) \frac{d\sigma(\vec{Q} - \vec{q})}{d\Omega} d\vec{q} \qquad (30)$$

A common problem in the analysis of x-ray data is the accurate determination of $R(\vec{q})$. Ideally, a perfect $\delta$-function scatterer (e.g., a silicon single crystal) is placed in the sample position and the resolution is measured at the relevant momentum transfers. The problem is that it is, in general, not possible to find a suitable test sample. A number of approaches are therefore adopted. For studies of critical fluctuations in the vicinity of a phase transition, it is common to take the scattering below $T_N$ to be long-range ordered, i.e., resolution limited. Fits to scans through this Bragg peak can then be used to parameterize the resolution function and in fitting the experimental data, following Equation 30. In other cases, nearby Bragg peaks can be used, again as a lower bound on the resolution function (the intrinsic resolution is equal to or better than the apparent widths of the Bragg peaks). The longitudinal resolution width (i.e., along the momentum transfer) is controlled by the monochromator and analyzer, if present, or the angular acceptance of the detector, if not. For a given monochromator-analyzer combination, the longitudinal resolution is a minimum at the nondispersive condition, i.e., when the scattering vectors of the monochromator, sample, and analyzer are equal, and in alternating directions. For a beamline with a double crystal monochromator and analyzer, the in-plane resolution functions are often Lorentzian squared functions. The out-of-plane resolution is determined by collimating slits and can be approximated by a triangular function. In such a case, with the assumption that the resolution function is separable

$$R(\vec{q}) = \left(\frac{1}{w_x^2 + q_x^2}\right)^2 \left(\frac{1}{w_y^2 + q_y^2}\right)^2 \left(1 - \left|\frac{q_z}{w_z}\right|\right) \qquad (31)$$

where the $x$ and $y$ directions are transverse and longitudinal to the momentum transfer, and $q_z$ is perpendicular to the scattering plane. For example, in a study of the critical fluctuations on holmium, Thurston et al. (1994) employed a Ge(111) analyzer and double bounce Ge(111) monochromator, and obtained resolution halfwidths of $2.9 \times 10^{-4}$, $4.5 \times 10^{-4}$, and $4.3 \times 10^{-3} \text{Å}^-$ in the transverse, longitudinal, and out-of-plane directions, respectively, at $Q = 1.92 \text{Å}^{-1}$.

Approximating the out-of-plane resolution with a triangular function, as in Equation 31, has the advantage that the integration in the $q_z$ direction may be performed analytically, for certain cross-sections, reducing the fitting of the data to a 2-D numerical integration and lessening the fitting time. Note that it is crucial that a full 3-D convolution procedure be carried out in fitting the data in order to obtain meaningful correlation lengths, even if the data themselves are only obtained in one direction.

Strictly, the cross-section is proportional to the dynamic structure factor, $S(\vec{Q}, \omega)$, where $\omega$ is the energy transfer of the scattering process. However, typically the energy resolution of an x-ray experiment is relatively broad (e.g., 1 to 10 eV) compared to the relevant magnetic energy scales, and all the magnetic fluctuations are integrated over, reducing $S(\vec{Q}, \omega)$ to the static structure factor, $S(\vec{Q})$. This is the quasi-elastic approximation. In turn, this structure factor may be expressed as

$$\frac{d\sigma}{d\Omega} = S(\vec{Q}) \propto \langle f(\vec{Q})f(-\vec{Q}, t = 0)\rangle = \langle |f(\vec{Q})|^2\rangle \qquad (32)$$

i.e., the instantaneous correlation function. For non-resonant magnetic scattering, this may be directly related to a particular moment-moment correlation function for a given scattering geometry. In the $MnF_2$ example (see Principles of the Method), for which $\vec{L} = 0$, then for $\sigma$ incident polarization, $S(\vec{Q}) = \langle \vec{S}_\perp(\vec{Q})\vec{S}_\perp(-\vec{Q})\rangle$, where $\vec{S}_\perp$ is the component of the spin perpendicular to the scattering plane, and we have ignored terms of order $\sin^2\theta$ in the scattering amplitude. For resonant scattering, the situation is more complex, (see e.g., Luo et al., 1993; Hill and McMorrow, 1996), though for first-order dipole scattering, and $\sigma$ incident polarization, $f(\vec{Q}) \propto \hat{z}(\vec{Q}) \cdot \hat{k}_f$, the projection of the moment direction along the scattered photon direction.

The form of $S(\vec{Q})$ depends on the system under study. Thermal critical fluctuations, for example, are expected to give a Lorentzian-like behavior

$$S(\vec{Q}) \sim \frac{k_B T_\chi(T)}{1 + (Q - Q_0)^2/\kappa^2} \qquad (33)$$

where $Q_0$ is the Bragg position, and $\kappa = 1/\xi$ is the inverse correlation length. Fits to the temperature dependence of the parameters, $\chi = \chi_0((T - T_N)/T_N)^{-\gamma}$ and $\xi = \xi_0((T - T_N)/T_N)^{-\nu}$ allow for the determination of the critical exponents $\gamma$ and $\nu$. The order parameter exponent, $\beta$, is determined from fits to the temperature dependence of the integrated intensity below $T_N$, i.e., $I = I_0((T_N - T)/T_N)^{2\beta}$. Conversely, static random disorder may give rise to Lorentzian-squared lineshapes. One of the strengths of high-resolution x-ray scattering is the ability to distinguish between such behavior, even at relatively long length scales, e.g., close to a phase transition. This ability has revealed new phenomena; for example the critical fluctuations immediately above a magnetic phase transition have been shown to exhibit two length scales (Thurston et al., 1993, 1994).

## SAMPLE PREPARATION

The range of experimental conditions and techniques described in this unit make it very difficult to make any useful general statements about what is required by way of sample preparation. However, perhaps the most important point to emphasize is that the better the sample is characterized before any synchrotron beamtime, the higher the chances of success. Particularly important properties to be characterized include the sample orientation and crystallographic quality, and the magnetic properties, through, for example bulk magnetization measurements to obtain transition temperatures, magnetic anisotropies etc. (see MAGNETROMETRY, THERMOMAGNETIC ANALYSIS and TECHNIQUES TO MEASURE MAGNETIC DOMAIN STRUCTURES). For reflection experiments, surface (and near-surface) quality can be all-important. Careful polishing of the appropriate surface (ideally one with a surface normal parallel to the direction of interest in reciprocal space) can provide large rewards in terms of signal to noise. However, the appropriate polishing method is extremely sample dependent, with electro-polishing or the use of diamond paste (with perhaps a 1 mm grit size) among the most common techniques used (see SAMPLE PREPARATION FOR METALLOGRAPHY).

## PROBLEMS

### Other Sources of Weak Scattering

One of the biggest problems in x-ray magnetic scattering experiments is distinguishing the weak magnetic scattering from other possible sources of scattering that may produce peaks coincident with it. In general, there is no single test that is always definitive, and it is necessary to use a combination of tests, together with prior knowledge of the system. Possible sources of scattering are discussed under the following subheadings.

**Multiple Scattering.** This is the scattering that results from two (or more) successive Thomson scattering events. It occurs when a second reciprocal lattice point, $\vec{q}_2$, intercepts the Ewald sphere, in addition to the point of interest, $\vec{q}_1$. Then double scattering of the form $\vec{q}_2 + \vec{q}_3$ occurs, where $\vec{q}_3 = \vec{q}_1 - \vec{q}_2$ is another allowed reciprocal lattice vector. It is a particular problem in systems for which the magnetic unit cell is the same size as the chemical unit cell. At synchrotrons, multiple scattering count rates can exceed $10^4$ photons per sec. The problem can be worse in crystals of lower crystallographic quality, or in lower-resolution configurations for which the requirements for making two successive Bragg reflections are less stringent. It is also aggravated at higher incident x-ray energies, because the radius of the Ewald sphere is larger and the probability of more than one reciprocal lattice point intercepting it becomes correspondingly larger. There are two possible means of eliminating multiple scattering. First,

by rotating the Ewald sphere around the momentum transfer, $\vec{q}_1$, thus moving the $\vec{q}_2$ lattice point off the Bragg condition, while leaving the scattering vector unchanged. Typically, only small rotations are required and the scattering geometry relative to the moment directions need not be changed significantly. The second alternative is to vary the incident photon energy. This alters the radius of the Ewald sphere, sweeping it through reciprocal space, thus moving off the $\vec{q}_2$ lattice point. Such a tactic is not an option for resonant experiments.

Unless there is a structural phase transition associated with the magnetic ordering, the multiple scattering is likely to be temperature independent.

**Higher Harmonic Contamination of the Beam.** This was already discussed in the section on beamlines. It can produce scattering at non-integer, but commensurate reciprocal, lattice points, but may be readily identified by the energy of the scattered photons, even with a NaI detector. It is typically temperature independent.

**Magnetoelastic Scattering.** This is charge scattering that arises from a lattice modulation induced by the magnetic order. A magnetoelastic coupling can occur when, for example, there is a distance dependence to the exchange interaction, though other mechanisms are possible (see e.g., McMorrow et al., 1999). A magnetization density wave can then induce a displacement wave in the lattice with an amplitude which can depend only on the relative change in magnitude of the magnetization wave. In such a situation, it is straightforward to show (e.g., Lovesey and Collins, 1996) that for a transverse magnetization density wave, a distortion is induced at twice the wave-vector (i.e., half the wavelength) with an amplitude that varies as the magnetization squared. The diffracted intensity then varies as the moment to the fourth power. Thus, although the scatterings from both modulations will go to zero at the same temperature, they will have very different temperature dependences, and may be distinguished in this manner. In addition, of course, the charge scattering does not rotate the polarization state of the photon.

In general, more complicated modulations are possible, and the magnetoelastic scattering need not occur at twice the wave-vector of the magnetic scattering. Such is the case for the spin structure in holmium, for example (Gibbs et al., 1988, 1991). Below $T_N = 132$ K, holmium forms a basal plane spiral. At low temperatures, the magnetic wave-vector, $\tau_m$, exhibits a tendency to lock into rational fractions of the c axis reciprocal lattice vector, $c* = 2\pi/c$. This is a result of a bunching (in pairs) of the moments in each successive ab plane about the six easy axes of magnetization, 60° apart. These pairs are known as doublets. If only a single spin (layer) is associated with a given easy axis, then this is known as a singlet, or spin slip. Ordered arrays of such spin slips give rise to the magnetic wave-vectors observed. In addition, at each spin slip, the phase of the magnetic modulation takes a jump in going from layer to layer and thus the magneto-elastic coupling is changed. This can give rise to a lattice distortion, with a new periodicity, $\tau_l$, where $\tau_l = 12\tau_m - 2$ (Bohr et al., 1986). In general then, the magnetoelastic scattering can appear close to the magnetic scattering, and can even be

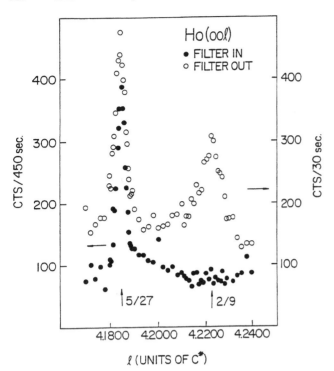

**Figure 10.** Magnetic and charge scattering in holmium. Data were taken at 7847 eV (off resonance). The open circles are recorded with no polarization analyzer and exhibit two peaks, at $\tau_m = 5/27$ and $\tau_I = 2/9$. With the polarization analyzer in, and set to record the $\sigma \rightarrow \pi$ scattering processes, only the $\tau_m$ peak is observed (closed circles). This is consistent with $\tau_m$ being magnetic scattering, and $\tau_I$ arising from charge scattering due to a lattice distortion induced by the magnetic order (Gibbs et al., 1985).

coincident with it (e.g., if $\tau_m = 2/11$). Polarization analysis is therefore required to distinguish the scattering arising from these two modulations, as illustrated in Figure 10, from Gibbs et al. (1985). The magnetic scattering observed at 5/27 is observed both with and without the analyzer. Conversely, the 2/9 peak is not seen in the $\sigma \rightarrow \pi$ channel and therefore corresponds to charge scattering. This peak is broader than the magnetic peak because of disorder in the spin slip structure, which is not reflected in the magnetic spiral.

Finally, there may be other couplings induced by the magnetic order that give rise to charge scattering. For example, in non-resonant magnetic scattering studies of $Fe_{0.75}Co_{0.25}TiO_3$, Harris et al. (1997) observed a quadratic increase in the scattering at the antiferromagnetic Bragg position (1,1,2.5), as a function of applied field, $H$. This effect was not seen in neutron data, and was attributed to a magneto-elastic distortion of the same wave-vector as the antiferromagnetic order. This can arise from terms in the free energy of the form $\rho_s M_s M$, where $\rho_s$ is the staggered charge density, $M_s$ the staggered magnetization, and $M$ the uniform magnetization. If the cost of the distortion due to the elastic energy, $1/2\rho_s^2$, is included and the total minimized with respect to $\rho_s$, then it follows that $\rho_s \sim M_s M$. The intensity of the scattering from the charge density is then $I_{charge} \sim (M_s M)^2$ and that from the magnetic density, $I_{mag} \sim M_s^2$. Therefore, in this case, the two intensities have the same temperature dependence. How-

ever, in an applied field, the uniform magnetization will grow linearly (for small fields) and so the charge scattering will increase as $H^2$ (note that for $H = 0$, the charge scattering should go to zero). In addition, polarization analysis should be able to distinguish between scattering mechanisms in this case. The magnetoelastic scattering was not seen in the neutron diffraction experiments because of the weakness of the coupling between the neutrons and the charge density waves, relative to the strength of the magnetic interaction.

### Near-Surface Effects

With the exception of high energy non-resonant scattering, all x-ray magnetic scattering experiments are performed in a reflection geometry, and penetration depths vary from perhaps $\sim 10$ µm in some non-resonant experiments to $\sim 1000$ Å, or less, in some resonant cases. While this surface sensitivity can be a boon, it can also cause varying degrees of complications in studying "bulk-like" behavior. The near-surface region can differ from the bulk in strain field densities, stochiometry, and different lattice connectivities (due to surface roughness), in addition to the intrinsic surface effects related to the termination of the lattice. Such effects can in turn alter the magnetic behavior of the near-surface region. This can be relatively benign, e.g., producing a single-domain state in cases where there would otherwise be a random distribution amongst energetically equivalent wave-vectors (e.g., Hill et al., 1995a; D. Watson et al., 1996), or more severe, introducing entirely new phenomena, such as the second length scale in magnetic critical fluctuations (e.g., Thurston et al., 1993, 1994; G. Watson et al., 1996a) or altering the observed magnetic state entirely (e.g., Hill et al., 1993). To date, little systematic work has been performed to assess the effect of surface preparation on "bulk" magnetic behavior and there is no general prescription for the correct surface preparation for a given x-ray experiment: It is something that must be approached on a case-by-case basis, and is beyond the scope of this unit. The purpose of this section is simply to emphasize the problem and to encourage careful characterization of the near-surface region for each experiment.

### ACKNOWLEDGMENTS

This work was supported by the U.S. Department of Energy, Division of Materials Science under contract no. DE-AC02-98CH10886.

### LITERATURE CITED

Bartolomé, F., Tonnerre, J. M., Sève, L., Raoux, D., Chaboy, J., García, L. M., Krisch, M., and Kao, C.-C. 1997. Identification of quadrupolar excitation channels at the L₃ edge of rare-earth compounds. *Phys. Rev. Lett.* 79:3775–3778.

Blume, M. 1985. Magnetic scattering of x-rays. *J. Appl. Phys.* 57:3615–3618.

Blume, M. 1994. Magnetic effects in anomalous dispersion. *In* Resonant Anomalous X-ray Scattering. (G. Materlik, C.

Sparks, and K. Fischer, eds.). pp. 513–528. Elsevier Science Publishing, New York.

Blume, M. and Gibbs, D. 1988. Polarization dependence of magnetic x-ray scattering. *Phys. Rev. B* 37:1779–1789.

Bohr, J., Gibbs, D., Moncton, D. E., and D'Amico, K. L. 1986. Spin slips and lattice modulations in holmium: A magnetic x-ray scattering study. *Physica A* 140:349–358.

Brown, G. and Lavendar, W. 1991. Synchrotron radiation spectra. *In* Handbook of Synchrotron Radiation, Vol. 3 (G. S. Brown and D. E. Moncton, eds.). pp. 37–61. North Holland Publishing, New York.

Brückel, Th., Lippert, M., Köhler, Th., Schneider, J. R., Prandl, W., Rilling, V., and Schilling, M. 1996. The non-resonant magnetic x-ray scattering cross-section of $MnF_2$. Part I. Medium x-ray energies from 5 to 12 keV. *Acta Crystallogr.* A52:427–437.

Brunel, M. and de Bergevin, F. 1991. Magnetic scattering. *In* Handbook of Synchrotron Radiation, Vol. 3 (G. S. Brown and D. E. Moncton, eds.). pp. 535–564. Elsevier Science Publishing, New York.

Carra, P. and Thole, B. T. 1994. Anisotropic x-ray anomalous diffraction and forbidden reflections. *Rev. Mod. Phys.* 66:1509–1515.

Carra, P., Thole, B. T., Altarelli, M., and Wang, X. 1993. X-ray circular dichroism and local magnetic fields. *Phys. Rev. Lett.* 70:694–697.

Collins, S. P., Laundy, D., and Guo, G. H. 1993. Spin and orbital magnetic x-ray diffraction in $HoFe_2$. *J. Phys. Condens. Matter* 5:L637–642.

Cromer, D. T. and Liberman, D. A. 1981. Anomalous dispersion calculations near to and on the long-wavelength side of an absorption edge. *Acta Cryst.* A37:267–68.

de Bergevin, F. and Brunel, M. 1981a. Diffraction of x-rays by magnetic materials. I. General formulae and measurements on ferro- and ferrimagnetic compounds. *Acta Crystallogr.* A37:314–324.

de Bergevin, F. and Brunel, M. 1981b. Diffraction by magnetic materials. II. Measurements on antiferromagnetic $Fe_2O_3$. *Acta Crystallogr.* A37:324–331.

de Bergevin, F., Brunel, M., Galéra, R. M., Vettier, C., Elkïm, E., Bessière, M., and Lefèbvre, S. 1992. X-ray resonant scattering in the ferromagnet, CoPt. *Phys. Rev. B* 46:10772–10776.

Detlefs, C., Goldman, A. I., Stassis, C., Canfield, P. C., Cho, B. K., Hill, J. P., and Gibbs, D. 1996. Magnetic structure of $GdNi_2B_2C$ by resonant and non-resonant x-ray scattering. *Phys. Rev. B* 53:6355–6361.

Detlefs, C., Islam, A. H. M. Z., Goldman, A. I., Stassis, C., Canfield, P. C., Cho, B. K., Hill, J. P., and Gibbs, D. 1997. Determination of magnetic moment directions using x-ray resonant exchange scattering. *Phys. Rev. B* 55:R680–683.

Everitt, B. A., Salamon, M. B., Park, B. J., Flynn, C. P., Thurston, T. R., and Gibbs, D. 1995. X-ray magnetic scattering from nonmagnetic Lu in a Dy/Lu alloy. *Phys. Rev. Lett.* 75:3182–3185.

Fernandez, V., Vettier, C., de Bergevin, F., Giles, C., and Neubeck, W. 1998. Observation of orbital moment in NiO. *Phys. Rev. B*57:7870–7876.

Ferrer, S., Fajardo, P., de Bergevin, F., Alvarez, J., Torrelles, X., van der Vegt, H. A., and Elgens, V. H., 1996. Resonant surface magnetic x-ray diffraction from $Co_3Pt(111)$. *Phys. Rev. Lett.* 77:747–750.

Ferrer, S., Alvarez, J., Lundgren, E., Torrelles, X., Fajardo, P., and Borscherini, F. 1997. Surface x-ray diffraction from Co/Pt(111) ultra thin films and alloys: Structure and magnetism. *Phys. Rev. B* 56:9848–9857.

Gibbs, D. 1992. X-ray magnetic scattering. *Synchr. Radiat. News* 5:18–23.

Gibbs, D., Moncton, D. E., D'Amico, K. L., Bohr, J., and Grier, B. H. 1985. Magnetic x-ray scattering studies of holmium using synchrotron radiation. *Phys. Rev. Lett.* 55:234–237.

Gibbs, D., Harshmann, D. R., Isaacs, E. D., McWhan, D. B., Mills, D., and Vettier, C. 1988. Polarization and resonant properties of magnetic x-ray scattering on holmium. *Phys. Rev. Lett.* 61:1241–1244.

Gibbs, D., Blume, M., Harshmann, D. R., and McWhan, D. B. 1989. Polarization analysis of magnetic x-ray scattering. *Rev. Sci. Instrum.* 60:1655–1660.

Gibbs, D., Grübel, G., Harshmann, D. R., Isaacs, E. D., McWhan, D. B., Mills, D., and Vettier, C. 1991. Polarization and resonant studies of x-ray magnetic scattering in holmium. *Phys. Rev. B* 61:1241–1244.

Goldman, A. I., Mohanty, K., Shirane, G., Horn, P. M., Greene, R. L., Peters, C. J., Thurston, T. R., and Birgeneau, R. J. 1987. Magnetic x-ray scattering measurements on $MnF_2$. *Phys. Rev. B* 36:5609–5612.

Hamrick, M. D. 1994. Magnetic and chemical effects in x-ray resonant exchange scattering in rare earths and transition metal compounds. Ph.D. thesis, Rice University, Houston, Tex.

Hannon, J. P., Trammell, G. T., Blume, M., and Gibbs, D. 1988. X-ray resonant exchange scattering. *Phys. Rev. Lett.* 61:1245–1249, *Phys. Rev. Lett.* 62:2644–2647.

Harris, Q. J., Feng, Q., Lee, Y. S., Birgeneau, R. J., and Ito, A. 1997. Random fields and random anisotropies in the mixed Ising-XY magnet $Fe_xCo_{1-x}TiO_3$. *Phys. Rev. Lett.* 78:346–349.

Hill, J. P. and McMorrow, D. F. 1996. X-ray resonant exchange scattering: Polarization dependence and correlation functions. *Acta Crystallogr.* A52:236–244.

Hill, J. P., Feng, Q., Birgeneau, R. J., and Thurston, T. R. 1993. Magnetic x-ray scattering study of random field effects in $Mn_{0.75}Zn_{0.25}F_2$. *Z. Phys. B* 92:285–305.

Hill, J. P., Helgesen, G. H., and Gibbs, D. 1995a. X-ray scattering study of charge and spin density waves in chromium. *Phys. Rev. B* 51:10336–10344.

Hill, J. P., Vigliante, A., Gibbs, D., Peng, J. L., and Greene, R. L. 1995b. Observation of x-ray magnetic scattering in $Nd_2CuO_4$. *Phys. Rev. B* 52:6575–6580.

Hill, J. P., Kao, C.-C., and McMorrow, D. F. 1996. K-edge resonant x-ray magnetic scattering from a transition metal oxide: NiO. *Phys. Rev. B* 55:R8662–8665.

Hill, J. P., Feng, Q., Harris, Q., Birgeneau, R. J., Ramirez, A. P., and Cassanho, A. 1997. Phase transition behavior in the random field antiferromagnet $Fe_{0.5}Zn_{0.5}F_2$. *Phys. Rev. B* 55:356–369.

Hirano, K., Ishikawa, T., and Kikuta, S. 1993. Perfect crystal phase plate retarders. *Nucl. Instrum. Methods A* 336:343–353.

Hirano, K., Ishikawa, T., and Kikuta, S. 1995. Development and application of phase plate retarders. *Rev. Sci. Instrum.* 66:1604–1609.

Isaacs, E. D., McWhan, D. B., Peters, C., Ice, G. E., Siddons, D. P., Hastings, J. B., Vettier, C., and Vogt, O. 1989. X-ray resonant exchange scattering in UAs. *Phys. Rev. Lett.* 63:1671–1674.

Isaacs, E. D., McWhan, D. B., Keliman, R. N., Bishop, D. J., Ice, G. E., Zschack, P., Gaulin, B. D., Mason, T. E., Garret, J. D., and Buyers, W. J. L. 1990. X-ray magnetic scattering in antiferromagnetic $URu_2Si_2$. *Phys. Rev. Lett.* 65:3185–3188.

Kao, C.-C., Hastings, J. B., Johnson, E. D., Siddons, D. P., Smith, G. C., and Prinz, G. A. 1990. Magnetic resonance exchange scattering at the iron $L_2$ and $L_3$ edges. *Phys. Rev. Lett.* 65:373–376.

Kao, C.-C., Chen, C. T., Johnson, E. D., Hastings, J. B., Lin, H. J., Ho, G. H., Meigs, G., Brot, J.-M., Hulbert, S. L., Idzerda, Y. U., and Vettier, C. 1994. Dichroic interference effects in circularly polarized soft x-ray resonant magnetic scattering. *Phys. Rev. B* 50:9599–9602.

Langridge, S., Lander, G. H., Bernhoeft, N., Stunault, A., Vettier, C., Grübel, G., Søutter, C., de Bergevin, F., Nuttall, W. J., Stirling, W. G., Mattenberger, K., and Vogt, O. 1997. Separation of the spin and orbital moments in antiferromagnetic UAs. *Phys. Rev. B* 55:6392–6398.

Langridge, S. Paixão, J. A., Bernhoeft. N., Vettier, C., Lander, G. H., Gibbs, D., Sørensen, S. Aa, Stunault, A., Wermeille, D, and Talik, E. 1999. Changes in $5d$ band polarization in rare-earth coumpounds. *Phys. Rev. Lett.* 82:2187–2190.

Laundry, D., Collins, S., and Rollason, A. J. 1991. Magnetic x-ray diffraction from ferromagnetic iron. *J. Phys. Condens. Matter* 3:369–372.

Lippert, M., Brückel, T., Kohler, T., and Schneider, J. R. 1994. High resolution bulk magnetic scattering of high energy synchrotron radiation. *Europhys. Lett.* 27:537–541.

Lovesey, S. W. and Collins, S. P. 1996. X-ray Scattering and Absorption by Magnetic Materials. Oxford University Press, New York.

Luo, J., Trammell, G. T., and Hannon, J. P. 1993. Scattering operator for elastic and inelastic resonant x-ray scattering. *Phys. Rev. Lett.* 71:287–290.

Mackay, J. F., Teichert, C., Savage, D. E., and Lagally, M. G. 1996. Element specific magnetization of buried interfaces probed by diffuse x-ray resonant magnetic scattering. *Phys. Rev. Lett.* 77:3925–3928.

Mannix, D., de Camargo, P. C., Giles, C., de Oliveira, A. J. A., Yokaichiya, F., and Vettier, C. 2001. The chromium spin density wave: Magnetic x-ray scattering studies with polarization analysis. *Eur. Phys. J. B* 20(1):19–25.

McMorrow, D. F., Gibbs, D., and Bohr, J. 1999. X-ray scattering studies of the rare earths. *In* Handbook on the Physics and Chemistry of the Rare Earths, vol. 26 (K. A. Gschneidner Jr. and L. Eyring, eds.).. Elsevier, North-Holland, Amsterdam.

McWhan, D. B. 1998. Synchrotron radiation as a probe of actinide magnetism. *J. Alloys Compounds* 271–273:408:413.

McWhan, D. B., Vettier, C., Isaacs, E. D., Ice, G. E., Siddons, P., Hastings, J. B., Peters, C., and Vogt, O. 1990. Magnetic x-ray scattering study of uranium arsenide. *Phys. Rev. B* 42:6007–6017.

McWhan, D. B., Isaacs, E. D., Carra, P., Shapiro, S. M., Thole, B. T., and Hoshino, S. 1993. Resonant magnetic x-ray scattering study of mixed valence TmSe. *Phys. Rev. B* 47:8630–8633.

Namikawa, K., Ando, M., Nakajima, T., and Kawata, H. 1985. X-ray resonance magnetic scattering. *J. Phys. Soc. Jpn.* 54:4099–4102.

Neubeck, W., Vettier, C., de Bergevin, F., Yakhov, F., Mannix, D., Bengone, O., Alouani, M., and Barbier, A. 2001. Probing the $4p$ electron-spin polarization in NiO. *Phys. Rev. B* 63:134430–134431.

Pengra, D. B., Thoft, N. B., Wulff, M., Feidenhans'l, R., and Bohr, J. 1994. Resonance-enhanced magnetic x-ray diffraction from a rare-earth alloy. *J. Phys. Condens. Matter* 6:2409–2422.

Robinson, I. K. 1991. Surface crystallography. *In* Handbook of Synchrotron Radiation, Vol. 3 (G. S. Brown and D. E. Moncton, eds.). pp. 221–266. North Holland Publishing, New York.

Sanyal, M. K., Gibbs, D., Bohr, J., and Wulff, M. 1994. Resonance magnetic x-ray scattering study of erbium. *Phys. Rev. B* 49:1079–1085.

Schülke, W. 1991. Inelastic scattering by electronic excitations. *In* Handbook of Synchrotron Radiation, Vol. 3 (G. S. Brown and D. E. Moncton, eds.). pp. 565–637. North Holland Publishing, New York.

Shen, Q., Shastri, S., and Finkelstein, K. D. 1995. Stokes polarimetry for x-rays using multiple beam diffraction. *Rev. Sci. Instrum.* 66:1610–1613.

Strempfer, J., Brückel, Th., Rütt, U., Schneider, J. R., Liss, K.-D., and Tschentscher, Th. 1996. The non-resonant magnetic x-ray cross-section of $MnF_2$. Part 2. High energy x-ray diffraction at 80 keV. *Acta Crystallogr.* A52:438–449.

Strempfer, J., Rütt, U., and Jauch, W. 2001. Absolute spin magnetic moment of $FeF_2$ from high-energy photon diffraction. *Phys. Rev. Lett.* 86:3152–3155.

Sutter, C., Grübel, G., Vettier, C., de Bergevin, F., Stunault, A., Gibbs, D., and Giles, C. 1997. Helicity of magnetic domains in holmium studied with circularly polarized x rays. *Phys. Rev. B* 55:954–959.

Tang, C. C., Stirling, W. G., Lander, G. H., Gibbs, D., Herzog, W., Carra, P., Thole, B. T., Mattenberger, K., and Vogt, O. 1992. Resonant magnetic scattering in a series of uranium compounds. *Phys. Rev. B* 46:5287–5297.

Thole, B. T., Carra, P., Sette, F., and van der Laan, G. 1992. X-ray circular dichroism as a probe of orbital magnetization. *Phys. Rev. Lett.* 68:1943–1947.

Thurston, T. R., Helgesen, G., Gibbs, D., Hill, J. P., Gaulin, B. D., and Shirane, G. 1993. Observation of two length scales in the magnetic critical fluctuations of holmium. *Phys. Rev. Lett.* 70:3151–3154.

Thurston, T. R., Helgesen, G., Hill, J. P., Gibbs, D., Gaulin, B. D., and Simpson, P. 1994. X-ray and neutron scattering measurements of two length scales in the magnetic critical fluctuations of holmium. *Phys. Rev. B* 49:15730–15744.

Tonnerre, J. M., Séve, L., Raoux, D., Soullié, G., Rodmacq, B., and Wolfers, P. 1995. Soft x-ray resonant magnetic scattering from a magnetically coupled Ag/Ni multilayer. *Phys. Rev. Lett.* 75:740–743.

van Veenendaal, M., Goedkoop, J. B., and Thole, B. T. 1997. Branching ratios of the circular dichroism at rare earth $L_{2,3}$ edges. *Phys. Rev. Lett.* 78:1162–1165.

Vettier, C., McWhan, D. B., Gyorgy, E. M., Kwo, J., Buntschuh, B. M., and Batterman, B. W. 1986. Magnetic x-ray scattering study of interfacial magnetism in a Gd-Y superlattice. *Phys. Rev. Lett.* 56:757–760.

Vigliante, A., von Zimmerman, M., Schneider, J. R., Frello, T., Andersen, N. H., Madsen, J., Buttrey, D. J., Gibbs, D., and Tranquada, J. M. 1997. Detection of charge scattering associated with stripe order in $La_{1.775}Sr_{0.225}NiO_4$ by hard x-ray diffraction. *Phys. Rev. B* 56:8248–8251.

Vigliante, A., Christensen, M. J., Hill, J. P., Helgesen, G., Sorensen, A. Aa., McMorrow, D. F., Gibbs, D., Ward, R. C. C., and Wells, M. R. 1998. Interplay between structure and magnetism in $Ho_xPr_{1-x}$ alloys II: Resonant magnetic scattering. *Phys. Rev. B* 57:5941.

Watson, D., Forgan, E. M., Nuttall, W. J., Stirling, W. G., and Fort, D. 1996. High resolution magnetic x-ray diffraction from neodymium. *Phys. Rev. B* 53:726–730.

Watson, D., Nuttall, W. J., Forgan, E. M., Perry, S., and Fort, D. 1998. Refinement of magnetic structures with x rays: Nd as a test case. *Phys. Rev. B Condensed Matter* 57(14):R8095–R8098.

Watson, G., Gaulin, B. D., Gibbs, D., Lander, G. H., Thurston, T. R., Simpson, P. J., Matzke, H. J., Wong, S., Dudley, M., and Shapiro, S. M. 1996a. Origin of the second length scale found above $T_N$ in $UO_2$. *Phys. Rev. B* 53:686–698.

Watson, G., Gibbs, D., Lander, G. H., Gaulin, B. D., Berman, L. E., Matzke, H. J., and Ellis, W. 1996b. X-ray scattering study of the magnetic structure near the (001) surface of $UO_2$. *Phys. Rev. Lett.* 77:751–754.

Watson, G., Gibbs, D., Lander, G. H., Gaulin, B. D., Berman, L. E., Matzke, H. J., and Ellis, W. 2000. Resonant x-ray scattering studies of the magnetic structure near the surface of an antiferromagnet. *Phys. Rev. B* 61:8966–8975.

## KEY REFERENCES

Als-Nielsen, J. and McMorrow, D. 2001. Elements of Modern X-Ray Physics. John Wiley & Sons, New York.

*Excellent introductory text for modern x-ray diffraction techniques.*

de Bergevin and Brunel, 1991. See above.

*Review of early experiments and non-resonant formalism.*

Lovesey and Collins, 1996. See above.

*Provides a general overview of theoretical background to magnetic x-ray scattering, and reviews some important experiments.*

McMorrow et al., 1999. See above.

*Excellent review article of the x-ray magnetic scattering work to date on the rare-earths, and rare-earth compounds. Includes review of the theoretical ideas and experiments.*

J. P. HILL
Brookhaven National
Laboratory
Upton, New York

# X-RAY MICROPROBE FOR FLUORESCENCE AND DIFFRACTION ANALYSIS

## INTRODUCTION

X-ray diffraction (see SYMMETRY IN CRYSTALLOGRAPHY, X-RAY POWDER DIFFRACTION and SURFACE X-RAY DIFFRACTION) and x-ray-excited fluorescence analysis are powerful techniques for the nondestructive measurement of crystal structure and chemical composition. X-ray fluorescence analysis is inherently nondestructive, with orders-of-magnitude lower power deposited for the same detectable limit as with fluorescence excited by charged particle probes (Sparks, 1980). X-ray diffraction analysis is sensitive to crystal structure with orders-of-magnitude greater sensitivity to crystallographic strain than electron probes (Rebonato et al., 1989; Chung and Ice, 1999). When a small-area x-ray microbeam is used as the probe, chemical composition ($Z > 14$), crystal structure, crystalline texture, and crystalline strain distributions can be determined. These distributions can be studied both at the surface of the sample and deep within the sample (Fig. 1). Current state-of-the-art can achieve an $\sim$1-$\mu$m diameter x-ray microprobe and an $\sim$0.1 $\mu$m diameter x-ray microprobe

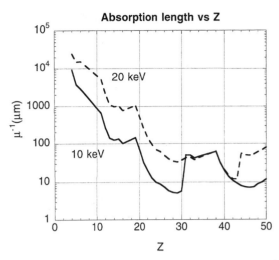

**Figure 1.** Absorption depth for 10- and 20-keV x rays as a function of elemental composition, $Z$.

has been demonstrated (Bilderback et al., 1994; Yun et al., 1999).

Despite their great chemical and crystallographic sensitivities, x-ray microprobe techniques have until recently been restricted by inefficient x-ray focusing optics and weak x-ray sources; x-ray microbeam analysis was largely superseded by electron techniques in the 1950s. However, interest in x-ray microprobe techniques has now been revived (Howells and Hastings, 1983; Ice and Sparks, 1984; Riekel, 1992; Thompson et al., 1992; Chevallier and Dhez, 1997; Hayakawa et al., 1990; APS Users Meeting Workshop, 1997) by the development of efficient x-ray focusing optics and ultra-high-intensity synchrotron x-ray sources (Buras and Tazzari, 1984; Shenoy et al., 1988). These advances have increased the achievable microbeam flux by more than 12 orders of magnitude (Fig. 2; also see Ice, 1997); the flux in a tunable 1 $\mu$m-diameter beam on a so-called third-generation synchrotron source such as the Advanced Photon Source (APS) can exceed the flux in a fixed-energy $mm^2$ beam on a conventional source. These advances place x-ray microfluorescence and x-ray microdiffraction analysis techniques among the most powerful techniques available for the nondestructive measurement of chemical and crystallographic distributions in materials.

This unit reviews the physics, advantages, and scientific applications of hard x-ray ($E > 3$ keV) microfluorescence and x-ray microdiffraction analysis. Because practical x-ray microbeam instruments are extremely rare, a special emphasis will be placed on instrumentation, accessibility, and experimental needs which justify the use of x-ray microbeam analysis.

### Competitive and Related Techniques

Despite their unique properties, x-ray microprobes are rare and the process of gaining access to an x-ray

**Tunable 20 keV Brilliance**

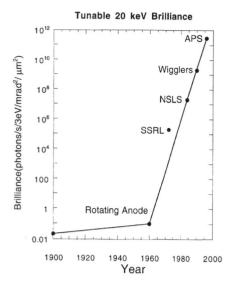

**Figure 2.** X-ray brilliance over the last 100 years shows an increase of more than 12 orders of magnitude since the use of hard x-ray synchrotron radiation sources began in the late 1960s. Both x-ray microdiffraction and x-ray microfluorescence have brilliance (photons/s/eV/mm$^2$/mrad$^2$) as the figure of merit (Ice, 1997).

microprobe can be difficult. For many samples, alternative techniques exist with far greater availability. Destructive methods such as laser ionization with mass spectrometry, as well as atom probe methods (Miller et al., 1996) can yield information on composition distributions. Atom probe measurements in particular can measure the atom-by-atom distribution in a small volume but require extensive sample preparation.

Auger spectroscopy (AUGER ELECTRON SPECTROSCOPY) and Rutherford backscattering techniques (HEAVY-ION BACK-SCATTERING SPECTROMETRY) are other possible methods for determining elemental distributions. Auger analysis is inherently surface-sensitive, whereas Rutherford back-scattering measurements can probe below the sample surface. Although these techniques are generally considered very sensitive, x-ray analysis can be even more sensitive.

The most directly comparable analysis techniques are charged particle microprobes such as electron or proton microprobes (Sparks, 1980; LOW-ENERGY ELECTRON DIFFRACTION and NUCLEAR REACTION ANALYSIS AND PROTON-INDUCED GAMMA RAY EMISSION). Whereas proton microprobes are almost as rare as x-ray microprobes, electron microprobes are widely used to excite fluorescence for chemical analysis. Electron microbeams are also used to measure crystallographic phase and texture, and strain resolution to $2 \times 10^{-4}$ has been demonstrated (Michael and Goehner, 1993). Advanced electron microbeams can deliver $10^{12}$ to $10^{15}$ electrons/$\mu$m$^2$ and can be focused to nanometer dimensions. Electron microbeams are available at many sites within the United States and around the world.

Sparks (1980) has compared the relative performance of x ray, proton, and electron microprobes for chemical analysis. He specifically compared the intrinsic ability of x ray and charged particle microprobes to detect trace elements in a matrix of other elements. He also compared their rela-

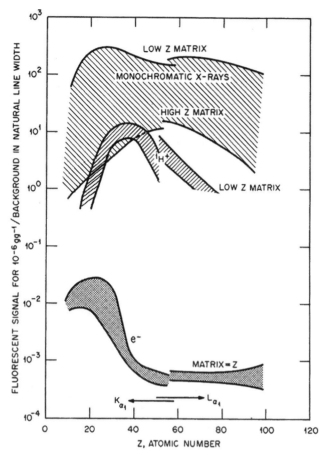

**Figure 3.** Comparison of the fluorescence signal-to-background ratio for various excitation probes at a concentration of $10^{-6}$ gg$^{-1}$ for an x-ray detection system with an energy resolution of the natural line width (Sparks, 1980).

tive abilities to resolve small-dimensioned features. In summary he finds that x rays have major advantages: (1) x rays are very efficient at exciting electrons in atoms, thereby creating inner-shell holes; (2) x-ray excitation produces very low backgrounds; (3) beam spreading with x rays is low; and (4) x-ray microprobe analysis requires minimum sample preparation.

A comparison of the signal-to-background ratio for various probes is shown in Figure 3. Monochromatic x-ray excitation produces the best fluorescence signal-to-background. Proton excitation produces signal-to-background ratios between that of x-ray- and electron-induced fluorescence. For low-$Z$ elements, proton microbeams can sometimes approach the signal-to-background of x-ray microbeams.

A rather direct comparison of the elemental sensitivity of various microbeam probes can be made by comparing their minimum-detectable limits (MDL). We adopt the MDL definition of Sparks (1980) for fluorescence analysis

$$C_{\mathrm{MDL}} = 3.29 \, C_{\mathrm{Z}} \frac{\sqrt{N_{\mathrm{b}}}}{N_{\mathrm{s}}} \qquad (1)$$

Here $C_{\mathrm{MDL}}$ is the minimum detectable limit, $C_{\mathrm{Z}}$ is the mass fraction in a calibrated standard, $N_{\mathrm{b}}$ is the

**Table 1. Estimated MDL/s from (Sparks, 1980), Scaled to $10^{12}$ Monochromatic Photons/s/$\mu m^2$**

| Probe | MDL (ppm/s)[a] |
|---|---|
| Proton | ~10 to 100 |
| Electron | ~5 to 30 |
| Filtered x ray | ~1 to 8 |
| Monochromatic x ray | ~0.005 to 0.08 |

[a]This comparison assumes that the matrix and trace element have similar $Z$ and that an advanced multielement solid-state detector is used where deadtime does not limit performance. With a low-$Z$ matrix or with an advanced crystal spectrometer, the MDL can be lower.

background counts beneath the fluorescence signal, and $N_s$ is the net counts at the fluorescence energy. Lowest MDL results when the ratio $N_s/\sqrt{N_b}$ is large (good signal-to-background and high flux), and when $N_s/C_Z$ is large (efficient inner-shell hole production and high flux).

Electron and proton microprobes cannot match the achievable MDL of an advanced x-ray microprobe; compared to x rays, the inner-shell hole production cross-section and signal-to-background of an electron probe are too low and the proton probe flux density on the sample is too low. An advanced x-ray microprobe with $10^{12}$ photons/s/$\mu m^2$ has about a $10^3$ lower MDL for most elements than a charged particle probe and can achieve the same MDL as electron probes with $10^4$ less energy deposited in the sample (Table 1).

For thin samples, the spatial resolution of electron probes is far better than that of either x-ray or proton microprobes. In terms of their practical spatial resolution for thick samples, however, x-ray microprobes are competitive or superior to charged particle probes. Although electron probes can be focused to nanometer dimensions, beam spreading in thick samples degrades their effective resolution. For example, in a thick Al sample, a nanometer electron probe spreads to an effective size of ~2-$\mu m$ diameter. In Cu and Au samples the same beam spreads to 1- and 0.4-$\mu m$ diameter, respectively (Goldstein, 1979; Ren et al., 1998). Proton microprobes can be made very small, but the fluxes are so low that few instruments exist with probe dimensions approaching 1-$\mu m$ diameter (Lindh, 1990; Doyle et al., 1991).

X-ray beams are now so intense that their flux density is approaching the maximum that can usefully be applied to most samples. For example, the estimated thermal rise of a thin target under an advanced x-ray microbeam is shown in Figure 4. Existing x-ray microbeams have highly monochromatic flux densities approaching $10^{12}$ photons/s/$\mu m^2$ and can go to $10^{13}$ to $10^{14}$ photons/s/$\mu m^2$ with larger band-pass optics. With $10^{14}$ photons/s/$\mu m^2$, the flux must be attenuated for most samples. Alternatively, the probe area can be decreased or the dwell time on the sample can be reduced to prevent sample melting.

Despite its many advantages, however, x-ray microbeam analysis remains an emerging field with very little available instrumentation. The effort required to gain access to x-ray microbeam facilities must therefore be weighed against the benefits. X-ray microfluorescence analysis becomes justified when the MDL from other techniques is inadequate. It may also be justified when the probe must penetrate the sample surface, when the analy-

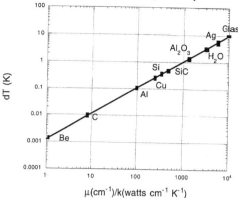

**Figure 4.** Thermal rise as a function of thermal conductivity $K$ and absorption coefficient, $\mu$ (Ice and Sparks, 1991). Note that existing x-ray microbeams have achieved $10^{12}$ photons/s/$\mu m^2$ and are anticipated to reach $10^{13}$–$10^{14}$ photons/s/$\mu m^2$ for some applications.

sis must be highly nondestructive, when the analysis must be quantitative, or when the measurement must be done in the presence of an air, water, or other low-$Z$ overlayer.

Average crystallographic grain orientations can be studied with large probes such as standard x-ray beams or neutron beams. Individual grain orientations can also be determined to ~1° with electron back-reflection measurements and strain resolution to $2 \times 10^{-4}$ has been reported (Michael and Goehner, 1993). However, large beam probes cannot study individual grain-grain correlations, and electron probes are predominantly surface sensitive, and in most cases have both strain and orientation precision two orders of magnitude worse than for x-ray microdiffraction. Microdiffraction analysis therefore becomes justified when the strain resolution $\Delta d/d$ must be better than $\sim 5 \times 10^{-4}$. Microdiffraction analysis is also justified for the study of crystallographic properties beneath the surface of a sample, for the measurement of texture in three dimensions or for nondestructive analysis of insulating samples where charge buildup can occur.

## PRINCIPLES OF THE METHOD

A typical x-ray microbeam experiment involves three critical elements: (1) x-ray condensing or aperturing optics on a high-brilliance x-ray source, (2) a high-resolution sample stage for positioning the sample, and (3) a detector system with one or more detectors (Fig. 5). The x-ray beam axis and focal position is determined by the optics of the particular arrangement. Different locations on the specimen are characterized by moving the specimen under the fixed x-ray microprobe beam. Details of the detector arrangement and the principles involved depend strongly on the particular x-ray microprobe and on whether elemental distributions or crystallographic information is to be collected.

Both x-ray microdiffraction and x-ray microfluorescence are brilliance (photons/s/eV/mm²/mrad²) limited (Ice, 1997). For x-ray microdiffraction, momentum

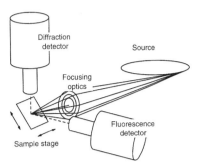

**Figure 5.** Key elements of an x-ray microbeam experiment.

transfer resolution is limited by spread in wavelength and angular divergence on the sample. Count rate is limited by flux per unit area. For x-ray microfluorescence, best signal-to-background occurs when the x-ray bandwidth $\Delta E/E$ is $\leq 3\%$ (Sparks, 1980). Spatial resolution for thick samples degrades when the divergence of the beam exceeds ~10 mrad. The principles of x-ray microfluorescence and x-ray microdiffraction analysis are briefly outlined below.

### X-Ray Microfluorescence Analysis

The unique advantages of x-ray probes arise from the fundamental interaction of x rays with matter. Below the electron-positron pair production threshold, the interaction of x rays with matter is dominated by three processes: photoabsorption (photoelectric effect), elastic scattering, and inelastic (Compton) scattering (Veigele et al., 1969). Of these three processes, photoabsorption has by far the largest cross-section in the 3- to 100-keV range.

Photoabsorption and Compton scattering are best understood in terms of a particle-like interaction between x rays and matter (Fig. 6). The quantized energy of an x-ray photon excites an electron from a bound state to an unbound (continuum) state, while the photon momentum is transferred either to the atomic nucleus (photoeffect) or

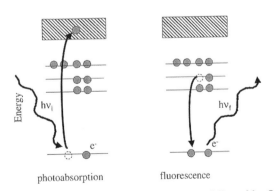

**Figure 6.** Schematic of x-ray photoabsorption followed by fluorescence. An x-ray photon of energy $h\nu_i$ is absorbed by the atom, which ejects an electron from an inner shell of the atom. The atom fills the electron hole by emitting an x ray with an energy $h\nu_f$; $h\nu_f$ is characteristic to the atom and has an energy equal to the energy difference between the initial and final hole-state energies. Alternatively the atom fills the inner-shell hole by emitting an energetic electron (Auger effect; Bambynek et al., 1972 also see AUGER ELECTRON SPECTROSCOPY).

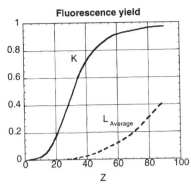

**Figure 7.** Fit to experimental fluorescence yields for $K$ and $L$ holes. (Bambynek et al., 1972.)

to the electron (Compton scattering). X-ray fluorescence is the name given to the elementally distinct or "characteristic" x-ray spectrum that is emitted from an atom as an inner-shell hole is filled.

**Fluorescence Yields.** An inner-shell hole can also be filled by nonradiative mechanisms (Auger and Coster-Kronig effects; Bambynek et al., 1972). Here the singly ionized atom fills the inner shell hole with a higher-energy electron and emits an energetic (Auger/Coster-Kronig) electron with a characteristic energy that is determined by the initial and final energy of the atom. The fraction of holes that are filled by x-ray fluorescence decay is referred to as the fluorescence yield. In general, nonradiative processes become increasingly likely as the initial-hole binding energy decreases (Fig. 7). We note that for $K$ holes with binding energies above 5 keV, the x-ray fluorescence yields are more than 20%. For $L$ holes with binding energies greater than 5 keV, all fluorescence yields exceed 10%. The fluorescence yields of deep inner-shell holes can exceed 90%.

**Characteristic Radiation.** The characteristic x-ray energies emitted when the initial hole decays by fluorescence serve as a "fingerprint" for the element and are quite distinct. Fluorescence spectra are labeled according to the electron hole being filled and the strength of the decay channel. For example, as shown in Figure 8, $K\alpha_1$ fluorescence arises when an $L_{III}$ ($2P_{3/2}$) electron fills a $K$ hole.

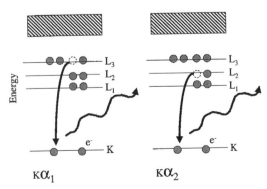

**Figure 8.** Fluorescence decay channels for $K$ holes.

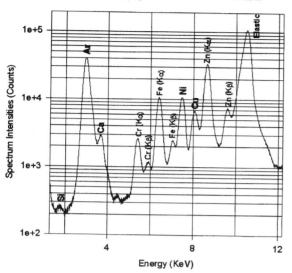

**Figure 9.** Fluorescence spectra from the SiC shell of an advanced nuclear fuel particle. The trace elements in the sample emit characteristic x-ray lines when excited with x rays.

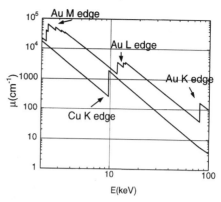

**Figure 10.** Photoabsorption cross-section for Cu and Au.

This transition is the strongest fluorescence decay channel for $K$ holes. A similar nomenclature is used for L holes, etc.

Because chemical effects on inner-shell wavefunctions are small, x-ray absorption cross-sections, fluorescence yields, and characteristic x-ray spectra are virtually unchanged by the sample environment except very near threshold. However, the measured-fluorescence signal can strongly depend on absorption and secondary excitation due to the sample matrix.

A typical spectrum from a low-$Z$ matrix with trace elements is shown in Figure 9. The characteristic lines have a natural bandwidth of a few eV that is smeared by the energy resolution of the detector. Even in this case where the trace elements are nearby in the periodic table, the fluorescence signature of each element is distinct. Crystal spectrometers with higher energy resolution can be used in more complicated cases with overlapping $L$ lines to resolve nearby fluorescence lines.

**Photoabsorption Cross-Sections.** To a first approximation, only x-rays with sufficient energy to excite an electron above the Fermi level (above the occupied electron states) can create an inner-shell hole. As a consequence, the photoabsorption cross-section for x-rays has thresholds that correspond to the energy needed to excite an inner-shell electron into unoccupied states. As shown in Figure 10, the photoabsorption cross-section exhibits a characteristic saw-tooth pattern as a function of x-ray energy. Highest x-ray efficiency for the creation of a given hole is just above its absorption edge energy. Maximum elemental sensitivity with minimum background results when the x-ray microprobe energy is tuned just above the absorption edge of the element of interest (Sparks, 1980). Because x rays are highly efficient at creating inner-shell holes, x-ray excited fluorescence has a very high signal-to-noise ratio (Fig. 3) and the energy deposited in the sample is low for a given signal.

**Micro-XAFS.** The saw-tooth pattern of Figure 10 shows the typical energy dependence of x-ray absorption cross-sections over a wide energy range, but does not include lifetime broadening of the inner shell hole, the density of unfilled electron states near the Fermi energy, or the influence of photoelectron backscattering. These various processes lead to fine structure in the photoabsorption cross-section which can be used to determine the valence state of an element, its local neighbor coordination, and bond distances. Near-edge absorption fine structure (NEXAFS) is particularly sensitive to the valence of the atom. Extended x-ray absorption fine structure (EXAFS) is sensitive to the near-neighbor coordination and bond distance. Fluorescence measurements have the best signal-to-background ratio for XAFS of trace elements. It is therefore possible to use XAFS techniques with an x-ray microprobe to determine additional information about the local environment of elements within the probe region. More detail about XAFS techniques is given in XAFS SPECTROSCOPY.

**Penetration Depth.** The effective penetration depth of a fluorescence microprobe depends on the energy of the incident and fluorescent x rays, the composition of the sample, and the geometry of the measurement. As shown in Figure 10, x-ray absorption decreases with an $\sim E^{-3}$ power dependence between absorption edges. For a low-$Z$ matrix, the penetration depth of an x-ray microfluorescence beam can be tens of millimeters into the sample, while for a high-$Z$ matrix the penetration can be only a few microns (Fig. 1).

With a thick sample, the fluorescence signal and effective depth probed depends on the total scattering angle and on the asymmetry of the incident-to-exit beam angles (Sparks, 1980). To a first approximation, the fluorescence signal is independent of total scattering angle but depends on the asymmetry between the incident angle $\psi$ and the exit angle $\phi$ (Fig. 11).

$$C_Z \propto \left( \mu_i + \mu_f \frac{\sin \psi}{\sin \phi} \right) \qquad (2)$$

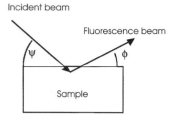

**Figure 11.** Depth penetrated by a fluorescence microprobe depends on the total absorption cross-section of the incident and fluorescence radiation and on the incident and exit angles with respect to the sample surface.

Here $\mu_i$ is the linear absorption coefficient for the incident beam and $\mu_f$ is the total absorption coefficient for the fluorescence beam. The approximation of Equation 2 is only valid for uniformly smooth sample surfaces. Where sample granularity and surface roughness are large, the fluorescence signal decreases as the glancing angle decreases (Campbell et al., 1985; Sparks et al., 1992). The effective depth probed depends both on the total scattering angle $\psi + \phi$ and on the asymmetry between $\psi$ and $\phi$.

The characteristic depth of an x-ray microprobe is given by

$$x = \left( \frac{\mu_i}{\sin \psi} + \frac{\mu_f}{\sin \phi} \right) \qquad (3)$$

Shallow depth penetration can be achieved with glancing angle and asymmetric geometries. With smooth surfaces, even greater surface sensitivity can be achieved by approaching or achieving total external reflection from the surface (Brennan et al., 1994). With total external reflection, the surface sensitivity can approach 1 nm (10 Å) or better (Fig. 12).

**Backgrounds.** Background signal is generated under fluorescence lines by various scattering and bremsstrahlung processes. With a white or broad bandpass incident beam, elastically scattered x-rays including those with the same energy as the fluorescent line can be directly scattered into the detector. This scattering can be greatly reduced by operating at a $2\theta$ scattering angle of $90°$ in the plane of the synchrotron storage ring where linear beam polarization inhibits elastic and Compton scattering. A

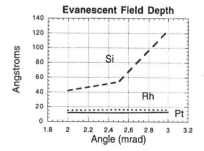

**Figure 12.** Evanescent wave depth of 10 keV x rays into Si, Rh, or Pt as a function of glancing angle.

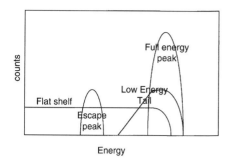

**Figure 13.** Schematic of low-energy tails in a solid state detector.

much better way to reduce background is through the use of a monochromatic x-ray beam. As shown in Figure 2, monochromatic x-ray beams produce the best signal-to-background because the background under a fluorescence peak must arise from multiple scattering events, bremsstrahlung from photoelectrons, or other cascade processes.

**Detectors.** Two kinds of detectors are used for observing fluorescence. Solid-state detectors are most widely used because they are efficient and can simultaneously detect many elements in the sample. A state-of-the-art solid-state detector with a 1-cm$^2$ active area has $\sim$130-eV resolution at 5.9 keV with a 5000 counts-per-second (cps) counting rate. Much higher counting rates are possible by compromising the energy resolution and by using multiple detector arrays. For example, 30-element arrays with 500,000 cps counting rates/element can achieve 15,000,000 cps. One drawback with solid-state detectors is additional background for trace elements, which is introduced by high intensity peaks in the spectra. As shown in Figure 13, the response of a solid-state detector to an x ray of a single energy typically has both short- and long-range low-energy tails due to insufficient charge collection (Cambell, 1990). These can be the dominant contribution to background under a trace element.

Wavelength dispersive spectrometers can also be used to measure fluorescence (Ice and Sparks, 1990; Koumelis et al., 1982; Yaakobi and Turner, 1979). These detectors have much poorer collection efficiency but much better energy resolution (lower background) and are not paralyzed by scattering or by fluorescence from the major elements in the sample. Because wavelength dispersive detectors only count one fluorescent energy at a time, they are not count-rate limited by fluorescence or scattering from the major elements in the sample matrix.

### X-Ray Microdiffraction Analysis

X-ray microdiffraction is sensitive to phase (crystal structure), texture (crystal orientation), and strain (unit cell distortion) (Ice, 1987; Rebonato et al., 1989). Diffraction (Fig. 14) is best understood in terms of the wave-like nature of x rays. Constructive and destructive interference from x-ray scattering off the charge-density distribution varies the x-ray scattering efficiency as a function of angle and wavelength. This so-called diffraction pattern can be Fourier transformed to recover charge-density

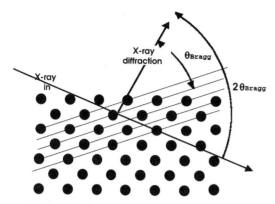

**Figure 14.** X-ray diffraction from a crystal becomes large at Bragg angles where the crystal lattice spacing, $d$, the wavelength, $\lambda$, and the angle of the incident beam with respect to the crystal lattice planes, $\theta_{Bragg}$, satisfy Bragg's law (Equation 4).

information. Although the basic diffraction process is identical for all diffraction probes (e.g., x rays, electrons, or neutrons), x rays have three very favorable attributes for the characterization of crystal structure: (1) the wavelength is similar to the atomic spacing of matter; (2) the cross-section is sufficiently low that multiple scattering effects are often small; and (3) the cross-section of elastic scattering is a large fraction of the total interaction cross-section of x rays, which contributes to low noise. In addition, x-ray scattering contrast can be adjusted by tuning near to x-ray absorption edges (Materlik et al., 1994).

Strong diffraction occurs when the incident beam satisfies Bragg's law

$$2d \sin \theta = n\lambda \qquad (4)$$

Here $d$ is a crystal lattice spacing, $\lambda$ is the x-ray wavelength, and $\theta$ is the so-called "Bragg angle" between the incident beam and the crystal plane. X-ray microdiffraction can yield detailed information about the sample unit cell and its orientation. If the x-ray wavelength is known, the angle between the incident and an intense exit beam ($2\theta$) determines the spacing, $d$. The relative orientation of crystals (mosaic spread or texture) can be determined by observing the rotation angles of a sample needed to maximize scattered intensity. These two methods are useful for studies of polycrystalline materials that have a strong preferred orientation.

For unknown crystal orientation, the unit-cell parameters and crystallographic orientation of a single crystal can be determined from the x-ray energy and the angles of three noncolinear reflections (Busing and Levy, 1967). In standard crystallography, reflections are found by rotation of the sample under the beam. With microdiffraction x-ray measurements on polycrystalline samples, however, the imprecision of mechanical rotations will cause the sample to translate relative to the beam on a micron scale. In addition, for complex polycrystalline samples with many grains, the penetration of the x-ray beam into the specimen ensures that the sample volume (and therefore microstructure) will vary as the sample rotates. The changing grain illumination makes a standard crystallographic

solution impossible. These two problems can be overcome by the use of Laue diffraction with white or broad-bandpass x-ray beams. With Laue diffraction, no sample rotations are required. Development is in progress to automatically index the overlapping reflections of up to 10 crystals (Marcus et al., 1996; Chung, 1997; Chung and Ice, 1998; Wenk et al., 1997).

X-ray microdiffraction also allows for three-dimensional imaging of crystal structure as demonstrated in some first experiments (Stock et al., 1995; Piotrowski et al., 1996). For these measurements, the sample-to-detector distance is changed and the origin of the reflecting crystal along the microprobe beam is determined by ray tracing. Spatial resolution of less than 2 $\mu$m along the microbeam direction has recently been demonstrated (Larson et al., 1999).

## PRACTICAL ASPECTS OF THE METHOD

### Sources

Only second- and third-generation synchrotron sources have sufficient x-ray brilliance for practical x-ray microprobe instrumentation. Worldwide, there are only three third-generation sources suitable for hard x-ray microprobes: the 6-GeV European Synchrotron Radiation Facility in Grenoble France (Buras and Tazzari, 1984), the 7-GeV Advanced Photon Source at Argonne, Illinois (Shenoy et al., 1998), USA, and the 8-GeV Spring-8 under construction in Japan. Although third-generation sources are preferred, x-ray microprobe work can also be done on second-generation sources like the National Synchrotron Light Source (NSLS), Brookhaven, New York. A list of contacts at synchrotron radiation facilities with x-ray microbeam instrumentation is given in Table 2.

### Optics

**Tapered Capillaries.** The development of intense synchrotron x-ray sources, with at least 12 orders of magnitude greater brilliance than conventional x-ray sources (Fig. 2), has revived interest in x-ray optics. At least three microbeam-forming options have emerged with various strengths and weaknesses for experiments (Ice, 1997). Tapered capillary optics have produced the smallest x-ray beams (Stern et al., 1988; Larsson and Engström, 1992; Hoffman et al., 1994). Bilderbach et al. (1994) have reported beams as small as 50 nm full width at half-maximum (FWHM). This option appears to be the best for condensing beams below 0.1 $\mu$m. One concern with capillary optics, however, is their effect on beam brilliance. Ray tracing and experimental measurements have found that the angular divergence following a capillary has a complex annular distribution. This distribution arises from roughness inside the capillary and from the nonequal number of reflections of different rays as they are propagated along the capillary.

**Zone Plates.** Hard x-ray zone plates are a rapidly emerging option for focusing synchrotron radiation to $\mu$m dimensions (Bionta et al., 1990; Yun et al., 1992). This

**Table 2. Worldwide Dedicated X-ray Microbeam Facilities**[a]

| Facility | Beamline | hv (keV) | Spot Size (μm²) | $\Delta E/E$ | Total Flux (photons/s) | Local Contact |
|---|---|---|---|---|---|---|
| APS | 7-ID | 5–20 | 0.5 | $2 \times 10^{-4}$ to $1 \times 10^{-1}$ | $10^{8}$–$10^{11}$ | Walter Lowe *Lowe@mhattcat.howard.edu* |
| APS | 2-ID-CD | 5–20 | ~0.03 | $2 \times 0^{-4}$ | $5 \times 10^{10}$ | Barry Lai *Lai@aps.anl.gov* |
| APS | 13-ID | 5–25 | ~0.7 | $2 \times 10^{-4}$ to $1 \times 10^{-2}$ | $1 \times 10^{13}$ | Steve Sutton *Sutton@cars.uchicago.edu* |
| ESRF | ID-13 | 6–16 | ~40 | $2 \times 10^{-4}$ | $2 \times 10^{11}$ | Christian Riekel *riekel@esrf.fr* |
| ESRF | ID-30 | 6–20 | ~4 | — | — | |
| ESRF | ID-22 | 4–35 | ~2 | $2 \times 10^{-4}$ | $10^{9}$–$10^{12}$ | Anatoly Snigirev *snigirev@esrf.fr* |
| ALS | 10.3.2 | 5–12 | ~0.6 | $2 \times 10^{-4}$ | — | Howard Padmore *Padmore@lbl.gov* |
| ALS | 10.3.1 | 6–12 | ~2 | $5 \times 10^{-2}$ | $2 \times 10^{10}$ | Scott McHugo *samchugo@lbl.gov* |
| LNLS | — | 2–14 | 100 | $2 \times 10^{-4}$ white | $3 \times 10^{7}$ $3 \times 10^{9}$ | Helio C.N. Tolentino *helio@lnls.br* |
| CHESS | B2 | 5–25 | 0.001 | $2 \times 10^{-2}$ | $10^{6}$ | Don Bilderbach *dhb2@cornell.edu* |
| Photon Factory | BL4-A | 5–15 | 25 | $2 \times 10^{-4}$ $5 \times 10^{-2}$ | $10^{8}$ $10^{10}$ | A. Iida |
| Hasylab | L | 4–80 | 9 | — | — | Thomas Wroblewski *wroblewt@mail.desy.de* |
| Hasylab | BW-1 | 10 | 1 | $1 \times 10^{-2}$ | $6 \times 10^{7}$ | Thomas Wroblewski *wroblewt@mail.desy.de* |
| DCI LURE | D15 | 6,10,14 | ~1–100 | $2 \times 10^{-2}$ | $10^{5}$–$10^{7}$ | P. Chevallier *chevallier@lure.u-psud.fr* |
| SSRL | — | — | — | — | — | J. Patel *Patel@ssrl01.Slac.stanford.edu* |
| NSLS | X16C | 5–20 | 4 μm² | White | $10^{10}$ | M. Marcus |
| NSLS | X26A | 5–20 | ~10–200 μm² | $10^{-4}$ white | $7 \times 10^{7}$– $7 \times 10^{10}$ | Steve Sutton *Sutton@cars.uchicago.edu* |

[a]The top four facilities are on hard x-ray third-generation sources. The ALS x-ray microprobe is on a third-generation VUV ring with a small emittance beam which acts like a high-performance second generation ring for hard x-rays. The remaining beamlines are on second-generation rings.

option appears especially promising for focusing monochromatic radiation ($\Delta E/E = \sim 10^{-4}$). These devices are simple to align, allow good working distance between the optics and the sample, and have already achieved submicron spots. Although zone plates are inherently chromatic, they can in principle be used with tunable radiation by a careful translation along the beam direction. State-of-the-art zone plates provide the most convenient optics for monochromatic experiments even though their focusing efficiency has not reached the 40% to 60% efficiency promised by more advanced designs.

**Kirkpatrick-Baez Mirrors.** Kirkpatrick-Baez (KB) mirrors provide a third highly promising option for focusing synchrotron radiation (Yang et al., 1995; Underwood et al., 1996). A Kirkpatrick-Baez mirror pair consists of mirrors that condense the beam in orthogonal directions. Both multilayer and total-external-reflection mirrors have been used for focusing synchrotron radiation. Multilayer mirrors appear most suitable for fluorescence measurements with a fixed wide-band-pass beam and where large divergences can be accepted. Total-external-reflection mirrors

appear to offer the best option for focusing white beams to μm dimensions. The key challenge with KB mirrors is achieving low figure and surface roughness with elliptical surfaces. There are numerous parallel efforts currently underway to advance mirror figuring for advanced KB focusing schemes.

**Refractive Lenses.** In addition to the focusing schemes mentioned above, there are two new options that have recently emerged from experiments at the European Synchrotron Radiation Facility (ESRF). These are compound refractive lenses and Bragg-Fresnel optics. Compound refractive lenses are very interesting because they are relatively easy to manufacture (Snigirev et al., 1996). Estimates of their theoretical efficiency, however, indicate that they cannot compete with the theoretical efficiency of KB mirrors or zone plates.

**Bragg-Fresnel Optics.** Bragg-Fresnel optics are also an interesting option used both at LURE (Chavallier et al., 1995, 1996) and the ESRF (Aristov, 1988). With Bragg-Fresnel optics, the phase-contrast steps are

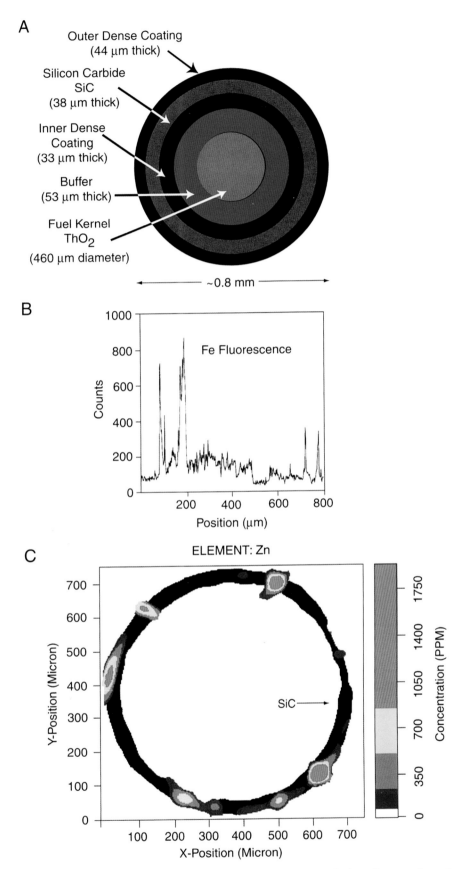

**Figure 15.** (**A**) Schematic of TRISO fuel element. (**B**) The Fe fluorescence during a line scan through a SiC shell shows a complicated spatial distribution with sharp features less than 1 μm wide. (**C**) Zn distribution in a SiC shell after reconstruction from x-ray microfluorescence data (Naghedolfeizi et al., 1998). The spatial resolution was limited because time restricted the number of steps.

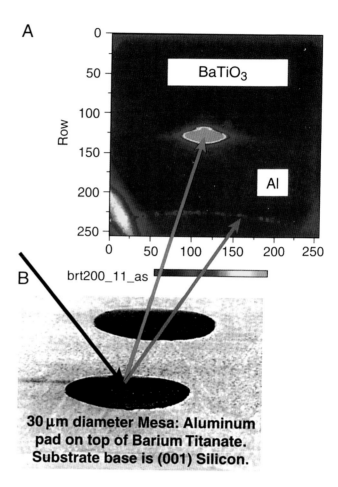

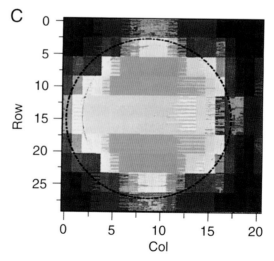

A

BaTiO₃

Al

brt200_11_as

30 µm diameter Mesa: Aluminum pad on top of Barium Titanate. Substrate base is (001) Silicon.

C

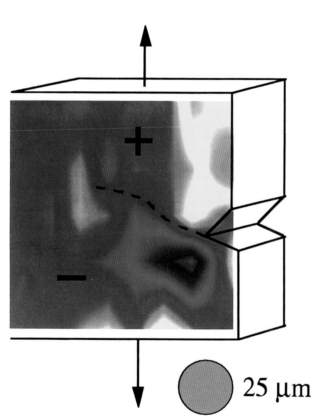

+

−

25 µm

**Figure 16.** Strain map in a Mo single crystal (Rebonato et al., 1989) showing $\Delta d/d$ for the (211) planes when the crystal is pulled. A region of contraction (below the dashed line) and a region of elongation (above the dashed line) are observed. The reduction in d beneath the dashed line arises from the orientation of the observed Bragg plane with respect to the free surface. The probe size of ~25 µm was barely small enough to detect the strain features.

**Figure 17.** (**A**) Mesas of BaTiO₃ show interesting poling behavior. (**B**) Schematic of microdiffraction process with x-ray microdiffraction image from an advanced thin film. The CCD image shows the single-crystal reflection and the powder-like image from the Al overlayer. (**C**) Strain map across the mesa. The strain map has a range of only $5 \times 10^{-4}$ and an uncertainty of ~$1 \times 10^{-5}$.

lithographically etched into a multilayer or monolithic Si substrate. This offers a rugged coolable substrate and serves to simultaneously focus and monochromatize the beam. Some of the alignment simplicity of zone plates is lost with these devices because the beam is deflected, but they have the advantage that the 0th order (direct beam) is spatially removed from the focus and can therefore be easily stopped.

## MICROBEAM APPLICATIONS

Microprobe analysis has already been applied to many problems with second- and third-generation x-ray beams (Jones and Gordon, 1989; Rebonato et al., 1989; Langevelde et al., 1990; Thompson et al., 1992; Wang et al., 1997). Studies include measurement of strain and texture in integrated circuit conduction paths (Wang et al., 1996, 1997), the measurement of buried trace elements in dissolution reactions (Perry and Thompson, 1994), and determination of chemistry in small regions. Three simple examples are given to illustrate possible applications.

### Example 1: Trace Element Distribution in a SiC Nuclear Fuel Barrier

TRISO-coated fuel particles contain a small kernel of nuclear fuel encapsulated by alternating layers of C and SiC as shown in Figure 15A. The TRISO-coated fuel particle is used in an advanced nuclear fuel designed for passive containment of the radioactive isotopes. The SiC layer provides the primary barrier for radioactive elements in the kernel. The effectiveness of this barrier layer under adverse conditions is critical to containment.

Shell coatings were evaluated to study the distribution and transport of trace elements in the SiC barrier after being subjected to various neutron fluences (Naghedolfeizi et al., 1998). The C buffer layers and nuclear kernels of the coated fuel were removed by laser drilling through the SiC and then leaching the particle in acid (Myers et al., 1986). Simple x-ray fluorescence analysis can detect the presence of trace elements, but does not indicate their distribution. Trace elements in the SiC are believed to arise at least in part as daughter products from the fission process; lower trace-element concentrations are found in unirradiated samples. The radial distribution of these elements in the SiC shells can be attributed to diffusion of elements in the kernel due to thermal and radiation-enhanced diffusion. Other elements in the shells may originate in the fabrication of the TRISO (three-layer coatings of pyrolytic carbon and SiC) particles.

Linear x-ray microprobe scans were made through the SiC shell. X-ray fluorescence is an ideal tool for this work because it is nondestructive (no spread of contamination), it is sensitive to heavy elements in a low-$Z$ matrix, and it provides a picture of the elemental distribution. Results of a simple line scan through a leached shell are shown in Figure 15B. This scan was made with 2-μm steps and an ~1-μm probe. As seen in Figure 15B, very localized Fe-rich regions <1-μm broad are observed throughout the shell. This behavior is typical of other trace elements observed in the shell.

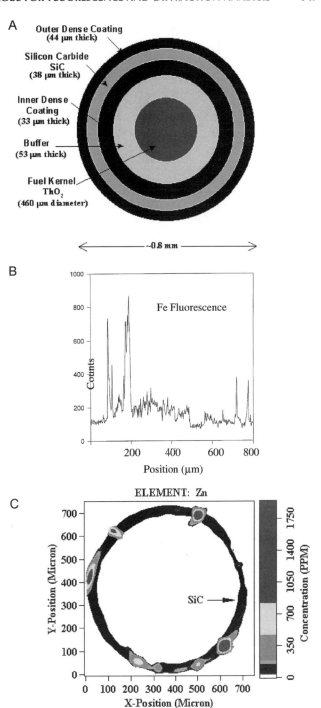

**Figure 15.** (**A**) Schematic of TRISO fuel element. (**B**) The Fe fluorescence during a line scan through a SiC shell shows a complicated spatial distribution with sharp features less than 1 μm wide. (**C**) Zn distribution in a SiC shell after reconstruction from x-ray microfluorescence data (Naghedolfeizi et al., 1998). The spatial resolution was limited because time restricted the number of steps. **See color figure.**

The spatial distribution can be further investigated by extending the x-ray microprobe analysis to x-ray microfluorescence tomography (Biosseau, 1986; Naghedolfeizi et al., 1998). Although x-ray fluorescence tomography is

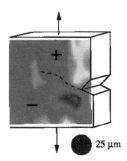

**Figure 16.** Strain map in a Mo single crystal (Rebonato et al., 1989) showing $\Delta d/d$ for the (211) planes when the crystal is pulled. A region of contraction (below the dashed line) and a region of elongation (above the dashed line) are observed. The reduction in d beneath the dashed line arises from the orientation of the observed Bragg plane with respect to the free surface. The probe size of $\sim$25 µm was barely small enough to detect the strain features. **See color figure.**

inherently time-consuming, the method yields three-dimensional distributions of trace elements. For example, the Zn distribution in a plane of the SiC shell is illustrated in Figure 15C. Note that the measurements are quantitative.

### Example 2: Change in the Strain Distribution Near a Notch During Tensile Loading

Early experiments with second-generation sources demonstrated some of the features of x-ray microprobe–based diffraction: good strain resolution, ability to study strain in thick samples, ability to study dynamics of highly strained samples, and the ability to distinguish lattice dilation from lattice rotations. For example, a measurement of differential strain induced by pulling on a notched Mo single crystal (Rebonato et al., 1989) shows lattice dilation above the notch and a lattice contraction below the notch (Fig. 16). This behavior would be totally masked with a topographic measurement because of the high density of initial dislocations. In addition, the measurement allowed the lattice rotations to be separated from the lattice dilations. Finally, the measurements are quantitative, and even in this early experiment yielded strain sensitivity $\Delta d/d$ of $\sim$5 $\times$ 10$^{-4}$.

### Example 3: Strain Distribution in an Advanced Ferroelectric Sample

BaTiO$_3$ has a tetragonal structure in the ferroelectric state. For a single-crystal thin film deposited on Si, the tetragonal axis can lie either in the plane of the thin film or normal to the surface. The direction of the tetragonal axis is referred to as the "poled" direction. In general, the poling is in the plane of the surface, but electrical measurements indicate that the poled state of the BaTiO$_3$ can be switched if the size of the BaTiO$_3$ film is small enough. X-ray microdiffraction provides a means to study the distribution of poled BaTiO$_3$ in small circular pads. The measurement sampled regions with $\sim$1 $\times$ 10 µm$^2$ spatial resolution. In the experiment, an x-ray microbeam was

scanned across monolithic films of the BaTiO$_3$ film (reference film) and across small $\sim$60-µm diameter mesasurements (Fig. 17A). The diffraction pattern from the film was recorded on a charge-coupled device (CCD) x-ray detector.

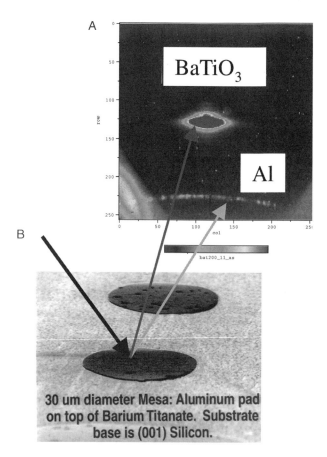

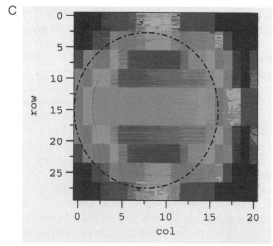

**Figure 17.** (**A**) Mesas of BaTiO$_3$ show interesting poling behavior. (**B**) Schematic of microdiffraction process with x-ray microdiffraction image from an advanced thin film. The CCD image shows the single-crystal reflection and the powder-like image from the Al overlayer. (**C**) Strain map across the mesa. The strain map has a range of only 5 $\times$ 10$^{-4}$ and an uncertainty of $\sim$1 $\times$ 10$^{-5}$. **See color figure.**

A typical diffraction pattern, shown in Figure 17B, illustrates many of the powerful attributes of microdiffraction. The $BaTiO_3$ 002/200 Bragg peak appears as a lenticular intense pattern in the CCD image even though it is covered with a 200-nm cap of aluminum. The strain in the thin film can be measured to $\Delta d/d = \sim 1 \times 10^{-5}$. This is sufficient to easily resolve the $\sim 1\%$ difference in the lattice parameter between 002 or 200 poled $BaTiO_3$. Texture, particle size, and strain of the powder-like aluminum overlayer can be inferred directly from the 111 and 200 Debye rings as seen in the lower part of Figure 17B. Even better strain resolution can be obtained by measurement of higher order reflections.

This example clearly indicates the need for an x-ray diffraction microprobe. Small spot size is required to study the distribution of poled material near small dimensioned features of the film. Good strain resolution is required to differentiate between the various poling options. A penetrating probe is required to measure the film in an unaltered state below a 200-nm cap of aluminum.

## METHOD AUTOMATION

X-ray microfluorescence and x-ray microdiffraction data collection at synchrotron sources occurs remotely in a hostile, high-radiation environment. Experiments are located inside shielded hutches that protect the researcher from the x rays near the experiment, but which restrict "hands-on" manipulation of the experiment. Fluorescence and simple diffraction data are typically collected by step-scanning the sample under the beam. A record is made of the fluorescence spectrum or the diffraction pattern from the sample at hundreds to thousands of positions. Fluorescence data collection and analysis is typically performed using specially designed one-of-a-kind programs. Some recent efforts have been directed to incorporate commercial fluorescence data analysis and collection software, but no standard has yet evolved. The general-purpose program SPECT has however been adopted on several beamlines worldwide for general purpose data collection and is used on a number of beamlines worldwide for collection of both microdiffraction and microfluorescence data.

Because the field has been revolutionized by new sources and instrumentation, there is increased urgency to further automate and standardize the data collection of both microfluorescence and microdiffraction analysis. The actual data collection program can be quite simple for both microfluorescence and microdiffraction. However, data analysis can be complicated, especially for microdiffraction. Improved computer processing and storage power is certain to make a major impact on the way data is collected, analyzed and stored. For example, to avoid the storage of huge data files, fluorescence spectra are often stored as "regions of interest." This procedure works well in many cases, but can lose important information for overlapping characteristic lines. A better solution is to record complete spectra or to fit the spectra "on the fly" to separate overlapping peaks. The availability of large storage devices and faster computers make these intrinsically superior options increasingly attractive.

Until recently, there has been little or no specialized software for x-ray microdiffraction. With monochromatic microdiffraction experiments, a key problem is the sphere of confusion of the diffractometer. Recent efforts have resulted in the development of automated methods of mapping the sphere of confusion of a diffractometer, to greatly reduce the experimental uncertainty of the relative beam/sample position (Noyan et al., 1999).

With microdiffraction experiments, a key challenge is automated indexing of Laue reflections. This quite general problem appears feasible for wide-band-pass x-ray beams and for samples where the unit cell is known (but not the grain orientation or strain) (Marcus et al., 1996; Chung, 1997; Wenk et al., 1997; Chung and Ice, 1999). Recent advances in specialized computer software for pattern recognition and microdiffraction data analysis has now demonstrated the ability to automatically separate the overlapping Laue patterns from 5 to 20 grains and to automatically determine strain to $2 \times 10^{-5}$ to $2 \times 10^{-4}$ and to localize the diffracting grain position along the incident beam to $\sim 2~\mu m$. The key elements of the so called ORDEX (Oak Ridge inDEXing) program used for automated indexing of overlapping Laue patterns on beamline 7 of the Advanced Photon Source have been described by Chung and Ice (1999). Parallel software development efforts are also underway at the European Synchrotron Radiation Facility and at the Advanced Light Source.

## SAMPLE PREPARATION

One major advantage of x-ray microprobe analysis is the minimal sample preparation required. In general, samples need to be small due to the short working distances between the focusing optics and the focal plane. Samples should also be mounted on low background-producing materials to reduce x-ray scatter and fluorescence (e.g., Mylar or Kapton for fluorescence and single-crystal Si for diffraction). In most cases, no additional preparation is required, and samples can be run in air, under water, or coated with low-Z barriers for radioactive containment or to prevent oxidation. Of course, what the x-ray microbeam measures is the elemental and crystallographic distributions in the sample as it is mounted. If, for example, a sample is machined, its surface will include a heavily damaged and chemically contaminated layer that may or may not be of interest to the experiment. In addition, because of the sensitivity of x-ray microfluorescence to trace elements, particular care must be taken during sample mounting to prevent contamination of the sample. For example, samples should be handled with low-Z plastic tweezers to avoid metal contamination from metallic tweezers.

Surface roughness and variations in sample thickness can also complicate the data analysis. In general, the analysis of fluorescence data is easier if the sample is homogeneous, smooth, and of uniform thickness. If the sample is very thick (semi-infinite) or very thin (negligible absorption), the data analysis is particularly easy. For most microdiffraction experiments to date, relative reflectivities are sufficient, so absorption corrections are not critical to the success of the experiment.

To aid in sample throughput, some microbeam instruments have mounting systems that allow the sample to be mounted on an optical microscope remote from the x-ray microprobe. The position of the sample with respect to a fiducial in the remote microscope holder can be accurately reproduced on the x-ray microbeam positioning stage. Features of interest can be identified off line and their coordinates noted for later x-ray microcharacterization.

## DATA ANALYSIS AND INITIAL INTERPRETATION

### Microfluorescence

For many samples, x-ray microflurorescence can yield rather direct information about trace element distributions. X-ray fluorescence data can be placed on an absolute scale if the approximate composition of the sample is understood and the incident flux known/measured. With a solid-state detector, the major components of the sample are readily identified for $Z > 12$. The composition can be used to estimate the linear absorption coefficient and therefore the sample volume probed by the x-ray beam. If the matrix is known, the absorption can be calculated from standard tables or can be found on various web sites. Because of uncertainties in the absorption coefficients of x-rays, first-principles methods are only good to ~5% (Sparks, 1980; Lachance and Claisse, 1995). More precise absolute measurements can be obtained with prepared standards. Standard samples can be obtained from the National Institutes of Standards and Technology (NIST) or by careful dilution of trace elements in a matrix of known composition.

Initial fluorescence data analysis can be made by simply selecting a region of interest that incorporates the fluorescence lines of interest and nearby regions for background subtraction. The elemental concentration in the probe volume can be approximated from the estimated linear absorption coefficients of the incident and fluorescence beams and the net counts in the line. Sparks (1980) gives a general formula for the mass concentration of element $z$, $C_z$, as a function of the experimental geometry

$$C_z = \frac{I_Z 4\pi R^2 (\mu_{S,0} + \mu_{S,i}\sin\psi/\sin\phi)\rho_S}{P_0 D_Z \sigma_{zij}\langle 1 - \exp\{1[(\mu_{S,0} + \mu_{S,i}\sin\psi/\sin\phi)/\rho_S]\rho_S T \sin\psi\}\rangle} \quad (5)$$

Here $R$ is the specimen-to-detector distance, $I_Z$ is the fluorescence intensity per area at the detector, $\mu_{s,0}/\rho_s$ and $\mu_{s,i}/\rho_s$ are the sample mass absorption coefficients for the incident and fluorescence radiation, $P_0$ is the incident power, $D_Z$ is an absorption correction for attenuation between the sample and the detector, $\sigma_{zij}$ is the fluorescence cross-section for the exciting energy, and $\rho_s T$ is the mass per unit area of the sample.

The precision of the measurement can be improved by fitting the fluorescence lines with linear (assumed shape) or nonlinear least-squares techniques. All fluorescence lines have a natural Lorenztian shape which is a consequence of the lifetime of the inner-shell holes that give rise to the fluorescence. For most measurements, however, the Lorenztian shape is dominated by the Gaussian resolution of the wavelength- or energy-dispersive spectrometer. Fluorescence lines are therefore typically fit to Gaussian or Voight functions. The Lorentzian tail of a fluorescence line is mainly of importance for estimating the background it contributes under a weak line.

### Microdiffraction

With a monochromatic x-ray beam, relative texture and strain information can be obtained rather directly from measured microdiffraction scattering angle and the x-ray wavelength (see X-RAY POWDER DIFFRACTION, XAFS SPECTROSCOPY and RESONANT SCATTERING TECHNIQUES). X-ray microdiffraction can measure absolute strain to better than 1 part in $10^5$, but is often used for high-precision measurements of strain or rotation differences, rather than absolute measurements. If the sample grain size is sufficiently small compared to the probe volume, standard powder diffraction methods can be used to determine the lattice constants of the sample (X-RAY POWDER DIFFRACTION). However, it is useful to utilize a position-sensitive area detector to collect a large fraction of the Debye ring for data analysis. Each powder ring can be fit to optimize the statistical determination of the $d$ spacing. The shape and intensity distribution of each ring also can be used to help determine the local texture and strain tensor of the sample. Standard samples are particularly useful for putting strain measurements on an absolute scale.

With white-beam microdiffraction, each grain intercepted by the beam produces a Laue pattern. This ensures that the deviatoric strain tensor information can be recovered for each grain, but complicates the analysis, since many grains simultaneously contribute to the Laue spectra. Often grains from various layers can be identified according to characteristics of the layer (i.e., intense sharp peaks from large grains, fuzzy weak peaks from small grains). Whatever the process, the key step is to identify which reflections are associated with a single grain and to index the reflections for each grain. This problem has now been solved for many systems with an automated indexing program that can simultaneously index ~5 to 20 grains or subgrains (Chung and Ice, 1999). With this program, the detector sample alignment is first calibrated using a perfect crystal of Ge, then the program automatically determines the grain orientation and strain. Typically, the orientation can be determined to better than 0.01° and the strain can be determined to less than 1 part in $10^4$. This precision already is better than for virtually all other probes, and can be further refined in some directions by moving the detector back both to localize the origin of the diffracted beam and also to gain angular precision in the measurement.

## PROBLEMS

As described previously, the two key problems of x-ray microbeam analysis are the limited number of useful sources and the difficulty of fabricating efficient x-ray optics. In addition to these two problems, there are several annoying problems that complicate the use of x-ray microbeam techniques. One difficulty with x-ray microbeam experiments is identification of the x-ray beam position on the sample. Because of the small beam size, even intense x-ray microbeams are difficult to view on fluorescent screens with visible light optics, and even when they are visualized, the transfer from beam position on a fluorescence screen to beam position on a sample can be difficult. For some experiments, it is necessary to place markers on the sample to help in locating the x-ray beam. For example, a cross-hair of a fluorescing heavy metal can be used to locate the beam position with respect to an interesting region.

Another annoying problem with x-ray microbeam analysis is the difficulty of monitoring the absolute beam intensity on the sample. Because the distance from the focusing optics to the focus is typically short, there is little room to install a transmission beam monitor. Even when such a monitor is installed, great care must be taken to avoid contamination of the monitor signal due to backscatter from the sample.

The problem of backscatter contamination into a transmission monitor is but one example of a general class of shielding problems that arise due to the proximity of sample, detector, and optics. In general, great care is required to reduce parasitic backgrounds associated with the beam path through x-ray optics and any air path to the sample. Typically, scatter from beam-defining slits and upstream condensing optics can swamp a CCD detector unless care is taken to shield against such direct scattering. The short working distance between the optics and the sample also require care to avoid collisions.

## ACKNOWLEDGMENTS

Research sponsored in part by the Laboratory Directed Research and Development Program of the Oak Ridge National Laboratory and the Division of Material Sciences, U.S. Department of Energy under contract DE-AC05-96OR22464 with Lockheed Martin Energy Research Corporation. Research carried out in part on beamline 10.3.1 at the ALS and on beamline 2ID at the APS. Both facilities are supported by the U.S. Department of Energy.

## LITERATURE CITED

APS Users Meeting Workshop. 1997. Technical report: Making and using very small X-ray beams. APS, Argonne, Ill.

Aristov, V. V. 1988. X-ray Microscopy II. pp. 108–117. Bogorodski Pechatnik Publishing Company, Chernogolovka, Moscow Region.

Bambynek, W., Crasemann, B., Fink, R. W., Freund, H. U., Mark, H., Swift, C. D., Price, R. E., and Venugopala Rao, P. 1972. X-ray fluorescence yields, Auger and Coster-Kronig transition probabilties. *Rev. Mod. Phys.* 44:716–813.

Bilderback, D. H., Hoffman, S. A., and Thiel, D. J. 1994. Nanometer spatial resolution achieved in hard X-ray imaging and Laue diffraction experiments. *Science* 263:201–202.

Bionta, R. M., Ables, E., Clamp, O., Edwards, O. D., Gabriele, P. C., Miller, K., Ott, L. L., Skulina, K. M., Tilley, R., and Viada, T. 1990. Tabletop X-ray microscope using 8 keV zone plates. *Opt. Eng.* 29:576–580.

Boisseau, P. 1986. Determination of three dimensional trace element distributions by the use of monochromatic X-ray microbeams. Ph.D. dissertation. Massachusetts Institute of Technology, Cambridge, Mass.

Brennan, S., Tompkins, W., Takaura, N., Pianetta, P., Laderman, S. S., Fischer-Colbrie, A., Kortright, J. B., Madden, M. C., and Wherry, D. C. 1994. Wide band-pass approaches to total-reflection X-ray fluorescence using synchrotron radiation. *Nucl. Instrum. Methods A* 347:417–421.

Buras, B. and Tazzari, S. 1984. European Synchrotron Radiation Facility Report of European Synchrotron Radiation Project. Cern LEP Division, Geneva, Switzerland.

Busing, W. R. and Levy, H. A. 1967. Angle calculations for 3- and 4-circle X-ray and neutron diffractometers. *Acta Crystallogr.* 22:457–464.

Cambell, J. L. 1990. X-ray spectrometers for PIXE. *Nucl. Instrum. Methods B* 49:115–125.

Campbell, J. R., Lamb, R. D., Leigh, R. G., Nickel, B. G., and Cookson, J. A. 1985. Effects of random surface roughness in PIXE analysis of thick targets. *Nucl. Instrum. Methods B* 12:402–412

Chevallier, P. and Dhez, P. 1997. Hard X-ray microbeam: Production and application. *In* Accelerator Based Atomic Physics Techniques and Applications (S. M. Shafroth and J. Austin, eds.). pp. 309–348. AIP Press, New York.

Chevallier, P., Dhez, P., Legrand, F., Idir, M., Soullie, G., Mirone, A., Erko, A., Snigirev, A., Snigireva, I., Suvorov, A., Freund, A., Engstrom, P., Als. Nielsen, J., and Grubel, A. 1995. Microprobe with Bragg-Fresnel multilayer lens at ESRF beam line. *Nucl. Instum. Methods A* 354:584–587

Chevallier, P., Dhez, P., Legrand, F., Irko, A., Agafonov, Y., Panchenko, L. A., and Yakshin, A. 1996. The LURE-IMT photon microprobe. *J. Trace Microprobe Techniques* 14:517–540.

Chung, J.-S. 1997. Automated indexing of wide-band-pass Laue images. 1997. APS Users Meeting Workshop on Making and Using Very Small X-ray Beams. APS, Argonne, Ill.

Chung, J.-S. and Ice, G. E. 1998. Automated indexing of Laue images from polycrystalline materials. MRS Spring Symposium, San Francisco, Calif.

Chung, J. S. and Ice, G. E. 1999. Automated indexing for texture and strain measurement with broad-bandpass X-ray microbeams. *J. Appl. Phys.* 86:5249–5256.

Doyle, B. L., Walsh, D. S., and Lee, S. R. 1991. External micro-ion-beam analysis (X-MIBA). *Nucl. Instrum. Methods B* 54:244–257.

Goldstein. J. I. 1979. Principles of thin film x-ray microanalysis. *In* Introduction to Analytical Electron Microscopy (J. J. Hern, J. I. Goldstein, and D. C. Joy, eds.). pp. 83–120. Plenum, New York.

Hayakawa, S., Gohshi, Y., Iida, A., Aoki, S., and Ishikawa, M. 1990. X-ray microanalysis with energy tunable synchrotron x-rays. *Nucl. Instrum. Methods B* 49:555.

Hoffman, S. A., Thiel, D. J., and Bilderback, D. H. 1994. Applications of single tapered glass capillaries-submicron X-ray imaging and Laue diffraction. *Opt. Eng.* 33:303–306.

Howells, M. R. and Hastings, J. B. 1983. Design considerations for an X-ray microprobe. *Nucl. Instrum. Methods* 208:379–386.

Ice, G. E. 1987. Microdiffraction with synchrotron radiation. *Nucl. Instrum. Methods B* 24/25:397–399.

Ice, G. E. 1997. Microbeam-forming methods for synchrotron radiation. *X-ray Spectrom.* 26:315–326.

Ice, G. E. and Sparks, C. J., Jr. 1984. Focusing optics for a synchrotron X-radiation microprobe. *Nucl. Instrum. Methods* 222:121–127.

Ice, G. E. and Sparks, C. J. 1990. Mosaic crystal X-ray spectrometer to resolve inelastic background from anomalous scattering experiments. *Nucl. Inst. Methods* A291:110–116.

Ice, G. E. and Sparks, C. J. 1991. MICROCAT proposal to APS. APS, Argonne, Ill.

Jones, K. W. and Gordon, B. M. 1989. Trace element determinations with synchrotron-induced X-ray emission. *Anal. Chem.* 61:341A–356A.

Koumelis, C. N., Londos, C. A., Kavogli, Z. I., Leventouri, D. K., Vassilikou, A. B., and Zardas, G. E. 1982. *Can. J. Phys.* 60:1241–1246.

Lachance, G. R. and Claisse, F. 1995. Quantitative X-ray Fluorescence Analysis: Theory and Application. John Wiley & Sons, New York.

Langevelde, F. V., Tros, G. H. J., Bowen, D. K., and Vis, R. D. 1990. The synchrotron radiation microprobe at the SRS, Daresbury (UK) and its applications. *Nucl. Instrum. Methods B* 49:544–550.

Larson, B. C., Tamura, N., Chung, J.-S., Ice, G. E., Budai, J. D., Tischler, J. Z., Yang, W., Weiland, H., and Lowe, W. P. 1999. 3-D measurement of deformation microstructure in Al(0.2%) Mg using submicron resolution white X-ray microbeams. MRS Fall Symposium Proceedings. Materials Research Society, Warrendale, Pa.

Larsson, S. and Engström, P. 1992. X-ray microbeam spectroscopy with the use of capillary optics. *Adv. X-ray Anal.* 35:1019–1025.

Lindh, U. 1990. Micron and submicron nuclear probes in biomedicine. *Nucl. Instrum. Methods B* 49:451–464.

Marcus, M. A., MacDowell, A. A., Isaacs, E. D., Evans-Lutterodt, K., and Ice, G. E. 1996. Submicron resolution X-ray strain measurements on patterned films: Some hows and whys. *Mater. Res. Soc. Symp.* 428:545–556.

Materlik, G., Sparks, C. J., and Fischer, K. 1994. Resonant Anomalous X-ray Scattering. North Holland Publishers, Amsterdam, The Netherlands.

Michael, J. R. and Goehner, R. P. 1993. Crystallographic phase identification in scanning electron microscope: Backscattered electron Kikuchi patterns imaged with a CCD-based detector. *MSA Bull.* 23:168–175

Miller, M. K., Cerezo, A., Hetherington, M. G., and Smith, G. D. W. 1996. Atom Probe Field Ion Microscopy. Oxford Science Publications, Oxford.

Myers, B. F., Montgomery, F. C., and Partain, K. E. 1986. The transport of fission products in SiC. Doc. No. 909055, GA Technologies. General Atomics, San Diego.

Naghedolfeizi, M., Chung, J.-S., and Ice, G. E. 1998. X-ray fluorescence microtomography on a SiC nuclear fuel ball. *Mater. Res. Soc. Symp.* 524:233–240.

Noyan, I. C., Kaldor, S. K., Wang, P.-C., Jordan-Sweet, J. 1999. A cost-effective method for minimizing the sphere-of-confusion error of x-ray microdiffractometers. *Rev. Sci. Inst.* 70:1300–1304.

Perry, D. L. and Thompson, A. C. 1994. Synchrotron induced X-ray fluorescence microprobe studies of the copper-lead sulfide solution interface. *Am. Chem. Soc. Abstr.* 208:429.

Piotrowski, D. P., Stock, S. R., Guvenilir, A., Haase, J. D., and Rek, Z. U. 1996. High resolution synchrotron X-ray diffraction tomography of large-grained samples. *Mater. Res. Soc. Symp. Proc.* 437:125–128.

Rebonato, R., Ice, G. E., Habenschuss, A., and Bilello, J. C. 1989. High-resolution microdiffraction study of notch-tip deforma-

tion in Mo single crystals using X-ray synchrotron radiation. *Philos. Mag. A* 60:571–583.

Ren, S. X, Kenik, E. A., Alexander. K. B., and Goyal, A. 1998. Exploring spatial resolution in electron back-scattered diffraction (EBSD) experiments via Monte Carlo simulation. *Microsc. Microanal.* 4:15–22.

Riekel, C. 1992. Beamline for microdiffraction and micro small-angle scattering. *SPIE* 1740:181–190.

Riekel, C., Bosecke. P., and Sanchez del Rio, M. 1992. 2 high brilliance beamlines at the ESRF dedicated to microdiffraction, biological crystallography and small angle scattering. *Rev. Sci. Instrum.* 63:974–981.

Shenoy, G. K., Viccaro, P. J., and Mills, D. M. 1988. Characteristics of the 7 GeV advanced photon source: A guide for users. ANL-88-9 Argonne National Laboratory, Argonne, Ill.

Snigirev, A., Kohn, V., Snigireva, I., and Legeler, B. 1996. A compound refractive lens for focusing high-energy X-rays. *Nature (London)* 384:49–51.

Snigirev, A., Snigireva, I., Bosecke, P., Lequien, S., and Schelokov, I. 1997. High energy X-ray phase contrast microscopy using a circular Bragg-Fresnel lens. *Opt. Commun.* 135:378–384.

Sparks, C. J., Jr. 1980. X-ray fluorescence microprobe for chemical analysis. *In* Synchrotron Radiation Research (H. Winick and S. Doniach, eds.). pp. 459–512. Plenum, New York.

Sparks, C. J., Kumar, R., Specht, E. D., Zschack, P., Ice, G. E., Shiraishi, T., and Hisatsune, K. 1992. Effect of powder sample granularity on fluorescent intensity and on thermal parameters in X-ray diffraction rietveld analysis. *Adv. X-ray Anal.* 35:57–62.

Stern, E. A., Kalman, Z., Lewis, A., and Lieberman, K. 1988. Simple method for focusing X-rays using tapered capillaries. *Appl. Opt.* 27:5135–5139.

Stock, S. R., Guvenilir, A., Piotrowski, D. P., and Rek, Z. U. 1995. High resolution synchrotron X-ray diffraction tomography of polycrstalline samples. *Mater. Res. Soc. Symp.* 375:275–280.

Thompson, A. C., Chapman, K. L., Ice, G. E., Sparks, C. J., Yun, W., Lai, B., Legnini, D., Vicarro, P. J., Rivers, M. L., Bilderback, D. H., and Thiel, D. J. 1992. Focusing optics for a synchrotron-based X-ray microprobe. *Nucl. Instrum. Methods A* 319:320–325.

Underwood, J. H., Thompson, A. C., Kortright, J. B., Chapman, K. C., and Lunt, D. 1996. Focusing x-rays to a 1 μm spot using elastically bent, graded multilayer coated mirrors. *Rev. Sci. Instrum. Abstr.* 67:3359.

Veigele, W. J., Briggs, E., Bates, L., Henry. E. M., and Bracewell, B. 1969. X-ray Cross Section Compilation from 0.1 keV to 1 MeV. Kaman Sciences Report No. DNA 2433F.

Wang, P. C., Cargill, G. S., Noyan, I. C., Liniger, E. G., Hu, C. K., and Lee, K. Y. 1996. X-ray microdiffraction for VLSI. *Mater. Res. Soc. Symp. Proc.* 427:35.

Wang, P. C., Cargill G. S., III, Noyan, I. C., Liniger, E. G., Hu, C. K., and Lee, K. Y. 1997. Thermal and electromigration strain distributions in 10 μm-wide aluminum conductor lines measured by x-ray microdiffraction. *Mater. Res. Soc. Symp. Proc.* 473:273.

Wenk, H. R., Heidelbach, F., Chadeigner, D., and Zontone, F. 1997. Laue orientation imaging. *J. Synchrotron Rad.* 4:95–101.

Wilkinson, A. J. 1997. Methods for determining elastic strains from electron backscattering diffraction and electron channeling patterns. *Mater. Sci. Tech.* 13:79.

Yaakobi, B. and Turner, R. E. 1979. Focusing X-ray spectrograph for laser fusion experiments. *Rev. Sci. Inst.* 50:1609–1611.

Yang, B. X., Rivers, M., Schildkamp, W., and Eng, P. J. 1995. Geo-CARS microfocusing Kirkpatric-Baez bender development. *Rev. Sci. Instrum.* 66:2278–2280.

Yun, W. B., Lai, B., Legnini, D., Xiao, Y. H., Chrzas, J., Skulina, K. M., Bionta, R. M., White, V., and Cerrina, F. 1992. Performance comparison of hard X-ray zone plates fabricated by two different methods. *SPIE* 1740:117–129.

## KEY REFERENCES

Boisseau, 1986. See above.

*Most complete discussion of x-ray microfluorescence tomography with some early examples.*

Chevallier and Dhez, 1997. See above.

*Recent overview of x-ray microbeam science and hardware.*

Chung and Ice, 1999. See above.

*Derives the mathematical basis for white-beam microdiffraction measurements of the deviatoric and absolute strain tensor and includes a description of the ORDEX program, which automatically indexes multiple overlapping Laue patterns.*

Ice, 1997. See above.

*Recent overview of x-ray focusing optics for x-ray microprobes with methods for comparing the efficiency of various x-ray focusing optics.*

Sparks, 1980. See above.

*Quantitative comparison of x-ray microfluorescence analysis to electron and proton microprobes.*

GENE E. ICE
Oak Ridge National Laboratory
Oak Ridge, Tennessee

# X-RAY MAGNETIC CIRCULAR DICHROISM

## INTRODUCTION

X-ray magnetic circular dichroism (XMCD) is a measure of the difference in the absorption coefficient ($\mu_c = \mu^+ - \mu^-$), with the helicity of incident circularly polarized light parallel ($\mu^+$) and antiparallel ($\mu^-$) to the local magnetization direction of the absorbing material. These measurements are typically made at x-ray energies spanning an absorption edge, as illustrated in Figure 1, where a deep atomic core electron is photoexited into the valence band. Because the binding energy of these core electrons is associated with specific elements, XMCD can be used to separately measure the magnetic contributions from individual constituents in a complex material, by tuning the incident beam energy to a specific edge. Furthermore, different absorption edges of the same element provide information on the magnetic contributions of different kinds of valence electrons within a single element.

XMCD measurements yield two basic types of magnetic information, both of which are specific to the element and orbital determined by the choice of measured absorption edge. First, the strength of the dichroic signal is proportional to the projection of the magnetic moment along

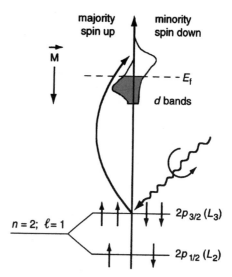

**Figure 1.** Schematic illustration of the energy bands and transitions involved in XMCD measurements. A circularly polarized photon excites a photoelectron from a spin-orbit-split core state to a spin-polarized valence band.

the incident x-ray beam direction. Thus, the signal can be used to determine magnetization direction for a sample in remanence or changes in the sample magnetization upon varying the applied magnetic field or temperature. Secondly, XMCD can be used to separate the orbital $\langle L_z \rangle$ and spin $\langle S_z \rangle$ contributions to the magnetic moment. This is possible through sum rules (Thole et al., 1992; Carra et al., 1993) that relate these contributions to the integrated dichroic signal. It is this elemental and orbital specificity and the ability to deconvolve the magnetic moment into its orbital and spin components that makes XMCD a unique tool for the study of magnetic materials.

In discussing x-ray spectra, one must refer quite frequently to the x-ray absorption edges; the terminology for labeling these edges should therefore be made clear from the outset of this discussion. The common convention has been to denote the principal quantum number ($n = 1$, 2, 3, 4, . . .) of the core state by $K, L, M, N, . . .$ and the total angular momentum ($s_{1/2}$, $p_{1/2}$, $p_{3/2}$, $d_{3/2}$, etc.) of the core state by 1, 2, 3, 4, . . . . Therefore, an edge referred to as the $L_3$ edge would denote an absorption threshold with an initial $2p_{3/2}$ core state. Table 1 lists the elements and edges for which XMCD measurements have been taken along with the corresponding transition, energy range, and reference to a typical spectrum. Most XMCD work in the soft x-ray regime ($E < 2000$ eV) has centered on the $3d$ transition metal (TM) $L_{2,3}$ edges and rare earth (RE) $M_{4,5}$ edges. These edges probe the largely localized TM $3d$ and RE $4f$ states, which are the primary magnetic electrons in almost all magnetic materials. The relatively large magnetic moments of these states lead to large dichroic effects in this energy regime. Hard x-ray measurements ($E > 2000$ eV) have focused on the TM $K$ edge and the RE $L_{2,3}$ edges, which probe much more diffuse bands with small magnetic moments leading to smaller dichroic effects. Although the edges probed and thus the final states are different for these regimes, the physics and

**Table 1. Representative Sample of the Elements and Edges Measured Using XMCD**

| Elements | Edges | Orbital Character of Transitions | Energy Range (keV) | Reference(s) Showing Typical Spectra |
|---|---|---|---|---|
| 3d (Mn-Cu) | $K$ | $s \rightarrow p$ | 6.5–9.0 | Stahler et al. (1993) |
| 3d (V-Cu) | $L_{2,3}$ | $p \rightarrow d$ | 0.5–1.0 | Rudolf et al. (1992); Wu et al. (1992) |
| 4d (Rd, Pd) | $L_{2,3}$ | $p \rightarrow d$ | 3.0–3.5 | Kobayashi et al. (1996) |
| 4f (Ce-Ho) | $M_{4,5}$ | $d \rightarrow f$ | 0.9–1.5 | Rudolf et al. (1992); Schillé et al. (1993) |
| 4f (Ce-Lu) | $L_{2,3}$ | $p \rightarrow d$ | 5.7–10.3 | Fischer et al. (1992); Lang et al. (1992) |
| 5d (Hf-Au) | $L_{2,3}$ | $p \rightarrow d$ | 9.5–13.7 | Schütz et al. (1989a) |
| 5f (U) | $M_{4,5}$ | $d \rightarrow f$ | 3.5–3.7 | Finazzi et al. (1997) |

analysis of the spectra is essentially the same, and they differ only in the techniques involved in spectrum collection.

This unit will concentrate on the techniques employed in the hard x-ray regime, but considerable mention will be made of soft x-ray results and techniques because of the fundamental importance in magnetism of the final states probed by edges in this energy regime. Emphasis will be given to the methods used for the collection of XMCD spectra rather than to detailed analysis of specific features in the XMCD spectra of a particular edge. More detailed information on the features of XMCD spectra at specific edges can be found in other review articles (Krill et al., 1993; Stahler et al., 1993; Schütz et al., 1994; Stöhr and Wu, 1994; Pizzini et al., 1995).

### Competitive and Related Techniques

Many techniques provide information similar to XMCD, but typically they tend to give information on the bulk macroscopic magnetic properties of a material rather than microscopic element–specific information. It is apparent that the basic criteria for choosing XMCD measurements over other techniques are the need for this element-specific information or for a separate determination of the orbital and spin moments. The need to obtain this information must be balanced against the requirements that XMCD measurements must be performed at a synchrotron radiation facility, and that collection of spectra can take several hours. Therefore, the amount of parameter space that can be explored via XMCD experiments is restricted by the amount of time available at the beamline.

The methods most obviously comparable to XMCD are the directly analogous laser-based methods such as measurement of Faraday rotation and magneto-optical Kerr effect. These techniques, while qualitatively similar to XMCD, are considerably more difficult to interpret. The transitions involved in these effects occur between occupied and unoccupied states near the Fermi level, and therefore the initial states are extended bands that are much more difficult to model than the atomic-like core states in XMCD measurements. In fact, recent efforts have focused on using XMCD spectra as a basis for understanding the features observed in laser-based magnetic

measurements. Another disadvantage of these measurements is that they lack the element specificity provided by XMCD spectra, because they involve transitions between all the conduction electron states. A final potential drawback is the surface sensitivity of the measurements.

Because the size of the dichroic signal scales with sample magnetization, XMCD spectra can also be used to measure relative changes in the magnetization as a function of temperature or applied field. This information is similar to that provided by vibrating sample or SQUID magnetometer measurements. Unlike a magnetometer, which measures the bulk magnetization for all constituents, dichroism measures the magnetization of a particular orbital of a specific element. Therefore, XMCD measurements can be used to analyze complicated hysteresis behavior, such as that encountered in magnetic multilayers, or to measure the different temperature dependencies of each component in a complicated magnetic compound. That these measurements are generally complementary to those obtained with a magnetometer should nevertheless be stressed, since it is frequently necessary to use magnetization measurements to correctly normalize the XMCD spectra taken at fixed field.

The size of the moments and their temperature behavior can also be ascertained from magnetic neutron or nonresonant magnetic x-ray diffraction. These two techniques are the only other methods that can deconvolve the orbital and spin contributions to the magnetic moment. For neutron diffraction, separation of the moments can be accomplished by measuring the magnetic form factor for several different momentum transfer values and fitting with an appropriate model, while for non-resonant magnetic x-ray scattering the polarization dependence of several magnetic reflections must be measured (Blume and Gibbs, 1988; Gibbs et al., 1988). The sensitivity of both these techniques, however, is very limited and thus measurements have been restricted to demonstration experiments involving the heavy rare earth metals with large orbital moments ($\mu_{orb} \sim 6 \, \mu_B$). XMCD on the other hand provides orbital moment sensitivity down to $\sim 0.005 \, \mu_B$ (Samant et al., 1994). Furthermore, for compound materials these measurements will not be element specific, unless a reflection is used for which only one atomic species produces constructive interference. This is usually only possible for simple compounds and generally not possible for multi-layered materials. As diffractive measurements, these techniques have an advantage over XMCD in that they are able to examine antiferromagnetic as well as ferromagnetic materials. But, the weakness of magnetic scattering relative to charge scattering for nonresonant x-ray scattering makes the observation of difference signals in ferromagnetic compounds difficult.

A diffractive technique that does provide the element specificity of XMCD is x-ray resonant exchange scattering (XRES; Gibbs et al., 1988). XRES measures the intensity enhancement of a magnetic x-ray diffraction peak as the energy is scanned through a core hole resonance. This technique is the diffraction analog of dichroism phenomena and involves the same matrix elements as XMCD. In fact, theoretical XRES calculations reduce to XMCD at the limit of zero momentum transfer. At first glance, this would seem to put XMCD measurements at a disadvantage to XRES measurements, because the former is just a simplification of the later. It is this very simplification, however, that makes possible the derivation of the sum rules relating the XMCD spectrum to the orbital and spin moments. Most XRES measurements in the hard x-ray regime have been limited to incommensurate magnetic Bragg reflections, which do not require circularly polarized photons. In these cases, XRES measurements are more analogous to linear dichroism, for which a simple correlation between the size of the moments and the dichroic signal has not been demonstrated. Lastly, even though the XRES signal is enhanced over that obtained by nonresonant magnetic scattering, it is still typically lower by a factor of $10^4$ or more than the charge scattering contributions. Separating signals this small from the incoherent background is difficult for powder diffraction methods, and therefore XRES, and magnetic x-ray scattering in general, have been restricted to single-crystal samples.

Spin-polarized photoemission (SPPE) measures the difference between the emitted spin-up and spin-down electrons upon the absorption of light (McIlroy et al., 1996). SPPE spectra can be related to the spin polarization of the occupied electron states, thereby providing complimentary information to XMCD measurements, which probe the unoccupied states. The particular occupied states measured depend upon the incident beam energy, UV or soft x-ray, and the energy of the photoelectron. Soft x-ray measurements probe the core levels, thereby retaining the element specificity of XMCD but only indirectly measuring the magnetic properties, through the effect of the exchange interaction between the valence band and the core-level electrons. UV measurements, on the other hand, directly probe the partially occupied conduction electron bands responsible for the magnetism, but lose element specificity, which makes a clear interpretation of the spectra much more difficult. The efficiency of spin-resolving electron detectors for these measurements, however, loses 3 to 4 orders of magnitude in the analysis of the electron spin. This inefficiency in resolving the spin of the electron has prevented routine implementation of this technique for magnetic measurements. A further disadvantage (or advantage) of photoemission techniques is their surface sensitivity. The emitted photoelectrons typically come only from the topmost 1 to 10 monolayers of the sample. XMCD, on the other hand, provides a true bulk characterization because of the penetrating power of the x-rays. Magnetic sensitivity in photoemission spectra has also been demonstrated using circularly polarized light (Baumgarten et al., 1990). This effect, generally referred to as photoemission MCD, should not be confused with XMCD, because the states probed and the physics involved are quite different. While this technique eliminates the need for a spin-resolving electron detector, it does not provide a clear relationship between measured spectra and size of magnetic moments.

## PRINCIPLES OF THE METHOD

XMCD measures the difference in the absorption coefficient over an absorption edge upon reversing the sample magnetization or helicity of the incident circularly polarized photons. This provides magnetic information that is specific to a particular element and orbital in the sample. Using a simple one-electron theory it is easy to demonstrate that XMCD measurements are proportional to the net moment along the incident beam direction and thus can be used to measure the variations in the magnetization of a particular orbital upon application of external fields or change in the sample temperature. Of particular interest has been the recent development of sum rules which can be used to separate the orbital and spin contributions to the magnetic moment. XMCD measurements are performed at synchrotron radiation facilities, with the specific beamline optics determined by the energy regime of the edge of interest.

### Basic Theory of X-Ray Circular Dichroism

Although a number of efforts had been made over the years to characterize an interplay between x-ray absorption and magnetism, it was only in 1986 that the first unambiguous evidence of magnetization-sensitive absorption was observed in a linear dichroism measurement at the $M_{4,5}$ edges of Tb metal (Laan et al., 1986). This measurement was quickly followed by the first observation of XMCD at the $K$ edge of Fe and, a year later, of an order-of-magnitude-larger effect at the $L$ edges of Gd and Tb (Schütz et al., 1987, 1988). Large XMCD signals directly probing the magnetic $3d$ electrons were then discovered in the soft x-ray regime at the $L_{2,3}$ edge of Ni (Chen et al., 1990).

The primary reason for the lack of previous success in searching for magnetic contributions to the absorption signal is that the count rates, energy tunability, and polarization properties required by these experiments make the use of synchrotron radiation sources absolutely essential. Facilities for the production of synchrotron radiation, however, were not routinely available until the late 1970s. The mere availability of these sources was clearly not sufficient, because attempts to observe XMCD had been made prior to 1988 without success (Keller, 1985). It was only as the stability of these sources increased that the systematic errors which can easily creep into difference measurements were reduced sufficiently to allow the XMCD effect to be observed.

The basic cause of the enhancement of the dichroic signal that occurs near an absorption edge can be easily understood using a simple one-electron model for x-ray absorption (see Appendix A). While this model fails to explain all of the features in most spectra, it does describe the fundamental interaction that leads to the dichroic signal, and more complicated analyses of XMCD spectra are typically just perturbations of this basic model. In this crude treatment of x-ray absorption, the transition probabilities between spin-orbit-split core states and the spin-polarized final states are calculated starting from Fermi's golden rule. The most significant concept in this analysis is that the fundamental cause of the XMCD signal is the spin-orbit splitting of the core state. This splitting causes the difference between the matrix elements for the left or right circularly polarized photons. In fact, the presence of a spin-orbit term is a fundamental requirement for the observation of any magneto-optical phenomena regardless of the energy range. Without it no dichroic effects would be seen.

Specifically, the example illustrated in Appendix A demonstrates that, for an initial $2p_{1/2}$ ($L_2$) core state, left (right) circularly polarized photons will make preferential transitions to spin-down (-up) states. The magnitude of this preference is predicted to be proportional to the difference between the spin-up and spin-down occupation of the probed shell and as such the net moment on that orbital. Therefore the measured signal scales with the local moment and can be used to determine relative changes in the magnetization upon applying a magnetic field or increasing the sample temperature. Furthermore, the XMCD signal for an $L_3$ edge is also predicted to be equal and opposite that observed at an $L_2$ edge. This $1:-1$ dichroic ratio is a general rule predicted for any pair of spin-orbit-split core states [i.e., $2p_{1/2,\ 3/2}$ ($L_{2,3}$), $3d_{3/2,\ 5/2}$ ($M_{4,5}$), etc.].

Although this simple model describes the basic underlying cause of the XMCD effect, it proves far too simplistic to completely explain most dichroic spectra. In particular, this model predicts no dichroic signal from $K$-edge measurements, because the initial state at this edge is an $s$ level with no spin-orbit coupling. Yet it is at this edge that the first (admittedly weak) XMCD effects were observed. Also, while experimental spectra for spin-orbit-split pairs have been found to be roughly in the predicted $1:-1$ ratio, for most materials they deviate considerably. To account for these discrepancies, more sophisticated models need to incorporate the following factors, which are neglected in the simple one-electron picture presented.

1. The effects of the spin-orbit interaction in the final states.

2. Changes in the band configuration due to the presence of the final state core hole.

3. Spin-dependent factors in the radial part of the matrix elements.

4. Contributions from higher-order electric quadrupole terms, which can contribute significantly to the XMCD signal for certain spectra.

The first two of these factors are significant for all dichroic spectra, while the later two are particularly relevant to RE $L$-edge spectra.

First consider the effects of the spin-orbit interaction in the final states. The inclusion of this factor can quickly explain the observed dichroic signal at TM $K$-edge spectra (i.e., repeat the example in Appendix A with initial states and final states inverted). This observation of a $K$-edge dichroic signal illustrates the extreme sensitivity of the XMCD technique, because not only do $4p$ states probed by this edge posses a relatively small spin moment ($\sim 0.2\,\mu_B$), but the substantial overlap with neighboring states results in a nearly complete quenching of the orbital moment. This orbital contribution ($\sim 0.01\,\mu_B$), however, is nonzero, and thus a nonzero dichroic signal is observed.

The influence of the final state spin-orbit interaction also explains the deviations from the predicted $1:-1$ ratio observed at $L$ and $M$ edges. The spin-orbit term tends to enhance the $L_3-$ ($M_5-$) edge dichroic signal at the expense of the $L_2$ ($M_4$) edge. In terms of the simple one-electron model presented, the spin-orbit interaction effectively breaks the degeneracy of the $m_l$ states, resulting in an enhancement of the XMCD signal at the $L_3$ edge. An example of this enhancement is shown in Figure 2, which plots XMCD spectra obtained at the Co $L$ edges for metallic Co and for a Co/Pd multilayer (Wu et al., 1992). The large enhancement of the $L_3$-edge dichroic signal indicates that the multilayer sample possesses an enhanced Co $3d$ orbital moment compared to that of bulk Co.

A quantitative relationship between the degree of this enhancement and the strength of the spin-orbit coupling is expressed in terms of sum rules that relate the integrated dichroic ($\mu_c$) and total ($\mu_o$) absorptions over a

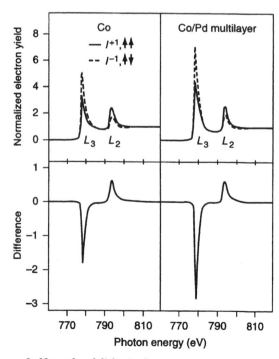

**Figure 2.** Normal and dichroic absorption at the Co $L_{2,3}$ edges for Co metal and Co/Pd multilayers. Courtesy of Wu et al. (1992).

spin-orbit-split pair of core states to the orbital and spin moments of that particular final state orbital (Thole et al., 1992; Carra et al., 1993). The integrated signal rather than the value at a specific energy is used in order to include all the states of a particular band. This has the added benefit of eliminating the final state core-hole effects, thereby yielding information on the ground state of the system. Although the general expressions for the sum rules can appear complicated, at any particular edge they reduce to extremely simple expressions. For instance, the sum rules for the $L_{2,3}$ edges simplify to the following expressions:

$$\frac{\int_{L_3+L_2} \mu_c dE}{\int_{L_3+L_2} 3\mu_0 dE} = \frac{\langle L_z \rangle /2}{(10 - n_{occ})}, \quad (1)$$

$$\frac{\int_{L_3} \mu_c dE - 2\int_{L_2} \mu_c dE}{\int_{L_3+L_2} 3\mu_0 dE} = \frac{\langle S_z \rangle + 7\langle T_z \rangle}{(10 - n_{occ})} \quad (2)$$

Here $\langle L_Z \rangle$, $\langle S_Z \rangle$, and $\langle T_Z \rangle$ are the expectation values for the orbital, spin, and spin magnetic dipole operators in Bohr magnetons [$\mathbf{T} = \frac{1}{2}(\mathbf{S} - 3\hat{\mathbf{r}}(\hat{\mathbf{r}} \cdot \mathbf{S}))$] and $n_{occ}$ is the occupancy of the final-state band (i.e., $10 - n_{occ}$ corresponds to the number of final states available for the transition). The value of $n_{occ}$ must usually be obtained via other experimental methods or from theoretical models of the band occupancy, which can lead to uncertainties in the values of the moments obtained via the sum rules. To circumvent this, it is sometimes useful to express the two equations above as a ratio:

$$\frac{\int_{L_3+L_2} \mu_c dE}{\int_{L_3} \mu_c dE - 2\int_{L_2} \mu_c dE} = \frac{\langle L_z \rangle}{2\langle S_z \rangle + 14\langle T_z \rangle} \quad (3)$$

This yields an expression that is independent of the shell occupancy and also eliminates the need to integrate over the total absorption, which can also be a source of systematic error in the measurement. Comparison with magnetization measurements can then be used to obtain the absolute value of the orbital moment. Applying the sum rules to the spectra shown in Figure 2 yields values of 0.17 and 0.24 $\mu_B$ for the orbital moments of the bulk Co and Co/Pd samples, respectively. This example again illustrates the sensitivity of the XMCD technique, because the clearly resolvable differences in the spectra correspond to relatively small changes in the size of the orbital component.

Several assumptions go into the derivation of the sum rules, which can restrict their applicability to certain spectra; the most important of these is that the two spin-orbit split states must be well resolved in energy. This criterion means that sum rule analysis is restricted to measurements involving deep core states and is generally not applicable to spectra taken at energies below ~250 eV. Moreover, the radial part of the matrix elements is assumed to be independent of the electron spin and identical at both edges of a spin-orbit-split pair. This is typically not the case for the RE $L$ edges, where the presence of the open $4f$ shell can introduce some spin dependence (Konig et al., 1994).

Also, the presence of the magnetic dipole term $\langle T_Z \rangle$ in the these expressions poses problems for determining the exact size of the spin moment or the orbital-to-spin ratio. The importance of the size of this term has been a matter of considerable debate (Wu et al., 1993; Wu and Freeman, 1994). Specifically, it has been found that, for TM $3d$ states in a cubic or near-cubic symmetry, the $\langle T_Z \rangle$ term is small and can be safely neglected. For highly anisotropic conditions, however, such as those encountered at interfaces or in thin films, $\langle T_Z \rangle$ can become appreciable and therefore would distort measurements of the spin moment obtained from the XMCD sum rules. For RE materials, $\langle T_Z \rangle$ can be calculated analytically for the $4f$ states, but an exact measure of the relative size of this term for the $5d$ states is difficult to obtain.

## PRACTICAL ASPECTS OF THE METHOD

The instrumentation required to collect dichroic spectra naturally divides XMCD measurements into two general categories based on the energy ranges involved: those in the soft x-ray regime, which require a windowless UHV-compatible beamline, and those in the hard x-ray regime, which can be performed in nonevacuated environments. The basic elements required for an XMCD experiment, however, are similar for both regimes: i.e., a source of circularly polarized x-rays (CPX), an optical arrangement for monochromatizing the x-ray beam, a magnetized sample, and a method for detecting the absorption signal. The main differences between the two regimes, other than the sample environment, lie in the methods used to detect the XMCD effect.

### Circularly Polarized X-Ray Sources

In evaluating the possible sources of CPX, the following quantities are desirable; high circular polarization rate ($P_c$), high flux ($I$), and the ability to reverse the helicity. For a given source, though, the experimenter must sometimes sacrifice flux in order to obtain a high rate of polarization, or vice versa. Under these circumstances, it should be kept in mind that the figure of merit for circularly polarized sources is $P_c^2 I$, because it determines the amount of time required to obtain measurements of comparable statistical accuracy on different sources (see Appendix B). Laboratory x-ray sources have proved to be impractical for XMCD measurements because they offer limited flux and emit unpolarized x-rays. The use of a synchrotron radiation source is therefore essential for performing XMCD experiments. Three different approaches can be used to obtain circularly polarized x-rays from a synchrotron source: using (1) off-axis bending magnet radiation, (2) a specialized insertion device, such as an elliptical multipole wiggler, or (3) phase retarders based on perfect crystal optics.

In a bending magnet source, a uniform circular motion by the relativistic electrons or positrons in the storage ring is used to generate synchrotron radiation. This radiation, when observed on the axis of the particle orbit, is purely linearly polarized in the orbital plane ($\sigma$ polarization).

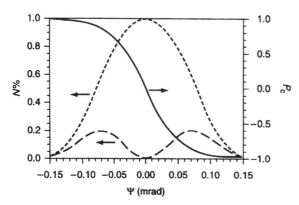

**Figure 3.** Horizontally (short dashes) and vertically (long dashes) polarized flux along with circular polarization rate (solid) from an APS bending magnet at 8.0 keV.

Appreciable amounts of photons polarized out of the orbital plane ($\pi$ polarization) are only observed off-axis (Jackson, 1975). These combine to produce a CPX source because the $\sigma$- and $\pi$-polarized photons are exactly $\delta = \pm \pi/2$ radians out of phase. The sign of this phase difference depends on whether the radiation is viewed from above or below the synchrotron orbital plane. Therefore, the off-axis synchrotron radiation will be elliptically polarized, with a helicity dependent on the viewing angle and a degree of circular polarization given by

$$P_c = \frac{2E_\sigma E_\pi}{|E_\sigma|^2 + |E_\pi|^2} \sin \delta, \qquad (4)$$

where $E_\sigma$ and $E_\pi$ are the electric field amplitudes of the radiation. An example of this is shown in Figure 3, which plots the $\sigma$ and $\pi$ intensities along with the circular polarization rate as a function of viewing angle for 8.0 keV x-rays at an Advanced Photon Source bending magnet (Shenoy et al., 1988).

Although simple in concept, obtaining CPX in this fashion does have some drawbacks. The primary one is that the off-axis position required to get appreciable circular polarization reduces the incident flux by a factor of 5 to 10. Furthermore, measurements taken at the sides of the synchrotron beam where the intensity changes more rapidly are particularly sensitive to any motions of the particle beam. Moreover, the photon helicity cannot be changed easily because this requires moving the whole experimental setup vertically. This movement can result in slightly different Bragg angles incident on the monochromator, thereby causing energy shifts that would have to be compensated for. Attempts have been made to overcome this by using slits, which define beams both above and below the orbital plane simultaneously in order to make XMCD measurements. These efforts have had limited success, however, since they are particularly sensitive to sample inhomogeneity (Schütz et al., 1989b).

Standard planar insertion devices are magnetic arrays placed in the straight sections of synchrotron storage rings to make the particle beam oscillate in the orbital plane. Because each oscillation produces synchrotron radiation, a series of oscillations greatly enhances the emitted flux over that produced by a single bending magnet source. These devices produce linearly polarized light on axis, but the off-axis radiation of a planar device, unlike that of a bending magnet, is not circularly polarized, because equal numbers of right- and left-handed bends in the particle orbit produce equal amounts of left- and right-handed circular polarization, yielding a net helicity of zero. Specialized insertion devices for producing circularly polarized x-rays give the orbit of the particle beam an additional oscillation component out of the orbital plane (Yamamoto et al., 1988). In this manner, the particle orbit is made to traverse a helical or pseudo-helical path to yield a net circular polarization on axis of these devices. When coupled with a low-emittance synchrotron storage ring, these devices can provide a high flux with a high degree of polarization. The main disadvantage of these devices is that their high cost has made the beamlines dedicated to them rather rare. Also, to preserve the circular polarization through monochromatization, specialized optics are required, particularly for lower energies (Malgrange et al., 1991).

Another alternative for producing CPX is using x-ray phase retarders. Phase retarders employ perfect crystal optics to transform linear to circular polarization by inducing a $\pm\pi/2$ radian phase shift between equal amounts of incoming $\sigma$- and $\pi$-polarized radiation. (Here $\sigma$ and $\pi$ refer to the intensities of the radiation polarized in and out of the scattering plane of the phase retarder, and should not be confused with the radiation emitted from a synchrotron source. Unfortunately, papers on both subjects tend to use the same notation; it is kept the same for this paper in order to conform with standard practice.) Equal $\sigma$ and $\pi$ intensities incident on the phase retarder are obtained by orienting its plane of diffraction at a 45° angle with respect to the synchrotron orbital plane. As the final optical element before the experiment, it offers the greatest degree of circular polarization incident on the sample. Thus far phase-retarding optics have only been developed for harder x-rays, with soft x-ray measurements restricted to bending-magnet and specialized-insertion-device CPX sources.

Typically materials in the optical regime exhibit birefringence over a wide angular range; in the x-ray regime, however, materials are typically birefringent only at or near a Bragg reflection. For the hard x-ray energies of interest in XMCD measurements, a transmission-phase retarder, shown in Figure 4, has proved to be the most suitable type of phase retarder (Hirano et al., 1993). In this

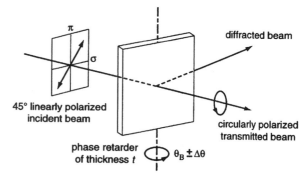

**Figure 4.** Schematic of a transmission phase retarder.

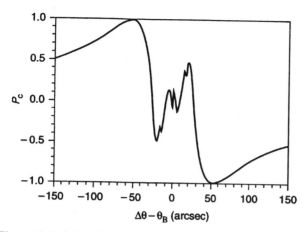

**Figure 5.** Calculated degree of circular polarization of the transmitted beam for a 375-μm-thick diamond (111) Bragg reflection at 8.0 keV.

phase retarder, a thin crystal, preferably of a low-$Z$ material to minimize absorption, is deviated ($\Delta\theta \sim 10$ to 100 arcsec) from the exact Bragg condition, and the transmitted beam is used as the circularly polarized x-ray source. Figure 5 plots the predicted degree of circular polarization in the transmitted beam as a function of the off-Bragg position for a 375-μm-thick diamond (111) crystal and 8.0 keV incoming x-ray beam. Note that, for a particular crystal thickness and photon energy, either helicity can be obtained by simply reversing $\Delta\theta$. Further, when used with a low-emittance source, a high degree of polarization ($P_c > \sim 0.95$) can be achieved with a transmitted beam intensity of $\sim 10\%$ to 20% of the incident flux. This flux loss is comparable to that encountered using the off-axis bending magnet radiation. The phase retarder, however, provides easy helicity reversal and can be used with a standard insertion device to obtain fluxes comparable to or greater than those obtained from a specialized device (Lang et al., 1996). The main drawback of using a phase retarder as a CPX source is that the increased complexity of the optics can introduce a source of systematic errors. For instance, if the phase retarder is misaligned and does not precisely track with the energy of the monochromator, the $\pm\Delta\theta$ movements of the phase retarder will not be symmetric about the Bragg reflection. This leads to measurements with different $P_c$ values and can introduce small energy shifts in the transmitted beam.

### Detection of XMCD

The most common method used to obtain XMCD spectra is to measure the absorption by monitoring the incoming ($I_o$) and transmitted ($I$) fluxes. The absorption in the sample is then given by:

$$\mu_o t = \ln\left(\frac{I_o}{I}\right) \tag{5}$$

The dichroism measurement is obtained by reversing the helicity or magnetization of the sample and taking the difference of the two measurements ($\mu_c = \mu^+ - \mu^-$).

Typically the thickness of the sample, $t$, is left in as a proportionality factor, because its removal requires an absolute determination of the transmitted flux at a particular energy. Some experimenters choose to remove this thickness factor by expressing the dichroism as a ratio between the dichroic and normal absorption, $\mu_c/\mu_o$. Absorption not due to resonant excitation of the measured edge, however, is usually removed by normalizing the pre-edge of the $\mu_o$ spectra to zero; therefore, expressing the XMCD signal in terms of this ratio tends to accentuate features closer to the absorption onset.

Rather than taking the difference of Eq. 5 for two magnetizations, the dichroic signal is also frequently expressed in terms of the asymmetry ratio of the transmitted fluxes only. This asymmetry ratio is related to the dichroic signal by Eq. 6.

$$\frac{I^+ - I^-}{I^+ + I^-} = \frac{e^{-\mu^+ t} - e^{-\mu^- t}}{e^{-\mu^+ t} + e^{-\mu^- t}} = \frac{e^{0.5\mu_c t} - e^{-0.5\mu_c t}}{e^{0.5\mu_c t} + e^{-0.5\mu_c t}}$$
$$= -\tanh(0.5\mu_c t) \cong -f 0.5\mu_c t \tag{6}$$

The last approximation can be made, because reasonable attenuations require $\mu_o t \sim 1$ to 2 and $\mu_c < \sim 0.05\,\mu_o t$. The factor $f$ is introduced in an *ad hoc* fashion to account for the incomplete circular polarization in the incident beam and the finite angle ($\theta$) between the magnetization and beam directions, $f = P_c \cos\theta$. In some cases, the factor $f$ also includes the incomplete sample magnetization ($M'$) relative to the saturated moment at $T = 0$ K. If the experimenter is interested in obtaining element specific magnetizations as a function of field or temperature, however, the inclusion of this factor would defeat the purpose of the measurement.

This type of measurement is illustrated in Figure 6, which shows a plot of a hysteresis measurement taken using the dichroic signal at the Co and Fe $L_3$ edges of a Co/Cu/Fe multilayer (Chen et al., 1993; Lin et al., 1993). In this compound the magnetic anisotropy of the Co layer is much greater than that of the Fe layer. Thus, in an applied magnetic field, the directions of Fe moments reverse before those of Co moments. By monitoring the strength of the dichroic signal as a function of applied field for the Fe and Co $L_3$ edges, the hysteresis curves of each constituent were traced out. Similarly, XMCD can also be used to measure the temperature variation of the magnetization of a specific constituent in a sample. This is demonstrated in Figure 7 (Rueff et al., 1997), which shows the variation of the dichroic signal at the Gd $L$ edges as a function of temperature for a series of amorphous GdCo alloys. These XMCD measurements provide orbital- and element-specific information complementary to that obtained from magnetometer measurements.

### Measurement Optics

Two basic optical setups, shown in Figure 8, are used for the measurement of x-ray attenuations. The scanning monochromator technique (Fig. 8a) selects a small energy range ($\Delta E \sim 1$ eV) to impinge on the sample, and the full

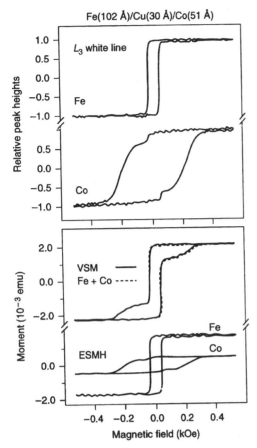

Fe(102 Å)/Cu(30 Å)/Co(51 Å)

**Figure 6.** Hysteresis measurements of the individual magnetizations of Fe and Co in a Fe/Cu/Co multilayer obtained by measuring the $L_3$ edge XMCD signal (top), along with data obtained from a vibrating sample magnetometer (VSM) and a least-squares fit to the measured Fe and Co hysteresis curves (bottom). Courtesy of Chen et al. (1993).

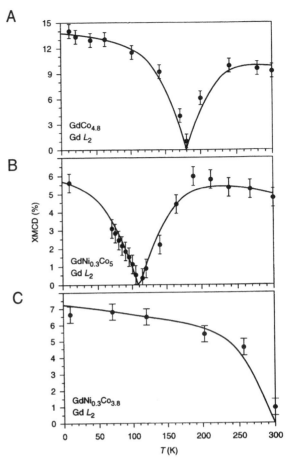

**Figure 7.** Variation of the Gd $L_2$-edge XMCD signal as a function of temperature for various GdNiCo amorphous alloys. Courtesy of Rueff et al. (1997).

systematic errors that can be introduced as a result of the decay or movements of the particle beam.

In the hard x-ray regime, because of the high intensity of the synchrotron radiation, the incident ($I_o$) and

spectrum is acquired by sequentially stepping the monochromator through the required energies. In the dispersive monochromator technique (Fig. 8b), a broad range of energies ($\Delta E \sim 200$ to $1000$ eV) is incident on the sample simultaneously and the full spectrum is collected by a position-sensitive detector. Both of these setups are common at synchrotron radiation facilities where they have been widely used for EXAFS measurements. Their conversion to perform XMCD experiments usually requires minimal effort. The added aspects are simply use of CPX, using the methods discussed above, and a magnetized sample.

Although the scanning monochromator technique can require substantial time to step through the required beam energies, the simplicity of data interpretation and compatibility with most synchrotron beamline optics have made it by far the most common method utilized for XMCD measurements. Typically, these measurements are taken by reversing the magnetization (helicity) at each energy step, rather than taking a whole spectrum, then changing the magnetization (helicity) and repeating the measurement. The primary reason for this is to minimize

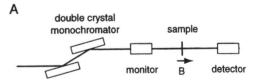

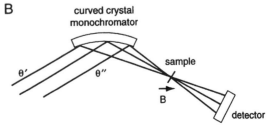

**Figure 8.** Experimental setups for the measurement of the absorption. (**A**) Scanning monochromator (double crystal); (**B**) dispersive monochromator (curved crystal).

attenuated ($I$) beam intensities are usually measured using gas ionization chambers. For these detectors, a bias (500 to 2000 V) is applied between two metallic plates and the x-ray beam passes though a gas (generally nitrogen) flowing between them. Absorption of photons by the gas creates ion pairs that are collected at the plates, and the induced current can then be used to determine the x-ray beam intensity. Typical signal currents encountered in these measurements range from $\sim 10^{-9}$ A for bending magnet sources to $\sim 10^{-6}$ A for insertion-device sources. While these signal levels are well above the electronic noise levels, $\sim 10^{-14}$ A, produced by a state-of-the-art current amplifier, other noise mechanisms contribute significantly to the error. The most common source of noise in these detectors arises from vibrations in the collection plates and the signal cables. Vibrations change the capacitance of the detection system, inducing currents that can obscure the signal of interest—particularly for measurements made on a bending magnet source, where the signal strength is smaller. This error can be minimized by rigidly mounting the ion chamber and minimizing the length of cable between the ion chamber and the current amplifier. Even when these effects are minimized, however, they still tend to be the dominant noise mechanism in XMCD measurements. This makes it difficult to get an *a priori* estimate of the error based on the signal strength. Therefore, the absolute error in an XMCD measurement is frequently obtained by taking multiple spectra and computing the average and variance in the data.

Using a scanning monochromator, one can also indirectly measure the effective absorption of a sample by monitoring the fluorescence signal. Fluorescence radiation results from the filling of the core hole generated by the absorption of an x-ray photon. The strength of the fluorescence will thus be proportional to the number of vacancies in the initial core state and therefore to the absorption. An advantage of this technique over transmittance measurements is that fluorescence arises only from the element at resonance, thereby effectively isolating the signal of interest. In a transmittance measurement, on the other hand, the resonance signal from the absorption edge sits atop the background absorption due to the rest of the constituents in the sample.

Fluorescence measurements require the use of an energy-sensitive detector to isolate the characteristic lines from the background radiation of elastically and Compton scattered radiation. Generally Si or Ge solid-state detectors have been used for fluorescence detection, since they offer acceptable energy resolution ($\Delta E \sim 200$ eV) and high efficiencies. The count rate offered by these solid-state detectors, however, is typically restricted to <500 kHz. Fluorescence signals this strong are easily obtained even using bending magnet sources; thus, the detector becomes the rate-limiting factor for data acquisition. Considering that a typical XMCD difference signal for a RE $L$ edge is $\sim 0.5\%$, obtaining a 10% accuracy for this difference requires a total count of $\sim 4 \times 10^6$. Thus, acquiring this accuracy over the entire scanned range, which may contain $\sim 100$ points, can require data acquisition times on the order of several hours, irrespective of the CPX source utilized.

A further complication encountered with fluorescence data is that the signal must be corrected for the self absorption of the sample. For a thick sample, this correction is given by:

$$\mu_o(E) = \frac{I_f}{I_o} \sim \frac{\mu_o(E)\sin\theta}{\mu_T(E)/\sin\theta + \mu_T(E_f)/\sin\varphi} \tag{7}$$

where $\mu_T(E)$ is the total absorption at energy $E$, $E_f$ is the fluorescence energy, $\theta$ is the entrance angle of the incident x-rays, and $\phi$ is the exit angle of the fluorescent x-rays. From the count rates involved and the process needed to deconvolve the true absorption from the measured fluorescence signal, it is clear that this technique is at a disadvantage compared to a straightforward transmission measurement. Fluorescence measurements, however, are the preferred technique when examining dilute systems or thin magnetic films on thick substrates. For these samples, $\mu_T(E)$ is nearly constant, so the correction in Eq. 7 becomes negligible. Further, the difference signals in transmission arising from dilute constituents are minimal; thus, they benefit greatly from isolation of absorption only due to the atomic species of interest achieved with the fluorescence technique.

Another indirect measure of the absorption spectra used for XMCD is the measurement of the total electron yield from a sample. In this method, an electrical lead is attached directly to the sample, and a metallic enclosure with an entrance hole for the beam surrounds it. The enclosure is kept at a small bias ($\sim 10$ V) with respect to the sample, effectively turning the sample itself into an ionizing detector. This technique is quite similar to fluorescence measurements, in that, instead of probing the radiative decay of the core hole, it probes the decay of the core hole via Auger transitions. The main contribution to the total yield signal, however, does not arise directly from the Auger electrons themselves or even from the ejected photoelectrons, but rather from the secondary inelastically scattered electrons. Despite the complications encountered with secondary electron processes, it has been shown that the total yield signal obtained is proportional to the x-ray absorption (Gudat and Kunz, 1972). To minimize possible systematic errors, the incident beam intensity is monitored using a similar detection method. A partially transmitting wire mesh is placed in the incident beam, again surrounded by a metal enclosure with an applied bias, and the induced current measured to determine $I_o$.

Electron yield measurements are the detection method most often utilized in the soft x-ray regime, because the photon attenuation lengths at these energies ($\sim 100$ Å) make transmittance measurements difficult. Furthermore, the probability of fluorescence decay relative to Auger decay of the core hole becomes small below $\sim 2000$ eV. The main drawback (or advantage) of this technique is its high surface sensitivity, since nearly all the Auger electrons originate within $\sim 50$ Å of the surface.

A particularly novel use of electron yield detection in XMCD has been imaging of magnetic domains (Stöhr et al., 1993). In this technique, a photoelectron microscope is used to detect the secondary electrons from a particular

**A**

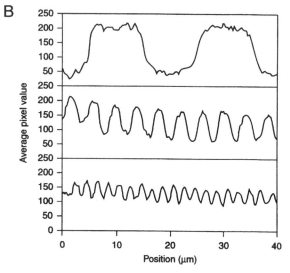

**B**

**Figure 9.** Image of the magnetic domains in Co/Pt multilayer obtained from XMCD signal detected using a photoelectron microscope (**A**). Also shown are integrated traces of the image (**B**). Reprinted with permission from Stöhr et al. (1993). Copyright 1993 American Association for the Advancement of Science.

spot on the sample, thereby obtaining a two-dimensional absorption profile (Tonner et al., 1994). By tuning the energy to an absorption edge with a strong XMCD signal and utilizing CPX, magnetization-sensitive contrast in the resultant image is obtained. This is demonstrated in Figure 9, which shows just such an image taken at the Co $L$ edges of a CoPtCr film. In this particular material, the magnetic domains align themselves perpendicular to the plane of the sample. Therefore, taking the difference in the signal measured with either helicity produces a picture of the magnetic domains of the sample. Figure 9 is a combination of the images taken at both the $L_2$ and $L_3$ edges. Domain sizes down to $10 \, \mu m \times 0.5 \, \mu m$ are clearly visible. The uniqueness of this measurement over the same information available from other techniques is again the element specificity of XMCD. Thus the degree of magnetic alignment of individual constituents can be measured.

An alternative to the scanning monochromator is the use of energy dispersive optics (Baudelet et al., 1991). In a dispersive arrangement a curved crystal is used to monochromatize the beam. The angle that the incident beam makes with the diffraction planes ($\theta' \rightarrow \theta''$) varies across the crystal, and thus a range of energies ($E' \rightarrow E''$) are diffracted (Fig. 8b). The dispersion of the monochromator crystal is such that the photon energies vary approximately linearly with angle, focusing to a point ($\sim 100 \, \mu m$) $\sim 3$ to $4 \, m$ from the monochromator, where the sample is generally placed. A position-sensitive detector placed behind the sample can then be used to collect the full spectrum. There are several advantages to using dispersive optics for collecting XMCD data. The foremost is that acquiring the spectra requires no motorized motions, eliminating the time required to scan the monochromator and thereby greatly enhancing data collection rates. Another advantage is that the small focal spot reduces errors caused by sample nonuniformity. The disadvantage of the technique is that the data analysis is in general more complicated than for data obtained using a scanning monochromator. The charge-coupled device detectors generally used for these measurements possess an inherent readout noise that must be carefully subtracted. This subtraction can be further complicated by structure that may be introduced into this noise over time by the x-ray beam; thus, the noise must be monitored periodically throughout the data collection process by collecting background spectra. Furthermore, the $I_o$ spectrum may not be uniform with energy, as a result of variations in the reflectivity of the monochromator crystal or post-monochromator mirror optics; therefore, an image of the incident beam must also be taken. This technique has almost exclusively been used with off-axis bending-magnet CPX sources, although recently the applicability of phase retarders with dispersive optics has been demonstrated (Giles et al., 1994).

## Sample Magnetization

Once the CPX source and optics have been chosen, the sample must be magnetized. The analysis presented in Appendix A assumed that the magnetization and photon helicity were collinear. A more detailed calculation demonstrates that for dipolar transitions the magnitude of the XMCD signal should scale with the cosine of the angle between the magnetization and beam. Therefore, in order to maximize the possible signal, the magnetic field should be as close to the beam direction as possible. Problems arise, however, in applying a field exactly parallel to the beam direction. Demagnetization effects make thin foils extremely difficult to magnetize normal to the foil plane and thus they are usually magnetized in plane. In a transmittance measurement the x-ray beam must be incident at some angle to the plane of the sample, which results in a finite angle between the beam and the field directions ($\sim 25° \rightarrow 45°$). Even when a solenoid coaxial with the incident x-ray beam is used to provide the magnetic field, the sample plane should not be perpendicular to the solenoid axis but rather at some angle. In this arrangement,

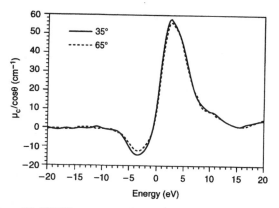

**Figure 10.** XMCD spectra at the Dy $L_3$ edge of a $Dy_{0.4}Tb_{0.6}$ alloy for different angles between the incident beam and magnetization directions, showing the contributions arising from quadrupole transitions. Courtesy of Lang et al. (1995).

unless the solenoid generates a very high magnetic field ($|B| > 1$ T), the direction of the local magnetization direction will generally lie along the sample plane rather than along the field direction.

Varying the angle that the beam makes with the sample magnetization can also be used as a means of identifying features in the dichroic spectra that do not arise from dipolar transitions which scale with the cosine. Figure 10 shows the XMCD signal taken for two different angles at the Dy $L_3$ edge in a $Dy_{0.4}Tb_{0.6}$ alloy (Lang et al., 1995). These spectra are normalized to the cosine of the angle between the beam and the magnetization directions; therefore, the deviation in the lower feature indicates nondipolar-like behavior. This feature has been shown to arise from quadrupolar transitions to the $4f$ states and is a common feature in all RE $L_{2,3}$-edge XMCD spectra (Wang et al., 1993). Although quadrupolar effects in the absorption spectrum are smaller than dipolar effects by a factor of ~100, their size becomes comparable in the difference spectrum due to the strong spin-polarization of the RE $4f$ states. The presence of quadrupolar transitions can complicate the application of sum rule to the analysis of these spectra but also opens up the possibility of obtaining magnetic information on two orbitals simultaneously from measurements at only one edge.

## METHOD AUTOMATION

Acquisition of XMCD spectra requires a rather high degree of automation, because the high intensities of x-ray radiation encountered at synchrotron sources prevent any manual sample manipulation during data acquisition. The experiment is normally set up inside a shielded room and all experimental equipment operated remotely through a computer interface. These interfaces are typically already installed on existing beamlines and the experimenter simply must ensure that the equipment that they are considering for the experiment is compatible with the existing hardware. Note that the time required to enter and exit the experimental enclosure is not minimal. Therefore, computerized motion control of as many of the experimen-

tal components (i.e., magnet, sample, slits, etc.) as possible is desirable.

The typical beamline equipment used in XMCD data acquisition includes motorized stages, various NIM (nuclear instrumentation module electronic standard) electronic components, power supplies, and cryostats with associated temperature controllers. The data acquisition system for the collection of the XMCD spectra should be fully automated. The computer should step through the required energies if a scanning monochromator is used, alternate the magnet field or helicity, and record the detector counts. In addition, since obtaining enough statistical accuracy in full XMCD spectra frequently requires repeated scanning for a substantial time, the software should also collect multiple scans and save the data without manual intervention.

## DATA ANALYSIS AND INITIAL INTERPRETATION

The first thing to realize when taking an initial qualitative look at an XMCD spectrum is that these measurements probe the unoccupied states, whose net polarization is opposite that of the occupied states. Thus, the net difference in the XMCD signal reflects the direction of the so-called minority spin band, because it contains the greater number of available states. It is therefore easy to make a sign error for the dichroic spectra, and care should be taken when determining the actual directions of the moments in a complicated compound. To prevent confusion, it is essential to measure the direction of the magnetic field and the helicity of the photons used from the outset of the experiment.

If XMCD is being used to measure changes in the magnetization of the sample, the integrated dichroic signal should be used rather than the size of the signal at one particular energy for two reasons in particular. The integrated signal is less sensitive to long-term drifts in the experimental apparatus than are measurements at one nominal energy. And, systematic errors are much more apparent in the entire spectrum than at any particular energy. For instance, the failure of the spectrum to go to zero below or far above the absorption edge would immediately indicate a problem with the data collection, but would not be detected by measurement at a single energy.

When analyzing spectra taken from different compounds, one should note that a change in the magnitude of the dichroic signal at just one edge does not necessarily reflect a change in the size or direction of the total magnetic moment. Several factors complicate this, some of which have already been mentioned. For example, the development of an orbital moment for the spectra shown in Figure 2 greatly enhanced the signal at the $L_3$ edge but diminished the signal at the $L_2$ edge. Therefore, although the size of the total moment remains relatively constant between the two samples, the dichroic signal at just one edge shows substantial changes that do not necessarily reflect this. The total moment in this case must be obtained through the careful application of both sum rules or comparison with magnetometer measurements. The strength of the TM $K$-edge dichroism also tends not to

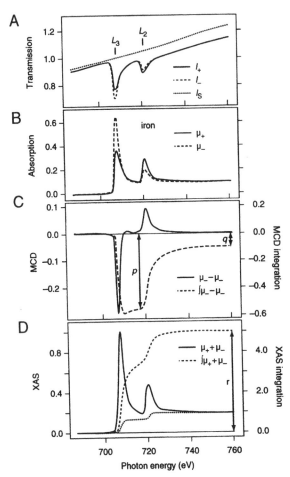

**Figure 11.** Steps involved in the sum rule analysis of the XMCD spectra demonstrated for a thin layer of metallic Fe. (**A**) Transmitted intensity of parylene substrate and substrate with deposited sample. (**B**) Calculated absorption with background subtracted. (**C**) Dichroic spectra and integrated dichroic signal. (**D**) Absorption spectra and edge-jump model curve along with integrated signal. Courtesy of Chen et al. (1995).

absorption not due to the particular element of interest. In this example, this consists of separately measuring and subtracting the substrate absorption (Fig. 11a), leaving only the absorption from the element of interest (Fig. 11b). In compound materials, however, there are generally more sources of absorption than just the substrate material, and another method of background removal is necessary. The common method has been subtraction from the signal of a $\lambda^3$ function fit to the pre-edge region of the absorption spectrum.

The next step in the application of the sum rules is determining the integrated intensities of the dichroic signal. The easiest method is to plot the integral of the dichroic spectrum as a function of the scanning energy (Fig. 11c). The integrated values over each edge can then be directly read at the positions where the integrated signal becomes constant. In the figure, $p$ corresponds to the signal from the $L_3$ edge alone and $q$ corresponds to the integrated value over both edges. Although there is some uncertainty in the appropriate value of the $p$ integral, since the dichroic signal of the $L_3$ edge extends well past the edge, choosing the position closest to the onset of the $L_2$ edge signal minimizes possible errors. This uncertainty diminishes for edges at higher energies where the spin-orbit splitting of the core levels becomes more pronounced.

The final step in the sum rule analysis is obtaining the integral of the absorption spectrum (Fig. 11d). At first glance, it seems that this integral would be nonfinite because the absorption edges do not return to zero at the high-energy end of the spectrum. The sum rule derivation, however, implicitly assumes that the integral includes only absorption contributions from the final states of interest (i.e., the $3d$ states in this example). It is well known that the large peaks in the absorption spectrum just above the edge, sometimes referred to as the "white line," are due to the strong density of states of $d$ bands near the Fermi energy, while the residual step-like behavior is due to transitions to unbound free electron states. To deconvolve these two contributions, a two-step function is introduced to model the absorption arising from the transitions to the free electron states. The absorption-edge jumps for this function are put in a 2:1 ratio, expected because of the 2:1 ratio in the number of initial-state electrons. The steps are also broadened to account for the lifetime of the state and the experimental resolution. The integral $r$ shown in Fig 11d is then obtained by determining the area between the absorption curve and the two-step curve. Once the integrals $r$, $p$, and $q$ are determined, they can easily be put into Eqs. 1, 2, and 3 to obtain the values of the magnetic moments.

## SAMPLE PREPARATION

XMCD measurements in the hard x-ray regime require little sample preparation other than obtaining a sample thin enough for transmission measurements. Measurements in the soft x-ray regime, on the other hand, tend to require more complicated sample preparation to account for UHV conditions and small attenuation lengths. Hard x-ray XMCD measurements therefore tend to be favored if

reflect the size of the moment for different materials. This is particularly true for the Fe $K$-edge spectra which demonstrate a complete lack of correlation between the size of the total Fe moment and the strength of the dichroic signal, although better agreement has been found for Co and Ni (Stahler et al., 1993). In addition, RE $L_{2,3}$-edge spectra show a strong spin dependence in the radial matrix elements that can in some cases invert the relationship between the dichroic signal and the size of the moment (Lang et al., 1994). These complications notwithstanding, the dichroic spectra at one edge and within one compound should still scale with changes in the magnetization due to applied external fields or temperature changes.

If the XMCD signal is not complicated by the above effects and both edges of a spin-orbit-split pair are measured, the sum rules may be applied. The steps involved in the implementation of the sum rules are illustrated in Figure 11 for a thin Fe metal film deposited on a parylene substrate (Chen et al., 1995). The first step in sum rules analysis is removal from the transmission signal of all

the same type of information (i.e., relative magnetization) can be obtained using edges at the higher energies. The incident photon fluxes typically do not affect the sample in any way, so repeated measurements can be made on the same sample without concern for property changes or degradation.

Attenuation lengths (i.e., the thickness at which $I/I_o \approx 1/e$) for hard x-rays are generally on the order of $\sim 2$ to $10 \, \mu m$. While it is sometimes possible to roll pure metals to these thicknesses, the rare earth and transition metal compounds generally studied in XMCD are typically very brittle, so that rolling is impractical. The sample is therefore often ground to a fine powder in which the grain size is significantly less than the x-ray attenuation length ($\sim 1 \, \mu m$). This powder is then spread on an adhesive tape, and several layers of tape combined to make a sample of the proper thickness. Most of the absorbance in these tapes is usually due to the adhesive rather than the tape itself; thus, a commercial tape brand is as effective as a specialized tape such as Kapton that minimizes x-ray absorbance. An alternative to tape is to mix the powder with powder of a light-Z material such as graphite and make a disk of appropriate thickness using a press. A note of warning: the fine metallic powders used for these measurements, especially those containing RE elements, can spontaneously ignite. Contact with air should therefore be kept to a minimum by using an oxygen-free atmosphere when applying the powder and by capping the sample with a tape overlayer during the measurements.

The techniques used to detect the XMCD signal in the soft x-ray energy regime are generally much more surface sensitive. Therefore, the surfaces must be kept as clean as possible and arrangements made to clean them in situ. Alternatively, a capping layer of a nonmagnetic material can be applied to the sample to prevent degradation. Surface cleanliness is also a requirement for the total electron-yield measurements; that is, the contacts used to measure the current should make the best possible contact to the material without affecting its properties. For transmittance measurements in this regime, the thinness of the sample attenuation lengths ($\sim 100 \, \mathring{A}$) makes supporting the sample impractical. Thus, samples are typically deposited on a semitransparent substrate such as parylene.

## PROBLEMS

There are many factors that can lead to erroneous XMCD measurements. These generally fall into two categories, those encountered in the data collection and those that arise in the data analysis. Problems during data collection tend to introduce distortions in the shape or magnitude of the spectra that can be difficult to correct in the data analysis; thus, great care must be taken to minimize their causes and be aware of their effects.

Many amplitude-distorting effects encountered in XMCD measurements, and in x-ray absorption spectroscopy in general, come under the common heading of "thickness effects." Basically these are systematic errors that diminish the relative size of the observed signal due to leakage of noninteracting x-rays into the measurement.

Leakage x-rays can come from pinholes in the sample, higher-energy harmonics in the incoming beam, or the monochromator resolution function containing significant intensity below the absorption-edge energy. While these effects are significant for all x-ray spectroscopies, they are particularly important in the near-edge region probed by XMCD.

First, to minimize pinhole effects, the sample should be made as uniform as possible. If the sample is a powder, it should be finely ground and spread uniformly on tape, with several layers combined to obtain a more uniform thickness. All samples should be viewed against a bright light and checked for any noticeable leakages. The sample uniformity should also be tested by taking absorption spectra at several positions on the sample before starting the XMCD measurements. Besides changes in the transmitted intensity, the strength and definition of sharp features just above the absorption edge in each spectrum provide a good qualitative indication of the degree of sample nonuniformity.

Another source of leakage x-rays is the harmonic content of the incident beam. The monochromator used for the measurement will pass not only the fundamental wavelength but multiples of it as well. These higher-energy photons are unaffected by the absorption edges and will cause distorting effects similar to those caused by pinholes. These harmonics are an especially bothersome problem with third-generation synchrotron sources because the higher particle-beam energies found at these sources result in substantially larger high-energy photon fluxes. Harmonics in the beam are generally reduced by use of an x-ray mirror. By adjusting the reflection angle of the mirror, near-unit reflectivity can be achieved at the fundamental energy, while the harmonic will be reduced by several orders of magnitude. Another common way of eliminating harmonics has been to slightly deviate the angle of the second crystal of a scanning monochromator (Fig. 8a) so that it is not exactly parallel to the first. The monochromator (said to be "detuned") then reflects only a portion of the beam reflected by the first crystal, typically $\sim 50\%$. The narrower reflection width of the higher-energy photons makes harmonics much more sensitive to this detuning. Thus while the fundamental flux is reduced by a factor of $\sim 2$, the harmonic content of the beam will be down by several orders of magnitude. Eliminating harmonics in this manner has a distinct disadvantage for XMCD measurements, however, because it also reduces the degree of circular polarization of the monochromatized beam. Therefore, monochromator detuning should not be used with off-axis bending-magnet or specialized-insertion-device CPX sources.

A third important consideration in minimizing thickness effects is the resolution of the monochromator. After passing through the monochromator, the beam tends to have a long-tailed Lorentzian energy distribution, meaning that a beam nominally centered around $E_o$ with a resolution of $\Delta E$ will contain a substantial number of photons well outside this range. When $E_o$ is above the edge, the bulk of the photons are strongly attenuated, and thus those on the lower end of the energy distribution will constitute a greater percentage of the transmitted intensity.

Typically a monochromator with a resolution well below the natural width of the edge is used to minimize this effect. This natural width is determined by the lifetime of the corresponding core state under study. For instance, a typical RE $L$-edge spectrum with a core-hole broadening of ~4 eV is normally scanned by a monochromator with ~1 eV resolution. Increasing the resolution, however, typically increases the Bragg angle of the monochromator, which reduces the circular polarization of the transmitted beam.

While all the methods for reducing the thickness effects mentioned above should be used, the primary and most efficient way to minimize these effects is to adjust the sample thickness so as to minimize the percentage of leakage photons. This is accomplished by optimizing the change in the transmitted intensity across the edge. It should be small enough to reduce the above thickness effect errors yet large enough to permit observation of the signal of interest. It has been found in practice that sample thicknesses ($t$) of ~1 attenuation length ($\mu t \sim 1$) below the absorption edge and absorption increases across the edge of <1.5 attenuation lengths ($\Delta \mu t < 1.5$) minimize the amplitude distortions.

Another source error in the XMCD measurement is sharp structure arising from spurious reflections in the monochromator or phase retarder, if used. These effects are usually called "glitches" because they have a small energy width and typically affect one data point. Generally the monochromator or phase plate must be adjusted by rotation about its diffraction vectors to move the offending glitch out of the energy range of interest. If this is not possible, an algorithm must be incorporated into the data reduction to omit the offending point. Similar glitch effects occur during data collection if data acquisition at one point started before a scanning monochromator had had time to settle into the relaxed position, causing the data point for one magnetization (helicity) to be taken under different conditions than the other. This can become particularly bothersome if the monochromator has some sort of feedback mechanism to maintain stability. These type of glitches can be minimized by introducing a 0.5- to 1.0-s delay between motor motions and triggering of the detectors.

The electromagnets used to magnetize samples are also a frequent cause of systematic errors in XMCD spectra. The efficiency of a gas ionization detector can be affected by the presence of a magnetic field. The resultant induced nonlinearities in these detectors can be as great as or greater than the dichroic signal itself. Therefore, for measurements taken by switching the magnetization of the sample, the detectors should be placed well away from the magnets. If this is not possible, the detectors or the magnet should be shielded with a high-permeability metal foil to minimize stray magnetic fields. Similarly, other detectors not directly sensitive to magnetic fields can nevertheless be indirectly influenced by the effects that the electromagnet power supplies can have on their electronic components. Strong electromagnets draw substantial currents, which can induce voltage drops in the circuit that powers them. To minimize cross-talk between the detector electronics and the power supply, they should be connected to different power sources. It is important to note that this does not simply mean lines on different circuit breakers, but rather entirely different, isolated lines. This problem can be particularly bothersome for solid-state detectors in which the switching of a magnet power supply can induce a change in the gain of a spectroscopy amplifier, thereby shifting the position of a fluorescence line or other measured signal.

Besides the errors encountered in the acquisition of the data, several errors can be introduced in the analysis of the spectra, especially in implementing the sum rules. Foremost among these are uncertainties in the values of the beam polarization and the sample magnetization. Both these factors scale directly with the size of the dichroic signal, and any uncertainty in them directly correlates with the size of the error of a magnetic moment obtained from XMCD. The rate of circular polarization is generally calculated from known parameters since it is difficult to measure directly. Because these calculations make assumptions about storage-ring parameters and the degree of perfection of the beamline optics, the error in the values for the polarization rate tend to be quite large and frequently dominate the error in a measurement of the absolute size of the magnetic moment. For measurements of the relative changes in the XMCD spectra, however, this uncertainty is eliminated since the polarization does not change. The magnetization of the sample is usually obtained from magnetometer measurements, but for a multiconstituent sample with different temperature dependencies estimation of each sublattice magnetization can prove difficult.

Systematic errors can also occur at each step in the implementation of the sum rules. As noted in the discussion of Figure 11, the values of the integrals $p$, $q$, and $r$ may vary slightly depending on the energy chosen to extract the integrated value. The greatest uncertainty in the sum rule analysis, however, arises in the integration of the absorption signal, which involves two steps, each of which can introduce uncertainties into the measurement. The first step is subtraction of the background absorption and the second (which causes a greater uncertainty) is modeling of the absorption from transitions to free electron states with a two-step function. Although the two-step function itself should be an adequate model of the transitions involved, the error arises in the placement of the inflection point of the two steps. This has generally been chosen to be at the maximum in the absorption spectrum, but this choice is rather arbitrary. A way around introducing this error is to use Eq. 3 to determine an orbital to spin ratio rather than the absolute moments themselves.

## ACKNOWLEDGMENTS

The author would like to thank Professor A.I. Goldman for offering suggestions on and proofreading this work, as well as Dr. C.T. Chen and Dr. J. Stöhr for providing copies of their figures for publication herein. This work was supported in part by the U.S. Department of Energy, Office of Basic Energy Sciences, under contract No. W-31-109-ENG-38.

# LITERATURE CITED

Baudelet, F., Dartyge, E., Fontaine, A., Brouder, C., Krill, G., Kappler, J. P., and Piecuch, M. 1991. Magnetic properties of neodymium atoms in Nd-Fe multilayers studied by magnetic x-ray dichroism on Nd $L_{II}$ and Fe $K$ edges. Phys. Rev. B 43:5857–5866.

Baumgarten, L., Schneider, C. M., Petersen, H., Schäfers, F., and Kirschner, J. 1990. Magnetic x-ray dichroism in core-level photoemission from ferromagnets. Phys. Rev. Lett. 65:492–495.

Blume, M. and Gibbs, D. 1988. Polarization dependence of magnetic x-ray scattering. Phys. Rev. B 37:1779-1789.

Brouder, C. 1990. Angular dependence of x-ray absorption spectra. J. Phys. Condens. Matter 2:701–738.

Brouder, C. and Hikam, M. 1991. Multiple-scattering theory of magnetic x-ray circular dichroism. Phys. Rev. B 43:3809–3820.

Carra, P., Thole, B. T., Altarelli, M., and Wang, X. 1993. X-ray circular dichroism and local magnetic fields. Phys. Rev. Lett. 70:694–697.

Chen, C. T., Sette, F., Ma, Y., and Modesti, S. 1990. Soft-x-ray magnetic circular dichroism at the $L_{q2,3}$ edges of nickel. Phys. Rev. B 42:7262–7265.

Chen, C. T., Idzerda, Y. U., Lin, H.-J., Meigs, G., Chaiken, A., Prinz, G. A., and Ho, G. H. 1993. Element-specific magnetic hysteresis as a means for studying heteromagnetic multilayers. Phys. Rev. B 48:642–645.

Chen, C. T., Idzerda, Y. U., Lin, H.-J., Smith, N. V., Meigs, G., Chaban, E., Ho, G. H., Pellegrin, E., and Sette, F. 1995. Experimental confirmation of the x-ray magnetic circular dichroism sum rules for iron and cobalt. Phys. Rev. Lett. 75:152–155.

Finazzi, M., Sainctvit, P., Dias, A.-M., Kappler, J.-P., Krill, G., Sanchez, J.-P., Réotier, P. D.d., Yaouanc, A., Rogalev, A., and Goulon, J. 1997. X-ray magnetic circular dichroism at the U $M_{4,5}$ absorption edges of UFe$_2$. Phys. Rev. B 55:3010–3014.

Fischer, P., Schütz, G., Scherle, S., Knülle, M., Stahler, S., and Wiesinger, G. 1992. Experimental study of the circular magnetic x-ray dichroism in Ho-metal, Ho$_2$Fe$_5$O$_{12}$, Ho$_2$Fe$_{23}$ and Ho$_2$Co$_{17}$. Solid State Commun. 82:857–861.

Gibbs, D., Harshman, D. R., Isaacs, E. D., McWhan, D. B., Mills, D., and Vettier, C. 1988. Polarization and resonance properties of magnetic x-ray scattering in Ho. Phys. Rev. Lett. 61:1241–1244.

Giles, C., Malgrange, C., Goulon, J., Bergevin, F.d., Vettier, C., Dartyge, E., Fontaine, A., Giorgetti, C., and Pizzini, S. 1994. Energy-dispersive phase plate for magnetic circular dichroism experiments in the x-ray range. J. Appl. Cryst. 27:232–240.

Gudat, W. and Kunz, C. 1972. Close similarity between photoelectric yield and photoabsorption spectra in the soft-x-ray range. Phys. Rev. Lett. 29:169–172.

Hirano, K., Ishikawa, T., and Kikuta, S. 1993. Perfect crystal x-ray phase retarders. Nucl. Instr.Meth. A336:343–353.

Jackson, J. D. 1975. Classical Electrodynamics. John Wiley & Sons, New York.

Keller, E. N. 1985. Magneto X-ray Study of a Gadolinium-Iron Amorphous Alloy. Ph.D. thesis, University of Washington.

Kobayashi, K., Maruyama, H., Iwazumi, T., Kawamura, N., and Yamazaki, H. 1996. Magnetic circular x-ray dichroism at Pd $L_{2,3}$ edges in Fe-Pd alloys. Solid State Commun. 97:491–496.

Konig, H., Wang, X., Harmon, B. N., and Carra, P. 1994. Circular magnetic x-ray dichroism for rare earths. J. Appl. Phys. 76:6474–6476.

Krill, G., Schillé, J. P., Sainctavit, J., Brouder, C., Giorgetti, C., Dartyge, E., and Kappler, J. P. 1993. Magnetic dichroism with synchrotron radiation. Phys. Scripta T49:295–301.

Laan, G. v. d., Thole, B. T., Sawatzky, G. A., Goedkoop, J. B., Fluggle, J. C., Esteva, J.-M., Karnatak, R., Remeika, J. P., and Dabkowska, H. A. 1986. Experimental proof of magnetic x-ray dichroism. Phys. Rev. B 34:6529–6531.

Lang, J. C., Kycia, S. W., Wang, X., Harmon, B. N., Goldman, A. I., McCallum, R. W., Branagan, D. J., and Finkelstein, K. D. 1992. Circular magnetic x-ray dichroism at the erbium $L_3$ edge. Phys. Rev. B 46:5298–5302.

Lang, J. C., Wang, X., Antropov, V. P., Harmon, B. N., Goldman, A. I., Wan, H., Hadjipanayis, G. C., and Finkelstein, K. D. 1994. Circular magnetic x-ray dichroism in crystalline and amorphous GdFe$_2$. Phys. Rev. B 49:5993–5998.

Lang, J. C., Srajer, G., Detlefs, C., Goldman, A. I., Konig, H., Wang, X., Harmon, B. N., and McCallum, R. W. 1995. Confirmation of quadrupolar transitions in circular magnetic x-ray dichroism at the dysprosium $L_{III}$ edge. Phys. Rev. Lett. 74:4935–4938.

Lang, J. C., Srajer, G., and Dejus, R. J. 1996. A comparison of an elliptical multipole wiggler and crystal optics for the production of circularly polarized x-rays. Rev. Sci. Instrum. 67:62–67.

Lin, H.-J., Chen, C. T., Meigs, G., Idzerda, Y. U., Chaiken, A., Prinz, G. A., and Ho, G. H. 1993. Element-specific magnetic hysteresis measurements, a new application of circularly polarized soft x-rays. SPIE Proc. 2010:174–180.

Malgrange, C., Carvalho, C., Braicovich, L., and Goulon, J. 1991. Transfer of circular polarization in Bragg crystal x-ray monochromators. Nucl. Instr. Meth. A308:390–396.

McIlroy, D. N., Waldfried, C., Li, D., Pearson, J., Bader, S. D., Huang, D.-J., Johnson, P. D., Sabiryanov, R. F., Jaswal, S. S., and Dowben, P. A. 1996. Oxygen induced suppression of the surface magnetization of Gd(0001). Phys. Rev. Lett. 76:2802–2805.

Pizzini, S., Fontaine, A., Fiorgetti, C., Dartyge, E., Bobo, J.-F., Piecuch, M., and Baudelet, F. 1995. Evidence for the spin polarization of copper in Co/Cu and Fe/Cu multilayers. Phys. Rev. Lett. 74:1470–1473.

Rudolf, P., Sette, F., Tjeng, L. H., Meigs, G., and Chen, C. T. 1992. Magnetic moments in a gadolinium iron garnet studied by soft-x-ray magnetic circular dichroism. J. Mag. Mag. Matr. 109:109–112.

Rueff, J. P., Galéra, F. M., Pizzini, S., Fontaine, A., Garcia, L. M., Giorgetti, C., Dartyge, E., and Baudelet, F. 1997. X-ray magnetic circular dichroism at the Gd $L$ edges in Gd-Ni-Co amorphous systems. Phys. Rev. B 55:3063–3070.

Samant, M. G., Stöhr, J., Parkin, S. S. P., Held, G. A., Hermsmeier, B. D., Herman, F., Schilfgaarde, M. v., Duda, L.-C., Mancini, D. C., Wassdahl, N., and Nakajima, R. 1994. Induced spin polarization in Cu spacer layers in Co/Cu multilayers. Phys. Rev. Lett. 72:1112–1115.

Schillé, J. P., Sainctavit, P., Cartier, C., Lefebvre, D., Brouder, C., Kappler, J. P., and Krill, G. 1993. Magnetic circular x-ray dichroism at high magnetic field and low temperature in ferrimagnetic HoCo$_2$ and paramagnetic Ho$_2$O$_3$. Solid State Commun. 85:787–791.

Schütz, G., Wagner, W., Wilhelm, W., Kienle, P., Zeller, R., Frahm, R., and Materlik, G. 1987. Absorption of circularly polarized x-rays in iron. Phys. Rev. Lett. 58:737–740.

Schütz, G., Knulle, M., Wienke, R., Wilhelm, W., Wagner, W., Kienle, P., and Frahm, R. 1988. Spin-dependent photoabsorption at the $L$-edges of ferromagnetic Gd and Tb metal. Z. Phys. B 73:67–75.

Schütz, G., Wienke, R., Wilhelm, W., Wagner, W., Kienle, P., Zeller, R., and Frahm, R. 1989a. Strong spin-dependent absorption at the $L_{2,3}$-edges of $5d$-impurities in iron. *Z. Phys. B* 75:495–500.

Schütz, G., Frahm, R., Wienke, R., Wilhelm, W., Wagner, W., and Kienle, P. 1989b. Spin-dependent $K$- and $L$-absorption measurements. *Rev. Sci. Instrum.* 60:1661–1665.

Schütz, G., Fischer, P., Atternkofer, K., Knulle, M., Ahlers, D., Stahler, S., Detlefs, C., Ebert, H., and Groot, F. M. F. d. 1994. X-ray magnetic circular dichroism in the near and extended absorption edge structure. *J. Appl. Phys.* 76:6453–6458.

Shenoy, G. K., Viccaro, P. J., and Mills, D. M. 1988. Characteristics of the 7-GeV Advanced Photon Source: A Guide for Users. Argonne National Laboratory ANL-88–9.

Smith, N. V., Chen, C. T., Sette, F., and Mattheiss, L. F. 1992. Relativistic tight-binding calculations of x-ray absorption and magnetic circular dichroism at the $L_2$ and $L_3$ edges of nickel and iron. *Phys. Rev. B* 46:1023–1032.

Stahler, S., Schutz, G., and Ebert, H. 1993. Magnetic $K$ edge absorption in the $3d$ elements and its relation to local magnetic structure. *Phys. Rev. B* 47:818–826.

Stöhr, J., Wu, Y., Hermsmeier, B. D., Samant, M. G., Harp, G. R., Koranda, S., Dunham, O., and Tonner, B. P. 1993. Element-specific magnetic microscopy with circularly polarized x-rays. *Science* 259:658–661.

Stöhr, J. and Wu, Y. 1994. X-ray magnetic circular dichroism: Basic concepts and theory for $3d$ transition metal atoms. *In* New Directions in Research with 3rd Generation Soft X-ray Synchrotron Radiation Sources (A. S. Schlachter and F. J. Wuilleumier, eds.).. pp. 221–251. Kluwer Academic Publishers, Dordrecht, Netherlands.

Thole, B. T., Carra, P., Sette, F., and Laan, G. v. d. 1992. X-ray circular dichroism as a probe of orbital magnetization. *Phys. Rev. Lett.* 68:1943–1946.

Tonner, B. P., Dunham, D., Shang, J., O'Brien, W. L., Samant, M., Weller, D., Hermsmeier, B. D., and Stöhr, J. 1994. Imaging magnetic domains with the x-ray dichroism photoemission microscope. *Nucl. Instr. Meth.* A347:142–147.

Wang, X., Leung, T. C., Harmon, B. N., and Carra, P. 1993. Circular magnetic x-ray dichroism in the heavy rare-earth metals. *Phys. Rev. B* 47:9087–9090.

Wu, Y., Stöhr, J., Hermsmeier, B. D., Samant, M. G., and Weller, D. 1992. Enhanced orbital magnetic moment on Co atoms in Co/Pd multilayers: A magnetic circular x-ray dichroism study. *Phys. Rev. Lett.* 69:2307–2310.

Wu, R., Wang, D., and Freeman, A. J. 1993. First principles investigation of the validity and range of applicability of the x-ray magnetic circular dichroism sum rule. *Phys. Rev. Lett.* 71:3581–3584.

Wu, R. and Freeman, A. J. 1994. Limitation of the magnetic-circular-dichroism spin sum rule for transition metals and importance of the magnetic dipole term. *Phys. Rev. Lett.* 73:1994–1997.

Yamamoto, S., Kawata, H., Kitamura, H., Ando, M., Saki, N., and Shiotani, N. 1988. First production of intense circularly polarized hard x-rays from a novel multipole wiggler in an accumulation ring. *Phys. Rev. Lett.* 62:2672–2675.

## KEY REFERENCES

Lovesey, S. W. and Collins, S. P. 1996. X-ray Scattering and Absorption by Magnetic Materials. Oxford University Press, New York.

*Provides an overall theoretical picture of the interaction of x-rays with magnetic materials. Chapter 5 is devoted to XMCD measurements and includes several excellent examples of recent experimental work.*

Koningsberger, D. C. and Prins, R. 1988. X-ray Absorption: Principles, Applications, Techniques of EXAFS, SEXAFS, and XANES. John Wiley & Sons, New York.

*Although this is primarily devoted to EXAFS measurements, the methods for and problems encountered with the acquisition of data are also applicable to XMCD measurements. Also useful is the theoretical work presented on XANES spectra, which cover the same energy range as XMCD, although without polarization analysis.*

Chen et al., 1995. See above.

Idzerda, Y. U., Chen, C. T., Lin, H.-J., Meigs, G., Ho, G. H., and Kao, C.-C. 1994. Soft x-ray magnetic circular dichroism and magnetic films. *Nucl. Motr. Meth.* 347:134–141.

*These two references provide excellent discussion of the various experimental limitations encountered when applying the XMCD sum rules.*

Stöhr and Wu, 1994. See above.

*A good overall theoretical treatment of XMCD, with emphasis on the interpretation of 3d L edge spectra.*

## APPENDIX A:
## ONE-ELECTRON PICTURE OF DICHROISM

This appendix provides a simple theory to explain the origin of dichroic signals. In this theory, the x-ray absorption is treated within the one-electron picture and the electric dipole approximation is assumed to be valid. The explanation below follows the same reasoning as put forth by Brouder and Hikam (1991) and Smith et al. (1992), and the reader should refer to these papers for a more detailed explanation of all the edges.

Using the first-order term in Fermi's golden rule, the x-ray absorption cross-section due to dipole transitions may be written as follows (Brouder, 1990):

$$\sigma(\varepsilon, \omega) = 4\pi^2 \alpha \hbar \omega \sum_{f, m_j, \beta} |\langle f|\varepsilon^\beta \cdot \boldsymbol{r}|jm_j\rangle|^2 \delta(E_f - E_i - h\omega) \quad (8)$$

where the $\alpha$ is the fine-structure constant.

Here the common bracket notation is followed in which the expression above represents an integral over the spacial and spin dimensions of the system and $|jm_j\rangle$ and $|f\rangle$ represent the initial- and final-state wave functions involved in the transition. Because the initial atomic core states are strongly spin-orbit coupled, the wave functions are expressed in terms of the total angular momentum $j$. As an example of an absorption calculation, consider the above expression for an $L_2$ edge with initial $2p_{1/2}$ core state. By the dipole selection rules, the final states accessible from this edge would be either the s or d states. As a general rule, however, the transitions to the $\ell - 1$ states are suppressed by 1 to 2 orders of magnitude compared to the $\ell + 1$ states due to the negligible overlap of the former with the initial level. Thus to a very good approximation only transitions to the $d$ states need be considered. Neglecting any spin-orbit coupling in the final states, the final $d$-band states can be written as the product of spatial and spin wave functions:

$$\langle f| = \varphi_{\ell m}(r)\langle s| \sim 2m_\ell \langle\uparrow(\downarrow)| \qquad (9)$$

where $m = -2, -\ell, \ldots, 2$ are the magnetic quantum numbers and $\uparrow$ and $\downarrow$ refer to the spin-up and spin-down states. The electric dipole operator $(\varepsilon \cdot r)$ for left (right) circularly polarized photons can be expressed in terms of spherical harmonics by:

$$\varepsilon^\pm \cdot \mathbf{r} = \pm\left[\frac{4\pi}{3}\right]^{\frac{1}{2}} rY_{1\pm1} \qquad (10)$$

Here the sign $(+,-)$ in the polarization vector (spherical harmonic) refers to (left-, right-) handed circular polarization. To obtain matrix elements involving three spherical harmonics, the initial $|jm_j\rangle$ states ($m_j = \pm1/2$ for an $L_2$ edge) can be expanded in terms $|lm_l\rangle$ of wave functions using Clebsch-Gordan coefficients:

$$\left|\frac{1}{2}-\frac{1}{2}\right\rangle = \phi_{1\frac{1}{2}}(r)\left[\sqrt{\frac{1}{3}}Y_{10}|\downarrow\rangle - \sqrt{\frac{2}{3}}Y_{1-1}|\uparrow\rangle\right] \qquad (11)$$

$$\left|\frac{1}{2}\frac{1}{2}\right\rangle = \phi_{1\frac{1}{2}}(r)\left[-\sqrt{\frac{1}{3}}Y_{10}|\uparrow\rangle - \sqrt{\frac{2}{3}}Y_{11}|\downarrow\rangle\right] \qquad (12)$$

If we assume that the spin functions are orthogonal and the radial parts of the wave functions are independent of spin, we can define quantities $\rho^\uparrow$ and $\rho^\downarrow$ that incorporate all the constants and the radial matrix elements. Here $\rho^\uparrow(\rho^\downarrow)$ represents a transition to a spin-up (-down) state and, to a first-order approximation, can be considered proportional to the spin-up (-down) spin density of unoccupied states. Taking a sum on initial and final states, Eq. 8 for left circularly polarized light $(\varepsilon^+)$ reduces to:

$$\sigma_{L_2}^+ \sim \frac{1}{3}|\langle21|11|10\rangle|^2\rho^\downarrow + \frac{2}{3}|\langle20|11|1-1\rangle|^2\rho^\uparrow$$
$$+ \frac{1}{3}|\langle21|11|10\rangle|^2\rho^\uparrow + \frac{2}{3}|\langle22|11|11\rangle|^2\rho^\downarrow \qquad (13)$$

The integrals denote the product of three spherical harmonics of index $|lm_l\rangle$. Here all the null matrix elements of terms with spin opposite to the factors in Eqs. 11 and 12 above have been omitted, and the selection rule $m_f = m_i + 1$ for the integration of three spherical harmonics has been used. Upon performing the integrations and normalizing, Eq. 13 reduces to:

$$\sigma_{L_2}^+ \sim \frac{3}{5}\rho^\downarrow + \frac{12}{5}\rho^\uparrow + \frac{3}{5}\rho^\uparrow + \frac{2}{5}\rho^\downarrow = \rho^\uparrow + 3\rho^\downarrow \qquad (14)$$

Thus left circularly polarized x-rays will be three times more likely to make a transition to a spin-down state than a spin-up state. Repeating the above steps for right-handed circular polarization yields exactly the opposite result,

$$\sigma_{L_2}^- \sim 3\rho^\uparrow + \rho^\downarrow \qquad (15)$$

Therefore, right circularly polarized light will make preferential absorption to spin-up states. The same procedure can be repeated for the $L_3$ edge with an initial $2p_{3/2}$ state to obtain the following results for left and right circular polarization, respectively:

$$\sigma_{L_3}^+ \sim 5\rho^\uparrow + 3\rho^\downarrow, \quad \sigma_{L_3}^- \sim 3\rho^\uparrow + 5\rho^\downarrow \qquad (16)$$

Here the assumption has been made that the $2p_{1/2}$ and $2p_{3/2}$ radial matrix elements are essentially equal.

$$\rho_{1/2}^{\uparrow(\downarrow)} = \rho_{3/2}^{\uparrow(\downarrow)} = \rho^{\uparrow(\downarrow)} \qquad (17)$$

Using Eqs. 14, 15, and 16, the following expressions for the XMCD at $L_{2,3}$ edges can be written as shown in Eq. 18.

$$\mu_c(L_2) \sim \sigma^+ - \sigma^- \sim -2(\rho^\uparrow - \rho^\downarrow),$$
$$\mu_c(L_3) \sim \sigma^+ - \sigma^- \sim 2(\rho^\uparrow - \rho^\downarrow) \qquad (18)$$

Therefore XMCD is, to a first-order approximation, proportional to the spin density of unoccupied states, and the ratio of the signal from the $L_{2,3}$ spin-orbit split states is $1:-1$.

## APPENDIX B:
## FIGURE OF MERIT FOR CPX

The figure of merit for CPX sources ($P_c^2I$) can be obtained by expressing the measured signal as a sum of terms arising from charge and magnetic effects,

$$I^\pm \sim \sigma_c I + \sigma_m I, \quad \sigma_c \gg \sigma_m \qquad (19)$$

Here $I^\pm$ indicates the measured intensities taken with opposite helicities (or magnetizations), $I$ is the incoming beam intensity, and $\sigma_c$ and $\sigma_m$ are the charge and magnetic cross-sections. This expression applies not only to XMCD measurements but more generally to other techniques that utilize CPX for magnetic measurements (i.e., magnetic Compton scattering and nonresonant magnetic x-ray scattering). For ferromagnetic materials, the magnetic cross-section depends linearly on $P_c$ (or the magnetization), and thus can be separated out, $\sigma_m \sim P_c\sigma_m'$ making the difference-to-sum ratio

$$\frac{I^+ - I^-}{I^+ + I^-} \sim \frac{2P_c\sigma_m'I}{2\sigma_cI} \sim P_c\frac{\sigma_m'}{\sigma_c} \qquad (20)$$

Therefore the measured signal depends linearly on the degree of circular polarization. The percentage error in this quantity, which is the quantity to be minimized in any experimental measurement, is obtained by adding the errors in the numerator and denominator in quadrature:

$$\left(\Delta\%\left(\frac{I^+ - I^-}{I^+ + I^-}\right)\right)^2 = \left(\frac{\sqrt{I^+ - I^-}}{I^+ + I^-}\right)^2 + \left(\frac{\sqrt{I^+ - I^-}}{I^+ - I^-}\right)^2 \qquad (21)$$

The first term above will always be much smaller than the second, and thus can be neglected, yielding:

$$\Delta\% \left(\frac{I^+ - I^-}{I^+ + I^-}\right) \cong \frac{\sqrt{I^+ + I^-}}{I^+ - I^-} \sim \frac{\sqrt{2\sigma_c I}}{2P_c\sigma_m I} \sim \frac{1}{P_c\sqrt{I}} \qquad (22)$$

Therefore, the minimum error in the measurement is achieved by maximizing $P_c \cdot \sqrt{I}$. Although this quantity is sometimes referred to as a figure of merit, the experimenter is generally more interested in the amount of time required to perform a measurement on one source compared to another. The time in this quantity is implicitly included in the flux, and therefore to obtain a source comparison for comparable times the quantity above must be squared, yielding a figure of merit of $P_c^2 I$.

J. C. LANG
Advanced Photon Source
Argonne National Laboratory
Argonne, Illinois

# X-RAY PHOTOELECTRON SPECTROSCOPY

## INTRODUCTION

X-ray photoelectron spectroscopy (XPS) uses x rays of a characteristic energy (wavelength) to excite electrons from orbitals in atoms. The photoelectrons emitted from the material are collected as a function of their kinetic energy, and the number of photoelectrons collected in a defined time interval is plotted versus kinetic energy. This results in a spectrum of electron counts (number per second) versus electron kinetic energy (eV). Peaks appear in the spectrum at discrete energies due to emission of electrons from states of specific binding energies (orbitals) in the material. The positions of the peaks identify the chemical elements in the material. Peak areas are proportional to the number of orbitals in the analysis volume and are used to quantify elemental composition. The positions and shapes of the peaks in an XPS spectrum can also be analyzed in greater detail to determine the chemical state of the constituent elements in the material, including oxidation state, partial charge, and hybridization.

X-ray photoelectron spectroscopy is widely applied to all types of solids, including metals, ceramics, semiconductors, and polymers, in many forms, including foils, fibers, and powders. It has also been used to obtain spectra of gas-phase compounds. In general, it is a nondestructive method of analysis. When applied to solids, XPS is a surface-sensitive technique. The nominal analysis depth is on the order of 1 to 10 nm (10 to 100 monolayers). Surface sensitivity can be increased by collecting the emitted photoelectrons at to glancing angles to the surface. Typical spatial resolutions on a sample surface are on the order of 1 to 5 mm, with so-called small-spot systems having spatial resolutions as low as 25 μm. Analysis times range from under 5 min to identify the chemistry and composition of a sample to 1 hr or more for characterizing chemical states of trace elements.

The best sensitivity of XPS for quantifying elemental composition in solids is on the order of 0.1 at. %. All elements with atomic number greater than three can be detected. The detection sensitivity varies for each element, with some elements requiring greater concentrations to reach a nominal detection threshold in a reasonable analysis time. Relative uncertainties in atomic compositions are typically $\sim\pm 10\%$ for samples with homogeneous compositions throughout the sampling volume. This error can be reduced by analyzing standards to calibrate the sensitivity factors of the instrument for each element. Materials with heterogeneous distributions of elements represent a greater challenge to obtaining accurate compositions with XPS. Samples with contaminant or other thin film overlayers will always be found to have attenuated concentration values for the supposed bulk material, and samples exposed to air before analysis typically show accentuated carbon and oxygen concentrations. Atomic concentrations reported from XPS analysis should be understood to be indicative of those from a hypothetical sample of homogeneous composition throughout the sampling volume, unless otherwise stated.

Thin film samples can be analyzed with XPS to determine film thickness. Typical methods of analysis involve measuring composition either as a function of sputter depth into the sample (sputter profiling) or as a function of electron emission angle (angle-resolved XPS). The former method requires the use of a noble-gas ion-etching process to sputter away successive layers of the sample, and is therefore destructive. Sputter profiling is limited in film thickness only by the patience of the user and the robustness of the instrumentation. Angle-resolved XPS is limited to films with a thickness that is on the order of the sampling depth of XPS (1 to 10 nm). Both techniques have other limitations, primarily involving surface roughness for angle-resolved XPS and both surface roughness and film thickness for sputter profiling.

The chemical state information obtained with XPS includes oxidation states or hybridization states for chemical bonds of the elements. Typical examples include determining whether a constituent element is or is not oxidized in a metal alloy or characterizing whether carbon is present as a carbonyl, ether, or ester linkage in a polymer. Application of peak fitting routines is generally required to distinguish peaks from overlapping oxidation or hybridization states. Assignment of chemical state using XPS is not without ambiguities, because other factors besides chemical environment shift the peaks in an XPS spectrum. The use of references with known composition and chemistry can usually help resolve these ambiguities.

The primary limitation of XPS is the need for ultrahigh vacuum conditions during analysis. This generally limits the type of material to those with a low vapor pressure ($<10^{-8}$ mbar) at room temperature and limits the sample size to that which will fit through the introduction ports on the vacuum chamber. Some compounds, such as polymers, can also degrade under the x-ray flux. Another limitation in interpretation of spectra from insulator and semiconductor samples arises from sample charging. This is an artifact that can be corrected either during or after data acquisition, although the extent of the correction is often not straightforward to determine.

For those wishing to find further information on XPS, a discussion of the available literature is provided at the end of this unit (see Key References).

### Competitive and Related Techniques

A common technique that is sometimes considered to compete with XPS for quantitative chemical analysis of materials surfaces is Auger electron spectroscopy (AES; AUGER ELECTRON SPECTROSCOPY). By comparison to XPS, AES is more surface sensitive, with sampling depths typically below 1 nm. Sub-micron spatial resolutions are also possible with AES. In this regard, AES is often the preferred technique for compositional analysis of surfaces in metallurgy. The disadvantage of AES in comparison to XPS is that AES must be done on conductive materials because it uses electrons as the excitation source; it cannot analyze ceramic or polymeric materials as well as XPS can, if at all. In terms of sample damage, AES is also more destructive of ceramic or polymeric materials and molecular layers adsorbed on surfaces. Both XPS and AES have comparable accuracies for quantitative analysis of elemental composition. The sensitivity of AES is better for light elements. Finally, XPS is generally considered as being able to provide much richer information about the chemical state of the elements in the material, although reports have shown exceptions to this (Sekine et al., 1996). This often makes XPS a better technique than AES for analysis of the chemistry of fundamental processes occurring at or on surfaces, such as chemical bonding, adhesion, or corrosion. The two techniques actually complement each other very closely, especially for analytical determination of elemental compositions at the surface of a material. Although the terminology ESCA was introduced for and has become synonymous primarily with XPS because of the resolution of the elemental chemical state information it provides, AES can also be considered as an electron spectroscopy for chemical analysis (ESCA), especially in cases were it is more widely used as an analytical tool for quantitative surface analysis than is XPS.

Another common technique that competes with XPS for analysis of elemental composition is energy-dispersive spectroscopy (EDS; see ENERGY-DISPERSIVE SPECTROMETRY) or energy-dispersive x-ray analysis (EDXA). The EDS technique is commonly available with scanning electron microscopes. By comparison to XPS, EDS has better spatial resolution on the surface but probes deeper into the bulk of the material, on the order of microns. No chemical state (oxidation state) information about the elements can be obtained with EDS. Finally, XPS has a better (elemental) resolution of the lighter elements.

A technique that is closely related to XPS is ultraviolet photoelectron spectroscopy (UPS; see ULTRAVIOLET PHOTOELECTRON SPECTROSCOPY). In UPS, the source is ultraviolet light rather than x rays. The principles of the two techniques are practically the same; both are based on photoemission. Because the UPS source has a lower energy than typical XPS sources, UPS can only probe electrons

in the valence band energy region of a material. The sensitivity of UPS to changes in the valence band structure of an element is greater than that of XPS. Therefore, UPS is often a used for determination of band structures of materials, especially semiconductors.

Other techniques of interest in relation to XPS are x-ray adsorption fine structure (XAFS; see XAFS SPECTROSCOPY), extended XAFS (EXAFS), and near-edge XAFS (NEXAFS). These techniques use x-rays of variable energy from synchrotron sources, whereas in XPS, the source energy (wavelength) remains constant. The information obtained from XAFS and NEXAFS pertains more to the fundamental nature of the chemical bond for a specific element in a material rather than the composition of the material. The average coordination number (number of bonds) surrounding a specific atom (element) in a material can be determined with EXAFS.

### PRINCIPLES OF THE METHOD

X-ray photoelectron spectroscopy is based on the principle of photoemission. The process is illustrated in Figure 1 for electrons in atomic orbitals. The principles are nearly the same for molecular orbitals and free electron or valence bands. A photon (x ray) of frequency $\nu = c/\lambda$, where $\lambda$ is its wavelength, has a kinetic energy of $h\nu$. As this photon passes through a material, it can interact with the electrons in the material. Absorption of the x ray by an atom can lead to an electronic excitation of the atom. The photon absorption process conserves energy, leaving an electron in the atom to a first approximation with a kinetic energy (KE or $E_K$) that is equal to the energy of the incident photon less the initial binding (electronic ground state) energy (BE or $E_B$) of the electron, $E_K = h\nu - E_B$. Under proper circumstances, when the electron does not scatter

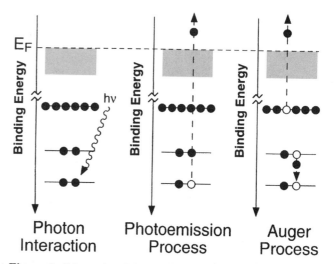

**Figure 1.** Schematics of the excitation process leading to photoemission of a core electron and the subsequent Auger relaxation process that can occur. Core-level electrons are solid circles, holes are open circles, and the valence band is a shaded box. The position of the Fermi energy is shown by the dotted line at $E_F$.

back, it acquires a sufficient velocity to escape from the material. In most common XPS instruments, electrostatic or electromagnetic analyzers focus the escaping electrons (photoelectrons) at a specific KE from the sample onto an electron detector. The number of electrons that arrive at the detector is counted for a certain time interval, and the electron counting is subsequently repeated at a new KE value. Alternatively, the number of electrons at the detector is counted concurrently while sweeping the energy selection range of the analyzer. Either method produces a spectrum of number of electrons emitted from the sample per unit time (electron counts per time, or $I_i^e$) versus the KE of the electron. The KE scale of the spectrum is converted to a BE scale using energy conservation. Peaks appear in the spectrum at energies where electronic excitation has occurred.

The electronic excitation by the photon leaves behind a hole (a site of positive charge). How this hole is subsequently filled is important in the overall energy balance, as discussed in the section on high-resolution scans. One important relaxation process that can occur is the Auger process. This is shown in the third panel in Figure 1. An upper lying (lower BE) electron drops into the hole, perhaps from the valence band as shown. The energy released by this process causes ejection of a second electron called the Auger electron. The Auger process is independent of how the initial hole was created. When electrons rather than x-rays are used to excite the initial core electron, the method is called Auger electron spectroscopy (AES), as detailed in AUGER ELECTRON SPECTROSCOPY. Auger electron peaks will also appear in XPS spectra, and their analysis in this case is considered under the acronym, x-ray induced AES (XAES). The position of an XAES peak relative to a corresponding photoemission peak is used to define a characteristic parameter called the Auger parameter. The Auger parameter is affected by the chemistry of the material. Further consideration of the use of the Auger parameter is given elsewhere (Briggs and Seah, 1994).

Not all photons that enter a material are successful at causing a photoemission event. The probability of such an event occurring is represented by the photoelectron cross-section, $\sigma$. Photoelectron cross-sections provide a start in determining the flux, $J_i^e$, of emitted electrons from a given orbital (or band) for a given x-ray flux $F$ and given number of orbitals $n$ present in the material. The relationship is $J_i^e \propto F\sigma n$, where the electron flux $J_i^e$ has units of electrons/m$^2$/sec, $F$ has units of photons/m$^2$/sec, and $n$ is the number of orbitals per unit area viewed by the analyzer (orbitals/m$^2$). Although the cross-section has explicit units of area (m$^2$), it contains implicit units of (number of photoelectrons)/(number of photons)(number of orbitals), as can be derived from the flux equation).

### The XPS Survey Spectrum

Using the above information, we can prepare a schematic of what might be expected for an XPS spectrum from a pure material, Au, under ideal conditions. A survey, or wide-range scan covers the majority or all of the levels that can be excited by the x-ray energy being used. In

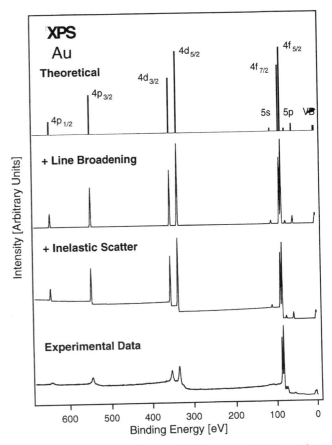

**Figure 2.** A series of XPS spectra for Au, generated from a theoretical standpoint and incorporating two factors associated with the overall photoemission process from a solid. An experimental XPS spectrum of Au is included for comparison.

what follows, we will use a scale of BE rather than KE. Further consideration of how the x-ray source energy influences the BE scan range and spectrum shape is presented in the section on practical aspects of XPS.

Binding energies for electrons in Au can be obtained from standard handbooks (Lide, 1993; Moulder et al., 1995), and values of $\sigma$ have also been calculated (Scofield, 1976). Using this information, we can generate a theoretical XPS survey spectrum in the BE range 0 to 650 eV for an assumed ideal, monochromatic x-ray source (with an as yet unspecified energy). The result is presented in the uppermost plot in Figure 2. It is essentially a series of delta function lines appearing at the respective BE values for each orbital. The heights of the lines are scaled in proportion to $\sigma$ for each orbital.

One noticeable feature of the theoretical spectrum is a splitting of $p$ through $f$ orbitals into two peaks. This is due to coupling between the electron spin and orbital angular momenta for these orbitals. The $s$ orbitals do not split because they have no orbital angular momentum. One of two methods can be applied to determine the magnitude of this coupling, $j$-$j$ coupling or L-S (Russell-Sanders) coupling. The former predominates for atomic numbers below $\sim$20. In $j$-$j$ coupling, the magnitude of total angular momentum for a given electron is the vector sum of its spin and orbital angular momenta. Since $s$ orbitals have no angular momentum, electrons in spin-up or spin-down configurations have the same magnitude of total angular momentum. In $p$ orbitals, vector coupling of spin-up or spin-down states with the $\pm 1$ orbital angular momentum vectors gives total angular momentum of either 1/2 (from $|1-1/2|$ or $|-1+1/2|$) or 3/2 (from $|1+1/2|$ or $|-1-1/2|$). These two spin-orbit couplings have different binding energies in the magnetic field of the nucleus, leading to the splitting of the levels. In $L$-$S$ coupling, the total angular momentum of is determined by summing (vectorially) the total spin angular momentum ($S$) and the total orbital angular momentum ($L$) after the electron is removed. Similar splitting of orbitals arises. The nomenclature used in XPS to designate the spin-orbit splitting is applied as though $j$-$j$ coupling predominates and is adopted from spectroscopy (rather than from x-ray notation). It uses principal quantum number, orbital designation ($s$, $p$, $d$, $f$), and $|1 \pm s|$. For example, the $3p_{3/2}$ peak designates electrons from $p$ orbital of the $n = 3$ level with a total angular momentum of either $|1+1/2|$ (spin-up in an "upwardly" aligned $p$ orbital) or $|-1-1/2|$ (spin down in a "downwardly" aligned $p$ orbital).

The width of the spin-orbit splitting depends on the magnitude of the coupling between electron spins and orbital angular momentum. This coupling is proportional to $\langle 1/r^3 \rangle$, the expectation value of $1/r^3$, where $r$ is the radius of the orbital. In practical terms, for a given element, the width of the spin-orbit splitting decreases with increasing principal quantum number ($4p$ levels are split more than $5p$ levels), and with increasing angular momentum quantum number ($4p$ levels are split more than $4f$ levels). The strength of the spin-orbit splitting also increases with increasing atomic number because the nuclear charge increases. For example, the experimentally determined $4f$ level splittings of the series Ir–Pt–Au (increasing atomic sizes) are ~3.0 eV, 3.4 eV, and 3.7 eV. The splitting of valence band levels increases within a homologous series going down a period—for example Cu $2p$ (20 eV), Ag $3p$ (30 eV), and Au $4p$ (100 eV)—because the increase due to an increase in nuclear charge overrides the decrease due to an increase in principal quantum number. The spin-orbit split levels have different relative occupancies, roughly in proportion to the respective degeneracies of the levels. Thus, $p$ 1:2, $d$ 2:3, and $f$ 3:4. These relative occupancies are included in the theoretical spectrum in Figure 2 by means of the relative cross-sections.

When the energy of the excitation source $h\nu$ is greater than the work function of the sample, electrons from valence bands (VB) will also be emitted from the sample. This is the principle behind the photoelectric effect. The valence bands are the topmost bonding levels with energies just above the Fermi energy ($E_F$). In the representative model spectrum for Au, the VB has been modeled as a band of finite width, just above 0 eV in BE, the nominal position of $E_F$. Excitation from the VB levels is typically better examined with UPS because the excitation cross-sections are greater. Both XPS and UPS provide useful information about the chemical state of a material, with XPS is the only appropriate method of the two for investigations of changes in core levels and UPS applicable to the valence-band region. Studies of valence bands during gas

adsorption and reaction phenomena on single crystal surfaces benefit by using both XPS and UPS. The two techniques are also useful for studying the valence band chemistry of polymeric materials.

Photoemission of electrons from solids will not produce the delta function responses shown in Figure 2 for the theoretical spectrum. Even under the assumption of ideal sources and analyzers, we must account for finite lifetime broadening effects. These effects are tied to the time needed for the ion created by the XPS process to relax back to its ground state, neutral atom condition. Heisenberg's uncertainty principle specifies $\Delta E_t \geq \hbar/2\Delta t_l$, where $\Delta E_l$ is the expected broadening of the emission feature due to a lifetime uncertainty of $\Delta t_l$. Shorter lifetimes lead to larger energy spreads in the photoemission distribution for the respective level. Core and valence band hole lifetimes are on the order of $10^{-14}$ to $10^{-16}$ sec, leading to lifetime broadenings on the order of 0.05 to as much as 5 eV. In general, lifetime broadening will be greater for photoemission from lower-lying (higher BE) levels than from upper-level (valence) levels, because the lower lying holes relax in a shorter time. As discussed in the section on high-resolution spectra, ions created in solids by the photoemission process will also relax differently from those created in gas-phase compounds, leading to a shifting of the apparent BE and to wider photoemission peaks from solids as compared to those from gases. Gas phase photoelectron peaks can have peak widths that reach the theoretical thermal broadening limit of $kT$ (Boltzmann's constant times temperature), or ~0.025 eV at room temperature.

Lifetime broadening was simulated by convolving a Gaussian with a 2 eV half-width with the theoretical spectrum in Figure 2. This does not cover the full range of broadening effects, primarily because the true broadening is not constant for all levels. Lifetime broadened photoemission lineshapes may also contain some Lorentzian character. This is discussed further below (see Data Analysis and Initial Interpretation).

One remaining factor can be illustrated for photoemission from solids using the survey scan range shown in Figure 2. As a photoelectron leaves a solid, it may lose energy to its surroundings. At this point, we will only consider inelastic scattering of the exciting photoelectron, as shown in Figure 3. The scattering occurs through interaction of the outgoing electron with phonon (vibrational) modes of the material. Those electrons that lose too

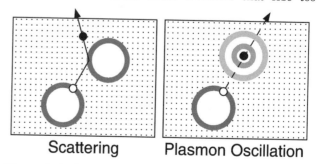

**Figure 3.** A pictorial representation of electron scatter and the bulk plasmon loss processes that can occur as an electron travels through a solid.

much energy through inelastic collisions will not leave the solid and are of no concern. Those electrons that lose energy yet still can leave the solid are collected at lower KE values than those at which they started. They contribute to an increase in the background on the higher BE side of the main photoemission peak and are not considered part of the primary photoemission process.

In general, the background due to inelastically scattered electrons extends over a range as much as 500 eV in BE above a primary photoelectron peak (which is up to 500 eV in KE below the $E_K$ of the photoelectron). The shape of the loss signal over this range of energy depends on the material being analyzed and the initial $E_K$ of the photoelectron. Each photoemission peak represents an additional source of electrons that can scatter as they leave the solid. Therefore, inelastic scattering leads to an overall increase in the background signal with increasing BE in an XPS spectrum.

Electrons will also scatter inelastically in multiple collisions as they travel through the solid. Because this is a cumulative process, the number of inelastically scattered electrons begins to increase again at low KE values. A peak in number of multiply scattered electrons is reached ~20 to 50 eV in $E_K$. This peak is called the secondary electron peak. Secondary electron emission decreases to zero at a value of zero $E_K$. The consequence of the secondary electron peak is that the background intensity in an XPS spectrum will begin to increase dramatically as BE approaches within ~100 eV of the source energy ($h\nu$). This will be shown in spectra presented in the section on practical aspects of the technique.

A simulated electron loss curve was convolved with the lifetime-broadened spectrum in Figure 2. The shape of the energy loss curve was simulated to give a resulting spectrum with a comparable background to an experimental spectrum around the Au $4f$ peaks. The actual shape of the inelastic loss spectrum and the secondary background can be measured with electron energy loss spectroscopy (EELS; see SCANNING TRANSMISSION ELECTRON MICROSCOPY: Z-CONTRAST IMAGING). The hypothetical loss spectrum did not account for the secondary electron peak. The result, and the experimental spectrum for Au, are included in Figure 2.

Lifetime broadening and inelastic electron scattering, when convolved with a theoretically generated spectrum of delta functions, can been seen to account for a significant amount of the overall shape of a survey scan in XPS. Some of the remaining differences between the generated and experimental spectra in Figure 2 arise from factors that depend on the source and analyzer. These factors are discussed below (see Practical Aspects of the Method), where the utility of survey scans is also considered. Further differences between generated and experimental spectra become apparent when we focus on the exact shape of a specific photoelectron peak, the background immediately near this peak, and the peak position. These factors are considered below.

## The XPS High-Resolution Spectrum

A high-resolution XPS spectrum is one that defines a narrow energy region, nominally 10 to 20 eV, around a photo-

emission peak. High-resolution spectra are taken to obtain improved quantitative analysis for elements in the material as well as chemical state information about the elements. In the following, we concentrate on the photoemission principles that define the shape of the peak in a high-resolution scan. How high-resolution spectra are analyzed to determine the composition of a material will be discussed below (see Data Analysis and Initial Interpretation).

The chemical state of an element affects the position and shape of its photoemission peak. The chemical state is an initial-state effect, in that it is defined before the photoemission process occurs. Unfortunately, the chemical state of an atom is not the only factor involved in shifting or broadening an XPS peak. Factors that occur after the photoemission process can also influence the apparent BE of the photoelectron. These factors are called final-state effects. Final-state effects will now be considered.

**Final-State Effects.** Final state effects can be classified into five types, plasmon loss, Auger excitation, relaxation, shake-up or shake-off, and multiplet splitting. All but the first one are concerned with how the core hole, that is left immediately after the photoemission process, is filled. They all shift, and in some cases broaden, the XPS peak.

The plasmon loss process is illustrated in Figure 3. It is a loss process that is similar to inelastic scattering discussed above, and it also leads to an increase in background intensity in survey scans. It occurs when the outgoing photoelectron interacts with the free electrons in a solid material (often called the free electron gas). Unlike what is traditionally considered under the term inelastic scattering, however, the plasmon loss process is quantized and does not extend over a broad energy range below the principal photoemission peak.

Two types of plasmon loss processes are recognized, bulk and surface. In bulk plasmon loss, the outgoing photoelectron loses energy by interacting via collective oscillations with the free electron gas in the bulk of material. These oscillations will be at resonance at the so-called bulk plasmon frequency, $\omega_p$, for the free electron gas. Therefore, the probability that an exiting photoelectron loses energy will be greatest for loss energies defined approximately by $\Delta E_{loss} = h\omega_p$. Values of $\omega_p$ can be readily calculated for simple free electron gas models for various materials (Ibach and Lüth, 1990). They range from 3 to 20 eV and decrease with decreasing free electron density (in correspondence with the decrease in $\omega_p$). The increasing magnitude of $\omega_p$ is what causes ceramics (and polymers) to be transparent just above or in the infrared range (the lowest energy), semiconductors to become transparent near (lower energy) ultraviolet frequencies, and metals to be opaque to high frequencies (energies) of radiation (up to x-ray frequencies). Surface plasmon losses occur when the outgoing photoelectron interacts with surface plasmon states in the material. The corresponding energies are generally lower than those for the respective bulk plasmon loss.

The net effect of the (bulk or surface) plasmon-loss process is the creation of a new peak or peaks at higher apparent BE (lower $E_K$) than the main XPS peak. The loss peak

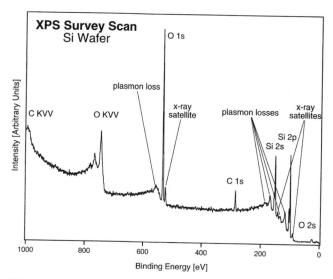

**Figure 4.** A survey scan from a Si wafer showing plasmon loss peaks that appear above the main Si peaks and O 1s peak. X-ray satellites are also labeled.

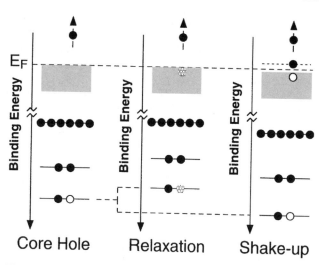

**Figure 5.** Schematics of the relaxation and shake-up processes in solids and their effects on the position of the core level created by the photoemission event.

will be closer to the primary photoelectron peak in ceramics than in metals. Because the plasmon-loss process is quantized, peaks can also appear at integer multiples of $\omega_p$. This will give rise to a series of peaks at equally spaced increments in BE above the primary XPS peak. The intensity of these multiple plasmon-loss peaks decreases with distance from the primary peak. An example is shown for Si in Figure 4. The Si 2p peak has multiple plasmon loss peaks. The position of a plasmon loss peak relative to a primary photoemission peak will depend on the type of material. The loss peak will be furthest away (toward higher BE) from the primary peak in materials with high free-electron densities, such as metals. It will be closest to the main peak in ceramics and polymers. Plasmon loss peaks are broader and generally lower in intensity than the primary XPS peak. They can be distinct in survey scans as well as in high-resolution spectra, as seen in Figure 4. As with inelastically scattered (background) electrons, bulk and surface plasmon loss peaks are not included in the calculations of the total number of electrons emitted from a given orbital for a given element in the sample. In other words, they are not considered in calculating atomic concentrations of elements.

The Auger excitation process is shown in the third panel of Figure 1. It is a means of removing the initial photoemission core hole. The Auger excitation process leaves the original atom in a doubly charged state. This energy reconfiguration will affect the measured BE of the outgoing photoelectron. The consequences are such that the energy gained by filling the initial core hole with an electron is subsequently lost by creation of the second core hole, for a balance that is nearly zero.

The relaxation process is a partial filling of the positive charge in the initial hole. This is represented in Figure 5 by a partial shading of the initial core hole after the photoemission event. The relaxation process can occur via a rehybridization of the atom, in which case it is an intraatomic process. This is illustrated in the figure by a removal

of partial charge from the valence band. Relaxation can also occur by extraatomic (also called interatomic) processes, in particular through a process equivalent to diffusion of the localized core hole throughout the valence band of the material.

Relaxation processes typically occur instantaneously in comparison to the lifetime of the core hole and always lead to a decrease in apparent BE for the outgoing photoelectron. This is shown in Figure 5 by the shift of the orbital positions toward the Fermi energy due to the relaxation process. A useful if somewhat naive analogy is to consider relaxation as leaving the atom less positively charged than immediately after the electron is ejected. The outgoing electron senses this as an increase in negative charge, and is correspondingly repelled to higher $E_K$. Higher $E_K$ translates to lower apparent BE in the XPS spectrum.

The magnitude of the decrease in BE due to relaxation can be on the order of several electron volts. It is distinctly apparent in comparison of gas-phase and solid-phase XPS spectra from compounds. Relaxation is also one reason why work functions of bulk materials are lower than the first ionization energies of the corresponding isolated (gas phase) atoms. Relaxation shifts in photoemission peaks also manifest themselves in XPS spectra from small particles versus bulk solids, although only at particle sizes approaching the scale of nanoclusters. This can occur with samples such as metal particles on heterogeneous catalysts or isolated clusters embedded in a different matrix material. Finally, the magnitude of the relaxation shift can be different for atoms in different chemical environments.

Relaxation occurs to the same extent for all atoms in a material to the extent that the atoms are in the same chemical and structural state. It therefore does not typically result in a new feature (peak) in an XPS spectrum. It may only be apparent in comparing of the position of a photoemission peak from one sample to another. As noted above, it is most prominent when comparing spectra from gas phase and bulk materials or from particles of different sizes or in different matrices. In materials having clusters of different sizes or particles in different chemical states, variations in the extent of relaxation due to these chemical or structural inhomogeneities can lead to relatively minor peak-broadening effects.

The fourth final-state effect is either the shake-up or shake-off process. This is shown in the third panel in Figure 5. The process involves rehybridization of the ion in such a way as to excite an upper-lying (valence band) electron. In the shake-up process, illustrated in Figure 5, the electronic excitation leads to promotion of a valence-band electron, for example, from the highest occupied molecular orbital (HOMO), to an unoccupied state above the Fermi energy (the lowest unoccupied molecular orbital or LUMO). In the shake-off process, the electron is promoted to a higher virtual state, although it does not necessarily leave the vicinity of the material (in which case, it could be considered an Auger electron).

Both shake-up and shake-off processes shift the BE of the initial hole to higher values. One view of why this is so is shown in Figure 5. The creation of a localized, occupied state above the initial $E_F$ of the material causes an upward shift of $E_F$. This is equivalent to shifts in $E_F$ that occur when semiconductors are doped. The upward shift in $E_F$ causes the increase in $E_B$ of all electrons in states below $E_F$. Other factors, such as changes in orbital energies during the rehybridization process, are also involved.

Unlike relaxation, which leads to an overall increase in the background of a spectrum, the shake-up or shake-off processes lead to the appearance of a new peak, called a satellite peak (not to be confused with satellite lines due to the x-ray source). The position of the satellite peak can be up to a few electron volts in BE above the primary photoelectron peak. In some cases, the intensity of the satellite peak can be comparable to that of the main peak. The satellites can also be turned off or on by the chemical state of the element. An example of this is seen in high-resolution spectra of the Cu $2p_{3/2}$ peaks for unoxidized and oxidized foils shown in Figure 6. Satellites are

only weakly visible above the main peak for the sputtered foil. The spectrum from the oxidized foil clearly shows two shake-off satellites at ~7 and 10 eV above the main Cu $2p_{3/2}$ peak.

Because the shake-up and shake-off processes occur after the photoemission process, the electrons that appear as satellite peaks in the XPS spectra should actually be considered as an integral part of the photoemission event. In particular, with reference to the equation $I_i^e \propto F\sigma n$, this means the electrons contributing to the satellite peaks should also be counted to determine the total number of orbitals, $n$, that produced the photoelectrons. In simpler terms, satellite peaks must generally be considered in all calculations of atomic concentrations. We will return to this point in considering data analysis of high-resolution spectra for atomic concentrations. Looking ahead, it suggests that acquisition of high-resolution spectra should always favor a wider range on the high BE side of a principal photoemission peak in order to cover potential changes in the appearance of satellites that may be present.

The concluding final state effect for consideration is multiplet splitting. Multiplet splitting occurs when the hole created by the photoemission process leaves behind an unpaired electron. The basis for the process is similar to that leading to the spin-orbit splitting of $p$, $d$, and $f$ levels discussed in conjunction with survey scans (see discussion of The XPS Survey Spectrum). Multiplet spitting is especially important for photoemission from filled $s$ levels, since photoemission always leaves behind an unpaired electron. The unpaired electron can couple with the nuclear and remaining electron spins as either a spin-up or a spin-down state. These two couplings will have different energies, and thereby shift the position of the orbital. The magnitude of the splitting is on the order of a few electron volts. It can be turned off and on by the chemical state of the atom, as with satellites. The width of the multiplet splitting will be largest for valence $s$ levels, and it decreases in magnitude with increasing BE.

**Initial-State Effects.** Initial-state effects influence the BE of the orbital before the photoemission process occurs. Three types of initial state effects are recognized: chemical state shifts, inhomogeneous matrix broadening, and charging shifts. The first two are considered below. Charging is an artifact that will be dealt with in the section on problems associated with XPS (see Problems).

Chemical state information is typically what is sought from a high-resolution XPS scan. The spectra in Figure 6 show that we can clearly distinguish between the chemical states of Cu and CuO, for example. The shapes of the Cu $2p$ peaks and their positions are distinctly different for Cu in the neutral (0) and charged (+2) oxidation states. The enhancement of the shake-up satellite adds a distinct signature to the peak shape. Differentiation of chemical state information is unfortunately not this straightforward for all elements, as the background information on final state effects already suggests.

The chemical state of an atom depends on the type of bond it forms: metallic, covalent, or ionic. Dispersive or hydrogen bonding also plays a role, as does the

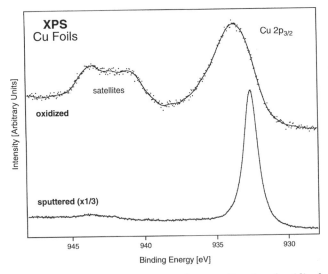

**Figure 6.** High-resolution spectra from sputtered and oxidized Cu foils showing how the intensity of the Cu $2p_{3/2}$ satellite peaks can be influenced by chemical state.

hybridization in covalently bonded systems. These factors affect the partial charge on the atom. In metals, atoms are expected to be neutral (have an oxidation state of zero). In covalent materials, atoms can have partial charges due to electronegativity differences as well as differences in hybridization. At extremes of electronegativity difference, ionic bonding leads to consideration of the formal oxidation state (charge) of the atom.

Although bonding occurs almost exclusively through valence electrons, the $E_B$ of core levels are also affected, as seen clearly in Figure 6, where Cu $2p$ peaks are shown for $Cu^0$ and $Cu^{2+}$ states. This can be simply illustrated based on the picture of the electrons in orbitals as charges contained on hollow, conductive shells. The core electrons are contained spatially on shells within the outer valence shells. The work needed to remove the inner shell (core) electrons is affected proportionally by changes in the potentials on the valence shell. In atoms that acquire a partial negative valence charge, the core levels will shift to lower $E_B$ because less work is needed. The core levels in atoms that become more positively charged will correspondingly shift to higher $E_B$. The analogy presented to explain peak shifts due to relaxation effects is also useful to consider here. An outgoing photoelectron will sense an increase in negative valence charge on an atom as it leaves the atom. This causes the outgoing photoelectron to gain KE in comparison to one that has been ejected from a neutral atom. This leads to a decrease in the measured BE of a photoelectron from a negatively charged ion relative to one from the same orbital in a neutral atom. The opposite shift (toward higher BE values) occurs for electrons leaving ions of greater positive charge.

The chemical shift due to a change in valence charge can be expressed as

$$\Delta E_B = k\Delta\left(\frac{q}{r}\right) + \sum_{j \neq i} k \frac{\Delta q_i}{r_{ij}} \tag{1}$$

where $\Delta E_B$ is the chemical state shift, $k\Delta q$ is the change in valence charge on the atom, and $r$ is the radius of the valence orbital. An increase in negative charge is considered as a positive value of $\Delta q$. The summation accounts for interaction with charges from surrounding atoms. This model considers the core level as being confined within a uniformly charged shell of radius $r$. The radius of the valence orbital will also change as the valence charge changes. Neglecting this effect for small changes in charge, we can replace the valence radius by its expectation (average) value, $\langle r_2 \rangle$. We can also neglect the second term when $\Delta q$ is small, since the $r_{ij}$ values will be larger than $\langle r_2 \rangle$. This leads to

$$\Delta E_B \approx k\left(\frac{\Delta q}{\langle r_2 \rangle}\right) \tag{2}$$

A consequence of this is that for the same amount of valence charge shift the chemical state shift in photoelectron $E_B$ will be larger for atoms of smaller radius. Alternatively, less valence charge perturbation is needed to give the same shift in $E_B$ in smaller atoms. We expect therefore

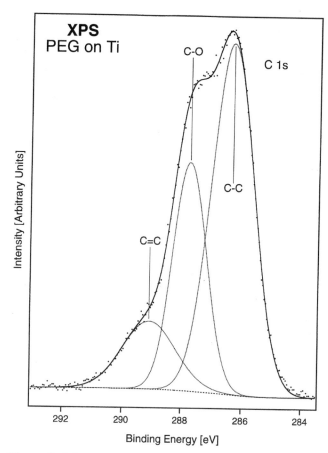

**Figure 7.** A C $1s$ peak from poly(ethylene glycol) (PEG) adsorbed on a Ti surface. The peak has been resolved into components for the various hybridization states.

to be better able to distinguish subtle changes in chemical state for light elements (with smaller atomic radii) as opposed to heavy elements (with larger atomic radii). This has practical applications in analysis of polymer materials. Although changes in valence charge due to rehybridization can be relatively small, shifts in the peak positions of core levels for carbon, oxygen, and nitrogen often readily observable for different hybridizations. An example of this is shown in Figure 7. After deconvolving the various overlapping peaks, we can clearly see contributions from carbon in aliphatic, ester, and carbonyl hybridization states in the C $1s$ peak from the polymer material.

We may want to compare shifts in $E_B$ values for an element in two different materials. In this case, Equation 1 becomes

$$E_{B_2} - E_{B_1} = k\left(\frac{\Delta q_2}{\langle r_2 \rangle} - \frac{\Delta q_1}{\langle r_1 \rangle}\right) + (V_2 - V_1) \tag{3}$$

where $E_{B_i}$ is the binding energy of a given orbital in chemical state $i$. The first difference term accounts for differences in valence charge between the two atoms, and the $V_i$ terms are abbreviations of the summation term in Equation 1. They essentially account for differences in Madelung potentials for the atom in the two different states.

**Tabel 1. Peak Shifts Measured Going from a Pure Metal to an Oxide**[a]

| Compounds | Orbital | $\Delta$BE (eV) | $R$ (nm) |
|---|---|---|---|
| Fe/FeO | Fe $2p_{3/2}$ | 2.9 | 0.126 |
| Cu/CuO | Cu $2p_{3/2}$ | 1.8 | 0.128 |
| Os/OsO$_2$ | Os $4f_{7/2}$ | 1.3 | 0.135 |
| Pt/PtO | Pt $4f_{7/2}$ | 3.2 | 0.139 |
| Pt/Pt(OH)$_2$ | Pt $4f_{7/2}$ | 1.6 | 0.139 |

[a] The corresponding atomic radius of the metal atom is included.

Madelung potentials at a localized atom will change if the matrix surrounding the atom contains ionic or charged constituents and the crystalline symmetry of the matrix changes as viewed from the atom.

The effects of both valence potential and Madelung potential terms may cancel each other. This can lead to counterintuitive peak shifts based solely on considerations of chemical bonding. An example is found by comparison of the position for the Cu $2p_{3/2}$ peak in Cu (932.7 eV), Cu$_2$O (932.4 eV), and CuO (933.8 eV). The Cu $2p$ peak shifts first to slightly <u>lower</u> BE values going from Cu to Cu$_2$O, then shifts to higher BE values with the bonding of one more oxygen atom. The initial reversal in $E_B$ of the peak shift going from Cu to Cu$_2$O may be due to differences in the Madelung potential term in Equation 3 or to differences in relaxation effects for the metal and metal oxide matrices.

Further examples of peak shifts for elements in different chemical states are given in Table 1 (Krause and Ferreira, 1975). The shift in BE going through the sequence Fe/FeO$\rightarrow$Cu/CuO$\rightarrow$Os/OsO is increasing with decreasing atomic (or ionic) size (Fe$\rightarrow$Cu$\rightarrow$Os), suggesting an overriding role due to differences in the valence charges. By comparison, in the Fe/FeO$\rightarrow$Pt/PtO sequence, BE increases with decreasing atomic size (Fe$\rightarrow$Pt), suggesting a greater role for the Madelung terms. Finally, the BE values in the Fe/FeO$\rightarrow$Pt/Pt(OH)$_2$ sequence are again more in line with a trend of decreasing shift with increasing atomic size (Fe$\rightarrow$Pt).

The second prominent initial-state effect, inhomogeneous matrix broadening, arises because not all atoms are in exactly equivalent bonding configurations in a material. The BE of an orbital will differ from atom to atom because of localized differences either in composition or structure surrounding each atom. This causes variations in the chemical potential or Madelung potential at each site. A specific orbital will therefore have a value of $E_B$ specific to its atom.

In practice, the shifts in $E_B$ of atomic orbitals due to micro-scale inhomogeneities in otherwise homogeneous materials are well below the detection limits of standard XPS instruments. Since the number of micro-inhomogeneities in a material is expected to be distributed as a Gaussian about a mean type, we expect to see peak broadening in the XPS high-resolution spectra rather than peak shifts. An example of this might be expected in comparisons of XPS spectra from a material with a glassy structure versus a spectrum from the same material in its single-crystal or polycrystalline state.

## PRACTICAL ASPECTS OF THE METHOD

The practical aspects of XPS will be covered by first considering the instrumentation, including the source, analyzer, and detector. We will then consider unique aspects of the x-ray interaction with the sample for applications to obtain quantitative composition, chemical state, and film thickness information. Limitations of the method with regard to determining these parameters are discussed below (see Problems). The final part of this section covers important practical aspects related to the use of XPS for analysis of the four basic types of materials: metals, semiconductors, ceramics, and polymers.

### Instrumentation

Facilities for XPS analysis consist primarily of an x-ray source, electron analyzer, and electron detector. This is all housed in an ultrahigh vacuum (UHV) chamber with nominal operating pressures below $\sim 10^{-8}$ mbar. One reason for UHV conditions is that mean free paths of the photoelectrons are inversely proportional to gas pressure. At room temperature and atmospheric pressure, the mean free path of an electron in air is only hundreds of nanometers. It is tens of kilometers at $10^{-8}$ mbar. Significant gas ionization can also occur for electrons traveling in gases above $\sim 10^{-4}$ mbar pressure. Another reason UHV conditions are often important during XPS analysis is to keep the sample surface clean. The impact rate of gas molecules with a surface is also inversely proportional to gas pressure. A common method of estimating coverage is to consider the general rule of thumb that 1 cm$^2$ of an initially pristine surface can be covered in 1 sec by a monolayer of absorbed gas when put under a pressure of $10^{-6}$ mbar of gas. Note that this rule of thumb is obtained from the historically important unit of dose called the Langmuir, defined using the pressure unit Torr as 1 liter $= 10^{-6}$ Torr sec/cm$^2$, where 1 Torr $= 1.33$ mbar. Suppose that analysis of 1 cm$^2$ of surface is to take 1 hr. To avoid having the analysis area completely covered by a monolayer of absorbed gas in this time requires that we work at gas pressures below $3 - 4 \times 10^{-10}$ mbar. Although typical survey scans usually take only a few minutes with modern instrumentation, high-resolution spectra can require an hour or more in some cases, and XPS can be sensitive enough to detect the growth of the monolayer in this time. The final reason for using UHV conditions is that laboratory x-ray sources will be damaged at higher pressures, especially when the background gas contains significant concentrations of water vapor or oxygen (see GENERAL VACCUM TECHNIQUES for details on achieving and maintaining UHV conditions).

**Sources.** Laboratory x-ray sources consist of an electron filament in close proximity to a metal surface. The filament is typically a tungsten wire that is heated by passing a current through it. A high voltage, on the order of 10 to 20 kV, is applied between the filament and the metal to accelerate electrons from the filament to the metal. The current that flows between the filament and the target metal (the anode) is the emission current. Electrons strike the metal

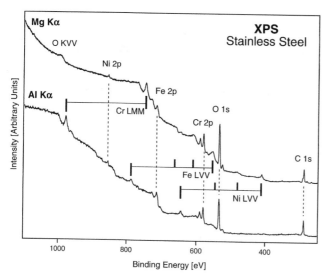

**Figure 8.** Spectra measured on a stainless steel alloy using either Mg Kα or Al Kα radiation. The shifts of the Auger lines are noted, while the photoemission peaks are seen to stay in one place on the binding energy scale. The background energy between the two spectra is also different at high binding energy due to the onset of the secondary loss peak.

anode and, in deaccelerating, produce x rays. Two types of x rays are produced, Bremsstrahlung ("braking radiation") and emission lines. Bremsstrahlung is a broad background underlying the emission lines and is independent of the choice of metal. Because it is low in intensity and broad in energy, Bremsstrahlung has no utility for XPS. The frequency (energy) of the principal emission line from a laboratory x-ray source is characteristic of the anode metal. The two most frequently used laboratory anodes in XPS are Al and Mg, with primary Kα emission lines at 1486.6 eV and 1253.6 eV, respectively. Other metals are sometimes used to have lower or higher excitation energies. Many laboratory XPS systems have a dual-anode arrangement that permits selection of one or another type of anode metal. The advantages of this arrangement are shown in Figure 8.

The peaks from AES electrons will always appear at constant $E_K$ regardless of the excitation source. Their peak positions will therefore shift on the BE scale as a function of excitation source energy. This is clearly seen in Figure 8, where the AES peaks have shifted by 233 eV to higher BE when Al Kα is used as the excitation source rather than Mg Kα. The true photoemission peaks remain at the same position in BE. In cases where Auger peaks overlap with principal photoemission peaks, having a second x-ray source of a different excitation energy can sometimes help resolve any ambiguities in peak assignments.

The background intensity also increases more sharply at higher BE values for the XPS spectrum taken with an Mg x-ray source compared to that using an Al source. This is due to the onset of the secondary electron peak. For the Mg source, the increase in secondary electron background should start at ∼1150 eV (1253 to 100), whereas for the Al source, it should start at ∼1380 eV. This means that an Al source can be used to probe core

levels at higher values of BE than a Mg source can. This is the same principle that limits UPS to valence band spectra.

Dual x-ray sources (Mg and Al) are typically used in conjunction with one another to shift Auger peaks out of the way of primary photoemission peaks. X-ray anodes with lower excitation energy, such as Ti at 395.3 eV, are also useful. The cross-section for photoemission of electrons from low $E_B$ orbitals is sometimes greater at lower excitation energy, leading to greater sensitivity. In other words, a lower amount (concentration) of an element is needed in a material to give a specific signal intensity in an XPS spectrum when its cross-section is higher. The tradeoff to this is, as the excitation energy of the x-ray source decreases, fewer of the higher atomic number elements can be detected.

Principal emission lines from laboratory x-ray anodes are not monochromatic. A lineshape analysis shows they have a characteristic spreading about the principal emission energy. The x-ray half-width of Al anodes is ∼0.85 eV and of Mg is ∼0.7 eV, for example. This leads to further broadening of the photoemission peaks in the XPS spectra beyond lifetime broadening effects. The total peak broadening due to lifetime and x-ray source effects will be $\Delta E_T^2 = \Delta E_1^2 + \Delta E_s^2$, where $\Delta E_s$ is the source broadening. For a state with an intrinsic width of 0.10 eV, its XPS peak measured with MgKα will have a width of 0.71 eV. Laboratory x-ray sources also have x-ray satellite emission lines, typically at higher energy from the main emission line. Both Al and Mg have a number of such secondary emission lines (Krause and Ferreira, 1975). Their intensities are generally ∼10% or lower than that from the primary x-ray emission line from the source. The secondary emission lines from an x-ray source lead to additional photoemission peaks at lower BE in the XPS spectra. These peaks are called x-ray satellites. For example, Kβ lines appear in the XPS spectra in Figure 4 at BE values ∼7 eV lower than each of the principal peaks. X-ray satellite peaks should not be confused with the satellite peaks that are characteristic of final state shake-up and shake-off processes. The former appear at lower BE values than the principal photoemission peak and the latter at higher BE values.

In operating laboratory x-ray sources, one characteristically defines any two of the parameters of applied voltage ($V$), filament emission current ($I_F$), or x-ray power ($P$) to define the operating characteristics of the source. They are related as $P = I_F V$. For a given x-ray source geometry, setting any two of these parameters defines the flux of x rays to the sample.

Most XPS systems in use today are equipped with an x-ray monochromator between the x-ray source and the sample. A typical x-ray monochromator is essentially an array of x-ray diffracting elements (quartz crystal strips for example) positioned to diffract the principal emission line from a conventional x-ray source onto the sample. Systems using a monochromator have a number of distinct advantages over conventional laboratory x-ray sources. First, the monochromator removes the x-ray satellite lines and significantly reduces Bremsstrahlung background radiation, leading to cleaner spectra, especially in survey scan mode. Secondly, an x-ray monochromator reduces the broadening of the x-ray source line significantly, making the resolution of overlapping features in high-resolution XPS spectra an easier task. Finally, a monochromator reduces the potential for sample damage, in one case because of lower overall x-ray flux and in another because inadvertent bombardment by electrons through damaged x-ray windows is no longer possible. Although use of a monochromator decreases the net x-ray flux to the sample, the output flux of modern x-ray sources and the sensitivity of electron detectors have improved to the point where this loss is overridden by the improvement in signal quality that is obtained. For these reasons, monochromatic x-ray sources are to be preferred for XPS analysis in all cases where a choice is offered.

Synchrotron radiation is also a viable source of x-ray radiation. Because the use of synchrotron radiation involves a movable monochromator between the beam line and the analysis system, the advantages are its tunability through a range of x-ray energies, high flux, narrow linewidth, low background intensity, and absence of satellites. The disadvantages are that one must schedule time on a synchrotron facility and be adept at the use of such facilities. Collaboration with research groups doing comparable studies is encouraged at the outset. The tunability of the x-ray energy from synchrotron sources is often exploited for x-ray absorption fine structure (XAFS; see XAFS SPECTROSCOPY) or near-edge XAFS (NEXAFS) studies. Both techniques are related to XPS. They look at the intensity of photoemitted electrons from a sample at a well-defined BE as a function of the energy of the incident x-ray. Further information is provided in XAFS SPECTROSCOPY.

**Analyzers.** Two types of analyzers are widely used for XPS analysis, the cylindrical mirror analyzer (CMA) and the hemispherical sector analyzer (HSA), shown in the schematic in Figure 9. Historically, the HSA pre-dates the CMA in design (Palmberg et al., 1969; Siegbahn et al., 1969a,b; Palmberg, 1975). The CMA is popular for AES systems, and further details of its configuration and operation can be found elsewhere (Palmberg et al., 1969; Palmberg, 1975). The HSA has become the norm for XPS

analysis and will be considered exclusively in the following discussion.

The HSA is recognized on XPS systems by its distinctive clam-shell appearance. This hemispherical portion of the analyzer serves as an electron monochromator. It is typically preceded in the electron path by a series of electrostatic lenses and is followed by an electron detector.

Photoelectrons leaving the sample are collected by the lens system. The frontmost lens is at sample potential (ground) to provide a field-free region between the analyzer and sample. The lenses focus the electrons spatially to the front aperture of the HSA and may provide image magnification. In principle, an image of the aperture is focused by the lenses onto the sample, and the bounds of this image define the analysis area. Retarding fields in front and back of the lenses serve as energy selectors. These retarding fields are swept in potential (voltage) to determine the cutoff of electron $E_K$ values that reach the analyzer. In doing this, the lenses retard the electrons in KE; they do not perform any further energy discrimination. Electrons reach the analyzer with a full range of $E_K$ values above that selected by the sweep energy on the lenses.

The function of the HSA is to select electrons of a particular $E_K$ to pass to the detector. The optimal $E_K$ for an electron to pass through an HSA (or CMA) is called the pass energy, $E_p$. It is defined for an HSA of inner and outer radii $R_i$ and $R_o$, respectively, according to

$$E_p = K_s \Delta V$$
$$K_s = \left( \frac{R_o}{R_i} - \frac{R_i}{R_o} \right)^{-1} \qquad (4)$$

where $K_s$ is a spectrometer constant and $\Delta V$ is the potential difference between the inner and outer hemispheres (outer minus inner potential). For a given set of voltages (potentials) on the inner and outer shells of an ideal HSA of a specific size, only electrons at $E_p$ in KE will pass through to the detector. Those electrons with KE higher than $E_p$ are lost to the outer hemisphere of the HSA, and those that travel too slowly are lost to the inner hemisphere.

All hemispherical sector analyzers allow electrons with $E_K$ values that are not exactly at $E_p$ to reach the detector. This is determined solely by the geometry of the analyzer. The range of $E_K$ values that are passed, $E_p \pm \Delta E_A$, describes the absolute energy resolution, $\Delta E_A$, of the HSA. The primary geometric parameter of concern in establishing $\Delta E_A$ is the acceptance angle, $\alpha$, of electrons at the entrance slit of the HSA. Electrons with $E_K$ equal to $E_p$ that enter the HSA normal to its entrance slit ($\alpha = 0$) will always be transmitted; this defines the ideal case. To increase the number of electrons that can be transmitted through the HSA, we can also allow it to transmit electrons with $E_K$ equal to $E_p$ that enter at an angle $\alpha$ from the normal to the entrance slit. This is determined in practice by specifying the lens and aperture designs according to electromagnetic principles, which typically requires expansion of the governing equations for electron travel in a Taylor series about $\alpha$. The optimal design for an HSA is to be focusing to second order in $\alpha$.

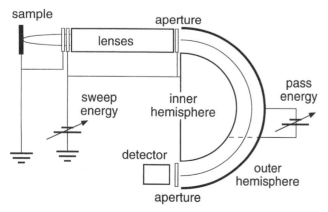

**Figure 9.** A schematic representation of the key parts of a hemispherical sector analyzer (HSA).

When this second-order focusing constraint is applied, a limit on $\Delta E_A$ is obtained. It is found to be a function of the entrance slit width $W$ and geometric radius of the optimal path through the analyzer $R_p$ in the form

$$\frac{\Delta E_A}{E_p} \propto \sqrt{W/R_p} \qquad (5)$$

where $\Delta E_A/E_p$ is the relative resolution. An optimally designed HSA is therefore not perfectly monochromatic; it introduces peak broadening of the order of $\Delta E_A$. The total peak broadening in an experimental XPS spectrum due to lifetime, source, and analyzer effects is therefore $\Delta E_T^2 = \Delta E_l^2 + \Delta E_s^2 + \Delta E_A^2$. Analyzers with larger values of $R_p$ or smaller values of $E_p$ will give smaller values of $\Delta E_A$, meaning better absolute energy resolution. In practice, typical XPS systems with an HSA are designed with $\Delta E_A/E_p \sim 0.1$. Specialty XPS instruments exist with far better relative resolutions, in which case the diameters of the hemispheres are significantly larger.

The transmission function of the HSA is an important factor to consider for quantitative analysis with XPS (Nöller et al., 1974). It defines how well the lenses and analyzer transmit an electron that leaves the sample at a given $E_K$. Three cases have been recognized: (a) where the lenses define the transmission, (b) where the lenses define the transmission only along one plane perpendicular to the sample along the lens axis, and (c) where the analyzer defines the transmission. In general, for XPS analysis using an HSA, the HSA defines the transmission except at low $E_K$ (high $E_B$).

The HSA can typically be operated in one of two modes. In one, the analyzer transmission function is held constant. The transmission function of the HSA varies as $1/E_K$ and as $1/\sqrt{E_p}$. In fixed analyzer transmission mode, the value of $E_p$ is varied to give a constant transmission function over the entire sweep range of KE in the spectrum. The pass energy increases with increasing KE in the spectrum. Because $\Delta E_A/E_p$ depends only on the analyzer geometry, the absolute resolution $\Delta E_A$ correspondingly varies with KE in the scan. In practice, this leads to broader peaks at low BE (high KE, where $\Delta E_A$ is large) compared to high BE. Constant transmission function mode can be used for AES analysis with an HSA. Peak widths in AES are much broader than those in XPS, and the variation in $\Delta E_A$ as a function of $E_p$ is correspondingly not as important. In addition, having a low transmission function at low KE in the scan significantly reduces the overall size of the secondary background signal (below $\sim 50$ eV in KE). While constant analyzer transmission mode may have some utility for XPS survey scans, it is to be avoided for high-resolution XPS scans and in all cases where accurate elemental compositional information is desired.

The more common mode of operation of an HSA for XPS analysis is to hold $E_p$ constant while the input sweep energy is varied. In practice, this is done by retarding the electrons through the lenses so that electrons leaving the sample at the desired EK arrive at the entrance slit to the analyzer with EK equal to $E_p$. Because $E_p$ is constant throughout a scan, $\Delta E_A$ is constant over the entire BE range of the XPS sweep. Constant pass energy mode is the one to use for high-resolution scans where constant absolute energy resolution is essential in order to obtain reliable chemical state information. In addition, for automated analysis, elemental compositions must be determined from scans taken with the XPS instrument in constant pass energy mode.

In constant pass energy mode, the transmission function of an HSA typically varies approximately as $1/E_K$, which is the KE of the electrons at the entrance to the lens. This means that even if the flux of electrons emitted from a sample is constant over the entire KE range, the apparent intensity of electrons measured with an HSA in constant $E_p$ mode will decrease with decreasing BE. In experimental spectra, peaks at high BE will be accentuated by the HSA relative to those at low BE. This effect will be considered again in later discussions regarding quantitative analysis using XPS, especially for ARXPS.

At a given value of KE in a scan, the transmission also varies as $1/\sqrt{E_p}$, so that lower values of $E_p$ will give better absolute energy resolution and higher analyzer transmission. This latter point can lead to confusion because, in practice, when $E_p$ is decreased, absolute resolution ($\Delta E_A$) increases but the measured signal intensity decreases (suggesting that transmission function decreases). The resolution of this apparent contradiction is to realize that all electrons that exit the HSA are counted as part of the signal at the given KE in the scan, regardless of what their $E_K$ actually was. In other words, the intensity that is measured at the detector as belonging to a signal at $E_K$, $I^m(E_K)$, is actually the integrated signal between $E_K \pm \Delta E_A$

$$I^m(E_K) \propto \int_{E_K - E_A}^{E_K + E_A} J^e(E) dE \qquad (6)$$

As analyzer pass energy is decreased, the relative fraction of electrons that are transmitted by the analyzer (the transmission function) increases (in proportion to $\sqrt{E_p}$). However, the integration range in the above equation decreases (in proportion to $E_p$). Because the latter effect dominates, measured intensity decreases.

Survey scans are typically taken at a higher pass energy where the signal intensity $I^m(E_K)$ is large but the absolute energy resolution $\Delta E_A$ is poor. The objective is to obtain a high signal-to-noise ratio in a relatively short time; the objective is not primarily to resolve overlapping peaks. This was the case for the experimental spectrum in Figure 2. High-resolution scans are generally run at lower $E_p$ to obtain peak resolutions approaching the inherent lifetime and source broadening limit. This means that high-resolution scans typically require longer acquisition times to get comparable signal-to-noise ratios.

Hemispherical analyzers for XPS analysis are more frequently being designed to be spatially focusing. This is because x-ray flux cannot be easily focused, and we often want to analyze only specific spatial regions on a sample. One way to achieve spatial focusing with an HSA is to insert mechanical apertures with various diameter entrance slits into the electron path through the lens, typically at a point just prior to the analyzer. The image of this aperture is focused back to the sample, and smaller apertures select smaller spots on the sample. While this method has, in principle, no lower limit in object size at the sample, it is limited by the sensitivity of the detector systems. The nominal spot sizes that can be obtained using such mechanical aperture systems are on the order of millimeters, with hundreds of microns being possible only with extremely long (tens of hours) signal acquisition time.

Small-spot, or imaging XPS refers to systems that have electrostatic or electromagnetic lenses that spatially focus the analyzer entrance aperture to a defined spot at the sample (Seah and Smith, 1988; Drummond et al., 1991). No mechanical apertures are used, and the (mechanical) entrance slit width is always a constant size. Small-spot systems in use today can obtain spatial resolutions at the sample on the order of a hundred microns routinely, with acquisition times well within an hour for high-resolution scans. Lower limits in analysis spot size at the sample are currently ~25 μm.

Another advance in analyzer and electrostatic lens design has led to XPS systems that can scan the analysis spot over the sample while acquiring data. Scanning XPS systems are typically coupled to small-spot analysis capabilities. This allows acquisition of chemical state maps over a sample. In this case, the intensity of photoelectrons at specific BE values is plotted versus spatial position. Within the spatial resolution limit of the instrument, scanning XPS can be used to determine such information as the uniformity of coatings on surfaces.

**Detectors.** Three types of electron detectors are common on XPS systems: electron multipliers, channeltrons, and channel plate arrays. The electron multiplier functions in a manner similar to an amplifier tube, often using multiple amplifier stages connected in series. The channeltron is a smaller, single-component system. The channel plate array is a circular, wafer-shaped detector with a multitude of narrow channels that each function, in a sense, as a separate channeltron. All three detectors have gain factors on the order of $10^3$ to $10^6$, meaning that one electron in will produce $10^3$ to $10^6$ electrons out. All of the detectors run at high voltage, often as much as 5 kV. They all require secondary amplification stages external to the detector to boost the output signal further. The external amplifiers are typically connected as closely as possible to the output stage of the detector to reduce concurrent amplification of circuit noise.

An increasingly common arrangement in XPS systems is to use multiple electron detectors arranged for position-sensitive detection. This makes use of the fact that electrons leave the exit aperture of the analyzer at different angles depending on their $E_K$. Multiple detectors can be configured to collect and process these electrons simultaneously, thereby increasing the signal-to-noise ratios obtained for comparable acquisition times with single-detector schemes.

For the electron multiplier or channeltron, position sensitivity is realized by positioning multiple detectors at key positions clustered around the exit aperture. At any

instant, as one detector positioned off-center from the aperture is collecting electrons at $E_p + \Delta E_A$, the on-axis detector(s) collect electrons at $E_p$, and an opposing detector to the first collects electrons at $E_p - \Delta E_A$. With channel plate arrays, the entire channel plate detector can be hard-wired to become a position-sensitive detector. The positions of electrons arriving on diametrically opposing sides of the circular array are typically sensed and recorded by a four-element resistive network attached in a four-fold symmetrical pattern to the outer edge of the channel plate.

**Electron Detection.** The final consideration with regard to instrumentation in XPS is how the $E_K$ of the photoelectron is sensed. This is not the same as how the analyzer selects for a particular KE of incoming photoelectrons. The latter is done by applying a retarding field to the lens system, as discussed above. Photoelectrons are counted when they successfully travel through the lens and analyzer systems because they have the appropriate $E_K$. However, photoelectrons have a different KE as seen from the reference frame of the sample versus the reference frame of the analyzer. The reason for this is shown schematically in Figure 10.

Photoelectrons leave the sample with a $E_K$ determined by their initial $E_B$, the excitation energy $h\nu$, and the work function of the sample, $\phi_s$. The work function of the sample is the difference between the Fermi energy of the sample $E_F$ and the vacuum level. The vacuum level at the sample is where the photoelectrons have zero $E_K$ <u>as seen by the sample</u>.

When the photoelectrons reach the detector, they will have traveled through an analyzer with an overall work function of $\phi_A$. The analyzer work function is a constant that is independent of all intervening potentials (voltages) between the sample and the detector. A constraint on a properly configured electron analysis

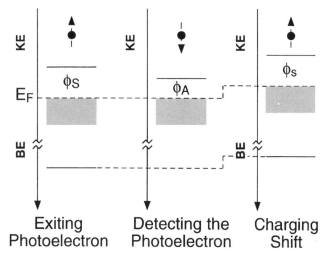

**Figure 10.** A schematic representation of the translation of electron kinetic energy (KE) to binding energy (BE), with consideration of the work functions of the sample and analyzer. How sample charging affects the considerations is represented by the panel on the far right.

instrument is that the Fermi energy of the sample and the system are equal, and electrons are "detected" (counted) when they reach the Fermi energy of the system. Because the work functions of the sample and analyzer are generally different, the electron will appear to have either a greater KE (as shown in Fig. 10) or lower KE from the perspective of the analyzer. The perspective of the analyzer, not the sample, is the one that counts in sensing the $E_K$ of the electron as it passes through the analyzer.

Combining all of the above considerations, we can write a series of equations that lead to the formal relationship between the $E_B$ of the electron in the sample and the $E_K$ of the electron as it is sensed by the analyzer, where the S and A subscripts denote sample and analyzer, respectively.

$$E_{KS} = h\nu - (E_B + \phi_s)$$
$$E_{KA} = E_{KS} - (\phi_A - \phi_S)$$
$$E_{KA} = h\nu - E_B - \phi_A$$
$$E_B = h\nu - E_{KA} - \phi_A$$

The final equation above states that we must know the work function of the analyzer to know the absolute values of $E_B$ for electrons in samples. In practice, the analyzer work function is a calibration constant for the instrument. It is determined by aligning peak positions from a sample with those of a reference. The typical calibration sample is Au, with the position of the Au $4f_{7/2}$ peaks being set at 84.0 eV. The issue of instrument calibration will be addressed more thoroughly in the section on problems with the technique.

As an example of the above equations in practice, consider a C $1s$ electron with $E_B = 284.6$ eV excited by Mg K$\alpha$ radiation (1253.6 eV) from a sample with a work function of 5.0 eV and measured through an analyzer with a work function of 4.0 eV. The electron will appear to have $E_{KS} = 1253.6 - (284.6 + 5.0) = 964.0$ eV from the sample. It will appear to have $E_{KA}$ 965.0 eV to the analyzer and will be recorded at this value if KE is used as the scale on the XPS spectrum. In a properly calibrated spectrometer, the $E_B$ of this electron will be determined to be $1253.6 - 965.0 - 4.0 = 284.6$ eV.

**Maintenance.** The importance of maintaining a well-calibrated instrument for any analytical technique cannot be overemphasized. The sections that follow will highlight how instrument parameters affect quantitative results in XPS analysis. Both compositional (peak areas) and chemical (peak positions) information can be unknowingly distorted by improper or inattentive maintenance and calibration of an XPS instrument. A number of articles are available that specifically discuss calibration procedures in greater detail. In particular, the series of articles by Seah and others is highly recommended for consideration of instrument parameters that affect determination of compositional information with XPS (Anthony and Seah, 1984a,b; Seah et al., 1984; Seah, 1993). A corresponding set from Anthony and Seah is recommended for discussion of calibration of energy scales (which affect chemical information; Anthony and Seah, 1984a,b). A summary article

pertaining to both AES and XPS was also published (Seah, 1985). Those who are involved in daily operation of XPS systems are strongly encouraged to apply the procedures set forth in these articles as part of their instrument maintenance and calibration routines. Laboratory experiments can also be designed around these articles as a method of introducing graduate students to the rigor needed for accurate and reproducible quantitative analysis of peak parameters with XPS.

### Applications

In analysis of solid materials, the two primary uses of XPS are to determine elemental compositions and to characterize the chemical states of elements. Other applications include determining film thickness or composition profiles at the surface of a material, either through nondestructive or destructive means, and characterizing molecular orientation for adsorbates on surfaces.

The discussion below considers the practical factors that are involved in determining compositions and chemistry using XPS. The use of XPS for determining film thickness and molecular orientation in absorbed layers is also outlined. Finally, a brief review of materials-specific considerations is presented.

**Elemental Compositions.** The basis behind the use of XPS to determine elemental compositions is provided above (see Principles of the Method). A relatively more rigorous development is given in what follows, using what has been learned from both the principles of the technique and practical aspects of the instrumentation. Those interested in further details and alternate derivations are referred to the reference list (see Key References) and to Jablonski and Powell (1993). The formulation below starts from the perspective of an ideal analyzer viewing an ideal sample at an angle that is fixed relative to the surface plane. Nonideal factors introduced by practical analyzers will then be introduced. Considerations of nonideal samples will be covered at the end.

The flux of electrons emitted from a given orbital for a given element, i, in a sample, $J_i^e$ (electrons/m$^2$-sec), for a given x-ray flux F (photons/m$^2$-sec) can be written as

$$J_i^e = F\sigma n_i \tag{7}$$

where $\sigma$ (in units of m$^2$) is the cross-section and $n_i$ (in units of 1/m$^2$) is the number of orbitals per unit area over which the x-ray beam impinges. We can recast this expression in terms of elemental (atomic) concentrations in a bulk solid by accounting for the volume sampled. For the electron KE values of interest in XPS, x rays penetrate much further into a sample (on the order of microns) than the electrons can escape. Therefore, the total flux of electrons from a bulk sample is limited by the probability of the electron escaping. As discussed above (see Principles of the Method), inelastic scattering and plasmon loss processes are the two main factors that remove photoelectrons from consideration as part of a primary photoelectron peak, where the former dominates. In analogy with scattering processes in ideal gases, the escape probability can be represented by a mean free path, $\lambda_i$ (in units of m), for the electrons to travel before scattering. The escape probability decays exponentially with distance traveled relative to $\lambda_i$. Therefore

$$J_i^e = F\sigma \int_0^\infty C_i(z)\, \exp(-z/\lambda_i)dz \qquad (8)$$

Assuming a homogeneous composition $C_i$ (in units of atoms/m$^3$) of element i throughout the material, this equation becomes

$$J_i^e = F\sigma\lambda_i C_i \qquad (9)$$

This represents the total photoemission flux from orbital $i$ in a homogeneous sample.

A wide range of references are available on the subject of electron mean free paths in materials (Tanuma et al., 1988, 1991a,b, 1996; Marcu et al., 1993). Electron mean free paths increase sharply with decreasing KE at low $E_K$ because the quantum-mechanical wavelength of the electron increases, and this reduces the inelastic scattering efficiency. They increase as $\sqrt{E_K}$ at high KE because the electrons essentially travel further between scattering events. The minimum mean free path is ~0.2 to 0.3 nm in the range of 30 to 100 eV. Mean free paths are also dependent on the type of material, its compositional homogeneity, crystal structure, and microstructure. Variations in electron mean free paths by factors of 2 to 3 are not uncommon from material to material for the same photoelectron $E_K$ from a specific orbital in a given element.

Once the electrons are emitted from the sample, they must pass through the analyzer and be detected. This gives rise to considerations of nonidealities in the instrument, in particular of the analysis system. The relationship between emitted flux $J_i^e$ and measured intensity $I_i^m$ (number per unit time) can be expressed succinctly as

$$I_i^m = f_s f_n \theta_s TDA J_i^e \qquad (10)$$

where $A$ is the sampling area on the surface, $D$ is the transmission coefficient of the detector, $T$ is the transmission coefficient of the lens and analyzer, $\theta_s$ is a collection efficiency factor for the frontmost lens of the spectrometer, $f_n$ is a response function for the lens, analyzer, and detector, and $f_s$ is a response function for the system electronics. The five coefficients other than sampling area generally depend on $E_K$ of the incoming photoelectron (as viewed by the analyzer). They range between zero and unity. How they depend on $E_K$ cannot always be exactly determined. This introduces errors in the consideration of atomic concentration, as discussed below (see Data Analysis and Initial Interpretation).

The transmission function of the analyzer has been discussed in the previous section and is typically the most prominent factor considered in the above equation. The detector response function $D$ defines the gain of the detector as a function of the $E_K$ of the electron at the detector. It is typically important to consider only when the pass energy varies, which should not occur during quantitative XPS analysis. In most practical cases, it is not possible or

even really necessary to determine $D$ separately from the value of $T$.

The angular efficiency factor $\theta_s$ is an abbreviated form of an integral over the solid angle of view of the analyzer, often called the analyzer acceptance angle. It accounts for any potential dependence of photoemission intensity from a solid on the analysis angle, often called the electron take-off angle. In some cases, this factor is also abbreviated as $d\sigma/d\Omega$ and termed the differential cross-section as a function of measurement angle. Even for pure elements as solids, photoemission intensity does indeed depend on emission angle, and therefore on take-off angle, especially for single crystal substrates. This is due both to the angular dependence of photoemission from orbitals of nonspherical shape as well as to diffraction phenomena. The angular efficiency factor is for the most part only important in analysis using XPS when considering spectra from highly ordered (single crystal) materials as a function of take-off angle or when considering variations in intensity as a function of analyzer acceptance angle. The former is typically a far more widely performed measurement than the latter.

The two response-function terms, $f_n$ and $f_s$, account for non-idealities in the instrument and the electronics respectively. Basically, these terms account for temporal variations in such things as stray magnetic field strengths within the analyzer or fluctuations in voltage precision and stability as a function of sweep energy. Typically, they are also not readily separable from the value of $T$.

The overall transmission function of an analyzer, $T' = f_n f_s \theta TD$, can be determined by using ideal electron sources with known emission behavior as a function of $E_K$. A typical methodology might involve the use of an electron gun and analysis of the intensity of elastically scattered electrons from a sample reaching the detector as a function of electron $E_K$. An alternate method is to compare a measured spectrum $I^m(E_K)$ from one instrument to a spectrum that is known to represent the expected emission spectrum $J^e(E_K)$ from a well-defined sample. Overall instrument response functions are useful for fundamental studies into the behavior of an XPS instrument. They are typically not recorded as part of the everyday operation of the system. As will be seen below, the behavior is instead lumped into one factor, the sensitivity factor $S_i$, that is provided in tabular form by the instrument manufacturer.

Substituting the electron flux equation for a homogeneous sample into the above equations leads to a relationship between measured electron intensity and the concentration of an element in a homogeneous material.

$$I_i^m = F\sigma f_a AT\lambda_i C_i \qquad (11)$$

where the angular efficiency as well as the detector and instrument response functions have been lumped into the factor $f_a$. The factor $AT$, the product of the viewing area of the analyzer and its transmission function, is called the analyzer étendue, $G$. As mentioned previously, for an HSA operating in <u>constant</u> pass energy mode under conditions where the analyzer defines the transmission function, $G$ varies approximately as $1/E_K$.

The practical applications of XPS for determining atomic concentrations start from the above equation. It is typically presented in a far more abbreviated form, namely

$$I_i^m = FS_iC_i \tag{12}$$

where $S_i$ is the instrument-dependent sensitivity factor for orbital i in an element. In practice, sensitivity factors should be provided by instrument vendors for each orbital in all elements. In most cases, sensitivity factors are only provided for the most prominent orbitals. As seen from the above discussion, they are not directly transferable from one instrument to another, since they contain instrument response and transmission functions. Also, sensitivity factors are confined to a particular sample-analyzer geometry. Finally, they do not take into account nonidealities in the sample, especially in regard to composition and structure. Despite all these apparent limitations, the use of tabulated sensitivity factors for an instrument has become the accepted practice for determining atomic concentrations with XPS, as discussed below (also see Data Analysis and Initial Interpretation).

Equation 12 can be used in three ways. The first is to determine absolute atomic (mole) fractions, $x_i$ for an element in a material. In this regard, we can write

$$x_i = \frac{I_i^m/S_i}{\sum I_i^m/S_i} \tag{13}$$

showing in principle that only sensitivity factors and measured intensities are required to determine atomic concentrations of a material. This equation is based on analysis of a material with a homogeneous composition throughout. This restriction will be relaxed below in the discussion on other applications of XPS.

To obtain absolute accuracy in atomic compositions using the above equation, we must have values of $S_i$ that were obtained (a) from a sample of similar if not identical atomic composition and structure to the one we are analyzing and (b) from a sample-analyzer configuration equivalent to that used to determine $S_i$. While the second condition can generally be met without much difficulty, meeting the first condition can be difficult if not impossible. Therefore, the above equation can be considered an approximation for almost all real-case systems. Given the extent of variations that are possible in $\lambda_i$ and $\theta_i$ from one material to another, relative inaccuracies in $x_i$ can be as much as ±20% using the above equation. A worthwhile note from this discussion is that absolute atomic fractions determined from XPS will be accurate to better than ~ ±1% only if they have been derived after a painstakingly thorough calibration of the XPS instrument concerned. This issue is separate from the precision of the measurements and will be highlighted again below (see Data Analysis and Initial Interpretation).

A far more useful form of Equation 12 (Briggs and Seah, 1994) can be derived by considering analysis to obtain "relative" atomic concentrations. Then, we can write

$$\frac{x_i}{x_j} = \frac{I_i^m/S_i}{I_j^m/S_j} = \frac{S_j}{S_i}\left(\frac{I_i^m}{I_j^m}\right) = R_{ji}\left(\frac{I_i^m}{I_j^m}\right) \tag{14}$$

where $R_{ji}$ is the relative sensitivity factor of element (orbital) j to element i. Be aware in applications of this form whether you are using $R_{ji}$ or $R_{ij}$, which are related as $R_{ji} = 1/R_{ij}$. Relative sensitivity factors are far less dependent on variations in sample composition and structure, as well as instrument configuration, than absolute concentrations. Changes in these factors tend to cancel each other in taking the ratio. Therefore, while uncertainties in absolute concentrations determined using XPS can be high, the uncertainties in relative atomic concentrations are in many cases bounded primarily by errors in measurement precision (repeatability). This often means that relative atomic concentrations determined with XPS can be compared more reliably from sample to sample and from system to system than absolute atomic concentrations.

The final form of Equation 12 (Briggs and Seah, 1994) that is of practical use arises when taking samples of known (and uniform) atomic composition for comparison to samples of unknown (and uniform) composition with similar micro- and molecular structure. In this case, the formulation can be recast to give

$$\frac{x_i}{x_{io}} = \left(\frac{I_i^m}{I_{io}^m}\right)\left(\frac{S_{io}}{S_i}\right) \tag{15}$$

where the "o" subscript indicates the sample with known composition. A particularly useful form of this equation for illustration purposes and applications arises when the known material is pure and we are comparing with binary (two-element) materials. The equation can then be rewritten as

$$\frac{x_A}{x_B}\left(\frac{x_{Ao}}{x_{Bo}}\right) = \left(\frac{S_B/S_{Bo}}{S_A/S_{Ao}}\right)\left(\frac{I_A^m/I_{Ao}^m}{I_B^m/I_{Bo}^m}\right)$$
$$\frac{x_a}{x_B} = f_{BA}^m\left(\frac{I_A^m/I_{Ao}^m}{I_B^m/I_{Bo}^m}\right) \tag{16}$$

where $F_{BA}^m$ is known as the matrix factor for the binary system. The matrix factor accounts primarily for the changes produced by putting element B in a matrix of element A. It is not always unity. To a first approximation, it can be taken as a ratio of the square root of the molar specific volumes of the elements

$$F_{BA}^m \approx \sqrt{\frac{\hat{V}_B}{\hat{V}_A}} \approx \left(\frac{R_B}{R_A}\right)^{3/2} \tag{17}$$

or the corresponding atomic radii for similar crystal systems. To use this equation, we would first measure intensities for the principal XPS peaks from pure materials, $I_{Ao}^m$ and $I_{Bo}^m$. Then, intensities from samples of known composition, $I_A^m$ and $I_B^m$ would be measured to calculate the matrix factor for the binary system. Thereafter, alloy composition can be calculated directly from measured peak intensities of unknown binary alloys. This equation is limited primarily to homogeneous alloy systems.

Care must be used in reporting atomic compositions determined with XPS. They must be understood to be accurate only for a sample that has a uniform composition throughout the entire sampling volume. This is rarely the case with real-world samples. Concentrations can vary dramatically across an analysis area, for example when analysis is done across grain boundaries in metals or across copolymer regions in polymers. One must keep in mind that even the best spatial resolution of ~25 µm obtainable with small-spot XPS is often larger than spatial variations in composition across the surfaces of most real-world samples. Conventional XPS systems can have analysis areas at the sample as large as a few mm². In cases where the sample is known to have lateral variations in composition, the atomic concentrations determined with XPS will generally be spatially averaged values.

Concentrations also vary with sample depth. In this regard, a key parameter is the average sampling depth in comparison to the depth over which the composition varies. The average sampling depth depends on the mean free path $\lambda_i$ of the electrons. For example, most samples quickly become covered by an overlayer of carbonaceous and water contamination upon exposure to air. This overlayer, if thicker than $\lambda_i$, will mask the true composition of the underlying material, and anomalously high values of atomic concentration for oxygen and carbon will result. Other examples of potential problems that can arise in determining compositions when using XPS spectra from samples with nonuniform composition profiles have been pointed out elsewhere (Tougaard, 1996).

Further discussion on how to extract information about compositional variations with sample depth is given in the following section titled Other Applications. Analysis of XPS spectra to obtain values for $I_i^m$ is also considered (see Data Analysis and Initial Interpretation). Methods for preparing samples to reduce the contaminant overlayer are given in the section on sample preparation.

**Chemical State.** Obtaining chemical state information from an XPS spectrum can often be as important as obtaining elemental composition. Any application that must consider the relative amounts of different oxide, valence, or hybridization states of an element will be interested in using XPS to determine chemical state. Specific examples of this for different families of materials (metals, ceramics, semiconductors, polymers, and composites) are found in almost any journal that publishes studies from XPS (see Key References).

The BE of a photoemission peak is affected by the chemical state of the element (see Principles of the Method). When an atom acquires a partial positive charge due to bonding or rehybridization, we expect a shift to higher BE values. Correspondingly, a shift to lower BE values is expected when an atom becomes negatively charged. Variations in the chemical state of atoms throughout the material can also affect the width of the photoemission peak.

As also pointed out (see Principles of the Method), other factors besides the chemical state of the element influence both the BE and width of a photoemission peak. The Madelung potential term arises when the lattice structure about the atom changes. The extent of intra- or extra-atomic relaxation may also change. This means, a peak shift to higher or lower BE values cannot always be interpreted as arising solely from a change in chemical state.

In practice, one should therefore not view XPS as an unambiguous means of identifying unknown chemical states of elements. The better approach is to have initial information regarding the bonding of the element in the material. This information should then serve as a guide to select the main factors that could cause peak shifts or peak broadening for photoemission from the sample. Other factors that lead to peak shifts and broadening are considered below (see Problems). Finally, an enormous number of reference sources are available that contain XPS peak positions and even reference spectra from compounds (see Key References). The reader is also encouraged in this regard to search the Internet. These reference texts and pertinent literature on the sample being analyzed should be consulted for fuller insight into peak assignments.

For further discussion on resolving overlapping peaks via peak fitting procedures and on deconvolving instrument response functions and source broadening effects from high-resolution XPS spectra (see Data Analysis).

**Other Applications.** Two other applications of XPS deserve mention. Both are extensions of the use of XPS to determine atomic concentrations. They apply in cases where composition varies with sample depth, $C_i(z)$. The first is the use of what is commonly termed sputter profiling. The second is called angle-resolved XPS (ARXPS). Both methods rely on the limitation that sampling depth in XPS is only on the order of the electron mean free path, $\lambda_i$.

Sputter profiling is a destructive technique. Details are provided in the reference from Briggs and Seah (1994). It is useful for mapping any form of concentration profile as a function of depth. In sputter profiling, the sample is bombarded with high-energy ions, typically $Ar^+$. This etches away layers of the sample. Sputtering rates can be controlled to remove on the order of sub-monolayers of material with careful calibration of the sputter gun and its configuration relative to the sample. Sputter depth profiling consists of performing a series of analysis-sputter cycles on a sample. Sputter rates and times during each cycle are set by the user. Typical results from sputter profiling are plots of $I_i^m$, $x_i$, or peak positions versus sample depth. Sample depth is obtained from sputter time by calibrating the sputter rate. This is done by using a layered sample with a known thickness. Oxide layers on Si wafers and oxide layers on Ta are most common and can be readily obtained through commercial vendors.

A number of factors limit the utility of sputter profiling with XPS (Hofmann, 1993; Wetzig et al., 1997), including sample surface roughness (Gunter et al., 1997) and compositional heterogeneity. Both contribute to nonuniform sputtering rates across a sample surface, especially in systems where sputtering is done at an incident angle that is not normal to the sample surface. A common method of overcoming this limitation, at least for rough sample surfaces, is to rotate the sample under the sputter gun. Another limit on sputter profiling is the inconsistency between sampling depth and sputter depth per cycle. Generally, the sputter depth per analysis cycle is set to be lower than the sampling depth (mean free path of the escaping electrons). This leads to the effect that $I_i^m$ will start to decrease or increase before the depth is reached where concentration actually increases or decreases. This problem is essentially due to a convolution of the true concentration profile $C_i(z)$ with functions representing the effective sampling depth during analysis $f_d(z)$ as well as any non-uniformities in sputtering depth $f_n(z)$, giving a measured concentration profile $C_i^m = C_i(z) \oplus f_d(z) \oplus f_n(z)$. A number of methods exist for handling the deconvolution of $f_d(z)$ and $f_n(z)$ from $C_i^m$ in a simple or rigorous manner (Blaschuk, 2001; Hofmann, 2001; Lesch et al., 2001). Ion bombardment can also cause remixing at the sputter crater of components in a heterogeneous sample, and segregation of specific bulk elements to surfaces and interfaces can occur (Hofmann, 1992). These factors will lead to false composition readings during sputter profiling. Finally, care must be taken during sputter profiling to make sure that sputtering is done uniformly over an area that is larger than the analysis area.

With the exception of nonuniform sputtering rates, the above limitations are less important in sputter profiling during AES analysis. The analysis area and sampling depth are both much smaller in AES than they are in XPS. Therefore, in cases where it can be used, AES is generally preferred for sputter profiling, especially when only composition (not chemical state information) is desired. The only other concern prior to sputter profiling is to determine whether any of the AES or XPS peaks of interest overlap. This can sometimes rule out or favor one technique over the other. For example, N peaks overlap with Ti peaks in AES but not in XPS, so nitrided Ti samples would be better analyzed with depth profiling using XPS.

In ARXPS, compositions are measured as a function of sampling angle relative to the surface plane, often called the analyzer take-off angle. For samples with layered compositions, changing the analyzer take-off angle results in changes in measured intensity, $I_i^m$, because emitted flux, $J_i^e$, varies from elements within the different layers. Measurements of $I_i^m$ as a function of take-off angle can be used to determine values for layer thickness. Other applications include the determination of molecular orientation in ordered adlayers and crystal packing symmetry in lattices. Two extensive reviews by C. Fadley of the principles and applications of ARXPS are recommended as a starting point for further details (Fadley, 1976, 1984).

One case is useful to present as an illustration of the methodology. This is an example for a bilayer material, such as occurs when a coating, oxide, or contaminant layer covers a surface. An illustration of the set up is shown in Figure 11. The overlayer has a thickness, $d$; the outmost surface and boundary interface are assumed to be flat on the microscale, and the boundary between the overlayer and the underlying substrate is abrupt (a step function).

Consideration of this situation starts with modification of Equation 8 to account for a change in the effective path length the electrons must travel to escape. The path length is dependent on the take-off angle, $\theta$, of the photoelectrons. We define take-off angle relative to the surface plane, so that glancing angle analysis occurs as $\theta$ approaches zero (some texts define take-off angle with respect to the surface normal). Under these conditions, for a fixed value of

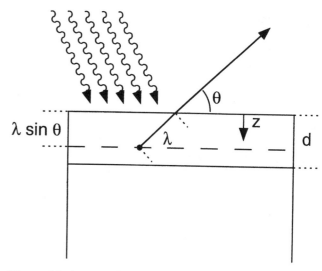

**Figure 11.** An example of a simple system that illustrates the application of angle resolved XPS. An overlayer of thickness $d$ is being analyzed at a take-off angle $\theta$ with respect to the surface plane.

$\theta$ and a semi-infinite sample, we can write Equation 8 for the overlayer o and underlying substrate s to be

$$J_o^e = F\sigma_o \int_0^d C_o(z)\, \exp(-z/\lambda_o \sin\theta)dz \tag{18}$$

$$J_s^e = F\sigma_s \int_d^\infty c_s(z)\, \exp(-z/\lambda_s \sin\theta)dz \tag{19}$$

We assume the overlayer and substrate have uniform composition throughout. We can then obtain representative equations for the measured intensity of photoelectrons from an orbital in the overlayer or in the substrate as

$$I_o^m = F\sigma_o f_{a,o} G_o C_o \lambda_o (1 - \exp(-d/\lambda_o \sin\theta)) \tag{20}$$

$$I_s^m = F\sigma_s f_{a,s} G_s C_s \lambda_s \exp(-d/\lambda_s \sin\theta) \tag{21}$$

Note that, factors dealing with the behavior of the instrument, $f_a$ and $G$, as well as the cross-section $\sigma$ depend on $E_K$ of the electron and the orbital being considered, respectively.

A ratio of measured intensity for a photoemission peak in the overlayer $I_o^m$ to that from the underlying substrate $I_s^m$ has the form

$$\frac{I_o^m}{I_s^m} = \left(\frac{\sigma_o}{\sigma_s}\right)\left(\frac{G_o}{G_s}\right)\left(\frac{\lambda_o(1 - \exp(-d/\lambda_o \sin\theta))}{\lambda_s\exp(-d/\lambda_s \sin\theta)}\right)\left(\frac{C_o}{C_s}\right) \tag{22}$$

In this equation, the parameter sought is $d$, the overlayer thickness. This can be determined if we obtain $I_o^m/I_s^m$ as a function of $\theta$, and know all the other parameters. In practice, the ratios of cross-sections and transmission functions are usually taken as unity to simplify the expression. The concentration ratios are either set equal to unity as well or are estimated from materials properties. An example of the use of this equation is provided below (see Data Analysis).

Photoemission intensity for one peak from an overlayer can also be compared relative to the intensity from same peak for the pure sample of the element concerned using the above equations. This leads to

$$\frac{I_o^m}{I_p^m} = \frac{\lambda_o(1 - \exp(-d/\lambda_o \sin\theta))}{\lambda_p \sin\theta}\left(\frac{\bar{V}_p}{\bar{V}_o}\right)x_o \qquad (23)$$

where $x_o$ is the mole fraction of the element in the overlayer and the V]$_i$ values are the molar specific volumes of the element in the overlayer or pure material.

Angle-resolved XPS can also be used to determine the bonding angle of a molecular adsorbate on a surface. Consider a linear-chain molecule that can bond in either an up or down configuration on a surface, and imagine that one end is tagged with a different element than the other. If we form a monolayer on a microscopically flat surface, analysis with XPS at glancing take-off angle will accentuate the signal from the element that is at the outermost position in the absorbed layer. This has applications for molecular coatings and polymers where information about the chemistry of the exposed surface relative to that for the underlying surface layer is of utmost importance. On microscopically flat surfaces, ARXPS can often be used to distinguish between perpendicular and horizontal configurations of molecules. It can also be used to determine the bond angle of molecular adsorbates in certain cases where the adsorbate has well-defined long-range order. These types of angle-resolved XPS measurements should be confirmed using other techniques.

Surface roughness complicates the analysis with ARXPS. In addition, the angular resolution of the data can never be any better than the acceptance angle of the front lens on the analyzer. Most commercial systems have an acceptance angle of 15° to 20°. This may not present problems for film thickness measurements, but is too wide for sophisticated ARXPS measurements of bond angles for molecular adsorbates. Some HSA-equipped XPS instruments have variable magnifications for the lenses. The acceptance angle of the analyzer will always be larger at higher magnifications. Both surface roughness and an overly wide analyzer acceptance angle lead to a fall off of intensity from theoretically expected values as take-off approaches glancing angles.

## Materials-Specific Considerations

Almost all of the previously mentioned journals (see Introduction) present articles on XPS as applied to all classes of materials. The scope ranges from fundamental (*Surf. Sci.* or *Surf. Interface Sci.*) to more applied (*Appl. Surf. Sci.* or *J. Vac. Sci. Technol.*). The use of XPS for analysis of biomaterials surfaces is also becoming more widely represented in *Colloid Interface Sci.* and *J. Biomed. Mater. Research.*

With regard to metals, the ASM Handbook (ASM Handbook Committee, 1986) is a reasonable starting point for practical applications of XPS in metallurgy. Analysis of metals that are used in routine applications is often not straightforward with XPS. Most metals will have oxide or contaminant overlayers on their surfaces, and surface segregation is also common in alloys. Structural variations, such as grains, precipitates, or other microconstitu-

ents can also be present, especially in multi-component metal alloys. These factors will all affect the measured intensity of a photoelectron peak $I_i^m$ by introducing nonidealities into the equations that were derived primarily for analysis of homogeneous samples. In addition, as seen in Figure 8, the more elements that are in a metal sample, the greater is the chance that Auger peaks will interfere with XPS peaks in the spectrum. This can make the determination of composition and chemical state difficult. Finally, on metals with oxide layers, the oxide layer can act as an electrical insulator, leading to peak shifts due to differential charging (see Problems).

As mentioned above (see Competitive and Related Techniques), better spatial and depth resolution are typically obtained using AES. This may be preferred for analysis of metals, especially for compositional analysis across grains or within two-phase regions. Because of the larger sampling depth and spot size of XPS, compositions will be averaged in these cases. The distinct advantage of XPS over AES for analysis of polycrystalline metals is that the oxidation states of metals are generally much easier to resolve with XPS, though this is not always the case (Sekine et al., 1996). Assignment of peak shifts to changes in metal oxidation state can be complicated when multiple oxidation states exist. Peak fitting is generally required to resolve the overlapping features. The use of sputtering to remove the outermost contamination layer, as well as glancing angle analysis to become more sensitive to surface compositions, can often help.

Analysis of semiconductors is relatively routine with XPS. The overriding concern is generally to remove the outermost contamination layer. With extrinsic semiconductors, sensitivity factors for the dopant concerned should be compared to those for the intrinsic material, keeping in mind that lower sensitivity factors mean greater amounts (concentrations) of dopant will be needed to give the same signal-to-noise ratio in an XPS spectrum. The mean free paths of the relevant photoelectrons in the material should be considered when analyzing for dopant compositions that may vary with depth. Sample charging can occur in XPS of semiconductors. In this regard, though, XPS presents fewer problems than AES.

With ceramics, XPS is generally the only spectroscopic laboratory method that can be used to determine (surface) composition and chemical state information. Analysis with AES is typically not feasible because of sample charging. Even with XPS, ceramic samples charge considerably, leading to peak shifts.

In regard to analysis of polymers with XPS, the articles by D.T. Clark are good, though somewhat dated, starting points (Clark and Feast, 1978; Clark, 1979). Analysis of polymers with XPS is a common method of determining both concentrations and hybridization states of elements in the material. As shown in Figure 7, XPS can easily resolve certain hybridization states of carbon. It can also resolve variations in functional groups attached to the carbon. In most cases, shifts of the C 1$s$ peak can be assigned solely to changes in the electronegativity of the surrounding atoms. The primary difficulty with polymer samples is charging. Some polymer samples are also damaged by high x-ray flux.

Metal and semiconductor single-crystal samples are often used as model systems for analysis with XPS. Quantification of the rates of chemical reaction processes and determination of the fundamental nature of chemical bonds formed by adsorbates to surfaces are two areas that benefit by the use of XPS on such model systems. Single-crystal metals and semiconductors have also been used to develop much of the theoretical basis behind the shape of the XPS spectrum. Angle-resolved XPS also benefits by using microscopically if not atomically flat surfaces.

## METHOD AUTOMATION

All XPS systems in operation today are run by computers, with many modern systems being entirely automated after introduction of the sample. The user generally has to define a number of parameters to control the instrument. On older-generation instruments, some of the parameters may be input manually. In most cases, those parameters that are input manually are not supposed to vary from run to run.

Starting with the source, after selecting an anode type (in systems with more than one anode), one typically defines any two of three operating parameters of x-ray source voltage, emission current, or power. This defines the type of x ray (its primary energy $h\nu$) and the flux $F$. We are typically not interested in knowing the absolute value of $F$ for the x-ray source, only in knowing that it is the same value from analysis to analysis. Most XPS instruments also allow the user to select an analysis area on the sample, either using manual apertures or through computer settings to control focus voltages to the lenses of the analyzer. Modern systems also allow the user to position the sample in different locations and possibly different orientations under the analyzer using computer control.

The scan parameters are set next. Among these parameters are: the sweep starting and/or ending energy, $BE_{start}$ and $BE_{end}$; sweep width, $\Delta BE_{sweep}$; step size, $\delta BE_{step}$, or number of points per sweep $n_{pps}$; dwell time at each step, $t_{dwell}$, or the total time per sweep, $t_{sweep}$; and finally the number of scans, $n_{scans}$, or total analysis time $t_{total}$. These parameters are related as listed below.

$$\Delta BE_{sweep} = |BE_{end} - BE_{start}|$$
$$\Delta BE_{step} = \Delta BE_{sweep}/n_{pps}$$
$$t_{sweep} = n_{pps}\, t_{dwell}$$
$$t_{total} = t_{sweep}\, n_{scans}\, n_{sweeps}$$

Not all of these parameters are used by all systems. Some are set automatically by the system depending on the type of scan (survey or high-resolution, for example). In this regard, the remaining input parameter is the pass energy $E_p$.

In addition to automated data acquisition, most instruments with computer control can be programmed to cycle through a series of spectra and determine peak parameters. This is generally done during sputter profiling or angle-resolved XPS. The instrument automatically shows profiles of composition as a function of sputter time (or depth) or take-off angle.

Automation of the data acquisition sequence leads to problems when the results deviate from those expected or when the system has problems during operation. The former case arises, for example, when samples have different degrees of charging, shifting the peaks out of preprogrammed sweep ranges. In some cases, trying to abort or break an established acquisition routine can lead to the loss of the settings, or worse, loss of data, especially on older-generation instruments. The user is therefore cautioned to be well aware of the potential variations that could arise before automating data acquisition.

Automation of data analysis does not typically present as many problems. The user can always repeat the analysis using different parameters. Unfortunately, some older XPS systems with automated data analysis may not allow this luxury, so caution is again warranted. Also, one should always do spot checks on the analysis results after an automated analysis. The computer is not an "all-knowing" instrument, and exceptions to the standard methods of data analysis can always arise.

## DATA ANALYSIS AND INITIAL INTERPRETATION

Analysis of an XPS spectrum starts by determining parameters for the photoemission peaks. The three primary parameters that define a photoemission peak in XPS are its area $A_{pi}$, position $BE_i$, and full-width-at-half-maximum (FWHM) $\Delta w_i$. Additional consideration may be given to peak shape, as discussed below (see Peak Integration and Fitting). To obtain accurate values for these peak parameters, we have to process the XPS spectrum correctly. This means applying methodologies that are well grounded in the principles of the method. Data analysis may be preceded or followed by application of methodologies to improve the appearance of a spectrum. Finally, interpretation of the results should keep in mind the uncertainties that arise during data analysis.

The following discussion first presents the methods used to determine peak parameters from XPS spectra. This is followed by a discussion of methods that can be used to improve the appearance of an XPS spectrum. We then consider how to interpret the results, with particular emphasis on uncertainty analysis. The concluding presentation in this section offers an example of the analysis of ARXPS data to determine the film thickness of an oxide layer on a Si wafer.

### Analysis

The most routine analysis of an XPS spectrum involves identifying the peaks with their corresponding elements. Tables of peak positions are available to help with this task. Most XPS systems also have software libraries of peak positions that will automatically mark elements in a spectrum. This automation is only as good as the software routines for locating peaks; it should not be taken as a definitive indication of the presence of an element, and it may miss small peaks on noisy backgrounds. The best method of determining unambiguously that an element is present is to cross-correlate for all peaks that should appear for that particular element. For example, if a peak at 100 eV is identified as the Si 2$p$, a corresponding 2$s$ peak should be found at ~150 eV with the appropriately scaled height. Auger peaks should also be present if the scan range covers them.

We generally also want to obtain elemental compositions from an XPS spectrum. This requires quantitative analysis of the spectrum. Quantitative analysis for compositions is based on background subtraction and peak integration. In cases where more than one chemical state is present for a given element, we may need to fit the peaks with components. Finally, we may want to quantify the positions and half-widths of peaks in order to compare changes from sample to sample. The practical aspects associated with these procedures are discussed in the sections immediately below.

**Background Subtraction.** The area of a photoemission peak is determined by subtracting an appropriate background and integrating (numerically) under the resultant curve. The background underlies the true spectral features, as illustrated in Figure 2. The types of background shapes that can be used range from a constant to a complex shape and are based to some degree on theoretical considerations of the inelastic scattering processes that generate a background in XPS spectra. The choice of a background shape can affect the resulting peak parameters. Although this effect is often below the precision (repeatability) of the measurements from sample to sample, proper attention should be paid to the application of appropriate background subtraction methods with XPS spectra. This is especially true in cases where peak fitting is to be done and where subtle changes in peak parameters, especially peak position or FWHM, are important to resolve.

A constant or linear background removes a straight line from under a spectrum, typically bringing the lowest value(s) in the spectrum to a value of zero. The constant background removes a constant offset throughout the spectrum. The higher-BE side of a primary photoemission peak will generally have a greater intensity due to inelastically scattered electrons and potentially to satellite peaks. A linear background rather than a constant one would therefore appear to be a better choice in most cases. This is the case for both spectra in Figure 6. The intensities on the left and right sides of the peaks in Figure 7 are nearly equal, so a constant offset could be removed. In both of these cases though, better choices for background shapes exist. In practice, a constant or linear background should only be applied as an approximation when the background intensity on the left and right sides of the photoemission features under consideration are nearly identical or have identical slopes.

A widely applied background shape with some degree of theoretical support is the integrated or Shirley background (Shirley, 1972). It is based on the principle that background intensity $I_i^b$ (BE) at a given BE is proportional to the total intensity of photoelectrons at higher $E_B$. This is similar to the nature of the inelastic scattering process that leads to a significant portion of the background in XPS. The intensity of scattered (background) electrons increases at lower KE (higher BE) away from a photoemis-

sion peak in proportion to the intensity of photoelectrons at higher $E_K$ (lower $E_B$). The Shirley background is derived analytically from a formulation similar to

$$I_i^b(\mathrm{KE}) = C \int_{\mathrm{KE}}^{\infty} (I_i^m(E) - I_i^b(E))dE \qquad (24)$$

The integration is best represented in KE space, since higher $E_K$ electrons contribute to the background at a given KE. The value $C$ is essentially a normalization constant to bring the background into alignment at the endpoints. This function can be readily coded into XPS analysis software and is therefore routinely available with the instrument or as part of off-line data analysis software. The above equation must be solved iteratively using numerical integration, and has no user-defined parameters other than the start and end ranges of the background. Some older XPS instruments may only use a simpler single-pass numerical integration for increased processing speed.

Disagreements exist about the accuracy of the Shirley background routine. It has been shown to distort peak parameters (peak positions) in certain critical cases. In the experience of the author, the potential errors introduced by the use of the iterative Shirley background versus a supposedly more accurate background shape are usually well within the systematic errors of determining peak parameters for most routine XPS experiments. Only when a high degree of accuracy is essential during peak fitting should the user generally have to consider other (more accurate) background shapes.

The Shirley background is only appropriate for fitting as a baseline to high-resolution peaks. The algorithm does not normally converge well over the wide BE range of a survey scan. Application of the Shirley background also leads to problems in cases where the signal intensity increases or decreases dramatically on either side of a photoemission peak. This happens when a photoemission peak of interest sits as a shoulder on a sloping background from a significantly larger peak. The resulting complex background shape is a convolution of the overlapping peak intensity and the normal (inelastic) background. Off-line processing routines exist to handle such complex background shapes. Whenever possible, data acquisition should encompass a full range of all overlapping peaks to avoid this problem.

The Tougaard background shape for XPS has the strongest basis in theory (Tougaard, 1988). It is derived on first-principles consideration of the intensity of inelastically scattered electrons as a function of decreasing KE away from the $E_K$ of the primary photoelectron. The resultant analytical expression in its general form is an integral over peak intensity that has two materials dependent fitting parameters, $B$ and $C$.

$$I_i^o(\mathrm{KE}) \approx C \int_{\mathrm{KE}}^{\infty} \frac{E - \mathrm{KE}}{\left[ B + (E - \mathrm{KE})^2 \right]^2} I_i^m(E)dE \qquad (25)$$

The fitting parameters relate to the nature of the inelastic scattering in the material of concern and are usually left to be determined by measurements with techniques

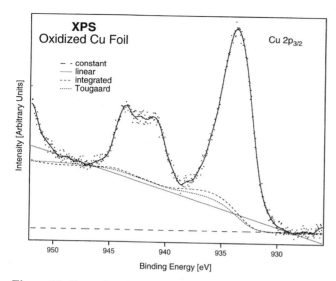

**Figure 12.** Examples of constant, linear, and integrated (Shirley) types of background shapes fit to a Cu $2p_{3/2}$ peak and the companion satellite for CuO.

such as EELS (see SCANNING TRANSMISSION ELECTRON MICROSCOPY: Z-CONTRAST IMAGING). Since this degree of rigor is beyond the scope of interest for typical XPS users, the Tougaard background is usually used with predefined "universal" values of fitting parameters, in particular $B \approx 1643$ eV$^2$. The value of $C$ is primarily a normalization constant to fit the background at the ends of the spectrum under consideration. The resulting background shape is similar to the Shirley background.

Examples of a constant, linear, iterative integrated (Shirley), and Tougaard backgrounds applied to the CuO peak shown earlier are given in Figure 12. This figure clearly shows the potential for error that can arise when a proper background is not used prior to further processing of a spectrum. The constant background is inappropriate for determination of peak area. Compare the linear, integrated, and Tougaard backgrounds. The resulting integrated area under the entire Cu $2p_{3/2}$ peak may be different in the three cases; however the extent of the differences is likely to be within the reproducibility of the measurements.

**Peak Integration and Fitting.** Once a suitable background is removed, $A_{pi}$ can be found by numerical integration. The units of $A_{pi}$ are typically counts-eV/sec. This is directly proportional to measured intensity $I_i^m$ in the corresponding equations above (see Practical Aspects of the Method). The additional factor of eV cancels in the calculation of concentration or concentration ratio.

Broad peaks may be deconvolved into (fitted by) more than one component peak to help resolve overlapping peaks. The spectrum in Figure 7 shows resolution of the broad photoemission feature into three component peaks. The following guidelines apply when doing peak fitting: (1) peak fitting should always be done after removing an appropriate background; (2) the shape of the component peak matters; (3) the initial conditions used to start the peak fit should be best approximations to the expected

chemistry of the sample; and (4) peak parameters should be appropriately constrained during the fitting process.

With regard to background subtraction, many peak-fitting algorithms in use today for analysis of XPS spectra do not include the capability to automatically fit an appropriate background while optimizing fit parameters for component peaks. In such cases, the appropriate background must be carefully chosen and removed beforehand to obtain consistent, accurate results from peak fitting. Note in Figure 12 that differences are to be expected in the final results for peak fitting when different backgrounds are used, because the initial peak shapes will be different. In particular, small peaks that are shoulders on larger peaks can be accentuated or lost merely by shifting the position or changing the shape of the background that is removed. Calibration of the combined baseline subtraction and peak-fitting methodologies with samples of known chemistry and composition is recommended. Examples where simultaneous fitting of baselines and peaks was done specifically to obtain detailed information about sample chemistry have been reported (Castle et al., 1990; Castle et al., 2000; Salvi et al., 2001).

A large number of peak shapes can be used to fit to a feature in an XPS spectrum. All peak shapes have at least three parameters: peak height, peak FWHM (or half-width), and peak position. Peak half-width is half of the FWHM value. Additional parameters are used to mix fractional amounts of the different peak shapes and to define such factors as asymmetry or tail height and extent. The doublet is a peak shape that is particular to XPS spectra. It is used to fit to the multiplet split $p$, $d$, or $f$ peaks, and it is essentially defined as two peaks separated by a certain distance.

The Appendix provides further information about peak shapes commonly used to fit XPS spectra. This information should clarify how the parameters are used to define the peak shape. In confronting the wide range of peak shapes, the first-time XPS user is cautioned to proceed conservatively. The Gaussian and Gaussian-Lorentzian mixed peak shapes (and their doublets) provide adequate representation of XPS peak shapes for almost all routine analysis work. The peak shapes with greater degrees of freedom should be used only with a clear understanding of the physical significance behind their application. In other words, do not choose a peak shape with more free parameters than can be reasonably justified within the confines of the chemistry and physics of the system with the goal of covering all potential contingencies. Rather, expand the degrees of freedom (number of peak-fit parameters) from a minimum to a larger set when no other options appear adequate to cover the case being examined.

During peak fitting, any or all of the parameters that define a specific component peak are allowed to vary until an optimal resultant peak shape, compared to the input raw spectrum, is obtained. The computational methods involved in optimizing the parameters are beyond the scope of this unit. Suffice it to say that local and global optima exist in the space of the fit parameters. Therefore, do not use more peaks than absolutely justified, and always start with peak parameters that provide the best representation of the chemistry as known from the sample. Starting with too many peaks or with parameters that are too far from the expected values could lead to an end result that is far removed from physical reality. This problem can be avoided in almost all cases in a straightforward manner. At the start, use a minimum number of peaks and peak parameters in order to obtain a resultant "fit" spectrum that represents the raw data well at the outset. This may require iterative, manual adjustments of the initial peak parameters. With practice, this step will become a matter of good practice during peak fitting.

In correspondence with setting the initial peak parameters correctly, allowing too many parameters to vary during the optimization can lead the optimization routine astray. The fitting may fall into an unrealistic local minimum or, in some cases, fail to converge properly. Most peak-fitting routines allow the user to constrain the peak parameters either to stay within certain bounds or to stay constant during peak fitting. As the number of potential peak-fitting parameters increases, the number of parameters that are constrained should also increase.

In conclusion, with regard to peak fitting to XPS spectra, let the expected chemistry of the material be the guide in defining and accepting the peak parameters, instead of letting the fitting results determine the chemistry. Further information about the steps involved during peak fitting of XPS spectra is available in a flowchart and comparative study (Crist, 1998). In addition, reports of methods that have been used to validate algorithms for peak fitting to XPS spectra (Seah and Brown, 1998) should be consulted when user-defined routines are developed or used. Finally, as a way to define the level of confidence in the methods being used during peak fitting, standardized XPS test data have been developed for peak fitting (Conny and Powell, 2000).

**Peak Position and Half-Width.** The remaining peak parameters, $BE_i$ and $\Delta w_i$, are determined by the position of the peak maximum and the width at half maximum height, respectively. Unlike peak area, they can routinely be determined by eye from XPS spectra. Be careful in such cases. A common mistake made in setting the positions of components in an overlapping feature by eye is to set the peak positions too close together. Underlying component peaks, when properly resolved using peak fitting, will end up being further apart than the result obtained "by hand." This is especially true in cases where one or more of the peaks is a shoulder on a larger peak. Consider in particular the spectrum in Figure 7. Look carefully at where the component peak positions actually are, according to the peak fit components, in comparison to where an otherwise untrained observer might set them "by eye" on the spectrum of the raw data.

In conjunction with the above, before calculating the FWHM of a peak "by hand," remember to consider the peak height after removing a background. This is done by drawing a baseline by hand, typically a sloping, linear background (not just a constant baseline). Someone doing an analysis of an XPS spectrum by hand for the first time is apt to make a common mistake with overlapping peaks by drawing two separate, linear baselines, one under each overlapping peak. While this may be an appro-

priate method for such techniques as infrared spectroscopy, where the baseline shape can be dramatically curved throughout a narrow spectral range, it is entirely inappropriate in XPS spectra, where the baseline extends smoothly under all peaks associated with the spectral feature. Review again the sloping linear baseline shown in Figure 6 as a reference; the slope is not disjoint at the junction of the two satellite peaks.

Depending on the computuation package used for peak fitting, peak parameters may be selected through autodetection routines. One method to locate a peak is by the position of maxima in a residual spectrum. The residual spectrum is determined as the difference between the original spectrum and the summation envelope of peaks already fit to the original spectrum. The FWHM for the peak in the original data can also be estimated by the peak width at half the height of the residual spectrum peak. Another method is by scanning derivatives of a spectrum for minima and zero crossings. Assymetries in the shapes of the derivatives can help determine whether closely overlapping peaks exist in a spectrum that otherwise appears to be a single peak. Numerical differentiation is inherently a smoothing process that requires good S/N in the original spectrum to avoid distoration of the derivatives (smoothing) or loss of signal (increase in noise). Either method may have difficulties resolving peaks that closely overlap. New methods that involve wavelet transformation may also prove useful in autodetection of peaks in a spectrum (Ehrentreich et al., 1998) and for improving the fitting process (Zhang et al., 2000).

**Principal Component Analysis.** Principal component analysis (PCA) is a strong multivariate statistical data analysis tool applied to a variety of fields from science to engineering (Rummel, 1990). It is a potent technique in the arsenal of chemometrics (Malinowski and Howery, 1980) and surface spectroscopies (Solomon, 1987). It has been applied to resolve problems in XPS (Sastry, 1997; Artyushkova and Fulghum, 2001; Richie et al., 2001; Gilbert et al., 1982; Simmons et al., 1999; Balcerowska and Siuda, 1999; Clumpson, 1997; Holgado et al., 2000) and the related surface spectroscopic technique AES (Balcerowska et al., 1999; Passeggi, 1998). Applications of PCA to XPS include reduction of noise (Sastry, 1997), resolution (deconvolution) of overlapped spectra (Gilbert et al., 1982), resolution of XPS spectra into surface and bulk components (Simmons et al., 1999), subtraction of inelastic background signals (Balcerowska and Siuda, 1999), and detection of principal components (Clumpson, 1997; Holgado et al., 2000).

In mathematical terminology, PCA transforms the space of data from a set of possibly correlated variables into a smaller number of uncorrelated variables. This is expressed simply in matrix format as

$$|S| + |W|\Lambda|P| \qquad (26)$$

where $|S|$ is data (as a matrix), $|\Lambda|$ is a weightings matrix, $|L|$ is an eigenvalue matrix, and $|P|$ is the matrix of principal components. The eigenvalue matrix is a square diagonal matrix that is obtained from the data through singular value decomposition, the principal component matrix has the same dimensions as the data matrix, and the weightings determine how each principal component (scaled by its eigenvalue) is factored to obtain each part of the data. The eigenvalues $\lambda_i$ in $|\Lambda|$ are arranged in order from largest to smallest across the diagonal. This means, components in $|P|$ are arranged in order from most to least significant in contributing to $|S|$. The statistical significance of each component in $|P|$ can also be determined through indicator functions.

In terms of spectroscopy, PCA transforms a set of spectra from a system into a set of independent, representative spectra for the system (the principal spectra). The principal spectra can serve to regenerate all of the features in any of the original spectra. The data matrix $|S|$ contains the experimental spectra ($m$ spectra of $n$ data points each), all from the same region (BE range in XPS). The principal component matrix $|P|$ contains spectra ($m$ spectra of $n$ data points each) in order starting from one that contributes the most to $|S|$. Peaks that appear consistently throughout spectra in $|S|$ will appear in the first principal spectrum in $|P|$. The least significant principal spectrum in $|P|$ generally will only be noise because the magnitude of noise contributes stochastically throughout all spectra in $|S|$.

The utility of PCA in XPS is primarily to decompose high-resolution spectra. Recall that high-resolution spectra are typically taken to determine oxidation or hybridation states of an element, and the desired information is traditionally obtained using peak fitting. Principal component analysis can be considered to have three advantages over peak fitting. Peak fitting often requires that certain parameters of component peaks be well-defined and constrained to obtain realistic results, whereas PCA requires no *a priori* or constrained information about the principal spectra. Peak fitting typically requires subjective input from the user during the analysis, whereas PCA is entirely objective in its analysis. Finally, peak fitting requires significantly more time to analyze one spectrum than PCA takes to return results from a set of spectra. To its disadvantage for applications in XPS, PCA does not return results that can be directly and unambiguously related to specific information about the chemistry of the system. For example, PCA often returns principal spectra that show negative going peaks and therefore would otherwise appear to be entirely unrealistic.

The number of spectra $m$ provided in $|S|$ is important in order to be able to perform proper PCA. It should be significantly (from a statistical sense) larger than the number of significant principal spectra expected in $|P|$. If $m$ is too small, then all spectra in $|P|$ will appear to be equally significant in contributing to $|S|$. As a guide, $m$ should usually be at least a factor of three or more times larger than the number of significant spectra expected from $|P|$. For XPS, the number of significant spectra expected from $|P|$ can be taken as the number of component peaks that will otherwise be fit to a spectrum in $|S|$ with peak fitting. For peak fitting, the minimum value of $m$ is the number of samples with different types of chemistry. To increase this minimum $m$ to a more reliable value for PCA, spectra can be acquired from different samples of the same type,

from different regions of the same sample, or even from different orientations (analyzer acceptance angles) of the same region on a sample.

In summary, peak fitting and PCA are complementary and supplementary tools to analyze XPS spectra. Peak fitting requires subjectivity and time to analyze the data (fit peaks to a spectrum), whereas PCA requires subjectivity to interpret the results (determine the physical significance of the principal spectra) and time to obtain the data. The results from PCA are obtained objectively and can guide the otherwise subjective analysis done with peak fitting. For peak fitting results to be statistical reliable, the number of component peaks fit to any spectrum in $|S|$ cannot exceed the number of significant principal spectra determined by PCA to be in $|P|$. In XPS, when more component peaks are used to peak fit than can be supported by PCA, the chemistry of the samples has been unjustifiably represented. Alternatively, when fewer components are used to peak fit than number of significant components found by PCA, certain chemical states of the element being considered have been neglected.

**Appearance.** When signal quality is poor, as can be the case in spectra from elements at low concentrations, improving the appearance of an XPS spectrum is sometimes considered an important part of data analysis and interpretation. Methods that increase the signal-to-noise ratio, $S/N$, are applied with the goal of enhancing the results. The use of signal smoothing procedures to improve the precision of quantitative results, particularly elemental compositions as well peak positions and half-widths, must be approached with care. While signal smoothing can be a useful tool, over-reliance on smoothing is a mistake.

The first concern in data acquisition should be to obtain a spectrum with sufficient $S/N$ to obtain reliable quantitative information. The $S/N$ ratio is, in practical terms, the height of a photoemission peak after subtracting a background divided by the root mean square of the intensity of the random noise at a location somewhere near the photoemission peak. The random noise is a stochastic process proportional to a deviation about a mean signal. Increasing the acquisition time in a data channel proportionally decreases the variance of a signal about the mean. Noise therefore increases as the square root of acquisition time. Signal increases proportionally with acquisition time. Therefore $S/N$ increases as the square root of total acquisition time. The formulations provided in the previous section can be used to determine the corresponding relationships to scan, sweep, and dwell time as well as number of scans or sweeps. The value of $S/N$ increases as the square root of any one of these when all others are held constant. In practice, $S/N$ ratios of $\sim$2 to 3 are necessary lower bounds for determining peak parameters. Spectra with such low values of $S/N$ are generally not suitable, however, to determine peak parameters reliably. A more reasonable $S/N$ is above $\sim$10. The raw data in Figure 7 has a value of $S/N$ value of $\sim$75 to 80, while that for CuO in Figure 6 has a value of $\sim$20 to 25. To obtain a comparable value of $S/N$ for these two spectra, we would have to acquire the latter spectrum (Figure 6) for a 16- to 25-fold longer acquisition time.

Smoothing routines are used when data acquisition has failed to obtain sufficient S/N. Smoothing algorithms are, however, not without their problems. Those that depend on n-point averaging methods are notorious for changing peak parameters. Although this change may be below the level of uncertainty in measuring the peak parameters, a change will occur. The FWHM of a peak will be affected in all cases. For overlapping peaks, peak positions will shift, and asymmetrical peaks will change shape. A theoretically more rigorous smoothing algorithm is performed in the time (Fourier) domain using appropriately defined Fourier filter functions. This has been shown to distort peak shapes less. Finally, the advent of algorithms to perform noise reduction in chemical spectroscopy using PCA (Sastry, 1997) or wavelet analysis (Jetter et al., 2000; Shunsuke, 2000; Harrington, 1998) may offer a choice of a better post-processing method to improve the S/N in XPS spectra without distorting peak parameters.

Smoothing is to be avoided in cases where rigorous and exacting quantitative analysis of XPS spectra is desired. Nothing substitutes in this case for acquiring more data to increase S/N in the raw spectrum. Peak fitting is also best done to raw (noisy) data. To a first order, the optimization algorithms essentially "do not care" about the level of random noise in a spectrum. Smoothing of noisy spectra is appropriate to provide the viewer with a guide to follow the otherwise noisy signal. This is what has been done in Figure 6. The author strongly recommends that, when smoothing is used as an integral method of the data analysis and interpretation, both smoothed and raw (noisy) spectra always be shown simultaneously in the published XPS spectra.

One primary objective of high-resolution analysis with XPS is to obtain spectra that are not distorted by line shape broading due to the x-ray source or analyzer. On systems without a monochromator, computer software methods exist to deconvolve the line shape of the x-ray source. They depend traditionally on iterative routines or operate in the Fourier domain (Jansson, 1984). They all require accurate representations of the line shapes of the x-ray source, as published elsewhere (Krause and Ferreira, 1975). They are not without their problems. Spectra with low S/N or with inadvertent "spikes" will not deconvolve well. Methods to deconvolve x-ray lineshapes are falling out of favor, especially as monochromatic x-ray sources become more popular. Alternative (neither Fourier or iterative) deconvolution techniques however are being used successfully to improve peak resolution. They include the maximum entropy method (Pia and McIntyre, 2001; Pia and McIntyre, 1999; Pratt et al., 1998; Splinter and McIntyre, 1998; Schnydera et al., 2001) and PCA (Gilbert et al., 1982; Simmons et al., 1999). These methods appear to avoid some of the problems with the traditional Fourier or interative deconvolution methods and can be used to remove broadening effects from the analyzer.

Finally, some older XPS systems without a monochromator offer a software option to remove x-ray satellite lines. While this can be useful in survey scans, it has little utility in high-resolution scans. The scan range can be set to avoid the x-ray satellite. If an x-ray satellite interferes with a principal photoemission peak and accurate quantitative analysis is desired, the better option is to use an XPS system with a monochromatic x-ray source.

### Interpretation

The results obtained from XPS are a list of the elements present, the compositions, and the positions and half-widths of their associated peaks. How this information is to be interpreted is an important aspect of the proper use of XPS. The first section below deals with interpretation of the compositions from XPS. This is followed by a discussion about defining chemical states.

**Compositions.** As pointed out in the section on practical aspects of the technique, compositions determined by XPS are to be interpreted as arising from homogeneous samples. This leads to problems when the sample composition is not uniform. In particular, most samples analyzed immediately after insertion into the vacuum system will contain a layer of carbonaceous material, adsorbed water, and oxides depending on the type of material. The contaminant layer can be a few monolayers thick in some cases. This approaches the mean free path of most low KE electrons. Attenuation of peak intensities from the underlying material is therefore to be expected. The consequence of this is that atomic concentrations will have relative uncertainties approaching $\pm 10\%$ or more in most practical cases.

Consider a rather simple example. A monolayer of (graphitic) carbon on an iron surface will have thickness of $\sim 0.1$ nm. The prominent Fe 2p line appears at $\sim 706$ eV in BE. For Mg K$\alpha$ radiation, this corresponds to $\sim 550$ eV in KE, for a mean free path of $\sim 1$ nm. The intensity of electrons from in the Fe 2p peak will be attenuated by a monolayer of carbon to $\sim 91\%$ at 90° take-off angle, 87% at 45°, and 56% at 10°. Based on Equation 15 and assuming sensitivity factors do not change, the same sample of otherwise pure Fe with only a uniform monolayer of carbon on its surface would therefore appear to have compositions of 9%, 13%, and 44% carbon at the respective take-off angles.

Most contaminant layers are thicker than a monolayer. Doubling the thickness of the layer nearly doubles the respective carbon compositions. For this reason, unexpectedly high values of atomic concentrations for carbon and oxygen on samples just inserted into the vacuum system are not unusual. Values as high as 50% to 60% carbon due to a contaminant overlayer are not uncommon, even on supposedly well cleaned surfaces. Metals (other than Au) will also always be covered by an oxide layer. Methods that can be used to reduce this contaminant overlayer are discussed in the section on sample preparation.

The C 1s peak appears at $\sim 284$ to 286 eV in BE. All photoemission peaks at higher BE than the C 1s line will likely have lower mean free paths. The concentrations of the corresponding elements will therefore be attenuated relative to carbon. The peak intensities for those elements below the C 1s peak in BE will be attenuated less.

An important objective of XPS analysis for sample composition should also be to estimate the uncertainties in composition values. A thorough guide to uncertainty analysis of data can be found elsewhere (Taylor, 1997).

Uncertainties in atomic concentrations determined from XPS generally contain two contributions, random (statistical) uncertainties and measurement errors. Statistical uncertainties are reduced by repeating the measurements multiple times. Certainly, more than one sample should be analyzed to obtain valid concentrations for the sample. A general guide in numerical data analysis is that at least ten measurements of a value are needed to begin to have a sample size that statistically approaches the behavior of a true representative set for the system. Small-sample statistics must otherwise be applied. In this regard, methods to report on the confidence level for values and to propagate uncertainties in values through calculations should be followed. For any composition value reported, one should, at the minimum, always report an average, the calculated standard deviation, and the number of samples used to obtain the value.

The extent of uncertainty in composition that arises due to measurement errors in an XPS spectrum is considered below for illustration. First, assume an ideal sample, where the composition is homogeneous throughout. For any one given measurement made to determine concentration with XPS, the relative uncertainty in absolute concentration scales in proportion to the relative uncertainty in measuring the peak area and the sensitivity factor. The peak area $A_i$ and element sensitivity factor $S_i$ enter in the calculation of atomic composition $x_i$ (atom or mole percent) as a ratio $K_i = A_i/S_i$. The errors in measuring peak areas and element sensitivity factors may be correlated because they are measured using the same devices in both cases. For such situations, an upper boundary on the relative uncertainty in composition due to measure errors can be established as

$$\frac{\Delta x_i}{x_i} \leq \frac{\Delta K_i}{K_i} + \frac{\sqrt{\Sigma \Delta K_j^2}}{x_i K_i} \quad (27)$$

where $\Delta K_i^2 = K_i^2 [(\Delta A_i/A_i)^2 + (\Delta S_i/S_i)]^2$. Note that the terms in the above equation would be added in quadrature if the measurement errors for area and for sensitivity factor respectively were uncorrelated from all other values in their sets. The above formulation is still valid to establish a conservative upper boundary for the influence of measurement errors on composition. Relative errors in $S_i$ values can be as much as $\pm 10\%$, especially when the instrument has not been calibrated. The greater the number of elements considered, the greater the summation over all values of relative uncertainty in sensitivity factor will become. Therefore, the relative uncertainty in any one composition will increase as more components are considered. Relative errors in peak areas depend on how the measurement was made (by hand or by computer, for example) and on the choice of background. They can be reduced by making sure that the absolute peak area, $A_i$, is as large as possible. The effects of excessive noise on a spectrum complicate matters and will generally act to increase the value of $\Delta A_i/ = A_i$.

With an assumption that sensitivity factors are accurate, ($\Delta S_i/S_i = 0$), $\Delta K_i/K_i = \Delta A_i/A_i$, and assuming all

peaks areas are measured to the same degree of relative uncertainty, $f_a = \Delta A/A$, and recalling for any two elements that $K_A/K_B = x_A/x_B$, the above equation can be recast to

$$\frac{\Delta x_i}{x_i} \leq f_a \left[ 1 + \frac{\sqrt{\Sigma x_j^2}}{x_i^2} \right] \quad (28)$$

This shows that, all else being equal, components with the lowest composition $x_i$ will have the largest relative uncertainty in their composition $\Delta x_i/x_i$. It also shows that, as more components are incorporated in a calculation of composition, the relative uncertainty in the absolute composition of any one component $\Delta x_i/x_i$ increases. The above equation also can be applied for the case where peak areas are measured with infinite accuracy and only sensitivity factors have measurement errors, in which case $f_a$ is the relative error in sensitivity factor. Relative uncertainties in concentration ratios for any two components A and B are expressed as

$$R_{AB} = \left( \frac{x_A}{x_B} \right) \leq \frac{\Delta K_A}{K_A} + \frac{\Delta K_B}{K_B} \quad (29)$$

When either sensitivity factors or peak areas are considered infinitely accurate, this expression becomes

$$R_{AB} \leq f_{a,A} + f_{a,B} \quad (30)$$

where $f_{a,i}$ is the relative error in determining the peak area or sensitivity factor for component $i$. An important point of this equation is that, to a first approximation, uncertainties in the ratios of component compositions are not affected by the number of components being considered or the absolute compositions of the components. For this reason, composition ratios from XPS can be considered more significant (less uncertain) than absolute concentrations.

Order of magnitude numerical calculations can serve to illustrate the above formulations. First consider how increasing the number of components in a system increases the relative uncertainty in the composition of each component. When either sensitivity factors or areas are considered accurate, for systems with $n$ components of equal composition (ideal stoichiometric compounds), the relative uncertainty on any one composition becomes $f_a(1 + n\sqrt{n})$, where $f_a$ is the relative uncertainty in either sensitivity factor or peak area. The multiplier to $f_a$ is 3.8, 6, 9, and 12 for systems of two to five components of equal composition, respectively. Specifically, for a 1:1 stoichiometric binary material such as CuO or NaCl, when the relative error in measuring peak areas or elemental sensitivity factors is 5% ($f_a = 0.05$), the relative uncertainty in measuring the composition of either component will be $\pm 19\%$. Therefore, the composition for either component in such a binary could only be measured to an accuracy $x = 0.50 \pm 0.10$ (atom fraction) using XPS. For this case, XPS measurements of compositions between 40 atomic percent and 60 atomic percent could not be distinguished as statistically different. By further example, for a five-component material of equal compositions measured with a 5% relative error on peak areas or sensitivity factors, the absolute composition of any one component could only be measured to a precision of $0.20 \pm 0.12$ atomic percent. Incorporation of the errors for both area and sensitivity factors in the above formulation only serves to increase the overall inaccuracy in measuring the true value of composition. These examples illustrate clearly why absolute atomic concentrations reported from XPS should typically be considered to have a relative uncertainty (imprecision) of at least $\pm 10\%$ due to measurement errors. Calibration or otherwise clear validation of the sensitivity factors for the spectrometer being used and affirmation of stringent care in measuring peak areas should typically be sought before accepting reports of atomic concentrations with relative uncertainties lower than $\pm 10\%$.

A corresponding example for the concentration ratio $R_{AB}$ can be developed. When either peak areas or sensitivity factors are accurate, the relative error in any value of $R_{AB}$ is the sum of relative errors in the individual terms. For the above systems of two to five components, in cases where $f_a = 0.05$ for all components, the value of $R_{AB}$ will be 0.10, or 10%. For the binary CuO, the ratio of copper to oxygen can be measured at best to an accuracy of $1.0 \pm 0.1$. For a five-component system of equal compositions throughout, the ratio of any two compositions can be measured at best to an accuracy of $1.0 \pm 0.1$. The relative error in measuring a concentration ratio does not depend on the number of components included in the analysis. This illustration can serve to help save analysis time. Suppose that changes in concentrations of only two components are of overriding importance in the analysis of a multicomponent system. Only XPS spectra from those two components will need to be obtained. Although absolute concentrations will not be possible to determine, the ratio of concentrations will be a reliable indicator of the behavior of the system.

**Chemical State Information.** In regard to elemental and chemical state identification, reference spectra have been published. One standard for reference spectra from elements are the handbook from Perkin-Elmer Corporation (Moulder et al., 1995). A reference book has also recently been made available on polymers (Kratos Analytical, 1997). Further reference libraries of XPS spectra are available from commercial vendors such as XPS International (*http://www.xpsdata.com/*); the journal *Surface Science Spectra* publishes experimental XPS spectra taken under well-defined conditions (sample type and treatment and instrument parameters) as references for other researchers.

Unambiguous interpretation of shifts and broadening of primary photoemission peaks in XPS spectra should be based on systematic elimination of the various factors that contribute to these effects. The first factor to consider and eliminate is the potential for sample charging. This artifact has been mentioned already and will be discussed in detail in the section on problems with the technique. The following discussion is based on the assumption that sample charging does not occur or that peak shifting or broadening effects due to sample charging have been eliminated or, more likely, appropriately accounted for. Consideration is based first on comparison of XPS spectra from a "bulk" material for a before and after situation—before a treatment and after the treatment. Overlayers are considered separately. Comparison of XPS spectra from one experimental system to another, in order to find similarities in the chemistry of the materials, is then considered.

For the same instrument conditions, relatively small amounts of peak broadening without a commensurate peak shift or appearance of a new peak is indicative of a movement toward greater heterogeneity in the sample chemistry or structure. The effects may go hand in hand. A change in structure may also be associated with subtle changes in the type of bonding between constituent elements in the material. These effects are often only worth pursuing in detail for samples with otherwise well defined chemistry and structure.

Peak broadening to such an extent that a new peak appears in close conjunction with the initial peak usually indicates a change in the chemistry in a portion of the material. The new peak may overlap with the initial peak so that peak fitting is required to resolve it. As long as some indication of the initial peak is present, assignment of the new peak to chemical state effects can usually be done with confidence. Note that changes in structure without changes in chemistry of a bulk material can also give rise to a new peak. Such changes are not as prevalent in materials.

When a peak shifts or significantly changes shape such that the initial peak disappears, the chemistry or structure (or both chemistry and structure) of a bulk material has changed. In some cases, these factors cannot be separated. Reference XPS spectra should be consulted. The changes may be obvious, as shown in the comparison of the Cu $2p$ peaks in Figure 6. In cases where reference spectra are not available, the chemistry and structure of the material are best probed with other techniques that provide supplemental information (e.g., x-ray diffraction for structure).

Thin film overlayers have unique properties that make interpretation of changes in their XPS spectra more difficult. Oxide layers on metals and other comparable insulator-like overlayers are prone to charging, especially as their thickness increases. Molecular overlayers on surfaces can show peak shifts due to changes in the degree of extra-atomic relaxation in the layer, and changes in the shapes of satellite peaks are often associated with changes in chemistry. Angle-resolved XPS is one method that can be used to accentuate "bulk" versus near surface behavior. Where possible, thin film adsorbates should also be analyzed with techniques such as attenuated total reflectance Fourier transform infrared (ATR-IR) spectroscopy, glancing angle FTIR, EELS, AES (AUGER ELECTRON SPECTROSCOPY), surface Raman (RAMAN SPECTROSCOPY OF SOLIDS), or thin film x-ray diffraction (XRD; X-RAY MICROPROBE FOR FLUORESCENCE AND DIFFRACTION ANALYSIS) to confirm proposed changes in chemistry or structure.

In conclusion, analysis of chemical state information using XPS is rarely without ambiguities in complex

**Table 2. Peak Areas, Element Sensitivity Factors, and Atomic Compositions from Analysis of Si Wafer Using XPS**

| Orbital | Peak Area (Counts-eV/sec) | Sensitivity Factor | Atomic Concentration (Atomic %) |
|---------|---------------------------|--------------------|---------------------------------|
| O 1s    | 94960                     | 0.711              | 42                              |
| C 1s    | 11964                     | 0.296              | 13                              |
| Si 2p   | 48366                     | 0.339              | 45                              |

systems. The technique should not be viewed as a definitive one. Analysis of the material with as many techniques as appropriate to the system at hand, and comparison to published reference spectra or spectra from similar materials, is often highly recommended.

**Example.** The example presented here concerns analysis of an Si wafer for composition, chemical state information, and oxide film thickness. A survey scan, obtained with a PHI 5400 system using Mg radiation at 15 kV and 325 W at a take-off angle of 45° and pass energy of 89.5 eV, is shown in Figure 4. The main photoemission peaks and Auger KVV transitions are labeled. Plasmon loss peaks predominate for both the O 1s and Si peaks. The x-ray satellite peaks are also visible for these lines. Peak areas, sensitivity factors, and compositions determined from the survey scan are given in Table 2.

High-resolution scans of the Si 2p peak taken with a pass energy of 8.95 eV at three different take-off angles are shown in Figure 13. Two distinct states of Si are apparent. The peak at higher BE is from SiO$_2$, and the lower BE state is from Si. The SiO$_2$ peak is from the oxide layer on top of the wafer. The relative intensities of the two peaks change as a function of take-off angle. At glancing angle ($\Theta = 10°$), the peak from SiO$_2$ predominates because analysis is more surface sensitive. The asymmetry in the Si peak for the Si state, clearly visible at 45° and 90° take-

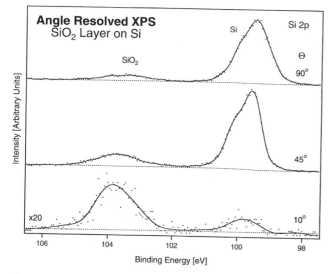

**Figure 13.** High-resolution scans of the Si 2p peak from a Si wafer at three different take-off angles. The oxide and neutral states of Si are clearly visible.

off angles, is due to the close overlap of the Si 2p$_{3/2}$ and 2p$_{1/2}$ peaks.

The areas under the two peaks were determined at each take-off angle by numerical integration (trapezoid rule) with computer software. The ratios of the SiO$_2$/Si peak areas were 4.71 (10°), 0.248 (45°), and 0.125 (90°). This data can be used in Equation 22 to determine the thickness of the oxide film. In this case, the cross-sections and mean free paths for the overlayer and substrate peaks can be taken to be the same since they are both the same orbital. Because an HSA was used for analysis, the ratio of étendue values is an inverse ratio of the KE of the peaks, which in this case is

$$\frac{G_o}{G_s} = \frac{E_{Ks}}{E_{Ko}} \approx \frac{(1253.6 - 99.4)}{(1253.6 - 103.4)} \approx 1.0 \qquad (31)$$

Equation 22 therefore reduces to

$$\frac{I_o^m}{I_s^m} \approx \frac{(1 - \exp(-d/\lambda\sin\theta))}{\exp(-d/\lambda\sin\theta)} \left(\frac{C_o}{C_o}\right) \qquad (32)$$

This can be rewritten as

$$\frac{d}{\lambda} = \ln\left[\frac{(A_o/A_s)}{(C_o/C_s)} + 1\right]\sin\theta \qquad (33)$$

We can estimate the ratio of atomic concentrations for Si in the oxide layer and the underlying substrate to a first approximation. This can be done by using the densities and atomic concentrations in the layers

$$C_o/C_s = (x_{si}\rho_o M_s/\rho_s M_o) \qquad (34)$$

where $x_{Si}$ is the mole fraction of Si in the oxide (0.33 for SiO$_2$), $\rho_i$ is the mass density of i, and $M_i$ is the molar mass. If the mass densities of Si and SiO$_2$ are taken to be equal, the value of $C_o/C_s$ is ~0.16. Using the area ratios from above, the ratio of film thickness to mean free path is determined to be 0.59 (10°), 0.66 (45°), and 0.58 (90°). The average ratio is 0.61 ± 0.04. Escape depths for the Si 2p peak are ~2.0 to 3.0 nm. Using an average of 2.5 nm, the oxide film thickness is found to be 1.5 ± 0.2 nm.

Before reviewing a more rigorous approach to analysis of the above ARXPS data, we can consider where mistakes are typically made in the above simple analysis. The first common mistake is to proceed without confirming that the ratios of cross-sections, mean free paths, analyzer étendues, and concentrations are unity. In the above calculation, since closely spaced peaks from the same orbital were used, all of the ratios but that for the atomic concentration could be safely approximated as unity. Had the ratio of concentrations also been taken as unity, the film thickness to mean free path ratios would have been determined to be 0.30 (10°), 0.16 (45°), and 0.12 (90°). This would have led to a reported oxide film thickness of 0.5 ± 0.2 nm. The significant difference between this value and the above, due to a failure to account for concentration differences, should be noted. When ARXPS data from two different peaks are used, the other ratios will not

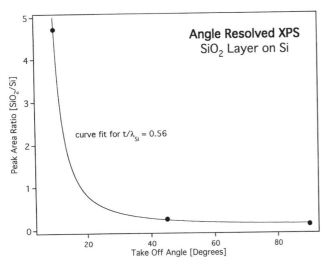

**Figure 14.** Results from nonlinear curve fitting of the peak area ratios to the equation for angle-resolved XPS. The solid line shows the fit for the given ratio of film thickness to electron mean free path.

necessarily be unity either. A second common mistake is to use ARXPS data from one take-off angle and apply Equation 22 in a significantly reduced form

$$d/\lambda = \ln[(A_o/A_s) + 1] \qquad (35)$$

This equation can only be used for data taken at 90° take-off angle under situations where the above mentioned ratios can confidently be set to unity. As seen by the above sample calculations, gross inaccuracies can be expected in all other cases.

A better approach to the determination of film thickness from ARXPS data is to use the full compliment of data for non-linear curve fitting to Equation 22. The ratios of terms can either be determined or left lumped as a single fitting parameter. An example of the results using this method is shown in Figure 14. The three area ratios were fit to Equation 22 using a non-linear $\chi^2$ optimization routine. The ratios of cross-sections, étendues, and concentrations were lumped into a single parameter, the mean free paths from the oxide and substrate were taken to be equal, and the other fitting parameter was the ratio of film thickness to electron mean free path. The results show good agreement with a fit to $d/\lambda$ of 0.56. The lumped parameter was close to unity. As the figure shows, ARXPS data should be taken near glancing take-off angles for the best sensitivity to film thickness. The data above ~45° contribute far less to the shape of the fitting curve. The confidence in the fitting would be improved by having data at 10° increments from 10° to 90°. Deviations that arise in fitting ARXPS data will also be more apparent at low take-off angles, as discussed in the section on problems with XPS.

## SAMPLE PREPARATION

Materials surfaces exposed to atmospheric conditions will contain a layer of contaminants, primarily adsorbed water and hydrocarbons. One objective of sample preparation prior to analysis with XPS could be to remove this layer without affecting the chemistry of the underlying material. For samples with intentionally applied molecular adsorbate layers, treatments that remove the contaminant layer may also remove the adsorbed layer, so that special precautions may be needed. Sample preparation should also focus on obtaining a representative cross-section of the material for analysis. When the extent of spatial variations in composition across a sample are expected to be larger than the analysis area at the sample surface, the recommended practice is to analyze over multiple regions on the sample surface. Finally, the optimal place for preparing a sample is often in the same vacuum chamber as the XPS analysis.

Because of constraints on the sample introduction ports of typical XPS vacuum chambers, samples are usually restricted to be a certain size. Typical sample sizes are ~3 to 5 mm thick and small enough to fit in a circle of 1 to 3 cm diameter (depending on the XPS instrument). Some newer XPS systems have been designed to handle larger size samples, up to 10 to 15 cm diameter (for example, for the Si wafer industry). The geometry of the sample is sometimes also restricted. Flat samples (like a hockey puck) are ideal in all systems. Curved surfaces and crevices present problems because either the x-ray source or the analyzer can be "shadowed" by the sample. Analysis cannot be done on hidden inner surfaces of samples.

Samples must be low out-gassing materials so that the pumps on the XPS chamber can maintain a suitable vacuum during sample analysis. This generally means they should have vapor pressures well below $10^{-8}$ mbar. Some XPS systems include a cold stage that cools the sample during analysis, potentially to as low as liquid $N_2$ temperatures. This can be beneficial in reducing the vapor pressure of otherwise high-vapor-pressure materials. Polymer samples are notorious for their outgassing problems. Small samples and thin films are best to use in this case, to minimize the total amount of material that outgasses. Porous materials and powders often present the same problems with extensive outgassing. They can often be analyzed if they have been well outgassed beforehand. This can often be done using a sample-preparation chamber that is attached to the main vacuum chamber containing the XPS instrumentation. In certain cases where ultraclean vacuum conditions in the main XPS analysis chamber are essential (with pressures below $10^{-9}$ mbar), samples such as polymers and powders that can potentially outgas a great deal are not permitted into the main chamber for XPS analysis. The policies on this vary from laboratory to laboratory.

For bulk-like sample analysis with XPS, the first step in sample preparation is to select a representative sample of the material being considered. Homogeneity of sample composition and structure are important factors. Samples with heterogeneities in composition or structure, such as grains or phase microstructures in metals or phase domains in copolymers, can be analyzed successfully either by analyzing over a larger area relative to the size of the grains or domains or by selectively analyzing over a statistically meaningful number of well-defined smaller areas across the sample surface. Samples with thin film overlayers should be considered as candidates for sputter profiling or angle-resolved XPS.

Cleaning samples for bulk-like analysis does not typically require a significant effort unless the material is corroded or dirty. Sample surfaces can generally be washed and degreased by standard means using solvents such as water, methanol, or acetone. The solvents should be low in their content of residual contaminants—distilled, deionized water and research grade solvents are recommended. Cleaning the sample with an aqueous detergent wash and distilled water rinse in an ultrasonic cleaner has more recently become a recommended final step. In addition, the use of an ozone-generating UV lamp to reactively remove hydrocarbons is becoming more widely adopted. Polishing is not always necessary; however rough or pitted surfaces tend to entrap contaminants more readily than smooth surfaces. If polishing is done, extra care should be given to making sure to clean the polishing compound from the sample surface.

Prior to XPS analysis for bulk-like information, the thin outermost contaminant layer on a sample can often be removed by reactive ion etching. Sputter cleaning to remove the first few surface layers on a sample is often a routine part of the XPS analysis procedure. Excessively rough or pitted samples will not sputter-clean adequately. Insulator and polymer samples also present problems. These materials generally charge while being sputtered. This can cause uneven sputter rates across the surface. Polymers can also degrade during sputtering.

Preparation of samples with thin-film, adsorbed layers is generally done using well-defined recipes specific to each system. When ultraclean surfaces are a requirement and sputter cleaning is not a viable cleaning method, well-defined sample-preparation methods using ultraclean chemicals can be essential. Even the simple task of rinsing a sample in a beaker of solvent subjects its surface to the potential for further contamination. Under normal conditions, the air-liquid interface of solvents in beakers collects a layer of adsorbed contaminants, and these contaminants can transfer to a sample surface when it is passed through the air-liquid interface. As illustrated above (see Practical Aspects of the Method), even a monolayer of contaminants attenuates the underlying signal intensity, leading to inaccurate results.

Some XPS facilities are equipped with a stage where the sample can be heated while it is in the vacuum. This can be used to desorb chemically adsorbed overlayers. Some XPS systems also have sample-preparation chambers where the samples can be exposed to reactive gases (oxygen). This can be used to react off contaminant layers. Finally, XPS systems can also be equipped with all manner of tools within the vacuum system to cleave, cut, notch, or slice a sample. This is the best method to obtain a pristine surface.

Preparation of powder samples for XPS analysis presents unique problems. In almost all cases, free-standing powders are not put into the main vacuum system for XPS analysis. Should the powders spill, the potential for damage to vacuum components inside the chamber, especially the pumping units, is far too high. Therefore, pow-

ders must be mounted securely to a holder in some manner. Common techniques for mounting powders include the use of double-sided tape, mounting wax, and soft metal foils. Double-sided tape is usually of a special quality that is used extensively in vacuum systems because of its low outgassing properties. Powders can also be pressed or sprinkled into thermoset or thermoplastic mounting waxes that are subsequently allowed to cure or harden. Finally, soft metal foils, in particular indium or gold, are often used to mount powders. The powder is pressed firmly into the metal. Typically, only a single layer or two of particles that are uniformly distributed across the sample surface is necessary to get a reasonable signal. In cases where the powder is somewhat self-adherent when compacted, a thicker layer can often be formed on the tape, wax, or metal foil. In all of these cases, the resulting mounted powder sample should be tapped firmly while its surface is in a vertical orientation in order to remove all loosely adhering particles.

Particles may also be mounted as free-standing samples after being tightly compacted into free-standing pellets. The pellets should be stable to the routine stresses involved in handling and mounting. They should not crumble easily. When pellets are used, they can also be mounted in sample holders that have wells to hold the pellet. Finally, powders have also been mounted by compacting them somewhat and then containing them in a specially designed cup-like holder under a wire gauze. This requires the same care to contain loose particles from spilling as all the other methods.

One disadvantage of mounting particles in any form other than pellets is that the mounting material will inevitably appear in the XPS spectrum. Double-sided tape contains carbon, silicon (adhesives), and oxygen. So do most mounting waxes. The indium foil will have an oxide coating. This problem must be handled on a case-by-case basis. Another problem is that powders in any form are notorious for having a carbonaceous overlayer, and sputter cleaning is not usually a viable option. Finally, porous powders can have problems with outgassing, and XPS will not provide a representative analysis of materials within pores due to limitations on electron escape depth.

## SPECIMEN MODIFICATION

Under proper conditions, XPS typically does not modify the sample. Polymers are, however, sensitive to photoinduced reactions during XPS analysis. In cases where the photoemission yield or secondary electron background is high, damage to polymers may occur by electron-induced chemical degradation. Photostimulated desorption of gaseous adlayers on surfaces can also occur for all materials. Finally, if electron beam neutralizers are used to reduce sample charging, the electron current can cause the sample to degrade.

## PROBLEMS

The most significant problem in XPS analysis is sample charging. As photoelectrons are emitted from the sample, a continual current (sample current) must flow from ground. Insulating and even semiconducting samples also have measurable capacitance in addition to resistance properties. Oxide layers on metals behave in the same way. This leads to a build-up of charge on these samples, typically on the surface being analyzed. The charge build-up is always positive, leading to a shift of the photoelectrons to lower KE and therefore higher BE. The shift is uniform across the entire XPS spectrum by as much as 2 to 3 eV. Inhomogeneous samples can also have differential charging across the sample surface. This leads to peak broadening, since the extent of charging is not uniform across the sampled areas.

Three methods are available to take care of sample charging. One method draws on experiences in SEM and AES. The sample surface is coated with a thin (sub-monolayer) film of conductive metal, typically Au. This method requires a metal-deposition system that can coat samples. Such systems are commonly found at scanning electron microscopy (SEM) facilities.

A second method is to supply additional electrons to the sample from a secondary source. This is usually done using an electron gun or an electron-emitting filament. The electrons that are supplied are flooded at a very low $E_K$, below ~50 eV. There are afew problems with this method. First, even at low $E_K$, electrons are damaging to some classes of materials, particularly polymers and sometimes ceramics. Secondly, the exact amount of electron flood current to supply is dependent on the sample type and surface condition. Trial and error experiments can be required to find the flood current that exactly compensates for the positive charge build-up on a surface. Changing the sample or sample orientation relative to the flood gun will change the current needed. Finally, when sample charging is inhomogeneous, an electron flood gun will not reduce peak broadening. Despite these limitations, electron flood guns are commonly used to correct for sample charging in many XPS systems.

The third method is to post-process the spectrum. Since charging shifts the spectrum uniformly, post-correction requires shifting all of the peaks back (to lower BE values) by a uniform amount. The problem with this method is that a reference is needed to align the peak positions. Without prior knowledge of the exact oxidation states of all principal peaks in the spectrum, the amount of charging correction shift to apply is entirely arbitrary. The most common standard internal reference to compensate for charging is to align the main C 1s peak to a value for graphitic carbon, 284.6 eV. Two problems arise with this. First, not all carbon is graphitic in nature. The C 1s peak position is particularly sensitive to changes in hybridization state, as seen in Figure 7. This means that the internal reference state may not be correct for the sample at hand. The second problem is that even when graphitic carbon is present, differential charging and the presence of other states of carbon will broaden the C 1s peak. If this broadening is asymmetric, the C 1s peak position will only be exact if it is determined by rigorous peak fitting procedures.

None of the above methods to correct for sample charging are reliable in all situations. When reporting peak positions from samples that may have charging shifts, one potentially better method of comparing chemical differences from sample to sample is to report relative peak positions. Define one peak that does not change position from sample to sample as an $E_B$ standard for the set of experiments. Report all other $E_B$ values relative to this standard peak. Changes in offsets from the reference peak should be a better indication of any possible changes in chemical state or final state effects. This is comparable to what is done in determining Auger parameters, where photoemission peak positions are calculated relative to respective XAES peaks.

As reported above (see Practical Aspects of the Method), the accuracy of the energy calibration of the XPS instrument should always be confirmed. Three common errors arise: offset, sloping, and nonlinear scaling. Offset errors are typically due to an incorrect setting of the analyzer work function. This will cause all of the peak positions to be off by a constant amount, toward either lower or higher BE. Offset errors are corrected by calibrating the work function of the instrument using the peak position of a well-established standard, typically the Au $4f_{7/2}$ line. Sloping errors typically arise when the electronics for the instrument have linear deviations from their required output voltages. This will cause peaks on one side of the calibration peak to shift to higher BE values and those on the other to shift to lower values, for example. This error is corrected by calibrating the instrument electronics for linearity across a wide BE range using at least two appropriately spaced reference peak positions. Finally, nonlinear scaling errors arise due to nonlinearities in the electronics. They will cause the BE of peaks to fall on a curved rather than straight line in a plot of measured BE versus expected BE.

One caution in operating XPS systems with conventional x-ray sources is to always confirm the integrity of the x-ray source window. The window is a thin metal foil at the end of the x-ray source between the electron filament and the sample. This x-ray window is thin enough to be nearly transparent to x rays, yet opaque to electrons. A typical x-ray window is Al. Small pinholes in the window will allow electrons from the x-ray filament to bombard the sample. This leads to an increase in the overall background of the spectrum without increasing the overall signal. The flood of high-KE electrons can also damage the sample, particularly polymers. The x-ray window evaporates over time and should be replaced as needed.

A second problem with laboratory x-ray sources arises as the anode degrades with time. The active metal coating (Mg or Al) on the anode (which is typically a block of Cu) evaporates with time or will oxidize if high water vapor or oxygen partial pressures are continually present. This problem leads to a significant decrease in S/N and signal relative to background intensity in the XPS spectrum. In systems with dual anodes (Mg/Al being most common), the metals from the two separate anodes can diffuse across the boundary separating them. This will cause the appearance of ghost peaks in survey scans. For example, when a Mg anode becomes contaminated with Al, every photoemission peak in a spectrum taken with the Mg anode will have a "ghost" peak at a value that is 233 eV higher in BE. The relative height of the ghost to the main peak will depend on the extent of contamination. To check for ghost peaks when using a Mg/Al dual anode system, look for the most intense photoemission peak, then look either 233 eV higher or lower in BE, depending on the original anode, to determine whether cross-contamination has occurred.

Finally, a concern that must be kept in mind when analyzing ARXPS data is that film roughness and analyzer acceptance angle will both act to degrade the quality of the results. The analyzer acceptance angle limits the angular resolution and lowest take-off angle that can be measured. Commercial instruments typically have half angles of acceptance at the input lens on the order of 10°; therefore measurements below ∼10° in take-off angle will show a significant decrease in signal intensity, and angular steps less than ∼10° will not add any significantly new information. Instruments are available with half angles of acceptance as low as 2°. Film roughness will also cause a deviation from Equation 22. Measured peak intensity ratios will be lower than the expected values at low take-off angles, and can be higher than expected at near normal take-off angles. Film thickness values estimated using Equation 22 that are less than the anticipated mean surface roughness of the film should be suspect.

## SUMMARY

The overall protocol for XPS analysis is generally to define the sample, determine the preparation and cleaning methods, mount and pre-pump the sample, load and align it, set up the analysis parameters, and obtain the XPS spectrum. After the sample has been prepared and cleaned outside of the vacuum system, the rest of the procedure can take anywhere from 15 min to a day to complete. Outgassing of the sample in a prep-vacuum preparation chamber is usually essential for porous or low-vapor-pressure materials.

A typical sequence of automated sample analysis on a modern XPS instrument proceeds after inserting a sample or set of samples. The user then defines, for each sample, the number of spatial regions and their locations, the number of acquisition sets, and the parameters for each set. High-resolution scans can be taken in sequence within a data acquisition set. This mode is called multiplex analysis on some instruments. Sputter profiling or angle-resolved analysis adds additional parameters for each acquisition set, such as the duty cycle of the sputter gun or the take-off angles for analysis. On modern instruments and with a well-established routine, a user can in principle load a set of samples, program the XPS instrument for data acquisition and analysis, and walk away.

## ACKNOWLEDGMENTS

The bulk of this unit was written against a background of questions that arose during a graduate lecture course on surface spectroscopic techniques taught at the University of Alabama in Huntsville in the spring of 1998. The input of the class was beneficial in defining the scope of this tutorial. The ARXPS data were obtained by a group of students as part of the laboratory assignment for the course. Critical reviews of the manuscript by M. A. George and J. C. Gregory are greatly appreciated.

## LITERATURE CITED

Anthony, M. T. and Seah, M. P. 1984a. XPS: Energy calibration of electron spectrometers. 1. An absolute, traceable energy calibration and the provision of atomic reference line energies. *Surf. Interface Anal.* 6:95–106.

Anthony, M. T. and Seah. M. P. 1984b. XPS: Energy calibration of electron spectrometers. 2. Results of an interlaboratory comparison. *Surf. Interface Anal.* 6:107–115.

Artyushkova, K. and Fulghum, J. 2001. Identification of chemical components in xps spectra and images using multivariate statistical analysis methods. *Journal of Electron Spectroscopy and Related Phenomena* 121:33–55.

ASM Handbook Committee (ed.) 1986. Metals Handbook, Vol. 10: Materials Characterization, 9th ed. American Society for Metals, Metals Park, Ohio.

Balcerowska, G. and Siuda, R. 1999. Inelastic background subtraction from a set of angle-dependent XPS spectra using PCA and polynomial approximation. *Vacuum Surface Engineering, Surface Instrumentation and Vacuum Technology* 54:195–199.

Balcerowska, G., Bukaluk, A., Seweryn, J. and Rozwadowski, M. 1999. Grain boundary diffusion of Pd through Ag thin layers evaporated on polycrystalline Pd in high vacuum studied by means of AES, PCA, and FA. *Vacuum Surface Engineering, Surface Instrumentation and Vacuum Technology* 54:93–97.

Blaschuk, A. G. 2001. Comparison of concentration profiles obtained by ion sputtering and nondestructive layer-by-layer analysis. *Metallofiz. Noveishie Tekhnol.* 23:255–271.

Briggs, D. and Seah, M. P. (eds.). 1994. Practical Surface Analysis: Auger and X-ray Photoelectron Spectroscopy, Vol. 1, 2nd ed. John Wiley & Sons, New York.

Bruhwiler, P. A. and Schnatterly, S. E. 1988. Core threshold and charge-density waves in alkali metals. *Phys. Rev. Lett.* 61:357–360.

Castle, J. E., Ruoru, K., and Watts, J. F. 1990. Additional in-depth information obtainable from the energy loss features of photoelectron peaks. *Corrosion Sci.* 30:771–798.

Castle, J. E., Chapman-Kpodo, H., Proctor, A., and Salvi, A. M. 2000. Curve-fitting in xps using extrinsic and intrinsic background structure. *Journal of Electron Spectroscopy and Related Phenomena* 106:65–80.

Clark, D. T. 1979. Structure, bonding, and reactivity of polymer surfaces studies by means of ESCA. *In* Chemistry and Physics of Solid Surfaces, Vol. 2 (R. Vanselow, ed.) pp. 1–51. CRC Press, Boca Raton, Fla.

Clark, D. T. and Feast, W. J. (eds.). 1978. Polymer Surfaces. John Wiley & Sons, New York.

Clumpson, P. 1997. Using principal components with chemical discrimination in surface analysis. *Spectra in Quantitative Analysis II 13*. VAM Bulletin (*http://www.vam.org.uk/ news/ news_bulletin.asp*).

Conny, J. M. and Powell, C. J. 2000. Standard test datafor estimating peak parameter errors in xps: I, II, and III. *Surface and Interface Analysis* 29.

Crist, V. 1998. Advanced peak fitting of monochromatic xps spectra. *J. Surf. Anal.* 4:428–434.

Doniach, S. and Sunjic, M. 1970. Many electron singularity in x-ray photoemission and x-ray line spectra from metals. *J. Phys. C* 3.

Drummond, I. W., Ogden, L. P., Street, F. J., and Surman, D. J. 1991. Methodology, performance, and application of an imaging x-ray photoelectron spectrometer. *J. Vacuum Sci. Technol.* A9:1434–1440.

Ehrentreich, F., S. Nikolov, G., Wolkenstein, M., and Hutter, H. 1998. The wavelet trans-form: A new preprocessign method for peak recognition of infrared spectra. *Mikrochimica Acta* 128:241–250.

Ertl, G. and Küppers, J. 1985. Low Energy Electrons and Surface Chemistry. Springer-Verlag, Weinheim, Germany.

Fadley, C. S. 1976. Solid state and surface analysis by means of angular-dependent x-ray photoelectron spectroscopy. *Prog. Surf. Sci.* 11:265–343.

Fadley, C. S. 1984. Angle-resolved x-ray photoelectron spectroscopy. *Prog. Surf. Sci.* 16:275–388.

Feldman, L. C. and Mayer, J. W. 1986. Fundamentals of Surface and Thin Film Analysis. North-Holland Publishing, New York.

Gilbert, R., Llewellyn, J., Swartz, W., and Palmer, J. 1982. Application of factor analysis to the resolution of overlapping XPS spectra. *Appl. Spectrosc.* 36:428–430.

Gunter, P. L. J., Gijzeman, O. L. J., and Niemantsverdriet, J. W. 1997. Surface roughness effects in quantitative XPS: Magic angle for determining overlayer thickness. *Appl. Surf. Sci.* 115:342–346.

Harrington, P. 1998. Different discrete wavelet transforms applied to denoising analytical data. *Journal of Chemical Information in Computer Science* 36:1161–1170.

Hofmann, S. 1992. Cascade mixing limitations in sputter profiling. *J. Vacuum Sci. Technol.* B10:316–322.

Hofmann, S. 1993. Approaching the limits of high resolution depth profiling. *Applied Surf. Sci.* 70-71:9-19

Hofmann, S. 2001. Profile reconstruction in sputter profiling. *Thin Solid Films* 398-399:336–342

Holgado, J., Rafael, A., and Guillermo, M. 2000. Study of CeO 2 XPS spectra by factor analysis: Reduction of $CeO_2$. *Applied Surface Science* 161:301.

Ibach, H. and Lüth, H. 1990. Solid State Physics: An Introduction to Theory and Experiment. Springer-Verlag, New York.

Jablonski, A. and Powell, C. J. 1993. Formalism and parameters for quantitative surface analysis by Auger electron spectroscopy and x-ray photoelectron spectroscopy. *Surf. Interface Anal.* 20:771–786.

Jansson, P. A. 1984. Deconvolution with Applications in Spectroscopy. Academic Press, Orlando, Fla.

Jetter, K., Depczynski, U., Molt, K., and Niemoller, A. 2000. Principles and applications of wavelet transformation to chemometrics. *Anal. Chim. Acta* 420:169–180.

Kratos Analytical. 1997. High-Resolution XPS Spectra of Polymers. Kratos Analytical, Chestnut Ridge, N.Y.

Krause, M. O. and Ferreira, J. G. 1975. K x-ray emission spectra of Mg and Al. *J. Phys. B: Atomic Mol. Phys.* 8:2007–2014.

Lesch, N., Aretz, A., Pidun, M. Richter, S., and Karduck, P. 2001. Application of sputter-assisted EPMA to depth profiling analysis. *Mikrochimica Acta* 132:377–382.

Lide, D. R. (ed.) 1993. Handbook of Chemistry and Physics, 73rd ed. CRC Press, Ann Arbor, Mich.

Malinowski, E. and Howery, D. 1980. Factor Analysis in Chemistry. John Wiley & Sons, New York.

Marcu, P., Hinnen, C., and Olefjord, I. 1993. Determination of attenuation lengths of photoelectrons in aluminum and aluminum oxide by angle-dependent x-ray photoelectron spectroscopy. *Surf. Interface Anal.* 20:923–929.

Moulder, J. F., Stickle, W. F., Sobol, P. E., and Bomben, K. D. 1995. Handbook of X-ray Photoelectron Spectroscopy. Perkin-Elmer, Eden Prairie, Minn.

Nöller, H. G., Polaschegg, H. D., and Schillalies, H. 1974. A step towards quantitative electron spectroscopy measurements. *J. Electron Spectrosc. Related Phenomena* 5:705–723.

Palmberg, P. W. 1975. A combined ESCA and Auger spectrometer. *J. Vacuum Sci. Technol.* 12:379–384.

Palmberg, P. W., Bohn, G. K., and Tracy, J. C. 1969. High sensitivity auger electron spectrometer. *Appl. Phys. Lett.* 15:254–255.

Passeggi, M. 1998. Auger electron spectroscopy analysis of the first stages of thermally stimulated oxidation of GaAs(100). *Appl. Surf. Sci.* 133:65–72.

Pia, H. and McIntyre, N. S. 1999. High resolution xps studies of thin film gold-aluminum alloy structures. *Surface Science* 421:L171-176.

Pia, H. and McIntyre, N. S. 2001. High-resolution valence band xps studies of thin film au-al alloys. *Journal of Electron Spectroscopy and Related Phenomena* 119:9–33.

Pratt, A. R., McIntyre, N. S., and Splinter, S. J. 1998. Deconvolution of pyrite, marcasite and arsenopyrite xps spectra using the maximum entropy method. *Surface Sci.* 396:266.

Richie, R., Oswald, S., and Wetzig, K. 2001. Xps and factor analysis for investigation of sputter-cleaned surface of metal (Re, Ir, Cr)-silicon thin films. *Appl. Surf. Sci.* 179:316–323.

Rummel, J. 1990. Applied Factor Analysis. Northwestern University Press, Evanston, Ill.

Salvi, A. M., Decker, F., Varsano, F, and Speranza, G. 2001. Use of xps for the study of cerium-vanadium (electrochromic) mixed oxides. *Surface and Interface Analysis* 31:255–264.

Sastry, M. 1997. Application of principal component analysis to X-ray photoelectron spectroscopy: The role of noise in the spectra. *Journal of Electron Spectroscopy and Related Phenomena* 83:143–150.

Schnydera, B., Alliata, D., Kotza, R., and Siegenthaler, H. 2001. Electrochemical intercalation of perchlorate ions in hopg: An sfm/lfm and xps study. *Applied Surface Science* 173:221–232.

Scofield, J. H. 1976. Hartree-Slater subshell photoionization cross-sections at 1254 and 1487 eV. *J. Electron Spectrosc. Related Phenomena* 8:129–137.

Seah, M. P. 1985. Measurement: AES and XPS. *J. Vacuum Sci. Technol.* A3:1330–1337.

Seah, M. P. 1993. XPS reference procedure for the accurate intensity calibration of electron spectrometers—results of a BCR intercomparison co-sponsored by the VAMAS SCA TWP. *J. Electron Spectrosc. Related Phenomena* 20:243–266.

Seah, M. P and Brown, M. T. 1998. Validation and accuracy of software for peak synthesis in xps. *Journal of Electron Spectroscopy and Related Phenomena* 95:71–93.

Seah, M. P. and Smith, G. C. 1988. Concept of an imaging XPS system. *Surf. Interface Anal.* 11:69–79.

Seah, M. P., Jones, M. E., and Anthony, M. T. 1984. Quantitative XPS: The calibration of spectrometer intensity-energy response functions. 2. Results of interlaboratory measurements for commercial instruments. *Surf. Interface Anal.* 6:242–254.

Sekine, T., Ikeo, N., and Nagasawa, Y. 1996. Comparison of AES chemical shifts with XPS chemical shifts. *Appl. Surf. Sci.* 100:30–35.

Shirley, D. A. 1972. High-resolution x-ray photoemission spectrum of the valence band of gold. *Phys. Rev. B* 5:4709–4714.

Shunsuke, M. 2000. Application of spline wavelet transformation to the analysis of extended energy-loss fine structure. *J. Electron Microsc.* 49:525–529.

Siegbahn, K. 1985. Photoelectron Spectroscopy: Retrospects and Prospects. Royal Society of London.

Siegbahn, K., Nordling, C., Fahlman, A., Nordberg, R., Hamrin, K., Hedman, J., Hohansson, G., Bergmark, T., Karlsson, S. E., Lindgren, I., and Lindberg, B. 1969a. ESCA: Atomic, molecular, and solid structure studied by means of electron spectroscopy. *Nova Acta Regiae Soc. Uppsala* 4:20.

Siegbahn, K., Nordling, C., Johansson, G., Hedman, J., Heden, P. F., Hamrin, R., Gelius, U., Bergmark, T., Werme, L. O., Manne, R., and Baer, Y. 1969b. ESCA Applied to Free Molecules. North-Holland Publishing, Amsterdam.

Simmons, G., Angst, D., and Klier, K. 1999. A self modeling approach to the resolution of XPS spectra into surface and bulk components. *Journal of Electron Spectroscopy and Related Phenomena* 105 :197–210.

Solomon, J. 1987. Factor analysis in surface spectroscopies. *Thin Solid Films* 154:11–20.

Splinter, S. J. and McIntyre, N. S. 1998. Resolution enhancement of x-ray photoelectron spectra by maximum entropy deconvolution. *Surface and Interface Analysis* 26:195–203.'

Tanuma, S., Powell, C. J., and Penn, D. R. 1988. Calculations of electron inelastic mean free paths I. Data for 27 elements and four compounds over the 200-2000 eV range. *Surf. Interface Anal.* 11:577.

Tanuma, S., Powell, C. J., and Penn, D. R. 1991a. Calculations of electron inelastic mean free paths. II. Data for 27 elements over the 50-2000 eV range. *Surf. Interface Anal.* 17:911–926.

Tanuma, S., Powell, C. J., and Penn, D. R. 1991b. Calculations of electron inelastic mean free paths. III. Data for 15 inorganic compounds over the 50-2000 eV range. *Surf. Interface Anal.* 17:927–939.

Tanuma, S., Ichimura, S., and Yoshihara, K. 1996. Calculations of effective inelastic mean free paths in solids. *Surf. Interface Anal.* 100:47–50.

Taylor, J. R. 1997. An Introduction to Error Analysis, 2nd ed. University Science Books, Sausalito, Calif.

Tougaard, S. 1988. Quantitative analysis of the inelastic background in surface electron spectroscopy. *Surf. Interface Anal.* 11:453–472.

Tougaard, S. 1996. Quantitative XPS: Non-destructive analysis of surface nanostructures. *Appl. Surf. Sci.* 100:1–10.

Wetzig, K., Baunack, S., Hoffmann, V., Oswald, S, and Prassler, F. 1997. Quantitative depth profiling of thin layers. *Fresenius J. Anal. Chem.* 358:25–31.

Zhang, X. Q., Zheng, J. B., and Gao, H. 2000. Comparison of wavelet transform and fourier self-deconvolution (FSD) and wavelet FSC for curve fitting. *Analyst* 125:915–919.

# KEY REFERENCES

ASM Handbook Committee, 1986. See above.

*Covers use of XPS for metallurgical applications.*

Barr, T. L. 1994. Modern ESCA. CRC Press, New York.

*A good overview of XPS, with descriptions of important photoemission phenomena from a quantum mechanical perspective. gives a more rigorous quantum mechanical presentation of certain photoemission processes in a manner that complements the information given in Briggs and Seah (1994). Also provides an important historical perspective, especially regarding the contributions of Siegbahn to the development of the experimental aspects of the technique that has become known as electron spectroscopy for chemical analysis (ESCA; Siegbahn et al., 1969a,b; Siegbahn, 1985). Recommended tor reader who is approaching XPS for the first time.*

Briggs, D. 1998. Surface analysis of polymers by xps and static sims. *In* Cambridge Solid State Science Series (D. R. Clarke, S. Suresh, and I. M. Ward, eds.).. Cambridge University Press, Cambridge, U.K.

*An in-depth treatment of the instrumentation, physical basis, and applications of XPS (and static secondary ion mass spectroscopy) with a specific focus on polymers. The book includes details from five case studies, including analysis of copolymer, biopolymer, and electropolymer surfaces.*

Briggs, D. and Beamson, G. 1992. High Resolution XPS of Organic Polymers: The Scientia ESCA300 Database. John Wiley & Sons, New York.

*This reference includes high resolution XPS spectra of over 100 organic polymers recorded at nearly intrinsic signal line widths. It contains a full spread of complete data, describes the database and the procedures used, and lists other relevant information.*

Briggs and Seah, 1994. See above.

*Perhaps the most comprehensive discussion of all significant practical aspects of XPS. Also covers Auger electron spectroscopy (treated separately in AUGER ELECTRON SPECTROSCOPY). Unattributed information in this unit is drawn primarily from this reference.*

Crist, B. V. 1991. Handbook of Monochromatic XPS Spectra: Polymers and Polymers Damaged by X-Rays. John Wiley & Sons, New York.

Crist, B. V. 2000a. Handbook of Monochromatic XPS Spectra: Semiconductors. John Wiley & Sons, New York.

Crist, B. V. 2000b. Handbook of Monochromatic XPS Spectra: The Elements and Native Oxides. John Wiley & Sons, New York.

*This recent three volume set by B. Vincent Crist includes spectra from native oxides, polymers, and semiconductors. Each volume contains an introductory section giving extensive details of the instrument, sample, experiment, and data processing. All spectra in the volumes are collected in a self consistent manner, all peaks in the survey spectra are fully annotated, and each high-energy resolution spectrum is peak-fitted with detailed tables containing binding energies, FWHM values, and relative percentages of the component.*

Feldman, L. C. and Mayer, J. W. 1986. Fundamentals of Surface and Thin Film Analysis. North-Holland Publishing, New York.

*Covers use of XPS for thin film applications.*

Ghosh, P. K. 1983. Introduction to Photoelectron Spectroscopy, Chemical Analysis, Vol. 67. John Wiley & Sons, New York.

*Though dated, covers the theoretical aspects of XPS with important and illustrative examples from gas phase systems. Provides a good treatment of XPS and the associated photoemission phenomena. Recommended tor reader who is approaching XPS for the first time.*

Kratos Analytical, 1997. See above.

*Handbook which is an essential source for reference XPS spectra of polymers.*

Moulder et al., 1995. See above.

*Handbook which is an essential source for reference XPS spectra of the elements*

Walls, J. M. 1989. Methods of Surface Analysis: Techniques and Applications. Cambridge University Press, Cambridge, U.K..

*Recommended for reader who is approaching XPS for the first time. Provides a well-rounded overview of the technique as a complement to the detail offered in Briggs and Seah (1994).*

Brundle, C. R. and Baker, A. D. 1978. Electron Spectroscopy— Theory, Technique, and Applications. Academic Press, London.

Carlson, T. A. 1975. Photoelectron and Auger Spectroscopy. Plenum Press, New York.

Watts, J. F. 1990. An Introduction to Surface Analysis by Electron Spectroscopy. Oxford Science Publications, Oxford.

Woodruff, D. P. and Delchar, T. A. (eds.). 1988. Modern Techniques in Surface Science. Cambridge University Press, Cambridge, U.K.

*The above four references include text and serial chapters that have been written over the past 25 years above XPS; some are no longer in print.*

Brundle, C. R., Evans, J. C. A., and Wilson, S. (eds.). 1992. Encyclopedia of Materials Characterization: Surfaces, Interfaces, Thin Films. Butterworth-Heinemann, Boston.

Cahn, R. W. and Lifshin, E. (eds.). 1993. Concise Encyclopedia of Materials Characterization, 1st ed. Pergamon Press, New York.

Sibilia, J. P. (ed.) 1988. A Guide to Materials Characterization and Chemical Analysis. VCH Publishers, New York.

Vickerman, J. C. 1997. Surface Analysis: The Principal Techniques. John Wiley & Sons, New York.

Wachtman, J. B. and Kalman, Z. H. 1993. Characterization of Materials. Butterworth-Heinemann, Stoneham, Mass.

Walls, J. M. and Smith, R. (eds.). 1994. Surface Science Techniques. Elsevier Science Publishing, New York.

*The above six publications are encyclopedic references that generally provide less depth on the specific topic of XPS but may provide a better perspective on its relation to other spectroscopic and analytical techniques. Walls and Smith (1994), Vickerman (1997), Brundle et al. (1992), and Cahn and Lifshin (1993) provide, potentially, the best overview. Sibilia (1988) is somewhat more limited in its coverage, and Wachtman and Kalman (1993) is relatively less practical (and correspondingly more abstract) in its style.*

Christmann, K. 1991. Introduction to Surface Physical Chemistry, Topics in Physical Chemistry. Springer-Verlag, New York.

Drummond et al., 1991. See above

Gunter et al., 1997. See above.

Prutton, M. 1994. Introduction to Surface Physics. Oxford University Press, New York.

Somorjai, G. A. 1994. Introduction to Surface Chemistry and Catalysis. John Wiley & Sons, New York.

*The above five publications include a discussion section on XPS while providing a comprehensive background on surface chemistry or surface physics.*

*Articles pertaining to the principles and applications of XPS are frequently presented in the journals Applied Surface Science, Colloid and Surface Science, Journal of Vacuum Science & Technology, Surface and Interface Science, and Surface Science.*

## INTERNET RESOURCES

http://srdata.nist.gov/xps/

*This Web page at the National Institute of Standards and Technology (NIST) is particularly worth citing. It contains positions of XPS and XAES peaks for elements in various compounds as well as plot (Wagner plots) of the Auger parameter for elements in various compounds.*

## APPENDIX: PEAK SHAPES

In general, a peak shape is named for the analytical function that defines it. Common peak shapes include the Gaussian, Lorentzian, and Voigt. Mixtures of these, such as a Gauss + Lorentz, and variations of these, such as an assymetric Lorentz or assymetric Voigt, are also widely used. For fitting to XPS spectra, special peak shapes such as a Gauss + tails (Briggs and Seah, 1994) or the Doniach-Sunjic (Bruhwiler and Schnatterly, 1988; Doniach and Sunjic, 1970) have also been developed. Some peak shape functions can be derived directly from a first-principles analysis of the response expected from the spectroscopic event (excitation of a photoelectron). These include the Gaussian and Lorentzian functions. Alternatively, analytical functions for peak shapes such as the Gaussiann + tails include formulations from semi-empirical analysis of experimental data.

Peak shapes can be classified in many ways. One method is according to the number of parameters needed to define the shape analytically. In this respect, at least four classes of peak shapes are commonly used to fit peaks in chemical spectroscopies, starting from those that require three parameters to those that must use six. Each class may contain more than one peak shape, but all peak shapes within a given class use the same number of parameters to define them.

The principal parameters used in an analytical function to define a peak shape in spectroscopy are the peak height, half-width or full-width at half-maximum, position, area, assymetry or shape factor, and mixing ratio. Certain types of XPS peaks are also defined by relative parameters from another peak, such as relative height or offset (relative position). This happens for spin-split doublets (for example, $p_{3/2}$ and $p_{1/2}$ peaks) and for satellites (shake-up and loss peaks). Finally, peak fitting can be done using an experimental spectrum or set of spectra rather than analytical peak shapes. Such reference style peaks require only two parameters to fit to an experimental spectrum: their height and their offset relative to the spectrum.

The above discussion suggests that XPS peak shapes can be categorized globally into three styles, single peaks, relative peaks (such as spin-split doublets), and reference peaks (using a full experimental spectrum). The first two styles are obtained from analytical functions and will be called singlets and ntuplets for the number of peaks they define in a spectrum. Reference-style peaks are derived from experimental data and will not be considered in any further detail in the discussion below. In principle, an ntuplet-style peak uses $n$ times as many parameters as its comparably shaped singlet, where $n$ is the number of ntuplets. Rather than expand the number of peak classes beyond four to accommodate this (potentially unlimited) expansion of parameters, classification of a peak shape is restricted to the number of principal parameters it requires. Singlet and ntuplet peak styles correspondingly then each contain only four different peak classes. As an example, a Gaussian singlet style peak and a Gaussian doublet-style peak (an ntuplet style with two peaks) are in the same peak class because they both use the same number of principal parameters (three as shown below). The doublet will require an additional set of three relative parameters to define shape of the second (relative) peak.

The four peak classes of XPS peak shapes are labeled A through D below. For each class, examples of analytical functions are given for singlet style peak shapes. A representative doublet style is given for a Gaussian peak shape (the first peak shape in class A). Where possible, the formula are given in a format suitable for input into a symbolic math or graphing package for programming purposes. In all of the functions, $S_{ps}$ is the analytical signal for the peak shape with a name that is abbreviated as ps, $x$ is the energy scale (assumed to be in eV binding energy in XPS), $w$ is the full width of the peak at exactly half of its height (full-width at half-maximum or FWHM in eV), $p$ is the peak position (eV), and $a$ is the peak area (counts-eV/sec). Other parameters are defined as needed.

The classification scheme above and the functions provided below are by no means unique or unequivocal.

### Class A (3 Parameters)

Two of the peak shapes in this class, Gaussian and Lorentzian are the most commonly used in spectroscopy.

**Gaussian.** Many processes that lead to a signal in chemical spectroscopy are random in nature and result statistically in a Gaussian signature. In statistical analysis, a normal probability curve is a Gaussian function defined by its mean value or position μ and its standard deviation, σ, which is measured as the half-width of the Gaussian peak at a distance of about 61% of the peak height. The normal probability curve also has an integrated area of unity.

The normal probability function is not useful for fitting peaks to spectroscopic data. Almost all spectroscopic peaks have areas that are not identically equal to unity, and $w$ (the FWHM or the width at 50% of the height) of a spectroscopic peak an experimental spectrum is more easily measured than σ (the width of the corresponding Gaussian function at about 61% of its height). The Gaussian functions given below are therefore defined in terms of spectroscopic parameters, especially $w$. As a point of reference for a Gaussian function, the FWHM is $w = 2\sigma\sqrt{2\ln(2)}$. Also, for any form of the Gaussian, the relationship between $h$, σ, $w$, and $a$ is $a = 2\sqrt{2\pi}h\sigma = \sqrt{\pi\ln(2)}hw$.

Three parameters define a normal probability Gaussian curve, μ, σ, and an area of unity. Therefore $p$ and any two parameters from the set of $h$, $w$, and $a$ are needed to define a spectroscopic Gaussian peak. Correspondingly, three different Gaussian shape peak functions exist in Class A depending on which two parameters are chosen in addition to $p$.

**Position/height/full-width-half-maximum.** A singlet style peak is represented as

$$S_{GPHF} = h\exp\left(-4\frac{\ln(2)(x-p)^2}{w^2}\right) \quad (36)$$

For this peak, the area is

$$a = hw\sqrt{\pi}/2\sqrt{\ln(2)} \quad (37)$$

An example of a doublet style where both peaks have the same Gaussian shape is given below. In the function, $D_{DGHF}$ is the signal, $r_h$ is the relative height, $o$ is the offset, and $r_w$ is the relative fwhm of the doublet peak

$$D_{DPHE} = h\exp\left(-4\frac{\ln(2)(x-p)^2}{w^2}\right)$$
$$+ r_h h\exp\left(-4\frac{\ln(2)(x-p-o)^2}{(r_w w)^2}\right) \quad (38)$$

**Position/area/full-width-half-maximum**

$$S_{GPAF} = \left(\frac{ea\sqrt{\ln(2)}}{w\sqrt{\pi}}\right)\exp\left(-4\frac{\ln(2)(x-p)^2}{w^2}\right) \quad (39)$$

The height of this peak is

$$h = 2a\sqrt{\ln(2)}/w\sqrt{\pi} \quad (40)$$

**Position/height/area**

$$S_{GPHA} = h\exp\left(-2\frac{h^2\pi(x-p)^2}{a^2}\right) \quad (41)$$

The FWHM of this peak is

$$w = 2a\sqrt{\ln(2)}/h\sqrt{\pi} \quad (42)$$

**Lorentzian.** The Lorentzian peak shape is also fundamental to chemical spectroscopy. It is often considered a more natural line shape for chemical spectroscopy than the Gaussian function because the broadening it displays is characteristic of the lifetime broadening found in quantized energy states. The Lorentzian function gives a broader peak at the base than does the Gaussian. Two variations of this peak shape exist:

**Position/height/full-width-half-maximum**

$$S_{LPHF} = h\left/\left(4\frac{(x-p)^2}{w^2}+1\right)\right. \quad (43)$$

The area of this peak is $a = \pi w/2$.

**Position/height/area**

$$S_{\text{LPHA}} = h \bigg/ \left( \frac{(x-p)^2 \pi^2}{a^2} + 1 \right) \tag{44}$$

The FWHM of this peak is $w = 2a/\pi$.

### Class B (4 Parameters)

These peak shapes include a fourth parameter that is either $f_a$—a *peak shape assymetry factor*—or $f_g$, a fractional Gaussian. In these peak shapes, $h$ and $w$ do not always define the actual height and half-width of the peak as measured in the resultant curve.

**Gaussian + Lorentzian.** This is a combination of the two Class A peak shapes. The $f_g$ value ranges from 0 (pure Lorentzian) to 1 (pure Gaussian). Two different variations can be formed

$$S_{\text{G+L(A)}} = f_g S_{\text{GPHF}} + (1-f_g) S_{\text{LPHF}} \tag{45}$$

$$S_{\text{G+L(B)}} = f_g S_{\text{GPHA}} + (1-f_g) S_{\text{LPHA}} \tag{46}$$

**Voigt.** The Voigt function has the form

$$S_{\text{V}} = \left( \frac{1}{\alpha} \right) \int_{-\infty}^{\infty} \frac{h \exp(-y^2)}{w_{\text{L}}^2 + ((x-p)/w_{\text{G}} - y)^2} \, dy \tag{47}$$

$$\alpha = \int_{-\infty}^{\infty} \frac{\exp(-y^2)}{w_{\text{L}}^2 + y^2} \, dy \tag{48}$$

where $w_{\text{L}}$ and $w_{\text{G}}$ are Lorentzian and Gaussian components of the FWHM of the peak. While an analytical solution for $\alpha$ exists, the remaining integral cannot be solved analytically. The Voigt peak shape is therefore typically obtained numerically. One method is to use a lookup function fVoigt($p_1, f_a$), where $f_a$ is a shape function and the value of $p_1$ is defined below

$$S_{\text{V}} = \text{f Voigt}(p_1, f_a) \tag{49}$$

$$p_1 = 2\sqrt{\ln(2)} \frac{(x-p)}{w} \tag{50}$$

The function f Voigt returns a value after numerical interpolation from a table. The shape factor $f_a$ varies from 0 to infinity. Larger values give flatter, broader peaks (less Gaussian in shape).

**Assymetric Lorentzian.** This form of the Lorentzian gives a skew to the higher $x$ side of the peak as the assymetry factor $f_a$ varies from 0 to 1. A value of 0 gives a symmetrical Lorentzian, as can be shown by analytical geometry. As $f_a$ increases from 0, the location of the maximum of the peak shifts toward greater $x$ and the value of $h$ no longer denotes the true peak height. For $f_a$ above 0.5, the function

returns to a symmetrical, flat, broad peak shape centered at $p$

$$S_{\text{AL}} = \left( \frac{h}{\alpha} \right) \cos\left( -f_a + (1-f_a)\arctan\left( 2\frac{x-p}{w} \right) \right) \tag{51}$$

$$\alpha = 2\sqrt{\frac{(x-p)^2}{w^2} + 1} \tag{52}$$

The area of the peak when $f_a = 0$ is $a = \pi w/2$ and otherwise goes to infinity as $f_a$ goes to unity.

**Doniach-Sunjic.** This peak shape has been used to fit the $d$ and $f$ levels of transition metals (Bruhwiler and Schnatterly 1988; Doniach and Sunjic 1970). The actual height, FWHM, and position of the resultant principle peak are only equal to the input values when the assymetry factor $f_a = 0$. As $f_a$ increases, the peak becomes assymetrical at higher $x$ values above $p$ and the maximum in the peak shifts toward higher values of $x$ away from $p$

$$S_{\text{DS}} = \frac{h}{\alpha} \cos\left( \frac{\pi f_a}{2} + (1-f_a)\arctan(p-x,w) \right) w^{1-f_a} \tag{53}$$

$$\alpha = ((p-x)^2 + w^2)^{(1-f_a)/2}$$
$$\times \cos\left( \frac{\pi f_a}{2} + (1-f_a)\arctan(0,w) \right) \tag{54}$$

In the above expression, arctan $(y, x)$ computes the principal value of the argument of the complex number $x + iy$ so that $-\pi < arctan(y, x) \leq \pi$. When $f_a$ is unity, the Doniach-Sunjic peak shape is undefined (the function returns a horizontal line at a value of $h$). The area of this peak is $a = \pi h w$ for $f_a = 0$ and goes to infinity as $f_a$ goes to unity.

### Class C (5 Parameters)

The simplest peaks in this class are just combinations of the Class A peak shapes and the assymetric Class B peak shapes. One additional parameter $f_g$ defines the mixing ratio.

#### Gaussian + assymetric Lorentzian

$$S_{\text{G+AL}} = f_g S_{\text{GPHF}} + (1-f_g) S_{\text{AL}} \tag{55}$$

### Class D (6 Parameters)

These peak shapes include parameters to define tails. Tails are assymetric extensions on one side of the peak. In XPS spectroscopy, tails are used to model the increase in intensity on the high binding energy side of the peak due to inelastic scattering.

**Gaussian + Tails.** Three additional parameters define the shape of the tail. They are $c_t$—constant tail height fraction—$e_t$—exponential tail height—and $fc$—fraction of constant tail (versus exponential tail). The tail function signal

$S_{\mathrm{TF}}$ is set to zero on the low $x$ side of the peak position $p$ using the logical test $(x \geq p)$.

$$S_{\mathrm{TF}} = F_c c_t + (1 - f_c) \exp\left(-\frac{|x - p|}{e_t}\right)(x \geq p) \quad (56)$$

$$S_{\mathrm{GPHF+T}} = S_{\mathrm{GPHF}}(1 - S_{\mathrm{TF}}) + h S_{\mathrm{TF}} \quad (57)$$

JEFFREY J. WEIMER
University of Alabama at
Huntsville
Huntsville, Alabama

# SURFACE X-RAY DIFFRACTION

## INTRODUCTION

### Surface X-ray Diffraction

Surface x-ray diffraction allows the determination of atomic structures of ordered crystal surfaces in an analogous way to x-ray crystallography for the determination of three-dimensional crystal structures. In doing this, crystallographic parameters such as lattice constants and Debye-Waller factors (thermal vibration amplitudes) are also accessed, so surface x-ray diffraction should be considered as a valid method for measuring these quantities too. Surface x-ray diffraction also probes surface morphological properties, such as roughness and facet formation, and allows the investigation of their thermodynamic and kinetic aspects, such as the study of phase transitions in surfaces.

Because x rays probe deep inside matter, buried interfaces can be treated in exactly the same manner as surfaces, so throughout this unit the words *surface* and *interface* will be used synonymously. In fact, some of the best applications of the method are for buried interfaces, where complementary electron-based probes such as low-energy electron diffraction (LEED; LOW-ENERGY ELECTRON DIFFRACTION and reflection high-energy electron diffraction cannot reach.

Surface x-ray diffraction is not a laboratory technique. While there are one or two exceptions of installations in individual laboratories, the majority are associated with national or international facilities. Even the very early experiments carried out by Eisenberger and Marra (1981) used synchrotron radiation. By now, facility-based science is already rather common and will become more so in the future, as it has considerable economies of scale and advantages of centralized operational safety. The practice of surface x-ray diffraction should necessarily be considered as a facility-based operation, and it will be assumed that the reader will have access as a "user" to one of these facilities. This is usually handled by the submission of a proposal to be judged on the importance of the science to be carried out. Access is then usually free of charge.

Because surface x-ray diffraction is facility based, it is not a technique for which one can purchase the equipment piecemeal from a vendor. Every surface x-ray diffraction installation is slightly different, but most were derived from and are based on the description that follows, which is the X16A facility of the National Synchrotron Light Source (NSLS). Others, such as the Advanced Photon Source (APS, Argonne), the European Synchrotron Radiation Facility (ESRF, Grenoble), Hamburger Synchrotronstrahlungslabor (HASYLAB, Hamburg), Daresbury Laboratory, Photon Factory (PF, Tsukuba), and Laboratoire pour l'Utilisation de Rayonnement Synchrotron (LURE, Orsay), use basically the same control and analysis programs and procedures as those described here.

## Competitive and Related Techniques

The need to travel to a facility to make surface x-ray diffraction measurements can be inconvenient, to say the least, so it is important to discuss its strengths and weaknesses compared with other techniques. Scanning probe methods, such as scanning tunneling microscopy (STM, SCANNING TUNNELING MICROSCOPY and atomic force microscopy (AFM), are extensively used in surface science. These imaging methods are unbeatable for detecting singular events on surfaces, and the "microscopist's eye" will often detect a wide variety of local behavior. Thus STM has been very important in understanding instabilities of surfaces, step properties, and island morphologies during growth. When "average" quantities are needed for understanding general behavior, such as thermodynamics, the averaging must be carried out explicitly; diffraction methods will access ensemble-average quantities directly and will filter out the exceptions. This is particularly important when studying the dynamical behavior associated with phase transitions, where the correlations within a moving arrangement of atoms can still be detected by diffraction. At their limit, scanning probe microscopies can reach atomic resolution, but they always detect the outermost features on a surface and will not see below the surface. Diffraction methods can attain atomic positions to an accuracy of 0.01 Å routinely, while scanning microscopies can only achieve that in the surface-normal direction under certain circumstances.

The direct competitors with surface x-ray diffraction are thus the various forms of electron diffraction (Electron Techniques), which provide information of an analogous form: LEED, reflection high-energy electron diffraction (RHEED), and transmission electron diffraction (TED). Since electrons interact more strongly with matter than do x rays, there are some important differences. First, electron diffraction is dynamical and requires a nonlinear theory to explain the intensity of the diffraction seen. This has been accomplished by deriving a careful theoretical description of the electron as it interacts with the sample in the case of LEED and to a lesser extent for RHEED. After 20 years of development, the method has reached the point where structures can be refined as well as just calculated, and thermal vibrations can now be included. With x rays, the simpler kinematical description can be used, which makes the calculations very easy, so that powerful refinement procedures can be followed. Transmission electron diffraction is near to this kinematical limit also, and the linear approximation has been successfully used to obtain surface structures. The potential user can make a choice between a hard experiment and an easy analysis in the case of x rays or between an easy experiment and a hard analysis in the case of electron diffraction.

Second, for the same reason, the penetration of x rays is significantly larger than electrons. The sensitivity of LEED and RHEED is limited to about four atomic layers, and the accuracy becomes reduced in the lower layers. To examine interfaces, LEED has been used, but it requires physical removal of half of the sample. X rays are not limited in this way and have many applications for interfaces: an important example is the electrochemical interface, which can be studied in situ. Transmission electron diffraction falls in between: samples must be thinned to 100 to 1000 Å but can still contain interfaces. The small differences in accuracy that can be obtained in surface structure determination by the different diffraction techniques can be attributed to the accessible range of reciprocal space. X-ray diffraction and TED tend to have higher accuracy parallel to the surface, because the largest component of momentum transfer is usually in that direction. Low-energy electron diffraction is a backscattering technique, which tends to give more accurate coordinates in the normal direction.

It used to be said that the momentum resolution of x rays was an unmatchable asset, but this has become less important in recent years. First, the spot profile analysis LEED (SPA-LEED) method was developed by Sheithauer et al. (1986) as well as the spot profile analysis RHEED (SPA-RHEED) method by Müller and Henzler (1995). These offer momentum resolution, which is comparable to x-ray diffraction, and are commercially available from Omicron. Surface-phase transitions can now be studied as well with LEED or RHEED as with x-ray diffraction. Second, the merit of probing correlations extending over large distances in real space has diminished with the widespread availability of STM and AFM, which routinely scan the same range.

## PRINCIPLES OF THE METHOD

Surface is a generic word that is often used as an antonym to bulk. Deviations from physical properties of bulk matter are often attributed to (i.e., blamed on) "surface effects." The study of surfaces is therefore a productive way of understanding these deviations. The study of surfaces always starts with their structure, and that is what is probed with surface x-ray diffraction. Typical examples of surfaces, generalized to include internal interfaces, are shown in Figure 1.

The general subject of diffraction will not be reviewed here since it appears elsewhere in this volume and also is covered in several excellent textbooks (James, 1950; Warren, 1969; Guinier, 1963; Lipson and Cochran, 1966; Woolfson, 1997). It is assumed that the reader is comfortable with the construction of a three-dimensional (3D) reciprocal space and the idea that the diffraction from a crystal is localized to a set of specific "Bragg" points forming a reciprocal lattice. Reciprocal space is spanned by the vectorial

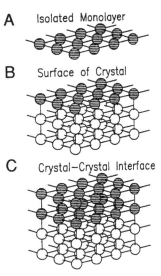

**Figure 1.** Entities that can be studied with surface x-ray diffraction. (**A**) isolated monolayer (2D crystal). (**B**) Same as A but connected to a crystalline substrate. (**C**) Same as A but at the interface between two different crystals. From Robinson and Tweet (1992).

quantity **q**, which is the momentum transfer of the diffraction experiment. The axes of reciprocal space are chosen to lie along simple directions within the reciprocal lattice. The coordinates are then normalized so that the Bragg points appear at simple integers, called Miller indices, *hkl*. This construction of reciprocal space was introduced in SYMMETRY IN CRYSTALLOGRAPHY and will be assumed henceforth.

In surface x-ray diffraction, it is conventional to choose a coordinate system in reciprocal space that places the third index of the momentum transfer, $q_z$, perpendicular to the surface. This is usually compatible with the Miller index notation because the naturally occurring surfaces are usually close-packed planes of atoms, which have small-integer Miller indices according to the usual construction. This convention often requires use of an unusual setting of the bulk crystal unit cell, e.g., hexagonal for a cubic (111) surface or tetragonal for a cubic (110) surface. The convention is exactly the same as that used in LEED; see LOW-ENERGY ELECTRON DIFFRACTION. Another common terminology is to separate the components of momentum transfer, **q**, parallel and perpendicular to the surface by writing $\mathbf{q} = (q_P, q_z)$. Here $q_P$ strictly represents both in-plane components. The reason for this separation is that a surface is a highly anisotropic object with completely different properties in the two directions.

### Diffraction from Surfaces

The most important characteristic of diffraction from surfaces is illustrated as a 3D sketch in Figure 2, which corresponds roughly to Figure 1 transformed into reciprocal space. Bulk crystal diffraction is concentrated at points of reciprocal space because of the 3D periodicity of the crystal. A surface or interface has reduced dimensionality, so its diffraction is no longer confined to a point but extends continuously in the direction perpendicular to the plane of

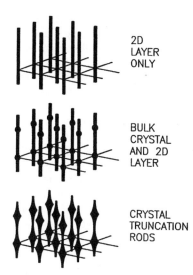

**Figure 2.** Schematic of the diffraction from the objects in Figure 1. The 3D points of reciprocal space are indicated as circles; 2D rods of reciprocal space are indicated as bars. From Robinson and Tweet (1992).

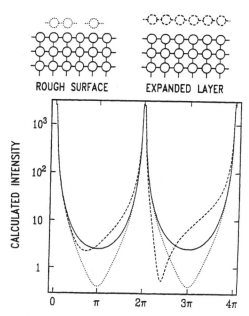

**Figure 3.** Calculated CTRs for an ideally terminated surface (solid curve), a surface with modified top-layer spacing (dashed), and a rough surface (dotted). The intensity is plotted as a function of perpendicular momentum transfer, $q_z$, in normalized units.

the surface or interface. These lines of diffraction, sharp in two in-plane directions and diffuse in the perpendicular direction, are called rods. The diffraction pattern of a two-dimensional (2D) crystal, as illustrated in Figure 2, is a 2D lattice of rods in reciprocal space.

A surface rarely exists as an isolated entity (Fig. 1A) but is normally intimately connected to a substrate of bulk crystalline material (Fig. 1B) or materials (Fig. 1C). The diffraction is then a superposition of 3D and 2D features, points and rods, as shown. However, the periodicities of the two lattices are the same (by construction), and so their diffraction patterns will be intimately related. This causes the lateral positions of the features to align so that the intensity distributions merge together. The 2D rods line up between the 3D points, and the two features simply merge together. The resulting object in reciprocal space, illustrated in the bottom panel of Figure 2, is called a crystal truncation rod (CTR). An example of an experimental study of a CTR is given below (see Practical Aspects of the Method).

Crystal truncation rods have been treated theoretically as extensions of bulk diffraction by Andrews and Cowley (1985), Afanas'ev and Melkonyan (1983), and Robinson (1986), and their connection to the surface structure has been derived. The sensitivity to the surface is found to vary smoothly along the rod. Near the divergence of the intensity at the 3D Bragg points, the intensity of the CTR depends only on the bulk structure, through its bulk structure factor. According to the kinematic theory of diffraction, the intensity follows a $(q_z - G_z)^{-2}$ law near each bulk Bragg peak at position $\mathbf{G} = (G_P, G_z)$. In the more precise dynamical theory discussed by Afanas'ev and Melkonyan (1983), this nonphysical divergence at $q_z = G_z$ becomes finite and follows a Darwin curve instead, but the asymptotic behavior remains the same as in the kinematic theory.

The connection between the intensity distribution along a CTR and the corresponding surface structure is

illustrated by a simple calculation in Figure 3. The full curve is for an ideally terminated simple-cubic lattice of atoms; its functional form is just $1/(\sin q_z)^2$, whose divergence at Bragg points $q = 2n\pi$ is clearly visible. The dashed curve corresponds to an outward displacement of a single layer of atoms at the surface; the intensity curve near the Bragg peaks, where there is little surface sensitivity, is barely changed, but the intensity at the CTR minimum is strongly modified. Finally, the dotted curve is for a rough surface, modeled by random omission of a fraction of the atoms in the top layer; here, again, the biggest effect is at the CTR minimum, this time with a symmetric drop of the intensity curve.

A simple rule of thumb is that the measurement is the most surface sensitive at the position on the CTR that is furthest from the bulk Bragg peaks, where the intensity is also weakest. This rule applies not only to the structure itself but also to the sensitivity to fluctuations in the surface, e.g., its roughness. In performing an experiment, one is frequently faced with a choice between surface sensitivity and signal level, since one quantity trades off directly with the other.

## Measurement of Surface X-ray Diffraction

The surface x-ray diffraction signal can be measured with a standard diffractometer. For reasons explained below (see Practical Aspects of the Method), it is common to use a diffractometer with extra degrees of freedom. Synchrotron radiation is highly desirable as it results in convenient signal levels around hundreds to thousands of counts per second. The primary responsibility of the diffractometer is to select $\mathbf{q}$ by suitable choice of its angles while keeping the beam always on the center of the sample.

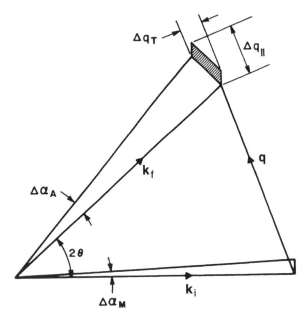

**Figure 4.** Construction of the resolution function within the scattering plane of a diffractometer. From Robinson (1990b).

Measurements then consist of scans of the **q** vector through the various features of the surface x-ray diffraction.

The shape and size of the resulting scans depend on the nature of the diffraction features being measured, e.g., due to the degree of order in the surface. The results also depend on the resolution function of the instrument, which is always convolved with the true diffraction pattern. The resolution function is controllable to a certain extent by instrumental parameters, particularly slits that define the beam divergence in different directions. The usual construction is shown in Figure 4, which is based on the vector construction of **q** in terms of the incident and exit wave vectors $\mathbf{k}_i$ and $\mathbf{k}_f$: $\mathbf{q} = \mathbf{k}_i - \mathbf{k}_f$. The variation of $\mathbf{k}_i$ is represented by the angular width of directions allowed by the entrance slits and monochromator, $\Delta\alpha_M$, while the variation of $\mathbf{k}_f$ is defined by $\Delta\alpha_A$. The resulting resolution function is the shaded parallelogram at the top of the picture. Simple geometry shows that its dimensions parallel and perpendicular to the vector **q** are given by Moncton and Brown (1983):

$$\Delta q_{\parallel} = k(\Delta\alpha_M + \Delta\alpha_A)\cos\theta$$
$$\Delta q_T = k(\Delta\alpha_M + \Delta\alpha_A)\sin\theta \qquad (1)$$

where $k = 2\pi/\lambda$ is the length of both wave vectors and $\theta$ is half the diffraction angle $2\theta$.

Note that this picture applies to the scattering plane of the diffractometer, spanned by the $2\theta$ angle of the diffractometer, and the notations $\Delta q_P$ and $\Delta q_T$ refer to parallel with (or radial) and transverse to the direction of **q**. The scattering plane is not the same as the surface plane of the sample, except in the grazing incidence geometry defined below, which is often used for precisely that reason. The third dimension of the resolution function,

denoted $\Delta q_{\perp}$, is also defined by slits, this time in the direction perpendicular to the scattering plane; the resolution function then becomes a parallelogram-based prism by simple extension of Figure 4 to 3D. This out-of-plane resolution is often the largest of the three because, in the grazing incidence geometry, the diffraction features are extended rods in that direction, and a considerable enhancement of intensity is thereby attained.

Many setups, including the one at X16A, gain additional flexibility by dividing the $\Delta q_{\perp}$ range into narrower sections by use of a linear position-sensitive detector (PSD). The x-ray counts recorded by the PSD are assigned a "position" along the detector by the height of its output pulse. This is accumulated in a multichannel analyzer (MCA) and then read out in bins, each corresponding to a section of $\Delta q_{\perp}$. The total counts in each bin are saved in the output file, to allow the data to be subsequently broken down into different $q_z$ positions with a correspondingly narrower $\Delta q_{\perp}$. Since surface x-ray diffraction in the grazing incidence geometry should give features that are broad in $q_z$, all the bins should rise and fall together in intensity and give independent structure factors at different $q_z$ values. However, if a powder grain or multiple scattering glitch happens to accidently satisfy the diffraction condition during some surface measurement, it will be recorded in only one MCA bin and can be suppressed.

### Surface Crystallographic Measurements

A subset of the general class of surface x-ray diffraction measurements is that pertaining to atomic structure determination alone. Here it is assumed that the surface under investigation is well enough ordered that the widths of all diffraction features are limited by the resolution function alone. This means that, in this case, any disorder that happens to be present is undetectable, and the surface is indistinguishable from ideal. It has been shown by Robinson (1990b) that in this limit the integrated intensity of the diffraction features is independent of the (unseen) disorder and so is representative of the crystallographic structure factor alone. Measurements of structure factors can then be analyzed in terms of the atomic structure of the surface, in a close analogy to bulk x-ray crystallography described by Woolfson (1997) or Lipson and Cochran (1966).

If the disorder is visible in the form of peak broadening that is not too severe, it may still be possible to reach the ideal limit by deliberately worsening the resolution function. While it is often hard to control $\Delta\alpha_M$ since this is a function of the beamline optics, $\Delta\alpha_A$ can be opened up significantly by opening the $2\theta$ detector slits. In all cases, the test for this desirable situation is that all diffraction features are resolution limited in their width. There are other limits in which the structure factor can be reliably measured if the peaks are significantly broader than the resolution function, then their peak intensity is representative of the square of the structure factor, provided a different, but calculable, correction is used.

The measurements needed for crystallography are most easily obtained by rocking scans of the principal diffractometer angle $\theta$. In the detailed protocols below, the name of the θ angle known to the usual computer software is TH; these names will be cross-referenced where appropriate. This ensures that the narrow direction of the resolution function sweeps roughly perpendicular to the peak. If the peak is slightly misaligned, such a scan will usually catch it either earlier or later and still give the correct structure factor; this will not work if the misalignment is too severe, say more than two peak widths away, because, in general, there would be components of the misalignment in all possible directions. The use of a rocking scan is also prescribed by the definition of integrated intensity, the quantity coupled to the structure factor, as explained by Robinson (1990b) (see Appendix B). This corrects for all kinds of disorder, including mosaic spread, provided this is not too severe. It is also the most reliable way to estimate the background underlying the peak.

### Grazing Incidence

It was explained (see Diffraction from Surfaces, above) that the use of grazing incidence orients the resolution function in a favorable direction for surface x-ray diffraction. This means that the incident beam, the exit beam, or both make a very small angle with the surface to be measured. This presents a special challenge in the diffractometry which is discussed below (see Practical Aspects of the Method). It also has other consequences.

The diffraction under grazing incidence conditions is no longer strictly kinematical because such a beam will partly reflect from the surface and also undergo a slight refraction as it crosses to the interior of the crystal. The effect, described by classical optics by Born and Wolf (1975), is small until the incidence angle $\alpha_i$ (or exit angle $\alpha_f$) becomes comparable with the critical angle for total external reflection, $\alpha_C$, which is of order a few tenths of a degree for most situations. When $\alpha_i < \alpha_C$, the beam is completely reflected from the sample and only an evanescent wave continues inside. This can be used to considerable advantage when extreme surface sensitivity is desired, since the typical penetration depth is of the order of 100 Å. Fortunately, Vineyard (1982) introduced a sound theoretical basis called the distorted-wave Born approximation for working with the effects of the critical angle. This approximation provides the grazing incidence method with great power.

The distorted-wave Born approximation has an additional consequence for surface x-ray diffraction: the intensity of the diffraction becomes modified under grazing incidence conditions. Born and Wolf (1975) show that the effect can be reduced to a simple transmission function for the intensity, $|T(\alpha)|^2$, as a function of either $\alpha_i$ or $\alpha_f$, which accounts for the refraction effects. An example of calculations of $|T(\alpha)|^2$ is given in the upper panel of Figure 5. As can be seen, there is a potential factor of 4 enhancement for both the incident and exit beams. The practical drawbacks with the factor of 16 are that it is only attainable when the collimation of both beams is severely restricted and the alignment is very difficult to achieve. The lower panel of Figure 5 shows the effect of the distortion on the $\alpha_f$ profile of an in-plane bulk Bragg peak, which in the absence of the refraction effect would lie at $\alpha_f = 0$. Careful examination of the experimental curve can be used to iden-

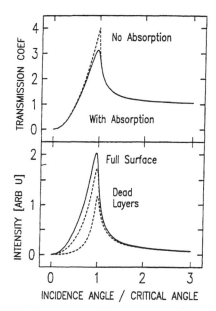

**Figure 5.** Calculations of $|T(\alpha)|^2$ and the profile of an in-plane diffraction peak for gold at $\lambda = 1.5$ Å. The dashed curves show the effect of 5 Å and 20 Å of inactive dead layers of gold present above the crystalline part of the sample. Adapted from Dosch et al. (1991).

tify whether the crystal diffracts from all layers up to its surface or whether there are inactive "dead" layers present, as shown.

## PRACTICAL ASPECTS OF THE METHOD

An overview of the generic vacuum diffractometer configuration is given in Figure 6. This instrument, which operates at beamline X16A at NSLS, is the most widely used design. All of the instrumental functions described in this section are generic to all installations of surface x-ray diffraction, although some specific details of the X16A instrument are given for clarity of illustration. Further machine-dependent details can be found below in the detailed protocols in the Appendices.

The overall size of the instrument in Figure 6 is ~2 m long by 1 m wide and it resides at the beamline endstation. The instrument slides on air bearings, which allows it to be slid quickly into and out of the beamline hutch at the beginning and end of each experimental run. The configuration consists of an ultrahigh-vacuum (UHV) chamber (left side) coupled to a precision diffractometer. These two components will be described separately.

### Vacuum System

On the left side, looking towards the x-ray source, lies the main vacuum chamber, which is used for sample preparation. It has turbomolecular (TMP), ion, and sublimation pumps (See GENERAL VACCUM TECHNIQUES) located in a cross below the level of the main chamber. The TMP and ion pumps have gate valves that allow them to be isolated, for protection against power failure for the TMP and to

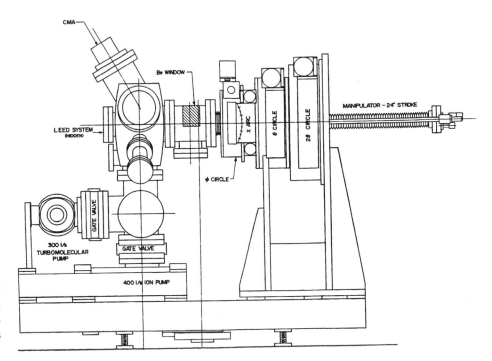

**Figure 6.** Overview of the vacuum diffractometer configuration used at the X16A facility at NSLS by Fuoss and Robinson (1984). Other installations vary slightly from this design.

reduce exposure to certain gases (e.g., Ar for sputtering and $O_2$ for dosing) for the ion pump. Numerous flanges are pointing at the sample, allowing the installation of various instruments for sample preparation and diagnosis. Samples are prepared by combinations of sputtering using an ion-bombardment gun, annealing by use of filaments, dosing with gases controlled by leak valves, and evaporation of substances using dosers. Dosers include commercial Knudsen cells and electron-beam evaporators, heated crucibles of molten materials, simple filaments coated with metal films, and getter-type alkali sources. Diagnosis is by means of LEED, Auger electron spectroscopy (AES; see AUGER ELECTRON SPECTROSCOPY), and residual gas analysis (RGA). Most conventional surface science preparations can be carried out using combinations of these tools.

Most importantly, the capability of performing an x-ray diffraction experiment in situ is provided in the chamber. X rays can enter and leave the chamber through a semicylindrical beryllium window, shown in the center of Figure 6. The window was brazed onto the vacuum wall at high temperature, so it is completely UHV compatible. It has 0.5 mm wall thickness so it transmits at least 95% of x rays with energies above 8 keV. In the diffraction position, the sample lies at the extreme right-hand side of the window with its flat face, upon which the surface has been prepared, facing left. The sample is rigidly connected to the diffractometer to the right. A manipulator wand located behind the diffractometer allows the sample to be detached from its rigid diffractometer mount and carried under UHV to the various instruments in the preparation chamber. There is a bellows and rotating seal coupling between the diffractometer and the chamber, which allows the passage of the diffractometer motions. These last essential features were specifically designed for the surface x-ray diffractometer and so will be discussed next.

Because preparation of $10^{-10}$ Torr pressure is time consuming, a loadlock is used to allow simple manipulations of samples, thermocouples, and heaters to be carried out without venting the main UHV chamber. The basic principle of operation is shown in Figure 7. For example, load locking is very useful for batch processing a series of samples or for introducing samples that cannot be baked at 150°C. Most instruments for surface x-ray diffraction provide some sort of load lock for rapid sample turnaround. Specific details for the operation of the X16A vacuum system are given later in the detailed protocol (see Appendix A), including the load-lock procedure, the venting procedure, and the bakeout procedure.

### Sample Manipulator

Once it is retracted, the manipulator moves entirely with the sample on the diffractometer. A spring-loaded kinematical mount comprising three balls resting in 3 V-grooves 120° apart ensures that any flexure of the long arm of the manipulator is decoupled from the sample, which is thereby allowed to rest against the diffractometer $\phi$ axis. To carry the sample into the preparation chamber, the manipulator is extended by collapsing the 0.8-m-long bellows using a motor and gear on a threaded shaft. Electrical connections to the sample traverse the full length of the manipulator from its accessible rear flange. The last 75 mm of the shaft is illustrated in Figure 7. It is crammed with intricate features that give it great versatility: a 90° elbowlike flexure driven by a push-rod running down the length of the manipulator that allows the sample to face the LEED and AES; a spring-loaded decoupling joint associated with the kinematic mount (not shown); a vacuum sealing flange penetrated by electrical and thermal feed-throughs for the load-lock system described below; and at the end 25 mm of clear space that allows the user to assem-

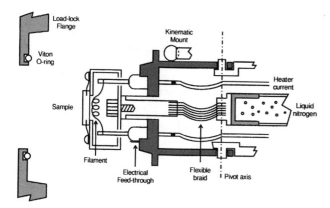

**Figure 7.** Detail of the manipulator head shown in cross-section. The pivot allows the sample to point at 90° to the main axis to be presented to surface characterization instruments. The viton seal of the load lock, through which the sample is passed, is shown on the left. The mating surface is halfway down the manipulator head and contains vacuum feedthroughs for cooling and electrical connections to the sample. All parts to the left of this are accessible to the user during the load-lock procedure.

ble his or her sample with the appropriate heaters, coolers, and sensors.

The rotating seal design is shown in Figure 8. It allows the passage of rotations across the vacuum wall using a rigid shaft that has neither play nor backlash. It achieves this with three concentric sliding seals made of Teflon between which differential vacua are applied, the first stage at $10^{-3}$ Torr with a roughing pump, the second at $10^{-6}$ torr with a TMP pump. If any seal momentarily leaks

during rotation, the pressure burst on the UHV side is minimized by the use of these differentials.

### Five-Circle Diffractometer

The diffractometer geometry is illustrated with the aid of Figure 9. This corresponds to a "five-circle" geometry, which was invented for surface diffraction by Vlieg et al. (1987). The standard names given to the angle variables describing the diffractometer setting are the same as the names of the axes in the figure. The principal axes are called (confusingly) $\theta$ and $2\theta$ due to their association with the two-circle diffractometer setting describing Bragg's law. The symbols $\phi$ and $\chi$ are Euler angles of the sample with respect to the main horizontal ($\theta$) axis. Finally $\alpha$ is the rotation of the entire instrument about the vertical axis. Historically at X16A, $\alpha$ was an afterthought, added to the instrument when it was found that the original four axes were too restrictive in their coverage of reciprocal space in the out-of-plane direction. Later instrument designs have incorporated $\alpha$ axes as a matter of course.

The convention for the zero settings of the axes is defined as follows: $\alpha = 0$ when the incident beam is perpendicular to the main horizontal axis. The remaining definitions all correspond to $\alpha = 0$: $2\theta = 0$ (dial = 270 for the X16A instrument) when the incident beam strikes the center of the detector; $\theta = 0$ (dial = 270) when the $\chi$ axis lies exactly along the incident beam direction; $\chi = 0$ when $\theta$ and $\phi$ are collinear; and $\phi = 0$ is arbitrary and usually chosen to be the same as the mechanical dial setting. The names used by the usual computer program for these angular settings are TTH, TH, PHI, CHI, and ALP, as discussed below (see Appendix C).

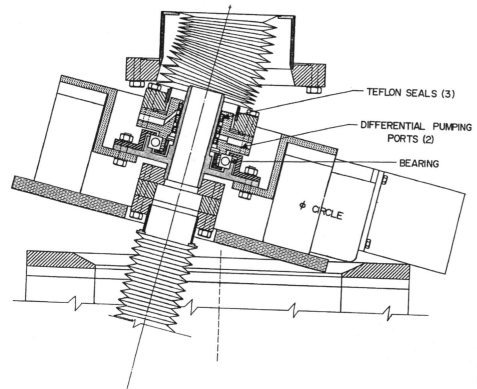

TEFLON SEALS (3)

DIFFERENTIAL PUMPING PORTS (2)

BEARING

$\phi$ CIRCLE

**Figure 8.** Cross-sectional view of the bellows and rotating vacuum seal that couple the rigid motions of the diffractometer to the sample in UHV. From Fuoss and Robinson (1984).

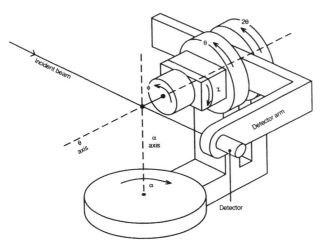

**Figure 9.** Isometric view of the five-circle diffractometer. The names of the axes and their positive directions of turning are indicated.

The additional degrees of freedom over the four circles of a nonspecialized diffractometer are needed for two reasons: (1) the additional constraints on the diffractometer setting due to grazing incidence or exit conditions and (2) the limited range of the angles due to potential collisions with the vacuum chamber.

The $\chi$ axis is particularly affected by the second constraint since it is limited to $\pm 0°$ at X16A by the coupling bellows and also by the manipulator. There is also a "six-circle" geometry due to Abernathy (1993) with an extra angular out-of-plane degree of freedom for the detector, called GAMMA, a "Z-axis" geometry due to Bloch (1985) that is a subset of six-circle geometry without the $\phi$ and $\chi$ motions, and also a "2+2" geometry used for surface diffraction by Evans-Lutterodt and Tang (1995).

**Laser Alignment**

Many of the typical measurement methods of surface x-ray diffraction employ the grazing angle geometry (see Principles of the Method) on the incident beam, the exit beam, or both. It is therefore crucial that the physical orientation of the sample's surface be well known. To allow for the general case (rather commonly occurring) that the surface orientation is not a simple crystallographic orientation, perhaps due to miscut, it is important that the optical surface orientation be defined independently of the crystallographic orientation. This step is known as "laser alignment."

A laser beam is directed roughly down the main ($\theta$) axis of the instrument and reflected from the polished sample surface and allowed to strike a screen located behind the laser. If the $\theta$ axis is rotated, the reflected spot will precess in a circular orbit. The center of this circle is the correct laser reflection angle if the sample face were exactly perpendicular to the axis. The beam can then be directed onto this center by adjustment of the diffractometer $\phi$ and $\chi$ angles until the laser spot remains stationary upon rotating $\theta$. These settings, called fphi and fchi in the detailed protocol below (see Appendix C), are then a

unique representation of the sample's optical orientation and are entered into the diffractometer control program so that the incident and exit angles can be precisely controlled.

**Sample Crystallographic Alignment**

Sample alignment is normally achieved by finding the diffractometer settings for diffraction from a small number of bulk Bragg peaks. If the lattice parameters are known, a complete orientation matrix is specified by the positions of just two Bragg peaks using the method of Busing and Levy (1967). If the lattice parameters are not known, it is necessary to specify the positions of at least three noncoplanar Bragg peaks. The orientation and calculation of arbitrary settings will be made by the control program once this information has been furnished.

The usual procedure for locating the Bragg peaks is to manually vary the diffractometer angles to maximize the diffracted intensity. This is less straightforward with a five- or six-circle diffractometer than with the more standard four-circle instrument, simply because there are more degrees of freedom in which the user can get lost. A reflection has 3 degrees of freedom, so the five-circle diffractometer has two motions too many. In particular, the detector moves on two axes, one of which ($\alpha$) also moves the sample: there is no longer a simple connection between Bragg's law and the detector motion. Here are some guidelines on the suggested method of finding peaks:

1. Set up a "dummy" orientation matrix (OM) with the best known values of the lattice parameters (see Appendix C). Two mutually perpendicular in-plane reflections can be specified as having the laser orientation angles. From here, the setting of a suitable bulk out-of-plane Bragg peak can be calculated. In-plane peaks are hard to find because both incident and exit beams must be exactly grazing; out-of-plane peaks have a much wider choice of possible settings. Set the diffractometer at that position.

2. The desired reflection can be found by rotating the sample about its surface normal using either $\phi$ or $\theta$. The peaks are so strong with synchrotron radiation that the motors can be searched at top speed; the reflection will "flash" on the detector oscilloscope or rate meter. There is usually sufficient thermal diffuse scattering (TDS) for a peak to be found in this way, even when it is several degrees away from the line that is searched.

3. Ride up the TDS until the center of the Bragg peak is found. Diffracted beam attenuators will be needed at this stage. Centering can be quickly and automatically carried out with "line-up" (lup) scans of a single motion (see Appendix C), after which the program calculates the center of mass and sends the motion there.

4. When using a PSD with an entrance window that is wide along the $q_z$ direction, the peak may appear at any position along the length ($q_z$). Since the diffractometer is aligned to the center point of the detector, the contents of the PSD in the MCA must be viewed

(function key F1 or F2) and the peak made to walk to the center. Alternatively, a region of interest should be defined about the center and only the counts in this region used for reflection centering.

5. Once a peak is centered, it is better to rotate the dummy matrix manually to the corresponding orientation. This way both defining reflections are rotated together. If merely one of the reflections is overwritten, the resulting two-peak orientation matrix may be very unrealistic and will make it hard to find the second peak. Alternatively, the second dummy reflection can be chosen to coincide with the surface normal direction.

Once two peaks are found, the orientation matrix is complete and can be used to find any further points in reciprocal space. There will always be small errors of both crystal and diffractometer alignment, which may accumulate and result in poor prediction of further Bragg peaks. There are two ways of avoiding this problem:

a. Use alignment peaks that are most relevant to the features to be measured. For example, a good alignment for measuring a CTR would be to choose two Bragg peaks through which the rod passes.

b. Use multiple-peak alignment. Diffractometer alignment errors can be partially compensated by using the crystallographic unit cell parameters as free variables. An excellent orientation matrix will always be obtained by using a least-squares refinement of the best lattice passing through many Bragg peaks, especially if they are at large diffraction angles.

## DATA ANALYSIS AND INITIAL INTERPRETATION

The following summarizes the different ways to present surface x-ray diffraction data to express one or other sensitivities or strengths of the technique. Different material properties of a given surface or interface would require entirely different kinds of measurement to be made. Each subsection lists a different form of presentation of the data that the reader might encounter in the published literature, then describes the typical measurement involved, and finally cross-references the typical diffractometer scans (Appendix C) and typical data analysis steps (Appendix B) that might have been followed. At the end of the section, a worked example is given.

### Rod Profiles

Crystallographic measurements referred to above (see Principles of the Method) are used for surface and interface structure determination. The same procedures apply to CTRs or superstructure reflections. The general form of the data is to convey the magnitude of the structure factor along a line of points in reciprocal space, running along a line perpendicular to the surface. Theta-rocking scans (see Appendix C) are taken at a range of $q_z$ values or Miller index $L$. These are numerically integrated and background subtracted, e.g., using the program PEAK (see Appendix B). For reliability, it is essential to measure multiple reflections related by the surface symmetry. When these symmetry equivalents are merged together, e.g., using the program AVE (Appendix B), their reproducibility is thereby determined and then used to specify the experimental error in the structure factor magnitude. Further details, including the equations defining the error estimate are given by Robinson (1990b).

The data are usually plotted as smooth curves against the perpendicular momentum transfer, $L$. Since the structure factor is a continuous function, enough $L$ points must be measured to be sure that the profiles are smoothly varying and that all oscillatory features are resolved. This last caution is particularly apt in the case of thin films, which have many closely spaced thickness fringes; in this case, some researchers choose to simply scan along the $q_z$ ridge of intensity in a "direct rod scan" and then to offset the $\theta$ angle and repeat the scan to estimate the background.

The rod profiles are then fit to the functional form of the structure factor of an atomic model of the surface, as outlined above (see Principles of the Method). In the kinematical limit, both for CTRs and superstructure rods, this is given by a complex linear sum over atoms representing the superposition of atomic form factors. Closed-form expressions are needed for the form of the CTRs to account for the infinite sum. The surface roughness must also be accounted for at this stage. The program ROD, written by E. Vlieg and described in Coppens et al. (1992) (see Appendix B), is a widely used computer program for this purpose. Atomic models of bulk and surface regions are specified in parametric form in a very general format. The parameters of the model are then refined by a least-squares minimization to give the atomic coordinates on the model.

It should be noted that surfaces are very different from bulk matter in the specification of atomic models. Each layer of the structure is expected to have a different Debye-Waller factor representing its atomic vibration amplitude. These Debye-Waller factors usually increase significantly in magnitude in passing from the bulk outward in the structure. Another common feature of surfaces is partial occupancy of lattice sites. It is also common to find more than one site, each partially occupied, for a structure with disorder. All of these situations are handled by the ROD program (see Appendix B). Finally, surface atoms are prime candidates for anharmonicity, since they have neighbors on one side but not the other. Surface anharmonicity has been analyzed in the pioneering work of Meyerheim et al. (1995).

### Reflectivity

The specular reflectivity is the special case of the CTR emanating from the origin of reciprocal space. While it can be measured in the way described in the previous section, this can also be achieved with the simple two-circle method if the sample is mounted perpendicular to the $\theta$ axis. It is customary to plot reflectivity data as the square of the reflection coefficient versus total momentum transfer, $q_z$. This gives a normalization to the data different from the crystallographic standard described above: an ideally terminated surface would have an intensity dropping as $q_z^{-4}$ in these units, according to the Fresnel law of reflection. To emphasize the differences, the reflectivity divided by the (calculated) Fresnel reflectivity is often what is plotted. There is a correction for the total external reflection region at small angle, which can be understood with a simple dynamical treatment.

Als-Nielsen (1987) has derived a "master formula" for the analysis of reflectivity data in terms of the spatial derivative of the electron density profile of the surface. When $q_z$ is small, the resolution is insufficient for atoms or atomic layers to be located, so the electron density is parameterized as a series of slabs of variable density connected by error function shaped transition regions of variable width. These parameters are then refined by least squares to fit the data and obtain the electron density profile explicitly.

Reflectivity is the method of choice for measuring the sharpness of the layer profiles of artificial multilayer structures that are chemically modulated by varying their composition during deposition. Thick multilayers of electron-dense materials require a full dynamical treatment for correct analysis, in which transmission and reflection coefficients of each layer interact self-consistently. The theory was worked out by Parratt (1954) and has been implemented by Wormington et al. (1992). It is also commercially available as a software product called REFS from Bede Scientific.

### Surface Diffuse Scattering

Surface roughness can be measured quantitatively using the CTR profiles, as mentioned above (see Principles of the Method). It is also measured routinely by the reflectivity method above. In either case, the intensity that is missing from the CTR or the reflectivity, resulting in a lower signal, reappears as diffuse scattering nearby. This was clearly explained in terms of the lateral correlations associated with the roughness in an important paper by Sinha et al. (1988).

The surface diffuse scattering can be measured most easily near the origin of reciprocal space by means of wide $\theta$ scans (see Appendix C). Such scans have several characteristic features that have simple interpretations: in the center there appears the CTR or the specular reflectivity as a sharp spike; at the two ends of the scan lie the "Yoneda wings," where the diffuse intensity rises by the $|T(\alpha)|^2$ transmission function as $\alpha_i \approx \alpha_C$ or $\alpha_f \approx \alpha_C$; in between lies the surface diffuse scattering. Multilayer samples are found to have rich surface diffuse scattering patterns that provide information about the correlations between the positions of the various layers making up the multilayer.

### Reciprocal Lattice Mapping

In some surfaces undergoing phase transitions or morphological changes, there is valuable information in just the diffraction peak positions, widths, and intensities. For example, surface diffraction is extremely sensitive to the appearance of new periodicities in surfaces, and this has been exploited extensively. The measurement consists of recording wide scans, e.g., using the *iscan* and *jscan* described below (see Appendix C), along simple directions of reciprocal space. Such curves would be plotted as a function of temperature, time, or other control variables. Sometimes two-dimensional scans (*kscan*) are also made to map out all the identifying features. A crystal analyzer, which provides much higher **q**-resolution, may be needed to separate finely spaced features.

The materials properties of interest that can be derived from such measurements concern grain shapes and sizes and the degree to which these are strained. The positions of Bragg peaks in any diffraction experiment determine the lattice parameters of the sample crystal; e.g., high-resolution x-ray diffraction is probably the most accurate of all and can achieve 1 part in $10^5$ using the Bond method. Such measurements provide information about strain and its distribution within the sample. Similarly, accurate measurements of diffraction line widths, through the Scherrer formula, determine the size of the diffracting grain. In both these situations it is routine to use a careful lineshape analysis to extract the most reliable sample parameters. The semi-automatic program ANA described below (see Appendix B) provides a battery of possible lineshapes that apply to different kinds of diffracting objects or distributions of objects.

### Grazing Incidence Measurements

Grazing incidence measurements fall into two categories. The first is a simple extension to all of the above techniques to determine materials properties as a function of depth inside a sample. Grazing incidence of exit conditions is then employed to vary the penetration depth into the bulk, which can be calculated reliably from the known refractive index and absorption values of the material at the x-ray energy used. The distribution of depths sampled is always exponential, so it is never possible to select a particular depth of interest; for this purpose it would be necessary to use a nonlinear technique such as ion scattering (see Chapter 12). Dosch (1992) has suggested the use of Laplace transforms to deconvolve the depth information.

Since the incident and exit angles always have to be under control in a surface x-ray diffraction experiment, it is a routine extension of the normal procedure to deliberately control them. For example, with the diffractometer control program "super," the penetration depth can be chosen by setting the target value of the incident or exit angle, $\alpha_i$ or $\alpha_f$ (B or B2 in Appendix C). A useful extension is to make a *vscan* at a fixed reciprocal lattice position as a function of $\alpha_i$ or $\alpha_f$.

The second category of grazing incidence measurement is to use the PSD to record an entire $\alpha_f$ profile in a single measurement. This method was pioneered for the study of surface-phase transitions in a depth-resolved way by Dosch (1987, 1992). It has the advantage that diffractometer motions are not required during measurement of small changes with temperature. More recently the method has been improved by Salditt et al. (1996) to record and analyze diffuse scattering from multilayer samples as a function of exit angle $\alpha_f$ in a PSD by making wide $\theta$ scans at fixed $\alpha_i$.

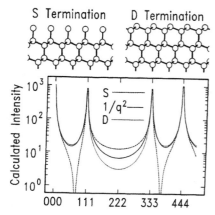

**Figure 10.** Top: the possible S and D terminations of the Si(111) surface. Bottom: calculated specular reflectivity for the two terminations according to the CTR theory and by superposition of $1/q^2$ intensity distributions.

## Worked Example: the Si(111) Surface

When the bulk crystal has a sufficiently simple structure, in the case of all primitive crystal structures, or the NaCl structure, there is only one possible way to cut the lattice to reveal a surface. However, in this case, the CTR presented above (see Principles of the Method) is technically indistinguishable from a simple superposition of $1/q^2$ intensity tails around each Bragg peak, as discussed by Robinson (1990a). Whenever the crystal has a multiple-atom basis that allows for more than one possible termination, the shape of the CTRs depends on which termination exists and is then clearly distinguishable from the superposition of $1/q^2$ intensity tails.

Such is the case of Si(111), where two possible terminations can exist, as illustrated in Figure 10. The double-layer (D) termination occurs when bonds perpendicular to the surface are cut; the single-layer (S) termination occurs when we break the bonds inclined at 19° to the surface, as shown. The density of dangling bonds is three times greater for the latter case, so it might be considered less likely to occur. The calculated CTRs are plotted for the two cases in the lower half of Figure 10. Up to a factor-of-10 difference in the two structure factors is seen to occur.

In reciprocal space, the specular reflectivity line is the direction with the momentum transfer, **q**, entirely perpendicular to the surface. This corresponds exactly to the zeroth-order CTR diffraction and so can be thought of as a CTR; the analysis of these data is particularly simple because only the vertical components of the atomic coordinates enter, by virtue of the section-projection theorem of Fourier transforms. The question, then, is how is the real Si(111) surface, prepared in UHV, terminated? This is complicated by the fact that it is reconstructed with 7×7 translational symmetry.

## Example of Measurements from Si(111)7 × 7

Figure 11 shows the measurements of the specular CTR of Si(111)7×7 using an improved data collection method. An x-ray PSD was placed so that it cuts across the diffracted beam. The background is then measured simultaneously

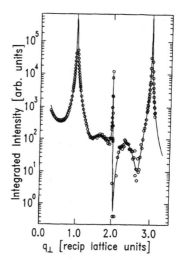

**Figure 11.** Measured x-ray reflectivity of a UHV-prepared Si(111) sample using a PSD. The solid line is the least-squares fit to a model based on the DAS model for the 7 × 7 reconstructed surface due to Takayanagi et al. (1985). From Robinson et al. (1994).

with the integrated intensity of the diffracted beam and can be immediately subtracted. It is clear from the data that neither the S nor D simple termination adequately describes the surface. The figure also shows the result of a least-squares fit to the data, which passes through the points respectably well. The model used was the dimer-adatom-stacking fault (DAS) model for the Si(111)7 × 7 reconstructed surface due to Takayanagi et al. (1985). The heights of the atoms and the vibration amplitudes in the top four layers were adjusted for a good fit using the program ROD described in Appendix B.

Figure 11 also shows an additional sharp feature in the center of the scan at the position of the specular 222 Bragg reflection. This is indeed the specular 222 Bragg reflection, which is forbidden in diffraction from the ideal diamond lattice but is not perfectly extinguished for two reasons suggested by Keating et al. (1971):

1. The vibrations of the ions in the lattice are not spherically symmetric but tend to bulge out along the tetrahedral directions (four of the eight {111} directions) opposite to each bond; the anisotropy is along different directions for the two atoms in the basis, meaning they are crystallographically inequivalent.

2. The electrons forming the $sp^3$-hybridized bonds between the Si atoms are not spherically symmetric either but have a tetrahedral shape, again meaning the two atoms are crystallographically inequivalent. This affects 4 of the 14 electrons in the Si form factor so it is a relatively large effect.

It was shown many years ago by Keating et al. (1971) that contribution 1, which is temperature dependent, is dominant in neutron diffraction, while contribution 2 is more important in x-ray diffraction experiments. We therefore included extra bond charges in our model to fit this 222 peak in our data. It is orders of magnitude weaker than the bulk diffraction but still stronger than the surface component of the CTR. A particularly noteworthy feature is the interference between the tails of the 222 and the CTR.

## PROBLEMS

This section is a critical appraisal of the weaknesses of the surface x-ray diffraction technique. It might be used in evaluating published work, e.g., in searching for alternative explanations of results obtained with the method.

### Surface Crystallographic Refinement

Crystallography has an inherent "phase problem." It is the structure factor amplitudes that are measured, not the phases. The solution of an atomic structure amounts to testing hypothetical models for agreement with the measurements. It is assumed that a good fit will only occur for one model, which therefore finds unique values for the missing phase information.

The difficulty with this logic is that it depends upon the quality and quantity of the data. For good, numerous data, experience has shown that there is usually a unique solution. In fact, many researchers are frustrated by the situation that they cannot find a model that agrees with their data! This is a heuristic support for the notion of uniqueness. "Direct methods" of structure solution, which exist for small-molecule crystallography (Woolfson, 1997), are being developed for surfaces but are not yet ready for widespread use.

The problem arises when there are insufficient data for the solution to be determined. The accepted rule of thumb (Lipson and Cochran, 1966) is for the structure to be overdetermined by a factor of 3:1. This means that for every structural degree of freedom in the model, there must be three independent measurements. It is presumed that nonuniqueness of the solution would be extremely unlikely when three separate measurements must all agree with each other for every parameter determined. For this to work, the counting must be done honestly: we will consider first the question of counting model freedoms and then that of counting the data.

It is possible to tie structural parameters together to reduce the degrees of freedom, but this always requires assumptions or constraints on the possible structures that might exist. Chemical constraints such as bond lengths and angles are usually safe, but sometimes unorthodox chemistry can take place at surfaces. This is often done through a Keating model (Keating, 1966) where a penalty for nonstandard bond lengths and angles is added to the $\chi^2$ functional used for least-squares minimization. Another common constraint is to collect the less-sensitive parameters, such as Debye-Waller factors, into groups applying to several related atoms. Yet another constraint is to enforce exponential decay of relaxation parameters from the surface toward the bulk; this amounts to assuming the form of the solution of the Poisson equation describing continuum elasticity in the material. In each case, if the hypothesis leading to the constraint is questionable, then the conclusions drawn are questionable.

A similar logic concerns counting the data. This relates also to their accuracy, which can be measured in the way described above (see Practical Aspects of the Method). The situation is more complicated for surface than for bulk crystallography because the data are in the form of "rod profiles," which are continuous functions of $L$. The amount of information in a given rod profile can be counted, but this must be done carefully: every distinct maximum and minimum in the rod counts as one independent data point. Whether or not an extremum is distinct is closely tied to the size of the error bars of the measurement.

In principle, all error analysis should be handled through the definition of the normalized $\chi^2$ functional, which is weighted by a factor $1/(N - P)$, where $N$ and $P$ are the number of data and number of parameters that we have just defined. The danger is that, while $P$ is usually correct, $N$ is frequently overestimated because it counts separately all the data points along each rod, rather than just the number of extrema. Nonspecialists and specialists alike quote a small goodness of fit $\chi^2$ (near 1.0) value as an endorsement of the correctness of a structure. This is nonsense! A small $\chi^2$ simply means that the data have been exhausted and the model is sufficiently detailed; it does not necessarily mean it is right. An easy way to obtain a small $\chi^2$ is to measure very few data!

In examining published work, there are certain indications that the refinement of the structure had been difficult. Debye-Waller factors are sometimes omitted because they refine to negative values when included. Negative Debye-Waller factors are unphysical and so usually indicate some other deficiency in the model. Another bad sign in a structural model is when many atoms with partial occupancy have been included. Partial occupancy is common on surfaces, because of their nature: when a crystal is cut, it must have a boundary, but not necessarily one with an ordered structure.

### Other Surface Crystallography Problems

Regarding the question of the quality of structure factor data, some aspects are specific to the case of surfaces. Apart from gross misalignments, there are subtle ways that additional errors can be introduced. When grazing incidence is used, the slightest variation of $\alpha_i$ from one measurement to the next causes a different value of the $|T(\alpha)|^2$ prefactor and hence the observed structure factor. In an extreme case, misalignment can cause $\alpha_i$ to become negative and the beam to disappear below the horizon. Sample curvature can aggravate the problem, leading to a different distribution of $\alpha_i$ values along different incidence azimuths. The only safe way to avoid the problem is to measure symmetry-equivalent reflections and record the data reproducibility.

When CTR data are included, there are several dangers associated with measuring very close to bulk Bragg peaks. Highly structured background is commonly found and must be carefully subtracted. The problems are compounded when the out-of-plane detector resolution is opened up to improve counting rates. There can be a significant distortion of the data from the $L$ variation of the intensity across the detector slit. The safest strategy is to avoid including data near the Bragg peaks, except where absolutely necessary.

Another problem can arise from unavoidable beam harmonics. The detectors are not always capable of discriminating $\lambda/3$ (and $\lambda/4$, etc.) contributions completely. This means that any measurement at or near an integer fraction of a bulk Bragg peak can give a strong "glitch." For simplicity, it is common to avoid all measurements at small-integer fractions of $L$.

A different source of possible misconception concerns symmetry. There are many cases of published work where the surface structure has lower symmetry than that of the underlying bulk crystal. For example, all reconstructed surfaces have lower translational symmetry than the bulk, and this is accepted for chemical reasons. The general rule is that a surface structure should have as high a symmetry as possible, compatible with the symmetry of its parent bulk. Only when this is shown to be impossible should the symmetry be broken. Then one symmetry element at a time should be removed from the model, since each symmetry that is lost can greatly increase the number of degrees of freedom.

A common source of confusion is that the symmetry of a surface is not immediately obvious from the symmetry of the diffraction pattern. Domain formation is common on surfaces, where differently oriented regions nucleate in different locations. It is usual to average the intensity contributions from each of the orientational domains, rather than the structure factor. This amounts to the assumption that the domains are far apart, beyond the coherence length of the measurement. The opposite limit of microdomains within the coherence length is usually apparent from a broadening of the peak lineshapes. This can be handled by mixed structures with partial occupancy once the peaks have been integrated correctly to account for their enlarged widths.

### Lineshape Analysis

The purpose of analyzing lineshapes is to extract parameters that relate to materials properties so that they can be tracked as the sample is varied in some way. It is usually the peak position and/or width that is of interest. Typically the fitting is done directly to the output data stream in a semi-automatic way, e.g., using the program ANA described in the detailed protocol in Appendix B. If the curve that is used for fitting has the wrong shape, systematic errors will be introduced that may or may not have consequences. For this reason, a battery of lineshapes is usually available, sometimes with several adjustable parameters. A good example of a generally useful lineshape is the Lorentzian function raised to a power:

$$p(x) = A[(x - B)^2 + C^2]^{-D} \qquad (2)$$

where $A$, $B$, $C$, and $D$ are adjustable parameters ($B$ and $C$ are the position and width most often sought). The danger here is that of parameter coupling or nonorthogonality, particularly between parameters $C$ and $D$. If $D$ is just allowed to vary from one fit to the next, and then later ignored, unwarranted changes in $C$ may occur that are not really present in the data. Here, the solution is to find a value of $D$ that applies to most of the data, then fix the parameter for fitting the data series to look for the trends.

A different situation applies to the handling of background, which is seen as a nonvanishing intensity in the extension of the tails of a peak. The background is due to unwanted additional scattering sources within the sample or the instrument. Usually the background is subtracted by extrapolation. The background can have a profound effect on the stability of fitting lineshapes such as the Lorentzian function raised to a power above, especially if $D$ is the parameter of interest. The best advice here is to measure a lot of background extending far away from the peak; this has a surprisingly beneficial effect in constraining the fits of functions with long tails such as this.

The best procedure for measuring diffuse scattering is again to control and understand the background. By its nature, there is no perfect way to separate diffuse scattering from background. The most difficult situation is that of small-angle scattering near the direction of the incident beam, which applies to the nonspecular reflectivity region for surfaces discussed above (see Data Analysis and Interpretation). Here, the potential sources of background extend all the way back up the beamline to the entrance slits and beyond. A common hazard in diffuse scattering measurements is the tails of the resolution function, which appear to be CTRs pointing in apparently unusual directions. In fact, they are the CTRs of the crystals in the monochromator or analyzer rather than those of the sample!

## DETAILED PROTOCOLS

The exact procedure will vary slightly from installation to installation, but the general principles will be the same. To be specific, we discuss here the procedure for the X16A surface diffractometer at NSLS. The generalization to other installations will usually be obvious. Details of the vacuum procedure are given in Appendix A, the data analysis programs in Appendix B, and the diffractometer control program in Appendix C.

### Alignment of the Beamline

The X16A beamline, like many others used for diffraction, consists of a grazing incidence toroidal mirror and double-crystal monochromator. It is the responsibility of the NSLS facility staff to help the user get to the point of operating the beamline. This requires verification of all safety-related hardware on a "safety checklist" and that the condition of all cooling and vacuum systems be "green" before the safety shutter will open.

The mirror focusing is 1:1 and so should produce a focal spot on the sample ~1 mm wide by 0.5 mm high. The focal length is fixed and requires the correct incidence angle of 5.65 mrad, which fixes the final height of the focused beam. It is important to ensure that the height is correct by using a focusing screen near the sample position and a TV camera with a close-up lens. The beam image is a U-shaped

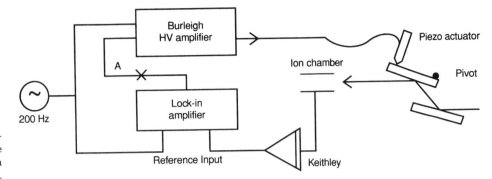

**Figure 12.** Monochromator feedback circuit. Point A is where the feedback loop should be broken in order to detune the monochromator.

"smile" when the incidence angle is too low and an inverted smile when the incidence angle is too high. The locus of focal images traces out an X shape on the screen upon tilting the mirror, with the optimum focus at the intersection of the X. A tilted smile means the mirror is not steered correctly and requires rotation about the vertical axis. The mirror bend should then be adjusted for the smallest vertical profile, and the beam position should be scanned to verify it is not being cut by any nearby aperture edges.

The double-crystal monochromator is controlled by five manual adjustments. The energy is set by a stepping motor on the back of the tank that is driven by a manual pulser. This can also be controlled by pulses from the beamline computer if needed. The energy reading is a dial counter, which is converted to kilo-electron-volts on a chart in the beamline log book. The $\theta$ and $\chi$ angles of each crystal are controlled by stepper motors driven by a shaft-encoder knob and selected by one of the four red "engage" buttons. The four angles are also read out by absolute encoders. It is very important not to move these in-vacuum motors far from their starting positions because there are no limit switches. After changing energy, it is necessary to bring the two crystals parallel again by adjustment of $\theta_2$ until a signal is seen in the ion chamber. Side-to-side steering of the beam position is achieved by $\chi_1$, which is generally used to bring the beam onto the sample (see below).

The monochromator feedback circuit is shown in Figure 12. The second crystal is modulated at 200 Hz by a piezoactuator. When its rocking curve is correctly centered on that of the first crystal, the lock-in amplifier reads zero output; if the angle is misset, a positive or negative error signal is generated that corrects the angle. The loop gain and integration time may have to be adjusted for correct performance. To detune the monochromator (e.g., to remove harmonics in the beam), the loop must be broken at point $A$ and the crystal angle set manually.

### Alignment of the Diffractometer

The intersection of the five axes of the diffractometer is at its center, and this must be positioned accurately on the beam's focal spot. This must be correct to within 100 μm. The beam direction must also be perpendicular to the $\theta$ and $\alpha$ axes to within 0.1° and 0.3°, respectively. Because a surface x-ray diffractometer has a sample in a UHV

environment also at its center, this presents a special challenge.

The vertical position is set by four screw jacks under the diffractometer table. These are manually or stepping-motor driven in two pairs, front and back, to allow the table to be tilted until the beam is perpendicular to the $\alpha$ axis. The tilt is equal to the beam's known 11.3 mrad angle to the horizontal and preset using a spirit level and rulers. The horizontal position is set coarsely using surveying marks and finely by moving the beam with the $\chi_2$ monochromator adjustment. The perpendicularity of the theta axis is determined by the choice of the zero of $\alpha$.

The essential alignment tool is an "off-axis pinhole" that is an L-shaped bracket with a 1-mm pinhole drilled through it. This bolts onto the $\phi$ circle of the diffractometer so that it wraps around the outside of the UHV chamber and presents itself to the incoming or outgoing beam, as shown in Figure 13. The distance from the pinhole to the mounting surface is accurately machined so that it corresponds to the manufactured radius of the $\chi$ arcs. Because of its offset from the center, the position of the pinhole with respect to the beam is scanned by moving the $\phi$ and $\chi$ axes of the diffractometer.

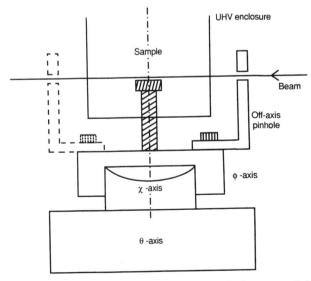

**Figure 13.** Off-axis pinhole designed to mark the center of the diffractometer without breaking the vacuum. The dashed position is reached by a 180° rotation of the $\theta$ axis.

Because the sample is blocking the true center position so that the x-ray beam cannot pass, the alignment must be carried out at a controlled location just above it. This is done by tilting the diffractometer $\chi$ by a known amount (usually $1°$) from its calibrated zero and moving the beam with the $\chi_2$ (monochromaor) to correspond. The center position of an arbitrarily placed starting beam is determined in $(\phi, \chi)$ space with the pinhole on one side of the sample and then again on the other side. An exact $180°$ rotation of $\theta$ takes the pinhole from front to back, as shown in Figure 13. Half the difference between the two $\phi$ values multiplied by the lever arm $L$ is the error in the table height, which is thereby corrected. Half the difference between the two $\chi$ values is the deviation of the beam from perpendicularity with the $\theta$ axis, which defines the zero of $\alpha$.

Once these corrections have been made, the pinhole is exactly aligned with the diffractometer center, offset horizontally to avoid the sample. The beam, passing through the pinhole, can then be used to set precisely the vertical incident and exit beam collimation slits. If the pinhole is then removed and the beam moved until it grazes across the sample at the center, the horizontal slits can be set too. This is also the opportunity to set up the channel definitions of the PSD (if used), by marking off known distances from the calibrated center line.

The centering of the $\alpha$ axis with respect to the other four is also adjustable, owing to its historical origin as an auxiliary sample motion. The adjustment is by means of a motorized linear translation across the beam and slotted screw-holes in the direction parallel to the beam. Previous settings of these adjustments are recorded in the logbook, and these should be used as starting values. The alignment should not be needed often as the performance does not depend critically upon it; the usual symptom of misalignment is a loss of intensity at large values of perpendicular momentum transfer. It is carried out using the off-axis pinhole once this has been centered exactly on the beam following the procedure above. Then $\theta$ is rotated by exactly $90°$ so that the pinhole is pointing straight up, thereby marking a point exactly above the intersection point of the axes, as can be seen in Figure 9. The center of the $\alpha$ axis is then brought to intersect this marked point by using the translations. This is most easily performed with a high-magnification TV camera hanging down from the hutch roof: when $\alpha$ is rotated by hand, the pinhole position should be invariant.

### Angle Calculations

The principal purpose of the diffractometer control program is to make the angle calculations that map the reciprocal crystal coordinates onto the angle setting of the diffractometer. This calculation depends on the mode of operation of the instrument, which is the "five-circle" mode for the installation discussed here. The program also contains the software drivers for the axis motors corresponding to the diffractometer angles so that these can be moved by the program. It also generates the most useful kinds of scans automatically and keeps track of the orientation matrix. Its other major task is to produce the output

files that contain the results of the experiments organized in a readable format. Here we make reference to the diffractometer control program "super," but many of the basic principles are common to other programs, e.g., the commercial "spec" program used elsewhere and available from Certified Scientific Software. Details of the command structure of "super" are included in Appendix C.

Angle calculations in the five-circle mode are described in a paper by Vlieg et al. (1987). The diffractometer has 5 degrees of freedom in the most general five-circle mode (alm = 2) and 4 degrees of freedom when alm = 1 (see Appendix C, Basic Super Variables). Three of these are needed to select a point of measurement in (3D) reciprocal space; one more is used to constrain the incidence or exit angle (see bem in Appendix C, Basic Super Variables). The 5th degree of freedom is used to help control the resolution function, bearing in mind that the diffraction features to be measured are highly elongated into "rods" along the direction perpendicular to the surface: the program arranges for the crystallographic $c^*$ axis to lie in a horizontal plane. When the diffraction rods (assumed parallel to $c^*$) are horizontal, the vertical high-resolution direction lies exactly across them.

The constraints on the five-circle diffractometer will allow inclination of the incident beam (ALP), but not the exit beam (GAMMA), from the sample plane. Because of the limits on CHI, the sample must lie within some degrees from perpendicular to the main TTH/TH axis. A consequence of these constraints is that most reciprocal space is accessible only in the bem = 2 mode (see Appendix C, Basic Super Variables) in which grazing exit conditions are maintained. This has the advantage that the PSD can then record the exit beam transmission profile for all reciprocal lattice settings. The exit angle (set by the variable B2) arises from setting TH near TTH + $90°$ and tilting CHI to a value near the negative of B2, as can be seen with the aid of Figure 9. The exact values depend on how close the "laser alignment" (see Practical Aspects of the Method) angles are to the ideal flat setting, fchi = 0. For this reason the $\theta$ axis should be expected to work roughly in the range $90° < \text{TH} < 150°$ and the limits should be set to accept this range, e.g., $-30° < \text{TH} < 150°$.

### ACKNOWLEDGMENTS

It is a pleasure to thank the various people who have helped develop this technique and the X16A facility over the past 15 years. Paul Fuoss helped develop the beamline and original surface diffractometer hardware. Elias Vlieg invented the five-circle mode and wrote the data analysis programs. Robert Fleming wrote the diffractometer control program. Alastair MacDowell developed the manipulator design. During the design and construction stages, the following people worked very hard: Warren Waskiewicz, Laura Norton, Steve Davey, and Jason Stark. In later years the input and contributions of the following people have been invaluable: Ken Evans-Lutterodt, Rolf Schuster, Peter Eng, Peter Bennett, and Don Walko. Finally, the vision and foresight of the Bell-Labs management to finance the whole project, at a time when it was unclear that it would work, deserves great commendation.

## LITERATURE CITED

Abernathy, D. 1993. An X-ray Scattering Study of the Si(113) Surface: Structure and Phase Behavior. Ph.D. dissertation, Massachusetts Institute of Technology.

Afanas'ev, A. M. and Melkonyan, M. K. 1983. *Acta Crystallogr. A* 39:207.

Als-Nielsen, J. 1987. Solid and liquid surfaces studied by synchrotron X-ray diffraction. *In* Structure and Dynamics of Surfaces II (W. Schommers and P. von Blanckenhagen, eds.). pp. 181-222. Springer-Verlag, Berlin.

Andrews, S. R. and Cowley, R. A. 1985. *J. Phys. C* 18:6247.

Bloch, J. M. 1985. *J. Appl. Crystallogr.* 18:33.

Born, M. and Wolf, E. 1975. Principles of Optics. Pergamon Press, Elmsford, N.Y.

Busing, W. R. and Levy, H. R. 1967. *Acta Crystallogr.* 22:457.

Coppens, P., Cox, D., Vlieg, E., and Robinson, I. K. 1992. Synchrotron Radiation Crystallography. Academic Press, London.

Dosch, H. 1987. *Phys. Rev. B* 35:2137.

Dosch, H. 1992. Critical Phenomena at Surfaces and Interfaces: Evanescent X-ray and Neutron Scattering. Springer-Verlag, Heidelberg.

Dosch, H., Mailander, L., Reichert, H., Peisl, J., and Johnson, R.L. 1991. *Phys. Rev. B* 43:13172.

Eisenberger, P. and Marra, W. C. 1981. *Phys. Rev. Lett.* 46:1081.

Evans-Lutterodt, K. and Tang, M. T. 1995. Angle calculations for a "2 + 2" surface X-ray diffractometer. *J. Appl. Crystallogr.* 28:318–326.

Fuoss, P. H. and Robinson, I. K. 1984. Apparatus for X-ray diffraction in ultra-high vacuum. *Nucl. Instr. Methods A* 222:171.

Guinier, A. 1963. X-ray Diffraction. W.H. Freeman, New York.

James, R. W. 1950. The Optical Principles of the Diffraction of X-rays. G. Bell and Sons, London.

Keating, D., Nunes, A., Batterman, B., and Hastings, J. 1971. *Phys. Rev. B* 4:2472.

Keating, P. N. 1966. *Phys. Rev.* 145:637.

Lipson, H. and Cochran, W. 1966. The Determination of Crystal Structures. Cornell University Press, Ithaca, N.Y.

Meyerheim, H. L., Moritz, W., Schulz, H., Eng, P. J., and Robinson, I. K. 1995. Anharmonic thermal vibrations observed by surface X-ray diffraction for Cs/Cu(001). *Surf. Sci.* 333:1422–1429.

Moncton, D. E. and Brown G. S. 1983. *Nucl. Instr. Methods* 208:579.

Müller, B. and Henzler, M. 1995. SPA-RHEED—A novel method in reflection high-energy electron diffraction with extremely high angular and energy resolution. *Rev. Sci. Instrum.* 66:5232–5235.

Parratt, L. G. 1954. *Phys. Rev.* 95:359.

Robinson, I. K. 1986. Crystal truncation rods and surface roughness. *Phys. Rev. B* 33:3830.

Robinson, I. K. 1990a. *Faraday Discuss. R. Soc. Chem.* 89:208.

Robinson, I. K. 1990b. Surface crystallography. *In* Handbook on Synchrotron Radiation, Vol. III (D.E. Moncton and G.S. Brown, eds.). pp. 221–266. Elsevier/North-Holland, Amsterdam.

Robinson, I. K. and Tweet, D. J. 1992. Surface X-ray diffraction. *Rep. Prog. Phys.* 55:599–651.

Robinson, I. K., Eng, P. J., and Schuster, R. 1994. Origin of the surface sensitivity in surface X-ray diffraction. *Acta Phys. Pol. A* 86:513.

Salditt, T., Lott, D., Metzger, T. H., Peisl, J., Vignaud, G., Legrand, J. F., Grübel, G., Høghøi, P., and Schärpf, O. 1996. Characterization of interface roughness in W/Si multilayers by high resolution diffuse X-ray scattering. *Physica B* 221:13–17.

Sheithauer, U., Meyer, G., and Henzler, M. 1986. *Surf. Sci.* 178:441.

Sinha, S. K., Sirota, E. B., Garoff, S., and Stanley, H. B. 1988. X-ray and neutron scattering from rough surfaces. *Phys. Rev. B* 38:2297.

Takayanagi, K., Tanishiro, Y., Takahashi, S., and Takahashi, M. 1985. *Surf. Sci.* 164:367.

Vineyard, G. H.. 1982. *Phys. Rev. B* 26:4146.

Vlieg, E., van der Veen, J. F., Macdonald, J. E., and Miller, M. 1987. Angle calculations for a five-circle diffractometer used for surface X-ray diffraction. *J. Appl. Crystallogr.* 20:330–337.

Warren, B. E. 1969. X-ray Diffraction. Addison-Wesley, Reading, Mass.

Woolfson, M. M. 1997. An Introduction to X-ray Crystallography. Cambridge, Cambridge University Press.

Wormington, M., Bowen, D. K., and Tanner, B. K. 1992. Principles and performance of a PC-based program for simulation of grazing incidence X-ray reflectivity profiles. *Mater. Res. Soc. Symp. Proc.* 238:119–124.

## KEY REFERENCES

Als-Nielsen, 1987. See above.

*Describes the principls of x-ray reflectivity from an experimental and theoretical point of view.*

Dosch, 1992. See above.

*Covers the use of evanescent x-ray waves to probe the depth dependence of phase transitions near surfaces.*

Robinson, 1990b. See above.

*Describes surface structural analysis from a crystallographic perspective.*

Sinha et al., 1988. See above.

*Describes phenomenological models of surface roughness and derives their signatures in diffraction.*

## APPENDIX A: VACUUM PROCEDURE

In order that samples have a chance to remain clean for the duration of an experiment, surface science experiments are usually carried out at pressures of $10^{-10}$ Torr or below, called the ultrahigh-vacuum (UHV) regime. Under these conditions, according to the kinetic theory of gases, an atom on the surface will be impacted by a residual gas atom from the vacuum about every 30 to 60 min. The vacuum protocols described here apply to the X16A facility at NSLS and are rather specific in this section. At other facilities, the principles will be the same, but the details will be a little different. In the latter case, the information provided here will enable the reader to assess the level of complexity involved in carrying out experiments under UHV.

## Load-Lock Procedure

Because preparation of $10^{-10}$ Torr pressure is time consuming (see below), a load-lock procedure has been developed that allows simple manipulation of samples, thermocouples, and heaters to be carried out without venting the main UHV chamber. The principle of operation is shown in Figure 7. Load locking is very useful for batch processing a series of samples, e.g., or for introducing samples that cannot stand to be baked at 150°C.

To load a sample, first ensure that the loading chamber is under vacuum. Its turbopump should have been at full speed for at least 10 min. Then open the gate valve that separates the loading chamber from the main UHV chamber. The pressure may rise into the $10^{-8}$ Torr range.

Set the diffractometer to the precalibrated angle settings recorded in the logbook or use the preassigned macro definition, called "load." The manipulator is now exactly in line with the sealing surface and can be extended fully until the motor will not move further. As always, when moving the manipulator, it is imperative to watch the progress so as to avoid unexpected collisions. The centering can be verified by observing the final sealing approach through the load-lock window.

Once sealed, the loading chamber is vented with air by (1) depowering the turbopump, (2) isolating the roughing pump (it is not necessary to turn it off), (3) waiting for the pump to slow down significantly (~2 min), and (4) slowly unscrewing its black venting screw while observing the pressure in the main chamber. If the pressure rises, immediately reverse the procedure and reseat the sample.

If the procedure goes well, the UHV conditions of the main chamber will be retained by the electrical and thermal feedthroughs in the design of the manipulator head. The loading chamber can then be unbolted to allow changing of or modification to the sample.

The reverse procedure is to bolt the loading chamber back over the finished sample, open the valve to the roughing pump, and start up the turbopump. Anywhere from 5 to 30 min after the pump reaches full speed, the manipulator can be retracted into the main chamber and the gate valve closed. The pressure may rise momentarily into the $10^{-7}$-torr range before dropping to $10^{-10}$ again. The fresh sample and filaments will now need to be outgassed by heating.

## Venting Procedure

When more serious vacuum interventions are required, the entire chamber will need to be vented. The procedure here is similar to that described above, except there are more precautions.

First, turn off all filaments inside the main chamber and disconnect their power cables to avoid accidental activation. The ion gauge is protected and so may be left on. Close the gate valve to the ion pump if this is not to be vented also; otherwise turn off the power here as well.

Manually override the protection circuit to the turbopump gate valve, so that it does not close when the pump is turned off. This is done by powering the solenoid directly. Then the main chamber is vented with nitrogen gas supplied to its vent valve as follows:

1. Depower the turbopump.
2. Isolate the roughing pump (it is not necessary to turn it off).
3. Wait for the pump to slow down significantly (~2 min).
4. Switch on its venting circuit on the turbo control panel.

The pressure has equilibrated with the atmosphere when the bellows become limp. Overpressurization is avoided by a check valve in the nitrogen feed.

## Bakeout Procedure

While the main chamber is open, the following routine maintenance may need to be carried out:

1. Replace the sublimation pump filaments.
2. Check the body of the manipulator for shorts or frayed wiring.
3. Recharge evaporation sources.
4. Clean any evaporated deposits from the interior of the windows.

The vacuum restart procedure is the reverse of the venting procedure given above (see Venting Procedure). The next step is the bakeout, which requires considerable disassembly of the diffraction parts of the experiment in order that they do not obstruct the ovens or get unnecessarily hot themselves. The following settings are advised:

1. The variables TTH, TH, CHI, and ALP should be set to zero (270 on dials of TH and TTH). The motor power can now be turned off.
2. The PSD high voltage should be turned off before moving it for storage.
3. Flight path vacuum should be turned off at its Nupro valve. The entire diffractometer TTH arm can now be dismounted and stored.
4. The turbopump gate valve should be in the protect mode (i.e., connected to its controller).

It is necessary to heat all of the chamber walls to 150°C for ~12 hr. This is achieved with ~1000 W of power distributed as follows:

1. Snap together the toggles that hold together the oven walls, which encase the main chamber. Any gaps can be filled with aluminum foil.
2. Lay the heater tape along the length of the manipulator and encase it with its clamshell oven. Any excess tape can be wrapped around the exposed feedthroughs.
3. Place the heater tape around the bellows/seal (between the diffractometer and the chamber).
4. Apply the following power settings: 110 V to each of two oven heaters and one ion pump heater, 60 V to heater tape around chi-bellows/seal; set the manipulator to 110 V if the tape is small (100 W) or to 70 V if the tape is large (300 W).

To shut down the bakeout, it is adequate to follow these quick steps:

1. Turn off all power.
2. About 30 to 60 min later, partially open roof of oven and unwrap the manipulator.
3. Another 30 to 60 min later, finish dismounting the oven.

The expected performance of the pressure will be in the $10^{-7}$-torr range at the start, rising to $10^{-6}$ when hot and eventually dropping to $10^{-7}$ at the end of the bake. After the heat is turned off, the pressure will fall steadily and reach $10^{-10}$ Torr in ~12 h.

## APPENDIX B: DATA ANALYSIS PROGRAMS

A suite of data analysis programs have been written to perform the various steps of data analysis. The plot program ANA is the most generally useful, while the full set of four programs are needed for a complete surface crystallographic analysis. They were all written by Elias Vlieg and are much more widely used than just at X16A, so their description here will be generally applicable. The programs are designed to work with the data file structure of "super" (Appendix C) but also accept input in other formats, including "spec."

The programs are interactive with the user. The commands to all programs are either keyed in at a prompt or redirected from a macro (*.mac) file. Each entered line is a continuous string of commands and arguments delimited by spaces. The commands are mostly English or Dutch verbs and may be highly abbreviated, usually to a single letter. This is made possible by the use of a hierarchical "menu" structure that defines the logical sequence of allowed commands. Each command is followed immediately by its arguments (separated by spaces) or by a <CR> to be prompted for the arguments. For example, the setup commands specific to plotting in ANA are unreachable at the main level but are reached by typing *p(lot)*. At any level the command *h(elp)* will provide a list of available commands and *l(ist)* will provide current values and status.

Upon exit from any of the programs (except AVE), an executable macro file is generated that records the current options that have been defined during the last session. This file is automatically executed when the program is restarted so that the previous configuration is restored. In the following descriptions the commands are defined by their full names with the minimum abbreviation indicated using the syntax *p(lot)*.

### Plot program ANA

This program maintains five data buffers into which spectra can be read. It uses the same buffers for generated fitting curves, so these can be superimposed. A variety of read formats are supported.

The *s(et)* menu is for setting parameters and modes. The *r(ead)* submenu is used to define filenames and col-

umn numbers for input. The *f(it)* submenu sets the mode of fitting, e.g., the weighting scheme.

The *r(ead)* menu is for reading in a spectrum and is followed by a format type, a scan number or filename, and a destination buffer number. Example formats are: *s(uper)*, *c(olumn)*, *x(y)*, and *(s)p(ec)*.

The *o(perate)* menu is for manipulating spectra. Mathematical operations are available through the *ad(d)*, *su(m)*, *mu(ltiply)*, and *sh(ift)* commands. Another functions is *l(ump)* for combining adjacent groups of data points within a spectrum. The function *me(rge)* combines two spectra, accounting for monitor counting time and removing duplicate data points; *w(rap)* removes 360° cuts in data; and *d(elete)* removes selected data points while *c(ut)* removes data points above or below a certain specified value.

The *p(lot)* menu makes graphs on the screen that can accumulate multiple entries for later printing or conversion to other image formats. A wide selection of *c(urve)*, *s(ymbol)*, and *li(nestyle)* options are supported. Curves can be *p(lot)*ed fresh or else *o(verlay)*ed and *ax(es)* and *te(xt)* can be supplied and modified. Logarithmic and linear scales can be chosen with, e.g., *ylog* and *xlin*. The program uses a commercial package called "graphi-c" for its plotting functions, and all file conversions of graphical output (including printing) must be handled through its internal mechanisms: hitting the space bar (a few times) allows access to the print menu of "graphi-c", and the simplest printout is activated by *l* for large or *m* for medium size.

The *f(it)* menu performs least-squares fitting to a library of functions. The command is followed by a code for the fit function required, which is assembled from *g(aussian)*, *l(orentzian)*, *v(oigt)*, *p(owerlaw)*, or *k(ummer)* components using commas as delimiters. Finally, the input and output spectrum numbers must be given. For example, the command "f g,l,l 1 2" will fit the sum of a gaussian and two lorentzians to the data in spectrum 1 and put the fit curve in spectrum 2. A two-parameter linear background is always assumed in addition to the chosen functions. Once inside the fit menu, the options are to *g(uess)* starting values of the parameters, specify a *v(alue)* by hand, and then *f(ix)* or *l(oosen)* these values. Finally, *r(un)* executes the least-squares fit and lists the parameter values.

A useful shortcut is the *a(utoplot)* menu with only two commands, *d(ata)* and *f(it)*. This generates a generic format plot with useful text information superimposed, such as the fitted values of the parameters.

### Integration Program PEAK

This is a batch program for converting rscans and jscans (see Appendix C, Important Scan Types) into structure factors. It is assumed that the scans cut directly across the desired diffraction features whose intensity is then related to the square of the structure factor; misalignment will, of course, lead to underestimation of the structure factor. It is also assumed that the scan reaches far enough that the background can be estimated by linear interpolation between its two ends. The "stretching" feature of "super" (see Appendix C, Important Scan Types) is designed to facilitate this.

The program is also responsible for four standard corrections to the data:

1. area correction, due to the changes in the active area of the sample as the diffractometer angles change, introduced by Robinson (1990b);
2. Lorentz factor correction, to convert an angular scan into an integral in reciprocal space, as in Warren (1969);
3. polarization correction, as in James (1950); and
4. monitor and stepsize normalization to account for different counting times and scanning rates that may have been used for different parts of the data.

Under the c(olumn) menu, it is necessary that all the diffractometer angles be identified by their column sequence number in the scan file. These must not change from one scan to the next.

The remaining parameters are entered at the command line. The le(ft) and ri(ght) numbers of points are to be considered as background; all points in between are integrated numerically to a total intensity. The s(cantype) can be r(ocking) or i(ndex) and will affect the Lorentz factor; all other scans found in the batch input will be ignored. For the area correction, the incident beam is taken to have dimensions wi vertically by hi horizontally, the sample has width ws, and the vertical exit beam slit is we.

The r(un) command followed by starting and ending scan numbers will perform the integration. Two output files are generated. The *.inf file has a line of information about each scan found, such as the date, background, and peak levels. The *.pk file contains the scan number hkl, the structure factor, and the structure factor error as the first three columns.

### Averaging Procedure AVE

This procedure uses the *.pk output of PEAK to locate and compare all the symmetry equivalents. It determines an overall estimate of the average systematic error, $\varepsilon$, assumed to be a constant fraction of each individual structure factor value. The value of $\varepsilon$ is taken to be a quality factor ($R$ value on structure factors) in assessing the data, as discussed in Robinson (1990b). The program then passes through the data a second time to generate a weighted average value (file *.wgt) of each inequivalent structure factor with an enlarged error bar that combines its statistical errors input with its overall systematic error estimated in the first pass.

The *.ave output listing is very useful for finding bad data. It lists together all reflections that are equivalent according to the specified symmetry. Any that fall out of line are flagged with warnings indicating they should be checked. They are identified by scan number in the listing. This is also a useful way of testing the symmetry of the data, if this is unknown from context, since the program can be run with different preassigned choices of assumed symmetry.

The s(et) menu has submenus for defining c(olumns), s(ymmetry) from the list of all 17 possible plane groups from p1 to p6mm, and a(veraging) parameters and modes. Under a(veraging) is the c(utoff) parameter, which is the number of times that a structure factor must be larger than its statistical error for it to be used in the estimation of $\varepsilon$. The default value (number of standard deviations) is $2\sigma$, and this should be increased to $10\sigma$ if the data are not strongly affected by counting statistics.

### Fitting Procedure ROD

This is a large refinement program for fitting a structural model to crystallographic data. It is specific to surface diffraction in that it calculates CTR rather than bulk diffraction structure factors. It differentiates the "bulk" and "surface" parts of the structure and includes a simple description of surface roughness that is usually necessary to get a good fit. The bulk cell is periodic and invariant; it only contributes to the CTRs. The atoms in the surface cell can be selectively refined in position, occupancy, and Debye-Waller factor. It is important that the z coordinates in the surface be a smooth continuation from those of the bulk.

The input files needed for ROD are as follows. All are expected to have a header line that is ignored on input. The second line of the first two files must contain the six unit cell parameters in real space, a, b, and c (in angstroms) and $\alpha$, $\beta$, and $\gamma$ (in degrees). The program checks that both are the same:

1. bulk unit cell coordinates, *.bul;
2. surface unit cell coordinates, *.sur (fixed), or *.fit (parameterized);
3. structure factor data, *.dat (renamed from *.wgt above); and
4. model parameters, *.par (optional).

In order that small data sets can be compatible with large models, ROD allows a very flexible parameterization scheme. The same displacement, occupancy, and DW parameters can be assigned to more than one atom. This is very convenient for structures that contain repeated motifs or different symmetries of components within the whole unit cell. In fact, the program does not handle symmetry at all, since this can always be built into the model; all structures are assumed to have triclinic, $1 \times 1$ unit cells.

### APPENDIX C: SUPER COMMAND FORMAT

Commands have a variable number of arguments, depending upon the specific application. Arguments follow on the same line as the command and are separated by spaces or commas. Most commands will prompt the user for the required arguments if no arguments are given. Multiple commands may be placed on one line provided they are separated by a semi-colon.

Values of important control parameters are stored as "variables." To obtain the current value of a variable, type its name followed by =. To change the value stored in a variable, type the new value after the =, e.g., as = 1.65. Rudimentary algebra is also allowed, e.g., as = bs * 3. Space characters are required as delimiters around the mathematical operation.

A manual of allowed commands will be available at the beamline. A list of all valid "super" commands can be obtained by typing "help" at the "super" prompt or using the F8 key. Pressing F5 lists the orientation matrix and the values of string variables. The values of all other variables are listed using F6.

## Basic Super Commands

**mv**    Move motors (four arguments: TTH, TH, PHI, CHI).

**ct**    Count (one optional argument = preset).

**br**    Go to position defined by $H$, $K$, and $L$ (three arguments).

**wh**    Print current $H$, $K$, and $L$ and values of the motor angles (no arguments).

**ca**    Calculate motor angles from $H$, $K$, and $L$ (three arguments).

**ci**    Calculate $H$, $K$, and $L$ from motor angles (five arguments).

**end**    Exit the program.

## Basic Super Variables

**as**    Crystal reciprocal lattice parameter $a^*$

**bs**    Crystal reciprocal lattice parameter $b^*$

**cs**    Crystal reciprocal lattice parameter $c^*$

**al**    Crystal reciprocal lattice angle $\alpha^*$

**be**    Crystal reciprocal lattice angle $\beta^*$

**ga**    Crystal reciprocal lattice angle $\gamma^*$

**wv**    Incident wave vector ($2\pi/\lambda$)

**nobs**    Number of alignment reflections used to calculate OM

**h1, k1, l1, h2, k2, ...**    HKLs of orientation reflections

**t1, u1, p1, c1, a1, u2, p2, ...**    2-Theta, theta, phi, chi, and alpha of orientation reflections

**B**    Target value of incidence angle, $\alpha_i$

**B2**    Target value of exit angle, $\alpha_f$

**alm**    Mode of calculating ALP for five-circle geometry (alm = 1 means ALP is frozen at its current value; alm = 2 means ALP is calculated)

**bem**    Mode of constraining incidence angle for five-circle geometry (bem = 1 means *incidence* angle $\alpha_i$ is fixed at target value B; (bem = 2 means *exit* angle $\alpha_f$ is fixed at target value B2; bem = 3 means incidence and exit angles are made equal)

**fchi**    CHI angle of optical surface orientation (laser alignment)

**fphi**    PHI angle of optical surface orientation (laser alignment)

## Orientation Matrix

There are two ways of achieving the sample orientation, depending on whether the reciprocal lattice parameters are known or unknown. In the more commonly used "lattice parameters known" mode, the six reciprocal lattice parameters are entered manually using syntax name = value. The names of the six reciprocal lattice constants are as, bs, cs, al, be, and ga. Two orientation reflections are required with h, k, l, TTH, TH, PHI, CHI, and ALP specified for each reflection. The meaning of the angles is slightly different in different modes of calculating the orientation matrix. We refer here only to the five-circle mode, which is set by the command *frz 5*.

Individual variables in the orientation matrix can be changed using syntax name = value. The variable names in this case consist of a single letter and a number (e.g., t1, t2, t3, ...). The number refers to the number of the orientation reflection. The letters are assigned as follows for the five-circle mode: t = 2-theta, u = theta, p = phi, c = chi, a = alpha, h,k,l = Miller indices, g = gamma (fixed out-of-plane detector position), and z = zeta (fixed sample offset angle).

The orientation reflections may also be entered using the *or* (orient) command. To enter the parameters for the current diffractometer position, type *or* followed by the number of the reflection (in this case 1 or 2) and the hkl. One may also manually enter all variables on one line by typing

or # h k l TTH TH PHI CHI ALP

The orientation matrix is a general $3 \times 3$ matrix with 9 degrees of freedom described by Busing and Levy (1967). Six degrees of freedom are constrained by the reciprocal lattice parameters, 2 more come from the first orientation reflection (the two polar angles), and the final degree of freedom comes from the second orientation reflection (the second reflection specifies the azimuthal rotation about the first vector). The direction of the first orientation reflection, called the primary reflection, will always be reproduced exactly. The angles of the second reflection may not necessarily be consistent with the lattice parameters, and consequently the program may not reproduce the angles of the secondary reflection exactly. The *sor n m* command allows reflections n and m to be swapped.

The "lattice parameters unknown" mode is set by toggling the *om* command. Then the six reciprocal lattice parameters are determined directly from the five angles specified for each of the orientation reflections (TTH, TH, PHI, CHI, and ALP). To define the nine matrix elements of the orientation matrix, at least three non-coplanar reflections must be used, so nobs = 3. If nobs3, the best matrix is constructed from the full list of nobs reflections using a least-squares fit and will usually improve as more reflections are added. Reciprocal lattice parameters that are consistent with the three observed reflections will then be calculated. In general, the reciprocal lattice parameters will correspond to a triclinic cell with interfacial angles nearly equal to those expected from the symmetry. As guide to selecting suitable reflections, an error is calculated for each, which is the distance (in reciprocal angstroms) of the reflection from the reciprocal lattice point derived from the orientation matrix. The orientation parameters may be entered manually or by using the *or* command.

### Important Scan Types

**ascan**   Angle scan—Scan four diffractometer motors TTH, TH, PHI, CHI.

**mscan**   Motor scan—Scan up to three arbitrary motors.

**iscan**   Index scan—Input three starting HKLs and three delta HKLs for the increments between points.

**jscan**   Centered index scan—Like iscan except scan is *centered* on the given starting hkl value.

**kscan**   Centered two-dimensional index scan—Like jscan it is *centered* on the given starting hkl value. Two sets of delta HKL step sizes are given.

**rscan**   Rocking scan—Input HKL value, motor number (or hkl) and one delta value for the increment. The scan moves either one motor or one hkl component and is centered at the given HKL value.

**lup**   Line-up scan—Input motor number and delta angle. Similar to a rocking scan, but the scan is centered at the current position of the motors. At the end of the scan, the motors move to the peak position, calculated as the center of gravity of the counts measured.

**fpk**   Find peak scan—Like lup, except fewer parameters are required and the data are not saved on disk. After the scan, the motors moved to the peak calculated from the first moment.

**vscan**   Variable scan—Input HKL, variable name, and starting value. The scan increments the value of the variable and moves the motors to the given HKL.

**vlup**   Centered vscan—Input HKL, variable name, delta, and npts. The scan increments the value of the variable and moves the motors to the given HKL. The scan is centered on the current value of the variable beforehand, which is reset to the center of gravity afterward.

**pkup**   Peakup reflection—The scan does a series of lup scans centered at each orientation reflection. At the end of the series, the angles of the orientation reflection are updated to reflect the new peak position.

The scan types *rscan* and *jscan* embody automatic "stretching"' of the background, whereby the step size at the beginning and end of the scan is increased by a factor of 3. The *lup* and *vlup* scans are very useful during the alignment stages of the experiment because of their auto-centering feature; they can also be used to make automatic realignments during an unattended batch procedure. All scans generate a video display of the data as they accumulate, which can be viewed at any time using the function keys: F1 (F2) to see the spectrum of the multichannel analyzer as it accumulates and F3 (F4) to see a histogram of the recorded counts on a linear (log) scale.

A scan may be executed by typing the command name followed by a list of arguments. To obtain help on the command format, type the name of the scan without arguments and the program will prompt you. Any scan can be interrupted during execution by ^C and continued by typing *co*. A series of commands (including scans) can be executed by using either the *sc* command or the *ex* command.

To use the *sc* command, one first has to enter a list of commands into the scan table. The list can contain almost any command executable from the command line. A scan list can be created in three ways. The most direct method is to use the *ds* (define scans) command. Alternatively, one can use the *es* (edit scans) command to edit the scan table directly. The command *gs* filename (get scans) will read a new scan table file. Once the scan table has been entered, a list of commands or scans can be executed with the *sc* command, which accepts a series of numbers in arbitrary order corresponding to the position of the command in the scan table. Before execution begins, the program does a "dry run" to check for setting calculation errors and limit problems. If the execution is halted, the *co* command will resume execution.

A series of commands or scans can also be executed with the *ex* command. This command diverts program input to a script file and executes the commands in order. There is no limit to the number of commands in the file. In contrast to *sc*, the *ex* command does not do a dry run before execution begins. If a problem is encountered during execution, the program skips to the next command in the list.

I. K. ROBINSON
University of Illinois
Urbana, Illinois

# X-RAY DIFFRACTION TECHNIQUES FOR LIQUID SURFACES AND MONOMOLECULAR LAYERS

## INTRODUCTION

X-ray and neutron scattering techniques are probably the most effective tools employed to determine the structure of liquid interfaces on molecular-length scales. These are not different in principle from conventional x-ray diffraction techniques that are usually applied to three-dimensional crystals, liquids, solid surfaces etc. However, special diffractometers that enable scattering from fixed horizontal surfaces are required to carry out the experiments. Indeed, systematic studies of liquid surfaces had not begun until the introduction of the first liquid surface reflectometer (Als-Nielsen and Pershan, 1983).

A basic property of a liquid-gas interface is the length scale over which the molecular density changes from the bulk value to that of the homogeneous gaseous medium. Molecular size and capillary waves, which depend on surface tension and gravity, are among the most important factors that shape the density profile across the interface and the planar correlations (Evans, 1979; Braslau et al., 1988; Sinha et al., 1988). In some instances the topmost layers of liquids are packed differently than in the bulk, giving rise to layering phenomena at the interface. Monolayers of compounds different than the liquid can be spread at the gas-liquid interface, and are termed Langmuir monolayers (Gaines, 1966; Swalen et al., 1987). The spread compound might "wet" the liquid surface to form a film of homogeneous thickness or cluster to form an inhomogeneous rough surface. The x-ray reflectivity (XR) technique allows one to determine the electron density across such interfaces, from which the molecular density and the total thickness can be extracted. The grazing angle diffraction (GID) technique is commonly used to determine lateral arrangements and correlations of the topmost layers at interfaces. GID is especially efficient in cases where surface crystallization of the liquid or spread monolayers occurs. Both techniques (XR and GID) provide structural information that is averaged over macroscopic areas, in contrast to scanning probe microscopies (SPMs), where local arrangements are probed (see, e.g., SCANNING TUNNELING MICROSCOPY). For an inhomogeneous interface, the reflectivity is an incoherent sum of reflectivities, accompanied by strong diffuse scattering, which, in general, is difficult to interpret definitively and often requires complementary techniques to support the x-ray analysis. Therefore, preparation of well-defined homogeneous interfaces is a key to a more definitive and straightforward interpretation.

### Competitive and Related Techniques

Although modern scanning probe microscopies (SPMs) such as scanning tunneling microscopy (STM; SCANNING TUNNELING MICROSCOPY; Binning and Rohrer, 1983) and atomic force microscopy (AFM; Binning, 1992) rival x-ray scattering techniques in probing atomic arrangements of solid surfaces, they have not yet become suitable techniques for free liquid surfaces (but see X-RAY DIFFRACTION TECHNIQUES FOR LIQUID SURFACES AND MONOMOLECULAR LAYERS). The large fluctuations due to the two-dimensional nature of the liquid interface, high molecular mobility, and the lack of electrical conductivity (which is necessary for STM) are among the main obstacles that make it difficult to apply these techniques to gas-liquid interfaces. In dealing with volatile liquids, inadvertent deposition or wetting of the probe by the liquid can occur, which may obscure the measurements. In addition, the relatively strong interaction of the probe with the surface might alter its pristine properties. For similar and other reasons, electron microscopy (see Section 11a) and electron diffraction techniques (LOW-ENERGY ELECTRON DIFFRACTION), which are among the best choices for probing solid surfaces, are not suitable to most liquid surfaces, and in particular aqueous interfaces. On the other hand, visible light microscopy techniques such as Brewster angle microscopy (BAM; Azzam and Bashara, 1977; Henon and Meunier, 1991; Hönig and Möbius, 1991) or fluorescence microscopy (Lösche and Möhwald, 1984) have been used very successfully in revealing the morphology of the interface on the micrometer-length scale. This information, in general, is complementary to that extracted from x-ray scattering. These techniques are very useful for characterizing inhomogeneous surfaces with two or more distinct domains, for which XR and GID results are usually difficult to interpret. However, it is impossible to determine the position of the domains with respect to the liquid interface, their thicknesses, or their chemical nature. Ellipsometry (ELLIPSOMETRY) is another technique that exploits visible light to allow determination of film thickness on a molecular length scale; it assumes that one knows the refractive

index of the substrate and of the film (Azzam and Bashara, 1977; Ducharme et al., 1990). Either of these values might be different from the corresponding bulk value, and therefore difficult to determine.

In the following sections, theoretical background to the x-ray techniques is presented together with experimental procedures and data analysis concerning liquid surfaces. This unit is intended to provide a basic formulation which can be developed for further specific applications. Several examples of these techniques applied to a variety of problems are presented briefly to demonstrate the strengths and limitations of the techniques. It should be borne in mind that the derivations and procedures described below are mostly general and can be applied to solid surfaces, and, vice versa, many results applicable to solid surfaces can be used for liquid surfaces. X-ray reflectivity from surfaces and GID have been treated in recent reviews (Als-Nielsen and Kjaer, 1989; Russell, 1990; Zhou and Chen, 1995).

### PRINCIPLES OF THE METHOD

We assume that a plane harmonic wave of frequency $\omega$ and wave-vector $\mathbf{k}_0$ (with electric field, $\mathbf{E} = \mathbf{E}_0 e^{i\omega t - i k_0 \cdot r}$) is scattered from a distribution of free electrons, with a number density $N_e(\mathbf{r})$. Due to the interaction with the electric field of the x-ray wave, each free electron experiences a displacement proportional to the electric field, $\mathbf{X} = (-e/m\,\omega^2)\mathbf{E}$. This displacement gives rise to a polarization $\mathbf{P}(\mathbf{r})$ distribution vector

$$\mathbf{P}(\mathbf{r}) = N_e(\mathbf{r})e\mathbf{X} \qquad (1)$$

in the medium. For the sake of convenience we define the scattering length density (SLD), $\rho(\mathbf{r})$, in terms of the classical radius of the electron $r_0 = e^2/4\pi\varepsilon_0 mc^2 = 2.82 \times 10^{-13}$ cm as follows

$$\rho(\mathbf{r}) = N_e(\mathbf{r})r_0 \qquad (2)$$

The polarization then can be written as

$$\mathbf{P}(\mathbf{r}) = \frac{-N_e(\mathbf{r})e^2}{\omega^2 m_e}\mathbf{E} = -\frac{4\pi\varepsilon_0}{k_0^2}\rho(\mathbf{r})\mathbf{E} \qquad (3)$$

The scattering length density (or the electron density) is what we wish to extract from reflectivity and GID experiments and relate it to atomic or molecular positions at liquid interfaces. The displacement vector $\mathbf{D}$ can now be constructed as follows

$$\mathbf{D} = \varepsilon_0\mathbf{E} + \mathbf{P}(\mathbf{r}) = \varepsilon(\mathbf{r})\mathbf{E} \qquad (4)$$

where $\varepsilon(\mathbf{r})$ is the permittivity of the medium, usually associated with the refractive index $n(\mathbf{r}) = \sqrt{\varepsilon(\mathbf{r})}$. To account for absorption by the medium we introduce a phenomenological factor $\beta$ that we calculate from the linear absorption coefficient $\mu$ (given in tables in Wilson, 1992) as follows: $\beta = \mu/(2k_0)$. Then the most general permittivity for x-rays becomes

$$\varepsilon(\mathbf{r}) = \varepsilon_0\left[1 - \frac{4\pi}{k_0^2}\rho(\mathbf{r})\right] + 2i\beta \qquad (5)$$

**Table 1. Electron Number Density, SDL, Critical Angles and Momentum Transfers, and Absorption Term for Water and Liquid Mercury**

| | $N_e(e/\text{Å}^3)$ | $\rho_s(\text{Å}^{-2}\times 10^{-5})$ | $Q_c(\text{Å}^{-1})$ | $\alpha_c(deg.)for$ $\lambda=_{1.5404}\text{Å}$ | $\beta(\times 10^{-8})$ |
|---|---|---|---|---|---|
| $H_2O$ | 0.334 | 0.942 | 0.02176 | 0.153 | 1.2 |
| Hg | 3.265 | 9.208 | 0.06803 | 0.478 | 360.9 |

Typical values of the SLD ($\rho$) and the absorption term ($\beta$) for water and liquid mercury are listed in Table 1. In the absence of true charges in the scattering medium (i.e., a neutral medium) and under the assumption that the medium is nonmagnetic (magnetic permeability $\mu = 1$) the wave equations that need to be solved to predict the scattering from a known SLD can be derived from the following Maxwell equations (Panofsky and Phillips, 1962)

$$\nabla \cdot \mathbf{D} = 0 \qquad \nabla \cdot \mathbf{H} = 0$$
$$\nabla \times \mathbf{E} = -\partial \mathbf{H}/\partial t \quad \nabla \times \mathbf{H} = \partial \mathbf{D}/\partial t \qquad (6)$$

Under the assumption of harmonic plane waves, $\mathbf{E} = \mathbf{E}_0 e^{i\omega t - i\mathbf{k}\cdot\mathbf{r}}$, the following general equations are obtained from Equation 6

$$\nabla^2 \mathbf{E} + [k_0^2 - 4\pi\rho(\mathbf{r})]\mathbf{E} = -\nabla(\nabla \ln \varepsilon(\mathbf{r}) \cdot \mathbf{E}) \qquad (7)$$
$$\nabla^2 \mathbf{H} + [k_0^2 - 4\pi\rho(\mathbf{r})]\mathbf{H} = -\nabla \ln \varepsilon(\mathbf{r}) \times (\nabla \times \mathbf{H}) \qquad (8)$$

In some particular cases the right hand side of Equations 7 and 8 is zero. We notice then, that the term $4\pi\rho(\mathbf{r})$ in the equation plays a role similar to that of a potential $V(\mathbf{r})$ in wave mechanics. In fact, for most practical cases the right hand side of Equation 7 and Equation 8 can be approximated to zero; thus the equation for each component of the fields resembles a stationary wave equation. In those cases, general mathematical tools, such as the Born approximation (BA) and the distorted wave Born approximation (DWBA; see DYNAMICAL DIFFRACTION and SURFACE X-RAY DIFFRACTION) can be used (Schiff, 1968).

### Reflectivity

In reflectivity experiments a monochromatic x-ray beam of wavelength $\lambda$ [wavevector $k_0 = 2\pi/\lambda$ and $\mathbf{k_i} = (0, k_y, k_z)$] is incident at an angle $\alpha_i$ on a liquid surface and is detected at an outgoing angle $\alpha_r$ such that $\alpha_i = \alpha_r$, as shown in Figure 1, with a final wave vector $\mathbf{k_f}$. The momentum transfer is defined in terms of the incident and reflected beam as follows

$$\mathbf{Q} = \mathbf{k_i} - \mathbf{k_f} \qquad (9)$$

where in the reflectivity case $\mathbf{Q}$ is strictly along the surface normal, with $Q_z = 2k_0\sin\alpha = 2k_z$.

#### Single, Ideally Sharp Interface: Fresnel Reflectivity. Sol-
Solving the scattering problem exactly for the ideally sharp interface, although simple, is very useful for the derivation of more complicated electron density profiles across interfaces. The wavefunctions employed are also

essential for inclusion of dynamical effects when dealing with non-specular scattering, i.e., GID and diffuse scattering. The following relates to the case of an s-polarized x-ray beam (see the Appendix at the end of this unit for a similar derivation in the case of the p-polarized x-ray beam).

For a stratified medium with an electron density that varies along one direction, $z$, $\rho(\mathbf{r}) = \rho(\mathbf{z})$, assuming no absorption, i.e., $\beta = 0$, an s-type polarized x-ray beam with the electrical field parallel to the surface (along the $x$ axis; see Fig. 1) obeys the stationary wave equation as derived from Equation 7 and is simplified as follows

$$\nabla^2 E_x + [k_0^2 - V(z)]E_x = 0 \qquad (10)$$

with an effective potential $V(z) = 4\pi\rho(z)$. The general solution to Equation 2 is then given by

$$E_x = E(z)e^{ik_y y} \qquad (11)$$

where the momentum transfer along $y$ is conserved when the wave travels through the medium, leading to the well-known Snell's rule for refraction. Inserting Equation 11 in

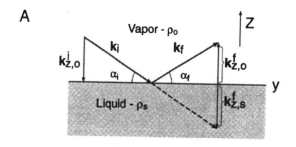

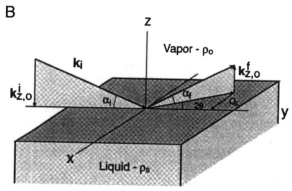

**Figure 1.** The geometry of incident and scattered beam in (**A**) specular reflectivity and (**B**) non-specular scattering experiments.

Equation 10 leads to a one-dimensional wave equation through a potential $V(z)$

$$\frac{d^2E}{dz^2} + [k_z^2 - V(z)]E = 0 \qquad (12)$$

The simplest case of Equation 12 is that of an ideally flat interface, separating the vapor phase and the bulk scattering length density $\rho_s$, at $z = 0$. The solution of Equation 12 is then given by

$$E(z) = \begin{cases} e^{ik_{z,0}z} + r(k_{z,s})e^{-ik_{z,0}z} & z \geq 0 \text{ in gas} \\ t(k_{z,s})e^{ik_{z,s}z} & z \leq 0 \text{ in liquid} \end{cases} \qquad (13)$$

where

$$k_{z,s} = \sqrt{k_{z,0}^2 - 4\pi\rho_s} = \sqrt{k_{z,0}^2 - k_c^2} \qquad (14)$$

with $k_c \equiv 2\sqrt{\pi\rho_s}$. By applying continuity conditions to the wavefunctions and to their derivatives at $z = 0$, the Fresnel equations for reflectance, $r(k_{z,s})$, and transmission, $t(k_{z,s})$, are obtained

$$r(k_{z,s}) = \frac{k_{z,0} - k_{z,s}}{k_{z,0} + k_{z,s}}, \qquad t(k_{z,s}) = \frac{2k_{z,0}}{k_{z,0} + k_{z,s}} \qquad (15)$$

The measured reflectivity from an ideally flat interface, $R_F$, is usually displayed as a function of the momentum transfer $Q_z = k_{z,0} + k_{z,s} \approx 2k_{z,0}$, and is given by

$$R_F(Q_z) = |r(k_{z,s})|^2 \qquad (16)$$

Below a critical momentum transfer, $Q_c \equiv 2k_c \equiv 4\sqrt{\pi\rho_s}$, $k_{z,s}$ is an imaginary number and $R_F(Q_z) = 1$; thus, total external reflection occurs. Notice that whereas the critical momentum transfer does not depend on the x-ray wavelength, the critical angle for total reflection does, and it is given by $\alpha_c \approx \lambda\sqrt{\rho_s/\pi}$. Typical values of critical angles for   x-rays of wavelength $\lambda_{CuK\alpha} = 1.5404$ Å are listed in Table 1. For $Q_z \gg Q_c$, $R_F(Q_z)$ can be approximated to a form that is known as the Born approximation

$$R_F(Q_z) \sim \left(\frac{Q_c}{2Q_z}\right)^4 \qquad (17)$$

This form of the reflectivity at large $Q_z$ values is also valid for internal scattering, i.e., reflectivity from liquid into the vapor phase. However, total reflection does not occur for the internal reflectivity case. Calculated external and internal reflectivity curves from an ideally flat surface, $R_F$, displayed versus momentum transfer (in units of the critical momentum transfer) are shown in Figure 2A. Both reflectivities converge at large momentum transfer where they can be both approximated by Equation 17. The dashed line in the same figure shows the approximation $(Q_c/2Q_z)^4$, which fails in describing the reflectivity close to the critical momentum transfer.

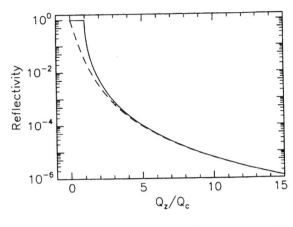

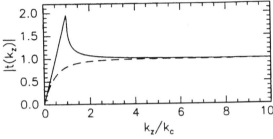

**Figure 2.** Calculated reflectivity curves for external (solid line) and internal (dashed line) scattering from an ideally flat interface versus momentum transfer given in units of the critical momentum transfer, $Q_c = 4(\pi\rho_s)^{1/2}$. The dotted line is kinematical approximation $(Q_c/2Q_z)^4$. The lower panel shows the amplitude of the wave in the medium for external (solid line) and external reflection (dashed line).

The photon transmission at a given $k_{z,s}$ is given by

$$T(k_{z,s}) = |t(k_{z,s})|^2 \frac{\text{Re}(k_{z,s})}{\text{Re}(k_{z,0})} \qquad (18)$$

where the ratio on the right hand side accounts for the flux through the sample. In the case of external reflection, and for values of $k_{z,0}$ that are smaller than $k_c$, the real part of $k_{z,s}$ is zero, and there is no transmission, whereas above the critical angle, $k_{z,s}$ is real, and the transmission is given by

$$T(k_{z,s}) = \frac{4k_{z,0}k_{z,s}}{(k_{z,0} + k_{z,s})^2} \quad \text{for} \quad k_{z,0} > k_c \qquad (19)$$

and the conservation of photons is fulfilled in the scattering process

$$T(k_{z,s}) + R(k_{z,s}) = 1 \qquad (20)$$

In Figure 2B the transmission amplitude $|t(k_{z,s})|$ for external (solid line) and for internal (dashed line) reflections are shown. This amplitude modulates non-specular scattering processes at the interface as will be discussed later in this unit.

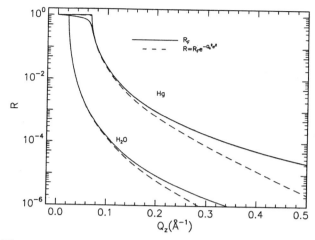

**Figure 3.** Calculated reflectivities from $H_2O$ and liquid mercury (Hg) showing the effects of absorption and surface roughness. The absorption modifies the reflectivity near the critical momentum transfer for mercury with insignificant effect on the reflectivity from $H_2O$. The dashed line shows the calculated reflectivity from the same interfaces with root mean square surface roughness, $\sigma = 3$ Å.

The effect of absorption on the reflectivity can be incorporated by introducing $\beta$ into the generalized potential in Equation 11, so that

$$k_{z,s} = \sqrt{k_{z,0}^2 - k_c^2 + 2i\beta} \qquad (21)$$

is used in the Fresnel equations (Equation 15). Calculated reflectivities from water and liquid mercury, demonstrating that the effect of absorption is practically insignificant for the former, yet has the strongest influence near the critical angle for the latter, are shown in Figure 3.

**Multiple Stepwise and Continuous Interfaces.** On average, the electron density across a liquid interface is a continuously varying function, and is a constant far away on both sides of the interface, as shown in Figure 4.

The reflectivity for a general function $\rho(z)$ can be then calculated by one of several methods, classified into two major categories: dynamical and kinematical solutions. The dynamical solutions (see DYNAMICAL DIFFRACTION) are in general more exact, and include all the features of the scattering, in particular the low-angle regime, close to the critical angle where multiple scattering processes occur. For a finite number of discrete interfaces, exact solutions can be obtained by use of standard recursive (Parratt, 1954) or matrix (Born and Wolf, 1959) methods. These methods can be extended to compute, with very high accuracy, the scattering from any continuous potential by slicing it into a set of finite layers but with a sufficient number of interfaces. On the other hand, the kinematical approach (see KINEMATIC DIFFRACTION OF X-RAYS) neglects multiple scattering effects and fails in describing the scattering at small angles.

1. *The matrix method.* In this approach the scattering length density with variation over a characteristic length $d_t$ is sliced into a histogram with $N$ interfaces.

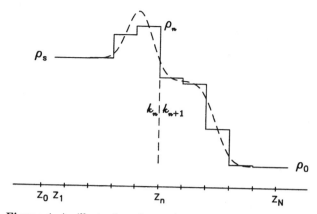

**Figure 4.** An illustration of a continuous scattering length density, sliced into a histogram.

The matrix method is practically equivalent to the Parratt formalism (Born and Wolf, 1959; Lekner, 1987). For each interface, the procedure described previously for the one interface is applied. Consider an arbitrary interface, $n$, separating two regions of a sliced SLD (as in Fig. 4), with $\rho_{n-1}$, and $\rho_n$ at position $z = z_n$, with the following wavefunctions

$$\begin{array}{ccc} \rho_{n-1} & & \rho_n \\ R_{n-1,n}e^{-ik_{n-1}z} \quad \leftarrow & \Big| & \leftarrow \quad R_{n,n+1}e^{-ik_nz} \\ T_{n-1,n}e^{-ik_{n-1}z} \quad \rightarrow & & \rightarrow \quad T_{n,n+1}e^{ik_nz} \\ & z = z_n & \end{array} \qquad (22)$$

where

$$k_n \equiv \sqrt{k_{z,0}^2 - 4\pi\rho_n} \qquad (23)$$

The effect of absorption can be taken into account as described earlier. For simplicity, the subscript $z$ is omitted from the component of the wavevector so that $k_{z,n} = k_n$. The solution at each interface in terms of a transfer matrix, $M_n$, is given by

$$\begin{pmatrix} T_{n-1,n} \\ R_{n-1,n} \end{pmatrix} = \begin{pmatrix} e^{-i(k_{n-1}-k_n)z_n} & r_n e^{-i(k_{n-1}+k_n)z_n} \\ r_n e^{i(k_{n-1}+k_n)z_n} & e^{i(k_{n-1}-k_n)z_n} \end{pmatrix} \begin{pmatrix} T_{n,n+1} \\ R_{n,n+1} \end{pmatrix} \qquad (24)$$

where

$$r_n = \frac{k_{n-1} - k_n}{k_{n-1} + k_n} \qquad (25)$$

is the Fresnel reflection function through the $z_n$ interface separating the $\rho_{n-1}$ and $\rho_n$ SLDs. The solution to the scattering problem is given by noting that beyond the last interface, (i.e., in the bulk), there is a transmitted wave for which only an arbitrary amplitude of the form $\begin{pmatrix} 1 \\ 0 \end{pmatrix}$ can be assumed (i.e., the reflectivity is normalized to the incident beam anyway). The effect of all interfaces is calculated as follows

$$\begin{pmatrix} T_{0,1} \\ R_{0,1} \end{pmatrix} = (M_1)(M_2) \dots (M_n) \dots (M_{N+1}) \begin{pmatrix} 1 \\ 0 \end{pmatrix} \qquad (26)$$

with

$$r_{N+1} = \frac{k_N - k_s}{k_N + k_s} \quad (27)$$

in the $M_{N+1}$ matrix given in terms of the substrate $k_s$. The reflectivity is then given by the ratio

$$R(Q_z) = \left| \frac{R_{0,1}}{T_{0,1}} \right|^2 \quad (28)$$

Applying this procedure to the one-box model of thickness $d$ with two interfaces yields

$$R(Q_z \equiv 2k_s) = \left| \frac{r_1 + r_2 e^{i2k_s d}}{1 + r_1 r_2 e^{i2k_s d}} \right|^2 \quad (29)$$

Figure 5 shows the calculated reflectivities from a flat liquid interface with two kinds of films (one box) of the same thickness $d$ but with different scattering length densities, $\rho_1$ and $\rho_2$. The reflectivities are almost indistinguishable when the normalized

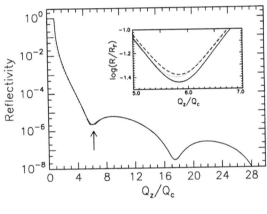

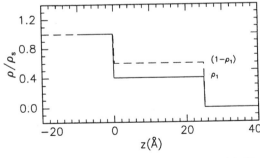

**Figure 5.** Calculated reflectivities for two films with identical thicknesses but with two distinct normalized electron densities, $\rho_1$ (solid line) and $1 - \rho_1$ (dashed line), and corresponding calculated reflectivities using the dynamical approach ($\sigma = 0$). The two reflectivities are almost identical except for a minute difference near the first minimum (see arrow in figure). The Born approximation (dotted line) for the two models yields identical reflectivities. The inset shows the normalized reflectivities near the first minimum. As $Q_z$ is increased, the three curves converge. This is the simplest demonstration of the phase problem, i.e., the nonuniqueness of models where two different potentials give the same reflectivities.

SLDs ($\rho_i/\rho_s$) of the films are complementary to one ($\rho_1/\rho_s + \rho_2/\rho_s = 1$), except for a very minute difference near the first minimum. In the kinematical method described below, the two potentials shown in Figure 5 yield identical reflectivities.

The matrix method can be used to calculate the exact solution from a finite number of interfaces, and it is most powerful when used with computers by slicing any continuous scattering length density into a histogram. The criteria for determining the optimum number of slices to use is based on the convergence of the calculated reflectivity at a point where slicing the SLD into more boxes does not change the calculated reflectivity significantly.

2. *The kinematical approach.* The kinematical approach for calculating the reflectivity is only applicable under certain conditions where multiple scattering is not important. It usually fails in calculating the reflectivity at very small angles (or small momentum transfers) near the critical angle. The kinematical approach, also known as the Born approximation, gives physical insight in the formulation of $R(Q_z)$ by relating the Fresnel normalized reflectivity, $R/R_F$, to the Fourier transform of spatial changes in $\rho(z)$ across the interface (Als-Nielsen and Kjaer, 1989) as discussed below.

As in the dynamical approach, $\rho(z)$ is sliced so that

$$k(z) = \sqrt{k_{z,0}^2 - 4\pi\rho(z)} \quad \frac{dk}{dz} = -\frac{2\pi}{k(z)}\frac{d\rho}{dz} \quad (30)$$

and the reflectance across an arbitrary point $z$ is given by

$$r(z) = \frac{k(z + \Delta z) - k(z)}{k(z + \Delta z) + k(z)} \approx -\frac{4\pi}{4k(z)^2}\frac{d\rho}{dz}dz$$
$$\approx (Q_c/2Q_z)^2 \frac{1}{\rho_s}\frac{d\rho}{dz}dz \quad (31)$$

In the last step of the derivation, $r(z)$ was multiplied and divided by $\rho_s$, the SLD of the subphase, and the identity $Q_c^2 \equiv 16\pi\rho_s$ was used. Assuming no multiple scattering, the reflectivity is calculated by integrating over all reflectances at each point, $z$, with a phase factor $e^{iQ_z z}$ as follows

$$R(Q_z) = R_F(Q_z)\left|\frac{1}{\rho_s}\int\left(\frac{d\rho}{dz}\right)e^{iQ_z z}dz\right|^2 = R_F(Q_z)|\Phi(Q_z)|^2 \quad (32)$$

where $\Phi(Q_z)$ can be regarded as the generalized structure factor of the interface, analogous to the structure factor of a unit cell in 3-D crystals. This formula also can be derived by using the Born approximation, as is shown in the following section.

As an example of the use of Equation 32 we assume that the SLD at a liquid interface can be approximated by a sum of error functions as follows

$$\rho(z) = \rho_0 + \sum_{j=1}^{N} \frac{(\rho_j - \rho_{j-1})}{2}\left[1 + \text{erf}\left(\frac{z - z_j}{\sqrt{2}\sigma_j}\right)\right] \quad (33)$$

where $\rho_0$ is the SLD of the vapor phase and $\rho_N = \rho_s$. Using Equation 32 the reflectivity is given by

$$R(Q_z) = R_F(Q_z) \left| \sum_j \left( \frac{\rho_j - \rho_{j-1}}{\rho_s} \right) e^{-\frac{(Q_z \sigma_j)^2}{2}} e^{iQ_z z_j} \right|^2 \qquad (34)$$

Assuming one interface at $z_1 = 0$ with surface roughness $\sigma_1 = \sigma$, the Fresnel reflectivity, $R_F(Q_z)$, is simply modified by a Debye-Waller-like factor

$$R(Q_z) = R_F(Q_z) e^{-(Q_z \sigma)^2} \qquad (35)$$

The effect of surface roughness on the reflectivities from water and from liquid mercury surfaces, assuming Gaussian smearing of the interfaces, is shown by the dashed lines in Figure 3. Braslau et al. (1988) have demonstrated that the Gaussian smearing of the interface due to capillary waves in simple liquids is sufficient in modeling the data, and that more complicated models cannot be supported by the x-ray data.

Applying Equation 34 to the one-box model discussed above (see Fig. 5), and assuming conformal roughness, $\sigma_j = \sigma$, the calculated reflectivity in terms of SLD normalized to $\rho_s$ is

$$R(Q_z) = R_F(Q_z)[(1-\rho_1)^2 + \rho_1^2 + 2\rho_1(1-\rho_1)\cos(Q_z d)]e^{-(Q_z \sigma)^2} \qquad (36)$$

In this approximation, the roles of the normalized SLD of the one box, $\rho_1$, and of the complementary model, $\rho_2 = 1 - \rho_1$, are equivalent. This demonstrates that the reflectivities for both models are mathematically identical. This is the simplest of many examples where two or more distinct SLD models yield identical reflectivities in the Born approximation. When using the kinematical approximation to invert the reflectivity to SLD, there is always a problem of facing a nonunique result. For discussion of ways to distinguish between such models see Data Analysis and Initial Interpretation.

In some instances, the scattering length density can be generated by several step functions that are smeared with one Gaussian (conformal roughness $\sigma_j = \sigma$), representing different moieties of the molecules on the surface. The reflectivity can be calculated by using a combination of the dynamical and the kinematical approaches (Als-Nielsen and Kjaer, 1989). First, the exact reflectivity from the step-like functions ($\sigma = 0$) is calculated using the matrix method, $R_{dyn}(Q_z)$, and the effect of surface roughness is incorporated by multiplying the calculated reflectivity with a Debye-Waller-like factor as follows (Als-Nielsen and Kjaer, 1989)

$$R(Q_z) = R_{dyn}(Q_z) e^{-(Q_z \sigma)^2} \qquad (37)$$

## Non-Specular Scattering

The geometry for non-specular reflection is shown in Figure 1B. The scattering from a 2-D system is very weak, and enhancements due to multiple scattering processes at the interface are taken advantage of. As is shown in Figure 1B the momentum transfer **Q** has a finite component parallel to the liquid surface ($\mathbf{Q}_\perp \equiv \mathbf{k}_\perp^i - \mathbf{k}_\perp^f$), enabling determination of lateral correlations in the 2-D plane. Exact calculation of scattering from surfaces is practically impossible except for special cases, and the Born approximation (BA; Schiff, 1968) is usually applied. When the incident beam or the scattered beam are at grazing angles (i.e., near the critical angle), multiple scattering effects modify the scattering, and these can be accounted for by a higher-order approximation known as the distorted wave Born approximation (DWBA). The features due to multiple scattering at grazing angles provide evidence that the scattering processes indeed occur at the interface.

**The Born Approximation.** In the BA for a general potential $V(\mathbf{r})$ the scattering length amplitude is calculated as follows (Schiff, 1968)

$$F(\mathbf{Q}) = \frac{1}{4\pi} \int V(\mathbf{r}) e^{i\mathbf{Q}\cdot\mathbf{r}} d^3\mathbf{r} \qquad (38)$$

where, in the present case, $V(\mathbf{r}) = 4\pi\rho(\mathbf{r})$. From the scattering length amplitude, the differential cross-section is calculated as follows (Schiff, 1968)

$$\frac{d\sigma}{d\Omega} = |F(\mathbf{Q})|^2 = \int [\rho(\mathbf{0})\rho(\mathbf{r})] e^{i\mathbf{Q}\cdot\mathbf{r}} d^3\mathbf{r} \qquad (39)$$

where $[\rho(0)\,\rho(\mathbf{r})] \equiv \int[\rho(\mathbf{r}'-\mathbf{r})\rho(\mathbf{r}')]d^3\mathbf{r}'$ is the density-density correlation function. The measured reflectivity is a convolution of the differential cross-section with the instrumental resolution, as discussed below and in the literature (Schiff, 1968; Braslau et al., 1988; Sinha et al., 1988).

The scattering length density, $\rho$, for a liquid-gas interface can be described as a function of the actual height of the surface, $z(x,y)$, as follows

$$\rho(\boldsymbol{\mu}, z) = \begin{cases} \rho_s & \text{for} \quad z < z(\boldsymbol{\mu}) \\ 0 & \text{for} \quad z > z(\boldsymbol{\mu}) \end{cases} \qquad (40)$$

where $\boldsymbol{\mu} = (x, y, 0)$ is a 2-D in-plane vector. The height of the interface $z$ is also time and temperature dependent due to capillary waves, and therefore thermal averages of $z$ are used (Buff et al., 1965; Evans, 1979). Inserting the SLD (Equation 40) in Equation 39 and performing the integration over the $z$ coordinate yields

$$F(\mathbf{Q}_\perp, Q_z) = \frac{\rho_s}{iQ_z} \int e^{i[\mathbf{Q}_\perp \cdot \boldsymbol{\mu} + Q_z z(\boldsymbol{\mu})]} d^2\boldsymbol{\mu} \qquad (41)$$

where $\mathbf{Q}_\perp = (Q_x, Q_y, 0)$ is an in-plane scattering vector. This formula properly predicts the reflectivity from an ideally flat surface, $z(x,y) = 0$ within the kinematical approximation

$$F(\mathbf{Q}_\perp, Q_z) = \frac{4\pi^2 \rho_s}{iQ_z} \delta^{(2)}(\mathbf{Q}_\perp) \qquad (42)$$

with a 2-D delta-function ($\delta^{(2)}$) that guarantees specular reflectivity only. The differential cross-section is then given by

$$\frac{d\sigma}{d\Omega} = \pi^2 \left(\frac{Q_c^2}{4Q_z}\right)^2 \delta^{(2)}(\mathbf{Q}_\perp) \qquad (43)$$

where $Q_c^2 \equiv 16\pi\rho_s$. This is the general form for the Fresnel reflectivity in terms of the differential cross-section $d\sigma/d\Omega$, which is defined in terms of the flux of the incident beam on the surface. In reflectivity measurements, however, the scattered intensity is normalized to the intensity of the incident beam, and therefore the flux on the sample is angle-dependent and is proportional to $\sin\alpha_i$. In addition, the scattered intensity is integrated over the polar angles $\alpha_f$ and $2\theta$ with $k_0^2 \sin\alpha_f d\alpha_f d(2\theta) = dQ_x dQ_y$. Correcting for the flux and integrating

$$R_{\mathrm{F}}(Q_z) \approx \sigma_{\mathrm{tot}}(Q_z) = \iint \left(\frac{d\sigma}{d\Omega}\right) \frac{dQ_x dQ_y}{4\pi^2 k_0^2 \sin\alpha_i \sin\alpha_f} = \left(\frac{Q_c}{2Q_z}\right)^4 \qquad (44)$$

as approximated from the exact solution, given in Equation 17.

Taking advantage of the geometrical considerations above, the differential cross-section to the reflectivity measurement can be readily derived in the more general case of scattering length density that varies along $z$ only, (i.e., $\rho(z)$). In this case, Equation 38 can be written as

$$\frac{d\sigma}{d\Omega} = 4\pi^2 \delta^{(2)}(\mathbf{Q}_\perp) \left|\int \rho(z) e^{iQ_z z} dz\right|^2 \qquad (45)$$

If we normalize $\rho(z)$ to the scattering length density of the substrate, $\rho_s$, and use a standard identity between the Fourier transform of a function and its derivative, we obtain

$$\frac{d\sigma}{d\Omega} = \pi^2 \left(\frac{Q_c^2}{4Q_z}\right)^2 \left|\int \frac{1}{\rho_s} \frac{d\rho(z)}{dz} e^{iQ_z z} dz\right|^2 \delta^{(2)}(\mathbf{Q}_\perp) \qquad (46)$$

which, with the geometrical corrections, yields Equation 32.

Thermal averages of the scattering length density under the influence of capillary waves and the assumption that the SLD of the gas phase is zero can be approximated as follows (Buff et al., 1965; Evans, 1979)

$$\rho(\mathbf{r}) \sim \frac{\rho_s}{2} \left(1 + \mathrm{erf}\left[\frac{z}{\sqrt{2}\sigma(\mu)}\right]\right) \qquad (47)$$

where $\sigma(\mu)$ is the height-height correlation function.

Inserting Equation 47 into Equation 39, and integrating over $z$, results in the differential cross-section

$$\frac{d\sigma}{d\Omega} \sim \frac{Q_c^4}{Q_z^2} \int e^{i\mathbf{Q}_\perp \cdot \mu - Q_z^2 \sigma^2(\mu)} d^2\mu \qquad (48)$$

and assuming isotropic correlation function in the plane yields (Sinha et al., 1988)

$$\frac{d\sigma}{d\Omega} \sim \frac{Q_c^4}{Q_z^2} \int \mu J_0(Q_\perp \mu) e^{-Q_z^2 \sigma^2(\mu)} d\mu \qquad (49)$$

where $J_0$ is a Bessel function of the first kind. This expression was used by Sinha et al. (1988) to calculate the diffuse scattering from rough liquid surfaces with a height-height density correlation function that diverges logarithmically due to capillary waves (Sinha et al., 1988; Sanyal et al., 1991).

**Distorted Wave Born-Approximation (DWBA).** Due to the weak interaction of the electromagnetic field (x-rays) with matter (electrons), the BA is a sufficient approach to the scattering from most surfaces. However, as we have already encountered with the reflectivity, the BA fails (or is invalid) when either the incident beam or the scattered beam is near the critical angle where multiple scattering processes take place. The effect of the bulk on the scattering from the surface can be accounted for by defining the scattering length density as a superposition of two parts, as follows

$$\rho(\mathbf{r}) = \rho_1(z) + \rho_2(\mu, z) \qquad (50)$$

Here $\rho_1(z)$ is a step function that defines an ideally sharp interface separating the liquid and gas phases at $z = 0$, whereas the second term, $\rho_2(\mu, z)$, is a quasi–two-dimensional function in the sense that it has a characteristic average thickness, $d_c$, such that $\lim_{z \to \pm d_c/2} \rho_2(\mu, z) = 0$. It can be thought of as film-like and is a detailed function with features that relate to molecular or atomic distributions at the interface. Although the definition of $\rho_2$ may depend on the location of the interface, ($z = 0$) in $\rho_1$, the resulting calculated scattering must be invariant for equivalent descriptions of $\rho(\mathbf{r})$. In some cases $\rho_2$ can be defined as either a totally external or totally internal function with respect to the liquid bulk, (i.e., $\rho_1$). In other cases, especially when dealing with liquid surfaces, it is more convenient to locate the interface at some intermediate point coinciding with the center of mass of $\rho_2$ with respect to $z$. The effect of the substrate term $\rho_1(z)$ on the scattering from $\rho_2$ can be treated within the distorted wave Born approximation (DWBA) by using the exact solution from the ideally flat interface (see Principles of the Method) to generate the Green function for a higher-order Born approximation (Rodberg and Thaler, 1967; Schiff, 1968; Vineyard, 1982; Sinha et al., 1988). The Green function in the presence of an ideally flat interface, $\rho_1$, replaces the free particle Green function that is commonly used in the Born approximation. The scattering amplitude in this case is given by (Rodberg and Thaler, 1967; Schiff, 1968)

$$F_{\mathrm{DWBA}}(\mathbf{Q}) = F_{\mathrm{F}}(Q_z) + F_2(\mathbf{Q}) = -i\pi Q_z r_{\mathrm{F}}(Q_z)$$
$$+ \int \tilde{\chi}_{k'}^*(\mathbf{r}) \rho_2(\mathbf{r}) \chi_k(\mathbf{r}) d\mathbf{r} \qquad (51)$$

where the exact Fresnel amplitude $F_F(Q_z)$ is written in the form of a scattering amplitude so that Equation 51 reproduces the Fresnel reflectivity in the absence of $\rho_2$.

The exact solution of the step function $\rho_1$, $\chi_k(\mathbf{r})$, is given by

$$\chi_k(\mathbf{r}) = e^{i k_\perp^i \cdot \mu} \begin{cases} e^{i k_{z,0}^i z} + r^i(k_{z,s}^i) e^{-i k_{z,0}^i z} & \text{for } z > 0 \\ t^i(k_{z,s}^i) e^{i k_{z,s}^i z} & \text{for } z < 0 \end{cases} \tag{52}$$

and the $\tilde{\chi}_k^*(\mathbf{r})$ is the time-reversed and complex conjugate solution of an incident beam with $-\mathbf{k}_i$,

$$\tilde{\chi}_k^*(\mathbf{r}) = e^{-i k_\perp^f \cdot \mu} \begin{cases} e^{-i k_{z,0}^f z} + r^f(k_{z,0}^f) e^{i k_{z,0}^f z} & \text{for } z > 0 \\ t^f(k_{z,0}^f) e^{-i k_{z,0}^f z} & \text{for } z < 0 \end{cases} \tag{53}$$

In Equation 53, it is assumed that the scattered beam is detected only for $z > 0$, i.e., above the liquid surface. The notation for transmission and reflection functions indicate scattering of the wave from the air onto the subphase and vice-versa according to

$$k_{z,s}^i = \sqrt{(k_{z,0}^i)^2 + k_c^2} \tag{54}$$

and

$$k_{z,0}^f = \sqrt{(k_{z,s}^f)^2 + k_c^2} \tag{55}$$

respectively. In the latter case, total reflectivity does not occur except for the trivial case $k_{z,s}^f = 0$, and no enhancement due to the evanescent wave is expected. In this approximation, the final momentum transfer in the $Q_z$ direction is a superposition of momentum transfers from $\rho_2$ (the film) and from the liquid interface, $\rho_1$. For instance, there could be a wave scattered with $Q_z = 0$ with respect to $\rho_2$ but reflected from the surface with a finite $Q_z$. This is due to multiple scattering processes between $\rho_2$ and $\rho_1$ and, therefore, in principle the amplitude of the detected beam contains a superposition of different components of the Fourier transform of $\rho_2(q_z)$ and interference terms between them. Detailed analysis of Equation 51 can be very cumbersome for the most general case, and is usually dealt with for specific cases of $\rho_2$. Assuming that the scattering from the film is as strong for $q_z$ as for $-Q_z$ (as is the case for an ideal 2-D system with equal scattering along the rod, i.e., $\rho_2$ is symmetrical under the inversion of $z$), we can write $\tilde{\rho}_2(\mathbf{Q}_\perp, Q_z) \approx \tilde{\rho}_2(\mathbf{Q}_\perp, -Q_z)$. In this simple case, the cross-section can be approximated as follows (Vineyard, 1982; Sinha et al., 1988; Feidenhas'l, 1989)

$$\frac{d\sigma}{d\Omega} = |t^i(k_{z,s}^i) \tilde{\rho}_2(\mathbf{Q}_\perp, Q_z') t^f(k_{z,s}^f)|^2 + |t^i(k_{z,s}^i) \tilde{\rho}_2(\mathbf{Q}_\perp, Q_z'') t^f(k_{z,0}^f)|^2 \tag{56}$$

where $\mathbf{Q}_\perp \equiv \mathbf{k}_\perp^i - \mathbf{k}_\perp^f$ and $Q_z' = k_{z,0}^i - k_{z,s}^f$ and $Q_z'' = k_{z,s}^i - k_{z,0}^f$. Notice that the transmission functions modulate the scattering from the film ($\rho_2$), and in particular they give rise to enhancements as $k_{z,s}^i$ and $k_{z,s}^f$ are scanned around the critical angle as depicted in Figure 2B. Also,

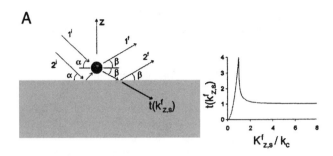

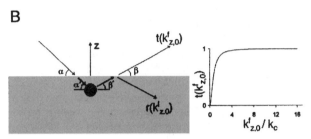

**Figure 6.** Illustration of wave paths for exterior (**A**) and interior (**B**) scatterer near a step-like interface. In both cases the scattering is enhanced by the transmission function when the angle of the incidence is varied around the critical angle. However, due to the asymmetry between external and internal reflectivity, the rod scan of the final beam modulates the scattering differently, as is shown on the right-hand side in each case.

it is only by virtue of the $z$ symmetry of the scatterer that such enhancements occur for an exterior film. From this analysis we notice that there will be no enhancement due to the transmission function of the final wave for an interior film.

To examine the results from the DWBA method, we consider scattering from a single scatterer near the surface. The discussion is restricted to the case where the detection of the scattered beam is performed in the vapor phase only. The scatterer can be placed either in the vacuum ($z > 0$) or in the liquid (see Fig. 6). When the particle is placed in the vacuum there are two major relevant incident waves: a direct one from the source, labeled $1^i$ in Figure 6A, and a second one reflected from the surface before scattering from $\rho_2$, labeled $2^i$. Assuming inversion symmetry along $z$, both waves scatter into a finite $\mathbf{Q}_\perp$ with similar strengths at $Q_z$ and $-Q_z$, giving rise to an enhancement near the critical angle if the incident beam is near the critical angle. Another multiple scattering process that gives rise to enhancement at the critical angle is one in which beam $1^i$ does not undergo a change in the momentum transfer along $z$ ($Q_z^{\text{film}} \approx 0$) before scattering from the liquid interface. The effect of these processes gives rise to enhancements if either the incident beam or the reflected beam are scanned along the $z$ direction. Slight modifications of the momentum transfer along the $z$ direction, such as

$$Q_z' = k_z^i + \sqrt{(k_z^f)^2 - k_c^2} \tag{57}$$

are neglected in the discussion above. The effective amplitude from the scatterer outside the medium is given by the following terms

$$|e^{iQ_z z} + e^{iQ'_z z} r(k^f_{z,s})| \approx |t(k^i_{z,s})| \qquad (58)$$

where the approximation is valid since the phase factors can be neglected, and $Q_z \approx Q'_z$. At large angles the reflectivity is negligible and the transmission function approaches $t(k_{z,s}) \approx 1$. Similar arguments hold for the outgoing wave. Neglecting the small changes in the momentum transfer due to dynamical effects, the transmission function modulates the scattering as is shown in solid line in Figure 2B.

The scattered wave from a particle that is embedded in the medium is different due to the asymmetry between external and internal reflection from the liquid subphase. The wave scattered from the particle rescatters from the liquid-gas interface. Upon traversing the liquid interface, the index of refraction increases from $n = 1 - (2\pi/k_0^2)\rho$ to 1, and no total internal reflection occurs, as discussed earlier; thus there is no evanescent wave in the medium. The transmission function for this wave is given by $t(-k^i_{z,s})$, like that of a wave emanating from the liquid interface into the vapor phase. In this case, the transmission function is a real function for all $k^i_{z,s}$ and does not have the enhancements around the critical angle (as shown in Fig. 6B), with zero intensity at the horizon.

**Grazing Incidence Diffraction (GID), and Rod Scans.** In some instances, ordering of molecules at liquid interfaces occurs. Langmuir monolayers spread at the gas-water interface usually order homogeneously at high enough lateral pressures (Dutta et al., 1987; Kjaer et al., 1987, 1994). Surface crystallization of $n$-alkane molecules on molten alkane has been observed recently (Wu et al., 1993b; Ocko et al., 1997). In these cases, $\rho_2$ is a periodic function in $x$ and $y$, and can be expanded as a Fourier series in terms of the 2-D reciprocal lattice vectors $\tau_\perp$ as follows

$$\rho_2(\mu, z) = \sum_{\tau_\perp} F(\tau_\perp, z) e^{i\tau_\perp \cdot \mu} \qquad (59)$$

Inserting Equation 59 in Equation 56 and integrating yields the cross-section for quasi–2-D Bragg reflection at $\mathbf{Q}_\perp = \tau_\perp$

$$\frac{d\sigma}{d\Omega} \sim P(\mathbf{Q})|t(k^i_{z,s})|^2 \langle |F(\tau_\perp, Q_z)|^2 \rangle \mathrm{DW}(Q_\perp, Q_z)|t^f(k^f_{z,s})|^2 \delta(\mathbf{Q} - \tau_\perp) \qquad (60)$$

where $P(\mathbf{Q})$ is a polarization correction and the 2-D unit cell structure factor is given as a sum over the atomic form factors $f_j(Q)$ with appropriate phase

$$F(\tau_\perp, Q_z) = \sum_j f_j(Q) e^{i\tau \cdot r_j + Q_z z_j} \qquad (61)$$

The structure factor squared is averaged for multiplicity due to domains and weighted for orientation relative to the surface normal. The ordering of monolayers at the air-water interface is usually in the form of 2-D powder consisting of crystals with random orientation in the plane. From Equation 60 we notice that the conservation of momentum expressed with the delta-function allows for observation of the Bragg reflection at any $Q_z$. A rod scan can be performed by varying either the incident or reflected beam, or both. The variation of each will produce some modulation due to both the transmission functions and to the average molecular structure factor along the z-axis. The Debye-Waller factor (see KINEMATIC DIFFRACTION OF X-RAYS), $\mathrm{DW}(\mathbf{Q}_\perp, Q_z)$, which is due to the vibration of molecules about their own equilibrium position with time-dependent molecular displacement $\mathbf{u}(t)$ is given by

$$\mathrm{DW}(\mathbf{Q}_\perp, Q_z) \sim e^{-(C_\perp Q_\perp^2 \langle \mathbf{u}_\perp^2 \rangle + Q_z^2 \sigma^2)} \qquad (62)$$

The term due to capillary waves on the liquid surface is much more dominant than the contribution from the in-plane intrinsic fluctuations. The Debye-Waller factor in this case is an average over a crystalline size and might not reflect surface roughness extracted from reflectivity measurements, where it is averaged over the whole sample.

## PRACTICAL ASPECTS OF THE METHOD

The minute sizes of interfacial samples on the sub-microgram level, combined with the weak interaction of x-rays with matter, result in very weak GID and reflectivity (at large $Q_z$) signals that require highly intense incident beams, which are available at x-ray synchrotron sources (see SURFACE X-RAY DIFFRACTION). A well prepared incident beam for reflectivity experiments at a synchrotron (for example, the X22B beam-line at the National Synchrotron Light Source at Brookhaven National Laboratory; Schwartz, 1992) has an intensity of $10^8$ to $10^9$ photons/ sec, whereas, for a similar resolution, an 18-kW rotating anode generator produces $10^4$ to $10^5$ photons/sec. Although reflectivity measurements can be carried out with standard x-ray generators, the measurements are limited to almost half the angular range accessible at synchrotron sources, and they take hours to complete compared to minutes at the synchrotron. GID experiments are practically impossible with x-ray generators, since the expected signals (2-D Bragg reflections, for example) normalized to the incident beam are on the order of $10^{-8}$ to $10^{-10}$.

### Reflectivity

X-ray reflectivity and GID measurements of liquid surfaces are carried out on special reflectometers that enable manipulation of the incident as well as the outgoing beam. A prototype liquid surface reflectometer was introduced for the first time by Als-Nielsen and Pershan (1983). In order to bring the beam to an angle of incidence $\alpha_i$ with respect to the liquid surface, the monochromator is tilted by an angle $\chi$ either about the axis of the incident beam (indicated by $\chi_1$ in Fig. 7) or about the axis normal to the reciprocal lattice wave vector of the monochromator, $\tau_0(\chi_2)$. Figure 7 shows the geometry that is used to deflect

**A**

Side View

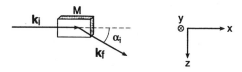

Top View

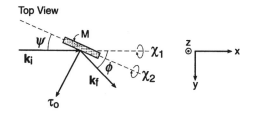

**B**

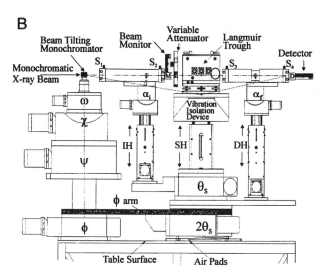

**Figure 7.** (**A**) Monochromator geometry to tilt a Bragg reflected beam from the horizon on a liquid surface. Two possible tilting configurations about the primary beam axis and about an axis along the surface of the reflecting planes are shown. (**B**) A side view diagram of the Ames Laboratory Liquid Surfaces Diffractometer at the 6-ID beam line at the Advanced Photon Source at Argonne National Laboratory.

the beam from the horizontal onto the liquid surface at an angle, $\alpha_i$, by tilting the monochromator. At the Bragg condition, the surface of the monochromator crystal is at an angle $\psi$ with respect to the incoming beam. Tilting over the incident beam axis is like tracing the Bragg reflection on the Debye-Scherer cone so that the $\psi$ axis remains fixed, with a constant wavelength at different tilting angles. The rotated reciprocal lattice vector and the final wave vector in this frame are given by

$$\tau_0 = \tau_0(-\sin \psi, \cos \psi \cos \chi_1, \cos \psi \sin \chi_1)$$

$$\mathbf{k}_f = k_0(\cos \alpha_i \cos \phi, \cos \alpha_i \sin \phi, \sin \alpha_i) \quad (63)$$

where $\phi$ is the horizontal scattering angle. The Bragg conditions for scattering are given by

$$\mathbf{k}_i + \tau_0 = \mathbf{k}_f; \qquad |\mathbf{k}_f| = k_0 \quad (64)$$

Using Equations 63 and 64, the following relations for the monochromator axes are obtained

$$\sin \psi = \frac{\tau_0}{2k_0}$$

$$\sin \chi_1 = \frac{k_0}{\tau_0} \cos \psi \sin \alpha_i \quad (65)$$

$$\cos \phi = \left(1 - \frac{\tau_0^2}{2k_0^2}\right) \Big/ \cos \alpha_i$$

and we notice that the monochromator angle $\psi$ is independent of $\alpha_i$. However, the scattering angle $\phi$ has to be modified as $\alpha_i$ is varied. This means that the whole reflectometer arm has to be rotated. Similarly, for the configuration where the monochromator is tilted over the axis normal to $\tau_0$ we get

$$\sin \psi = \frac{\tau_0}{2k_0\cos \chi_2}$$

$$\sin \chi_2 = \frac{k_0}{\tau_0} \sin \alpha_i \quad (66)$$

$$\cos \phi = \left(1 - \frac{\tau_0^2}{2k_0^2}\right) \Big/ \cos \alpha_i$$

From these relations, the conditions for a constant wavelength operation for any angle of incidence, $\alpha_i$, can be calculated and applied to the reflectometer. Here, unlike the previous mode, deflection of the beam to different angles of incidence requires both the adjustment of $\psi$ and of $\phi$ in order to maintain a constant wavelength. If $\psi$ is not corrected in this mode of operation, the wavelength varies as $\chi$ is varied. This mode is sometimes desirable, especially when the incident beam hitting the monochromator consists of a continuous distribution of wavelengths around the wavelength at horizontal scattering, $\chi_2 = 0$. Such continuous wavelength distribution exists when operating with x-ray tubes or when the tilting monochromator is facing the white beam of a synchrotron. Although the variation in the wavelength is negligible as $\chi_2$ is varied, without the correction of $\psi$, the exact wavelength and the momentum transfer can be computed using the relations in Equation 65 and Equation 66. In both modes of monochromator tilting, the surface height as well as the height of the slits are adjusted with vertical translations.

Figure 7, panel B, shows a side-view diagram of the Ames Laboratory Liquid Surfaces Diffractometer at the 6-ID beam line at the Advanced Photon Source at Argonne National Laboratory. A downstream Si double-crystal monochromator selects a very narrow-bandwidth energy ($\sim$1 eV) beam (in the range 4 to 40 keV) from an undulator-generated beam. The highly monochromatic beam is deflected onto the liquid surface to any desired angle of incidence, $\alpha_i$, by the beam-tilting monochromator; typically Ge(111) or Ge(220) crystals are used. To simplify its description, the diffractometer can be divided into two main stages, with components that may vary slightly depending on the details of the experiment. In the first stage, the incident beam on the liquid surface is optimized. This part consists of the axes that adjust the beam tilting monochromator ($\psi, \phi, \chi$), incident beam slits ($S_i$), beam monitor, and attenuator. The $\omega$ axis, below the monochromator crystal, is adjusted in the initial alignment process to ensure that the tilting axis $\chi$ is well defined. For each angle of incidence $\alpha_i$, the $\psi$, $\phi$, $\chi$ and the height of the incident beam arm (IH) carrying the $S_1$, $S_2$ slits, are adjusted. In addition, the liquid surface height (SH) is brought to the intersecting point between the incident beam and the detector arm axis of rotation $2\theta_s$. In the second stage of the diffractometer, the intensity of the scattered beam from the surface is mapped out. In this section, the angles $\alpha_f$ and $2\theta_s$ and the detector height (DH) are adjusted to select and detect the outgoing beam. The two stages are coupled through the $\phi$-arm of the diffractometer. In general, $\theta_s$ is kept idle because of the polycrystalline nature of monolayers at liquid surfaces. Surface crystallization of alkanes proceeds with the formation of a few large single crystals, and the use of $\theta_s$ is essential to orienting one of these crystals with respect to the incoming beam. The scattering from liquid metals is complicated by the fact that their surfaces are not uniformly flat. For details on how to scatter from liquids with curved surfaces see Regan et al. (1997).

For aqueous surfaces, the trough (approximate dimensions $\sim 120 \times 270 \times 5$ mm$^3$) is placed in an airtight aluminum enclosure that allows for the exchange of the gaseous environment around the liquid surface. To get total reflectivity from water below the critical angle (e.g., $\alpha_c = 0.1538°$ and $0.07688°$ at 8 keV and 16 keV respectively), the footprint of the beam has to be smaller than the specimen surface. A typical cross-section of the incident beam is $0.5 \times 0.1$ mm$^2$ with approximately 1010 to 1011 photons per second (6-ID beamline). At about $0.8\alpha_c$, the footprint of the beam with a vertical size (slit $S_2$) of about 0.1 mm is $\sim$47 and $\sim$94 mm at 8 keV and 16 keV, respectively, compared to 120 mm liquid-surface dimension in the direction of the primary beam. Whereas a 0.1 mm beam size is adequate for getting total reflectivity at 8 keV (exploiting about half of the liquid surface in the beam direction), a 0.05-mm beam size is more appropriate at 16 keV. This vertical beam size (slit $S_2$) can be increased at larger incident angles to maintain a relatively constant beam footprint on the surface.

The alignment of the diffractometer encompasses two interconnected iterated processes. First, the angles of the first stage ($\alpha_i$, $\psi$, $\phi$, $\chi$, and $\omega$) are optimized and set so that the x-ray flux at the monitor position is preserved upon deflection of the beam (tracking procedure). Second, the beam is steered so that it is parallel to the liquid surface. It should be emphasized that the beam, after the tracking procedure, is not necessarily parallel to the liquid surface. In this process, reflectivities from the liquid surface at various incident angles are employed to define the parallel beam, by adjustment of the angles and heights of the diffractometer. Then, repeatedly, the first and second processes are iterated until convergence is achieved (i.e., until corrections to the initial positions of motors are smaller than the mechanical accuracies).

The divergence of the monochromatic incident beam on the surface is determined by at least two horizontal slits located between the sample and the source. One of these

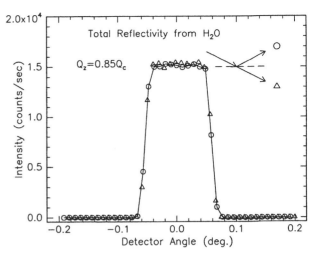

**Figure 8.** Superposition of the reflected beam (circles) below the critical angle and direct beam (triangles), demonstrating total reflectivity of x-rays from the surface of water. Severe surface roughness reduces the intensity and widens the reflected signal. A reduction from total reflectivity can also occur if the slits of the incident beam are too wide, so that the beam-footprint is larger than the surface sample.

slits is usually located as close as possible to the sample, and the other as close as possible to the source. These two slits determine the resolution of the incident beam. By deflecting the beam from the horizontal, the shape of the beam changes, and that may change incident beam intensity going through the slits; therefore the use of a monitor right after the slit in front of the sample is essential for the absolute determination of the reflectivity. The size of the two slits defining the incident beam is chosen in such a way that the footprint of the beam is much smaller than the width of the reflecting surface, so that total reflectivity occurs. Figure 8 shows the reflected beam and the direct beam from a flat surface of water, demonstrating total reflectivity at $Q_z = 0.85 Q_c$. In this experiment the detector slit is wide open at $\sim$10 times the opening of the sample slit. As is demonstrated, the effect of absorption is negligible for water, and roughness is significantly reduced by damping surface waves. The damping can be achieved by reducing the height of the water film to $\sim$0.3 mm and placing a passive as well as an active antivibration unit underneath the liquid sample holder, suppressing mechanical vibrations (Kjaer et al., 1994).

### Non-Specular Scattering: GID, Diffuse Scattering, and Rod Scans

X-ray GID measurements are performed at angles of incidence below the critical angle $\approx 0.9\alpha_c$. Operating with the incident beam below the critical angle enhances the signal from the surface with respect to that of the bulk by creating an evanescent wave in the medium that is exponentially decaying according to

$$E(z) = t(k_{z,s})e^{-z/\Lambda} \qquad (67)$$

where

$$\frac{1}{\Lambda} = \sqrt{k_c^2 - k_{z,0}^2} \qquad (68)$$

For water at $k_z \approx 0.9k_c$, $\Lambda \sim 100$ Å.

As illustrated in Figure 1, the components of the momentum transfer for GID are given by

$$\begin{aligned} Q_z &= k_0(\sin\alpha_i + \sin\alpha_f) \\ Q_x &= k_0(\cos\alpha_i - \cos\alpha_f)\cos 2\theta \qquad (69) \\ Q_y &= k_0\cos\alpha_f\sin 2\theta \end{aligned}$$

In most cases, the 2-D order on liquid surfaces is powder-like, and the lateral scans are displayed in terms of $Q_\perp$ which is given by

$$Q_\perp = k_0\sqrt{\cos^2\alpha_i + \cos^2\alpha_f - 2\cos\alpha_i\cos\alpha_f\cos 2\theta} \qquad (70)$$

To determine the in-plane correlations, the horizontal resolution of the diffractometer can be adjusted with a Soller collimator consisting of vertical absorbing foils stacked together between the surface and the detector. The area that is probed at each scattering angle $2\theta$ is proportional to $S_0/\sin 2\theta$, where $S_0$ is the area probed at $2\theta = \pi/2$. The probed area must be taken into account in the analysis of a GID scan that is performed over a wide range of angles.

Position-sensitive detectors (PSD) are commonly used to measure the intensity along the 2-D rods. It should be kept in mind that the intensity along the PSD is not a true rod scan of a Bragg reflection at a nominal $Q_\perp$ because of the variation in $Q_\perp$ as $\alpha_f$ is varied, as is seen in Equation 69.

## DATA ANALYSIS AND INITIAL INTERPRETATION

The task of finding the SLD from a reflectivity curve is similar to that of finding an effective potential for Equation 7 from the modulus of the wave-function. Direct inversion of the scattering amplitude to SLD is not possible except for special cases when the BA is valid (Sacks, 1993, and references therein). If the modulus and the phase are known, they can be converted by the method of Gelfand-Levitan-Marchenko (Sacks, 1993, and references therein) to SLD (GLM method). However, in reflectivity experiments, the intensity of the scattered beam alone is measured, and phase information is lost.

Step-like potentials have been directly reconstructed recently by retrieving the phase from the modulus—i.e., reflectivity—and then using the GLM method (Clinton, 1993; also Sacks, 1993, and references therein). Model-independent methods which are based on optimization of a model to reflectivity, without requiring any knowledge of the chemical composition of the SLD at the interface, were also developed recently (Pedersen, 1992; Zhou and Chen, 1995). Such models incorporate a certain degree of objectivity. These methods are based on the kinematical and the dynamical approaches for calculating the reflectiv-

ity. One method (Pedersen, 1992) uses indirect Fourier transformation to calculate the correlation function of $d\rho/dz$, which is subsequently used in a square-root deconvolution model to construct the SLD model. Zhou and Chen (1995), on the other hand, developed a groove tracking method that is based on an optimization algorithm to reconstruct the SLD using the dynamical approach to calculate the reflectivity at each step.

The most common procedure to extract structural information from reflectivity is the use of standard nonlinear least squares refinement of an initial SLD model. The initial model is defined in terms of a P-dimensional set of independent parameters, $\mathbf{p}$, using all the information available in guessing $\rho(z, \mathbf{p})$. The parameters are then refined by calculating the reflectivity ($R[Q_z^i\mathbf{p}]$) with the tools described earlier, and by minimizing the $\chi^2(\mathbf{p})$ quantity

$$\chi^2(\mathbf{p}) = \frac{1}{N-P}\sum_{i=1}\left[\frac{R_{\exp}(Q_z^i) - R(Q_z^i, \mathbf{p})}{\varepsilon(Q_z^i)}\right]^2 \qquad (71)$$

where $\varepsilon(Q_z^i)$ is the uncertainty of the measured reflectivity, $R_{\exp}(Q_z^i)$, and $N$ is the number of measured points. The criteria for a good fit can be found in Bevington (1968). Uncertainties of a certain parameter can be obtained by fixing it at various values and for each value refining the rest of the parameters until $\chi^2$ is increased by a factor of at least $1/(N-P)$.

The direct methods and model-independent procedures of reconstruction SLD do not guarantee uniqueness of the potential—i.e., there can be multiple SLD profiles that essentially yield the same reflectivity curve, as discussed with regard to Figure 5, for example. The uniqueness can be achieved by introducing physical constraints that are incorporated into the parameters of the model. Volume, in-plane density of electrons, etc., are among such constraints that can be used. Applying such constraints is discussed briefly below; see Examples (also see Vaknin et al., 1991a,b; Gregory et al., 1997). These constraints reduce the uncertainties and make the relationship of the SLD to the actual molecular arrangement apparent. In the dynamical approach, no two potentials yield exactly the same reflectivity, although the differences between two models might be too small to be detected in an experiment.

An experimental method for solving such a problem was suggested by Sanyal et al. (1992) using anomalous x-ray reflectivity methods. Two reflectivity curves from the same sample are measured with two different x-ray energies, one below and one above an absorption edge of the substrate atoms, thereby varying the scattering length density of the substrate. Subsequently the two reflectivity curves can be used to perform a direct Fourier reconstruction (Sanyal et al., 1992), or refinement methods can be used to remove ambiguities. This method is not efficient when dealing with liquids that consist of light atoms, because of the very low energy of the absorption edge with respect to standard x-ray energies. Another way to overcome the problem of uniqueness is by performing reflectivity experiments on similar samples with x-rays and with neutrons. In addition, the SLD, $\rho(z)$ across the interface can be changed significantly, in neutron scattering experiments, by chemical exchange of isotopes that change $\rho(z)$, but maintain the same structure (Vaknin et al., 1991b; Penfold and Thomas, 1990). The reflectivities (x-ray as well as neutrons) can be fitted to one structural model that is defined in terms of geometrical parameters only, calculating the SLDs from scattering lengths of the constituents and the geometrical parameters (Vaknin et al., 1991b,c).

### Examples

Since the pioneering work of Als-Nielsen and Pershan (1983), x-ray reflectivity and GID became standard tools for the characterization of liquid surfaces on the atomic length scales. The techniques have been exploited in studies of the physical properties of simple liquids (Braslau et al., 1988; Sanyal et al., 1991; Ocko et al., 1994), Langmuir monolayers (Dutta et al., 1987; Kjaer et al., 1987, 1989, 1994; Als-Nielsen and Kjaer, 1989; Vaknin et al., 1991b), liquid metals (Rice et al., 1986; Magnussen et al., 1995; Regan et al., 1995), surface crystallization (Wu et al., 1993a,b, 1995; Ocko et al., 1997), liquid crystals (Pershan et al., 1987), surface properties of quantum liquids (Lurio et al., 1992), protein recognition processes at liquid surfaces (Vaknin et al., 1991a, 1993; Lösche et al., 1993), and many other applications. Here, only a few examples are briefly described in order to demonstrate the strengths and the limitations of the techniques. In presenting the examples, there is no intention of giving a full theoretical background of the systems.

**Simple Liquids.** The term simple liquid is usually used for a monoatomic system governed by van der Waals–type interactions, such as liquid argon. Here, the term is extended to include all classical dielectric (nonmetallic) liquids such as water, organic solvents (methanol, ethanol, chloroform, etc.), and others. One of the main issues regarding dielectric liquids is the determination of the average density profile across the interface, $N(z)$. This density is the result of folding the intrinsic density $N_I(z)$ of the interface due to molecular size, viscosity, and compressibility of the fluid with density fluctuations due to capillary waves, $\Delta N_{CW}(z)$. The continuous nature of the density across the interface due to capillary waves was worked out by Buff, Lovett, and Stillinger (BLS; Buff et al., 1965) assuming that $N_I(z)$ is an ideal step-like function. The probability for the displacement is taken to be proportional to the Boltzmann factor, $e^{-\beta U(z)}$, where $U$ is the free energy necessary to disturb the surface from equilibrium state—i.e., $z(x,y) = 0$—and $\beta = 1/k_B T$ where $k_B$ is Boltzmann's constant. The free energy of an incompressible and nonviscous liquid surface consists of two terms; a surface tension ($\gamma$) term, which is proportional to the changes in area from the ideally flat surface and a gravitational term as follows

$$U = \int (\gamma[\sqrt{1 + |\nabla z|^2} - 1] + \frac{1}{2} m_s g z^2) d^2\mu$$
$$\approx \frac{1}{2} \int (\gamma|\nabla z|^2 + m_s g z^2) d^2\mu \quad (72)$$

where $m_s$ is the mass density of the liquid substrate. By using standard Gaussian approximation methods, Buff et al. (1965) find that

$$U(z) \sim \frac{z^2}{2\sigma_0^2} \quad (73)$$

Convolution of the probability with a step-like function, representing the intrinsic density of the liquid surface yields the following density function

$$N(z) = N_s \text{erfc}\left(\frac{z}{\sqrt{2}\sigma}\right) \quad (74)$$

with a form similar to the one given in Equation 47. The average surface roughness at temperature $T$, is then given by

$$\sigma_{CW}^2 = \frac{k_B T}{2\pi\gamma} \ln\left(\frac{L}{a_0}\right) \quad (75)$$

where $a_0$ is a molecular diameter and $L$ is the size of the surface. Notice the logarithmic divergence of the fluctuations as the size of the surface increases, as expected of a 2-D system (Landau and Lifshitz, 1980). This model was further refined by assuming that the intrinsic profile has a finite width (Evans, 1979). In particular if the width due to the intrinsic profile is also expressed by a Gaussian then, the effective surface roughness is given by

$$\sigma_{eff}^2 = \sigma_I^2 + \sigma_{CW}^2 \quad (76)$$

and the calculated reflectivity is similar to Equation 35 for an interface that is smeared like the error function

$$R_{CW} = R_F(Q_z)e^{-\sigma_{eff}^2 Q_z^2} \quad (77)$$

Figure 9 shows the reflectivity from pure water measured at the synchrotron (D. Vaknin, unpub. observ.) where it is shown that, using Equation 77 for fitting the reflectivity data is satisfactory. This implies that the error function type of density profile (BLS model) for the liquid interface is sufficient. Only the surface roughness parameter, $\sigma$, is varied to refine the fit to the data ($\sigma = 2.54$ Å). This small roughness value depends on the attenuation of capillary waves by minimizing the depth of the water to $\sim$0.3 mm by placing a flat glass under the water (Kjaer et al., 1994). The validity of a Gaussian approximation of $N(z)$ (BLS model) was examined by various groups and for a variety of systems (Braslau et al., 1988; Sanyal et al., 1991; Ocko et al., 1994). Ocko et al. (1994) have measured the reflectivity of liquid alkanes over a wide range of temperatures, verifying that the surface roughness is of the form given in Equations 75 and 76.

Experimentally, the reflectivity signal at each $Q_z$ from a rough interface is convoluted with the resolution of the spectrometer in different directions. The effect of the resolution in $Q_z$ can be calculated analytically or convoluted numerically. For simplicity, we consider that the resolution functions can be approximated as a Gaussian with a

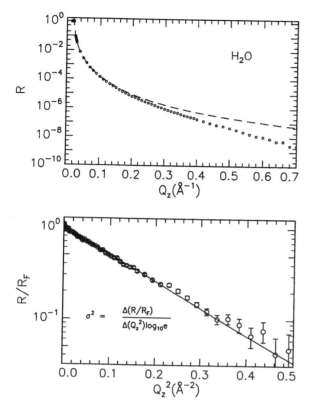

**Figure 9.** Experimental reflectivity from the surface of water. The dashed line is the calculated Fresnel reflectivity from an ideally flat water-interface, $R_F$. The normalized reflectivity versus $Q_z^2$ is fitted to a the form $R/R_F = e^{-(Q_z\sigma)^2}$, demonstrating the validity of the capillary-wave model (Buff et al., 1965).

width of $\Delta Q_z$ along $Q_z$ can be taken as $e^{-Q_z^2/\Delta Q_z^2}$ with appropriate normalization factor (Bouwman and Pederesen, 1996). The resolution, $\Delta Q_z$, is $Q_z$-dependent as the angles of incidence and scattering are varied (Ocko et al., 1994). However, if we assume that around a certain $Q_z$ value the resolution is a constant and we measure $\sigma_{exp}$, the convolution of the true reflectivity with the resolution function yields the following relation

$$\frac{1}{\sigma_{exp}^2} \approx \frac{1}{\sigma_{eff}^2} + \Delta Q_z^2 \qquad (78)$$

from which the effective roughness can be extracted as follows

$$\sigma_{eff} \approx \frac{\sigma_{exp}}{\sqrt{1 - \sigma_{exp}^2 \Delta Q_z^2}} \qquad (79)$$

Thus, if the resolution is infinitely good, i.e., with a $\Delta Q_z = 0$, the measured and effective roughness are the same. However, as the resolution is relaxed, the measured roughness gets smaller than the effective roughness. The effect of the resolution on the determination of true surface roughness was discussed rigorously by Braslau et al. (1988).

Diffuse scattering from liquid surfaces is practically inevitable, due to the presence of capillary waves. Calcula-

tion of the scattering from disordered interfaces of various characteristics are treated in Sinha et al. (1988). In the Born approximation, true specular scattering from liquid surfaces exists only by virtue of the finite cutoff length of the mean-square height fluctuations (Equation 75). In other words, the fluctuations due to capillary waves diverge logarithmically, and true specular reflectivity is observed only by virtue of the finite instrumental resolution. The theory for the diffuse scattering from fractal surfaces and other rough surfaces was developed in Sinha et al. (1988).

**Langmuir Monolayers.** A Langmuir monolayer (LM) is a monomolecular amphiphilic film spread at the air-water interface. Each amphiphilic molecule consist of a polar head group (hydrophilic moiety) and a nonpolar tail, typically hydrocarbon (hydrophobic) chains (Gaines, 1966; Swalen et al., 1987; Möhwald, 1990). Typical examples are fatty acids, lipids, alcohols, and others. The length of the hydrocarbon chain can be chemically varied, affecting the hydrophobic character of the molecule. On the other hand, the head group can be ionic, dipolar, or it may have with a certain shape that might attract specific compounds present in the aqueous solution. One important motivation for studying LMs is their close relationship to biological systems. Membranes of all living cells and organelles within cells consist of lipid bilayers interpenetrated with specific proteins, alcohols, and other organic compounds that combine to give functional macromolecules which determine transport of matter and energy through them. It is well known that biological functions are structural, and structures can be determined by XR and GID. In addition, delicate surface chemistry can be carried out at the head-group interface with the molecules in the aqueous solution. From the physics point of view, the LM belongs to an important class of quasi–2-D system, by means of which statistical models that depend on the dimension of the system can be examined.

In this example, results from a simple lipid, dihexadecyl hydrogen phosphate (DHDP), consisting of a phosphate head group ($PO_4^-$) and its two attached hydrocarbon chains, are presented. Figure 10A displays the normalized reflectivity of a DHDP monolayer at the air-water interface at a lateral pressure of 40 mN/m. The corresponding electron density profile is shown in the inset as a solid line. The profile in the absence of surface roughness ($\sigma = 0$) is displayed as a dashed line. The bulk water subphase corresponds to $z < 0$, the phosphate headgroup region is at $0 \geq z \geq 3.4$ Å, and the hydrocarbon tails are at the $3.4$ Å $\geq z \geq 23.1$ Å region. As a first-stage analysis of the reflectivity, a model SLD with minimum number of boxes, $i = 1, 2, 3 \ldots$, is constructed. Each box is characterized by a thickness $d_i$ and an electron density $N_{e,i}$, and one surface roughness, $\sigma$ for all interfaces. Refinement of the reflectivity with Equation 37 shows that the two-box model is sufficient. In order to improve the analysis, we can take advantage of information we know of the monolayer, i.e., the constituents used and the molecular area determined from the lateral-pressure versus molecular area isotherm. If the monolayer is homogeneous and not necessarily ordered, we can assume an average area per

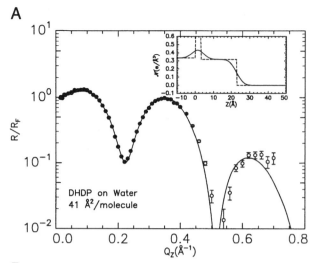

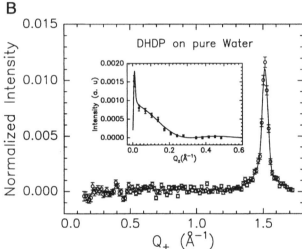

**Figure 10.** (**A**) Normalized x-ray reflectivity from dihexadecyl phosphate (DHDP) monolayer at the air-water interface with best-fit electron density, $N_e$, shown with solid line in the inset. The calculated reflectivity from the best model is shown with a solid line. The dashed line in the inset shows the box model with no roughness $\sigma = 0$. (**B**) A diffraction from the same monolayer showing a prominent 2-D Bragg reflection corresponding to the hexagonal ordering of individual hydrocarbon chains at $Q_\perp^B = 1.516$ Å$^{-1}$. The inset shows a rod scan from the quasi-2D Bragg reflection at $Q_\perp^B$, with a calculated model for tilted chains denoted by solid line (see text for more details).

molecule at the interface $A$, and calculate the electron density of the tail region as follows

$$\rho_{\text{tail}} = N_{e,\text{tail}} \, r_0/(Ad_{\text{tail}}) \qquad (80)$$

where $N_{e,\text{tail}}$ is the number of electrons in the hydrocarbon tail and $d_{\text{tail}}$ is the length of the tail in the monolayer. The advantage of this description is two-fold: the number of independent parameters can be reduced, and constraints on the total number of electrons can be introduced. However, in this case, the simple relation $\rho_{\text{head}} = N_{e,\text{phosphate}}/(Ad_{\text{head}})$ is not satisfactory, and in order to get a reasonable fit, additional electrons are necessary in the head-group region. These additional electrons can be associated

with water molecules that interpenetrate the head-group region, which is not densely packed. The cross-section of the phosphate head group is smaller than the area occupied by the two hydrocarbon tails, allowing for water molecules to penetrate the head group region. We therefore introduce an extra parameter $N_{\text{H}_2\text{O}}$, the number of water molecules with ten electrons each. The electron density of the head group region is given by

$$\rho_{\text{head}} = (N_{e,\text{phosphate}} + 10N_{\text{H}_2\text{O}})/(Ad_{\text{head}}) \qquad (81)$$

This approach gives a physical insight into the chemical constituents at the interface. In modeling the reflectivity with the above assumptions, we can either apply volume constraints or, equivalently, examine the consistency of the model with the literature values of closely packed moieties. In this case the following volume constraint can be applied

$$V_{\text{headgroup}} = A \, d_{\text{head}} = N_{\text{H}_2\text{O}}V_{\text{H}_2\text{O}} + V_{\text{phosphate}} \qquad (82)$$

where $V_{\text{H}_2\text{O}} \approx 30$ Å$^2$ is known from the density of water. The value of $V_{\text{phosphate}}$ determined from the refinement should be consistent within error with known values extracted from crystal structures of salt phosphate (Gregory et al., 1997). The value of $V_{\text{phosphate}}$ determined from the refinement should be consistent within error with known values extracted from crystal structures of salt phosphate (Gregory et al., 1997).

Another parameter that can be deduced from the analysis is the average tilt angle, $t$, of the tails with respect to the surface. For this the following relation is used

$$d_{\text{tail}}/l_{\text{tail}} = \cos t \qquad (83)$$

where $l_{\text{tail}}$ is the full length of the extended alkyl chain evaluated from the crystal data for alkanes (Gregory et al., 1997). Such a relation is valid under the condition that the electron density of the tails when tilted is about the same as that of closely packed hydrocarbon chains in a crystal, $\rho_{\text{tail}} \approx 0.32 \, e/\text{Å}^3 r_0$, as observed by Kjaer et al. (1989). Such a tilt of the hydrocarbon tails would lead to an average increase in the molecular area compared to the cross-section of the hydrocarbon tails ($A_0$)

$$A_0/A = \cos t \qquad (84)$$

Gregory et al. (1997) found that at lateral pressure $\pi = 40$ mN/m, the average tilt angle is very close to zero ($\approx 7 \pm 7°$), and extracted an $A_0 \approx 40.7$ Å$^2$ compared with a value of 39.8 Å$^2$ for closely packed crystalline hydrocarbon chains. The small discrepancy was attributed to defects at domain boundaries.

The GID for the same monolayer is shown in Figure 10B, where a lowest-order Bragg reflection at $1.516$ Å$^{-1}$ is observed. This reflection corresponds to the hexagonal ordering of the individual hydrocarbon chains (Kjaer et al., 1987, 1989) with lattice constant $d = 4.1144$ Å, and molecular area per chain $A_{\text{chain}} = 19.83$ Å$^2$. Note that in DHDP the phosphate group is anchored to a pair of hydrocarbon chains with molecular area $A = 39.66$ Å$^2$, and it is surprising that ordering of the head group with a larger unit cell (twice that of the hydrocarbon unit cell) is not observed, as is evident in Figure 10B. Also shown in the inset of Figure 10B is a rod scan of the Bragg reflection. To model the rod scan in terms of tilted chains, the procedure developed in Kjaer et al. (1989) is followed. The structure factor of the chain can be expressed as

$$F_{\text{chain}}(\mathbf{Q}'_{\perp}, Q'_z) = F(Q'_{\perp}) \frac{\sin(lQ'_z/2)}{(lQ'_z/2)} \qquad (85)$$

where $F(Q'_{\perp})$ is the in-plane Fourier transform of the cross-section of the electron density of chain, weighted with the atomic form factors of the constituents. The second term accounts for the length of the chain, and is basically a Fourier transform of a one-dimensional aperture of length $l$. If the chains are tilted with respect to the surface normal (in the $y - z$ plane) by an angle $t$, the $\mathbf{Q}'$ should be rotated as follows

$$\begin{aligned}
Q'_x &= Q_x \cos t + Q_z \sin t \\
Q'_y &= Q_y \qquad\qquad\qquad\qquad (86) \\
Q'_z &= -Q_x \sin t + Q_z \cos t
\end{aligned}$$

Applying this transformation to the molecular structure factor, Equation 85, and averaging over all six domains (see more details in Kjaer et al., 1989) with the appropriate weights to each tilt direction, we find that at 40 mN/m the hydrocarbon chains are practically normal to the surface, consistent with the analysis of the reflectivity.

In recent Brewster Angle Microscopy (BAM) and x-ray studies of $C_{60}$-propylamine spread at the air-water interface (see more details on fullerene films; Vaknin, 1996), a broad in-plane GID signal was observed (Fukuto et al., 1997). The GID signal was analyzed in terms of a 2-D radial distribution function that implied short-range positional correlations extending to only few molecular distances. It was demonstrated that the local packing of molecules on water is hexagonal, forming a 2-D amorphous solid.

**Surface Crystallization of Liquid Alkanes.** Normal alkanes are linear hydrocarbon chains $(CH_2)_n$ terminating with $CH_3$ groups similar to fatty acids and lipids. The latter compounds, in contrast to alkanes, possess a hydrophilic head group at one end. Recent extensive x-ray studies of pure and mixed liquid alkanes (Wu et al., 1993a,b, 1995; Ocko et al., 1997) reveal rich and remarkable properties near their melting temperature, $T_f$. In particular, a single crystal monolayer is formed at the surface of an isotropic liquid bulk up to $\approx 3°C$ above $T_f$ for a range of hydrocarbon number $n$. The surface freezing phenomenon exists for a wide range of chain lengths, $16 \geq n \geq 50$. The molecules in the ordered layer are hexagonally packed and show three distinct ordered phases: two rotator phases, one with the molecules oriented vertically ($16 \geq n \geq 30$), and the other tilted toward nearest neighbors ($30 \geq n \geq 44$). The third phase ($44 \geq n$) orders with the molecules tilted towards next-nearest neighbors. In addition to the 2-D Bragg reflections observed in the GID studies, reflectivity curves from the same monolayers were found to be consistent with a one-box model of densely packed hydrocarbon chains, and a thickness that corresponds to slightly tilted chains. This is an excellent demonstration of a case where no other technique but the x-ray experiments carried out at a synchrotron could be applied to get the detailed structure of the monolayers. Neutron scattering from this system would have yielded similar information; however, the intensities available today from reactors and spallation sources are smaller by at least a factor of $10^5$ counts/sec for similar resolutions, and will not allow observation of any GID signals above background levels.

**Liquid Metals.** Liquid metals, unlike dielectric liquids, consist of the classical ionic liquid and quantum free-electron gas. Scattering of conduction electrons at a step-like potential (representing the metal-vacuum interface) gives rise to quantum interference effects and leads to oscillations of the electron density across the interface (Lang and Kohn, 1970). This effect is similar to the Friedel oscillations in the screened potential arising from the scattering of conduction electrons by an isolated charge in a metal. By virtue of their mobility, the ions in a liquid metal can in turn rearrange and conform to these oscillations to form layers at the interface, not necessarily commensurate with the conduction electron density (Rice et al., 1986). Such theoretical predictions of atomic layering at surfaces of liquid metals have been known for a long time, and have only recently been confirmed by x-ray reflectivity studies for liquid gallium and liquid mercury (Magnussen et al., 1995; Regan et al., 1995). X-ray reflectivities of these liquids were extended to $Q_z \sim 3\text{Å}^{-1}$, showing a single peak that indicates layering with spacing on the order of atomic diameters. The exponential decay for layer penetration into the bulk of Ga (6.5Å) was found to be larger than that of Hg ($\sim 3$Å). Figure 11 shows a peak in the reflectivity of liquid Ga under in situ UHV oxygen-free surface cleaning (Regan et al., 1995, 1997). The normalized reflectivity was fitted to a model scattering length density shown in Figure 11B, of the following oscillating and exponentially decaying form (Regan et al., 1995)

$$\rho(z)/\rho_s = \text{erf}[(z - z_0)/\sigma] + \theta(z)A \sin (2\pi z/d)e^{-z/\xi} \qquad (87)$$

where $\theta(z)$ is a step function, $d$ is the inter-layer spacing, $\xi$ is the exponential decay length, and $A$ is an amplitude. Fits to this model are shown in Figure 11 with $d = 2.56$ Å, $\xi = 5.8$ Å. The layering phenomena in Ga showed a strong temperature dependence. Although liquid Hg exhibits layering with a different decay length, the reflectivity at small momentum transfers, $Q_z$, is significantly different than that of liquid Ga, indicating fundamental differences in the surface structures of the two metals. The layering phenomena suggest in-plane correlations that might be different than those of the bulk, but had not been observed yet with GID studies.

## SPECIMEN MODIFICATION AND PROBLEMS

In conducting experiments from liquid surfaces, the experimenter faces problems that are commonly encountered in other x-ray techniques. A common nuisance in

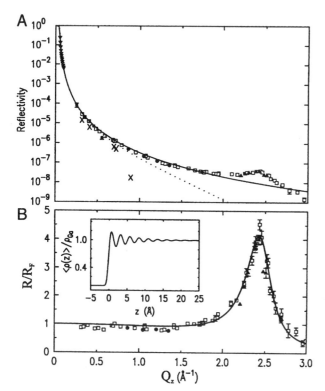

**Figure 11.** (**A**) Measured reflectivity for liquid gallium (Ga). Data marked with X were collected prior to sample cleaning whereas the other symbols correspond to clean surfaces (for details see Regan et al., 1995). Calculated Fresnel reflectivity from liquid Ga surface convoluted with a surface roughness due to capillary waves ($\sigma = 0.82$ Å), and the atomic form factor for Ga is denoted with a solid line. (**B**) The normalized reflectivity, with a solid line that was calculated with the best fit by an exponentially decaying sine model shown in the inset (courtesy of Regan et al., 1995).

an undulator beam line is the presence of high harmonic components that accompany the desired monochromatic beam. These high harmonics can skew the data as they affect the counting efficiency of the detector, can damage the specimen or the detector, and can increase undesirable background. The higher harmonic components can be reduced significantly (but not totally eliminated) by inserting a mirror below the critical angle for total reflection, or by detuning the second crystal of the double-crystal monochromator (XAFS SPECTROSCOPY). However, there are also problems that are more specific to reflectivity and GID in general, and to liquid surfaces in particular. Such problems can arise at any stage of the study, initially during the alignment of the diffractometer, or subsequently during data collection, and also in the final data analysis.

Poor initial alignment of the diffractometer will eventually result in poor data. Accurate tracking, namely small variation (5% to 15%) in the count rate of the monitor as the incident angle $\alpha_i$ is varied (workable range of $\alpha_i$ is 0° to 10°) is a key to getting reliable data. Although seemingly simple, one of the most important factors in getting a good alignment is the accurate determination of relevant distances in the diffractometer. For the specific design in

Figure 7B, these include the distances from the center of the tilting monochromator to the center of the IH elevator and to the sample SH, or the distances from the center of the liquid surface to the detector ($S_4$) and to the elevator (DH). These distances are used to calculate the translations of the three elevators (IH, SH, and DH) for each angle. Although initial values measured directly with a ruler can be used, effective values based on the use of the x-ray beam yield the best tracking of the diffractometer.

Radiation damage to the specimen is a common nuisance when dealing with liquid surfaces. Many of the studies of liquid surfaces and monolayers involve investigations of organic or biomaterials that are susceptible to chemical transformations in general and, in particular, in the presence of the intense synchrotron beam. Radiation damage is of course not unique to monolayers on liquid surfaces; other x-ray techniques that involve organic materials (protein, polymer, liquid crystals, and others) face similar problems. Radiation damage to a specimen proceeds in two steps. First, the specimen or a molecule in its surroundings is ionized (by the photoelectric, Auger, or Compton effects) or excited to higher energy levels (creating radicals). Subsequently, the ionized/ excited product can react with a nearby site of the same molecule or with a neighboring molecule to form a new species, altering the chemistry as well the structure at the surface. In principle, recovery without damage is also possible, but there will be a recovery time involved. The remedies that are proposed here are in part specific to liquid surfaces and cannot be always fulfilled in view of the specific requirements of an experiment. To minimize the primary effects (photoelectric, Auger, and Compton scattering) one can employ one or all of the following remedies. First, the exposure time to the x-ray beam can be minimized. For instance, while motors are still moving to their final positions, the beam can be blocked. Reduced exposure can be also achieved by attenuating the flux on the sample to roughly match it to the expected signal, so that the full intense beam is used only for signals with cross-sections for scattering that require it. That, of course, requires rough knowledge of signal intensity, which is usually the case in x-ray experiments. Monolayers on liquid surfaces are inevitably disposable and it takes a few of them to complete a study, so that in the advanced stage of a study many peak intensities are known. Another approach to reducing the effect of the primary stage is by operating at high x-ray energies. It is well known that the cross-section for all the primary effects is significantly reduced with the increase of x-ray energy. If the experiment does not require a specific energy, such as in resonance studies (anomalous scattering), it is advantageous to operate at high x-ray energies. However, operation at higher x-ray energies introduces technical difficulties. Higher mechanical angular resolutions, and smaller slits, are required in order to achieve reciprocal space resolutions comparable to those at lower energies. As discussed above, slit $S_2$ at 16 keV has to be set at about 50 μm width, which cannot be reliably achieved with variable slits. Finally, reducing the high harmonic component in the x-ray beam will also reduce the primary effect of the radiation.

Reducing the effects of secondary processes also depends on the requirements of a specific experiment. Air surrounding the sample has probably the worst effect on the integrity of an organic film at the liquid interface. The x-ray radiation readily creates potent radicals in air (such as monatomic oxygen) that are highly diffusive and penetrant and can interact with almost any site of an organic molecule. Working in an He environment or under vacuum can significantly reduce this source of radiation damage. Another approach to consider for reducing the secondary effect of radiation is scattering at low temperatures. It is well documented in protein crystallography that radiation damage is significantly reduced at liquid nitrogen temperatures. However, such low temperatures are not a choice when dealing with aqueous surfaces, and variations in temperature can also lead to dramatic structural transition in the films which may alter the objectives of the study.

The liquid substrate, even water, can create temporary radicals that can damage the monolayer: in particular, the head group region of lipids. Water under intense x-ray radiation can give many reactive products such as $H_2O_2$ or monatomic oxygen that can readily interact with the monolayer. Thus, some radiation damage, with an extent that may vary from sample to sample, is inevitable, and fresh samples are required to complete a study. Moving the sample underneath the footprint is a quick fix in that regard, assuming that the radiation damage is mostly localized around the illuminated area. To accomplish that, one can introduce a translation of the trough (perpendicular to the incident beam direction at $\alpha_i = 0$) to probe different parts of the surface. It should be noted that for lipid monolayers, a common observation suggests that radiation damage is much more severe at lower in-plane densities than for closely packed monolayers.

Another serious problem concerning scattering from surfaces is background radiation that can give count rates comparable to those expected from GID or rod scan signals. Background can be classified into two groups, one due to room background and the other due to the sample and its immediate surroundings. Although it is very hard to locate the sources of room background, it is important to trace and block them, as they reduce the capability of the diffractometer. The specimen can give unavoidable background signal due to diffuse or incoherent scattering from the liquid substrate, and this needs to be accounted for in the interpretation of the data. An important source of background is the very immediate environment of the liquid surface that does not include the sample but is included in the scattering volume of the beam: worst of all is air. Working under vacuum is not an option with monolayers, and therefore such samples are kept under an He environment. Air scattering in the trough can give rise to background levels that are at least two or three orders of magnitude higher than the expected signal from a typical 2-D Bragg reflection in the GID.

As discussed previously (see Data Analysis and Initial Interpretation), reflectivity data can give ambiguous SLD values, although that rarely happens. More often, however, the reflectivity is overinterpreted, with details in the SLD that cannot be supported by the data. The reflectivity from aqueous surfaces can be at best measured up to a momentum transfer $Q_z$ of $\sim 1$ Å$^{-1}$, which, roughly speaking, in an objective reconstruction of the SLD, should give uncertainties on the order of $2\pi/Q_z \sim 6$ Å. It is only by virtue of complementary knowledge on the constituents of the monolayer that these uncertainties can be lowered to about 1 to 2 Å. Another potential pitfall for overinterpretation lies in the fact that in the GID experiments only a few peaks are observed, and without the knowledge on the 3-D packing of the constituents it is difficult to interpret the data unequivocally.

## ACKNOWLEDGMENTS

The author would like to thank Prof. P. S. Pershan for providing a copy of Fig. 11 for this publication. Ames Laboratory is operated by Iowa State University for the U.S. Department of Energy under Contract No. W-7405-Eng-82. The work at Ames was supported by the Director for Energy Research, Office of Basic Energy Sciences.

## LITERATURE CITED

Als-Nielsen, J. and Kjaer, K. 1989. X-ray reflectivity and diffraction studies of liquid surfaces and surfactant monolayers. *In* Phase Transitions in Soft Condensed Matter (T. Riste and D. Sherrington, eds.).. pp. 113–138. Plenum Press, New York.

Als-Nielsen, J. and Pershan, P. S. 1983. Synchrotron X-ray diffraction study of liquid surfaces. *Nucl. Instrum. Methods* 208:545–548.

Azzam, R. M. A. and Bashara, N. M. 1977. Ellipsometry and Polarized Light. North-Holland Publishing, New York.

Bevington, P. R. 1968. Data Reduction and Error Analysis. McGraw-Hill, New York.

Binnig, G. 1992. Force microscopy. *Ultramicroscopy* 42–44:7–15.

Binnig, G. and Rohrer, H., 1983. Scanning tunneling microscopy. *Surf. Sci.* 126:236–244.

Born, M. and Wolf, E., 1959. Principles of Optics. MacMillan, New York.

Bouwman, W. G. and Pedersen, J. S., 1996. Resolution function for two-axis specular neutron reflectivity. *J. Appl. Crystallogr.* 28:152–158.

Braslau, A., Pershan, P. S., Swislow, G., Ocko, B. M., and Als-Nielsen, J. 1988. Capillary waves on the surface of simple liquids measured by X-ray reflectivity. *Phys. Rev. A* 38:2457–2469.

Buff, F. P., Lovett, R. A., and Stillinger, Jr., F. H. 1965. Interfacial density profile for fluids at the critical region. *Phys. Rev. Lett.* 15:621–623.

Clinton, W. L. 1993. Phase determination in X-ray and neutron reflectivity using logarithmic dispersion relations. *Phys. Rev. B* 48:1–5.

Ducharme, D., Max, J.-J., Salesse, C., and Leblanc, R. M. 1990. Ellipsometric study of the physical states of phosphotadylcholines at the air-water interface. *J. Phys. Chem.* 94:1925–1932.

Dutta, P., Peng, J. B., Lin, B., Ketterson, J. B., Prakash, M. P., Georgopoulos, P., and Ehrlich, S. 1987. X-ray diffraction studies of organic monolayers on the surface of water. *Phys. Rev. Lett.* 58:2228–2231.

Evans, R. 1979. The nature of the liquid-vapor interface and other topics in the statistical mechanics of non-uniform, classical fluids. *Adv. Phys.* 28:143–200.

Feidenhas'l, R. 1989. Surface structure determination by X-ray diffraction. *Surf. Sci. Rep.* 10:105–188.

Fukuto, M., Penanen, K., Heilman, R. K., Pershan, P. S., and Vaknin, D. 1997. $C_{60}$-propylamine adduct monolayers at the gas/water interface: Brewster angle microscopy and x-ray scattering study. *J. Chem. Phys.* 107:5531.

Gaines, G. 1966. Insoluble Monolayers at Liquid Gas Interface. John Wiley & Sons, New York.

Gregory, B. W., Vaknin, D., Gray, J. D., Ocko, B. M., Stroeve, P., Cotton, T. M., and Struve, W. S. 1997. Two-dimensional pigment monolayer assemblies for light harvesting applications: Structural characterization at the air/water interface with X-ray specular reflectivity and on solid substrates by optical absorption spectroscopy. *J. Phys. Chem. B* 101:2006–2019.

Henon, S. and Meunier, J. 1991. Microscope at the Brewster angle: Direct observation of first-order phase transitions in monolayers. *Rev. Sci. Instrum.* 62:936–939.

Hönig, D. and Möbius, D. 1991. Direct visualization of monolayers at the air-water interface by Brewster angle microscopy. *J. Phys. Chem.* 95:4590–4592.

Kjaer, K., Als-Nielsen, J., Helm, C. A., Laxhuber, L. A., and Möhwald, H. 1987. Ordering in lipid monolayers studied by synchrotron x-ray diffraction and fluorescence microscopy. *Phys. Rev. Lett.* 58:2224–2227.

Kjaer, K., Als-Nielsen, J., Helm, C. A., Tippman-Krayer, P., and Möhwald, H. 1989. Synchrotron X-ray diffraction and reflection studies of arachidic acid monolayers at the air-water interface. *J. Phys. Chem.* 93:3202–3206.

Kjaer, K., Als-Nielsen, J., Lahav, M., and Leiserowitz, L. 1994. Two-dimensional crystallography of amphiphilic molecules at the air-water interface. *In* Neutron and Synchrotron Radiation for Condensed Matter Studies, Vol. III (J. Baruchel, J.-L. Hodeau, M. S. Lehmann, J.-R. Regnard, and C. Schlenker, eds.). pp. 47–69. Springer-Verlag, Heidelberg, Germany.

Landau, L. D. and Lifshitz, E. M. 1980. Statistical Physics. p. 435. Pergamon Press, Elmsford, New York.

Lang, N. D. and Kohn, W. 1970. Theory of metal surfaces: Charge density and surface energy. *Phys. Rev. B1* 4555–4568.

Lekner, J. 1987. Theory of Reflection of Electromagnetic Waves and Particles. Martinus Nijhoff, Zoetermeer, The Netherlands.

Lösche, M. and Möhwald, H. 1984. Fluorescence microscope to observe dynamical processes in monomolecular layers at the air/water interface. *Rev. Sci. Instrum.* 55:1968–1972.

Lösche, M., Piepenstock, M., Diederich, A., Grünewald, T., Kjaer, K., and Vaknin, D. 1993. Influence of surface chemistry on the structural organization of monomolecular protein layers adsorbed to functionalized aqueous interfaces. *Biophys. J.* 65:2160–2177.

Lurio, L. B., Rabedeau, T. A., Pershan, P. S., Silvera, I. S., Deutsch, M., Kosowsky, S. D., and Ocko, B. M., 1992. Liquid-vapor density profile of helium: An X-ray study. *Phys. Rev. Lett.* 68:2628–2631.

Magnussen, O. M., Ocko, B. M., Regan, M. J., Penanen, K., Pershan, P. S., and Deutsch, M. 1995. X-ray reflectivity measurements of surface layering in mercury. *Phys. Rev. Lett.* 74 4444–4447.

Möhwald, H. 1990. Phospholipid and phospholipid-protein monolayers at the air/water interface. *Annu. Rev. Phys. Chem.* 41:441–476.

Ocko, B. M., Wu, X. Z., Sirota, E. B., Sinha, S. K., and Deutsch, M., 1994. X-ray reflectivity study of thermal capillary waves on liquid surfaces. *Phys. Rev. Lett.* 72:242–245.

Ocko, B. M., Wu, X. Z., Sirota, E. B., Sinha, S. K., Gang, O., and Deutsch, M., 1997. Surface freezing in chain molecules: I. Normal alkanes. *Phys. Rev. E* 55:3166–3181.

Panofsky, W. K. H. and Phillips, M. 1962. Classical Electricity and Magnetism. Addisson-Wesley, Reading, Mass.

Parratt, L. G. 1954. Surface studies of solids by total reflection of X-rays. *Phys. Rev.* 95:359–369.

Pedersen, J. S. 1992. Model-independent determination of the surface scattering-length-density profile from specular reflectivity data. *J. Appl. Crystallogr.* 25:129–145.

Penfold, J. and Thomas, R. K., 1990. The application of the specular reflection of neutrons to the study of surfaces and interfaces. *J. Phys. Condens. Matter* 2:1369–1412.

Pershan, P. S., Braslau, A., Weiss, A. H., and Als-Nielsen, 1987. Free surface of liquid crystals in the nematic phase. *Phys. Rev. A* 35:4800–4813.

Regan, M. J., Kawamoto, E. H., Lee, S., Pershan, P. S., Maskil, N., Deutsch, M., Magnussen, O. M., Ocko, B. M., and Berman, L. E. 1995. Surface layering in liquid gallium: An X-ray reflectivity study. *Phys. Rev. Lett.* 75:2498–2501.

Regan, M. J., Pershan, P. S., Magnussen, O. M., Ocko, B. M., Deutch, M., and Berman, L. E. 1997. X-ray reflectivity studies of liquid metal and alloy surfaces. *Phys. Rev. B* 55:15874–15884.

Rice, S. A., Gryko, J., and Mohanty, U. 1986. Structure and properties of the liquid-vapor interface of a simple metal. *In* Fluid Interfacial Phenomena (C. A. Croxton, ed.) pp. 255–342. John Wiley & Sons, New York.

Rodberg, L. S. and Thaler R. M. 1967. Introduction to the Quantum Theory of Scattering. Academic Press, New York.

Russell, T. P. 1990. X-ray and neutron reflectivity for the investigation of polymers. *Mater. Sci. Rep.* 5:171–271.

Sacks, P. 1993. Reconstruction of step like potentials. *Wave Motion* 18:21–30.

Sanyal, M. K., Sinha, S. K., Huang, K. G., and Ocko, B. M. 1991. X-ray scattering study of capillary-wave fluctuations at a liquid surface. *Phys. Rev. Lett.* 66:628–631.

Sanyal, M. K., Sinha, S. K., Gibaud, A., Huang, K. G., Carvalho, B. L., Rafailovich, M., Sokolov, J., Zhao, X., and Zhao, W. 1992. Fourier reconstruction of the density profiles of thin films using anomalous X-ray reflectivity. *Europhys. Lett.* 21:691–695.

Schiff, L. I. 1968. Quantum Mechanics, McGraw-Hill, New York.

Schwartz, D. K., Schlossman, M. L., and Pershan, P. S. 1992. Re-entrant appearance of the phases in a relaxed Langmuir monolayer of the tetracosanoic acid as determined by X-ray scattering. *J. Chem. Phys.* 96:2356–2370.

Sinha, S. K., Sirota, E. B., Garof, S., and Stanely, H. B. 1988. X-ray and neutron scattering from rough surfaces. *Phys. Rev. B* 38:2297–2311.

Swalen, J. D., Allra, D. L., Andrade, J. D., Chandross, E. A., Garof, S., Israelachvilli, J., Murray, R., Pease, R. F., Wynne, K. J., and Yu, H. 1987. Molecular monolayers and films. *Langmuir* 3:932–950.

Vaknin, D., Als-Nielsen, J., Piepenstock, M., and Lösche, M. 1991a. Recognition processes at a functionalized lipid surface observed with molecular resolution. *Biophys. J.* 60:1545–1552.

Vaknin, D., Kjaer, K., Als-Nielsen J., and Lösche, M. 1991b. A new liquid surface neutron reflectometer and its application to the study of DPPC in a monolayer at the air/water interface. *Makromol. Chem. Macromol. Symp.* 46:383–388.

Vaknin, D., Kjaer, K., Als-Nielsen, J., and Lösche, M. 1991c. Structural properties of phosphotidylcholine in a monolayer at the air/water interface: Neutron reflectivity study and reexamination of X-ray reflection experiments. *Biophys. J.* 59:1325–1332.

Vaknin, D., Kjaer, K., Ringsdorf, H., Blankenburg, R., Piepenstock, M. Diederich, A., and Lösche, M. 1993. X-ray and neutron reflectivity studies of a protein monolayer adsorbed to a functionalized aqueous surface. *Langmuir* 59:1171–1174.

Vaknin, D. 1996. $C_{60}$-amine adducts at the air-water interface: A new class of Langmuir monolayers. *Phys. B* 221:152–158.

Vineyard, G. 1982. Grazing-incidence diffraction and the distorted-wave approximation for the study of surfaces. *Phys. Rev. B* 26:4146–4159.

Wilson, A. J. C. (eds.). 1992. International Tables For Crystallography Volume C. Kluwer Academic Publishers, Boston.

Wu, X. Z., Ocko, B. M., Sirota, E. B., Sinha, S. K., Deutsch, M., Cao, B. H., and Kim, M. W. 1993a. Surface tension measurements of surface freezing in liquid normal-alkanes. *Science* 261:1018–1021.

Wu, X. Z., Sirota, E. B., Sinha, S. K., Ocko, B. M., and Deutsch, M. 1993b. Surface crystallization of liquid normal-alkanes. *Phys. Rev. Lett.* 70:958–961.

Wu, X. Z., Ocko, B. M., Tang, H., Sirota, E. B., Sinha, S. K., and Deutsch, M. 1995. Surface freezing in binary mixtures of alkanes: New phases and phase transitions. *Phys. Rev. Lett.* 75:1332–1335.

Zhou, X. L. and Chen, S. H. 1995. Theoretical foundation of X-ray and neutron reflectometry. *Phys. Rep.* 257:223–348.

## KEY REFERENCES

Als-Neilsen and Kjaer, 1989. See above.

*An excellent, self-contained, and intuitive review of x-ray reflectivity and GID from liquid surfaces and Langmuir monolayers by pioneers in the field of liquid surfaces.*

Braslau et al., 1988. See above.

*The first comprehensive review examining the properties of the liquid-vapor interface of simple liquids (water, carbon tetrachloride, and methanol) employing the reflectivity technique. The paper provides many rigorous derivations such as the Born approximation, the height-height correlation function of the surface, and surface roughness due to capillary waves. Discussion of practical aspects regarding resolution function of the diffractometer and convolution of the resolution with the reflectivity signal is also provided.*

Sinha et al., 1988. See above.

*This seminal paper deals with the diffuse scattering from a variety of rough surfaces. It also provides a detailed account of the diffuse scattering in terms of the distored-wave Born approximation (DWBA). It is relevant to liquid as well as to solid surfaces.*

## APPENDIX:
## P-POLARIZED X-RAY BEAM

A p-polarized x-ray beam has a magnetic field component that is parallel to the stratified medium (along the $x$ axis, see Fig. 1), and straightforward derivation of the wave equation (Equation 7) yields

$$\frac{d}{dz}\left(\frac{dB}{\varepsilon dz}\right) + [k_z^2 - V(z)]B = 0 \tag{88}$$

By introducing a dilation variable $Z$ such that $dZ = \varepsilon dz$, Equation 88 for $B$ can be transformed to a form similar to Equation 12

$$\frac{d^2 B}{dZ^2} = \left[\frac{k_z^2 - V(z)}{\varepsilon}\right]B = 0 \tag{89}$$

The solution of Equation 89 for an ideally flat interface in terms of $r_p(k_{z,s})$ and $t_p(k_{z,s})$ is then simply given by

$$r_p(k_{z,s}) = \frac{k_{z,0} - k_{z,s}/\varepsilon}{k_{z,0} + k_{z,s}/\varepsilon}, \quad t_p(k_z, s) = \frac{2k_{z,0}}{k_{z,0} + k_{z,s}/\varepsilon} \tag{90}$$

The critical momentum transfer for total external reflectivity of the p-type x-ray beam is $Q_c = 2k_c = \sqrt{4\pi\rho_s}$, identical to the one derived for the s-type wave. Also, for $2k_z^B \gg Q_z \gg Q_c$, ($k_z^B$ is defined below), $R_F(Q_z)$ can be approximated as

$$R_F(Q_z) \sim \left(\frac{Q_c}{2Q_z}\right)^4 \left(\frac{2}{1+\varepsilon}\right)^2 \tag{91}$$

The factor on the right hand side is equal to 1 for all practical liquids, and thus the Born approximation is basically the same as for the s-polarized x-ray beam (Equation 17). The main difference between the s-type and p-type waves occurs at larger angles near a Brewster angle that is given by $\theta_B = \sin^{-1}(k_z^B/k_0)$. At this angle, total transmission of the p-type wave occurs ($r_p(k_z, s) = 0$). Using Equations 14 and 90, $k_z^B$ can be derived

$$\frac{k_z^B}{k_0} = \frac{1}{\sqrt{2 - 4\pi\rho_s/k_0^2}} \tag{92}$$

The Brewster angle for x-rays is then given by $\theta_B = \sin^{-1}(k_z^B/k_0) \approx \pi/4$. This derivation is valid for solid surfaces, including crystals, where the total transmission effect of the p-polarized wave at a Bragg reflection is used to produce polarized and monochromatic x-ray beams.

DAVID VAKNIN
Iowa State University
Ames, Iowa

# ELECTRON TECHNIQUES

## INTRODUCTION

This part describes how electrons are used to probe the microstructures of materials. These methods are arguably the most powerful and flexible set of tools available for materials characterization. For the characterization of structures internal to materials, electron beam methods provide capabilities for determining crystal structure, crystal shapes and orientations, defects within the crystals, and the distribution of atoms within these individual crystals. For characterizing surfaces, electron methods can determine structure and chemistry at the level of the atomic monolayer.

The methods in this part use electrons with kinetic energies spanning the range from 1 to 1,000,000 eV. The low energy range is the domain of scanning tunneling microscopy (STM). Since STM originates with quantum mechanical tunneling of electrons through potential barriers, this method differs from the others in this part, which employ electron beams incident on the material. Progressively higher energies are used for the electron beams of low energy electron diffraction (LEED), Auger spectrometry, reflection high energy electron diffraction (RHEED), scanning electron microscopy (SEM), and transmission electron microscopy (TEM). Electron penetration through the solid increases strongly over this broad energy range. STM, Auger, and LEED techniques are used for probing the monolayer of surface atoms (although variants of STM can probe sub-surface structures). On the other hand, the penetration capability of high energy electrons makes TEM primarily a bulk technique. Nevertheless, even with TEM it is often unrealistic to study samples having thicknesses much greater than a fraction of a micron.

Since electron beams can be focused with high accuracy, the incident beam can take various forms. A tightly focused beam, rastered across the sample, is one useful form. The acquisition of various signals in synchronization with this raster pattern provides a spatial map of the signal. Signals can be locations of characteristic x-ray emission, secondary-electron emission, or electron energy losses, to name but three. A plane wave is another useful form of the incident electron beam. Plane wave illumination is typically used for image formation with conventional microscopy and diffraction. It turns out that for complementary optical designs, the same diffraction effects in the specimen that produce visible features in images will occur with either a point- or plane-illumination mode. There are, however, advantages to one mode of operation versus another, and instruments of the scanning type and imaging type are typically used for different types of measurements. Atomic resolution imaging, for example, can be performed with a plane wave illumination method in high resolution electron microscopy (HREM), or with a probe mode in Z-contrast imaging. A fundamental difference between these techniques is that the diffracted waves interfere coherently in the case of HREM imaging, and the phase information of the scattering is used to make interference patterns in the image. On the other hand, the operation of the scanning transmission electron microscope for Z-contrast imaging makes use of incoherent imaging, where the intensities of the scatterings from the individual atoms add independently to the image.

Electron scattering from a material can be either elastic or inelastic. In general, elastic scattering is used for imaging measurements, whereas inelastic scattering provides for spectroscopy. (Electron beam techniques are enormously versatile, however, so there are many exceptions to this generality.) Chemical mapping, for example, makes images by use of inelastic scattering. For chemical mapping, the energy loss of the electron itself may be used to obtain the chemical information, as in electron energy loss spectrometry (EELS). Alternatively, the subsequent radiations from atoms ionized by the inelastic scattering may be as signals containing the chemical information. A component of the characteristic atomic x-ray spectrum, measured with an energy dispersive x-ray spectrometer (EDS), is a particularly useful signal for making chemical maps with a scanning electron microscope. The intensity of selected Auger electrons emitted from the excited atoms provides the chemical mapping capability of scanning Auger microscopy. Auger maps are highly surface-sensitive, since the Auger electrons lose energy as they traverse only a short distance through the solid.

The chapters in this part enumerate some requirements for sample preparation, but it is remarkable that any one of these techniques permits studies on a large variety of samples. All classes of materials (metals, ceramics, polymers, and composites) have been studied by all of these electron techniques. Except for some problems with the lightest (or heaviest) elements in the periodic table, the electron beam techniques have few compositional limitations. There are sometimes problems with how the sample may contaminate the vacuum system that is integral to all electron beam methods. This is especially true for the surface techniques that employ ultrahigh vacuums.

When approaching a new problem in materials characterization, electron beam methods (especially SEM), provide some of the best reward/effort ratios of any materials characterization technique. Many, if not all, of the techniques in this part should therefore be familiar to everyone interested in materials characterization. Some of the methods are available through commercial laboratories on a routine basis, such as SEM, EDS and Auger spectroscopy. TEM services can sometimes be found at companies offering asbestos analysis. Manufacturers of electron beam instruments may also provide analysis services at reasonable rates. All the electron methods of this part are available at materials research laboratories at universities and national laboratories. Preliminary assessments of the applicability of an electron method to a materials problem can often be made by contacting someone at a local institution. These institutions usually have established policies for fees and service for outside users. For value in capital investment, perhaps a scanning electron microscope with an energy dispersive spectrometer is a best buy. The surface techniques are more specialized, and the expense and maintenance of their ultrahigh vacuum systems can be formidable. TEM is also a technique that cannot be approached casually, and some contact with an expert in the field is usually the best way to begin.

BRENT FULTZ

# SCANNING ELECTRON MICROSCOPY

## INTRODUCTION

The scanning electron microscope (SEM) is one of the most widely used instruments in materials research laboratories and is common in various forms in fabrication plants. Scanning electron microscopy is central to microstructural analysis and therefore important to any investigation relating to the processing, properties, and behavior of materials that involves their microstructure. The SEM provides information relating to topographical features, morphology, phase distribution, compositional differences, crystal structure, crystal orientation, and the presence and location of electrical defects. The SEM is also capable of determining elemental composition of micro-volumes with the addition of an x-ray or electron spectrometer (see ENERGY-DISPERSIVE SPECTROMETRY and AUGER ELECTRON SPECTROSCOPY) and phase identification through analysis of electron diffraction patterns (see LOW-ENERGY ELECTRON DIFFRACTION). The strength of the SEM lies in its inherent versatility due to the multiple signals generated, simple image formation process, wide magnification range, and excellent depth of field.

Lenses in the SEM are not a part of the image formation system but are used to demagnify and focus the electron beam onto the sample surface. This gives rise to two of the major benefits of the SEM: range of magnification and depth of field in the image. Depth of field is that property of SEM images where surfaces at different distances from the lens appear in focus, giving the image three-dimensional information. The SEM has more than 300 times the depth of field of the light microscope. Another important advantage of the SEM over the optical microscope is its high resolution. Resolution of 1 nm is now achievable from an SEM with a field emission (FE) electron gun. Magnification is a function of the scanning system rather than the lenses, and therefore a surface in focus can be imaged at a wide range of magnifications from $3\times$ up to $150,000\times$. The higher magnifications of the SEM are rivaled only by the transmission electron microscope (TEM) (see TRANSMISSION ELECTRON MICROSCOPY and SCANNING TRANSMISSION ELECTRON MICROSCOPY: Z-CONTRAST IMAGING), which requires the electrons to penetrate through the entire thickness of the sample. As a consequence, TEM sample preparation of bulk materials is tedious and time consuming, compared to the ease of SEM sample preparation, and may damage the microstructure. The information content of the SEM and TEM images is different,

with the TEM image showing the internal structure of the material.

Due to these unique features, SEM images frequently appear not only in the scientific literature but also in the daily newspapers and popular magazines. The SEM is relatively easy to operate and affordable and allows for multiple operation modes, corresponding to the collection of different signals. The following sections review the SEM instrumentation and principles, its capabilities and applications, and recent trends and developments.

## PRINCIPLES OF THE METHOD

### Signal Generation

The SEM electron beam is a focused probe of electrons accelerated to moderately high energy and positioned onto the sample by electromagnetic fields. The SEM optical column is utilized to ensure that the incoming electrons are of similar energy and trajectory. These beam electrons interact with atoms in the specimen by a variety of mechanisms when they impinge on a point on the surface of the specimen. For inelastic interactions, energy is transferred to the sample from the beam, while elastic interactions are defined by a change in trajectory of the beam electrons without loss of energy. Since electrons normally undergo multiple interactions, the inelastic and elastic interactions result in the beam electrons spreading out into the material (changing trajectory from the original focused probe) and losing energy. This simultaneous energy loss and change in trajectory produces an interaction volume within the bulk (Fig. 1). The size of this interaction volume can be estimated by Monte Carlo simulations, which incorporate probabilities of the multiple possible elastic and inelastic interactions into a calculation of electron trajectories within the specimen.

The signals resulting from these interactions (e.g., electrons and photons) will each have different depths within the sample from which they can escape due to their unique

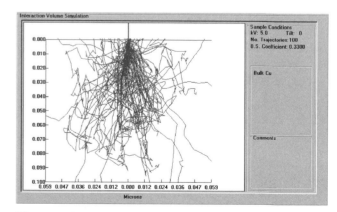

**Figure 1.** A Monte Carlo simulation of electron beam interaction with a bulk copper target at 5 keV shows the interaction volume within the specimen. The electron trajectories are shown, and the volume is symmetric about the beam due to the normal incidence angle (derived from Goldstein et al., 1992; Monte Carlo simulation using Electron Flight Simulator, Small World, D. Chernoff).

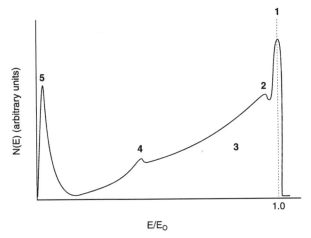

**Figure 2.** Distribution of relative energies ($E/E_0$; $E_0$ = incident beam energy) of electrons ejected from a surface by an incident electron beam (not to scale). The peak (1) is the elastic peak, BSEs that have lost no energy in the specimen. Slightly lower in energy is the plasmon loss region (2). The BSEs (3) are spread over the entire energy range from 0 eV to the energy of the incident beam. The characteristic Auger electron peak (4) is usually a small peak superimposed on the backscattered curve. Secondary electrons emitted from specimen atoms are responsible for the large peak at low energy (5).

physical properties and energies. For example, a secondary electron (SE) is a low-energy (2- to 5-eV) electron ejected from the outer shell of a sample atom after an inelastic interaction. These low-energy electrons can escape the surface only if generated near the surface. Thus we have an "interaction volume" in which beam electrons interact with the specimen and a "sampling volume" from which a given signal escapes the solid, which is some fraction of the interaction volume. It is this sampling volume and the signal distribution within it that determine the spatial resolution of the technique (Joy, 1984). We can thus expect different spatial resolutions for the various types of signals generated in the SEM.

Backscattered electrons (BSEs) are electrons from the incident probe that undergo elastic interactions with the sample, change trajectory, and escape the sample. These make up the majority of electrons emitted from the specimen at high beam voltage, and their average energy is much higher than that of the SEs (Fig. 2). The depth from which BSEs escape the specimen is dependent upon the beam energy and the specimen composition, but >90% generally escape from less than one-fifth the beam penetration depth. The intensity of the BSE signal is a function of the average atomic number ($Z$) of the specimen, with heavier elements (higher $Z$ samples) producing more BSEs. It is thus a useful signal for generating compositional images, in which higher $Z$ phases appear brighter than lower $Z$ phases. The BSE intensity and trajectory are also dependent upon the angle of incidence between the beam and the specimen surface. The topography, or physical features of the surface, is then imaged by using these properties of the BSE signal to generate BSE topographic images. Due to the relatively high energy of the

BSE signal, the sampling volume (sample depth) is greater than that of SEs.

Secondary electrons are due to inelastic interactions, are low energy (typically 2 to 5 eV), and are influenced more by surface properties than by atomic number. The SE is emitted from an outer shell of a specimen atom upon impact of the incident electron beam. The term "secondary" thus refers to the fact that this signal is not a scattered portion of the probe, but a signal generated within the specimen due to the transfer of energy from the beam to the specimen. In practice, SEs are arbitrarily defined as those electrons with <50 eV energy (Fig. 2). The depth from which SEs escape the specimen is generally between 5 and 50 nm due to their low energy. Secondary electrons are generated by both the beam entering the specimen and BSEs as they escape the specimen; however, SE generation is concentrated around the initial probe diameter. The SE sampling volume is therefore smaller than the BSE and provides for a high-resolution image of the specimen surface. Secondary electron intensity is a function of the surface orientation with respect to the beam and the SE detector and thus produces an image of the specimen morphology. The SE intensity is also influenced by chemical bonding, charging effects, and BSE intensity, since BSE-generated SEs are a significant part of the SE signal.

Characteristic x rays and Auger electrons are generated by inelastic interactions of the probe electrons in which an inner shell electron is emitted from a specimen atom (ionization). Following the creation of a hole in an inner shell, the atom relaxes by a transition in which an outer shell electron fills the inner shell (relaxation). Since the outer shell electron was at a higher energy level, this relaxation results in a release of energy from the atom (emission). This emission may be in the form of a photon (x ray) or may go into the emission of an electron (the Auger electron, AUGER ELECTRON SPECTROSCOPY) from the atom. These two processes therefore compete, since a single ionization can result in either an x ray or an Auger electron. The x-ray energy would then equal the difference in binding energy of the two electron levels within the atom and is thus characteristic of that element. The Auger electron generated from this ionization is emitted from an outer shell. It would thus have a kinetic energy equal to that of the x ray minus the binding energy of the outer shell from which it is emitted. The Auger sample depth (1 to 3 nm) is much less than the x-ray sample depth (0.1 to 100 μm) since electrons have a higher probability of interaction with the specimen than photons (x rays) of similar energy, and energy loss interactions will render them non-characteristic. X rays may interact with the specimen through absorption, in which the entire energy is absorbed by the specimen, but do not undergo partial energy loss interactions. These signals can be used for imaging as well as the identification of elements within the sample volume (see ENERGY-DISPERSIVE SPECTROMETRY and AUGER ELECTRON SPECTROSCOPY). It is common for the SEM in a materials laboratory to include an x-ray spectrometer as one of the detectors, and scanning Auger microanalysis combines the SEM with an electron spectrometer. Electron energy loss spectroscopy is a transmission electron technique, since multiple energy loss interactions occur for a single beam electron in the bulk samples characteristic of SEM and deconvolution of these energy losses is not possible.

Other signals originating from electron beam interactions with the sample include plasmons, phonons, bremsstrahlung radiation (noncharacteristic x rays), and cathodoluminescence. Cathodoluminescence is the emission of light from a material under the electron beam. The wavelength of the light is a function of the material's band gap energy, since the generation involves the production and recombination of electron-hole pairs. Cathodoluminescence is therefore useful in the characterization of semiconductors, especially in cases where a contactless, high-spatial-resolution method is needed (Yacobi and Holt, 1986).

The sample may also absorb a significant portion of the electron beam current, and the specimen current image can thus provide the same information as the combined BSE and SE signals (Newbury, 1976). Charge collection scanning electron microscopy with the specimen as the detector can also be used to study the electron-beam-induced current/ conductivity (EBIC), which has been used to determine carrier lifetime, diffusion length, defect energy levels, and surface recombination velocities and to locate *p-n* junctions and recombination sites (Leamy, 1982).

### Image Formation

Image construction in the SEM is accomplished by mapping intensity of one of these signals (usually SE and/or BSE) from the specimen onto a viewing screen or film (Fig. 3). There is a point-to-point transfer of this intensity information, since the SEM scan generator simultaneously drives an electron beam across the surface of the specimen and an electron beam on the viewing cathode-ray tube (CRT) or recording device. The intensity information from the detector is translated to a gray scale, with higher signal intensity being brighter on the display. The image is therefore an array of pixels, with each pixel defined by $x$ and $y$ spatial coordinates and a gray value proportional to the signal intensity. Magnification of the image is defined as the length of the scan line on the monitor or recording device divided by the length of the scan line on the specimen. Magnification is therefore independent of the lenses. Magnification is adjusted by changing the size of the area scanned on the specimen while the monitor or film size is held constant. Thus a smaller area scanned on the sample will produce a higher magnification. This is a major strength of SEM imaging, and with optimized beam conditions and focus, the image magnification can be changed through its entire range without loss of image quality. A typical magnification range for the SEM is $10\times$ to $100,000\times$; however, the higher magnifications are dependent upon acquiring sufficient resolution.

The region on the specimen from which information is transferred to a single pixel of the image is called a picture element. The size of the picture element is determined by the length of the scan on the specimen divided by the number of pixels in a line of the image. Picture element therefore refers to a position on the specimen, and pixel refers to

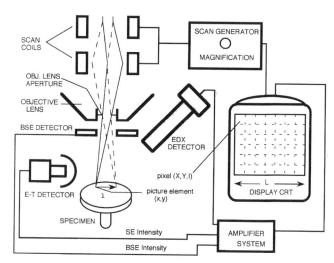

**Figure 3.** Schematic of the image formation system of the SEM. The electron beam is shown at two positions in time, scanned in the $X$ direction to define the distance $l$ on the specimen surface. Reproduction of this scan on the CRT, with a width $L$, gives a magnification $M = L/l$. The picture element is an area on the specimen surface, its location given by $(x, y)$. Signal from this area (SE, BSE, x ray, or other) is transferred to the CRT and converted to an intensity value for pixel $(X, Y, I)$. This simple image formation process is highly versatile in that the magnification is only a function of the scan length on the sample, and any signal may be used to provide the intensity variation, thus producing an image. (E-T, Everhart-Thornley.)

the corresponding portion of the recorded image. The intensity of the signal is recorded for a single pixel of the image while the beam is addressed to the center of a single picture element. Consider a digital image that is 10 cm on a side and $100\times$ magnification. The image represents an area on the sample that is 1 mm (1000 µm) wide, and the scale for the image is 1 cm = 100 µm. The picture element width is 1000 µm (the length of the scan on the specimen) divided by the number of pixels in the digital image. A common digital resolution for the SEM is $1024 \times 1024$, and thus the picture element width for the $100\times$ image would be 0.98 µm. The picture element width would be 0.098 µm at $1000\times$ and 0.0098 µm (9.8 nm) at $10,000\times$. The SEM image will appear in focus if the sampling volume is smaller in diameter than this picture element size. If the probe diameter or resultant sampling volume is larger than the picture element, blurring of the image occurs due to overlap of information from neighboring picture elements. This results in loss of resolution. When magnification is increased, the picture element size is reduced, and this overlap of information can result. The sampling volume is therefore a limiting factor on resolution for the SEM.

The SEM image conveys three-dimensional information due to the depth of field in the image. The depth of field depends upon the electron beam divergence angle, which is defined by the diameter of the objective lens aperture and the distance between the specimen and the aperture (Fig. 4). This angle is exceptionally small compared to that in a light microscope and results in a small change

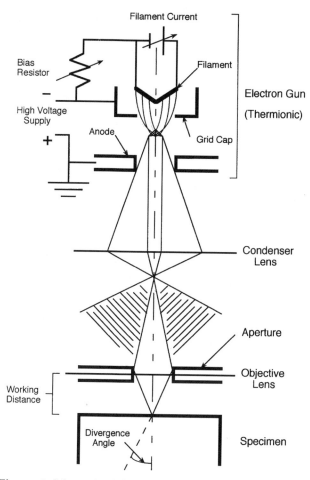

**Figure 4.** Schematic of the probe-forming system of the SEM. The thermionic electron gun is a triode system, with the initial crossover of electrons from the filament due to the field created by the filament, grid cap, and anode. This initial beam diameter is reduced by the condenser and objective lenses. A portion of the beam (shaded in the schematic) will not pass through the final aperture. Increasing the strength of the condenser lens will decrease the probe diameter but will also decrease the probe current. The objective lens will produce a smaller probe diameter when the working distance is reduced (the lens is stronger) and provide higher resolution. A longer working distance will reduce the divergence angle and therefore increase depth of field in the image.

## Resolution

The resolving power of the SEM is dependent upon the sampling volume, since this sampling volume is the portion of the specimen from which the signal originates. It was stated under Image Formation above that the image appears in focus if the sampling volume is smaller than the picture element. This is also a statement of the resolution; if the sampling volume is smaller than the feature of interest, then the feature can be resolved, or distinguished from its surroundings. This simple model of resolution works well for most situations, but for higher magnifications the actual signal distribution within the sampling volume becomes critical, since the signal density is not uniform across the sampling volume. The sampling volume is a function of the probe (energy and diameter) and the specimen (composition and orientation). Resolution is therefore not an instrument constant, and it varies significantly with application. Microscope manufacturers quote a resolution value that is determined under ideal conditions on a specimen produced specifically for the purpose of testing high resolution. These values are used to characterize the instrument rather than to predict results expected from a given specimen. It is the job of the microscopist to prepare the sample and control the instrument in such a way as to provide the optimum or maximum resolution.

Modern SEMs with field emission electron guns (FESEMs) are capable of resolutions near 1 nm on appropriate samples using the SE signal. Tungsten filament instruments can routinely obtain resolutions of 3 to 4 nm on appropriate specimens. Samples that require special conditions, such as low voltage or low dosage, and samples that are inherently poor signal generators due to their composition may fall far short of this resolution. The signal from these specimens can often be enhanced by depositing a thin film of metal such as chromium or gold onto the surface. This increases contrast in the SE signal and allows for higher resolution imaging. Operating conditions that favor one signal, for instance, the high beam currents and energies used for x-ray analysis, can also reduce image resolution. High-resolution BSE images can be obtained by selecting only high-energy BSEs (low-energy-loss BSEs), since these have undergone few interactions with the sample and are thus from a smaller sampling volume.

## PRACTICAL ASPECTS OF THE METHOD

### SEM Instrumentation

The production of an electron probe requires an electron gun to emit and accelerate electrons and a series of lenses to de-magnify and focus the beam (Fig. 4). "Scanning" involves positioning of the probe on the specimen surface by electromagnetic (or electrostatic) deflection and simultaneously positioning a beam that carries signal intensity information onto a viewing device (camera or monitor). When the two beams are scanned in a regular pattern covering a rectangular area (raster), the signal output onto the viewing monitor becomes a map of signal intensity coming from the sample, and this "map" is the SEM image.

in diameter of the beam with change in distance (depth) from the lens. This means that the sampling volume can likewise remain small over this range of depth. All points will appear in focus where the sampling volume diameter is smaller than the picture element. Depth of field is increased by reducing the beam divergence angle, which can be done by increasing the working distance or decreasing the aperture diameter. The excellent depth of field in SEM images is as important to the usefulness of scanning electron microscopy as the resolution and provides the topographic/morphological information well known to anyone who has admired any SEM image. Depth of field is typically between 10% and 60% of the field width.

This can be done in an analog or digital manner. Scanning a small area on the sample and mapping the resultant signal intensity information onto a larger viewing surface result in magnification.

Three types of electron guns are in common use today: tungsten and lanthanum hexaboride thermionic electron guns, and the field emission gun. Tungsten and lanthanum hexaboride are used as filaments in thermionic electron guns, in which the filament (cathode) is heated by applying a current. Emission of electrons occurs at high temperatures and requires that the gun be operated in a high vacuum to avoid oxidation. The emitted electrons are accelerated to high energy by the application of a negative voltage to the cathode and a bias voltage to the grid cap with the anode grounded. This triode system accelerates the electrons and provides an initial focusing into a probe. Tungsten is popular for its low cost, ease of replacement, and high current yield, while lanthanum hexaboride provides a brighter source (higher current density) useful for higher resolution. In the field emission gun, a fine tip is placed near a grid and electrons escape the surface due to the high extracting field strength created by the first grid. A second grid provides the accelerating voltage. Field emission requires ultrahigh vacuum and is more expensive, but it provides much higher brightness with a resultant increase in resolution.

Scanning, positioning, and focusing of the electron beam after it leaves the electron gun is accomplished by the use of electromagnetic or electrostatic fields. A simple electromagnetic lens requires only a coil of wire wrapped concentrically around the optic axis. A current through this coil produces a field that deflects electrons toward the center of the lens. This field focuses the electrons into a smaller area, thus reducing the diameter of the electron beam produced by the electron gun. Lenses in modern instruments are far from simple, utilizing shroud and pole piece designs that maximize lens strength while minimizing lens aberrations (spherical, chromatic, and diffraction; see Goldstein et al., 1992).

The "lenses" in the SEM all combine to produce a fine probe diameter. The condenser lens (often a system of several lenses adjusted by a single control) is used as the primary means of reducing the beam diameter produced by the electron gun. The condenser lens is usually directly below the electron gun and above a series of apertures necessary to define a small probe diameter (Fig. 4). Increasing the strength of the condenser lens will decrease the probe diameter but will also decrease the probe current due to interception of part of the beam by these apertures. This decrease in probe current with reduction of the probe diameter is a serious limitation in the SEM, since signal intensity is a fraction of probe intensity. High resolution requires a small probe diameter as well as a signal of high enough intensity that the differences in intensity are within the detector's range. It is this limitation that makes the brightness of the electron gun such an important consideration. The control knob for the condenser lens system may be labeled "spot size," "beam current," "probe diameter," or various other similar descriptive terms referring to its dual but interconnected effect on the beam.

The SEM objective lens is in basic respects no different than the condenser lens, in that it reduces and focuses the probe. Its position in the instrument, however, is such that its primary purpose is to reduce the probe to the smallest diameter at the surface of the sample (Fig. 4), which results in the clearest, or "focused," image. In the SEM there is no image formed behind the lenses as in the optical microscope. The lenses act on the electron beam rather than on the image. The objective lens aperture, often called the probe-forming aperture, defines the convergence (divergence) angle of the probe, which is important for the depth of field. Adjustment of the objective lens does not have a large effect on beam current as adjustment of the condenser lens does.

Electromagnetic or electrostatic fields that are not concentric with the optic axis will deflect the electron beam. This is utilized to center the beam onto the optic axis (alignment) and to scan or position the beam on the sample surface. Alignment coils are commonly utilized to position the electron beam coming from the electron gun along the optic axis of the instrument. These coils may also be used to "blank" the beam by directing it away from the sample toward a grounded aperture. The scan coils are usually located near or within the objective lens assembly and are used for many of the operator-controlled functions of the SEM. The scan coils control the position of the beam on the sample and are used for scanning of the beam to produce an image, determining magnification of the image, electronic shifting of the imaged area, and positioning a probe for x-ray analysis, among other functions. A final component of the optical system, the stigmator, utilizes paired coils arranged around the optic axis to eliminate astigmatism. Astigmatism arises due to irregularities in the electron-lens fields, which can be caused by aperture irregularities or contamination. The operator corrects for this irregularity by providing an equal but opposite irregularity with the coils of the stigmator. This operation should be done in conjunction with focusing, since it is necessary for producing a small, symmetric probe diameter on the sample surface.

### Contrast and Detectors

Microstructure for most solids can be imaged in the SEM with minimum specimen preparation, although interpretation of the microstructure can be aided by careful attention to specimen preparation, such as the polishing and etching of metals (Fig. 5). The SEM offers a choice of detectors and operating parameters that can be taken advantage of to produce a variety of images showing microstructure. Contrast in the SEM image is the difference in intensity, or brightness, of the pixels that make up the image. The difference in intensity represents a difference in signal from corresponding picture elements on the specimen. This signal intensity is dependent upon the signal generated within the specimen and the type of detector used. It is therefore necessary to recognize both the beam-specimen interaction that produces the signal and the detection and amplification process that converts the signal intensity to image brightness. Contrast components of the signal electrons—their number, trajectory,

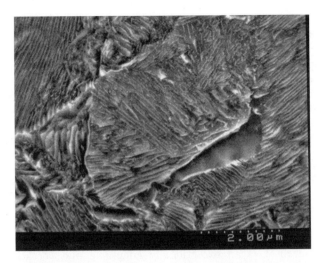

**Figure 6.** Secondary electron image of the internal structure of a dove eggshell. Note the bright edges of the fibers (collagen) and the differing brightness from crystal faces of the calcium carbonate.

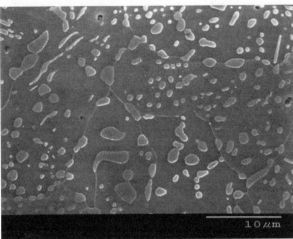

**Figure 5.** Microstructures of two steel samples of identical chemical composition (0.97% carbon by weight) but with different cooling and annealing histories. Scanning electron microscopy reveals the characteristic structure that makes the steel in the top image extremely hard and brittle, while the steel in the bottom image can be used for springs and is comparatively soft (Killick, 1996).

and energy—are a useful means to describe and understand how this information from the specimen is translated to an intensity value on the image. The number component is expressed by higher numbers of electrons from the specimen resulting in higher brightness on the image. Trajectory is equally important, since detectors have limited size and shape and therefore detect only a portion of the signal emitted. A high signal generated in the specimen that is directed away from the detector will result in a low detected signal. The energy component is also critical, since the ability to collect and detect the signal is a function of the signal energy. The combination of these three components (number, trajectory, and energy) creates a unique situation for each probe-specimen-detector combination. The most widely used images produced by SEM imaging are the SE image, the BSE compositional image, and the BSE topographic image. It is important to note that many images will have a

combination of contrast components that contribute to the final image and that the description here has summarized only the most common images.

The most popular and useful detector for observing surfaces and morphology is the Everhart-Thornley (ET) detector, composed of a scintillator, collector, light pipe, and photomultiplier tube. Electrons striking the positively biased scintillator are converted to a burst of photons that travel through the light pipe to the photomultiplier tube, where the signal is amplified. The collector is a grid or collar around the scintillator that is positively biased to draw the low-energy SEs toward the detector. The image produced from the ET detector with a positively biased collector is dominated by SE contrast, and this detector is often referred to as the SE detector. The SE image is dominated by surface features due to the shallow escape depth of SEs and the dependence of the SE signal on the incidence angle of the probe with the specimen (Fig. 6). Surfaces tilted away from the normal to the beam allow for more SE escape. A noticeable feature of the SE image is that edges appear bright. This is due to the escape of SEs from the top and side surfaces when the interaction volume intercepts these two surfaces. A surface facing away from the detector can still appear illuminated, since some portion of the SEs can be detected due to the bias on the collector. The traditional SEM instrument has an ET detector within the specimen chamber. The ET detector can also be placed within the optical column of the SEM, and with an alteration to the lens design, the SE signal can be trapped within the field of the objective lens and collected by the biased detector. The SE image from the ET detector, often coupled with enhancement of the signal by deposition of metal thin films on the specimen, results in high-resolution SEM images.

Common BSE detectors are the solid-state detector and the scintillator detector. These detectors are not responsive to low-energy electrons and therefore exclude the SE signal. Both detectors are designed such that they can be placed above the specimen, where the geometric collection

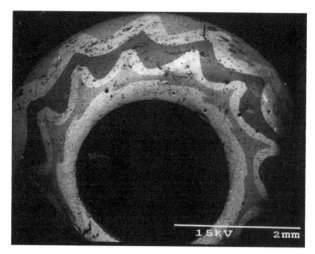

**Figure 7.** Backscattered electron image of a glass trade bead. The bands are different colored glasses. In this image, contrast is due to the different intensities of BSEs, which are due to differences in the average atomic number of the glasses. X-ray analysis coupled with this image provides a complete characterization of the specimen.

efficiency can be optimized to collect a high proportion of the BSE signal. The solid-state detector is based on the principle that high-energy electrons striking the silicon produce electron-hole pairs in the material. A bias on the detector separates the electrons and holes, and the resultant charge can be collected. The intensity of the charge is proportional to the energy of the incoming electrons (in silicon it takes $\sim$3.6 eV to produce one electron-hole pair) and the number of electrons striking the detector. The solid-state detector is usually an array of detectors that are symmetrically placed around the optic axis, and the signals can be mixed in additive or subtractive modes to enhance different contrast components. The scintillator detector relies on the interaction of BSEs with the detector to produce photons, similar to the ET detector. Unlike the ET detector, however, SEs are not detected due to their lower energy. The geometry or position of the detector itself can be altered to select for contrast components.

Both of these BSE detectors can be operated in two modes. In the first mode, placing the BSE detector above the specimen and symmetrically around the optic axis will eliminate the trajectory component of contrast by collecting all trajectories equally. This produces the BSE compositional image, in which the "number" component of contrast predominates and the signal intensity increases with increasing average atomic number of the sample (Fig. 7). This imaging strategy works best with flat, polished surfaces where information on the distribution of phases is easily interpreted. In the second mode, the trajectory of the signal can be accentuated by nonsymmetric detector geometry. This is accomplished in the solid-state detector by subtracting the signal from one part of the array from the other or by differential weighting of the signals from different sides of the specimen chamber. The scintillator detector can be partially removed from its position directly above the sample to create this same

enhancement of trajectory differences. The resulting BSE topographic image shows in high contrast minor irregularities on the surface. Figure 7 exhibits this topographic contrast, with surface pitting shown in sharp relief, superimposed upon a background of compositional contrast. Since the BSE trajectories are essentially straight due to their high energy, this image also shows "shadowing" where material between the detector and a surface region will block the signal reaching the detector. The ET detector can also be used to produce this BSE topographic image, but its small size and position away from the specimen leads to poor collection efficiency and therefore a noisy image.

## Special Techniques and Recent Innovations

The SEM is an exceptionally adaptable and versatile instrument, and many special application areas have been developed. Its unequaled depth of field can be enhanced by stereographic techniques, in which two images of the same area are recorded at different specimen tilt angles (Gabriel, 1985). The resulting stereopair can be viewed with the aid of filters or a stereo viewer to produce a three-dimensional image. Measurements of distances in three dimensions are possible if the initial geometry is recorded, and software programs are available to simplify the task. Metrology, particularly the measurement of critical dimensions in microelectronic devices, is possible with refined SEM scanning systems and appropriate standards and control. The SEM, by nature, is a bulk analysis technique, and many *in situ* experiments can be devised. Electronic testing in the SEM has the benefit of precise placement of the probe (beam), control of the beam voltage and/or current, and avoidance of mechanical damage and impedance loading associated with mechanical probes (Menzel and Buchanan, 1987). Physical testing of materials within the SEM chamber is also possible, with heating, cooling, strain, and fracture stages custom built for the microscope.

The diffraction of electrons by a crystalline sample can lead to electron channeling patterns in the SEM image (Joy et al., 1982; Newbury et al., 1986) or Kikuchi patterns in the BSE distribution (Adams et al., 1993; Michael and Goehner, 1993; Thaveeprungsriporn et al., 1994; Goehner and Michael, 1996). Electron diffraction techniques in the SEM can be used to identify phases, determine their orientation, and study crystal damage and defects (Dingley, 1981).

Variable-pressure SEM is a technique in which the specimen chamber pressure is controlled independently of the SEM column (Mathieu, 1996). The introduction of air (or other gases) into the specimen chamber neutralizes charging on the surface of insulating materials, allowing them to be imaged and analyzed (with decreased resolution) without having to deposit a conducting thin film (Fig. 8). The higher chamber pressures also allow for the observation of materials with high vapor pressures, including biological material, without extensive sample preparation to remove volatile components. A special form of variable-pressure scanning electron microscopy is referred to as environmental scanning electron microscopy, in which

**Figure 8.** Variable-pressure SEM image of an abrasive. The abrasive is composed of an epoxy resin, silicon nitride particles, and ceramic balloons. The ability to control the pressure in the chamber of the SEM (40 Pa for this image) allows for operation in an atmosphere capable of eliminating charge buildup on the specimen. This specimen is also a problem for conventional SEM due to outgassing of the epoxy and ceramic balloons. (Image by David Johnson, University of Arizona.)

chamber atmosphere can also be altered (Danilatos, 1988, 1993).

The FESEM deserves special mention due to its enormous impact on the field. The field emission (FE) electron gun allows for the creation of an exceptionally bright (small with high-current-density) electron beam. This in itself has made scanning electron microscopy competitive with transmission electron microscopy for materials characterization in the nanometer range (Fig. 9). The FE source is often coupled with a special lens design such

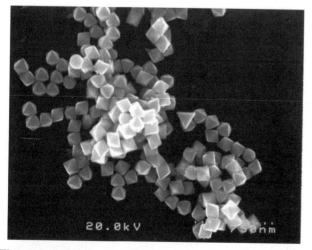

**Figure 9.** Field emission SEM image of gold particles in dichromic glass taken at 400× magnification. The image is at the upper range of magnification for an SEM with a thermionic electron gun but a normal range for the FESEM. The particle geometry and size are apparent, and also some irregularities on the surfaces and where two particles join. (Image by Toshinori Kokubu and G. Chandler, University of Arizona.)

**Figure 10.** Low-voltage image of PZT on platinum. The low accelerating voltage used here (2 kV) gives excellent surface contrast of this thin material. (Image by Mike Orr and G. Chandler, University of Arizona.)

that the specimen is in the field of the lens and the detector is within the column rather than the sample chamber. The FESEM also has the ability to produce a small probe diameter at low voltage, opening the way to many applications that were difficult with thermionic electron guns (Greenhut and Friel, 1997). Low-voltage scanning electron microscopy has the advantage of revealing more details of the surface since there is less penetration of the beam into the specimen (Fig. 10) and is less damaging to the material. The SE yield from the specimen is a function of voltage, and at lower beam voltages the specimen reaches a point at which emission of secondary and backscattered electrons equals the beam current. This is a unique voltage for each material at which there is no specimen charging, and the surface can be imaged without a conductive coating. This phenomenon has also been used as a voltage contrast mechanism in materials such as copolymers and ceramic composites that were traditionally difficult to observe with the SEM.

## METHOD AUTOMATION

The SEM used as a research tool typically requires the operator to make many value judgments during a viewing session. There are many situations, however, where automated routines are useful. Instruments with automatic electron gun start-up, automatic focus and astigmatism correction, and automatic contrast and brightness may be the choice for situations in which repetitive sampling is necessary, such as in quality control. Systematic sampling, in which a motorized stage positions the specimen below the beam or the beam is moved to positions on the specimen, is available and useful for precise, repetitive analysis. Automatic search routines are available that can identify points of interest with the SEM image and return to those areas for compositional analysis with a spectrometer. The SEM is commonly fitted with automated sample handling in industrial settings such as microelectronic fabrication plants. Instrument manufacturers offer extensive options for automated specimen preparation, handling, manipulation, sampling, instrument operation, and data collection, presentation, storage, processing, and analysis.

## DATA ANALYSIS AND INITIAL INTERPRETATION

Interpretation of SEM images is dependent upon the geometry of the probe, specimen, and detector. The specimen appears to the viewer as if looking down the optic axis, with the illumination coming from the detector. This is most clearly seen in the BSE topographic image but is also important for the SE image. The most common imaging mode utilizes SEs for information on the surface morphology (Fig. 6). This SE image is often an extremely complex convolution of contrast-producing factors but is amazing for its ease of interpretation. Surfaces at different angles to the probe produce different brightness levels and edges appear brighter. "Holes" are darker while raised areas are brighter. The image generally looks like a black-and-white optical view of the surface.

The BSE signal can also be used to show surface topography but is perhaps more often used to show atomic number contrast, in which areas of higher average atomic number produce more BSEs, and thus appear brighter, than lower atomic number regions. Figure 7 shows this contrast clearly on a glass trade bead. The bead was produced by wrapping layers of different colored glass around a core, forming the pattern seen in the image. The contrast in this image is not due to color differences but is produced by different intensities of electrons backscattering from the layers with different average atomic number.

The SE image is sensitive to changes in potential on (or near) the surface of the specimen. This can create problems when viewing insulators, but it can also be used to enhance features of the specimen. In Figure 11, the probe energy was set to create a buildup of electrons within an insulating layer below the surface of the specimen. This charged area then emits more SEs than surrounding, grounded material and appears brighter. The same example can be given to demonstrate that features below the surface can contribute to the image contrast. The operator can control the depth that the beam penetrates by changing the energy of the electron probe through changing voltage.

Specimens with multiple phases or grains, twinning, and other microstructure not associated with composition may exhibit contrast due to causes such as differences in crystal orientation or magnetic domain. These contrast mechanisms are superimposed upon the SE or BSE contrast, and careful attention to sample preparation and instrument set-up is necessary for interpretation (Newbury et al., 1986).

Imaging with the SEM has benefited greatly from the incorporation of digital techniques. Multiple low-dosage frames can be averaged to produce a low-noise image on specimens that are sensitive to electron dosage. Digital filtering techniques bring out details in the signals, which can vary over many orders of magnitude, and are not

**Figure 11.** Secondary electron image of microelectronic device. The accelerating voltage of the electron beam was set such that a large portion of the beam electrons were absorbed within an insulating material below the surface. The resulting buildup of charge creates a potential that causes an increased emission of SEs, thus providing a way to highlight the structure. (Image by Ward Lyman, University of Arizona.)

visible when converted directly to a visible gray scale as in conventional photography. The independent collection and storage of multiple signals (SE, BSE, and x ray) from a digital array can greatly enhance the productivity and utility of the SEM and allows for postsampling manipulation and study of the results. Digital images from the SEM can be analyzed with numerous commercially available and free-ware software products.

## SAMPLE PREPARATION

Surface topology, such as in fractography, may require no sample preparation if the specimen can withstand the low chamber pressure and electron beam bombardment of the SEM. The microstructure may be greatly enhanced, however, by such methods as polishing and selective etching of the surface, as with the metals in Figure 5. Most SEMs have sample chambers with limited dimensions, and the specimen must be affixed to a stage holder for orientation and manipulation within the chamber. Conductive adhesives with low vapor pressure or mechanical devices are used to mount the specimen. Special chambers and stages are available to suit most needs.

The most common form of sample preparation for the SEM is the deposition of a metal thin film onto the specimen surface. Vacuum evaporation and ion sputtering of metals are common methods of depositing these thin films (Goldstein et al., 1992). The metal thin film provides electrical conductivity, enhances the signal (if a higher-$Z$ metal is used), and may add strength to the specimen. If x-ray spectroscopy is coupled with scanning electron microscopy, the metal film may be replaced with a carbon thin film or deposited to a thickness that is not detectable by the spectrometer. The deposition of very thin (1-nm) films of metal may be necessary to obtain the highest resolution from an FESEM.

Materials with volatile or high-vapor-pressure components, such as biological specimens, must be prepared for viewing in the SEM. Sample preparation is often tedious and extensive, and special protocols are required to preserve the material's structure while eliminating the unwanted components. Elimination of water from biological specimens, for example, may employ freeze drying, critical point drying, or chemical means. Cryo-microscopy, in which a liquid nitrogen- or liquid helium-cooled stage is employed, is an option for especially challenging materials. Polymers often require some protection from the beam, and staining techniques similar to those employed for biological materials can increase contrast in the image.

Specimens such as living organisms, museum pieces, or very large objects can be surface replicated using polymer films or dental casting resins if they cannot be placed within the SEM chamber. These materials will closely replicate the surface features of the specimen and, after hardening, can be coated with a metal thin film for SEM viewing.

Extensive expertise on sample preparation is available through the Microscopy Society of America Listserver (see Internet Resources).

## SPECIMEN MODIFICATION

Scanning electron microscopes vary widely in the specimen chamber pressure but are traditionally in the high-vacuum range due to requirements of the electron optics. This limits materials that can be analyzed using the SEM; however, extensive sample preparation protocols have been developed for this purpose. The variable-pressure and environmental SEMs are designed to allow operation at higher pressures (still vacuum) and can therefore eliminate some sample preparation.

The electron beam can deposit significant energy within the specimen, and damage may occur due to heating, buildup of electrical charge, or breaking of bonds. Sample preparation, in particular coating with thin metal films, can eliminate many of these problems. Electron beam lithography in integrated circuit production is an example of where this damage is used to produce a pattern in a thin polymer film. Controlled scanning of the beam over the surface produces a pattern within the film. It is then exposed to a mild etch, so that the regions "damaged" by the electron beam etch at a different rate than regions not exposed, and the pattern becomes a mask for subsequent deposition procedures.

Specimens observed in the SEM often become contaminated with a thin film of carbon deposited on the area being scanned with the beam. This is often explained as being due to oil vapor from the vacuum system entering the specimen chamber. This will occur even in oil-free systems if the specimen itself has carbon compounds on its surface. Under electron beam bombardment, these molecules are broken down into constituent species, and while oxygen and nitrogen are removed by the vacuum system, the carbon remains on the surface. Migration of carbon compounds toward the area under electron beam bombardment creates a buildup of carbon.

## PROBLEMS

The sense of perspective described above (see Data Analysis and Interpretation) can be critical for specimens with simple surface geometry, since topography can appear inverted to some observers if the image is rotated 180°. Three-dimensional images produced by viewing stereo-pairs also require that the operator control this geometry of probe, specimen, and detector and in addition record sample tilt and tilt axis alignment.

Image quality in the SEM usually refers to the amount of noise, or graininess, of the image. There is a trade-off in SEM imaging between resolution and image quality in which higher resolution is only obtained by accepting noisier images. This situation is due to the SEM optics described earlier (see Practical Aspects of the Method), in which a stronger condenser lens results in a smaller probe diameter and a decreased probe current. The decrease in probe current will result in a decrease in signal being generated. Noise is inherent in any electronic system, and decreasing the signal will decrease the signal-to-noise ratio. Thus, when one decreases the probe diameter to obtain higher resolution, the signal must be amplified more, and the noise is apparent in the image. This results in grainy images. The use of higher-brightness electron guns, especially the FE sources, makes dramatic improvements in the image quality at higher magnifications by increasing the current density (brightness) of the probe.

The scan length on the specimen can be calculated from two measurements and used to provide an estimate of the magnification. Working distance is the distance between the specimen and the objective lens. Assuming the image is in focus, the current supplied to the objective lens can be sampled to estimate the working distance. A similar reading of the scanning coil current will give an estimate of the scan deflection angle. The length of the scan line on the sample can then be calculated from this angle and the working distance. The magnification readout on most instruments is a calculated value derived from these two measurements, and instruments vary significantly in their accuracy. This problem can easily be overcome by the use of standards to calibrate magnification. Images are also subject to distortions due to irregular scanning, variations in the working distance due to sample topography or tilt, and the projection of the scan (angular) onto a flat surface (film). The projection distortion lessens as magnification increases (scan angle decreases), and distortions are normally apparent only when viewing highly symmetrical samples such as grids. Use of the SEM for critical dimension measurement therefore requires special instruments and techniques.

Traditionally, SEM images were produced at high voltages, where probe electrons could build up a significant charge on insulating materials. This surface charge can manifest itself in many ways in the image. A negatively charged surface usually appears brighter due to increased emission of SEs from the region. The positioning of the beam on the surface may be altered by a nonuniform charge, creating a distorted image. Beam energy is altered by specimen bias from charge buildup and will affect imaging and spectroscopy. Particles on the specimen surface that collect charge can produce a local field that deflects the SE signal away from the detector, resulting in a "shadowed" region. Sporadic discharge of electrons from a charged surface may create bright dots and lines on the image that are not associated with the sample topography. Due to these many detrimental effects of charging, specimens are often coated with thin films of heavy metals (or carbon if x-ray analysis is desired) to provide a conductive path to ground. Newer instruments, particularly the FES-EM, have the ability to produce images at low accelerating voltages. Specimen charging can be reduced or eliminated with control of the accelerating voltage, and positive charging is possible when the intensity of secondary and backscattered electron emission exceeds the probe current. Constructive use of sample charging is used to create contrast in materials such as ceramics and microelectronics.

Contamination on the surface of the specimen plays a critical role in SE images, since the signal is influenced most by the surface characteristics. The most troublesome contamination is the buildup of carbon on the surface. Carbon compounds from oil vapor in the vacuum pumps, exposure to atmosphere, specimen handling, and solvents used to clean the specimen can migrate toward the area that the beam is striking. The high-energy electrons from the beam can break bonds within these materials, producing a carbon buildup on the surface and gases that are easily removed by the vacuum pumps. It is common to see this carbon buildup on the surface of the specimen while viewing the live image. This contaminant layer obscures and blurs the surface detail, and the image becomes darker, since much of the signal is within this low-$Z$ material on the surface. Contamination becomes worse at lower voltages. Careful attention to sample preparation and handling, rapid signal collection, and the use of clean vacuum techniques (see GENERAL VACCUM TECHNIQUES) will combat contamination problems.

## LITERATURE CITED

Adams, B. L., Wright, S. I., and Kunze, K. 1993. Orientation imaging: The emergence of a new microscopy. *Met. Trans.* 24A:819–831.

Danilatos, G. D. 1988. Foundation of environmental scanning electron microscopy. *Adv. Electron. Electron Phys.* 71:109–250.

Danilatos, G. D. 1993. Introduction to the ESEM instrument. *Microsc. Res. Tech.* 25:354–361.

Dingley, D. J. 1981. A comparison of diffraction techniques for the SEM. *SEM/1981* 4:273–286.

Gabriel, B. L. 1985. SEM: A Users Manual for Materials Science. American Society for Metals, Menlo Park, Ohio.

Goehner, R. P. and Michael, J. R. 1996. Phase identification in a scanning electron microscope using backscattered electron Kikuchi patterns. *J. Res. NIST* 101:301-308.

Goldstein, J. I., Newbury, D. E., Echlin, P., Joy, D. C., Romig, A. D., Lyman, C. E., Fiori, C., and Lifshin, E. 1992. Scanning Electron Microscopy and X-Ray Microanalysis. Plenum, New York.

Greenhut, V. A. and Friel, J. J. 1997. Application of advanced field emission scanning electron microscopy to ceramic materials. *USA Microsc. Anal.* March:15–17.

Joy, D. C. 1984. Beam interactions, contrast, and resolution in the SEM. *J. Microsc.* 136:241–258.

Joy, D. C., Newbury, D. E., and Davidson, D. L. 1982. Electron channeling patterns in the scanning electron microscope. *J. Appl. Phys.* 53(8):R81–R119.

Killick, D. 1996. Optical and electron microscopy in material culture studies. *In* Learning from Things (W. D. Kingery, ed.). pp. 204–230. Smithsonian Institution Press, Washington.

Leamy, H. J. 1982. Charge collection scanning electron microscopy. *J. Appl. Phys.* 53(6):R51–R80.

Mathieu, C. 1996. Principles and applications of the variable pressure scanning electron microscope. *USA Microsc. Anal.* Sept.:15–16.

Menzel, E. and Buchanan, R. 1987. Noncontact testing of integrated circuits using an electron beam probe. *J. SPIE* 795:188–200.

Michael, J. R. and Goehner, R. P. 1993. Crystallographic phase identification in the scanning electron microscope: Backscattered electron Kikuchi patterns imaged with a CCD-based detector. *MSA Bull.* 23:168–175.

Newbury, D. 1976. The utility of specimen current imaging in the SEM. *SEM/1976* 1:111–120.

Newbury, D. E., Joy, D. C., Echlin, P., Fiori, C. E., and Goldstein, J. I. 1986. Advanced Scanning Electron Microscopy and X-Ray Microanalysis. Plenum, New York.

Thaveeprungsriporn, V., Mansfield, J. F., and Was, G. S. 1994. Development of an economical backscattering diffraction system for an environmental scanning electron microscope. *J. Mater. Res.* 9:1887–1894.

Yacobi, B. G. and Holt, D. B. 1986. Cathodoluminescence scanning electron microscopy of semiconductors. *J. Appl. Phys.* 59(4): R1–R24.

## KEY REFERENCES

Goldstein et al., 1992. See above.

*Provides a very complete coverage of basics and written for a wide audience. It includes chapters on instrumentation, basic imaging principles, electron optics, and sample preparation. The authors have extensive experience in teaching electron microscopy to thousands of students and have addressed the novice as well as providing more advanced sections.*

Newbury et al., 1986. See above.

*A companion volume to the previous textbook and provides in-depth coverage of specific topics. Topics include modeling of electron beam-sample interactions, microcharacterization of semiconductors, electron channeling contrast, magnetic contrast, computer-aided imaging, specimen coating, biological specimen preparation, cryomicroscopy, and alternative microanalytical techniques.*

Lyman, C. E., Newbury, D. E., Goldstein, J. I., Williams, D. B., Romig, A. D., Armstrong, J. T., Echlin, P., Fiori, C. E., Joy, D. C., Lifschin, E., and Peters, K.-R. 1990. Scanning Electron Microscopy, X-Ray Microanalysis, and Analytical Electron Microscopy. A Laboratory Workbook. Plenum, New York.

*A detailed laboratory workbook with questions, problems, and solutions to the problems given for each of the exercises.*

*Although written as a guide for a course, this is a helpful reference for those who will be operating the instrument, as it provides detailed, step-by-step instructions for obtaining specific results from the instrument. Exercises include basic imaging, measurement of beam parameters, image contrast, stereomicroscopy, BSE imaging, transmitted electron imaging, low-voltage, high-resolution, SE signal components, electron channeling, magnetic contrast, voltage contrast, electron beam-induced current, environmental scanning electron microscopy, and computer-aided imaging.*

## INTERNET RESOURCES

The Microscopy and Microanalysis WWW server site: http://www.amc.anl.gov/

*This site is hosted by Argonne National Laboratory and is a complete guide to the Internet for microscopy. The site is also home of the Microscopy Society of America listserver, providing a forum for novice and expert microscopists to ask questions of their colleagues. Other links at this site include the MSA software library, ftp sites, a list of upcoming meetings, conferences and workshops, and lists of national and international www sites. This provides access to educational, government, commercial, and society resources.*

## APPENDIX: GUIDELINES FOR SELECTING AN SEM

The selection of a SEM can be quite involved, with ten companies offering new instruments, and some of these having as many as 15 models from which to choose. The market is highly competitive, and a thorough search by the purchaser will be very beneficial. Many options may be available from companies specializing in custom equipment rather than from the instrument manufacturer, and a potential purchaser should also contact these vendors for information. A quick look at the Microscopy Society of America web pages and a few days on their list server should give one a feeling for the variety of options available. Table 1 lists some features that a buyer should check, although it is not possible to include all options, adaptations, add-on systems, and interfaces available. The list of criteria allows for a comparison between microscopes, and the buyer must use some ranking system for the individual items, such as imperative, useful, a nice feature, or unnecessary. A cost analysis should also be conducted, and the many laboratories that offer SEM services are an alternative to purchase. The determination of employee needs, training, maintenance and repair costs, and instrument reliability/lifetime should be a part of consideration to purchase and are not covered here. A list such as the one below must be customized for each individual purchase, and performance of the microscope on real samples should always be the final test. Most manufacturers will arrange for a demonstration of their product, and the buyer should prepare well for this demonstration, including bringing in your most problematic specimens. Some general notes on the major systems of the SEM are included here for background information on the criteria.

## Vacuum System and Specimen Handling

The vacuum system for the SEM is designed to meet the requirements of the electron gun and probe-forming system, while the buyer must decide if the system is compatible with requirements of the specimen. Traditional SEM pressures are on the order of $1 \times 10^{-6}$ torr, with the electron gun and specimen chamber kept at the same pressure. Modern instruments may offer differential pumping systems such that ultrahigh vacuum can be obtained for a field emission source, or low-vacuum "environmental" chambers can be maintained for the specimen chamber, while other portions of the system are held at traditional high-vacuum levels. The cleanliness of the vacuum system and component gases may be equally important to a particular application as the pressure. Oil-free pumping systems, special sample introduction systems, inert gas back-filling, or environmental chambers with control of gas composition are examples of the options available.

The specimen-handling system is closely related to the vacuum system and should therefore be designed/chosen with the specimen and the vacuum in mind. Most instruments are equipped with specimen stages that are adequate for manipulating a specimen within the chamber but will not provide the fine control necessary for many special techniques. Stage options are available for eucentric geometry, large samples, *in situ* experiments, fine movement, cryo-microscopy, nonstandard detection systems, and other special requirements. Introduction of a specimen into the instrument can also be accomplished in a variety of ways, from manually opening the entire chamber to completely automated handling and transfer from one instrument/device to another under, e.g., vacuum or at low temperature.

## Electron Gun

Three electron gun types are commonly available today. The tungsten filament thermionic source is inexpensive, requires little in the way of vacuum systems, produces a stable beam current, and is appropriate for most routine SEM imaging. The lanthanum hexaboride thermionic source offers higher brightness than the tungsten source, with resulting higher resolution. It requires lower operating pressures, and although it has a longer lifetime, replacement cost is high. The field emission electron gun is much more expensive but produces the brightest and smallest probe diameter and thus the highest resolution. The FE electron gun is also the choice for low-voltage applications.

## Imaging System

The manufacturer will usually quote a resolution for the microscope that is a "best case" situation and is normally obtained using the SE signal on a high-contrast, standard specimen. The resolution of an image from the SEM is dependent on the beam parameters (accelerating voltage, beam diameter), the signal and detector used (SE, BSE, x ray), and the characteristics of the specimen (composition, structure, preparation). It is therefore imperative

**Table 1. Some Criteria for Evaluation of a Scanning Electron Microscope**

Vacuum system
Types of vacuum pumps
Chamber operating pressure(s)
Pump-down time/sample exchange time
Contamination rate (on your sample)
Vacuum meter/status indicator light
Safety interlock to high-voltage supply
Valve closure on power failure
Column isolation valve
Gun isolation valve
Differential pumping
Inert gas back-filling of chamber
Pump filters or venting system
Sample introduction system/airlock
Chamber/stage
   Maximum specimen size
   Range of movement ($X,Y,Z$)
   Range of rotation and tilt
   Eucentric point or points
   *In situ* capabilities
Vacuum feed-through for external controls
   Location and number of ports
   Energy dispersive spectrometry (EDS)
      angle/working distance
   Simultaneous BSE/SE/EDS
   Automation
   Vibration isolation
Electron gun and probe-forming system
Type of gun
Brightness
Accelerating voltage range/steps
Minimum probe diameter (specify working distance,
   probe current, accelerating voltage)
Probe current range
Special requirements/maintenance needs
Adjustable grid cap
Adjustable anode
Gun alignment (mechanical/electromagnetic)
Column liner
Column aperture
Aperture alignment (mechanical/electromagnetic)
Current-measuring device
Beam stability (keV and current)
Imaging system
   Image quality (your sample)
Magnification range/steps
Scan speeds (view/photo)
Digital resolution (view/photo/store)
Video out/peripheral devices
Waveform monitor for photo settings
Filament image
Channeling patterns
Viewing screen(s)
Scanning mode options
Electromagnetic image shift
Spot mode
Selected area
Line scan
Dual image/split screen
Dual magnification

Scan rotation
Image data display
Micrometer marker/ranges and format
Data entry (text on image)
Automatic features (manual override available)
Focus
   Stigmation
   Start-up
   Gun bias
   Contrast/brightness
Image modes and detectors
   SE detector(s) and position
SE resolution (kV/working distance)
   BSE detector
BSE resolution (kV/working distance)
      Tested resolution (on your samples)
         SE (kV/working distance)
         BSE (kV/working distance)
      X ray (kV/element line)
BSE $Z$ resolution (atomic number)
Specimen current amplifier
Additional options and customizing
Live image processing
   Gamma correction
   Y-modulated image
   Differential filter
   Dynamic focus
   Tilt correction
Camera system and film type(s)
Digital image storage system
Computer control/interface
Other
   Training in operation and maintenance
   Installation
Warranty
   Service contract
      Response time
      Parts covered
      Phone consultation
   Parts availability
   Guarantee of resolution/performance
   Other instruments in area
   Documentation, instructions, schematics
   Tool kit
   Internal diagnostics
   Ease of routine maintenance/repair
Ease and comfort of use
Environmental needs/stability
   Cooling water
   Gases
   Power
   Vibration
   Fields
   Temperature/humidity
EDS compatibility
   Apertures
   Working distance/detector geometry
Compatibility with other systems/instruments
Availability of other options

that the instrument be tested with "real" specimens if resolution is a critical factor in the selection.

All modern instruments will offer a wide array of capabilities to the user and will meet basic needs. Criteria such as magnification range and scanning speeds are important to consider, but the operator's preference and attention to most common usage should guide choices. Several types of detector are available and may be offered as options by the manufacturer or as an add-on system from a third party. Probably the most rapidly changing portion of the SEM is the system for recording, storing, and presenting images. Many laboratories are currently using a dual-system approach, with traditional photography and digital imaging available on the instrument. There are no standards for software control, digital image format, image data format, or printing. The SEM control and recording systems are rapidly evolving, and one should pay close attention to possible interfacing and compatibility problems with existing or desired equipment, computers, and software.

GARY W. CHANDLER
SUPAPAN SERAPHIN
University of Arizona
Tucson, Arizona

# TRANSMISSION ELECTRON MICROSCOPY

## INTRODUCTION

Transmission electron microscopy (TEM) is the premier tool for understanding the internal microstructure of materials at the nanometer level. Although x-ray diffraction techniques generally provide better quantitative information than electron diffraction techniques, electrons have an important advantage over x rays in that they can be focused using electromagnetic lenses. This allows one to obtain real-space images of materials with resolutions on the order of a few tenths to a few nanometers, depending on the imaging conditions, and simultaneously obtain diffraction information from specific regions in the images (e.g., small precipitates) as small as 1 nm. Variations in the intensity of electron scattering across a thin specimen can be used to image strain fields, defects such as dislocations and second-phase particles, and even atomic columns in materials under certain imaging conditions. Transmission electron microscopy is such a powerful tool for the characterization of materials that some microstructural features are defined in terms of their visibility in TEM images.

In addition to diffraction and imaging, the high-energy electrons (usually in the range of 100 to 400 keV of kinetic energy) in TEM cause electronic excitations of the atoms in the specimen. Two important spectroscopic techniques make use of these excitations by incorporating suitable detectors into the transmission electron microscope.

1. In energy-dispersive x-ray spectroscopy (EDS), an x-ray spectrum is collected from small regions of the specimen illuminated with a focused electron probe using a solid-state detector. Characteristic x rays of each element are used to determine the concentrations of the different elements present in the specimen (Williams and Carter, 1996). The principles behind this technique are discussed in details (ENERGY-DISPERSIVE SPECTROMETRY).

2. In electron energy loss spectroscopy (EELS), a magnetic prism is used to separate the electrons according to their energy losses after having passed through the specimen (Egerton, 1996). Energy loss mechanisms such as plasmon excitations and core-electron excitations cause distinct features in EELS. These can be used to quantify the elements present as well as provide information about atomic bonding and a variety of other useful phenomena.

In scanning transmission electron microscopy (STEM), a focused beam of electrons (typically <1 nm in diameter) is scanned in a television-style raster pattern across the specimen, as in a scanning electron microscope (SCANNING ELECTRON MICROSCOPY). In synchronization with the raster scan, emissions resulting from the interaction of the electron beam with the specimen are collected, such as x rays or secondary or backscattered electrons, to form images. Electrons that pass through the specimen can also be detected to form images that are similar to conventional TEM images. An annular detector can be used to collect the scattered transmitted electrons, which leads to $Z$-contrast imaging (discussed in SCANNING TRANSMISSION ELECTRON MICROSCOPY: Z-CONTRAST IMAGING). The STEM mode of operation is particularly useful for spectroscopic analysis, since it permits the acquisition of a chemical map of the sample typically with a resolution of a few nanometers. For example, one can make an image of the distribution of Fe in a sample by recording, in synchronization with the raster pattern, either the emission from the sample of Fe $K\alpha$ x rays (with the EDS spectrometer) or transmitted electrons with energy losses greater than that of the Fe $L$ edge (with the EELS spectrometer). The STEM mode of operation is different from the conventional TEM mode in that the objective lens is operated in tandem with the illumination lens system to assist in the formation of a focused electron probe on the specimen (Keyse et al., 1998).

A fully equipped transmission electron microscope has the capability to record the variations in image intensity across the specimen using mass thickness or diffraction contrast techniques, to reveal the atomic structure of materials using high-resolution (phase-contrast) imaging or $Z$-contrast (incoherent) imaging, to obtain electron diffraction patterns from small areas of the specimen using a selected-area aperture or a focused electron probe, and to perform EELS and EDS measurements with a small probe. Additional lenses can be installed in conjunction with an EELS spectrometer to create an energy filter, enabling one to form energy-filtered TEM images (EFTEM—Krivanek et al., 1987; Reimer, 1995). These images enable mapping of the chemical composition of a specimen with nanometer spatial resolution. A block diagram of such a transmission electron microscope is shown in Figure 1.

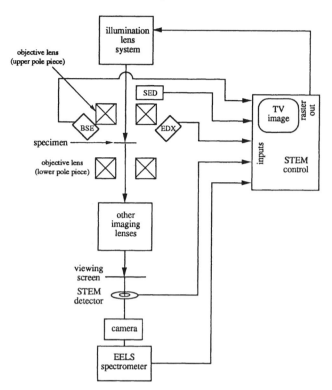

**Figure 1.** Typical transmission electron microscope with STEM capability. It is also possible to perform scanning electron microscopy (SEM) in a STEM using backscattered electron (BSE) and secondary electron detectors (SEDs) located above the specimen.

In addition to the main techniques of (1) conventional imaging, (2) phase-contrast imaging, (3) Z-contrast imaging, (4) electron diffraction, (5) EDS, and (6) EELS, in TEM many other analyses are possible. For example, when electrons pass through a magnetic specimen, they are deflected slightly by Lorentz forces, which change direction across a magnetic domain wall. In a method known as Lorentz microscopy, special adjustments of lens currents permit imaging of these domain walls (Thomas and Goringe, 1979). Phase transformations and microstructural changes in a specimen can be observed directly as the specimen is heated, cooled, or deformed in the microscope using various specimen stages (Butler and Hale, 1981). Differential pumping can be used to allow the introduction of gases into the microscope column surrounding a thin foil, making it possible to follow chemical reactions in TEM *in situ*. Many of these techniques can be performed at spatial resolutions of a few tenths of a nanometer. The possibilities are almost endless, and that is why TEM continues to be an indispensible tool in materials research. Since it is not possible to cover all of these techniques, this unit focuses on the theory and practice of conventional TEM electron diffraction and imaging techniques, including some examples of high-resolution TEM (HRTEM). A few imaging and diffraction techniques offer the resolution and versatility of TEM and are competitive or complementary in some respects.

Transmission electron microscopy is not able to readily image single point defects in materials. In contrast, the field ion microscope (FIM) can be used to study vacancies and surface atoms and therefore extends the available resolution to atomic dimensions (Miller and Smith, 1989). When combined with a mass spectrometer, the FIM becomes an atom probe (APFIM), easily capable of compositional analysis of regions ~1 nm wide and now approaching atom-by-atom analysis of local areas of samples. In the FIM, a high potential is applied to a very fine pointed specimen. A low pressure of inert gas is maintained between the specimen and imaging screen/spectrometer. Positively charged gas ions generated by the process of field ionization are used to produce images of the atoms on the surface of the specimen. Successive atom layers of material may be ionized and removed from the specimen surface by the process of field evaporation, enabling the three-dimensional (3D) structure of the material to be imaged in atomic detail and also to provide the source of ions for mass spectrometry.

Another instrument capable of imaging the internal structure of materials is the scanning acoustic microscope (SAM). In this microscope, a lens is used to focus the acoustic emission from an ultrasonic transducer onto a small region of a specimen through a coupling medium such as water. The same lens can be used to collect the acoustic echo from defects below the surface of the specimen. Either the lens or the specimen is scanned mechanically to form a two-dimensional (2D) image, the resolution of which depends on the wavelength of the acoustic wave in the specimen and thus on the frequency of the ultrasound and the velocity of sound in the specimen. Since the depth to which sound waves can penetrate decreases as their frequency increases, the choice of frequency depends on a compromise between resolution and penetration. Typical values might be a resolution of 40 to 100 μm at a depth of 5 mm below the surface for 50-MHz frequency, or a resolution of 1 to 3 μm at a similar depth for a frequency of 2 GHz. The resolution is clearly not as good as in TEM although the depth penetration is greater.

Lastly, x-ray diffraction and microscopy are alternative techniques to TEM and offer the advantage of greater penetration through materials. The main obstacle to high spatial resolution in both techniques is the difficulty of focusing x-ray beams, since x rays are uncharged and cannot be focused by electromagnetic or electrostatic lenses. While x-ray diffraction is commonly used for characterization of bulk samples in materials science and can be used to determine the atomic structures of materials (McKie and McKie, 1986), x-ray microscopy is not common, largely due to the better resolution and versatility of TEM.

## PRINCIPLES OF THE METHOD

Here we develop the theoretical basis necessary to understand and quantify the formation of diffraction patterns and images in TEM. The theory developed is known as the "kinematical theory" of electron diffraction, and the contrast that arises in TEM images due to electron diffraction is "diffraction contrast." These concepts are then utilized in examples of the method and data analyses.

## Structure Factor and Shape Factor

In both real space and in reciprocal space, it is useful to divide a crystal into parts according to the formula (Fultz and Howe, 2000)

$$\mathbf{r} = \mathbf{r}_g + \mathbf{r}_k + \delta_{\mathbf{r}_{g,k}} \qquad (1)$$

For a defect-free crystal the atom positions $\mathbf{R}$ are given by

$$\mathbf{R} = \mathbf{r}_g + \mathbf{r}_k \qquad (2)$$

where the lattice is one of the 14 Bravais lattice types (see SYMMETRY IN CRYSTALLOGRAPHY) and the basis is the atom group associated with each lattice site (Borchardt-Ott, 1995). Here we calculate the scattered wave $\psi(\Delta\mathbf{k})$ for the case of an infinitely large, defect-free lattice with a basis:

$$\psi(\Delta\mathbf{k}) = \sum_{\mathbf{R}} f_{at}(\mathbf{R})\exp(-i2\pi\,\Delta\mathbf{k}\cdot\mathbf{R}) \qquad (3)$$

where the scattered wave vector $\Delta\mathbf{k}$ is defined as the difference between the diffracted wave vector $\mathbf{k}$ and the incident wave vector $\mathbf{k}_0$ (refer to Fig. 2), or

$$\Delta\mathbf{k} = \mathbf{k} - \mathbf{k}_0 \qquad (4)$$

and $f_{at}(\mathbf{R})$ is the atomic scattering factor for electrons from an atom. We decompose our diffracted wave into a lattice component and a basis component with

$$\psi(\Delta\mathbf{k}) = \sum_{(\mathbf{r}_g+\mathbf{r}_k)} f_{at}(\mathbf{r}_g+\mathbf{r}_k)\exp[-i2\pi\,\Delta\mathbf{k}\cdot(\mathbf{r}_g+\mathbf{r}_k)] \qquad (5)$$

Since the atom positions in all unit cells are identical, $f_{at}(\mathbf{r}_g+\mathbf{r}_k)$ cannot depend on $\mathbf{r}_g$, so $f_{at}(\mathbf{r}_g+\mathbf{r}_k) = f_{at}(\mathbf{r}_k)$, and thus

$$\psi(\Delta\mathbf{k}) = \sum_{\mathbf{r}_g}\exp(-i2\pi\,\Delta\mathbf{k}\cdot\mathbf{r}_g)\sum_{\mathbf{r}_k} f_{at}(\mathbf{r}_k)\exp(-i2\pi\,\Delta\mathbf{k}\cdot\mathbf{r}_k) \qquad (6)$$

or, simplifying,

$$\Psi(\Delta\mathbf{k}) = \mathcal{S}(\Delta\mathbf{k})\mathcal{F}(\Delta\mathbf{k}) \qquad (7)$$

In writing Equation 7, we have given formal definitions to the two summations in Equation 6. The first sum, which is over all the lattice sites of the crystal (all unit cells), is known as the shape factor $\mathcal{S}$. The second sum, which is over the atoms in the basis (all atoms in the unit cell), is known as the structure factor $\mathcal{F}$. The notation $(\Delta\mathbf{k})$ is used to indicate the dependence of these terms on $\Delta\mathbf{k}$. The decomposition of the diffracted wave into the shape factor and the structure factor parallels the decomposition of the crystal into a lattice plus a basis.

Calculation of the structure factor $\mathcal{F}(\Delta\mathbf{k})$ for a unit cell is discussed in detail in KINEMATIC DIFFRACTION OF X RAYS and in standard books on diffraction (Schwartz and Cohen, 1987) and is not developed further here. Because in TEM we examine thin foils often containing small particles with different shapes, it is useful to examine the shape factor $\mathcal{S}(\Delta\mathbf{k})$ in Equations 6 and 7 in further detail. The shape factor $\mathcal{S}(\Delta\mathbf{k})$ is not very interesting for an infinitely large crystal where it becomes a set of delta functions centered at the various values of $\Delta\mathbf{k}$, where $\Delta\mathbf{k} = \mathbf{g}$ ($\mathbf{g}$ is a reciprocal lattice vector), but it is interesting for small crystals, which give rise to various spatial distributions of the diffracted electron intensity. The full 3D expression for the kinematical diffracted intensity due to the shape factor is given as

$$\mathcal{S}^*\mathcal{S}(\Delta\mathbf{k}) = \frac{\sin^2(\pi\,\Delta k_x a_x N_x)}{\sin^2(\pi\,\Delta k_x a_x)}\frac{\sin^2(\pi\,\Delta k_y a_y N_y)}{\sin^2(\pi\,\Delta k_y a_y)}\frac{\sin^2(\pi\,\Delta k_z a_z N_z)}{\sin^2(\pi\,\Delta k_z a_z)} \qquad (8)$$

where $a_x$, $a_y$, and $a_z$ are the magnitudes of the primitive translation vectors of the unit cell expressed along orthonormal $x$, $y$, and $z$ axes; $N_x$, $N_y$, and $N_z$ are the number of unit cells along the same axes, i.e., $a_x N_x = t_x$, the crystal thickness along the $x$ direction and similarly for $y$ and $z$; and $\Delta k_x$, $\Delta k_y$, and $\Delta k_z$ are the components of $\Delta\mathbf{k}$ expressed along the $x$, $y$, and $z$ axes. Equation 8 is valid for a crystal shaped as a rectangular prism.

The function in Equation 8 becomes large when the denominator goes to zero. This occurs when the argument of the sine function is equal to $\pi$ or to some integral multiple of it, expressed (in the $x$ direction only) as

$$\Delta k_x a_x = \text{integer} \qquad (9)$$

Since similar conditions are expected for $y$ and $z$, this condition requires that $\Delta\mathbf{k}$ is a reciprocal lattice vector $\mathbf{g}$. In other words, the kinematical intensity $\mathcal{S}^*\mathcal{S}$ is large when the Bragg condition is exactly satisfied. Since the denominator varies slowly with respect to the numerator, we can make the following approximation, which is valid near the center of the main peaks (expressed in only one direction):

$$\mathcal{S}^*\mathcal{S}(\Delta\mathbf{k}) \cong \frac{\sin^2(\pi\,\Delta k\,aN)}{(\pi\,\Delta k\,a)^2} \qquad (10)$$

This function describes an envelope of satellite peaks situated near the main Bragg diffraction peaks. By examining

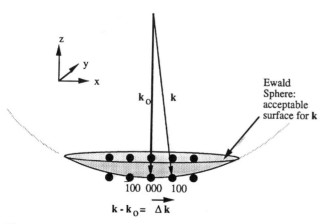

**Figure 2.** Ewald sphere construction showing incident $\mathbf{k}_0$ and scattered $\mathbf{k}$ wave vectors joined tail to tail and definition of $\Delta\mathbf{k}$.

the numerator, we see that the positions of the satellite peaks get closer to the main peak in proportion to $1/Na$, and the position of the first minimum in the intensity is located on either side of the main peak at the position $\Delta k = 1/Na$. Similarly, the widths of the main peaks and satellite peaks also decrease as $1/Na$. Thus, large crystal dimensions in real space lead to sharp diffracted intensities in reciprocal space, and vice versa (Hirsch et al., 1977; Thomas and Goringe, 1979).

### Ewald Sphere Construction

The Bragg condition for diffraction, $\Delta k = g$, where $g$ is a reciprocal lattice vector and $\Delta k$ is any possible scattered wave vector, can be implemented in a geometrical construction due to Ewald (McKie and McKie, 1986). The Ewald sphere depicts the incident wave vector $k_0$ and all possible $k$ for the diffracted waves. The tip of the wave vector $k_0$ is always placed at a point of the reciprocal lattice that serves as the origin. To obtain $\Delta k = k - k_0$, we would normally reverse the direction of $k_0$ and place it tail to head with the vector $k$, but in the Ewald sphere construction in Figure 2, we draw $k$ and $k_0$ tail to tail. The vector $\Delta k$ is the vector from the head of $k_0$ to the head of $k$. If the head of $k$ touches any reciprocal lattice point, the Bragg condition ($\Delta k = g$) is satisfied and diffraction occurs. In elastic scattering, the length of $k$ equals the length of $k_0$, since there is no change in wavelength (magnitude of the wave vector). The tips of all possible $k$ vectors lie on a sphere whose center is at the tails of $k$ and $k_0$. By this construction, one point on this sphere always touches the origin of the reciprocal lattice. Whenever another point on the Ewald sphere touches a reciprocal lattice point, the Bragg condition is satisfied and diffraction occurs. As illustrated in Figure 2, the Bragg condition is approximately satisfied for the (100) and ($\bar{1}$00) diffraction spots.

The Ewald sphere is strongly curved for x-ray diffraction because $|g|$ (typically 5 to 10 $nm^{-1}$) is comparable to $|k_0|$. Electron wave vectors, on the other hand, are much longer than the spacings in the reciprocal lattice (100-keV electrons have a wavelength of 0.0037 nm). So, for high-energy electrons, the Ewald sphere is approximately a plane, and consequently, $\Delta k$ is nearly perpendicular to $k_0$. In practice, the diffracted intensity distribution, such as that in Equation 8, is located in a finite volume around the reciprocal lattice points, so $\Delta k$ need not equal $g$ exactly in order for diffraction to occur. The shape factor intensity $S^*S(\cdot k)$ serves in effect to broaden the reciprocal lattice points. With a crystal oriented near a zone axis, i.e., along a direction parallel to the line of intersection of a set of crystal planes, the Ewald sphere passes through many of these small volumes around the reciprocal lattice points, and many diffracting spots are observed. Figure 3 provides an example of Ewald sphere constructions for two orientations of the specimen with respect to $k_0$, together with their corresponding selected-area diffraction (SAD) patterns. Figure 3 is drawn as a 2D slice (the $x$-$z$ plane) of Figure 2. The crystal on the right is oriented precisely along a zone axis, but the crystal on the left is not.

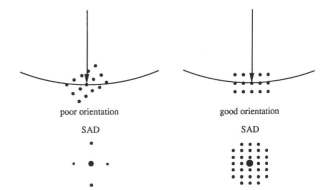

**Figure 3.** Two orientations of reciprocal lattice with respect to Ewald sphere and corresponding SAD patterns.

### Deviation Vector and Deviation Parameter

From the previous discussion, it is apparent that the diffracted intensity observed in an electron diffraction pattern depends on exactly how the Ewald sphere cuts through the diffracted intensity around the Bragg position. To locate the exact position of intersection, we introduce a new parameter, the deviation vector $s$, defined as

$$g = \Delta k + s \tag{11}$$

where $g$ is a reciprocal lattice vector and $\Delta k$ is the diffraction vector whose end lies on the Ewald sphere ($\Delta k = k - k_0$). For high-energy electrons, the shortest distance between the Ewald sphere and a reciprocal lattice point $g$ is parallel to the $z$ direction, so we often work with only the magnitude of $s$, i.e., $|s| = s$, referred to as the deviation parameter. We choose a sign convention for $s$ that is convenient when we determine $s$ by measuring the positions of Kikuchi lines (Kikuchi, 1928). Positive $s$ means that $s$ points along positive $z$. (By convention, $z$ points upward toward the electron gun.) Figure 4 shows that $s$ is positive when the reciprocal lattice point lies inside the Ewald sphere and negative when the reciprocal lattice point lies outside the sphere.

The parameter $s$ is useful because it is all that we need to know about the diffraction conditions to calculate the kinematical shape factor. Using Equation 11 for $\Delta k$ in Equation 7 yields

$$S(\Delta k) = \sum_{r_g} \exp(-i2\pi g \cdot r_g)\exp(+i2\pi s \cdot r_g) \tag{12}$$

$$\mathcal{F}(\Delta k) = \sum_{r_k} f_{at}(r_k, \Delta k)\exp(-i2\pi g \cdot r_k)\exp(+i2\pi s \cdot r_k) \tag{13}$$

$$\mathcal{F}(\Delta k) \cong \sum_{r_k} f_{at}(r_k, \Delta k)\exp(-i2\pi g \cdot r_k) = \mathcal{F}(g) \text{ or } \mathcal{F}_g \tag{14}$$

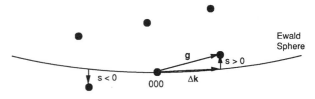

**Figure 4.** Convention for defining deviation vector $s$ and deviation parameter $s$, from Ewald sphere to reciprocal lattice spot $g$.

where we made the last approximation for the structure factor $\mathcal{F}(\mathbf{g})$ because $\mathbf{s} \cdot \mathbf{r}_k$ is small when the unit cell is small. Since $\mathbf{g}$ is a reciprocal lattice vector, $\exp(-i2\pi\mathbf{g} \cdot \mathbf{r}_g) = 1$ for all $\mathbf{r}_g$, which simplifies the shape factor to

$$S(\Delta\mathbf{k}) = S(\mathbf{s}) \sum_{\mathbf{r}_g} \exp(+i2\pi\mathbf{s} \cdot \mathbf{r}_g) \qquad (15)$$

and the diffracted wave $\psi(\mathbf{g}, \mathbf{s})$, which is a function of the reciprocal lattice vector and deviation parameter, becomes

$$\psi(\mathbf{g},\mathbf{s}) = \mathcal{F}_g \sum_{\mathbf{r}_g} \exp(+i2\pi\mathbf{s} \cdot \mathbf{r}_g) \qquad (16)$$

The shape factor depends only on the deviation vector $\mathbf{s}$ and not on the particular reciprocal lattice vector $\mathbf{g}$. When we substitute for the components of $\mathbf{s}$ and $\mathbf{r}_g$ along the $x$, $y$, and $z$ axes, we obtain an equation for the shape factor intensity that is similar to Equation 8:

$$S * S(\mathbf{s}) = \frac{\sin^2(\pi s_x a_x N_x)}{\sin^2(\pi s_x a_x)} \frac{\sin^2(\pi s_y a_y N_y)}{\sin^2(\pi s_y a_y)} \frac{\sin^2(\pi s_z a_z N_z)}{\sin^2(\pi s_z a_z)} \quad (17)$$

For high-energy electron diffraction, we can make the following simplifications: (1) the deviation vector is very nearly parallel to the $z$ axis, so $s_z$ is simply equal to $s$; (2) the denominator is given as $(\pi s_z)^2$; and (3) the quantity $a_z N_z$ is the crystal thickness $t$. Ignoring the widths along $x$ and $y$ of the diffracting columns, a useful expression for the shape factor intensity is

$$S * S(\mathbf{s}) = \frac{\sin^2(\pi s t)}{(\pi s)^2} \qquad (18)$$

Combining Equations 14 and 18 then gives the resulting diffracted intensity for the vector $\mathbf{g}$:

$$\mathbf{I}_g = |\psi(\mathbf{g},\mathbf{s})|^2 = |\mathcal{F}(\mathbf{g})|^2 \frac{\sin^2(\pi s t)}{(\pi s)^2} \qquad (19)$$

This intensity depends only on the diffracting vector $\mathbf{g}$ and the deviation parameter $s$. Below we present many examples of how $\mathbf{I}_g$ depends on the deviation parameter $s$ and the sample thickness $t$. The dependence of $\mathbf{I}_g$ on the position $(x,y)$ of the diffracting column provides diffraction contrast in bright-field (BF) or dark-field (DF) images. In a two-beam condition, where only the incident beam and one diffracting beam have appreciable intensity, the intensities of the transmitted beam $\mathbf{I}_0$ and diffracted beam $\mathbf{I}_g$ are complementary, and with the incident intensity normalized to 1, we have the relationship

$$\mathbf{I}_0 = 1 - \mathbf{I}_g \qquad (20)$$

The kinematical diffraction theory developed above is valid when the intensity of the diffracted beam is much less than the transmitted beam, or

$$\mathbf{I}_g \ll \mathbf{I}_0 \qquad (21)$$

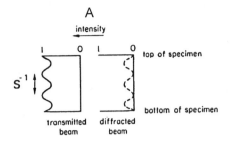

A

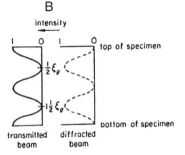

B

**Figure 5.** Extinction distance and amplitudes of transmitted and diffracted beams for two-beam condition in relatively thick crystal (**A**) under kinematical conditions with $s \gg 0$ and (**B**) under dynamical conditions with $s = 0$. (After Edington, 1974.)

When this condition is not satisfied, one must resort to the dynamical theory of electron diffraction (DYNAMICAL DIFFRACTION; Hirsch et al., 1977; Williams and Carter, 1996). Equations 19 through 21 tell us that, for a foil of constant thickness $t$, we expect the intensities of the transmitted and diffracted beams to vary with depth in the foil and the period of depth variation is approximately $s^{-1}$. This behavior is illustrated schematically in Figure 5A. The larger is the deviation parameter, the smaller is the period of oscillation and vice versa.

### Extinction Distance

A more general way of writing Equation 19 is

$$\mathbf{I}_g = \left(\frac{\pi}{\xi_g}\right)^2 \frac{\sin^2(\pi s t)}{(\pi s)^2} \qquad (22)$$

where the extinction distance $\xi_g$ is given by the expression

$$\xi_g = \frac{\pi V \cos\theta}{\lambda \mathcal{F}_g} \qquad (23)$$

and $V$ is the volume of the unit cell, $\theta$ is the Bragg angle for the diffraction vector $\mathbf{g}$, $\lambda$ is the electron wavelength, and $\mathcal{F}_g$ is the structure factor for the diffracting vector $\mathbf{g}$. Note that $\xi_g$ increases with increasing order of diffracting vectors because $\mathcal{F}_g$ decreases as $\theta$ increases.

The quantity $\xi_g$ defined in Equation 23 is an important length scale in the kinematical and dynamical theories of electron diffraction. The extinction distance is defined as twice the distance in the cystal over which 100% of the incident beam is diffracted when $s = 0$ and the Bragg condition is satisfied exactly. The magnitude of $\xi_g$ depends on

**Table 1. Extinction Distances $\xi_g$ (nm) for fcc Metals with 100-kV Electrons (from Hirsch et al., 1977)**

| Diffracting Plane | Al | Ni | Cu | Ag | Pt | Au | Pb |
|---|---|---|---|---|---|---|---|
| 111 | 55.6 | 23.6 | 24.2 | 22.4 | 14.7 | 15.9 | 24.0 |
| 200 | 67.3 | 27.5 | 28.1 | 25.5 | 16.6 | 17.9 | 26.6 |
| 220 | 105.7 | 40.9 | 41.6 | 36.3 | 23.2 | 24.8 | 35.9 |
| 311 | 130.0 | 49.9 | 50.5 | 43.3 | 27.4 | 29.2 | 41.8 |
| 222 | 137.7 | 52.9 | 53.5 | 45.5 | 28.8 | 30.7 | 43.6 |
| 400 | 167.2 | 65.2 | 65.4 | 54.4 | 34.3 | 36.3 | 50.5 |
| 331 | 187.7 | 74.5 | 74.5 | 61.1 | 38.5 | 40.6 | 55.5 |
| 420 | 194.3 | 77.6 | 77.6 | 63.4 | 39.8 | 42.0 | 57.2 |
| 422 | 219.0 | 89.6 | 89.7 | 72.4 | 45.3 | 47.7 | 63.8 |
| 511 | 236.3 | 98.3 | 98.5 | 79.2 | 49.4 | 51.9 | 68.8 |
| 333 | 236.3 | 112.0 | 112.6 | 90.1 | 55.8 | 58.7 | 77.2 |
| 531 | 279.8 | 119.6 | 120.6 | 96.4 | 59.4 | 62.6 | 82.2 |
| 600 | 285.1 | 122.1 | 123.2 | 98.4 | 60.6 | 63.8 | 83.8 |
| 442 | 285.1 | 122.1 | 123.2 | 98.4 | 60.6 | 63.8 | 83.8 |

the atomic form factors; the stronger the scattering, the shorter is $\xi_g$. Table 1 shows some values of $\xi_g$ for different diffracting planes in pure metals with a face-centered cubic (fcc) crystal structure (Hirsch et al., 1977). Notice how $\xi_g$ increases with the indices of the diffracting vectors *hkl* and decreases with increasing atomic number. The values of $\xi_g$ generally range from a few tens to a few hundred nanometers.

The extinction distance defines the depth variation of the transmitted and diffracted intensities in a crystal when the diffraction is strong. The extinction distance $\xi_g$ applies to the dynamical situation where $s = 0$, as illustrated in Figure 5B, not just the kinematical situation where $s \gg 0$. In the development of the dynamical theory of electron diffraction, the dependence of the diffracted intensity on the deviation parameter and sample thickness is much the same as in Equation 22 with the following modification: The extinction distance in Equation 23 is made dependent on $s$ by transforming $\xi_g$ into an effective extinction distance, $\xi_{geff}$,

$$\xi_{geff} = \frac{\xi_g}{\sqrt{1 + s^2 \xi_g^2}} \tag{24}$$

and another quantity known as the effective deviation parameter is defined as

$$s_{eff} = \sqrt{s^2 + \xi_g^{-2}} \tag{25}$$

Equation 24 shows that the effective extinction distance $\xi_{geff} = \xi_g$ when $s = 0$, but $\xi_{geff}$ decreases with increasing deviation from the exact Bragg position. With large deviations, $s \gg 1/\xi_g$, and $\xi_{geff} = s^{-1}$, so the kinematical result is recovered for large $s$.

**Diffraction Contrast from Lattice Defects**

The following are important variables in the diffraction contrast from defects in crystals:

$\mathcal{F}(\mathbf{g}) \equiv$ structure factor of unit cell
$t \equiv$ specimen thickness
$\Delta\mathbf{k} \equiv$ diffraction vector
$\mathbf{g} \equiv$ reciprocal lattice vector
$\mathbf{s} \equiv$ deviation vector ($\mathbf{g} = \Delta\mathbf{k} + \mathbf{s}$)
$\mathbf{r} \equiv$ actual atom centers
$\mathbf{R} \equiv$ atom centers in a perfect crystal ($\mathbf{R} = \mathbf{r}_g + \mathbf{r}_k$, where $\mathbf{r}_g$ refers to the lattice and $\mathbf{r}_k$ to the basis)
$\delta\mathbf{r} \equiv$ displacements of atoms off the ideal atom centers ($\mathbf{r} = \mathbf{R} + \delta\mathbf{r}$).

Note that the actual atom centers $\mathbf{r}$ as defined above are the locations of the atoms off of the atom positions in a perfect crystal $\mathbf{R}$ that are given by the lattice (plus basis) points, i.e., the ideal mathematical positions. Spatial variations in these variables (e.g., an $x$ dependence) can produce diffraction contrast in an image. Examples include the following:

$\mathcal{F}$: $d\mathcal{F}/dx$ causes chemical (compositional) contrast,
$t$: $dt/dx$ causes thickness contours,
$\mathbf{g}$: $d\mathbf{g}/dx$ causes bend contours,
$\mathbf{s}$: $d\mathbf{s}/dx$ causes bend contours, and
$\delta\mathbf{r}$: $d\delta\mathbf{r}/dx$ causes strain contrast.

Up to this point, we have only considered diffraction occurring from perfect crystals. We now consider the displacements in atom positions $\delta\mathbf{r}$ caused by strain fields around defects in a crystal. These displacements represent the time-averaged position of the atoms during electron scattering. We decompose $\mathbf{r}$ into components from the lattice vectors, basis vectors, and distortion vectors as

$$\mathbf{r} = \mathbf{r}_g + \mathbf{r}_k + \delta\mathbf{r} \tag{26}$$

and we use the familiar expression for our diffracting vector as the difference of a reciprocal lattice vector and a deviation vector:

$$\Delta\mathbf{k} = \mathbf{g} - \mathbf{s} \tag{27}$$

By using Equations 26 and 27 and noting that $\mathbf{r} = \mathbf{R} + \delta\mathbf{r}$, we can rewrite Equation 3 as

$$\psi(\Delta\mathbf{k}) = \sum_{\mathbf{r}} f_{\text{at}}(\mathbf{r})\exp[-i2\pi(\mathbf{g} - \mathbf{s}) \cdot (\mathbf{r}_g + \mathbf{r}_k + \delta\mathbf{r})] \quad (28)$$

We then arrange Equaton 28 into a product of a structure factor $\mathcal{F}(\Delta\mathbf{k})$ and something like a shape factor, as we did previously. In doing so, we assume that the distortion $\delta\mathbf{r}$ is the same for all atoms in a particular unit cell,

$$\psi(\Delta\mathbf{k}) = \sum_{\mathbf{r}_g} \exp[-i2\pi(\mathbf{g} - \mathbf{s}) \cdot (\mathbf{r}_g + \delta\mathbf{r})]$$
$$\times \sum_{\mathbf{r}_k} f_{\text{at}}(\mathbf{r}_k)\exp[-i2\pi(\mathbf{g} - \mathbf{s}) \cdot \mathbf{r}_k] \quad (29)$$

When there are only a few atoms in the unit cell, we can neglect the factor of $\exp(i2\pi\mathbf{s} \cdot \mathbf{r}_k)$. We then identify the second sum in Equation 29 as the familiar structure factor of the unit cell $\mathcal{F}(\Delta\mathbf{k})$, and rewriting Equation 29 gives

$$\psi(\Delta\mathbf{k}) = \sum_{\mathbf{r}_g} \exp[-i2\pi(\mathbf{g} \cdot \mathbf{r}_g + \mathbf{g} \cdot \delta\mathbf{r} - \mathbf{s} \cdot \mathbf{r}_g - \mathbf{s} \cdot \delta\mathbf{r})]\mathcal{F}(\Delta\mathbf{k})$$
$$(30)$$

First, note that the product $\mathbf{g} \cdot \mathbf{r}_g$ is an integer, so it yields a factor of unity in the exponential. Second, because both $\mathbf{s}$ and $\delta\mathbf{r}$ are small, their product is negligible compared to the other two terms in the exponential in Equation 30. Thus, one obtains

$$\psi_g - F(\mathbf{g}) \sum_{\mathbf{r}_g} \exp[-i2\pi(\mathbf{g} \cdot \delta\mathbf{r} - \mathbf{s} \cdot \mathbf{r}_g)] \quad (31)$$

It is convenient and appropriate to substitute $\mathcal{F}(\mathbf{g})$ for $\mathcal{F}(\Delta\mathbf{k})$ in Equation 31, since we use all of the diffracted intensity around a specific reciprocal lattice point in forming a DF image, as defined in the next section. Equation 31 is useful for determining the diffraction contrast from defects such as dislocations, stacking faults, and precipitates. Such defects possess strain fields, and it is these strain fields that produce diffraction contrast in TEM images. Equation 31 shows that, in addition to the local atomic displacements $\delta\mathbf{r}$, the defect image is also controlled by the deviation vector $\mathbf{s}$.

The diffraction condition $\mathbf{g} \cdot \delta\mathbf{r} = 0$ is particularly important for the study of defects with diffraction contrast. The condition $\mathbf{g} \cdot \delta\mathbf{r} = 0$ is a null-contrast condition. For this condition, $\delta\mathbf{r}$ lies in the diffracting plane, and the interplanar spacing $d$ (and thus $|\mathbf{g}|$) is therefore unaltered. The condition $\mathbf{g} \cdot \delta\mathbf{r}$ is the condition for zero diffraction contrast originating from the displacement $\delta\mathbf{r}$. Even when $\mathbf{g} \cdot \delta\mathbf{r}$ is not exactly zero, the magnitude of $\mathbf{g} \cdot \delta\mathbf{r}$ must be sufficient to change the local intensity from the background level so that contrast is visible in the image. This typically requires an intensity change of ~10%. A rule of thumb is that if $\mathbf{g} \cdot \delta\mathbf{r} \leq \frac{1}{3}$, there is no visible contrast associated with $\delta\mathbf{r}$. This criterion is adequate for the analysis of diffracton contrast from many defects of interest in materials, such as dislocations, dislocation loops, precipitates,

stacking faults, domain boundaries, and grain and interphase boundaries, and its use is illustrated below (see Data Analysis and Initial Interpretation).

## PRACTICAL ASPECTS OF THE METHOD

### Bright-Field/Dark-Field Imaging

A ray diagram for making an image with a conventional TEM (CTEM) containing two lenses is shown in Figure 6B. In this diagram, the intermediate lens is focused on the image plane of the objective lens. We assume the illumination system provides rays (electrons) that travel straight down the microscope (parallel to the optic axis) before hitting the specimen. In Figure 6A, all transmitted and diffracted rays leaving the specimen are combined to form an image at the viewing screen, much as in a conventional optical microscope. In this simple mode of imaging, the specimen generally shows little contrast because all of the diffracted intensity reaches the viewing screen. By tracing the individual rays in Figure 6A, one can see how each point in the back focal plane of the objective lens contains rays from all parts of the specimen. Not all of the rays in the back focal plane are therefore required to form an image. An image can be formed with only those rays passing through one point in the back focal plane. What distinguishes the points located in the back focal plane is that all rays entering a given point have been scattered by the specimen into the same angle. By positioning an objective aperture at a specific location in the back focal plane, an image will be made with only those electrons that have been diffracted by a specific angle. (A similar principle applies in EFTEM, except that an aperture is used to allow electrons of only a certain energy loss to form the image.) This defines two basic imaging modes, which are also illustrated in Figure 6:

1. When the objective aperture is positioned to pass only the transmitted (undiffracted) electrons, a BF image is formed, as illustrated in Figure 6B.
2. When the objective aperture is positioned to pass only some diffracted electrons, a DF image is formed, as illustrated in Figure 6C.

The best way to form a DF image is to tilt the incident illumination on the specimen by an angle equal to the angle of the particular diffraction used for making the DF image, as illustrated in Figure 6C. On the back focal plane, the position of the transmitted beam was tilted into the position of the diffracted spot on the left, and the diffracted spot on the right is now used to form the DF image. Notice how these diffracted rays remain near the optic axis. This minimizes the detrimental effect of spherical aberration on the image resolution (see below). The following section illustrates how one can visualize tilting either the specimen or the electron beam in the TEM mode using the Ewald sphere construction discussed previously.

In most CTEM studies of crystalline materials, features in the image originate primarily from diffraction contrast.

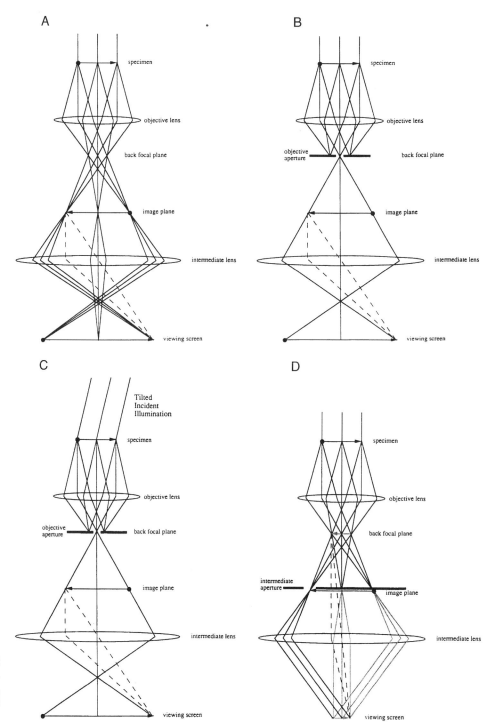

**Figure 6.** (**A**) Image mode for TEM containing objective and intermediate lens (**B**) BF imaging mode, (**C**) axial DF imaging mode, and (**D**) SAD mode in TEM.

Diffraction contrast is the variation in the diffracted intensity of particular electrons across the specimen, observed by inserting an objective aperture into the path of the beam. Upon doing so, features in the image become far more visible; without the objective aperture, the image tends to be gray and featureless. The physical reason that the diffraction contrast of a BF (or DF) image (Fig. 6B or C) is so much better than that of an apertureless image (Fig. 6A) is as follows: When there is a large

intensity in the diffracted beams, there is a large complementary loss of intensity in the transmitted beam, so either the BF or DF image alone will show strong diffraction contrast. Without the objective aperture, however, the diffracted intensity recombines with the transmitted intensity at the viewing screen. This recombination eliminates the observed diffraction contrast.

For thick specimens, an apertureless image mainly shows contrast caused by incoherent scattering in the sample. This generic type of contrast is called mass-thickness contrast because it increases with the atomic mass (squared) and the thickness of the material (Reimer, 1993). Mass-thickness contrast is particularly useful in biology, where techniques have been developed for selectively staining the different cell organelles with heavy elements to increase their mass-thickness contrast. In the case of very thin specimens (say $<\sim$10 nm), if a large objective aperture is used (or no aperture at all, as in Fig. 6A) and the transmitted and diffracted electron beams (waves) are allowed to recombine (interfere), the resulting image contrast depends on the relative phases of the beams involved; this mode of imaging is often called phase-contrast imaging. If the microscope resolution is less than the atomic spacings in the sample and both the sample and microscope conditions are optimized, phase-contrast imaging can be used to resolve the atomic structures of materials. This forms the basis of HRTEM.

In the microscope, typical objective apertures range from $\sim$0.5 to 20 μm in diameter. The apertures are moveable with high mechanical precision and can be positioned around selected diffraction spots in the back focal plane of the objective lens. The practice of positioning an objective aperture requires changing the operating mode of the microscope to the diffraction mode (described in the next section). In the diffraction mode, the images of both the diffraction pattern and the aperture are visible on the viewing screen, and the objective aperture can then be moved until it is in the desired position. Once the objective aperture is positioned properly, the microscope is switched back into the image mode, and either a DF or a BF image is formed.

### Tilting Specimens and Tilting Electron Beams

Many problems in the geometry of diffraction can be solved by employing an Ewald sphere and a reciprocal lattice. When working problems, it is important to remember the following:

1. The Ewald sphere and the reciprocal lattice are connected at the origin of the reciprocal lattice. Tilts of either the Ewald sphere or the reciprocal lattice are performed about this pivot point.

2. The reciprocal lattice is affixed to the crystal (for cubic crystals the reciprocal lattice directions are along the real-space directions). Tilting the specimen is represented by tilting the reciprocal lattice by the same angle and in the same direction.

3. The Ewald sphere surrounds the incident beam and is affixed to it. Tilting the direction of the incident beam is represented by tilting the Ewald sphere by the same amount.

These three facts are useful during practical TEM work. It is handy to think of the viewing screen as a section of the Ewald sphere, which shows a disk-shaped slice of the reciprocal lattice of the specimen. When we tilt the sample, the Ewald sphere and the viewing screen stay fixed, but the different points on the reciprocal lattice of the sample move into the viewing screen. For small tilts of the

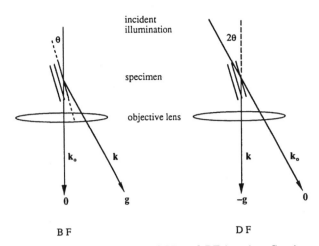

**Figure 7.** Procedures for axial BF and DF imaging. Specimen remains stationary as beam is tilted 2θ to center $-\mathbf{g}$ diffracted spot on optic axis.

specimen, the diffraction pattern does not move on the viewing screen, but the diffraction spots change in intensity. Alternatively, when we tilt the incident beam, we rotate the transmitted beam and the Ewald sphere. We can think of this operation as moving the disk-shaped viewing screen around the surface of the Ewald sphere. The diffraction spots therefore move in a direction opposite to the tilt of the incident beam, and the diffraction pattern moves as though the central beam were pivoting about the center of the sample.

### Tilted Illumination and Diffraction: How to Do Axial DF Imaging

To form DF images with good resolution, it is necessary to ensure that the image-forming diffracted rays travel straight down the optic axis. This requires that the direction of the incident electrons ($\mathbf{k}_0$) be tilted away from the optic axis by an angle 2θ, as shown in Figure 7. After tilting the illumination, there is a difference in the positions and intensities of the diffraction spots for the BF and DF modes. By tilting the illumination, we in fact tilt the Ewald sphere about the origin of the reciprocal lattice, as shown in Figure 8. Seen on the viewing screen, we tilt the transmitted beam into the former position of the diffraction spot $\mathbf{g}$. In doing so, the diffraction spot $\mathbf{g}$ moves far from the optic axis, but the diffraction spot $-\mathbf{g}$ becomes active, and its rays travel along the optic axis. This procedure seems counterintuitive, so why did we use it? The answer is that if the active diffraction spot $\mathbf{g}$ is tilted into the center of the viewing screen, the diffraction spot $\mathbf{g}$ becomes weak, and the diffraction spot $3\mathbf{g}$ becomes strong. Since the diffraction spot $\mathbf{g}$ is weak, it would be difficult to use for making a DF image. This latter procedure can be useful to improve resolution in DF analysis of defects and is referred to as weak-beam dark-field (WBDF) imaging (Cockayne et al., 1969).

### Selected-Area Diffraction

Figure 6D is a ray diagram for making a diffraction pattern with the simplified two-lens transmission electron

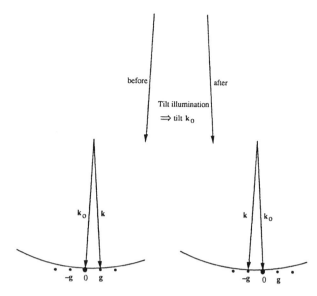

**Figure 8.** Procedure to set up axial DF image as in Figure 7. As seen on viewing screen, transmitted beam is moved into former position of diffracted spot **g** and −**g** diffracted spot becomes intense.

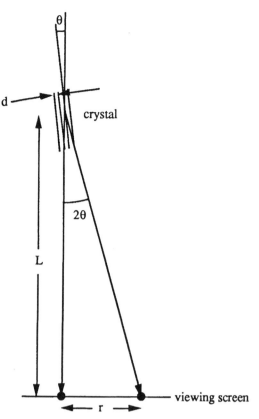

**Figure 9.** Geometry for electron diffraction and definition of camera length $L$.

microscope. The intermediate lens is now focused on the back focal plane of the objective lens. The transmitted beam and all of the diffracted beams are imaged on the viewing screen. In this configuration, a second aperture, called an intermediate (or selected-area) aperture, can be positioned in the image plane of the objective lens, and this provides a means of confining the diffraction pattern to a selected area of the specimen. This technique of SAD is usually performed in the following way. The specimen is first examined in image mode until a region of interest is found (the tip of the solid arrow in the image plane in Fig. 6D). The intermediate aperture is then inserted and positioned around this feature. Owing to spherical aberration, it may be necessary to slightly underfocus the objective lens to ensure that the SAD pattern comes from the region of interest (Hirsch et al., 1977; Thomas and Goringe, 1979). The microscope is then switched into the diffraction mode. The SAD pattern that appears on the viewing screen originates from the area selected in the image mode (the tip of the solid arrow in the image plane). Selected-area diffraction can be performed on regions a micrometer or so in diameter, but spherical aberration of the objective lens limits the technique to regions not much smaller than this. For diffraction work from smaller regions, it is necessary to use small-probe techniques such as convergent-beam electron diffraction (CBED).

We can use the separation of the diffraction spots on the viewing screen to determine interplanar spacings in crystals, and this can be used to help determine the lattice parameter of a crystal or to identify an unknown phase. To do this, we need to know the camera length of the microscope. Consider the geometry of the SAD pattern in Figure 9, which shows the camera length $L$ that is

characteristic of the optics of the microscope. Bragg's law is given as

$$2d \sin \theta = n\lambda \qquad (32)$$

where $d$ is the interplanar spacing, $\lambda$ is the electron wavelength, and $n$ is an integer. The value of $\theta$ is on the order of a degree for low-order diffraction spots from most materials using 100-keV electrons ($\lambda = 0.0037$ nm). For such small angles,

$$\sin \theta \sim \tan \theta \sim \frac{1}{2} \tan(2\theta) \qquad (33)$$

and from the geometry in Figure 9,

$$\tan 2\theta = \frac{r}{L} \qquad (34)$$

where $r$ is the separation of the diffraction spots on the viewing screen. If we substitute Equation 34 into Bragg's law (Equation 32) and arrange terms, we obtain

$$rd = \lambda L \qquad (35)$$

Equation 35 is referred to as the camera equation. It allows one to determine an interplanar spacing $d$ by measuring the separation of diffraction spots $r$ in the transmission electron microscope (or on a photographic negative). To do this, we need to know the product $\lambda L$, known as the camera constant (in units of nanometers-centimeters), and its approximate value ($\pm 5\%$) can be found on a console display of a modern transmission electron microscope. For higher precision, one should perform a calibration of the camera constant using a standard such as an evaporated Au film and insert the unknown specimen into the microscope, keeping the lens conditions fixed while adjusting the specimen focus using the height ($Z$) control on the goniometer. Better still is to evaporate the film directly on part of the unknown specimen, or if the unknown is a precipitate in a matrix, for example, the matrix material can be used for calibration, as illustrated below (see Data Analysis and Initial Interpretation).

### Indexing Diffraction Patterns

The reciprocal lattices of materials are three dimensional, so their diffraction patterns are three dimensional. Diffraction data from TEM are obtained as near-planar sections through diffraction space. The magnitude of the diffraction vector, $|\Delta\mathbf{k}|$, is obtained from the angle between the transmitted and diffracted beams. The interpretation of these data is simplified because the large electron wave vector provides an Ewald sphere that is nearly flat. This allows the convenient approximation that the diffraction pattern is sampling the diffraction intensities from a planar section of the reciprocal lattice. Two degrees of orientational freedom are required for the sample in TEM. These degrees of freedom are typically obtained with a double-tilt specimen holder, which provides two tilt axes oriented perpendicular to each other. Modern TEMs provide two methods for obtaining diffraction patterns from individual crystallites. One method is SAD, which is useful for obtaining diffraction patterns from regions as small as 0.5 μm in diameter. The second method is CBED (or microdiffraction/nanodiffraction), in which a focused electron beam is used to obtain diffraction patterns from regions as small as 1 nm. Both techniques provide a 2D pattern of diffraction spots, which can be highly symmetrical when a single crystal is oriented precisely along a zone axis orientation, as on the right in Figure 3.

We now describe how to index the planar sections of single-crystal diffraction patterns; i.e., label the individual diffraction spots with their appropriate values of $h$, $k$, and $l$. It is helpful to imagine that the single-crystal diffraction pattern is a picture of a plane in the reciprocal space of the material. (See SYMMETRY IN CRYSTALLOGRAPHY for more details on point groups.) Indexing begins with the identification of the transmitted beam, or the (000) diffraction spot. This is usually the brightest spot in the center of the diffraction pattern. Next, we need to index two independent (i.e., noncolinear) diffraction spots nearest to the (000) spot. Once these two vectors are determined, we can make linear combinations of them to obtain the positions and indices of all the other diffraction spots. To complete the indexing of a diffraction pattern, we also specify the normal to the plane of the spot pattern; this normal is the zone axis. By convention, the zone axis points toward the electron gun (i.e., upward in most transmission elec-

tron microsopes). The indexing of a diffraction pattern is not unique. If a crystal has high symmetry, so does its reciprocal lattice (although these symmetries may be different). A high symmetry leads to a multiplicity of different, but correct, ways to index a given diffraction pattern. For example, a vector cube axis can be a [100], [010], or [001] vector. Once the orientation of the diffraction pattern is specified, it is important to index the pattern consistently.

Most of the work in indexing diffraction patterns involves measuring angles and distances between diffraction spots and then comparing these measurements to geometrical calculations of angles and distances. When indexing a diffraction pattern, one must remember that structure factor rules eliminate certain diffraction spots. For consistency, one must also satisfy the "right-hand rule," which is given by the vector cross-product relation: $\mathbf{x} \times \mathbf{y} \parallel \mathbf{z}$. The procedures are straightforward for low-index zone axes of simple crystal structures but become increasingly difficult for crystal structures with low symmetry and for high-index zone axes, where many different combinations of interplanar spacings and angles provide diffraction patterns that look similar. In these cases, a computer program to calculate such diffraction patterns is extremely helpful (see Method Automation; also see Data Analysis and Initial Interpretation).

The procedure to index a diffraction pattern is illustrated by example below. Suppose we need to index the diffraction pattern in Figure 10A and we know that it is from an fcc crystal. The first step is to consult a reference and find an indexed diffraction pattern that looks like the one in Figure 10A. In this method, we guess the zone axis and its diffraction pattern. This method is most useful when the diffraction pattern shows an obvious symmetry, such as a square or hexagonal array of spots for a cubic crystal. With experience, one eventually remembers the symmetries for common fcc and body-centered cubic (bcc) diffraction patterns shown in Table 2. Without turning immediately to published SAD patterns, however, we first note that the pattern in Figure 10A is less symmetrical than those in Table 2. Nevertheless, we note that the

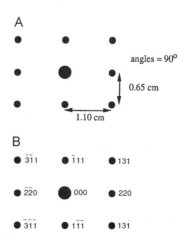

**Figure 10.** (**A**) An fcc diffraction pattern ready for indexing and (**B**) successful indexing of diffraction pattern in (**A**).

**Table 2. Some Symmetrical Diffraction Patterns**

| Zone Axis | [100] | [110] | [111] |
|---|---|---|---|
| Symmetry | square | rectangular $1 : \sqrt{2}$ for bcc (almost hexagonal for fcc) | hexagonal |

density of spots seems reasonably high, so we expect that we have a fairly low-order zone axis. The lowest order zone axes are

[100] [110] [111] [210] [211] [310]
[200] [220] [222]
[300]

Since the [200], [220], and [222] vectors point along the directions of [100], [110], and [111], we need only consider the lower index [100], [110], and [111] directions as candidate zone axes. We eliminate the first three zone axes because the pattern does not have the required symmetry as listed in Table 2. The diffraction pattern from the [310] zone axis is not rectangular, so the [210] and [211] patterns seem most promising.

If we knew the camera constant $\lambda L$, it would be appropriate to work with absolute distances for the spot spacings. In this case, we could rearrange the camera equation $rd = \lambda L$ (Equation 35) to obtain the measured distance $r$ of a diffraction spot from the transmitted beam by

$$r = \frac{\lambda L}{a} \sqrt{h^2 + k^2 + l^2} \qquad (36)$$

Here we work with relative spacings instead. Equation 36 shows that the ratio of the spot distances must equal the ratio of $\sqrt{h^2 + k^2 + l^2}$. We first measure the spacings to the vertical and horizontal spots as reference distances (0.65 and 1.10 cm, respectively, from Fig. 10A originally). We then seek the ratios of $\sqrt{h^2 + k^2 + l^2}$ from the allowed $(hkl)$ of an fcc crystal by making a list of possible distances, as shown in Table 3, and looking for two diffraction spots, preferably low-order ones, whose spacings are in the ratio of $0.65/1.10 = 0.591$. One can find by trial and error that $\sqrt{3}/\sqrt{8} = 0.61$. This ratio corresponds to the (111) and (220) diffraction spots, which seem promising candidates

**Table 3. List of Allowed fcc $(hkl)$ Reflections and Their Relative Interplanar Spacings**

| Allowed fcc $(hkl)$ | $\sqrt{h + k^2 + l^2}$ | Relative Spacing |
|---|---|---|
| (111) | $\sqrt{3}$ | $= 1.732$ |
| (200) | $\sqrt{4}$ | $= 2.000$ |
| (220) | $\sqrt{8}$ | $= 2.828$ |
| (311) | $\sqrt{11}$ | $= 3.317$ |
| (222) | $\sqrt{12}$ | $= 3.464$ |
| (400) | $\sqrt{16}$ | $= 4.000$ |
| (331) | $\sqrt{19}$ | $= 4.359$ |
| (420) | $\sqrt{20}$ | $= 4.472$ |
| (422) | $\sqrt{24}$ | $= 4.899$ |

for further work. Note also that the diffraction pair (200) and (311), the pair (200) and (222), and the pair (220) and (422) also have good ratios of their spot spacing, but they contain higher order reflections.

We need to choose specific vectors in the $\langle 111 \rangle$ and $\langle 220 \rangle$ families that provide the correct angles in the diffraction pattern. There are a number of ways to do this; here we use $[\bar{1}11]$ and $[220]$. Note that $[\bar{1}11] \cdot [220] = 0$, so these spots are consistent with the observed 90° angles. It turns out that we can eliminate two of our other three candidate pairs of diffraction spots—the pair (200) and (311) and the pair (200) and (222)—because no vectors in their families are at 90° angles. (For nonperpendicular normalized vectors, we find the angle between them by taking the arccosine of their dot product.) Now we complete the diffraction pattern by labeling the other diffraction spots by vector addition, as illustrated in Figure 10B. Finally, we get the zone axis of the crystal from the vector cross-product:

$$[220] \times [\bar{1}11] = (2 - 0)\hat{x} + (0 - 2)\hat{y} + (2 - -2)\hat{z} = [2\bar{2}4] = [1\bar{1}2] \qquad (37)$$

The astute reader may wonder what happened to the candidate pair of (220) and (422) reflections, which also have good ratios of their spot spacings, and a 90° angle is formed by [220] and $[2\bar{2}4]$. We could have gone ahead and constructed a candidate diffraction pattern with these diffraction vectors. The zone axis is

$$[220] \times [2\bar{2}4] = (8 - 0)\hat{x} + (0 - 8)\hat{y} + (-4 - 4)\hat{z} = [8\bar{8}\bar{8}] = [1\bar{1}\bar{1}] \qquad (38)$$

This should seem suspicious, however, because a $\langle 111 \rangle$ zone axis provides a diffraction pattern with hexagonal symmetry (Table 2), quite unlike the rectangular symmetry of Figure 10. These {220} diffracted spots make a hexagonal pattern around the transmitted beam. Once the zone axis is identified, it is important to check again all expected diffraction spots and ensure that the diffraction pattern accounts for all of them. Once we have done so, it is clear that the (220) and (422) spots are inappropriate for indexing the pattern in Figure 10.

Having gone through the exercise of indexing the diffraction pattern in Figure 10, one can appreciate how tedious the practice could be for low-symmetry patterns with nonorthogonal vectors. As mentioned above, several excellent computer programs are available to help simplify the task (but do not trust them blindly). It is also important to note that the eye is able to judge distances to ~0.1 mm, particularly with the aid of a 10× calibrated magnifier, so the measurement accuracy of spot spacings in a diffraction pattern is typically about that good. A diffraction spot 10 mm from the center of the pattern yields a measurement error of 1%. For the highest accuracy in determining spot spacings, it is often preferable to measure the distance between sharper, higher order spots and then divide by the number of spots separating them (plus 1). Photographic printing can distort spot spacings, so measurements should be performed directly on negatives or the photographic enlarger should be checked for possible distortions.

## Kikuchi Lines and Specimen Orientation

**Electron Diffuse Scattering.** In thick specimens, features are seen in electron diffraction patterns in addition to the usual Bragg diffraction spots and their fine structure. Inelastic scattering contributes a diffuse background to the diffraction pattern, even at moderate specimen thicknesses. More interestingly, intersecting sets of straight lines appear on top of the diffuse background. These are called Kikuchi lines (Kikuchi, 1928). They may be either bright or dark and are straight and regularly arranged. We can explain the existence of Kikuchi lines by a combination of two electron-scattering processes, the first one inelastic, followed by elastic Bragg diffraction. Although the first electron-scattering process is inelastic, a 100-keV electron does not lose too much energy, typically < 30 eV, owing to plasmon excitation, so the magnitude of its wave vector is nearly unaffected. Even for an energy loss of 1 keV, a 100-keV electron will undergo a change in wave vector of only ~0.5%. The direction of the inelastically scattered electron deviates from the direction of the incident beam, but usually only by a small amount. There is a distribution of directions for inelastically scattered electrons, but it is important to note that this distribution is peaked along the incident direction, as illustrated at the top of Figure 11A. A diffuse background in the diffraction pattern originates from the inelastically scattered electrons that are not Bragg diffracted but are forward peaked, as shown at the bottom in Figure 11A. This diffuse background has a higher intensity for thicker specimens, at least until electron transparency is lost.

**Origin of Kikuchi Lines.** Kikuchi lines are features in the inelastic background that show the crystallographic structure of the sample. To create Kikuchi lines, a few of the inelastically scattered electrons must undergo a second scattering, which is an elastic Bragg diffraction event. It is important to recognize that most of the inelastically scattered electrons have lost relatively little energy and undergo a minimal change in their wavelength $\lambda$. Of these electrons, some may be diverted by the first inelastic scattering event so they become incident at a Bragg angle to some crystal planes. These electrons can then be Bragg diffracted.

The two rays drawn in Figure 11B (labeled "two inelastically scattered rays") are special ones, being rays in the plane of the paper that are oriented properly for Bragg diffraction by the crystal planes (*hkl*). Notice the two directions for the electrons coming out of the crystal, labeled $K_{hkl}$ and $\mathbf{K}_{\bar{h}\bar{k}\bar{l}}$ in Figure 11B. The beam $\mathbf{K}_{hkl}$ consists of electrons that were Bragg diffracted out of the forward-scattered inelastic beam plus those electrons that were scattered inelastically at a modest angle but not diffracted. The other beam $\mathbf{K}_{\bar{h}\bar{k}\bar{l}}$ consists of the forward-scattered electrons plus those electrons that were Bragg diffracted into the forward direction (after first having been inelastically scattered by a modest angle). The important point is that the forward beam is stronger, so there are more electrons lost by secondary Bragg diffraction from the forward beam, $\mathbf{K}_{\bar{h}\bar{k}\bar{l}}$, than vice versa. Electron intensity in the diffuse background is therefore taken from the beam $\mathbf{K}_{\bar{h}\bar{k}\bar{l}}$ and

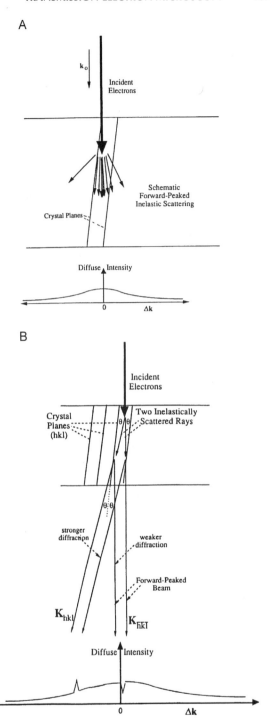

**Figure 11.** (**A**) Top: Electron paths through specimen. Bottom: Diffuse intensity recorded in diffraction pattern from forward-peaked inelastic scattering. (**B**) Origin of Kikuchi lines. Top: Electron paths through specimen for electrons subject to inelastic scattering followed by secondary Bragg diffraction. Bottom: Pattern of scattered intensity showing sharp modulations in diffuse intensity caused by Bragg diffraction.

added to the beam $\mathbf{K}_{hbl}$. As shown at the bottom of Figure 11B, the diffuse background has a deficit of intensity on the right and an excess of intensity on the left. The excess and deficit beams are separated by the angle $2\theta$.

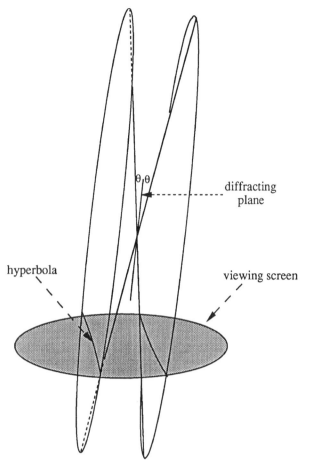

**Figure 12.** Intersection of Kossel cones with viewing screen of TEM.

Excesses and deficits in the inelastic background intensity can be produced in any direction that is at the Bragg angle to the crystal planes. In three dimensions, the set of these rays form pairs of Kossel cones, which are oriented symmetrically about the diffracting planes. All lines from the vertices of these cones make the Bragg angle θ with respect to the diffracting plane. In Figure 12, note how the minimum angle between the surfaces of the two cones is 2θ. The intersections of these two cones with the viewing screen of the microscope are nearly hyperbolas. Since θ is so small and the viewing screen is so far away, these hyperbolas appear as straight lines. The line closer to the transmitted beam is darker than the background (the deficit line); the other line is brighter (the excess line).

Because Kikuchi lines originate from inelastically scattered electrons, their intensity increases with the diffuse background in the diffraction pattern. Kikuchi lines are found to be weak or nonexistent in very thin regions of a specimen where the diffraction spot pattern is without much diffuse background. As one translates a specimen in TEM to observe thicker regions, the Kikuchi lines become more prominent, often becoming more observable than the diffraction spots themselves. The Kikuchi lines decrease in intensity when the sample becomes so thick that it is no longer electron transparent. Lastly, we note that by the explanation of Figure 11, there should be no

excess or deficit bands in the symmetric case where the Kikuchi lines straddle the transmitted beam. Kikuchi lines are in fact observed, however, showing that our simple explanation of their origin is not adequate. A much more complicated explanation based on dynamical diffraction theory with absorption is necessary to fully account for the intensities of Kikuchi lines (Hirsch et al., 1977; Williams and Carter, 1996). Regardless of this, we can use the crystallographic symmetry of the Kikuchi lines in the TEM analysis of crystals.

**Indexing Kikuchi Lines.** For a specific set of crystal planes (hkl), the complement of the two vertex angles of the Kossel cones is 2θ (see Fig. 12). This is the same angle as between the transmitted beam and the (hkl) diffracted beam. On the viewing screen, the separation between the two Kikuchi lines will therefore be the same as the separation of the (hkl) diffraction spot from the (000) spot. We can index the Kikuchi lines by measuring their separations in much the same way as we index diffraction spots. For two different pairs of Kikuchi lines, from the planes $(h_1k_1l_1)$ and $(h_2k_2l_2)$, their separations between excess and deficit lines $p_1$ and $p_2$ are in the ratio

$$\frac{p_1}{p_2} = \frac{\sqrt{h_1^2 + k_1^2 + l_1^2}}{\sqrt{h_2^2 + k_2^2 + l_2^2}} \tag{39}$$

Figure 13 shows ratios of $\sqrt{6}$, 2, and $\sqrt{2}$ for indexed (211), (200), and (110) Kikuchi line pairs. The angles between the Kikuchi line pairs can be measured with precision. High-energy electrons have very obtuse vertex angles for their Kossel cones (small 2θ); it is useful to think of the Kossel cones as extensions of the crystal planes onto the viewing screen. At the same time it is useful to think of

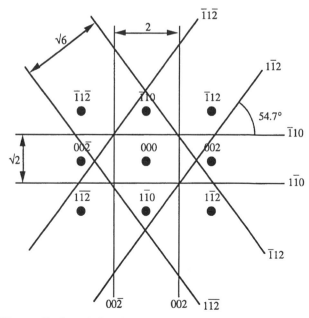

**Figure 13.** Some indexed Kikuchi lines for bcc diffraction pattern on a[110] zone axis.

the diffraction spots as representing the normals to the diffracting planes. It is then clear that the angles between intersecting Kikuchi line pairs will be the same as the angles between their corresponding diffraction spots, at least so long as the Kikuchi lines are not too far from the center of the viewing screen. These angles are helpful for indexing Kikuchi lines in the same way that the angles between pairs of diffraction spots were useful for indexing diffraction patterns above. For example, the angle $\phi$ between the $(1\bar{1}2)$ and $(1\bar{1}0)$ Kikuchi lines is given as

$$\phi = \arccos\left(\frac{1}{\sqrt{6}}(1\bar{1}2)\cdot\frac{1}{\sqrt{2}}(1\bar{1}0)\right) = 54.7° \quad (40)$$

**Specimen Orientation and Deviation Parameter.** Kikuchi lines are useful for precise determination of the specimen orientation in TEM. Recall that when we tilt the specimen, we tilt its reciprocal lattice. Tilts of the reciprocal lattice with respect to a stationary Ewald sphere do not cause any substantial changes in the positions of the diffraction spots, but the individual spots increase or decrease in intensity. Conversely, the positions of the Kikuchi lines vary sensitively with the specimen tilt and make an excellent tool for accurately determining specimen orientation. We see from Figure 12 that as the diffracting plane is tilted, the Kossel cones move by exactly the same angle. The Kikuchi lines behave as though they were affixed to the bottom of the crystal. Usually, we use a rather long camera length for diffraction work, so there is significant movement of the Kikuchi lines on the viewing screen when the specimen is tilted.

Figure 14 shows how the Kikuchi lines can be used to determine the sign and magnitude of the deviation parameter $s$, which quantifies how accurately the Bragg condition is satisfied for a particular reflection $\mathbf{g}$ (discussed above). This in turn determines the diffraction contrast and appearance of BF and DF TEM images. As illustrated in Figure 14, the distance between the Kikuchi lines of order $\mathbf{g}$ is $r$, where

$$r = 2\theta L \quad (41)$$

This is the same distance $r$ between the 000 and $\mathbf{g}$ ($hkl$) diffraction spots given in Equation 36, and $L$ is the camera length shown in Figure 9.

Consider first the special situation where the Kikuchi lines intersect the transmitted beam and the diffraction spot $\mathbf{g}$ exactly, as shown in Figure 14A. This special situation corresponds to the exact Bragg condition because the transmitted beam is oriented at the angle $\theta$ with respect to the diffracting planes. In this special case, $\mathbf{s} = 0$. Now tilt the crystal counterclockwise into the arrangement on the right. The angle of tilt is

$$\phi = \frac{x}{L} \quad (42)$$

where $x$ is the distance between the diffracted spot and its corresponding bright Kikuchi line. Since the Kikuchi lines move as though they are affixed to the crystal, the bright

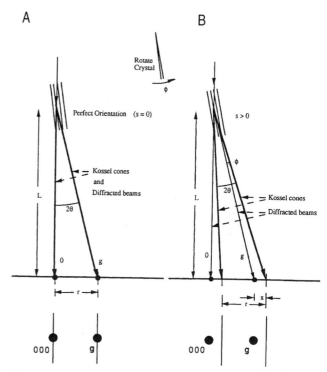

**Figure 14.** Geometry of crystal rotation and position of Kikuchi lines with respect to diffraction spots. Bottom portion of each figure illustrates positions of Kikuchi lines and diffraction spots as seen on viewing screen. (**A**) Intersection of Kikuchi lines with **g** vector occurs when specimen is at exact Bragg orientation for that **g** vector. (**B**) When crystal in (**A**) is rotated counterclockwise by angle $\phi$, Kikuchi lines move right, in the same sense as crystal rotation.

Kikuchi line moves to the right of the spot, as illustrated in Figure 14B. When we rotate the crystal by the angle $\phi$, we also rotate the reciprocal lattice with respect to the Ewald sphere by the same amount, as shown in Figure 15. By using Figure 15, we can find the relation between the magnitudes of **s** and **g**:

$$\phi = \frac{s}{g} \quad (43)$$

where $g$ is the magnitude of $\mathbf{g}$, i.e., $|\mathbf{g}|$. Combining Equations 42 and 43 yields

$$\frac{s}{g} = \frac{x}{L} \quad \text{or simply} \quad s = \frac{gx}{L} \quad (44)$$

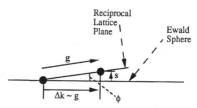

**Figure 15.** Relationship between deviation vector **s** and rotation of crystal by angle $\phi$ (as in Fig. 14).

where $x$ is the distance from the bright Kikuchi line to the corresponding spot **g**. Note that it is possible to eliminate the camera length from Equation 44 by inserting Equation 41 for $L$, so the magnitude of **s** is given by

$$s = g2\theta\frac{x}{r} \tag{45}$$

Equation 45 shows how one can obtain a value for the deviation parameter $s$ from the position of the Kikuchi lines with respect to the diffraction spots. Because $\theta$ (in radians) is small, we can determine very small values of $s$ with accuracy. In electron diffraction, typical units for both $s$ and $g$ are reciprocal nanometers.

**The Sign of s.** We say that $s > 0$ if the excess Kikuchi line lies outside its corresponding diffraction spot **g**. In this case, the reciprocal lattice point lies inside the Ewald sphere, as was the case in Figures 14B and 15. Alternatively, we say that $s < 0$ if the excess Kikuchi line lies inside its corresponding diffraction spot, and we say that $s = 0$ when the Kikuchi line runs exactly through its corresponding diffraction spot, e.g., Figure 14A. The convention is that $s$ points from the Ewald sphere to the reciprocal lattice point, and $s$ is positive if **s** points upward along the positive $z$ axis. This is consistent with the relationship $\mathbf{g} = \Delta\mathbf{k} + \mathbf{s}$ given previously in Equation 11, and shown in the Ewald sphere construction in Figure 4.

### Lens Defects and Resolution

Important performance figures in TEM are the smallest spatial features that can be resolved in a specimen, or the smallest probes that can be formed on a sample. Electromagnetic lenses have aberrations that limit their performance, and to understand resolution, we must first understand lens aberrations. We can then consider the cumulative effect of all lens defects on the performance of a TEM.

**Spherical Aberration.** Spherical aberration distorts the focus of off-axis rays. The further the ray deviates from the optic axis, the greater is its error in focal length. All magnetic lenses have a spherical aberration coefficient that is positive so that those rays furthest from the optic axis are focused most strongly. The angle of illumination into the lens is defined as the aperture angle $\alpha$; in paraxial imaging conditions, $\alpha$ is small. For reference, we define the true (or Gaussian) image plane as the image plane for paraxial imaging conditions. Spherical aberration causes an enlargement of the image of a point $P$ to a distance $QQ'$, as illustrated in Figure 16. The minimum enlargement of point $P$ occurs just in front of $QQ'$ and is termed the disk of least confusion. For a magnetic lens, the diameter $d_s$ of the disk of least confusion caused by spherical aberration corresponds to a distance $d_s$ on the specimen (Hirsch et al., 1977; Reimer, 1993; Williams and Carter, 1996):

$$d_s = 0.5C_s\alpha^3 \tag{46}$$

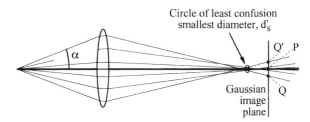

**Figure 16.** Lens with positive spherical aberration, i.e., positive $C_s$, forming disk of confusion with diameter $d_s$.

where $C_s$ is the spherical aberration coefficient (usually 1 to 2 mm) and $\alpha$ is the semiangle of convergence of the electron beam.

**Chromatic Aberration.** Magnetic lenses are not achromatic. Electrons entering the lens along the same path but having different energies have different focal lengths. The spread in focal lengths is proportional to the spread in energy of the electrons. There are two main sources of this energy distribution. First, the electron gun does not produce monochromatic electrons. Typically, $<\pm1$ eV of energy spread can be attributed to irregularities of the high-voltage supply. There is also a thermal distribution of electron energies as the electrons are thermionically emitted from a hot filament, which contributes on the order of $\pm0.1$ eV of energy spread. With high beam currents, there are also Coulomb repulsions of the electrons at the condenser cross-over. This contributes an energy spread of $\pm1$ eV through a phenomenon known as the Boersch effect. The second main cause of nonmonochromaticity of electrons is the specimen itself. Inelastic scatterings of the high-energy electrons by plasmon excitations are a common way for electrons to lose 10 to 20 eV. Thin specimens will help minimize the blurring of TEM images caused by chromatic aberration. The disk of least confusion for chromatic aberration has a diameter $d_c$ at the specimen,

$$d_c = \frac{\Delta E}{E}C_c\alpha \tag{47}$$

where $\Delta E/E$ is the fractional variation in electron beam voltage and $C_c$ is the chromatic aberration coefficient ($\sim1$ mm).

**Diffraction from Apertures.** Diffraction of the electron wave from the edge of an aperture contributes a disk of confusion corresponding to a distance at the specimen of $d_d$:

$$d_d = \frac{0.61\lambda}{\alpha_{OA}} \tag{48}$$

where $\lambda$ is the electron wavelength and $\alpha_{OA}$ is the aperture angle of the objective lens. Equation 48 is the classic Rayleigh criterion that sets the resolution in light optics (Smith and Thomson, 1988). In essence, Equation 48 states that when the intensity between two point (Gaussian) sources of light reaches 0.81 of the maximum intensity of the sources, they can no longer be resolved.

**Astigmatism.** Astigmatism occurs when the focusing strength of a lens varies with angle $\theta$ around the optic axis, again leading to a spread of focus and a disk of least confusion. Two lenses of the transmission electron microscope require routine correction for astigmatism. To obtain a circular probe on the specimen, we must correct for astigmatism in the second condenser lens. Similarly, objective lens astigmatism blurs the image and degrades resolution, and it is important to correct astigmatism of the objective lens. Unlike spherical aberration, it is possible to use stigmators to correct for astigmatism. This correction can in fact be performed so well that astigmatism has a negligible effect on image resolution.

A stigmator in a modern transmission electron microscope is a pair of magnetic quadrupole lenses arranged one above the other and rotated with respect to each other. Correction of objective lens astigmatism is one of the more difficult skills to learn in electron microscopy. This correction is particularly critical in HRTEM, where the image detail depends on the phases of the beams and hence on the symmetry of the magnetic field of the objective lens about the optic axis. The astigmatism correction is tricky because three interdependent adjustments are needed: (1) main focus (adjustment of objective lens), (2) adjustment of $x$ stigmator, and (3) adjustment of $y$ stigmator. These adjustments must be performed in an iterative manner using features in the image as a guide. The procedure to accomplish this is a bit of an art and a matter of personal preference. A holey carbon film is an ideal specimen for practicing this correction. When the objective lens is overfocused (strong current) or underfocused (weak current) with respect to the Gaussian image plane, dark and bright Fresnel fringes, respectively, appear around the inside of a hole.

**Resolution.** Here we collect the results of lens defects (besides astigmatism) described above to obtain a general expression for the resolution of the transmission electron microscope for its two important modes of operation. In one case, we are concerned with the smallest probe that we can put on a specimen. In high-resolution imaging, we are concerned with the smallest feature in the specimen that can be resolved. We begin by obtaining an expression for the minimum probe diameter on a specimen, $d_0$.

If no aberrations were present in the optics of an electron microscope, the minimum probe size at the specimen could be readily calculated. The probe diameter $d_0$ is related to the total probe current $I_p$ by the expression (Reimer, 1993)

$$I_p = \frac{1}{4}\pi d_0^2 j_0 \tag{49}$$

where $j_0$ is the current density (in amperes per square centimeter) in the probe and $I_p$ is $j_0$ times the area of the probe. Conservation of gun brightness on the optic axis implies that

$$j_0 = \pi\beta\alpha_p^2 \tag{50}$$

where $\alpha_p$ is the semiangle of beam convergence of the electron probe on the specimen and $\beta$ is the brightness, defined as current density per solid angle. Substituting the second expression into the first and solving for $d_0$ yield

$$d_0 = \frac{\sqrt{4I_p/\beta}}{\pi\alpha_p} = \frac{C_0}{\alpha_p} \tag{51}$$

So for a given probe current $I_p$, small values of the probe diameter $d_0$ are obtained by increasing the brightness $\beta$ and semiangle of convergence $\alpha_p$. Because of lens aberrations, however, $\alpha_p$ has a maximum value, and $\beta$ is limited by the design of the electron gun.

A general expression for the minimum probe size and image resolution can be obtained by summing in quadrature all diameters of the disks of least confusion, $d_s$, $d_c$, $d_d$, and $d_0$, as

$$d_p^2 = d_s^2 + d_c^2 + d_d^2 + d_0^2 \tag{52}$$

Substituting the diameters of these disks of least confusion from Equations 46, 47, 48, and 51 yields

$$d_p^2 = \frac{C_0^2 + (0.61\lambda)^2}{\alpha_p^2} + 0.25C_s^2\alpha_p^6 + \left(\alpha_p C_c \frac{\Delta E}{E}\right)^2 \tag{53}$$

For a thermionic gun with a large probe current, $C_0 \gg \lambda$ and the contributions of $d_d$ and $d_c$ can be neglected. In this case, the diameters $d_0$ and $d_s$ superpose to produce a minimum probe diameter $d_{min}$ at an optimum aperture angle $\alpha_{opt}$ for a constant $I_p$. The optimum aperture angle is found by setting $(d/d\alpha_p)d_p = 0$, giving

$$\alpha_{opt} = \left(\frac{4}{3}\right)^{1/8}\left(\frac{C_0}{C_s}\right)^{1/4} \tag{54}$$

and substitution in Equation 53 yields

$$d_{min} = \left(\frac{4}{3}\right)^{3/8} C_0^{3/4} C_s^{1/4} \cong 1.1 C_0^{3/4} C_s^{1/4} \tag{55}$$

In the case of image resolution in TEM (or for a field emission gun where the total probe current is lower so that $C_0 \ll \lambda$), the contributions of $d_0$ and $d_c$ can be neglected. Superposition of the remaining terms again yields a minimum that is less than with $d_0$. In this case, $\alpha_{opt}$ and $d_{min}$ are given by

$$\alpha_{opt} = 0.9\left(\frac{\lambda}{C_s}\right)^{1/4} \tag{56}$$

$$d_{min} = 0.8\lambda^{3/4}C_s^{1/4} \tag{57}$$

These expressions can be used to calculate the optimum aperture angle and the resolution limit (or minimum probe size) of a transmission electron microscope. (Note that the same equations with slightly different values of the leading constants, which are all near 1, are found in the literature depending on the exact expressions used for the quantities in Equation 52.) In a modern field emission gun TEM, it is common to achieve spatial resolutions <0.2 nm and probe sizes of 0.5 nm for a 200-keV microscope with $C_s = 1.0$ mm. It is important to note that these quantities represent the instrumental resolution limits and are well-defined, calculated quantities. The actual resolution that one can achieve for any given sample depends on a variety of specimen parameters, such as the sample thickness, the diffracting conditions, and how localized scattering is in the sample, so that the image resolution actually achieved in a typical BF or DF image is usually one or two orders of magnitude larger than the instrument resolution. For example, the width of the dark contrast associated with an edge dislocation in a BF image is $\sim \xi_g/3$ (Hirsch et al., 1977), or 10 nm wide for low-index reflections in metals (refer to Table 2).

Detailed alignment procedures for the transmission electron microscope are provided by the instrument manufacturers. In addition, the book by Chescoe and Goodhew (1984) and Chapters 5 to 9 in Williams and Carter (1996) explain the basic TEM components and their alignment procedures.

## METHOD AUTOMATION

Transmission electron microscopy traditionally has been operator intensive with little computer automation except during postmicroscope analysis of data or when the microscope is operated in the scanning (STEM) mode. This trend is changing rapidly since the evolution of digital charge-coupled diode (CCD) cameras and faster computers now makes it possible to acquire and process images in real time on the microscope (Egerton, 1996; Williams and Carter, 1996). Companies such as Gatan offer several different types of CCD cameras and software packages such as Autotuning and DigitalMicrograph, which allow the user to correct objective lens astigmatism and process images on the microscope. Some TEM manufacturers have hardware/software that can be installed to step a probe across the sample to perform x-ray analyses across an interface or to automatically measure the spot spacings and angles in a diffraction pattern on the screen, for example. The specimen stages in TEM can now be computer controlled, allowing the user to input or recall particular specimen positions and/or tilts in the microscope. In addition, considerable effort is being made to establish remote microscopy centers in national electron microscope laboratories (Voelkl et al., 1998). Microscopes are becoming automated in many other ways, and it is likely that this trend will continue, although the operator will still need to plan the desired investigation.

## DATA ANALYSIS AND INITIAL INTERPRETATION

The theories underlying electron diffraction and imaging were outlined (see Principles of the Method) and practical considerations associated with implementation of the theory in TEM were described (see Practical Aspects of the Method). The present section uses several examples to

illustrate how we can analyze and interpret typical data from TEM. It should be emphasized that a variety of computer programs are available to calculate, process, and analyze diffraction patterns and images.

### Effect of Shape Factor

One of the most important aspects in our discussion of the shape factor in Equations 8 to 10 was that the width of diffracted intensity in reciprocal space, i.e., the SAD pattern, varies inversely with the dimensions of a crystal in real space. This shape factor effect is often observed in diffraction patterns from multiphase materials containing small plate-shaped or rod-shaped precipitates, as illustrated below.

Figure 17 shows an example of the shape factor effect in an Al-Cu alloy, where very thin Cu-rich precipitates form in the Al-rich matrix. These precipitates, called Guinier-

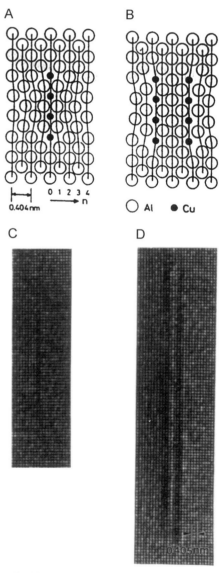

**Figure 17.** (**A**) Illustration of GP(1) zone, (**B**) illustration of GP(2) zone, (**C**) HRTEM image of GP(1) zone in Al-Cu, and (**D**) HRTEM image of GP(2) zone in Al-Cu. (From Chang, 1992.)

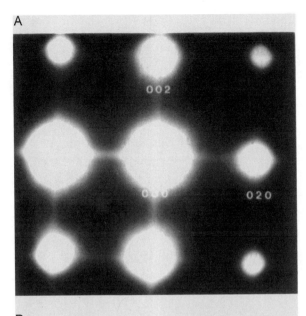

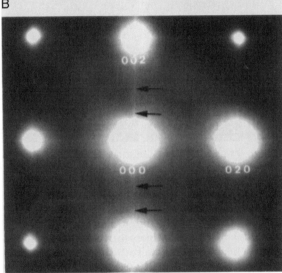

**Figure 18.** SAD pattern from sample containing (**A**) GP(1) zones and (**B**) GP(2) zones in Al-Cu alloy. (From Rioja and Laughlin, 1977.)

Figure 18A, the streaks from the single layer of Cu atoms are practically continuous along the two $\langle 100 \rangle$ directions in the plane of the figure because the precipitate is essentially a monolayer of Cu atoms, as in Figures 17A and C. In Figure 18B, the streaks along $\langle 100$ are no longer continuous but have maxima at $1/4\langle 100 \rangle$ positions in the diffraction pattern. This periodicity arises because the Cu planes in the GP(2) zone are spaced four $\{100\}$ planes apart, as illustrated in Figures 17B and D. (Note that the precipitate reflections still appear as streaks because these plates are very thin.) This illustrates the important point that for every real-space periodicity in a specimen, there is a corresponding reciprocal space intensity, and diffraction patterns and corresponding images should be carefully compared to interpret microstructures.

### Variations in Specimen Thickness and Deviation Parameter

As described below (see Sample Preparation), many TEM specimens are thinned to perforation using techniques such as electropolishing or ion milling, so that the specimens are thin foils that vary in thickness from the edge of the perforation (hole) inward. In such specimens, it is common to observe thickness fringes running parallel to the edge of the hole, particularly when the specimen is oriented in a strong two-beam diffracting condition with $s = 0$ for a particular **g**. These fringes are a result of the variation in diffracted intensity **Ig** (and $\mathbf{I}_0$) in Equation 22 as a function of $t$ for constant $s$, i.e., for a fixed specimen orientation. Figures 19A and B show complementary BF and DF images of thickness fringes in Al, taken by placing a small objective aperture around the optic axis and tilting the beam to obtain the DF image, as illustrated in Figures 6 through 8. Note that the intensity in the BF and DF images is complementary, as expressed by Equation 20. For a constant value of $s$, the intensity varies periodically with $t$, becoming zero each time the product $st$ in Equation 22 is an integer, as illustrated schematically in Figure 5. Thus, the fringes in a DF image are dark for a thickness $t = n/s$, where $n$ is an integer and bright for $t = (n + \frac{1}{2})/s$; the reverse is true for a BF image. We can determine the magnitude of **s** from the position of the Kikuchi line in a diffraction pattern (refer to Equation 45 and Fig. 14). If we also know the value of $\xi_g$ (Table 1), we can estimate the sample thickness anywhere in the foil based on the position of the thickness fringes. If $s = 0$ in a BF image, the foil is $\frac{1}{2}\xi_g$ thick in the middle of the first dark fringe, $\frac{3}{2}\xi_g$ in the middle of the second dark fringe, etc. (Fig. 19C); the reverse is true for the DF image. It is possible to calculate the periodicity of the fringes for quantitative comparison with experimental images such as those in Figure 19 using Equations 22 to 25. If the sample is tilted in the microscope, the thickness fringes remain parallel to the edge of the foil, but their spacing changes because $s$ is changing.

Preston (GP) zones, lie on the $\{100\}$ planes and have a thin disk shape. When the disks have a thickness of only an atom or two, the broadened diffraction spots appear as streaks rather than as discrete spots, and two types of GP zones are found. As illustrated in Figures 17A and B, a GP(1) zone consists of a single layer of Cu atoms that have substituted for Al on a $\{100\}$ plane, while a GP(2) zone contains two such Cu layers separated by three $\{100\}$ planes of Al atoms. Experimental HRTEM images of these two types of GP zones, taken along a $\langle 100 \rangle$ matrix direction in an Al–4 wt% Cu alloy, are shown in Figures 17C and D (Chang, 1992). The atomic columns in the high-resolution images are white and the Cu-rich layers appear as darker planes in the images.

Figure 18 shows two diffraction patterns from samples that were aged to contain GP(1) and GP(2) zones. In

Transmission electron microscopy foils of ductile materials such as metal alloys often bend during specimen preparation or handling, so that bend contours are observed in the TEM foil. These intensity fringes are a result of the variation in diffracted intensity **Ig** (and $\mathbf{I}_0$) in Equation 22 as a function of $s$ for constant $t$, i.e., changing specimen

A

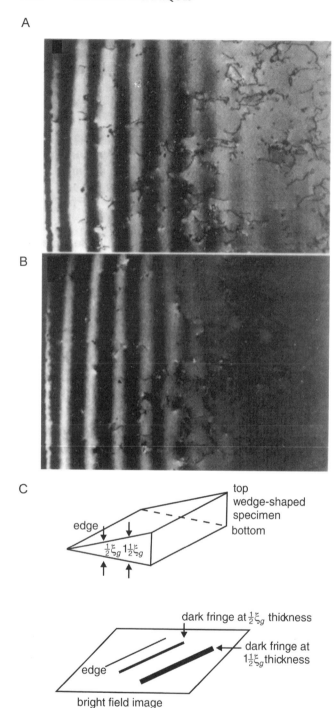

B

C

Figure 19. Complementary (**A**) BF and (**B**) DF images of thickness fringes in Al taken under dynamical conditions with $s = 0$, and (**C**) schematic showing position of fringes in BF image relative to wedge-shaped specimen. Foil edge is on left. (From Edington, 1974.)

A                    B

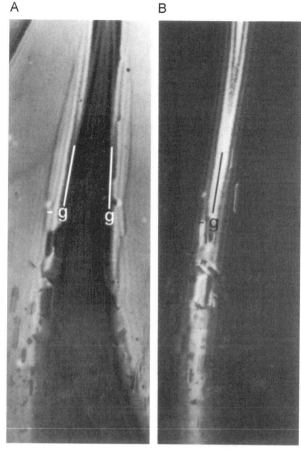

Figure 20. Complementary (**A**) BF and (**B**) DF images of bend contour, with thickness constant horizontally across images. Left side of bend contour in (**A**) is bright in (**B**). (From Thomas and Goringe, 1979.)

orientation for a fixed thickness. Figure 20 shows complementary BF and DF images of bend contours in a metal foil taken under conditions similar to those in Figure 19. The specimen thickness is approximately constant horizontally across the field of view, and the specimen is slightly curved (bent) about a vertical axis in the center of the images. Thus, for a constant value of $t$, the intensity varies periodi-

cally with $s$ horizontally across the images. However, unlike thickness fringes where $t$ continually increases, $s$ varies from negative to zero to positive as the specimen bends through orientation space. The DF image in Figure 20B shows only the left contour as a bright line for the diffracting planes $-\mathbf{g}$. Diffracted intensity $\mathbf{I}_{-g}$ has a symmetric profile about the maximum at $s = 0$, indicated by a line in the figure. In contrast to the thickness fringes in Figure 19, the corresponding BF image in Figure 20A is not symmetric across $-g$ (or $\mathbf{g}$) or complementary to the DF image, for reasons that can be accounted for by the dynamical theory of electron diffraction. Examining the BF image in Figure 20A shows that the crystal is generally darker between $\mathbf{g}$ and $-\mathbf{g}$ when $s < 0$ than when $s > 0$, with the darkest line of intensity lying approximately at $s = 0$, again indicated by lines in the figure. Unlike thickness fringes, bend contours can lie along any direction with respect to the edge of the foil and generally move across the sample when it is tilted.

### Use of Bright-Field, Dark-Field, and Selected-Area Diffraction as Complementary Techniques

Using BF/DF imaging and SAD as complementary techniques to understand the microstructures of materials is one

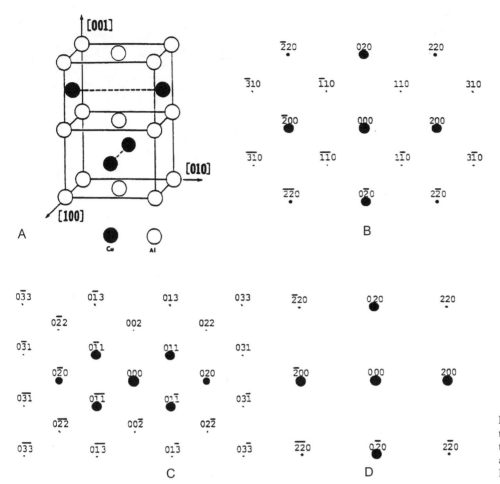

**Figure 21.** (**A**) Crystal structure of the θé phase, (**B**) calculated diffraction pattern for θé phase along [001] and (**C**) along [100], and (**D**) calculated diffraction pattern for [100] Al.

of the most important aspects of CTEM. Here we illustrate the use of these techniques to understand the microstructure of an Al-Cu alloy (the same alloy described in the discussion of shape factors above) heat treated to form $\theta'$ precipitate plates on the {100} planes in the fcc Al-rich matrix. The $\theta'$ plates have the tetragonal crystal structure illustrated in Figure 21A with a composition of $Al_2Cu$. The [001] axis of the $\theta'$ plates contains the fourfold axis while the [100] and [010] axes of the $\theta'$ phase have twofold rotation axes and are crystallographically indistinguishable. Diffraction patterns calculated for the $\theta'$ phase along the [001] and [100] zone axes are shown in Figures 21B and C, and a calculated [100] diffraction pattern for Al is shown in Figure 21D.

The $\theta'$ phase forms in the Al-rich matrix such that the [001] axis of $\theta'$ aligns along each of the three possible ⟨100⟩ directions in the matrix. When the sample is viewed along a ⟨100⟩ matrix direction, one variant of the $\theta'$ plates is seen face-on, while the other two variants are seen edge-on. Figure 22A shows a BF TEM image of all three variants of the $\theta'$ precipitates, and Figure 22B shows a corresponding SAD pattern, taken with a large aperture containing all three variants of precipitates. The square pattern of bright spots is from the Al-rich matrix, which is the majority phase, while the weaker spots, particularly noticeable in the middle of the square pattern of Al spots, arise from the three variants of $\theta'$ plates. A schematic of the diffraction pattern that distinguishes the three variants of spots from the $\theta'$ plates is shown in Figure 22C.

It is possible to determine which precipitate spots in the SAD pattern correspond to each variant by forming DF images with the precipitate spots. This procedure is illustrated in Figure 22D, where a small objective aperture was centered on the optic axis and the electron beam was tilted (see Practical Aspects of the Method) to move the precipitate spot indicated by an arrow in Figures 22B and C into the aperture. As a result, only the vertical plates in Figure 22D appear bright, verifying that the reflection came from that variant. It is also possible to determine which precipitate spots correspond to a particular variant of $\theta'$ by placing a selected-area aperture around one variant and recording its corresponding SAD pattern. This procedure is illustrated in Figures 22E and F, where an aperture placed around the vertical plate reveals that its spots form the rectangular diffraction pattern outlined in Figure 22F. This pattern is identical to the calculated pattern for $\theta'$ in Figure 21C. This procedure can be repeated on the other variants to account for all of the precipitate spots in Figure 22A.

It is also possible to determine the lattice parameters of the $\theta'$ phase using the Al-rich matrix spots for calibration and Equation 35. For example, Al has a lattice parameter

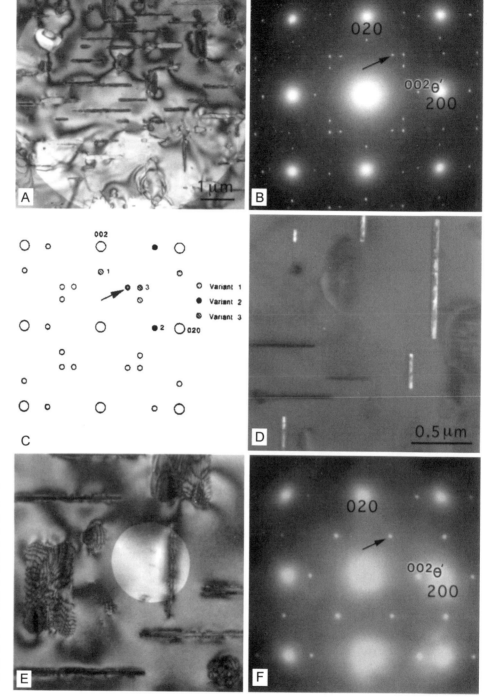

**Figure 22.** (**A**) BF TEM image of $\theta'$ plates viewed along $a\langle 100 \rangle$ Al direction; (**B**) corresponding SAD pattern containing weak spots from all three variants of $\theta'$ plates and stronger spots from Al matrix (labeled); (**C**) schematic of diffraction pattern in (**B**) with arrow indicating which precipitate spot was used for axial DF imaging; (**D**) corresponding DF image showing bright, vertical $\theta'$ plates; (**E**) SAD placed around vertical $\theta'$ plate (revealed by double exposure); and (**F**) corresponding SAD pattern, similar to that in Figure 21C.

of 0.40496 nm so that the {200} interplanar spacing $d_{\{200\}} = 0.2025$ nm. The {200} spots in Figure 22B have a spacing $r = 21.0$ mm on the original image. Substituting these values for $r$ and $d_{\{200\}}$ into Equation 35 gives $\lambda L = 0.425$ nm·cm. The [002] spots of the $\theta'$ phase (labeled in Fig. 22F) have a spacing $r = 14.8$ mm, which, when substituted into Equation 35 with $\lambda L$ then gives $d_{(002)} = 0.575$ nm. This is 0.8% different than the reported value for the $c$ lattice parameter of the $\theta'$ phase of 0.58 nm (Weatherly and Nicholson, 1968). Since the [200] spots of $\theta'$ phase coin-

cide with the {200} spots of the Al matrix, the $a$ lattice parameter of $\theta'$ must be approximately the same as that of Al, or 0.40496 nm. This good matching of lattice parameters between the two phases is why the phase forms as plates on the {100} planes of Al.

### Use of $\mathbf{g} \cdot \delta\mathbf{r} = 0$ in Defect Analysis

The same $\theta'$ plates discussed above can be used to illustrate the application of the $\mathbf{g} \cdot \delta\mathbf{r}$ criterion for defect

A

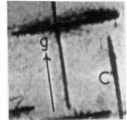

B

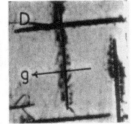

C

**Figure 23.** Burgers vector determination for misfit dislocations A, B, C, and D on flat faces of θ′ plates in Al-Cu alloy aged 1 day at 275°C. P represents dislocations around periphery of plates and Q complex dislocations arising from agglomeration of two plates: (**A**) **g** = [200], dislocations A visible with enlargement showing dislocations C; (**B**) **g** = [020], dislocations B visible with enlargement showing dislocations D; (**C**) **g** = [220], all dislocations visible. (From Weatherly and Nicholson, 1968.)

analysis in TEM (see Diffraction Contrast from Lattice Defects, above). Figure 23 shows the three variants of θ′ plates viewed along a ⟨100⟩ matrix direction at fairly high magnification (Weatherly and Nicholson, 1968). Four types of dislocation arrays are visible in the BF TEM images.

1. Complex arrays with no simple Burgers vector (for a definition of the Burgers vector, refer to Hull and Bacon, 1984) that seem to be associated with the agglomeration of two or more θ′ plates (Q in Fig. 23).

2. Dislocations associated with the periphery of the θ′ plate (P in Fig. 23). These frequently show a characteristic double or triple image and a line of no contrast that is typical of pure edge dislocations having a Burgers vector parallel to the electron beam (Hirsch et al., 1977). The Burgers vector of these dislocations is $a$[001] for a precipitate plate lying parallel to (001); these dislocations accommodate the large misfit (~4.3%) associated with the

peripheral interface of the plate and the $c$ direction of the θ′ lattice.

3. Dislocations associated with the flat face of the θ′ plate (A and B in Fig. 23). These are affected by the contours of the θ′ plates but lie predominantly in the [100] and [010] directions for a plate on (001). It is shown below that the Burgers vectors of these dislocations are $a$[100] and $a$[010].

4. Dislocations associated with plastic deformation of the foil (either accidental or deliberate). They tend to lie in the [110] or [110] directions for a θ′ plate on (001) and have Burgers vectors of the type $a/2$[110].

Here we concentrate on the genuine misfit dislocations associated with the flat interface of the θ′ plate, type 3. The Burgers vector analysis that follows is thus confined to dislocations lying away from the periphery of the plates.

The four sets of dislocations A, B, C, and D are marked in Figure 23. If the dislocations are discrete loops wrapped around the θ′ plates, an analysis in a [001] foil gives a complete determination. The results in Figure 23 show that when $\mathbf{g} = [\bar{2}00]$, dislocations A and C are visible (Fig. 23A); when $g = [0\bar{2}0]$, dislocations B and D are visible (Fig. 23B); and when $\mathbf{g} = [\bar{2}20]$, all four loops are visible (Fig. 23C). Applying the $\mathbf{g} \cdot \delta\mathbf{r} = 0$ criterion (which becomes the $\mathbf{g} \cdot \mathbf{b} = 0$ criterion since the displacement $\delta\mathbf{r} = \mathbf{b}$, where $\mathbf{b}$ is the Burgers vector) for dislocation invisibility to the network AB, it must consist of dislocations with Burgers vectors in directions $\pm[h0l]$ and $\pm[h0l]$, where $h$, $k$, and $l$ are undetermined. The θ′ plate with the dislocations marked D lies on (100) and, by analogy with the θ′ plate on (001), must be covered by a network of dislocations with Burgers vectors in directions $\pm[hk0]$ and $\pm[h0l]$. The dislocations marked D have Burgers vectors in directions $\pm[hk0]$ because they are visible when $\mathbf{g} = [0\bar{2}0]$. Because dislocations D are invisible when $\mathbf{g} = [\bar{2}00]$, $l = 0$. Hence the network AB for a θ′ plate on (001) has Burgers vectors in directions $\pm[100]$ and $\pm[010]$. There are no contrast effects associated with the interface to suggest that the Burgers vectors of the dislocations are not lattice translation vectors, and hence the most likely Burgers vectors are $\pm a$[100] and $\pm a$[010] for a (001) plate. The dislocations are mainly edge in character since their line direction is perpendicular to the Burgers vectors and are mainly accommodating a small misfit (<1%) in the plane of the flat interface.

### Kikuchi Lines and the Effect of Deviation Parameter on Defect Contrast

One of the most important uses of the deviation parameter $s$ is for obtaining and quantifying high-quality TEM images of defects in materials. Figure 24 shows three diffraction conditions with (a) $s = 0$, (b) $s < 0$, and (c) $s > 0$ and the corresponding BF images of dislocations in Al, respectively. Note that the best images are formed when $s > 0$, because the dislocations display sharp, strong contrast above a bright background. The magnitude of $\mathbf{s}$ must be determined to calculate images of such dislocations (Head et al., 1973).

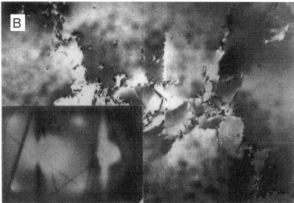

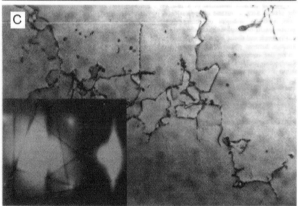

**Figure 24.** Changes in diffraction patterns and corresponding images of dislocations with changes in $s$: (**A**) $s = 0$, (**B**) $s60; 0$, and (**C**) $s > 0$. (From Edington, 1974.)

## SAMPLE PREPARATION

The ideal TEM specimen is usually a thin, parallel-sided foil that is stable in the electron microscope as well as during storage. A gradually tapered wedge is often obtained in practice, and this too is quite useful. The TEM foil should be representative of the bulk material being studied and thus unaltered by sample preparation. The exact thickness of the specimen depends on the type of analysis desired and may range from a few nanometers for HRTEM to tens of nanometers for CTEM imaging. The surface condition is important, particularly for very thin specimens. It is difficult to meet all of these requirements, but the quality

of the results depends on the specimen quality, so specimen preparation is an important aspect of TEM. The range of possible specimen preparation techniques is enormous, and the discussion below is intended to provide an overview of some of the more common techniques available to materials scientists. More comprehensive discussions of specimen preparation procedures are provided by Edington (1974), Goodhew (1984), and Williams and Carter (1996).

There are usually two stages in the preparation of thin specimens from bulk material. The first is to prepare, often by cutting and grinding, a thin slice of the material that is a few millimeters in diameter and a tenth of a millimeter or so thick. The main consideration in the first stage is to avoid using a coarse cutting technique so that damage is introduced throughout the specimen. A diamond wafer saw is often used to cut thin specimens of metals and ceramics, and SiC paper with successively finer grit (e.g., down to 600 grit) is used to grind these materials to the required thickness. The second stage involves thinning this specimen to electron transparency for which several techniques are available. Four of the most common—electropolishing, chemical thinning, ion milling, and $Ga^+$ ion beam thinning—are briefly described next.

### Electropolishing and Chemical Thinning

The most common technique for thinning electrically conductive materials such as metal alloys is electropolishing. The principle behind the method is that the specimen is made the anode in an electrolytic cell so that when current is passed through the cell, the solution dissolves the alloy and deposits it on the cathode. The main considerations in electropolishing are the composition, temperature and flow of the electrolyte across the sample, and the potential difference applied between the sample (anode) and cathode. Under the right conditions, the foil becomes smoother and thinner until perforation is achieved near the center of the specimen. The technique is fairly simple, but sophisticated electropolishing units that allow the operator to carefully control the polishing variables and terminate polishing immediately upon perforation are commercially available. The commercial units are designed to accommodate 3-mm-diameter disks, which is the size required for most TEM specimen holders.

A major limitation of electropolishing is that it cannot be used on nonconducting materials such as ceramics and semiconductors. Chemical thinning, using mixtures of acids without an applied potential, is frequently used for these materials. Both electropolishing and chemical thinning share the advantage that the thinning rate is high, so they are quick and generally nondamaging, although it is possible to preferentially leach out elements or introduce hydrogen into some materials, so as with any technique, one must exercise some caution when utilizing these procedures.

### Ion Milling and Focused $Ga^+$ Ion Beam Thinning

It is common to thin specimens to electron transparency by bombarding them with energetic ions. The ions knock atoms off the surface in a process called sputtering, gradually thinning the material. The most common ions used are $Ar^+$, and these are accelerated toward the sample with an applied potential ranging from 4 to 6 keV at angles ranging from 5° to 20° from the sample surface, i.e., at a glancing angle. Usually two ion guns are used so that the sample can be thinned from both sides simultaneously, and the sample is often cooled with liquid nitrogen to temperatures of $\sim-180°$C to avoid excessive heating by the ion beams. The samples are usually rotated during thinning to prevent preferential sputtering along one direction. The milling rate is usually only a few micrometers per hour, so compared to electropolishing, sample preparation is much more time consuming. Therefore, a dimpler is often used to grind the center of the sample to a few tens of micrometers in thickness to reduce the ion milling time. Ion milling is very useful for thinning nonconducting samples, for thinning materials that consist of phases that electropolish at much different rates, for thinning cross-sectional samples of thin films on substrates such as semiconductor and multilayer materials, or for cleaning the surfaces of samples prepared by other methods. Ion milling produces an amorphous layer on the surfaces of most materials, which can be detrimental to some analyses.

A focused ion beam (FIB) microscope is becoming popular for preparing thin membranes for TEM examination (Basile et al., 1992). A FIB microscope is similar to a scanning electron microscope (SCANNING ELECTRON MICROSCOPY) except that it uses electrostatic lenses to focus a $Ga^+$ ion beam on the specimen instead of electromagnetic lenses and electrons. The $Ga^+$ ions are typically accelerated toward the specimen at 30 keV so that rapid sputtering of the sample occurs. The $Ga^+$ ion probe can be focused to a size as small as 10 nm, and a computer can be used to raster the beam across the sample in a variety of patterns, so that it is possible to fabricate very specific specimen geometries. This technique can be used on both conducting and nonconducting materials, and it is particularly useful for preparing cross-sectional samples of semiconductors and thin films. There is evidence that FIB milling can produce a damaged layer on the surfaces of materials (Susnitzky and Johnson, 1998), but the extent of this behavior and its relation to specimen and microscope parameters remain to be fully determined.

Another instrument that is particularly useful for thinning semiconductor materials to electron transparency is a tripod polisher (Klepeis et al., 1988). The tripod polisher, so called because it has three micrometer feet, is used to hold the specimen while it is mechanically thinned to electron transparency on a polishing wheel using diamond lapping films. This "wedge technique" can be used to examine particular depths in a sample, such as a microelectronic device. Even if not used for final sample preparation, the use of a tripod polisher can dramatically reduce the time for the final thinning step.

### Ultramicrotomy

An ultramicrotome utilizes a fixed diamond knife to slice a thin section from a specimen, which is typically $\sim 0.5 \times 0.5$ mm in cross-section and can be mounted in epoxy for support. The resulting slices are collected in a liquid-filled trough and mounted on copper grids before being inserted into the transmission electron microscope. Ultramicrotomy is widely used to prepare samples of polymers for TEM and can also be used for other materials such as powders, metals, and composites. Slicing generally damages the material (e.g., introducing dislocations into metal alloys), but chemical information is retained, so ultramicrotomy is well suited for preparing thin sections for analytical techniques such as EDS in TEM. Although it only takes a few minutes to obtain a slice from a specimen, considerable expertise is required to mount the specimen and obtain a thin section, so ultramicrotomy can be a fairly time consuming preparation method.

### Replication

A thin film of carbon can be deposited on the surface of a specimen, typically by striking an arc between two carbon rods in vacuum. The carbon layer conforms to the shape of the surface and is allowed to build up to a few tens of nanometers in thickness. The carbon film is then scored into squares a few millimeters wide, removed from the surface by etching, and collected on a copper grid for examination in the transmission electron microscope. If the sample contains highly dispersed phases that can be left in relief on the surface by etching, the carbon film can be deposited on the surface, and the phases are then embedded in the carbon film when it is removed from the surface by further etching. This specimen is called an extraction replica. Since the particles are present in the replica, they can be analyzed by diffraction or analytical techniques without interference from the surrounding matrix. A common alternative to carbon media is cellulose acetate, which can be deposited on the sample as a sheet with solvent.

### Dispersion

One final technique that is simple and works well for small particles such as catalysts or brittle ceramics that can be ground to a fine powder is to disperse a small quantity of the material in ethanol by sonication and then place a few drops of the dispersion on a carbon support film mounted on a grid. When the solution evaporates, the particles cling to the support film and can be examined in the transmission electron microscope. Depending on the material, it may be preferable to use a solvent other than ethanol or a grid material other than copper, but the procedure is generally the same.

### SPECIMEN MODIFICATION

Inelastic collisions of electrons with a specimen often produce the undesirable effect of beam damage. The damage that affects the structure and/or chemistry of the material depends mainly on the beam energy. Damage usually occurs through one of two main mechanisms: (1) radiolysis, in which inelastic scattering breaks chemical bonds in the material, or (2) knock-on damage, in which atoms are displaced off their sites, creating point defects.

The first mechanism generally decreases with increasing beam energy whereas the second increases with increasing beam energy. It is important to remember that, under certain conditions, in TEM it is possible to damage any material. Sometimes this is useful as it allows one to simulate various irradiation conditions or perform in situ studies, but more often it produces unwanted side effects. Some of the more important aspects of beam damage are summarized below. This is followed by a brief discussion of intentional specimen modification in TEM during in situ experiments.

Specimen heating can occur in TEM. Temperature is difficult to measure experimentally because the temperature is affected by many variables, such as specimen thickness, thermal conductivity, surface condition, contact with the holder, beam energy, and beam current. Hobbs (1979) has performed calculations on the effects of beam current and thermal conductivity on specimen temperature. The results of these calculations indicate that beam heating is generally negligible for good thermal conductors such as metals but substantial for insulating materials. To minimize beam heating, one can (1) operate at the highest accelerating potential to reduce the cross-section for inelastic scattering, (2) cool the specimen to liquid nitrogen or liquid helium temperatures in a cooling holder, (3) coat the specimen with a conducting film, (4) use low-dose imaging techniques, or (5) all of these techniques (Sawyer and Grubb, 1987).

Polymers are highly sensitive to the process of radiolysis in TEM. Electrons usually cause the main polymer chain to break up or side groups to break off, leading to breakdown or crosslinking of the polymer, respectively. In the first case, the polymer continually loses mass under irradiation, whereas in the second case, the polymer ultimately ends up as a mass of carbon. If the polymer is crystalline, radiation damage causes a loss of crystallinity, which can be measured experimentally by the loss of diffraction contrast in the image or the reduction in intensity of diffraction spots in the diffraction pattern and the appearance of an amorphous pattern. Radiolysis can also occur in ionic and covalent materials such as ceramics and minerals. This often results in the formation of new compounds or the amorphization of the original material, processes that can be observed in TEM. Radiolysis is not affected by heat transfer considerations, so the main ways to minimize this process are to increase the accelerating voltage and minimize the specimen thickness and exposure to the electron beam.

Metals are mainly affected by knock-on damage in the transmission electron microscope. Knock-on damage is directly related to the beam energy and the atomic weight of the atoms (Hobbs, 1979). Accelerating potentials $> \sim 200$ kV can damage light metals such as aluminum, and the effect on heavier elements continually increases as the beam energy increases above this value. Knock-on damage usually manifests itself by the formation of small vacancy or interstitial clusters and dislocation loops in the specimen, which can be observed by diffraction contrast. The main way to avoid displacment damage is to use lower accelerating voltages. Knock-on damage can also occur in polymers and minerals where there is generally a trade-off

between knock-on damage and radiolysis. Since knock-on damage creates vacancies and interstitials, it can be used to study the effects of irradiation in situ.

Another important aspect of specimen modification in TEM is the ability to perform in situ studies to directly observe the behavior of materials under various conditions (Butler and Hale, 1981). This potentially useful effect for electron irradiation studies was mentioned above. In addition, specimen holders designed to heat, cool, and strain (and combinations of these) a TEM sample in the microscope are available commercially from several manufacturers. It is also possible to introduce gases into the transmission electron microscope to observe chemical reactions in situ by using apertures and differential pumping in the microscope. Once they have equilibrated, most commercial TEM holders are sufficiently stable that it is possible to perform EDS and EELS on foils to obtain compositional information in situ (Howe et al., 1998). To reproduce the behavior of bulk material in TEM, it is often desirable to perform in situ experiments in high-voltage TEM (e.g., 1 MeV) so that foils on the order of 1 µm thick can be used. Using thicker foils can be particularly important for studies such as straining experiments. As with most techniques, one needs to be careful when interpreting in situ data, and it is often advisable to compare the results of in situ experiments with parallel experiments performed in bulk material. For example, if we want to examine the growth behavior of precipitates during in situ heating in TEM, we might compare the size of the precipitates as a function of time and temperature obtained in situ in the transmission electron microscope with that obtained by bulk aging experiments. The temperature calibration of most commercial holders seems to be reliable to $\pm 20°$C, and calibration specimens need to be used to obtain greater temperature accuracy. One problem that can arise during heating experiments, for example, is that solute may preferentially segregate to the surface. Oxidation of some samples during heating can also be a problem. In spite of all these potential difficulties, in situ TEM is a powerful technique for observing the mechanisms and kinetics of reactions in solids and the effects of electron irradiation on these phenomena.

## PROBLEMS

A number of limitations should be considered in the TEM analysis of materials, including but not limited to (1) the sampling volume, (2) image interpretation, (3) radiation damage, (4) specimen preparation, and (5) microscope calibration. This section briefly discusses some of these factors. More thorough discussion of these topics can be found in Edington (1974), Williams and Carter (1996), and other textbooks on TEM (see Literature Cited).

It is important to remember that in TEM only a small volume of material is observed at high magnification. If possible, it is important to examine the same material at lower levels of resolution to ensure that the microstructure observed in TEM is representative of the overall specimen. It is often useful to prepare the TEM specimen by more than one method to ensure that the microstructure was not altered during sample preparation. As discussed above (see Specimen Modification), many materials damage under a high-energy electron beam, and it is important to look for signs of radiation damage in the image and diffraction pattern. In addition, as shown above (see Data Analysis and Initial Interpretation), image contrast in TEM varies sensitively with diffracting conditions, i.e., the exact value of the deviation parameter $s$. Therefore, one must carefully control and record the diffracting conditions during imaging in order to quantify defect contrast.

It is important to remember that permanent calibration of TEM features such as image magnification and camera length, which are typically performed during installation, are only accurate to within $\pm 5\%$. If one desires higher accuracy, then it is advisable to perform *in situ* calibrations using standards. A variety of specimens are available for magnification calibration. The most commonly used are diffraction grating replicas for low and intermediate magnifications ($100\times$ to $200,000\times$) and direct lattice images of crystals for high magnifications ($>200,000\times$). Similarly, knowledge of the camera constant in Equation 35 greatly simplifies indexing of diffraction patterns and is essential for the identification of unknown phases. The accuracy of calibration depends on the experiment, but an error of $\pm 1\%$ can be achieved relatively easily, but this can be improved to $\pm 0.1\%$ if care is used in operation of the microscope and measurement of the diffraction patterns. An example of this calibration was given above (see Data Analysis and Initial Interpretation), and it is common to evaporate a metal such as Au directly onto a sample or to use a thin film as a standard. Permanent calibrations are normally performed at standard lens currents that must be subsequently reproduced during operation of the microscope for the calibration to be valid. This is accomplished by setting the lens currents and using the height adjustment ($z$ control) on the goniometer to focus the specimen. Other geometric factors that can introduce errors into the magnification of diffraction patterns are discussed by Edington (1974). There are also a number of other calibrations that are useful in TEM, including calibration of (1) the accelerating voltage, (2) specimen drift rate, (3) specimen contamination rate, (4) sense of specimen tilt, (5) focal increments of the objective lens, and (6) beam current, but these are not detailed here.

## ACKNOWLEDGMENTS

The authors are grateful for support of this work by the National Science Foundation, JMH under Grant DMR-9630092 and BTF under Grant DMR-9415331.

## LITERATURE CITED

Basile, D. P., Boylan, R., Hayes, K., and Soza, D. 1992. FIBX-TEM—Focussed ion beam milling for TEM sample preparation. *In* Materials Research Society Symposium Proceedings, Vol. 254 (R. Anderson, B. Tracy, and J. Bravman, eds.). pp. 23–41. Materials Research Society, Pittsburgh.

Borchardt-Ott, W. 1995. Crystallography, 2nd ed. Springer-Verlag, New York.

Butler, E. P. and Hale, K. F. 1981. Dynamic Experiments in the Electron Microscope, Vol. 9. *In* Practical Methods in Electron Microscopy (A. M. Glauert, ed.). North-Holland, New York.

Chang, Y.-C. 1992. Crystal structure and nucleation behavior of (111) precipitates in an AP-3.9Cu-0.5Mg-0.5Ag alloy. Ph.D. thesis, Carnegie Mellon University, Pittsburgh.

Chescoe, D. and Goodhew, P. J. 1984. The Operation of the Transmission Electron Microscope. Oxford University Press, Oxford.

Cockayne, D. J. H., Ray, I. L. F., and Whelan, M. J. 1969. Investigation of dislocation strain fields using weak beams. *Philos. Mag.* 20:1265–1270.

Edington, J. W. 1974. Practical Electron Microscopy in Materials Science, Vols. 1-4. Macmillan Philips Technical Library, Eindhoven.

Egerton, R. F. 1996. Electron Energy-Loss Spectroscopy in the Electron Microscope, 2nd ed. Plenum Press, New York.

Fultz, B. and Howe, J. M. 2000. Transmission Electron Microscopy and Diffractometry of Materials. Springer-Verlag, Berlin.

Goodhew, P. J. 1984. Specimen Preparation for Transmission Electron Microscopy of Materials. Oxford University Press, Oxford.

Head, A. K., Humble, P., Clarebrough, L. M., Morton, A. J., and Forwood, C. T. 1973. Computed Electron Micrographs and Defect Identification. North-Holland, Amsterdam, The Netherlands.

Hirsch, P. B., Howie, A., Nicholson, R. B., Pashley D. W., and Whelan, M. J. 1977. Electron Microscopy of Thin Crystals, 2nd ed. Krieger, Malabar.

Hobbs, L. W. 1979. Radiation effects in analysis of inorganic specimens by TEM. *In* Introduction to Analytical Electron Microcopy (J. J. Hren, J. I. Goldstein, and D. C. Joy, eds.) pp. 437–480. Plenum Press, New York.

Howe, J. M., Murray, T. M., Csontos, A. A., Tsai, M. M., Garg, A., and Benson, W. E. 1998. Understanding interphase boundary dynamics by *in situ* high-resolution and energy-filtering transmision electron microscopy and real-time image simulation. *Microsc. Microanal.* 4:235–247.

Hull, D. and Bacon, D. J. 1984. Introduction to Dislocations, 3rd ed. (see pp. 17–21). Pergamon Press, Oxford.

Keyse, R. J., Garratt-Reeed, A. J., Goodhew, P. J., and Lorimer, G. W. 1998. Introduction to Scanning Transmission Electron Microscopy. Springer-Verlag, New York.

Kikuchi, S. 1928. Diffraction of cathode rays by mica. *Jpn. J. Phys.* 5:83–96.

Klepeis, S. J., Benedict, J. P., and Anderson, R. M. 1988. A grinding/ polishing tool for TEM sample preparation. *In* Specimen Preparation for Transmission Electron Microscopy of Materials (J. C. Bravman, R. M. Anderson, and M. L. McDonald, eds.). pp. 179–184. Materials Research Society, Pittsburgh.

Krivanek, O. L., Ahn, C. C., and Keeney, R. B. 1987. Parallel detection electron spectrometer using quadrupole lenses. *Ultramicroscopy* 22:103–116.

McKie, D. and McKie, C. 1986. Essentials of Crystallography, p 208. Blackwell Scientific Publications, Oxford.

Miller, M. K. and Smith, G. D. W. 1989. Atom Probe Microanalysis: Principles and Applications to Materials Problems. Materials Research Society, Pittsburgh.

Reimer, L. 1993. Transmission Electron Microscopy: Physics of Image Formation and Microanalysis, 3rd ed. Springer-Verlag, New York.

Reimer, L. (ed.). 1995. Energy-Filtering Transmission Electron Microscopy. Springer-Verlag, Berlin.

Rioja, R. J. and Laughlin, D. E. 1977. The early stages of GP zone formation in naturally aged Al-4 wt pct Cu alloys. *Metall. Trans.* 8A:1257–1261.

Sawyer, L. C. and Grubb, D. T. 1987. Polymer Microscopy. Chapman and Hall, London.

Schwartz, L. H. and Cohen, J. B. 1987. Diffraction from Materials, 2nd ed. Springer-Verlag, New York.

Smith, F. G. and Thomson, J. H. 1988. Optics, 2nd ed. John Wiley & Sons, Chichester.

Susnitzky, D. W. and Johnson, K. D. 1998. Focused ion beam (FIB) milling damage formed during TEM sample preparation of silicon. *In* Microscopy and Microanalysis 1998 (G. W. Bailey, K. B. Alexander, W. G. Jerome, M. G. Bond, and J. J. McCarthy, eds.) pp. 656–667. Springer-Verlag, New York.

Thomas, G. and Goringe, M. J. 1979. Transmission Electron Microscopy of Metals. John Wiley & Sons, New York.

Voelkl, E., Alexander, K. B., Mabon, J. C., O'Keefe, M. A., Postek, M. J., Wright, M. C., and Zaluzec, N. J. 1998. The DOE2000 Materials MicroCharacterization Collaboratory. *In* Electron Microscopy 1998, Proceedings of the 14th International Congress on Electron Microscopy (H. A. Calderon Benavides and M. Jose Yacaman, eds.) pp. 289–299. Institute of Physics Publishing, Bristol, U.K.

Weatherly G. C. and Nicholson, R. B. 1968. An electron microscope investigation of the interfacial structure of semi-coherent precipitates. *Philos. Mag.* 17:801–831.

Williams, D. B. and Carter, C. B. 1996. Transmission Electron Microscopy: A Textbook for Materials Science. Plenum Press, New York.

**KEY REFERENCES**

Edington, 1974. See above.

*Reprinted edition available from Techbooks, Fairfax, Va. Filled with examples of diffraction and imaging analyses.*

Fultz and Howe, 2000. See above.

*An integrated treatment of microscopy and diffraction, with emphasis on principles.*

Hirsch et al., 1977. See above.

*For many years, the essential text on CTEM.*

Reimer, 1993. See above.

*Excellent reference with emphasis on physics of electron scattering and TEM.*

Shindo, D. and Hiraga, K. 1998. High Resolution Electron Microscopy for Materials Science. Springer-Verlag, Tokyo.

*Provides numerous high-resolution TEM images of materials.*

Williams and Carter, 1996. See above.

*A current and most comprehensive text on modern TEM techniques.*

**INTERNET RESOURCES**

http://www.amc.anl.gov

*An excellent source for TEM information on the Web in the United States. Provides access to the Microscopy ListServer and a Software Library as well as a connection to the Microscopy & Microanalysis FTP Site and Libraries plus connections to many other useful sites.*

http://cimewww.epfl.ch/welcometext.html

*A similar site based at the Ecole Polytechnique Federale de Lausanne in Switzerland that contains software and a variety of electron microscopy information.*

http://www.msa.microscopy.com

*Provides access to up-to-date information about the Microscopy Society of America, affiliated societies, and microscopy resources that are sponsored by the society.*

http://rsb.info.nih.gov/nih-image

*Public domain software developed by National Institutes of Health (U.S.) for general image processing and manipulation. Available from the Internet by anonymous ftp from zippy.nimh.nih. gov or on floppy disk from National Technical Information Service, 5285 Port Royal Rd., Springfield, VA 22161, part number PB93-504868.*

JAMES M. HOWE
University of Virginia
Charlottesville, Virginia

BRENT FULTZ
California Institute of
Technology
Pasadena, California

## SCANNING TRANSMISSION ELECTRON MICROSCOPY: Z-CONTRAST IMAGING

### INTRODUCTION

As its name suggests, the scanning transmission electron microscope is a combination of the scanning electron microscope and the transmission electron microscope. Thin specimens are viewed in transmission, while images are formed serially by the scanning of an electron probe. In recent years, electron probes have become available with atomic dimensions, and, as a result, atomic resolution images may now be achieved in this instrument. The nature of the images obtained in scanning transmission electron microscopy (STEM) can differ in significant ways from those formed by the more widespread conventional transmission electron microscopy (CTEM). The key difference lies in their modes of image formation; the STEM instrument can be configured for almost perfect incoherent imaging whereas CTEM provides almost perfect coherent imaging. The latter technique is generally referred to as high-resolution electron microscopy (HREM), though both methods now provide atomic resolution.

The difference between coherent and incoherent imaging was first discussed over one hundred years ago in the context of light microscopy by Lord Rayleigh (1896). The difference depends on whether or not permanent phase relationships exist between rays emerging from different parts of the object. A self-luminous object results in perfect incoherent imaging, as every atom emits independently, whereas perfect coherent imaging occurs if the entire object is illuminated by a plane wave, e.g., a point source at infinity. Lord Rayleigh noted the factor of 2 improvement in resolution available with incoherent

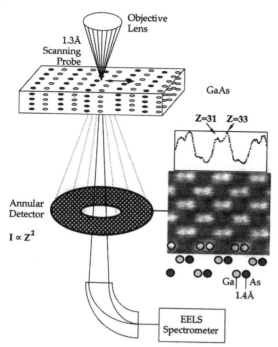

**Figure 1.** Schematic showing $Z$-contrast imaging and atomic resolution electron energy loss spectroscopy with STEM. The image is of GaAs taken on the 300-kV STEM instrument at Oak Ridge National Laboratory, which directly resolves and distinguishes the sublattice, as shown in the line trace.

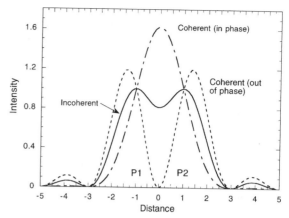

**Figure 2.** Image intensity for two point objects, P1 and P2, illuminated coherently in phase, 180° out of phase, and with incoherent illumination. (After Lord Rayleigh, 1896.)

imaging and also the lack of artifacts caused by interference phenomena that could be mistaken for real detail in the object. Of particular importance in the present context, he appreciated the role of the condenser lens (1896, p. 175): "It seems fair to conclude that the function of the condenser in microscopic practice is to cause the object to behave, at any rate in some degree, as if it were self-luminous, and thus to obviate the sharply-marked interference bands which arise when permanent and definite phase relationships are permitted to exist between the radiations which issue from various points of the object." A large condenser lens provides a close approximation to perfect incoherent imaging by ensuring a range of optical paths to neighboring points in the specimen.

A hundred years later, STEM can provide these same advantages for the imaging of materials with electrons. A probe of atomic dimensions illuminates the sample (see Fig. 1), and a large annular detector is used to detect electrons scattered by the atomic nuclei. The large angular range of this detector performs the same function as Lord Rayleigh's condenser lens in averaging over many optical paths from each point inside the sample. This renders the sample effectively self-luminous; i.e., each atom in the specimen scatters the incident probe in proportion to its atomic scattering cross-section. With a large central hole in the annular detector, only high-angle Rutherford scattering is detected, for which the cross-section depends on the square of the atomic number ($Z$); hence this kind of microscopy is referred to as $Z$-contrast imaging.

The concept of the annular detector was introduced by Crewe et al. (1970), and spectacular images of single heavy

atoms were obtained (see, e.g., Isaacson et al., 1979). In the field of materials, despite annular detector images showing improved resolution (Cowley, 1986a) and theoretical predictions of the lack of contrast reversals (Engel et al., 1974), it was generally thought impossible to achieve an incoherent image at atomic resolution (Cowley, 1976; Ade, 1977). Incoherent images of thick crystalline materials were first reported by Pennycook and Boatner (1988), and the reason for the preservation of incoherent characteristics despite the strong dynamical diffraction of the crystal was soon explained (Pennycook et al., 1990), as described below.

An incoherent image provides the most direct representation of a material's structure and, at the same time, improved resolution. Figure 2 shows Lord Rayleigh's classic result comparing the observation of two point objects with coherent and incoherent illumination. Each object gives rise to an Airy disc intensity distribution in the image plane, with a spatial extent that depends on the aperture of the imaging system. The two point objects are separated so that the first zero in the Airy disc of one coincides with the central maximum of the other, a condition that has become known as the Rayleigh resolution criterion. With incoherent illumination, there are clearly two peaks in the intensity distribution and a distinct dip in between; the two objects are just resolved, and the peaks in the image intensity correspond closely with the positions of the two objects. With coherent illumination by a plane-wave source (identical phases at the two objects), there is no dip, and the objects are unresolved. Interestingly, however, if the two objects are illuminated 180° out of phase, then they are always resolved, with the intensity dropping to zero half-way between the two. Unfortunately, this desirable result can only be achieved in practice for one particular image spacing (e.g., by illuminating from a particular angle), and other spatial frequencies will have different phase relationships and therefore show different contrast, a characteristic that is generic to coherent imaging. Note also that the two peaks are significantly displaced from their true positions.

In an incoherent image, there are no fixed phase relationships, and the intensity is given by a straightforward

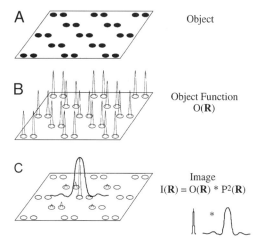

**Figure 3.** Schematic showing incoherent imaging of a thin specimen with STEM: (**A**) monolayer raft of Si$\langle 110 \rangle$; (**B**) the $Z$-contrast object function represents the high-angle scattering power localized at the atomic sites; (**C**) illumination of the sites for a 1.26-Å probe located over one atom in the central dumbbell. As the probe scans, it maps out the object function, producing an incoherent image.

convolution of the electron probe intensity profile with a real and positive specimen object function, as shown schematically in Figure 3. With Rutherford scattering from the nuclei dominating the high-angle scattering, the object function is sharply peaked at the atomic positions and proportional to the square of the atomic number. In the figure, a monolayer raft of atoms is scanned by the probe, and each atom scatters according to the intensity in the vicinity of the nucleus and its high-angle cross-section. This gives a direct image with a resolution determined by the probe intensity profile. Later it will be shown how crystalline samples in a zone axis orientation can also be imaged incoherently. These atomic resolution images show similar characteristics to the incoherent images familiar from optical instruments such as the camera; in a $Z$-contrast image, atomic columns do not reverse contrast with focus or sample thickness. In Figure 1, columns of Ga can be distinguished directly from columns of As simply by inspecting the image intensity. Detailed image simulations are therefore not necessary.

This unit focuses on $Z$-contrast imaging of materials at atomic resolution. Many reviews are available covering other aspects of STEM, e.g., microanalysis and microdiffraction (Brown, 1981; Pennycook, 1982; Colliex and Mory, 1984; Cowley, 1997). Unless otherwise stated, all images were obtained with a 300-kV scanning transmission electron microscope, a VG Microscopes HB 603U with a 1.26-Å probe size, and all spectroscopy was performed on a 100-kV VG Microscopes HB 501UX STEM.

### Competitive and Related Techniques

The absence of phase information in an incoherent image is an important advance; it allows the direct inversion from the image to the object. In coherent imaging, the structural relationships between different parts of the object are encoded in the phase relationships between the various diffracted beams. These are lost when the intensity is recorded, giving rise to the well-known phase problem. Despite this, much effort has been expended in attempts to measure the phase relationships between different diffracted beams in order to reconstruct the object (e.g., Coene et al., 1992; Orchowski et al., 1995; Möbus, 1996; Möbus and Dehm, 1996).

However, electrons interact with the specimen potential, which is a real quantity. There is no necessity to involve complex quantities; in an incoherent image, there are no phases and so none can be lost. Information about the object is encoded in the image intensities, and images may be directly inverted to recover the object. As seen in Figure 2, intensity maxima in an incoherent image are strongly correlated with atomic positions, so that often this inversion can be done simply by eye, exactly as we interpret what we see around us in everyday life. With the additional benefit of strong $Z$ contrast, structures of defects and interfaces in complex materials may often be determined directly from the image.

In a phase-contrast image, the contrast changes dramatically as the phases of the various diffracted beams change with specimen thickness or objective lens focus, which means that it is much more difficult to determine the object uniquely. It is often necessary to simulate many trial objects to find the best fit. The situation is especially difficult in regions such as interfaces or grain boundaries, where atomic spacings deviate from those in the perfect crystal, which can also cause atomic columns to reverse contrast. If one does not think of the correct structure to simulate, obviously the fit will be spurious. Unexpected phenomena such as the formation of new interfacial phases or unexpected atomic bonding are easily missed. Such phenomena are much more obvious in an incoherent image, which is essentially a direct image of atomic structure. As an example, Figure 4 shows the detection of van der Waals bonding between an epitaxial film and its substrate from the measurement of film-substrate atomic separation (Wallis et al., 1997). At 3.2 Å, this is significantly larger than bond lengths in covalently bonded semiconductors, which are typically in the range of 2.3 to 2.6 Å.

Closely related to STEM, scanning electron microscopy (SEM) gives an image of the surface (or near-surface)

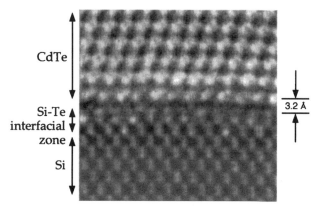

**Figure 4.** $Z$-contrast image of an incommensurate CdTe(111)/ Si(100) interface showing a 3.2-Å spacing between film and substrate indicating van der Waals bonding.

region of a bulk sample, whereas STEM gives a transmission image through a thin region of the bulk, which by careful preparation we hope is representative of the original bulk sample. The information from the two microscopes is therefore very different, as is the effort required for sample preparation. Whereas in STEM we detect the bright field and dark field transmitted electrons to reveal the atomic structure, in SEM we detect secondary electrons, i.e., low-energy electrons ejected from the specimen surface. The resolution is limited by the interaction volume of the beam inside the sample, and atomic resolution has not been achieved. Probe sizes are therefore much larger, ∼1 to 10 nm. Secondary electron images show topographic contrast with excellent depth of field. Backscattered primary electrons can also be detected to give a Z-contrast image analogous to that obtained with the STEM annular detector, except at lower resolution and contrast. Other signals such as cathodoluminescence or electron beam-induced conductivity can be used to map optical and electronic properties, and x-ray emission is often used for microanalysis. These signals can also be detected in STEM (Pennycook et al., 1980; Pennycook and Howie, 1980; Fathy and Pennycook, 1981), except that signals tend to be weaker because the specimen is thin.

Atomic resolution images of surfaces are produced by scanning tunneling microscopy (STM) and atomic force microscopy (AFM). Here, the interaction is with the surface electronic structure, although this may be influenced by defects in underlying atomic layers to give some subsurface sensitivity. In scanning probe microscopy, it is the valence electrons that give rise to an image; it is not a direct image of the surface atoms, and interpretation must be in terms of the surface electronic structure. The STEM Z-contrast image is formed by scattering from the atomic nuclei and is therefore a direct structure image. Valence electrons are studied in STEM through electron energy loss spectroscopy (EELS) and can be investigated at atomic resolution using the Z-contrast image as a structural reference, as discussed below. Scanning transmission electron microscopy can study a wider range of samples than STM or AFM as it has no problem with rough substrates and can tolerate substrates that are quite highly insulating.

Data from EELS provide information similar to that from x-ray absorption spectroscopy. The position and fine structure on absorption edges give valuable information on the local electronic structure. The EELS data give such information from highly localized regions, indeed, from single atomic columns at an interface (Duscher et al., 1998a). Instead of providing details of bulk electronic structure, it provides information on how that structure is modified at defects, interfaces, and grain boundaries and insight into the changes in electronic, optical, and mechanical properties that often determine the overall bulk properties of a material or a device.

The detailed atomic level characterization provided by STEM, with accurate atomic positions determined directly from the Z-contrast image and EELS data on impurities, their valence, and local band structure, represents an ideal starting point for theoretical studies. This is particularly valuable for complex materials; it would be impractical to explore all the possible configurations of a complex extended defect with first-principles calculations, even with recent advances in computational capabilities. As the complexity increases, so the number of required trial structures grows enormously, and experiment is crucial to cut the possibilities down to a manageable number. Combining experiment and theory leads to a detailed and comprehensive picture of complex atomistic mechanisms, including equilibrium structures, impurity or stress-induced structural transformations, and dynamical processes such as diffusion, segregation, and precipitation. Recent examples include the observation that As segregates at specific sites in a Si grain boundary in the form of dimers (Chisholm et al., 1998) and that Ca segregating to an MgO grain boundary induces a structural transformation (Yan et al., 1998).

## PRINCIPLES OF THE METHOD

### Comparison to Coherent Phase-Contrast Imaging

The conventional means of forming atomic resolution images of materials is through coherent phase-contrast imaging using plane-wave illumination with an (approximately) parallel incident beam (Fig. 5A). The objective aperture is behind the specimen and collects diffracted beams that are brought to a focus on the microscope screen where they interfere to produce the image contrast. The electrons travel from top to bottom in the figure. Not shown are additional projector lenses to provide higher magnification. In Figure 5B, the optical path of the STEM is shown, with the electrons traveling from bottom to top. A point source is focused into a small probe by the objective lens, which is placed before the specimen. Not shown are the condenser lenses (equivalent to projector lenses in the CTEM) between the source and the objective lens to provide additional demagnification of the source. Transmitted electrons are then detected through a defined angular range. For the small axial collector aperture shown, the two microscopes have identical optics, apart from the fact that the direction of electron propagation is reversed. Since image contrast in the electron microscope is dominated by elastic scattering, no energy loss is involved and time reversal symmetry applies. With equivalent apertures, the image contrast is independent of the direction of electron propagation, and the two microscopes are optically equivalent: the STEM bright-field image will be the same image, and be described by the same imaging theory, as that of a conventional TEM with axial illumination. This is the principle of reciprocity, which historically was used to predict the formation of high-resolution lattice images in STEM (Cowley, 1969; Zeitler and Thomson, 1970).

For phase-contrast imaging, the axial aperture (illumination aperture in TEM, collection aperture in STEM) must be much smaller than a typical Bragg diffraction angle. If $\beta$ is the semiangle subtended by that aperture at the specimen, then the transverse coherence length in the plane of the specimen will be of order $\lambda/\beta$, where $\lambda$ is the electron wavelength. Coherent illumination of

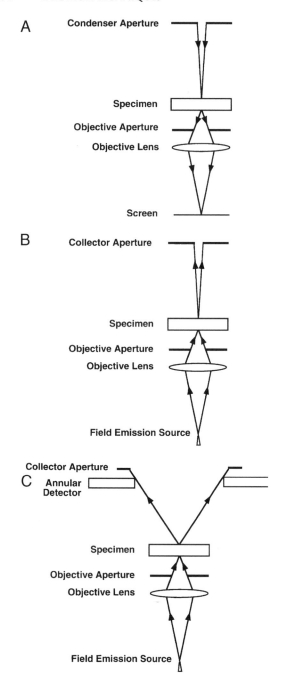

**Figure 5.** Schematics of the electron-optical arrangement in (**A**) CTEM and (**B**) STEM with a small axial detector. Note the direction of the electron propagation is reversed in the two microscopes. (**C**) A large-angle bright-field detector or annular dark-field detector in STEM gives incoherent imaging.

neighboring atoms spaced a distance $d$ apart requires $d \ll \lambda/\beta$, or $\beta \ll \lambda/d \approx 2\theta_B$, where $\theta_B$ is the Bragg angle corresponding to the spacing $d$. Similarly, if the collection aperture is opened much wider than a typical Bragg angle (Fig. 5C), then the transverse coherence length at the specimen becomes much less than the atomic separation $d$, and the image becomes incoherent in nature, in precisely the same way as first described by Lord Rayleigh. Now the annular detector can be seen as the complementary dark-field

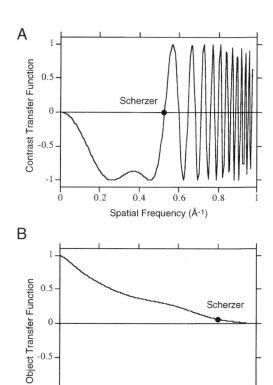

**Figure 6.** Contrast transfer functions for a 300-kV microscope with an objective lens with $C_s = 1$ mm; (**A**) coherent imaging conditions; (**B**) incoherent imaging conditions. Curves assume the Scherzer (1949) optimum conditions shown in Table 1: (**A**) defocus $-505$ Å; (**B**) defocus $-438$ Å, aperture cutoff $0.935$ Å$^{-1}$.

version of the Rayleigh case; conservation of energy requires the dark-field image $I = 1 - I_{BF}$, where $I_{BF}$ is the bright-field image and the incident beam intensity is normalized to unity. In practice, it is more useful to work in the dark-field mode for weakly scattering objects. If $I \ll 1$, the incoherent bright-field image shows weak contrast on a large background due to the unscattered incident beam. The dark-field image avoids this background, giving greater contrast and better signal-to-noise ratio.

The difference between coherent and incoherent characteristics is summarized in Figure 6, where image contrast is plotted as a function of spatial frequency in a microscope operating at an accelerating voltage of 300 kV. In both modes some objective lens defocus is used to compensate for the very high aberrations of electron lenses, and the conditions shown are the optimum conditions for each mode as defined originally by Scherzer (1949). In the coherent imaging mode (Fig. 6A), the contrast transfer function is seen to oscillate rapidly with increasing spatial frequency (smaller spacings). Spatial frequencies where the transfer function is negative are imaged in reverse contrast; those where the transfer function crosses zero will be absent. For this reason, atomic positions in distorted regions such as grain boundaries may reverse contrast or be absent from a phase-contrast image. In the incoherent mode (Fig. 6B), a smoothly

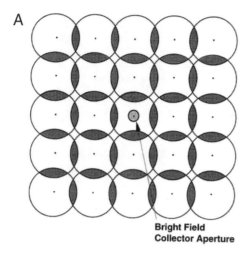

**Bright Field
Collector Aperture**

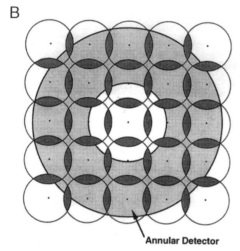

**Annular Detector**

**Figure 7.** Diffraction pattern in the detector plane from a simple cubic crystal of spacing such that the angle between diffracted beams is greater than the objective aperture radius. (**A**) An axial bright-field detector shows no contrast. (**B**) Regions of overlapping discs on the annular detector produce atomic resolution in the incoherent image.

that all image contrast arises through interference; as the probe is scanned across the atomic planes, only the regions of overlap between the discs change intensity as the phases of the various reflections change (Spence and Cowley, 1978).

For the crystal spacing corresponding to Figure 7A, coherent imaging in CTEM or STEM using a small axial aperture will not detect any overlaps and will show a uniform intensity with no contrast. The limiting resolution for this mode of imaging requires diffraction discs to overlap on axis, which means they must be separated by less than the objective aperture radius. Any detector that is much larger than the objective aperture will detect many overlapping regions, giving an incoherent image. Clearly, the limiting resolution for incoherent imaging corresponds to the discs just overlapping, i.e., to a crystal spacing such that the diffracted beams are separated by the objective aperture diameter (as opposed to the radius), double the resolution of the coherent image formed by an axial aperture. The approximately triangular shape of the contrast transfer function in Figure 6B results from the simple fact that a larger crystal lattice produces closer spaced diffraction discs with more overlapping area, thus giving more image contrast.

In practice, of course, one would tend to use larger objective apertures in the phase-contrast case to improve resolution. The objective aperture performs a rather different role in the two imaging modes; in the coherent case the objective aperture is used to cut off the oscillating portion of the contrast at high spatial frequencies, and the resolution is directly determined by the radius of the objective aperture, as mentioned before. In the incoherent case, again it is clear that the available resolution will be limited to the aperture diameter, but there is another consideration. To take most advantage of the incoherent mode requires that the image bear a direct relationship to the object, so that atomic positions correlate closely with peaks in the image intensity. This is a rather more stringent condition and necessitates some loss in resolution. The optimum conditions worked out long ago by Scherzer (1949) for both coherent and incoherent imaging conditions are shown in Figure 5 and summarized in Table 1. The Scherzer resolution limit for coherent imaging is now only 50% poorer than for the incoherent case.

The role of dynamical diffraction is also very different in the two modes of imaging. As will be seen later, dynamical diffraction manifests itself in a Z-contrast image as a columnar channeling effect, not by altering the incoherent nature of the Z-contrast image, but simply by scaling the columnar scattering intensities. In coherent imaging, dynamical diffraction manifests itself through contrast

decaying positive transfer is obtained (referred to as an object transfer function to distinguish from the coherent case), which highlights the lack of contrast reversals associated with incoherent imaging. It is also apparent from the two transfer functions that incoherent conditions give significantly improved resolution.

To demonstrate the origin of the improved resolution available with incoherent imaging, Figure 7 shows the diffraction pattern formed on the detector plane in the STEM from a simple cubic crystal with spacing $d$. Because the illuminating beam in the STEM is a converging cone, each Bragg reflection appears as a disc with diameter determined by the objective aperture. Electrons, unlike light, are not absorbed by a thin specimen but are only diffracted or scattered. The terms are equivalent, diffraction being reserved for scattering from periodic assemblies of atoms when sharp peaks or discs are seen in the scattered intensity. The periodic object provides a good demonstration of Abbé theory (see, e.g., Lipson and Lipson, 1982)

**Table 1. Optimum Conditions for Coherent and Incoherent Imaging**

| Parameter | Coherent Imaging | Incoherent Imaging |
|---|---|---|
| Resolution limit | $0.66 C_S^{1/4} \lambda^{3/4}$ | $0.43 C_S \lambda^{1/4} \lambda^{3/4}$ |
| Optimum defocus | $-1.155 (C_S \lambda)^{1/2}$ | $-(C_S \lambda)^{1/2}$ |
| Optimum aperture | $1.515 (\lambda/C_S)^{1/4}$ | $1.414 (\lambda/C_S)^{1/4}$ |

reversals with specimen thickness, as the phases between various interfering Bragg reflections change. These effects are also nonlocal, so that atom columns adjacent to interfaces will show different intensity from columns of the same composition far from the interface. Along with the fact that atom columns can also change from black to white by changing the objective lens focus, these effects make intuitive interpretation of coherent images rather difficult. Of course, one usually has a known material next to an interface that can help to establish the microscope conditions and local sample thickness, but these characteristics of coherent imaging make it very difficult to invert an image of an interface without significant prior knowledge of its likely structure. Typically, one must severely limit the number of possible interface structures that are considered, e.g., by assuming an interface that is atomically abrupt, with the last monolayer on each side having the same structure as the bulk. Although some interfaces do have these characteristics, it is also true that many, and perhaps most, do not have such simple structures. In the case of the $CoSi_2/Si(111)$ interface, this procedure gives six possible interface models. At interfaces made by ion implantation, however, different interface structures were seen by $Z$-contrast imaging (Chisholm et al., 1994). There are now a large number of examples where interface structures have proved to be different from previous models: in semiconductors (Jesson et al., 1991; McGibbon et al., 1995; Chisholm and Pennycook, 1997), superconductors (Pennycook et al., 1991; Browning et al., 1993a), and ceramics (McGibbon et al., 1994, 1996; Browning et al., 1995).

If we apply reciprocity to the STEM annular detector, it is clear that an equivalent mode exists in CTEM using high-angle hollow-cone illumination. In practice, it has not proved easy to achieve such optics. In any case, a much higher incident beam current will pass through the specimen than in the equivalent STEM geometry, because the illumination must cover an angular range much greater than a typical Bragg angle, whereas the STEM objective aperture is of comparable size to a Bragg angle. Beam damage to the sample would therefore become a serious concern.

Another advantage of the STEM geometry that should be emphasized is that the detection angle can be increased without limit, up to the maximum of 180° if so desired. Lord Rayleigh's illumination was limited to the maximum aperture of the condenser lens, so that at the limit of resolution (with condenser and objective apertures equal), significant partial coherence exists in an optical microscope. In STEM, by increasing the inner angle of the annular detector, the residual coherence can be reduced well below the limiting resolution imposed by the objective aperture. It is therefore a more perfect form of incoherent imaging than fixed-beam light microscopy. However, increasing detector angles will lead to reduced image intensity due to the rapid fall-off in atomic scattering factor, and a backscattered electron detector is really not practical. With high-energy electrons, the full unscreened Rutherford scattering cross-section is seen above a few degrees scattering angle, so higher scattering angles bring no advantage. A useful criterion for the minimum detector

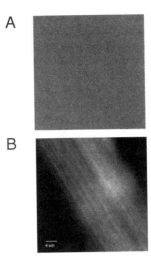

**Figure 8.** (**A**) Bright-field phase-contrast STEM image of an iodine intercalated bundle of carbon nanotubes showing lattice fringes. (**B**) $Z$-contrast image taken simultaneously showing intercalated iodine.

aperture $\theta_i$ to achieve incoherent imaging of two objects separated by $\Delta R$ is

$$\theta_i = 1.22\lambda/\Delta R \qquad (1)$$

where the detected intensity varies by <5% from the incoherent expectation (Jesson and Pennycook, 1993). In this case the Airy disc coherence envelope of the detector (illumination) aperture is half the width of that of the probe-forming (imaging) aperture. This condition therefore corresponds to separating the coherence envelopes by double the Rayleigh criterion.

Another benefit of STEM is that the $Z$-contrast image and the conventional HREM image are available simultaneously. However, phase-contrast images in STEM tend to be substantially noisier than those recorded in TEM using a charge-coupled device (CCD) camera or photographic plate, because STEM is a serial imaging instrument that is far less efficient than the parallel recording capabilities of TEM. An example of simultaneous bright-field and $Z$-contrast imaging of a bundle of iodine-intercalated carbon nanotubes is shown in Figure 8. Because of their cylindrical form, only a few atomic layers are parallel to the electron beam. Nevertheless, lattice fringes from the tubes are seen clearly in the phase-contrast image because it is tuned to the spacing expected and filters out the uniform background near-zero spatial frequency due to all other atoms. On the other hand, the $Z$-contrast image is sensitive to the absolute numbers of atoms under the beam. Here, where there is no significant dynamical diffraction, the $Z$-contrast image can be considered as an image of projected mass thickness. It shows no detectable contrast from the tubes themselves, but the iodine intercalation is now clearly visible. Thus the two kinds of images are highly complementary in this case.

Figure 9 shows a Rh catalyst cluster supported on $\gamma$-alumina (Pennycook et al., 1996). In this case, the bright-field image is dominated by phase contrast from

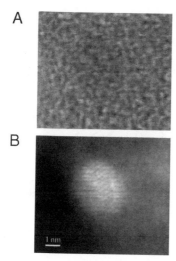

**Figure 9.** (A) Bright-field and (B) Z-contrast images of an Rh catalyst particle on γ-alumina. The particle is barely detectable in the bright-field image due to the phase contrast of the support, but the dark-field image reveals its atomic structure, internal twins, and external facets.

the carbon film ($Z = 6$) used to support the sample, whereas the Rh particle ($Z = 45$) is clearly visible in the Z-contrast image. Of course, if pieces of γ-alumina overlapping holes in the carbon film were chosen, the phase-contrast image would not be so obscured, but nevertheless, bright-field imaging of small metal clusters becomes very difficult for particles $<\sim1$ nm in size, due to the inevitable coherent interference effects from the support and the lack of Z contrast (Datye and Smith, 1992). The ultimate example of Z-contrast imaging is the detection of single Pt atoms on γ-alumina shown in Figure 10 (Nellist and Pennycook, 1996). Here, again, the two images are very complementary. The orientation of the γ-alumina support can be deduced from the bright-field image, while single atoms, dimers, and trimers are detectable in the Z-contrast image. Spacings and angles between the Pt atoms are constrained to match the atomic spacings in the γ-alumina surface, suggesting the possible adsorption sites shown in the schematic.

## Probe Formation

An electron of wavelength λ passing at an angle θ through an objective lens with spherical aberration coefficient $C_s$ set to a defocus of $\Delta f$ experiences a phase shift γ given by

$$\gamma = \frac{\pi}{\lambda}\left(\Delta f \theta^2 + \frac{1}{2}C_s\theta^4\right) \quad (2)$$

The amplitude distribution $P(\mathbf{R} - \mathbf{R_0})$ of the STEM probe at position $\mathbf{R_0}$ on the specimen is obtained by integrating all possible pathways through the objective lens, as defined by the objective aperture,

$$P(\mathbf{R} - \mathbf{R_0}) = \int_{\substack{\text{objective} \\ \text{aperture}}} e^{i[\mathbf{K}\bullet(\mathbf{R}-\mathbf{R_0})+\gamma(\mathbf{K})]}d\mathbf{K} \quad (3)$$

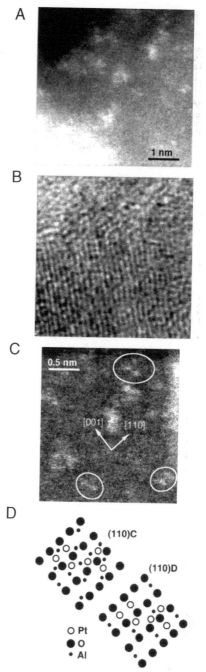

**Figure 10.** (A) Z-contrast image of Pt clusters supported on γ-Al₂O₃. (B) Bright-field image showing strong {222} fringes of Al₂O₃. (C) Higher magnification Z-contrast image after high- and low-pass filtering to enhance contrast revealing a Pt trimer and two dimers (circled). (D) The spacings and angles between Pt atoms match the orientation of the support, suggesting likely configurations on the two possible (110) surfaces.

where $|\mathbf{K}| = \chi\theta$ is the transverse component of the incident electron wave vector $\chi = 2\pi/\lambda$. The intensity distribution is given by $P^2(\mathbf{R} - \mathbf{R_0})$. A focal series is shown in Figure 11 for a 300-kV STEM instrument with $C_s = 1$ mm and a Scherzer optimum objective aperture of 9.3 mrad. Notice how the intensity profiles are not symmetric about the Scherzer defocus value of −430 Å. This

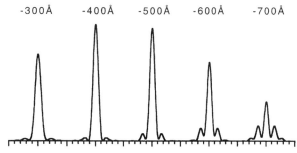

**Figure 11.** Probe intensity profiles for a 300-kV STEM instrument with $C_s = 1$ mm and a Scherzer optimum objective aperture of 9.4 mrad.

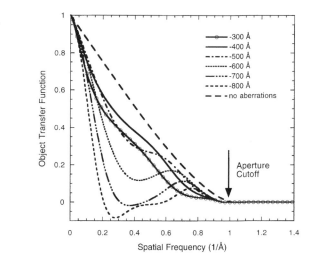

is because the entire probe is (ideally) a single electron wavepacket, all angles fully coherent with each other. The probe is best thought of as a converging, phase-aberrated spherical wave. At a defocus below the Scherzer optimum, the probe is close to Gaussian in nature, while at high defocus values, the probe develops a well-defined "tail," a subsidiary maximum around the central peak. Although the width of the central peak is significantly reduced, actually dropping to below 1 Å at a defocus of $-500$ Å, now over half the total intensity is in the tails. This gives rise to significant false detail in the image, which makes intuitive interpretation no longer possible, as shown by the corresponding simulated images of Si$\langle 110 \rangle$ in Figure 12.

The object transfer function for incoherent imaging is the Fourier transform of the probe intensity profile. Figure 13A shows transfer functions corresponding to the focal series of Figures 11 and 12. Also shown is the ideal transfer function without any aberration. These curves show how increasing the defocus enhances the high spatial frequencies but reduces the transfer at lower frequencies. In all cases the transfer reaches zero at the cutoff defined by the aperture. If the aperture size is increased, the transfer function can be extended even further. Figure 13B shows transfer functions obtained with a 13-mrad objective aperture, corresponding to an aperture cutoff of 0.74 Å. At low defocus values there is little transfer at the high spatial frequencies, and an intuitive image is expected but with reduced contrast compared to the optimum aperture.

Experimental verification that the resolution is enhanced under such conditions is seen in Figure 14, which shows images of Si$\langle 110 \rangle$ obtained with a 17-mrad objective aperture (Nellist and Pennycook, 1998a). Under optimum defocus the dumbbells are well resolved (Fig. 14A) with maxima close to the atomic positions, the expected intuitive image. The Fourier transform of the image intensity, Figure 14B shows the spatial frequencies

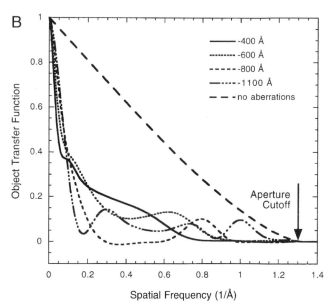

**Figure 13.** Object transfer functions for (**A**) the optimum condition in Figures 11 and 12; (**B**) extended transfer obtained with an oversized objective aperture of 13 mrad, though contrast at lower spatial frequencies is reduced.

**Figure 12.** Simulated images of Si(110) corresponding to the probe profiles in Figure 11.

being transferred, which includes the {004} reflection at a spacing of 1.36 Å. On increasing the objective lens defocus (Fig. 14C), the image is no longer intuitive, but many additional spots are seen in the Fourier transform (Fig. 14D). The {444} spot at 0.78 Å is the highest resolution yet achieved in any electron microscope. This result demonstrates very clearly the improved resolution available with incoherent imaging, as first pointed out by Lord Rayleigh.

### Incoherent Scattering

Because the probe is necessarily a coherent spherical wave, incoherent imaging must come by ensuring incoherent scattering. Of course, as mentioned earlier, all imaging

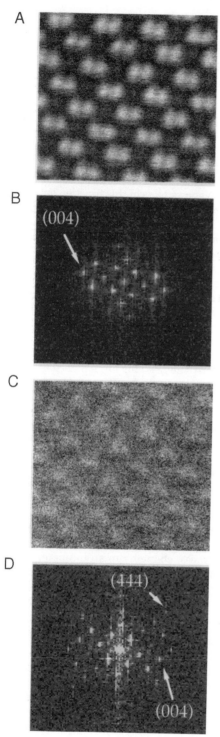

**Figure 14.** Images and corresponding Fourier transforms of Si<110> obtained with a 17-mrad objective aperture (Nellist and Pennycook, 1998a); (**A, B**) at Scherzer defocus passing the {004} spacing and resolving the dumbbell; (**C, D**) using a high underfocus giving increased transfer at high spatial frequencies, including the {444} spacing at 0.78 Å.

imaging can be achieved with a large detector to any desired degree, as shown in Appendix A. A simple justification is that, depending on the projected potential, a monolayer crystal changes only the phase of the electron wave passing through it. Clearly, if only the phase of the spherical wave is altered inside the crystal, the intensity profile will remain the same as that of the incident probe. Each nucleus will scatter in proportion to the intensity of the probe in its vicinity, and incoherent imaging is obtained as shown schematically in Figure 3. The image intensity $I(\mathbf{R})$ is given by a convolution of the probe intensity with the object function $O(\mathbf{R})$,

$$I(\mathbf{R}) = O(\mathbf{R}) * P^2(\mathbf{R}) \qquad (4)$$

In the simple case of a monolayer crystal, the object function is just an array of atomic cross-sections that can be approximated as delta functions at the atomic positions.

A typical Bragg angle for a high-energy electron is only ~10 mrad, or ~0.5°, so that high-angle scattering corresponds to only a few degrees. This means that path differences between atoms separated in the transverse plane are much greater than between atoms separated along the beam direction, as shown in Figure 15. Atoms spaced by $x$ and $z$ in the transverse and longitudinal directions, respectively, when viewed from direction $\theta$, have phase differences of $x \sin \theta$ and $z(1 - \cos \theta)$, respectively. For small scattering angles these are approximately $x\theta$ and $z\theta^2/2$; the latter is much smaller, showing how longitudinal coherence is more difficult to break with the detector alone. Clearly, if the Rayleigh detector is sufficient to ensure incoherence in the $x$ direction, there is no justification for believing that atoms separated by a similar distance in the

is based on interference, and incoherent characteristics arise when averaging a large number of optical paths of different lengths. For a monolayer raft of weakly scattering atoms, it can be shown rigorously how incoherent

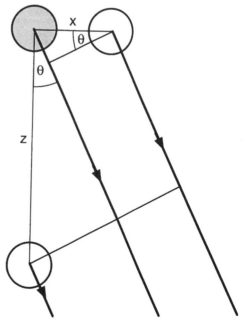

**Figure 15.** Path differences between scattering from atoms separated by $x$ and $z$ in the transverse and longitudinal directions, respectively.

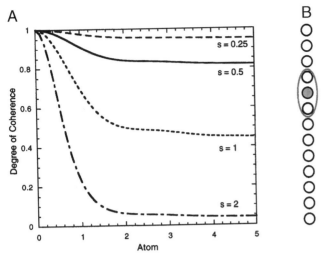

**Figure 16.** (**A**) Degree of coherence between an atom at the origin and neighboring atoms along a column using a phonon model of thermal vibrations. (**B**) With increasing separation the degree of coherence drops rapidly, defining an effective coherence envelope.

indeed transverse modes of phonons with wave vectors in the $z$ direction that are most effective in breaking the coherence along a column (Jesson and Pennycook, 1995). The analysis showed further that the coherence of two atoms separated along the $z$ direction by $m$ unit cells is

$$W_m = \exp\left\{ 2Ms^2 \left[ \frac{\mathrm{Si}(\pi m)}{\pi m} - 1 \right] \right\} \qquad (5)$$

where $\mathrm{Si}(x)$ is the sine integral function, $M$ is the usual Debye-Waller factor, and $s = \theta/2\lambda$ is the scattering vector. Figure 16A shows plots of this function for different values of $s$ showing how the coherence rapidly reduces as the separation of atoms along the column increases. For large separations the degree of coherence approaches the limiting value $e^{-2M}$, which is the Einstein value for the strength of coherent reflections in his model of independently vibrating atoms. This model, though convenient, does ignore the correlation between near neighbors seen in the figure. The phonon model shows clearly that atoms close together scatter with greater coherence than those far apart, leading to the concept of a longitudinal coherence volume, shown schematically in Figure 16B.

Therefore, the need for large detector angles to ensure *inter*column incoherence (transverse incoherence) will automatically break the *intra*column coherence leading to longitudinal incoherence also. The change in the thickness dependence of the image intensity with increasing annular detector angle is largely governed by longitudinal coherence, changing from predominantly coherent at low angles to predominantly incoherent at large angles. This is illustrated in Figure 17 for a column of Rh atoms 2.7 Å apart illuminated by 300-kV electrons. With low detection angles the scattering is almost entirely coherent, becoming almost entirely incoherent at high detector angles. Intermediate angles exhibit an initial coherent dependence with thickness, changing to an incoherent dependence as the column becomes significantly longer than the correlation length.

$z$ direction will scatter independently. Indeed, the incoherence in the beam direction comes from another effect, the unavoidable thermal vibrations of the atoms themselves.

Thermal vibrations act to randomize the phase differences between atoms at different heights in the column by inducing transverse atomic displacements. Consider the two atoms separated by $z$ to have an additional instantaneous relative displacement due to thermal motion of $u_z$ and $u_x$ in the longitudinal and transverse directions, respectively. The resulting path differences are $(z + u_z)$ $(1 - \cos\theta)$ and $u_x \sin\theta$. Clearly, the longitudinal displacements $u_z$ have relatively little effect as they are much smaller than the atomic separation $z$, but the transverse displacements $u_x$ are significant. A detailed analysis using a phonon model of thermal vibrations shows that it is

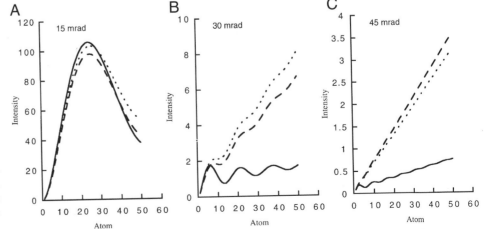

**Figure 17.** Thickness dependence of the scattering from a column of Rh atoms spaced 2.7 Å apart along the beam direction for (**A**) low, (**B**) medium, and (**C**) high scattering angles. The coherent thickness oscillations at low angles are almost completely suppressed at high angles.

Finally, it should be mentioned that a significant fraction of the thermal displacements of the atoms are due to zero-point fluctuations that will not disappear on cooling the sample. Thus, it should not be assumed that the scattering will become coherent on cooling, although the required detection angles will be increased somewhat.

## Dynamical Diffraction

The above analysis makes clear how at high scattering angles thermal vibrations effectively destroy the coherence between atoms in a column. Given the transverse incoherence due to the annular detector, we may therefore think of each atom as scattering independently with an atomic cross-section σ, the electron equivalent of the self-luminous object. Now all that remains is to determine how the probe propagates through the thick crystals typically encountered in materials science studies, i.e., what is the effective illumination of each atom in a column? Electrons interact strongly with matter, undergoing multiple scattering, which is referred to as dynamical diffraction. With amorphous materials, or crystals in a random orientation, the multiple scattering is uncorrelated and leads to beam broadening. In a crystal oriented to a low-order zone axis, the situation is very different since the scattering is strongly correlated, often leading to a periodic behavior with increasing sample thickness. Phase-contrast images usually exhibit periodic contrast changes or reversals. One of the most surprising features of Z-contrast images from zone axis crystals is the lack of any apparent change in the form of the image with increasing sample thickness (Pennycook and Jesson, 1990, 1991). The Z-contrast image shows no strong oscillatory behavior, just reducing contrast with increasing thickness. The 300-kV STEM instrument at Oak Ridge National Laboratory still resolves the dumbbell spacing of 1.36 Å in a Si⟨110⟩ crystal 1000 Å thick.

The explanation of this remarkable behavior requires a quantum mechanical analysis of the probe propagation. Dynamical diffraction effects can be conveniently described in a Bloch wave formulation, where an incident plane-wave electron is described as a superposition of Bloch states, each propagating along the zone axis with a different wave vector. It is the interference between these Bloch states that leads to dynamical diffraction effects, thickness fringes in a diffraction contrast image, and contrast reversals in a phase-contrast lattice image. Figure 18 shows the first six Bloch states for Si⟨110⟩, which are seen to take the form of molecular orbitals about the atomic strings. Usually, the wave function inside the crystal can be well represented with just a few strongly excited Bloch states, and it is their propagation with different wave vectors through the crystal that leads to depth-dependent dynamical diffraction effects. Note in particular the 1s states, which are located over the atomic columns, and overlap little with neighboring columns. They are the most localized states, and we will find they are responsible for the Z-contrast image. These 1s states exist around every atomic column, and it is because of their localization that we can think of the Z-contrast image as a column-by-

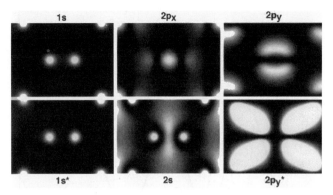

**Figure 18.** Intensities of the first six Bloch states in Si(110) with their molecular orbital assignments. The 1s states are located around the Si atomic columns.

column image of the crystal. They are also the basis for atomic column spectroscopy, as described under Atomic Resolution Spectroscopy, below. All the other states are much less localized and overlap significantly with neighboring columns, as is clear from the figure. These delocalized states are responsible for the nonlocal effects in phase-contrast imaging, which makes it necessary to simulate images with supercells. Such states are not specific to a particular type of column but are strongly dependent on surrounding columns. They are the reason that atomic columns at an interface in a phase-contrast image may appear different to those in the bulk.

We can now better appreciate the action of the high-angle detector. The 1s Bloch states are the most highly *localized* in real space, and therefore the most *extended* in reciprocal space. Only these states have transverse momentum components that extend to the high-angle detector. So the detector acts as an efficient Bloch state filter, with only one type of state per column involved in Z-contrast imaging. All states have low-order Fourier components, and so depending on their excitation, they all interfere in a low-angle detector such as used for axial bright-field imaging. This is how the Z-contrast image can show no interference phenomena while simultaneously the bright-field image shows strong interference effects, such as thickness fringes or contrast reversals.

The high-angle scattering comes entirely from the 1s states, specifically from their sharp peaks at the atomic columns. The detector therefore acts rather like a detector placed inside the sample at each atomic nucleus. In fact, the high-angle components of 1s states scale as $Z^2$, as would be anticipated on the basis of Rutherford scattering. However, dynamical diffraction does make its presence felt because not all the incident beam excites 1s states, and the total amplitude inside the crystal must be described as a sum over Bloch states $j$ propagating with different wave vectors $k_z^j$ along the z direction. The 1s states are located over the deepest part of the projected potential and so have the highest kinetic energy and the largest $k_z^j$. All the other less localized states have very similar $k_z^j$ values, and so in thin crystals they all propagate approximately in phase through the thickness z. Therefore, it is a good

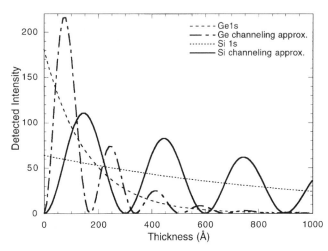

**Figure 19.** Intensity of coherent scattering reaching the annular detector from Si and Ge$\langle 110 \rangle$ using the channeling approximation, Equation 6. Parameters: Si, $\xi = 300$ Å, $\mu^{1s} = 0.00048$; Ge, $\xi = 169$ Å, $\mu^{1s} = 0.0032$.

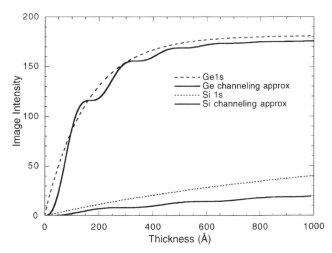

**Figure 20.** Intensity of incoherent scattering reaching the annular detector from Si and Ge $\langle 110 \rangle$ using the channeling approximation, Equation 7. Parameters are the same as Figure 19.

approximation to consider just two components to the electron wave function, the $1s$ state $\phi^{1s}$ propagating with wave vector $k_z^{1s}$ and a term $1 - \phi^{1s}$ propagating at an average $k_z^0$. This is referred to as a channeling approximation (Pennycook and Jesson, 1992). The beating between these two components occurs with an extinction distance $\xi = 2\pi/(k_z^{1s} - k_z^0)$, and the total intensity scattered to high angles from a depth $z$ in the dynamical wave field is given by

$$I \propto Z^2 \varepsilon^{\mathrm{av}^2} [1 - \cos(2\pi z/\xi)] e^{-2\mu^{1s}z} \qquad (6)$$

Here, $Z^2$ represents the high-angle scattering cross-section of the column; $\varepsilon^{\mathrm{av}}$ is the $1s$ state excitation averaged over the angular range of the objective aperture. The term in square brackets represents the dynamical oscillations with depth $z$, and $\mu^{1s}$ represents the $1s$ state absorption coefficient that damps the oscillations. Figure 19 shows a plot of this function for Si and Ge in the $\langle 110 \rangle$ orientation, compared to the $1s$ state intensities alone.

Experimental images do not show these strong depth oscillations because, as mentioned earlier, most of the scattering reaching the detector is thermal diffuse scattering. This is generated incoherently from the dynamical wave field described in Equation 6 and is therefore given by the integral of Equation 6 over the crystal thickness $t$. This results in a thickness-dependent columnar object function (Nellist and Pennycook, 1999)

$$O(R, t) \propto \frac{Z^2 \varepsilon^{\mathrm{av}^2}}{2\mu^{1s}(\xi^2 \mu^{1s^2} + \pi^2)}$$
$$\times \left[ \pi^2 (1 - e^{-2\mu^{1s}t}) - \xi^2 \mu^{1s^2} e^{-2\mu^{1s}t} \left( 1 - \cos\frac{2\pi t}{\xi} \right) \right.$$
$$\left. - \pi\xi\mu^{1s} e^{-2\mu^{1s}t} \sin\frac{2\pi t}{\xi} \right] \qquad (7)$$

As shown in Figure 20, the thickness integration has removed the strong dynamical oscillations. Again we see

that the thermal vibrations are important in breaking the coherence through the thickness $t$ and giving an image intensity that is increasing with thickness. The limiting value is again proportional to $Z^2$, allowing very simple image quantification. The form of the curve is in good agreement with both experimental observations and multislice simulations for a high detection angle (Loane et al., 1992; Hillyard et al., 1993; Anderson et al., 1997; Hartel et al., 1996). Under these conditions, this channeling approximation is very useful for simulating images and saves enormously on computer time.

For lower detection angles the other states become more important, and a fraction of the coherent exit-face wave function given by Equation 6 must be added. The image contrast begins to show more oscillatory dependence on thickness, and the incoherent characteristics are progressively lost. Other states also become important if the atomic columns are no longer straight but are bent due to the presence of a defect or impurity. Then transitions occur between Bloch states, which is the origin of diffraction contrast imaging in CTEM, and strain contrast effects will also be seen in the annular detector signal, as discussed under Strain Contrast, below. A similar case is that of a single heavy impurity atom at a depth $z$, which will sample the oscillating wave field of Figure 19 and show depth-dependent contrast (Loane et al., 1988; Nakamura et al., 1997).

Nevertheless, in most cases of high-resolution imaging, the $Z$-contrast image represents a thickness integrated signal, and the oscillatory behavior is strongly suppressed. This is in marked contrast to a phase-contrast image that reflects the exit-face wave function and therefore does show an oscillatory dependence on sample thickness. The $Z$-contrast image is a near-perfect approximation to an ideal incoherent image. It removes the contrast reversals with objective lens focus, the proximity effects due to non-local Bloch states, and finally, the depth-dependent oscillations due to dynamical diffraction. The Bloch state description of the imaging process shows clearly the physical origin of the incoherent characteristics, despite the presence of strong dynamical scattering.

### Atomic Resolution Spectroscopy

Scanning transmission electron microscopy provides a sound basis for atomic resolution analysis using x-ray fluorescence or EELS. To date, atomic resolution has been demonstrated only for EELS because of the much lower detection efficiency for x rays (Browning and Pennycook, 1995; Pennycook et al., 1997; Duscher et al., 1998a). Using the incoherent Z-contrast image, a stationary probe can be centered over a chosen column by maximizing the annular detector intensity. Provided that incoherent optics are also used for the EELS, i.e., a large collector aperture, the EELS signal from that column will also be maximized at the same probe position, giving an atom column resolved analysis.

Because it is an inner shell electron that is the scatterer rather than the nucleus, the spatial resolution of the spectral information may not be quite as high as that of the Z-contrast image, an effect referred to as delocalization. For inner shell excitations, however, the width of the object function is less than an angstrom for most elements, as shown in Figure 21 (Rafferty and Pennycook, 1999). Core loss EELS therefore allows individual atomic columns to be analyzed, as shown in Figure 22. The Z-contrast image and spectra from specific atomic columns show the variation in both concentration and valence of Mn in a doped $SrTiO_3$ grain boundary (Duscher et al., 1998a). The highest peak intensity comes from the pair of columns labeled 2, which in the undoped boundary are Ti sites. A much lower concentration is seen at position 3, which represents the Sr sites. The ratio of the two Mn peaks, or white lines, indicates valence and suggests a change from 4+ in the bulk to 3+ at the boundary. Data of this nature are particularly valuable for linking the structure of the grain boundary to its electrical activity.

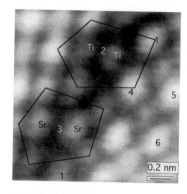

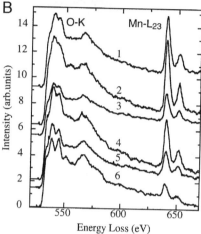

**Figure 22.** (**A**) Z-contrast image from a Mn-doped $SrTiO_3$ grain boundary with EELS spectra from individual atomic columns; (**B**), showing differences in composition and Mn valence at the grain boundary.

With the phase-contrast image of CTEM, it is not so simple to accurately illuminate an individual column. Neither is it practical to form an energy-filtered image at atomic resolution except from low-loss electrons that are intrinsically delocalized, in which case the contrast is due to elastic scattering, but the chemical information is nonlocal. In CTEM the objective lens is behind the specimen (Fig. 5A), so that the collection angle for EELS imaging is restricted to the objective aperture size. Furthermore, the energy loss electrons will suffer chromatic aberration on passing through the lens to form an image. For the energy-filtered image to show atomic resolution, the energy window selected by the filter must be kept small. Core loss cross-sections are orders of magnitude lower than elastic scattering cross-sections that determine the signal-to-noise ratio of the Z-contrast image. Attempting to form an image from core loss electrons in CTEM will therefore result in a very noisy image, probably with insufficient statistics to show atomic resolution. Increasing the incident beam current to compensate will increase the chances of beam damage. The information would be more efficiently gathered in the STEM mode by illuminating the chosen column, using the high-intensity Z-contrast image as a reference, and collecting the transmission EELS spectrum column by column. The objective lens is before the specimen in STEM, so that

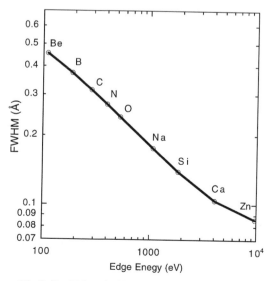

**Figure 21.** Full-width at half-maxima of EELS object functions for K-shell excitations by 300-kV incident electrons.

chromatic aberration no longer degrades the spatial resolution, and a large collector aperture can be used to collect a large fraction of the inelastic scattering and partially compensate for the low cross-section.

This combination of the $Z$-contrast image and EELS at atomic resolution is very powerful. It is particularly useful for analyzing elements such as oxygen, which are too light to be seen directly in the $Z$-contrast image (McGibbon et al., 1994; Dickey et al., 1998). In the high-temperature superconductors, a pre-edge feature on the oxygen $K$ edge is directly related to the concentration of holes that are responsible for superconductivity. This feature can be used to map hole concentrations at a spatial resolution below the superconducting coherence length (Browning et al., 1993b).

## PRACTICAL ASPECTS OF THE METHOD

Besides the need for a high-resolution objective lens, a critical requirement for a probe of atomic dimensions is a field emission source. Without the high brightness such sources offer, there is insufficient current in the probe to focus and stigmate the image and insufficient current for an atomic resolution analysis. The cold field emission source is preferred for EELS because of its smaller energy spread compared to the Schottky-type source, whereas the latter is preferred for general TEM work because it is capable of a higher current mode and therefore gives less noisy images at low magnification when searching the specimen. It is easier to search the sample in the CTEM mode, and once an interesting region is found, the microscope can be switched to the STEM mode for $Z$-contrast imaging.

Formally, atomic-sized probes were only available in dedicated STEM instruments, but in recent years, conventional TEM has dramatically improved its probe-forming capabilities. It is now possible to obtain true atomic resolution $Z$-contrast capability in a conventional TEM column (James et al., 1998; James and Browning, 1999).

To adjust the STEM instrument, a shadow image or ronchigram is very useful (Cowley 1979, 1986b; Lin and Cowley, 1986). Formed by stopping the STEM probe and observing the intensity in the diffraction plane, it is a powerful means to align the microscope, correct astigmatism, and tilt a specific area precisely on axis. The objective lens strength is used as a nonlinear magnification control. With the probe focused away from the specimen plane, a broad area of sample is illuminated, and the diffraction screen shows a shadow image. This will also contain diffraction information in the form of Kikuchi lines, as different parts of the image are illuminated by electrons at different angles in the defocused probe (see Fig. 23). As the probe is focused on the specimen, the magnification increases, until at focus (infinite magnification) a ronchigram is formed that is an angular map of the objective lens aberrations. A very thin region of sample is required, ideally an amorphous edge or glue, and the optical axis can be seen clearly. Astigmatism can be adjusted by making the phase-contrast features circular, and stability of all lenses and alignments can be checked by wobbling and counting fringes as the probe is displaced. This procedure

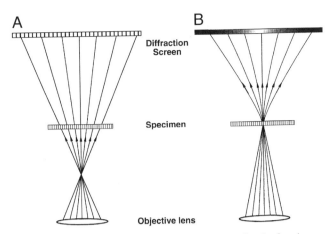

**Figure 23.** Schematic showing the formation of a shadow image or ronchigram by a stationary probe in the STEM instrument. (**A**) A defocused probe gives a shadow image containing simultaneous image and diffraction information. (**B**) focused probe produces a ronchigram.

gives an accurate alignment and precise position of the objective lens optic axis about which the objective aperture and all detectors are centered. If the microscope has a stage capable of adjusting the sample height, then the shadow image is a useful method to set each sample at the same height and therefore at the same objective lens excitation. In this way all alignments and astigmatism settings are constant and all that remains is to tilt the sample on axis.

With the microscope now adjusted, the scan is turned on and $Z$-contrast images can be collected. Unless there is a large change in height from area to area, the astigmatism should not need adjustment, and any changes in the image will most likely reflect changes in sample tilt. The $Z$-contrast image loses contrast more rapidly than the phase-contrast image with specimen tilt. The first indication is an asymmetry in the image, progressing to a pronounced streaking along the direction of tilt. This can be distinguished from astigmatism because it will not change direction when the objective lens is changed from underfocus to overfocus. Fine focus can be judged by eye, remembering the general characteristics shown in Figure 12, that on weakening the lens (underfocusing), the image contrast first increases, then goes through a maximum at the optimum Scherzer focus, and finally reduces but acquires increasing fine detail, until the image disappears into noise. It should also be remembered that slowing the scan will reduce the image noise quite significantly, although specimen drift becomes a limitation. One can of course also use the conventional phase-contrast image to stigmate and focus, although the noise in a scanned image makes it more difficult to judge minimum image contrast than with conventional TEM.

It should also be realized that in extremely thick regions of sample, the $Z$-contrast image may show contrast reversals, not in the atomic resolution image (usually the region is so thick that an atomic resolution image is no longer possible), but in the relative brightness of materials of different $Z$. In a thick region, the bright-field image has long disappeared, and this effect can be confusing when searching the sample for the interface of interest. It is caused by multiple elastic scattering that takes electrons off the detector to even higher angles. In the limit there is no more transmission, and obviously the higher $Z$ material will be the first to approach this limit and therefore appear less bright than a material of lower $Z$.

Finally, the EELS detector needs to be of the highest efficiency possible, which means it should be a cooled CCD parallel detection system (see, e.g., McMullan et al., 1990). Because damage is a more serious concern in EELS, it is also essential to have a time-resolved capability in the system, to allow a series of spectra to be recorded in succession so that they can be compared for signs of damage. If none are seen, they can subsequently be summed to improve statistics. Usually it is in the spectral fine structure that damage effects are first observed, long before the image shows signs of amorphization.

## DATA ANALYSIS AND INITIAL INTERPRETATION

### Retrieval of the Object Function

The only specimen information lost in recording an incoherent image is at high spatial frequencies, due to the convolution with a probe of finite width. No phase information is lost, as in the recording of a coherent image, because there is no phase information in an incoherent image. It might therefore be assumed that, to retrieve the object, all that would be required would be to deconvolve the probe from the image. Unfortunately this is not useful, as illustrated in Figure 24 (Nellist and Pennycook, 1998b), because there is insufficient information at high frequencies. Although the raw image data in Figure 24A can be Fourier filtered to remove the noise (Fig. 24B), the result of the deconvolution in part C is to produce artifacts between the columns. These arise because, lacking any information at high spatial frequencies, one is forced to cut off the transfer abruptly near the maximum spatial frequency present in the original image. To avoid such artifacts, a slower cutoff can be imposed on the image, but obviously this would then degrade the resolution. In fact, it is not useful to attempt to improve upon the incoherent transfer function in this way, and an alternative method is required to reconstruct the missing high-frequency information.

For atoms far apart, it is reasonable to locate the maximum of each image feature, but this procedure does not work near the limit of resolution because it does not take any account of the probe profile. In Si$\langle 110 \rangle$, for example, pairs of columns are spaced by distances comparable to the probe size, and the peak image intensity is displaced outward by a few tenths of an angstrom depending on defocus. Maximum entropy (Gull and Skilling, 1984) is a method to accurately account for the effects of the convolution. No prior knowledge concerning the nature of the image is assumed, except that it is incoherent. It does not assume that the object is comprised of atoms, but, given a specified probe profile, it reconstructs the most likely object function consistent with the image data. It produces an object with the least amount of structure needed to account for the image intensity, an object with

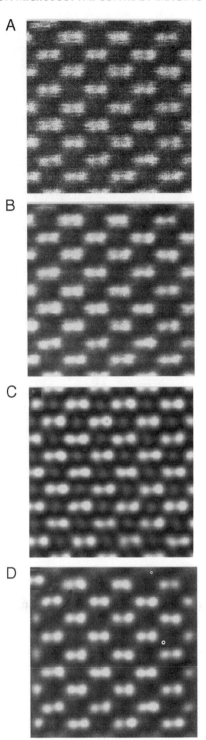

**Figure 24.** (**A**) Raw $Z$-contrast image of Si $\langle 110 \rangle$; (**B**) low-pass filtered image to reduce noise; (**C**) deconvolution of the probe function leads to artifacts between the columns; (**D**) maximum entropy retrieves the correct object, giving a reconstructed image free of artifacts.

maximum entropy. Figure 24D shows the reconstructed Si image, obtained by convolving the maximum-entropy object with the probe profile, where the artifacts between the columns are no longer present.

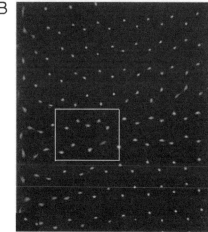

**Figure 25.** (**A**) $Z$-contrast image of a threading dislocation in a GaN thin film grown on sapphire, viewed along the $\langle 0001 \rangle$ direction. (**B**) The 8-fold structure of the core is clear from the maximum-entropy object.

An example of this method applied to a dislocation core in GaN is shown in Figure 25 (Xin et al., 1998). The maximum-entropy object function consists of fine points, the best fit atomic positions, and clearly reveals the core structure of the threading dislocation. The accuracy of the method is easily checked by measuring spacings far from the dislocation core, here, as found typically, $\sim\pm0.1$ Å for individual columns. More detailed descriptions of the maximum-entropy approach applied to $Z$-contrast images and its accuracy and comparison with alternative schemes are given elsewhere (McGibbon et al., 1999; Nellist and Pennycook, 1998b).

### Strain Contrast

Elastic strain fields due to impurities, point defects, or extended defects will affect the image by disturbing the channeling of the probe, even if they do not significantly affect the columnar scattering cross-section. Dislocations are visible because they induce transitions between Bloch states. This is the usual mechanism of diffraction contrast

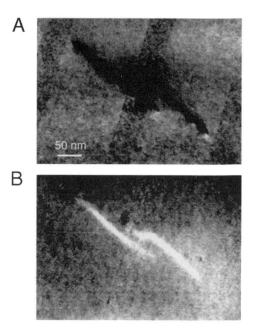

**Figure 26.** (**A**) STEM bright-field and (**B**) annular dark-field diffraction contrast images of inclined dislocations in a thick Si/Si(B) superlattice.

in CTEM images, but clearly, if the $1s$-type Bloch states are involved, dislocations will also be visible in the annular detector image. In general, transitions may occur into or out of the $s$ states, depending on the depth in the crystal, giving oscillatory contrast from inclined dislocations, as seen in Figure 26 (Perovic et al., 1993a,b). Even dislocations that are viewed end-on may show strain contrast due to the transverse relaxations of the atomic positions that occur near the sample surface. For this reason, grain boundaries, which are closely spaced arrays of dislocations, often appear brighter or darker than the matrix. Strain contrast is relatively long range compared to the lattice parameter and can be removed by Fourier filtering if desired. As shown below, strain contrast depends strongly on detector angle and can be distinguished from compositional changes ($Z$ contrast) by comparing images taken with different inner detector angles.

From the discussion of the loss of coherence through atomic vibrations, it is clear that in a zone axis crystal transverse displacements comparable to the atomic vibration amplitude will significantly affect the scattering at high angles. The diffuse scattering cross-section per atom $\sigma$ depends on scattering angle and temperature through the atomic form factor $f$ and the Debye-Waller factor $M^T$:

$$\sigma = f^2[1 - \exp(-2M^T s^2)] \tag{8}$$

where $s$ is the scattering angle, $M^T = 8\pi^2 u_T^2$, and $u_T^2$ is the mean-square thermal vibration amplitude of the atom (Hall and Hirsch, 1965). In the presence of static random atomic displacements, assuming a Gaussian distribution of strain with a mean-square static displacement of $u_s^2$, the atomic scattering cross-section will be modified to (Hall et al., 1966)

$$\sigma^s = f^2 \{1 - \exp[-2(M^T + M^S)s]\} \qquad (9)$$

where $M^s = 8\pi u_s^2$. It is clear from the form of these expressions that both tend to the full atomic scattering cross-section $\sigma = f^2$ at a sufficiently high scattering angle. At lower angles where the Debye-Waller factor is significant, static strains comparable to the thermal vibration amplitude may lead to a significantly enhanced scattering cross-section. Figure 27 shows images of a thermally grown Si-SiO$_2$ interface. The bright-field phase-contrast image (Fig. 27A) shows dark contrast that could be due to a number of effects, such as strain, thickness variation, bending of the crystal, or a combination of these mechanisms. Figure 27B shows an incoherent dark-field image collected simultaneously using a low (25-mrad) inner radius for the annular detector. Now there is a bright band near the interface indicating additional scattering.

With this image alone, this additional scattering could be due either to strain or to the presence of some heavy impurity atoms. However, when the inner detector angle is increased further, to 45 mrad, the bright line disappears (Fig. 27C), showing that the contrast cannot be due to the presence of heavy impurity atoms that would still give increased scattering. The contrast must therefore be due to a static strain effect. Intensity profiles across the two dark-field images are shown in Figure 28A. Taking their ratio normalizes the channeling effect of each column and gives the additional atomic displacement due to strain, as shown in Figure 28B (Duscher et al., 1998b). This is seen to decrease exponentially from the incoherent interface, as would be expected for a uniform array of misfit dislocations.

Strain effects are also commonly seen at grain boundaries and can be distinguished from true $Z$ contrast in the same way, by comparing images at different detector

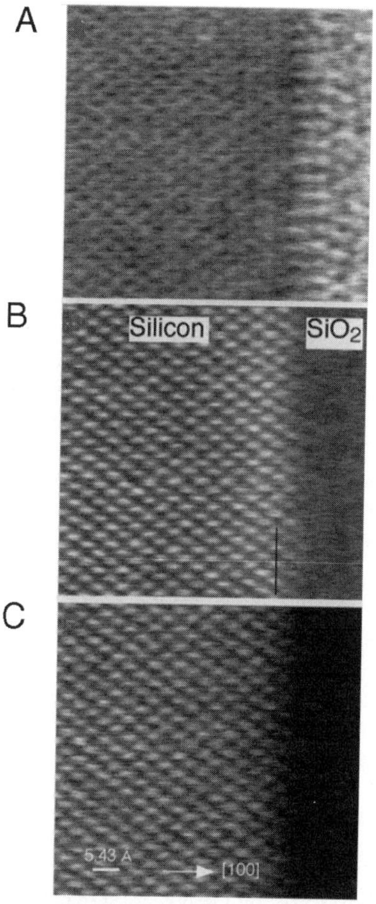

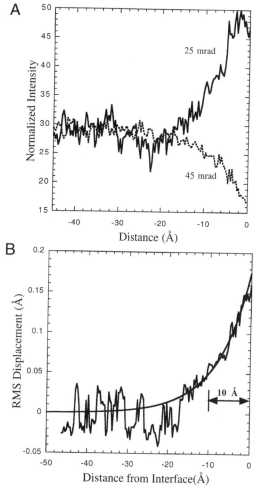

**Figure 28.** (**A**) Intensity profiles across the $Z$-contrast images of Figure 27. The high-angle profile shows the dechanneling effect near the interface, which can be used to normalize the profile obtained with a lower detector angle. (**B**) Root-mean-squared atomic displacement due to static strain induced by the Si-SiO$_2$ interface and an exponential decay fitted to the data. Profile ends at the vertical line shown in Figure 27.

**Figure 27.** STEM images of a Si-SiO$_2$ interface: (**A**) bright-field phase-contrast image; (**B**) $Z$-contrast image with 25 mrad inner detector angle showing strain contrast. The vertical line marks the last Si plane used for strain profiling. (**C**) $Z$-contrast image with 45 mrad inner detector angle.

angles. A strain effect will appear bright at low detector angles, changing to dark at high detector angles. True segregation of heavy impurities will appear bright at both detector angles.

## SAMPLE PREPARATION

Specimen preparation requirements for STEM are essentially the same as for HREM—a similar sample thickness, but generally the thinner the better to avoid problems with projection. However, some differences are required for working with small probes. In particular, the crystal surface must be as free of damage as possible. Ion milling leaves a thin amorphous layer, which broadens the probe before it reaches the crystal surface below. As the resolution is directly determined by the probe size, this leads to loss of contrast and perhaps no atomic resolution at all, whereas it might hardly be noticed in a phase-contrast image. Low-voltage final milling is essential for STEM, and samples made by cleaving or chemical means will give the best results. Nevertheless, most of the images shown in this unit were obtained from samples that had been ion milled. The amorphous layer increases the intensity fluctuations between identical columns as the probe profile is affected by the randomly located atoms above each column.

## SPECIMEN MODIFICATION

Damage is usually expected to be a major concern with the high brightness of a small probe, but it should be remembered that for a coherent probe at the electron optical limit, a high demagnification is used and probe currents are in the region of 10 to 100 pA, small compared with the nanoampere currents used typically in microanalysis. There is only one electron in the column at a time, and some excitations, such as electron-hole pairs, may have time to decay before the next electron arrives. For this reason, many damage processes have a threshold beam current, and generally damage by an atomic resolution probe is not a limiting factor for imaging. Nevertheless, when imaging grain boundaries, it is sometimes observed that the structure at the boundary becomes amorphous after a few scans (Maiti et al., 1999). In STEM, because the only area illuminated is that scanned by the beam, it is possible to "walk" the probe down the boundary, taking images from a fresh area each time. In EELS studies damage is a more serious concern, because of the longer exposure times needed for data collection and also because the light elements are the first to suffer displacement damage.

## PROBLEMS

Perhaps the main problem with the Z-contrast technique has been the lack of available instruments except in selected centers. Now that dedicated instruments are no longer required and the technique is being incorporated into the conventional TEM column, it should see more widespread use.

Because the STEM instrument is serial, with data recorded pixel by pixel, those familiar with CTEM will find a number of differences. One difference is certainly the noisy appearance of the raw Z-contrast image, although the atomic positions can be determined through maximum-entropy analysis to an accuracy of $\pm 0.1$ Å, comparable to that of the best conventional methods. Reconstructed images are less noisy, although the noise in the original data now appears as random displacements of the "noise free" atomic columns.

Sample drift also appears in a very different manner than in CTEM. Drift blurs the image in CTEM, whereas in STEM it distorts the image. One can view this as an advantage or disadvantage, but the result is that STEM images often show angular distortion or wavy atomic planes due to specimen drift. This makes it difficult to determine rigid shifts across grain boundaries by the usual methods. One would usually rotate the scan direction so that the fast scan is across the boundary, giving the best measurement of expansion or contraction. One can also compare images taken at different rotations.

Vacuum is a critical requirement for STEM studies, as hydrocarbons are attracted to the small probe and can build up rapidly during high-magnification viewing or analysis when the probe scans only a small area. Sometimes this problem can be alleviated by "flooding," irradiating the sample at the maximum available current by removing all apertures and reducing the demagnification. Obviously, besides polymerizing the hydrocarbons, this may also damage the sample. Usually the vacuum in modern instruments is sufficiently good that the specimen itself becomes the major source of carbon contamination. Use of oil-free specimen preparation techniques is useful, particularly if samples are sensitive to carbon, as in the case of the high-temperature superconductors. Plasma cleaning can be useful, or even a gentle bake in air using an oven or by placing the specimen on a light bulb can be very effective in eliminating contamination. Only those hydrocarbons that adsorb strongly to the sample at room temperature are a problem; lighter species evaporate in the vacuum while heavier species are immobile. Hence, only a moderate temperature is required.

## ACKNOWLEDGMENTS

The author is grateful to their colleagues P. D. Nellist, D. E. Jesson, M. F. Chisholm, N. D. Browning, Y. Yan, Y. Xin, B. Rafferty, G. Duscher, and E. C. Dickey for research collaborations. This research was supported by Lockheed Martin Energy Research Corp., under Department of Energy Contract No. DE-AC05-96OR22464, and by appointments to the ORNL Postdoctoral Research Associates Program administered jointly by ORNL and ORISE.

## LITERATURE CITED

Ade, G. 1977. On the incoherent imaging in the scanning transmission electron microsocpe. *Optik* 49:113–116.

Anderson, S. C., Birkeland, C. R., Anstis, G. R., and Cockayne, D. J. H. 1997. An approach to quantitative compositional profiling at near atomic resolution using high-angle annular dark-field imaging. *Ultramicroscopy* 69:83–103.

Brown, L. M. 1981. Scanning transmission electron microscopy: Microanalysis for the microelectronic age. *J. Phys. F* 11:1–26.

Browning, N. D., Buban, J. P., Nellist, P. D., Norton, D. P., Chisholm, M. F., and Pennycook, S. J. 1998. The atomic origins of reduced critical currents at [001] tilt grain boundaries in YBa$_2$Cu$_3$O$_{7-\delta}$ thin films. *Physica C* 294:183–193.

Browning, N. D., Chisholm, M. F., Nellist, P. D., Pennycook, S. J., Norton, D. P., and Lowndes, D. H. 1993b. Correlation between hole depletion and atomic structure at high-angle grain boundaries in YBa$_2$Cu$_3$O$_{7-\delta}$. *Physica C* 212:185–190.

Browning, N. D., Chisholm, M. F., and Pennycook, S. J. 1993a. Atomic-resolution chemical analysis using a scanning transmission electron microscope. *Nature* 366:143–146.

Browning, N. D. and Pennycook, S. J. 1995. Atomic-resolution electron energy loss spectroscopy in the scanning transmission electron microscope. *J. Microsc.* 180:230–237.

Browning, N. D., Pennycook, S. J., Chisholm, M. F., McGibbon, M. M., and McGibbon, A. J. 1995. Observation of structural units at symmetric [001] tilt boundaries in SrTiO$_3$. *Interface Sci.* 2:397–423.

Chisholm, M. F., Maiti, A., Pennycook, S. J., and Pantelides, S. T. 1998. Atomic configurations and energetics of arsenic impurities in a silicon grain boundary. *Phys. Rev. Lett.* 81:132–135.

Chisholm, M. F. and Pennycook, S. J. 1997. Z-contrast imaging of grain-boundary core structures in semiconductors. *Mater. Res. Soc. Bull.* 22:53–57.

Chisholm, M. F., Pennycook, S. J., Jebasinski, R., and Mantl, S. 1994. New interface structure for A-type CoSi$_2$/Si(111). *Appl. Phys. Lett.* 64:2409–2411.

Coene, W., Janssen, G., Op de Beeck, M., and Van Dyck, D. 1992. Phase retrieval through focus variation for ultra-resolution in field-emission transmission electron microscopy. *Phys. Rev. Lett.* 69:3743–3746.

Colliex, C. and Mory, C. 1984. Quantitative aspects of scanning transmission electron microscopy. *In* Quantitative Electron Microscopy (J. N. Chapman and A. J. Craven, eds.). pp. 149–216. Scottish Universities Summer School in Physics, Edinburgh.

Cowley, J. M. 1969. Image contrast in a transmission scanning electron microscope. *Appl. Phys. Lett.* 15:58–59.

Cowley, J .M. 1976. Scanning transmission electron microscopy of thin specimens. *Ultramicroscopy* 2:3–16.

Cowley, J. M. 1979. Adjustment of a STEM by use of shadow images. *Ultramicroscopy* 4:413–418.

Cowley, J. M. 1986a. Principles of image formation. *In* Principles of Analytical Electron Microscopy (J. J. Hren, J. I. Goldstein, and D. C. Joy, eds.). pp. 343–368. Plenum, New York.

Cowley, J. M. 1986b. Electron diffraction phenomena observed with a high resolution STEM instrument. *J. Electron Microsc. Techne.* 3:25–44.

Cowley, J. M. 1997. Scanning transmission electron microscopy. *In* Handbook of Microscopy (S. Amelinckx, D. van Dyck, J. van Landuyt, and G. van Tendeloo, eds.). pp. 563–594. VCH Publishers, Weinheim, Germany.

Crewe, A. V., Wall, J., and Langmore, J. 1970. Visibility of single atoms. *Science* 168:1338–1340.

Datye, A. K. and Smith, D. J. 1992. The study of heterogeneous catalysts by high-resolution transmission electron microscopy. *Catal. Rev. Sci. Eng.* 34:129–178.

Dickey, E. C., Dravid, V. P., Nellist, P. D., Wallis, D. J., and Pennycook, S. J. 1998. Three-dimensional atomic structure of NiO-ZrO$_2$(cubic) interfaces. *Acta Metall. Mater.* 46:1801–1816.

Duscher, G., Browning, N. D., and Pennycook, S. J. 1998a. Atom column resolved electron energy loss spectroscopy. *Phys. Stat. Solidi (a)* 166:327–342.

Duscher, G., Pennycook, S. J., Browning, N. D., Rupangudi, R., Takoudis, C., Gao, H-J., and Singh, R. 1998b. Structure, composition and strain profiling of Si/SiO$_2$ interfaces. *In* Characterization and Metrology for ULSI Technology, Conf. Proc. 449 (D. G. Seiler, A. C. Diebold, W. M. Bullis, T. J. Shaffer, R. McDonald, and E. J. Walters, eds.). pp. 191–195. American Institute of Physics, Woodbury, N.Y.

Engel, A., Wiggins, J. W., and Woodruff, D. C. 1974. A comparison of calculated images generated by six modes of transmission electron microscopy. *J. Appl. Phys.* 45:2739–2747.

Fathy, D. and Pennycook, S. J. 1981. STEM Microanalysis of Impurity Precipitates and Doping in Semiconductor Devices, Conf. Ser. 60 (M. J. Goringe, ed.) pp. 243–248. Institute of Physics, Bristol and London.

Gull, S. F. and Skilling, J. 1984. Maximum entropy methods in image processing. *IEE Proc.* 131F:646–659.

Hall, C. R. and Hirsch, P. B. 1965. Effect of thermal diffuse scattering on propagation of high energy electrons through crystals. *Proc. R. Soc. A* 286:158–177.

Hall, C. R., Hirsch, P. B., and Booker. 1966. Effects of point defects on absorption of high energy electrons passing through crystals. *Philos. Mag.* 14:979–989.

Hartel, P., Rose, H., and Dinges, C. 1996. Conditions and reasons for incoherent imaging in STEM. *Ultramicroscopy* 63:93–114.

Hillyard, S., Loane, R. F. and Silcox, J. 1993. Annular dark-field imaging: Resolution and thickness effects. *Ultramicroscopy* 49:14–25.

Isaacson, M. S., Ohtusuki, M., and Utlaut, M. 1979. Electron microscopy of individual atoms. *In* Introduction to Analytical Electron Microscopy (J. J. Hren, J. I. Goldstein, and D. C. Joy, eds.). pp. 343–368. Plenum, New York.

James, E. M. and Browning, N. D. 1999. Practical aspects of atomic resolution imaging and analysis in STEM. *Ultramicroscopy.* 78:125–139.

James, E. M., Browning, N. D., Nicholls, A. W., Kawasaki, M., Xin, Y., and Stemmer, S. 1998. Demonstration of atomic resolution Z-contrast imaging in a JEOL-2010F scanning transmission electron microscope. *J. Electron Microsc.* 47:561–574.

Jesson, D. E. and Pennycook, S. J. 1993. Incoherent imaging of thin specimens using coherently scattered electrons. *Proc. R. Soc. (London) A* 441:261–281.

Jesson, D. E. and Pennycook, S. J. 1995. Incoherent imaging of crystals using thermally scattered electrons. *Proc. R. Soc. (London) A* 449:273–293.

Jesson, D. E., Pennycook, S. J., and Baribeau, J.-M. 1991. Direct imaging of interfacial ordering in ultrathin (Si$_m$Ge$_n$)$_p$ superlattices. *Phys. Rev. Lett.* 66:750–753.

Lin, J. A. and Cowley, J. M. 1986. Calibration of the operating parameters for an HB5 STEM instrument. *Ultramicroscopy* 19:31–42.

Lipson, S. G. and Lipson, H. 1982. Optical Physics, 2nd ed. Cambridge University Press, Cambridge.

Loane, R. F., Kirkland, E. J., and Silcox, J. 1988. Visibility of single heavy atoms on thin crystalline silicon in simulated annular dark-field STEM images. *Acta Crystallogr.* A44:912–927.

Loane, R. F., Xu, P., and Silcox, J. 1992. Incoherent imaging of zone axis crystals with ADF STEM. *Ultramicroscopy* 40:121–138.

Maiti, A., Chisholm, M. F., Pennycook, S. J., and Pantelides, S. T. 1999. Vacancy-induced stuctural transformation at a Si grain boundary. *Appl. Phys. Lett.* 75:2380–2382.

McGibbon, M. M., Browning, N. D., Chisholm, M. F., McGibbon, A. J., Pennycook, S. J., Ravikumar, V., and Dravid, V. P. 1994. Direct determination of grain boundary atomic structure in SrTiO$_3$. *Science* 266:102–104.

McGibbon, M. M., Browning, N. D., McGibbon, A. J., and Pennycook, S. J. 1996. The atomic structure of asymmetric [001] tilt boundaries in SrTiO$_3$. *Philos. Mag. A* 73:625–641.

McGibbon, A. J., Pennycook, S. J., and Angelo, J. E. 1995. Direct observation of dislocation core structures in CdTe/GaAs(001). *Science* 269:519–521.

McGibbon, A. J., Pennycook, S. J., and Jesson, D. E. 1999. Crystal structure retrieval by maximum entropy analysis of atomic resolution incoherent images. *J. Microsc.* 195:44–57.

McMullan, D., Rodenburg, J. M., Murooka, Y., and McGibbon, A. J. 1990. Parallel EELS CCD detector for a VG HB501 STEM. *Inst. Phys. Conf. Ser* 98:55–58.

Möbus, G. 1996. Retrieval of crystal defect structures from HREM images by simulated evolution. I. Basic technique. *Ultramicroscopy* 65:205–216.

Möbus, G. and Dehm, G. 1996. Retrieval of crystal defect structures from HREM images by simulated evolution. II. Experimental image evaluation. *Ultramicroscopy* 65:217–228.

Nakamura, K., Kakibayashi, H., Kanehori, K., and Tanaka, N., 1997. Position dependence of the visibility of a single gold atom in HAADF-STEM image simulation. *J. Electron Microsc.* 1:33–43.

Nellist, P. D. and Pennycook, S. J. 1996. Direct imaging of the atomic configuration of ultra-dispersed catalysts. *Science* 274:413–415.

Nellist, P. D. and Pennycook, S. J. 1998a. Sub-Ångstrom resolution TEM through under-focussed incoherent imaging. *Phys. Rev. Lett.* 81:4156–4159.

Nellist, P. D. and Pennycook, S. J. 1998b. Accurate structure determination from image reconstruction in ADF STEM. *J. Microsc.* 190:159–170.

Nellist, P. D. and Pennycook, S. J. 1999. Incoherent imaging using dynamically scattered coherent electrons. *Ultramicroscopy.* 78:111–124.

Orchowski, A., Rau, W. D., and Lichte, H. 1995. Electron holography surmounts resolution limit of electron microscopy. *Phys. Rev. Lett.* 74:399–401.

Pennycook, S. J. 1982. High resolution electron microscopy and microanalysis. *Contemporary Phys.* 23:371–400.

Pennycook, S. J. 1995. Z-contrast electron microscopy for materials science. *In* Encyclopedia of Advanced Materials (D. Bloor, R. J. Brook, M. C. Flemings, and S. Mahajan, eds.). pp. 2961–2965. Pergamon Press, Oxford.

Pennycook, S. J. and Boatner, L. A. 1988. Chemically sensitive structure imaging with a scanning transmission electron microscope. *Nature* 336:565–567.

Pennycook, S. J., Brown, L. M., and Craven, A. J. 1980. Observation of cathodoluminescence at single dislocations by STEM. *Philos. Mag.* 41:589–600.

Pennycook, S. J., Chisholm, M. F., Jesson, D. E., Norton, D. P., Lowndes, D. H., Feenstra, R., Kerchner, H. R., and Thomson, J. O. 1991. Interdiffusion, growth mechanisms, and critical currents in YBa$_2$Cu$_3$O$_{7-x}$/PrBa$_2$Cu$_3$O$_{7-x}$ superlattices. *Phys. Rev. Lett.* 67:765–768.

Pennycook, S. J. and Howie, A. 1980. Study of single electron excitations by electron microscopy; II cathodoluminescence image contrast from localized energy transfers. *Philos. Mag.* 41:809–827.

Pennycook, S. J. and Jesson, D. E. 1990. High-resolution incoherent imaging of crystals. *Phys. Rev. Lett.* 64:938–941.

Pennycook, S. J. and Jesson, D. E. 1991. High-resolution Z-contrast imaging of crystals. *Ultramicroscopy* 37:14–38.

Pennycook, S. J. and Jesson, D. E. 1992. Atomic resolution Z-contrast imaging of interfaces. *Acta Metall. Mater.* 40:S149–S159.

Pennycook, S. J., Jesson, D. E., and Nellist, P. D. 1996. High angle dark field STEM for advanced materials. *J. Electron Microsc.* 45:36–43.

Pennycook, S. J., Jesson, D. E., Nellist, P. D., Chisholm, M. F., and Browning, N. D. 1997. Scanning transmission electron microscopy: Z-contrast. *In* Handbook of Microscopy (S. Amelinckx, D. van Dyck, J. van Landuyt, and G. van Tendeloo, eds) pp. 595–620. VCH Publishers, Weinheim, Germany.

Perovic, D. D., Howie, A., and Rossouw, C. J. 1993a. On the image contrast from dislocations in high-angle annular dark-field scanning transmission electron microscopy. *Philos. Mag. Lett.* 67: 261–277.

Perovic, D. D., Rossouw, C. J. and Howie, A. 1993b. Imaging inelastic strains in high-angle annular dark field scanning transmission electron microscopy. *Ultramicroscopy* 52:353–359.

Rafferty, B. and Pennycook, S. J. 1999. Towards atomic column-by-column spectroscopy. *Ultramicroscopy* 78:141–152.

Lord Rayleigh 1896. On the theory of optical images with special reference to the microscope. *Philos. Mag.* 42(5):167–195.

Scherzer, O. 1949. The theoretical resolution limit of the electron microscope. *J. Appl. Phys.* 20:20–29.

Spence, J. C. H. and Cowley, J. M. 1978. Lattice imaging in STEM. *Optik* 50:129–142.

Wallis, D. J., Browning, N., Sivananthan, S., Nellist, P. D., and Pennycook, S. J. 1997. Atomic layer graphoepitaxy for single crystal heterostructures. *Appl. Phys. Lett.* 70:3113–3115.

Xin, Y., Pennycook, S. J., Browning, N. D., Nellist, P. D., Sivananthan, S., Omnès, F., Beaumont, B., Faurie, J.-P., and Gibart, P. 1998. Direct observation of the core structures of threading dislocations in GaN. *Appl. Phys. Lett.* 72:2680–2682.

Yan, Y., Chisholm, M. F., Duscher, G., Maiti, A., Pennycook, S. J., and Pantelides, S. T. 1998. Impurity induced structural transformation of a MgO grain boundary. *Phys. Rev. Lett.* 81:3675–3678.

Zeitler, E. and Thomson, M. G. R. 1970. Scanning transmission electron microscopy. *Optik* 31:258–280, 359–366.

## KEY REFERENCES

Bird, D. M. 1989. Theory of zone axis electron diffraction. *J. Electron. Microsc. Tech.* 13:77–97.

*A good introduction to the theory of dynamical diffraction using Bloch states.*

Keyse, R. J., Garret-Redd, A. J., Goohhew, P. J., and Lorimer, G. W. 1998. Introduction to Scanning Transmission Electron Microscopy. Bios Scientific Publishers, Oxford.

*A general introduction to STEM.*

Pennycook, S. J. 1992. Z-contrast transmission electron microscopy: Direct atomic imaging of materials. *Annu. Rev. Mater. Sci.* 22:171–195.

*A simple description of Z-contrast imaging and a number of early applications.*

Pennycook, S. J. and Jesson, D. E. 1992. High-resolution imaging in the scanning transmission electron microscope. *In* Proceedings of the International School on Electron Microscopy in Materials Science (P. G. Merli and M. Vittori Antisari, eds.). pp. 333–362. World Scientific, Singapore.

*Includes a detailed discussion on breaking the coherence of the imaging process through detector geometry and thermal vibrations.*

Pennycook, S. J., Jesson, D. E., Chisholm, M. F., Browning, N. D., McGibbon, A. J., and McGibbon, M. M. 1995. Z-contrast imaging in the scanning transmission electron microscope. *J. Microsc. Soc. Am.* 1:231–251.

*A nonmathematical description of Z-contrast imaging and a number of applications to semiconductors and superconductors.*

Pennycook et al., 1997. See above.

*Provides a more detailed theoretical treatment of incoherent imaging with elastic, thermal, and inelastic scattering and a survey of applications to materials.*

## APPENDIX A:
## INCOHERENT IMAGING OF WEAKLY SCATTERING OBJECTS

For a very thin specimen, effects of probe dispersion and absorption may be ignored, and the scattered amplitude $y_s$ in the direction $\mathbf{K}_f$ is obtained immediately from the first Born approximation (Pennycook et al., 1997),

$$\psi_s(\mathbf{K}_f) = \frac{m}{2\pi\hbar^2} \int e^{-i\mathbf{K}_f \cdot \mathbf{R}} V(\mathbf{R}) P(\mathbf{R} - \mathbf{R_0}) d\mathbf{R} \qquad (10)$$

where $V(\mathbf{R})$ is the specimen potential integrated along the beam direction, called the projected potential. Integrating the scattered intensity $|\psi_s|^2$ over all final states $\mathbf{K}_f$ and using the identity

$$\int e^{-i\mathbf{K}_f \cdot (\mathbf{R} - \mathbf{R'})} d\mathbf{K}_f = (2\pi)^2 \delta(\mathbf{R} - \mathbf{R'}) \qquad (11)$$

the total scattered intensity is given as

$$I(\mathbf{R_0}) = \int O(\mathbf{R}) P^2(\mathbf{R} - \mathbf{R_0}) d\mathbf{R} \qquad (12)$$

$$= O(\mathbf{R}) * P^2(\mathbf{R_0}) \qquad (13)$$

a convolution of the probe intensity profile $P^2(\mathbf{R})$ with an object function $O(\mathbf{R})$ given by

$$O(\mathbf{R}) = \left(\frac{\chi}{2E}\right)^2 V^2(\mathbf{R}) \qquad (14)$$

where $\sigma = \chi/2E$ is the interaction constant and $E$ is the beam energy. Therefore, provided *all* scattered electrons could be collected, we see immediately that incoherent imaging would be obtained with a resolution controlled by the incident probe intensity profile.

For many years it was considered that incoherent imaging at atomic resolution was impossible in principle because much of the total scattered intensity occurs at small angles, in the same angular range as the unscattered

beam (e.g., Ade, 1977). There is no way to distinguish unscattered electrons from elastically scattered electrons, which was referred to as the hole-in-the-detector problem. However, realizing that the Rayleigh bright-field detector is equivalent to the annular dark-field detector in Figure 5, it is clear that the problem does not arise in the dark-field case and perfect incoherent imaging may be achieved.

## APPENDIX B:
## STEM MANUFACTURERS

### Dedicated STEM Instruments

Conventional phase-contrast and diffraction contrast modes are available, but all images are recorded in the scanning mode. Instruments have cold field emission guns and are typically equipped with EELS and x-ray analysis capabilities.

VG Microscopes: No longer commercially available, but often can be found in user centers. The entire microscope is bakeable, resulting in a clean vacuum.

Hitachi: http://www.hii.hitachi.com/prdem.htm
TEM/STEM Instruments
JEOL: http://www.jeol.com/navbar/tembar.html
Hitachi: http://www.hii.hitachi.com/prdem.htm
Philips: http://www.feic.com/tecnai/main.htm

The Philips and JEOL instruments have a Schottky source. Hitachi instruments have a cold field emission source. Note that manufacturers specify resolution in different ways. The best comparison is to find the objective lens spherical aberration coefficient and use the definitions of resolution given in Table 1.

S. J. PENNYCOOK
Oak Ridge National Laboratory
Oak Ridge Tennessee

# SCANNING TUNNELING MICROSCOPY

## INTRODUCTION

Scanning tunneling microscopy (STM) is a technique for the determination of the structure of surfaces, with spatial resolution on the angstrom scale. In an oversimplified description, STM allows imaging of the surface of conductive materials down to the atomic level.

The technique is based on the measurement and control of the current of electrons tunneling between a sharp stylus (hereafter called the tip) and the sample surface. During imaging, the tip is separated a few angstroms (3 to 10 Å) from the surface, while it is being rastered with the help of piezoelectric transducers. The area scanned is tens to thousands of angstroms on a side. STM has the advantage over other microscopies that its operation does not require any special environmental conditions. Imaging can be performed in vacuum, at high pressures (including air), and in liquids. The choice of the environment is determined mostly by the requirements of the sample and its surface condition.

Acquisition of STM images is relatively simple with modern instruments. Most of the experimental time is spent on sample preparation and tip conditioning in the appropriate environment (ultrahigh vacuum, electrochemical cell, or controlled atmosphere). Tip conditioning and noise reduction can sometimes add considerable time, depending on factors such as the nature of the experiment and reactivity of the sample. This time depends also on the type of information being sought. A medium-resolution (0.5 to 1 Å in $z$, i.e., normal to the surface) large-area image of a few hundred angstroms is relatively easy to obtain. In such a case, however, similar information can be obtained with atomic force microscopy (AFM), which is easier to use. It is in the atomic-resolution area that STM excels over any other surface microscopy, because atom-to-atom corrugations <0.5 Å—and sometimes <0.1 Å, as in the case of close-packed metal surfaces—need to be resolved. Therefore, the requirements for both instrument and sample preparation are substantially more stringent.

STM was discovered at the beginning of the 1980s and has since experienced an explosive development in both instrumentation and applications (Olgetree and Salmeron, 1990; Wiesendanger, 1994). It has given rise to numerous related techniques generally referred to as scanning probe microscopies (SPM). The most popular of these is AFM. The related technique known as scanning electrochemical microscopy is covered in detail elsewhere (SCANNING ELEC-TROCHEMICAL MICROSCOPY).

In considering the ultimate resolution obtainable with the STM, it is important to consider the role of tip geometry. The schematic drawing at the bottom of Figure 1 illustrates the basic point: on flat parts of the sample, where one tip atom is responsible for the tunneling, STM reaches its maximum performance. On rough surfaces, however, the tunneling point on the tip apex changes as it moves over sharp corners and changing slopes. The image is then a convolution of the tip and surface shapes, and the final resolution is determined by the tip radius and profile. STM is thus not the best technique to probe, for example, highly corrugated surfaces, deep-patterned surfaces, or powdered and porous materials.

### Complementary and Competitive Techniques

Atomic force microscopy (AFM) is a technique that has many characteristics in common with STM, although the physical processes involved are quite different. In AFM, there is also a sharp tip but it is mounted on a very flexible cantilever. As in the STM, it is rastered over the surface by means of piezoelectric transducers. Tip-surface interaction forces are sensed in AFM by the deflection of the lever, instead of the tunneling current. AFM is therefore ideal for studies of insulating materials that are not directly accessible to STM imaging.

The resolution of AFM in its most usual operation modes (contact, friction, or tapping) is not truly atomic, as is the case with STM. The forces of interaction produce a contact spot that is several tens of angstroms in diameter, depending on the applied load. Thus atomic-size point defects are not observed in AFM. However, new non-

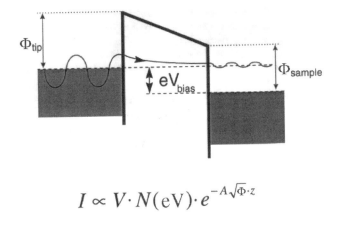

$$I \propto V \cdot N(\text{eV}) \cdot e^{-A\sqrt{\Phi} \cdot z}$$

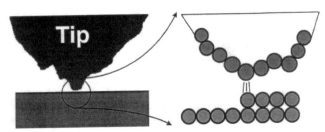

**Figure 1.** Top: Schematic energy diagram of the tip and sample in an STM experiment. Electrons tunnel from the negatively-biased tip (left) to the sample (right). Only the electrons within the eV range of states in the tip participate in the tunneling. The tunnel barrier is related to the work function $\Phi$ of the material involved. Bottom: Schematic of the shape of a tip near a flat surface. At the atomic scale shown on the right, the most protruding atom concentrates the tunnel current. Secondary images can be obtained if another protruding atom can enter within tunneling range in surface irregularities.

contact imaging techniques using resonant-lever modulation are being developed, and true atomic resolution, i.e., the visualization of point defects, has been demonstrated (Giessibl, 1995; Kitamura and Iwatsuki, 1995; Sugawara et al., 1995). These developments should be incorporated in commercial instruments soon.

Because contact with the (inert) tip is part of the standard operation of the AFM, the technique is less sensitive to chemical instabilities, and to some forms of mechanical vibration noise, than STM. For the routine examination of samples with a resolution of a few nanometers in $x$-$y$ and angstroms in $z$, AFM is clearly advantageous over STM.

Electrons in the far field regime, as opposed to tunneling in the near field, are used in transmission electron microscopy (TEM) to provide angstrom-resolved projections of bulk atomic rows in crystalline samples (see SCAN-NING TRANSMISSION ELECTRON MICROSCOPY: Z-CONTRAST IMAGING). TEM has been used on several occasions in an edge-on configuration to resolve surface atomic features. The sample requirements for TEM are quite strict, as samples must be thinned down to a few hundred angstroms to allow transmission of the electron beam. TEM is excellent, for example, for use in studies of bulk phases of crystalline

**Table 1. Comparison of Various Microscopies**

| Technique[a] | Probe/Principle | Environment | Sample Requirements | Resolution (Å) | |
|---|---|---|---|---|---|
| | | | | $xy$ | $z$ |
| STM | Electron tunneling | Air, vacuum, liquid | Flat, conductor | ~1–2 | ~$10^{-2}$ |
| AFM | Forces: attractive, repulsive contact, noncontact | Air, vacuum, liquid | Flat, insulator, conductor | ~10 | 0.1 |
| TEM | Electron transmission | High vacuum | Solid, ~100 Å thin | ~2 | — |
| SEM | Secondary electrons | High vacuum | Solid, conductive | 20 | — |
| SAM | Auger emission | Ultrahigh vacuum | Conductive | 100 | — |
| SIMS | Ion sputtering | Ultrahigh vacuum | Most samples | 1000 | — |

[a]Abbreviations: STM, scanning tunneling microscopy; AFM, atomic force microscopy; TEM, transmission electron microscopy; SEM, scanning electron microscopy; SAM, scanning Auger microscopy; SIMS, scanning secondary ion mass spectroscopy.

precipitates, in metallurgy or in small particles supported on amorphous or crystalline materials. It is also used extensively in studies of biomaterials supported on carbon films. A thin coating (~50 Å) of heavy metal is usually necessary to provide contrast.

Scanning electron microscopy (SEM) is based on the measurement of the secondary electron yield of conductive substrates (see SCANNING TRANSMISSION ELECTRON MICROSCOPY: Z-CONTRAST IMAGING). This yield changes both as a function of composition and local surface slope. The spatial resolution of SEM is determined by the spot size of the electron beam (~20 Å in the best instruments), and by the diffusion of the secondary electrons before exiting the sample. The SEM operates in vacuum and the best results are obtained with conductive substrates. Combined with x-ray emission detection for elemental composition determination, it is a powerful technique for material studies.

At lower resolutions, a number of other techniques should be considered: scanning Auger microscopy (SAM; see AUGER ELECTRON SPECTROSCOPY), scanning secondary ion mass spectroscopy (SIMS), and small-spot x-ray photoelectron spectroscopy (XPS). These techniques offer the great advantage of spectroscopic information in addition to imaging. Table 1 is a compendium of several commonly used techniques for real-space imaging.

## PRINCIPLES OF THE METHOD

A schematic view of the operation of the STM is shown in Figure 1. On top is an energy diagram of the sample and tip. The wavefunction of a tunneling electron is represented by the wavy lines inside the solids connected by a decaying line in the intervening gap. The negative bias applied to the tip causes a net flow of electrons to the sample. The electrons leave occupied levels in an energy band of width eV between the Fermi levels of tip and surface, with $V$ being the bias voltage, and tunnel through the potential barrier separating the two materials to occupy empty levels at the sample side. The reverse situation occurs if the bias is positive. Both the height of the tunneling barrier and the separation between the two surfaces enter into the exponential dependence of the current

$$I \propto V \cdot N(eV) \cdot e^{-A\sqrt{\Phi} \cdot z} \tag{1}$$

where $A$ is a number of order 1 with units $(eV)^{-1/2} \, \mathring{A}^{-1}$, $\Phi$ the tunneling barrier height in eV and $z$ the separation in Å. The strong $z$ dependence of $I$ is at the origin of the very high $z$-distance resolution of the STM. In properly designed microscopes, where the noise level is determined by the Johnson noise of the preamplifier gain resistance, the $z$ resolution can be 1/1000 Å. The in-plane resolution is determined, as illustrated at the bottom of Figure 1, by the atomic arrangement at the tip apex. In the optimal case of a single-atom-terminated tip, it is the electronic "size" of this atom (~1 to 2 Å) that determines the $x$-$y$ resolution. The nature of the tip (geometry and chemical composition) is paramount in determining the quality of the images. Except in special cases, the nature of the tip is unknown, and the operation of the STM to date is still subject to some degree of empiricism. This empiricism is the art of tip preparation, for which no well-proven recipes exist yet. In practice, however, atomically resolved images are often obtained due to the exponential dependence of $I$ on $z$, which naturally selects the most protruding atom as the imaging element.

Equation 1 is the well-known quantum mechanical result that relates the tunneling probability of electrons through a one-dimensional potential energy barrier of width $z$ (the tip-sample separation) and height $\Phi$. The value of $\Phi$ is related to the work functions of the tip and sample materials and is due to the overlap of the potential energy curves of the electrons beyond the surface atoms. Its value is therefore on the order of a few electron volts.

Typical values of $I$, $V$, $z$ and $\Phi$ are $I = \sim 1$ nA, $V = \sim 10$ mV to 3 V, $z = \sim 5$ to 15 Å, and $\Phi = \sim 4$ to 5 eV. Equation 1 is valid in the limit of low applied voltages ($V \ll \Phi$), and so it can be viewed as the relation between bias current and gap resistance $R$. The value of $R$ is the most practical way to characterize the tunneling gap, since it can be directly measured ($R = V/I$).

STM images are typically acquired in a wide range of $R$ values, from MΩ to GΩ and even TΩ. To put the value of $R$ in perspective, it should be compared with the quantum resistance unit $2e^2/h$, which is 12.9 kΩ and observed in single-atom contacts. Between this value and a few tens of MΩs, the risk of strong tip-surface interactions, leading to such problems as sample indentation or tip and surface modifications through atom transfer, is considerable. In surfaces that have low chemical activity or that are covered with passivating layers, contact imaging is sometimes possible in a friction-like regime without permanent modification of the tip. This type of imaging, usually by interposed atoms or molecules between the tip and surface, is not uncommon, even if often not explicitly known to many experimenters.

A clear signature of contact STM imaging is the value of the barrier height $\Phi$, obtained in $I$ versus $z$ measurements. In these, the tip is advanced toward or away from the surface while the current $I$ is measured. The slope of log $I$ versus $z$ should be $\sim 2$ Å$^{-1}$, corresponding to a value of $\Phi = \sim 4-5$ eV for clean, true tunneling gaps. This corresponds to a change of about one decade of current per angstrom. Values in the range of a few tenths of an electron volt and below are signatures of "contact" imaging. An example of these two situations is shown in Figure 2, which corresponds to a nominally clean Pt tip and a Pd substrate (Behler et al., 1997a). The explanation for the low barrier height is the elastic deformation of tip and sample due to the contact forces through interposed material in the gap. Because of these elastic deformations, the change in gap width is always less than the change in tip or sample position calculated from the voltages applied to the piezo transducers.

The discussion below (see Data Analysis and Initial Interpretation) contains additional information on the physical nature of the tunneling process and its relation to the actual image contrast.

## PRACTICAL ASPECTS OF THE METHOD

### Acquisition of STM Images

As mentioned above, the tip (or sample) scans over areas that range from a few Å to several μm in $x$ and $y$, while an electronic feedback control keeps the tunnel current constant. The acquisition of images is done electronically by means of computers interfaced with an STM controller using data conversion cards. The number of data points is variable, but $128 \times 128$, $512 \times 512$, or $1024 \times 1024$ pixels per image are common. On each pixel, the value of $z$ is stored as a 12- or 16-bit number. The resulting data files are 200 kb to 1 Mb in size.

Many variant modes of operation are used, depending on the experiment. Besides the "topographic" mode in which $z(x,y)$ profiles are acquired at constant $I$, one can also acquire $I(x,y)$ profiles at constant $z$. The latter is the so-called "current" mode and is useful in fast scanning of atomically flat surfaces. Here the feedback control response time is slow relative to the scanning time, so that only the average value of $I$ is kept constant.

More specialized operation modes include multiple bias imaging, in which interlaced curves at different preset values of $V$ are acquired, and the current image tunneling spectroscopy (CITS) mode, in which the current corresponding to a few selected $V$ values is collected at each pixel during a brief interruption of the feedback control (Hamers, 1992). These two imaging modes have been used principally in investigations of semiconductor surfaces. At selected points of the surface, complete $I(V)$ curves can be obtained for a more detailed study of the band structure (Feenstra et al., 1987). This type of experiment is fraught with difficulties, however, since at the interesting voltages (those of a few volts) that are necessary to access states in the conduction and valence bands, the strong electric field at the tip can easily modify its structure, making reproducibility problematic in many cases. The large information content of the images acquired in these modes can only be displayed in multiple images (one for each value of $V$) or by post-analysis of the $I(V)$ curves to extract the density-of-states information.

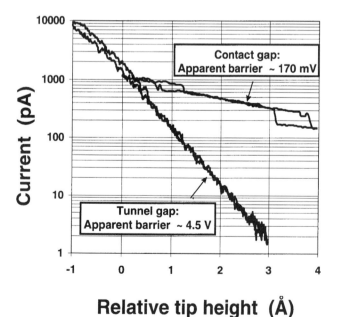

**Figure 2.** Current versus distance for a tunneling junction of a Pt tip on a Pd(111) crystal surface in ultrahigh vacuum. A true vacuum tunneling produces a change of current of roughly a decade per Å, and corresponds to an apparent barrier height of 4.5 eV. A contact gap, characteristic of dirty junctions, gives a much smaller current decay and apparent barrier heights of less than 1 eV. From Behler et al. (1997a).

### Types of Materials and Problems

The study of semiconductor surfaces is one area where STM has been applied most intensely. Once clean and well-prepared, semiconductor surfaces are relatively easy to image. Large surface reconstructions occur in many of these materials, due to the strong directionality of the tetrahedral covalent bonds, coupled with the tendency to minimize the number of broken bonds. This produces rather large corrugations, from 0.2 to 1 Å in $z$, and relatively large unit cells in the $x$-$y$ plane, which makes it possible to obtain good atomic resolution images even in instruments with modest stability performance. In addition, semiconductor surfaces tend to be less reactive than metals.

Metal surfaces are another important class of materials studied with STM. The applications include surface chemistry, i.e., the study of chemisorbed structures, and electrochemistry. These topics are covered in detail in many recent reviews (Ogletree and Salmeron, 1990). Here we will comment only on some peculiarities in imaging these materials, which should be kept in mind when considering and planning an STM experiment. The most important characteristic of metal surfaces is that their corrugation is substantially smaller than that of semiconductors. This is due to the delocalized nature of the metallic bond. The conduction electrons that are responsible for the tunneling signal are spread over the unit cell, rather than spatially concentrated in covalent bonds. The resulting electronic corrugation is always small, particularly in the close-packed surfaces—i.e., the (111) planes of fcc metals, the (110) planes in bcc metals, and (0001) or basal planes in hcp metals, which are also the most widely studied because of their stability. Corrugations ranging from 0.01 to 0.2 Å are typical. The more open surfaces tend to have more corrugation and are then easier to resolve atomically.

Adsorbates can change the corrugation substantially, since in many cases they give rise to localized, covalent-type bonds. The apparent height of isolated atoms or molecules can vary from a few tenths of an angstrom to several angstroms. Imaging of isolated adsorbates, however, is not an easy task even for strongly bound elements. The reason is the high diffusion rate, at room temperature, of the adatoms. For example, the binding energy of sulfur on Re is very high, ~4.5 eV for the (0001) plane (Dunphy et al., 1993). However, its diffusion energy barrier is ~0.7 eV, which implies a residence time of 100 ms on a site at room temperature. Thus, during scanning, the S adatoms are changing positions continuously, making imaging problematic. More weakly bound adsorbates, e.g., CO molecules, might therefore appear invisible in the STM images because of rapid diffusion. In all these situations, sample cooling is necessary, sometimes to very low temperatures. The pioneering work of Eigler and his collaborators demonstrated the great advantages of low temperature STM for fundamental studies (Eigler and Schweizer, 1990; Zeppenfeld et al., 1992; Heller et al., 1994). As a result, the design and construction of cryogenic STMs has recently increased.

Imaging while the sample is at high temperature imposes strict demands on the design of the microscope head, because of large thermal drifts (Curtis et al., 1997; Kuipers et al., 1993; Kuipers et al., 1995). In vacuum, good shielding, together with careful choice of construction materials with the appropriate expansion coefficients, have made it possible to obtain images up to several hundreds of degrees Celsius (McIntyre et al., 1993). Under high pressures (a few torr to several atmospheres), convection in the gas phase adds to the problem of thermal drifts, and limitations in the highest temperatures are to be expected.

Similar considerations apply to metals in an electrochemical cell, although heating and cooling of the sample is not feasible except in a very narrow temperature window. Therefore, imaging of diluted or isolated adsorbates is difficult or impossible.

Oxides tend to be insulating, and therefore the surfaces of bulk oxide materials cannot be easily imaged by STM. Exceptions include the cases of small bandgap oxides and of oxides that can be made conductive by defect doping. Such is the case for $TiO_2$ (Rohrer et al., 1990; Onishi and Iwasawa, 1994; Sander and Engel, 1994; Murray et al., 1995), $UO_2$ (Castell et al., 1996), $Fe_3O_4$ (Jansen et al., 1996), and others. Nonconductive oxides, in the form of very thin layers, have also been studied by deposition or growth onto metal substrates.

Biological material can, in principle, be imaged with STM. However, the normally poor electrical conductivity of these materials, and the need to adsorb them on flat conductive substrates, impose severe constrictions. As a result, today these materials are preferentially studied with AFM techniques.

Other applications of STM include the study of epitaxial growth of metals and oxides on metals and the mapping of superconducting and normal-conducting areas in superconductors by using a bias voltage above or below the superconducting gap (Hess et al., 1989; Hess et al., 1990a,b). Recently, there has been increasing use of STM as a microfabrication (Quate, 1989) or atomic manipulation tool (Eigler and Schweizer, 1990; Stipe et al., 1997; Bartels et al., 1998).

## METHOD AUTOMATION

A major component of any STM instrument is the software for data acquisition and analysis. With the help of such software, color-coded, picture-like graphic displays of the images can be produced. Top views, as well as three-dimensional representations, including artificial light shading, are commonly used to enhance contrast, as in the example of Figure 3. The selection of the color or gray-scale mapping is arbitrary and left to the subjective taste of the operator. Image processing features (e.g., smoothing, filtering, and flattening) are common, as well as analysis capabilities for extracting cursor profiles along selected paths over the image. Other features included in most software packages are Fourier analysis and height histograms, among others. Today's sophisticated programs have rendered the task of analysis and data display relatively easy, and have made possible the beautiful images of atomic-surface landscapes (see the examples in Figure 3) that are currently produced in many STM laboratories.

Complete automation of an STM experiment is not common for either commercial or home-made instruments. The nature of the STM experiment itself is not easily conducive to automation. This is because the atomic structure of the surfaces under study is a delicate function of preparation and environmental conditions. Exceptions are the most chemically inert surfaces such as graphite, $MoS_2$, and others of limited intrinsic interest, except

## Sulfur rings on Re(0001)

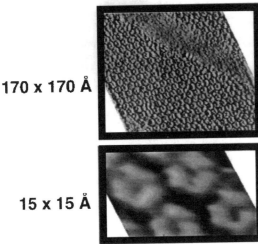

170 x 170 Å

15 x 15 Å

## Sulfur layer on Molybdenum (001)

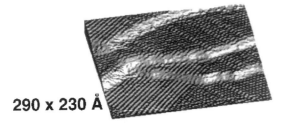

290 x 230 Å

**Figure 3.** Examples of STM images obtained in vacuum. The top image was obtained on a Re(0001) surface covered by a saturation layer of sulfur. Sulfur is seen forming nearly hexagonal rings. Notice the blurred area corresponding to a monoatomic step. The radius of the tip determines the resolution in this case. The middle image shows a magnified view of the atomic arrangement. The bottom image was obtained on a Mo(001) surface, again with a saturation layer of sulfur. Sulfur atoms are seen forming two domains at a 90°angle. Steps 2-, 3- and 4-atoms high separate the terraces.

when used as substrates for other adsorbed material. Thus, it is not possible in general to simply insert a sample into a sample holder and have the instrument position the sample and acquire atomically resolved images. The approach of the tip from a large distance (centimeters or millimeters) down to a few angstroms from the surface is risky and often accompanied by some instability when nearing the "in range" or tunneling position that leads to atom exchanges and tip or surface modification. STM is thus an operator-intensive technique, as it requires considerable experience and exercise of scientific judgment for its successful realization.

## DATA ANALYSIS AND INITIAL INTERPRETATION

The fundamental question in image interpretation is, of course, the relation between the measured topography, i.e., the $z(x,y)$ profile and the actual "height" or position of the atomic features. Contrast and symmetry are largely

determined by the bonding orbitals of the atoms or molecules being imaged, and by the chemical nature of the tip termination. In its simplest interpretation, the $z(x,y)$ profile corresponds to the electronic corrugation of the surface at a constant integrated density of states between the Fermi levels of the tip and sample (see diagram in Figure 1). The latent capability of the STM to provide both topographic and electronic maps of the surface cannot therefore be fully realized without a theory of the tunneling process.

While an exact theory does not yet exist, several approximations have been useful in interpreting the images. The first one came from Tersoff and Hamann (1985), who assumed no interaction between the tip and surface. The tunneling current in this model, for a tip with electron states of $s$ symmetry, is proportional to the density of states near the Fermi edge at the position of the tip center (considered to be terminated in a sphere). Other theoretical treatments have also been developed and can be found in recent reviews (Güntherodt and Wiesendanger, 1993). A recent one, called electron scattering quantum chemistry (ESQC), deserves special mention because of its relative simplicity and good results (Sautet and Joachim, 1988; Cerda et al., 1997). It is based on the combination of an exact treatment of the electron transport through the tunneling gap with approximate quantum chemistry methods (e.g., extended Hückel) to calculate the Hamiltonian matrix elements. It does not require the neglect of tip-surface interactions, and the structure of the tip is explicitly taken into account in the calculation of the image. It has been used in chemisorbed systems with remarkable success.

The strong electronic contribution to the image contrast sometimes gives rise to counterintuitive observations. For example, O atoms are usually imaged as holes (lower

## Ga sublattice

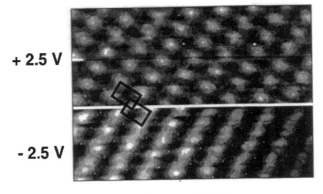

+ 2.5 V

- 2.5 V

## As sublattice

**Figure 4.** STM image of the cleavage face of an $n$-type GaAs sample. The bias applied to the sample was switched from +2.5 to −2.5 V near the center of the image. At positive bias, electrons tunnel from the tip states to empty states of the GaAs. These states are due to empty Ga dangling bonds, which have a lobe localized nearly atop the Ga atoms. At negative bias, electrons leave As dangling bonds, which are also localized near the As atoms. The two images represent the sublattices of these two atoms, which are offset in the manner indicated by the two boxes representing the surface unit cell.

tunneling probability) rather than protrusions (Kopatzki and Behm, 1991; Wintterlin et al., 1988; Sautet, 1997). Interference, when the tip overlaps several adsorbate atoms, can give rise to strong effects. These effects are manifested sometimes in a contrast distribution inside the unit cell that does not reflect the number of maxima expected from the atomic composition. For example, S on Mo(001) forms a c(2 × 4) structure, with 3 S atoms in the unit cell and yet, only one maximum is observed in the images (see Fig. 3), giving the apparent impression that only one S atom is present (Sautet et al., 1996).

Electronic effects in the STM images are particularly strong for semiconductor surfaces, due to the existence in these materials of energy gaps and surface states close to the Fermi level that are filled or empty. This is the case of GaAs(110), for example, where a filled dangling bond located on the As atom is imaged at negative sample bias while an empty dangling bond, located at the Ga atom position, is imaged at the opposite polarity (Feenstra and Stroscio, 1987). This is shown in Figure 4.

## SAMPLE PREPARATION

The preparation of samples for STM is not different from that in other microscopies. In UHV-STM (ultrahigh vacuum STM), samples are prepared following well established surface-science methods. These include such techniques as sputtering, annealing, characterization by Auger electron spectroscopy (AES; see AUGER ELECTRON SPECTROSCOPY), or low-energy electron diffraction (LEED; see LOW-ENERGY ELECTRON DIFFRACTION), whenever appropriate. Air or high-pressure environments require special procedures, sometimes in combination with sample preparation in environmental chambers. In electrochemical cells equipped with STM, cleaning is part of the oxidation-reduction cycles and can be tested by well-established electrochemical voltammograms (see SCANNING ELECTROCHEMICAL MICROSCOPY).

It is the preparation of tips that is a distinct procedure of STM. Ideally single-atom-terminated tips are desired. However, short of performing field ion microscopy examination prior to or after imaging, no technique or procedure has been developed that meets universal standards of characterization. Tip preparation is thus, regrettably, a technique that is still largely empirical and varies from laboratory to laboratory.

The most widely used tip materials are W and Pt. The first is easily prepared using electrochemical etching methods with alkaline solutions (KOH or NaOH) to thin, cut, and sharpen a wire. Tips with radii from a few tens to a few hundreds of angstroms can thus be produced. If the wire is a single crystal of (111) orientation, single-atom-terminated tips of defined structure can, in principle, be prepared (Kuk, 1992). Pt alloyed with Ir or Rh is also a popular tip material. Such Pt alloys might be preferable in applications with reactive atmospheres and in air, where W tips can oxidize and more easily produce insulating layers. Pt tips can be prepared by electrochemical etching using molten (∼300°C) mixtures of 3:1 $NaNO_3$:NaCl as the electrolyte bath. The sharpened tips usually are subjected to additional conditioning in vacuum, through heating by resistive or electron-beam methods and/or electron field emission on a sacrificial substrate. The purpose of these final treatments is to remove contaminant layers that can diffuse to the tip apex during imaging.

Unfortunately, even these procedures are no guarantee of a "good" tip, which is defined as one that produces stable, well resolved images. Deterioration and accidental changes occur by brief contacts (induced by vibrational noise) with the sample, by capture of contaminant atoms or molecules that form unstable bonds with the tip, and by loss of material. The cleaner and better controlled the sample and environment, the higher is the success rate of tip preparation and utilization. Often, lost performance can be restored by application of voltage pulses to the tip (>1 to 2 V), or by intentional contact with the sample. In the end, as previously pointed out, tip preparation is still an empirical procedure in most STM operations, and performance is decided in a somewhat subjective manner by the operator. For example, tips may be considered "good" and one-atom terminated when sharp and stable images with symmetric "atoms" are obtained. Satisfaction of that criterion, however, does not guarantee that "good" images acquired with different single-atom-terminated tips will have reproducible heights of the atoms, with $z$ variations of a few tenths of an angstrom possible.

## PROBLEMS

The immediate signature of a defective STM is the observation of an unstable tunneling gap. This is visible in the high noise level in the $z$ height or the tunneling current signals. The cause may vary from the trivial—such as broken wires, cracked piezos, depolarization, or poor mechanical isolation between head and/or sample and the rest of the apparatus—to the more subtle and problematic, related to uncontrolled chemical processes occurring at the tip or sample. The tunneling gap, which must be stable within a small fraction of an angstrom, is subject to several instabilities. These instabilities can excite mechanical resonances in the head and sample support, which in turn can introduce a phase change in the feedback response. When this change becomes larger than 90°, positive feedback occurs and the tip oscillation is then driven instead of suppressed by the electronic control.

A chemically unstable gap can occur as a result of the continuous transfer of atoms or molecules at the surface or at the tip apex in dirty samples or tips. Sample and tip cleaning, as well as improvement of the environmental conditions, are then in order.

An excellent test of gap stability and a good troubleshooting exercise that should be common practice with students and technicians using STM is the analysis of the harmonic content of the current signal. This analysis gives a good deal of information about the sources and origin of the noise. It is performed by the application of fast Fourier transform (FFT) techniques, which are (or should be) a common feature of most data acquisition and software systems. The feedback control signal (usually the $z$ height, in the form of a voltage applied to the $z$ piezo scanner) can also be analyzed in this way, but it responds only within the frequency range of the feedback bandwidth, which is rather small. Thus, at low gain, the $z$ response of an unstable gap has the appearance of a saw tooth—the large error signal charges the feedback capacitor in the integrator at a nearly linear rate until saturation. At high gain, the feedback control–gap circuit can enter into resonance and cause the tip to oscillate against the surface. If the frequency of the offending noise is within the feedback bandwidth, the gain can be adjusted to suppress the noise in either $z$ or $I$, but not both, since this does not solve the problem.

The best way to analyze the noise is to look at the FFT of the tunnel current, with the feedback at low gain or disabled. A clean, stable and well behaved gap should show no peaks in the crucial region of 1 to 500 Hz, since this is the region covered by the feedback response time and is also the region of typical scanning frequencies. An example of FFT analysis is shown in Figure 5, obtained

with a beetle-type STM head developed by Besocke and Wolf (Frohn et al., 1989). This head consists of three piezo tubes attached perpendicularly to a disk, forming an equilateral triangle. The free end of each of these piezos moves when voltages are applied, allowing it to displace and rotate. A fourth piezo tube, at the center of the disk, supports the tip and can be used for scanning. The top curve corresponds to a noisy gap, with a $z$ noise of a few angstroms, and is due to a loose sample holder. Loose parts tend to enhance the noise in the low-frequency region, below a few tens of hertz. Other poor conditions can derive from such factors as wires touching the chamber or a cracked piezo. The bottom curve shows the FFT of the same head after fastening all the parts that were loose. Peaks at the frequency of the power line (60 Hz or 50 Hz) are due mostly to ground loops and can be eliminated. The FFT spectrum will also show high frequency peaks due to the natural resonances of the STM scanner head, which should be greater than 1 kHz (the higher the better.) In the example of Figure 5, they are above 4.5 kHz. There are several studies published about this subject that the reader can consult for additional details (Behler et al., 1997b).

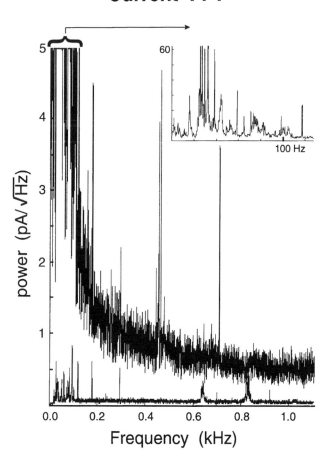

**Figure 5.** Example of Fourier transform analysis of the tunneling current for a beetle-type STM tip. The top curve was obtained in a highly noisy situation. In this case, the noise was due to the loose condition of the sample holder. Low-frequency noise is dominant, in the region 0 to 50 Hz (see top right inset). This noise couples strongly with the scanning rate at similar frequencies. Rattling of the entire head, which is simply resting on the sample holder in this design, is also excited and produces the peaks at ~500 and ~700 Hz. The spectrum below this was obtained after tightly securing the sample holder. The overall noise decreased by almost 3 orders of magnitude.

### ACKNOWLEDGMENTS

The author acknowledges the support and help of various members of the STM group at LBNL, especially M. Rose and D.F. Ogletree for discussions and the data used in this part. This work was supported by the Director, Office of Energy Research, Office of Basic Energy Sciences, Materials Science Division of the U.S. Department of Energy under Contract No. DE-AC03-76SF00098.

### LITERATURE CITED

Bartels, L., Meyer, G., Rieder, K.-H., Velic, D., et al. 1998. Dynamics of electron-induced manipulations of individual CO molecules on Cu(111). *Phys. Rev. Lett.* 80:2004–2007.

Behler, S., Rose, M. K., Dunphy, J. C., Olgetree, D. F., Salmeron, M., and Chapelier, C. 1997a. A scanning tunneling microscope with continuous flow cryostat sample cooling. *Rev. Sci. Instrum.* 68:2479–2485.

Behler, S., Rose, M. K., Ogletree, D. F., and Salmeron, M. 1997b. Method to characterize the vibrational response of a beetle type scanning tunneling microscope. *Rev. Sci. Instrum.* 68:124–128.

Castell, M. R., Muggelberg, C., Briggs, G. A. D., and Goddard, D. T. 1996. Scanning tunneling microscopy of the $UO_2/(111)$ surface. *J. Vac. Sci. Technol. B* 14:1173–1175.

Cerda, J., Van Hove, M. A., Sautet, P. and Salmeron, M. 1997. Efficient method for the simulation of STM images. I. Generalized Green-function formalism. *Phys. Rev. B* 56:15885–15889.

Curtis, R., Mitsui, T., and Ganz, E. 1997. An ultrahigh vacuum high speed scanning tunneling microscope. *Rev. Sci. Instrum.* 68:2790–2796.

Dunphy, J. C., Sautet, P., Ogletree, D. F., Daboussi, O., and Salmeron, M. 1993. Scanning-tunneling-microscopy study of the surface diffusion of sulfur on Re(0001). *Phys. Rev. B* 47:2320–2328.

Eigler, D. M. and Schweizer, E. K. 1990. Positioning single atoms with a scanning tunneling microscope. *Nature* 344:524–526.

Feenstra, R. M. and Stroscio, J. A. 1987. Tunneling spectroscopy of the GaAs(110) surface. *J. Vac. Sci. Technol. B* 5:923–929.

Feenstra, R. M., Stroscio, J. A., and Fein, A. P. 1987. Tunneling spectroscopy of the Si(111)2*1 surface. *Surf. Sci.* 181:295–306.

Frohn, J., Wolf, J. F., Besocke, K. H., and Teske, M. 1989. Coarse tip distance adjustment and positioner for a scanning tunneling microscope. *Rev. Sci. Instrum.* 60:1200.

Giessibl, F. J. 1995. Atomic resolution of the silicon (111)-(7*7) surface by atomic force microscopy. *Science* 267:68–71.

Güntherodt, H.-J. and Wiesendanger, R. (eds.).. 1993. Scanning Tunneling Microscopy, Vol III. Springer-Verlag, Heidelberg.

Hamers, R. J. 1992. STM on semiconductors. *In* Scanning Tunneling Microscopy, Vol. I (H.-J. Güntherodt and R. Wiesendanger, eds.).. pp. 83–129. Springer-Verlag, Heidelberg.

Heller, E. J., Crommie, M. F., Lutz, C. P., and Eigler, D. M. 1994. Scattering and absorption of surface electron waves in quantum corrals. *Nature* 369:464–466.

Hess, H. F., Robinson, R. B., Dynes, R. C., Valles, J. M., and Waszczak, J. V. 1989. Scanning-tunneling-microscope observation of the Abrikosov flux lattice and the density of states near and inside a fluxoid. *Phys. Rev. Lett.* 62:214–216.

Hess, H. F., Robinson, R. B., Dynes, R. C., Valles, J. M., and Waszczak, J. V. 1990a. Spectroscopic and spatial characterization of superconducting vortex core states with a scanning tunneling microscope. *J. Vac. Sci. Technol. A* 8:450–454.

Hess, H. F., Robinson, R. B., and Waszczak, J. V. 1990b. Vortex-core structure observed with a scanning tunneling microscope. *Phys. Rev. Lett.* 64:2711–2714.

Jansen, R., van Kempen, H., and Wolf, R. M. 1996. Scanning tunneling microscopy and spectroscopy on thin Fe$_3$O$_4$(110) films on MgO. *J. Vac. Sci. Technol. B* 14:1173–1175.

Kitamura, S. and Iwatsuki, M. 1995. Observation of 7*7 reconstructed structure on the silicon (111) surface using ultrahigh vacuum noncontact atomic force microscopy. *Jpn J. Appl. Phys.* 34:L145–L148.

Kopatzki, E. and Behm, R. J. 1991. STM imaging and local order of oxygen adlayers on Ni(100). *Surf. Sci.* 245:255–262.

Kuipers, L., Hoogeman, M. S., and Frenken, J. W. M. 1993. Step dynamics on Au(110) studied with a high-temperature, high-speed scanning tunneling microscope. *Phys. Rev. Lett.* 71:3517–3520.

Kuipers, L., Loos, R. W. N., Neerings, H., ter Horst, J. et al. 1995. Design and performance of a high-temperature, high-speed scanning tunneling microscope. *Rev. Sci. Instrum.* 66:4557–4565.

Kuk, Y. 1992. STM on metals. *In* Scanning Tunneling Microscopy, Vol. I (H.-J. Güntherodt and R. Wiesendanger, eds.).. pp.17–37. Springer-Verlag, Berlin.

McIntyre, B. J., Salmeron, M., and Somorjai, G. A. 1993. A variable pressure/temperature scanning tunneling microscope for surface science and catalysis studies. *Rev. Sci. Instrum.* 64:687–691.

Murray, P. W., Condon, N. G., and Thornton, G. 1995. Na absorption sites on TiO$_2$/(110)-1*2 and its 2*2 superlattice. *Surf. Sci.* 323:L281–L286.

Ogletree, D. F. and Salmeron, M. 1990. Scanning tunneling microscopy and the atomic structure of solid surfaces. *Prog. Solid State Chem.* 20:235–303.

Onishi, H. and Iwasawa, Y. 1994. Reconstruction of TiO$_2$(110) surface: STM study with atomic-scale resolution. *Surf. Sci.* 313:L783–L789.

Quate, C. F. 1989. Surface modification with the STM and the AFM. *In* Scanning Tunneling Microscopy and Related Methods (R. J. Behm, N. Garcia, and H. Rohrer, eds.).. p. 281. Kluwer Academic Publishers, Norwell, Mass.

Rohrer, G., Henrich, V. E., and Bonnell, D. A. 1990. Structure of the reduced TiO$_2$/(110) surface determined by scanning tunneling microscopy. *Science* 250:1239–1241.

Sander, M. and Engel, T. 1994. Atomic level structure of TiO$_2$[110] as a function of surface oxygen coverage. *Surf. Sci.* 302:L263–268.

Sautet, P. 1997. Atomic absorbate identification with the STM: A theoretical approach, *Surf. Sci.* 374:406–417.

Sautet, P., Dunphy, J. C., and Salmeron, M. 1996. The Origin of STM contrast differences for inequivalent S atoms on a Mo(100) surface. *Surf. Sci.* 364:335–344.

Sautet, P. and Joachim, C. 1988. Electronic transmission coefficient for the single-impurity problem in the scattering-matrix approach. *Phys. Rev. B* 38:12238–12247.

Stipe, B. C. Rezaei, M. A., Ho, W., Gao, S., Persson, M., and Lundqvist, B. I. 1997. Single-molecule dissociation by tunneling electrons. *Phys. Rev. Lett.* 78:4410–4413.

Sugawara, Y., Ohta, M., Ueyama, H., and Morita, S. 1995. Defect motion on an InP(110) surface observed with noncontact atomic force microscopy. *Science* 270:1646–1648.

Tersoff, J. and Hamann, D. R. 1985. Theory of the scanning tunneling microscope. *Phys. Rev. B* 31:805–813.

Wiesendanger, R. 1994. Scanning Probe Microscopy and Spectroscopy: Methods and Applications. Cambridge University Press, New York.

Wintterlin, J., Brune, H., Hofer, H., and Behm, R. J. 1988. Atomic scale characterization of oxygen absorbates on Al(111) by scanning tunneling microscopy. *Appl. Phys. A* 47:99–102.

Zeppenfeld, P., Lutz, C. P., and Eigler, D. M. 1992. Manipulating atoms and molecules with a scanning tunneling microscope. *Ultramicroscopy* 42–44:128–133.

## KEY REFERENCES

Cerda et al. 1997. See above.

*A modern implementation of the ESQC is described.*

Güntherodt, H.-J. and Wiesendanger, R. 1994. Scanning Tunneling Microscopy, Vols. I and II. Springer-Verlag, Heidelberg.

*Contains an excellent collection of review papers covering most aspects of STM.*

Wiesdanger, 1994. See above.

*Contains a good description of the state of the STM field at the time of its publication.*

## INTERNET RESOURCES

http://www.omicron-instruments.com

http://www.RHK-TECH.com

*Web sites of two STM manufacturers, describing scanning microscopes and containing links.*

MIQUEL SALMERON
Lawrence Berkeley National Laboratory
Berkeley, California

# LOW-ENERGY ELECTRON DIFFRACTION

## INTRODUCTION

Low-energy electron diffraction (LEED) is a powerful method for determining the geometric structure of solid surfaces. The phenomenon was first observed by Davisson and Germer in 1927 and provided the earliest direct experimental proof of the wavelike properties of electrons. LEED has since evolved into one of the most powerful and widespread tools of the surface scientist. It is similar to x-ray diffraction (XRD) in the type of information that it provides, and XRD will be frequently referred to for analogy throughout this unit. The most obvious difference is that a beam of electrons, rather than x rays, is used. Since electrons have a mean free path (or attenuation length) measured in angstroms (Å) as opposed to microns (μ) for x rays, LEED is particularly sensitive to surface geometry.

LEED is a structural technique that can provide essentially two levels of information. In both cases, it is most commonly used to determine the structure of a solid surface when the bulk structure of the material is already known by other means (e.g., XRD). It is possible, and indeed straightforwardly simple, to use LEED to determine both the absolute dimensions of the surface unit cell and the unit cell symmetry. One can easily deduce that, for example, a surface has a two-dimensional (2D) unit cell twice as large as that of the bulk; this often allows a reasonable guess at the true structure. Sometimes it is sufficient to know that a particular phase is present without knowing the details of the structure. For example, the $(2 \times 4)$ reconstruction of GaAs (nomenclature explained below) has been found to be optimal for device growth (Farrow, 1995). Determination of the exact atomic positions is more difficult, but quantitative experiments can elucidate this second level of information. Sophisticated calculations, generally run on a workstation, can provide atomic coordinates with a typical precision of $\pm 0.05$ Å, which is generally more than adequate to determine the adsorption site of a molecule or the atomic positions in a reconstructed surface.

Many different surfaces can be examined by LEED, and their terminology must be understood. We assume familiarity with the basic crystal structures [face-centered cubic (fcc), zincblende, etc.]. A particular crystal face is specified by its Miller indices. These are of the form $(hkl)$, and the ideal bulk termination can be determined in the following way: place one corner of the unit cell at the origin of a Cartesian coordinate system and construct a plane that intersects the axes at the reciprocal of the respective Miller index (see SYMMETRY IN CRYSTALLOGRAPHY). In the general case, this plane would pass through the points $(1/h, 0, 0)$, $(0, 1/k, 0)$, and $(0, 0, 1/l)$. A Miller index of zero indicates that the plane is parallel to that axis. Further examples can be found in any solid-state textbook (e.g., Ashcroft and Mermin, 1976; Kittel, 1986). It is conceptually useful to regard a surface as being composed of stacked planes of atoms, each parallel to the topmost surface plane (see SYMMETRY IN CRYSTALLOGRAPHY).

The structure of an actual surface is usually very different from an ideal termination of the bulk. The low-Miller-

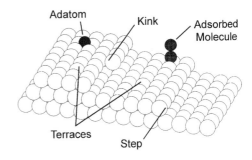

**Figure 1.** A model of the most common surface features.

index surfaces—(100), (110), (111)—are relatively flat, but higher-index surfaces generally have a large number of steps and often kinks as well. Adatoms and adsorbed molecules also appear in many systems. Figure 1 illustrates some of these features. LEED is sensitive to any such structures that recur at regular intervals, that is, with measurable periodicity.

LEED is a mature area of surface science and it is useful to consider the state of the art in this field. A great deal of work has been done on clean metal and semiconductor surfaces, and much of this is considered "done"; current work in this area consists mostly of refinements. Most low-index clean metal surfaces are found to merely relax the spacing between the topmost layers. Higher-index metal surfaces tend to be less closely packed and more prone to reconstruction. Clean semiconductor surfaces almost always reconstruct. Adsorbed atoms and molecules are sometimes found to lift reconstructions, and they often modify the dimensions and orientation of the surface unit cell themselves. Several atoms and simple molecules (H, O, S, CO, $C_2H_4$) have been investigated on low-Miller-index surfaces of semiconductors (Si, GaAs) and most transition metals. An important direction of current research is the growth of ordered crystalline epitaxial films for a variety of purposes. They serve as models for catalytic systems (Boffa et al., 1995), model magnetic structures, or avenues for studying otherwise intractable systems (Roberts et al; 1998), to name a few examples. Other recent advances seek to extend the power of the technique to systems with disorder in geometry (Döll et al., 1997; Starke et al., 1996) or composition (Gauthier et al., 1985; Crampin et al., 1991). Because of the time-consuming trial-and-error computations involved in a structural determination, there is also interest in developing a LEED-based holographic technique for rapid adsorption site determination (Saldin, 1997).

### Complementary and Related Techniques

Information similar or complementary to that provided by LEED can be obtained by other techniques. For example, XRD provides the same information as LEED, but without surface selectivity. Similar surface-specific diffraction information can be obtained from such methods as photoelectron diffraction (Woodruff and Bradshaw, 1994; Fadley et al., 1995), small-angle x-ray diffraction (SAXD; SURFACE X-RAY DIFFRACTION), x-ray standing waves (XSW;

Woodruff and Delchar, 1994, pp. 97-104), Rutherford back-scattering spectroscopy (RBS; Van Der Veen, 1985), and reflection high-energy electron diffraction (RHEED). The great advantage of LEED over these other diffraction methods is its ease of use and the simplicity of the instrumentation required. X-ray standing wave experiments require a monochromatic x-ray source of a brilliance only to be found at a synchrotron, and small-angle x-ray scattering requires very precise control over sample alignment. RBS needs a mega-electron-volt (MeV) Van de Graaff generator or an ion implantation machine, equipment only found at specialized facilities, to create the needed beam of high-energy ions (50 keV to 2 MeV $H^+$ or $He^+$) to probe the surface. In contrast, the majority of all ultrahigh-vacuum (UHV) surface analysis chambers are typically fitted with a relatively simple set of LEED optics, and sample alignment of $\sim0.5°$ precision is easily attained with standard equipment. RHEED is relatively simple to implement, but it suffers from the disadvantage that full structural determinations are far from routine, and it is generally applicable only in a qualitative mode. Nevertheless, it finds wide application particularly in the study of semiconductor growth because it allows for easy *in situ* study of MBE growth processes. Under some circumstances, RBS, also known as high-energy ion scattering (HEIS), can be used to infer surface structure. This is generally done by channeling ions between the aligned rows of atoms along major crystal axes in the substrate lattice and looking for scattering from displaced surface atoms that block those channels. By channeling along several axes, displaced surface and often second larger atoms can be located by triangulation. This has the advantage of being a real-space technique, rather than a diffraction-based reciprocal-space technique and can distinguish foreign adatoms under available conditions.

Methods that provide information complementary to LEED are of greater interest. The chief weakness of LEED is shared with all other diffraction techniques: it requires single-crystal samples and selectively views only the well-ordered portions of the sample. Neither disordered regions nor intimate details of atomic motion can generally be examined, except in an average sense. For a complete structural picture, the combination of LEED with microscopy techniques is especially powerful; scanning tunneling microscopy (STM; SCANNING TUNNELING MICROSCOPY) and atomic force microscopy (AFM) are especially useful in this regard. This has been addressed in a recent review (Van Hove et al., 1997). Another useful technique is SEXAFS, which is merely surface extended x-ray absorption fine structure (EXAFS) analysis (Koningsberger and Prins, 1988). This technique is a local probe of structure and provides a radial distribution function rather than crystallographic data. It is possible to use SEXAFS to determine structures just as in LEED, but its real utility is its applicability to disordered systems where diffraction cannot be done. SEXAFS also suffers from reliance on synchrotron light sources. Electron microscopy (EM; SCANNING ELECTRON MICROSCOPY, TRANSMISSION ELECTRON MICROSCOPY, SCANNING TRANSMISSION ELECTRON MICROSCOPY: Z-CONTRAST IMAGING) is useful for investigating structures on larger scales, and sophisticated EM

instruments can provide atomic resolution as well. Unlike LEED, neither SEXAFS nor EM requires a flat sample in order to work. However, it should be stressed that an inherent problem in all microscopy techniques is the question of representative sampling. LEED rapidly examines a sample of macroscopic proportions ($\sim1$ to 10 mm$^2$, the spot size of the electron beam).

Finally, LEED is a purely structural technique. Information about bond energies or strengths is simply not available, except insofar as correlations can be drawn with changes in bond lengths. In general, dynamic information is also not available. Temperature-dependent measurements can determine the mean amplitudes of isotropic thermal vibrations, and some efforts have been made to investigate anisotropic vibrations, but this work is in its early stages (Löffler et al., 1995). Certainly, none of this can compete with, for example, femtosecond laser spectroscopic methods for true molecular dynamic information.

## PRINCIPLES OF THE METHOD

As the name implies, LEED operates on the basis of diffraction of low-energy electrons, typically in the range of 20 to 400 eV (1 eV = $1.602 \times 10^{-19}$ J) from solid surfaces. Electrons in this energy range have wavelengths close to an angstrom and thus diffract from crystalline arrays of atoms. Electrons in this energy range are also very surface sensitive. Electrons can act as quantum mechanical waves with a De Broglie wavelength given by $\lambda(\text{Å}) = \sim \sqrt{150.4/E(\text{eV})}$, so energies in the range 20 to 400 eV are typically used to diffract from crystalline solids with lattice constants of typically several angstroms (1 = $10^{-10}$ m). This technique is surface sensitive because electrons in this energy range have a short inelastic mean free path ($\sim3$ to 20 Å) in all known materials. Seah and Dench (1979) have fit an analytic expression to the available experimental data that is plotted in Figure 2. This "universal curve" can be qualitatively understood by recognizing that the dominant mechanism for inelastic scattering is loss of energy to a plasmon. At lower energies, insufficient

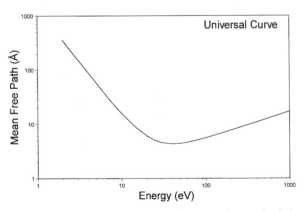

**Figure 2.** The "universal curve" of the mean free path of electrons in a solid, as a function of electron kinetic energy. (After Seah and Dench, 1979.) The mean free path represents a distance over which the flux of electrons drops to $1/e$ of its original value. The kinetic energy is that measured in vacuum.

energy exists to efficiently excite a plasmon, and at higher energies, the electron may well escape the solid before any excitation occurs. Once the electrons escape the solid, their mean free path in air requires pressures below $\sim 10^{-4}$ Torr for typical operating distances. Lower pressures are generally required for most experiments because of the need to keep samples clean; ultrahigh-vacuum conditions ($\leq 10^{-9}$ Torr) are generally recommended.

LEED is distinguished from more conventional diffraction techniques in that only the surface layers contribute to the diffraction pattern. This has consequences in the types of pattern that are seen. In conventional x-ray diffraction, for example, the real-space three-dimensional (3D) crystalline lattice generates a 3D lattice in reciprocal space (cf. X-RAY TECHNIQUES and SYMMETRY IN CRYSTALLOGRAPHY). The Laue condition for diffraction gives rise to the construction of an Ewald sphere that may or may not intersect another point on the reciprocal lattice. Put plainly, for arbitrary conditions of wavelength (or energy) and incident angle, diffraction will not necessarily occur from a 3D crystal. In practice, this means it is necessary to examine the reflection of x rays from a single crystal at many different angles in order to observe all possible diffraction spots. (Those familiar with typical experiments might notice that x-ray diffraction from a single crystal commonly is seen at all orientations; however, all spots are not seen at the same time. The common appearance of several spots under arbitrary conditions is usually due to deviations of the experiment from "ideal" conditions— finite divergence of the x-ray beam, polychromatic x-ray source.) Diffraction from a rigorously 2D crystal, however, guarantees that the diffraction condition will be met (above some critical energy). To see how this is so, examine the reciprocal lattice vectors, $\mathbf{b}_1$, $\mathbf{b}_2$, and $\mathbf{b}_3$ of a real-space lattice with lattice vectors $\mathbf{a}_1$, $\mathbf{a}_2$, and $\mathbf{a}_3$. The relation is

$$\mathbf{b}_1 = \frac{\mathbf{a}_2 \times \mathbf{a}_3}{\mathbf{a}_1 \bullet (\mathbf{a}_2 \times \mathbf{a}_3)} \qquad \mathbf{b}_2 = \frac{\mathbf{a}_3 \times \mathbf{a}_1}{\mathbf{a}_1 \bullet (\mathbf{a}_2 \times \mathbf{a}_3)}$$
$$\mathbf{b}_3 = \frac{\mathbf{a}_1 \times \mathbf{a}_2}{\mathbf{a}_1 \bullet (\mathbf{a}_2 \times \mathbf{a}_3)} \tag{1}$$

The periodicity of a 2D lattice in the direction normal to the lattice plane can be considered infinite. Thus, let $\mathbf{a}_3 = d\mathbf{n}$, where $\mathbf{n}$ is a unit vector normal to the surface plane. As $d \to \infty$ (the limit of infinite interplanar separation), the magnitude of $\mathbf{b}_3$ approaches zero, resulting in continuous rods, rather than points, in reciprocal space. This means that, once the Ewald sphere is sufficiently large (for sufficiently short wavelength), diffraction will always be observed for arbitrary orientation of the incident angle. That is, one is guaranteed to always see spots on the LEED screen for a sufficiently large energy of the electron beam.

The assumption of a rigorously 2D lattice is somewhat incorrect as a model for real 2D diffraction. Real crystals do have periodicity in the direction normal to the surface, but that periodicity terminates abruptly at the surface. This termination manifests as the rods in reciprocal space becoming "lumpy," and diffraction in certain directions being favored. This is realized in the LEED pattern as variations of the intensities of spots with the kinetic energy of the electrons. It is precisely this variation that allows extraction of detailed crystallographic parameters. Certain spots in the pattern will occasionally vanish to zero intensity, but these are isolated instances of angle and beam energy, and the spot will otherwise be generally visible. It happens to turn out that there are almost always conditions under which all of the spots can be seen.

**Qualitative Observations**

There are essentially two levels of information that can be extracted from a LEED experiment: the qualitative and the quantitative. Qualitatively, a LEED experiment will determine the symmetry of a surface structure. Consider, e.g., Figure 3. Shown in part A is the diffraction pattern obtained from a square 2D lattice [perhaps the (100) plane of a face- or body-centered cubic (bcc) metal] along with the corresponding real-space structure. Part B shows the diffraction pattern due to a so-called (2 × 2) pattern of adsorbates on the substrate crystal (notation explained below). One such real-space structure is also illustrated in B; adsorbates have been arbitrarily placed atop the atoms.

Also shown in Figure 3 is the indexing usually applied to the diffracted beams. The specularly reflected beam is placed at the center of the coordinate system, and two of the beams with the lowest value of $|\mathbf{k}|$ are selected as the (10) and (01) beams. Other beams are labeled according to the coordinate system generated. Negative indices are generally indicated by a bar over the appropriate number. When an adsorbate or reconstruction exists and induces extra spots, the most common method of indexing assigns fractional values to the new spots.

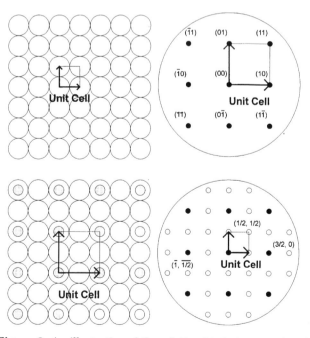

**Figure 3.** An illustration of the relationship between real and reciprocal space. In (**A**), we see a clean surface of square symmetry, with atoms represented as circles. In (**B**), we see the same surface with a p(2 × 2) unit cell. The translational symmetry has been lowered by an adsorbate arbitrarily placed atop an atom. Typical spot indexing is indicated by a label above the appropriate beam.

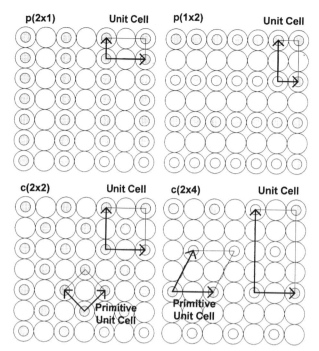

**Figure 4.** Some simple, single-domain real-space structures on a substrate of square symmetry. The Wood notation for each is given above the structure. These real-space models correspond to the diffraction patterns in Figure 5.

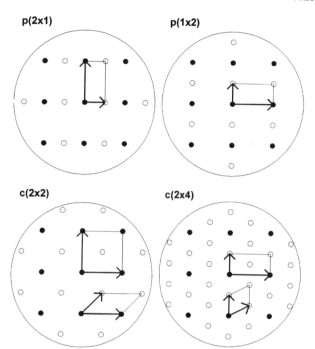

**Figure 5.** Some simple, single-domain reciprocal space structures on a substrate of square symmetry. These diffraction patterns correspond to the real-space models in Figure 4. Note that the p(2 × 1) and p(1 × 2) structures are not incorrectly reversed; the long dimension in real space is the short dimension in reciprocal space. The centered structures, c(2 × 2) and c(2 × 4), have two unit cells indicated. The lower one is the primitive unit cell, while the upper one is a larger cell that more obviously preserves the symmetry of the real-space unit cell.

The notation used above, the so-called Wood notation (after Wood et al., 1963), relates the periodicity of the overlayer to that of the substrate. It is commonly found that an overlayer adopts lattice vectors that are integral multiples of the substrate unit vectors. For example, in the (2 × 2) case above, the overlayer has a unit cell that is twice as large in both dimensions as the substrate unit cell. Figure 4 illustrates, again on a "square" lattice substrate, a p(2 × 1), a p(1 × 2), a c(2 × 2), and a c(2 × 4) structure. Their diffraction patterns are illustrated in Figure 5. Figures 6 and 7 illustrate the real-space lattice and diffraction patterns, respectively, for a p(1 × 2), p(2 × 2), $(\sqrt{3} \times \sqrt{3})R30°$, and c(2 × 4) overlayer on a surface of hexagonal symmetry [e.g., the (111) plane of an fcc metal]. Notice that all of the "c" structures have an adsorbate at the center of the unit cell as well as at the corners. If one examines the square c(2 × 2) structure more closely, one can also construct a smaller unit cell that is square, contains only one adsorbate, but is rotated 45° with respect to the substrate. The sides of this square are $\sqrt{2}$ times longer than the dimensions of the substrate unit cell, so this cell can be described as $(\sqrt{2} \times \sqrt{2})R45°$ as well.

More formally, Wood's notation can be described as $j[(b_1/a_1) \times (b_2/a_2)]R\theta$, where $j$ is either p or c, $a_n$ and $b_n$ are substrate and overlayer unit cell lengths, respectively, and $\theta$ is the angle through which the cell must be rotated to produce the overlayer unit cell. The notation "$R\theta$" indicates that the unit cell of the substrate must be rotated by an angle $\theta$ to give unit cell vectors parallel to those of the overlayer. Typically, $R\theta$ is omitted when $\theta$ is zero. The preceding "p" signifies a primitive unit cell, while "c"

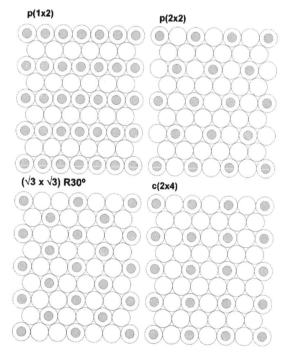

**Figure 6.** Some simple, single-domain real-space structures on a substrate of pseudo-hexagonal symmetry. The Wood notation for each is given above the structure. These real-space models correspond to the diffraction patterns in Figure 7.

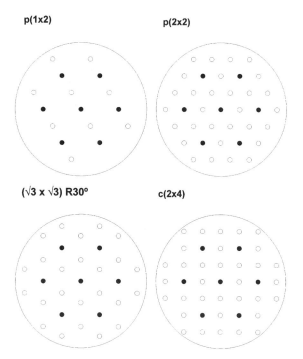

**p(1x2)**    **p(2x2)**

**(√3 x √3) R30°**    **c(2x4)**

**Figure 7.** Some simple, single-domain reciprocal space structures on a substrate of square symmetry. These diffraction patterns correspond to the real-space models in Figure 6.

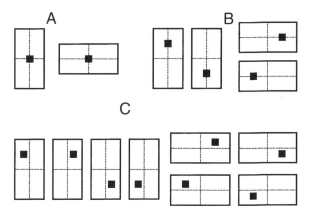

**Figure 8.** The various possible domains for a (1 × 2) unit cell on a substrate of square symmetry [e.g., (100) face of a bcc or fcc lattice]. Two domains are possible in (**A**), where the unit cell has the maximum possible symmetry, including two mirror planes and a 2-fold rotation axis. In (**B**), four possible unit cells have only one mirror plane. In (**C**), eight unit cells are possible with no internal symmetry. These unit cells can be distinguished by quantitative LEED structural analysis, but their diffraction patterns are qualitatively identical with that shown in the upper right of Figure 9.

represents a "centered" unit cell with a lattice point at the center of the cell identical to the ones at the corners. The p is sometimes omitted from descriptions of structures as well, although its omission does not always mean that p is implied. It should be apparent that in all cases, a centered cell can be decomposed into smaller primitive cells, but the c usage survives—if only because it is intuitively easier to recognize centered cells than their primitive counterparts.

It should likewise be apparent that structures are possible that cannot be described using Wood's simple notation. When the angle between the unit vectors of the overlayer is different from the angle between the unit vectors of the substrate, the two unit cells cannot be related by a simple expansion and/or rotation. In such cases, matrix notation must be used. The matrix used is the most general relation between the substrate unit vectors, $\mathbf{a}_1$ and $\mathbf{a}_2$, and the overlayer unit vectors, $\mathbf{b}_1$ and $\mathbf{b}_2$ (expressed in terms of a common basis):

$$\begin{pmatrix} \boldsymbol{b}_1 \\ \boldsymbol{b}_2 \end{pmatrix} = \begin{pmatrix} m_{11} & m_{12} \\ m_{21} & m_{22} \end{pmatrix} \begin{pmatrix} \boldsymbol{a}_1 \\ \boldsymbol{a}_2 \end{pmatrix} \qquad (2)$$

This has the advantage of always being applicable, but the disadvantage of being more cumbersome.

Another important point concerns the existence of multiple domains of a surface structure. In general, a particular overlayer geometry is energetically equivalent when operated on by one of the symmetry elements of the substrate. If the symmetry of the overlayer is lower than that of the substrate, several domains are possible. For example, on a bcc (100) face, two distinguishable

domains of a (1 × 2) structure exist, which can be termed (1 × 2) and (2 × 1), as seen in Figure 4. The bcc (100) surface has a 4-fold rotation axis and four mirror planes, while the (1 × 2) cell has at most a 2-fold axis and two mirror planes. The importance of this is that at least two domains of the (1 × 2) will be present on any real sample. Figure 8 illustrates all of the possible domains of a (1 × 2) superstructure on a bcc (100) substrate. The fewer symmetry elements possessed by the unit cell, the more possible domains will be present. In cases of lower symmetry, the large number of domains will only be observed in the quantitative structure of the I(E) spectra, and the qualitative appearance of the diffraction pattern is indistinguishable from a unit cell with the maximum possible symmetry. Most domains turn out to be larger than the coherence length (see below) but smaller than the typical beam spot size of ~1 mm. This means that all possible domains will be illuminated by the electron beam, and diffraction from all of them will appear on the LEED screen. As a result, the pattern observed from a low-symmetry structure on the LEED screen may not be a Bravais lattice, but a superposition of such lattices. Figure 9 illustrates the patterns typically seen from some common multidomain structures.

### Quantitative Measurements

Quantitative crystallographic information is also accessible. Atomic positions can be determined to within several hundredths of an angstrom by modeling the intensities of the diffraction spots observed. These intensity versus beam energy curves are typically called I(E) curves (as well as "I-V curves," referring to the intensity and electron beam voltage). In principle, the formalism is similar to

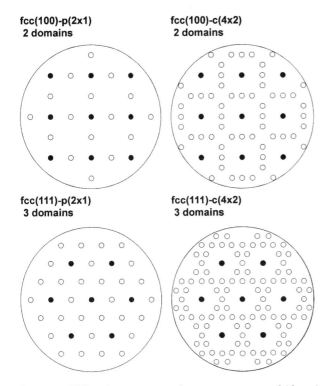

**fcc(100)-p(2x1)**
**2 domains**

**fcc(100)-c(4x2)**
**2 domains**

**fcc(111)-p(2x1)**
**3 domains**

**fcc(111)-c(4x2)**
**3 domains**

**Figure 9.** Diffraction patterns of some common multidomain structures. Note that the fcc(111)-p(2 × 1) structure is indistinguishable from the fcc(111)-p(2 × 2) structure in Figure 7. However, the fcc(100)-p(2 × 1) can be distinguished from the p(2 × 2) structure on the same substrate seen in Figure 3.

x-ray diffraction, but important differences exist owing primarily to the effect of multiple scattering. Because of the low cross-section for x-ray scattering, one may assume that an x ray scatters only once before exiting a crystal, but the cross-section for electron scattering approaches the geometric cross-section of an atom. As a result, multiple scattering is a common occurrence and its effects cannot be ignored. A practical result of this is that an x-ray diffraction pattern can be solved, in principle, by a direct Fourier transform of the pattern. However, a great deal of information is convolved in the process of multiple scat-

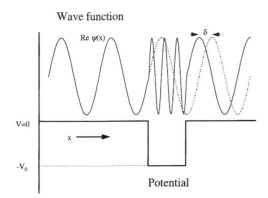

Wave function

Re $\psi(x)$

$\delta$

$V=0$

$x$

$-V_0$

Potential

**Figure 10.** A simple model illustrating the phenomenon of a phase shift. A plane wave passing over a square well will change its wavelength in the region of the well. After exiting the well, its phase will have advanced by an amount indicated by $\delta$.

tering, and the best that can be achieved is comparison of experimental data with a reference calculation. As an illustration, consider that the wave function of an electron undergoes a phase shift when scattering from an atom, modeled as a square well in Figure 10. This phase shift is indistinguishable from a phase shift due to the electron traversing a longer distance, and this ambiguity denies the possibility of directly determining a lattice spacing as one might do, for example, from x-ray diffraction data. Hence, peaks in the $I(E)$ curves cannot be fit by a simple Bragg law, and sophisticated multiple scattering calculations must be performed.

## PRACTICAL ASPECTS OF THE METHOD

Performing the actual LEED experiment is reasonably straightforward. It requires an ultrahigh-vacuum chamber fitted with a single-crystal sample, electron gun, display system, and provisions for intensity measurements. These will be discussed in turn.

The sample is usually a single crystal of some electrically conducting material. Doped semiconductors can also be used easily. The sample must usually be cleaned of surface contaminants prior to use, and one typically needs facilities such as an ion bombardment gun and sample heating capabilities to anneal out defects caused by ion bombardment. Sample cleanliness is usually monitored by such techniques as Auger electron spectroscopy (AES; AUGER ELECTRON SPECTROSCOPY) or x-ray photoelectron spectroscopy (XPS). Since thermal vibrations are known to diminish the intensity of all diffraction peaks, LEED experiments should be done with the sample as cold as is practical. Typically, liquid nitrogen cooling is used during the recording of the pattern, even if higher temperatures are required to prepare the desired surface structure. Another important consideration is that the diffraction pattern is typically viewed from behind the sample, resulting in a need to reduce the bulk of the manipulator as much as possible for maximum viewability (the advent of so-called "rear-view" optics has relaxed this requirement on some systems).

Another requirement for the sample manipulator is the ability to rotate the sample around a minimum of two perpendicular axes. This is necessary to achieve the normal incidence condition useful for quantitative work, although this is not strictly required for qualitative data. The reason for positioning the crystal normal to the incoming electron beam is 2-fold: (1) normal incidence is the easiest angle to specify exactly, so that high angular precision can be ensured; quantitative calculations require a precision of half a degree or better; and (2) normally, incident electron beams also maximize the symmetry of the resulting diffracted beams; making full use of symmetry is a great advantage when doing a quantitative analysis, as will be discussed later. Since finding normal incidence is a nontrivial task, two methods of checking for normal incidence are described briefly. Both depend on the fact that, for a normally incident beam of electrons, the intensities of the diffracted beams have the same symmetries as the crystal plane being investigated. For example, an fcc(100) face has a 4-fold symmetry axis and four mirror planes, and an fcc(111) plane has a 3-fold symmetry axis and three mirror planes. The simplest method is useful only for highly symmetric surfaces—at normal incidence the intensities of symmetrically equivalent beams are equal. For example, four symmetrically equivalent spots in the diffraction pattern from an fcc(100) surface, say the (10) set of beams [e.g., as distinct from the (11) or (20)], should have equal intensity. If the left pair of spots were brighter than the right pair of spots, some sort of left/ right angular adjustment would need to be made. The other method is to allow beams with identical values of $|\Delta \mathbf{k}_{\parallel}|$ to approach the edge of the screen, which is generally a cylindrically symmetric section of a sphere. These beams should all arrive at the edge of the screen at the same time, as the energy of the incident beam is varied.

The electron guns typically employed are widely used and only a few brief comments are warranted. Since LEED examines back-reflected electrons, the gun is usually mounted inside the display apparatus. As a result, only electrostatic deflection is used to reduce the effect of stray fields on the diffraction pattern. The filament is usually tungsten, although sometimes $LaB_6$ has been used for its low work function. One problem that can arise is that the light emitted from the hot filament can make the fluorescent screen of the display system difficult to see. For this reason, the filament is typically mounted off of the axis of the gun and the emitted electrons deflected onto the axis. Another consideration is that the beam current is not usually constant as the accelerating voltage is increased. This is important in quantitative measurements because increased beam current will result in increased intensity in the diffracted beams; this can artificially influence the position of peaks in the $I(E)$ curves. There are various ways to control this, but the simplest method is to directly measure the beam current as a function of energy and correct the $I(E)$ curves accordingly.

The display system must provide for measurement of the intensities of diffracted beams. The earliest methods for doing so employed a Faraday cup that could be moved to the proper position. It is currently much more popular to use a fluorescent screen held at high voltage (3 to 7 kV), which glows when struck by electrons with brightness proportional to current. A typical modern apparatus is shown in Figure 11. The hemispherical phosphorescent screen is preceded by a series (two, three, or four) of concentric hemispherical wire meshes or grids, with four grids being most common. The sample is ideally placed at the center of curvature of the instrument. The grid closest to the sample is generally held at ground/earth potential, as is the sample, so that diffracted beams propagate through a field-free region between the two. The second grid is held at a potential a few volts below that of the beam to energy select out all inelastically scattered electrons. In a four-grid system, the second and third grids are held at the same potential to improve the quality of energy selection, and the fourth grid is grounded to isolate the energy selection potential from the high voltage placed on the phosphorescent screen. Two grids are generally sufficient for LEED work, but the four-grid system offers improved energy selection. This is useful because the same apparatus is often also

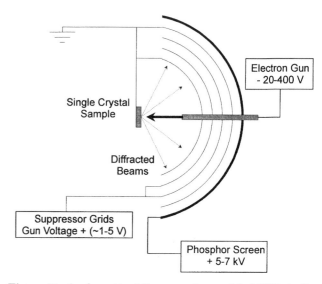

**Figure 11.** A schematic of the apparatus used in LEED studies. Voltages are indicated on the diagram. Most older instruments view the screen from behind the sample, but newer rear-view optics view the screen from the right side of the diagram.

used for AES measurements, for which energy selection is crucial.

Several methods have been employed for the direct measurement of spot intensities. The earliest methods involved the use of a Faraday cup that was moved to intersect the diffracted electron beam. This presented difficulties in part because it is difficult to measure currents of the low-energy electrons employed in LEED. Later refinements introduced the postacceleration to a phosphorescent screen described above. A spot photometer could be used to track each diffracted beam and directly measure the $I(E)$ spectrum. However, this process was found to be slow, owing to the need to manually readjust the photometer at each new energy value. Moreover, this method requires that separate acquisitions be made of every spot in the pattern. A method that is currently in wider use involves photographing the screen at regular intervals as the beam energy is increased. This allows for parallel measurement of all visible spots, and delays measurement of intensities until after the rapid data acquisition. This was pioneered by Stair et al. (1975) with a Polaroid camera, but a digital camera interfaced to a computer is more popular currently. In the authors' experience, a common 35-mm camera designed for personal use is not usually sufficient to capture the low light levels emitted by typical fluorescent screens, and a more sensitive device must be employed. The authors successfully used both a manually shuttered Polaroid camera (for rapid development and evaluation) and a computer-controlled CCD camera.

It is important to know the boundaries of LEED's capabilities. As a rule of thumb, LEED interrogates an ordered region of the surface ~100 Å across, and is sensitive to ~5% to 10% of a monolayer. The limitation on lateral resolution is a typical number for most instruments, and it is determined by deviations from the "ideal" beam—a monochromatic plane wave described as $A$

exp($i\mathbf{k} \bullet \mathbf{r}$), with well-defined energy, $\hbar^2|\mathbf{k}|^2/2m_e$, wave vector $\mathbf{k}$ and position in space $\mathbf{r}$. The two important factors characterizing deviations are the finite energy spread of the electron beam, deviating from monochromaticity, and the finite angular divergence of said beam, owing to the filament typically being an imperfect plane wave source. Qualitatively, these imperfections will give rise to unpredictable lateral variations in the phase of the beam; the lateral distance at which the phase relationship becomes arbitrary is generally termed the instrument response function, or coherence length. The coherence length is generally around a few hundred angstroms; an estimate is given in Appendix A.

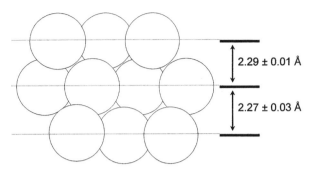

**Figure 12.** A side view of the Pt(111) surface along the $\{1\bar{1}0\}$ direction. (After Materer et al., 1995.) The bulk interlayer distance is 2.2650 Å.

## METHOD AUTOMATION

It is possible and desirable, although not strictly necessary, to acquire LEED data using a computer. The bare minimum requirements for qualitative data can be met with a simple camera. However, as noted above, a common 35-mm camera designed for personal use may not prove sufficient due to the low light level of the phosphorescent screen. Quantitative data requires detailed measurements of spot intensities. It is recommended that a camera interfaced to a computer be capable of at least 8 bits of intensity resolution with adjustable gain and dark level.

Two approaches to automated data acquisition are generally used. One is a procedure whereby a single spot is tracked by a camera while the energy is scanned. This has the advantage of directly measuring the desired $I(E)$ curve in real time. Another popular method is to use a camera to photograph the entire LEED screen at particular energies (e.g., 2-eV steps) and perform an off-line measurement of the intensities later. This method has the advantage of photographing all spots at once for maximum efficiency and not requiring particular effort to track individual spots. Since measuring all of the spots in parallel is also more rapid, instrument drift is minimized. We typically implement the latter method with a CCD camera, a personal computer with a frame-grabber expansion card, and our own software for image acquisition.

## DATA ANALYSIS AND INITIAL INTERPRETATION

Two different sorts of data analysis are appropriate to LEED experiments: qualitative and quantitative. Knowledge of the symmetry of the diffraction pattern can yield powerful restrictions on the atomic geometry of the surface. These prove invaluable in reducing the number of candidate structures examined by calculations. The two different forms of analysis will be examined in turn.

### Qualitative Analysis

The qualitative features of LEED are the easiest and most commonly applied. As illustrated above, the diffraction pattern reveals a great deal of information about the symmetry of the overlayer structure. This is often enough to make a reasonable guess at a structure or to help deduce

a coverage. However, it is dangerous to make blind assumptions about coverage without corroborating information. A less obvious qualitative feature is the $I(E)$ spectra themselves—these can be used as a "fingerprint" of a particular surface structure in much the same way that vibrational spectra are unique to a particular molecule.

This process is perhaps best illustrated by the following examples: Pt(111)-(1 × 1), Si(100)-(2 × 1), β-SiC(100)-(2 × 1), Pt(111)-p(2 × 2)-C₂H₃, and Ni(100)-(2 × 2)-2C. These structures can be seen in Figures 12 to 16, respectively. All of these structures have been solved by quantitative LEED techniques, and are examined in light of what qualitative information could be deduced. They also illustrate some of the more important points to bear in mind when investigating a structure.

The Pt(111) surface (Fig. 12) has the simplest possible overlayer structure (Materer et al., 1995). The (1 × 1) translational symmetry prohibits almost all lateral reconstructions, and the most reasonable structures to consider involve simply relaxations of interlayer spacings. In principle, a shift of the topmost layer of surface atoms to a different registry [at bridge sites or hexagonal close-packed (hcp) hollows instead of fcc hollows] is possible, but it has not been observed in any system examined to date. This structure is also in line with other known metal surface structures; the (111) plane of an fcc metal is the most dense, and lateral reconstructions are rare. Some (100) faces have been known to reconstruct [e.g., the Pt(100)-hex reconstruction; Van Hove et al., 1981], and (110) faces commonly exhibit reconstructions [e.g., Ir(110)-(2 × 1); Chan and Van Houe, 1986]. For bcc metals, the (110) surface is the most dense and the (111) is the least dense, and the propensity for reconstruction varies accordingly.

The Si(100) structure (Fig. 13; Holland et al., 1984) is simple in that no adsorbate is involved, merely a reconstruction of the substrate. The major question is the geometry of the topmost layers. Semiconductor surface reconstructions can be understood as driven to minimize the number of "dangling bonds." While metals are understood to have nearly isotropic bonding, semiconductors tend to have bonds that are more covalent in nature and highly directional. Consequently, the simple octet rule of "four bonds" from organic chemistry also applies well to silicon surfaces. Si(100) eliminates its "dangling bonds"

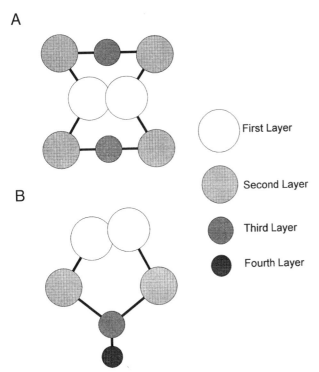

A

B

First Layer

Second Layer

Third Layer

Fourth Layer

**Figure 13.** A top (**A**) and side view (**B**) of a single unit cell of the Si(100)-(2 × 1) reconstruction. The details of this structure are still a matter of debate, but the most recent evidence agrees that the surface dimer is tilted. Reconstructions have been found penetrating as far as the fifth atomic layer (Holland et al., 1984.)

by forming dimers, which halves the number of unoccupied bonds.

The SiC(100) structure (Fig. 14; Powers, et al., 1992) is remarkably similar to the Si(100) structure, which is an example of the extent to which chemical analogy is useful. In the bulk, the (100) planes of SiC are composed of alternating layers of pure Si or pure C composition. However, one should also consider the possibility that chemical mixing will occur at the surface. It turns out that in this case, independent evidence (from Auger spectroscopy) was available to indicate that the face was Si terminated, reducing the number of possible structures to be considered. This illustrates that it is generally advisable to obtain as much information as possible about a surface structure so that some possibilities can be ruled out.

The p(2 × 2) ethylidyne (CH₃–C≡) structure on Pt(111) (Fig. 15) is representative of adsorbate-induced structures (Starke et al., 1993). The LEED pattern would naively seem to suggest a coverage of 0.25 monolayers (ML), with one molecule adsorbed per unit cell. However, models with 0.75 or 0.50 ML coverage can also be constructed consistent with this symmetry, and one is urged not to leap too rapidly to such a conclusion. Another complication of this system is that a p(2 × 2) structure on an fcc(111) surface is indistinguishable from three rotationally equivalent domains of a p(2 × 1) structure. This ambiguity requires that both models be considered, and in fact, there was a debate in the literature at one time as to which was correct before a general agreement was reached. Part of the data

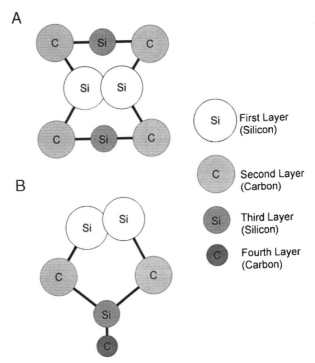

A

B

First Layer (Silicon)

Second Layer (Carbon)

Third Layer (Silicon)

Fourth Layer (Carbon)

**Figure 14.** A top (**A**) and side view (**B**) of a single unit cell of the SiC(100)-(2 × 1) reconstruction. (After Powers et al., 1992.) Note the correspondence with Figure 13.

that resolved this ambiguity was a careful calibration of the coverage. The measured coverage of 0.25 ML is consistent with a p(2 × 2) but not a p(1 × 2) structure. Another interesting point is that the calculated structure based on quantitative LEED only indicates the positions of the carbon atoms. Hydrogen atoms are very light, and hence poor scatterers of electrons. It is generally considered impossible to locate hydrogen atoms by LEED, except indirectly by their effects on the positions of heavier atoms.

The case of carbon on Ni(100) (Fig. 16; Gauthier et al., 1991) is interesting because the apparent structure, which resembles a p(2 × 2), leads one to the wrong conclusion upon simplistic assessment. Instead of a coverage of 0.25 or 0.75 ML, it turns out to be 0.50 ML. A critical clue is found in the systematic extinctions of certain spots in the LEED pattern. The (0, $n/2$) and ($n/2$, 0) fractional-order spots all vanish when the diffraction pattern is viewed with the electron beam at normal incidence to the crystal. This is characteristic of glide-plane symmetry elements, and these elements place severe restrictions on the possible structures. It is important to note that if the electron beam is not at normal incidence, the symmetry of the beam with respect to the glide planes is broken and the extinguished spots reappear.

### Quantitative Analysis

Quantitative structural determinations represent the pinnacle of what LEED can currently achieve. A full tutorial in the theory, programming, and use of LEED computer codes is beyond the scope of this unit, and one is referred to Pendry (1974), Clarke (1985), Van Hove and Tong

A

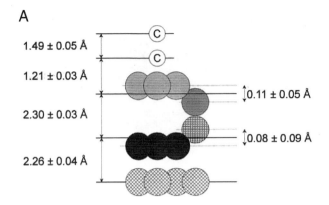

B

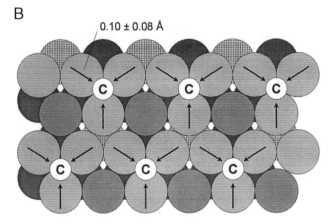

**Figure 15.** A side (**A**) and top view (**B**) of the structure of ethylidyne on platinum, Pt(111)-p(2 × 2)-C$_2$H$_3$. The hydrogen atoms have been omitted since the original investigation (Starke et al., 1993) was not able to locate them; this might have been due to either poor scattering characteristics of the light atoms or dynamic motion that was averaged out in the diffraction experiment. The distances shown are not to scale and are exaggerated to display some of the subtle atomic displacements detected.

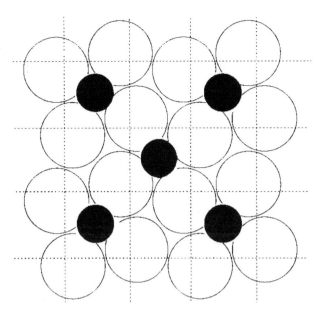

**Figure 16.** A top view of the Ni(100)-(2 × 2)-2C structure. Carbon atoms are represented as small black circles, and the glide planes are indicated by dotted lines.

(1979), and Van Hove et al. (1986) for the details. However, the outline and important elements of the theory will be described.

The objective of a LEED calculation is to compute the intensities of diffracted beams as a function of the beam energy. These intensities are evaluated for a particular geometry and compared with the experimentally measured data. If the agreement is not satisfactory, the intensities are recomputed for a different trial geometry, and this process of trial and error is repeated until a best agreement is found.

The scattering of LEED electrons is modeled quantum mechanically as reflection from the topmost layers of the surface. The electron beam is generally modeled as a plane wave, $A \exp(i\mathbf{k} \bullet \mathbf{r})$, where $\mathbf{k}$ is complex. The surface is typically modeled with a muffin-tin, one-electron potential. The calculation is typically organized in a hierarchical fashion by determining the scattering of electrons from atoms, organizing those atoms into 2D layers, and finally evaluating the scattering between successive layers. An atomic-like potential for the surface atoms is embedded in and matched to a constant muffin-tin potential, which

also includes an imaginary component to simulate the damping of the electron wave by inelastic scattering. The real part of the potential is a calculational device, and it has only limited physical significance, although qualitatively one would expect its value to change for a particular system in correlation with work function changes. The energy of a LEED electron inside the crystal, between the ion cores, is

$$
\frac{\hbar^2 k^2}{2m_e} = \frac{\hbar^2}{2m_e}(k_r + ik_i)^2
$$
$$
= \frac{\hbar^2}{2m_e}(k_r^2 - k_i^2) + i\frac{\hbar^2}{m_e}k_r k_i = E + V_{0r} + iV_{0i} \quad (3)
$$

thus,

$$
\frac{\hbar}{2m_e}(k_r^2 - k_i^2) = E + V_{0r} \quad \text{and} \quad \frac{\hbar^2}{m_e}k_r k_i = V_{0i} \quad (4)
$$

where $V_{0r}$ is the real part of the potential, $V_{0i}$ is the imaginary part, $k_r$ is the real part of the complex magnitude of the wave vector of the electron, $k_i$ is the imaginary part, and $E$ is the electron kinetic energy outside of the crystal. Typically, $V_{0i}$ is ~1 to 5 eV, and $V_{0r}$ is generally used as a fitting parameter. An estimate of $V_{0i}$ can be obtained from the width of the observed peaks; approximately, $\Delta E_{FWHM} \approx V_{0i}$ (Van Hove et al., 1986, p. 119). LEED electrons are not terribly sensitive to valence electrons, so some liberties can be taken with the details of the atomic part of the potential. Most commonly, one uses the results of a band structure calculation as the source of the atomic potentials, such as those of Moruzzi et al. (1978). However, LEED is sufficiently sensitive to valence electrons that simple free atom calculations do not give good results (Van Hove, 1986, p. 121). Further details of the calculations are given in Appendix B.

In order to actually perform a structural search, some quantitative measure of the goodness of fit of two sets of $I(E)$ curves must be devised. This is typically done by means of an $R$ factor, as is used in x-ray crystallography. The simplest criterion one might imagine is a squared difference, as is employed in a least-squares type of fit. Many other $R$ factors have been proposed and used as well. One particularly popular one was suggested by Pendry (1980), who recognized that peak positions contain most of the important information and chose to emphasize position at the expense of intensity. In part, this is useful because peak positions are determined primarily by geometric factors, but the factors affecting intensity include nonstructural parameters, such as thermal motion, that introduce artificial biases. Pendry's $R$ factor ($R_P$) is based on the logarithmic derivative of the intensity, $L = (I)^{-1}(dI/dE)$, evaluated as part of a $Y$ function, $Y = L/(1 + V_{0i}^2 L^2)$ so as to avoid singularities near zero intensity. The $R$ factor is defined as

$$R_P = \frac{\int (Y_{\text{expt}} - Y_{\text{theo}})^2 \, dE}{\int (Y_{\text{expt}}^2 + Y_{\text{theo}}^2) \, dE} \qquad (5)$$

The values for Pendry's $R$ factor range from 0 to 1, with 0 being a perfect correlation between theory and experiment and 1 being no correlation.

Another major advance has been the development by Rous and Pendry (1989) of Tensor LEED, a perturbative method that greatly accelerates structural searches. This method computes a tensor (hence the name) that describes the effect on the LEED matrices of small geometric displacements. Since the full dynamical calculations of $I(E)$ spectra are the most time-consuming step in a fit, this method accelerates a search by allowing a single reference calculation to cover a wide range of parameter space. Instead of requiring an inefficient grid search of parameter space at, say, 0.05-Å intervals, Tensor LEED calculations allow all structures within a certain radius of convergence to be rapidly tested. This radius has been estimated as ~0.3 to 0.4 Å (Rous, 1992), although more conservative estimates are often recommended. In any event, it is wise to check the results of a Tensor LEED calculation by starting a new reference calculation at the last minimum, followed by a new minimization, and perhaps iterating this once or twice.

An important consideration is what kind of experimental data set is required for these calculations. This is usually estimated by examining the total data range for each of the contributing $I(E)$ curves and summing them together to obtain a cumulative energy range. Cumulative ranges of at least 1000 eV are recommended, and more is always better, especially when many parameters are to be fit. A method of estimating the uncertainty in a structural determination was developed by Pendry (1980), and it scales inversely with the cumulative energy range. The sampling of points in experimental spectra should be at least every 3 eV, and 1 to 2 eV is preferable, because peaks in $I(E)$

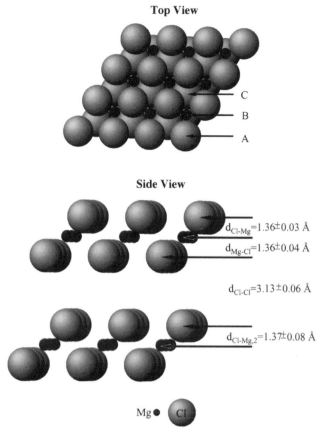

**Figure 17.** Top view (top) of the $\alpha$-MgCl$_2$(0001) surface with labels indicating the lateral stacking of the ions. Ionic radii were reduced by 15% for clarity. Side view (bottom) of the $\alpha$-MgCl$_2$ (0001) surface with the surface termination on top and with the ionic radii reduced by 50%. Refined interlayer spacings are labeled with there respective error bars. Bulk interlayer spacing values are $d_{\text{Mg–Cl,bulk}} = d_{\text{Cl–Mg,bulk}} = 1.33$ Å and $d_{\text{Cl–Cl,bulk}} = 3.22$ Å. The labels A, B, C in the top view indicate the different layers in the side view.

spectra are generally ~4 to 5 eV wide. This brief, but comprehensive, overview of the theoretical aspects involved in a detailed structural solution provides the reader with adequate insight to critically examine the current LEED literature.

The analysis of the MgCl$_2$ multilayer surface structure (Roberts et al., 1998) will be used as an example of a typical LEED publication. Figure 17 illustrates the objective of a LEED $I(E)$ analysis—the surface structure itself. The use of top and side views familiarizes the reader with the crystallography of MgCl$_2$. In this example, the hexagonal surface symmetry is clearly evident in the top view (Fig. 17 top) along with the relative lateral positions of the atoms. Additionally, it is correctly assumed that the lateral coordinates were held at their bulk values. This is indicated by the absence of any explicit labeling of atomic positions. The side view (Fig. 17 bottom) shows the layered alternating stacking of the Mg and Cl ions along with the interlayer spacings, the only structural parameters varied, in the refined region with their respective error bars. The

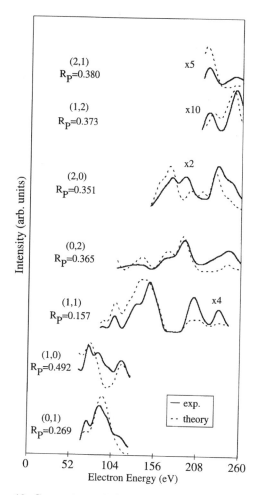

**Figure 18.** Comparison of the theoretical (dashed lines) and experimental (solid lines) $I(E)$ curves for the fully optimized α-$MgCl_2$(0001) surface. All beams used in the calculation are plotted along with their indices.

late to the experimental intensities, since the reason for the inequity between these intensities is directly related to the nature of the Pendry $R$ factor. This $R$ factor is sensitive to the shape of the $I(E)$ curve with respect to energy and not to their intensities. Consequently, a good fit is determined if the minima and maxima of the theoretical curves occur at the same energies as in the experimental curves. The comments and questions that arose from this specific example are common to all LEED-determined structures and provide a more profound understanding of the analysis than could be obtained from just looking at the overall $R$ factor.

Since these calculations are computationally expensive, the tractability of the problem is always a consideration. The difficulty of calculations scales primarily with the number of atoms in a unit cell, usually as the square or cube of that number. Some examples of recent calculations follow. A study of Pt(111)-p(2 × 2)-NO and Ni(111)-c(4 × 2)-2 NO required calculation times of 4 to 5 h for one geometry on an IBM RISC 6000 (Materer et al., 1994). On the same computer, a study of Rh(110)-(2 × 1)-2 O required 1.5 h to calculate a single geometry, and 200 s to complete a search of nearby structures using a Tensor LEED algorithm (Batteas et al., 1995). One benchmark structure is the Si(111)-(7 × 7) reconstruction (Tong et al., 1988), which takes ~1 day on an IBM RISC 6000. A (1 × 1) cell of $MgCl_2$, a simple structure (Roberts et al., 1998), took ~15 min of computational time on a DEC Alpha, using current LEED codes utilizing all benefits afforded by the high symmetry. If symmetry is ignored, the same full dynamical calculation takes ~1 h. For comparison, the DEC Alpha used is observed to be roughly 10 to 20 times faster than the quoted IBM machine. The Si(111)-(7 × 7) structure takes ~1 h to compute on the DEC Alpha, using all available symmetries.

## SAMPLE PREPARATION

The basic requirements of a sample are that it be flat, well ordered, and able to reflect diffracted electrons. Samples are typically polished to at least an optically fine finish (1.0 μm or better polishing paste) and mounted in a UHV chamber. Preparation of clean, well-ordered samples for UHV surface science work is somewhat of an art unto itself. Beyond the advice given by Musket et al. (1982), the best that can be said is that one should experiment for oneself. Generally, cycles of noble gas ion bombardment (e.g., 0.5 to 3.0 keV $Ar^+$), annealing (usually 500 to 2000°C, depending on the material), and gas treatments (usually $H_2$ or $O_2$) are used. Ion bombardment can be thought of as "atomic sandblasting" and is good for removing gross contaminants no greater than a few microns in thickness. If contaminants are thick enough to be visible to the naked eye, other methods of cleaning should be used. It is sometimes useful to "momentum match" the ion used to the type of contaminant expected (e.g., $Ar^+$ for S, $Ne^+$ for C), but the relative cost of the various rare gases is more often the deciding factor. Annealing heals the damage caused by ion bombardment, and it also allows diffusion of impurities to take place. It is not uncommon for a new crystal to

spacings below the refined region are again assumed to have bulklike separations because no explicit values are given for these parameters.

The accompanying figure to the structural illustration of all finished calculations is the comparison of the experiment and theoretical $I(E)$ curves (Fig. 18). Although the goodness of fit is quantified by the overall $R$ factor (here $R_P = 0.32$), a visual inspection of the experimental data allows one to determine if the researcher made the correct choice in the number of refinable structural parameters. For $MgCl_2$, the total experimental energy range of 658 eV is smaller than for most data sets, because of the material's extreme sensitivity to the electron beam. This small data set limited the optimization to only four interlayer spacings. Additional structural parameters would have resulted in refined values with very large uncertainties, although the overall $R$ factor would be lowered due to the fact that more variables were used to fit the experimental $I(E)$ curves. When inspecting the $I(E)$ curves, the quality of the theoretical fit to the data should not be judged by how well the theoretical beam intensities corre-

require several weeks to clean for the first time, while the near-surface region is depleted of contaminants. Subsequent cleaning occurs far quicker, however, and it can usually be accomplished in ~1 h on a daily basis. Chemical treatments are useful in especially difficult cases, if an impurity can be volatilized by reaction (e.g., burning carbon to CO and $CO_2$). Not all of these steps are necessarily required or even useful in every case.

## SPECIMEN MODIFICATION

The only modification to the specimen typically observed is possible electron beam damage. Most of the time, this is not a serious concern, especially when fairly heavy atoms are involved (e.g., reconstructed metals or semiconductors). Damage has more commonly been observed as a result of more energetic beams than those used for LEED (e.g., during Auger electron spectroscopy; see AUGER ELECTRON SPECTROSCOPY), but it is a serious issue for insulating overlayers in some cases. In a recent study of ice multilayers on Pt(111) (Materer et al., 1997), electron beam damage was so severe that a LEED experiment could only be done by use of special Digital LEED equipment (Ogletree et al., 1992).

## PROBLEMS

There are several kinds of difficulties that experimenters often face. The most basic is that of finding an ordered structure in a given adsorbate-substrate system. This is the most difficult to address because the specifics vary from one system to another. One must begin with a substrate that is itself well ordered. The spots on the display system should be sharp, nearly as narrow as the incident beam, and the space between the spots should have minimal background intensity. Unless the substrate reconstructs, the pattern seen should form a simple Bravais lattice corresponding to the face of the crystal being studied. If the sample is polycrystalline, one may see either a uniformly bright background or many spots that cannot be reconciled with a single Bravais lattice, perhaps without any obvious pattern at all. The second alternative can usually be distinguished from an unknown substrate reconstruction by the fact that even a complex multiple-domain reconstruction will generally give a pattern with all or most of the bulk rotational and mirror-plane symmetry elements of this crystal plane when a normal incidence electron beam is used. Polycrystalline samples tend to lack recognizable symmetry elements.

Once the substrate crystallinity is assured, a common approach is to adsorb the desired overlayer at as cold a temperature as practical, and begin inspecting the LEED pattern. Frequently, one must warm the sample somewhat to anneal the structure into place. At very cold temperatures, molecules frequently adsorb wherever they strike the surface, and thermal energy must be supplied to overcome kinetic barriers to diffusion so that the well-ordered thermodynamic ground state can be attained. The annealing temperature must be kept low enough that the adsorbate not desorb from the surface. The results of a temperature programmed desorption (TPD) experiment are thus frequently useful in designing a "recipe" for sample preparation (see Woodruff and Delchar, 1994, for more about TPD). Another, more subtle problem is that the adsorbate might itself rearrange on the surface. For example, ethylene adsorbed on Pt(111) at room temperature gives a p($2 \times 2$) LEED pattern, but it was not obvious until some time later that the ordered species was ethylidyne, $CH_3-C\equiv$, rather than ethylene itself. Since these transformations are typically thermally driven, one again would wish to use the minimum temperature necessary to anneal an overlayer to avoid side reactions of the adsorbate.

One of the best indicators of reliable data is a close correspondence between the $I(E)$ curves of symmetrically equivalent beams. For example, at normal incidence, the (10), (01), ($\bar{1}0$), and ($0\bar{1}$) beams are all symmetrically equivalent on an fcc(100) surface. These curves are usually averaged together to reduce noise, but they should look very similar to begin with. Another delicate factor is the integration of intensities. In the photographic method, one integrates the intensity of the spot profile found within a particular window as the total intensity of the spot, since the spot is of finite dimensions. One should take care that the integration windows do not overlap or include intensity from other spots. A less obvious precaution is to normalize the $I(E)$ spectra to the intensity of the primary electron beam. The beam current will generally change as a function of beam voltage, and this can artificially influence the position of peak maxima. Electron beam damage sometimes occurs, as mentioned above. This can be detected by making repeated measurements in ascending and descending order of electron energy. If the two spectra look different, damage has occurred.

Although obtaining an ordered structure represents the greatest challenge to surface crystallographers, finding the sample position where the incident beam is normal to the crystal is always tedious, especially if the overlayer and/or substrate is susceptible to damage from the electron beam. The procedure previously presented for checking normal incidence does have one pitfall—small residual electromagnetic fields inside the chamber can cause deflections in the diffracted beams particularly at low energies. This effect can be circumvented by using higher-order beams to confirm normal incidence of the electron beam on the sample.

## ACKNOWLEDGMENTS

The authors gratefully acknowledge support from the U.S. Department of Energy through the Lawrence Berkeley National Laboratory.

## LITERATURE CITED

Ashcroft, N. W. and Mermin, N. D. 1976. Solid State Physics. W. B. Saunders, Philadelphia.

Batteas, J. D., Barbieri, A., Starkey, E. K., Van Hove, M. A., and Somorjai, G. A. 1995. The Rh(110)-p2mg($2 \times 1$)-2O surface structure determined by Automated Tensor LEED—structure changes with oxygen coverage. *Surf. Sci.* 339:142–150.

Beeby, J. L. 1968. The diffraction of low-energy electrons by crystals. *J. Phys. C* 1:82–87.

Boffa, A. B., Galloway, H. C., Jacobs, P. W, Benitez, J. J., Batteas, J. D., Salmeron, M., Bell, A. T., and Somorjai, G. A. 1995. The growth and structure of titanium oxide films on Pt(111) investigated by LEED, XPS, ISS, and STM. *Surf. Sci.* 326:80–92.

Chan, C. M. and Van Hove, M. A. 1986. Confirmation of the missing-row model with three-layer relaxations for the reconstructed Ir(110)-(1 × 2) surface. *Surf. Sci.* 171:226–238.

Clarke, L. J. 1985. Surface Crystallography—An Introduction to Low-Energy Electron Diffraction. Wiley, New York.

Crampin, S. and Rous, P. J. 1991. The validity of the average *t*-matrix approximation for low energy electron diffraction from random alloys. *Surf. Sci.* 244:L137–L142.

Davisson, C. and Germer, L. H. 1927. Diffraction of electrons by a crystal of nickel. *Phys. Rev.* 30:705–740.

Döll, R., Gerken, C. A., Van Hove, M. A., and Somorjai, G. A. 1997. Structure of disordered ethylene adsorbed on Pt(111) analyzed by diffuse LEED: Asymmetrical di-sigma bonding favored. *Surf. Sci.* 374:151–161.

Fadley, C. S., Van Hove, M. A., Hussain, Z., and Kaduwela, A. P. 1995. Photoelectron diffraction—New dimensions in space, time and spin. *J. Electron Spectrosc. Relat. Phenom.* 75:273–297.

Farrow, R. F. C. 1995. Molecular Beam Epitaxy. Noyes Publications, Park Ridge, N.J.

Gauthier, Y., Baudoing-Savois, R., Heinz, K., and Landskron, H. 1991. Structure determination of p4g Ni(100)-(2 × 2)-C by LEED. *Surf. Sci.* 251/252:493–497.

Gauthier, Y., Joly, Y., Baudoing, R., and Rundgren, J. 1985. Surface-sandwich segregation on nondilute bimetallic alloys: $Pt_{50}Ni_{50}$ and $Pt_{78}Ni_{22}$ probed by low-energy electron diffraction. *Phys. Rev. B* 31:6216–6218.

Holland, B. W., Duke, C. B., and Paton, A. 1984. The atomic geometry of Si(100)-(2 × 1) revisited. *Surf. Sci. Lett.* 140:L269–L278.

Kittel, C. 1986. Introduction to Solid State Physics, 6th ed. Wiley, New York.

Koningsberger, D. C. and Prins, R. 1988. X-Ray Absorption: Principles, Applications, Techniques of EXAFS, SEXAFS and XANES. Wiley, New York.

Löffler, U., Muschiol, U., Bayer, P., Heinz, K., Fritzsche, V., and Pendry, J. B. 1995. Determination of anisotropic vibrations by Tensor LEED. *Surf. Sci.* 331-333:1435–1440.

Materer, N., Barbieri, A., Gardin, D., Starke, U., Batteas, J. D., Van Hove, M. A., and Somorjai, G. A. 1994. Dynamical LEED analyses of the Pt(111)-p(2 × 2)-NO and the Ni(111)-c(4 × 2)-2NO structures—substrate relaxation and unexpected hollow-site adsorption. *Surf. Sci.* 303:319–332.

Materer, N., Starke, U., Barbieri, A., Döll, R., Heinz, K., Van Hove, M. A., and Somorjai, G. A. 1995. Reliability of detailed LEED structural analyses: Pt(111) and Pt(111)-p(2 × 2)-O. *Surf. Sci.* 325:207–222.

Materer, N., Starke, U., Barbieri, A., Van Hove, M. A., Somorjai, G. A., Kroes, G. J., and Minot, C. 1997. Molecular surface structure of ice(0001): Dynamical low-energy electron diffraction, total-energy calculations and molecular dynamics simulations. *Surf. Sci.* 381:190–210.

Morruzi, V. L., Janak, J. F., and Williams, A. R. 1978. Calculated Electronic Properties of Metals. Pergamon Press, Elmsford, N.Y.

Musket, R. G., McLean, W., Colmenares, C. A., Makowiecki, D. M., and Siekhaus, W. J. 1982. Preparation of atomically clean surfaces of selected elements: A review. *Appl Surf. Sci.* 10:143–207.

Ogletree, D. F., Blackman, G. S., Hwang, R. Q., Starke, U., Somorjai, G. A., and Katz, J. E. 1992. A new pulse counting low-energy electron diffraction system based on a position sensitive detector. *Rev. Sci. Instrum.* 63:104–113.

Pendry, J. B. 1974. Low Energy Electron Diffraction. Academic Press, New York.

Pendry, J. B. 1980. Reliability factors for LEED calculations. *J. Phys. C* 13:937–944.

Powers, J. M., Wander, A., Van Hove, M. A., and Somorjai, G. A. 1992. Structural analysis of the β-SiC(100)-(2 × 1) surface reconstruction by Automated Tensor LEED. *Surf. Sci. Lett.* 260:L7–L10.

Roberts, J. G., Gierer, M., Fairbrother, D. H., Van Hove, M. A., and Somorjai, G. A. 1998. Quantitiative LEED analysis of the surface structure of a $MgCl_2$ thin film grown on Pd(111). *Surf. Sci.* 399:123–128.

Rous, P. J. 1992. The tensor LEED approximation and surface crystallography by low-energy electron diffraction. *Prog. Surf. Sci.* 39:3–63.

Rous, P. J. and Pendry, J. B. 1989. The theory of tensor LEED. *Surf. Sci.* 219:355–372.

Saldin, D. K. 1997. Holographic crystallography for surface studies: A review of the basic principles. *Surf. Rev. Lett.* 4:441–457.

Seah, M. P. and Dench, W. A. 1979. Quantitative electron spectroscopy of surfaces: A standard data base for electron inelastic mean free paths in solids. *Surf. Interface Anal.* 1:2–11.

Stair, P. C., Kaminska, T. J., Kesmodel, L. L., and Somorjai, G. A. 1975. New rapid and accurate method to measure low-energy-electron-diffraction beam intensities: The intensities from the clean Pt(111) crystal face. *Phys. Rev. B.* 11:623–629.

Starke, U., Barbieri, A., Materer, N., Van Hove, M. A., and Somorjai, G. A. 1993. Ethylidyne on Pt(111)—Determination of adsorption site, substrate relaxation, and coverage by automated tensor LEED. *Surf. Sci.* 286:1–14.

Starke, U., Pendry, J. B., and Heinz, K. 1996. Diffuse low-energy electron diffraction. *Prog. Surf. Sci.* 52:53–124.

Tong, S. Y., Huang, H., and Wei, C. M. 1988. Low-energy electron diffraction analysis of the Si(111)-(7 × 7) structure. *J. Vac. Sci. Technol.* 6:615–624.

Tong, S. Y. and Van Hove, M. A. 1977. Unified computation scheme of low-energy electron diffraction—the combined-space method. *Phys. Rev. B* 16:1459–1467.

Van Der Veen, J. F. 1985. Ion beam crystallography of surfaces and interfaces. *Surf. Sci. Rep.* 5:199–288.

Van Hove, M. A., Cerda, J., Sautet, P., Bocquet, M.-L., and Salmeron, M. 1997. Surface structure determination by STM vs. LEED. *Prog. Surf. Sci.* 54:315–329.

Van Hove, M. A., Koestner, R. J., Stair, P. C., Biberian, J. P., Kesmodel, L. L., Bartos, I., and Somorjai, G. A. 1981. The surface reconstructions of the (100) crystal faces of iridium, platinum and gold. I. Experimental observations and possible structural models. *Surf. Sci.* 103:189–217.

Van Hove, M. A. and Tong, S. Y. 1979. Surface Crystallography by LEED—Theory, Computation and Structural Results. Springer-Verlag, Berlin.

Van Hove, M. A., Weinberg, W. H., and Chan, C.-M. 1986. Low Energy Electron Diffraction—Experiment, Theory and Surface Structure Determination. Springer-Verlag, Berlin.

Wood, E. A. 1963. Vocabulary of surface crystallography. *J. Appl. Phys.* 35:1306–1312.

Woodruff, D. P. and Bradshaw, A. M. 1994. Adsorbate structure determination on surfaces using photoelectron diffraction. *Rep. Prog. Phys.* 57:1029–1080.

Woodruff, D. P. and Delchar, T. A. 1994. Modern Techniques of Surface Science, 2nd ed. Cambridge University Press, Cambridge.

## KEY REFERENCES

Pendry, 1974. See above.

Clarke, 1985. See above.

*Two excellent introductions to the field, both very readable by the beginner. Both emphasize the background necessary to LEED calculations, while still describing other qualitative aspects.*

Van Hove et al., 1986. See above.

*The more recent of Van Hove's books, which provides a solid exposition of the theory needed to perform structural calculations. The Barbieri/Van Hove software package is available from Michel Van Hove at Lawrence Berkeley Laboratory and is highly recommended, as its use will save a great deal of effort in programming one's own computer codes.*

Woodruff and Delchar, 1994. See above.

*An excellent overview of all of the major techniques used in modern surface science, with enough information for both the newcomer and the experienced user looking to broaden his or her scope.*

*Another good book on the same topic as the previous reference.*

Rivere, J. C. 1990. Surface Analytical Techniques. Clarendon Press, New York.

## APPENDIX A:
## ESTIMATION OF COHERENCE LENGTH

The following quantitative estimate is after Pendry (1974): variations in the wave vector parallel and perpendicular to the beam can be estimated from the dot- and cross-product, respectively, between $\mathbf{k}$ and $\Delta\mathbf{k}$. The parallel deviations can be described as

$$|\mathbf{k}|^{-2}\overline{(\Delta\mathbf{k}\bullet\mathbf{k})^2} = |\mathbf{k}|^2\,\overline{(\Delta E)^2}/(4E^2) \tag{6}$$

since

$$\hbar|\mathbf{k}|^2 = 2m_e E \tag{7}$$

where $\overline{(\Delta E)^2}$ is the mean-square spread in energy. Perpendicular variations can be described as

$$|\mathbf{k}|^{-2}\overline{|\Delta\mathbf{k}\times\mathbf{k}|^2} = \overline{(\Delta\theta)^2}|\mathbf{k}|^2 \tag{8}$$

where $\overline{(\Delta\theta)^2}$ is the mean-square angular spread of the beam. On the surface, two points separated by a distance $\mathbf{l}$ differ in phase by $\mathbf{l}\bullet\mathbf{k}$, and the mean-square deviation from this phase is given by

$$\overline{(\Delta\phi)^2} = \overline{(\Delta\theta)^2}\frac{1}{2}|\mathbf{k}|^2|\mathbf{l}|^2\sin^2(\alpha) + \overline{(\Delta E)^2}(2E)^{-2}|\mathbf{k}|^2|\mathbf{l}|^2\cos^2(\alpha) \tag{9}$$

where $\alpha$ is the angle between $\mathbf{l}$ and $\mathbf{k}$. When the mean-square phase deviation approaches $\pi^2$, all phase relationship has vanished. The coherence length becomes

$$l_c = \frac{2\pi|\mathbf{k}|}{[2\overline{(\Delta\theta)^2}\sin^2(\alpha) + \overline{(\Delta E/E)^2}\cos^2(\alpha)]^{1/2}} \tag{10}$$

Choosing, e.g., $\Delta\theta \cong 0.001$ radians, $\Delta E \cong 0.2$ eV, $E = 150$ eV and $\alpha = 45°$ gives

$$l_c \cong 500\text{Å} \tag{11}$$

## APPENDIX B:
## DETAILS OF CALCULATIONS

The details of the atomic scattering event are only important at the level of a "black box"—knowing what goes in and comes out. Consequently, knowledge of the wave function of the electron beam within the atom is unnecessary; knowledge of the amount by which the phase of the scattered wave was advanced is all that is useful or required, and this describes the scattering completely. These phase shifts, $\delta_l$ are components of the atomic scattering $t$ matrix, which has the form

$$t_l = \frac{\hbar^2}{2m_e}\left(\frac{1}{k_0}\right)\sin(\delta_l)\exp(i\delta_l) \tag{12}$$

where $k_0 = |\mathbf{k}|$, $\hbar \equiv h/2\pi$ (where $h$ is Planck's constant), and $m_e$ is the mass of the electron. Note that it has been found convenient to represent scattering from an assumed spherical atom by a spherical wave formalism. This necessitates decomposing the incident plane waves into spherical partial waves for purposes of the scattering event. In a practical calculation, one cannot carry out the spherical wave expansion to infinite orders of precision, so one may approximate the maximum required angular momentum, $l$ by $l_{\max} \approx k_0 \times r_{\mathrm{mt}}$, where $r_{\mathrm{mt}}$ is the muffin-tin radius of the atom in question. Of course, this is only a rule of thumb, and one should check that the inclusion of higher-order terms does not change the results of a calculation.

Once the atomic scattering parameters have been determined, an atomic layer is constructed. Usually, it is simple to divide the crystal into layers, although sometimes it is necessary to create a "composite layer" of inequivalent atoms at either the same or similar heights; this is most frequently true for the topmost layer and for closely spaced planes. Scattering matrices are determined for each plane, $M_{\mathrm{gg}}^{\pm\pm}$, which relate the amplitudes of a scattered wave with momentum $k_{\mathrm{g}}^{\pm}$ to that of an incident wave with momentum $k_{\mathrm{g}}^{\pm}$. In the previous notation, "+" is taken to signify a wave propagating into the crystal while "−" denotes an outgoing wave, and $k_{\mathrm{g}}$ is the momentum of an electron having experienced parallel momentum transfer of $\mathbf{g}$, one of the reciprocal lattice vectors of the crystal. A matrix inversion formalism can be used for this calculation, using a spherical wave representation for each of the subplanes of a composite layer. This approach was developed by Beeby (1968) and generalized by Tong and Van Hove (1977) for use with all beams. The following are defined in atomic units $\hbar = m_e = 1$ in a spherical wave representation [$L = (l,m)$] directly after Van Hove and Tong (1979):

$t_l^i = \frac{1}{2k_0}\exp(i\delta_l)\sin(\delta_l)$: Scattering $t$-matrix for a single atom in subplane $i$.

$\tau_{LL'}^i$: Scattering matrix that contains all scattering paths within subplane $i$.

$T_{LL'}^i$: Scattering matrix that includes all those scattering paths within the composite layer that end at subplane $i$.

$G_{ll'}^{ji}$: Structural propagator describing all unscattered propagations from atoms in subplane $i$ to atoms in subplane $j$.

The diffraction matrix is defined as

$$M_{gg}^{\pm\pm} = -\frac{16\pi^2 i}{Ak_{g'\perp}}\sum_{LL'}Y_L(k_{g'}^{\pm})\sum_{i=1}^{N}\{\exp[i(\pm k_g^{\pm}\mp k_{g'}^{\pm})\cdot r_i]T_{LL'}^i\}$$
$$\times Y_{L'}^{8}(k_{g'}^{\pm}) + \delta_{g'g}\delta_{\pm\pm} \quad (13)$$

with

$$G_{Ll'}^{ji} = \hat{G}_{LL'}^{ji}\exp[-ik_g^{\pm}\cdot(r_j - r_i)] \quad (14)$$

$G_{LL'}^{ji}$ may be expressed as a sum over reciprocal lattice points with

$$\hat{G}_{Ll'}^{ji} = \frac{16\pi^2 i}{A}\sum_{g}\frac{\exp[ik_g^{\pm}\cdot(r_j - r_i)]}{k_g^{\pm}}Y_L^*(k_g^{\pm})Y_{L'}(k_g^{\pm}) \quad (15)$$

where $Y_{lm}$ is a spherical harmonic, $A$ is the unit cell area, and $r_i$ and $r_j$ are the positions of arbitrary reference atoms in subplanes $i$ and $j$, respectively. The subplane $\tau$ matrix is

$$\tau_{LL'}^i = [(I - t^iG^{ii})^{-1}]_{LL'}t_l^i \quad (16)$$

where $I$ is the multiplicative identity matrix. According to Beeby (1968), the matrices $T^1, \ldots, T^N$ are obtained directly from the equation

$$\begin{pmatrix} T^1 \\ T^2 \\ \vdots \\ T^N \end{pmatrix} = \begin{pmatrix} I & -\tau^1G12 & \ldots & -\tau^1G1N \\ -\tau^2G21 & I & \ldots & -\tau^2G2N \\ \vdots & \vdots & & \vdots \\ -\tau^NG^{N1} & -\tau^NG^{N2} & \ldots & I \end{pmatrix}^{-1}\begin{pmatrix} \tau^1 \\ \tau^2 \\ \vdots \\ \tau^N \end{pmatrix} \quad (17)$$

This completes the calculation of $M_{g'g}^{\pm\pm}$.

The final step is to calculate the interlayer scattering, which is typically done using the renormalized forward scattering (RFS) method developed by Pendry (1974). The RFS method takes advantage of the physical fact that forward scattering is much more probable than backscattering (backscattered flux is typically 1% to 5% incident flux), and the reflectivity is expanded in terms of the number of reflections experienced by the electron wave. Note that only an odd number of reflections will contribute to any observed diffraction beam. This method can be implemented by allowing the transmitted (forward scattered) waves to propagate through the crystal until their amplitude decays to a negligible value due to inelastic effects. The backscattered waves are then propagated back through the crystal, and the exiting amplitudes become the first-order result. The second order is obtained by allowing each of the backscattered waves to reflect twice more and exit the crystal, and so on for higher orders. This is continued until convergence is attained, when the reflected amplitude changes negligibly from a previous pass. This method tends to converge because inelastic damping prohibits very high-order waves from surviving. The formalism makes use of two iterative expressions for the interlayer amplitudes, one for penetration

$$a_{(i)g}^{new} = \sum_{g'}[M_{gg'}^{++}P_{g'}^{+(i-1)}a_{(i-1)g} + M_{gg'}^{+-}P_{g'}^{-(i)}a_{(i)g'}] \quad (18)$$

and one for emergence (Van Hove et al., 1986)

$$a_{(i)g}^{new} = \sum_{g'}[M_{gg'}^{--}P_{g'}^{-(i+1)}a_{(i+1)g'} + M_{gg'}^{-+}P_{g'}^{+(i)}a_{(i)g}] \quad (19)$$

where $P_g^{\pm -i}$ are plane wave propagators between appropriate reference points on successive layers and $a_{(i)g}^{new}$ is constantly overwriting $a_{(i)g}$. RFS typically requires 12 to 15 layers and 3 to 4 orders to achieve convergence (Van Hove et al., 1986, p. 174). This method also requires that layer spacings be greater than $\sim 1$ Å, or too many plane waves are required for convergence. In these cases, alternative methods such as layer doubling must be employed.

CRAIG A. GERKEN
GABOR A. SOMORJAI
University of California at
Berkeley and Lawrence
Berkeley National Laboratory
Berkeley, California

# ENERGY-DISPERSIVE SPECTROMETRY

## INTRODUCTION

Energy-dispersive x-ray spectrometry (EDS) is a technique for measuring the intensity of x-ray emission as a function of the energy of the x-ray photons (Fitzgerald et al., 1968; Goldstein et al., 1992). The excitation source for the x-rays can be energetic electrons, photons, or ions, and the target can be solid, liquid, or gas, although solid targets are the norm in virtually all materials science applications. The measured x-ray intensity can be related to the concentration (i.e., mass or atomic fraction) for each element present by performing physical/empirical matrix corrections to account for the interelement modification of the generated x radiation caused by (1) primary radiation stopping power and ionization effects, (2) attenuation of secondary characteristic x-rays during their passage through matter, and (3) any inefficiencies in detection. In this chapter, various aspects of the process of obtaining high-quality x-ray spectra, the interpretation of spectra, and the accurate extraction of x-ray intensities from spectra will be considered.

Materials analysis techniques that employ EDS include the following.

*Electron excitation*: (1) electron probe x-ray microanalysis (EPMA)/analytical scanning electron microscopy (ASEM), where EDS complements wavelength-dispersive x-ray spectroscopy (WDS); (2) analytical electron microscopy (AEM).

*Photon excitation:* X-ray fluorescence (XRF).

*Ion excitation: particle-induced x-ray emission (PIXE).*

This chapter will concentrate on EDS performed in electron-beam instruments (Williams et al., 1995). The general principles of EDS x-ray detection, processing, display, and spectral manipulation are similar for all analytical techniques regardless of the excitation source. The procedures for quantitative analysis depend in detail on the excitation process. Quantitative x-ray microanalysis by electron excitation is considered in detail in the section on Electron Probe Microanalysis. Those aspects of quantitative x-ray microanalysis that are particular to EDS spectral measurement will be considered.

**Measurement Challenge.** All elements except H and He produce characteristic x rays. Because of self-absorption in the target and absorption in the components of the spectrometer/detector, the practical minimum photon energy that is accessible by any analytical spectrometry technique is ~0.1 keV, which excludes measurement of Li (0.052 keV). Measurement of all elements beginning with the Be K-shell (0.108 keV) is possible, with heavy elements typically detected with their M- or L-shell lines (with an upper bound of ~15 keV for the L-shell). For certain applications, photon energies as high as 100 keV may be of interest (e.g., K-lines of any element excited with the AEM operating at electron energies of 200 keV or higher). The ideal form of x-ray spectrometry would measure photons (1) in the range 100 eV to 100 keV with high-energy resolution to avoid spectral peak interferences and to reduce the contribution of background processes to the peaks; (2) with true parallel detection of different photon energies to maximize measurement efficiency and to improve the utility for qualitative analysis; or (3) if serial detection of photons is the only possibility, then with a short measurement time constant to minimize coincidence events; and (4) with no added artifacts in the spectrum. These factors will be considered for semiconductor detector-based EDS systems in current use, and for other types of EDS detectors now under development for future application to analytical systems.

## PRINCIPLES OF THE METHOD

The basic principles of the measurement process for the semiconductor energy-dispersive x-ray spectrometer are illustrated in Figure 1A, and a schematic diagram of a generic semiconductor EDS system is shown in Figure 1B (Goldstein et al., 1992). An x ray with energy $E_v$ is absorbed photoelectrically by an atom in a semiconductor crystal (Si or Ge) creating a photoelectron with energy $E_v - E_c$, where $E_c$ is the critical ionization energy (binding energy) for the shell (e.g., $E_K = 1.84$ keV for the Si K-shell) and leaving the absorbing atom in an excited, ionized state. Consider a detector fabricated with Si. The Si atom will subsequently de-excite through electron shell transitions that will cause the subsequent emission of

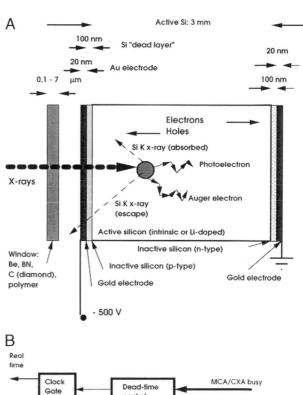

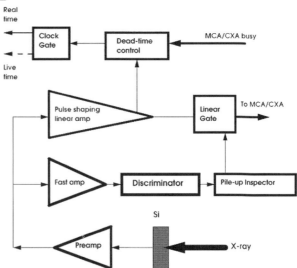

**Figure 1.** (**A**) The measurement process for the semiconductor energy-dispersive x-ray spectrometer. (**B**) Schematic diagram of a generic semiconductor EDS spectrometry system.

either an Auger electron (e.g., Si KLL) or a characteristic x-ray (Si Kα, $E = 1.74$ keV or Si Kβ, $E = 1.83$ keV). If an x-ray is emitted it may be reabsorbed in a photoelectric event with a lower binding energy shell (e.g., Si L-shell) with the emission of another photoelectron and subsequently another Auger electron. This cascade of energetic electron generation within the semiconductor is completed within picoseconds. The energetic electrons subsequently scatter inelastically with a very short mean free path (tens of nanometers or less) while traveling in the crystal, giving up their energy to crystal electron and phonon excitations. In particular, electrons from the filled valence level in the semiconductor crystal can be promoted to the empty conduction band where they are free to move under an applied electric field, leaving behind a positive "hole" in the valence band, which can also move. By placing a bias potential (500 to 1000 V) across the faces of the crystal, the free electrons and holes can be separated before they recombine. Creation of an electron-hole pair in silicon requires ~3.6 eV, so that a 3.6-keV photon will create ~1000 electron-hole pairs. The charge carriers drift out of the detector in a time of hundreds of nanoseconds. Measuring the collected charge gives a value proportional to the energy of the original photon. This is the critical measurement in x-ray EDS. A detailed description of the amplification and charge measurement electronics of the EDS system is beyond the scope of this unit. It is critical that a high-gain, low-noise amplification process be employed, since the detection of a photon involves measuring a deposited charge of ~$10^{-16}$ C and this measurement must be completed in ~50 μs (Williams et al., 1995).

An EDS system consists of several key components, Figure 1B: (1) A semiconductor crystal detector, typically housed with a field-effect transistor (FET) preamplifier in a cryostat cooled with liquid nitrogen to maintain a temperature of ~100 K or lower at the detector to minimize thermal noise. (Some systems use electrical Peltier cooling and operate at higher temperature and noise to avoid the need for liquefied gas.) (2) A high-voltage power supply to provide the detector bias (~ 500 V). (3) A slow main amplifier, which takes the pulses from the preamplifier and provides further amplification and pulse shaping. (4) A fast pulse inspection function, which monitors the charge state of the detector as a function of time and is used to reduce pileup events. (5) A "deadtime" correction function, which compensates for the time when the detector is processing a pulse and unavailable for other pulses by adding time to the clock to achieve the specified "live" accumulation time. (6) A computer-assisted analysis system [multichannel analyzer (MCA) or computer x-ray analyzer (CXA)] for spectrum display and manipulation. With the rapid evolution of computer control of measurement systems, many of these functions may take place within the computer control system, and in modern systems virtually all are controlled by software commands. Moreover, digital pulse processing, in which the entire preamplifier waveform is digitized by a high-speed circuit for subsequent processing, is emerging as a means to improve pulse handling and deadtime correction (Mott and Friel, 1995).

The EDS x-ray spectrum presented to the analyst by the computer-assisted analyzer consists of a histogram of channels calibrated in energy units containing counts corresponding to the detection of individual photons. An example of an EDS spectrum of a pure element, titanium, excited with an incident beam energy of 20 keV is shown in Figure 2A and expanded vertically in 2B. Consider this spectrum in terms of the measurement challenge posed above. The semiconductor EDS detector fulfills the first requirement that a wide energy range be accessible. Photons in the range 100 to 20 keV can be readily measured with a Si EDS detector, as shown in the logarithmic display in Figure 2C, and with decreasing efficiency, photon energies as high as 30 keV can be measured. The range can be extended to 100 keV with the use of Ge as a higher-density detector. It is this "energy-dispersive" character that provides the greatest value in practical analytical x-ray spectrometry, because it enables access to the entire periodic table (except H, He, and Li) with every spectral measurement. A complete qualitative analysis can (and should!) be performed at every sample location analyzed. With EDS measurements, constituents should not be missed unless they suffer severe interference from another peak or are present at concentrations below the limit of detection for the measurement conditions chosen.

Concerning the second criterion, the resolution (i.e., width) of x-ray peaks measured by EDS is relatively coarse and is limited by the statistical nature of the charge generation and collection process. A typical figure of merit used to define the resolution of a system is the width of the Mn Kα peak (chosen because of the use of the $^{55}$Fe radioactive source, which decays by electron capture with subsequent emission of Mn K-radiation). The resolution of a typical Si EDS is measured as the full width of the peak at half the peak intensity (full width half-maximum, FWHM) and is typically ~130 to 140 eV at Mn Kα (5890 eV) for Si detectors operating in the optimum resolution (i.e., long integration time constant) or a fractional width of 2.3%. This compares to the natural peak width of ~ 2 eV; i.e., the line is broadened by a factor of ~75. The Ge EDS can achieve an improved resolution of ~120 eV. The resolution is energy dependent and is described approximately by the following expression (Fiori and Newbury, 1978):

$$\mathrm{FWHM} = [2.5(E - E_{\mathrm{ref}}) + \mathrm{FWHM}_{\mathrm{ref}}^2]^{0.5} \qquad (1)$$

where the reference peak is usually Mn Kα. This relatively poor energy resolution leads to a significant number of peak overlaps in practical analysis situations (e. g., S K, Mo L, and Pb M), and also constrains the limit of detection (concentration) when a significant background process is present, as in the case of the electron-excited bremsstrahlung (continuum) radiation.

Considering the third criterion for parallel collection, when the accumulation of an EDS spectrum is observed, the EDS appears to provide parallel collection across the entire energy range. Actually, the term "energy dispersive" in the name of the technique is traditional and not strictly correct. No dispersion in the classic spectroscopic sense actually occurs in the measurement of the x-ray spectrum with EDS. [In the older literature, the more descriptive term "nondispersive" was often applied

(Heinrich, 1981).] The photons are detected individually, and the EDS detection process can only accommodate one photon at a time since the entire volume of the monolithic semiconductor crystal is involved. While a photon of any energy within the range specified above can be detected at any given time, detection is a serial process, so that if a second photon enters the detector while the first is being processed, the measurement of one or both may be corrupted. Because it takes time for the charge to drift out of the detector and an even longer time to measure that

charge accurately, the EDS detection process is paralyzable. As more and more photons enter the detector, the output count rate first increases with increasing input count rate, but eventually reaches a maximum and decreases with any further increase in the input count rate.

The final criterion for defect-free spectra is also not satisfied for semiconductor EDS, because of the formation of certain parasitic peaks during the detection process (escape and sum peaks), the nonuniform efficiency with

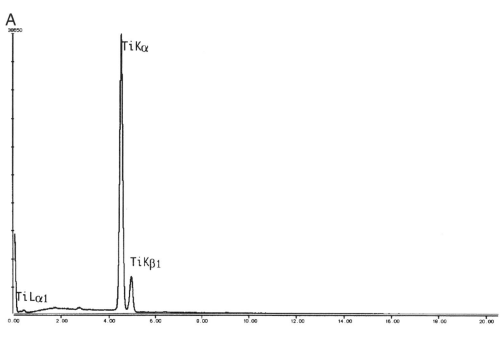

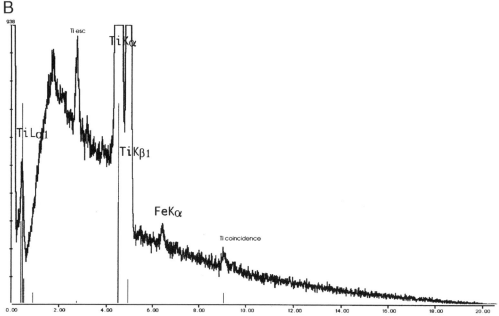

**Figure 2.** EDS spectrum of a pure element, titanium, excited with an incident beam energy of 20 keV. (**A**) Vertical (linear) scale set to highest peak; (**B**) vertical (linear) scale expanded to show bremsstrahlung background and artifact peaks due to coincidence and escape from the detector; (**C**) logarithmic scale.

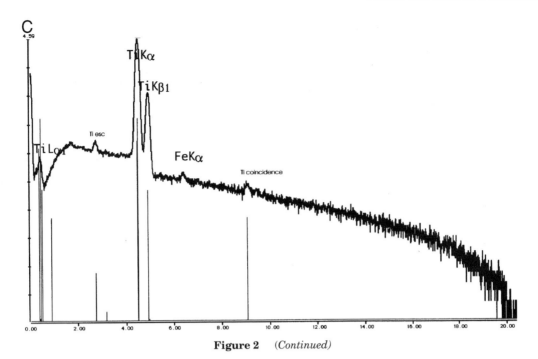

**Figure 2**    (*Continued*)

photon energy, and the fundamental distortion imposed upon the spectrum by the relatively coarse resolution function.

1. *Escape Peaks.* The escape peak arises because during the photon capture process within the detector, the de-excitation of the Si (or Ge) atom leads to the emission of a Si Kα photon (or Ge Lα and Ge Kα) in ~10% of the events. A material is relatively transparent to its own characteristic x rays, so the range of Si Kα in Si is much greater than that for the Si photoelectrons and Si Auger electrons, and moreover, the Si Kα x-ray does not lose significant energy by inelastic scattering within the detector. While the Si Kα x ray can be absorbed photoelectrically in an interaction with a Si L-shell electron, there is a significant chance that it will escape the detector, which robs the photon being measured of 1.74 keV of energy. These deficient pulses form a parasitic peak displaced by 1.74 keV below the parent peak, as noted in the spectrum of pure titanium shown in Figures 2B and C. The relative size of the escape peak depends on the average depth of capture of the primary photon within the detector crystal, so that peaks with energies just above the silicon K edge (1.84 keV) show the largest escape peaks. For P Kα (2.015 keV), the escape peak is ~2% of the parent peak intensity, while for photons above 5 keV, the escape peak becomes less significant, e.g., 0.5% for Fe Kα.

2. *Sum Peaks.* The pulse-inspection circuitry constantly monitors the rate of change of charge flow from the detector. After a photon arrival is detected and the measurement process is initiated, any subsequent photon detection during the pulse processing time changes the rate of charge flow from the detector and triggers a decision to retain or exclude the initial photon measurement in process. This pulse inspection circuit has a finite time resolution, so that it is possible for two photons to enter the detector so close in time that they cannot be distinguished. In such a case, their energies are added, creating a coincidence or "sum" peak. It is possible for any two photons to coincide, but it is only the high-abundance peaks that can give rise to noticeable spectral defects, as marked in Figures 2B and C. Coincidence peaks become progressively greater in relative height as the input count rate increases. For system deadtime below ~30%, the coincidence effect will be negligible, especially if the target contains several major elements so that the effective peak count rate is reduced compared to that for a pure element. Note that for equivalent counting rates, sum peaks become relatively more significant as the photon energy decreases, because the charge pulse produced by a low-energy photon is closer to the fundamental noise level and the coincidence exclusion circuit is therefore less efficient at distinguishing photon events. For low-energy photons (<1 keV), the pileup inspector may not function at all.

3. *Nonuniform Detection Efficiency.* Any photon in the energy range 100 to ~15 keV that actually enters the active volume of the detector is captured with near unit efficiency. Photons >15 keV in energy are increasingly lost with increasing energy due to penetration through a Si detector, while for a Ge detector, useful efficiency remains out to 100 keV. However, photons below ~4 keV suffer significant loss due to absorption in the spectrometer components: aluminum reflective coatings, window materials, gold surface electrodes, and silicon "dead" (actually,

A
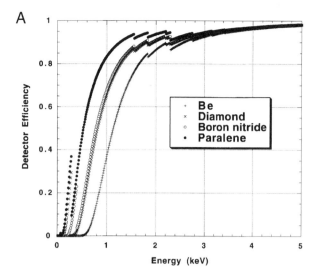

B
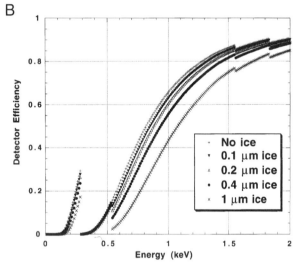

**Figure 3.** (**A**) Detector efficiency for various windows: boron nitride (0.5 μm), diamond (0.4 μm), beryllium (7.6 μm), and paralene (0.3 μm) (Al window coating = 0.02 μm; Au electrode = 0.01 μm; Si dead layer = 0.03 μm). The absorption edge at 0.28 keV is due to the presence of carbon in some of the windows. Edges due to the Au electrode and the Si dead layer can also be seen in the figure. (**B**) Detector efficiency for a diamond (0.4 μm) thin window showing the effect of ice buildup.

partially active) layers, as well as pathological layers due to possible contamination on the window, and ice on the detector. Efficiency curves are presented in Figure 3A for various windows (Be, diamond, boron nitride, and polymer) on a Si detector with a 20-nm gold electrode, a 30-nm silicon dead layer, and a 20-nm Al reflective coating. The effect of ice buildup on the detector efficiency (diamond window) is shown in Figure 3B.

4. *Spectrum Distortion*. The action of the detector resolution function on the spectrum is illustrated in Figure 4, where the "true" spectrum calculated theoretically (using NIST-NIH Desktop Spectrum Analyzer; Fiori et al., 1992) following absorption in the

specimen is compared with the actual EDS spectrum after all stages of the measurement process. The broadening of the peaks from "line" spectra (the characteristic peaks are shown as single lines because the true peak width is actually less than the 10-eV channel width) into Gaussian peaks 10 to 15 channels wide is evident. Less evident is the effect of spectrometer broadening on background distort ion, where the absorption edges are broadened. The parasitic escape and pileup peaks also contribute to spectrum distortion.

## PRACTICAL ASPECTS OF THE METHOD

### Optimizing EDS Collection

To obtain the best possible x-ray spectra for subsequent analysis, the analyst must carefully choose and/or adjust several aspects of the operation of the EDS system.

**Resolution/Count Rate Range.** The effective resolution of an EDS system depends on the time spent measuring each pulse (the "shaping time"). A tradeoff can be made between the maximum acceptable count rate and the resolution. Operating at the best resolution limits the maximum input count rate to ≤3 kHz. If the x-ray peaks of interest are well separated, it may be acceptable to operate with poorer resolution to obtain a higher acceptable counting rate, 10 kHz or higher. Whatever the choice, it is critical that the analyst be consistent when collecting a series of spectra, especially if measurements are taken over a period of time and compared to digitally archived standards.

**Energy Calibration.** The charge deposited in the detector must be accurately converted to the equivalent energy in a linear fashion. At the beginning of each measurement campaign, the user should check that the calibration (peak position) is correct within ±10 eV using known x-ray peaks from standards. Accurate calibration is critically important for both qualitative and quantitative analysis. Adjustments of the zero and coarse/fine gain may be necessary via hardware or software settings to achieve accurate calibration. Ideally, the test peaks should span the analytical range, e.g., 100 to 12 keV. A typical first choice for this task is pure copper, which provides peaks at 0.93 keV (Cu Lα), 8.04 keV (Cu Kα), and 8.68 keV (Cu Kβ). Note that a prudent analyst will also check selected lines at intermediate energies [e.g., Si Kα (1.74 keV), Ti Kα (4.51 keV), and Fe Kα (6.40 keV)], as well as higher-energy lines [e.g., Pb Lα (10.549 keV)]. It is commonly observed that even with a linear calibration established above 1 keV, the energy response may be nonlinear for photons below 1 keV. Due to the effects of incomplete charge collection for low-energy photons that are absorbed near the front electrode of the crystal, the peak position (and shape) may deviate significantly. Calibrating with a peak in this range is likely to introduce a nonlinearity for intermediate energies. The best procedure is to calibrate with the low reference peak chosen at 1 keV or above and then to check the position of important low-energy peaks, e.g., C K (0.282 keV) and O K (0.523 keV).

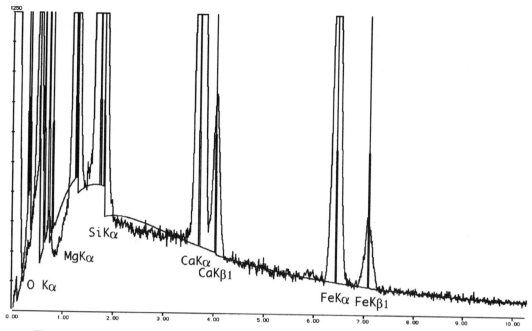

**Figure 4.** Comparison of experimentally measured spectrum (noisy) with theoretical spectrum as generated in the target (smooth). Specimen, NIST Standard Reference Material K-411 glass [Mg (0.08885); Si (0.254); Ca (0.111); Fe (0.112); O (0.424) mass fraction]; beam energy, 20 keV.

**Deadtime Correction.** The basis of quantitative analysis methods is the measurement of the x-ray intensity produced with a known dose (i.e., beam current × integration time). However, because the EDS has a paralyzable deadtime, the effective integration time would depend on the count rate if a simple clock (real) time were used for the measurement. The deadtime correction circuit automatically determines how long the detector is busy processing pulses and therefore unavailable to detect the arrival of another photon, and then adds on additional time units to the specified time of accumulation to compensate for this deadtime. The performance of the deadtime correction function can be checked by noting that all signals scale linearly with the beam current. An accurate current meter should be used to measure the beam current in a Faraday cup. Choosing a fixed "live time" (i.e., deadtime-correction applied), spectra should be collected from a pure-element standard (e.g., Cu or Fe) with progressive increases in the beam current. The x-ray counts in any peak (or background region) should increase linearly with the beam current to indicated system deadtime of ≥50%. Note that operation with the dose rate chosen to keep the deadtime <30% is advisable, since at higher deadtime artifacts such as coincidence peaks and pileup distortions in the background will occur.

## METHOD AUTOMATION

Computers play an extensive role in energy-dispersive x-ray spectrometry, and in fact the advance of laboratory computerization has been a driving force in the incorporation of dedicated automation systems in electron microscope/EDS laboratories throughout the history of the technique (Williams et al., 1995). In current instrumentation, the computer plays a role at all stages of EDS detector control, digital pulse processing, spectrum display, qualitative and quantitative analysis, and archiving of spectral data. The multichannel analyzer (MCA) has given way to the computer-assisted x-ray analyzer (CXA), which is capable of controlling both the full functions of the scanning electron microscope (SEM) or analytical electron microscope (AEM) and the EDS system for unattended, automatic operation. Spectra can be collected from an operator-specified list of predetermined sites on the specimen, or in the most advanced systems, objects of a certain class, such as particles, can be automatically detected and located in electron images and then analyzed according to a specified protocol.

## DATA ANALYSIS AND INITIAL INTERPRETATION

### Qualitative Analysis

Qualitative analysis is the identification of the spectral peaks present and their assignment to specific elements. While state-of-the-art computer-assisted analysis systems will automatically identify peaks, a prudent analyst will always check on the suggested solutions to see if they agree with the analyst's assessment. This is particularly true for the low intensity peaks, which might correspond to minor (e.g., 1 to 10 wt.%) and trace (e.g., <1 weight percent) constituents, but which might also arise as minor family members, escape peaks, or sum peaks of high-concentration constituents. Automatic qualitative analysis usually operates on the basis of a look-up table in which the detected peak is compared with an energy database and then all possible elements within a specified range, e.g., ±50 eV, are reported. These automatic results should only be considered as a guide, and the analyst must take the responsibility to confirm or deny each suggested possibility with manual qualitative analysis. One possible strategy for manual qualitative analysis is described below (Fiori and Newbury, 1978).

Before attempting manual qualitative analysis, it is critical that the analyst first determine the reliability of the database of x-ray information available in the computer-assisted analysis system in use. Certain minor lines in the L and M families (especially L1 and $M_2N_4$) produce low-relative-intensity peaks but yet are well separated in energy from the main family lines and therefore are readily visible. Such lines may not be included in the database and consequently will not appear in the "KLM" line markers for the display. If those lines are unavailable in the database, it is almost certain that they will be misidentified and assigned to another element. To evaluate the situation, the analyst should examine spectra measured from pure heavy metals (e.g., Ta, Au, Pb) to test the particular implementation of KLM markers as well as the automatic qualitative-analysis procedure against complex families of x-ray peaks. An example of a complete treatment of the lines is shown for the bismuth L and M families in Figure 5. The relative weights for the L- and M-family lines may also be of questionable value, and should only be considered as a crude guide. In addition, the analyst should test the response of the automatic qualitative analysis procedure against escape peaks and sum peaks with pure-element spectra, e.g., Al, Si, and Ti.

The basis for robust qualitative analysis is the identification, when possible, of multiple lines for each element and the careful location and notation of all minor-family-member and artifact peaks (Fiori and Newbury, 1978). To improve confidence in the identification of light elements that only produce a single line in the low photon energy range (<3 keV), it is necessary to carefully exclude possible L- and M-line interferences from heavy elements. Finally, the identification of low-intensity peaks as arising from minor and trace constituents can only be made with confidence if all minor family members from high-intensity parent peaks have been located and properly assigned, as well as the corresponding escape and sum peaks. Adequate peak statistics must be obtained, which may require additional spectrum accumulation for identification of minor and trace constituents.

1. First identify the high-intensity peaks that arise from major constituents. The analyst should start at the upper end of the spectrum (>5 keV) and work down in photon energy because the members of each x-ray family are most widely separated above 5 keV and are almost always resolvable by semiconductor EDS, thus giving greater confidence to elemental assignments by providing two or more peaks for identification.

2. Once a tentative assignment has been made to a peak, all other family members must be located, e.g., Kα-Kβ; Lα-Lβα-Lγ-Lη-Ll; Mα-Mβ-Mγ-Mζ-$M_2N_4$. Both energy position and relative height (weights of lines) are important in making an assignment.

3. For x-ray peaks above 5 keV, if a K family is identified, then a lower-energy L family of the same element will also be present and must be located and identified. Similarly, if an L family above 5 keV is identified, there must be a corresponding M family at lower energy.

4. The escape peaks and sum peaks corresponding to each high-intensity peak must be located and marked. Note that both the escape and sum peaks increase in relative significance as the parent peak energy decreases.

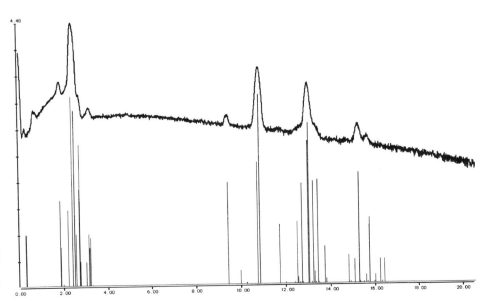

**Figure 5.** EDS spectrum of bismuth (logarithmic scale), showing KLM markers depicting the full M and L families. Beam energy, 25 keV.

5. Low-atomic-number elements ($Z < 15$) will produce a single unresolved K peak, since the Kα-Kβ peaks become progressively closer in energy and the relative Kα/Kβ ratio increases to more than 50:1 as the atomic number decreases. For fluorine and below, only one K peak exists. Thus, light elements can only be identified with a single peak, which reduces the confidence with which such assignments can be made. This is especially true for photon energies <1 keV, where x-ray absorption in the specimen significantly reduces the measured peak height relative to the higher-energy portion of the spectrum. Thus, care in identifying the higher-atomic number-elements that produce low-energy L- and M-family members is critical.

6. After the spectrum has been examined for major constituents and all possible peaks located and marked, the remaining low-intensity peaks can be assigned to minor and trace constituents following the same strategy. Note that because of the lower intensity of the peaks produced by minor and trace constituents, it may not be possible to locate and identify more than one peak per element, thus reducing the confidence with which an assignment can be made. If it is necessary to increase the confidence level that a minor or trace constituent is present, a longer accumulation time may be required to improve the counting statistics and reveal additional peaks against the statistical fluctuations in the background.

7. A final question should always be considered: What minor/trace elements could be obscured by the high-intensity peaks of the major constituents, e.g., Ti-K and Ba-L, Ta-M, and Si-K?

## Quantitative Analysis

The first step in achieving accurate quantitative analysis is the extraction of the characteristic x-ray intensities from the EDS spectrum (Heinrich and Newbury, 1991; Goldstein et al., 1992; Williams et al., 1995). The relatively poor resolution in EDS results in low spectral peak-to-background ratios and frequent severe peak interferences. Careful spectral deconvolution is a critical step in any quantification scheme. Several mathematical tools are used in various combinations in different deconvolution strategies.

### Background Removal

Characteristic peaks can be separated from background by two approaches: modeling and filtering.

In background modeling, a physical model for the background parameterized in terms of primary radiation energy, x-ray energy, and specimen composition is used to predict the background intensity as a function of energy, constrained to match the experimental spectrum at two or more energy values. Figure 6 shows the results of a background fit using the background model of Small et al. (1987) to a multielement glass (NIST K309). An excellent fit is found across the full energy range, with the only significant deviation occurring below the oxygen K line (0.523 keV). When background modeling the spectrum from an unknown specimen, the composition is initially estimated as part of a quantitative analysis procedure and an iteration procedure is followed as the specimen composition is refined. An accurate background fit is most critical for minor and trace constituents. High-intensity peaks are less sensitive to the background, so that the estimate/iteration approach is effective for unknowns.

In background filtering, a mathematical algorithm that acts as a frequency filter is applied to the spectrum (Goldstein et al., 1992). Transformed to frequency space, the peaks in the spectrum reside in the high-frequency component, while the background represents the low-frequency component. These components can be separated using a top-hat filter, which acts by transforming the contents of

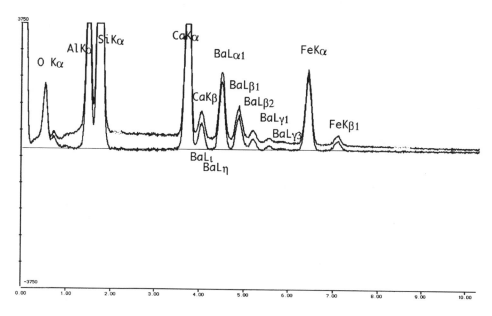

**Figure 6.** Result of a background fit to a complex specimen. NIST glass K309 [Al (0.079 mass fraction); Si (0.187); Ca (0.107); Fe (0.105); Ba (0.134); O (0.387)]. Beam energy, 20 keV.

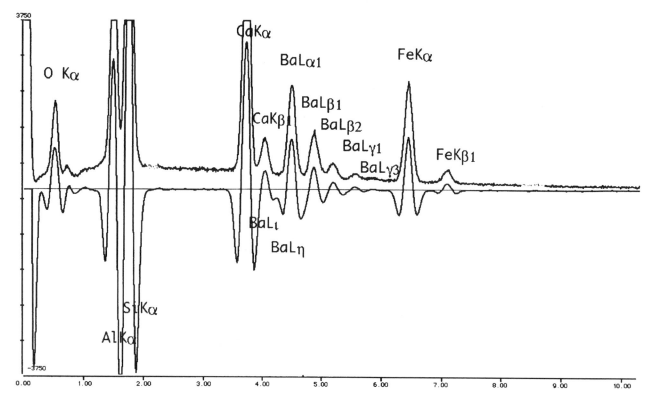

**Figure 7.** Result of the application of a top-hat digital filter to a complex specimen. NIST glass K309 [Al (0.079 mass fraction); Si (0.187); Ca (0.107); Fe (0.105); Ba (0.134); O (0.387)]. Beam energy, 20 keV.

a channel with a sum of scaled values from adjacent channels. Figure 7 shows the effect of a top-hat filter on the K309 spectrum.

### Deconvolution of Peak Overlaps

Most commercial software systems utilize peak fitting by the method of multiple linear least squares (MLLS). In MLLS, a peak reference is first prepared that contains the full peak shape for an element separated from the background and without interference from other elements. Ideally this reference is measured on the same EDS system that is used to measure the unknown. The reference contains the measured response of the particular EDS for all peaks in the family in the energy range of interest. The reference shows the true response of the EDS to each peak, including any deviations from the ideal Gaussian shape, such as incomplete charge collection, which is a function of the particular detector. To deconvolve an unknown, references for all elements expected in a specified energy region of interest are collectively fit to the unknown peak bundle. The goodness of fit between the synthesized spectrum and the real spectrum is determined on a channel-by-channel basis, using a statistical criterion such as the chi-squared value. The only variable is the peak amplitude for each component, so combinations of the references can be synthesized by linear superposition. An example of MLLS fit to the Sr-L, W-M region of a $SrWO_4$ spectrum is shown in Figure 8, which contains both the original spectrum and the background remaining

after peak stripping. Because linear mathematics is used, MLLS fitting can be very fast, requiring <1 sec per spectral region of interest. It is critical that the MLLS references be appropriate to the EDS system in use in terms of calibration and peak shape (resolution). Any deviation in system response will impact the accuracy of an MLLS deconvolution. It is extremely useful to be able to inspect the spectrum after MLLS peak stripping to detect failures in the fitting. A fit to the Ti $K\alpha,\beta$ peaks in $SrTiO_3$ with a good peak reference derived from $TiO_2$ measured in the same sequence is shown in Figure 9A. Both the original spectrum and the residuals following peak stripping are shown. The effect of a 10-eV (one-channel) shift in the position of the recorded spectrum relative to the reference is shown in Figure 9B, and a 20-eV difference in resolution (unknown broader than reference) is shown in Figure 9C. The value of examining the residuals following any spectrum modification operation such as stripping is well illustrated by these examples.

Sequential simplex fitting is an alternative approach to peak deconvolution that does not require prior determination of peak references (Fiori et al., 1981). The simplex approach requires an assumption of the peak shape, taken as a Gaussian for EDS x-ray peaks. A test spectrum is synthesized for the elements being fit, and the correspondence to the measured spectrum is compared on a channel-by-channel basis. The simplex is a nonlinear fitting procedure, and can simultaneously be fit for three variables for each peak: position (energy calibration), width (resolution), and amplitude. Thus, simplex fitting can respond

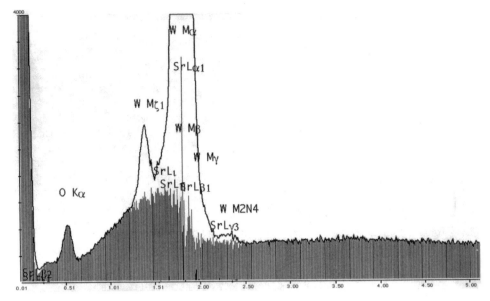

**Figure 8.** Multiple linear least-squares stripping of the W M family and the Sr L family in SrWO4. Beam energy, 20 keV. Solid black trace, original spectrum; gray spectrum, background remaining after stripping.

to instability in the EDS system. Because of the more complex, nonlinear calculations, the simplex procedure is significantly slower, by a factor of 10 or more, than the MLLS approach.

### Matrix Corrections

Once characteristic x-ray intensities have been extracted from spectra of unknowns and standards, quantitative analysis proceeds in the same way for EDS as for wavelength-dispersive x-ray spectrometry (WDS) in electron probe microanalysis. The central paradigm of quantitative electron-excited x-ray microanalysis with WDS or EDS is the measurement of the unknown and standards under identical conditions of beam energy, spectrometer efficiency, known electron dose, and spectral deconvolution. The first step is the determination of the ratio of intensities for the same x-ray peak in the unknown and the standard, which to a first approximation is equal to the ratio of the elemental concentrations in weight fractions (Castaing, 1951):

$$k = I_{\mathrm{unk}}/I_{\mathrm{std}} \sim C_{\mathrm{unk}}/C_{\mathrm{std}} \qquad (2)$$

Note that in this ratio the efficiency of the detector cancels quantitatively because the same x-ray peak is measured for the unknown and for the standard.

The concentration ratio is not exactly equal to the $k$ ratio because of the action of matrix (interelement) effects (Goldstein et al., 1992). Matrix effects arise because of compositionally dependent differences between the sample and standard in the following processes.

1. **_Electron Scattering._** The backscattering of high-energy electrons increases with the atomic number of the target (and therefore varies depending on

composition). Backscattering reduces the ionization power of the beam electrons.

2. **_Electron Stopping._** The loss of energy from the beam electrons due to inelastic scattering depends upon the composition. The efficiency of the ionization depends upon the ratio of the electron energy to the critical excitation energy of the atomic shell of interest. Differences in electron stopping with composition modify the ionization power.

3. **_X-ray Absorption._** Characteristic x rays are produced over a range of depth in the target and must travel through the solid to reach the detector. X rays are subject to photoelectric absorption, and this absorption depends on the path length and the mass-absorption coefficients, which are compositionally dependent.

4. **_Secondary X-Ray Fluorescence._** X rays that are higher in energy than the excitation energy of a particular atomic shell can be photoelectrically absorbed through interaction with that shell, leading to subsequent emission of the shell's characteristic x ray. This secondary fluorescence can be induced by higher-energy characteristic x rays or by continuum (bremsstrahlung) x rays.

These four effects can be expressed as a series of multiplicative correction factors that convert the $k$ ratio of intensities to the equivalent ratio of weight concentrations.

$$C_{\mathrm{unk}}/C_{\mathrm{std}} = kZAF \qquad (3)$$

The $Z$, $A$, and $F$ factors can be derived from first-principles physical calculations (such as Monte Carlo electron-trajectory simulations), but because of the difficulty in determining certain critical parameters with adequate accuracy, the matrix factors are often based at least partially upon experimental measurements of the required data.

Since the $Z$, $A$, and $F$ factors depend upon the composition, which is initially unknown, a starting estimate of the composition is obtained by normalizing the set of measured $k$ values for all elements in the specimen:

$$C_{i,n} = i_i / \Sigma k \qquad (4)$$

From this initial estimate of composition, a set of correction factors is calculated and the corresponding $k$ values are calculated and compared with the measured $k$ values. This procedure is iterated until convergence is achieved, which is generally within three iterations. Note that for oxidized systems, the low energy of the oxygen charac-

teristic x ray (0.523 keV) leads to severe absorption in the target and in the detector components, and consequently a large uncertainty in the calculated absorption correction. It is thus common practice to include oxygen by the method of assumed stoichiometry, where based upon the analysis of the cations, an appropriate amount of oxygen is added according to the assumed cation valences.

An example of a quantitative analysis of a microhomogeneous standard reference material (SRM 482; NIST) is listed in Table 1. The SRM values are listed along with the value determined by EDS analysis. The standards were pure elements, and matrix corrections were calculated with the NIST ZAF procedure (Myklebust et al., 1979; Fiori et al., 1992). Note that the relative errors are defined as

$$\text{rel err } (\%) = 100\% \times (\text{measured} - \text{true})/\text{true} \qquad (5)$$

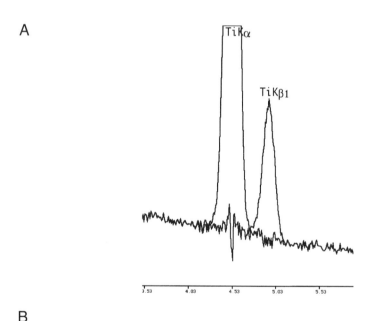

A

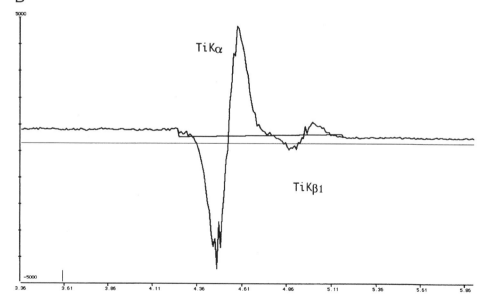

B

**Figure 9.** (**A**) Multiple linear least-squares fit to the titanium Kα-Kβ peaks with good peak references. (**B**) Effect of a one-channel (10-eV) shift between the reference and the experimental spectrum on multiple linear least-squares fitting. Specimen, Ti; beam energy, 20 keV. (**C**) Effect of a 20-eV difference in resolution between the reference and the experimental spectrum on multiple linear least-squares fitting. Specimen, Ti; beam energy, 20 keV.

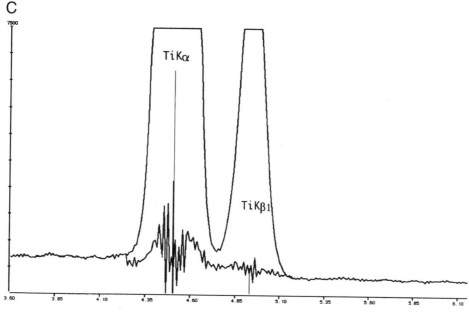

**Figure 9**   *(Continued)*

In no case is the error >1.6% relative. This is comparable to the errors that are encountered with WDS analysis of the same materials.

When large numbers of analyses of known materials spanning a large fraction of the periodic table are performed, the distribution of analytical errors can be determined. For such studies, spectra are recorded with a large number of counts to reduce the statistical variation to a negligible level compared to the systematic errors. An example of this distribution for the case of binary metallic alloys analyzed with pure-element standards is shown in Figure 13A. This distribution predicts that in 95% of the analyses performed, the relative errors are less than ±5%. For EDS analysis, this distribution is expected for concentrations greater than 0.05 mass fraction (or 5% by weight) and with minimal peak interference. As the concentration is reduced below a mass fraction of 0.05, the uncertainty in the background correction becomes more significant, and the level of error will depend strongly on the photon energy.

### Standardless Analysis

An increasing trend in recent years in performing quantitative electron probe x-ray microanalysis with energy-dispersive x-ray spectrometry has been the substitution of "standardless" methods in place of the traditional approach of measuring standards containing the elements of interest on the same analytical instrument under the same excitation and detection conditions. In standardless methods, the standard intensities necessary for quantification are either calculated from first principles, considering all aspects of x-ray generation, propagation through the solid target, and detection ("true standardless"); or else they are derived from a suite of experimental measurements performed remotely and adjusted for the characteristics of the local instrument actually used to measure the unknowns ("fitted standards"). With either route to obtaining "standard intensities," the resulting $k$ values (intensity of the unknown/intensity of the standard) are then subjected to matrix corrections with one of the usual approaches—e.g., ZAF or $\phi(\rho z)$.

The apparent advantages of standardless analysis are considerable. Instrument operation can be extremely simple. There is no need to know the beam current, and indeed, it is not even necessary for the beam current to be stable during the spectrum accumulation, a real asset for instruments such as the cold-field emission gun SEM where the beam current can be strongly time dependent. Moreover, the detector solid angle is of no consequence to

**Table 1. Electron-Excited EDS X-Ray Microanalysis of SRM 482 (Au-Cu)**[a]

| Cu SRM | Cu Conc | Rel Err % | Au SRM | Au Conc | Rel Err % | Total |
|--------|---------|-----------|--------|---------|-----------|-------|
| 0.198 | 0.198 | 0 | 0.801 | 0.790 | −1.4 | 0.988 |
| 0.396 | 0.399 | 0 | 0.603 | 0.594 | −1.6 | 0.993 |
| 0.599 | 0.605 | 1.0 | 0.401 | 0.402 | 0.1 | 1.007 |
| 0.798 | 0.797 | −0.1 | 0.200 | 0.199 | −1.2 | 0.996 |

[a] Values in mass fractions. Conc,, concentration; rel err, relative error; SRM, standard reference material.

the quantitative procedure. Both the beam-current uncertainty and the detector solid-angle uncertainty are effectively hidden by normalizing the analytical total to a predetermined value, e.g., unity when all constituents are measured. When the spectrum has been accumulated to the desired level of statistical precision, the analyst needs to specify only the beam energy and x-ray takeoff angle, and list of elements to be quantified (or this list can be derived from an automatic qualitative analysis). The software then proceeds to calculate the composition directly from the spectrum, as described below, and in the resulting output report the concentration is often specified to 3 or 4 significant figures. The measurement precision for each element, calculated from the integrated peak and background counts, is also reported. The precision is really only limited by the patience of the analyst and the stability of the specimen under electron bombardment, so that relative precision values below 1% (1 σ) can be readily achieved, even for minor constituents. While such excellent measurement precision is invaluable when sequentially comparing different locations on the same specimen, the precision is no indication of accuracy. When standardless analysis procedures are tested to produce error histograms like that in Figure 13A, the errors are found to be substantially larger (Newbury et al., 1995; Newbury, 1999).

### First-Principles Standardless Analysis

Calculating an equivalent standard intensity from first principles requires solution of Equation 6 (Goldstein et al., 1992):

$$I_{std} = \left[ (\rho N_0/A)\omega \int_{E_0}^{E_c} \frac{Q}{(dE/ds)} dE \right] Rf(\chi)\varepsilon(E)_v \quad (6)$$

where the terms in brackets represent the excitation function: $\rho$ is the density, $N_0$ is Avogadro's number, $A$ is the atomic weight, $\omega$ is the fluorescence yield, $Q$ is the ionization cross-section, $dE/ds$ is the stopping power, $E$ is the electron energy, $E_0$ is the incident-beam energy, and $E_c$ is the critical excitation energy. The other terms correct for the loss of x-ray production due to electron backscattering $(R)$, the self-absorption of x rays propagating through the solid $[f(\chi)]$, and the efficiency of the detector, $\varepsilon(E_v)$, as a function of photon energy, $E_v$. It is useful to consider the confidence with which each of these terms can be calculated.

**Excitation.** There are three critical terms in the excitation function: the ionization cross-section, fluorescence yield, and stopping power.

*Ionization cross-section.* Several parameterizations of the K-shell ionization cross-section are plotted in Figure 10. The variation among these choices exceeds 25%. While it is not possible to say that any one of these is correct, it is certain that they cannot all be correct. Moreover, because of the continuous energy loss in a solid target, the cross-section must be integrated from $E_0$ to $E_c$, through the peak in $Q$ and the rapid decrease to $U = 1$

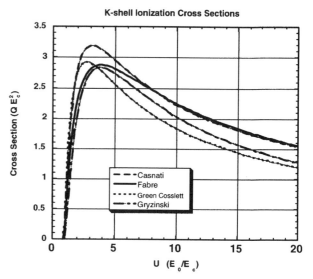

**Figure 10.** Ionization cross-sections (Casnati, S4; Fabre, S5; Green-Cosslett, S6; and Gryzinski, S7) as a function of overvoltage $U$, as formulated by different authors.

(where $U$ is overvoltage $U = E_0/E_c$). This region of the cross-section is poorly characterized, so that it is difficult to choose among the cross section formulations based on experimental measurements. The situation for L- and M-shell cross-sections is even more unsatisfactory.

*Fluorescence yield.* Figure 11 plots various experimental determinations of the K-shell fluorescence yield. Again, a variation of >25% exists for many elements. The situation for L- and M-shell transitions is even less certain.

*Stopping power.* The classic Bethe formulation of the stopping power becomes inaccurate at low beam energies (<5 keV), and eventually with decreasing energy becomes

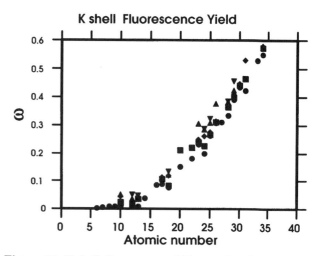

**Figure 11.** K-shell fluorescence yield as a function of atomic number. Symbols indicate measurements by different workers, as derived from the compilation in Fink et al. (1966).

physically unrealistic with a sign change. The accuracy of the stopping power matters for calculating Equation 5, because the cross-section must be integrated down to $E_c$, which for low-energy x-rays (e.g., C, N, O, F) involve electron energies in this regime. Several authors (e.g., Heinrich and Newbury, 1991) have suggested modifications to the Bethe formulation to correct for the low-beam-energy regime. Unfortunately, the base of experimental measurements necessary to select the best choice is just being developed.

**Backscatter Loss.** The backscatter loss correction factor $R$ was initially formulated based upon experimental measurements of the total backscatter coefficient and the differential backscatter coefficient with energy (Heinrich, 1981). While careful, extensive measurements of total backscatter were available in the literature, the database of the differential backscatter coefficient with energy, a much more difficult experimental measurement, was limited to a few elements and was available at only one emergence angle. The development of advanced Monte Carlo simulations has permitted the rigorous calculation of $R$ over all scattering angles and energy losses so that this factor is now probably known to an accuracy within a few percent across the periodic table and the energy range of interest (Heinrich and Newbury, 1991).

**X-Ray Self-Absorption.** The self-absorption of x rays in the hypothetical standard is calculated based upon the distribution in depth of x-ray production, a parameter that has been extensively studied by experimental measurement of layered targets and by Monte Carlo electron-trajectory simulation. The formulation of the absorption factor due to Heinrich and Yakowitz, as used in the NIST ZAF matrix correction, is employed (Heinrich, 1981). Fortunately, the absorption correction is generally small for the x rays of a pure element, so that at least for higher-energy characteristic x rays, e.g., photon energies >3 keV, there is little error in this factor. However, the self-absorption increases both as photon energy decreases and as incident electron energy increases, so that the error in calculating the intensity emitted from an element emitting low-energy photons, such as carbon, can be significant.

**Detector Efficiency.** The last term in Equation 5 is one of the most difficult with which to deal. In the traditional $k$ value approach, the detector efficiency cancels quantitatively in the intensity ratio and can be ignored because the same x-ray peak is measured for the unknown and the standard under identical (or at least accurately reproducible) spectrometer conditions. When standardless analysis is performed, this cancellation can not occur because x-ray peaks of different energies are effectively being compared, and therefore accurate knowledge of the detector efficiency becomes critical. Detector efficiency is mainly controlled by absorption losses in the window(s) and detector structure. The expression for detector efficiency, Equation 7, consists of a multiplicative series of absorption terms for each com-

ponent: detector window (denoted "win" in the subscripted terms in Equation 6), gold surface electrode ("Au"), semiconductor ("Si" or "Ge") dead layer ("DL"; actually a partially active layer below the electrode and the source of incomplete charge phenomena), and a transmission term for the detector thickness. Additionally, for most practical measurement situations there may be pathological absorption contributions from contamination on the detector crystal, usually arising from ice buildup due to pinhole leaks in the window or support ("ice"), and from contamination on the detector window usually deposited as hydrocarbons from the microscope environment ("con").

$$\varepsilon(E_v) = \exp[(-\mu/\rho)_{win}\rho_{win}win + (-\mu/\rho)_{Au}\rho_{Au}Au \\ + (-\mu/\rho)_{Si}\rho_{Si}SiDL + (-\mu/\rho)_{con}\rho_{con}t_{con} \\ + (-\mu/\rho)_{ice}\rho_{ice}t_{ice}\}]^{*}[1 - \exp(\mu/\rho)_{Si}\rho_{Si}t_{Si}] \quad (7)$$

In Equation 7, the mass absorption coefficients are those appropriate to the photon energy, $E_v$, of interest. An example of the detector efficiency as a function of photon energy for several window materials is shown in Figure 3A. The choice of the window material has a strong effect on the detector efficiency for photon energies <3 keV, and accurate knowledge of the window and detector parameters is vital for accurate interelement efficiency correction across the working range of the detector, typically 100 to 12 keV. The pathological change in the detector efficiency with the accumulation of ice is illustrated in Figure 13B. The buildup of ice and other contaminants and the resulting loss in efficiency is referred to as "detector aging." Detector aging can result in significant loss of low-energy photons (<3 keV) relative to higher-energy photons (3 to 12 keV). Detector aging due to ice buildup can often be reversed by following the manufacturer's recommended procedure for warming the detector.

**Fitted-Standards Standardless Analysis**

The fitted-standards technique is the more widely used approach for implementing standardless analysis on commercial computer-assisted EDS analyzer systems. In the fitted standards technique, a suite of pure element standards covering K-, L-, and M-family x-rays is measured at one or more beam energies on an electron-beam instrument equipped with an EDS detector whose efficiency is known from independent measurements or at least can be estimated from knowledge of the detector construction. An example of such a measurement for a portion of the K-series from pure elements measured at 20 keV is shown in Figure 12A. In the fitted standards technique, missing elements can be calculated by mathematically fitting the available peaks and interpolating (e.g., in Fig. 12A, the intensity for gallium could be estimated by fitting the smoothly varying data and interpolating). From the smooth change in peak height with atomic number seen in Figure 12A, such an interpolation should be possible with reasonable accuracy. The situation is not as satisfactory in the L and M families, as illustrated in Figures 12B and 12C, because the cross-section/fluorescence yield

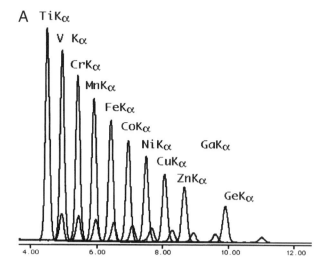

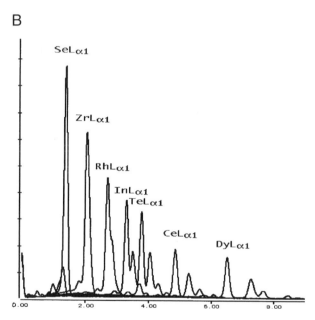

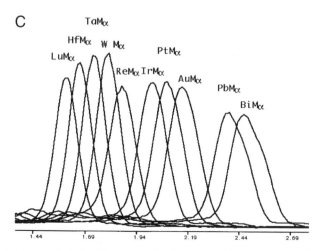

**Figure 12.** (**A**) K-family peaks from transition elements; $E_0$, 20 keV. (**B**) L-family peaks; $E_0$, 20 keV. (**C**) M-family peaks; $E_0$, 20 keV.

product is a much more complicated function of the atomic number.

If the analysis must be performed at a beam energy other than that of the spectral database, then Equation 6 must be used to shift the fitted standards intensities appropriately. Similarly, if a different EDS detector is used, the detector efficiency must be corrected for the differences in efficiency between the two detectors using Equation 7. The fitted-standards standardless procedure is expected to be more accurate than the "first-principles" standardless procedure because it is tied to actual experimental measurements that directly incorporate the effects of the cross-section/fluorescence yield product, at least over the range of the elements actually measured.

### Testing the Accuracy of Standardless Analysis

The accuracy of standardless analysis procedures has been tested by carrying out analyses on microhomogeneous materials of known composition (Newbury et al., 1995): NIST Microanalysis Standard Reference Materials (SRM; Newbury, 1999), NIST Microanalysis Research Materials (RM), stoichiometric binary compounds (e.g., III-V compounds such as GaAs and II-VI compounds such as SrTe), and other materials such as ceramics, alloys, and minerals for which compositions were available from independent chemical analysis and for which microhomogeneity could be established (Newbury et al., 1995). Compositions were carefully chosen to avoid serious spectral overlaps (as in, e.g., PbS and $MoS_2$). Light elements such as boron, carbon, nitrogen, oxygen, and fluorine were also eliminated from consideration because of large errors due to uncertainties in mass absorption coefficients. In oxidized systems, the oxygen was calculated by means of assumed stoichiometry, but the resulting oxygen values were not included in the error histograms because of their dependence on the cation determinations.

Figure 13B shows an error histogram for the first-principles standardless analysis procedure embedded in the National Institute of Standards and Technology–National Institutes of Health Desktop Spectrum Analyzer (DTSA) x-ray spectrometry software engine. The error distribution shows symmetry around 0% error, but in comparing this distribution with that for the conventional standards/ ZAF procedure shown in Figure 13A, the striking fact is that the error bins are 10 times wider for first-principles standardless analysis. Thus, the 95% error range is $\sim \pm 50\%$ relative rather than $\pm 5\%$ relative.

The error distribution for a commercial standardless procedure based upon the fitted-standards approach is shown in Figure 13C. This distribution is narrower than the first principles standardless approach, but the error bins are still 5 times wider than those of the conventional standards/ZAF procedure, so that the 95% error range is $\pm 25\%$ compared to $\pm 5\%$. It must be emphasized that this distribution represents a test of only one of the many implementations of standardless analysis in commercial software systems and that more extensive testing is needed.

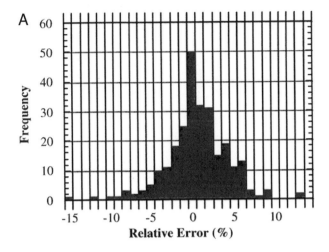

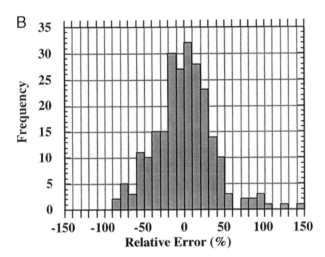

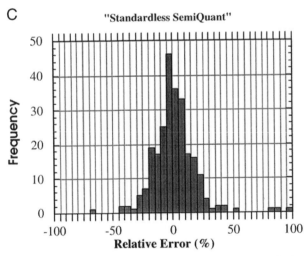

**Figure 13.** (**A**) Distribution of errors observed in the analysis of binary metal alloys against pure element standards (Heinrich-Yakowitz binary data ZAF; Heinrich, 1981). Beam energy, 20 keV. (**B**) Error distribution for the first-principles standardless analysis procedure embedded in the NIST-NIH Desktop Spectrum Analyzer x-ray spectrometry software system. Beam energy, 20 keV. (**C**) Error distribution for a fitted-standards standardless analysis procedure embedded in a commercial x-ray spectrometry software system (beam energy, 20 keV).

## Using Standardless Analysis

Given the broad width of these error distributions for standardless analysis, it is clear that reporting composition values to 3 or 4 significant figures can be very misleading in the general case. At the same time, the error distributions show that there are significant numbers of analyses for which the relative errors are acceptably small, <10%. Usually, these analyses involve elements of similar atomic number (e.g., Cr, Fe, and Ni in stainless steel) for which the x-ray peaks are of similar energy and are therefore excited and measured with similar efficiency. Users of standardless analysis must be wary that any confidence obtained in such analyses does not extend beyond those particular compositions. The most significant errors are usually found when elements must be measured involving a mix of K-, L-, and M-shell x rays. The errors in the tails of the distributions are so large that the utility of concentration values obtained in this fashion is very limited.

Standardless analysis does have a legitimate value. If a microhomogeneous material with a known composition similar to the unknown specimen of interest is available, then the errors due to standardless analysis can be assessed and included with the report of analysis. For example, if the Fe-S system is to be studied, then analyzing the minerals pyrite ($FeS_2$) and troilite (FeS) would serve as a good test to challenge the standardless analysis procedure.

The analyst should never attempt to estimate relative concentrations merely by inspecting a spectrum. There are simply too many complicated physical effects of relative excitation, absorption, and detection efficiency to allow such a quantitative detail to be obtained from a casual inspection of a spectrum. Standardless analysis incorporates enough of the corrections to allow a sensible classification of the constituents of the specimen into broad categories, e.g.:

    major: (>10 wt.%)

    minor: (1 to 10 wt.%)

    trace: (<1 wt.%)

In the absence of a known material to test a standardless analysis procedure, it is recommended that these broad classification categories be used instead of numerical concentration values, which may imply far more apparent accuracy than is justified and which may lead to a loss of confidence in quantitative electron-probe microanalysis when an independent test is conducted.

**Limits of Detection with EDS X-Ray Microanalysis.** The ultimate limit of detection is determined by statistical variance of the background counts within the peak region of interest. As such, the concentration limit of detection ($C_{DL}$) for a peak that does not suffer from interference from a nearby high-intensity peak depends on the peak-to-background ratio (Ziebold, 1967):

$$C_{DL} \geq 3.3a/[n\tau P(P/B)]^{0.5} \qquad (8)$$

where $a$ is a factor relating the concentration to the measured intensity ratio and is generally close to unity, $n$ is

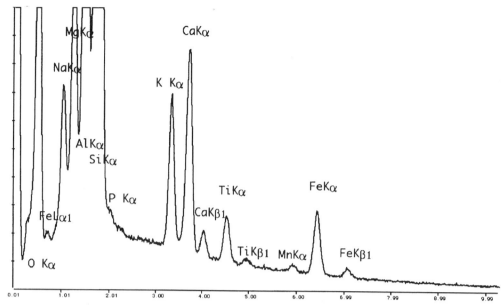

**Figure 14.** EDS spectrum of NIST glass K961; beam energy, 15 keV. Composition: O (0.470); Na (0.0297); Mg (0.0301); Al (0.0582); P (0.0022); K (0.0249); Ca (0.0357); Ti (0.012); Mn (0.0032); Fe (0.035) (mass fraction).

the number of replicate measurements, $\tau$ is the measurement time, and $P$ and $B$ are the peak and background count rates, respectively. While the detection limit can be lowered by using longer counting times, the appearance of this term within the square root limits the improvement. For practical counting conditions ($\tau < 1000$ s) and deadtime ($< 30\%$), the limit of detection for EDS varies from concentration levels of 0.0005 to 0.002 (mass fraction) depending on the analytical line and the matrix. An example of a spectrum showing minor and trace elements is presented in Figure 14. Note the obvious detection of Mn (0.0032 mass fraction), while P (0.0022) suffers severe interference from the major Si peak.

## SAMPLE PREPARATION

The ideal form of the specimen for x-ray analysis with any type of primary excitation (electrons, photons, or ions) is a flat, mechanically polished surface with topography reduced below 50 nm (Goldstein et al., 1992). The requirement for flatness arises because of the strong absorption of x rays, particularly for photon energies $<4$ keV and increasing in severity as the photon energy decreases. Low-energy x rays, e.g., C-K, N-K, and O-K, are particularly strongly absorbed. Even with a flat surface, 90% or more of the x rays generated within the target are absorbed along the exit path toward the EDS detector. The effect of surface topography is to introduce an uncontrolled geometric variable that can modify the x-ray spectrum independent of the composition, which in conventional analysis is assumed to be the only variable. An example of the magnitude of the effect of anomalous absorption is shown in Figure 15, which shows two spectra obtained from a "123" $YBa_2Cu_3O_7$ high-$T_c$ superconductor. The spectrum obtained with the beam placed on the

polished surface shows significant x-ray counts at all photon energies, while the spectrum obtained in the crack shows significant attenuation of the low-energy photons, which is caused by the additional path length through the solid. Note that if only the "in-crack" spectrum were available, it would not be readily apparent that the anomalous absorption situation existed. The low-energy photon peaks from Y-L, Cu-L, and O-K can easily be detected, but their intensities are not representative of the polished material, and if quantification were attempted, the concentrations calculated from these peaks would be below the correct value by a large factor.

## SPECIMEN MODIFICATION

Generally, EDS performed with electron bombardment does not lead to significant changes for most specimen compositions, such as metal alloys, ceramics, glasses, and minerals. Important exceptions occur, however, where charged ionic species can move under the influence of charge injected by the beam and for certain classes of specimens for which radiation damage can result in the generation and loss of volatile components (Joy et al., 1986; Goldstein et al., 1992).

### Ionic Migration

Insulating materials can accumulate charge under electron bombardment, leading to the development of local fields strong enough to deflect the incident beam. To minimize the effects of charging on the primary beam, a thin ($\sim$ 10- to 20-nm) conducting coating such as carbon is typically applied to the surface of the specimen and this layer is connected to an electrical ground. However, beam electrons passing through this layer can accumulate within

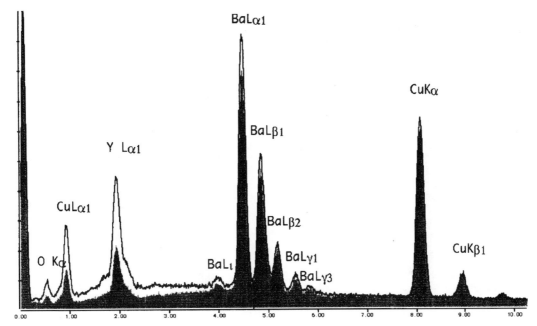

**Figure 15.** Comparison of spectra obtained on polished surface (trace) and in crack (solid gray). Specimen: $YBa_2Cu_3O_7$ high-$T_c$ superconductor. Beam energy, 25 keV.

the specimen and create a significant internal electric field. Sodium, for example, is known to be capable of motion under such fields, which can be detected as a time dependency in the x-ray signal. The degree of sodium migration depends on the specimen composition.

### Loss of Volatile Components

Some specimen compositions are sensitive to radiation damage, which can produce broken chemical bonds or free radicals, for example. Biological and organic specimens are particularly prone to the loss of water and other volatile components under electron bombardment. Even after chemical substitution to stabilize such specimens, significant modification may occur during electron bombardment. To correct for such mass loss in the quantification of minor and trace constituents, Hall (1968) developed a method of normalizing the characteristic peaks to a band of high-energy bremsstrahlung radiation that scaled with the total mass of the electron-excited volume.

## PROBLEMS

Energy-dispersive x-ray spectrometry systems are capable of stable, long-term operation with a high degree of reproducibility. In the next section operational quality-assurance protocols will be described. However, occasional deviations will occur, even in an otherwise properly operating EDS system, and it is the responsibility of the user to recognize such pathological conditions. The careful user of EDS will develop the ability to recognizing such deviations. One of the best ways to do this is to record and archive spectra from high-purity elements and simple binary compounds and carefully study the forms of the spectra. Modern computer-assisted x-ray analyzers provide ready means for comparing spectra both visually and

mathematically. An archive of reference spectra obtained on the local instrument is invaluable for rapid comparison with spectra obtained under current operating conditions.

### Pathological Conditions

**Ground Loops.** The EDS detector-amplifier chain operates at extremely high gain and may therefore be sensitive to other sources of electromagnetic radiation. Improperly shielded cables may act as antennae, introducing spectral degradation through resolution loss, background distortion, false peaks, and other artifacts. Another source of interference arises from "ground loops"—AC currents flowing between two points at nominal ground potential. It is critical to prevent electrical contact between the EDS detector snout and the microscope column. A high value of electrical resistance should be maintained to isolate the EDS from the microscope. The EDS (as well as microscope power supplies) should all be connected to a common high-quality ground. An example of spectra obtained with and without a ground loop operating is shown in Figure 16. Note the high and complex background when the ground loop is present. It is highly recommended that when a new EDS system is being installed, the EDS performance first be checked on the bench (i.e, with the EDS system isolated from the microscope) with a radioactive source (e.g., $^{55}Fe$). Upon installation on the microscope, the EDS system should be checked again with the radioactive source placed in the microscope specimen position. Ideally, these two spectra should be identical. Any degradation in EDS performance (resolution, deadtime, etc.) is probably due to ground loops and should be corrected by careful cable shielding and routing to avoid proximity to power supplies, etc. If no degradation is found, the microscope should then be fully activated with the vacuum system in full operation and the electron beam on but blanked to prevent it from hitting the specimen (the

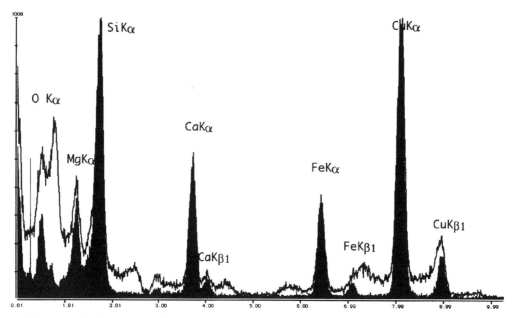

**Figure 16.** Pathological artifact: Action of ground loop in a low-resistance path between the EDS and the microscope chassis distorting the spectrum. Trace, spectrum with ground-loop active; solid filled areas, spectrum with ground loop eliminated.

[55]Fe source). A third radioactive source spectrum should be obtained. If this spectrum is identical to the previous two conditions, then the EDS system is free of ground loops.

**Light Leakage.** Under electron bombardment, certain specimens (especially minerals, ceramics, and some semiconductors) emit light, a process called cathodoluminescence. The vacuum-isolation window of the EDS is coated with a thin (~20-nm) aluminum layer to reflect visible light. This layer may not provide sufficient attenuation of the light, or there may be leakage through pinholes in the coating. Depending on the light intensity, the spectrum may undergo energy shift (miscalibration) or resolution degradation, and in extreme cases the peaks may undergo enormous distortion, as illustrated in Figure 17 for ZnS. Other sources of light can include inspection lights and the infrared source chamber TV camera. These sources should be shut off during EDS operation. A symptom of light leakage into the EDS detector is anomalously high deadtime, which can reach 100% (no x-ray photons processed) if the light source is strong enough. Note that after the EDS detector is exposed to a high level of light, it may take several minutes to return to stable operation

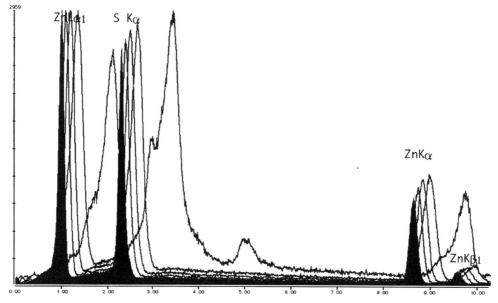

**Figure 17.** Pathological artifact: Effect of cathodoluminescent light leakage into detector. Specimen, ZnS; beam energy, 20 keV. Solid filled areas, reference ZnS; traces, increasing cathodoluminescence output achieved by increasing scanned area at fixed current.

and can be subject to severe spectral artifacts during this recovery.

**Ice Accumulation.** Figure 3B shows the calculated decrease in low-energy detection efficiency with the build-up of ice. To monitor this phenomenon in practice, a pure nickel target can provide a spectrum in which the Ni Kα radiation (7.477 keV) is sufficiently energetic that it is unaffected by detector windows and ice layers, while the low-energy Ni Lα (0.849 keV) suffers severe absorption passing through ice, since it is located above the O K edge (0.531 keV). A plot of the Ni Lα/Ni Kα ratio is an effective monitor of detector performance. When the Ni Lα/Ni Kα ratio has fallen by a significant percentage (e.g., arbitrarily 10%, to be determined by the user's established quality-assurance plan) relative to the performance of the new detector, the EDS manufacturer's protocol for conditioning the detector should be followed scrupulously.

**Distortions of Low-Energy Peaks.** Low-energy photons, less than 1 keV, are measured with charge pulses that are very close in magnitude to the system noise. Several artifacts are observed in this energy range even when the detector is performing optimally for photons above 1 keV in energy. Peaks are distorted from the ideal Gaussian shape by the effects of incomplete charge collection on the low- energy side. This effect depends entirely on the detector construction and cannot be modified. The pulse-pileup inspector is virtually ineffective in this energy range. Pulse pileup effects can be severe, with the result that the peak position and shape may change with input count rate, which can be controlled by the analyst. Both peak shift and peak distortion at high deadtime (50% versus 8%) can be seen in the carbon spectra shown in Figure 18.

**Stray Radiation.** The operating environment of the EDS system on an electron microscope can be severe. Beam electrons will backscatter from the specimen, and additionally may scatter off microscope components such as apertures and stage materials, especially in a high-energy analytical electron microscope. The EDS is equipped with an "electron trap" (actually a permanent magnet deflector) to prevent energetic electrons (whose prevalence actually exceeds the rate of x-ray production by a large factor) from reaching the detector and dominating or distorting the spectrum. However, these remote sources of electrons can create anomalous contributions to the x-ray spectrum. The EDS is also commonly equipped with a collimator to restrict the contributions of x rays from the remote electron scattering, but typically such collimation is limited to an area with linear dimensions of millimeters centered around the on-axis beam impact point on the specimen. Electrons not in the focused beam that strike the specimen within this collimation acceptance disk will contribute to the spectrum. To assess the magnitude of stray electron contributions, it is highly recommended that an "in-hole" spectrum be obtained. For the solid specimen SEM, a "Faraday cup" should be constructed using materials not present in the microscope construction. A polished block of titanium is a good choice. Into this metal block should be drilled a blind hole and over this hole a microscope aperture with a small opening (<100 μm) should be placed. If possible, the aperture metal should be different from that actually used for apertures in the microscope. Three spectra should then be recorded for the same live time: (1) beam placed on the titanium; (2) beam placed on the aperture; and (3) beam placed in the blind hole. Ideally, the in-hole spectrum should have no counts for either characteristic or bremsstrahlung x rays. It is likely that a small component of x rays representing nonbeam electrons striking the aperture and/or metal block will be detected. This contribution should be <1% of the intensity recorded when the beam is placed on these materials.

Note that this in-hole procedure detects only electrons not in the focused beam. In normal operation with the beam striking the specimen, significant numbers of backscattered electrons will be generated that will strike the polepiece above the specimen and can rescatter as an uncollimated radiation source raining down on the specimen. To examine this effect, a second specimen should be fabricated consisting of microscopic particles or fibers embedded in a silver-doped conducting epoxy and polished. Choose a particle or fiber whose diameter exceeds the expected interaction volume by at least a factor of 4: e.g., at least 20-μm diameter if the electron x-ray range is 5 μm. With the beam placed on the center on the particle, which should have a composition that does not contain silver, examine a spectrum accumulated with at least 200,000 counts in the full energy range. Any silver radiation must arise from rescattered backscattered electrons. This rescattering component might be reduced if the polepiece can be shielded with a plate consisting of a light element, such as carbon. It is very important never to use carbon paint.

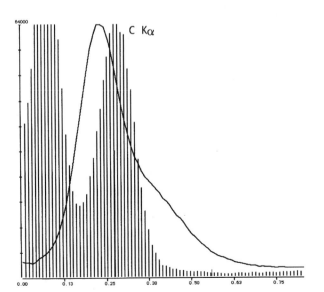

**Figure 18.** Pathological artifact: Peak shift induced by count-rate effects. Barred spectrum taken at low deadtime (8%); line spectrum taken at high deadtime (50%). Beam energy, 15 keV.

**Electron Penetration.** Energetic electrons that penetrate the detector window and strike the detector will be measured as if they were x-ray photons, leading to anomalous spectral background shape (Fiori and Newbury, 1978). To avoid or at least minimize this problem, a permanent-magnet deflector is usually placed in the collimator assembly. During very-high-beam-energy operation in the AEM (100 keV or more), electron penetration into the detector may still occur, especially during surges in beam current that can happen during changes in the operating parameters of the electron lenses, especially under computer automation. Such electron penetration may actually damage the EDS detector.

## PROTOCOLS

Measurement protocols have been partially described as part of the previous sections, e.g., the recommendations for testing a new EDS detector system for ground loops, stray radiation, etc., and the recommendations for testing qualitative-analysis procedures. In this section, protocols for quality assurance in operating an EDS system will be summarized.

Quality-assurance operations for an EDS system operating on an electron microscope takes place at several time intervals: once per session, weekly, and long term.

### Session

At the start of each measurement session, the analyst must go through a checklist of operational parameters to ensure that the system is operating in the configuration that is required. This is especially true if the facility has multiple users. In the following discussion, please note that terms are used which are as generic as possible to describe the functions. There is no standardization for these terms, and when the historical range of EDS systems is considered, these functions may be implemented with controls ranging from hardware to full software selection and control.

The key elements on this checklist are as follows.

**Resolution (Count-Rate Range).** The longer the time spent processing a pulse, the more precisely the energy can be determined, and thus the narrower the peak and the better the resolution. Selection of this parameter (e.g., specified as "shaping time," "amplifier time constant," "count rate range," and "resolution") is made depending on the analytical problem. Generally, one will wish to accumulate as many x-ray counts as possible in the available counting time, so it is desirable to operate at the highest allowable count rate and then select a beam current/detector solid angle to achieve that count rate, recognizing that the EDS will operate at the poorest resolution. However, if the elements of interest produce interfering peaks, especially if an element of interest is present as a minor or trace constituent whose peak is close to the intense peak of a major constituent, then the poor-resolution performance at high count-rate range may preclude achieving an adequate detection limit, and it may be necessary to operate at high resolution and a lower limiting count rate. Whatever the choice, the analyst must be consistent when quantitative procedures are followed with archived standards.

**EDS Spectrum Channel Width and Number.** The width of the individual channel of the EDS histogram can be selected, as well as the number of channels. Typical width choices are 5, 10, 20 and 40 eV. It is important to have as many channels as possible to define the x-ray peak, so a choice of 10 or 5 eV is desirable. If possible, it is also important to record the entire x-ray spectrum up to the beam energy, at least in the SEM case where $E_0$ is typically between 5 and 30 keV. Thus, it is desirable to record 2048 or 4096 channels. Given the continuing drastic decreases in the cost of computer memory and mass storage, it is advisable to record and archive the complete spectrum from every measurement location.

**Calibration.** The calibration should be checked on a specimen with characteristic peaks in the 1- to 2-keV and 8- to 10-keV ranges, e.g., Cu Kα (8.04 keV) and Cu Lα (0.928 keV). The calibration should be established within 10 eV of the reference value. After these endpoints have been fixed, several intermediate peaks should be checked, e.g., Si (1.740 keV), Ti (4.508 keV), and Fe (6.400 keV). Noting that the low-energy photon peaks (<1 keV) are likely to be out of calibration, it is useful to record pure carbon (0.282 keV) and an oxygen-containing target, e.g., $Al_2O_3$ (O = 0.523 keV).

**Specimen Position.** The first condition to define the detector geometric efficiency (solid angle) is the specimen position. In electron beam systems with an optical microscope, the specimen can be accurately positioned because of the shallow depth of focus of the optical microscope. If only SEM imaging is available, then to position the specimen relative to the EDS in terms of the working distance, the objective lens current can be used to select a particular lens strength. By using a large final aperture to minimize the depth of focus, the specimen position can then be selected with the z motion (along the beam) of the stage.

**Detector Solid Angle.** The absolute x-ray intensity depends upon the electron dose (current × live time), the detector quantum efficiency, and the detector solid angle. Many detectors are capable of mechanical motion along the axis of the takeoff angle. The solid angle $\Omega$ of collection depends on the detector area $A$ (as collimated) and the detector-to-specimen distance, $r$:

$$\Omega = A/r^2 \tag{9}$$

Because of the strong dependence of the solid angle on the detector-to-specimen distance, it is critical to have a method to accurately reset the detector position. Typically a ruler scale is provided for coarse adjustments, but this may not be adequate for the most accurate work. If the beam current and specimen position can be set accurately and reproducibly, fine adjustment of the solid angle can be based upon the absolute intensity measured in a high-energy x-ray peak, e.g., Cu Kα.

### Weekly

The performance of the deadtime correction circuit should be checked by using the beam current (as measured in a

Faraday cup) to independently control the input count rate. The output count in a defined live time should scale with the electron dose (beam current) at least up to an indicated deadtime of 50%, and preferably to even higher deadtime. Detector aging should be monitored with a spectrum of Ni.

### Long Term

EDS systems are capable of a high degree of stability provided they are continuously maintained under power. When powered up from a cold start, several hours are typically needed to establish thermal equilibrium in all electronic circuits and thus achieve a stable operating condition. It is therefore not advisable to power the EDS system down as part of routine operation but rather to maintain power at all times. When such a shutdown must be carried out, it is important to reestablish thermal equilibrium before resuming measurement operations. All parameters then should be checked, and adjusted if necessary: resolution, calibration, and deadtime correction.

## LITERATURE CITED

Castaing, R. 1951. Application of Electron Probes to Metallographic Analysis. Ph.D. dissertation, University of Paris.

Fink, R. W., Jopson, R. C., Mark, H., and Swift, C. D. 1966. Atomic fluorescence yields. *Rev. Mod. Phys.* 38:513–540.

Fiori, C. E., Myklebust, R. L., and Gorlen, K. E. 1981. Sequential simplex: A procedure for resolving spectral interference in energy dispersive X-ray spectrometry. *In* Energy Dispersive X-ray Spectrometry (K. F. J. Heinrich, D. E. Newbury, R. L. Myklebust, and C. E. Fiori, eds.). pp. 233–272. National Bureau of Standards Publication 604, U.S. Government Printing Office, Washington, D.C.

Fiori, C. E. and Newbury, D. E. 1978. Energy dispersive x-ray spectrometry in the scanning electron microscope. *Scanning Electron Microsc.* 1:401–422.

Fiori, C. E., Swyt, C. R., and Myklebust, R. L. 1992. Desktop Spectrum Analyzer. Office of Standard Reference Data, National Institute of Standards and Technology, Gaithersburg, Md.

Fitzgerald, R., Keil, K., and Heinrich, K. F. J. 1968. Solid-state energy-dispersion spectrometer for electron-microprobe X-ray analysis. *Science* 528:159–160.

Goldstein, J. I., Newbury, D. E., Echlin, P., Joy, D. C., Romig, A. D. Jr., Lyman, C. E., Fiori, C., and Lifshin, E. 1992. Scanning Electron Microscopy and X-ray Microanalysis. Plenum, New York.

Hall, T. 1968. Some aspects of the microprobe analysis of biological specimens. *In* Quantitative Electron Probe Microanalysis (K. F. J. Heinrich, ed.) pp. 269–299. U.S. Dept. of Commerce, Washington, D.C.

Heinrich, K. F. J. 1981. Electron Beam X-ray Microanalysis. Van Nostrand–Reinhold, New York.

Heinrich, K. F. J. and Newbury, D. E. (eds.). 1991. Electron Probe Quantitation. Plenum, New York.

Joy, D. C., Romig, A. D. Jr., and Goldstein, J. I. 1986. Principles of Analytical Electron Microscopy. Plenum, New York.

Mott, R. B. and Friel, J. J. 1995. Improved EDS performance with digital pulse processing. *In* X-ray Spectrometry in Electron Beam Instruments (D. B. Williams, J. I. Goldstein, and D. E. Newbury, eds.). pp. 127–157. Plenum, New York.

Myklebust, R. L., Fiori, C. E., and Heinrich, K. F. J. 1979. FRAME C: A Compact Procedure for Quantitative Energy-Dispersive Electron Probe X-ray Analysis. National Bureau of Standards Technical Note 1106, U.S. Dept. of Commerce, Washington, D.C.

Newbury, D. E. 1999."Standardless" quantitative electron-excited x-ray microanalysis by energy-dispersive spectrometry: what is its proper role? *Microscop. Microanalys.* 4:585–597.

Newbury, D. E., Swyt, C. R., and Myklebust, R. L. 1995. 'Standardless" quantitative electron probe microanalysis with energy-dispersive x-ray spectrometry: Is it worth the risk? *Anal. Chem.* 67:1866–1871.

Small, J. A., Leigh, S. D., Newbury, D. E., and Myklebust, R. L. 1987. Modeling of the bremsstrahlung radiation produced in pure-element targets by 10-40 keV electrons. *J. Appl. Phys.* 61:459–469.

Williams, D. B., Goldstein, J. I., and Newbury, D. E. (eds.). 1995. X-ray Spectrometry in Electron Beam Instruments. Plenum, New York.

Ziebold, T. O. 1967. Precision and sensitivity in electron microprobe analysis. *Anal. Chem.* 39:858–861.

## KEY REFERENCES

Goldstein et al., 1992. See above.

*Comprehensive reference to x-ray microanalysis procedures for SEM.*

Heinrich and Newbury, 1991. See above.

*Comprehensive treatment of leading topics in microanalysis research.*

Joy et al., 1986. See above.

*Comprehensive reference to x-ray microanalysis procedures for AEM.*

Williams et al., 1995. See above.

*Comprehensive treatment of leading topics in x-ray spectrometry.*

DALE E. NEWBURY
National Institute of Standards
and Technology
Gaithersburg, Maryland

# AUGER ELECTRON SPECTROSCOPY

## INTRODUCTION

Auger electron spectroscopy (AES) is a powerful technique for determining the elemental composition of the few outermost atomic layers of materials. Surface layers often have a composition that is quite different from the bulk material, due to contamination, oxidation, or processing. In the most commonly used form of AES, a specimen is bombarded with electrons having an energy between 3 and 30 keV, resulting in the ejection of core-level electrons from atoms to a depth up to ~1 µm. The resulting core vacancy can be filled by an outer-level electron, with the excess energy being used to emit an x ray (the principle of electron probe microanalysis) or another electron (the principle of Auger electron spectroscopy) from the atom. This emitted electron is called an Auger electron and is named after Pierre Auger who first observed such events in a cloud chamber in the 1920s (Goto, 1995). AES is a surface-sensitive technique, due to the strong inelastic scattering of low-energy electrons in specimens. Auger electrons from only the outermost few atomic layers are emitted from the specimen without energy loss, and contribute to the peaks in a spectrum. Auger electrons that have lost energy in escaping from the specimen will appear as an additional background signal at lower kinetic energies. Obviously, hydrogen and helium cannot be detected, as three electrons are needed for the Auger process. The Auger electron kinetic energies are characteristic of the material, and the measurement of their kinetic energies is used to identify the elements that produce these Auger electrons. As the Auger electron kinetic energies depend on the binding energies of the electron levels involved, changes in surface chemistry (such as oxidation) can produce detectable shifts in Auger kinetic energies, thereby providing useful information about the surface chemistry as well. Auger electrons emitted from the specimen will appear as peaks on a continuous background of secondary electrons and backscattered electrons. Secondary electrons are usually defined as the electron background below 50 eV, and backscattered electrons are those in the background with energies from 50 eV up to the incident beam energy. The concentrations of elements detected can be determined from the intensities of the Auger peaks. An Auger spectrum is usually a plot of the number of electrons detected as a function of kinetic energy, but sometimes it is displayed as the first derivative of the number of electrons emitted, to enhance the visibility of the Auger electrons and to suppress the continuous background of secondary and backscattered electrons. Electron beams can be focused to small diameters, thereby allowing the composition of small areas ($\sim$50 nm and below) on a surface to be determined. This is often called point analysis. The electron beam can be defocused or rastered over small areas to reduce possible electron beam damage where this might be a problem. Analysis can also be performed at preselected points or areas on the specimen. The electron beam can alternatively be scanned in a straight line across part of the specimen surface and Auger data can be acquired as a function of beam position, resulting in what is called an Auger line scan. Auger maps can also be measured, showing the variation in elemental composition (and concentration) across a region of the surface; this is referred to as scanning Auger microscopy (SAM). Variation in composition with depth can be determined by depth profiling, which is usually accomplished by continuously removing atomic layers by sputtering with inert gas ions while monitoring the Auger signals from the newly created surfaces. For most elements, the detection limit with AES is between 0.1 and 1 atom %.

### Competitive and Related Techniques

Methods for surface analysis that are alternatives to AES include x-ray photoelectron spectroscopy (XPS), which is also often called electron spectroscopy for chemical analysis (ESCA); low-energy ion scattering (LEIS), which is also called ion scattering spectroscopy (ISS); and secondary ion mass spectrometry (SIMS). XPS/ESCA uses soft x rays to eject electrons from atoms. Compared with AES, the interpretation of XPS/ESCA spectra is generally simpler; it is more easily quantifiable and easier to use when studying insulators, and produces less beam damage. The detection limit of XPS/ESCA is similar to that of AES, but the highest spatial resolution is a few micrometers (as opposed to $\sim$50 nm with AES). Auger peaks will also appear in XPS spectra. In some cases, energetic ions can also produce Auger electrons. ISS is the most surface sensitive of these techniques; low-energy inert gas ion beams measure just the outermost atomic layer; but it is not generally useful for identifying unknown elements at the surface due to its relatively poor specificity. Of these techniques, SIMS has the highest detectability as well as high elemental specificity; it is also capable of detecting atomic parts per million and below. In SIMS, the ions that are sputtered from the surface are detected in a mass spectrometer. Static SIMS is used for identifying the surface composition, whereas dynamic SIMS is used for depth profiling. SIMS can detect hydrogen as well as isotopes of the elements, which AES cannot. SIMS can be quantified for simple, well-characterized specimens, but is not routinely quantifiable for complex surface compositions. A variation of SIMS is sputtered neutral mass spectrometry (SNMS), where the neutral species that are emitted by sputtering are post-ionized, and then measured with a mass spectrometer.

Surface analysis systems cost several hundred thousand U.S. dollars. Systems can be stand-alone (e.g., AES only) or can incorporate more than one technique (such as AES and XPS). Analysis is performed in a vacuum chamber at pressures typically of the order of $10^{-18}$ Pa. A vacuum is needed since electron beams are involved, and ultrahigh vacuum ($< 10^{-7}$ Pa) is required so the surface composition does not change during analysis. There are several manufacturers of Auger systems, and manufacturers usually offer more than one model. Manufacturers include JEOL (*http://www.jeol.com*), Omicron Associates (*http://www.omicron-instruments.com*), Physical Electronics (*http://www.phi.com*), Staib Instruments (*http://www.staib-instruments.com*), and VG Scientific (*http://www.vgscientific.com*). These companies, and many others, also manufacture components such as electron sources, electron kinetic energy analyzers, vacuum systems, and systems for computerized data acquisition and processing, which can be assembled to make a customized system. There are several commercial companies that provide analytical services using AES, and costs are typically 1000 to 3000 U.S. dollars per day. Some commercial companies in the United States are Anderson Materials Evaluation (*http://www.andersonmaterials.com*), Charles Evans & Associates (*http://www.cea.com*), Geller Micro Analytical Lab (*http://www.gellermicro.com*), and Physical Electronics (*http://www.phi.com*). Samples can usually be mounted and inserted into the analysis chamber in $\sim$30 min. Simple analysis can be performed within another 30 min, whereas depth profiling and Auger mapping can take longer.

Surface topography can be measured using optical microscopy (OPTICAL MICROSCOPY and REFLECTED-LIGHT

OPTICAL MICROSCOPY); scanning electron microscopy (SEM; SCANNING ELECTRON MICROSCOPY); scanning tunneling microscopy (STM; SCANNING TUNNELING MICROSCOPY); and atomic force microscopy (AFM). Scanning Auger microscopy systems have SEM capability, and this is used for locating regions on the surface of the specimen to be studied.

## PRINCIPLES OF THE METHOD

An example of the Auger process is shown in Figure 1. With AES, the electron energy levels are denoted by the x-ray notation, K, L, M, etc., corresponding to the principal quantum number 1, 2, 3, etc. Subscripts are based on the various combinations of the orbital angular momentum 0, 1, 2, 3, etc. and the electron spin $+1/2$ or $-1/2$, and have the values 1, 2, 3, 4, 5, etc. The K level is not given a subscript. The L levels are $L_1$, $L_2$, and $L_3$, the M levels are $M_1$, $M_2$, $M_3$, $M_4$, and $M_5$, etc. Sometimes the energy levels are close in energy and not resolvable in the measurement, and are then designated as $L_{2,3}$, $M_{2,3}$, $M_{4,5}$, etc. In the example shown in Figure 1, the K and L levels in the atom are shown to be fully occupied by electrons (Fig. 1, panel A). In this example, the initial ionization occurs in the K level (Fig. 1, panel B). Following relaxation and the emission of an Auger electron, the atom has two vacancies in outer levels, the $L_2$ and the $L_3$ levels in this example (Fig. 1, panel C). This process can be thought of as consisting of the following processes. (1) An electron drops down from the $L_2$ level to fill the K level vacancy. (2) Energy equal to the difference in binding energies between the K and the $L_2$ levels is produced. (3) This energy is sufficient to remove an electron from the $L_3$ level of the atom, with the excess energy being the energy of the Auger electron. The three atomic levels involved in Auger electron production are used to designate the Auger transition. The first letter designates the shell containing the initial vacancy and the last two letters designate the shells containing electron vacancies created by Auger emission. The Auger electron produced in the example shown in Figure 1 is therefore referred to as a $KL_2L_3$ Auger electron. When a bonding electron is involved, the letter V is sometimes

used (e.g., KVV and LMV). Coupling terms for the electronic orbital angular momentum and spin momentum may also be added where known, e.g., $L_3M_{4,5}M_{4,5}$; $^1D$. More complicated Auger processes can also occur; e.g., if an atom is doubly ionized before Auger emission, it will be triply charged after emission (Grant and Haas, 1970). In such cases, the Auger transition can be designated by separating the initial and final states by a dash, e.g., LL-VV would be used for an atom that is doubly ionized in the L shell before Auger emission, and that uses two electrons from the valence shell for the Auger emission. When an Auger relaxation process involves an electron from the same principal shell as the initial vacancy, e.g., $L_1L_2M$, it is sometimes referred to as a Coster-Kronig transition. If both electrons involved in relaxation are from the same principal shell as the initial vacancy (e.g., $N_5N_7N_7$), it can be called a super Coster-Kronig transition. In some cases, such as MgO, electrons from neighboring atoms can also be involved in the Auger process and are referred to as cross transitions (Janssen et al., 1974). After the initial Auger electron is produced, further Auger relaxations will occur to fill the electron vacancies, resulting in a cascade of Auger electron emissions as vacancies are filled by outer electrons.

The energy of the Auger electron will depend on the binding energies, $E_b$, of the levels involved (and relaxation effects) and not on the energy of the incident beam. In its simplest form, the kinetic energy, $E_k$, of the Auger electron $KL_2L_3$ is given approximately by

$$E_k(KL_2L_3) \approx E_b(K) - E_b(L_2) - E_b(L_3) \qquad (1)$$

This equation neglects contributions from the interaction energy between the holes in the $L_2$ and $L_3$ levels, as well as intra-atomic and extra-atomic relaxation. Therefore, each element (except hydrogen and helium) will have a characteristic spectrum of Auger electrons, and this forms the basis for qualitative analysis. The intensity of the Auger electrons emitted forms the basis for quantitative analysis. Since binding energies (and relaxation effects) depend on the chemical environment of atoms, information about the chemical environment can often be obtained by studying changes in the kinetic energies of the Auger electrons, which are typically a few electron volts (Haas and Grant, 1969; Haas et al., 1972a,b). Auger transitions involving valence electrons can also exhibit changes in line shape (Haas and Grant, 1970; Haas et al., 1972b). In solids, after an Auger electron is produced, it has to escape from the specimen and pass through the energy analyzer to be measured. All elements except hydrogen and helium can produce Auger electrons, and each has at least one major Auger transition with a kinetic energy below 2000 eV. In AES systems, the electron energy analyzers typically measure electron kinetic energies up to 2000 or 3000 eV. The dominant Auger transitions are KLL, LMM, MNN, etc., with combinations of the outermost levels producing the most intense peaks. These Auger electrons have inelastic mean free paths of only a few atomic layers (Seah and Dench, 1979), and make AES a surface-sensitive technique. Auger electrons produced deeper in

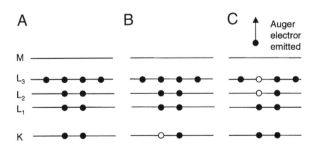

**Figure 1.** Schematic diagram illustrating the $KL_2L_3$ Auger process. Atom showing: (**A**) electrons present in filled K and L levels before an electron is removed from the K level; (**B**) after removal of an electron from the K level; and (**C**) following the Auger process, where a $KL_2L_3$ Auger electron is emitted. In (c), one L-level electron fills the K vacancy and the other L-level electron is ejected due to the energy available on filling the K level.

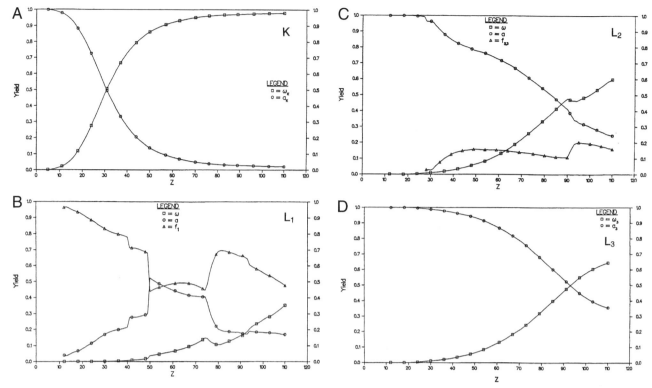

**Figure 2.** Examples of fluorescent yield and Auger yield as a function of atomic number. Fluorescent yield (boxes), non-Coster-Kronig Auger yield (circles), and Coster-Kronig Auger yield (triangles) for: (**A**) K, (**B**) $L_1$, (**C**) $L_2$, and (**D**) $L_3$ levels as a function of atomic number, $Z$. Auger electron energy analyzers usually measure kinetic energies up to ~3 keV. For this energy range, K levels in atoms up to atomic number 19 and $L_{2,3}$ levels up to atomic number 50 can be ionized; in these atoms, the fluorescent yields for these levels are much lower than the combined Auger yields. Data from Krause, 1979. (Reprinted with permission.)

the specimen will lose energy in traveling to the surface and appear as a background at lower kinetic energy.

A competing process for Auger electron emission from atoms is x-ray emission. Following the creation of the initial core-level vacancy, a characteristic x ray can be emitted (instead of an Auger electron) when an outer level electron fills this vacancy. The probability for relaxation by Auger emission (called the Auger yield) is much higher than for x-ray emission (called the fluorescent yield) for the energies usually measured in AES (Krause, 1979). Examples of fluorescent yield and Auger yield as a function of atomic number for the K, $L_1$, $L_2$, and $L_3$ levels are shown in Figure 2 (note that Coster-Kronig Auger yields are identified separately from the non-Coster-Kronig Auger yields in this figure).

## PRACTICAL ASPECTS OF THE METHOD

As mentioned earlier, most AES work is done with electron-beam excitation. Tungsten filaments, lanthanum hexaboride cathodes, and field emitter tips are all available as electron sources. Lanthanum hexaboride and field emitters have high brightness and are used for high-spatial-resolution instruments. Electron beams can be focused to small areas for point analysis, and rastered in one direction for Auger line scans or in two directions for

Auger mapping. Some materials are susceptible to electron beam damage, and care must be exercised in studying them (see Specimen Modification). Insulating specimens can also charge electrically during analysis, resulting in a shift in the measured Auger kinetic energy. Such charging can usually be reduced and/or stabilized using low-energy electron flood guns to replace the lost charge.

A number of electron energy analyzers can be used for AES, the two most popular being the cylindrical mirror analyzer and the spherical sector analyzer. Specimens must be positioned accurately, particularly for single-pass cylindrical mirror analyzers where the measured kinetic energy depends somewhat on specimen position. Positioning is also important in sputter depth profiling where the electron beam, the ion beam, and the specimen must all be aligned at the focus of the analyzer. The electrons transmitted through the analyzer are usually amplified by one or more channel electron multipliers (or microchannel plates) and measured using pulse counting, voltage-to-frequency conversion, or phase-sensitive detection coupled with electron beam or analyzer modulation.

Because of the surface sensitivity of AES, measurements should be made in an ultra-high-vacuum system (~$10^{-7}$ Pa and below) to prevent contamination of the specimens under study. Sputter depth profiling at higher background pressures can be tolerated if the sputter rate is sufficiently high that contamination from the background gases does not occur. Commercial systems usually include devices for specimen manipulation and treatment (e.g., specimen heating, cooling, and fracture). For many specimens, some treatment, such as inert gas sputtering to remove surface contaminants, might be necessary before meaningful analysis can be made. Some specimens, even though they might appear clean to the eye, may have a thin contaminant layer precluding the detection of any Auger signal from the material of interest. Specimen handling is very important to avoid any contamination of the specimen. Specimens are usually inserted into the analysis chamber through a vacuum interlock.

Auger line scans give the distribution of elements in a straight line across the surface, and Auger maps provide the distribution of elements within a selected area on the surface. The distribution of elements with depth can be obtained by depth profiling. Analysis is rapid, from a few minutes to a few hours depending on the information required, and several specimens can be examined in one day.

## Types of Auger Spectra

Auger spectra can be acquired in several ways, but the methods fall into two main categories, namely, (1) the measurement of the total electron signal including Auger electrons and the background from secondary and backscattered electrons, or (2) the measurement of the derivative of this signal using a lock-in amplifier in order to suppress the slowly varying background. Both types of spectra are measured as a function of the kinetic energy of the electrons leaving the sample. Backgrounds can also be suppressed by taking a derivative of the total electron signal obtained in (1) above with a computer. Examples of such spectra are shown in Figure 3, and were taken from a used Cu gasket. Figure 3A is often referred to as a "direct spectrum," and the Auger electrons appear superimposed on the continuous background of secondary and backscattered electrons. This spectrum was obtained in the pulse-counting mode, and the Auger peaks from Cu, Cl, C, N, and O are identified. A large part of the increasing background with increasing kinetic energy is due to the way Auger spectra are usually measured—i.e., with the analyzer energy resolution set at a constant percentage of the measured kinetic energy of the emitted electrons. This results in a continuous, linear change in the measured energy resolution across the spectrum, and therefore the enhancement of the kinetic energy distribution of electrons entering the analyzer, $N(E)$, by a multiple of the kinetic energy, $E$. This is why the ordinate is usually labeled $E*N(E)$ in Auger spectra. Figure 3B is referred to as a "derivative spectrum", where the background is reduced and the Auger features are enhanced; note especially the improved N signal. This spectrum was obtained by differentiation of the spectrum in Figure 3A with a computer, and then cutting off some of the large signal at low kinetic energy. The ordinate is then labeled $d[E*N(E)]/dE$.

## Specimen Alignment

The specimen is usually moved to the analyzer focus and approximate analysis position using an optical method

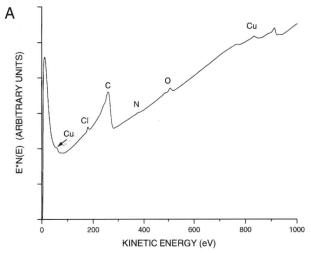

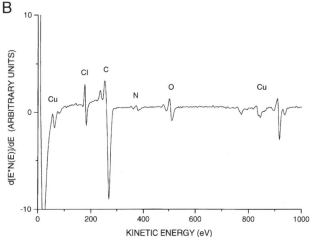

**Figure 3.** Auger spectra of a used Cu gasket showing: (**A**) the direct spectrum and (**B**) the derivative spectrum. Spectra were obtained using a 40-nA, 5-keV electron beam for excitation, and a hemispherical analyzer with a 1-eV step interval for analysis. The spectrum took 18 min to acquire.

such as a previously calibrated video camera. The electron gun is then switched on, and precise positioning of the specimen is accomplished by using the scanning electron microscopy (SEM) capability of the system. The principles of SEM are discussed in SCANNING ELECTRON MICROSCOPY, and images are obtained from the signal variation with beam position produced by either secondary electrons, backscattered electrons, or specimen current to ground. The analyzer focus can also be checked by maximizing the signal detected by the electron energy analyzer at the same energy as the electron beam, typically 1, 2, or 3 keV. This signal arises from elastically backscattered primary electrons.

## Qualitative Analysis

Elements present at the surface are identified from the energies of the Auger peaks in the spectra. An element (or compound) can be considered positively identified if the peak energies, peak shapes, and the relative signal strengths (for multiple peaks) coincide with a standard reference spectrum of the element (or compound). The procedure starts with the most intense peak in a spectrum and is repeated until all peaks have been identified (Fig. 3). Peak identification depends on an increase in signal intensity above the slowly varying background, and if a computer program is used for peak identification it compares the kinetic energy of the peak with a reference library of associated peaks. If no element lies within a certain energy range from the measured peak, it is not identified by the program. Such problems can arise if the kinetic energy scale is not calibrated, or if the measured kinetic energies shift due to specimen charging. In some cases, peaks can also be misidentified with such software, particularly when different elements have Auger peaks with similar kinetic energies, and the user needs to be aware of this limitation in qualitative analysis. Handbooks of Auger spectra are available and are very useful for helping the novice identify (or confirm) peaks in spectra (Sekine et al., 1982; Childs et al., 1995). The reported Auger peak energies vary by a few electron volts between handbooks, but such variations between handbooks or laboratories should be eliminated in the future, now that energy calibration procedures for AES have been provided for both direct and derivative spectra (Seah et al., 1990). The relative intensities of peaks at widely different kinetic energies will also vary between instruments (and the handbooks) due to instrument design and operating conditions. Therefore, reference spectra should be measured in the same instrument used for the unknown. This problem of intensity variation with kinetic energy can also be reduced by measuring direct spectra of Cu, Ag, and Au in a given instrument, allowing intensity-energy relationships to be derived by comparing such spectra with standard reference AES spectra of Cu, Ag, and Au (Smith and Seah, 1990).

Sets of Auger spectra from some of the elements (Mn, Fe, Co, Ni, and Cu) in the first row of transition elements are shown in direct form in Figure 4, and in derivative form in Figure 5. These spectra were taken from argon-ion sputtered metal foils. The dominant peaks at higher kinetic energy (500 to 1000 eV) are the $L_{2,3}MM$ series, and those at low kinetic energy (40 to 70 eV) are from $M_{2,3}VV$ transitions. The peaks near 100 eV are from $M_1VV$ transitions. Note the rich structure in the $L_{2,3}MM$ series due to the various final M shell vacancies, and how these sets of peaks are similar to each other but move to higher kinetic energy with increasing atomic number. Spectra from other elements will show similar trends in their LMM Auger spectra, as will KLL and MNN transitions. The low-kinetic-energy $M_{2,3}VV$ Auger peaks from the transition metals in Figures 4 and 5 shift by smaller amounts than the $L_{2,3}MM$ peaks, and are not usually used for elemental analysis. The types of Auger spectra shown in these two figures are often referred to as survey spectra and are taken with relatively poor energy resolution and relatively large energy steps so as to optimize the signal-to-noise for a relatively short data acquisition time. Survey spectra are usually the first spectra taken in an analysis, as they provide an overall indication of the surface composition, and are typically taken over an energy range from a few electron volts to 1000 or 2000 eV.

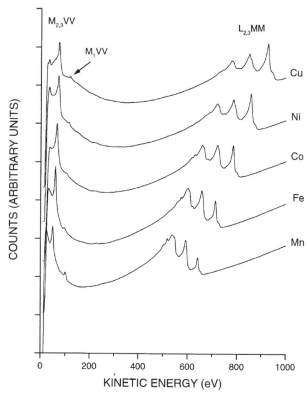

**Figure 4.** Auger spectra of Mn, Fe, Co, Ni, and Cu taken from argon-sputtered foils, shown in the direct mode. These spectra were obtained with a 40-nA, 5-keV electron beam, and a hemispherical analyzer with a 1-eV step interval. Each spectrum took <2 min to acquire. The Fe, Co, Ni, and Cu spectra are offset for clarity.

The derivative spectra are normally used for peak identification because the Auger features are enhanced in these spectra. The data can also be taken over a longer time to increase the detectability, and higher energy resolution spectra can be obtained if such information is needed. The Auger spectra displayed in Figures 4 and 5 were taken using a hemispherical analyzer with an energy resolution of 0.6% of the kinetic energy and 1.0-eV energy steps applied to the analyzer; each took less than 2 min to acquire with a 40-nA, 5-keV electron beam.

### Chemical Effects

Besides identifying the elements present at surfaces, AES can often provide useful information about the chemical environment of surface atoms. Such chemical effects can affect the measured Auger spectrum in a number of ways, e.g., the kinetic energy at which an Auger peak occurs can change, the energy distribution of the Auger electrons can change, or the energy loss structure (see Problems in reference to plasmon loss peaks) associated with Auger peaks can change.

Energy shifts are expected to occur whenever there is charge transfer from one atom to another. Thus, changes in composition of metallic components in metal alloys would not be expected to produce measurable changes in

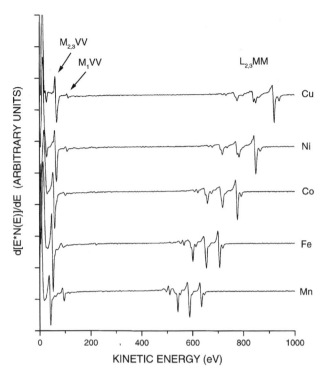

**Figure 5.** Auger spectra of Mn, Fe, Co, Ni, and Cu taken from argon-sputtered foils, shown in the derivative mode. Derivatives were taken from the direct spectra shown in Figure 4. Note how the spectral features are enhanced in the derivative mode, compared to the direct mode (Fig. 4). The spectra were obtained with a 40-nA, 5-keV electron beam, and each took <2 min to acquire. The spectra are offset for clarity.

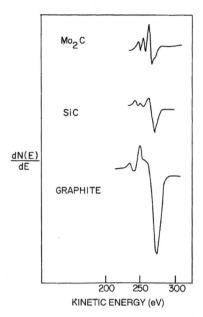

**Figure 6.** Carbon KVV Auger spectra from molybdeum carbide, silicon carbide, and graphite, showing the different Auger line shapes. Metal carbides usually exhibit the fairly symmetric, three-peak structure of molybdenum carbide, whereas surface contamination from organics usually resembles graphite.

the Auger energies (for core levels) of the components. However, submonolayer quantities of oxygen adsorbed on clean metal surfaces can produce measurable changes in the metal Auger peaks, the shift (from a few tenths of an eV to >1 eV) increasing with coverage (Haas and Grant, 1969). For bulk metal oxides, shifts in the metal Auger peaks are of the order of 10 eV. Changes in the line shape of Auger spectra can also occur due to changes in bonding, particularly when one or two valence electrons are involved in the relaxation process (Haas et al., 1972b). Changes in lineshape of transitions involving valence electrons are usually accompanied by significant shifts in Auger energies as well (Grant and Haas, 1970). Examples of changes in the carbon KVV Auger line shape for different chemical states of carbon are shown in Figure 6, and are due to differences in the density of states in the valence band.

## Sensitivity and Detectability

AES measures the composition of the several outermost atomic layers. The sensitivity of the technique should be limited by the shot noise in the electron beam used for Auger electron production. This can be checked by measuring the noise in the direct spectrum, which should follow the square root of the counts detected. Obviously, the longer the time allowed for taking data, the better the detection

limit. Also, the sensitivity factor (see Data Analysis) varies for different elements, and for different Auger transitions of the same element. Problems in detectability can also arise due to peak overlap, and this is discussed further below (see Data Analysis). Typical detection limits are in the range of 0.1 to 1 atom %. Remember that H and He are not detected.

## Auger Line Scans

With Auger line scans, Auger data are acquired as a function of position across a straight line on the specimen. The maximum length of the line depends on the acceptance area of the analyzer and the deflection capability of the electron gun. The specimen itself is imaged in the SEM mode, and the analysis line is then selected on this image. Also selected is a region (or regions) of the Auger energy spectrum from which to derive the signal that is monitored as the scan progresses.

An SEM image of a Cu grid over a Pd foil, together with a selected line scan position (the white horizontal line near the center), are shown in Figure 7. There is a small vertical displacement between the Cu grid and the Pd foil. The specimen was cleaned by argon-ion bombardment at several azimuths to eliminate shadowing. The Cu and Pd Auger line scans are shown in Figure 8. The Auger line scan signals in this case were the difference between the Auger spectral intensity at the Cu LMM and Pd MNN Auger peak maxima and the backgrounds at the high-kinetic-energy sides of each peak, in the direct spectra. The Cu line scan (solid line) is rather simple, with the signal intensity being high at the Cu grid, as expected. The Pd signal (dashed line) is more complex, being zero at the Cu grid (as

**Figure 7.** SEM image of a slightly elevated Cu grid over a Pd foil, together with a selected line scan position shown as the white horizontal line near the center. This image was obtained from secondary electrons, using a 5-keV electron beam in the Auger system.

expected), and at a maximum in the region between the Cu grid except for a region to the left of the grid where the signal decreases to near zero. This decrease in signal is due to shadowing of part of the Pd signal by the Cu grid (i.e., blocking Auger electrons from reaching the analyzer), as there is a small gap between the grid and the Pd foil. The specimen was mounted horizontally in the vacuum chamber; the incident electron beam was in the vertical plane that included the line scan, and at an angle of 45° to the horizontal from the left of the image. The analyzer was at an angle of 45° to the horizontal, toward the top of the image.

Several Auger survey spectra are shown in Figure 9 for different positions of the electron beam as it approached

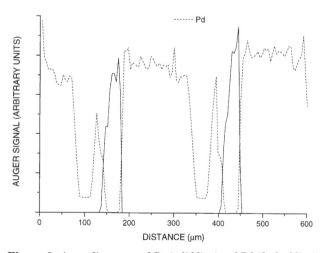

**Figure 8.** Auger line scans of Cu (solid line) and Pd (dashed line) taken where indicated by the horizontal white line in Figure 7. The Auger line scan signals were, respectively, the difference between the intensity at the Cu LMM or Pd MNN Auger peak maxima and the backgrounds at the high kinetic energy sides of each peak, in the direct spectra. Each scan took 6 min to acquire, using a 40-nA, 5-keV electron beam.

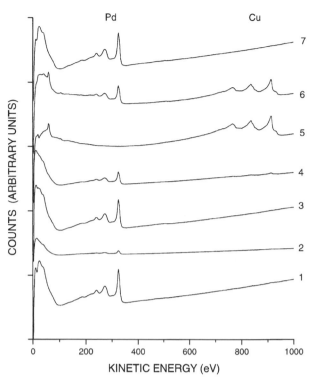

**Figure 9.** Seven Auger survey spectra in the direct mode, taken at different positions of the electron beam as it approached and crossed the Cu grid at the line scan position shown in Figure 7. The spectra 2 to 7 have been offset for clarity, but were not normalized. Spectra were acquired with a 40-nA, 5-keV electron beam, and each spectrum took ~1 min to acquire.

and crossed the elevated Cu grid. Spectrum number 1 was obtained when the electron beam was on the Pd foil, to the left of the grid as displayed in Figure 7. The Auger peaks between 150 and 350 eV are from Pd. Spectrum 2 was obtained when the electron beam was still on the Pd but underneath the Cu grid, where most of the Pd Auger electrons were blocked by the grid from reaching the analyzer. Spectrum 3 was obtained when the electron beam had passed beneath the elevated Cu grid, but the beam was still not on the grid. Spectrum 4 was obtained when the beam moved partly onto the side of the grid (but shadowing reduced the Cu Auger intensity). Spectrum 5 was obtained when the beam was actually on top of the Cu grid (maximum Cu intensity). Spectrum 6 when obtained when the beam had moved partly off the Cu grid, and spectrum 7 was obtained when the beam was on the Pd foil once more. From these spectra, it can be seen that the initial signal drop in the Pd Auger line scans occurs when the electron beam is not yet on the Cu grid, and is due to shadowing by the grid and not to surface contamination. Other problems in Auger line scans due to topography (not shadowing) and electron backscattering are discussed below (see Data Analysis).

## Auger Maps

With Auger maps, Auger data are acquired as a function of position within a defined area on the specimen. The

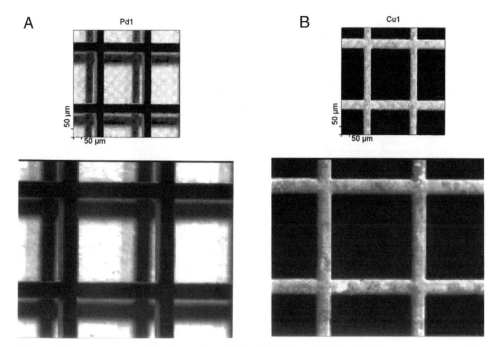

**Figure 10.** Auger maps of the Cu grid on Pd for, (**A**) Pd, and (**B**) Cu. Bright areas correspond to high signal intensity, and darker areas to lower intensity. These maps were taken with a 40-nA, 5-keV electron beam, with a 128×128 point spatial resolution. Each map took ∼25 min to acquire.

maximum dimensions of the area that can be mapped depend on the acceptance area of the analyzer and the deflection capability of the electron gun. The specimen itself is imaged in the SEM mode, and the analysis area is then selected on this image.

Pd and Cu Auger maps of the Cu grid on Pd foil are shown in Figure 10, where the bright areas correspond to higher Auger intensity. The Cu Auger map on the right is rather simple, with the grid being displayed quite clearly. The Pd map is more complex as shadowing also comes into play (see discussion of Auger Line Scans). The region of the Cu grid itself is black, but there is another image of the grid in gray due to the shadowing of this part of the Pd foil by the Cu grid, resulting in a reduction of the Pd Auger signal from this part of the foil.

Auger maps can be displayed in different ways, such as a selected number of gray levels, on a thermal scale, in pseudocolor, or in a selected color for comparing with a map of a different element in the same map area. Auger line scans for each element mapped can also be displayed across a specified line on the appropriate map. As in the case of Auger line scans, corrections of Auger intensity due to topography and electron backscattering are often required (see Data Analysis).

### Depth Profiling

In AES, depth profiling is usually accomplished by inert gas ion bombardment to remove successive layers of material from the specimen. The removal of atoms by ion bombardment is also referred to as sputtering. Auger measurements are made either continuously while simultaneously sputtering, or sequentially with alternating cycles of sputtering and analysis. Ar or Xe ions at an energy of a few kilo-electron volts are usually used for sputtering. Typical sputter rates are of the order of 10 nm/min. It is important that the ion beam is correctly aligned to the analysis area on the specimen, and the ion beam is often rastered over a small area of the specimen to ensure that the electron beam used for Auger analysis strikes the flat-bottomed crater formed by the rastering.

Auger survey scans could be taken in depth profiling, but a technique called multiplexing is normally used instead. With multiplexing, several energy spectral windows are preselected, with each one encompassing an Auger peak of the element of interest (Sickafus and Colvin, 1970). This procedure saves time, as only signals from energy regions of interest are measured. This will also provide a higher density of data points in the depth profile, but means that other elements that might be present below the original surface could go undetected.

An Auger sputter depth profile of $SiO_2$ on Si is shown in Figure 11. The Auger transitions selected were O KLL, Si LMM, and Si KLL. Depth profiles can be plotted with peak height intensities in direct spectra, or with peak-to-peak heights in derivative spectra. The depth profile shown here was obtained using the derivative spectra. The ion beam was rastered over a 2-mm × 2-mm area on the specimen. Auger measurements were made sequentially with sputtering, and the abscissa is plotted as sputter time. Two scans of each Auger transition were obtained before sputtering (one of these is plotted as a negative time). Different compounds sputter at different rates, and it is often not a trivial matter to convert sputter time to depth. Note that near the interface the O signal decreases towards zero, and the Si LMM signal increases

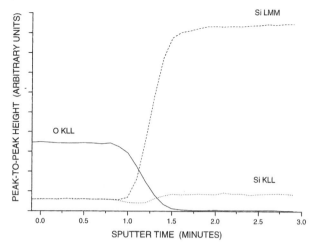

**Figure 11.** An Auger sputter depth profile of $SiO_2$ on Si. The Auger transitions selected were O KLL (solid line), Si LMM (dashed line), and Si KLL (dotted line). The profile was obtained by sequentially taking Auger spectra and sputtering. Two scans of each Auger transition were obtained before sputtering (one of these is plotted as a negative time). The sample was sputtered for 0.1 min between measurements. The Auger spectra were taken with a 40-nA, 5-keV electron beam. A 3-keV $Ar^+$ ion beam, rastered over an area 2 mm×2 mm, was used for sputtering.

**Figure 13.** Auger line scans for the O KLL (solid line) and Si LMM (dashed line) Auger transitions taken across the crater edge shown in Figure 12b, using Auger peak heights above background in the direct spectra. The line scans were taken with a 40-nA, 5-keV electron beam, and each scan took ~1 min to acquire.

significantly. The Si KLL signal undergoes a different behavior with an apparent dip at the interface and a small increase in intensity in the pure Si substrate. For discussion of these signal variations, see Data Analysis.

Depth profiles can also be obtained by conducting an Auger line scan across the "crater wall" formed by sputtering (Taylor et al., 1976). An SEM image of the sputtered area for the $SiO_2$ layer on Si is shown in Figure 12A, with the left-hand crater wall shown expanded in Figure 12B. The Auger line scan was obtained across 400 μm of the left-hand crater wall, as shown in Figure 12B. The Auger line scans for the O KLL and Si LMM Auger transitions are shown in Figure 13 using peak height above background in the direct spectra. Line

scans across crater walls are also useful if there was a problem with the alignment of the electron and ion beams, since the Auger line scans are performed on the crater wall after sputtering is finished. In the case of very thin films (up to ~10 nm thickness) where Auger peaks from the substrate can also be detected without sputtering, depth information can be obtained by angle-resolved measurements (the specimen is tilted in relation to the analyzer to change the angle of emission). Other variations on depth profiling include mechanical methods for specimen preparation such as angle lapping (Tarng and Fisher, 1978), where a flat sample is polished at a small angle (~0.5°) to the original surface to expose the depth distribution of elements, and ball cratering (Walls et al., 1979), where a spherical ball grinds a crater into the sample to expose the depth distribution. These two techniques are

**Figure 12.** An SEM image of the sputtered area for the $SiO_2$ layer on Si, following the acquisition of the depth profile, is shown in (**A**). The left-hand crater edge is shown expanded in (**B**), and the horizontal white line in (**B**) is the region that was used for crater edge profiling. The SEM images were taken with (**A**) 2-keV and (**B**) 5-keV electron beams.

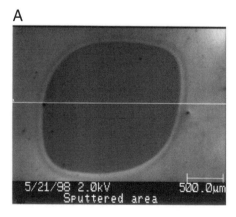

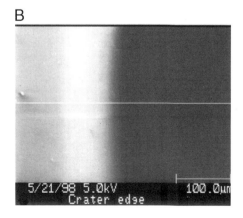

typically used where analysis is required to depths >1 μm. The depth resolution in sputter depth profiling is typically several percent of the depth sputtered, but the depth resolution can be improved by rotating the specimen during sputtering (Zalar, 1985).

## METHOD AUTOMATION

All modern AES systems operate under computer control, and many repetitive functions can be carried out by the computer.

Specimen introduction is semiautomatic: the specimen insertion chamber can be evacuated automatically, the gate valve between the insertion chamber and the analysis chamber can be pneumatically opened and closed, and the specimen manipulator in the analysis chamber can be parked in the transfer position by computer control. Some AES systems also allow transferability of specimen location information, on a platen obtained prior to insertion, to the specimen manipulator in the analysis chamber.

The electron gun used for producing Auger electrons in the specimen is also controlled by the computer. As electron beam energy is changed, the lens parameters required for focusing, beam alignment in the column, astigmatism correction, etc., are all set to predetermined values. The beam current can also be set to preselected values. Different regions of interest on the specimen surface can also be selected and stored in the computer, so these different regions can be examined in turn automatically during an analysis, e.g., survey spectra could be obtained from the different regions, or individual depth profiles could be obtained from each region during one sputtering operation.

The electron energy analyzer also operates under computer control, although any physical apertures in the analyzer are usually selected manually. The energy resolution of the analyzer is usually controlled by computer, as is the collection and display of the output signal. The kinetic energy range for acquiring data, the kinetic energy step interval, the dwell time at each energy step for acquiring data, and the number of sweeps, among other settings, are all controllable with the computer. The computer also controls the acquisition of Auger line scans and Auger maps.

The computer is also used for data processing. This ranges from simple procedures such as differentiation of spectra to rather complex procedures such as linear least-squares fitting and factor analysis of data. This will be discussed further in the Data Analysis and Initial Interpretation section.

## DATA ANALYSIS AND INITIAL INTERPRETATION

Qualitative analysis was discussed earlier (see Practical Aspects of the Method); there are, however, some very useful data analysis methods that need to be mentioned in this section, namely, spectrum subtraction, linear least-squares fitting, and factor analysis. Also, these techniques have all been used to enhance the analysis and interpretation of spectra and to improve detectability.

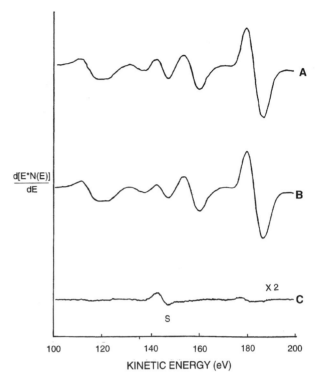

**Figure 14.** An example of spectrum subtraction: (**A**) the derivative Auger spectrum from a Mo surface, (**B**) the spectrum from pure Mo, and (**C**) the result after subtracting (**B**) from (**A**) to minimize the Mo peak near 180 eV. Note the recovery of an S Auger signal after subtraction.

Spectrum subtraction is a rather simple procedure whereby a spectrum from one element is subtracted from that of another to eliminate overlap problems (Grant et al., 1975a). In this procedure, the spectra must be aligned to allow for any shifts due to chemical effects or specimen charging. If the line shapes are different due to chemical effects, the subtraction will not completely remove the features that are due to the subtracted element. An example of spectrum subtraction is shown in Figure 14, where the S Auger signal (C) is recovered from an Auger spectrum of a Mo surface (A) by subtracting the spectrum from a clean Mo surface (B). The S is then more easily quantified as well.

The different behavior of the Si LMM and Si KLL intensities in the depth profile of $SiO_2$ on Si (Fig. 11) is due to chemical effects that occur in Si Auger peaks in $SiO_2$ when compared with pure Si. This depth profile was obtained using derivative Auger data, and the actual Si LMM and KLL spectra used are shown in Figure 15, panels A and B, respectively. Note that in each case the spectra cluster around two different line shapes, which are due to the different Si chemistries in Si and $SiO_2$. The LMM spectrum from $SiO_2$ has its main structure ~65 to 75 eV, whereas that from Si is ~80 to 100 eV. Also, the peak-to-peak heights of these LMM transitions increase quite markedly in going from the oxide to the Si substrate. Both of these large differences are due to

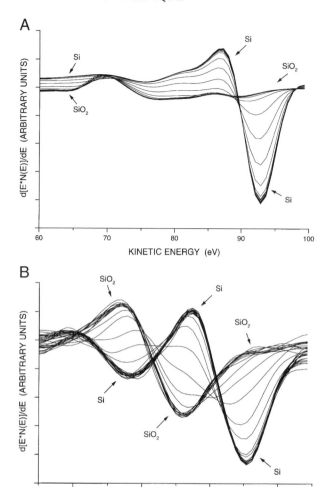

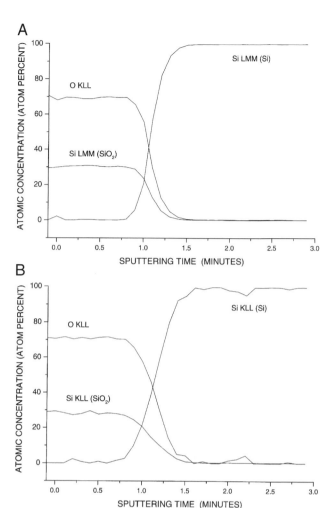

**Figure 15.** The Si (**A**) LMM, and (**B**) KLL derivative Auger spectra obtained at different depths during the depth profile of $SiO_2$ on Si shown in Figure 11. Note how in each case the spectra cluster around two different line shapes, corresponding to the Auger transitions in $SiO_2$ and Si. The few intermediate spectra are from the region near the $SiO_2$/Si interface.

**Figure 16.** The sputter depth profiles of $SiO_2$ on Si obtained after applying linear least-squares processing to (**A**) the derivative Si LMM Auger spectra, and (**B**) the derivative Si KLL spectra to separate the Si signals from the oxide and the substrate. The concentrations were calculated using the manufacturer-supplied elemental sensitivity factors for Si and O.

involvement of valence electrons in the LMM transitions. For the KLL transitions, a well defined chemical shift occurs between the oxide and the substrate (the oxide being at lower kinetic energy), and the main peak intensity is more symmetrical in the oxide. Also present in these spectra are a few spectra from the interfacial region, but it is not obvious whether these spectra are linear combinations of the spectra from the oxide and from pure Si, or are line shapes corresponding to a different chemical state of Si. However, it can be seen that the reason for the dip in the Si KLL profile at the interface is a reduction in the peak-to-peak height of the Si KLL signal near the interface since both KLL peaks will be present side-by-side in this region, and not necessarily a decrease in Si concentration. Numerical methods such as linear least-squares fitting (Stickle and Watson, 1992) and factor analysis (Gaarenstroom, 1981) can be used to separate overlapping spectra, or the different chemical components in spectra such as in

the $SiO_2$ on Si sputter depth profile in Figure 11. These processing programs are incorporated into numerical software packages, and have been included in some surface analysis processing software. Examples of use of the linear least-squares method to separate the two different chemical states of Si in the sputter depth profile of $SiO_2$ on Si are shown in Figure 16A for the derivative Si LMM and O KLL peaks, and in Figure 16B using the derivative Si KLL and O KLL peaks. Note in both examples how the Si Auger signal has been separated into the two different chemical states of Si, namely $SiO_2$ and pure Si, and that there is no unusual decrease in total silicon concentration at the interface, as indicated from the raw KLL data in Figure 11. The atomic concentrations were calculated using the manufacturer-supplied elemental sensitivity factors for Si and O (see discussion of Quantitative Analysis). Similar results were obtained by selecting two-component fits in factor analysis.

## Quantitative Analysis

The most commonly used method for quantitative analysis of Auger spectra is that of sensitivity factors. With sensitivity factors, the atomic fractional concentration of element i, $C_i$, is often calculated from

$$C_i = \frac{I_i/I_i^\infty}{\sum_j I_j/I_j^\infty} \qquad (2)$$

where $I_i$ is the measured Auger intensity of element i in the unknown, $I_i^\infty$ is the corresponding intensity from the pure element i, and the summation is over one term for each element in the unknown (remember that H and He are not detected, and are not included in this calculation). The Auger intensities should all be measured under the same experimental conditions to achieve reliable results. Sometimes, sensitivity factors are measured in the same instrument relative to a particular element such as silver, and the method is then referred to as using relative sensitivity factors. Unfortunately, most people do not measure sensitivity factors in their own instruments but rely on published values, which can differ by more than a factor of 2. Sensitivity factors between elements can vary by as much as a factor of 100 across the periodic table (Sekine et al., 1982; Childs et al., 1995).

Further improvements to Equation 2 can also be made by including corrections for atom size, electron inelastic mean free path, and electron backscatter (Seah, 1983). The atom size adjusts for density, the inelastic mean free path adjusts for the analysis depth, and backscattering adjusts for the increase in Auger intensity that occurs when energetic electrons are scattered back toward the surface and have another chance to produce Auger electrons in the atoms that are closer to the surface. In a homogeneous binary alloy of elements A and B, the atomic concentration ratio of A ($C_A$) to B ($C_B$) is then given by

$$\frac{C_A}{C_B} = F_{AB} \frac{I_A/I_A^\infty}{I_B/I_B^\infty} \qquad (3)$$

where

$$F_{AB} = \left(\frac{1 + r_A(E_A)}{1 + r_B(E_A)}\right)\left(\frac{a_B}{a_A}\right)^{3/2} \qquad (4)$$

and $F_{AB}$ is nearly constant over the entire AB compositional range. $r_i(E_i)$ is the backscattering factor for element i for an electron with kinetic energy $E_i$, and $a_i$ is the size of atoms in element i (Seah, 1983). These corrections are straightforward and can improve quantitative analysis, but unfortunately are rarely made.

Auger currents cannot be measured directly as they are superimposed on background currents due to secondary and backscattered electrons. Also, there is current on the low kinetic energy side of Auger peaks due to inelastic losses suffered by Auger electrons in escaping from the specimen being analyzed, and this background must be removed in the measurement. Some work has been conducted to measure the areas under Auger peaks, but this is not done routinely in quantitative analysis due to the

difficulty in removing the background. Peak height measurements (direct spectra) or peak-to-peak height measurements (derivative spectra) are normally used for quantitative analysis, and each method has its own sets of sensitivity factors.

Peak height measurements are often used for Auger line scans and Auger maps, as data acquisition times can be minimized by collecting data at just two or three energy values for each element. The sloping background above the peaks should be extrapolated to the energy of the peak maximum to measure peak height, although this is not always done. This can be done using a simple algorithm if measurements are made at three energies, namely at the peak maximum, at the background just above the peak, and at a higher background energy equally spaced from the central energy. Problems in quantitative analysis for Auger line scans or maps, due to specimen topography and changes in electron backscattering coefficient, can be satisfactorily corrected in peak height data by dividing the Auger intensity by the background on the high-energy side of the peak (Prutton et al., 1983).

Peak-to-peak height measurements from derivative spectra are the most commonly used method for quantitative analysis in AES, and this is the method normally used for survey scans and for depth profiles. The peak-to-peak height method has met with a lot of success, particularly in studying metals and their alloys. The accuracy of such measurements is typically ±20%. More accurate quantitative analysis can be achieved if standards with concentrations that are close to those of the actual specimens are used, thereby reducing errors due to variations in electron backscatter and inelastic mean free path. Large errors can occur if the standards are not measured using the same analysis equipment as for the unknown. For example, different instruments from the same manufacturer have shown a fivefold variation in the intensity ratios for high- and low-energy Auger peaks (Baer and Thomas, 1986).

Problems also arise in quantitative analysis, particularly when using peak-to-peak heights, if Auger lineshape changes occur due to a change in chemistry. For example, for oxygen adsorption on nickel, the concentration of oxygen can be in error by a factor of two if peak-to-peak height measurements are used, even though the oxygen Auger line shape changes are subtle (Hooker et al., 1976).

Improvements in quantitative analysis can be made by measuring the intensity/energy dependency of each instrument for Cu, Ag, and Au for the set of conditions normally used for analysis (e.g., different analyzer energy resolutions), and comparing the results with standard reference spectra (see NPL under Standards Sources section). This will provide the intensity response function for each experimental condition, allow accurate transferability of results between instruments, and allow the performance of a particular instrument to be monitored over time.

### Standards Sources

There are several standards for AES that have been developed by national standards organizations. Standards for depth profiling are available from the United States National Institute of Standards and Technology (NIST), and from the National Physical Laboratory (NPL) in the United Kingdom. Software for calibration of Auger electron spectrometers is available from NPL. Standard terminology, practices, and guides have been developed by the American Society for Testing and Materials (ASTM). Details are listed in the Appendix.

## SAMPLE PREPARATION

Since AES is a surface-sensitive technique, the handling, preparation, mounting, and analysis of a sample is of paramount importance. For example, a fingerprint on a specimen can be sufficient contamination to mask any signal from the actual surface of interest. If the specimen has been contaminated in some way, cleaning prior to mounting might be required. Cleaning with acetone, followed by rinsing in methanol, is often used to remove soluble contaminants. It must be remembered that any solvent used for cleaning might also leave a contaminant residue. Loose contaminating particles can be removed using a jet of nitrogen gas. If the contaminant is a thin layer, it might be possible to remove the contamination by low-energy inert gas ion sputtering in the analysis chamber. Sputtering can also cause its own problems, as elements sputter at different rates and their relative concentrations at the surface can therefore change. Sample handling is described in detail in ASTM Guides E1078 and E1829 (see the Appendix).

Size limitations on specimens depend on the particular analysis chamber. Specimen diameters up to several centimeters can usually be accommodated, but height limitations are often ∼1 cm when vacuum-interlocked sample insertion systems are used. Some AES systems are designed to handle 300-mm wafers. If larger samples are to be routinely studied, customized insertion systems can be designed. Samples might need to be cut from a larger piece or sectioned for analysis. If this is done, care should be exercised to minimize possible contamination. If information is needed to a depth greater than ∼1 μm, samples might be sectioned or ball cratered (see discussion of Depth Profiling).

Powder-free gloves should always be used when mounting specimens, and the tools used should be clean and nonmagnetic. Samples are usually mounted on Al or stainless steel holders, and are held in place with small screws or clips. Sometimes, samples are mounted with colloidal Ag or colloidal graphite. Powders are often mounted on conducting adhesive tape.

Samples prone to outgassing can be prepumped in an auxiliary vacuum system, or in the introduction chamber, to remove volatile species. However, cross-contamination between samples might occur.

When in the analysis chamber, specimens can sometimes be prepared by fracture, cleavage, or scribing. In some cases, it may be possible to remove contaminants by heating if subsequent studies on clean surfaces are required, such as on how specific gases react with a clean surface.

Once in the analysis chamber, specimens can be moved by a manipulator and imaged with a microscope or video camera (initially) and then with the electron beam (as in an SEM), to locate the regions to be analyzed.

If samples are electrically insulating, they are sometimes mounted beneath a metal mask with a small opening or beneath a metal grid to minimize charging by placing the incident electron beam close to the metal. Tilting the sample to change the angle of incidence of the electron beam can help control charging. Changing the beam energy, reducing the beam current density, and using a low-energy electron flood gun are also effective in controlling charging.

## SPECIMEN MODIFICATION

Since an energetic electron beam is used for Auger analysis, and since it is a surface-analysis technique, specimens can sometimes be modified by the beam. A manifestation of this would be that the surface composition varies with time of exposure to the beam (Pantano and Madey, 1981).

Specimen modification can occur from heating by the beam, where species might diffuse to the surface from the bulk, diffuse across the surface, or desorb from the surface. The electron beam itself can also cause desorption (e.g., fluorine desorption), it can assist the adsorption of molecules from the vacuum environment onto the surface,

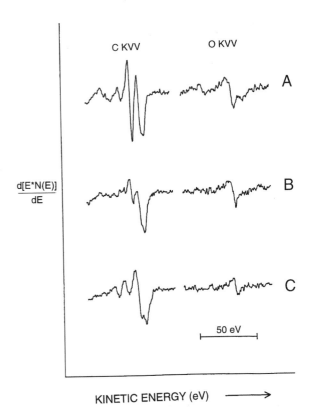

**Figure 17.** C and O KVV derivative Auger spectra from (**A**) CO molecularly adsorbed on Ni, (**B**) after exposure to a 1.5-keV electron beam with a current density of ∼1 μA/mm² for 10 min, and (**C**) after exposure to the beam for 40 min.

and it can decompose species at the surface. Electron beam modification can be reduced by decreasing the beam energy, by decreasing the current density at the surface by defocusing the beam, or by rastering the beam over a certain area. If the specimen is laterally homogeneous, electron beam rastering can always be used. An example of electron beam decomposition is shown in Figure 17, where it can be seen that a low-energy beam (1 keV) with a current density of ~1 μA/mm$^2$ produces marked changes, in a few minutes, in the C KVV and O KVV Auger spectra for CO molecularly adsorbed on Ni. In fact, some C and O is desorbed from the surface, and the remaining species are converted into nickel carbide and oxide (Hooker and Grant, 1976).

## PROBLEMS

Problems in conducting an Auger analysis can occur from many sources. These include specimen modification from the electron beam and electrical charging of insulators. Problems can also be encountered in data reduction, such as peak overlap, misidentification of Auger peaks, and the possible presence of ionization loss peaks, plasmon loss peaks, and ion-excited Auger peaks in the spectra. Since ionization loss peaks, plasmon loss peaks, and ion-excited Auger peaks have not been discussed earlier, their effects on the interpretation of spectra will be discussed here.

### Ionization Loss Peaks

When an electron is ejected from an atomic shell (having a binding energy $E_b$) by an electron from the incident beam (having a kinetic energy $E_p$), the energy remaining will be $(E_p - E_b)$. This remaining energy is shared between the ejected electron and the incident electron, and these electrons will also appear in the measured energy distribution, $N(E)$. This means that there will be a background of electrons in $N(E)$ from this process up to a kinetic energy of $(E_p - E_b)$, resulting in a small reduction in the intensity of $N(E)$ above this energy. This small step in $N(E)$ is sometimes observed in derivative spectra, and is called an ionization loss peak (Bishop and Riviere, 1970). Its source can be readily confirmed by changing the incident beam energy, as the feature will move in the $N(E)$ distribution by the change in beam energy (remember that Auger peaks do not change their kinetic energy when the electron beam energy is changed). These ionization loss peaks can be useful in studying surfaces, for example, where overlap of Auger peaks might make detection difficult. They can also be used to study changes in surface chemistry, as any shift in binding energy of the energy level in the atom will be reflected as a change in kinetic energy of the ionization loss peak. An example of ionization loss peaks is shown in Figure 18, for the Si $L_1$ and $L_{2,3}$ levels in SiO$_2$. In the case of conductors, ionization loss peaks might overlap plasmon loss peaks (see Plasmon Loss Peaks section), for levels with low binding energy.

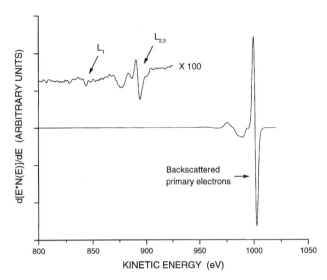

**Figure 18.** Ionization loss peaks in the derivative mode for the $L_1$ and $L_{2,3}$ levels in SiO$_2$. The spectrum was taken with a 100-nA, 1000-eV electron beam in <10 min. The symmetrical peak at 1000 eV is from backscattered primary electrons.

### Plasmon Loss Peaks

An electron traveling through a conducting material can interact with the free electrons in the material and lose a specific amount of energy, resulting in an additional peak separated in energy by what is called the plasmon energy (Klemperer and Shepherd, 1963). Several orders of bulk plasmons can occur, and they will appear as a series of equally separated peaks of decreasing intensity on the low-kinetic-energy side of the Auger peak. Besides these bulk plasmons, the presence of the surface results in additional plasmons, called surface plasmons, that are associated with the back-scattered primary electrons and with each bulk plasmon peak. Higher-order surface plasmons are very weak and not observed directly in the loss of spectrum (Pardee et al., 1975). An example of plasmon peaks is shown in Figure 19A for 1-keV backscattered primary electrons from an Al surface, where the bulk (B) and surface (S) plasmons have energies of 15.3 and 10.3 eV, respectively. This backscattered spectrum was taken under high energy resolution, and typical Auger peaks and their corresponding plasmon loss peaks will be broader, as can be seen for the Al KLL Auger spectrum in Figure 19B. The Al KL$_1$L$_{2,3}$ Auger electrons near 1336 eV have their own plasmons, and the KL$_1$L$_{2,3}$ Auger peak also overlaps with plasmon peaks from Al KL$_{2,3}$L$_{2,3}$ Auger electrons, which have a peak near 1386 eV. Note that the plasmon peaks are equally spaced and decrease in intensity as they move away from the Auger peaks. Aluminum oxide would not have any plasmon peaks. The presence of plasmons will change the shape of an Auger peak, and take electrons away from the Auger peak, making quantitative analysis somewhat more difficult.

### Ion-Excited Auger Peaks

Core electron levels for some elements can be ionized by low-energy (1 to 5 keV) inert gas ion bombardment,

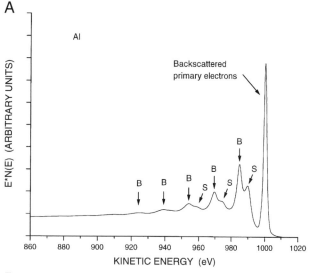

A

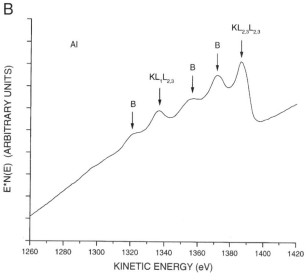

B

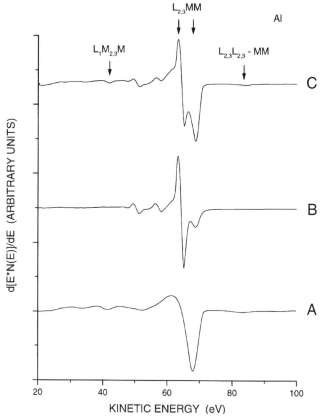

**Figure 20.** Derivative LMM Auger spectra, (**A**) excited by a 5-keV electron beam, (**B**) excited by a 3-keV Ar$^+$ ion beam, and (**C**) excited by both electron and ion beams together, as might occur in a sputter depth profile through Al.

**Figure 19.** Plasmon peaks (bulk B, surface S) associated with (**A**) a backscattered primary electron beam at 1000 eV from Al, and (**B**) the Al KL$_1$L$_{2,3}$ (near 1337 eV) and KL$_{2,3}$L$_{2,3}$ (near 1386-eV) Auger peaks from Al obtained with a 5-keV electron beam. For Al, the bulk and surface plasmons have energies of 15.3 and 10.3 eV, respectively. The surface plasmons in (**B**) are obscured by the broad bulk plasmon peaks (reflecting the broad Auger peaks).

resulting in the emission of Auger electrons from these atoms (Haas et al., 1974). Only those energy levels in an atom of the sample that can be promoted to a much lower binding energy by the incident ion during its interaction with the atom can be ionized (the relatively heavy ion cannot transfer enough energy to an electron to ionize core shell electrons). The atom is ionized in a core shell if an electron is transferred to the incident ion from a promoted core shell energy level during impact, and if it remains transferred after the collision. The ionized level then returns to its higher binding energy, and Auger electron emission can then occur. Typical levels that can be ionized by this method are the L$_{2,3}$ levels of Mg, Al, and Si, and additional Auger intensity might occur on depth profiling

through layers containing these elements. The L$_1$ level of Mg, Al, and Si cannot be ionized, as this level is not promoted during collision with the incoming ion. During ion bombardment, Auger emission can occur from sputtered specimen ions as well as from the specimen. The spectrum produced by the ion beam will be different from that produced by an electron beam in two ways (1) the spectrum produced by ion impact will not contain all the transitions produced by electron impact, since not all levels can be promoted (e.g., the L$_1$ levels in these atoms), and (2) sharp, atomic-like Auger transitions will occur from sputtered ions as well (Grant et al., 1975b). A comparison of electron- and ion-excited Auger emission from Al is shown in Figure 20. The spectrum produced by electron excitation, Figure 20A, shows the main L$_{2,3}$MM transition at 68 eV, and the L$_1$L$_{2,3}$M and L$_{2,3}$L$_{2,3}$-MM transitions at 41 and 83 eV, respectively. The spectrum produced by Ar$^+$ excitation, Figure 11d.20B, shows the L$_{2,3}$MM transition at 68 eV from the bulk Al, together with some overlapping symmetric Auger peaks from sputtered Al ions, but the L$_1$L$_{2,3}$M and L$_{2,3}$L$_{2,3}$-MM transitions are not seen. The L$_1$ level is not promoted on impact with an Ar$^+$ ion, so the L$_1$L$_{2,3}$M transition cannot occur. Also, double initial ionization of the L$_{2,3}$ level does not occur on ion impact.

Composite spectra, such as the one shown in Figure 20C, can be obtained during simultaneous electron and ion bombardment of Al. Note that the symmetric, ion-excited peaks from sputtered ions are sharper than the electron-excited peaks, and will therefore appear relatively larger in derivative spectra than in direct spectra.

## LITERATURE CITED

Baer, D. R. and Thomas, M. T. 1986. A technique for comparing Auger electron spectroscopy signals from different spectrometers using common materials. *J. Vac. Sci. Technol. A* 4:1545–1550.

Bishop, H. E. and Riviere, J. C. 1970. Characteristic ionization losses observed in Auger electron spectroscopy. *Appl. Phys. Lett.* 16:21–23.

Childs, K. D., Carlson, B. A., LaVanier, L. A., Moulder, J. F., Paul, D. F., Stickle, W. F., and Watson, D. G. 1995. Handbook of Auger electron spectroscopy. Physical Electronics, Inc., Eden Prairie, Minn.

Gaarenstroom, S. W. 1981. Principal component analysis of Auger line shapes at solid-solid interfaces. *Appl. Surf. Sci.* 7:7–18.

Goto, K. 1995. Historical Auger electron spectroscopy. I. Works of P. Auger and other pioneers. *J. Surf. Anal.* 1:328–341.

Grant, J. T. and Haas, T. W. 1970. Auger electron spectroscopy of Si. *Surf. Sci.* 23:347–362.

Grant, J. T., Hooker, M. P., and Haas, T. W. 1975a. Spectrum subtraction techniques in Auger electron spectroscopy. *Surf. Sci.* 51:318–322.

Grant, J. T., Hooker, M. P., Springer, R. W., and Haas, T. W. 1975b. Comparison of Auger spectra of Mg, Al, and Si excited by low-energy electron and low-energy argon-ion bombardment. *J. Vac Sci. Technol.* 12:481–484.

Haas, T. W. and Grant, J. T. 1969. Chemical shifts in Auger electron spectroscopy from the initial oxidation of Ta(110). *Phys. Lett.* 30A:272.

Haas, T. W. and Grant, J. T. 1970. Chemical effects on the KLL Auger electron spectrum from surface carbon. *Appl. Phys. Lett.* 16:172–173.

Haas, T. W., Grant, J. T., and Dooley, G. J. 1972a. On the identification of adsorbed species with Auger electron spectroscopy. *In* Adsorption-Desorption Phenomena (F. Ricca, ed.) pp. 359–368. Academic Press, New York.

Haas, T. W., Grant, J. T., and Dooley, G. J. 1972b. Chemical effects in Auger electron spectroscopy. *J. Appl. Phys.* 43:1853–1860.

Haas, T. W., Springer, R. W., Hooker, M. P., and Grant, J. T. 1974. Ion excited Auger spectra of aluminum. *Phys. Lett.* 47A:317–318.

Hooker, M. P. and Grant, J. T. 1976. Auger electron spectroscopy studies of CO on Ni(110)–spectral line shapes and quantitative aspects. *Surf. Sci.* 55:741–746.

Hooker, M. P., Grant, J. T., and Haas, T. W. 1976. Some aspects of an AES and XPS study of the adsorption of $O_2$ on Ni. *J. Vac. Sci. Technol.* 13:296–300.

Janssen, A. P., Schoonmaker, R. C., Chambers, A., and Prutton, M. 1974. Low energy Auger and loss electron spectra from magnesium and its oxide. *Surf. Sci.* 45:45–60.

Klemperer, O. and Shepherd, J. P. G. 1963. Characteristic energy losses of electrons in solids. *Adv. Phys.* 12:355–390.

Krause, M. O. 1979. Atomic radiative and radiationless yields for K and L shells. *J. Phys. Chem. Ref. Data* 8:307–327.

Pantano, C. G. and Madey, T. E. 1981. Electron beam effects in Auger electron spectroscopy. *Appl. Surf. Sci.* 7:115–141.

Pardee, W. J., Mahan, G. D., Eastman, D. E., Pollak, R. A., Ley, L., McFeely, F. R., Kowalczyk, S. P., and Shirley, D. A. 1975. Analysis of surface- and bulk-plasmon contributions to x-ray photoemission spectra. *Phys. Rev. B* 11:3614–3616.

Prutton, M., Larson, L. A., and Poppa, H. 1983. Techniques for the correction of topographical effects in scanning Auger microscopy. *J. Appl. Phys.* 54:374–381.

Seah, M. P. 1983. Quantification of AES and XPS. *In* Practical Surface Analysis (D. Briggs and M. P. Seah, eds.)., Chapter 5. John Wiley & Sons, Chichester, U.K.

Seah, M. P. and Dench, W. A. 1979. Quantitative electron spectroscopy of surfaces: A standard data base for electron inelastic mean free paths in solids. *Surf. Interface Anal.* 1:2–11.

Seah, M. P., Smith, G. C., and Anthony, M. T. 1990. AES: Energy calibration of electron spectrometers. I. An absolute, traceable energy calibration and the provision of atomic reference line energies. *Surf. Interface Anal.* 15:293–308.

Sekine, T., Nagasawa, Y., Kudoh, M., Sakai, Y., Parkes, A. S., Geller, J. D., Mogami, A., and Hirata, K. 1982. Handbook of Auger Electron Spectroscopy. JEOL Ltd., Tokyo.

Sickafus, E. N. and Colvin, A. D. 1970. A mutichannel monitor for repetitive Auger electron spectroscopy with application to surface composition changes. *Rev. Sci. Instrum.* 41:1349–1354.

Smith, G. C. and Seah, M. P. 1990. Standard reference spectra for XPS and AES: Their derivation, validation and use. *Surf. Interface Anal.* 16:144–148.

Stickle, W. F. and Watson, D. G. 1992. Improving the interpretation of x-ray photoelectron and Auger electron spectra using numerical methods. *J. Vac. Sci. Technol. A* 10:2806–2809.

Tarng, M. L. and Fisher, D. G. 1978. Auger depth profiling of thick insulating films by angle lapping. *J. Vac. Sci. Technol.* 15:50–53.

Taylor, N. J., Johannessen, J. S., and Spicer, W. E. 1976. Crater-edge profiling in interface analysis employing ion beam etching and AES. *Appl. Phys. Lett.* 29:497–499.

Walls, J. M., Hall, D. D., and Sykes, D. E. 1979. Composition-depth profiling and interface analysis of surface coatings using ball cratering and the scanning Auger microprobe. *Surf. Interface Anal.* 1:204–210.

Zalar, A. 1985. Improved depth resolution by sample rotation during Auger electron spectroscopy depth profiling. *Thin Solid Films* 124:223–230.

## KEY REFERENCES

Annual Book of ASTM Standards, Vol. 03.06. American Society for Testing and Materials, West Conshohocken, Pa.

*This volume includes the ASTM Standards that have been developed for surface analysis.*

Briggs, D. and Seah, M. P. (eds.). 1990. Practical Surface Analysis, 2nd ed., Vol. 1. Auger and X-ray Photoelectron Spectroscopy. John Wiley & Sons, Chichester, U.K.

*This is perhaps the most popular book on Auger electron spectroscopy, and is available in paperback.*

Ferguson, I. 1989. Auger Microprobe Analysis. Adam Hilger, Bristol, U.K.

*This book provides an overall view of Auger electron spectroscopy, with many historical references. It also has a lot of information on the scanning aspects of the measurement.*

## INTERNET RESOURCES

See the Appendix for additional internet resources.

http://sekimori.nrim.go.jp/

*Web site of the Surface Analysis Society of Japan. This site has access to a Common Data Processing System that is being developed, and to Auger spectra from many of the elements in the periodic table. The spectra can be viewed on line, and members of the Society can download files. Membership information is available online.*

http://www.vacuum.org/

*The American Vacuum Society is the main professional society for those working in Auger electron spectroscopy in the United States. This page has links to its scientific journals, scientific meetings, a buyers guide, and to the International Union for Vacuum Science, Technique, and Applications.*

http://www.uwo.ca/ssw/

*Surface Science Western home page. A surface science mailing list for users in surface science (mainly AES and XPS) has been set up by Surface Science Western at the University of Western Ontario, and information about the mailing list can be accessed from this page. An archive of messages is also kept.*

## APPENDIX:
## SOURCES OF STANDARDS FOR AUGER ELECTRON SPECTROSCOPY

### National Institute of Standards and Technology (NIST)

Standard Reference Materials Program, Building 202, Room 204, Gaithersburg, Md., 20899. Telephone: (301) 975-6776; Fax: (301) 948-3730; e-mail: *srminfo@nist.gov*; WWW: *http://ts.nist.gov/ts/htdocs/230/232/232.htm*.

The following two standards from NIST are for calibrating equipment used to measure sputtered depth and erosion rates in surface analysis.

*SRM 2135c:* Nickel/chromium thin-film depth profile standard.

*SRM 2136:* Chromium/chromium oxide thin-film depth profile standard.

### National Physical Laboratory (NPL)

Surfaces and Interfaces Section, CMMT, National Physical Laboratory, Queens Road, Teddington, Middlesex TW11 0LW, U.K. Telephone: +44 181 943 6620; Fax:

+44 181 943 6453; e-mail: *sjs@npl.co.uk*; WWW: *http://www.npl.co.uk/npl/cmmt/sis/refmat.html*.

*CRM261:* Tantalum pentoxide depth profile reference materials, consisting of four 30-nm-thickness samples and four 100-nm-thickness samples of $Ta_2O_5$ on Ta, each ~10 mm×5 mm in size.

*SCAA87:* Copper, silver, and gold reference materials for intensity and energy calibration of Auger electron spectrometers.

*AES intensity calibration software:* Used to calculate the intensity response (or transmission) function of the spectrometer over a kinetic energy range of 20 to 2500 eV.

*PC138 software:* To check that data files can be saved in compliance with the VAMAS (Versailles Project on Advanced Materials and Standards) Standard Data Transfer Format.

### American Society for Testing and Materials (ASTM)

100 Barr Harbor Drive, West Conshohocken, Pa. 19428-2959. Telephone: (610) 832-9500; Fax: (610) 832-9555; e-mail: *gcolling@astm.org*; WWW: *http://www.astm.org/COMMIT/*.

There are currently 13 standards covering terminology, practices and guides published by ASTM, that relate to Auger electron spectroscopy. Some of the most useful documents are as follows.

*E673:* Standard Terminology Relating to Surface Analysis.

*E827:* Standard Practice for Elemental Identification by Auger Electron Spectroscopy.

*E983:* Standard Guide for Minimizing Unwanted Electron Beam Effects in Auger Electron Spectroscopy.

*E984:* Standard Guide for Identifying Chemical Effects and Matrix Effects in Auger Electron Spectroscopy.

*E995:* Standard Guide for Background Subtraction Techniques in Auger Electron and X-ray Photoelectron Spectroscopy.

*E1078:* Standard Guide for Specimen Preparation and Mounting in Surface Analysis.

*E1127:* Standard Guide for Depth Profiling in Auger Electron Spectroscopy.

*E1829:* Standard Guide for Handling Specimens Prior to Surface Analysis.

John T. Grant
University of Dayton
Dayton, Ohio

# ION-BEAM TECHNIQUES

## INTRODUCTION

Ion Beam Analysis (IBA) involves the use of a well-defined *Beam* of energetic *Ions* for the purpose of *Analyzing* materials. In general, it is the combined attributes of the incident ions' energy, mass, and mode of interaction with the target that distinguish each of the analytic techniques described in this chapter. The range of possible ion-atom interactions that can occur in a target, and the results of that interaction that produces the detected signal, give each technique specific advantages under specific circumstances. In the units of this chapter, IBA techniques are described which utilize ion species spanning nearly the entire periodic table, ion energies which extend from several electron volts to several million electron volts, and detected signals consisting of nuclei, electrons, $\gamma$ rays, x rays, optical photons, collected electrical charge, and even logic states in digital circuits.

One feature of IBA that distinguishes it from most other analytical techniques is the incredible wealth of signals that are available for detection. A simple beam of protons incident on a target offers the experimenter a "detection menu" consisting of: (1) back-scattered protons, (2) $\alpha$ particles formed by nuclear reactions (if the energy is high enough), (3) $\gamma$ rays caused by these same reactions or Coulomb excitation, (4) x rays produced by inner atomic shell ionization followed by radiative decay, (5) secondary ions emitted by sputtering, (6) secondary electrons ionized near the surface, (7) photons produced by the ionolumenescence process, (8) electron-hole pairs collected from within biased semiconductor targets, and on and on in glutinous profusion. Of course we are not limited to just protons—one can accelerate deuterons, tritons, $^3$He, $^4$He, $^6$Li, $^7$Li, etc. Given all of the possible signals from all of the possible beams at all of the possible energies, a virtually mind-boggling range of measurements is possible. Moreover, owing to the interplay of ion, energy, and target composition combined with the capability to finely focus the beam, the experimenter can preferentially probe different regions of a sample at times in three dimensions. In the simplest of cases, for example, low-energy ions interact only with atoms that are near the surface, while high-energy ions permit measurements of the composition deep within a specimen. Thus, a modern IBA lab can usually provide a way to analyze virtually any isotope of any element, at any position, inside any sample.

The connecting thread that binds the broad range of analytical techniques presented in this chapter may not be immediately obvious, due to the widely disparate energies and ion species that are employed to probe for specific elements in specific targets. Yet, the relationships among the different IBA techniques becomes more apparent when each is plotted in a "phase space" defined by the projectile ion, energy, and the electronic and nuclear stopping powers for a specific target (i.e., the rates at which an ion loses energy in a sample due to electronic and nuclear collisions). In such a phase space of IBA techniques, shown in Figure 1, a coordinate system of projectile atomic number (specifically for H, He, C, Si, Mo and Au) is plotted versus ion energy/mass (logarithmic in E/amu) within the framework of a log-log plot of nuclear stopping power $[S(n)]$ versus electronic stopping power $[S(e)]$, where the stopping power units are MeV/(mg/cm$^2$). This plot has been generated for the "favorite son" of IBA—silicon—as the target. Aside from the qualitative rendering of the relationships among the IBA techniques, such plots also readily provide quantitative comparisons of the stopping powers of all ions across nearly all energies in a given target.

The different IBA techniques are mapped into this space by assigning each technique that region of the phase space representing its typical ion and energy parameters. Beginning with the least utilized region, i.e., that of relativistic energies, the region labeled "REL" applies to energies greater than $\sim$100 MeV/amu—approximately half the speed of light. The region labeled NRA (nuclear reaction analysis) is next. Its low energy boundary is set by the height of the Coulomb barriers for collisions between ions and Si; above this barrier, the nuclei can collide permitting the possibility of nuclear reactions. Below the Coulomb barrier is the elastic scattering region characterized by RBS (Rutherford back-scattering) for low-Z projectiles, and ERD (elastic recoil detection) for high-Z projectiles. Note that it is in this region that the electronic (and total) stopping power is maximum, and therefore the highest depth resolution IBA depth profiling is done with ions in this category. This is also the region where proton-induced x-ray emission (PIXE) is performed. The low energy boundary for this region occurs where electron screening effects become important. At still lower energies, MEIS (medium energy ion scattering) and HIBS (heavy ion back-scattering spectrometry) come into use. HIBS denotes the use of higher-Z projectiles with larger stopping powers and high sensitivity to trace elements. MEIS also takes advantage of very large stopping powers and the enormous Coulomb scattering cross-sections which tend to vary with the inverse of the projectile energy squared. Because of this, MEIS techniques also have very good depth resolution and, at times, extremely high sensitivity. We have defined this region's lower energy boundary using a velocity equal to approximately one-half the Bohr velocity. Ions with energies less than the Bohr velocity become neutralized very easily because their velocity is almost exactly matched to that of a target's outer-shell electrons which can then be easily captured by the ion. This marks the onset of the LEIS (low energy ion scattering) regime; owing to its low energy, LEIS is highly surface sensitive. When inert-gas ions are used, LEIS can provide an analysis of the outermost atomic layer in a material. Moreover, since in this regime the nuclear stopping power is a maximum, SIMS (secondary ion mass spectrometry) is performed at these energies with heavier ions that have correspondingly large sputtering yields. Finally, for ions with energies of less than 100 eV, we leave the realm of

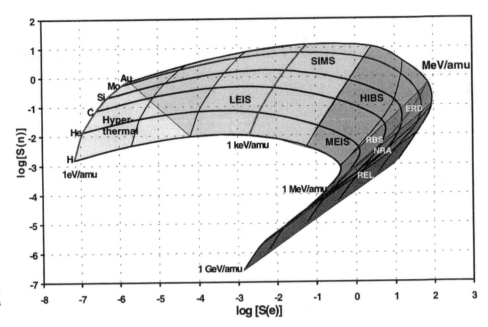

**Figure 1.** IBA Regimes for Si Target Incident Ions H-Au Energies from 1 eV to 1 GeV/amu.

binary collisions and enter a regime called hyperthermal, where ion-solid interactions begin to include both multibody kinematic and chemical interactions.

Common to each of the IBA techniques described in this chapter is the use of the particle accelerator—a device that the public most commonly associates with the "atom smashers" used in the early part of the 20th century to probe the structure of the atom, and later, the nucleus. Our dependence on this tool is somehow fitting given that virtually all of the techniques discussed in the following units can trace their development to the pioneering work performed by the great physicists of that era—Rutherford, Compton, Bethe, Bohr, etc. This legacy becomes apparent by simply noting that in performing any of these analyses, we are usually repeating a classic experiment, albeit with a different emphasis and better equipment, that was once performed by one of the giants of modern physics. Today's IBA techniques are the direct descendants of investigations into the nature of the atom by Lord Rutherford using back-scattered α particles, the structure of the electronic orbitals by Niels Bohr using characteristic x-ray spectra, and even the nature of stellar energy generation by Hans Bethe using the cross-sections of ion-induced nuclear reactions. When performing IBA, it is sometimes hard not to actually feel the presence of one of these legends on whose shoulders we stand.

As the reader delves into each of the following chapters, the underlying similarities in these techniques will repeatedly appear. This is not surprising, since the basis of all ion-atom scattering kinematics is contained in the simple central force problem first solved by Kepler, (and perfected for IBA purposes by Minnesota Fats and Willie Mosconi!). Likewise, in a large fraction of these techniques, the scattering cross-sections are given by the famous Rutherford equation. Thus, while it is the unique nature of the collision physics within each ion/energy regime that lends

each IBA technique its particular utility for a given measurement, it is the shared traits of ion-solid interactions which unite these techniques into a single scientific field which has been the useful servant of much of today's science and technology.

BARNEY L. DOYLE
KEVIN M. HORN

# HIGH-ENERGY ION BEAM ANALYSIS

## INTRODUCTION

It is very hard to identify the earliest application of high-energy ion beam analysis (IBA), which was a natural outgrowth of accelerator-based atomic and nuclear physics that started in the 1940s with the development of particle accelerators. In these early experiments, the composition of the target to be studied was well known, and it was the nuclear scattering or reaction cross-section that was measured. It would not be surprising if one of the very first nuclear physics pioneers turned this emphasis around and used their beams to measure the compositional uniformity of an unknown sample. Perhaps the earliest known proposal for IBA was that of E. Rubin in 1949 (Rubin, 1949, 1950) at the Stanford Research Institute, which spelled out the principle of Rutherford backscattering spectrometry (RBS). Another very early application of IBA was performed by Harald Enge at MIT (Enge et al., 1952). Enge used mega-electron-volt deuterons from one of Van de Graaff's first accelerators with a magnetic spectrometer to measure the constituents of aluminized Formvar in perhaps the earliest materials analysis application of nuclear reaction analysis (NRA). It is interesting to note that the

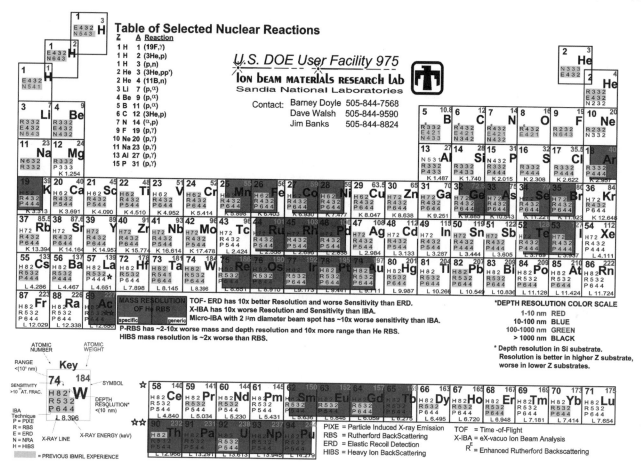

**Figure 1.** Ion beam analysis table of the elements.

use of magnetic spectrometers to obtain ultrahigh depth resolution is now common, and Dollinger et al. (1995) have obtained single atomic layer resolution by using an Enge Q3D magnetic spectrometer for IBA.

High-Energy IBA is generally thought to apply when the ion beam has energy greater than 0.5 MeV. There are several attributes of high-energy IBA, including (1) the ability to measure virtually any element in the periodic table, at times with isotopic specificity; (2) the ability to determine concentration as a function of depth; (3) that trace element sensitivity is not uncommon; and (4) that the technique is nondestructive with respect to the chemical makeup of the sample.

The periodic table in Figure 1 provides a convenient overview of high-energy IBA capabilities.

## ION BEAM ANALYSIS PERIODIC TABLE

The purpose of the periodic table is to quickly provide information on the sensitivity, depth of analysis, and depth resolution of most modern ion beam analysis techniques in a single, easy-to-use format. One can find an inter-active version of this table at: *http://www.sandia.gov/ 1100/1111/Elements/tablefr.htm.*

### Key: Technique, Sensitivity, and Depth Resolution

The information is organized in a periodic table to facilitate its use, as most readers may have only little knowledge of IBA but will know what element they want to detect. A key to the information included in this table is found in the lower left for the element W, or tungsten. The first letter indicates the technique. The numbers that follow the technique label are logarithmic scales for sensitivity, analysis range, and depth resolution, respectively. For example (using the key), H (HIBS) can be used to measure W with a sensitivity better than $10^{-8}$ atomic fraction to a depth range greater than $10^2$ nm. Likewise, R (RBS) can measure W with a sensitivity of $10^{-5}$ atomic fraction to a depth of $10^3$ nm with a depth resolution better than $10^2$ nm, and P (PIXE) can measure W down to $10^{-6}$ atomic fraction to a range and depth resolution of $10^4$ nm.

This sensitivity and depth information has been calculated based on available ion-atom interaction cross-sections and assumes that (1) 100 counts are measured, (2) the integrated incident beam current is 10 particle microcoulombs, and (3) the largest practical detector solid angles are used. These calculations ignore backgrounds and other interferences, but IBA scientists routinely use integrated incident beams of greater than 10 μC, and therefore the sensitivities cited are actually rather conservative. In many cases, actual sensitivities are an order of magnitude better. The depth resolutions listed are also very conservative and should be thought of as upper limits.

## Mass Resolution

The shaded regions in the table are used to indicate the mass resolution range for He beam RBS. For the first four rows of the table (through Ar), the resolution is adequate to separate all adjacent elements except specific cases of mass interference, which are shaded (e.g., Mn and Fe cannot be separated because their mass differs by only 1 amu). For the remainder of the table, no adjacent elements are separable, and the lightly shaded "generic" mass resolution range is characteristic for an entire row of the periodic table (e.g., the shading of Ru-Rh-Pd indicates that for row 5 of the periodic table the mass resolution is not adequate to separate Ru from Pd but could separate Ru from Ag, and so on for other elements in this row).

High-energy IBA includes elastic scattering, nuclear reactions, and particle-induced X-ray emission (PIXE). We next provide brief descriptions of the techniques that fall under these headings.

## ELASTIC SCATTERING (see ELASTIC ION SCATTERING FOR COMPOSITION ANALYSIS)

### Rutherford Backscattering Spectrometry

Rutherford backscattering spectrometry (Chu et al., 1978; Feldman and Mayer, 1986) has matured into one of the most important analytical techniques for the elemental characterization of materials and is used routinely in material science applications in many laboratories around the world. Typically, in conventional RBS, H or He ions, at mega-electron-volt energies, are elastically scattered from a nucleus in a sample and detected in a backward direction in a Si surface barrier detector. For a given detection angle, the kinematics of the elastic scattering is well known, and the energy transferred in the collision can be determined from the Rutherford theory. The scattering kinematics and the energy loss of the ion entering and leaving the sample are well understood. Therefore, the energy of the scattered ion provides information about the mass of the scattering center as well as the depth in the sample. The RBS information may provide the elemental composition and the depth distributions for single- or multi-layered samples at sensitivities as low as a few parts per million. Sensitivities are much better for heavier elements on lighter substrates, although elements with atomic numbers, $Z$, from 4 to 94 have been detected. Light elements in heavy matrices are much more difficult to analyze. Depth resolutions may vary from 1 to 100 nm. Analysis depth ranges are 0.1 μm to tens of micrometers, and

sensitivities vary from 0.1 at. % to ~10 ppm, depending on the atomic number of the trace element.

### Heavy-Ion Backscattering

Heavy-ion backscattering spectrometry (HIBS) (Doyle et al., 1989; Weller et al., 1990; Knapp and Banks, 1993) employs heavier incident ions. This increases the elastic scattering cross-section, which is proportional to $(Z_1 Z_2/E)^2$, and increases the sensitivity for detection as well as the mass resolution. Here, $Z_1$ and $E$ are the atomic number and energy, respectively, of the incident ion and $Z_2$ is the atomic number of the target atom. Because of the heavier mass of the incident ion, the depth analysis range is reduced to 0.01 to 0.1 μm while the sensitivity is increased to 0.1 to 1 ppm.

### Elastic Recoil Detection Analysis

Elastic recoil detection analysis (ERDA) (L'Ecuyer et al., 1976; Doyle and Peercy, 1979; Green and Doyle, 1986; Barbour and Doyle, 1995; Dollinger et al., 1995) uses a high-energy, heavy incident ion to recoil light target elements from a sample in the forward direction. Since RBS is not suitable when $M_1 \geq M_2$, where $M_1$ is the mass of the incident ion and $M_2$ is the mass of the target element, ERDA is especially useful for light elements such as H through F. Elastic recoil detection analysis depends on the ability to distinguish between incident ions scattered in the forward direction and light recoiling atoms. Since the heavier incident ions have a larger energy loss than the lighter recoil target atoms, a thin foil is sometimes used to block the incident ions from the Si surface barrier detector. This technique may be used with a time-of-flight detector to achieve much higher depth resolution (1 nm) and even permit the separation of different isotopes of light elements. Sensitivities of 10 to 1000 ppm have been obtained. The glancing angle limits the depth analysis range to ~1 μm.

## NUCLEAR REACTION ANALYSIS (see NUCLEAR REACTION ANALYSIS AND PROTON-INDUCED GAMMA RAY EMISSION)

Nuclear reaction analysis (Bird, 1989; Moore et al., 1975) and resonant nuclear reaction analysis (RNRA) (Vizkelethy, 1995) are very element specific analysis techniques in which nucleus $a$ bombards target $A$ resulting in a light product $b$ and residual nucleus $B$, $A(a,b)B$. Nonresonant nuclear reactions (NRA) may occur when the energy of the incident nucleus is above a threshold value for the reaction. Nuclear reaction analysis allows bulk measurements of target nuclei distributed throughout a sample. The sensitivity of NRA depends upon the cross-section for the nuclear reaction. Resonant NRA may be used to depth profile the concentration of target nuclei below the surface in a sample. For both NRA and RNRA, the concentration of nuclei is then used to determine elemental concentration from isotopic compositions of the sample. Depending upon the resonant nuclear reaction employed, the technique may be used to profile from the surface to a few micrometers in depth with a resolution of a few to hundreds of nanometers. For example, commonly used resonant nuclear reactions for H detection are $^{1}\text{H}(^{15}\text{N},\alpha\gamma)^{12}\text{C}$ (resonance at 6.385 MeV) and $^{1}\text{H}(^{19}\text{F},\alpha\gamma)^{16}\text{O}$ (resonance at 6.418 MeV). Depending upon the cross-section for the nuclear reaction, sensitivities may vary from 1 to 1000 ppm.

## PARTICLE-INDUCED X-RAY EMISSION
### (see PARTICLE-INDUCED X-RAY EMISSION)

Particle- (or equivalently proton-) induced X-Ray emission is a multielement, nondestructive analysis technique with sensitivities from major element down to parts-per-million (ppm) trace element, depending upon the elements of interest (Rickards and Ziegler, 1975; Mitchell and Ziegler, 1977; Johansson and Campbell, 1988). With PIXE, a low-atomic-number ($Z_1$) incident ion beam up to a few mega-electron-volts in energy bombards a sample producing atomic excitations of the elements ($Z_2$) present in the sample. Here, $Z_2$ is the atomic number of the target element. Subsequently, characteristic K, L, or M X rays are emitted from the excited elements and are counted to determine the concentrations of the trace elements. Up to 30 elements can be analyzed simultaneously.

Since, the target ionization cross-sections are proportional to $Z_1^2/Z_2^4$, the number of target element X rays decreases with increasing target atomic number for a given electronic shell. The target ionization cross-section increases with higher incident ion energy $E$ until the velocity of the ion matches the orbital velocity of the electron being ionized and then decreases as $\sim 1/E$. Typically, Si(Li) detectors with Be entrance windows allow detection of elements of atomic number from 12 to 92. Windowless Si(Li) detectors allow elements down to carbon to be analyzed. Sensitivities of 1 to 100 ppm atomic (ppma) have been achieved. Analysis ranges are 1 μm to tens of micrometers.

## SUMMARY

The units that follow provide the details required to select the optimum high-energy IBA technique for specific applications. The IBA periodic table in this introduction clearly portrays the interrelationships of these units.

## LITERATURE CITED

Barbour, J. C. and Doyle, B. L. 1995. Elastic recoil detection: ERD. *In* Handbook of Modern Ion Beam Materials Analysis (J. R. Tesmer, M. Nastasi, J. C. Barbour, C. J. Maggiore, and J. W. Mayer, eds.). pp. 83–138. Materials Research Society, Pittsburgh, Pa.

Bird, J. R. 1989. Nuclear reactions. *In* Ion Beams for Materials Analysis (J. R. Williams and J. R. Bird, eds.). pp. 149–207. Academic Press, Marrickville, Australia.

Chu, W.-K., Mayer, J. W, and Nicolet, M.-A. 1978. Backscattering Spectrometry. Academic Press, New York.

Dollinger, G., Bergmaier, A., Faestermann, T., and Frey, C. M. 1995. High-resolution depth profile analysis by elastic recoil detection with heavy ions. *Fresenius J. Anal. Chem.* 353:311–315.

Doyle, B. L. and Peercy, P. S. 1979. Technique for profiling H-1 with 2.5-MeV van de Graaff accelerators. *Appl. Phys. Lett.* 34:811–813.

Doyle, B. L., Knapp, J. A., and Buller, D. L. 1989. Heavy-ion back-scattering spectrometry (HIBS): An improved technique for trace-element detection. *Nucl. Instrum. Methods* B42:295–297.

Enge, H., Buechner, W. W., and Sperduto, A. 1952. Magnetic analysis of the Al$^{27}$(d,p)Al$^{28}$ reaction. *Phys. Rev.* 88:963–968.

Feldman, L. C. and Mayer, J. W. 1986. Fundamentals of Surface and Thin Film Analysis. North-Holland, New York.

Green, P. F. and Doyle, B. L. 1986. Silicon elastic recoil detection studies of polymer diffusion: Advantages and disadvantages. *Nucl. Instrum. Methods* B18:64–70.

Johansson, S. A. E and Campbell, J. L. 1988. PIXE: A Novel Technique for Elemental Analysis. John Wiley & Sons, Chichester.

Knapp, J. A. and Banks, J. C. 1993. Heavy-ion backscattering spectrometry for high sensitivity. *Nucl. Instrum. Methods* B79:457–459.

L'Ecuyer, J., Brassard, C., Cardinal, C., Chabbal, J., and Deschenes, L. 1976. Elastic recoil detection. *J. Vac. Sci. Technol.* 14:492.

Mitchell, I. V. and Ziegler, J. F. 1977. Ion induced X-rays. *In* Ion Beam Handbook for Material Analysis (J. W. Mayer and E. Rimini, eds.). pp. 311–484. Academic Press, New York.

Moore, J. A., Mitchell, I. V., Hollis, M. J., Davies, J. A., and Howe, L. M. 1975. Detection of low-mass impurities in thin films using MeV heavy-ion elastic scattering and coincidence detection techniques. *J. Appl. Phys.* 46:52–61.

Rickards, J. and Ziegler, J. F. 1975. Trace element analysis of water samples using 700 keV protons. *Appl. Phys. Lett.* 27:707–709.

Rubin, E. 1949. Chemical analysis by proton scattering. *Bull. Am. Phys Soc.* December 1949.

Rubin, E. 1950. Chemical analysis by proton scattering. *Phys. Rev.* 78:83.

Vizkelethy, G. 1995. Nuclear reaction analysis: Particle-particle reactions. *In* Handbook of Modern Ion Beam Materials Analysis (J. R. Tesmer, M. Nastasi, J. C. Barbour, C. J. Maggiore, and J. W. Mayer, eds.). pp. 139–165. Materials Research Society, Pittsburgh, Pa.

Weller, M. R., Mendenhall, M. H., Haubert, P. C., Döbeli, M., and Tombrello, T. A. 1990. Heavy ion Rutherford backscattering. *In* High Energy and Heavy Ion Beams in Materials Analysis (J. R. Tesmer, C. J. Maggiore, M. Nastasi, J. C. Barbour, and J. W. Mayer, eds.). pp. 139–151. Materials Research Society, Pittsburgh, Pa.

BARNEY DOYLE
Sandia National Laboratories
Albuquerque, New Mexico

FLOYD DEL MCDANIEL
University of North Texas
Denton, Texas

## ELASTIC ION SCATTERING FOR COMPOSITION ANALYSIS

### INTRODUCTION

Composition analyses for virtually all of the elements on the periodic table can be performed through a combination of elastic ion-scattering techniques: Rutherford backscattering spectrometry (RBS), elastic recoil detection (ERD), and enhanced-cross-section backscattering spectrometry (EBS). This unit will review the basic concepts used in elastic ion scattering for composition analysis and give examples of their use. Elastic ion scattering is also used for structure determination utilizing techniques such as ion channeling (see, e.g., Feldman et al., 1982). Composition determination using ion-beam analysis (IBA) falls into two categories: composition-depth profiling and determination of the total integrated composition of a thin region. Unknown species in a sample may be identified using elastic ion-scattering techniques, but in general, analyses involving characteristic x rays or γ rays are more accurate for determining the presence of unknown constituents. For elastic ion scattering, if the ion beam and ion energy are chosen to have sufficiently high energy loss in the sample, then a composition profile is determined as a function of depth by collecting one ion-scattering spectrum. To represent accurately the composition profile as a function of depth in the sample, an accurate knowledge of the sample's volume density is required because IBA techniques are insensitive to density. If the density of the sample is unknown, then the depth in the sample is given in terms of an areal density: depth times volume density. When the incident ion beam is very light or the ion energy very high, then the composition as a function of depth may not be resolved fully, in which case the average composition of the layer of interest is determined by integrating the signal. In this way, the number of atoms per unit area (areal density) is determined from the integrated signal. Table 1 lists typical examples of composition analyses using different elastic ion-scattering techniques with different ion beams and ion-beam energies.

The term elastic refers to the fact that the target nucleus is unchanged after the scattering event (except for kinetic energy), and therefore inelastic conversion between mass and energy or excitation of the nucleus to a higher energy state is absent. Energy and momentum are conserved, and in the center-of-mass frame of reference, the position of the center of mass remains unchanged. The term Rutherford is used to indicate that the scattering cross-section σ is predominantly a result of Coulomb scattering, and therefore σ can be described by the Rutherford formula (Rutherford, 1911). Elastic ion-scattering techniques are often thought of as nondestructive because they do not require activation of the sample, sputtering of the sample, or any other change of the sample to determine the composition of the sample. However, the ion beam does interact with the sample by breaking bonds, exciting electrons, producing phonons, and possibly creating displacements. Therefore, techniques that are sensitive to these properties (e.g., electrical and optical measurements or x-ray diffraction) should be performed prior to IBA.

Now consider a general composition analysis experiment in which a flux of particles is incident upon a sample. These particles can be electrons, ions, or photons that are scattered elastically or absorbed in the sample through energy loss of the particle. If the particle causes an electronic transition in the target atoms, then the energy of emitted radiation is characteristic of that atom and is used to identify the atom. The number of atoms in a sample is determined from knowledge of the type of interaction between the incident particle and the target atom and the probability that an interaction will occur. Each type of interaction has a different cross-section, which is the effective area for an interaction to occur between one of the incident particles and one of the target atoms. The probability for an interaction is then given by the cross-section times the areal density. This is a general description for many types of composition analysis techniques, including elastic ion scattering, and a single equation describes the quantitative composition analysis of a material: $Y = Q\sigma N\tau\Omega\eta$, where $Y$ is the ion-scattering yield, $Q$ is the number of incident particles, $\sigma$ is the cross-section for a given process, $N$ is the atomic density, $\tau$ is the layer thickness, $\Omega$ is the detector solid angle, and $\eta$ is the detector efficiency. The yield is a measured quantity, and thus all of the details needed to determine $N$, the composition, reduce to a determination of the other variables in the yield equation. The rest of this unit will describe how these other variables are measured or calculated from measured

**Table 1. Elastic Ion-Scattering Techniques for Typical Composition Analysis**

|  | Energy (MeV) | θ or φ | Elements Analyzed | Sensitivity (at.%) | Analysis Range (μm) | Depth Resolution (μm) | Comments |
|---|---|---|---|---|---|---|---|
| He RBS | 1–3.5 | 164 | Na to U | 0.001–0.1 | ~1.5 | 0.03 | |
| H RBS[a] | 1–2 | 164 | Na to U | 0.01–1.0 | ~10 | 0.08 | |
| He ERD | 2–3 | 164 | H and D | 0.1 | ~1 (polymers) | 0.1 | Non-Rutherford cross-sections |
| Si ERD | 12–16 | 30 | H to Li | 0.05 | ~0.4 | 0.02 | Range for H |
|  | 18–28 | | H to N | 0.1 | ~0.4 | 0.03 | Range for N |
|  | 28–32 | | H to N and O | 0.2 | ~0.2 | 0.05 | Range for O |
| He EBS | 3.5 | 164 | C | 0.01 | ~0.5 | 0.03 | Depth profile |
|  | 3.5 | | N | 0.01 | ~0.5 | 0.03 | Depth profile |
|  | 8.7 | | O | 0.01 | ~7 | 0.05 | Depth profile |
| H EBS[a] | 2.5 | 164 | O | 0.1 | ~4 | 0.2 | Depth profile |
|  | 4.4 | | S | 0.1 | ~1 | | Areal density |

[a]*CAUTION:* Risks possible radiation exposure to people from prompt nuclear reactions or activation of the samples. Consult your health physicist before doing high-energy proton irradiation.

quantities. In addition, IBA techniques provide depth profiles because the energy loss of the incident and scattered particles are well characterized through analytical formulas (Ziegler, 1980, 1987; Ziegler et al., 1985; Tesmer et al., 1995), and therefore the energy of these particles is described well along the path of the particles. The yield is then calculated as a function of depth in the sample because the energy at each depth in the sample is well known.

Elastic ion scattering methods using both RBS and ERD are simple and fast for quantitative elemental depth profiling in thin films ($\leq 1$ µm thick) for every element on the periodic table. Elastic recoil detection is often used for profiling elements with atomic number $Z \leq 8$. Sometimes enhanced-cross-section RBS is used to profile elements in the range $5 \leq Z \leq 8$, but RBS is often used for profiling elements with $Z \geq 8$. Once the system parameters are calibrated, this technique is standardless, and the spectra are turned into concentration profiles within minutes. The many attributes of RBS analysis are also shared by ERD analysis, and therefore collection of ERD spectra can be done on the same accelerator and in the same analysis chamber as RBS. The analysis of the data is fully analytical and generally independent of bonding in the sample. This is a tremendous benefit over other techniques that depend upon extensive comparison to standards and also on sputter rates that may vary from sample to sample. Finally, conventional RBS and ERD give adequate depth resolution (tens of nanometers) and sensitivity ($\sim 0.1$ at.%) for many types of thin-film analyses performed in materials science. If greater depth resolution or greater sensitivity is required, then more exotic detection schemes can be used [e.g., time-of-flight (TOF) detectors, which produce a depth resolution an order-of-magnitude greater than that for conventional surface barrier detectors].

Before considering the principles of the IBA method, it is useful to place some practical limitations on the conventional use of IBA in comparison to other common materials analysis techniques. Composition analysis with approximately mega-electron-volt ions, in a rather standard IBA configuration, is less surface sensitive than techniques such as Auger electron spectroscopy (see AUGER ELECTRON SPECTROSCOPY) or x-ray photoelectron spectroscopy (XPS). Further, XPS can be used to identify changes in bonding even when the composition is not changing. For determination of the material microstructure, transmission electron microscopy (TEM) should be considered (see TRANSMISSION ELECTRON MICROSCOPY and SCANNING TRANSMISSION ELECTRON MICROSCOPY: Z-CONTRAST IMAGING), and if local compositional analysis of regions $\sim 10$ nm in size is needed, then TEM combined with energy-dispersive x-ray spectroscopy (EDS) should be considered (see ENERGY-DISPERSIVE SPECTROMETRY). Ion beam analysis is insensitive to film density, and therefore accurate measurement of film thickness is best made by other techniques, such as ellipsometry, x-ray reflectivity, or simple step-height measurements from profilometry. Finally, for trace element analysis (ppm or less), techniques such as secondary ion mass spectroscopy (SIMS) or nuclear reaction analysis (NRA; see NUCLEAR REACTION ANALYSIS AND PROTON-INDUCED GAMMA RAY EMISSION) should be considered. In any case, thin-film composition analysis and monitoring reactions and interdiffusion of thin films are performed very effectively using RBS, ERD, and EBS, especially when complimented by the other techniques listed here.

## PRINCIPLES OF THE METHOD

This unit will describe the typical IBA experiment using Si surface barrier detectors (SBDs), although several types of IBA techniques have been developed in which the detector is changed to improve the depth resolution or elemental sensitivity. Once the principles are understood for conventional IBA, these principles are extended easily for the case of specialized detectors, and the reader is referred to the literature for those cases (see, e.g., Thomas et al., 1986; Wielunski et al., 1986; Bird and Williams, 1989; Gebauer et al., 1990; Whitlow, 1990; Tesmer et al., 1995; and the Additional References for Special Detectors and Detection Schemes following the Literature Cited section). Rutherford backscattering was first done by Geiger and Marsden (1913), but this technique waited until the 1960s for wider recognition (outside the nuclear physics community) that heavy element analysis on lighter substrates could be performed using RBS (e.g., see Turkevich, 1961). One of the most useful modern IBA techniques for easy depth profiling of light elements is ERD, which was first introduced in 1976 by L'Ecuyer et al. (1976).

The physical concepts governing RBS and ERD are identical, and through the use of a computer, their spectra are easily manipulated to obtain full quantitative analyses. However, even before manipulation of the spectra to obtain a quantitative analysis, it is useful to learn to interpret the spectra in order to understand the data as it is being collected. Therefore, we will begin with a description of the concepts common to RBS and ERD, and then demonstrate how to read the spectra.

Both RBS and ERD depend on only four physical concepts: (1) the kinematic factor describes the energy transfer in an elastic two-body collision; (2) the differential scattering cross-section gives the probability for the scattering event to occur; (3) the stopping powers give the average energy loss of the projectile and scattered atoms as they traverse the sample, thereby establishing a depth scale; and (4) the energy straggling gives the statistical fluctuation in the energy loss. By applying these four physical concepts, the spectra are transformed into a quantitative concentration profile as a function of depth. The main difference in the treatment of ERD data and RBS data is in the calculation of stopping powers: the incident projectile and recoil atoms are different for ERD data, but these incident and detected ion species are the same for RBS data. As a result of the similarity of these two techniques, the analytical expressions describing the backscattering process and the elastic recoil process will be derived in parallel.

A comparison of RBS and ERD is shown in Figure 1 with $\sim 1$-MeV/amu projectiles: 1-MeV protons are used to profile Si in RBS analysis and 28-MeV Si ions are used to profile H in ERD analysis. At the top of the figure the experimental setup is shown schematically. This setup is

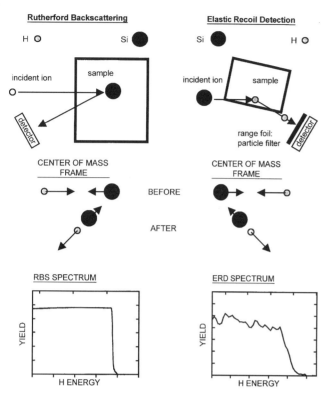

**Figure 1.** Schematic comparison of RBS and ERD scattering geometries in the laboratory (top) and center-of-mass (middle) frames of reference. The larger filled circle represents the Si atom and the smaller gray circle represents the H atom. In the center-of-mass frame, the scattering events for RBS and ERD are equivalent. The representative spectra for the collection of scattered H atoms for both RBS and ERD are shown at bottom.

essentially the same for RBS and ERD but with the addition of a range foil for the ERD detector. In fact, RBS and ERD can be set up in the same vacuum chamber by positioning one detector in the backscattering direction and another detector with range foil in the forward-scattering position and by giving the sample holder the rotation capability to allow for grazing incidence of the projectile and exit of the recoiled atoms. The incident projectiles lose energy as they enter the sample and kinematically scatter hydrogen atoms (recoil H atoms for ERD and backscatter H atoms for RBS), and then the H atoms lose energy as they traverse the sample out to a particle detector. In both analyses, the atom entering the detector is H, but the incident ion projectiles differ. The energy loss in the range foil must be calculated when analyzing the ERD data. The differential Rutherford scattering cross-section ($d\sigma/d\Omega$) for the recoil scattering event is ~1.7 b/sr while the differential Rutherford scattering cross-section for the backscattering event is ~0.26 b/sr. The middle portion of Figure 1 schematically shows the scattering events in the center-of-mass reference frame in which RBS and ERD appear as symmetrical scattering events. The yield of H as a function of detected H energy is shown in the bottom portion of the figure for the RBS Si spectrum and the ERD H spectrum.

Both spectra in Figure 1 give the yield of detected H particles as a function of the H energy, but the important

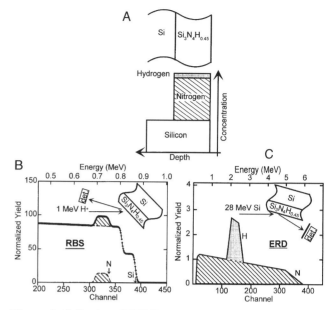

**Figure 2.** Collection of RBS (**B**) and ERD (**C**) spectra from a 300-nm-thick $Si_3N_4H_{0.45}$ film on Si, shown schematically (**A**), with the concentration profile shown in the middle (depth increasing to the left). The incident projectiles are 1-MeV protons for RBS and 28-MeV Si ions for ERD.

point to note is that for RBS the detected particle is the same species as the incident ion, whereas for ERD the detected particle is coming from the sample. In fact, for samples containing several light elements, many of these elements can be detected simultaneously in a single ERD spectrum. In this latter case, the detected particle energy is that for several different recoiled particles causing several overlapping ERD spectra for each detected particle. Schematically, an example is shown in Figure 2 of overlapping spectra for both RBS and ERD data representative of a 300-nm-thick $Si_3N_4$ layer containing 6 at.% H on a Si substrate. The elemental concentration profiles are given (Fig. 2A), one on top of the other, as stacked bars in which the height of each bar is the concentration in atomic percent and the width of the bar corresponds to depth in the sample. Simulated RBS and ERD spectra from this sample are shown in Figure 2B and C in which the signals from the different elements overlap. The signals are shaded in order to show the contributions coming from the different elements in the sample. Also, the contribution to the RBS spectrum (data points shown as plus symbols) from N alone is shown as a dashed line between channels 300 and 350. Therefore, just as for the bar graph in Figure 2A, the individual contributions to the spectra from the different elements are stacked one on top of the other. These spectra would appear on a multichannel analyzer (MCA) as the yield versus channel, and then the top energy axis is determined from an energy calibration of the channel axis. Note that a signal for H does not appear in the RBS spectrum because the incident protons are not backscattered by the H in the sample, and a signal for Si does not appear in the ERD spectrum because the Si in the sample is not scattered effectively and the incident Si is stopped in a range foil.

The height of each peak in the spectra corresponds to a concentration for that element as function of depth, which is correlated to the energy axis. The high-energy edge (front edge) of a given peak is a result of scattering from the surface of the sample and the depth scale increases to the left (decreasing energy). First, note that the front-edge position is representative of the mass of the target atom in the scattering event. For RBS, the greater the target mass, the greater is the backscattered proton energy, and therefore the increasing energy axis in this spectrum also represents increasing target mass. Similarly, the energy axis in the ERD spectrum is correlated to surface scattering for different target masses, but this correlation is made more complicated by the use of the range foil, as is discussed below, and the surface scattering for a heavy element is not always detected at a higher energy than for a lighter mass element. The surface scattering positions for each detected element are marked in the spectra of Figure 2 with the symbol for that element. The energy axis is often calibrated relative to the measured channels by using the front edge of peaks (surface scattering) for several elements present at the surface of the sample and by calculating the energy for surface scattering. Although the N has the highest concentration in this film, the height of the Si signal in the RBS spectrum and the height of the H signal in the ERD spectrum are both greater than the height of the N as a result of the difference in cross-sections and effective stopping powers relative to those for N. The height of the Si signal in the RBS spectrum below channel 350 is a result of proton scattering from Si in the substrate (100 at.% Si), and the relative decrease in the Si signal between channels 350 and 400 is a result of proton scattering from Si in the silicon nitride film with a decreased Si concentration (40 at.% Si).

The energy scale of the spectra is directly related to depth scale for each element through knowledge of the stopping power for the incident ion and the different scattered (detected) atoms. The width of the N signal in the RBS spectrum corresponds to the width of the 300-nm-thick silicon nitride film, and similarly the width of the Si signal from the $Si_3N_4H_{0.45}$ film corresponds to a depth of 300 nm. In fact, the width of these two signals are equal because the scattered particle in both cases is a proton, and therefore only one type of stopping power is involved in determining the depth scale. In contrast, the width of the N and H signals in the ERD spectrum both correspond to 300 nm of $Si_3N_4H_{0.45}$, but because the scattered particle is a H atom in one case and a N atom in the other case, two different stopping powers are involved in determining the energy-to-depth-scale conversions. Thus, each element profiled in an ERD spectrum has its own depth scale, which complicates the comparison of yields from different species as a function of depth during data collection. To accurately determine the composition-depth profiles from the RBS and ERD data, the spectra need to be separated for each element and the yield converted to atomic density N, as described under Equations for IBA (see Data Analysis and Initial Interpretation).

In contrast to RBS, where light masses produce a signal at low energies with a low yield and heavy masses produce a signal at high energies with high yield, the energy of the

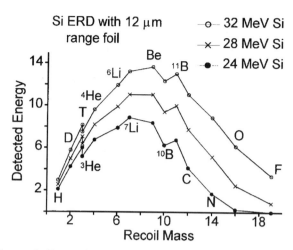

**Figure 3.** Detected energy calculated for recoiled atoms as a function of recoil mass for 24-, 28-, and 32-MeV Si ERD using a 12-µm Mylar range foil. For these calculations, the Si ions were incident upon the samples at an angle of 75° and the detector was at a scattering angle of 30°.

detected particle, $E_d$, and the yield of the ERD signals for the different masses depend strongly on the stopping power and thickness of the range foil. A heavier mass may or may not be detected at a higher energy than a light mass depending on the choice of projectile, projectile energy, type of range foil, and thickness of the range foil. Figure 3 shows $E_d$ for recoiled atoms as a function of mass of the recoiled atom for 24-, 28-, and 32-MeV Si ion beams incident at 75° with a scattering angle of 30° and using a 12-µm Mylar range foil. If no range foil were used for these calculations, then the detected recoil energy would steadily increase with recoil mass. These curves show that the detected energy depends strongly on the increased stopping power in the range foil for the heavier recoiled atoms (amu $\geq 9$). Mass resolution is the ability to separate in the spectra along the detected-energy axis the scattering signal from different target atoms. For surface scattering, $E_d$ in RBS analyses depends solely on the change in kinematic scattering as a function of the target mass, whereas in ERD analyses $E_d$ depends on the energy imparted to a recoiled atom and its stopping in the range foil. The kinematic factor $K$ is the ratio of the scattered particle energy to the projectile energy for an elastic collision. Consequently, the selectivity for different masses in RBS analyses is determined by $K$, while the mass resolution in ERD analyses is determined by $K$ and the effect of stopping in the range foil, as shown by the curves in Figure 3. For RBS the mass resolution is improved for a given ion beam by increasing the ion energy, and this is often true for ERD, but not always. For example, Figure 3 shows the ability to resolve B from He in a Si ERD spectrum is greatly improved by increasing the Si ion energy from 24 to 32 MeV, causing the difference $|E_d(B) - E_d(He)|$ to increase from 0.25 to 3.1 MeV; however, the same increase in Si ion energy causes $|E_d(He) - E_d(N)|$ to decrease from 5.1 to 0.73 MeV.

Depth resolution is the ability to separate, in the spectrum along the energy axis, the signal coming from scattering events at different depths in the sample. Therefore, the smallest resolvable detected energy $\delta E_d$ determines the smallest resolvable depth interval $\delta x$. Experimentally, the energy width $\delta E_d$ is taken from the energies corresponding to 12% and 88% of full signal height for an abrupt change in sample composition (e.g., at the sample surface). An example of the difference in energy width is demonstrated in Figure 2, in which the width of the front edge (measured between the 12% and 88% yields) for the H signal is less than for the N signal in the ERD spectrum, and even though the energy-to-depth conversions for the H and N signals differ significantly, the depth resolution for the H signal is better than that for the N. The depth resolution for both RBS and ERD is improved by using lower energy projectiles or using heavier projectiles. For example, most thin-film RBS analyses are performed using $^4$He ion beams rather than protons because of the increased depth resolution and increased cross-section. Also, $^4$He ion beams, like beams of $^1$H, are relatively easy to produce from simple ion sources.

The description given above for ERD analysis requires the use of a range foil in front of an SBD. However, one rare example of ERD analysis using a Si SBD without using a range foil is when a very heavy projectile is used with moderate-to-light-mass substrates. For this case, if the geometry and beam are chosen such that the recoil scattering angle $\phi$ is $> \sin^{-1}(M_{\text{substrate}}/M_{\text{incident ion}})$, where $M$ is the mass of the particle, then the incident ion is not scattered into the detector and no range foil is required. This condition is particularly useful for high-depth-resolution ERD analysis with a Au ion.

## PRACTICAL ASPECTS OF THE METHOD

### Typical Experimental Setup

Now we describe the apparatus used in a conventional RBS or ERD experiment. Rather than give an exhaustive treatment, this discussion is intended to allow a first-time practitioner to become familiar with the purpose of different parts of the apparatus before entering the laboratory. A more detailed treatment is found in Tesmer et al. (1995), Bird and Williams (1989), Chu et al. (1978), and Mayer and Rimini (1977).

First, the ion beam is generated from a plasma or sputtered target (Fig. 4). These ions are extracted at low energy from the ion source and accelerated to millions of electron volts. A magnetic field is used to bend the ions into a specific direction along the analysis beam line (evacuated tube), and by separating the ions according to their mass-to-charge ratio for a given energy, the proper ion beam is selected for the IBA technique of interest. The ion beam is then steered onto a sample that is in an ion-scattering vacuum chamber. A variety of IBA experiments may be performed in an ion-scattering chamber, depending on sample manipulation capabilities and detector geometry. In addition, the chamber may be configured to perform *in situ* materials science experiments such as thermal treatments and gas exposures. The first variable to be determined in the yield equation is the detector solid angle $\Omega$, which is measured with the use of a standard

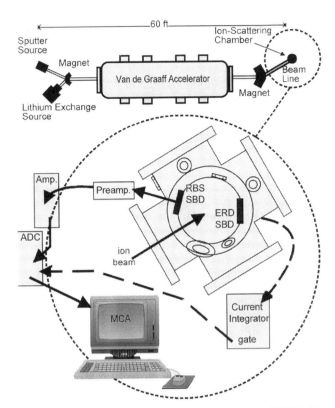

**Figure 4.** Typical experimental setup for RBS and ERD. Ion-beam analysis facilities are generally <100 ft in length, but this depends on the size of the accelerator being used. The expanded view, at bottom, schematically shows the ion beam scattering chamber and the electronics used to measure the target current and signal detected in an SBD (ADC = analog-to-digital converter).

sample if all other parameters are known. However, most ion-scattering chambers are designed with fixed dimensions; therefore the solid angle can be calculated from simple geometry:

$$\Omega = \int_{\text{angle subtended by detector}} \sin \phi \, d\phi \, d\varepsilon \approx \frac{\text{detector area}}{z^2} \quad (1)$$

where $\varepsilon$ is the stopping cross-section and $z$ is the sample-to-detector distance.

The next variable to be calculated for the yield equation is $Q$, the number of incident projectiles, which is determined from the target current and the projectile charge. The target current may be measured either in front of the ion-scattering chamber on a rotating wheel containing slots to allow the ion beam to enter the chamber or directly from the sample of interest. In either case, care is taken to accurately measure all of the incident charge by suppressing the escape of secondary electrons from the target. One solution is to electrically isolate the scattering chamber from ground and then measure the total current on the chamber relative to ground. This technique effectively treats the entire chamber as a Faraday cup for accurate charge collection (see Chapter 12 in Tesmer et al., 1995). To ensure that low-energy secondary electrons are not lost from the chamber in the direction of the incident beam, a suppression bias of $-500$ to $-1000$ V is applied to a wire mesh or metal aperture, which is isolated from the chamber and ground, just in front of the ion beam entrance to the chamber. The ion beam is collimated with movable slits or steered to avoid scattering from the aperture along its path to the sample. A more complete discussion of Faraday cup designs is found in Tesmer et al. (1995). The target current is then sent into a current integrator to measure the total charge incident upon the sample during the data collection, as shown in Figure 4. Generally, the amount of charge is preset on the integrator (typically to a value between 1 and 50 µC for most experiments) such that when the preset value is reached, a signal is sent to gate off the data collection. The length of time or amount of charge collected for an experiment depends on the counting statistics desired. The value needed for $Q$ is the number of particles (not the charge), and therefore the measured charge in microcoulombs is divided by $1.6022 \times 10^{-13}$ µC per ionic charge and by the ion's charge state to obtain $Q$ (e.g., charge state 2 for $He^{2+}$, or 5 for $Si^{5+}$).

The signal measured in RBS or ERD comes from scattered atoms collected in an SBD (Fig. 4). For RBS, the backscattered beam is the same as the incident beam, while for ERD the forward-scattered beam may be composed of many atom species with mass less than that of the incident beam. Also, the incident beam may be forward scattered into the detector for ERD, and therefore a stopping foil (the range foil) is used to stop heavy species such as the incident beam from entering the detector. The number of incident projectiles that scatter into the direction of the detector is generally much greater than the number of recoiled atoms; therefore the foil is needed to prevent damage to the detector and to minimize the background signal from the primary beam.

The SBD is a reverse-biased surface barrier diode (Mayer and Lau, 1990) that contains a depletion region large enough to stop the highest energy particle to be detected. The depletion region for $n$-type Si is established through the application of a positive bias typically <130 V. As the detected particle traverses the depletion region, the particle loses energy, and that energy lost to electron excitation gives rise to electron-hole pair production. The electrons and holes are swept to opposite terminals of the diode and provide a pulse of current. The detected particle loses its energy quickly enough so that all of the electrons and holes created by this particle are collected in a single pulse of current. Therefore, the greater the energy of the detected particle, the more electron-hole pairs are produced, and the larger the current pulse is generated at the diode terminals. This current pulse is then sent to a charge-sensitive preamplifier to produce a signal proportional to the total charge. This signal is then sent to a spectroscopy amplifier, which outputs an approximately Gaussian signal proportional in amplitude to the size of the input pulse. The unipolar output signal from the amplifier is then sent to an analog-to-digital converter (ADC) to create a digital signal corresponding to the maximum amplitude of each pulse. A useful exercise at the beginning of each experiment is to examine on an oscilloscope the signal coming from the bipolar output of the amplifier. This signal should be examined as the bias voltage is applied to the detector in order to examine the noise reduction obtained for the optimum detector bias voltage. As the data are collected, different amplitude Gaussian peaks are seen on the oscilloscope corresponding to different energies of the detected particles. The gain of the amplifier is adjusted such that all of the most energetic particles will appear at a voltage less than the maximum input voltage for the ADC (typically 8- to 10-V maximum signal).

The digital signal coming from the ADC is recorded as a data point in a channel on an MCA in which the channel number is proportional to the amplitude of the digital signal. This creates a histogram (spectrum) of data in which the channel number represents the height of detected pulses and the yield for a given channel equals the number pulses detected for that pulse height. Each new pulse-height signal that is collected is then added to and stored in the spectrum as one additional count for that channel. This type of data acquisition is known as pulse-height analysis (PHA). Thus, the yield recorded on the MCA as a function of channel number corresponds to the yield of scattered particles with a detected energy $E_d$ in which each channel number corresponds to a value of $E_d$. As mentioned above, the gate signal from the current integrator (used to measure the incident beam current) is used to turn off the ADC in order to stop the data collection. The remainder of this unit will review how the RBS and ERD spectra are converted from yield to composition and provide concrete examples.

### Other Detectors and Detection Geometries

So far this unit has concentrated on conventional ERD analysis geometries, but another possible geometry for ERD is the transmission mode. For transmission ERD, the sample must be thinner than the range of the recoiled atom to be profiled. The projectile beam usually impinges on the sample at or near normal incidence, and the detector is placed at a recoil scattering angle of 0°. Additional range foils are used to stop the high-intensity projectile beam or the sample itself may be sufficient to stop the projectile. The main advantage of transmission ERD is increased sensitivity, by as much as 2 orders of magnitude, in comparison to conventional reflection geometry ERD. This increase in sensitivity stems from several factors: (1) since the analysis range scales with $\cos \Theta_1$, more material is probed with transmission ERD; (2) larger solid angles are used without significant kinematic broadening in transmission ERD since $dK/d\phi = 0$ at $\phi = 0°$ and finally, (3) the background caused by surface H is reduced considerably in the transmission mode due to smaller multiple scattering cross-sections. Wielunski et al. (1986) demonstrated a one- atom ppm sensitivity for detecting H in Ni using 4- to 6-MeV He projectiles in the transmission mode.

The scattering dynamics of ERD and RBS are quite similar, especially when viewed in the center-of-mass frame of reference. Some of these similarities are good, such as the ease of acquiring both types of spectra; but other shared aspects are actually bad, such as the built-in mass-depth ambiguity that complicates the interpretation of both ERD and RBS energy spectra. This ambiguity results because the energy of the backscattered or recoiled ions depends both on the mass of the target atom and on the depth in the sample where the scattering occurred. Further, in the case of ERD, a mass ambiguity for scattering from the surface of a sample due to the energy loss suffered in the range foil also exists. Avoiding this ambiguity is nearly impossible in RBS because the detected ion is always the same as the incoming ion. However, in ERD, the recoil ion energy and mass are measured independently either by employing an $E - \Delta E$ particle telescope (Gebauer et al., 1990) or by combining a measurement of the ion's flight time (i.e., velocity) to that of total energy (Thomas et al., 1986; Whitlow, 1990). This latter technique is referred to as time of flight. An additional benefit of TOF techniques is that the depth resolution is improved considerably, and the use of a TOF detector is beneficial to increase depth resolution for both RBS and ERD analyses. This improvement in depth resolution results because the timing resolution involved with TOF is generally much better than the energy resolution involved with SBDs. Time-of-flight detection involves the simultaneous measurement of both the velocity and total energy of the detected atoms. The detected atom velocity is determined by measuring the time $t$ required to pass along a preset flight path of length $L$, and then the detected energy is given by $E_d = ML^2/2t^2$.

## DATA ANALYSIS AND INITIAL INTERPRETATION

### Equations for IBA

The analytical expressions that relate the observed RBS and ERD energy spectra (the yield equation) to target atom type and concentration with depth are quite similar and will be developed here in brief. A more exhaustive treatment in the development of these equations is found in Chapters 4 and 5 of Tesmer et al. (1995), Chapter 3 of Chu et al. (1978), Brice (1973), Foti et al. (1977), and Doyle and Brice (1988). The geometries for RBS and ERD kinematic scattering in the laboratory reference frame are given in Figure 5, in which the incident beam and the directions of detection are coplanar. In this figure, for simplicity, the RBS and ERD geometries are combined, and the atom configurations before a scattering event are shown with cross-hatched circles and the atom configurations after a scattering event are shown schematically with an open circle for RBS and solid circles for ERD. However, the backscattering angle $\theta$ for RBS and forward-scattering angle $\phi$ for ERD are not necessarily coplanar, and the following development is made independent of that representation in the figure. The scattering angles are both measured relative to the incident beam direction. An ion beam with energy $E_0$, mass $M_1$, and atomic number $Z_1$ is incident onto a target atom with mass $M_2$ and atomic number $Z_2$. For RBS, the projectile is typically protons or He ions and $M_1 < M_2$, while the projectile for ERD varies widely from He ions to Au ions and $M_1 > M_2$. Through conservation of energy and momentum, the energy of the scat-

**KINEMATICS: LAB FRAME**

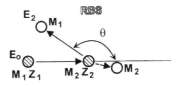

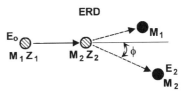

**Figure 5.** Schematic representation of two different elastic collision events occurring between a projectile of mass $M_1$, atomic number $Z_1$, and energy $E_0$ and a target of mass $M_2$ (atomic number $Z_2$) that is at rest before the collision. For RBS (open circle), the incident ion is backscattered with an energy $E_2$ at an angle $\theta$ measured relative to the incident beam direction. For ERD (filled circles), the target mass is recoiled forward at an angle $\phi$ with an energy $E_2$.

tered atom, $E_2$, is related to $E_0$ by the kinematic factor $K$ such that

$$E_2 = KE_0 \qquad (2)$$

and

$$K = \left( \frac{\sqrt{M_2^2 - M_1^2 \sin^2 \theta} + M_1 \cos \theta}{M_1 + M_2} \right)^2 \qquad (3)$$

for RBS or

$$K = \frac{4 M_1 M_2 \cos^2 \phi}{(M_1 + M_2)^2} \qquad (4)$$

for ERD.

Equations 2 to 4 relate the energy before and after the scattering event, independent of depth in the sample.

The following development is based on a slab analysis in which the variables needed for the yield equation are evaluated at energies corresponding to uniform increments of depth in the sample. Figure 6 gives the geometry for the ion-scattering processes occurring at a depth $x$ in the sample measured along the sample normal. This figure describes either the forward-scattering process using a range foil with thickness $X^{(0)}$ or the backscattering process without a range foil $[X^{(0)} = 0]$. The projectile ion beam is incident upon the sample at an angle $\Theta_1$ measured from the surface normal to the beam direction, and the detected beam is collected in an SBD at angle $\Theta_2$ measured from the surface normal. In an experiment, the scattering angles are fixed by the apparatus while $\Theta_1$ and $\Theta_2$ vary depending on the sample tilt. In Figure 6, $\Theta_1$ and $\Theta_2$ are coplanar and related to the scattering angle by $\theta = \pi - (\Theta_1 + \Theta_2)$ for

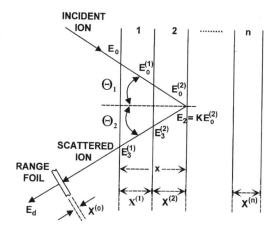

**Figure 6.** Schematic configuration for an elastic ion-scattering experiment used to describe either RBS or ERD analysis in which the sample is divided into $n$ layers of thickness $X^{(n)}$) for calculations. The superscripts in parentheses refer to layer number and the subscripts on $E$ refer to the type of atom: $0$ = incident ion, $2$, $3$ = scattered ion. The parameters are incident ion energy $E_0$, incident and exit angles $\Theta_1$ and $\Theta_2$ (measured relative to the sample normal), the energy ($E_2$) of the scattered atom at depth $x$, the energy of the scattered atom ($E_3$) as it traverses the sample, and the energy detected in an SBD ($E_d$).

RBS or $\phi = \pi - (\Theta_1 + \Theta_2)$ for ERD [note that $\Theta_1 + \Theta_2 > \pi/2$ for ERD]; however, the formalism developed below does not require that $\Theta_1$ be coplanar with $\Theta_2$, and in a general experimental apparatus they may not be coplanar. The incident projectile enters the sample with energy $E_0$ and loses energy ($dE/dx$) along its path in the sample, and at a depth $x$ the projectile has an energy $E_0^{(2)}$ just prior to a scattering event. Thus, the initial energy of the detected atom after the scattering event is $E_2 = KE_0^{(2)}$, and after traversing the sample this atom exits with energy $E_3^{(1)}$. For ERD, the recoiled atom is then incident normal to a range foil, which is assumed to be parallel to the detector surface, and this atom then loses more energy in the range foil. For either RBS or ERD, the atom is then detected with energy $E_d$ in an SBD ($E_d = E_3^{(1)}$ for RBS).

The yield $Y$ of detected atoms in energy channel $E_d$ with a channel width $E_d$ (in kilo-electron-volts perchannel) is given by

$$Y(E_d) = \frac{QN(x)\sigma(E_0^{(n)})\Omega\,\delta E_d}{\cos\Theta_1 dE_d/dx} \tag{5}$$

where $Q$ is the number of incident projectiles, $N(x)$ is the atomic density of the target atom at depth $x$, $\sigma(E_0^{(n)})$ is the average scattering cross-section in laboratory coordinates at an energy $E_0^{(n)}$, $\Omega$ is the detector solid angle, and $dx$ is the increment of depth at $x$ corresponding to an increment of energy $dE_d$. The quantities $Q$, $\Omega$, and $\Theta_1$ are fixed by the experimental setup. A defining aperture is usually placed in front of the detector in ERD to maximize the depth resolution, which is degraded by the trajectory of the particles entering the detector at different scattering angles. Rigorously, $\Omega$ is the integral of the differential surface area element in spherical coordinates, but it is often calibrated

by the yield from a standard sample (generally a H profile in ERD or a Si substrate profile in RBS). Care should be taken to avoid ion beam channeling in a crystalline sample when using the spectrum's surface edge height to calibrate a detector solid angle. Typical solid angles are ~5 msr for ERD analyses and 1 msr $\leq \Omega \leq$ 10 msr for RBS analyses. To transform the yield to a concentration profile, analytical expressions for $\sigma(E_0^{(n)})$ and $dE_d/dx$ must be determined.

First, the cross-sections for many instances are governed by Coulomb scattering, as modeled by Rutherford (1911), and expressed in the laboratory reference frame as (Marion and Young, 1968)

$$\frac{d\sigma_R(E_0^{(n)}, \theta)}{d\Omega} = \left(\frac{Z_1 Z_2 e^2}{4E_0^{(n)}}\right)^2 \left(\frac{4}{\sin^4\theta}\right)$$
$$\times \frac{\sqrt{1 - [(M_1/M_2)\sin\theta]^2} + \cos\theta)^2}{\sqrt{1 - [(M_1/M_2)\sin\theta]^2}} \tag{6}$$

for RBS and

$$\frac{d\sigma_R(E_0^{(n)}, \phi)}{d\Omega} = \left(\frac{Z_1 Z_2 e^2}{4E_0^{(n)}}\right)^2 \left(\frac{4}{\cos^3\phi}\right)\frac{(M_1 + M_2)^2}{(M_2)^2} \tag{7}$$

for ERD, where $(e^2/E)^2 = 0.020731$ b for $E = 1$ MeV ($1\,b = 1 \times 10^{-24}$ cm$^2$). In practice, the differential cross-section is averaged over a finite volume $\Omega$ and given simply as $\sigma_R$. For those cases in which the cross-section is non-Rutherford (see Problems), the simplest solution for the thin-film approximation is to multiply the $\sigma_R$ in Equations 6 and 7 by a scaling factor $f$ in order to model the true cross-section, i.e., $\sigma = f\sigma_R$. It is important to note the functional dependencies of $\sigma_R$ in Equations 6 and 7.

1. Both cross-sections are inversely proportional to the square of the projectile energy, and therefore $Y(E_d)$ increases as $(1/E_0^{(n)})^2$. Thus, the projectile energy should be decreased in order to increase the yield or sensitivity for a given element.

2. Since atomic mass is proportional to atomic number, the RBS $\sigma_R$ is proportional to $(Z_1)^2$, and the ERD $\sigma_R$ is $\sim (Z_1)^2$ for $M_1 \approx M_2$ and $\sim (Z_1)^4$ for $M_1 \gg M_2$. Therefore the yield for a given element is increased by using heavier projectiles.

3. The RBS $\sigma_R$ is proportional to $(Z_2)^2$, and the ERD $\sigma_R$ is $\sim (Z_2)^2$ for $M_1 \approx M_2$; therefore the scattering probability is greater for heavy targets than for light targets. The yield in RBS is greater for heavy targets than for light targets, but the yield in ERD also depends upon the stopping powers for the recoiled atom. For $M_1 \gg M_2$, the ERD $\sigma_R$ is less dependent upon $Z_2$.

4. Both cross-sections are axially symmetric with respect to the incident beam direction and independent of $\Theta_1$ and $\Theta_2$. Thus the sample can be tilted and rotated to change the ion beam path lengths without changing the angular dependence in $\sigma_R$. For RBS,

the $(1/\sin\theta)^4$ dependence increases the yield as the backscattering angle is increased while the $\cos\Theta_1$ term in the yield equation tends to cancel the $1/\cos^3\phi$ increase in the ERD cross-section.

As described above (see Introduction), the areal density (AD) is the integrated concentration $N$ over the layer thickness $\tau$. From the yield Equation 5,

$$\text{AD} = \int_0^\tau N(x)\,dx = \frac{\cos\Theta_1}{Q\Omega}\int_0^\tau \frac{S^{\text{eff}}}{\sigma\,\delta E_{\text{d}}}Y(E_{\text{d}})\,dx$$
$$= \frac{\cos\Theta_1}{Q\Omega}\int_{E_{\text{d}}(0)}^{E_{\text{d}}(x)}\frac{Y(E_{\text{d}})}{\sigma}\frac{dE_{\text{d}}}{\delta E_{\text{d}}} \qquad (8)$$

where $S^{\text{eff}}$ is the effective stopping power $(=dE_{\text{d}}|dx|_x)$. Equation 8 shows that the effective stopping power term is eliminated by integration when determining the areal density. In practice, the final integral of Equation 8 is evaluated as a summation; and for a surface peak, the cross-section $\sigma_0$ is evaluated at the specific energy $E_0$:

$$\text{AD} = \frac{\cos\Theta_1}{Q\Omega\sigma_0}\sum_{\text{channel}[E_{\text{d}}(0)]}^{\text{channel}[E_{\text{d}}(x)]}Y(\text{channel}) \qquad (9)$$

Expressions for $\sigma(E_0^{(n)}$ and $dE_{\text{d}}/dx$ are needed to convert the yield to a concentration profile, and these expressions are derived in terms of the energies $E_0^{(n)}$ and stopping powers $(dE/dx)$ evaluated at $E_0^{(n)}$ corresponding to uniform increments of depth in the sample. A simple method to calculate the apparent energy loss in the detected beam as a function of depth is to divide the sample into a series of slabs of equal thickness, and then create a table of the corresponding energies and stopping powers for each layer or slab of the material, e.g., as shown schematically in Figure 6. The energy and depth relationships are expressed by the equations given below using the following convention for superscripts and subscripts. The subscripts refer to the atom type, except for the notation given above for the angles $\Theta_1$ and $\Theta_2$, and the superscripts refer to the layer of the target (1, 2, ...) or to the range foil (superscript 0). For example, the projectile energy and stopping powers are $E_0^{(n)}$ and $S_{\text{p}}^{(n)}$, respectively; the energy for the detected atom is either $E_2^{(n)}$, or $E_3^{(n)}$ and the stopping power is $S_{\text{r}}^{(n)}$ for recoiled atoms and $S_{\text{b}}^{(n)}$ for backscattered atoms. The electronic energy lost by a particle as it traverses a slab containing more than one type of atom is approximated using the principle of additivity of stopping cross-sections first given by Bragg and Kleeman (1905). The stopping cross-section $\varepsilon \equiv (1/N)(dE/dx)$ is the proportionality factor between the amount of energy lost in a thin slab and the areal density of that slab. The energy loss in a material composed of several atom species is the sum of the losses for each species weighted according to the composition. This is known as Bragg's rule, and for a material with formula unit $A_\alpha B_\beta$,

$$\varepsilon^{A_\alpha B_\beta} = \alpha\varepsilon^A + \beta\varepsilon^B \quad\text{and}\quad S = N^{A_\alpha B_\beta}\varepsilon^{A_\alpha B_\beta} \qquad (10)$$

The length of path traveled by the projectile crossing the $n$th slab with thickness $X^{(n)}$ is $X^{(n)}/\cos\Theta_1$, and the

energy lost by the projectile is $E_0^{(n)} - E_0^{(n-1)}$. For $X^{(n)}$ sufficiently small, $S_{\text{p}}^{(n)}$ changes little across the slab and is approximated by the value at energy $E_0^{(n-1)}$ such that:

$$E_0^{(n)} = E_0^{(n-1)} - \frac{X^{(n)}}{\cos\Theta_1}S_{\text{p}}^{(n)}(E_0^{(n-1)}) \qquad (11)$$

Similarly, the energy $E_0^{(n)}$ for the detected particle as a function of depth in the sample is written in terms of a recursion relation for the energy from the previous slab boundary, or in terms of the kinematic relationship Equation 2 just after a scattering event. For an RBS scattering event occurring in slab $n$,

$$E_3^{(n)} = KE_0^{(n)} - \frac{X^{(n)}}{\cos\Theta_2}S_{\text{b}}^{(n)}(KE_0^{(n)}) \qquad (12)$$

and for an RBS event occurring in slab $j > n$,

$$E_3^{(n)} = E_3^{(n+1)} - \frac{X^{(n)}}{\cos\Theta_2}S_{\text{b}}^{(n)}(E_3^{(n+1)}) \qquad (13)$$

Equations 12 and 13 for ERD have a similar form and are given below (see Appendix B). In practice, the depth in the sample is related to the detected energy measured in a spectrum, and therefore the following equations relate the detected energy for the $n$th slab, $E_{\text{d}}^{(n)}$, to the thickness $X^{(n)}$:

$$E_{\text{d}}^{(n)} = K\left[E_0 - \frac{X^{(n)}}{\cos\Theta_1}\sum_{j=1}^n S_{\text{p}}^{(j)}(E_0^{(j-1)})\right]$$
$$- \frac{X^{(n)}}{\cos\Theta_2}\left[S_{\text{b}}^{(n)}(KE_0^{(n)}) + \sum_{j=n}^2 S_{\text{b}}^{(j-1)}(E_3^{(j)})\right] \qquad (14)$$

for RBS and

$$E_{\text{d}}^{(n)} = K\left[E_0 - \frac{X^{(n)}}{\cos\Theta_1}\sum_{j=1}^n S_{\text{p}}^{(j)}(E_0^{(j-1)})\right]$$
$$- \frac{X^{(n)}}{\cos\Theta_2}\left[S_{\text{r}}^{(n)}(KE_0^{(n)}) + \sum_{j=n}^2 S_{\text{r}}^{(j-1)}(E_3^{(j)})\right] - \Delta E_{\text{foil}} \qquad (15)$$

for ERD. Equations 14 and 15 are very similar with the change of notation to indicate the species of the detected atom [backscattered (b) vs. recoil (r)] and with the inclusion of stopping by the recoiled atom in the range foil for ERD calculations. The energy lost by the recoil atom in traversing the foil ($\Delta E_{\text{foil}}$) should be calculated from a slab analysis of the foil containing NF slabs of thickness $\Delta x^{(0)}$ (i.e., $\text{NF} = X^{(0)}/\Delta x^{(0)}$) as

$$\Delta E_{\text{foil}} = \Delta x^{(0)}\sum_{j=1}^{\text{NF}}S_{\text{r}}^{(0)}(E_{\text{f}}^{(j-1)}) \qquad (16)$$

where the recoil atom stopping power is evaluated at the energy $E_{\text{f}}^{(j-1)}$ before traversing the $j$th slab (e.g., $E_{\text{f}}^{(0)} =$

$E_3^{(1)}$). Finally, the effective stopping power for the deleted particle at a given projectile energy, $dE_d/dx$, is given as

$$\left. \frac{dE_d}{dx} \right|_n = [S]_{p,b}^{(n)} \frac{S_b(E_d)}{S_b(KE_0^{(n)})} \qquad (17)$$

for RBS and

$$\left. \frac{dE_d}{dx} \right|_n = [S]_{p,r}^{(n)} \prod_n \qquad (18)$$

for ERD, where the energy loss factors $[S]$ corresponding to a backscattering or recoil event at energy $E_0^{(n)}$ are

$$[S]_{p,b}^{(n)} = \frac{KS_p^{(n)}(E_0^{(n)})}{\cos \Theta_1} + \frac{S_b^{(n)}(KE_0^{(n)})}{\cos \Theta_2} \qquad (19)$$

for RBS and

$$[S]_{p,r}^{(n)} = \frac{KS_p^{(n)}(E_0^{(n)})}{\cos \Theta_1} + \frac{S_r^{(n)}(KE_0^{(n)})}{\cos \Theta_2} \qquad (20)$$

for ERD. The product term $\prod_n$ in Equation 18 for ERD analysis is given by

$$\prod_n = \frac{S_r^{(n)}(E_3^{(n)})}{S_r^{(n)}(E_2)} \frac{S_r^{(n-1)}(E_3^{(n-q)})}{S_r^{(n-1)}(E_3^{(n)})} \cdots \frac{S_r^{(0)}(E_d)}{S_r^{(0)}(E_3^{(1)})} \qquad (21)$$

which is similar to the product term $S_b(E_d)/S_b(KE_0^{(n)})$ for RBS analysis. Definitions for variables and the equations used for IBA are summarized below (see Appendices A and B).

Several methods can be followed to determine composition profiles from RBS and ERD spectra using the equations given above. In general, an iterative process is used in which the stopping powers, energies, and cross-sections for each slab layer are calculated from an estimated composition. A spectrum is simulated and compared to the data or the data are directly converted into a concentration-depth profile. Based on these results, the composition or slab thicknesses are changed and the energies, $S$, and $\sigma$ are recalculated for another simulation or conversion of the data to a profile until a convergence between the estimated composition and results are obtained. This type of approach is used in RBS analysis programs such as RUMP (Doolittle, 1986, 1994). The method of directly converting the data to a concentration profile by scaling the data with cross-sections and $dE_d/dx$ as a function of depth is known as spectral scaling. In this approach, a table is made containing the sample depth, $E_0^{(n)}$, $\sigma(E_0^{(n)})$, $E_d$, and $dE_d/dx$ as a function of slab layer. The data are then scaled one channel at a time to produce a concentration-depth profile by interpolating between values in the table, and the energy scale is transformed to depth while the yield scale converts to concentration. The yield equation 5 is thereby converted to appear as though the cross-section

and effective stopping power are constant through the sample. Analytically, the scaling is written as

$$\begin{aligned} N^{(n)}(x) &= \frac{\cos \Theta_1}{Q\Omega \, \delta E_d} \frac{dE_d/dx|_n}{\sigma_R(E_0^{(n)})} Y^{(n)}(E_d) \\ &= \frac{\cos \Theta_1}{Q\Omega \, \delta E_d} \frac{S_0^{eff}}{\sigma_0} \left[ \frac{\sigma_R(E_0)}{\sigma_R(E_0^{(n)})} \frac{dE_d/dx|_n}{dE_d/dx|_0} Y^{(n)}(E_d) \right] \end{aligned} \qquad (22)$$

where $S_0^{eff}$ and $\sigma_0$ are evaluated at the incident energy $E_0$. The cross-section scaling factor for Rutherford cross-sections is $\sigma_R(E_0)/\sigma_R(E_0^{(n)}) = [(E_0^{(n)})/E_0]^2$, and the stopping power scaling factor, $(dE_d/dx|_n)/(dE_d/dx|_0)$, is determined by interpolating from tables. Examples of spectral scaling and slab analysis of RBS and ERD spectra will be given below.

## PROBLEMS

### Non-Rutherford Cross-Sections

Deviations from the Rutherford scattering cross-sections given in Equations 6 and 7 may result from Coulomb barrier penetration, which occurs most often for low-$Z$ projectiles at high energies, or from excess screening effects, which occur most often for high-$Z$ projectiles at low energies. At high energies, resonances occur in the backscattering cross-sections, and those resonances for which $\sigma$ is enhanced relative to $\sigma_R$ are useful for enhanced sensitivity of light elements with ion backscattering (EBS). These resonances occur because the incident particle has sufficient energy to penetrate the Coulomb potential barrier and probe the nuclear potential, which produces an attractive, strong, short-range force. The Coulomb potential produces a relatively weak, long-ranged repulsive force. At high energies, the projectile may penetrate into the nucleus of the target atom such that the two nuclei combine for a finite time as a single excited-state "compound nucleus." The width of a given resonance depends on the lifetime of the excited state in the compound nucleus, and for large resonance widths or energies with overlapping resonances, $\sigma$ is relatively constant with energy in order to permit EBS depth profiling. Bozoian et al. (1990), Bozoian (1991a,b), and Hubbard et al. (1991) modeled the case of Coulomb barrier penetration and determined the energies $E_{NR}$, where the cross-sections may become non-Rutherford by 4%. In the lab reference frame their formulation of this energy is given as (Bozoian et al., 1990)

$$E_{NR}(MeV) \approx (0.12 \pm 0.01)Z_2 - (0.5 \pm 0.1) \qquad (23)$$

for proton backscattering with $160° < \theta(lab) < 180°$,

$$E_{NR}(MeV) \approx (0.25 \pm 0.01)Z_2 + (0.4 \pm 0.2) \qquad (24)$$

for $^4$He backscattering with $160° < \theta(lab) < 180°$, and

$$E_{NR}(MeV) \approx \frac{-1.1934(Z_1 Z_2)(M_1 + M_2)}{M_1^{4/3} M_2 \ln(0.001846 Z_1 M_2 / Z_2)} \qquad (25)$$

for ERD with $\phi = 30°$. Furthermore, Andersen et al. (1980) modeled the case where screening effects cause significant deviations from the Rutherford formula, and in the laboratory reference frame for 1% screening effects (i.e., $\sigma/\sigma_R = 0.99$), the non-Rutherford energy boundary is given by

$$E_{NR}(\text{MeV/amu}) = 99V_{LJ}\frac{M_1 + M_2}{M_1 M_2} \qquad (26)$$

where

$$V_{LJ} = 48.73Z_1Z_2(Z_1^{2/3} + Z_2^{2/3})^{1/2} \qquad (27)$$

is the Lenz-Jensen potential in electron volts.

The use of a He ion beam for ERD analysis is nearly always in an energy range where penetration of the Coulomb barrier occurs, and the use of a Au ion beam for ERD is nearly always in an energy range where significant screening effects occur (for current IBA accelerator technology). A quick method to determine the shape of α particle and proton backscattering cross-sections at high energies is to examine the nuclear data sheet of the compound nucleus for the given nuclear reaction (see Internet Resources). As an example, the nuclear reaction $^{16}O(\alpha,\alpha)^{16}O$ for high-energy α particles backscattering from the $^{16}O$ nucleus creates a compound nucleus of $^{20}Ne$. Figure 7 shows a portion of the data sheet (Tilley et al., 1998) for $^{20}Ne$ at the top and a magnified section of the $^{16}O(\alpha,\alpha)^{16}O$ cross-section at the bottom. This figure shows that sharp resonances occur at laboratory energies (in mega-electron-volts) of 2.52, 3.04, 3.08, 3.37, 3.89, 4.9,..., and a broader resonance occurs from ~8.35 to 8.85 MeV, which is particularly useful for oxygen depth profiling. Even though the shape of the cross-section is shown here, the actual value of an enhanced cross-section should be determined from a well-characterized standard for the energy and scattering angle used in a given experiment.

### Radiation Hazards

When the incident beam energy exceeds the Coulomb barrier of the target, nuclear reactions other than elastic scattering are possible. High-energy proton scattering can produce unwanted γ or neutron radiation exposures to workers, even along sections of the beam line away from the sample chamber. This concern is particularly significant for irradiation of light elements, but it is also important for elements such as Cu. Neutrons and γ rays are the main prompt radiation hazard while activation of the sample produces a longer lasting radiation hazard (of particular importance for high-energy proton beams). Activated samples can be easily controlled to avoid accidental radiation exposure of workers as well as satisfy regulatory agencies, but a knowledge of possible activating nuclear reactions is necessary for this purpose. Novices should seek help from a qualified radiation-health physicist to assess the hazards when using a high-energy proton beam.

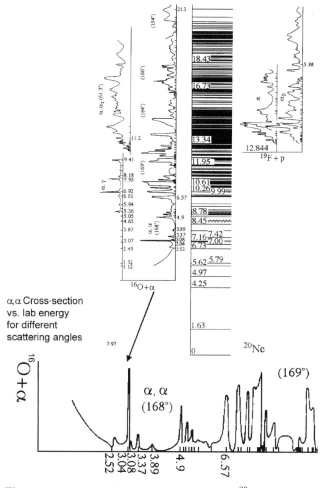

**Figure 7.** A portion of the nuclear data sheet for $^{20}Ne$ is shown at top, and a magnified view of the $^{16}O(\alpha,\alpha)^{16}O$ cross-section from this data sheet is shown at bottom.

### Enhanced-Cross-Section Backscattering Spectrometry

Rutherford backscattering spectrometry was shown above to be highly useful for thin-film composition analysis. However, one of the disadvantages of RBS is its relative insensitivity to light elements, particularly lighter elements on heavy element substrates. Enhanced-cross-section backscattering with He ions offers sensitivity for light elements such as B, C, N, and O, and EBS with protons offers increased sensitivity for these light elements as well as even heavier elements such as S. The disadvantages of EBS as opposed to RBS is that the depth resolution for the higher energy EBS technique is generally less than that for RBS and the cross-section does not obey the Rutherford equations 6 and 7. Therefore $\sigma_{EBS}$ must be calibrated from a standard. The decreased sensitivity to light elements in RBS results directly from the fact that $\sigma_R$ is proportional to $(Z_1Z_2/E)^2$. Furthermore, RBS analysis is complicated by the fact that both H and He projectiles show deviations from $\sigma_R$ for light elements at relatively low energy, and in the low-energy regime where low-$Z$ element cross-sections are still Rutherford, the depth of analysis is rather small (<0.5 μm) and the background from heavier substrates is high.

The analytical formulas for EBS analysis are identical to those for RBS analysis with a change in the value of the cross-section. The elastic scattering cross-sections for high-energy α particles and protons were some of the first cross-sections measured with the emergence of modern particle accelerators and nuclear physics in the mid-20th century. The elements with the most useful α high-energy cross-sections are [11]B, [12]C, [14]N, and [16]O, while elements out to [32]S have been analyzed using high-energy proton cross-sections but with much less depth resolution. One difficulty found for EBS analysis is when the matrix contains moderate- to low-Z elements such as Si or Al, which have sharp resonances in their cross-sections, producing wild fluctuations in the backscattering signal from the matrix. These fluctuations make a determination of the light atom concentration nearly impossible, as is the case for oxygen profiling from $SiO_2$ or $Al_2O_3$ using 8.7-MeV $^4He$ ion EBS analysis.

## EXAMPLES

Several examples are now given to demonstrate mass resolution, energy resolution, and depth of analysis for backscattering and ERD as a function of ion energy and ion species. Also, examples of how to convert spectra to composition-depth profiles for both RBS and ERD analyses are demonstrated. Figure 8 shows an RBS spectrum (+) for 2.2-MeV He$^+$ ions incident upon a film of $La_xSr_yCo_zO_3$ grown on a $LaAlO_3$ substrate. The spectrum is plotted from channel 15 to channel 615 similar to how it would appear when collected as data on an MCA. An MCA gives yield in counts, but for convenience the yield in Figure 8

**Table 2. Energy Scale for RBS Spectrum in Figure 8**

| Element | Surface Scattering Energy (MeV) | Substrate Scattering Energy (MeV) |
|---|---|---|
| O | 0.8069 | (0.620) |
| Al | (1.224) | 1.040 |
| Co | 1.685 | |
| Sr | 1.839 | |
| La | 1.965 | 1.780 |

is normalized by the number of incident particles ($Q = 20$ pμC), the solid angle ($\Omega = 6.95$ msr), and the channel energy width (3.314 keV). [Note that 20 pμC (particle micro- coulombs) is equal to $(20 \times 10^{-6}/1.602 \times 10^{-19})$ particles.] The sample was tilted 10° in order to decrease channeling in the substrate, and the detector was mounted at a scattering angle $\theta = 164°$. The top (energy) axis was determined using the surface scattering energies for O, Co, and La. These energies are given in Table 2 along with the detected energies for scattering from O, Al, and La at the film-substrate interface. Also, the surface scattering positions are marked with arrows in the figure. The Al surface energy is shown in parentheses in the table because Al is only present in the substrate, and therefore the Al edge energy is decreased from 1.224 to 1.040 MeV. The O substrate edge energy is marked in parentheses because the signal from oxygen in the substrate is small and difficult to differentiate from the background. Therefore, the O substrate energy given in Table 2 is based on the relative position of the O surface energy and the energy width of the layer as determined from the La signal.

Figure 8 shows a spectrum from a relatively simple sample, a thin film with a uniform composition (≈190 nm thick). Yet, the spectrum contains many overlapping peaks, which makes composition-depth profiling of a single element difficult because of the variation in the background signal due to the other elements in the film and the substrate. In these circumstances, a simulation of the spectrum often suffices to determine the composition of the film. A simulation of Rutherford backscattering from a $La_{0.85}Sr_{0.15}Co_{0.95}O_3$ film with an areal density of $1.6 \times 10^{18}$ atoms/cm$^2$ is shown in this figure as a solid line using the analysis program RUMP (Doolittle, 1994). The oxygen content was determined from the reduced height of the individual metallic constituents in the film and later confirmed with greater accuracy using 8.7-MeV He$^{2+}$ EBS. The height of the simulated substrate signal confirmed the values previously measured for $Q$ and $\Omega$. Deviations from the measured values occur for large dead times on the ADC or if the incident ion channels into the substrate; however, the agreement between the values given above and those determined from the simulation is better than 3%. At energies below ~0.6 MeV, the data and the simulation begin to deviate substantially as a result of multiple scattering in the substrate, which was not taken into account in the simulation. The layer thickness is given as areal density in the simulation because the exact density of the film is unknown, but if a density of $0.85 \times 10^{23}$ atoms/cm$^3$ is assumed, then the film thickness is 188 nm. For more simple spectra in which

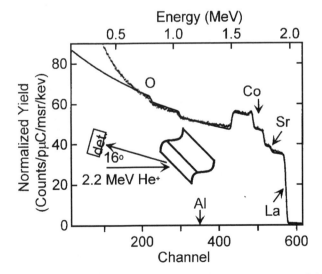

**Figure 8.** Backscattering spectrum (+) and simulation (solid line) obtained from 2.2-MeV He$^+$ ions scattered off $La_xSr_yCo_zO_3$/$LaAlO_3$ at 164° relative to the incident beam direction, with $Q = 20$ pμC, $\Omega = 6.95$ msr, $\Theta_1 = 10°$, and channel energy width 3.314 keV/channel.

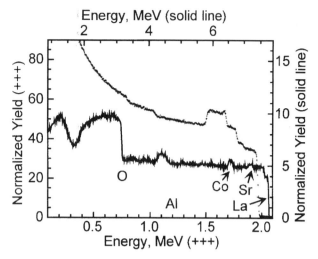

**Figure 9.** Spectra for backscattering off the same sample as in Figure 8. The 2.2-MeV He$^+$ spectra is reproduced here ($+$) in addition to a spectrum (solid line) obtained from 8.7-MeV He$^{2+}$ ion scattering at 164°, with $Q = 18$ pµC, $\Omega = 6.95$ msr, $\Theta_1 = 45°$, and channel energy width 7.82 keV/channel. The left and bottom axes are for the low-energy RBS spectrum, and the right and top axes are for the high-energy EBS spectrum.

peaks of the individual elements are easily separated from a background signal, each peak is integrated to determine the areal density for that element and thus provide the relative areal densities for the composition of the layer. Also, if the ion energy were increased for analysis of this sample, then the mass resolution would increase and the individual peaks would separate in the spectrum.

A He ion energy of 8.7 MeV was used to analyze the same sample as given in Figure 8, and the spectrum is shown as a solid line in Figure 9, with the normalized yield plotted on the right axis and the energy plotted along the top axis. The parameters used for this spectrum were $Q = 18$ pµC, $\Omega = 6.95$ msr, channel energy width 7.82 keV/channel, $\Theta_1 = 45°$, and $\theta = 164°$. The sample was tilted 45° to make the apparent thickness increase for increased depth resolution. For comparison, the 2.2-MeV He$^+$ RBS spectrum ($+$) is plotted versus energy along the bottom axis and normalized yield on the left axis. To put these spectra on the same figure in a readable manner, two separate sets of axes were used because the energy and yield scales differ considerably between the spectra. In fact, it is this difference in the energy scale that gives rise to the separation of signals from different masses. The surface energies for 8.7-MeV He EBS are given in Table 3 along

**Table 3. Energy Scale for EBS Spectrum in Figure 9 (Solid Curve)**

| Element | Surface Scattering Energy (MeV) | Substrate Scattering Energy (MeV) |
|---------|---------------------------------|-----------------------------------|
| O | 3.192 | 3.03 |
| Al | (4.841) | 4.68 |
| Co | 6.663 | |
| Sr | 7.272 | |
| La | 7.770 | 7.61 |

with the detected energies for scattering from O, Al, and La at the film-substrate interface, and the surface scattering positions for the EBS spectrum are marked with arrows in Figure 9. A comparison of the surface energies in Tables 2 and 3 show that the mass resolution, which is given by the difference in energy between surface edges, is greater for 8.7-MeV He scattering than for 2.2-MeV He scattering by a factor of 4. At the same time, the depth resolution is decreased by using a higher energy ion beam. Depth resolution $\delta x$ is given by the resolution measured along the detected energy axis divided by the effective stopping power for the detected atom: $\delta x = E_d/(dE_d/dx)$. The front edges for the peaks shows that $E_d \sim 2$ times smaller for the low-energy spectrum than for the high-energy spectrum, and the effective stopping power is greater for the low-energy spectrum than for the high-energy spectrum. Thus, the depth resolution is considerably better at low energies than at high energies.

The high-energy ion beam with the lower stopping cross-section does afford a greater depth of analysis that is useful for thick samples. The depth of analysis can be estimated from stopping and range computer codes such as TRIM (Ziegler, 1987), and as a practical limit the maximum depth is between one-tenth and one-third the projected range. Moreover, the greatest advantage to using higher energy ion beams in an EBS analysis is demonstrated by the increased yield from O relative to La. The yield from the low-energy backscattering is considerably greater than the yield from the high-energy backscattering because of a $1/E^2$ dependence for the Rutherford cross-section. At 8.7 MeV, the He scattering from both Sr and La obeys the Rutherford formula, and therefore the yield is less than at 2.2 MeV, but the O cross-section is $22\sigma_R$ at 8.7 MeV ($\theta = 164°$, $\Omega = 6.95$ msr), and therefore the signal-to-background noise for the O signal is improved at the higher energy. The Co and Al also have non-Rutherford $\sigma$ at 8.7 MeV, but $\sigma(\text{Co}) < \sigma_R(\text{Co})$ and the $\sigma(\text{Al})$ varies considerably. The rapid variation in the Al cross-section for 8.7-MeV He EBS adds uncertainty to the background below the O signal. The addition of sharp resonance background peaks below an O signal becomes much worse for Si, Al, or Al$_2$O$_3$ substrates.

A sample more complicated than the example just given can still yield a composition-depth profile as shown in the next analysis. Figure 10 is an RBS spectrum ($+$) collected from a thick film of Al deposited on a Si substrate in such a manner that the oxygen content in the film varies greatly with depth. A 2.8-MeV He$^+$ ion beam was used for this analysis to avoid sharp resonance peaks from Al or Si and to give a compromise between depth of analysis and depth resolution. Also, $\sigma(\text{O})$ was calibrated previously at this energy and for this detector geometry such that $\sigma(\text{O})/\sigma_R(\text{O}) = 1.25$. The energy scale for the spectrum in Figure 10 was calibrated from another sample such that the channel energy width is 3.11 keV/channel, $\Theta_1 = 5°$, $Q = 20$ pµC, $\Omega = 6.95$ msr, and $\theta = 164°$. The surface energies for O, Al, and Si are marked in the figure at energies 1.027, 1.558, and 1.595 MeV, respectively. To simulate this spectrum fairly well, the film was divided into six separate layers in which the thickness and composition were allowed to vary, as shown in Table 4 with the layer number

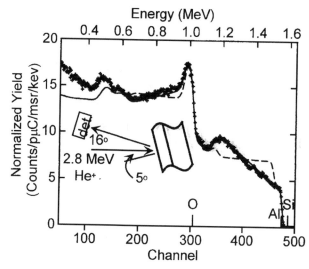

**Figure 10.** An RBS spectrum (+) and simulations (solid line and dashed line) for 2.8-MeV He$^+$ backscattering from an AlO$_x$/Si sample with a greatly varying oxygen content. The scattering angle is 164°, $\Theta_1 = 5°$, $Q = 20$ p$\mu$C, and $\Omega = 6.95$ msr.

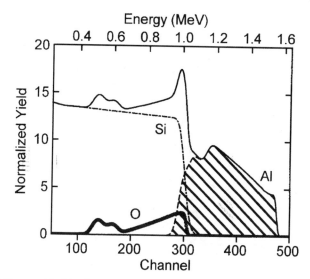

**Figure 11.** Simulated RBS spectrum (solid line) and the contributions to this simulation from the individual elements: Si (dash-dotted line), O (open circles), and Al (dashed line delineating the cross-hatched region). The simulation is for 2.8-MeV He$^+$ backscattering as described in Figure 10.

increasing in depth. Again, the thickness is given as a real density because the volume density of each layer is not known for such a complicated sample. The thicker layers were divided into two or more equal-thickness sublayers to increase the accuracy of the energy and stopping power calculations as a function of depth. Layer 2 was simulated as ten equal-thickness sublayers such that the composition across layer 2 graded linearly, as shown This simulation is presented in Figure 10 as a solid line in addition to a simulation (dashed line) assuming that layer 2 has a constant average composition Al$_{0.675}$O$_{0.325}$. The dashed-line simulation fits the data poorly. For such a complicated sample, the simulated sample structure in Table 4 may not be unique, but it will serve well to calculate the stopping powers and energies as a function of depth, and then the spectrum is scaled to give the composition-depth profile from the data, as demonstrated below. Further, the figure shows that the number of sample subdivisions used in the calculation is quite important, as determined from the comparison of simulations for layer 2.

The linearly graded sample simulation is given again for clarity as a solid line in Figure 11, and the individual contributions to the simulation are presented as a dash-dotted line for Si, open circles for O, and a dashed line

with a cross-hatched region for Al. The contributions to the RBS spectrum from all three elements overlap in the spectral region from ~0.9 to 1.1 MeV, and this overlap produces a sharp peak at 1 MeV. The RBS signal from each element is separated from the background in the spectrum by subtracting the individual contributions to the simulation. Figures 12 and 13 show the RBS signals from the O (open circles) and the Al (+) after the simulated Si and Al (or O) signals were subtracted from the data. The solid lines are the contributions to the simulation for that element. The yields in Figures 12 and 13 were then scaled with cross-sections and stopping powers as a function of energy to give the concentration as a function of depth (in atoms per square centimeter). These scaled data are

**Table 4. Sample Structure Used for Simulation in Figure 10**

| Layer | Thickness (10$^{15}$ atoms/cm$^2$) | Composition |
|---|---|---|
| 1 | 800 (with 2 sublayers) | Al$_{0.4}$O$_{0.6}$ |
| 2 | 4900 (with 10 sublayers) | Linearly graded from Al$_{0.4}$O$_{0.6}$ to Al$_{0.95}$O$_{0.05}$ |
| 3 | 200 | Al$_{0.95}$O$_{0.05}$ |
| 4 | 1000 (with 2 sublayers) | Al$_{0.7}$O$_{0.3}$ |
| 5 | 200 | Al$_{0.95}$O$_{0.05}$ |
| 6 | 1000 (with 2 sublayers) | Al$_{0.6}$O$_{0.4}$ |

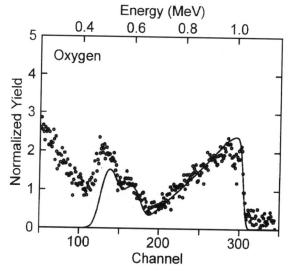

**Figure 12.** Oxygen signal (open circles) from the RBS spectrum of Figure 10 after the background Si and Al signals were subtracted. The simulated O signal is shown as a solid line.

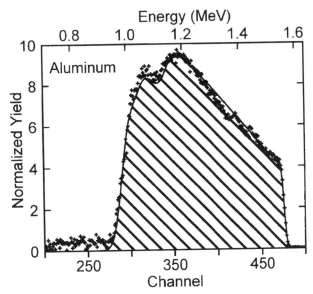

**Figure 13.** Aluminum signal (+) from the RBS spectrum of Figure 10, after the background Si and O signals were subtracted. The simulated Al signal is shown as a solid line delineating a cross-hatched region.

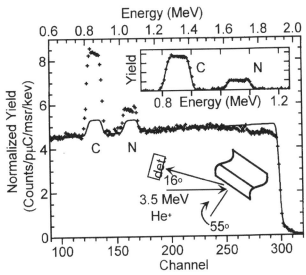

**Figure 15.** An RBS spectrum (+) and simulation based on Rutherford scattering (solid line) for 3.5-MeV He$^+$ backscattering from a $CN_x$/Si sample. The scattering angle is 164°, $\Theta_1 = 55°$, $Q = 20$ p$\mu$C, and $\Omega = 6.95$ msr. The inset shows the C and N signals above background overlayed by a simulation (solid line) based on enhanced scattering cross-sections ($5.6\sigma_R$ for C and $2\sigma_R$ for N).

plotted in Figure 14 with the depth (in micrometers) calculated from an assumed average volume density of $0.831 \times 10^{23}$ atoms/cm$^3$ (about half way between Al$_2$O$_3$ and Al). Note that the depth scale is plotted with depth increasing to the right.

The previous analysis was given as an RBS example because the Al and Si cross-sections obeyed the Rutherford formula and the O cross-section was very nearly Rutherford. Still, it shows that there is no clear boundary between naming a technique as RBS or EBS, and convention allows that RBS is used as the acronym for backscattering analysis independent of the cross-section formulation. Also, $\sigma$

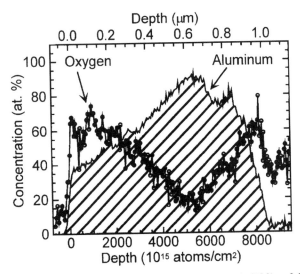

**Figure 14.** Oxygen (open circles) and aluminum (solid line delineating a cross-hatched region) concentration-vs.-depth profiles determined from the data given in Figures 12 and 13.

was non-Rutherford for O at a relatively low He$^+$ ion energy, showing that a check is always required for the $\sigma$ calibration involving He scattering from elements with $Z \leq 9$.

Another example of useful enhanced-scattering cross-sections is given by analysis of a $CN_x$ film in which both the C and N cross-sections are non-Rutherford. An example of a high-energy backscattering spectrum (+) taken from a 44-nm-thick $C_{0.65}N_{0.35}$ sample deposited on a Si substrate is shown in Figure 15. The analysis beam was 3.5-MeV He$^+$ incident at 55° from the sample normal and $\theta = 164°$. This angle of incidence was chosen to increase the depth resolution. The surface energy for C and N are $\sim 0.9$ and 1.1 MeV, respectively. (The ERD analysis showed that the sample contains less than 2% H.)

A simulated spectrum for a film with composition $C_{0.65}N_{0.35}$ and using Rutherford cross-sections is shown in Figure 15 as a solid line. The Si cross-section is Rutherford at this energy, and thus a small amount of channeling in the Si substrate is evident in the data from channel 250 to channel 300, but channeling does not affect the height of the C and N signals because this $C_{0.65}N_{0.35}$ film is amorphous. The carbon cross-section was calibrated from a pure C film on Si, and the N cross-section was calibrated from an amorphous Si$_3$N$_4$ film on Si. The following enhancements to $\sigma$ were determined for this energy and scattering geometry: carbon $\sigma = 5.6\sigma_R$, and nitrogen $\sigma = 2\sigma_R$. The C and N signals above background (+) are shown inset at the top, with the simulated signal (solid line) determined using these factors for the enhanced cross-sections. The depth resolution in these signals is 9 to 10 nm.

Great care should be taken when choosing a carbon standard. A sample that contains a considerable amount of H has a significantly different height for the C signal

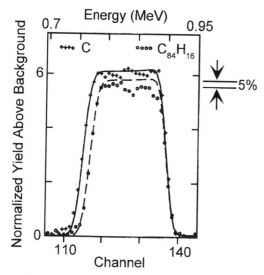

**Figure 16.** Carbon signal above background for RBS spectra from a pure carbon sample (+) and a $C_{84}H_{16}$ sample (open circles). The spectra were obtained using 3.5-MeV $He^+$ with $\theta = 164°$; $\Theta_1 = 55°$ and $\Theta = 45°$, respectively; $Q = 20$ pμC; and $\Omega = 6.95$ msr. Each spectrum (solid line and dashed line, respectively) was simulated based on the enhanced scattering cross-section given in Figure 15.

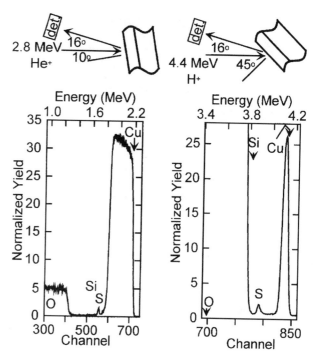

**Figure 17.** Helium (left) and H (right) backscattering spectra from a sulfidized Cu layer on a Si substrate. The 2.8-MeV $He^+$ spectrum was collected using $\theta = 164°$, $\Theta_1 = 10°$, $Q = 10$ pμC, and $\Omega = 6.95$ msr. The 4.4-MeV $H^+$ spectrum was collected using $\theta = 164°$, $\Theta_1 = 45°$, $Q = 10$ pμC, and $\Omega = 6.95$ msr.

than a sample that is pure C. Figure 16 shows the C signal above background for a C standard (+) containing <0.05 at.% H and for a $C_{0.84}H_{0.16}$ sample (open circle). If the $C_{0.84}H_{0.16}$ sample were used for determining $\sigma$ without considering the H content, then a value of only $5\sigma_R$ would be obtained, but the low-H-content standard has a C signal that is $5.6\sigma_R$, as demonstrated by the solid line. Thus the knowledge of the H content of the standard is important to avoid errors (~10%) when calibrating the cross-section. Also, the stopping powers of C and C-H samples are often corrected for bonding effects that influence the composition analysis and thickness determination. A correction of ~5% must be applied to the simulation (dashed line) in order to obtain the proper height on the C signal from the $C_{0.84}H_{0.16}$ sample. This correction then causes an ~5% difference in the thickness of the film determined using RBS. The bonding effects on the stopping power add an increased uncertainty to the composition analysis of a $CN_x$ film containing H because the nature of the bonding is often unknown and its effects on the stopping power are also unknown, and therefore the inherent limit to the accuracy of the composition analysis is ~5%.

An example of enhanced RBS comparing $H^+$ and $He^+$ backscattering is demonstrated by the composition analysis of sulfur on a heavy element substrate. Analysis of sulfur on Si or Al substrates is relatively easy with He[5] RBS because S is heavier than Si and Al, but analysis of a thin layer containing S on bulk Cu is more difficult. The spectra shown in Figure 17 are for 2.8-MeV $He^+$ and 4.4-MeV $H^+$ backscattering from a thin $CuS_x$ film on a Cu layer on a Si substrate. The $\Theta_1$ for each analysis are as indicated in the figure. For this case of He backscattering, the small S peak is just separated from the low-energy side of the Cu signal because the Cu layer was only 325 nm thick. However, a

thicker Cu layer would cause the S signal to be lost on top of a large Cu background signal. Aldridge et al. (1968) examined the high-energy backscattering cross-sections for $\alpha$ particles incident on $^{32}S$ up to an energy of 17.5 MeV. The absence of enhanced cross-sections for $(alpha, \alpha)$ scattering necessitates the use of proton scattering for thick Cu substrates. Proton scattering cross-sections for the $^{32}S(p,p)^{32}S$ reaction at moderate energies are larger than $\sigma_R$ (Abbandanno et al., 1973), and similar to the example given above for the $^{16}O(\alpha, \alpha)^{16}O$ reaction, a flat-top cross-section exists for the $^{32}S(p,p)^{32}S$ reaction in the energy range $\approx 4.3$ to $\approx 4.5$ MeV. A CdS standard was used to calibrate the sulfur cross-section for 4.4-MeV incident protons scattering at an angle of 164° such that $\sigma = 6.55\sigma_R$.

Analysis of the proton scattering S and Cu peaks in Figure 17 give the following composition in areal densities: $0.45 \times 10^{17}$ $S/cm^2$ and $2.79 \times 10^{18}$ $Cu/cm^2$. Although the sample was tilted to 45° relative to the incident beam in order to increase depth resolution, the sulfidized layer remained unresolved; therefore only areal densities are determined rather than a composition-depth profile. In fact, the loss of depth resolution is one of the difficulties encountered when using proton RBS rather than $\alpha$ RBS, as shown in this figure. Nevertheless, the enhanced S cross-section gives an enhanced yield relative to the Cu signal and good counting statistics for the areal density, yielding an uncertainty of ±10%.

Finally, focus is placed on one sample analyzed with different ion beam species and energies in order to demonstrate

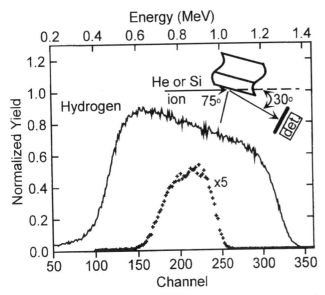

**Figure 18.** The ERD spectra collected from a ≈300-nm-thick $Si_3N_4$ layer on Si. The only detected recoil atom for these spectra is H. The 16-MeV Si ERD spectrum is shown as a solid line, and the 2.8-MeV He ERD spectrum is magnified 5 times and shown as plus symbols. For both spectra, $\phi = 30°$, $\Theta_1 = 75°$, $Q = 4$ p$\mu$C, and $\Omega = 5.26$ msr.

ERD analysis. Figure 18 shows two ERD spectra collected from the same $Si_3N_4$ film on a Si substrate using either 2.8-MeV He$^+$ ions (+) or 16-MeV Si$^{3+}$ ions (solid line) incident at 75° relative to the sample normal. For greater visibility, the yield for the He ERD spectrum is magnified 5 times in this figure, even though the recoil cross-section for $^4$He

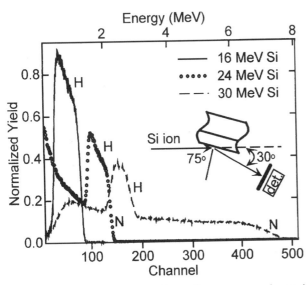

**Figure 19.** The ERD spectra from the same sample as in Figure 18 collected using three different ion energies: 16-MeV Si (solid line), 24-MeV Si (open circles), and 30-MeV Si (dashed line). The higher energy ERD spectra have the signal from recoiled H atoms sitting on the signal from recoiled N atoms. For all three spectra, $\phi = 30°$, $\Theta_1 = 75°$, and $\Omega = 5.26$ msr. For the 16- and 24-MeV Si spectra, $Q = 4$ p$\mu$C, whereas $Q = \Omega$ p$\mu$C for the 30-MeV spectrum.

recoiling $^1$H is ~2.5 times greater than Rutherford. The recoil cross-section for $^{28}$Si recoiling $^1$H is Rutherford for 16- to 30-MeV Si. Both spectra were collected at a scattering angle $\phi = 30°$ and using a 12-$\mu$m Mylar range foil in front of the SBD. The only signal visible in these spectra arises from recoiled H atoms because all other species are either not recoiled or stopped in the range foil. This figure demonstrates that the yield and depth resolution are greatly increased by using a heavier incident projectile. Nevertheless, He ERD is often done (Green and Doyle, 1986) because either a He ion beam causes less radiation damage in the sample (e.g., polymers) or the heavy ion sources and high energies required for the heavy ion ERD are unavailable on a given accelerator.

A comparison of ERD spectra collected from this same silicon nitride sample for different incident Si ion energies is shown in Figure 19. Each spectrum was collected for $\Theta_1 = 75°$ and $\phi = 30°$ using a 12-$\mu$m Mylar range foil. This figure shows that increasing the incident Si ion energy above 16 MeV decreases the yield for H and introduces a signal from the recoiled N that can now get through the Mylar foil. The front edges of the N and H nearly overlap for the 24-MeV Si ERD spectrum but are well separated for the 30-MeV Si ERD spectrum. As a result of recoil atom stopping in the range foil, this variation in mass resolution does not follow a simple analytical formula and is best determined by experiment or using a computer code such as SERDAP (Barbour, 1994) to determine the expected surface energies. Further, the N signal from the entire thickness of the film was observed using 30-MeV Si ERD, whereas only a portion of the N signal was observed using 24-MeV Si ERD. Thus the depth of analysis is increased by using a higher energy ion beam. The H signals are easily separated from the background signals using a linear fit of the background. These signals were then scaled by the cross-sections and stopping powers as a function of depth for the 16- and 30-MeV Si ERD spectra. The H profiles obtained from this spectral scaling, shown in Figure 20, demonstrate the increased depth

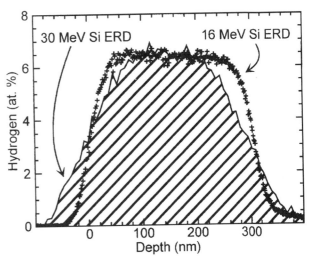

**Figure 20.** The H concentration-vs.-depth profiles determined from the 16-MeV Si (+) and 30-MeV Si (solid line) ERD data given in Figure 19. The N background was subtracted with a linear fit of the N signal under the H signal for the 30-MeV Si ERD data.

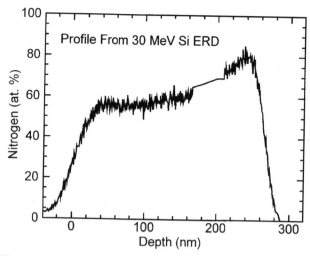

**Figure 21.** The N concentration-vs.-depth profile determined from the N signal for the 30-MeV Si ERD data given in Figure 19. The straight-line profile from ~160 to 210 nm has no visible statistical noise because of the linear fit used to eliminate the H signal sitting on top of the N signal.

resolution obtained with the lower energy ion beam. The depth and concentration scales were determined using the volume density of $0.903 \times 10^{23}$ atoms/cm$^3$ for Si$_3$N$_4$.

The N profile was obtained only for the higher energy IBA and is given in Figure 21 using the same assumed volume density as in Figure 20. Comparing Figures 20 and 21 shows that the depth resolution for N is better at a given Si energy than for H as a result of the greater stopping by N than by H in the range foil. Comparison of these two figures also gives a check on the accuracy of the nitrogen stopping power in the Mylar range foil. Independently, the H and N depth profiles yield approximately the same Si$_3$N$_4$ film thickness, and therefore the stopping power used for N is acceptable. The uncertainty in the N stopping power may, however, cause inaccurate scaling of the N concentration at depth. The N profile in Figure 21 should remain constant with depth but appears to increase beyond ~160 nm. This increase may result from inaccurately calculating the N stopping power, from multiple scattering in the sample, or from a pulse-height defect effect called the nuclear deficit (i.e., some ion energy is deposited in the detector as atomic displacements rather than through the process of producing electron-hole pairs). Further work is needed to accurately determine the N stopping power in Mylar for the N energy range from 1 to 8 MeV.

## LITERATURE CITED

Abbondanno, U., Lagonegro, M., Pauli, G., Poliani, G., and Ricci, R. A. 1973. Isospin-forbidden analogue resonances in $^{33}$Cl, II: Levels of $^{33}$Cl and higher $T = 3/2$ resonances. *Nuovo Cimento* 13A:321.

Aldridge, J. P., Crawford, G. E., and Davis, R. H., 1968. Optical-model analysis of the $^{32}$S$(\alpha, \alpha)^{32}$S elastic scattering from 10.0 MeV to 17.5 MeV. *Phys. Rev.* 167:1053.

Andersen, H. H., Besenbacher, F., Loftager, P., and Moller, W., 1980. Large-angle scattering of light ions in the weakly screened Rutherford region. *Phys. Rev. A* 21:1891.

Barbour, J. C. 1994. SERDAP computer program. Sandia National Laboratories, Dept. 1111, Albuquerque, New Mex.

Bird, J. R. and Williams, J. S. (eds.). 1989. Ion Beams for Materials Analysis. Academic Press, San Diego.

Bozoian, M. 1991a. Threshold of non-Rutherford nuclear cross-sections for ion beam analysis. *Nucl. Instrum. Methods B* 56/57:740.

Bozoian, M. 1991b. Deviations from Rutherford backscattering for $Z = 1, 2$ projectiles. *Nucl. Instrum. Methods B* 58:127.

Bozoian, M., Hubbard, K. M., and Nastasi, M. 1990. Deviations from Rutherford-Scattering cross-sections. *Nucl. Instrum. Methods B* 51:311.

Bragg, W. H. and Kleeman, R. 1905. The $\alpha$ particles of radium and their loss of range in passing through various atoms and molecules. *Philos. Mag.* 10:318.

Brice, D. K. 1973. Theoretical analysis of the energy spectra of back-scattered ions. *Thin Solid Films* 19:121.

Chu, W.-K., Mayer, J. W., and Nicolet, M.-A. 1978. Backscattering Spectrometry. Academic Press, New York.

Doolittle, L. R. 1986. Algorithms for the rapid simulation of Rutherford backscattering spectra. *Nucl. Instrum. Methods B* 9:344.

Doolittle, L. R. 1994. Computer code "RUMP." Computer Graphics Services, Ithaca, N.Y.

Doyle, B. L. and Brice, D. K. 1988. The analysis of elastic recoil detection data. *Nucl. Instrum. Methods B* 35:301.

Feldman, L. C., Mayer, J. W., and Picrauz, S. T. 1982. Materials Analysis by Ion Channeling. Academic Press, New York.

Foti, G., Mayer, J. W., and Rimini, E. 1977. Backscattering spectrometry. In Ion Beam Handbook for Material Analysis (J. W. Mayer and E. Rimini, eds.). p. 22. Academic Press, New York.

Gebauer, B., Fink, D., Goppelt, P., Wilpert, M., and Wilpert, Th. 1990. A multidimensional ERDA spectrometer at the VICKSI heavy ion accelerator. In High Energy and Heavy Ion Beams in Materials Analysis (J. R. Tesmer, C. J. Maggiore, M. Nastasi, J. C. Barbour, and J. W. Mayer, eds.). p. 257. Materials Research Society, Pittsburgh.

Geiger, H. and Marsden, E. 1913. The laws of deflexion of $\alpha$ particles through large angles. *Philos. Mag.* 25:604.

Green, P. F. and Doyle, B. L. 1986. Silicon elastic recoil detection studies of polymer diffusion: Advantages and disadvantages. *Nucl. Instrum. Methods B* 18:64.

Hubbard, K. M., Tesmer, J. R., Nastasi, M., and Bozoian, M. 1991. Measured deviations from Rutherford backscattering cross sections using Li-ion beams. *Nucl. Instrum. Methods B* 58:121.

L'Ecuyer, J., Brassard, C., Cardinal, C., Chabbal, J., Deschenes, L., Labrie, J. P., Terrault, B., Martel, J. G., and St.-Jacques, R. 1976. An accurate and sensitive method for the determination of the depth distribution of light elements in heavy materials. *J. Appl. Phys.* 47:381.

Marion, J. B. and Young, F. C. 1968. Nuclear Reaction Analysis. North-Holland Publishing, New York.

Mayer, J. W. and Lau, S. S. 1990. Electronic Materials Science: For Integrated Circuits in Si and GaAs. Macmillan, New York.

Mayer, J. W. and Rimini, E. (eds.). 1977. Ion Beam Handbook for Materials Analysis. Academic Press, New York.

Rutherford, E. 1911. The scattering of $\alpha$ and $\beta$ particles by matter and the structure of the atom. *Philos. Mag.* 21:669.

Tesmer, J. R., Nastasi, M., Barbour, J. C., Maggiore, C. J., and Mayer, J. W. (eds.). 1995. Handbook of Modern Ion Beam Materials Analysis. Materials Research Society, Pittsburgh.

Thomas, J. P., Fallavier, M., and Ziani, A., 1986. Light elements depth-profiling using time-of-flight and energy detection of recoils. Nucl. Instrum. Methods B 15:443.

Tilley, D. R., Cheves, C. M., Kelley, J. H., Raman S., and Weller, H. R. 1998. Energy levels of light nuclei A = 20. Nucl. Phys. A 636(3):249–364.

Turkevich, A. L., 1961. Chemical analysis of surfaces by use of large-angle scattering of heavy charged particles. Science 134:672.

Whitlow, H. J. 1990. Mass and energy dispersive recoil spectrometry: A new quantitative depth profiling technique for microelectronic technology. In High Energy and Heavy Ion Beams in Materials Analysis (J. R. Tesmer, C. J. Maggiore, M. Nastasi, J. C. Barbour, and J. W. Mayer, eds.). p. 73. Materials Research Society, Pittsburgh.

Wielunski, L., Benenson, R., Horn, K., and Lanford, W. A. 1986. High sensitivity hydrogen analysis using elastic recoil detection. Nucl. Instrum. Methods B 15:469.

Ziegler, J. F. 1980. Handbook of Stopping Cross-Sections for Energetic Ions in All Elements. Pergamon Press, New York.

Ziegler, J. F., 1987. RSTOP computer code from TRIM-91. IBM Research Center, Yorktown Heights, N.Y.

Ziegler, J. F., Biersack, J. P., and Littmark, U., 1985. The Stopping and Range of Ions in Solids. Pergamon Press, Elmsford, N.Y.

## ADDITIONAL REFERENCES FOR SPECIAL DETECTORS AND DETECTION SCHEMES

Bowman, J. D. and Heffner, R. H. 1978. A novel zero time detector for heavy ion spectroscopy. Nucl. Instrum. Methods 148:503.

Chu, W.-K. and Wu, D.T., 1988. Scattering recoil coincidence spectroscopy. Nucl. Instrum. Methods B 35:518.

Gossett, C. R. 1986. Use of a magnetic sector spectrometer to profile light elements by elastic recoil detection. Nucl. Instrum. Methods B 15:481.

Hofsass, H. C., Parich, N. R., Swanson, M. L., and Chu, W.-K. 1990. Depth profiling of light elements using elastic recoil coincidence spectroscopy (ERCS). Nucl. Instrum. Methods B 45:151.

Kraus, R. H., Vieira, D. J., Wollnik, H., and Wouters, J. M. 1988. Large-area fast-timing detectors developed for the TOFI spectrometer. Nucl. Instrum. Methods A 264:327.

Nagai, H., Hayashi, S., Aratani, M., Nozaki, T., Yanokura, M., Kohno, I., Kuboi, O., and Yatsurugi, Y., 1987. Reliability, detection limit and depth resolution of the elastic recoil measurement of hydrogen. Nucl. Instrum. Methods B 28:59.

Whitlow, H. J. 1990. Time of flight spectroscopy methods for analysis of materials with heavy ions: A tutorial. In High Energy and Heavy Ion Beams in Materials Analysis (J. R. Tesmer, C. J. Maggiore, M. Nastasi, J. C. Barbour, and J. W. Mayer, eds.). p. 243. Materials Research Society, Pittsburgh.

Zebelman, A. M., Meyer, W. G., Halbach, K., Poskanzer, A. M., Sextro, R. G., Gabor, G., and Landis, D. A., 1977. A zero-time detector utilizing isochronous transport of secondary electrons. Nucl. Instrum. Methods 141:439.

## KEY REFERENCES

Bird and Williams, 1989. See above.

*Good general overview of ion beam analysis techniques.*

Chu et al., 1978. See above.

*Tutorial-style book for learning RBS.*

Tesmer et al., 1995. See above.

*Tutorial handbook for learning ion-beam analysis.*

## INTERNET RESOURCES

http://www.sandia.gov/1100/ibanal.html

http://corpbusdev.sandia.gov/Facilities/Descriptions/beam-materials. htm

*The home page of the Ion Beam Materials Research Laboratory at Sandia National Laboratories*

http://www.tunl.duke.edu/nucldata/groupinfo.html

*A web site for nuclear data sheets maintained by Nuclear Data Group, Triangle Univesities Nuclear Laboratory (TUNL)*

http://fai.idealibrary.com:80/cgi-bin/fai.idealibrary.com_8011/iplogin/toc/ds

*A web site for nuclear data sheets maintained by Academic Press.*

## APPENDIX A: GLOSSARY OF TERMS AND SYMBOLS

| | |
|---|---|
| AD | Areal density (number of atoms per unit area) |
| $dE/dx$ | Energy loss of a particle (generally as a result of electronic stopping) |
| $dE_d/dx$ | Effective stopping power for the detected particle at a given projectile energy (an incremental change in $E_d$ corresponds to an incremental change in depth, $dx$, at $x$) |
| $dx$ | Increment of depth at $x$ corresponding to an increment of energy $dE_d$ |
| $E_0$ | Energy of the incident projectile particle |
| $E_0'$ | Projectile energy at a depth $x'$ |
| $E_0^{(n)}$ | Projectile energy incident upon slab $n + 1$ |
| $E_3^{(n)}$ | Energy of the backscattered atom (or recoiled atom for ERD) as it emerges from slab $n$ |
| $E_d$ | Energy of detected particle |
| $E_d^{(n)}$ | Detected energy for the $n$th slab |
| $E_f^{(j-1)}$ | Energy in foil before recoil atom traverses the $j$th slab |
| $E_{NR}$ | Energy where $\sigma$ becomes non-Rutherford |
| $K$ | Kinematic factor (ratio of the scattered particle energy to the projectile energy for an elastic collision) |
| $L$ | Length of light tube |
| $M_1$ | Mass of the incident projectile particle |

| | |
|---|---|
| $M_2$ | Mass of a target atom involved in a scattering event |
| $N$ | Volume density of atoms in the target |
| NF | Number of slabs in foil |
| $Q$ | Number of incident projectiles (determined from a target current and the projectile charge) |
| $S^{\mathrm{eff}}$ | Effective stopping power |
| $S_{\mathrm{p}}$ | Projectile particle stopping power (energy loss), $= dE/dx$ projectile |
| $S_{\mathrm{r}}$ ($S_{\mathrm{b}}$) | Recoil (backscattered) particle stopping power (energy loss) |
| $[S]_{\mathrm{p,r}}$ or $[S]_{\mathrm{p,b}}$ | Energy loss factor corresponding to a recoil (backscattering) event at energy $E$ |
| $t$ | Layer thickness; also time |
| $V_{\mathrm{LJ}}$ | Lenz-Jensen potential |
| $x$ | Distance (depth) in the sample, measured along the sample normal |
| $X^{(0)}$ | Thickness of range foil |
| $X^{(n)}$ | Thickness of slab (sublayer) $n$ within the sample |
| $Y(E_{\mathrm{d}})$ | Yield (height) of detected particles at a channel corresponding to energy $E_{\mathrm{d}}$ |
| $z$ | Sample-to-detector distance |
| $Z_1$ | Atomic number of the incident projectile particle |
| $Z_2$ | Atomic number of a target atom involved in a scattering event |
| $\delta E_{\mathrm{d}}$ | Energy width of a channel in the multi-channel analyzer |
| $\delta x$ | Smallest resolvable depth interval |
| $\Delta E_{\mathrm{foil}}$ | Energy lost by a recoil atom in traversing the range foil |

| | |
|---|---|
| $\Delta x^{(0)}$ | Incremental slab thickness for range foil containing $X^{(0)}/\Delta x^{(0)}$ slabs |
| $\varepsilon$ | Stopping cross-section, $\equiv (1/N)(dE/dx)$ |
| $\eta$ | Detector efficiency |
| $\theta$ | Backscattering angle measured in the laboratory frame of reference |
| $\Theta_1$ | Incident angle, measured between the sample normal and the incident projectile direction |
| $\Theta_2$ | Detection angle, measured between the sample normal and the scattered particle direction |
| $\sigma$ | Average differential scattering cross-section (also known as the scattering cross-section) |
| $\sigma_0$ | Scattering cross-section of surface peak |
| $\sigma_{\mathrm{R}}$ | Rutherford scattering cross-section |
| $\tau$ | Layer thickness |
| $\phi$ | Forward (recoil) scattering angle measured in the laboratory frame of reference |
| $\Omega$ | Detector solid angle |

## APPENDIX B:
## RBS AND ERD EQUATIONS

Figure 22 shows equations applicable to both RBS and ERD, whereas Figure 23 shows equations applicable to either RBS or ERD. Although these analytical expressions appear complicated, their implementation through the use of a computer is simple and fast.

$$\Omega = \int_{\substack{\text{area subtended} \\ \text{by detector}}} \sin\phi\, d\phi\, d\varepsilon \approx \frac{\text{detector area}}{z^2} \quad (z = \text{sample-to-detector distance}) \tag{1}$$

$$Y(E_{\mathrm{d}}) = \frac{QN(x)\sigma(E_0^{(n)})\Omega\delta E_{\mathrm{d}}}{\cos\Theta_1\, dE_{\mathrm{d}}/dx} \tag{5}$$

$$\mathrm{AD} = \int_0^\tau N(x)(dx) = \frac{\cos\Theta_1}{Q\Omega}\int_0^\tau \frac{S^{\mathrm{eff}}}{\sigma\delta E_{\mathrm{d}}}Y(E_{\mathrm{d}})dx = \frac{\cos\Theta_1}{Q\Omega}\int_{E_{\mathrm{d}}(0)}^{E_{\mathrm{d}}(x)} \frac{Y(E_{\mathrm{d}})}{\sigma}\frac{dE_{\mathrm{d}}}{\delta E_{\mathrm{d}}} \tag{8}$$

$$\mathrm{AD} = \frac{\cos\Theta_1}{Q\Omega\sigma_0}\sum_{\mathrm{channel}[E_{\mathrm{d}}(0)]}^{\mathrm{channel}[E_{\mathrm{d}}(x)]} Y(\text{channel}) \tag{9}$$

Bragg's rule:  $\varepsilon^{\mathrm{A}}\alpha^{\mathrm{B}}\beta = \alpha\varepsilon^{\mathrm{A}} + \beta\varepsilon^{\mathrm{B}}$  and  $S = N^{\mathrm{A}}\alpha^{\mathrm{B}}\beta\varepsilon^{\mathrm{A}}\alpha^{\mathrm{B}}\beta$ \hfill (10)

$$E_0^{(n)} = E_0^{(n-1)} - \frac{X^{(n)}}{\cos\Theta_1}S_{\mathrm{p}}^{(n)}(E_0^{(n-1)}) \tag{11}$$

Spectral scaling:  $N^{(n)}(x) = \dfrac{\cos\Theta_1}{Q\Omega\,\delta E_{\mathrm{d}}}\dfrac{dE_d/dx|_n}{\sigma_{\mathrm{R}}(E_0^{(n)})}Y^{(n)}(E_{\mathrm{d}}) = \dfrac{\cos\Theta_1}{Q\Omega\delta E_{\mathrm{d}}}\dfrac{S_0^{\mathrm{eff}}}{\sigma_0}\left[\dfrac{\sigma_{\mathrm{R}}(E_0)}{\sigma_{\mathrm{R}}(E_0^{(n)})}\dfrac{dE_{\mathrm{d}}/dx|_n}{dE_{\mathrm{d}}/dx|_0}Y^{(n)}(E_{\mathrm{d}})\right]$ \hfill (22)

**Figure 22.** Equations applicable to both RBS and ERD.

| RBS | Equation(s) | ERD |
|---|---|---|

$$K = \left( \frac{\sqrt{M_2^2 - M_1^2 \sin^2 \theta} + M_1 \cos\theta}{M_1 + M_2} \right)^2$$

3,4

$$K = \frac{4 M_1 M_2 \cos^2 \phi}{(M_1 + M_2)^2}$$

$$Y^{(n)}(E_d) = \frac{Q \Omega N^{(n)}(x) \sigma(E_0^{(n)}, \theta) \delta E_d}{\cos\Theta_1 \, dE_d/dx|_n}$$

5

$$Y_r^{(n)}(E_d) = \frac{Q \Omega N_r^{(n)}(x) \sigma_r(E_0^{(n)}, \phi) \delta E_d}{\cos\Theta_1 \, dE_d/dx|_n}$$

$$\frac{d\sigma_R(E_0^{(n)}, \theta)}{d\Omega} = \left( \frac{Z_1 Z_2 e^2}{4 E_0^{(n)}} \right)^2 \left( \frac{4}{\sin^4 \theta} \right)$$

6,7

$$\frac{d\sigma_R(E_0^{(n)}, \phi)}{d\Omega} = \left( \frac{Z_1 Z_2 e^2}{4 E_0^{(n)}} \right)^2 \left( \frac{4}{\cos^3 \phi} \right) \frac{(M_1 + M_2)^2}{(M_2)^2}$$

$$\times \frac{(\sqrt{1 - [(M_1/M_2)\sin\theta]^2} + \cos\theta)^2}{\sqrt{1 - [(M_1/M_2)\sin\theta]^2}}$$

For a scattering event occurring in slab $n$:

$$E_3^{(n)} = K E_0^{(n)} - \frac{X^{(n)}}{\cos\Theta_2} S_b^{(n)}(K E_0^{(n)})$$

12

For a scattering event occurring in slab $n$:

$$E_3^{(n)} = K E_0^{(n)} - \frac{X^{(n)}}{\cos\Theta_2} S_r^{(n)}(K E_0^{(n)})$$

For a scattering event occurring in slab $j > n$:

$$E_3^{(n)} = E_3^{(n+1)} - \frac{X^{(n)}}{\cos\Theta_2} S_b^{(n)}(E_3^{(n+1)})$$

13

For a scattering event occurring in slab $j > n$:

$$E_3^{(n)} = E_3^{(n+1)} - \frac{X^{(n)}}{\cos\Theta_2} S_r^{(n)}(E_3^{(n+1)})$$

$$E_d^{(n)} = K \left[ E_0 - \frac{X^{(n)}}{\cos\Theta_1} \sum_{j=1}^{n} S_p^{(j)}(E_0^{(j-1)}) \right]$$
$$- \frac{X^{(n)}}{\cos\Theta_2} \left[ S_b^{(n)}(K E_0^{(n)}) + \sum_{j=n}^{2} S_b^{(j-1)}(E_3^{(j)}) \right]$$

14,15

$$E_d^{(n)} = K \left[ E_0 - \frac{X^{(n)}}{\cos\Theta_1} \sum_{j=1}^{n} S_p^{(j)}(E_0^{(j-1)}) \right]$$
$$- \frac{X^{(n)}}{\cos\Theta_2} \left[ S_r^{(n)}(K E_0^{(n)}) + \sum_{j=n}^{2} S_r^{(j-1)}(E_3^{(j)}) \right] - \Delta E_{foil}$$

16

$$\Delta E_{foil} = \Delta x^{(0)} \sum_{j=1}^{NF} S_r^{(0)}(E_f^{(j-1)})$$

$$\left. \frac{dE_d}{dx} \right|_n = [S]_{p,b}^{(n)} \frac{S_b(E_d)}{S_b(K E_0^{(n)})}$$

17,18

$$\left. \frac{dE_d}{dx} \right|_n = [S]_{p,r}^{(n)} \prod_n$$

$$[S]_{p,b}^{(n)} = \frac{K S_p^{(n)}(E_0^{(n)})}{\cos\Theta_1} + \frac{S_b^{(n)}(K E_0^{(n)})}{\cos\Theta_2}$$

19,20

$$[S]_{p,r}^{(n)} = \frac{K S_p^{(n)}(E_0^{(n)})}{\cos\Theta_1} + \frac{S_r^{(n)}(K E_0^{(n)})}{\cos\Theta_2}$$

21

$$\prod_n = \frac{S_r^{(n)}(E_3^{(n)})}{S_r^{(n)}(E_2)} \frac{S_r^{(n-1)}(E_3^{(n-1)})}{S_r^{(n-1)}(E_3^{(n)})} \cdots \frac{S_r^{(0)}(E_d)}{S_r^{(0)}(E_3^{(1)})}$$

**Figure 23.** Equations specific to either RBS or ERD.

J. C. BARBOUR
Sandia National Laboratories
Albuquerque, New Mexico

# NUCLEAR REACTION ANALYSIS AND PROTON-INDUCED GAMMA RAY EMISSION

## INTRODUCTION

Nuclear reaction analysis (NRA) and proton- (particle-) induced gamma ray emission (PIGE) are based on the interaction of energetic (from a few hundred kilo-electron-volt to several mega-electron-volt) ions with light nuclei. Every nuclear analytical technique that uses nuclear reactions has a unique feature—isotope sensitivity. Therefore, it is insensitive to matrix effects, and there is much less interference than in methods where signals from different elements overlap. In NRA the nuclear reaction produces charged particles, while in PIGE the excited nucleus emits gamma rays. Sometimes the charged particle emission and the gamma ray emission occur simultaneously, such as in the $^{19}F(p,\alpha\gamma)^{16}O$ reaction. Both NRA and PIGE measure the concentration and depth distribution of elements in the surface layer (few micrometers) of the sample. Both techniques are limited by the available nuclear reactions. Generally, they can be used for only light elements, up to calcium. This is the basis for an important property of these methods: the nuclear reaction technique is one of the few analytical techniques that can quantitatively measure hydrogen profiles in solids. Since these techniques are sensitive to the nuclei in the sample, they are unable to provide information about the chemical states and bonds of the elements in the sample. For the same reason they cannot provide information about the microscopic structure of the sample. Combined with channeling, NRA or PIGE can provide information about the location of the measured element in a crystalline lattice.

The sensitivity and depth resolution of these methods depend on the specific nuclear reaction. The sensitivity typically varies between 10 and 100 ppm while the depth resolution can be as good as a few nanometers or as large as hundreds of nanometers. The lateral resolution of the method depends on the size of the bombarding ion beam. Good nuclear microprobes are currently in the few-micrometer range. Also, using nuclear microprobes, an elemental/isotopic image of the sample can be recorded. Both NRA and PIGE are nondestructive, although some materials might be lattice damaged after long, high-current bombardment.

Although NRA and PIGE are quantitative, in most cases standards have to be used. Depending on the shape of the particular nuclear cross-section, nonresonant or resonant depth profiling can be used. The resonant profiling (in most cases PIGE uses resonances but there are a few charged particle reactions that have usable resonance) typically can give very good resolution, but the measurement takes much longer than in nonresonant profiling; therefore, the probability of inducing changes in the sample by the ion beam is higher.

These techniques require a particle accelerator capable of accelerating ions up to several mega-electron-volts. This limits their availability to laboratories dedicated to these methods or that have engaged in low-energy nuclear physics in the past (i.e., they have a particle accelerator that is no longer adequate for modern nuclear physics experiments because of its low energy but is quite suitable for nuclear analytical techniques).

This unit will concentrate on the specific aspects of these two nuclear analytical techniques and will not cover the details of the ion-solid interaction or the general aspects of ion beam techniques (e.g., stopping power, detection of ions). These are described in detail elsewhere in this part (see ELASTIC ION SCATTERING FOR COMPOSITION ANALYSIS and MEDIUM-ENERGY BACKSCATTERING AND FORWARD-RECOIL SPECTROMETRY).

## Competitive and Complementary Techniques

Practically every analytical technique that measures the concentration and depth profile of elements in the top few micrometers of a solid competes with NRA and PIGE. However, either most of these techniques are not isotope sensitive or their isotope resolution is not very good. The competing techniques can be divided into two groups: ion beam techniques and other techniques. The ion beam techniques that compete with NRA and PIGE are Rutherford backscattering spectrometry (RBS; see ELASTIC ION SCATTERING FOR COMPOSITION ANALYSIS), elastic recoil detection (ERD or ERDA; see ELASTIC ION SCATTERING FOR COMPOSITION ANALYSIS), proton-induced x-ray emission (PIXE; see PARTICLE-INDUCED X-RAY EMISSION), secondary ion mass spectroscopy (SIMS), medium-energy ion scattering (MEIS; see MEDIUM-ENERGY BACKSCATTERING AND FORWARD-RECOIL SPECTROMETRY), and ion scattering spectroscopy (ISS; see HEAVY-ION BACKSCATTERING SPECTROMETRY).

Rutherford backscattering spectrometry and ERD can be considered special cases of NRA in which the nuclear reaction is just an elastic scattering. The main advantage of RBS and ERD is that they are able to see almost all elements in the periodic table (with the obvious exception in RBS when the element to be detected is lighter than the bombarding ions). This can be a disadvantage when a light element has to be measured in a heavy matrix (e.g., the measurement of oxygen in tantalum by RBS). This is the typical case when NRA can solve the problem but RBS cannot. In ERD, since the forward recoiled atoms have to be mass analyzed, the measurement becomes more complicated than in NRA, requiring more sophisticated (and more expensive) instruments. The resolution and sensitivity achieved when using RBS, ERD, and NRA are on the same order, although when using very heavy ions, the sensitivity and depth resolution can be an order of magnitude better in ERD than in NRA. Using heavy ions presents other problems, but these are beyond the scope of this unit (see HEAVY-ION BACKSCATTERING SPECTROMETRY). For more discussion see Green and Doyle (1986) and Davies et al. (1995).

Proton-induced x-ray emission (see PARTICLE-INDUCED X-RAY EMISSION) can detect most elements, except H and He. Also, the detection of x rays from low-$Z$ elements (below Na) requires a special, windowless detector. Another drawback of PIXE is that it generally cannot provide depth information.

Secondary ion mass spectroscopy can be used for most of the analyses performed in NRA, including hydrogen profiling. Its sensitivity and depth resolution are superior to those of NRA. The main advantage of NRA over SIMS is that while NRA is a nondestructive technique, SIMS depth profiling destroys the sample. Also, the depth scale with SIMS depends on accurate knowledge of sputtering rates for a specific sample.

Although MEIS and ISS can be considered competing techniques, their probing depth is much smaller (which also means much better depth resolution) than that of NRA.

Among the non−ion beam techniques, Auger electron spectroscopy (AES, see AUGER ELECTRON SPECTROSCOPY), x-ray photoelectron spectroscopy (XPS), x-ray fluorescence (XRF, see X-RAY MICROPROBE FOR FLUORESCENCE AND DIFFRACTION ANALYSIS), and neutron activation analysis (NAA) should be mentioned.

Auger electron spectroscopy and XPS not only provide concentration information but also are sensitive to the chemical states of the elements; therefore, they give information about chemical bonds. Since both methods get the information from the first few nanometers of the sample, to measure a depth profile, layers of the sample have to be removed (usually by sputtering); therefore, the sample is destroyed. X-ray fluorescence can provide concentration information about elements heavier than Na (again, it depends on the x-ray detector) but it cannot measure the depth profile. The only technique listed that has the same isotope sensitivity as NRA is NAA. However, NAA is a bulk technique and does not provide any depth information.

Most of the above-mentioned techniques are complementary techniques to NRA, and in many cases are, used concurrently with NRA. [It is quite common in ion beam analysis (IBA) laboratories for an ion-scattering chamber to have an AES or XPS spectrometer mounted on it.] The most frequent combination is RBS and NRA, since they are closely related and use the same equipment. Whereas, NRA can measure light elements in a heavy matrix but cannot see the matrix itself (at least not directly), RBS is very sensitive and has a good mass resolution for heavy elements but cannot see a small amount of some light element in a heavy matrix.

## PRINCIPLES OF THE METHOD

Although NRA and PIGE are similar to other high-energy ion beam techniques (see ELASTIC ION SCATTERING FOR COMPOSITION ANALYSIS and PARTICLE-INDUCED X-RAY EMISSION), here we will discuss only the principles specific to NRA and PIGE.

Both NRA and PIGE measure the prompt reaction products from nuclear reactions. The yield (number of detected particles, $\gamma$ rays) provides information about the concentration of elements, and the energy of the detected charged particles provides information about the depth distribution (depth profile) of the elements. Depending on whether the reaction cross-section has a sharp resonance or not, the methods are distinguished as either resonant or nonresonant (e.g., PIGE uses only the resonant method).

When a high-energy ion beam is incident on a target, the ions slow down in the material and either undergo elastic scattering (RBS, ERDA) or induce a nuclear reaction at various depths. In a nuclear reaction, a compound nucleus is formed and almost immediately a charged particle (p, d, $^3$He, and $^4$He) and/or a $\gamma$ photon is emitted. Many (p, $\alpha$) reactions have an associated $\gamma$ photon. The emitted charged particle/photon then leaves the target and is detected by an appropriate detector. The energy of the charged particle depends on the angle of the incident ion and the emitted particle, the kinetic energy of the incident particle, and the $Q$ value of the reaction, where $Q$ is the energy difference between the initial and final nuclear states. The energy of the emitted $\gamma$ photon is determined by the energy structure of the compound nucleus (for the formulas, see Appendix). The detected number of emitted particles is proportional to the number of incident particles, the solid angle of the detector, the concentration of the atoms participating in the nuclear reaction, and the cross-section of the reaction. The charged reaction products will lose energy on their way out from the target; therefore, their energy will carry information about the depth where the nuclear reaction occurred. Since the energy of the $\gamma$ photons does not change while they travel through the target, they do not provide depth information.

### Nonresonant Methods

**Overall Near-Surface Content.** If the cross-section changes slowly in the vicinity of the $E_0$ bombarding energy, the absolute value of nuclei per square centimeters

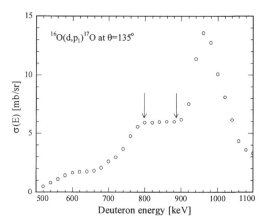

**Figure 1.** Cross-section of the $^{16}O(d, p_1)^{17}O$ reaction; $\theta$ is the scattering angle (Jarjis, 1979).

can be determined in thin layers independent of the concentration profile and the other components of the target. Assuming that the incident ions lose $\Delta E$ energy in the thin layer and the reaction cross-section $\sigma(E) \approx \sigma(E_0)$ for $E_0 > E > E - \Delta E$, the number of particles detected is

$$Y = \frac{Q_C \Omega \sigma(E_0) Nt}{\cos \alpha_{in}} \tag{1}$$

where $Q_C$ is the collected incident ion charge, is the solid angle of the detector, $\alpha_{in}$ is the angle between the direction of the incident beam and the surface normal of the target, and $Nt$ is the number of nuclei per square centimeter. The spectrum of emitted particle would contain a peak with an area $Y$. An example of such a cross-section is shown in Figure 1. The $^{16}O(d, p_1)^{17}O$ reaction has a plateau between 800 and 900 keV, as indicated by arrows in the figure. This reaction is frequently used to determine oxygen content of thin layers or thickness of surface oxide layers up to several hundreds of nanometers.

**Depth Profiling.** When the thickness of the sample becomes larger and the cross-section cannot be considered constant, the spectrum becomes the convolution of the concentration profile, the energy resolution of the incident and the detected beam, and the cross-section. A typical scattering geometry is shown in Figure 2. Although the figure shows only backward geometry, forward geometry is used in certain cases as well. The depth scale can be calculated by determining the energy loss of the incident ions

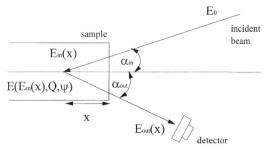

**Figure 2.** Typical scattering geometry used in NRA experiments.

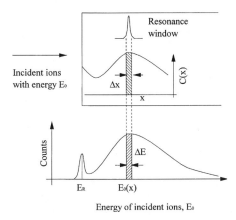

**Figure 3.** Principle of resonant depth profiling.

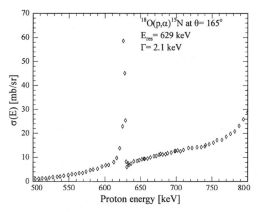

**Figure 4.** Cross-section of the $^{18}$O(p, $\alpha$)$^{15}$N reaction around the 629-keV resonance; $\theta$ is the scattering angle (Amsel and Samuel, 1967).

before they induce the nuclear reaction and the energy loss of the reaction product particles:

$$E_{\text{out}}(x) = E(E_{\text{in}}(x),\ Q,\ \psi) - \int_0^{x/\cos\alpha_{\text{out}}} S_{\text{out}}(E)\,dx \quad (2)$$

$$E_{\text{in}}(x) = E_0 - \int_0^{x/\cos\alpha_{\text{in}}} S_{\text{in}}(E)\,dx \quad (3)$$

where $x$ is the depth at which the nuclear reaction occurs, $E_0$ is the energy of the incident ions, $Q$ is the $Q$ value of the reaction, $\psi$ is the angle between the incident and detected beams, $\alpha_{\text{in}}$ and $\alpha_{\text{out}}$ are angles of the incident and detected beams to the surface normal of the sample, and $S_{\text{in}}$ and $S_{\text{out}}$ are the stopping powers of the incident ions and the reaction products. The term $E(E_{\text{in}}(x),Q,\psi)$ is the energy of the reaction product calculated from Equation 7. (Also see Fig. 7.)

In most cases, to get quantitative results, a nonlinear fit of the spectrum to a simulated spectrum is necessary (Simpson and Earwaker, 1984; Vizkelethy, 1990; Johnston, 1993). For a discussion of energy broadening, see the Appendix.

### Resonant Depth Profiling

In case of sharp resonances in the cross-section, most particles/$\gamma$ photons come from a very narrow region in the target that is equal to the energy-loss-related width of the resonances. As the energy of the incident beam increases, the incident ions slow to the resonance energy at deeper and deeper depth; therefore, the thin layer from which the reaction products are coming lies deeper and deeper. In this way a larger depth can be probed and a depth profile can be measured. The principle of the method is illustrated in Figure 3.

The depth $x$ and the energy $E_0$ of the incident ions are related through the equation

$$E_0(x) = E_{\text{R}} + \int_0^{x/\cos\alpha_{\text{in}}} S_{\text{in}}(E)\,dx \quad (4)$$

where $E_{\text{R}}$ is the resonance energy. In the figure $C(x)$ is the concentration of the element to be detected. Figure 4 shows

the cross-section of the $^{18}$O(p, $\alpha$)$^{15}$N reaction. The resonance at 629 keV is widely used in $^{18}$O tracing experiments where oxygen movement is studied. This is an excellent example of how useful NRA and PIGE are, since these methods are sensitive to only $^{18}$O and not to the other oxygen isotopes. The measured spectrum ("excitation curve") is the convolution of the concentration profile and the energy spread of the detection system. To extract the depth profile from the excitation function, we should deconvolve it with the depth-dependent energy spread or fit the excitation function to a simulation of it. The theory and simulation of narrow resonances are discussed in detail in Maurel (1980), Maurel et al. (1982), and Vickridge (1990).

### PRACTICAL ASPECTS OF THE METHOD

### Equipment

The primary equipment used in NRA and PIGE is an electrostatic particle accelerator, which is the source of the high-energy ion beam. The accelerators used for IBA are either Cockroft-Walton or van de Graaff type (or some slight variation of them, such as pelletron accelerators) with terminal voltage in the range of a few million volts. When higher energies are needed or heavy ions have to be accelerated, tandem accelerators are used. In general, the requirements for the accelerators used in NRA and PIGE are similar to those of any other IBA technique. In addition, to perform resonance profiling, the accelerators must have better energy resolution and stability than are necessary for RBS or ERDA and be capable of changing the beam energy easily. Usually, accelerators with slit feedback energy stabilization systems are satisfactory. Another NRA-specific requirement is needed because of the deuteron-induced nuclear reactions. When deuterons are accelerated, neutrons are generated mainly from the D(d, n) $^3$He reaction. Therefore, if deuteron-induced nuclear reactions are used, additional shielding is necessary to protect personnel against neutrons. Many laboratories using ion beam techniques do not have this shielding.

Most NRA and PIGE measurements take place in vacuum, with the exception of a few extracted beam experiments. The required vacuum, i.e., >10$^{-3}$ Pa, depends on the particular application; it usually ranges from $10^{-3}$ to $10^{-7}$ Pa. The main problem caused by bad vacuum is hydrocarbon layer formation on the sample under irradiation, which can interfere with the measurement. Having a load-lock system attached to the scattering chamber makes the sample changing convenient and fast.

In most cases, standard surface barrier or ion-implanted silicon particle detectors are used to detect charged particles. These detectors usually have an energy resolution around 10 to 12 keV, which is adequate for NRA. When better energy resolution is required, electrostatic or magnetic spectrometers or time-of-flight (TOF) techniques can be used.

In PIGE, NaI, BGO (bismuth germanate), Ge(Li), or HPGE (high-purity germanium) detectors are used to detect γ rays. The NaI and BGO detectors are used when high efficiency is needed (measuring low concentrations, or the cross-section is too small) using well-isolated resonances. The NaI detectors were used mainly before BGO detectors became available, but currently BGO detectors are preferred since their efficiency is higher for a given size (Kuhn et al., 1990) and have a better signal-to-background ratio. The Ge(Li) or HPGE detectors are used when there are interfering peaks in the spectrum that cannot be resolved using other detectors.

To process the signals from the detectors, standard NIM (nuclear instrument module) or CAMAC (computer-automated measurement and control) electronics modules, such as amplifiers, single-channel analyzers (SCA), and analog-to-digital converters (ADCs), are used. In most cases, the ADC is on a card in a computer or connected to a computer through some interface (e.g., GPIB, Ethernet). Therefore, the spectrum can be examined while it is being collected and the data can be evaluated immediately after the measurement.

### Filtering of Unwanted Particles (NRA)

From Figure 2 it is obvious that not only the reaction products but also all the backscattered particles will reach the detector. Since the $Q$ value of the most frequently used reactions is positive, the backscattered particle will have less energy than the reaction products; therefore, the signal from the backscattered particles will not overlap in the spectrum. [The exceptions are the (p, α) reactions with low $Q$ value. Due to the higher stopping power of the α particles than that of the protons, the energy of an α particle coming from a deeper layer can be the same as the energy of the proton coming from another layer.] However, the cross-section of the Rutherford scattering is usually much larger than the cross-section of the nuclear reaction we want to use. Therefore, much less reaction product would be detected in a unit time than backscattered particles. Since every detection system has finite dead time, the number of backscattered particles would limit the maximum incident ion current, so the NRA measurement would take a very long time. To make the measurement time reasonable, the backscattered particles should be filtered out.

The simplest method to get rid of the backscattered particles is to use an absorber foil, since the energy of the reaction products is higher than the energy of the backscattered particles. The energy of the reaction products after the absorber foil is

$$E_{abs}(x) = E_{out}(x) - \int_0^{x_{abs}} S_{abs}(E)\, dx \qquad (5)$$

where $x_{abs}$ and $S_{abs}$ are the thickness and stopping power of the absorber foil and $E_{out}(x)$ is from Equation 2. Choosing an absorber foil thickness such that $E_{abs}$ is larger than the energy range of the backscattered particles will eliminate the unwanted particles. Usually Mylar or aluminum foils are used as absorbers. The main disadvantage of the method is the poor depth resolution due the large energy spread (straggling) in the absorber foil.

There are more sophisticated methods to filter the backscattered particles. Here we briefly list them without detailed discussion:

1. Electrostatic or magnetic deflection (or the use of electrostatic or magnetic spectrometers) gives much better resolution than the absorber method but is complicated, time consuming, and expensive (spectrometers). For applications see Möller (1978), Möller et al. (1977), and Chaturvedi et al. (1990).

2. The TOF technique, a standard technique used in nuclear physics to distinguish particles, can be used to select the reaction products only. This technique requires sophisticated electronics and a two-dimensional multichannel analyzer.

3. The thin-detector technique is used when protons and α peaks overlap in a spectrum. Since the stopping power of protons is much smaller than the stopping power of α particles, the protons would lose only the fraction of their energies in a thin detector while the α particles stop completely in it. Thus the α particles will be separated from the mix. This technique uses either "$dE/dx$ detectors," which are quite expensive, or low-resistivity detectors, with low bias voltage. Recently the use of low-resistivity detectors was studied in detail by Amsel et al. (1992).

The TOF and thin-detector techniques have the disadvantage that although they select the reaction product, the large flux of backscattered particles still reach and can damage the detector.

### Energy Scanning (for Resonance Depth Profiling)

To do resonance depth profiling, the energy of the incident ion beam has to be changed in small steps. After acceleration, the ion beam goes through an analyzing magnet that selects the appropriate energy. Generally, the energy is changed by adjusting the terminal voltage and the magnetic field. Automatic energy-scanning systems have been developed using electrostatic deflection plates after the analyzing magnet by Amsel et al. (1983) and Meier and Richter (1990). Varying the voltage on these plates, the magnetic field is kept constant and the terminal voltage is changed by the slit feedback system. This system allows energy scanning up to several hundred kilo-electron-volts.

## Background, Interference, and Sensitivity

Since NRA and PIGE are both sensitive to specific isotopes, the background, interference from the sample matrix, and sensitivity, all depend on which nuclear reaction is used. When the nonresonant technique is used, the spectrum is generally background free. Interference from elements in the matrix is possible, especially in the deuteron-induced reaction case. An element-by-element discussion of the possible interferences can be found in Vizkelethy (1995) for NRA and in Hirvonen (1995) for PIGE.

In resonant depth profiling, there are two sources of background. One is natural background radiation (present only for PIGE). The effect of this background can be taken into account by careful background measurements or can be minimized by using appropriate active and passive shielding of the detector (Damjantschitsch et al., 1983; Kuhn et al., 1990; Horn and Lanford, 1990). The second background source is also present if the resonance is not an isolated resonance that sits on top of a non-resonant cross-section. In this case, the background is not constant and depends on the concentration profile of the element we want to measure. To extract a reliable depth profile from the excitation curve, a non-linear fit to a simulation is necessary.

As mentioned above, the sensitivity depends on the cross-section used and the composition of the sample. Generally the sensitivities of NRA and PIGE are on the order of 10 to 100 ppm. Lists of sensitivities for special applications can be found in Bird et al. (1974) and Hirvonen (1995).

## Cross-Sections and $Q$ values

The most important data in NRA and PIGE are the cross-sections, the $Q$ values of the reactions, and the energy level diagrams of nuclei. The most recent compilation of $Q$ values and energy level diagrams can be found for $A = 3$ in Tilley et al. (1987); $A = 5, \ldots, 10$ in Ajzenberg-Selove (1988); $A = 11, 12$ in Ajzenberg-Selove (1990); $A = 13, \ldots, 15$ in Ajzenberg-Selove (1991); $A = 16, 17$ in Tilley et al. (1993); f$A = 18, 19$ in Tilley et al. (1995); $A = 18, \ldots, 20$ in Ajzenberg-Selove (1987); and $A = 21, \ldots, 44$ in Endt and van der Leun (1978). Data of frequently used cross-sections can be found in Foster et al. (1995), and an extensive list of (p, $\gamma$) resonances can be found in Hirvonen and Lappalainen (1995).

The data are also available on the Internet (see Internet Resources). The energy levels are available from Lund Nuclear Data Service (LNDS) and from Triangle Universities Nuclear Laboratories (TUNL). An extensive database of nuclear cross-sections is available from the National Nuclear Data Center, NNDC) and from the T-2 Nuclear Information Services, although these serve more the needs of nuclear physicists than people working in the field of IBA. Nuclear reaction cross-sections used in IBA are available from the Sigmabase together with many (p, p) and ($\alpha,\alpha$) cross-sections.

## Usage of Standards

There are two basic reasons to use standards with NRA and PIGE. First, the energy of the bombarding beam is determined by reading the generating voltmeter (GVM) of the accelerator or by reading the magnetic field of the analyzing magnet. Usually these readings are not absolute and might shift with time. Therefore, in depth profiling it is necessary to use standards to determine the energy of the resonance (with respect to the accelerator readings) and the energy spread of the beam.

Second, since the yield is proportional to the collected charge, the cross-section, and the solid angle of the detector, to make an absolute measurement, the absolute values of these three quantities must be known precisely. In most cases, these values are not available. Ion bombardment always causes secondary electron emission. Imperfect suppression of the secondary electrons falsifies the current (and therefore the collected charge) measurement. It is not easy to design a measurement setup that can measure current with high absolute precision, but high relative precision and reproducibility are easy to achieve. Also, the solid angle of the detector and the absolute cross-section are usually not known precisely. Using reference targets that compare the reference yield to the yield from the unknown sample can eliminate most of these uncertainties.

A good reference target must satisfy the following requirements: (1) have high lateral uniformity; (2) be thick enough to provide sufficient yield in reasonable time but thin enough not to cause significant change in the cross-section; (3) have a standard that is amorphous (to avoid accidental channeling); (4) be stable in air, in vacuum, and under ion bombardment; and (5) be highly reproducible.

A detailed discussion of reference targets used in nuclear microanalysis can be found in Amsel and Davies (1983).

# METHOD AUTOMATION

Automation of acquisition of NRA spectra is the same as in RBS. The only task that can be automated is changing the sample. Since most of the measurements are done in high vacuum, there is a need for some mechanism to change samples without breaking the vacuum. If the scattering chamber is equipped with a load-lock mechanism, the sample change is easy and convenient and does not require automation. Without the load lock, usually large sample holders that can hold several samples are used. In this case stepping motors connected to a computer can be used to move from one sample to another.

In PIGE or resonant depth profiling, an automatic energy scanning system synchronized with data acquisition electronics is desired. Sophisticated energy scanning systems have been developed by Amsel et al. (1983) and Meier and Richter (1990). An alternative to these methods is the complete computer control of the accelerator and the analyzing magnet.

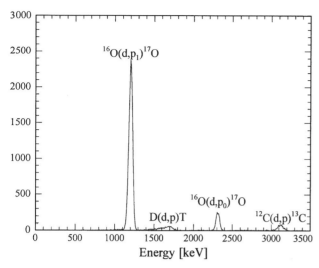

**Figure 5.** Measured spectrum from an 834-keV deuteron beam on 100 nm $SiO_2$ on Si.

## DATA ANALYSIS AND INITIAL INTERPRETATION

### Overall Surface Content (Film Thickness)

As mentioned above, when the cross-section can be considered flat across the thickness of the film, the overall surface content can be determined easily. Figure 5 shows a spectrum from an 834-keV deuteron beam incident on 100 nm $SiO_2$ on bulk Si. The $^{16}O(d, p)^{17}O$ reaction has two proton groups that show up as two separate peaks in the spectrum. Since the $(d, p_1)$ cross-section has a plateau around 850 keV, we need to consider only the $p_1$ peak.

Using Equation 1, we can easily calculate the number of oxygen atoms per square centimeter. In most cases reference standards are used (thin $Ta_2O_5$ layers on Ta backing or thin $SiO_2$ layer on Si); therefore, the exact knowledge of the detector solid angle, cross-section, and absolute value of the collected charge is not necessary. If the reference sample contains $Nt_{ref}$ oxygen atoms per square centimeter, the number of oxygen atoms per square centimeter in the unknown sample is

$$Nt = Nt_{ref} \frac{Y}{Y_{ref}} \frac{Q_C^{ref}}{Q_C} \qquad (6)$$

where $Q_C^{ref}$ and $Q_C$ are collected charges and $Y_{ref}$ and $Y$ are the peak areas of the $p_1$ protons for the reference standard and for the unknown sample, respectively. There are two peaks in the spectrum that need further explanation. The peak at higher energies is the result of the $^{12}C(d, p)^{13}C$ reaction from the hydrocarbon layer deposited during ion bombardment. The broad peak between the $p_0$ and $p_1$ peaks is due to the $D(d, p)T$ reaction. Since this particular sample was used as a reference standard, during the measurements a considerable amount of deuterium had been implanted into it and is now a detectable component of the target.

### Nonresonant Depth Profiling

When the cross-section cannot be considered constant, the spectrum becomes the convolution of the concentration

profile with the energy broadening of the incident ions and the detected reaction particles. The interpretation is not that simple now. The energy-depth conversion can be calculated using Equations 2 and 3. If the cross-section changes only slowly, the spectrum shows the main features of the concentration profile, but this cannot be considered as quantitative analysis. To extract the concentration profile from the spectrum, the spectrum has to be simulated and the assumed concentration profile should be changed until an acceptable agreement between the simulation and the measurement is reached. Several computer programs are available that can do the simulations and fits, e.g., ANALRA (Johnston, 1993) and SEN-RAS (Vizkelethy, 1990). These programs are available from the Sigmabase (see Internet Resources).

### Resonant Depth Profiling

In resonant depth profiling, the spectrum is usually not recorded, but the particles or $\gamma$ photons are counted in an energy window. The data that carry the depth profile information is the excitation function, i.e., the number of counts vs. the incident energy. The excitation function is the convolution of the resonance cross-section, the energy profile of the incident beam as a function of depth, and the concentration profile. Qualitative features of the concentration profile can be deduced from the excitation function. Quantitative concentration profiles can be obtained by simulating the excitation function and using least-squares fitting to match it to the measured excitation functions. [More detailed discussion can be found in Hirvonen (1995).] Also, several computer programs have been developed for profiling purposes, most recently in Smulders (1986); Lappalainen (1986), and Rickards (1991). The SPACES program (Vickridge and Amsel, 1990) was developed especially for high-resolution surface studies using very narrow resonance. This program is also available from the Sigmabase.

A simple example is shown in Figure 6. In this experiment an $^{18}O$ profile was measured in YBaCuO using the already mentioned $^{18}O(p, \alpha)^{15}N$ reaction. From the spectrum

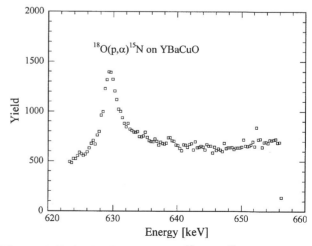

**Figure 6.** Excitation function of the $^{18}O(p, \alpha)^{15}N$ reaction measured on an $^{18}O$-enriched YBaCuO sample (Cheang-Wong, 1992).

we can deduce qualitatively that there is an $^{18}O$ enrichment on the surface. Using Equation 6, we can estimate the thickness of the enriched region (width of the peak divided by the stopping power of proton around the resonance energy), which is about 20 nm. Using simulation of the excitation curve and a nonlinear fit to the measured excitation function gives a 15-nm-thick layer enriched to 30% $^{18}O$ with 3.5% constant volume enrichment.

A recent review of the computer simulation methods in IBA can be found in Vizkelethy (1994).

## SPECIMEN MODIFICATION

Although NRA and PIGE are nondestructive techniques, certain modifications to the samples can occur due to the so-called beam effect. Since a high-energy ion beam is bombarding the sample surface during the measurement, considerable heat transfer occurs. A 1-MeV ion beam with 1-$\mu$A/cm$^2$ current (not unusual in NRA and PIGE) means 1-W/cm$^2$ heat is transferred to the sample. This can be a serious problem in biological samples, and the measurement can lead to the destruction of the samples. In good heat-conducting samples, the heat does not cause a big problem, but diffusion and slight structural changes can occur. Poor heat-conducting samples will suffer local overheating that can induce considerable changes in the composition of the sample.

Apart from the heat delivered to the sample, the large measuring doses can cause considerable damage to single-crystal samples. It is especially significant in heavy-ion beams, such as $^{15}N$, which is used to profile hydrogen using the $^{1}H(^{15}N, \alpha\gamma)^{12}C$ reaction. In this case significant hydrogen loss can occur during the analysis. Heavy ions also can cause ion beam mixing and sputtering.

To avoid these problems, the beam should be spread out over a large area on the sample if high current must be used or the current should be kept at the minimum acceptable level.

## PROBLEMS

Several factors can lead to false results. Most of them can be minimized with careful design of the measurement.

A frequently encountered problem is the inability to precisely measure the current. The secondary electrons emitted from the sample can falsify the current measurement and thus measurement of the collected charge. To minimize the effect of secondary electron emission, several techniques can be used:

1. A suppression cage around the sample holder will minimize the number of escaping electrons, but it is not possible to build a perfect suppression cage. A much simpler technique is to apply positive bias to the sample, but this alternative cannot be used for every sample.

2. Isolating the scattering chamber from the rest of the beamline and allowing its use as a Faraday cup is also a good solution, but in many cases the chamber is connected electrically to other equipment that cannot be isolated.

3. There are different transmission Faraday cup designs where the current is measured not on the sample but in a separate Faraday cup that intercepts the beam in front of the sample several times per second.

4. As described above (see Practical Aspects of the Method), standards can be used to accurately measure current.

Another problem is hydrocarbon layer deposition during measurement. This can cause serious problems when narrow resonances are used for depth profiling. The incident ions will lose energy in the hydrocarbon layer that has an unknown thickness. This can lead to a false depth scale if the energy loss is significant in the hydrocarbon layer. Since the hydrocarbon layer formation is slower in higher vacuum, the best way to minimize its effect is to maintain good vacuum (see GENERAL VACCUM TECHNIQUES). Cold traps in the beamline and in the scattering chamber can significantly reduce hydrocarbon layer formation. Another way to minimize the effect of the carbon layer in case of resonance profiling (when many measurements have to be made on the same sample) is to frequently move the beam spot to a fresh area.

Interference from the elements of the matrix can be another problem. This is a potential danger in deuteron-induced reactions. Many light elements have (d, p) or (d, $\alpha$) reactions, and the protons and $\alpha$ particles can overlap in the spectrum. Careful study of the kinematics and the choice of appropriate absorber foil thickness can solve this problem, although in certain cases the interference cannot be avoided. In PIGE, when there are interfering reactions, the use of a Ge(Li) or HPGE detector can help. The better resolution of germanium detectors allows the separation of the $\gamma$ peaks from the different reactions, but at the cost of lower efficiency.

A special problem arises in single-crystal samples—accidental channeling. When the direction of the incident ions coincides with one of the main axes of the crystal, the ions are steered to these channels and the reaction yield becomes lower than it would otherwise be. (Also there is a difference between the energy loss of ions in a channel and in a random direction, which can affect the energy-to-depth conversion.) The solution is to tilt the samples to 5° to 7° with respect to the direction of the beam. (There are cases when channeling is preferred. Using NRA or PIGE combined with channeling, the lattice location of atoms can be determined.)

Electrically nonconducting samples will become charged under ion bombardment up to voltages of several kilovolts, and this can modify the beam energy. This can be especially detrimental when narrow (few hundred electron volts wide) resonances are used for depth profiling. There are several solutions to avoid surface charging: (1) A thin conducting coating (usually carbon) on the surface of the sample can reduce the charging significantly. (2) Supplying low-energy electrons from a hot filament will neutralize the accumulated positive charge on the sample surface.

A problem can arise from the careless use of the fitting programs. As in any deconvolution problem, there is usually no unique solution determined by the measured yield curve only. (One example is the fact that these deconvolutions tend to give an oscillating depth profile, which is in most cases obviously incorrect.) Extra boundary conditions and assumptions for the profile are needed to make the solution acceptable.

## LITERATURE CITED

Ajzenberg-Selove, F. 1987. Energy levels of light nuclei $A = 18-20$. *Nucl. Phys.* A475:1–198.

Ajzenberg-Selove, F. 1988. Energy levels of light nuclei $A = 5-10$. *Nucl. Phys.* A490:1.

Ajzenberg-Selove, F. 1990. Energy levels of light nuclei $A = 11-12$. *Nucl. Phys.* A506:1–158.

Ajzenberg-Selove, F. 1991. Energy levels of light nuclei $A = 13-15$. *Nucl. Phys.* A523:1–196.

Amsel, G., d'Artemare, E., and Girard., E. 1983. A simple, digitally controlled, automatic, hysteresis free, high precision energy scanning system for van de Graaff type accelerators. *Nucl. Instrum. Methods.* 205:5–26.

Amsel, G. and Davies, J. A. 1983. Precision standard reference targets for microanalysis with nuclear reactions. *Nucl. Instrum. Methods* 218:177–182.

Amsel, G., Pászti, F., Szilágyi, E., and Gyulai, J. 1992. p, d, and $\alpha$ particle discrimination in NRA: Thin, adjustable sensitive zone semiconductor detectors revisited. *Nucl. Instrum. Methods* B63:421–433.

Amsel, G. and Samuel, D. 1967. Microanalysis of the stable isotopes of oxygen by means of nuclear reactions. *Anal. Chem.* 39:1689–1698.

Bird, J. R., Campbell, B. L., and Price, P. B. 1974. Prompt nuclear analysis. *Atomic Energy Rev.* 12:275–342.

Blatt, J. M. and Weisskopf, V. 1994. Theoretical Nuclear Physics. Dover Publications, New York.

Chaturvedi, U. K., Steiner, U., Zak, O., Krausch, G., Shatz, G., and Klein, L. 1990. Structure at polymer interfaces determined by high-resolution nuclear reaction analysis. *Appl. Phys. Lett.* 56:1228–1230.

Cheang-Wong, J. C., Ortega, C., Siejka, J., Trimaille, I., Sacuto, A., Balkanski, M., and Vizkelethy, G. 1992. RBS analysis of thin amorphous YBaCuO films: Comparison with direct determination of oxygen contents by NRA. *Nucl. Instrum. Methods* B64:169–173.

Damjantschitsch, H., Weiser, G., Heusser, G., Kalbitzer, S., and Mannsperger, H. 1983. An in-beam-line low-level system for nuclear reaction γ-rays. *Nucl. Instrum. Methods* 218:129–140.

Davies, J. A., Lennard, W. N., and Mitchell, I. V. 1995. Pitfalls in ion beam analysis. *In* Handbook of Modern Ion Beam Analysis (J. R. Tesmer and M. Nastasi, eds.). pp. 343–363. Materials Research Society, Pittsburgh, Pa.

Endt, P. M. and van der Leun, C. 1978. Energy levels of $A = 21-44$ nuclei. *Nucl. Phys.* A310:1–752.

Foster, L., Vizkelethy, G., Lee, M., Tesmer, J. R., and Nastasi, M. 1995. Particle-particle nuclear reaction cross sections. *In* Handbook of Modern Ion Beam Materials Analysis (J.R. Tesmer and M. Nastasi, eds.). pp. 549–572. Materials Research Society, Pittsburgh, Pa.

Green, P. F. and Doyle, B. L. 1986. Silicon elastic recoil detection studies of polymer diffusion: Advantages and disadvantages. *Nucl. Instrum. Methods* B18:64–70.

Hirvonen, J.-P. 1995. Nuclear reaction analysis: Particle-gamma reactions. *In* Handbook of Modern Ion Beam Materials Analysis (J. R. Tesmer and M. Nastasi, eds.) pp. 167–192. Materials Research Society, Pittsburgh, Pa.

Hirvonen, J.-P. and Lappalainen, R. 1995. Particle-gamma data. *In* Handbook of Modern Ion Beam Materials Analysis (J. R. Tesmer and M. Nastasi, eds.). pp. 573–613. Materials Research Society, Pittsburgh, Pa.

Horn, K. M. and Lanford, W. A. 1990. Suppression of background radiation in BGO and NaI detectors used in nuclear reaction analysis. *Nucl. Instrum. Methods* B45:256–259.

Jarjis, R. A. 1979. Internal Report, University of Manchaster, U.K.

Johnston, P. N. 1993. ANALRA—charged particle nuclear analysis software for the IBM PC. *Nucl. Instrum. Methods* B79:506–508.

Kuhn, D., Rauch, F., and Baumann, H. 1990. A low-background detection system using a BGO detector for sensitive hydrogen analysis with the $^1H(^{15}N,a\gamma)^{12}C$ reaction. *Nucl. Instrum. Methods* B45:252–255.

Lappalainen, R. 1986. Application of the NRB method in range, diffusion, and lifetime measurements, Ph.D. Thesis, University of Helsinki, Finland.

Maurel, B. 1980. Stochastic Theory of Fast Charged Particle Energy Loss. Application to Resonance Yield Curves and Depth Profiling. Ph.D. Thesis, University of Paris, France.

Maurel, B., Amsel, G., and Nadai, J. P. 1982. Depth profiling with narrow resonances of nuclear reactions: Theory and experimental use. *Nucl. Instrum. Methods* 197:1–14.

Meier, J. H. and Richter, F. W. 1990. A useful device for scanning the beam energy of a van de Graaff accelerator. *Nucl. Instrum. Methods* B47:303–306.

Möller, W. 1978. Background reduction in D($^3$He, α)H depth profiling experiments using a simple electrostatic deflector. *Nucl. Instrum. Methods* 157:223–227.

Möller, W., Hufschmidt, M., and Kamke, D. 1977. Large depth profile measurements of D, $^3$He, and $^6$Li by deuterium induced nuclear reactions. *Nucl. Instrum. Methods* 140:157–165.

Rickards, J. 1991. Fluorine studies with a small accelerator. *Nucl. Instrum. Methods.* B56/57:812–815.

Simpson, J. C. B. and Earwaker, L. G. 1984. A computer simulation of nuclear reaction spectra with applications in analysis and depth profiling of light elements. *Vacuum* 34:899–902.

Smulders, P. J. M. 1986. A deconvolution technique with smooth, non-negative results. *Nucl. Instrum. Methods* B14:234–239.

Tilley, D. R., Weller, H. R., and Hassian, H. H. 1987. Energy levels of light nuclei $A = 3$. *Nucl. Phys.* A474:1–60.

Tilley, D. R., Weller, H. R., and Cheves, C. M. 1993. Energy levels of light nuclei $A = 16-17$. *Nucl. Phys.* A564:1–183.

Tilley, D. R., Weller, H. R., Cheves, C. M., and Chesteler, R. M. 1995. Energy levels of light nuclei $A = 18-19$. *Nucl. Phys.* A595:1–170.

Vickridge, I. C. 1990. Stochastic Theory of Fast Ion Energy Loss and Its Application to Depth Profiling Using Narrow Nuclear Resonances. Applications in Stable Isotope Tracing Experiment for Materials Science. PhD Thesis, University of Paris, France.

Vickridge, I. C. and Amsel, G. 1990. SPACES: A PC implementation of the stochastic theory of energy loss for narrow resonance profiling. *Nucl. Instrum. Methods* B45:6–11.

Vizkelethy, G. 1990. Simulation and evaluation of nuclear reaction spectra. *Nucl. Instrum. Methods* B45:1–5.

Vizkelethy, G. 1994. Computer simulation of ion beam methods in analysis of thin films. *Nucl. Instrum. Methods* B89:122–130.

Vizkelethy, G. 1995. Nuclear reaction analysis: Particle-particle reactions. *In* Handbook of Modern Ion Beam Materials Analysis (J. R. Tesmer and M. Nastasi, eds.). pp. 139–165. Materials Research Society, Pittsburgh, Pa.

## KEY REFERENCES

Amsel, G. and Lanford, W. A. 1984. Nuclear reaction technique in materials analysis. *Ann. Rev. Nucl. Part. Sci.* 34:435–460.

*Excellent review paper of RNA.*

Deconnick, G. 1978 Introduction to Radioanalytical Chemistry. Elsevier, Amsterdam.

*Excellent reviews of the methods.*

Feldman, L. C. and Mayer, J. W. 1986. Fundamentals of Surface and Thin Film Analysis. North-Holland Publishing, New York. See chapter on "Nuclear techniques: Activation analysis and prompt radiation analysis," pp. 283–310.

*Good discussion of the fundamental physics involved in NRA.*

Hirvonen, 1995. See above.

*Detailed discussion of PIGE with several worked-out examples.*

Lanford, W. A. 1995. Nuclear reactions for hydrogen analysis. *In* Handbook of Modern Ion Beam Materials Analysis (J. R. Tesmer and M. Nastasi, eds.). pp. 193–204. Materials Research Society, Pittsburgh, Pa.

*Detailed discussion of hydrogen detection with NRA.*

Peaisach, M. 1992. Nuclear reaction analysis. *In* Elemental Analysis by Particle Accelerators (Z. B. Alfassi and M. Peisach, eds.). pp. 351–383. CRC Press, Boca Raton, Fla.

*General discussion of NRA with valuable references.*

Vizkelethy, 1995. See above.

*Detailed discussion of NRA with worked-out examples and discussion of useful reactions.*

## INTERNET RESOURCES

Lund Nuclear Data Service, *http://nucleardata.nuclear.lu.se/nucleardata*

National Nuclear Data Center, *http://www.nndc.bnl.gov/*

Sigmabase, *http://ibaserver.physics.isu.edu/sigmabase*

T-2 Nuclear Information Service, *http://t2.lanl.gov/*

Triangle Universities Nuclear Laboratory, *http://www.tunl.duke.edu/NuclData*

## APPENDIX

### Energy Relations in Nuclear Reactions

In a charged particle reaction, the energies of the emitted particles in the laboratory coordinate system are (see Fig. 7):

$$E_{\text{light}} = B\left(\cos\psi \pm \sqrt{\frac{D}{B} - \sin^2\psi}\right)^2 (E_0 + Q) \qquad (7)$$

$$E_{\text{heavy}} = A\left(\cos\varphi \pm \sqrt{\frac{C}{A} - \sin^2\varphi}\right)^2 (E_0 + Q) \qquad (8)$$

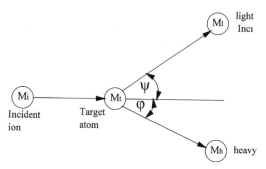

**Figure 7.** Reaction kinematics.

where $E_0$ is the energy of the incident ions; $Q$ is the $Q$ value of the reaction; and $A$, $B$, $C$, and $D$ are given as

$$\begin{aligned}
A &= \frac{M_1 M_4}{(M_1 + M_2)(M_3 + M_4)}\frac{E_0}{E_0 + Q} \\[4pt]
B &= \frac{M_1 M_3}{(M_1 + M_2)(M_3 + M_4)}\frac{E_0}{E_0 + Q} \\[4pt]
C &= \frac{M_2 M_3}{(M_1 + M_2)(M_3 + M_4)}\left(1 + \frac{M_1 Q}{M_2(E_0 + Q)}\right) \\[4pt]
D &= \frac{M_2 M_4}{(M_1 + M_2)(M_3 + M_4)}\left(1 + \frac{M_1 Q}{M_2(E_0 + Q)}\right)
\end{aligned} \qquad (9)$$

where $M_1$, $M_2$, $M_3$, and $M_4$ are the masses of the incident ion, the target nucleus, the lighter reaction product, and the heavier product, respectively.

In Equation 7 only plus signs are used unless $B > D$, in which case $\psi_{\text{max}} = \sin^{-1}\sqrt{D/B}$, and in Equation 8 only plus signs are used unless $A > C$, in which case $\varphi_{\text{max}} = \sin^{-1}\sqrt{C/A}$. The emitted particle then loses energy along its path until it leaves the target. The energy loss of ions inward and outward is described by Equations 8 and 9.

### Energy Spread (Depth Resolution)

Equations 2 and 3 are true only on the average. Since the energy loss is a stochastic process, the ions, after passing a certain thickness, will have an energy distribution rather than a sharp energy. This phenomenon is called straggling. There are several models to describe the energy straggling; in most cases, the simplest one, Bohr straggling, is satisfactory. [One exception is the case of very narrow resonances, which is described in details in Maurel (1980) and Vickridge (1990).]

There are other factors that cause broadening of the energy in the detected spectrum. The following contribute to the energy spread of detected particles in NRA: (1) initial energy spread of the incident beam, (2) straggling of the incident ions, (3) multiple scattering, (4) geometric spread due to the finite acceptance angle of the detector and to the finite beam size, (5) straggling of the outgoing particles, (6) energy resolution of the detector, and (6) energy straggling in the absorber foil.

A detailed treatment of most of these factors can be found in Vizkelethy (1994). In PIGE and resonant depth profiling with NRA, only a few contribute to the depth resolution. Since the energy of the detected particles or $\gamma$

rays is not used to extract concentration profile information, only the initial energy spread, the straggling of the incident beam, and the resonance width determine the depth resolution. (The contribution from the multiple scattering is negligible since then normal incidence is used.)

### Calculation of the Excitation Function for Resonance Depth Profiling

The excitation function at an $E_0$ incident energy is

$$Y(E_0) \propto \sigma(E_0) * h(E_0) * \int_0^\infty c(x)g(E_0;x)\,dx \qquad (10)$$

where $\sigma(E_0)$ is the reaction cross-section, $h(E_0)$ is the initial energy distribution of the ion beam, $c(x)$ is the concentration profile of the atoms to be detected, $g(E;x)$ is the probability that an ion loses $E$ energy when penetrating a thickness $x$, and the asterisk denotes convolution. [This convolution is a very complicated calculation; for details see Maurel (1980) and Vickridge (1990).] When the resonance is not extremely narrow, the initial energy spread of the ions is not too small, and the profile is not measured very close to the surface (the energy loss is large enough that the straggling can be considered Gaussian); the function $g(E;x)$ can then be approximated with a Gaussian:

$$g(E;x) = \frac{1}{\sqrt{2\pi s^2(x)}} e^{-[E-E(x)]^2/2s^2(x)} \qquad (11)$$

where $E(x)$ is the same as $E_{in}(x)$ in Equation 3 and $s(x)$ is the root-mean-square energy straggling at $x$. If the resonance cross-section is given by a Breit-Wigner resonance (Blatt and Weiskopf, 1994),

$$\sigma(E_0) = \sigma_0 \frac{\Gamma^2/4}{\Gamma^2/4 + (E-E_R)^2} \qquad (12)$$

where $\sigma_0$ is the strength, $\Gamma$ is the width, and $E_R$ is the energy of the resonance, then the straggling, the Doppler-broadening, and the initial energy spread can be combined into a single Gaussian with depth-dependent width. Evaluating Equation 10, the excitation function is

$$Y(E_0) \propto \int_0^\infty c(x)\sigma_0 \frac{\Gamma}{\sqrt{8}S(x)} \operatorname{Re} w\left(\frac{E_0-E_R}{\sqrt{2}S(x)} - i\frac{\Gamma}{\sqrt{8}S(x)}\right) dx \qquad (13)$$

where $\operatorname{Re} w(z)$ is the real part of the complex error function and $S(x)$ the width of the combined Gaussian.

GYÖRGY VIZKELETHY
Idaho State University
Pocatello, Idaho

# PARTICLE-INDUCED X-RAY EMISSION

## INTRODUCTION

Particle-induced x-ray emission (PIXE) is an elemental analysis technique that employs a mega-electron-volt energy beam of charged particles from a small electrostatic accelerator to induce characteristic x-ray emission from the inner shells of atoms in the specimen. The accelerator can be single ended or tandem. Most PIXE work is done with 2- to 4-MeV proton beams containing currents of a few nanoamperes within a beam whose diameter is typically a few millimeters; a small amount of work with deuteron and helium beams has been reported, and use of heavier ion beams has been explored. The emitted x rays are nearly always detected in the energy-dispersive mode using a Si(Li) spectrometer. Although wavelength-dispersive x-ray spectrometry would provide superior energy resolution, its low geometric efficiency is a significant disadvantage for PIXE, in which potential specimen damage by heating limits the beam current and therefore the emitted x-ray intensity. In the proton microprobe, magnetic or electrostatic quadrupole lenses are employed to focus the beam to micrometer spot size; this makes it possible to perform micro-PIXE analysis of very small features and, also, by sweeping this microbeam along a preselected line or over an area of the specimen, to determine element distributions in one or two dimensions in a fully quantitative manner. Most proton microprobes realize a beam spot size smaller than a few micrometers at a beam current of at least 0.1 nA. In trace element analysis or imaging by micro-PIXE, the high efficiency of a Si(Li) detector, placed as close to the specimen as possible, is again mandatory for x-ray collection. To date, PIXE and micro-PIXE have most often been used to conduct trace element analysis in specimens whose major element composition is already known or has been measured by other means, such as electron microprobe analysis. However, both variants are capable of providing major element analysis for elements of atomic number down to $Z = 11$.

## Competing Methods

In the absence of beam focusing, conventional PIXE is an alternative to x-ray fluorescence analysis (XRFA; see X-RAY MICROPROBE FOR FLUORESCENCE AND DIFFRACTION ANALYSIS) but requires more complex equipment. Micro-PIXE is a powerful and seamless complement to electron microprobe analysis (EPMA), a longer established microbeam x-ray emission method based on excitation by kilo-electron-volt electron beams; detection limits of a few hundred parts per million in EPMA compare with limits of a few parts per million in micro-PIXE. While the electron microprobe is ubiquitous, there are only about 50 proton microprobes around the world, but many of these can provide beam time on short notice. Conducted with the highly polarized x radiation from a synchroton storage ring equipped with insertion devices such as wigglers and undulators, XRFA now provides a strong competitor to micro-PIXE in terms of both spatial resolution (a few micrometers) and detection limits (sub−parts per million). The x-ray spectrum from an undulator is highly collimated, extremely intense, and nearly monochromatic; the exciting x-ray energy can be tuned to lie just above the absorption edge energy of the element of greatest interest, where the photoabsorption cross-section is highest, thereby optimizing the detection limit. However, there are only a handful of synchrotron facilities, and beam time is at a premium. A much cheaper method of obtaining an x-ray microbeam is through the use of a conventional fine-focus x-ray tube and a nonimaging optical system based upon focusing capillaries (Rindby, 1993). Total-reflection XRFA (Klockenkamper et al., 1992) offers similar detection limits, but this method is restricted to very thin films deposited on a totally reflecting substrate and lacks the versatility of PIXE and XRFA as regards different specimen types. One of PIXEs merits relative to XRFA is the possibility of deploying other ion beam analysis techniques simultaneously or sequentially; these include Rutherford backscattering spectroscopy (see MEDIUM-ENERGY BACKSCATTERING AND FORWARD-RECOIL SPECTROMETRY and HEAVY-ION BACKSCATTERING SPECTROMETRY), nuclear reaction analysis (NUCLEAR REACTION ANALYSIS AND PROTON-INDUCED GAMMA RAY EMISSION), ionoluminescence, and scanning transmission ion microscopy (STIM).

Optical spectroscopy methods based on emission (AES—atomic emission spectroscopy) and absorption (AAS—atomic absorption spectroscopy) are the workhorses of trace element analysis in materials that can be reduced to liquid form or dissolved in a liquid and then atomized. The inductively coupled plasma (ICP) has become the method of choice for inducing light emission, and ICP-AES attains detection limits of 1 to 30 ng/mL of liquid; the limits referred to the original material depend on the dilution factor and typically reach down to 0.1 ppm. Interference effects among optical lines are more complex than is the case in PIXE with x-ray lines. Use of mass spectrometry (ICP-MS) reduces interferences, provides a more complete elemental coverage than ICP-AES, and has much less element-to-element variation in sensitivity than does ICP-AES; its matrix effects are less straightforward to handle than those of PIXE. Overall, PIXE is more versatile as regards specimen type and preparation than ICP-AES and ICP-MS and has the advantages of straightforward matrix correction, smoothly varying sensitivity with atomic number, and high accuracy. But for conventional bulk analysis, the highly developed optical and mass spectrometry methods will frequently be more accessible and entirely satisfactory and provide excellent detection limit without the need for an accelerator. For particular specimen types such as aerosol particulates and awkwardly shaped specimens (e.g., in archaeology) requiring nondestructive analysis, PIXE has unique advantages. The ability to handle awkwardly shaped specimens by extracting the proton beam through a thin window into the laboratory milieu is particularly valuable in archaeometric applications.

The nondestructive nature, high spatial resolution, parts per million detection limits, and high accuracy of micro-PIXE together with the simultaneous deployment of STIM and backscattering make it very competitive with other microprobe methods. It has been applied intensively in the analysis of individual mineral grains and in zoning phenomena within these, individual fly-ash particles, single biological cells, and thin tissue slices containing features such as Alzheimer's plaques (Johansson et al., 1995). Secondary ion mass spectrometry (SIMS) often provides better detection limits but is destructive. The combination of laser ablation microprobe with inductively coupled plasma (ICP) excitation and mass spectrometry offers resolution of some tens of micrometers and detection limits as low as 0.5 ppm; however, the ablation basis of this method renders matrix effects and standardization more complex than in micro-PIXE.

Detailed accounts of the fundamentals and of many applications can be found in two recent books on PIXE (Johansson and Campbell, 1988; Johansson et al., 1995), the first of which provides considerable historical perspective. The present offering is an overview, including some typical applications. The proceedings of the triennial international conferences on PIXE provide both an excellent historical picture and an account of recent developments and new applications.

## PRINCIPLES OF THE METHOD

The most general relationship between element concentrations in a specimen and the intensity of detected x rays of each element is derived for the simple case of $K$ x rays (see Appendix A). The $L$ and $M$ x rays may be dealt with in similar manner, with the added complexity that in these cases ionization occurs in three and five subshells, respectively, and there is the probability of vacancies being transferred by the Coster-Kronig effect to higher subshells prior to the x-ray emission occurring. In principle, Equation 6 could be used to perform standardless analysis; this would involve a complete reliance on the underlying theory and the database and on the assumed properties of the x-ray detector. At the other extreme, by employing standards that very closely mimic the major element (matrix) composition of a specimen, the PIXE analyst working on trace elements could completely avoid dependence upon theory and rely only upon calibration curves relating x-ray yield to concentration. In practice, there is much practical merit in adopting an intermediate stance. We recommend a fundamental parameter approach using the well-understood theories of the interaction of charged particle beams and photon beams with matter and of characteristic x-ray emission from inner shell vacancies but also relying on a small set of standards that need bear only limited resemblance to the specimens at hand.

In this approach the equation used to derive element concentrations from measured x-ray intensities is

$$Y(Z) = Y_1(Z)H\varepsilon_Z t_Z C_Z Q \qquad (1)$$

where $Y(Z)$ is the measured intensity of x rays in the principal line of element $Z$ (concentration $C_Z$), $Y_1(Z)$ is the theoretically computed intensity per unit beam charge per unit detector solid angle per unit concentration, $H$ is an instrumental constant that subsumes the detector solid angle and any calibration factor required to convert an indirect measurement of beam charge to units of microcoulombs, $Q$ is the direct or indirect measure of beam charge, $t_Z$ is the x-ray transmission fraction through any absorbers (see Practical Aspects of the Method) that are deliberately interposed between specimen and detector, and $\varepsilon_Z$ is the detector's intrinsic efficiency. It is also straightforward to calculate the contribution of secondary x rays fluoresced when proton-induced x rays are absorbed in the specimen and to correct for these. The various atomic physics quantities required are obtained from databases that have been developed by fitting appropriately parameterized expressions (in part based on theory and in part semiempirical) to either compilations of experimental data or theoretically generated values. The $H$ value is determined via Equation 1 by use of appropriate standards or standard reference materials. It is implicit in the direct use of Equation 1 that the major elements have already been identified and that their concentrations are known a priori. This permits computation of the integrals that involve matrix effects (slowing down of the protons and attenuation of the x rays); Equation 1 then provides the trace element concentrations. However, if the x-ray lines of the major elements are observed in the spectrum, Equation 1 may be solved in an iterative manner to provide both major and trace element concentrations.

The method is fully quantitative, and its accuracy is determined in part by the accuracy of the database. The theoretically computed x-ray intensity uses ionization cross-sections and stopping powers for protons, x-ray mass attenuation coefficients, x-ray fluorescence yields and Coster-Kronig probabilities for the various subshells, and relative intensities of the lines within each x-ray series (e.g., $K$, $L_1$, ...) emitted by a given element. For ionization cross-sections, the choice is among the hydrogenic model calculations of Cohen and Harrigan (1985, 1989) or of Liu and Cipolla (1996) based on the so-called ECPSSR model of Brandt and Lapicki (1981), the equivalent but more sophisticated ECPSSR treatment of Chen and Crasemann (1985, 1989) based on self-consistent field wave functions, or experimental cross-section compilations such as those of Paul and Sacher (1989) for the $K$ shell and Orlic et al. (1994) for the $L$ subshells. Proton stopping powers (i.e., rate of energy loss as a function of distance traveled in an element) are usually taken from one of the various compilations of Ziegler and his colleagues, which are summarized in a report by the International Commission on Radiation Units and Measurements (1993). A variety of mass attenuation coefficient schemes ranging from theoretical to semiempirical have been used, and the XCOM database provided by the National Institute of Science and Technology (NIST; Berger and Hubbell, 1987) appears to us to be admirably suited. The relative x-ray intensities within the $K$ and $L$ series are invariably taken from the calculations of Scofield (1974a,b, 1975) and from fits to these by Campbell and Wang (1989), although corrections must be made for the $K$ x rays of the elements $21 < Z < 30$ where configuration mixing effects arising from the open $3d$ subshell cause divergences; relative intensities for the $M$ series are given by Chen and Crasemann (1984). Atomic fluorescence and Coster-Kronig probabilities may be taken from the theoretical calculations of Chen et al. (1980a,b, 1981, 1983) or from critical assessments of experimental data such as those of Krause (1979) and Bambynek (see Hubbell, 1989). Campbell and Cookson (1983) have assessed how the various uncertainties in all these quantities are transmitted into accuracy estimates for PIXE, but the best route for assessing overall

accuracy remains the analysis of known reference materials.

The intrinsic efficiency of the Si(Li) x-ray detector enters Equation 1 and so must be known. This efficiency is essentially unity in the x-ray energy region between 5 and 15 keV. At higher energies $\varepsilon_Z$ falls off because of penetration through the silicon crystal (typically 3 to 5 mm thick). At lower energies, $\varepsilon_Z$ falls off due to attenuation in the beryllium or polymer vacuum window, the metal contact, and internal detector effects such as the escape of the primary photoelectrons and Auger electrons created in x-ray interactions with silicon atoms and loss of the secondary ionization electrons due to trapping and diffusion. The related issues of detector efficiency and resolution function (lineshape) are dealt with below (see Appendix B).

A basic assumption of the method is that the element distribution in the volume analyzed is homogeneous. The presence of subsurface inclusions, for example, would negate this assumption. Of course, micro-PIXE can be used to probe optically visible inhomogeneities such as zoning and grain boundary phenomena in minerals and to search for optically invisible inhomogeneities in, for example, air particulate deposits from cascade impactors. It can then be used to select homogeneous regions for analysis. Given (1) homogeneity and (2) knowledge of the major element concentrations, PIXE or micro-PIXE analysis is fully quantitative, with the accuracy and the detection limits determined by the type of specimen.

A great deal of PIXE work is conducted on specimens that are so thin that the proton energy is scarcely altered in traversing the specimen and there is negligible attenuation of x rays. Films of fine particulates collected from the ambient air by sampling devices are an example. In such cases, a similar approach may be taken with concentration $C_Z$ replaced in Equation 1 by the areal density of elements in the specimen.

The main engineering advance responsible for PIXE was the advent of the Si(Li) x-ray detector in the late 1960s. Micro-PIXE was made possible by the development of high-precision magnetic quadrupole focusing lenses in the 1970s. The acceptance of PIXE and other accelerator-based ion beam techniques stimulated advances in accelerator design that have resulted in a new generation of compact, highly stable electrostatic accelerators provided primarily for ion beam analysis.

## PRACTICAL ASPECTS OF THE METHOD

Most PIXE work is done with nanoampere currents of 2- to 4-MeV protons transmitted in a vacuum beam line to the specimen. However, some laboratories extract the beam into the laboratory through a thin Kapton window to deal with unwieldy or easily damaged specimens such as manuscripts or objects of art (Doyle et al., 1991).

As with EPMA, most of the technical work for the user lies in specimen preparation (see Sample Preparation). Conduct of PIXE analysis requires an expert accelerator operator to steer and focus the beam onto the specimen.

As indicated above (see Principles of the Method), the intrinsic efficiency of the Si(Li) detector as a function of x-ray energy is an important practical aspect. This function may be determined from, e.g., the manufacturer's data on crystal thickness and the contact and window thickness. More accurate determinations of these quantities may be effected via methods outlined below (see Appendix B). The lineshape or resolution function of the detector is required (see Data Analysis and Initial Interpretation) for the interpretation of complex PIXE spectra with their many overlapping lines. Guidance in this direction is also provided below (see Appendix B).

Specimens are generally classified as thin or thick. Thin specimens are those that cause negligible reduction in the energy of the protons, and they are usually deposited on a substrate of trace-element-free polymer film; examples include microtome slices of tissue, residue from dried drops of fluid containing suspended or dissolved solids, or films of atmospheric particulate material collected by a sampling device. Thick specimens are defined as those having sufficient areal density to stop the beam within the specimen, and they may be self-supporting or may reside on a substrate. As indicated below (see Appendix A), their analysis requires full attention to the effect of the matrix in slowing down the protons and in attenuating the excited characteristic x rays. The x-ray production decreases rapidly with depth but is significant for some 10 to 30 μm. Examples of thick specimens include mineral grains, geological thin sections, metallurgical specimens, and archaeological artefacts such as jewelry, bronzes, and pottery. Specimens between these limiting cases are referred to as having intermediate thickness, and their thickness must be known if matrix effects are to be accurately handled. In early PIXE work, many intermediate specimens were treated as if they were thin; i.e., matrix effects were neglected. However, it is now customary to correct for these, using, for example, an auxiliary proton scattering or transmission measurement to determine specimen thickness and major element composition.

With specimens that stop the beam, current and charge measurements are usually accomplished very simply by having the specimen electrically insulated from the chamber and by connecting it to a charge integrator. If thick specimens are not conducting, they must be coated with a very thin layer of carbon to prevent charging; neglecting this results in periodic spark discharge, which in turn causes intense bremsstrahlung background in the x-ray spectrum. Secondary electrons are emitted from the specimen, potentially causing an error of up to 50% in beam charge determination. These electrons must be returned to the specimen by placing a suitable electrode or grid close by, at a negative potential of typically 100 V. In thin specimens, the beam is transmitted into a graphite-lined Faraday cup, also with an electron suppressor, and the charge is integrated.

Electronic dead time effects must be accounted for. After detection of an event, there is a finite processing time in the electronic system, during which any subsequent events will be lost. A dead time signal from the pulse processor may be used to gate off the charge integrator and effect the necessary correction. Alternatively, an on-demand beam deflection system may be used. Here, the proton beam passes between two plates situated about 1 m upstream of the specimen and carrying equal voltages; detection of an x ray triggers a unit that grounds one plate as rapidly as possible, thereby deflecting the beam onto a tantalum collimator. The beam is restored when electronic processing is complete, so no corrections for dead time are required. This approach has two further advantages. The first is that specimen heating effects are reduced by removing the beam when its presence is not required. The second is that pile-up of closely spaced x-ray events is substantially decreased, thereby removing undesired artefacts from the spectrum.

The finite counting rate capacity of a Si(Li) detector and its associated pulse processor and multichannel pulse height analyzer demand that unnecessary contributions in the x-ray spectrum be minimized and as much of the capacity as possible be used for the trace element x rays of interest. Most thin-specimen analyses in the atmospheric or biological science context are performed with a Mylar absorber of typically 100 to 200 μm thick employed to reduce the intense bremsstrahlung background that lies below 5 keV in the spectrum and whose intensity increases rapidly toward lower energy. With thick metallurgical, geological, or archaelogical specimens, x-ray lines of major elements dominate the spectrum, making it necessary to reduce their intensities with an aluminum absorber of thickness typically 100 to 1000 μm. For example, in silicate minerals, the $K$ x rays of the major elements Na, Mg, Al, and Si occupy most of the counting rate capability; an aluminum foil of thickness 100 μm reduces the intensity of these x rays to a level such that they are still present at useful intensity in the spectrum, but the x rays from trace elements of higher atomic number can now be seen, with detection limits approaching the 1-ppm level. The thickness of such filters must be accurately determined.

As a thin-specimen example, Figure 1 shows the PIXE spectrum of an air particulate standard reference material. Proton-induced x-ray emission has proven extremely powerful in aerosol analysis, and special samplers have been devised to match its capabilities (Cahill, 1995). The low detection limits facilitate urban studies with short sampling intervals in order to determine pollution patterns over a day. At the other extreme, in remote locations, robust samplers of the IMPROVE type (Eldred et al., 1990) run twice weekly for 24-h periods at sites around the world in a project aimed at assembling continental and global data on visibility and particulate composition (Eldred and Cahill, 1994). The low detection limits match PIXE perfectly with cascade impactors, where the sample is divided according to particle diameter. Figure 2 (Maenhaut et al., 1996) presents measured detection limits for two cascade impactors that are well suited to PIXE in that they create on each impaction stage a deposit of small diameter that may be totally enveloped by a proton beam of a few millimeters in diameter. Now, PIXE appears in about half of all analyses of atmospheric aerosols in the standard aerosol journals.

Figure 3 shows an example spectrum from an "almost thin" specimen where an auxiliary technique is used to determine thickness and major element composition so that matrix corrections may be applied in the data reduction. The specimen is a slice of plant tissue, and the

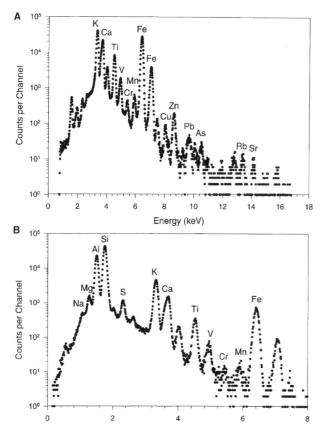

**Figure 1.** PIXE spectra of the BCR-128 fly ash reference standard, measured at Guelph with two x-ray detectors: **(A)** 8-μm window, no absorber; **(B)** 25-μm window, 125-μm Mylar absorber. Proton energy was 2.5 MeV.

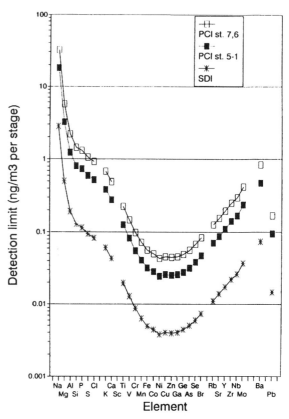

**Figure 2.** Detection limits for the PIXE International cascade impactor (PCI) and a small-deposit low-pressure impactor (SDI). Note that the detection limit depends upon the impactor stage. From Maenhaut et al. (1996) with permission of Elsevier Science.

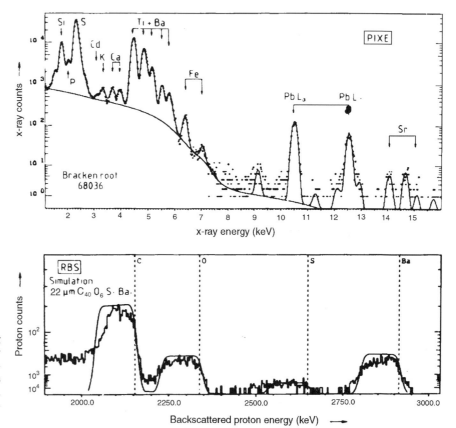

**Figure 3.** PIXE and proton backscattering energy spectra from a focal deposit of heavy metals in a plant root (Watt et al., 1991). Proton energy was 3 MeV. The continuous curves represent a fit (PIXE case) and a simulation (BS case). Reproduced by permission of Elsevier Science.

auxiliary technique is proton backscattering, whose results, acquired simultaneously, are in the lower panel of the figure. Many PIXE systems employ an annular silicon surface barrier detector upstream from the specimen and situated on the beam axis so that the beam passes through the annulus. This affords a detection angle close to 180° for scattered particles, which results in optimum energy resolution. This combination has proved to be a powerful means to determine major element concentrations in a large range of biological applications. Another powerful technique ancillary to micro-PIXE in the analysis of thin tissue slices is STIM. Here, a charged particle detector situated downstream from the specimen at about 15° to 25° to the beam axis records the spectrum of forward-scattered particles that reflects energy losses from both nuclear scattering and ion-electron collisions. This technique is effective in identifying structures such as individual cultured cells on a substrate, individual aerosol particles, and lesions in medical tissue. It has been used to identify neuritic plaques in brain tissue from Alzheimer's disease patients (Landsberg et al., 1992), thereby eliminating the need for immunohistochemical staining and so removing a sample preparation step that has the potential for contamination. Simultaneous recording of STIM, backscatter, and PIXE spectra and two-dimensional images has become a standard approach with biological specimens whose structure remains unaffected by preparation.

Micro-PIXE elegantly complements electron probe microanalysis in the trace element analysis of mineral grains using the same specimens and extending detection limits down to the parts per million level (Campbell and Czamanske, 1998). One major niche is the study of precious and other metals in sulfide ores. Figure 4 shows spectra of three iron-nickel sulfides (pentlandites) from a Siberian ore deposit; the large dynamic range between major elements such as iron (over 30% concentration in this case) and trace elements at tens to hundreds of parts

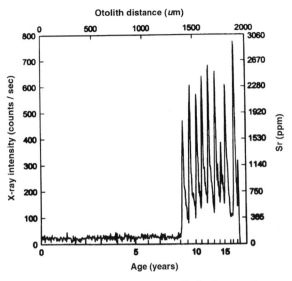

**Figure 5.** Strontium concentration profile in the growth zones of an otolith from an Arctic char collected from the Jayco River in the Northwest Territories of Canada (Babaluk et al., 1997). A 30-min micro-PIXE linescan was conducted using a $10 \times 10$-mm beam of 3-MeV protons. Reproduced by permission of the Arctic Institute of Canada.

per million concentration is typical of the earth science application area, and it necessitates use of absorbing filters (several hundred micrometers of aluminum) to depress the intensities of the intense x rays of the lighter major elements. Another important area is the study of trace elements in silicate minerals, where PIXE has aided in the development of geothermometers and fingerprinting approaches that are proving immensely powerful in, for example, assessing kimberlites for diamond potential (Griffin and Ryan, 1995).

Micro-PIXE is also widely used in trace element studies of zoned minerals, where the zoning contains information on the history of the mineral. Figure 5 shows a recently developed zoning application in calcium carbonate of biological origin—the otolith of an Arctic fish (Babaluk et al., 1997). The oscillatory behavior of Sr content in annual growth rings reflects annual migration of the fish from a freshwater lake to the open ocean, starting at age 9 years in the example shown. Micro-PIXE is now being applied in population studies for stock management.

Many more examples, including exotic ones in the fields of art and archaeometry, may be found in the book by Johansson et al. (1995). A unique example of nondestructive analysis is the use of micro-PIXE to identify use of different inks in individual letters in manuscripts. An example of a partly destructive analysis is the extraction of a very narrow core from a painting and the subsequent scanning of a microbeam along its length to identify and characterize successive paint layers.

In the absence of interfering peaks, the detection limit for an element is determined by the intensity of the continuous background on which its principal x-ray peak is superimposed in the measured spectrum. This background is due mainly to secondary electron bremsstrahlung,

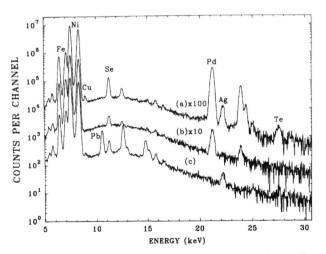

**Figure 4.** PIXE spectra of three Siberian pentlandites (Czamanske et al., 1992). The spectra were recorded using 3-MeV protons and an aluminum filter of 352 mm thickness. Element concentrations (in ppm) are (*a*) Se 261, Pd 2540, Ag 112, Te 54; (*b*) Se 80, Pd 132; (*c*) Se 116, Pd< 5, Ag 34, Te 36, Pb 1416.

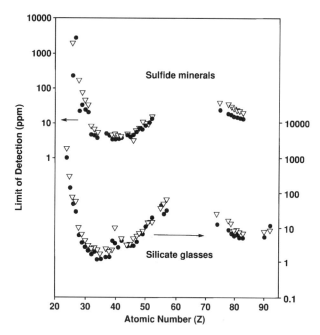

**Figure 6.** Micro-PIXE detection limits measured at Guelph for sulfide minerals and silicate glasses. Upper plot (left scale): triangles, pentlandites (350-μm Al filter); dots, pyrrhotite (250-mm Al filter); 10 mC of 3-MeV protons in a 5 × 10-mm spot. Lower plot (right scale): triangles, BHVO-1 basalt fused to glass; dots, rhyolitic glass RLS-158; 2.5 μC of 3-MeV protons in 5 × 10-μm spot; Al filter 250 mm.

whose intensity diminishes rapidly with increasing photon energy, but it can be augmented by gamma rays from nuclear reactions in light elements such as Na and F; these produce an essentially flat contribution to the background, visible at higher x-ray energies beyond the lower energy region where the bremsstrahlung dominates. The usual criterion is that a peak is detectable if its integrated intensity ($I$) exceeds a three-standard-deviation ($3\sqrt{B}$) fluctuation of the underlying background intensity ($B$). The desirability of a thin substrate in the case of thin specimens is obvious, since the substrate contributes only background to the spectrum. The ratio $I/3\sqrt{B}$ increases as the square root of the collected charge, suggesting that detection limits will be minimized by maximizing beam current and measuring time. Beam current is often limited by specimen damage, and measuring time may be limited by cost. Detection limits should be quoted for a given beam charge and detector geometry. The $Z$ dependence of detection limits is U shaped, as shown in the examples of Figure 6, and this reflects the theoretical x-ray intensity expression that is included in Equation 1. For many thick specimen types, major elements contribute intense peaks and both pile-up continua and pile-up peaks to the spectrum, and these can locally worsen detection limits considerably; electronic pile-up rejection or on-demand beam deflection, which minimize such artefacts, are therefore necessary parts of a PIXE system.

The hazards of the technique are those involved with operation of any small accelerator operating with low beam currents. While radiation fields just outside the spe-

cimen chamber are very small, steps are necessary to shield radiation from beam-defining slits and from the accelerator itself. Special precautions (Doyle et al., 1991) are necessary if the beam is extracted through a window into the laboratory to analyze large or fragile specimens.

## METHOD AUTOMATION

Specimen chambers usually accommodate many specimens, and computer-controlled stepping motors, either inside or outside the vacuum, are employed to expose these sequentially to the beam and to control all other aspects of data acquisition, e.g., charge measurement and selection of absorbing filter. In the case of micro-PIXE, the coordinates for each point analysis, linescan, or area map may be recorded by optical examination at the outset, permitting subsequent unsupervised data acquisition. Human intervention is only needed to break vacuum and insert new specimen loads into the analysis chamber, reevacuate, and input the parameters of the measurement (e.g., integrated charge per specimen). However, the accelerator itself is rarely automated, and a technical expert is required to steer and focus the beam prior to the analysis.

## DATA ANALYSIS AND INITIAL INTERPRETATION

On the assumption that the volume sampled by the beam is homogeneous, the method is fully quantitative. Prior to analysis of measured spectra, checks are necessary to ensure that the expected detector energy resolution was maintained and that counting rate was in the proper domain.

The main step in data analysis is a nonlinear least-squares fit of a model energy-dispersive x-ray spectrum to the measured spectrum. The software requires the database mentioned earlier to describe the x-ray production process, a representation (Gaussian with corrections, as outlined below; see Appendix B) of the peaks in the energy-dispersed spectrum, and a means of dealing with continuous background; the latter may be fitted by an appropriate expression or removed by applying an appropriate algorithm prior to the least-squares fitting. Accurate knowledge of the properties of the detector and of any x-ray absorbers introduced between specimen and detector is necessary.

The quantities determined by the fit are the heights, and therefore the intensities, of the principal x-ray line of each element present in the spectrum. The intensities of all the remaining lines of each element are then set by that of the element's main line together with the relative x-ray intensity ratios provided by the database; these ratios are adjusted to reflect the effects of both absorber transmission and detector efficiency. For thick specimens, the database intensity ratios are also adjusted to reflect the effects of the matrix. This last adjustment is straightforward in a trace element determination in a known matrix; the matrix corrections to the ratios remain fixed through the iterations of the fit. The more complex case of major element measurement will be discussed below. Given the very large number of x-ray lines present, the number of pile-up combinations can run into hundreds, and these double and triple peaks must also be modeled. A sorting process is used to eliminate pile-up combinations that are too weak to be of importance.

The final intensities provided by the fit for the principal lines of each element present are corrected for secondary fluorescence contributions, for pulse pile-up, and dead time effects (if necessary), prior to conversion to concentrations using Equation 1.

There is a linear relationship between the x-ray energy and the corresponding channel number of the peak centroid in the spectrum, and there is a linear relationship between the variance of the Gaussian peaks and the corresponding x-ray energy. The four system calibration parameters inherent in these relationships may be fixed or variable in the fitting process. The latter option has merit in that it accounts for small changes due to electronic drifts and does not require that the system be preset in any standard way.

If the major element concentrations are known *a priori*, then the second step, the conversion of these peak intensities to concentrations, is accomplished directly from Equation 1 or a variant thereof using the instrumental constant $H$, which is determined either using standards or from known major elements in the specimen itself. If the major elements have to be determined by the PIXE analysis, then Equation 1 has to be solved by an iterative process, starting with some estimate of concentrations and repeating the least-squares fit until consistency is achieved between the concentrations utilized and generated by Equation 1.

Various software packages are available for the above tasks (Johansson et al., 1995). They differ mainly in the databases that they adopt and in their approaches to handling the continuous background component of the spectra. The first option for the background is to add an appropriate expression to the peak model, thus describing the whole spectrum in mathematical terms, and then determine the parameter values in the background expression via the fitting process; the most common choice to describe the background has been an exponential polynomial of order up to 6 for the electron bremsstrahlung plus a linear function for the gamma ray component. The first of these two expressions must be modified to describe filter transmission and detector efficiency effects; the second needs no modification, because high-energy gamma rays are not affected by the absorbers and windows. While one particular form of expression may cope well, in an empirical sense, with a given type of specimen, there is little basis for assuming that the expressions described will be universally satisfactory. This has led to approaches that involve removing the background continuum by mathematically justifiable means. The simplest such approach, developed in the electron microprobe analogue, involves convoluting the spectrum with the "top-hat" filter shown in Figure 7. Such a convolution reduces a linear background to zero and will therefore be effective if the continuum is essentially linear within the dimensions of the filter. These dimensions are prescribed by Schamber (1977) as $UW = 1$ FWHM and $LW = 0.5$ FWHM, where FWHM is the full width at half-maximum of the resolution function at the

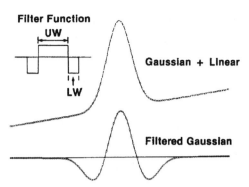

**Figure 7.** Top-hat filter and its effect on a Gaussian peak with linear background: UW and LW are the widths of the central and outer lobes.

x-ray energy specified. A more sophisticated multistep approach to continuum removal uses a peak-clipping algorithm to remove all peaks from the measured spectrum, and Ryan et al. (1988) have developed a version of this that is configured to cope with the very large dynamic range in peak heights that are encountered in PIXE spectra. First, the measured spectrum is smoothed by a "low-statistics digital filter" of variable width; at each channel of the spectrum this width is determined by the intensity in that channel and its vicinity. As the filter moves from a valley toward a peak, the smoothing interval is reduced to prevent the peak intensity from influencing the smoothing of the valley region. In a second step, a double-logarithmic transformation of channel content is applied, i.e.,

$$z = \ln[\ln(y + 1)] \qquad (2)$$

in order to compress the dynamic range. The third step is a multipass peak clipping, in which each channel content $z(c)$ is replaced by the lesser of $z(c)$ and

$$\bar{z} = \frac{1}{2}[z(c + w) + z(c - w)] \qquad (3)$$

where the scan width $w$ is twice the FWHM corresponding to channel $c$. After no more than 24 passes, during which $w$ is reduced by a factor of 2, a smooth background remains. The final step is to transform this background back through the inverse of Equation 2. The accuracy of this so-called SNIP (statistics-insensitive nonlinear peak-clipping algorithm) approach has been demonstrated in many PIXE analyses of geochemical reference materials (Ryan et al., 1990).

The VAX software package GEOPIXE (Ryan et al., 1990) is designed for thick specimens encountered in geochemistry and mineralogy and can also provide accurate elemental analysis of subsurface fluid inclusions (Ryan et al., 1995). It is installed in various micro-PIXE laboratories concerned with geochemical applications. It takes the SNIP approach to dealing with the background continuum. The widely used PC package GUPIX (Maxwell et al., 1989, 1995) deals with thin-, thick-, intermediate-, and multiple-layer specimens, employing the simple top-hat filter method to strip background. GUPIX offers the option, useful in mineralogy, of analyzing in terms of oxides rather than elements; this enables the user to compare the sum of generated oxide concentrations to 100% and draw appropriate conclusions. GUPIX also allows the user to include one "invisible" element in the list of elements to be included in the calculation, requiring the sum of concentrations of all "visible" elements and the invisible element to sum to 100%; this can be used to determine oxygen content in oxide minerals or sulfide content in sulfide minerals when the oxygen and sulfur $K$ x rays are not observable in the spectrum (as is usually the case). The GUPIX code can provide analysis and thickness determination of multilayer film structures, provided that the elements whose concentrations in a given layer are to be determined are not also present at unknown concentrations in any other layer. GUPIX is presently being extended to cope with analysis using deuteron and helium beams (Maxwell et al., 1997).

The above has been concerned with PIXE analysis for element concentrations in a fixed area on the specimen. In imaging applications, the beam is scanned over the specimen, and a record is built of the coordinates and channel number of each recorded x-ray event. From this record, a one- or two-dimensional image may be reconstructed using the spectrum intensity in preset "windows" to represent amounts of each element in a semiquantitative fashion. Obviously, such an approach can encounter errors due to peak overlap, but these may be dealt with by appropriate nonlinear least-squares spectrum fitting upon conclusion of the analysis. It is, however, desirable to transform observed window intensities into element concentrations in real time, updating as the data accumulate. Such a dynamic analysis method has been developed by Ryan and Jamieson (1993), and it enables on-line accumulation of PIXE maps that are inherently overlap resolved and background subtracted.

Returning now to direct analysis, standard reference materials of a similar nature to each specimen type of interest should be analyzed to demonstrate, e.g., accuracy and detection limits, which are determined in part by the measurement and in part by the fitting procedure. There are many examples of the accurate analysis of standard reference materials (SRMs). Table 1 shows the results of PIXE analysis by Maenhaut (1987) of the European Community fly-ash reference material used to simulate exposed filters from air particulate samplers. Agreement between measured and certified values is excellent, except for the elements of very low atomic number; these light elements tend to occur in larger soil-derived particles, and the data treatment used did not account for particle size effects. Table 2 summarizes a PIXE study of eight biological standard reference materials prepared for PIXE analysis by two different methods; one method (A) involved direct deposition of freeze-dried powder on the polymer substrate, and the other (D) involved Teflon bomb digestion of the powder followed by deposition and drying of aliquots. Table 3 shows micro-PIXE analyses of geochemical reference standards from three laboratories that specialize in earth science applications of micro-PIXE; these are the Commonwealth Scientific and Industrial Research Organization (CSIRO) in Australia, the University of Guelph

**Table 1. Analysis by PIXE of Thin Films (BCR 128) Containing Certified Fly Ash (BCR 38)[a]**

| Element | Reference Value[a,b] | PIXE Value[c] | PIXE/ Reference[d] |
|---|---|---|---|
| Mg[e] | $9.45 \pm 0.34$[f] | $8.3 \pm 1.1$ | $0.88 \pm 0.12$ |
| Al[e] | $127.5 \pm 4.9$[f] | $104.0 \pm 1.6$ | $0.82 \pm 0.01$ |
| Si[e] | $227 \pm 12$[f] | $176.0 \pm 2.1$ | $0.78 \pm 0.01$ |
| P[e] | | $1.60 \pm 0.15$ | |
| S[e] | $3.89 \pm 0.13$[f] | $4.17 \pm 0.03$ | $1.07 \pm 0.01$ |
| K[e] | $34.1 \pm 1.6$[f] | $30.6 \pm 0.7$ | $0.90 \pm 0.02$ |
| Ca[e] | $13.81 \pm 0.63$[f] | $13.35 \pm 0.16$ | $0.97 \pm 0.01$ |
| Ti[e] | | $5.07 \pm 0.12$ | |
| V | $334 \pm 23$ | $340 \pm 2$ | $1.02 \pm 0.07$ |
| Cr | $178 \pm 12$ | $161 \pm 16$ | $0.90 \pm 0.09$ |
| Mn | $479 \pm 16$ | $454 \pm 10$ | $0.95 \pm 0.02$ |
| Fe | $33.8 \pm 0.7$ | $32.9 \pm 0.5$ | $0.97 \pm 0.01$ |
| Ni | $194 \pm 26$ | $192 \pm 8$ | $0.99 \pm 0.04$ |
| Cu | $176 \pm 9$ | $169 \pm 8$ | $0.96 \pm 0.05$ |
| Zn | $581 \pm 29$ | $572 \pm 18$ | $0.98 \pm 0.03$ |
| Ga | | $56.2 \pm 4.5$[g] | |
| Ge | | $16.3 \pm 4.7$[g] | |
| As | $48.0 \pm 2.3$ | $49 \pm 11$[g] | $1.02 \pm 0.23$ |
| Br | | $29 \pm 7$[g] | |
| Rb | | $183 \pm 18$[g] | |
| Sr | | $187 \pm 18$[g] | |
| Zr | | $169 \pm 23$[g] | |
| Pb | $262 \pm 11$ | $225 \pm 26$[g] | $0.86 \pm 0.10$ |

[a] Concentrations are given in micrograms per gram unless indicated otherwise.
[b] Reference values and associated errors are for the certified fly ash (BCR 38) unless indicated otherwise.
[c] PIXE data are averages and standard deviations, based on the analysis of five thin-film samples, unless indicated otherwise.
[d] Values were obtained by dividing both the PIXE result and its associated error by the reference concentration.
[e] Concentration in mililligrams per gram.
[f] Concentration and standard deviation obtained by averaging round-robin values for the fly ash.
[g] Result derived from the sum spectrum, obtained by summing the PIXE spectra of the five films analyzed; the associated error is the error from counting statistics.

in Canada, and the National Accelerator Center at Faure in South Africa; details of these analyses are given by Campbell and Czamanske (1998). In such studies, the accuracy of the PIXE technique is usually assessed by comparing the mean element concentrations over several replicate samples (or spots on the same sample in the micro-PIXE case). In these and other such studies, the accuracy is typically a few percent when the concentration is well above the detection limit. The matter of reproducibility or precision is discussed below (see Problems).

## SAMPLE PREPARATION

In thin-specimen work, it is straightforward to deposit thin films of powder or fluid (which is then dried) or microtome slices of biological tissue onto a polymer substrate. In aerosol specimens, the particulate material is deposited on an appropriate substrate, usually Nuclepore or Teflon filters, by the air-sampling device. Reduction of original bulk

**Table 2. Comparison between Certified Values and PIXE Data[a] for 16 Elements in 8 Reference Materials**

| Element | Number of Certified Values | Average Absolute % Difference Relative to Certified Value[b] | |
|---|---|---|---|
| | | Method A | Method D |
| K | 7 | 2.1 (7) | 5.7 (5) |
| Ca | 7 | 6.6 (7) | 51 (7) |
| Cr | 5 | [17 (4)] | [19 (4)] |
| Mn | 7 | 3.3 (6) | 3.1 (6) |
| Fe | 8 | 5.9 (8) | 3.7 (8) |
| Ni | 2 | [19 (2)] | [25 (2)] |
| Cu | 8 | 5.5 (8) | 3.0 (8) |
| Zn | 6 | 2.3 (6) | 4.0 (6) |
| As | 6 | 7.8 (2) | 6.2 (2) |
| Se | 3 | [0.6 (1)] | [15 (1)] |
| Br | 1 | 0.7 (1) | |
| Rb | 7 | 2.0 (7) | 5.6 (7) |
| Sr | 6 | 2.6 (5) | 3.1 (5) |
| Mo | 3 | [0.6 (1)] | [1.4 (1)] |
| Ba | 1 | | |
| Pb | 7 | 4.4 (4) | 3.0 (4) |

[a]With standard deviation from counting statistics <15%.
[b]The numbers in parentheses designate how many values the averages are based upon; brackets are used to indicate that all PIXE data for the certified element have standard deviation values in the range 5% to 15%.
*Source*: From Maenhaut et al., 1987. Reproduced by permission of Springer-Verlag.

material to a fluid or powder draws on the standard repertoire of grinding, ashing, and digestion techniques. Powdered material may be suspended in a liquid that is pipetted onto the substrate. Similarly, an aliquot of liquid containing dissolved material may be allowed to dry on the substrate. However, nonuniform deposits left from drying liquid drops may result in poorer than optimal reproducibility. The beam diameter is kept less than the specimen diameter to avoid problems of nonuniform specimen at the edges. If specimens are thin enough, there are no charging problems, but on occasion graphite coating has been necessary.

With thick specimens, a polished surface must be presented to the beam. If the specimen is an insulator, it must be carbon coated to prevent charge buildup. Geological specimens are prepared for micro-PIXE precisely as they are for electron probe microanalysis. The options are thin sections (30 to 50 µm) or multiple grains "potted" in epoxy resin.

## SPECIMEN MODIFICATION

Overly large energy deposition per unit area may cause heating damage to specimens. This is observed as loss of matter from biological specimens or as crazing of mineral grains, and its avoidance is largely a matter of experience. Obviously the risk of damage increases with finer beam diameters. Volatile elements can be lost from biological materials, e.g., some compounds of sulfur, chlorine, and bromine. A guideline for tolerable beam current densities on thin biological specimens is 5 nA/mm$^2$ *in vacuo*

**Table 3. Micro-PIXE Analyses of Fused Rock Standards at Three Laboratories[a]**

| | W-2 (CSIRO)[b] | | BHVO-1 (Guelph)[c] | | BCR-1 (Faure)[d] | |
|---|---|---|---|---|---|---|
| | Measured | Nominal | Meassured | Nominal | Measured | Nominal |
| K | 6130 ± 500 | 5200 | nm | — | nm | — |
| Ca | 8.1 ± 0.3% | 78% | nm | — | nm | — |
| Ti | 6450 ± 100 | 6350 | nm | — | nm | — |
| V | 289 ± 40 | 262 | nm | — | nm | — |
| Cr | nd | 93 | nm | — | nm | — |
| Mn | 1270 ± 40 | 1260 | 1319 ± 25 | 1300 | 1358 ± 219 | 1400 |
| Fe | 7.5% | 7.5% | 8.62% | 8.64% | 9.5% | 9.5% |
| Ni | 64 ± 8 | 70 | 119 ± 6 | 121 | 17 ± 5 | 13 |
| Cu | 45 ± 4 | 103 | 160 ± 3 | 136 | 14 ± 2 | 18 |
| Zn | 74 ± 5 | 77 | 113 ± 3 | 105 | 137 ± 8 | 125 |
| Ga | 20 ± 1.4 | 20 | 22 ± 2 | 21 | 21 ± 1 | 22 |
| As | 1.4 ± 0.6 | (1.2) | nd | — | nd | — |
| Rb | 20 ± 0.8 | 20 | 9 ± 1 | 11 | 43 ± 2 | 47 |
| Sr | 195 ± 6 | 194 | 398 ± 3 | 403 | 330 ± 8 | 330 |
| Y | 19 ± 1 | 24 | 23.5 ± 0.5 | 27.6 | 32 ± 4 | 39 |
| Zr | 100 ± 3 | 94 | 170 ± 3 | 179 | 183 ± 5 | 190 |
| Nb | 5.7 ± 0.7 | (8) | 18 ± 1 | 19 | 13 ± 2 | 16 |
| Ba | 179 ± 15 | 182 | 119 ± 17 | 139 | 652 ± 21 | 680 |
| Pb | nd | (9) | nd | — | 16 ± 3 | 15 |
| Th | 2.1 ± 1.2 | 2.2 | nd | — | nd | — |

[a] Measured concentration in ppm (except Fe in wt.%); nominal concentration in ppm (except Fe in wt.%); brackets signify that the nominal value is only an estimate, as opposed to a formal recommendation. nm, not measured; nd, not detected.
[b] Concentrations from a sum spectrum from 10 spots each having 3-µC charge.
[c] Mean concentrations from spectra of 10 spots, each having 2.5-µC charge.
[d] Concentrations from a sum spectrum from 6 spots, each having 1-µC charge.

(Maenhaut, 1990), but it is advisable to conduct tests on a particular specimen type. In geological materials, e.g., volcanic glass, mobility of sodium results in significant analytical error. Careful design is needed when analyzing certain works of art and archaeologic specimens; glass, ceramics, and paper may be discolored (Malmqvist, 1995), and the intrinsic merit of in-air PIXE analysis, which provides continuous cooling of the specimen, is obvious.

## PROBLEMS

Lack of homogeneity in the volume sampled by the ion beam is a potential problem, especially in mineralogical work, where a subsurface grain may be present but invisible just under the region being analyzed or microinclusions may be present in an optically homogeneous crystallite.

Spectrum artefacts may be interpreted as element peaks by the spectrum-fitting software, leading to erroneous detection of elements that are not present. Peak tailing effects on the predominantly Gaussian response function, double and triple pile-up peaks, and escape peaks (which arise when the x-ray interaction causes creation of a silicon $K$ x ray that escapes from the detector) must be dealt with rigorously in the code, and this requires precharacterization of the lineshape as a function of x-ray energy. Overlaps between major x-ray lines of neighboring elements can cause error, and this places demands upon the accuracy of the x-ray physics portion of the database. Inadequate description of the continuous background underlying peaks of low intensity is another source of error. As a general rule, a fitting code should be directed to fit only those elements that visual inspection suggests are contributing to the spectrum; given an extensive list of elements that is not limited by inspection, fitting codes will happily generate small concentrations that represent nothing more than minor spectral artifacts such as imperfect description of tailing.

Precision for a given specimen type is generally determined by replicate analyses that provide a standard deviation to accompany each mean concentration, as in the example of Table 3. Analysis of the same region will test the reproducibility of all aspects except the specimen itself; analysis of different spots will include any specimen inhomogeneity in the variance that is determined. Usually, if the reduced chi-squared of the fit is good, the standard deviation for a suite of replicates will be close to that expected from the counting statistics of a single measurement; if it is in excess, this constitutes evidence for sample inhomogeneity. Comparison of mean concentrations determined in this manner for standard reference materials with formally accepted or recommended concentration values provides estimates of systematic errors; these tend to derive from database errors (e.g., in x-ray attenuation coefficients or relative x-ray intensities) that affect the accuracy of the fit.

## LITERATURE CITED

Babaluk, J. A., Halden, N. M., Reist, J. D., Kristofferson, A. H., Campbell, J. L., and Teesdale, W. J. 1997. Evidence for non-anadromous behavior of Arctic char from Lake Hazen, Ellesmere Island, Northwest territories, Canada, based on scanning proton microprobe analysis of otolith strontium distribution. *Arctic* 50:224–233.

Berger, M. J. and Hubbell, J. H. 1987. Photon cross-sections on a personal computer. National Bureau of Standards Report NBSIR 87–3597.

Brandt, W. and Lapicki, G. 1981. Energy-loss effect in inner-shell Coulomb ionization by heavy charged particles. *Phys. Rev. A* 23:1717–1729.

Cahill, T. A. 1995. Compositional analysis of atmospheric aerosols. *In* Particle-Induced X-ray Emission Spectrometry (S. A. E. Johansson, J. L. Campbell, and K. G. Malmqvist, eds.). pp. 237–312. John Wiley & Sons, New York.

Campbell, J. L. 1996. Si(Li) detector response and PIXE spectrum fitting. *Nucl. Instrum. Methods* B109/110:71–78.

Campbell, J. L. and Cookson, J. A. 1983. PIXE analysis of thick targets. *Nucl. Instrum. Methods* B3:185–197.

Campbell, J. L. and Czamanske, G. C. 1998. Micro-PIXE in earth science. *Rev. Econ. Geol.* 7:169–185.

Campbell, J. L. and Wang, J.-X. 1989. Interpolated Dirac-Fock values of $L$-subshell x-ray emission rates including overlap and exchange effects. *At. Data Nucl. Data Tables* 43:281–291.

Chen, M. H. and Crasemann, B. 1984. $M$ x-ray emission rates in Dirac-Fock approximation. *Phys. Rev. A* 30:170–176.

Chen, M. H. and Crasemann, B. 1985. Relativistic cross-sections for atomic $K$ and $L$ ionization by protons, calculated from a Dirac-Hartree-Slater model. *At. Data Nucl. Data Tables* 33:217–233.

Chen, M. H. and Crasemann, B. 1989. Atomic $K$-, $L$- and $M$-shell cross-sections for ionization by protons: A relativistic Hartree-Slater calculation. *At. Data Nucl. Data Tables* 41:257–285.

Chen, M. H., Crasemann, B., and Mark, H. 1980a. Relativistic $K$-shell Auger rates, level widths and fluorescence yields. *Phys. Rev. A* 21:436–441.

Chen, M. H., Crasemann, B., and Mark, H. 1980b. Relativistic $M$-shell radiationless transitions. *Phys. Rev. A* 21:449–453.

Chen, M. H., Crasemann, B., and Mark, H. 1981. Widths and fluorescence yields of atomic $L$ vacancy states. *Phys. Rev. A* 24:177–182.

Chen, M. H., Crasemann, B., and Mark, H. 1983. Radiationless transitions to atomic $M_{1,2,3}$ shells: Results of relativistic theory. *Phys. Rev. A* 27:2989–2993.

Cohen, D. D. and Harrigan, M. 1985. $K$ and $L$ shell ionization cross-sections for protons and helium ions calculated in the ECPSSR theory. *At. Data Nucl. Data Tables* 33:256–342.

Cohen, D. D. and Harrigan, M. 1989. $K$-shell and $L$-shell ionization cross-sections for deuterons calculated in the ECPSSR theory. *At. Data Nucl. Data Tables* 41:287–338.

Czamanske, G. C., Kunilov, V. E., Zientek, M. L., Cabri, L. J., Likhachev, A. P., Calk, L. C., and Oscarson, R. L. 1992. A proton microprobe analysis of magmatic sulfide ores from the Noril'sk-Talnakh district, Siberia. *Can. Mineralogist* 30:249–287.

Doyle, B. L., Walsh, D. S., and Lee, S. R. 1991. External micro-ion-beam analysis. *Nucl. Instrum. Methods* B54:244–257.

Eldred, R. A. and Cahill, T. A. 1994. Trends in elemental concentrations of fine particles at remote sites in the U.S.A. *Atmos. Environ.* 5:1009–1019.

Eldred, R. A., Cahill, T. A., Wilkinson, L. K., Feeney, P. J., Chow, J. C., and Malm, W. C. 1990. Measurement of fine particles and their chemical components in the IMPROVE/NPS networks. *In*

Visibility and Fine Particles (C. V. Mathai, ed.). pp. 187–196. Air and Waste Management Association, Pittsburgh, Pa.

Griffin, W. L. and Ryan, C. G. 1995. Trace elements in indicator minerals: Area selection and target evaluation in diamond exploration. *J. Geochem. Exploration* 53:311–337.

Hubbell, J. H. 1989. Bibliography and current status of *K, L* and higher shell fluorescence yields for computations of photon energy-absorption coefficients. National Institute of Standards and Technology Report NISTIR 89–4144, U.S. Department of Commerce.

International Commission on Radiation Units and Measurements (ICRUM). 1993. Stopping Powers and Ranges for Protons and Alpha Particles. ICRUM, Bethesda, Md.

Johansson, G. I. 1982. Modifications of the HEX program for fast automatic resolution of PIXE spectra. *X-ray Spectrom.* 11:194–200.

Johansson, S. A. E. and Campbell, J. L. 1988. PIXE: A Novel Technique for Elemental Analysis. John Wiley & Sons, Chichester.

Johansson, S. A. E., Campbell, J. L. and Malmqvist, K. G. 1995. Particle-Induced X-ray Emission Spectrometry. John Wiley & Sons, New York.

Klockenkamper, R., Knoth, J., Prange, A., and Schwenke, H. 1992. Total reflection x-ray fluorescence spectroscopy. *Anal. Chem.* 64:1115A–1121A.

Krause, M. O. 1979. Atomic radiative and radiationless yields for *K* and *L* shells. *J. Phys. Chem. Ref. Data* 8:307–327.

Landsberg, J. P., McDonald, B., and Watt, F. 1992. Absence of aluminium in neuritic plaque cores in Alzheimer's disease. *Nature (London)* 360:65–68.

Larsson, N. P.-O., Tapper, U. A. S., and Martinsson, B. G. 1989. Characterization of the response function of a Si(Li) detector using an absorber technique. *Nucl. Instrum. Methods* B43:574–580.

Liu, Z. and Cipolla, S. J. 1996. A program for calculating *K, L* and *M* cross-sections from ECPSSR theory using a personal computer. *Comp. Phys. Commun.* 97:315–330.

Maenhaut, W. 1987. Particle-induced x-ray emission spectrometry: An accurate technique in the analysis of biological, environmental and geological samples. *Anal. Chim. Acta* 195:125–140.

Maenhaut, W. 1990. Multi-element analysis of biological materials by particle-induced x-ray emission. *Scan. Microscy.* 4: 43–62.

Maenhaut, W., Hillamo, R., Makela, T., Jaffrezo, J.-L., Bergin, M. H., and Davidson, C. I. 1996. A new cascade impactor for aerosol sampling with subsequent PIXE analysis. *Nucl. Instrum. Methods* B109/110:482–487.

Maenhaut, W., Vandenhaute, J., and Duflour, H. 1987. Applicability of PIXE to the analysis of biological reference materials. *Fresenius Z. Anal. Chem.* 326:736–738.

Malmqvist, K. G. 1995. Application in art and archaeology. *In* Particle-Induced X-Ray Emission Spectrometry (S. A. E. Johansson, J. L. Campbell, and K. G. Malmqvist, eds.). pp. 367–417. John Wiley & Sons, New York.

Maxwell, J. A., Teesdale, W. J., and Campbell, J. L. 1989. The Guelph PIXE software package. *Nucl. Instrum. Methods* B43:218–230.

Maxwell, J. A., Teesdale, W. J., and Campbell, J. L. 1995. The Guelph PIXE software package II. *Nucl. Instrum. Methods* B95:407–421.

Maxwell, J. A., Teesdale, W. J., and Campbell, J. L. 1997. Private communication.

Orlic, I., Sow, C. H., and Tang, S. M. 1994. Semiempirical formulas for calculation of *L*-subshell ionization cross sections. *Int. J. PIXE* 4:217–230.

Paul, H. and Sacher, J. 1989. Fitted empirical reference cross-sections for *K*-shell ionization by protons. *At. Data Nucl. Data Tables* 42:105–156.

Rindby, A. 1993. Progress in X-ray microbeam spectroscopy. *X-Ray Spectrom.* 22:187–191.

Ryan, C. G., Clayton, E., Griffin, W. L., Sie, S. H., and Cousens, D. R. 1988. SNIP—a statistics-sensitive background treatment for the quantitative analysis of PIXE spectra in geoscience applications. *Nucl. Instrum. Methods* B34:396–402.

Ryan, C. G., Cousens, D. R., Sie, S. H., Griffin, W. L., and Suter, G. F. 1990. Quantitative PIXE microanalysis of geological material using the CSIRO proton microprobe. *Nucl. Instrum. Methods* B47:55–71.

Ryan, C. G., Heinrich, C. A., van Achterberg, E., Ballhaus, C., and Mernagh, T.P. 1995. Microanalysis of ore-forming fluids using the scanning proton microprobe. *Nucl. Instrum. Methods* B104:182–190.

Ryan, C. G. and Jamieson, D. N. 1993. Dynamic analysis: On-line quantitative PIXE microanalysis and its use in overlap-resolved elemental mapping. *Nucl. Instrum. Methods* B77: 203–214.

Schamber, F. H. 1977. A modification of the non-linear least-squares fitting method which provides continuum suppression. *In* X-Ray Fluorescence Analysis of Environmental Samples (T. G. Dzubay, ed.). pp. 241–248. Ann Arbor Science, Ann Arbor, Mich.

Scofield, J. H. 1974a. Exchange corrections of *K* x-ray emission rates. *Phys. Rev. A* 9:1041–1049.

Scofield, J. H. 1974b. Hartree-Fock values of *L* x-ray emission rates. *Phys. Rev. A* 10:1507–1510.

Scofield, J. H. 1975. Erratum. *Phys. Rev. A* 12:345.

Watt, F., Grime, G., Brook, A. J., Gadd, G. M., Perry, C. C., Pearce, R. B., Turnau, K., and Watkinson, S. C. 1991. Nuclear microscopy of biological specimens. *Nucl. Instrum. Methods* B54:123–143.

## KEY REFERENCES

Johansson and Campbell, 1988. See above.

*The first book on PIXE, providing extensive practical detail on all aspects and covering the early history.*

Johansson et al., 1995. See above.

*With the technique at maturity, this book provides the basics and presents a very wide range of examples of PIXE and micro-PIXE analysis in biology, medicine, earth science, atmospheric science, and archaeometry.*

*Nuclear Instruments and Methods in Physics Research*, vols. 142 (1977), 181 (1981), B3 (1984), B22 (1987), B49 (1990), B75 (1993), B109/110 (1996), B150 (1999).

*Proceedings of the seven international conferences on PIXE, they reflect its development and increasing sophistication.*

## APPENDIX A:
## RELATIONSHIP BETWEEN X-RAY INTENSITIES AND CONCENTRATIONS

Consider the general case of a proton beam incident with energy $E_0$ at angle $\alpha$ to the specimen normal and a detector at angle $\Theta_{T_0}$ to the specimen surface; the specimen thickness is $t$. To an excellent approximation, the proton travels in a straight line, its path length being $t/\cos\alpha$, and emerges from the specimen with energy $E_f$. The proton's energy profile ($E$) along its path (direction $x$) is given by the stopping power of the matrix $S_M$:

$$S_M(E) = \rho^{-1}\frac{dE}{dx} \qquad (4)$$

which is the concentration-weighted sum of the stopping powers of the major (matrix) elements comprising the specimen, whose density is $\rho$.

We consider only $K$ x-ray production. The treatment for $L$ and $M$ x rays is similar, but the existence of subshells ($L_1$ to $L_3$, $M_1$ to $M_5$) renders matters more complex. The ionization cross-section $\sigma_Z(E)$ for a constituent element Z (concentration $C_Z$, atomic mass $A_Z$) decreases along the proton path as the energy decreases.

When an ionization occurs, the probability of $K$ x-ray emission is the fluorescence yield $\omega_Z$, and the fraction of the $K$ x rays appearing in the principal line ($K\alpha_1$) is $b_Z$. The x rays from each successive element of the path have a transmission probability $T_Z(E)$ for reaching the detector through the overlying matrix,

$$T_Z(E) = \exp\left[-\left(\frac{\mu}{\rho}\right)_{Z,M}\frac{\cos\alpha}{\sin\Theta_{To}}\int_{E_0}^{E}\frac{dE}{S_M(E)}\right] \qquad (5)$$

in which the matrix mass attenuation coefficient $(\mu/\rho)_{Z,M}$ is the concentration-weighted sum of the mass attenuation coefficients $(\mu/\rho)_{Z,M}$ of the major (or matrix) elements.

Integration along the proton track gives the total x-ray intensity due to a total incident beam charge $Q$ as

$$Y(Z) = \frac{(N_{av}/e)\omega_Z b_Z t_Z \varepsilon_Z(\Omega/4\pi)}{A_Z}QC_Z\int_{E_0}^{E_f}\frac{\sigma_Z(E)T_Z(E)}{S_M(E)}dE \qquad (6)$$

in which $N_{av}$ is Avogadro's number, $e$ is the electronic charge, $t_Z$ represents the transmission of the x rays through any absorbing filter used, $\Omega/4\pi$ is the solid angle fraction subtended by the detector, and $\varepsilon_Z$ is the detector's intrinsic efficiency.

For a thick target ($E_f = 0$), it follows that

$$Y(Z) = Y_1(Z)H\varepsilon_Z t_Z C_Z Q \qquad (7)$$

where $Y_1(Z)$ is the yield computed from the PIXE database per steradian per microcoulomb per unit concentration (ppm) and $H$ is an instrumental constant that subsumes both detector solid angle and any beam change calibration factor.

For a thin target ($E_f \sim E_0$), $T_Z(E) \sim 1$, $\sigma_Z(E)$ becomes $\sigma_Z(E_0)$, and the integral in Equation 7 reduces to $t/\cos a$. We then have

$$Y(Z) = Y_1(Z)H\varepsilon_Z t_Z m_a(Z)Q \qquad (8)$$

where $Y_1(Z)$ is the computed x-ray yield per microcoulomb per unit areal density of element Z per steradian, $H$ is the same instrumental constant, and $m_a(Z)$ is the areal density of element Z.

Specimens of intermediate thickness and multiple-layer cases are dealt with by straightforward extensions of this formalism.

## APPENDIX B:
## RESPONSE FUNCTION AND EFFICIENCY OF SI(LI) DETECTOR

A hypothetically perfect Si(Li) detector would provide a delta function response in the spectrum to a monoenergetic x ray. In reality, x rays have an intrinsic Lorentzian energy distribution whose width is small compared to the Gaussian broadening that arises from the statistics of charge formation and from the electronic noise. The main feature of the lineshape is therefore a Gaussian. Partial escape from the front surface of photoelectrons and Auger electrons created in the silicon causes a flat shelf extending down to low energy. Some detectors, but not all, exhibit a truncated shelf whose origin is likely due to diffusion of thermalized ionization electrons out of the front surface. Most detectors show a quasi-exponential tail that may be due in part to this out-diffusion and includes the effects of Lorentzian broadening but may also arise in part from other origins. The mathematical model for a given peak may be assembled from these features, which are shown in Figure 8. Various authors have determined the values of the parameters of these components (e.g., height, width, and truncation energy) as a function of x-ray energy by using monochromatized x rays (Campbell, 1996). PIXE analysts have recourse to an adequate, albeit less accurate, characterization by fitting PIXE spectra of pure single elements. The KMM, KLL, and KLM radiative Auger satellites present in $K$ x-ray spectra also contribute to apparent quasi-exponential tailing features and therefore have to be allowed for in this process. An elegant method for dealing with this complication through use of multiple absorbers is given by Larsson et al. (1989).

The detectors's intrinsic efficiency, i.e., the counts in the Gaussian peak relative to the number of x-ray photons incident on the detector window, is given by

$$\varepsilon_i = \frac{(\exp[-\sum_i^4\mu_i t_i])f_{CCE}f_E[1-\exp(-\mu_{Si}D)]}{(1+z/d)^2} \qquad (9)$$

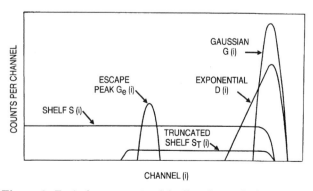

**Figure 8.** Typical components of the line shape of a Si(Li) detector.

where

$$z = \frac{1 - [\exp(-\mu_{Si}D)(1 + \mu_{Si}D)]}{\mu_{Si}[1 - \exp(-\mu_{Si}D)]} \quad (10)$$

and $D$ is detector thickness, $d$ the distance from specimen to detector surface, and $\mu_{Si}$ the linear attenuation coefficient of incident x rays in silicon. The first factor in the numerator of Equation 9 reflects x-ray attenuation in, successively, the vacuum window (beryllium or plastic), any ice layer on the cooled detector surface, the metal contact (most often gold), and any silicon dead layer that acts as a simple absorber. The second factor describes x-ray interactions that happen near the front surface of the active silicon and from which charge carriers escape via the processes described above and any others; the resulting signals fall in the shelves or the exponential features and not in the Gaussian. The factor $f_E$ describes loss to the "escape peak," which occurs when a silicon $K$ x ray, resulting from deexcitation after a photoelectric interaction, escapes from the crystal. The fourth factor describes penetration through the silicon. The reader will recall that the detector's solid angle was incorporated into Equation 1. That is the solid angle defined by the front surface of the silicon. Such a definition of solid angle is correct for only low-energy x rays that interact at the front of the crystal. As x-ray energy increases, the effective interaction depth is greater and the effective solid angle changes. The denominator of Equation 9 incorporates this nonnegligible effect in the intrinsic efficiency.

The detector manufacturer usually provides the thicknesses of the window, the contact, and the crystal. The presence of an ice layer is difficult to determine, but a steady decrease in efficiency at low energies betrays its presence; a detector that incorporates a heater for ice removal is advantageous. Detailed spectra measurements on pure elements are necessary to elucidate the correction for charge loss, which varies greatly among detectors. The term $f_E$ has been determined accurately by various authors, e.g., Johansson (1982).

To determine the distance from specimen to detector crystal, one requires the distance of the crystal inside the window, which the manufacturer provides only approximately. This may be determined by recording the $K$ x-ray spectrum from a $^{55}$Fe radionuclide point source placed at accurately determined positions along the axis of the detector and invoking the inverse square law to describe its distance dependence.

The crystal thickness can be determined by constructing an efficiency curve using radionuclide point source standards and by least-squares fitting the model of Equation 9 to this curve with $D$ as a variable. Suitable nuclides emitting x rays in the energy region 5 to 60 keV and enabling efficiency determination at accuracies of 1% to 3% are supplied by the Physikalisches-Technisches Institut in Braunschweig, Germany, and by other national laboratories charged with responsibility for standards.

JOHN L. CAMPBELL
University of Guelph
Guelph, Ontario, Canada

# RADIATION EFFECTS MICROSCOPY

## INTRODUCTION

Radiation effects microscopy is a general name that includes both ion-beam-induced charge (IBIC) microscopy and single-event upset (SEU) imaging. Ion-beam-induced charge microscopy was initially developed as a means of imaging the distribution of buried $pn$ junctions in microelectronic devices (Breese et al., 1992) and dislocation networks in semiconductors (Breese et al., 1993a). More recently it has found a wider range of applications, e.g., in the study of solar cells (Beckman et al., 1997), segmented silicon detectors (R.A. Bardos, pers. comm.), and chemical vapor deposition (CVD) diamond films (Manfredotti et al., 1995, 1996). Many other applications are reviewed in the literature (Breese et al., 1996), and Doyle et al. (1997) is a good source of up-to-date applications.

Ion-beam-induced charge microscopy uses a beam of mega-electron-volt light ions, mainly protons or helium, to create electron-hole pairs in semiconductor material. This energy transfer from the incident ions to the material causes them to lose energy and eventually stop in the material at a well-defined distance called the ion range. The number of these electron or hole charge carriers that are collected at a junction within the sample is measured and displayed as a function of position of the focused beam within the scanned area. This process is analogous to electron-beam-induced current (EBIC) microscopy (Leamy, 1982), which uses a focused kilo-electron-volt electron beam, and optical-beam-induced current (OBIC) microscopy (Acciarri et al., 1996), which uses a focused laser beam to study similar problems in semiconductor technology. They are also used to image dislocations and inversion layers in semiconducting materials and depletion regions in devices and to give quantitative measurements of the minority carrier diffusion length (Wu and Wittry, 1978; Chi and Gatos, 1979). A description of different EBIC imaging modes as well as more detailed accounts of the theory and technical aspects of image generation is given elsewhere (Holt et al., 1974; Leamy, 1982; Piqueras et al., 1996).

As microelectronic device features continually shrink, the amount of charge that defines the different logic levels becomes smaller. This means that the charge generated by the passage of ionizing radiation through the device is more likely to cause a change in its logic state, called a soft upset, or a SEU. This phenomenon is particularly important for satellite-based devices, which are exposed to a high flux of ionizing cosmic radiation. It is a serious problem in the design of high-density semiconductor memories (May and Woods, 1979). Much work has been done using unfocused high-energy heavy-ion beams from large tandem accelerators (Knudson and Campbell, 1983), pulsed lasers (Buchner et al., 1988), and pulsed electron beams as sources of ionizing radiation to study upset mechanisms. However, the lack of spatially resolved information led to difficulties in determining which part of the device caused the upsets. Single-event upset analyses with collimated heavy-ion beams revealed the benefits of spatially resolving the results (Geppert et al., 1991;

Nashiyama et al., 1993). This prompted the development of SEU microscopy at Sandia (Doyle et al., 1992; Horn et al., 1992; Sexton et al., 1993a). This has many similarities to IBIC, which is usually performed in conjunction with SEU microscopy to image the device components present within the area analyzed for upsets. The two techniques use almost all the same nuclear microprobe hardware and electronics. Whereas IBIC mainly uses mega-electron-volt light-ion beams for analysis, SEU analysis uses heavier ions, such as carbon or silicon, to create a plasma density of electron-hole charge carriers high enough to cause upsets (McLean and Oldham, 1982; Knudson and Campbell, 1983).

There is no displacement damage in silicon using kilo-electron-volt electrons for EBIC, and a low-power laser beam such as Ar or He-Ne used for OBIC also produces no material damage. In comparison, mega-electron-volt ions generate defects in semiconductors. This is a major drawback in IBIC and SEU imaging as it imposes a limitation on the maximum number of ions that can be used to form an image, and care must be taken that the material property that is being measured is not altered beyond what can be accounted for during data collection. The effects of ion-induced damage on the measured charge pulse height spectrum is outlined in Specimen Modification for IBIC and are more fully described by Breese et al. (1996) and for SEU analysis by Sexton et al. (1995).

A major advantage of IBIC and SEU microscopy over competing electron beam techniques is their ability to give high-spatial-resolution analysis in thick layers. This is due to the different shapes of the charge carrier generation volume of mega-electron-volt ions compared with kilo-electron-volt electrons, which are compared in Figure 1. The measured carrier generation volume is shown in the form of concentration contours. The axes are normalized to the particle range. For kilo-electron-volt electrons the generation volume is approximately spherical. The lateral extent of the generation volume can be reduced by lowering the electron beam energy, but at the expense of further decreasing the analytical depth. Because of the large lateral extent of the generation volume of the electron beam within the silicon, the high spatial resolution of the electron beam on the sample surface is degraded, so deeply buried areas cannot be imaged with high spatial resolution. For 3-MeV protons there is little lateral scattering of the proton beam in the top few micrometers, and most occurs close to the end of range, giving a teardrop-shaped carrier generation volume. If the ion range is much greater than both the depletion depth and the diffusion length of the sample (which is usually the case with mega-electron-volt protons), then carriers generated deep within the sample, where there is significant lateral scattering, will not be measured.

The main advantages of IBIC microscopy over electron and optical methods are therefore a large penetration depth and a high spatial resolution for analysis of buried features within microelectronic devices (Breese et al., 1994; Kolachina et al., 1996) or semiconductor wafers. The main advantage of SEU microscopy is its ability to give spatially resolved information on which components within microelectronic devices are responsible for changes

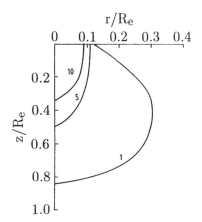

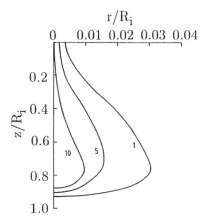

**Figure 1.** (**A**) Carrier generation contours for 10-keV electrons and 3 MeV protons in silicon. The lateral distance $r$ and the depth $z$ are plotted as fractions of (A) the electron range $R_e$ and (**B**) the ion range $R_i$. The numbers 10, 5, and 1 refer to the intensity of the individual generation contour. Courtesy of Breese et al. (1992). Copyright 1992 American Institute of Physics.

in the device logic states due to the passage of high-energy ionizing radiation. A practical disadvantage of both forms of radiation effects microscopy is that they are only available in conjunction with nuclear microprobes located on megavolt particle accelerators (Grime and Watt, 1984; Watt and Grime, 1987; Breese et al., 1996). This makes them expensive compared with competing electron and laser beam techniques, and they are only available in laboratories that specialize in ion beam analysis, of which there are however a considerable number.

## PRINCIPLES OF THE METHOD

The basic semiconductor theory relevant for IBIC and SEU imaging is reviewed here. Other sources (Sze, 1981; Kittel, 1986) are recommended for in-depth discussions of these and other theoretical aspects.

## Semiconductor Theory

Ionizing radiation such as mega-electron-volt ions, kilo-electron-volt electrons, or high-frequency photons can create mobile charge carriers in semiconducting materials by transferring enough energy to the atoms to move valence electrons to the conduction band, leaving behind a positively charged hole. The average energy ($E_{eh}$) needed to create this electron-hole pair does not depend on the type of ionizing radiation and is constant for a given material (Klein, 1968). Charge carriers produced by ionizing radiation diffuse randomly through the semiconductor lattice, and if it contains no electric field, they become trapped and recombine until all excess carriers are removed.

If the semiconductor contains an electric field, the charge carriers are separated in the field region, and this charge flow can be measured in an external circuit. Because of the large number of charge carriers produced by individual mega-electron-volt ions, a charge pulse from each incident ion can be detected above the thermal noise level. The use of mega-electron-volt ions differs in this respect from electron and optical methods such as EBIC and OBIC, in which variations about a steady-state injection level produced by a large incident flux of ionizing radiation is detected rather than individual charge pulses, since these would not be resolved from the thermal noise of the sample.

Typical sample geometries used for IBIC microscopy are shown in Figure 2. The electric field is usually provided by $pn$ junctions in microelectronic devices or by a Schottky

barrier for the analysis of semiconductor wafers. Microelectronic devices typically consist of a semiconductor substrate with a patterned array of $pn$ junctions at the substrate surface, which comprise the different transistors and other electronic components present (Sze, 1988). Above the semiconductor surface there are usually thick, patterned layers of insulating and metal tracks that make up the interconnecting device layers, with a thick passivation layer over the device surface. The total thickness of all the surface layers present can be up to 10 μm. As mentioned above, this is difficult to penetrate with good spatial resolution using an electron beam, and a laser beam suffers high attenuation in passing through any metallization layers, so analysis of small underlying junctions is very limited.

For both IBIC and SEU microscopy a focused beam current of ~1 fA of mega-electron-volt ions is incident on the front face of the sample, and each ion generates charge carriers along its trajectory. The number of charge carriers generated by each incident ion is measured for IBIC microscopy using a contact from the relevant depletion layer at the front face and a contact to the rear surface, each connected to a charge-sensitive preamplifier. This gives an output voltage that is amplified and measured by the data acquisition computer of the nuclear microprobe.

### Quantitative Interpretation of the IBIC Pulse Height

A simple one-dimensional theory used to calculate the IBIC pulse height as a function of ion type and energy, minority carrier diffusion length, and surface and depletion layer thicknesses is described in the literature (Breese, 1993), and only a brief summary is given here. The ion-induced charge pulse height is given here in units of kilo-electron-volts. For example, a measured charge pulse height of 1 MeV from a 2-MeV ion means that half of the beam energy has generated charge carriers that contribute to the measured charge pulse height, and the other 1 MeV has been lost in passing through the surface layers or through carrier recombination within the substrate.

The total measured charge pulse height of Π kilo-electron-volts can be calculated by integrating the rate of ion electronic energy loss $dE/dz$ between the semiconductor surface ($z = 0$) and the ion range, i.e., mean penetration depth ($z = R_i$):

$$\Pi = \int_0^{z_d} \frac{dE}{dz} dz + \int_{z_d}^{R_i} \frac{dE}{dz} \exp\left[-\frac{z - z_d}{L}\right] dz \qquad (1)$$

where $L$ is the minority carrier diffusion length in the sample. The first term is the contribution from the charge carriers generated within the depletion region of the collecting junction, which is $z_d$ micrometers thick. The second term is the contribution from the charge carriers diffusing to the collecting junction from the substrate. Figure 3 shows the calculated charge pulse height for protons and [4]He ions for different values of $L$, based on Equation 1. For a short diffusion length, the charge pulse height reaches a maximum value at a certain proton energy and then decreases as the proton energy is raised further. This is because at a low beam energy there is a high rate of

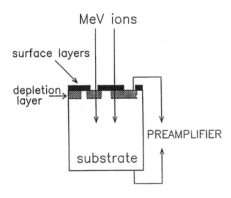

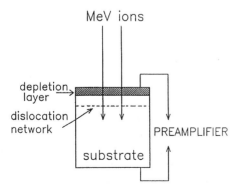

**Figure 2.** Schematic of geometry used for IBIC analysis of (**A**) microcircuits and (**B**) semiconductor and insulator wafers.

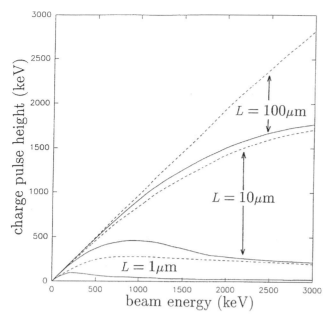

**Figure 3.** Variation of the charge pulse height with protons (solid lines) and ⁴He ions (dashed lines) versus energy for diffusion lengths $L = 1, 10, 100$ μm. Courtesy of Breese et al. (1993). Copyright 1993 American Institute of Physics.

energy loss close to the surface where the generated charge carriers can be measured. As the beam energy is raised further, there is a gradual reduction in the rate of energy loss at the surface, and most carriers are generated too deep to be measured. Similar behavior is also found for kilo-electron-volt electrons (Wu and Wittry, 1978).

### IBIC Topographical Contrast

It is important to know how the IBIC image contrast changes as a function of changes in surface layer thickness in order to correctly interpret the distribution of charge pulses (Breese et al., 1995). Figure 4A shows the rate of electronic energy loss with distance traveled for 2-MeV protons and 3-MeV ⁴He ions. Since they have the same range of ~12 μm as 3-MeV ⁴He ions, 38-keV electrons have been included, so that the charge pulse height from these different charged particles with the same range can be compared.

Figure 4B shows the charge pulse height resulting from a sample consisting of a depletion layer thickness of 1 μm and a substrate diffusion length $L = 6$ μm, with increasing surface layer thickness for the same incident charged particles. The maximum slope of the charge pulse height variation with surface layer thickness occurs close to the end of range, i.e., between 10 and 12 μm, for 3-MeV ⁴He ions, whereas the charge pulse height for 2-MeV protons for similar thickness surface layers changes at a rate that is a hundred times slower. Thus, mega-electron-volt ⁴He ions can produce IBIC images that are highly sensitive to device topography. In comparison, similar energy protons produce IBIC images that are almost independent of device topography, in which the contrast just depends on the distribution of the underlying pn junctions, as will be demonstrated in Figure 10.

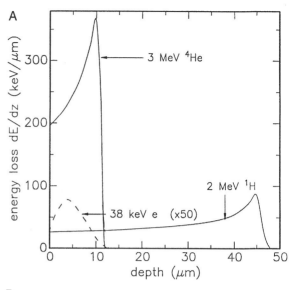

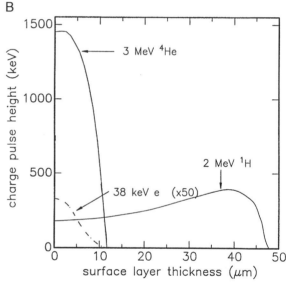

**Figure 4.** (**A**) Electronic energy loss with depth for 2-MeV protons, 3-MeV He ions, and 38-keV electrons (dashed line with the energy loss is multiplied by 50). (**B**) Charge pulse height resulting from a depletion layer thickness of 1 μm and a diffusion length of 6 μm as a function of increasing surface layer thickness for the same charged particles. Courtesy of Breese et al. (1995). Copyright 1995 American Institute of Physics.

The charge pulse height from 38-keV electrons is shown on the same vertical scale as for mega-electron-volt ions to compare the sensitivity of EBIC to changes in surface layer thickness (in practice, EBIC uses a steady-state large incident beam current). The maximum slope of the resultant charge pulse height variation is typically between that measured with protons and with mega-electron-volt ⁴He ions, showing that IBIC can be made either sensitive or insensitive to topographical contrast whereas EBIC does not have this flexibility. Optical methods also lack this flexibility owing to strong attenuation by metallization layers.

This section has briefly described a basic model of ion-induced charge collection, which provides some insight into how the measured charge pulses vary with beam energy and surface layer thickness. There are two routes to obtaining better quantitative understanding for more detailed analysis and interpretation. First, better analytical models of ion-induced charge diffusion and collection are being developed (Donolato et al., 1996; Nipoti et al., 1998). These take into account two-dimensional lateral charge diffusion effects, which should result in better quantitative understanding of how the proximity of *pn* junctions, grain boundaries, and dislocations affect IBIC image contrast. The second route to acquiring a better insight into charge collection mechanisms is with detailed computer simulations. These seem to be the best path to a quantitative understanding of three-dimensional charge collection for use with IBIC and SEU microscopy, as described above.

## SEU Microscopy

The passage of a high-energy, heavy ion through a semiconductor can produce a region with a very high density of excess charge carriers along the ion path because of the ion's high rate of electronic energy loss. The dense plasma created around heavy ions is a reasonable experimental analogy for the passage of strongly ionizing cosmic radiation passing through device memories in satellite-based microelectronic devices, which is why this has become an important research area. If the carrier density reaches a critical level, then enough charge might be collected at a sensitive device junction to change the device logic state, causing a corruption of the information stored, i.e., a SEU has occurred. Figure 5 gives a schematic of the SEU imaging process within a nuclear microprobe. A focused heavy-ion beam is scanned over the chip, and the information stored within the device is sampled at each beam position within the scanned area, so that an image can be constructed showing the upset probability at each location.

If a heavy ion passes through a depletion region of a device memory (Fig. 6), the resultant dense plasma forma-

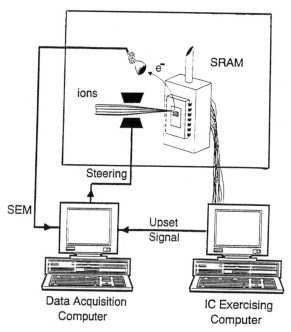

**Figure 5.** Schematic of apparatus used for SEU analysis. Courtesy of Doyle et al. (1992).

tion can distort its shape, giving rise to the phenomenon of charge funneling (McLean and Oldham, 1982). The carrier density can reach up to $10^{20}$ cm$^{-3}$ along the path of a heavy ion, which is much greater than the typical device substrate doping density. As a consequence, the junction depletion layer in the vicinity of the ion path is quickly neutralized, and the high electric field in the junction is screened by electrons being drawn off at the electrode. The electric field associated with the junction is reduced in size and becomes elongated in the direction of the ion path, and Figure 6C shows the distorted equipotential line associated with this effect. The effect of this charge funnel is to give a much larger measured charge pulse than would otherwise occur by charge drift and diffusion alone. The charge is also much more localized around the

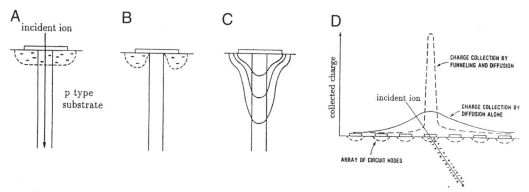

**Figure 6.** Schematic of charge funneling showing (**A**) a heavy ion hitting a junction, (**B**) the depletion region being neutralized by the resultant plasma column, and (**C**) equipotential lines from the junction being extended down. (**D**) Resultant charge collection profile over a circuit array, with enhanced charge collection due to charge funneling. Adapted from McLean et al. (1982). Reprinted in modified form. Copyright 1982 IEEE.

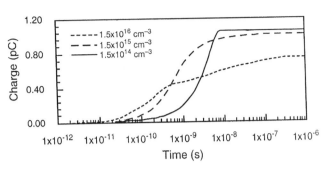

**Figure 7.** Simulated charge transients generated by a 100-MeV Fe ion in three $n^+/p$ silicon diodes with substrate doping densities shown in the top left. Courtesy of Dodd et al. (1994).

struck junction, as less charge diffuses away to be absorbed by other junctions, resulting in a much increased susceptibility of the device to SEUs.

Charge-funneling effects are also important to IBIC as well as for SEU analysis because they should be taken into account when calculating the charge pulse height due to heavy ions if the ionization carrier density is much greater than the substrate density. Sophisticated two- and three-dimensional computer codes (Dodd et al., 1994; Dodd, 1996) can be used to model the effects of SEU and charge-funneling effects in order to gain a detailed understanding of the basic mechanisms that can upset devices and to distinguish funneling from the effects of charge drift and diffusion.

An example is a heavily doped ($5 \times 10^{20}$ cm$^{-3}$) $n^+$ diffusion in a $p$-type silicon substrate with different doping concentrations (Dodd et al., 1994). The passage of a 100-MeV Fe ion through this system was modeled to investigate under what conditions it could upset the device. Figure 7 shows the simulated charge collection transients from one quadrant of the symmetrical device area for three different substrate doping densities. The total measured amount of charge decreases with increasing substrate doping density, because it is less likely that significant additional amounts of charge will be collected via the funneling effect. For the lightly doped substrates, nearly all the charge carriers are collected by a charge-funneling process, so there is little extra charge to be collected by diffusion after the funnel collapses $\sim$10 ns after the passage of the ion through the device. For the more heavily doped substrate of $1.5 \times 10^{16}$ cm$^{-3}$, the funnel collapses 400 ps after ion impact, leaving a considerable fraction of charge carriers free to diffuse through the device. Some carriers diffuse to the collecting junction and others recombine, so less charge is measured in total.

### Further Advances

An interesting development that should further enhance the capabilities of IBIC and SEU analysis is the production of time-resolved, or transient, IBIC images. Presently being developed by the Sandia microprobe group (Schone, 1998), this enables those charge components from the depletion region and from the substrate to be measured directly, being distinguished by their different charge col-

lection time scales. The ion-induced transient response is measured with a timing resolution of tens of picoseconds, as has already been done with unfocused ion beams (Wagner et al., 1988). The experimental procedure is to store the digitized ion-induced current transient along with the $(x,y)$ position coordinates of the beam. From the stored data, the shape of the current transient can be studied as a function of position as well as time, enabling direct measurements of the critical charge needed to cause SEUs. These also help to test numerical three-dimensional simulations of charge diffusion and charge funneling.

A similar system has been built by the Melbourne microprobe group (Laird et al., 1999). The installation of a cold stage on this system capable of cooling samples to liquid helium temperature should enable transient IBIC pulses to be studied as a function of temperature, in a similar manner to deep level transient spectroscopy; see DEEP-LEVEL TRANSIENT SPECTROSCOPY (Lang, 1974).

## PRACTICAL ASPECTS OF THE METHOD

The basic layout used for IBIC analysis of microcircuits and Schottky barrier samples is shown in Figure 2. The IBIC pulses are measured using standard charged particle detection electronics, similar to those used for other ion beam analysis techniques. Each pulse is fed to a low-noise charge-sensitive preamplifier whose output voltage is proportional to the measured number of carriers generated by individual ions.

Charge-sensitive preamplifiers are ideal for use in IBIC experiments because they integrate the induced charge on a feedback capacitor. A charge-sensitive preamplifier typically has an open-loop gain of $\sim 10^4$ so that it appears as a large capacitance to the sample, rendering the gain insensitive to changes in the sample capacitance. Details of preamplifier design, pulse shaping, and methods of noise reduction are described in many texts (e.g., Bertolini and Coche, 1968; England, 1974). The preamplifier output voltage $V_o$, is given as

$$V_o = \frac{1000 \Pi e}{E_{eh} C} \tag{2}$$

where $\Pi$ is the measured charge pulse height in kiloelectron-volts, given by Equation 1. With a feedback capacitance $C = 1$ pF, a typical preamplifier output pulse size is 44 mV per mega-electron-volts of energy of the incident ion. The small output pulses from the preamplifier are fed into an amplifier, which gives an output voltage of $\sim$1 V. This is then fed into the data acquisition computer and an image is generated showing either variations in the average measured charge pulse height or the intensity of counts from different "windows" of the spectrum at each pixel within the scanned area.

A beam current of $\sim$1 fA (i.e., $\sim$6000 ions/s) is typically used for both IBIC and SEU microscopy since the maximum data acquisition rate available with most nuclear microprobes is less than 10 kHz. To produce a focused spot containing such a small current, a much larger beam current of 100 pA is first focused in a conventional

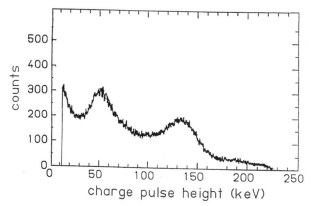

**Figure 8.** IBIC pulse height spectrum from the $300 \times 300$ μm$^2$ device area shown in Figure 9, produced with 3-MeV protons. The preamplifier was connected between the data pin and ground.

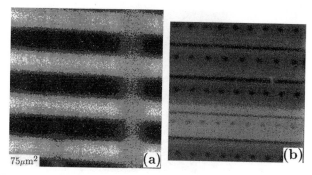

**Figure 10.** (**A**) $75 \times 75$ μm$^2$ average pulse height IBIC image of the central region in Figure 9 using 3-MeV protons. (**B**) 2.3-MeV $^4$He IBIC images of the same area, under the same conditions, with a scan size of $40 \times 40$ μm$^2$. Reprinted from (Breese et al. 1994) with permission. Copyright 1994 American Institute of Physics.

manner (Breese et al., 1996). The object and divergence slits are then closed until the remaining current is a few thousand ions per second, as measured by a semiconductor detector placed in the path of the ion beam. This procedure ensures that the sample is not irradiated with a high ion dose prior to analysis. The sample is then moved onto the beam axis, and a large-area IBIC image is used to identify and position the region of interest. With SEU analysis, damage effects are even more critical since heavier ions are used. Therefore, initial sample positioning with SEU microscopy is also best carried out using IBIC with light ions, since they will cause much less damage prior to SEU analysis.

### IBIC Examples

A typical IBIC charge pulse height spectrum measured from a $300 \times 300$ μm$^2$ area of an extended programmable read-only-memory (EPROM) chip using a focused 3-MeV proton beam is shown in Figure 8. The charge pulses in this case are up to five times larger than the thermal noise level, which occupies the lower end of the spectrum. The lower input threshold of the amplifier should be raised to a level such that very few noise pulses are measured or else the high noise level will saturate the data acquisition system and the ion-induced charge pulses will not be measured. However, if the threshold is raised too far, then the smallest charge pulses will not be measured, demonstrating the problem of a poor signal-to-noise ratio, which must always be borne in mind in IBIC experiments. This may also influence the choice of incident ion used for the IBIC

analysis, since heavier ions usually give larger charge pulses (see Fig. 4).

In this example the device is an $n$-type metal-oxide-semiconductor (NMOS) memory chip. The area analyzed contains two output driver field effect transistors and also two transistors comprising the input buffer. The metallization consists of a 1-μm-thick layer of Al(1% Si). Over the surface of the device is an ~1.5-μm-thick SiO$_2$ passivation layer, with a total surface layer thickness up to 4 μm. Figure 9 shows three IBIC images of this $300 \times 300$ μm$^2$ area, but with different preamplifier connections and device pin voltages in each case. In each case, spectra similar to that shown in Figure 8 are measured, and IBIC images showing the average charge pulse height at each pixel are produced. The differing contrasts are explained in detail elsewhere (Breese et al., 1992), but they demonstrate here that IBIC can generate images showing the different distributions of active components buried several micrometers beneath the surface of the functioning device.

Figure 10 shows IBIC images of the central region of Figure 9c with the device under the same operating conditions. Figure 10 shows a $75 \times 75$ μm$^2$ image generated using 3-MeV protons and a $40 \times 40$ μm$^2$ image generated with 2.3-MeV $^4$He ions. Figure 10b shows strong topographical contrast where the circles along the lengths of the horizontally running metallized source and drain regions can be seen. These are not visible in Figure 10a, which only shows IBIC contrast due to the layout of the underlying

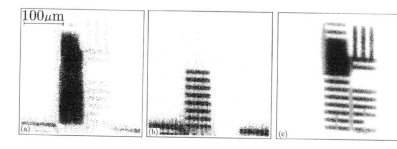

**Figure 9.** Three $300 \times 300$ μm$^2$ IBIC images of the same device area. The preamplifier connections are (**A**) between the supply voltage pin, with $V_{cc} = +5$ V and ground. The output driver voltage $V_{o=+5}$ V, and the data pin is not connected here. (**B**) Same as (A) except the data pin is now also connected to a different preamplifier. (**C**) Measured between the data pin and ground but with other transistors on.

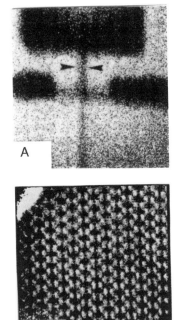

**Figure 11.** (**A**) $20 \times 20~\mu m^2$ IBIC image of a GaAs device, between the $p$- and $n$-type contacts. A 0.8-$\mu$m-wide depletion region is indicated. (**B**) $65 \times 65~\mu m^2$ IBIC image of a DRAM. Reprinted from Breese et al. (1993b) with permission from Elsevier Science B.V., Amsterdam, The Netherlands.

$pn$ junctions. This demonstrates the insensitivity of IBIC to device topography using mega-electron-volt protons, as calculated in Figure 4B. The maximum signal-to-noise level in Figure 10b is ~50, whereas in the proton IBIC images of this same area it was ~5, so in this case the use of heavier ions has greatly increased the pulse sizes.

### Other IBIC Examples

An IBIC image of a GaAs high-electron-mobility transistor is shown in Figure 11a. The surface of the device was patterned using a focused $Be^+$ ion beam, which formed $p$-type material and resulted in the formation of a field effect transistor with a very narrow gate depletion region that was confined both laterally and in depth. The IBIC images were generated using contacts between the $p$- and $n$-type material, with a 10-$\mu$m-thick Al foil placed between the focused 3-MeV protons and the device surface. On the IBIC image, a 0.8-$\mu$m-wide depletion region is indicated by arrows; this high spatial resolution resulted from the low lateral straggling of mega-electron-volt protons through thick surface layers.

Figure 11b shows an IBIC image of a $65 \times 65~\mu m^2$ area of the memory array in a 4-Mbit dynamic random-access memory (DRAM) device. Smaller charge pulses were measured from the light-colored honeycomb structure of 1-$\mu$m-wide trench cells that comprise parts of the individual memory cells. To prepare a sample suitable for IBIC microscopy, all the device surface layers were removed using hydrofluoric acid, leaving just the $p$-type substrate and the $n$-doped trench walls. Electrical contacts were

then produced by depositing a thin gold layer onto the back surface to form an ohmic contact and a thin aluminum layer onto the front surface to form a Schottky contact in the geometry shown in Figure 2B. This device also formed part of an IBIC study of methods of compensating for the effects of beam-induced damage in IBIC images, as described by Breese et al. (1993b).

## DATA ANALYSIS AND INTERPRETATION

### SEU analysis of SRAMs

An example of the SEU analysis of radiation-hardened static random-access memory (SRAM) devices is given here. An SEU image generated using a low-intensity beam of 24-MeV $Si^{6+}$ ions scanning over a $40 \times 40~\mu m^2$ area of the device is shown in Figure 12a, and the schematic of the device layout within this area is shown in Figure 12b. The SEU image was produced by interrogating the device logic state as a function of irradiation with the focused ion beam at each pixel within the scanned area. The SEU image shows variations in the upset probability within this area as darker shading at the center and lighter shading toward the edges of the upset-generating region.

An IBIC image of the same area generated using 2.4-MeV $^4$He ions is shown in Figure 12c. The ion impact locations that give rise to upsets within this scanned area can now be directly identified by comparing the SEU and IBIC images with the device layout. In this case, the upset

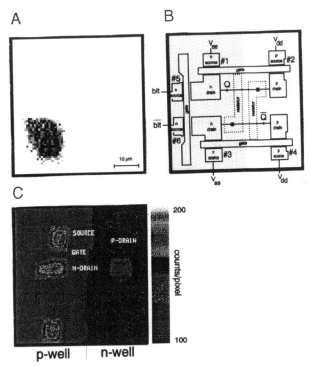

**Figure 12.** (**A**) $40 \times 40~\mu m^2$ SEU image of the SRAM device. (**B**) Schematic of the device layout in this area and (**C**) IBIC image from the same area. Courtesy of Horn et al. (1993). Reprinted from Sexton et al. (1993b) with permission from Elsevier Science B.V., Amsterdam, The Netherlands.

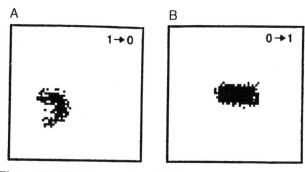

**Figure 13.** SEU images of an SRAM memory cell showing the logic state dependence: (**A**) with logic 1 initially stored in the memory cell and (**B**) with logic state 0 initially stored. Courtesy of Horn et al. (1993).

area was localized around an *n*-channel transistor drain. In the IBIC image, a large charge pulse was measured from the *n* drains within the *p* well of the device. Within the *p* well, a large charge pulse was also measured from the *n* source, but from the SEU image this is shown to be insensitive to upsets. There is a weaker region of charge collection within the *n*-well region on the right of the IBIC image, corresponding to a *p* drain. A detailed explanation of the observed IBIC image contrast in terms of the device layout is given in the literature (Horn et al., 1993; Sexton et al., 1993b), where the behavior observed in the SEU image was correlated with the observed IBIC contrast. It was postulated that regions of larger than expected charge pulses could arise from a "shunt" effect (Kreskovsky and Grubin, 1986) caused by the high ion-induced carrier density along the ion path.

Single-event upset microscopy also gives the opportunity to measure logic-state-dependent upsets. Figure 13 shows an SEU image from the same SRAM device structure for each of two different initial logic states. As can be seen, the upsets occur in different positions, and the sensitive volume is of a different shape and size, depending on the initial value stored. This observation is typical of the additional information to be gained by the ability to spatially resolve the locations of SEUs.

## SPECIMEN MODIFICATION

The main damage mechanism of mega-elextron-volt ions in semiconductors is the creation of vacancy/interstitial pairs, called Frenkel defects. This is caused by direct collisions between the incident ions and the lattice nuclei, in which a sufficient amount of energy is imparted to the nucleus to displace it from its lattice site. Their primary effect for IBIC and SEU microscopy is a reduction in the minority carrier diffusion length, because the defects act as trapping and recombination centers. Since many charge pulses should ideally be measured at each pixel to reduce the statistical noise in IBIC and SEU images, ion-induced damage is a major drawback as it limits the beam dose, which can be used to produce an image before the image contrast greatly changes.

Prior to the development of IBIC, it was demonstrated that a nuclear microprobe could be used to spatially resolve ion beam damaged areas in a *pn* junction by measuring the charge pulses from the focused beam (Angell et al., 1989). Since the subsequent development of IBIC as an analytical technique, the effects of ion-induced damage and methods of compensating for its effects have been studied extensively (Breese et al., 1993b, 1995).

A study into the observed damage occurring during SEU analysis (Sexton et al., 1995) investigated whether the observed damage effects with heavy ions were due to the ionizing energy loss component, i.e., to possible charge buildup in insulating layers, or the nonionizing component, i.e., to displacement effects. Devices were irradiated with X-rays to model the effects of ionizing radiation, and it was found that no significant changes were observed in subsequent SEU behavior. In comparison, identical devices irradiated with large doses of 30-MeV copper ions underwent significant changes in SEU behavior after a certain dose, with more components becoming sensitive to upsets with cumulative beam dose. It was postulated that this was due to a reduction of the critical charge threshold for upset produced by beam damage effects of the nonionizing component of the energy loss.

Using the same simple theory to characterize charge collection as described in Principles of the Method, a rough guide to the behavior of IBIC pulses with cumulative beam dose can be derived. The calculated rate of charge pulse height reduction for different energy protons and $\alpha$-particles is shown in Figure 14. In brief, the charge pulse height reduces with cumulative beam dose because of a reduction in the minority carrier diffusion length. The reduction is faster for heavier ions because they introduce more defects and thus decrease the diffusion length more rapidly.

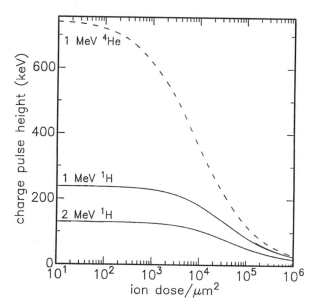

**Figure 14.** Calculated charge pulse height with cumulative ion irradiation with 1- and 2-MeV protons (solid lines) and 1-MeV $^4$He ions (dashed lines). An initial diffusion length of 5 $\mu$m is assumed. Except where indicated, reprinted from Breese (1993) with permission. Copyright 1993 American Institute of Physics.

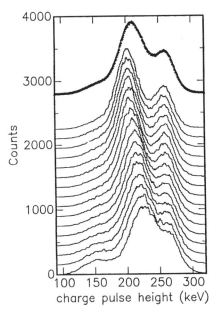

**Figure 15.** Charge pulse height spectra from a $80 \times 80\ \mu m^2$ area produced using 2-MeV $^4$He ions. The measured charge pulse height spectra with cumulative sequential dose increments of 9 ions/$\mu m^2$ are offset vertically by a fixed arbitrary amount for clarity. The uppermost spectrum is the cumulative spectrum from which the sequential spectra were extracted. Reprinted from Breese et al. (1995) with permission. Copyright 1995 American Institute of Physics.

Therefore, although heavier ions give a larger measured IBIC pulse than similar energy light ions, this advantage might be outweighed by the more rapid rate of damage. This is another consideration when choosing the optimum type of ion for use in IBIC or SEU analysis.

The best type of data acquisition system to use for IBIC and SEU microscopy is an event-by-event system (O'Brien et al., 1993), which stores the measured charge pulses along with their position coordinates for subsequent processing and analysis. The specific advantage of this is that the measured charge pulse height data set can be sliced into sequential dose increments so that the evolution of the charge pulse height spectrum and image contrast from different parts of the scanned area can be examined with the cumulative ion dose.

Figure 15 shows the IBIC pulse height spectra from a $80 \times 80\ \mu m^2$ area of a different microelectronic device. Here 2-MeV $^4$He ions are used, and the resultant spectrum is shown at the start of data collection and also after fixed sequential dose increments. It can be seen that different parts of the charge pulse height spectrum behave in different ways; the left-hand peak decreases in size with increasing beam dose, whereas the right-hand peak remains the same size. The explanation for this type of behavior is given below. Here it suffices to say that even though the charge pulse height in adjacent pixels may vary differently with cumulative ion dose, i.e., the contrast may change, the charge pulses can still be used to generate an IBIC image when an event-by-event acquisition system is used.

## Minimizing the Effects of Ion-Induced Damage

Ion-induced damage of the semiconductor substrate decreases the measured amount of charge because the increased recombination of the slowly diffusing charge carriers causes a reduction in the carrier diffusion length, as described below. The mechanism by which the amount of charge measured from the depletion region reduces with ion-induced damage is more complex. Charge carriers generated in the depletion region have a much lower recombination probability than those generated in the substrate because of the associated electric field. This accelerates the carriers in opposite directions and reduces the probability of them becoming trapped and recombining at any defects present.

Figure 16 shows a schematic of the final locations of ions with a fixed energy penetrating through an increasingly thick surface layer into a narrow depletion region and then into the substrate. For a thin surface layer, the ions penetrate deep into the substrate, and most of the measured charge is due to carriers diffusing from the substrate to the collecting junction. Since the distribution of the nonionizing energy loss along the ion trajectory follows a similar trend as the electronic energy loss, most of the defects are created in the substrate, where they have a large effect on the measured charge pulse height. For a thicker surface layer, all the ions may be stopped in the depletion layer at the semiconductor surface. The maximum defect generation rate also occurs here, where they have a much smaller effect on the measured charge pulse height since charge carriers are much less likely to be trapped here. For a still thicker surface layer (c), most of the ions are stopped in the surface layers and only a few penetrate into the depletion region. If the ion does not

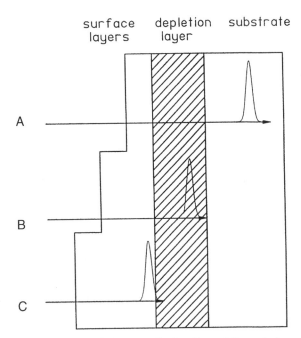

**Figure 16.** Schematic showing the locations of the end-of-range defects (approximated as Gaussian distributions) for increasing surface layer thicknesses from (**A**) to (**C**).

generate enough charge carriers to be measured above the noise level, then it will not be detected.

The different behavior of the two charge peaks in Figure 15 is due to the effect shown in Figure 16. The left-hand peak is due to ions that penetrate into the substrate and so reduces in height rapidly, whereas the right-hand peak is due to ions that are stopped in depletion regions present.

## PROBLEMS

Two experimental problems commonly encountered during IBIC experiments are difficulties in producing measurable charge pulses and the associated large noise level present from the microelectronic device or semiconductor wafer.

### Absence of Observed Charge Pulses

It is important to test the electrical contacts to the IBIC or SEU sample before irradiating them in the microprobe. Otherwise there is a high probability of unintentionally damaging the sample by mistaking the absence of measured IBIC pulses for the absence of a beam irradiating it. First the sample's sensitivity to light should be checked by measuring the current flowing through it. Light generates charge carriers in a similar manner to ions, so on most types of samples there should be a large increase in the measured current with light shining on them. Next the sample in the shielded microprobe chamber should be connected and a determination made as to whether the noise level measured with an oscilloscope (with the chamber lights off) is small enough to resolve charge pulses. Excessively large noise level might be due to poor barrier preparation or wrong connections to the microelectronic device. A further wise step is to check that charge pulses generated by α-particles from a test source can be resolved from the noise level. If not, then the use of heavier ions to produce larger charge pulses should be considered. Also, the amplifier polarity should be set so as to give positive pulses into the data acquisition system—a commonly made mistake.

### Methods of Noise Reduction

If the noise level in an SEU or IBIC sample is too high, then it is difficult to sufficiently resolve the pulses. Great care must therefore be taken to reduce the noise level as much as possible before the start of data collection.

The noise level typically encountered with IBIC and SEU samples is ~1 mV at the preamplifier output for microcircuit devices. Noise in charge-sensitive preamplifiers has three main sources: the input field effect transistor, the input capacitance, and the preamplifier resistance. The noise contribution from the input capacitance increases at a typical rate of 15 to 20 eV/pF, so any excess capacitance should be removed. Leads between the preamplifier and the sample should be as short as possible since the capacitance increases with lead length, and ideally the preamplifier should be mounted inside the target chamber to be as close as possible to the sample. For IBIC analysis of semiconductor wafers, the area of the Schottky barrier or the number of pins connected to the preamplifier should be as small as possible to minimize the total active area, as this reduces the capacitance. The leads should be well screened, and the sample should be isolated from earth loops, circulating currents, and radio frequency pickup from other components of the microprobe electronics. For SEU analysis, which typically requires a larger number of feedthroughs and connections to the device in the microprobe chamber in order to check its logic state after every incident ion, it is even more important that no external noise is introduced that might affect the device.

The noise level in the measured charge pulse height spectrum also depends on the amplifier time constant. A typical value of 1 μs is frequently used for the best signal-to-noise ratio. The optimum value depends on both the current flowing in the sample and its capacitance. Decreasing the time constant too much can decrease the measured charge pulse height because not all the trapped carriers may have detrapped, so it may be necessary to make a compromise between the best signal-to-noise ratio and quantifying the diffusion length from the measured charge pulse height. Once all excess input capacitance to the preamplifier has been eliminated, the remaining measured noise level is dominated by thermal generation of charge carriers. Cooling the material reduces the thermally generated noise level. The charge pulse height spectra measured from Schottky barriers for IBIC microscopy have been very noisy to date, because large-area barriers have been used in order to manually attach a thin wire to the front surface using silver paint. Since both the capacitance and the thermally generated noise contributions increase with the barrier area, this should ideally be as small as possible.

## LITERATURE CITED

Acciarri, M., Binetti, S., Garavaglia, M., and Pizzini, S. 1996. Detection of junction failures and other defects in silicon and III–V devices using the LBIC technique in lateral configuration. *Mat. Sci. Eng.* B42:208–212.

Angell, D., Marsh B. B., Cue, N., and Miao, J. W. 1989. Charge collection microscopy: Imaging defects in semiconductors with a positive ion microbeam. *Nucl. Instrum. Methods* B44:172–178.

Beckman, D. R., Saint, A., Gonon, P., Jamieson, D. N., Prawer, S., and Kalish, R. 1997. Spatially resolved imaging of charge collection efficiency in polycrystalline CVD diamond by the use of ion beam induced current. *Nucl. Instrum. Methods* B130:518–523.

Bertolini, G. and Coche, A. 1968. Semiconductor Detectors. North-Holland, Amsterdam.

Breese, M. B. H. 1993. A theory of ion beam induced charge collection. *J. Appl. Phys.* 74(6):3789–3799.

Breese, M. B. H., Grime, G. W., and Watt, F. 1992. Microcircuit imaging using ion beam induced charge. *J. Appl. Phys.* 72(6):2097–2105.

Breese, M. B. H, King, P. J. C., Grime, G. W., and Wilshaw, P. R. 1993a. Dislocation imaging using ion beam induced charge. *Appl. Phys. Lett.* 62:3309–3311.

Breese, M. B. H., Grime, G. W., and Dellith, M. 1993b. The effects of ion induced damage on IBIC images. *Nucl. Instrum. Methods* B77:332–338.

Breese, M. B. H., Laird, J. S., Moloney, G. R., Saint, A., and Jamieson, D. N. 1994. High signal to noise ion beam induced charge images. *Appl. Phys. Lett.* 64(15):1962–1964.

Breese, M. B. H., Saint, A., Sexton, F. W., Schone, H., Doyle, B. L., Laird, J. S., and Legge, G. J. F. 1995. Optimisation of ion beam induced charge microscopy for the analysis of integrated circuits. *J. Appl. Phys.* 77(8):3734–3741.

Breese, M. B. H., Jamieson, D. N., and King, P. J. C. 1996. Materials Analysis Using a Nuclear Microprobe. John Wiley & Sons, New York.

Buchner, S., Knudson, A. R., Kang, K., Campbell, A. B. 1988. Charge collection from focused pico-second laser pulses. *IEEE Trans. Nucl. Sci.* NS-35:1517–1522.

Chi, J. C. and Gatos, H. C. 1979. Determination of dopant concentration diffusion lengths and lifetime variations in silicon by scanning electron microscopy. *J. Appl. Phys.* 50(5):3433–3440.

Dodd, P. E. 1996. Device simulation of charge collection and single-event upset. *IEEE Trans. Nucl. Sci.* NS-43(2):561–575.

Dodd, P. E., Sexton, F. W., and Winokur, P. S. 1994. Three-dimensional simulation of charge collection and multiple-bit upset in Si devices. *IEEE Trans. Nucl. Sci.* NS-41(6):2005–2017.

Donolato, C., Nipoti, R., Govoni, D., Egeni, G. P., Rudello, V., and Rossi, P. 1996. Images of grain boundaries in polycrystalline silicon solar cells by electron and ion beam induced charge collection. *Mat. Sci. Eng.* B42:306–310.

Doyle, B. L., Horn, K. M., Walsh, D. S., and Sexton, F. W. 1992. Single event upset imaging with a nuclear microprobe. *Nucl. Instrum. Methods* B64:313–320.

Doyle, B. L., Maggiore, C. J., and Bench. G. 1997. *In* Proceedings of the Fifth International Conference on Nuclear Microprobe Technology and Applications (H.H. Andersen and L. Rehn, eds.). pp. 1–750. Elsevier/North-Holland, Amsterdam.

England, J. B. A. 1974. Techniques in Nuclear Structure Physics. MacMillan Press, London.

Geppert, L. M., Bapst, U., Heidel, D. F., and Jenkins, K. A. 1991. Ion Microbeam Probing of Sense Amplifiers to Analyse Single Event Upsets in a CMOS DRAM. *IEEE Trans. Solid-State Circuits* 26(2):132–134.

Grime, G. W. and Watt, F. 1984. Beam Optics of Quadrupole Probe-forming Systems. Adam Hilger, Bristol.

Holt, D. B., Muir, M. D., Grant, P. R., and Boswarva, I. M. 1974. Quantitative Scanning Electron Microscopy. Academic Press, London.

Horn, K. M., Doyle, B. L., Walsh, D. S., and Sexton, F. W. 1992. Nuclear microprobe imaging of single event upsets. *IEEE Trans. Nucl. Sci.* 39(1):7–12.

Horn, K. M., Doyle, B. L., Sexton, F. W., Laird, J. S., Saint, A., Cholewa, M., and Legge, G. J. F. 1993. Ion beam induced charge collection (IBICC) microscopy of ICs: Relation to single event upsets (SEUs). *Nucl. Instrum. Methods* B77:355–361.

Kittel, C. 1986. Introduction to Solid State Physics. John Wiley & Sons, Singapore.

Klein, C. A. 1968. Bandgap dependence and related features of radiation ionisation energies in semiconductors. *J. Appl. Phys.* 39:2029–2038.

Knudson, A. R. and Campbell, A. B. 1983. Investigation of soft upsets in integrated circuit memories and charge collection and charge collection in semiconductor structures by the use of an ion microbeam. *Nucl. Instrum. Methods* 218:625–631.

Kolachina, S., Ong, V. K. S., Chan, S. H., Phang, J. C. H., Osipowicz, T., and Watt, F. 1996. Unconnected junction contrast in ion beam induced charge microscopy. *Appl. Phys. Lett.* 68(4):532–534.

Kreskovsky, J. P. and Grubin, H. L. 1986. Numerical simulation of charge collection in two- and three-dimensional silicon diodes—a comparison. *Solid-State Electron.* 29(5):505–518.

Laird, J. S., Bardos, R. A., Jagadish, C., Jamieson, D. N., and Legge, G. J. F. 1999. Scanning ion deep level transient spectro-scopy. *Nucl. Instrum. Methods* B158:464–469.

Lang, D. V. 1974. Deep level transient spectroscopy: A new method to characterize traps in semiconductors. *J. Appl. Phys.* 45(7):3023–3032.

Leamy, H. J. 1982. Charge collection scanning electron microscopy. *J. Appl. Phys.* 53:R51-R80.

Manfredotti, C., Fizzotti, F., Vittone, E., Boero, M., Polesello, P., Galassini, S., Jaksic, M., Fazinic, S., and Bogdanovic, I. 1995. IBIC Investigations on CVD diamond. *Nucl. Instrum. Methods* B100:133–140.

Manfredotti, C., Fizzotti, F., Vittone, E., Boero, M., Polesello, P., Galassini, S., Jaksic, M., Fazinic, S., and Bogdanovic, I. 1996. Study of physical and chemical inhomogeneities in semiconducting and insulating materials by a combined use of micro-PIXE and micro-IBIC. *Nucl. Instrum. Methods* B109/110:555–562.

May, T. C. and Woods, M. H. 1979. Alpha-particle-induced soft errors in dynamic memories. *IEEE Trans. Electron Devices* ED-26:2–9.

McLean, F. B. and Oldham, T. R. 1982. Charge funneling in n- and p-type Si substrates. *IEEE Trans. Nucl. Sci.* NS-29:2018–2023.

Nashiyama, I., Hirao, T., Kamiya, T., Yutoh, H., Nishijima, T., and Sekiguti, H., 1993. Single-event current transients induced by high energy ion microbeams. *IEEE Trans. Nucl. Sci.* NS-40(6):1935–1940.

Nipoti, R., Donolato, C., Govoni, D., Rossi, P., Egeni, E., and Rudello, V. 1998. A study of He$^+$ ion induced damage in silicon by quantitative analysis of collection efficiency data. *Nucl. Instrum. Methods.* B138:1340–1344.

O'Brien, P. M., Moloney, G., O'Connor, A., and Legge, G. J. F. 1993. A versatile system for the rapid collection, handling, and graphics of multidimensional data. *Nucl. Instrum. Methods* B77:52–55.

Piqueras, J., Fernandez, P., and Mendez, B. 1996. *In* Proceedings of the Fourth Workshop on Beam Injection Assessment of Defects in Semiconductors (M. Balkanski, H. Kamimura, and S. Mahajan, eds.). pp. 1–310. Elsevier/North-Holland, Amsterdam.

Schone, H., Walsh, D. S., Sexton, F. W., Doyle, B. L., Dodd, P. E., Aurand, J. F., Flores, R. S., and Wing, N. 1998. Time-resolved ion beam induced charge collection in microelectronics. *IEEE Trans. Nuc. Sci.* 45(6):2544–2549.

Sexton, F. W., Horn, K. M., Doyle, B. L., Laird, J. S., Cholewa, M., Saint, A., and Legge, G. J. F. 1993a. Relationship between IBICC imaging and SEU in CMOS ICs. *IEEE Trans. Nucl. Sci.* NS-40:1787–1794.

Sexton, F. W., Horn, K. M., Doyle, B. L., Laird, J. S., Saint, A., Cholewa, M., and Legge, G. J. F. 1993b. Ion beam induced charge collection imaging of CMOS ICs. *Nucl. Instrum. Methods* B79:436–442.

Sexton, F. W., Horn, K. M., Doyle, B. L., Shaneyfelt, M. R., and Meisenheimer, T. L. 1995. Effects of ion damage on IBICC and SEU imaging. *IEEE Trans. Nucl. Sci.* 42(6):1940–1947.

Sze, S. M. 1981. Physics of Semiconductor Devices. John Wiley & Sons, New York.

Sze, S. M. 1988. VLSI Technology. McGraw-Hill, Singapore.

Wagner, R. S., Bordes, N., Bradley, J. M., Maggiore, C., Knudson, A. R., and Campbell, A. B. 1988. Alpha-, boron-, silicon- and iron-ion induced current transients. *IEEE Trans. Nucl. Sci.* NS-35(4):1578–1584.

Watt, F. and Grime, G. W. 1987. Principles and Applications of High Energy Ion Microbeams. Adam Hilger, Bristol.

Wu, C. J. and Wittry, D. B. 1978. Investigation of minority carrier diffusion lengths by electron bombardment of Schottky barriers. *J. Appl. Phys.* 49(5):2827–2836.

## KEY REFERENCES

Breese et al., 1996. See above.

*A comprehensive description of IBIC analysis with many examples. Explains in detail the one-dimensional theory of IBIC. Describes other SEU work.*

Doyle et al., 1997. See above.

*Supplies the most up-to-date papers on both technological aspects and applications of IBIC and SEU microscopy.*

Mark B. H. Breese
National University of Singapore
Kent Ridge, Singapore

# TRACE ELEMENT ACCELERATOR MASS SPECTROMETRY

## INTRODUCTION

Accelerator mass spectrometry (AMS; Jull et al., 1997) is a relatively new analytical technique that is being used in over 40 laboratories worldwide for the measurements of cosmogenic long-lived radioisotopes, as a tracer in biomedical applications, and for trace element analysis of stable isotopes. The use of particle accelerators, along with mass spectrometric methods, has allowed low-concentration measurements of small-volume samples. The use of AMS to directly count ions, rather than measure radiation from slow radioactive decay processes in larger samples, has resulted in sensitivities of one part in $10^{15}$ (Elmore and Phillips, 1987; Jull et al., 1997). The ions are accelerated to mega-electron-volt energies and can be detected with 100% efficiency in particle detectors. The improved sensitivities and smaller sample sizes have resulted in a wide variety of applications in anthropology, archaeology, astrophysics, biomedical sciences, climatology, ecology, geology, glaciology, hydrology, materials science, nuclear physics, oceanography, sedimentology, terrestrial studies, and volcanology, among others (Jull et al., 1997). Accelerator mass spectrometry has opened new areas of research in the characterization of trace elements in materials.

This unit will focus on the technique of trace element accelerator mass spectrometry (TEAMS) and its applications in the analysis of stable isotopes in materials. This technique has been labeled by different groups as TEAMS (McDaniel et al., 1992; Datar et al., 1996, 1997a,b; Grabowski et al., 1997; McDaniel et al., 1998a,b), accelerator secondary-ion mass spectrometry (SIMS; Purser et al., 1979; Anthony et al., 1994a,b; Ender et al., 1997a,b,c,d), secondary-ion accelerator mass spectrometry (SIAMS;

Ender et al., 1997a,b; Massonet et al., 1997), super-SIMS (Anthony et al., 1986), atomic mass spectrometry (Anthony et al., 1990b; McDaniel et al., 1990a), AMS (Anthony et al., 1985; Wilson et al., 1997b), and microbeam AMS (Sie and Suter, 1994; Sie et al., 1997a,c; Niklaus et al., 1997; Sie et al., 1999).

Trace element AMS for stable isotopes is very similar to SIMS, which is one of the most sensitive techniques for determination of elemental composition in materials. As with SIMS, TEAMS is a destructive technique that may be used in a static mode for bulk measurements of elemental composition or may be used in a dynamic mode to profile elemental composition as a function of depth in the material. As with SIMS, TEAMS employs a positively charged, primary-ion beam at energies of 3 to 30 keV to sputter secondary ions, atoms, and molecules from the surface or near surface of a material. Of course, most of the particles sputtered from the surface are electrically neutral and cannot be analyzed at all unless one uses a laser or some other means to ionize the neutrals, e.g., sputter-initiated resonance ionization spectrometry (SIRIS). With SIMS, either positive or negative secondary ions may be extracted from the surface through a few kilovolts and then magnetically analyzed to determine composition. This is an advantage because some atoms more readily form positive ions while others more readily form negative ions (Wilson et al., 1989). A disadvantage of SIMS is that both atomic and molecular ions are produced and, in many cases, the molecular mass is nearly the same as the atomic mass of interest. For example, $^{30}SiH$ has the same nominal mass as $^{31}P$. Operating SIMS in a high-mass-resolution mode can reduce the molecular interferences by reducing the slit openings into the magnetic spectrometer, but this is at the cost of reduced signal strength and therefore reduced sensitivity (Wilson et al., 1989).

Trace element AMS employs a tandem electrostatic accelerator with the center terminal at a positive potential. The secondary ions from the sample, which are injected into the tandem accelerator, must be negatively charged to be accelerated to the terminal. At the terminal, the atomic and molecular secondary ions from the sample are stripped of a number of electrons in a gas or foil stripper. The electron stripping causes molecular break-up through a Coulomb explosion into elemental ions, which are then accelerated to higher energy and exit the accelerator. If the diatomic or triatomic molecular ion is stripped to a $3^+$ or greater charge state, almost all molecules break up into elemental constituents (Weathers et al., 1991). With TEAMS, the resulting atomic ions, which are in a number of different charge states with correspondingly different energies, are extracted from the high-energy end of the tandem accelerator. The atomic ions are then magnetically analyzed for momentum/charge ($mv/q$) and electrostatically analyzed for energy/charge ($E/q$), which together give mass/charge ($m/q$) discrimination. Here, $m$ is the mass, $v$ is the velocity, $E$ is the kinetic energy, and $q$ is the charge of the ion. To eliminate atomic interferences from possible break-up fragments that have the same $m/q$ (e.g., $^{28}Si^{2+}$ from $Si_2$ break-up vs. $^{56}Fe^{4+}$), the total energy of the ion is measured in a surface barrier detector or ionization chamber.

Sensitivities for TEAMS measurements vary for different elements. These sensitivities depend on (1) the ability to form a negative atomic ion or a negatively charged molecule that can be accelerated to the terminal and then broken apart; (2) the charge state used for the measurements since the charge state ($\geq 3^+$) must be chosen to remove molecular interferences; and (3) transmission of the ion from the sample through the accelerator and beamlines to the detector. By calibrating the secondary-ion yield against reference standards and by measuring the depth of the sputtered crater, a quantified depth profile can be obtained. The sensitivity or detection limit depends on the amount or volume of material available to analyze. The sensitivity can be much better for bulk measurements, where the element of interest is uniformly dispersed throughout the material. Then, the sample can be run almost indefinitely until good statistics have been obtained. Typically, sensitivities of 0.1 part per billion (ppb) can be obtained in the depth-profiling mode (McDaniel et al., 1998a,b; Datar et al., 2000) and 5 parts per trillion (ppt) in the bulk analysis mode. Theoretical sensitivities of 1 ppt are expected under ideal conditions. Because sensitivities are so high, it is sometimes difficult to obtain suitable reference standards.

### Complementary, Competitive, and Alternative Methods

There are a number of complementary trace analysis techniques of lower sensitivity and lower cost than TEAMS. The logical choice for TEAMS measurements is as a complement to other measurements provided by photon, electron, proton, or heavier ions. Depending upon the concentrations of the trace elements of interest in the sample, one may want to use one of the lower cost methods which can have sensitivities from major-element down to sub-ppb. Since TEAMS can be used for both bulk and depth-profiling measurements, some of the techniques discussed below are primarily for bulk analysis [inductively coupled plasma mass spectrometry and neutron activation analysis (ICPMS, NAA)], and others may be used for depth profile analysis (SIMS, SIRIS). Other competitive techniques that involve high-energy ion beam analysis (IBA) are described in the section introduction (HIGH-ENERGY ION BEAM ANALYSIS).

### Secondary-Ion Mass Spectrometry.

A complementary, alternative method to TEAMS, and one of the most powerful techniques for the characterization of materials, is SIMS (Benninghoven et al., 1987; Wilson et al., 1989; Lu, 1997; Gillen et al., 1998). In fact, the front end of TEAMS is like a SIMS system. In the next few paragraphs, the operating principles and analytical capabilities of SIMS are reviewed both as an introduction to TEAMS and as a competitive and complementary technique. An excellent SIMS tutorial is provided at the web site of Charles Evans and Associates: http://www.cea.com/tutorial.htm.

With SIMS, a positive primary-ion beam, which can be $O_2^+$, $O^+$, $Cs^+$, $Ar^+$, $Xe^+$, $Ga^+$, or any number of other species, typically 3 to 30 keV in energy, sputters secondary ions from the surface of a material. The bombarding primary ion produces atomic and molecular secondary ions and neutrals of the sample material, resputtered primary ions, and electrons and photons. Typically, $O_2^+$ is used if the secondary ions are more likely to be electropositive, and $Cs^+$ is typically used if the secondary ions are more likely to be electronegative. The choice of sputter ion can cause a change in secondary-ion yield of several orders of magnitude. Some elements do not form negative ions (e.g., N, Mg, Mn, Zn, Cd, Hg, and the noble gases), and many elements form negative ions very weakly. Of course, most of the particles sputtered from the surface are electrically neutral and cannot be analyzed at all unless one uses a laser or some other means to ionize the neutrals. Sputter-initiated resonance ionization spectrometry is such a method. The small fraction that are ionized depend strongly upon the matrix from which they are sputtered. The sputtering process is not just with the surface layer but consists of the implantation of the primary species into the sample at depths of 1 to 10 nm and the removal of the atoms or molecules from the surface by the energy loss of the primary ion through cascade collisions. Typical sputter rates in SIMS vary between 0.5 and 5 nm/s and depend upon beam intensity, sample material, and crystal orientation. The secondary ions, which are extracted through a potential of a few kilovolts and mass analyzed, are representative of the composition of the sample being analyzed. The sputter yield is the ratio of the number of atoms sputtered to the number of primary ions impinging on the sample and can range from 5 to 15. An advantage of SIMS is that secondary-ion yields can be measured over a dynamic range from matrix atoms to ppb.

The SIMS detection limits for most trace element are $1 \times 10^{12}$ to $1 \times 10^{16}$ atoms/cm$^3$. Factors that affect sensitivity and detection limits are ionization efficiencies or relative sensitivity factors (RSFs; Wilson et al., 1989); dark currents from electron multiplier detectors, which limits the minimum count rate; vacuum system elements and elements from SIMS components, which produce background counts; and mass interferences. Because both atomic and molecular ions are sputtered from the surface and analyzed with SIMS, the detection sensitivity for some elements is limited by molecular interferences. For example, in Si, $^{56}$Fe is masked by $(^{28}\text{Si})_2$; $^{31}$P is masked by $^{30}$SiH, $^{29}$SiH$_2$, and $^{28}$SiH$_3$; and $^{32}$S is masked by $(^{16}\text{O})_2$, among others. However, the molecular species do allow chemical information to be extracted from a sample in some cases.

The SIMS instruments may use magnetic sector, radio frequency (RF) quadrupole, or time-of-flight (TOF) mass analyzers. The secondary ions must also be energy analyzed. The magnetic sector mass analyzer separates the secondary ions by their momentum/charge ($mv/q$). Because of the high extraction voltage used for secondary ions, magnetic sector analyzers have high mass transmission. Magnetic sector analyzers may be used in a high-mass-resolution mode to separate molecular masses that interfere with an atomic mass of interest. This is usually accomplished at the expense of signal strength and therefore reduced sensitivity. RF quadrupole mass analyzers are smaller, more compact, and easier to use. A mass spectrum is produced by applying RF and direct current electric fields to a quadrupole rod assembly. Quadrupole SIMS instruments have loss of transmission and mass resolution with increasing mass and also cannot operate in a high-mass-resolution mode. The TOF mass analyzers, which are used with pulsed secondary-ion beams, have high transmission and mass resolution. TOF-SIMS provides excellent surface sensitivity, and many commercial instruments couple this with submicrometer primary optics for analysis.

The SIMS data may be taken in either a bulk analysis or a depth-profiling mode. Bulk analysis allows scanning of the mass spectrometer while a sample is being sputtered away. Bulk analysis may also provide greater sensitivity since a larger volume of material is available for analysis. Bulk measurements are mainly useful for uniformly doped impurities or for surface analysis. Depth profiling allows the concentration of impurities to be determined as a function of depth beneath the surface. One or more masses are monitored as the mass spectrometer is sequentially switched. For the mass of interest, the detector signal comes from increasingly greater depths in the sample. Accurate depth profiles require uniform bombardment of the sample analysis area (i.e., a flat bottom crater) with no contribution from the crater walls, adjacent sample surfaces, or nearby instrument surfaces. Usually the primary-ion beam, which may range in size from 1 μm to hundreds of micrometers, is rastered over a region of the sample. Only data from the center part of the scan are retained for analysis to reduce contributions from crater walls.

Secondary-ion yields for elements can vary by many orders of magnitude for a given material or matrix and even from matrix to matrix. It is therefore difficult to quantify SIMS results without using calibration standards. Wilson et al. (1989) have provided tables of RSFs for many elements for both $O_2^+$ and $Cs^+$ bombardment of different substrate materials, which help quantify SIMS data.

**Sputter-Initiated Resonance Ionization Spectrometry.** Sputter-initiated resonance ionization spectrometry is a relatively young and powerful method for selectively ionizing and analyzing materials (Winograd et al., 1982; Pellin et al., 1987, 1990; Bigelow et al., 1999, 2001). As with TEAMS and SIMS, SIRIS also removes atoms and molecules from a sample by sputtering with ions at kiloelectron-volt energies. Again, the majority of the sputtered atoms and molecules are neutral and only a small fraction is ionized. This fraction depends strongly upon the matrix of the material. TEAMS and SIMS mass analyze this ionized fraction. SIRIS uses lasers to ionize the much larger sputtered neutral fraction for a particular species with essentially 100% efficiency and then detects the postionized atoms. With SIRIS, two or more lasers are used to first resonantly excite an atom and then ionize it from the excited state. By suitably tuning the lasers, only one atomic species is ionized at a time. Therefore, even without subsequent mass analysis, SIRIS is highly selective and efficient. The yield ratio of atoms detected to atoms sputtered can approach 0.1, which may be several orders of magnitude better than SIMS or TEAMS. The SIRIS technique has been demonstrated to have bulk detection limits as low as 100 ppt (Pellin et al., 1990). Because ≅80% of the

sputtered atoms originate from the surface monolayer of the sample and ≅95% from the top two surface monolayers (Hubbard et al., 1989), SIRIS has excellent surface sensitivity. Because of the high useful yield, measurements may be made while sputtering sub-monolayer target thicknesses, making SIRIS essentially nondestructive. However, SIRIS may also be used to measure atoms below the surface if the primary-ion beam is allowed to sputter the sample for a longer time. One limitation of SIRIS is that it is very element specific, and multielement work may require several laser schemes.

**Neutron Activation Analysis.** Other mature methods that are readily available and provide rapid and economical, multielement, ppb-level detection of trace elements include NAA (Kruger, 1971). While NAA has excellent sensitivity in some cases (e.g., $^{197}Au$ in Si), the sensitivity of NAA is limited by the neutron capture cross sections, which vary from element to element. In addition, NAA may be primarily useful for Si since irradiation of other compound semiconductor materials like GaAs and HgCdTe will produce higher background radiation from the additional substrate elements.

**Inductively Coupled Plasma Mass Spectrometry.** Inductively coupled plasma mass spectrometry is a sensitive bulk materials analysis technique ($10^{14}$ atoms/cm$^3$ or 10 ppb; Houk, 1986; Baumann and Pavel, 1989). Essentially the entire sample is vaporized, and the resulting plasma is analyzed by mass spectrometry. However, there may be a concern about incomplete elemental dissolution and contamination during sample preparation.

**Neutron Activation–Accelerator Mass Spectrometry.** Neutron activation (NA) followed by AMS has been used to study light impurities in Si (Elmore et al., 1989; Gove et al., 1990). Neutron activation is used to produce long-lived radioisotopes, which generally have lower concentrations in the environment. After activation, AMS can provide more sensitive background-free measurements of the neutron activation products.

For example, neutron activation to convert stable $^{35}Cl$ to the long-lived radioisotope $^{36}Cl$ followed by conventional AMS has been used to study both the diffusion of Cl in Si (Datar et al., 1995) and the impurity Cl concentration in Si. This gets around the problem of having to ensure source cleanliness since the ambient $^{36}Cl$ has an abundance ratio of $10^{-15}$ vis-à-vis stable Cl. This was especially useful in the Rochester measurements (Datar et al., 1995) since the ion source used for the measurements was heavily contaminated with stable Cl. This technique can be used in principle for any isotope that produces a suitable, relatively long lived isotope by thermal neutron activation. The sensitivity of the measurement can be increased by increasing the total thermal neutron flux incident on the samples.

**Choosing the Proper Technique.** From the above discussion, the basic criteria for choosing TEAMS over other techniques should be based upon the need to determine elemental concentrations in a sample for those elements that need high sensitivity or have a molecular interference. Because of the cost and difficulty of a TEAMS measurement, TEAMS should not be used for all elements if the element can be analyzed by an alternative and less expensive method such as Rutherford backscattering spectrometry (RBS; see ELASTIC ION SCATTERING FOR COMPOSITION ANALYSIS), particle-induced X-ray emission (PIXE; see PARTICLE-INDUCED X-RAY EMISSION), or nuclear reaction analysis (NRA; see NUCLEAR REACTION ANALYSIS AND PROTON-INDUCED GAMMA RAY EMISSION). In some cases, chemical information may be needed for a sample, and SIMS would be more useful. For some elements, molecular interferences are not a problem, and SIMS would be easier and less expensive. In addition, for low-mass-resolution applications, SIMS may give lower detection limits due to higher ionization rates with $O_2^+$ primary-ion bombardment.

## PRINCIPLES OF THE METHOD

### The Trace Element Accelerator Mass Spectrometry Facility

The TEAMS facility consists of three main components: (1) the ultraclean ion source for the generation of negative secondary ions from a material by primary-ion sputtering, much like in SIMS; (2) the tandem accelerator, which accelerates the secondary ions to mega-electron-volt energies and strips electrons to remove molecular interferences; and (3) the detector system, which includes the high-energy beam transport line with magnetic and electrostatic analysis, and the particle total energy detection system. Different laboratories around the world are employing a number of different approaches to TEAMS, and attempts will be made to contrast these different approaches in this unit. However, the following description of TEAMS will be primarily the University of North Texas (UNT) facility.

### Ultraclean Ion Source for Negatively Charged Secondary-Ion Generation

The ion source is designed to produce secondary ions from the sample while minimizing contaminant ions produced from the ion source hardware and from previous samples. The ion source is discussed in detail under Practical Aspects of the Method.

### Secondary-Ion Acceleration and Electron-Stripping System

With the sample biased at $-13$ kV, the negatively charged secondary ions are extracted from the chamber at the energy of 13 keV and passed through a 45° electrostatic analyzer (ESA). The ESA discriminates against energy/charge ($E/q$) with $E/\Delta E = 75$ for a 2-mm slit width. The ESA passes all masses originating from the sample surface that have the correct extraction energy. Secondary ions originating from other surfaces in the ion source chamber are prevented from being passed down the extraction beamline and into the tandem accelerator. Since the negative ion yield is at most 1% to 2%, a typical secondary-ion beam after the ESA due to 900 nA of 23-keV $Cs^+$ (10 keV from Cs oven plus 13 keV from negative sample bias) is

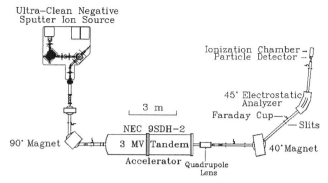

**Figure 1.** The TEAMS facility in the Ion Beam Modification and Analysis Laboratory at the University of North Texas.

$\cong$50 to 60 nA of $^{28}Si$ through a 2-mm slit opening. After the ESA, the negative ion beam is passed through a 90° dipole magnet and momentum/charge analyzed ($mv/q$) with $m/\Delta m = 500$ for a 2-mm slit width. A mass scan after the 90° magnet would be equivalent to a SIMS analysis and would include molecular as well as atomic negative ions. After the 90° magnet, the $^{28}Si$ secondary-ion beam is $\cong$12 to 15 nA. Figure 1 shows the TEAMS system at UNT.

The 13-keV negative ions are injected into the tandem accelerator and accelerated to the terminal, which typically is operated at 1.7 MV for TEAMS. For an ion with charge $1^-$, injection voltage $U = 13$ kV, and terminal voltage $V = 1.7$ MV, the ion energy is given by

$$E = e(U + V) \tag{1}$$

The higher energy negative ions (1.713 MeV) are then efficiently stripped of one or more electrons in the terminal by an $N_2$ gas or carbon foil stripper. The gas or foil stripper is sufficiently thick to produce atomic and molecular ions in a distribution of positive charge states $q$ (Betz, 1972; Wiebert et al., 1994). All of the charge states produced will be accelerated again from the terminal down the high-energy side of the tandem accelerator, producing a number of different energy ions. The $1^+$ ions will gain another 1.7 MeV of energy; the $2^+$ ions will gain another 3.4 MeV of energy, and so on for each subsequent charge state $q$. The mass $m$ of the ion exiting the accelerator may be less than the mass $m_0$ of the ion injected (if dissociation occurs in the terminal) or equal to the injected mass if no dissociation occurs. The total energy of the ion of mass $m$ after the accelerator is given by

$$E = e(m/m_0)(U + V) + eqV \tag{2}$$

where the first term is the residual energy from the dissociated ion in the terminal and the second term is the post-stripping energy. This energy neglects the energy loss of the ion in the stripper gas, which can vary from 0.1 keV for H to 7.0 keV for Si and must be taken into account in subsequent analysis (Arrale et al., 1991). The $N_2$ gas pressure may be used to optimize the production of the charge state of interest. For 1.7 MV terminal voltage, the most populated charge state is $3^+$ for masses in the range from 20 to 60 amu. The charge state chosen after acceleration and for further analysis depends upon possible interferences with other $m/q$ ions, mainly from matrix elements in the sample and upon the charge state needed for molecular break-up. Molecules in a $1^+$ or $2^+$ charge state will still exist and will be accelerated to the higher energies. There are several molecules that are known to be at least metastable in a $3^+$ charge state. While most of these molecular ions are larger clusters or organic molecules (Burdick et al., 1986; Saunders, 1989), a few are relatively small trimers $CS_2$ and $CSe_2$ and the dimer $S_2$ (Morvay and Cornides, 1984). The small dimers $B_2$ and AlO have been observed to remain in a $3^+$ metastable state for many microseconds, enough time to travel all the way from the terminal to the detector (Weathers et al., 1991; Kim et al., 1995). Normally, ions in a $3^+$ charge state are chosen to ensure that there are no molecular interferences. Occasionally, for some elements, a charge state of $4^+$ or $5^+$ might be more desirable. The rule of thumb is that $m/q$ should be a noninteger.

For elements that do not form negative ions or do so very poorly, the sensitivity for atomic detection would be very low or zero. However, for these cases, a negative molecular ion may be injected into the tandem accelerator of the form $(MX)^{1-}$, where M is the matrix atom or some other atom present in the sample (e.g., Si) and X is the impurity atom of interest. After accelerating the molecule to the terminal and breaking it apart by electron stripping, the atomic fragment of interest, $X^{q+}$, may be analyzed, where $q$ is the charge state of the ion. Therefore, essentially, any element in the periodic table may be analyzed by the TEAMS technique. Indeed, the sensitivity for detection of each element depends upon a number of things, such as negative ion formation probability, matrix composition, sample contamination, surface oxidation or covering, charge state chosen, and transmission through the accelerator.

### High-Energy Beam Transport, Analysis, and Detection System

After acceleration to mega-electron-volt energies and break-up of molecular interferences, the total ion beam in all charge states is 20 to 30 nA of $^{28}Si$. The atomic ions in the chosen charge state are selected by magnetic analysis at $40°$ and by electrostatic analysis at $45°$, as seen in Figure 1. After the magnetic and electrostatic analysis, the $^{28}Si^{3+}$ beam is $\cong 6$ nA and $\cong 5$ nA, respectively. The magnetic rigidity is given by $Br_b = mv/q$ with $m/\Delta m \cong 300$ for a 2-mm slit width, and the electrostatic rigidity is given by $\mathcal{E}r_e = 2E/q$ with $E/\Delta E \cong 250$. Here, $B$ and $\mathcal{E}$ are the magnetic and electric field strengths, respectively, and $r_b$ and $r_e$ are the radii of curvature in the magnetic and electric fields, respectively. Combining these two analyses gives $m/q$. To eliminate atomic interferences from possible break-up fragments that have the same $m/q$ (e.g., $^{28}Si^{2+}$ from the break-up of $Si_2$ and $^{56}Fe^{4+}$), the total energy of the ion is measured in a surface barrier detector or ionization chamber.

For bulk measurements of uniformly distributed impurities or dopants, the beam spot may be spread out to $\sim 2$ mm in diameter to increase count rate and the sample

may be counted for longer time periods. For depth-profiling measurements, only data collected from the center $\cong 12\%$ of the rastered area (plus the width of the sputter beam) are gated into the pulse height analyzer. Data from the remaining $\cong 88\%$ of the rastered area, which includes the secondary ions sputtered from the crater wall, are discarded. It is still possible and even likely that some of the material comprising the $\cong 88\%$ could be resputtered back onto the sample center $\cong 12\%$ region, producing a contamination of the sample and reduced sensitivity.

The TEAMS system is automated and under computer control. The software to operate and control the TEAMS system is a combination of in-house and LabView programs at UNT. The software allows magnetic and electrostatic analyzers to scan $mv/q$ and $E/q$, respectively, to a precision of 1:10,000 and to make either bulk or depth-profiling measurements. The magnetic and electrostatic analyzers are calibrated by analyzing known masses, usually from different matrix elements or molecules. Both atomic and molecular ions and atomic fragments are typically used for calibrating a wide range of magnetic ($mv/q$) and electric ($E/q$) rigidities. The software also operates Faraday cups, viewers, detectors, quadrupoles, steerers, and beam attenuators. Transmission through the entire TEAMS system is $\cong 1\%$ to 8%.

### Future Possible Advances in TEAMS

Future possible advances in TEAMS include the development of a tandem accelerator specially designed to directly couple to a SIMS source (with suitably modified injection optics) and the possible development of an efficient charge exchange canal to enable conversion of positive ions to negative ions. The Naval Research Laboratory (NRL) group (Grabowski et al., 1997) is constructing a TEAMS system by attaching a SIMS instrument to a tandem accelerator. Future improvements would include a submicrometer primary-ion beam and a means for secondary electron imaging of features on a sample. The Australia group (Sie et al., 1998) and the Zurich group (Ender et al., 1997c) have already developed microbeam TEAMS systems. Also on the horizon are other techniques to increase ionization probabilities such as laser postionization of neutrals sputtered from a sample (e.g., SIRIS; Winograd et al., 1982; Pellin et al., 1987, 1990; Bigelow et al., 1999, 2001). For some applications, the truly atomic aspects of TEAMS may be used to advantage, for example, in profiles through layered films.

### PRACTICAL ASPECTS OF THE METHOD

The uniqueness of TEAMS is that it combines the two techniques of AMS and SIMS and can thereby be used to measure very low levels of stable isotopes of almost any element in a variety of different matrices corresponding to different materials (Rucklidge et al., 1981, 1982; Wilson et al., 1991, 1997a,b; Ender et al., 1997a-c; Massonet et al., 1997; McDaniel et al., 1998a,b; Datar et al., 2000). TEAMS has the capability to achieve ppt detection limits for some elements in Si, which is considerably better than SIMS. TEAMS analysis is similar to SIMS. Perhaps, the largest benefit of TEAMS results from the passage of the secondary ions through a tandem accelerator where electrons are stripped from the molecular ion, resulting in a Coulomb explosion of the molecule into its elemental components. With TEAMS, after molecular break-up, magnetic ($mv/q$), electrostatic ($E/q$), and total energy ($E$) analyses are performed that uniquely identify the elemental ion. The mega-electron-volt energies of the secondary ions allow nuclear physics particle detectors to be used with essentially 100% detection efficiency because of the reduction in detector noise and scattered particles. The sensitivity of TEAMS is reduced because only one charge state ion is analyzed from the distribution of charge states produced during the electron-stripping process in the tandem accelerator. Since the most dominant charge state is usually chosen for analysis, the reduction in sensitivity is usually by only a factor of 3. The current depth-profiling detection limits at UNT of $\cong 0.1$ ppb atomic are primarily due to ion source contamination and suboptimal secondary-ion transmission through the accelerator and analysis system (McDaniel et al., 1998a,b). Removal of source contamination and improvements in system transmission should allow ppt-level detection of stable elements in the depth-profiling mode. Bulk detection limits are lower because more material is available to analyze just by counting longer. Other TEAMS facilities around the world have similar detection limits (see section on existing TEAMS facilities in operation or under construction below).

The TEAMS technique can be used to easily discriminate against molecular interferences. However, it cannot easily discriminate against isobaric interferences, since at relatively low energies the ionization chamber detector cannot provide significant $dE/dx$ information. Thus, in the case of more than one isotope with possible isobaric interferences, the isotope without any interferences should be chosen for analysis.

## Ultraclean Ion Source Design Details

The ion source is designed to reduce sample contamination from elements in the ion source hardware and from chemically similar elements in the Cs and cross-contamination from other or previously analyzed samples (commonly called source memory effects). To reduce sample contamination, ion source components exposed to the primary-ion beam may be coated with ultraclean Si or some other pure material. The ion optics should be designed to reduce scattering onto the sample (Kirchhoff et al., 1994). The primary-ion beam is Cs$^+$, which readily provides electrons for the production of negatively charged atomic and molecular ions. The Cs$^+$ may be produced by a surface ionization ion source with an energy spread of approximately $kT = 0.1$ eV, where $k = 1.38 \times 10^{-23}$ JK$^{-1}$ is the Boltzmann constant, and $T$ is the temperature in K. For nominal energies of 10 keV incident upon the sample, the fractional energy spread is approximately $\Delta E/E = 10^{-5}$. The sample is held at a negative bias of 11 to 30 kV to allow the secondary ion produced to have enough energy to be injected into the accelerator and transmitted through the entire

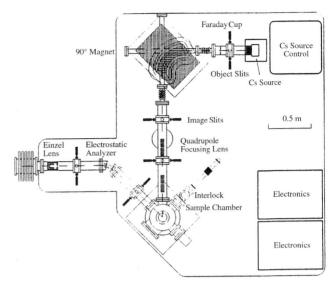

**Figure 2.** The TEAMS ultraclean negative sputter ion source. Shown are the Cs oven, the 90° magnet for preanalysis of the Cs sputter beam, sample chamber with sample load interlock, and 45° electrostatic analyzer in the secondary-ion beamline. Not shown are the eight Einzel focusing lenses, the six octupole steerers, and Cs beam scanner, which are electrostatic and therefore mass independent. Also, not shown is the aperture plate with laser drilled holes from 2 mm down to 0.2 mm, which allows the beam object size to be varied for microbeam production. Because of the many components for focusing and steering the primary Cs beam and the secondary ions from the sample, the entire source is under computer control at the source and in the control room.

TEAMS system. The lower sample bias allows better depth resolution in the sample. For the sample biased at $-13$ kV, the total primary-ion impact energy is 23 keV. While the Cs$^+$ ion beam may be very monoenergetic, it is not very pure. The primary Cs$^+$ ion beam should also be magnetically preanalyzed ($mv/q$) before striking the sample to reduce contamination of the sample (Kirchhoff et al., 1994). The ion source is normally capable of depth profiling with a microbeam primary ion as well as bulk impurity measurements in materials. For depth profiling, the last optical element before the sample is usually used to raster the primary-ion microbeam over a selected region of the sample. Secondary-ion data are then accumulated only from the center (12% to 20%) region of the crater sputtered by the primary ion to reduce wall contributions. It is necessary to have a microscope in the sample chamber to observe and monitor the beam spot on the sample.

Figure 2 shows the main components of the ultraclean ion source developed at the Ion Beam Modification and Analysis Laboratory (IBMAL) at the UNT (Kirchhoff et al., 1994). The intensity of the ion source is determined by the brightness of the primary Cs source, which is a UNT-modified, model 133 Cs source available from HVEC-Europa. The modifications are in the extraction and focusing optics and steering fields and have resulted in an increase in source brightness. Using computer control and optimization algorithms, the output parameters of the source can be rapidly optimized. The maximum half-angle of divergence of the Cs beam from the source was measured to be ~9 mrad. The laboratory emittance of the source was determined to be $\cong 7 \times 10^{-4}$ π-cm-mrad. For the 10-keV Cs ions, this corresponds to a normalized emittance of $\cong 7 \times 10^{-5}$ π-cm-mrad (MeV)$^{1/2}$. This parameter indicates that, in the absence of aberrations, the theoretical spot size, without significant current loss, is $\cong 40$ μm in diameter. In reality, the smallest spot size attained thus far at UNT is $\cong 60$ μm in diameter. The use of reducing apertures can further reduce the spot size at the expense of intensity. This beam spot is continuously variable up to $\cong 2$ mm in diameter for use in bulk analysis. Sie et al. (1998) and Ender et al. (1997c) have developed primary-ion microbeams that are on the order of 1 μm to a few micrometers in diameter.

The 10-keV Cs$^+$ ion beam is magnetically analyzed with a double-focusing 90° electromagnet to remove other alkali metals and Mo and Fe. These elements are produced from the components in the Cs and the ion gun. The ion source magnetic spectrometer helps to remove ambiguity and chance of contamination. Therefore, the Cs beam striking the sample is $^{133}$Cs. Contamination has been shown to be well below 1 ppt after magnetic filtering.

Following the magnetic spectrometer, the primary ions enter an electrostatic zoom lens column that forms an intense microprobe. For depth-profiling measurements, the Cs$^+$ is scanned across the sample in a spiral digital meander pattern. The sample region scanned is ~1 mm × 1 mm in area and is divided into a 128 × 128 pixel integrated image, which is regenerated for each frame. The scanning system is digitally controlled from a digital input-output (I/O) card resident in a personal computer. Because of the number of active focusing and steering elements in the primary-ion beamline and the secondary-ion beamline, the ion source operation components are under computer control.

A microbeam may be formed from the momentum-analyzed $^{133}$Cs ion beam with a combination of two quadrupole/octupole quadruplets. The first quadrupole is presently designed to produce a stigmatic image with a magnification $M_{Q1} = -1/50$ or less. This demagnification is followed by a long-working-distance quadrupole/octupole quadruplet objective to produce an elliptical spot at near-unity magnification in the vertical direction and a magnification of ~0.707 in the horizontal direction. Therefore, when the beam impinges on the sample surface at near 45°, the spot will appear almost circular. In practice, the strong horizontal electric field between the extraction plate and the sample bends the primary beam so that the angle of impact is $\cong 25°$ for $\cong -20$-keV sample bias. The final optical element in the primary column of the ion source is a two-stage, eight-electrode dipole element that produces a translation of the beam of ±4 mm without introducing any variation in the angle of impact upon the sample. This feature is important when sputtering a uniformly deep crater for depth profiling. The dipole element scans the ion beam in a raster pattern. By correlating electronically the position of the primary-ion beam spot and the detection of the secondary ion at the other end of the apparatus, one can build an image of impurities distributed across the surface of the sample. The lateral resolution of the image, however, is limited by the beam spot size, which is presently $\cong 60$ μm at the UNT facility. By

acquiring the secondary ions versus time and knowing the rate at which the ion beam erodes the surface of the sample, one can build a profile of the impurity distribution versus depth. Therefore, in principle, a three-dimensional impurity map is possible. The limitation of available impurity ions in a sputtered volume will set a lower limit on the size of the beam spot for a given concentration of impurity atom. For example, for a typical Cs sputtering rate of ~2 Å/s ($2 \times 10^{-4}$ μm/s) and a 100 × 100 μm analyzed area, the sample volume of 2 μm$^3$ would contain $10^{11}$ Si atoms. One impurity atom detected in 1 s in this volume would correspond to a sensitivity of $10^{-11}$, or 10 ppt. Higher impurity concentrations would require a smaller sample volume while lower impurity concentrations would require a larger sample volume.

Because of the developments in SIMS instruments that allow for a primary-ion microbeam and secondary electron imaging of a sample, some laboratories are using existing SIMS instruments to inject into the tandem accelerator and analysis system for the TEAMS facility (e.g., Grabowski et al., 1997). To improve the depth resolution of the data, the primary-ion energy and the sample bias should be as low as possible, while the sample bias should still be high enough to ensure adequate transmission through the remainder of the accelerator and beam transport system.

Secondary electron images permit accurate and rapid positioning of the primary Cs$^+$ ion beam on the sample. The sample is also viewed by a long-working-distance microscope with a field of view equal to the sample size ($\cong 3$ mm) and a resolution of a few micrometers. A remote high-resolution TV monitor views the optical image. The position of the sample is adjusted by a four-axis manipulator ($x$-$y$-$z$-$\theta$) to allow three translations and one rotation about the secondary ion axis which is perpendicular to the sample surface.

Throughout the ion source, all first surfaces, which can be "seen" by the primary or secondary ions, are Si. The use of ultrapure and ultraclean Si on the first surfaces of the focusing and steering lenses, slits, Faraday cups, and apertures reduces the contamination of the sample from ion-wall scattering. These specialized components may be fabricated from relatively inexpensive materials (Kirchhoff et al., 1994). To further reduce sample contamination, the vacuum system is all metal sealed and has a base pressure, without baking, of $\cong 2 \times 10^{-9}$ torr. The sample holder, which will hold up to 19 individual samples 3 mm in diameter, is admitted and removed from the sample chamber via a sample interlock and manipulator. Samples are positioned on the sample holder and prepared for analysis in a class 100, portable clean room.

For samples that are insulating, a charge compensation system is required. This may be an electron gun that sprays electrons upon the sample to reduce sample charging in much the same way as with a standard SIMS instrument.

### Calibration of TEAMS Magnetic and Electrostatic Analyzers

With TEAMS, in most cases, one is attempting to measure a trace quantity of an impurity element in a matrix, as

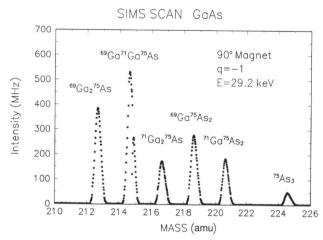

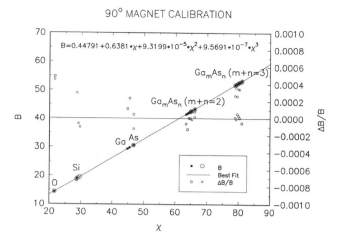

**Figure 3.** SIMS mass scan of the 90° preinjection magnet showing the characteristic signature spectrum of the trimer molecules formed in a charge state $q = 1^-$ from a GaAs matrix sample in the ion source. The peak heights are related to the natural abundances of the $^{69}$Ga (60.1%) and $^{71}$Ga (39.9%) isotopes. For each mass in the spectrum, a magnetic rigidity $Br_b$ is determined, which is used to build a calibration file for the mass 212- to 225-amu region. Atomic and molecular ions of different masses are used for other mass regions.

**Figure 4.** Graph of the 90° preinjection magnet calibration file showing magnetic field strength $B$ versus $\chi_B = Br_b = mv/q$, the magnetic rigidity, which is dependent on the $m$, $E$, and $q$ of the analyzed ion. The calibration file was determined from spectra like the one shown in Figure 3. The right-hand axis is a measure of the goodness of fit, the residual $\Delta B/B$, and is less than 0.0006 for all $\chi$. The calculated fit, which is given at the top of the graph, is interpolated to find magnetic field strengths for other masses.

either a bulk elemental mass scan or an elemental depth profile of one or two elements. The small concentration of the impurity element requires that the TEAMS magnetic and electric fields used for analysis be precisely calibrated. The TEAMS system is calibrated by measuring mass scans of elements of known concentrations, i.e., the elements of the matrix for the sample being measured. Figure 3 shows a characteristic signature spectrum of the trimers (e.g., $(^{69}$Ga$^{71}$Ga$^{75}$As$)^{1-}$) that occurs with the 90° preinjection magnet mass scan of a GaAs sample. The calibrations are run sequentially through the 90° preinjection magnet, the 40° postacceleration magnet, and the 45° ESA for a fixed terminal voltage. A calibration file is constructed for each magnet, and the ESA from the isotopic mass and the magnetic ($\chi_B = Br_b = mv/q$) or electrostatic ($\chi_{\mathcal{E}} = \mathcal{E}r_e = 2E/q$) rigidity. Atomic and molecular ions as well as dissociated ions may be used for calibration. After the calibration files have been constructed, cubic polynomial fits to the data for each magnet were found to be necessary while a straight-line fit was found to be sufficient for the ESA. Figure 4 shows a representative calibration for the 90° preinjection magnet. The 40° postacceleration magnet (HVEC) is similarly calibrated. The electrostatic voltage for the ESA is plotted versus the electrostatic rigidity, $\chi_{\mathcal{E}} = \mathcal{E}r_e = 2E/q$, for different known masses. The calculated fits are interpolated to set magnetic and electric field strengths for future mass analyses. The secondary ions from the sample are analyzed through the entire spectrometer under computer control. A calibration precision of 0.001 in $\Delta B/B$ and 0.002 in $\Delta V/V$ is required for preservation of isotopic ratios, which is the true test of an accurate calibration of the system. For example, for elements with more than one stable isotope, the stable isotopes should appear in the correct proportions in a mass scan, as shown in Figure 3.

### Existing TEAMS Facilities in Operation or Under Construction

The characteristics of a number of TEAMS facilities are discussed below. The information was provided to the author by personnel from the different TEAMS laboratories. (See the Internet Resources for contact information.)

**Paul Scherrer Institute (PSI)/ETH Zurich Accelerator SIMS Laboratory.** The PSI/ETH tandem accelerator facility serves as a national and international center for AMS. This very sensitive method is currently the leading technique for the detection of long-lived radionuclides such as $^{10}$Be, $^{14}$C, $^{26}$Al, $^{36}$Cl, and $^{129}$I at isotopic ratios between $10^{-10}$ and $10^{-15}$. These isotopes are intensively studied as natural tracers in earth sciences and in climate history research. In addition to radionuclide AMS, the PSI/ETH Zurich laboratory is actively involved in TEAMS research.

The TEAMS facility in Zurich includes a sputter chamber with an ideal extraction of secondary ions into the already existing beamline of the AMS facility in Zurich. At the same time, the chamber design keeps the contamination due to sputtering of secondary ions in the vicinity of the sample as low as possible. This was achieved with the extraction geometry shown in Figure 5.

The finely focused Cs$^+$ primary-ion beam is produced with a commercial Cs431 SIMS ion source from ATOMIKA in Munich, Germany (Wittmaack, 1995). It can produce beam currents from 0.2 to 800 nA with corresponding beam spot sizes of 1 to 150 μm. The Cs beam is analyzed with an $E \times B$ filter and sent through a 1° electrostatic deflector to remove impurities and neutral components and to keep contamination from the primary beam as low as possible. Also, the primary beam can be scanned over the sample to perform imaging of trace elements on the sample surface.

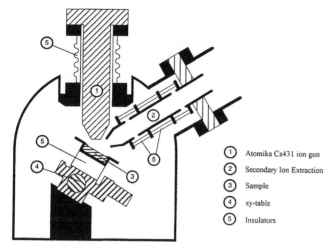

**Figure 5.** Schematic cross section through the TEAMS-sputtering chamber at the AMS facility in Zurich.

① Atomika Cs431 ion gun
② Secondary Ion Extraction
③ Sample
④ xy-table
⑤ Insulators

Using specific extraction voltages on the electrodes in the vicinity of the sample, the positive secondary ions are kept on the sample as much as possible and negative ions are extracted as much as possible so that they cannot be accelerated toward electrodes and produce tertiary ions and neutrals that would contaminate the sample. However, since this process cannot be suppressed completely, all the electrodes of the source have been coated with a 20-μm-thick layer of gold so that contamination of the sample due to the sputtering processes is kept low for all elements except gold (Döbeli et al., 1994; Ender et al., 1997a–c).

The extracted negative ions are deflected into the main beamline of the AMS facility and analyzed as in routine measurements. The EN6-Van de Graaff tandem accelerator is usually run at a terminal voltage of 5 MV, and a two-electrode gas ionization detector is used for the separation of isobaric interferences (Ender, 1997).

The facility has been tested with respect to detection limits, imaging, and depth profiling on trace elements in Si wafers. Table 1 and Figure 6 show a summary of the results for the TEAMS facility in Zurich for the bulk detection of trace elements in Si (Ender et al., 1997d).

Ongoing projects on the TEAMS facility are the measurement of isotopic ratios of trace elements in sedimen-

**Table 1. Bulk Detection Limits for Trace Elements in Si Measured at PSI-ETH Zurich**[a]

| Element | Atomic Ratio | Atomic Density (atoms/cm$^3$) |
|---|---|---|
| B | $1 \times 10^{-10}$ | $5 \times 10^{12}$ |
| Al | $7 \times 10^{-11}$ | $3.5 \times 10^{12}$ |
| P | $2 \times 10^{-10}$ | $1 \times 10^{13}$ |
| Fe | $2 \times 10^{-9}$ | $1 \times 10^{14}$ |
| Ni | $2 \times 10^{-9}$ | $1 \times 10^{14}$ |
| Cu | $2 \times 10^{-9}$ | $1 \times 10^{14}$ |
| As | $9 \times 10^{-12}$ | $4.5 \times 10^{11}$ |
| Sb | $2 \times 10^{-10}$ | $1 \times 10^{13}$ |

[a] The bulk detection limits are given as atomic ratios of trace element to Si and as atomic densities.

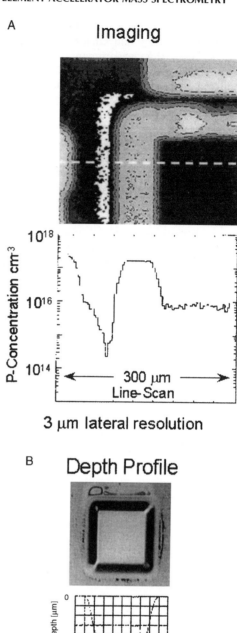

**A**  Imaging

3 μm lateral resolution

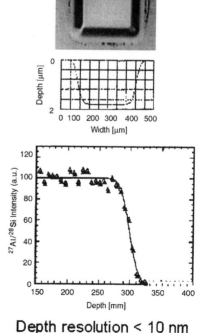

**B**  Depth Profile

Depth resolution < 10 nm

**Figure 6.** (**A**) Lateral resolution and (**B**) depth resolution of the TEAMS facility in Zurich.

tary layers and extraterrestrial matter with the goal to solve problems in geology and nuclear astrophysics. The main interest lies in the measurement of the heavy platinum group elements (PGEs). With the finely focused Cs beam, one can measure samples as small as ∼200 ng. This also might have potential for the analysis of radioisotopes (Maden et al., 1999, 2000).

**University of Toronto IsoTrace Laboratory.** The University of Toronto is well known as one of the first laboratories to do radionuclide AMS. The existing AMS facility has also been used for a number of years for *in situ* TEAMS measurements of Au, Ag, and PGEs in mineral grains and meteorites (Rucklidge et al., 1981, 1982; Wilson et al., 1990; Rucklidge et al., 1990a,b; Kilius et al., 1990; Wilson et al., 1995; Rucklidge et al., 1992; Wilson et al., 1997a–c; Wilson, 1998; Sharara et al., 1999; Litherland et al., 2000, 2001). The "heavy-element" program developed at IsoTrace had its beginnings in the fertile period of collaborative research at the University of Rochester, circa 1977 to 1981, which saw trial analyses of PGEs such as Ir and Pt in pressed, powdered targets of rock and mineral samples using a broad-beam Cs$^+$ source on the tandem accelerator. Further development at IsoTrace was slow but sped up in the late 1980s with the evolution of a multitarget sample-mount format that involved the drilling of small (4-mm-wide) cylindrical "minicores" of samples and reference materials that were then mounted in sets of 12 in a 25-mm circular holder and polished to a fine finish, permitting mineralogical or metallographic tests prior to analysis. It should be noted that the sample chamber and 25 × 25-mm micrometer-resolution stage came first, in 1984, to combine unrestrained three-dimensional mobility and adequate sample-viewing optics, essential features for analysis of targets such as rocks. Rocks are almost always heterogeneous mixtures of one or more mineral species of variable shapes and grain sizes. The initial work conducted on the new stage was again of the pressed-powder variety, but the move to solid targets promoted a spate of measurements and rapid progress from circa 1989 to 1997, as summarized below. The principal limitations of the technology (aside from questions of cost and availability) that have thus far restricted adoption of these methods are perceived to be (1) the need to use conducting targets, whereas common rock-forming minerals are mostly good insulators, and (2) the broad Cs$^+$ beam size (maximum range may be 250 to 1500 μm, but commonly ∼1000 μm), which limits application to relatively coarse grained samples. If any group succeeds in going beyond these constraints, which are common to $^{14}$C-oriented accelerator laboratories, the next question will probably return to quantitation and the development of reference materials. In the course of this work, IsoTrace has addressed this by fabricating "fire assay" beads based on dissolution of well-characterized samples into a conducting nickel sulfide matrix.

Possible applications of TEAMS are almost limitless if the questions of surface charge build-up and spatial resolution can be addressed at a facility with a flexible analytical repertoire. Analytical capability evolved at IsoTrace due to the efforts of the late Linas Kilius and the other

**Table 2. Standard Bulk Detection Levels (±30%) for Some Trace Elements for Analysis Times Less Than 200 s at the University of Toronto$^a$**

| Element | Bulk Detection Level (ppb) | Atoms/cm$^3$ |
|---------|---------------------------|--------------|
| Ru | 4 | $1.2 \times 10^{14}$ |
| Rh | 0.5 | $1.5 \times 10^{13}$ |
| Pd | 4 | $1.1 \times 10^{14}$ |
| Ag | 0.05 | $1.4 \times 10^{12}$ |
| Os | 20 | $3.2 \times 10^{14}$ |
| Ir | 0.25 | $3.9 \times 10^{12}$ |
| Pt | 0.1 | $1.5 \times 10^{12}$ |
| Au | 0.005 | $8 \times 10^{10}$ |

$^a$ If required, lower detection levels can be obtained with increased analysis time, limited by the availability of material.

laboratory staff but did not exist in the very first incarnation of the laboratory, a straightforward $^{14}$C system. Although test measurements were made on semiconductors, most trace element data at IsoTrace have involved conducting minerals and metals in rocks, meteorites, and archaeological artifacts. In addition, a large majority of the measurements have centered on just eight elements, the precious metals, namely Au and Ag, and the six PGEs (Pt, Ir, and Rh, which form negative ions readily, and Pd, Ru, and Os, for which sensitivity is significantly lower). Typical bulk detection levels achieved in short (e.g., 10 to 200 s) counting times are given in Table 2. Approximately 40 elements have been the subject of basic research at IsoTrace (see the short essay on heavy element research on the web page), but the precious metals have dominated the applications work.

It is important to note that in-house research has emphasized the importance of collecting TEAMS data as part of a broad spectrum of research on a topic. This ideally includes collection of samples in the field, classic petrographic (microscopic) studies, and sometimes whole-rock analyses of bulk samples plus collection of mineral-chemical data for major to trace elements using a full range of methods—some combination of electron, proton, and ion microprobe analyses as well as the broad-beam TEAMS.

The PGEs and Au typically behave as siderophile or chalcophile (iron- or sulfur-loving) metals, present at low-ppb levels or below in the earth's crust but at much higher levels in many meteorites and (by inference) in the cores of the differentiated terrestrial planets. A factor of roughly 1000 is necessary to enrich typical crystal rocks from ppb levels to the ppm grades, which constitute ore in most hard-rock gold and platinum mines today. The TEAMS of precious metals has largely been directed to the following materials: common Fe-Cu-Ni-As sulfides, Fe-Ti-Cr oxides, copper (both natural native copper and refined metal), and the common Fe-Ni-Co metal phases in iron meteorites and graphite. These materials have been analyzes in three broad contexts:

1. Distribution of the metals among coexisting ore minerals in various classes of mineral deposit. All the precious metals may occur both in concentrated form (rare minerals such as Au, PtAS$_2$, and PdSb)

and "invisibly" within the crystal structures of common ore minerals at levels from <1 ppb to 100 ppm or more—much more in the case of Ag.

2. Chemistry of iron meteorites, where the PGE and Au must occur dispersed among a generally small number of reduced phases: Ni-Fe alloys, phosphides, carbides, FeS, and graphite.

3. Provenance of archaeological metals. Thus, smelted Cu of European derivation has been shown by instrumental neutron activation analysis (INAA), AMS, and humble petrographic methods to be distinct from the purer native copper recovered for perhaps seven millennia by the inhabitants of the Great Lakes region. As with natural materials, samples must be characterized by context and microscopic properties prior to analyses, which may include $^{14}$C dating of suitable components.

In terms of trace element work, the most important research at IsoTrace since 1997 has been method and equipment development by graduate students (see web site). In the geological context, the work of Ilia Tomski and Jonathan Doupé promises to broaden the range of available analytical strategies, while Jenny Krestow's project addresses the fundamental problem of analyzing insulating targets, including common minerals such as olivine and quartz.

**Technical University Munich Secondary Ion AMS Facility.** The Munich Tandem Accelerator Laboratory at the Technical University Munich, Garching, Germany, has developed a radionuclide AMS facility and has recently developed an ultraclean ion source for stable isotope impurities (Massonet et al., 1997).

At the Munich accelerator laboratory an ultraclean injector has been built that is used for SIAMS. It utilizes a magnetically analyzed Cs sputter beam and a large-volume ultrahigh vacuum target chamber. In the first stage apertures, electrodes, and target shielding were made of copper; in the second stage these parts were gold plated; and in the third stage the target aperture and the target shielding were made of high-purity Si. In the third stage, the best results were obtained. See, e.g., Fe and Cu in Table 3. The other elements were measured with copper- or gold-plated parts. The work was the doctoral thesis of Stefan Massonet (Massonet, 1998). Currently the AMS system is being used for the measurement of depth profiles. Table 3 presents bulk detection limits for trace elements in Si.

**CSIRO Heavy Ion Analytical Facility (HIAF), Sydney, Australia.** Another specialized TEAMS system is the AUSTRALIS (AMS for Ultra Sensitive Trace Element and Isotopic Studies) system, a microbeam TEAMS system that utilizes a 3-MV Tandetron at CSIRO HIAF, Sydney, Australia (Sie and Suter, 1994; Sie et al., 1997a–c, 1998, 1999; Sie et al., 2000a,b, 2001). Commissioned in 1997, the system is aimed at enabling *in situ* microanalysis of geological samples for radiogenic and stable isotope data

**Table 3. Bulk Detection Limits for Trace Elements in Si Measured with the UltraClean Injector at the Munich Accelerator Laboratory with SIAMS**[a]

| Element | Atomic Ratio | Atomic Density (atoms/cm$^3$) |
|---|---|---|
| B | $1.1 \times 10^{-10}$ | $5.5 \times 10^{12}$ |
| N | $8.2 \times 10^{-8}$ | $4.1 \times 10^{16}$ |
| O | $1.2 \times 10^{-6}$ | $6.1 \times 10^{15}$ |
| Na | $2.6 \times 10^{-6}$ | $1.3 \times 10^{17}$ |
| Al | $6.4 \times 10^{-7}$ | $3.2 \times 10^{16}$ |
| P | $1.7 \times 10^{-8}$ | $8.5 \times 10^{14}$ |
| Cl | $9.0 \times 10^{-6}$ | $4.5 \times 10^{17}$ |
| K | $4.4 \times 10^{-8}$ | $2.2 \times 10^{15}$ |
| Ti | $4.6 \times 10^{-6}$ | $2.3 \times 10^{17}$ |
| V | $2.8 \times 10^{-8}$ | $1.4 \times 10^{15}$ |
| Cr | $3.8 \times 10^{-7}$ | $1.9 \times 10^{16}$ |
| Fe[b] | $2.0 \times 10^{-10}$ | $1.0 \times 10^{13}$ |
| Co | $1.3 \times 10^{-8}$ | $6.6 \times 10^{14}$ |
| Ni | $4.8 \times 10^{-7}$ | $2.4 \times 10^{16}$ |
| Cu[b] | $6.5 \times 10^{-10}$ | $3.3 \times 10^{13}$ |
| Ga | $7.0 \times 10^{-8}$ | $3.5 \times 10^{15}$ |
| Ge | $1.9 \times 10^{-8}$ | $9.3 \times 10^{14}$ |
| As | $4.6 \times 10^{-9}$ | $2.3 \times 10^{14}$ |
| Se | $1.8 \times 10^{-10}$ | $8.8 \times 10^{12}$ |

[a] The bulk detection limits are given as atomic ratios of trace element to Si and as atomic densities.
[b] Target aperture and target shielding made of high-purity Si.

free from molecular and isobaric interferences. The microbeam ion source, developed from a HICONEX source, produces a 30-μm-diameter Cs$^+$ beam routinely and includes a high-magnification sample viewing system in the reflected geometry, facilitating sample positioning and tuning of the primary beam. In measurements of Pb and S isotopes, high-precision isotope ratios of better than 1:1000 have been achieved for geochronological applications. The high precision is made possible by a fast bouncing system at both the low- and high-energy ends to counter the effect of instabilities in the ion source and beam transport system. Measurements were conducted at a terminal voltage of 1.5 to 2 MV. Lead isotopes are measured as Pb$^{4+}$ ions from PbS$^-$ injected ions. Sulfur isotopes are measured as S$^{2+}$ or S$^{3+}$ ions from injected S$^-$ ions. The trace element detection capability is demonstrated in measurements of precious metals, the PGEs, and Au from an assortment of geological and meteoritic samples. Detection sensitivity as low as sub-ppb has been obtained for Au in sulfides, which will be of great benefit in studies of ore deposit mineralogy and mineral processing. The Australia group has also demonstrated the facility for Os isotope measurements in meteorite samples, opening up the possibility of widespread use of the Re-Os system in exploration programs.

**Naval Research Laboratory TEAMS Facility.** One of the most ambitious programs is at the NRL, Washington, DC (Grabowski et al., 1997). A large portion of the novel TEAMS facility at the NRL is now installed. When completed, this facility will provide for parallel mass analysis over a broad mass range for conducting and insulating samples and offer 10 μm lateral image resolution, depth profiling, and sensitivity down to tens of ppt of trace impurities. The facility will use a modified commercial SIMS system as the source of secondary ions to provide these capabilities. This ion source is being installed. After the ion source, the facility uses a Pretzel magnet (Knies et al., 1997a) to act as a unique recombinator to simultaneously transmit from 1 to 200 amu ions but attenuate intense matrix-related beams. Following acceleration, a single charge state is selected by a $3°$ electrostatic bend; then the selected ions are electrostatically analyzed for energy/charge $(E/q)$ by a 2.2-m-radius, $30°$ spherical ESA with $E/\Delta E \cong 800$. Finally, a split-pole mass spectrograph with a 1.5-m-long focal plane provides parallel analysis over a broad mass range $(m_{max}/m_{min} = 8)$ with high mass resolution $(m/\Delta m \cong 2500)$. Micro-channel-plate-based position-sensitive detector modules are used to populate the focal plane.

Currently, vacuum and beam optics hardware are in place, and testing is underway with a multicathode ion source in place of the SIMS ion source. This more intense source simplifies diagnostic testing and initial research efforts. All twelve position-sensitive detector modules for the focal plane of the spectrograph have been received, and their testing is progressing well. Programmatically, the NRL is actively involved in the study of gas hydrates present under the ocean floor, which includes plans to analyze cycling between various carbon pools present there. Since $^{14}$C analysis is an important part of this work, the TEAMS facility has been modified to include a switching ESA to choose between the new multicathode ion source and the SIMS ion source for injection into the Pretzel magnet. It is anticipated that operations can be easily switched between standard radioisotope AMS and TEAMS efforts. Preliminary $^{14}$C analysis is currently underway.

### University of North Texas Ion Beam Modification and Analysis Laboratory TEAMS Facility.

The UNT TEAMS facility has been described in detail in other parts of this unit and will only be briefly summarized here. The UNT TEAMS facility is a dedicated trace element AMS that is part of the IBMAL.

The facility is characterized by an ultraclean ion source with magnetic $(mv/q)$ analysis of the primary Cs$^+$ sputter beam to remove contaminants. The Cs$^+$ beam may be raster scanned over the sample for depth-profiling measurements with a spot size of 60 to 90 μm or enlarged to ∼2 mm diameter for bulk impurity measurements. The samples are analyzed in a high-vacuum chamber with all components that could be hit by the beam coated with ultrapure Si to reduce sample contamination. The secondary ions extracted from the sample are electrostatically analyzed at $45°$, magnetically analyzed at $90°$, and injected into the 3-MV National Electrostatics tandem accelerators. After the accelerator, the energetic secondary ions are magnetically analyzed at $40°$ and electrostatically analyzed at $45°$. The secondary ions are detected in a Faraday cup for matrix ions and in a surface barrier detector or ionization chamber for trace impurity ions.

The TEAMS system has been used mainly to demonstrate depth-profiling capability for a number of trace elemental species in a variety of semiconductor materials,

**Table 4. Detection Limits in Si in Depth-Profiling Mode at the University of North Texas**[a]

| Element | Matrix | Depth-Profiling Detection Limits (atoms/cm$^3$) |
|---|---|---|
| B | Si | $4 \times 10^{14}$ |
| F | Si | $2 \times 10^{13}$ |
| P | Si | $3 \times 10^{13}$ |
| Ni | Si | $1 \times 10^{14}$ |
| Cu | Si | $2 \times 10^{14}$ |
| Zn | Si | $1 \times 10^{15}$ |
| As | Si | $2 \times 10^{14}$ |
| Mo | Si | $1 \times 10^{15}$ |
| Sb | Si | $1 \times 10^{15}$ |
| Se | Si | $1 \times 10^{12}$ |
| As | Ge$_x$Si$_{1-x}$ | $1 \times 10^{15}$ |

[a] The depth window is ∼50 Å.

e.g., Si, GeSi, GaAs, GaN, and SiC (Datar et al., 1997a,b; McDaniel et al., 1998a,b; Datar et al., 2000). This is noteworthy because depth profile measurements are the primary concern of the semiconductor industry. Depth profile measurements can also be compared directly with SIMS results.

The TEAMS facility can be operated in a bulk analysis mode or depth-profiling mode. The bulk analysis mode is generally more sensitive, since one can count the impurity ions for a longer time. In the depth-profiling mode, there is a limited amount of material available at a particular sputtering depth. Table 4 gives the detection limits in Si in the depth-profiling mode in atoms per cubic centimeter.

### Engineering Advances That Made TEAMS Possible

Some of the advances that made TEAMS feasible include voltage-stable tandem accelerators, specially designed low-contamination ion sources, the use of all-electrostatic focusing and steering elements, double-focusing spherical electrostatic energy analyzers, the development of fast-switching magnet power supplies, the development of nuclear detectors for mega-electron-volt energy ions. The operation and voltage control of tandem accelerators allows the production of energetic ions with less energy uncertainty, thereby enabling magnetic and electrostatic analysis.

### Historical Evolution of TEAMS

Accelerator mass spectrometry was first used by Alvarez and Cornog (1939a,b) to measure natural abundances of $^3$He. It was reincarnated when Muller (1977) suggested that a cyclotron could be used for detection of $^{14}$C, $^{10}$Be, and other long-lived radioisotopes. Independently in 1977, the University of Rochester group (Purser et al., 1977) demonstrated that AMS could be used to separate $^{14}$C from the isobar $^{14}$N because of the instability of the $^{14}$N$^-$ ion. Later, Nelson et al. (1977) at McMaster University and Bennett et al. (1977) at the University of Rochester reported accelerator measurements of natural $^{14}$C.

Purser et al. (1979) reported that accelerator SIMS is a very promising technique for the detection of trace elements in ultrapure materials. A lot of the early work done at Rochester and Toronto concentrated on the *in situ* TEAMS measurements of PGEs, Au, and Ag in mineralogical and geological samples (Rucklidge et al., 1981, 1982). Since then, the Toronto group has continued this TEAMS research in mineral grains and meteorites (Wilson et al., 1997a,b). The Rochester group measured quantitative TEAMS depth profiles of N and Cl in semiconductor-grade Si wafers (Elmore et al., 1989; Gove et al., 1990).

Anthony and Thomas (1983) first suggested the use of AMS for the detection of stable isotopes present as contaminants or impurity dopants in ultrapure semiconductor materials. Preliminary measurements were made on the carbon-dating AMS facility at the University of Arizona (Anthony et al., 1985, 1986). Anthony and Donahue (1987) showed that AMS could help provide solutions for semiconductor problems. Following this lead, a TEAMS system dedicated to broad-band stable isotope measurements over the entire range of the periodic table was designed and constructed at IBMAL at the UNT in collaboration with Texas Instruments (Anthony et al., 1989, 1990a,b, 1991, 1994a,b; Matteson et al., 1989, 1990; McDaniel et al., 1990a,b, 1992, 1993, 1994, 1995, 1998a,b; Kirchhoff et al., 1994; Datar et al., 1996, 1997a,b, 2000; Zhao et al., 1997). Other groups soon followed with modifications of existing radionuclide AMS facilities to perform stable isotope measurements at PSI/ETH Zurich (Döbeli et al., 1994; Ender, 1997; Ender et al., 1997a-d) and the Technical University Munich (Massonet et al., 1997) or construction of new facilities at CSIRO in Australia (Sie and Suter, 1994; Niklaus et al., 1997; Sie et al., 1997a–c) and at the NRL (Grabowski et al., 1997; Knies et al., 1997a,b).

## METHOD AUTOMATION

Due to the extreme complexity of the ion source at UNT, computer control is essential for steering and focusing all elements in both primary- and secondary-ion beamlines. By systematically varying the control voltage and measuring the beam current, the optimum control voltage can be determined for each parameter. Ideally, if all beamlines are properly aligned, the steering voltages should be zero. Practically, as long as the steering voltages are near zero, near-optimum tuning can be assumed. The accelerator, beamline, and analyzing elements are also under computer control and can be set to the appropriate settings by keyboard entry. Figure 7 shows a block diagram of the computer control and data acquisition system at UNT.

For depth profile measurements, one simply clicks on the depth profile icon. A pop-up menu enables one to enter all the necessary parameters, and the computer automatically acquires the raw depth profile data. The operator determines the endpoint and stops data acquisition. The raw data are stored in a file of the operator's choice. Sample changes have to be done manually. The measurement should not be left unattended. For bulk measurements, the computer control program is also used, except that the entries in the pop-up menu are different. Due to the

## TEAMS Computer Control Schematic

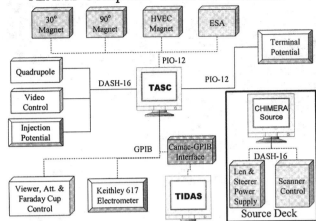

**Figure 7.** Block diagram of the TEAMS computer control and data acquisition system at UNT. The ion source, accelerator, and data acquisition system are controlled by three personal computers. The tandem accelerator system control (TASC) computer controls the magnets, electrostatic analyzer, quadrupole, terminal potential, ion source injection potential, viewers, Faraday cups, and attenuators and is connected to the CAMAC modular data-handling system crate through a GPIB interface. The TI data acquisition system (TIDAS) is connected to a CAMAC crate through a GPIB interface. Detector signals from either the ionization chamber or Si particle detector are passed into the CAMAC crate amplifier and stored in the TIDAS. The voltages (or currents) for the ion source Cs source bias, 90° magnet, eight Einzel lenses, six *x-y* steerers, Cs beam scanner, sample bias, extraction bias, and 45° ESA are controlled by the ion source computer, which can be operated in the accelerator control room or at the source.

unique nature of the computer control system, most software was initially developed in-house and later converted to a LabView-based system.

## DATA ANALYSIS AND INITIAL INTERPRETATION

### Bulk Impurity Measurements

For bulk measurements, where a constant, very low level concentration is to be measured, the acquisition time can be increased and the tuning can be done on each sample and not just the blank, to increase transmission. Furthermore, if a standard sample with a constant low-level impurity doping is available, additional fine tuning of the TEAMS system can be done, leading to enhanced impurity transmission with a corresponding increase in sensitivity. This procedure is essentially the same as radioisotope AMS measurements, where following the use of a stable isotope as a pilot beam, the final fine tuning is done using a known standard. Bulk analysis measurements are identical to that used for isotope ratio measurements in AMS. The matrix ion count rate is measured briefly as a current, while the impurity count rate is measured for a longer time in a particle detector or ionization chamber. The time difference is because there are fewer impurity ions compared to matrix ions. By repeatedly cycling between the two measurements, good statistics can be obtained for both matrix

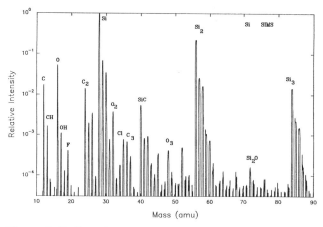

**Figure 8.** Bulk analysis SIMS mass scan of a heavily contaminated Si sample. All peak intensities are relative to the $^{28}$Si peak, which is set equal to 1.0.

and impurity ions. Elmore et al. (1984) has given a detailed exposition of the technique.

The TEAMS bulk measurements can also provide a quick mass scan of elements present in a sample in much the same way as SIMS, except the molecular interferences can be suppressed. For example, Figures 8 to 10 show representative bulk analysis mass scans of a heavily contaminated Si sample over an extended mass range of 10 to 90 amu. All isotopic relative intensities are normalized to the $^{28}$Si peak, which is set equal to 1.0 in Figures 8 and 9. The pre-tandem-acceleration analysis shown in Figure 8 represents a conventional negative ion SIMS spectrum with contributions to many mass regions from molecular ions. The atomic and molecular ions are extracted from the sample, pass out of the ion source chamber through the 45° electrostatic analyzer and the 90° magnetic spectrometer, and are measured as a current in a Faraday cup, as shown in Figure 1. Detection before the accelerator is limited to current measurements because of the high currents, but molecular ions are still present. For example,

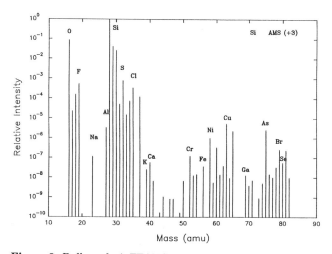

**Figure 9.** Bulk analysis TEAMS mass scan of the same sample as in Figure 8. All peak intensities are relative to $^{28}$Si, which is set equal to 1.0.

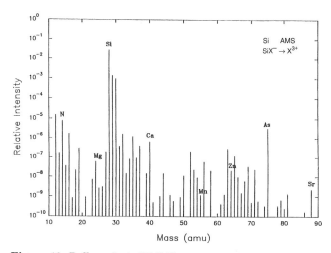

**Figure 10.** Bulk analysis TEAMS mass scan of the same sample similar to Figure 9, except a molecular mass was chosen to accelerate to the terminal of the accelerator. The signals have been normalized to the $^{28}$Si signal in Figure 9 for direct comparison.

some of the molecular ion interferences shown in the SIMS mass scan in Figure 8 are given in Table 5 with their intensity relative to $^{28}$Si$^-$. In many cases the molecular ion intensity is much greater than an atomic ion at the same mass.

Figure 9 is a TEAMS bulk analysis mass scan of the same Si sample after passage through the tandem accelerator and high-energy magnetic ($mv/q$) and electrostatic ($E/q$) analysis with charge state $q = 3^+$ and total energy ($E$) detection. The relative intensities of the isotopes shown are normalized to the intensity of the $^{28}$Si$^{3+}$ peak, which is set equal to 1.0. Most of the lower intensity measurements shown require direct ion counting in a particle detector as opposed to higher intensity current measurements. Counting time was 5 s per mass interval for all measurements or ~400 s for the entire mass scan. Since all isotopic ions analyzed were in a $q = 3^+$ charge state, interfering molecular ions have been rejected by breaking up the molecules in the terminal of the accelerator. In addition to removal of the molecular interferences, detection of ions in high-energy particle detectors has removed detector background noise as well as most $m/q$ interferences. Signals from break-up products that still exhibit $m/q$ interferences at masses 36, 42, 45, 48, 51, 57, and 84 are not shown in Figure 9. Elements with zero electron affinity (e.g., Mg, Mn, Zn) will not form negative ions, and the TEAMS mass scan does not show any signal for these masses. As seen by comparison of the relative intensities in Figures 8 and 9 and Table 5, the molecular interferences are reduced, in some cases, by many orders of magnitude. For example, the relative intensity of ($^{28}$Si)$_2$ at mass 56 ($^{56}$Fe) is reduced by almost seven orders of magnitude in the TEAMS data compared to the SIMS data. The $^{28}$Si$^{12}$C at mass 40 ($^{40}$Ca) is reduced by almost five orders of magnitude in the TEAMS data compared to the SIMS data. The molecular peaks ($^{29}$Si)$_2$, $^{29}$Si$^{28}$Si$^1$H, etc. at mass 58 ($^{58}$Ni) are reduced by over four orders of magnitude in the TEAMS data compared to the SIMS data.

**Table 5. Molecular Ion Impurities in SIMS Mass Scan in Figure 8.1 Relative to $^{28}Si^-$ Intensity and Atomic Ion Intensities from TEAMS Mass Scan in Figure 9 Relative to $^{28}Si^{3+}$ Intensity**[a]

| Molecular Ion | Mass (amu) | Intensity/$^{28}Si^-$ (Figure 8) | Intensity/$^{28}Si^{3+}$ (Figure 9) | Elemental Ion |
|---|---|---|---|---|
| $^{16}O^1H$ | 17 | $1.2 \times 10^{-3}$ | $3.0 \times 10^{-5}$ | $^{17}O$ |
| $(^{12}C)_2$ | 24 | $1.6 \times 10^{-2}$ | $<10^{-10}$ | $^{24}Mg$ |
| $^{30}Si^1H$ | 31 | $7.0 \times 10^{-4}$ | $7.0 \times 10^{-5}$ | $^{31}P$ |
| $(^{16}O)_2$ | 32 | $4.0 \times 10^{-3}$ | $9.0 \times 10^{-4}$ | $^{32}S$ |
| $^{28}Si^{12}C$ | 40 | $6.0 \times 10^{-3}$ | $7.0 \times 10^{-8}$ | $^{40}Ca$ |
| $^{24}Mg^{28}Si$, etc | 52 | $5.0 \times 10^{-4}$ | $1.3 \times 10^{-7}$ | $^{52}Cr$ |
| $(^{28}Si)_2$ | 56 | $2.5 \times 10^{-1}$ | $4.0 \times 10^{-8}$ | $^{56}Fe$ |
| $^{28}Si^{29}Si^1H$, $^{29}Si_2$, etc. | 58 | $2.0 \times 10^{-2}$ | $1.0 \times 10^{-6}$ | $^{58}Ni$ |
| $^{28}SiO_2$, $^{30}Si_2$, etc. | 60 | $7.0 \times 10^{-4}$ | $4.0 \times 10^{-7}$ | $^{60}Ni$ |
| $(^{28}Si)_2{}^{16}O$ | 72 | $1.8 \times 10^{-4}$ | $<10^{-10}$ | $^{72}Ge$ |

[a] Break-up of the molecular interferences reduces the intensities by orders of magnitude in many cases.

Because of the removal of the molecular interferences in the TEAMS data, the correct isotopic abundances are found for O, Si, Cl, Ni, and Cu, which indicates that calibration of the magnetic and electrostatic analyzers is correct.

Even though all the molecular ions are dissociated in the stripping canal of the tandem accelerator, there still may be some $m/q$ interferences. During a mass scan, the preacceleration magnet is tuned for a charge state $q = 1^-$, while the postacceleration magnet is tuned for the charge state of interest, which is $q = 3^+$ for Figure 9. Even though direct molecular interferences are removed, molecular fragments may still cause an ambiguity. For example, $((^{12}C)_n)^-$ molecules can be produced from the vacuum system during the sputtering process (as seen in Figure 8). Any $((^{12}C)_3)^-$ molecular ions will dissociate after stripping but will result in a large production of $^{12}C^+$ ions. The $^{12}C^+$ ions at mass 12 have the same $m/q$ as the atomic ions at mass 36 (e.g., $^{36}S^{3+}$) and therefore will pass into the final total energy detector. The $^{12}C^-$ ions created during the sputtering process will not be an interference for mass 36 because it will not pass through the preacceleration magnet. Even though the total energy spectroscopy will separate the peaks, the tails from high levels of $^{12}C$ may lead to saturation of the detector and incorrect results. Several of the masses with known interferences described above (36, 42, 45, 48, 51, 57, and 84) are not included in the mass scan in Figure 9. These interferences can be removed by analyzing a different charge state and therefore a different $m/q$.

Some elements do not form negative ions directly, e.g., N, Mg, Mn, and Zn. In these cases, molecular ions of the form SiX-, where X represents the element of interest, are injected into the accelerator. The break-up products $X^{3+}$ are accelerated down the high-energy side of the accelerator, magnetically and electrostatically analyzed, and detected in the total energy detector. Figure 10 shows such a TEAMS spectrum in which the intensity of each signal has been normalized to the maximum $^{28}Si$ signal in Figure 9 for direct comparison. Several ions that were not detectable in the TEAMS mass scan in Figure 9 are now measured in the correct isotopic abundances, e.g., N,

Mg, Mn, and Zn. Signals from break-up products at masses 36, 42, 45, 48, 51, 57, and 84 are not removed in Figure 10.

### Data Analysis for Bulk Measurements

This procedure for bulk analysis provides a relative normalization of the impurity element to the amount of matrix element present in the sample, which is well known. Of course, this procedure does not account for the differences in negative ion formation probabilities of different elements or molecules or charge state fractionation of the ion in the terminal of the accelerator. In fact, some elements will not form a negative ion at all (e.g., N, Mg, Mn, Zn). For these elements that do not form a negative ion, in almost all cases, a molecule with this element can be found that will form a negative ion, therefore allowing analysis of essentially every element in the periodic table. A distribution of charge states is produced during electron stripping in the terminal of the accelerator. Because the magnetic ($mv/q$) and electrostatic ($E/q$) analyses pass only one charge state from this distribution, only a fraction of the original ions from the sample are analyzed and detected. Fortunately, the $q = 3^+$ charge state most used in the analysis is the most populated for masses in the range of ~30 to 60 amu when stripped at an energy of ~1.5 to 3 MeV. In addition, this procedure requires that the Cs sputter beam be constant over the length of the mass scan and that the calibration of the magnetic and electrostatic analyzers be accurate over the mass range of the scan. The calibration is normally checked before and after the measurement, as discussed earlier with Figure 3, by running a mass scan of the masses present for the matrix elements to see the characteristic signature spectrum.

Mass scans over a range of masses are useful to confirm isotopic natural abundances of an element of interest as well as to check for interferences that may be present at a particular mass. In the TEAMS analysis beamline, the UNT facility has a pneumatically controlled Faraday cup followed by a Si charged particle detector that may be inserted into the beam or withdrawn and then an ionization chamber. For a mass scan as shown in Figure 9 or 10, the magnetic and electrostatic fields are set to the correct values to pass the initial mass of interest with the Faraday cup in the beam path. If the current measured in the Faraday cup for that mass is below a preset value, the Faraday cup is removed from the beam path and the ions of that mass are detected in the Si detector or ionization chamber for a preset time. If not, then the mass is measured as a current in the Faraday cup as Coulombs per second and converted to particles per second, taking into account the charge state of the ion. After completing the preset measuring time, the magnetic and electrostatic fields are incremented to the next mass of interest, and a measurement is made for that mass, and so on for the entire mass range of interest. Usually the time of measurement is of the order of a few seconds for each mass value. Counting for a longer time period can reduce the minimum background count and therefore improve the detection limit. For example, for a 5-s count at each mass value, the minimum detectable count rate is 1 count in 5 s or 0.2 counts/s (cps). For a 100-s count at each mass value, the minimum detectable count rate is 1 count in 100 s or 0.01 cps, which is 20 times more sensitive.

## Depth-Profiling Impurity Measurements

To measure a depth profile of Cu in Si, for example, the key parameter is the terminal voltage on the accelerator. The terminal voltage should be chosen such that the most abundant charge state for Cu is $3^+$ or higher. There is some flexibility in the choice of impurity ion species injected into the accelerator. For some elements, the ion $MX^-$, where M is a matrix atom and X is an impurity atom, is more prolific than the ion $X^-$. As far as possible, then the most prolific negative ion should be chosen. The RSF tables in Wilson et al. (1989) provide a good guide to the best possible negative ion to be injected. Care must be exercised in the choice of the positive charge state selected for postaccelerator analysis to avoid $m/q$ degeneracies with the matrix ion(s) since it can swamp the detector. For example, during $^{58}Ni$ measurements, charge state $4^+$ has the same $m/q$ as $^{29}Si^{2+}$ and the Ni signal can be swamped in the detector by the tail of the Si signal, even though they are at different energies. The $m/q$ interferences from non-matrix species can be separated by energy in the detector, as the count rate is usually not very high. If the impurity concentration to be measured is high as for a dopant, then a less abundant charge state can be chosen to avoid swamping the detector. However, in the rest of the present discussion, it will be assumed that one is interested in only measuring very low concentration levels, which is the strength of TEAMS.

As previously mentioned, the nuclear-type particle detector has essentially zero dark current because the ion is at mega-electron-volt energies. Thus, the lowest counting rate of the detector is determined by the acquisition time for that particular data point in the depth profile, which in turn is determined by the volume of the available material. For example, with a sputtering rate of 2 Å/s, an acquisition time of 20 s implies that the measurement represents an average concentration for a slice of material 40 Å thick. The lowest measurable concentration in 20 s is

0.05 cps (one count in the 20 s). By increasing the acquisition time to 50 s, the lowest measured concentration now corresponds to 0.02 cps, but now it represents an average over 100 Å. However, one must remember that the sputtering process in the ion source causes depth profile broadening due to the penetration of the Cs ions into the sample. This effect is of the order of 100 Å (Vandervorst and Shepherd, 1987). Therefore, the Cs penetration depth limits the ultimate depth sensitivity.

For the example of a Cu depth profile in a Si sample, the experimental parameters were as follows:

Sample bias: 11.17 keV. The lower the sample bias, the better, as it reduces the Cs penetration depth (Vandervorst and Shepherd, 1987).

Secondary ions injected into the tandem accelerator: $^{63}Cu^-$, $^{28}Si^-$.

Terminal voltage: 1.7 MV.

Ions transported to the end of the beamline: 6.8 MeV $^{63}Cu^{3+}$, 6.8 MeV $^{28}Si^{3+}$.

Cu ions were measured in an ionization chamber with an acquisition time of 20 s, and Si ion current was measured in a Faraday cup just in front of the detector.

The tuning was done using a nominally blank pure Si sample. The focusing and steering elements were adjusted to get maximum $^{28}Si^{3+}$ current in the final Faraday cup. It was assumed that the analyzing elements (the magnets and the electrostatic analyzer) were already calibrated. This calibration is periodically checked. For depth profile measurements, one has to specify the isotopes to be measured along with the charge states and the acquisition time. In the case of $m/q$ interferences, the total energy gate width in the ionization chamber has to be specified also. Typically, in depth profile measurements, one sequentially cycles between a matrix ion count rate measurement and an impurity isotope count rate measurement. Switching the system between different masses involves dead time since the magnets cannot be switched instantaneously. By suitable choice of matrix charge state, this dead time can be reduced. The matrix ion count rate measurement serves as a basis for normalization of unknown depth profiles to standard implanted depth profiles as well as serving as a check on system stability (very much similar to both AMS and SIMS).

Before measuring an unknown depth profile, a standard implant is measured. Thus, for the Cu profile, after the blank measurement, an implanted standard was measured. This serves as an additional check on the measurement process. Finally, the unknown sample was measured with exactly the same parameters as the standard implant. Figure 11 shows an example of a Cu impurity depth profile measurement for a 10-keV, $1 \times 10^{16}$ atoms/cm$^2$ As-implanted Si wafer. The Cu concentration decreases from the surface but then shows an unexpected peak around 0.1 μm into the wafer before finally tailing off at a concentration level of $4 \times 10^{14}$ atoms/cm$^3$, which is a factor of 6 lower than the SIMS measurement. A nominally blank sample of Si is run after measurements of an unknown sample to get an estimate of cross-contamination from other samples and also to make sure that running the unknown sample has not contaminated the source.

## As implant at 10 keV 1e16

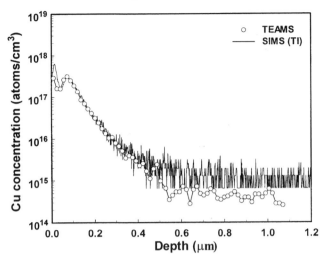

**Figure 11.** TEAMS Cu impurity depth profile in a 10-keV, $1 \times 10^{16}$ atoms/cm$^2$ $^{75}$As implant in Si. Also shown is a SIMS depth profile for comparison. The TEAMS and SIMS profiles are in good agreement until the SIMS detection limit is reached at ∼0.4 µm in depth. After 0.4 µm, TEAMS is about a factor of 6 more sensitive than the SIMS measurements.

Figures 12 and 13 show TEAMS Cu impurity concentrations as a function of depth in As-implanted Si wafers. Figure 12 shows an impurity Cu depth profile for a 40-keV, $1 \times 10^{16}$ atoms/cm$^2$ As implant. Figure 13 shows an impurity Cu depth profile for an 80-keV, $1 \times 10^{16}$ atoms/cm$^2$ As implant. The Cu concentrations in both spectra decrease from the surface peak but then increase again to form a deeper secondary peak. The depth of the secondary peak scales with the As implant energy. SIMS measurements

## As implant at 40 keV 1e16

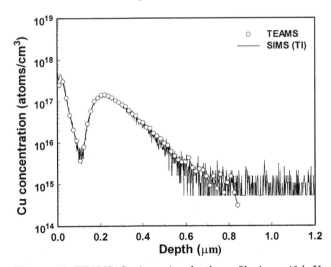

**Figure 12.** TEAMS Cu impurity depth profile in a 40-keV, $1 \times 10^{16}$ atoms/cm$^2$ $^{75}$As implant in Si. Also shown is a SIMS depth profile for comparison. The TEAMS and SIMS profiles are in good agreement until the SIMS detection limit for Cu is reached.

## As implant at 80 keV 1e16

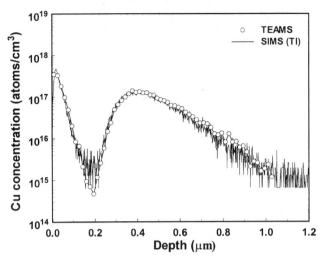

**Figure 13.** TEAMS Cu impurity depth profile in an 80-keV, $1 \times 10^{16}$ atoms/cm$^2$ $^{75}$As implant in Si. Also shown is a SIMS depth profile for comparison. The TEAMS and SIMS profiles are in good agreement. TEAMS can be seen to be more sensitive than the SIMS measurements in the dip in the spectrum at ∼0.2 µm.

were done later to confirm the TEAMS data. The SIMS depth profile data are overlaid on the TEAMS data and the agreement is excellent. During data acquisition, both isotopes of Cu were measured with TEAMS, and the isotope ratios were found to be in agreement with their natural abundances. This is an indication that the Cu was not implanted along with the As and that the two Cu isotopes were not implanted separately. Only one mass can be implanted at a time.

### Data Analysis for Depth-Profiling Measurements

The raw data are measurements of count rate versus time. After analysis, the samples are removed from the ion source and the crater depths are measured with a surface profilometer. The measured depth gives a straightforward time-to-depth conversion assuming a constant sputtering rate, which is monitored via the matrix ion count rate. The count rate data are normalized to concentration (atoms per cubic centimeter). Normalization can be accomplished by scaling the peak count rate of the standard implant to the known concentration as well as by integrating the area under the standard profile and obtaining a yield factor. These methods are independent of each other. They have been described in detail by Wilson et al. (1989), Datar (1994), and Datar et al. (1997a, 1997b). The section on calibration of data acquisition below also has a sample demonstration of data normalization for an implant standard sample. The final depth profile is achieved by multiplying both axes of the raw data plot with the respective normalization factors. Typically, after a series of unknowns, another standard is run with the unknowns being sandwiched between standard samples. The final data are referred to the standards and the nominal blank. A nominally blank sample is run after measurements of an unknown sample to get an estimate of cross-contamination from other samples and to make sure that running the unknown sample has not contaminated the source. Float-zone Si is a good blank sample.

Standard samples for depth-profiling measurements are implanted samples with known doses and known energies. However, any sample where the depth profile is known beforehand can be used as a standard. For bulk measurements, standard samples should be uniformly doped with the impurity of interest. Standards may be cross-checked in a number of TEAMS laboratories.

**Interpretation of the Data.** The final depth profile has exactly as much information as a SIMS profile, and in general, data interpretation has to take into account the same factors. Due to the extreme sensitivity, accuracy can only be compared to a known standard. Absolute measurements without reference to a standard are not possible. Also, for bulk measurements, one can only refer the data to a known standard in a manner similar to conventional AMS.

**Calibration of Data Acquisition.** Assuming a constant sputtering rate (checked by monitoring the Cs current before and after the measurement as well as by periodically measuring the matrix element current during the measurement), the time-to-depth conversion is simply done by measuring the depth of the crater after measurement using a profilometer. Converting the count rate to concentration is more involved. The SIMS RSFs cannot be used due to charge state fractionation as well as possible variations in transmission efficiency. In due course, TEAMS RSFs can be developed that take into account these factors as well. At present, calibration is done using implant standards. Assuming the depth profile is measured to exhaustion, the ratio of the integrated area under the depth profile to the implant dose gives the "yield" of the system. This information is used to normalize the unknown profile. The yield factor can also be used to normalize bulk impurity concentration measurements in the absence of uniformly doped standards. The concentration normalization factor (CNF) is given by Datar (1994) as

$$CNF = \frac{\text{implanted dose (atoms / cm}^2)}{\text{area under profile} \times \text{sputtering rate (cm/s)}} \quad (3)$$

Then, the concentration at a given depth is given by

$$\text{Concentration} = \text{cps} \times \text{CNF} \quad (4)$$

For example, Figure 14A shows a raw depth profile for a 100-keV, $2 \times 10^{14}$ atoms/cm$^2$, $^{60}$Ni implant in Si. The total number of Ni ions measured (obtained by integrating the area under the cps versus the time profile) was $4.287 \times 10^6$. The sputtering rate (obtained by measuring the depth of the crater after the measurement) was 2.6 Å/s. Since the implanted dose was $2 \times 10^{14}$ atoms/cm$^2$,

$$CNF = \frac{2 \times 10^{14} \text{ atoms/cm}^2}{4.287 \times 10^6 \text{ atoms} \times 2.6 \times 10^{-8} \text{ cm/s})}$$
$$= 1.82 \times 10^{15} \text{ s/cm}^3 \quad (5)$$

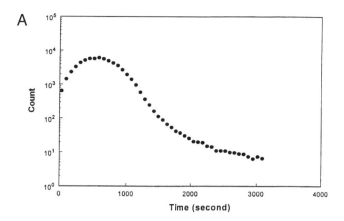

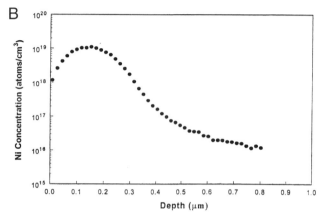

**Figure 14.** (**A**) Measured depth profile for a 100-keV, $2 \times 10^{14}$ atoms/cm$^2$, $^{60}$Ni implant in Si. (**B**) Normalized depth profile for Figure 14A showing concentration versus depth.

This means that a count rate of 1 cps corresponds to a concentration of $1.82 \times 10^{15}$ atoms/cm$^3$. The detection limit is the least count multiplied by the normalization factor. Figure 14B shows the normalized depth profile as concentration (atoms per cubic centimeter) versus depth (micrometers). The two graphs are identical; just the $x$ and $y$ axes have been renormalized.

For the unknown sample, which has been run under the same conditions, the count rate numbers can be multiplied by the CNF to convert the depth profile into a concentration profile. Thus, the normalization is critically dependent upon the use of implant standards for depth profile measurements. For bulk measurements, an even better standard would be samples with known constant impurity levels. But basically, as in radioisotope AMS, the final numbers should always be referred to the standard used.

## SAMPLE PREPARATION

Extreme care has to be taken during sample preparation to avoid surface particulate impurities. For depth profile measurements, it is desirable that the sample be flat and at least somewhat conductive. Typically, a TEAMS sample is a small square at least $4 \times 4$ mm. The samples are loaded into the sample holder in a class 100 clean room. Care is taken to ensure that different kinds of samples do not come in contact with each other, essentially the same procedures as with SIMS.

## SPECIMEN MODIFICATION

Trace element AMS is a destructive technique, and the samples cannot be reused for other measurements. In addition, the sputtering process itself affects the depth distribution of the impurity being measured due to ion beam mixing. Here, the sputtering ion actually drives the impurity further into the sample. This effect is more pronounced for higher Cs impact energies and causes depth profile broadening. This is the primary reason to attempt to use the lowest Cs impact energy possible that allows good transmission through the accelerator and analysis beamline. At UNT, the Cs impact energy is the sum of the 10 keV from the Cs oven and the energy from the negative sample bias, which has been operated as high as 30 kV, for a total impact energy of 40 keV. Recently, the sample bias has been reduced to 11 kV, which makes the Cs impact energy 21 keV, without significant loss of transmission through the entire system. The ranges of 21- and 40-keV Cs ions in Si are $\sim$178 Å and $\sim$266 Å, respectively. Therefore, 40-keV Cs ions penetrate Si $\sim$1.5 times as far as 21-keV Cs ions.

## PROBLEMS

Accidental contamination from particulates, elements in the vacuum system, and from sputtering neighboring samples is a hazard. Running samples in order from low concentration to high concentration if approximate concentration levels are known can minimize contamination. Sometimes, due to improper sample positioning, the transmission is unusually low. The matrix ion current measurement is usually a good indicator of this problem.

In general, sample positioning with respect to the extraction optics is the most critical factor affecting the reproducibility of the data. The ion source at UNT is being modified to improve the reproducibility in sample positioning.

Destroying a particle detector is possible by placing too much ion current on it due to unexpected matrix-based signal. Use of an ionization chamber for tune-up and initial measurements essentially solves this problem. The surface barrier particle detector can then be used to make higher resolution measurements.

A long measurement can take as long as an hour. Accelerator terminal voltage drift has to be monitored since this can affect transmission. Once again, periodically running standard samples or measuring matrix currents can be used to monitor the terminal voltage.

Over time, the analyzing magnet calibrations may change and will have to be recalibrated. Periodic measurements of known samples are the best safeguard against erroneous readings due to a number of different effects.

## ACKNOWLEDGMENTS

The author would like to acknowledge the many funding sources that were required for the development of the TEAMS facility in the Ion Beam Modification and Analysis Laboratory (IBMAL) at the University of North Texas (UNT), including the State of Texas Coordinating Board Advanced Technology Program, the University of North Texas, Texas Instruments (TI), the Office of Naval Research, the National Science Foundation, and the Robert A. Welch Foundation. The TEAMS system at UNT would not have been possible without the technical support of Mark Anthony, Roy Beavers, and Tommy Bennett at TI.

The author would like to thank the many colleagues that have worked with him over the years on the design and development of the TEAMS system at UNT. Jerry Duggan, Sam Matteson, Duncan Weathers, Dwight Maxson, Jimmy Cobb, and Bobby Turner of UNT have been involved with the TEAMS development from the beginning. Graduate students and postdoctoral research scientists that have worked long and hard on TEAMS include Danny Marble, Joe Kirchhoff, György Vizkelethy, Sameer Datar, Yong-Dal Kim, Zhiyong Zhao, Steve Renfrow, Elaine Roth, Eugene Reznik, Baonian Guo, Karim Arrale, and David Gressett.

The author has benefited from many discussions with numerous people around the world that are experts in this field, including Soey Sie, Martin Suter, Ken Grabowski, Graham Hubler, David Knies, Barney Doyle, Ken Purser, Drew Evans, David Elmore, Harry Gove, Liam Kieser, Graham Wilson, Eckehart Nolte, Colin Maden, Hans-Arno Synal, John Vogel, Jay Davis, Doug Donahue, and Greg Norton. I especially thank Sameer Datar, who helped write parts of this unit and for his critical reading, and my wife Linda McDaniel for grammatical corrections. I particularly want to thank Mark Anthony for his belief in this project, for his continued support over the past few years, and for reading and commenting on the manuscript.

## LITERATURE CITED

Alvarez, L. W. and Cornog, R. 1939a. $^3$He in helium. *Phys. Rev.* 56:379–379.

Alvarez, L. W. and Cornog, R. 1939b. Helium and hydrogen of mass 3. *Phys. Rev.* 56:613–613.

Anthony, J. M. and Donahue, D. J. 1987. Accelerator mass spectrometry solutions to semiconductor problems. *Nucl. Instrum. Methods Phys. Res.* B29:77–82.

Anthony, J. M. and Thomas, J. 1983. Frontier analysis of stable isotopes present as contaminants or impurity dopants in ultrapure semiconductor materials. *Nucl. Instrum. Methods* 218:463–467.

Anthony, J. M., Donahue, D. J., Jull, A. J. T., and Zabel, T. H. 1985. Detection of semiconductor dopants using accelerator mass spectrometry. *Nucl. Instrum. Methods Phys. Res.* B10/11:498–500.

Anthony, J. M., Donahue, D. J., and Jull, A. J. T. 1986. Super SIMS for ultra-sensitive impurity analysis. *Mat. Res. Soc. Symp. Proc.* 69:311–316.

Anthony, J. M., Matteson, S., McDaniel, F. D., and Duggan, J. L. 1989. Accelerator mass spectrometry at the University of North Texas. *Nucl. Instrum. Methods Phys. Res.* B40/41: 731–733.

Anthony, J. M., Matteson, S., Marble, D. K., Duggan, J. L., McDaniel, F. D., and Donahue, D. J. 1990a. Applications of accelerator mass spectrometry to electronic materials. *Nucl. Instrum. Methods Phys. Res.* B50:262–266.

Anthony, J. M., Matteson, S., Duggan, J. L., Elliott, P., Marble, D. K., McDaniel, F. D., and Weathers, D. L. 1990b. Atomic mass spectrometry of materials. *Nucl. Instrum. Methods Phys. Res.* B52:493–497.

Anthony, J. M., Beavers, R. L., Bennett, T. J., Matteson, S., Marble, D. K., Weathers, D. L., McDaniel, F. D., and Duggan, J. L. 1991. Materials characterization using accelerator mass spectrometry. *Nucl. Instrum. Methods Phys. Res.* B56/57: 873–876.

Anthony, J. M., Kirchhoff, J. F., Marble, D. K., Renfrow, S. N., Kim, Y. D., Matteson, S., and McDaniel, F. D. 1994a. Accelerator based secondary ion mass spectrometry for impurity analysis. *J. Vac. Sci. Tech.* A12:1547–1550.

Anthony, J. M., McDaniel, F. D., and Renfrow, S. N. 1994b. Molecular free SIMS for semiconductors. *In* Diagnostic Techniques for Semiconductor Materials and Devices (D. K. Schroder, J. L. Benton, and P. Rai-Choudhury, eds.). Proceedings of the Electrochemical Society 94:349–354.

Arrale, A. M., Matteson, S., McDaniel, F. D., and Duggan, J. L. 1991. Energy loss corrections for MeV ions in tandem accelerator stripping. *Nucl. Instrum. Methods Phys. Res.* B56/57:886–888.

Baumann, H. and Pavel, J. 1989. Determination of trace impurities in electronic grade quartz: Comparison of inductively coupled plasma mass spectroscopy with other analytical techniques. *Mikrochim. Acta [Wien]* 111:413–422.

Bennett, C. L., Beukens, R. P., Clover, M. R., Gove, H. E., Liebert, R. B., Litherland, A. E., Purser, K. H., and Sondheim, W. E. 1977. Radiocarbon dating using electrostatic accelerators: Negative ions the key. *Science* 198:508–510.

Benninghoven, A., Rüdenauer, F.G., and Werner, H. W. 1987. Secondary Ion Mass Spectrometry: Basic Concepts, Instrumental Aspects, Applications, and Trends. John Wiley & Sons, New York.

Betz, H. D. 1972. Charge state and charge-changing cross sections of fast heavy ions penetrating through gaseous and solid media. *Rev. Mod. Phys.* 44:465–507.

Bigelow, A. W., Li, S. L., Matteson, S., and Weathers, D. L. 1999. Sputter-initiated resonance ionization spectroscopy at the University of North Texas. *AIP Conf. Proc.* 475:569–572.

Bigelow, A. W., Li, S. L., Matteson, S., and Weathers, D. L. 2001. Measured energy and angular distributions of sputtered neutral atoms from a Ga-In eutectic alloy target. *AIP Conf. Proc.* 576:108–111.

Burdick, G. W., Appling, J. R., and Moran, T. F. 1986. Stable multiply charged molecular ions. *J. Phys. B* 19:629–641.

Datar, S. A. 1994. The Diffusion of Ion-Implanted Cl in Si. Ph.D. thesis, University of Rochester.

Datar, S. A., Gove, H. E., Teng, R. T. D., and Lavine, J. P. 1995. AMS studies of the diffusion of Cl in Si wafers. *Nucl. Instrum. Methods* B99:549–552.

Datar, S. A., Renfrow, S. N., Anthony, J. M., and McDaniel, F. D. 1996. Trace-element accelerator mass spectrometry: A new technique for low-level impurity measurements in semiconductors. *Mat. Res. Soc. Symp. Proc.* 406:395–400.

Datar, S. A., Renfrow, S. N., Guo, B. N., Anthony, J. M., Zhao, Z. Y., and McDaniel, F. D. 1997a. TEAMS depth profiles in semiconductors. *Nucl. Instrum. Methods Phys. Res.* B123:571–574.

Datar, S. A., Zhao, Z. Y., Renfrow, S. N., Guo, B. N., Anthony, J. M., and McDaniel, F. D. 1997b. High sensitivity impurity measurements in semiconductors using trace-element accelerator mass spectrometry (TEAMS). *AIP Conf. Proc.* 392:815–818.

Datar, S. A., Wu, L., Guo, B. N., Nigam, M., Necsoiu, D., Zhai, Y. J., Smith, D. E., Yang, C., El Bouanani, M., Lee, J. J., and McDaniel, F. D. 2000. High sensitivity measurement of implanted As in the presence of Ge in $Ge_xSi_{1-x}$/Si layered alloys using trace element accelerator mass spectrometry. *Appl. Phys. Lett.* 77:3974–3976.

Döbeli, M., Nebiker, P. W., Suter, M., Synal, H.-A., and Vetterli, D. 1994. Accelerator SIMS for trace element detection. *Nucl. Instrum. Methods* B85:770–774.

Elmore, D. and Phillips, F. M. 1987. Accelerator mass spectrometry for measurement of long-lived radioisotopes. *Science* 236:543–550.

Elmore, D., Conard, N., Kubik, P. W., and Fabryka-Martin, J. 1984. Computer controlled isotope ratio measurements and data analysis. *Nucl. Instrum. Methods* B5:233–237.

Elmore, D., Hossain, T. Z., Gove, H. E., Hemmick, T. K., Kubik, P. W., Jiang, S., Lavine, J. P., and Lee, S. T. 1989. Depth profiles of nitrogen and Cl in pure materials through AMS of the neutron activation products $^{14}C$ and $^{36}Cl$. *Radiocarbon* 31:292–297.

Ender, R. M. 1997. Analyse von Spurenelementen mit Beschleuniger-Sekundärionen-Massenspektrometrie. Dissertation, Eidgenössiche Technische Hochschule, Zurich.

Ender, R. M., Döbeli, M., Suter, M., and Synal, H.-A. 1997a. Accelerator SIMS at PSI/ETH Zurich. *Nucl. Instrum. Methods Phys. Res.* B123:575–578.

Ender, R. M., Döbeli, M., Suter, M., and Synal, H.-A. 1997b. Characterization of the accelerator SIMS setup at PSI/ETH Zurich. *AIP Conf. Proc.* 392:819–822.

Ender, R. M., Döbeli, M., Nebiker, P., Suter, M., and Synal, H.-A., Vetterli, D., 1997c. A new setup for accelerator SIMS. *Proc. SIMS* 10:1018 –1022.

Ender R. M., et al. 1997d. *Applications of Accelerator SIMS.* PSI Annual Report Annex IIIA, p. 43.

Gillen, G., Lareau, R., Bennett, J., and Stevie, F. 1998. Secondary Ion Mass Spectrometry, SIMS XI. John Wiley & Sons, New York.

Gove, H. E., Kubik, P. W., Sharma, P., Datar, S., Fehn, U., Hossain, T. Z., Koffer, J., Lavine, J. P., Lee, S.-T., and Elmore, D. 1990. Applications of AMS to electronic and silver halide imaging research. *Nucl. Instrum. Methods Phys. Res.* B52:502–506.

Grabowski, K., Knies, D. L., Hubler, G. K., and Enge, H. A. 1997. A new accelerator mass spectrometer for trace element analysis at the Naval Research Laboratory. *Nucl. Instrum. Methods Phys. Res.* B123:566–570.

Houk, R. S. 1986. Spectrometry of inductively coupled plasmas. *Anal. Chem.* 58:97A–105A.

Hubbard, K. M., Weller, R. A., Weathers, D. L., and Tombrello, T. A. 1989. The angular distribution of atoms sputtered from a Ga-In eutectic alloy target. *Nucl. Instrum. Methods* B36: 395–403.

Jull, A. J. T., Beck, J. W., and Burr, G. S. 1997. *Nucl. Instrum. Methods Phys. Res.* B123:1–612.

Kilius, L. R., Baba, N., Garwan, M. A., Litherland, A. E., Nadeau, M.-J., Rucklidge, J. C., Wilson, G. C., and Zhao, X.-L., 1990. AMS of heavy ions with small accelerators. *Nucl. Instrum. and Methods in Phys. Res.* B52:357–365.

Kim, Y. D., Jin, J. Y., Matteson, S., Weathers, D. L., Anthony, J. M., Marshall, P., and McDaniel, F. D. 1995. Collision-induced interaction cross sections of 1-7 MeV $B_2$ ions incident on an $N_2$ gas target. *Nucl. Instrum. Methods Phys. Res.* B99:82–85.

Kirchhoff, J. F., Marble, D. K., Weathers, D. L., McDaniel, F. D., Matteson, S., Anthony, J. M., Beavers, R. L., and Bennett, T. J. 1994. Fabrication of Si-based optical components for an ultra-clean accelerator mass spectrometry ion source. *Rev. Sci. Instrum.* 65:1570–1574.

Knies, D. L., Grabowski, K. S., Hubler, G. K., and Enge, H. A. 1997a. 1-200 amu tunable Pretzel magnet notch-mass-filter and injector for trace element accelerator mass spectrometry. *Nucl. Instrum. Methods Phys. Res.* B123:589–593.

Knies, D. L., Grabowski, K. S., Hubler, G. K., Treacy Jr., D. J., DeTurck, T. M., and Enge, H. A. 1997b. Status of the Naval Research Laboratory trace element accelerator mass spectrometer: Characterization of the pretzel magnet. *AIP Conf. Proc.* 392:783–786.

Kruger, P. 1971. Principles of Neutron Activation Analysis. Wiley-Interscience, New York.

Litherland, A. E., Beukens, R. P., Doupe, J., Kieser, W. E., Krestow, J., Rucklidge, J. C., Tomski, I., Wilson, G. C., and Zhao, X.-L. 2000. Progress in accelerator mass spectrometry (AMS) research at IsoTrace. *Nucl. Instrum. Methods Phys. Res.* B172:206–210.

Litherland, A. E., Doupe, J. P., Tomski, I., Krestow, J., Zhao, X.-L., Kieser, W. E., and Beukens, R. P. 2001. Ion preparation systems for atomic isobar reduction in accelerator mass spectrometry. *AIP Conf. Proc.* 576:390–393.

Lu, S. 1997. Secondary ion mass spectrometry (SIMS) applications on characterization of RF npn bipolar polySi emitter process. *AIP Conf. Proc.* 392:823–826.

Maden, C., Döbeli, M., and Hoffman, B. 1999. Investigation of platinum group elements with accelerator SIMS. PSI Annual Report, Vol. 1, p. 199 (available at *http://ihp-power1.ethz.ch/IPP/tandem/Annual/1999.html*).

Maden, C., Frank, M., Suter, M., Kubik, P. W., and Döbeli, M. 2000. Investigation of natural $^{10}Be/Be$ ratios with accelerator mass spectrometry. PSI Annual Report, Vol. 1, p. 164 (available at *http://ihp-power1.ethz.ch/IPP/tandem/Annual/2000.html*).

Massonet, S., Faude, Ch., Nolte, E., and Xu, S. 1997. AMS with stable isotopes and primordial radionuclides for material analysis and background detection. *AIP Conf. Proc.* 392:795–797.

Massonet, S. 1998. Beschleunigermassenspektrometrie mit stabilen Isotopen und primordialen radionukliden. Ph.D Thesis, Technical University, Munich

Matteson, S., McDaniel, F. D., Duggan, J. L., Anthony, J. M., Bennett, T. J., and Beavers, R. L. 1989. A high resolution electrostatic analyzer for accelerator mass spectrometry. *Nucl. Instrum. Methods Phys. Res.* B40/41:759–761.

Matteson, S., Marble, D. K., Hodges, L. S., Hajsaleh, J. Y., Arrale, A. M., McNeir, M. R., Duggan, J. L., McDaniel, F. D., and Anthony, J. M. 1990. Molecular-interference-free accelerator mass spectrometry. *Nucl. Instrum. Methods Phys. Res.* B45:575–579.

McDaniel, F. D., Matteson, S., Weathers, D. L., Marble, D. K., Duggan, J. L., Elliott, P. S., Wilson, D. K., and Anthony, J. M. 1990a. The University of North Texas atomic mass spectro-

metry facility for detection of impurities in electronic materials. *Nucl. Instrum. Methods Phys. Res.* B52:310–314.

McDaniel, F. D., Matteson, S., Marble, D. K., Hodges, L. S., Arrale, A. M., Hajsaleh, J. Y., McNeir, M. R., Duggan, J. L., and Anthony, J. M. 1990b. Molecular dissociation in accelerator mass spectrometry for trace impurity characterization in electronic materials. *In* Proceedings of the High Energy and Heavy Ion Beams in Materials Analysis Workshop, Albuquerque, NM, June 14–16, 1989 (J. R. Tesmer, C. J. Maggiore, M. Nastasi, J. C. Barbour, and J. W. Mayer, eds.). pp. 269–278. Materials Research Society, Pittsburgh, PA.

McDaniel, F. D., Matteson, S., Weathers, D. L., Duggan, J. L., Marble, D. K., Hassan, I., Zhao, Z. Y., and Anthony, J. M. 1992. Radionuclide dating and trace element analysis by accelerator mass spectrometry. *J. Radioanal. Nucl. Chem.* 160:119–140.

McDaniel, F. D., Matteson, S., Anthony, J. M., Weathers, D. L., Duggan, J. L., Marble, D. K., Hassan, I., Zhao, Z. Y., Arrale, A. M., and Kim, Y. D. 1993. Trace element analysis by accelerator mass spectrometry. *J. Radioanal. Nucl. Chem.* 167:423–432.

McDaniel, F. D., Anthony, J. M., Kirchhoff, J. F., Marble, D. K., Kim, Y. D., Renfrow, S. N., Grannan, E. C., Reznik, E. R., Vizkelethy, G., and Matteson, S. 1994. Impurity determination in electronic materials by accelerator mass spectrometry. *Nucl. Instrum. Methods Phys. Res.* B89:242–249.

McDaniel, F. D., Anthony, J. M., Renfrow, S. N., Kim, Y. D., Datar, S. A., and Matteson, S. 1995. Depth profiling analysis of semiconductor materials by accelerator mass spectrometry. *Nucl. Instrum. Methods.* B99:537–540.

McDaniel, F. D., Datar, S. A., Guo, B. N., Renfrow, S. N., Matteson, S., Zoran, V. I., Zhao, Z. Y., and Anthony, J. M. 1998a. A method to reduce molecular interferences in SIMS: Trace-element accelerator mass spectrometry (TEAMS). *In* Proceedings of the Eleventh Inter. Conf. On Secondary Ion Mass Spectrometry, Orlando, FL, September 7–12, 1997 (G. Gillen, R. Lareau, J. Bennett, and F. Stevie, eds.). pp. 167–170. John Wiley & Sons, New York.

McDaniel, F. D., Datar, S. A., Guo, B. N., Renfrow, S. N., Zhao, Z. Y., and Anthony, J. M. 1998b. Low-level copper concentration measurements in Si wafers using trace-element accelerator mass spectrometry (TEAMS). *Appl. Phys. Lett.* 72:3008–3010.

Morvay, L. and Cornides, I. 1984. Triply charged ions of small molecules. *Int. J. Mass Spectrom. Ion Process.* 62:263–268.

Muller, R. A. 1977. Radioisotope dating with a cyclotron. *Science* 196:489–494.

Nelson, D. E., Korteling, R. G., and Stott, W. R. 1977. Carbon-14: Direct detection at natural concentrations. *Science* 198:507–508.

Niklaus, Th.R., Sie, S. H., and Suter, G. F. 1997. Australis: A microbeam AMS beamline. *AIP Conf. Proc.* 392:779–782.

Pellin, M. J., Young, C. E., Calaway, W. F., Burnett, J. W., Jorgensen, B., Schweitzer, E. L., and Gruen, D. M. 1987. Sensitive, low damage surface analysis using resonance ionization of sputtered atoms. *Nucl. Instrum. Methods* B18:446–451.

Pellin, M. J., Young, C. E., Calaway, W. F., Whitten, J. E., Gruen, D. M., Blim, J. D., Hutcheon, I. D., and Wasserburg, G. J. 1990. Secondary neutral mass spectrometry using 3-colour resonance ionization: Osmium detection at the ppb level and ion-detection in Si at the less than 200 ppt level. *Philos. Trans. R. Soc. Lond.* A333:133–146.

Purser, K. H., Liebert, R. B., Litherland, A. E., Beukens, R. P., Gove, H. E., Bennett, C. L., Clover, M. E., and Sondheim,

W. E. 1977. An attempt to detect stable atomic nitrogen (−) ions from a sputter source and some implications of the results for the design of tandems for ultra-sensitive carbon analysis. *Rev. Appl. Phys.* 12:1487–1492.

Purser, K. H., Litherland, A. E., and Rucklidge, J. C. 1979. Secondary ion mass spectrometry using dc accelerators. *Surf. Interface Anal.* 1:12–19.

Rucklidge, J. C., Evensen, N. M., Gorton, M. P., Beukens, R. P., Kilius, L. R., Lee, H. W., Litherland, A. E., Elmore, D., Gove H. E., and Purser, K. H. 1981. Rare isotope detection with tandem accelerators. *Nucl. Instrum. Methods* 191:1–9.

Rucklidge, J. C., Gorton, M. P., Wilson, G. C., Kilius, L. R., Litherland, A. E., Elmore, D., and Gove, H. E. 1982. Measurement of Pt and Ir at sub-ppb levels using tandem accelerator mass spectrometry. *Can. Mineral.* 20:111–119.

Rucklidge, J. C., Wilson, G. C., and Kilius, L. R. 1990a. AMS advances in the geosciences and heavy-element analysis. *Nucl. Instrum. Methods Phys. Res.* B45:565–569.

Rucklidge, J. C., Wilson, G. C., and Kilius, L. R. 1990b. In situ trace element determination by AMS. *Nucl. Instrum. Methods Phys. Res.* B52:507–511.

Rucklidge, J. C., Wilson, G. C., Kilius, L. R., and Cabri, L. J. 1992. Trace element analysis of sulfide concentrates from Sudbury by accelerator mass spectrometry. *Can. Mineral.* 30:1023–1032.

Saunders, W. A. 1989. Charge exchange and metastability of small multiply charged gold clusters. *Phys. Rev. Lett.* 62: 1037–1040.

Sharara, N. A, Wilson, G. C., and Rucklidge, J. C. 1999. Platinum-group elements and gold in Cu-Ni mineralized peridotite at Gabbro Akarem, Eastern Desert, Egypt. *Can. Mineral.* 37:1081–1097.

Sie, S. H. and Suter, G. F. 1994. A microbeam AMS system for mineralogical applications. *Nucl. Instrum. Methods Phys. Res.* 92:221–226.

Sie, S. H., Niklaus, T. R., and Suter, G. F. 1997a. Microbeam AMS: Prospects of new geological applications. *Nucl. Instrum. Methods Phys. Res.* B123:112–121.

Sie, S. H., Niklaus, T. R., and Suter, G. F. 1997b. The AUSTRALIS microbeam ion source. *Nucl. Instrum. Methods Phys. Res.* B123:558–565.

Sie, S. H., Niklaus, T. R., and Suter, G. F. 1997c. Prospects of new geological applications with a microbeam-ams. *AIP Conf. Proc.* 392:799–802.

Sie, S. H., Niklaus, T. R., Suter, G. F., and Bruhn, F. 1998. A microbeam Cs source for accelerator mass spectrometry. *Rev. Sci. Instrum.* 69:1353–1358.

Sie, S. H., Sims, D. A., Bruhn, F. Suter, G. F., and Niklaus, T. R. 1999. Microbeam AMS for trace element and isotopic studies. *AIP Conf. Proc.* 475:648–651.

Sie, S. H., Sims, D. A., Suter, G. F., Cripps, G. C., Bruhn, F., and Niklaus, T. R. 2000a. Precision Pb and S isotopic ratio measurements by microbeam AMS. *Nucl. Instrum. Methods* B172:228–234.

Sie, S. H., Sims, D. A., Niklaus, T. R., and Suter, G. F. 2000b. A fast bouncing system for the high energy end of AMS. *Nucl. Instrum. Methods* B172:268–273.

Sie, S. H., Sims, D. A., and Suter, G. F. 2001. Microbeam AMS measurements of PGE and Au trace and osmium isotopic ratios. *AIP Conf. Proc.* 576:399–402.

Vandervorst, W. and Shepherd, F. R. 1987. Secondary ion mass spectrometry profiling of shallow implanted layers using quadrupole and magnetic sector instruments. *J. Vac. Sci. Technol.* A5:313–320.

Weathers, D. L., McDaniel, F. D., Matteson, S., Duggan, J. L., Anthony, J. M., and Douglas, M. A. 1991. Triply-ionized B$_2$ molecules from a tandem accelerator. *Nucl. Instrum. Methods Phys. Res.* B56/57:889–892.

Wiebert, A., Erlandsson, B., Hellborg, R., Stenström, K., and Skog, G. 1994. The charge state distributions of carbon beams measured at the Lund pelletron accelerator. *Nucl. Instrum. Methods Phys. Res.* B89:259–261.

Wilson, G. C. 1998. Economic applications of accelerator mass spectrometry. *Rev. Econ. Geol.* 7:187–198.

Wilson, G. C., Rucklidge, J. C., and Kilius, L. R. 1990. Sulfide gold content of skarn mineralization at Rossland, British Columbia. *Econ. Geol.* 85:1252–1259.

Wilson, G. C., Kilius, L. R., and Rucklidge, J. C. 1995. Precious metal contents of sulfide, oxide and graphite crystals: Determinations by accelerator mass spectrometry. *Econ. Geol.* 90:255–270.

Wilson, G. C., Rucklidge, J. C., Kilius, L. R., Ding, G.-J., and Zhao, X.-L. 1997a. Trace-element analysis of mineral grains using accelerator mass spectrometry—from sampling to interpretation. *Nucl. Instrum. Methods Phys. Res.* B123:579–582.

Wilson, G. C., Rucklidge, J. C., Kilius, L. R., Ding, G.-J., and Cresswell, R. G. 1997b. Precious metal abundances in selected iron meteorites: In-situ AMS measurements of the six platinium-group elements plus gold. *Nucl. Instrum. Methods Phys. Res.* B123:583–588.

Wilson, G. C., Pavlish, L. A., Ding, G.-J., and Farquhar, R. M. 1997c. Textural and in-situ analytical constraints on the provenance of smelted and native archaeological copper in the Great Lakes region of eastern North America. *Nucl. Instrum. Methods Phys. Res.* B123:498–503.

Wilson, R. G., Stevie, F. A., and Magee, C. W. 1989. Secondary Ion Mass Spectrometry: A Practical Handbook for Depth Profiling and Bulk Impurity Analysis. John Wiley & Sons, New York.

Winograd, N., Baxter, J. P., and Kimock, F. M. 1982. Multiphonon resonance ionization of sputtered neutrals: A novel approach to materials characterization. *Chem. Phys. Lett.* 88:581–584.

Wittmaack, K. 1995. Small-area depth profiling in a quadrupole based SIMS instrument. *Int. J. Mass Spectrom Ion Process.* 143:19.

Zhao, Z. Y., Mehta, S., Angel, G., Datar, S. A., Renfrow, S. N., McDaniel, F. D., and Anthony, J. M. 1997. Use of accelerator mass spectrometry for trace element detection. *In* Proceedings of the Eleventh International Conference on Ion Implantation Technology, Austin, TX, June 16–21, 1996, Vol. 11 (E. Ishidida, S. Banerjee, S. Mehta, T. C. Smith, M. Current, L. Larson, A. Tasch, and T. Romig, eds.). pp. 131–134. IEEE, Piscataway, NJ.

## KEY REFERENCES

Benninghoven et al., 1987. See above.

*An excellent reference for SIMS.*

Datar et al., 1997a. See above.

*Discusses TEAMS techniques to depth profile impurities in materials.*

Datar et al., 2000. See above.

*A reference to using TEAMS to measure As in the presence of Ge in* $Ge_xSi_{1-x}$ *layers where a number of mass interferences exist.*

Kirchhoff et al., 1994. See above.

*Discusses how to construct an ultraclean ion source for TEAMS measurements.*

McDaniel et al., 1998b. See above.

*A reference to TEAMS measurements of low-level Cu impurities in Si.*

Sie et al., 1998. See above.

*Discusses a microbeam ion source for TEAMS.*

Wilson et al., 1989. See above.

*A complete reference to SIMS measurement techniques and contains many useful tables and figures to help determine negative ion formation probabilities and relative sensitivity factors.*

## INTERNET RESOURCES

http://mstd.nrl.navy.mil/6370/ams.html

*The Naval Research Laboratory, Washington, DC. Currently constructing a TEAMS facility by mating a SIMS instrument to a tandem accelerator. The NRL facility features parallel injection of a range of masses with a pretzel magnet and detection using a split-pole spectrograph. The contact person is Dr. Kenneth E. Grabowski. Email: grabowski@nrl.navy.mil.*

http://www.ams.physik.tu-muenchen.de

*The AMS group at the Technical University Munich. Performs radionuclide AMS and has recently built an ultraclean ion source for stable isotope impurities for materials analysis. The project leader is Dr. Eckehart Nolte. Email: eckehart_nolte@physik.tu-muenchen.de.*

http://www.cea.com/tutorial.htm

*Charles Evans and Associates, specialists in materials characterization. Provides an excellent SIMS tutorial.*

http://www.phys.ethz.ch/ipp/tandem

*The AMS facility at PSI/ETH Ion Beam Physics Institute of Particle Physics HPK, ETH Hönggerberg, CH-8093, Zurich, Switzerland. Mainly involved in radionuclide AMS and has built a separate ion source for TEAMS measurements. Dr. Martin Suter is project leader. Email: suter@particle.phys.ethz.ch.*

http://www.physics.utoronto.ca/~isotrace

*The IsoTrace Laboratory in the Department of Physics at the University of Toronto, Toronto, Canada. One of the first laboratories to make radionuclide AMS measurements and also to make stable isotope measurements of the platinum group elements, Au, and Ag in mineral grains and meteorites. The project leader is Dr. W. E. "Liam" Kieser. Email: liam.kieser@ utoronto.ca.*

http://www.phys.unt.edu/ibmal

*The TEAMS facility at the Ion Beam Modification and Analysis Laboratory, Department of Physics, University of North Texas, Denton, Texas. Used to routinely analyze bulk materials as well as depth profiles of impurities. Dr. Floyd Del McDaniel is the project leader of the TEAMS research at UNT. Email: mcdaniel@unt.edu.*

http://www.syd.dem.CSIRO.AU/research//hiaf/AUSTRALIS

*The AUSTRALIS project at CSIRO, Sydney, Australia. An AMS system with a microprobe ion source. This system is designed primarily for in situ microanalysis of geological samples. The web site is case sensitive. Dr. Soey Sie is project leader at CSIRO in Australia. Email: soey.sie@dem.csiro.au.*

## APPENDIX A:
## GLOSSARY OF TERMS

| | |
|---|---|
| AMS | Accelerator mass spectrometry |
| Charge state | The amount of charge on an ion. Usually refers to positive ions since only singly charged negative ions have been observed. Obviously, the maximum charge state of an elemental ion corresponds to its atomic number. |
| Class-100 clean room | A room with atmospheric particulate concentration of particles greater than 0.5 μm diameter not exceeding 100 particles/ft$^3$ (or 3500 particles/m$^3$) |
| CNF | Concentration normalization factor |
| cps | Counts per second |
| Depth profile | Plot of elemental concentration as a function of depth from the surface |
| Detection limit | Lowest meaningful concentration level that can be measured |
| ICPMS | Inductively coupled plasma mass spectrometry |
| Ionization chamber | An instrument used to measure the energy of a particle. It is basically a parallel-plate capacitor filled with a gas. Energetic particles entering the chamber lose energy by ionizing the gas. The amount of ionization produced is proportional to the energy. It is a very sensitive instrument capable of detecting single particles. |
| ion source | Produces ions for injection into accelerator |
| INAA | Instrumental neutron activation analysis |
| NAA | Neutron activation analysis |
| NA-AMS | Neutron activation followed by accelerator mass spectroscopy |
| NRA | Nuclear reaction analysis |
| PGEs | Platinum group elements (Ru, Rh, Pd, Os, Ir, Pt) |
| PIXE | Particle-induced X-Ray emission |
| ppb | parts per billion (1/10$^9$) |
| ppm | parts per million (1/10$^6$) |
| ppt | parts per trillion (1/10$^{12}$) |
| Primary ion | Ion incident on sample surface that sputters the sample. Typically for TEAMS it is Cs$^+$. |
| RBS | Rutherford backscattering spectroscopy |
| RSFs | Relative sensitivity factors |
| Secondary ion | Ion sputtered from the sample by the primary ion. Can be positive or negative. Neutral atoms or molecules can also be sputtered from the sample surface. |
| Sensitivity | Lowest measurable concentration or detection limit in atoms per cubic centimeters |
| SIMS | Secondary-ion mass spectrometry |
| SIRIS | Sputter-initiated resonance ionization spectrometry |
| Solid-state or surface barrier | A semiconductor detector that works in a manner similar to an ionization |

detector    chamber but is substantially more compact. Energetic particles incident on the detector create electron-hole pairs in the depletion region of a p-n junction. The electrons and holes are swept off in opposite directions by an applied electric field, constituting a current pulse that generates a signal. The signal strength is roughly proportional to the energy of the incident particles.

Tandem accelerator    Electrostatic accelerator based on the Van De Graaff principle, with a high positive potential at the center terminal. The center terminal is charged to a high potential by charges carried from ground by a belt or a chain. The accelerating column is enclosed in a pressure vessel filled with $SF_6$ or other insulating gases.

TEAMS    Trace element accelerator mass spectrometry or trace element atomic mass spectrometry

TOF-SIMS    Time of flight–secondary ion mass spectrometry

## APPENDIX B:
## INSTRUMENT SPECIFICATIONS AND SUPPLIERS

1. Cs gun: General Ionex, model 133 surface ionization Cs source. This gun is no longer made. Very similar guns are sold by Peabody Scientific, Atomika.

2. Cs analyzing magnet: Made by Magnecoil, Peabody, MA. Supplied by National Electrostatics Corp.

3. Cs Analyzing Magnet Power supply, TCR 20T125: Made by Electronic Measurements.

4. Sample chamber: Made to specification by Huntington Mechanical Laboratories, 1040 L'Avenida, Mountain View, CA 94043. Chamber features four-axis manipulator and load-lock for sample changing.

5. Electron flood gun for charge compensation; model EFG-8 with EGPS-8 power supply: Kimball Physics, 311 Kimball Hill Rd. Wilton NH, 03086.

6. 45° secondary-ion electrostatic analyzer: In-house fabrication.

7. 90° analyzing magnet: Supplied by National Electrostatics Corp.

8. 3MV, electrostatic, tandem pelletron accelerator, model 9SDH-2: Part of the accelerator beamline, supplied by National Electrostatics Corp., 7540 Graber Rd. Box 620310, Middleton, WI 53562-0310.

9. 40° bending magnet: HVEE Europa BV PO Box 99, 3800 AB Amersfoort, The Netherlands. Mass-Energy product $(mE/q^2) \sim 60$.

10. 45° double-focusing spherical electrostatic analyzer: In-house fabrication.

11. Si surface-barrier particle detector: Ortec, 801 S. Illinois Ave., Oak Ridge, Tenn. 37831-0895.

12. Total energy detector ionization chamber: In-house development.

13. Data acquisition and control software: Developed in-house using LabWindows and LabView 4.1 (National Instruments) for PCs using a CAMAC crate with a GPIB interface.

FLOYD DEL McDANIEL
Ion Beam Modification and
Analysis Laboratory
University of North Texas
Denton, Texas

# INTRODUCTION TO MEDIUM-ENERGY ION BEAM ANALYSIS

## INTRODUCTION

The earliest applications of ion beam techniques for materials analysis, typified by the work of Tollestrup et al. (1949), were carried out with mega-electron-volt beams in the course of basic research in nuclear physics, and the mega-electron-volt beam techniques (described in Section 12a of this chapter) still form the indispensable core of the subject. In the broadest sense, the importance of these techniques lies in the fact that the information they yield depends upon nuclear properties and interactions and is, therefore, relatively insensitive to perturbation by local properties such as the chemical environment and electronic structure. The relevant nuclear physics and scattering theory are well known experimentally and often theoretically, and in consequence, important parameters such as interaction cross-sections are known from first principles (PARTICLE SCATTERING). This is in sharp contrast to the situation for some other important techniques (e.g., secondary ion mass spectrometry) where the fundamental measurement interactions may be very vulnerable to the unique idiosyncrasies of a specimen. It is essential to stress the unity of ion beam analysis even as one seeks to differentiate the relatively newer medium-energy techniques from their more mature high-energy predecessors.

First it is important to understand what is meant by "medium energy." There are two answers, one historical and one based upon the physics of scattering. Physically, ion beam techniques may be termed medium energy if the energy of the beam is low enough that the deviation of the scattering cross-section from the unscreened Rutherford value is greater than $\sim 1\%$. In addition, the energy must be high enough to assume both that the beam interacts with the target by binary elastic collisions (except when it is traveling in a channeling direction) and that most backscattered particles remain ionized. (The latter constraint differentiates medium- from low-energy ion scattering, where it is assumed that the ionized component of backscattered particles is strongly correlated with scattering events at the specimen's surface.) For practical purposes, medium energy encompasses ion beam energies from a few tens of to a few hundreds of kilo-electron-volts, and even a bit higher for heavy ions being used in forward-recoil work. Below this range, ion scattering becomes increasingly surface specific. Above it lies the full range of conventional elastic scattering and nuclear reaction analysis.

Historically, medium energy has been defined by technology. The field of ion beam analysis owes its ubiquity, if not its existence, to a single device, the silicon surface-barrier particle detector (see the review by McKenzie, 1979, and references therein). A surface-barrier detector is essentially a large-area Schottky-barrier diode. It has approximately unit efficiency for detecting particle radiation, produces an output signal that is highly linear with the energy of the particle striking it, and is remarkably inexpensive to produce and operate. Along with the lead pencil and the digital compact disk, it is an example of a technology that, from its inception, was so elegant in design, so well suited to its purpose, and so demonstrably superior in function and economics to competitive alternatives that it at once revolutionized a field of activity. However, surface-barrier detectors experience diminished relative energy resolution when used for $\alpha$ particle radiation below $\sim$500 keV. As a result, for years almost all ion beam analysis has been conducted using projectiles with energy greater than this.

At the low end of the energy scale, a technique called low-energy ion scattering (known both by the acronym LEIS and, in early literature, as ISS, for ion surface scattering) was used to study surface structure. Typically, this work was done by pulsing the accelerator beam and using an electron multiplier with exposed cathode to detect scattered ions. Formidable technical problems with producing adequately short beam pulses, as well as other fundamental issues, placed a practical upper limit on this technique of 20 to 30 keV. Thus, the historic range of medium energy began above LEIS and ranged up to the domain of the surface-barrier detector.

As is the case for high-energy ion beam analysis, most of the tools needed for medium-energy ion beam analysis have been developed and perfected for nuclear research and are widely available commercially. This is notably not the case for spectrometers, which still fall largely in the class of research instrumentation and are available, but less readily so.

The units that follow describe some medium-energy ion beam analytical techniques that are optimized for different problems. Time-of-flight medium-energy backscattering and forward-recoil spectrometry, described in MEDIUM-ENERGY BACKSCATTERING AND FORWARD-RECOIL SPECTROMETRY, are quite similar to the high-energy techniques of Section 12a but offer increased depth resolution, sensitivity, and surface specificity at the expense of total analyzable depth and ease of use. Forward-recoil measurement of hydrogen at medium energies can be particularly effective, especially when it is important to measure hydrogen and other light elements simultaneously.

Heavy-ion backscattering spectrometry, described in HEAVY-ION BACKSCATTERING SPECTROMETRY, is a variant of time-of-flight backscattering that is optimized for the highest possible sensitivity, specifically for the measurement of ultra-low-level metallic contaminants on device-grade Si wafer surfaces. Along with total-reflection x-ray fluorescence spectrometry, it is the definitive method currently available for this measurement.

Medium-energy ion scattering, which uses a toroidal electrostatic spectrometer to analyze both energy and angle of scattered particles, is the oldest of the medium-energy techniques and is a definitive method for the precise lattice location of surface atoms by a detailed analysis of channeling and blocking.

If you are reading this, you are probably trying to decide if a medium-energy ion beam technique is right for your problem. If you have not done so, you should first familiarize yourself with the general principles of ion scattering in PARTICLE SCATTERING. If you can solve your problem with one of the high-energy techniques described in Section 12a, then you should probably do so. Medium-energy techniques are appropriate for specialized problems and circumstances as described in the pages that follow.

## LITERATURE CITED

McKenzie, J. M. 1979. Development of the semiconductor radiation detector. *Nucl. Instrum. Methods* 162:49–73.

Tollestrup, A. V., Fowler, W. A., and Lauritsen, C. C., 1949. Energy release in beryllium and lithium reactions with protons. *Phys. Rev.* 76:428–450.

ROBERT A. WELLER
Vanderbilt University
Nashville, Tennessee

# MEDIUM-ENERGY BACKSCATTERING AND FORWARD-RECOIL SPECTROMETRY

## INTRODUCTION

Medium-energy backscattering spectrometry is a variation of Rutherford backscattering spectrometry (RBS) (ELASTIC ION SCATTERING FOR COMPOSITION ANALYSIS) using ions with energies ranging from a few tens to a few hundreds of kilo-electron-volts. The ion beam, frequently $He^+$, is directed onto a specimen and the number of backscattered ions per incident ion is measured as a function of scattered-ion energy at one or more angles relative to the beam direction. The fundamental physical process being used is that of a binary collision between the nuclei of an ion in the beam and a near-stationary atom in the target. In such a collision, the energy of the backscattered ion is determined uniquely by its original energy, the scattering angle (which is the angle between the ion's trajectories before and after the collision), and the ratio of the masses. Consequently, a measurement of the energy of a backscattered ion at a specific angle may be used to infer the mass of the target atom with which it collided. The number of backscattered ions is proportional to the collision cross-section and the total number of available scattering centers. Since collision cross-sections are known *a priori* in this energy range, the number of backscattered ions with a given energy is a direct measure of the number of atoms of the corresponding mass in the target.

The method achieves additional richness as the result of a second physical phenomenon, the gradual loss of kinetic energy through collisions as an ion penetrates through matter. This is measured by stopping power (also called linear energy transfer), which is the average energy lost, typically in electron volts per unit of length (measured in atoms per square centimeter) of trajectory in the target. If an ion passes through the surface of a specimen and experiences a collision at some depth within it, it will lose energy on both the inward and outward legs of its trajectory. Consequently, a second factor controlling the energy of backscattered ions is the depth within the target at which the collision occurs. As a result, it is usually possible to extract quantitative information about the distribution of specific species within the near-surface region of the specimen. There are a number of excellent reviews of ion penetration in the literature that include both theoretical background and algorithms for computing stopping power (Ziegler et al., 1985; Rauhala, 1995; Nastasi et al., 1996). For the purposes of this unit, it is sufficient to note that for medium-energy ions the stopping power is dominated by collisions between the nucleus of the projectile and the electrons of the target, so-called electronic stopping, which results in relatively uniform loss of energy with minimal angular divergence of the beam. Widely accepted semiempirical procedures are available for computing electronic stopping power.

Additional information can often be derived from backscattering spectra for specimens that are crystalline or multilayer structures with some crystalline and other amorphous layers by exploiting the phenomenon of particle channeling (Feldman et al., 1982; Swanson, 1995). At medium energies, the angular widths of particle channels, which vary as the inverse square root of particle energy, are of the general order of a degree. This can be exploited through precise control of target orientation in order to obtain information about target crystallinity or additional information about target structure or to deliberately suppress scattering from one region of a specimen in order to emphasize that from another. Channels are so wide at medium energies that accidental channeling occurs frequently and can adversely affect the precision of measurements.

Backscattering spectrometry is most formidable for the analysis of higher atomic mass constituents of a sample, because the spectral features attributable to lower-atomic-mass surface constituents such as carbon or oxygen will occur at the same energy as features attributable to higher-atomic-mass constituents such as silicon located more deeply in the sample. Thus, a heavier substrate produces a background that complicates the analysis of light elements on the surface. Forward-recoil spectrometry, particularly when done by time of flight, addresses this situation.

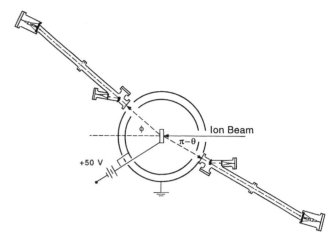

**Figure 1.** Schematic of the vacuum chamber of a medium-energy ion scattering apparatus showing the target, a forward-scattering and backscattering spectrometer, and their relative orientations with respect to the incident ion beam. The target is biased at +50 V and is situated in a large Faraday cup.

Forward-recoil and backscattering spectrometry, as the names imply, differ primarily in the geometry of the basic ion beam–target collision (Fig. 1). In medium-energy forward-recoil spectrometry, in the laboratory the detector of scattered ions is located at a direction that is typically ~30° with respect to the original direction of the ion beam. In binary collisions between beam ions and atoms of the target, the latter are forbidden from having laboratory scattering angles exceeding 90°, so that the features of backscattering spectra are all attributable to scattered ions from the beam. In forward-recoil spectrometry, both target and beam species are present. The optimum situation for light-element detection is achieved by a combination of time-of-flight spectrometry and the use of a beam projectile with higher atomic mass than the target elements of interest. When a heavy projectile strikes a light target, the latter moves away at a speed that, in the laboratory, can be nearly twice that of the former. As a result, at the fixed angle of the detector, of the common light elements hydrogen will be moving most swiftly, followed by carbon, nitrogen, and so on. By measuring the time that these particles require to traverse a fixed distance, they may be sorted by element in a spectrum devoid of the background that would be present in either a backscattering spectrum (where hydrogen would be unobservable) or a forward-recoil measurement using an energy-dispersive detector. Medium-energy forward-recoil spectrometry is among the most sensitive techniques available for the quantitative measurement of surface hydrogen, particularly when it is important to measure other light elements, such as carbon, nitrogen, or oxygen, simultaneously.

Particle detection technology is an extremely important issue in medium-energy ion beam analysis. The remainder of this unit will focus on time-of-flight spectrometry optimized for high-resolution measurements. Other methods, including magnetic and electrostatic spectrometers, pulsed-beam accelerator systems, and time-of-flight spectrometry optimized for other measurements, will be discussed briefly (see Complementary and Alternative Techniques, below), along with several other methods for depth profiling, trace-element identification, and hydrogen measurement. Throughout, special attention will be paid to the relationship between medium-energy backscattering and conventional high-energy RBS.

## Complementary and Alternative Techniques

Time-of-flight medium-energy backscattering is a variation of conventional (high-energy) RBS that has been optimized for high-resolution depth profiling of the first few tens of nanometers of the surface of multilayer planar structures. When devising an ion beam analytical strategy for a specific specimen, it is best to determine first if high-energy ($\approx$2-MeV) backscattering will provide satisfactory information. If it will, then considerations of cost, availability, and ease of interpretation of the data all argue that it should be used. Medium-energy backscattering should be considered if the region of interest is within a few tens of nanometers of the surface and if the depth resolution of conventional RBS appears to be inadequate. Medium-energy backscattering can routinely achieve near-surface depth resolutions better by about a factor of 4 than high-energy RBS with a surface-barrier detector. For the special case of profiling oxygen in $SiO_2$ films on Si, one should also consider medium-energy proton scattering using a toroidal electrostatic analyzer.

In general, it is always best to use an ion beam technique for elemental depth profiling, if one is available. This is because absolute cross-sections are known, and therefore measurements are quantitative, and because ion beam techniques typically look at the specimen in depth. An alternative procedure is to combine a highly surface-specific technique with the gradual erosion of the specimen by either chemical erosion or sputtering. Examples would include secondary ion mass spectrometry (SIMS) and sputter Auger profiling (see AUGER ELECTRON SPECTROSCOPY). SIMS is extremely sensitive to trace elements. When it is performed in conjunction with slow near-threshold sputtering of a sample, it is termed "static" and can be used for elemental depth profiling by observing secondary ion yields as a function of duration of erosion. This procedure can produce excellent depth profiles of many elements simultaneously, but like all variations of SIMS, absolute values are not readily obtained. In sputter Auger electron spectrometry, one or more specific Auger lines are monitored as a function of erosion. Both static SIMS and sputter Auger require an independent measurement of erosion to assign an absolute depth scale. With ion beam techniques, the depth scale follows from a knowledge of the stopping power.

Medium-energy forward-recoil spectrometry is particularly well suited to measure surface hydrocarbons, especially when they are on metallic or other high-atomic-mass substrates. Alternative techniques include high-energy forward recoil (Barbour and Doyle, 1995; Tirira et al., 1996) and, for specific cases such as hydrogen, nuclear reaction analysis. At mega-electron-volt energies, forward-recoil spectrometry has been carried out both by time-of-flight spectrometry and by the use of surface-

barrier detectors in conjunction with thin foils that function essentially as filters to reject forward-scattered beam ions (Barbour and Doyle, 1995). Whenever possible, all forward recoil, regardless of beam energy, should be carried out using time-of-flight techniques. This is because velocity and not energy is the appropriate natural parameter for dispersive analysis. Medium-energy forward-recoil spectrometry should be considered when very high sensitivity is a goal of the experiment. It may also be appropriate when a specimen is very fragile and must receive the minimum possible radiation dose consistent with analysis. Since damage cross-sections often vary as $E_0^{-1}$ while scattering cross-sections vary as $E_0^{-2}$, the lowest energy suitable for analysis is optimum. Medium-energy forward-recoil measurements can be made on thin organic foils with as few as $10^{11}$ total ions striking the target (Arps and Weller, 1995).

The preferred method for hydrogen measurement on and near the surfaces of materials is by nuclear resonant reaction analysis using $^1H(^{15}N, \alpha\gamma)^{12}C$ or $^1H(^{19}F, \alpha\gamma)^{16}O$ (PARTICLE-INDUCED X-RAY EMISSION). However, neither of these methods is capable of simultaneously returning information about other components of the specimen. Medium-energy forward-recoil spectrometry provides information about all low-atomic-mass constituents at once and, therefore, can be used in situations where the picture obtained by nuclear reaction analysis is incomplete. Medium-energy forward-recoil spectrometry has been shown to be capable of achieving a depth resolution and sensitivity for hydrogen of $\sim$6 nm and $10^{13}$ cm$^{-2}$, respectively (Arps and Weller, 1996). SIMS is also used successfully for hydrogen measurement, but is not as predictable as nuclear reaction or forward-recoil analysis because the cross-section for secondary hydrogen ion production is a function of the state of the surface and is not well known.

## PRINCIPLES OF THE METHOD

The definitive treatment of backscattering spectrometry is the book by Chu et al. (1978). The subject has been reviewed recently by Leavitt et al. (1995) and also in ELASTIC ION SCATTERING FOR COMPOSITION ANALYSIS of this work, which describes conventional mega-electron-volt methods. What follows is a brief review of the basic principles of backscattering spectrometry with particular emphasis on those aspects in which medium-energy techniques differ from their high-energy counterparts. Additional material specific to forward-recoil spectrometry is also provided. The basic geometry of a medium-energy backscattering experiment with a time-of-flight spectrometer is shown in Figure 1.

When a particle with energy $E_0$ and mass $M_1$ strikes a stationary target particle with mass $M_2 > M_1$ and is observed to be scattered by an angle $\theta$ in the laboratory, then the energy of the scattered particle will be

$$E_1 = K \cdot E_0 = E_0 \left( \frac{\{x \cdot \cos(\theta) + [1 - x^2\sin^2(\theta)]^{1/2}\}^2}{(1 + x^2)^2} \right) \quad (1)$$

where $x \equiv M_1/M_2$. This expression defines the kinematic factor $K$, which relates the incident energy of a particle to its energy after the collision. This is the fundamental relationship by which the mass $M_2$ of a surface constituent is obtained from the measured backscattered energy $E_1$.

When an ion beam is normally incident on a surface covered by a thin layer of thickness $t$, the observed number of backscattering events $Y_i$ attributable to the $i$th constituent is given by the relationship

$$Y_i = \eta(K \cdot E_0) \cdot Q \cdot (N_i \cdot t) \cdot \Omega \cdot \sigma_i(E_0, \theta) \qquad (2)$$

where $N_i$ is the density of the $i$th component in atoms per cubic centimeter, $Q$ is the total number of incident projectiles, $\Omega$ is the solid angle of the spectrometer, and $\sigma_i$ is the differential cross-section for the collision (PARTICLE SCATTERING). Strictly speaking, one should integrate the cross-section over the solid angle subtended by the detector, but in practice the solid angle is usually so small that the approximation of Equation 2 is satisfactory. The remaining term, $\eta(K \cdot E_0)$, is the spectrometer quantum efficiency, which is the probability that a particle incident upon the detector will result in a measurable event (Weller et al., 1994). The efficiency is a function of the energy of the backscattered particle and, implicitly, of the particle species. The presence of the efficiency term in Equation 2 is one of the features that distinguishes medium-energy backscattering from its higher-energy analog. At mega-electron-volt energies, the particle detector of choice is the silicon surface barrier detector. The quantum efficiency of surface barrier detectors is so near unity that it is almost universally ignored, and linearity of the output signal with particle energy is also assumed (Lennard et al., 1986). As a result, $N_i t$ can be directly inferred from a measurement of $Y_i$, $Q$, and $\Omega$, the latter being, at least in principle, a simple exercise in geometry. This, of course, supposes a knowledge of $\sigma_i$, but this is known from theory for mega-electron-volt scattering to be the Rutherford cross-section, given in the laboratory reference frame (Darwin, 1914; Chu et al., 1978) by

$$\sigma_R(E_0, \theta) = \left(\frac{Z_1 Z_2 e^2}{2 E_0}\right)^2 \left(\frac{\{\cos(\theta) + [1 - x^2\sin^2(\theta)]^{1/2}\}^2}{\sin^4(\theta)[1 - x^2\sin^2(\theta)]^{1/2}}\right) \qquad (3)$$

Here $Z_1$ and $Z_2$ are the atomic numbers of the projectile and the $i$th target constituent, respectively, and $e^2$ is the square of the electron's charge, which may be conveniently expressed as $1.44$ eV$\cdot$nm.

In medium-energy work, both the unit efficiency and Rutherford cross-section approximations fail, adding significantly to the complexity of interpreting medium-energy backscattering data. It is perhaps some consolation, however, that they only appear as a product $\eta \cdot \sigma$ (see Equation 2) and therefore can be treated as a single entity. Discussion of the efficiency will be postponed until after the discussion of the general features of a time-of-flight spectrometer (see Practical Aspects of the Method, Spectrometer Efficiency). The scattering cross-section will be considered first.

A general rule for the range of validity of the Rutherford cross-section has been given by Davies et al. (1995). They assert that the above expression should be within $\pm 4\%$ between the limits of $0.03 Z_1 \cdot Z_2^2$ keV and $0.3 Z_1 \cdot Z_2^{2/3}$ MeV. Above this limit, nuclear reactions begin to occur. Below it, in the range of medium-energy collisions, the presence of electrons significantly screens the collision. For the common case of a He$^+$ particle beam, the lower threshold for silicon, $Z_2 = 14$, is $\sim 12$ keV, while for gold, $Z_2 = 79$, it is $\sim 370$ keV. The departure of the cross-section from the Rutherford value at low energies has been studied extensively (L'Ecuyer et al., 1979, and references therein) and found to be well described by the Lenz-Jensen (Andersen et al., 1980) screened Coulomb potential. For reference purposes, algorithms have been developed (Mendenhall and Weller, 1991) that are sufficiently accurate and efficient to permit the classical cross-section for this potential to be computed in real time. However, in view of the uncertainty in the spectrometer efficiency, the following simple expression for the ratio of the screened to the Rutherford cross-section proposed by Andersen et al. (1980), which is typically accurate to a few percent, is recommended for most practical calculations:

$$\frac{\sigma}{\sigma_R} = \left(1 + \frac{48.73 Z_1 Z_2 (1 + x)(Z_1^{2/3} + Z_2^{2/3})^{1/2}}{E_0(\text{eV})}\right)^{-1} \qquad (4)$$

where, as indicated, $E_0$ is expressed in electron volts, and the mass ratio $x$ has been defined above. Figure 2A shows the ratio of the Lenz-Jensen screened cross-section to the Rutherford cross-section for He$^+$ ions on Si, Cu, and Au. Figure 2B, which shows a comparison of the approximation of Equation 4 with the exact Lenz-Jensen cross-section for the extreme case of He$^+$ on gold, serves to illustrate the quality of the approximation.

Figure 3 presents a graph of the stopping power of He$^+$ ions in Si, Cu, and Au in the range of Rutherford and medium-energy backscattering. As described above, stopping power is the physical phenomenon that gives backscattering spectrometry its depth-profiling ability. From this graph alone, one would conclude that, because the stopping power is smaller, medium-energy depth profiles would, in principle, contain less information. However, the loss in intrinsic depth differentiation is more than made up by the gains in spectrometer resolution made possible by exchanging a surface barrier detector for a time-of-flight spectrometer.

Backscattering data consist of spectra made by producing a histogram of the number of backscattered particles observed within energy bins of equal width. Figures 4A–D contain computer-simulated backscattering spectra as they would be observed with a surface barrier detector (Figs. 4A–C, $E_0$ equals 2 MeV, 1 MeV, and 500 keV, respectively) and a time-of-flight spectrometer (Fig. 4D, $E_0 = 250$ keV). In all cases the specimen is assumed to be a device-grade silicon wafer covered with 20 nm of SiO$_2$ (density 2.22 g/cm$^3$), 10 nm of TiN$_{0.8}$ (molecular density same as stoichiometric TiN), 15 nm of TiN (density 4.94 g/cm$^3$), and a trace contamination of W on the surface with an areal density of $4 \times 10^{14}$ cm$^{-2}$. In all cases, the He$^+$

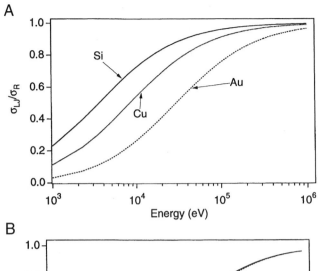

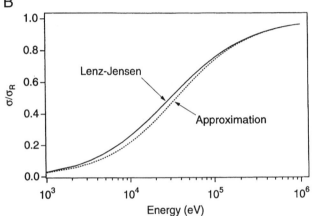

**Figure 2.** Medium-energy scattering cross-sections. (**A**) The ratio of the Lenz-Jensen cross-section to the Rutherford cross-section, as a function of energy, for He$^+$ scattering from several target nuclei at a laboratory angle of 150°. (**B**) A comparison of the ratio of the Lenz-Jensen cross-section to the Rutherford cross-section and the approximation to this ratio suggested by Andersen et al. (1980). The approximation is given here as Equation 4. The comparison is for He$^+$ scattering from Au at 150° in the laboratory.

beam is incident normally on the surface and the scattering angle is 150°. The amount of deposited charge is 100 μC and the solid angle of both detectors is 1 msr. Both spectrometers are assumed to have unit efficiency as a function of energy. The resolution of the surface barrier detector is taken to be 13 keV, which is typical of commercial devices when new. The resolution of the medium-energy system is more complicated since it is energy dependent, but it has been taken to be the experimentally determined resolution of a typical time-of-flight spectrometer, expressed in the energy domain. For the simulation, all isotopes of all the elements present have been included explicitly, as has energy straggling, using the results of Yang et al. (1991). Stopping power has been computed using the methods of Ziegler et al. (1985) and cross-sections using the algorithms of Mendenhall and Weller (1991). It is assumed that the spectra have been taken in such a way that channeling is not significant. In practice, this is probably best accomplished by aligning the beam a few degrees from a known channel or by rotating the target

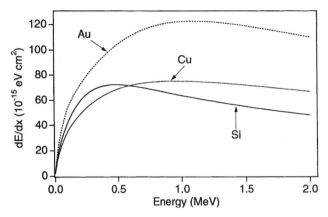

**Figure 3.** The stopping power of He ions in Si, Cu, and Au as a function of energy. These were computed as described in Ziegler et al. (1985).

continuously during the measurement. The plots have been scaled vertically to mimic data taken with a multi-channel analyzer adjusted for a conversion gain of 1 keV per channel for the surface barrier spectra and 250 eV per channel for the time-of-flight spectrum. The effects of counting statistics have been omitted for clarity. These figures help to illustrate graphically the relative merits of mega-electron-volt and kilo-electron-volt backscattering.

The small isolated peak at the right of each of the spectra shown in Figures 4A–D is attributable to the trace contamination of W on the sample surface. To quantify the surface concentration, one would simply use Equation 2 with the total yield taken as the integrated number of counts comprising the peak. Because the W is not distributed in depth but instead is concentrated on the surface, the width of the peak is a measure of the system resolution. For the surface barrier spectra this energy resolution is approximately constant as a function of energy. For the time-of-flight spectrum the resolution is poorer at larger energies. For the spectra of Figures 4A and B the large isolated peak is attributable to Ti, and its width is also narrow. However, it shows some broadening relative to the W peak. In Figure 4C it is clearly apparent that the Ti peak has width, but no structure is visible. In all these spectra, the total Ti can be easily measured by the same procedure used for quantifying W.

The pronounced step feature near the center of the spectra of Figures 4A–D is attributable to silicon in the SiO$_2$ layer and in the silicon substrate, and is referred to as the silicon edge. Because this target is thick, the silicon signal occupies the full range of the spectrum below (to the left of) the edge. Notice the two small peaks visible above the silicon background in Figure 4A. These are attributable to nitrogen and oxygen, with the oxygen peak being at higher energy. For beam energies of 1 and 2 MeV these are fully resolved, but at 0.5 MeV, the surface barrier detector is unable to separate these peaks.

The spectrum of Figure 4D shows the view of the specimen, again omitting statistical fluctuations, that would be obtained by time-of-flight medium-energy backscattering. (See Practical Aspects of the Method for a method to derive an energy spectrum from raw time-of-flight data.) The

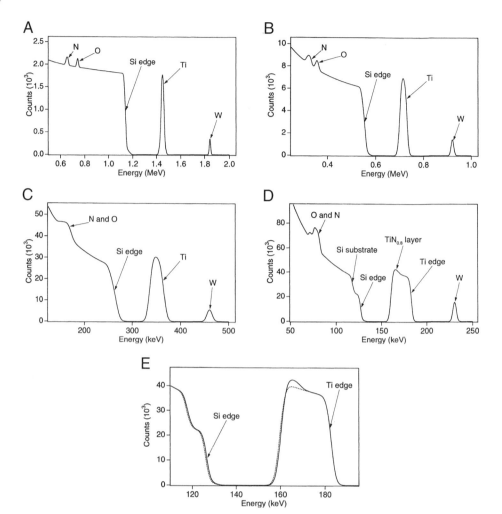

**Figure 4.** A computer simulation comparing backscattering at various energies using a surface barrier detector (**A–C**) and a time-of-flight spectrometer (**D–E**). The target is a TiN film, with a nitrogen-deficient region, on $SiO_2$ on Si. There is a trace of W contamination on the surface. (**E**) An expanded view of the Ti peak from (**D**) along with the result (dotted line) that would be expected from a fully stoichiometric TiN film, demonstrating that the nitrogen-deficient region is clearly discernible.

important points to notice are that the Ti peak now clearly contains structure to the extent that it is higher on the left than the right side, and that the silicon edge now contains definite evidence of a step. From these observations alone, it is possible to conclude that the TiN layer does not have uniform composition and that the oxygen is associated with silicon in the form of an $SiO_2$ film whose thickness can be inferred from the spectrum. This additional insight has been achieved at the expense of the loss of resolution of the nitrogen and oxygen features, which are now fully merged. This is demonstrated clearly in Figure 4E, which compares the Ti peak with a portion partially depleted in N against a simulation of the same target with a single stoichiometric layer of TiN 25 nm thick.

The method of inferring the stoichiometry of a thin film from backscattering data is the same for both medium-energy and high-energy backscattering and is described at length by Chu et al. (1978) and by Leavitt et al. (1995) as well as in ELASTIC ION SCATTERING FOR COMPOSITION ANALYSIS. Similarly, procedures have been devised to infer depth profiles from measured backscattering data. It is important to emphasize, however, that the problem of inverting a backscattering spectrum to produce a target structure does not, in general, have a unique solution

and is further complicated by statistical fluctuations in the data, small- and large-angle multiple scattering of the beam and scattered ions, channeling, uncertainties in density, and the surface topography of the specimen, in addition to factors such as energy straggling and measurement resolution, which have been simulated in Figures 4A–E.

It is generally true that the more *a priori* knowledge that one has about a specimen, the more precisely one can infer its structure. For example, in the semiconductor industry, where fabricating thin-film structures is the *sine qua non* of integrated circuit manufacture, engineers ordinarily know the order and approximate composition of individual layers in a multilayer thin-film structure. In this case, it is possible, working from approximate starting values, to use a simulation procedure such as the one used to produce Figures 4A–E in a closed-loop algorithm to find the values of parameters such as film thickness and composition that provide the best fit to measured data in the sense of $\chi^2$ minimization or some other measure of goodness of fit. For mega-electron-volt data there are computer algorithms in the published literature to implement such procedures (Doolittle, 1984, 1985, 1990) as well as proprietary computer programs available from commercial sources (e.g., Strathman, 1990). (See Leavitt et al., 1995, for a more complete list of published simulation algorithms.) However, as of this writing, these have not yet been adapted to explicitly handle medium-energy backscattering. The general problem of inversion of backscattering spectra continues to be an area of active research as the advance of computer power continues to widen the range of methods that can be undertaken practically. The goal of this research, which is to extract all statistically significant information contained in backscattering spectra, is still elusive, although important progress is being made (Barradas et al., 1997).

## PRACTICAL ASPECTS OF THE METHOD

The previous section discussed issues relating to medium-energy ion scattering for the most part without reference to any specific spectrometric technique. This is appropriate because there are a number of methods, including electrostatic and magnetic spectrometers, pulsed-beam and foil-based time-of-flight spectrometers, and even cryogenic solid-state detectors, by which the experiments can be performed, and novel variations frequently appear in the literature. Each method has advantages and disadvantages and continues to benefit from time to time by advances in technology. The choice of which method to use is not abstract, but is closely related to the nature of the problem at hand. The discussion in this section will be restricted to a particular style of foil-based time-of-flight spectrometer that has a straightforward design and an attractive set of operating characteristics. Foil-based time-of-flight spectrometers were originally introduced for heavy-ion spectrometry in nuclear physics (see Pfeffer et al., 1973, and Betts, 1979, and references therein) and later adapted for backscattering spectrometry using mega-electron-volt heavy ions (Chevarier et al., 1981). They were first applied to medium-energy ion beam analysis by Mendenhall and Weller (1989, 1990). Several time-of-flight configurations have been reviewed critically by Whitlow (1990).

### Time-of-Flight Spectrometry

An energy-dispersive particle spectrometer such as a silicon surface-barrier detector functions by converting the energy of a kinetic particle to an approximately proportional quantity of electric charge. A time-of-flight spectrometer functions by measuring the time that it takes for a particle to traverse a fixed distance. This requires two signals representing a "start" and a "stop" that are associated with, but possibly offset by a fixed time from, the beginning and end of the traversal. In time-of-flight spectrometry, the issue is how best to obtain a start signal, since a stop signal can always be produced by simply collecting the particle in any of several detectors. In pulsed-beam time of flight, the start signal is derived by knowing that all ions of the beam arrive at the target in a narrow pulse. In foil-based time of flight, a very thin foil, usually of carbon, is placed at the beginning of the measured flight path. This foil is so thin that an ion hitting it passes through with minimal loss of energy and perturbation of its trajectory.

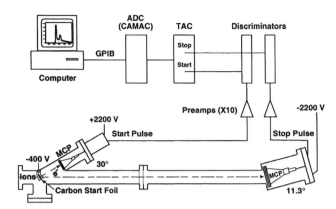

**Figure 5.** Principal components of a medium-energy time-of-flight spectrometer system. This system is computer controlled and CAMAC based (Weller, 1995a). The angle of the stop detector is chosen to optimize resolution as described in the text.

As it exits the foil, the ion produces secondary electrons with significant (but not unit) probability, and these are accelerated away and detected to produce a start signal. A time-of-flight spectrometer of this type is shown in Figure 5.

### Spectrometer Efficiency

The characteristics of the spectrometer enter into the determination of the concentrations of constituents via the energy-dependent efficiency term $\eta$ in Equation 2. It is possible to understand the physical origin of this efficiency by considering individually the mechanisms by which events may be lost. For example, while the perturbation of the trajectory by the start foil is small, it is not zero, and a predictable number of events will be lost because small-angle multiple scattering deflects some particles enough to miss the stop detector (see Fig. 5). Similarly, since the production of a start pulse requires secondary electrons, and since the production of secondary electrons is inherently statistical, there will be instances in which a particle will pass through the start foil without creating a start pulse. Additional processes that may disrupt the normal sequence of events in the spectrometer include scattering of ions or electrons from various meshes, the intrinsic dead area on the surface of microchannel plates, and the statistical nature of the development of pulses in these detectors. The cumulative effect of these various events has been summarized in a model of time-of-flight spectrometer performance that is, in principle, capable of predicting the energy-dependent performance of a time-of-flight spectrometer from first principles (Weller et al., 1994). However, because many input parameters to the model, such as secondary electron emission yields, are highly variable and dependent upon vacuum conditions and age and composition of the components, among other things, the model is best used in conjunction with measurements of efficiency, so that these parameters need not be known *a priori*. This semiempirical approach is demonstrated in Figure 6, which uses experimentally determined parameters along with the equations of Weller et al. (1994) to obtain efficiency curves

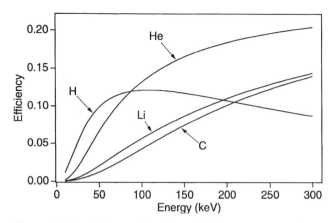

**Figure 6.** The efficiency of the time-of-flight spectrometer shown in Figure 5 for hydrogen, helium, and carbon ions as a function of energy, computed using empirical parameters in conjunction with the formalism described by Weller et al. (1994).

for a time-of-flight spectrometer like that shown in Figure 5. Note that the curve for H, unlike the others, has a maximum in the region of interest. This is attributable to the dependence of the secondary electron yield on electronic stopping power. It is important to emphasize that the efficiency of a spectrometer for a given ion is a property of the measuring device and not of the target being measured. Thus, while it is necessary to use a standard for the initial calibration of the instrument, it is not necessary to employ standards that mimic the properties of individual targets in order to achieve quantitative results. In this sense, the method is to be contrasted with SIMS, where the characteristics of a surface may have a strong impact on the relative intensities of the signals from various surface constituents.

### Sensitivity

In general, elastic scattering spectrometries are more sensitive at lower incident beam energies, all other things being equal, because the collision cross-section decreases with increasing energy (see Equation 3). Thus, medium-energy backscattering and forward-recoil spectrometry enjoy an intrinsic advantage of as much as two orders of magnitude when compared with their high-energy counterparts. However, a portion of this is forfeited in practice because of the efficiency of the spectrometer, as discussed in the previous paragraph. See HEAVY-ION BACKSCATTERING SPECTROMETRY for a discussion of a system that has been optimized for extremely high sensitivity.

The trace-element sensitivity of medium-energy backscattering is limited by two factors, the gradual sputter removal of the constituent of interest and multiple scattering of the beam within the target. If, instead of scattering once at an angle $\theta$ in order to enter the spectrometer, a particle scatters twice at an angle $\theta/2$, then its energy will be considerably higher. Ordinarily, these multiple collisions (including all possible pairs of events that sum to an angle $\theta$) are sufficiently rare that they are ignored in analyzing backscattering data. However, for heavy trace elements such as, for example, iron on silicon, multiple

scattering in the silicon substrate can lead to the production of counts in the region of the iron peak that is kinematically forbidden to binary collisions with individual substrate atoms. These counts constitute a background that limits the signal-to-noise ratio of true iron counts. This multiple-scattering background, which has been studied extensively by Brice (1992), is most significant for heavy-ion backscattering (HEAVY-ION BACKSCATTERING SPECTROMETRY).

For light-ion backscattering, as for example with He$^+$, the predominant factor controlling sensitivity is sputtering (Weller, 1993). Sputtering, the erosion of a material by ion impacts, typically results in from $10^{-3}$ to as many as 10 or more surface atoms being removed for each ion impact (Yamamura and Tawara, 1996). In the absence of sputtering, a trace-element backscattering measurement could proceed until all atoms of the element of interest are involved in a collision. With sputtering, it is convenient to define the sensitivity by the constraint that the fractional amount of the trace element removed is equal to the fractional uncertainty that is acceptable in the final answer. The only difficulty with this is that the *a priori* sputtering yields of trace elements in a foreign host are not well known (Pedersen et al., 1996), so that plausible assumptions about these yields must be made. Doing so results in the following formula for the minimum areal concentration $\rho$ of a surface constituent that can be measured with a fractional uncertainty $\varepsilon$ (Weller, 1993):

$$\rho = \frac{1 + \sqrt{1 + 8\varepsilon^2 B}}{2\varepsilon^3} \cdot \frac{Y_s}{\rho_s A_s \Omega \sigma \eta} \quad (5)$$

where $A_s$ is the area of the target that is irradiated by the beam and $B$ is the number of background counts in the region of the spectrum that contains valid counts. The quantities $\sigma$ and $\eta$ are the cross-section and spectrometer efficiency, respectively, and $\Omega$ is the spectrometer's solid angle. This expression is based upon the assumption that the probability $P$ that a beam ion removes a target atom is related to the (dimensionless) sputtering yield of the substrate $Y_s$ through the relation

$$P \approx Y_s \cdot \frac{\rho}{\rho_s} \quad (6)$$

where $\rho_s$ is taken to be the areal density of substrate atoms in the region from which sputtered particles are drawn. This is typically the first two or three atomic layers of the target (Dumke et al., 1983). Equation 6 expresses the assumption that trace elements are sputtered randomly from the surface with a probability equal to their fractional representation in the volume from which sputtered particles can emerge. Estimates of sputtering yields may be obtained using the equations or tables of Yamamura and Tawara (1996). For trace quantities of copper on silicon analyzed by He$^+$ with a general-purpose spectrometer like that shown in Figure 5, Equation 6 predicts a sensitivity of $\sim 10^{11}$ cm$^{-2}$ in the absence of background ($B = 0$). However, achieving this figure would require exceptionally long measuring times. For trace elements with concentrations below $\sim 10^{13}$ cm$^{-2}$, a system optimized for sensitivity should be used (HEAVY-ION BACKSCATTERING SPECTROMETRY).

## Resolution

The energy resolution of a spectrometer is important because it is a direct measure of the depth resolution of backscattering spectrometry. It is important to emphasize, however, that resolution is not a unique concept. Most often, in analogy to the Rayleigh criterion familiar from optics, two isolated peaks in a backscattering spectrum are said to be resolved when their separation is equal to the full width at half-maximum (FWHM) of the peaks. In this case, any additional separation results in an observable bimodal structure. This definition certainly captures the essence of the physics for cases, such as superlattices, where the objective is to discern changes in the concentration of an element as a function of depth. However, there is another closely related concept, that of layer thickness, for which this definition of resolution results in an unduly pessimistic estimate of the quality of the data. In this second type of measurement, one is typically interested in small variations in the thickness of a layer rather than in the details of its structure. Essentially, the problem is that of locating the edge of a spectral feature, which is approximately equivalent to locating the centroid of an isolated peak. This can typically be done with an uncertainty of $\approx 10\%$ of a standard deviation, $\sigma$, of the peak, as compared with the FWHM, which is $2.35\sigma$. Thus, it is possible to measure thickness variations that are more than an order of magnitude smaller than the standard definition of resolution would suggest. There is no contradiction here, simply an inconsistency in what exactly is meant by "resolution." In the paragraphs that follow, the usual Rayleigh concept will be adopted, with the understanding that it may not be appropriate in all circumstances.

If $\delta E$ is the uncertainty (FWHM) from all sources in the measurement of a backscattered particle's energy, then the corresponding uncertainty in depth of the target atom is (Chu et al., 1978)

$$\delta x = \frac{\delta E}{[S]} \tag{7}$$

where the denominator is known as the energy loss or stopping power factor and is given by

$$[S] = \left[ K\left(\frac{dE}{dx}\right)_{in} \cdot \frac{1}{\cos(\theta_1)} + \left(\frac{dE}{dx}\right)_{out} \cdot \frac{1}{\cos(\theta_2)} \right] \tag{8}$$

Here $K$ is the kinematic factor defined above, $dE/dx$ is the stopping power, and $\theta_1$ and $\theta_2$ are angles made with the target surface normal by the inward and outward trajectories of the particle, respectively (Leavitt et al., 1995). As a particle penetrates a target, the uncertainty in its energy grows as a result of straggling, which is the intrinsic uncertainty associated with the statistics of energy loss (Yang et al., 1991). Thus, depth resolution is always best at the surface of a target, where it is limited by the uncertainty of the beam energy $E_0$ and the properties of the spectrometer.

Three primary design characteristics govern the resolution of a time-of-flight spectrometer such as is shown in Figure 5: minor differences in the length of the flight paths of different ions, straggling in the carbon start foil, and nonuniformity of the foil either in thickness or density. Of these, foil nonuniformity is the most serious (McDonald et al., 1999). An additional variation due to the natural kinematic broadening associated with the variation of $\theta$ across the finite acceptance solid angle of the spectrometer is also significant (Equation 1), as is a charge-exchange process that produces small parasitic features offset from the true value by plus or minus the value of the start foil bias, or $\sim 300$ eV (Weller et al., 1996).

The velocity of a 100-keV He$^+$ ion is $\approx 2.2$ mm/ns. As a result, it is necessary to design a spectrometer so that most trajectories differ by no more than a millimeter or so. This is the reason for the relative angles between the start and stop detectors in Figure 5. If all ions entered the spectrometer along parallel trajectories, then the offset angles measured from the spectrometer axis should be equal. However, for particles emerging from a point source, there is another angle that produces minimum path length dispersion, which is dependent upon the distance between the target and the start foil and the length of the drift section. Interestingly, a further adjustment of this angle conveys an advantage unique to time-of-flight methods. Ions with larger scattering angles move more slowly as a result of the natural variation of the kinematic factor $K$ with angle (Equation 1). As a result, it is possible to adjust the relative angles of the start and stop detectors to minimize kinematic dispersion from the finite solid angle of the spectrometer. This adjustment, although it amounts to only about a degree, produces a clear improvement in performance for the ion for which it is optimized (Weller et al., 1996).

Considering fundamental physical processes such as straggling and limitations imposed by geometry, time-of-flight spectrometers like that shown in Figure 5 should have a resolution of <1%, or perhaps 1 keV for a 150-keV ion (McDonald et al., 1999). With favorable target orientation (see Equation 8), this should produce a depth resolution of $\approx 1$ nm at the target surface. This is approximately a factor of 2 better than has been achieved in the laboratory because of the microscopic roughness of available carbon start foils, but still somewhat inferior to results obtainable in proton backscattering using high-resolution electrostatic analyzers. It compares quite favorably with results obtained using surface barrier detectors simply because of the improved resolution (Williams and Möller, 1978).

## Instrumentation

The electronic instrumentation needed to implement a time-of-flight system is standard (Weller, 1995a) and available from commercial vendors. The essential components are the start and stop detectors, timing discriminators, and time-to-amplitude or time-to-digital converters. The fundamental, active, particle-detecting elements in time-of-flight spectrometers are microchannel plates (Fig. 5). These are high-density arrays of microscopic, continuous-dynode electron multipliers fabricated in millimeter-thick glass wafers (Wiza, 1979). New microchannel plates operated at rated bias can produce pulses that are a significant fraction of a volt. However, they are so fast that most general-purpose oscilloscopes cannot display them.

With proper signal-handling techniques (Weller, 1995a, and references cited therein), it should be possible to couple the start and stop pulses directly to timing discriminators, whose function it is to produce uniform time markers that are insensitive to pulse amplitude. However, in practice, it may be necessary to add a stage of voltage amplification between the microchannel plates and timing discriminators. Amplifiers with bandwidth of at least 1 GHz are needed for this purpose. The timing discriminator outputs are used as the definitive start and stop signals are fed to a timing module.

Two styles of timing modules are in common use. One, called a time-to-amplitude converter (TAC), produces an output pulse whose height is proportional to the time difference between the start and stop signal. This is very convenient because the resulting linear pulse may be directly substituted for the amplified and shaped output of a surface-barrier detector. In a system of this type, the pulse from the TAC goes to a multichannel pulse-height analyzer that completes the analysis by producing a histogram of the number of pulses as a function of amplitude. This scheme may sound laboriously indirect, but it is capable of achieving resolutions of ≈50 ps when configured for medium-energy scattering and, until very recently, was clearly superior to directly counting ticks of a clock (Porat, 1973). The alternative time-to-digital converter (TDC) scheme in its purist form does, indeed, directly count the ticks of a clock between the start and stop signals. In this approach, the number of clock ticks is used as a pointer into a memory bank (a histogramming memory) and one count is recorded for each valid interval measurement. With countable clock frequencies now in the tens of gigahertz, producing correspondingly high resolution, this method is a practical alternative to time-to-amplitude conversion. However, it is worth noting that, at the time of this writing, commercial modules called TDCs are often, in reality, TACs and pulse analog-to-digital converters packaged as a unit.

## Examples

Figures 7A–C show the result of measuring an $SiO_2$ thin film nominally 13.9 nm thick grown on crystalline Si by medium-energy backscattering using a beam of 270-keV $He^+$ ions. Figure 7A shows the raw time-of-flight spectrum as accumulated. In Figure 7B, this spectrum has been transformed to the energy domain using the procedure described below. Figure 7C contains the final version of this spectrum, which has been corrected for the efficiency of the spectrometer. These data were taken with the specimen oriented so that the beam is aligned with the $\langle 110 \rangle$ direction of the crystalline Si. This causes the number of counts from the bulk Si region to be suppressed, while Si and O counts from the amorphous $SiO_2$ layer are unaffected. An analysis of the oxygen peak reveals that the oxide is ~13.1 nm thick, in good agreement with the nominal thickness. The analysis of the Si peak is more complicated because it contains constituents from both the $SiO_2$

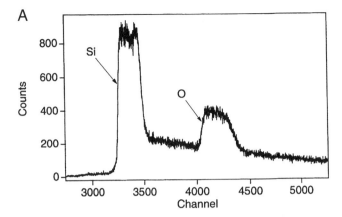

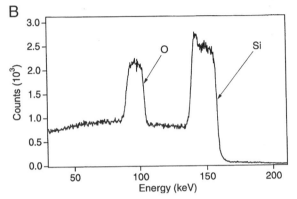

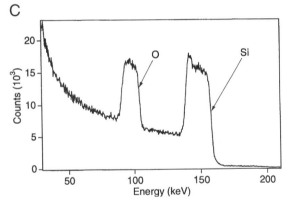

**Figure 7.** Time-of-flight medium-energy backscattering spectrum of a 13-nm $SiO_2$ film on Si. The ion beam was $He^+$ at 270 keV. The beam is channeled in the $\langle 110 \rangle$ direction of the Si substrate in order to accentuate the Si and O in the amorphous oxide layer. The scattering angle was $150°$. (**A**) Raw data showing number of counts versus channel in the multichannel analyzer. Each channel has a width of 122 ps. (**B**) The data rendered in the energy domain using the transformation described in the text. (**C**) The energy spectrum corrected for spectrometer efficiency using the efficiency function plotted in Figure 6. Note in each case, but particularly in the energy spectra, that the channeling surface peak attributable to the first few atomic layers of Si in the substrate is clearly visible on the low-energy side of the Si spectral feature.

layer and the first few layers of the crystalline Si. This channeling surface peak is discussed by Feldman et al. (1982) and Swanson (1995).

Figure 8 shows the result of measuring the surface of an Si wafer, as delivered by the vendor, for hydrocarbon

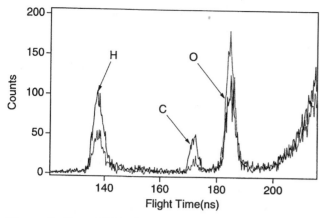

**Figure 8.** Time-of-flight forward-recoil spectrum of Si wafer showing surface H, C, and O before and after cleaning with ozone. The solid line is the spectrum of the as-delivered wafer. The spectrum shown as a dotted line was made after cleaning. The ozone exposure removed H and C but increased the thickness of the surface $SiO_2$ layer as indicated by the increased O signal. The ion beam was $Ar^{3+}$ at 810 keV.

contamination using medium-energy forward-recoil spectrometry. The beam was 810 keV $Ar^{3+}$. The spectrum reveals three distinct peaks that are attributable to surface H, C, and O. While the hydrogen could have been measured by, for example, nuclear reaction analysis (Lanford, 1995; also see NUCLEAR REACTION ANALYSIS AND PROTON-INDUCED GAMMA RAY EMISSION), and the oxygen separately by backscattering, this measurement has revealed both in a way that makes it straightforward to determine the relative areal densities of the three constituents simultaneously.

### Safety Considerations

Medium-energy backscattering spectrometry poses no special laboratory safety hazards. General safety considerations are the same as for most classes of surface analytical equipment. The most serious concern is shock hazard from any of several high-voltage sources that are needed to carry out the experiments. Ion beam irradiation by kilo-electron-volt ions is essentially nondestructive for most targets and produces no lingering effect to alter the intrinsic toxicity of a specimen. Particle accelerators used in the technique are often capable of producing x rays, especially when operated improperly, and must therefore be supported by appropriate radiation safety staff under the laws of the jurisdiction in which they are located.

A subtle safety issue in this regard is less well known. When a highly insulating material such as silica is irradiated with kilo-electron-volt ions, the surface can become charged to many tens of kilovolts. Eventually, the surface will break down electrically and a significant spark will occur in the vacuum chamber. These sparks are easily capable of producing measurable pulses of x rays that may sometimes even be observed outside the vacuum system, especially through windows. Although it is quite unlikely that this source would ever generate enough x rays to be an issue for personal safety, it could affect the results of

radiation surveys and might also impact the need for shielding or placement of the target chamber in the laboratory.

## METHOD AUTOMATION

In principle, the acquisition of backscattering spectra may be highly automated. A full range of NIM and CAMAC modular electronics is available commercially, and multichannel pulse-height analysis is now usually done by computer (Weller, 1995b). Specialized equipment, such as goniometers for channeling studies, may easily be synchronized with data collection under computer control. However, medium-energy backscattering is still a specialized technique, and production-scale systems are not available as they are for conventional mega-electron-volt Rutherford backscattering.

The analysis of backscattering data is, in general, not highly automated, at least not in the sense of, for example, gamma ray spectroscopy, where peak search-and-fit algorithms have been developed that are capable of nearly independently extracting the identity and intensity of all the gamma rays in a spectrum. Although algorithms have existed for many years to assist in the reduction of backscattering data (Doolittle, 1984, 1985, 1990) and new methods continue to be explored that further reduce the need for operator intervention (Barradas et al., 1997), human judgment still plays a role in most ion scattering data analysis. These principles are discussed next (see Data Analysis and Interpretation), along with considerations that are specific to the handling of medium-energy ion scattering data.

## DATA ANALYSIS AND INTERPRETATION

### Backscattering Data

The steps for analyzing time-of-flight medium-energy backscattering data are, in order, to convert the spectrum from the time to the energy domain, to apply a correction for the efficiency of the spectrometer as a function of energy, and finally to extract information on elemental concentrations and depth profiles using techniques identical to those used for high-energy Rutherford backscattering spectra. In the final stage, it is essential to use cross-sections and stopping powers that are appropriate for the energy range.

The process of converting a time-of-flight spectrum to an equivalent representation in the energy domain is often called rebinning, since it represents a sorting of collected counts from equal-width time bins to equal-width energy bins. Both time-of-flight and energy spectra are essentially differential quantities, since they are histograms of the number of events whose defining parameter, time or energy, falls within a given narrow range defined by a single channel. Thus, to convert from a time-domain spectrum $P_t(t)$ to an equivalent energy-domain spectrum $P_E(E)$ requires use of the fundamental relationship

$$P_E(E) = P_t(t(E)) \left| \frac{dt(E)}{dE} \right| \qquad (9)$$

where $t$ is the flight time that corresponds with backscattered energy $E$, and the parameters are related through the mass $M_1$ of the backscattered particle and the drift length $d$ of the spectrometer by

$$E = \frac{1}{2} M_1 \left( \frac{d}{t} \right)^2 \qquad (10)$$

Although Equation 9 can be used directly in order to perform the conversion, it is not the preferred method. A more direct procedure is to use the integral form of this equation:

$$\int_0^E P_E(E')dE' = \int_{t(T)}^\infty P_t(t')dt' \qquad (11)$$

To apply Equation 11, first it is necessary to create an integral time spectrum by iteratively summing adjacent channels of the time-of-flight spectrum from the longest flight time (rightmost, highest array index) to the shortest. This can be accomplished on an array in place. At the $N-1$ channel of an N-channel spectrum, the value of channel $N-1$ is added to the value of channel N and the sum is put in location $N-1$. This is repeated with $N$ replaced by $N-1$ iteratively until N is zero. This is the integral time spectrum. To obtain an integral energy spectrum, an array is created and each location is filled with the value of the integral time spectrum at the time corresponding to the various equally spaced energies. It will, of course, rarely be the case that an energy and corresponding time value both fall exactly on channel boundaries in their respective spectra, so normally one must interpolate linearly between two values in the integral time spectrum to obtain the energy value. After the integral energy spectrum has been created, the process must be inverted by subtracting adjacent channels to recover a differential energy spectrum. This too can be done on an array in place. This conversion from a time to an energy spectrum has the virtue that it very accurately preserves total numbers of counts through the transformation and is very economical to compute.

It is important to issue a caution at this point. It is generally assumed in the analysis of ion scattering data that the variance associated with each channel of a multichannel spectrum is the Poisson value, that is, that the variance equals the number of counts and *the uncertainties of adjacent channels are uncorrelated*. This is true in raw time-of-flight spectra but may not be true in the corresponding rebinned energy spectra. The problem lies with the widths of the channels. For proper statistics, the width of time channels must be less than the time-equivalent width of the highest-energy channel in the energy spectrum. If this condition fails, then it is clear from the above description of rebinning that, in the high-energy region of the spectrum, channel values will be created by extracting two or more energy values by interpolation between the same two points in the integral time spectrum. This introduces strong correlation into adjacent channel values and changes the statistical character of the data. It is good practice to choose the width of time bins so that, even at the highest energies of interest, each energy bin corresponds to two or more time bins. In this way, the statistics of rebinned time-of-flight spectra will be practically equivalent to those of energy spectra that have been measured directly by an energy-dispersive spectrometer.

After time-of-flight spectra have been rebinned to the energy domain, they should be corrected for spectrometer efficiency as described above (see Practical Aspects of the Method, Spectrometer Efficiency). This consists of a straightforward channel-by-channel multiplication of the energy spectrum by a value appropriate to the ion and energy. It is important to emphasize again that this correction is completely independent of the specimen and only a function of spectrometer characteristics. As such, it does not introduce sample-to-sample variation. It does, however, change the statistics of the spectrum, since in any given channel statistical uncertainly is based upon the actual number of measured counts and not the corrected number of counts. Strictly speaking, one should also include at this point an estimate of error produced by the correction itself. With careful measurements, however, this can be made quite small. Also, the efficiency correction produces its greatest effect in the lower-energy portion of the spectrum, where its impact on the appearance of a spectrum is large but where there are seldom features of quantitative interest.

After rebinning and efficiency correction, time-of-flight spectra, along with their associated arrays of variances, may be treated with the tools and techniques that have been developed for elastic scattering spectrometry with energy-dispersive spectrometers (ELASTIC ION SCATTERING FOR COMPOSITION ANALYSIS).

### Forward-Recoil Data

The rebinning process described above is of limited usefulness for one-dimensional forward-recoil data. Since forward-recoil spectra contain events attributable to different species, it is only possible to do rebinning of a one-dimensional spectrum sensibly in a spectral region known to contain only one species of ion, such as hydrogen or carbon. Furthermore, in forward recoil, time is the proper distinguishing characteristic, the one with the greatest power of differentiation of species. Thus, with the exception of near-surface hydrogen profiling, one-dimensional forward-recoil spectra should be used directly to infer properties of the specimen.

The mass of surface constituents can be identified by their energy $E_2$, computed using Equation 10, along with the recoil kinematic factor $K_R$:

$$E_2 = K_R \cdot E_0 = \frac{4M_1 M_2}{(M_1 + M_2)^2} \cos^2(\phi) \qquad (12)$$

where $\phi$ is the angle of the recoil measured with respect to the initial beam direction. Absolute concentrations of low-atomic-mass surface constituents can be computed directly using Equation 2 along with spectrometer efficiencies, the peak areas (total counts), and the forward-recoil cross-section $\sigma_r$, given here in terms of the center-of-mass scattering angle $\theta_c = \pi - 2\phi$ and center-of-mass cross-section

$\sigma_c(\theta_c)$ by a standard kinematic transformation (e.g., Weller, 1995b):

$$\sigma_r(\phi) = 4\sigma_c[\theta_c(\phi)]\cos(\phi) = 4\sigma_c(\pi - 2\phi)\cos(\phi) \quad (13)$$

Most previous research on forward-recoil spectrometry has used high-energy ions and surface-barrier detectors in conjunction with range foils. These are self-supporting thin films placed between the specimen and the detector to act as filters to the passage of certain ions. Procedures have been developed to analyze these data (Barbour and Doyle, 1995) but are of limited applicability for medium-energy time-of-flight work.

The most advanced technology for forward-recoil spectrometry combines time-of-flight with energy spectrometry in an experiment in which both parameters are measured for each event (Tirira et al., 1996). This gives rise to two-dimensional histograms with distinct spectral features that correspond to specific elements. This method works best for heavy ions with energies of many (often tens of) mega-electron-volts, and as a result, application of this method has been limited to a relatively small number of laboratories. Multiparameter forward-recoil spectrometry at medium energies has not been well studied.

## PROBLEMS

The definitive enumeration of the myriad of ways that ion scattering measurements can go astray has been compiled by Davies et al. (1995) and should be read by all serious students and practitioners of the field. Issues important for medium-energy time-of-flight work are emphasized here.

For quantitative measurements, it is essential to be able to integrate the beam charge accurately, since it is the measure of the total number of incident ions (Equation 2). This is done with a specially designed charge-collection structure known as a Faraday cup (England, 1974; Davies et al., 1995). Nature confounds the casual designer of a Faraday cup with a dizzying array of processes that complicate what would seem to be a straightforward measurement task. Currents used in time-of-flight medium-energy ion beam analysis range from a maximum of a few tens of nanoamperes, when analyzing for trace elements on a low-atomic-mass substrate, to <1 nA for forward-recoil measurements of hydrogen. Making a 1% measurement in the latter case could be compromised by a mere millivolt of stray voltage (or instrumental offset) across a resistance of 100 M$\Omega$. More likely to cause trouble, however, are processes intrinsic to ion beam experiments such as secondary electron emission at the target or ion beam neutralization through charge exchange with the residual vacuum. With a good design and careful attention to detail (Davies et al., 1995), charge integration at the level of 5% should be easily achieved.

With good charge integration, the efficiency of the time-of-flight spectrometer can be measured. The simplest approach is to obtain a target with a known areal density of a heavy trace element such as gold at the surface and to measure this target with different incident beam energies so that the range of backscattered energies covers the range of interest in general measurements. The effectiveness of this approach is dependent upon the accuracy of beam-current integration. The quality of the measurement can be improved by using a target with two isolated peaks that are well separated in energy. The beam energy is then raised or lowered in such a way that, through successive runs, the low-energy peak moves to the position formerly occupied by the high-energy peak, or the reverse. This maintains an internal check of charge integration and provides data to normalize from run to run. In this way, a curve of spectrometer efficiency as a function of energy can be produced from the calculated cross-section and reliably interpolated between measured points, using either a semiempirical model (Weller et al., 1994) or convenient analytic functions of energy. Since spectrometer efficiency must be measured for every ion to achieve the highest-accuracy results, it is sufficient to use the approximate screened cross-section given by Equation 4.

At medium energies, the critical angles for channeling in crystalline targets are very large (Swanson, 1995). This can be of very great help in suppressing background counts so that, for example, the oxygen content of a thin oxide on Si can be measured more accurately (Fig. 7C). However, accidental channeling can distort a spectrum and create a misleading impression about the structure of the sample. It is good practice to make critical measurements at several angles in order to be able to check for this kind of distortion. It is even better to continuously rotate the sample during a measurement to approximate the appearance of an amorphous target.

Another effect that varies strongly with target orientation is loss of resolution due to surface roughness. Ion scattering measurements at all energies work best on planar targets. Moreover, depth resolution can be enhanced by tilting the target so that ion trajectories, and consequently energy loss, are maximized in layers of interest. Detailed studies by Williams and Möller (1978) suggest that for 2-MeV ions the best depth resolution can be obtained using grazing incidence and exit angles of 5° to 10°. However, at these extreme angles, even minor surface imperfections can produce large variations in the cumulative energy loss of ions on different trajectories. For this reason, discretion should be exercised when using target angle to enhance resolution, and under normal circumstances, trajectories closer than <15° to the plane of the target should be used with considerable caution. It is possible to obtain ~2-nm resolution using conventional RBS with grazing exit-angle geometry provided the target surface is (atomically) smooth. Similar resolution can be obtained using medium-energy backscattering without the need for grazing exit angles.

Finally, it is important to keep in mind that the beam that is striking the target may not be what it is thought to be. This is because most accelerators use magnets to select a specific beam from the many that are produced by the ion source. The relevant parameter, known as magnetic rigidity, is $(2M \cdot E/q^2)^{1/2}$, where $M$, $E$, and $q$ are, respectively, the mass, energy, and charge of the ions in the beam. In medium-energy accelerators, $E$ is almost always equal to $q \cdot V$, where $V$ is the accelerating voltage, so that any time two ions have the same $M/q$, they cannot be distinguished. The most important case for medium-energy backscattering is that of $He^{2+}$ and $H_2^+$. The omnipresence of the hydrogen molecular ion makes the use of doubly charged helium, to obtain higher energies, a dangerous proposition. This mass interference taken together with the difficulty of producing $He^{2+}$ in the first place argues strongly against the use of this ion for routine analytical work.

## ACKNOWLEDGMENTS

The author would like to thank Kyle McDonald and Len C. Feldman for helpful comments during the preparation of this unit. The primary support for the development of time-of-flight medium-energy ion beam analysis has been provided by the U.S. Army Research Office under contracts DAAL 03-92-G-0037, DAAH 04-94-G-0148, and DAAH 04-95-1-0565.

## LITERATURE CITED

Andersen, H. H., Besenbacher, F., Loftager, P., and Möller, W. 1980. Large-angle scattering of light ions in the weakly screened Rutherford region. *Phys. Rev. A* 21:1891–1901.

Arps, J. H. and Weller, R. A. 1995. Time-of-flight elastic recoil analysis of ion-beam modified nitrocellulose thin films. *Nucl. Instrum. Methods Phys. Res. B* 100:331–335.

Arps, J. H. and Weller, R. A. 1996. Determination of hydrogen sensitivity and depth resolution of medium-energy, time-of-flight, forward-recoil spectrometry. *Nucl. Instrum. Methods Phys. Res. B* 119:527–532.

Barbour, J. C. and Doyle, B. L. 1995. Elastic recoil detection: ERD. *In* Handbook of Modern Ion Beam Materials Analysis (J. R. Tesmer and M. Nastasi, eds.). pp. 83–138. Materials Research Society, Pittsburgh.

Barradas, N. P., Jeynes, C., and Webb, R. P. 1997. Simulated annealing analysis of Rutherford backscattering data. *Appl. Phys. Lett.* 71:291–293.

Betts, R. R. 1979. Time-of-flight detectors for heavy ions. *Nucl. Instrum. Methods* 162:531–538.

Brice, D. K. 1992. Screened Rutherford multiple scattering estimates for heavy ion backscattering applications. *Nucl. Instrum. Methods Phys. Res. B* 69:349–360.

Chevarier, A., Chevarier, N., and Chiodelli, S. 1981. A high resolution spectrometer used in MeV heavy ion backscattering analysis. *Nucl. Instrum. Methods* 189:525–531.

Chu, W. K., Mayer, J. W., and Nicolet, M.-A. 1978. Backscattering Spectrometry. Academic Press, New York.

Darwin, C. G. 1914. Collision of $\alpha$ particles with light atoms. *Philos. Mag.* 27:499–506.

Davies, J. A., Lennard, W. N., and Mitchell, I. V. 1995. Pitfalls in ion beam analysis. *In* Handbook of Modern Ion Beam Materials Analysis (J. R. Tesmer and M. Nastasi, eds.). pp. 343–363. Materials Research Society, Pittsburgh.

Doolittle, L. R. 1984. Algorithms for the rapid simulation of Rutherford backscattering spectra. *Nucl. Instrum. Methods Phys. Res. B* 9:344–351.

Doolittle, L. R. 1985. A semiautomatic algorithm for Rutherford backscattering analysis. *Nucl. Instrum. Methods Phys. Res. B* 15:227–231.

Doolittle, L. R., 1990. High energy backscattering analysis using RUMP. *In* High Energy and Heavy Ion Beams in Materials Analysis: Proceedings High Energy and Heavy Ion Beams in Materials Analysis Workshop (J. R. Tesmer, C. J. Maggiore, M. Nastasi, J. C. Barbour, and J. W. Mayer, eds.). pp. 175–182. Materials Research Society, Pittsburgh.

Dumke, M. F., Tombrello, T. A., Weller, R. A., Housley, R. M., and Cirlin, E. H. 1983. Sputtering of the gallium-indium eutectic alloy in the liquid phase. *Surf. Sci.* 124:407–422.

England, J. B. A. 1974. Techniques in Nuclear Structure Physics. John Wiley & Sons, New York.

Feldman, L. C., Mayer, J. W., and Picraux, S. T. 1982. Materials Analysis by Ion Channeling. Academic Press, New York.

Lanford, W. A. 1995. Nuclear reactions for hydrogen analysis. *In* Handbook of Modern Ion Beam Materials Analysis (J. R. Tesmer and M. Nastasi, eds.). pp. 193–204. Materials Research Society, Pittsburgh.

Leavitt, J. A., McIntyre, Jr., L. C., and Weller, M. R. 1995. Backscattering spectrometry. *In* Handbook of Modern Ion Beam Materials Analysis (J. R. Tesmer and M. Nastasi, eds.). pp. 37–81. Materials Research Society, Pittsburgh.

L'Ecuyer, J., Davies, J. A., and Matsunami, N. 1979. How accurate are absolute Rutherford backscattering yields? *Nucl. Instrum. Methods* 160:337–346.

Lennard, W. N., Geissel, H., Winterbon, K. B., Phillips, D., Alexander, T. K., and Forster, J. S. 1986. Nonlinear response of Si detectors for low-z ions. *Nucl. Instrum. Methods Phys. Res. A* 248:454–460.

McDonald, K., Weller, R. A., and Liechtenstein, V. Kh. 1999. Quantitative evaluation of the determinants of resolution in time-of-flight spectrometers for medium energy ion beam analysis. *Nucl. Instrum. Methods Phys. Res. B* 152:171–181.

Mendenhall, M. H. and Weller, R. A. 1989. A time-of-flight spectrometer for medium energy ion scattering. *Nucl. Instrum. Methods Phys. Res. B* 47:193–201.

Mendenhall, M. H. and Weller, R. A. 1990. Performance of a time-of-flight spectrometer for thin film analysis by medium energy ion scattering. *Nucl. Instrum. Methods Phys. Res. B* 40/41:1239–1243.

Mendenhall, M. H. and Weller, R. A. 1991. Algorithms for the rapid computation of classical cross sections for screened Coulomb collisions. *Nucl. Instrum. Methods Phys. Res. B* 58:11–17.

Nastasi, M., Mayer, J. W., and Hirvonen, J. K. 1996. Ion-Solid Interactions: Fundamentals and Applications. Cambridge University Press, Cambridge.

Pedersen, D., Weller, R. A., Weller, M. R., Montemayor, V. J., Banks, J. C., and Knapp, J. A. 1996. Sputtering and migration of trace quantities of transition metal atoms on silicon. *Nucl. Instrum. Methods Phys. Res. B* 117:170–174.

Pfeffer, W., Kohlmeyer, B., and Schneider, W. F. W. 1973. A fast zero-time detector for heavy ions using the channel electron multiplier. *Nucl. Instrum. Methods* 107:121–124.

Porat, D. I. 1973. Review of sub-nanosecond time-interval measurements. *IEEE Trans. Nucl. Sci.* NS-20(5):36:51.

Rauhala, E. 1995. Energy loss. *In* Handbook of Modern Ion Beam Materials Analysis (J. R. Tesmer and M. Nastasi, eds.). pp. 3–19. Materials Research Society, Pittsburgh.

Strathman, M. D. 1990. SCATT. *In* High Energy and Heavy Ion Beams in Materials Analysis: Proceedings High Energy and Heavy Ion Beams in Materials Analysis Workshop (J. R. Tesmer,

C. J. Maggiore, M. Nastasi, J. C. Barbour, and J. W. Mayer, eds.). pp. 183–188. Materials Research Society, Pittsburgh.

Swanson, M. L. 1995. Channeling. *In* Handbook of Modern Ion Beam Materials Analysis (J. R. Tesmer and M. Nastasi, eds.). pp. 231–300. Materials Research Society, Pittsburgh.

Tirira, J., Serruys, Y., and Trocellier, P. 1996. Forward Recoil Spectrometry. Plenum Press, New York.

Weller, R. A. 1993. Instrumental effects on time-of-flight spectra. *Nucl. Instrum. Methods Phys. Res. B* 79:817–820.

Weller, R. A. 1995a. Instrumentation and laboratory practice. *In* Handbook of Modern Ion Beam Materials Analysis (J. R. Tesmer and M. Nastasi, eds.). pp. 301–341. Materials Research Society, Pittsburgh.

Weller, R. A. 1995b. Scattering and reaction kinematics. *In* Handbook of Modern Ion Beam Materials Analysis (J. R. Tesmer and M. Nastasi, eds.). pp. 412–416. Materials Research Society, Pittsburgh.

Weller, R. A., Arps, J. H., Pedersen, D., and Mendenhall, M. H. 1994. A model of the intrinsic efficiency of a time-of-flight spectrometer for keV ions. *Nucl. Instrum. Methods Phys. Res. A* 353:579–582.

Weller, R. A., McDonald, K., Pedersen, D., and Keenan, J. A. 1996. Analysis of a thin, silicon-oxide, silicon-nitride multilayer target by time-of-flight medium energy backscattering. *Nucl. Instrum. Methods Phys. Res. B* 118:556–559.

Whitlow, H. J. 1990. Time of flight spectroscopy methods for analysis of materials with heavy ions: A tutorial. *In* High Energy and Heavy Ion Beams in Materials Analysis: Proceedings High Energy and Heavy Ion Beams in Materials Analysis Workshop (J. R. Tesmer, C. J. Maggiore, M. Nastasi, J. C. Barbour, and J. W. Mayer, eds.). pp. 243–256. Materials Research Society, Pittsburgh.

Williams, J. S. and Möller, W. 1978. On the determination of optimum depth-resolution conditions for Rutherford backscattering analysis. *Nucl. Instrum. Methods* 157:213–221.

Wiza, J. L. 1979. Microchannel plate detectors. *Nucl. Instrum. Methods* 162:587–601.

Yamamura, Y. and Tawara, H., 1996. Energy dependence of ion-induced sputtering yields from monatomic solids at normal incidence. *Atomic Data Nucl. Data Tables* 62:149–253.

Yang, Q., O'Connor, D. J., and Wang, Z. 1991. Empirical formulae for energy loss straggling of ions in matter. *Nucl. Instrum. Methods Phys. Res. B* 61:149–155.

Ziegler, J. F., Biersack, J. P., and Littmark, U. 1985. The Stopping and Range of Ions in Solids. Pergamon Press, New York.

## KEY REFERENCES

Chu et al., 1978. See above.

*The definitive work on conventional mega-electron-volt backscattering with surface barrier detectors.*

Feldman, L. C. and Mayer, J. W. 1986. Fundamentals of Surface and Thin Film Analysis. North-Holland, Elsevier Science Publishing, New York.

*Introductory text containing excellent descriptions of many surface and thin-film analytical techniques.*

Gibson, W. M. and Teitelbaum, H. H. 1980. Ion-Solid Interactions, a Comprehensive Bibliography, Vols. 1–3. INSPEC, Institution of Electrical Engineers, London.

*Exhaustive compilation of the literature of the field of ion beam interactions with solids up to approximately the date of its publication.*

Knoll, G. F. 1989. Radiation Detection and Measurement, 2nd ed. John Wiley & Sons, New York.

*A standard text on all aspects of radiation detection.*

Nastasi et al., 1996. See above.

*Excellent text reviewing fundamentals of ion beam interactions with materials.*

*Nuclear Instruments and Methods in Physics Research*, Section B. Elsevier Science Publishers, Amsterdam.

*In recent years a very significant portion of the literature of the field of ion beam interactions with materials has been published in this journal. Much additional material may be found in Section A of the journal and in the original journal Nuclear Instruments and Methods from which the modern series are derived.*

Tesmer, J. R. and Nastasi, M. (eds.). 1995. Handbook of Modern Ion Beam Materials Analysis. Materials Research Society, Pittsburgh.

*Detailed treatment of all aspects of ion beam analysis with numerous tables and tutorial material. The standard reference.*

Tirira et al., 1996. See above.

*A recent, comprehensive treatment of mega-electron-volt forward-recoil spectrometry.*

ROBERT A. WELLER
Vanderbilt University
Nashville, Tennessee

# HEAVY-ION BACKSCATTERING SPECTROMETRY

## INTRODUCTION

Heavy-ion backscattering spectrometry (HIBS) is a new technique for nondestructively analyzing ultratrace levels of impurities on the surface of very pure substrates. Although any high-$Z$ contaminant ($Z \geq 18$) on a pure low-$Z$ substrate can be measured, recent practical applications for HIBS have focused on measuring contaminants found on silicon wafers used in semiconductor manufacturing. Therefore, discussions of HIBS will focus in this area. The technique is based on the same principles described elsewhere (see ELASTIC ION SCATTERING FOR COMPOSITION ANALYSIS and MEDIUM-ENERGY BACKSCATTERING AND FORWARD-RECOIL SPECTROMETRY) for high-energy Rutherford backscattering spectrometry (RBS) and medium-energy ion backscattering spectrometry. The approach was invented and patented at Sandia National Laboratories and led to the construction of a HIBS User Facility, built in collaboration with SEMATECH member companies and Vanderbilt University. The facility was opened in June 1995 for use by U.S. industry, national laboratories, and universities in conducting contamination studies and is located at Sandia.

The motivation for HIBS is to detect metallic contamination at levels significantly below those that can be detected by RBS or medium-energy ion backscattering, both of which have a limit of $\sim 1 \times 10^{13}$ atoms/cm$^2$ for near-surface impurities. By the year 2001, as reported in the National Technology Roadmap for Semiconductors

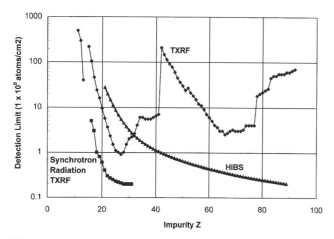

**Figure 1.** HIBS vs. total-reflection x-ray fluorescence (TXRF) detection limits on pure silicon. TXRF limits are shown using traditional x-ray and newer synchrotron radiation sources.

(Semiconductor Industry Association, 1997), contamination control in the microelectronics industry will require tools that can measure $\sim 1 \times 10^{10}$ atoms/cm$^2$ for transition metals in starting material. A sensitivity curve, shown in Figure 1, demonstrates how HIBS meets this requirement on silicon. The detection limit for well-separated elements on a clean Si surface ranges from $\sim 6 \times 10^9$ atoms/cm$^2$ for Fe to $\sim 3 \times 10^8$ atoms/cm$^2$ for Au, without the use of vapor

phase decomposition (VPD) to preconcentrate the impurities. Using VPD would improve the sensitivity by at least an order of magnitude.

The HIBS technique can be illustrated by following the ion trajectory shown in Figure 2. An 120-keV C$^+$ ion beam is typically used, with the beam focused onto an $\sim 3$-mm$^2$ spot at the sample surface. The relatively large beam spot size, which reduces sputtering and sample heating, allows the use of a higher beam current ($\sim 20$ nA). Ion beam effects such as sputtering, sample heating, and ion beam enhanced diffusion are discussed later. It should be noted that sputtering, although limiting the ultimate sensitivity of HIBS, is not as serious a limitation as once thought (Pedersen et al., 1996). Carbon ions backscattered from the sample are collected by a time-of-flight detector array having a large solid angle maximized for high sensitivity. The screening foil in each detector has a thickness chosen to filter out carbon ions scattered from the substrate but to pass ions scattered from heavier atoms, such as impurities on the surface. Measuring the yield versus flight time of the backscattered carbon ions allows both the mass and areal density of the near-surface impurities to be determined from basic physical backscattering concepts.

Other surface analysis techniques for detection of trace element contamination are briefly discussed next, with the strengths and weaknesses of each compared to those of HIBS. A detailed discussion of the basic theoretical and practical aspects of HIBS will then be provided.

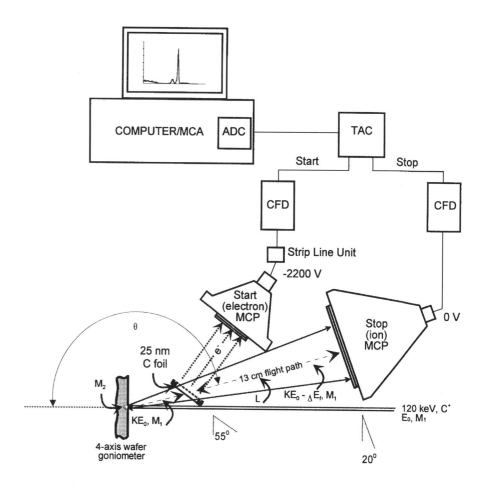

**Figure 2.** Typical HIBS time-of-flight detector with associated data acquisition instrumentation. Backscattering kinematics are also illustrated.

## Competitive, Complementary, and Alternative Techniques

A number of other techniques for trace element analysis are available, with total-reflection x-ray fluorescence (TXRF) and secondary-ion mass spectroscopy (SIMS) being the most widely used. All the techniques differ from HIBS in one very important aspect: they require standard samples for quantitative analysis. Since HIBS is a standardless measurement, one of its primary applications is to calibrate standard samples for use by the other analysis techniques. Another difference is that all but TXRF are destructive; that is, the sample may not be reused or subjected to an alternative analysis in the same area. The HIBS technique is nondestructive for further wafer analysis, but the wafer cannot be returned to production without further processing.

Total-reflection x-ray fluorescence using traditional x-ray sources is both a competitive and a complementary technique. Recently, a version of TXRF using synchrotron radiation has been introduced with the potential for higher sensitivity to many elements than can be achieved with any other method. Neither HIBS nor TXRF require sample preparation, and both provide spatial information about the impurities on the test sample. For an accurate measurement, TXRF requires an optically flat surface, such as the front side of a wafer, whereas HIBS can also measure rough surfaces, such as the backside of a wafer. Typically, TXRF is used to measure only the transition metals, with an anode change required to measure high-$Z$ metals, while HIBS surveys all elements with $Z \geq 18$ in a single experiment. As shown in Figure 1, TXRF has relatively poor sensitivity for high-$Z$ metallic contamination ($Z > 30$), so HIBS complements TXRF for measuring high-$Z$ contamination. On the other hand, TXRF, using traditional x-ray sources, has better mass resolution than HIBS and is used extensively by the semiconductor industry, being the tool of choice when measuring production wafers. Mass resolution in HIBS ranges from $\sim$2 amu for Fe to $\sim$20 amu for Pb, but its survey capability and accuracy make it an excellent tool for precise quantification of known contaminants. The analysis depth for TXRF is $\sim$5 nm on an unpatterned bare silicon wafer or $SiO_2$, whereas HIBS measures contamination in samples to a depth of $\sim$10 nm. Even when measuring the same wafer for the concentration of contaminants, TXRF instruments may report widely varying areal densities. Because of its high accuracy and standardless configuration, HIBS has been used to nondestructively verify the areal density of contamination on TXRF standards and calibrate TXRF tools (Werho et al., 1997; Banks et al., 1998). High concentrations of one element can cause interference in detecting other elements, for both TXRF and HIBS. The TXRF instruments have been highly developed, providing automated wafer exchange, data acquisition, and analysis. The HIBS system at Sandia is partially automated, but it requires up to 90 min per point when analyzing for $\sim 1 \times 10^9$ atoms/cm$^2$ levels, while TXRF measurements take $\sim$15 min per point for the same level.

Secondary-ion mass spectrometry has been used for many years and is a valuable tool for many problems. Based on the sputter erosion of the surface being analyzed, SIMS is inherently a depth profiling technique with good bulk sensitivity ($<$1 ppm). However, its sensitivity to a thin surface contaminant is not as good as other techniques and its sensitivity to various elements can vary by orders of magnitude in the same substrate. Even the sensitivity for a single element varies with different substrates.

Destructive alternative techniques such as inductively coupled plasma mass spectroscopy (ICP-MS), graphite furnace atomic absorption mass spectroscopy (GFAA-MS), glow discharge mass spectroscopy (GD-MS), and a variation of TXRF called microsample x-ray analysis (MXA) are used to measure contamination, mostly in liquids, to the ppt level. They can be used, indirectly, to measure surface contamination by concentrating the elements through dissolution of the sample surface using VPD. The VPD technique exposes the surface of a wafer to an acidic HF vapor for a set time, and then the etched products are collected by a droplet of ultraclean deionized water rolled around the surface. Finally, the liquid droplet is analyzed by one of the above techniques. The droplet can also be placed on a separate clean low-$Z$ film and allowed to dry for x-ray analysis (or HIBS). Collection efficiencies between 95% and 99% have been reported using the VPD process, except for Cu, for which it is $\sim$20%. There is some concern about the use of this technique as wafer sizes increase. All the techniques using VPD are labor intensive; however, automated VPD equipment has recently been introduced and could increase throughput. The availability of these techniques and their exceptional sensitivity, ranging to $1 \times 10^8$ atoms/cm$^2$, make them competitive.

## PRINCIPLES OF THE METHOD

The four principles of traditional RBS on which HIBS is based (see MEDIUM-ENERGY BACKSCATTERING AND FORWARD-RECOIL SPECTROMETRY) are (1) the kinematic factor $K$, which describes the energy transfer between a projectile and target nucleus in an elastic collision; (2) the scattering cross-section $\sigma$, defined as the probability of a scattering event occurring; (3) the stopping cross-section $\varepsilon$, which determines the energy loss of a projectile as it undergoes interactions with target nuclei and electrons; and (4) the energy straggling $\Omega_{Bohr}$, i.e., a statistical fluctuation in the energy loss of a projectile.

The yield equation $Y = Q\sigma Nt\Omega_D\eta$ gives the yield of backscattered particles as a function of the number of ions ($Q$) hitting the target layer, the atomic density ($N$) and thickness ($t$) of the layer, and the probability ($\sigma$) of the projectile ions being backscattered from the layer into a detector with a finite solid angle ($\Omega_D$) and detection efficiency ($\eta$).

It follows from the yield equation that if the factors $Q$, $\sigma$, $\Omega_D$, and $\eta$ are increased, then the corresponding yield increase allows for the detection of increasingly lower areal densities, where the areal density is the product of the target atomic density and thickness, $Nt$, normally expressed in atoms per square centimeter. Detection of increasingly lower areal densities is commonly referred to as increasing the sensitivity.

Although HIBS is based on the same fundamental concepts as RBS and medium-energy ion backscattering, the emphasis is placed on maximizing sensitivity to trace elements on the surface of a lower $Z$ substrate. This increased sensitivity is accomplished in two ways. First, since the backscattering cross-section ($\sigma$) scales as $Z^2/E^2$, where $Z$ and $E$ are the atomic number and energy of the analysis beam, the HIBS scattering yield from all elements is dramatically increased by using a high-$Z$ analysis beam with a relatively low energy, such as $C^+$ at 120 keV.

Of course, this increase in yield also increases the backscattering from the substrate, so a means of eliminating the ions scattered from substrate atoms is needed. This is accomplished by the second, key aspect of the technique: a screening foil incorporated in the design of the time-of-flight detectors eliminates essentially all ions backscattered from the substrate while still allowing those backscattered from heavy impurities on the surface to be detected. The foil is already required in the time-of-flight detector for producing a start pulse for the timing. By making the foil somewhat thicker than required, only ions backscattered with energies above a given threshold are passed through the foil to the second part of the detector, thus eliminating most of the ions backscattered from the substrate (such as silicon), which would otherwise comprise an overwhelmingly large background in the spectrum. It should be noted that simply using electronic discrimination of this background cannot be used because the microchannel plates would have to process the recoiled ions from substrate atoms, saturating the detectors.

Some ions scattered from the substrate will still be passed through the foil because of straggling in the foil and multiple small-angle scattering in the substrate, so further improvements can be made by use of ion channeling. When allowed by the sample configuration, aligning the analyzing ion beam with the crystalline channels in the substrate will further reduce the residual background from the substrate, helping to increase sensitivity.

Further improvements in sensitivity are obtained in HIBS by using multiple time-of-flight detectors, each with a large solid angle ($\Omega_D$), and a relatively large ($\sim$3-mm$^2$) beam spot. The large beam allows higher beam current and longer counting time, increasing the number of projectile ions ($Q$) while minimizing sputter erosion of the sample. Although the efficiency of each time-of-flight detector is lower than that of the surface barrier detector (SBD) commonly used in RBS measurements, the gain in mass resolution offsets the lower efficiency. Another advantage of an ion beam analysis such as HIBS is that measurement of areal density of surface impurities on a target is largely insensitive to chemical bonding or electronic configuration in the target (Feldman and Mayer, 1986; Ziegler and Manoyan, 1988).

Depth information about the sample is contained in the energy loss of the projectile ion and can be extracted using the known stopping cross-section ($\epsilon$), with straggling ($\Omega_{Bohr}$) limiting the resolution of this depth information. The energy loss is due to two mechanisms: projectile interactions with electrons in the target, referred to as electronic stopping, and small angle scattering with target atoms, called nuclear stopping. Since HIBS uses a low-energy

beam of carbon ions, depth information is limited to very near the surface, and thus the technique is generally not useful for depth profiling.

**Basic Theory**

The fundamentals of HIBS are based on the four basic physical concepts of traditional RBS, as previously noted and found in several excellent reference books (e.g., Chu et al., 1978; Feldman and Mayer, 1986; Tesmer et al., 1995). These references discuss RBS fundamentals using a He ion beam and an SBD to obtain a yield-versus-energy spectrum. Early HIBS experiments also used an SBD (Knapp et al., 1994a). However, limitations of the SBD, discussed later, caused its discontinued use. The HIBS technique now uses the carbon foil-based time-of-flight detection scheme, which is described in the literature (Chevarier et al., 1981; Mendenhall and Weller, 1989; Dobeli et al., 1991; Knapp et al., 1994b; Anz et al., 1995).

The four basic Rutherford backscattering physical concepts are kinematic factor, scattering cross-section, stopping cross-section, and straggling, with the information gained being that of mass identity, atomic composition, depth, and the limits to mass and depth resolution, respectively. Equations based on these concepts allow analysis of the raw HIBS backscattering data.

Referring to Figure 2, when a projectile ion with initial energy $E_0$ and mass $M_1$ undergoes a simple elastic collision with a stationary atom $M_2$ at the surface of the substrate, the backscatter energy $E_1$ at an angle $\theta$ in the laboratory frame of reference can be determined from the kinematic factor, defined as $K \equiv E_1/E_0$, where $M_2 > M_1$. Thus the backscatter energy is

$$E_1 = KE_0 = E_0 \left[ \frac{\{x\cos(\theta) + [1 - x^2\sin^2(\theta)]^{1/2}\}}{1 + x} \right]^2 \quad (1)$$

where $x \equiv M_1/M_2$. Note that the backscattered particle energy $E_1$ is only dependent on the initial energy $E_0$, the ratio of the projectile and target masses, $M_1/M_2$, and the laboratory scattering angle $\theta$. After backscattering, the particle energy can also be described by the product of its mass and velocity, $E_1 = \frac{1}{2}M_1V_1^2$, where the projectile velocity can be found from $V_1 = L/\tau$ by measuring the time $\tau$ to travel the fixed path length $L$ the ion traverses in the detector. These basic equations allow a determination of the mass of the atom from which the ion was backscattered, which is the general principle on which the time-of-flight technique is based.

The carbon projectile loses energy when penetrating and exiting the target and when passing through the carbon foil. Carbon ions that are not immediately backscattered from surface contaminant atoms penetrate the surface of the target, lose a finite amount of energy, are backscattered, and then lose more energy on their exit path, with the total energy loss denoted as $\Delta E_s$. Many of the ions, of course, are not backscattered and come to rest in the sample. (The probability of backscattering, or the scattering cross-section, will be discussed later.) The backscattered ions are collimated, to reduce stray ion scattering, and then pass through the carbon foil, losing additional energy $\Delta E_f$. Most energetic ions that pass through the carbon foil eject electrons that are accelerated into the start microchannel plate (MCP) and create a start timing pulse $\tau_0$. As described earlier, the cumulative energy losses occur through electronic and nuclear stopping, with the losses used to advantage by stopping most projectile ions scattered from the substrate in the carbon foil, reducing unnecessary counts from these scattering events. However, due to straggling in the foil and multiple scattering in the substrate, the ions backscattered from silicon can never be eliminated completely.

Energy lost in a layer can be calculated from the stopping power of the material in the layer. This loss is given as $\Delta E_s = [\varepsilon]Nt$, where $[\varepsilon]$ is the stopping cross-section factor, $N$ is the atomic density of the layer, and $t$ is the thickness of the layer. The stopping cross-section is defined as $\varepsilon \equiv (1/N)(dE/dx)$, where $dE/dx$ is the stopping power or change in energy per unit distance. The stopping cross-section typically has units expressed as $eV/10^{15}$ atoms/cm$^2$. The stopping cross-section factor is given by the formula (Chu et al., 1978)

$$[\varepsilon] = \left[ \frac{K}{\cos\theta_1}\varepsilon_{in} + \frac{1}{\cos\theta_2}\varepsilon_{out} \right] \quad (2)$$

where $\theta_1$ and $\theta_2$ are the angles that the projectile ion makes with respect to the surface normal as it enters and leaves the target and $\varepsilon_{in}$ and $\varepsilon_{out}$ are the stopping cross-sections along the inward and outward paths of the projectile ion, respectively. Although the low-energy carbon projectile ions lose a small amount of energy while penetrating and exiting the target, the loss is neglected in HIBS analyses because the surface energy loss $KE_0$ is approximately equal to the total loss $\Delta E_s$.

The energy of the ions escaping the substrate or the carbon foil exhibits a statistical energy distribution referred to as straggling. This straggling limits both the energy (and therefore the time) and mass resolution of the system. Straggling is commonly described using Bohr's equation,

$$\Omega_{Bohr}^2 = 4\pi Z_1^2 e^4 N Z_2 t \quad (3)$$

which shows that for a projectile ion with atomic number $Z_1$ passing through a layer with atomic density $N$, atomic number $Z_2$, and thickness $t$, the projectile ion energy straggling ($\Omega_{Bohr}$) increases with the square root of the foil electron areal density, $NZ_2t$. Straggling exhibits a projectile ion velocity dependence, not shown in Bohr's equation, and therefore, modified versions of Bohr's theory are now used (Lindhard and Scharff, 1953; Bonderup and Hvelplund, 1971; Chu, 1976; and Besenbacher et al., 1980). It should be noted that no theory exists for slow heavy ions in a light matrix. Bohr's equation predicts a symmetrical Gaussian distribution of energy equal to $2.355\,\Omega_{Bohr}$ at full width at half-maximum (FWHM) due to projectile ion interactions with electrons in the foil. For most backscattering spectra, straggling appears as the integral of the Gaussian distribution and is described by a complementary error function. Empirically, this is seen as a slope

in the backscattered energy edge for each target mass, with the energy distribution of $2.355\,\Omega_{Bohr}$ at FWHM corresponding to the 12% to 88% range of the slope.

After the ion traverses the start foil and produces the start timing signal, it continues its flight until being stopped by the ion MCP, where it produces a stop signal $\tau_{stop}$. After signal processing, a time-to-amplitude converter is used to give a digital pulse, with a height proportional to the time difference between the start and stop timing signal, $\tau_{stop} - \tau_0$. These signals are binned in a multichannel analyzer, giving a yield-versus-time spectrum. The yield for the $i$th element, $Y_i$, is dependent on several factors, according to the equation

$$Y_i = \eta(KE_0 - \Delta E_f)Q(N_it)\Omega_D\sigma_i(E_0,\theta) \quad (4)$$

where $\eta(KE_0 - \Delta E_f)$ is the efficiency of the detector at the energy of the projectile ion after passing through the foil, $Q$ is the total number of projectiles hitting the target, $N_i$ is the atomic density in atoms per cubic centimeter of the $i$th contaminant element, $t$ is the thickness of the contaminant layer, $\Omega_D$ is the detector solid angle, and $\sigma_i(E_0,\theta)$ is the screened cross-section for the collision with the $i$th element. Determining the yield is complicated by the energy dependence of the detector efficiency and the reduction of the scattering cross-section due to electron screening as discussed in detail by elsewhere (see MEDIUM-ENERGY BACKSCATTERING AND FORWARD-RECOIL SPECTROMETRY). However, the yield can be determined with reasonable accuracy by a computer analysis algorithm, described later. The screened scattering cross-section (Anderson et al., 1980) is given in the equation

$$\sigma_i = \sigma_R \left( 1 + \frac{0.04873\,Z_1Z_2(Z_1^{2/3}Z_2^{2/3})^{1/2}(1+x)}{E_0} \right)^{-1} \quad (5)$$

where $E_0$ is the initial projectile ion energy, $Z_1$ and $Z_2$ are the projectile and the $i$th target element atomic numbers, respectively, and $x \equiv M_1/M_2$. This value differs by only a few percent from the more precise Lenz-Jensen screened cross-section (see MEDIUM-ENERGY BACKSCATTERING AND FORWARD-RECOIL SPECTROMETRY). The Rutherford cross-section $\sigma_R$ in the laboratory frame of reference is given (Darwin, 1914; Chu et al., 1978) as

$$\sigma_R(E_0,\theta) = \left( \frac{Z_1Z_2e^2}{2E_0} \right)^2 \left( \frac{\{\cos(\theta) + [1 - x^2\sin^2(\theta)]^{1/2}\}^2}{\sin^4(\theta)[1 - x^2\sin^2(\theta)]^{1/2}} \right) \quad (6)$$

where $E_0, Z_1, Z_2$, and $x$ are described in Equation 5, $\theta$ is the laboratory angle through which the projectile ion is scattered, and $e^2$ is the square of the electron's charge, which may be expressed as $1.4398 \times 10^{-10}$ keV·cm.

## PRACTICAL ASPECTS OF THE METHOD

The first HIBS experiments used an SBD. However, limitations of the SBD (e.g., low count rate leading to pulse

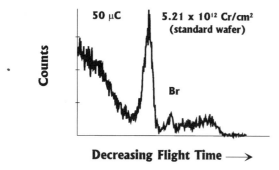

**Figure 3.** An important use of HIBS is to calibrate other trace element analysis tools. The HIBS measurements on Cr (above), Fe, and Ni contamination standard wafers were used to calibrate TXRF instruments at Motorola.

pileup, degraded energy resolution due to electronic noise, and radiation damage from high-$Z$ ions) warranted changing to a time-of-flight detection scheme using a carbon foil. Since the time-of-flight HIBS detectors give a yield-versus-flight-time spectrum, it looks similar to one acquired by RBS, with spectra showing increasing mass numbers to the right (this assumes that flight time is shown decreasing to the right). Typical spectra can be seen in Figure 3 and Figure 4. Portions of the spectra in Figure 4 show the physical phenomena that limit the mass resolution and sensitivity of the time-of-flight technique.

### Time-of-Flight Spectrometer

A schematic of the time-of-flight spectrometer used in HIBS is shown in Figure 2 and consists of a collimator, a thin carbon foil (~25 nm) placed in the path of the back-scattered carbon projectiles, an electron accelerating grid, and two MCPs. Secondary electrons are produced

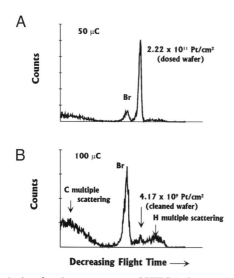

**Figure 4.** Another important use of HIBS is for contamination-free semiconductor manufacturing research. Texas Instruments used HIBS analyses to determine the dose of Pt contaminant on a wafer (**A**) and to show the efficacy of a $HNO_3$:HF (95:5) mix to clean the wafer (**B**). Scales for (**A**) and (**B**) are identical. Spectra background (**B**) is due to multiple scattering of the C analysis beam in the Si substrate and H recoiled from the wafer's surface.

when the projectile passes through the foil, with the number ejected based on statistics. These electrons are then accelerated by the grid into the first MCP to provide a start timing pulse. As discussed previously, the foil also serves as a filter to eliminate most of the carbon ions scattered from the substrate, suppressing the substrate background. Carbon ions that pass through the foil continue their flight until being stopped by the second MCP, where they produce a stop timing pulse.

### Sensitivity and Mass Resolution

A number of factors influence the detection limits and mass resolution that can be achieved with HIBS: detector efficiency, choice of beam species and energy, sputtering, and background due to multiple scattering and random coincidences (Knapp et al., 1996). The detection limits can be reduced in a number of ways: increasing the solid angle by using multiple detectors, using a larger beam spot, channeling the incident beam along a crystal axis of the substrate, and matching the choice of beam to the screening foil thickness. All of these were incorporated in the design of the HIBS system at Sandia. Of course, sputtering of the sample surface by the analysis beam is the ultimate limit to the statistics that can be obtained, and hence the sensitivity. Beam effects due to sample heating are considered negligible. Sputtering and ion beam–enhanced diffusion can be potential problems when measuring surface contaminants in a metal-silicon system because erosion and diffusion can cause a decrease in the backscattering signal. If it is suspected that sputtering or beam-enhanced diffusion might be a problem, the areal density as a function of beam flux can be measured to determine if the problem exists and, if it does, extract the starting areal density. A study using a large $N^+$ ion flux to measure metallic surface contaminants (Fe, Cu, Mo, W, and Au on silicon) showed that sputtering is not a significant limitation to the sensitivity of HIBS for these elements and that diffusion was limited to <10 nm (Pedersen et al., 1996).

We define the detection limit as $DL = 2 \times$ (background)$^{0.5} \times$ (calibration factor) where the background is taken under the full width of the peak for the element of interest. A number of alternative definitions could be used, and many are discussed by Currie (1960). Heavy-ion backscattering spectrometry cannot be considered to have a "well-known blank" (a sample with no contaminants, to be measured as a background), since we have found that no Si wafer is completely free of heavy-element contamination at levels HIBS can detect. However, since the peaks due to contaminants in a spectrum are generally surrounded on each side by wide regions of well-defined background, the background under the peaks can be fit fairly well and its standard deviation reduced to a low level. In this case, the criteria of Currie (1960) would be $DL = 1.64 \times$ (background)$^{0.5} \times$ (calibration factor), or 18% lower than our definition.

The efficiency of the MCPs for detecting electrons, ions, and neutral particles in the sub-100-keV energy range is not well known, nor is the electron-producing efficiency of the carbon screening foil. The MCPs also have a count rate limitation that may in turn limit the amount of beam current that can be used. Straggling in the foil is also an important effect, dominating the observed time resolution and limiting the mass resolution that can be achieved, through broadening of the peak widths. The mass resolution of the Sandia HIBS system, with its present 5-$\mu$g/cm$^2$ carbon foil, is about 1 amu for Ti-V, 3 amu for Cu-Zn, and 18 to 20 amu for the heaviest elements. That is, for elements near Cu, two peaks of equal magnitude can be separated if the masses of the two elements differ by 3 amu or more.

Scattering in the foil is also a serious consideration for the detector sensitivity, since it results in some ions exiting at angles that do not intercept the ion MCP, reducing the effective solid angle of the detector. This effect is stronger at lower energy, offsetting the gains in overall sensitivity that can be obtained by going to lower beam energy. The problem has been illustrated by TRIM (a code applied to the transport of ions in matter; Ziegler and Biersack, 1995) calculations of the distribution of trajectories of $N$ particles passing through a foil at two different energies. The percentage of particles that remain in the acceptance angle of the ion MCP is much lower at low energy: the simulation shows 85% of the 160-keV particles reach the MCP, but only 22% do so at 40 keV.

The effect of foil scattering will vary with beam species and energy and of course will be different for different foil thicknesses. For a given foil, measurements of the detector efficiency will show that some ions are detected more efficiently than others, primarily because of the relative amounts of foil scattering, but also due to details of electron production efficiency and MCP detection efficiency. An analytical model of the efficiency of time-of-flight ion detectors of this general configuration has been developed elsewhere (Weller et al., 1994).

Background in the spectra is the major limit to the sensitivity that can be achieved. We have found that background can come from several sources: scattering of ions and electrons around the chamber, scattering from the substrate "leaking" through the foil, and beam contamination posing major problems. The first problem can be minimized by proper containment of the beam and shielding of the detectors from unwanted ions and electrons. The second problem is due not just to straggling in the foil but also to multiple scattering of the ion beam in the substrate: for a very low number of incoming ions, multiple scattering events can steer the ions back out of the substrate with higher than expected energy, making the particles indistinguishable from those scattered from heavy atoms on the surface. This low-level background cannot be eliminated, but it can be minimized by channeling of the analysis beam.

While using N$^+$ and O$^+$ ion beams for HIBS analysis, we also identified beam contamination by NH$_2$ and CH$_2$ and forward scattering of H as sources of background in the Sandia HIBS system, often at such low levels that they had remained unnoticed in an earlier HIBS system with less sensitivity. To minimize beam contamination problems, we settled on a C$^+$ beam for all high-sensitivity measurements. Using the Sandia HIBS system with three detectors and 25-nm foils, we have observed background

levels implying detection limits to impurities on Si ranging monotonically from $\sim 6 \times 10^9$ atoms/cm$^2$ for Fe to $\sim 3 \times 10^8$ atoms/cm$^2$ for Au.

## Instrumentation

Figre 2 shows a typical time-of-flight detector and associated instrumentation used for HIBS measurements. The detector and instruments are commercially available and consist of MCP assemblies for the start and stop MCPs, a strip line unit for the start MCP, constant fraction discriminators (CFDs), a time-to-amplitude converter (TAC), an analog-to-digital converter (ADC) board, and a multichannel analyzer (MCA). The start MCP uses a 20-mm-diameter chevron and the stop MCP uses a 40-mm-diameter chevron. Since HIBS uses three time-of-flight detectors to maximize the detector solid angle, three each of the units mentioned above are needed, with the exception of the CFDs and MCA. Two CFDs (with quad inputs) can be used to process all start and stop timing signals, and the MCA has the capability of processing data from all three detectors. The strip line unit is used to block the $-2200$-V output level of the electron MCP and allow the timing signal to pass while providing impedance matching to reduce signal reflections. A fast oscilloscope (bandwidth $\geq 400$ MHz) is needed to monitor the timing pulse outputs from the MCP assemblies and to set up the CFDs.

A CFD that provides a zero-crossing timing pulse output for use by the TAC module gives better time resolution than the leading-edge type. A short coaxial cable, whose length is determined experimentally, must be used to set the zero-crossover point and adjust the CFD walk, which can be labor intensive. At least one manufacturer (Ortec) has recently offered a picosecond timing discriminator that does not require this cable but processes only one signal.

Recent advances in MCP assembly technology (Galileo) have resulted in low-profile versions being made available, with a tenfold height reduction, which could allow a more compact HIBS system to be designed. This new generation of MCPs has an extended dynamic range, a higher gain, a shorter output pulse width, and lower noise. These features could allow a higher count rate, greater efficiency, and better mass resolution. These low-profile MCPs also have integral resistor biasing of the plates and feature a quick-change module for replacement of the plates, reducing both installation and maintenance time.

A small 200-keV accelerator (Peabody Scientific, model PS-250) using a Duoplasmatron gas source is used to obtain the C$^+$ projectile ions measured by the HIBS instrumentation described above. However, implanters or other accelerators with good energy stability ($\sim$250 eV), a high-resolution mass analyzing magnet ($\leq$0.5 amu), and other features similar to the Peabody accelerator would also work well.

## Examples

Measurements by two major semiconductor manufacturers, Motorola and Texas Instruments, serve as examples of using HIBS to solve integrated circuit manufacturing problems. Motorola has used the HIBS facility to help solve problems in the calibration of TXRF instruments (Werho et al., 1997; Banks et al., 1998). The TXRF instruments are used for detecting trace levels of transition metals introduced by semiconductor processing procedures and equipment. These instruments are typically calibrated using the manufacturer's calibration data. This independent calibration has led to inaccuracies and a resulting lack of interchangeability in TXRF data. To illustrate the problem caused by independent calibration, the average error in measuring the same contamination standard wafers by different laboratory sites during a round robin conducted at Motorola showed errors in excess of 350%. To solve this problem, HIBS was used to nondestructively verify the areal density of a contaminant on commercially obtained contamination standard wafers used to calibrate these instruments. Figure 3 shows a typical spectrum taken on a Cr standard wafer. The HIBS values measured on this Cr-contaminated wafer, as well as other standard wafers contaminated with Fe and Ni, generally verified the vendor's value to within ±10%. The HIBS data were then used to create a universal sensitivity factor curve for recalibration of the TXRF instruments used in the study. This recalibration generally reduced the error between TXRF instruments to < ±10%. Using this universal method for calibrating TXRF instruments has led to improved quality control during integrated circuit (IC) manufacturing. The calibration method has international applicability, since the lack of interchangeability of data between TXRF instruments is a problem that has been reported worldwide (Hockett et al., 1992).

Texas Instruments has used HIBS technology to evaluate the handling and processing of new materials that can be used in the manufacture of ultralarge-scale integrated (ULSI) dynamic random-access memories (DRAMs). Texas Instruments used the HIBS facility at Sandia National Laboratories in the development of cleaning processes for the removal of high-$Z$ metallic contamination (Banks et al., 1998). To decrease the size of capacitors used in ULSI DRAMs, Texas Instruments has been studying the use of a high-$k$ dielectric (where $k$ is the relative dielectric constant), barium-strontium-titanate (BST) between platinum electrodes (Pt/BST/Pt). Because of the concern over cross-contamination of wafers and processing tools with these materials, several questions needed to be answered, including how effective their present cleaning methods were at removing these heavy metals. To begin answering this question, wafers were uniformly dosed with the contaminants and the areal density was measured. The wafers were then subjected to various cleaning methods, the results of one of which is shown in Figure 4, and measured again. This comparison shows the efficacy of the cleaning process and demonstrates the excellent sensitivity of HIBS.

### Safety Considerations

The two primary safety hazards associated with HIBS measurements are exposure to (1) high voltage and (2) x radiation. These hazards are mitigated through the use of stringent safeguards. A physical barrier must be set up around the accelerator with an appropriate door interlock used to disconnect power to the high-voltage supply in the event that high voltage is inadvertently left on when attempting to enter the accelerator enclosure. In addition, it is a good practice to use a radiation monitor interlocked with the high-voltage supply, shutting the supply off when radiation above acceptable limits is detected. Another feature that should be incorporated in the accelerator enclosure design is to provide an OSHA-approved grounding rod for discharging high voltages that might exist on various isolated accelerator components. The accelerator should always be grounded before conducting maintenance or repairs. Finally, easily accessible panic switches should be located both inside and outside an enclosure, providing for emergency shutdown of the accelerator. All applicable regulations of local, regional, and national authorities should be adhered to for installation and operation of the accelerator. General radiological guidelines for conducting ion beam measurements and a listing of regulating agencies can be found in Tesmer et al. (1995).

### METHOD AUTOMATION

Sample positioning and data collection in HIBS can be semiautomated so that little operator intervention is needed during data acquisition, with the operator only needing to make minor adjustments of the ion accelerator parameters. In the HIBS system at Sandia, motor control hardware and software using the IEEE-488 protocol are used in conjunction with Microsoft Visual Basic® to control a four-axis sample goniometer. This control is needed so that a wafer (or sample) may be positioned at precise $x$ and $y$ coordinates and also allow channeling. The software automatically converts $x$ and $y$ coordinates (given in millimeters) into polar coordinates for translation by the stepping motors. To mimic the spot size of TXRF instruments, HIBS steps the beam spot ($\sim$3 mm$^2$) around in a $\sim$1-cm$^2$ pattern. This allows direct comparison of the areal density collected and reported by TXRF instruments to that of HIBS. This large pattern also further reduces the sputtering of contaminants being measured.

The raw time-of-flight data are collected by three ADC boards installed in a personal computer, with the three spectra being processed by a commercially available MCA software package (Oxford) that interfaces with the boards. Data are normally acquired until a preset amount of charge has been collected on a sample. An automated beam stop incorporated into the beam line blocks the beam after reaching a preset condition. With this minimal degree of automation, data collection can proceed while data are being analyzed.

### DATA ANALYSIS AND INITIAL INTERPRETATION

The HIBS data analysis provides information about the identity and concentration of heavy-metal surface impurities on very pure substrates. Computer algorithms, derived from fundamental equations and discussed below, are used to determine areal density and mass for a given peak in the time-of-flight spectrum. It should be noted that this spectrum is the sum of three separate but simultaneously acquired spectra from the HIBS detectors. Spectra manipulation, such as shifting and scaling prior to the analysis, is not discussed. However, this spectra manipulation and subsequent analysis have been semi-automated with software written in Microsoft Visual Basic®. Copies of the program, called HIBWIN© (Knapp, 1995), can be obtained by contacting the author at Sandia National Laboratories.

### Areal Density

To find the areal density, Equation 4 can be used. To use this equation, however, the detector efficiency $\eta$, which is a function of the backscattered ion energy, must be determined empirically (Weller, 1993). For the HIBS system at Sandia, a standard target consisting of three elements, Fe, Ag, and Pt, evaporated onto a clean silicon substrate is used to determine $\eta$. The concentration of each element is approximately one monolayer so that standard RBS can be used to calibrate the standard. Using Equation 4, it follows that the areal density of an unknown element $i$ can be found from

$$Y_i = \eta(KE_0 - \Delta E_f)Q(N_i t)\Omega_D\sigma_i(E_0, \theta) \qquad (7)$$

and for a known standard,

$$Y_s = \eta(KE_0 - \Delta E_f)Q(N_s t)\Omega_D\sigma_s(E_0, \theta) \qquad (8)$$

so that, solved simultaneously, the two equations give

$$N_i t = \frac{Y_i}{Q\sigma_i(E_0, \theta)}f \qquad (9)$$

where $f$ is the factor $1/(a + bE + cE^2)$. The coefficients of this second-order polynomial equation are determined from the known areal densities (from RBS measurements) and HIBS yields for Fe, Ag, and Pt measured on the standard by substituting them into the equation

$$N_n t = \frac{Y_n}{Q\sigma_n(E_0, \theta)}(a + bE_n + cE_n^2) \qquad (10)$$

where $n = 1, 2, 3$ for Fe, Ag, and Pt, respectively. The values needed for the energy $E_n$ of the carbon particle after scattering from the standard elements and passing through the range foil can be approximated from the equation (Doyle et al., 1989)

$$E_n = [KE_0^{(1-a)} - C_1]^{1/(1-a)} \qquad (11)$$

where $n = 1, 2, 3$ and the terms $K$ and $E_0$ have been described. The exponential terms $1 - a$ and $1/(1 - a)$ allow a fit of the $E$-versus-mass curve, and $C_1$ is an energy constant proportional to the thickness of the carbon foil. Once the unknown coefficients, exponential, and constant terms from Equations 10 and 11 are found, subsequent areal densities for other elements are easily determined.

### Background and Sensitivity

From the previous discussion on HIBS sensitivity, it follows that the sensitivity is found by substituting $2\sqrt{\text{Bkg}}$ for the yield in Equation 9. The value of the background is found by performing a fit to the spectral background using the equation

$$y = de[-a(x - c_2)]^m + be[-a(x - c_1)]^m \qquad (12)$$

The variable $d$ is found from the average height of the spectral background in the high-velocity (fast-time) part of the spectrum and $b$ from the average height in the slow-time portion of the spectrum. A best fit to the background is obtained in the software by the user inputting a value for $m$ and setting four markers at strategic locations on the spectral background to define the variables $d$ and $b$ as well as the exponential terms involving $c_1$ and $c_2$.

### Mass

When performing the areal density calibration, as discussed above, the flight times for a carbon ion backscattered from the standard masses Fe, Ag, and Pt can also be determined by using the beam energy $E_n$ after passing through the foil, calculated from Equation 11 and using $V_n = (L/\tau_n)$. The velocity is found from

$$V_n = \left(\frac{2E_n(\text{keV})}{M_1(\text{amu})}k\right)^{1/2} \qquad (13)$$

where $E_n$ and $M_1$ were previously described and $k = 9.648 \times 10^{14}$ for units of centimeters per second, $L = 13$ cm, and $n$ is 1, 2, 3 for Fe, Ag, and Pt, respectively. Knowing the time-to-amplitude converter time range, the number of channels in a spectrum, and the calculated flight times from the standard masses allows calibration of the time/channel. The peak channel position for each standard element is found by calculating the peak median (counts on either side of the peak are equal). The calibration is then used for subsequent time-to-mass conversions.

## SAMPLE PREPARATION

The HIBS measurements require no sample preparation; however, special precautions should be taken to prevent any extraneous contamination. Unpacking and insertion of samples for HIBS analyses into the chamber loadlock should be done in at least a class 10 environment. The loadlock is required to prevent damaging the thin foils used by the HIBS detectors. A small footprint class 10 minienvironment is ideal for space-limited laboratories and only requires wearing clean room gloves and a laboratory coat. Samples or wafers should only be handled with low-Z tweezers or a wafer wand.

## PROBLEMS

There are four main areas where the potential for misapplication or interpretive errors might occur when using the

HIBS technique. First, HIBS is an ultratrace element surface analytical technique, with the depth of analysis being ~10 nm. Use for analysis of samples with impurities extending well below the surface is difficult. Other techniques more suited for trace element depth profiling, such as SIMS, are discussed under Competitive, Complementary, and Alternative Techniques, above. Second, HIBS mass resolution is relatively poor and mass dependent, as compared to TXRF and other techniques, with HIBS mass resolution ranging from ~2 amu for Fe to ~20 amu for Pb. This can lead to interpretive errors, for example, in assigning a mass to a single high-Z peak or overlapping peaks that are close in mass, as discussed under Sensitivity and Mass Resolution, above. Also, as with other techniques, high concentrations of one element can interfere in the detection of other elements. Third, sputtering and ion beam-enhanced diffusion can be potential problems when measuring surface comtaminants in a metal-silicon system because erosion and diffusion can cause a decrease in the backscattering signal and reduced sensitivity. However, these effects do not appear to be a significant problem for the HIBS technique (Pedersen et al., 1996). If it is suspected that sputtering or beam-enhanced diffusion might be a problem, the areal density as a function of beam flux can be measured to determine if the problem exists and, if it does, extract the starting areal density. Fourth, background in the spectra is a major limit to the sensitivity that can be achieved with the HIBS technique, which ranges from $\sim 6 \times 10^9$ atoms/cm$^2$ for Fe to $\sim 3 \times 10^8$ atoms/cm$^2$ for Au when measured on very pure substrates. Careful design eliminates many factors contributing to this background, as discussed under Sensitivity and Mass Resolution, above. However, multiple scattering of the ion beam in the sample and forward scattering of H on the sample surface cannot be entirely eliminated. Multiple scattering is greatly reduced by channeling of the analysis beam, leaving H scattered by the ion beam into the detectors as a problem. This is discussed in more detail under Sensitivity and Mass Resolution and is shown in the spectrum of Figure 4B. The analysis software used for HIBS, discussed above (see Data Analysis and Initial Interpretation), was written to minimize errors introduced by background.

## ACKNOWLEDGMENTS

The authors would like to thank B. L. Doyle for initiating the HIBS concept and providing technical oversight during the research and development phases of HIBS. Sandia National Laboratories is a multiprogram laboratory operated by Sandia Corporation, a Lockheed Martin Company, for the U.S. Department of Energy under contract no. DE-AC04-94AL85000.

## LITERATURE CITED

Andersen, H. H., Besenbacher, F., Loftager, P., and Möller, W. 1980. Large-angle scattering of light ions in the weakly screened Rutherford region. *Phys. Rev. A* 21:1891–1901.

Anz, S. J., Felter, T. E., Daley, R. S., Roberts, M. L., Williams, R. S., and Hess, B. V. 1995. Recipes for high resolution time-of-flight detectors. Sandia Report, SAND94–8251, Albuquerque, N.M.

Banks, J. C., Doyle, B. L., Knapp, J. A., Werho, D., Gregory, R. B., Anthony, M., Hurd, T. Q., and Diebold, A. C. 1998. Using heavy ion backscattering spectrometry (HIBS) to solve integrated circuit manufacturing problems. *Nucl. Instrum. Methods B* 136/138:1223–1228.

Besenbacher, F., Andersen, J. U., and Bonderup, E. 1980. Straggling in energy loss of energetic hydrogen and helium ions. *Nucl. Instrum. Methods* 168:1–15.

Bonderup, E. and Hvelplund, P. 1971. Stopping power and energy straggling for swift protons. *Phys. Rev. A* 4:562.

Chu, W. K. 1976. Calculation of energy straggling for protons and helium ions. *Phys. Rev. A* 13:2057.

Chevarier, A., Chevarier, N., and Chiodelli, S., 1981. A high resolution spectrometer used in MeV heavy ion backscattering analysis. *Nucl. Instrum. Methods* 189:525–531.

Currie, L. A. 1960. *Anal. Chem.* 40:586.

Darwin, C. G. 1914. Collision of α particles with light atoms. *Philos. Mag.* 27:499–506.

Döbeli, M., Haubert, P. C., Livi, R. P., Spicklemire, S. J., Weathers, D. L., and Trombrello, T. A. 1991. A time-of-flight detector for heavy ion RBS. *Nucl. Instrum. Methods* B56/57:764.

Doyle, B. L., Knapp, J. A., and Buller, D. L. 1989. Heavy ion backscattering spectrometry (HIBS)—an improved technique for trace element detection. *Nucl. Instrum. Methods* B42:295–297.

Feldman, L. C. and Mayer, J. W. 1986. Fundamentals of Surface and Thin Film Analysis. Prentice-Hall, Englewood Cliffs, N.J.

Hockett, R. S., Ikeda, S., and Taniguchi, T., 1992. TXRF round robin results. *In* Proceedings of the Second International Symposium on Cleaning Technology in Semiconductor Device Manufacturing (J. Ruzyllo and R. E. Novak, eds.), pp. 324–337. Electrochemical Society, Pennington, N.J.

Knapp, J. A. 1995. HIBWIN© computer software for HIBS analysis. Sandia National Laboratories, Dept. III, Albuquerque, N. Mex.

Knapp, J. A., Banks, J. C., and Doyle, B. L. 1994a. Time-of-flight detector for heavy ion backscattering spectrometry. Sandia National Laboratories, publication SAND94–0391.

Knapp, J. A., Banks, J. C., and Doyle, B. L., 1994b. Time-of-flight heavy ion backscattering spectrometry. *Nucl. Instrum. Methods* B85:20–23.

Knapp, J. A., Brice, D. K., and Banks, J. C. 1996. Trace element sensitivity for Heavy Ion Backscattering Spectrometry. *Nucl. Instrum. Methods* B108:324–330.

Lindhard, J. and Scharff, M. 1953. Approximation method in classical scattering by screened Coulomb fields. *Mat. Fys. Medd. Dan. Vid. Selsk.* 27(15).

Mendenhall, M. H. and Weller, R. A. 1989. A time-of-flight spectrometer for medium energy ion scattering. *Nucl. Instrum. Methods* B47:193–201.

Pedersen, D., Weller, R. A., Weller, M. R., Montemayor, V. J., Banks, J. C., and Knapp, J. A. 1996. Sputtering and migration of trace quantities of transition metal atoms on silicon. *Nucl. Instrum. Methods* B117:170–174.

Semiconductor Industry Association. 1997. The National Technology Roadmap for Semiconductors: Technology Needs. SEMATECH, Inc., San Jose, Calif.

Weller, R. A. 1993. Instrumental effects on time-of-flight spectra. *Nucl. Instrum. Methods* B79:817–820.

Weller, R. A., Arps, J. H., Pedersen, D., and Mendenhall, M. H. 1994. A model of the intrinsic efficiency of a time-of-flight spectrometer for keV ions. *Nucl. Instrum. Methods* A353:579–582.

Werho, D., Gregory, R. B., Schauer, S., Liu, X., Carney, G., Banks, J. C., Knapp, J. A., Doyle, B. L., and Diebold, A. C. 1997. Calibration of reference materials for total-reflection X-ray fluorescence analysis by heavy ion backscattering spectrometry. *Spectrochim. Acta B* 52:881–886.

Ziegler, J. F. and Biersack, J. P. 1995. Transport of Ions in Matter (version TRIM-95) software. IBM-Research, Yorktown, N.Y.

Ziegler, J. F. and Manoyan, J. M., 1988. The stopping ions in compounds. *Nucl. Instrum. Methods B* 35:215.

## KEY REFERENCES

Chu, W. K., Mayer, J. W., and Nicolet, M-A. 1978. Backscattering Spectrometry. Academic Press, New York.

*Definitive reference for traditional MeV RBS using surface barrier detectors, with the fundamentals applicable to HIBS.*

Feldman and Mayer, 1986. See above.

*Excellent introductory text for ion and electron beam techniques used for surface and thin film analyses.*

Tesmer, J. R., and Nastasi, M. (eds.). 1995. Handbook of Modern Ion Beam Materials Analysis. Materials Research Society, Pittsburgh, Pa.

*Provides an invaluable overview of ion beam analysis fundamentals and techniques, including sections on traditional RBS and medium-energy ion beam analysis, from which the HIBS technique is derived.*

JAMES C. BANKS
JAMES A. KNAPP
Sandia National Laboratories
Albuquerque, New Mexico

# NEUTRON TECHNIQUES

## INTRODUCTION

This part and its supplements explore the wide range of applications of elastic and inelastic neutron scattering to the study of materials. Neutron scattering provides information complementary to several other techniques found in this volume. Perhaps most closely related are the x-ray scattering methods described in X-RAY TECHNIQUES. The utility of both x-ray and neutron scattering methods in investigations of atomic scale structure arises from the close match of the wavelength of these probes to typical interatomic distances (a few Ångstroms). The differences between these two techniques bear some further discussion.

The absorption of neutrons by most elemental species is generally quite weak, while the x-ray absorption cross-sections, at typical energies (around 8 keV), are much larger. Therefore, neutron scattering measurements probe bulk properties of the system while x-rays generally sample only the first few microns of the bulk. The principal x-ray scattering interaction in materials involves the atomic electrons, so the scattering power for x rays is proportional to the electron density and, therefore, scales with the atomic number, $Z$, of elements in the sample. Neutron scattering lengths, on the other hand, do not exhibit any simple scaling relationship with $Z$ and can vary significantly between neighboring elements in the periodic table. For this reason, neutron and x-ray diffraction measurements are often combined to provide elemental contrast in complicated systems.

The principal neutron scattering interactions involve both the nuclei of constituent elements and the magnetic moment of the outer electrons. Indeed, the cross-section for scattering from magnetic electrons is of the same order as scattering from the nuclei, so this technique is of great utility in studying magnetic structures of magnetic materials. Indeed, neutron scattering is generally the probe of choice for the determination of magnetic structure and microscopic investigations of magnetic properties of materials.

Finally, thermal neutron energies are typically on the order of a few meV to tens of meV, an energy scale that is comparable to many important elementary excitations in solids. Inelastic neutron scattering measures the gain or loss in energy of the scattered neutron as compared to the incident neutron energy. The energy loss or gain corresponds to energy transferred to or from the system in the form of excitations. Therefore inelastic neutron scattering has become a critical probe of elementary excitations including phonon and magnon dispersion in solids.

Facilities for neutron scattering are found at both nuclear reactors and, more recently, at accelerator-based spallation neutron sources. The steady flux at thermal energies at reactor sources is well-suited to instruments such as triple-axis spectrometers, which can probe elementary excitations at a particular point in reciprocal space. Spallation sources are, by design, best suited to time-of-flight studies of inelastic scattering because of the inherent time structure of the source.

ALAN I. GOLDMAN

## NEUTRON POWDER DIFFRACTION

### INTRODUCTION

#### Historical Aspects of the Method

The first powder diffraction studies on simple materials such as iron metal using x rays were done independently by Debye and Scherrer (1916) in Germany and by Hull (1917) in the United States. For a long time x-ray powder diffraction was primarily used for qualitative purposes such as phase identification and the assessment of crystallinity. The earliest structure determinations using only powder diffraction data was demonstrated when Zachariasen (1949) solved the structures of α- and β-$UF_5$ by an intuitive trial-and-error approach. An important subsequent step was taken when Zachariasen and Ellinger (1963) solved the monoclinic structure of β-plutonium using manual direct-method procedures (see X-RAY POWDER DIFFRACTION). The use of neutrons for such powder diffraction work has advanced significantly over the past 50 years.

The first neutron powder diffractometer was built at Argonne National Laboratory in 1945 (Zinn, 1947). The first generation of neutron powder diffractometers were subsequently built almost simultaneously by Wollan and Shull (1948) at Oakridge National Laboratory, Hurst and co-workers (1950) at Chalk River, Canada, and Bacon et al. (1950) at Harwell, England. They were the prototype of the so-called constant-wavelength angle-dispersive two-axis diffractometer whose basic features have not changed over the last 50 years. What has changed and is largely responsible for the rapid development of neutron powder diffraction was the increased flux of neutrons made available by the controlled fission processes of uranium. The highest neutron fluxes at a nuclear research reactor source can be found at the Institute Laue-Langevin (ILL) in Grenoble, France, with its 60-MW high-flux reactor (HFR). High-flux neutron sources allow the use of tighter neutron beam collimation and larger "take-off" angles, thereby increasing the resolution of diffraction experiments by reducing the line widths of Bragg reflections. Today, the required mechanical precision to move a massive detector bank is routinely somewhere between 0.01° and 0.02° in 2θ. The size and type of the detector also have undergone dramatic changes. Instead of the $BF_3$ counters used in the early days, today either position-sensitive detectors based on microstrip technology (Convert et al., 1997) or multi-counter arrays typically with up to 64 individual $^3He$ counters are state-of-the-art. And last but not least, improvements in large, vertically focusing monochromators with a reproducible aniso-tropic mosaic microstructure yielding well-defined spectral peak shapes have allowed the potential of high-resolution neutron powder diffraction to be fully developed (Axe et al., 1994).

Investigations of the magnetic susceptibility of materials such as MnO led to the concept of antiferromagnetism proposed by Néel (1932). The first experimental proof of this concept was provided by the powder neutron scattering experiments of Shull and Smart (1949). They demonstrated that, by cooling below the transition temperature, new Bragg diffraction peaks ("magnetic peaks") appeared that arise because the magnetic unit cell is twice the size of the "chemical" unit cell.

The seminal achievement of Rietveld in the late 1960s opened the door for refinements of neutron and x-ray powder diffraction patterns of complex structures with up to 50 atoms in the asymmetric unit cell (Rietveld, 1967, 1969). In the early 1970s Rietveld refinements were still performed predominantly to refine neutron powder diffraction data due to the simpler peak shape of the Bragg reflections. More complex peak shapes were developed subsequently for x-ray and especially synchrotron powder diffraction data. Since then, the attempt to solve more and more complex structures from powder diffraction data alone (*ab initio*) by using and adapting the tools developed for single-crystal crystallography was pursued by a steadily growing community of powder diffraction users. Today, structure solution and refinement are no longer the exclusive domain of single-crystal diffraction. The availability of highly collimated monochromatic neutron and x-ray beams, the development of the neutron time-of-flight powder diffraction technique used at neutron spallation sources, and the advances in instrumentation and computational methods make it possible to tackle the problems of *ab initio* structure determination in an almost routine and systematic manner (Cheetham, 1986; Cheetham and Wilkinson, 1992; see KINEMATIC DIFFRACTION OF X RAYS and DYNAMICAL DIFFRACTION). In fact, powder diffraction is the only tool capable of obtaining crystallographic information from samples such as catalytically active zeolites and low- and high-temperature phases where single crystals are simply not available. At the same time, information concerning the mesoscopic properties of materials such as texture, particle size, and stacking faults can also be obtained. The continuing efforts made in unraveling the structural details of superconducting oxides and the recently discovered oxides exhibiting colossal magnetoresistance have demonstrated the indisputable usefulness of high-resolution neutron powder diffraction. Modern powder diffraction using neutrons, x rays, and synchrotron x-ray radiation has developed from a qualitative method some 20 years ago to a quantitative method now used to detect new phases and determine their atomic, magnetic, and mesoscopic structure as well as their volume fractions if they are present in a mixture.

Neutron powder diffraction is a complementary technique to x-ray powder diffraction and electron diffraction. The greater penetration depth of neutrons, the fact that the neutron-nucleus interaction is a point scattering process implying no variation of the nuclear scattering length with scattering angle, the independence of the scattering cross-section from the number of electrons ($Z$) of an element, and therefore the stronger interaction of neutrons with "light" elements such as oxygen and hydrogen, and its isotope specificity as well as its interaction with unpaired electrons ("magnetic scattering") make neutrons a unique and indispensable probe for structural condensed matter physics and chemistry. However, x-ray powder diffraction and, in particular, synchrotron x-ray powder diffraction can investigate samples many thousand times smaller and have an intrinsic resolution at least one order of magnitude better than the best neutron powder diffractometer. A major drawback for neutron scattering is that it can presently only be done at large facilities (research reactors and spallation sources), whereas laboratory-based x-ray scattering equipment provides the same flux much more conveniently. Current governmental policies will not permit the construction or upgrade of present-day research reactor sources in the United States. The future of U.S. neutron scattering therefore relies solely on accelerator-based spallation sources. Electron diffraction is a very powerful tool to obtain both real and reciprocal space images of minute amounts of powders. In combination with other supplementary techniques such as solid-state nuclear magnetic resonance (NMR), which allows, e.g., the determination of Wyckoff multiplicities and interatomic distances by exploiting the nuclear Overhauser effect (NOE), physicists, chemists, and materials scientists have a powerful arsenal to elucidate condensed matter structures.

## PRINCIPLES OF THE METHOD

### The Neutron—A Different Kind of Probe

Independent of the specific interaction probes such as electrons, x rays, neutrons, or positrons have with matter, there is a general formalism for scattering phenomena upon which we will rely in the following. We distinguish two types of scattering. In elastic scattering, the probing particle is deflected without energy loss or gain. In inelastic scattering, the probing particle loses or gains energy. This energy gain or loss is called energy transfer. In both cases the probe is scattered by an angle $2\theta$ and the scattering event is defined by the scattering vector $\mathbf{Q} = \mathbf{K'} - \mathbf{K}$ (Bragg's law), where $\mathbf{K}$ and $\mathbf{K'}$ are the wave vectors before and after diffraction. In the elastic case, $\mathbf{Q} = 4\pi \sin \theta / \lambda$, with $\lambda$ being the wavelength. The scattering vector $\mathbf{Q}$ multiplied by $h/2\pi$ is called the momentum transfer, where $h$ is Planck's constant. If the probing particles interact with the scattering centers in such a way that the scattered waves have a phase relationship with each other, the case is referred to as coherent scattering and interference is possible between scattered waves with different amplitudes and phases. Diffraction or Bragg scattering is the simplest form of coherent scattering. If the probes interact in an independent, random-phase-related manner among the different scattering centers, the case is referred to as incoherent scattering. No interference is possible in the incoherent scattering case. The intensities rather than the amplitudes originating from the different scattering centers only add up. This may simply increase the background of a diffraction experiment or be used to provide information about a specific scattering species at different times and positions.

The neutron is an invaluable probe for condensed matter research. When neutrons interact elastically with matter, there are important differences from x rays.

Neutrons do not primarily interact with the electrons but with the nuclei of atoms. Due to the short range of nuclear forces, the size of these scattering centers are on the order of $10^5$ times smaller than the distances between them. The nucleus is effectively a point scatterer, and the neutron is therefore scattered isotropicially. The interaction potential is a delta function, called Fermi's pseudo-potential, and can be expressed as

$$V(r) = \frac{2\pi h^2}{m} b\delta(r) \tag{1}$$

$h$ being Planck's constant and $m$ the mass of the neutron. The Fermi length or scattering length $b$ has the dimension of a length and is often given in femtometers. It can be expressed using the Breit-Wigner (1936) equation

$$b = R - \frac{\Gamma}{2kE_r} \tag{2}$$

where $R$ is the nuclear radius, $k$ the Boltzman constant, $E_r$ a resonant energy for the neutron-nucleus system, and $\Gamma$ the width of the nuclear energy level.

In contrast, x rays are not scattered isotropically since the size of the "electron cloud" is comparable to the wavelengths used to probe them. As a result, one observes a decrease of the atomic form factor and therefore of the Bragg intensities at high scattering angles. In the case of neutrons, however, the scattering length $b$ is constant at all scattering angles. The interaction of the neutron with the nucleus is weak, especially when compared to electrons involved in very strong electrostatic potential interactions or x rays interacting via their electromagnetic radiation field with electrons. The energy of a 1-Å-wavelength neutron is 82 meV, whereas the energy of a 1-Å x-ray photon is 12 keV. A 0.037-Å electron has an energy of 100 keV. An often-neglected experimental consequence of this weak neutron-nucleus interaction is that neutrons have a very high penetration depth into matter, typically on the order of centimeters. Neutrons are a true bulk probe. The penetration depth of x rays (typically 0.5 mm) and electrons (typically 100 Å) is low enough that one has to take into consideration the possibility of near-surface effects, especially when the sample contains high-$Z$ atoms, $Z$ being the atomic number. The linear absorption coefficients in Al for neutrons and x rays both with a wavelength of 1.79 Å (i.e., comparable to a lattice spacing) are 0.014 and 212 cm$^{-1}$, respectively. In other words, the intensity of the neutron beam is reduced by half after going through roughly 50 cm of Al, whereas for x rays, it is already halved after going through 0.003 cm. This is why, from an experimentalist's point of view, sample environments (e.g., cryostats, magnets, pressure cells) are a lot easier to engineer and use in neutron than in x-ray scattering experiments. However, even with the highest currently available neutron fluxes, one will only be able to work with monochromatic beams that have fluxes of only $\sim10^8$ neutrons/cm$^2$/s impinging on the sample. This is equivalent to what an ordinary sealed x-ray tube generates. Typical fluxes at synchrotron sources are more than 6 orders of magnitude higher. As a result of this intrinsically low flux, the typical neutron powder diffraction experiment requires gram quantities of material in order to obtain data with good statistics within a reasonable amount of time. This is in marked contrast to conventional and synchrotron x-ray powder diffraction experiments where milligram quantities are sufficient.

Another important difference from x rays is that the interaction of the neutron with the nucleus does not vary systematically with the atomic number $Z$. Thus, in certain cases, low-$Z$ elements such as deuterium ($Z = 1$, $b = 6.671$ fm) or oxygen ($Z = 8$, $b = 5.80$ fm) interact more strongly with neutrons than do high-$Z$ elements such as Ce ($Z = 58$, $b = 4.84$ fm) or W ($Z = 74$, $b = 4.86$ fm). In oxides, the scattering contribution of the oxygen to the Bragg reflections is in general higher with neutrons than with x rays, especially when high-$Z$ cations are present. This is of tremendous advantage when investigating oxide structures, where the subtleties of oxygen displacements and tilts of oxygen coordination polyhedra are structural signatures of materials with significantly different physical properties. This oxygen-sensitive scattering power combined with the fact that there is no "fall-off" of the scattering length at high scattering angles when scattering neutrons led to the enormous number of neutron powder diffraction studies performed over the last decades on oxides and in particular the high-temperature cuprate superconductors.

Furthermore, even neighboring elements can have very different scattering lengths such as Mn ($Z = 25$, $b = -3.73$ fm), Fe ($Z = 26$, $b = 9.45$ fm), and Co ($Z = 27$, $b = 2.5$ fm). Without the need to perform x-ray diffraction experiments at various wavelengths and to use anomalous dispersion to distinguish between Mn and Fe or between Fe and Co, one neutron measurement can establish a particular cation distribution. This is very important when investigating the magnetic properties in solid solutions of, e.g., $Fe_{3-x}Co_xO_4$ oxides with the spinell structure. The distribution of Fe and Co among the tetrahedral and octahedral sites of the oxide structure determines the magnetism in these compounds.

In another illustrative example, the compound NaMnFeF$_6$ can be described as an ordered distribution of Na$^+$, Mn$^{2+}$, and Fe$^{3+}$ cations located in the hexagonal close-packed (hcp) lattice of fluorine. X rays have no problem distinguishing between Na$^+$ and the Mn$^{2+}$ or Fe$^{3+}$ sites. However, there are two possible ordering schemes for Mn and Fe in the space group P321 where there are three Wyckoff positions available for these metals: 1$a$, 2$d$, and 3$f$. The difference between the model where Mn is located in 3$f$ and Fe in 1$a$ and 2$d$ and the model where Fe is located in 3$f$ and Mn in 1$a$ and 2$d$ is striking. This remarkable difference between the two models for the cation distribution is shown in Figure 1.

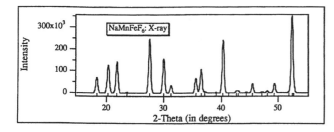

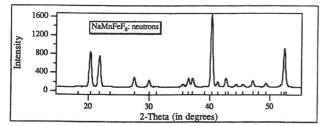

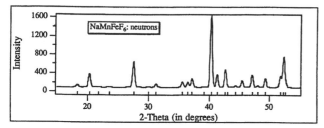

**Figure 1.** Powder diffraction pattern of $NaMnFeF_6$ calculated for two different ordering models. Top: X-ray diffraction pattern: cation ordering is not distinguishable since $Fe^{3+}$ and $Mn^{2+}$ have the same form factor far from the absorption edge. Middle: Neutron powder diffraction pattern calculated with Mn in $3f$ and Fe in $1a$ and $2d$. Bottom: Neutron powder diffraction pattern calculated with Mn in $1a$ and $2d$ and Fe in $3f$.

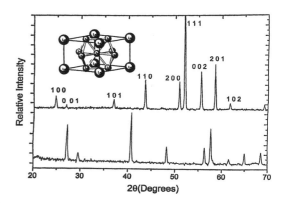

**Figure 2.** Neutron powder diffraction pattern of $LaNi_{3.55}Co_{0.75}$ $Al_{0.3}Mn_{0.4}$ ($\lambda = 1.9$ Å) and $La(^{58}Ni_{0.376}\ ^{62}Ni_{-0.623})_{3.55}Co_{0.75}$ $Al_{0.3}Mn_{0.4}$ ($\lambda = 2.1$ Å).

The neutron-nucleus interaction depends on the atomic number $Z$ as well as on the atomic weight. Isotopes, which have the same $Z$ but different weight, can vary substantially in their neutron scattering length. The neutron-nucleus interaction also allows for a negative scattering length $b$ (e.g., H, Mn, $^{62}Ni$, $^{48}Ti$). This corresponds to a phase change of p radians during the scattering process.

Exploiting these last two properties of the neutron-nucleus interaction allows contrast variation experiments to be done by altering the scattering length $b$ of a given atomic species by isotope substitution. The most known example is hydrogen ($b = -3.74$ fm) and deuterium ($b = 6.674$ fm). Varying the ratio of, e.g., $H_2O/D_2O$ changes the "visibility" of water in diffraction experiments. This technique is extensively used in small-angle neutron scattering (SANS). In general, neutron powder diffraction experiments are performed with deuterated samples due to the large incoherent scattering of hydrogen. However, if only small amounts of hydrogen are present, hydrogenated samples can be used, as will be discussed below.

In certain favorable cases the net elastic scattering contribution of certain elements can be reduced to zero by varying the isotopic ratios within the sample: Ni has three isotopes with positive scattering lengths and one with a negative scattering length ($^{58}Ni$, $b = 14.4$ fm; $^{60}Ni$,

$b = 2.8$ fm; $^{61}Ni$, $b = 7.60$ fm; and $^{62}Ni$, $b = -8.7$ fm); Ti has two isotopes, namely, $^{48}Ti$ with $b = -6.08$ fm and $^{50}Ti$ with $b = 6.18$ fm. Certain members of intermetallic alloys with the $AB_5$ structure are replacing the cadmium electrode in nickel cadmium batteries due to their benign environmental impact and high-energy density. The $AB_5$ structure contains planes of A and B cations in the basal plane and only B cations in the midplane, as shown in the inset of Figure 2. To reduce corrosion and thus enhance the cycle life of the battery while maintaining good energy storage capacity, commercial electrodes have the stoichiometry $LaNi_{3.55}Co_{0.75}Mn_{0.4}Al_{0.3}$. However, due to the high cost of cobalt, current research efforts are focused on reducing its amount to make the battery economically competitive. It is therefore important to know the location of the cobalt within the alloy. When using metals with isotopes present in their natural abundance, the scattering is dominated by nickel, and only very little stems from cobalt since nickel has a scattering length of 10.3 fm, which is four times higher than that of cobalt ($b = 2.5$ fm). If, however, one makes an alloy using a mixture of the isotopes $^{58}Ni$ and $^{62}Ni$ in the respective proportions 37.6% to 62.3%, the picture changes dramatically; nickel no longer contributes to the Bragg scattering, and cobalt dominates the scattering from the $B_5$ sublattice. The tremendous changes in the diffraction pattern are shown in Figure 2, where the first nine Bragg reflections of an alloy containing metals with a natural isotopic abundance are compared to one containing the zero-scattering "isotopic alloy" $^{58}Ni_{0.376}\ ^{62}Ni_{0.623}$. Rietveld refinements of this data set together with data of samples with natural isotope abundance revealed that cobalt prefers to be located in the midplane and not in the basal plane of this structure.

Another important application of neutron scattering is possible because neutrons have a magnetic moment of $1.9132\mu_n$ that can interact via dipole-dipole interactions with the magnetic moments of unpaired electrons. The magnitudes of magnetic and nuclear scattering are of the same order of magnitude, but now the unpaired electron density can no longer be approximated as a point scatterer, as is done for the nucleus. Magnetic scattering of neutrons is not isotropic, and angle-dependent form factors similar

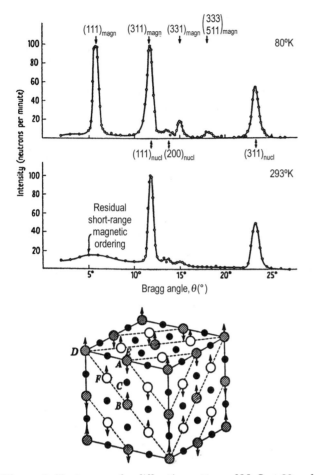

**Figure 3.** Neutron powder diffraction pattern of MnO at 80 and 293 K. The low-temperature pattern reveals extra antiferromagnetic reflections that can be indexed within a magnetic unit cell twice the size of the chemical unit cell. The magnetic unit cell below the Curie point is shown as an inset.

to the ones in x-ray scattering are observed (see MAGNETIC NEUTRON SCATTERING).

As mentioned above, the theory of antiferromagnetism was developed by Néel (1932), but it took 17 years until neutron powder diffraction provided its first experimental proof. Shull and Smart (1949) measured the neutron powder diffraction pattern of MnO at 80 and 293 K, which is below and above its antiferromagnetic phase transition at 120 K. As shown in Figure 3, the pattern at 80 K revealed the appearance of additional diffraction peaks, which could only be accounted for by a doubling of the "chemical" cell of MnO. The first reflection is thus the 111 of the "magnetic" unit cell. Subsequent experiments by Shull et al. (1951) revealed that the intensity of this peak decreases and becomes zero as one approaches the phase transition. This was the first observation of magnetic Bragg scattering, and the observed magnetic intensities were compared with model calculations provided by Néel (1948). The model has a magnetic unit cell made up of eight chemical unit cells. These four sublattices are arranged so that they have sheets of atoms with their spin pointing in the same direction parallel to (111) planes. These sheets of ferromagnetically coupled spins are

coupled antiferromagnetically along (111). This antiferromagntic coupling between second nearest neighbors is achieved via a superexchange mediated by the oxygen. This was postulated by Kramers (1934). Figure 3 depicts the magnetic unit cell. A model that gives equally good agreement with the intensities of the magnetic Bragg reflections is one where the magnetic moments of the neighboring atoms in the individual sublattices vary randomly from crystallite to crystallite; in other words, the sublattices are not coupled. If the sublattices are coupled, the structure is no longer cubic but rhombohedral. This was confirmed by low-temperature x-ray measurements by Tombs and Rooksby (1950)—a historical first to illustrate the valuable complementary use of x-ray and neutron scattering. In the rhombohedral symmetry, the four pairs of reflections that make up the {111} magnetic Bragg reflections are no longer equivalent. Only the (111) and (−1−1−1) planes will give rise to magnetic intensities. The (−11−1) plane has as many atoms with their magnetic moment pointing in one way as in the opposite and thus will contribute no magnetic intensity to the {111} reflection. The orientation of the magnetic moment parallel to the [100] axis is not arbitrary. If one calculates the intensity of the {111} magnetic Bragg reflection with the magnetic moments perpendicular to (111), no magnetic intensity results since only the perpendicular component of the magnetic moment with respect to the scattering vector contributes to magnetic Bragg scattering. If one would align the magnetic moments along the (111) sheets, an exceedingly strong magnetic intensity results. That distinction between these various models is a direct consequence of the reduction of symmetry from cubic to rhombohedral.

The neutron also has a spin angular momentum of $\frac{1}{2}$. One can separate neutrons into beams of spin-up and spin-down neutrons having equal moments of $\pm\frac{1}{2}\hbar$ (where $\hbar = h/2\pi$), respectively. This is done by, e.g., using the (111) reflection of the Heusler alloy $Cu_2MnAl$, whose nuclear and magnetic scattering contributions compensate each other exactly for one of the two spin states. These neutrons are referred to as polarized neutrons and can be used to distinguish between nuclear and magnetic scattering since only the perpendicular component of the atomic magnetic moment with respect to the scattering vector will contribute to magnetic scattering. Polarized neutron scattering is severely hampered and has difficulties to live up to its potential since the available present-day sources are all too weak to provide high-flux polarized neutron beams using current techniques: polarized neutron scattering is currently a truly signal-limited technique. Further details concerning this neutron diffraction technique are found in Williams (1988).

## PRACTICAL ASPECTS OF THE METHOD

### The Angle-Dispersive Constant-Wavelength Diffractometer

This is the most familiar experimental set-up for neutron powder diffraction experiments and resembles the standard laboratory-based x-ray powder diffractometer. In a nuclear reactor, neutrons are created by controlled fission of $^{235}$U. Neutrons released by nuclear fission have a kinetic energy of ~5 MeV. A so-called moderator made in most cases of either $H_2O$ or $D_2O$ surrounds the reactor core. Its purpose is to "slow down" the neutrons via inelastic neutron-proton or neutron-deuterium collisions. Using the wave-particle formalism and de Broglie's equation $\lambda = h/mv$, with $v$ being the velocity, the energy of a neutron can be expressed as

$$E(\text{meV}) = 0.08617T = 5.227v^2 = \frac{81.81}{\lambda^2} \qquad (3)$$

where the temperature of the moderator $T$ is in kelvins, the velocity $v$ is in kilometers per second, and the wavelength $\lambda$ is in angstroms. If water is used ($T = 300$ K), the mean wavelength is 1.78 Å and the neutrons have an energy of 25.8 meV. This moderation process is rather imperfect and results in a broad distribution of wavelengths available for experiments. If the moderator is sufficiently thick, the neutrons will have a Maxwellian energy distribution, their average kinetic energy being $\frac{3}{2}kT$, where $k$ is the Boltzmann constant. If one selects a monochromatic beam with a certain wavelength resolution $\Delta\lambda/\lambda$ by using a single-crystal monochromator and the scattering angle $2\theta$ is varied by stepping the detector, then the measurement mode is referred to as angle dispersive. The reflectivity of a perfect single crystal is low due to the small angular misalignment of the mosaic blocks—the so-called mosaic spread. In neutron powder diffraction, germanium is the natural choice for monochromation, especially since it has the diamond structure in which lattice planes whose Miller indices ($hkl$) are all odd are systematically absent. Therefore, $\lambda/2$ contamination will not occur from Bragg scattering. However, the mosaic spread has to be increased to make monochromators that can provide a reasonable flux. Recently, a significant advancement was reported by Axe et al. (1994). Instead of "squashing" single crystals, thin wafers are plastically deformed and then reassembled into composites. The reproducible creation of a spatially homogeneous but anisotropic mosaic spread led to a well-defined neutron energy profile and symmetrical spectral peak shapes, which are crucial for high-resolution neutron powder diffraction. The angle at which the neutrons are reflected off their flight path by the monochromator toward the sample is called the take-off angle. The higher this take-off angle, the better the angular resolution $\Delta d/d$ will be. Differentiating Bragg's law, one obtains

$$\frac{\Delta\lambda}{\lambda} = \frac{\Delta d}{d} + \Delta\theta \cot\theta \qquad (4)$$

The resolution is given by

$$\frac{\Delta d}{d} = \Delta\theta \cot\theta \qquad (5)$$

with $\Delta\theta$ being the mosaic spread of the monochromator reflecting at an angle $\theta$, the take-off angle. To obtain a high angular resolution (small $\Delta d/d$), a large take-off

angle has to be chosen. However, the intensity also depends on $\Delta/\lambda\lambda$ and will thus decrease as the take-off angle is increased. This is why only in high-flux reactors can the gain in resolution be afforded without paying the penalty of unreasonably small fluxes at the samples (resulting in prohibitive long counting times) in order to achieve good signal-to-noise ratios. The resolution of an angle-dispersive set-up depends on the narrow band $\Delta/\lambda\lambda$ selected by the monochromator and is further enhanced by using Soller collimators before the monochromator and detector and optionally between the monochromator and sample. Soller collimators (Soller, 1924) consist of a number of thin sheets bound together to form a series of long and narrow channels. These sheets are painted with a neutron-absorbing paint containing gadolinium to absorb diverging neutrons. The resolution of an angle-dispersive neutron powder diffractometer is further discussed in Appendix A.

**Time-of-Flight Diffractometers**

For didactic reasons, most arguments and examples that follow will be made using the constant-wavelength or angle-dispersive diffractometers found at research reactors. However, there is another type of neutron source, the spallation source and associated with it a second type of diffractometer, the time-of-flight diffractometer. The spallation source relies on a process whereby highly accelerated particles such as protons with energies of ~800-MeV impact on a target made of uranium or lead. Thereby neutrons are "kicked out" of the nuclei. The experimental requirements for diffraction at a spallation source are different. The time structure of the neutron beam created by the proton beam on the target is exploited in these measurements. In the time-of-flight experiment the detector is located at a fixed value of $2\theta$ while the sample is irradiated by a pulsed, "white" neutron beam containing a very large wavelength distribution $\Delta/\lambda\lambda$. All wavelengths created in the spallation process are then moderated and impinge onto the sample without being selected by a monochromator. A coarse wavelength selection is achieved by placing choppers in the beam. These are rotating discs of neutron-absorbing material with an opening cut into it. By varying the speed, one can select neutrons with different speeds and therefore energies and wavelengths. For a given reflection, the Bragg condition $\lambda = 2d\sin\theta$ is satisfied at a specific wavelength. This wavelength can be determined by measuring the time it takes the neutron to reach the detector after hitting the sample. The resolution of a time-of-flight powder diffractometer is given by

$$\frac{\Delta Q}{Q} = \left[ \left(\frac{\Delta t}{t}\right)^2 + \left(\frac{\Delta L}{L}\right)^2 + \left(\frac{\cot\theta}{\Delta\theta}\right)^2 \right]^{1/2} \qquad (6)$$

where $t$ is the time of flight of the neutrons and $\Delta t$ its uncertainty mainly due to the uncertainty in the moderation time, $L$ the flight path length and $\Delta L$ its uncertainty due to a finite target and moderator size. The angle $\theta$ is the scattering angle and $\Delta\theta$ is the angle subtended at the source by the active area of the detector. From this equation, one can deduce that the highest resolution is obtained using a long flight path $L$ and a detector close to $\theta = 90°$, referred to as the backscattering position. As an example, the high-resolution neutron powder diffractometer located at the spallation source ISIS at the Rutherford Laboratory in the United Kingdom has a 96-m-long flight path and a detector in the backscattering position. If one would use only a detector array covering a small angular range close to the backscattering position, the largest $d$ spacing using a neutron pulse with a wavelength distribution $\lambda_1 < \lambda < \lambda_2$ that could be observed would be $\lambda_2/2$. At the $2\theta = 90°$ position, a $d$ spacing of $\lambda_2\sqrt{2}$ can be observed. One has to keep this in mind because the observation of high $d$ spacings is of crucial importance for the determination and refinement of magnetic structures as well as for indexing unknown structures. Each time neutrons are created in a spallation process, typically every 20 ms, a complete diffraction pattern is collected. Successive pulses have to be accumulated to improve the signal-to-noise ratio. The resolution curve of a time-of-flight diffractometer is almost constant with scattering angle $2\theta$. Thus, minute lattice constant deviations can be resolved from high-order reflections without suffering the instrumental broadening that angle-dispersive instruments suffer beyond the minimum at the take-off angle (see Appendix A). A disadvantage of time-of-flight neutron powder diffraction is that wavelength-dependent corrections for the incident intensity, absorption, extinction, and detector efficiencies have to be made and the peak shape is highly asymmetrical. Despite these tedious corrections, sophisticated software packages exist and the time-of-flight method is used as routinely as the angle-dispersive procedure. It appears that in the future new neutron sources will probably be spallation sources rather than reactors. Time-of-flight neutron powder data have been used to determine structures from powder data alone (Cheetham, 1986). David et al. (1993) used high-resolution time-of-flight neutron powder diffraction data collected as the spallation source ISIS to study in detail the structure of the soccer ball–shaped molecule $C_{60}$ as a function of temperature. Subtle aspects of an orientational glass transition and precursor effects of the order-disorder transition at 260 K were observed. Detailed analysis revealed that, despite the lower precision when compared to single-crystal measurements and the resulting larger errors, time-of-flight neutron powder diffraction data can be used to reproduce all the essential details found in a single-crystal synchrotron radiation experiment. This illustrates the remarkable amount of information that can be extracted using state-of-the-art neutron time-of-flight data and data analysis.

## SAMPLE PREPARATION

Samples should resemble as much as possible an ideal random powder consisting of particles with sizes between 1 and 5 μm. The size distribution should be as smooth and Gaussian as possible. If larger crystallites dominate the sample, the measured intensities will have severe and nonsystematic deviations from the "true" values based on

crystallography. The intensities of certain $d$ spacings are then no longer evenly distributed on the Debye-Scherrer cones but rather are clustered in spots depending on the orientation of the bigger crystallites. There are no corrections that can be applied to these type of nonrandom samples. If too small particles are present, line-broadening effects will occur that can be corrected for (see The particle size effect). The minute amount of sample present in the beam and the highly collimated x-ray beam in a synchrotron x-ray powder diffraction experiment will exacerbate problems with nonrandomness. If the nonrandomness is systematic, as in samples with preferred orientation, it can be corrected for in Rietveld refinements. Recently, Wessels et al. (1999) used preferred orientation to solve structures from powder data alone. In neutron powder diffraction experiments, the larger size of the beam, its larger divergence, and the required gram quantities of sample to obtain useful signal-to-noise ratios within reasonable time result in a better sampling and averaging of the powder sample. Therefore, fewer problems with nonrandomness are encountered. If, however, a low-temperature structure of a compound is investigated that is a liquid or a gas at room temperature, it is advisable to cool and reheat the sample several times in order to obtain a "good" powder sample. The cooling time might become a crucial parameter since cooling too slow might result in crystallites that are too large, and therefore not all crystallite orientations will be equally represented.

## DATA ANALYSIS AND INITIAL INTERPRETATION

### Positions of the Bragg Reflections

For constant-wavelength diffractometers the $d$ spacings of a crystalline sample are obtained from the measured angular position $2\theta$ using Bragg's law:

$$\lambda = 2d \sin\theta \tag{7}$$

In a time-of-flight neutron powder diffractometer, the $d$ spacings can be obtained by converting the time-of-flight $t$ via

$$d = \frac{ht}{2mL\sin\theta} \tag{8}$$

where the detectors are located so that $L\sin\theta$ is a constant and $m$ is the mass of the neutron.

The obtained $d$ spacings allow the determination of the lattice constants with an accuracy of a few hundredths of angstroms in the case of neutrons. This accuracy is crucial when attempting to use lattice constants to determine phases in mixtures, index unknown phases, or determine thermal expansion coefficients and residual stress in materials. Accurate $d$ spacings can only be obtained by carefully aligning the diffractometer, thus minimizing systematic errors (zero shift of the detector in $2\theta$), and using the proper peak shape when fitting the diffraction peaks. An adequate correction to the asymmetry due to axial divergence is especially important (see Peak Asymmetry Due to Axial Divergence). Figure 4 shows a diffraction pattern

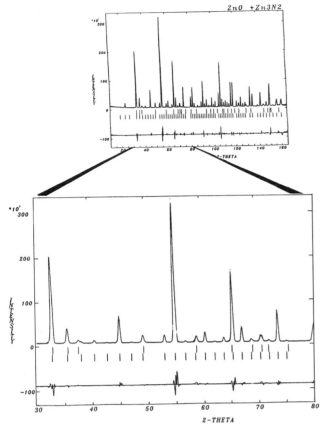

**Figure 4.** Neutron powder diffraction pattern of $Zn_3N_2$ containing ZnO as an impurity. The first row of tick marks below the pattern indicates the position of the Bragg reflections stemming from $Zn_3N_2$ [$a = b = 3.255(1)$ Å, $c = 5.2112(1)$ Å, space group $P63mc$], the second from cubic ZnO [$a = 9.7839(2)$ Å, space group $Ia3$].

of a sample of $Zn_3N_2$ that contains ZnO as an impurity phase. It is apparent that due to the superposition of the Bragg reflections of these two phases, only a high angular resolution allows the separation of these two phases.

### The Peak Shape

The diffraction profile function or peak shape describes the intensity distribution about the Bragg position $2\theta$. Ultimately, the width of Bragg reflections is determined by the Darwin width. For an ideal crystal, Darwin (1914) found that the width of a reflection depends only on the structure factor and the number of interfering $d$ spacings. Its finite width is due to the weakening of the primary beam inside the crystal due to multiple Bragg reflection—also called primary extinction. In addition to the Darwin width, various effects broaden the diffraction lines. The most commonly used quantifier for the width of a reflection is the full width at half-maximum (FWHM). The experimentally observed peak shape in the case of a constant-wavelength diffractometer is the result of the convolution of two components: (1) the optical characteristics of a diffractometer such as the wavelength distribution $\Delta\lambda/\lambda$ diffracted off the monochromator at a given take-off angle $\theta_m$ and the various collimators (see

Appendix A) and (2) the actual physical contributions of the sample to the profile.

### Instrumental Contributions to the Peak Shape

The resolution of a constant-wavelength neutron powder diffractometer can be calculated according to a theory first developed by Caglioti et al. (1958). The most important result of this simple theory is that the neutron peak shapes are to a very good first approximation Gaussian:

$$\Omega(2\theta) = \frac{2}{\text{FWHM}} \left( \frac{\ln 2}{\pi} \right)^{1/2} \exp\left( \frac{-4 \ln 2 (2\theta - 2\theta_0)^2}{\text{FWHM}^2} \right) \quad (9)$$

The intensity distribution $\Omega(2\theta)$ about their Bragg position $2\theta_0$ is fully described by the FWHM. This is a very good approximation for low- and medium-resolution neutron powder diffractometers, and the rapid development of the Rietveld method was largely due to this simple mathematical description of the peak shape. For high-resolution neutron powder diffraction data and especially laboratory-based and synchrotron x-ray powder diffraction data, more sophisticated peak shapes had to be developed. In general, the resolution curve of a neutron powder diffractometer that describes the angular variation of the FWHM of the Bragg reflections can be given according to Cagliotti et al. as a simple expression (see Appendix A):

$$\text{FWHM}^2 = U \tan^2 \theta + V \tan \theta + W \quad (10)$$

However, there are significant deviations from this simple description that are especially apparent in high-resolution neutron powder diffraction.

### Peak Asymmetry Due to Axial Divergence

An adequate peak-shape function is of utmost importance in the analysis of crystalline structures when using the Rietveld profile refinement technique. This is especially true for the low-angle region of a constant-wavelength high-resolution neutron powder diffraction pattern where severe asymmetries due to axial divergence occur. The appropriate asymmetry correction in these low-angle $2\theta$ regions might determine success or failure when using, e.g., auto-indexing routines to determine the unit-cell dimensions of unknown structures. With a nondivergent point source and a truly randomized powder, the radiation scattered at a given Bragg reflection will lie on the surface of a Debye-Scherrer cone with semiangle $2\theta$. The detector itself lies on the surface of a cylinder with its axis parallel to the $2\theta$ axis of the diffractometer. The intersection of the Debye-Scherrer cone with the detector cylinder is an ellipse. The center of this ellipse is at $0°$ for Bragg reflections with $2\theta < 90°$ and at $180°$ for reflections with a $2\theta > 90°$. The further from $90°$, the more the ellipticity will increase. As Figure 5 shows, this departure leads to a peak asymmetry because diffracted neutrons from the ends of the intercepted part of the Debye-Scherrer cone will hit the detector on the side of the peak closer to the center of the ellipse first. This is known as axial divergence. Increasing the height of the detector and the size of

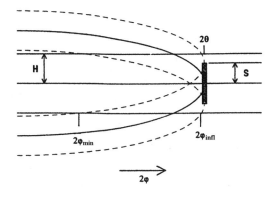

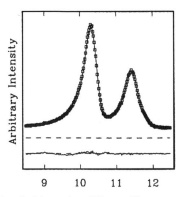

**Figure 5.** Band of intensity, diffracted by a sample with height $2S$, as seen by a detector with opening $2H$ and a detector angle $2\varphi$ moving in the detector cylinder. For angles below $2\varphi_{min}$, no intensity is detected. For angles between $2\varphi_{infl}$ and $2\theta$, the entire sample can be seen by the detector. Observed and calculated profiles are for neutron powder diffraction peaks of a zeolite sample measured at the instrument HRNPD at Brookhaven National Laboratory.

the sample will lead to greater counting rates and shorter measurement times—an important factor for high-resolution neutron powder diffraction due to the intrinsically weaker sources available in neutron scattering, as discussed earlier. However, these combined measures to increase the counting rates will also increase the asymmetry. The integrated intensities of affected reflections are biased since a larger part of the Debye-Scherrer ring is collected at lower angles. Rietveld (1969) introduced a refineable asymmetry factor providing a purely empirical correction. A widespread and commonly used empirical correction is the split–Pearson VII function described by Brown and Edmonds (1980) and Toraya (1986). Another approach is to take into account the essential optical features of the diffraction optics to correct for the asymmetry. Cooper and Sayer (1975) introduced an asymmetry correction for neutron powder diffraction based on the calculation of the resolution function in reciprocal space. However, this correction was not widely used due to the difficulties of deriving asymmetric profile functions from the source, sample, and detector contribution. Howard (1982) used the approach of Eastabrook (1952) and applied it to neutron diffraction with a point source and sample as well as a finite detector height. The correction term contains one variable parameter depending on the ratio of the detector

height to the distance between sample and detector. This is currently incorporated into the Rietveld refinement programs LHPM (Hill and Howard, 1987), DBW (Wiles and Young, 1981), and older versions of GSAS (Larson and Von Dreele, 1990) and works quite well for the predominantly Gaussian peak shapes occurring in neutron powder diffraction. A more rigorous approach was advanced by van Laar and Yelon (1984). Their correction includes a finite sample size and detector height when the latter is bigger and a weight function takes into account edge effects when the detector first intercepts the Debye-Scherrer cone up to the point where it "sees" the whole sample. Finger et al. (1994) generalized the van Laar and Yelon approach, and it is currently encoded in the latest versions of GSAS. The latest generation of high-resolution neutron powder diffractometers has benefited tremendously from this correction. These diffractometers use large, vertically focusing monochromators and have detector apertures with axial divergences of several degrees. The asymmetry of low-angle peaks from materials with large unit cells such as zeolites can be described without using any free parameters.

### Mesoscopic Properties Influencing the Peak Shape

When using a "real" diffractometer, the measured resolution will always be worse than the design specifications. Besides neutron optical effects stemming from nonideal collimators and monochromators that degrade the flux and resolution, there are so-called mesoscopic properties of the sample that influence the peak shape. It is important to account for these structural variations at length scales of tens to thousands of angstroms in high-resolution neutron powder diffraction data sets. We will mainly concentrate on four of them: the particle size effect, microstrain, extinction, and stacking faults.

**The Particle Size Effect.** The particle size effect is due to coherent diffraction emanating from finite domain sizes within the grains of a powder. Kinematical diffraction theory assumes an infinite lattice. In this case, the Bragg reflections are $\delta$ functions. The smaller the particle from which diffraction emanates, the more this $\delta$ function will be "smeared out." The broadening of all points in reciprocal space is uniform. Using Bragg's law, one can derive that

$$\frac{\Delta d}{d^2} = -\frac{\Delta \theta}{\tan \theta} = -\frac{\Delta(2\theta)\cos \theta}{\lambda} \tag{11}$$

This reveals that particle size broadening $\Delta(2\theta)$ varies with $1/\cos \theta$ in $2\theta$ space. Analysis of size broadening is often performed using the Scherrer equation:

$$\Delta(2\theta) = \frac{K\lambda}{A \cos \theta} \tag{12}$$

where $A$ is the thickness of the coherently diffracting domain and $K$ is a dimensionless crystal shape constant close to unity, which is also referred to as the Scherrer constant.

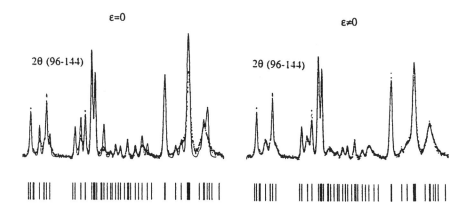

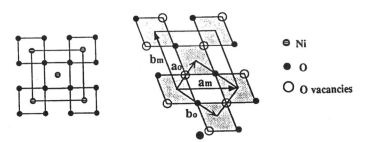

**Figure 6.** Portions of the neutron powder diffraction pattern of $La_{1.5}Sr_{0.5}NiO_{3.6}$ showing the effect of microstrain due to the partial ordering of the oxygen vacancies as depicted in the lower part. The fit depicted in the left part does not take microstrain into account.

**Microstrain Broadening.** Microstrain $\varepsilon$ is due to a local variation $\Delta d/d$ of the average $d$ spacing. These variations can be due to external stresses leading to anisotropic strain in the crystallites, lattice defects, or local compositional fluctuations described by a constant $\varepsilon = \Delta d/d$. In contrast to particle size broadening, the effect of strain varies in reciprocal space. Differentiating Bragg's law with respect to the diffraction angle $\theta$, one obtains

$$d\theta = -\tan\theta \frac{dd_{hkl}}{d_{hkl}} \qquad (13)$$

In this case, we have the compositional fluctuations as constant and not the size $A$, and therefore

$$\Delta\theta = -\tan\theta \frac{\Delta d}{d} \qquad (14)$$

This term is no longer wavelength dependent. Therefore, if we change wavelength, the width of a line (corrected for instrumental contribution!) should not vary in the case of only microstrain broadening. The variations in $d$ spacings $\Delta d/d$ manifest themselves in a broadening of the diffraction line centered at $2\theta$:

$$\Delta(2\theta) = -2\varepsilon\tan\theta \qquad (15)$$

These two mesoscopic properties of materials, particle size and strain, are increasingly studied due to their important consequences for the physical properties of mat-

erials such as resistivity, magnetism, and the occurrence of phase transitions. Medarde and Rodriguez-Carvajal (1997) investigated the oxygen vacancy ordering in $La_{2-x}Sr_xNiO_{4-\delta}(0 \leq x \leq 0.5)$. In $La_2NiO_4$, the crystallographic structure is locally orthorhombic. However, the local fluctuations are correlated in such a way that their macroscopic average is tetragonal. This is due to the stress induced by the size mismatch between $NiO_2$ and $La_2O_2$ layers. The $(hhl)$ reflections are narrow, whereas the $(h0l)$ and $(0kl)$ reflections reveal an increasing broadening with increasing values of $h$ and $k$ (Rodriguez-Carvajal et al., 1991). When substituting $La^{3+}$ by $Sr^{2+}$, the broadening diminishes since the size mismatch between the $NiO_2$ and $(La,Sr)_2O_2$ layers is reduced. Above a critical $x \sim; 0.135$, the oxygen vacancies order and give rise to an extra broadening of reflections. This microstrain created by ordered oxygen vacancies along the orthorhombic $a$ axis corresponds to ~1.7% of its total length. Figure 6 shows two Rietveld refinements of $La_{1.5}Sr_{0.5}NiO_{3.6}$, one of them refining a model for the microstrain, the other assuming no microstrain.

Microstrain and particle size effects can be separated by their $q$ dependency when both are present in a material. There are various techniques to extract these types of mesoscopic properties from powder diffraction data. The most commonly used is based on the Williamson-Hall plot. Combining the Scherrer equation with the equation describing microstrain broadening leads to

$$\text{FWHM}_{corr} \frac{\cos\theta}{\lambda} = \frac{1}{A} + \frac{C\varepsilon\sin\theta}{\lambda} \qquad (16)$$

**ZnO-B: size & strain broadening**

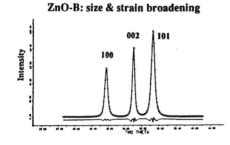

**ZnO-A: size broadening only**

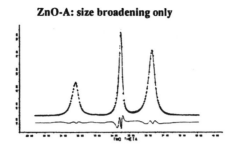

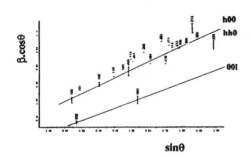

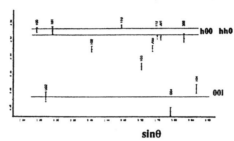

**Figure 7.** Williamson-Hall plots for two samples of ZnO prepared in different conditions: sample A is strain free but made of small ansiotropic crystallites. Sample B reveals both particle size and microstrain effect. (Adapted from Langford and Louër 1986.)

where $FWHM_{corr}$ is corrected for instrumental resolution and $C$ varies with $\varepsilon$ and the FWHM. The Scherrer constant $K$ is assumed to be unity. When plotting $FWHM_{corr} \cos\theta$ of Bragg reflections versus $\sin\theta$, one can extract the domain size $A$ as the $y$ intercept and $\varepsilon$ from the slope. Both $T$ and $\varepsilon$ can be anisotropic. One can therefore determine the crystalline shape and the direction of microstrain in crystallites. Figure 7 compares two samples of ZnO, one of them revealing broadening due to particle sizes, the other where both a broadening due to particle size and strain occur. The Williamson-Hall plots for both cases show how powerful this simple analysis can be.

Extinction is defined as the reduction of the ideal kinematical intensity due to depletion of the primary beam and rescattering of the diffracted beam. Sabine et al. (1988) demonstrated that in time-of-flight neutron powder diffraction primary extinction can be substantial when highly crystalline samples are measured. The extinction coefficient $E$ is a function of the scattering angle and can be expressed as

$$E = E_L \cos^2\theta + E_B \sin^2\theta \qquad (17)$$

$$E_L = 1 - \frac{x}{2} + \frac{x^2}{4} - \frac{5x^3}{48} + \frac{7x^4}{192} \qquad (18)$$

$$E_B = \frac{1}{(1+x)^{1/2}} \qquad (19)$$

$$x = (KN_c \lambda FD)^2 \qquad (20)$$

Here, $D$ is a refinable parameter and $K$ a shape factor that is unity for a cube and $\frac{3}{4}$ for a sphere of diameter $D$; $N_c$ is the number of unit cells, $F$ the structure factor, and $\lambda$ the wavelength.

**Stacking Faults.** Due to the high angular resolutions achievable with modern synchrotron and neutron powder diffractometers, stacking faults are more and more notice-able. The diffracted intensity from a sample containing stacking disorder is the weighted incoherent sum of diffraction pattern emanating from each crystallite orientation and defect arrangement. Any single stacking fault can be included into the calculations in a rather straightforward fashion, but the number of permutations of stacking faults increases the computational effort substantially. Stacking faults were originally discovered in cold-worked metals. They represent disorder in the stacking sequence of layers along a particular direction. In the face-centered cubic (fcc) structure layers of atoms can be seen stacking along the [111] direction, whereas in the hcp structure the layers stack along [001]. The possible atomic layer positions in the hcp lattice is ABABAB... along [001], and in the fcc lattice it is ABCABCABCABC... along [111]. A so-called deformation fault occurs in the fcc structure when the sequence ABCABCABC is altered to, e.g., ABCACABCA This deformation fault corresponds to a shift of all layers after the fault (bold **C**) to the right. A twin fault in the fcc structure is represented by a sequence change from ABCABCABC to ABCABACBA. The stacking sequence is reversed to the left and right of the fault. Twin faults in fcc structures along [111] lead to diffraction line broadening, whereas deformation faults lead to line broadening as well as line positional shifts. This can be understood in a simple qualitative picture: if one writes down three possible deformation fault sequences and compares them to the original sequence, one observes that all A, B, and C layers following the fault (in bold) are now closer to the original A* layer than in the unfaulted fcc sequence:

A* B C A C A
A* B C A B **A**
A* B C **B** C A
A* B C A B C

This change of the translational periodicity leads to a broadening due to local variations and a peak shift. If one does the same for possible twin-fault structures, one observes that the A, B, and C layers following the fault (in bold) are shifted forward or backward or remain in the same sequence position as the unfaulted fcc sequence, leading only to an increased dispersion along $c$ resulting in peak broadening:

A* B C A **B** A
A* B C **A** C B
A* B C A B **C**
A* B C A B C

An exemplary study by Berliner and Werner (1986) showed that the structure of lithium at 20 K has a so-called 9R structure with the nine-layer repeating sequence ABCBCACAB, which is highly faulted below 20 K. Their stacking fault analysis revealed deformation stacking-fault probabilities of 7%.

Wilson (1942, 1943) developed a difference equation method, where the probability of the occurrence of a given layer is related to the probability that it occurred in previous layers. Hendricks and Teller (1942) generalized this approach by using correlation probability matrices including effects of the nearest-neighbor layer-layer correlations and different stacking vectors.

Cowley (1976a,b, 1981) came up with another approach by analyzing single-layer terms in the Patterson function and separating intensity contributions into "no-fault" terms, which were summed as geometric series, and "fault" terms, which were summed explicitly. This led to quicker convergence in low-fault cases. Recently, Treacy et al. (1991) implemented a general recursion algorithm to calculate the diffraction pattern with coherent planar faults. Their DIFFaX program is rather easy to apply to even complex structures such as zeolite beta (Newsam et al., 1988) and intergrowths.

Delmas and Tessier (1997) correlated the broadening observed in $Ni(OH)_2$ to stacking faults. The occurrence of fcc domains in the hcp of oxygen can be related to the electrochemical behavior of this material used as electrodes in batteries and explains unexpected Raman bands. The work by Roessli et al. (1993) emphasizes the fact that when diffracting with neutrons magnetic stacking faults can also occur. In their low-temperature measurements of $HoBa_2Cu_3O_7$, the magnetic superstructure reflections show remarkably different widths and shapes: asymmetric, e.g., $0, \frac{1}{2}, \frac{1}{2}$; broad but symmetric, $0, \frac{1}{2}, \frac{3}{2}$; and narrow and resolution limited, $1, \frac{1}{2}, \frac{1}{2}$. They can be explained by an interplay of the $q$ dependence of the magnetic form factor and finite magnetic correlations on the order of 30 Å along the $c$ axis. The latter is the equivalent to magnetic stacking faults along the $c$ axis.

## The Rietveld Method

The Rietveld method makes use of the fact that the peak shapes of Bragg reflections can be described analytically and the variation of their widths (FWHM) with the scattering angle $2\theta$ can be expressed as a convolution of optical characteristics depending on the diffractometer (take-off angle of the monochromator and collimation) and sample-related effects such as strain and particle size broadening effects, as discussed above.

This allows least-squares refinements of crystal structure parameters without explicitly extracting structure

factors or integrated intensities. For details with respect to reliability factors and estimated standard deviations, see Appendices A and B. Less frequently, one finds studies relying on a two-stage approach where individual intensities are extracted (Jansen et al., 1988).

The Rietveld algorithm fits the observed diffraction pattern using as variables the optical characteristics and the structural parameters. A function

$$M = \sum w_i (y_i - y_{c,i})^2 \qquad (21)$$

is minimized, where $w_i = 1/\sigma_i^2$ is the weight assigned to an individually measured intensity $y_i$ at step $i$, $\sigma$ is the variance assigned to the observation $y_i$, and $y_{c,i}$ is the calculated intensity at step $i$. The calculated intensities $y_{c,i}$ are obtained by adding the contributions stemming from the contributions of the overlapping Bragg reflections to the background $y_{b,i}$:

$$y_{c,i} = y_{b,i} + \sum_i s_\phi \sum j_{\phi k} \mathrm{Lp}_{\phi k} O_{\phi k} |F_{\phi k}|^2 \Omega_{i\phi k} \qquad (22)$$

The first sum adds all the various contributions from different phases $\phi$; the second sum adds all contributions from different reflections $k$. The scale factor $S_\phi$ is proportional to the volume fraction of the phase, $j_{\phi k}$ is the multiplicity of the $k$th reflection, $\mathrm{Lp}_{\phi k}$ is the Lorentz factor, $O_{\phi k}$ is a factor describing any mesoscopic effects (particle size, strain, preferred orientation), $F_{\phi k}$ is the nuclear and magnetic structure factor including the displacement contributions, and $\Omega_{i\phi k}$ describes the peak profile function of reflection $k$. A well-defined and symmetrical peak profile function that can be parameterized by a small number of variables and whose $2\theta$ dependency can be described by Cagliotti et al. (1958) equation or similar functions is advantageous. It will result in fewer correlations between the so-called machine parameters and the structural parameters (positional coordinates and displacement parameters). A not well-defined peak shape leads to "unstable" least-squares refinements and larger errors of the refined structural parameters. The density of reflections increases tremendously with increasing $2\theta$, and the quality of the Rietveld fit relies on the separation of close $d$ spacings at higher scattering angles. This separation will benefit from the highest resolution available at high scattering angles and a well-understood peak shape. Furthermore, as shown above, certain mesoscopic effects manifest themselves as deviations of the peak shape at high scattering angles.

## Quantitative Phase Analysis

In the absence of severe absorption, the intensity diffracted by a crystalline phase is proportional to the amount of irradiated material. The Rietveld method can be used to extract the structures and volume fractions of individual phases when phase mixtures are present (Hill and Howard, 1987; Bish and Howard, 1988). One just has to sum over the contributions of all phases present represented by their individual scale factor proportional to the volume fraction of the phase. When measuring in the Debye-Scherrer geometry, the scale factor for each phase is proportional to

$$a_j \propto \frac{m_j}{(ZMV_c)_j} \qquad (23)$$

with $M$ being the mass per formula unit of the phase, $Z$ the number of formula units per unit cell, and $V_c$ the volume of the unit cell. By constraining the sum of the weight fractions to unity, one can determine the relative weight of any component of a mixture. This is an implicit constraint since one assumes no amorphous phase to be present. Adding an internal standard to the sample allows determination of absolute weight fractions. Even in severe overlap one can determine the weight fractions of phase mixtures. The high-resolution neutron powder diffraction pattern of the $Zn_3N_2$ and $ZnO$ mixture depicted in Figure 4 was refined as a two-phase mixture revealing that the sample contains 10% $ZnO$ as an impurity phase. Any refinements that did not take into account $ZnO$ resulted in poor fits and unreliable $Zn_3N_2$ parameters.

### Partial Structure Determination

In many structural investigations using neutron powder diffraction, a large portion of the crystallographic structure is already known but the location of a few atoms remains elusive. When solving and refining structures using x-ray powder diffraction data, often low-$Z$ elements cannot be located with high enough precision and accuracy. Then neutron powder diffraction becomes essential to fully determine and understand the structure. This is the case when hydrogen or lithium is present as extra framework atoms in zeolites or silicates (Paulus et al., 1990; Vogt et al., 1990). The location of organic molecules in zeolites opened up an important application for neutron powder diffraction. In such cases the known fragment of the structure is refined against the diffraction data using the Rietveld refinement technique, and a difference Fourier map based on the difference between experimentally observed and calculated structure factors of the known fragment using the calculated phases is constructed:

$$\Delta(hkl) = |F_{obs}(hkl)| - |F_{calc}(hkl)| \qquad (24)$$

This map will in many cases reveal the positions of the missing atoms as peaks. Rotella et al. (1982) used this approach to locate the deuterium atoms in $DTaWO_6$. In certain cases even nondeuterated samples can be investigated: Harrison et al. (1995) measured the neutron powder diffraction pattern of $(NH_4)_2(WO_3)_3SeO_3$ and using difference Fourier techniques were able to determine the four hydrogen atoms as negative peaks (due to their negative scattering length). One hydrogen reveals larger displacement parameters than the other three. Examining the structure revealed that this was the only hydrogen not involved in hydrogen bonding. The interpretation of difference Fourier maps is not always as straightforward as described above due to noise introduced by series termination errors (Prince, 1994). An alternative to locate atoms, especially extra framework cations in zeolites, is the max-

imum-entropy technique (Papoular and Cox, 1995). This technique is based on Bayesian analysis and the use of entropy maximization in crystallography (Bricogne, 1991). The use in powder diffraction was pioneered by Antoniadis et al. (1990).

### Ab Initio Structure Determination from Power Data

The development of high-resolution neutron powder diffractometers with their high angular $2\theta$ resolution resulted in a dramatic reduction of peak overlap in the neutron powder diffraction pattern. Although solving structures from powder diffraction data alone is by no means as straightforward as when using single-crystal diffraction data, more and more structures of important materials are determined this way. Many times a combination of electron, x-ray, and neutron diffraction is used. The complexities of structures being solved is impressive: Morris et al. (1992) solved and refined the structure of $Ga_2(HPO_3)_3 \cdot 4H_2O$, a novel framework structure with 29 atoms in the asymmetric unit cell and 171 structural parameters by combining the use of synchrotron x-ray and neutron powder diffraction data. From high-resolution neutron powder diffraction data alone, Buttrey et al. (1994) determined the structure of the high-temperature polymorph of $Bi_2MoO_6$. This monoclinic structure has 36 atoms in the asymmetric unit cell and was refined using 146 structural parameters. In another structure determination using synchrotron x-ray and neutron powder diffraction data, Morris et al. (1994) solved the structure of $La_3Ti_5Al_{15}O_{37}$ with 60 atoms in the asymmetric unit and 183 structural parameters. The successive steps of such an ab initio structure determination are presented in the following paragraphs.

**Indexing.** Before intensities can be extracted, one has to determine the unit cell and space group. Using single peak-fitting routines, individual $d$ spacings are determined. It is very important in this step that systematic errors are eliminated as much as possible. In neutron powder diffraction data, the two most serious errors that will determine success or failure when indexing are (1) the low-angle asymmetry due to axial divergence (see above) and (2) the zero-point shift of the detector. One way to determine the zero point of the detector is to mix the sample with a known standard such as silicon and thus "sacrifice" a certain amount of material for indexing purposes. Another approach is to determine the zero-point shift in a separate experiment with a known standard since the mechanical reproducibility is in general <0.01° with state-of-the-art equipment such as absolute encoders. If possible, one should also attempt to use larger wavelengths, if available, since the peak overlap will be further reduced in $2\theta$, as can be seen from Bragg's law. However, any wavelength with $\lambda/2$ components should be avoided. Any presence of even minute impurity phases will make the task of indexing the powder pattern more difficult and in many cases impossible, especially when dealing with low-symmetry structures. The pattern of the $Zn_3N_2$-$ZnO$ mixture shown in Figure 4 shows the effect an impurity phase would have. Peak splitting and slight shoulders in the low-angle 2θ region could lead to the wrong assumption that the structure actually has a lower symmetry. The significant better angular resolution of synchrotron x-ray powder diffraction data due to almost ten times smaller FWHMs can be of great help if one runs into problems with large unit cells and low-symmetry space groups. Once reliable $d$ spacings have been extracted, several computer programs for indexing are available. They are based on various strategies ranging from trial and error, deductive determination, and semiexhaustive to exhaustive search algorithms. Certain programs work better for low-symmetry space groups, others for high-symmetry ones. An overview of these techniques is given by Shirley (1984). In general, indexing programs will give the user more than one solution each associated with a certain quality factor value. The most common quality factor used is $M_{20}$ and is defined as

$$M_{20} = \frac{Q_{\mathrm{calc}}^{\max}}{2N_{\mathrm{calc}}\langle\delta\rangle} \quad (25)$$

using the first 20 peak positions where $Q = 1/(d_{hkl})^2$, $N_{\mathrm{calc}}$ is the number of potentially observable calculated lines up to the last indexed line, and $\langle\delta\rangle$ is the absolute mean difference in $Q$ between experimental and calculated peak positions. As a general rule of thumb, any solution with an $M_{20} < 10$ is suspicious, whereas solutions with $M_{20} > 20$ are generally correct. Another common "quality factor" is $F_N$:

$$F_N = \frac{N}{\langle|\Delta(2\theta)|\rangle N_{\mathrm{possible}}} \quad (26)$$

In this case, $N_{\mathrm{possible}}$ is the number of possible peak positions out to the $N$th observed peak.

The most popular autoindexing programs are ITO (Visser, 1969), TREOR (Werner et al., 1985), and DIVCOL (Louër and Vargas, 1982).

The space group can be determined the "old fashioned" way by comparing the various extinction rules listed, e.g., in the International Tables (IUC, 1992). Certain program suites provide little search routines for this purpose (e.g., Byrom and Lucas, 1991).

**The Extraction of Individual Structure Factors.** Once the cell parameters and the space group are known, a full cell-constrained profile refinement can be performed. The individual intensities are treated as variables together with the unit cell parameters, the zero-point shift, and other "machine parameters" such as the $U$, $V$, and $W$ coefficients (Appendix A), the asymmetry parameter, and any other parameters describing the peak shape. Cell-constrained refinements were first proposed by Pawley (1981). However, in the original procedure all peaks that occurred within a given angular range were generated from the starting cell parameters and then their intensities were refined in a least-squares procedure together with the machine parameters and the cell constants. These led to large correlations among the parameters and ill conditioning of the least-squares fit, often resulting in diver-

gence. The use of the individual intensities in the least-squares refinement resulted in enormous normal equations and the major cause of instabilities (see Basics of Crystallographic Refinements). A computationally more stable algorithm was proposed by J.C. Taylor and first implemented by LeBail et al. (1988). The integrated intensities of the individual reflections are not used in the least-squares process but are instead calculated by iteration starting from crudely estimated intensities. However, due to this iterative process, the errors assigned to the individual intensities are no longer correct in the sense that they represent the counting statistics. These are crucial for the subsequent step in an *ab initio* structure determination. Therefore, according to Pawley, a cell-constrained refinement should be used as a last step to obtain correct errors. Furthermore, the errors of completely overlapping reflection should be assigned with the method derived by David (1987).

**Structure Solution.** All known standard structure solution techniques can be applied (Christensen et al., 1985, 1989; Cheetham, 1986; Cheetham and Wilkinson, 1991). Often trial-and-error attempts based on a good knowledge of crystal chemistry are successful. The two main methods are Patterson methods and direct methods. Patterson methods are based on the use of the autocorrelation function:

$$P(x,y,z) = \frac{1}{V^2} \sum\sum\sum F_{hkl}^2 \cos 2\pi(hx + ky + lz) \quad (27)$$

They are easy to calculate but often difficult to interpret since they represent interatomic vectors rather than atomic positions. If heavy scatterers are present, they will dominate the Patterson maps. Therefore, the success rate with x-ray data sets is higher than with neutron data sets, where the scattering power is in general more equally distributed among the atoms. Structures are frequently solved more easily using direct methods based on probabilistic relations between intensities and phases of the structure factors when neutron data are available. Other less-frequently used approaches to solve structures from powder diffraction only are (1) the MEDIC algorithm based on Patterson map deconvolution (Rius and Miravitles, 1988), (2) methods based on statistical mechanics such as simulated annealing using the $R$ factor as a cost function (Semenovskaya et al., 1985), (3) image reconstruction techniques (Johnson and Davide, 1987), (4) the increasingly popular maximum-entropy approach (David, 1990, Gilmore et al., 1993), and (5) a recently very promising genetic algorithm for fitting trial structures against measured powder diffraction data (Shankland et al., 1997).

**Time-Dependent Neutron Powder Diffraction**

The combination of powerful neutron sources and the development of fast detectors allows diffraction experiments to be recorded within a few minutes and in some cases even in a few seconds. We can thus observe rapid changes. With the new high-intensity diffractometer D20

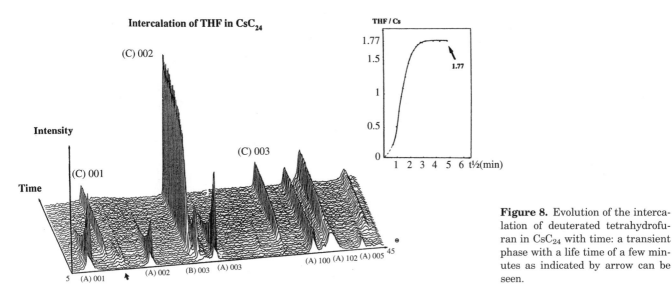

**Intercalation of THF in CsC$_{24}$**

**Figure 8.** Evolution of the intercalation of deuterated tetrahydrofuran in CsC$_{24}$ with time: a transient phase with a life time of a few minutes as indicated by arrow can be seen.

at the ILL (Convert et al., 1997) and the GEM project at the spallation source ISIS in the United Kingdom, neutron powder diffraction will open up the "time domain" for chemists and material scientists in an unprecedented manner. The temporal information gained during a chemical or physical process can be very important and opens up the study of chemical reactions by identifying intermediate crystalline species as well as the study of phase transformations in general. The notion of a static chemical structure "frozen in space" is being replaced by the dynamics of an evolving system. Time-dependent diffraction experiments are performed by externally perturbing a sample and then monitoring the relaxation of this sample back toward equilibrium. The time it takes to perform a single measurement will be shorter than the relaxation time of the process. The external perturbation can be temperature, pressure, magnetic, or electric field. It can be applied once or in a cyclic fashion. Two distinctions can be made with respect to the observed processes: First, in nonreversible processes time dependence is monitored by recording data sequentially. Second, in reversible processes the relaxation time of the process can be much shorter than the time it takes to record a pattern with good statistics. In such an experiment the perturbation time is divided into different time slices, and the data are recorded and accumulated over many cycles. Reversible processes in the range of 10 ms were observed using a special variable-time-delay electronics at the D20 prototype by Convert et al. (1990).

The use of position-sensitive detectors (PSDs) in thermal neutron scattering (Convert and Forsyth, 1983) in the late 1970s triggered these investigations. The low-resolution high-intensity diffractometer D1B at the ILL was increasingly being used by chemists and material scientists to identify intermediate and transient crystalline species during chemical reactions and phase transformations. Pannetier (1986, 1988) performed the groundbreaking work in this field. Besides qualitative studies, time-dependent studies allow in a quantitative way the determination of the kinetic laws governing the trans-

formations. This is done by measuring the volume fraction of each phase as a function of time. By monitoring the normalized intensities of a few strong reflections, one can determine a time-dependent volume fraction $\alpha(t) = I(t) - I(0)/I(\infty) - I(0)$ for an appearing and $\alpha(t) = I(\infty) - I(t)/I(\infty) - I(0)$ for a disappearing phase. The temporal evolution of $\alpha$ can be described in very simple terms of nucleation and growth (Christian, 1975) using the so-called Avrami-Erofeev equation: $a(t) = 1 - \exp[-k_T(t)^n]$, where the exponent $n$ is a constant related to the dimensionality of nucleation and growth [$3 \leq n \leq 4$ for a three-dimensional (3D) process] and $k_T(t)$ the kinetic constant of the reaction depending on the temperature.

Examples in this field are the pressure-induced NaCl-to-CsCl type structural transformation in RbI (Yamada et al., 1984) and the solid-gas reaction of tetrahydrofuran (THF) vapor with the second-stage graphite intercalation compound CsC$_{24}$. The latter reaction takes place over a few hours at room temperature. The temporal evolution of this chemical system is shown in Figure 8. Goldman et al. (1988) identified the distinct steps in this reaction: CsC$_{24}$ vanishes rapidly as an intermediate CsC$_{24}$(THF)$_x$ with $x \approx 1$ evolves. This intermediate phase, indicated by the arrow in Figure 8, has a lifetime on the order of minutes until the final first-stage intermediate CsC$_{24}$(THF)$_{\sim 1.8}$ appears. There were no indications from the bulk properties that this structural intermediate exists.

Neutron powder diffraction studies, combined with the unique capabilities of neutrons to alter the scattering lengths by the right mixture of isotopes, was shown earlier to be a very powerful tool, especially in site-disordered alloys. In time-dependent studies, this can be used as a "marker" for the specific element during transformations, as the following example illustrates: the discovery of quasicrystalline Al-Mn alloys with long-range icosahedral symmetry (Schechtman et al., 1984) led to many intriguing questions. One of them was what happens when these quasicrystals transform to "normal" crystalline alloys? Substituting part of the Mn ($b_{\mathrm{coh}} = -3.73$ fm) with various amounts of Fe ($b_{\mathrm{coh}} = 9.54$ fm) allows again the formation of a "zero-scattering" composition, namely $Mn_{0.72}Fe_{0.28}$ in the transition metal sublattice. Comparing $Al_{85}Si(Mn_{0.72}$ $Fe_{0.28})_{14}$ with $Al_{85}SiMn_{14}$ revealed diffuse scattering at the onset of crystallization of the latter but not in the case of the zero-scattering alloy. It appears that the transformation is an order-disorder transition involving the transition metal sublattice and not the Al subnetwork (Pannetier et al., 1987).

Another increasingly popular time-dependent structural investigation is the *in situ* electrochemical experiment. Using electrochemical cells directly in the neutron beam, Latoche et al. (1992) investigated the structural changes taking place in a $LaNi_{4.5}Al_{0.5}Dx$ electrode during charge and discharge. Phases with out-of-equilibrium cell parameters were observed and related to the charge-discharge rate. In the near future more and more studies using high-intensity neutron powder diffractometers to probe the time dependency of chemical processes will become feasible due to new and powerful machines currently under construction or close to completion.

## PROBLEMS

### Basics of Crystallographic Refinements

An increasing sophistication in data collection and correction procedures for absorption, extinction, preferred orientation, strain, particle size, and asymmetry due to axial divergence has substantially enhanced the quality of data available nowadays from modern neutron powder diffractometers. However, the misuse of Rietveld refinement programs, perhaps through failure to grasp the principles, sometimes leads to poor refinements of high-quality data. With increasing complexity (large unit cell, low symmetry) of the investigated structures, the need to use restraints and constraints in the refinement to limit parameter space and achieve convergence is becoming increasingly important. The following is an attempt to shed some light onto the mystery of modern refinement strategies.

Any structural refinement program for x-ray, neutron, or electron diffraction data obtained from gases, liquids, or powder or single crystal will provide a model for the pattern of electron or nuclear density describing the atomic structure. We have well-established and proven theories of the interaction of x rays, neutrons (mainly kinematical scattering), and electrons (dynamical scattering) with the electrons (in the case of x rays and electrons) and nuclei (in the case of neutrons) of the investigated matter. Given high-quality observations of scattering amplitudes and estimates of phases, we can compute electron or nuclear density by using a 3D Fourier summation of structure factors. Crystallography is about defining and refining these models.

Detailed references for a full mathematical background of least-squares techniques can be found in Press (1986) and Prince (1994). Here is the "poor man's version":

We have a mathematical expression for $Y_c$ (calculated intensities) as a function of variables $x_i$ (positional and displacement parameters) whose values we wish to determine. We have experimental values $Y_0$ from our diffraction

experiment. Expanding $Y_c$ as a Taylor series in $x_i$, we determine shifts $\delta x$ zthat will be applied to the initial values of $x$ to improve $Y_c$, thus minimizing the difference between $Y_0$ and $Y_c$:

$$Y_{old} + \left(\frac{\partial Y}{\partial x_1}\right)\delta x_1 + \left(\frac{\partial Y}{\partial x_2}\right)\delta x_2 \cdots = Y_{new} \quad (28)$$

By minimizing,

$$Y_0 - Y_c = \sum \left(\frac{\partial Y}{\partial x_i}\right)\delta x_i \quad (29)$$

we express all individual equations in matrices as

$$\mathbf{A}\,\delta x = \mathbf{y} \quad (30)$$

Multiplying both sides by $\mathbf{A}'$, the transpose of $\mathbf{A}$, we obtain the so-called normal equations:

$$\mathbf{H}\,\delta x = \mathbf{A}'\mathbf{y} = \mathbf{r} \quad (31)$$

The normal matrix $\mathbf{H}$ is square and made up of terms $h_{ij} = \Sigma(\partial Y_m/\partial x_i)\,(\partial Y_m/\partial x_j)$ and $\mathbf{r}$ is a vector of terms

$$r_i = \sum \left(\frac{\partial y_m}{\partial x_i}\right)(Y_{obs} - Y_{calc}) \quad (32)$$

Solving these equations for $\delta x$ leads to

$$\delta x = \mathbf{H}^{-1}\mathbf{r} \quad (33)$$

As is standard practice in crystallography, the least-squares errors, the estimated standard deviations, are obtained from the diagonal elements of the variance-covariance matrix and the goodness-of-fit index $S$:

$$\sigma_i = S_{wp}A^{-1/2} \quad (34)$$

with $S_{wp} = [\sum w_i(y_{i,obs} - y_{i,calc})^2/(N_0 - N_{par})^2]^{1/2}$, where $N_0$ is the number of profile points and $N_{par}$ the number of parameters used. Prince (1981) has criticized this procedure. A recent summary of this highly controversial and statistically unsound procedure is given by Post and Bish (1989). The underlying assumption that a proper weighting scheme is being used is questionable. As a "rule of thumb," the calculated estimated standard deviations tend to underestimate the "true" error by a factor of 2 to 3 for positional parameters and even more in the case of displacement parameters and site occupancies (see Appendix C).

The most used refinement algorithm is a weighted least-squares algorithm that minimizes $\chi^2$. For more details on statistics, see Prince (1994). This assumes that the distribution of the random variable is due to counting statistics and is Gaussian. If the errors of the observations are independent and normally distributed, one obtains the correct estimates of the parameters. If this is no longer true due to, e.g., low counting statistics, other better suited methods should be used. In this case, a Poisson distribution might better represent the experimental errors and the maximum-likelihood method used (Antoniadis et al., 1990). Currently, one Rietveld program (FULLPROF) has this option incorporated. With good counting statistics, maximum-likelihood and minimum-$\chi^2$ methods give the same results. The maximum-likelihood method is an iteratively reweighted least-squares fit. The weights are determined by the fit. This is in marked difference to the weighted least-squares fit where the weights determine the fit.

## Constraints

Constraints and restraints are both assumptions made about the solutions to crystallographic problems. In the case of a restraint, the solution is more tightly restrained to satisfy the assumption the greater our confidence is in the assumption. When using a constraint, we insist that this assumption is obeyed. In the early literature (Rollet, 1965), restraints were often referred to as "slack constraints." We can subdivide constraints into explicit and implicit constraints. Implicit constraints are implied to be without appreciable errors, and our model will be implicitly constrained by the fact that we are using these corrections and assumptions. Examples of implicit constraints are the corrections for Lorentz polarization, absorption corrections, the scattering lengths used in neutron diffraction, and the form factors used in x-ray or magnetic neutron scattering. The assumption that no significant amorphous phase is present in a phase mixture is also an implicit constraint when refining multiphase mixtures and attempting to determine weight fractions. Any errors in these implicit constraints will lead to errors in the final model that are not accounted for by the estimated standard deviations of the refined parameters. In neutron powder diffraction refinements, these considerations are generally applicable to refinements where the choice of form factors will crucially influence the refined magnetic moment, or for instance an incorrect absorption correction will bias the nonharmonic expansion of the anisotropic displacement parameters. To extract physically meaningful information from displacement parameters, one has to deal with implicit constraints in an appropriate manner. Explicit constraints are choices we make in the course of a refinement. The peak shape is an important but often overlooked one. The choice of space groups is an obvious one. In many refinements, it is difficult to decide if a structure is better described in a high-symmetry space group than in a lower symmetry space group with more variables that are highly correlated and lead to an unstable refinement. One solution to this problem is to refine in the lower space group but apply symmetry constraints between certain atoms related by pseudomirror planes. Using the same parameters but assigning opposite signs to the two atoms will achieve this. Coupling the parameters of these atom pairs leads to only one pair of least-squares parameters, thus reducing correlations and leading to a well-converging refinement. Thus, although the structure is described in a lower symmetry space group, the use of explicit constraints introduces symmetry elements from a higher symmetry space group. Another "classical" pitfall in refinements is the need to impose a constraint when

describing the model in certain space groups where the origin is not fixed by symmetry such as $P1$, $Pc$, and $P2$. If there is no fixed origin, the normal matrix will be singular. The solution is to fix a certain atom. This is done by not refining, e.g., the $z$ coordinate in space group $P2$. However, this position will not have an estimated standard deviation and all estimated standard deviations of the $z$ components of the other atoms will therefore be biased. Using the variance-covariance matrix is the appropriate way to extract the correct estimated standard deviations in such a case (Prince, 1994). Another problem often occurring is site disorder. This can be the case of two different atomic species occupying or partially occupying the same site. In both cases, the total occupancy can be an explicit constraint. Many problems in refinements are concerned with the displacement parameters. They are often loosely referred to as "temperature factors." The question arises if a given atom has an isotropic or anisotropic displacement parameter. This choice will influence the bond distances to the neighboring atoms. Sometimes, even anisotropic displacement parameters are not the solution to the problem. One might have to describe rigid units of the structure by using a TLS model (Prince, 1994). The T stands for translation, the L for libration, and the S for screw movement of the rigid group. The right choice of the rigid unit and explicit constraints (e.g., all translational components of the fragments are the same) is crucial and will influence the final $R$ factors (see Appendix B). Often, rigid-body constraints are used to lead the refinement in the right direction. Certain structural coordination polyhedra such as octahedra or tetrahedra are initially constrained to a rigid unit with an "ideal" geometry. The individual atomic parameters are replaced by a total of six parameters, three defining a rotation, three a translation. The significant reduction of parameters from $3n$ to 6 for $n$ atoms reduces the complexity of the normal matrix substantially and generally leads to a smoother refinement. In the initial states of an attempt to solve the structure of $SF_6$, Cockcroft and Firch (1988) applied such chemical constraints to preserve the $SF_6$ octahedra in refinements against high-resolution neutron powder diffraction data. This resulted in a smooth convergence to a model, which was then the starting model for further refinements.

## Restraints

Restraints, or "slack or soft constraints" as they are referred to in the older literature (Waser, 1963; Rollet, 1965), are ways of influencing the refinement by introducing other observables $Y_{obs}$. These new observables are then subject to the same mathematical treatment as the "real" experimental observables (intensities). They are added into the normal matrix and the **r** vector. Additional variables can be as simple as a dummy atom needed to restrain planarity in a benzene molecule, a nonbonding restraint to avoid overly unphysical distances during the refinement, or the restraints with respect to refining combined x-ray and neutron diffraction data sets. It is important to consider the weights applied to these supplemental observations or subsidiary conditions, as they are also sometimes called, when refining with restraints. If there are only few restraints, then we can assign a desired estimated standard deviation to them and scale them to the weighted residual for the intensities (Rollet, 1965). The most widely used restraint is the bond restraint:

$$(x_1 - x_2)^2 + (y_1 - y_2)^2 + (z_1 - z_2)^2 = d^2 \qquad (35)$$

The value for $d$ is generally taken from the literature. Six parameters are coupled. In the refinement $d^2 = \Delta x' \, \mathbf{G} \, \Delta x$, with $\mathbf{G}$ being the metric tensor. Differentiating leads to

$$\frac{\partial d}{\partial x_1} = \frac{g_{11}\Delta x + g_{12}\Delta y + g_{13}\Delta z}{d} \qquad (36)$$

and

$$\frac{\partial d}{\partial x_2} = \frac{g_{11}\Delta x + g_{12}\Delta y + g_{13}\Delta z}{d} \qquad (37)$$

However, note that the position within the unit cell is not fixed. If used alone, this will lead to a singular matrix. For fairly rigid units, this type of constraint is quite appropriate since the assigned estimated standard deviation is symmetric. When trying to mimic, for example, van der Waals interactions, one might want to use an "asymmetric" estimated standard deviation. Using a "penalty function" can achieve this by calculating an energy using a van der Waals potential involving the restraints. This takes into account the fact that when dealing with van der Waals interactions, you accept longer but not shorter interactions as physically sensible.

## LITERATURE CITED

Antoniadis, A., Berruyer, J., and Filhol, A. 1990. Maximum likelihood method in powder diffraction refinements. *Acta Crystallogr. A* 46:692–711.

Axe, J. D., Cheung, S., Cox, D. E., Passell, L., Vogt, T., and Bar-Ziv, S.-B. 1994. Composite germanium monochromators for high resolution neutron powder diffraction applications. *J. Neutron Res.* 2(3):85.

Bacon, G. E., Smith, J. A. G., and Whitehead, C. D. 1950. *J. Sci. Instrum.* 27:330.

Baerlocher, C. 1986. *Zeolites* 6:325.

Baerlocher, C. 1993. Restraints and constraints in Rietveld refinement. *In* The Rietveld Method (R. A. Young, ed.) pp. 186–196. Oxford University Press, New York.

Bendall, P. J., Fitch, A. N., and Fender, B. E. F. 1983. The structure of $Na_2UCl_6$ and $Li_2UCl_6$ from mulitiphase powder neutron profile refinement. *J. Appl. Crystallogr.* 16:164.

Berar, J. F. and Lelann, P. 1991. E. S. D.'s and estimated probable error obtained in Rietveld refinements with local correlations. *J. Appl. Crystallogr.* 24:1–5.

Berliner, R. and Werner, S. A. 1986. Effect of stacking faults on diffraction. The structure of lithium metal. *Phys. Rev. B* 34:3586–3603.

Bish, D. L. and Howard, S. A. 1988. Quantitative phase analysis using the Rietveld method. *J. Appl. Crystallogr.* 21:86–91.

Breit, G. and Wigner, E. 1936. *Phys. Rev.* 49:519.

Bricogne, G. 1991. Maximum entropy and the foundations of direct methods. *Acta Crystallogr. A* 47:803–829.

Brown, A. and Edmonds, J. W. 1980. *Adv. X-ray Anal.* 23:361–374.

Buttrey, D. J., Vogt, T., Wildgruber, U., and Robinson, W. R. 1994. The high temperature structure of $Bi_2MoO_6$. *J. Solid State Chem.* 111:118.

Byrom, P. G. and Lucas, B. W. 1991. POWABS: A computer program for the automatic determination of reflection conditions in powder diffraction patterns. *J. Appl. Crystallogr.* 24:70–72.

Cagliotti, G., Paoletti, A., and Ricci, F. P. 1958. Choice of collimators for a crystal spectrometer for neutron diffraction. *Nucl. Instrum. Methods* 3:223.

Cheetham, A. K. 1986. Structure determination by powder diffraction. *Mater. Sci. Forum* 9:103.

Cheetham, A. K., David, W. I. F., Eddy, M. M., Jakeman, R. J. B., Johnson, M. W., and Torardi, C. C. 1986. Crystal structure determination by powder neutron diffraction at the spallation source ISIS. *Nature (London)* 320:46–48.

Cheetham, A. K. and Wilkinson, A. P. W. 1991. Structure determination and refinement with synchrotron X-ray powder diffraction data. *J. Phys. Chem. Solids* 52:1199–1208.

Cheetham, A. K. and Wilkinson, A. P. 1992. Synchrotron X-ray and neutron diffraction studies in solid state chemistry. *Angew. Chem. Int. Ed. Engl.* 31:1557–1570.

Christensen, A. N., Cox, D. E., and Lehmann, M. S. 1989. A crystal structure determination of $PbC_2O_4$ from synchrotron X-ray and neutron powder diffraction data. *Acta Chem. Scand.* 43:19–25.

Christensen, A. N., Lehmann, M. S., and Nielsen, M. 1985. Solving crystal structures from powder diffraction data. *Aust. J. Phys.* 38:497–505.

Christian, J. W. 1975. The Theory of Transformations in Metals and Alloys. Pergamon Press, Oxford.

Cockcroft, J. K. and Firch, A. N. 1988. The solid phases of sulphur hexafluoride by powder neutron diffraction. *Z. Kristallogr.* 184:123–145.

Convert, P., Berneron, M., Gandelli, R., Hansen, T., Oed, A., Rambaud, A., Ratel, J., and Torregrossa, J. 1997. A large high counting rate one-dimensional position sensitive detector: The D20 banana. *Physica B* 234–236:1082.

Convert, P. and Forsyth, J. B. 1983. Position-Sensitive Detection of Thermal Neutrons. Academic Press, San Diego.

Convert, P., Hock, R., and Vogt, T. 1990. High-speed neutron diffraction—A feasibility study. *Nucl. Instrum. Methods A* 292:731–733.

Cooper, M. J. and Sayer, J. P. 1975. The asymmetry of neutron powder diffraction peaks. *J. Appl. Crystallogr.* 8:615–618.

Cowley, J. M. 1976a. Diffraction by crystals with planar faults. I. General theory. *Acta Crystallogr. A* 34:83–87.

Cowley, J. M. 1976b. Diffraction by crystals with planar faults. II. Magnesium fluorogermanate. *Acta Crystallogr. A* 32:88–91.

Cowley, J. M. 1981. Diffraction Physics. North-Holland Publishing, New York (see pp. 388–400).

Darwin, G. G. 1914. *Philos. Mag.* 27:315.

David, W. I. F. 1987. The probabilistic determination of intensities of completely overlapping reflections in powder diffraction patterns. *J. Appl. Crystallogr.* 20:316–319.

David W. I. F. 1990. Extending the power of powder diffraction for structure determination. *Nature (London)* 346:731.

David, W. I. F., Ibberson, R. M., and Matsuo, T. 1993. High resolution neutron powder diffraction: A case study of the structure of $C_{60}$. *Proc. R. Soc. London A* 442:129–146.

Debye, P. and Scherrer, P. 1916. *Z. Phys.* 17:277–283.

Delmas, C. and Tessier, C. 1997. Stacking faults in the structure of nickel hydroxides: A rationale of its high electrochemical activity. *J. Mater. Chem.* 7:1439–1443.

Eastabrook, J. N. B. 1952. *J. Appl. Phys.* 3:349–352.

Finger, L. W., Cox, D. E., and Jephcoat, A. P. J. 1994. A correction for powder diffraction peak asymmetry due to axial divergence. *Appl. Crystallogr.* 27:892–900.

Gilmore, G., Shankland, K., and Bricogne, G. 1993. Applications of maximum entropy method to powder diffraction and electron crystallography. *Proc. R. Soc. London A* 442:97–111.

Goldman, M., Pannetier, J., Beguin, F., and Gonzalez, F. 1988. *Synthetic Metals* 23:133.

Harrison, W. T. A., Dussack, L. L., Vogt, T., and Jacobson, A. J. 1995. Synthese, crystal structures and properties of new layered tungsten(VI)-containing materials based on the hexagonal-$WO_3$ structure: $M_2(WO_3)_3SeO_3$, M = $NH_4$, Rb, Cs). *J. Solid State Chem.* 120:112.

Hendricks, S. and Teller, E. 1942 X-ray interference in partially ordered layer lattices. *J. Chem. Phys.* 10:147–167.

Hewat, A. W. 1973. UKAEA, Harwell, United Kingdom, Research Group Report R7350.

Hill, R. 1992. Rietveld refinement round robin. I. Analysis of standard X-ray and neutron data for $PbSO_4$. *J. Appl. Crystallogr.* 25:589–610.

Hill, R. J. and Flack, H. D. 1987. The use of the Durbin-Watson d-statistics in Rietveld analysis. *J. Appl. Crystallogr.* 20:356–361.

Hill, R. J. and Howard, C. J. 1987. Quantitative phase analysis fron neutron powder diffraction data using the Rietveld method. *J. Appl. Crystallogr.* 20:467–474.

Howard, C. J. 1982. The approximation of asymmetric neutron powder diffraction peaks by sums of gaussians. *J. Appl. Crystallogr.* 15:615–620.

Hull, A. W. 1917. *Phys. Rev.* 9:84.

Hurst, D. G., Pressesky, A. J., and Tunnicliffe, P. R. 1950. *Rev. Sci. Instrum.* 21:705.

International Union of Crystallography (IUC). 1992. International Tables of Crystallography. Dordrecht, Boston and London.

Jansen, E., Schäfer, W., and Will, G. 1988. Profile fitting and the two-stage method in neutron powder diffractometry for structure and texture analysis. *J. Appl. Crystallogr.* 21:228–239.

Johnson, M. W. and Davide, W. I. F. 1987. Rutherford Appleton Lab, United Kingdom, Report RAL-87-059.

Kramers, H. A. 1934. *Physica* 1:182.

Langford, J. B. and Louër, D. 1986. *Powder Diffraction* 1:211.

Larson, A. C. and Von Dreele, R. B. 1990. Unpublished GSAS Program Manual, Los Alamos National Lab.

Latoche, M., Pecheron-Guyan, A., Chabre, Y., Poinsignon, C., and Pannetier, J. 1992. Correlations between the structural and thermodynamic properties of $LaNi_5$ type hydrides and their electrode performances. *J. Alloys Compounds* 189:59–65.

LeBail, A., Duroy, H., and Fourquet, J. L. 1988. *Ab-initio* structure determination of $LiSbWO_6$ by X-ray powder diffraction. *Mater. Res. Bull.* 23:447–452.

Louër, D. and Vargas, R. 1982. Indexation automatique des diagrammes des poudres per dichotomies successives. *J. Appl. Crystallogr.* 15:542.

Medarde, M. and Rodriguez-Carvajal, J. 1997. Oxygen vacancy ordering in $La_{2-x}Sr_xNiO_{4-d}$ ($0 \leq x \leq 0.5$): The crystal structure and defects investigated by neutron diffraction. *Z. Phys. B* 102:307–315.

Morris, R. E., Harrisson, W. T. A., Nicol, J. M., Wilkinson, A. P., and Cheetham, A. K. 1992. Determination of complex structures by combined neutron and synchrotron X-ray powder diffraction. *Nature (London)* 359:519–522.

Morris, R. E., Owen, J. J., Stalick, J. K., and Cheetham, A. K. 1994. Determination of complex structures from powder diffraction data: The crystal structure of $La_3Ti_5Al_{15}O_{37}$. *J. Solid State Chem.* 111:52–57.

Néel, L. 1932. *Ann. Phys.* 17:64.

Néel, L. 1948. *Ann. Phys.* 3:137.

Newsam, J. M., Treacy, M. M. J., Koetsier, W. T., and deGruyter, C. B. 1988. Structural characterization of zeolite beta. *Proc. R. Soc. London A* 420:375–405.

Pannetier, J. 1986. Time-resolved neutron powder diffraction. *Chem. Scr.* 26A:131–139.

Pannetier, J. 1988. Real time neutron powder diffraction. *In* Chemical Crystallography with Pulsed Neutrons and Synchrotron X-rays (M. A. Carrondo and G. A. Jeffrey, eds.) p. 313. D. Reidel Publishing, Oxford.

Pannetier, J., Dubois, J. M., Janot, C., and Bilde, A. 1987. *Philos. Mag. B* 55:435–457.

Papoular, R. J. and Cox, D. E. 1995. Model-free search for extra framework cations in zeolites using powder diffraction. *Europhys. Lett.* 32:337–342.

Paulus, H., Fuess, H., Müller, G., and Vogt, T. 1990. The crystal structure of β-quartz type $HALSi_2O_6$. *N. Jb. Miner. Mh.* 5:232–240.

Pawley, G. S. 1981. Unit cell refinement from powder diffraction data. *J. Appl. Crystallogr.* 14:357–361.

Post, J. E. and Bish, D. L. 1989. Modern powder diffraction. *Rev. Miner.* 20:277.

Press, W. H. 1986. Numerical Recipies—The Art of Scientific Computing. Cambridge University Press, Cambridge.

Prince, E. 1981. *J. Appl. Crystallogr.* 14:157.

Prince, E. 1994. Mathematical Techniques in Crystallography and Materials Science. Springer-Verlag, Heidelberg.

Rietveld, H. M. 1967. *Acta Crystallogr.* 22:151–152.

Rietveld, H. M. 1969. A profile refinement for nuclear and magnetic structures. *J. Appl. Crystallogr.* 2:65.

Rius, J. and Miravitles, C. 1988. Determination of crystal structures with large known fragments directly from measured x-ray powder diffraction intensities. *J. Appl. Crystallogr.* 21:224–227.

Rodriguez-Carvajal, J. 1993. Recent advances in magnetic structure determination by neutron powder diffraction. *Physica B* 192:55–69.

Rodriguez-Carvajal, J., Fernandez-Diaz, M. T., and Martinez, J. L. 1991. *J. Phys. Condensed Matter* 3:3215.

Roessli, B., Fischer, P., Staub, U., Zolliker, M., and Furrer, A. 1993. Combined electronic-nuclear magnetic ordering of the $Ho^{3+}$ ions and magnetic stacking faults in the high-Tc superconductor $HoBa_2Cu_3O_7$. *Europhys. Lett.* 23:511–515.

Rollett, J. S. 1965. Computing Methods in Crystallography. Pergamon Press, Elmsford, N.Y.

Rotella, F. J., Jorgensen, J. D., Biefeld, R. M., and Morosin, B. 1982. Location of deuterium sites in the defect pyrochlore $DTaWO_6$ from neutron powder diffraction data. *Acta Crystallogr. B* 38:1697.

Sabine, T. M., Von Dreele, R. B., and Jorgensen, J. E. 1988. *Acta Crystallogr. A* 44:274.

Schechtman, D., Blech, I., Gratias, D., and Cahn, J. W. 1984. Metallic phase with long-range orientational order and no translational symmetry. *Phys. Rev. Lett.* 53:1951.

Semenovskaya, S. U., Khatchaturayan, K. A., and Khatchaturayan, A. L. 1985. Statistical mechanics approach to the structure determination of a crystal. *Acta Crystallogr. A* 41:268–276.

Shankland, K., David, W. I. F., and Csoka, T. 1997. Crystal structure determination from powder diffraction data by the application of a genetic algorithm. *Z. Kristallogr.* 112:550–552.

Shirley, R. 1984. Measurements and analysis of powder data from single solid phases. *In* Methods and Applications in Crystallographic Computing (S. R. Hall and T. Ashida, eds.) pp. 414–437. Clarendon Press, Oxford.

Shull, C. G. and Smart, J. S. 1949. *Phys. Rev.* 76:1256.

Shull, C. G., Strauser, W. A., and Wollan, E. O. 1951. *Phys. Rev.* 83:333.

Soller, W. A. 1924. A new precision X-ray spectrometer. *Phys. Rev.* 24:158–167.

Tombs, N. C. and Rooksby, H. P. 1950. *Nature (London)* 165:442.

Toraya, H. J. 1986. Whole-powder-pattern fitting without reference to a structural model: Application to X-ray powder diffraction spectra. *Appl. Crystallogr.* 19:940–947.

Treacy, M. M. J., Newsam, J. M., and Deem, M. W. 1991. A general recursion method for calculating diffracted intensities from crystals containing planar faults. *Proc. R. Soc. London* 433:499–520.

van Laar, B. and Yelon, W. B. 1984. The peak in neutron powder diffraction. *J. Appl. Crystallogr.* 17:47–54.

Visser, J. 1969. A fully automatic program for finding the unit cell from powder data. *J. Appl. Crystallogr.* 2:89.

Vogt, T., Paulus, H., Fuess, H., and Muller, G. 1990 The crystal structure of $HAlSi_2O_6$ with a keatite-type framework. *Z. Kristallogr.* 190:7–18.

Waser, J. 1963. Least-squares refinement with subsidiary conditions. *Acta Crystallogr.* 16:1091–1094.

Werner, P.-E., Eriksson, L., and Westdahl, M. J. 1985. TREOR, a semi-exhaustive trial-and-error powder indexing program for all symmetries. *J. Appl. Crystallogr.* 18:367.

Wessells, T., Baerlocher, C., and McCusker, L. B. 1999. Single-crystal-like diffraction from polycrystalline materials. *Science* 284:477.

Wiles, D. B. and Young, R. A. 1981. A new computer program for Rietveld analysis of X-ray powder diffraction patterns. *J. Appl. Crystallogr.* 14:149–151.

Williams, W. G. 1988. Polarized Neutrons. Clarendon Press, Oxford.

Wilson, A. J. C. 1942. Imperfections in the structure of cobalt II. Mathematical treatment of proposed structures. *Proc. R. Soc. London A* 180:277–285.

Wilson, A. J. C. 1943. The reflection of X-rays from the 'antiphase nuclei' of $AuCu_3$. *Proc. R. Soc. London A* 181:360–368.

Wollan, E. O. and Shull, C. G. 1948. *Phys. Rev.* 73:830.

Woodward, P. M., Sleight, A. W., and Vogt, T. 1995. Structure refinement of triclinic tungsten trioxide. *J. Phys. Chem. Solids* 56:1305–1315.

Yamada, Y., Hamaya, N., Axe, J. D., and Shapiro, S. M. 1984. Nucleation, growth, and scaling in a pressure-induced first-order phase transition: RbI. *Phys. Rev. Lett.* 53:1665.

Zachariasen, W. H. 1949. *Acta Crystallogr.* 2:296.

Zachariasen, W. H. and Ellinger, F. H. 1963. *Acta Crystallogr.* 16:369.

Zinn, W. H. 1947. *Phys. Rev.* 71:752.

## KEY REFERENCES

Bish, D. L. and Post, J. E. 1989. Modern Powder Diffraction, Reviews in Mineralogy, Vol. 20. The Mineralogical Society of America, Washington, D.C.

*Covers neutron powder diffraction as well as x-ray and synchrotron powder diffraction. Contains valuable information with respect to experimental procedures and data analysis.*

Young, R. A. 1993. The Rietveld Method. International Union of Crystallography. Oxford University Press, New York.

*Provides an excellent introduction to modern Rietveld refinement techniques and neutron powder diffraction using pulsed sources and reactor-based instruments. The 15 parts are written by acknowledged researchers who cover the field in theoretical as well as practical aspects. Highly recommended.*

## APPENDIX A:
## OPTICS OF A CONSTANT WAVELENGTH DIFFRACTOMETER

The Cagliotti coefficients $U$, $V$, and $W$ are functions of the collimators $\alpha_1$, $\alpha_2$, and $\alpha_3$, of the mosaic spread $\beta$ of the monochromator crystals and the monochromator scattering angle ("take-off" angle) $\theta_m$:

$$\text{FWHM} = (U \tan^2 \theta + V \tan \theta + W)^{1/2} \qquad (38)$$

where

$$U = \frac{4(\alpha_1^2 \alpha_2^2 + \alpha_1^2 \beta^2 + \alpha_2^2 \beta^2)}{\tan 2\theta_m (\alpha_1^2 + \alpha_2^2 + 4\beta^2)} \qquad (39)$$

$$V = -\frac{4\alpha_2^2 (\alpha_1^2 + 2\beta^2)}{\tan \theta_m (\alpha_1^2 + \alpha_2^2 + 4\beta^2)} \qquad (40)$$

$$W = \frac{\alpha_1^2 \alpha_2^2 + \alpha_1^2 \alpha_3^2 + \alpha_2^2 \alpha_3^2 + 4\beta^2(\alpha_2^2 + \alpha_3^2)}{\alpha_1^2 + \alpha_2^2 + 4\beta^2} \qquad (41)$$

and with a minimum of the FWHM at

$$\left( W - \frac{V^2}{4U} \right)^{1/2} \qquad (42)$$

Here, $\alpha_1$, $\alpha_2$, $\alpha_3$ are the beam divergences of the in-pile, monochromator-sample, and sample-detector collimators, respectively. This FWHM can be rewritten as

$$\text{FWHM} = \frac{N}{(\alpha_1^2 + \alpha_2^2 + 4\beta^2)^{1/2}} \qquad (43)$$

where

$$N = \left( \frac{\alpha_1^2 \alpha_2^2 + \alpha_1^2 \alpha_3^2 + \alpha_2^2 \alpha_3^2 + 4\beta^2(\alpha_2^2 + \alpha_3^2)}{(\alpha_1^2 + \alpha_2^2 + 4\beta^2) - 4a\alpha_2^2(\alpha_1^2 + 2\beta^2) + 4a^2(\alpha_1^2 \alpha_2^2 + \alpha_1^2 \beta^2 + \alpha_2^2 \beta^2)} \right)^{1/2} \qquad (44)$$

$$a = \frac{\tan \theta}{\tan \theta_m} \qquad (45)$$

$\theta$, being the Bragg angle. An important result is that the FWHM decreases with $N$ and $N$ becomes smaller the closer $\theta$ is to the take-off angle $\theta_m$. The best resolution of a neutron powder diffractometer will be near the take-off position. After the take-off position, the FWHM increases considerably. For high-resolution purposes, one will attempt to use a take-off angle as high as possible. The diffractometer D2B at the ILL has a $\theta_m$ of 135°, and the HRNPD at Brookhaven National Laboratory has a $\theta_m$ of 120°.

## APPENDIX B:
## RELIABILITY FACTORS

One of the most debated areas of Rietveld refinement is of the the so-called reliability factors, often called "$R$ factors," which assess the quality of the model with respect to the data. In a least-squares refinement, one minimizes $R$. Therefore, the choice of $R$ is nontrivial. The two most-used $R$ factors in a Rietveld refinement are the profile $R$ factor $R_p$ and the weighted profile $R$ factor $R_{wp}$:

$$R_p = \frac{\sum |y_i - y_{i,\mathrm{calc}}|}{\sum y_i} \tag{46}$$

$$R_{wp} = \left(\frac{m}{\sum w_i y_I^2}\right)^{1/2} \tag{47}$$

where $M = \Sigma w_i(y_i - y_{i,\mathrm{calc}})^2$ is the actual function the refinement routine is minimizing and $w_i = 1/\sigma_i^2$ the weight assigned to the observed intensity $y_i$ at step $i$, $\sigma_i$ being the variance of observation $y_i$ and $y_{i,\mathrm{calc}}$ the calculated intensity at the $i$th step of the pattern. Another important $R$ factor is

$$R_{\mathrm{Bragg}} = \frac{\sum |I_i - I_{i,\mathrm{calc}}|}{\sum I_i} \tag{48}$$

which is a reliability factor commonly used in refinements of integrated intensity data. The term $I_i$ is the integrated intensities at point i.

The best $R$ factor to judge the agreement between experimental data and the structural model is $R_{\mathrm{Bragg}}$. However, $R_{\mathrm{Bragg}}$ is deduced with the help of a model and is therefore biased in favor of the model. This is different when using model-free structure refinement techniques such as maximum-entropy techniques. The $R_{\mathrm{Bragg}}$ is also rather insensitive to inadequate modeling of the peak shape and the presence of peaks stemming from other phases. This is not the case with $R_{wp}$, which depends very much on how well the profile function used describes the peak shape, especially the low-angle asymmetry correction. Therefore, $R_{wp}$ is less sensitive to the structural parameters than $R_{\mathrm{Bragg}}$. Another problem with $R_{wp}$ is that it can be misleadingly small if the refined background is high since it is easier to get a good fit of a slowly varying background than of the Bragg reflections that contain the structural information. This is an important consideration when measuring materials containing hydrogen, since the incoherent scattering will raise the background. Even

minute residual hydrogen in deuterated compounds or materials synthesized from hydrogen-containing precursors will result in a higher background.

Other $R$ factors are

$$\chi^2 = \frac{M}{N - P + C} \tag{49}$$

$$R_e = \frac{R_{wp}}{\sqrt{\chi^2}} \tag{50}$$

where $N$, $P$, and $C$ are the numbers of data points, refined parameters, and constraints, respectively.

The frequently encountered Durbin-Watson statistic parameters $d$ and $Q$ (Hill and Flack, 1987) measure the correlation between adjacent residuals:

$$d = \frac{\sum [w_i(y_i - y_{i,\mathrm{calc}}) - w_{i-1}(y_{i-1} - y_{i-1,\mathrm{calc}})]^2}{\sum w_1^2(y_i - y_{i,\mathrm{calc}})^2} \tag{51}$$

Serial correlation is tested at the 99.9% confidence level by comparing $d$ to $Q$:

$$Q = 2\left(\frac{N-1}{N-P} - \frac{3.0902}{(N+2)^{1/2}}\right) \tag{52}$$

If $d < Q$, there is a positive serial correlation, which means successive residuals will have the same sign. If $Q < d < 4 - Q$, there is no correlation between neighboring points, and if $d > 4 - Q$, there is a negative correlation. The ideal value for $d$ is 2.00. If the calculated and observed profile functions do not match well, there will be a strong correlation of the residuals of neighboring points and the Durbin-Watson $d$ value will be far from 2.00.

## APPENDIX C:
## ESTIMATED STANDARD DEVIATIONS

The question of what is an independent observation $N$ is not only for purists, the reason being that $N$ is used to calculate the estimated standard deviation (esd):

$$\mathrm{esd}_i = \left(\frac{M_{ii}^{-1} \sum w_i(y_i - y_{i,\mathrm{calc}})^2}{N - P + C}\right)^{1/2} \tag{53}$$

where $M_{ii}^{-1}$ is the diagonal in the inverted least-squares matrix, $P$ the number of refined variables, and $C$ the total number of constraints. In Rietveld refinements one uses the number of points measured and not the number of individual reflections $I_{hkl}$ as in single-crystal refinements. The estimated standard deviations are not the probable experimental error; rather they are the minimal possible probable error arising from random errors alone. There are certain systematic errors (e.g., peak shape, background) that arise from implicit constraints that are not considered in the estimated standard deviations. In other words, the experimental reproducibility will only then reflect the estimated standard deviations when the random errors ("counting statistics") are the dominant errors.

Berar and Lelann (1991) showed that the underestimation of the estimated standard deviations is due to serial correlations as assessed by the Durbin-Watson statistics and modified the summation to adjust the estimated standard deviations. An important problem arises when reflections completely overlap and therefore the error associated with the individual intensity has to be assigned. Using a probabilistic method, David (1987) derived a method to deal with these cases.

## APPENDIX D:
## COMPUTER PROGRAMS FOR RIETVELD ANALYSIS

The best source for up-to-date Rietveld refinements programs and other useful programs for powder diffraction are available on the Commission for Powder Diffraction's homepage *www.za.iucr.org/iucr-top/comm/cpd/rietveld. html*. This site provides access to new and updated program versions and should be consulted by anyone looking for powder diffraction–related software.

The first Rietveld refinement code was written in ALGOL by H. M. Rietveld (1969). Hewat (1973) modified the code and converted it to FORTRAN. PROF, as it became known, was the standard program in the early days to analyze neutron powder pattern. A further offspring of the original Rietveld program was MPROF, which featured a modification to incorporate two phase refinements (Bendall et al., 1983).

**GSAS** Generalized Structure Analysis System, written by Allen C. Larson and Robert B. von Dreele, LANSCE, MS-H805, Los Alamos National Laboratory, Los Alamos, NM 87545. This is one of the most used Rietveld refinement programs in the world, even though, as its name implies, it is a multipurpose program suite. It runs under OPENVMS/VAX, OPENVW/ALPHA, Silicon Graphics IRIX, Ultrix, HPUX, and DOS. This program allows the refinement of fixed wavelength as well as time-of-flight data. It furthermore allows the combined refinement of various powder and single- crystal data sets as well as neutron and x-ray diffraction data and has all the "mesoscopic" features incorporated such as particle size and microstrain as well as a recently added feature for the refinement of texture using spherical harmonics.

Website: *http://www.mist.lansce.lanl.gov*

Author: *vondreele@lanl.gov*

**FULLPROF** Written by Juan Rodriguez-Carvajal (unpublished). It is also originally based on DBW3.2 but incorporates, among many extensions, the capability to refine microstrain and particle size in a generalized manner. It runs on Vax machines under VMS, on Suns, IBM Riscs, HP Unix, and Silicon Graphics as well as PCs. This program is especially useful for refining magnetic structures. The formula for the magnetic structure factor and its derivatives with respect to the parameters was programmed in a general fashion. The Fourier coefficients can be given either as standard crystallographic or spherical components. The program generates satellite reflections of up to 24 propagation vectors per phase and refines

their components in reciprocal lattice units (Rodriguez-Carvajal, 1993). This program allows the user to perform many nonstandard refinements because of user-supplied executables that can be linked to the FULLPROF library. FULLPROF also allows refinements using the maximum-likelihood technique.

Internet address of author: *juan@bali.scalay.cea.fr*

UNIX-anonymous ftp areas where FULLPROF is stored: *www.bali.scalay.cea.fr/pub/diverse/fullp/freadme ftp://charybde.saclay.cea.fr/pub/divers/fullprof98*

**LHPM** Written by Hill and Howard. It is based on Wiles and Young's DBW3.2 program and features an improved peak asymmetry, absorption correction, multiphase refinement, and a Voigt peak shape (Hill and Howard, 1987).

Anonymous ftp: *atom.ansto.gov.au/pub/physics/ neutron/rietveld*

Author: *bah@atom.ansto.gov.au* (Brett Hunter)

**RIETAN** Written by F. Izumi. The program runs on a UNIX platform. It is also a modern Rietveld refinement program that allows the refinement of incommensurate structures as well as superstructures. Furthermore, it also allows the choice of least-squares algorithms (Gauss-Newton, modified Marquardt, and conjugate direction) at the beginning of each refinement cycle, which is very helpful to assure smoother convergence especially in the early stages of a refinement.

Anonymous ftp: *nirim.go.jp/pub/sci/rietan*

Author: *izumi@nirim.go.ip*

**XRS-82** Written by C. Baerlocher (1986). It is based on the X-Ray 72 system. This code is particularly interesting for scientist working on the structures of zeolites since it has elaborate constraints and restraints with various stereochemical aspects as well as a learned profile function (Baerlocher, 1993).

**PROFIL** Written by J. K. Cockcroft. It is a modern Rietveld refinement that was not based on any older versions of code but written completely from scratch. It is a very fast and easy-to-use program resembling SHELX in the way its free format input file is constructed. The output files are very detailed and contain very useful crystallographic information, such as conversion of the various displacement parameters, distances, and angles with estimated standard deviations, correlation matrices, and resolution curves converted in FWHM versus $2\theta$. This is a program suited for the nonspecialist. It allows the focus to remain on crystallography. Various output files feed into other programs. It also contains a full-profile-refinements option based on LeBails algorithm, which generates input files for use in SHELX-82 direct methods and Patterson techniques.

Anonymous ftp: *gordon.cryst.bbk.ac.uk*

It should be pointed out that refining neutron powder diffraction data with different Rietveld refinement programs leads to differences that sometimes exceed the determined errors. This was shown by Woodward et al. (1995) when comparing the results of refinements of the same data sets of triclinic and monoclinic $WO_3$ using the three Rietveld refinement programs GSAS, Rietan, and

PROFIL. Subtle differences in peak shapes, the different low-angle asymmetry corrections for axial divergence, the way the background is accounted for, and the way errors are calculated lead to these discrepancies. In another study, Hill (1992) reported on the results of a Rietveld round robin. The main conclusion was again that the refinement of neutron data showed a rather narrow distribution for the coordinates and anisotropic displacement parameters about the single-crystal values and that the errors are substantially smaller than those of interrefinement variations.

THOMAS VOGT
Brookhaven National
Laboratory
Upton, New York

# SINGLE-CRYSTAL NEUTRON DIFFRACTION

## INTRODUCTION

Single-crystal neutron diffraction measures the elastic Bragg reflection intensities from crystals of a material, the structure of which is the subject of investigation. A single crystal is placed in a beam of neutrons produced at a nuclear reactor or at a proton accelerator-based spallation source. Single-crystal diffraction measurements are commonly made at thermal neutron beam energies, which correspond to neutron wavelengths in the neighborhood of 1 Å. For high-resolution studies requiring shorter wavelengths ($\sim$0.3 to 0.8 Å), a pulsed spallation source, or a high-temperature moderator (a "hot source") at a reactor may be used. When complex structures with large unit-cells are under investigation, as is the case in structural biology, a cryogenic-temperature moderator (a "cold source") may be employed to obtain longer neutron wavelengths ($\sim$4 to 10 Å). A monochromatic beam technique is generally used at reactors, which normally operate as continuous wave (CW) neutron sources. At spallation sources, which normally are pulsed, a time-of-flight Laue method with a broad range of neutron wavelengths is generally used, and the neutron wavelength is determined from the time-of-flight for neutrons to reach the detector. The type of moderator is again generally optimized for the particular application, in a manner similar to what is the common practice at reactors.

A single-crystal neutron diffraction analysis will determine the crystal structure of the material, typically including its unit cell and space group, the positions of the atomic nuclei and their mean-square displacements, and relevant lattice site occupancies. Because the neutron possesses a magnetic moment, the magnetic structure of the material can be determined as well from the magnetic contribution to the Bragg intensities (see MAGNETIC NEUTRON SCATTERING).

Instruments for single-crystal diffraction (single-crystal diffractometers or SCDs) are generally available among the facilities at the major neutron-scattering centers. Beam time on many of these instruments is available

through a proposal mechanism. For a listing of neutron SCD instruments and their corresponding facility contacts, see Table 1.

## Complementary and Related Techniques

A number of techniques give information on crystal structure complementary to that provided by single-crystal neutron diffraction. Some of the most important of these are listed below along with brief descriptions. For additional information on these techniques, and their advantages and disadvantages, see the corresponding chapters of *Characterization of Materials* in which they are described. It is worth noting that neutron diffraction is a very expensive proposition, first of all involving the often demanding and time-consuming preparation of large, high-quality crystals (at least 1 mm$^3$ in volume). Then it is generally necessary to travel to a neutron scattering research center to perform the measurements, which typically require from several days to several weeks. One therefore should be certain that neutron diffraction is the technique of choice for solving the problem at hand— otherwise one of the complementary methods should be considered first.

Single-crystal x-ray diffraction is the method most directly complementary to single-crystal neutron diffraction. The chief difference between the two methods is that, while neutron diffraction images the nuclear scattering density in the crystal, x-ray diffraction images the electron-density distribution. Neutron diffraction therefore has important advantages in studies where hydrogen or other light atoms must be located, or where isotopic substitutions may be of interest, and of course for magnetic structures as well. Magnetic structures can also be studied using x-rays, although there the effect is much smaller, so that synchrotron radiation is generally required. In the special case where the x-ray energy is close to an absorption edge of a magnetic center in the sample, resonance enhancement of the magnetic scattering can produce effects comparable in magnitude to those observed with neutrons. X-ray and neutron diffraction measurements can be combined to good effect in studies seeking detailed information about the valence electron-density distribution in a material, particularly when hydrogen is present (Coppens, 1997).

Powder neutron (NEUTRON POWDER DIFFRACTION) and powder x-ray (X-RAY POWDER DIFFRACTION) diffraction are also powerful techniques for determining crystal structure. For many important classes of materials single crystals may be difficult or even impossible to obtain. Powder diffraction methods have the advantage of wide applicability and generally allow for much more rapid data collection than do single-crystal methods. This makes powder methods ideal for studies of materials over ranges of temperature and pressure, for example, in studies investigating phase transitions, and for in situ real-time studies. The chief disadvantage of powder methods stems from the more limited amount of information obtained in the diffraction pattern, which generally places an upper limit on the complexity of structures that can be satisfactorily treated. Progress in this regard continues at a rapid

**Table 1. Neutron SCD Instruments and Facility Website Contacts**

| Facility | Source Type | Location | Instrument Name | Instrument Characteristics[a] | Proposal Cycle | Website URL |
|---|---|---|---|---|---|---|
| BENSC | Reactor | Hahn-Meitner Institute, Berlin, Germany | E5 | Monochromatic thermal beam | Semiannual | http://www.hmi.de/bensc |
| FRJ-2 | Reactor | Jülich Research Center, Jülich, Germany | SV28 | Monochromatic thermal beam | Consult web site | http://www.kfa-juelich. de/iff/Institute/ins/ Broschuere_NSE/ |
| HIFAR | Reactor | ANSTO, Lucas Heights, NSW, Australia | 2TANA | Monochromatic thermal beam | Consult web site | http://www.ansto.gov.au/ natfac/hifar.html |
| HFIR | Reactor | Oak Ridge National Laboratory, Oak Ridge, Tenn., U.S.A. | HB-2A | Monochromatic thermal beam | Continuous | http://www.ornl.gov/hfir |
| HFR | Reactor | Institut Laue-Langevin, Grenoble, France | D9 | Monochromatic hot beam | Semiannual | http://www.ill.fr |
| | | | D10 | Monochromatic 4-circle and triple axis | | |
| | | | D19 | Monochromatic thermal beam | | |
| | | | LADI | Laue with imaging plate Weissenberg geometry | | |
| IBR-2 | Reactor | Frank Laboratory of Nuclear Physics, Dubna, Russia | DN2 | Monochromatic thermal beam | Consult web site | http://nfdfn.jinr.dubna.su |
| IPNS | Spallation | Argonne National Laboratory, Argonne, IL, U.S.A. | SCD | Time-of-flight Laue | Semiannual | http://www.pns.anl.gov |
| ISIS | Spallation | Rutherford Appleton Laboratory, Chilton, Oxon, U.K. | SXD | Time-of-flight Laue | Semiannual | http://www.isis.rl.ac.uk |
| JRR-3 | Reactor | Japan Atomic Energy Research Institute, Tokai-mura, Japan | BIX-III | Monochromatic thermal beam with imaging plate. Oscillation geometry | Consult web site | http://www.jaeri.go.jp/ english/ |
| KENS | Spallation | KEK Laboratory, Tsukuba, Japan | FOX | Time-of-flight Laue | Annual | http://neutron-www. kek.jp/ |
| LANSCE | Spallation | Los Alamos National Laboratory, Los Alamos, N.M., U.S.A. | SCD | Time-of-flight Laue | Consult web site | http://www.lansce. lanl.gov |
| | | | PCS | Time-of-flight Laue protein crystallogra phy station | — | http://www.lansce. lanl.gov |
| NRU | Reactor | Chalk River Laboratories, Chalk River, Ontario, Canada | E3 | Monochromatic thermal beam | Continuous | http://neutron.nrc.ca |
| Orphée | Reactor | Laboratoire Léon Brillouin, Saclay, France | 5C2 | Monochromatic hot beam | Semiannual | http://www-llb.cea.fr |
| R2 | Reactor | Studsvik Neutron Research Laboratory, Nyköping, Sweden | SXD | Monochromatic thermal beam | Every 4 months | http://www.studsvik. uu.se |
| SINQ | CW Spallation | Paul Scherer Institute, Villigen, Switzerland | TriCS | Monochromatic thermal beam | Semiannual | http://www.psi.ch |

[a]The instruments are 4-circle diffractometers unless otherwise noted.

pace, however, so that structures with on the order of 100 independent atoms in the crystallographic asymmetric unit may now be considered as suitable candidates for powder studies. Recently, powder-diffraction studies of proteins have even been undertaken, fitting highly constrained structure models to the data (Von Dreele, 1999). Both powder and single-crystal x-ray techniques have been enormously enhanced by the availability of synchrotron x-ray sources, with their high-intensity beams. At the third-generation synchrotron sources, x-ray intensities may be as much as nine or ten orders of magnitude larger than for the most currently available intense neutron beams.

Electron diffraction (Section 11c) is another complementary technique for crystal-structure determination—particularly when studying surfaces and interfaces, where low-energy electron diffraction (LOW-ENERGY ELECTRON DIFFRACTION) is especially powerful, or when only microcrystals are available. Progress in dynamical computations that take multiple scattering into account (n-beam dynamical diffraction computations) is extending the power of electron diffraction for the solution and refinement of crystal structures.

Nuclear magnetic resonance (NUCLEAR MAGNETIC RESONANCE IMAGING) is also a valuable technique for investigating the structure of materials. To cite just one application, high-field solid-state NMR can yield accurate distances between hydrogen atoms and—particularly in materials with high thermal motion or disorder—results may sometimes be more reliable than for comparable interatomic distances obtained from neutron diffraction.

## Scope of This Unit

This unit outlines the basic principles of single-crystal neutron diffraction and provides examples to illustrate the range of materials that can be studied. Sufficient practical details are included, it is hoped, to help a non-expert get started with the technique and locate an appropriate neutron diffraction facility. A listing of the instruments that are available at various neutron sources is included in Table 1.

For readers desiring additional background, the classic monograph by Bacon (1975) provides an excellent starting point. Wilson (1999) has recently published an overview that includes a comprehensive survey of neutron diffraction results on molecular materials. For those wishing to investigate the subject in more detail, Marshall and Lovesey (1971) provide an exhaustive theoretical treatment of all aspects of neutron scattering.

## PRINCIPLES OF THE METHOD

Single-crystal neutron diffraction measures elastic, coherent scattering, i.e., the Bragg reflection intensities (see X-RAY AND NEUTRON DIFFUSE SCATTERING MEASUREMENTS). Although the measurements and their interpretation are generally quite straightforward, and amount to a standard crystal-structure analysis analogous to what is done with x rays, neutron structure determinations are still rela-

tively uncommon. This is so because of the limited number of neutron crystallography facilities and the need to obtain large samples in order to observe adequate reflection intensities (see Complementary and Related Techniques). Thus, for example, the contents of a recent release of the Cambridge Structural Database of organic and organometallic crystal structures (Allen and Kennard, 1993) consist of less than 1% neutron-determined structures (the overwhelming majority of the remaining 99% are x-ray-determined structures).

### Basic Theory of Single-Crystal Neutron Diffraction

The nuclear scattering density in a crystal may be expressed in terms of the neutron structure factors, $F(h,k,l)$, by the familiar relationship

$$\rho(x,y,z) = \Sigma F(h,k,l)\exp\{-2\pi i(hx + ky + lz)\} \quad (1)$$

where the Fourier summation (Equation 1) runs over the Miller indices $h,k,l$, and the coordinates $x,y,z$ are expressed in terms of fractions of the unit-cell translations $\mathbf{a}, \mathbf{b}, \mathbf{c}$. The nuclear scattering density is simply the atomic probability density at any point multiplied by the corresponding neutron scattering amplitude.

As noted above, neutrons are scattered by the atomic nuclei in a crystal, while x-rays are scattered by the electrons. Because the nuclear radius is $\sim 10^4$ times smaller than the neutron wavelengths that are used in diffraction, the neutron scattering amplitude is a constant independent of the momentum transfer, $Q = 4\pi(\sin\theta/\lambda)$. For x rays, on the other hand, the scattering amplitude, also called the scattering factor or form factor, falls off with increasing $Q$ due to destructive interference of the scattering from different regions of the electron-density cloud. Each isotope has a unique neutron scattering amplitude, but these values lie within the same order of magnitude across the entire periodic table, as illustrated in Figure 1. For x-rays, the scattering amplitude increases with the atomic number of atoms, so that diffraction is dominated by contributions from the high-Z atoms. For this reason, neutron diffraction has important advantages in studies where hydrogen or other light atoms must be located, or

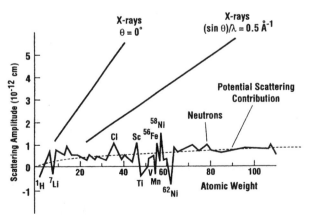

**Figure 1.** Neutron scattering amplitudes as a function of atomic weight. Adapted from Bacon (1975).

where isotopic substitutions may be of interest, as has been mentioned earlier. Further, because neutron diffraction images the nuclear scattering density in the crystal, it provides an unbiased estimate of the mean nuclear position. X-ray diffraction studies generally assume that the nucleus is located at the centroid of its electron-density cloud. This can be a poor assumption, e.g., where hydrogen atoms form covalent bonds. For example, C-H bond distances determined by x-ray diffraction are systematically shortened by ∼0.1 Å due to the effects of bonding on the hydrogen $1s$ electron-density distribution.

It is, of course, not possible to obtain the scattering-density function $\rho(x, y, z)$ directly from the observed diffraction intensities, because of the crystallographic phase problem. In practice, an initial estimate of the neutron structure factor phases is usually obtained from a prior x-ray diffraction study. Probabilistic direct methods of phase determination developed for x-ray structure determination have, however, also been shown to work in the neutron case, e.g., for organic and organometallic crystals (Broach et al., 1979). Direct methods seem to work in practice in spite of the existence of negative scattering density whenever hydrogen is present, which actually contradicts one of the underlying assumptions of direct methods. Phase-determination methods based on anomalous dispersion can also be applied in neutron diffraction for special cases involving crystals that contain one of a handful of highly absorbing elements, including Li, B, Cd, Sm, and Gd (see, e.g., Koetzle and Hamilton, 1975, and references cited therein).

Neutron structure analysis and refinement generally utilizes standard kinematical diffraction theory. According to kinematical theory (see KINEMATIC DIFFRACTION OF X RAYS), the Bragg reflection intensity $I(h,k,l)$ is given by the relationship

$$I(h, k, l) \propto (i_o k \lambda^2 V_c / V^2) |F(h, k, l)|^2 \quad (2)$$

where $i_o$ is the incident neutron flux, $V_c$ is the crystal volume, $V$ is the unit-cell volume, and $|F(h, k, l)|^2$ is the squared structure factor. The relationship in Equation 2 holds for the monochromatic beam case. For the Laue case, $i_o$ is a function of $\lambda$ reflecting the neutron flux spectrum incident on the sample. Equation 2 omits the Lorenz term, which must be introduced to correct for the geometrical effect caused by scanning reflections to obtain integrated Bragg intensities.

The structure factor is

$$F(h, k, l) = \Sigma a_i b_i \exp(-W_i) \exp\{2\pi i(hx + ky + lz)\} \quad (3)$$

where $a_i$ is the atomic site occupancy, $b_i$ is the neutron scattering amplitude, and $W_i$ is the Debye-Waller factor, and the summation in Equation 3 runs over $n$ atoms in the crystallographic unit cell.

From Equations 2 and 3, it is apparent that the diffracted intensity will fall off as the unit-cell volume increases. In order to achieve sufficient diffracted intensity, it is therefore necessary to grow larger single crystals for materials with larger unit cells. The use of longer neu-

tron wavelengths is also highly desirable in this case, e.g., by performing the experiment at an instrument located on a cold moderator beamline.

The Debye-Waller factor $W_i$ takes into account the reduction in diffracted intensity from interference produced by smearing of the scattering density due to thermal motion in the sample. Equation 4 is the isotropic approximation

$$W_i = \exp\{-U_i Q^2 / 2\} = \exp\{-8\pi^2 U_i (\sin^2\theta / \lambda^2)\} \quad (4)$$

where $U_i$ is the atomic mean-square displacement (adp). It is often the practice to introduce an anisotropic harmonic model in which second-order tensors are employed to describe the adps. In that case, the Debye-Waller factor is given by

$$\begin{aligned} W_i = \exp\{-2\pi^2 (&U_{11} a^{*2} h^2 + U_{22} b^{*2} k^2 \\ &+ U_{33} c^{*2} l^2 + 2U_{12} \boldsymbol{a}^* \bullet \boldsymbol{b}^* hk + 2U_{13} \boldsymbol{a}^* \bullet \boldsymbol{c}^* hl \\ &+ 2U_{23} \boldsymbol{b}^* \bullet \boldsymbol{c}^* kl \end{aligned} \quad (5)$$

where $\boldsymbol{a}^*$, $\boldsymbol{b}^*$, and $\boldsymbol{c}^*$ are reciprocal lattice vectors. Higher-order tensor descriptions may be used to model anharmonic effects.

Neutron diffraction is often the method of choice in studies aimed at a precise description of thermal motion and/or disorder because it is easier to determine the Debye-Waller factor in the neutron case where the atomic scattering amplitude is a constant independent of $Q$ (see above), and there is therefore minimal correlation between the Debye-Waller factor and the site occupancy. In contrast, the x-ray atomic form factor may assume a $Q$ dependence quite similar to that of the Debye-Waller factor, leading to high correlations in the least-squares minimization approach to structure refinement.

That the neutron scattering amplitude is independent of $Q$ confers an additional advantage not always well appreciated. When diffraction measurements are made at cryogenic temperatures, thereby lowering the adp parameters, neutron Bragg intensities will exhibit a much reduced fall-off with $Q$ compared to the x-ray case. This results in substantially improved precision in the atomic positions.

The standard kinematical theory outlined above ignores the effects of absorption and extinction in real crystals. Absorption is often quite minimal except in the case of crystals that contain hydrogen, where the effective absorption due to hydrogen incoherent scattering can be quite appreciable. The measured neutron intensities can be corrected for absorption using the method of Gaussian integration (Busing and Levy, 1957) or by employing an analytical technique in which the crystal is described in terms of Vornoi polyhedra (Templeton and Templeton, 1973).

The use of large crystals often requires that extinction corrections be included during neutron structure refinement to obtain a satisfactory fit to the strong reflections. A number of formalisms have been developed for this purpose. Perhaps the most widely used is that of Becker and Coppens (1975), which in its most general implementation treats both primary and secondary extinction.

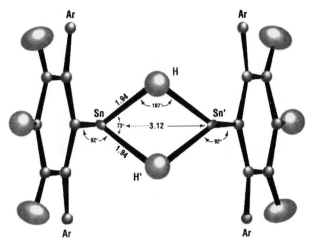

**Figure 2.** The molecular structure of a tin(II) hydride complex determined by a neutron diffraction study on IPNS SCD at 20 K (Koetzle et al., 2001). Estimated standard deviations in bond distances and angles are less than 0.01 Å and 0.1°, respectively.

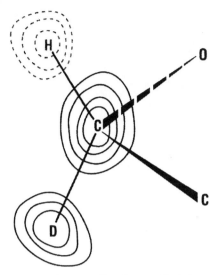

**Figure 3.** Contoured neutron Fourier map for (+)-neopentyl-1-d-alcohol (S) with positive nuclear scattering density at C and D shown as solid contours and negative nuclear scattering density at H shown as dashed contours (Yuan et al., 1994).

## PRACTICAL ASPECTS OF THE METHOD

The initial applications of single-crystal neutron diffraction to the study of molecular structure were carried out at first generation research reactors around 1950. For additional information in this regard, see, e.g., the studies of $KHF_2$ (Peterson and Levy, 1952) and ice (Peterson and Levy, 1953, 1957) at Oak Ridge National Laboratory and the study of $KH_2PO_4$ (KDP) at Harwell (Bacon and Pease, 1953, 1955). KDP was also studied independently at about the same time at Oak Ridge (Peterson et al., 1953, 1954). These studies provided an important demonstration of the advantages of neutron diffraction for determining the positions of hydrogen atoms and for studying hydrogen bonding.

Single-crystal methods are being employed at modern reactor and spallation neutron sources to study the structures of a broad range of materials. For example, the tin hydride complex shown in Figure 2 was investigated (Koetzle et al., 2001) using the SCD spectrometer at the Intense Pulsed Neutron Source (IPNS) at Argonne National Laboratory. Collection of Bragg intensity data

at a temperature of 20 K for this triclinic crystal with 111 independent atoms, unit-cell volume $V = 1986$ Å$^3$, and crystal volume $V_c = 15$ mm$^3$ required 11 days on SCD. Low-temperature studies of this quality allow full, unconstrained refinement of the structure model incorporating anisotropic Debye-Waller factors and readily yield metal-hydrogen bond distances with a precision of better than 0.01 Å. The amount of beam time used by this experiment is quite representative at present at IPNS for this type of organometallic crystal.

Figure 3 shows an application of Fourier methods to single-crystal neutron diffraction data obtained at the Brookhaven National Laboratory High Flux Beam Reactor (HFBR; note that the HFBR neutron source ceased operation in 1999). In this site-specific isotope labeling study of a chiral organic alcohol (Yuan et al., 1994), the neutron Fourier map readily gives the location of the deuterium label that was introduced in an enzymatic hydrogenation reaction.

Figure 4 shows the results of a study of the variation with temperature of the anharmonic thermal motion in the cubic form of ZnS (zincblende; Moss et al., 1980,

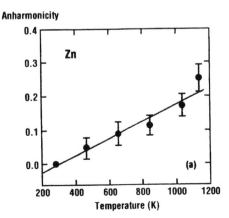
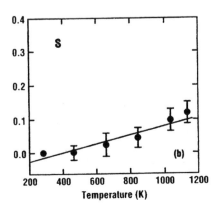

**Figure 4.** Anharmonicity of atomic displacements in zincblende (ZnS; m.p. 1973 K; cubic phase) as a function of temperature (Moss et al., 1980, 1983).

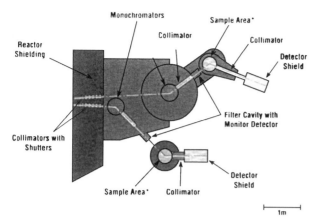

**Figure 5.** Schematic plan of the SCD spectrometers formerly installed at the dual thermal beam port H6 at the Brookhaven High-Flux Beam Reactor (HFBR).

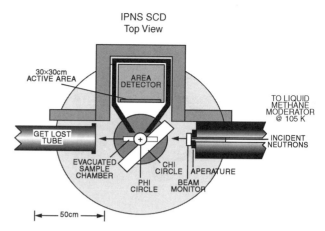

**Figure 6.** Schematic plan of the SCD spectrometer installed on a liquid methane moderator at the Argonne Intense Pulsed Neutron Source (IPNS).

1983). This high-resolution study, also carried out on one of the single-crystal diffractometers at the HFBR, used a relatively short neutron wavelength of 0.83 Å and successfully modeled the higher-order contributions to the Debye-Waller factor for a zincblende sample drawn from the collections of the Smithsonian Institution, Washington, D.C.

### Neutron Sources and Instrumentation

As mentioned earlier, neutron sources fall into two categories: nuclear reactors, which normally operate as CW neutron sources, and spallation sources, which normally are pulsed. A pulsed nuclear reactor (IBR-30) operates at the Frank Laboratory of Nuclear physics in Dubna, Russia, and a CW spallation source (SINQ) operates at the Paul Scherer Institute in Villigen, Switzerland. A listing of neutron scattering centers offering single-crystal diffraction facilities that will host users from other organizations is given in Table 1.

Figure 5 schematically illustrates a representative SCD set-up at a reactor. Using this facility, which was formerly installed on a dual thermal beam port at the HFBR at Brookhaven, neutron beams in the wavelength range 0.8 to 1.7 Å could be extracted using a variety of monochromator crystals. Common choices for monochromators include Be, Si, Ge, Cu, and pyrolitic graphite. The incident beam horizontal divergence could be selected to match the mosaic spread of the sample using a Soller slit collimator, and a circular aperture was employed to limit the beam diameter to approximately twice the dimensions of the sample to minimize the background. Samples were mounted on one of the two four-circle diffractometers with Eulerian cradles that provided for full orientation control. The samples were generally centered in the incident beam using the observed setting angles for an azimuthal reflection. This enabled a precise centering to be carried out with the sample inside a cryostat or other specialized environment. Integrated Bragg intensities were recorded by scanning the samples in steps about the vertical ($\omega$) axis, counting for a preset incident-beam monitor

count to correct for any variation in the incident neutron flux from the reactor over time. The addition of position-sensitive area detectors has significantly enhanced the performance of many of the SCDs currently in operation at reactors (see discussion of Detectors, below). In particular, the recent emergence of neutron imaging-plate detectors and the development of improved glass scintillation detectors have been important advances.

Figure 6 is a schematic view of the SCD at Argonne's IPNS, which serves as an illustration of an instrument at a spallation neutron source. This diffractometer (Schultz, 1987), which is installed at a beam port on a liquid methane cryogenic moderator, operates by the time-of-flight Laue method (see X-RAY AND NEUTRON DIFFUSE SCATTERING MEASUREMENTS). The full spectrum of neutrons from the moderator is allowed to impinge on the sample, and the wavelength is determined by the time-of-flight required for neutrons to reach the detector, i.e., $\lambda = ht/ml$, where $h$ is Planck's constant, $t$ is the neutron time-of-flight, $m$ is the neutron mass, and $l$ is the source-to-detector distance.

The Eulerian cradle mount is used to orient the sample for a series of intensity data collection histograms, recording $t$ and the position for each neutron arriving at the area detector. Counts are accumulated for a preset number of accelerator pulses. The IPNS proton accelerator operates at 30 Hz and a proton beam power of 7 kW. A depleted uranium spallation target is employed at present. With the geometry shown in Figure 6, a complete hemisphere of reciprocal space can be explored while collecting a total of 45 data histograms. In contrast to monochromatic beam methods, where the sample must be rotated to scan through Bragg reflections in order to obtain integrated intensities, in the Laue method the sample is held stationary while counting is in progress. This significantly simplifies the experiment and may be of particular advantage, for example, when a special sample environment is required or when carrying out rapid reciprocal-lattice surveys, e.g., while searching for phase transitions over a range of temperature and/or pressure.

In general, each neutron scattering center has developed its own instrumentation, and the design of each SCD spectrometer is unique in its details. From the prospective experimenter's point of view, however, most SCDs probably will look quite similar in the sense that they will feature an Eulerian cradle sample mount that offers the possibility of deploying cryostats, furnaces, pressure cells, etc., to obtain a wide range of sample environments. As was mentioned earlier, a successful experiment requires production of a large, well-formed single crystal. Generally, the minimum satisfactory sample size falls in the range 1 to 10 mm$^3$, with larger samples required for larger unit cells. These large samples are required because even at the current most powerful spallation neutron source, the 180 kW ISIS at the Rutherford Appleton Laboratory in the U.K., or at the most powerful high-flux reactor sources, the 58 MW High Flux Reactor (HFR) at the Institut Laue-Langevin in Grenoble, France, and the 85 MW High Flux Isotope Reactor (HFIR) at Oak Ridge National Laboratory in the U.S.A., the neutron flux on sample at the SCDs is several orders of magnitude reduced compared to the flux from a conventional, sealed-tube laboratory x-ray source.

Bent crystal focusing monochromators (see, e.g., Tanaka et al., 1999b, 2000) have been installed at a number of reactor SCD instruments. Significant intensity enhancement has been achieved in this way, in some cases approaching one order of magnitude.

### Detectors

The first generation of SCD instruments at reactors utilized single-channel gas proportional counters filled with $BF_3$ or $^3He$. These detectors capture neutrons by the reactions $^{10}B + ^1n = ^7Li + ^4He + 2.8$ MeV, and $^3He + ^1n = ^1H + ^3H + 0.76$ MeV, respectively. The thermal neutron capture cross-section of $^3He$ is ~1.5 times that of $^{10}B$, and for this reason $^3He$ counters are usually lighter and smaller than $BF_3$ counters.

Modern neutron SCD instruments generally include two-dimensional position-sensitive detectors, allowing for parallel data collection. Area detectors are, in any event, required for the time-of-flight Laue instruments at pulsed spallation sources. Neutron position-sensitive detectors are of three types: gas-filled proportional counters, generally using $^3He$; scintillation counters, such as the $^6Li$-glass detector on the SCD at IPNS (see Fig. 6); and imaging-plate detectors, which utilize a Gd converter. Neutron imaging plates (Niimura et al., 1994, Tazaki et al., 1999) have the important advantage of being flexible, which facilitates covering a large solid angle. Powerful new SCDs utilizing imaging plates and oscillation/Weissenberg geometry have been developed since 1990 (LADI, Wilkinson and Lehmann, 1991; Cipriani et al., 1995; BIX, Tanaka et al, 1999a). Imaging plates are integrating detectors and are therefore, unfortunately, not suitable for use on time-of-flight Laue instruments.

### Future Outlook

The application of neutron scattering in general, including single-crystal diffraction, has always been constrained by the limited number of research facilities available. With the advent of the 2 MW Spallation Neutron Source (SNS), now under construction at Oak Ridge and scheduled to begin operation in 2006, the neutron scattering community will have at its disposal a state-of-the-art facility of unprecedented power and intensity, with approximately an order of magnitude increase over the power currently delivered by ISIS. Planning for major new spallation facilities is at an advanced stage in Europe, with the European Spallation Source (ESS) project, and also in Japan (JHF project). The SNS is expected to make possible a broad range of new SCD applications. For example, it should be possible to do the following.

1. Use much smaller single crystals, perhaps approaching normal x-ray sized samples (~0.01 mm$^3$ in volume), which will greatly expand the range of materials that can be investigated.
2. Obtain complete SCD data sets in minutes, as opposed to the days or weeks that are presently required. This should enable the study of many interesting processes in real time. To cite just one example of interest, it should be possible to investigate solid-state photochemical reactions where the mechanisms can be probed by following site-specific isotope labels (Ohgo et al., 1996, 1997).
3. Carry out extensive parametric studies, including investigating materials over a broad range of temperatures and pressures, which, in practice, at existing facilities, can only be done conveniently with powders. This would be extremely important for many materials applications, e.g., to understand the response of hydrogen-bonded ferroelectric, ferroelastic, and nonlinear optical (NLO) materials to temperature and pressure changes.

A critical challenge facing the neutron scattering community is how to replace the facilities at the current generation of research reactors. The majority of these facilities were commissioned in the 1950s and 1960s, and some are nearing the end of their projected lifetimes. Two new research reactor facilities are currently under development, at Garching, Germany, where the 20 MW FRM-II is scheduled to be operational in 2002, and at ANSTO, Lucas Heights, Australia, where a 20-MW replacement for the HIFAR reactor is scheduled to be completed in 2005.

### DATA ANALYSIS AND INITIAL INTERPRETATION

As discussed above (see Principles of the Method), initial values of the neutron structure factor phases are usually calculated based on a structure model obtained from a prior x-ray diffraction study. Although the neutron phases can be obtained by direct methods (see above), the x-ray study provides an efficient structure solution and serves the important dual purpose of allowing a check for problems, such as twinning, that might make it difficult to obtain a high-quality neutron structure. Most neutron scattering facilities strongly recommend that prospective SCD users have an x-ray structure of their material in hand before embarking on their neutron measurements.

Each SCD instrument generally provides its users with its own specially adapted suite of programs to handle data reduction, i.e., for Bragg peak indexing, integration of peak intensities, absorption correction, and merging of equivalent reflections. For Laue data, equivalent reflections are generally not merged, since they may be collected at different neutron wavelengths and their extinction corrections may therefore differ one from another. For information on data reduction software, the reader should consult the facility websites listed in Table 1. These websites generally identify the responsible instrument scientists who can advise prospective users on what to expect.

Refinement of neutron structures may be carried out with a standard crystallographic program package. Program systems in common use at neutron scattering centers include GSAS (*http://www.ncnr.nist.gov/programs/crystallography/software/gsas.html*; Larson and Von Dreele, 2000) and SHELX (*http://shelx.uni-ac.gwdg.de/SHELX/*; Sheldrick, 1997).

## SAMPLE PREPARATION

As has been noted above, single-crystal neutron diffraction requires a large, high-quality crystal, generally at least $1 \text{ mm}^3$ in volume. There usually is no need for the crystal to be deuterated. This contrasts with the situation for neutron powder diffraction (see NEEUTRON POWDER DIFFRACTION), where gram-quantities of sample generally are required, and the material must be completely deuterated to avoid prohibitively high background from incoherent scattering of hydrogen. Single-crystal samples are normally mounted on an aluminum or vanadium pin with a suitable adhesive. Vanadium is sometimes used to reduce the background because of its extremely small neutron scattering amplitude. Aluminum will usually suffice, however, so long as the diameter of the pin is substantially smaller than that of the crystal. The personnel responsible for the facility's SCD instrument can advise as to the proper type of pin to use. Often the most suitable course is to bring unmounted samples to the neutron scattering center and to use the local laboratory facilities for final sample preparation.

Provision for on-site sample-handling facilities is frequently an important factor to consider when choosing a neutron scattering center. For example, prospective users with air-sensitive samples will want to be certain that facilities exist for handling and mounting their crystals under an inert atmosphere. The availability of well-equipped conventional laboratory facilities adjacent to the neutron source can critically affect the outcome of an experiment.

## PROBLEMS

The most common reason for failure of a single-crystal neutron diffraction experiment is poor crystal quality.

Ideally, smaller crystals from the same crystallization batch as the neutron samples will be examined first by x-ray diffraction, to assure their quality and to ascertain that the crystals scatter to adequate resolution (in $Q$). Optical examination of the samples under a polarizing microscope frequently may help to reject twinned specimens and agglomerates.

Once a neutron sample has been chosen, a careful assessment should be made of its mosaic spread along the principal crystal axes. At reactor instruments, this is usually done using $\omega$ scans, while at spallation-source instruments it is accomplished by examining preliminary data histograms that are collected to determine the sample orientation. A crystal with noticeably split diffraction profiles often will give unsatisfactory results. It is also important to avoid overlap in neighboring reflections that may result in high mosaic spreads, i.e., greater than about $2°$ in $\omega$, particularly for crystals with large unit cells.

For proper estimation of intensities, it is critical that position-sensitive detectors be calibrated to remove effects of nonuniform response and to estimate detector dead-time losses, if any. At spallation-source instruments, the detector calibration must include allowance for the wavelength dependence of detector efficiency. Detector calibrations generally will have been carried out by the instrument scientist, placing an incoherent scatterer such as a vanadium rod in the sample position, and the resulting corrections will be applied automatically during data reduction.

Background corrections are also important, especially when data are acquired using a position-sensitive detector. A very successful and widely used technique that employs a variable background mask adjusted to maximize $I/\sigma(I)$, where $I$ is the integrated intensity and $\sigma(I)$ its estimated standard deviation (Lehmann and Larsen, 1974), has been adapted to treat such data.

As noted above, the ideal kinematical expression for the Bragg reflection intensity given in Equation 2 ignores the effects of absorption and extinction that are present in real crystals. In particular, extinction is often important in neutron diffraction because of the requirement for large, well-formed crystals. Zachariasen (1967) gave an early treatment of extinction. Coppens and Hamilton (1970) extended the Zachariasen formalism to treat crystals with anisotropic extinction. Becker and Coppens (1974a,b, 1975) made further improvements, while Thornley and Nelmes (1974) treated the case of highly anisotropic extinction. For single-crystal neutron studies, it is important to have access to a crystallographic least-squares structure refinement program that allows for extinction. This should not pose a serious problem, however, since many modern refinement programs include the popular Becker-Coppens extinction formalism. When extinction is particularly severe, high correlations may occur in the least-squares refinement between extinction parameters and some of the adps, or more uncommonly even with the atomic positions. In these cases, special care should be taken, e.g., by employing restraints or damping factors, and by carefully testing the various extinction models to determine which one best fits the observed intensity data.

An absorption correction should always be applied to the observed Bragg intensities before attempting to treat extinction in the least-squares refinement. It is preferable to measure the crystal, and index its faces, before beginning data collection. In practice, however, many samples are approximated as spheres, e.g., when the material is air-sensitive and therefore difficult to measure. Fortunately, absorption is usually quite modest for neutrons and actually rarely exceeds around 50%. An exception might be for crystals containing highly absorbing elements such as Li, B, Cd, Sm, and Gd, where it also may be important to include anomalous dispersion corrections. For many samples, however, the majority of the absorption will result from the reduction in the effective incident-beam intensity due to incoherent scattering of hydrogen. Howard et al. (1987) have determined the wavelength dependence of the hydrogen incoherent scattering cross-section, which is needed for proper absorption correction of time-of-flight Laue data from spallation sources.

## ACKNOWLEDGMENTS

The author would like to thank Dr. Arthur Schultz for reading a draft of this manuscript and providing comments. Work at Argonne and Brookhaven National Laboratories was supported by the U.S. Department of Energy, Office of Basic Energy Sciences, under Contracts W-31-109-ENG-38 and DE-AC2-98CH10886, respectively. Additional financial support was provided to the author by NATO under grant PST.CLG.976225.

## LITERATURE CITED

Allen, F. H. and Kennard, O. 1993. 3D search and research using the Cambridge structural database. *Chem. Des. Autom. News* 8:31–37.

Bacon, G. E. 1975. Neutron Diffraction (Third Edition). Oxford University Press, London.

Bacon, G. E. and Pease, R. S. 1953. A neutron diffraction study of potassium dihydrogen phosphate by Fourier synthesis. *Proc. Roy. Soc.* A220:397–421.

Bacon, G. E. and Pease, R. S. 1955. A neutron-diffraction study of the ferroelectric transition of potassium dihydrogen phosphate. *Proc. Roy. Soc.* A230:359–381.

Becker, P. J. and Coppens, P. 1974a. Extinction within the limit of validity of the Darwin transfer equations. I. General formalisms for primary and secondary extinction and their application to spherical crystals. *Acta Cryst.* A30:148–153.

Becker, P. J. and Coppens, P. 1974b. Extinction within the limit of validity of the Darwin transfer equations. II. Refinement of extinction in spherical crystals of $SrF_2$ and LiF. *Acta Cryst.* A30:148–153.

Becker, P. J. and Coppens, P. 1975. Extinction within the limit of validity of the Darwin transfer equations. III. Non-spherical crystals and anisotropy of extinction. *Acta Crystallogr.* A31:417–425.

Broach, R. W., Schultz, A. J., Williams, J. M., Brown, G. M., Manriquez, J. M., Fagan, P. J., and Marks, T. J. 1979. Molecular structure of an unusual organoactinide hydride complex determined solely by neutron diffraction. *Science* 203:172–174.

Busing, W. R. and Levy, H. A. 1957. High speed computation of the absorption correction for single crystal diffraction measurements. *Acta Crystallogr.* 10:180–187.

Cipriani, F., Castagna, J.-C., Lehmann, M. S., and Wilkinson, C. 1995. A large image-plate detector for neutrons. *Physica B* 213–214:975–977.

Coppens, P. 1997. X-ray Charge Densities and Chemical Bonding. Oxford University Press, London.

Coppens, P. and Hamilton, W. C. 1970. Anisotropic extinction corrections in the Zachariasen approximation. *Acta Cryst.* A26:71–83.

Howard, J. A. K., Johnson, O., Schultz, A. J., and Stringer, A. M. 1987. Determination of the neutron absorption cross-section for hydrogen as a function of wavelength with a pulsed neutron source. *J. Appl.Cryst.* 20:120–22.

Koetzle, T. F. and Hamilton, W. C. 1975. Neutron diffraction study of NaSmEDTA.8H$_2$O: An evaluation of methods of phase determination based on three-wavelength anomalous dispersion data. *In* Anomalous Scattering (S. Ramaseshan and S.C. Abrahams, eds.). pp. 489–502. Munksgaard, Copenhagen.

Koetzle, T. F., Henning, R., Schultz, A. J., Power, P. P., Eichler, B. E., Albinati, A., and Klooster, W. T. 2001. Neutron Diffraction Study of the First Known Tin(II) Hydride Complex, [2,6-Trip$_2$C$_6$H$_3$Sn($\mu$-H)]$_2$, Trip = 2,4,6-tri-isopropylphenyl. XX European Crystallographic Meeting, Krakow, Poland, 2001, Abstract O.M2.05.

Larson, A. C. and Von Dreele, R. B. 2000. General Structure and Analysis System (GSAS). Los Alamos National Laboratory Report LAUR 86-7481, Los Alamos, New Mexico.

Lehmann, M. S. and Larsen, F. K. 1974. A method for location of the peaks in step-scan-measured Bragg reflections. *Acta Cryst.* A31:580–584.

Marshall, G. W., and Lovesey, S. W. 1971. Theory of Thermal Neutron Scattering. Oxford University Press, London.

Moss, B., McMullan, R. K., and Koetzle, T. F. 1980. Temperature dependence of thermal vibrations in cubic ZnS: A comparison of anharmonic models. *J. Chem. Phys.* 73:495–508.

Moss, B., Roberts, R. B., McMullan, R. K., and Koetzle, T. F. 1983. Comment on "temperature dependence of thermal vibrations in cubic ZnS: A comparison of anharmonic models." *J. Chem. Phys.* 78:7503–7505.

Niimura, N., Karasawa, Y., Tanaka, I., Miyahara, J., Takahashi, K., Saito, H., Koizumi, S., and Hidaku, M. 1994. An imaging plate neutron detector. *Nucl. Instr. Meth. Phys. Res.* A349:521–525.

Ohgo, Y., Ohashi, Y., Klooster, W. T., and Koetzle, T. F. 1996. Direct observation of hydrogen-deuterium exchange reaction in a cobaloxime crystal by neutron diffraction. *Chem. Lett.* 445-446, 579.

Ohgo, Y., Ohashi, Y., Klooster, W. T., and Koetzle, T. F. 1997. Analysis of hydrogen-deuterium exchange reaction in a crystal by neutron diffraction. *Enantiomer* 2:241–248.

Peterson, S. W. and Levy, H. A. 1952. A single crystal neutron diffraction determination of the hydrogen position in potassium bifluoride. *J. Chem. Phys.* 20:704–707.

Peterson, S. W. and Levy, H. A. 1953. A single-crystal neutron diffraction study of heavy ice. *Phys. Rev.* 92:1082.

Peterson, S. W. and Levy, H. A. 1957. A single-crystal neutron diffraction study of heavy ice. *Acta Crystallogr.* 10:70–76.

Peterson, S. W., Levy, H. A., and Simonsen, S. H. 1953. Neutron diffraction study of tetragonal potassium dihydrogen phosphate. *J. Chem. Phys.* 21:2084–2085.

Peterson, S. W., Levy, H. A., and Simonsen, S. H. 1954. Neutron diffraction study of the ferroelectric modification of potassium dihydrogen phosphate. *Phys. Rev.* 93: 1120-1121.

Schultz, A. J. 1987. Pulsed neutron single-crystal diffraction. *Trans. Am. Crystallogr. Assoc.* 23:61–69.

Sheldrick, G. M. 1997. SHELX-97. Program for the Refinement of Crystal Structures using Single Crystal Diffraction Data. University of Göttingen, Germany.

Tanaka, I., Ahmed, F. U., and Niimura, N. 2000. Application of a stacked elastically bent perfect Si monochromator with identical and different crystallographic planes for single crystal and powder neutron diffractometry. *Physica B* 283:195–298.

Tanaka, I., Kurihara, K., Haga, Y., Minezaki, Y., Fujiwara, S., Kumazawa, S., and Niimura, N. 1999a. An upgraded neutron diffractometer (BIX-$I_M$) for macromolecules with a neutron imaging plate. *J. Phys. Chem. Solids* 60:1623–1626.

Tanaka, I., Niimura, N., and Mikula, P. 1999b. An elastically bent silicon monochromator for a neutron diffractometer. *J. Appl. Crystallogr.* 32:525–529.

Tazaki, S., Neriishi, K., Takahashi, K., Etoh, M., Karasawa, Y., Kumazawa, S., and Niimura, N. 1999. Development of a new type of imaging plate for neutron detection. *Nucl. Instr. Meth. Phys. Res.* A424:20–25.

Templeton, D. H. and Templeton, L. K. 1973. Am. Crystallogr. Assoc. Meeting Abstracts, Abstract E10, Storrs, Connecticut.

Thornley, F. R. and Nelmes, R. J. 1974. Highly anisotropic extinction. *Acta Cryst.* A30:748–757.

Von Dreele, R. B. 1999. Combined Rietveld and stereochemical restraint refinement of a protein crystal structure. *J. Appl. Crystallogr.* 32:1084–1089.

Wilkinson, C. and Lehmann, M. S. 1991. Quasi-Laue neutron diffractometer. *Nucl. Instr. Meth. Phys. Res.* A310:411–415.

Wilson, C. C. 1999. Single Crystal Neutron Diffraction from Molecular Materials. World Scientific, Singapore.

Yuan, H. S. H., Stevens, R. C., Bau, R., Mosher, H. S., and Koetzle, T. F. 1994. Determination of the molecular configuration of (+)-Neopentyl-1-d alcohol by neutron and x-ray diffraction analysis. *Proc. Natl. Acad. Sci. U.S.A.* 91:12872–12876.

Zachariasen, W. H. 1967. A general theory of x-ray diffraction in crystals. *Acta Cryst.* 23:558–564.

## KEY REFERENCES

Bacon, 1975. See above.

*A classic treatment of the principles and practice of neutron diffraction. Part 4 is devoted to a description of experimental techniques.*

Marshall and Lovesey, 1971. See above.

*Exhaustive treatment of all aspects of neutron scattering, including inelastic and quasielastic scattering, as well as diffraction. Sometimes referred to as "the bible" of neutron scattering.*

Stout, G. H. and Jensen, L. H. 1989. X-ray Structure Determination: A Practical Guide, 2nd ed. John Wiley & Sons, New York.

*An excellent basic crystallography text outlining the techniques used in crystal structure analysis.*

Willis, B. T. M. and Pryor, A. W. 1975. Thermal Vibrations in Crystallography. Cambridge University Press.

*Provides a comprehensive overview of the use of crystallographic techniques to analyze thermal motion and disorder in solids.*

Wilson, 1999. See above.

*A compendium describing the current state-of-the-art and giving a review of the literature including many recent results in neutron crystal structure analysis.*

THOMAS F. KOETZLE
Brookhaven National
Laboratory
Upton, New York and
Argonne National Laboratory
Argonne, Illinois

# PHONON STUDIES

## INTRODUCTION

Until the beginning of this century, the models used to understand the properties of solids were based on the assumption that atomic nuclei are fixed at their equilibrium positions. These models have been very successful in explaining many of the low-temperature properties of solids, and especially those of metals, whose properties are determined to a large extent by the behavior of the conduction electrons. Many well-known properties of solids (such as sound propagation, specific heat, thermal expansion, conductivity, melting, and superconductivity), however, cannot be explained without considering the vibrational motion of the nuclei around their equilibrium positions. The study of lattice dynamics—or of phonons, which are the quanta of the vibrational field in a solid—was initiated by the seminal papers of Einstein (1907, 1911), Debye (1912), and Born and von Kármán (1912, 1913). Initially, most of the theoretical and experimental studies were devoted to the study of phonons in crystalline solids, as the name lattice dynamics implies. At present, the field encompasses the study of the dynamical properties of solids with defects, surfaces, and amorphous solids. It is impossible in a single unit to present even a brief overview of this field. The discussion that follows will therefore be limited to a brief outline of the neutron scattering techniques used for the study of phonons in crystalline solids. Additional information can be found in many books devoted to this subject (see Bibliography). Particularly useful is the four-volume book *Dynamical Properties of Solids* edited by Horton and Maradudin (1974), and several parts in *Neutron Scattering*, Vol. 23 of *Methods of Experimental Physics*, edited by Sköld and Price (1986).

Indirect information about the lattice dynamical properties of solids can be obtained from a variety of macroscopic measurements. For instance, in some cases, the phonon contribution to the specific heat or resistivity can be easily separated from other contributions and can provide important information about the lattice dynamical properties of a solid. In addition, measurements of sound velocities are particularly useful, since they provide the slopes of the acoustic phonon branches. Since such macroscopic measurements are relatively easy and inexpensive to make in a modern materials science laboratory, they should, in principle, be performed before one undertakes a detailed phonon study by spectroscopic techniques.

Direct measurement of the frequencies of long-wavelength optical phonons can be obtained by Raman scattering and infrared spectroscopy (Cardona, 1975–1991; Burstein, 1967). These experiments provide essential information for solids containing several atoms per unit cell and should be performed, if possible, before one undertakes a time-consuming and expensive detailed neutron scattering study. Also, for simple solids, information about the phonon dispersion curves can be obtained from x-ray diffuse scattering experiments.

Presently, the most powerful technique for the study of phonons is the inelastic scattering of thermal neutrons. The technique directly determines the dispersion relation, that is, the relationship between the frequency and propagation vector of the phonons. Its power was demonstrated by Brockhouse and Stewart (1955), who, following a suggestion put forward by Placzek and Van Hove (1954), measured the dispersion relation of aluminum.

## PRINCIPLES OF THE METHOD

Given the potential seen by the nuclei of a solid, the description of their motion is reduced to the study of the oscillations of a system of particles around their equilibrium positions, which is a well-known problem in mechanics. The basic problem here is to determine the potential seen by the nuclei. If we assume that the electrons follow nuclear motion adiabatically, the effective potential $\phi(R)$ seen by the nuclei is written as,

$$\phi(R) = V(R) + V_e(R) \qquad (1)$$

where $R$ collectively denotes the nuclear positions, $V(R)$ is the nuclear interaction energy, and $V_e(R)$ is the average electronic energy determined with the nuclei fixed at $R$.

For a given potential, the motion of the nuclei can be immediately determined if we assume that the displacements of the nuclei from their equilibrium positions are small compared to their separation (the harmonic approximation). This approximation is very good, except for solids consisting of light nuclei, such as solid $^4$He. If the harmonic approximation is valid, it is well known from mechanics that any nuclear motion can be considered as a superposition of a number of monochromatic waves, the normal vibrational modes or characteristic vibrations of the system. For a system containing $N$ nuclei, the number of normal modes is $3N - 6$, the number of vibrational degrees of freedom of the system; for a solid $3N - 6 \cong 3N$, since the number of nuclei in this case is very large. The frequencies of the normal modes are obtained by diagonalizing the force constant matrix, a relatively simple problem in the case of a molecule. In an amorphous solid, one is faced with the formidable problem of diagonalizing a $3N \times 3N$ matrix. With present advances in computer technology, such direct diagonalizations are becoming possible and play an increasingly important role in the study of the vibrational properties of amorphous solids.

The translational symmetry of crystalline solids considerably simplifies the determination of the frequencies of their normal modes. In this case, the problem is reduced to the determination of the motion of the atoms contained in a unit cell. The normal modes are propagating vibrational waves with the propagation wave vector $\mathbf{q}$ determined by the crystal geometry. If the unit cell contains $r$ nuclei, their frequencies $\omega_j(\mathbf{q})$ are obtained by diagonalizing a $3r \times 3r$ force constant matrix, which is a more tractable problem than the one encountered in amorphous solids. The relation

$$\omega = \omega_j(\mathbf{q}) \qquad (j = 1, 2, \ldots, 3r) \qquad (2)$$

between the frequency and wave vector is called the dispersion relation. It consists of $3r$ branches labeled by the index $j$. Among these branches, there are three acoustic branches whose frequencies approach zero at the long-wavelength limit ($\mathbf{q} \to 0$). In this limit, the dispersion of the acoustic branches is linear, $\omega = v_j\mathbf{q}$, where $v_j$ is the appropriate sound velocity in the solid and the propagating vibrational waves in the solid are simply sound waves. The remaining $3r - 3$ branches tend to a finite frequency as $\mathbf{q} \to 0$ and are called optic branches, since some of these vibrational modes interact with light. For each wave vector of a branch, the motion of the nuclei in the cell is determined by solving the equations of motion. The pattern of motion in the cell is specified by the polarization vectors $\mathbf{e}_d^j(\mathbf{q})(d = 1, 2, \ldots, r)$, which provide the direction of motion of the nuclei in the cell. A vibrational mode is called longitudinal or transverse if the polarization vectors are parallel and perpendicular, respectively, to the propagation vector $\mathbf{q}$. The allowed values of the propagation vector $\mathbf{q}$ are determined by the boundary conditions on the surface of the crystal. The bulk properties of the crystal are not influenced by the specific form of the boundary conditions. For convenience, periodic boundary conditions are usually adopted. Because of the periodicity of the crystal, all physically distinct values of $\mathbf{q}$ can be obtained by restricting the allowed values of $\mathbf{q}$ in one of the primitive cells of the reciprocal lattice. Actually, instead of the primitive cell of the reciprocal lattice, it is more convenient to use the first Brillouin zone, which is the volume enclosed by planes that are the perpendicular bisectors of the lines joining a reciprocal lattice point to its neighboring points.

In the classical description outlined above, any nuclear motion can be considered as the superposition of propagating waves with various propagation vectors and polarizations. The analogy with the classical description of the electromagnetic field should be noted. As in the case of the electromagnetic field, the quantum description is easily obtained. In the harmonic approximation, the motions of the nuclei can be decoupled by a canonical transformation to normal coordinates. By this transformation, the nuclear Hamiltonian is reduced to a sum of Hamiltonians corresponding to independent harmonic oscillators, and the quantization of the vibrational field is reduced to that of the harmonic oscillator. Thus, instead of the classical wave description, we have the quantum description in terms of quanta, called phonons, that propagate through the lattice with definite energy and momentum. The energy $E_j(\mathbf{q})$ and momentum $\mathbf{p}$ of the phonon are related to the frequency and wave vector of the corresponding vibrational wave by the well-known relations

$$E_j(\mathbf{q}) = \hbar\omega_j(\mathbf{q}) \qquad \mathbf{p} = \hbar\mathbf{q} \qquad (3)$$

where $\hbar = h/2\pi$ is Planck's constant.

The free motion of noninteracting phonons corresponds to the free propagation of the monochromatic waves in the harmonic approximation. The phonons have finite lifetimes, however, since anharmonic interaction as well as interaction with other elementary excitations in the solid are always present. At thermal equilibrium, brought about by the above mentioned interactions, the average number of phonons, $n_j$, is given by the Bose-Einstein relation

$$n_j = \frac{1}{\exp(E_j/k_\mathrm{B}T) - 1} \qquad (4)$$

where $k_\mathrm{B}$ is Boltzmann's constant.

Notice that at high temperatures $(E_j/k_\mathrm{B}T \ll 1)$ the number of phonons is proportional to the temperature and inversely proportional to their energy.

From the measured phonon frequencies, one can evaluate the frequency distribution or phonon spectrum, which can be used to calculate the thermodynamic properties of the solid, such as the lattice specific heat. The frequency distribution, $Z(\omega)$, is defined so that $Z(\omega)\Delta\omega$ is the fraction of vibrational frequencies in the interval between $\omega$ and $\omega + \Delta\omega$.

$$Z(\omega) = \frac{1}{3N}\sum_{j,\mathbf{q}}\delta[\omega - \omega_j(\mathbf{q})] \qquad (5)$$

where $\delta$ is Dirac's delta function.

As mentioned earlier, the most detailed information about the lattice dynamical properties of solids is obtained by studying the scattering of thermal neutrons by a crystalline specimen of solid. This is because the energy and wave vector of thermal neutrons are of the same order as those of the normal vibrations in a solid.

The scattering cross-section consists of an incoherent and a coherent part. The most detailed information about the lattice dynamical properties of the solid is obtained by measuring the coherent one-phonon inelastic neutron scattering cross-section from a single crystal. In such studies, incoherent and multiphonon processes are of importance only because their contribution to the background scattering may complicate observation of the one-phonon coherent processes.

The coherent cross-section for scattering of the neutron to a final state characterized by a wave vector $\mathbf{k}_1$ (and energy $E_1 = \hbar k_1/2m$) due to the creation or annihilation of a single phonon of frequency $\omega_j(\mathbf{q})$ is essentially (see, e.g., Chapter 1 of Sköld and Price, 1986) given by

$$\frac{k_1}{k_0}\frac{\left[n_j(\mathbf{q}) + \frac{1}{2} \pm \frac{1}{2}\right]}{\omega_j(\mathbf{q})}|F(\mathbf{Q})|^2\delta[E_1 - E_0 \pm \hbar\omega_j(\mathbf{q})]\delta(\mathbf{Q}\mp\mathbf{q} - \tau) \qquad (6)$$

In this equation,

$$\mathbf{Q} = \mathbf{k}_0 - \mathbf{k}_1 \qquad (7)$$

is the scattering vector; $[n_j(\mathbf{q}) + \frac{1}{2} \pm \frac{1}{2}]$ is the population factor with $n_j(\mathbf{q})$ given by Equation 4, the upper and lower signs correspond to phonon creation and phonon annihilation, respectively; the delta functions assure conservation of energy and momentum in the scattering process

$$\mathbf{Q} = \mathbf{q} + \tau \qquad E_1 - E_0 = \mp\hbar\omega_j(\mathbf{q}) \qquad (8)$$

and the inelastic structure factor $F(\mathbf{Q})$ is defined by

$$F(\mathbf{Q})\sum_d M_d^{-1/2}b_d\mathbf{e}_d^j(\mathbf{q})\cdot\mathbf{Q}e^{i\mathbf{Q}\cdot\mathbf{d}}e^{-W_d} \qquad (9)$$

with the sum extending over the atoms of the unit cell. In this equation, $M_d$ is the mass of the $d$th atom in the unit cell, $b_d$ its coherent neutron scattering length, $\mathbf{d}$ its position vector, and $e^{-W_d}$ the Debye-Waller factor.

## PRACTICAL ASPECTS OF THE METHOD

The study of the coherent one-phonon inelastic neutron scattering from a single crystal is the most powerful technique for detailed investigation of the lattice dynamics of crystalline solids. Simple inspection of the coherent one-phonon neutron scattering cross-section (Equation 6) provides valuable information for the practical implementation of the technique.

The cross-section is inversely proportional to the phonon frequency and proportional to the population factor $(n_j + \frac{1}{2} \pm \frac{1}{2})$. Thus, for the low-lying modes $(\hbar\omega/kT < 1)$, the intensity is inversely proportional to the square of the phonon frequency. The population factor is $n_j$ for processes with neutron energy gain (phonon annihilation) and $(n_j + 1)$ for processes with neutron energy loss (phonon creation). Since $n_j(\mathbf{q})$ rapidly decreases with increasing $\hbar\omega/kT$, experiments with neutron energy loss are generally preferred. The $k_1/k_0$ factor in the cross-section, which favors energy-gain processes, cannot outweigh the gain in intensity due to the population factor for the energy-loss processes. Measurements with neutron energy gain, however, can be very useful in checking, whenever necessary, the results obtained by energy-loss measurements. Also, in experiments using neutrons of very low-incident energies (e.g., 5 meV), one is restricted in most cases to the observation of energy-gain processes. It should also be noted that because of the population factor, the intensity of the observed neutron groups usually increases with increasing sample temperature. Actually, for the low-lying modes $[(\hbar\omega/kT) \ll 1]$, the intensity of the neutron groups is proportional to the sample temperature. Because of this gain in intensity with increasing temperature, most detailed studies of the dispersion curves have been performed at room or higher temperatures. The decrease in the Debye-Waller factor $e^{-2W}$ with increasing temperature, with few exceptions, only moderates the increase of the intensity with increasing temperature. At high temperatures, however, the increase in background scattering and the changes in the shape and width of the observed neutron groups, resulting from anharmonic effects, may considerably complicate the measurements.

The intensity of a neutron group corresponding to a certain mode depends on the crystal structure of the material through the square of the inelastic structure factor. It can be seen that there are definite advantages to performing the measurements, whenever possible, in mirror symmetry planes—such as the (110) and (001) planes in the case of cubic crystals. In fact, if the neutron scattering vector $\mathbf{Q}$ lies in a mirror plane, in principle no one-phonon scattering can be observed from modes polarized perpendicularly to the mirror plane. This considerably simplifies the task of identifying the modes with polarization vectors lying in the plane. In addition, the frequencies and polarization vectors of these modes vary slowly in the vicinity of and perpendicular to the mirror plane. As a result, the vertical collimation of the incident and scattered neutron beams can usually be much more relaxed than the horizontal collimation, thus obtaining more intensity without any significant loss in the accuracy of the measurements. The polarization dependence of the inelastic structure factor is exploited in practice to differentiate between various normal modes and to adopt the best experimental arrangement for their detection. For instance, if in a certain direction the modes are by symmetry purely longitudinal or transverse, the experimental configuration can be arranged so that the intensity is maximized for either the longitudinal or the transverse modes.

To illustrate these elementary considerations, let us consider the simple case of a monatomic face-centered cubic (fcc) structure like Cu. In this case, the modes along the [100], [110], and [111] directions are, by symmetry, purely longitudinal or transverse. Also, by symmetry, the transverse branches along the [100] and [111] directions are degenerate. The transverse branches along the [110] direction, on the other hand, are not degenerate and usually are denoted by $T_2$ (polarization parallel to the cube edge) and $T_1$ (polarization parallel to a face diagonal). Most of the measurements of the dispersion curves in these directions can be obtained by studying the scattering in the (110) mirror plane of the crystal. With the crystal in this orientation, all the longitudinal branches, the transverse branches along [111] and [100], and the $T_2$ [110] branch can be obtained. Typical scattering geometries for the detection of longitudinal and transverse [100] modes in the (1$\bar{1}$0) scattering plane are indicated in Figure 1. The $T_1$ (110) branch can be observed by measurements in the (100) plane of the crystal. With the crystal in this orientation, measurements can be made along the [1$\xi$0] direction as well. These modes, usually designated $\Lambda$ and $\Pi$, are not purely longitudinal or transverse. The dispersion curves of Cu are given in Figure 2. Notice that the measurements along the [110] direction were extended beyond the zone boundary to point $X$, where the $L$[110] branch is degenerate with $T$[100] and $T_2$[110] is degenerate with the $L$[100].

In simple cases, such as that of Cu, the symmetry of the structure determines the polarization vectors in high-symmetry directions and allows selection of the proper experimental arrangements. In more complicated struc-

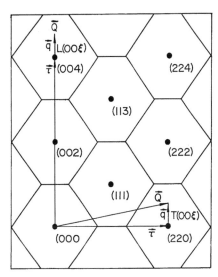

**Figure 1.** Neutron scattering amplitudes as a function of atomic weight. Adapted from Bacon (1975).

tures with unit cells containing several atoms, planning of experiments is difficult, if not impossible, without some prior knowledge regarding the inelastic structure factor. In these cases, one proceeds by adopting a Born–von Kármán model with forces extending to a couple of nearest neighbors. By assigning reasonable values to the force constants (using, e.g., information available about the lattice dynamics of the system, such as measured values of the elastic constants), the structure factors can be evaluated. These calculated inelastic structure factors can then be used to plan the experimental measurements. As soon as some data have been obtained, the force constants can be readjusted, and a more realistic set of intensities can be calculated to serve as a guide for the selection of experimental conditions.

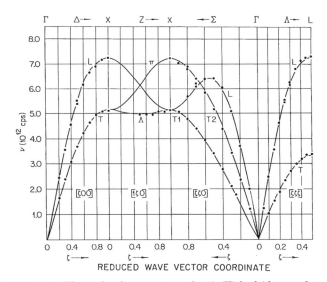

**Figure 2.** The molecular structure of a tin(II) hydride complex determined by a neutron diffraction study on IPNS SCD at 20 K (Koetzle et al., 2001). Estimated standard deviations in bond distances and angles are less than 0.01 Å and 0.1°, respectively.

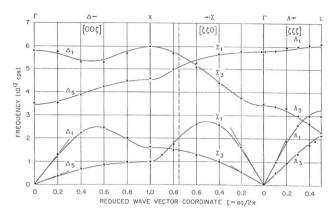

**Figure 3.** Contoured neutron Fourier map for (+)-neopentyl-1-d-alcohol (S) with positive nuclear scattering density at C and D shown as solid contours and negative nuclear scattering density at H shown as dashed contours (Yuan et al., 1994).

To illustrate this point, let us consider AgCl, which is a slightly more complicated case than that of Cu. This ionic crystal has the sodium chloride structure, and the Bragg intensities with even and odd Miller indices are proportional, respectively, to $(b_{Ag} + b_{Cl})^2$ (strong reflections) and $(b_{Ag} - b_{Cl})^2$ (weak reflections). By symmetry, the high-symmetry-direction branches are purely longitudinal or transverse. At $\Gamma$, the two atoms in the unit cell move in phase and out of phase, respectively, for the acoustic and optical branches. It is easily seen by using Equation 9 that, close to $\Gamma$, measurements for the acoustic and optical branches should be performed around strong and weak reflections, respectively—a rule of thumb applicable to systems with two atoms per unit cell. This simple rule is only valid close to the zone center, however. In this particular experiment, the measurements away from $\Gamma$ were planned using shell-model calculations of the inelastic structure factor. The measured dispersion curves of AgCl are given in Figure 3. Notice the similarity of the acoustic branches to those of the fcc structure, in particular the degeneracies at the point $X$.

The inelastic structure factor, and therefore the intensity, increases as $Q^2$, although this increase is moderated at large $Q$ values by the Debye-Waller factor. This decrease of the Debye-Waller factor is not usually significant. The case of quantum crystals, such as $^4$He, is a notable exception. In these crystals, the displacements of the atoms from their equilibrium positions are so large that the decrease with increasing $Q$ of the Debye-Waller factor makes the observation of phonons impossible except around a few reflections with relatively small Miller indices. Thus, with some exceptions, such as the case of $^4$He mentioned above, the measurements must be performed around reflections with large Miller indices.

Once the experimental conditions have been defined, following the general principles just outlined, the intensity of the observed neutron groups is determined by the intensity of the neutron source and the resolution of the instrument. Since the intensities vary inversely as a relatively high power ($\sim$4) of the overall resolution of the instrument, one must adopt a compromise between high resolution and

reasonable intensity. With the neutron fluxes available in present-day sources, the energy resolution for a typical experiment is of the order of 1%, a relatively poor resolution compared to those ordinarily obtained in light-scattering experiments. There are essentially two methods allowing the determination of the neutron energy with such precision: (1) Bragg scattering from a single crystal (used as monochromator or analyzer) and (2) measuring the time of flight of the neutrons, chopped into pulses, over a known distance. A large number of instruments have been designed utilizing the first, the second, or a combination of these methods.

The choice of a particular instrument depends on the type of problem to be investigated. For single crystals, the region of interest in ($\mathbf{Q}\omega$) space is considerably reduced by the symmetry of the crystal. Most of the relevant information regarding the lattice dynamics can be obtained by measuring a relatively small number of phonon frequencies at points or along lines of high symmetry of reciprocal space. For this type of experiment, one needs to perform a variety of rather specific scans in ($\mathbf{Q}\omega$) space. It is generally accepted that, for these types of measurements, the triple-axis spectrometer has a definite advantage over other instruments. For incoherent scattering experiments, on the other hand, time-of-flight instruments are more convenient. These instruments are also preferred in coherent scattering experiments involving the measurement of a large number of frequencies in a relatively extended region of ($\mathbf{Q}\omega$) space (as, e.g., in amorphous solids and complex compounds with a very large primitive unit cell).

The arguments presented here are mainly valid for instruments installed at nuclear reactors. Time-of-flight instruments are, of course, the natural machines to use with a pulsed source. Several time-of-flight machines, such as the multianalyzer crystal instrument and the constant-$Q$ spectrometer, are presently being used at pulsed sources for coherent inelastic scattering experiments. Of these instruments, only the multianalyzer crystal instrument can reach the versatility of the triple-axis spectrometers (operating at continuous neutron sources) for coherent neutron scattering studies. Since the various instruments and their characteristics are described in detail in recent reviews (Dolling, 1974; Windsor, 1981, 1986), we present here only a few remarks regarding the use of the triple-axis spectrometer.

**Triple-Axis Spectrometry**

The design of present-day triple-axis spectrometers has changed little from that originally proposed by Brockhouse (1961). A schematic diagram of such an instrument is shown in Figure 4. A monochromatic beam of neutrons, obtained by Bragg reflection (through an angle $2\theta_M$) from a monochromator crystal, is scattered by the sample (through an angle $\phi$) and the energy of this scattered beam is determined by Bragg scattering (through an angle $2\theta_A$) from an analyzer crystal. The orientation of the sample is defined by the angle $\chi$ between a certain crystal axis and the incident beam. The overall resolution of the instruments is governed by the collimators $C_1$, $C_2$, $C_3$, and $C_4$

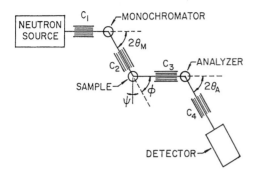

**Figure 4.** Anharmonicity of atomic displacements in zincblende (ZnS; m.p. 1973 K; cubic phase) as a function of temperature (Moss et al., 1980, 1983).

and the mosaic spreads of the crystals (monochromator, analyzer, and sample).

In some designs, no collimator ($C_1$) between the source and the monochromating crystal is used, since the natural collimation of the incident beam due to the relatively large distance between source and monochromator may be sufficient for most practical applications. Some horizontal collimation, however, is desirable, especially if it is easily adjustable, since it provides additional flexibility and reduces the overall background (for most experiments, a horizontal collimation before the monochromator between $40'$ and $1°$ is sufficient). Since the distances between the sample and, respectively, the monochromator and the analyzer crystal are relatively short, it is necessary to introduce collimators ($C_2$, $C_3$) in order to restrict the horizontal divergence of the neutrons incident on and scattered from the specimen. In typical experiments, the horizontal collimation of the incident and scattered beam is $10'$ to $40'$. In most experiments no collimator ($C_4$) is used between the analyzer and the counter; a coarse ($40'$ to $2°$) horizontal collimation can, however, reduce background if this is a serious problem in the measurements. Since most of the measurements in coherent neutron scattering experiments are performed in mirror planes, usually no vertical collimation is necessary. If, however, it is suspected that in a particular experiment scattering by modes off the horizontal scattering plane has been observed, the experiment should be repeated with some vertical collimation.

A low-sensitivity counter (usually a $^{235}$U fission chamber) is placed before the sample to monitor the flux of neutrons incident on the sample. This monitor counter is normally used to control the counting times, so that no corrections are necessary for any changes in the incident neutron flux during the experiment. A second monitor counter is usually placed after the sample to detect Bragg scattering from the specimen, which may give rise to a spurious neutron group by incoherent scattering from the analyzer crystal. The signal counter is usually a high-efficiency (80% to 90%) $^{10}$BF$_3$ gas detector. Recently, the more compact $^3$He gas detectors have replaced the $^{10}$BF$_3$ counters in several installations.

The intensity obtained in an experiment depends critically on the monochromator and analyzer crystals used in the spectrometer. In principle, the nuclei of the material used for a good monochromator (or analyzer) should have negligible absorption and incoherent scattering and fairly large coherent scattering cross-sections. Unfortunately, it is not always possible to obtain large single crystals of sufficient mosaic spread to obtain high reflectivity for all materials with favorable neutron characteristics. There are many methods of improving the reflectivities and other characteristics of the crystals used as monochromators (or analyzers), the most common being the mechanical distortion of the crystal at room or high temperatures. For details on this subject, we refer to the review by Freund (1976). The most useful monochromator (and analyzer) at present is pyrolytic graphite, which is ordered only along the $c$ axis. In typical experiments, the (002) reflection of pyrolytic graphite is used at energies below $\sim 50$ meV. At higher energies the Bragg scattering angles become quite small [because of the relatively large lattice constant ($c = 6.71$ Å) of pyrolytic graphite]. This results in relatively poor energy resolution (since from Bragg's law $\Delta E = 2E \cot \theta_M \Delta \theta_M$). In addition, at these higher energies, the reflectivity decreases because of the scattering by parasitic reflections. Thus, at higher energies Be [usually reflecting from the (002) planes] is often used as a monochromator and/or analyzer in typical experiments. In practice, it is highly desirable to have many good monochromator crystals available in order to optimize intensity and resolution in a particular experiment. In most neutron scattering research centers, pyrolytic graphite as well as Be and (distorted) Ge crystals are available to be used as monochromators (or analyzers).

One of the most important advantages of the triple-axis spectrometer is its versatility. Highly specific scans along selected lines of the $(\mathbf{Q}\omega)$ space can be performed under various experimental conditions, since both the neutron energy and the instrumental resolution can be easily adjusted during the experiments. There are several modes of operation. Irrespective of the mode of operation, however, the experiment determines $\omega$ as a function of $\mathbf{Q}$ (or $\mathbf{q}$). In the experimental plane, the energy and momentum conservation conditions determine the three unknowns, $\omega$ and the two components of $\mathbf{Q}$ along two axes in the scattering plane. Since in a crystal spectrometer there are four independently variable parameters ($E_0$, $E_1$, $\phi$, and $\psi$), there are evidently a large number of ways in which to perform an experimental scan. Usually $E_1$ or $E_0$ is kept fixed during a particular scan.

The triple-axis spectrometer can, like a time-of-flight machine, be operated in the so-called conventional mode. In this mode $\mathbf{k}_0$ (and $E_0$) remain fixed, and the energies of the neutrons scattered at a fixed angle are analyzed by changing the setting of the analyzer crystal. As the magnitude (but not the direction) of $\mathbf{k}_1$ (and $E_1$) is varied, the extremity of the scattering vector moves through reciprocal space causing $\mathbf{q}$ to change, and neutron groups are observed if the energy and momentum conservation conditions are fulfilled. The disadvantage of this method is that it cannot be used to determine phonon frequencies at specified values of the wave vector $\mathbf{q}$, except by interpolation between values obtained in a series of experiments.

To determine the phonon frequencies at preselected values of $\mathbf{q}$, the constant-$\mathbf{Q}$ method is used. The ease with which measurements with constant $\mathbf{Q}$ (and $\mathbf{q}$) can be performed with a triple-axis spectrometer is one of its main advantages over time-of-flight methods. The principle of this method, first introduced by Brockhouse (1960, 1961), is very simple. The choice of $\mathbf{q}$ and the reciprocal lattice point around which the scan is to be performed fix the components, say $Q_x$, $Q_y$, of the vector $\mathbf{Q}$, which is to remain constant during the scan. For instance, by assuming that $E_1$ is to be kept constant during the scan, a range of values for $E_0$ (or equivalently, $2\,\theta_M$) is chosen using a guess for the energy of the phonon. For each of these values of $E_0$, the appropriate values of $\phi$ and $\psi$ are obtained from the momentum conservation conditions. If a neutron group is observed in the scan, its center will determine the frequency of a phonon with propagation vector $\mathbf{q}$. Similarly, the constant-$\mathbf{Q}$ scan can be performed keeping the incident neutron energy $E_0$ fixed. In most cases it is preferable to keep $E_1$ fixed, since in a constant-$E_0$ scan the variation in the sensitivity of the analyzer spectrometer as the scattered neutron energy varies during the scan can distort the observed neutron groups. In a constant-$E_1$ scan, no correction is needed for the $k_1/k_0$ factor in the cross-section since $k_1$ is fixed and $k_0$ is canceled by the $1/v_0$ sensitivity of the monitor counter. Another important advantage of the constant-$\mathbf{Q}$ method is that integrated intensities can be obtained and interpreted more easily than with other methods. This can be easily seen by momentarily neglecting the finite resolution of the instrument. The theoretical intensity is then obtained by simply integrating the theoretical cross-section over the phonon modes observed during an experimental scan. This matter is not trivial (Waller and Fröman, 1952), since $E_0$, $E_1$, $k_0$, and $k_1$, which appear in the delta functions, are related. For a general scan, the result of the integration is to multiply the expression preceding the delta functions by $|\mathbf{J}_j|^{-1}$, where the Jacobian $\mathbf{J}_j$ depends on the slope of the dispersion curve under study. This dependence on the slope of the branch being studied complicates the analysis of the measured intensities. For a constant-$\mathbf{Q}$ scan, on the other hand, $\omega_j(\mathbf{q})$ is independent of the energy transfer, since $\mathbf{q}$ is constant, and the integration is trivial, $\mathbf{J}_j = 1$.

In addition to the conventional and constant-$\mathbf{Q}$ scans, a variety of other scans can be programmed with a triple-axis spectrometer. One of the most useful is the constant-energy-transfer scan. In this type of scan, $E_0$ and $E_1$ are kept fixed and the angles $\phi$ and $\psi$ are varied so that the measurement is performed with constant energy transfer along a predetermined line in reciprocal space. This type of scan is preferred over a constant-$\mathbf{Q}$ scan for very steep dispersion curves or whenever there is a sharp dip in a dispersion curve—as, e.g., in the study of the longitudinal phonon frequencies in the vicinity of $\mathbf{q} = \frac{2}{3}[111]$ of some body-centered cubic (bcc) metals.

Finally, a few comments should be made regarding the resolution of a triple-axis spectrometer and its focusing properties. Assume that the instrument has been set to detect neutrons scattered by the sample with momentum transfer $\mathbf{Q}_0$ and energy transfer $\hbar\omega_0$. Because of the finite collimation of the neutron beams and the finite mosaic spreads of the crystals (monochromator, analyzer, and sample), there is also a finite probability for the instrument to detect neutrons experiencing momentum transfer $\mathbf{Q}$ and energy transfer $\hbar\omega$ differing from $\mathbf{Q}_0$ and $\hbar\omega_0$ by $\Delta\mathbf{Q}$ and $\hbar\Delta\omega$, respectively. This probability is usually denoted by $R(\mathbf{Q} - \mathbf{Q}_0, \omega - \omega_0)$ and defines the resolution or transmission function of the instrument. The intensity observed for a particular setting of the spectrometer $(\mathbf{Q}_0, \omega_0)$ is then simply the convolution of the one-phonon coherent scattering cross-section with the resolution function around $\mathbf{Q}_0, \omega_0$:

$$I(\mathbf{Q}_0, \omega_0) = \int R(\mathbf{Q} - \mathbf{Q}_0, \omega - \omega_0)\sigma(\mathbf{Q}, \omega)d\mathbf{Q}\,d\omega \qquad (10)$$

The course of any particular scan can then be visualized as the stepwise motion of the resolution function through the dispersion surface. If one assumes that the transmission functions of the collimators and single crystals (monochromator, analyzer, and sample) are Gaussian distributions, a relatively simple analytical expression for the resolution of the instrument can be obtained (Cooper and Nathans, 1967; Mller et al., 1970),

$$R(\mathbf{Q} - \mathbf{Q}_0, \omega - \omega_0) = R_0 \exp\left\{-\frac{1}{2}\sum_{k,l=1}^{4} M_{kl}x_k x_l\right\} \qquad (11)$$

where $x_i = \Delta Q_i (i = 1, 2, 3)$, $x_4 = \Delta\omega$, and $R_0$ and $M_{kl}$ are rather complicated expressions specified by the collimations and the mosaic spreads of the crystals. In writing this expression, $x_1$ was taken along $\mathbf{Q}_0$, and $x_3$ was vertical. The surface over which $R = R_0/2$ is represented by

$$\sum_{k,l=1}^{4} M_{kl}x_k x_l = 1.386 \qquad (12)$$

and is referred to as the resolution ellipsoid. Since the experiment is performed in a plane that is usually horizontal, in practically all cases, one is only concerned with the three-dimensional ellipsoid in the $\Delta Q_1 = \Delta Q_\parallel, \Delta Q_2 = \Delta Q_\perp, \Delta\omega$ space (where $\Delta Q_\parallel$ and $\Delta Q_\perp$ denote the components of $\Delta\mathbf{Q}$ parallel and perpendicular to $\mathbf{Q}_0$, respectively). Typically, the resolution ellipsoid is quite elongated along $\Delta\omega$, and it is more elongated along $\Delta Q_\parallel$ than along $\Delta Q_\perp$. The elongated nature of the resolution ellipsoid is responsible for the focusing characteristics of the spectrometer: the width of the observed neutron groups will depend on the orientation of the long axes of the ellipsoid with respect to the dispersion surface. Clearly (see Fig. 5), the sharpest neutron group will be obtained if the experimental configuration is such that the long axis of the resolution ellipsoid is parallel to the dispersion surface. Because of the shape of the ellipsoid, focusing effects are more pronounced for transverse acoustic phonons (with $\Delta_q\omega$ perpendicular to $\mathbf{Q}_0$) than for longitudinal phonons (with $\Delta_q\omega$ parallel to $\mathbf{Q}_0$). In the experiments, the transverse phonons are always measured with a focused configuration. Sometimes, it may also be useful to measure longitudinal phonons not in a purely longitudinal configuration ($\mathbf{Q} \parallel \mathbf{q}$) but in a configuration favorable for focusing. To determine the focused configuration for a given

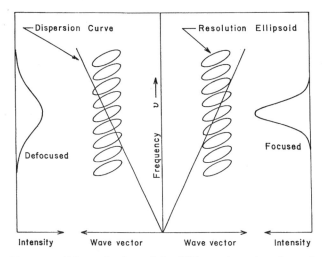

**Figure 5.** Schematic plan of the SCD spectrometers formerly installed at the dual thermal beam port H6 at the Brookhaven High-Flux Beam Reactor (HFBR).

instrument, one usually measures the neutron groups corresponding to a low-lying transverse acoustic phonon with **Q** taken, in the clockwise and counterclockwise sense, respectively, from a selected reciprocal lattice point. The sharpest of the two neutron groups (see Fig. 6) indicates which of the two configurations is focused, and thus establishes in which direction (clockwise or counterclockwise) from a reciprocal point the propagation vector of a transverse phonon (with $\omega$ increasing with increasing **q**) should be taken to exploit focusing.

Computer programs for the evaluation of the resolution function are available in all neutron scattering centers. In many studies (e.g., the determination of intrinsic phonon widths), such detailed calculations of the resolution function are essential. For the planning of most experiments, however, simple estimates of the resolution are sufficient since they provide the information needed for the proper choice of the instrumental parameters. The contribution to the energy resolution of the spectrometer of each collimator and crystal involved can be evaluated, and an estimate of the overall energy resolution can be obtained by taking the square root of the sum of the squares of these contributions. Each contribution, $\Delta\omega$, to the energy resolution is obtained (Stedman and Nilsson, 1966), from

$$\Delta\omega = \Delta\mathbf{k} \cdot \left[\frac{\hbar}{m}\mathbf{k} - \nabla_q\omega(\mathbf{q})\right] \quad (13)$$

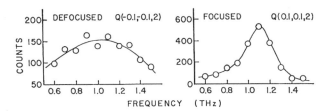

**Figure 6.** Schematic plan of the SCD spectrometer installed on a liquid methane moderator at the Argonne Intense Pulsed Neutron Source (IPNS).

In this equation, **k** stands for the average neutron wave vectors ($\mathbf{k}_0$ for the incident or $\mathbf{k}$, or the scattered beam) and $\Delta\mathbf{k}$ for the spread around these average values introduced by the collimators and finite mosaic spreads of the monochromator and analyzer crystal. To be more specific, let us assume that the monochromator is set to reflect neutrons of wave vector $\mathbf{k}_0$. As a result of the finite collimation $\alpha$ of the beam and the mosaic width $\beta$ of the monochromator, the distribution of incident neutron wave vectors around $\mathbf{k}_0$ has a width $\Delta\mathbf{k}_0 = \Delta\mathbf{k}_{0c} + \Delta\mathbf{k}_{0\eta}$ (where $\Delta\mathbf{k}_{0c}$ from the collimation and $\Delta\mathbf{k}_{0\eta}$ from the mosaic width of the monochromator). It is easily shown, using Bragg's law, that $\Delta\mathbf{k}_{0c}(|\Delta\mathbf{k}_{0c}| = \alpha k_0\theta_M)$ is parallel to the reflecting planes of the monochromator, whereas $\Delta\mathbf{k}_{0\eta}(|\Delta\mathbf{k}_{0\eta}| = \beta k_0 \cot\theta_M)$ is along $\mathbf{k}_0$. Notice that the distribution has a pronounced elongation along $\Delta\mathbf{k}_{0c}$, since $|\Delta\mathbf{k}_{0c}|/|\Delta\mathbf{k}_{0\eta}|$ is typically between 2.5 and 5, a feature alluded to earlier in this section. It can be seen from Equation 13 that to minimize this contribution to the energy resolution, the experimental arrangement should be such that the vector $[(\hbar\mathbf{k}_0/m) - \nabla_q\omega]$ is approximately normal to the monochromator reflecting planes, and this is the focusing condition at the monochromator. When this condition is approximately satisfied, the spread of incident energy around $E_0$ due to the distribution of incident wave vectors (first term in the right-hand side of Equation 13) approximately matches the spread in $\omega$ (second term of Equation 13) brought about by the spread in **q** due to $\Delta k_0$. Similar considerations apply to the scattered beam. In particular, the configuration of the analyzer spectrometer can be chosen to satisfy the focusing condition: $[(\hbar\mathbf{k}_0/m) - \nabla_q\omega]$ parallel to the normal to the analyzer reflecting planes. By using such simple considerations, the optimum focusing configuration can be adopted and a reasonable estimate of the resolution of the instrument can be obtained. The resolution widths calculated by this simple method are accurate to within ~15%, which is sufficient for typical experiments.

## METHOD AUTOMATION

Presently, all neutron scattering spectrometers used for the study of phonons are computer operated. By simple commands, the user can direct the instrument to perform any given scan or a series of scans with different parameters. Of course, software limits must be set so that the spectrometer does not assume a configuration incompatible with the experimental setup. In a triple-axis spectrometer, it is recommended to use limit-switches, in addition to the software limits, to restrict the motion of the monochromator drum, counter, and sample table within the limits set by a particular experimental setup. This is particularly important if the sample is in a cryostat, dilution refrigerator, or superconducting magnet.

In practically all neutron scattering spectrometers, the data can be printed and/or displayed graphically. In addition, in most spectrometers the user can perform a preliminary analysis of the collected data by using the computer programs available at the instrument site.

## DATA ANALYSIS

Over the last 35 years, considerable progress has been made toward understanding of the connection between phonon measurements and the crystalline and electronic structure of solids. Until the pioneering work of Toya (1958), on the dispersion curves of Na, the measured dispersion curves of solids were analyzed in terms of simple phenomenological models based on the Born–von Kármán formalism.

As mentioned earlier, the vibration frequencies of a crystalline solid, in the harmonic approximation, are determined by a $3r \times 3r$ force constant (or dynamical) matrix, which depends on the forces between the nuclei. In a Born–von Kármán model, only forces extending to a few nearest neighbors are introduced. The number of independent force constants one needs to introduce is considerably reduced by the crystal symmetry and the requirement of translational and rotational invariance of the solid as a whole. The force constants introduced are considered as adjustable parameters to be determined by fitting to the experimental data. With few exceptions, however, Born–von Kármán models with a few adjustable parameters cannot adequately describe the measured dispersion curves. This is because of the different ways with which the electrons screen and modify the internuclear Coulomb forces in the various classes of materials. By making reasonable assumptions about the response of the electronic medium, however, a large number of phenomenological models have been advanced to explain the lattice dynamical properties of various classes of solids. A detailed description of these phenomenological models can be found in many books (Born and Huang, 1954; Bilz and Kress, 1979; Hardy and Karo, 1979; Brüesch, 1982; Venkataraman et al., 1975; Horton and Maradudin, 1974) and review articles (Hardy, 1974; Cochran, 1971; Sinha, 1973).

Phenomenological models are extremely useful since they can be used to calculate the phonon density of states and the lattice specific heat. Also, they are used to calculate the inelastic structure factors that are invaluable in planning experimental measurement of the dispersion curves of solids containing several atoms per unit cell. The main drawback of phenomenological models is that in most cases the parameters introduced to characterize the electronic response to the nuclear motion have no clear physical interpretation. In many cases, it is difficult, if not impossible, to understand how the results are related to the detailed electronic structure of materials, as provided by energy-band calculations.

Following the work of Toya (1958) on the dispersion curves of Na, the lattice dynamics of simple metals was developed within the framework of the pseudopotential theory of metals (Harrison, 1966; Heine, 1968; Cohen and Heine, 1968; Heine and Weaire, 1968). For most of the other classes of materials, however, it was impossible until the mid-1970s to relate the measured dispersion curves to the electronic properties of the materials. Progress in computational techniques has presently made possible first-principles calculations. The most direct approach describes the electronic response to the nuclear motion by a dielectric matrix (or electronic susceptibility). In practical applications, however, one is faced with the problem of inverting this dielectric matrix, a rather formidable problem. Because of this problem, only few calculations have been performed to date with this direct (or inversion) method.

Various methods have been proposed to overcome the computational problems of the dielectric function formalism of the general theory. The basic idea is to separate the band structure contribution to the dynamical matrix. One of the most commonly used methods is that put forward by Varma and Weber (1977, 1979). In this approach, the dynamical matrix is reduced to

$$D = D_{\mathrm{sr}} + D_2 \tag{14}$$

where $D_{\mathrm{sr}}$ is determined by relatively short-range forces and $D_2$ is the main band structure contribution. In practical applications the only calculation needed is that of $D_2$, since $D_{\mathrm{sr}}$ is parametrized by using short-range force constants. Many phonon anomalies observed in the dispersion curves of metals have been explained by this method.

Finally, it should be emphasized that presently it is possible to evaluate phonon frequencies directly and to compare them with the experimental results. These *ab initio* or frozen phonon calculations (Chadi and Martin, 1976; Wendel and Martin, 1978, 1979; Yin and Cohen, 1980; Kunc and Martin, 1981; Harmon et al., 1981; Yin and Cohen, 1982; Kunc and Martin, 1982; Ho et al., 1982) are particularly powerful in relating observed phonon anomalies in the dispersion curves to the electronic properties of the solid. To illustrate the principle of the method, consider a perfect (monatomic) crystal and another obtained from the first by displacing the atoms to the positions they assume instantaneously in the presence of a lattice wave of wave vector $\mathbf{q}$ and frequency $\omega(\mathbf{q})$. Clearly, the second lattice is obtained from the perfect lattice by displacing the atoms, at the equilibrium positions $\mathbf{l}$ by

$$\Delta \mathbf{R}_l = \mathbf{u}_0 \cos(\mathbf{q} \cdot \mathbf{l}) \tag{15}$$

where $\mathbf{u}_0$ defines the amplitude and direction of the displacement. The energy difference per atom, $\Delta E$, between the two lattices is simply the potential energy per atom for the "frozen" phonon mode,

$$\Delta E = \frac{1}{N} \sum_l \frac{1}{2} M \omega^2(\mathbf{q}) |\delta \mathbf{R}_l|^2 \tag{16}$$

which can be written as

$$\Delta E = \frac{1}{2} M \omega^2(\mathbf{q}) \left[ \frac{1}{N} |\mathbf{u}_0|^2 \sum_l \cos^2(\mathbf{q} \cdot \mathbf{l}) \right] \tag{17}$$

For a zone boundary phonon [$\mathbf{q} = (\tau/2)$], the expression in brackets is simply $|\mathbf{u}_0|^2$, whereas for arbitrary $\mathbf{q}$, it is $\frac{1}{2} |\mathbf{u}_0|^2$. To obtain the phonon frequency, the energy difference $\Delta E$ is evaluated as a function of the displacement $|\mathbf{u}_0|$ by calculating the total energies of the distorted and perfect crystal. The phonon frequency is then obtained from the curvature of the energy-difference versus displacement curve.

The method is used to obtain frequencies of phonon modes with wave vector $\mathbf{q}$ such that $n\mathbf{q} = \tau$ (where $n$ is an integer), since only in this case is the distorted crystal also periodic (with a larger real-space cell), and its total energy can be calculated by the same techniques as those used to obtain the total energy of the perfect crystal. The computational times increase, of course, as the number of atoms in the larger cell of the distorted crystal increases. Accurate calculations of total energies for crystal structures with 10 to 15 atoms per cell are presently feasible.

Frozen phonon calculations have been performed for simple metals and semiconductors, as well as transition metals. In general, the agreement between calculated and experimentally determined frequencies is excellent. In a slightly different approach (Kunc and Martin, 1982; Yin and Cohen, 1982), the phonon dispersion curves along high symmetry directions can be obtained by calculating the force constants between atomic layers perpendicular to the direction of wave propagation. In these calculations, a supercell containing a sufficient number of layers is adopted and the interlayer force constants are obtained from the Hellmann-Feynman forces acting on individual atomic layers when a certain layer is displaced along the direction of wave propagation.

Probably the most appealing aspect of this method is that it provides not only phonon frequencies, but also additional information for comparison with the experimental results: lattice constants, bulk moduli, and elastic constants (obtained by subjecting the perfect crystal to a homogeneous deformation). Also, by studying the departure from a parabola of the energy versus displacement curves, one can obtain information about anharmonic effects. In addition, extra minima in the energy versus displacement curves may indicate instability of the lattice toward a certain phase transformation. It should be pointed out, however, that the calculations are valid essentially at $T = 0$, a fact that is sometimes overlooked in comparing the calculations to the experimental results.

In practice, the analysis and interpretation of the results is facilitated considerably if good quality data are collected. Given a sample of good quality this can be achieved by minimizing the background, optimizing the resolution, and adopting sufficient counting times for satisfactory counting statistics. The phonon frequencies, obtained by fitting the phonon peaks, are assigned to the various branches using Born–von Kármán model calculations. The dispersion curves are then fitted to those given by various theoretical or phenomenological models. A preliminary analysis of the results can be made as the experiment is in progress by making use of the programs available in practically all neutron scattering facilities.

## SAMPLE PREPARATION

A few remarks should be made regarding the samples used for phonon measurements. From the expression of the inelastic structure factor, it is evident that light elements with relatively large coherent cross-sections are favored for coherent inelastic scattering experiments. Actually, the small mass of $^4$He made possible the measurements of the dispersion curves of this quantum crystal. If the incoherent cross-section that contributes to the background is not sufficiently small, it may also complicate the measurements. In such cases the use of single isotopes, as in materials containing Ni, can be very helpful. The use of appropriate isotopes may actually be essential if the material to be studied has a significant neutron absorption cross-section (much larger than $\sim$100 b). For instance, the use of nonabsorbing isotopes made possible the study of the dispersion curves of Cd (Chernyshov et al., 1979; Dorner et al., 1981). In such extreme cases, however, one must seriously examine the possibility of obtaining the desired information by an x-ray diffuse scattering experiment instead (see X-RAY AND NEUTRON DIFFUSE SCATTERING MEASUREMENTS). For instance, such measurements have been performed on V, which, because of its low coherent scattering length ($0.05 \times 10^{-12}$ cm), is difficult to study with standard coherent inelastic scattering techniques.

For coherent neutron scattering experiments, it is almost essential to have a single-crystal specimen whose size is dictated by the material properties and the intensity of the neutron source. For most materials, information about lattice dynamics can be obtained in a high-flux source on samples as small as 0.05 cm$^3$. In general, for detailed studies, one needs specimens of the order of 1 cm$^3$ or larger. If it is available, a single crystal of high perfection (mosaic spread 0.2° should be used. In high-temperature or -pressure experiments, as well as in studies where the sample is brought through a phase transformation, however, the experimentalist may have little, if any, control of the mosaic spread of the sample. In such cases, measurements may have to be performed on crystals with mosaic spreads as large as 5°. Difficulties arising from loss in experimental resolution and Bragg-phonon scattering processes in such cases are discussed in the Problems section.

When possible, high-purity samples should be used, since impurities as well as defects increase the background scattering. It should be noted, however, that impurity contents as high as 1000 ppm have, in most cases, very little effect on the dispersion curves of the material being studied. Significant hydrogen content, on the other hand, can complicate the experiments considerably. In addition to its large incoherent cross-section, which considerably increases the background, hydrogen can precipitate as a hydride at a certain temperature and cause a large increase in the mosaic of the sample. This is particularly troublesome in studies of lattice dynamics as a function of temperature.

As mentioned earlier, most of the detailed coherent neutron scattering studies have been performed on single-crystal specimens. Recently, however, it has been demonstrated (Buchenau et al., 1981) that considerable information about the lattice dynamics of materials can be obtained by studies of the coherent neutron scattering by polycrystalline samples. In the case of Ca, the dispersion curves obtained by this technique were found to be in good agreement with measurements performed on single-crystal specimens (Stassis et al., 1983). For materials for which relatively large single crystals are not available, study of the coherent scattering by a polycrystalline sample may prove to be extremely valuable.

## SPECIMEN MODIFICATION

In general, the scattering of thermal neutrons is a nondestructive experimental technique. However, after irradiation practically all samples become to some extent radioactive. It is therefore important before performing an experiment to estimate the activity of the samples after irradiation and to avoid, if possible, the use of elements that upon irradiation become long-lived radioactive isotopes. After the experiment is completed, users are required to have the samples checked by the institutional radiation safety office (Health Physics). In many cases the sample is stored for several days or weeks to allow its activity to decay to admissible levels before it is shipped to the user.

## PROBLEMS

Collection of data on a triple-axis spectrometer is complicated by the observation of peaks arising from higher-order contamination. If a monochromator is set to reflect neutrons of energy $E_0$ from the $(hkl)$ plane, it is also set to reflect in the same direction neutrons of energy $n^2 E_0$ from the $(nh\ nk\ nl)$ plane. Similarly, if the analyzer is set to reflect neutrons of energy $E_1$, it will also reflect in the direction of the counter neutrons of energy $m^2 E_1$. It is easily seen that, as a result, various spurious neutron groups may be observed. One of the most common is due to the elastic incoherent scattering by the sample, in the direction of the analyzer, of the $n$th-order component of the incident beam. Clearly, if $n$ and $m$ are such that $n^2 E_0 = m^2 E_1 = m^2(E_0 \pm \hbar\omega)$, a spurious neutron group will be observed that could be attributed to the creation (or annihilation) of a phonon of energy $\hbar\omega$. If such a problem is suspected, one can repeat the measurement using a sufficiently high $E_1$ (or $E_0$) so that this process is not energetically possible for any values of $n$ and $m$. Quite generally, by measurements at different energies one can avoid ambiguities arising from spurious neutron groups due to higher-order contamination. This problem is serious, however, in studies of complex systems, and various methods are used to minimize it.

The simplest method to deal with the problem of order contamination is to choose the incident energy close to the maximum of the reactor spectrum so that the higher-order contamination in the incident beam arises from the low-flux region of the spectrum. In most experiments, however, one needs lower energies for better energy resolution. It is, therefore, highly desirable to decrease the flux of higher-energy neutrons incident on the monochromator crystal. This can be achieved by placing a single crystal of high perfection before the monochromator. The choice of the best filter for this purpose is restricted by the difficulty of obtaining large, highly perfect, single crystals of the desired material. Quartz, silicon, and sapphire have been

used with considerable success. Of these, silicon and quartz are used at liquid nitrogen temperatures and have similar characteristics. Sapphire seems to be more effective and can be used at room temperature without significant loss of efficiency. Note that mechanical velocity selectors and curved neutron guides can also be used for the same purpose. For instance, an S-shaped neutron guide is installed between the cold source and the IN-12 triple-axis spectrometer of the high flux reactor at the Institut Laue-Langevin.

Even when the higher-energy neutron flux incident on the monochromator has been minimized as described above, additional discrimination of the higher-order contamination is desirable in a large number of experiments. This is usually achieved by using appropriate filters. Pyrolitic graphite is probably the most useful, and it is used routinely in typical experiments. It is particularly efficient in a narrow energy range around 13.7 and 14.7 meV. Pyrolitic graphite filters are also quite efficient at 30.5 and 42 meV. The filter, typically several inches thick, is oriented so that the $c$ axis is along the beam and is positioned in front of either the analyzer or the sample (depending on whether the experiment is performed at fixed scattered or incident neutron energy). For experiments at low energies ($\leq 5$ meV), polycrystalline filters such as Be are frequently used. In this case, one exploits the fact that no coherent elastic scattering occurs if the neutron wavelength exceeds twice the largest spacing between planes in the material. Finally, it should be mentioned that one can eliminate a specific higher-order contamination by appropriate choice of the analyzer or monochromator. For instance, one of the most troublesome processes is the incoherent elastic scattering by the sample of the primary beam, in the direction of the analyzer, followed by second-order scattering from the analyzer ($n = 1, m = 2$). This process can be eliminated by using Ge (or Si) reflecting from the (111) planes as analyzer, since for all practical purposes the (222) reflection is forbidden for diamond-structure crystals.

Since we discussed in some detail how to deal with the spurious neutron groups arising from higher-order contamination, it is appropriate now to mention three other processes that may introduce similar problems. The first process, alluded to earlier in this section, involves Bragg scattering of the primary incident beam (of energy $E_0$) by the sample in the direction of the scattered beam followed by incoherent scattering by the analyzer in the direction of the counter. In the second process, neutrons with energy $E_1$ in the incident beam may be Bragg reflected from the sample and then from the analyzer into the detector. In both cases, a relatively high count rate will be observed in the second monitor, which is placed behind the sample. The problem is then easily eliminated by a slight change of experimental conditions. The third case involves a double-scattering process in the sample: coherent one-phonon scattering followed or preceded by Bragg scattering. This process is particularly troublesome since it gives rise to neutron groups corresponding to modes that in principle should not be observed under the experimental conditions of the measurement. For instance, quite often one observes in a longitudinal scan, neutron groups corresponding to transverse modes. In principle, this problem can be practically eliminated by using a single crystal of small mosaic spread. In experiments with crystals of large mosaic spreads (as is often the case in high-pressure or -temperature studies), this double-scattering process can complicate the measurements considerably.

Finally, it should be mentioned that radiation damage of the sample is not usually a problem because the sample is exposed to the relatively low intensity and low energy beam reflected by the monochromator. A notable exception is the cast of some classes of biological materials.

## LITERATURE CITED

Bilz, H. and Kress, W. 1979. Phonon Dispersion Relations in Insulators. Springer-Verlag, Berlin and New York.

Born, M. and Huang, K. 1954. Dynamical Theory of Crystal Lattices. Oxford University Press, London.

Born, M. and von Kàrmàn, T. 1912. *Z. Phys.* 14:65.

Born, M. and von Kàrmàn, T. 1913. *Z. Phys.* 13:297.

Brockhouse, B. N. 1960. *Bull. Am. Phys. Soc.* 5:462.

Brockhouse, B. N. 1961. *In* Inelastic Scattering of Neutrons in Solids and Liquids, p. 113. IAEA, Vienna.

Brockhouse, B. N. and Stewart, A. T. 1955. *Phys. Rev.* 100:756.

Brüesch, P. 1982. Phonons: Theory and Experiments. Springer-Verlag, Berlin and New York.

Buchenau, U., Schober, H. R., and Wagner, R. 1981. *J. Phys. (Orsay, France)* 42:C6–365.

Burstein, E. 1967. *In* Phonons and Phonon Interactions (T. A. Bak, ed.) p. 276. W. A. Benjamin, New York.

Cardona, M. and Güntherodt, G. 1975–1991. Light Scattering in Solids. Vols. 1–6, Springer-Verlag, Berlin.

Chadi, J. D. and Martin, R. M. 1976. Solid State Commun. 19:643.

Chernyshov, A. A., Pushkarev, V. V., Rumpantsev, A. Y., Dorner, B., and Pynn, R. 1979. *J. Phys. F* 9:1983.

Cochran, W. 1971. *Crit. Rev Solid State Sci.* 2:1.

Cohen, M. L. and Heine, V. 1968. *Solid State Phys.* 24:38 [and references cited therein].

Cooper, M. J. and Nathans, R. 1967. *Acta Crystallogr.* 23:357.

Debye, P. 1912. *Ann. Phys. (Leipzig)* 39:789.

Dolling, G. 1974. *In* Dynamical Properties of Solids (G. K. Horton and A. A. Maradudin, eds.) Vol. 1, p. 541. North-Holland, Amsterdam.

Dorner, B., Chernysov, A.A., Pushkarev, V.V., Rumyantsev, A.Y., and Pynn, R. 1981. J. Phys. F 11:365.

Einstein, A. 1907. *Ann. Phys. (Leipzig)* 22:180.

Einstein, A. 1911. *Ann. Phys. (Leipzig)* 35:679.

Freund, A. 1976. *In* Proceedings of the Conference on Neutron Scattering. Gatlinburg, Tenn. p. 1143.

Hardy, J. R. 1974. *In* Dynamical Properties of Solids (G. K. Horton and A. A. Maradudin, eds.) Vol. 7, p. 157. North-Holland, Amsterdam.

Hardy, J. R. and Karo, A. M. 1979. The Lattice Dynamics and Statics of Alkali Halide Crystals. Plenum, New York.

Harmon, B. N., Weber, W., and Hamann, D. R. 1981. *J. Phys. (Orsay, France)* 42:C6–628.

Harrison, W. A. 1966. Pseudopotentials in the Theory of Metals. Benjamin, New York.

Heine, V. 1968. *Solid State Phys.* 24:1 [and references cited therein].

Heine, V. and Weaire, D. 1968. *Solid State Phys.* 24:250 [and references cited therein].

Ho, K.-M., Fu, C.-L, Harmon, B. N., Weber, W., and Hamann, D. R. 1982. *Phys. Rev. Lett.* 49:673.

Horton, G. K. and Maradudin, A. A. (eds.) 1974. Dynamical Properties of Solids, Vols. 1 to 4. North-Holland, Amsterdam.

Kunc, K. and Martin, R. M. 1981. *J. Phys. (Orsay, France)* 42:C6–649.

Kunc, K. and Martin, R. M. 1982. *Phys. Rev. Lett.* 48:406.

Mller, H. Bjerrum, and Nielsen, M. 1970. *In* Instrumentation for Neutron Inelastic Scattering Research. p. 49. IAEA, Vienna.

Nicklow, R. M. Gilat, G., Smith, H. G., Raubenheimer, L. J., and Wilkinson, M. K. 1967. Phys. Rev. 164:922.

Placzek, G. and Van Hove, L. 1954. Phys. Rev. 93:1207.

Sinha, S. K. 1973. *Crit. Rev. Solid State Sci.* 4:273.

Sköld, K. and Price, D. L. (eds.) 1986. Neutron Scattering (Methods of Experimental Physics, Vol. 23). Academic, New York.

Stassis, C., Zarestky, J., Misemer, D. K., Skriver, H. L., and Harmon, B. N. 1983. *Phys. Rev. B* 27:3303.

Stedman, R. and Nilsson, G. 1966. *Phys. Rev.* 145:492.

Toya, T. 1958. *J. Res. Inst. Catal. Hokkaido Univ.* 6:183.

Varma, C. M. and Weber, W. 1977. *Phys. Rev. Lett.* 39:1094.

Varma, C. M. and Weber, W. 1979. *Phys. Rev B* 19:6142.

Venkataraman, G., Feldkamp, L. A., and Shani, V. C. 1975. Dynamics of Perfect Crystal. MIT Press, Cambridge, Mass.

Vijayaraghavan, P. R., Nicklow, R. M., Smith, H. G., and Wilkinson, M. K. 1970. *Phys. Rev. B* 1:4819.

Waller, I. and Fröman, P. O. 1952. *Ark. Fys.* 4:183.

Wendel, H. and Martin, R. M. 1978. *Phy. Rev. Lett.* 40:950.

Wendel, H. and Martin, R. M. 1979. *Phys. Rev. B* 19:5251.

Windsor, C. G. 1981. Pulsed Neutron Scattering. Taylor & Francis, Halsted Press, New York.

Windsor, C. G. 1986. *In* Methods of Experimental Physics (K. Sköld and D. L. Price, eds.) Vol. 23, p. 197. Academic, New York.

Yin, M. T. and Cohen, M. L. 1980. *Phys. Rev. Lett.* 45:1004.

Yin, M. T. and Cohen, M. L. 1982. *Phys. Rev. D* 25:4317.

Yin, M. T. and Cohen, M. L. 1982. *Phys. Rev. B* 26:3259 [and references cited therein].

## KEY REFERENCES

Born and Huang, 1954. See above.

*Although quite old, still the best introduction to the subject of lattice dynamics.*

Venkataraman, Feldkamp, and Sahni, 1975. See above.

*A relatively up-to-date treatment of lattice dynamics. It contains a detailed outline of the Born-von Kármán formalism.*

Sköld and Price (eds.), 1986. See above.

*An excellent source for information regarding phonons in various materials. The first part is a concise introduction to neutron scattering.*

## BIBLIOGRAPHY

Bilz, H. and Kress, W. 1979. Phonon Dispersion Relations in Insulators. Springer-Verlag, Berlin and New York.

Born, M. and Huang, K. 1954. Dynamical Theory of Crystal Lattices. Oxford University Press, London.

Brillouin, L. 1953. Wave Propagation in Periodic Structures. Dover, New York.

Brüesch, P. 1982. Phonons: Theory and Experiments. Springer-Verlag, Berlin and New York.

Choquard, Ph. 1967. The Anharmonic Crystal. Benjamin, New York.

Cochran, W. 1973. The Dynamics of Atoms in Crystals. Arnold, London.

Decius, J. C. and Hexter, R. M. 1977. Molecular Vibrations in Crystals. McGraw-Hill, New York.

Donovan, B. and Angress, J. F. 1971. Lattice Vibrations. Chapman & Hall, London.

Dorner, B. 1982. Coherent Inelastic Neutron Scattering in Lattice Dynamics. Springer-Verlag, Berlin and New York.

Hardy, J. R. and Karo, A. M. 1979. The Lattice Dynamics and Statics of Alkali Halide Crystals. Plenum, New York.

Horton, G. K. and Maradudin, A., eds. 1974. Dynamical Properties of Solids, Vols. 1 to 4. North-Holland, Amsterdam.

Ludwig, W. 1967. Recent Developments in Lattice Theory. Springer-Verlag, Berlin and New York.

Maradudin, A. A., Ipatova, I. P., Montroll, E. W., and Weiss, G.H. 1971. Theory of Lattice Dynamics in the Harmonic Approximation. In Solid States Physics, Suppl. 3, 2nd Ed. Academic, New York.

Reissland, J. A. 1973. The Physics of Phonons. Wiley, New York.

Venkataraman, G., Feldkamp, L. A., and Sahni, V. C. 1975. Dynamics of Perfect Crystals. MIT Press, Cambridge, Mass.

C. STASSIS
Iowa State University
Ames, Iowa

# MAGNETIC NEUTRON SCATTERING

## INTRODUCTION

The neutron is a spin-$\frac{1}{2}$ particle that carries a magnetic dipole moment of $-1.913$ nuclear magnetons. Magnetic neutron scattering then originates from the interaction of the neutron's spin with the unpaired electrons in the sample, either through the dipole moment associated with an electron's spin or via the orbital motion of the electron. The strength of this magnetic dipole-dipole interaction is comparable to the neutron-nuclear interaction, and thus there are magnetic cross-sections that are analogous to the nuclear ones, which reveal the structure and dynamics of materials over wide ranges of length scale and energy. Magnetic neutron scattering plays a central role in determining and understanding the microscopic properties of a vast variety of magnetic systems—from the fundamental nature, symmetry, and dynamics of magnetically ordered materials—to the elucidation of the magnetic characteristics essential in technological applications.

One traditional role of magnetic neutron scattering has been the measurement of magnetic Bragg intensities in the magnetically ordered regime. Such measurements can be used to determine the spatial arrangement and directions of the atomic magnetic moments, the atomic magnetization density of the individual atoms in the material, and the value of the ordered moments as a function of thermodynamic parameters such as temperature, pressure, and applied magnetic field. These types of measurements can be carried out on single crystals, powders, thin films, and artificially grown multilayers, and often the information collected can be obtained by no other experimental technique. For magnetic phenomena that occur over length scales that are large compared to atomic distances, the technique of magnetic small angle neutron scattering (SANS) can be applied, in analogy to structural SANS. This is an ideal technique to explore domain structures, long wavelength oscillatory magnetic states, vortex structures in superconductors, and other spatial variations of the magnetization density on length scales from 1 to 1000 nm. Another specialized technique is neutron reflectometry, which can be used to investigate the magnetization profile in the near-surface regime of single crystals, as well as the magnetization density of thin films and multilayers, in analogy with structural reflectometry techniques. This particular technique has enjoyed dramatic growth during the last decade due to the rapid advancement of atomic deposition capabilities.

Neutrons can also scatter inelastically, to reveal the magnetic fluctuation spectrum of a material over wide ranges of energy ($\sim 10^{-8} - 1\,\mathrm{eV}$) and over the entire Brillouin zone. Neutron scattering plays a truly unique role in that it is the only technique that can directly determine the complete magnetic excitation spectrum, whether it is in the form of the dispersion relations for spin wave excitations, wave vector and energy dependence of critical fluctuations, crystal field excitations, magnetic excitons, or moment fluctuations. In the present overview we will discuss some of these possibilities.

## Competitive and Related Techniques

Much of the information that can be obtained from neutron scattering, such as ordering temperatures and ferromagnetic moments, can and should be compared with magnetic measurements of quantities such as specific heat, susceptibility, and magnetization (MAGNETISM AND MAGNETIC MEASUREMENTS). Indeed it is often highly desirable to have measurements available from these traditional techniques before undertaking the neutron measurements, as a way of both characterizing the particular specimen being used and to be able to focus the neutron data collection to the range of temperature, applied field, etc., of central interest. Thus these complementary measurements are highly desirable.

For magnetic diffraction, the technique that is directly competitive is magnetic x-ray scattering. The x-ray magnetic cross-sections are typically $\sim 10^{-6}$ of those of charge scattering, but with the high fluxes available from synchrotron sources x-ray scattering from these cross-sections is readily observable. The information that can be obtained is also complementary to neutrons in that neutrons scatter from the total magnetization in the system, whether it originates from spin or orbital motions, while for x rays the spin and orbital contributions can be distinguished. X rays also have the advantage that the wavelength can be tuned to an electronic resonance, which can increase the cross-section by as much as $\sim 10^4$, and thus it can be an element-specific technique. However, the limited number of resonances available restricts the materials that can be investigated, while the high absorption cross-sections and the generally more stringent experimental needs and specimen requirements are disadvantages. Each x-ray photon also carries $\sim 10^6$ times the energy that a neutron carries, and the combination of this with the brightness of a synchrotron source can cause rather dramatic changes in the physical or chemical properties of the sample during the investigation. Neutrons are a much gentler probe, but the low flux relative to synchrotron sources also means that larger samples are generally needed. Both techniques must be carried out at centralized facilities. Magnetic x-ray diffraction is a relatively young technique, though, and its use is expected to grow as the technique develops and matures. However, neutron scattering generally will continue to be the technique of choice when investigating new materials.

For the investigation of spin dynamics, there are a number of experimental techniques that can provide important information. Mössbauer spectroscopy (MÖSSBAUER SPECTROMETRY), muon spin precession, and perturbed angular correlation of nuclear radiation measurements can reveal characteristic time scales, as well as information about the site symmetry of the magnetism and the ordered moment as a function of temperature. Raman and infrared (IR) spectroscopy can reveal information about the long-wavelength (Brillouin zone–center) excitations (RAMAN SPECTROSCOPY OF SOLIDS), and the magnon density of states via two-magnon processes. Microwave absorption can determine the spin wave excitation spectrum at very long wavelengths (small wave vectors **q**), while nuclear magnetic resonance (NUCLEAR MAGNETIC RESONANCE IMAGING) can give important information about relaxation rates. However, none of these techniques can explore the full range of dynamics that neutrons can, and in this regard there is no competitive experimental technique for measurements of the spin dynamics.

## PRINCIPLES OF THE METHOD

### Magnetic Diffraction

The cross-section for magnetic Bragg scattering (Bacon, 1975) is given by

$$I_M(\mathbf{g}) = C \left( \frac{\gamma e^2}{2mc^2} \right)^2 \mathbf{M}_g A(\theta_B) |F_M(\mathbf{g})|^2 \tag{1}$$

where $I_M$ is the integrated intensity for the magnetic Bragg reflection located at the reciprocal lattice vector, **g**, the neutron-electron coupling constant enclosed in large parentheses evaluates as $-0.27 \times 10^{-12}$ cm, $C$ is an instrumental constant that includes the resolution of the measurement, $A(\theta_B)$ is an angular factor that depends on the method of measurement (e.g., sample angular rotation, or

$\theta : 2\theta$ scan), and $\mathbf{M}_g$ is the multiplicity of the reflection (for a powder sample). The magnetic structure factor $F_M(\mathbf{g})$ is given in the general case (Blume, 1961) by

$$F_M(\mathbf{g}) = \sum_{j=1}^{N} e^{i g \cdot r_j} \hat{\mathbf{g}} \times [\mathbf{M}_j(\mathbf{g}) \times \hat{\mathbf{g}}] e^{-W_j} \tag{2}$$

where $\hat{\mathbf{g}}$ is a unit vector in the direction of the reciprocal lattice vector **g**, $M_j(\mathbf{g})$ is the vector form factor of the $j$th ion located at $\mathbf{r}_j$ in the unit cell, $W_j$ is the Debye-Waller factor (see MÖSSBAUER SPECTROMETRY) that accounts for the thermal vibrations of the $j$th ion, and the sum is over all (magnetic) atoms in the unit cell. The triple cross product originates from the vector nature of the dipole-dipole interaction of the neutron with the electron. A quantitative calculation of $M_j(\mathbf{g})$ in the general case involves evaluating matrix elements of the form $\langle \pm | 2S e^{i g \bullet R} + O | \pm \rangle$, where $S$ is the (magnetic) spin operator, $O$ is the symmetrized orbital operator introduced by Trammell (1953), and $|\pm\rangle$ represents the angular momentum state. This can be quite a complicated angular-momentum computation involving all the electron orbitals in the unit cell. However, usually the atomic spin density is *collinear*, by which we mean that at each point in the spatial extent of the electron's probability distribution, the *direction* of the atomic magnetization density is the same. In this case, the direction of $M_j(\mathbf{g})$ does not depend on **g**, and the form factor is just a scalar function, $f(\mathbf{g})$, which is simply related to the Fourier transform of the magnetization density. The free-ion form factors have been tabulated for essentially all the elements. If one is familiar with x-ray diffraction, then it is helpful to note that the magnetic form factor for neutron scattering is analogous to the form factor for charge scattering of x rays, except that for x rays it corresponds to the Fourier transform of the total charge density of all the electrons, while in the magnetic neutron case it is the transform of the "magnetic" electrons, which are the electrons whose spins are unpaired. Recalling that a Fourier transform inverts the relative size of objects, the magnetic form factor typically decreases much more rapidly with **g** than for the case of x-ray charge scattering, since the unpaired electrons are usually the outermost ones of the ion. This dependence of the scattering intensity on $f(\mathbf{g})$ is a convenient way to distinguish magnetic cross-sections from nuclear cross-sections, where the equivalent of the form factor is just a constant (the nuclear coherent scattering amplitude $b$).

If, in addition to the magnetization density being collinear, the magnetic moments in the ordered state point along a unique direction (i.e., the magnetic structure is a ferromagnet, or a simple $+ - + -$ type antiferromagnet), then the square of the magnetic structure factor simplifies to

$$|F_M(\mathbf{g})|^2 = \langle 1 - (\hat{\mathbf{g}} \bullet \hat{\boldsymbol{\eta}})^2 \rangle \left| \sum_j \eta_j \langle \mu_j^z \rangle f_j(\mathbf{g}) e^{-W_j} e^{i g \bullet r_j} \right|^2 \tag{3}$$

where $\hat{\boldsymbol{\eta}}$ denotes the (common) direction of the ordered moments and $\eta_j$ the sign of the moment ($\pm 1$), $\langle \mu_j^2 \rangle$ is the average value of the ordered moment in thermodynamic equilibrium at $(T, H, P, \ldots)$, and the orientation factor $\langle 1 - (\hat{\mathbf{g}} \bullet \hat{\boldsymbol{\eta}})^2 \rangle$ represents an average over all possible domains. If the magnetic moments are the same type, then this expression further simplifies to

$$|F_{\mathrm{M}}(\mathbf{g})|^2 = \langle 1 - (\hat{\mathbf{g}} \bullet \hat{\boldsymbol{\eta}})^2 \rangle \langle \mu^z \rangle^2 f^2(\mathbf{g}) e^{-2W_j} \left| \sum_j \eta_j e^{i\mathbf{g}\bullet\mathbf{r}_j} \right|^2 \quad (4)$$

We see from these expressions that neutrons can be used to determine several important quantities: the location of magnetic atoms in the unit cell and the spatial distribution of their magnetic electrons, as well as the dependence of $\langle \mu^z \rangle$ on temperature, field, pressure, or other thermodynamic variables, which is directly related to the order parameter for the phase transition (e.g., the sublattice magnetization). Often the preferred magnetic axis, $\hat{\boldsymbol{\eta}}$, can also be determined from the relative intensities. Finally, the scattering can be put on an absolute scale by internal comparison with the nuclear Bragg intensities $I_{\mathrm{N}}$ from the same sample, given by

$$I_{\mathrm{N}}(\mathbf{g}) = C M_{\mathbf{g}} A(\theta_{\mathrm{B}}) |F_{\mathrm{N}}(\mathbf{g})|^2 \quad (5)$$

with

$$|F_{\mathrm{N}}(\mathbf{g})|^2 = \left| \sum_j b_j e^{-W_j} e^{i\mathbf{g}\bullet\mathbf{r}_j} \right|^2 \quad (6)$$

Here $b_j$ is the coherent nuclear scattering amplitude for the $j$th atom in the unit cell, and the sum is over all atoms in the unit cell. Often the nuclear structure is known accurately and $F_{\mathrm{N}}$ can be calculated, whereby the saturated value of the magnetic moment can be obtained.

### Subtraction Technique

There are several ways that magnetic Bragg scattering can be distinguished from the nuclear scattering from the structure. Above the magnetic ordering temperature, all the Bragg peaks are nuclear in origin, while as the temperature drops below the ordering temperature the intensities of the magnetic Bragg peaks rapidly develop, and for unpolarized neutrons the nuclear and magnetic intensities simply add. If these new Bragg peaks occur at positions that are distinct from the nuclear reflections, then it is straightforward to distinguish magnetic from nuclear scattering. In the case of a ferromagnet, however, or for some antiferromagnets that contain two or more magnetic atoms in the chemical unit cell, these Bragg peaks can occur at the same position. One standard technique for identifying the magnetic Bragg scattering is to make one diffraction measurement in the paramagnetic state well above the ordering temperature, and another in the ordered state at the lowest temperature possible, and then subtract the two sets of data. In the paramagnetic

state, the (free ion) diffuse magnetic scattering is given (Bacon, 1975) by

$$I_{\mathrm{Para}} = \frac{2}{3} C \left( \frac{\gamma e^2}{2mc^2} \right)^2 \mu_{\mathrm{eff}}^2 f(\mathbf{Q})^2 \quad (7)$$

where $\mu_{\mathrm{eff}}$ is the effective magnetic moment $\{ = g[J(J+1)]^{1/2}$ for a free ion$\}$. This is a magnetic incoherent cross-section, and the only angular dependence is through the magnetic form factor $f(\mathbf{Q})$. Hence this scattering looks like "back-ground." There is a sum rule on the magnetic scattering in the system, though, and in the ordered state most of this diffuse scattering shifts into the coherent magnetic Bragg peaks. A subtraction of the high-temperature data (Equation 7) from the data obtained at low temperature (Equation 1) will then yield the magnetic Bragg peaks, on top of a deficit (negative) of scattering away from the Bragg peaks due to the disappearance of the diffuse paramagnetic scattering in the ordered state. On the other hand, usually none of the nuclear cross-sections change significantly with temperature, and hence drop out in the subtraction. A related subtraction technique is to apply a large magnetic field in the paramagnetic state, to induce a net (ferromagnetic-like) moment. The zero-field (nuclear) diffraction pattern can then be subtracted from the high-field pattern to obtain the induced-moment diffraction pattern.

### Polarized Beam Technique

When the neutron spins that impinges on a sample has a well-defined polarization state, then the nuclear and magnetic scatterin that originates from the sample interferes *coherently*, in contrast to being separate cross-sections like Equations 1 and 5, where magnetic and nuclear intensities just add. A simple example is provided by the magnetic and nuclear scattering from elemental iron, which has the body-centered cubic (bcc) structure and a (saturated) ferromagnetic moment of $2.2\mu_{\mathrm{B}}$ at low temperature. It is convenient to define the magnetic scattering amplitude $p$ (Bacon, 1975) as

$$p = \left( \frac{\gamma e^2}{2mc^2} \right) \langle \mu \rangle f(\mathbf{g}) \quad (8)$$

and then the structure factor for the scattering of polarized neutrons is given by

$$F(\mathbf{g}) = (b \pm p) e^{-W_j} (1 + e^{i\pi(h+k+l)}) \quad (9)$$

where the $\pm$ indicates the two polarization states of the incident neutrons. The phase factor in parentheses originates from the sum over the unit cell, which in this case is just two identical atoms located at the origin $(0,0,0)$ and the body-centered position $(1/2,1/2,1/2)$, and $(hkl)$ are Miller indices (SYMMETRY IN CRYSTALLOGRAPHY). This gives the familiar selection rule that for a bcc lattice $h + k + l$ must be an even integer for Bragg reflection to occur. The smallest nonzero reciprocal lattice vector is then the $(1,1,0)$, and for this reflection ($\mathbf{g} = 3.10$ Å$^{-1}$) the magnetic scattering amplitude is as follows:

$$p = (0.27 \times 10^{-12} \text{ cm})(2.2)(0.59) = 0.35 \times 10^{-12} \text{ cm} \quad (10)$$

The coherent nuclear scattering amplitude for iron is $b = 0.945 \times 10^{-12}$ cm and is independent of the Bragg reflection. The flipping ratio $R$ is defined as the ratio of the intensities for the two neutron polarizations; for this reflection we have

$$R = \left(\frac{0.945 + 0.35}{0.945 - 0.35}\right)^2 = 4.74 \quad (11)$$

Note that, for a different Bragg reflection, the only quantity that changes is $f(\mathbf{g})$, so that a measurement of the flipping ratio at a series of Bragg peaks ($hkl$) can be used to make precision determinations of the magnetic form factor, and thus by Fourier transform the spatial distribution of the atomic magnetization density. It also should be noted that if $b \approx p$, then the flipping ratio will be very large, and this is in fact one of the standard methods employed to produce a polarized neutron beam in the first place.

In the more general situation when the magnetic structure is not a simple ferromagnet, polarized beam diffraction measurements with polarization analysis of the scattered neutrons must be used to establish unambiguously which peaks are magnetic, which are nuclear, and more generally to separate the magnetic and nuclear scattering at Bragg positions where there are both nuclear and magnetic contributions. The polarization analysis technique as applied to this problem is in principle straightforward; complete details can be found elsewhere (Moon et al., 1969; Williams, 1988). Nuclear coherent Bragg scattering never causes a reversal, or spin-flip, of the neutron spin direction upon scattering. Thus the nuclear peaks will only be observed in the non-spin-flip scattering geometry. We denote this configuration as $(+ +)$, where the neutron is incident with up spin, and remains in the up state after scattering. Non-spin-flip scattering also occurs if the incident neutron is in the down state, and remains in the down state after scattering [denoted $(- -)$]. The magnetic cross-sections, on the other hand, depend on the relative orientation of the neutron polarization $\mathbf{P}$ and the reciprocal lattice vector $\mathbf{g}$. In the configuration where $\mathbf{P} \perp \mathbf{g}$, half the magnetic Bragg scattering involves a reversal of the neutron spin [denoted by $(- +)$ or $(+ -)$], and half does not. Thus, for the case of a purely magnetic reflection the spin-flip $(- +)$ and non-spin-flip $(+ +)$ intensities should be equal in intensity. For the case where $\mathbf{P} \parallel \mathbf{g}$, all the magnetic scattering is spin-flip. Hence for a pure magnetic Bragg reflection the spin-flip scattering should be twice as strong as for the $\mathbf{P} \perp \mathbf{g}$ configuration, while ideally no non-spin-flip scattering will be observed. The analysis of these cross-sections can be used to unambiguously identify nuclear from magnetic Bragg scattering.

## Inelastic Scattering

Magnetic inelastic scattering plays a unique role in determining the spin dynamics in magnetic systems, as it is the only probe that can directly measure the complete magnetic excitation spectrum. Typical examples are spin wave dispersion relations, critical fluctuations, crystal field excitations, and moment/valence fluctuations.

As an example, consider identical spins $S$ on a simple cubic lattice, with a coupling given by $-JS_i \cdot S_j$ where $J$ is the (Heisenberg) exchange interaction between neighbors separated by the distance $a$. The collective excitations for such an ensemble of spins are magnons (or loosely termed "spin waves"). If we have $J > 0$ so that the lowest energy configuration is where the spins are parallel (a ferromagnet), then the magnon dispersion along the edge of the cube (the [100] direction) is given by $E(q) = 8JS$ $[\sin^2(qa/2)]$. At each wave vector $\mathbf{q}$, the magnon energy is different, and a neutron can interact with the system of spins and either create a magnon at $(\mathbf{q}, E)$, with a concomitant change of momentum and loss of energy of the neutron, or conversely destroy a magnon with a gain in energy. The observed change in momentum and energy for the neutron can then be used to map out the magnon dispersion relation. Neutron scattering is particularly well suited for such inelastic scattering studies since neutrons typically have energies that are comparable to the energies of excitations in the solid, and therefore the neutron energy changes are large and easily measured. The dispersion relations can then be measured over the entire Brillouin zone (see, e.g., Lovesey, 1984).

Additional information about the nature of the excitations can be obtained by polarized inelastic neutron scattering techniques, which are finding increasing use. Spin wave scattering is represented by the raising and lowering operators $S\pm = S_x \pm iS_y$, which cause a reversal of the neutron spin when the magnon is created or destroyed. These "spin-flip" cross-sections are denoted by $(+ -)$ and $(- +)$. If the neutron polarization $\mathbf{P}$ is parallel to the momentum transfer $\mathbf{Q}$, $\mathbf{P} \parallel \mathbf{Q}$, then spin angular momentum is conserved (as there is no orbital contribution in this case). In this experimental geometry, we can only create a spin wave in the $(- +)$ configuration, which at the same time causes the total magnetization of the sample to decrease by one unit ($1\mu_B$). Alternatively, we can destroy a spin wave only in the $(+ -)$ configuration, while increasing the magnetization by one unit. This gives us a unique way to unambiguously identify the spin wave scattering, and polarized beam techniques in general can be used to distinguish magnetic from nuclear scattering in a manner similar to the case of Bragg scattering.

Finally, we note that the magnetic Bragg scattering is comparable in strength to overall magnetic inelastic scattering. However, all the Bragg scattering is located at a single point in reciprocal space, while the inelastic scattering is distributed throughout the (three-dimensional) Brillouin zone. Hence, when actually making inelastic measurements to determine the dispersion of the excitations, one can only observe a small portion of the dispersion surface at any one time, and thus the observed inelastic scattering is typically two to three orders of magnitude less intense than the Bragg peaks. Consequently these are much more time-consuming measurements, and larger samples are needed to offset the reduction in intensity. Of course, a successful determination of the dispersion relations yields a complete determination of the fundamental atomic interactions in the solid.

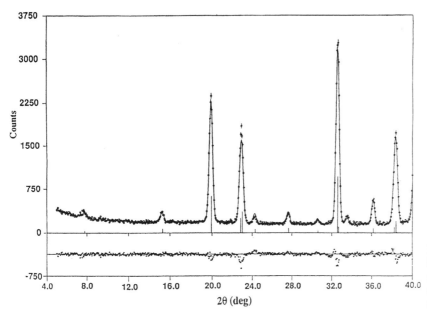

**Figure 1.** Calculated (solid curve) and observed intensities at room temperature for a powder of antiferromagnetically ordered $YBa_2Fe_3O_8$. The differences between calculated and observed are shown at the bottom. (Huang et al., 1992.)

## PRACTICAL ASPECTS OF THE METHOD

### Diffraction

As an example of Bragg scattering, a portion of the powder diffraction pattern from a sample of $YBa_2Fe_3O_8$ is shown in Figure 1 (Huang et al., 1992). The solid curve is a Rietveld refinement (Young, 1993) of both the antiferromagnetic and crystallographic structure for the sample. From this type of data, we can determine the full crystal structure; lattice parameters, atomic positions in the unit cell, site occupancies, etc. We can also determine the magnetic structure and value of the ordered moment. The results of the analysis are shown in Figure 2; the crystal structure is identical to the structure for the $YBa_2Cu_3O_7$ high-$T_C$ cuprate superconductor, with the Fe replacing the Cu, and the magnetic structure is also the same as has been observed for the Cu spins in the oxygen-reduced ($YBa_2Cu_3O_6$) semiconducting material.

Experimentally, we can recognize the magnetic scattering by several characteristics. First, it should be temperature-dependent, and the Bragg peaks will vanish above the ordering temperature. Figure 3 shows the temperature dependence of the intensity of the peak at a scattering angle of $19.5°$ in Figure 1. The data clearly reveal a phase transition at the Néel temperature for $YBa_2Fe_3O_8$ of 650 K (Natali Sora et al., 1994); the Néel temperature is where long-range antiparallel order of the spins first occurs. Above the antiferromagnetic phase transition, this peak completely disappears, indicating that it is a purely magnetic Bragg peak. A second characteristic is that the magnetic intensities become weak at high scattering angles (not shown), as $f(\mathbf{g})$ typically falls off strongly with increasing angle. A third, more elegant, technique is to use polarized neutrons. The polarization technique can be used at any temperature, and for any material, regardless of whether or not it has a crystallographic distortion (e.g., via magnetoelastic interactions) associated with the magnetic transition. It is more involved and time-consuming

experimentally, but yields an unambiguous identification and separation of magnetic and nuclear Bragg peaks.

First consider the case where $\mathbf{P} \parallel \mathbf{g}$, which is generally achieved by having a horizontal magnetic field, which must also be oriented along the scattering vector. In this geometry, all the magnetic scattering is spin-flip, while the nuclear scattering is always non-spin-flip. Hence for a magnetic Bragg peak the spin-flip scattering should be twice as strong as for the $\mathbf{P} \perp \mathbf{g}$ configuration (vertical

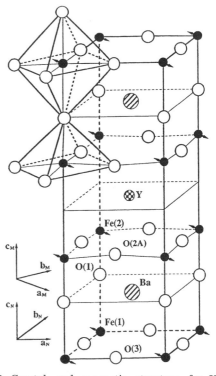

**Figure 2.** Crystal and magnetic structure for $YBa_2Fe_3O_8$ deduced from the data of Figure 1. (Huang et al., 1992.)

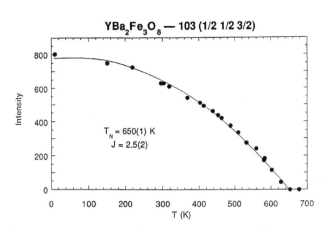

**Figure 3.** Temperature dependence of the intensity of the magnetic reflection found at a scattering angle of 19.5° in Figure 1. The Néel temperature is 650 K. (Natali et al., 1994.)

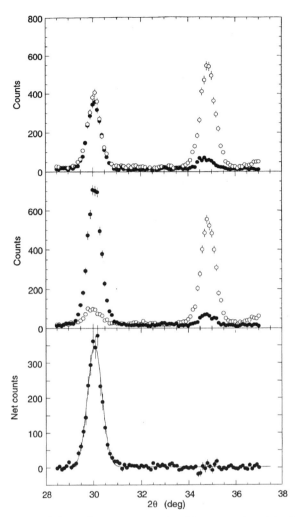

**Figure 4.** Polarized neutron scattering. The top portion of the figure is for $\mathbf{P} \perp \mathbf{g}$, where the open circles show the non-spin-flip scattering and the filled circles are the observed scattering in the spin-flip configuration. The low angle peak has equal intensity for both cross-sections, and thus is identified as a pure magnetic reflection, while the ratio of the $(++)$ to $(-+)$ scattering for the high-angle peak is 11, the instrumental flipping ratio. Hence this is a pure nuclear reflection. The center portion of the figure is for $\mathbf{P} \perp \mathbf{g}$ and the bottom portion is the subtraction of the $\mathbf{P} \perp \mathbf{g}$ spin-flip scattering from the data for $\mathbf{P} \| \mathbf{g}$. Note that in the subtraction procedure all background and nuclear cross-sections cancel, isolating the magnetic scattering. (Huang et al., 1992.)

field), while the nuclear scattering is non-spin-flip scattering and independent of the orientation of $\mathbf{P}$ and $\mathbf{g}$. Figure 4 shows the polarized beam results for the same two peaks, at scattering angles (for this wavelength) of 30° and 35°; these correspond to the peaks at 19.5° and 23° in Figure 1. The top section of the figure shows the data for the $\mathbf{P} \perp \mathbf{g}$ configuration. The peak at 30° has the identical intensity for both spin-flip and non-spin-flip scattering, and hence we conclude that this scattering is purely magnetic in origin, as inferred from Figure 3. The peak at 35°, on the other hand, has strong intensity for $(+ +)$, while the intensity for $(- +)$ is smaller by a factor of 1/11, the instrumental flipping ratio in this measurement. Hence, ideally there would be no spin-flip scattering, and this peak is identified as a pure nuclear reflection. The center row shows the same peaks for the $\mathbf{P} \| \mathbf{g}$ configuration, while the bottom row shows the subtraction of the $\mathbf{P} \perp \mathbf{g}$ spin-flip scattering from the $\mathbf{P} \| \mathbf{g}$ spin-flip scattering. In this subtraction procedure, instrumental background, as well as all nuclear scattering cross-sections, cancel, isolating the magnetic scattering. We see that there is magnetic intensity only for the low angle position, while no intensity survives the subtraction at the 35° peak position. These data unambiguously establish that the 30° peak is purely magnetic, while the 35° peak is purely nuclear. This simple example demonstrates how the technique works; obviously it would play a much more critical role in cases where it is not clear from other means what is the origin of the peaks, such as in regimes where the magnetic and nuclear peaks overlap, or in situations where the magnetic transition is accompanied by a structural distortion. If needed, a complete "magnetic diffraction pattern" can be obtained and analyzed with these polarization techniques.

In cases where there is no significant coupling of the magnetic and lattice systems, on the other hand, the subtraction technique can also be used to obtain the magnetic diffraction pattern (see, e.g., Zhang et al., 1990). This technique is especially useful for low-temperature magnetic phase transitions where the Debye-Waller effects can be

safely neglected. Figure 5 shows the diffraction patterns for $Tm_2Fe_3Si_5$, which is an antiferromagnetic material that becomes superconducting under pressure (Gotaas et al., 1987). The top part of the figure shows the diffraction pattern obtained above the antiferromagnetic ordering temperature of 1.1 K, where just the nuclear Bragg peaks are observed. The middle portion of the figure shows the low-temperature diffraction pattern in the ordered state, which contains both magnetic and nuclear Bragg peaks, and the bottom portion shows the subtraction, which gives the magnetic diffraction pattern.

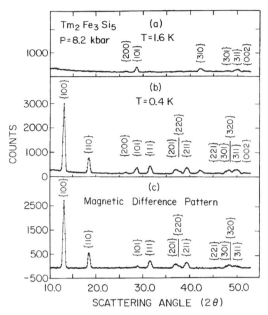

**Figure 5.** Diffraction patterns for the antiferromagnetic superconductor $Tm_2Fe_3Si_5$. The top part of the figure shows the nuclear diffraction pattern obtained above the antiferromagnetic ordering temperature of 1.1 K, the middle portion of the figure shows the low-temperature diffraction pattern in the ordered state, and the bottom portion shows the subtraction of the two, which gives the magnetic diffraction pattern. (Gotaas et al., 1987.)

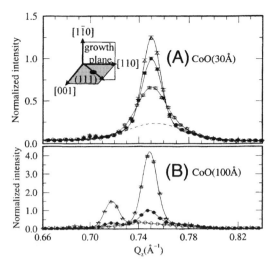

**Figure 7.** Neutron diffraction scans of the (111) reflection along the [001] growth axis direction (scattering vector $\mathbf{Q}_z$) for (**A**) $[Fe_3O_4/(100 \text{ Å})CoO(30 \text{ Å})]_{50}$ and (**B**) $[Fe_3O_4/(100 \text{ Å})CoO(100 \text{ Å})]_{50}$ superlattices, taken at 78 K in zero applied field. The closed circles, open circles, and triangles indicate data taken after zero-field cooling (initial state), field cooling with an applied field ($H = 14$ kOe) in the [110] direction, and field cooling with an applied field in the [110] direction, respectively. The inset illustrates the scattering geometry. The dashed lines indicate the temperature- and field-independent $Fe_3O_4$ component (Ijiri et al., 1998).

Another example of magnetic Bragg diffraction is shown in Figure 6. Here we show the temperature dependence of the intensity of an antiferromagnetic Bragg peak on a single crystal of the high-$T_C$ superconductor ($T_C \approx 92$ K) $ErBa_2Cu_3O_7$ (Lynn et al., 1989). The magnetic interactions of the Er moments in this material are highly anisotropic, and this system turns out to be an ideal two-dimensional (2D) (planar) Ising antiferromagnet; the solid curve is Onsager's exact solution to the $S = \frac{1}{2}$, 2D Ising model (Onsager, 1944), and we see that it provides an excellent representation of the experimental data.

A final diffraction example is shown in Figure 7, where the data for two $Fe_3O_4/CoO$ superlattices are shown (Ijiri et al., 1998). The superlattices consist of 50 repeats of 100-Å thick layers of magnetite, which is ferrimagnetic, and either 30- or 100-Å thick layers of the antiferromagnet CoO. The superlattices were grown epitaxially on single-crystal MgO substrates, and thus these may be regarded as single-crystal samples. These scans are along the growth direction ($Q_z$), and show the changes in the magnetic scattering that occur when the sample is cooled in an applied field, versus zero-field cooling. These data, together with polarized beam data taken on the same samples, have elucidated the origin of the technologically important exchange biasing effect that occurs in these magnetic superlattices.

It is interesting to compare the type and quality of data that are represented by these three examples. The powder diffraction technique is quite straightforward, both to obtain and analyze the data. In this case, typical sample sizes are ~1–20 g, and important and detailed information can be readily obtained with such sample sizes in a few hours of spectrometer time, depending on the particulars of the problem. The temperature dependence of the order parameter in $ErBa_2Cu_3O_7$, on the other hand, was obtained on a single crystal weighing only 31 mg. Note that the statistical quality of the data is much better than for the powder sample, even though the sample is more than 2 orders of magnitude smaller; this is because it is a single crystal and all the scattering is directed into a single peak, rather than scattering into a powder-diffraction ring. The final example was for $Fe_3O_4/CoO$

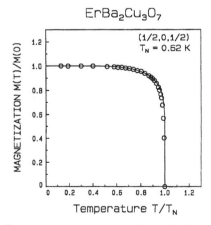

**Figure 6.** Temperature dependence of the sublattice magnetization for the Er spins in superconducting $ErBa_2Cu_3O_7$, measured on a single crystal weighing 31 mg. The solid curve is Onsager's exact theory for the 2D, $S = \frac{1}{2}$, Ising model (Lynn et al., 1989).

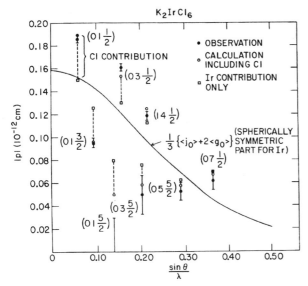

**Figure 8.** Comparison of the observed and calculated magnetic form factor for the cubic $K_2IrCl_6$ system, showing the enormous directional anisotropy along the spin direction ([001] direction) compared with perpendicular to the spin direction. The contribution of the Cl moments along the spin direction, arising from the covalent bonding, is also indicated; no net moment is transferred to the Cl ions that are perpendicular to the spin direction. (Lynn et al., 1976.)

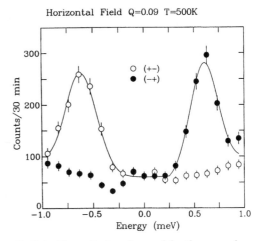

**Figure 9.** Spin-flip scattering observed for the amorphous Invar $Fe_{86}B_{14}$ isotropic ferromagnetic system in the $\mathbf{P} \parallel \mathbf{Q}$ configuration. Spin waves are observed for neutron energy gain ($E < 0$) in the $(+-)$ cross-section, and for neutron energy loss ($E > 0$) in the $(-+)$ configuration. (Lynn et al., 1993.)

superlattices, where the weight of the superlattices that contribute to the scattering is $\sim$1 mg. Thus it is clear that interesting and successful diffraction experiments can be carried out on quite small samples.

The final example in this section addresses the measurement of a magnetic form factor, in the cubic antiferromagnetic $K_2IrCl_6$. The Ir ions occupy a face-centered cubic (fcc) lattice, and the $5d$ electrons carry a magnetic moment that is covalently bonded to the six Cl ions located octahedrally around each Ir ion. One of the interesting properties of this system is that there is charge transfer from the Ir equally onto the six Cl ions. A net spin transfer, however, only occurs for the two Cl ions along the direction in which the Ir spin points. This separation of spin and charge degrees of freedom leads to a very unusual form factor that is highly anisotropic, as shown in Figure 8 (Lynn et al., 1976). Note the rapid decrease in the form factor as one proceeds along the ($c$ axis) spin direction, from the $(0,1,1/2)$, to the $(0,1,3/2)$, then to the $(0,1,5/2)$ Bragg peak, which in fact has an unobservable intensity. In this example, $\sim$30% of the total moment is transferred onto the Cl ions, which is why these covalency effects are so large in this compound. It should be noted that, in principle, x-ray scattering should be able to detect the charge transferred onto the (six) Cl ions as well, but this is much more difficult to observe because it is a very small fraction of the total charge. In the magnetic case, in contrast, it is a large percentage effect and has a different symmetry than the lattice, which makes the covalent effects much easier to observe and interpret.

## Inelastic Scattering

There are many types of magnetic excitations and fluctuations that can be measured with neutron scattering techniques, such as magnons, critical fluctuations, crystal field excitations, magnetic excitons, and moment/valence fluctuations. To illustrate the basic technique, consider an isotropic ferromagnet at sufficiently long wavelengths (small $q$). The spin wave dispersion relation is given by $E_{sw} = D(T)q^2$, where $D$ is the spin wave "stiffness" constant. The general form of the spin wave dispersion relation is the same for all isotropic ferromagnets, a requirement of the (assumed) perfect rotational symmetry of the magnetic system, while the numerical value of $D$ depends on the details of the magnetic interactions and the nature of the magnetism. One example of a prototypical isotropic ferromagnet is amorphous $Fe_{86}B_{14}$. Figure 9 shows an example of polarized beam inelastic neutron scattering data taken on this system (Lynn et al., 1993). These data were taken with the neutron polarization $\mathbf{P}$ parallel to the momentum transfer $\mathbf{Q}(\mathbf{P} \parallel \mathbf{Q})$, where we should be able to create a spin wave only in the $(-+)$ configuration, or destroy a spin wave only in the $(+-)$ configuration. This is precisely what we see in the data—for the $(-+)$ configuration the spin waves can only be observed for neutron energy loss scattering ($E > 0$), while for the $(+-)$ configuration, spin waves can only be observed in neutron energy gain ($E < 0$). This behavior of the scattering uniquely identifies these excitations as spin waves.

Data like these can be used to measure the renormalization of the spin waves as a function of temperature, as well as to determine the lifetimes as a function of wave vector and temperature. An example of the renormalization of the "stiffness" constant $D$ for the magnetoresistive oxide $Tl_2Mn_2O_7$ is shown in Figure 10 (Lynn et al., 1998). Here, the wave vector dependence of the dispersion relation has been determined at a series of $q$'s, and the stiffness parameter is extracted from the data. The variation in the stiffness parameter is then plotted, and indicates a

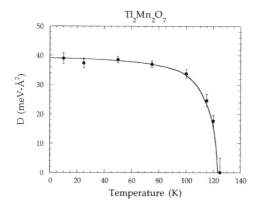

**Figure 10.** Temperature dependence of the spin wave stiffness $D(T)$ in the magnetoresistive $Tl_2Mn_2O_7$ pyrochlore, showing that the spin waves renormalize as the ferromagnetic transition is approached. (Lynn et al., 1998.) Below $T_C$ the material is a metal, while above $T_C$ it exhibits insulator behavior.

smooth variation with a ferromagnetic transition temperature of 123 K. These measurements can then be compared directly with theoretical calculations. They can also be compared with other experimental observations, such as magnetization measurements, whose variation with temperature originates from spin wave excitations.

Finally, these types of measurements can be extended all the way to the zone boundary on single crystals. Figure 11 shows an example of the spin wave dispersion relation for the ferromagnet $La_{0.85}Sr_{0.15}MnO_3$, a system that undergoes a metal-insulator transition that is associated with the transition to ferromagnetic order (Vasiliu-Doloc et al., 1998). The top part of the figure shows the dispersion relation for the spin wave energy along two high-symmetry directions, and the solid curves are fits to a simple nearest-neighbor spin-coupling model. The overall trend of the data is in reasonable agreement with the model, although there are some clear discrepancies as well, indicating that a more sophisticated model will be needed in order to obtain quantitative agreement. In addition to the spin wave energies, though, information about the intrinsic lifetimes can also be determined, and these linewidths are shown in the bottom part of the figure for both symmetry directions. In the simplest type of model, no intrinsic spin wave linewidths at all would be expected at low temperatures, while we see here that the observed linewidths are very large and highly anisotropic, indicating that an itinerant-electron type of model is more appropriate for this system.

## METHOD AUTOMATION

Neutrons for materials research are produced by one of two processes; fission of $U^{235}$ in a nuclear reactor, or by the spallation process where high energy protons from an accelerator impact on a heavy-metal target like W and explode the nuclei. Both techniques produce high-energy (MeV) neutrons, which are then thermalized or sub-thermalized in a moderator (such as heavy water) to produce a Maxwellian spectrum of thermal or cold neu-

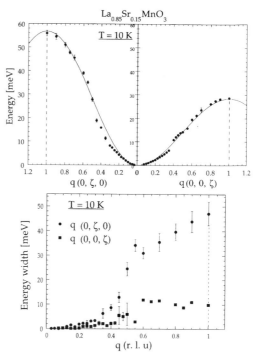

**Figure 11.** Ground state spin wave dispersion along the (0,1,0) and (0,0,1) directions measured to the zone boundary for the magnetoresistive manganite $La_{0.85}Sr_{0.15}MnO_3$. The solid curves are a fit to the dispersion relation for a simple Heisenberg exchange model. The bottom part of the figure shows the intrinsic linewidths of the excitations. In the standard models, intrinsic linewidths are expected only at elevated temperatures. The large observed linewidths demonstrate the qualitative inadequacies of these models. (Vasiliu-Doloc et al., 1998.)

trons, with energies in the meV range. Thus all neutron scattering instrumentation is located at centralized national facilities, and each facility generally has a wide variety of instrumentation that has been designed and constructed for the specific facility, and is maintained and scheduled by facility scientists. Generally, any of these instruments can be used to observe magnetic scattering, and with the number and diversity of spectrometers in operation it is not practical to review the instrumentation here; the interested user should contact one of the facilities directly. The facilities themselves operate for periods of weeks at a time continuously, and hence, since the early days of neutron scattering all the data collection has been accomplished by automated computer control.

## SAMPLE PREPARATION

As a general rule, there is no particular preparation that is required for the sample, in the sense that there is no need, for example, to polish the surface, cleave it under high vacuum, or perform similar procedures. Neutrons are a deeply penetrating bulk probe, and hence generally do not interact with the surface. There are two exceptions to this rule. One is when utilizing polarized neutrons in the investigation of systems with a net magnetization, where a rough surface can cause the variations in the local magnetic field that depolarize the neutrons. In such cases, the surface may need to be polished. The second case is for neutron reflectometry, where small-angle mirror reflection is used to directly explore the magnetization profile of the surface and interfacial layers. In this case, the samples need to be optically flat over the full size of the sample (typically 1 to 10 cm$^2$), and of course the quality of the film in terms of, e.g., surface roughness and epitaxial quality becomes part of the investigation.

For magnetic single crystal diffraction, typical sample sizes should be no more than a few mm$^3$ in order to avoid primary and secondary extinction effects; if the sample is too large, extinction will always limit the accuracy of the data obtained. Since the magnetic Bragg intensity is proportional to the square of the ordered moment, any extinction effects will depend on the value of $\langle \mu_z \rangle$ at a particular temperature and field, while the ideal size of the sample can vary by three to four orders of magnitude, depending on the value of the saturation moment.

Powder diffraction requires gram-sized samples, generally in the range of 1 to 20 g. The statistical quality of the data that can be obtained is directly related to the size, so in principle there is a direct trade-off between sample size and the time required to collect a set of diffraction data at a particular temperature or magnetic field. Sample sizes smaller than 1 g can also be measured, but since the neutron facilities are heavily oversubscribed, in practice a smaller sample size translates into fewer data sets, so an adequate-sized sample is highly desirable.

The cross-sections for inelastic scattering are typically two to three orders of magnitude smaller than elastic cross-sections, and therefore crystals for inelastic scattering typically must be correspondingly larger in comparison with single-crystal diffraction. Consequently, for both powder diffraction and inelastic scattering, the general rule is the bigger the sample the better the data. The exception is if one or more of the elements in a material has a substantial nuclear absorption cross-section, in which case the optimal size of the sample is then determined by the inverse of this absorption length. The absorption cross-sections are generally directly proportional to the wavelength of the neutrons being employed for a particular measurement, and the optimal sample size then depends on the details of the spectrometer and the neutron wavelength(s). For the absorption cross-sections of all the elements, see Internet Resources. In some cases, particular isotopes can be substituted to avoid some large absorption cross-sections, but generally these are expensive and/or are only available in limited quantity. Therefore isotopic substitution can be employed only for a few isotopes that are relatively inexpensive, such as deuterium or boron, or in scientifically compelling cases where the cost can be justified.

## DATA ANALYSIS AND INITIAL INTERPRETATION

One of the powers of neutron scattering is that it is a very versatile tool and can be used to probe a wide variety of magnetic systems over enormous ranges of length scale ($\sim$0.1 to 10000 Å) and dynamical energy ($\sim$10 neV to 1 eV). The instrumentation to provide these measurement capabilities is equally diverse, but data analysis software is generally provided for each instrument to perform the initial interpretation of the data. Moreover, the technique is usually sufficiently fundamental and the interpretation sufficiently straightforward that the initial data analysis is often all that is needed.

## PROBLEMS

Neutron scattering is a very powerful technique, but in general it is a flux-limited, and this usually requires a careful tradeoff between instrumental resolution and signal. There can also be unwanted cross-sections that appear and contaminate the data, and so one must of course exercise care in collecting the data and be vigilant in its interpretation. If a time-of-flight spectrometer is being employed, for example, there may be frame overlap problems that can mix the fastest neutrons with the slowest, and give potentially spurious results. Similarly, if the spectrometer employs one or more single crystals to monochromate or analyze the neutron energies, then the crystal can reflect higher-order wavelengths [since Bragg's law is $n\lambda = 2d \sin(\theta)$] which can give spurious peaks. Sometimes these can be suppressed with the use of filters, but more generally it is necessary to vary the spectrometer conditions and retake the data in order to identify genuine cross-sections from spurious ones. This identification process can consume considerable beam time.

Another problem can be encountered in identifying the magnetic Bragg scattering in powders using the subtraction technique. If the temperature is sufficiently low, the nuclear spins may begin to align with the electronic spins, particularly if there is a large hyperfine field. Significant polarization can occur at temperatures of a few degrees kelvin and lower, and the nuclear polarization can be confused with electronic magnetic ordering. Another problem can be encountered if the temperature dependence of the Debye-Waller factor is significant, or if there is a structural distortion associated with the magnetic transition, such as through a magnetoelastic interaction. In this case, the nuclear intensities will not be identical, and thus will not cancel correctly in the subtraction. It is up to the experimenter to decide if this is a problem, within the statistical precision of the data. A problem can also occur if there is a significant thermal expansion, where the diffraction peaks shift position with temperature. By significant, we mean that the shift is noticeable in comparison with the instrumental resolution employed. Again it is up to the experimenter to decide if this is a problem. In both these latter cases, though, a full refinement of the combined magnetic and nuclear structures in the ordered phase can be carried out. Alternatively, polarized beam techniques can be employed to unambiguously separate the magnetic and nuclear cross-sections. Finally, if one uses the subtraction technique in the method where a field is applied, the field can cause the powder particles to reorient if there is substantial crystalline magnetic anisotropy in the sample. This preferred orientation will remain when the field is removed, and will be evident in the nuclear peak intensities, but must be taken into account.

Finally, we remark about the use of polarized neutrons. Highly polarized neutron beams can be produced by single-crystal diffraction from a few special magnetic materials, from magnetic mirrors, or from transmission of the beam through a few specific nuclei where the absorption cross-section is strongly spin dependent. The choices are limited, and consequently most spectrometers are not equipped with polarization capability. For instruments that do have a polarized beam option, generally one has to sacrifice instrumental performance in terms of both resolution and intensity. Polarization techniques are then typically not used for many problems in a routine manner, but rather are usually used to answer some specific question or make an unambiguous identification of a cross-section, most often after measurements have already been carried out with unpolarized neutrons.

## ACKNOWLEDGMENTS

I would like to thank my colleagues, Julie Borchers, Qing Huang, Yumi Ijiri, Nick Rosov, Tony Santoro, and Lida Vasiliu-Doloc, for their assistance in the preparation of this unit.

There are vast numbers of studies in the literature on various aspects presented here. The specific examples used for illustrative purposes were chosen primarily for the author's convenience in obtaining the figures, and his familiarity with the work.

## LITERATURE CITED

Bacon, G. E. 1975. Neutron Diffraction, 3rd ed. Oxford University Press, Oxford.

Blume, M. 1961. Orbital contribution to the magnetic form factor of $Ni^{++}$. Phys. Rev. 124:96–103.

Gotaas, J. A., Lynn, J. W., Shelton, R. N., Klavins, P., and Braun, H. F. 1987. Suppression of the superconductivity by antiferromagnetism in $Tm_2Fe_3Si_5$. Phys. Rev. B 36:7277–7280.

Huang, Q., Karen, P., Karen, V. L., Kjekshus, A., Lynn, J. W., Mighell, A. D., Rosov, N., and Santoro, A. 1992. Neutron powder diffraction study of the nuclear and magnetic structures of $YBa_2Cu_3O_8$, Phys. Rev. B 45:9611–9619.

Ijiri, Y., Borchers, J. A., Erwin, R. W., Lee, S.-H., van der Zaag, P. J., and Wolf, R. M. 1998. Perpendicular coupling in exchange-biased $Fe_3O_4/CoO$ superlattices. Phys. Rev. Lett. 80:608–611.

Lovesey, S. W. 1984. Theory of Neutron Scattering from Condensed Matter, Vol. 2. Oxford University Press, New York.

Lynn, J. W., Clinton, T. W., Li, W.-H., Erwin, R. W., Liu, J. Z., Vandervoort, K., and Shelton, R. N. 1989. 2D and 3D magnetic order of Er in $ErBa_2Cu_3O_7$. Phys. Rev. Lett. 63:2606–2610.

Lynn, J. W., Rosov, N., and Fish, G. 1993. Polarization analysis of the magnetic excitations in Invar and non-Invar amorphous ferromagnets. J. Appl. Phys. 73:5369–5371.

Lynn, J. W., Shirane, G., and Blume, M. 1976. Covalency effects in the magnetic form factor of Ir in $K_2IrCl_6$. Phys. Rev. Lett. 37:154–157.

Lynn, J. W., Vasiliu-Doloc, L., and Subramanian, M. 1998. Spin dynamics of the magnetoresistive pyrochlore $Tl_2Mn_2O_7$. Phys. Rev. Lett. 80:4582–4586.

Moon, R. M., Riste, T., and Koehler, W. C. 1969. Polarization analysis of thermal neutron scattering. Phys. Rev. 181:920–931.

Natali, S. I., Huang, Q., Lynn, J. W., Rosov, N., Karen, P., Kjekshus, A., Karen, V. L., Mighell, A. D., and Santoro, A. 1994. Neutron powder diffraction study of the nuclear and magnetic structures of the substitutional compounds $(Y_{1-x}Ca_x)$ $Ba_2Fe_3O_{8+\delta}$. Phys. Rev. B 49:3465–3472.

Onsager, L. 1944. Crystal statistics. I. A two-dimensional model with an order-disorder transition. Phys. Rev. 65:117–149.

Trammell, G. T. 1953. Magnetic scattering of neutrons from rare earth ions. Phys. Rev. 92:1387–1393.

Vasiliu-Doloc, L., Lynn, J. W., Moudden, A. H., de Leon-Guevara, A. M., and Revcolevschi, A. 1998. Structure and spin dynamics of $La_{0.85}Sr_{0.15}MnO_3$. Phys. Rev. B 58:14913–14921.

Williams, G. W. 1988. Polarized Neutrons. Oxford University Press, New York.

Young, R. A. 1993. The Rietveld Method. Oxford University Press, Oxford.

Zhang, H., Lynn, J. W., Li, W-H., Clinton, T. W., and Morris, D. E. 1990. Two- and three-dimensional magnetic order of the rare earth ions in $RBa_2Cu_4O_8$. Phys. Rev. B 41:11229–11236.

## KEY REFERENCES

Bacon, 1975. See above.

*This text is more for the experimentalist, treating experimental procedures and the practicalities of taking and analyzing data. It does not contain some of the newest techniques, but has most of the fundamentals. It is also rich in the history of many of the techniques.*

Balcar, E. and Lovesey, S. W. 1989. Theory of Magnetic Neutron and Photon Scattering. Oxford University Press, New York.

*More recent work that specifically addresses the theory for the case of magnetic neutron as well as x-ray scattering.*

Lovesey, 1984. See above.

*This text treats the theory of magnetic neutron scattering in depth. Vol. 1 covers nuclear scattering.*

Moon et al., 1969. See above.

*This is the classic article that describes the triple-axis polarized beam technique, with examples of all the fundamental measurements that can be made with polarized neutrons. Very readable.*

Price, D. L. and Sköld, K. 1987. Methods of Experimental Physics: Neutron Scattering. Academic Press, Orlando, Fla.

*A recent compendium that covers a variety of topics in neutron scattering, in the form of parts by various experts.*

Squires, G. L. 1978. Thermal Neutron Scattering. Cambridge University Press, New York.

*This book is more of a graduate introductory text to the subject of neutron scattering.*

Williams, 1988. See above.

*This textbook focuses on the use of polarized neutrons, with all the details.*

Young, 1993. See above.

*This text details the profile refinement technique for powder diffraction.*

## INTERNET RESOURCES

### Magnetic Form Factors

http://papillon.phy.bnl.gov/form.html

### Numerical Values of the Free-Ion Magnetic Form Factors

http://www.ncnr.nist.gov/resources/n-lengths/

*Values of the coherent nuclear scattering amplitudes and other nuclear cross-sections.*

J. W. LYNN
University of Maryland
College Park, Maryland

# INDEX